Recommended Regimen and Indications for Routine Vaccinations

Routine Schedules for Children and Adults											Shaded bars indicate that a dose can be given once during this time frame
Vaccine	**Birth**	**2 Months**	**4 Months**	**6 Months**	**12 Months**	**15 Months**	**18 Months**	**4–6 Years**	**11–12 Years**	**Adults**	**Comments**
Mixed vaccines											
Diphtheria, Tetanus, Pertussis[1] (DTaP)		DTaP	DTaP	DTaP	DTaP one dose			DTaP	Td or T		Td is tetanus/diphtheria; T is tetanus alone; either one should be given as booster every 10 years.
Measles,[2] Mumps, Rubella (MMR)					MMR			MMR or MMR			First dose varies with disease incidence; booster given at either 4–6 or 11–12 years.
Haemophilus influenzae type b (Hib)		Hib	Hib	Hib	Hib						Schedule depends upon source of vaccine; given with DTaP as TriHIBit
Single vaccines											
Poliovirus (IPV)[3]		IPV	IPV	IPV	one dose			IPV			Similar schedule to DTaP; injected vaccine
Hepatitis B (HB)	HB option 1		HB option 2	HB option 3							Option depends upon condition of infant; 3 doses given
Chickenpox (CPV)					CPV one dose					CPV two doses	Cannot be given to children <1 year old
Pneumococcus Vaccine (PV)		PV	PV	PV	PV						Used to protect children against otitis media

[1]DTaP. The diphtheria–tetanus–acellular pertussis vaccine has replaced the DTP.

[2]Measles vaccine (Attenuvax) can be given alone to children during epidemics or to adults immunized before 1970.

[3]IPV—inactivated polio vaccine—is now indicated as a safer alternative to the oral polio vaccine.

Used in Cases of Specific Risk Due to Occupational or Other Exposure

Vaccine	Group Targeted
Hepatitis B (Recombivax)	Health care personnel; people exposed through life-style
Hepatitis A (Havrix)	Children 2–14 years who live in areas of high prevalence
Pneumococcus (Pneumovax)	Elderly patients, children with sickle-cell anemia
Influenza, polio, tuberculosis (BCG)	Hospital, laboratory, health care workers
Rabies, plague, Lyme disease	People whose jobs involve contact with animals (veterinarians, forest rangers); known or suspected exposure to rabid animal; living in areas of high incidence
Cholera, hepatitis B, hepatitis A, measles, yellow fever, meningococcal meningitis, polio, rabies, typhoid, plague	Travelers to endemic regions, including military recruits (varies with geographic destination)

Foundations in

Microbiology

FOURTH EDITION

Foundations in Microbiology

FOURTH EDITION

Kathleen Park Talaro
Pasadena City College

Arthur Talaro
Pasadena City College

Boston   Burr Ridge, IL   Dubuque, IA   Madison, WI   New York   San Francisco   St. Louis
Bangkok   Bogotá   Caracas   Kuala Lumpur   Lisbon   London   Madrid   Mexico City
Milan   Montreal   New Delhi   Santiago   Seoul   Singapore   Sydney   Taipei   Toronto

McGraw-Hill Higher Education

*A Division of The **McGraw-Hill** Companies*

FOUNDATIONS IN MICROBIOLOGY, FOURTH EDITION

Published by McGraw-Hill, a business unit of The McGraw-Hill Companies, Inc., 1221 Avenue of the Americas, New York, NY 10020. Copyright © 2002, 1999, 1996 by The McGraw-Hill Companies, Inc. All rights reserved. No part of this publication may be reproduced or distributed in any form or by any means, or stored in a database or retrieval system, without the prior written consent of The McGraw-Hill Companies, Inc., including, but not limited to, in any network or other electronic storage or transmission, or broadcast for distance learning.

Some ancillaries, including electronic and print components, may not be available to customers outside the United States.

 This book is printed on recycled, acid-free paper containing 10% postconsumer waste.

International   1 2 3 4 5 6 7 8 9 0 QPH/QPH 0 9 8 7 6 5 4 3 2 1
Domestic   1 2 3 4 5 6 7 8 9 0 QPH/QPH 0 9 8 7 6 5 4 3 2 1

ISBN 0–07–232042–7
ISBN 0–07–112275–3 (ISE)

Publisher: *James M. Smith*
Senior developmental editor: *Jean Sims Fornango*
Associate marketing manager: *Tami Petsche*
Project manager: *Rose Koos*
Senior production supervisor: *Laura Fuller*
Designer: *K. Wayne Harms*
Cover/interior designer: *Mary Sailer*
Cover image: *Juergen Berger, Max-Planck Institute/Science Photo Researchers*
Senior photo research coordinator: *Lori Hancock*
Photo research: *Connie Mueller*
Supplement producer: *Brenda A. Ernzen*
Media technology producer: *Lori Welsh*
Compositor: *GTS Graphics, Inc.*
Typeface: *10/12 Times Roman*
Printer: *Quebecor World Hawkins,TN*

The cover photograph is a colorized electronmicrograph of a single *Salmonella* cell, trailing its long flagella behind it like streamers. This bacterium commonly resides in the intestine of cattle, poultry, and rodents, and is the second most common cause of food infection in the United States.

The credits section for this book begins on page C-1 and is considered an extension of the copyright page.

Library of Congress Cataloging-in-Publication Data

Talaro, Kathleen P.
        Foundations in microbiology / Kathleen Park Talaro, Arthur Talaro. — 4th ed.
            p.   cm.
        Includes index.
        ISBN 0–07–232042–7 — ISBN 0–07–112275–3 (ISE)
        1. Microbiology. 2. Medical microbiology. I. Talaro, Arthur. II. Title.

QR41.2 .T35   2002
579—dc21   2001045264
                                                                CIP

INTERNATIONAL EDITION ISBN 0–07–112275–3
Copyright © 2002. Exclusive rights by The McGraw-Hill Companies, Inc., for manufacture and export. This book cannot be re-exported from the country to which it is sold by McGraw-Hill. The International Edition is not available in North America.

This fourth edition is dedicated to two pioneering science textbooks that first introduced a new approach to the study of biology and microbiology: Life, *authored by George Gaylord Simpson, and* The Microbial World, *by Roger Stanier, Michael Doudoroff, and Edward Adelberg. These textbooks presented their subjects in a fashion that integrated structure, function, and diversity as a fascinating and thoroughly readable story of the living world. They both profoundly influenced the teaching and organization of these subjects for generations of budding scientists. We count ourselves among the students who were inspired by these textbooks to continue a lifelong pursuit of further learning and enlightenment.*

Brief Contents

Contents

CHAPTER **10**   Genetic Engineering: A Revolution in Molecular Biology   286

CHAPTER **11**   Physical and Chemical Control of Microbes   318

CHAPTER **12**   Drugs, Microbes, Host—The Elements of Chemotherapy   348

CHAPTER 13   Microbe–Human Interactions: Infection and Disease   380

CHAPTER 14   The Nature of Host Defenses   416

CHAPTER 15   The Acquisition of Specific Immunity and Its Applications   446

CHAPTER **19** The Gram-Positive Bacilli
of Medical Importance 574

CHAPTER **20** The Gram-Negative Bacilli
of Medical Importance 602

CHAPTER **21** Miscellaneous Bacterial Agents
of Disease 628

Preface

Perspectives on Microbiology

It has been nearly ten years since the first edition of this text was published, a decade marked by extensive discoveries and developments related to the science of microbiology. In fact, the total amount of information on this subject has doubled and possibly tripled during this relatively short time. Dealing with such an abundance of new information has, at times, been overwhelming. But this degree of enrichment has only served to reinforce the far-reaching importance of the subject matter. One has only to pick up a newspaper to be struck by daily reminders of microbiology's impact, whether it be emerging diseases, the roles of viruses in cancer, the development of new vaccines, drugs, and bioengineered organisms, or the use of microbes to clean up toxic wastes. Thanks to technologies that really originated with microbiologists, we now have detailed genetic maps of hundreds of microbes, plants, and animals, including humans. These discoveries, in turn, have spawned entirely new sciences and applications and an explosion of new discoveries. So, as we look back over these few years, one idea that rings even truer than ever is an observation made about 120 years ago by the renowned microbiologist Louis Pasteur:

> "Life would not long remain possible in the absence of microbes."

Looking ahead to the future, microbiology will continue to dominate biology, medicine, ecology, and industry for many years to come. Clearly, the more you learn about this subject, the better prepared you will be for personal and professional challenges, and to make decisions as a citizen of the world.

Emphasis of *Foundations in Microbiology*

The primary goals of this textbook are to:

- involve you in the relevance and excitement of microbiology.
- help you understand and appreciate the natural roles, structure, and functions of microorganisms.
- continue building your knowledge and facilitating your ability to apply the subject matter.
- encourage skills that make you a lifelong learner in the subject.

Microbiology is an inherently valuable and useful discipline that offers an intimate view of an invisible world. We have often felt that certain areas of the subject should be taught at the high school, junior high, and even elementary levels, so that knowledge of microbes and their importance becomes second nature from an early age. The orientation of this textbook continues to be a presentation that is understandable to students of diverse backgrounds. We hope to promote interest in this fascinating subject, and to share our sense of excitement and awe for it. We hope our involvement in the subject, our love of language, and our fun with analogies, models, and figures are so contagious that they stimulate your interest and catch your imagination.

Like all technical subjects, microbiology contains a wide array of facts and ideas that will become part of your growing body of knowledge. One of the ways to fulfill the goals of the textbook is to concentrate on understanding concepts—important, fundamental themes that form a framework for ideas and words. Most of these concepts are laid out like links in a chain of information, each leading you to the next level. As you continue to progress through the book, you can branch out into new areas, refine your knowledge, make important connections, and develop sophistication with the subject. Most chapters are structured with two levels of coverage—a general one that provides an overall big picture, and a more specific one that fills in the details of the topics.

The order and style of our presentation are similar to those of the previous edition, but as in past editions, we have included extensive revisions. We realize that many courses do not have extra time to cover every possible topic, and so we embarked on this edition with the goals of updating and simplifying content where necessary or possible, improving illustrations, streamlining and balancing the coverage, and editing for currency, accuracy, and clarity. We extensively updated figures and statistics, and introduced pertinent events and major discoveries of the past three years. We have added approximately 20 new figures and 50 new photographs, and have revised about half of the figures. Although the basic book plan is similar to that of the last edition, it has been redesigned with a new color scheme, chapter opening page, table structure, and boxed reading organization.

Despite the amount of new information being generated every year, we have aimed to cover both traditional and new developments in microbiology without adding to the length of the book. We have streamlined the disease chapters by removing some material on diagnosis and laboratory tests, and we have balanced the first chapter to emphasize the highly beneficial nature of microbes. In addition, we have added a series of "overview" statements at the opening of each chapter. These statements replace the outline, which is still available in the contents section. The chapter capsules have been converted to an outline format, with more concise summaries, and new questions have been added to most chapters.

A Note to Students

This book has been selected as part of a course that will prepare you for a career in the health or natural sciences. The information it contains is highly technical and provides a foundation for the practical, hands-on work and critical thinking that are an integral part of many science-based professions. You will need to understand concepts such as cell structure, physiology, disinfection, drug actions, genetics, pathogenesis, transmission of diseases, and immunology, just to name a few. Like all science courses, this type of course will require prior preparation, background, and significant time for study. You will need to develop a working knowledge of terminology and definitions, and learn the "how and why" of many phenomena. Like all learning, the study of microbiology can be a lifelong discovery experience that makes you a well-informed person who can differentiate fact from fiction and make well-reasoned interpretations and decisions.

FACTS ABOUT LEARNING STYLES

We assimilate information in several ways, including visual, auditory, or some combination of these. According to William Glasser, the retention of information can be quantified as follows:

> We remember about
> > 10% of what we read,
> > 20% of what we hear,
> > 30% of what we see,
> > 50% of what we see and hear,
> > 70% of what is discussed with others,
> > 80% of what we experience personally,
> > 95% of what we teach to someone else.

With this background in mind, what are some of the ways you can maximize your learning? First, you will want to develop consistent study habits, preferably having some contact with the material every day. Many students highlight key portions in a chapter as they read, but such passive activity may use up valuable time and energy without involving your emotions. You will retain far more information if you engage your mind with the words and ideas. This might include writing marginal notes to yourself, questioning yourself on understanding, and outlining only the most significant points as you read.

Another strategy for active learning is to write questions and answers on index cards to use as a portable review and self-quiz. The benefits of this are twofold: first, it uses muscular activity (writing) and second, it requires you to think about the material.

Making models is another valuable technique for setting down memories. This could include making "mental maps" or flow diagrams of how various ideas interrelate, or the order of steps in a process.

Since the highest levels of learning occur within a group setting, it is highly desirable to collaborate in study groups or with a tutor. Be aware that teaching uses all of the sensory and motor parts of the brain, which is why it is the most effective pathway to learning. In the group setting, you can take the role of a teacher by asking questions, explaining ideas, giving definitions, and drawing diagrams.

Another factor that contributes to successful study is the realization that the brain is not a tireless sponge that can "soak up" information without rest. We now know that a chemical messenger in the part of the brain that regulates memory must be regenerated about every 30–45 minutes. Any information studied when the messenger is inactive will not be placed into memory. This explains why trying to "cram" a lot of information in a long marathon of studying is relatively ineffective. The best learning takes place in short bursts with frequent breaks. Even if you have to study over a longer stretch, you should relax for a few moments, take a walk, or involve your mind in some activity that doesn't require intense thought. Spending an hour every day with flash cards is a far more effective way of learning than trying to absorb three chapters of material in a single marathon session.

RESOURCES

The text features several resources to help you in your studies.

Vocabulary, Glossary, and Index
The study of microbiology will immerse you in a rich source of terminology. No one expects the beginner to learn all of these new terms immediately, but an enhanced vocabulary will certainly be essential to understand, speak, and write this new language. To assist you in building vocabulary, the principal terms appear in boldface or italics and are defined or used in context. For terms marked by an asterisk, pronunciation and derivation information is given in a footnote at the bottom of the page. As a rule, speaking a word will help you spell it, and learning its origin will help you understand its meanings and those of related words.

The glossary is expanded in this edition to include definitions of all the boldface and italicized terms used in the text. The index is also detailed enough to serve as a rapid locator of terms and subject matter.

Chapter Checkpoints and Chapter Capsules with Key Terms
Sometimes the amount of factual information in a chapter can make it difficult to see the "forest for the trees." A beneficial strategy at such times is to pause and review important points before continuing to the next topic. Throughout each chapter, we have included three to six brief summaries called *Chapter Checkpoints* that concisely state the most important ideas under a major heading, and provide you with a quick recap of what has been covered

to that point. Many instructors assign these as a guide for study and review.

At the end of each chapter, the major content of the chapter is condensed into short summaries called *Chapter Capsules with Key Terms*. These summaries take the form of an outline, with key terms placed in context with their associated topics. Capsules can be used as both a quick review and preview of the chapter.

The subject matter in this text is basic, but that doesn't mean it is simple, or that it is merely a review of information you have had in a prior biology course. Microbiology is, after all, a specialized area of biology with its own orientation and emphasis. There is more information presented here than can be covered in a single course, so be guided by your instructor's reading assignments and study guide.

Question Section

Each chapter concludes with an extensive question section intended to guide and supplement your study and self-testing. The number and types of questions are diverse so that your instructor can assign questions for desired focus and emphasis. Due to space constraints, the text contains answers only to multiple-choice and selected matching questions (see appendix E). The *multiple-choice* type of objective question is commonly used in class testing and standardized exams, and is a quick way to assess your grasp of chapter content. *Matching* questions have a list of words and a list of numbered descriptions that are meant to correlate. The *concept* questions direct you to review the chapter by composing complete answers that cover essential topics and use correct terminology. *Critical-thinking* questions challenge you to use scientific thinking, analysis, and problem solving. They require that you find relationships, suggest plausible explanations, and apply these concepts to real-world situations. By their nature, most of these questions allow more than one interpretation and do not have a predetermined, "correct" answer.

FINAL NOTE OF ENCOURAGEMENT

One of life's little truths is that you get out of any endeavor what you put into it. Therefore, the more time you spend in serious study, the more you will learn. This will lead to a pride in mastery, greater skill in discovery, and the thrill of learning that is almost like being a microbiological detective!

Acknowledgments

A textbook is a collaboration that takes on a life of its own. No single person can take full credit for its final form. The one thing that we all agree upon, whether author, reviewer, or editor, is that we want it to be the best possible microbiology book we can create. The authors have been fortunate to have an exceptional editing and production team from McGraw-Hill for this edition. The person most responsible for keeping us on track and focused on our goals is Jean Sims Fornango, our Developmental Editor. Her contributions run the gamut from careful synopsis of the book pedagogy and detailed proofreading, to soothing our concerns and twisting our arms. She meets every challenge with good humor, insight, and professionalism, and we feel truly fortunate to have been in her capable hands. We also value our relationship with our publisher, James M. Smith, for his can-do attitude and support for meaningful and long overdue additions and improvements to the book. We also enjoyed collaborating with the able production team, including Rose Koos, Lori Hancock, Wayne Harms, and Connie Mueller.

Valuable support has also come from reviewers who shared their expertise in several specialized areas of microbiology. We would like to express sincere appreciation to Robert White, Lundy Pentz, Harry Kesler, Leland Pierson, Hugh Pross, and Valeria Howard for their detailed analyses of the chapters on chemistry, metabolism, genetics, drug therapy, and immunology. Many thanks also to Louis Giacinti, Jackie Butler, and Joseph Jaworski for their valued contributions and suggestions for improving several chapters. We would like to thank our many student readers and instructors from around the country for their kind and informative e-mails. You are the unsung heroes of textbook publishing.

We value the support and feedback from colleagues and students at Pasadena City College. In particular, we would like to recognize Barry Chess, a good friend and a talented microbiologist who can navigate his students through the most challenging areas of the subject with ease and humor; and Terry Pavlovitch, an able and creative biologist, who shares her love of teaching with us. We also owe a debt of gratitude to Mary Timmer, our proficient lab technician, whose attendance to the demands of a very busy microbiology laboratory have freed us to devote time to writing and conceptualizing illustrations. Over the past 30 years, countless fine students here at Pasadena City College have literally served as the "test lab" for shaping and refining the content of the book. It has been a wonderful side effect of teaching microbiology to watch our students grow and become friends and associates. Abigail Bernstein deserves special mention. She has been by Kathy's side as a tutor, lab assistant, and friend, and is a budding microbiologist. Abigail has the difficult job of being the "front" woman who works tirelessly answering questions and helping students use the book.

It takes about a year and a half to complete a textbook revision, during which time the manuscript is edited, reedited, and then edited again. All alterations are carefully spell-checked and proofread by the author, editors, and a number of others from the production staff. The figures are scrutinized for accuracy in labeling and composition. Unfortunately, even in these days of computerized cross-checks, some errors can still slip through. We appreciate knowing about errors you detect or critiques you may have regarding text content, figures, and boxed material, and encourage you to share any ideas you have for changes and improvements. We can be reached through the McGraw-Hill Company (www.mcgraw-hill.com) or by e-mail at ktalaro@aol.com.

We have enjoyed a superb team of reviewers for the fourth edition who were both formative and informative members of the team. They have been a significant source of suggestions about content, order, depth, organization, and readability. So, too, have they lent their microscopic precision for screening the accuracy and soundness of the science. They have been there for us for nearly 18 years, keeping us on our toes and contributing in hundreds of ways to this ongoing project. We couldn't do it without them.

REVIEWERS

Fourth Edition
Kevin Anderson, *Mississippi State University*
Cheryl K. Blake, *Indian Hills Community College*
Bruce Bleakley, *South Dakota State University*
Harold Bounds, *University of Louisiana*
Brenda Breeding, *Oklahoma City Community College*
Karen Buhrer, *Tidewater Community College*
Charles Denny, *University of South Carolina*
Richard Fass, *Ohio State University*
Denise Friedman, *Hudson Valley Community College*
Bernard Frye, *University of Texas*
Louis Giacinti, *Milwaukee Area Technical College*
Ted Gsell, *University of Montana*
Herschel Hanks, *Collin County Community College*
Ann Heise, *Washtenaw Community College*
Valeria Howard, *Bismarck State College*
Harold Kessler, *Lorain County Community College*
George Lukasic, *University of Florida*
Sarah MacIntire, *Texas Women's University*
Lundy Pentz, *Mary Baldwin College*
Hugh Pross, *Queen's University*
Leland Pierson, III, *University of Arizona*
Ken Slater, *Utah Valley State College*
Edward Simon, *Purdue University*
Robert A. Smith, *University of the Sciences–Philadelphia*
Kristine M. Snow, *Fox Valley Technical College*
Cynthia V. Sommer, *University of Wisconsin–Milwaukee*
Linda Harris Young, *Motlow State Community College*
Robert White, *Dalhousie University*

Second/Third Editions
Rodney P. Anderson, *Ohio Northern University*
Robert W. Bauman, Jr., *Amarillo College*
Leon Benefield, *Abraham Baldwin Agricultural College*
Lois M. Bergquist, *Los Angeles Valley College*
L. I. Best, *Palm Beach Community College–Central Campus*
Bruce Bleakley, *South Dakota State University*
Kathleen A. Bobbitt, *Wagner College*
Jackie Butler, *Grayson County College*
R. David Bynum, *SUNY at Stony Brook*
David Campbell, *St. Louis Community College–Meramec*
Joan S. Carter, *Durham Technical Community College*
Barry Chess, *Pasadena City College*
John C. Clausz, *Carroll College*
Margaret Elaine Cox, *Bossier Parish Community College*
Kimberlee K. Crum, *Mesabi Community College*
Paul A. DeLange, *Kettering College of Medical Arts*
Michael W. Dennis, *Montana State University–Billings*

William G. Dolak, *Rock Valley College*
Robert F. Drake, *State Technical Institute at Memphis*
Mark F. Frana, *Salisbury State University*
Elizabeth B. Gargus, *Jefferson State Community College*
Larry Giullou, *Armstrong State College*
Safawo Gullo, *Abraham Baldwin Agricultural College*
Christine Hagelin, *Los Medanos College*
Geraldine C. Hall, *Elmira College*
Heather L. Hall, *Charles County Community College*
Theresa Hornstein, *Lake Superior College*
Anne C. Jayne, *University of San Francisco*
Patricia Hilliard Johnson, *Palm Beach Community College*
Patricia Klopfenstein, *Edison Community College*
Jacob W. Lam, *University of Massachusetts–Lowell*
James W. Lamb, *El Paso Community College*
Hubert Ling, *County College of Morris*
Andrew D. Lloyd, *Delaware State University*
Marlene McCall, *Community College of Allegheny County*
Joan H. McCune, *Idaho State University*
Gordon A. McFeters, *Montana State University*
Karen Mock, *Yavapai College*
Jacquelyn Murray, *Garden City Community College*
Robert A. Pollack, *Nassau Community College*
Judith A. Prask, *Montgomery College*
Leda Raptis, *Queen's University*
Carol Ann Rush, *La Roche College*
Andrew M. Scala, *Dutchess Community College*
Caren Shapiro, *D'Youville College*
Linda M. Sherwood, *Montana State University*
Lisa A. Shimeld, *Crafton Hills College*
Cynthia V. Sommer, *University of Wisconsin–Milwaukee*
Donald P. Stahly, *University of Iowa*
Terrence Trivett, *Pacific Union College*
Garri Tsibel, *Pasadena City College*
Leslie S. Uhazy, *Antelope Valley College*
Valerie Vander Vliet, *Lewis University*
Frank V. Veselovsky, *South Puget Sound Community College*
Katherine Whelchel, *Anoka-Ramsey Community College*
Vernon L. Wranosky, *Colby Community College*
Dorothy M. Wrigley, *Mankato State University*

First Edition
Shirley M. Bishel, *Rio Hondo College*
Dale DesLauriers, *Chaffey College*
Warren R. Erhardt, *Daytona Beach Community College*
Louis Giacinti, *Milwaukee Area Technical College*
John Lennox, *Penn State, Altoona Campus*
Glendon R. Miller, *Wichita State University*
Joel Ostroff, *Brevard Community College*
Nancy D. Rapoport, *Springfield Technical Community College*
Mary Lee Richeson, *Indiana University; Purdue University at Fort Wayne*
Donald H. Roush, *University of North Alabama*
Pat Starr, *Mt. Hood Community College*
Pamela Tabery, *Northampton Community College*

Guided Tour

Foundations for Success!

Everything you need to master microbiology—clear presentation of principles, strong links between principles and applications, great learning tools to tie it all together. Provides a greater understanding of the place of microbial populations in the scheme of life than ever before!

Overview

Each chapter opens with a vignette that states the relevance of the chapter focus and a bulleted list that outlines the main themes of the chapter.

From Atoms to Cells:
A Chemical Connection

CHAPTER 2

In laboratories all over the world, sophisticated technology is being developed for a wide variety of scientific applications. Refinements in molecular biology techniques now make it possible to routinely identify microorganisms, detect genetic disease, diagnose cancer, sequence the genes of organisms, break down toxic wastes, synthesize drugs and industrial products, and genetically engineer microorganisms, plants, and animals. A common thread that runs through new technologies and hundreds of traditional techniques is that, at some point, they involve chemicals and chemical reactions. In fact, if nearly any biological event is traced out to its ultimate explanation, it will invariably involve atoms, molecules, reactions, and bonding.

It is this relationship between the sciences that makes a background in chemistry necessary to biologists and microbiologists. Students with a basic chemistry background will enhance their understanding of and insight into microbial structure and function, metabolism, genetics, drug therapy, immune reactions, and infectious disease. This chapter has been organized to promote a working knowledge of atoms, molecules, bonding, solutions, pH, and biochemistry and to build foundations to later chapters. It concludes with an introduction to cells and a general comparison of procaryotic and eucaryotic cells as a preparation for chapters 4 and 5.

Chapter Overview

- The understanding of living cells and processes is enhanced by a knowledge of chemistry.
- The structure and function of all matter in the universe is based on atoms.
- Atoms have unique structures and properties that allow chemical reactions to occur.
- Atoms contain protons, neutrons, and electrons in combinations to form elements.
- Living things are composed of approximately 25 different elements.
- Elements interact to form bonds that result in molecules and compounds with different characteristics than the elements that form them.
- Atoms can show variations in charge and polarity.
- Atoms and molecules undergo chemical reactions such as oxidation/reduction, ionization, and dissolution.
- The properties of carbon have been critical in forming macromolecules of life such as proteins, fats, carbohydrates, and nucleic acids.
- The nature of macromolecule structure and shape dictates its functions.
- Cells carry out fundamental activities of life, such as growth,

A molecular probe machine, called an ion microprobe, designed to analyze the isotopes found in very tiny samples of meteors, ancient rocks, and fossil samples. Chemists have used this device to determine the age of certain rocks found in Greenland (3.85 billion years old) and whether the sample may have come from a living thing.

metabolism, reproduction, synthesis, and transport, that are all essentially chemical reactions on a grand scale.

Atoms, Bonds, and Molecules: Fundamental Building Blocks

The universe is composed of an infinite variety of substances existing in the gaseous, liquid, and solid states. All such tangible materials that occupy space and have mass are called **matter**. The

26

Visual Learning

Extensively revised and updated art program.
Numerous overview figures help students master
important principles. Over 60 new photos.

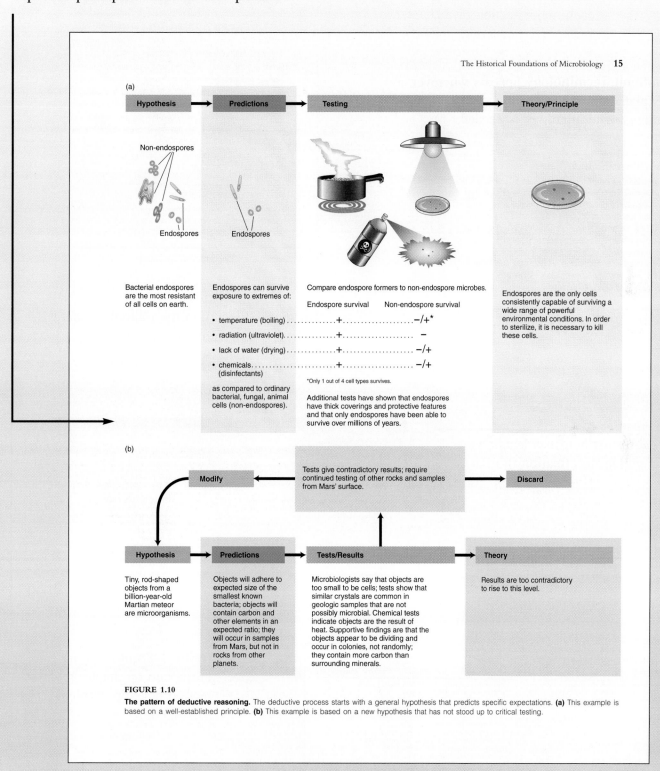

(a)

| Hypothesis | Predictions | Testing | Theory/Principle |

Non-endospores

Endospores

Endospores

Bacterial endospores are the most resistant of all cells on earth.

Endospores can survive exposure to extremes of:

Compare endospore formers to non-endospore microbes.

Endospore survival   Non-endospore survival

- temperature (boiling) . . . . +. . . . . −/+*
- radiation (ultraviolet). . . . . +. . . . . −
- lack of water (drying) . . . . +. . . . . −/+
- chemicals. . . . . +. . . . . −/+
  (disinfectants)

*Only 1 out of 4 cell types survives.

as compared to ordinary bacterial, fungal, animal cells (non-endospores).

Additional tests have shown that endospores have thick coverings and protective features and that only endospores have been able to survive over millions of years.

Endospores are the only cells consistently capable of surviving a wide range of powerful environmental conditions. In order to sterilize, it is necessary to kill these cells.

(b)

| Modify | Tests give contradictory results; require continued testing of other rocks and samples from Mars' surface. | Discard |

| Hypothesis | Predictions | Tests/Results | Theory |

Tiny, rod-shaped objects from a billion-year-old Martian meteor are microorganisms.

Objects will adhere to expected size of the smallest known bacteria; objects will contain carbon and other elements in an expected ratio; they will occur in samples from Mars, but not in rocks from other planets.

Microbiologists say that objects are too small to be cells; tests show that similar crystals are common in geologic samples that are not possibly microbial. Chemical tests indicate objects are the result of heat. Supportive findings are that the objects appear to be dividing and occur in colonies, not randomly; they contain more carbon than surrounding minerals.

Results are too contradictory to rise to this level.

FIGURE 1.10

The pattern of deductive reasoning. The deductive process starts with a general hypothesis that predicts specific expectations. **(a)** This example is based on a well-established principle. **(b)** This example is based on a new hypothesis that has not stood up to critical testing.

Student-Friendly Learning Tools

Chapter Checkpoints highlight the main themes of each major section of a chapter.

CHAPTER CHECKPOINTS

Metabolism includes all the biochemical reactions that occur in the cell. It is a self-regulating complex of interdependent processes that encompasses many thousands of chemical reactions.

Anabolism is the energy-requiring subset of metabolic reactions, which synthesize large molecules from smaller ones.

Catabolism is the energy-releasing subset of metabolic reactions, which degrade or break down large molecules into smaller ones.

Enzymes are proteins that catalyze all biochemical reactions by forming enzyme-substrate complexes. The binding of the substrate by an enzyme makes possible both bond-forming and bond-breaking reactions, depending on the pathway involved. Enzymes may utilize cofactors as carriers and activators.

Enzymes are classified and named according to the kinds of reactions they catalyze.

To function effectively, enzymes require specific conditions of temperature, pH, and osmotic pressure.

Enzyme activity is regulated by processes of feedback inhibition, induction, and repression, which, in turn, respond to availability of substrate and concentration of end products, as well as to other environmental factors.

A Note on Terminology appears wherever an explanation of the variations or meanings of terminology is needed.

A Note on Terminology

The word spore can have more than one usage in microbiology. It is a generic term that refers to any tiny compact cells that are produced by vegetative or reproductive structures of microorganisms. Spores can be quite variable in origin, form, and function. The bacterial type discussed here is called an endospore, because it is produced inside a cell. It functions in *survival,* not in reproduction, because no increase in cell numbers is involved in its formation. In contrast, the fungi produce many different types of spores for both survival and reproduction (see Chapter 5).

*sporangium (spor-anj'-yum) L. *sporos,* and Gr. *angeion,* vessel.

Running Glossary in the footnotes assures student understanding of terminology.

Chapter Capsule in an outline format helps students review the most important information in each chapter.

282   CHAPTER 9 Microbial Genetics

CHAPTER CAPSULE WITH KEY TERMS

I. Genes and the Genetic Material
A. **Genetics** is the study of **heredity,** and the **genome** is the sum total of genetic material of a cell.
B. A **chromosome** is composed of DNA in all organisms; **genes** are specific segments of the elongate DNA molecule. Genes code for polypeptides and proteins that become enzymes, antibodies, or structures in the cell.

II. Gene Structure and Replication
A. A **gene** consists of DNA, a double helix formed from linked **nucleotides** composed of a **phosphate, deoxyribose sugar,** and a **nitrogen base—purine** or **pyrimidine.**
B. The backbone of the molecule is formed of **antiparallel** strands of repeating deoxyribose sugar-phosphate units that are linked together by the base-pairing of adenine with thymine and cytosine with guanine. The order of base pairs in DNA constitutes the genetic code. The very long DNA molecule must be highly coiled to fit into the cell.
C. Pairing ensures the accuracy of the copying of DNA synthesis or **replication.**
   1. Replication is **semiconservative** and requires enzymes such as **helicase, DNA polymerase, ligase,** and **gyrase.**
   2. These components in conjunction with the chromosome being duplicated constitute a **replicon.** The unzipped strands of DNA function as templates. Synthesis proceeds along two

transcript requires splicing to delete stretches that correspond to introns.

IV. The Genetics of Animal Viruses
A. Genomes of viruses can be linear or circular; segmented or not; made of double-stranded (ds) DNA, single-stranded (ss) DNA, ssRNA, or dsRNA.
B. In general, DNA viruses replicate in the nucleus, RNA viruses in the cytoplasm.
C. Retroviruses synthesize dsDNA from ssRNA.
D. The DNA of some viruses can be silently integrated into the host's genome. Integration by **oncogenic** viruses can lead to **transformation** of the host cell into an immortal cancerous cell.
E. RNA viruses have strand polarity (positive- or negative-sense genome) and double-strandedness.

V. Regulation of Genetic Function
A. Protein synthesis and metabolism are regulated by **gene induction** or **repression,** as controlled by an **operon.**
B. An operon is a DNA unit of **regulatory genes** (made up of **regulators, promoters,** and **operators**) that controls the expression of **structural genes** (which code for enzymes and structural peptides).
   1. **Inducible operons** such as the **lactose operon** are normally *off* but are turned *on* by a lactose inducer.

Critical-Thinking Questions in the end-of-chapter review section develop problem-solving skills.

CRITICAL-THINKING QUESTIONS

1. A simple test you can do to demonstrate the coiling of DNA in bacteria is to open a large elastic band, stretch it taut, and twist it. First it will form a loose helix, then a tighter helix, and finally, to relieve stress, it will twist back upon itself. Further twisting will result in a series of knotlike bodies; this is how bacterial DNA is condensed.

2. Knowing that retroviruses operate on the principle of reversing the direction of transcription from RNA to DNA, propose a drug that might possibly interfere with their replication.

3. Using the piece of DNA in concept question 14, show a deletion, an insertion, a substitution, and an inversion. Which ones are frameshift mutations? Are any of your mutations nonsense? Missense? (Use the universal code to determine this.)

4. Using figure 9.14 and table 9.5, go through the steps in mutation of a codon followed by its transcription and translation that will give the end result in silent, missense, and nonsense mutations.

5. Explain the principle of "wobble" and find four amino acids that are encoded by wobble bases (figure 9.14). Suggest some benefits of this phenomenon to microorganisms.

6. Suggest a reason for having only one strand of DNA serve as a source of useful genetic information. What could be some possible functions of the coding strand?

7. The enzymes required to carry out transcription and translation are themselves produced through these same processes. Speculate which may have come first in evolution—proteins or nucleic acids—and explain your choice.

Focus on the Big Picture

Special-interest essays expand students' horizons and understanding of a broad range of topics. Additional topics are available on the website. **Internet Search Topics** at the end of every chapter provide research and problem-solving opportunities.

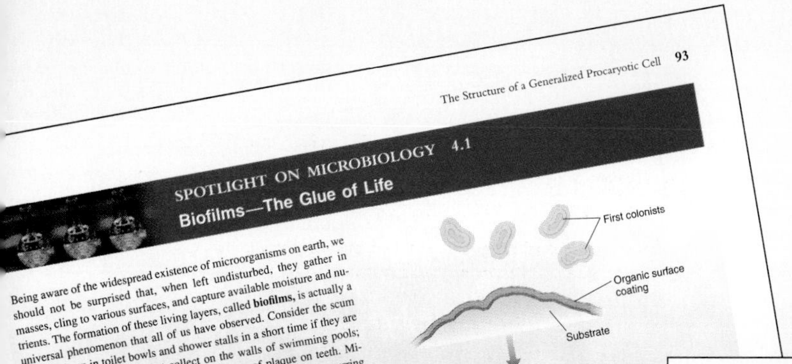

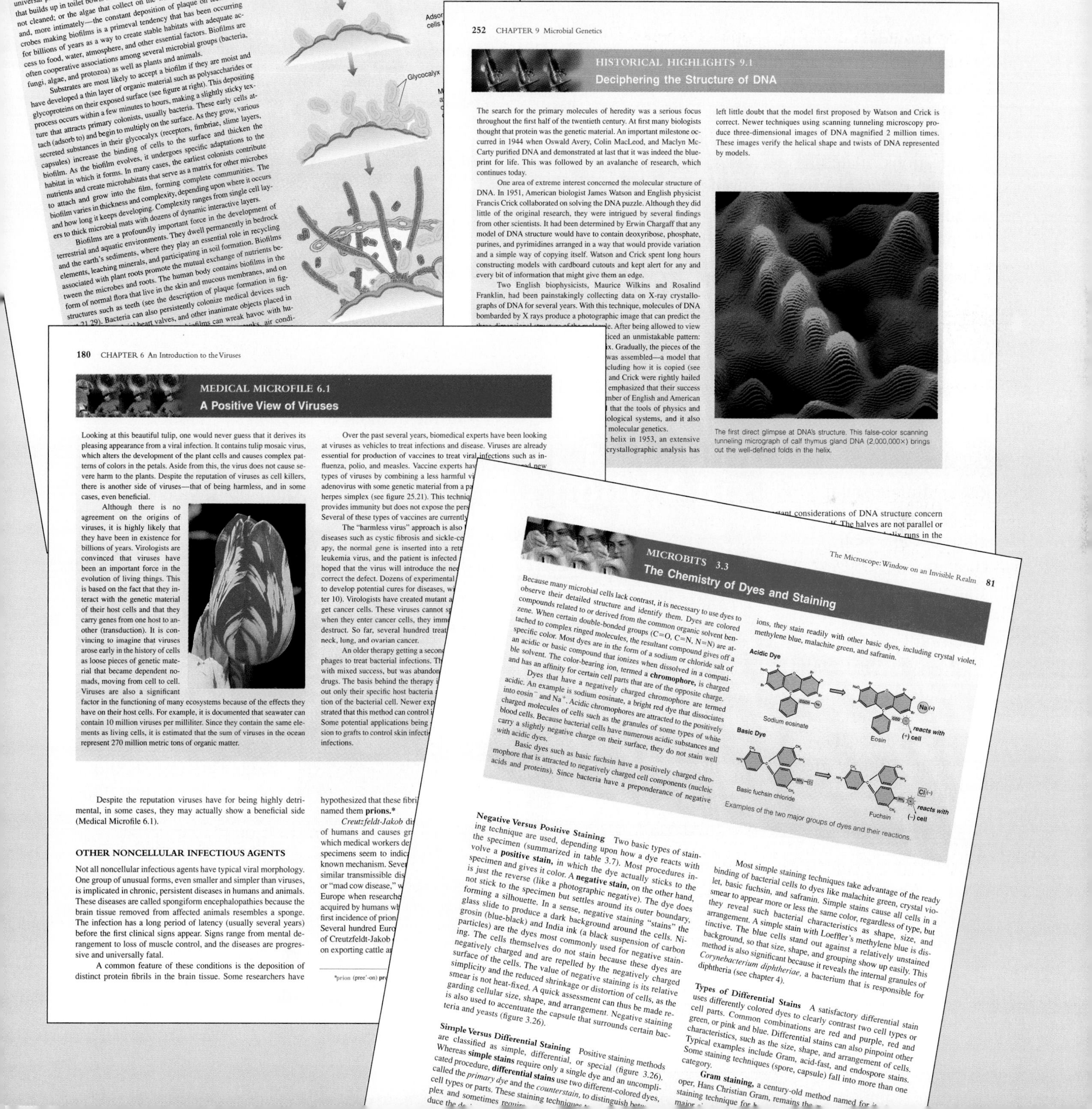

The Structure of a Generalized Procaryotic Cell **93**

SPOTLIGHT ON MICROBIOLOGY 4.1
Biofilms—The Glue of Life

Being aware of the widespread existence of microorganisms on earth, we should not be surprised that, when left undisturbed, they gather in masses, cling to various surfaces, and capture available moisture and nutrients. The formation of these living layers, called **biofilms**, is actually a universal phenomenon that all of us have observed. Consider the scum that builds up in toilet bowls and shower stalls in a short time if they are not cleaned; or the algae that collect on the walls of swimming pools; and, more intimately—the constant deposition of plaque on teeth. Microbes making biofilms is a way to create stable habitats with adequate access to food, water, atmosphere, and other essential factors. Biofilms are often cooperative associations among several microbial groups (bacteria, fungi, algae, and protozoa) as well as plants and animals.

Substrates are most likely to accept a biofilm if they are moist and have developed a thin layer of organic material such as polysaccharides or glycoproteins on their exposed surface (see figure at right). This depositing process occurs within a few minutes to hours, making a slightly sticky texture that attracts primary colonists, usually bacteria. These early cells attach (adsorb) to and begin to multiply on the surface. As they grow, various secreted substances in their glycocalyx (receptors, fimbriae, slime layers, capsules) increase the binding of cells to the surface and thicken the biofilm. As the biofilm evolves, it undergoes specific adaptations to the habitat in which it forms. In many cases, the earliest colonists contribute nutrients and create microhabitats that serve as a matrix for other microbes to attach and grow into the film, forming complete communities. The biofilm varies in thickness and complexity, depending upon where it occurs and how long it keeps developing. Complexity ranges from single cell layers to thick microbial mats with dozens of dynamic interactive layers.

Biofilms are a profoundly important force in the development of terrestrial and aquatic environments. They dwell permanently in bedrock and the earth's sediments, where they play an essential role in recycling elements, leaching minerals, and participating in soil formation. Biofilms associated with plant roots promote the mutual exchange of nutrients between the microbes and roots. The human body contains biofilms in the form of normal flora that live in the skin and mucous membranes, and on structures such as teeth (see the description of plaque formation in figures 21.29). Bacteria can also persistently colonize medical devices placed in the body and heart valves. Biofilms can wreak havoc with hu-

First colonists

Organic surface coating

Substrate

Adsorb cells

Glycocalyx

---

252 CHAPTER 9 Microbial Genetics

HISTORICAL HIGHLIGHTS 9.1
Deciphering the Structure of DNA

The search for the primary molecules of heredity was a serious focus throughout the first half of the twentieth century. At first many biologists thought that protein was the genetic material. An important milestone occurred in 1944 when Oswald Avery, Colin MacLeod, and Maclyn McCarty purified DNA and demonstrated at last that it was indeed the blueprint for life. This was followed by an avalanche of research, which continues today.

One area of extreme interest concerned the molecular structure of DNA. In 1951, American biologist James Watson and English physicist Francis Crick collaborated on solving the DNA puzzle. Although they did little of the original research, they were intrigued by several findings from other scientists. It had been determined by Erwin Chargaff that any model of DNA structure would have to contain deoxyribose, phosphate, purines, and pyrimidines arranged in a way that would provide variation and a simple way of copying itself. Watson and Crick spent long hours constructing models with cardboard cutouts and kept alert for any and every bit of information that might give them an edge.

Two English biophysicists, Maurice Wilkins and Rosalind Franklin, had been painstakingly collecting data on X-ray crystallographs of DNA for several years. With this technique, molecules of DNA bombarded by X rays produce a photographic image that can predict the three-dimensional structure of the molecule. After being allowed to view [...] noticed an unmistakable pattern: [...]ix. Gradually, the pieces of the [...] was assembled—a model that [...] including how it is copied (see [...] and Crick were rightly hailed [...] emphasized that their success [...] number of English and American [...] that the tools of physics and [...]logical systems, and it also [...] molecular genetics.

[...] helix in 1953, an extensive [...] crystallographic analysis has

left little doubt that the model first proposed by Watson and Crick is correct. Newer techniques using scanning tunneling microscopy produce three-dimensional images of DNA magnified 2 million times. These images verify the helical shape and twists of DNA represented by models.

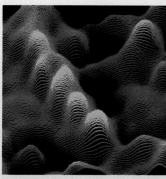

The first direct glimpse at DNA's structure. This false-color scanning tunneling micrograph of calf thymus gland DNA (2,000,000×) brings out the well-defined folds in the helix.

[...]ant considerations of DNA structure concern [...] The halves are not parallel or [...]lix runs in the

---

180 CHAPTER 6 An Introduction to the Viruses

MEDICAL MICROFILE 6.1
A Positive View of Viruses

Looking at this beautiful tulip, one would never guess that it derives its pleasing appearance from a viral infection. It contains tulip mosaic virus, which alters the development of the plant cells and causes complex patterns of colors in the petals. Aside from this, the virus does not cause severe harm to the plants. Despite the reputation of viruses as cell killers, there is another side of viruses—that of being harmless, and in some cases, even beneficial.

Although there is no agreement on the origins of viruses, it is highly likely that they have been in existence for billions of years. Virologists are convinced that viruses have been an important factor in the evolution of living things. This is based on the fact that they interact with the genetic material of their host cells and that they carry genes from one host to another (transduction). It is convincing to imagine that viruses arose early in the history of cells as loose pieces of genetic material that became dependent nomads, moving from cell to cell. Viruses are also a significant factor in the functioning of many ecosystems because of the effects they have on their host cells. For example, it is documented that seawater can contain 10 million viruses per milliliter. Since they contain the same elements as living cells, it is estimated that the sum of viruses in the ocean represent 270 million metric tons of organic matter.

Over the past several years, biomedical experts have been looking at viruses as vehicles to treat infections and disease. Viruses are already essential for production of vaccines to treat viral infections such as influenza, polio, and measles. Vaccine experts have [...] new types of viruses by combining a less harmful vi[...] adenovirus with some genetic material from a pa[...] herpes simplex (see figure 25.21). This techniq[...] provides immunity but does not expose the pers[...] Several of these types of vaccines are currently [...]

The "harmless virus" approach is also [...] diseases such as cystic fibrosis and sickle-ce[...] apy, the normal gene is inserted into a retr[...] leukemia virus, and the patient is infected [...] hoped that the virus will introduce the nee[...] correct the defect. Dozens of experimental [...] to develop potential cures for diseases, w[...] ter 10). Virologists have created mutant a[...] get cancer cells. These viruses cannot sp[...] when they enter cancer cells, they imm[...] destruct. So far, several hundred treat[...] neck, lung, and ovarian cancer.

An older therapy getting a secon[...] phages to treat bacterial infections. Th[...] with mixed success, but was abandon[...] drugs. The basis behind the therapy i[...] out only their specific host bacteria a[...] tion of the bacterial cell. Newer exp[...] strated that this method can control [...] Some potential applications being [...] sion to grafts to control skin infecti[...] infections.

Despite the reputation viruses have for being highly detrimental, in some cases, they may actually show a beneficial side (Medical Microfile 6.1).

OTHER NONCELLULAR INFECTIOUS AGENTS

Not all noncellular infectious agents have typical viral morphology. One group of unusual forms, even smaller and simpler than viruses, is implicated in chronic, persistent diseases in humans and animals. These diseases are called spongiform encephalopathies because the brain tissue removed from affected animals resembles a sponge. The infection has a long period of latency (usually several years) before the first clinical signs appear. Signs range from mental derangement to loss of muscle control, and the diseases are progressive and universally fatal.

A common feature of these conditions is the deposition of distinct protein fibrils in the brain tissue. Some researchers have

hypothesized that these fibri[...] named them **prions.** *

Creutzfeldt-Jakob di[...] of humans and causes gr[...] which medical workers de[...] specimens seem to indic[...] known mechanism. Sever[...] similar transmissible dis[...] or "mad cow disease," w[...] Europe when research[...] acquired by humans wh[...] first incidence of prion[...] Several hundred Euro[...] of Creutzfeldt-Jakob [...] on exporting cattle a[...]

*prion (pree´-on) pr[...]

The Microscope: Window on an Invisible Realm **81**

MICROBITS 3.3
The Chemistry of Dyes and Staining

Because many microbial cells lack contrast, it is necessary to use dyes to observe their detailed structure and identify them. Dyes are colored compounds related to or derived from the common organic solvent benzene. When certain double-bonded groups (C=O, C=N, N=N) are attached to complex ringed molecules, the resultant compound gives off a specific color. Most dyes are in the form of a sodium or chloride salt of an acidic or basic compound that ionizes when dissolved in a compatible solvent. The color-bearing ion, termed a **chromophore,** is charged and has an affinity for certain cell parts that are of the opposite charge.

Dyes that have a negatively charged chromophore are termed acidic. An example is sodium eosinate, a bright red dye that dissociates into eosin⁻ and Na⁺. Acidic chromophores are attracted to the positively charged molecules of cells such as the granules of some types of white blood cells. Because bacterial cells have numerous acidic substances and carry a slightly negative charge on their surface, they do not stain well with acidic dyes.

Basic dyes such as basic fuchsin have a positively charged chromophore that is attracted to negatively charged cell components (nucleic acids and proteins). Since bacteria have a preponderance of negative

ions, they stain readily with other basic dyes, including crystal violet, methylene blue, malachite green, and safranin.

Acidic Dye

Sodium eosinate → Eosin reacts with (−) cell

Basic Dye

Basic fuchsin chloride → Fuchsin reacts with (−) cell

Examples of the two major groups of dyes and their reactions.

Negative Versus Positive Staining
Two basic types of staining technique are used, depending upon how a dye reacts with the specimen (summarized in table 3.7). Most procedures involve a **positive stain,** in which the dye actually sticks to the specimen and gives it color. A **negative stain,** on the other hand, is just the reverse (like a photographic negative). The dye does not stick to the specimen but settles around its outer boundary, forming a silhouette. In a sense, negative staining "stains" the glass slide to produce a dark background around the cells. Nigrosin (blue-black) and India ink (a black suspension of carbon particles) are the dyes most commonly used for negative staining. The cells themselves do not stain because these dyes are negatively charged and are repelled by the negatively charged surface of the cells. The value of negative staining is its relative simplicity and the reduced shrinkage or distortion of cells, as the smear is not heat-fixed. A quick assessment can thus be made regarding cellular size, shape, and arrangement. Negative staining is also used to accentuate the capsule that surrounds certain bacteria and yeasts.

Simple Versus Differential Staining
Positive staining methods are classified as simple, differential, or special (figure 3.26). Whereas **simple stains** require only a single dye and an uncomplicated procedure, **differential stains** use two different-colored dyes, called the *primary dye* and the *counterstain,* to distinguish between cell types or parts. These staining techniques tend to be more complex and sometimes require additional steps, including the [...] duce the de[...]

Most simple staining techniques take advantage of the ready binding of bacterial cells to dyes like malachite green, crystal violet, basic fuchsin, and safranin. Simple stains cause all cells in a smear to appear more or less the same color, regardless of type, but they reveal such bacterial characteristics as shape, size, and arrangement. A simple stain with Loeffler's methylene blue is distinctive. The blue cells stand out against a relatively unstained background, so that size, shape, and grouping show up easily. This method is also significant because it reveals the internal granules of *Corynebacterium diphtheriae,* a bacterium that is responsible for diphtheria (see chapter 4).

Types of Differential Stains
A satisfactory differential stain uses differently colored dyes to clearly contrast two cell types or cell parts. Common combinations are red and purple, red and green, or pink and blue. Differential stains can also pinpoint other characteristics, such as the size, shape, and arrangement of cells. Typical examples include Gram, acid-fast, and endospore stains. Some staining techniques (spore, capsule) fall into more than one category.

Gram staining, a century-old method named for its developer, Hans Christian Gram, remains the [...] staining technique for [...] major [...]

Supplements

Study Guide

The study guide to accompany *Foundations in Microbiology*, 4e, was prepared by Jackie Butler, *Grayson County College,* Dennison, TX.

The guide provides:
- study objectives and chapter overviews
- test-taking strategies
- crossword puzzles
- multiple-choice questions, critical-thinking questions, matching exercises, and pathway mapping problems to reinforce the concepts in each section
- answers to the objective questions

Multimedia

Microbes in Motion 3 CD-ROM   **Free with the text**
Interactive, easy-to-use general microbiology CD-ROM helps students explore and understand microbial structure and function through audio, video, animations, illustrations, and text. The CD-ROM is appropriate for any microbiology course. CD-ROM icons throughout the book direct the student to text-related material on the CD-ROM. The CD-ROM is compatible with both Windows and Macintosh systems.

Online Learning Center (Student Resources)
Passcard is **Free** with the text.
This online resource provides student access to interactive study tools, including terminology flash cards, interactive quizzes, web links to related topics, supplemental readings, and more.

HyperClinic 2 CD-ROM
From the authors of *Microbes in Motion*. Students evaluate realistic case studies that include patient histories and descriptions of signs and symptoms. Animations, videos, and interactive exercises explore all the avenues of clinical microbiology. Allied health students may analyze the results of physician-ordered clinical tests to reach a diagnosis. Medical students can evaluate a case study scenario, and then decide which clinical samples should be taken and which diagnostic test should be run. More than 200 pathogens are profiled, 105 case studies presented, and 46 diagnostic tests covered.

Instructor Resources

A full complement of instructor resources includes:

- 450 four-color transparencies
- Visual Resource Library CD-ROM with jpg files suitable for use with PowerPoint.
- computerized test-bank CD-ROM compatible with either Macintosh or PC systems.
- Online Learning Center (Instructor Resources) with correlation guides for the text and media, integration guide to coordinate organism topics with basic principle topics, and Online Learning Center material compatible with either Web CT or Blackboard.

Visit www.mhhe.com/talaro4

Foundations in
Microbiology

FOURTH EDITION

CHAPTER 1

The Main Themes of Microbiology

The earth is an amazingly nurturing environment for microorganisms. Altshough it has been fancifully nicknamed the "blue planet" or the "water planet," the earth is truly the "planet of the microbes." They are the dominant organisms living in most natural environments, and they are woven tightly through the cycles of all living things. Because of this fact, microbiologists are accustomed to finding extraordinary microbes in unusual places. In the fall of 2000, scientists from Pennsylvania unearthed a microbe that made even the most seasoned microbiologists take notice. They were able to isolate and grow a living bacterium that had been lying dormant and protected in a salt crystal for about 250 million years. This creature was alive even before the time of the dinosaurs, during the Permian period of geologic time. Its source was deep in an underground cavern near Carlsbad, New Mexico. The microbiologists on the project dated it by nearby fossils, and are now taking a close look at its characteristics and genetics. One secret to its longevity is that this bacterium, like its modern relatives, forms very resistant endospores. This finding has prompted a dramatic revision in our ideas about the nature of life and longevity.

Clearly, microorganisms pervade our lives in both an everyday, mundane sense and in a far wider view. We wash our clothes with detergents containing microbe-produced enzymes, eat food that derives flavor from microbial action, and, in many cases, even eat microorganisms themselves. We are vaccinated with altered microbes to prevent diseases that are caused by those very same microbes. We treat various medical conditions with drugs produced by microbes; we dust our plants with insecticides of microbial origin; and we use microorganisms as tiny factories to churn out various industrial chemicals and plastics. We depend upon microbes for many facets of life—one might say even for life itself.

No one can emerge from a microbiology course without a changed view of the world and of themselves.

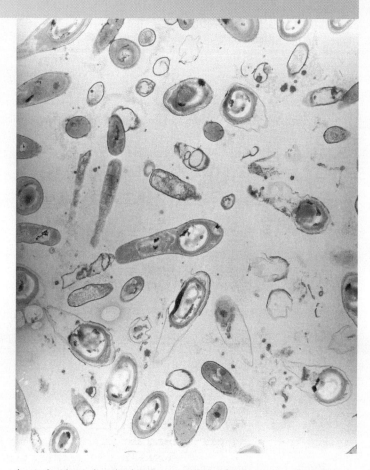

An ancient bacterium that has been awakened from a quarter of a billion years' sleep. What secrets does it have to share?

Chapter Overview

- Microorganisms, also called microbes, are organisms that require a microscope to be readily observed.
- In terms of numbers and range of distribution, microbes are the dominant organisms on earth.
- Major groups of microorganisms include bacteria, algae, protozoa, fungi, parasitic worms, and viruses.
- Microbiology involves study in numerous areas involving cell structure, function, genetics, immunology, biochemistry, epidemiology, and ecology.
- Microorganisms have developed complex interactions with other organisms and the environment.
- Microorganisms are essential to the operation of the earth's ecosystems, as photosynthesizers, decomposers, and recyclers.
- Humans use the versatility of microbes to make improvements in industrial production, agriculture, medicine, and environmental protection.
- The beneficial qualities of microbes are in contrast to the many infectious diseases they cause.
- Microbiologists use the scientific method to develop theories and explanations for microbial phenomena.
- The history of microbiology is marked by numerous significant discoveries and events in microscopy, culture techniques, and other methods of handling or controlling microbes.
- Microorganisms are the oldest organisms, having evolved over the 4 billion years of earth's history to the modern varieties we now observe.
- Microbes are classified into groups according to evolutionary relationships, provided with standard scientific names, and identified by specific characteristics.
- Microorganisms can be classified by means of general categories called domains and cell types (procaryotes and eucaryotes).

The Scope of Microbiology

Microbiology is a specialized area of **biology*** that deals with living things ordinarily too small to be seen without magnification. Such **microscopic*** organisms are collectively referred to as **microorganisms,*** **microbes,*** or several other terms, depending upon the purpose. Some people call them germs or bugs in reference to their role in infection and disease, but those terms have other biological meanings and perhaps place undue emphasis on the disagreeable reputation of microorganisms. Other terms that are encountered in our study are **bacteria, viruses, fungi, protozoa, algae,** and **helminths;** these microorganisms are the major biological groups that microbiologists study. The very nature of microorganisms makes them ideal subjects for study. They often are more accessible than **macroscopic*** organisms because of their relative simplicity, rapid reproduction, and adaptability, which is the capacity of a living thing to change its structure or function in order to adjust to its environment.

Microbiology is one of the largest and most complex of the biological sciences because it deals with many diverse biological disciplines. In addition to studying the natural history of microbes, it also deals with every aspect of microbe-human and microbe-environmental interactions. These interactions include genetics, metabolism, infection, disease, drug therapy, immunology, genetic engineering, industry, agriculture, and ecology. The subordinate branches that come under the large and expanding umbrella of microbiology are presented in table 1.1.

Microbiology has numerous practical uses in industry and medicine. Some prominent areas that are heavily based on applications in microbiology are as follows:

Immunology studies the system of body defenses that protects against infection. It includes *serology,* a discipline that looks for the products of immune reactions in the blood and tissues and aids in diagnosis of infectious diseases by that means, and *allergy,* the study of hypersensitive responses to ordinary, harmless materials (see chapters 14, 15, 16, and 17).

Public health microbiology and **epidemiology** aim to monitor and control the spread of diseases in communities. The principal U.S. and global institutions involved in this concern are the United States Public Health Service (USPHS) with its main agency, the Centers for Disease Control and Prevention (CDC) located in Atlanta, Georgia, and the World Health Organization (WHO), the medical limb of the United Nations (see chapter 13). The CDC collects information on disease from around the United States and publishes it in a weekly newsletter called the *Morbidity and Mortality Weekly Report.* (Visit www.cdc.gov/mmwr/ for the most current report.)

Food microbiology, dairy microbiology, and **aquatic microbiology** examine the ecological and practical roles of microbes in food and water (see chapter 26).

Agricultural microbiology is concerned with the relationships between microbes and crops, with an emphasis on improving yields and combating plant diseases.

Biotechnology includes any process in which humans use the metabolism of living things to arrive at a desired product, ranging from bread making to gene therapy (see chapters 10 and 26).

Industrial microbiology is concerned with the uses of microbes to produce or harvest large quantities of substances such as beer, vitamins, amino acids, drugs, and enzymes (see chapters 7 and 26).

Genetic engineering and **recombinant DNA technology** involve techniques that deliberately alter the genetic makeup of organisms to mass-produce human hormones and other drugs, create totally novel substances, and develop organisms with unique methods of synthesis and adaptation. This is the most powerful and rapidly growing area in modern microbiology (see chapter 10).

Each of the major disciplines in microbiology contains numerous subdivisions or specialties that in turn deal with a specific subject area or field. In fact, many areas of this science have become so specialized that it is not uncommon for a microbiologist to spend his or her whole life concentrating on a single group or type of microbe, biochemical process, or disease. On the other hand, rarely is one person a single type of microbiologist, and most can be classified in several ways. There are, for instance, bacterial physiologists who study industrial processes, molecular biologists who focus on the genetics of viruses, mycologists doing research on agricultural pests, epidemiologists who are also nurses, and dentists who specialize in the microbiology of gum disease.

Studies in microbiology have led to greater understanding of many theoretical biological principles. For example, the study of microorganisms established universal concepts concerning the chemistry of life (see chapters 2 and 8), systems of inheritance (see chapter 9), and the global cycles of nutrients, minerals, and gases (see chapter 26).

The Impact of Microbes on Earth: Small Organisms with a Giant Effect

The most important knowledge that should emerge from a microbiology course is the profound influence microorganisms have on all aspects of the earth and its residents (figure 1.1). For billions of years, microbes have extensively shaped the development of the earth's habitats and the evolution of other life forms. It is understandable that scientists searching for life on other planets first look for signs of microorganisms.

Microbes can be found nearly everywhere, from deep in the earth's crust, to the polar ice caps and oceans, to the bodies of plants and animals. Being mostly invisible, the actions of microorganisms are usually not as obvious or familiar as those of larger

---

***biology** Gr. *bios,* life, and *logos,* to study. The study of organisms.

***microscopic** (my″-kroh-skaw′-pik) Gr. *mikros,* small, and *scopein,* to see.

***microorganism** (my-kroh″-or′-gun-izm)

***microbe** (my′-krohb) Gr. *mikros,* small, and *bios,* life.

***macroscopic** (mak″-roh-skaw′-pik) Gr. *macros,* large, and *scopein,* to see. Visible with the naked eye.

TABLE 1.1

Branches of Microbiology

Science	Area of Study	Chapter Reference
Bacteriology	The bacteria—the smallest, simplest single-celled organisms	4
Mycology	The fungi, a group of organisms that includes both microscopic forms (molds and yeasts) and larger forms (mushrooms, puffballs)	5, 22
Protozoology	The protozoa—animal-like and mostly single-celled organisms	5, 23
Virology	Viruses—minute, noncellular particles that parasitize cells	6, 24, 25
Parasitology	Parasitism and parasitic organisms—traditionally including pathogenic protozoa, helminth worms, and certain insects	5, 23
Phycology or Algology	Simple aquatic organisms called algae, ranging from single-celled forms to large seaweeds	5
Microbial Morphology	The detailed structure of microorganisms	4, 5, 6
Microbial Physiology	Microbial function (metabolism) at the cellular and molecular levels	7, 8
Microbial Taxonomy	Classification, naming, and identification of microorganisms	1, 4, 5
Microbial Genetics, Molecular Biology	The function of genetic material and the biochemical reactions of cells involved in metabolism and growth	9, 10
Microbial Ecology	Interrelationships between microbes and the environment; the roles of microorganisms in the nutrient cycles of soil, water, and other natural communities	7, 26

FIGURE 1.1

The world of microbes. Whether on land, sea, air, or deep in the earth's crust, microbes sustain a living support network for all the earth's habitats.

plants and animals. They make up for their small size by occurring in large numbers and living in places that many other organisms cannot survive. Above all, they play central roles in the earth's landscape that are essential to life.

MICROBIAL INVOLVEMENT IN ENERGY AND NUTRIENT FLOW

Microbes are deeply involved in the flow of energy and food through the earth's ecosystems.[1] At the producer end of this range is **photosynthesis,** the formation of food using energy derived from the sun. Photosynthetic microorganisms, including algae and cyanobacteria, account for more than 50% of the earth's photosynthesis (figure 1.2*a*). In addition to serving as the basis for the food chains in the ocean and fresh water, these microorganisms also contribute the majority of oxygen to the atmosphere.

Another process that helps keep the earth in balance is the process of biological **decomposition** and nutrient recycling. Decomposition involves the breakdown of dead matter and wastes into simple compounds that can be directed back into the natural cycles of living things (figure 1.2*b*). If it were not for multitudes of bacteria and fungi, many chemical elements would become locked up and unavailable to organisms. In the long-term scheme of things, microorganisms are the main forces that drive the structure and content of the soil, water, and atmosphere.

- Decomposers play strategic and often very specific roles in the cycling of elements such as nitrogen, sulfur, phosphorus, and carbon between the living and nonliving environment. The very temperature of the earth is regulated by "greenhouse gases," such as carbon dioxide and methane, that create an insulation layer in the atmosphere and help retain heat. Much of this gas is produced by microbes living in the environment and the digestive tracts of animals.
- Recent estimates propose that, based on weight and numbers, up to 50% of all organisms exist within and beneath the earth's crust in sediments, rocks, and even volcanoes. These are mostly primitive microbes that can survive high temperatures and nutrient extremes. It is increasingly evident that this enormous underground community of microbes is a significant influence on weathering, mineral extraction, and soil formation (figure 1.2*c*).
- Microbes have also developed symbiotic[2] associations with other organisms that are highly beneficial to the participating members. Bacteria and fungi live in complex associations with plants that assist the plants in obtaining nutrients and water and may protect them against disease (figure 1.2*d*). Microbes form similar interrelationships with animals, notably represented by the stomach of cattle, which harbor a rich assortment of bacteria to digest the complex carbohydrates of the animals' diets. Other microbes become normal flora[3] that serve as barriers to infectious agents.

---

1. Ecosystems are any interactions that occur between living organisms and their environment.

2. Symbiosis is a partnership between organisms that benefits at least one of them.

3. Flora are microbes that normally inhabit the bodies of animals.

(a) Summer pond with a thick mat of algae—a rich photosynthetic community.

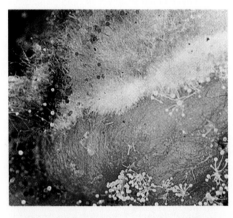

(b) A fruit being decomposed by a common soil fungus.

(c) Hydrothermal vents deep in the ocean present a hostile habitat that teems with unusual microorganisms.

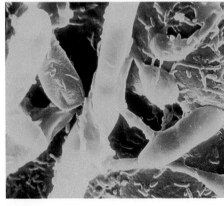

(d) A high-magnification view of plant roots reveals a clinging growth of fungi and bacteria.

FIGURE 1.2
Microbial habitats.

APPLICATIONS USING MICROORGANISMS: VERSATILE CHEMICAL MACHINES

It is clear that microorganisms have monumental importance to the earth's operation. It is this very same diversity and versatility that also makes them excellent candidates for solving human problems. By accident or choice, humans have been using microorganisms for thousands of years to improve life and even to mold civilizations. The use of microbes to create products is the science of biotechnology. For the most part, this technology relies on the chemical reactions of microorganisms[4] to produce many types of foods and manufactured materials through a process called fermentation[5] (figure 1.3a). Yeasts, a type of microscopic fungi, supply the necessary reactions to make bread, alcoholic beverages, and vitamins. Some specialized bacteria have unique capacities to ferment milk products, pickle foods, and even to mine precious metals from raw minerals (figure 1.3b). Microbes are also employed to synthesize drugs (antibiotics and hormones) and to mass-produce enzymes for industry and amino acids for health supplements (figure 1.3a).

Genetic engineering is a newer area of biotechnology that manipulates the genetics of microbes, plants, and animals for the purpose of creating new products and genetically modified organisms (figure 1.3c). One powerful technique for designing new organisms is termed **recombinant DNA.** This technology makes it possible to deliberately alter DNA[6] and to switch genetic material from one organism to another. Bacteria and fungi were some of the first microorganisms to be genetically engineered, because they are so adaptable to changes in their genetic makeup. Recombinant DNA technology has unlimited potential in terms of medical, industrial, and agricultural uses. Microbes can be engineered to synthesize desirable proteins such as drugs, hormones, enzymes, and physiological substances.

Among the genetically unique organisms that have been designed by bioengineers are bacteria that contain a natural pesticide, viruses that serve as vaccines, pigs that produce human hemoglobin, and plants that do not ripen too rapidly (figure 1.3d). The techniques also extend to the characterization of human genetic material and diseases.

Another way of tapping into the unlimited potential of microorganisms is the relatively new science of **bioremediation.*** This process involves the introduction of microbes into the environment to restore stability or to clean up toxic pollutants. Bioremediation is required to control the massive levels of pollution from industry and modern living. Microbes have a surprising capacity to break down chemicals that would be harmful to other organisms. Agencies and companies have developed microbes to handle oil spills and detoxify sites contaminated with heavy metals, pesticides, and other chemical wastes (figure 1.3e). The solid waste disposal industry is interested in developing methods for degrading the tons of garbage in landfills, especially human-made plastics and paper products. One form of bioremediation that has been in use for some

4. These chemical reactions are also called metabolism.

5. Large-scale processes in industry, using microbes as tiny factories.

6. DNA, or deoxyribonucleic acid, the chemical substance that comprises the genetic material of organisms.

* **bioremediation** (by'-oh-ree-mee-dee-ay"-shun) *bios,* life; *re,* again; *mederi,* to heal. The use of biological agents to remedy environmental problems.

(a) Microbes as synthesizers. A large complex fermentor manufactures drugs and enzymes using microbial metabolism.

(b) An aerial view of a copper mine looks like a giant quilt pattern. The colored patches are various stages of bacteria extracting metals from the ore.

(c) Workers in a clean biotechnology lab isolate genes for possible testing.

(d) Genetically engineered tomatoes have genes manipulated to slow ripening and increase flavor and nutritional content.

(e) A bioremediation platform placed in a river for the purpose of detoxifying the water containing industrial pollutants.

FIGURE 1.3
Microbes at work.

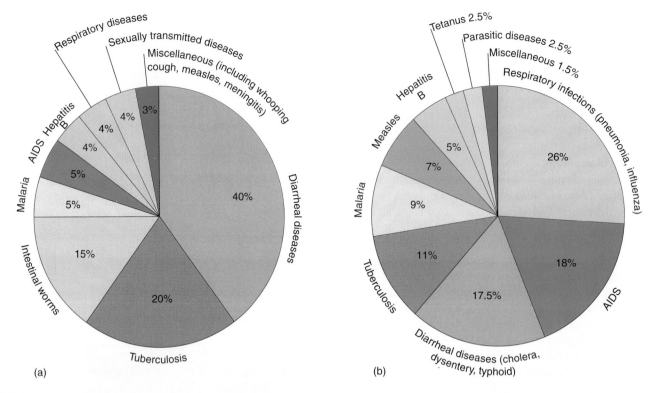

FIGURE 1.4

Global infectious disease statistics. Infectious disease statistics rank the major causes of **(a)** morbidity (rate of disease) and **(b)** mortality (deaths). A large number of diseases can be treated with drugs or prevented altogether with vaccination and improvements in health care and sanitation.

Source: World Health Organization, 1999; most recent data available.

time is the treatment of water and sewage. Since clean freshwater supplies are rapidly dwindling worldwide, it will become even more important to find ways to reclaim polluted water.

INFECTIOUS DISEASES AND THE HUMAN CONDITION

One of the most fascinating aspects of the microorganisms with which we share the earth is that, despite all of the benefits they provide, they also contribute significantly to human misery as **pathogens.*** Humanity is plagued by nearly 2,000 different microbes that can infect the human body and cause various types of disease. Infectious diseases still devastate human populations worldwide, despite significant strides in understanding and treating them. The most recent estimates from the World Health Organization (WHO) point to a total of 10 billion new infections across the world every year (figure 1.4a). There are more infections than people because many people acquire more than one infection. Infectious diseases are also the most common cause of death in much of humanity, and they still kill about one-third of the U.S. population. The worldwide death toll is about 13 million people, and many of these diseases are preventable by drugs and vaccines (figure 1.4b). Those hardest hit are residents in countries where access to adequate medical care is lacking. One-third of the earth's inhabitants live on less than $1 per day, are malnourished, and are not fully immunized.

Adding to the overload of infectious diseases, we are also witnessing an increase in the number of new (emerging) and older (reemerging) diseases (Spotlight on Microbiology 1.1). AIDS, hepatitis C, and viral encephalitis are examples of recently identified diseases that cause severe mortality and morbidity and are currently on the rise. To somewhat balance this trend, there have also been some advances in eradication of diseases such as polio, measles, leprosy, and certain parasitic worms. The WHO is currently on a global push to vaccinate children against the most common childhood diseases.

It is significant that, in addition to known infectious diseases, many other diseases are suspected of having a microbial origin. A connection has been established between certain cancers and viruses, between diabetes and the Coxsackie virus, and between schizophrenia and a virus called the borna agent. Diseases as disparate as multiple sclerosis, obsessive compulsive disorder, and coronary artery disease have been linked to chronic infections with microorganisms.

A further health complication from infectious diseases is the increasing number of patients with weakened defenses that are kept alive for extended periods. They are subject to infections by common microbes that are not pathogenic to healthy people. There is also an increase in microbes that are resistant to drugs. It appears that even with the most modern technology available to us, microbes still have the "last word," as the great French microbiologist Louis Pasteur observed.

---

*pathogens (path′-oh-jenz) Gr. *pathos*, disease, and *gennan*, to produce. Disease-causing agents.

SPOTLIGHT ON MICROBIOLOGY 1.1
Infectious Diseases in the Global Village

Eradicating infectious diseases and arresting their spread have long been goals of medical science. There is no doubt that advances in detection, treatment, and prevention have gradually reduced the numbers of such diseases, but most of these decreases have occurred in more developed countries. In less developed countries, however, infections still account for over 40% of deaths.

From the standpoint of infectious diseases, the earth's inhabitants serve as a collective incubator for old and new diseases. Newly identified diseases that are becoming more prominent are termed **emerging**. Table 1.A lists 20 of the most common emerging diseases that have been diagnosed over a span of 30 years. Some of them were associated with a specific geographic site (Ebola fever), whereas others were spread across all continents (AIDS). Older diseases that have been known for hundreds of years and are increasing in occurrence are termed **reemerging**. Among the diseases currently experiencing a resurgence are tuberculosis, influenza, malaria, cholera, and hepatitis B. Many factors play a part in emergence, but fundamental to all emerging diseases is the formidable capacity of microorganisms to respond and adapt to alterations in the individual, community, and environment.

What events in the world cause emerging diseases? Dr. David Satcher, director of the Centers for Disease Control and Prevention (CDC), states simply that "organisms changed and people changed." Among the most profound influences are disruptions in the human population, such as crowding or immigration. For an excellent example of this effect, we have only to look at AIDS, which began as a focus of infection in remote African villages and was transported out of the region through immigration and tourism. These factors caused the disease to spread rapidly, so that by 2001, it had become the second most common cause of death worldwide (see figure 1.4). Another population factor is an increase in the number of people who are susceptible to infections. Some countries have a high percentage of young people at risk. These children lack immunization or are malnourished, both of which increase their susceptibility to common childhood and diarrheal diseases.

A number of prominent emerging diseases are associated with changing methods in agriculture and technology. The mass production and packing of food increases the opportunity for large outbreaks, especially if foods are grown in fecally contaminated soils or are eaten raw or poorly cooked. In the past several years, dozens of food-borne outbreaks caused by emerging pathogens have occurred. Epidemics have been associated with the bacterium *Escherichia coli* 0157:H7 in fresh vegetables, fruits, and meats; hundreds of thousands of people are exposed to the cholera bacteria in seafood and to *Salmonella* bacteria in eggs and milk. Even municipal water supplies have spread water-borne protozoa such as *Cryptosporidium* and *Giardia* that slipped past the usual water treatment systems. An unusual protein-infectious agent (prion) was spread to consumers of beef in some parts of Europe, where it was associated with a disease known as *spongiform encephalopathy*.

Other influences on emergent diseases are fluctuations in ecology, climate, animal migration, and human travel. Warming in some regions has increased the spread of mosquitoes that carry dengue fever or encephalitis viruses. There is considerable concern about the migration of tropical diseases such as malaria and West Nile fever into northern climates.

The encroachment of humans into wild habitats has opened the way to *zoonotic** pathogens. A zoonosis is an infection indigenous to animals that can be transmitted to humans. It is important to realize that many microbes are not specific to their host, and many of them have mutated to become more virulent. A recent study found that 79% of emerging human pathogens originate from animals. Examples of host switching are the African outbreaks of monkeypox and Ebola fever (ostensibly from contact with wild monkeys), a Malaysian outbreak of the Nipah virus that is harbored in fruit bats, and hantavirus disease from exposure to rodents.

Increased personal freedom and opportunities for travel favor the rapid dispersal of microbes. A person may become infected and be home for several days before symptoms appear. Pathogens can literally be transmitted around the globe in a short time. Because of this potential, we can no longer separate the world into "them" and "us." Health authorities from every country must be constantly vigilant to prevent another crisis like AIDS and to keep common diseases in check through vaccination and medication.

*zoonotic (zoh-naw'-tik) Gr. *zoion*, animal, and *nosis*, disease.

TABLE 1.A
Prominent Emerging Diseases over a Span of 30 Years

Microbe	Source	Disease	Year of Emergence
Lassa fever virus	Rodents	Lassa fever	1969
Monkeypox virus	Monkeys, rodents	Monkeypox	1970
Marburg virus	Monkeys?	Hemorrhagic fever	1975
Ebola virus	Monkeys?	Hemorrhagic fever	1977
Legionella bacteria	Water	Legionnaire's disease	1977
Human T-cell virus	Humans	Lymphoma, leukemia	1980
Staphylococcus bacteria	Humans	Toxic shock syndrome	1981
Human immuno-deficiency virus	Humans	AIDS	1981
Escherichia coli bacteria	Cattle	Hemolytic uremic syndrome	1982
Borrelia bacteria	Ticks	Lyme disease	1982
Helicobacter bacteria	Humans	Peptic ulcers	1983
Hepatitis C virus	Humans	Hepatitis, liver diseases	1989
Hantavirus	Mice	Respiratory distress syndrome	1993
Hendra virus	Bats	Encephalitis	1994
Cryptosporidium protozoa	Water	Diarrhea, enteric disease	1994
Ehrlichia bacteria	Ticks	Arthritis, muscle pain	1995
New variant CJD prion	Cattle	Creutzfeldt-Jakob disease	1996
Nipah virus	Bats	Encephalitis, pneumonia	1998
Arenavirus	Woodrats	Respiratory distress	1999
West Nile virus	Mosquitoes	Encephalitis	1999

- Microorganisms are defined as "living organisms too small to be seen with the naked eye." Among the members of this huge group of organisms are bacteria, fungi, protozoa, algae, viruses, and parasitic worms.

- Microorganisms live nearly everywhere and have impact on many biological and physical activities on earth.

- There are many kinds of relationships between microorganisms and humans; most are beneficial, but some are harmful.

- The scope of microbiology is incredibly diverse. It includes basic microbial research, research on infectious diseases, study of prevention and treatment of disease, environmental functions of microorganisms, and industrial use of microorganisms for commercial, agricultural, and medical purposes.

- In the last 120 years, microbiologists have identified the causative agents for most of the infectious diseases. In addition, they have discovered distinct connections between microorganisms and diseases whose causes were previously unknown.

- Microorganisms: We have to learn to live with them because we cannot live without them.

The General Characteristics of Microorganisms

CELLULAR ORGANIZATION

Two basic cell lines have appeared during evolutionary history. These lines, termed **procaryotic*** **cells** and **eucaryotic*** **cells,** differ primarily in the complexity of their cell structure (figure 1.5*a*).

In general, procaryotic cells are smaller than eucaryotic cells, and they lack special structures such as a nucleus and **organelles.*** Organelles are small membrane-bound cell structures that perform specific functions in eucaryotic cells. These two cell types and the organisms that possess them (called procaryotes and eucaryotes) are covered in more detail in chapters 2, 4, and 5.

All procaryotes are microorganisms, but only some eucaryotes are microorganisms. The bodies of most microorganisms consist of either a single cell or just a few cells (figure 1.5*c, d*). Because of their role in disease, certain animals such as helminth worms and insects, many of which can be seen with the naked eye, are also considered in the study of microorganisms. Even in its seeming simplicity, the microscopic world is every bit as complex and diverse as the macroscopic one. There is no doubt that microorganisms also outnumber macroscopic organisms by a factor of several million.

---

*procaryotic (proh″-kar-ee-ah′-tik) Gr. *pro,* before, and *karyon,* nucleus.

*eucaryotic (yoo″-kar-ee-ah′-tik) Gr. *eu,* true or good, and *karyon,* nucleus. Sometimes spelled prokaryotic and eukaryotic.

*organelle (or′-gan-el″) L., little organ.

A NOTE ON VIRUSES

Viruses are subject to intense study by microbiologists. They are small particles that exist at a level of complexity somewhere between large molecules and cells (figure 1.5*b*). Viruses are much simpler than cells; they are composed essentially of a small amount of hereditary material wrapped up in a protein covering. Some biologists refer to viruses as parasitic particles; others consider them to be very primitive organisms. One thing is certain—they are highly dependent on a host cell's machinery for their activities.

MICROBIAL DIMENSIONS: HOW SMALL IS SMALL?

When we say that microbes are too small to be seen with the unaided eye, what sorts of dimensions are we talking about? This concept is best visualized by comparing microbial groups with the larger organisms of the macroscopic world and also with the molecules and atoms of the molecular world (figure 1.6). Whereas the dimensions of macroscopic organisms are usually given in centimeters (cm) and meters (m), those of most microorganisms fall within the range of micrometers (μm) and to a lesser extent, nanometers (nm) and millimeters (mm). The size range of most microbes extends from the smallest viruses, measuring around 20 nm and actually not much bigger than a large molecule, to protozoans measuring 3 to 4 mm and visible with the naked eye.

LIFE-STYLES OF MICROORGANISMS

The majority of microorganisms live a free existence in habitats such as soil and water, where they are relatively harmless and often beneficial. A free-living organism can derive all required foods and other factors directly from the nonliving environment. Some microorganisms require interactions with other organisms. One such group, termed **parasites,** are harbored and nourished by other living organisms, called **hosts.** A parasite's actions cause damage to its host through infection and disease. Most microbial parasites are some type of bacterium, fungus, protozoan, worm, or virus. Although parasites cause important diseases, they make up only a small proportion of microbes. As we shall see later in the chapter, a few microorganisms can exist on either free-living or parasitic levels.

- Excluding the viruses, there are two types of microorganisms: procaryotes, which are small and lack a nucleus and organelles, and eucaryotes, which are larger and have both a nucleus and organelles.

- Viruses are not cellular and are therefore called particles rather than organisms. They are included in microbiology because of their small size and close relationship with cells.

- Most microorganisms are measured in micrometers, with two exceptions. The helminths are measured in millimeters, and the viruses are measured in nanometers.

- Contrary to popular belief, most microorganisms are harmless, free-living species that perform vital functions in both the environment and larger organisms. Comparatively few species are agents of disease.

(a) Cell Types

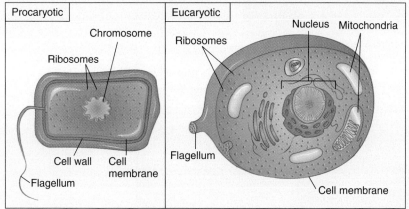

| Procaryotic | Eucaryotic |

Chromosome

Ribosomes

Ribosomes

Nucleus   Mitochondria

Cell wall   Cell membrane

Flagellum

Flagellum

Cell membrane

Microbial cells are of the small, relatively simple procaryotic variety (left) or the larger, more complex eucaryotic type (right).

(b) Virus Types

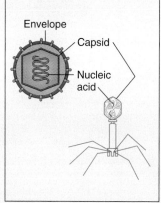

Envelope

Capsid

Nucleic acid

Viruses are tiny particles, not cells, that consist of genetic material surrounded by a protective covering. Shown here are a human virus (top) and bacterial virus (bottom).

(c) Examples of procaryotic organisms

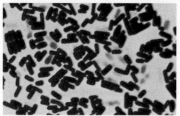

Rod-shaped bacteria, *Clostridium,* found in soil

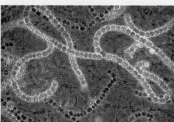

Nostoc, a cyanobacterium that lives in fresh water

(d) Examples of eucaryotic organisms

The stalked protozoan *Vorticella* is shown in feeding mode. These free-living eucaryotes are common in pond water.

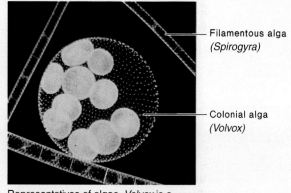

Filamentous alga (*Spirogyra*)

Colonial alga (*Volvox*)

Representatives of algae. *Volvox* is a large, complex colony composed of smaller colonies (spheres) and cells (dots). *Spirogyra* is a filamentous alga composed of elongate cells joined end to end.

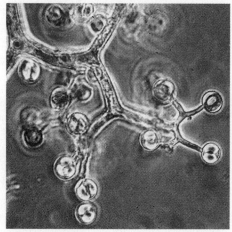

Example of a fungus; shown here is the mold *Thamnidium* displaying its sac-like reproductive vessels.

FIGURE 1.5

The organization of living things and viruses.

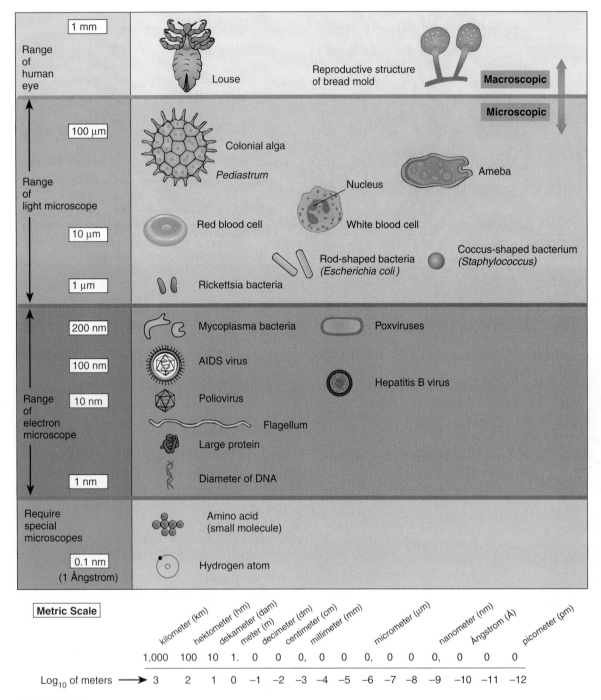

FIGURE 1.6

The size of things. Common measurements encountered in microbiology and a scale of comparison from the macroscopic to the microscopic, molecular, and atomic. Most microbes encountered in our studies will fall between 100 µm and 10 nm in overall dimensions. The microbes shown are more or less to scale within size zone but not between size zones.

The Historical Foundations of Microbiology

If not for the extensive interest, curiosity, and devotion of thousands of microbiologists over the last 300 years, we would know little about the microscopic realm that surrounds us. Many of the discoveries in this science have resulted from the prior work of men and women who toiled long hours in dimly lit laboratories with the crudest of tools. Each additional insight, whether large or small, has added to our current knowledge of living things and processes. This treatment of the early history of microbiology will summarize the prominent discoveries made in the past 300 years: microscopy, the rise of the scientific method, and the development of medical microbiology, including the germ theory and the origins of modern microbiological techniques. See table B.1 in appendix B, which summarizes some of the pivotal events in microbiology, from its earliest beginnings to the present. Additional historical vignettes are integrated throughout this text to emphasize the developmental stages of modern microbiology.

FIGURE 1.7

An oil painting of Antonie van Leeuwenhoek (1632–1723) sitting in his laboratory. J.R. Porter and C. Dobell have commented on the unique qualities Leeuwenhoek brought to his craft: "He was one of the most original and curious men who ever lived. It is difficult to compare him with anybody because he belonged to a genus of which he was the type and only species, and when he died his line became extinct."

THE DEVELOPMENT OF THE MICROSCOPE: "SEEING IS BELIEVING"

It is likely that from the very earliest history, humans noticed that when certain foods spoiled they became inedible or caused illness, and yet other spoiled foods did no harm and even had enhanced flavor. Indeed, several centuries ago, there was already a sense that diseases such as the black plague and smallpox were caused by some sort of transmissible matter. But the causes of such phenomena were vague and obscure because the technology to study them was lacking. Consequently, they remained cloaked in mystery and regarded with superstition—a trend that led even well-educated scientists to believe in spontaneous generation (Historical Highlights 1.2). True awareness of the widespread distribution of microorganisms and some of their characteristics was finally made possible by the development of the first microscopes. These devices revealed microbes as discrete entities sharing many of the characteristics of larger, visible plants and animals. Several early scientists fashioned magnifying lenses, but their microscopes lacked the optical clarity needed for examining bacteria and other small, single-celled organisms. The first careful and exacting observations awaited the clever single-lens microscope hand-fashioned by Antonie van Leeuwenhoek, a Dutch linen merchant and self-made microbiologist (figure 1.7). The original purpose of the microscopes was to examine cloth for flaws, but Leeuwenhoek turned them to other uses as well.

Leeuwenhoek's wide-ranging investigations included observations of tiny organisms he called *animalcules* (little animals), blood, and other human tissues (including his own tooth scrapings), insects, minerals, and plant materials. He constructed more than 250 small, powerful microscopes that could magnify up to 300 times (figure 1.8).

Considering that he had no formal training in science and that he was the first person ever to faithfully record this strange new world, his descriptions of bacteria and protozoa were astute and precise. Because of Leeuwenhoek's extraordinary contributions to microbiology, he is known as the father of bacteriology and protozoology.

From the time of Leeuwenhoek, microscopes evolved into more complex and improved instruments with the addition of refined lenses, a condenser, finer focusing devices, and built-in light sources. The prototype of the modern compound microscope, in use from about the mid-1800s, was capable of magnifications of 1,000 times or more. Even our modern laboratory microscopes are not greatly different in basic structure and function from those early microscopes. The technical characteristics of microscopes and microscopy are the major focus of chapter 3.

THE ESTABLISHMENT OF THE SCIENTIFIC METHOD

A serious impediment to the development of true scientific reasoning and testing was the tendency of early scientists to explain natural phenomena by a mixture of belief, superstition, and argument. The development of an experimental system that answered questions objectively and was not based on prejudice marked the beginning of true scientific thinking. These ideas gradually crept into the consciousness of the scientific community during the 1600s. The general approach taken by scientists to explain a certain natural phenomenon is called the **scientific method.** A primary aim of this method is to formulate a **hypothesis,** a tentative explanation to account for what has been observed or measured. A good hypothesis must be capable of being either supported or discredited by careful, systematic observation or experimentation. For example, the statement that "microorganisms cause diseases" can be experimentally determined by the tools of science, but the statement that "diseases are caused by evil spirits" cannot.

The two types of reasoning that are commonly applied separately or in combination to develop and support hypotheses are **induction** and **deduction.** In the **inductive approach,** a scientist first accumulates specific data or facts and then formulates a general hypothesis that accounts for those facts (figure 1.9). The inductive approach asks, "Are various observed events best explained by this hypothesis or by another one?" In the **deductive approach,** a scientist constructs a hypothesis, tests its validity by outlining particular events that are predicted by the hypothesis, and then performs experiments to test for those events (figure 1.10). The deductive process states: "If the hypothesis is valid, then certain specific events can be expected to occur."

Natural processes have numerous physical, chemical, and biological factors, or *variables,* that can hypothetically affect their outcome. To account for these variables, scientists design experiments: (1) to thoroughly test for or measure the consequences of each possible variable and (2) to accompany each variable with one or more *control groups.* A control group is designed exactly as the test group but omits only that variable being tested; thus, it may serve as a basis of comparison for the test group. The reasoning is that, if a certain experimental finding occurs only in the test group and not in the control group, the finding must be due to the variable being tested and not to some uncontrolled, untested factor that is not part of the hypothesis (figure 1.11).

For thousands of years, people believed that certain living things arose from vital forces present in nonliving or decomposing matter. This ancient belief, known as **spontaneous generation,** was continually reinforced as people observed that meat left out in the open soon "produced" maggots, that mushrooms appeared on rotting wood, that rats and mice emerged from piles of litter, and other similar phenomena. Though some of these early ideas seem quaint and ridiculous in light of modern knowledge, we must remember that, at the time, mysteries in life were accepted, and the scientific method was not widely practiced.

Even after single-celled organisms were discovered during the mid-1600s, the idea of spontaneous generation continued to exist. Some scientists assumed that microscopic beings were an early stage in the development of more complex ones.

Redi's Experiment

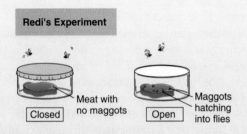

Meat with no maggots — Closed

Maggots hatching into flies — Open

Jablot's Experiment

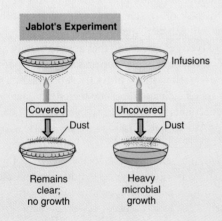

Infusions

Covered | Uncovered

Dust | Dust

Remains clear; no growth

Heavy microbial growth

Over the subsequent 200 years, scientists waged an experimental battle over the two hypotheses that could explain the origin of simple life forms. Some tenaciously clung to the idea of **abiogenesis,*** which embraced spontaneous generation. On the other side were advocates of **biogenesis** saying that living things arise only from others of their same kind. There were serious proponents on both sides, and each side put forth what appeared on the surface to be plausible explanations of why their evidence was more correct. Gradually, the abiogenesis hypothesis was abandoned, as convincing evidence for biogenesis continued to mount. The following series of experiments were among the most important in finally tipping the balance. Among the important variables to be considered in challenging the hypotheses were the effects of nutrients, air, and heat and the presence of preexisting life forms in the environment. One of the first people to test the spontaneous generation theory was Francesco Redi of Italy. He conducted a simple experiment in which he placed meat in a jar and covered it with fine gauze. Flies gathering at the jar were blocked from entering and thus laid their eggs on the outside of the gauze. The maggots subsequently developed without access to the meat, indicating that maggots were the offspring of flies and did not arise from some "vital force" in the meat. This and related experiments laid to rest the idea that more complex animals such as insects and mice developed through abiogenesis, but it did not convince many scientists of the day that simpler organisms could not arise in that way.

Frenchman Louis Jablot reasoned that even microscopic organisms must have parents, and his experiments with infusions (dried hay steeped in water) supported that hypothesis. He divided an infusion that had been boiled to destroy any living things into two containers: a heated container that was closed to the air and a heated container that was freely open to the air. Only the open vessel developed microorganisms, which he presumed had entered in air laden with dust. Regrettably, the validation of biogenesis was temporarily set back by John Needham, an Englishman who did similar experiments using mutton gravy. His results were in conflict with Jablot's because both his heated and unheated test containers teemed with microbes. Unfortunately, his experiments were done before the concepts of heat-resistant endospores and true methods of sterilization were understood and widely known.

*abiogenesis (ah-bee″-oh-jen-uh-sis) Gr. *a*, not, *bios*, living, and *gennan*, to produce.

A lengthy process of experimentation, analysis, and testing eventually leads to conclusions that either support or refute the hypothesis. If experiments do not uphold the hypothesis—that is, if it is found to be flawed—the hypothesis or some part of it is rejected; it is either discarded or modified to fit the results of the experiment. If the hypothesis is supported by the results from the experiment, it is not (or should not be) immediately accepted as fact. It then must be tested and retested. Indeed, this is an important guideline in the acceptance of a hypothesis. The results of the experiment must be published and then repeated by other investigators.

In time, as each hypothesis is supported by a growing body of data and survives rigorous scrutiny, it moves to the next level of acceptance—the **theory.** A theory is a collection of statements, propositions, or concepts that explains or accounts for a natural event. A theory is not the result of a single experiment repeated over and over again, but is an entire body of ideas that expresses or explains many aspects of a phenomenon. It is not a fuzzy or weak speculation, as is

Additional experiments further defended biogenesis. Franz Shultze and Theodor Schwann of Germany felt sure that air was the source of microbes and sought to prove this by passing air through strong chemicals or hot glass tubes into heat-treated infusions in flasks. When the infusions again remained devoid of living things, the supporters of abiogenesis claimed that the treatment of the air had made it harmful to the spontaneous development of life.

Georg Schroeder and Theodor Van Dusch followed up these studies. They did not treat the air with heat or chemicals but passed it through cotton wool to filter out microscopic organisms. Again, no microbes grew in the infusions. Although all these experiments should have finally laid to rest the arguments for spontaneous generation, they did not.

Then, in the mid-1800s, the acclaimed microbiologist Louis Pasteur entered the arena. He had recently been studying the roles of microorganisms in the fermentation of beer and wine, and it was clear to him that these processes were brought about by the activities of microbes introduced into the beverage from air, fruits, and grains. The methods he used to discount abiogenesis were simple yet brilliant.

Pasteur repeated the experiments using cotton filters to trap dust from air and observed tiny objects (probably spores) in the filters. He also observed that these same filters would initiate growth in previously sterile broths. To further clarify that air and dust were the source of microbes, he filled flasks with broth and fashioned their openings into elongate, swan-neck–shaped tubes. The flasks' openings were freely open to the air but were curved so that gravity would cause any airborne dust particles to deposit in the lower part of the necks. He heated the flasks to sterilize the broth and then incubated them. As long as the flask remained intact, the broth remained sterile, but if the neck was broken off so that dust fell directly down into the container, microbial growth immediately commenced.

Pasteur summed up his findings, "For I have kept from them, and am still keeping from them, that one thing which is above the power of man to make; I have kept from them the germs that float in the air, I have kept from them life."

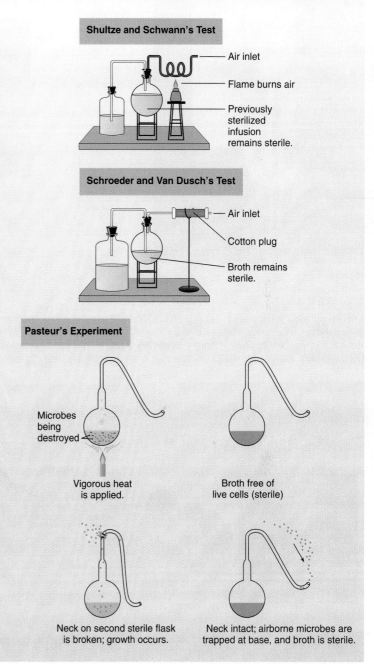

Shultze and Schwann's Test

Air inlet

Flame burns air

Previously sterilized infusion remains sterile.

Schroeder and Van Dusch's Test

Air inlet

Cotton plug

Broth remains sterile.

Pasteur's Experiment

Microbes being destroyed

Vigorous heat is applied.

Broth free of live cells (sterile)

Neck on second sterile flask is broken; growth occurs.

Neck intact; airborne microbes are trapped at base, and broth is sterile.

sometimes the popular notion, but a viable declaration that has stood the test of time and has yet to be disproved by serious scientific endeavors. Often, theories develop and progress through decades of research and are added to and modified by new findings. At some point, evidence of the accuracy and predictability of a theory is so compelling that the next level of confidence is reached and the theory becomes a **law,** or principle. For example, although we still refer to the germ *theory* of disease, so little question remains that microbes can cause disease that it has clearly passed into the realm of law.

Science and its hypotheses and theories must progress along with technology. As advances in instrumentation allow new, more detailed views of living phenomena, old theories may be reexamined and altered and new ones proposed. But scientists do not take the stance that theories are ever absolutely proved. The characteristics that make scientists most effective in their work are curiosity, open-mindedness, skepticism, creativity, cooperation, and readiness to revise their views of natural processes as new discoveries are made (Medical Microfile 1.3).

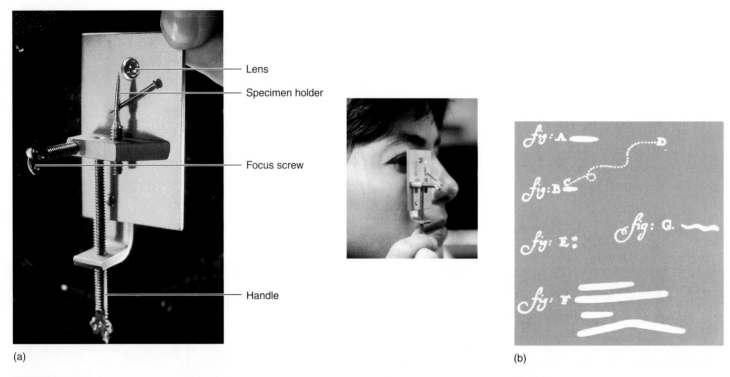

(a)

(b)

FIGURE 1.8

Leeuwenhoek's microscope. (a) A brass replica of a Leeuwenhoek microscope and how it is held. **(b)** Examples of bacteria drawn by Leeuwenhoek. He keenly observed, "I discovered living creatures in rain water which had stood but a few days in a new earthen pot. This invited me to view this water with great attention, especially those little animals appearing to me ten thousand times less than those which may be perceived in the water with the naked eye." This is probably the first observation of bacteria.

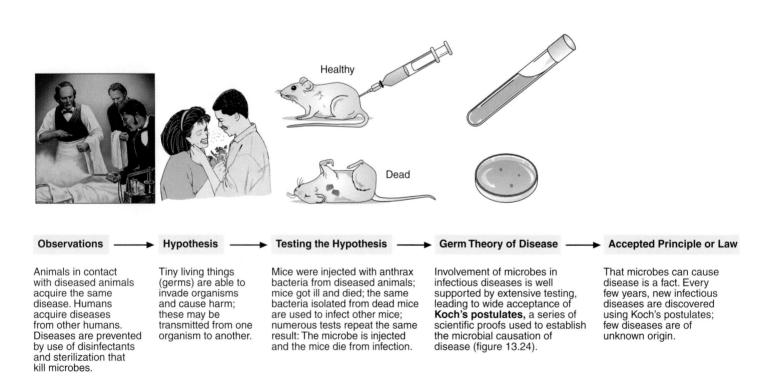

Observations	Hypothesis	Testing the Hypothesis	Germ Theory of Disease	Accepted Principle or Law
Animals in contact with diseased animals acquire the same disease. Humans acquire diseases from other humans. Diseases are prevented by use of disinfectants and sterilization that kill microbes.	Tiny living things (germs) are able to invade organisms and cause harm; these may be transmitted from one organism to another.	Mice were injected with anthrax bacteria from diseased animals; mice got ill and died; the same bacteria isolated from dead mice are used to infect other mice; numerous tests repeat the same result: The microbe is injected and the mice die from infection.	Involvement of microbes in infectious diseases is well supported by extensive testing, leading to wide acceptance of **Koch's postulates,** a series of scientific proofs used to establish the microbial causation of disease (figure 13.24).	That microbes can cause disease is a fact. Every few years, new infectious diseases are discovered using Koch's postulates; few diseases are of unknown origin.

FIGURE 1.9

Induction and the germ theory. The inductive process proceeds from specific observations to a general hypothesis. This example presents the classic events in developing the germ theory of disease.

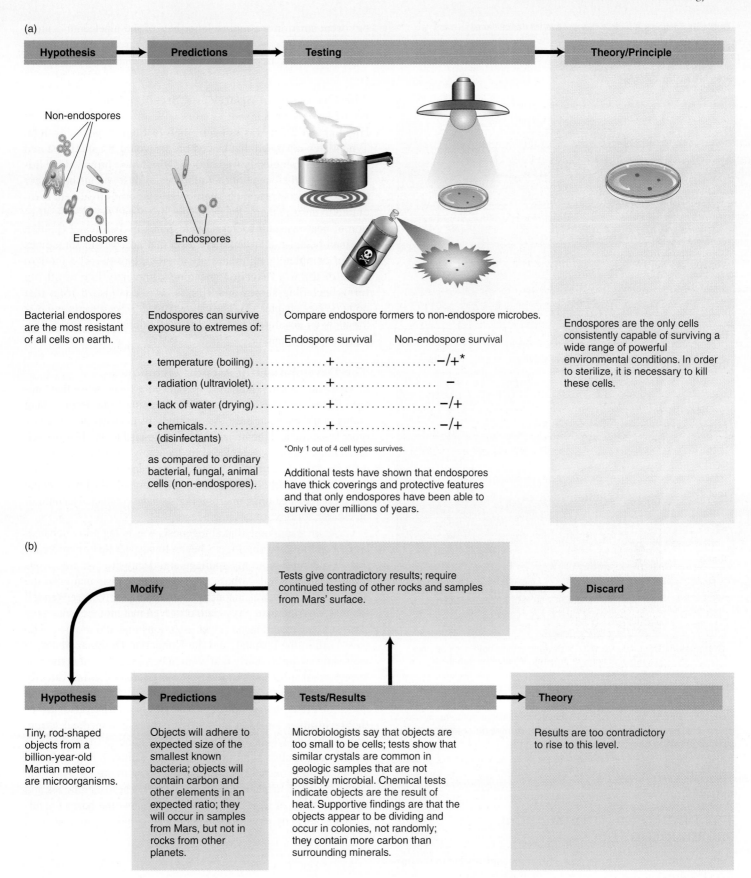

FIGURE 1.10

The pattern of deductive reasoning. The deductive process starts with a general hypothesis that predicts specific expectations. **(a)** This example is based on a well-established principle. **(b)** This example is based on a new hypothesis that has not stood up to critical testing.

Subject: Testing the factors responsible for dental caries

Hypothesis: Dental caries (cavities) involve dietary sugar or microbial action or both.

Variables:

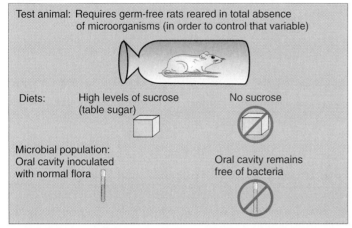

Test animal: Requires germ-free rats reared in total absence of microorganisms (in order to control that variable)

Diets: High levels of sucrose (table sugar)   No sucrose

Microbial population: Oral cavity inoculated with normal flora   Oral cavity remains free of bacteria

Experimental Protocol:

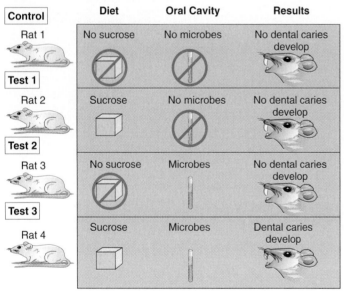

Control	Diet	Oral Cavity	Results
Rat 1	No sucrose	No microbes	No dental caries develop
Test 1 Rat 2	Sucrose	No microbes	No dental caries develop
Test 2 Rat 3	No sucrose	Microbes	No dental caries develop
Test 3 Rat 4	Sucrose	Microbes	Dental caries develop

Conclusion: Dental caries will not develop unless both sucrose and microbial action are present. What other variables were not controlled?

FIGURE 1.11

Variables. Any factor that can affect the experimental outcome is called a variable, and each combination of variables must be controlled while the hypothesis is being tested.

THE DEVELOPMENT OF MEDICAL MICROBIOLOGY

Early experiments on the sources of microorganisms led to the profound realization that microbes are everywhere: Not only are air and dust full of them, but the entire surface of the earth, its waters, and all objects are exposed to them. This discovery led to immediate applications in medicine, thus the seeds of medical microbiology were sown in the mid to latter half of the nineteenth century with the introduction of the germ theory of disease and the resulting use of sterile, aseptic, and pure culture techniques.

The Discovery of Spores and Sterilization

Following Pasteur's inventive work with infusions (see Historical Highlights 1.2), it was not long before English physicist John Tyndall demonstrated that heated broths would not spoil if stored in chambers completely free of dust. His studies provided the initial evidence that some of the microbes in dust and air have very high heat resistance and that particularly vigorous treatment is required to destroy them. Later, the discovery and detailed description of heat-resistant bacterial endospores by Ferdinand Cohn, a German botanist, clarified the reason that heat would sometimes fail to completely eliminate all microorganisms. The modern sense of the word **sterile,*** meaning completely free of all life forms including spores and viruses, was established from that point on (see chapter 11). The capacity to sterilize objects and materials is an absolutely essential part of microbiology, medicine, dentistry, and industry.

The Development of Aseptic Techniques

From earliest history, humans experienced a vague sense that "unseen forces" or "poisonous vapors" emanating from decomposing matter could cause disease. As the study of microbiology became more scientific and the invisible was made visible, the fear of such mysterious vapors was replaced by the knowledge and sometimes even the fear of "germs." About 120 years ago, the first studies by Robert Koch clearly linked a microscopic organism with a specific disease. Since that time, microbiologists have conducted a continuous search for disease-causing agents.

At the same time that abiogenesis was being hotly debated, a few budding microbiologists began to suspect that microorganisms could cause not only spoilage and decay but also infectious diseases. It occurred to these rugged individualists that even the human body itself was a source of infection. Dr. Oliver Wendell Holmes, an American physician, observed that mothers who gave birth at home experienced fewer infections than did mothers who gave birth in the hospital, and the Hungarian Dr. Ignaz Semmelweis showed quite clearly that women became infected in the maternity ward after examinations by physicians coming directly from the autopsy room. The English surgeon Joseph Lister took notice of these observations and was the first to introduce **aseptic* techniques** aimed at reducing microbes in a medical setting and preventing wound infections. Lister's concept of asepsis was much more limited than our modern precautions. It mainly involved disinfecting the hands and the air with strong antiseptic chemicals, such as phenol, prior to surgery. These techniques and the application of heat for sterilization became the bases for microbial control by physical and chemical methods, which are still in use today.

---

*sterile (stair'-il) Gr. *steira*, barren.

*aseptic (ay-sep'-tik) Gr. *a*, no, and *sepsis*, decay or infection. These techniques are aimed at reducing pathogens and do not necessarily sterilize.

MEDICAL MICROFILE 1.3
The Serendipity of the Scientific Method: Discovering Drugs

The discoveries in science are not always determined by the strict formulation and testing of a formal hypothesis. Quite often, they involve serendipity* and the luck of being in the right place and time, followed by a curiosity and willingness to change the direction of an experiment. This is especially true in the field of drug discoveries. The first antibiotic, penicillin, was discovered in the late 1920s by Dr. Alexander Fleming, who found a mold colony growing on a culture of bacteria that was wiping out the bacteria. He isolated the active ingredient that eventually launched the era of antibiotics. The search for new drugs to treat infections and cancer has been a continuous focus since that time. Even though the detailed science of testing a drug and working out its chemical structure and action require sophisticated scientific technology, the first and most important part of discovery often lies in a keen eye and an open mind.

In 1987, Dr. Michael Zasloff, a physician and molecular biologist, was doing research in gene expression, using African clawed frogs as a source of eggs. After performing surgery on the frogs and routinely placing them back in a nonsterile aquarium, he was surprised to notice that most of the time the frogs did not get infected or die. If the animal had been a mammal such as a mouse, it would probably not have survived the nonsterile surgery. This led him to conclude that the frog's skin must provide some form of natural protection. He observed that when the skin was stimulated by injury or irritants, it formed a thick white coating in a few moments that reminded him of a self-made "bandage" over the wound. He took a section of skin and extracted the components that were responsible for killing the microbes. His tests showed that they were small proteins called peptides, which he named *magainins,* after the Hebrew word for shield. Within 6 months of these findings, Dr. Zasloff made the deci-

sion to completely change the subject of his research and started up a new biotechnology company (Magainin Pharmaceuticals) to explore the therapeutic potential for magainins as well as other frog peptides.

The initial tests on this new class of drugs would indicate that they do indeed destroy a variety of bacteria as well as fungi, protozoa, and viruses. Although they are toxic to human cells too, this makes them a possible candidate for cancer treatment. Currently the drugs are being synthesized and tested in the lab for effectiveness and safety. Dr. Zasloff's intriguing observation and subsequent experiments had the impact of opening up a whole new area of biology: isolating antimicrobic peptides from multicellular organisms. Additional studies have shown that these compounds are widespread among amphibians, fish, birds, mammals, and plants. A number of companies are involved in developing applications for animal peptides. This discovery has been well timed, since resistance among microorganisms to traditional drugs is a continuing problem.

An African clawed frog responding to an irritant on its back first forms spots and then a thick opaque blotch of protective chemicals.

---

*serendipity Making useful discoveries by accident.

The Discovery of Pathogens and the Germ Theory of Disease

Two ingenious founders of microbiology, Louis Pasteur of France (figure 1.12) and Robert Koch of Germany (figure 1.13), introduced techniques that are still used today. Pasteur made enormous contributions to our understanding of the microbial role in wine and beer formation. He invented pasteurization and completed some of the first studies showing that diseases could arise from infection. These studies, supported by the work of other scientists, became known as the **germ theory of disease.** Pasteur's contemporary, Koch, established *Koch's postulates,* a series of proofs that verified the germ theory and could establish whether an organism was pathogenic and which disease it caused (see figure 1.9). About 1875, Koch used this experimental system to show that anthrax was caused by a bacterium called *Bacillus anthracis.* So useful were his postulates that the causative agents of 20 other diseases were discovered between 1875 and 1900, and even today, they are the standard for identifying pathogens.

FIGURE 1.12

Louis Pasteur (1822–1895), one of the founders of microbiology, viewing a sample. Few microbiologists can match the scope and impact of his contributions to the science of microbiology.

FIGURE 1.13

Robert Koch (1843–1910), intent at his laboratory workbench and surrounded by the new implements of his trade: Petri plates, tubes, and flasks filled with media; smears of bacteria; and bottles of stains.

Numerous exciting technologies emerged from Koch's prolific and probing laboratory work. During this golden age of the 1880s, he realized that study of the microbial world would require separating microbes from each other and growing them in culture. It is not an overstatement to say that he and his colleagues invented most of the techniques that are described in chapter 3: inoculation, isolation, media, maintenance of pure cultures, and preparation of specimens for microscopic examination. Other highlights in this era of discovery are presented in later chapters on microbial control (see chapter 11) and vaccination (see chapter 16).

Taxonomy: Organizing, Classifying, and Naming Microorganisms

Students just beginning their microbiology studies are often dismayed by the seemingly endless array of new, unusual, and sometimes confusing names for groups and specific types of microorganisms. Learning microbial **nomenclature*** is very much like learning a new language, and occasionally its demands may be a bit overwhelming. But paying attention to proper microbial names is just like following a baseball game or a movie plot: You cannot tell the players apart without a program! Your understanding and appreciation of microorganisms will be greatly improved by learning a few general rules about how they are named.

The formal system for organizing, classifying, and naming living things is **taxonomy.*** This science originated more than 250 years ago when Carl von Linné (Linnaeus; 1701–1778), a Swedish botanist, laid down the basic rules for taxonomic categories, or **taxa.*** Von Linné realized early on that a system for recognizing and defining the properties of living things would prevent chaos in scientific studies by providing each organism with a unique name and an exact "slot" in which to catalogue it. This classification would then serve as a means for future identification of that same organism and permit workers in many biological fields to know if they were indeed discussing the same organism. The von Linné system has served well in categorizing the 2 million or more different types of organisms that have been discovered since that time.

The primary concerns of taxonomy are classification, nomenclature, and identification. These three areas are interrelated and

---

*nomenclature (noh′-men-klay″-chur) L. *nomen,* name, and *clare,* to call. A system of naming.

*taxonomy (tacks-on″-uh-mee) Gr. *taxis,* arrangement, and *nomos,* name.

*taxa (tacks′-uh) sing. taxon.

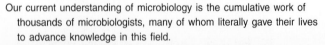

CHAPTER CHECKPOINTS

Our current understanding of microbiology is the cumulative work of thousands of microbiologists, many of whom literally gave their lives to advance knowledge in this field.

The microscope made it possible to see microorganisms and thus to identify their widespread presence, particularly as agents of disease.

Antonie van Leeuwenhoek is considered the father of bacteriology and protozoology because he was the first person to produce precise, correct descriptions of these organisms using microscopes he made himself.

The theory of spontaneous generation of living organisms from "vital forces" in the air was disproved once and for all by Louis Pasteur.

The scientific method is a process by which scientists seek to explain natural phenomena. It is characterized by specific procedures that either support or discredit an initial hypothesis.

Knowledge acquired through the scientific method is rigorously tested by repeated experiments by many scientists to verify its validity. A

collection of valid hypotheses is called a theory. A theory supported by much data collected over time is called a law.

Scientific truth changes through time as new research brings new information. Scientists must be able and willing to change theory in response to new data.

Medical microbiologists developed the germ theory of disease and introduced the critically important concept of aseptic technique to control the spread of disease agents.

Koch's postulates are the cornerstone of the germ theory of disease. They are still used today to pinpoint the causative agent of a specific disease.

Louis Pasteur and Robert Koch were the leading microbiologists during the golden age of microbiology (1875–1900). Each had his own research institute.

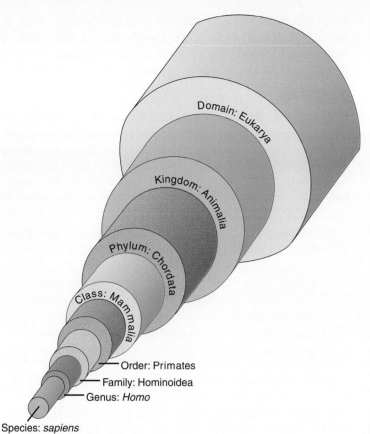

FIGURE 1.14

Classification scheme. The levels in classification from domain to species operate like a set of nesting boxes. Humans are the example here.

Domain: Eukarya (All eucaryotic cells)

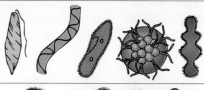

Kingdom: Protista (Protozoa and algae)

Phylum: Ciliophora (Only protozoa with cilia)

Class: Oligohymenophorea (Single cells with regular rows of cilia; rapid swimmers)

Order: Hymenostomatida (Elongate oval cells)

Family: Parameciidae (Cells rotate while swimming)

Genus: *Paramecium* (Pointed, cigar-shaped cells with an oral groove)

Species: *caudatum* (Cells pointed at one end)

FIGURE 1.15

Sample taxonomy. A common species of protozoan, *Paramecium caudatum*, traced through its taxonomic series. Note the gradual narrowing of the members, proceeding from general to specific levels.

play a vital role in keeping a dynamic inventory of the extensive array of living things. *Classification* is the orderly arrangement of organisms into groups, preferably in a format that shows evolutionary relationships. *Nomenclature* is the process of assigning names to the various taxonomic rankings of each microbial species. *Identification* is the process of discovering and recording the traits of organisms so that they may be placed in an overall taxonomic scheme. A survey of some general methods of identification appears in chapter 3.

THE LEVELS OF CLASSIFICATION

The main taxa, or groups, in a classification scheme are organized into several descending ranks, beginning with **domain,** which is a giant, all-inclusive category based on a unique cell type, and ending with a **species,*** the smallest and most specific taxon. All the members of a domain share only one or few general characteristics, whereas members of a species are essentially the same kind of organism—that is, they share the majority of their characteristics. The taxa between the top and bottom levels are, in descending order: **kingdom, phylum*** or **division,[7] class, order, family,** and **genus.*** Thus, each domain can be subdivided into a series of kingdoms, each kingdom is made up of several phyla, each phylum contains several classes, and so on. Because taxonomic schemes are to some extent artificial, certain groups of organisms do not exactly fit into the eight taxa. In that case, additional levels can be imposed immediately above (super) or below (sub) a taxon, giving us such categories as superphylum and subclass.

To illustrate the fine points of this system, we compare the taxonomic breakdowns of a human (figure 1.14) and a protozoan (figure 1.15). Humans and protozoa belong to the same domain (Eukarya) but are placed in different kingdoms. To emphasize just how broad the category kingdom is, ponder the fact that we belong to the same kingdom as sponges. Of the several phyla within this kingdom, humans belong to the Phylum Chordata (notochord-bearing animals), but even a phylum is rather all-inclusive, considering that humans share it with other vertebrates as well as creatures called sea squirts. The next level, Class Mammalia, narrows

---

*species (spee′-sheez) L. *specere,* kind. In biology, this term is always in the plural form.

7. The term *phylum* is used for protozoa and animals; the term *division* is used for bacteria, algae, plants, and fungi.

*phylum (fy′-lum) pl. phyla (fye′-luh) Gr. *phylon,* race.

*genus (jee′-nus) pl. genera (jen′-er-uh) L. birth, kind.

the field considerably by grouping only those vertebrates that have hair and suckle their young. Humans belong to the Order Primates, a group that also includes apes, monkeys, and lemurs. Next comes the Family Hominoidea, containing only humans and apes. The final levels are our genus, *Homo* (all races of modern and ancient humans), and our species, *sapiens* (meaning wise). Notice that for both the human and the protozoan, the categories become less inclusive and the individual members more closely related. Other examples of classification schemes are provided in sections of chapters 4 and 5 and in several later chapters.

It would be well to remember that all taxonomic **hierarchies*** are based on the judgment of scientists with certain expertise in a particular group of organisms and that not all other experts may agree with the system being used. Consequently, no taxa are permanent to any degree; they are constantly being revised and refined as new information becomes available or new viewpoints become prevalent. Because this text does not aim to emphasize details of taxonomy, we will usually be concerned with only the most general (kingdom, phylum) and specific (genus, species) levels.

ASSIGNING SPECIFIC NAMES

Many larger organisms are known by a common name suggested by certain dominant features. For example, a bird species might be called a red-headed blackbird or a flowering species a black-eyed Susan. Some species of microorganisms (especially pathogens) are also called by informal names, such as the gonococcus (*Neisseria gonorrhoeae*) or the tubercle bacillus (*Mycobacterium tuberculosis*), but this is not the usual practice. If we were to adopt common names such as the "little yellow coccus"* or the "club-shaped diphtheria bacterium,"* the terminology would become even more cumbersome and challenging than scientific names. Even worse, common names are notorious for varying from region to region, even within the same country. A decided advantage of standardized nomenclature is that it provides a universal language, thereby enabling scientists from all countries on the earth to freely exchange information.

The method of assigning the **scientific,** or **specific name** is called the **binomial (two-name) system of nomenclature.** The scientific name is always a combination of the generic (genus) name followed by the species name. The generic part of the scientific name is capitalized, and the species part begins with a lowercase letter. Both should be italicized (or underlined if italics are not available), as follows:

Saccharomyces cerevisiae

Because other taxonomic levels are not italicized and consist of only one word, one can always recognize a scientific name. An organism's scientific name is sometimes abbreviated to save space, as in *S. cerevisiae,* but only if the genus name has already been

stated. The source for nomenclature is usually Latin or Greek. If other languages such as English or French are used, the endings of these words are revised to have Latin endings. In general, the name first applied to a species will be the one that takes precedence over all others. An international group oversees the naming of every new organism discovered, making sure that standard procedures have been followed and that there is not already an earlier name for the organism or another organism with that same name. The inspiration for names is extremely varied and often rather imaginative. Some species have been named in honor of a microbiologist who originally discovered the microbe or who has made outstanding contributions to the field. Other names may designate a characteristic of the microbe (shape, color), a location where it was found, or a disease it causes. Some examples of specific names, their pronunciations, and their origins are:

- *Saccharomyces cerevisiae* (sak′-air-oh″-my-seez sair′-uh-vis″-ee-ee) Gr. *sakcharon,* sugar, *mykes,* fungus, and L. *cerevisia,* beer. The common yeast used in making beer, wine, and bread.
- *Haemophilus aegypticus* (hee′-mah-fil-us ee-jip′-tih-kus) Gr. *haema,* blood, *philos,* to love, and Egypt, the country. The causative agent of pinkeye.
- *Pseudomonas tomato* (soo′-doh-mon′-us toh-may′-toh) Gr. *pseudo,* false, *monas,* unit, and *tomato,* the fruit. A bacterium that infects the common garden tomato.
- *Campylobacter jejuni* (cam-pee′-loh-bak-ter jee-joo′-neye) Gr. *kampylos,* curved, *bakterion,* little rod, and *jejunum,* a section of intestine. One of the most important causes of intestinal infection worldwide.
- *Lactobacillus sanfrancisco* (lak″-toh-bass-ill′-us san-fran-siss′-koh) L. *lacto,* milk, and *bacillus,* little rod. A bacterial species used to make sourdough bread.
- *Vampirovibrio chlorellavorus* (vam-py′-roh-vib-ree-oh klor-ell-ah′-vor-us) F. *vampire;* L. *vibrio,* curved cell; *Chlorella,* a genus of green algae; and *vorus,* to devour. A small, curved bacterium that sucks out the cell juices of *Chlorella.*
- *Giardia lamblia* (jee-ar′-dee-uh lam′-blee-uh) for Alfred Giard, a French microbiologist, and Vilem Lambl, a Bohemian physician, both of whom worked on the organism, a protozoan that causes a severe intestinal infection.

THE ORIGIN AND EVOLUTION OF MICROORGANISMS

Earlier we indicated that taxonomists prefer to use a system of classification that shows the degree of relatedness of organisms, one that places closely related organisms into the same categories. This pattern of organization, called a *natural* or *phylogenetic system,* often uses selected observable traits to form the categories.

A phylogenetic system is based on the concept of evolutionary relationships among types of organisms. **Evolution*** is an important theme that underlies all of biology, including microbiology. From its simplest standpoint, evolution states that living things

---

***hierarchy** (hy′-ur-ar-kee) L. *hierarchia,* levels of power. Things arranged in the order of rank.

***Micrococcus luteus** (my″-kroh-kok′-us loo′-tee-us) Gr. *micros,* small, and *kokkus,* berry; L. *luteus,* yellow.

****Corynebacterium diphtheriae*** (kor-eye″-nee-bak-ter″-ee-yum dif′-theer-ee-eye) Gr. *coryne,* club, *bacterion,* little rod, and *diphtheriae,* the causative agent of the disease diphtheria.

***evolution** (ev-oh-loo′-shun) L. *evolutio,* to roll out.

change gradually through hundreds of millions of years and that these evolvements are expressed in various types of structural and functional changes through many generations. The process of evolution is selective: Those changes that most favor the survival of a particular organism or group of organisms tend to be retained, and those that are less beneficial to survival tend to be lost. Space does not permit a detailed analysis of evolutionary theories, but the occurrence of evolution is supported by a tremendous amount of evidence from the fossil record and from the study of **morphology,*** **physiology,*** and **genetics** (inheritance). Evolution accounts for the millions of different species on the earth and their adaptation to its many and diverse habitats.

Evolution is founded on two preconceptions: (1) that all new species originate from preexisting species and (2) that closely related organisms have similar features because they evolved from common ancestral forms. Usually, evolution progresses toward greater complexity, and evolutionary stages range from simple, primitive forms that are close to an ancestral organism to more complex, advanced forms. Although we use the terms *primitive* and *advanced* to denote the degree of change from the original set of ancestral traits, it is very important to realize that all species presently residing on the earth are modern, but some have arisen more recently in evolutionary history than others.

The evolutionary patterns of organisms are often drawn as a family tree, with the trunk representing the main ancestral lines and the branches showing offshoots into specialized groups of organisms. This sort of arrangement places the more ancient groups at the bottom and the more recent ones at the top. The branches may also indicate origins, how closely related various organisms are, and an approximate timescale for evolutionary history (figures 1.16 and 1.17).

SYSTEMS OF PRESENTING A UNIVERSAL TREE OF LIFE

The first phylogenetic trees of life were constructed on the basis of just two kingdoms (plants and animals). In time, it became clear that certain organisms did not truly fit either of those categories, so a third kingdom for simpler organisms that lacked tissue differentiation (protists) was recognized. Eventually, when significant differences became evident even among the protists, Robert Whittaker proposed a fourth kingdom for the bacteria and a fifth one for the fungi.

Although biologists have found the system of five kingdoms and two basic cell types to be a valuable method of classification, recent studies in molecular biology have provided a more accurate view of the relationships and origins of cells. It has been determined that certain types of molecules in cells, called small ribosomal ribonucleic acid (rRNA), provide a "living record" of the evolutionary history of an organism. Analysis of this molecule in procaryotic and eucaryotic cells indicates that certain unusual cells called archaea (originally archaebacteria) are so different from the other two groups that they should be included in a separate super-

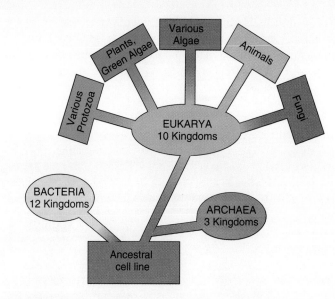

FIGURE 1.16

Woese system. A system for representing the origins of cell lines and major taxonomic groups as proposed by Carl Woese and colleagues. They propose three distinct cell lines placed in superkingdoms called domains. The first primitive cells, called progenotes, were ancestors of both lines of procaryotes (Domains Bacteria and Archaea), and the Archaea emerged from the same cell line as eucaryotes (Domain Eukarya). Some of the traditional kingdoms are still present with this system (see figure 1.17). Protozoa and some algal groups (called various algae here) are lumped into general categories.

kingdom. Those same studies have also revealed that the cells of archaea, though procaryotic in nature, are actually more closely related to eucaryotic cells than to bacterial cells (see table 4.7). To reflect these relationships, Carl Woese and George Fox have proposed a system that assigns all organisms to one of three domains, each described by a different type of cell (figure 1.16). The procaryotic cell types are placed in the Domains **Archaea** and **Bacteria.** Eucaryotes are all placed in the Domain **Eukarya.** It is believed that these three superkingdoms arose from an ancestor most similar to the archaea. This new system is still undergoing analysis and somewhat complicates the presentation of organisms in that it disposes of some traditional groups, although many of the traditional kingdoms still work within this framework (animals, plants, and fungi). The original Kingdom Protista is now a collection of protozoa and algae that exist in several separate kingdoms (see chapter 5). This new scheme will not greatly affect our presentation of most microbes, because we will be discussing them at the genus or species level. It is also an important truism that our methods of classification reflect our current understanding and are constantly changing as new information is uncovered.

In the interest of balance, we will also present the traditional Whittaker system of classification (figure 1.17). This system places all living things in one of five basic kingdoms: (1) the Procaryotae or Monera, (2) the Protista, (3) the Myceteae or Fungi, (4) the Plantae, and (5) the Animalia. The simple, single-celled organisms at the base of the family tree are in the **Kingdom Procaryotae** (also called **Monera**). Because only those organisms with procaryotic

---

*morphology (mor-fol′-oh-jee) Gr. *morphos,* form, and *logos,* to study. The study of organismic structure.

*physiology (fiz″-ee-ol′-oh-jee) Gr. *physis,* nature. The study of the function of organisms.

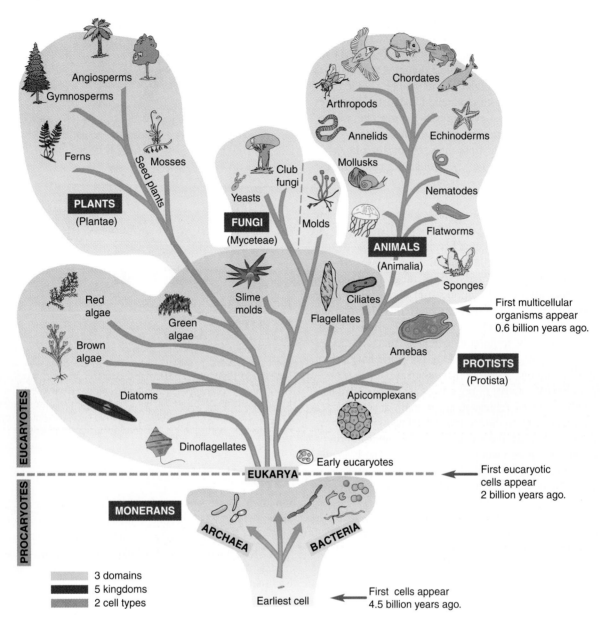

FIGURE 1.17

Traditional Whittaker system of classification. Kingdoms are based on cell structure and type, the nature of body organization, and nutritional type. Bacteria and Archaea (monerans) have procaryotic cells and are unicellular. Protists have eucaryotic cells and are mostly unicellular. They can be photosynthetic (algae), or they can feed on other organisms (protozoa). Fungi have eucaryotic cells and are unicellular or multicellular; they have cell walls and are not photosynthetic. Plants have eucaryotic cells, are multicellular, have cell walls, and are photosynthetic. Animals have eucaryotic cells, are multicellular, do not have cell walls, and derive nutrients from other organisms.

After Dolphin, Biology Lab Manual, *4th ed., Fig. 14.1, p. 177, McGraw-Hill Companies.*

cells are placed in it, the nature of cell structure is the main defining characteristic for this kingdom. It includes all of the microorganisms commonly known as **eubacteria,*** cells with typical procaryotic cell structure, and the **archaebacteria,*** cells with atypical cell structure that live in extreme environments (high salt and temperatures). Primitive procaryotes were the earliest cells to appear on the earth (see figure 4.30) and were the original ancestors of both more advanced bacteria and eucaryotic organisms. The Kingdom Procaryotae is a large and complex group that is surveyed in more detail in chapter 4.

The other four kingdoms contain organisms composed of eucaryotic cells. Their probable origin from procaryotic cells is discussed in Spotlight on Microbiology 5.1. The **Kingdom Protista*** contains mostly single-celled microbes that lack more complex levels of organization, such as tissues. Its members include both the

---

*eubacteria (yoo″-bak-ter′-ee-uh) Gr. *eu*, true, and *bakterion*, little rod. All bacteria besides the archaebacteria.

*archaebacteria (ark″-ee-uh-bak-ter′-ee-uh) Gr. *archaios*, ancient. The same name as archaea.

*Protista (pro-tiss′-tah) Gr. *protos*, the first.

microscopic algae, defined as independent photosynthetic cells with rigid walls, and the protozoans, animal-like creatures that feed upon other live or dead organisms and lack cell walls. More information on this group's taxonomy is given in chapter 5. The **Kingdom Myceteae*** contains the fungi, single- or multi-celled eucaryotes that are encased in cell walls and absorb nutrients from other organisms (see chapter 5). With the exception of certain infectious worms and arthropods, the final two kingdoms, **Animalia** and **Plantae,** are

---

***Myceteae** (my-cee′-tee-eye) Gr. *mycos*, the fungi.

generally not included in the realm of microbiology because most are large, multicellular organisms with tissues, organs, and organ systems. In general, animals move freely and feed on other organisms, whereas plants grow in an attached state and exhibit a nutritional scheme based on photosynthesis. It is possible to integrate the two systems as shown in figure 1.17.

Please note that viruses are *not* included in any of the classification or evolutionary schemes, because they are not cells and their position cannot be given with any confidence. Their special taxonomy is discussed in chapter 6.

CHAPTER CHECKPOINTS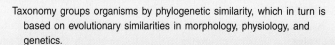

Taxonomy is the formal filing system scientists use to classify living organisms. It puts every organism in its place and makes a place for every living organism.

The taxonomic system has three primary functions: classification, nomenclature, and identification of species.

The eight major taxa, or groups, in the taxonomic system are (in descending order): domain, kingdom, phylum or division, class, order, family, genus, and species.

The binomial system of nomenclature describes each living organism by two names: genus and species.

Taxonomy groups organisms by phylogenetic similarity, which in turn is based on evolutionary similarities in morphology, physiology, and genetics.

Evolutionary patterns show a treelike branching from simple, primitive life forms to complex, advanced life forms.

The Woese-Fox classification system places all eucaryotes in the Domain (Superkingdom) Eukarya and subdivides the procaryotes into the two Domains Archaea and Bacteria.

The Whittaker five-kingdom classification system places all bacteria in the Kingdom Procaryotae and subdivides the eucaryotes into Kingdoms Protista, Myceteae, Animalia, and Plantae.

CHAPTER CAPSULE WITH KEY TERMS

I. **Microbiology** is the study of **bacteria, viruses, fungi, protozoa,** and **algae,** which are collectively called **microorganisms,** or **microbes.** In general, microorganisms are **microscopic** and, unlike **macroscopic** organisms, which are readily visible, they require magnification to be adequately observed or studied.

II. Microbes live in most of the world's habitats and are indispensable for normal, balanced life on earth. They play many roles in the functioning of the earth's ecosystems. Most organisms are free-living, but a few are **parasites.**

   A. Microbes are involved in nutrient production and energy flow. Algae and certain bacteria trap the sun's energy to produce food through **photosynthesis.**

   B. Other microbes are responsible for the breakdown and recycling of nutrients through **decomposition.** Microbes are essential to the maintenance of the air, soil, and water.

III. Microbes have been called upon to solve environmental, agricultural, and medical problems.

   A. **Biotechnology** applies the power of microbes toward the manufacture of industrial products, foods, and drugs.

   B. Microbes form the basis of **genetic engineering** and **recombinant DNA** technology, which alter genetic material to produce new products and modified life forms.

   C. With **bioremediation,** microbes are used to clean up pollutants and wastes in natural environments.

IV. Nearly 2,000 microbes are **pathogens** that cause **infectious diseases.** Infectious diseases result in high levels of mortality and morbidity. Many infections are **emerging,** meaning that they are newly identified pathogens gaining greater prominence. Many older diseases are also increasing.

V. The simplicity, growth rate, and adaptability of microbes are some of the reasons that microbiology is so diverse and has branched out into many subsciences and applications. Important subsciences include **immunology, epidemiology, public health, food, dairy, aquatic,** and **industrial** microbiology.

VI. **Important Historical Events**

   A. Microbiology as a science is about 200 years old. Hundreds of contributors have provided discoveries and knowledge to enrich our understanding.

   B. With his simple microscope, Leeuwenhoek discovered organisms he called animalcules. As a consequence of his findings and the rise of the **scientific method,** the notion of **spontaneous generation,** or **abiogenesis,** was eventually abandoned for **biogenesis.** The scientific method applies **inductive** and **deductive** reasoning to develop rational **hypotheses** and **theories** that can be tested. Principles that withstand repeated scrutiny become **law** in time.

   C. Early microbiology blossomed with the conceptual developments of **sterilization, aseptic techniques,** and the **germ theory of disease.**

VII. **Characteristics and Classification of Microorganisms**

   A. Organisms can be described according to their **morphology** and **physiology.** The **genetics** of organisms reveals an ancestral **evolutionary** relationship among these kingdoms.

   B. Cells of **eucaryotic** organisms contain a nucleus, but those of **procaryotic** organisms do not.

   C. **Taxonomy** is a hierarchy scheme for the **classification, identification,** and **nomenclature** of organisms, which are grouped in categories called **taxa,** based on features ranging from general to specific.

1. Starting with the broadest category, the taxa are **domain, kingdom, phylum** (or **division**), **class, order, family, genus,** and **species.** Organisms are assigned **binomial scientific names** consisting of their genus and species names.

2. The latest classification scheme for living things is based on the genetic structure of their ribosomes. The Woese-Fox system recognizes three domains: **Archaea,** simple procaryotes that live in extremes; **Bacteria,** typical procaryotes; and **Eukarya,** all types of eucaryotic organisms.

3. An alternative classification scheme uses a simpler five-kingdom organization: **Kingdom Procaryotae (Monera),** containing the **eubacteria** and the **archaebacteria; Kingdom Protista,** containing primitive unicellular microbes such as algae and protozoa; **Kingdom Myceteae,** containing the fungi; **Kingdom Animalia,** containing animals; and **Kingdom Plantae,** containing plants.

MULTIPLE-CHOICE QUESTIONS

Select the correct answer from the answers provided. For questions with blanks, choose the combination of answers that most accurately completes the statement.

1. Which of the following is not considered a microorganism?
   a. alga
   b. bacterium
   c. protozoan
   d. mushroom

2. Which microorganism(s) is/are responsible for photosynthesis in the aqueous environment?
   a. bacteria
   b. algae
   c. cyanobacteria
   d. both b and c

3. Which process involves the deliberate alteration in an organism's genetic material?
   a. bioremediation
   b. biotechnology
   c. decomposition
   d. recombinant DNA

4. A prominent difference between procaryotic and eucaryotic cells is the
   a. larger size of procaryotes
   b. lack of pigmentation in eucaryotes
   c. presence of a nucleus in eucaryotes
   d. presence of a cell wall in procaryotes

5. Which of the following parts was absent from Leeuwenhoek's microscopes?
   a. focusing screw
   b. lens
   c. specimen holder
   d. condenser

6. Abiogenesis refers to the
   a. spontaneous generation of organisms from nonliving matter
   b. development of life forms from preexisting life forms
   c. development of aseptic technique
   d. germ theory of disease

7. A hypothesis can be defined as
   a. a belief based on knowledge
   b. knowledge based on belief
   c. a scientific explanation that is subject to testing
   d. a theory that has been thoroughly tested

8. Which early microbiologist was most responsible for developing sterile laboratory techniques?
   a. Louis Pasteur
   b. Robert Koch
   c. Carl von Linné
   d. John Tyndall

9. Which scientist is most responsible for finally laying the theory of spontaneous generation to rest?
   a. Joseph Lister
   b. Robert Koch
   c. Francesco Redi
   d. Louis Pasteur

10. The process of observing an event and then constructing a hypothesis to explain it involves
    a. inductive reasoning
    b. deductive reasoning
    c. a controlled experiment
    d. guesswork

11. When a hypothesis has been thoroughly supported by long-term study and data, it is considered
    a. a law
    b. a speculation
    c. a theory
    d. proved

12. Which is the correct order of the taxonomic categories, going from most specific to most general?
    a. domain, kingdom, phylum, class, order, family, genus, species
    b. division, domain, kingdom, class, family, genus, species
    c. species, genus, family, order, class, phylum, kingdom, domain
    d. species, family, class, order, phylum, kingdom

13. By definition, organisms in the same _____ are more closely related than are those in the same _____ .
    a. order, family
    b. class, phylum
    c. family, genius
    d. phylum, division

14. Which of the following are procaryotic?
    a. bacteria
    b. archaea
    c. protists
    d. both a and b

15. Order the following items by size, using numbers: 1 = smallest and 8 = largest.
    ____ AIDS virus   ____ worm
    ____ ameba   ____ coccus bacterium
    ____ rickettsia   ____ white blood cell
    ____ protein   ____ atom

CONCEPT QUESTIONS

These questions are suggested as a *writing-to-learn* experience. For each question, compose a one- or two-paragraph answer that includes the factual information needed to completely address the question. Discuss the concepts in a sequence that allows you to present the subject using clear logic and correct terminology

1. Explain the important contributions microorganisms make in the earth's ecosystems.

2. Describe five different ways in which humans exploit microorganisms for our benefit.

3. Identify the groups of microorganisms included in the scope of microbiology, and explain the criteria for including these groups in the field.

4. Briefly identify the subdivisions of microbiology and tell what is studied in each. What do the following microbiologists study: algologist, epidemiologist, biotechnologist, ecologist, virologist, and immunologist?

5. Why was the abandonment of the spontaneous generation theory so significant? Using the scientific method, describe the steps you would take to test the theory of spontaneous generation.

6. a. Explain how inductive reasoning and deductive reasoning are similar and different.
   b. What are variables and controls?
   c. Look at figure 1.11 and answer the question at the bottom of the figure.

7. a. Differentiate between a hypothesis and a theory.
   b. Is the germ theory of disease really a law, and why?

8. a. Differentiate between taxonomy, classification, and nomenclature.
   b. What is the basis for a phylogenetic system of classification?
   c. What is a binomial system of nomenclature, and why is it used?
   d. Give the correct order of taxa, going from most general to most specific. A mnemonic (memory) device for recalling the order is *Darling King Phillip Came Over For Good Spaghetti.*

9. a. Construct a table that compares cell types and places them into domains and kingdoms. In which kingdoms do we find microorganisms?
   b. Compare the new domain system with the five-kingdom system. Does the newer system change the basic idea of procaryotes and eucaryotes? What is the third cell type?

CRITICAL-THINKING QUESTIONS

Critical thinking is the ability to reason and solve problems using facts and concepts. It requires you to apply information to new or different circumstances, to integrate several ideas to arrive at a solution, and to perform practical demonstrations as part of your analysis. These questions can be approached from a number of angles, and in most cases, they do not have a single correct answer.

1. What do you suppose the world would be like if there were cures for all infectious diseases and a means to destroy all microbes? What characteristics of microbes will prevent this from ever happening?

2. a. Where do you suppose the "new" infectious diseases come from?
   b. Name some factors that could cause older diseases to show an increase in the number of cases.
   c. Comment on the sensational ways that some tabloid media portray infectious diseases to the public.

3. Add up the numbers of deaths worldwide from infectious diseases (figure 1.4). Look up each disease in the index and see which ones could be prevented by vaccines or treated with drugs. How many do you think could have been prevented by modern medicine?

4. Correctly label the types of microorganisms in the drawing at right, using basic characteristics featured in the chapter.

5. What events, discoveries, or inventions were probably the most significant in the development of microbiology and why?

6. List the major variables in abiogenesis outlined in Historical Highlights 1.2 and explain how each was tested and controlled by the scientific method.

7. Can you develop a scientific hypothesis and means of testing the cause of stomach ulcers? (Is it caused by an infection? By too much acid? By a genetic disorder?)

8. Construct the scientific name of a newly discovered species of bacterium, using your name, a pet's name, a place, or a unique characteristic. Be sure to use proper notation and endings.

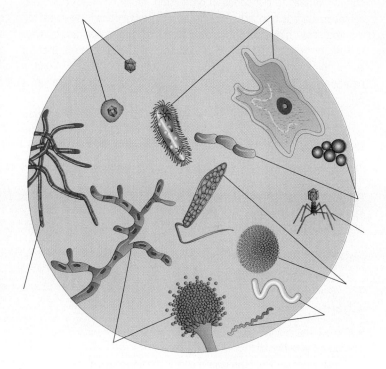

9. Archaea are found in hot, sulfuric, acidic, salty habitats, much like the early earth's conditions. Postulate on the origins of life, especially as it relates to the archaea.

INTERNET SEARCH TOPICS

1. Access a search engine on the World Wide Web under the heading *emerging diseases.* Adding terms like WHO and CDC will refine your search and take you to several appropriate websites. List the top 10 emerging diseases in the United States and worldwide.

2. Locate websites that discuss the ancient sporeformer isolated from a cavern in New Mexico. Determine the exact methods used in its isolation and what characteristics allowed it to survive.

From Atoms to Cells:
A Chemical Connection

I n laboratories all over the world, sophisticated technology is being developed for a wide variety of scientific applications. Refinements in molecular biology techniques now make it possible to routinely identify microorganisms, detect genetic disease, diagnose cancer, sequence the genes of organisms, break down toxic wastes, synthesize drugs and industrial products, and genetically engineer microorganisms, plants, and animals. A common thread that runs through new technologies and hundreds of traditional techniques is that, at some point, they involve chemicals and chemical reactions. In fact, if nearly any biological event is traced out to its ultimate explanation, it will invariably involve atoms, molecules, reactions, and bonding.

It is this relationship between the sciences that makes a background in chemistry necessary to biologists and microbiologists. Students with a basic chemistry background will enhance their understanding of and insight into microbial structure and function, metabolism, genetics, drug therapy, immune reactions, and infectious disease. This chapter has been organized to promote a working knowledge of atoms, molecules, bonding, solutions, pH, and biochemistry and to build foundations to later chapters. It concludes with an introduction to cells and a general comparison of procaryotic and eucaryotic cells as a preparation for chapters 4 and 5.

A molecular probe machine, called an ion microprobe, designed to analyze the isotopes found in very tiny samples of meteors, ancient rocks, and fossil samples. Chemists have used this device to determine the age of certain rocks found in Greenland (3.85 billion years old) and whether the sample may have come from a living thing.

Chapter Overview

- The understanding of living cells and processes is enhanced by a knowledge of chemistry.
- The structure and function of all matter in the universe is based on atoms.
- Atoms have unique structures and properties that allow chemical reactions to occur.
- Atoms contain protons, neutrons, and electrons in combinations to form elements.
- Living things are composed of approximately 25 different elements.
- Elements interact to form bonds that result in molecules and compounds with different characteristics than the elements that form them.
- Atoms can show variations in charge and polarity.
- Atoms and molecules undergo chemical reactions such as oxidation/reduction, ionization, and dissolution.
- The properties of carbon have been critical in forming macromolecules of life such as proteins, fats, carbohydrates, and nucleic acids.
- The nature of macromolecule structure and shape dictates its functions.
- Cells carry out fundamental activities of life, such as growth,

metabolism, reproduction, synthesis, and transport, that are all essentially chemical reactions on a grand scale.

Atoms, Bonds, and Molecules: Fundamental Building Blocks

The universe is composed of an infinite variety of substances existing in the gaseous, liquid, and solid states. All such tangible materials that occupy space and have mass are called **matter.** The

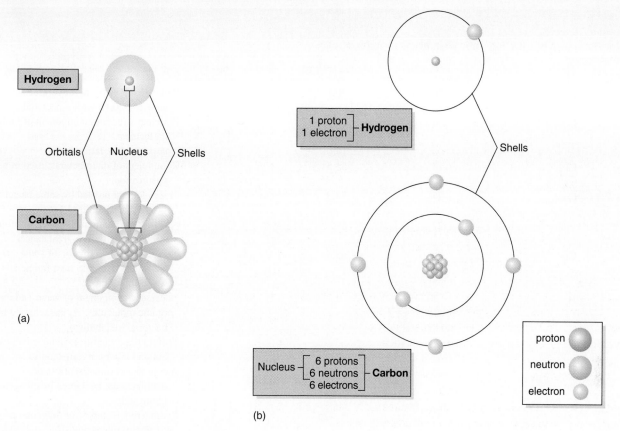

FIGURE 2.1

Models of atomic structure. (a) Three-dimensional models of hydrogen and carbon that approximate their actual structure. The nucleus is surrounded by electrons in orbitals that occur in levels called shells. Hydrogen has just one shell and one orbital. Carbon has two shells and four orbitals; the shape of the outermost orbitals is paired lobes rather than circles or spheres. **(b)** Simple models of the same atoms make it easier to show the numbers and arrangements of shells and electrons, and the numbers of protons and neutrons in the nucleus.

organization of matter—whether air, rocks, or bacteria—begins with individual building blocks called atoms. An **atom*** is defined as a tiny particle that cannot be subdivided into smaller substances without losing its properties. Even in a science dealing with very small things, an atom's minute size is striking; for example, an oxygen atom is only 0.0000000013 mm (0.0013 nm) in diameter, and one million of them in a cluster would barely be visible to the naked eye.

Although scientists have not directly observed the detailed structure of an atom, the exact composition of atoms has been well established by extensive physical analysis using sophisticated instruments. In general, an atom derives its properties from a combination of subatomic particles called **protons** (p^+), which are positively charged; **neutrons** (n^0), which have no charge (are neutral); and **electrons** (e^-), which are negatively charged. The relatively larger protons and neutrons make up a central core, or **nucleus,** that is surrounded by one or more electrons (figure 2.1). The nucleus makes up the larger mass (weight) of the atom, whereas the electron region accounts for the greater volume. To get a perspective on pro-

portions, consider this: If an atom were the size of a football stadium, the nucleus would be about the size of a marble! The stability of atomic structure is largely maintained by: (1) the mutual attraction of the protons and electrons (opposite charges attract each other) and (2) the exact balance of proton number and electron number, which causes the opposing charges to cancel each other out. At least in theory then, isolated intact atoms do not carry a charge.

DIFFERENT TYPES OF ATOMS: ELEMENTS AND THEIR PROPERTIES

All atoms share the same fundamental structure. All protons are identical, all neutrons are identical, and all electrons are identical. But when these subatomic particles come together in specific, varied combinations, unique types of atoms called **elements** result. Each element has a characteristic atomic structure and predictable chemical behavior. To date, 92 naturally occurring elements have been described, and 18 have been produced artificially by physicists. By convention, an element is assigned a distinctive name with an abbreviated shorthand symbol. Table 2.1 lists some of the elements common to biological systems, their atomic characteristics, and some of the natural and applied roles they play.

---

*atom (at′-um) Gr. *atomos,* not cut.

TABLE 2.1

The Major Elements of Life and Their Primary Characteristics

Element	Atomic Symbol*	Atomic Number	Atomic Weight	Ionized Form**	Significance in Microbiology
Calcium	Ca	20	40.1	Ca^{++}	Part of outer covering of certain shelled amebas; stored within bacterial spores
Carbon	C	6	12.0	—	Principal structural component of biological molecules
		6	14.0	—	Isotope used in dating fossils
Chlorine	Cl	17	35.5	Cl^-	Component of disinfectants; used in water purification
Cobalt	Co	27	58.9	Co^{++}, Co^{+++}	Trace element needed by some bacteria to synthesize vitamins
		27	60	—	An emitter of gamma rays; used in food sterilization; used to treat cancer
Copper	Cu	29	63.5	Cu^+, Cu^{++}	Necessary to the function of some enzymes; Cu salts are used to treat fungal and worm infections
Hydrogen	H	1	1	H^+	Necessary component of water and many organic molecules; H_2 gas released by bacterial metabolism
		1	3	—	Tritium has 2 neutrons; radioactive; used in clinical laboratory procedures
Iodine	I	53	126.9	I^-	A component of antiseptics and disinfectants; contained in a reagent of the Gram stain
		53	131, 125		Radioactive isotopes for diagnosis and treatment of cancers
Iron	Fe	26	55.8	Fe^{++}, Fe^{+++}	Necessary component of respiratory enzymes; some microbes require it to produce toxin
Magnesium	Mg	12	24.3	Mg^{++}	A trace element needed for some enzymes; component of chlorophyll pigment
Manganese	Mn	25	54.9	Mn^{++}, Mn^{+++}	Trace elements for certain respiratory enzymes
Nitrogen	N	7	14.0	—	Component of all proteins and nucleic acids; the major atmospheric gas
Oxygen	O	8	16.0	—	An essential component of many organic molecules; molecule used in metabolism by many organisms
Phosphorus	P	15	31	—	A component of ATP, nucleic acids, cell membranes; stored in granules in cells
		15	32	—	Radioactive isotope used as a diagnostic and therapeutic agent
Potassium	K	19	39.1	K^+	Required for normal ribosome function and protein synthesis; essential for cell membrane permeability
Sodium	Na	11	23.0	Na^+	Necessary for transport; maintains osmotic pressure; used in food preservation
Sulfur	S	16	32.1	—	Important component of proteins; makes disulfide bonds; storage element in many bacteria
Zinc	Zn	30	65.4	Zn^{++}	An enzyme cofactor; required for protein synthesis and cell division; important in regulating DNA

*Based on the Latin name of the element. The first letter is always capitalized; if there is a second letter, it is always lowercased.
**A dash indicates an element that is usually found in combination with other elements, rather than as an ion.

SPOTLIGHT ON MICROBIOLOGY 2.1
Searching for Ancient Life with Isotopes

Determining the age of the earth and the historical time frame of living things has long been a priority of biologists. Much evidence comes from fossils, geologic sediments, and genetic studies, yet there has always been a need for an exacting scientific reference for tracing samples back in time, possibly even to the beginnings of the earth itself. One very precise solution to this problem comes from patterns that exist in isotopes. The isotopes of an element have the same basic chemical structure, but over billions of years, they have come to vary slightly in the number of neutrons. For example, carbon has 3 isotopes: C12, predominantly found in living things; C13, a less common form associated with nonliving matter; and C14, a radioactive isotope. All isotopes exist in relatively predictable proportions in the earth, solar system, and even universe, so that any variations from the expected ratios would indicate some other factor besides random change.

Isotope chemists use giant machines called microprobes to analyze the atomic structures in fossils and rock samples (see chapter opening photo). These amazing machines can rapidly sort and measure the types and amounts of isotopes, which reflect a sample's age and possibly its origins. The accuracy of this method is such that it can be used like an "atomic clock." It was recently used to verify the dateline for the origins of the first life forms, using 3.85 billion-year-old sediment samples from Greenland. Testing indicated that the content of C12 in the samples was substantially higher than the amount in inorganic rocks, and it was concluded that living cells must have accumulated the C12. This finding shows that the origin of life was 400 million years earlier than the previous estimates.

In a separate study, some ancient Martian meteorites were probed to determine if certain microscopic rods could be some form of microbes (figure 1.10b). By measuring the ratios of oxygen isotopes in carbonate ions (CO_3^{-2}), chemists were able to detect significant fluctuations in the isotopes from different parts of the same meteorite. Such differences would most likely be caused by huge variations in temperature or other extreme environments that are incompatible with life. From this evidence, they concluded that the tiny rods were not Martian microbes.

THE MAJOR ELEMENTS OF LIFE AND THEIR PRIMARY CHARACTERISTICS

The unique properties of each element result from the numbers of protons, neutrons, and electrons it contains, and each element can be identified by certain physical measurements.

Each element is assigned an **atomic number (AN)** based on the number of protons it has. The atomic number is a valuable measurement because an element's proton number does not vary, and knowing it automatically tells you the usual number of electrons (recall that a neutral atom has an equal number of protons and electrons). Another useful measurement is the **mass[1] number (MN),** equal to the number of protons and neutrons. If one knows the mass number and the atomic number, it is possible to determine the numbers of neutrons by subtraction. Hydrogen is a unique element because its common form has only one proton, one electron, and no neutron, making it the only element with the same atomic and mass number.

Isotopes are variant forms of the same element that differ in the number of neutrons and thus have different mass numbers. These multiple forms occur naturally in certain proportions. Carbon, for example, exists primarily as carbon 12 with 6 neutrons (MN = 12); but a small amount (about 1%) is carbon 13 with 7 neutrons and carbon 14 with 8 neutrons. Although isotopes have virtually the same chemical properties, some of them have unstable nuclei that spontaneously release energy in the form of radiation. Such *radioactive isotopes* play a role in a number of research and medical applications. Because they emit detectable signs, they can be used to trace the position of key atoms or molecules in chemical reactions, they are tools in diagnosis and treatment, and they are even applied in sterilization procedures (see ionizing radiation in chapter 11). Another application of isotopes is in dating fossils and other ancient materials (Spotlight on Microbiology 2.1). An element's **atomic weight** is the average of the mass numbers of all its isotopic forms (table 2.1).

Electron Orbitals and Shells

The structure of an atom can be envisioned as a central nucleus surrounded by a "cloud" of electrons that constantly rotate about the nucleus in pathways (see figure 2.1). The pathways, called **orbitals,** are not actual objects or exact locations, but represent volumes of space in which an electron is likely to be found. Electrons occupy energy **shells,** proceeding from the lower-level energy electrons nearest the nucleus to the higher-energy electrons in the farthest orbitals.

Electrons fill the orbitals and shells in *pairs,* starting with the shell nearest the nucleus. The first shell contains one orbital and a maximum of 2 electrons; the second shell has four orbitals and up to 8 electrons; the third shell with 9 orbitals can hold up to 18 electrons; and the fourth shell with 16 orbitals contains up to 32 electrons. The number of orbitals and shells and how completely they are filled depends on the numbers of electrons, so that each element will have a unique pattern. For example, helium (AN = 2) has only a filled first shell of 2 e$^-$; oxygen (AN = 8) has a filled first shell and a partially filled second shell of 6 e$^-$; and magnesium (AN = 12) has a filled first shell, a filled second one, and a third shell that fills only one orbital, so is nearly empty. As we will see, the chemical properties of an element are controlled mainly by the distribution of electrons in the outermost shell. Figures 2.1 and 2.2 present various simplified models of atomic structure and electron maps.

---

1. Mass refers to the amount of matter that a particle contains. The proton and neutron have almost exactly the same mass, which is about 1.7×10^{-24} g, or 1 dalton.

Helium (He)

First Shell

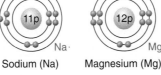

Carbon (C)   Nitrogen (N)   Oxygen (O)

First and Second Shells

Sodium (Na)   Magnesium (Mg)   Phosphorus (P)   Sulfur (S)   Chlorine (Cl)

First, Second, and Third Shells

FIGURE 2.2

Electron orbitals and shells. Models of several elements show how the shells are filled by electrons as the atomic numbers increase (numbers noted inside nuclei). Electrons tend to appear in pairs, but certain elements have incompletely filled outer shells. Chemists depict elements in shorthand form (red Lewis structures) that indicate only the valence electrons, since these are the electrons involved in chemical bonds.

CHAPTER CHECKPOINTS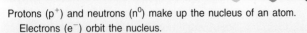

Protons (p$^+$) and neutrons (n^0) make up the nucleus of an atom. Electrons (e$^-$) orbit the nucleus.

All elements are composed of atoms but differ in the numbers of protons, neutrons, and electrons they possess.

Elements are identified by *atomic weight*, or *mass*, or by *atomic number*.

Isotopes are varieties of one element that contain the same number of protons but different numbers of neutrons.

The number of electrons in an element's outermost orbital (compared with the total number possible) determines its chemical properties and reactivity.

Bonds and Molecules

Most elements do not exist naturally in pure, uncombined form but are bound together as molecules and compounds. A **molecule*** is a distinct chemical substance that results from the combination of two or more atoms. Some molecules such as oxygen (O_2) and nitrogen gas (N_2) consist of atoms of the same element. Molecules that are combinations of two or more *different* elements are termed **compounds.** Compounds such as water (H_2O) and biological molecules (proteins, sugars, fats) are the predominant substances in living systems. When atoms bind together in molecules, they lose the properties of the atom and take on the properties of the combined substance. In the same way that an atom has an atomic weight, a molecule has a molecular weight (MW), which is calculated from the sum of all of the atomic weights of the atoms it contains.

The **chemical bonds** of molecules and compounds result when two or more atoms share, donate (lose), or accept (gain) electrons (figure 2.3). The number of electrons in the outermost shell of

an element is known as its **valence.*** The valence determines the degree of reactivity and the types of bonds an element can make. Elements with a filled outer orbital are relatively stable because they have no extra electrons to share with or donate to other atoms. For example, helium has one filled shell, with no tendency either to give up electrons or to take them from other elements, making it a stable, inert (nonreactive) gas. Elements with partially filled outer orbitals are less stable and are more apt to form some sort of bond. Many chemical reactions are based on the tendency of atoms with unfilled outer shells to gain greater stability by achieving, or at least approximating, a filled outer shell. For example, an atom such as oxygen that can accept 2 additional electrons will bond readily with atoms (such as hydrogen) that can share or donate electrons. We explore some additional examples of the basic types of bonding in the following section.

In addition to reactivity, the number of electrons in the outer shell also dictates the number of chemical bonds an atom can make. For instance, hydrogen can bind with one other atom, oxygen can bind with up to two other atoms, and carbon can bind with four (see figure 2.13).

COVALENT BONDS AND POLARITY: MOLECULES WITH SHARED ELECTRONS

Covalent (cooperative valence) **bonds** form between atoms with valences that suit them to sharing electrons rather than to donating or receiving them. A simple example is hydrogen gas (H_2), which consists of two hydrogen atoms. A hydrogen atom has only a single electron, but when two of them combine, each will bring its electron to orbit about both nuclei, thereby approaching a filled orbital (2 electrons) for both atoms and thus creating a **single covalent bond** (figure 2.4*a*). Covalent bonding also occurs in oxygen gas (O_2), but with a difference. Because each atom has 2 electrons to share in this molecule, the combination creates two pairs of shared

---

*molecule (mol′-ih-kyool) L. *molecula,* little mass.

*valence (vay′-lents) L. *valentia,* strength. A measure of atomic binding capacity.

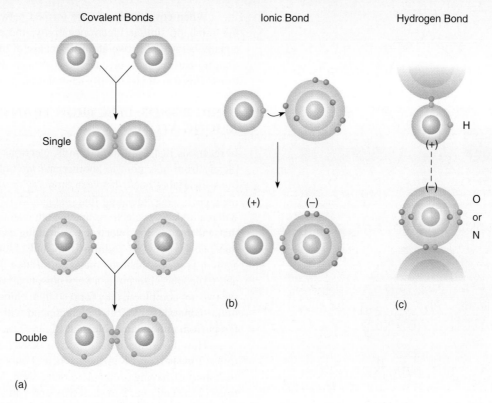

Covalent Bonds   Ionic Bond   Hydrogen Bond

Single

Double

(a)   (b)   (c)

FIGURE 2.3

General representation of three types of bonding. (a) Covalent bonds, both single and double. **(b)** Ionic bond. **(c)** Hydrogen bond. Note that hydrogen bonds are represented in models and formulas by dotted lines, as shown in **(c)**.

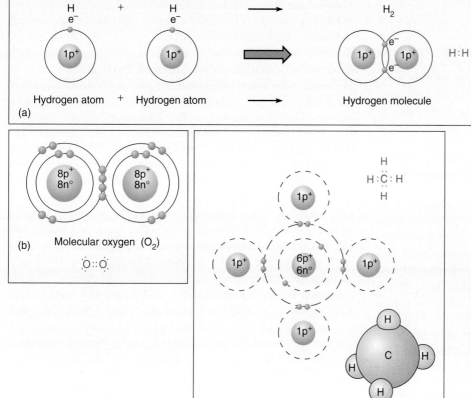

(a)

Hydrogen atom + Hydrogen atom ⟶ Hydrogen molecule

(b)   Molecular oxygen (O₂)

(c)   Methane (CH₄)

FIGURE 2.4

Examples of molecules with covalent bonding.

(a) A hydrogen molecule is formed when two hydrogen atoms share their electrons and form a single bond. **(b)** In a double bond, the outer orbitals of two oxygen atoms overlap and permit the sharing of 4 electrons (one pair from each) and the saturation of the outer orbital for both.

(c) Simple, working, and three-dimensional models of methane. Note that carbon has 4 electrons to share and hydrogens each have one, thereby completing the shells for all atoms in the compound.

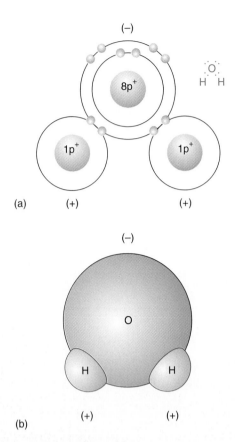

FIGURE 2.5

Polar molecule. (a) Simple models and **(b)** a three-dimensional model of a water molecule indicate the polarity, or unequal distribution, of electrical charge, which is caused by the pull of the shared electrons toward the oxygen side of the molecule.

electrons, also known as a **double covalent bond** (figure 2.4*b*). The majority of the molecules associated with living things are composed of single and double covalent bonds between the most common biological elements (carbon, hydrogen, oxygen, nitrogen, sulfur, and phosphorus), which are discussed in more depth in chapter 7. A slightly more complex pattern of covalent bonding is shown for methane gas (CH₄) in figure 2.4*c*.

Other effects of bonding result in differences in polarity. When atoms of different electronegativity form covalent bonds, the electrons are not shared equally and may be pulled more toward one atom than another. This pull causes one end of a molecule to assume a partial negative charge and the other end to assume a partial positive charge. A molecule with such an asymmetrical distribution of charges is termed **polar** and has positive and negative poles. Observe the water molecule shown in figure 2.5 and note that, because the oxygen atom is larger and has more protons than the hydrogen atoms, it will tend to draw the shared electrons with greater force toward its nucleus. This unequal force causes the oxygen part of the molecule to express a negative charge (due to the electrons' being attracted there) and the hydrogens to express a positive charge (due to the protons). The polar nature of water plays an extensive role in a number of biological reactions, which are discussed later. Polarity is a significant property of many large molecules in living systems and greatly influences both their reactivity and their structure.

When covalent bonds are formed between atoms that have the same or similar electronegativity, the electrons are shared equally between the two atoms. Because of this balanced distribution, no part of the molecule has a greater attraction for the electrons. This sort of electrically neutral molecule is termed **nonpolar.**

IONIC BONDS: ELECTRON TRANSFER AMONG ATOMS

In reactions that form **ionic bonds,** electrons are transferred completely from one atom to another and are not shared. These reactions invariably occur between atoms with valences that complement each other, meaning that one atom has an unfilled shell that will readily accept electrons and the other atom has an unfilled shell that will readily lose electrons. A striking example is the reaction that occurs between sodium (Na) and chlorine (Cl). Elemental sodium is a soft, lustrous metal so reactive that it can burn flesh, and molecular chlorine is a very poisonous yellow gas. But when the two are combined, they form sodium chloride[2] (NaCl)—the familiar nontoxic table salt—a compound with properties quite different from either parent element (figure 2.6).

How does this transformation occur? Sodium has 11 electrons (2 in shell one, 8 in shell two, and only 1 in shell three), so it is 7 short of having a complete outer shell. Chlorine has 17 electrons (2 in shell one, 8 in shell two, and 7 in shell three), making it 1 short of a complete outer shell. These two atoms are very reactive with one another, because a sodium atom will readily donate its single electron and a chlorine atom will avidly receive it. (The reaction is slightly more involved than a single sodium atom's combining with a single chloride atom (Microbits 2.2), but this complexity does not detract from the fundamental reaction as described here.) The outcome of this reaction is not many single, isolated molecules of NaCl but rather a solid crystal complex that interlinks millions of sodium and chloride ions (figure 2.6*b*).

Ionization: Formation of Charged Particles

Molecules with intact ionic bonds are electrically neutral, but they can produce charged particles when dissolved in a liquid called a solvent. This phenomenon, called **ionization,** occurs when the ionic bond is broken and the atoms dissociate (separate) into unattached, charged particles called **ions*** (figure 2.7). To illustrate what imparts a charge to ions, let us look again at the reaction between sodium and chlorine. When a sodium atom reacts with chlorine and loses one electron, the sodium is left with one more proton than electrons. This imbalance produces a positively charged sodium ion (Na⁺). Chlorine, on the other hand, has gained one electron and now has one more electron than protons, producing a negatively charged ion (Cl⁻). Positively charged ions are termed **cations,*** and negatively charged ions are termed **anions.*** (A good mnemonic device is to think of the "t" in cation as a plus (+) sign and the first "n" in anion as a negative (−) sign.) Substances such

---

2. In general, when a salt is formed, the ending of the name of the negatively charged ion is changed to -*ide*.

*ion (eye′-on) Gr. *ion,* going.

*cation (kat′-eye-on) An ion that migrates toward the negative pole, or cathode, of an electrical field.

*anion (an′-eye-on) An ion that migrates toward the positive pole, or anode.

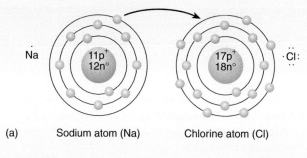

(a)   Sodium atom (Na)   Chlorine atom (Cl)

(b) Na :Cl:

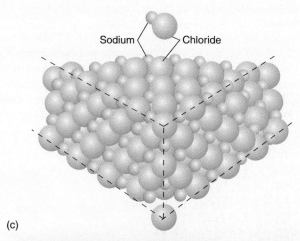

Sodium   Chloride

(c)

(d)

FIGURE 2.6

Ionic bonding between sodium and chlorine. (a) When the two elements are placed together, sodium loses its single outer orbital electron to chlorine, thereby filling chlorine's outer shell. **(b)** Simple model of ionic bonding. **(c)** Sodium and chloride ions form large molecules, or crystals, in which the two atoms alternate in a definite, regular, geometric pattern. **(d)** Note the cubic nature of NaCl crystals at the macroscopic level.

as salts, acids, and bases that release ions when dissolved in water are termed **electrolytes** because their charges enable them to conduct an electrical current. Owing to the general rule that particles of like charge repel each other and those of opposite charge attract each other, we can expect ions to interact electrostatically with other ions and polar molecules. Such interactions are important in many cellular chemical reactions, in the formation of solutions, and in the reactions microorganisms have with dyes. The transfer of electrons from one molecule to another constitutes a significant mechanism by which biological systems store and release energy (see Microbits 2.2).

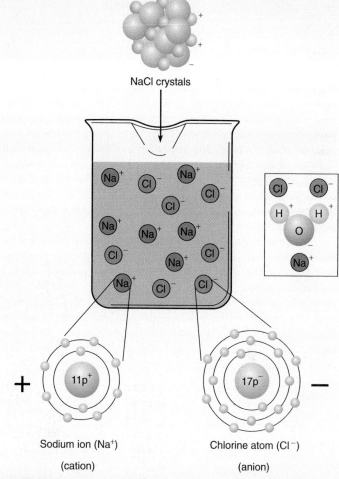

NaCl crystals

Sodium ion (Na⁺)   Chlorine atom (Cl⁻)

(cation)   (anion)

FIGURE 2.7

Ionization. When NaCl in the crystalline form is added to water, the ions are released from the crystal as separate charged particles (cations and anions) into solution. (See also figure 2.11.)

Hydrogen Bonding Some types of bonding involve neither sharing, losing, nor gaining electrons but instead are due to attractive forces between nearby molecules or atoms. One such bond is a **hydrogen bond,** a weak type of bond that forms between a hydrogen covalently bonded to one molecule and an oxygen or nitrogen atom on the same molecule or on a different molecule. Because hydrogen in a covalent bond tends to be positively charged, it will attract a nearby negatively charged atom and form an easily disrupted bridge with it. This type of bonding is usually represented in molecular models with a dotted line. A simple example of hydrogen bonding occurs between water molecules (figure 2.8). More extensive hydrogen bonding is partly responsible for the structure and stability of proteins and nucleic acids (see figure 2.22*b* and 2.25).

Chemical Shorthand: Formulas, Models, and Equations

The atomic content of molecules can be represented by a few convenient **formulas.** We have already been exposed to the molecular formula, which concisely gives the atomic symbols and the number of the elements involved in subscript (CO_2, H_2O). More complex molecules such as glucose ($C_6H_{12}O_6$) can also be symbolized this

MICROBITS 2.2
Redox: Electron Transfer and Oxidation-Reduction Reactions

The metabolic work of cells, such as synthesis, movement, and digestion, revolves around energy exchanges and transfers. The management of energy in cells is almost exclusively dependent on chemical rather than physical reactions because most cells are far too delicate to operate with heat, radiation, and other more potent forms of energy. The outer-shell electrons are readily portable and easily manipulated sources of energy. It is in fact the movement of electrons from molecule to molecule that accounts for most energy exchanges in cells. Fundamentally, then, a cell must have a supply of atoms that can gain or lose electrons if they are to carry out life processes.

The phenomenon in which electrons are transferred from one atom or molecule to another is termed an **oxidation** and **reduction** (shortened to **redox**) **reaction.** Although the term *oxidation* was originally adopted for reactions involving the addition of oxygen, the term oxidation can include any reaction causing electron release, regardless of the involvement of oxygen. By comparison, reduction is any reaction that causes an atom to receive electrons. All redox reactions occur in pairs. To analyze the phenomenon, let us again review the production of NaCl, but from a different standpoint. Although it is true that these atoms form ionic bonds, the chemical combination of the two is also a type of redox reaction.

When these two atoms react to form sodium chloride, a sodium atom gives up an electron to a chlorine atom. During this reaction, sodium is oxidized because it loses an electron, and chlorine is reduced because it gains an electron. To take this definition further, an atom or molecule, such as sodium, that can donate electrons and thereby reduce another molecule is a **reducing agent;** one that can receive extra electrons and thereby oxidize another molecule is an **oxidizing agent.** You may find this concept easier to keep straight if you think of redox agents as partners: The one that gives its electrons away is oxidized; the partner that receives the electrons is reduced. (A mnemonic device to keep track of this is *LEO* says *GER*. "Lose Electrons Oxidized; Gain Electrons Reduced."

Redox reactions are essential to many of the biochemical processes discussed in chapter 8. In cellular metabolism, electrons alone can be transferred from one molecule to another as described here, but sometimes oxidation and reduction occur with the transfer of hydrogen atoms (which are a proton and an electron) from one compound to another.

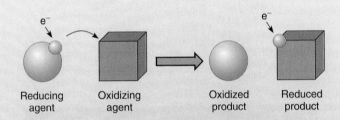

Simplified diagram of the exchange of electrons during an oxidation-reduction reaction.

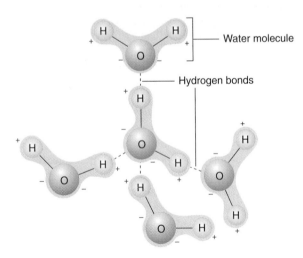

Water molecule

Hydrogen bonds

FIGURE 2.8

Hydrogen bonding in water. Because of the polarity of water molecules, the negatively charged oxygen end of one water molecule is weakly attracted to the positively charged hydrogen end of an adjacent water molecule.

way, but this formula is not unique, since fructose and galactose also share it. Molecular formulas are useful, but they only summa-

rize the atoms in a compound; they do not show the position of bonds between atoms. For this purpose, chemists use structural formulas illustrating the relationships of the atoms and the number and types of bonds (figure 2.9). Other structural models present the three-dimensional appearance of a molecule, illustrating the orientation of atoms (differentiated by color codes) and the molecule's overall shape (figure 2.10).

The printed page tends to make molecules appear static, but this picture is far from correct, because molecules are capable of changing through chemical reactions. For ease in tracing chemical exchanges between atoms or molecules, and to derive some sense of the dynamic character of reactions, chemists use shorthand **equations** containing symbols, numbers, and arrows to simplify or summarize the major characteristics of a reaction. Molecules entering or starting a reaction are called **reactants,** and substances left by a reaction are called **products.** In most instances, summary chemical reactions do not give the details of the exchange, in order to keep the expression simple and to save space.

In a **synthesis*** **reaction,** the reactants bond together in a manner that produces an entirely new molecule (reactant A plus reactant B yields product AB). An example is the production of sulfur

---

*synthesis (sin′-thuh-sis) Gr. *synthesis,* putting together.

(a)
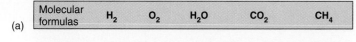

(b)

(c)
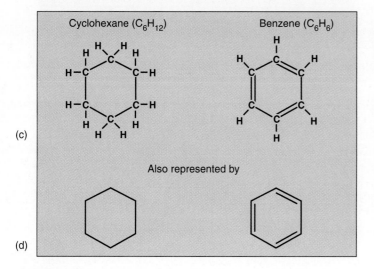

(d)

FIGURE 2.9

Comparison of molecular and structural formulas. (a) Molecular formulas provide a brief summary of the elements in a compound. **(b)** Structural formulas clarify the exact relationships of the atoms in the molecule, depicting single bonds by a single line and double bonds by two lines. **(c)** In structural formulas of organic compounds, cyclic or ringed compounds may be completely labeled, or **(d)** they may be presented in a shorthand form in which carbons are assumed to be at the angles and attached to hydrogens. See figure 2.14 for structural formulas of three sugars with the same molecular formula, $C_6H_{12}O_6$.

FIGURE 2.10

Three-dimensional, or space-filling, models of (a) water, (b) carbon dioxide, and (c) glucose. By convention, the red atoms are oxygen, the white ones hydrogen, and the black ones carbon.

dioxide, a by-product of burning sulfur fuels and an important component of smog:

$$S + O_2 \rightarrow SO_2$$

Some synthesis reactions are not such simple combinations. When water is synthesized, for example, the reaction does not really involve one oxygen atom combining with two hydrogen atoms, because elemental oxygen exists as O_2 and elemental hydrogen exists as H_2. A more accurate equation for this reaction is:

$$2H_2 + O_2 \rightarrow 2H_2O$$

The equation for reactions must be balanced—that is, the number of atoms on one side of the arrow must equal the number on the other side to reflect all of the participants in the reaction. To arrive at the total number of atoms in the reaction, multiply the prefix number by the subscript number; if no number is given, it is assumed to be 1.

In **decomposition reactions,** the bonds on a single reactant molecule are permanently broken to release two or more product molecules. One example is the resulting molecules when large nu-

trient molecules are digested into smaller units; a simpler example can be shown for the common chemical hydrogen peroxide:

$$2H_2O_2 \rightarrow 2H_2O + O_2$$

During **exchange reactions,** the reactants trade portions between each other and release products that are combinations of the two. This type of reaction occurs between acids and bases when they form water and a salt:

$$AB + XY \rightleftharpoons AX + BY$$

The reactions in biological systems can be **reversible,** meaning that reactants and products can be converted back and forth. These reversible reactions are symbolized with a double arrow, each pointing in opposite directions, as in the exchange reaction above. Whether a reaction is reversible depends on the proportions of these compounds, the difference in energy state of the reactants and products, and the presence of catalysts (substances that increase the rate of a reaction). Additional reactants coming from another reaction can also be indicated by arrows that enter or leave at the main arrow:

$$\begin{array}{c} CD \quad\; C \\ \diagdown\!\!\nearrow \\ X + Y \rightarrow XYD \end{array}$$

SOLUTIONS: HOMOGENEOUS MIXTURES OF MOLECULES

A **solution** is a mixture of one or more substances called **solutes** uniformly dispersed in a dissolving medium called a **solvent.** An important characteristic of a solution is that the solute cannot be

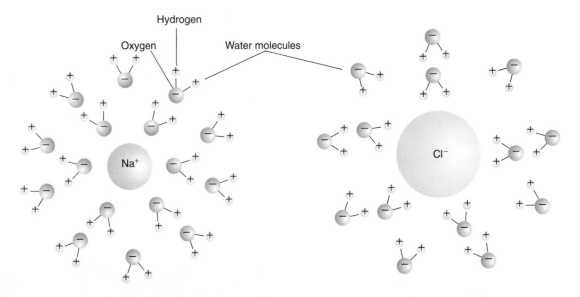

FIGURE 2.11

Hydration spheres formed around ions in solution. In this example, a sodium cation attracts the negatively charged region of water molecules, and a chloride anion attracts the positively charged region of water molecules. In both cases, the ions become covered with spherical layers of specific numbers and arrangements of water molecules.

separated by filtration or ordinary settling. The solute can be gaseous, liquid, or solid, and the solvent is usually a liquid. Examples of solutions are salt or sugar dissolved in water and iodine dissolved in alcohol. In general, a solvent will dissolve a solute only if it has similar electrical characteristics as indicated by the rule of solubility, expressed simply as "like dissolves like." For example, water is a polar molecule and will readily dissolve an ionic solute such as NaCl, yet a nonpolar solvent such as benzene will not dissolve NaCl.

Water is the most common solvent in natural systems, having several characteristics that suit it to this role. The polarity of the water molecule causes it to form hydrogen bonds with other water molecules, but it can also interact readily with charged or polar molecules. When an ionic solute such as NaCl crystals is added to water, it is **dissolved,** thereby releasing Na^+ and Cl^- into solution. Dissolution occurs because Na^+ is attracted to the negative pole of the water molecule and Cl^- is attracted to the positive pole; in this way, they are drawn away from the crystal separately into solution. As it leaves, each ion becomes **hydrated,** which means that it is surrounded by a sphere of water molecules (figure 2.11). Molecules such as salt or sugar that attract water to their surface are termed **hydrophilic.*** Nonpolar molecules, such as benzene, that repel water are considered **hydrophobic.*** A third class of molecules, such as the phospholipids in cell membranes, are considered *amphipathic** because they have both hydrophilic and hydrophobic properties.

Because most biological activities take place in aqueous (water-based) solutions, the concentration of these solutions can be very important (see chapter 7). The **concentration** of a solution expresses the amount of solute dissolved in a certain amount of solvent. It can be calculated by weight, volume, or percentage. A common way to calculate percentage of concentration is to use the weight of the solute, measured in grams (g), dissolved in a specified volume of solvent, measured in milliliters (ml). For example, dissolving 3 g of NaCl in 100 ml of water produces a 3% solution; dissolving 30 g in 100 ml produces a 30% solution; and dissolving 3 g in 1,000 ml (1 liter) produces a 0.3% solution. A solution with a small amount of solute and a relatively greater amount of solvent (0.3%) is considered dilute or weak. On the other hand, a solution containing significant percentages of solute (30%) is considered concentrated or strong.

A common way to express concentration of biological solutions is by its molar concentration, or *molarity* (M). A standard molar solution is obtained by dissolving one *mole,* defined as the molecular weight of the compound in grams, in 1 L (1,000 ml) of solution. To make a 1 M solution of sodium chloride, we would dissolve 58 g of NaCl to give 1 L of solution; a 0.1 M solution would require 5.8 g of NaCl in 1 L of solution.

ACIDITY, ALKALINITY, AND THE pH SCALE

Another factor with far-reaching impact on living things is the concentration of acidic or basic solutions in their environment. To understand how solutions develop acidity or basicity, we must look again at the behavior of water molecules. Hydrogens and oxygen tend to remain bonded by covalent bonds, but in certain instances, a single hydrogen can break away as the ionic form (H^+), leaving the remainder of the molecule in the form of an OH^- ion. The H^+ ion is positively charged because it is essentially a hydrogen ion that has lost its electron; the OH^- is negatively charged because it remains in possession of that electron. Ionization of water is constantly occurring, but in pure water containing no other ions,

---

***hydrophilic** (hy-droh-fil′-ik) Gr. *hydros,* water, and *philos,* to love.

***hydrophobic** (hy-droh-fob′-ik) Gr. *phobos,* fear.

***amphipathic** (am′-fy-path′-ik) Gr. *amphi,* both.

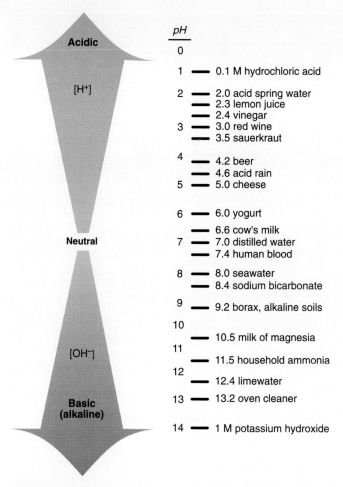

FIGURE 2.12

The pH scale. Shown are the relative degrees of acidity and basicity and the approximate pH readings for various substances.

pH
- 0
- 1 — 0.1 M hydrochloric acid
- 2 — 2.0 acid spring water
      — 2.3 lemon juice
      — 2.4 vinegar
- 3 — 3.0 red wine
      — 3.5 sauerkraut
- 4 — 4.2 beer
      — 4.6 acid rain
- 5 — 5.0 cheese
- 6 — 6.0 yogurt
      — 6.6 cow's milk
- 7 — 7.0 distilled water
      — 7.4 human blood
- 8 — 8.0 seawater
      — 8.4 sodium bicarbonate
- 9 — 9.2 borax, alkaline soils
- 10
      — 10.5 milk of magnesia
- 11 — 11.5 household ammonia
- 12 — 12.4 limewater
- 13 — 13.2 oven cleaner
- 14 — 1 M potassium hydroxide

Acidic $[H^+]$ — Neutral — Basic (alkaline) $[OH^-]$

TABLE 2.2

Hydrogen Ion and Hydroxyl Ion Concentrations at a Given pH

Moles/L of Hydrogen Ions	Logarithm	pH	Moles/L of OH⁻
1.0	10^{-0}	0	10^{-14}
0.1	10^{-1}	1	10^{-13}
0.01	10^{-2}	2	10^{-12}
0.001	10^{-3}	3	10^{-11}
0.0001	10^{-4}	4	10^{-10}
0.00001	10^{-5}	5	10^{-9}
0.000001	10^{-6}	6	10^{-8}
0.0000001	10^{-7}	7	10^{-7}
0.00000001	10^{-8}	8	10^{-6}
0.000000001	10^{-9}	9	10^{-5}
0.0000000001	10^{-10}	10	10^{-4}
0.00000000001	10^{-11}	11	10^{-3}
0.000000000001	10^{-12}	12	10^{-2}
0.0000000000001	10^{-13}	13	10^{-1}
0.00000000000001	10^{-14}	14	10^{-0}

H^+ and OH^- are produced in equal amounts, and the solution remains neutral. By one definition, a solution is considered **acidic** when a component dissolved in water (acid) releases excess hydrogen ions[3] (H^+); a solution is **basic** when a component releases excess hydroxyl ions (OH^-), so that there is no longer a balance between the two ions.

To measure the acid and base concentrations of solutions, scientists use the **pH scale,** a graduated numerical scale that ranges from 0 (the most acidic) to 14 (the most basic). This scale is a useful standard for rating relative acidity and basicity; use figure 2.12 to familiarize yourself with the pH readings of some common substances. It is not an arbitrary scale but actually a mathematical derivation based on the negative logarithm (reviewed in appendix B) of the concentration of H^+ ions in moles per liter (symbolized as $[H^+]$) in a solution, represented as:

$$pH = -\log[H^+]$$

Acidic solutions have a greater concentration of H^+ than OH^-, starting with pH 0, which contains 1.0 moles $H^+/1$. Each of the subsequent whole-number readings in the scale changes in $[H^+]$ by a tenfold reduction, so that pH 1 contains [0.1 moles $H^+/1$], pH 2 contains [0.01 moles $H^+/1$], and so on, continuing in the same manner up to pH 14, which contains [0.00000000000001 moles $H^+/1$]. These same concentrations can be represented more manageably by exponents: pH 2 has a $[H^+]$ of 10^{-2} moles, and pH 14 has a $[H^+]$ of 10^{-14} moles (table 2.2). It is evident that the pH units are derived from the exponent itself. Even though the basis for the pH scale is $[H^+]$, it is important to note that, as the $[H^+]$ in a solution decreases, the $[OH^-]$ increases in direct proportion. At midpoint—pH 7, or neutrality—the concentrations are exactly equal and neither predominates, this being the pH of pure water previously mentioned.

In summary, the pH scale can be used to rate or determine the degree of acidity or basicity (also called alkalinity) of a solution. On this scale, a pH below 7 is acidic, and the lower the pH, the greater the acidity; a pH above 7 is basic, and the higher the pH, the greater the basicity. Incidentally, although pHs are given here in even whole numbers, more often, a pH reading exists in decimal form; for example, pH 4.5 or 6.8 (acidic) and pH 7.4 or 10.2 (basic). Because of the damaging effects of very concentrated acids or bases, most cells operate best under neutral, weakly acidic, or weakly basic conditions (see chapter 7).

Aqueous solutions containing both acids and bases may be involved in **neutralization** reactions, which give rise to water and other neutral by-products. For example, when equal molar solutions of hydrochloric acid (HCl) and sodium hydroxide (NaOH, a base) are mixed, the reaction proceeds as follows:

$$HCl + NaOH \rightarrow H_2O + NaCl$$

Here the acid and base ionize to H^+ and OH^- ions, which form water, and other ions, Na^+ and Cl^-, which form sodium chloride. Any product other than water that arises when acids and bases react is called a **salt.** Many of the organic acids (such as lactic and

---

3. Actually, it forms a hydronium ion (H_3O^+), but for simplicity's sake, we will use the notation of H^+.

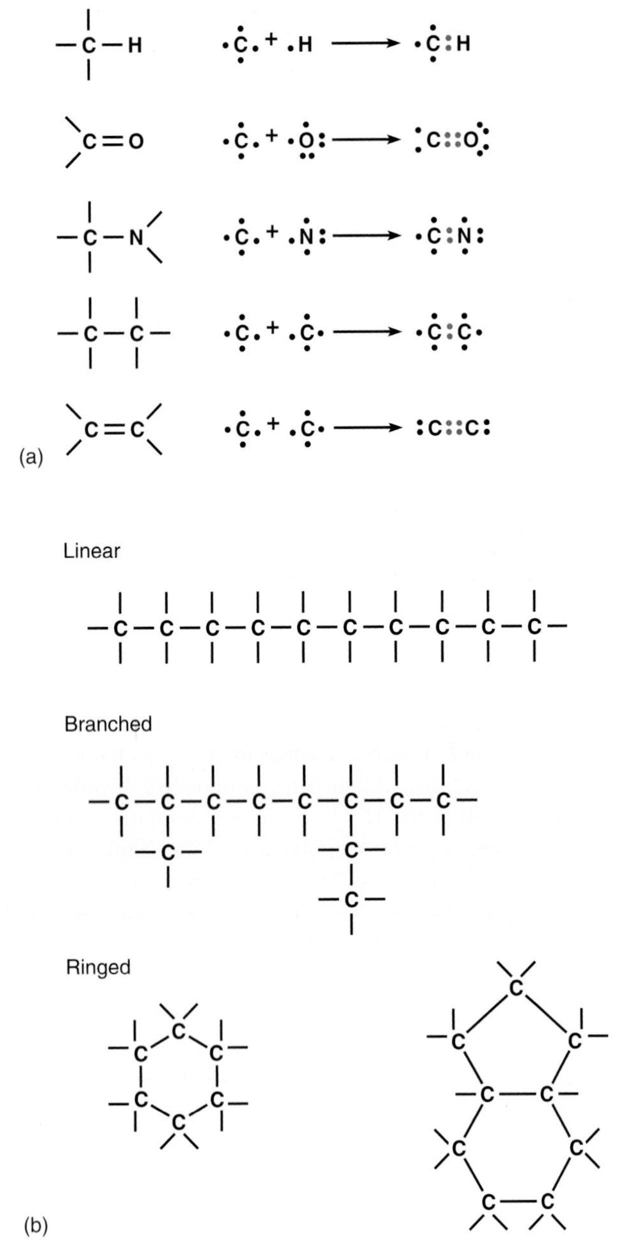

(a)

Linear

Branched

Ringed

(b)

FIGURE 2.13

The versatility of bonding in carbon. In most compounds, each carbon makes a total of four bonds. **(a)** Both single and double bonds can be made with other carbons, oxygen, and nitrogen; single bonds are made with hydrogen. Simple electron models show how the electrons are shared in these bonds. **(b)** Multiple bonding of carbons can give rise to long chains, branched compounds, and ringed compounds, many of which are extraordinarily large and complex.

TABLE 2.3 Representative Functional Groups and Classes of Organic Compounds		
Formula of Functional Group	Name	Class of Compounds
R* — O — H	Hydroxyl	Alcohols, carbohydrates
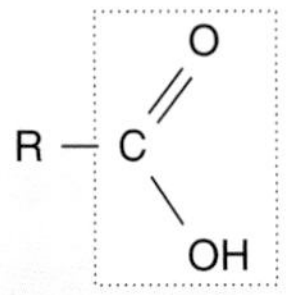 R — C (=O) (OH)	Carboxyl	Fatty acids, proteins, organic acids
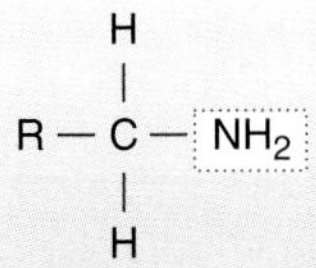 R — C(H)(H) — NH₂	Amino	Proteins, nucleic acids
R — C (=O) (O—R)	Ester	Lipids
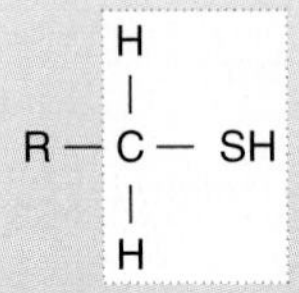 R — C(H)(H) — SH	Sulfhydryl	Cysteine (amino acid), proteins
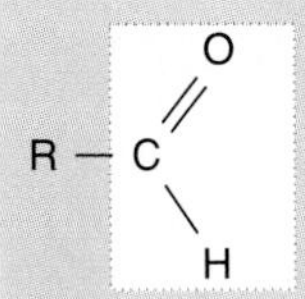 R — C (=O) — H	Carbonyl, terminal end	Aldehydes, polysaccharides
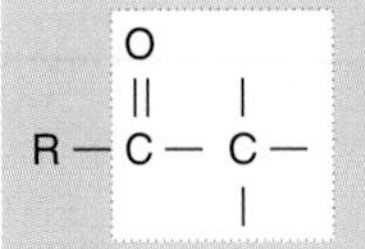 R — C(=O) — C —	Carbonyl, internal	Ketones, polysaccharides
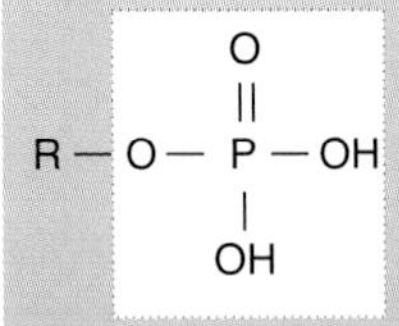 R — O — P(=O)(OH) — OH	Phosphate	DNA, RNA, ATP

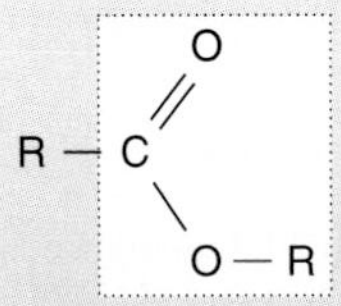

The R designation on a molecule is shorthand for residue, and its placement in a formula indicates that what is attached at that site varies from one compound to another.

THE CHEMISTRY OF CARBON AND ORGANIC COMPOUNDS

So far, our main focus has been on the characteristics of atoms, ions, and small, simple substances that play diverse roles in the structure and function of living things. These substances are often lumped together in a category called **inorganic* chemicals.** Examples of inorganic chemicals include $NaCl$ (sodium chloride), $Mg_3(PO_4)_2$ (magnesium phosphate), $CaCO_3$ (calcium carbonate), and CO_2 (carbon dioxide). In reality, however, most of the chemi-

succinic acids) that function in **metabolism*** are available as the acid and the salt form (such as lactate, succinate), depending on the conditions in the cell (see chapter 8).

---

*metabolism (muh-tab′-oh-lizm) A general term referring to the totality of chemical and physical processes occurring in the cell.

*inorganic (in-or-gan′-ik) Any chemical substances that do not contain both carbon and hydrogen.

TABLE 2.4

Macromolecules and Their Functions

Macromolecule	Description/Basic Structure	Examples/Functions
Carbohydrates		
Monosaccharides	3–7-carbon sugars	Glucose, fructose / Sugars involved in metabolic reactions; building block of disaccharides and polysaccharides
Disaccharides	Two monosaccharides	Maltose (malt sugar) / Composed of two glucoses; an important breakdown product of starch
		Lactose (milk sugar / Composed of glucose and galactose
		Sucrose (table sugar) / Composed of glucose and fructose
Polysaccharides	Chains of monosaccharides	Starch, cellulose, glycogen / Cell wall, food storage
Lipids		
Triglycerides	Fatty acids + glycerol	Fats, oils / Major component of cell membranes; storage
Phospholipids	Fatty acids + glycerol + phosphate	Membranes
Waxes	Fatty acids, alcohols	Mycolic acid / Cell wall of mycobacteria
Steroids	Ringed structure (not a polymer)	Cholesterol, ergosterol / Membranes of eucaryotes and some bacteria
Proteins	Amino acids	Enzymes; part of cell membrane, cell wall, ribosomes, antibodies / Metabolic reactions; structural components

Nucleic acids	Pentose sugar + phosphate + nitrogenous base
	  Purines; adenine, guanine
	  Pyrimidines: cytosine, thymine, uracil	

Deoxyribonucleic acid (DNA)	Contains deoxyribose sugar and thymine, not uracil	Chromosomes; genetic material of viruses / Inheritance
Ribonucleic acid (RNA)	Contains ribose sugar and uracil, not thymine	Ribosomes; mRNA, tRNA / Expression of genetic traits

cal reactions and structures of living things occur at the level of more complex molecules, termed **organic* chemicals.** These are defined as molecules that contain a basic framework of the elements carbon and hydrogen. Organic molecules vary in complexity from the simplest, methane (CH_4; figure 2.4c), which has a molecular weight of 16, to certain antibody molecules (produced by an immune reaction) that have a molecular weight of nearly 1,000,000 and are among the most complex molecules on earth.

The role of carbon as the fundamental element of life can best be understood if we look at its chemistry and bonding patterns. The valence of carbon makes it an ideal atomic building block to form the backbone of organic molecules; it has 4 electrons in its outer orbital to be shared with other atoms (including other carbons) through covalent bonding. As a result, it can form stable chains containing thousands of carbon atoms and still has bonding sites available for forming covalent bonds with numerous other atoms. The bonds that carbon forms are linear, branched, or ringed, and it can form four single bonds, two double bonds, or one triple bond (figure 2.13). The atoms with which carbon is most often associated in organic compounds are hydrogen, oxygen, nitrogen, sulfur, and phosphorus.

FUNCTIONAL GROUPS OF ORGANIC COMPOUNDS

One important advantage of carbon's serving as the molecular skeleton for living things is that it is free to bind with an unending array of other molecules. These special molecular groups or accessory molecules that bind to organic compounds are called **functional groups.** Functional groups help define the chemical class of certain groups of organic compounds and confer unique reactive properties on the whole molecule (table 2.3). Because each type of functional group behaves in a distinctive manner, reactions of an organic compound can be predicted by knowing the kind of functional group or groups it carries. Many synthesis, decomposition, and transfer reactions rely upon functional groups such as R—OH or R—NH_2. The **—R** designation on a molecule is shorthand for residue, and its placement in a formula indicates that the group attached at that site varies from one compound to another.

CHAPTER CHECKPOINTS

Covalent bonds are chemical bonds in which electrons are shared between atoms. Equally distributed electrons form nonpolar covalent bonds, whereas unequally distributed electrons form polar covalent bonds.

*organic (or-gan'-ik) Gr. *organikos,* instrumental.

Ionic bonds are chemical bonds in which the outer electron shell either donates or receives electrons from another atom so that the outer shell of each atom is completely filled.

Hydrogen bonds are weak chemical bonds that form between covalently bonded hydrogens and either oxygens or nitrogens on different molecules.

Chemical energy is generated by the movement of electrons from one atom or molecule to another. Chemical equations express movement of energy in chemical reactions such as synthesis or decomposition reactions.

Solutions are mixtures of solutes and solvents that cannot be separated by filtration or settling.

The pH, ranging from a highly *acidic* solution to highly *basic* solution, refers to the concentration of hydrogen ions. It is expressed as a number from 0 to 14.

Biologists define organic molecules as those containing both carbon and hydrogen.

Carbon is the backbone of biological compounds because of its ability to form single, double, or triple covalent bonds with itself and many different elements.

Functional (R) groups are specific arrangements of organic molecules that confer distinct properties, including chemical reactivity, to organic compounds.

Macromolecules: Superstructures of Life

The compounds of life fall into the realm of **biochemistry.** Biochemicals are organic compounds produced by (or components of) living things, and they include four main families: carbohydrates, lipids, proteins, and nucleic acids (table 2.4). The compounds in these groups are assembled from smaller molecular subunits, or building blocks, and because they are often very large compounds, they are termed **macromolecules.** All macromolecules except lipids are formed by **polymerization,** a process in which repeating subunits termed **monomers*** are bound into chains of various lengths termed **polymers.*** For example, proteins (polymers) are composed of a chain of amino acids (monomers) (see figure 2.22a). The large size and complex, three-dimensional shape of macromolecules enables them to function as structural components, molecular messengers, energy sources, enzymes (biochemical catalysts), nutrient stores, and sources of genetic information. In the following section and in later chapters, we will consider numerous concepts relating to the roles of macromolecules in cells.

CARBOHYDRATES: SUGARS AND POLYSACCHARIDES

The term **carbohydrate** originates from the way that most members of this chemical class resemble combinations of carbon and water. Although carbohydrates can be generally represented by the formula $(CH_2O)_n$, in which n indicates the number of units of this combination of atoms, some carbohydrates contain additional

atoms of sulfur or nitrogen. In molecular configuration, the carbons form chains or rings with two or more hydroxyl groups and either an aldehyde or a ketone group, giving them the technical designation of *polyhydroxy aldehydes* or *ketones* (figure 2.14).

Carbohydrates exist in a great variety of configurations. The common term **sugar** (*saccharide*)* refers to a simple carbohydrate such as a monosaccharide or a disaccharide that has a sweet taste. A **monosaccharide** is a simple polyhydroxy aldehyde or ketone molecule containing from 3 to 7 carbons; a **disaccharide** is a combination of two monosaccharides; and a **polysaccharide** is a polymer of five or more monosaccharides bound in linear or branched chain patterns (figure 2.14). Monosaccharides and disaccharides are specified by combining a prefix that describes some characteristic of the sugar with the suffix **-ose.** For example, **hexoses** are composed of 6 carbons, and **pentoses** contain 5 carbons. **Glucose** (Gr. sweet) is the most common and universally important hexose; **fructose** is named for fruit (one of its sources); and xylose, a pentose, derives its name from the Greek word for wood. Disaccharides are named similarly: **lactose** (L. milk) is an important component of milk; **maltose** means malt sugar; and **sucrose** (Fr. sugar) is common table sugar or cane sugar.

The Nature of Carbohydrate Bonds

The subunits of disaccharides and polysaccharides are linked by means of **glycosidic bonds,** in which carbons (each is assigned a number) on adjacent sugar units are bonded to the same oxygen atom like links in a chain (figure 2.15). For example, maltose is formed when the number 1 carbon on a glucose bonds to the oxygen on the number 4 carbon on a second glucose; sucrose is formed when glucose and fructose bind oxygen between their number 1 and number 2 carbons; and lactose is formed when glucose and galactose connect by their number 1 and number 4 carbons. In order to form this bond, one carbon gives up its OH group and the other (the one contributing the oxygen to the bond) loses the H from its OH group. Because a water molecule is produced, this reaction is known as **dehydration synthesis,** a process common to most polymerization reactions (see proteins, page 47). Three polysaccharides (starch, cellulose, and glycogen) are structurally and biochemically distinct, even though all are polymers of the same monosaccharide—glucose. The basis for their differences lies primarily in the exact way the glucoses are bound together, which greatly affects the characteristics of the end product (figure 2.16). The synthesis and breakage of each type of bond requires a specialized catalyst called an enzyme (see chapter 8).

The Functions of Polysaccharides

Polysaccharides typically contribute to structural support and protection and serve as nutrient and energy stores. The cell walls in plants and many microscopic algae derive their strength and rigidity from **cellulose,** a long, fibrous polymer (figure 2.16a). Because of this role, cellulose is probably one of the most common organic substances on the earth, yet it is digestible only by certain bacteria, fungi, and protozoa. These microbes, called decomposers, play an essential role in breaking down and recycling plant materials (see figure 7.2). Some bacteria secrete slime layers of a glucose polymer

---

*monomer (mahn′-oh-mur) Gr. *mono,* one, and *meros,* part.

*polymer (pahl′-ee-mur) Gr. *poly,* many; also the root for polysaccharide and polypeptide.

*saccharide (sak′-uh-ryd) Gr. *sakcharon,* sweet.

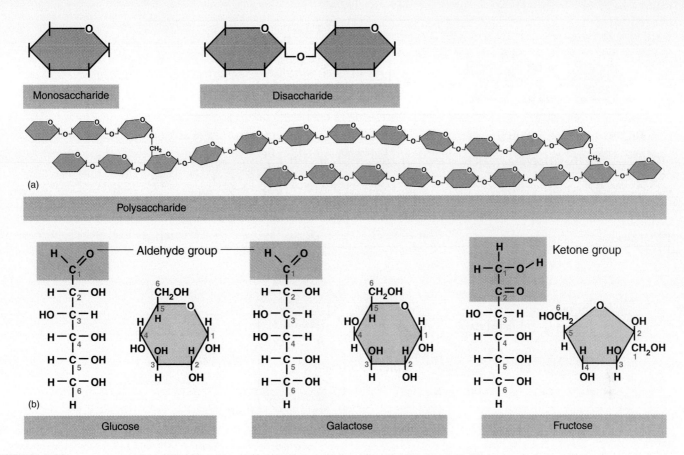

FIGURE 2.14

Common classes of carbohydrates. (a) Major saccharide groups, named for the number of sugar units each contains. **(b)** Three hexoses with the same molecular formula and different structural formulas. Both linear and ring models are given. The linear form indicates aldehyde and ketone groups, although in solution the sugars exist in the ring form. Note that the carbons are numbered so as to keep track of reactions within and between monosaccharides.

called *dextran*. This substance causes a sticky layer to develop on teeth that leads to plaque (see figure 4.12).

Other structural polysaccharides can be conjugated (chemically bonded) to amino acids, nitrogen bases, lipids, or proteins. **Agar,** an indispensable polysaccharide in preparing solid culture media, is a natural component of certain seaweeds. It is a complex polymer of galactose and sulfur-containing carbohydrates. The exoskeletons of certain fungi contain **chitin,** a polymer of glucosamine (a sugar with an amino functional group). **Peptidoglycan*** is one special class of compounds in which polysaccharides (glycans) are linked to peptide fragments (a short chain of amino acids). This molecule provides the main source of structural support to the bacterial cell wall. The cell wall of gram-negative bacteria also contains **lipopolysaccharide,** a complex of lipid and polysaccharide responsible for symptoms such as fever and shock (see chapters 4 and 13).

The outer surface of many cells has a delicate "sugar coating" composed of polysaccharides bound in various ways to proteins (the combination is called mucoprotein or glycoprotein). This structure, called the **glycocalyx,*** functions in attachment to other cells or as a site for *receptors*—surface molecules that receive and respond to external stimuli. Small sugar molecules account for the differences in human blood types, and carbohydrates are a component of large protein molecules called antibodies. Some viruses have glycoproteins on their surface with which they bind to and invade their host cells.

Polysaccharides are usually stored by cells in the form of glucose polymers such as **starch** (figure 2.16*b*) or **glycogen,** but only organisms with the appropriate digestive enzymes can break them down and use them as a nutrient source. Because a water molecule is required for breaking the bond between two glucose molecules, digestion is also termed **hydrolysis.*** Starch is the primary storage food of green plants, microscopic algae, and some fungi; glycogen (animal starch) is a stored carbohydrate for animals and certain groups of bacteria.

LIPIDS: FATS, PHOSPHOLIPIDS, AND WAXES

The term **lipid,** derived from the Greek word *lipos,* meaning fat, is not a chemical designation, but an operational term for a variety of substances that are not soluble in polar solvents such as water

*peptidoglycan (pep-tih-doh-gly'-kan).

*glycocalyx (gly"-koh-kay'-lix) Gr. *glycos,* sweet, and *calyx,* covering.

*hydrolysis (hy-drol'-uh-sis) Gr. *hydros,* water, and *lyein,* to dissolve.

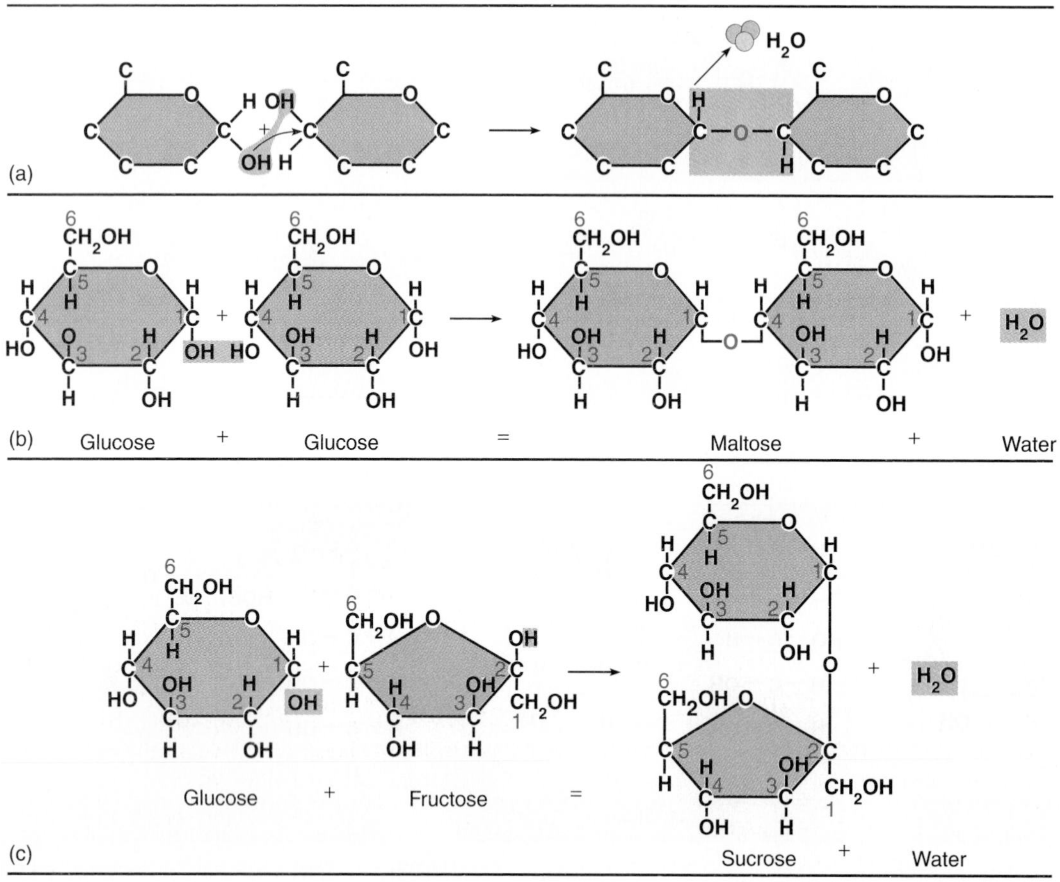

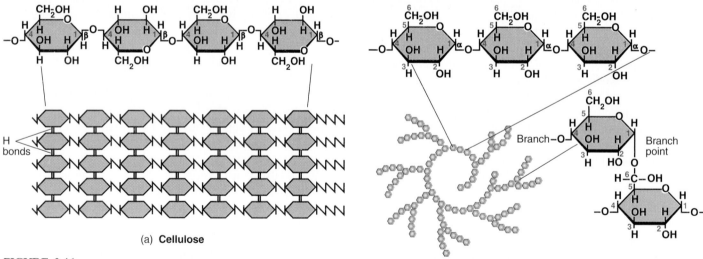

FIGURE 2.15

Glycosidic bond. (a) General scheme in the formation of a glycosidic bond by dehydration synthesis. **(b)** Formation of the 1,4 bond between two α glucoses to produce maltose and water. **(c)** Formation of the 1,2 bond between glucose and fructose to produce sucrose and water.

(a) **Cellulose**

(b) **Starch**

FIGURE 2.16

Polysaccharides. (a) Cellulose is composed of β glucose bonded in 1,4 bonds that produce linear, lengthy chains of polysaccharides that are H-bonded along their length. This is the typical structure of wood and cotton fibers. **(b)** Starch is also composed of glucose polymers, in this case α glucose. The main structure is amylose bonded in a 1,4 pattern, with side branches of amylopectin bonded by 1,6 bonds. The entire molecule is compact and granular.

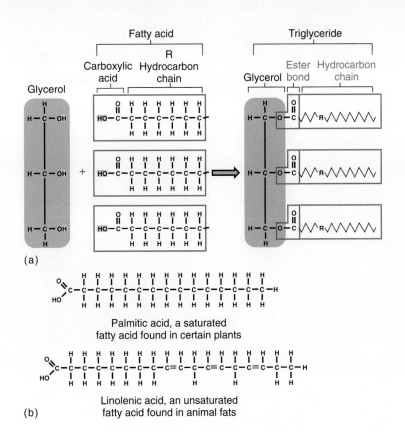

Palmitic acid, a saturated
fatty acid found in certain plants

Linolenic acid, an unsaturated
fatty acid found in animal fats

(b)

FIGURE 2.17

Synthesis and structure of a triglyceride.
(a) Because a water molecule is released at
each ester bond, this is another form of
dehydration synthesis. The jagged lines and
R symbol represent the hydrocarbon chains of
the fatty acids, which are commonly very long.
(b) Structural formulas for saturated and
unsaturated fatty acids.

(recall that oil and water do not mix) but will dissolve in nonpolar solvents such as benzene and chloroform. This property occurs because the substances we call lipids contain relatively long or complex C—H (hydrocarbon) chains that are nonpolar and thus hydrophobic. The main groups of compounds classified as lipids are triglycerides, phospholipids, steroids, and waxes.

Important storage lipids are the **triglycerides,** a category that includes fats and oils. Triglycerides are composed of a single molecule of glycerol bound to three fatty acids (figure 2.17). **Glycerol** is a 3-carbon alcohol with three OH groups that serve as binding sites, and **fatty acids** are long-chain hydrocarbon molecules with a carboxyl group (COOH) at one end that is free to bind to the glycerol. The bond that forms between the —OH group and the —COOH is defined as an **ester bond.** The hydrocarbon portion of a fatty acid can vary in length from 4 to 24 carbons and, depending on the fat, it may be saturated or unsaturated. If all carbons in the chain are single-bonded to 2 other carbons and 2 hydrogens, the fat is saturated; if there is at least one C=C double bond in the chain, it is unsaturated. The structure of fatty acids is what gives fats and oils (liquid fats) their greasy, insoluble nature. In general, solid fats (such as beef tallow) are more saturated, and oils (or liquid fats) are more unsaturated. In most cells, triglycerides are stored in long-term concentrated form as droplets or globules. When the ester linkage is acted on by digestive enzymes called lipases, the fatty acids and glycerol are freed to be used in metabolism. Fatty acids are a superior source of energy, yielding twice as much per gram as other storage molecules (starch). Soaps are K^+ or Na^+ salts of fatty acids whose qualities make them excellent grease removers and cleaners (see chapter 11).

Membrane Lipids

A class of lipids that serves as a major structural component of cell membranes is the **phospholipids.** Although phospholipids also contain glycerol and fatty acids, they have some significant differences from triglycerides. Phospholipids contain only two fatty acids attached to the glycerol, and the third glycerol binding site holds a phosphate group. The phosphate is in turn bonded to an alcohol,[4] which varies from one phospholipid to another (figure 2.18a). These lipids have a hydrophilic region from the charge on the phosphoric acid–alcohol "head" of the molecule and a hydrophobic region that corresponds to the long, uncharged "tail" (formed by the fatty acids). When exposed to an aqueous solution, the charged heads are attracted to the water phase, and the nonpolar tails are repelled from the water phase (figure 2.18b). This property causes lipids to naturally assume single and double layers (bilayers), which contribute to their biological significance in membranes. When two single layers of polar lipids come together to form a double layer, the outer hydrophilic face of each single layer will orient itself toward the solution, and the hydrophobic portions will become immersed in the core of the bilayer. The structure of lipid bilayers confers characteristics on membranes such as selective permeability and fluid nature (Microbits 2.3).

Miscellaneous Lipids

Steroids are complex ringed compounds commonly found in cell membranes and animal hormones. The best known of these is the sterol (meaning a steroid with an OH group) called **cholesterol**

---

4. Alcohols are hydrocarbons containing OH groups.

FIGURE 2.18

Phospholipids—membrane molecules.
(a) A complex model of a single molecule of a phospholipid. The phosphate-alcohol head lends a charge to one end of the molecule; its long, trailing hydrocarbon chain is uncharged. **(b)** The behavior of phospholipids in water-based solutions causes them to become arranged **(1)** in single layers called micelles, with the charged head oriented toward the water phase and the hydrophobic nonpolar tail buried away from the water phase, or **(2)** in double-layered phospholipid systems with the hydrophobic tails sandwiched between two hydrophilic layers.

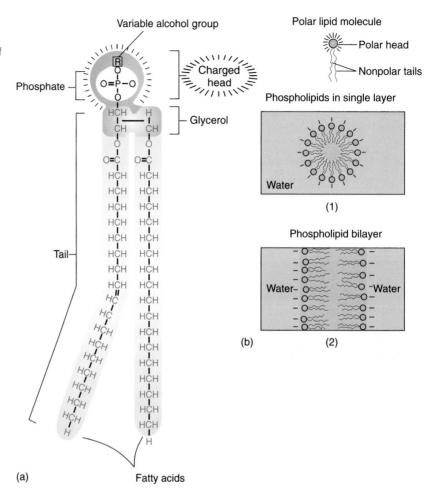

FIGURE 2.19

Formula for cholesterol, an alcoholic steroid that is inserted in some membranes. Cholesterol can become esterified with fatty acids at its OH group, imparting a polar quality similar to that of phospholipids.

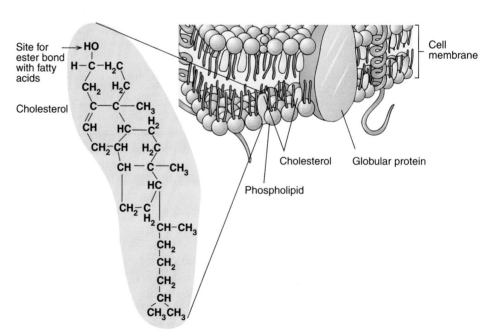

MICROBITS 2.3
Membranes: Cellular Skins

The word **membranes** appears frequently in descriptions of cells in this chapter and in chapters 4 and 5. The word itself describes any lining or covering, including such multicellular structures as the mucous membranes of the body. From the perspective of a single cell, however, a membrane is a thin, double-layered sheet composed of lipids such as phospholipids and sterols (averaging about 40% of membrane content) and protein molecules (averaging about 60%). The primary role of membranes is as a **cell membrane** that completely encases the cytoplasm. Membranes are also components of eucaryotic organelles such as nuclei, mitochondria, and chloroplasts, and they appear in internal pockets of certain procaryotic cells. Even some viruses, which are not cells at all, can have a membranous protective covering.

Cell membranes are so thin—on the average, just 0.0070 μm (7 nm) thick—that they cannot actually be seen with an optical microscope. Even at magnifications made possible by electron microscopy (500,000×), very little of the precise architecture can be visualized, and a cross-sectional view has the appearance of railroad tracks. Following detailed microscopic and chemical analysis, S. J. Singer and C. K. Nicholson proposed a simple and elegant theory for membrane structure

called the **fluid mosaic* model.** According to this theory, a membrane is a continuous bilayer formed by lipids that are oriented with the polar lipid heads toward the outside and the nonpolar heads toward the center of the membrane. Embedded at numerous sites in this bilayer are various-sized globular proteins. Some proteins are situated only at the surface; others extend fully through the entire membrane. The configuration of the inner and outer sides of the membrane can be quite different because of the variations in protein shape and position.

Membranes are dynamic and constantly changing because the lipid phase is in motion and many proteins can migrate freely about, somewhat as icebergs do in the ocean. This fluidity is essential to such activities as engulfment of food and discharge or secretion by cells. The structure of the lipid phase provides an impenetrable barrier to many substances. This property accounts for the **selective permeability** and capacity to regulate transport of molecules. It also serves to segregate activities within the cell's cytoplasm. Membrane proteins function in receiving molecular signals (receptors), in binding and transporting nutrients, and in acting as enzymes, topics to be discussed in chapters 7 and 8.

---

*mosaic (moh-zay′-ik) An intricate design made up of many small fragments.

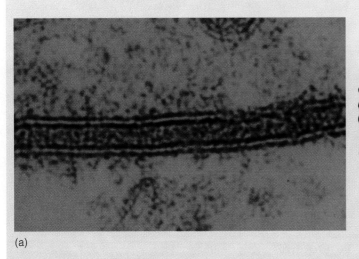

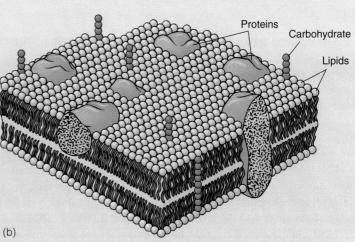

(a)   (b)

Extreme magnification of **(a)** a cross section of a cell membrane, which appears as double tracks. **(b)** A generalized version of the fluid mosaic model of a cell membrane indicates a bilayer of lipids with globular proteins embedded to some degree in the lipid matrix. This structure explains many characteristics of membranes, including flexibility, solubility, permeability, and transport.

(figure 2.19). Cholesterol reinforces the structure of the cell membrane in animal cells and in an unusual group of cell-wall-deficient bacteria called the mycoplasmas (see chapter 4). The cell membranes of fungi also contain a sterol, called ergosterol. *Prostaglandins* are fatty acid derivatives found in trace amounts that function in inflammatory and allergic reactions, blood clotting, and smooth muscle contraction. Chemically, a *wax* is an ester formed between a long-chain alcohol and a saturated fatty acid. The resulting material is typically pliable and soft when warmed but hard and water-resistant when cold (paraffin, for example). Among

living things, fur, feathers, fruits, leaves, human skin, and insect exoskeletons are naturally waterproofed with a coating of wax. Bacteria that cause tuberculosis and leprosy produce a wax (wax D) that repels ordinary stains and contributes to their pathogenicity.

PROTEINS: SHAPERS OF LIFE

The predominant organic molecules in cells are **proteins,** a fitting term adopted from the Greek word *proteios,* meaning first or prime. To a large extent, the structure, behavior, and unique qualities of

TABLE 2.5

Twenty Amino Acids and Their Abbreviations

Acid	Abbreviation	Characteristic of R Groups*
Alanine	Ala	NP
Arginine	Arg	+
Asparagine	Asn	P
Aspartic acid	Asp	–
Cysteine	Cys	P
Glutamic acid	Glu	–
Glutamine	Gln	P
Glycine	Gly	P
Histidine	His	+
Isoleucine	Ile	NP
Leucine	Leu	NP
Lysine	Lys	+
Methionine	Met	NP
Phenylalanine	Phe	NP
Proline	Pro	NP
Serine	Ser	P
Threonine	Thr	P
Tryptophan	Trp	NP
Tyrosine	Tyr	P
Valine	Val	NP

*NP, nonpolar; P, polar; +, positively charged; –, negatively charged.

Amino Acid	Structural Formula

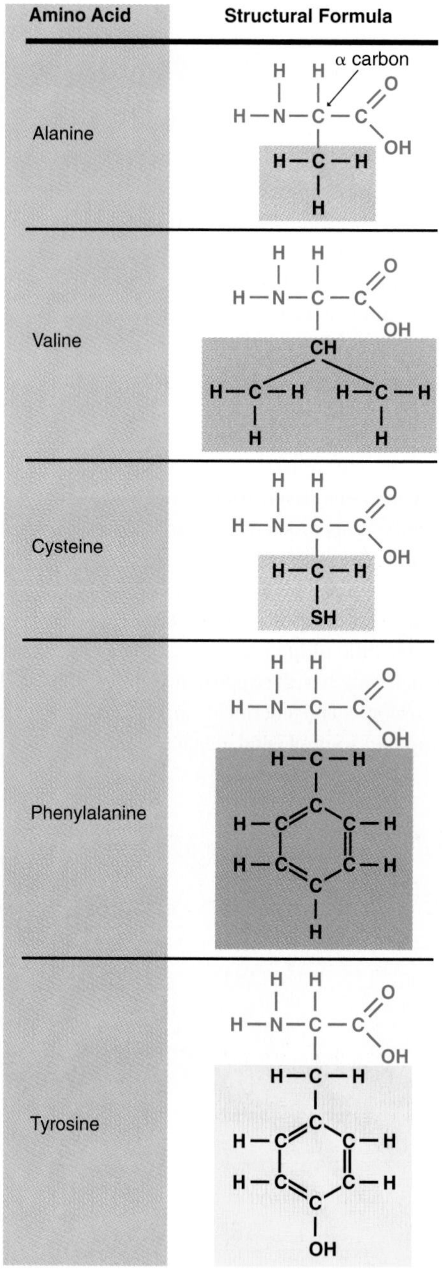

FIGURE 2.20

Structural formulas of selected amino acids. The basic structure common to all amino acids is shown in blue and the variable group, or R group, is placed in a colored box. Note the variations in structure of this reactive component.

each living thing are a consequence of the proteins they contain. To best explain the origin of the special properties and versatility of proteins, we must examine their general structure. The building blocks of proteins are **amino acids,** which exist in 20 different naturally occurring forms (table 2.5). Various combinations of these amino acids account for the nearly infinite variety of proteins. Amino acids have a basic skeleton consisting of a carbon (called the α carbon) linked to an amino group (NH_2), a carboxyl group (COOH), a hydrogen atom (H), and a variable R group. The variations among the amino acids occur at the R group, which is different in each amino acid and imparts the unique characteristics to the molecule and to the proteins that contain it (figure 2.20). A covalent bond called a **peptide bond** forms between the amino group on one amino acid and the carboxyl group on another amino acid. As a result of peptide bond formation, it is possible to produce molecules varying in length from two amino acids to chains containing thousands of them.

Various terms are used to denote the nature of compounds containing peptide bonds. **Peptide*** usually refers to a molecule composed of short chains of amino acids, such as a dipeptide (two amino acids), a tripeptide (three), and a tetrapeptide (four) (figure 2.21). A **polypeptide** contains an unspecified number of amino acids, but usually has more than 20, and is often a smaller subunit of a protein. A protein is the largest of this class of compounds and usually contains a minimum of 50 amino acids. It is common for the terms *polypeptide* and *protein* to be used interchangeably,

though not all polypeptides are large enough to be considered proteins. In chapter 9 we see that protein synthesis is not just a random connection of amino acids; it is directed by information provided in DNA.

Protein Structure and Diversity

The reason that proteins are so varied and specific is that they do not function in the form of a simple straight chain of amino acids (called the primary structure). A protein has a natural tendency to assume more complex levels of organization, called the secondary,

*peptide (pep'-tyd) Gr. *pepsis*, digestion.

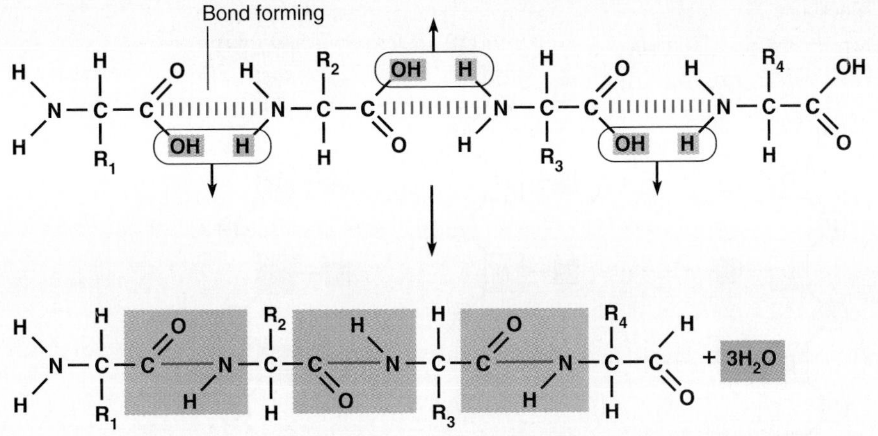

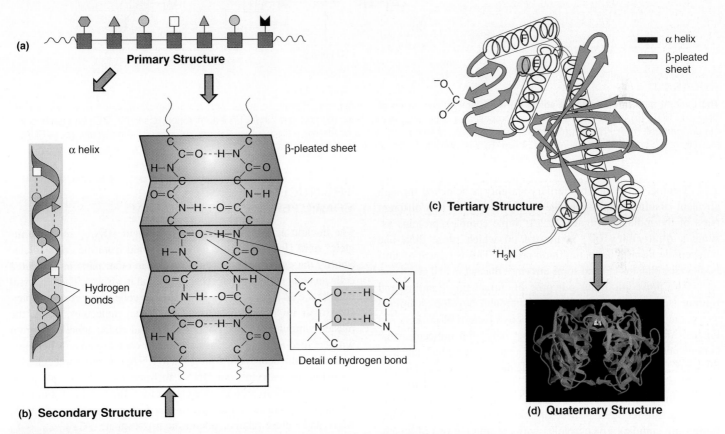

FIGURE 2.21

The formation of peptide bonds in a tetrapeptide.

FIGURE 2.22

🔘 **Stages in the formation of a functioning protein. (a)** Its primary structure is a series of amino acids bound in a chain. **(b)** Its secondary structure develops when the chain forms hydrogen bonds that fold it into one of several configurations such as an alpha helix or beta-pleated sheet. Some proteins have several configurations in the same molecule. **(c)** A protein's tertiary structure is due to further folding of the molecule into a three-dimensional mass that is stabilized by hydrogen, ionic, and disulfide bonds between functional groups. The letters and arrows denote the order and direction of the folded chain. **(d)** The quaternary structure exists only in proteins that consist of more than one polypeptide chain. Shown here is a computer model of the nitrogenase iron protein, with the two polypeptide chains arranged symmetrically.

tertiary, and quaternary structures (figure 2.22). The **primary (1°) structure** is more correctly described as the type, number, and order of amino acids in the chain, which varies extensively from protein to protein. The **secondary (2°) structure** arises when various functional groups exposed on the outer surface of the molecule interact by forming hydrogen bonds. This interaction causes the

amino acid chain to twist into a coiled configuration called the α *helix* or to fold into an accordion pattern called a β-*pleated sheet*. Some proteins contain both types of secondary configurations. Proteins at the secondary level undergo a third degree of torsion called the **tertiary (3°) structure** created by additional bonds between functional groups. In proteins with the sulfur-containing amino

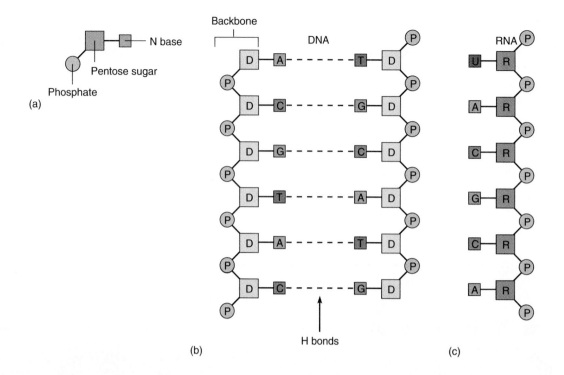

FIGURE 2.23

The general structure of nucleic acids. (a) A nucleotide, composed of a phosphate, a pentose sugar, and a nitrogen base, is the monomer of both DNA and RNA. **(b)** In DNA, the polymer is composed of alternating deoxyribose (D) and phosphate (P) with nitrogen bases (A, T, C, G) attached to the deoxyribose. Two of these polynucleotide strands are oriented so that the bases are paired across the central axis of the molecule. **(c)** In RNA, the polymer is composed of alternating ribose (R) and phosphate (P) attached to nitrogen bases (A, U, C, G), but it is only a single strand.

acid **cysteine,*** considerable tertiary stability is achieved through covalent **disulfide bonds** between sulfur atoms on two different parts of the molecule (figure 2.22c). Some complex proteins assume a **quaternary (4°) structure,** in which more than one polypeptide forms a large, multiunit protein. This is typical of antibodies (see chapter 15) and some enzymes that act in cell synthesis.

The most important outcome of intrachain[5] bonding and folding is that each different type of protein develops a unique shape, and its surface displays a distinctive pattern of pockets and bulges. As a result, a protein can react only with molecules that complement or fit its particular surface features like a lock and key. Such a degree of specificity can provide the functional diversity required for many thousands of different cellular activities. **Enzymes** serve as the catalysts for all chemical reactions in cells, and nearly every reaction requires a different enzyme (see chapter 8). **Antibodies** are complex glycoproteins with specific regions of attachment for bacteria, viruses, and other microorganisms; certain bacterial toxins (poisonous products) react with only one specific organ or tissue; and proteins embedded in the cell membrane have reactive sites restricted to a certain nutrient. Some proteins function as *receptors* to receive stimuli from the environment. The functional three-dimensional form of a protein is termed the *native state,* and if it is disrupted by some means, the protein is said to be *denatured.* Such agents as heat, acid, alcohol, and some disinfectants disrupt the stabilizing intrachain bonds and cause the molecule to become nonfunctional (see figure 11.4).

THE NUCLEIC ACIDS: A CELL COMPUTER AND ITS PROGRAMS

The nucleic acids, **deoxyribonucleic*** **acid (DNA)** and **ribonucleic*** **acid (RNA),** were originally isolated from the cell nucleus. Shortly thereafter, they were also found in other parts of nucleated cells, in cells with no nuclei (bacteria), and in viruses. The universal occurrence of nucleic acids in all known cells and viruses emphasizes their important roles as informational molecules. DNA, the master computer of cells, contains a special coded genetic program with detailed and specific instructions for each organism's heredity. It transfers the details of its program to RNA, operator molecules responsible for carrying out DNA's instructions and translating the DNA program into proteins that can perform life functions. For now, let us briefly consider the structure and some functions of DNA, RNA, and a close relative, adenosine triphosphate (ATP).

Both nucleic acids are polymers of repeating units called **nucleotides,*** each of which is composed of three smaller units: a *nitrogen base,* a *pentose* (5-carbon) sugar, and a *phosphate* (figure 2.23a). The nitrogen base is a cyclic compound that comes in two forms: *purines* (two rings) and *pyrimidines* (one ring). There are two types of purines—**adenine (A)** and **guanine (G)**—and three types of pyrimidines—**thymine (T), cytosine (C),** and **uracil (U)** (figure 2.24). A characteristic that differentiates DNA from RNA is that DNA contains all of the nitrogen bases except uracil, and RNA contains all of the nitrogen bases except thymine. The nitrogen base

---

5. **Intra**chain means within the chain; **inter**chain would be between two chains.

*cysteine (sis'-tuh-yeen) Gr. *kystis,* sac. An amino acid first found in urine stones.

*deoxyribonucleic (dee-ox"-ee-ry"-boh-noo-klay'-ik).

*ribonucleic (ry"-boh-noo-klay'-ik) It is easy to see why the abbreviations are used!

*nucleotide (noo'-klee-oh-tyd) From nucleus and acid.

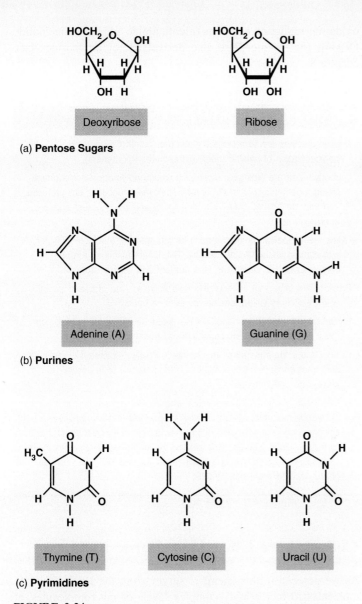

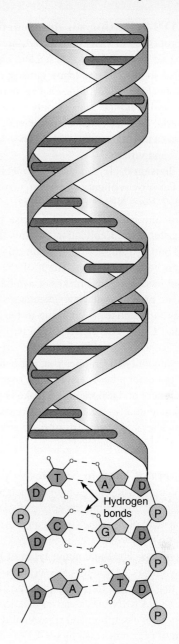

FIGURE 2.24

The sugars and nitrogen bases that make up DNA and RNA.
(a) DNA contains deoxyribose, and RNA contains ribose. **(b)** A and G
purines are found in both DNA and RNA. **(c)** C pyrimidine is found in
both DNA and RNA, but T is found only in DNA, and U is found only
in RNA.

FIGURE 2.25

A structural representation of the double helix of DNA. Shown are
the details of hydrogen bonds between the nitrogen bases of the two
strands.

is covalently bonded to the sugar *ribose* in RNA and *deoxyribose*
(because it has one less oxygen than ribose) in DNA. Phosphate
(PO_4^{-3}), a derivative of phosphoric acid (H_3PO_4), provides the final covalent bridge that connects sugars in series. Thus, the backbone of a nucleic acid strand is a chain of alternating phosphate-
sugar-phosphate-sugar molecules, and the nitrogen bases branch
off the side of this backbone (see figure 2.23*b,c*).

The Double Helix of DNA

DNA is a huge molecule formed by two very long polynucleotide
strands linked along their length by hydrogen bonds between complementary pairs of nitrogen bases. The pairing of the nitrogen
bases occurs according to a predictable pattern: Adenine ordinarily
pairs with thymine, and cytosine with guanine. The bases are
attracted in this way because each pair shares oxygen, nitrogen, and

hydrogen atoms exactly positioned to align perfectly for hydrogen
bonds (figure 2.25).

For ease in understanding the structure of DNA, it is sometimes compared to a ladder, with the sugar-phosphate backbone representing the rails and the paired nitrogen bases representing the
steps. Owing to the manner of nucleotide pairing and stacking of the
bases, the actual configuration of DNA is a *double helix* that looks
somewhat like a spiral staircase (see figures 2.25 and 9.4). As is true
of protein, the structure of DNA is intimately related to its function.
DNA molecules are usually extremely long, a feature that satisfies
a requirement for storing genetic information in the sequence of
base pairs the molecule contains. The hydrogen bonds between pairs
can be disrupted when DNA is being copied, and the fixed complementary base pairing is essential to maintain the genetic code.

Making New DNA: Passing on the Genetic Message

The biological properties of cells and viruses are ultimately programmed by a master code composed of nucleic acids. This code is in the form of DNA in all cells and many viruses; other viruses are based on RNA alone. Regardless of the exact genetic program, both cells and viruses will continue to exist only if they can duplicate their genetic material and pass it on to subsequent generations. Figure 2.26 summarizes the main steps in this process and how it differs between cells and viruses.

During its division cycle, the cell has a mechanism for making a copy of its DNA by **replication,*** using the original strand as a pattern (figure 2.26*a*). Note that replication is guided by the double-stranded nature of DNA and the precise pairing of bases that create the master code. Replication requires the separation of the double strand into two single strands by an enzyme that helps to split the hydrogen bonds along the length of the molecule. This event exposes the base code and makes it available for copying. Free nucleotides are used to synthesize matching strands that complement the bases in the code by adhering to the pairing requirements of A–T and C–G. The end result is two separate double strands with the same order of bases as the original molecule. Viruses have a related mode of replication, but being genetic parasites, they must first enter a host cell and take over the cell's regular synthetic machinery and program it to make copies of the virus nucleic acid and, eventually, create whole viruses (figure 2.26*b*). The details of nucleic acid chemistry of cells and viruses will be covered in greater detail in chapter 9.

RNA: Organizers of Protein Synthesis

Like DNA, RNA consists of a long chain of nucleotides. However, RNA is a single strand containing ribose sugar instead of deoxyribose and uracil instead of thymine (see figure 2.23). Several functional types of RNA are formed using the DNA template through a replication-like process. The three major types of RNA are important for protein synthesis. Messenger RNA (mRNA) is a copy of a gene from DNA that provides the order and type of amino acids in a protein; transfer RNA (tRNA) is a carrier that delivers the correct amino acids for protein assembly; and ribosomal RNA (rRNA) is a major component of ribosomes (see figure 4.19). More information on these important processes is presented in chapter 9.

ATP: The Energy Molecule of Cells

A relative of RNA involved in an entirely different cell activity is **adenosine triphosphate (ATP).** ATP is a nucleotide containing adenine, ribose, and three phosphates rather than just one (figure 2.27). It belongs to a category of high-energy compounds (also including guanosine triphosphate, GTP) that give off energy when the bond is broken between the second and third (outermost) phosphate. The presence of these high-energy bonds makes it possible for ATP to release and store energy for cellular chemical reactions. Breakage of the bond of the terminal phosphate releases energy to do cellular work and also generates adenosine diphosphate (ADP). ADP can be converted back to ATP when the third phosphate is restored, thereby serving as an energy depot. Carriers for

oxidation-reduction activities (nicotinamide adenine dinucleotide [NAD], for instance) are also derivatives of nucleotides (see chapter 8).

CHAPTER CHECKPOINTS

Macromolecules are very large organic molecules (polymers) built up by polymerization of smaller molecular subunits (monomers).

Carbohydrates are biological molecules whose polymers are monomers linked together by glycosidic bonds. Their main functions are protection and support (in organisms with cell walls) and also nutrient and energy stores.

Lipids are biological molecules such as fats that are insoluble in water and contain special ester linkages. Their main functions are cell components, cell secretions, and nutrient and energy stores.

Proteins are biological molecules whose polymers are chains of amino acid monomers linked together by peptide bonds.

Proteins are called the "shapers of life" because of the many biological roles they play in cell structure and cell metabolism.

Protein shape determines protein function. Shape is dictated by amino acid composition and by the pH and temperature of the protein's immediate environment.

Nucleic acids are biological molecules whose polymers are chains of nucleotide monomers linked together by phosphate–pentose sugar covalent bonds. Double-stranded nucleic acids are linked together by hydrogen bonds. Nucleic acids are information molecules that direct cell metabolism and reproduction. Nucleotides such as ATP also serve as energy transfer molecules in cells.

Cells: Where Chemicals Come to Life

As we proceed in this chemical survey from the level of simple molecules to increasingly complex levels of macromolecules, at some point we cross a line from the realm of lifeless molecules and arrive at the fundamental unit of life called a **cell.**[6] A cell is indeed a huge aggregate of carbon, hydrogen, oxygen, nitrogen, and many other atoms, and it follows the basic laws of chemistry and physics, but it is much more. The combination of these atoms produces characteristics, reactions, and products that can only be described as **living.**

FUNDAMENTAL CHARACTERISTICS OF CELLS

The bodies of living things such as bacteria and protozoa consist of only a single cell, whereas those of animals and plants contain trillions of cells. Regardless of the organism, all cells have a few common characteristics. They tend to be spherical, polygonal, cubical, or cylindrical, and their protoplasm (internal cell contents) is encased in a cell or cytoplasmic membrane (see Microbits 2.3). They have chromosomes containing DNA and ribosomes for protein synthesis, and they are exceedingly complex in function. Aside from

---

*replication (reh″-plih-kay′-shun) A process of making an exact copy of something.

6. The word *cell* was originally coined from an Old English term meaning "small room" because of the way plant cells looked to early microscopists.

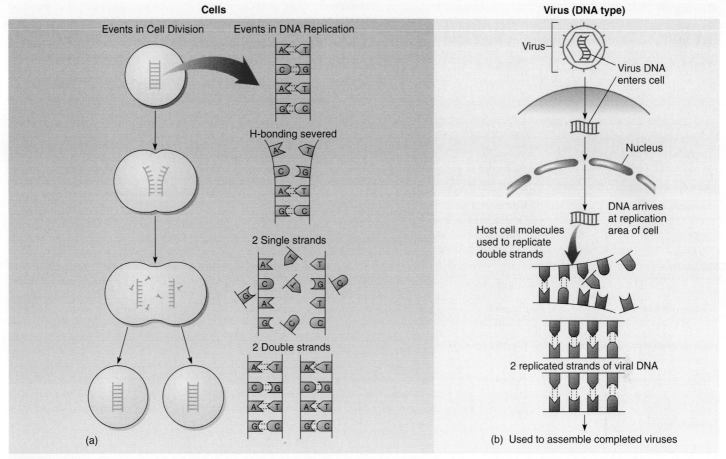

Cells

Events in Cell Division   Events in DNA Replication

H-bonding severed

2 Single strands

2 Double strands

(a)

Virus (DNA type)

Virus

Virus DNA enters cell

Nucleus

DNA arrives at replication area of cell

Host cell molecules used to replicate double strands

2 replicated strands of viral DNA

(b) Used to assemble completed viruses

FIGURE 2.26

Simplified view of DNA replication in cells and viruses. (a) The DNA in the cell's chromosome must be duplicated as the cell is dividing. This duplication is accomplished through the separation of the double DNA strand into two single strands. New strands are then synthesized using the original strands as guides to assemble the correct complementary new bases. **(b)** In DNA viruses, the DNA is inserted into its host cell and enters the replication area. Here it is separated into two strands and copied by the machinery of the host cell. Thousands of copies of viral nucleic acid will become part of the completed viruses later released by the infected host cell.

FIGURE 2.27

The structural formula of an ATP molecule, the chemical form of energy transfer in cells. The wavy lines that connect the phosphates represent bonds that release large amounts of energy when broken.

these few similarities, most cell types fall into one of two fundamentally different lines (discussed in chapter 1): the small, seemingly simple procaryotic cells and the larger, structurally more complicated eucaryotic cells.

Eucaryotic cells are found in animals, plants, fungi, and protists. They contain a number of complex internal parts called organelles that perform useful functions for the cell involving growth, nutrition, or metabolism. By convention, organelles are defined as cell components that perform specific functions and are enclosed by membranes. Organelles also partition the eucaryotic cell into smaller compartments. The most visible organelle is the nucleus, a roughly ball-shaped mass surrounded by a double membrane that contains the DNA of the cell. Other organelles include the Golgi apparatus, endoplasmic reticulum, vacuoles, and mitochondria (table 2.6).

TABLE 2.6

A General Comparison of Procaryotic and Eucaryotic Cells and Viruses*

Function or Structure	Characteristic	Procaryotic Cells	Eucaryotic Cells	Viruses**
Genetics	Nucleic acids	+	+	+
	Chromosomes	+	+	−
	True nucleus	−	+	−
	Nuclear envelope	−	+	−
Reproduction	Mitosis	−	+	−
	Production of sex cells	+/−	+	−
	Binary fission	+	+	−
Biosynthesis	Independent	+	+	−
	Golgi apparatus	−	+	−
	Endoplasmic reticulum	−	+	−
	Ribosomes	+***	+	−
Respiration	Enzymes	+	+	−
	Mitochondria	−	+	−
Photosynthesis	Pigments	+/−	+/−	−
	Chloroplasts	−	+/−	−
Motility/locomotor structures	Flagella	+/−***	+/−	−
	Cilia	−	+/−	−
Shape/protection	Cell wall	+***	+/−	−
	Capsule	+/−	+/−	−
	Spores	+/−	+/−	−
Complexity of function		+	+	+/−
Size (in general)		0.5–3 μm	2–100 μm	< 0.2 μm

*+ means most members of the group exhibit this characteristic; − means most lack it; +/− means only some members have it.

** Viruses cannot participate in metabolic or genetic activity outside their host cells.

*** The procaryotic type is functionally similar to the eucaryotic, but structurally unique.

Procaryotic cells are possessed only by the bacteria and archaea. Sometimes it may seem that procaryotes are the microbial "have-nots" because, for the sake of comparison, they are described by what they lack. They have no nucleus or other organelles. This apparent simplicity is misleading, because the fine structure of procaryotes is complex. Overall, procaryotic cells can engage in nearly every activity that eucaryotic cells can, and many can function in ways that eucaryotes cannot.

PROCESSES THAT DEFINE LIFE

To lay the groundwork for a detailed coverage of cells in chapters 4 and 5, this section provides an overview of cell structure and function and introduces the primary characteristics of life. The biological activities or properties that help define and characterize cells as living entities are: (1) growth, (2) reproduction and heredity, (3) metabolism, including cell synthesis and the release of energy, (4) movement and/or irritability, (5) the capacity to transport substances into and out of the cell, and (6) cell support, protection, and storage mechanisms. Although eucaryotic cells have specific organelles to perform these functions, procaryotic cells must rely on a few simple, multipurpose cell components. As indicated in chapter 1, viruses are not cells, are not generally considered living things, and show certain signs of life only when they invade a host cell. Table 2.6 indicates their relative simplicity compared with cells.

Reproduction: Bearing Offspring

A cell's **genome,*** its complete set of genetic material, is composed of elongate strands of DNA. The DNA is packed into discrete bodies called **chromosomes,*** In eucaryotic cells, the chromosomes are located entirely within a nuclear membrane. Procaryotic DNA occurs in a special type of circular chromosome that is not enclosed by a membrane of any sort.

Living things devote a portion of their life cycle to producing offspring that will carry on their particular genetic line for many generations. In **sexual reproduction,** offspring are produced through the union of sex cells from two parents. In **asexual[7] reproduction,** offspring originate through the division of a single parent cell into two daughter cells. Sexual reproduction occurs in most eucaryotes, and eucaryotic cells also reproduce asexually by several processes. One is a type of cell division called *binary fission,* a simple process in which the cell splits equally in two. Many eucaryotic cells engage in **mitosis,*** an orderly division of chromosomes that

---

7. *Asexual* refers to the absence of sexual union. (The prefix *a* or *an* means not or without.)

*genome (jee′-nohm) A combination of the words *gene* and *chromosome.* Refers to an organism's entire set of hereditary factors.

*chromosome (kro′-moh-sohm) Gr. *chroma,* colored, and *soma,* body. The name is derived from the fact that chromosomes stain readily with dyes.

*mitosis (my-toh′-sis) Gr. *mitos,* thread, and *osis,* a condition. Often the term *mitosis* is used synonymously with eucaryotic cell division.

usually accompanies cell division (see figure 5.6). In contrast, procaryotic cells reproduce primarily by binary fission. They have no mitotic apparatus, nor do they reproduce by typical sexual means.

Metabolism: Chemical and Physical Life Processes

Cells synthesize proteins using hundreds of tiny particles called **ribosomes.** In eucaryotes, ribosomes are dispersed throughout the cell or inserted into membranous sacs known as the **endoplasmic reticulum** (see figure 5.8). Procaryotes have smaller ribosomes scattered throughout the protoplasm, since they lack an endoplasmic reticulum. Eucaryotes generate energy by chemical reactions in the **mitochondria,** whereas procaryotes use their cell membrane for this purpose. Photosynthetic microorganisms (algae and some bacteria) trap solar energy by means of pigments and convert it to chemical energy in the cell. Algae (eucaryotes) have compact, membranous bundles called **chloroplasts,** which contain the pigment and perform the photosynthetic reactions. Photosynthetic reactions and pigments of procaryotes do not occur in chloroplasts, but in specialized areas of the cell membrane.

Irritability or Motility

All cells have the capacity to respond to chemical, mechanical, or light stimuli. This quality, called **irritability,** helps cells adapt to the environment and obtain nutrients. Although not present in all cells, true **motility,** or self-propulsion, is a notable sign of life. Eucaryotic cells move by one of the following locomotor organelles: cilia, which are short, hairlike appendages; flagella, which are longer, whiplike appendages; or pseudopods, fingerlike extensions of the cell membrane. Motile procaryotes move by means of unusual, propeller-like flagella unique to bacteria or by special fibrils that produce a gliding form of motility. They have no cilia or pseudopods.

Protection and Storage

Many cells are supported and protected by rigid **cell walls,** which prevent them from rupturing while also providing support and shape. Among eucaryotes, cell walls occur in plants, microscopic algae, and fungi, but not in animals or protozoa. The majority of procaryotes have cell walls, but they differ in composition from the eucaryotic varieties. As protection against depleted nutrient sources, many microbes store nutrients intracellularly. Eucaryotes store nutrients in membranous sacs called vacuoles, and procaryotes concentrate them in crystals called granules or inclusions.

Transport: Movement of Nutrients and Wastes

Cell survival depends on drawing nutrients from the external environment and expelling waste and other metabolic products from the internal environment. This two-directional transport is accomplished in both eucaryotes and procaryotes by the cell membrane. This membrane, described in Microbits 2.3, has a very similar structure in both eucaryotic and procaryotic cells. Eucaryotes have an additional organelle, the **Golgi apparatus,** that assists in sorting and packaging molecules for transport and removal from the cell.

Table 2.6 summarizes the differences and similarities among procaryotes, eucaryotes, and viruses. You will probably want to review this table after you have completed chapters 4 and 5.

CHAPTER CHECKPOINTS

As the atom is the fundamental unit of matter, so is the cell the fundamental unit of life.

All true cells contain biological molecules that carry out the processes that define life: metabolism and reproduction; these two basic processes are supported by the functions of irritability and motility, protection, storage, and transport.

The cell membrane is of critical importance to all cells because it controls the interchange between the cell and its environment.

CHAPTER CAPSULE WITH KEY TERMS

I. **Basic Properties of Atoms and Elements**
   A. Atomic Structure
      1. All **matter** in the universe is composed of minute particles called **atoms**—the simplest form of matter not divisible into a simpler substance by chemical means. Atoms are composed of smaller particles called **protons** (p^+), positive in charge; **neutrons** (n^0), uncharged; and **electrons** (e^-), negative in charge. Protons and neutrons comprise the **nucleus** of an atom, and electrons move about the nucleus in **orbitals** arranged in **shells.**
      2. Atoms that differ in numbers of protons, neutrons, and electrons are **elements.** Elements can be described by **mass number** (the total number of protons and neutrons in the nucleus) and **atomic number** (the number of protons alone), and each is known by a distinct name and symbol. An unreacted atom is neutral because electron charge cancels out proton charge. Elements may exist in variant forms called **isotopes,** which differ in the number of neutrons.
      3. Electrons fill the orbitals in pairs: The first orbital (shell) holds 2 e^-, the second shell can hold 8 e^- (4 pairs), and the third shell can hold 18 e^- (9 pairs). The electron number of each element dictates orbital filling; the outermost shell becomes the focus of reactivity and bonding. In general, atoms with a filled outermost shell are less reactive than atoms with unfilled ones.
   B. Chemical Bonds and Molecules
      1. Atoms interact to form **chemical bonds** and **molecules.** Member atoms of molecules may be the same element or different elements; if the member elements are different, the substance is a **compound.**
      2. The type of bond is dictated by the electron makeup (**valence**) of the outer orbitals of the atoms.
      3. With **covalent bonds,** the electrons are shared and orbit within the entire molecule. This molecule can be **polar,** due to an imbalance of charge, or **nonpolar.**
      4. With **ionic bonds,** an atom with a low number of valence electrons loses them to an atom that has a nearly filled outer orbital. When ionic bonds are broken, the atoms **ionize** into

charged particles called **ions.** Ions that have lost electrons and have a positive charge are **cations** (Na^+), and those that have gained electrons and have a negative charge are **anions** (Cl^-). Ions of the opposite charge are attracted to each other, and those with the same charges repel each other.

    5. **Hydrogen bonds** are caused by weak attractive forces between covalently bonded hydrogen and negatively charged oxygen or nitrogen on the same or nearby molecules.

C. Redox Reactions

    Chemicals may participate in a transfer of electrons, called an **oxidation-reduction** (redox) reaction, between pairs of atoms or molecules. **Oxidation** is a reaction in which electrons are released, and **reduction** is a reaction in which these same electrons are received. Any atom or molecule that donates electrons to another atom or molecule is a **reducing agent,** and one that picks up electrons is an **oxidizing agent.**

D. Chemical Formulas, Models, and Equations

    1. **Chemical formulas** may be presented as simple molecular summaries of the atoms or expressed as structural formulas that give the details of bonding.

    2. **Chemical equations** summarize a chemical reaction by showing the reactants (starting chemicals) and the products (end results), and can indicate **synthesis** reactions, **decomposition** reactions, **exchange** reactions, and reversible reactions.

E. Solutions, Acids, Bases, and pH

    1. A **solution** is a combination of a solid, liquid, or gaseous chemical termed a **solute** dissolved in a liquid medium called a **solvent.** Water-soluble solutes are either charged or polar. The dissolved solute will become **hydrated** due to electrostatic attraction. Chemicals that attract water are **hydrophilic;** those that repel it are **hydrophobic.** The **concentration** of a solution is defined as the amount of solute dissolved in a given amount of solvent.

    2. Acidity and basicity of solutions are represented by the **pH scale,** based on a standardized range of **hydrogen ion concentration** $[H^+]$; an **acid** is a solution that contains a concentration of $[H^+]$ greater than 0.0000001 moles/liter (10^{-7}), and a **base** is a solution that contains a concentration of $[H^+]$ below that amount. When a solution contains exactly that $[H^+]$, it is considered neutral (pH 7). The working pH scale ranges from the most acidic reading of 0 to the most basic (alkaline) reading of 14. Acidic and basic ions in aqueous solutions may interact to form water and a salt through **neutralization.**

II. Organic Chemistry

A. **Organic compounds** contain some combination of carbon and hydrogen covalently bonded. Carbon can also bond with other carbons and with oxygen, nitrogen, and phosphorus to create the diversity of biological molecules. Organic compounds constitute the most important molecules in the structure and function of cells. **Inorganic compounds** are composed of some combination of atoms other than carbon and hydrogen.

B. **Functional groups** are special accessory molecules that bind to carbon and provide the diversity and reactivity seen in organic compounds.

C. Biochemistry and Macromolecules

    1. Organic compounds that function in living things are **biochemicals.** Many of them are **macromolecules** (very large compounds). Levels of structure are **monomers** (single units) and **polymers** (long chains of single units) assembled by **polymerization.**

    2. **Carbohydrates** are composed of carbon, hydrogen, and oxygen (CH_2O) and contain aldehyde or ketone groups.

    3. **Monosaccharides (glucose)** are the simplest sugars. When two sugars are joined by the $-OH$ group and carbon through **dehydration synthesis,** a **glycosidic bond** occurs.

    4. **Disaccharides** are composed of two monosaccharides **(lactose, sucrose). Polysaccharides** are chains of five or more monosaccharides; examples are **cellulose, peptidoglycan, starch,** and **glycogen.** Di- and polysaccharides are digested by specific enzymes that break the bond through **hydrolysis.**

    5. **Lipids** are not soluble in water and other polar solvents due to their nonpolar, hydrophobic chains. **Triglycerides,** including fats and oils, consist of a **glycerol** molecule bonded to 3 **fatty acid** molecules. They are important storage lipids. **Phospholipids** are composed of a glycerol bound to two fatty acids and a phosphoric acid–alcohol group; the molecule has a hydrophilic head and long hydrophobic fatty acid chains; they form single or double lipid layers in the presence of water and are important constituents of cell membranes.

    6. **Proteins** are highly complex macromolecules assembled from 20 different subunits called **amino acids (aa).** Amino acids are combined in a certain order by **peptide bonds:**

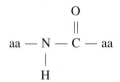

    a. A **peptide** is a short chain of aa's; a dipeptide has two, and a tripeptide has three; a **polypeptide** is usually 20–50 aa's; a **protein** contains more than 50 aa's. Larger proteins predominate in cells.

    b. The chain of amino acids is a protein's **primary structure.** Interactions between functional groups result in additional levels of structure. Hydrogen bonds within the chain fold it first into a helix or sheet called the **secondary structure.** This folds again, forging stronger **disulfide bonds** on nearby cysteines and producing a three-dimensional **tertiary structure.** Proteins composed of two or more polypeptides exist in a **quaternary** state. Variations in shape provide specificity and give rise to the diversity in enzymes, antibodies, and structural proteins.

    7. **Nucleic Acids**

    a. **Deoxyribonucleic acid (DNA)** and **ribonucleic acid (RNA)** are very complex molecules that carry, express, and pass on the genetic information of all cells and viruses. Their basic building block is a nucleotide, composed of a **nitrogen base,** a **pentose sugar,** and a **phosphate.** Nitrogen bases are the ringed compounds: **adenine (A), cytosine (C), thymine (T), guanine (G), and uracil (U);** pentose sugars may be deoxyribose or ribose. The basic design is a polynucleotide, with the sugars linking up in an alternating series with phosphates to make a backbone, and the bases branching off the sugars.

    b. DNA contains **deoxyribose** sugar, has all of the bases except uracil, and occurs as a double-stranded helix with the bases hydrogen-bonded in pairs between the helices; the pairs mate according to the pattern A–T and C–G. DNA is the master code for a cell's life processes and must be transmitted to offspring. During cell division, it is **replicated** by separation of the double strand into two single strands, which are used as a template to form two new double strands. RNA contains **ribose** sugar, has all of the bases except thymine, and is a single-stranded molecule. It expresses the DNA code into proteins.

8. **Adenosine triphosphate (ATP)** is a nucleotide involved in the transfer and storage of energy in cells. It contains adenine, ribose, and three phosphates in a series. Splitting off the last phosphate in the triphosphate releases a packet of energy that may be used to do cell work.

III. **Introduction to Cell Structure**
   A. **Cells** are huge aggregates of macromolecules organized to carry out complex processes described as **living.** All organisms consist of cells, which fall into one of two types: **procaryotic,** which are small, structurally simple cells that lack a **nucleus** and other **organelles;** and **eucaryotic,** larger cells with a nucleus and organelles, found in plants, animals, fungi, and protozoa. **Viruses** are not cells and are not generally considered living because they cannot function independently.

B. Organisms demonstrate several essential qualities of life.
   1. **Growth** and **reproduction** involves producing offspring asexually (with one parent) or sexually (with two parents).
   2. **Metabolism** refers to the chemical reactions in cells, including the synthesis of proteins on ribosomes and the release of energy (ATP).
   3. **Motility** originates from special locomotor structures such as flagella and cilia; **irritability** is responsiveness to external stimuli.
   4. **Protective** external structures include capsules and cell walls; nutrient **storage** takes place in compact intracellular masses.
   5. **Transport** involves conducting nutrients into the cell and wastes out of the cell.

C. **Membranes** surround the cytoplasm and may occur internally in organelles. The cell membrane is a continuous ultrathin bilayer of lipids studded with proteins that controls cell **permeability** and transport. Proteins also serve as sites of recognition and cell communication.

MULTIPLE-CHOICE QUESTIONS

1. The smallest unit of matter with unique characteristics is
   a. an electron   c. an atom
   b. a molecule   d. a proton

2. The _____ charge of a proton is exactly balanced by the _____ charge of a (an) _____ .
   a. negative, positive, electron
   b. positive, neutral, neutron
   c. positive, negative, electron
   d. neutral, negative, electron

3. Electrons move around the nucleus of an atom in pathways called
   a. shells   c. circles
   b. orbitals   d. rings

4. Which part of an element does not vary in number?
   a. electron   c. proton
   b. neutron   d. all of these vary

5. The number of electrons of a neutral atom is automatically known if one knows the
   a. atomic number
   b. atomic weight
   c. number of orbitals
   d. valence

6. Elements
   a. exist in 92 natural forms
   b. have distinctively different atomic structures
   c. vary in atomic weight
   d. a and b are correct
   e. a, b, and c are correct

7. If a substance contains two or more elements of different types, it is considered
   a. a compound   c. a molecule
   b. a monomer   d. organic

8. Bonds in which atoms share electrons are defined as _____ bonds.
   a. hydrogen   c. double
   b. ionic   d. covalent

9. What kind of bond would you expect potassium to form with chlorine?
   a. ionic   c. polar
   b. covalent   d. nonpolar

10. When a compound carries a positive charge on one end and a negative charge on the other end, it is said to be
   a. ionized   c. polar
   b. hydrophilic   d. oxidized

11. Hydrogen bonds can form between _____ adjacent to each other.
   a. two hydrogen atoms
   b. two oxygen atoms
   c. a hydrogen atom and an oxygen atom
   d. negative charges

12. Ions with the same charge will be _____ each other, and ions with opposite charges will be _____ each other.
   a. repelled by, attracted to
   b. attracted to, repelled by
   c. hydrated by, dissolved by
   d. dissolved by, hydrated by

13. An atom that can donate electrons during a reaction is called
   a. an oxidizing agent
   b. a reducing agent
   c. an ionic agent
   d. an electrolyte

14. In a solution of NaCl and water, NaCl is the _____ and water is the _____ .
   a. acid, base   c. solute, solvent
   b. base, acid   d. solvent, solute

15. A substance that releases H^+ into a solution
   a. is a base   c. is an acid
   b. is ionized   d. has a high pH

16. A solution with a pH of 2 _____ than a solution with a pH of 8.
   a. has less H^+
   b. has more H^+
   c. has more OH^-
   d. is less concentrated

17. Fructose is a type of
    a. disaccharide
    b. monosaccharide
    c. polysaccharide
    d. amino acid

18. Bond formation in polysaccharides and polypeptides is accompanied by the removal of a
    a. hydrogen atom
    b. hydroxyl ion
    c. carbon atom
    d. water molecule

19. The monomer unit of polysaccharides such as starch and cellulose is
    a. fructose
    c. ribose
    b. glucose
    d. lactose

20. A phospholipid contains
    a. three fatty acids bound to glycerol
    b. three fatty acids, a glycerol, and a phosphate
    c. two fatty acids and a phosphate bound to glycerol
    d. three cholesterol molecules bound to glycerol

21. Proteins are synthesized by linking amino acids with _____ bonds.
    a. disulfide
    c. peptide
    b. glycosidic
    d. ester

22. The amino acid that accounts for disulfide bonds in the tertiary structure of proteins is
    a. tyrosine
    c. cysteine
    b. glycine
    d. serine

23. DNA is a hereditary molecule that is composed of
    a. deoxyribose, phosphate, and nitrogen bases
    b. deoxyribose, a pentose, and nucleic acids
    c. sugar, proteins, and thymine
    d. adenine, phosphate, and ribose

24. What is meant by DNA replication?
    a. duplication of the sugar-phosphate backbone
    b. matching of base pairs
    c. formation of the double helix
    d. the exact copying of the DNA code into two new molecules

CONCEPT QUESTIONS

1. How are the concepts of an atom and an element related? What causes elements to differ?

2. a. How are mass number and atomic number derived? What is the atomic weight?
   b. Using data in table 2.1, give the electron number of nitrogen, sulfur, calcium, phosphorus, and iron.
   c. What is distinctive about isotopes of elements, and why are they important?

3. a. How is the concept of molecules and compounds related?
   b. Compute the molecular weight of oxygen and methane.

4. a. Why is an isolated atom neutral?
   b. Describe the concept of the atomic nucleus, electron orbitals, and shells.
   c. What causes atoms to form chemical bonds?
   d. Why do some elements not bond readily?
   e. Draw the atomic structure of magnesium and predict what kinds of bonds it will make.

5. Distinguish between the general reactions in covalent, ionic, and hydrogen bonds.

6. a. Which kinds of elements tend to make covalent bonds?
   b. Distinguish between a single and a double bond.
   c. What is polarity?
   d. Why are some covalent molecules polar and others nonpolar?
   e. What is an important consequence of the polarity of water?

7. a. Which kinds of elements tend to make ionic bonds?
   b. Exactly what causes the charges to form on atoms in ionic bonds?
   c. Verify the proton and electron numbers for Na^+ and Cl^-.
   d. Differentiate between an anion and a cation.
   e. What kind of ion would you expect magnesium to make, on the basis of its valence?

8. Differentiate between an oxidizing agent and a reducing agent.

9. Why are hydrogen bonds relatively weak?

10. a. Compare the three basic types of chemical formulas.
    b. Review the types of chemical reactions and the general ways they can be expressed in equations.

11. a. Define solution, solvent, and solute.
    b. What properties of water make it an effective biological solvent, and how does a molecule like NaCl become dissolved in it?
    c. How is the concentration of a solution determined?
    d. What is molarity? Tell how to make a 1M solution of $Mg_3(PO_4)_2$ and a 0.1 M solution of $CaSO_4$.

12. a. What determines whether a substance is an acid or a base?
    b. Briefly outline the pH scale.
    c. How can a neutral salt be formed from acids and bases?

13. a. What atoms must be present in a molecule for it to be considered organic?
    b. What characteristics of carbon make it ideal for the formation of organic compounds?
    c. What are functional groups?
    d. Differentiate between a monomer and a polymer.
    e. How are polymers formed?
    f. Name several inorganic compounds.

14. a. What characterizes the carbohydrates?
    b. Differentiate between mono-, di-, and polysaccharides, and give examples of each.
    c. What is a glycosidic bond, and what is dehydration synthesis?
    d. What are some of the functions of polysaccharides in cells?

15. a. Draw simple structural molecules of triglycerides and phospholipids to compare their differences and similarities.
    b. What is an ester bond?
    c. How are saturated and unsaturated fatty acids different?
    d. What characteristic of phospholipids makes them essential components of cell membranes?
    e. Why is the hydrophilic end of phospholipids attracted to water?

16. a. Describe the basic structure of an amino acid.
    b. What makes the amino acids distinctive, and how many of them are there?
    c. What is a peptide bond?
    d. Differentiate between a peptide, a polypeptide, and a protein.
    e. Explain what causes the various levels of structure of a protein molecule.
    f. What functions do proteins perform in a cell?

17. a. Describe a nucleotide and a polynucleotide, and compare and contrast the general structure of DNA and RNA.
    b. Name the two purines and the three pyrimidines.
    c. Why is DNA called a double helix?
    d. What is the function of RNA?
    e. What is ATP, and what is its function in cells?

18. a. Outline the general structure of a cell, and describe the characteristics of cells that qualify them as living.
    b. Why are viruses not considered living?
    c. Compare the general characteristics of procaryotic and eucaryotic cells.
    d. What are cellular membranes, and what are their functions?
    e. Explain the fluid mosaic model of a membrane.

CRITICAL-THINKING QUESTIONS

1. The "octet rule" in chemistry helps predict the tendency of atoms to acquire or donate electrons from the outer shell. It says that those with fewer than 4 tend to donate electrons and those with more than 4 tend to accept additional electrons; those with exactly 4 can do both. Using this rule, determine what category each of the following elements falls into: N, S, C, P, O, H, Ca, Fe, and Mg. (You will need to work out the valence of the atoms.)

2. Predict the kinds of bonds that occur in ammonium (NH_3), phosphate (PO_4), disulfide (S−S), and magnesium chloride ($MgCl_2$). (Use simple models such as those in figure 2.3.)

3. Work out the following problems:
   a. What is the number of protons in helium?
   b. Will an H bond form between $H_3C—CH=O$ and H_2O? Why or why not?
   c. Draw the following molecules and determine which are polar: Cl_2, NH_3, CH_4.
   d. What is the pH of a solution with a concentration of 0.00001 moles/ml (M) of H^+?
   e. What is the pH of a solution with a concentration of 0.00001 moles/ml (M) of OH^-?

4. a. Describe how hydration spheres are formed around cations and anions.
   b. Which substances will be expected to be hydrophilic and hydrophobic, and what makes them so?
   c. Distinguish between polar and ionic compounds, using your own words.

5. In what way are carbon-based compounds like children's Tinker Toys or Lego blocks?

6. Is galactose an aldehyde or a ketone sugar?

7. a. How many water molecules are released when a triglyceride is formed?
   b. How many peptide bonds are in a tetrapeptide?

8. a. Use pipe cleaners to help understand the formation of the 2° and 3° structures of proteins.
   b. Note the various ways that your pipe cleaner structure can be folded and the diversity of shapes that can be formed.

9. a. Looking at figure 2.25, can you see why adenine forms hydrogen bonds with thymine and why cytosine forms them with guanine?
   b. Show on paper the steps in replication of the following segment of DNA:

$$\begin{array}{l} ATGTTCCCGATCGGC \\ |\;|\;|\;|\;|\;|\;|\;|\;|\;|\;|\;|\;|\;|\;| \\ TACAAGGGCTAGCCG \end{array}$$

10. A useful mnemonic (memory) device for recalling the major characteristics of life is: *Giant Rats Have Many Colored Teeth*. Can you list some of the characteristics for which each letter stands?

INTERNET SEARCH TOPIC

Go to a search engine and explore the topic of isotopes and dating ancient rocks. How can isotopes be used to determine if rocks contain evidence of life?

Tools of the Laboratory:
The Methods for Studying Microorganisms

Every year in the United States, hundreds of outbreaks of foodborne illness are reported to public health authorities. Because such episodes could potentially travel through the population, epidemiologists are under pressure to determine, as rapidly as possible, the agent involved, the source of contaminated food, and how the illness was acquired. To gather information on microorganisms that may be involved, infectious disease specialists have developed some remarkable techniques and tools: special media for isolating the microbe, microscopes for observing it, and numerous biochemical and genetic tests. Such refined technology is so sensitive that it can uncover a pathogen lying hidden in samples that often contain large numbers of bacteria that are not involved in disease. It has made it possible to detect *Salmonella* bacteria in ice cream and eggnog, hepatitis A virus in strawberries, *Escherichia coli* 0157:H7 bacteria in meat and produce, and a veritable "cafeteria" of other common food-borne agents.

The concerns surrounding food-borne disease have led the Department of Agriculture to create new food preparation guidelines. All fresh meats and poultry must include instructions for safe handling, including thorough cooking and clean kitchen techniques (see Microbits 20.3). The same agency is also reevaluating the techniques used in the slaughterhouse for determining the safety of food. A final precautionary note on microbes and food emphasizes that their small size and invisibility are serious impediments to their control. It is impossible to judge food fitness and safety on the basis of macroscopic appearance alone.

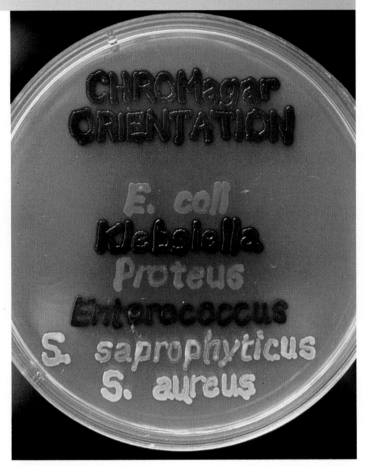

A state-of-the-art medium developed for culturing and identifying the most common urinary pathogens. CHROMagar Orientation™ uses color-forming reactions to distinguish at least seven species and permits rapid identification and treatment.

Chapter Overview

- A driving force of microbiology has been to find ways to visualize and handle microorganisms.
- Microbes are managed and characterized by implementing the Five I's—inoculation, incubation, isolation, inspection, and identification.
- Cultures are made by removing a sample from a desired source and placing it in containers of media.
- Media can be varied in chemical, physical, and functional purposes, depending on the intention.
- Growth and isolation of microbes leads to pure cultures that permit the study and testing of single species.
- Cultures can be used to provide information on microbial morphology, biochemistry, and genetic characteristics.
- Unknown, invisible samples can become known and visible.
- The microscope is a powerful tool for magnifying and resolving cells and their parts.
- Microscopes exist in several forms, using light, radiation, and electrons to form images.
- Specimens and cultures are prepared for study in fresh (live) or fixed (dead) form.
- Staining procedures highlight cells and allow them to be described and identified.

Methods of Culturing Microorganisms—The Five I's

Biologists studying large organisms such as animals and plants can, for the most part, immediately see and differentiate their experimental subjects from the surrounding environment and from one another. In fact, they can use their senses of sight, smell, hearing, and even touch to detect and evaluate identifying characteristics and to keep track of growth and developmental changes. Because microbiologists cannot rely as much as other scientists on senses other than sight, they are confronted by some unique problems. First, most habitats (such as the soil and the human mouth) harbor microbes in complex associations, so it is often necessary to separate the species from one another. Second, to maintain and keep track of such small research subjects, microbiologists usually have to grow them under artificial conditions. A third difficulty in working with microbes is that they are invisible and widely distributed, and undesirable ones can be introduced into an experiment and cause misleading results. These impediments motivated the development of techniques to control microbes and their growth, primarily sterile, aseptic, and pure culture techniques.[1]

Microbiologists use five basic techniques to manipulate, grow, examine, and characterize microorganisms in the laboratory: inoculation, incubation, isolation, inspection, and identification (the Five I's; figure 3.1). Some or all of these procedures are performed by microbiologists, whether the beginning laboratory student, the researcher attempting to isolate drug-producing bacteria from soil, or the clinical microbiologist working with a specimen from a patient's infection. These procedures make it possible to handle and maintain microorganisms as discrete entities whose detailed biology can be studied and recorded.

INOCULATION: PRODUCING A CULTURE

To cultivate, or **culture,*** microorganisms, one introduces a tiny sample (the **inoculum**) into a container of nutrient **medium*** (pl. media), which provides an environment in which they multiply. This process is called **inoculation.*** The observable growth that appears in or on the medium is known as a culture. The nature of the sample being cultured depends on the objectives of the analysis. Clinical specimens for determining the cause of an infectious disease are obtained from body fluids (blood, cerebrospinal fluid), discharges (sputum, urine, feces), or diseased tissue. Other samples subject to microbiological analysis are soil, water, sewage, foods, air, and inanimate objects. Procedures for proper specimen collection are discussed on page 535.

ISOLATION: SEPARATING ONE SPECIES FROM ANOTHER

Certain isolation techniques are based on the concept that if an individual bacterial cell is separated from other cells and provided adequate space on a nutrient surface, it will grow into a discrete mound of cells called a **colony** (figure 3.2). Because it was formed from a single cell, a colony consists of just that one species and no other. Proper isolation requires that a small number of cells be inoculated into a relatively large volume or over an expansive area of medium. It generally requires the following materials: a medium that has a relatively firm surface (see agar in "Physical States of

Media," page 62), a **Petri plate** (a clear, flat dish with a cover), and inoculating tools. In the **streak plate method,** a small droplet of culture or sample is spread over the surface of the medium according to a pattern that gradually thins out the sample and separates the cells spatially over several sections of the plate (figure 3.3a,b). Because of its effectiveness and economy of materials, the streak plate is the method of choice for most applications.

In the **loop dilution,** or **pour plate, technique,** the sample is inoculated serially into a series of cooled but still liquid agar tubes so as to dilute the number of cells in each successive tube in the series (figure 3.3c,d). Inoculated tubes are then plated out (poured) into sterile Petri plates and are allowed to solidify (harden). The end result (usually in the second or third plate) is that the number of cells per volume is so decreased that cells have ample space to grow into separate colonies. One difference between this and the streak plate method is that in this technique some of the colonies will develop in the medium itself and not just on the surface.

With the **spread plate technique,** a small volume of liquid, diluted sample is pipetted onto the surface of the medium and spread around evenly by a sterile spreading tool (sometimes called a "hockey stick"). Like the streak plate, cells are pushed into separate areas on the surface so that they can form individual colonies (figure 3.3e,f).

Before we continue to cover information on the Five I's, we will take a side trip to look at media in more detail.

MEDIA: PROVIDING NUTRIENTS IN THE LABORATORY

A major stimulus to the rise of microbiology 100 years ago was the development of techniques for growing microbes out of their natural habitats and in pure form in the laboratory. This milestone enabled the close examination of a microbe and its morphology, physiology, and genetics. It was evident from the very first that for successful cultivation, each microorganism had to be provided with all of its required nutrients in an artificial medium.

Nutritional requirements of microbes vary from a few very simple inorganic compounds to a complex list of specific inorganic and organic compounds. This tremendous diversity is evident in the types of media that can prepared. At least 500 different types of media are used in culturing and identifying microorganisms. Culture media are contained in test tubes, flasks, or Petri plates, and they are inoculated by such tools as loops, needles, pipettes, and swabs. Media are extremely varied in nutrient content and consistency and can be specially formulated for a particular purpose. Culturing parasites that cannot grow on artificial media (all viruses and certain bacteria) requires cell cultures or host animals (Microbits 3.1). For an experiment to be properly controlled, sterile technique is necessary. This means that the inoculation must start with a sterile medium and inoculating tools with sterile tips must be used. Measures must be taken to prevent introduction of nonsterile materials such as room air and fingers directly into the culture.

Types of Media

Media can be classified on three primary levels: (1) physical form, (2) chemical composition, and (3) functional type (table 3.1). Most media discussed here are designed for bacteria and fungi, though algae and some protozoa can be propagated in media.

---

1. *Sterile* relates to the complete absence of viable microbes; *aseptic* relates to prevention of infection; *pure culture* refers to growth of a single species of microbe.

   * **culture** (kul′-chur) Gr. *cultus*, to tend or cultivate. It can be used as a verb or a noun.

   * **medium** (mee′-dee-um) pl. media; L., middle.

   * **inoculation** (in-ok″-yoo-lay′-shun) L. *in*, and *oculus*, bud.

An Overview of Major Techniques Performed by Microbiologists to Locate, Grow, Observe, and Characterize Microorganisms.

Specimen Collection:
Nearly any object or material can serve as a source of microbes. Common ones are body fluids and tissues, foods, water, or soil. Specimens are removed by some form of sampling device. This may be a swab, syringe, or a special transport system that holds, maintains, and preserves the microbes in the sample.

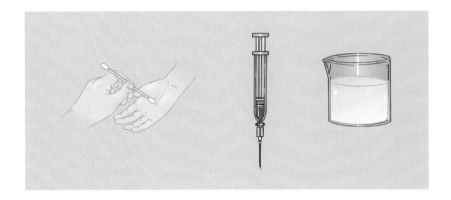

A GUIDE TO THE FIVE I'S: How the Sample Is Processed and Profiled

1. Inoculation:
During inoculation, the sample is placed into a container of sterile medium that provides microbes with the appropriate nutrients to sustain growth. Inoculation involves using a sterile tool to spread the sample out on the surface of a solid medium or to introduce the sample into a flask or tube. Selection of media with specialized functions can improve later steps of isolation and identification. Some microbes may require a live organism (animal, egg) as the inoculation medium.

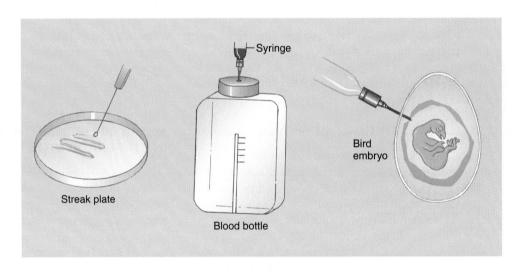

2. Incubation:
An incubator can be used to adjust the proper growth conditions of a sample. Setting the optimum temperature and gas content promotes multiplication of the microbes over a period of hours, days, and even weeks. Incubation produces a culture—the visible growth of the microbe in the medium.

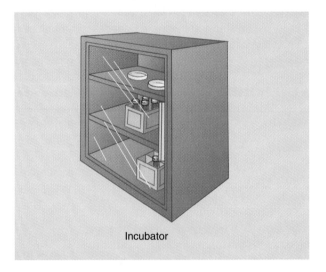

FIGURE 3.1

A summary of the general laboratory techniques carried out by microbiologists. It is not necessary to perform all the steps shown or to perform them exactly in this order, but all microbiologists participate in at least some of these activities. In some cases, one may proceed right from the sample to inspection, and in others, only inoculation and incubation on special media are required.

3. Isolation:

The end result of inoculation and incubation is isolation of the microbe in macroscopic form. The isolated microbes take the form of separate colonies (discrete mounds of cells) on solid media, or turbidity in broths. Further isolation, also known as subculturing, involves taking a tiny bit of growth and inoculating an additional culture of it. This is one way to make a pure culture that contains only a single species of microbe.

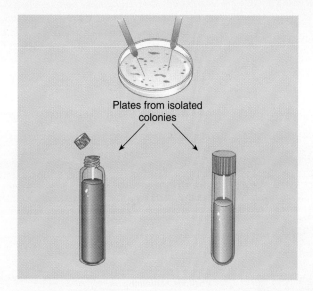

Plates from isolated colonies

4. Inspection:

The cultures are observed macroscopically for obvious growth characteristics (color, texture, size) that could be useful in analyzing the specimen contents. Slides are made to assess microscopic details such as cell shape, size, and motility. Staining techniques may be used to gather specific information on microscopic morphology.

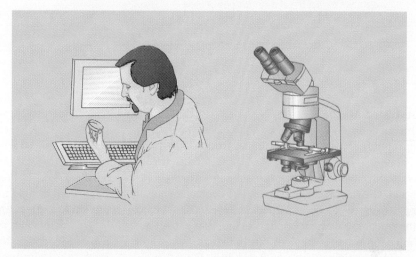

5. Identification:

A major outcome is to pinpoint an isolate down to the level of species. Summaries of accumulated data are used to develop profiles of the microbe or microbes isolated from the sample. Information can include relevant characteristics already taken during inspection or additional tests that further describe and differentiate the nature of microbes isolated. Other types of specialized tests include biochemical tests to determine metabolic activities specific to the microbe, immunologic tests, and genetic analysis.

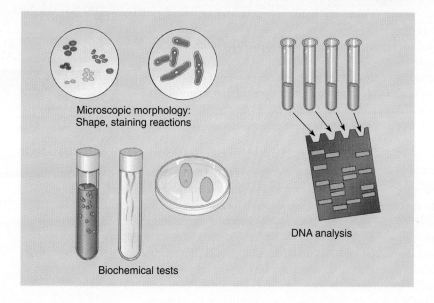

Microscopic morphology: Shape, staining reactions

Biochemical tests

DNA analysis

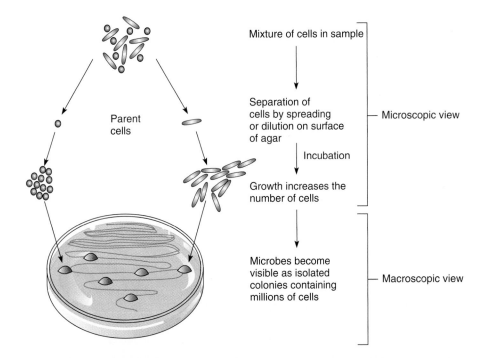

Parent cells

Mixture of cells in sample

Separation of cells by spreading or dilution on surface of agar

Incubation

Growth increases the number of cells

— Microscopic view

Microbes become visible as isolated colonies containing millions of cells

— Macroscopic view

FIGURE   3.2

Isolation technique. Stages in the formation of an isolated colony, showing the microscopic events and the macroscopic result. Separation techniques such as streaking can be used to isolate single cells. After numerous cell divisions, a macroscopic, clinging mound of cells, or a colony, will be formed. This is a relatively simple yet successful way to separate different types of bacteria in a mixed sample.

Physical States of Media

Liquid media are defined as water-based solutions that do not solidify at temperatures above freezing and that tend to flow freely when the container is tilted (figure 3.4). These media, termed broths, milks, or infusions, are made by dissolving various solutes in distilled water. Growth occurs throughout the container and can then present a dispersed, cloudy or particulate appearance. A common laboratory medium, nutrient broth, contains beef extract and peptone dissolved in water. Methylene blue milk and litmus milk are opaque liquids containing whole milk and dyes. Fluid thioglycollate is a slightly viscous broth used for determining patterns of growth in oxygen.

At ordinary room temperature, **semisolid media** exhibit a clotlike consistency (figure 3.5) because they contain an amount of solidifying agent (agar or gelatin) that thickens them but does not produce a firm substrate. Semisolid media are used to determine the motility of bacteria and to localize a reaction at a specific site. Both motility test medium and SIM contain a small amount (0.3–0.5%) of agar. The medium is stabbed carefully in the center and later observed for the pattern of growth around the stab line. In addition to motility, SIM can test for physiological characteristics used in identification (hydrogen sulfide production and indole reaction).

Solid media provide a firm surface on which cells can form discrete colonies (see figure 3.3) and are advantageous for isolating and subculturing bacteria and fungi. They come in two forms: liquefiable and nonliquefiable. **Liquefiable solid media,** sometimes called reversible solid media, contain a solidifying agent that is thermoplastic: Its physical properties change in response to temperature. By far the most widely used and effective of these agents is **agar,*** a complex polysaccharide isolated from the red alga *Gelidium.* The benefits of agar are numerous. It is solid at room

temperature, and it melts (liquefies) at the boiling temperature of water (100°C). Once liquefied, agar does not resolidify until it cools to 42°C, so it can be inoculated and poured in liquid form at temperatures (45°–50°C) that will not harm the microbes or the handler. Agar is flexible and moldable, and it provides a basic framework to hold moisture and nutrients, though it is not itself a digestible nutrient for most microorganisms.

Any medium containing 1% to 5% agar usually has the word *agar* in its name. Nutrient agar is a common one. Like nutrient broth, it contains beef extract and peptone, as well as 1.5% agar by weight. Many of the examples covered in the section on functional

TABLE 3.1		
Three Categories of Media Classification		
Physical State (Medium's Normal Consistency)	**Chemical Composition (Type of Chemicals Medium Contains)**	**Functional Type (Purpose of Medium)***
1. Liquid	1. Synthetic (chemically defined)	1. General purpose
2. Semisolid		2. Enriched
3. Solid (can be converted to liquid)	2. Nonsynthetic (not chemically defined)	3. Selective
4. Solid (cannot be liquefied)		4. Differential
		5. Anaerobic growth
		6. Specimen transport
		7. Assay
		8. Enumeration

* *Some media can serve more than one function. For example, a medium such as brain-heart infusion is general purpose and enriched; mannitol salt agar is both selective and differential; and blood agar is both enriched and differential.*

---

* agar (ah′-gur) The Malay word for this seaweed product is *agar-agar.*

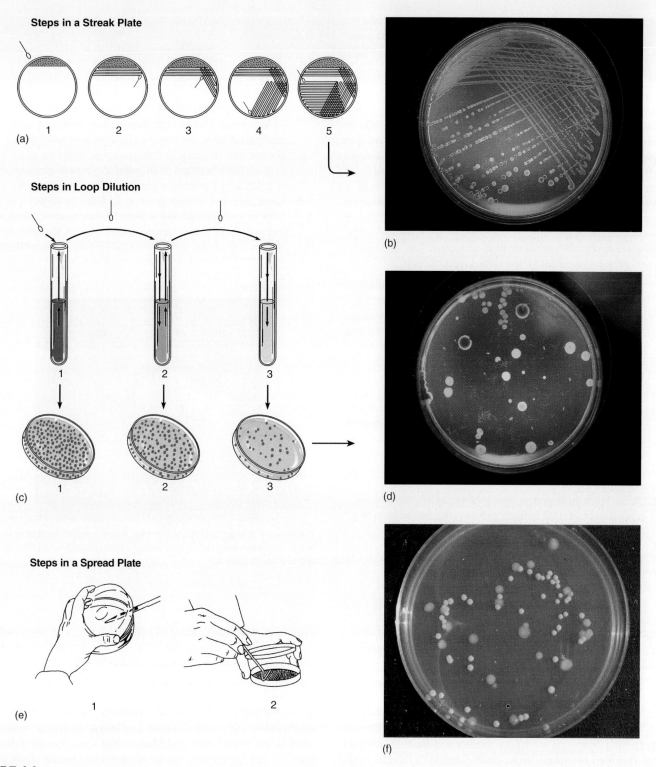

Steps in a Streak Plate

(a) 1   2   3   4   5

(b)

Steps in Loop Dilution

1   2   3

(c) 1   2   3

(d)

Steps in a Spread Plate

1   2

(e)

(f)

FIGURE 3.3

Methods for isolating bacteria. (a) Steps in a quadrant streak plate and **(b)** resulting isolated colonies of bacteria. **(c)** Steps in the loop dilution method and **(d)** the appearance of plate 3. **(e)** Spread plate and **(f)** its result.

categories of media contain agar. Although **gelatin** is not nearly as satisfactory as agar, it will create a reasonably solid surface in concentrations of 10% to 15% (but it probably will not remain solid). Agar and gelatin media are illustrated in figure 3.6.

Nonliquefiable solid media have less versatile applications than agar media because they do not melt. They include materials such as rice grains (used to grow fungi), cooked meat media (good for anaerobes), and potato slices; all of these media start out solid

MICROBITS   3.1
Animal Inoculation: "Living Media"

A great deal of attention has been focused on the uses of animals in biology and medicine. Animal rights activists are vocal about practically any experimentation with animals and have expressed their outrage quite forcefully. Certain kinds of animal testing may seem trivial and unnecessary, but many times it is absolutely necessary to use animals bred for experimental purposes, such as guinea pigs, mice, chickens, and even armadillos. Such animals can be an indispensable aid for studying, growing, and identifying microorganisms. One special use of animals involves inoculation of the early life stages (embryos) of birds. Vaccines for influenza are currently produced in duck embryos. The following is a summary of major rationales for live animal inoculation:

1. Animal inoculation is an essential step in testing the effects of drugs and the effectiveness of vaccines before they are administered to humans. It makes progress toward prevention, treatment, and cure possible without risking the lives of humans.

2. Researchers develop animal models for evaluating new diseases or for studying the cause or process of a disease. Koch's postulates is a series of proofs to determine the causative agent of a disease and requires a controlled experiment with an animal that can develop a typical case of the disease. So far, there has not been a completely successful model for HIV infection and human AIDS, but a similar disease in monkeys called simian AIDS has clarified some aspects of human AIDS and is serving as a model for vaccine development.

3. Animals are an important source of antibodies, antisera, antitoxins, and other immune products that can be used in therapy or testing.

4. Animals are sometimes required to determine the pathogenicity or toxicity of certain bacteria. One such test is the mouse neutralization test for the presence of botulism toxin in food. This test can help identify even very tiny amounts of toxin and thereby can avert outbreaks of this disease. Occasionally, it is necessary to inoculate an animal to distinguish between pathogenic or nonpathogenic strains of *Listeria* or *Candida* (a yeast).

5. Some microbes will not grow on artificial media but will grow in a suitable animal and can be recovered in a more or less pure form. These include animal viruses, the spirochete of syphilis, and the leprosy bacillus (grown in armadillos).

The nude or athymic mouse has genetic defects in hair formation and thymus development. It is widely used to study cancer, immune function, and infectious diseases.

---

and remain solid after heat sterilization. Other solid media containing egg and serum start out liquid and are permanently coagulated or hardened by moist heat.

Chemical Content of Media

Media whose compositions are chemically defined are termed **synthetic.** Such media contain pure organic and inorganic compounds that vary little from one source to another and have a molecular content specified by means of an exact formula. Synthetic media come in many forms. Some media, such as minimal media for fungi, contain nothing more than a few essential compounds such as salts and amino acids dissolved in water. Others contain a variety of defined organic and inorganic chemicals (table 3.2). Such standardized and reproducible media are most useful in research and cell culture when the exact nutritional needs of the test organisms are known. If even one component of a given medium is not chemically definable, the medium belongs in the next category.

Complex, or **nonsynthetic, media** contain at least one ingredient that is *not* chemically definable—not a simple, pure compound and not representable by an exact chemical formula. Most of these substances are extracts of animals, plants, or yeasts, including such materials as ground-up cells, tissues, and secretions. Examples are blood, serum, and meat extracts or infusions. Infusions are high in vitamins, minerals, proteins, and other organic nutrients. Other nonsynthetic ingredients are milk, yeast extract, soybean digests, and peptone. Peptone is a partially digested protein, rich in amino acids, that is often used as a carbon and nitrogen source. Nutrient broth, blood agar, and MacConkey agar, though different in function and appearance, are all nonsynthetic media. They present a rich mixture of nutrients for microbes that have complex nutritional needs.

A specific example can be used to compare what differentiates a synthetic medium from a nonsynthetic one. Both synthetic *Euglena* medium (table 3.2) and nonsynthetic nutrient broth contain amino acids. But *Euglena* medium has three known amino acids in known amounts, whereas nutrient broth contains amino acids (in peptone) in variable types and amounts. Pure inorganic salts and organic acids are added in precise quantities for *Euglena*, whereas those components are provided by undefined beef extract in nutrient broth.

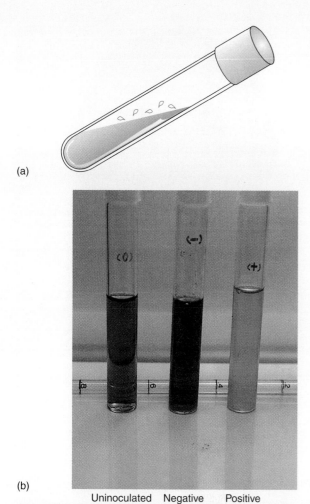

(a)

(b)

Uninoculated   Negative   Positive

FIGURE 3.4

Sample liquid media. (a) Liquid media tend to flow freely when the container is tilted. **(b)** *Enterococcus faecalis* broth, a selective medium for identifying this species. On the left (0) is a clear, uninoculated broth. The broth in the center is growing without a color change (−), and on the right is a broth with growth and color change (+).

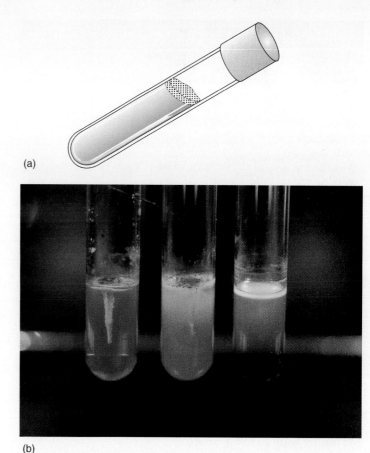

(a)

(b)

FIGURE 3.5

Sample semisolid media. (a) Semisolid media have more body than liquid media but less body than solid media. They do not flow freely and have a soft, clotlike consistency. **(b)** Sulfur indole motility (SIM) medium. The growth patterns that develop in this medium can be used to determine various characteristics. The tube on the left shows no motility; the tube in the center shows motility; the tube on the right shows both motility and production of hydrogen sulfide gas (black precipitate).

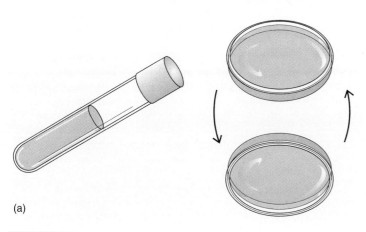

(a)

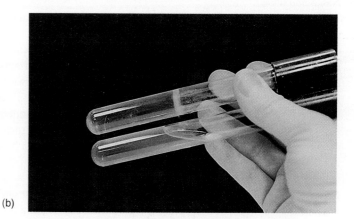

(b)

FIGURE 3.6

Solid media that are reversible to liquids. (a) Media containing 1–5% agar-agar are solid enough to remain in place when containers are tilted or inverted. They are liquefiable by heat but generally not by bacterial enzymes. **(b)** Nutrient gelatin contains enough gelatin (12%) to take on a solid consistency. The top tube shows it as a solid. The bottom tube indicates a result when microbial enzymes digest the gelatin and liquefy it.

TABLE 3.2

Medium for the Growth and Maintenance of the Green Alga *Euglena*

Glutamic acid (aa)	6 g
Aspartic acid (aa)	4 g
Glycine (aa)	5 g
Sucrose (c)	30 g
Malic acid (oa)	2 g
Succinic acid (oa)	1.04 g
Boric acid	1.14 mg
Thiamine hydrochloride (v)	12 mg
Monopotassium phosphate	0.6 g
Magnesium sulfate	0.8 g
Calcium carbonate	0.16 g
Ammonium carbonate	0.72 g
Ferric chloride	60 mg
Zinc sulfate	40 mg
Manganese sulfate	6 mg
Copper sulfate	0.62 mg
Cobalt sulfate	5 mg
Ammonium molybdate	1.34 mg

Note: These ingredients are dissolved in 1,000 ml of water.
aa, amino acid; c, carbohydrate; oa, organic acid; v, vitamin; g, gram; mg, milligram.

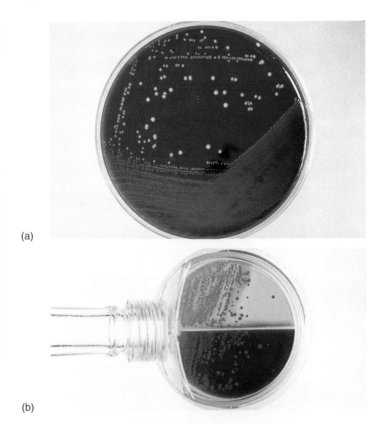

(a)

(b)

FIGURE 3.7

Examples of enriched media. (a) Blood agar growing *Enterococcus faecalis*, a common fecal bacterium. **(b)** Chocolate agar (lower half of plate), a medium that gets its brown color from heated blood, not from chocolate. It is commonly used to culture the fastidious gonococcus *Neisseria gonorrhoeae.* (The upper plate chamber contains MacConkey agar with lactose-fermenting bacteria.)

Media to Suit Every Function

Microbiologists have many types of media at their disposal, with new ones being devised all the time. Depending upon what is added, a microbiologist can fine-tune a medium for nearly any purpose. As a result, only a few species of bacteria or fungi cannot yet be cultivated artificially. Media are used for primary isolation, to maintain cultures in the lab, to determine biochemical and growth characteristics, and for numerous other functions.

General-purpose media are designed to grow as broad a spectrum of microbes as possible. As a rule, they are nonsynthetic and contain a mixture of nutrients that could support the growth of pathogens and nonpathogens alike. Examples include nutrient agar and broth, brain-heart infusion, and trypticase soy agar (TSA). TSA contains partially digested milk protein (casein), soybean digest, NaCl, and agar.

An **enriched medium** contains complex organic substances such as blood, serum, hemoglobin, or special **growth factors** (specific vitamins, amino acids) that certain species must have in order to grow. Bacteria that require growth factors and complex nutrients are termed **fastidious.*** Blood agar, which is made by adding sterile sheep, horse, or rabbit blood to a sterile agar base (figure 3.7a) is widely employed to grow fastidious streptococci and other pathogens. Pathogenic *Neisseria* (one species causes gonorrhea) are grown on Thayer-Martin medium or chocolate agar, which is made by heating blood agar (figure 3.7b).

Selective and Differential Media. Some of the cleverest and most inventive media belong to the categories of selective and differential media (figure 3.8). These media are designed for special

TABLE 3.3

Selective Media, Agents, and Functions

Medium	Selective Agent	Used For
Mueller tellurite	Potassium tellurite	Isolation of *Corynebacterium diphtheriae*
Enterococcus faecalis broth	Sodium azide Tetrazolium	Isolation of fecal enterococci
Phenylethanol agar	Phenylethanol chloride	Isolation of staphylococci and streptococci
Tomato juice agar	Tomato juice, acid	Isolation of lactobacilli from saliva
MacConkey agar	Bile, crystal violet	Isolation of gram-negative enterics
Salmonella/Shigella (SS) agar	Bile, citrate, brilliant green	Isolation of *Salmonella* and *Shigella*
Lowenstein-Jensen	Malachite green dye	Isolation and maintenance of *Mycobacteria*
Sabouraud's agar	pH of 5.6 (acid)	Isolation of fungi— inhibits bacteria

---

* **fastidious** (fass-tid'-ee-us) L. *fastidium*, loathing or disgust.

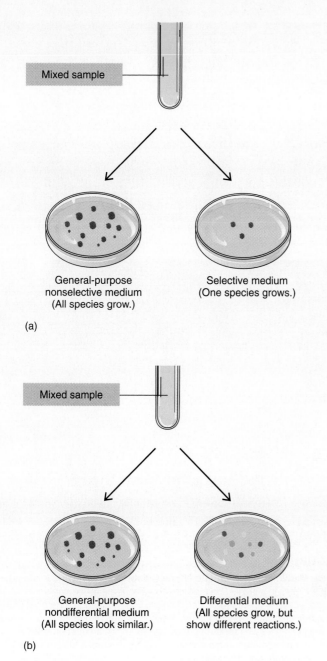

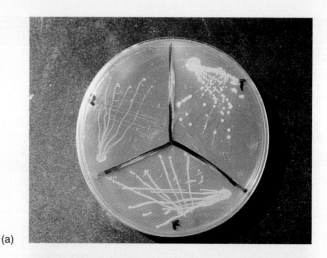

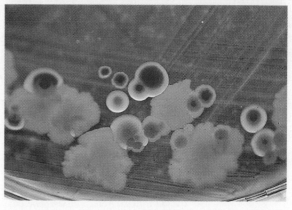

(a)

(b)

FIGURE 3.9

Example of selective media. **(a)** Mannitol salt agar is used to isolate members of the genus *Staphylococcus*. It is selective for this genus because its members can grow in the presence of 7.5% sodium chloride, whereas many other species are inhibited by this high concentration. It contains a dye that also pinpoints those species of *Staphylococcus* that produce acid from mannitol and turn the phenol red dye to a bright yellow. **(b)** MacConkey agar differentiates between lactose-fermenting bacteria (indicated by a pink-red reaction in the center of the colony) and lactose-negative bacteria (indicated by an off-white colony with no dye reaction).

General-purpose
nonselective medium
(All species grow.)

Selective medium
(One species grows.)

(a)

General-purpose
nondifferential medium
(All species look similar.)

Differential medium
(All species grow, but
show different reactions.)

(b)

FIGURE 3.8

Comparison of selective and differential media with general-purpose media. **(a)** The same mixed sample containing five different species is streaked onto plates of general-purpose nonselective medium and selective medium. Note the results. **(b)** Another mixed sample containing three different species is streaked onto plates of general-purpose nondifferential medium and differential medium. Note the results.

microbial groups, and they have extensive applications in isolation and identification. They can permit, in a single step, the preliminary identification of a genus or even a species.

A **selective medium** (table 3.3) contains one or more agents that inhibit the growth of a certain microbe or microbes (A, B, C) but not others (D) and thereby encourages, or *selects*, microbe D and allows it to grow. Selective media are very important in primary isolation of a specific type of microorganism from samples containing

dozens of different species—for example, feces, saliva, skin, water, and soil. They hasten isolation by suppressing the unwanted background organisms and favoring growth of the desired ones.

Mannitol salt agar (MSA) (figure 3.9*a*) contains a concentration of NaCl (7.5%) that is quite inhibitory to most human pathogens. One exception is the genus *Staphylococcus,* which grows well in this medium and consequently can be amplified in very mixed samples. Bile salts, a component of feces, inhibit most gram-positive bacteria while permitting many gram-negative rods to grow. Media for isolating intestinal pathogens (MacConkey agar, eosin methylene blue [EMB] agar) contain bile salts as a selective agent (figure 3.9*b*). Dyes such as methylene blue and crystal violet also inhibit certain gram-positive bacteria. Other agents that have selective properties are antimicrobic drugs and acid. Some selective media contain strongly inhibitory agents to favor the growth of a pathogen that would otherwise be overlooked because of its low

TABLE 3.4

Differential Media

Medium	Substances That Facilitate Differentiation	Differentiates Between
Blood agar	Intact red blood cells	Types of hemolysis
Mannitol salt agar	Mannitol, phenol red, and 7.5% NaCl	Species of *Staphylococcus* NaCl also inhibits the salt-sensitive species
Eosin methylene blue (EMB) agar	Eosin, methylene blue, lactose, and bile	*Escherichia coli,* other lactose fermenters, and lactose nonfermenters Dyes and bile also inhibit gram-positive bacteria
Spirit blue agar	Spirit blue dye and oil	Bacteria that use fats from those that do not
Urea broth	Urea, phenol red	Bacteria that hydrolyze urea to ammonia
Sulfur indole motility (SIM)	Thiosulfate, iron	H₂S gas producers from nonproducers
Triple-sugar iron agar (TSIA)	Triple sugars, iron, and phenol red dye	Fermentation of sugars, H₂S production
XLD agar	Lysine, xylose, iron, thiosulfate, phenol red	*Enterobacter, Escherichia, Proteus, Providencia, Salmonella,* and *Shigella*
Birdseed agar	Seeds from thistle plant	*Cryptococcus neoformans* and other fungi

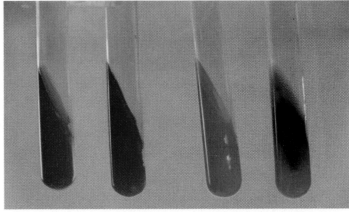

(a)

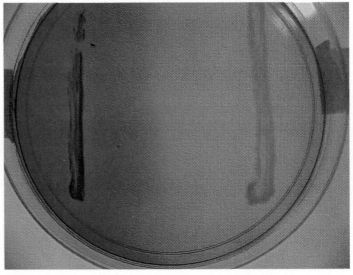

(b)

FIGURE 3.10

Media that differentiate characteristics. (a) Triple-sugar iron agar (TSIA) in a slant. This medium contains three fermentable carbohydrates, phenol red to indicate pH changes, and a chemical (iron) that indicates H₂S gas production. Reactions (from left to right) are: a tube in which no acid was produced (red); a tube in which acid production (yellow) has occurred in the bottom (butt) of the slant; a tube in which acid has been produced throughout the slant; and a tube in which H₂S has been produced, forming a black precipitate. **(b)** A plate of spirit blue agar helps differentiate bacteria that can digest fat (lipase producers) from those that cannot. A positive reaction (left) causes the blue indicator dye to develop a dark blue color on the colony; a negative result (right) is pale yellow or white.

numbers in a specimen. Selenite and brilliant green dye are used in media to isolate *Salmonella* from feces, and sodium azide is used to isolate enterococci from water and food (see EF broth, figure 3.4*b*).

Differential media grow several types of microorganisms and are designed to display visible differences among those microorganisms. Differentiation shows up as variations in colony size or color, in media color changes, or in the formation of gas bubbles and precipitates (table 3.4). These variations come from the type of chemicals these media contain and the ways that microbes react to them. For example, when microbe X metabolizes a certain substance not used by organism Y, then X will cause a visible change in the medium and Y will not. The simplest differential media show two reaction types such as the use or nonuse of a particular nutrient or a color change in some colonies but not in others. Some media are sufficiently complex to show three or four different reactions (see the chapter opening photo and TSIA, figure 3.10*a*).

Dyes can be used as differential agents because many of them are pH indicators that change color in response to the production of

an acid or a base. For example, MacConkey agar contains neutral red, a dye that is yellow when neutral and pink or red when acidic. A common intestinal bacterium such as *Escherichia coli* that gives off acid when it metabolizes the lactose in the medium develops red to pink colonies, and one like *Salmonella* that does not give off acid remains its natural color (off-white). Spirit blue agar is used to detect the hydrolysis (digestion) of fats by lipase enzyme. Positive hydrolysis is indicated by the dark blue color that develops in colonies (figure 3.10*b*).

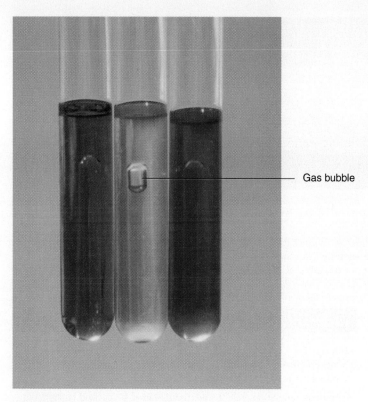

Gas bubble

FIGURE 3.11

Carbohydrate fermentation in broths. This medium is designed to show fermentation (acid production) and gas formation by means of a small, inverted Durham tube for collecting gas bubbles. The tube on the left is an uninoculated negative control; the center tube is positive for acid (yellow) and gas (open space); the tube on the right shows growth but neither acid nor gas.

Miscellaneous Media A **reducing medium** contains a substance (thioglycollic acid or cystine) that absorbs oxygen or slows the penetration of oxygen in a medium, thus reducing its availability. Reducing media are important for growing anaerobic bacteria or determining oxygen requirements (see figure 7.12). **Carbohydrate fermentation media** contain sugars that can be fermented (converted to acids) and a pH indicator to show this reaction (figures 3.9a and 3.11). Media for other biochemical reactions that provide the basis for identifying bacteria and fungi are presented in the second half of this book.

Transport media are used to maintain and preserve specimens that have to be held for a period of time before clinical analysis or to sustain delicate species that die rapidly if not held under stable conditions. Stuart's and Amies transport media contain salts, buffers, and absorbants to prevent cell destruction by enzymes, pH changes, and toxic substances, but will not support growth. **Assay media** are used by technologists to test the effectiveness of antimicrobial drugs (see chapter 12) and by drug manufacturers to assess the effect of disinfectants, antiseptics, cosmetics, and preservatives on the growth of microorganisms. **Enumeration media** are used by industrial and environmental microbiologists to count the numbers of organisms in milk, water, food, soil, and other samples.

CHAPTER CHECKPOINTS

Most microorganisms can be cultured on artificial media, but some can be cultured only in living tissue.

Artificial media are classified by their **physical state** as either liquid, semisolid, liquefiable solid, or nonliquefiable solid.

Artificial media are classified by their **chemical content** as either *synthetic* or *nonsynthetic,* depending on whether the exact chemical composition is known.

Artificial media are classified by their **function** as either general-purpose media or media with one or more specific purposes. Enriched, selective, differential, transport, assay, and enumerating media are all examples of media designed for specific purposes.

INCUBATION, INSPECTION, AND IDENTIFICATION

Once a container of medium has been inoculated, it is **incubated.*** This means it is placed in a temperature-controlled chamber or incubator to encourage multiplication. Although microbes have adapted to growth at temperatures ranging from freezing to boiling, the usual temperatures used in laboratory propagation fall between 20° and 40°C. Incubators can also control the content of atmospheric gases such as oxygen and carbon dioxide that may be required for the growth of certain microbes. During the incubation period (ranging from a day to several weeks), the microbe multiplies and produces growth that is observable macroscopically.[2] Microbial growth in a liquid medium materializes as cloudiness, sediment, scum, or color. A common manifestation of growth on solid media is the appearance of colonies, especially in bacteria and fungi. Colonies are actually large masses of clinging cells with distinctions in size, shape, color, and texture (see figure 4.12a,b).

In some ways, culturing microbes is analogous to gardening on a microscopic scale. Cultures are formed by "seeding" tiny plots (media) with microbial cells to grow into separate rows. Extreme care is taken to exclude weeds. A **pure culture** is a container of medium that grows only a single known species or type of microorganism (figure 3.12a). This type of culture is most frequently used for laboratory study, because it allows the systematic examination and control of one microorganism by itself. Instead of the term *pure culture,* some microbiologists prefer the term **axenic,*** meaning that the culture is free of other living things except for the one being studied. A standard method for preparing a pure culture is to **subculture,** or make a second-level culture from a well-isolated colony. A tiny bit of cells is transferred into a separate container of media and incubated (see figure 3.1, step 3).

A **mixed culture** (figure 3.12b) is a container that holds two or more *identified,* easily differentiated species of microorganisms, not unlike a garden plot containing both carrots and onions. A **contaminated*** **culture** (figure 3.12c) was once pure or mixed (and thus

---

2. For all intents and purposes, the macroscopic level is synonymous with the cultural level.

* **incubate** (in′-kyoo-bayt) Gr. *incubatus,* to lie in or upon.

* **axenic** (ak-zee′-nik) Gr. *a,* no, and *xenos,* stranger.

* **contaminated** (kon-tam′-ih-nay-tid) Gr. *con,* together, and L. *tangere,* to touch.

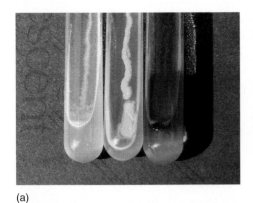

(a)

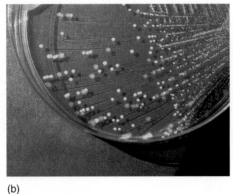

(b)

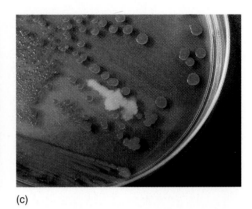

(c)

FIGURE 3.12

Various conditions of cultures. (a) Three tubes containing pure cultures of *Escherichia coli* (white), *Micrococcus luteus* (yellow), *and Serratia marcescens* (red). **(b)** A mixed culture of *M. luteus* and *E. coli* readily differentiated by their colors. **(c)** This plate of *S. marcescens* was overexposed to room air, and it has developed a large, white colony. Because this intruder is not desirable and not identified, the culture is now contaminated.

a known entity) but has since had **contaminants** (unwanted microbes of uncertain identity) introduced into it, like weeds into a garden. Because contaminants have the potential for causing disruption, constant vigilance is required to exclude them from microbiology laboratories, as you will no doubt witness from your own experience.

How does one determine what sorts of microorganisms have been isolated in cultures? Certainly the combination of microscopic and macroscopic appearance can be valuable in differentiating the smaller, simpler procaryotic cells from the larger, more complex eucaryotic cells. Appearance can be especially useful in identifying eucaryotic microorganisms to the level of genus or species because of their distinctive morphological features; however, bacteria are generally not identifiable by these methods because very different species may appear quite similar. For them, we must include techniques that characterize their cellular metabolism. These methods, called biochemical tests, can determine fundamental chemical characteristics such as nutrient requirements, products given off during growth, presence of enzymes, and mechanisms for deriving energy.

Several modern analytical and diagnostic tools that focus on genetic and molecular characteristics can detect the exact nature of microbial DNA. In the case of certain pathogens, further information on a microbe is obtained by inoculating a suitable laboratory animal. A profile is prepared by compiling physiological testing results with both macroscopic and microscopic traits. The profile then becomes the raw material used in final identification. In several subsequent chapters, we present more detailed examples of identification methods (see figure 4.28 for an example).

Maintenance and Disposal of Cultures

In most medical laboratories, the cultures and specimens constitute a potential hazard and will require immediate and proper disposal. Both steam sterilizing (see autoclave, chapter 11) and incineration (burning) are used to destroy microorganisms. On the other hand, many teaching and research laboratories maintain a line of stock cultures that represent "living catalogues" for study and experimentation. The largest culture collection can be found at the American Type Culture Collection in Rockville, Maryland, which maintains a voluminous array of frozen and freeze-dried fungal, bacterial, viral, and algal cultures.

CHAPTER CHECKPOINTS

The Five I's—inoculation, incubation, isolation, inspection, and identification—summarize the kinds of laboratory procedures used in microbiology.

Following **inoculation,** cultures are **incubated** at a specified temperature and time to encourage growth.

Isolated colonies that originate from single cells are composed of large numbers of cells clinging together.

A culture may exist in one of the following forms: A **pure culture** contains only one species or type of microorganism. A **mixed culture** contains two or more known species. A **contaminated culture** contains both known and unknown (unwanted) microorganisms.

During inspection, the cultures are examined and evaluated macroscopically and microscopically.

Microorganisms are identified in terms of their macroscopic, or colony, morphology; their microscopic morphology; and their biochemical reactions and genetic characteristics.

Microbial cultures are disposed of in two ways: steam sterilization or incineration.

The Microscope: Window on an Invisible Realm

Imagine Leeuwenhoek's excitement and wonder when he first viewed a drop of rainwater and glimpsed an amazing microscopic world teeming with unearthly creatures. Beginning microbiology students still experience this sensation, and even experienced microbiologists never forget their first view. The microbial existence is indeed another world, but it would remain largely uncharted without an essential tool: the microscope. Your efforts in exploring microbes will be more meaningful if you understand some essentials of **microscopy*** and specimen preparation.

---

* microscopy (mye-kraw´-skuh-pee) Gr. The science that studies microscope techniques.

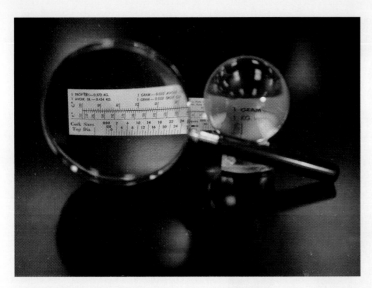

FIGURE 3.13

Effects of magnification. Demonstration of the magnification and image-forming capacity of clear glass "lenses." Given a proper source of illumination, this magnifying glass and crystal ball magnify a ruler two to three times.

MAGNIFICATION AND MICROSCOPE DESIGN

The two key characteristics of a reliable microscope are **magnification,** or the ability enlarge objects, and **resolving power,** or the ability to show detail.

A discovery by early microscopists that spurred the advancement of microbiology was that clear, glass spheres could act as a lens to **magnify*** small objects. Magnification in most microscopes results from a complex interaction between visible light waves and the curvature of the lens. When a beam or ray of light transmitted through air strikes and passes through the convex surface of glass, it experiences some degree of **refraction,*** defined as the bending or change in the angle of the light ray as it passes through a medium such as a lens (see figure 3.15). The greater the difference in the composition of the two substances the light passes between, the more pronounced is the refraction. When an object is placed a certain distance from the spherical lens and **illuminated*** with light, an optical replica, or **image,** of it is formed by the refracted light. Depending upon the size and curvature of the lens, the image appears enlarged to a particular degree, which is called its power of magnification and is usually identified with a number combined with × (read "times"). This behavior of light is evident if one looks through an everyday object such as a glass ball or a magnifying glass (figure 3.13). It is basic to the function of all **optical,** or **light, microscopes,** though many of them have additional features that define, refine, and increase the size of the image.

The first microscopes were **simple,** meaning they contained just a single magnifying lens and a few working parts. Examples of this type of microscope are a magnifying glass, a hand lens, and Leeuwenhoek's basic little tool shown in figure 1.8a. Among the refinements that led to the development of today's **compound microscope** were the addition of a second magnifying lens system, a lamp in the base to give off visible light and **illuminate** the specimen, and a special lens called the **condenser** that converges or focuses the rays of light to a single point on the object. The fundamental parts of a modern compound light microscope are illustrated in figure 3.14.

Principles of Light Microscopy

To be most effective, a microscope should provide adequate magnification, resolution, and clarity of image. Magnification of the object or specimen by a compound microscope occurs in two phases. The first lens in this system (the one closest to the specimen) is the **objective lens,** and the second (the one closest to the eye) is the **ocular lens,** or eyepiece (figure 3.15). The objective forms the initial image of the specimen, called the **real image.** When this image is projected up through the microscope body to the plane of the eyepiece, the ocular lens forms a second image, the **virtual image.** The virtual image is the one that will be received by the eye and converted to a retinal and visual image. The magnifying power of the objectives alone usually ranges from 4× to 100×, and the power of the ocular alone ranges from 10× to 20×. The **total power of magnification** of the final image formed by the combined lenses is a product of the separate powers of the two lenses:

Power of objective	×	Power of ocular	=	Total magnification
10× low power objective		10×	=	100×
40× high dry objective		10×	=	400×
100× oil immersion objective		10×	=	1,000×

Microscopes are equipped with a nosepiece holding three or more objectives that can be rotated into position as needed. The power of the ocular usually remains constant for a given microscope. Depending on the power of the ocular, the total magnification of standard light microscopes can vary from 40× with the lowest power objective (called the scanning objective) to 2,000× with the highest power objective (the oil immersion objective).

Resolution: Distinguishing Magnified Objects Clearly

Despite the importance of magnification in visualizing tiny objects or cells, an additional optical property is essential for seeing clearly. That property is **resolution,** or **resolving power.** Resolution is the capacity of an optical system to distinguish or separate two adjacent objects or points from one another. For example, at a certain fixed distance, the lens in the human eye can resolve two small objects as separate points just as long as the two objects are no closer than 0.2 mm apart. The eye examination given by optometrists is in fact a test of the resolving power of the human eye for various-sized letters read at a distance of 20 feet. Because microorganisms are extremely small and usually very close together, they will not be seen with clarity or any degree of detail unless the microscope's lenses can resolve them.

---

* magnify (mag′-nih-fye) L. *magnus,* great, and *ficere,* to make.

* refract, refraction (ree-frakt′, ree-frak′-shun) L. *refringere,* to break apart.

* illuminate (ill-oo′-mih-nayt) L. *illuminatus,* to light up.

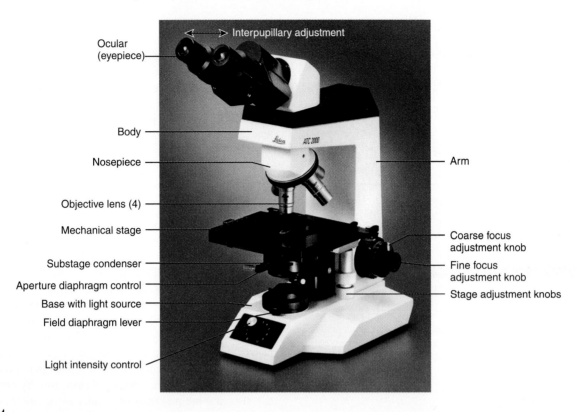

FIGURE 3.14

The parts of a student laboratory microscope. This microscope is a compound light microscope with two oculars (called binocular). It has four objective lenses, a mechanical stage to move the specimen, a condenser, an iris diaphragm, and a built-in lamp.

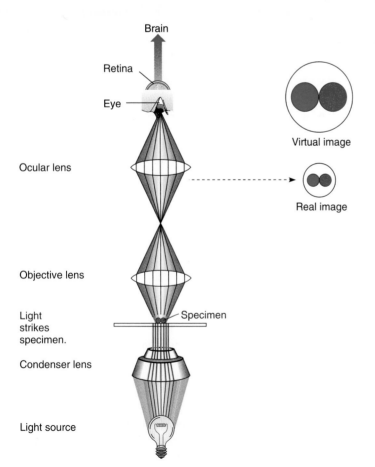

FIGURE 3.15

The pathway of light and the two stages in magnification of a compound microscope. As light passes through the condenser, it forms a solid beam that is focused on the specimen. Light leaving the specimen that enters the objective lens is refracted so that an enlarged primary image, the real image, is formed. One does not see this image, but its degree of magnification is represented by the lower circle. The real image is projected through the ocular, and a second image, the virtual image, is formed by a similar process. The virtual image is the final magnified image that is received by the retina and perceived by the brain. Notice that the lens systems cause the image to be reversed.

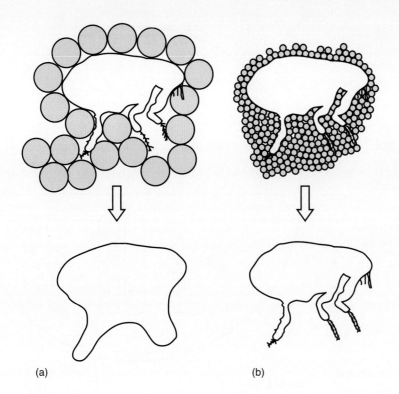

(a)   (b)

FIGURE 3.16

Effect of wavelength on resolution. A simple model demonstrates how the wavelength influences the resolving power of a microscope. Here an outline of a flea represents the object being illuminated, and two different-sized circles represent the wavelengths of light. In **(a),** the longer waves are too large to penetrate between the finer spaces and produce a fuzzy, undetailed image. In **(b),** shorter waves are small enough to enter small spaces and produce a much more detailed image that is recognizable as a flea.

A simple equation in the form of a fraction expresses the main determining factors in resolution:

$$\text{Resolving power (R.P.)} = \frac{\text{Wavelength of light in nm}}{2 \times \text{Numerical aperture of objective lens}}$$

From this equation it is evident that the resolving power is a function of the wavelength of light that forms the image, along with certain characteristics of the objective. The light source for optical microscopes consists of a band of colored wavelengths in the visible spectrum (see figure 11.7). The shortest visible wavelengths are in the violet-blue portion of the spectrum (400 nm), and the longest are in the red portion (750 nm). Because the wavelength must pass between the objects that are being resolved, shorter wavelengths (in the 400–500 nm range) will provide better resolution (figure 3.16). Some microscopes have a special blue filter placed over the lamp to limit the longer wavelengths of light from entering the specimen.

The other factor influencing resolution is the **numerical aperture,** a mathematical constant that describes the relative efficiency of a lens in bending light rays. Without going into the mathematical derivation of this constant, it is sufficient to say that each objective has a fixed numerical aperture reading that is determined by the microscope design and ranges from 0.1 in the lowest power lens to approximately 1.25 in the highest power (oil immersion) lens. The most important thing to remember is that a higher numerical aperture number will provide better resolution. In order for the oil immersion lens to arrive at its maximum resolving capacity, a drop of oil must be inserted between the tip of the lens and the specimen on the glass slide. Because oil has the same optical qualities as glass, it prevents refractive loss that normally occurs as peripheral light passes from the slide into the air; this property effectively increases the numerical aperture (figure 3.17). There is an absolute limitation to resolution in optical microscopes, which can be demonstrated by calculating the resolution of the oil immersion lens using a blue-green wavelength of light:

$$\text{R.P.} = \frac{500 \text{ nm}}{2 \times 1.25}$$

$$= 200 \text{ nm (or } 0.2 \text{ } \mu\text{m)}$$

In practical terms, this means that the oil immersion lens can resolve any cell or cell part as long as it is at least 0.2 μm in diameter, and that it can resolve two adjacent objects as long as they are at least 0.2 μm apart (figure 3.18). In general, organisms that are 0.5 μm or more in diameter are readily seen. This includes fungi and protozoa and some of their internal structures, and most bacteria. However, a few bacteria and most viruses are far too small to be resolved by the optical microscope and require electron microscopy (discussed later in this chapter). In summary then, the factor that most limits the clarity of a microscope's image is its resolving power. Even if a light microscope were designed to magnify several thousand times, its resolving power could not be increased, and the image it produced would simply be enlarged and fuzzy.

Other constraints to the formation of a clear image are the quality of the lens and light source and the lack of contrast in the specimen. No matter how carefully a lens is constructed, flaws remain. A typical problem is spherical aberration, a distortion in the image caused by irregularities in the lens, which creates a curved, rather than flat, image (see figure 3.13). Another is chromatic aberration, a rainbowlike image that is caused by the lens acting as a prism and separating visible light into its colored bands. Brightness and direction of illumination also affect image formation. Because too much light can reduce contrast and burn out the

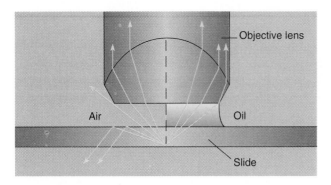

FIGURE 3.17

Workings of an oil immersion lens. To maximize its resolving power, an oil immersion lens (the one with highest magnification) must have a drop of oil placed at its tip. This forms a continuous medium to transmit a beam of light from the condenser to the objective and effectively increase the numerical aperture. Without oil, some of the peripheral light that passes through the specimen is scattered into the air or onto the glass slide; this scattering decreases resolution.

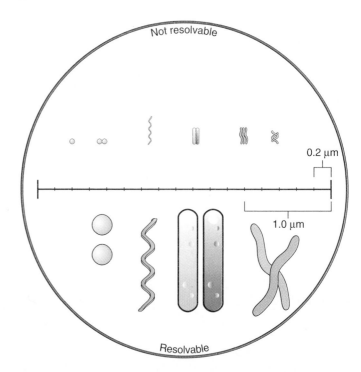

FIGURE 3.18

Effect of magnification. Comparison of cells that would not be resolvable versus those that would be resolvable under oil immersion at 1,000× magnification. Note that in addition to differentiating two adjacent things, good resolution also means being able to see an object clearly.

image, an iris diaphragm on most microscopes controls the amount of light entering the condenser. The lack of contrast in cell components is compensated for by using special lenses (the phase-contrast microscope) and by adding dyes.

VARIATIONS ON THE OPTICAL MICROSCOPE

Optical microscopes that use visible light can be described by the nature of their **field,** meaning the circular area viewed through the ocular lens. With special adaptations in lenses, condenser, and light sources, four special types of microscopes can be described: bright-field, dark-field, phase-contrast, and interference. A fifth type of optical microscope, the fluorescence microscope, uses ultraviolet radiation as the illuminating source. Each of these microscopes is adapted for viewing specimens in a particular way, as described in the next sections and summarized in parts of table 3.5.

Bright-Field Microscopy

The **bright-field microscope** is the most widely used type of light microscope. Although we ordinarily view objects like the words on this page with light reflected off the surface, a bright-field microscope forms its image when light is transmitted through the specimen. The specimen, being denser and more opaque than its surroundings, absorbs some of this light, and the rest of the light is transmitted directly up through the ocular into the field. As a result, the specimen will produce an image that is darker than the surrounding brightly illuminated field. The bright-field microscope is a multipurpose instrument that can be used for both live, unstained material and preserved, stained material. The bright-field image is compared with that of other microscopes in figure 3.19.

Dark-Field Microscopy

A bright-field microscope can be adapted as a **dark-field microscope** by adding a special disc called a *stop* to the condenser. The stop blocks all light from entering the objective lens except peripheral light that is reflected off the sides of the specimen itself. The re-

sulting image is a particularly striking one: brightly illuminated specimens surrounded by a dark (black) field (figure 3.19*b*). Some of Leeuwenhoek's more successful microscopes probably operated with dark-field illumination. The most effective use of dark-field microscopy is to visualize living cells that would be distorted by drying or heat or cannot be stained with the usual methods. It can outline the organism's shape and permit rapid recognition of swimming cells that might appear in dental and other infections (figure 3.20), but it does not reveal fine internal details.

Phase-Contrast and Interference Microscopy

If similar objects made of clear glass, ice, cellophane, or plastic were immersed in the same container of water, an observer would have difficulty telling them apart because they have similar optical properties. Internal components of a live, unstained cell also lack contrast and can be difficult to distinguish. But cell structures do differ slightly in density, enough that they can alter the light that passes through them in subtle ways. The **phase-contrast microscope** has been constructed to take advantage of this characteristic. This microscope contains devices that transform the subtle changes in light waves passing through the specimen into differences in light intensity. For example, denser cell parts such as organelles alter the pathway of light more than less dense regions (the cytoplasm). Light patterns coming from these regions will vary in contrast. The amount of internal detail visible by this method is greater than by either bright-field or dark-field methods. The phase-contrast microscope is most useful for observing intracellular struc-

TABLE 3.5

Comparison of Light Microscopes and Electron Microscopes

Characteristic	Light or Optical	Electron (Transmission)
Useful magnification	2,000×	1,000,000 or more
Maximum resolution	200 nm	0.5 nm
Image produced by	Visible light rays	Electron beam
Image focused by	Glass objective lens	Electromagnetic objective lenses
Image viewed through	Glass ocular lens	Fluorescent screen
Specimen placed on	Glass slide	Copper mesh
Specimen may be alive	Yes	No
Specimen requires special stains or treatment	Not always	Yes
Colored images produced	Yes	No

tures such as bacterial spores, granules, and organelles, as well as the locomotor structures of eucaryotic cells (figures 3.19c and 3.21a).

Like the phase-contrast microscope, the **differential interference contrast (DIC) microscope** provides a detailed view of unstained, live specimens by manipulating the light. But this microscope has additional refinements, including two prisms that add contrasting colors to the image and two beams of light rather than a single one. DIC microscopes produce extremely well-defined images that are vividly colored and appear three-dimensional (figure 3.21b).

Most optical microscopes have difficulty forming a clear image of cells at higher magnifications, because cells are often too thick for conventional lenses to focus all levels of the cell simultaneously. This is especially true of larger cells with complex internal structures. A newer type of microscope that overcomes this impediment is called the *confocal scanning optical microscope*. This microscope uses a beam of light to scan various depths in the specimen and deliver a sharp image focusing on just a single plane. It is thus able to capture a highly focused view at any level, ranging from the surface to the middle of the cell. It is versatile because it works on live or preserved, stained specimens and it does not require sectioning of the specimen.

Fluorescence Microscopy

The **fluorescence microscope** is a specially modified compound microscope furnished with an ultraviolet (UV) radiation source and a filter that protects the viewer's eye from injury by these dangerous rays. The name of this type of microscopy originates from the use of certain dyes (acridine, fluorescein) and minerals that show **fluorescence.** This means that the dyes emit visible light when bombarded by shorter ultraviolet rays. For an image to be formed, the specimen must first be coated or placed in contact with a source of fluorescence. Subsequent illumination by ultraviolet radiation

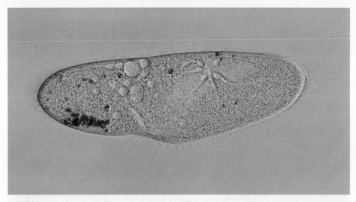

(a)

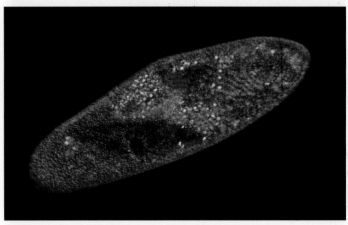

(b)

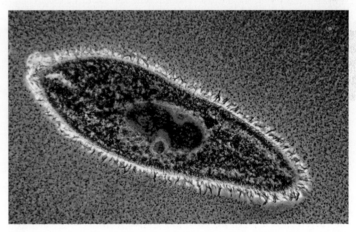

(c)

FIGURE 3.19

Three views of a basic cell. A live cell of *Paramecium* viewed with **(a)** bright-field (400×), **(b)** dark-field (400×), and **(c)** phase-contrast (400×). Note the difference in the appearance of the field and the degree of detail shown by each method of microscopy. Only in phase-contrast are the cilia (fine hairs) on the cells noticeable. Can you see the nucleus? The oral groove?

causes the specimen to give off light that will form its own image, usually an intense yellow, orange, or red against a black field.

Fluorescence microscopy has its most useful applications in diagnosing infections caused by specific bacteria, protozoans, and viruses. A staining technique with fluorescent dyes is commonly

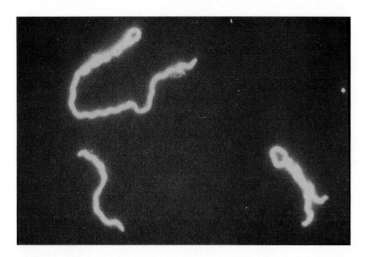

FIGURE 3.20

Dark-field photomicrograph of a spiral-shaped oral bacterium called *Treponema vincenti* (1,100×). This species lives in the space between the gums and the teeth and may be involved in a deteriorating infection called necrotizing gingivitis.

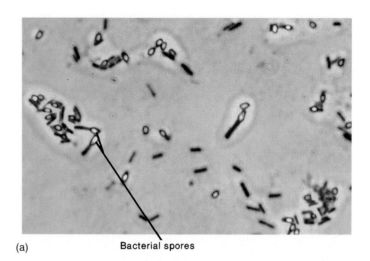

(a)   Bacterial spores

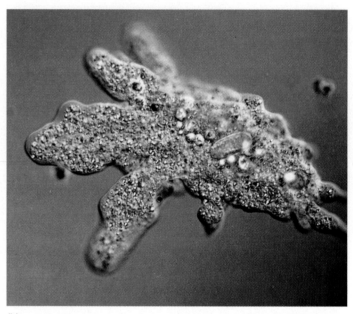

(b)

FIGURE 3.21

Visualizing internal structures. (a) Phase-contrast micrograph of a bacterium containing spores. The relative density of the sporers causes them to appear as bright, shiny objects against the darker cell parts (600×). **(b)** Differential interference micrograph of *Amoeba proteus*, a common protozoan. Note the outstanding internal detail, the depth of field, and the bright colors, which are not natural (160×).

used to detect *Mycobacterium tuberculosis* (the agent of tuberculosis) in patients' specimens (see figure 19.20). In a number of diagnostic procedures, fluorescent dyes are affixed to specific antibodies. These *fluorescent antibodies* can be used to detect the causative agents in such diseases as syphilis, chlamydiosis, trichomoniasis, herpes, and influenza (see figure 6.24*a*). A newer technology using fluorescent nucleic acid stains can differentiate between live and dead cells in mixtures (figure 3.22). A fluorescence microscope can be handy for locating microbes in complex mixtures because only those cells targeted by the technique will fluoresce.

ELECTRON MICROSCOPY

If conventional light microscopes are our windows on the microscopic world, then the electron microscope (EM) is our window on the tiniest details of that world. Although this microscope was originally conceived and developed for studying nonbiological materials such as metals and small electronics parts, biologists immediately recognized the importance of the tool and began to use it in the early 1930s. One of the most impressive features of the electron microscope is the resolution it provides.

Unlike the light microscope, the electron microscope forms an image with a beam of electrons that can be made to travel in wavelike patterns when accelerated to high speeds. These waves are 100,000 times shorter than the waves of visible light. Because resolving power is a function of wavelength, electrons have tremendous power to resolve minute structures. Indeed, it is possible to resolve atoms with an electron microscope, though the practical resolution for biological applications is approximately 0.5 nm. Because the resolution is so substantial, it follows that magnification can also be extremely high—usually between 5,000× and 1,000,000× for biological specimens and up to 5,000,000× in some applications. Its capacity for magnification and resolution makes the EM an invaluable tool for seeing the finest structure—or

ultrastructure—of cells and viruses. If not for electron microscopes, our understanding of biological structure and function would still be in its early theoretical stages.

In fundamental ways, the electron microscope is a derivative of the compound microscope. It employs components analogous to, but not necessarily the same as, those in light microscopy (figure 3.23). For instance, it magnifies in stages by means of two lens systems, and it has a condensing lens, a specimen holder, and focusing apparatus. Otherwise, the two types have numerous differences (table 3.6). An electron gun aims its beam through a vacuum to ring-shaped electromagnets that focus this beam on the specimen.

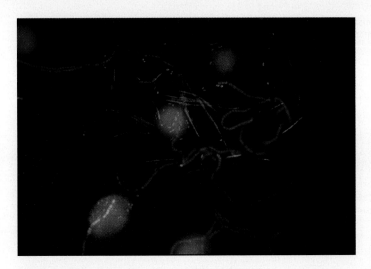

FIGURE 3.22

Fluorescent staining on a fresh sample of cheek scrapings from the oral cavity. Cheek epithelial cells are the larger unfocused red or green cells. Bacteria appearing here are streptococci (tiny spheres in long chains) and filamentous rods. This technique also indicates whether cells are alive or dead; live cells fluoresce green, and dead cells fluoresce red.

Specimens must be pretreated with chemicals or dyes to increase contrast and cannot be observed in a live state. The enlarged image is displayed on a viewing screen or photographed for further study rather than being observed directly through an eyepiece. Because images produced by electrons lack color, electron micrographs (a micrograph is a photograph of a microscopic object) are always shades of black, gray, and white. The color-enhanced micrographs used in this and other textbooks have computer-added color (yes, they are colorized!).

Two general forms of EM are the transmission electron microscope (TEM) and the scanning electron microscope (SEM) (see table 3.6). **Transmission electron microscopes** are the method of choice for viewing the detailed structure of cells and viruses. This microscope produces its image by transmitting electrons through the specimen. Because electrons cannot readily penetrate thick preparations, the specimen must be sectioned into extremely thin slices (20–100 nm thick) and stained or coated with metals that will increase image contrast. The darkest areas of TEM micrographs represent the thicker (denser) parts, and the lighter areas indicate the more transparent and less dense parts (figure 3.24). The TEM can also be used to produce negative images and shadow casts of whole microbes (see figure 6.1).

TABLE 3.6

Comparisons of Types of Microscopy

Microscope	Maximum Practical Magnification	Resolution	Important Features
Visible light as source of illumination			
Bright-field	2,000×	0.2 μm (200 nm)	Common multipurpose microscope for live and preserved stained specimens; specimen is dark, field is white; provides fair cellular detail
Dark-field	2,000×	0.2 μm	Best for observing live, unstained specimens; specimen is bright, field is black; provides outline of specimen with reduced internal cellular detail
Phase-contrast	2,000×	0.2 μm	Used for live specimens; specimen is contrasted against gray background; excellent for internal cellular detail
Differential interference	2,000×	0.2 μm	Provides brightly colored, highly contrasting, three-dimensional images of live specimens
Ultraviolet rays as source of illumination			
Fluorescent	2,000×	0.2 μm	Specimens stained with fluorescent dyes or combined with fluorescent antibodies emit visible light; specificity makes this microscope an excellent diagnostic tool
Electron beam forms image of specimen			
Transmission electron microscope (TEM)	1,000,000×	0.5 nm	Sections of specimen are viewed under very high magnification; finest detailed structure of cells and viruses is shown; used only on preserved material
Scanning electron microscope (SEM)	100,000×	10 nm	Scans and magnifies external surface of specimen; produces striking three-dimensional image

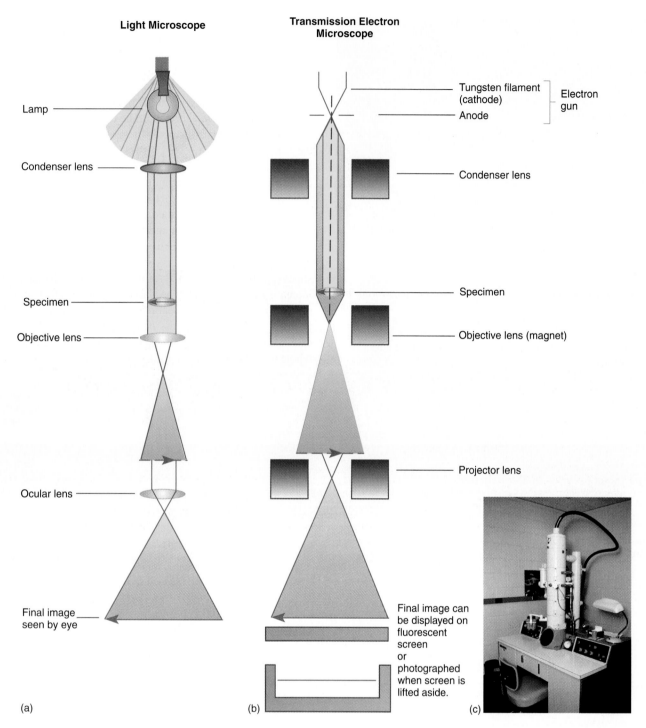

Light Microscope

Transmission Electron Microscope

Lamp

Condenser lens

Specimen

Objective lens

Ocular lens

Final image seen by eye

(a)

Tungsten filament (cathode)

Anode

Electron gun

Condenser lens

Specimen

Objective lens (magnet)

Projector lens

Final image can be displayed on fluorescent screen or photographed when screen is lifted aside.

(b)

(c)

FIGURE 3.23

Comparison of two microscopes: (a) the light microscope and **(b)** one type of electron microscope (EM; transmission type). These diagrams are highly simplified, especially for the electron microscope, to indicate the common components. Note that the EM's image pathway is actually upside down compared with that of a light microscope. **(c)** The EM is a larger machine with far more complicated working parts than most light microscopes.

The **scanning electron microscope** provides some of the most dramatic and realistic images in existence. This instrument is designed to create an extremely detailed three-dimensional view of anything from a fly's eye to AIDS viruses escaping their host cells. To produce this image, the SEM does not transmit electrons; it bombards the surface of a whole, metal-coated specimen with electrons while scanning back and forth over it. A shower of electrons deflected from the surface is picked up with great fidelity by a sophisticated detector, and the electron pattern is displayed as an image on a television screen. The contours of the specimens resolved

(a)

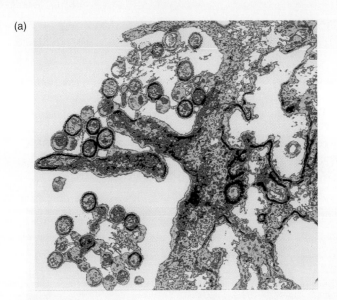

FIGURE 3.24

TEM. (a) Hantaviruses isolated from a patient with hantavirus respiratory syndrome are seen budding off the surface of a cell. The virus particles have been colorized with bright red nucleic acid strands and blue-green capsids. **(b)** TEM section through a cell of *Chlamydomonas,* a motile alga that lives in fresh water. At this magnification, the ultrastructure of several organelles is evident (30,000×).

(b)

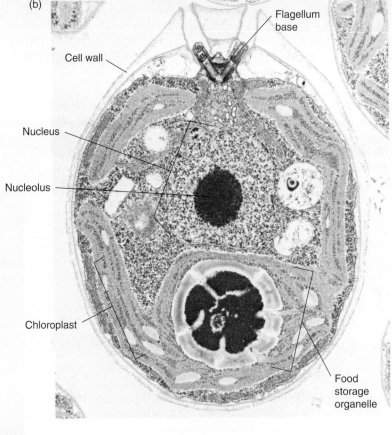

Flagellum base

Cell wall

Nucleus

Nucleolus

Chloroplast

Food storage organelle

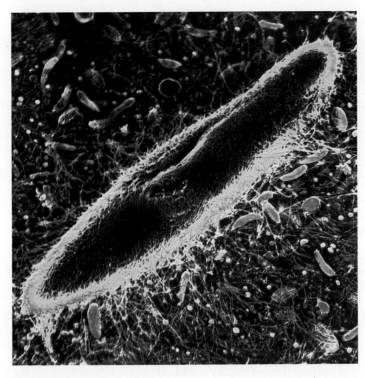

FIGURE 3.25

A false-color scanning electron micrograph (SEM) of *Paramecium,* covered in masses of fine hairs (100×). These are actually its locomotor and feeding structures—the cilia. Cells in the surrounding medium are bacteria that serve as the protozoan's "movable feast." Compare this with figure 3.20 to appreciate the outstanding three-dimensional detail shown by an SEM.

with scanning electron micrography are very revealing and often surprising. Areas that look smooth and flat with the light microscope display intriguing surface features with the SEM (figure 3.25). Improved technology has continued to refine electron microscopes and to develop variations on the basic plan. One of the most inventive relatives of the EM is the scanning probe microscope (Microbits 3.2).

PREPARING SPECIMENS FOR OPTICAL MICROSCOPES

A specimen for optical microscopy is generally prepared by mounting a sample on a suitable glass slide that sits on the stage between the condenser and the objective lens. The manner in which a slide specimen, or **mount,** is prepared depends upon: (1) the condition of the specimen, either in a living or preserved state; (2) the aims of the examiner, whether to observe overall structure, identify the microorganisms, or see movement; and (3) the type of microscopy available, whether it is bright-field, dark-field, phase-contrast, or fluorescence.

Fresh, Living Preparations

Live samples of microorganisms are placed in **wet mounts** or in **hanging drop mounts** so that they can be observed as near to their natural state as possible. The cells are suspended in a suitable fluid (water, broth, saline) that temporarily maintains viability and provides space and a medium for locomotion. A wet mount consists of a drop or two of the culture placed on a slide and overlaid with a cover glass. Although this type of mount is quick and easy to

MICROBITS 3.2

An Evolution in Resolution: Probing Microscopes

In the past, chemists, physicists, and biologists had to rely on indirect methods to provide information on the structures of the smallest molecules. But technological advances in the last decade have created a new generation of microscopes that "see" atomic structure by actually feeling it. *Scanning probe microscopes* operate with a minute needle tapered to a tip that can be as narrow as a single atom! This probe scans over the exposed surface of a material and records an image of its outer texture. These revolutionary microscopes have such profound resolution that they have the potential to image single atoms (but not subatomic structure yet) and to magnify 100 million times. The scanning tunneling microscope (STM) was the first of these microscopes. It uses a tungsten probe that hovers near the surface of an object and follows its topography while simultaneously giving off an electrical signal of its pathway, which is then imaged on a screen. The STM is used primarily for detecting defects on the surfaces of electrical conductors and computer chips composed of silicon, but it recently provided the first incredible close-up views of DNA, the genetic material (see Spotlight on Microbiology 9.1).

Another exciting new variant is the atomic force microscope (AFM), which gently forces a diamond and metal probe down onto the surface of a specimen like a needle on a record. As it moves along the surface, any deflection of the metal probe is detected by a sensitive device that relays the information to an imager. The AFM is very useful in viewing the detailed patterns of biological molecules.

These powerful new microscopes, along with tools that can move and position atoms, have spawned a field called *nanotechnology*—the science of the "small." Scientists in this area use physics, chemistry, biology, and engineering to explore and manipulate small molecules and

atoms. Working at these dimensions, they hope to create tiny molecular tools to miniaturize computers and other electronic devices. In the future, it may be possible to use microstructures to deliver drugs, analyze DNA, and treat disease.

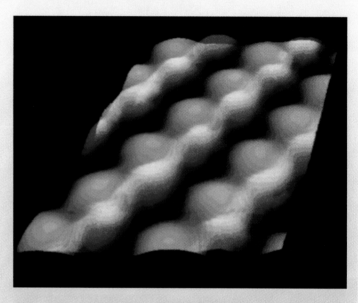

These chains of colored beads are actually atoms of the element gallium (blue) and arsenic (red) as imaged by a scanning tunneling microscope.

prepare, it has certain disadvantages. The cover glass can damage larger cells, and the slide is very susceptible to drying and can contaminate the handler's fingers. A more satisfactory alternative is the hanging drop preparation made with a special concave (depression) slide, a Vaseline adhesive or sealant, and a coverslip from which a tiny drop of sample is suspended (see figure 4.4). These types of short-term mounts provide a true assessment of the size, shape, arrangement, color, and motility of cells. Greater cellular detail can be observed with phase-contrast or interference microscopy.

Fixed, Stained Smears

A more permanent mount for long-term study can be obtained by preparing fixed, stained specimens. The **smear** technique, developed by Robert Koch more than 100 years ago, consists of spreading a thin film made from a liquid suspension of cells on a slide and air-drying it. Next, the air-dried smear is usually heated gently by a process called **heat fixation** that simultaneously kills the specimen and secures it to the slide. Another important action of fixation is to preserve various cellular components in a natural state with minimal distortion. Fixation of some microbial cells is performed with chemicals such as alcohol and formalin.

Like images on undeveloped photographic film, the unstained cells of a fixed smear are quite indistinct, no matter how great the magnification or how fine the resolving power of the microscope. The process of "developing" a smear to create contrast and make inconspicuous features stand out requires staining techniques. **Staining** is any procedure that applies colored chemicals called **dyes** to specimens. Dyes impart a color to cells or cell parts by becoming affixed to them through a chemical reaction. In general, they are classified as **basic (cationic) dyes,** which have a positive charge, or **acidic (anionic) dyes,** which have a negative charge. Because chemicals of opposite charge are attracted to each other, cell parts that are negatively charged will attract basic dyes and those that are positively charged will attract acidic dyes (table 3.7). Many cells, especially those of bacteria, have numerous negatively charged acidic substances and thus stain more readily with basic dyes. Acidic dyes, on the other hand, tend to be repelled by cells, so they are good for negative staining (discussed in the next section). The chemistry of dyes and their staining reactions is covered in more detail in Microbits 3.3.

MICROBITS 3.3
The Chemistry of Dyes and Staining

Because many microbial cells lack contrast, it is necessary to use dyes to observe their detailed structure and identify them. Dyes are colored compounds related to or derived from the common organic solvent benzene. When certain double-bonded groups (C=O, C=N, N=N) are attached to complex ringed molecules, the resultant compound gives off a specific color. Most dyes are in the form of a sodium or chloride salt of an acidic or basic compound that ionizes when dissolved in a compatible solvent. The color-bearing ion, termed a **chromophore,** is charged and has an affinity for certain cell parts that are of the opposite charge.

Dyes that have a negatively charged chromophore are termed acidic. An example is sodium eosinate, a bright red dye that dissociates into eosin⁻ and Na⁺. Acidic chromophores are attracted to the positively charged molecules of cells such as the granules of some types of white blood cells. Because bacterial cells have numerous acidic substances and carry a slightly negative charge on their surface, they do not stain well with acidic dyes.

Basic dyes such as basic fuchsin have a positively charged chromophore that is attracted to negatively charged cell components (nucleic acids and proteins). Since bacteria have a preponderance of negative ions, they stain readily with other basic dyes, including crystal violet, methylene blue, malachite green, and safranin.

Acidic Dye

Sodium eosinate → Eosin

reacts with (+) **cell**

Basic Dye

Basic fuchsin chloride → Fuchsin

reacts with (−) **cell**

Examples of the two major groups of dyes and their reactions.

Negative Versus Positive Staining   Two basic types of staining technique are used, depending upon how a dye reacts with the specimen (summarized in table 3.7). Most procedures involve a **positive stain,** in which the dye actually sticks to the specimen and gives it color. A **negative stain,** on the other hand, is just the reverse (like a photographic negative). The dye does not stick to the specimen but settles around its outer boundary, forming a silhouette. In a sense, negative staining "stains" the glass slide to produce a dark background around the cells. Nigrosin (blue-black) and India ink (a black suspension of carbon particles) are the dyes most commonly used for negative staining. The cells themselves do not stain because these dyes are negatively charged and are repelled by the negatively charged surface of the cells. The value of negative staining is its relative simplicity and the reduced shrinkage or distortion of cells, as the smear is not heat-fixed. A quick assessment can thus be made regarding cellular size, shape, and arrangement. Negative staining is also used to accentuate the capsule that surrounds certain bacteria and yeasts (figure 3.26).

Simple Versus Differential Staining   Positive staining methods are classified as simple, differential, or special (figure 3.26). Whereas **simple stains** require only a single dye and an uncomplicated procedure, **differential stains** use two different-colored dyes, called the *primary dye* and the *counterstain,* to distinguish between cell types or parts. These staining techniques tend to be more complex and sometimes require additional chemical reagents to produce the desired reaction.

Most simple staining techniques take advantage of the ready binding of bacterial cells to dyes like malachite green, crystal violet, basic fuchsin, and safranin. Simple stains cause all cells in a smear to appear more or less the same color, regardless of type, but they reveal such bacterial characteristics as shape, size, and arrangement. A simple stain with Loeffler's methylene blue is distinctive. The blue cells stand out against a relatively unstained background, so that size, shape, and grouping show up easily. This method is also significant because it reveals the internal granules of *Corynebacterium diphtheriae,* a bacterium that is responsible for diphtheria (see chapter 4).

Types of Differential Stains   A satisfactory differential stain uses differently colored dyes to clearly contrast two cell types or cell parts. Common combinations are red and purple, red and green, or pink and blue. Differential stains can also pinpoint other characteristics, such as the size, shape, and arrangement of cells. Typical examples include Gram, acid-fast, and endospore stains. Some staining techniques (spore, capsule) fall into more than one category.

Gram staining, a century-old method named for its developer, Hans Christian Gram, remains the most universal diagnostic staining technique for bacteria. It permits ready differentiation of major significant categories based upon the color reaction of the cells: **gram positive,** which stain purple, and **gram negative,** which stain pink (red). The Gram stain is the basis of several important bacteriological topics, including bacterial taxonomy, cell wall structure, and identification and diagnosis of infection; in some cases, it even guides the selection of the correct drug for an infection. Gram staining is discussed in greater detail in Historical Highlights 4.2.

The **acid-fast stain,** like the Gram stain, is an important diagnostic stain that differentiates acid-fast bacteria (pink) from non-acid-fast bacteria (blue). This stain originated as a specific method to detect *Mycobacterium tuberculosis* in specimens. It was determined that these bacterial cells have a particularly impervious outer wall that holds fast (tightly or tenaciously) to the dye (carbol fuchsin) even when washed with a solution containing acid or acid alcohol. This stain is used for other medically important mycobacteria such as the leprosy bacillus and for *Nocardia,* an agent of lung or skin infections.

The **endospore stain** (spore stain) is similar to the acid-fast method in that a dye is forced by heat into resistant bodies called spores or endospores (their formation and significance are discussed in chapter 4). This stain is designed to distinguish between spores and the cells that they come from (so-called **vegetative cells**). Of significance in medical microbiology are the gram-positive, spore-forming members of the genus *Bacillus* (the cause of anthrax) and *Clostridium* (the cause of botulism and tetanus)—

dramatic diseases of universal fascination that we consider in later chapters.

Special stains are used to emphasize certain cell parts that are not revealed by conventional staining methods. **Capsule staining** is a method of observing the microbial capsule, an unstructured protective layer surrounding the cells of some bacteria and fungi. Because the capsule does not react with most stains, it is often negatively stained with India ink, or it may be demonstrated by special positive stains. The fact that not all microbes exhibit capsules is a useful feature for identifying pathogens. One example is *Cryptococcus,* which causes a serious fungal infection in AIDS patients.

Flagellar staining is a method of revealing flagella, the tiny, slender filaments used by bacteria for locomotion (see chapter 4). Because the width of bacterial flagella lies beyond the resolving power of the light microscope, in order to be seen, they must be enlarged by depositing a coating on the outside of the filament and then staining it. This stain works best with fresh, young cultures, because flagella are delicate and can be lost or damaged on older cells. Their presence, number, and arrangement on a cell are taxonomically useful.

TABLE 3.7

Comparison of Positive and Negative Stains

	Positive Staining	Negative Staining
Appearance of cell	Colored by dye	Clear and colorless
Background	Not stained (generally white)	Stained (dark gray or black)

Dyes employed	Basic dyes:
	  Crystal violet
	  Methylene blue
	  Safranin
	  Malachite green
		Acidic dyes:
		  Nigrosin
		  India ink

Subtypes of stains	Several types:
	  Simple stain
	  Differential stains
	    Gram stain
	    Acid-fast stain
	    Spore stain
	  Special stains
	    Capsule
	    Flagella
	    Spore
	    Granules
	    Nucleic acid
		Few types:
		  Capsule
		  Spore (Dorner)

CHAPTER CHECKPOINTS

Magnification, resolving power, lens quality, and illumination source all influence the clarity of specimens viewed through the optical microscope.

The maximum resolving power of the optical microscope is 200 nm, or 0.2 μm. This is sufficient to see the internal structures of eucaryotes and the morphology of most bacteria.

There are five types of optical microscopes. Four types use visible light for illumination: bright-field, dark-field, phase-contrast, and interference microscopes. The fifth type, the fluorescence microscope, uses UV light for illumination, but it has the same resolving power as the other optical microscopes.

Electron microscopes (EM) use electrons, not light waves, as an illumination source to provide high magnification ($5,000\times$ to $1,000,000\times$) and high resolution (0.5 nm). Electron microscopes can visualize cell ultrastructure (TEM) and three-dimensional images of cell and virus surface features (SEM).

Specimens viewed through optical microscopes can be either alive or dead, depending on the type of specimen preparation, but all EM specimens are dead because they must be treated with metals for effective viewing.

Stains are important diagnostic tools in microbiology because they can be designed to differentiate cell shape, structure, and biochemical composition of the specimens being viewed.

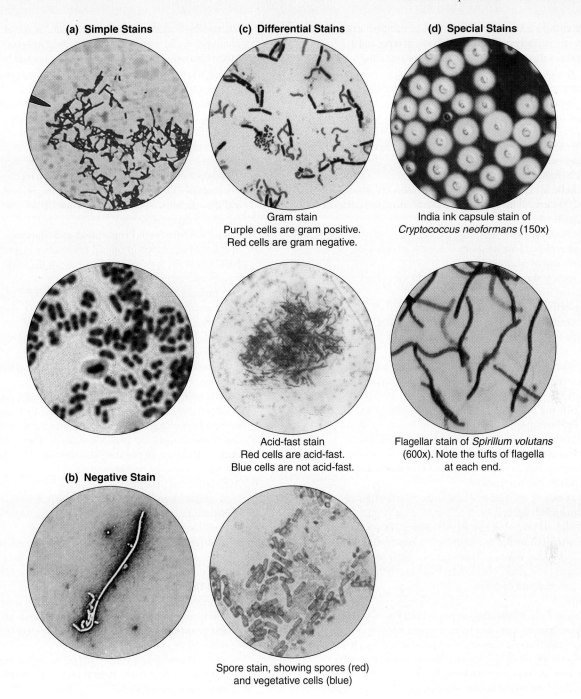

(a) Simple Stains

(c) Differential Stains

Gram stain
Purple cells are gram positive.
Red cells are gram negative.

(d) Special Stains

India ink capsule stain of
Cryptococcus neoformans (150x)

Acid-fast stain
Red cells are acid-fast.
Blue cells are not acid-fast.

Flagellar stain of *Spirillum volutans*
(600x). Note the tufts of flagella
at each end.

(b) Negative Stain

Spore stain, showing spores (red)
and vegetative cells (blue)

FIGURE 3.26

Types of microbiological stains. (a) Simple stains. **(b)** Negative stain. **(c)** Differential stains: Gram, acid-fast, and spore. **(d)** Special stains: capsule and flagellar.

CHAPTER CAPSULE WITH KEY TERMS

I. Culture Techniques

A. Microbiology as a science is very dependent on a number of specialized laboratory techniques. Laboratory steps routinely employed in microbiology are inoculation, incubation, isolation, inspection, and identification.

B. To prepare a **culture** (visible growth specimen), a **medium** (nutrient substrate) is **inoculated** (implanted, seeded) with a sample.

C. Methods of **isolation.** Cells are spread over a large area (**streak plate** and **spread plate**) or are diluted in a large volume (**pour plate**) so that individual cells are completely separated and can grow into **colonies.**

D. The cultures are **incubated, subcultured, observed** macroscopically and microscopically, and **identified** by means of morphological, physiological, genetic, and serological methods.

E. A **pure culture** consists of a single type of microbe; a **mixed culture** deliberately contains more than one type; and a **contaminated culture** is tainted with some intruding microbe.

II. **Artificial Media**

Artificial nutrient media vary according to their physical form, chemical characteristics, and purpose.

A. Physical Subgroups

They can be **liquid** (broth, milk), **semisolid,** or **solid,** depending upon the absence or the quantity of solidifying agent (usually **agar** and **gelatin**).

B. Chemical Subgroups

A **synthetic medium** is any preparation that is chemically defined, but a medium containing a poorly identified component is **nonsynthetic.**

C. Functional Subgroups

1. A **general-purpose medium** is used to grow a wide assortment of microbial types.

2. An **enriched medium** is supplemented with blood or tissue infusion to culture **fastidious** species.

3. A **selective medium** permits preferential growth of certain organisms in a mixture and inhibits others. Highly selective media are used expressly to favor the growth of one organism over another.

4. A **differential medium** distinguishes among different microbes by bringing out their variations in a particular reaction.

5. A **reducing medium** limits oxygen availability and is useful for cultivating anaerobes.

6. The ability to utilize sugars can be determined with **carbohydrate fermentation media.**

7. To convey fragile microbes, special **transport media** are needed to stabilize viability.

8. **Assay media** are used to evaluate the effectiveness of antimicrobial agents.

9. Environmental and industrial surveillance routinely call for **enumeration media** to determine the number of microbes in food, drinking water, soil, sewage, and other sources.

10. In certain instances, microorganisms have to be grown in animals and bird embryos.

III. **Microscopy**

A. **Optical,** or **light, microscopy** depends upon lenses that **refract** light rays, drawing them to a focus to produce a magnified **image.**

B. A **simple microscope** consists of a single magnifying lens, whereas a **compound microscope** relies on two lenses: the **ocular lens** and the **objective lens.** The objective lens is responsible for the **real image,** and the ocular lens forms the **virtual image.**

C. The **total power of magnification** is calculated from the product of the ocular and objective magnifying powers.

D. **Resolution,** or the **resolving power,** is a measure of a microscope's capacity to make clear images of very small objects. Resolution is improved with shorter wavelengths of illumination and with a higher **numerical aperture** of the lens.

E. Modifications in the lighting or the lens give rise to the **bright-field, dark-field, phase-contrast, interference,** and **fluorescence microscopes.**

F. Magnification and resolution in the **electron microscope** obey the same laws as in optical microscopes. Fundamental substitutions include electron emission for light, electromagnets for focusing the lens, and the proper apparatus to provide the necessary voltage and vacuum. Compared with optical microscopes and imagery, the **transmission electron microscope (TEM)** is analogous to the bright-field microscope, and the **scanning electron microscope (SEM)** corresponds to the dark-field microscope.

IV. **Techniques in Specimen Preparation and Staining**

A. Specimen preparation of optical microscopy is governed by the condition of the specimen, the purpose of the inspection, and the type of microscope being used.

B. **Wet mounts** and **hanging drop mounts** permit examination of the characteristics of live cells such as motility, shape, and arrangement.

C. **Stained, fixed** mounts are made by drying and heating a thin film of specimens called a **smear.**

D. To provide more detailed information on cells, a smear is stained with one or more dyes.

E. A **basic (cationic) dye** carries a positive charge, and an **acidic (anionic) dye** bears a negative charge, thus they selectively bind to oppositely charged surfaces. The surfaces of microbes are negatively charged and attract basic dyes. This is the basis of **positive staining.** In **negative staining,** the microbe repels the dye and it stains the background.

F. Applications of Staining

1. A **simple stain** uses just one dye, such as methylene blue, malachite green, crystal violet, basic fuchsin, or safranin. It is useful in determining cell morphology.

2. A **differential stain** requires a primary dye and a contrasting counterstain; consequently, the procedure is more involved. Classic examples of differential stains are the **Gram stain, acid-fast stain,** and **endospore stain.**

3. Differential stains have practical diagnostic significance. The ability of the Gram stain to distinguish **gram-positive** bacteria and **gram-negative** bacteria is useful in identification and choosing drug therapy.

The **acid-fast stain** is useful in diagnosing tuberculosis and leprosy. The **spore stain** is helpful in anthrax, botulism, and tetanus diagnosis.

4. **Special stains** are designed to bring out distinctive characteristics, as with the spore stain, capsule stain, and flagellar stain. Finding internal granules by means of a simple stain is valuable in diphtheria diagnosis.

MULTIPLE-CHOICE QUESTIONS

1. Which of the following is not one of the Five I's?
   a. inspection   d. incubation
   b. identification   e. inoculation
   c. induction

2. The term *culture* refers to the _____ growth of microorganisms in _____.
   a. rapid, an incubator
   b. macroscopic, media
   c. microscopic, the body
   d. artificial, colonies

3. A mixed culture is
   a. the same as a contaminated culture
   b. one that has been adequately stirred
   c. one that contains two or more known species
   d. a pond sample containing algae and protozoa

4. Agar is superior to gelatin as a solidifying agent because agar
   a. does not melt at room temperature
   b. solidifies at 75°C
   c. is not usually decomposed by microorganisms
   d. both a and c

5. The process that most accounts for magnification is
   a. a curved glass surface
   b. refraction of light
   c. illumination
   d. resolution

6. A subculture is a
   a. colony growing beneath the media surface
   b. culture made from a contaminant
   c. culture made in an embryo
   d. culture made from an isolated colony

7. Resolution is ____ with a longer wavelength of light.
   a. improved   c. not changed
   b. worsened   d. not possible

8. A real image is produced by the
   a. ocular   c. condenser
   b. objective   d. eye

9. A microscope that has a total magnification of 1,500× with the oil immersion lens has an ocular of what power?
   a. 150×   c. 15×
   b. 1.5×   d. 30×

10. The specimen for an electron microscope is always
    a. stained with dyes
    b. sliced into thin sections
    c. killed
    d. viewed directly

11. Motility is best observed with a
    a. hanging drop preparation
    b. negative stain
    c. streak plate
    d. flagellar stain

12. Bacteria tend to stain more readily with cationic (positively charged) dyes because bacteria
    a. contain large amounts of alkaline substances
    b. contain large amounts of acidic substances
    c. are neutral
    d. have thick cell walls

13. **Multiple Matching.** For each type of medium, select all descriptions that fit. For media that fit more than one description, briefly explain why this is the case.
    ____ mannitol salt agar
    ____ chocolate agar
    ____ MacConkey agar
    ____ nutrient broth
    ____ Sabouraud's agar
    ____ triple-sugar iron agar
    ____ *Euglena* agar
    ____ SIM medium
    ____ Stuart's medium

    a. selective medium
    b. differential medium
    c. chemically defined (synthetic) medium
    d. enriched medium
    e. general-purpose medium
    f. complex medium
    g. transport medium

CONCEPT QUESTIONS

1. a. Describe briefly what is involved in the Five I's.
   b. Name three basic differences between inoculation and contamination.

2. a. Name two ways that pure, mixed, and contaminated cultures are similar and two ways that they differ from each other.
   b. What must be done to avoid contamination?

3. a. Explain what is involved in isolating microorganisms and why it is necessary to do this.
   b. Compare and contrast three common laboratory techniques for separating bacteria in a mixed sample.
   c. Describe how an isolated colony forms.
   d. Explain why an isolated colony and a pure culture are not the same thing.

4. a. Explain the two principal functions of dyes in media.
   b. Differentiate among the ingredients and functions of enriched, selective, and differential media.

5. Differentiate between microscopic and macroscopic methods of observing microorganisms, citing a specific example of each method.

6. a. Contrast the concepts of magnification, refraction, and resolution.
   b. Briefly explain how an image is made and magnified

7. a. On the basis of the formula for resolving power, explain why a smaller R. P. value is preferred to a larger one.
   b. What does it mean in practical terms if the resolving power is 1.0 μm?
   c. What does it mean if the value is greater than 1.0 μm?
   d. What if it is less than 1.0 μm?
   e. What can be done to improve resolution?

8. a. Compare bright-field, dark-field, phase-contrast, and fluorescence microscopy as to field appearance, specimen appearance, light source, and uses.
   b. Of what benefit is the dark-field microscope if the field is black?

9. a. Compare and contrast the optical compound microscope with the electron microscope.
   b. Why is the resolution so superior in the electron microscope?
   c. What will you never see in an unretouched electron micrograph?
   d. Compare the way that the image is formed in the TEM and SEM.

10. Evaluate the following preparations in terms of showing microbial size, shape, motility, and differentiation: spore stain, negative stain, simple stain, hanging drop slide, and Gram stain.

11. a. Itemize the various staining methods, and briefly characterize each.
    b. For a stain to be considered a differential stain, what must it do?
    c. Explain what happens in positive staining to cause the reaction in the cell.
    d. Explain what happens in negative staining that causes the final result.

CRITICAL-THINKING QUESTIONS

1. Describe the steps you would take to isolate, cultivate, and identify a microbial pathogen from a urine sample. (Hint: Look at the Five I's.)

2. A certain medium has the following composition:

   Glucose   15 g
   Yeast extract   5 g
   Peptone   5 g
   KH_2PO_4 2 g
   Distilled water   1,000 ml

   a. To what chemical category does this medium belong?
   b. How could you convert *Euglena* agar (table 3.2) into a nonsynthetic medium?

3. a. Name four categories that blood agar fits into.
   b. Name four differential reactions that TSIA shows.
   c. Can you tell what functional kind of medium *Enterococcus faecalis* medium is?

4. a. What kind of medium might you make to selectively grow a bacterium that lives in the ocean?
   b. One that lives in the human stomach?
   c. What characteristic of dyes makes them useful in differential media?
   d. Why are intestinal bacteria able to grow on media containing bile?

5. a. When buying a microscope, what features are most important to check for?
   b. What is probably true of a $20 microscope that claims to magnify 1,000×?

6. How can one obtain 2,000× magnification with a 100× objective?

7. a. In what ways are dark-field microscopy and negative staining alike?
   b. How is the dark-field microscope like the scanning electron microscope?

8. Biotechnology companies have engineered hundreds of different types of mice, rats, pigs, goats, cattle, and rabbits to have genetic diseases similar to diseases of humans or to synthesize drugs and other biochemical products. They have patented these animals and sell them to researchers for study and experimentation.
   a. What do you think of creating new life forms just for experimentation?
   b. Comment or start a class discussion on the benefits, safety, and humanity of this trend.

9. This is a test of your living optical system's resolving power. Prop your book against a wall about 20 inches away and determine the line that is no longer resolvable by your eye.
   Source: Poem by Jonathan Swift.

So, Naturalists observe,

a flea has smaller

fleas that on him prey;

and these have smaller still

to bite 'em; and so proceed,

ad infinitum.

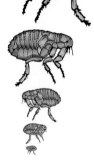

INTERNET SEARCH TOPICS

1. Use a search engine to find a description of confocal microscopes. How are they different from regular light microscopes?

2. Access information on the World Wide Web about nanotechnology and some of its applications.

CHAPTER 4

Procaryotic Profiles
The Bacteria and Archaea

S mall and deceptively simple, procaryotes are among nature's most abundant and ubiquitous microorganisms. If it were somehow possible to eradicate all bacteria in the world, humans would notice the effects immediately and, for a while, might find it a favorable change. We would not have to be as careful about preparing and refrigerating foods; plaque would no longer develop on our teeth; and there would be fewer cleaning chores around the house. Quite suddenly, the medical community's goal of eradicating certain infectious diseases, such as tuberculosis, cholera, and tetanus, could be a reality. But there are other considerations to take into account. We would also have to do without certain foods, like sauerkraut, yogurt, Swiss cheese, and sourdough bread. At first, this may seem a small price to pay, but are these slight inconveniences the only sacrifices to be expected? Within a few days, industrial processes that produce vitamins, drugs, and solvents would lie abandoned and useless, and most molecular biology research labs and biotechnology companies would be shut down. In a few months and years, humus containing dead animal and plant matter would build up and trap the very elements needed to sustain the living world. Clearly, bacteria play vital roles in all aspects of our existence. We literally can't live without them!

In order to explore the roles of bacteria in nutrition, genetics, drug therapy, infection, and microbial ecology, we must first understand several aspects of the structure and behavior of procaryotic cells. The primary topics to be covered in this chapter are elements of microscopic anatomy, physiology, identification, and classification and a survey of selected bacterial groups.

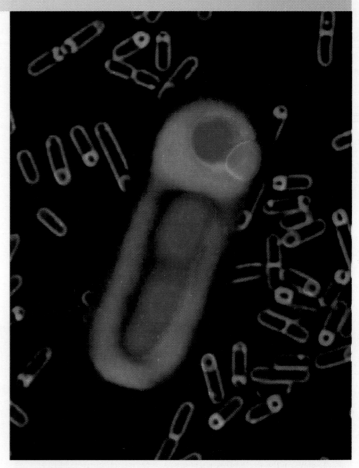

These biological "safety pins" are actually stages in endospore formation of *Bacillus subtilis,* stained with fluorescent proteins. The large red and blue cell is a vegetative cell in the early stages of sporulating. The developing spores are shown in green and orange.

Chapter Overview

- Procaryotic cells are the smallest, simplest, and most abundant cells on earth.
- Representative procaryotes include bacteria and archaea, both of which lack a nucleus and organelles but are functionally complex.
- The structure of bacterial cells is compact and capable of adaptations to a myriad of habitats.
- The cell is encased in an envelope that protects, supports, and regulates transport.
- Bacteria have special structures for motility and adhesion to the environment.
- Bacterial cells contain genetic material in a single chromosome, and ribosomes for synthesizing proteins.
- Bacteria have the capacity for reproduction, nutrient storage, dormancy, and resistance to adverse conditions.
- Shape, size, and arrangement of bacterial cells are extremely varied.
- Bacterial taxonomy and classification is based on their structure, metabolism, and genetics.
- Archaea are procaryotes related to eucaryotic cells that possess unique biochemistry and genetics.
- Archaea are adapted to the most extreme habitats on earth.

Procaryotic Form and Function: External Structure

The evolutionary history of procaryotic cells extends back at least 3.5 billion years (see figure 4.30). It is now generally thought that the very first cells to appear on the earth were a type of archaea possibly related to modern forms that live on sulfur compounds in geothermal ocean vents. The fact that these organisms have endured for so long in such a variety of habitats indicates a cellular structure and function that are amazingly versatile and adaptable. The general cellular organization of a procaryotic cell can be represented with the following flowchart:

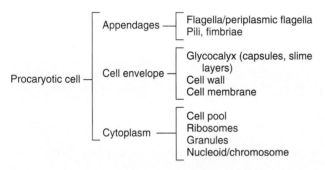

All bacterial cells invariably have a cell membrane, cytoplasm, ribosomes, and a chromosome; the majority have a cell wall and some form of surface coating or glycocalyx. Specific structures that are found in some, but not all, bacteria are flagella, pili, fimbriae, capsules, slime layers, and granules.

The Structure of a Generalized Procaryotic Cell

Bacterial cells appear featureless and two-dimensional when viewed with an ordinary microscope. Not until they are subjected to the scrutiny of the electron microscope and biochemical studies does their intricate and functionally complex nature become evident. The descriptions of bacterial structure, except where otherwise noted, refer to the **bacteria,*** a category of procaryotes with peptidoglycan in their cell walls. Figure 4.1 presents a three-dimensional anatomical view of a generalized (rod-shaped) bacterial cell. As we survey the principal anatomical features of this cell, we will perform a microscopic dissection of sorts, following a course that begins with the outer cell structures and proceeds to the internal contents.

---

*bacteria Formerly known as eubacteria (yoo′-bak-ter-ee-uh) Gr. *eu*, true, and *bakterion*, rod. Includes all procaryotes besides the archaea.

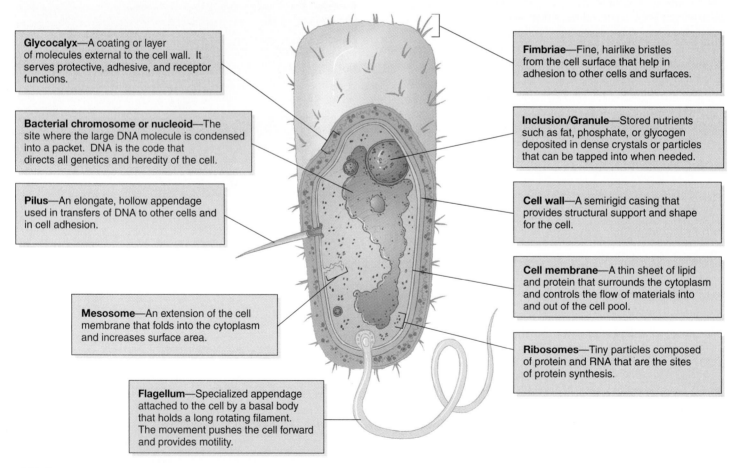

Glycocalyx—A coating or layer of molecules external to the cell wall. It serves protective, adhesive, and receptor functions.

Bacterial chromosome or nucleoid—The site where the large DNA molecule is condensed into a packet. DNA is the code that directs all genetics and heredity of the cell.

Pilus—An elongate, hollow appendage used in transfers of DNA to other cells and in cell adhesion.

Mesosome—An extension of the cell membrane that folds into the cytoplasm and increases surface area.

Flagellum—Specialized appendage attached to the cell by a basal body that holds a long rotating filament. The movement pushes the cell forward and provides motility.

Fimbriae—Fine, hairlike bristles from the cell surface that help in adhesion to other cells and surfaces.

Inclusion/Granule—Stored nutrients such as fat, phosphate, or glycogen deposited in dense crystals or particles that can be tapped into when needed.

Cell wall—A semirigid casing that provides structural support and shape for the cell.

Cell membrane—A thin sheet of lipid and protein that surrounds the cytoplasm and controls the flow of materials into and out of the cell pool.

Ribosomes—Tiny particles composed of protein and RNA that are the sites of protein synthesis.

FIGURE 4.1

Structure of a prokaryotic cell. Cutaway view of a typical rod-shaped bacterium, showing major structural features. Note that not all components are found in all cells.

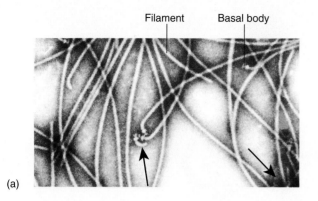

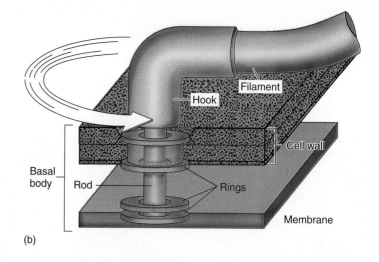

FIGURE 4.2

Structure of flagella. (a) Electron micrograph of basal bodies and filaments (at arrows) of bacterial flagella (66,000×). **(b)** Details of the basal body in a gram-negative cell. The hook, rings, and rod function together as a tiny device that rotates the filament 360°.

APPENDAGES: CELL EXTENSIONS

Several discrete types of accessory structures sprout from the surface of bacteria. These elongate **appendages*** are common but are not present on all species. Appendages can be divided into two major groups: those that provide motility (flagella and axial filaments) and those that provide attachments (fimbriae and pili).

Flagella—Bacterial Propellers

The procaryotic **flagellum*** is an appendage of truly amazing construction, certainly unique in the biological world. The primary function of flagella is to confer **motility,** or self-propulsion—that is, the capacity of a cell to swim freely through an aqueous habitat. The extreme thinness of a bacterial flagellum necessitates high magnification to reveal its special architecture, which occurs in three distinct parts: the filament, the hook (sheath), and the basal body (figure 4.2). The **filament,** a helical structure composed of proteins, is approximately 20 nm in diameter and varies from 1 to 70 nm in length. It is inserted into a curved, tubular hook. The hook is anchored to the cell by the basal body, a stack of rings firmly anchored through the cell wall, to the cell membrane. This arrangement permits the hook with its filament to rotate 360°, rather than undulating back and forth like a whip as was once thought. As the flagellum rotates in a counterclockwise direction, it causes the cell body to swim in a direct forward path (see figure 4.5).

One can generalize that all spirilla, about half of the bacilli, and a small number of cocci are flagellated. Flagella vary both in number and arrangement according to two general patterns: (1) In a *polar* arrangement, the flagella are attached at one or both ends of the cell. Three subtypes of this pattern are: **monotrichous,*** with a single flagellum; **lophotrichous,*** with small bunches or tufts of flagella emerging from the same site; and **amphitrichous,*** with flagella at both poles of the cell. (2) In a **peritrichous*** arrangement, flagella are dispersed randomly over the surface of the cell (figure 4.3). The type of arrangement has some bearing on the swimming speed of the bacterium. The speediest forms are polar, flagellated cells such as *Thiospirillum,* which can zip along at 5.2 mm/minute, and *Pseudomonas aeruginosa,* which can swim 4.4 mm/minute. Taking into account the small dimensions of these bacteria, such speeds are comparable to some protozoa and animals. Peritrichous rods such as *Escherichia coli* tend to swim at a relatively slower pace (1 mm/minute).

The presence of motility is one piece of information used in the laboratory identification or diagnosis of pathogens. Special stains or electron microscope preparations must be used to see arrangement, since flagella are too minute to be seen in live preparations with a light microscope. Often it is sufficient to know simply whether a bacterial species is motile. One way to detect motility is to stab a tiny mass of cells into a soft (semisolid) medium. Growth spreading rapidly through the entire medium is indicative of motility (see figure 3.5). Alternatively, cells can be observed microscopically with a hanging drop slide (figure 4.4). A truly motile cell will flit, dart, or wobble around the field, making some progress, whereas one that is nonmotile jiggles about in one place but makes no progress.

Fine Points of Flagellar Function   Flagellated bacteria can perform some rather sophisticated feats. They can detect and move in response to chemical signals—a type of behavior called **chemotaxis.*** Positive chemotaxis is movement of a cell in the direction of a favorable chemical stimulus (usually a nutrient); negative chemotaxis is movement away from a repellent (potentially harmful) compound.

---

*appendage (uh-pen´-dij) Any external projection of a body.

*flagellum (flah-jel´-em) pl. flagella; L., a whip.

*monotrichous (mah˝-noh-trik´-us) Gr. *mono,* one, and *tricho,* hair.

*lophotrichous (lo˝-foh-trik´-us) Gr. *lopho,* tuft or ridge.

*amphitrichous (am˝-fee-trik´-us) Gr. *amphi,* on both sides.

*peritrichous (per˝-ee-trik´-us). Gr. *peri,* around.

*chemotaxis (ke˝-moh-tak´-sis) pl. chemotaxes; Gr. *chemo,* chemicals, and *taxis,* an ordering or arrangement.

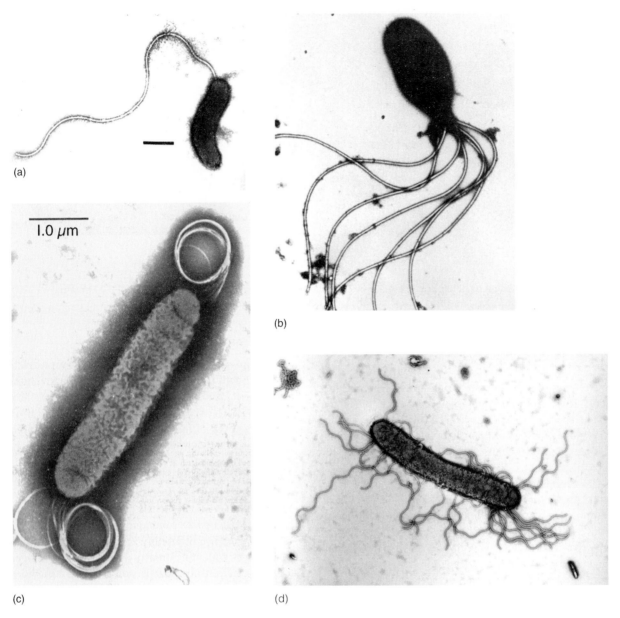

FIGURE 4.3

Electron micrographs depicting types of flagellar arrangements. (a) Monotrichous flagellum on the predatory bacterium *Bdellovibrio*.
(b) Lophotrichous flagella on *Vibrio fischeri*, a common marine bacterium (23,000×). **(c)** Unusual flagella on *Aquaspirillum* are amphitrichous (and lophotrichous) in arrangement and coil up into tight loops. **(d)** An unidentified bacterium discovered inside the cells of *Paramecium* exhibits peritrichous flagella.

(b) From Reichelt and Baumann, Arch. Microbiol. *94:283–330. © Springer-Verlag, 1973.*

The flagellum is effective in guiding bacteria through the environment primarily because the system for detecting chemicals is linked to the mechanisms that drive the flagellum. Located in the cell membrane are clusters of receptors[1] that bind specific molecules coming from the immediate environment. The attachment of sufficient numbers of these molecules transmits signals to the flagellum and sets it into rotary motion. If several flagella are present, they become aligned and rotate as a group (figure 4.5). As a flagellum rotates counterclockwise, the cell itself swims in a smooth linear direction toward the stimulus, called a **run.** Runs are interrupted at various intervals by **tumbles,** during which the flagellum

reverses direction and causes the cell to stop and change its course. It is believed that attractant molecules inhibit tumbles and permit progress toward the stimulus. Repellents cause numerous tumbles, allowing the bacterium to redirect itself away from the stimulus (figure 4.6). Some photosynthetic bacteria exhibit *phototaxis*, a type of movement in response to light rather than chemicals.

Periplasmic Flagella: Internal Flagella

Corkscrew-shaped bacteria called **spirochetes*** show an unusual, wriggly mode of locomotion caused by two or more long, coiled threads, the **periplasmic flagella** or *axial filaments*. A periplasmic

---

1. Cell surface molecules that bind specifically with other molecules.

*spirochete (spy′-roh-keet) Gr. *speira*, coil, and *chaite*, hair.

Top view

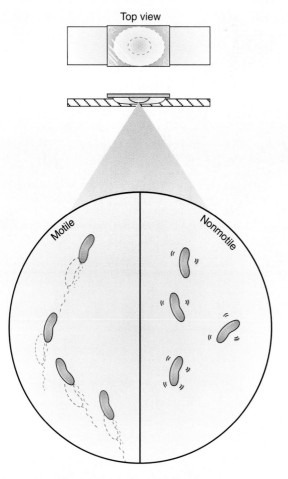

FIGURE 4.4

🔘 **Motility detection.** A hanging drop slide can be used to detect motility in bacteria by observing the microscopic behavior of the cell. In true motility, the cell swims and progresses from one point to another. It is assumed that a motile cell has one or more flagella even though the flagella are not visible. Nonmotile cells oscillate in the same relative space because of bombardment by molecules, a physical process called Brownian movement.

flagellum is a type of internal flagellum that is enclosed in the space between the cell wall and the cell membrane (figure 4.7). The filaments curl closely around the spirochete coils yet are free to contract and impart a twisting or flexing motion to the cell. This form of locomotion must be seen in live cells such as the spirochete of syphilis to be truly appreciated.

Appendages for Attachment and Mating

The structures termed **pilus*** and **fimbria*** both refer to bacterial surface appendages that provide some type of adhesion, but not locomotion. We will use the term *fimbriae* to refer to the shorter, numerous strands and the term *pili* to refer to the longer, sparser appendages.

Fimbriae are small, bristlelike fibers sprouting off the surface of many bacterial cells (figure 4.8). Their exact composition varies, but most of them contain protein. Fimbriae have an inherent ten-

---

*pilus (py′-lus) pl. pili; L., hair.

*fimbria (fim′-bree-ah) pl. fimbriae; L., a fringe.

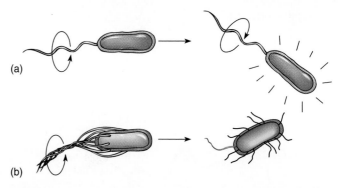

FIGURE 4.5

🔘 **The operation of flagella and the mode of locomotion in bacteria with polar and peritrichous flagella. (a)** In general, when a polar flagellum rotates in a counterclockwise direction, the cell swims forward. When the flagellum reverses direction and rotates clockwise, the cell stops and tumbles. **(b)** In peritrichous forms, all flagella sweep toward one end of the cell and rotate as a single group. During tumbles, the flagella lose coordination.

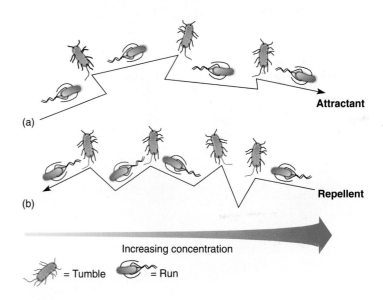

FIGURE 4.6

🔘 **Chemotaxis in bacteria.** The cell shows a primitive mechanism for progressing **(a)** toward positive stimuli and **(b)** away from irritants by swimming in straight runs or by tumbling. Bacterial runs allow straight, undisturbed progress toward the stimulus, whereas tumbles interrupt progress to allow the bacterium to redirect itself away from the stimulus after sampling the environment.

dency to stick to each other and to surfaces. They may be responsible for the mutual clinging of cells that leads to biofilms and other thick aggregates of cells on the surface of liquids and for the microbial colonization of inanimate solids such as rocks and glass (Spotlight on Microbiology 4.1). Some pathogens can colonize and infect host tissues because of a tight adhesion between their fimbriae and epithelial cells (figure 4.8*b*). For example, the gonococcus (agent of gonorrhea) invades the genitourinary tract, and *Escherichia coli* invades the intestine by this means. Mutant forms of these pathogens that lack fimbriae are unable to cause infections.

A pilus (also called a *sex pilus*) is an elongate, rigid tubular structure made of a special protein, *pilin*. So far, true pili have been

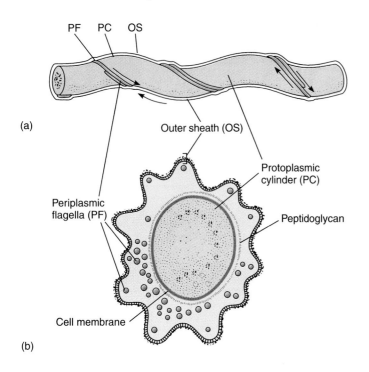

(a)

(b)

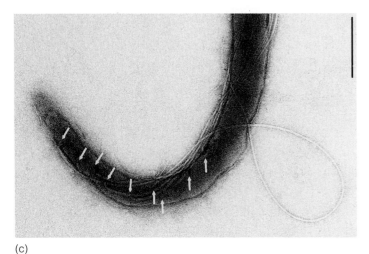

(c)

FIGURE 4.7

⊕ **The orientation of periplasmic flagella on the spirochete cell.**
(a) Longitudinal section. **(b)** Cross section. Contraction of the filaments imparts a spinning and undulating pattern of locomotion. **(c)** Electron micrograph captures the details of periplasmic flagella and their insertion points (arrows) in *Borrelia burgdorferi*. (Bar = 0.2 μm)

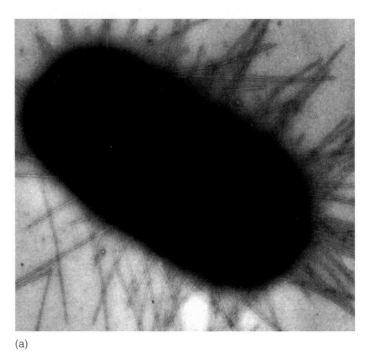

(a)

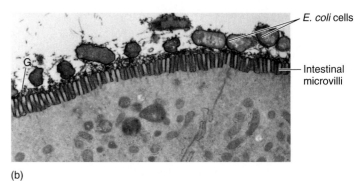

(b)

FIGURE 4.8

⊕ **Form and function of bacterial fimbriae. (a)** Single cell of pathogenic *Escherichia coli* is bristling with numerous stiff fibers called fimbriae (40,000×). **(b)** A row of *E. coli* cells tightly adheres by their fimbriae to the surface of intestinal cells (12,000×). This is how the bacterium clings and gains access to the body during an infection. (G = glycocalyx)

THE CELL ENVELOPE: THE OUTER WRAPPING OF BACTERIA

The majority of bacteria have a chemically complex external covering, termed the **cell envelope,** that lies outside of the cytoplasm. It is composed of three basic layers known as the glycocalyx, the cell wall, and the cell membrane (figure 4.10). The layers of the envelope are stacked one upon another and are often tightly bonded together like the outer husk and casings of a coconut. Although each envelope layer performs a distinct function, together they act as a single protective unit. The envelope is extensive and can account for one-tenth to one-half of a cell's volume.

The Bacterial Surface Coating, or Glycocalyx

The bacterial cell surface is frequently exposed to severe environmental conditions. The **glycocalyx** develops as a coating of macromolecules to protect the cell and, in some cases, help it adhere to its

found only on gram-negative bacteria, where they are involved primarily in a mating process between cells called **conjugation,*** which involves partial transfer of DNA from one cell to another (figure 4.9). A pilus from the donor cell unites with a recipient cell, thereby providing a cytoplasmic connection for making the transfer. Production of pili is controlled genetically, and conjugation takes place only between compatible gram-negative cells. The roles of pili and conjugation are further explored in chapter 9.

---

*conjugation (kon-joo-gay'-shun) L. *conjugatus,* linked together.

Being aware of the widespread existence of microorganisms on earth, we should not be surprised that, when left undisturbed, they gather in masses, cling to various surfaces, and capture available moisture and nutrients. The formation of these living layers, called **biofilms,** is actually a universal phenomenon that all of us have observed. Consider the scum that builds up in toilet bowls and shower stalls in a short time if they are not cleaned; or the algae that collect on the walls of swimming pools; and, more intimately—the constant deposition of plaque on teeth. Microbes making biofilms is a primeval tendency that has been occurring for billions of years as a way to create stable habitats with adequate access to food, water, atmosphere, and other essential factors. Biofilms are often cooperative associations among several microbial groups (bacteria, fungi, algae, and protozoa) as well as plants and animals.

Substrates are most likely to accept a biofilm if they are moist and have developed a thin layer of organic material such as polysaccharides or glycoproteins on their exposed surface (see figure at right). This depositing process occurs within a few minutes to hours, making a slightly sticky texture that attracts primary colonists, usually bacteria. These early cells attach (adsorb to) and begin to multiply on the surface. As they grow, various secreted substances in their glycocalyx (receptors, fimbriae, slime layers, capsules) increase the binding of cells to the surface and thicken the biofilm. As the biofilm evolves, it undergoes specific adaptations to the habitat in which it forms. In many cases, the earliest colonists contribute nutrients and create microhabitats that serve as a matrix for other microbes to attach and grow into the film, forming complete communities. The biofilm varies in thickness and complexity, depending upon where it occurs and how long it keeps developing. Complexity ranges from single cell layers to thick microbial mats with dozens of dynamic interactive layers.

Biofilms are a profoundly important force in the development of terrestrial and aquatic environments. They dwell permanently in bedrock and the earth's sediments, where they play an essential role in recycling elements, leaching minerals, and participating in soil formation. Biofilms associated with plant roots promote the mutual exchange of nutrients between the microbes and roots. The human body contains biofilms in the form of normal flora that live in the skin and mucous membranes, and on structures such as teeth (see the description of plaque formation in figure 21.29). Bacteria can also persistently colonize medical devices such as catheters, artificial heart valves, and other inanimate objects placed in the body (see figure 4.13). Invasive biofilms can wreak havoc with human-made structures such as cooling towers, storage tanks, air conditioners, and even stone buildings. Additional information on biofilms is found in chapter 26.

First colonists

Organic surface coating

Substrate

Adsorption of cells to surface

Glycocalyx

More permanent attachment of cells by means of slimes or capsules; growth of colonies

Mature biofilm with microbial community in complex matrix

environment. Glycocalyces differ among bacteria in thickness, organization, and chemical composition. Some bacteria are covered with a loose, soluble shield called a **slime layer** that evidently protects them from loss of water and nutrients (figure 4.11*a*). Other bacteria produce **capsules** of repeating polysaccharide units, of protein, or of both (figure 4.11*b*). A capsule is bound more tightly to the cell than a slime layer is, and it has a thicker, gummy consistency that gives a prominently sticky (mucoid) character to the colonies of most encapsulated bacteria (figure 4.12).

Specialized Functions of the Glycocalyx Capsules are formed by a few pathogenic bacteria, such as *Streptococcus pneumoniae* (a cause of pneumonia, an infection of the lung), *Haemophilus influenzae* (one cause of meningitis), and *Bacillus anthracis* (the cause of anthrax). Encapsulated bacterial cells generally have greater pathogenicity because capsules protect the bacteria against white blood cells called phagocytes. Phagocytes are a natural body defense that can engulf and destroy foreign cells through phagocytosis, thus preventing infection. A capsular coating

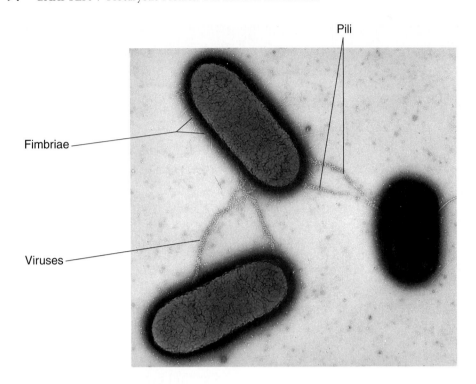

Pili

Fimbriae

Viruses

FIGURE 4.9

Three bacteria in the process of conjugating. Clearly evident are the sex pili forming mutual conjugation bridges between a donor (upper cell) and two recipients (two lower cells). Interesting points to observe are the fimbriae on the donor cell and the bacterial viruses (phages) forming tiny spots on the pili.

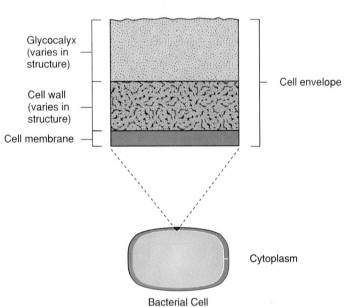

Glycocalyx (varies in structure)

Cell wall (varies in structure)

Cell membrane

Cell envelope

Cytoplasm

Bacterial Cell

FIGURE 4.10

The relationship of the three layers of the cell envelope.

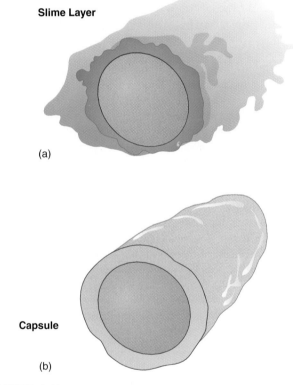

Slime Layer

(a)

Capsule

(b)

FIGURE 4.11

Bacterial cells sectioned to show the types of glycocalyces. **(a)** The slime layer is a loose structure that is easily washed off. **(b)** The capsule is a thick, structured layer that is not readily removed.

makes the bacteria too slippery for the phagocyte to capture or digest. By escaping phagocytosis, the bacteria are free to multiply and infect body tissues. Encapsulated bacteria that mutate to nonencapsulated forms usually lose their pathogenicity.

Other types of glycocalyces can be important in formation of biofilms. The thick, white plaque that forms on teeth comes in part from the surface slimes produced by certain streptococci in the oral cavity. This slime initially allows them to adhere to the teeth and provides a niche for other oral bacteria that, in time, can lead to dental disease. The glycocalyx of some bacteria is so highly adher-

ent that it is responsible for persistent colonization of nonliving materials such as plastic catheters, intrauterine devices, and metal pacemakers that are in common medical use (figure 4.13).

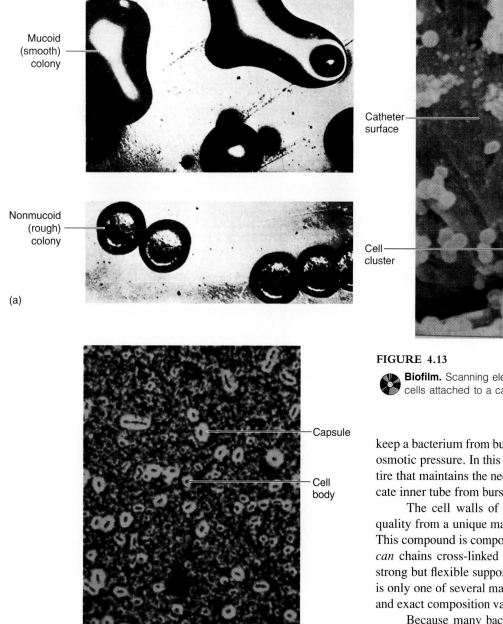

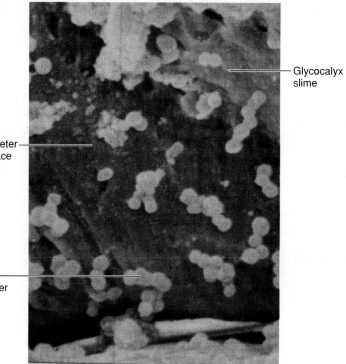

FIGURE 4.13

Biofilm. Scanning electron micrograph of *Staphylococcus aureus* cells attached to a catheter by a slime secretion.

FIGURE 4.12

Structure of emcapsulated bacteria. (a) The appearance of colonies composed of encapsulated cells (mucoid) compared with those lacking capsules (nonmucoid). Even at the macroscopic level, the slippery, gel-like character of the capsule is evident. **(b)** Staining reveals the microscopic appearance of a large, well-developed capsule.

STRUCTURE OF THE CELL WALL

Immediately below the glycocalyx lies a second layer, the **cell wall.** This structure accounts for a number of important bacterial characteristics. In general, it determines the shape of a bacterium, and it also provides the kind of strong structural support necessary to keep a bacterium from bursting or collapsing because of changes in osmotic pressure. In this way, the cell wall functions like a bicycle tire that maintains the necessary shape and prevents the more delicate inner tube from bursting when it is expanded.

The cell walls of most bacteria gain their relatively rigid quality from a unique macromolecule called **peptidoglycan (PG).** This compound is composed of a repeating framework of long *glycan* chains cross-linked by short peptide fragments to provide a strong but flexible support framework (figure 4.14). Peptidoglycan is only one of several materials found in cell walls, and its amount and exact composition vary among the major bacterial groups.

Because many bacteria live in aqueous habitats with a low solute concentration, they are constantly absorbing excess water by osmosis. Were it not for the strength and relative rigidity of the peptidoglycan in the cell wall, they would rupture from internal pressure. Understanding this function of the cell wall has been a tremendous boon to the drug industry. Several types of drugs used to treat infection (penicillin, cephalosporins) are effective because they target the peptide cross-links in the peptidoglycan, thereby disrupting its integrity. With their cell walls incomplete or missing, such cells have very little protection from **lysis*** (see figure 12.3). Some disinfectants (alcohol, detergents) also kill bacterial cells by damaging the cell wall. Lysozyme, an enzyme contained in tears and saliva, provides a natural defense against certain bacteria by hydrolyzing the bonds in the glycan chains and causing the wall to break down.

---

*lysis (ly′-sis) Gr., to loosen. A process of cell destruction, as occurs in bursting.

(a) The peptidoglycan
of a cell wall can be
presented as a
crisscross network
pattern similar to a
chain-link fence,
forming a single
massive molecule
that molds the
structure of the cell
into a tight box.

(b) An idealized view of
the molecular meshwork
of peptidoglycan. It
contains alternating
glycans (G and M)
bound together in long
strands. The G stands for
N-acetyl glucosamine,
and the M stands for
N-acetyl muramic acid.
A muramic acid
molecule binds to an
adjoining muramic acid
on a parallel chain by
means of a cross-linkage
of peptides.

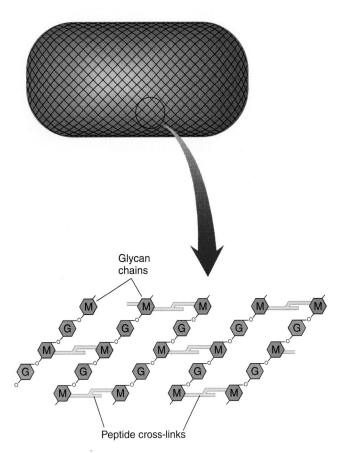

Glycan
chains

Peptide cross-links

(c) A more detailed view
of the connection
linking the muramic
acids. Tetrapeptide
chains branching off
the muramic acids are
connected by
interbridges
composed of another
chain of amino acids.
The types of amino
acids in the interbridge
can vary. It is this
linkage that provides
rigid yet flexible
support to the cell and
that may be targeted
by drugs like penicillin.

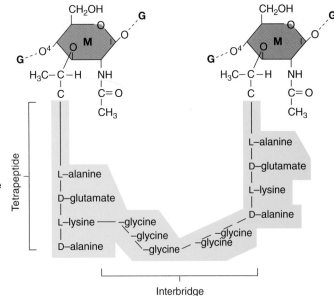

FIGURE 4.14

Structure of peptidoglycan in the cell wall.

Differences in Cell Wall Structure

More than a hundred years ago, long before the detailed anatomy of bacteria was even remotely known, a Danish physician named Hans Christian Gram developed a staining technique, the **Gram stain,** that delineates two generally different groups of bacteria (Historical Highlights 4.2). We now know that the contrasting staining reactions that occur with this stain are due entirely to some very fundamental differences in the structure of bacterial cell walls. The two major groups shown by this technique are the gram-positive bacteria and the gram-negative bacteria. Because the Gram stain does

HISTORICAL HIGHLIGHTS 4.2
The Gram Stain: A Grand Stain

In 1884, Hans Christian Gram discovered a staining technique that could be used to make bacteria in infectious specimens more visible. His technique consisted of a timed, sequential application of crystal violet (the primary dye), Gram's iodine (IKI, the mordant), an alcohol rinse (decolorizer), and a contrasting counterstain. The initial counterstain used was yellow or brown and later replaced by the red dye, safranin. Since that substitution, bacteria that stained purple are called gram positive, and those that stained red are called gram negative.

Although these staining reactions involve an attraction of the cell to a charged dye (see chapter 3), it is important to note that the terms *gram positive* and *gram negative* are not used to indicate the electrical charge of cells or dyes but whether or not a cell retains the primary dye-iodine complex after decolorization. There is nothing specific in the reaction of gram-positive cells to the primary dye or in the reaction of gram-negative cells to the counterstain. The different results in the Gram stain are due to differences in the structure of the cell wall and how it reacts to the series of reagents applied to the cells.

In the first step, crystal violet is attracted to the cells in a smear and stains them all the same purple color. The second and key differentiating step is the addition of the mordant—Gram's iodine. The mordant is a stabilizer that causes the dye to form large crystals in the peptidoglycan meshwork of the cell wall. Because the peptidoglycan layer in gram-positive cells is thicker, the entrapment of the dye is far more extensive in them than in gram-negative cells. Application of alcohol in the third step dissolves lipids in the outer membrane and removes the dye from the peptidoglycan layer and the gram-negative cells. By contrast, the crystals of dye tightly embedded in the peptidoglycan of gram-positive bacteria are relatively inaccessible and resistant to removal. Because gram-negative bacteria are colorless after decolorization, their presence is demonstrated by applying the counterstain safranin in the final step.

This century-old staining method remains the universal basis for bacterial classification and identification. It permits differentiation of four major categories based upon color reaction and shape: gram-positive rods, and gram-positive cocci, gram-negative rods, and gram-negative cocci (see table 4.6). The Gram stain can also be a practical aid in diagnosing infection and in guiding drug treatment. For example, gram staining a fresh urine or throat specimen can help pinpoint the possible cause of infection, and in some cases it is possible to begin drug therapy on the basis of this stain. Even in this day of elaborate and expensive medical technology, the Gram stain remains an important and unbeatable first tool in diagnosis.

	Microscopic Appearance of Cell		Chemical Reaction in Cell Wall (very magnified view)	
Step	Gram (+)	Gram (−)	Gram (+)	Gram (−)
1. Crystal violet				
			Both cell walls affix the dye	
2. Gram's iodine			Dye crystals trapped in wall	No effect of iodine
3. Alcohol			Crystals remain in cell wall	Cell wall partially dissolved, loses dye
4. Safranin (red dye)			Red dye has no effect	Red dye stains the colorless cell

Gram stain technique and theory.

not actually reveal the nature of these physical differences, we must turn to the electron microscope and to biochemical analysis.

The extent of the differences between gram-positive and gram-negative bacteria is evident in the physical appearance of their cell envelopes (figure 4.15). In gram-positive cells, a microscopic section resembles an open-faced sandwich with two layers: the thick outer cell wall, composed primarily of peptidoglycan, and the cell membrane. A similar section of a gram-negative cell envelope shows a complete sandwich with three layers: the cell wall, composed of an outer membrane and a thin layer of peptidoglycan, and the cell membrane. See table 4.1 for a further comparison of cell wall types.

The Gram-Positive Cell Wall The bulk of the gram-positive cell wall is a thick, homogeneous sheath of peptidoglycan ranging from 20 to 80 nm in thickness. It also contains tightly bound acidic polysaccharides, including teichoic acid and lipoteichoic acid (figure 4.16). Teichoic acid is a polymer of ribitol or glycerol and phosphate embedded in the peptidoglycan sheath. Lipoteichoic acid is similar in structure but is attached to the lipids in the plasma membrane. These molecules appear to function in cell wall maintenance and enlargement during cell division, and they also contribute to the acidic charge on the cell surface. In some cases, the cell wall of gram-positive bacteria is pressed tightly against the cell membrane with very little space between them, but in other cells, a

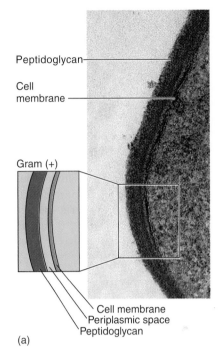

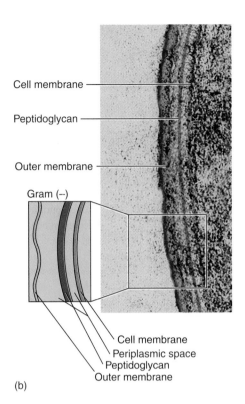

FIGURE 4.15

A comparison of the envelopes of gram-positive and gram-negative cells. (a) A photomicrograph of a gram-positive cell wall/membrane and an artist's interpretation of its open-faced sandwich–style layering with two layers. **(b)** A photomicrograph of a gram-negative cell wall/membrane and an artist's interpretation of its complete sandwich–style layering with three distinct layers.

TABLE 4.1

Comparison of Gram-Positive and Gram-Negative Cell Walls

Characteristic	Gram-Positive	Gram-Negative
Number of major layers	1	2
Chemical composition	Peptidoglycan Teichoic acid Lipoteichoic acid	Lipopolysaccharide Lipoprotein Peptidoglycan
Overall thickness	Thicker (20–80 nm)	Thinner (8–11 nm)
Outer membrane	No	Yes
Periplasmic space	Narrow	Extensive
Porin proteins	No	Yes
Permeability to molecules	More penetrable	Less penetrable

thin **periplasmic* space** is evident between the cell membrane and cell wall.

The Gram-Negative Cell Wall   The gram-negative cell wall is more complex in morphology because it contains an **outer membrane (OM),** has a thinner shell of peptidoglycan, and has an extensive space surrounding the peptidoglycan (figure 4.16). The outer membrane is somewhat similar in construction to the cell membrane, except that it contains specialized types of polysaccharides and proteins. The uppermost layer of the OM contains

lipopolysaccharide (LPS). The polysaccharide chains extending off the surface function as antigens and receptors. The innermost layer of the OM is another lipid layer anchored by means of lipoproteins to the peptidoglycan layer below. The outer membrane serves as a partial chemical sieve by allowing only relatively small molecules to penetrate. Access is provided by special membrane channels formed by *porin proteins* that completely span the outer membrane. The size of these porins can be altered so as to block the entrance of harmful chemicals, making them one defense of gram-negative bacteria against certain antibiotics.

The bottom layer of the gram-negative wall is a single, thin (1–3 nm) sheet of peptidoglycan. Although it acts as a somewhat rigid protective structure as previously described, its thinness gives gram-negative bacteria a relatively greater flexibility and sensitivity to lysis. There is a well-developed *periplasmic space* surrounding the peptidoglycan. This space is an important reaction site for a large and varied pool of substances that enter and leave the cell.

Practical Considerations of Differences in Cell Wall Structure
Variations in cell wall anatomy contribute to several other differences between the two cell types. The outer membrane contributes an extra barrier in gram-negative bacteria that makes them more impervious to some antimicrobic chemicals such as dyes and disinfectants, so they are generally more difficult to inhibit or kill than are gram-positive bacteria. One exception is for alcohol-based compounds, which can dissolve the lipids in the outer membrane and disturb its integrity. Treating infections caused by gram-negative bacteria often requires different drugs from gram-positive infections, especially drugs that can cross the outer membrane.

The cell wall or its parts can interact with human tissues and contribute to disease. The lipids have been referred to as *endotoxins* because they stimulate fever and shock reactions in gram-negative

---

*periplasmic (per″-ih-plaz′-mik) Gr. *peri*, around, and *plastos*, formed.

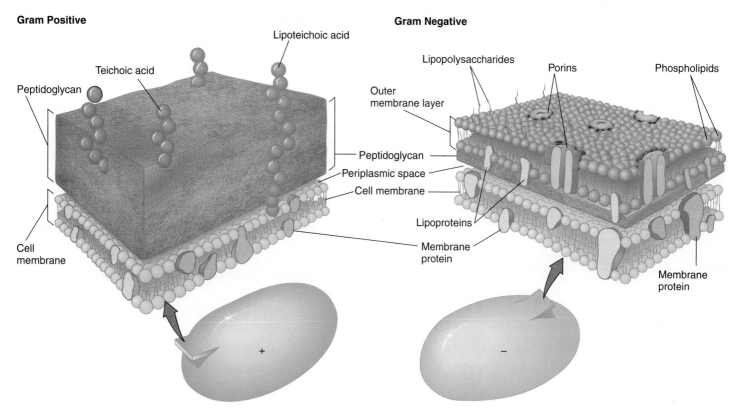

Gram Positive

- Peptidoglycan
- Teichoic acid
- Lipoteichoic acid
- Cell membrane

Gram Negative

- Lipopolysaccharides
- Porins
- Phospholipids
- Outer membrane layer
- Peptidoglycan
- Periplasmic space
- Cell membrane
- Lipoproteins
- Membrane protein
- Membrane protein

+

−

FIGURE 4.16

A comparison of the detailed structure of gram-positive and gram-negative cell walls.

infections such as meningitis and typhoid fever. Proteins attached to the outer portion of the cell wall of several gram-positive species, including *Corynebacterium diphtheriae* (the agent of diphtheria) and *Streptococcus pyogenes* (the cause of strep throat), also have toxic properties. The lipids in the cell walls of certain *Mycobacterium* species are harmful to human cells as well. Because most macromolecules in the cell walls are foreign to humans, they stimulate antibody production by the immune system (see chapter 15).

Nontypical Cell Walls

Several bacterial groups lack the cell wall structure of gram-positive or gram-negative bacteria, and some bacteria have no cell wall at all. Although these exceptional forms can stain positive or negative in the Gram stain, examination of their fine structure and chemistry shows that they do not really fit the descriptions for typical gram-negative or -positive cells. For example, the cells of *Mycobacterium* and *Nocardia* contain peptidoglycan and stain gram positive, but the bulk of their cell wall is composed of unique types of lipids. One of these is a very-long-chain fatty acid called *mycolic acid,* or cord factor, that contributes to the pathogenicity of this group (see chapter 19). The thick, waxy nature imparted to the cell wall by these lipids is also responsible for a high degree of resistance to certain chemicals and dyes. Such resistance is the basis for the **acid-fast stain** used to diagnose tuberculosis and leprosy. In this stain, hot carbol fuchsin dye becomes tenaciously attached (is held fast) to these cells so that an acid-alcohol solution will not remove the dye (see chapter 3).

Because they are from a more ancient and primitive line of procaryotes, the archaea exhibit unusual and chemically distinct cell walls. In some, the walls are composed almost entirely of polysaccharides, and in others, the walls are pure protein; but as a group, they all lack the true peptidoglycan structure described previously. Since a few archaea and all mycoplasmas (discussed in a later section of this chapter) lack a cell wall entirely, their cell membrane must serve the dual functions of support as well as transport.

Some bacteria that ordinarily have a cell wall can lose it during part of their life cycle. These wall-deficient forms are referred to as **L forms** or L-phase variants (for the Lister Institute, where they were discovered). L forms arise naturally from a mutation in the wall-forming genes, or they can be induced artificially by treatment with a chemical such as lysozyme or penicillin that disrupts the cell wall. When a gram-positive cell is exposed to either of these two chemicals, it will lose the cell wall completely and become a **protoplast,*** a fragile cell bounded only by a membrane that is highly subject to lysis (figure 4.17*a*). A gram-negative cell exposed to these same substances loses it peptidoglycan but retains its outer membrane, leaving a less fragile but nevertheless weakened **spheroplast*** (figure 4.17*b*). Evidence points to a role for L forms in certain infections (see chapter 21).

---

*protoplast (proh′-toh-plast) Gr. *proto,* first, and *plastos,* formed.

*spheroplast (sfer′-oh-plast) Gr. *sphaira,* sphere.

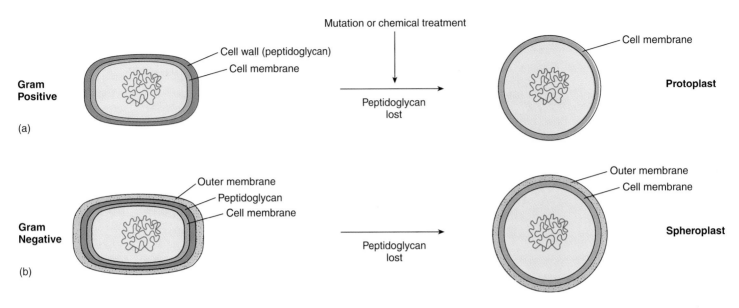

FIGURE 4.17

The conversion of walled bacterial cells to L forms: (a) gram-positive bacteria; **(b)** gram-negative bacteria.

CELL MEMBRANE STRUCTURE

Appearing just beneath the cell wall is the **cell,** or **cytoplasmic, membrane,** a very thin (5–10 nm), flexible sheet molded completely around the cytoplasm. Its general composition was described in chapter 2 as a lipid bilayer with proteins embedded to varying degrees (see Microbits 2.3). Bacterial cell membranes have this typical structure, containing primarily phospholipids (making up about 30–40% of the membrane mass) and proteins (contributing 60–70%). Major exceptions to this description are the membranes of mycoplasmas, which contain high amounts of sterols—rigid lipids that stabilize and reinforce the membrane—and the membranes of archaea, which contain unique branched hydrocarbons rather than fatty acids.

In some locations, the cell membrane forms internal pouches in the cytoplasm called **mesosomes*** (see figure 4.1). These are prominent in gram-positive bacteria but are harder to see in gram-negative bacteria because of their relatively small size. Mesosomes presumably increase the internal surface area available for membrane activities. Some of the proposed functions of mesosomes are to participate in cell wall synthesis and to guide the duplicated bacterial chromosomes into the two daughter cells during cell division (see figure 7.14).

Photosynthetic procaryotes such as cyanobacteria contain dense stacks of internal membranes that carry the photosynthetic pigments (see figure 4.34)

Functions of the Cell Membrane

Since bacteria have none of the eucaryotic organelles, the cell membrane provides a site for functions such as energy reactions, nutrient processing, and synthesis. A major action of the cell membrane is to regulate *transport,* that is, the passage of nutrients into the cell and the discharge of wastes. Although water and small un-

charged molecules can diffuse across the membrane unaided, the membrane is a *selectively permeable* structure with special carrier mechanisms for passage of most molecules (see chapter 7). The glycocalyx and cell wall can bar the passage of large molecules, but they are not the primary transport apparatus. The cell membrane is also involved in *secretion,* or the discharge of a metabolic product into the extracellular environment.

The membranes of procaryotes are an important site for a number of metabolic activities. Most enzymes of respiration and ATP synthesis reside in the cell membrane since procaryotes lack mitochondria (see chapter 8). Enzyme structures located in the cell membrane also help synthesize structural macromolecules to be incorporated into the cell envelope and appendages. Other products (enzymes and toxins) are secreted by the membrane into the extracellular environment.

CHAPTER CHECKPOINTS

Bacteria are the oldest form of cellular life. They are also the most widely dispersed, occupying every conceivable microclimate on the planet.

The appendages of bacteria provide motility (flagella), attachment (pili and fimbriae) and in some bacteria, a means of DNA transfer (sex pili).

Flagella vary in number and arrangement as well as in the type and rate of motion they produce.

The cell envelope is the outermost covering of bacteria. It consists of three basic layers: (1) the glycocalyx, (2) the cell wall, and (3) the cell membrane.

The composition of the procaryotic cell wall is used to classify bacteria into four major divisions: gram-positive bacteria, gram-negative bacteria, bacteria with no cell walls, and bacteria with chemically unique cell walls.

Gram-positive bacteria retain the crystal violet and stain purple. Gram-negative bacteria lose the crystal violet and stain red from the safranin counterstain.

*mesosome (mes′-oh-sohm) Gr. *mesos,* middle, and *soma,* body.

Gram-positive bacteria have thick cell walls of peptidoglycan and acidic polysaccharides such as teichoic acid, and they have a thin periplasmic space. The cell walls of gram-negative bacteria are thinner but contain an additional outer membrane and have a wide periplasmic space.

The bacterial cell membrane is typically composed of phospholipids and proteins, and it performs many metabolic functions as well as transport activities.

Bacterial Form and Function: Internal Structure

CONTENTS OF THE CELL CYTOPLASM

Encased by the cell membrane is a dense, gelatinous solution referred to as **cytoplasm,** which is another prominent site for many of the cell's biochemical and synthetic activities. Its major component is water (70–80%), which serves as a solvent for the **cell pool,** a complex mixture of nutrients including sugars, amino acids, and salts. The components of this pool serve as building blocks for cell synthesis or as sources of energy. The cytoplasm also contains larger, discrete cell masses such as the chromatin body, ribosomes, mesosomes, and granules.

Bacterial Chromosomes and Plasmids: The Sources of Genetic Information

The hereditary material of bacteria exists in the form of a single circular strand of DNA designated as the **bacterial chromosome.*** By definition, bacteria do not have a nucleus; that is, their DNA is not enclosed by a nuclear membrane but instead is aggregated in a dense area of the cell called the **nucleoid.*** The chromosome is actually an extremely long molecule of DNA that is tightly coiled around special basic protein molecules so as to fit inside the cell compartment (see Microbits 9.3). Arranged along its length are genetic units (genes) that carry information required for bacterial maintenance and growth. When exposed to special stains or observed with an electron microscope, chromosomes have a granular or fibrous appearance (figure 4.18).

Although the chromosome is the minimal genetic requirement for bacterial survival, many bacteria contain other, nonessential pieces of DNA called **plasmids.*** These tiny, circular extra-chromosomal strands can be free or integrated into the chromosome; they are duplicated and passed on to offspring. They are not essential to bacterial growth and metabolism, but they often confer protective traits such as resisting drugs and producing toxins and enzymes (see chapter 9). Because they can be readily manipulated in the laboratory and transferred from one bacterial cell to another, plasmids are an important agent in modern genetic engineering techniques.

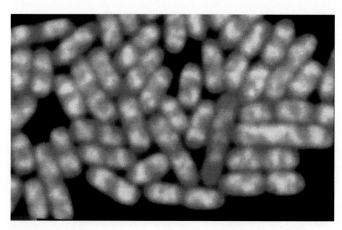

FIGURE 4.18

Chromosome structure. Fluorescent staining highlights the chromosomes of the bacterial pathogen *Salmonella enteriditis.* The cytoplasm is orange, and the chromosome fluoresces bright yellow. Some bacteria appear to have more than one chromosome because they are in the process of dividing.

Ribosomes: Sites of Protein Synthesis

A bacterial cell contains thousands of tiny, discrete units called **ribosomes.*** When viewed even by very high magnification, ribosomes show up as fine, spherical specks dispersed throughout the cytoplasm that often occur in chains (polysomes). They are also attached to the cell membrane. Chemically, a ribosome is a combination of a special type of RNA called ribosomal RNA, or rRNA (about 60%), and protein (40%). One method of characterizing ribosomes is by S, or Svedberg,[2] units, which rate the molecular sizes of various cell parts that have been spun down and separated by molecular weight and shape in a centrifuge. Heavier, more compact structures sediment faster and are assigned a higher S rating. Combining this method of analysis with high-resolution electron micrography has revealed that the procaryotic ribosome, which has an overall rating of 70S, is actually composed of two smaller subunits. The 30S unit looks something like a heart; the 50S unit bears a resemblance to a crown (figure 4.19). They fit together to form a miniature platform upon which protein synthesis is performed. We examine the more detailed functions of ribosomes in chapter 9.

Inclusions, or Granules: Storage Bodies

Most bacteria are exposed to severe shifts in the availability of food. During periods of nutrient abundance, they compensate by laying down nutrients intracellularly in **inclusion bodies,** or **inclusions,*** of varying size, number, and content. As the environmental source of these nutrients becomes depleted, the bacterial cell can mobilize its own storehouse as required. Some inclusion bodies enclose condensed, energy-rich organic substances, including glycogen and poly β-hydroxybutyrate (PHB), within special single-layered membranes (figure 4.20). A unique type of inclusion found in some aquatic bacteria are gas vesicles that provide buoyancy and flotation.

---

*chromosome (kroh'-mah-som) Gr. *chromato,* color, and *soma,* body.

*nucleoid (noo'-klee-oid) L. *nucis,* nut, and *oid,* form or like.

*plasmid (plaz'-mid) Gr. *plasma,* a form, and *id,* belonging to.

---

2. Named in honor of T. Svedberg, the Swedish chemist who invented the ultracentrifuge in 1926.

*ribosome (reye'-boh-sohm) Gr. *ribose,* a pentose sugar, and *soma,* body.

*inclusion (in-kloo'-zhun). Any substance held in the cell cytoplasm in an insoluble state.

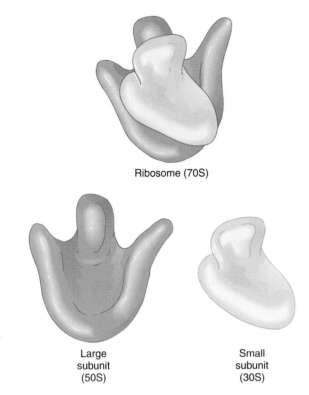

Ribosome (70S)

Large
subunit
(50S)

Small
subunit
(30S)

FIGURE 4.19

A model of a procaryotic ribosome, showing the small (30S)
and large (50S) subunits, both separate and joined.

Other inclusions, also called granules, contain crystals of inorganic
compounds and are not enclosed by membranes. Sulfur granules of
photosynthetic bacteria (see figure 4.35*b*) and polyphosphate granules of *Corynebacterium* (see figure 4.22) and *Mycobacterium* are
of this type. The latter represent an important source of building
blocks for nucleic acid and ATP synthesis. They have been termed
metachromatic* **granules** because they stain a contrasting color
(red, purple) in the presence of methylene blue dye.

BACTERIAL ENDOSPORES: AN EXTREMELY RESISTANT STAGE

Ample evidence indicates that the anatomy of bacteria helps them
adjust rather well to adverse habitats. But of all microbial structures, nothing can compare to the bacterial **endospore*** (or simply
spore) for withstanding hostile conditions and facilitating survival.
Endospores are dormant bodies produced by the gram-
positive genera *Bacillus, Clostridium,* and *Sporosarcina.* These
bacteria have a two-phase life cycle—a vegetative cell and an endospore (figure 4.21*a*). The vegetative cell is a metabolically active
and growing entity that can be induced by environmental conditions to undergo spore formation, or **sporulation.** Once formed, the
spore exists in an inert, resting condition that shows up prominently
in a spore or Gram stain (figure 4.21*b*). The chapter opening photo
also shows stages in this cycle, stained with differential fluorescent
dyes. Features of spores, including size, shape, and position in the
vegetative cell, are somewhat useful in identifying some species.

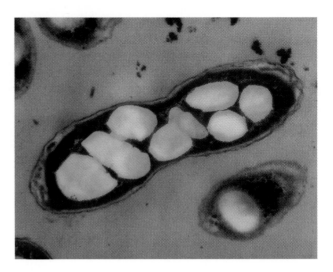

FIGURE 4.20

An example of a storage inclusion in a bacterial cell (32,500×).
Substances such as polyhydroxybutyrate can be stored in an insoluble,
concentrated form that provides an ample, long-term supply of that
nutrient.

Endospore Formation and Resistance

The depletion of nutrients, especially an adequate carbon or nitrogen
source, is the stimulus for a vegetative cell to begin spore formation.
Once this stimulus has been received by the vegetative cell, it undergoes a conversion to a committed sporulating cell called a
sporangium.* Complete transformation of a vegetative cell into a
sporangium and then into a spore requires 6 to 8 hours in most spore-
forming species. Table 4.2 illustrates some major physical and chemical events in this process. Bacterial endospores are the hardiest of all
life forms, capable of withstanding extremes in heat, drying, freezing,
radiation, and chemicals that would readily kill vegetative cells. Their
survival under such harsh conditions is due to several factors. The heat
resistance of spores has been linked to their high content of calcium
and *dipicolinic acid,* although the exact role of these chemicals is not
yet clear. We know, for instance, that heat destroys cells by inactivating proteins and DNA and that this process requires a certain amount
of water in the protoplasm. Because the deposition of calcium dipi-
colinate in the spore removes water and leaves the spore very dehydrated, it is less vulnerable to the effects of heat. It is also metabolically inactive and highly resistant to damage from further drying. The
thick, impervious cortex and spore coats also protect against radiation

---

A Note on Terminology

The word spore can have more than one usage in microbiology.
It is a generic term that refers to any tiny compact cells that are
produced by vegetative or reproductive structures of
microorganisms. Spores can be quite variable in origin, form,
and function. The bacterial type discussed here is called an
endospore, because it is produced inside a cell. It functions in
survival, not in reproduction, because no increase in cell
numbers is involved in its formation. In contrast, the fungi
produce many different types of spores for both survival and
reproduction (see Chapter 5).

---

*metachromatic (met″-uh-kroh-mah′-tik) Gr. *meta,* other, and *chromo,* color.

*endospore (en′-doh-spor) Gr. *endo,* inside, and *sporos,* seed.

*sporangium (spor-anj′-yum) L. *sporos,* and Gr. *angeion,* vessel.

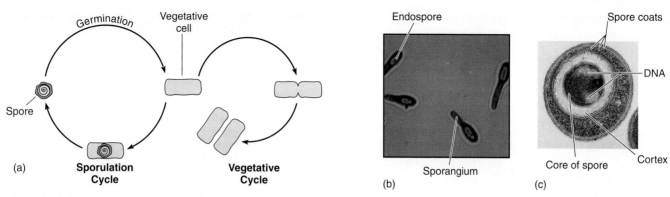

FIGURE 4.21

The general life cycle of a spore-forming bacterium. (a) Sporeformers can exist in an active, growing, vegetative state, or they can enter a dormant survival state through sporulation. Their actual fate depends upon the availability of nutrients. **(b)** A Gram stain of *Clostridium* shows the distinctive appearance of spores swelling up the end of the sporangium. **(c)** A cross section of a single spore reveals numerous layers.

and chemicals (figure 4.21*c*). The longevity of bacterial spores verges on immortality. One record describes the isolation of viable spores from a fossilized bee that was 25 million years old. More recently, microbiologists unearthed a viable spore from a 250-million-year-old salt crystal (see Chapter 1)! Initial analysis of this ancient microbe indicates it is a species of *Bacillus* that is genetically unique.

The Germination of Endospores

After lying in a state of inactivity for an indefinite time, spores can be revitalized when favorable conditions arise. The breaking of dormancy, or germination, happens in the presence of water and a specific chemical or environmental stimulus (germination agent). Once initiated, it proceeds to completion quite rapidly ($1\frac{1}{2}$ hours). Although the specific germination agent varies among species, it is generally a small organic molecule such as an amino acid or an inorganic salt. This agent stimulates the formation of hydrolytic (digestive) enzymes by the spore membranes. These enzymes digest the cortex and expose the core to water. As the core rehydrates and takes up nutrients, it begins to grow out of the spore coats. In time, it reverts to a fully active vegetative cell, resuming the vegetative cycle.

TABLE 4.2

General Stages in Endospore Formation

Stage		State of Cell	Process/Event
1		Vegetative cell	Cell in early stage of binary fission doubles chromosome.
2		Vegetative cell becomes **sporangium** in preparation for sporulation	One chromosome and a small bit of cytoplasm are walled off as a protoplast at one end of the cell. This **core** contains the minimum structures and chemicals necessary for guiding life processes. During this time, the sporangium remains active in synthesizing compounds required for spore formation.
3		Sporangium	The protoplast is engulfed by the sporangium to continue the formation of various protective layers around it.
4		Sporangium with **prospore**	Special peptidoglycan is laid down to form a cortex around the spore protoplast, now called the prospore; calcium and dipicolinic acid are deposited; core becomes dehydrated and metabolically inactive.
5		Sporangium with prospore	Three heavy and impervious protein spore coats are added.
6		Mature endospore	Endospore becomes thicker, and heat resistance is complete; sporangium is no longer functional and begins to deteriorate.
7		Free spore	Complete lysis of sporangium frees spore; it can remain dormant yet viable for thousands of years.
8		Germination	Addition of nutrients and water reverses the dormancy. The spore then swells and liberates a young vegetative cell.
9		Vegetative cell	Restored vegetative cell.

Medical Significance of Bacterial Spores

Although the majority of spore-forming bacteria are relatively harmless, several bacterial pathogens are sporeformers. In fact, some aspects of the diseases they cause are related to the persistence and resistance of their spores. *Bacillus anthracis* is the agent of anthrax, a skin and lung infection of domestic animals transmissible to humans. The genus *Clostridium* includes even more pathogens, including *C. tetani,* the cause of tetanus (lockjaw), and *C. perfringens,* the cause of gas gangrene. When the spores of these species are embedded in a wound that contains dead tissue, they can germinate, grow, and release potent toxins. Another toxin-forming species, *C. botulinum,* is the agent of botulism, a deadly form of food poisoning.

Because they inhabit the soil and dust, spores are a constant intruder where sterility and cleanliness are important. They resist ordinary cleaning methods that use boiling water, soaps, and disinfectants, and they frequently contaminate cultures and media. Hospitals and clinics must take precautions to guard against the potential harmful effects of spores in wounds. Spore destruction is a particular concern of the food-canning industry. Several endospore-forming species cause food spoilage or poisoning. Ordinary boiling (100°C) will usually not destroy such spores, so canning is carried out in pressurized steam at 120°C for 20 to 30 minutes. Such rigorous conditions will ensure that the food is sterile and free from viable bacteria.

TABLE 4.3

Comparison of the Two Spiral-Shaped Bacteria

	Spirilla	Spirochetes
Overall appearance	Rigid helix	Flexible helix
Mode of locomotion	Polar flagella; cells swim by rotating around like corkscrews; do not flex	Periplasmic flagella within sheath; cells flex; can swim by rotation or by creeping on surfaces
	1 to several flagella; can be in tufts	2 to 100 periplasmic flagella
Number of helical turns	Varies from 1 to 20	Varies from 3 to 70
Gram reaction (cell wall type)	Gram negative	Gram negative
Examples of important types	Most are harmless saprobes; one species, *Spirillum minor,* causes rat bite fever	*Treponema pallidum* cause of syphilis; *Borrelia* and *Leptospira,* important pathogens

CHAPTER CHECKPOINTS

The cytoplasm of bacterial cells serves as a solvent for materials (the cell pool) used in all cell functions.

The genetic material of bacteria is DNA. Genes are arranged in a single circular chromosome. Additional genes are carried on plasmids.

Bacterial ribosomes are dispersed in the cytoplasm in chains (polysomes) and are also embedded in the cell membrane.

Bacteria store nutrients in their cytoplasm in structures called inclusions. Inclusions vary in structure and the materials that are stored.

A few families of bacteria produce dormant bodies called endospores, which are the hardiest of all life forms, surviving for centuries.

The genera *Bacillus* and *Clostridium* are sporeformers, and both contain deadly pathogens.

Bacterial Shapes, Arrangements, and Sizes

For the most part, bacteria function as independent single-celled, or unicellular, organisms. Although it is true that an individual bacterial cell can live attached to others in colonies or other such groupings, each one is fully capable of carrying out all necessary life activities, such as reproduction, metabolism, and nutrient processing (unlike the more specialized cells of a multicellular organism).

Bacteria exhibit considerable variety in shape, size, and colonial arrangement. It is convenient to describe most bacteria by one of three general shapes as dictated by the configuration of the cell wall (figure 4.22). If the cell is spherical or ball-shaped, the bacterium is described as a **coccus.*** Cocci can be perfect spheres, but they also can exist as oval, bean-shaped, or even pointed variants. A cell that is cylindrical (longer than wide) is termed a **rod,** or **bacillus.*** There is also a genus named *Bacillus.* As might be expected, rods are also quite varied in their actual form. Depending on the bacterial species, they can be blocky, spindle-shaped, round-ended, long and threadlike (filamentous), or even clubbed or drumstick-shaped. When a rod is short and plump, it is called a **coccobacillus;** if it is gently curved, it is a **vibrio.*** A bacterium having the shape of a curviform or spiral-shaped cylinder is called a **spirillum,*** a rigid helix, twisted twice or more along its axis (like a corkscrew). Another spiral cell mentioned earlier in conjunction with periplasmic flagella is the **spirochete,** a more flexible form that resembles a spring. Refer to table 4.3 for a comparison of other features of the two helical bacterial forms. Because bacterial cells look rather two-dimensional and flat with traditional staining and microscope techniques, they are seen to best advantage with a scanning electron microscope that emphasizes their striking three-dimensional forms (figure 4.23).

It is rather common for cells of the same species to vary to some extent in shape and size. This phenomenon, called **pleomorphism*** (figure 4.24*a*), is due to individual variations in cell wall structure caused by nutritional or slight hereditary differences. For example, although the cells of *Corynebacterium diphtheriae* are generally considered rod-shaped, in culture they display variations

*coccus (kok′-us) pl. cocci (kok′-seye) Gr. *kokkos,* berry.

*bacillus (bah-sil′-lus) pl. bacilli (bah-sil′-eye) L. *bacill,* small staff or rod.

*vibrio (vib′-ree-oh) L. *vibrare,* to shake.

*spirillum (spy-ril′-em) pl. spirilla; L. *spira,* a coil.

*pleomorphism (plee″-oh-mor′-fizm) Gr. *pleon,* more, and *morph,* form or shape.

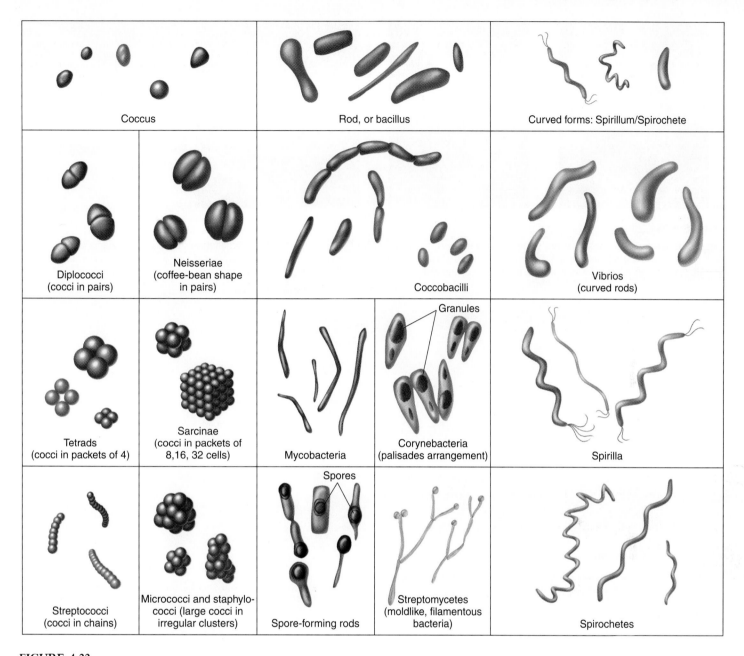

FIGURE 4.22

Bacterial shapes and arrangements. May not be shown to exact scale.

such as club-shaped, swollen, curved, filamentous, and coccoid. Pleomorphism reaches an extreme in the mycoplasmas, which entirely lack cell walls and thus display extreme variations in shape (see figure 4.33).

The cells of bacteria can also be categorized according to **arrangement,** or style of grouping (see figure 4.22). The main factors influencing the arrangement of a particular cell type are its pattern of division and how the cells remain attached afterward. The greatest variety in arrangement occurs in cocci, which can be single, in pairs **(diplococci),*** in **tetrads** (groups of four), in irregular

clusters (both **staphylococci*** and **micrococci),*** or in chains of a few to hundreds of cells **(streptococci).*** An even more complex grouping is a cubical packet of eight, sixteen, or more cells called a **sarcina** (sar´-sih-nah). These different coccal groupings are the result of the division of a coccus in a single plane, in two perpendicular planes, or in several intersecting planes; after division, the resultant daughter cells remain attached (figure 4.25).

*diplococci; Gr. diplo, double.

*staphylococci (staf″-ih-loh-kok´-seye) Gr. staphyle, a bunch of grapes.

*micrococci Gr. mikros, small.

*streptococci (strep″-toh-kok´-seye) Gr. streptos, twisted.

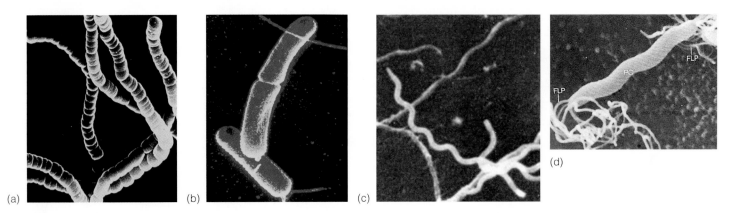

(a) (b) (c) (d)

FIGURE 4.23

Basic bacterial shapes in three dimensions with surface features. (a) Cocci in chains. (b) A rod-shaped bacterium (*Escherichia coli*) in a diplobacillus arrangement. (c) A spirochete (*Borrelia burgdorferi,* the cause of Lyme disease) is a long, thin cell with irregular coils and no external flagella. (d) A spirillum is thicker with a few even coils (PC) and external flagella (FLP). Can you tell what the flagellar arrangement is?

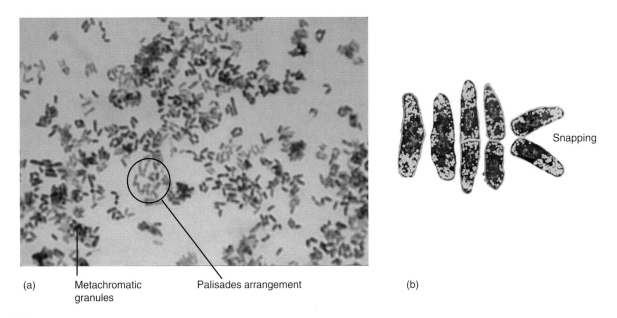

(a) Metachromatic granules Palisades arrangement (b) Snapping

FIGURE 4.24

Pleomorphism in *Corynebacterium.* (a) Cells occur in a great variety of shapes and sizes. This genus typically exhibits an unusual formation called a palisades arrangement that is caused by snapping, shown in (b). Close examination will also reveal darkly stained granules inside the cells.

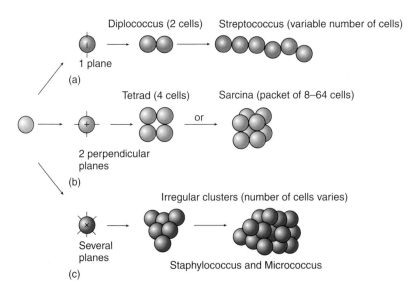

Diplococcus (2 cells) Streptococcus (variable number of cells)

1 plane

(a)

Tetrad (4 cells) Sarcina (packet of 8–64 cells)

or

2 perpendicular planes

(b)

Irregular clusters (number of cells varies)

Several planes

Staphylococcus and Micrococcus

(c)

FIGURE 4.25

Arrangement of cocci resulting from different planes of cell division. (a) Division in one plane produces diplococci and streptococci. **(b)** Division in two planes at right angles produces tetrads and packets. **(c)** Division in several planes produces irregular clusters.

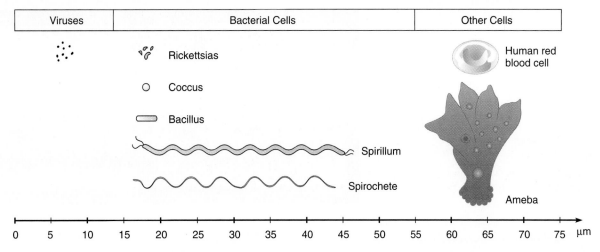

Viruses	Bacterial Cells	Other Cells

Rickettsias

Coccus

Bacillus

Spirillum

Spirochete

Human red blood cell

Ameba

0 5 10 15 20 25 30 35 40 45 50 55 60 65 70 75 μm

FIGURE 4.26

The dimensions of bacteria. The sizes of bacteria range from those just barely visible with light microscopy (0.2 μm) to those measuring a thousand times that size. Cocci measure anywhere from 0.5 to 3.0 μm in diameter; bacilli range from 0.2 to 2.0 μm in diameter and from 0.5 to 20 μm in length; vibrios and spirilla vary from 0.2 to 2.0 μm in diameter and from 0.5 to 100 μm in length. Spirochetes range from 0.1 to 3.0 μm in diameter and from 0.5 to 250 μm in length. Note the range of sizes as compared with eucaryotic cells and viruses. Comparisons are given as average sizes.

Bacilli are less varied in arrangement because they divide only in the transverse plane (perpendicular to the axis). They occur either as single cells, as a pair of cells with their ends attached (diplobacilli), or as a chain of several cells (streptobacilli). A **palisades*** arrangement, typical of the corynebacteria, is formed when the cells of a chain remain partially attached by a small hinge region at the ends. The cells tend to fold (snap) back upon each other, forming a row of cells oriented side by side (see figure 4.24*b*). The reaction can be compared to the behavior of boxcars on a jack-knifed train, and the result looks superficially like an irregular picket fence. Spirilla are occasionally found in short chains, but spirochetes rarely remain attached after division. Comparative sizes of typical cells are presented in figure 4.26.

CHAPTER CHECKPOINTS

Most bacteria have one of three general shapes: coccus (round), bacillus (rod), or spiral, based on the configuration of the cell wall. Two types of spiral cells are spirochetes and spirilla.

Shape and arrangement of cells are key means of describing bacteria. Arrangements of cells are based on the number of planes in which a given species divides.

Cocci can divide in many planes to form pairs, chains, packets, or clusters. Bacilli divide only in the transverse plane. If they remain attached, they form chains or palisades.

Bacteria range in size from the smallest rickettsias to the largest spiral forms.

Bacterial Identification and Classification Systems

Every speck of soil, dab of saliva, and droplet of pond water is teeming with a rich variety of bacteria. Although study of such mixed populations can be useful, in most laboratory studies, one must prepare pure cultures of the individual members of the population (see chapter 3). This is especially critical in medical labs, which are charged with gathering the characteristics of an infectious agent and identifying it so that proper treatment can be given. Isolation and laboratory growth are also necessary for discovery and verification of new species.

The methods that a microbiologist uses to identify bacteria to the level of genus and species fall into the main categories of morphology (microscopic and macroscopic), bacterial physiology or biochemistry, serological analysis, and genetic techniques. Data from a cross section of such tests can produce a unique profile of each bacterium. Final differentiation of any unknown species is accomplished by comparing its profile with the characteristics of known bacteria in tables, charts, and keys (see figure 20.8). Many of the identification systems are automated and incorporate computers to process data and provide a "best fit" identification. However, not all methods are used on all bacteria. A few bacteria can be identified by placing them in an automated machine that analyzes only the kind of fatty acids they contain; in contrast, some are identifiable by a Gram stain and a few physiological tests; others may require a diverse spectrum of morphological, biochemical, and genetic tests. The following list summarizes some of the general areas of bacteriological testing. (See also figure 3.1.)

METHODS USED IN BACTERIAL IDENTIFICATION

Microscopic Morphology Traits that can be valuable aids to identification are combinations of cell shape and size, Gram stain reaction, acid-fast reaction, and special structures, including endospores, granules, and capsules. Electron microscope studies can pinpoint additional structural features (such as the cell wall, flagella, pili, and fimbriae).

Macroscopic Morphology Appearance of colonies, including texture, size, shape, pigment, speed of growth and patterns of growth in broth and gelatin media.

***palisades** (pal′-ih-saydz) L. *pale,* a stake. A fence made of a row of stakes.

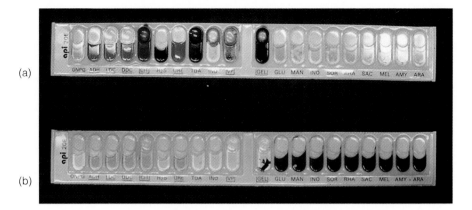

(a)

(b)

FIGURE 4.27

Rapid tests. The API 20E manual biochemical system for microbial identification. **(a)** Positive and **(b)** negative results.

Physiological/Biochemical Characteristics These have been the traditional mainstay of bacterial identification. Enzymes and other biochemical properties of bacteria are fairly reliable and stable expressions of the chemical identity of each species. Dozens of diagnostic tests exist for determining the presence of specific enzymes and to assess nutritional and metabolic activities. Examples include tests for fermentation of sugars; capacity to digest or metabolize complex polymers such as proteins and polysaccharides; production of gas; presence of enzymes such as catalase, oxidase, and decarboxylases; and sensitivity to antimicrobic drugs. Special rapid identification test systems that record the major biochemical reactions of a culture have streamlined data collection (figure 4.27).

Chemical Analysis Analyzing the types of specific structural substances that the bacterium contains, such as the chemical composition of peptides in the cell wall and lipids in membranes.

Serological Analysis Bacteria have surface and other molecules called antigens that are recognized by the immune system. One immune response to antigens is the production of molecules called antibodies that are designed to bind tightly to the antigens. This response is so specific that antibodies can be used as a means of identifying bacteria in specimens and cultures. Laboratory kits based on this technique are available for immediate identification of a number of pathogens.

Genetic and Molecular Analysis Examining the genetic material itself has revolutionized the identification and classification of bacteria.

    G + C base composition. The overall percentage of guanine and cytosine (the G + C content as compared with A + T content) in DNA is a general indicator of relatedness because it is a trait that does not change rapidly. Bacteria with a significant difference in G + C percentage are likely to be genetically distinct species or genera. For example, although superficially similar in Gram reaction, shape, and other morphological characteristics, *Escherichia* has a G + C base composition of 48–52% and *Pseudomonas* has a composition of 58–70%, indicating that they probably are not closely related. This technique is most applicable for clarifying the taxonomic position of a bacterium, but it is too nonspecific to be applicable as a precise identification tool.

DNA analysis using genetic probes. The exact order and arrangement of the DNA code is unique to each organism (see figure 2.26). With a technique called *hybridization,* it is possible to identify a bacterial species by analyzing segments of its DNA (figure 4.28). This requires small fragments of single-stranded DNA (or RNA) called **probes** that are known to be complementary to the specific sequences of DNA from a particular microbe. The test is conducted by extracting unknown test DNA from cells in specimens or cultures and binding it to special blotter paper. After several different probes have been added to the blotter, it is observed for visible signs that the probes have become fixed (hybridized) to the test DNA. The binding of probes onto several areas of the test DNA indicates close correspondence and makes positive identification possible (also discussed in chapter 10).

Nucleic acid sequencing and rRNA analysis. One of the most viable indicators of evolutionary relatedness and affiliation is comparison of the sequence of nitrogen bases in ribosomal RNA, a major component of ribosomes (figure 4.29). Ribosomes have the same function (protein synthesis) in all cells, and they tend to remain more or less stable in their nucleic acid content over long periods. Thus, any major differences in the sequence, or "signature," of the rRNA is likely to indicate some distance in ancestry. This technique is powerful at two levels: It is effective for differentiating general group differences (it was used to separate the three superkingdoms of life discussed in chapter 1), and it can be fine-tuned to identify at the species level (for example in *Mycobacterium* and *Legionella*). Elements of these and other identification methods are presented in more detail in chapters 10, 16, 20, and 21.

CLASSIFICATION SYSTEMS IN THE PROCARYOTAE

Classification systems serve both practical and academic purposes. They aid in differentiating and identifying unknown species in medical and applied microbiology. They are also useful in organizing bacteria and as a means of studying their relationships and origins. Since the classification was started around 200 years ago,

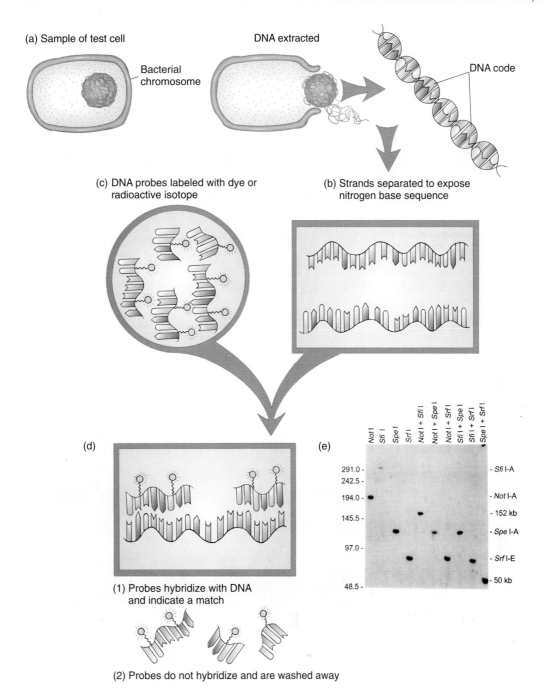

(a) Sample of test cell

Bacterial chromosome

DNA extracted

DNA code

(c) DNA probes labeled with dye or radioactive isotope

(b) Strands separated to expose nitrogen base sequence

(d)

(1) Probes hybridize with DNA and indicate a match

(2) Probes do not hybridize and are washed away

(e)

FIGURE 4.28

DNA hybridization using probes. Note that the scale is exaggerated for visibility. **(a)** DNA is extracted from an unknown specimen and processed. **(b)** DNA is separated into two strands and fixed in position. **(c)** Solutions containing probes of known base sequence are applied. The probes are labeled with tracer molecules that indicate where a reaction has taken place. **(d)** When probes contact DNA that has a complementary base sequence, they become affixed through base-pairing and will cause a visible reaction in a localized area of the test DNA (1). Probes that are not of the correct specificity will remain free and will not show up as a visible reaction (2). **(e)** A DNA hybridization result from the syphilis spirochete, *Treponema pallidum.* Dark bands indicate areas where a specific probe has reacted with *Treponema* DNA.

several thousand species of bacteria and archaea have been identified, named, and catalogued.

For years there has been intense interest in tracing the origins of and evolutionary relationships among bacteria, but doing so has not been an easy task. As a rule, tiny, relatively soft organisms do not form fossils very readily. Several times since the 1960s, however, scientists have discovered microscopic fossils of procaryotes that look very much like modern bacteria. Some of the rocks that contain these fossils have been dated back billions of years (figure 4.30). One of the questions that has plagued taxonomists is,

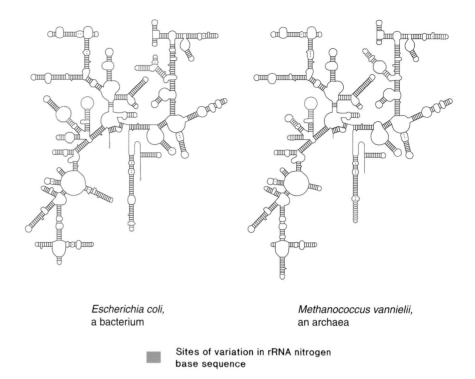

FIGURE 4.29

A modern molecular technique that identifies the cell type or subtype by analyzing the nitrogen base sequence of its ribosomal RNA. The segments of rRNA shown here are from the small subunit of *Escherichia coli,* a common intestinal bacterium, and *Methanococcus vannielii,* an archaea that inhabits the deep sediments of marshes. The shaded regions indicate areas of rRNA that differ significantly between bacteria (*Escherichia*) and archaea (*Methanococcus*).

Escherichia coli,
a bacterium

Methanococcus vannielii,
an archaea

Sites of variation in rRNA nitrogen base sequence

FIGURE 4.30

Ancient procaryotes. Electron micrograph of fossilized rod-shaped bacterial cells preserved in rock sediment. This specimen is more than 2 billion years old (24,000×).

What characteristics are the most indicative of closeness in ancestry? Early bacteriologists found it convenient to classify bacteria according to shape, variations in arrangement, growth characteristics, and habitat. However, as more species were discovered and as techniques for studying their biochemistry were developed, it soon became clear that similarities in cell shape, arrangement, and staining reactions do not automatically indicate relatedness. Even though the gram-negative rods look alike, there are hundreds of different species, with highly significant differences in biochemistry and genetics. If we attempted to classify them on the basis of Gram stain and shape alone, we could not assign them to a more specific level than class. Increasingly, classification schemes are turning to genetic and molecular traits that cannot be visualized under a microscope or in culture.

The most functional classification schemes make use of current knowledge to show natural relationships. Not only must

---

TABLE 4.4

Major Taxonomic Groups of Bacteria per *Bergey's Manual*

Division I. Gracilicutes: Gram-Negative Bacteria

Class I.   Scotobacteria: Gram-negative non-photosynthetic bacteria (examples in table 4.5)

Class II. Anoxyphotobacteria: Gram-negative photosynthetic bacteria that do not produce oxygen (purple and green bacteria)

Class III. Oxyphotobacteria: Gram-negative photosynthetic bacteria that evolve oxygen (cyanobacteria)

Division II. Firmicutes: Gram-Positive Bacteria

Class I.   Firmibacteria: Gram-positive rods or cocci (examples in table 4.6)

Class II. Thallobacteria: Gram-positive branching cells (the actinomycetes)

Division III. Tenericutes

Class I.   Mollicutes: Bacteria lacking a cell wall (the mycoplasmas)

Division IV. Mendosicutes

Class I.   Archaebacteria: Bacteria with atypical compounds in the cell wall and membranes

Source: Data from Bergey's Manual of Systematic Bacteriology, *9th ed. Williams & Wilkins Company, Baltimore, 1984.*

TABLE 4.5

Bergey's Manual* Taxonomic Rankings for *Borrelia Burgdorferi

Taxonomic Rank	Includes
Kingdom: Monera	All bacteria
Division: Gracilicutes	All bacteria with gram-negative cell wall
Class: Scotobacteria	Non-photosynthetic gram-negative bacteria
Order: Spirochaetales	Helical, flexible, motile bacteria (spirochetes)
Family: Spirochaetaceae	Helical, motile bacteria lacking hooked ends
Genus: *Borrelia*	Tiny, loose, irregularly coiled spirochetes
Species: *burgdorferi*	Causative agent of Lyme disease

---

they be flexible enough to add newly discovered species, but they must also complement a number of microbiology disciplines. In general, schemes for organizing bacteria can be phylogenetic, based on evolutionary relationships, or **phenetic,*** based on their morphology or biochemistry. One system of classification has not been permanent or universally accepted to date; indeed, most systems are in a state of flux as new information and methods of analysis become available. Of the current proposed systems, we present three: (1) the overall scheme of classification from the ninth edition of *Bergey's Manual of Systematic Bacteriology,* a manual of bacterial descriptions and classification published continuously since 1923 (table 4.4); (2) a method that groups bacteria by comparing rRNA sequence; and (3) a practical system that uses a few morphological and physiological traits to categorize the major, medically important bacterial families (see table 4.6).

The ninth edition of *Bergey's Manual* organizes the Kingdom Procaryotae into four major divisions. These somewhat natural divisions are based upon the nature of the cell wall. The **Gracilicutes*** have gram-negative cell walls and thus are thin-skinned; the **Firmicutes*** have gram-positive cell walls that are thick and strong; the **Tenericutes*** lack a cell wall and thus are soft; and the **Mendosicutes*** are the archaea (also called archaebacteria), primitive procaryotes with unusual cell walls and nutritional habits. The

first two divisions contain the greatest number of species. The 200 or so species that cause human and animal diseases can be found in four classes: the Scotobacteria, Firmibacteria, Thallobacteria, and Mollicutes. The system used in *Bergey's Manual* further organizes bacteria into subcategories such as classes, orders, and families, but these are not available for all groups. An example of the entire classification of one bacterial species is shown in table 4.5.

The classification scheme developed by analyzing the similarities in base sequence of rRNA revealed 11 distinct branches on the bacterial "tree" (figure 4.31). It just so happens that some groupings are similar to those used in Bergey's system (gram-positive bacteria), but others are genetically different enough to be placed into their own separate categories (spirochetes).

1. Gram-positive eubacteria: Selected representatives are *Bacillus, Clostridium, Mycobacterium, Staphylococcus, Actinomyces,* and the cell-wall-free mycoplasmas.

2. Gram-negative eubacteria (Proteobacteria) includes purple photosynthetic bacteria (*Chromatium*) and non-photosynthetic relatives represented by *Pseudomonas, Vibrio, Neisseria,* and the rickettsias.

3. Cyanobacteria: photosynthetic bacteria with chlorophyll *a* that evolve (give off) oxygen; includes *Oscillatoria* and *Spirulina.*

4. Spirochetes: flexible helical cells with periplasmic flagella such as *Treponema* and *Borrelia.*

5. Walled, budding bacteria that lack peptidoglycan in their cell walls: includes *Planctomyces.*

6. The *Bacteroides, Flavobacterium, Fusobacterium,* and *Cytophaga:* a mixed group morphologically and physiologically.

7. Chlamydias: unusual obligate parasites of vertebrates; lack ability to complete metabolism independently; lack peptidoglycan; one genus—*Chlamydia.*

8. Green sulfur bacteria: anaerobic bacteria that contain bacteriochlorophyll and use sulfur in metabolism; do not give off oxygen during photosynthesis; includes *Chlorobium.*

*****phenetic** (fuh-neh′-tik) Referring to the phenotype or expression of traits.

*****Gracilicutes** (gras″-ih-lik′-yoo-teez) L. *gracilus,* thin, and *cutis,* skin.

*****Firmicutes** (fer-mik′-yoo-teez) L. *firmus,* strong.

*****Tenericutes** (ten″-er-ik′-yoo-teez) L. *tener,* soft

*****Mendosicutes** (men-doh-sik′-yoo-teez) L. *mendosus,* to be false.

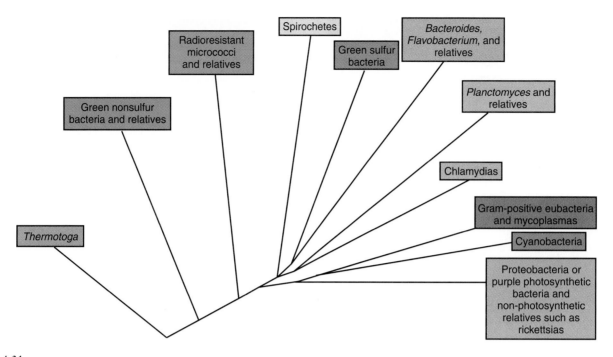

FIGURE 4.31

Separation chart for the eubacteria proposing phylogenetic relationships based on rRNA sequences. This scheme divides them into 11 genetically discrete groups. The branches suggest evolutionary origins, and branches that have greater similarities appear closer to each other.

Source: Data from C. R. Woese, Microbiol. Rev. 51(1987):221–71. American Society for Microbiology, Washington, D. C.

9. Green nonsulfur bacteria: filamentous, gliding, thermophilic, photosynthetic bacteria that contain bacteriochlorophyll, do not evolve oxygen; includes *Chloroflexus*.

10. Unique bacteria with extreme resistance to electromagnetic radiation: *Deinococcus*, gram-positive cocci, and *Thermus*, thermophilic rods.

11. Unusual thermophilic bacteria inhabiting hot oceanic vents: *Thermotoga*.

Ribosomal RNA analysis of the archaea detected three well-defined groups:

1. Representatives that synthesize methane from simple inorganic compounds (methanogens)

2. Extremely halophilic (salt-loving) archaea

3. Extremely thermophilic (heat-loving) archaea that use sulfur in their metabolism

The last section this chapter covers features of archaeal form, function, and ecology.

Many medical microbiologists prefer an informal working system that outlines the major families and genera (table 4.6). This system is more applicable for their purposes because it is restricted to bacterial disease agents, depends less on nomenclature, and is based on readily accessible morphological and physiological tests rather than on phylogenetic relationships. It also divides the bacteria into gram positive, gram negative, and those without cell walls and then subgroups them according to cell shape, arrangement, and certain physiological traits such as oxygen usage: *Aerobic* bacteria use oxygen in metabolism; *anaerobic* bacteria do not use oxygen in

metabolism; and facultative bacteria may or may not use oxygen. Further tests not listed on the table would be required to separate closely related genera and species. Many of these are included in later chapters on specific bacterial groups.

Species and Subspecies in Bacteria

Among most organisms, the species level is a distinct, readily defined, and natural taxonomic category. In animals, for instance, a species is a distinct type of organism that can produce viable offspring only when it mates with others of its own kind. This definition does not work for bacteria primarily because they do not exhibit a typical mode of sexual reproduction. They can accept genetic information from unrelated forms, and they can also alter their genetic makeup by a variety of mechanisms. Thus, it is necessary to hedge a bit when we define a bacterial species. Theoretically, it is a collection of bacterial cells, all of which share an overall similar pattern of traits, in contrast to other groups whose pattern differs significantly. Although the boundaries that separate two closely related species in a genus are in some cases very arbitrary, this definition still serves as a method to separate the bacteria into various kinds that can be cultured and studied. As additional information on bacterial genomes is discovered, it may be possible to define species according to specific combinations of genetic codes found only in a particular isolated culture.

Since the individual members of given species can show variations, we must also define levels within species (subspecies) called **strains** and **types.** A strain or variety of bacteria is a culture derived from a single parent that differs in structure or metabolism from other cultures of that species (also called biovars or morphovars).

TABLE 4.6

Medically Important Families and Genera of Bacteria, with Notes on Some Diseases★

I. Bacteria with gram-positive cell wall structure

Cocci in clusters or packets that are aerobic or facultative

Family Micrococcaceae: *Staphylococcus* (members cause boils, skin infections)

Cocci in pairs and chains that are facultative

Family Streptococcaceae: *Streptococcus* (species cause strep throat, dental caries)

Anaerobic cocci in pairs, tetrads, irregular clusters

Family Peptococcaceae: *Peptococcus, Peptostreptococcus* (involved in wound infections)

Spore-forming rods

Family Bacillaceae: *Bacillus* (anthrax), *Clostridium* (tetanus, gas gangrene, botulism)

Non-spore-forming rods

Family Lactobacillaceae: *Lactobacillus, Listeria* (milk-borne disease), *Erysipelothrix* (erysipeloid)

Family Propionibacteriaceae: *Propionibacterium* (involved in acne)

Family Corynebacteriaceae: *Corynebacterium* (diphtheria)

Family Mycobacteriaceae: *Mycobacterium* (tuberculosis, leprosy)

Family Nocardiaceae: *Nocardia* (lung abscesses)

Family Actinomycetaceae: *Actinomyces* (lumpy jaw), *Bifidobacterium*

Family Streptomycetaceae: *Streptomyces* (important source of antibiotics)

II. Bacteria with gram-negative cell wall structure

Family Neisseriaceae

Aerobic cocci

Neisseria (gonorrhea, meningitis), *Branhamella*

Aerobic coccobacilli

Moraxella, Acinetobacter

Anaerobic cocci

Family Veillonellaceae

Veillonella (dental disease)

Miscellaneous rods

Brucella (undulant fever), *Bordetella* (whooping cough), *Francisella* (tularemia)

Aerobic rods

Family Pseudomonadaceae: *Pseudomonas* (pneumonia, burn infections)

Miscellaneous: *Legionella* (Legionnaires' disease)

Facultative or anaerobic rods and vibrios

Family Enterobacteriaceae: *Escherichia, Edwardsiella, Citrobacter, Salmonella* (typhoid fever), *Shigella* (dysentery), *Klebsiella, Enterobacter, Serratia, Proteus, Yersinia* (one species causes plague)

Family Vibronaceae: *Vibrio* (cholera, food infection), *Campylobacter, Aeromonas*

Miscellaneous genera: *Chromobacterium, Flavobacterium, Haemophilus* (meningitis), *Pasteurella, Cardiobacterium, Streptobacillus*

Anaerobic rods

Family Bacteroidaceae: *Bacteroides, Fusobacterium* (anaerobic wound and dental infections)

Helical and curviform bacteria

Family Spirochaetaceae: *Treponema* (syphilis), *Borrelia* (Lyme disease), *Leptospira* (kidney infection)

Obligate intracellular bacteria

Family Rickettsiaceae: *Rickettsia* (Rocky Mountain spotted fever), *Coxiella* (Q fever)

Family Bartonellaceae: *Bartonella* (trench fever, cat scratch disease)

Family Chlamydiaceae: *Chlamydia* (sexually transmitted infection)

III. Bacteria with no cell walls

Family Mycoplasmataceae: *Mycoplasma* (pneumonia), *Ureaplasma* (urinary infection)

★*Details of pathogens and diseases in chapters 18, 19, 20, and 21.*

For example, there are pigmented and nonpigmented strains of *Serratia marcescens* and flagellated and nonflagellated strains of *Pseudomonas fluorescens*. A type is a subspecies that can show differences in antigenic makeup (serotype or serovar), in susceptibility to bacterial viruses (phage type), and in pathogenicity (pathotype).

CHAPTER CHECKPOINTS

A bacterial species can be properly identified only if it is grown in pure culture—that is, in isolation from all other forms of life.

Key traits that are used to identify a bacterial species include (1) morphology, (2) Gram stain or other stain characteristics, (3) presence of specialized structures, (4) macroscopic appearance of colonies, (5) biochemical reactions, and (6) nucleotide composition of both DNA and rRNA.

Bacteria are formally classified by phylogenetic relationships and phenotypic characteristics.

Medical identification of pathogens uses a more informal system of classification based on Gram stain, morphology, biochemical reactions, and metabolic requirements.

A bacterial species is loosely defined as a collection of bacterial cells that shares an overall similar pattern of traits different from other groups of bacteria.

Variant forms within a species (subspecies) include strains and types.

Survey of Procaryotic Groups with Unusual Characteristics

The bacterial world is so diverse that we cannot do complete justice to it in this introductory chapter. This variety extends into all areas of bacterial biology, including nutrition, mode of life, and behavior. Certain types of bacteria exhibit such unusual qualities that they deserve special mention. In this minisurvey, we will consider some medically important groups and some more remarkable representatives of bacteria living free in the environment that are ecologically important. Many of the bacteria mentioned here do not have the morphology typical of bacteria discussed previously, and in a few cases, they are vividly different (Spotlight on Microbiology 4.3).

UNUSUAL FORMS OF MEDICALLY SIGNIFICANT BACTERIA

Most bacteria are free-living or parasitic forms that can metabolize and reproduce by independent means. Two groups of bacteria—the rickettsias and chlamydias—have adapted to life inside their host cells, where they are considered **obligate intracellular parasites.** These unusual bacteria are covered in greater detail in chapter 21, but here we provide brief descriptions.

Rickettsias Rickettsias[3] are distinctive, very tiny, gram-negative bacteria (figure 4.32). Although they have a somewhat typical bacterial morphology, they are atypical in their life cycle and other adaptations. Most are pathogens that alternate between a mam-

malian host and blood-sucking arthropods,[4] such as fleas, lice, or ticks. Rickettsias cannot survive or multiply outside a host cell and cannot carry out metabolism completely on their own, so they are closely attached to their hosts. Several important human diseases are caused by rickettsias. Among these are Rocky Mountain spotted fever, caused by *Rickettsia rickettsii* (transmitted by ticks), and epidemic typhus, caused by *Rickettsia prowazekii* (transmitted by lice). An exceptionally resistant rickettsia, *Coxiella burnetti* (the cause of Q fever), is transmitted in air and dust by arthopods.

Chlamydias Bacteria of the genus *Chlamydia* are similar to the rickettsias in that they require host cells for growth and metabolism, but they are not closely related and are not transmitted by arthropods. Because of their tiny size and obligately parasitic lifestyle, they were at one time considered a type of virus. Later studies indicated that their structure was that of a gram-negative cell and that their mode of cell division (binary fission) was clearly procaryotic. Species that carry the greatest medical impact are *Chlamydia trachomatis,* the cause of both a severe eye infection (trachoma) that can lead to blindness and one of the most common sexually transmitted diseases; *Chlamydia psittaci,* the agent of ornithosis, or parrot fever, a disease of birds that can be transmitted to humans; and *Chlamydia pneumoniae,* an agent in lung infections (see chapter 21).

Mycoplasmas and Other Cell-Wall-Deficient Bacteria Mycoplasmas are bacteria that naturally lack a cell wall. Although other bacteria require an intact cell wall to prevent the bursting of the cell, the mycoplasmal cell membrane is stabilized by sterols and is resistant to lysis. These extremely tiny, pleomorphic cells are considered the smallest cells, ranging from 0.1 to 0.5 μm in size. They range in shape from filamentous to coccus or doughnut-shaped. They are *not* obligate parasites and can be grown on artificial media, although added sterols are required for the cell membranes of some species. Mycoplasmas are found in many habitats, including plants, soil, and animals. The most important medical species is *Mycoplasma pneumoniae* (figure 4.33), which adheres to the epithelial cells in the lung and causes an atypical form of pneumonia in humans.

FREE-LIVING NONPATHOGENIC BACTERIA

Photosynthetic Bacteria

The nutrition of most bacteria is heterotrophic, meaning that they derive their nutrients from other organisms. Photosynthetic bacteria, however, are independent cells that contain special light-trapping pigments and can use the energy of sunlight to synthesize all required nutrients from simple inorganic compounds. The two general types of photosynthetic bacteria are those that produce oxygen during photosynthesis and those that produce some other substance, such as sulfur granules or sulfates.

Cyanobacteria: Blue-Green Bacteria The cyanobacteria were called blue-green algae for many years and were grouped with the eucaryotic algae. However, further study verified that they are indeed bacteria with a gram-negative cell wall (Gracilicutes) and general procaryotic structure. These bacteria range in size from

---

3. Named for Howard Ricketts, a physician who first worked with these organisms and later lost his life to typhus.

4. An arthropod is an invertebrate with jointed legs, such as an insect, tick, or spider.

SPOTLIGHT ON MICROBIOLOGY 4.3
Redefining Bacterial Size

Many microbiologists believe we are still far from having a complete assessment of the bacterial world, mostly because the world is so large and bacteria are so small. This fact becomes evident in the periodic discoveries of exceptional bacteria that are reported in newspaper headlines. Among the most remarkable are giant and dwarf bacteria.

Big Bacteria Break Records
In 1985, biologists discovered a new bacterium living in the intestine of surgeonfish that at the time was a candidate for the *Guinness Book of World Records*. The large cells, named *Epulopiscium fishelsoni* ("guest at a banquet of fish"), measure around 100 μm in length, although some specimens were as large as 300 μm. This record was recently broken when marine microbiologist Heide Schultz discovered an even larger species of bacteria living in ocean sediments near the African country of Namibia. These gigantic cocci are arranged in strands that look like pearls and contain hundreds of golden sulfur granules, inspiring their name, *Thiomargarita namibia* ("sulfur pearl of Namibia") (see photo at right). The size of the individual cells ranges from 100 up to 750 μm (3/4 mm), and many are large enough to see with the naked eye. By way of comparison, if the average bacterium were the size of a mouse, *Thiomargarita* would be as large as a blue whale!

Closer study revealed that they are indeed procaryotic and have bacterial ribosomes and DNA, but that they also have some unusual adaptations to their life cycle. They live an attached existence embedded in sulfide sediments (H_2S) that are free of gaseous oxygen. They obtain energy through oxidizing these sulfides using dissolved nitrates (NO_3). Since the quantities of these substances can vary with the seasons, they must be stored in cellular depots. The sulfides are borne as granules in the cytoplasm, and the nitrates occupy a giant, liquid-filled vesicle that takes up a major proportion of cell volume. Because of their morphology and physiology, the cells can survive for up to three months without an external source of nutrients by tapping into their "storage tanks." These bacteria are found in such large numbers in the sediments that it is thought that they are essential to the ecological cycling of H_2S gas in this region, converting it to less toxic substances.

Miniature Microbes—The Smallest of the Small
At the other extreme, microbiologists are being asked to reevaluate the lower limits of bacterial size. Up until now it has been generally accepted that the smallest cells on the planet are some form of mycoplasma with

dimensions of 0.2 to 0.3 μm, which is right at the limit of resolution with light microscopes. A new controversy is brewing over the discovery of tiny cells that look like dwarf bacteria but are ten times smaller than mycoplasmas and a hundred times smaller than the average bacterial cell. These minute cells have been given the name **nanobacteria** or **nanobes** (Gr. *nanos,* one-billionth).

Nanobacteria-like forms were first isolated from blood and serum samples. The tiny cells appear to grow in culture, have cell walls, and contain protein and nucleic acids, but their size range is only from 0.05 μm to 0.2 μm. Similar nanobes have been extracted by minerologists studying sandstone rock deposits in the ocean at temperatures of 100°–170°C and deeply embedded in billion-year-old minerals. The minute filaments were able to grow and are capable of depositing minerals in a test tube. Many geologists are convinced that these nanobes are real, that they are probably similar to the first microbes on earth, and that they play a strategic role in the evolution of the earth's crust. Microbiologists tend to be more skeptical. It has been postulated that the minimum cell size to contain a functioning genome and reproductive and synthetic machinery is approximately 0.14 μm. They believe that the nanobes are really just artifacts or bits of larger cells that have broken free. It is partly for this reason that most bacteriologists rejected the idea that small objects found in a Martian meteor were microbes but were more likely caused by chemical reactions. Additional studies are needed to test this curious question of nanobes, and possibly to answer some questions about the origins of life on earth and even other planets.

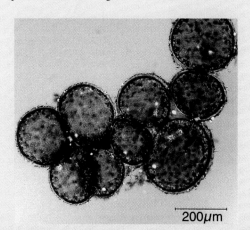

Thiomargarita namibia—giant cocci.

1 μm to 10 μm, and they can be unicellular or can occur in colonial or filamentous groupings (figure 4.34*b,c*). Some species occur in packets surrounded by a gelatinous sheath (figure 4.34*c*). A specialized adaptation of cyanobacteria are extensive internal membranes called **thylakoids,** * which contain granules of chlorophyll *a* and other photosynthetic pigments (figure 4.34*a*). They also have gas inclusions, which permit them to float on the water surface and increase their light exposure, and cysts that convert gaseous nitrogen (N_2) into a form usable by plants. This group is sometimes called

the blue-green bacteria in reference to their content of **phycocyanin*** pigment that tints some members a shade of blue, although other members are colored yellow and orange. Some representatives glide or sway gently in the water from the action of filaments in the cell envelope that cause wavelike contractions.

Cyanobacteria are very widely distributed in nature. They grow profusely in fresh water and seawater and are thought to be responsible for periodic blooms that kill off fish by depleting the

---

*thylakoid (theye′-luh-koid) Gr. *thylakon,* and *eidos,* form.

*phycocyanin (feye-koh′seye′-an-in) Gr. *phykos,* seaweed, and *cyan,* blue. A bluish pigment unique to this group.

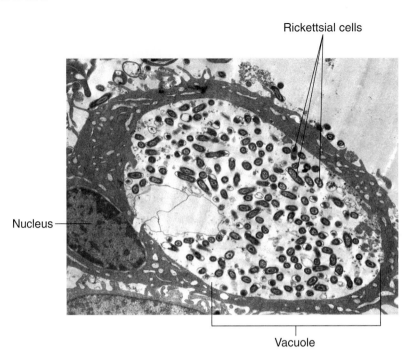

Rickettsial cells

Nucleus

Vacuole

FIGURE 4.32

Transmission electron micrograph of the rickettsia *Coxiella burnetii*, the cause of Q fever. Its mass growth inside a host cell has filled a vacuole and displaced the nucleus to one side.

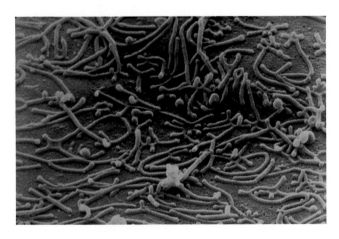

FIGURE 4.33

Scanning electron micrograph of *Mycoplasma pneumoniae* **(62,000×).** Cells like these that naturally lack a cell wall exhibit extreme pleomorphism.

available oxygen. Some members are so pollution-resistant that they serve as biological indicators of polluted water. Cyanobacteria inhabit and flourish in hot springs (see Spotlight on Microbiology 7.1) and have even exploited a niche in dry desert soils and rock surfaces. Some types of lichens are symbiotic associations between fungi and cyanobacteria that assist in the breakdown of rock and soil formation (see chapter 26).

Green and Purple Sulfur Bacteria The green and purple bacteria are also photosynthetic and contain pigments. They differ from the cyanobacteria in having a different type of chlorophyll called *bacteriochlorophyll* and by not giving off oxygen as a product of photosynthesis. They live in sulfur springs, freshwater lakes, and swamps that are deep enough for the anaerobic conditions they require yet where their pigment can still absorb wavelengths of light (figure 4.35*a*). These bacteria are named for their predominant colors, but they can also develop brown, pink, purple, blue, and orange coloration. They exist as single cells of many different shapes and frequently are motile. Both groups utilize sulfur compounds (H_2S, S) in their metabolism, and some can deposit intracellular granules of sulfur or sulfates (figure 4.35*b*).

Gliding, Fruiting Bacteria The **gliding bacteria** are a mixed collection of gram-negative bacteria that live in water and soil. The name is derived from the tendency of members to glide over moist surfaces. The gliding property evidently involves rotation of filaments or fibers just under the outer membrane of the cell wall. They do not have flagella. There are several morphological forms, including slender rods, long filaments, cocci, and some miniature, tree-shaped fruiting bodies. Probably the most intriguing and exceptional members of this group are the slime bacteria, or **myxobacteria** (figure 4.36). What sets the myxobacteria apart from other bacteria are the complexity and advancement of their life cycle. During this cycle, the vegetative cells swarm together and differentiate into a many-celled, colored structure called the fruiting body. The fruiting body is a survival structure that makes spores by a method very similar to that of certain fungi. These fruiting structures are often large enough to be seen with the unaided eye on tree bark and plant debris.

Appendaged Bacteria

The appendaged bacteria are quite varied in their structure and life cycles, but all of them produce an extended process of the cell wall in the form of a bud, a stalk, or a long thread (figure 4.37). The stalked bacteria live attached to the surface of objects in aquatic environments. One type can even grow in distilled water or tap water. The stalks evidently help them trap minute amounts of organic

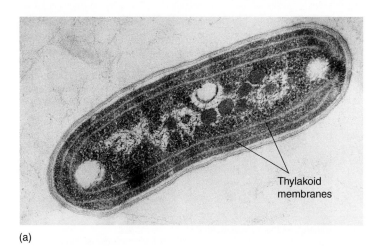

(a)

(b)

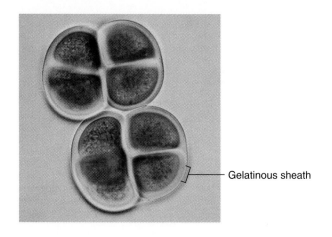

(c)

FIGURE 4.34

Structure and examples of cyanobacteria. (a) Electron micrograph of a cyanobacterial cell (80,000×) reveals folded stacks of membranes that contain the photosynthetic pigments and increase surface area for photosynthesis. **(b)** Two species of *Oscillatoria,* a gliding, filamentous form (100×). **(c)** *Chroococcus,* a colonial form surrounded by a gelatinous sheath (600×).

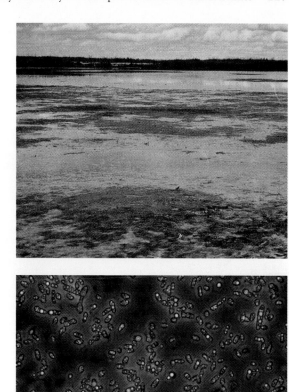

(a)

(b)

FIGURE 4.35

Behavior of purple sulfur bacteria. (a) Floating purple mats are huge masses of purple sulfur bacteria blooming in the Baltic Sea. Photosynthetic bacteria can have significant effects on the ecology of certain habitats. **(b)** Microscopic view of purple sulfur bacteria (*Chromatium vinosum*) carrying large, yellow sulfur granules internally.

materials present in the water. Budding bacteria reproduce entirely by budding; that is, they form a tiny bulb (bud) at the end of a thread. The bud then breaks off, enlarges, develops a flagellum, and swarms to another area to start its own cycle. These bacteria can also grow in very low-nutrient habitats.

ARCHAEA: THE OTHER PROCARYOTES

The discovery and characterization of novel procaryotic cells that have unusual anatomy, physiology, and genetics changed our views of microbial taxonomy and classification (see chapter 1). These single-celled, simple organisms, called **archaea*** are now considered a third cell type in a separate superkingdom (the Domain Archaea). We include them in this chapter because they are procaryotic in general structure and they do share many bacterial

---

*archaea (ark′-ee-uh) Gr. *arche,* beginning.

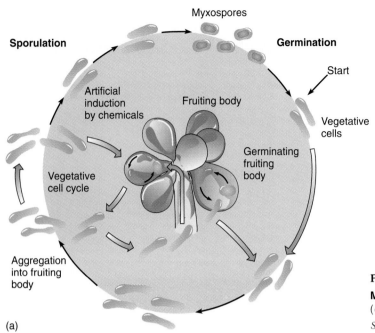

(a)

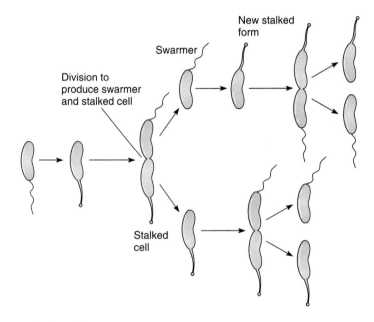

(b)

FIGURE 4.36

Myxobacterium. (a) The life cycle of a myxobacterium (*Chondromyces*). **(b)** A photograph of an actual mature fruiting body.

Source: From ASM News 63(August 1997):425. Photographer David Graham.

characteristics. Evidence is accumulating that they are actually more closely related to Domain Eukarya than to bacteria. For example, archaea and eucaryotes share a number of ribosomal RNA sequences that are not found in bacteria, and their protein synthesis and ribosomal subunit structures are similar. Table 4.7 outlines selected points of comparison of the three domains.

Among the ways that the archaea differ significantly from other cell types are that certain genetic sequences (CACACACCG) are found only in their rRNA (see figure 4.29), and that they have unique membrane lipids and cell wall construction. It is clear that the archaea are the most primitive of all life forms and are most closely related to the first cells that originated on the earth 4 billion years ago. The early earth is thought to have contained a hot, anaerobic "soup" with sulfuric gases and salts in abundance. The modern archaea still live in the remaining habitats on the earth that have these same ancient conditions—the most extreme habitats in nature. It is for this reason that they are often called extremophiles, meaning that they "love" extreme conditions in the environment.

Metabolically, the archaea exhibit nearly incredible adaptations to what would be deadly conditions for other organisms. These hardy microbes have adapted to multiple combinations of heat, salt, acid, pH, pressure, and atmosphere. Included in this group are methane producers, hyperthermophiles, extreme halophiles, and sulfur reducers.

Members of the group called *methanogens* can convert CO_2 and H_2 into methane gas (CH_4) through unusual and complex pathways. These archaea are common inhabitants of anaerobic swamp mud, the bottom sediments of lakes and oceans, and even the digestive systems of animals. The gas they produce collects in swamps and may become a source of fuel. Methane may also contribute to the "greenhouse effect," which maintains the earth's temperature and can contribute to global warming (see chapter 26).

Other types of archaea—the extreme halophiles—require salt to grow and may have such a high salt tolerance that they can multiply in sodium chloride solutions (36% NaCl) that would de-

FIGURE 4.37

Budding bacteria. The life cycle of *Caulobacter*, showing the development of the flagellated swarm cell and stalked phases.

stroy most cells (figure 4.38). They exist in the saltiest places on the earth—inland seas, salt lakes, salt mines, and salted fish. They are not particularly common in the ocean because the salt content is not high enough. Many of the "halobacteria" use a red pigment to synthesize ATP in the presence of light. These pigments are responsible for "red herrings," the color of the Red Sea, and the red color of salt ponds. An interesting type of halophile are the square bacteria—tiny packets or sheets (figure 4.38).

Archaea adapted to growth at very high temperatures are defined as hyperthermophilic (they love high temperatures). These or-

TABLE 4.7

Comparison of Three Cellular Domains

Characteristic	Bacteria	Archaea	Eukarya
Cell type	Procaryotic	Procaryotic	Eucaryotic
Chromosomes	Single, circular	Single, circular	Several, linear
Types of ribosomes	70S	70S but structure is more similar to 80S	80S
Contains unique ribosomal RNA signature sequences	+	+	+
Number of sequences shared with Eukarya	1	3	
Protein synthesis similar to Eukarya	−	+	
Presence of peptidoglycan in cell wall	+	−	−
Cell membrane lipids	Fatty acids with ester linkages	Long-chain, branched hydrocarbons with ether linkages	Fatty acids with ester linkages
Sterols in membrane	− (some exceptions)	−	+

(a)

(b)

FIGURE 4.38

Examples of halophiles. (a) A plate of milk salt agar growing the obligate halophile *Halobacterium salinarium*. This archaea has a bright pink pigment that increases upon exposure to light, and it requires at least 13% salt concentration to grow. **(b)** A packet of square bacteria. Not only are these unusual salt-loving cells square to rectangular in shape, but they are also extremely flat, giving the appearance of tiles.

ganisms flourish at temperatures between 80° and 105°C and cannot grow at 50°C. They live in volcanic waters and soils and submarine vents and are also often salt- and acid-tolerant as well. One member, *Thermoplasma,* lives in hot, acidic habitats in the waste piles around coal mines that regularly sustain a pH of 1 and a temperature of nearly 60°C. Researchers sampling sulfur vents in the deep ocean discovered thermophilic archaea flourishing at temperatures up to 250°C—150° above the temperature of boiling water! Not only were these archaea growing prolifically at this high temperature, but they were also living at 265 atmospheres of pressure. (On the earth's surface, pressure is about one atmosphere.) For additional discussion of the unusual adaptations of archaea, see chapter 7.

CHAPTER CHECKPOINTS

The rickettsias are a group of bacteria that are intracellular parasites, dependent on their eucaryote host for energy and nutrients. Most are pathogens that alternate between arthropods and mammalian hosts.

The chlamydias are also small, intracellular parasites that infect humans, mammals, and birds. They do not require arthropod vectors.

The mycoplasmas are bacteria that lack cell walls and therefore lack a definite cell shape. Although they are not obligate parasites, many require a eucaryote host and have specialized growth requirements when grown in culture.

Many bacteria are free-living, rather than parasitic. The photosynthetic bacteria, gliding bacteria, and appendaged bacteria encompass many subgroups that colonize specialized habitats, not other living organisms.

Archaea are another type of procaryotic cell that constitute the third domain of life. They exhibit unusual biochemistry and genetics that make them different from bacteria. Members are adapted to extreme habitats with high temperature, salt, pressure, or acid.

CHAPTER CAPSULE WITH KEY TERMS

THE PROCARYOTES: DOMAINS BACTERIA AND ARCHAEA

Names for Procaryotes: Bacteria, Archaea, Monera, Archaebacteria, Eubacteria, Schizomycetes (the last three are older)

I. **Overall Morphology**
   A. Procaryotic cells lack a nucleus, mitochondria, chloroplasts, or other membranous organelles.
   B. The outer covering is a **cell envelope** composed of **glycocalyx (slime layer, capsule), cell walls** (in most), and **cell membrane** with **mesosomes. Gram stain** differentiates two types of cells on the basis of wall structure: **Gram-positive** cells have a thick layer of **peptidoglycan; gram-negative** cells have an outer membrane and a thin layer of peptidoglycan. The diameter of most bacteria is $1-5$ μm.
   C. Special **appendages:** Bacterial **flagella** that provide motility are common; **fimbriae** are important in cell adhesion, and **pili** in genetic recombination.
   D. Genetic material is in a chromosome and **plasmids; ribosomes** synthesize proteins.
   E. Other special structures include **endospores** that are highly resistant survival cells and **inclusions** and **granules** for storing nutrients.
   F. Most bacteria are unicellular and form colonies on solid nutrients. Their shapes are distinct, including **cocci, bacilli, spirilla,** and **spirochetes.** Certain groups can assume long filaments; some show variations in shape, or **pleomorphism.**
   G. Various **arrangements** based upon mode of cell division are termed **diplo-, strepto-, staphylo-,** or **palisade.**

II. **Nutritional/Habitat Requirements**
   A. A large number of bacteria are heterotrophs that require an organic carbon source derived from dead organisms by absorption or from a live host (parasites). Several groups are capable of synthesizing nutrients using sunlight (photosynthetic).
   B. Bacteria can be aerobic, facultative, or anaerobic with respect to oxygen, and most exist between 10° and 40°C. Some species live in very cold habitats, even at subfreezing temperatures, and some at high temperatures (up to 250°C).

III. **Reproduction and Life Cycles**
   Bacteria generally have a single chromosome; reproduce primarily by asexual means, including budding and simple binary fission; and have no mitosis. Genetic exchange can occur through conjugation. The life cycle is complex in sporeformers, gliding bacteria, and appendaged bacteria.

IV. **Identification**
   Bacteria are identified by means of microscopic morphology (Gram stain, shape, special structures), macroscopic morphology, and biochemical, molecular, and genetic characteristics.

V. **Classification of Major Groups**
   One system divides the Kingdom Procaryotae into divisions based on the type of cell wall.

   A. *Gracilicutes:* The largest group; contains bacteria with gram-negative cell walls; includes several medically significant microbes, such as *Salmonella, Shigella,* and other intestinal pathogens, the **Rickettsias,** and the agents of gonorrhea and syphilis. The photosynthetic, gliding, sheathed, and appendaged bacteria are also in this group.
   B. *Firmicutes:* Gram-positive cells; examples of pathogens are in the genera *Streptococcus, Staphylococcus, Mycobacterium* (tuberculosis), and *Clostridium* (tetanus).
   C. *Tenericutes:* Cells without cell walls; primarily in the genus **Mycoplasma** (one type causes pneumonia).
   D. *Mendosicutes* (also **Archaea**)
   Unusual procaryotes different from bacteria in genetics and some structural features; they are adapted to extremes of habitat (salt, temperature, and nutrients). Halophiles live in salt concentrations of 10–30%; hyperthermophiles can live at temperatures of 60°–110°C or more. Methanogens live without oxygen and produce methane gas. No known medically important species.

VI. **Ecological Importance**
   A. Because procaryotes are highly adaptable as a group, they are found nearly everywhere: water, soil, air, dust, food, deep sea, plants, animals, swamps, hot springs, North and South Poles.
   B. They are important as decomposers of organic matter and play a role in the cycles of nitrogen, phosphorus, sulfur, and carbon.
   C. The cyanobacteria contribute oxygen to the atmosphere through photosynthesis.

VII. **Economic Importance**
   A. Bacteria are used in a number of industries, including food, drugs (production of antibiotics, vaccines, and other medicines), and biotechnology (manipulation of bacteria for making large quantities of hormones, enzymes, and other proteins).
   B. Bacteria are responsible for spoilage of food and vegetables. Many plant pathogens adversely affect the agricultural industry.

VIII. **Medical Importance**
   Bacteria are very common pathogens of humans. Approximately 200 species are known to cause disease in humans, and many normally inhabit the bodies of humans and other animals. Infections can be treated with antibiotics.

IX. **Unique Features**
   A. Bacteria and archaea are the smallest cells in existence.
   B. Bacteria and archaea have procaryotic structure, unique shapes and arrangements, and lack traditional sexual reproduction and mitosis.
   C. The cell wall of most contains peptidoglycan, and the cytoplasm of some contains mesosomes.
   D. Other unique features are a single chromosome and 70S ribosomes.

MULTIPLE-CHOICE QUESTIONS

1. Which of the following is not found in all bacterial cells?
   a. cell membrane   c. ribosomes
   b. a nucleoid   d. capsule

2. The major locomotor structures in bacteria are
   a. flagella   c. fimbriae
   b. pili   d. cilia

3. Pili are tubular shafts in ____ bacteria that serve as a means of ____.
   a. gram-positive, genetic exchange
   b. gram-positive, attachment
   c. gram-negative, genetic exchange
   d. gram-negative, protection

4. An example of a glycocalyx is
   a. a capsule   c. outer membrane
   b. pili   d. a cell wall

5. Which of the following is a primary bacterial cell wall function?
   a. transport   c. support
   b. motility   d. adhesion

6. Which of the following is present in both gram-positive and gram-negative cell walls?
   a. an outer membrane
   b. peptidoglycan
   c. teichoic acid
   d. lipopolysaccharides

7. Mesosomes are internal extensions of the
   a. cell wall   c. cell membrane
   b. chromosome   d. capsule

8. Metachromatic granules are concentrated crystals of ____ that are found in ____.
   a. fat, *Mycobacterium*
   b. dipicolinic acid, *Bacillus*
   c. sulfur, *Thiobacillus*
   d. PO$_4$, *Corynebacterium*

9. Bacterial endospores function in
   a. reproduction   c. protein synthesis
   b. survival   d. storage

10. A bacterial arrangement in packets of eight cells is described as a ____.
    a. micrococcus   c. tetrad
    b. diplococcus   d. sarcina

11. The major difference between a spirochete and a spirillum is
    a. presence of flagella
    b. a cell with coils
    c. the nature of motility
    d. size

12. Genetic analysis of bacteria would include
    a. fermentation testing
    b. ability to digest complex nutrients
    c. presence of oxidase
    d. G + C content

13. Which division of bacteria has a gram-positive cell wall?
    a. Gracilicutes   c. Firmicutes
    b. Archaea   d. Tenericutes

14. To which division of bacteria do cyanobacteria belong?
    a. Tenericutes   c. Firmicutes
    b. Gracilicutes   d. Mendosicutes

15. Archaea differ from bacteria on the basis of
    a. structure of envelope
    b. size
    c. the archaea having a nucleus
    d. type of locomotor structures

CONCEPT QUESTIONS

1. a. Name several general characteristics that could be used to define the procaryotes.
   b. Do any other microbial groups besides bacteria have procaryotic cells?
   c. What does it mean to say that bacteria are ubiquitous? In what habitats are they found? Give some general means by which bacteria derive nutrients.

2. a. Describe the structure of a flagellum and how it operates. What are the four main types of flagellar arrangement?
   b. How does the flagellum dictate the behavior of a motile bacterium? Differentiate between flagella and periplasmic flagella.
   c. List some direct and indirect ways that one can determine bacterial motility.

3. a. List the components of the cell envelope.
   b. Explain the position of the glycocalyx.
   c. What are the functions of slime layers and capsules?
   d. How is the presence of a slime layer evident even at the level of a colony?

4. a. Differentiate between pili and fimbriae.
   b. How do their structure differ?
   c. How do their functions differ?

5. a. Compare the cell envelopes of gram-positive and gram-negative bacteria.
   b. What function does peptidoglycan serve?
   c. To which part of the cell envelope does it belong?
   d. Give a simple description of its structure.

e. What happens to a cell that has its peptidoglycan disrupted or removed?
   f. What functions does the LPS layer serve?

6. a. What is the Gram stain?
   b. What is there in the structure of bacteria that causes some to stain purple and others to stain red?
   c. How does the precise structure of the cell walls differ in gram-positive and gram-negative bacteria?
   d. What other properties besides staining are different in gram-positive and gram-negative bacteria?
   e. What is the periplasmic space, and how does it function?
   f. What characteristics does the outer membrane confer on gram-negative bacteria?

7. a. List five functions that the cell membrane performs in bacteria.
   b. What are mesosomes and some of their possible functions?

8. a. Compare the composition of the bacterial chromosome (nucleoid) and plasmids.
   b. What are the functions of each?

9. a. What is unique about the structure of bacterial ribosomes?
   b. How do they function?
   c. Where are they located?

10. a. Compare and contrast the structure and function of inclusions and granules.
    b. What are metachromatic granules, and what do they contain?

11. a. Describe the vegetative stage of a bacterial cell.
    b. Describe the structure of an endospore, and explain its function.
    c. Describe the endospore-forming cycle.
    d. Explain why an endospore is not considered a reproductive body.
    e. Why are endospores so difficult to destroy?
12. a. Draw the three bacterial shapes.
    b. How are spirochetes and spirilla different?
    c. What is a vibrio? A coccobacillus?
    d. What is pleomorphism?
    e. What is the difference between the use of the term bacillus and the name *Bacillus*? *Staphylococcus* and staphylococcus?
13. a. Rank the size ranges in bacteria according to shape.
    b. Rank the bacteria in relationship to viruses and eucaryotic cell size.
14. a. What characteristics are used to classify bacteria?
    b. What are the most useful characteristics for categorizing bacteria into families?
    c. In what ways is ribosomal RNA an important method for differentiating bacteria and groups of organisms?

15. a. How is the species level in bacteria defined?
    b. Name at least three ways bacteria are grouped below the species level.
    c. In what ways are they important?
16. a. Describe at least two circumstances that give rise to L forms.
    b. How do L forms survive?
    c. In what ways are they important?
17. Name several ways in which bacteria are medically and ecologically important.
18. a. Explain the characteristics of archaea that indicate it is a unique domain of living things that is neither a bacterium nor a eucaryote.
    b. What leads microbiologists to believe the archaea are more closely related to eucaryotes than to bacteria?
    c. What is meant by the term *extremophile?* Describe some archaeal adaptations to extreme habitats.

CRITICAL-THINKING QUESTIONS

1. What would happen if one stained a gram-positive cell with safranin? A gram-negative cell with crystal violet? What would happen to the two types if the mordant were omitted?
2. What is required to kill endospores? How do you suppose archaeologists were able to date some spores as being 3,000 years old?
3. Using clay, demonstrate how cocci can divide in several planes and show the outcome of this division. Show how the arrangements of bacilli occur, including palisades.
4. Using a corkscrew and a spring to compare the flexibility and locomotion of spirilla and spirochetes, explain which group is represented by each of them.
5. Under the microscope, you see a rod-shaped cell that is swimming rapidly forward.
    a. What do you automatically know about that bacterium's structure?
    b. How would a bacterium use its flagellum for phototaxis?
    c. Can you think of another function of flagella besides locomotion?
6. Can you determine which probes in figure 4.28 successfully hybridized and which did not? Account for the reasons that some probes did not attach. What are the implications of the probe's attaching to the test DNA?
7. a. Name an unusual type of bacterium that lives in extreme habitats.
    b. What adaptations enable it to survive there?
8. a. Name a bacterium that has no cell walls.
    b. How is it protected from osmotic destruction?
9. a. Name a bacterium that is aerobic, gram positive, and spore-forming.
    b. What habitat would you expect this species to occupy?
10. Name a bacterium that is pleomorphic and has a palisades arrangement and metachromatic granules.

11. a. Name an acid-fast bacterium.
    b. What characteristics make this bacterium different from other gram-positive bacteria?
12. a. Name two main groups of obligate intracellular parasitic bacteria.
    b. Why can't these groups live independently?
13. Name a bacterium that lives in extremes of heat and salt.
14. Name a coccus that is gram positive and in chains.
15. Name a gram-negative diplococcus.
16. a. Name a bacterium that contains sulfur granules.
    b. What is the advantage in storing these granules?
17. a. Name a bacterium that uses chlorophyll to photosynthesize.
    b. Describe the two major groups of photosynthetic bacteria.
    c. How are they similar?
    d. How are they different?
18. a. What are some possible adaptations that the giant bacterium *Thiomargarita* has had to make because of its large size?
    b. If a regular bacterium were the size of an elephant, estimate the size of a nanobe at that scale.
19. Propose a hypothesis to explain how bacteria and archaea could have, together, given rise to eucaryotes.
20. Explain or illustrate exactly what will happen to the cell wall if the synthesis of the interbridge is blocked by penicillin. What if the glycan is hydrolyzed by lysozyme?
21. Ask your lab instructor to help you make a biofilm and examine it under the microscope. One possible technique is to suspend a glass slide in an aquarium for a few weeks, then carefully air-dry, fix, and stain it. Observe the diversity of cell types.

INTERNET SEARCH TOPICS

1. Go to a search engine and type in "Martian Microbes." Look for papers and information that support or reject the idea that fossil structures discovered in an ancient meteor from Mars could be bacteria. What are some of the reasons that microbiologists are skeptical of this possibility?

2. Search the Internet for information on nanobacteria. Give convincing reasons why these are or are not real organisms.

Eucaryotic Cells and Microorganisms

T he eucaryotic cell is a complex, compartmentalized unit that differs from the procaryotic cell by containing a nucleus and several other specialized structures called organelles. Although exact cell structures differ somewhat among the several groups of eucaryotic organisms, the eucaryotic cell is the typical cell of certain microbial groups (fungi, algae, protozoa, and helminth worms) as well as all animals and plants. In this chapter, we examine the overall structure and function of eucaryotic cells in preparation for later chapters that deal with related microbiological concepts, including metabolism, genetics, nutrition, drug therapy, immunology, and disease. Because of the tremendous variety of eucaryotic microorganisms and their practical importance in medicine, industry, and agriculture, this chapter will also cover the major characteristics of each group of eucaryotic microorganisms.

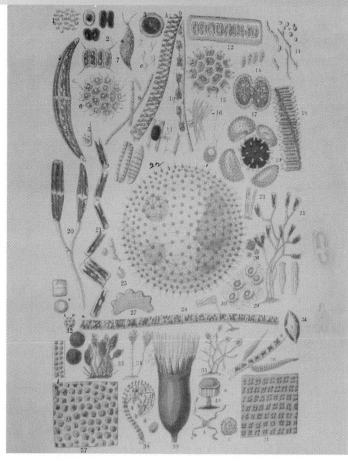

These drawings are from a 100-year-old book "Common Objects of the Microscope" that displays hundreds of well-known algae, molds, and yeasts found in water and soil samples.

Chapter Overview

- Eucaryotic cells are large complex cells divided into separate compartments by membrane-bound components called organelles.
- Major organelles—the nucleus, mitochondria, chloroplasts, endoplasmic reticulum, Golgi apparatus, and locomotor appendages— each serve an essential function to the cell, such as heredity, production of energy, synthesis, transport, and movement.
- Eucaryotic cells are found in fungi, protozoa, algae, plants, and animals, and they exhibit single-celled, colonial, and multicellular body plans.
- Fungi and protists (algae and protozoa) are the major kingdoms that contain eucaryotic microorganisms.
- Fungi are eucaryotes that feed on organic substrates, have cell walls, reproduce asexually and sexually by spores, and exist in macroscopic or microscopic forms.
- Most fungi are free-living decomposers that are beneficial to biological communities; some may cause infections in animals and plants.
- Microscopic fungi include yeasts with spherical budding cells and molds with elongate filamentous hyphae in mycelia.
- Algae are aquatic photosynthetic protists with rigid cell walls and chloroplasts containing chlorophyll and other pigments.
- Algae belong to several groups based on their type of pigments, cell wall, stored food materials, and body plan.
- Protozoa are protists that feed by engulfing other cells, lack a cell wall, usually have some type of locomotor organelle, and may form dormant cysts.
- Subgroups of protozoa vary in organelle of motility (flagella, cilia, pseudopods, nonmotile).
- Most protozoa are free-living aquatic cells that feed on bacteria and algae, and a few are animal parasites.
- The parasitic helminths are flatworms and roundworms that infect humans and other animals.
- Their existence imposes reduction in the development of organ systems in favor of the organs of reproduction.

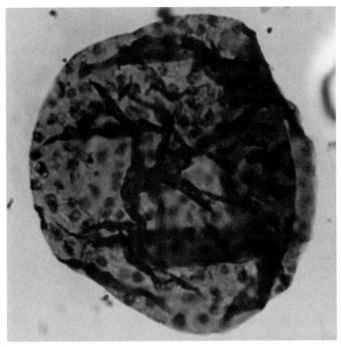

(a)

FIGURE 5.1

Ancient eucaryotic protists caught up in fossilized rocks. Both forms are called acritarchs. **(a)** An ornamented algalike cell found in Siberian shale deposits and dated from 850 million to 950 million years old. **(b)** A large, disclike cell bearing a crown of spines is from Chinese rock dated 590 million to 610 million years ago.

The Nature of Eucaryotes

Evidence from paleontology indicates that the first eucaryotic cells appeared on the earth approximately 2 billion years ago. Some fossilized cells that look remarkably like algae or protozoa appear in shale sediments from China and Australia that date from 800 million to 900 million years ago (figure 5.1). Biologists have discovered profound evidence to suggest that the eucaryotic cell evolved

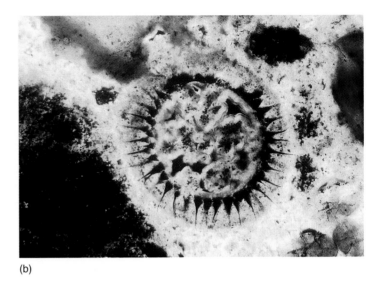

(b)

from procaryotic organisms by a process of intracellular *symbiosis** (Spotlight on Microbiology 5.1). It now seems clear that some of the **organelles** that distinguish eucaryotic cells originated from procaryotic cells that became trapped inside them. The structure of these first eucaryotic cells was so versatile that eucaryotic microorganisms soon spread out into available habitats and adopted greatly diverse styles of living.

The first primitive eucaryotes were probably single-celled and independent, but, over time, some forms began to aggregate, forming colonies and filaments. With further evolution, some of the cells within colonies became *specialized*, or adapted to perform a particular function advantageous to the whole colony, such as locomotion, feeding, or reproduction. Complex multicellular organisms evolved as individual cells in the organism lost the ability to survive apart from the intact colony. Although a multicellular organism is composed of many cells, it is more than just a disorganized assemblage of cells like a colony. Rather, it is composed of distinct groups of cells that cannot exist independently of the

---

*symbiosis (sim-beye-oh′-sis) Gr. *sym*, together, and *bios*, to live.

Structure Flowchart

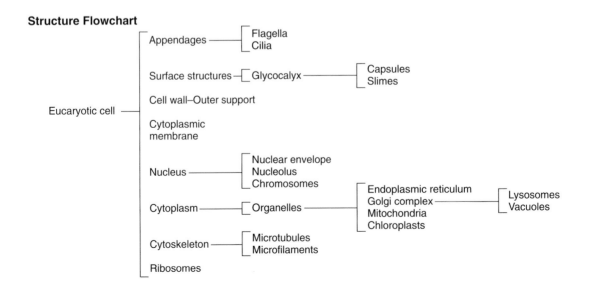

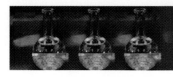

SPOTLIGHT ON MICROBIOLOGY 5.1
The Extraordinary Emergence of Eucaryotic Cells

For years, biologists have grappled with the problem of how a cell as complex as the eucaryotic cell originated. One of the most fascinating explanations is that of **endosymbiosis**,* which proposes that eucaryotic cells arose when a much larger procaryotic cell engulfed smaller bacterial cells that began to live and reproduce there rather than being destroyed. As the smaller cells took up permanent residence, they came to perform specialized functions for the larger cell, such as food synthesis and oxygen utilization, that enhanced the cell's versatility and survival. Over time, when the cells evolved into a single functioning entity, the relationship became obligatory. Although the theory of endosymbiosis has been greeted with some controversy, this phenomenon has been demonstrated in the laboratory with amebas infected with bacteria that gradually became dependent upon the bacteria for survival.

The biologist most responsible for validation of the theory of endosymbiosis is Dr. Lynn Margulis. Using modern molecular techniques, she has accumulated convincing evidence of the relationships between the organelles of modern eucaryotic cells and the structure of bacteria. In many ways, the mitochondrion of eucaryotic cells is something like a tiny cell within a cell. It is capable of independent division, contains a circular chromosome that has bacterial DNA sequences, and has ribosomes that are clearly procaryotic. Mitochondria also have bacterial membranes and can be inhibited by drugs that affect only bacteria. This link is so well established that recent phylogenetic trees of bacteria show mitochondria and chloroplasts as two of the bacterial branches.

Chloroplasts likely arose when endosymbiotic cyanobacteria provided their host cells with a built-in feeding mechanism. Evidence is seen in a modern flagellated protist that harbors specialized chloroplasts with cyanobacterial chlorophyll and thylakoids. Margulis also has convincing evidence that eucaryotic cilia and flagella are the consequence of endosymbiosis between spiral bacteria and the cell membrane of early eucaryotic cells.

It is tempting to envision the development of the eucaryotic cell through a fusion of archaeal and bacterial cells. The archaea would have served as a source of ribosomes and certain aspects of protein synthesis, and bacteria would have given rise to mitochondria and chloroplasts. The first eukarya probably resembled modern protists such as *Giardia* or *Plagiopyla*. Researchers are pursuing further genetic analysis of these microbes and their organelles in order to add further support to the theory.

*endosymbiosis (en′-doh-sym-bee-oh′-sis) Gr. *endo*, within; *syn*, together, and *bios*, life. An intimate association between two cells, in which one is living inside another.

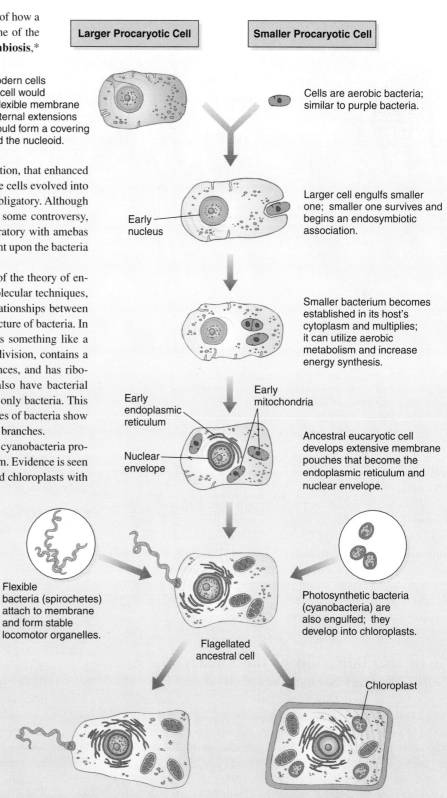

Larger Procaryotic Cell   **Smaller Procaryotic Cell**

No modern cells exist; cell would have flexible membrane and internal extensions that could form a covering around the nucleoid.

Cells are aerobic bacteria; similar to purple bacteria.

Early nucleus

Larger cell engulfs smaller one; smaller one survives and begins an endosymbiotic association.

Smaller bacterium becomes established in its host's cytoplasm and multiplies; it can utilize aerobic metabolism and increase energy synthesis.

Early endoplasmic reticulum

Early mitochondria

Nuclear envelope

Ancestral eucaryotic cell develops extensive membrane pouches that become the endoplasmic reticulum and nuclear envelope.

Flexible bacteria (spirochetes) attach to membrane and form stable locomotor organelles.

Photosynthetic bacteria (cyanobacteria) are also engulfed; they develop into chloroplasts.

Flagellated ancestral cell

Chloroplast

Protozoa, fungi, animals

Algae, higher plants

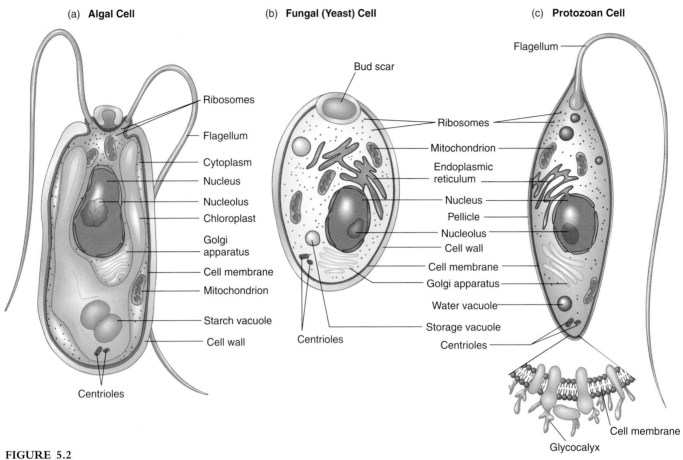

(a) Algal Cell

Ribosomes
Flagellum
Cytoplasm
Nucleus
Nucleolus
Chloroplast
Golgi apparatus
Cell membrane
Mitochondrion
Starch vacuole
Cell wall
Centrioles

(b) Fungal (Yeast) Cell

Bud scar
Ribosomes
Mitochondrion
Endoplasmic reticulum
Nucleus
Nucleolus
Cell wall
Cell membrane
Golgi apparatus
Storage vacuole
Centrioles

(c) Protozoan Cell

Flagellum
Ribosomes
Mitochondrion
Endoplasmic reticulum
Nucleus
Pellicle
Nucleolus
Cell membrane
Golgi apparatus
Water vacuole
Centrioles
Cell membrane
Glycocalyx

FIGURE 5.2

The structure of three representative eucaryotic cells. (a) *Chlamydomonas*, a unicellular motile alga. **(b)** A yeast cell (fungus). **(c)** *Peranema*, a flagellated protozoan.

rest of the body. The cell groupings of multicellular organisms that have a specific function are termed *tissues*, and groups of tissues make up *organs*.

Looking at modern eucaryotic organisms, we find examples of many levels of cellular organization. All protozoa, as well as numerous algae and fungi, live a unicellular or colonial existence. Truly multicellular organisms are found only among plants and animals and some of the fungi (mushrooms) and algae (seaweeds). Only certain eucaryotes are traditionally studied by microbiologists—primarily the protozoa, the microscopic algae and fungi, and animal parasites.

Form and Function of the Eucaryotic Cell: External Structures

The cells of eucaryotic organisms are so varied that no one member can serve as a typical example. Figure 5.2 presents the generalized structure of typical algal, fungal, and protozoan cells. The outline at the bottom of page 124 shows the organization of a eucaryotic cell. (Only motile algae contain all categories of eucaryotic organelles.)

In general, eucaryotic microbial cells have a cytoplasmic membrane, nucleus, mitochondria, endoplasmic reticulum, Golgi apparatus, vacuoles, cytoskeleton, and glycocalyx. A cell wall, locomotor appendages, and chloroplasts, are found only in some

groups. In the following sections, we cover the microscopic structure and functions of the eucaryotic cell. As with the procaryotes, we begin on the outside and proceed inward through the cell.

LOCOMOTOR APPENDAGES: CILIA AND FLAGELLA

Motility allows a microorganism to locate life-sustaining nutrients and to migrate toward positive stimuli such as sunlight; it also permits avoidance of harmful substances and stimuli. Locomotion by means of flagella or cilia is a common property of protozoa, many algae, and a few fungal and animal cells.

Although they share the same name, eucaryotic **flagella** are much different from those of procaryotes. The eucaryotic flagellum is thicker (by a factor of 10), structurally more complex, and covered by an extension of the cell membrane. A single flagellum is a long, sheathed cylinder containing regularly spaced hollow tubules—**microtubules**—that extend along its entire length (figure 5.3*a*). A cross section reveals nine pairs of closely attached microtubules surrounding a single central pair. This scheme, called the 9 + 2 arrangement, is a universal pattern of flagella and cilia. During locomotion, the adjacent microtubules slide past each other, whipping the flagellum back and forth. Although details of this process are too complex to discuss here, it involves expenditure of energy and a coordinating mechanism in the cell membrane. Flagella can move the cell by pushing it forward like a fishtail or by

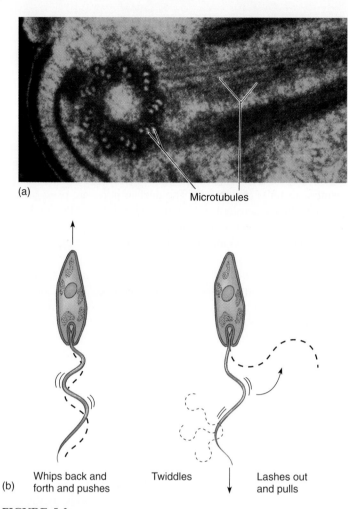

(a)

Microtubules

Whips back and forth and pushes | Twiddles | Lashes out and pulls

(b)

FIGURE 5.3

Longitudinal section through a flagellum, showing microtubules.
(a) On the left is a cross section (circular area) that reveals the typical 9 + 2 arrangement found in both flagella and cilia. **(b)** Locomotor patterns seen in flagellates.

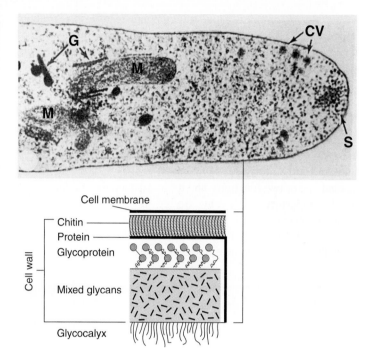

FIGURE 5.4

Glycocalyx structure. Cross section through the tip of a fungal cell to show the general structure of the cell wall and other features. Top: Photomicrograph. (S, growing tip; CV, coated vesicles; G, Golgi apparatus; M, mitochondrion.) Bottom: The cell wall is a thick, rigid structure composed of complex layers of polysaccharides and proteins.

pulling it by a lashing or twirling motion (figure 5.3*b*). The placement and number of flagella can be useful in identifying flagellated protozoa and certain algae.

Cilia* are very similar in overall architecture to flagella, but they are shorter and more numerous (some cells have several thousand). They are found only on a single group of protozoa and certain animal cells. In the ciliated protozoa, the cilia occur in rows over the cell surface, where they beat back and forth in regular oar-like strokes (see figure 5.30). Such protozoa are among the fastest of all motile cells. The fastest ciliated protozoon can swim up to 2,500 µm/s—a meter and a half per minute! On some cells, cilia also function as feeding and filtering structures.

SURFACE STRUCTURES: THE GLYCOCALYX

Most eucaryotic cells have a **glycocalyx,** an outermost boundary that comes into direct contact with the environment (see figure 5.2). This structure is usually composed of polysaccharides and appears as a network of fibers, a slime layer, or a capsule much like the glycocalyx of procaryotes. Because of its positioning, the glycocalyx

contributes to protection, adherence of cells to surfaces, and reception of signals from other cells and from the environment. We will explore the function of receptors that are part of the glycocalyx in chapters 14 and 15 (host defenses and immune interactions). The nature of the layer beneath the glycocalyx varies among the several eucaryotic groups. Fungi and most algae have a thick, rigid cell wall, whereas protozoa, a few algae, and all animal cells lack a cell wall and have only a cell membrane.

THE CELL WALL

The cell walls of the fungi and algae are rigid and provide structural support and shape, but they are different in chemical composition from procaryotic cell walls. Fungal cell walls have a thick, inner layer of polysaccharide fibers composed of chitin or cellulose and a thin outer layer of mixed glycans (figure 5.4). The cell walls of algae are quite varied in chemical composition. Substances commonly found among various algal groups are cellulose, pectin,[1] mannans,[2] and minerals such as silicon dioxide and calcium carbonate.

THE CELL MEMBRANE

The cell membrane of eucaryotic cells is a typical bilayer of lipids in which protein molecules are embedded. In addition to phospholipids, eucaryotic membranes also contain *sterols* of various kinds. Sterols are different from phospholipids in both structure and behavior (see figure 2.19). Their relative rigidity confers stability on

---

*cilia (sil′-ee-uh) sing. cilium; L. *siliam,* eyelid.

---

1. A polysaccharide composed of galacturonic acid subunits.

2. A polymer of the sugar known as mannose.

eucaryotic membranes. This strengthening feature is extremely important in cells that lack a cell wall. Cytoplasmic membranes of eucaryotes are functionally similar to those of procaryotes, serving as selectively permeable barriers in transport (see chapter 7). Unlike procaryotes, eucaryotic cells also contain a number of individual membrane-bound organelles that are extensive enough to account for 60% to 80% of their volume.

Survey of Major Organelles

Eucaryotic cells accomplish their cellular work with unique intracellular membrane-wrapped organelles. Each type of organelle carries out one or two functions, and it can partition the cell into many small compartments, thereby serving to organize and separate metabolic events. Functions such as synthesis and secretion often proceed in an assembly-line fashion, with each organelle in the series affecting the outcome of the final product or process.

CHAPTER CHECKPOINTS

Eucaryotes are cells with organelles and a nucleus compartmentalized by membranes. They might have originated from procaryote ancestors about 2 billion years ago. Eucaryotic cell structure enabled eucaryotes to diversify from single cells into a huge variety of complex multicellular forms.

The cell structures common to most eucaryotes are the cell membrane, a membrane-enclosed nucleus with nucleolus inside, vacuoles, mitochondria, endoplasmic reticulum, Golgi apparatus, and a cytoskeleton. Cell walls, chloroplasts, and locomotor organs are present in some eucaryote groups.

Microscopic eucaryotes use locomotor organs such as flagella or cilia for moving themselves or their food.

The glycocalyx is the outermost boundary of most eucaryotic cells. Its functions are protection, adherence, and reception of chemical signals from the environment or from other organisms. The glycocalyx is supported by either a cell wall or a cell membrane.

The cytoplasmic membrane of eucaryotes is similar in function to that of procaryotes, but it differs in composition, possessing sterols as additional stabilizing agents.

Eucaryotic cells have membrane-bound organelles, which are never present in procaryotes. Such compartmentalization allows a wide variety of cellular functions to occur in specialized areas of the cell.

Form and Function of the Eucaryotic Cell: Internal Structures

THE NUCLEUS: THE CONTROL CENTER

The **nucleus** is a compact sphere that is the most prominent organelle of eucaryotic cells. It is separated from the cell cytoplasm by an external boundary called a **nuclear envelope.** The envelope has a unique architecture. It is composed of two parallel membranes separated by a narrow space, and it is perforated with small, regularly spaced openings, or pores, formed at sites where the two membranes unite (figure 5.5). The nuclear pores are passageways through which macromolecules migrate from the nucleus to the cytoplasm and vice versa. The nucleus contains a matrix called the **nucleoplasm** and a granular mass, the **nucleolus,*** that can stain more intensely than the immediate surroundings because of its RNA content. The nucleolus is the site for ribosomal RNA synthesis and a collection area for ribosomal subunits. The subunits are transported through the nuclear pores into the cytoplasm for final assembly into ribosomes.

A prominent feature of the nucleoplasm in stained preparations is a network of dark fibers known as **chromatin*** because of its attraction for dyes. Analysis has shown that chromatin actually comprises the eucaryotic **chromosomes,** large units of genetic information in the cell. The chromosomes in the nucleus of most cells are not readily visible because they are long, linear DNA molecules bound in varying degrees to *histone*[3] proteins, and they are far too fine to be resolved as distinct structures without extremely high magnification. During **mitosis,*** however, when the duplicated chromosomes are separated equally into daughter cells, the chromosomes themselves finally become readily visible as discrete bodies (figure 5.6). This appearance arises when the DNA becomes highly condensed by forming coils and supercoils around the histones to prevent the chromosomes from tangling as they are separated into new cells.

---

3. A simple type of protein containing a high proportion of lysine and arginine.

*nucleolus (noo-klee´-oh-lus) pl. nucleoli; L. "a little nucleus."

*chromatin (kroh´-muh-tin) Gr. *chromos*, color.

*mitosis (my-toh´-sis) Gr. *mitos*, thread, and *osis*, a condition of.

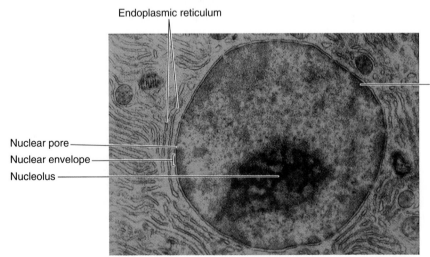

FIGURE 5.5

The nucleus. Electron micrograph section of an interphase nucleus, showing its most prominent features.

Endoplasmic reticulum

Chromatin

Nuclear pore
Nuclear envelope
Nucleolus

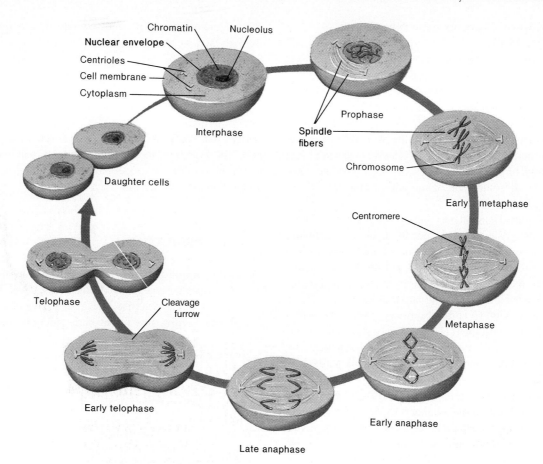

(a)

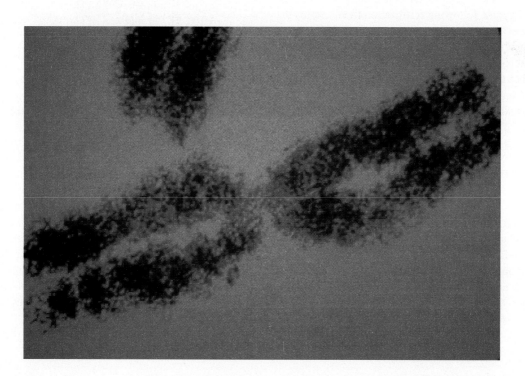

(b)

FIGURE 5.6

Changes in the cell and nucleus that accompany mitosis in a eucaryotic cell such as a yeast. (a) Before mitosis (at interphase), chromosomes are visible only as chromatin. As mitosis proceeds (early prophase), chromosomes take on a fine, threadlike appearance as they condense, and the nuclear membrane and nucleolus are temporarily disrupted. **(b)** By metaphase, the chromosomes are fully visible as X-shaped structures. The shape is due to duplicated chromosomes attached at a central point, the centromere. Spindle fibers attach to these and facilitate the separation of individual chromosomes during metaphase. Later phases serve in the completion of chromosomal separation and division of the cell proper into daughter cells.

The Relationship of Chromosome Number to Reproductive Mode

To ensure its continued existence, a eucaryotic cell maintains a specified number of chromosomes that is typical for its species. The set of chromosomes exist in the **haploid*** (single, unpaired) state or in the **diploid*** (matched pair) state. Most fungi, many algae, and some protozoa are examples of organisms whose cells occur in the haploid state during most of the life cycle. For example, the chromosome number of one type of ameba is 25, and one species of the fungus *Penicillium* possesses 5 chromosomes. Regardless of the chromosome state of the adult organism, all gametes (sex cells) are haploid.

The cells of animals and plants, and of some protozoa, fungi, and algae, spend the greater part of their life cycle in the diploid state. This state comes about when a female haploid gamete (egg) fuses with a male haploid gamete (sperm) and produces a double (diploid) set of chromosomes in the offspring during sexual reproduction. Examples are humans, with 46 chromosomes (23 pairs), a parasitic blood fluke, with 16 chromosomes (8 pairs), and a redwood tree, with 22 chromosomes (11 pairs). During normal cell division (asexual reproduction) of all eucaryotic organisms, the chromosome number, whether haploid or diploid, is maintained by mitosis (figures 5.6 and 5.7).

When it comes to sexual reproduction in diploid organisms, however, the diploid chromosome number must be reduced to haploid, because gametes must be haploid for proper fertilization. This reduction is accomplished through reduction division, or **meiosis*** (figure 5.7*b*). During this process, diploid cells in the sex organs undergo two sets of divisions that result in haploid sex cells. Formation of a diploid zygote following fertilization then restores the chromosome number characteristic of that organism. Another example of the function of meiosis occurs in the sexual reproductive cycles of haploid fungi. Normal mitosis alone can produce their gametes, but after these gametes unite, they form a diploid zygote. In order to restore the fungus to the haploid state, the diploid nucleus undergoes meiosis and produces haploid offspring (see figure 5.7*a*, step 2, and figure 5.20).

ENDOPLASMIC RETICULUM: A PASSAGEWAY IN THE CELL

The **endoplasmic reticulum*** **(ER)** is a microscopic series of tunnels used in transport and storage. In sections, it appears as parallel, flat pouches bounded by membranes. Two kinds of endoplasmic reticulum are the **rough endoplasmic reticulum (RER)** (figure 5.8) and the **smooth endoplasmic reticulum (SER)**. Electron micrographs show that the RER originates from the outer membrane of the nuclear envelope and extends in a continuous network through the cytoplasm, even out to the cell membrane. This architecture permits the spaces in the RER, or **cisternae,** to transport materials from the nucleus to the cytoplasm and ultimately to the cell's exterior. The RER appears rough because of large numbers of ribosomes partly attached to its membrane to act as secretory factories. Proteins synthesized on the ribosomes are shunted into the cavity of the reticulum and held there for later packaging and transport. The SER is a closed tubular network without ribosomes that functions in nutrient processing and in synthesis and storage of nonprotein macromolecules such as lipids.

GOLGI APPARATUS: A PACKAGING MACHINE

The **Golgi*** **apparatus,** also called the Golgi **complex** or body, is a discrete organelle consisting of a stack of several flattened, disc-shaped sacs called cisternae. These sacs have outer limiting membranes and cavities like those of the endoplasmic reticulum, but they do not form a continuous network (figure 5.9). This organelle is always closely associated with the endoplasmic reticulum both in its location and function. At a site where it meets the Golgi apparatus, the endoplasmic reticulum buds off tiny membrane-bound packets of protein called *transitional vesicles** that are picked up by the forming face of the Golgi apparatus. Once in the complex itself, the proteins are often modified by the addition of polysaccharides and lipids. The final action of this apparatus is to pinch off finished *condensing vesicles* that will be conveyed to organelles such as lysosomes or transported outside the cell as secretory vesicles (figure 5.10). The production of antibodies and receptors in human white blood cells involves this packaging and transport system (see chapter 15).

Nucleus, Endoplasmic Reticulum, and Golgi Apparatus: Nature's Assembly Line

As the keeper of the eucaryotic genetic code, the nucleus ultimately governs and regulates all cell activities. But, because the nucleus remains fixed in a specific cellular site, it must direct these activities through a structural and chemical network (figure 5.10). This network includes ribosomes, which originate in the nucleus, and the rough endoplasmic reticulum, which is continuously connected with the nuclear envelope. Initially, a segment of the genetic code of DNA is copied into RNA and passed out through the nuclear pores directly to the ribosomes on the endoplasmic reticulum. Here, specific proteins are synthesized from the RNA code and deposited in the lumen (space) of the endoplasmic reticulum. After being

---

***haploid** (hap′-loyd) Gr. *haplos*, simple, and *eidos*, form.

***diploid** (dip′-loyd) Gr. *di*, two, and *eidos*, form.

***meiosis** (my-oh′-sis) Gr. *meiosis*, diminution. A process that permits paired chromosome sets to be reduced to single sets.

***endoplasmic** (en″-doh-plas′-mik) Gr. *endo*, within, and *plasm*, something formed; reticulum (rih-tik′-yoo-lum) L. *rete*, net or network.

***Golgi** (gol′-jee) Named for C. Golgi, an Italian histologist who first described the apparatus in 1898.

***vesicle** (ves′-ik-l) L. *vesios*, bladder. A small sac containing fluid.

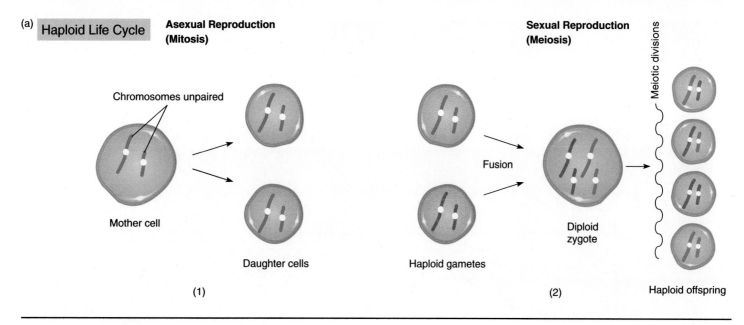

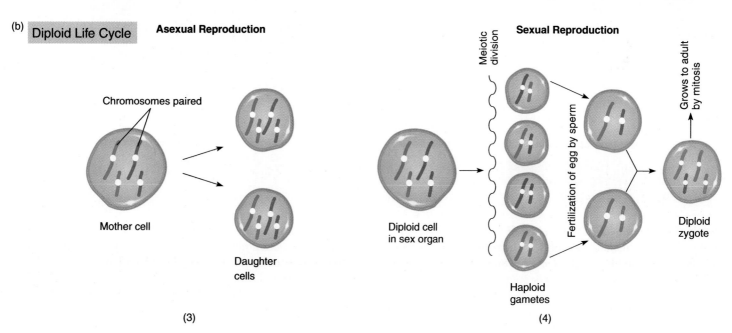

FIGURE 5.7

Schematic of haploid versus diploid life cycles. Shown are the roles of mitosis and meiosis in reproduction and maintaining an organism's chromosome number. **(a)** Cells with haploid life cycle (algae, fungi, some protozoa). Chromosomes are unpaired. (*1*) Asexual reproduction is accomplished through mitosis, which maintains the chromosome number in the daughter cells. (*2*) Sexual reproduction involves the fusion of specialized sex cells (already haploid), forming a diploid zygote. The haploid state in the offspring is restored by meiosis of the zygote. Note that reassortment of chromosomes can produce haploid offspring unlike either original gamete. **(b)** Cells with a diploid life cycle (animals, plants, some fungi, algae, and protozoa). Chromosomes occur in pairs. (*3*) Asexual reproduction through mitosis maintains the chromosome number and content. (*4*) In order to reproduce sexually, the normally diploid cells must produce haploid gametes through meiotic division. During fertilization, these haploid gametes (egg, sperm) form a zygote that restores the diploid number. Depending upon which gametes fuse, combinations of chromosomes unlike the original parents are possible in offspring.

transported to the Golgi apparatus, the protein products are chemically modified and packaged into vesicles that can be used by the cell in a variety of ways. Some of the vesicles contain enzymes to digest food inside the cell; other vesicles are secreted to digest materials outside the cell, and yet others are important in the enlargement and repair of the cell wall and membrane.

A **lysosome*** is one type of vesicle originating from the Golgi apparatus that contains a variety of enzymes. Lysosomes are involved in intracellular digestion of food particles and in protection against invading microorganisms. They also participate in

---

*lysosome (ly'-soh-sohm) Gr. *lysis,* dissolution, and *soma,* body.

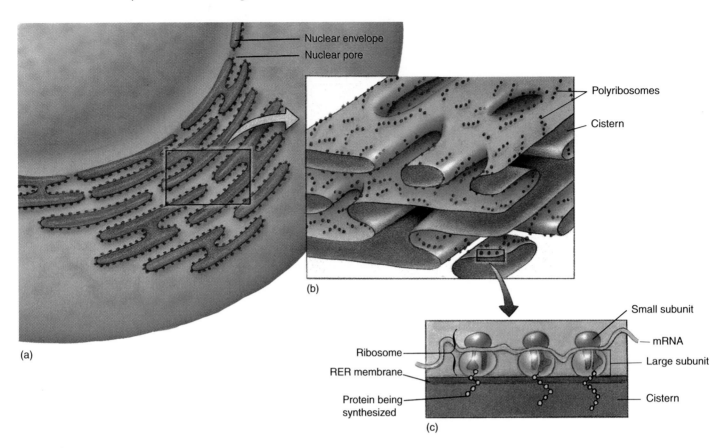

Nuclear envelope
Nuclear pore
Polyribosomes
Cistern
(b)
(a)
Small subunit
mRNA
Ribosome
Large subunit
RER membrane
Protein being synthesized
Cistern
(c)

FIGURE 5.8

The origin and detailed structure of the rough endoplasmic reticulum (RER). (a) Schematic view of the origin of the RER from the outer membrane of the nuclear envelope. **(b)** Three-dimensional projection of the RER. **(c)** Detail of the orientation of a ribosome on the RER membrane.

digestion and removal of cell debris in damaged tissue. Other types of vesicles include **vacuoles,*** which are membrane-bound sacs containing fluids or solid particles to be digested, excreted, or stored. They are formed in phagocytic cells (certain white blood cells and protozoa) in response to food and other substances that have been engulfed. The contents of a food vacuole are digested through the merger of the vacuole with a lysosome (figure 5.11). Other types of vacuoles are used in storing reserve food such as fats and glycogen. Protozoa living in freshwater habitats regulate osmotic pressure by means of contractile vacuoles, which regularly expel excess water that has diffused into the cell (see figure 5.29).

MITOCHONDRIA: ENERGY GENERATORS OF THE CELL

Although the nucleus is the cell's control center, none of the cellular activities it commands could proceed without a constant supply of energy, the bulk of which is generated in most eucaryotes by **mitochondria.*** When viewed with light microscopy, mitochondria

appear as round to elongate particles scattered throughout the cytoplasm. The internal ultrastructure reveals that a single mitochondrion consists of a smooth, continuous outer membrane that forms the external contour, and an inner, folded membrane nestled neatly within the outer membrane (figure 5.12a). The folds on the inner membrane, called **cristae,*** may be tubular, like fingers, or folded into shelflike bands.

The cristae membranes hold the enzymes and electron carriers of aerobic respiration. This is an oxygen-using process that extracts chemical energy contained in nutrient molecules and stores it in the form of high-energy molecules, or ATP. More detailed functions of mitochondria are covered in chapter 8. The spaces around the cristae are filled with a chemically complex fluid called the **matrix,*** which holds ribosomes, DNA, and the pool of enzymes and other compounds involved in the metabolic cycle. Mitochondria (along with chloroplasts) are unique among organelles in that they divide independently of the cell, contain circular strands of DNA, and have procaryotic-sized 70S ribosomes. These findings have prompted some intriguing speculations on their evolutionary origins (see Spotlight on Microbiology 5.1).

---

* vacuole (vak'-yoo-ohl) L. *vacuus,* empty. Any membranous space in the cytoplasm.

* mitochondria (my"-toh-kon'-dree-uh) sing. mitochondrion; Gr. *mitos,* thread, and *chondrion,* granule.

---

* cristae (kris'-te) sing. crista; L. *crista,* a comb.

* matrix (may'-triks) L. *mater,* mother or origin.

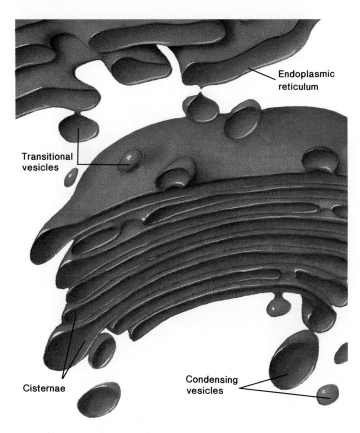

FIGURE 5.9

Detail of the Golgi apparatus. The flattened layers are cisternae. Vesicles enter the upper surface and leave the lower surface.

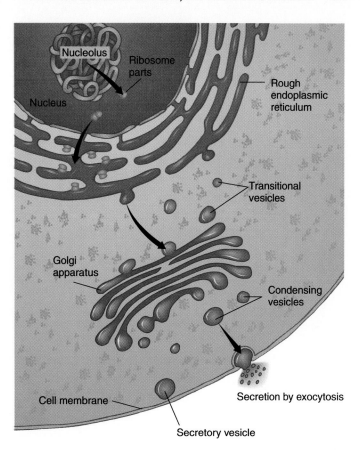

FIGURE 5.10

The transport process. The cooperation of organelles in protein synthesis and transport: nucleus → RER → Golgi apparatus → vesicles → secretion.

CHLOROPLASTS: PHOTOSYNTHESIS MACHINES

Chloroplasts* are remarkable organelles found in algae and plant cells that are capable of converting the energy of sunlight into chemical energy through photosynthesis. The photosynthetic role of chloroplasts makes them the primary producers of organic nutrients upon which all other organisms (except certain bacteria) ultimately depend. Another important photosynthetic product of chloroplasts is oxygen gas. Although chloroplasts resemble mitochondria, chloroplasts are larger, contain special pigments, and are much more varied in shape.

There are differences among various algal chloroplasts, but most are generally composed of two membranes, one enclosing the other. The smooth, outer membrane completely covers an inner membrane folded into small, disclike sacs called *thylakoids** that are stacked upon one another into *grana*. These structures carry the green pigment chlorophyll and sometimes additional pigments as well. Surrounding the thylakoids is a ground substance called the *stroma** (figure 5.13). The role of the photosynthetic pigments is to absorb and transform solar energy into chemical energy, which is then converted by reactions in the stroma to carbohydrates. We further explore some important aspects of photosynthesis in chapters 7 and 26.

THE CYTOSKELETON: A SUPPORT NETWORK

The cytoplasm of a eucaryotic cell is penetrated by a flexible framework of molecules called the cytoskeleton (figure 5.14). This framework appears to have several functions, such as anchoring organelles, providing support, and permitting shape changes and movement in some cells. The two main types of cytoskeletal elements are **microfilaments** and **microtubules.** Microfilaments are thin protein strands that attach to the cell membrane and form a network through the cytoplasm. Some microfilaments are responsible for movements of the cytoplasm, often made evident by the streaming of organelles around the cell in a cyclic pattern. Other microfilaments are active in *ameboid motion,* a type of movement typical of cells such as amebas and phagocytes that produces extensions of the cell membrane (pseudopods) into which the cytoplasm flows (see figure 5.29). Microtubules are long, hollow tubes that maintain the shape of eucaryotic cells without walls and transport substances from one part of a cell to another. The spindle fibers that play an essential role in mitosis are actually microtubules that attach to chromosomes and separate them into daughter cells. As indicated earlier, microtubules are also responsible for the movement of cilia and flagella.

---

*chloroplast (klor′-oh-plast) Gr. *chloros*, green, and *plastos*, to form.

*thylakoid (thy′-lah-koid) Gr. *thylakon*, a small sac, and *eidos*, form.

*stroma (stroh′-mah) Gr. *stroma*, mattress or bed.

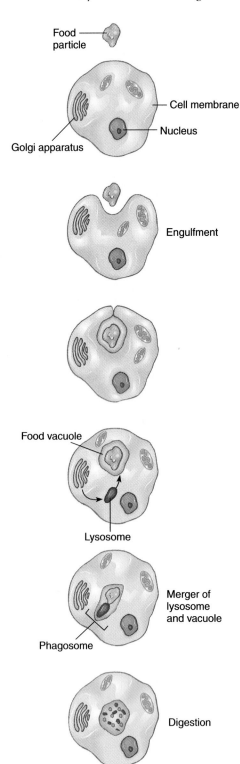

FIGURE 5.11

The origin and action of lysosomes in phagocytosis.

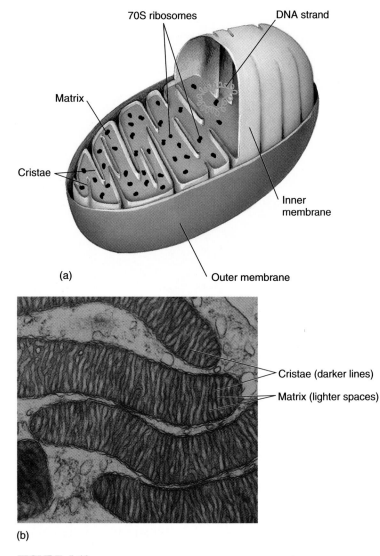

(a)

(b)

FIGURE 5.12

General structure of a mitochondrion. (a) in three-dimensional projection and **(b)** by electron microscopy. In most cells, mitochondria are elliptical or spherical, although in certain fungi, algae, and protozoa, they are long and filament-like.

RIBOSOMES: PROTEIN SYNTHESIZERS

In an electron micrograph of a eucaryotic cell, ribosomes are numerous, tiny particles that give a stippled appearance to the cytoplasm. Ribosomes are distributed in two ways: Some are scattered freely in the cytoplasm and cytoskeleton; others are intimately associated with the rough endoplasmic reticulum as previously described. Multiple ribosomes are often found arranged in short chains called polyribosomes (polysomes). The basic structure of eucaryotic ribosomes is similar to that of procaryotic ribosomes. Both are composed of large and small subunits of ribonucleoprotein (see figure 5.8). By contrast, however, the eucaryotic ribosome is the larger 80S variety that is a combination of 60S and 40S subunits. As in the procaryotes, eucaryotic ribosomes are the staging areas for protein synthesis.

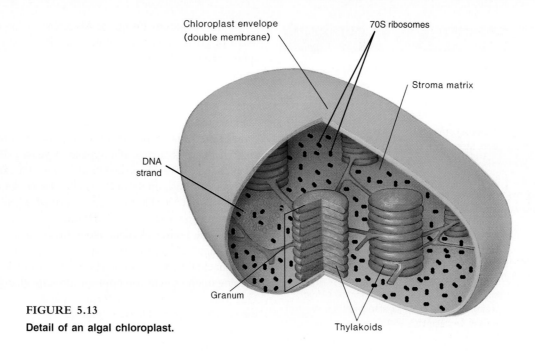

FIGURE 5.13

Detail of an algal chloroplast.

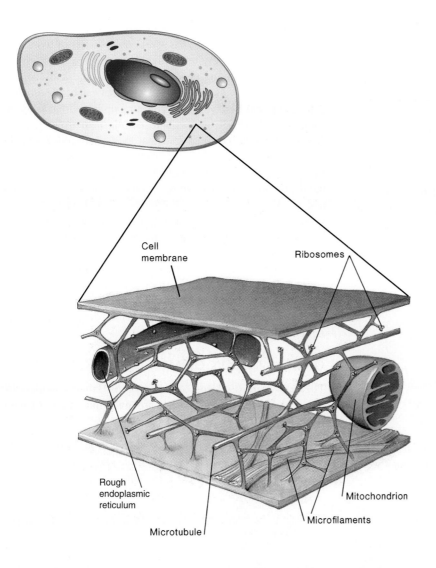

FIGURE 5.14

A model of the cytoskeleton. Depicted is the relationship between microtubules, microfilaments, and organelles.

Survey of Eucaryotic Microorganisms

With the general structure of the eucaryotic cell in mind, let us next examine the amazingly wide range of adaptations that this cell type has undergone. The following sections contain a general survey of the principal eucaryotic microorganisms—fungi, algae, protozoa, and parasitic worms—while simultaneously introducing elements of their structure, life history, classification, identification, and importance.

The Kingdom of the Fungi

The position of the **fungi*** in the biological world has been debated for many years. Although they were originally classified with the green plants (along with algae and bacteria), they were later separated from plants and placed in a group with algae and protozoa (the Protista). Even at that time, however, many *mycologists** were struck by several unique qualities of fungi that warranted their being placed into their own separate kingdom, and eventually they were.

---

*fungi (fun'-jy) sing. fungus; Gr. *fungos,* mushroom.

*mycologist (my-kol'-uh-jist) Gr. *mykes,* standard root for fungus. Scientists who study fungi.

The **Kingdom Fungi,** or Myceteae, is large and filled with forms of great variety and complexity. For practical purposes, the approximately 100,000 species of fungi can be divided into two groups: the *macroscopic fungi* (mushrooms, puffballs, gill fungi) and the *microscopic fungi* (molds, yeasts). Although the majority of fungi are either unicellular or colonial, a few complex forms such as mushrooms and puffballs do have a limited form of cellular specialization. Cells of the microscopic fungi exist in two basic morphological types: yeasts and hyphae. A **yeast** cell is distinguished by its round to oval shape and by its mode of asexual reproduction. It grows swellings on its surface called **buds.** Some species form a **pseudohypha,*** a chain of yeasts formed when buds remain attached in a row (figure 5.15). **Hyphae*** are long, threadlike cells found in the bodies of filamentous fungi, or **molds** (figure 5.16). Some fungal cells exist only in a yeast form; others occur primarily as hyphae; and a few, called *dimorphic,** can take either form, depending upon growth conditions, especially changing temperature. This variability in growth form is particularly characteristic of some pathogenic molds (see figure 22.1).

FUNGAL NUTRITION

All fungi are **heterotrophic.*** They acquire nutrients from a wide variety of organic materials called **substrates** (figure 5.17). Most fungi are **saprobes,*** meaning that they obtain these substrates from the remnants of dead plants and animals in soil or aquatic habitats. Fungi can also be parasites on the bodies of living animals or plants, although very few fungi absolutely require a living host. In general, the fungus penetrates the substrate and secretes enzymes that reduce it to small molecules that can be absorbed by the cells. Fungi have enzymes for digesting an incredible array of substances, including feathers, hair, cellulose, petroleum products, wood, rubber. It has been said that every naturally occurring organic material on the earth can be attacked by some type of fungus. Fungi are often found in nutritionally poor or adverse environments. In our laboratory, we occasionally discover fungi in mineral and weak acid solutions ("bottle imps") and on specimens preserved in formaldehyde. Various fungi thrive in substrates with high salt or sugar content, at relatively high temperatures, and even in snow and glaciers. Their medical and agricultural impact is extensive. A number of species cause **mycoses** (fungal infections) in animals, and thousands of species are important plant pathogens. Fungal toxins may cause disease in humans, and airborne fungi are a frequent cause of allergies and other medical conditions (Spotlight on Microbiology 5.2).

ORGANIZATION OF MICROSCOPIC FUNGI

The cells of most microscopic fungi grow in loose associations or colonies. The colonies of yeasts are much like those of bacteria in

---

*pseudohypha (soo"-doh-hy'-fuh) pl. pseudohyphae; Gr. *pseudo,* false, and *hyphe.*

*hypha (hy'-fuh) pl. hyphae (hy'-fee) Gr. *hyphe,* a web.

*dimorphic (dy-mor'-fik) Gr. *di,* two, and *morphe,* form.

*heterotrophic (het-ur-oh-tro'-fik) Gr. *hetero,* other, and *troph,* to feed. A type of nutrition that relies on an organic nutrient source.

*saprobe (sap'-rohb) Gr. *sapros,* rotten, and *bios,* to live. Also called saprotroph or saprophyte.

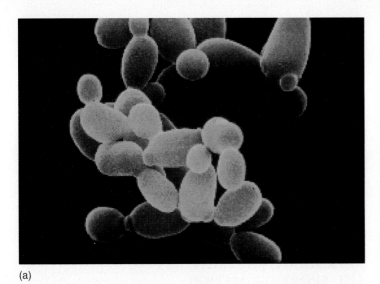

(a)

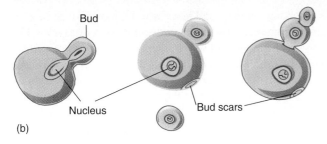

(b)

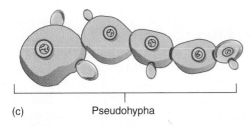

(c)   Pseudohypha

FIGURE 5.15

Microscopic morphology of yeasts. (a) Scanning electron micrograph of the brewer's, or baker's, yeast *Saccharomyces cerevisiae* (21,000×). **(b)** Formation and release of yeast buds. **(c)** Formation of pseudohypha (a chain of budding yeast cells).

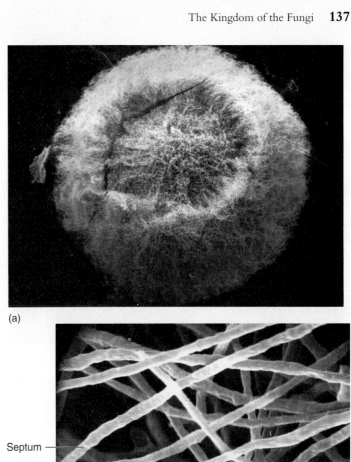

(a)

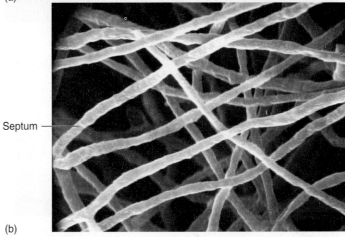

(b)

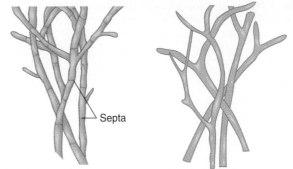

(c)

FIGURE 5.16

***Diplodia maydis*, a pathogenic fungus of corn plants.**
(a) Scanning electron micrograph showing a colony or mycelium (24×). **(b)** Close-up of hyphal structure (1,200×). **(c)** Basic structural types of hyphae.

that they have a soft, uniform texture and appearance. The colonies of filamentous fungi are noted for the striking cottony, hairy, or velvety textures that arise from their microscopic organization and morphology. The woven, intertwining mass of hyphae that makes up the body or colony of a mold is called a **mycelium*** (see figure 5.16).

Although hyphae contain the usual eucaryotic organelles, they also have some unique organizational features. In most fungi, the hyphae are divided into segments by cross walls, or **septa,*** a condition called **septate.** The nature of the septa varies from solid partitions with no communication between the compartments to partial walls with small pores that allow the flow of organelles and

*mycelium (my-see′-lee-um) pl. mycelia; Gr. *mykes,* fungus, and *helos,* nail.

*septa (sep′-tuh) sing. septum; L. *saepio,* to fence or wall.

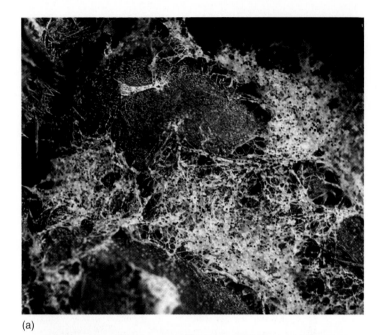

(a)

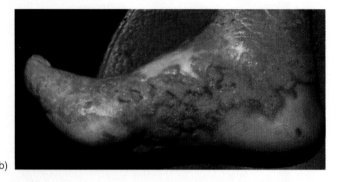

(b)

FIGURE 5.17

A small assortment of nutritional sources (substrates) of fungi.
(a) A fungal mycelium growing in soil and forest litter (black spheres are spores). **(b)** The skin of the foot infected by a soil fungus, *Fonsecaea pedrosoi.*

FIGURE 5.18

Functional types of hyphae using the mold Rhizopus as an example. (a) Vegetative hyphae are those surface and submerged filaments that digest, absorb, and distribute nutrients from the substrate. This species also has special anchoring structures called rhizoids. **(b)** Later, as the mold matures, it sprouts reproductive hyphae that produce asexual spores. **(c)** During the asexual life cycle, the dispersed mold spores settle on a usable substrate and send out germ tubes that elongate into hyphae. Through continued growth and branching, an extensive mycelium is produced. So prolific are the fungi that a single colony of mold can easily contain 5,000 spore-bearing structures. If each of these released 2,000 single spores and if every spore released from every sporangium on every mold were able to germinate, we would soon find ourselves in a sea of mycelia. Most spores do not germinate, but enough are successful to keep the numbers of fungi and their spores very high in most habitats.

nutrients between adjacent compartments. **Nonseptate** hyphae consist of one long, continuous cell *not* divided into individual compartments by cross walls. With this construction, the cytoplasm and organelles move freely from one region to another, and each hyphal element can have several nuclei.

Hyphae can also be classified according to their particular function. **Vegetative hyphae** (mycelia) are responsible for the visible mass of growth that appears on the surface of a substrate and penetrates it to digest and absorb nutrients. During the development of a fungal colony, the vegetative hyphae give rise to structures called **reproductive,** or **fertile hyphae,** which branch off vegetative mycelium. These hyphae are responsible for the production of fungal reproductive bodies called **spores.** Other specializations of hyphae are illustrated in figure 5.18.

REPRODUCTIVE STRATEGIES AND SPORE FORMATION

Fungi have many complex and successful reproductive strategies. Most can propagate by the simple outward growth of existing hyphae or by fragmentation, in which a separated piece of mycelium can generate a whole new colony. But the primary reproductive mode of fungi involves the production of various types of spores. Do not confuse fungal spores with the more resistant, nonreproductive bacterial spores. Fungal spores are responsible not only for multiplication but also for survival, producing genetic variation, and dissemination. Because of their compactness and relatively light weight, spores are dispersed widely through the environment by air, water, and living things. Upon encountering a favorable substrate, a spore will germinate and produce a new fungus colony in a very short time (figure 5.18).

The fungi exhibit such a marked diversity in spores that they are largely classified and identified by their spores and spore-forming structures. Although there are some elaborate systems for naming and classifying spores, we will present a basic overview of the principal types. The most general subdivision is based on the way the spores arise. **Asexual spores** are the products of mitotic division of a single parent cell, and **sexual spores** are formed

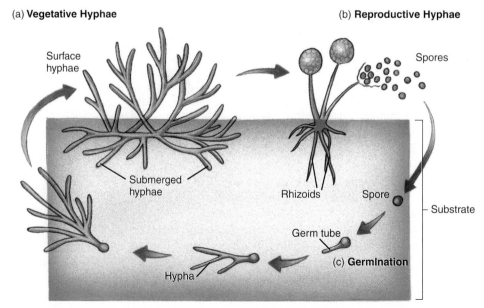

(a) **Vegetative Hyphae**

Surface hyphae

Submerged hyphae

(b) **Reproductive Hyphae**

Spores

Rhizoids

Spore

Germ tube

Substrate

(c) **Germination**

Hypha

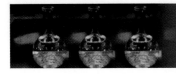

The Mother of All Fungi

Far from being microscopic, one of the largest organisms in the world is a massive basidiomycete growing in a forest in eastern Oregon. The mycelium of a single colony of *Armillaria ostoye* covers an area of 2,200 acres and stretches 3.5 miles across and 3 feet into the ground. Most of the fungus lies hidden beneath the ground, where it periodically fruits into edible mushrooms. Experts with the forest service took random samples over a large area and found that the mycelium is from genetically identical stock. They believe this fungus was started from a single mating pair of sexual hyphae around 2,400 years ago. Another colossal species from Washington state covers 1,500 acres and weighs around 100 tons. Both fungi penetrate the root structure of trees and spread slowly from tree to tree, sapping their nutrients and gradually destroying whole forests.

A Fungus in Your Future

Biologists are developing some rather imaginative uses for fungi as a way of controlling both the life and death of plants. Dutch and Canadian researchers studying ways to control a devastating fungus infection of elm trees (Dutch elm disease) have come up with a brand new use of an old method—they actually vaccinate the trees. Ordinarily, the disease fungus invades the plant vessels and chokes off the flow of water. The natural tendency of the elm to defend itself by surrounding and inactivating the fungus is too slow to save it from death. But treating the elm trees before they get infected helps them develop an immunity to the disease. Plants are vaccinated somewhat like humans and animals: nonpathogenic spores or proteins from fungi are injected into the tree over a period of time. So far it appears that the symptoms of disease and the degree of damage can be significantly reduced. This may be the start of a whole new way to control fungal pests.

At the other extreme, government biologists working for narcotic control agencies have unveiled a recent plan to use fungi to kill unwanted plants. The main targets would be plants grown to produce illegal drugs like cocaine and heroin in the hopes of cutting down on these drugs right at the source. A fungus infection (*Fusarium*) that wiped out 30% of the coca crop in Peru dramatically demonstrated how effective this might be. Since then, at least two other fungi that could destroy opium poppies and marijuana plants have been isolated. Purposefully releasing plant pathogens such as *Fusarium* into the environment has stirred a great deal of controversy. Critics in South America emphasize that even if the fungus appears specific to a particular plant, there is too much potential for it to switch hosts to food and ornamental plants and wreak havoc with the ecosystem. United States biologists who support the plan of using fungal control agents say that it is not as dangerous as massive spraying with pesticides, and that extensive laboratory tests have proved that the species of fungi being used will be very specific to the illegal drug plants and will not affect close relatives. Limited field tests will be started in the near future, paid for by a billion-dollar fund created by the U.S. government as part of its war on drugs.

Fungi, Fungi, Everywhere

The importance of fungi in the ecological structure of the earth is well founded. They are essential contributors to complex environments such as soil, and they play numerous beneficial roles as decomposers of organic debris and as partners to plants. Fungi also have great practical importance due to their metabolic versatility. They are productive sources of drugs (penicillin) to treat human infections and other diseases, and they are used in industry to ferment foods and synthesize organic chemicals.

The fact that they are so widespread also means that they frequently share human living quarters, especially in locations that provide ample moisture and nutrients. Often their presence is harmless and limited to a film of mildew on shower stalls or other moist environments. In some cases, depending on the amount of contamination and the type of mold, these indoor fungi can also give rise to various medical problems. Such common air contaminants as *Penicillium, Aspergillus, Cladosporium,* and *Stachybotrys* all have the capacity to give off airborne spores and toxins that, when inhaled, cause a whole spectrum of symptoms sometimes referred to as "sick building syndrome." The usual source of harmful fungi is the presence of chronically water-damaged walls, ceilings, and other building materials that have come to harbor these fungi. Such materials contain plant products that serve as a rich source of nutrients. People exposed to these houses or buildings report symptoms that range from skin rash, flulike reactions, sore throat, and headaches to fatigue, diarrhea, and even immune suppression. Because they can mimic other diseases, the association with indoor toxic fungi may be missed. Recent reports of sick buildings have been on the rise, affecting thousands of people, and some deaths have been reported in small children. The control of indoor fungi requires correcting the moisture problem, removing the infested materials, and decontaminating the living spaces. In the worst cases, the buildings have been burned down to prevent the spread of fungi to surrounding structures.

Armillaria (honey) mushrooms sprouting from the base of a tree in Oregon provide only a tiny hint of the massive fungus from which they arose.

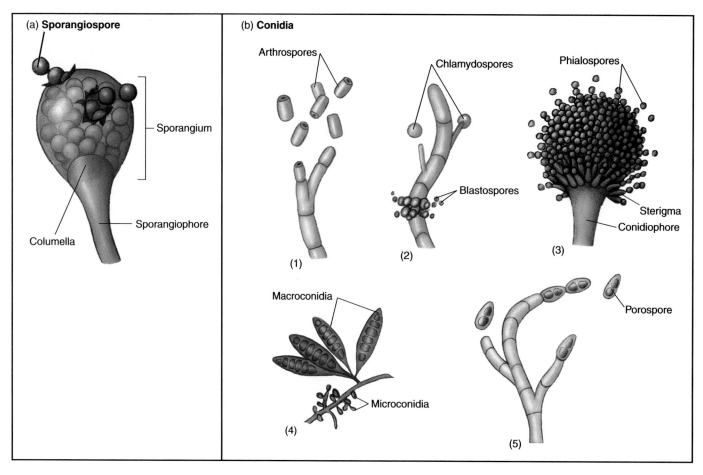

FIGURE 5.19

Types of asexual mold spores. (a) Sporangiospores (e.g., *Absidia*). **(b)** Conidia: (*1*) arthrospores (e.g., *Coccidioides*), (*2*) chlamydospores and blastospores (e.g., *Candida albicans*), (*3*) phialospores (e.g., *Aspergillus*), (*4*) macroconidia and microconidia (e.g., *Microsporum*), and (*5*) porospores (e.g., *Alternaria*).

through a process involving the fusing of two parental nuclei followed by meiosis.

Asexual Spore Formation

On the basis of the nature of the reproductive hypha and the manner in which the spores originate, there are two subtypes of asexual spore (figure 5.19):

1. **Sporangiospores** (figure 5.19*a*) are formed by successive cleavages within a saclike head called a **sporangium,** * which is attached to a stalk, the sporangiophore. These spores are initially enclosed but are released when the sporangium ruptures.
2. **Conidia** * (conidiospores) are free spores not enclosed by a spore-bearing sac (figure 5.19*b*). They develop either by the pinching off of the tip of a special fertile hypha or by the segmentation of a preexisting vegetative hypha. Conidia are the most common asexual spores, and they occur in the following forms:

arthrospore (ar′-thro-spor) Gr. *arthron,* joint. A rectangular spore formed when a septate hypha fragments at the cross walls.

chlamydospore (klam-ih′-doh-spor) Gr. *chlamys,* cloak. A spherical conidium formed by the thickening of a hyphal cell. It is released when the surrounding hypha fractures, and it serves as a survival or resting cell.

blastospore. A spore produced by budding from a parent cell that is a yeast or another conidium; also called a bud.

phialospore (fy′-ah-lo-spor) Gr. *phialos,* a vessel. A conidium that is budded from the mouth of a vase-shaped spore-bearing cell called a phialide or *sterigma,* leaving a small collar.

microconidium and *macroconidium.* The smaller and larger conidia formed by the same fungus under varying conditions. Microconidia are one-celled, and macroconidia have two or more cells.

porospore. A conidium that grows out through small pores in the spore-bearing cell; some are composed of several cells.

---

*sporangium (spo-ran′-jee-um) pl. sporangia; Gr. *sporos,* seed, and *angeion,* vessel.

*conidia (koh-nid′-ee-uh) sing. conidium; Gr. *konidion,* a particle of dust.

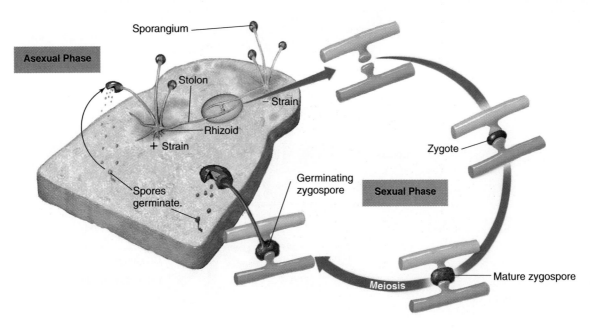

FIGURE 5.20

Formation of zygospores in *Rhizopus stolonifer*. Sexual reproduction occurs when two mating strains of hyphae grow together, fuse, and form a mature zygospore. Germination of the zygospore involves meiotic division and production of a haploid sporangium that looks just like the asexual one.

Sexual Spore Formation

If fungi can propagate themselves successfully with millions of asexual spores, what is the function of their sexual spores? The answer lies in important variations that occur when fungi of different genetic makeup combine their genetic material. Just as in plants and animals, this linking of genes from two parents creates offspring with combinations of genes different from that of either parent. The offspring from such a union can have slight variations in form and function that are potentially advantageous in the adaptation and survival of their species. Because fungal cells spend most of their life cycle in the haploid state, their cells already have a chromosome number compatible for sexual union. In general, during sexual reproduction, the haploid nuclei from two compatible parental cells fuse, forming a diploid nucleus that subsequently participates in sexual spore development (see figure 5.7*a*, step 2).

The majority of fungi produce sexual spores at some point. The nature of this process varies from the simple fusion of fertile hyphae of two different strains to a complex union of differentiated male and female structures and the development of special fruiting structures. We will consider the three most common sexual spores: zygospores, ascospores, and basidiospores. These spore types provide an important basis for classifying the major fungal divisions.

Zygospores* are sturdy diploid spores formed when hyphae of two opposite strains (called the plus and minus strains) fuse and create a diploid zygote that swells and becomes covered by strong, spiny walls (figure 5.20). When its wall is disrupted, and moisture and nutrient conditions are suitable, the zygospore germinates and forms a mycelium that gives rise to a sporangium. Meiosis of diploid cells of the sporangium results in haploid nuclei that de-

velop into sporangiospores. Both the sporangia and the sporangiospores that arise from sexual processes are outwardly identical to the asexual type, but because the spores arose from the union of two separate fungal parents, they are not genetically identical.

In general, haploid spores called **ascospores*** are created inside a special fungal sac, or **ascus** (pl. asci) (figure 5.21). Although details can vary among types of fungi, the ascus and ascospores are formed when two different strains or sexes join together to produce offspring. In many species, the male sexual organ fuses with the female sexual organ. The end result is a number of terminal cells, each containing a diploid nucleus. Through differentiation, each of these cells enlarges to form an ascus, and its diploid nucleus undergoes meiosis (often followed by mitosis) to form four to eight haploid nuclei that will mature into ascospores. A ripe ascus breaks open and releases the ascospores. Some species form an elaborate fruiting body to hold the asci (figure 5.21).

Basidiospores* are haploid sexual spores formed on the outside of a club-shaped cell called a **basidium** (figure 5.22). In general, spore formation follows the same pattern of two mating types coming together, fusing, and forming terminal cells with diploid nuclei. Each of these cells becomes a basidium, and its nucleus produces, through meiosis, four haploid nuclei. These nuclei are extruded through the top of the basidium, where they develop into basidiospores. Notice the location of the basidia along the gills in mushrooms, which are often dark from the spores they contain. It may be a surprise to discover that the fleshy part of a mushroom is actually a fruiting body designed to protect and help disseminate its sexual spores.

---

*zygospores (zy′-goh-sporz) Gr. *zygon*, yoke, to join.

*ascospores (as′-koh-sporz) Gr. *ascos*, a sac.

*basidiospores (bah-sid′-ee-oh-sporz) Gr. *basidi*, a pedestal.

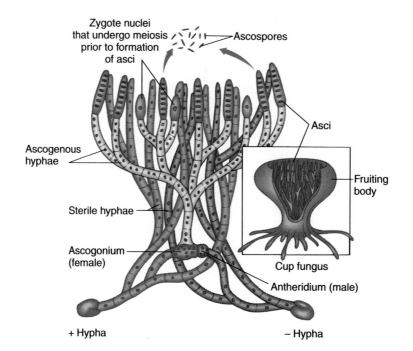

FIGURE 5.21

Production of ascospores in a cup fungus. Inset shows the cup-shaped fruiting body that houses the asci.

FUNGAL CLASSIFICATION

It is often difficult for microbiologists to assign logical and useful classification schemes to microorganisms while at the same time preserving their relationships. This difficulty is due to the fact that the organisms do not always perfectly fit the neat categories made for them, and even experts cannot always agree on the nature of the categories. The fungi are no exception, and there are several ways to classify them. For our purposes, we will adopt a classification scheme with a medical mycology emphasis, in which the Kingdom Fungi is subdivided into two subkingdoms, the Amastigomycota and the Mastigomycota. The Amastigomycota are common inhabitants of terrestrial habitats. Some of the characteristics used to separate them into subgroups are the type of sexual reproduction and their hyphal structure. The Mastigomycota are primitive filamentous fungi that live primarily in water and may cause disease in potatoes and grapes.

The following section outlines the four divisions of Amastigomycota, including major characteristics and important members.

Amastigomycota: Fungi That Produce Sexual and Asexual Spores (Perfect)

Division I—Zygomycota (also Phycomycetes) Sexual spores: zygospores; asexual spores: mostly sporangiospores, some conidia. Hyphae are usually nonseptate. If septate, the septa are complete. Most species are free-living saprobes; some are animal parasites. Can be obnoxious contaminants in the laboratory and on food and vegetables. Examples are mostly molds: *Rhizopus,* a black bread mold; *Mucor; Syncephalastrum; Circinella* (figure 5.23, asexual phase only).

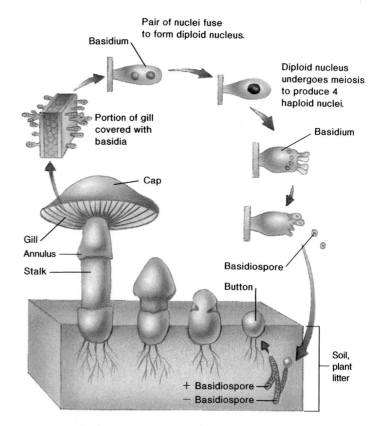

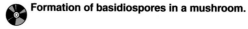

FIGURE 5.22

Formation of basidiospores in a mushroom.

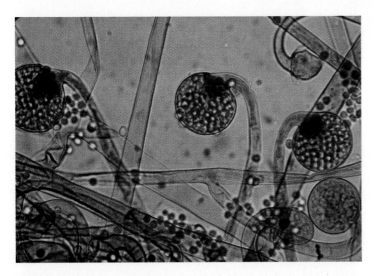

FIGURE 5.23

A representative Zygomycota, *Circinella*. Note the sporangia, sporangiospores, and nonseptate hyphae.

Division II—Ascomycota (also Ascomycetes)   Sexual spores: produce ascospores in asci, asexual spores: many types of conidia, formed at the tips of conidiophores. Hyphae with porous septa. Many important species. Examples: *Histoplasma,* the cause of Ohio Valley fever; *Microsporum,* one cause of ringworm (a common name for certain fungal skin infections that often grow in a ringed pattern); *Penicillium,* one source of antibiotics (figure 5.24, asexual phase only); and *Saccharomyces,* a yeast used in making bread and beer. Most of the species in this division are either molds or yeasts; it also includes many human and plant pathogens, such as *Pneumocystis carinii,* a pathogen of AIDS patients.

Division III—Basidiomycota (also Basidiomycetes)   Sexual reproduction by means of basidia and basidiospores; asexual spores: conidia. Incompletely septate hyphae. Some plant parasites and one human pathogen. Fleshy fruiting bodies are common. Examples: mushrooms, puffballs, bracket fungi, and plant pathogens called rusts and smuts. The one human pathogen, the yeast *Cryptococcus neoformans,* causes an invasive systemic infection in several organs, including the skin, brain, and lungs (see figure 22.26).

Amastigomycota: Fungi That Produce Asexual Spores Only (Imperfect)

From the beginnings of fungal classification, any fungus that lacked a sexual state was called "imperfect" and was placed in a catchall category, the Fungi Imperfecti, or Deuteromycota. A species would remain classified in that category until its sexual state was described (if ever). Gradually, many species of Fungi Imperfecti were found to make sexual spores, and they were assigned to the taxonomic grouping that best fit those spores. This system created a quandary, because several very important species, especially pathogens, had to be regrouped and renamed. In many cases, the older names were so well entrenched in the literature that it was easier to retain them and assign a new generic name for their sexual stage. Consequently, *Blastomyces* and *Histoplasma* are also known as *Ajellomyces* (sexual phase).

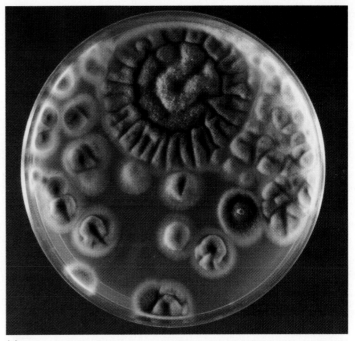

(a)

(b)

FIGURE 5.24

A common Ascomycota, *Penicillium*. (a) Macroscopic view of a typical blue-green colony. **(b)** Microscopic view shows the brush arrangement of phialospores (220×).

Division IV—Deuteromycota*   Asexual spores: conidia of various types. Hyphae septate. Majority are yeasts or molds, some dimorphic. Saprobes and a few animal and plant parasites. Examples: Several human pathogens were originally placed in this group, especially the imperfect states of *Blastomyces* and *Microsporum.* Other species are *Coccidioides immitis,* the cause of valley fever; *Candida albicans,* the cause of various yeast infections; and *Cladosporium,* a common mildew fungus (figure 5.25).

*Deuteromycota (doo′-ter-oh-my-koh″-tuh) Gr. *deutero,* second, and *mykes,* fungus.

Mycelium   Conidia

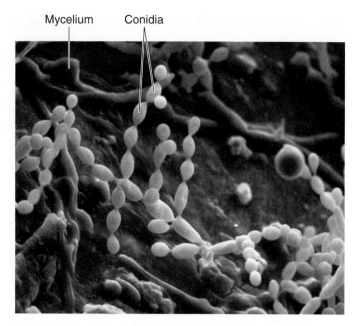

FIGURE 5.25

Mycelium and spores of a representative of Deuteromycota called *Cladosporium*. This black mold is one of the most common fungi in the world and a ubiquitous air contaminant (see Spotlight on Microbiology 5.2).

FUNGAL IDENTIFICATION AND CULTIVATION

Fungi are identified in medical specimens by first being isolated on special types of media and then being observed macroscopically and microscopically. Examples of media for cultivating fungi are cornmeal, blood, and Sabouraud's agar. The latter medium is useful in isolating fungi from mixed samples because of its low pH, which inhibits the growth of bacteria but not of most fungi. Because the fungi are classified into general groups by the presence and type of sexual spores, it would seem logical to identify them in the same way, but sexual spores are rarely if ever demonstrated in the laboratory setting. As a result, the asexual spore-forming structures and spores are usually used to identify organisms to the level of genus and species. Other characteristics that contribute to identification are hyphal type, colony texture and pigmentation, physiological characteristics, and genetic makeup (see figure 22.3).

THE ROLES OF FUNGI IN NATURE AND INDUSTRY

Nearly all fungi are free-living and do not require a host to complete their life cycles. Even among those fungi that are pathogenic, most human infection occurs through accidental contact with an environmental source such as soil, water, or dust. Humans are generally quite resistant to fungal infection, except for two main types of fungal pathogens: the true pathogens, which can infect even healthy persons, and the opportunistic pathogens, which attack persons who are already weakened in some way.

Mycoses (fungal infections) vary in the way the agent enters the body and the degree of tissue involvement (table 5.1). The list of opportunistic fungal pathogens has been increasing in the

TABLE 5.1

Major Fungal Infections of Humans

Degree of Tissue Involvement and Area Affected	Name of Infection	Name of Causative Fungus
Superficial (not deeply invasive)		
Outer epidermis	Pityriasis versicolor	*Malassezia furfur*
Epidermis, hair, and dermis can be attacked	Dermatophytosis, also called tinea or ringworm of the scalp, body, feet (athlete's foot), toenails	*Microsporum, Trichophyton,* and *Epidermophyton*
Mucous membranes, skin, nails	Candidiasis, or yeast infection	*Candida albicans*
Subcutaneous (invades just beneath the skin)		
Nodules beneath the skin; can invade lymphatics	Sporotrichosis	*Sporothrix schenckii*
Large fungal tumors of limbs and extremities	Mycetoma, madura foot	*Pseudoallescheria boydii, Madurella mycetomatis*
Systemic (deep; organism enters lungs; can invade other organs)		
Lung	Coccidioidomycosis (San Joaquin Valley fever)	*Coccidioides immitis*
	North American blastomycosis (Chicago disease)	*Blastomyces dermatitidis*
	Histoplasmosis (Ohio Valley fever)	*Histoplasma capsulatum*
	Cryptococcosis (torulosis)	*Cryptococcus neoformans*
Lung, skin	Paracoccidioidomycosis (South American blastomycosis)	*Paracoccidioides brasiliensis*

past few years because of newer medical techniques that keep compromised patients alive. Even so-called harmless species found in the air and dust around us may be able to cause opportunistic infections in patients who already have AIDS, cancer, or diabetes.

Fungi are involved in other medical conditions besides infections (see Spotlight on Microbiology 5.2). Fungal cell walls give off chemical substances that can cause allergies. The toxins produced by poisonous mushrooms can induce neurological disturbances and even death. The mold *Aspergillus flavus* (see figure 22.27) synthesizes a potentially lethal poison called aflatoxin,[4] which is the cause of a disease in domestic animals that have eaten grain infested with the mold and is also a cause of liver cancer in humans.

Fungi pose an ever-present economic hindrance to the agricultural industry. A number of species are pathogenic to field plants

---

4. From **aspergillus, flavus, toxin.**

TABLE 5.2
Summary of Algal Characteristics

Group	Organization	Primary Habitat	Cell Wall	Pigmentation	Ecology/Importance	Example(s)
Euglenophyta (euglenids)	Mainly unicellular; motile by flagella	Fresh water	None; pellicle instead	Chlorophyll, carotenoids, xanthophyll	Some are close relatives of Mastigophora	*Euglena*
Pyrrophyta (dinoflagellates)	Unicellular, dual flagella	Marine plankton	Cellulose or atypical wall	Chlorophyll, carotenoids	Cause of "red tide"	*Gonyaulax*
Chrysophyta (diatoms or golden-brown algae)	Mainly unicellular, some filamentous forms, unusual form of motility	Fresh water and marine	Silicon dioxide	Chlorophyll, fucoxanthin	Diatomaceous earth, major component of plankton	*Navicula,* other diatoms
Phaeophyta (brown algae— kelps)	Multicellular, vascular system, holdfasts	Marine, subtidal forests	Cellulose, alginic acid	Chlorophyll, carotenoids, fucoxanthin	Source of an emulsifier, alginate	*Fucus, Sargassum*
Rhodophyta (red seaweeds)	Multicellular	Marine, intertidal forests	Cellulose	Chlorophyll, carotenoids, xanthophyll, phycobilin	Source of agar and carrageenan, a food additive	*Gelidium*
Chlorophyta (green algae, grouped with plants)	Varies from unicellular, colonial, filamentous, to multicellular	Fresh water and salt water	Cellulose	Chlorophyll, carotenoids, xanthophyll	Precursor of higher plants	*Chlamydomonas, Spirogyra, Volvox*

such as corn and grain, and fungi also rot harvested fresh produce during shipping and storage. It has been estimated that as much as 40% of the yearly fruit crop is consumed not by humans but by fungi. On the beneficial side, however, fungi play an essential role in decomposing organic matter and returning essential minerals to the soil. They form stable associations with plant roots (mycorrhizae) that increase the ability of the roots to absorb water and nutrients. Industry has tapped the biochemical potential of fungi to produce large quantities of antibiotics, alcohol, organic acids, and vitamins. Some fungi are eaten or used to impart flavorings to food. The yeast *Saccharomyces* produces the alcohol in beer and wine and the gas that causes bread to rise. Blue cheese, soy sauce, and cured meats derive their unique flavors from the actions of fungi (see chapter 26).

The Protists

The algae and protozoa have been traditionally combined into the Kingdom Protista. The two major taxonomic categories of this kingdom are Subkingdom Algae and Subkingdom Protozoa. Although these general types of microbes are now known to occupy several kingdoms, it is still useful to retain the concept of a protist as any unicellular or colonial organism that lacks true tissues. This concept is especially helpful in our survey of major groups of microorganisms, because many algae are multicellular and macroscopic, and we cannot present an exhaustive coverage of all forms.

THE ALGAE: PHOTOSYNTHETIC PROTISTS

The **algae*** are a group of photosynthetic organisms usually recognized by their larger members, such as seaweeds and kelps. In addition to being beautifully colored and diverse in appearance, they vary in length from a few micrometers to 100 meters. Algae occur in unicellular, colonial, and filamentous forms, and the larger forms can possess tissues and simple organs. Table 5.2 lists the major characteristics of the various subgroups of algae.

Algal cells as a group contain all of the eucaryotic organelles. The most noticeable of these are the chloroplasts, which contain, in addition to the green pigment chlorophyll, a number of other pigments that create the yellow, red, and brown coloration of some groups (table 5.2). The chloroplasts can be of such unique character that they are used in identification (for example, *Spirogyra* in figure 5.26b). Most algal cells are enclosed by a complex cell wall, which accounts for the distinctive appearance of unicellular members such as diatoms (figure 5.26c) and dinoflagellates. The outermost structure in a group called the euglenids (for example, *Euglena* in figure 5.26a) is a thick, flexible membrane called a **pellicle.** Motility by flagella or gliding is common among the algae, and many members contain tiny light-sensitive areas (eyespots) that coordinate with the flagella to guide the cell toward the light it requires for photosynthesis.

*algae (al′-jee) sing. alga; L. seaweeds.

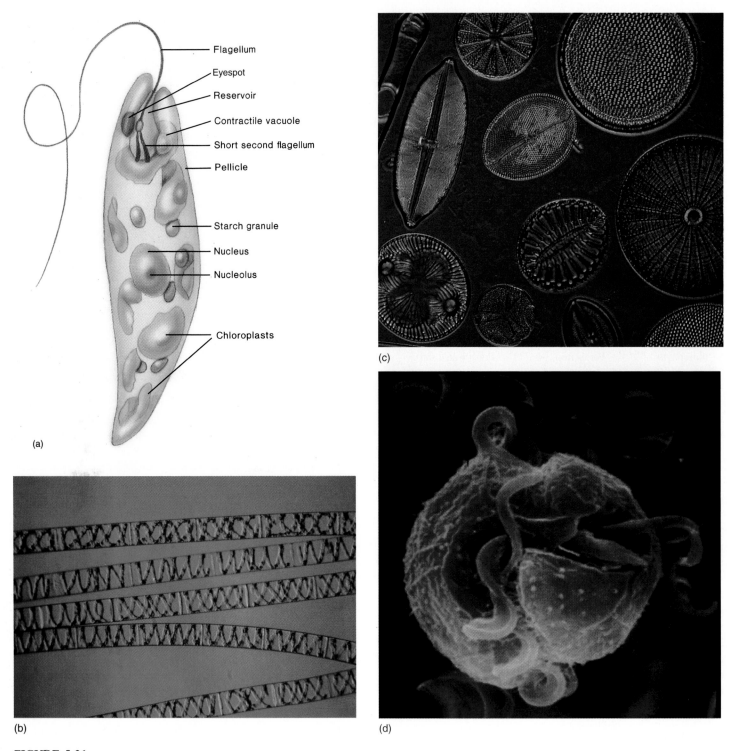

FIGURE 5.26

Representative microscopic algae. (a) *Euglena*, a unicellular motile example (150×). **(b)** *Spirogyra*, a colonial filamentous form with spiral chloroplasts. **(c)** A strew of beautiful chrysophyta—diatoms—shows the intricate and varied structure of their silica cell wall. **(d)** Zoospore of the dinoflagellate *Pfiesteria piscicida*. Although it is free-living, it is known to parasitize fish and release potent toxins that kill fish and sicken humans.

Algae are widespread inhabitants of fresh and marine waters. They are one of the main components of the large floating community of microscopic organisms called **plankton.** In this capacity, they play an essential role in the aquatic food web and produce most of the earth's oxygen. Other algal habitats include the surface of soil, rocks, and plants, and several species are even hardy enough to live in hot springs or snowbanks (see figure 7.10).

Algae are often described by common names such as green algae, brown algae, golden brown algae, and red algae in reference to their predominant color. A more technical system divides the microscopic algae into divisions or kingdoms based on the types of chlorophyll and other pigments, the type of cell covering, the nature of their stored foods, and genetic factors. The common names for these groups are: (1) Euglenophyta, or euglenids; (2) Pyrrophyta, or dinoflagellates; (3) Chrysophyta, or diatoms; (4) Phaeophyta, or brown algae; (5) Rhodophyta, or red seaweeds; and (6) Chlorophyta, or green algae.

Algae reproduce asexually through fragmentation, binary fission, and mitosis, and some produce motile spores. Their sexual reproductive cycles can be highly complex, with stages similar to those of fungi. The capacity of algae to photosynthesize is extremely important to the earth. They form the basis of aquatic food webs and play an essential role in the earth's oxygen and carbon dioxide balance. Some products of algae have industrial applications. For example, fossilized marine diatoms yield an abrasive powder called diatomaceous earth that is used in polishes, bricks, and filters, and certain seaweeds are a source of agar and algin used in microbiology, dentistry, and the food and cosmetics industries. We consider the role of algae in microbial ecology in chapter 26.

Animal tissues would be rather inhospitable to algae, so they are rarely infectious. One exception, however, is *Prototheca,* an unusual non-photosynthetic alga, which has been associated with skin and subcutaneous infections in humans and animals.

The primary medical threat from algae is due to a type of food poisoning caused by the toxins of certain marine **dinoflagellates.** During particular seasons of the year, the overgrowth of these motile algae imparts a brilliant red color to the water, which is referred to as a "red tide." When intertidal animals feed, their bodies accumulate toxins given off by the dinoflagellates that can persist for several months (see figure 26.20).

Paralytic shellfish poisoning is caused by eating exposed clams or other invertebrates. It is marked by severe neurological symptoms and can be fatal. Ciguatera is a serious intoxication caused by dinoflagellate toxins that have accumulated in fish such as bass and mackerel. Cooking does not destroy the toxin, and there is no antidote.

Recently, a U.S. East Coast outbreak of fish disease that spread to humans was traced to a remarkable dinoflagellate called *Pfiesteria piscicida.* This newly identified species occurs in at least 20 forms, including spores, cysts, and amebas (see figure 5.26*d*), that can release potent toxins. Both fishes and humans develop neurological symptoms and bloody skin lesions. The cause of the epidemic has been traced to nutrient-rich agricultural run-off water that promoted the sudden "bloom" of dinoflagellates which attacked and killed millions of fish.

BIOLOGY OF THE PROTOZOA

If a poll were taken to choose the most engrossing and vivid group of microorganisms, many people would choose the protozoa. Although their name comes from the Greek for "first animals," they are far from being simple, primitive organisms. The protozoa constitute a very large group (about 65,000 species) of creatures that although single-celled, have startling properties when it comes to movement, feeding, and behavior. Although most members of this group are harmless, free-living inhabitants of water and soil, a few species are parasites collectively responsible for hundreds of millions of infections of humans each year. Before we consider a few examples of important pathogens, let us examine some general aspects of protozoan biology.

Protozoan Form and Function

Most protozoan cells are single cells containing the major eucaryotic organelles except chloroplasts (see figure 5.2*c*). Their organelles can be highly specialized for feeding, reproduction, and locomotion. The cytoplasm is usually divided into a clear outer layer called the *ectoplasm* and a granular inner region called the *endoplasm.* Ectoplasm is involved in locomotion, feeding and protection. Endoplasm houses the nucleus, mitochondria, and food and contractile vacuoles. Some ciliates and flagellates[5] even have organelles that work somewhat like a primitive nervous system to coordinate movement. Because protozoa lack a cell wall, they have a certain amount of flexibility. Their outer boundary is a cell membrane that regulates the movement of food, wastes, and secretions. Cell shape can remain constant (as in most ciliates) or can change constantly (as in amebas). Certain amebas (foraminiferans) encase themselves in hard shells made of calcium carbonate. The size of most protozoan cells falls within the range of 3 to 300 μm. Some notable exceptions are giant amebas and ciliates that are large enough (3–4 mm in length) to be seen swimming in pond water.

Nutritional and Habitat Range   Protozoa are heterotrophic and usually require their food in a complex organic form. Free-living species scavenge dead plant or animal debris and even graze on live cells of bacteria and algae. Some species have special feeding structures such as oral grooves, which carry food particles into a passageway or gullet that packages the captured food into vacuoles for digestion. A remarkable feeding adaptation can be seen in the ciliate *Didinium,* which can easily devour another ciliate that is nearly its size (see figure 5.30*c*). Some protozoa absorb food directly through the cell membrane. Parasitic species live on the fluids of their host, such as plasma and digestive juices, or they can actively feed on tissues.

Although protozoa have adapted to a wide range of habitats, their main limiting factor is the availability of moisture. Their predominant habitats are fresh and marine water, soil, plants, and animals. Even extremes in temperature and pH are not a barrier to their existence; hardy species are found in hot springs, ice, and habitats with low or high pH. Many protozoa can convert to a resistant, dormant stage called a cyst.

Styles of Locomotion   Except for one group (the Apicomplexa), protozoa are motile by **pseudopods,*** **flagella,** or **cilia.** A few species have both pseudopods (also called pseudopodia) and flagella. Some unusual protozoa move by a gliding or twisting movement that does not appear to involve any of these locomotor structures. Pseudopods are blunt, branched, or long and pointed, depending on the particular species. As previously described, the flowing action of the pseudopods results in ameboid motion, and

---

5. The terms *ciliate* and *flagellate* are common names of protozoan groups that move by means of cilia and flagella.

*pseudopod (soo′-doh-pod) pl. pseudopods or pseudopodia; Gr. *pseudo,* false, and *pous,* feet.

pseudopods also serve as feeding structures in many amebas (see figure 5.29). The structure and behavior of flagella and cilia were discussed in the first section of this chapter. Flagella vary in number from one to several, and in certain species they are attached along the length of the cell by an extension of the cytoplasmic membrane called the *undulating membrane* (see figure 5.28). In most ciliates, the cilia are distributed over the entire surface of the cell in characteristic patterns. Because of the tremendous variety in ciliary arrangements and functions, ciliates are among the most diverse and awesome cells in the biological world. In certain protozoa, cilia line the oral groove and function in feeding; in others, they fuse together to form stiff props (cirri) that serve as primitive rows of walking legs (see figure 5.30).

Life Cycles and Reproduction Most protozoa are recognized by a motile feeding stage called the **trophozoite**** that requires ample food and moisture to remain active. A large number of species are also capable of entering into a dormant, resting stage called a **cyst** when conditions in the environment become unfavorable for growth and feeding. During *encystment,* the trophozoite cell rounds up into a sphere, and its ectoplasm secretes a tough, thick cuticle around the cell membrane (figure 5.27). Because cysts are more resistant than ordinary cells to heat, drying, and chemicals, they can survive adverse periods. They can be dispersed by air currents and may even be an important factor in the spread of diseases such as amebic dysentery. If provided with moisture and nutrients, a cyst breaks open and releases the active trophozoite.

The life cycles of protozoans vary from simple to complex. Several protozoan groups exist at all times in the trophozoite state. Many alternate between a trophozoite and a cyst stage, depending on the conditions of the habitat. The life cycle of a parasitic protozoon dictates its mode of transmission to other hosts. For example, the flagellate *Trichomonas vaginalis* causes a common sexually transmitted disease. Because it does not form cysts, it is more delicate, and must be transmitted by intimate contact between sexual partners. In contrast, intestinal pathogens such as *Entamoeba histolytica* and *Giardia lamblia* form cysts and are readily transmitted in contaminated water and foods.

All protozoa reproduce by relatively simple, asexual methods, usually mitotic cell division. Several parasitic species, including the agents of malaria and toxoplasmosis, reproduce asexually inside a host cell by multiple fission. Sexual reproduction also occurs during the life cycle of most protozoa. Ciliates participate in *conjugation,* a form of genetic exchange in which members of two different mating types fuse temporarily and exchange micronuclei. This process of sexual recombination yields new and different genetic combinations that can be advantageous in evolution.

Classification of Selected Medically Important Protozoa

Taxonomists have not escaped problems classifying protozoa. They, too, are very diverse and frequently frustrate attempts to generalize or place them in neat groupings. The most recent system

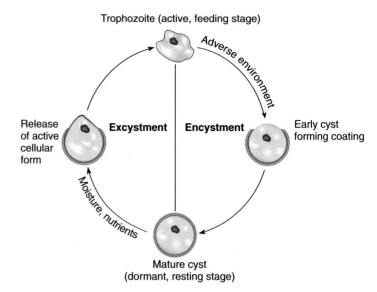

Trophozoite (active, feeding stage)

Adverse environment

Release of active cellular form

Excystment   **Encystment**

Early cyst forming coating

Moisture, nutrients

Mature cyst (dormant, resting stage)

FIGURE 5.27

The general life cycle exhibited by many protozoa. All protozoa have a trophozoite, but not all produce cysts.

places them in all five of the Kingdoms on the Eukarya tree (see figure 1.16), but this method may be more complex than is necessary for our survey. We will simplify this system by presenting four groups, based on method of motility, mode of reproduction, and stages in the life cycle, summarized as follows:

The Mastigophora (Flagellata)[6] Motility is primarily by flagella alone or by both flagellar and ameboid motion. Single nucleus. Sexual reproduction, when present, by syngamy; division by longitudinal fission. Several parasitic forms lack mitochondria and Golgi apparatus. Most species form cysts and are free-living; the group also includes several parasites. Some species are found in loose aggregates or colonies, but most are solitary. Members include: *Trypanosoma* and *Leishmania,* important blood pathogens spread by insect vectors: *Giardia,* an intestinal parasite spread in water contaminated with feces; *Trichomonas,* a parasite of the reproductive tract of humans spread by sexual contact (figure 5.28).

The Sarcodina[6] Cell form is primarily an ameba (figure 5.29). Major locomotor organelles are pseudopods, although some species have flagellated reproductive states. Asexual reproduction by fission. Two groups have an external shell; mostly uninucleate; usually encyst. Most amebas are free-living and not infectious; *Entamoeba* is a pathogen or parasite of humans; shelled amebas called foraminifera and radiolarians are responsible for chalk deposits in the ocean.

The Cilophora (Ciliata) Trophozoites are motile by cilia; some have cilia in tufts for feeding and attachment (figure 5.30); most develop cysts; have both macronuclei and micronuclei; division by

---

***trophozoite** (trof″-oh-zoh′-yte) Gr. *trophonikos,* to nourish, and *zoon,* animal.

6. Some biologists prefer to combine Mastigophora and Sarcodina into the phylum Sarcomastigophora. Because the algal group Euglenophyta has flagella, it may also be included in the Mastigophora.

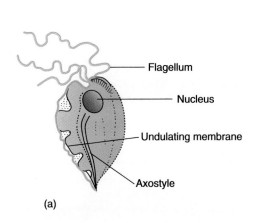

Flagellum

Nucleus

Undulating membrane

Axostyle

(a)

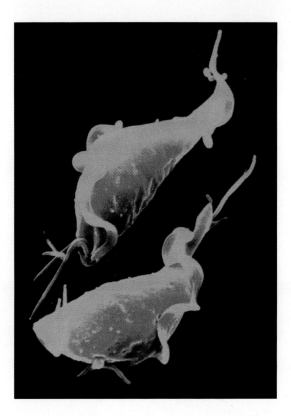

(b)

FIGURE 5.28

The structure of a typical mastigophoran, *Trichomonas vaginalis.* This genital tract pathogen, is shown in **(a)** a drawing and **(b)** a scanning electron micrograph.

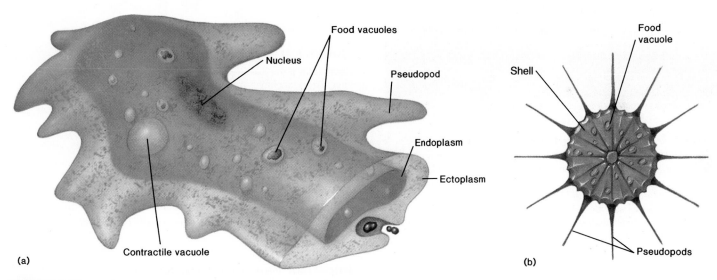

Food vacuoles

Nucleus

Pseudopod

Endoplasm

Ectoplasm

Contractile vacuole

(a)

Food vacuole

Shell

Pseudopods

(b)

FIGURE 5.29

Examples of sarcodinians. (a) The structure of an ameba. **(b)** Radiolarian, a shelled ameba with long, pointed pseudopods. Pseudopods are used in both movement and feeding.

transverse fission; most have a definite mouth and feeding organelle; show relatively advanced behavior. The majority of ciliates are free-living and harmless. One important pathogen, *Balantidium coli,* lives in vertebrate intestines and can infect humans.

The Apicomplexa (Sporozoa)   Motility is absent in most cells except male gametes. Life cycles are complex, with well-developed asexual and sexual stages. Sporozoa produce special sporelike cells

called **sporozoites*** (figure 5.31) following sexual reproduction, which are important in transmission of infections; most form thick-walled zygotes called oocysts; entire group is parasitic. *Plasmodium,* the most prevalent protozoan parasite, causes 100 million to 300 million cases of malaria each year worldwide. It is an intracellular parasite with a complex cycle alternating between humans and

*sporozoite (spor″-oh-zoh′-yte) Gr. *sporos,* seed, and *zoon,* animal.

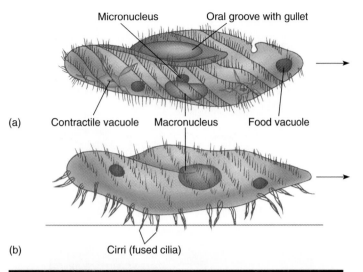

(a)

(b)

Cirri (fused cilia)

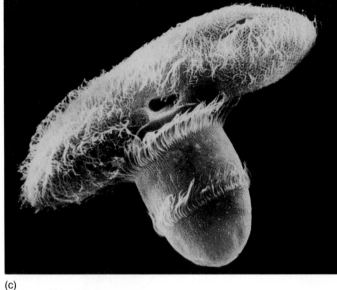

(c)

FIGURE 5.30

Selected ciliate representatives. (a) The structure of a typical representative, *Paramecium*. Cilia beat in coordinated waves, driving the cell forward and backward. Cilia are also used to sweep food particles into the gullet to form food vacuoles. (b) Fused cilia (cirri) of *Stylonychia* are used for walking over surfaces. (c) Scanning electron micrograph of a *Didinium* cell eating a *Paramecium* (1,400×). Note the number and distribution of cilia in each protozoon.

mosquitoes (see figure 23.11). *Toxoplasma gondii* causes an acute infection (toxoplasmosis) in humans, which is acquired from cats and other animals.

Protozoan Identification and Cultivation

The unique appearance of most protozoa makes it possible for a knowledgeable person to identify them to the level of genus and often species by microscopic morphology alone. Characteristics to consider in identification include the shape and size of the cell; the type, number, and distribution of locomotor structures; the presence of special organelles or cysts; and the number of nuclei. Medical specimens taken from blood, sputum, cerebrospinal fluid, fe-

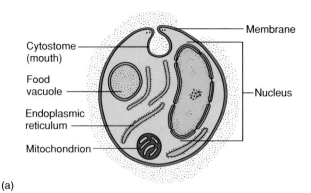

(a)

(b)

FIGURE 5.31

Sporozoan parasites. (a) General cell structure. Note the lack of specialized locomotor organelles. **(b)** Scanning electron micrograph of the sporozoite of *Cryptosporidium*, an intestinal parasite of humans and other mammals.

ces, or the vagina are smeared directly onto a slide and observed with or without special stains. Occasionally, protozoa are cultivated on artificial media or in laboratory animals for further identification or study.

Important Protozoan Parasites

Although protozoan infections are very common, they are actually caused by only a small number of species often restricted geographically to the tropics and subtropics. In this survey, we first introduce some concepts of the host-protozoan relationship and then look at examples that illustrate some of the main features of protozoan diseases. Details of several other medically important protozoa are given in chapter 23.

Parasitic protozoa are traditionally studied along with the helminths in the science of parasitology. Although a parasite is generally defined as an organism that obtains food and other requirements at the expense of a host, the range of host-parasite relationships can be very broad. At one extreme are the so-called good parasites, which occupy their host with little harm. An example is certain amebas that live in the human intestine and feed off organic matter there. At the other extreme are parasites (*Plasmodium, Trypanosoma*) that multiply in host tissues such as the blood or brain, causing severe damage and disease. Between these two extremes are parasites of varying pathogenicity, depending on their particular adaptations.

Most human parasites go through three general stages: (1) The microbe is transmitted to the human host from a source such as soil, water, food, other humans, or animals. (2) The microbe invades and multiplies in the host, producing more parasites that can infect other suitable hosts. (3) The microbe leaves the host in large numbers by a specific means and, to survive, must find and enter a new host. There are numerous variations on this simple theme. For instance, the microbe can invade more than one host species (alternate hosts) and undergo several changes as it cycles through these hosts, such as sexual reproduction or encystment. Some microbes are spread from human to human by means of **vectors,*** defined as animals such as insects that carry diseases. Others can be spread through bodily fluids and feces.

Pathogenic Flagellates: Trypanosomes   Trypanosomes are parasites belonging to the genus *Trypanosoma.** The two most important representatives are *T. brucei* and *T. cruzi,* species that are closely related but geographically restricted. *Trypanosoma brucei* occurs in Africa, where it causes approximately 10,000 new cases of sleeping sickness each year. *Trypanosoma cruzi,* the cause of Chagas disease,[7] is endemic to South and Central America, where it infects several million people a year. Both species have long, crescent-shaped cells with a single flagellum that is sometimes attached to the cell body by an undulating membrane (figure 5.32). Both occur in the blood during infection and are transmitted by blood-sucking vectors. We will use *T. cruzi* to illustrate the phases of a trypanosomal life cycle and to demonstrate the complexity of parasitic relationships.

The trypanosome of Chagas disease relies on the close relationship of a warm-blooded mammal and an insect that feeds on mammalian blood. The mammalian hosts are numerous, including dogs, cats, opossums, armadillos, and foxes. The vector is the *reduviid* bug,* an insect that is sometimes called the "kissing bug" because of its habit of biting its host at the corner of the mouth. Transmission occurs from bug to mammal and from mammal to bug, but usually not from mammal to mammal, except across the placenta during pregnancy. The general phases of this cycle are presented in figure 5.32.

The trypanosome trophozoite multiplies in the intestinal tract of the reduviid bug and is harbored in the feces. The bug seeks a host and bites the mucous membranes, usually of the eye, nose, or lips. As it fills with blood, the bug soils the bite with feces containing the trypanosome. Ironically, the victims themselves inadvertently contribute to the entry of the microbe by scratching the bite wound. The trypanosomes ultimately become established and mul-

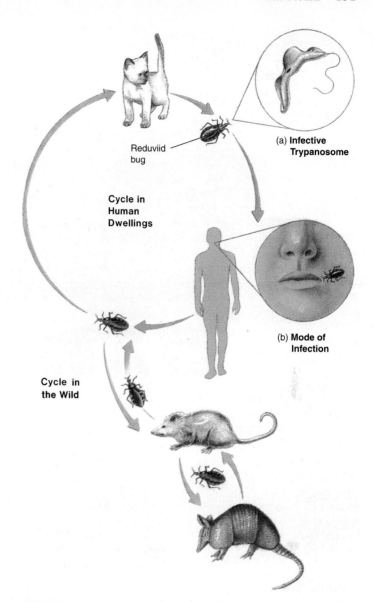

FIGURE 5.32

Cycle of transmission in Chagas disease. Trypanosomes (inset *a*) are transmitted among mammalian hosts and human hosts by means of a bite from the kissing bug (inset *b*).

tiply in muscle and white blood cells. Periodically, these parasitized cells rupture, releasing large numbers of new trophozoites into the blood. Eventually, the trypanosome can spread to many systems, including the lymphoid organs, heart, liver, and brain. Manifestations of the resultant disease range from mild to very severe, and include fever, inflammation, and heart and brain damage. In many cases, the disease has an extended course and can cause death.

Infective Amebas: Entamoeba   Several species of amebas cause disease in humans, but probably the most common disease is amebiasis, or amebic dysentery,* caused by *Entamoeba histolytica.* This

---

7. Named for Carlos Chagas, the discoverer of *T. cruzi.*

   ***vector** (vek′-tur) L. *vectur,* one who carries.

   ***Trypanosoma** (try″-pan-oh-soh′-mah) Gr. *rypanon,* borer, and *soma,* body.

   ***reduviid** (ree-doo′-vee-id) A member of a large family of flying insects with sucking, beaklike mouths.

***dysentery** (dis′-en-ter″-ee) Any inflammation of the intestine accompanied by bloody stools. It can be caused by a number of factors, both microbial and nonmicrobial.

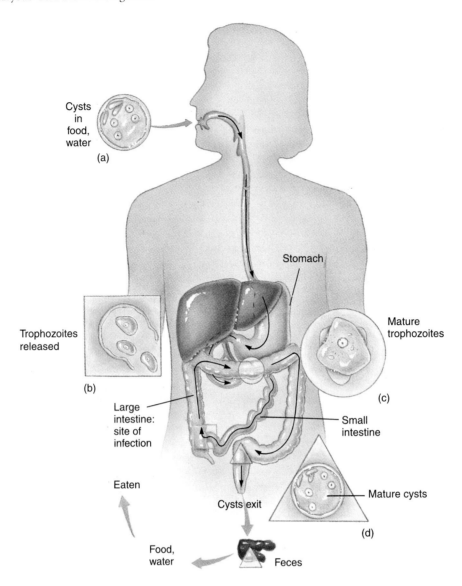

FIGURE 5.33

Stages in the infection and transmission of amebic dysentery. Arrows show the route of infection; insets show the appearance of *Entamoeba histolytica*. **(a)** Cysts are eaten. **(b)** Trophozoites (amebas) emerge from cysts. **(c)** Trophozoites invade the large intestinal wall. **(d)** Mature cysts are released in the feces, and may be spread through contaminated food and water.

microbe is widely distributed in the world, from northern zones to the tropics, and is nearly always associated with humans. Amebic dysentery is the fourth most common protozoan infection in the world. This microbe has a life cycle quite different from the trypanosomes in that it does not involve multiple hosts and a blood-sucking vector. It lives part of its cycle as a trophozoite and part as a cyst. Because the cyst is the more resistant form and can survive in water and soil for several weeks, it is the more important stage for transmission. The primary way that people become infected is by ingesting food or water contaminated with human feces.

Figure 5.33 shows the major features of the amebic dysentery cycle, starting with the ingestion of cysts. The viable, heavy-walled cyst passes through the stomach unharmed. Once inside the small intestine, the cyst germinates into a large multinucleate ameba that subsequently divides to form small amebas (the trophozoite stage). These trophozoites migrate to the large intestine and begin to feed and grow. From this site, they can penetrate the lining of the intestine and invade the liver, lungs, and skin. Common symptoms include gastrointestinal disturbances such as nausea, vomiting, and diarrhea, leading to weight loss and dehydration. Untreated cases with extensive damage to the organs experience a high death rate. The cycle is completed in the infected human when certain trophozoites in the feces begin to form cysts, which then pass out of the body with fecal matter. Knowledge of the amebic cycle and role of cysts has been a boon in controlling the disease. Important preventive measures include sewage treatment, curtailing the use of human feces as fertilizers, and adequate sanitation of food and water.

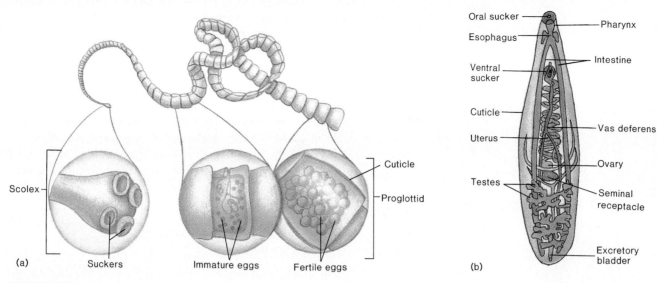

FIGURE 5.34

Parasitic flatworms. (a) A cestode (beef tapeworm), showing the scolex; long, tapelike body; and magnified views of immature and mature proglottids. **(b)** The structure of a trematode (liver fluke). Note the suckers that attach to host tissue and the dominance of reproductive and digestive organs.

The Parasitic Helminths

Tapeworms, flukes, and roundworms are collectively called **helminths,** from the Greek word meaning worm. Adult animals are usually large enough to be seen with the naked eye, and they range from the longest tapeworms, measuring up to about 23 m in length, to roundworms less than 1 mm in length. Nevertheless, they are included among microorganisms because the microscope is necessary to see the details of smaller helminths and to identify their eggs and larvae.

On the basis of morphological form, the two major groups of parasitic helminths are the **flatworms** (Phylum Platyhelminthes), with a very thin, often segmented body plan (figure 5.34), and the **roundworms** (Phylum Aschelminthes, also called **nematodes***), with an elongate, cylindrical, unsegmented body plan (figure 5.35). The flatworm group is subdivided into the **cestodes,*** or tapeworms, named for their long, ribbonlike arrangement, and the **trematodes,*** or flukes, characterized by flat, ovoid bodies. Not all flatworms and roundworms are parasites by nature; many live free in soil and water. Both here and in chapter 23, we concern ourselves with the medically important worms (see table 23.4).

GENERAL WORM MORPHOLOGY

All helminths are multicellular animals equipped to some degree with organs and organ systems. In parasites, the most developed organs are those of the reproductive tract, with some degree of reduction in the digestive, excretory, nervous, and muscular systems. In particular groups, such as the cestodes, reproduction is so dominant that the worms are reduced to little more than a series of flattened sacs filled with ovaries, testes, and eggs (see figure 5.34a). Not all worms have such extreme adaptations as cestodes, but most have a highly developed reproductive potential, thick cuticles for protection, and mouth glands for breaking down the host's tissue.

LIFE CYCLES AND REPRODUCTION

Many worms have complex life cycles that alternate between hosts. Reproduction of individuals is primarily sexual, involving the production of eggs and sperm in the same worm or in separate male and female worms. Fertilized eggs are usually released to the environment and are provided with a protective shell and extra food to aid their development into larvae. Even so, most eggs and larvae are vulnerable to heat, cold, drying, and predators and are destroyed or unable to reach a new host. To counteract this formidable mortality rate, certain worms have adapted a reproductive capacity that borders on the incredible: A single female *Ascaris*[8] can lay 200,000 eggs a day, and a large female can contain over 25,000,000 eggs at varying stages of development! If only a tiny number of these eggs makes it to another host, the parasite will have been successful in completing its life cycle.

A HELMINTH CYCLE: THE PINWORM

To illustrate a helminth cycle in humans, we will use the example of a roundworm, *Enterobius vermicularis,* the pinworm or seatworm.

---

*nematode (neem′-ah-tohd) Gr. *nemato,* thread, and *eidos,* form.

*cestode (sess′-tohd) L. *cestus,* a belt, and *ode,* like.

*trematode (treem′-a-tohd) Gr. *trema,* hole. Named for the appearance of having tiny holes.

---

8. *Ascaris* is a genus of parasitic intestinal roundworms.

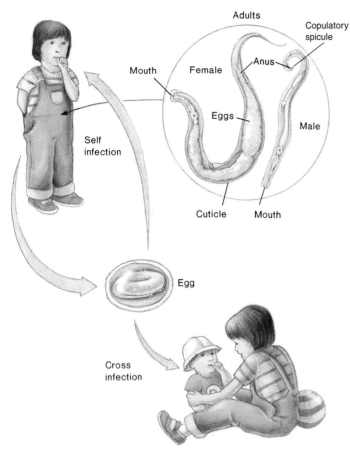

FIGURE 5.35

🔵 **The life cycle of the pinworm, a roundworm.** Eggs are the infective stage and are transmitted by unclean hands. Children frequently reinfect themselves and also pass the parasite on to others.

This worm causes a very common infestation of the large intestine (see figure 5.35). Worms range from 2 to 12 mm long and have a tapered, curved cylinder shape. The condition they cause, enterobiasis, is usually a simple, uncomplicated infection that does not spread beyond the intestine.

A cycle starts when a person swallows microscopic eggs picked up from another infected person by direct contact or by touching articles that person has touched. The eggs hatch in the intestine and then release larvae that mature into adult worms in the appendix (within about one month). There, male and female worms mate, and the female migrates out to the anus to deposit eggs, which cause intense itchiness that is relieved by scratching. Herein lies a significant means of dispersal: Scratching contaminates the fingers, which, in turn, transfer eggs to bedclothes and other inanimate objects. This person becomes a host and a source of eggs and can spread them to others in addition to reinfesting himself (see figure 5.35). Enterobiasis occurs most often among families and in other close living situations. Its distribution is worldwide among all

socioeconomic groups, but it seems to attack younger people more frequently than older ones.

HELMINTH CLASSIFICATION AND IDENTIFICATION

The helminths are classified according to their shape; their size; the degree of development of various organs; the presence of hooks, suckers, or other special structures; the mode of reproduction; the kinds of hosts; and the appearance of eggs and larvae. They are identified in the laboratory by microscopic detection of the adult worm or its larvae and eggs, which often have distinctive shapes or external and internal structures (see chapter 23). Occasionally, they are cultured in order to verify all of the life stages.

DISTRIBUTION AND IMPORTANCE OF PARASITIC WORMS

About 50 species of helminths parasitize humans. They are distributed in all areas of the world that support human life. Some worms are restricted to a given geographic region, and many have a higher incidence in tropical areas. This knowledge must be tempered with the realization that jet-age travel, along with human migration, are gradually changing the patterns of worm infections, especially of those species that do not require alternate hosts or special climatic conditions for development. The yearly estimate of worldwide cases numbers in the billions, and these are not confined to developing countries. A conservative estimate places 50,000,000 helminth infections in North America alone. The primary targets are malnourished children.

CHAPTER CHECKPOINTS

The eucaryotic microorganisms include the Fungi (Myceteae), the Protista (algae and protozoa), and the Helminths (Kingdom Animalia).

The Kingdom Fungi (Myceteae) is composed of non-photosynthetic haploid species with cell walls. The fungi are either saprobes or parasites, and may be unicellular, colonial, or multicellular. Forms include yeasts (unicellular budding cells) and molds (filamentous cells called hyphae). Their primary means of reproduction involves asexual and sexual spores.

The protists are mostly unicellular or colonial eucaryotes that lack specialized tissues. There are two major organism types: the Algae and the Protozoa. Algae are photosynthetic organisms that contain chloroplasts with chlorophyll and other pigments. Protozoa are heterotrophs that usually display some form of locomotion. Most are single-celled trophozoites, and many produce a resistant stage, or cyst.

The Kingdom Animalia has only one group that contains microscopic members. These are the helminths or worms. Parasitic members include flatworms and roundworms, that are able to invade and reproduce in human tissues.

CHAPTER CAPSULE WITH KEY TERMS

I. **Eucaryotic Structure**

A. Major organelles and other structural features include: Appendages (cilia, flagella), glycocalyx, cell wall, cytoplasmic (or cell) membrane, ribosomes, organelles (nucleus, nucleolus, endoplasmic reticulum, Golgi complex, mitochondria, chloroplasts, cytoskeleton, microfilaments, microtubules). A review comparing the major differences between eucaryotic and procaryotic cells is provided in table 2.6, page 52.

II. **The Kingdom Fungi (Myceteae)**

Common names of two particular types: macroscopic fungi (mushrooms, bracket fungi, puffballs) and microscopic fungi (yeasts, molds).

A. *Overall Morphology:* At the cellular (microscopic) level, typical eucaryotic cell, with thick **cell walls. Yeasts** are single cells that form **buds** and **pseudohyphae. Hyphae** are long, tubular filaments that can be septate or nonseptate and grow in a network called a **mycelium;** hyphae are characteristic of the filamentous fungi called **molds.** Some fungi (mushrooms) produce multicellular structures such as fleshy fruiting bodies. Fungal cells are largely **nonmotile,** except for some motile gametes.

B. *Nutritional Mode/Distribution:* All are **heterotrophic.** The majority are harmless **saprobes** living off organic **substrates** such as dead animal and plant tissues. A few are **parasites,** living on the tissues of other organisms, but none are obligate. Preferred temperature of growth is 20°–40°C. Distribution is extremely widespread in many habitats.

C. *Reproduction:* Primarily through **spores** formed on special **reproductive hyphae.** In **asexual reproduction,** spores are formed through budding, partitioning of a hypha, or in special sporogenous structures; examples are **conidia** and **sporangiospores.** In **sexual reproduction,** spores are formed following fusion of male and female strains and the formation of a sexual structure; sexual spores are one basis for classification.

D. *Major Groups:* The four main divisions among the terrestrial fungi, given with sexual spore type, are **Zygomycota (zygospores), Ascomycota (ascospores), Basidiomycota (basidiospores),** and **Deuteromycota** (no sexual spores).

E. *Importance:* Essential decomposers of plant and animal detritis in the environment with return of valuable nutrients to the ecosystem. Economically beneficial as sources of antibiotics; used in making foods and in genetic studies. Adverse impact: decomposition of fruits and vegetables; several fungi cause infections, or **mycoses;** some produce substances that are toxic if eaten.

III. **Microscopic Protists**

A. The Algae

Include plantlike protists, kelps, and seaweeds. Specific groups are the **euglenids, green algae, diatoms, dinoflagellates, brown algae,** and **red seaweeds.**

1. *Overall Morphology:* Contain **chloroplasts** with **chlorophyll** and other pigments; cell wall; may or may not have flagella. Microscopic forms are unicellular, colonial, filamentous; macroscopic forms are colonial and multicellular.

2. Nutritional Mode/Distribution: **Photosynthetic;** most are free-living in the aquatic environment, both fresh water and marine (common component of **plankton**).

3. *Major Groups:* Classification according to types of pigments and cell walls. Microscopic algae: Euglenophyta, Chlorophyta, Chrysophyta, Pyrrophyta: Macroscopic algae: Phaephyta (kelps) and Rhodophyta (seaweeds) are multicellular algae.

4. *Importance:* Algae provide the basis of the food web in most aquatic habitats, and they produce a large proportion of atmospheric O_2 through photosynthesis. Some are harvested as a source of cosmetics, food, and medical products. Dinoflagellates cause red tides and give off toxins that are harmful to humans and animals.

B. The Protozoa

Include animal-like **protists;** unicellular animals such as **amebas, flagellates, ciliates, sporozoa.**

1. *Overall Morphology:* Most are unicellular, colonies rare, no multicellular forms; most have locomotor structures, such as **flagella, cilia, pseudopods;** special feeding structures can be present; lack a cell wall; extreme variations in shape. Can exist in **trophozoite,** a motile, feeding stage, or **cyst,** a dormant resistant stage.

2. *Nutritional Mode/Distribution:* All are **heterotrophic.** Most are free-living in a moist habitat (water, soil); feed by engulfing other microorganisms and organic matter. A number of animal parasites; can be spread from host to host by insect **vectors.**

3. *Reproduction:* Asexual by binary fission and **mitosis,** budding; sexual by fusion of free-swimming gametes, conjugation.

4. *Major Groups:* Protozoa are subdivided into four groups based upon mode of locomotion and type of reproduction: **Mastigophora,** the flagellates, motile by flagella; **Sarcodina,** the amebas, motile by pseudopods; **Ciliophora,** the ciliates, motile by cilia; **Apicomplexa,** all parasites; motility not well developed; produce unique reproductive structures.

5. *Importance*: Ecologically important in food webs and decomposing organic matter. Medical significance: hundreds of millions of people are afflicted with one of the many protozoan infections (malaria, trypanosomiasis, amebiasis).

IV. **The Helminth Parasites**

Includes parasitic worms, tapeworms, flukes, nematodes.

A. *Overall Morphology:* Animal cells; multicellular; individual organs specialized for reproduction, digestion, movement, protection, though some of these are reduced.

B. *Nutritional/Reproductive Mode:* Parasitize host tissues; have mouthparts for attachment to or digestion of host tissues. Most have well-developed sex organs that produce eggs and sperm. Fertilized eggs go through larval period in or out of host body.

C. *Major Groups:*

1. **Flatworms** have highly flattened body; no definite body cavity; digestive tract a blind pouch; simple excretory and nervous systems. **Cestodes** (tapeworms) are long chains of segments; attach to host's intestine by hooked mouthpart. **Trematodes,** or flukes, are flattened, nonsegmented worms with sucking mouthparts.

2. **Roundworms (nematodes)** have round bodies, a complete digestive tract, a protective surface cuticle, spines and hooks on mouth; excretory and nervous systems poorly developed.

D. *How Transmitted and Acquired:* Through ingestion of larvae or eggs in food; from soil or water; some are carried by insect vectors.

E. *Importance:* Afflict billions of humans. Because these worms are so common, they have medical and economic impact of staggering proportions.

MULTIPLE-CHOICE QUESTIONS

1. Both flagella and cilia are found primarily in
   a. algae
   b. protozoa
   c. fungi
   d. both b and c

2. Features of the nuclear envelope include
   a. ribosomes
   b. a double membrane structure
   c. pores that allow communication with the cytoplasm
   d. b and c
   e. all of these

3. In general, if two haploid cells fuse, _____ will result.
   a. a germ cell
   b. a diploid zygote
   c. mitosis
   d. meiosis

4. The cell wall is found in which eucaryotes?
   a. fungi
   b. algae
   c. protozoa
   d. a and b

5. What is embedded in rough endoplasmic reticulum?
   a. ribosomes
   b. Golgi apparatus
   c. chromatin
   d. vesicles

6. Yeasts are _____ fungi, and molds are _____ fungi.
   a. macroscopic, microscopic
   b. unicellular, filamentous
   c. motile, nonmotile
   d. water, terrestrial

7. In general, fungi derive nutrients through
   a. photosynthesis
   b. engulfing bacteria
   c. digesting organic substrates
   d. parasitism

8. A hypha divided into compartments by cross walls is called
   a. nonseptate
   b. imperfect
   c. septate
   d. perfect

9. A conidium is a/an _____ spore, and a zygospore is a/an _____ spore.
   a. sexual, asexual
   b. free, endo
   c. ascomycete, basidiomycete
   d. asexual, sexual

10. Algae generally contain some types of
    a. spore
    b. chlorophyll
    c. locomotor organelle
    d. toxin

11. All protozoa have a
    a. locomotor organelle
    b. cyst stage
    c. pellicle
    d. trophozoite

12. The protozoan trophozoite is the
    a. active feeding stage
    b. inactive dormant stage
    c. infective stage
    d. spore-forming stage

13. All mature sporozoa are
    a. parasitic
    b. nonmotile
    c. carried by vectors
    d. both a and b

14. Helminth parasites reproduce with
    a. spores
    b. eggs and sperm
    c. mitosis
    d. cysts
    e. all of these

15. Mitochondria likely originated from
    a. archaea
    b. invaginations of the cell membrane
    c. purple bacteria
    d. cyanobacteria

16. **Single Matching.** Select the description that best fits the word in the left column.
    _____ diatom
    _____ *Rhizopus*
    _____ *Histoplasma*
    _____ *Cryptococcus*
    _____ euglenid
    _____ dinoflagellate
    _____ *Trichomonas*
    _____ *Entamoeba*
    _____ *Plasmodium*
    _____ *Enterobius*

    a. the cause of malaria
    b. single-celled alga with silica in its cell wall
    c. fungal cause of Ohio Valley fever
    d. the cause of amebic dysentery
    e. genus of black bread mold
    f. helminth worm involved in pinworm infection
    g. motile flagellated alga with eyespots
    h. a yeast that infects the lungs
    i. flagellated protozoan genus that causes an STD
    j. alga that causes red tides

CONCEPT QUESTIONS

1. Construct a chart indicating the major similarities and differences between procaryotic and eucaryotic cells.

2. a. Which kingdoms of the five-kingdom system contain eucaryotic microorganisms? How do unicellular, colonial, and multicellular organisms differ from each other?
   b. Give examples of each type.

3. a. Describe the anatomy and functions of each of the major eucaryotic organelles.
   b. How are flagella and cilia similar? How are they different?
   c. Compare and contrast the smooth ER, the rough ER, and the Golgi apparatus in structure and function.

4. Trace the synthesis of cell products, their processing, and their packaging through the organelle network.

5. a. Describe the detailed structure of the nucleus.
   b. Why can one usually not see the chromosomes?
   c. When are the chromosomes visible?
   d. What causes them to be visible?

6. a. Define mitosis and explain its function.
   b. What happens to the chromosome number during this process?
   c. How does a diploid organism remain diploid and a haploid organism remain haploid?

7. a. Define meiosis and explain its function.
   b. When does it occur in diploid organisms?
   c. In haploid organisms?
   d. How does it differ from mitosis?

8. Describe some of the ways that organisms use lysosomes.

9. For what reasons would a cell need a "skeleton?"

10. a. Differentiate between the yeast and hypha types of fungal cell.
    b. What is a mold?
    c. What does it mean if a fungus is dimorphic?

11. a. How does a fungus feed?
    b. Where would one expect to find fungi?

12. a. Describe the functional types of hyphae.
    b. Describe the two main types of asexual fungal spores and how they are formed.
    c. What are some types of conidia?
    d. What is the reproductive potential of molds in terms of spore production?
    e. How do mold spores differ from procaryotic spores?

13. a. Explain the importance of sexual spores to fungi.
    b. Describe the three main types of sexual spores, and show how each is formed by means of a simple diagram.

14. How are fungi classified? Give an example of a member of each fungus division and describe its structure and importance.

15. What is a mycosis? What kind of mycosis is athlete's foot? What kind is coccidioidomycosis?

16. What is a working definition of a "protist?"

17. a. Describe the principal characteristics of algae that separate them from protozoa.
    b. How are algae important?
    c. What causes the many colors in the algae?
    d. Are there any algae of medical importance?

18. a. Explain the general characteristics of the protozoan life cycle.
    b. Describe the protozoan adaptations for feeding.
    c. Describe protozoan reproductive processes.

19. a. Briefly outline the characteristics of the four protozoan groups.
    b. What is an important pathogen in each group?

20. a. Which protozoan group is the most complex in structure and behavior?
    b. In life cycle?
    c. What characteristics set the apicomplexa apart from the other protozoan groups?

21. a. Construct a chart that compares the four groups of eucaryotic microorganisms (fungi, algae, protozoa, helminths) in cellular structure.
    b. Indicate whether the group has a cell wall, chloroplasts, motility, or some other distinguishing feature.
    c. Include also the manner of nutrition and body plan (unicellular, colonial, filamentous, or multicellular).

CRITICAL-THINKING QUESTIONS

1. Suggest some ways that one would go about determining if mitochondria and chloroplasts are a modified procaryotic cell.

2. Give the common name of a eucaryotic microbe that is unicellular, walled, non-photosynthetic, nonmotile, and bud-forming.

3. Give the common name of a microbe that is unicellular, nonwalled, motile with flagella, and has chloroplasts.

4. Which group of microbes has long, thin pseudopods and is encased in a hard shell?

5. What general type of multicellular parasite is composed primarily of thin sacs of reproductive organs?

6. a. Name two parasites that are transmitted in the cyst form.
   b. How must a non-cyst-forming pathogenic protozoan be transmitted? Why?

7. You just found an old container of food in the back in your refrigerator. You open it and see a mass of multicolored fuzz. As a budding microbiologist, describe how you would determine what types of organisms are growing on the food.

8. Explain what factors could cause opportunistic mycoses to be a growing medical problem.

9. a. How are bacterial endospores and cysts of protozoa alike?
   b. How do they differ?

10. You have gone camping in the mountains and plan to rely on water present in forest pools and creeks for drinking water. Certain encysted pathogens often live in this type of water, but you do not discover this until you arrive at the campground. How might you treat the water to prevent becoming infected?

11. a. Explain the two levels of parasitism at work in Chagas disease.
    b. Do you suppose the trypanosome has a parasite, too?
    c. If so, could its parasite have a parasite?
    d. What does this tell you about the prevalence of parasitism and its success as a mode of life?
    e. What is a potential weakness in parasitism?

12. Can you think of a way to determine if a child is suffering from pinworms? Hint: Scotch tape is involved.

INTERNET SEARCH TOPICS

Use the World Wide Web to explore two topics:

1. The endosymbiotic theory of eucaryotic cell evolution. List data from studies that support this idea.

2. The medical problems caused by mycotoxins and "sick-building" syndrome.

An Introduction to the Viruses

The notion of viruses can inspire a sense of mystery and awe. At times we assign to them imaginary powers that they do not have, and yet, at other times, they seem to have powers beyond imagination. Despite success in eradicating viral diseases like smallpox and polio, humans continue to experience outbreaks of new viral infections every few months. In fact, viruses are the most prominent emerging microbes, accounting for over 50% of new outbreaks of infectious diseases worldwide.

Recently, infections with the West Nile virus were reported in parts of New England, resulting in several deaths and prompting a great deal of concern (chapter opening photo). This virus has somehow migrated from Africa or Asia and become established in American birds and mosquitoes. At about the same time, the Nipah virus jumped hosts from bats to pigs to humans, causing a deadly epidemic among agricultural workers in parts of Asia.

Meanwhile in southern California, virologists have detected high levels of viruses in the seawater near storm drains and sewage outfalls that are putting millions of beachgoers at risk for hepatitis A and other intestinal infections. Viruses are often blamed for unexplained illnesses and symptoms, and are usually considered the culprit when other infectious agents have been ruled out. They are tied intimately to their host cells, where they often linger around and even become part of their host's genetic material. Because of their roles in disease and genetics, it is very important to have a working knowledge of the basic characteristics of viruses. The primary aim of this chapter is to familiarize you with their many unique properties and to provide a survey of their structure, physiology, multiplication, and diversity.

A young man demonstrates a net designed to keep mosquitoes off. This was just one of several protective strategies used by residents of the Northeast to avoid exposure to the West Nile fever virus. Other measures were mass spraying of insecticides and staying indoors.

Chapter Overview

Viruses:

- Are a unique group of tiny infectious particles that are obligate parasites of cells.
- Do not exhibit the characteristics of life, but can regulate the functions of host cells.
- Infect all groups of living things and produce a variety of diseases.
- Are not cells but resemble complex molecules composed of protein and nucleic acid.
- Are encased in an outer shell or envelope and contain either DNA or RNA as their genetic material.
- Are genetic parasites that take over the host cell's metabolism and synthetic machinery.
- Can instruct the cell to manufacture new virus parts and assemble them.
- Are released in a mature, infectious form, followed by destruction of the host cell.
- May persist in cells, leading to slow progressive diseases and cancer.
- Are identified by structure, host cell, type of nucleic acid, outer coating, and type of disease.
- Are among the most common infectious agents, causing serious medical and agricultural impact.

The Search for the Elusive Viruses

The discovery of the light microscope made it possible to see firsthand the agents of many bacterial, fungal, and protozoan diseases. But the techniques for observing and cultivating these relatively large microorganisms were virtually useless for viruses. For many years, the cause of viral infections such as smallpox and polio was unknown, even though it was clear that the diseases were transmitted from person to person. The French bacteriologist Louis Pasteur was certainly on the right track when he postulated that rabies was caused by a "living thing" smaller than bacteria, and in 1884 he was able to develop the first vaccine for rabies. Pasteur also proposed the term **virus** (L. poison) to denote this special group of infectious agents.

The first substantial revelations about the unique characteristics of viruses occurred in the 1890s. First, D. Ivanovski and M. Beijerinck showed that a disease in tobacco was caused by a virus (tobacco mosaic virus). Then, Friedrich Loeffler and Paul Frosch discovered an animal virus that causes foot-and-mouth disease in cattle. These early researchers found that when infectious fluids from host organisms were passed through porcelain filters designed to trap bacteria, the filtrate still remained infectious. This result proved that an infection could be caused by a cell-free fluid containing agents smaller than bacteria and thus first introduced the concept of a *filterable virus*.

Over the succeeding decades, a remarkable picture of the physical, chemical, and biological nature of viruses began to take form. Years of experimentation were required to show that viruses were noncellular particles with a definite size, shape, and chemical composition. Like bacteria, they could be cultured in the laboratory. By the 1950s, virology had grown into a multifaceted discipline that promised to provide much information on disease, genetics, and even life itself.

The Position of Viruses in the Biological Spectrum

Viruses are a unique group of biological entities known to infect every type of cell, including bacteria, algae, fungi, protozoa, plants, and animals. Although the emphasis in this chapter is on animal viruses, much credit for our knowledge must be given to experiments with bacterial and plant viruses. The exceptional and curious nature of viruses prompts numerous questions, including: (1) Are they organisms; that is, are they alive? (2) What are their distinctive biological characteristics? (3) How can particles so small, simple, and seemingly insignificant be capable of causing disease and death? and (4) What is the connection between viruses and cancer? In this chapter, we address these questions and many others.

The unusual structure and behavior of viruses have led to debates about their connection to the rest of the microbial world. One viewpoint holds that viruses are unable to exist independently from the host cell, so they are not living things but are more akin to large, infectious molecules. Another viewpoint proposes that even though viruses do not exhibit most of the life processes of cells (discussed in chapter 2), they can direct them and thus are certainly more than inert and lifeless molecules. Depending upon the circumstances, both views are defensible. In applied virology, this debate has greater philosophical than practical importance because viruses are agents of disease and must be dealt with through control, therapy, and prevention, whether we regard them as living or not. In keeping with their special position in the biological spectrum, it is best to describe viruses as *infectious particles* (rather than organisms) and as either *active* or *inactive* (rather than alive or dead).

TABLE 6.1
Novel Properties of Viruses

- **Are obligate intracellular parasites of bacteria, protozoa, fungi, algae, plants, and animals.**
- **Ultramicroscopic size, ranging from 20 nm up to 450 nm (diameter).**
- **Are not cells; structure is very compact and economical.**
- **Do not independently fulfill the characteristics of life (see chapter 2).**
- **Are inactive macromolecules outside of the host cell and active only inside host cells.**
- **Are geometric; can form crystal-like masses.**
- **Basic structure consists of protein shell (capsid) surrounding nucleic acid core.**
- **Nucleic acid can be either DNA or RNA but not both.**
- **Nucleic acid can be double-stranded DNA, single-stranded DNA, single-stranded RNA, or double-stranded RNA.**
- **Molecules on virus surface impart high specificity for attachment to host cell.**
- **Multiply by taking control of host cell's genetic material and regulating the synthesis and assembly of new viruses.**
- **Lack enzymes for most metabolic processes.**
- **Lack machinery for synthesizing proteins.**

Viruses are different from their host cells in size, structure, behavior, and physiology. They are a type of **obligate intracellular parasites** that cannot multiply unless they invade a specific host cell and instruct its genetic and metabolic machinery to make and release quantities of new viruses. Because of this characteristic, viruses are capable of causing serious damage and disease. Other unique properties of viruses are summarized in table 6.1.

CHAPTER CHECKPOINTS	

Viruses are noncellular entities whose properties have been identified through technological advances in microscopy and tissue culture.

Viruses are infectious particles that invade every known type of cell. They are not alive, yet they are able to redirect the metabolism of living cells to reproduce virus particles.

Viral replication inside a cell usually causes death or disease of that cell.

The General Structure of Viruses

SIZE RANGE

As a group, viruses represent the smallest infectious agents (with some unusual exceptions to be discussed later in this chapter). Their size relegates them to the realm of the **ultramicroscopic.** This term means that most of them are so minute (< 0.2 μm) that an electron microscope is necessary to detect them or to examine their fine structure. They are dwarfed by their host cells: More than 2,000 bacterial viruses could fit into an average bacterial cell, and more than 50 million polioviruses could be accommodated by an average human cell. Animal viruses range in size

TABLE 6.2
Relative Sizes of Selected Cells, Viruses, and Molecules

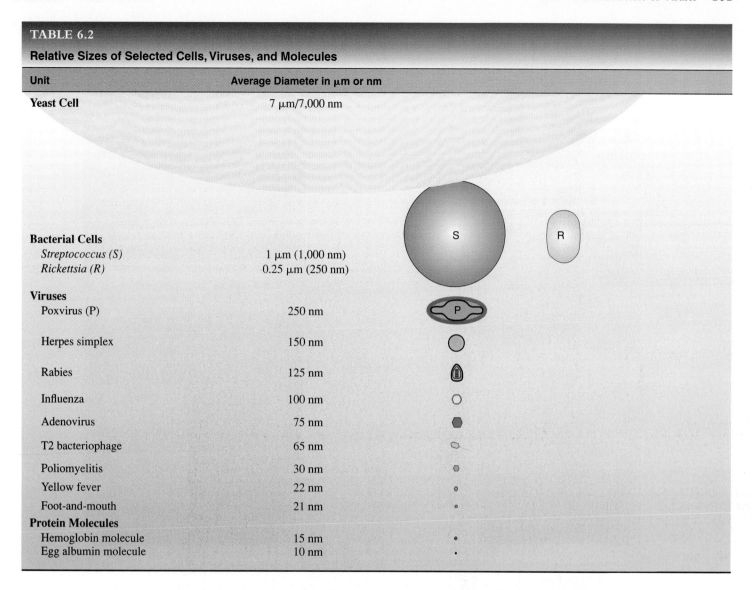

Unit	Average Diameter in μm or nm
Yeast Cell	7 μm/7,000 nm
Bacterial Cells	
Streptococcus (S)	1 μm (1,000 nm)
Rickettsia (R)	0.25 μm (250 nm)
Viruses	
Poxvirus (P)	250 nm
Herpes simplex	150 nm
Rabies	125 nm
Influenza	100 nm
Adenovirus	75 nm
T2 bacteriophage	65 nm
Poliomyelitis	30 nm
Yellow fever	22 nm
Foot-and-mouth	21 nm
Protein Molecules	
Hemoglobin molecule	15 nm
Egg albumin molecule	10 nm

from the small parvoviruses[1] (around 20 nm [0.02 μm] in diameter) to poxviruses[2] that are as large as small bacteria (up to 450 nm [0.4 μm] in length). Some cylindrical viruses are relatively long (800 nm [0.8 μm] in length) but so narrow in diameter (15 nm [0.015 μm]) that their visibility is still limited without the high magnification and resolution of an electron microscope. Table 6.2 compares the sizes of several viruses with procaryotic and eucaryotic cells and molecules.

Viral architecture is most readily observed through special stains in combination with electron microscopy (figure 6.1). Negative staining uses very thin layers of an opaque salt to outline the shape of the virus against a dark background and to enhance textural features on the viral surface. Internal details are revealed by positive staining of specific parts of the virus such as protein or nucleic acid. The *shadowcasting* technique attaches a virus preparation to a surface and showers it with a dense metallic vapor directed from a certain angle. The thin metal coating over the surface of the virus approximates its contours, and a shadow is cast on the unexposed side.

---

1. DNA viruses that cause respiratory infections in humans.

2. A group of large, complex viruses, including smallpox, that cause raised skin swellings called pox.

UNIQUE VIRAL CONSTITUENTS: CAPSIDS, NUCLEIC ACIDS, AND ENVELOPES

It is important to realize that viruses bear no real resemblance to cells and that they lack any of the protein-synthesizing machinery found in even the simplest cells. Their molecular structure is composed of regular, repeating subunits that give rise to their crystalline appearance. Indeed, many purified viruses can form large aggregates or crystals if subjected to special treatments (figure 6.2). The general plan of virus organization is the utmost in its simplicity and compactness. Viruses contain only those parts needed to invade and control a host cell: an external coating and a core containing one or more nucleic acid strands of either DNA or RNA. This pattern of organization can be represented with a flowchart:

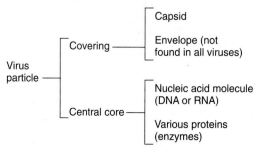

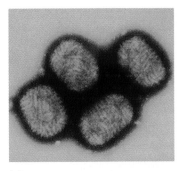

(a)

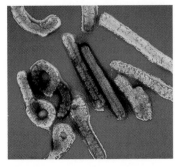

(b)

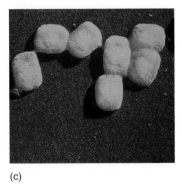

(c)

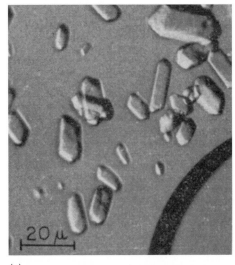

(a)

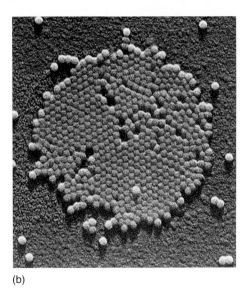

(b)

FIGURE 6.1

Methods of viewing viruses. (a) Negative staining of an orfvirus (a type of poxvirus), revealing details of its outer coat (122,000×). **(b)** Positive stain of the Ebola virus, a type of filovirus, so named because of its tendency to form long strands. Note the textured capsid. **(c)** Shadowcasting image of a vaccinia virus (17,500×).

FIGURE 6.2

The crystalline nature of viruses. (a) Light microscope magnification (1,200×) of purified poliovirus crystals. **(b)** Highly magnified (150,000×) electron micrograph of the capsids of this same virus, demonstrating their highly geometric nature.

All viruses have a protein **capsid,*** or shell, that surrounds the nucleic acid in the central core. Together the capsid and the nucleic acid are referred to as the **nucleocapsid** (figure 6.3). Members of 13 of the 20 families of animal viruses possess an additional covering external to the capsid called an **envelope,** which is actually a modified piece of the host's cell membrane (figure 6.3*b*). Viruses that consist of only a nucleocapsid are considered **naked viruses** (figure 6.3*a*). As we shall see later, the enveloped viruses also differ from the naked viruses in the way that they enter and leave a host cell.

The Viral Capsid: The Protective Outer Shell

When a virus particle is magnified several hundred thousand times, the capsid appears as the most prominent geometric feature (see figure 6.2*b*). In general, each capsid is constructed from identical subunits called **capsomers*** that are constructed from protein molecules. The capsomers spontaneously self-assemble into the finished capsid. Depending on how the capsomers are shaped and arranged, this binding results in two different types: helical and icosahedral.

*capsid (kap′-sid) L. *capsa*, box.

*capsomer (kap′-soh-meer) L. *capsa*, box, and *mer*, part.

(a) Naked Nucleocapsid Virus

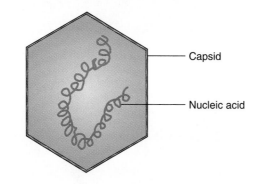

Capsid

Nucleic acid

(b) Enveloped Virus

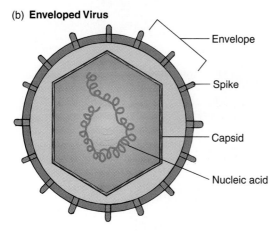

Envelope

Spike

Capsid

Nucleic acid

FIGURE 6.3

Generalized structure of viruses. **(a)** The simplest virus is a naked virus (nucleocapsid) consisting of a geometric capsid assembled around a nucleic acid strand or strands. **(b)** An enveloped virus is composed of a nucleocapsid surrounded by a flexible membrane called an envelope. The envelope usually has special receptor spikes inserted into it.

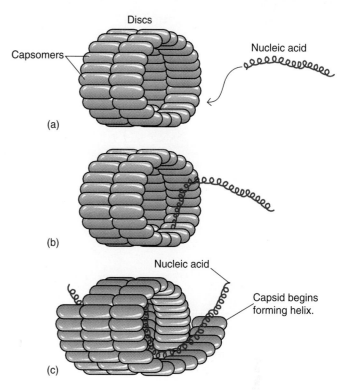

Discs

Capsomers

Nucleic acid

(a)

(b)

Nucleic acid

Capsid begins forming helix.

(c)

FIGURE 6.4

Assembly of helical nucleocapsids. (a) Capsomers assemble into hollow discs. **(b)** The nucleic acid is inserted into the center of the disc. **(c)** Elongation of the nucleocapsid progresses from both ends, as the nucleic acid is wound "within" the lengthening helix.

The simpler **helical capsids** have rod-shaped capsomers that bond together to form a series of hollow discs resembling a bracelet. During the formation of the nucleocapsid, these discs link with other discs to form a continuous helix into which the nucleic acid strand is coiled (figure 6.4). In electron micrographs, the appearance of a helical capsid varies with the type of virus. The nucleocapsids of naked helical viruses are very rigid and tightly wound into a cylinder-shaped package (figure 6.5*a,b*). An example is the *tobacco mosaic virus,* which attacks tobacco leaves. Enveloped helical nucleocapsids, on the other hand, are more flexible and tend to be arranged as a looser helix within the envelope (figure 6.5*c,d*). This type of morphology is found in several enveloped human viruses, including those of influenza, measles, and rabies.

The capsids of a number of major virus families are arranged in an **icosahedron***—a three-dimensional, 20-sided figure with 12 evenly spaced corners. The arrangements of the capsomers vary from one virus to another. Some viruses construct the

*icosahedron (eye″-koh-suh-hee′-drun) Gr. *eikosi,* twenty, and *hedra,* side. A type of polygon.

capsid from a single type of capsomer while others may contain several types of capsomers (figure 6.6). Although the capsids of all icosahedral viruses have this sort of symmetry, they can have major variations in the number of capsomers; for example, a poliovirus has 32, and an adenovirus has 240 capsomers. Individual capsomers can look either ring- or dome-shaped, and the capsid itself can appear spherical or cubical (figure 6.7). During assembly of the virus, the nucleic acid is packed into the center of this icosahedron, forming a nucleocapsid. Another factor that alters the appearance of icosahedral viruses is whether or not they have an outer envelope; contrast a papillomavirus (warts) and its naked nucleocapsid with herpes simplex (cold sores) and its enveloped nucleocapsid (figure 6.7).

The Viral Envelope

When **enveloped viruses** (mostly animal) are released from the host cell, they take with them a bit of its membrane system in the form of an **envelope** (see figure 6.19). Some viruses bud off the cell membrane; others leave via the nuclear envelope or the endoplasmic reticulum. Whichever avenue of escape, the viral envelope differs significantly from the host's membranes. In the envelope, some or all of the regular membrane proteins are replaced with special viral proteins. Some proteins form a binding layer between the envelope and capsid of the virus, and glycoproteins (proteins bound to a carbohydrate) remain exposed on the outside of the envelope. These

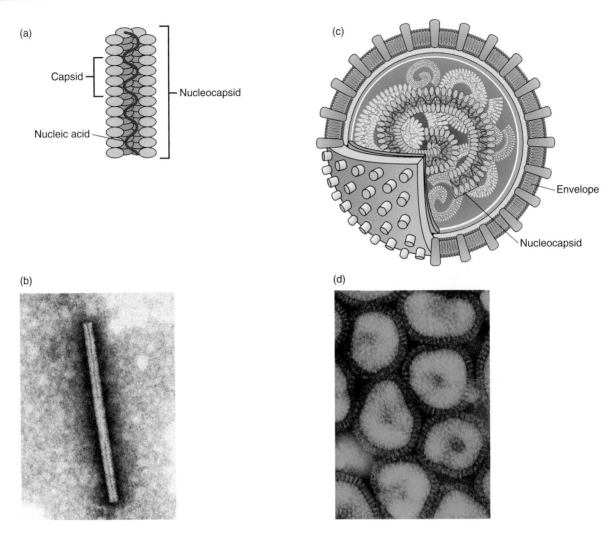

FIGURE 6.5

🔴 **Typical variations of viruses with helical nucleocapsids.** Naked helical virus (tobacco mosaic virus): **(a)** a schematic view and **(b)** a greatly magnified micrograph. Note the overall cylindrical morphology. Enveloped helical virus (influenza virus): **(c)** a schematic view and **(d)** an electron micrograph of the same virus (350,000×).

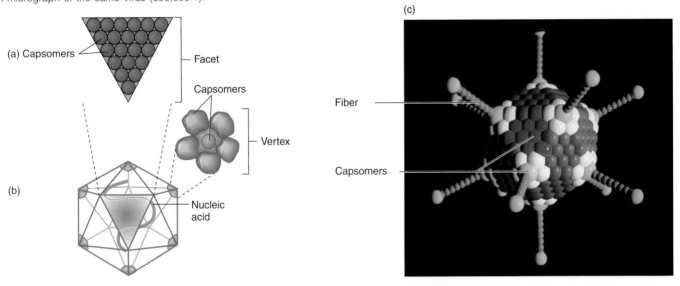

FIGURE 6.6

🔵 **The structure and formation of an icosahedral virus (adenovirus is the model). (a)** A facet or "face" of the capsid is composed of 21 identical capsomers arranged in a triangular shape. The vertices or "points" consist of 5 capsomers arranged with a single penton in the center. Other viruses can vary in the number, types, and arrangement of capsomers. **(b)** An assembled virus shows how the facets and vertices come together to form a shell around the nucleic acid. **(c)** A three-dimensional model of this virus shows fibers attached to the pentons.

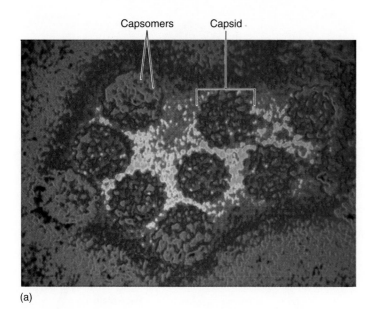

(a)

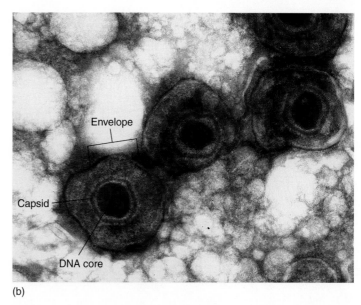

(b)

FIGURE 6.7

Two types of icosahedral viruses, highly magnified. (a) False-color micrograph of papillomaviruses with unusual, ring-shaped capsomers. **(b)** Herpesvirus, an enveloped icosahedron (300,000×).

protruding molecules, called **spikes** or **peplomers,*** are essential for the attachment of viruses to the next host cell. Because the envelope is more supple than the capsid, enveloped viruses are pleomorphic and range from spherical to filamentous.

Functions of the Viral Capsid/Envelope

The outermost covering of a virus is indispensable to viral function because it protects the nucleic acid from the effects of various enzymes and chemicals when the virus is outside the host cell. For example, the capsids of enteric (intestinal) viruses such as polio and hepatitis A are resistant to the acid- and protein-digesting enzymes of the gastrointestinal tract. Capsids and envelopes are also responsible for helping to introduce the viral DNA or RNA into a suitable host cell, first by binding to the cell surface and then by assisting in penetration of the viral nucleic acid (to be discussed in more detail later in the chapter). In addition, parts of viral capsids and envelopes stimulate the immune system to produce antibodies that can neutralize viruses and protect the host's cells against future infections (see chapters 15, 24, and 25).

Complex Viruses: Atypical Viruses

Two special groups of viruses, termed **complex viruses** (figure 6.8), are more intricate in structure than the helical, icosahedral, naked, or enveloped viruses just described. The **poxviruses** (including the agent of smallpox) are very large viruses that contain a DNA core but lack a regular capsid and have in its place several layers of lipoproteins and coarse surface fibrils. Another group of very complex viruses, the **bacteriophages,*** have a polyhedral

head, a helical tail, and fibers for attachment to the host cell. Their mode of multiplication is covered in a later section of this chapter. Figure 6.9 summarizes the morphological types of some common viruses.

Nucleic Acids: At the Core of a Virus

So far, one biological constant is that the genetic information of living cells is carried by nucleic acids (DNA, RNA). Viruses, although neither alive nor cells, are no exception to this rule, but there is a significant difference. Unlike cells, which contain both DNA and RNA, viruses contain either DNA or RNA *but not both*. Because viruses must pack into a tiny space all of the genes necessary to instruct the host cell to make new viruses, the size of a viral genome is quite small compared with that of a cell. It varies from four genes in hepatitis B virus to hundreds of genes in some herpesviruses. By comparison, the bacterium *Escherichia coli* has approximately 4,000 genes, and a human cell has approximately 40,000 genes.

In chapter 2 we learned that DNA usually exists as a double-stranded molecule and that RNA is single-stranded. Although most viruses follow this same pattern, a few exhibit distinctive and exceptional forms. Notable examples are the parvoviruses (the cause of erythema infectiosum[3]), which contain single-stranded DNA, and reoviruses (a cause of respiratory and intestinal tract infections), which contain double-stranded RNA. In all cases, these tiny strands of genetic material carry the blueprint for viral structure and functions. In a very real sense, viruses are **genetic parasites** because they cannot multiply until their nucleic acid has reached the internal habitat of the host cell. At the minimum, they must carry genes for synthesizing the viral capsid and genetic material, for regulating the actions of the host, and for packaging the mature virus.

---

*peplomer (pep'-loh-meer) Gr. *peplos*, envelope, and *mer*, part.

*bacteriophage (bak-teer'-ee-oh-fayj″) From *bacteria*, and Gr. *phagein*, to eat. These viruses parasitize bacteria.

---

3. A common childhood disease described in chapter 24.

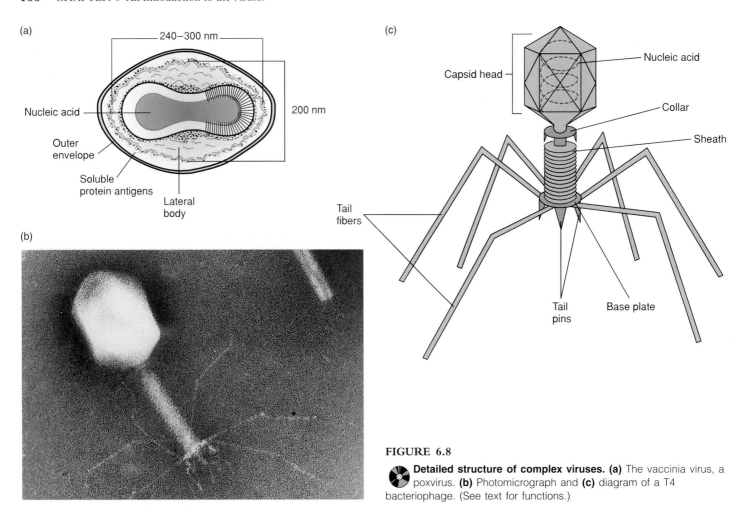

(a)

240–300 nm

200 nm

Nucleic acid

Outer envelope

Soluble protein antigens

Lateral body

(b)

(c)

Capsid head

Nucleic acid

Collar

Sheath

Tail fibers

Tail pins

Base plate

FIGURE 6.8

Detailed structure of complex viruses. (a) The vaccinia virus, a poxvirus. **(b)** Photomicrograph and **(c)** diagram of a T4 bacteriophage. (See text for functions.)

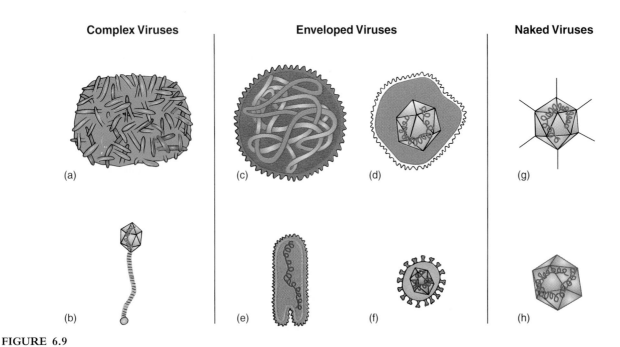

Complex Viruses

(a)

(b)

Enveloped Viruses

(c)

(d)

(e)

(f)

Naked Viruses

(g)

(h)

FIGURE 6.9

Complex viruses: (a) poxvirus, a large DNA virus; **(b)** flexible-tailed bacteriophage. Enveloped viruses: **(c)** mumps virus, an enveloped RNA virus with a helical nucleocapsid; **(d)** herpesvirus, an enveloped DNA virus with an icosahedral nucleocapsid; **(e)** rhabdovirus, an RNA virus with a bullet-shaped envelope; **(f)** HIV (AIDS), an RNA retrovirus with an icosahedral capsid. Naked viruses: **(g)** adenovirus, a DNA virus with fibers on the capsid; **(h)** papillomavirus, a DNA virus that causes warts.

Other Substances in the Virus Particle

In addition to the protein of the capsid, the proteins and lipids of envelopes, and the nucleic acid of the core, viruses can contain enzymes for specific operations within their host cell. They may come with pre-formed enzymes that are required for viral replication. Examples include *polymerases** that assemble DNA and RNA and replicases that copy RNA. The AIDS virus comes equipped with reverse transcriptase for synthesizing DNA from RNA. Viruses completely lack the genes for synthesis of metabolic enzymes. As we shall see, this deficiency has little consequence, because viruses have adapted to completely take over their hosts' metabolic resources. Some viruses can actually carry away substances from their host cell. For instance, arenaviruses pack along host ribosomes, and retroviruses "borrow" the host's tRNA molecules.

How Viruses Are Classified and Named

Although viruses are not classified as members of the kingdoms discussed in chapter 1, they are diverse enough to require their own classification scheme to aid in their study and identification. In an informal and general way, we have already begun classifying viruses—as animal, plant, or bacterial viruses; enveloped or naked viruses; DNA or RNA viruses; and helical or icosahedral viruses. These introductory categories are certainly useful in organization and description, but the study of specific viruses requires a more standardized method of nomenclature. For many years, the animal viruses were classified mainly on the basis of their hosts and the kind of diseases they caused. Newer systems for naming viruses take into account the actual nature of the virus particles themselves, with only partial emphasis on host and disease. The main criteria presently used to group viruses are structure, chemical composition, and similarities in genetic makeup.

A widely used scheme for classifying animal viruses first assigns them to one of two superfamilies, either those containing DNA or those containing RNA. DNA viruses can be further subdivided into six families, and RNA viruses into 13 families, for a total of 19 families of animal viruses. Virus families are given a name composed of a Latin root followed by *-viridae*. Characteristics used for placement in a particular family include type of capsid, nucleic acid strand number, presence and type of envelope, overall viral size, and area of the host cell in which the virus multiplies. Some virus families are named for their microscopic appearance (shape and size). Examples include *rhabdoviruses,** which have a bullet-shaped envelope, and *togaviruses,** which have a cloaklike envelope. Anatomical or geographic areas have also been used in naming. For instance, *adenoviruses** were first discovered in adenoids (one type of tonsil), and hantaviruses were originally isolated in an area in Korea called Hantaan. Viruses can also be named for their effects on the host. *Lentiviruses** tend to cause slow, chronic infec-

tions. Acronyms made from blending several characteristics include *picornaviruses,** which are tiny RNA viruses, and reoviruses (or **r**espiratory **e**nteric **o**rphan viruses), which inhabit the respiratory tract and the intestine and are not yet associated with any known disease state.

Each different type of virus is also assigned genus status according to its host, target tissue, and the type of disease it causes (table 6.3). Viral genera are denoted by a special Latinized root followed by the suffix *-virus* (for example, *Enterovirus* and *Hantavirus*). Because the use of standardized species names has not been widely accepted, the genus or common English vernacular names (for example, poliovirus and rabies virus) predominate in discussions of specific viruses in this text.

A Note on Terminology

Although the terms *virus* and *virus particle* are interchangeable, virologists find it convenient to distinguish between the various states in which a virus can exist. A fully formed, extracellular particle that is virulent (able to establish infection in a host) is called a **virion** (vir'-ee-on). Once its genetic material has entered a host cell, a virus may undergo a multiplication phase called the **lytic** (lih'-tik) **cycle** in which the host cell is disrupted to release more virions. Some viruses remain in an inactive, or **latent** (lay'-tunt), stage in which the virus does not lyse the host cell. This stage is called the **lysogenic phase.**

Modes of Viral Multiplication

Viruses tend to remain closely associated with their hosts. In addition to providing the viral habitat, the host cell is absolutely necessary for viral multiplication. The process of viral multiplication is an extraordinary biological phenomenon. Viruses have often been aptly described as minute parasites that appropriate the synthetic and genetic machinery of cells. The nature of this cycle dictates viral pathogenicity, transmission, the responses of the immune defenses, and human measures to control viral infections. From these perspectives, we cannot overemphasize the importance of a working knowledge of the relationship between viruses and their host cells.

The multiplication cycles of viruses are similar enough that virologists use certain viruses (such as bacteriophages and enveloped animal viruses) as general models. Although a given cycle occurs continuously and not in discrete steps, it is helpful to demarcate its major sequential events. These events are **adsorption,** a recognition process between a virus and host cell that results in virus attachment to the external surface of the host cell; **penetration,** entrance of the virion (either a whole virus or just its nucleic acid) into the host cell; **replication,** copying and expression of the viral genome at the expense of the host's synthetic

---

*polymerase (pol-im'-ur-ace) An enzyme that synthesizes a large molecule from smaller subunits.

*rhabdovirus (rab″-doh-vy′-rus) Gr. *rhabdo*, little rod.

*togavirus (toh″-guh-vy′-rus) L. *toga*, covering or robe.

*adenovirus (ad″-uh-noh-vy′-rus) G. *aden*, gland.

*lentivirus (len″-tee-vy′-rus) Gr. *lente*, slow. HIV, the AIDS virus, belongs in this group.

---

*picornavirus (py-kor″-nah-vy′-rus) Sp. *pico*, small, plus RNA.

*adsorption (ad-sorp′-shun) L. *ad*, to, and *sorbere*, to suck. The attachment of one thing onto the surface of another.

*replication (rep-lih-kay′-shun) L. *replicare*, to reply. To make an exact duplicate.

TABLE 6.3

Important Human Virus Families, Genera, Common Names, and Types of Diseases★

	Family	Genus of Virus	Common Name of Genus Members	Name of Disease
DNA Viruses				
	Poxviridae	*Orthopoxvirus*	Variola major and minor	Smallpox, cowpox
	Herpesviridae	*Simplexvirus*	Herpes simplex (HSV) 1 virus	Fever blister, cold sores
			Herpes simplex (HSV) 2 virus	Genital herpes
		Varicellovirus	Varicella zoster virus (VZV)	Chickenpox, shingles
		Cytomegalovirus	Human cytomegalovirus (CMV)	CMV infections
	Adenoviridae	*Mastadenovirus*	Human adenoviruses	Adenovirus infection
	Papovaviridae	*Papillomavirus*	Human papillomavirus (HPV)	Several types of warts
		Polyomavirus	JS virus (JCV)	Progressive multifocal leukoencephalopathy (PML)
	Hepadnaviridae	*Hepadnavirus*	Hepatitis B virus (HBV or Dane particle)	Serum hepatitis
	Parvoviridae	*Erythrovirus*	Parvovirus B19	Erythema infectiosum
RNA Viruses				
	Picornaviridae	*Enterovirus*	Poliovirus	Poliomyelitis
			Coxsackievirus	Hand-foot-mouth disease
		Hepatovirus	Hepatitis A virus (HAV)	Short-term hepatitis
		Rhinovirus	Human rhinovirus	Common cold, bronchitis
	Calciviridae	*Calicivirus*	Norwalk virus	Viral diarrhea, Norwalk virus syndrome
	Togaviridae	*Alphavirus*	Eastern equine encephalitis virus	Eastern equine encephalitis (EEE)
			Western equine encephalitis virus	Western equine encephalitis (WEE)
			Yellow fever virus	Yellow fever
			St. Louis encephalitis virus	St. Louis encephalitis
		Rubivirus	Rubella virus	Rubella (German measles)
	Flaviviridae	*Flavivirus*	Dengue fever virus	Dengue fever
			West Nile fever virus	West Nile fever
	Bunyaviridae	*Bunyavirus*	Bunyamwera viruses	California encephalitis
		Hantavirus	Sin Nombre virus	Respiratory distress syndrome
		Phlebovirus	Rift Valley fever virus	Rift Valley fever
		Nairovirus	Crimean–Congo hemorrhagic fever virus (CCHF)	Crimean–Congo hemorrhagic fever
	Filoviridae	*Filovirus*	Ebola, Marburg virus	Ebola fever
	Reoviridae	*Coltivirus*	Colorado tick fever virus	Colorado tick fever
		Rotavirus	Human rotavirus	Rotavirus gastroenteritis
	Orthomyxoviridae	*Influenza virus*	Influenza virus, type A (Asian, Hong Kong, and swine influenza viruses)	Influenza or "flu"
	Paramyxoviridae	*Paramyxovirus*	Parainfluenza virus, types 1–5	Parainfluenza
			Mumps virus	Mumps
		Morbillivirus	Measles virus	Measles (red)
		Pneumovirus	Respiratory syncytial virus (RSV)	Common cold syndrome
	Rhabdoviridae	*Lyssavirus*	Rabies virus	Rabies (hydrophobia)
	Retroviridae	*Oncornavirus*	Human T-cell leukemia virus (HTLV)	T-cell leukemia
		Lentivirus	HIV (human immunodeficiency viruses 1 and 2)	Acquired immunodeficiency syndrome (AIDS)
	Arenaviridae	*Arenavirus*	Lassa virus	Lassa fever
	Coronaviridae	*Coronavirus*	Infectious bronchitis virus (IBV)	Bronchitis
			Enteric corona virus	Coronavirus enteritis

★*See also tables 24.1 and 25.1 for additional information.*

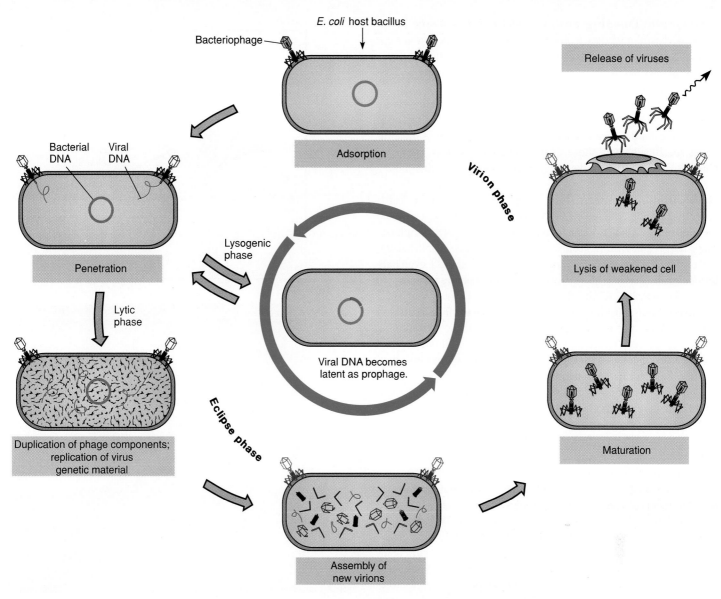

FIGURE 6.10

Events in the multiplication cycle of T-even bacteriophages. The cycle is divided into the eclipse phase (during which the phage is developing but is not yet infectious) and the virion phase (when the virus matures and is capable of infecting a host). (The text provides further details of this cycle. See figure 6.14 for the steps in the lysogenic phase.)

equipment, resulting in the production of the various virus components; **assembly** and **maturation** of these individual viral parts into whole, intact virions; and **release,** escape from the host cell of the active, infectious viral particles (figure 6.10). We present an overview of the genetic events of virus cycles here, but this topic is covered in greater detail in chapter 9. The following section compares and contrasts the multiplication stages for bacterial and human viruses.

THE MULTIPLICATION CYCLE IN BACTERIOPHAGES

When Frederick Twort and Felix d'Herelle discovered bacterial viruses in 1915, it first appeared that the bacterial host cells were being eaten by some unseen parasite, hence the name bacterio-

phage was used. Most bacteriophages (often shortened to *phage*) contain double-stranded DNA, though single-stranded DNA and RNA types exist as well. So far as is known, every bacterial species is parasitized by various specific bacteriophages. Probably the most widely studied bacteriophages are those of the intestinal bacterium *Escherichia coli*—especially the T-even (for even-numbered type) phages. Their complex structure has been previously described. They have an icosahedral capsid head containing DNA, a central tube (surrounded by a sheath), collar, base plate, tail pins, and fibers, which in combination make an efficient package for infecting a bacterial cell (see figure 6.8c). Momentarily setting aside a strictly scientific and objective tone, it is tempting to think of these extraordinary viruses as minute spacecrafts docking on an alien planet, ready to unload their genetic cargo.

Adsorption: Docking onto the Host Cell Surface

For a successful infection, a phage must meet and stick to a susceptible host cell. Adsorption takes place when certain molecules on the tail fibers bind to specific molecules (receptors) on the cell envelope of the host bacterium (see figure 6.10). Among the bacterial structures that serve as phage receptors are surface molecules of the cell wall, pili, and flagella. Attachment is a function of correct fit and chemical attraction; that is, the phage's tail molecules must interact specifically with those of the bacterial receptors. Once the phage is firmly affixed to the cell, it is positioned for penetration.

Bacteriophage Penetration: Entry of the Nucleic Acid

After adsorption, a phage is still on the outside of its host cell and remains inactive unless it can gain entrance to the cell's cytoplasm. The strong, rigid bacterial cell wall is quite an impenetrable barrier, and the entire virus particle is unable to cross it. But the T-even bacteriophages have an exquisite mechanism for injecting nucleic acid across this barrier and into the cell. Like a miniscule syringe, the sheath constricts, pushing the inner tube through the host's cell wall and membrane, forming a passageway for its nucleic acid into the interior of the cell (figure 6.11). Once inside, the viral nucleic acid alone can complete the multiplication cycle. The spent virus shell has fulfilled its principal functions—protection, adsorption, and injection of DNA—and it will remain attached outside to the cell wall fragment even after the cell has lysed.

The Bacteriophage Assembly Line

Entry of the nucleic acid into the cell results in sweeping changes in the bacterial cell's activities. Within a few minutes, the bacterium stops synthesizing its own molecules, and its metabolism shifts to the expression of genes on the viral nucleic acid strand. In T-even bacteriophages, the viral DNA redirects the genetic and metabolic activity of the cell, blocking the utilization of host DNA and ensuring that viral DNA is copied and used to synthesize new viral components (see replication of DNA in chapter 2). Expression of that phage's genetic information gives rise to the following viral molecules: (1) proteins to "seal" the cell (because the host cell is punctured by the virus during injection, and it is advantageous for the virus to repair the hole before the host cell's integrity is destroyed); (2) enzymes for copying the virus genome; (3) proteins that make up the capsid head and parts of the tail; and (4) enzymes that help weaken the cell wall so that the new phages can easily escape. This period of viral synthesis exploits the host's cytoplasmic nutrient resources, its ribosomes, and its energy supplies. The early stages of viral replication are known as *eclipse,* a period during which no mature virions can be detected within the host cell. This corresponds with the period of penetration, replication, and early assembly, and if the cell were disrupted at this time, no infective virus would be released.

As the host cell rapidly produces new phage parts, these parts spontaneously assemble into bacteriophages, similar to a mass production assembly line (figure 6.12). The first parts to assemble are the capsid and tail. Viral DNA is inserted into the capsid before the capsomers are completely joined and the collar and sheath unite with the tail pins. Mature phage particles are formed when the capsid and tail fit together and the fibers are anchored to the base plate.

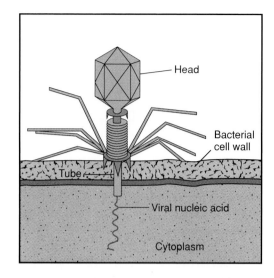

(a)

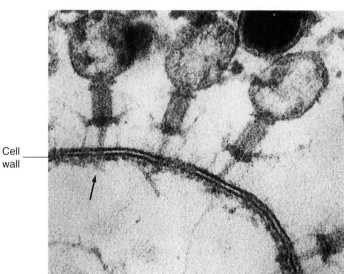

(b)

FIGURE 6.11

Penetration of a bacterial cell by a T-even bacteriophage. (a) After adsorption, the phage plate becomes embedded in the cell wall, and the sheath contracts, pushing the tube through the cell wall and releasing the nucleic acid into the interior of the cell. **(b)** Section through *Escherichia coli* with attached phages. Note that these phages have injected their nucleic acid through the cell wall and now have empty heads.

An average-sized *Escherichia coli* cell can contain up to 200 new phage units at the end of this period. Eventually, the host cell becomes so packed with viruses that it **lyses**—splits open—thereby liberating the mature virions (figure 6.13). This process is hastened by viral enzymes produced late in the infection cycle that digest the cell envelope, thereby weakening it. Upon release, the virulent phages can spread to other susceptible bacterial cells and begin a new cycle of infection.

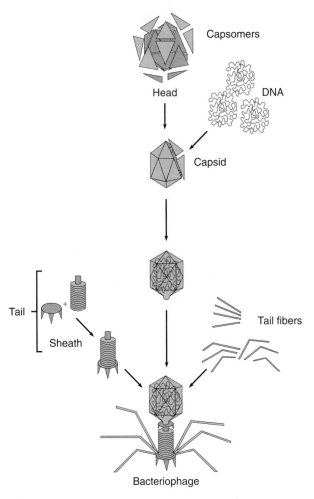

FIGURE 6.12

Bacteriophage assembly line. First the protein subunits (head, sheath, tail fibers) are synthesized by the host cell. A strand of viral nucleic acid also replicated by the host's machinery is inserted during the later stages of capsid formation. During final assembly, the prefabricated components fit together into whole parts and finally into the finished products.

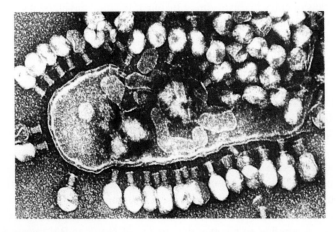

FIGURE 6.13

A weakened bacterial cell, crowded with viruses. The cell has ruptured and released numerous virions that can then attack nearby susceptible host cells. Note the empty heads of "spent" phages lined up around the ruptured wall.

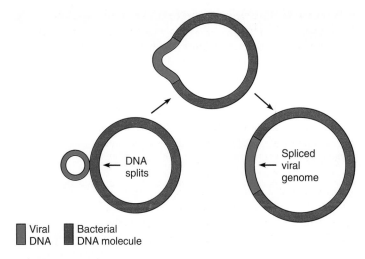

FIGURE 6.14

The lysogenic state in bacteria. A bacterial DNA molecule can accept and insert viral DNA molecules at specific sites on its genome. This additional viral DNA is duplicated along with the regular genome and can provide adaptive characteristics for the host bacterium.

Lysogeny: The Silent Virus Infection

The lethal effects of a virulent phage on the host cell present a dramatic view of virus-host interaction. Not all bacteriophages complete the lytic cycle, however. Special DNA phages, called **temperate*** phages, undergo adsorption and penetration in the bacterial host but are not replicated or released. Instead, the viral DNA enters an inactive **prophage*** state, in which it is inserted into the bacterial chromosome. This viral DNA will be retained by the bacterial cell and copied during its normal cell division so that the cell's progeny will also have the temperate phage DNA (figure 6.14). This condition, in which the host chromosome carries bacteriophage DNA, is termed **lysogeny.*** Because the viral genome is not expressed, the bacterial cells carrying temperate phages do not lyse, and they appear entirely normal. On occasion, the prophage in a lysogenic cell will be activated and progress directly into viral replication and the lytic cycle. Lysogeny is a less deadly form of parasitism than the full lytic cycle and is thought to be an advancement that allows the virus to spread without killing the host. In a later section and in chapters 9, 17, and 24, we describe a similar relationship that exists between certain animal viruses and human cells.

The cycle of bacterial viruses illustrates general features of viral multiplication in a very concrete and memorable way. It is fascinating to realize that viruses are capable of lying "dormant" in their host cells, possibly becoming active at some later time. Because of the intimate association between the genetic material of the virus and host, phages occasionally serve as transporters of bacterial

---

*temperate (tem'-pur-ut) A reduction in intensity.

*prophage (pro'-fayj) L. *pro,* before, plus phage.

*lysogeny (ly-soj'-uhn-ee) The potential ability to produce phage.

TABLE 6.4

Comparison of Bacteriophage and Animal Virus Multiplication

	Bacteriophage	Animal Virus
Adsorption	Precise attachment of special tail fibers to cell wall	Attachment of capsid or envelope to cell surface receptors
Penetration	Injection of nucleic acid through cell wall; no uncoating of nucleic acid	Whole virus is engulfed and uncoated, or virus surface fuses with cell membrane, nucleic acid is released
Replication and Maturation	Occurs in cytoplasm Cessation of host synthesis	Occurs in cytoplasm and nucleus Cessation of host synthesis
	Viral DNA or RNA is replicated and begins to function Viral components synthesized	Viral DNA or RNA is replicated and begins to function Viral components synthesized
Viral Persistence	Lysogeny	Latency, chronic infection, cancer
Exit from Host Cell	Cell lyses when viral enzymes weaken it	Some cells lyse; enveloped viruses bud off host cell membrane
Cell Destruction	Immediate	Immediate; delayed in some

genes from one bacterium to another and consequently can play a profound role in bacterial genetics. This phenomenon, called transduction, is one way that genes for toxin production and drug resistance are transferred between bacteria (see chapters 9 and 12).

MULTIPLICATION CYCLES IN ANIMAL VIRUSES

Now that we have a working concept of viral multiplication in bacteria, let us turn to the animal viruses for yet another model system. Although the basic cycle is very similar to that of bacteriophages, several significant differences exist between the host cells, their viruses, and the multiplication cycles (table 6.4). The general phases in the cycle of animal viruses are adsorption, penetration, uncoating, replication, assembly, and departure from the host cell. The length of the entire multiplication cycle varies from 8 hours in polioviruses to 36 hours in herpesviruses. See figure 6.15 for a comparison of the major phases of two types of animal viruses.

Adsorption and Host Range

Invasion begins when the virus encounters a susceptible host cell and adsorbs specifically to receptor sites on the cell membrane. The membrane receptors that viruses attach to are usually glycoproteins the cell requires for its normal function. For example, the rabies virus affixes to the acetylcholine receptor of nerve cells, and the human immunodeficiency virus (HIV or AIDS virus) attaches to the CD4 protein on certain white blood cells. The mode of attachment varies between the two general types of viruses. In enveloped forms such as influenza virus and HIV, glycoprotein spikes bind to the cell membrane receptors. Viruses with naked nucleocapsids (poliovirus, for example), possess surface molecules that adhere to cell membrane receptors (figure 6.16).

Because a virus can invade its host cell only through making an exact fit with a specific host molecule, the scope of hosts it can infect in a natural setting is limited. This limitation, known as the **host range,** may be as restricted as hepatitis B, which infects only liver cells of humans; intermediate like the poliovirus, which infects intestinal and nerve cells of primates (humans, apes, and monkeys); or as broad as the rabies virus, which can infect various cells of all mammals. Cells that lack compatible virus receptors are resistant to adsorption and invasion. This explains why, for example, human liver cells are not infected by the canine hepatitis virus and dog liver cells cannot host the human hepatitis A virus. It also explains why viruses usually have tissue specificities called *tropisms** for certain cells in the body. The hepatitis B virus targets the liver, and the mumps virus targets salivary glands. However, the fact that most viruses can be induced to infect cells that they would not infect naturally makes it possible to cultivate them in the laboratory.

Penetration/Uncoating of Animal Viruses

Animal viruses exhibit some impressive mechanisms for entering a host cell. Unlike bacteriophages, they have no mechanism to inject their nucleic acids. Instead, the flexible cell membrane of the host is penetrated by the whole virus or its nucleic acid (figure 6.17). In penetration by **endocytosis** (figure 6.17a), the entire virus is engulfed by the cell and enclosed in a vacuole or vesicle. When enzymes in the vacuole dissolve the envelope and capsid, the virus is said to be **uncoated,** a process that releases the viral nucleic acid into the cytoplasm. The exact manner of uncoating varies, but in most cases, the virus fuses with the wall of the vesicle. Another means of entry involves direct fusion of the viral envelope with the host cell membrane (as in influenza and mumps viruses) (figure 6.17b). In this form of penetration, the envelope merges directly with the cell membrane, thereby liberating the nucleocapsid into the cell's interior.

Replication and Maturation of Animal Viruses: Host Cell As Factory

The synthetic and replicative phases of animal viruses are highly regulated and extremely complex at the molecular level. They are too detailed to cover at this point. A more thorough coverage of viral genetics is included in chapters 9, 24, and 25. As with bacteriophages, the free viral nucleic acid exerts control over the host's synthetic and metabolic machinery. How this control proceeds will vary, depending on whether the virus is a DNA or an RNA virus. In general, the DNA viruses (except poxviruses) enter the host cell's

---

*tropism (troh′-pizm) Gr. *trope,* a turn. Having a special affinity for an object or substance.

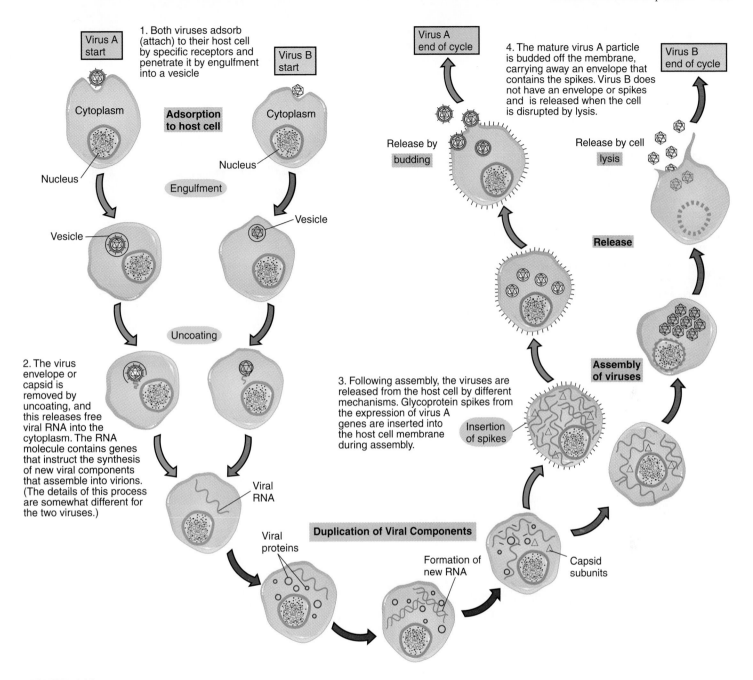

FIGURE 6.15

General features in the multiplication cycle of animal viruses. Two types of RNA viruses are compared: an enveloped virus (rubella virus) and a naked virus (poliovirus).

nucleus and are replicated and assembled there (see figure 9.19). With few exceptions (such as retroviruses), RNA viruses are replicated and assembled in the cytoplasm (see figure 9.20).

For an overview of the phases of duplication and assembly, we will use RNA viruses as a model. Almost immediately upon entry, the viral nucleic acid alters the genetic expression of the host and instructs it to synthesize the building blocks for new viruses. First, the RNA of the virus becomes a message for synthesizing viral proteins (translation). The viruses with *positive sense** RNA

molecules already contain the correct message for translation into proteins. Viruses with *negative sense* RNA molecules must first be converted into a positive sense message. Some viruses come equipped with the necessary enzymes for synthesis of viral components; others utilize those of the host.

In the next phase, new RNA is synthesized using host nucleotides. Proteins for the capsid, spikes, and viral enzymes are synthesized on the host's ribosomes using its amino acids. Toward the end of the cycle, mature virus particles are constructed from the growing pool of parts. In most instances, the capsid is first laid down as an empty shell that will serve as a receptacle for the nucleic acid

*sense The "sense" of the strand relates to its readability into a protein.

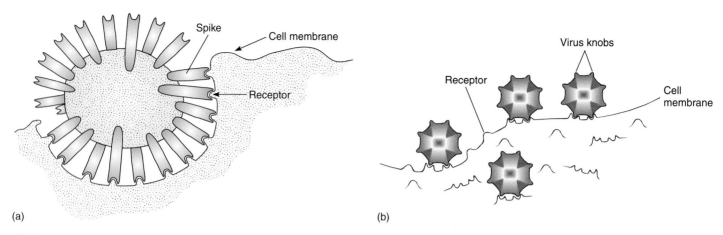

(a)

(b)

FIGURE 6.16

The mode by which animal viruses adsorb to the host cell membrane. (a) A virus with spikes. The configuration of the spike has a complementary fit for cell receptors. The process in which the virus lands on the cell and plugs into receptors is termed docking. **(b)** A virus with a naked capsid adheres to its host cell by nestling surface molecules on its capsid into the receptors on the host cell's membrane.

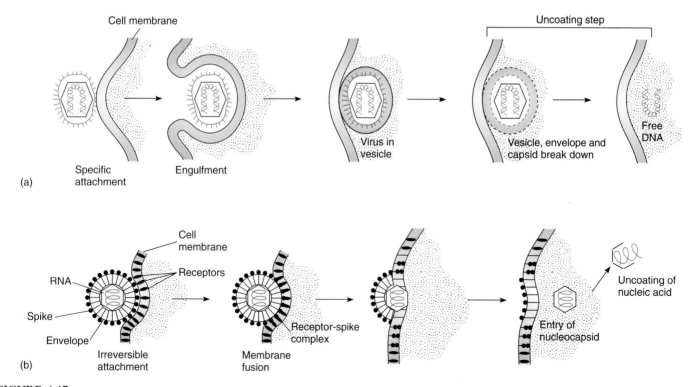

(a)

(b)

FIGURE 6.17

Two principal means by which animal viruses penetrate. (a) Endocytosis (engulfment) and uncoating of a herpesvirus. **(b)** Fusion of the cell membrane with the viral envelope (mumps virus).

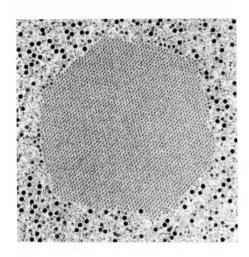

FIGURE 6.18

Nucleus of a cell, containing a crystalline mass of adenovirus (35,000×).

strand. Electron micrographs taken during this time show cells with masses of viruses, often in crystalline packets (figure 6.18). One important event in enveloped viruses is the insertion of viral spikes into the host's cell membrane so they can be picked up as the virus buds off with its envelope (see figures 6.15 and 6.19).

Release of Mature Viruses

To complete the cycle, assembled viruses leave their host in one of two ways. Nonenveloped and complex viruses that reach maturation in the cell nucleus or cytoplasm are released when the cell lyses or ruptures. Enveloped viruses are liberated by **budding** or **exocytosis**[4] from the membranes of the cytoplasm, nucleus, endoplasmic reticulum, or vesicles. During this process, the nucleocapsid binds to the membrane, which curves completely around it and forms a small pouch. Pinching off the pouch releases the virus with its envelope (figure 6.19). Budding of enveloped viruses causes

---

4. For enveloped viruses, these terms are interchangeable. They mean the release of a virus from an animal cell by enclosing it in a portion of membrane derived from the cell.

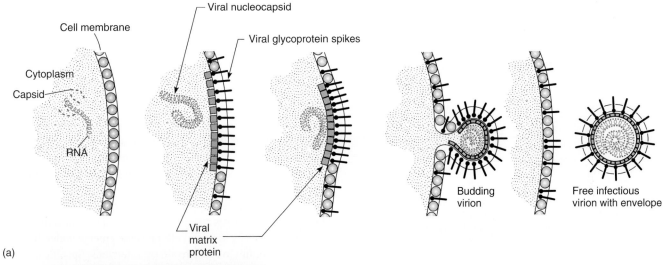

(a)

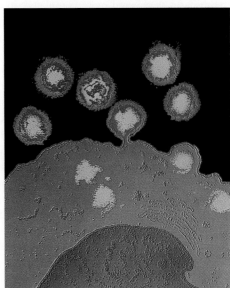

(b)

FIGURE 6.19

Process of maturation of an enveloped virus (parainfluenza virus). (a) As the virus is budded off the membrane, it simultaneously picks up an envelope and spikes. (b) AIDS viruses (HIV) leave their host white blood cell by budding off its surface. (Note early and mature stages.)

Normal cell   Giant cell

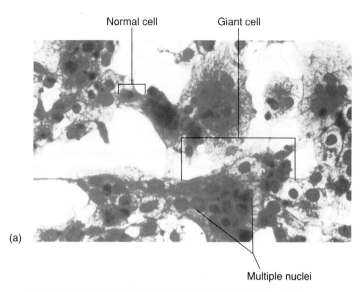

(a)

Multiple nuclei

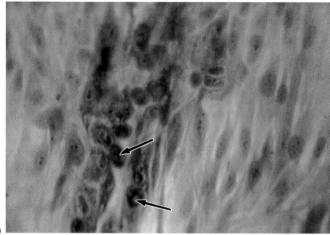

(b)

FIGURE 6.20

Cytopathic changes in cells and cell cultures infected by viruses. **(a)** Human epithelial cells infected by herpes simplex virus demonstrate multinucleate giant cells. **(b)** Human cells infected with cytomegalovirus. Note the inclusion bodies (arrows). Note also that both viruses disrupt the cohesive junctions between cells.

TABLE 6.5

Cytopathic Changes in Selected Virus-Infected Animal Cells

Virus	Response in Animal Cell
Smallpox virus	Cells round up; inclusions appear in cytoplasm
Herpes simplex	Cells fuse to form multinucleated giant cells; nuclear inclusions
Adenovirus	Clumping of cells; nuclear inclusions
Poliovirus	Cell lysis; no inclusions
Reovirus	Cell enlargement; vacuoles and inclusions in cytoplasm
Influenza virus	Cells round up; no inclusions
Rabies virus	No change in cell shape; cytoplasmic inclusions (Negri bodies)
HIV	Giant cells with numerous nuclei (multinucleate)

them to be shed gradually, without the sudden destruction of the cell. Regardless of how the virus leaves, most active viral infections are ultimately lethal to the cell because of accumulated damage. Lethal damages include a permanent shutdown of metabolism and genetic expression, destruction of cell membrane and organelles, toxicity of virus components, and release of lysosomes.

The number of viruses released by infected cells is variable, controlled by factors such as the size of the virus and the health of the host cell. About 3,000 to 4,000 virions are released from a single cell infected with poxviruses, whereas a poliovirus-infected cell can release over 100,000 virions. If even a small number of these virions happens to meet another susceptible cell and infect it, the potential for rapid viral proliferation is immense.

Damage to the Host Cell and Persistent Infections

The short- and long-term effects of viral infections on animal cells are well documented. **Cytopathic* effects** (CPEs) are defined as virus-induced damage to the cell that alters its microscopic appearance. Individual cells can become disoriented, undergo gross changes in shape or size, or develop intracellular changes (figure 6.20*a*). It is common to note *inclusion bodies,* or compacted masses of viruses or damaged cell organelles, in the nucleus and cytoplasm (figure 6.20*b*). Examination of cells and tissues for cytopathic effects is an important part of the diagnosis of viral infections. Table 6.5 summarizes some prominent cytopathic effects associated with specific viruses.

Although accumulated damage from a virus infection kills most host cells, some cells maintain a carrier relationship, in which the cell harbors the virus and is not immediately lysed. These so-called *persistent infections* can last from a few weeks to the rest of one's life. One of the more serious complications occurs with the measles virus. It may remain hidden in brain cells for many years, causing progressive damage and loss of function. Several viruses remain in a *chronic latent state,*[5] periodically becoming reactivated. Examples of this are herpes simplex viruses (fever blisters and genital herpes) and herpes zoster virus (chickenpox and shingles), which can go into latency in nerve cells and later emerge under the influence of various stimuli to cause recurrent infections. Specific damage that occurs in viral diseases is covered more completely in chapters 24 and 25.

Some animal viruses enter their host cell and permanently alter its genetic material, leading to cancer. These viruses are termed *oncogenic,* and their effect on the cell is called *transformation.* A startling feature of these viruses is that their nucleic acid is consolidated into the host DNA like a lysogenic phage. Transformed cells have an increased rate of growth; alterations in chromosomes; changes in the cell's surface molecules; and the capacity to divide for an indefinite period, unlike normal animal cells. Mammalian viruses capable of initiating tumors are called *oncoviruses.* Some of these are DNA viruses such as papillomavirus (genital warts are associated with cervical cancer), herpesviruses (Epstein-Barr virus causes Burkitt's lymphoma), and hepatitis B virus. Retroviruses are an unusual family of RNA viruses that can program the synthesis of DNA using their single-stranded RNA as a template. They can then

---

5. Meaning that they exist in an inactive state over long periods.

*cytopathic (sy″-toh-path′-ik) Gr. *cyto,* cell, and *pathos,* disease.

insert their DNA into a host chromosome, where it can remain for the life of the cell (see figure 25.16*a*). One type of retrovirus is HIV, the cause of AIDS. Other viruses related to HIV—HTLV I and II[6]—are involved in human cancers. These findings have spurred a great deal of speculation on the possible involvement of viruses in cancers whose cause is still unknown. Additional information on the connection between viruses and cancer is found in chapters 9, 17, 24, and 25.

CHAPTER CHECKPOINTS

Virus size range is from 20 nm to 450 nm (diameter).

Viruses are composed of an outer protein capsid enclosing either DNA or RNA plus a variety of enzymes. Some viruses also exhibit an envelope around the capsid.

Viral particles are called virions.

Viruses reproduce through five major events that turn a cell into a virus factory that synthesizes viral components, then assembles them into complete virion units. The new virus particles leave the cell host by lysing the cell or by budding.

Bacterial viruses, or bacteriophages, sometimes inadvertently incorporate bacterial DNA into the virion in place of viral DNA, thus providing a means of genetic transfer in bacteria.

Lysogeny is a condition in which viral DNA is inserted into the bacterial chromosome and remains inactive for an extended period. It is replicated right along with the chromosome every time the bacterium divides.

Animal viruses vary significantly from bacteriophage in their methods of adsorption, penetration, site of replication, and method of exit from host cells.

Animal viruses can persist in host tissues as chronic latent infections that can reactivate periodically throughout the host's life. Some persistent animal viruses are oncogenic.

Techniques in Cultivating and Identifying Animal Viruses

One problem hampering earlier animal virologists was their inability to propagate specific viruses routinely in pure culture and in sufficient quantities for their studies. Virtually all of the pioneering attempts at cultivation had to be performed in an organism that was the usual host for the virus. But this method had its limitations. How could researchers have ever traced the stages of viral multiplication if they had been restricted to the natural host, especially in the case of human viruses? Fortunately, systems of cultivation with broader applications were developed, including *in vivo** inoculation of laboratory-bred animals and embryonic

bird tissues and *in vitro** cell (or tissue) culture methods. Such use of substitute host systems permits greater control, uniformity, and wide-scale harvesting of viruses.

The primary purposes of viral cultivation are: (1) to isolate and identify viruses in clinical specimens; (2) to prepare viruses for vaccines; and (3) to do detailed research on viral structure, multiplication cycles, genetics, and effects on host cells.

USING LIVE ANIMAL INOCULATION

Specially bred strains of white mice, rats, hamsters, guinea pigs, and rabbits are the usual choices for animal cultivation of viruses. Invertebrates (insects) or nonhuman primates are occasionally used as well. Because viruses can exhibit some host specificity, certain animals can propagate a given virus more readily than others. Depending on the particular experiment, tests can be performed on adult, juvenile, or newborn animals. The animal is exposed to the virus by injection of a viral preparation or specimen into the brain, blood, muscle, body cavity, skin, or footpads.

USING BIRD EMBRYOS

An **embryo** is an early developmental stage of animals marked by rapid differentiation of cells. Birds undergo their embryonic period within the closed protective case of an egg, which makes an incubating bird egg a nearly perfect system for viral propagation. It is an intact and self-supporting unit, complete with its own sterile environment and nourishment. Furthermore, it furnishes several embryonic tissues that readily support viral multiplication.

Chicken, duck, and turkey eggs are the most common choices for inoculation. The egg must be injected through the shell, usually by drilling a hole or making a small window. Rigorous sterile techniques must be used to prevent contamination by bacteria and fungi from the air and the outer surface of the shell. The exact tissue that is inoculated is guided by the type of virus being cultivated and the goals of the experiment (figure 6.21).

Viruses multiplying in embryos may or may not cause effects visible to the naked eye. The signs of viral growth include death of the embryo, defects in embryonic development, and localized areas of damage in the membranes, resulting in discrete, opaque spots called **pocks** (a variant of pox). If a virus does not produce overt changes in the developing embryonic tissue, virologists have other methods of detection. Embryonic fluids and tissues can be prepared for direct examination with an electron microscope. Certain viruses can also be detected by their ability to agglutinate* red blood cells or by their reaction with an antibody of known specificity that will affix to its corresponding virus, if it is present.

USING CELL (TISSUE) CULTURE TECHNIQUES

The most important early discovery that led to cultivation of viruses was the development of a simple and effective way to grow populations of isolated animal cells in culture. These types of in

---

6. human T-cell lymphotropic viruses; cause types of leukemia.

*in vivo (in vee´-voh) L. *vivos*, life. Experiments performed in a living body.

*in vitro (in vee´-troh) L. *vitros*, glass. Experiments performed in test tubes or other artificial environments.

*agglutinate To form large masses or clumps.

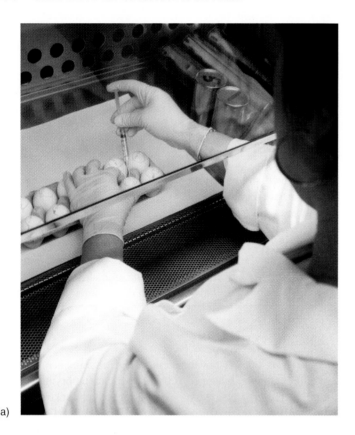

(a)

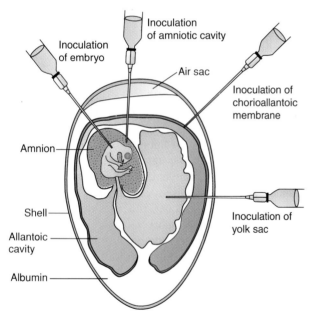

Inoculation
of embryo

Inoculation
of amniotic cavity

Air sac

Inoculation of
chorioallantoic
membrane

Amnion

Shell

Allantoic
cavity

Albumin

Inoculation of
yolk sac

(b)

FIGURE 6.21

Cultivating animal viruses in a developing bird embryo. (a) A technician inoculates fertilized chicken eggs with viruses in the first stage of preparing vaccines. This process requires the highest levels of sterile and aseptic precautions. **(b)** The shell is perforated using sterile techniques, and a virus preparation is injected into a site selected to grow the viruses. Targets include the allantoic cavity, a fluid-filled sac that functions in embryonic waste removal; the amniotic cavity, a sac that cushions and protects the embryo itself; the chorioallantoic membrane, which functions in embryonic gas exchange; the yolk sac, a membrane that mobilizes yolk for the nourishment of the embryo; and the embryo itself.

vitro cultivation systems are termed **cell culture** or **tissue culture.** (Although these terms are used interchangeably, cell culture is probably a more accurate description.) So prominent is this method that most viruses are propagated in some sort of cell culture, and much of the virologist's work involves developing and maintaining these cultures. Animal cell cultures are grown in sterile chambers with special media that contain the correct nutrients required by animal cells to survive. The cultured cells grow in the form of a *monolayer,* a single, confluent sheet of cells that supports viral multiplication and permits close inspection of the culture for signs of infection (figure 6.22).

Cultures of animal cells usually exist in the primary or continuous form. *Primary cell cultures* are prepared by placing freshly isolated animal tissue in a growth medium. The cells undergo a series of mitotic divisions to produce a monolayer. Embryonic, fetal, adult, and even cancerous tissues have served as sources of primary cultures. A primary culture retains several characteristics of the original tissue from which it was derived, but this original line generally has a limited existence. Eventually, it will die out or mutate into a line of cells that can grow continuously. These **cell lines** tend to have altered chromosome numbers, grow rapidly, and show changes in morphology, and they can be continuously subcultured, provided they are routinely transferred to fresh nutrient medium. One very clear advantage of cell culture is that a specific cell line can be available for viruses with a very narrow host range. Strictly human viruses can be propagated in one of several primary or continuous human cell lines, such as embryonic kidney cells, fibroblasts, bone marrow, and heart cells.

One way to detect the growth of a virus in culture is to observe degeneration and lysis of infected cells in the monolayer of cells. The areas where virus-infected cells have been destroyed show up as clear, well-defined patches in the cell sheet called **plaques*** (figure 6.22*c*). This same technique is used to detect and count bacteriophages, because they also produce plaques when grown in soft agar cultures of their host cells (bacteria). A plaque develops when the viruses released by an infected host cell radiate out to adjacent host cells. As new cells become infected, they die and release more viruses, and so on. As this process continues, the infection spreads gradually and symmetrically from the original point of infection, causing the macroscopic appearance of round, clear spaces that correspond to areas of dead cells.

CHAPTER CHECKPOINTS

Animal viruses must be studied in some type of host cell environment such as laboratory animals, bird embryos, or tissue cultures.

Tissue cultures are cultures of host cells grown in special sterile chambers containing correct types and proportions of growth factors using aseptic techniques to exclude unwanted microorganisms.

Virus growth in tissue culture is detected by the appearance of plaques.

*plaque (plak) Fr. *placke,* patch or spot.

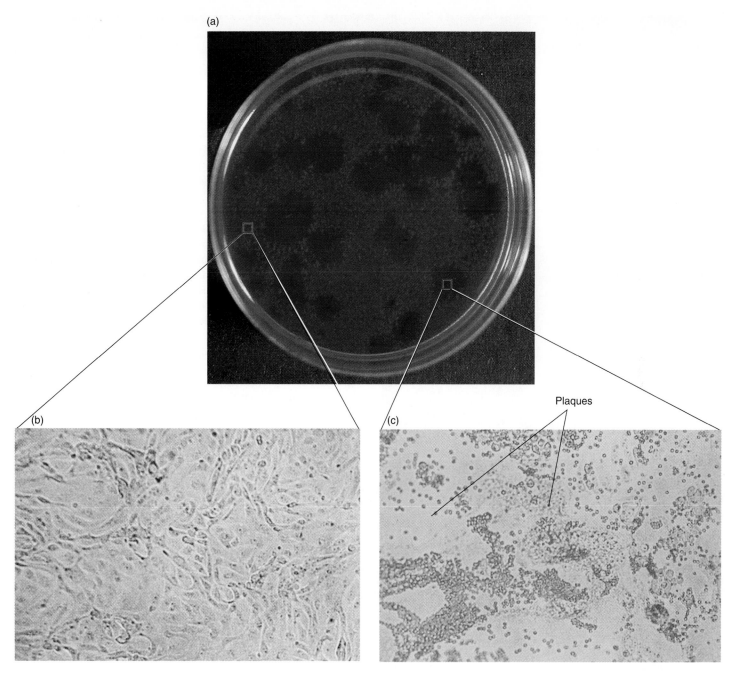

FIGURE 6.22

Appearance of normal and infected cell cultures. (a) Macroscopic view of a Petri dish containing a monolayer (single layer of attached cells) of monkey kidney cells. Clear spaces in culture indicate sites of virus growth (plaques). Microscopic views of **(b)** normal, undisturbed cell layer and **(c)** plaques, which consist of cells disrupted by viral infection.

Medical Importance of Viruses
[See also chapters 24 and 25]

The number of viral infections that occur on a worldwide basis is nearly impossible to predict accurately. Certainly, viruses are the most common cause of acute infections that do not result in hospitalization, especially when one considers widespread diseases such as colds, hepatitis, chickenpox, influenza, herpes, and warts. If one also takes into account prominent viral infections found only in certain regions of the world, such as West Nile fever, Rift Valley fever, and yellow fever, the total could easily exceed several billion cases each year. Although most viral infections do not result in death, some, such as rabies, AIDS, and Ebola infection have very high mortality rates, and others can lead to long-term debility (polio, neonatal rubella). Current research is focused on the possible connection of viruses to chronic afflictions of unknown cause, such as type I diabetes, multiple sclerosis, and various cancers.

MEDICAL MICROFILE 6.1
A Positive View of Viruses

Looking at this beautiful tulip, one would never guess that it derives its pleasing appearance from a viral infection. It contains tulip mosaic virus, which alters the development of the plant cells and causes complex patterns of colors in the petals. Aside from this, the virus does not cause severe harm to the plants. Despite the reputation of viruses as cell killers, there is another side of viruses—that of being harmless, and in some cases, even beneficial.

Although there is no agreement on the origins of viruses, it is highly likely that they have been in existence for billions of years. Virologists are convinced that viruses have been an important force in the evolution of living things. This is based on the fact that they interact with the genetic material of their host cells and that they carry genes from one host to another (transduction). It is convincing to imagine that viruses arose early in the history of cells as loose pieces of genetic material that became dependent nomads, moving from cell to cell. Viruses are also a significant

factor in the functioning of many ecosystems because of the effects they have on their host cells. For example, it is documented that seawater can contain 10 million viruses per milliliter. Since they contain the same elements as living cells, it is estimated that the sum of viruses in the ocean represent 270 million metric tons of organic matter.

Over the past several years, biomedical experts have been looking at viruses as vehicles to treat infections and disease. Viruses are already essential for production of vaccines to treat viral infections such as influenza, polio, and measles. Vaccine experts have also engineered new types of viruses by combining a less harmful virus such as vaccinia or adenovirus with some genetic material from a pathogen such as HIV and herpes simplex (see figure 25.21). This technique creates a vaccine that provides immunity but does not expose the person to the intact pathogen. Several of these types of vaccines are currently in development.

The "harmless virus" approach is also being used to treat genetic diseases such as cystic fibrosis and sickle-cell anemia. With gene therapy, the normal gene is inserted into a retrovirus, such as the mouse leukemia virus, and the patient is infected with this altered virus. It is hoped that the virus will introduce the needed gene into the cells and correct the defect. Dozens of experimental trials are currently underway to develop potential cures for diseases, with some successes (see chapter 10). Virologists have created mutant adenoviruses (ONYX) that target cancer cells. These viruses cannot spread among normal cells, but when they enter cancer cells, they immediately cause the cells to self-destruct. So far, several hundred treatment plans are proceeding for neck, lung, and ovarian cancer.

An older therapy getting a second chance involves use of bacteriophages to treat bacterial infections. This technique was tried in the past with mixed success, but was abandoned for more efficient antimicrobic drugs. The basis behind the therapy is that bacterial viruses would seek out only their specific host bacteria and would cause complete destruction of the bacterial cell. Newer experiments with animals have demonstrated that this method can control infections as well as traditional drugs. Some potential applications being considered are adding phage suspension to grafts to control skin infections and to intravenous fluids for blood infections.

Despite the reputation viruses have for being highly detrimental, in some cases, they may actually show a beneficial side (Medical Microfile 6.1).

OTHER NONCELLULAR INFECTIOUS AGENTS

Not all noncellular infectious agents have typical viral morphology. One group of unusual forms, even smaller and simpler than viruses, is implicated in chronic, persistent diseases in humans and animals. These diseases are called spongiform encephalopathies because the brain tissue removed from affected animals resembles a sponge. The infection has a long period of latency (usually several years) before the first clinical signs appear. Signs range from mental derangement to loss of muscle control, and the diseases are progressive and universally fatal.

A common feature of these conditions is the deposition of distinct protein fibrils in the brain tissue. Some researchers have hypothesized that these fibrils are the agents of the disease and have named them **prions.***

Creutzfeldt-Jakob disease afflicts the central nervous system of humans and causes gradual degeneration and death. Cases in which medical workers developed the disease after handling autopsy specimens seem to indicate that it is transmissible, but by an unknown mechanism. Several animals (sheep, mink, elk) are victims of similar transmissible diseases. Bovine spongiform encephalopathy, or "mad cow disease," was recently the subject of fears and a crisis in Europe when researchers found evidence that the disease could be acquired by humans who consumed contaminated beef. This was the first incidence of prion disease transmission from animals to humans. Several hundred Europeans developed symptoms of a variant form of Creutzfeldt-Jakob disease, leading to strict governmental controls on exporting cattle and beef products.

---

*prion (pree′-on) **pr**oteinacious **i**nfectious particle.

Whether prions actually represent an infectious agent is currently controversial. If continued study supports this hypothesis, the discovery that they are composed primarily of protein (no nucleic acid) will certainly revolutionize our ideas of what can constitute an infectious agent. One of the most compelling questions is just how a prion could be replicated, since all other infectious agents require some nucleic acid.

Other fascinating viruslike agents in human disease are defective forms called satellite viruses that are actually dependent on other viruses for replication! Two remarkable examples are the adeno-associated virus (AAV), which can replicate only in cells infected with adenovirus, and the delta agent, a naked strand of RNA that is expressed only in the presence of the hepatitis B virus and can worsen the severity of liver damage.

Plants are also parasitized by viruslike agents called **viroids** that differ from ordinary viruses by being very small (about one-tenth the size of an average virus) and being composed of only naked strands of RNA, lacking a capsid or any other type of coating. Viroids are significant pathogens in several economically important plants, including tomatoes, potatoes, cucumbers, citrus trees, and chrysanthemums.

DETECTION AND CONTROL OF VIRAL INFECTIONS

Life-threatening viral diseases such as AIDS, hantavirus pulmonary syndrome, influenza, and those that pose a serious risk to the fetus (rubella or cytomegalovirus) demand rapid detection and correct diagnosis so that appropriate therapy and other control measures can be prescribed. Viruses tend to be more difficult to identify than cellular microbes. Physicians often use the overall clinical picture of the disease (specific signs) to guide diagnosis, but exact virus identification from clinical specimens (blood, respiratory secretions, feces, throat, and vaginal secretions) is more satisfactory (figure 6.23*a,b*). One direct, rapid method is to examine a specimen for the presence of the virus or signs of cytopathic changes in cells or tissues (see herpesviruses, figure 24.9). This may involve immunofluorescence techniques or direct examination with an electron microscope (figure 6.23*c*). Samples can also be screened for the presence of indicator molecules (antigens) from the virus itself. A new technique that is rapidly becoming standard procedure is the polymerase chain reaction (PCR), which can detect and amplify even minute amounts of viral DNA or RNA in a sample (figure 6.23*e*). In certain infections, definitive diagnosis will require the use of cell culture, embryos, or animals, but this method can be time-consuming and slow to give results (figure 6.23*f*). Frequently, it is helpful to use a screening test to detect specific antibodies that indicate signs of virus infection in a patient's blood. It is the main test for HIV infection (figure 6.23*d*). Details of viral diagnosis are provided in chapters 10, 16, 24, and 25.

TREATMENT OF ANIMAL VIRAL INFECTIONS

The nature of viruses has at times been a major impediment to effective therapy. Because viruses are not bacteria, antibiotics aimed at disrupting procaryotic cells do not work on them. On the other hand, many antiviral drugs block virus replication by targeting the function of host cells, and can cause severe side effects. An example is azidothymidine (AZT), a drug used to treat AIDS, which has severe toxic side effects such as immunosuppression and anemia. In addition, most antiviral drugs prevent the virus from replicating or maturing and therefore must be capable of entering human cells to have an effect. Another compound that shows some potential for treating and preventing viral infections is a naturally occurring human cell product called *interferon* (see chapter 14). Vaccines that stimulate immunity are an extremely valuable tool but are available for only a limited number of viral diseases (see chapter 16).

CHAPTER CHECKPOINTS

Viruses are easily responsible for several billion infections each year. It is conceivable that many chronic diseases of unknown cause will be connected to viral agents.

Other noncellular agents of disease are the prions, which are not viruses at all, but protein fibers; viroids, extremely small lengths of protein-coated nucleic acid; and satellite viruses, which require larger viruses to cause disease.

Diagnosis of viral disease agents is done directly through culture, electron microscopy, immunofluorescence microscopy, and by the PCR technique. Indirect diagnosis is done by testing patient serum for host antibody to viral agents.

Viral infections are difficult to treat because the drugs that attack the viral replication cycle also cause serious side effects in the host.

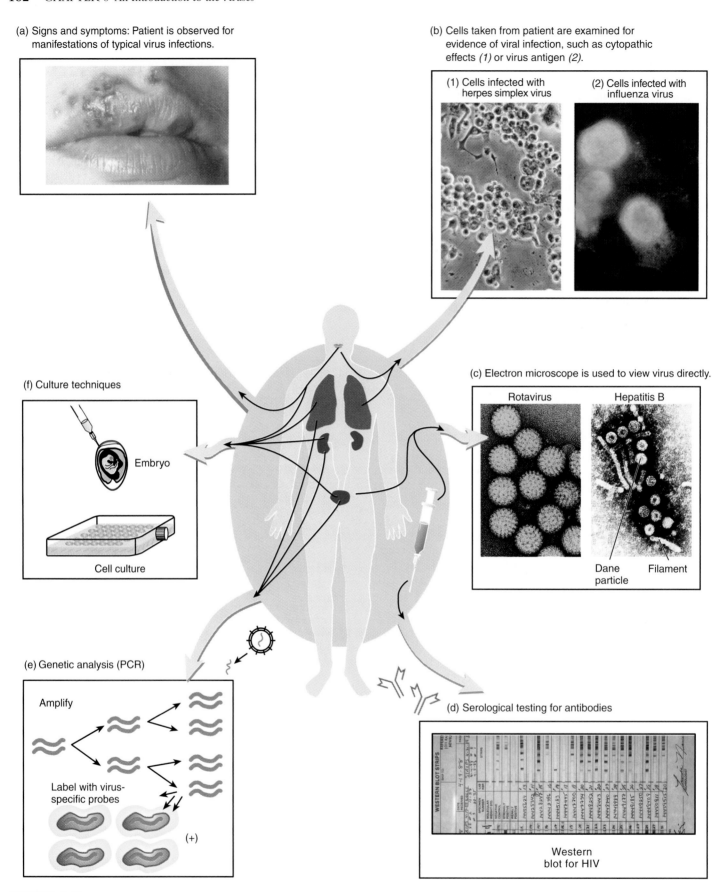

(a) Signs and symptoms: Patient is observed for manifestations of typical virus infections.

(b) Cells taken from patient are examined for evidence of viral infection, such as cytopathic effects *(1)* or virus antigen *(2)*.

(1) Cells infected with herpes simplex virus

(2) Cells infected with influenza virus

(c) Electron microscope is used to view virus directly.

Rotavirus

Hepatitis B

Dane particle

Filament

(f) Culture techniques

Embryo

Cell culture

(e) Genetic analysis (PCR)

Amplify

Label with virus-specific probes

(+)

(d) Serological testing for antibodies

Western blot for HIV

FIGURE 6.23

Summary of methods used to diagnose viral infections.

CHAPTER CAPSULE WITH KEY TERMS

Other terms used to describe viruses are general (virions, viral particles) and specific (bacteriophage, poxvirus, herpesvirus, etc.). They may also be named for host cell, disease, and morphology.

I. **Overall Morphology of Viruses**
   A. Viruses are infectious particles and not cells; lack protoplasm, organelles, locomotion of any kind; are large, complex molecules; can be **crystalline** in form. A virus particle is composed of a nucleic acid core (**DNA** or **RNA**, not both) surrounded by a geometric protein shell, or **capsid**; combination called a **nucleocapsid**; capsid is **helical** or **icosahedral** in configuration; many are covered by a membranous **envelope** containing viral protein **spikes; complex viruses** have additional external and internal structures.
   B. *Shapes/Sizes:* Cuboidal, spherical, cylindrical, brick- and bullet-shaped. Smallest infectious forms range from the largest poxvirus (0.45 mm or 450 nm) to the smallest viruses (0.02 mm or 20 nm).
   C. *Nutritional and Other Requirements:* Lack enzymes for processing food or generating energy; are tied entirely to the host cell for all needs (**obligate intracellular parasites**).

II. **Distribution/Host Range**
   Viruses are known to parasitize all types of cells, including bacteria, algae, fungi, protozoa, animals, and plants. Each viral type is limited in its **host range** to a single species or group, mostly due to specificity of adsorption of virus to specific host receptors.

III. **Classification**
   A. The two major types of viruses are **DNA** and **RNA viruses.** These are further subdivided into families, depending on shape and size of capsid, presence or absence of an envelope, whether double- or single-stranded nucleic acid, and antigenic similarities.
   B. **Animal viruses:** Common names of major DNA animal viruses include poxviruses (smallpox), herpesviruses, adenoviruses, papillomavirus (wart), hepatitis B virus, parvoviruses. RNA animal viruses include poliovirus, hepatitis A virus, rhinovirus (common cold), encephalitis viruses, yellow fever virus, rubella virus, influenza virus, mumps virus, measles virus, rabies viruses, and HIV (AIDS virus).
   C. Other noncellular infectious agents cause spongiform encephalopathies such as Creutzfeldt-Jakob disease in humans and bovine spongiform encephalopathy (mad cow disease) in cattle. The agents are probably protein particles called **prions.**

IV. **Multiplication Cycle**
   A. A unique method of multiplication occurs—no division is involved. Viruses make multiple copies of themselves through:
      1. attachment, or **adsorption,** to a host cell; they must have special binding molecules for receptors on the host cell;
      2. **penetration** of a cell by means of whole virus or only nucleic acid;

3. assuming control of the cell's genetic and metabolic components, thus
4. programming synthesis of new viral parts;
5. directing **assembly** of these parts into new virus particles, and
6. **releasing** complete, mature viruses.
   B. **Bacteriophages** are viruses that attack bacteria; they penetrate by injecting their nucleic acid; they are released as virulent phage upon **lysis** of the cell.
   C. Enveloped viruses penetrate by being engulfed into a vesicle. They are released by **budding** off with an envelope.
   D. Some viruses go into a **latent** or **lysogenic** phase in which they integrate into the DNA of the host cell and later may become active and produce a lytic infection.

V. **Method of Cultivation**
   The need for an intracellular habitat makes it necessary to grow viruses in living cells, either in the intact host animal, in bird embryos, or in isolated cultures of host cells (**cell culture**).

VI. **Identification**
   Viruses are identified by means of **cytopathic effects** in host cells, direct examination of viruses or their components in samples, analyzing blood for antibodies against viruses, performing genetic analysis of samples to detect virus nucleic acid, growing viruses in culture, and symptoms.

VII. **Importance of Viruses**
   A. *Medical:* Viruses attach to specific target hosts or cells. They cause a variety of infectious diseases, ranging from mild respiratory illness (common cold) to destructive and potentially fatal conditions (rabies, AIDS). Some viruses can cause birth defects and cancer in humans and other animals.
   B. *Agricultural:* Hundreds of cultivated plants and domestic animals are susceptible to viral infections, often with adverse economic and ecologic repercussions.
   C. *Research:* Because of their simplicity, viruses have become an invaluable tool for studying basic genetic principles.

VIII. **Unique Features**
   - Viruses exist at a level between living things and nonliving molecules.
   - They consist of a capsid and nucleic acid, either DNA or RNA, not both.
   - Lack metabolism and respiratory enzymes.
   - Multiply inside host cells using the assembly line of the host's synthetic machinery.
   - Are genetic parasites.
   - Can pass through fine filters.
   - Are ultramicroscopic and crystallizable.
   - Some viruses are enveloped, others are naked.
   - Some have unique nucleic acid structure (single-stranded DNA and double-stranded RNA).
   - Some animal viruses can become latent.

MULTIPLE-CHOICE QUESTIONS

1. A virus is a tiny infectious
   a. cell   c. particle
   b. living thing   d. nucleic acid

2. Viruses are known to infect
   a. plants   c. fungi
   b. bacteria   d. all organisms

3. The capsid is composed of protein subunits called
   a. spikes   c. virions
   b. protomers   d. capsomers

4. The envelope of an animal virus is derived from the ____ of its host cell.
   a. cell wall   c. glycocalyx
   b. cell membrane   d. receptors

5. The nucleic acid of a virus is
   a. DNA only
   b. RNA only
   c. both DNA and RNA
   d. either DNA or RNA

6. The general steps in a viral multiplication cycle are
   a. adsorption, penetration, replication, maturation, and release
   b. endocytosis, uncoating, replication, assembly, and budding
   c. adsorption, uncoating, duplication, assembly, and lysis
   d. endocytosis, penetration, replication, maturation, and exocytosis

7. A prophage is an early stage in the development of a/an
   a. bacterial virus   c. lytic virus
   b. poxvirus   d. enveloped virus

8. The nucleic acid of animal viruses enters the host cell through
   a. translocation   c. endocytosis
   b. fusion   d. all of these

9. In general, RNA viruses multiply in the cell _____, and DNA viruses multiply in the cell _____.
   a. nucleus, cytoplasm
   b. cytoplasm, nucleus
   c. vesicles, ribosomes
   d. endoplasmic reticulum, nucleolus

10. Enveloped viruses carry surface receptors called
    a buds   c. fibers
    b. spikes   d. sheaths

11. Viruses that persist in the cell and cause recurrent disease are considered
    a. oncogenic   c. latent
    b. cytopathic   d. resistant

12. Examples of cytopathic effects of viruses are
    a. inclusion bodies c. multiple nuclei
    b. giant cells   d. all of these

13. Viruses cannot be cultivated in
    a. tissue culture   c. live mammals
    b. bird embryos   d. blood agar

14. Clear patches in cell cultures that indicate sites of virus infection are called
    a. plaques   c. colonies
    b. pocks   d. prions

15. Label the parts of this virus. What type of virus might this be?

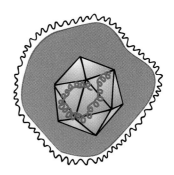

CONCEPT QUESTIONS

1. a. Describe 10 *unique* characteristics of viruses (can include structure, behavior, multiplication).
   b. After consulting table 6.1, what additional statements can you make about viruses, especially as compared with cells?

2. a. What does it mean to be an obligate intracellular parasite?
   b. What is another way to describe the sort of parasitism exhibited by viruses?

3. a. Characterize the viruses according to size range.
   b. What does it mean to say that they are ultramicroscopic?
   c. That they are filterable?

4. a. Describe the general structure of viruses.
   b. What is the capsid, and what is its function?
   c. How are the two types of capsids constructed?
   d. What is a nucleocapsid?
   e. Give examples of viruses with the two capsid types.
   f. What is an enveloped virus, and how does the envelope arise?
   g. Give an example of a common enveloped human virus.
   h. What are spikes, how are they formed, and what is their function?

5. a. What are bacteriophages, and what is their structure?
   b. What is a tobacco mosaic virus?
   c. How are the poxviruses different from other animal viruses?

6. a. Since viruses lack metabolic enzymes, how can they synthesize necessary components?
   b. What are some enzymes with which the virus is equipped?

7. a. How are viruses classified? What are virus families?
   b. How are generic and common names used?

   c. Looking at table 6.3, how many different viral diseases can you count?

8. a. Compare and contrast the main phases in the lytic multiplication cycle in bacteriophages and animal viruses.
   b. When is a virus a virion?
   c. What is necessary for adsorption?
   d. Why is penetration so different in the two groups?
   e. What is eclipse?
   f. In simple terms, what does the virus nucleic acid do once it gets into the cell?
   g. What processes are involved in assembly?

9. a. What is a prophage or temperate phage?
   b. What is lysogeny?

10. a. What dictates the host range of animal viruses?
    b. What are two ways that animal viruses penetrate the host cell?
    c. What is uncoating?
    d. Describe the two ways that animal viruses leave their host cell.

11. a. Describe several cytopathic effects of viruses.
    b. What causes the appearance of the host cell?
    c. How might it be used to diagnose viral infection?

12. a. What does it mean for a virus to be persistent or latent, and how are these events important?
    b. Briefly describe the action of an oncogenic virus.

13. a. Describe the three main techniques for cultivating viruses.
    b. What are the advantages of using cell culture?
    c. The disadvantages of using cell culture?

d. What is a disadvantage of using live intact animals or embryos?

e. What are a cell line and a monolayer?

f. How are plaques formed?

14. a. What is the principal effect of the agent of Creutzfeldt-Jakob disease?

b. How is the proposed agent different from viruses?

c. What are viroids?

15. Why are virus diseases more difficult to treat than bacterial diseases?

16. Circle the viral infections from this list: cholera, rabies, plague, cold sores, whooping cough, tetanus, genital warts, gonorrhea, mumps, Rocky Mountain spotted fever, syphilis, rubella, rat bite fever.

CRITICAL-THINKING QUESTIONS

1. a. What characteristics of viruses could be used to characterize them as life forms?

b. What makes them more similar to lifeless molecules?

2. a. Comment on the possible origin of viruses. Is it not curious that the human cell welcomes a virus in and hospitably removes its coat?

b. How do spikes play a part in the action of the host cell?

3. a. If viruses that normally form envelopes were prevented from budding, would they still be infectious?

b. If the RNA of an influenza virus were injected into a cell by itself, could it cause a lytic infection?

4. The end result of most viral infections is death of the host cell.

a. If this is the case, how can we account for such differences in the damage that viruses do (compare the effects of the cold virus with those of the rabies virus)?

b. Describe the adaptation of viruses that does not immediately kill the host cell and explain what its function might be.

5. a. Given that DNA viruses can actually be carried in the DNA of the host cell's chromosomes, comment on what this phenomenon means in terms of inheritance in the offspring.

b. Could some cancers be explained on the basis of inheritance of a latent virus?

6. You have been given the problem of constructing a building on top of a mountain. It will be impossible to use huge pieces of building material. Using the virus capsid as a model, can you think of a way to construct the building?

7. The AIDS virus attacks only specific types of human cells, such as certain white blood cells and nerve cells. Can you explain why a virus can enter some types of human cells but not others?

8. a. Consult table 6.3 to determine which viral diseases you have had and which ones you have been vaccinated against.

b. Which viruses would you investigate as possible oncoviruses?

9. One early problem in cultivating the AIDS virus was the lack of a cell line that would sustain indefinitely *in vitro,* but eventually one was developed. What do you expect were the stages in developing this cell line?

10. a. If you were involved in developing an antiviral drug, what would be some important considerations? (Can a drug "kill" a virus?)

b. How could multiplication be blocked?

11. a. Is there such a thing as a "good virus"?

b. Explain why or why not. Consider both bacteriophages and viruses of eucaryotic organisms.

12. Why is an embryonic or fetal viral infection so harmful?

13. How are computer program viruses analogous to real viruses?

INTERNET SEARCH TOPICS

1. Look up West Nile virus on the World Wide Web. Determine how the infection is acquired and the symptoms it causes. Find a site that theorizes how this virus got from Africa to the East Coast of the United States.

2. Find websites that discuss prions and prion-based diseases. What possible way do the prions replicate?

Elements of Microbial Nutrition, Ecology, and Growth

Carved into a hillside near Butte, Montana, lies the Berkeley Pit, an industrial body of water that stretches about one mile across and contains a volume of close to 30 billion gallons. This small lake was formerly the site of a copper pit mine abandoned in 1982 and left to fill up with water seeping out of local springs. Lying at the bottom of the pit was a massive layer of mining waste that had accumulated over many years of mining.

The gradual dissolution of the waste in the water transformed Berkeley Pit into a giant vat of concentrated chemicals so toxic that it was considered uninhabitable to living things. Among the substances found in abundance are heavy metals such as lead, cadmium, aluminum, copper, iron, and zinc, as well as arsenic and sulfides. The pit also has a high level of acidity—10,000 times more than a normal freshwater habitat. The greatest concern is that water from the pit will leak into local streams and contaminate groundwater, river drainages, and eventually the ocean. The Environmental Protection Agency has targeted the site for cleanup to avert a possible ecological disaster.

When scientists from the University of Montana began researching the composition of the water in the pit, they were startled by what they found. The samples showed signs of a well-established community of microorganisms that had taken hold over the 20 years the pit lay undisturbed. It included an array of very hardy bacteria, protozoa, algae, and fungi—nearly 50 species in all. The microbes evidently entered the pit from the water and air, and rather than being killed, they survived, grew, and spread into available *niches*.* In a relatively short time, some of them had actually evolved to depend on the toxic sludge for survival. An important observation made by the Montana researchers is that the microbes appear to be naturally detoxifying the water. They are currently investigating a way to adapt this "self-cleaning" technology to clear the system of toxic chemicals and in time convert the lake to a useful water reservoir.

This example should serve as a dramatic reminder of the overwhelming adaptability of microorganisms, which allows them to prevail in environments that would be toxic to all other living things (see Spotlight on Microbiology 7.1). There are millions of habitats on earth, both of natural and human origin. Microbes are exposed to a tremendous variety in nutrient sources, temperatures, gases, salt, pH, and radiation. With this theme in mind, this chapter will take a closer look at this adaptability in the light of nutrition, responses to environmental factors, transport, and growth.

The Berkeley Pit near Butte, Montana—a legacy of 45 years of mining—now teaching us about the survival capacity of the microbial world.

Chapter Overview

- Microbes exist in every known natural habitat on earth.
- Microbes show enormous capacity to adapt to environmental factors.
- Factors that have the greatest impact on microbes are nutrients, temperature, pH, amount of available water, atmospheric gases, light, pressure, and other organisms.
- Nutrition involves absorbing required chemicals from the environment for use in metabolism.
- Autotrophs can exist solely on inorganic nutrients, while heterotrophs require both inorganic and organic nutrients.
- Energy sources for microbes may come from light or chemicals.
- Temperature adaptations may be at cold, moderate, and hot temperatures.
- Oxygen and carbon dioxide are primary gases used in metabolism.
- Transport of materials by cells involves movement by passive or active mechanisms across the cell membrane.
- The water content of the cell versus its environment dictates the osmotic adaptations of cells.
- Microbes interact in a variety of ways with one another and with other organisms that share their habitats.
- The pattern of population growth in simple microbes is to double the number of cells in each generation.
- Growth rate is limited by lack of nutrients and buildup of waste products.

---

*niche (nitch) Fr. *nichier*, to nest. The role of an organism in its ecological setting.

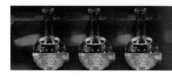

SPOTLIGHT ON MICROBIOLOGY 7.1
Life in the Extremes

Any extreme habitat—whether hot, cold, salty, acidic, alkaline, high pressure, arid, oxygen-free, or toxic—is likely to harbor microorganisms that have made special adaptations to their conditions. Although in most instances the inhabitants are archaea and bacteria, certain fungi, protozoans, and algae are also capable of living in harsh habitats. Microbiologists have termed such remarkable organisms **extremophiles.**

Hot and Cold

Perhaps the most extreme habitats are hot springs, geysers, volcanoes, and ocean vents, all of which support flourishing microbial populations. Temperatures in these regions range from 50°C to well above the boiling point of water, with some ocean vents even approaching 350°C. Many heat-adapted microbes are highly primitive archaea whose genetics and metabolism are extremely modified for this mode of existence. A unique ecosystem based on hydrogen sulfide–oxidizing bacteria exists in the hydrothermal vents lying along deep oceanic ridges (see Spotlight on Microbiology 7.5). Heat-adapted bacteria even plague home water heaters and the heating towers of power and industrial plants.

A large part of the earth exists at cold temperatures. Microbes settle and grow throughout the Arctic and Antarctic, and in the deepest parts of the ocean, in temperatures that hover near the freezing point of water (2°–10°C). Several species of algae and fungi thrive on the surfaces of snow and glaciers (see figure 7.10). More surprising still is a strategy of bacteria and algae adapted to the sea ice of Antarctica. Although the ice appears to be completely solid, it is honeycombed by various-sized pores and tunnels filled with liquid water. These frigid microhabitats harbor a virtual microcosm of planktonic life, including predators (fish and shrimp) that live on these algae and bacteria.

Salt, Acidity, Alkalinity

The growth of most microbial cells is inhibited by high amounts of salt; for this reason, salt is a common food preservative. Yet whole communities of salt-dependent bacteria and algae occupy habitats in oceans, salt lakes, and inland seas, some of which are saturated with salt (30%). Most of these microbes have demonstrable metabolic requirements for high levels of minerals such as sodium, potassium, magnesium, chlorides, or iodides. Because of their salt-loving nature, some species are pesky contaminants in salt-processing plants, pickling brine, and salted fish.

Highly acidic or alkaline habitats are not common, but acidic bogs, lakes, and alkaline soils contain a specialized microbial flora. A few species of algae and bacteria can actually survive at a pH near that of concentrated hydrochloric acid. They not only require such a low pH for growth, but particular bacteria (for example, *Thiobacillus*) actually help maintain the low pH by releasing strong acid.

Other Frontiers to Conquer

It was once thought that the region far beneath the soil and upper crust of the earth's surface was sterile. However, new work with deep core samples (from 330 m down) indicates a vast microbial population in these zones. Myriad bacteria, protozoa, and fungi exist in this moist clay, which is high in minerals and complex organic substrates. Even deep mining deposits two miles into the earth's crust harbor a rich assortment of bacteria. They thrive in mineral deposits that are hot (90°C) and radioactive. Many biologists believe these are very similar to the first ancient microbes to have existed on earth.

Numerous species have carved a niche for themselves in the depths of mud, swamps, and oceans, where oxygen gas and sunlight cannot penetrate. The predominant living things in the deepest part of the oceans (10,000 m or below) are pressure- and cold-loving microorganisms. Even parched zones in sand dunes and deserts harbor a hardy brand of microbes, and thriving bacterial populations can be found in petroleum, coal, and mineral deposits containing copper, zinc, gold, and uranium.

As a rule, a microbe that has adapted to an extreme habitat will die if placed in a moderate one. And, except for rare cases, none of the organisms living in these extremes are pathogens, because the human body is a hostile habitat for them.

(a)

(b)

(a) Cells of *Sulfolobus,* an archaean that lives in mineral deposits of hot springs and volcanoes. It can survive temperatures of about 90°C and acidity of pH 1.5. **(b)** Clumps of bacteria (dark matter) clinging to life on crystals of ice deep in the Antarctic sediments.

MICROBITS 7.2
Dining with an Ameba

An ameba gorging itself on bacteria could be compared to a person eating a bowl of vegetable soup, because its nutrient needs and uses are fundamentally similar to a human's. Most food is a complex substance that contains many different types of nutrients. Some smaller molecules such as sugars can be absorbed directly by the cell; larger food debris and molecules must be first be ingested and broken down into a size that can be absorbed. As nutrients are taken in, they add to a dynamic pool of inorganic and organic compounds dissolved in the cytoplasm. This pool will provide raw materials to be assimilated into the organism's own specialized proteins, carbohydrates, lipids, and other macromolecules used in growth and metabolism.

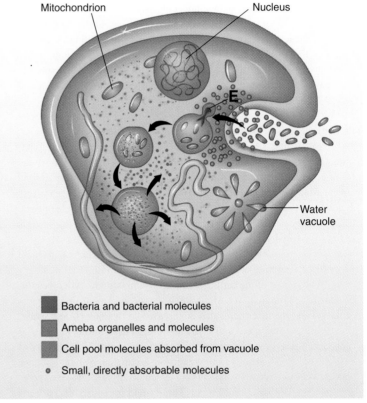

Steps in obtaining and incorporating nutrients. Food particles are phagocytosed into a vacuole that fuses with a lysozyme containing digestive enzymes (E). Smaller subunits of digested macromolecules are transported out of the vacuole into the cell pool and are used in the metabolic and synthetic activities of the cell.

■ Bacteria and bacterial molecules

■ Ameba organelles and molecules

■ Cell pool molecules absorbed from vacuole

• Small, directly absorbable molecules

Microbial Nutrition

Nutrition* is a process by which chemical substances called **nutrients** are acquired from the environment and used in cellular activities such as metabolism and growth. With respect to nutrition, microbes are not really so different from humans (Microbits 7.2). Bacteria living in mud on a diet of inorganic sulfur or protozoa digesting wood in a termite's intestine seem to show radical adaptations, but even these organisms require a constant influx of certain substances from their habitat. In general, all living things require a source of elements such as carbon, hydrogen, oxygen, phosphorus, potassium, nitrogen, sulfur, calcium, iron, sodium, chlorine, magnesium, and certain other elements.[1] But the ultimate source of a particular element, its chemical form, and how much of it the microbe needs are all points of variation (table 7.1). Any substance, whether in elemental or molecular form, that must be provided to an organism is called an **essential nutrient.** Once absorbed, nutrients are processed and transformed into the chemicals of the cell.

Two categories of essential nutrients are **macronutrients** and **micronutrients.** Macronutrients are required in relatively large quantities and play principal roles in cell structure and metabolism. Examples of macronutrients are proteins, carbohydrates, and other molecules that contain carbon, hydrogen, and oxygen. Micronutrients, or **trace elements,** such as manganese, zinc, and nickel are present in much smaller amounts and are involved in enzyme function and maintenance of protein structure. What constitutes a micronutrient can vary from one microbe to another and often must be determined in the laboratory. This determination is made by deliberately omitting the substance in question from a growth medium to see if the microbe can grow in its absence.

Another way to categorize nutrients is according to their carbon content. An **inorganic nutrient** is an atom or simple molecule that contains a combination of atoms other than carbon and hydrogen. The natural reservoirs of inorganic compounds are mineral deposits in the crust of the earth, bodies of water, and the atmosphere. Examples include metals and their salts (magnesium sulfate, ferric nitrate, sodium phosphate), gases (oxygen, carbon dioxide), and water (table 7.2). In contrast, the molecules of **organic nutrients** contain carbon and hydrogen atoms and are usually the products of living things. They range from the simplest organic molecule, methane (CH_4), to large polymers (carbohy-

---

1. A useful mnemonic device for keeping track of major elements required by most organisms is: **C HOPKINS Ca Fe, Na Cl Mg**ood. Read aloud, it becomes, See Hopkin's Cafe, NaCl (salt) M good! Boldfaced letters are short for the atomic symbol of that element.

*nutrition L. *nutrire,* nourishment.

TABLE 7.1

Nutrient Requirements and Sources for Two Ecologically Different Bacterial Species

	Thibacillus thioxidans	*Mycobacterium tuberculosis*
Habitat	Sulfur springs	Human lung
Nutritional Type	Chemoautotroph (sulfur-oxidizing bacteria)	Chemoheterotroph (parasite; cause of tuberculosis in humans)
Essential Element	**Provided Principally By**	
Carbon	Carbon dioxide (CO_2) in air	Minimum of L-glutamic acid (an amino acid), glucose (a simple sugar), albumin (blood protein), oleic acid (a fatty acid), and citrate
Hydrogen	H_2O, H^+ ions	Nutrients listed above, H_2O
Nitrogen	Ammonium (NH_4)	Ammonium, L-glutamic acid
Oxygen	Phosphate (PO_4), sulfate (SO_4), O_2	Oxygen gas (O_2) in atmosphere
Phosphorus	PO_4	PO_4
Sulfur	Elemental sulfur (S), SO_4	SO_4
Miscellaneous Nutrients		
Vitamins	None required	Pyridoxine, biotin
Inorganic salts	Potassium, calcium, iron, chloride	Sodium, chloride, iron, zinc, calcium, copper, magnesium
Principal energy source	Oxidation of sulfur (inorganic)	Oxidation of glucose (organic)

drates, lipids, proteins, and nucleic acids). The source of nutrients is extremely varied: Some microbes obtain their nutrients entirely from inorganic sources, and others require a combination of organic and inorganic sources (see table 7.1). Parasites capable of invading and living on the human body derive all essential nutrients from host tissues, tissue fluids, secretions, and wastes.

CHEMICAL ANALYSIS OF MICROBIAL CYTOPLASM

Examining the chemical composition of a bacterial cell can indicate its nutritional requirements. Table 7.3 lists the major contents of the intestinal bacterium *Escherichia coli*. Some of these components are absorbed in a ready-to-use form, and others must be synthesized by the cell from simple nutrients. Several important features of cell composition can be summarized as follows:

- Water content is the highest of all the components (70%).
- Proteins are the next most prevalent chemical.
- About 97% of the dry cell weight is composed of organic compounds.
- About 96% of the cell is composed of six elements (represented by CHNOPS).
- Bioelements are needed in the overall scheme of cell growth, but most of them are available to the cell as compounds and not as pure elements (table 7.3).
- A cell as "simple" as *E. coli* contains on the order of 5,000 different compounds, yet it needs to absorb only a few types of nutrients to synthesize this great diversity. These include $(NH_4)_2SO_4$, $FeCl_2$, NaCl, trace elements, glucose, KH_2PO_4, $MgSO_4$, $CaHPO_4$, and water.

TABLE 7.2

Principal Inorganic Reservoirs of Elements

Element	Inorganic Environmental Reservoir
Carbon	CO_2 in air; CO_3^{-2} in rocks and sediments
Oxygen	O_2 in air, certain oxides, water
Nitrogen	N_2 in air; NO_3^-, NO_2^-, NH_4^+ in soil and water
Hydrogen	Water, H_2 gas, mineral deposits
Phosphorus	Mineral deposits (PO_4^{-3}, H_3PO_4)
Sulfur	Mineral deposits, volcanic sediments (SO_4^{-2}, H_2S, S)
Potassium	Mineral deposits, the ocean (KCl, K_3PO_4)
Sodium	Mineral deposits, the ocean (NaCl, NaSi)
Calcium	Mineral deposits, the ocean ($CaCO_3$, $CaCl_2$)
Magnesium	Mineral deposits, geologic sediments ($MgSO_4$)
Chloride	The ocean (NaCl, NH_4Cl)
Iron	Mineral deposits, geologic sediments ($FeSO_4$)
Manganese, molybdenum, cobalt, nickel, zinc, copper, other micronutrients	Various geologic sediments

SOURCES OF ESSENTIAL NUTRIENTS

The elements that make up nutrients ultimately exist in an environmental inorganic reservoir of some type. These reservoirs serve not only as a permanent, longtime source of these elements but also can be replenished by the activities of organisms. In fact, as we shall see in chapter 26, the ability of microbes to keep elements cycling is essential to all life on the earth.

TABLE 7.3		
Analysis of the Chemical Composition of an *Escherichia coli* Cell		
	% Total Weight	**% Dry Weight**
Organic Compounds		
Proteins	15	50
Nucleic acids		
RNA	6	20
DNA	1	3
Carbohydrates	3	10
Lipids	2	Not determined
Miscellaneous	2	Not determined
Inorganic Compounds		
Water	70	
All others	1	3
Elements		
Carbon (C)		50
Oxygen (O)		20
Nitrogen (N)		14
Hydrogen (H)		8
Phosphorus (P)		3
Sulfur (S)		1
Potassium (K)		1
Sodium (Na)		1
Calcium (Ca)		0.5
Magnesium (Mg)		0.5
Chloride (Cl)		0.5
Iron (Fe)		0.2
Manganese (Mn), zinc (Zn), molybdenum (Mo), copper (Cu), cobalt (Co), zinc (Zn)		0.3

For convenience, this section on nutrients is organized by element. You will no doubt notice that some categories overlap and that many of the compounds furnish more than one element.

Carbon Sources

It seems worthwhile to emphasize a point about the *extracellular source* of carbon as opposed to the *intracellular function* of carbon compounds. Although a distinction is made between the type of carbon compound cells absorb as nutrients (inorganic or organic), the majority of carbon compounds involved in the normal structure and metabolism of all cells are organic.

A **heterotroph*** is an organism that must obtain its carbon in an organic form. Because organic carbon originates from the bodies of other organisms, heterotrophs are dependent on other life forms. Among the common organic molecules that can satisfy this requirement are proteins, carbohydrates, lipids, and nucleic acids. In most cases, these nutrients provide several other elements as well. Some organic nutrients available to heterotrophs already exist in a form that is simple enough for absorption (for example, mono-

saccharides and amino acids), but many larger molecules must be digested by the cell before absorption. Moreover, heterotrophs vary in their capacities to use various organic carbon sources. Some are restricted to a few substrates, whereas others (certain *Pseudomonas* bacteria, for example) are so versatile that they can metabolize more than 100 different substrates.

An **autotroph*** is an organism that uses CO_2, an inorganic gas, as its carbon source. Because autotrophs have the special capacity to convert CO_2 into organic compounds, they are not nutritionally dependent on other living things. In a later section, we enlarge on the topic of nutritional types as based on carbon and energy sources.

Nitrogen Sources

The main reservoir of nitrogen is nitrogen gas (N_2), which makes up about 79% of the earth's atmosphere. This element is indispensable to the structure of proteins, DNA, RNA, and ATP. Such nitrogenous compounds are the primary source for heterotrophs, but to be useful, they must first be degraded into their basic building blocks (proteins into amino acids; nucleic acids into nucleotides). Some bacteria and algae utilize inorganic nitrogenous nutrients (NO_3^-, NO_2^-, or NH_3). A small number of bacteria can transform N_2 into compounds usable by other organisms through the process of nitrogen fixation (see chapter 26). Regardless of the initial form in which the inorganic nitrogen enters the cell, it must first be converted to NH_3, the only form that can be directly combined with carbon to synthesize amino acids and other compounds.

Oxygen Sources

Because oxygen is a major component of organic compounds such as carbohydrates, lipids, and proteins, it plays an important role in the structural and enzymatic functions of the cell. Oxygen is likewise a common component of inorganic salts such as sulfates, phosphates, nitrates, and water. Free gaseous oxygen (O_2) makes up 20% of the atmosphere. It is absolutely essential to the metabolism of many organisms, as we shall see later in this chapter and in chapter 8.

Hydrogen Sources

Hydrogen is a major element in all organic compounds and several inorganic ones, including water (H_2O), salts ($Ca[OH]_2$), and certain naturally occurring gases (H_2S, CH_4, and H_2). These gases are both used and produced by microbes. Hydrogen performs the following overlapping roles in the biochemistry of cells: (1) maintaining pH, (2) forming hydrogen bonds between molecules, and (3) serving as the source of free energy in oxidation-reduction reactions of respiration (see chapter 8).

Phosphorus (Phosphate) Sources

The main inorganic source of phosphorus is phosphate (PO_4^{-3}), derived from phosphoric acid (H_3PO_4) and found in rocks and oceanic mineral deposits. Phosphate is a key component of nucleic acids and is thereby essential to the genetics of cells and viruses. Because it is found in the nucleotide ATP, it also serves in cellular energy transfers. Other phosphate-containing compounds are phos-

---

*heterotroph (het'-uhr-oh-trohf) Gr. *hetero*, other, and *troph*, to feed.

*autotroph (aw'-toh-trohf) Gr. *auto*, self, and *troph*, to feed.

pholipids in cell membranes and coenzymes such as NAD (see chapter 8). Phosphate can be so scarce in the environment that its rarity severely limits growth, which explains why bacteria such as *Corynebacterium* concentrate it in metachromatic granules.

Sulfur Sources

Sulfur is widely distributed throughout the environment in mineral form. Rocks and sediments (such as gypsum) can contain sulfate (SO_4^{-2}), sulfides (FeS), hydrogen sulfide gas (H_2S), and elemental sulfur (S). Sulfur is an essential component of some vitamins (vitamin B_1) and the amino acids cysteine (sis′-tee-een) and methionine (meth-eye′-oh-neen). Cysteine contributes to the shape and structural stability of proteins by forming unique linkages called disulfide bonds (see figure 2.22).

Other Nutrients Important in Microbial Metabolism

The remainder of important elements are mineral ions. **Potassium** is essential to protein synthesis and membrane function. **Sodium** is important for some types of cell transport. **Calcium** is a stabilizer of the cell wall and endospores of bacteria. It is also combined with carbonate (CO_3^{-2}) in the formation of shells by foraminiferans and radiolarians. **Magnesium** is a component of chlorophyll and a stabilizer of membranes and ribosomes. **Iron** is an important component of the cytochrome pigments of cell respiration. Zinc is an essential regulatory element for eucaryotic genetics. It is a major component of "zinc fingers"—binding factors that help enzymes adhere to specific sites on DNA. Copper, cobalt, nickel, molybdenum, manganese, silicon, iodine, and boron are needed in small amounts by some microbes but not others. A discovery with important medical implications is that metal ions can directly influence certain diseases by their effects on microorganisms. For example, the bacteria that cause gonorrhea and meningitis grow more rapidly in the presence of iron ions.

Growth Factors: Essential Organic Nutrients

Few microbes are as versatile as *Escherichia coli* in assembling molecules from scratch. Many fastidious bacteria lack the genetic and metabolic mechanisms to synthesize every organic compound they need for survival. An organic compound such as an amino acid, nitrogen base, or vitamin that cannot be synthesized by an organism and must be provided as a nutrient is a **growth factor.** For example, although all cells require 20 different amino acids for proper assembly of proteins, many cells cannot synthesize them all. Those that must be obtained from food are called **essential amino acids.** A notable example of the need for growth factors occurs in *Haemophilus influenzae,* a bacterium that causes meningitis and respiratory infections in humans. It can grow only when hemin (factor X), NAD (factor V), thiamine and pantothenic acid (vitamins), uracil, and cysteine are provided by another organism or a growth medium.

HOW MICROBES FEED: NUTRITIONAL TYPES

The earth's limitless habitats and microbial adaptations are rivaled by an elaborate menu of nutritional schemes. Fortunately, most organisms show consistent trends and can be described by a few general categories (table 7.4) and a few selected terms (see "A Note on Terminology"). The main determinants of a microbe's nutritional

TABLE 7.4

Nutritional Categories of Microbes by Carbon and Energy Source

Category	Carbon Source	Energy Source	Example
Autotroph	CO_2	**Nonliving Environment**	
Photoautotroph	CO_2	Sunlight	Photosynthetic organisms, such as algae, plants, cyanobacteria
Chemoautotroph	CO_2	Simple inorganic chemicals	Only certain bacteria, such as methanogens, vent bacteria
Heterotroph	**Organic**	**Other Organisms or Sunlight**	
Photoheterotroph	Organic	Sunlight	Nonsulfur bacteria
Chemoheterotroph	Organic	Metabolic conversion of the nutrients from other organisms	Protozoa, fungi, many bacteria, animals
Saprobe	Organic	Metabolizing the organic matter of dead organisms	Fungi, bacteria (decomposers)
Parasite	Organic	Utilizing the tissues, fluids of a live host	Various parasites and pathogens; can be bacteria, fungi, protozoa, animals

type are its sources of carbon and energy. In a previous section, microbes were defined as autotrophs, whose primary carbon source is inorganic carbon (CO_2), and heterotrophs, which are dependent on organic carbon compounds. In terms of energy source, microbes that photosynthesize are generally classified as **phototrophs,** and those that oxidize chemical compounds are **chemotrophs.** The terms for carbon and energy source are often merged into a single word for convenience (table 7.4). The categories described here are meant to describe only the major nutritional groups and do not include unusual exceptions.

Autotrophs and Their Energy Sources

Autotrophs derive energy from one of two possible nonliving sources: sunlight (photoautotrophs) and chemical reactions involving simple inorganic chemicals (chemoautotrophs). **Photoautotrophs** are photosynthetic; that is, they capture the energy of light rays and transform it into chemical energy that can be used in cell metabolism (Microbits 7.3). Because photosynthetic organisms (algae, plants, some bacteria) produce organic molecules that can be used by themselves and heterotrophs, they form the basis for most food webs. Their role as primary producers of organic matter is discussed in chapter 26.

A Note on Terminology

Much of the vocabulary for describing microbial adaptations is based on some common root words. These are combined in various ways that assist in discussing the types of nutritional or ecologic adaptations, as shown in this partial list:

Root	Meaning	Example of Use
troph-	Food, nourishment	Trophozoite—the feeding stage of protozoa
-phile	To love	Extremophile—an organism that has adapted to ("loves") extreme environments
-obe	To live	Microbe—to live "small"
hetero-	Other	Heterotroph—an organism that requires nutrients from other organisms
auto-	Self	Autotroph—an organism that does not need other organisms for food (obtains nutrients from a nonliving source)
photo-	Light	Phototroph—an organism that uses light as an energy source
chemo-	Chemical	Chemotroph—an organism that uses chemicals for energy, rather than light
sapro-	Rotten	Saprobe—an organism that lives on dead organic matter
halo-	Salt	Halophile—an organism that can grow in high-salt environments
thermo-	Heat	Thermophile—an organism that grows best at high temperatures
psychro-	Cold	Psychrophile—an organism that grows best at cold temperatures
aero-	Air (O_2)	Aerobe—an organism that uses oxygen in metabolism

Modifier terms are also used to specify the nature of an organism's adaptations. *Obligate* or *strict* refers to being restricted to a narrow niche or habitat, such as an obligate thermophile that requires high temperatures to grow. By contrast, *facultative** means not being so restricted, but being able to adapt to a wider range of metabolic conditions and habitats. A facultative halophile can grow with or without high salt concentration.

A significant type of bacteria called **chemoautotrophs** have an unusual nutritional adaptation that requires neither sunlight nor organic nutrients. Some microbiologists prefer to call them **lithoautotrophs** (rock feeders) in reference to their total reliance on inorganic minerals. These bacteria derive energy in diverse and rather amazing ways. In very simple terms, they remove electrons from inorganic substrates such as hydrogen gas, hydrogen sulfide,

sulfur, or iron and combine them with carbon dioxide and hydrogen. This reaction provides simple organic molecules and a modest amount of energy to drive the synthetic processes of the cell. Chemoautotrophic bacteria play an important part in recycling inorganic nutrients. For an example of chemoautotrophy and its importance to deep-sea communities, see Spotlight on Microbiology 7.5.

An interesting group of chemoautotrophs are *methanogens,** which produce methane (CH_4) from hydrogen gas and carbon dioxide (figure 7.1).

$$4H_2 + CO_2 \rightarrow CH_4 + 2H_2O$$

Methane, sometimes called "swamp gas," is formed in anaerobic, hydrogen-containing microenvironments of soil, swamps, mud, and even in the intestines of some animals. Many methanogens are archaea that live in extreme habitats such as ocean vents and hot springs, where temperatures reach up to 125°C. Methane can be harvested and used as an inexpensive energy source in certain industries. Biogas generators are devices primed with a mixed population of microbes (including methanogens) and fueled with various waste materials that can supply enough methane to drive a steam generator. Methane also plays a role as one of the greenhouse gases that is currently an environmental concern (see chapter 26).

Heterotrophs and Their Energy Sources

The majority of heterotrophic microorganisms are **chemoheterotrophs** that derive both carbon and energy from organic compounds. Processing these organic molecules by respiration or fermentation releases energy in the form of ATP. An example of chemoheterotrophy is *aerobic respiration,* the principal energy-yielding reaction in animals, most protozoa and fungi, and aerobic bacteria. It can be simply represented by the equation:

$$\text{Glucose } [(CH_2O)_n] + O_2 \rightarrow CO_2 + H_2O + \text{Energy (ATP)}$$

You might notice that this reaction is complementary to photosynthesis. Here, glucose and oxygen are reactants, and carbon dioxide is given off. Indeed, the earth's balance of both energy and metabolic gases is greatly dependent on this relationship. Chemoheterotrophic microorganisms belong to one of two main categories that differ in how they obtain their organic nutrients: **Saprobes**[2] are free-living microorganisms that feed primarily on organic detritus from dead organisms, and **parasites** ordinarily derive nutrients from the cells or tissues of a host.

Saprobic Microorganisms Saprobes occupy a niche as decomposers of plant litter, animal matter, and dead microbes. If not for the work of decomposers, the earth would gradually fill up with organic material, and the nutrients it contains would not be recycled. Most saprobes, notably bacteria and fungi, have a rigid cell wall and cannot engulf large particles of food. To compensate, they release enzymes to the extracellular environment and digest the food

---

*facultative (fak′-uhl-tay-tiv) Gr. *facult,* capability or skill, and *tatos,* most.

2. Synonyms are *saprotroph* and *saprophyte.* We prefer to use the terms *saprobe* and *saprotroph* because they are more consistent with other terminology and because the term *saprophyte* is a holdover from the time when bacteria and fungi were considered plants.

*methanogen (meth-an-oh-gen) From *methane,* a colorless, odorless gas, and *gennan,* to produce.

MICROBITS   7.3
Light-Driven Organic Synthesis

Two equations sum up the reactions of photosynthesis in a simple way. The first equation shows a reaction that results in the production of oxygen:

$$CO_2 + H_2O \xrightarrow[\text{by chlorophyll}]{\text{Sunlight absorbed}} (CH_2O)_{n*} + O_2$$

This oxygenic (oxygen-producing) type of photosynthesis occurs in plants, algae, and cyanobacteria (see figure 4.34). The function of chlorophyll is to capture light energy. Carbohydrates produced by the reaction can be used by the cell to synthesize other cell components, and they also become a significant nutrient for heterotrophs that feed on them. The production of oxygen is vital to maintaining this gas in the atmosphere.

A second equation shows a photosynthetic reaction that does not result in the production of oxygen:

$$CO_2 + H_2S \xrightarrow[\text{by bacteriochlorophyll}]{\text{Sunlight absorbed}} (CH_2O)_n + S + H_2O$$

This anoxygenic (no oxygen produced) type of photosynthesis is found in bacteria such as purple and green sulfur bacteria. Note that the type of chlorophyll (bacteriochlorophyll, a substance unique to this microbes), one of the reactants (hydrogen sulfide gas), and one product (elemental sulfur) are different from those in the first equation. These bacteria live in the anaerobic regions of aquatic habitats.

*$(CH_2O)_n$ is shorthand for a carbohydrate.

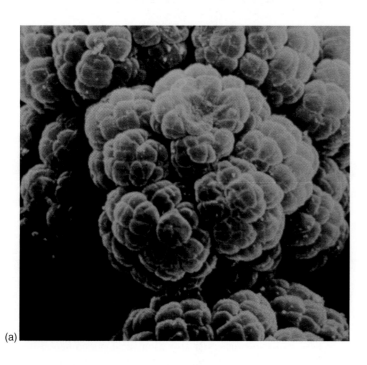

(a)

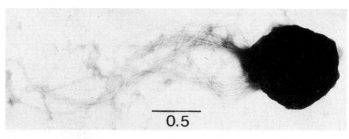

(b)

FIGURE 7.1

Methane-producing archaea. Members of this group are primitive procaryotes with unusual cell walls and membranes. **(a)** A small colony of *Methanosarcina*. **(b)** *Methanococcus junnuschii*, a motile archaea that inhabits hot vents in the seafloor and uses hydrogen gas as a source of energy.

particles into smaller molecules that can pass freely into the cell (figure 7.2). *Obligate saprobes* exist strictly on organic matter in soil and water and are unable to adapt to the body of a live host. This group includes many free-living protozoa, fungi, and bacteria. Apparently, there are fewer of these strict species than was once thought, and many supposedly nonpathogenic saprobes can infect a susceptible host. When a saprobe infects a host, it is considered a *facultative parasite*. Such an infection usually occurs when the host is compromised, and the microbe is considered an *opportunistic pathogen*. For example, although its natural habitat is soil and water, *Pseudomonas aeruginosa* frequently causes infections in hospitalized patients. The yeast *Cryptococcus neoformans* causes a severe lung and brain infection in AIDS patients, yet its natural habitat is the soil.

Parasitic Microorganisms Parasites live in or on the body of a host, which they usually harm to some degree. Parasites inclined to cause damage to tissues (disease) or even death are called **pathogens.** Parasites range from viruses to helminth worms, and they can live on the body (ectoparasites), in the organs and tissues (endoparasites), or even within cells (intracellular parasites, the most extreme type). Although there are several degrees of parasitism, the more successful parasites generally have no fatal effects and eventually evolve to a less harmful relationship with their host. *Obligate parasites* (for example, the leprosy bacillus and the syphilis spirochete) are unable to grow outside of a living host. Parasites that are less strict can be cultured artificially if provided with the correct nutrients and environmental conditions. Bacteria such as *Streptococcus pyogenes* (the cause of strep throat) and *Staphylococcus aureus* can grow on artificial media as saprobes.

Digestion in Bacteria and Fungi

Organic debris

(a) Walled cell is inflexible.

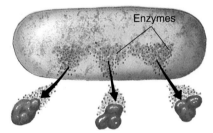

Enzymes

(b) Enzymes are transported across the wall.

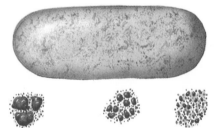

(c) Enzymes hydrolyze the bonds on nutrients.

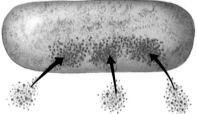

(d) Smaller molecules are transported into the cytoplasm.

FIGURE 7.2

Extracellular digestion in a saprobe with a cell wall (bacterium or fungus). **(a)** A walled cell is inflexible and cannot engulf large pieces of organic debris. **(b)** In response to a usable substrate, the cell synthesizes enzymes that are transported across the wall into the extracellular environment. **(c)** The enzymes hydrolyze the bonds in the debris molecules. **(d)** Digestion produces molecules small enough to be transported into the cytoplasm.

Obligate intracellular parasitism is an extreme but relatively common mode of life. Microorganisms that spend all or part of their life cycle inside a host cell include the viruses, a few bacteria (rickettsias, chlamydias), and certain protozoa (apicomplexa). Contrary to what one might think, the inside of a cell is not completely without hazards, and microbes must overcome some difficult challenges. They must find a way into the cell, keep from being destroyed, not destroy the host cell too soon, multiply, and find a way

to infect other cells. Intracellular parasites obtain different substances from the host cell, depending on the group. Viruses are the most extreme, parasitizing the host's genetic and metabolic machinery. Rickettsias are primarily energy parasites, and the malaria protozoan is a hemoglobin parasite.

TRANSPORT MECHANISMS FOR NUTRIENT ABSORPTION

A microorganism's habitat provides necessary nutrients—some abundant, others scarce—that must still be taken into the cell. Survival also requires that cells transport waste materials into the environment. Whatever the direction, transport occurs across the cell membrane, the structure specialized for this role. This is true even in organisms with cell walls (bacteria, algae, and fungi), because the cell wall is usually too nonselective to screen the entrance or exit of molecules. Three general types of transport are **passive transport,** which follows physical laws that are not unique to living systems and do not generally require direct energy input from the cell; **active transport,** which requires carrier proteins in the membranes of living cells and the expenditure of energy; and *endocytosis,* the movement of large masses of material across membranes (table 7.5).

Passive Transport: Diffusion, Osmosis

The driving force of passive transport is atomic and molecular movement—the natural tendency of atoms and molecules to be in constant random motion. The existence of this motion is evident in Brownian movement (see figure 4.4) of small particles suspended in liquid. It can be demonstrated by a variety of simple observations. A drop of perfume released into one part of a room is soon smelled in another part, or a lump of sugar in a cup of tea spreads through the whole cup without stirring. This phenomenon of molecular movement, in which atoms or molecules move in a gradient from an area of higher density or concentration to an area of lower density or concentration, is **diffusion*** (figure 7.3). Although passive transport requires molecules to be arranged in a gradient, diffusion remains an important way for cells to obtain freely diffusible materials (oxygen, carbon dioxide, and water) and to release wastes to the environment.

Diffusion of water through a selectively permeable membrane, a process called **osmosis,*** is also a physical phenomenon that is easily demonstrated in the laboratory with nonliving materials. It provides a model of how cells deal with various solute concentrations in aqueous solutions (figure 7.4). In an osmotic system, the membrane is *selectively,* or *differentially, permeable,* having passageways that allow free diffusion of water but can block certain other dissolved molecules. When this membrane is placed between solutions of differing concentrations and the solute is not diffusible (protein, for example), then under the laws of diffusion, water will diffuse at a faster rate from the side that has more water to the side that has less water. As long as the concentrations of the solutions differ, one side will experience a net loss of water and the other a net gain of water, until equilibrium is reached and the rate of diffusion is equalized.

***diffusion** (dih-few′-zhun) L. *dis,* apart, and *fundere,* to pour.

***osmosis** (oz-moh′-sis) Gr. *osmos,* impulsion, and *osis,* a process.

TABLE 7.5

Summary of Transport Processes in Cells

General Process	Nature of Transport	Examples	Description	Qualities
Passive	Energy expenditure is not required. Substances exist in a gradient and move from areas of higher concentration towards areas of lower concentration in the gradient.	Diffusion	A fundamental property of atoms and molecules that exist in a state of random motion	Nonspecific Brownian movement
		Osmosis	Diffusion of water molecules across a membrane barrier that is freely permeable to water but selectively permeable to other molecules	Direction depends on osmolarity of cell vs. habitat
		Facilitated diffusion	Molecule binds to a receptor in membrane and is carried across to other side	Molecule specific; transports both ways
Active	Energy expenditure is required. Molecules need not exist in a gradient. Rate of transport is increased. Transport may occur against a concentration gradient.	Carrier-mediated active transport	Atoms or molecules are pumped into or out of the cell by specialized receptors. Driven by ATP	Transports simple sugars, amino acids, inorganic ions (Na^+, K^+)
		Group translocation	Molecule is moved across membrane and simultaneously converted to a metabolically useful substance.	Alternate system for transporting nutrients (sugars, amino acids)
		Bulk transport	Mass transport of large particles, cells, and liquids by engulfment and vesicle formation	Includes endocytosis, exocytosis, pinocytosis

Be aware of an important point about the dynamics of osmosis and the solutions involved. Whereas the primary considerations are the concentration and direction of movement of water (solvent), the solute concentration is also involved, because it will dictate how much water is present to diffuse. For example, a more concentrated 50% solution contains 50 parts of solute and 50 parts of water. By comparison, a less concentrated 5% solution contains 5 parts of solute and 95 parts of water. When these two solutions are placed on opposite sides of a membrane, water will diffuse at a faster rate from the 5% solution into the 50% solution.

Osmosis in living systems is similar to the model shown in figure 7.4. Living membranes generally block the entrance and exit of larger molecules and permit free diffusion of water. Because most cells are surrounded by some free water, the amount of water entering or leaving has a far-reaching impact on cellular activities and survival. This osmotic relationship between cells and their environment is determined by the relative concentrations of the solutions on either side of the cell membrane (figure 7.5). Such systems can be compared using the terms isotonic, hypotonic, and hypertonic.

Under **isotonic*** conditions, the environment is equal in concentration to the cell's internal environment, and because diffusion of water proceeds at the same rate in both directions, there is no net change in cell volume. Isotonic solutions are the most stable environments for cells, because they are already in osmotic equilibrium with the cell. Parasites living in host tissues are most likely to be living in isotonic habitats.

Under **hypotonic*** conditions, the solute concentration of the external environment is lower than that of the cell's internal environment. Pure water provides the most hypotonic environment for cells because it has no solute. The net direction of osmosis is from the hypotonic solution into the cell, and cells without walls swell and can burst.

Hypertonic* conditions are also out of balance with the tonicity of the cell's cytoplasm, but in this case, the environment has a higher solute concentration than the cytoplasm. Because a

---

*isotonic (eye-soh-tahn´-ik) Gr. *iso*, same, and *tonos*, tension.

*hypotonic (hy-poh-tahn´-ik) Gr. *hypo*, under, and *tonos*, tension.

*hypertonic (hy-pur-tahn´-ik) Gr. *hyper*, under, and *tonos*, tension.

How Molecules Diffuse In Aqueous Solutions

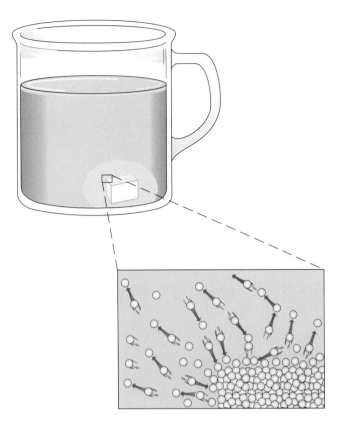

FIGURE 7.3

Diffusion of molecules in aqueous solutions. A high concentration of sugar exists in the cube at the bottom of the liquid. An imaginary molecular view of this area shows that sugar molecules are in a constant state of motion. Those at the edge of the cube diffuse from the concentrated area into more dilute regions. As diffusion continues, the sugar will spread evenly throughout the aqueous phase, and eventually there will be no gradient. At that point, the system is said to be in equilibrium.

hypertonic environment will force water to diffuse out of a cell, it is said to have high *osmotic pressure* or potential. The lethal effect of hypertonic solutions on microbes is the principle behind using concentrated salt and sugar solutions as preservatives for food.

Adaptations to Osmotic Variations in the Environment

Let us now see how specific microbes have adapted osmotically to their environments. In general, isotonic conditions pose little stress on cells, so survival depends on counteracting the adverse effects of hypertonic and hypotonic environments.

A bacterium and an ameba living in fresh pond water are examples of cells that live in constantly hypotonic conditions. The rate of water diffusing across the cell membrane into the cytoplasm is rapid and constant, and the cells would die without a way to adapt. The majority of bacterial cells compensate by having a cell wall that protects them from bursting even as the cytoplasm membrane be-

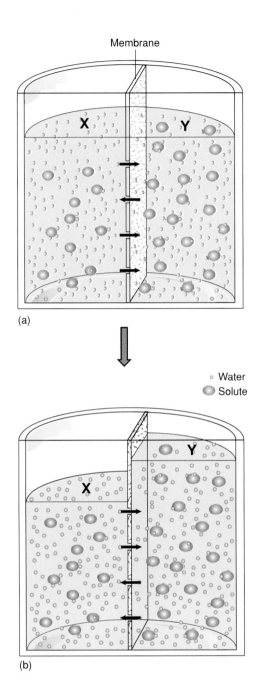

FIGURE 7.4

Osmosis, the diffusion of water through a selectively permeable membrane. **(a)** A membrane has pores that allow the ready passage of water but not large solute molecules from one side to another. Placement of this membrane between solutions of different solute concentrations (X = less concentrated and Y = more concentrated) results in a diffusion gradient for water. Water molecules undergo diffusion and move across the membrane pores in both directions. Because there is more water in solution X, the opportunity for a water molecule to successfully hit and go through a pore is greater for X than for Y. The result will be a net movement of water from X to Y. **(b)** The level of solution on the Y side rises as water continues to diffuse in. This process will continue until equilibration occurs and the rate of diffusion of water is equal on both sides.

Isotonic	Hypotonic	Hypertonic

Cells with Cell Wall

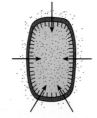

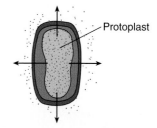
— Protoplast

Water concentration is equal inside and outside the cell, thus rates of diffusion are equal in both directions.

Net diffusion of water is into the cell; this swells the protoplast and pushes it tightly against the wall. Wall usually prevents cell from bursting.

Water diffuses out of the cell and shrinks the protoplast away from the cell wall; process is **plasmolysis.**

Cells Lacking Cell Wall

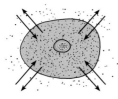

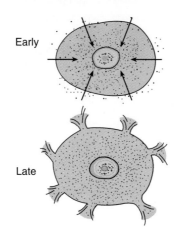

Early

Late

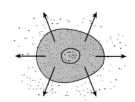

Early

Late

Rates of diffusion are equal in both directions.

Diffusion of water into the cell causes it to swell, and may burst it if no mechanism exists to remove the water.

Water diffusing out of the cell causes it to shrink and become distorted.

 Direction of net water movement.

FIGURE 7.5

Cell responses to solutions of differing osmotic content.

comes *turgid** from pressure. The ameba's adaptation is an anatomical and physiological one that requires the constant expenditure of energy. It has a water, or contractile, vacuole that siphons excess water back out into the habitat like a tiny pump (see figure 5.29).

A microbe living in a high-salt environment (hypertonic) has the opposite problem and must either restrict its loss of water to the environment or increase the salinity of its internal environment. Halobacteria living in the Great Salt Lake and the Dead Sea actually absorb salt to make their cells isotonic with the environment, thus they have a physiological need for a high-salt concentration in their habitats (see halophiles on page 118).

A form of passive transport called **facilitated diffusion** also requires a concentration gradient of the transported molecule and does not expend energy, but it is more specific than other passive systems. The only molecules to be transported are those that can bind to specialized membrane proteins. Examples of facilitated

diffusion include transport of sugars by some eucaryotic cells and transport of glycerol by bacteria. In the case of glucose transport, glucose binds to a permease in the membrane, which carries it through the membrane while simultaneously changing shape (figure 7.6).

Active Transport: Bringing in Molecules Against a Gradient

Free-living microbes exist under relatively nutrient-starved conditions and cannot rely completely on slow and rather inefficient passive transport mechanisms. To ensure a constant supply of nutrients and other required substances, microbes must capture those that are in extremely short supply and actively transport them into the cell. Features inherent in **active transport systems** are (1) the transport of nutrients against the natural diffusion gradient or in the same direction as the natural gradient but at a rate faster than by diffusion alone, (2) the presence of specific membrane proteins (permeases and pumps; figure 7.7a), and (3) the expenditure of energy. Examples of substances transported actively are monosaccharides,

*turgid (ter'-jid) A condition of being swollen or congested.

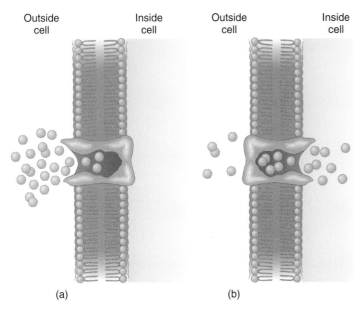

(a)   (b)

FIGURE 7.6

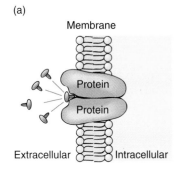 **Passive transport.** Facilitated diffusion involves the attachment of a molecule to a specific protein carrier. **(a)** Bonding of the molecule causes a conformational change in the protein that facilitates the molecule's passage across the membrane. **(b)** The membrane receptor opens into the cell and releases the molecule.

amino acids, organic acids, phosphates, and metal ions. Some freshwater algae have such efficient active transport systems that an essential nutrient can be found in intracellular concentrations 200 times that of the habitat.

An important type of active transport involves specialized **pumps,** which can rapidly carry ions such as K$^+$, Na$^+$, and H$^+$ across the membrane. This behavior is particularly important in mitochondrial ATP formation (see the discussion of oxidative phosphorylation in chapter 8) and protein synthesis. Another type of active transport, **group translocation,** couples the transport of a nutrient with its conversion to a substance that is immediately useful inside the cell (figure 7.7*b*). This method is used by certain bacteria to transport sugars (glucose, fructose) while simultaneously adding molecules such as phosphate that prepare them for the next stage in metabolism.

Endocytosis: Eating and Drinking by Cells

Some eucaryotic cells transport large molecules, particles, liquids, or even other cells across the cell membrane. Because the cell usually expends energy to carry out this transport, it is also a form of active transport. The substances transported do not pass physically through the membrane but are carried into the cell by **endocytosis.** * First the cell encloses the substance in its membrane, simultane-

---

*endocytosis (en″-doh-cy-toh′-sis) Gr. *endo,* in; *cyte,* cell; and *osis,* a process.

(a)

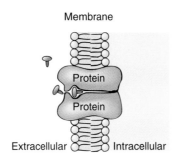

(b)

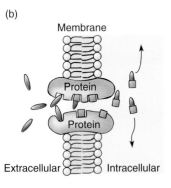

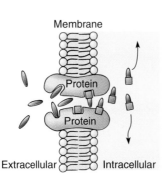

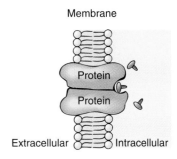

FIGURE 7.7

Active transport. In active transport mechanisms, energy is expended to transport the molecule across the cell membrane. **(a)** Molecular pumps. The membrane proteins (permeases) have attachment sites for essential nutrient molecules. As these molecules bind to the permease, they are pumped into the cell's interior through special membrane protein channels. Microbes have these systems for transporting various ions (sodium, iron) and small organic molecules. **(b)** In group translocation, the molecule is actively captured, but along the route of transport, it is chemically altered. By coupling transport with synthesis, the cell conserves energy.

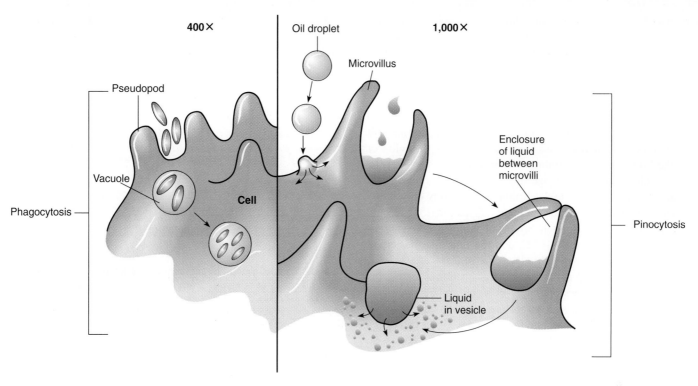

FIGURE 7.8

 Endocytosis (phagocytosis and pinocytosis). Solid particles are phagocytosed by large cell extensions called pseudopods, and they are pinocytosed into vesicles by very fine cell protrusions called microvilli. Oil droplets fuse with the membrane and are released directly into the cell.

ously forming a vacuole and engulfing it (figure 7.8). Amebas and certain white blood cells ingest whole cells or large solid matter by **phagocytosis.*** Liquids, such as oils or large molecules in solution, enter the cell through **pinocytosis.***

---

*phagocytosis (fag″-oh-cy-toh′-sis) Gr. *phagein,* to eat.

*pinocytosis (pin″-oh-cy-toh′-sis) Gr. *pino,* to drink.

CHAPTER CHECKPOINTS

Nutrition is a process by which all living organisms obtain substances from their environment to convert to metabolic uses.

Although the chemical form of nutrients varies widely, all organisms require six bioelements—carbon, hydrogen, oxygen, nitrogen, phosphorus, and sulfur—to survive, grow, and reproduce.

Nutrients are categorized by the amount required (macronutrients or micronutrients), by chemical structure (organic or inorganic), and by their importance to the organism's survival (essential or nonessential).

Microorganisms are classified both by the chemical form of their nutrients and the energy sources they utilize.

Nutrient requirements of microorganisms determine their respective niches in the food webs of major ecosystems.

Nutrients are transported into microorganisms by two kinds of processes: active transport that expends energy and passive transport that occurs independently of energy input.

The molecular size and concentration of a nutrient determine which method of transport is used.

Environmental Factors That Influence Microbes

Microbes are exposed to a wide variety of environmental factors in addition to nutrients. Microbial ecology focuses on ways that microorganisms deal with or adapt to such factors as heat, cold, gases, acid, radiation, osmotic and hydrostatic pressures, and even other microbes. Adaptation is a complex adjustment in biochemistry or genetics that enables long-term survival and growth. For most microbes, environmental factors fundamentally affect the function of metabolic enzymes. Thus, survival in a changing environment is largely a matter of whether the enzyme systems of microorganisms can adapt to alterations in their habitat. Incidentally, one must be careful to differentiate between growth in a given condition and tolerance, which implies survival without growth.

TEMPERATURE ADAPTATIONS

Microbial cells are unable to control their temperature and therefore assume the ambient temperature of their natural habitats. Their survival is dependent on adapting to whatever temperature variations are encountered in that habitat. The range of temperatures for microbial growth can be expressed as three *cardinal temperatures*. The **minimum temperature** is the lowest temperature that permits a microbe's continued growth and metabolism; below this temperature, its activities are inhibited. The **maximum temperature** is the highest temperature at which growth and metabolism can proceed. If the temperature rises slightly above maximum, growth will stop, but if it continues to rise beyond that point, the enzymes and nucleic acids will eventually become permanently inactivated and the cell will die. This is why heat works so well as an agent in microbial control. The **optimum temperature** covers a small range, intermediate between the minimum and maximum, which promotes the fastest rate of growth and metabolism (rarely is the optimum a single point).

Depending on their natural habitats, some microbes have a narrow cardinal range, others a broad one. Some strict parasites will not grow if the temperature varies more than a few degrees below or above the host's body temperature. For instance, the typhus rickettsia multiplies only in the range of 32°–38°C, and rhinoviruses (one cause of the common cold) multiply successfully only in tissues that are slightly below normal body temperature (33°–35°C). Other parasites are not so limited. Strains of *Staphylococcus aureus* grow within the range of 6°–46°C, and the intestinal bacterium *Enterococcus faecalis* grows within the range of 0°–44°C.

Another way to express temperature adaptation is to describe whether an organism grows optimally in a cold, moderate, or hot temperature range. The terms used for these ecological groups are psychrophile, mesophile, and thermophile (figure 7.9), respectively.

A **psychrophile** (sy′-kroh-fyl) is a microorganism that has an optimum temperature below 15°C and is capable of growth at 0°C. It is obligate with respect to cold and generally cannot grow above 20°C. Laboratory work with true psychrophiles can be a real challenge. Inoculations have to be done in a cold room because room temperature can be lethal to the organisms. Unlike most laboratory cultures, storage in the refrigerator incubates, rather than inhibits, them. As one might predict, the habitats of psychrophilic bacteria, fungi, and algae are snowfields (figure 7.10), polar ice, and the deep ocean. Rarely, if ever, are they pathogenic. True psychrophiles must be distinguished from *psychrotrophs* or *facultative psychrophiles* that grow slowly in cold but have an optimum temperature above 20°C. Bacteria such as *Staphylococcus aureus* and *Listeria monocytogenes* are a concern because they can grow in refrigerated food and cause food-borne illness.

The majority of medically significant microorganisms are **mesophiles** (mez′-oh-fylz), organisms that grow at intermediate temperatures. Although an individual species can grow at the extremes of 10°C or 50°C, the optimum growth temperatures (optima) of most mesophiles fall into the range of 20°–40°C. Organisms in this group inhabit animals and plants as well as soil and water in temperate, subtropical, and tropical regions. Most human pathogens have optima somewhere between 30°C and 40°C (human body temperature is 37°C). *Thermoduric* microbes, which can survive short exposure to high temperatures but are normally mesophiles, are common contaminants of heated or pasteurized foods (see chapter 11). Examples include heat-resistant cysts such as *Giardia* or sporeformers such as *Bacillus* and *Clostridium*.

A **thermophile** (thur′-moh-fyl) is a microbe that grows optimally at temperatures greater than 45°C. Such heat-loving microbes live in soil and water associated with volcanic activity and in habitats directly exposed to the sun. Thermophiles vary in heat requirements, with a general range of growth of 45°–80°C. Most eucaryotic forms cannot survive above 60°C, but a few thermophilic bacteria called hyperthermophiles, grow between 80°C and 110°C (currently thought to be the temperature limit endured by enzymes and cell structures). Strict thermophiles are so heat-

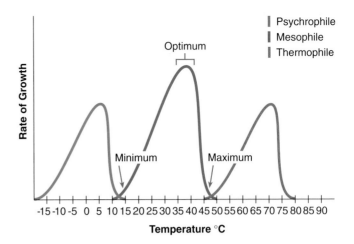

FIGURE 7.9

Ecological groups by temperature of adaptation. Psychrophiles can grow at or near 0°C and have an optimum below 15°C. As a group, mesophiles can grow between 10°C and 50°C, but their optima usually fall between 20°C and 40°C. Generally speaking, thermophiles require temperatures above 45°C and grow optimally between this temperature and 80°C. The cardinal temperatures are labeled for mesophiles. Note that the extremes of the ranges can overlap to an extent.

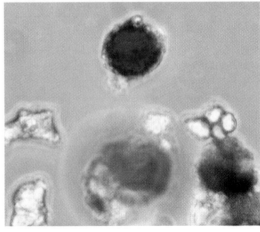

(a)

(b)

FIGURE 7.10

Red snow. (a) An early summer snowbank provides a perfect habitat for psychrophilic photosynthetic organisms like *Chlamydomonas nivalis.* **(b)** Microscopic view of this snow alga (actually classified as a "green" alga).

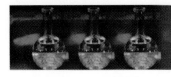

SPOTLIGHT ON MICROBIOLOGY 7.4
Cashing in on "Hot" Microbes

The smoldering thermal springs in Yellowstone National Park are more than just one of the geologic wonders of the world. They are also a hotbed of some of the most unusual microorganisms in the world. The thermophiles thriving at temperatures near the boiling point are the focus of serious interest from the scientific community. For many years, biologists have been intrigued that any living thing could function at such high temperatures. Such questions as these come to mind: Why don't they melt and disintegrate, why don't their proteins coagulate, and how can their DNA possibly remain intact?

One of the earliest thermophiles to be isolated was *Thermus aquaticus*. It was discovered by Thomas Brock in Yellowstone's Mushroom Pool in 1965 and was registered with the American Type Culture Collection. Interested researchers studied this species and discovered that it has extremely heat-stable proteins and nucleic acids, and its cell membrane does not break down readily at high temperatures. Later, an extremely heat-stable DNA-replicating enzyme was isolated from the species.

What followed is a riveting example of how pure research for the sake of understanding and discovery also offered up a key ingredient in a multimillion-dollar process. Developers of the polymerase chain reaction (PCR), a versatile tool for making multiple copies of DNA fragments, found that the technique would work only if they performed the test at temperatures around 65°–72°C. The mesophilic enzymes they tested were destroyed at such high temperatures, but the *Thermus* DNA-copying enzyme (Taq polymerase) functioned splendidly. The PCR technique became the basis for a variety of test procedures in forensics and gene detection and analysis.

Spurred by this remarkable success story, biotechnology companies have descended on Yellowstone, which contains over 10,000 hot springs, geysers, and hot habitats. These industries are looking to unusual bacteria and archaea as a source of "extremozymes," enzymes that oper-

Biotechnology researchers harvesting samples in Yellowstone National Park.

ate under high temperatures and acidity. So far, about a dozen other organisms with useful enzymes have been discovered. They may have applications in developing high-temperature fermentations, cleaning up toxic wastes, and organic syntheses.

This quest has also brought attention to questions such as: Who owns these microbes, and can their enzymes be patented? As of the year 2000, the Park Service has gotten a legal ruling that allows them to share in the profits from companies and to add that money to their operating budget. The U.S. Supreme Court has also ruled that a microbe isolated from natural habitats cannot be patented. Only the technology that uses the microbe can be patented.

---

tolerant that researchers may use an autoclave to isolate them in culture. Currently, there is intense interest in thermal microorganisms by biotechnology companies (Spotlight on Microbiology 7.4).

GAS REQUIREMENTS

The atmospheric gases that most influence microbial growth are O_2 and CO_2. Of these, oxygen gas has the greatest impact on microbial adaptation. Not only is it an important respiratory gas, but it is also a powerful oxidizing agent that exists in many toxic forms. In general, microbes fall into one of three categories: those that use oxygen and can detoxify it; those that can neither use oxygen nor detoxify it; and those that do not use oxygen but can detoxify it.

How Microbes Process Oxygen

As oxygen enters into cellular reactions, it is transformed into several toxic products. Singlet oxygen ($^1O_2$) is an extremely reactive molecule produced by both living and nonliving processes. No-

tably, it is one of the substances produced by phagocytes to kill invading bacteria (see chapter 14). The buildup of singlet oxygen and the oxidation of membrane lipids and other molecules can damage and destroy a cell. The highly reactive superoxide ion (O_2^-), peroxides (H_2O_2), and hydroxyls (OH^-) are other destructive metabolic by-products of oxygen. To protect themselves against damage, most cells have developed enzymes that go about the business of scavenging and neutralizing these chemicals. The complete conversion of superoxide ion into harmless oxygen requires a two-step process and at least two enzymes:

Step 1. $O_2^- + O_2^- + 2H^+ \xrightarrow{\text{Superoxide dismutase}} H_2O_2 \text{ (hydrogen peroxide)} + O_2$

Step 2. $H_2O_2 + H_2O_2 \xrightarrow{\text{Catalase}} 2H_2O + O_2$

In this series of reactions (essential for aerobic organisms), the superoxide ion is first converted to hydrogen peroxide and normal oxygen by the action of an enzyme called superoxide dismutase. Because hydrogen peroxide is also toxic to cells (it is a disinfectant and antiseptic), it will be degraded by the enzyme catalase into water and oxygen. If a microbe is not capable of dealing with toxic oxygen by these or similar mechanisms, it is forced to live in habitats free of oxygen.

With respect to oxygen requirements, several general categories are recognized. An **aerobe*** (aerobic organism) can use gaseous oxygen in its metabolism and possesses the enzymes needed to process toxic oxygen products. An organism that cannot grow without oxygen is an **obligate aerobe.** Most fungi and protozoa, as well as many bacteria (genera *Micrococcus* and *Bacillus*), have strict requirements for oxygen in their metabolism.

A **facultative anaerobe** is an aerobe that does not require oxygen for its metabolism and is capable of growth in the absence of oxygen. This type of organism metabolizes by aerobic respiration when oxygen is present, but in its absence, it adopts an anaerobic mode of metabolism such as fermentation. Facultative anaerobes usually possess catalase and superoxide dismutase. A large number of bacterial pathogens fall into this group (for example, gram-negative *enteric*** bacteria and staphylococci). A **microaerophile** (myk″-roh-air′-oh-fyl) does not grow at normal atmospheric tensions of oxygen but requires a small amount of it in metabolism. Most organisms in this category live in a habitat (soil, water, or the human body) that provides small amounts of oxygen but is not directly exposed to the atmosphere.

An **anaerobe** (anaerobic microorganism) lacks the metabolic enzyme systems for using oxygen in respiration. Because **strict,** or **obligate, anaerobes** also lack the enzymes for processing toxic oxygen, they cannot tolerate any free oxygen in the immediate environment and will die if exposed to it. Strict anaerobes live in highly reduced habitats, such as deep muds, lakes, oceans, and soil. Even though human cells use oxygen and oxygen is found in the blood and tissues, some body sites present anaerobic pockets or microhabitats where colonization or infection can occur. One region that is an important site for anaerobic infections is the oral cavity. Dental caries are partly due to the complex actions of aerobic and anaerobic bacteria, and most gingival infections consist of similar mixtures of oral bacteria that have invaded damaged gum tissues. Another common site for anaerobic infections is the large intestine, a relatively oxygen-free habitat that harbors a rich assortment of strictly anaerobic bacteria. Anaerobic infections can accompany abdominal surgery and traumatic injuries (gas gangrene and tetanus). Growing anaerobic bacteria usually requires special media, methods of incubation, and handling chambers that exclude oxygen (figure 7.11*a*).

Aerotolerant anaerobes do not utilize oxygen but can survive and grow to a limited extent in its presence. These anaerobes are not harmed by oxygen, mainly because they possess alternate mechanisms for breaking down peroxides and superoxide. Certain lactobacilli and streptococci use manganese ions or peroxidases to perform this task.

Determining the oxygen requirements of a microbe from a biochemical standpoint can be a very time-consuming process. Often it is illuminating to perform culture tests with reducing media

---

*aerobe (air′-ohb) Although the prefix means air, it is used in the sense of oxygen.

*enteric (en-terr′-ik) Gr. *enteron*, intestine. A family of bacteria that live in the large intestines of animals.

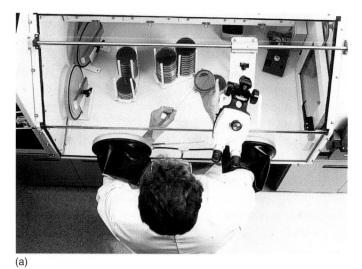

(a)

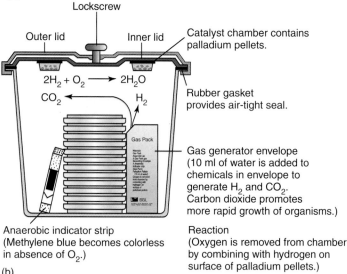

(b)

FIGURE 7.11

Culturing techniques for anaerobes. (a) A special anaerobic environmental chamber makes it possible to handle strict anaerobes without exposing them to air. It also has provisions for incubation and inspection in a completely O_2-free system. **(b)** The anaerobic jar, or CO_2 incubator system. To create an anaerobic environment, a packet is activated to produce hydrogen gas, and the chamber is sealed tightly. The gas reacts with available oxygen to produce water. Carbon dioxide can also be added to the system for growth of capnophiles.

(those that contain an oxygen-absorbing chemical). One such technique demonstrates oxygen requirements by the location of growth in a tube of fluid thioglycollate (figure 7.12).

Although all microbes require some carbon dioxide in their metabolism, *capnophiles*** grow best at a higher CO_2 tension than is normally present in the atmosphere. This becomes important in the initial isolation of some pathogens from clinical specimens, notably *Neisseria* (gonorrhea, meningitis), *Brucella* (undulant fever), and *Streptococcus pneumoniae*. Incubation is carried out in a CO_2 incubator that provides 3% to 10% CO_2 (see figure 7.11*b*).

---

*capnophile (kap′-noh-fyl) Gr. *kapnos*, smoke.

Demonstration of Oxygen Requirements

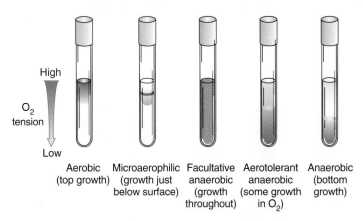

Aerobic (top growth) Microaerophilic (growth just below surface) Facultative anaerobic (growth throughout) Aerotolerant anaerobic (some growth in O$_2$) Anaerobic (bottom growth)

FIGURE 7.12

Use of thioglycollate broth to demonstrate oxygen requirements. Thioglycollate is a chemical that absorbs O$_2$ gas from the air. Oxygen at the top of the tube is dissolved in the medium and absorbed; its presence is indicated by the red dye resazurin. When a series of tubes is inoculated with bacteria that differ in O$_2$ requirements, the relative position of growth provides some indication of their adaptations to oxygen use.

EFFECTS OF pH

Microbial growth and survival are also influenced by the pH of the habitat. The pH was defined in chapter 2 as the degree of acidity or alkalinity (basicity) of a solution. It is expressed by the pH scale, a series of numbers ranging from 0 to 14. The pH of pure water (7.0) is neutral, neither acidic nor basic. As the pH value decreases toward 0, the acidity increases, and as the pH increases toward 14, the alkalinity increases. The majority of organisms live or grow in habitats between pH 6 and 8 because strong acids and bases can be highly damaging to enzymes and other cellular substances.

A few microorganisms live at pH extremes. Obligate *acidophiles* include *Euglena mutabilis,* an alga that grows in acid pools between 0 and 1.0 pH, and *Thermoplasma,* an archaea that lacks a cell wall, lives in hot coal piles at a pH of 1 to 2, and will lyse if exposed to pH 7. Because many molds and yeasts tolerate moderate acid, they are the most common spoilage agents of pickled foods. Alkalinophiles live in hot pools and soils that contain high levels of basic minerals (up to pH 10.0). Bacteria that decompose urine create alkaline conditions, since ammonium (NH$_4^+$, an alkaline ion) can be produced when urea (a component of urine) is digested. Metabolism of urea is one way that the ulcer bacterium *Helicobacter pylori* can neutralize the acidity of the stomach.

OSMOTIC PRESSURE

Although most microbes exist under hypotonic or isotonic conditions, a few, called **halophiles** (hay'-loh-fylz), live in habitats with a high solute concentration. *Obligate halophiles* such as *Halobacterium* and *Halococcus* inhabit salt lakes, ponds, and other hypersaline habitats. They grow optimally in solutions of 25% NaCl but require at least 9% NaCl (combined with other salts) for growth. These archaea have significant modifications in their cell walls and membranes and will lyse in hypotonic habitats. *Facultative halophiles* are remarkably resistant to salt, even though they do not normally reside in high-salt environments. For example, *Staphylococcus aureus* can grow on NaCl media ranging from 0.1% up to 20%. Although it is common to use high concentrations of salt and sugar to preserve food (jellies, syrups, and brines), many bacteria and fungi actually thrive under these conditions and are common spoilage agents. The term to describe microbes that withstand and grow at high osmotic pressures is **osmophile.**

MISCELLANEOUS ENVIRONMENTAL FACTORS

Various forms of electromagnetic radiation (ultraviolet, infrared, visible light) stream constantly onto the earth from the sun. Some microbes (phototrophs) can use visible light rays as an energy source, but non-photosynthetic microbes tend to be damaged by the toxic oxygen products produced by contact with light. Some microbial species produce yellow carotenoid pigments to protect against the damaging effects of light by absorbing and dismantling toxic oxygen. Other types of radiation that can damage microbes are ultraviolet and ionizing rays (X rays and cosmic rays). In chapter 11, we will see just how these types of energy are applied in microbial control.

Descent into the ocean depths subjects organisms to increasing hydrostatic pressure. Deep-sea microbes called **barophiles** exist under pressures that range from a few times to over 1,000 times the pressure of the atmosphere. These bacteria are so strictly adapted to high pressures that they will rupture when exposed to normal atmospheric pressure.

Because of the high water content of cytoplasm, all cells require water from their environment to sustain growth and metabolism. Water is the solvent for cell chemicals, and it is needed for enzyme function and digestion of macromolecules. A certain amount of water on the external surface of the cell is required for the diffusion of nutrients and wastes. Even in apparently dry habitats, such as sand or dry soil, the particles retain a thin layer of water usable by microorganisms. Dormant, dehydrated cell stages (for example, spores and cysts) tolerate extreme drying because of the inactivity of their enzymes.

ECOLOGICAL ASSOCIATIONS AMONG MICROORGANISMS

Up to now, we have considered the importance of nonliving environmental influences on the growth of microorganisms. Another profound influence comes from other organisms that share (or sometimes are) their habitats. In all but the rarest instances, microbes live in shared habitats, which give rise to complex and fascinating associations. Some associations are between similar or dissimilar types of microbes; others involve multicellular organisms such as animals or plants. Interactions can have beneficial, harmful, or no particular effects on the organisms involved; they can be obligatory or nonobligatory to the members; and they often

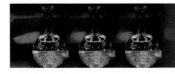

SPOTLIGHT ON MICROBIOLOGY 7.5
Life Together: Mutualism

A tremendous variety of mutualistic partnerships occur in nature. These associations gradually evolve over millions of years as the participating members come to rely on some critical substance or habitat that they share. One of the earliest such associations is thought to have resulted in eucaryotic cells (see Spotlight on Microbiology 5.1).

Protozoan cells often receive growth factors from symbiotic bacteria and algae that, in turn, are nurtured by the protozoan cell. One peculiar ciliate propels itself by affixing symbiotic bacteria to its cell membrane to act as "oars." These relationships become so obligatory that some amebas and ciliates require mutualistic bacteria for survival. This kind of relationship is especially striking in the complex mutualism of termites, which harbor protozoans specialized to live only inside them. The protozoans, in turn, contain endosymbiotic bacteria. Wood eaten by the termite gets processed by the protozoan and bacterial enzymes, and all three organisms thrive.

A view of a vent community based on mutualism and chemoautotrophy. The giant tube worm *Riftia* houses bacteria in its specialized feeding organ, the trophosome. Raw materials in the form of dissolved inorganic molecules are provided to the bacteria through the worm's circulation. With these, the bacteria produce usable organic food that is absorbed by the worm.

Symbiosis Between Microbes and Animals

Microorganisms carry on symbiotic relationships with animals as diverse as sponges, worms, and mammals. Bacteria and protozoa are essential in the operation of the rumen (a complex, four-chambered stomach) of cud-chewing mammals. These mammals produce no enzymes of their own to break down the cellulose that is a major part of their diet, but the microbial population harbored in their rumens does. The complex food materials are digested through several stages, during which time the animal regurgitates and chews the partially digested plant matter (the cud) and occasionally burps methane produced by the microbial symbionts.

Thermal Vent Symbionts

Another fascinating symbiotic relationship has been found in the deep hydrothermal vents in the seafloor, where geologic forces spread the crustal plates and release heat and gas. These vents are a focus of tremendous biological and geologic activity. Discoveries first made in the late 1970s demonstrated that the source of energy in this community is not the involve nutritional interactions. The following outline provides an overview of the major types of microbial associations:

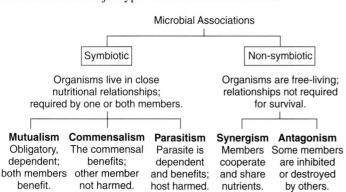

A general term used to denote a situation in which two organisms live together in a close partnership is **symbiosis,** * and the members are termed *symbionts*. Three main types of symbiosis occur. **Mutualism** exists when organisms live in an obligatory but mutually beneficial relationship. This association is rather common in nature because of the survival value it has for the members involved. Spotlight on Microbiology 7.5 gives several examples to illustrate this concept. In the other symbiotic relationships—commensalism and parasitism—the relationship tends to be unequal, meaning it benefits one member and not the other, and it can be obligatory.

*symbiosis (sim″-bye-oh′-sis) Gr. *syn*, together, and *bios*, to live.

sun, because the vents are too deep for light to penetrate (2,600 m). Instead, this ecosystem is based on a massive chemoautotrophic bacterial population that oxidizes the abundant hydrogen sulfide (H_2S) gas given off by the volcanic activity there. As the bottom of the food web, these bacteria serve as the primary producers of nutrients that service a broad spectrum of specialized animals.

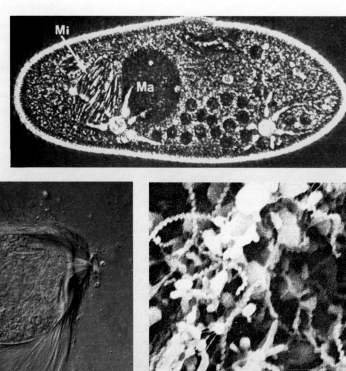

(a)

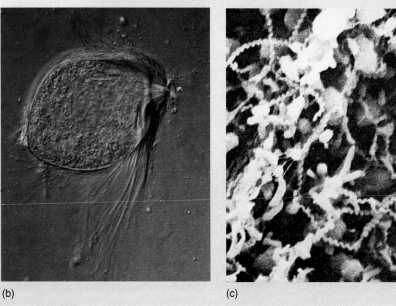

(b)   (c)

(a) The endosymbiotic bacteria of *Paramecium*. This familiar pond protozoan harbors a bacterium in its micronucleus (Mi) that swells it to 50 times its normal size and makes it as large as the macronucleus (Ma). **(b)** The real culprits in the final breakdown of wood are species of protozoa in a termite's gut. Here we see a trichonymph filled with tiny particles of wood (135×). Also note the fringe of flagella radiating over its body. **(c)** Scanning electron micrograph of microbial flora of a sheep's rumen (5,500×). This sample is extremely mixed, with pockets of rod- and spiral-shaped bacteria.

In a relationship known as **commensalism,*** the member called the **commensal** receives benefits, while its coinhabitant is neither harmed nor benefited. A classic commensal interaction between microorganisms called *satellitism* arises when one member provides nutritional or protective factors to the other. One example of nutritional satellitism is observed when one microbe provides a growth factor that another one needs (figure 7.13). Some microbes can break down a substance that would be toxic or inhibitory to another microbe. Relationships between humans and resident commensals that derive nutrients from the body are discussed in a later section.

In an earlier section, we introduced the concept of **parasitism** as a relationship in which the host organism provides the parasitic microbe with nutrients and a habitat. Multiplication of the parasite usually harms the host to some extent. As this relationship evolves, the host can even develop tolerance for or dependence on a parasite, at which point we call the relationship commensalism or mutualism.

Synergism* is an interrelationship between two or more free-living organisms that benefits them but is not necessary for their survival. Together, the participants cooperate to produce a result that none of them could do alone. This form of shared

---

*commensalism (kuh-men′-sul-izm) L. *com*, together, and *mensa*, table.

*synergism (sin′-ur-jizm) Gr. *syn*, together; *erg*, work; and *ism*, process.

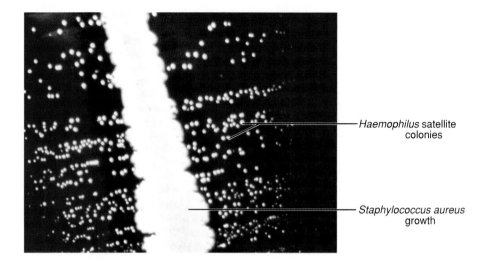

FIGURE 7.13

Satellitism, a type of commensalism between two microbes. In this example on blood agar, *Staphylococcus aureus* provides growth factors to *Haemophilus influenzae,* which grows as tiny satellite colonies near the streak of *Staphylococcus.* By itself, *Haemophilus* could not grow on blood agar.

Haemophilus satellite colonies

Staphylococcus aureus growth

metabolism can be viewed by the reaction:

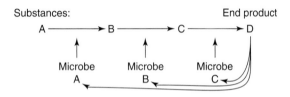

Substances: End product

Each microbe performs a specific action on a chemical in the series. The end product is useful to all three microbes.

An example of synergism is observed in the exchange between soil bacteria and plant roots (see chapter 26). The plant provides various growth factors, and the bacteria help fertilize the plant by supplying it with minerals. In synergistic infections, a combination of organisms can produce tissue damage that a single organism would not cause alone. Gum disease, dental caries, and gas gangrene involve mixed infections by bacteria interacting synergistically.

Antagonism* is an association between free-living species that arises when members of a community compete. In this interaction, one microbe secretes chemical substances into the surrounding environment that inhibit or destroy another microbe in the same habitat. The first microbe may gain a competitive advantage by increasing the space and nutrients available to it. Interactions of this type are common in the soil, where mixed communities often compete for space and food. *Antibiosis*—the production of inhibitory compounds called antibiotics—is actually a form of antagonism. Hundreds of naturally occurring antibiotics have been isolated from bacteria and fungi and used as drugs to control diseases (see chapter 12). *Bacteriocins* are another class of antimicrobial proteins that are toxic to bacteria other than the ones that produced them.

---

*antagonism (an-tag'-oh-nizm) Gr. *antagonistes,* an opponent.

*antibiosis (an"-tee-by-oh'-sis) Gr. *anti,* against, and *bios,* life.

*bacteriocin (bak-teer'-ee-oh-sin) Gr. *bakterion,* little rod, and *ios,* poison.

INTERRELATIONSHIPS BETWEEN MICROBES AND HUMANS

The human body is a rich habitat for symbiotic bacteria, fungi, and a few protozoa. Microbes that normally live on the skin, in the alimentary tract, and in other sites are called the *normal microbial flora* (see chapter 13). These residents participate in commensal, parasitic, and synergistic relationships with their human hosts. For example, certain bacteria living symbiotically in the intestine produce some vitamins, and species of symbiotic *Lactobacillus* residing in the vagina help maintain an environment that protects against infection by other microorganisms. Hundreds of commensal species "make a living" on the body without either harming or benefiting it. For example, many bacteria and yeasts reside in the outer dead regions of the skin; oral microbes feed on the constant flow of nutrients in the mouth; and billions of bacteria live on the wastes in the large intestine. Because the normal flora and the body are in a constant state of change, these relationships are not absolute, and a commensal can convert to a parasite by invading body tissues and causing disease.

CHAPTER CHECKPOINTS

The environmental factors that control microbial growth are temperature, pH, moisture, radiation, gases, and other microorganisms.

Environmental factors control microbial growth by their influence on microbial enzymes.

Three cardinal temperatures for a microorganism describe its temperature range and the temperature at which it grows best. These are the minimum temperature, the maximum temperature, and the optimum temperature.

Microorganisms are classified by their temperature requirements as psychrophiles, mesophiles, or thermophiles.

Most eucaryotic microorganisms are aerobic, but bacteria vary widely in their oxygen requirements from facultative to anaerobic.

Microorganisms live in associations with other species that range from mutually beneficial symbiosis to parasitism and antagonism.

A. A young cell at early phase of cycle.

B. A parent cell prepares for division by enlarging its cell wall, cell membrane, and overall volume. Midway in the cell, the wall develops notches that will eventually form the transverse septum, and the duplicated chromosome becomes affixed to a special membrane site.

C. The septum wall grows inward, and the chromosomes are pulled toward opposite cell ends as the membrane enlarges. Other cytoplasmic components are distributed (randomly) to the two developing cells.

D. The septum is synthesized completely through the cell center, and the cell membrane patches itself so that there are two separate cell chambers.

E. At this point, the daughter cells are divided. Some species will separate completely as shown here, while others will remain attached.

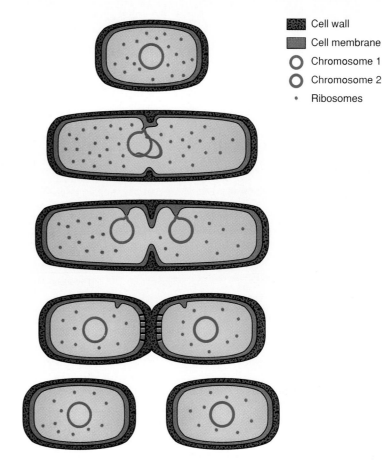

- Cell wall
- Cell membrane
- Chromosome 1
- Chromosome 2
- Ribosomes

FIGURE 7.14

 Steps in binary fission of a rod-shaped bacterium.

The Study of Microbial Growth

When microbes are provided with nutrients and the required environmental factors, they become metabolically active and grow. Growth takes place on two levels. On one level, a cell synthesizes new cell components and increases its size; on the other level, the number of cells in the population increases. This capacity for multiplication, increasing the size of the population by cell division, has tremendous importance in microbial control, infectious disease, and biotechnology. In the following sections, we will focus primarily on the characteristics of bacterial growth that are generally representative of single-celled microorganisms.

THE BASIS OF POPULATION GROWTH: BINARY FISSION

The division of a bacterial cell occurs mainly through **binary,** or **transverse, fission;** *binary* means that one cell becomes two, and *transverse* refers to the division plane forming across the width of the cell. During binary fission, the parent cell enlarges, duplicates its chromosome, and forms a central transverse septum that divides the cell into two daughter cells. This process is repeated at intervals by each new daughter cell in turn, and with each successive round of division, the population increases. The stages in this continuous process are shown in greater detail in figures 7.14 and 7.15.

THE RATE OF POPULATION GROWTH

The time required for a complete fission cycle—from parent cell to two new daughter cells—is called the **generation,** or **doubling, time.** The term *generation* has a similar meaning as it does in humans. It is the period between an individual's birth and the time of producing offspring. In bacteria, each new fission cycle or generation increases the population by a factor of 2, or doubles it. Thus, the initial parent stage consists of 1 cell, the first generation consists of 2 cells, the second 4, the third 8, then 16, 32, 64, and so on (figure 7.15). As long as the environment remains favorable, this doubling effect can continue at a constant rate. With the passing of each generation, the population will double, over and over again.

The length of the generation time is a measure of the growth rate of an organism. Compared with the growth rates of most other living things, bacteria are notoriously rapid. The average generation time is 30 to 60 minutes under optimum conditions. The shortest generation times average 5 to 10 minutes, and the longest generation times require days. For example, *Mycobacterium leprae,* the cause of leprosy, has a generation time of 10 to 30 days—as long as in some animals. Most pathogens have relatively short doubling times. *Salmonella enteritidis* and *Staphylococcus aureus,* bacteria that cause food-borne illness, double in 20 to 30 minutes, which is why leaving food at room temperature even for a short period has caused many a person to be suddenly stricken with an attack of food-borne

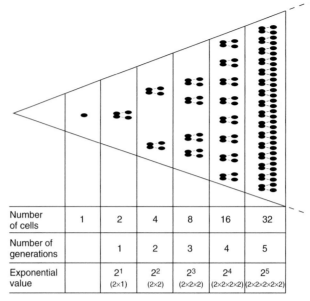

Number of cells	1	2	4	8	16	32
Number of generations		1	2	3	4	5
Exponential value		2^1 (2×1)	2^2 (2×2)	2^3 (2×2×2)	2^4 (2×2×2×2)	2^5 (2×2×2×2×2)

(a)

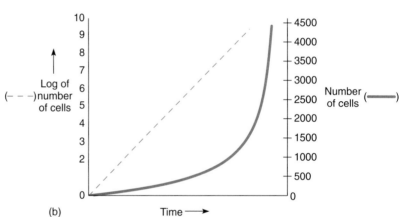

(b)

FIGURE 7.15

The mathematics of population growth.
(a) Starting with a single cell, if each product of reproduction goes on to divide in a binary fashion, the population doubles with each new division cycle or generation. This process can be represented by logarithms (2 raised to an exponent) or simple numbers. **(b)** Plotting the logarithm of the cells produces a straight line indicative of exponential growth, whereas plotting the cell numbers arithmetically gives a curved slope.

disease. In a few hours, a population of these bacteria can easily grow from a small number of cells to several million.

Figure 7.15 shows several quantitative characteristics of growth: (1) The cell population size can be represented by the number 2 with an exponent (2^1, 2^2, 2^3, 2^4); (2) the exponent increases by one in each generation; and (3) the number of the exponent is also the number of the generation. This growth pattern is termed **exponential.** Population growth expressed this way is **geometric,** with a constantly increasing slope. Because these populations often contain very large numbers of cells, it is useful to express them by means of exponents or logarithms (see appendix A). The data from a growing bacterial population are graphed by plotting the number of cells as a function of time. The cell number can be represented logarithmically or arithmetically. Plotting the logarithm number over time provides a straight line indicative of exponential growth. Plotting the data arithmetically gives a constantly curved slope. In general, logarithmic graphs are preferred because an accurate cell number is easier to read, especially during early growth phases.

Predicting the number of cells that will arise during a long growth period (yielding millions of cells) is based on a relatively simple concept. One could use the method of addition $2 + 2 = 4$; $4 + 4 = 8$; $8 + 8 = 16$; $16 + 16 = 32$, and so on, or a method of multiplication (for example, $2^5 = 2 \times 2 \times 2 \times 2 \times 2$), but it is easy to see that for 20 or 30 generations, this calculation could be very tedious. An easier way to calculate the size of a population over time is to use an equation such as:

$$N_f = (N_i)2^n$$

In this equation, N_f is the total number of cells in the population at some point in the growth phase, N_i is the starting number, the exponent n denotes the generation number, and 2^n represents the number of cells in that generation. If we know any two of the values, the other values can be calculated. Let us use the example of *Staphylococcus aureus* to calculate how many cells (N_f) will be present in an egg salad sandwich after it sits in a warm car for 4 hours. We will assume that N_i is 10 (number of cells deposited in the sandwich while it was being prepared). To derive n, we need to divide 4 hours (240 minutes) by the generation time (we will use 20 minutes). This calculation comes out to 12, so 2^n is equal to 2^{12}. Referring to a table on the powers of numbers or using a calculator, we find that 2^{12} is 4,096.

Final number (N_f) = 10 × 4,096
= 40,960 cells in the sandwich

This same equation, with modifications, is used to determine the generation time, a more complex calculation that requires knowing the number of cells at the beginning and end of a growth period. Such data are obtained through actual testing by a method discussed in the following section.

THE POPULATION GROWTH CURVE

In reality, a population of bacteria does not maintain its potential growth rate and does not double endlessly, because in most systems numerous factors prevent the cells from continuously dividing at their maximum rate. Quantitative laboratory studies indicate that a population typically displays a predictable pattern, or **growth curve,** over time. The method traditionally used to observe the population growth pattern is a viable count technique, in which the total number of live cells is counted over a given time period. In brief, this method entails (1) placing a tiny number of cells into a sterile liquid medium; (2) incubating this culture over a period of several hours; (3) sampling the broth at regular intervals during incubation; (4) plating each sample onto solid media; and (5) counting the number of colonies present after incubation. Microbits 7.6 gives the details of this process.

STAGES IN THE NORMAL GROWTH CURVE

The system of batch culturing described in Microbits 7.6 is *closed,* meaning that nutrients and space are finite and there is no mechanism for the removal of waste products. Data from an entire growth period of 3 to 4 days typically produce a curve with a series of phases termed the lag phase, the exponential growth (log) phase, the stationary phase, and the death phase (figure 7.16).

The **lag phase** is a relatively "flat" period on the graph when the population appears not to be growing or is growing at less than the exponential rate. Growth lags primarily because: (1) The newly inoculated cells require a period of adjustment, enlargement, and synthesis; (2) the cells are not yet multiplying at their maximum rate; and (3) the population of cells is so sparse or dilute that the sampling misses them. The length of the lag period varies somewhat from one population to another.

The cells reach the maximum rate of cell division during the **exponential growth (log) phase,** a period during which the curve increases geometrically. This phase will continue as long as cells have adequate nutrients and the environment is favorable.

At the **stationary growth phase,** the population enters a survival mode in which cells stop growing or grow slowly. The curve levels off because the rate of cell inhibition or death balances out the rate of multiplication. The decline in the growth rate is caused by depleted nutrients and oxygen, excretion of organic acids and other biochemical pollutants into the growth medium, and an increased density of cells.

As the limiting factors intensify, cells begin to die in exponential numbers (literally perishing in their own wastes), and they are unable to multiply. The curve now dips downward as the **death phase** begins. The speed with which death occurs depends on the rel-

ative resistance of the species and how toxic the conditions are, but it is usually slower than the exponential growth phase. Viable cells often remain many weeks and months after this phase has begun. In the laboratory, refrigeration is used to slow the progression of the death phase so that cultures will remain viable as long as possible.

Practical Importance of the Growth Curve

The tendency for populations to exhibit phases of rapid growth, slow growth, and death has important implications in microbial control, infection, food microbiology, and cultural technology. Antimicrobial agents such as heat and disinfectants rapidly accelerate the death phase in all populations, but microbes in the exponential growth phase are more vulnerable to these agents than are those that have entered the stationary phase. In general, actively growing cells are more vulnerable to conditions that disrupt cell metabolism and binary fission.

Growth patterns in microorganisms can account for the stages of infection (see chapter 13). Microbial cells produced during the exponential phase are far more numerous and virulent than those released at a later stage of infection. A person shedding bacteria in the early and middle stages of an infection is more likely to spread it to others than is a person in the late stages. The course of an infection is also influenced by the relatively faster rate of multiplication of the microbe, which can overwhelm the slower growth rate of the host's own cellular defenses.

Understanding the stages of cell growth is crucial for work with cultures. Sometimes a culture that has reached the stationary phase is incubated under the mistaken impression that enough nutrients are present for the culture to multiply. In most cases, it is unwise to continue incubating a culture beyond the stationary phase, because doing so will reduce the number of viable cells and the culture could die out completely. It is also preferable to do stains (an exception is the spore stain) and motility tests on young cultures, because the cells will show their natural size and correct reaction, and motile cells will have functioning flagella.

For certain research or industrial applications, closed batch culturing with its four phases is inefficient. The alternative is an automatic growth chamber called the *chemostat,* or continuous culture system. This device can admit a steady stream of new nutrients and siphon off used media and old bacterial cells, thereby stabilizing the growth rate and cell number. The chemostat is very similar to the industrial fermenters used to produce vitamins and antibiotics (see chapter 26). It has the advantage of maintaining the culture in a biochemically active state and preventing it from entering the death phase.

OTHER METHODS OF ANALYZING POPULATION GROWTH

Microbiologists have developed several alternative ways of analyzing bacterial growth qualitatively and quantitatively. One of the simplest methods for estimating the size of a population is through turbidometry. This technique relies on the simple observation that a tube of clear nutrient solution loses its clarity and becomes cloudy, or **turbid,** as microbes grow in it. In general, the greater the turbidity, the larger the population size, which can be measured by means of sensitive instruments (figure 7.17*b*).

MICROBITS 7.6
Steps in a Viable Plate Count—Batch Culture Method

A growing population is established by inoculating a flask containing a known quantity of sterile liquid medium with a few cells of a pure culture. The flask is incubated at that bacteria's optimum temperature and timed. The population size at any point in the growth cycle is quantified by removing a tiny measured sample of the culture from the growth chamber and plating it out on a solid medium to develop isolated colonies. This procedure is repeated at evenly spaced intervals (every hour for 24 hours).

Evaluating the samples involves a common and important principle in microbiology: One colony on the plate represents one cell or colony-forming unit (CFU) from the original sample. Because the CFU of some bacteria is actually composed of several cells (consider the clustered arrangement of *Staphylococcus,* for instance), using a colony count can underestimate the exact population size to an extent. This is not a serious problem because, in such bacteria, the CFU is the smallest unit of colony formation and dispersal. Multiplication of the number of colonies in a single sample by the container's volume gives a fair estimate of the total population size (number of cells) at any given point. The growth curve is determined by graphing the number for each sample in sequence for the whole incubation period (see figure 7.16).

Because of the scarcity of cells in the early stages of growth, some samples can give a zero reading even if there are viable cells in the culture. The sampling itself can remove enough viable cells to alter the tabulations, but since the purpose is to compare relative trends in growth, these factors do not significantly change the overall pattern.

Flask inoculated

Samples taken at equally spaced intervals (0.1 ml)

	60 min	120 min	180 min	240 min	300 min	360 min	420 min	480 min	540 min	600 min
Sample is diluted in liquid agar medium and poured or spread over surface of solidified medium										
Plates are incubated, colonies are counted	None									
Number of colonies (CFU) per 0.1 ml	0*	1	3	7	13	23	45	80	135	230
Total cell population in flask	0*	5,000	15,000	35,000	65,000	115,000	225,000	400,000	675,000	1,150,000

500 ml — 0.1 ml

* Zero CFUs only means that too few cells are present to be assayed.

The viable plate count, a technique for determining population size and rate of growth.

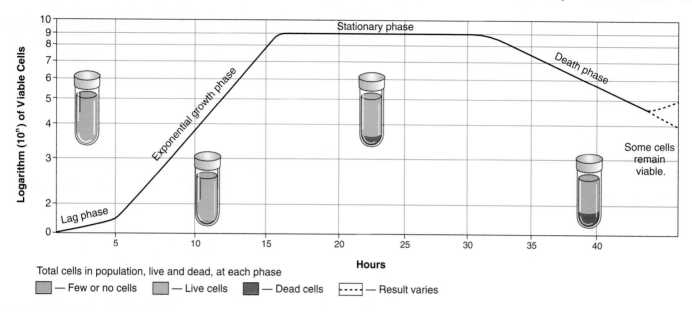

FIGURE 7.16

The growth curve in a bacterial culture. On this graph, the number of viable cells expressed as a logarithm (log) is plotted against time. See text for discussion of the various phases. Note that with a generation time of 30 minutes, the population has risen from 10 (10^1) cells to 1,000,000,000 (10^9) cells in only 16 hours.

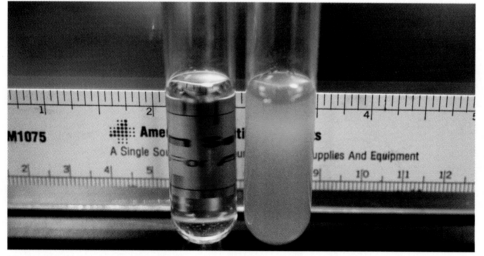

(a)

FIGURE 7.17

Turbidity measurements as indicators of growth. (a) Holding a broth to the light is one method of checking for gross differences in cloudiness (turbidity). The broth on the left is transparent, indicating little or no growth; the broth on the right is cloudy and opaque, indicating heavy growth. **(b)** The eye is not sensitive enough to pick up fine degrees in turbidity; more sensitive measurements can be made with a spectrophotometer. **(1)** A tube with no growth transmits more light and gives a higher reading. **(2)** In a tube with growth, the cells scatter the light so that transmittance is reduced and the reading is lower. This technique provides only a relative measure of growth; it cannot determine actual numbers or differentiate between dead and live cells.

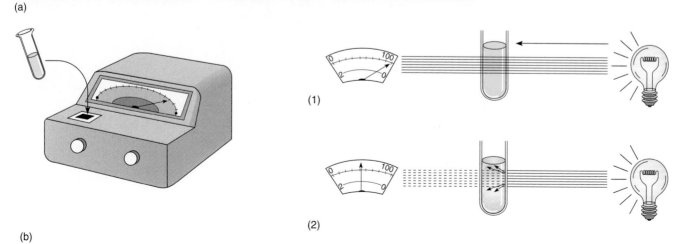

(b)

(1)

(2)

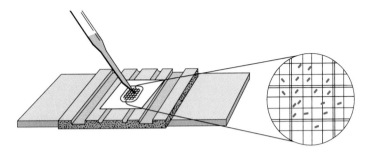

FIGURE 7.18

Direct microscopic count of bacteria. A small sample is placed on the grid under a cover glass. Individual cells, both living and dead, are counted. This number can be used to calculate the total count of a sample.

Enumeration of Bacteria

Turbidity readings are useful for evaluating relative amounts of growth, but if a more quantitative evaluation is required, the viable colony count described previously or some other enumeration (counting) procedure is necessary. The **direct,** or **total, cell count** involves counting the number of cells in a sample microscopically (figure 7.18). This technique, very similar to that used in blood cell counts, employs a special microscope slide (cytometer) calibrated to accept a tiny sample that is spread over a premeasured grid. The cell count from a cytometer can be used to estimate the total number of cells in a larger sample (for instance, of milk or water). One inherent inaccuracy in this method is that no distinction can be made between dead and live cells, both of which are included in the count.

Counting can be automated by sensitive devices such as the Coulter counter, which electronically scans a culture as it passes through a tiny pipette. As each cell flows by, it is detected and registered on an electronic sensor (figure 7.19a). A flow cytometer works on a similar principle, but in addition to counting, it can measure cell size and even differentiate between live and dead cells. When used in conjunction with fluorescent dyes and antibodies, it has been used to differentiate between gram-positive and gram-negative bacteria. It is being adapted for use as a rapid method to identify pathogens in patient specimens and to compare bacterial species on the basis of genetic differences such as guanine and cytosine content (figure 7.19b).

CHAPTER CHECKPOINTS

Microbial growth refers both to increase in cell size and increase in number of cells in a population.

The generation time is a measure of the growth rate of a microbial species. It varies in length according to environmental conditions.

Microbial cultures in a nutrient-limited environment exhibit four distinct stages of growth: the lag phase, the exponential growth (log) phase, the stationary phase, and the death phase.

Microbial cell populations show distinct phases of growth in response to changing nutrient and waste conditions.

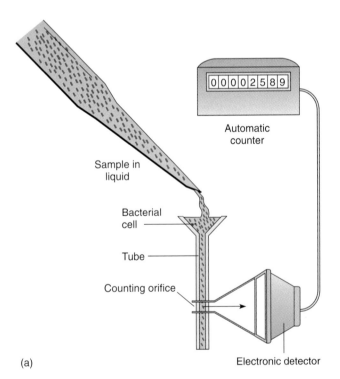

(a)

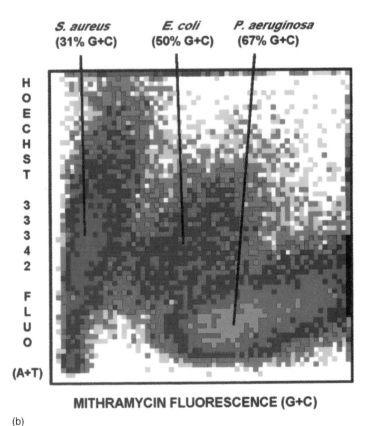

(b)

FIGURE 7.19

Coulter counter. (a) As cells pass through this device, they trigger an electronic sensor that tallies their numbers. **(b)** A variation on this machine, called a flow cytometer, can record the number, size, and types of cells by tagging them with fluorescent substances and passing them through a beam of light. Shown here is a fluorescence signature of three species, which are differentiated on the basis of guanine and cytosine percentages.

CHAPTER CAPSULE WITH KEY TERMS

I. **Microbial Nutrition, Ecology, and Growth**
  A. Source of Nutrients
    Nutrition consists of taking in chemical substances (nutrients) and assimilating and extracting energy from them.
    1. Substances required for survival are **essential nutrients**—usually containing the elements (C, H, N, O, P, S, Na, Cl, K, Ca, Fe, Mg). Essential nutrients are considered **macronutrients** (required in larger amounts) or **micronutrients** (trace elements required in smaller amounts—Zn, Mn, Cu). Nutrients are classed as either **inorganic** or **organic.**
    2. A **growth factor** is an organic nutrient (amino acid and vitamin) that cannot be synthesized and must be provided.

II. **Nutritional Categories**
  A. An **autotroph** depends on carbon dioxide for its carbon needs. If its energy needs are met by light, it is a **photoautotroph,** but if it extracts energy from inorganic substances, it is a **chemoautotroph.**
  B. A **heterotroph** acquires carbon from organic molecules. A **saprobe** is a decomposer that feeds upon dead organic matter, and a **parasite** feeds from a live host and usually causes harm. Disease-causing parasites are pathogens.

III. **Environmental Influences on Microbes**
  A. *Temperature:* An organism exhibits **optimum, minimum,** and **maximum temperatures.** Organisms that cannot grow above 20°C but thrive below 15°C and continue to grow even at 0°C are known as **psychrophiles. Mesophiles** grow from 10°C to 50°C, having temperature optima from 20°C to 40°C. The growth range of **thermophiles** is 45°C to 80°C.
  B. *Oxygen Requirements:* The ecological need for free oxygen (O_2) is based on whether a cell can handle toxic by-products such as superoxide and peroxide.
    1. **Aerobes** grow in normal atmospheric oxygen and have enzymes to handle toxic oxygen by-products. An aerobic organism capable of living without oxygen if necessary is a **facultative anaerobe.** An aerobe that prefers a small amount of oxygen but does not grow under anaerobic conditions is a **microaerophile.**
    2. **Strict (obligate) anaerobes** do not use free oxygen and cannot produce enzymes to dismantle reactive oxides. They are actually damaged or killed by oxygen. An **aerotolerant anaerobe** cannot use oxygen for respiration, yet is not injured by it.
  C. *Effects of pH:* Acidity and alkalinity affect the activity and integrity of enzymes and the structural components of a cell. Optimum pH for most microbes ranges approximately from 6 to 8. **Acidophiles** prefer lower pH, and **alkalinophiles** prefer higher pH.
  D. *Other Environmental Factors:* Electromagnetic radiation and barometric pressure affect microbial growth. A **barophile** is adapted to life under high pressure (bottom dwellers in the ocean, for example).

IV. **Transport Mechanisms**
  A. A microbial cell must take on nutrients from its surroundings by transporting them across the cell membrane.

B. **Passive transport** involves the natural movement of substances down a concentration gradient and requires no additional energy (**diffusion**).
  C. **Osmosis** is diffusion of water through a selectively permeable membrane. A form of passive transport that can move specific substances is **facilitated diffusion.**
  D. Osmotic changes that affect cells are **hypotonic** solutions, which contain a lower solute concentration, and **hypertonic** solutions, which contain a higher solute concentration. **Isotonic** solutions have the same solute concentration as the inside of the cell. A **halophile** thrives in hypertonic surroundings, and an **obligate halophile** requires a salt concentration of at least 15%, but grows optimally in 25%.
  E. In **active transport,** substances are taken into the cell by a process that consumes energy. In **group translocation,** molecules are altered during transport.
  F. **Phagocytosis** and **pinocytosis** are forms of active transport in which bulk quantities of solid and fluid material are taken into the cell.

V. **Microbial Interactions**
  A. Microbes coexist in varied relationships in nature.
    1. Types of **symbiosis** are **mutualism,** a reciprocal, obligatory, and beneficial relationship between two organisms, and **commensalism,** an organism receiving benefits from another without harming the other organism in the relationship. Parasitism occurs between a host and an infectious agent.
    2. **Synergism** is a mutually beneficial but not obligatory coexistence. **Antagonism** entails competition, inhibition, and injury directed against the opposing organism. A special case of antagonism is antibiotic production.

VI. **Microbial Growth**
  Microbes multiply by division.
  A. The splitting of a parent bacterial cell to form a pair of similar-sized daughter cells is known as **binary,** or **transverse, fission.**
  B. The duration of each division is called the **generation,** or **doubling, time.** A population theoretically doubles with each generation, so the growth rate is **exponential,** and each cycle increases in **geometric progression.**
  C. A **growth curve** is a graphic representation of a closed population over time. Plotting a curve requires an estimate of live cells, called a **viable count.** The initial flat period of the curve is called the **lag phase,** followed by the **exponential growth phase,** in which viable cells increase in logarithmic progression. Adverse environmental conditions combine to inhibit the growth rate, causing a plateau, or **stationary growth phase.** In the **death phase,** nutrient depletion and waste buildup cause increased cell death.
  D. Cell numbers can be counted directly by a microscope counting chamber, Coulter counter, or flow cytometer. Cell growth can also be determined by turbidometry and a total cell count.

MULTIPLE-CHOICE QUESTIONS

1. An organic nutrient essential to an organism's metabolism that cannot be synthesized itself is termed a/an
   a. trace element
   c. growth factor
   b. micronutrient
   d. essential nutrient

2. The source of the necessary elements of life is
   a. an inorganic environmental reservoir
   b. the sun
   c. rocks
   d. the air

3. An organism that can synthesize all its required organic components from CO_2 using energy from the sun is a
   a. photoautotroph
   b. photoheterotroph
   c. chemoautotroph
   d. chemoheterotroph

4. An obligate halophile requires high
   a. pH
   c. salt
   b. temperature
   d. pressure

5. Chemoautotrophs can survive on _____ alone.
   a. minerals
   c. minerals and CO_2
   b. CO_2
   d. methane

6. Which of the following statements is true for *all* organisms?
   a. they require organic nutrients
   b. they require inorganic nutrients
   c. they require growth factors
   d. they require oxygen gas

7. A pathogen would most accurately be described as a
   a. parasite
   c. saprobe
   b. commensal
   d. symbiont

8. Which of the following is true of passive transport?
   a. it requires a gradient
   b. it uses the cell wall
   c. it includes endocytosis
   d. it only moves water

9. A cell exposed to a hypertonic environment will _____ by osmosis.
   a. gain water
   c. neither gain nor lose water
   b. lose water
   d. burst

10. Active transport of a substance across a membrane requires
    a. a gradient
    b. the expenditure of ATP
    c. water
    d. diffusion

11. Environmental factors such as temperature and pH exert their effect on the _____ of microbial cells.
    a. membranes
    c. enzymes
    b. DNA
    d. cell wall

12. Psychrophiles would be expected to grow
    a. in hot springs
    b. on the human body
    c. at refrigeration temperatures
    d. at low pH

13. Superoxide ion is toxic to strict anaerobes because they lack
    a. catalase
    c. dismutase
    b. peroxidase
    d. oxidase

14. The time required for a cell to undergo binary fission is called the
    a. exponential growth rate
    b. growth curve
    c. generation time
    d. lag period

15. In a viable plate count, each _____ represents a _____ from the sample population.
    a. cell, colony
    c. hour, generation
    b. colony, cell
    d. cell, generation

16. During the _____ phase, the rate of new cells being added to the population has slowed down.
    a. stationary
    c. lag
    b. death
    d. exponential growth

CONCEPT QUESTIONS

1. Differentiate between micronutrients and macronutrients.
   a. What elements do the letters in the mnemonic device on page 188 stand for?
   b. Briefly describe the general function of CHNOPS in the cell.
   c. Define growth factors, and give examples of them.

2. Name some functions of metallic ions in cells

3. Compare autotrophs and heterotrophs with respect to the form of carbon-based nutrients they require.

4. Describe the nutritional strategy of two types of chemoautotrophs (lithotrophs) in the chapter.

5. Briefly fill in the following table:

	Source of Carbon	Usual Source of Energy	Example
Photoautotroph			
Photoheterotroph			
Chemoautotroph			
Chemoheterotroph			
Saprobe			
Parasite			

6. a. Compare and contrast passive and active forms of transport, using examples of what is being transported and the requirements for each.
   b. How are phagocytosis and pinocytosis similar? How are they different?

7. Compare the effects of isotonic, hypotonic, and hypertonic solutions on an ameba and on a bacterial cell. If a cell is in a hypotonic environment, what is the condition of its cytoplasm relative to the environment?

8. Look at the following diagrams and predict in which direction osmosis will take place. Use arrows to show the net direction of osmosis. Is one of these microbes a halophile? Which one?

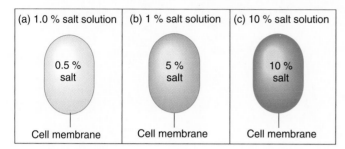

| (a) 1.0 % salt solution | (b) 1 % salt solution | (c) 10 % salt solution |

0.5 % salt — Cell membrane   5 % salt — Cell membrane   10 % salt — Cell membrane

9. Why are most pathogens mesophilic?

10. What are the ecological roles of psychrophiles and thermophiles?

11. What is the natural habitat of a facultative parasite? Of a strict saprobe?

12. Name three groups of obligate intracellular parasites.

13. Classify a human with respect to oxygen requirements.

14. a. What might be the habitat of an aerotolerant anaerobe?
   b. Where in the body are anaerobic habitats apt to be found?

15. Where do superoxide ions and hydrogen peroxide originate?

16. a. Define symbiosis and differentiate among mutualism, commensalism, synergism, parasitism, and antagonism, using examples.
   b. How are parasitism and antagonism similar and different?
   c. Are any of these relationships obligatory?

17. Explain the relationship between colony counts and colony-forming units. Why can one use the number of colonies as an index of population size?

18. Why is growth called exponential? What makes it a geometric progression?

19. Explain what is happening to the population at points A, B, C, and D in the following diagram.

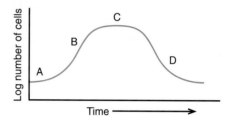

20. Why are we concerned with differentiating live versus dead cells in populations?

CRITICAL-THINKING QUESTIONS

1. a. Is there a microbe that could grow on a medium that contains only the following compounds dissolved in water: $CaCO_3$, $MgNO_3$, $FeCl_2$, $ZnSO_4$, and glucose? Defend your answer.
   b. Check the last entry in the cell composition summary on table 7.3. Are all needed elements present here? Where is the carbon?

2. Describe how one might determine the nutrient requirements of a microbe from Mars. If, after exhausting the nutrient schemes, it still does not grow, what other factors might one take into account?

3. a. Explain what ultimately determines whether a microorganism can adapt to a certain habitat.
   b. Give two examples of ways that microbes become modified to survive.

4. a. How can osmotic pressure and pH be used in preserving foods.
   b. How do they affect microbes?

5. a. How might one effectively treat anaerobic infections using gas?
   b. Explain how it would work.

6. How can you explain the observation that unopened milk will spoil even while refrigerated?

7. What would be the effect of a fever on a thermophilic pathogen?

8. Patients with acidotic diabetes are especially susceptible to fungal infections. Can you explain why?

9. Using the concept of synergism, can you describe a way to grow a fastidious microbe?

10. Describe a way to isolate an antibiotic-producing bacterium.

11. a. If an egg salad sandwich sitting in a warm car for 4 hours develops 40,960 bacterial cells, how many more cells would result with just one more hour of incubation?
    b. With 10 hours of incubation?
    c. What would the cell count be after 4 hours if the initial bacterial dose were 100?
    d. What do your answers tell you about using clean techniques in food preparation (other than aesthetic considerations)?

12. Why is an older culture needed for spore staining?

13. Should biotechnology companies be allowed to isolate and own microorganisms taken from the earth's habitats and derive profits from them? Why or why not?

INTERNET SEARCH TOPICS

1. Look up the term **extremophile** on a search engine. Find examples of microbes that have adapted to various extremes. Is there a species that exists in several extremes simultaneously?

2. Find a website that discusses the use of microbes living in high temperatures to bioremediate (clean up) the environment.

CHAPTER 8

Microbial Metabolism:
The Chemical Crossroads of Life

Cells maintain a complex internal habitat that is highly structured and controlled. They host a never-ending array of metabolic reactions that are required for functions such as nutrient processing, growth, and release of energy. Among the most remarkable features of cell metabolism are enzymes, specialized proteins that perform key roles in most of these reactions. Recently, biochemists clarified the actions of an enzyme called OMP decarboxylase, used to synthesize a molecule that is found universally in the genetic material of all organisms and viruses. They found that in the presence of the OMP enzyme, this reaction happens very rapidly—about 30 times per second. The chances of the reaction happening without the enzyme would take so long (78 million years) that it is essentially not a realistic possibility. It is clear that without this enzyme, life as we know it could never have evolved. This is but one example of the amazing cellular machinery that is ultimately responsible for the tremendous diversity in the earth's life forms. In addition, it shows us that the metabolic activities of cells often share common patterns. With these themes in mind, this chapter will cover some of the major unifying characteristics of metabolism, enzymes, the flow of energy, and the pathways that govern nutrient processing. Having this background will greatly benefit your understanding of future topics such as genetics, drug therapy, disinfection, biotechnology, diseases, and the identification of microorganisms.

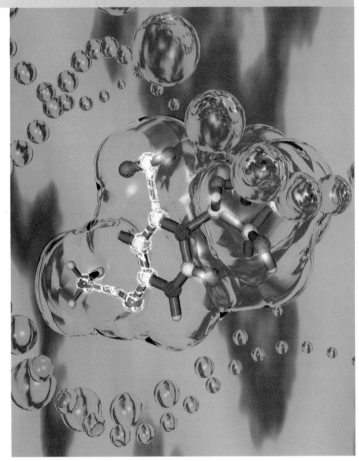

An artist's vision reveals the enzyme OMP decarboxylase. This enzyme is shown assisting in the formation of uracil, a component of RNA, by releasing carbon dioxide gas. It is one of the most efficient enzymes ever discovered.

Chapter Overview

- Cells are constantly involved in an orderly activity called metabolism that encompasses all of their chemical and energy transactions.
- Enzymes are essential metabolic participants that drive cell reactions.
- Enzymes are protein catalysts that speed up chemical processes by lowering the required energy.
- Enzymes have a specific shape tailored to perform their actions on a single type of molecule called a substrate.
- Enzymes derive some of their special characteristics from cofactors such as vitamins, and they show sensitivity to environmental factors.
- Enzymes are involved in activities that synthesize, digest, oxidize, and reduce compounds, and convert one substance to another.
- Enzymes are regulated by several mechanisms that alter the structure or synthesis of the enzyme.
- The energy of living systems resides in the atomic structure of chemicals that can be acted upon and changed.
- Cell energetics involves the release of energy that powers the formation of bonds.

- The energy of electrons is transferred from one molecule to another in coupled redox reactions.
- Electrons are transferred from substrates such as glucose to coenzyme carriers and ultimately captured in high-energy adenosine triphosphate (ATP).
- Cell pathways involved in extracting energy from fuels are glycolysis, the tricarboxylic acid cycle, and electron transport.
- The molecules used in aerobic respiration are glucose and oxygen, and the products are CO_2, H_2O, and ATP.
- Microbes participate in alternate pathways such as fermentation and anaerobic respiration.
- Cells manage their metabolites through linked pathways that have numerous functions and can proceed in more than one direction.

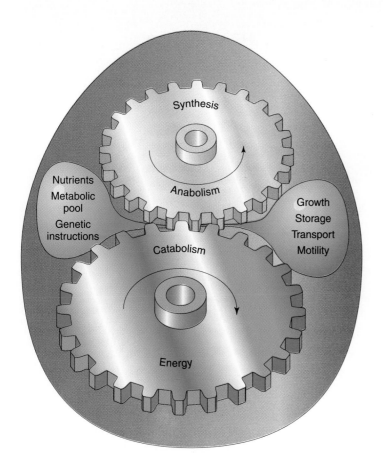

FIGURE 8.1

A marvelous metabolic machine. This simplified model summarizes cell metabolism. The chemical reactions in cells are interactive and highly balanced. Synthetic actions of anabolism work with the energy-producing actions of catabolism. They mutually drive each other and keep the basic actions of the cell constantly in play.

The Metabolism of Microbes

Metabolism, from the Greek term *metaballein,* meaning change, pertains to all chemical reactions and physical workings of the cell. Although metabolism entails thousands of different reactions, most of them fall into one of two general categories. **Anabolism,*** sometimes also called *biosynthesis,* is any process that results in synthesis of cell molecules and structures. It is a building and bond-making process that forms larger molecules from smaller ones, and it usually requires the input of energy. **Catabolism*** is the opposite, or complement, of anabolism. Catabolic reactions are degradative; they break bonds, convert larger molecules into smaller components, and often produce energy. The linking of anabolism to catabolism ensures the efficient completion of many thousands of cellular processes.

Metabolism is a self-regulatory process that maintains the stability of the cell (figure 8.1). Its actions provide a dynamic pool of chemical building blocks and give rise to enzymes and structural components of the cell. Metabolism of nutrients can extract energy in the form of adenosine triphosphate (ATP), or other high-energy compounds, that can be channeled into such processes as biosynthesis, transport, growth, and motility. As we will see, **metabolites**—compounds given off by the complex networks of metabolism—often serve multipurpose roles in the economy of the cell. Metabolism is highly organized and responsive to regulation.

It has built-in controls for reducing or stopping a process that is not in demand and for storing excess nutrients. The metabolic workings of the cell are indeed intricate and complex, but they are also elegant and efficient. It is this very organization that sustains life.

ENZYMES: CATALYZING THE CHEMICAL REACTIONS OF LIFE

A microbial cell could be viewed as a microscopic factory, complete with basic building materials, a source of energy, and a "blueprint" for running its extensive network of metabolic reactions. But the chemical reactions of life, even when highly organized and complex, cannot proceed without a special class of proteins called **enzymes.*** Enzymes are a remarkable example of **catalysts,*** chemicals that increase the rate of a chemical reaction without becoming part of the products or being consumed in the reaction. Do not make the mistake of thinking that an enzyme creates a reaction. Because of the great energy of some molecules, a reaction could occur spontaneously at some point even without an enzyme, but at a very slow rate (figure 8.2). A study of the enzyme urease shows that it increases the rate of the breakdown of urea by a factor of 100 trillion as compared to an uncatalyzed reaction. Because most uncatalyzed metabolic reactions do not occur fast enough to sustain cell processes, enzymes, which speed up the rate of reactions, are indispensable to life. Other major characteristics of enzymes are summarized in table 8.1.

---

* anabolism (ah-nab'-oh-lizm) Gr. *anabole,* a throwing up.

* catabolism (kah-tab'-oh-lizm) Gr. *katabole,* a throwing down.

---

* enzyme (en'-zyme) Gr. *en,* in, and *syme,* leaven. Named for catalytic agents first found in yeasts.

* catalyst (kat'-uh-list) Gr. *katalysis,* dissolution.

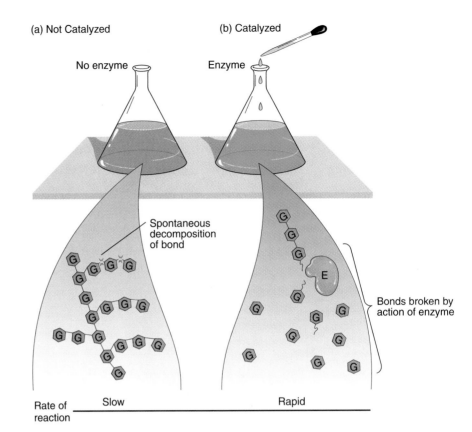

FIGURE 8.2

The effects of a catalyst. This example shows the breakdown of starch into glucose. **(a)** A few of the bonds will break in the absence of an enzyme, but it is a very slow process. **(b)** Addition of an enzyme (E) can speed up the reaction so that it happens very rapidly.

How Do Enzymes Work?

We have said that an enzyme speeds up the rate of a metabolic reaction, but just how does it do this? During a chemical reaction, reactants are converted to products by bond formation or breakage. A certain amount of energy is required to initiate every such reaction, which limits its rate. This resistance to a reaction, which must be overcome for a reaction to proceed, is measurable and is called the **energy of activation** (Microbits 8.1). In the laboratory, overcoming this initial resistance can be achieved by (1) increasing thermal energy (heating) to increase molecular velocity,

(2) increasing the concentration of reactants, or (3) adding a catalyst. In most living systems, the first two alternatives are not feasible, because elevating the temperature is potentially harmful and higher concentrations of reactants are not practical. This leaves only the action of catalysts, and enzymes fill this need efficiently and potently.

At the molecular level, an enzyme promotes a reaction by serving as a physical site upon which the reactant molecules, called **substrates,** can be positioned for various interactions. The enzyme is much larger in size than its substrate, and it presents a unique niche that fits only that particular substrate. Although an enzyme binds to the substrate and participates directly in changes to the substrate, it does not become a part of the products, is not used up by the reaction, and can function over and over again. Enzyme speed, defined as the number of substrate molecules converted per enzyme per second, is well documented. Speeds range from several million for catalase to a thousand for lactate dehydrogenase. To further visualize the roles of enzymes in metabolism, we must next look at their structure.

Enzyme Structure

The primary structure of all enzymes is protein (with some exceptions—Microbits 8.2), and they can be classified as simple or conjugated. **Simple** enzymes consist of protein alone, whereas **conjugated** enzymes (figure 8.3) contain protein and nonprotein molecules. A conjugated enzyme, sometimes referred to as a **holoenzyme,*** is a combination of a protein, now called the **apoenzyme,** and one or more **cofactors** (table 8.2). Cofactors are either organic molecules, called **coenzymes,** or inorganic ele-

TABLE 8.1
Checklist of Enzyme Characteristics

- Are composed of protein and may require cofactors
- Act as organic catalysts to speed up the rate of cellular reactions
- Lower the activation energy required for a chemical reaction to proceed (Microbits 8.1)
- Have unique characteristics such as shape, specificity, and function
- Enable metabolic reactions to proceed at a speed compatible with life
- Provide a reactive site for target molecules called substrates
- Are much larger in size than their substrates
- Associate closely with substrates but do not become integrated into the reaction products
- Are not used up or permanently changed by the reaction
- Can be recycled, thus function in extremely low concentrations
- Are limited by particular conditions of temperature and pH
- Can be regulated by feedback and genetic mechanisms

---

* holoenzyme (hol-oh-en′-zyme) Gr. *holos,* whole.

MICROBITS   8.1

Enzymes as Biochemical Levers

An analogy will allow us to envision the relationship of enzymes to the energy of activation. A large boulder sitting precariously on a cliff's edge contains a great deal of potential energy, but it will not fall to the ground and release this energy unless something disturbs it. It might eventually be disturbed spontaneously as the cliff erodes below it, but that could take a very long time. Moving it with a crowbar (to overcome the resistance of the boulder's weight) will cause it to tumble down freely by its own momentum. If we relate this analogy to chemicals, the reactants are the boulder on the cliff, the activation energy is the energy needed to move the boulder, the enzyme is the crowbar, and the product is the boulder at the bottom of the cliff. Like the crowbar, an enzyme permits a reaction to occur rapidly by lowering the energy obstacle. (Although this analogy concretely illustrates the concept of the energy of activation, it is imperfect in that the enzyme, unlike the person with the crowbar, does not actually add any energy to the system.)

This phenomenon can be represented by plotting the energy of the reaction against the direction of the reaction. The curve of reaction is shown in the accompanying graph. It is evident that there is an energy "hill" called the energy of activation (E_{act}) that must be overcome for the reaction to proceed. When a catalyst is present, it lowers the E_{act}. It does this by providing the reactants with an energy "shortcut" by which they can progress to the final state. Note that the products are the same final energy state with or without an enzyme.

Analogy demonstrating the influence of enzymes on chemical reactions. **(a)** A boulder can represent potential energy available for a chemical reaction. **(b)** Graph of chemical reaction, with and without an enzyme. Energy (called energy of activation) is required in both cases to convert a reactant molecule to products. But in an enzyme-catalyzed reaction, the enzyme significantly lowers this energy of activation and allows the reaction to proceed more readily and rapidly.

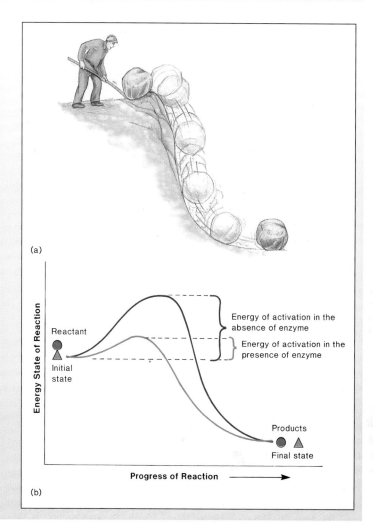

ments (metal ions). In some enzymes, the cofactor is loosely associated with the apoenzyme by noncovalent bonds; in others, it is linked by covalent bonds.

Apoenzymes: Specificity and the Active Site

Apoenzymes range in size from small polypeptides with about 100 amino acids and a molecular weight of 12,000 to large polypeptide conglomerates with thousands of amino acids and a molecular weight of over one million. Like all proteins, an apoenzyme exhibits levels of molecular complexity called the primary, secondary, tertiary, and, in larger enzymes, quaternary organization (figure 8.4). As we saw in chapter 2, the first three levels of structure arise when a single polypeptide chain undergoes an automatic folding process and achieves stability by forming disulfide and other types of bonds. Folding causes the surface of the apoenzyme to acquire three-dimensional surface features that result in the enzyme's specificity for substrates. The actual site where the substrate binds is a crevice or groove called the **active site,** or **catalytic site,** and

there can be from one to several such sites (figure 8.4). Each type of enzyme has a different primary structure (type and sequence of amino acids)[1], variations in folding, and unique active sites.

Enzyme-Substrate Interactions

For a reaction to take place, a temporary enzyme-substrate union must occur at the active site. There are several explanations for the process of attachment (figure 8.5). The specificity is often described as a "lock-and-key" fit in which the substrate is inserted into the active site's pocket. This is a useful analogy, but an enzyme is not a rigid lock mechanism. It is likely that the enzyme actually helps pull the substrate into the active site through slight changes in its shape. This is called an *induced fit* (figure 8.5d).

The bonds formed between the substrate and enzyme are weak and, of necessity, easily reversible. Once the enzyme-substrate

---

1. As specified by the genetic blueprint in DNA (see chapter 9).

MICROBITS 8.2
Unconventional Enzymes

The molecular reactions of cells are still an active and rich source of discovery, and new findings come along nearly every few months that break older "rules" and change our understanding. It was once an accepted fact that proteins were the only biological molecules that act as catalysts, until biologists found a novel type of RNA termed **ribozymes.** Ribozymes are associated with many types of cells and a few viruses, and they display some of the properties of protein catalysts, such as having a specific active site and interacting with a substrate. But these molecules are remarkable because their substrate is other RNA. Ribozymes are thought to be remnants of the earliest molecules on earth that could have served as both catalysts and genetic material. In natural systems, ribozymes are involved in self-splicing or cutting of RNA molecules during final processing of the genetic code (see chapter 9). Further research has shown that ribozymes can be designed to handle other kinds of activities, such as inhibiting gene expression. Several companies have developed and are testing ribozyme-based therapies for treating cancer and AIDS.

Another intriguing new type of enzyme comes to us from studies of the immune system. Using antibodies as their basic tool, immunologists have developed catalytic molecules called *abzymes*. Antibodies, like enzymes, are proteins that are very specific to their target molecules and attach tightly to them. They can be modified in the lab to create abzymes that are capable of forming a reversible complex with substrate, speeding up reactions, and participating in bond formation and breakage. Abzymes have an additional benefit—they may be fine-tuned to fit almost any substrate shape, and they can be manufactured through monoclonal antibody technology (see chapter 15). The potential of abzymes for use as drugs is also being explored by medicine and industry. Tests are currently underway on specific drugs and other treatments for cancer, infections, and medical problems such as blood clots.

Another focus of study that could lead to new applications is the search for *extremozymes,* enzymes from microbes that live under rigorous conditions of temperature, pH, salt, and pressure. Because these enzymes are adapted to working under such extreme conditions, they may well be useful in industrial processes that require work in heat or cold or very low pH. Several enzymes that can work under high temperature and low pH have been isolated and are either currently in use (PCR technique) or in development.

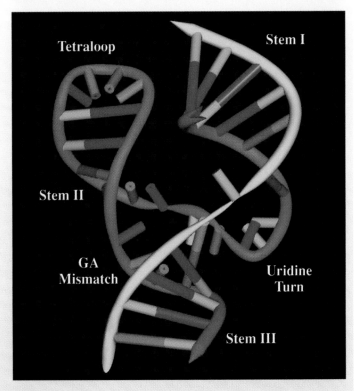

A "hammerhead" ribozyme consisting of a single RNA strand curved around to form an active site (indentation between tetraloop and stem I).

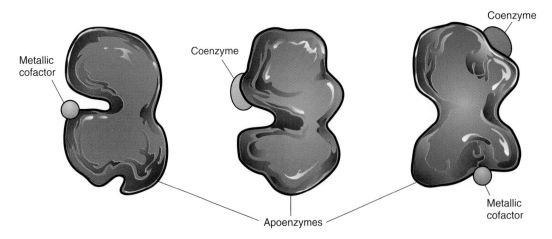

FIGURE 8.3

Conjugated enzyme structure. All have an apoenzyme (polypeptide or protein) component and one or more cofactors.

TABLE 8.2

Selected Enzymes, Catalytic Actions, Cofactors, and Functions

Enzyme	Action	Metallic Cofactor Required
Catalase	Breaks down hydrogen peroxide	Iron (Fe)
Oxidase	Adds electrons to oxygen	Iron, copper (Cu)
Hexokinase	Transfers phosphate to glucose	Magnesium (Mg)
Urease	Splits urea into an ammonium ion	Nickel (Ni)
Nitrate reductase	Reduces nitrate to nitrite	Molybdenum (Mo)
DNA polymerase complex	Synthesis of DNA	Zinc (Zn) and Mg

Enzyme	Coenzyme Involved	Vitamin Required	Metabolic Function of Coenzyme Complex
Pyruvate dehydrogenase	Coenzyme A	Pantothenic acid (B3)	Recognition site for binding substrate
Pyruvic dehydrogenase	Thiamine pyrophosphate	Thiamine (B1)	Transfer of aldehyde groups
Various dehydrogenases	Nicotinamide adenine dinucleotide (NAD)	Niacin	Carries electrons in redox reactions
Succinate reductase	Flavin adenine dinucleotide (FAD)	Riboflavin (B2)	Electron transfers in mitochondria
Various amino acid synthetases	Tetrahydrofolate	Folic acid	Transfer carbon groups in amino acids and nucleotides
Transaminases; decarboxylases	Pyridoxal phosphate	B6	Numerous conversions of amino acids

complex has formed, appropriate reactions occur on the substrate, often with the aid of a cofactor, and a product is formed and released. The enzyme can then attach to another substrate molecule and repeat this action. Although enzymes can potentially catalyze reactions in both directions, most examples in this chapter will depict them working in one direction only.

Cofactors: Supporting the Work of Enzymes

In chapter 7 we learned that microorganisms require specific metal ions called trace elements and certain organic growth factors. In many cases, the need for these substances arises from their roles as cofactors. The **metallic cofactors,** including iron, copper, magne-

sium, manganese, zinc, cobalt, selenium, and many others, participate in precise functions between the enzyme and its substrate. In general, metals activate enzymes, help bring the active site and substrate close together, and participate directly in chemical reactions with the enzyme-substrate complex.

Coenzymes are organic compounds that work in conjunction with an apoenzyme to perform a necessary alteration of a substrate. The general function of a coenzyme is to remove a functional group from one substrate molecule and add it to another substrate, thereby serving as a transient carrier of this group (figure 8.6). The specific activities of coenzymes are many and varied. In a later section of this chapter, we shall see that coenzymes carry and transfer hydrogen atoms, electrons, carbon dioxide, and amino groups. One of the most important components of coenzymes are **vitamins** (see table 8.2), which explains why vitamins are important to nutrition and may be required as growth factors for living things. Vitamin deficiencies prevent the complete holoenzyme from forming. Consequently, both the chemical reaction and the structure or function dependent upon that reaction are compromised.

Classification of Enzyme Functions

Enzymes are classified and named according to characteristics such as site of action, type of action, and substrate (Microbits 8.3).

Location and Regularity of Enzyme Action Enzymes exhibit several patterns of performance. After initial synthesis in the cell, **exoenzymes** are transported extracellularly, where they break down (hydrolyze) large food molecules or harmful chemicals. Examples of exoenzymes are cellulase, amylase, and penicillinase. By contrast, **endoenzymes** are retained intracellularly and function there. Most enzymes of the metabolic pathways are of this variety (figure 8.7).

In terms of their presence in the cell, enzymes are not all produced in equal amounts or at equal rates. Some, called **constitutive enzymes** (figure 8.7c), are always present and in relatively constant amounts, regardless of the amount of substrate. The enzymes involved in utilizing glucose, for example, are very important in metabolism and thus are constitutive. An **induced** (or inducible) **enzyme** (figure 8.7d) is not constantly present and is produced only when its substrate is present. Induced enzymes are present in amounts ranging from a few molecules per cell to several thousand times that many, depending on the metabolic requirement. This property of selective synthesis of enzymes prevents a cell from wasting resources by making enzymes that will not be used immediately. The induction of enzymes constitutes an important metabolic control discussed later in this section of the chapter and again in chapter 9.

Synthesis and Hydrolysis Reactions A growing cell is in a frenzy of activity, constantly synthesizing proteins, DNA, and RNA; forming storage polymers such as starch and glycogen; and assembling new cell parts. Such anabolic reactions require enzymes (ligases) to form covalent bonds between smaller substrate molecules. Also known as *condensation reactions,* synthesis reactions typically require ATP and always release one water molecule for each bond made (see figure 8.8a). Catabolic reactions involving energy transactions, remodeling of cell structure, and digestion of

Levels of Structure

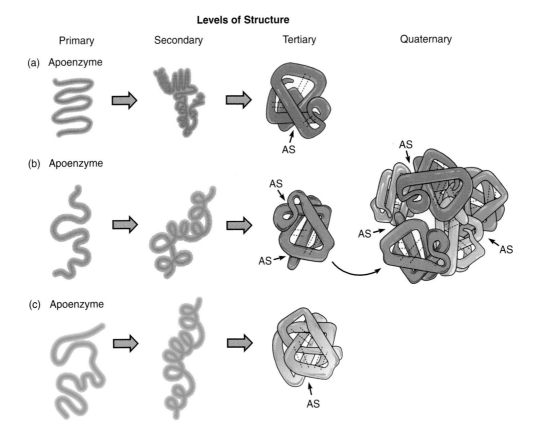

FIGURE 8.4

How the active site and specificity of the apoenzyme arise. As the polypeptide forms intrachain bonds and folds, it assumes a three-dimensional (tertiary) state with numerous surface features. Because each different polypeptide folds differently **(a, b, c),** each apoenzyme will have differently shaped active sites (AS). Some enzymes have more than one active site; others have sites to attach cofactors and regulatory compounds; and more complex enzymes have a quaternary structure consisting of several polypeptides bound by weak forces, as in **(b).**

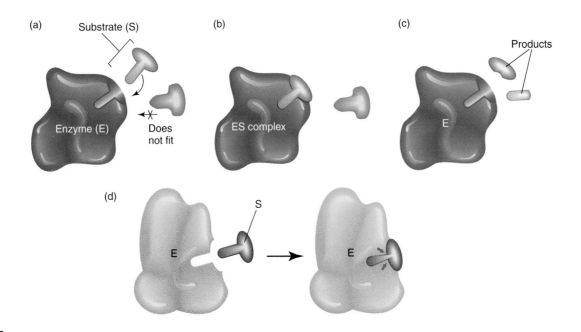

FIGURE 8.5

Enzyme-substrate reactions. (a) When the enzyme and substrate come together, the substrate (S) must show the correct fit and position with respect to the enzyme (E). **(b)** When the ES complex is formed, it enters a transition state. During this temporary but tight interlocking union, the enzyme participates directly in breaking or making bonds. **(c)** Once the reaction is complete, the enzyme releases the products. **(d)** The induced fit model proposes that the enzyme recognizes its substrate and adapts slightly to it so that the final binding is even more precise.

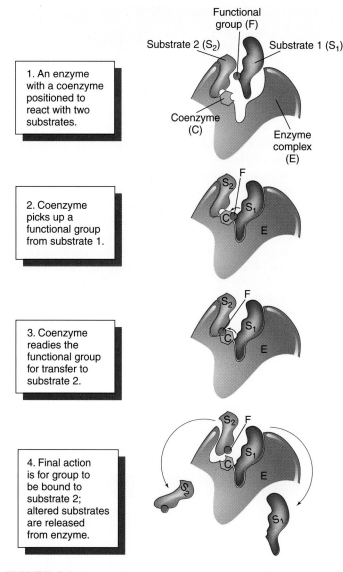

1. An enzyme with a coenzyme positioned to react with two substrates.

2. Coenzyme picks up a functional group from substrate 1.

3. Coenzyme readies the functional group for transfer to substrate 2.

4. Final action is for group to be bound to substrate 2; altered substrates are released from enzyme.

FIGURE 8.6

The carrier functions of coenzymes. Many enzymes use coenzymes to transfer functional groups from one substrate to another.

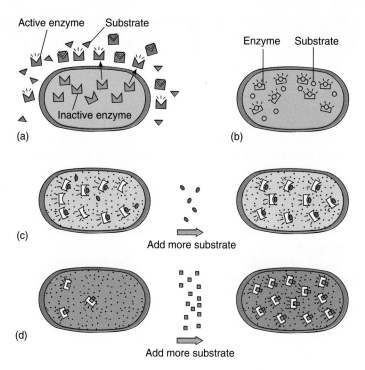

FIGURE 8.7

Types of enzymes, as described by their location of action and quantity. (a) Exoenzymes function extracellularly. **(b)** Endoenzymes function intracellularly. **(c)** Constitutive enzymes are present in relatively constant amounts in a cell. The addition of more substrate does not increase the numbers of these enzymes. **(d)** Induced enzymes are normally present in trace amounts, but their quantity can be increased a thousandfold by the addition of substrate.

macromolecules are also very active during cell growth. For example, digestion requires enzymes to break down substrates into smaller molecules so they can be used by the cell. Because the breaking of bonds requires the input of water, digestion is often termed a *hydrolysis* reaction* (figure 8.8*b*).

Transfer Reactions by Enzymes   Other enzyme-driven processes that involve the simple addition or removal of a functional group are important to the overall economy of the cell. Oxidation-reduction and other transfer activities are examples of these types of reactions.

Some atoms and compounds readily give or receive electrons and participate in oxidation (the loss of electrons) or reduction (the gain of electrons). The compound that loses the electrons is **oxidized,** and the compound that receives the electrons is **reduced.**

Such oxidation-reduction (redox) reactions are common in the cell and indispensable to the energy transformations discussed later in this chapter (see Microbits 2.2). Important components of cellular redox reactions are oxidoreductases, which remove electrons from one substrate and add them to another, and their coenzyme carriers, nicotinamide adenine dinucleotide (NAD; see figure 8.14) and flavin adenine dinucleotide (FAD).

Other enzymes play a role in the molecular conversions necessary for the economical use of nutrients by directing the transfer of functional groups from one molecule to another. For example, *aminotransferases* convert one type of amino acid to another by transferring an amino group (see figure 8.28); *phosphotransferases* participate in the transfer of phosphate groups and are involved in energy transfer; *methyltransferases* move a methyl (CH_3) group from substrate to substrate; and *decarboxylases* (also called carboxylases) catalyze the removal of carbon dioxide from organic acids in several metabolic pathways.

The Role of Microbial Enzymes in Disease   Many pathogens secrete unique exoenzymes that help them avoid host defenses or promote their multiplication in tissues. Because these enzymes contribute to pathogenicity, they are referred to as virulence factors, or toxins in some cases. *Streptococcus pyogenes* (a cause of throat and skin infections) produces a streptokinase that digests blood clots and apparently assists in invasion of wounds. *Pseudomonas aeruginosa,* a respiratory and skin pathogen, produces elastase and

---

* **hydrolysis** (hy-drol′-uh-sis) Gr. *hydro,* water, and *lysis,* setting free.

Most metabolic reactions require a separate and unique enzyme. Up to the present time, researchers have discovered and named over 5,000 of them. Given the complex chemistry of the cell and the extraordinary diversity of living things, many more enzymes probably remain to be discovered. A standardized system of nomenclature and classification was developed to prevent discrepancies.

In general, an enzyme name is composed of two parts: a prefix or stem word derived from a certain characteristic—usually the substrate acted upon or the type of reaction catalyzed, or both—followed by the ending *-ase.*

The system classifies the enzyme in one of the following six classes, on the basis of its general biochemical action: (1) *Oxidoreductases* transfer electrons from one substrate to another, and *dehydrogenases* transfer a hydrogen from one compound to another. (2) *Transferases* transfer functional groups from one substrate to another. (3) *Hydrolases* cleave bonds on molecules with the addition of water. (4) *Lyases* add groups to or remove groups from double-bonded substrates. (5) *Isomerases* change a substrate into its isomeric* form. (6) *Ligases* catalyze the formation of bonds with the input of ATP and the removal of water.

Each enzyme is also assigned a technical name that indicates the specific reaction it catalyses. These names are useful for biochemists but are often too lengthy for routine use. More often, a common name is derived from some notable feature such as the substrate or enzyme action, with *-ase* added to the ending. For example, *hydrogen peroxide oxidoreductase* is more commonly known as *catalase,* and *mucopeptide N-acetylmuramoylhydrolase* is called (thankfully) *lysozyme.* With this system, an enzyme that digests a carbohydrate substrate is a *carbohydrase;* a specific carbohydrase, *amylase,* acts on starch (amylose is a major component of starch). The enzyme *maltase* digests the sugar maltose. An enzyme that hydrolyzes peptide bonds of a protein is a *proteinase, protease,* or *peptidase,* depending on the size of the protein substrate. Some fats and other lipids are digested by *lipases.* DNA is hydrolyzed by *deoxyribonuclease,* generally shortened to *DNase.* A *synthetase* or *polymerase* bonds together many small molecules into large molecules. Other examples of enzymes are presented in table 8.A. (See also table 8.2).

---

*An isomer is a compound that has the same molecular formula as another compound but differs in arrangement of the atoms.

TABLE 8.A
A Sampling of Enzymes, Their Substrates, and Their Reactions

Common Name	Systematic Name	Enzyme Class	Substrates	Action
Cellulase	1,4 β-glycosidase	Hydrolase	Cellulose	Cleaves 1,4 β-glycosidic linkages
Lactase	β-D-galactosidase	Hydrolase	Lactose	Breaks lactose down into glucose and galactose
Penicillinase	Beta-lactamase	Hydrolase	Penicillin	Hydrolyzes beta-lactam ring
Lipase	Triacylglycerol acylhydrolase	Hydrolase	Triglycerides	Cleaves bonds between glycerol and fatty acids
DNA polymerase	DNA nucleotidyl-transferase	Transferase	DNA nucleosides	Synthesizes a strand of DNA using the complementary strand as a model
Hexokinase	ATP-glucose phosphotransferase	Transferase	Glucose	Catalyzes transfer of phosphate from ATP to glucose
Aldolase	Fructose diphosphate aldolase	Lyase	Fructose diphosphate	Catalyzes the conversion of the substrate to two 3-carbon fragments
Lactate dehydrogenase	Same as common name	Oxidoreductase	Pyruvic acid	Catalyzes the conversion of pyruvic acid to lactic acid
Oxidase	Cytochrome oxidase	Oxidoreductase	Molecular oxygen	Catalyzes the reduction (addition of electrons and hydrogen) to O_2

collagenase, which digest elastin and collagen. These increase the severity of certain lung diseases and burn infections. *Clostridium perfringens,* an agent of gas gangrene, synthesizes lecithinase C, a lipase that profoundly damages cell membranes and accounts for the tissue death associated with this disease. Not all enzymes digest tissues; some, such as penicillinase, inactivate penicillin and thereby protect a microbe from its effects.

The Sensitivity of Enzymes to Their Environment
The activity of an enzyme is highly influenced by the cell's environment. In general, enzymes operate only under the natural temperature, pH, and osmotic pressure of an organism's habitat. When enzymes are subjected to changes in these normal conditions, they tend to be chemically unstable, or **labile.** Low temperatures inhibit catalysis, and high temperatures denature the

(a) Condensation Reaction

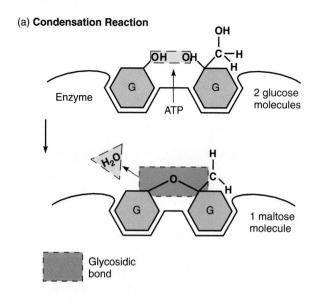

Enzyme

2 glucose molecules

ATP

H₂O

1 maltose molecule

Glycosidic bond

(b) Hydrolysis Reaction

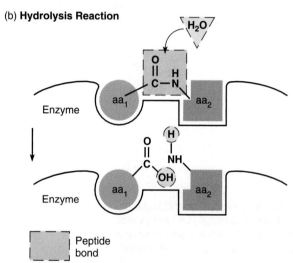

H₂O

Enzyme   aa₁   aa₂

Enzyme   aa₁   aa₂

Peptide bond

FIGURE 8.8

Examples of enzyme-catalyzed synthesis and hydrolysis reactions.
(a) Condensation reaction. Forming a glycosidic bond between two glucose molecules to generate maltose requires the removal of a water molecule and energy from ATP. **(b)** Hydrolysis reaction. Breaking a peptide bond between two amino acids requires a water molecule that adds an H and OH to the amino acids.

Multienzyme Systems

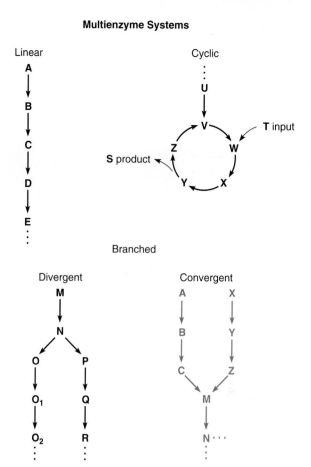

Linear

Cyclic

Branched

FIGURE 8.9

Patterns of metabolism. In general, metabolic pathways consist of a linked series of individual chemical reactions that produce intermediary metabolites and lead to a final product. These pathways occur in several patterns, including linear, cyclic, and branched. Anabolic pathways involved in biosynthesis result in a more complex molecule, each step adding on a functional group, whereas catabolic pathways involve the dismantling of molecules and can generate energy. Virtually every reaction in a series involves a specific enzyme.

---

apoenzyme. **Denaturation** is a process by which the weak bonds that collectively maintain the native shape of the apoenzyme are broken. This disruption causes extreme distortion of the enzyme's shape and prevents the substrate from attaching to the active site (see figure 11.4). Such nonfunctional enzymes block metabolic reactions and thereby can lead to cell death. Low or high pH or certain chemicals (heavy metals, alcohol) are also denaturing agents.

REGULATION OF ENZYMATIC ACTIVITY AND METABOLIC PATHWAYS

Metabolic reactions proceed in a systematic, highly regulated manner that maximizes the use of available nutrients and energy. The cell responds to environmental conditions by adopting those meta-

bolic reactions that most favor growth and survival. Because enzymes are critical to these reactions, the regulation of metabolism is largely the regulation of enzymes by an elaborate system of checks and balances. Before we examine some of these control systems, let us take a look at some general features of metabolic pathways.

Metabolic Pathways

Metabolic reactions rarely consist of a single action or step. More often, they occur in a multistep series or pathway, with each step catalyzed by an enzyme. An individual reaction is shown in various ways, depending on the purpose at hand (figure 8.9). The product of one reaction is often the reactant (substrate) for the next, forming a linear chain of reactions. Many pathways have branches that provide alternate methods for nutrient processing. Others take a cyclic

Negative Feedback

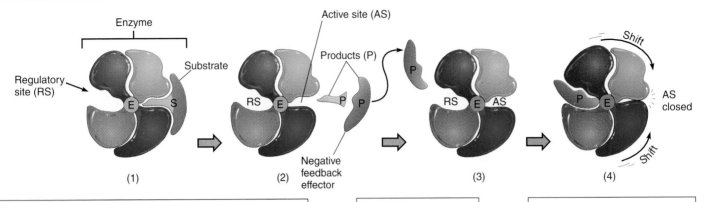

(1), (2) Allosteric enzymes have a quaternary structure with two different sites of attachment—the active site (AS) and the regulatory site (RS). The enzyme complex normally attaches to the substrate at the AS and releases products (P).

(3) One product can function as a negative-feedback effector by fitting into a regulatory site.

(4) The entrance of P into RS causes a confirmational shift of the enzyme that closes the active site. The enzyme cannot catalyze further reactions with substrate as long as the product is inserted in the regulatory site.

Competitive Inhibition

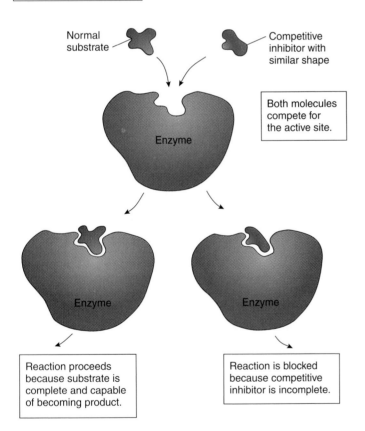

Normal substrate

Competitive inhibitor with similar shape

Both molecules compete for the active site.

Enzyme

Enzyme

Enzyme

Reaction proceeds because substrate is complete and capable of becoming product.

Reaction is blocked because competitive inhibitor is incomplete.

FIGURE 8.10

Examples of two common control mechanisms for enzymes.

form, in which the starting molecule is regenerated to initiate another turn of the cycle (for example, the TCA cycle; see figure 8.21). Pathways generally do not stand alone; they are interconnected and merge at many sites.

Every pathway has one or more enzyme pacemakers (usually the slowest enzyme in the series) that sets the rate of a pathway's progression. These enzymes respond to various control signals and, in so doing, determine whether a pathway proceeds. Regulation of pacemaker enzymes proceeds on two fundamental levels. Either the enzyme itself is directly inhibited or activated, or the amount of the enzyme in the system is altered (decreased or increased). Factors that affect the enzyme directly provide a means for the system to be finely controlled or tuned, whereas regulation at the genetic level (enzyme synthesis) provides a slower, less sensitive control.

Direct Controls on the Actions of Enzymes

Competitive Inhibition   In **competitive inhibition,** other molecules with a structure similar to the normal substrate can occupy the enzyme's active site. Although these molecular mimics have the proper fit for the site, they cannot be further acted upon. The attachment of a mimic effectively prevents an enzyme from attaching to its usual substrate and blocks its activity, output, and possibly the rest of that pathway (figure 8.10). This mechanism is common in natural metabolic pathways. It is also an important mode of action by some drugs for treating infections (see chapter 12).

Feedback Control   The term *feedback* is borrowed from the field of engineering, where it describes a process in which the product of a system is *fed back* into the system to keep it in balance. Feedback is positive or negative, depending upon whether the process is stimulated (+) or inhibited (−) by the product. **Negative feedback** is common in electrical appliances such as hot plates or furnaces with automatically controlled temperatures. These devices possess a means of controlling the temperature (a heat source) and a temperature sensor (thermostat) that is set to a desired point and regulates the turning on or off of the heat source. In the case of a hot plate, when the set temperature has been reached, the thermostat switches the heating elements off. A subsequent fall below the set temperature causes the heating element to be turned on again.

Many enzymatic reactions are controlled by negative feedback, whereby the end product being fed back into the system *negates* (cancels) an enzyme's activity:

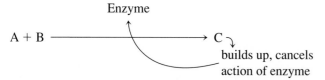

At high levels of substrate and scant levels of end product, the enzyme works freely. In time, the end product builds up to sufficient levels, and it stops the action of that enzyme. It usually targets the first enzyme in a metabolic pathway. The mechanisms of negative feedback—that is, how the product actually stops the action of an enzyme—can best be understood if we look at the allosteric behavior of enzymes.

We have previously shown that globular proteins of enzymes are bulky and possess numerous surface features and that one part of an enzyme's terrain (the active site) accommodates the substrate. **Allosteric* enzymes** have an additional **regulatory site** for the attachment of molecules other than substrate. This regulatory site usually exists on a different polypeptide unit of a quaternary structure. When an end-product molecule fits into this regulatory site, the enzyme's active site is so distorted that it can no longer bind to its substrate. Although this distortion does not denature the enzyme and is quite reversible, allosteric inhibitors can temporarily stop the action of that enzyme (figure 8.10). This mechanism is termed **feedback inhibition.** When at some point the end product is used up and more of it is needed, the enzyme will be released from inhibition and can resume catalysis.

Controls on Enzyme Synthesis

Controlling enzymes by controlling their synthesis is another effective mechanism, because enzymes do not last indefinitely. Some wear out, some are deliberately degraded, and others are diluted with each cell division. For catalysis to continue, enzymes eventually must be replaced. This cycle works into the scheme of the cell, where replacement of enzymes can be regulated according to cell demand. The mechanisms of this system are genetic in nature; that is, they require regulation of DNA and the protein synthesis machinery, topics we shall encounter once again in chapter 9.

Enzyme repression is a means to stop further synthesis of an enzyme somewhere along its pathway. As the level of the end product from a given enzymatic reaction has built to excess, the genetic apparatus responsible for replacing these enzymes is automatically suppressed (figure 8.11). The response time takes longer than feedback inhibition, but its effects are more enduring.

A response that resembles the inverse of feedback repression is **enzyme induction.** In this process, enzymes appear (are induced) only when suitable substrates are present; that is, the synthesis of enzyme is induced by its substrate (see figures 8.7*d* and 9.21). Both mechanisms are important genetic control systems in bacteria.

A classic model of enzyme induction occurs in the response of *Escherichia coli* to certain sugars. For example, if a particular strain of *E. coli* is inoculated into a medium whose principal carbon

FIGURE 8.11

One type of genetic control of enzyme synthesis: feedback/enzyme repression. The enzyme is synthesized continuously until enough product has been made, at which time the excess product reacts with a site on DNA that regulates the enzyme's synthesis, thereby inhibiting further enzyme production.

source is lactose, it will produce lactase to hydrolyze it into glucose and galactose. If the bacterium is subsequently inoculated into a medium containing only sucrose as a carbon source, it will cease synthesizing lactase and begin synthesizing sucrase. This response enables the organism to adapt to a variety of nutrients, and it also prevents a microbe from "spinning its wheels," making enzymes for which no substrates are present.

CHAPTER CHECKPOINTS

Metabolism includes all the biochemical reactions that occur in the cell. It is a self-regulating complex of interdependent processes that encompasses many thousands of chemical reactions.

Anabolism is the energy-requiring subset of metabolic reactions, which synthesize large molecules from smaller ones.

Catabolism is the energy-releasing subset of metabolic reactions, which degrade or break down large molecules into smaller ones.

Enzymes are proteins that catalyze all biochemical reactions by forming enzyme-substrate complexes. The binding of the substrate by an enzyme makes possible both bond-forming and bond-breaking reactions, depending on the pathway involved. Enzymes may utilize cofactors as carriers and activators.

Enzymes are classified and named according to the kinds of reactions they catalyze.

To function effectively, enzymes require specific conditions of temperature, pH, and osmotic pressure.

Enzyme activity is regulated by processes of feedback inhibition, induction, and repression, which, in turn, respond to availability of substrate and concentration of end products, as well as to other environmental factors.

---

* **allosteric** (al-oh-stair´-ik) Gr. *allos,* other, and *steros,* solid. Literally, "another space."

The Pursuit and Utilization of Energy

Energy is defined as the capacity to do work or to cause change. In order to carry out the work of an array of metabolic processes, cells require constant input and expenditure of some form of usable energy. Energy commonly exists in various forms: (1) thermal, or heat, energy from molecular motion; (2) radiant (wave) energy from visible light or other rays; (3) electrical energy from a flow of electrons; (4) mechanical energy from a physical change in position; (5) atomic energy from reactions in the nucleus of an atom; and (6) chemical energy present in the bonds of molecules. Cells are far too fragile to rely in any constant way on thermal or atomic energy for cell transactions. Except for certain chemoautotrophic bacteria, the ultimate source of energy is the sun, but only photosynthetic autotrophs can tap this source directly. This capacity of photosynthesis to convert solar energy into chemical energy provides both a nutritional and an energy basis for all heterotrophic living things. For the most part, only chemical energy can routinely operate cell transactions, and chemical reactions are the universal basis of cellular energetics.

In this section we will further explore the role of energy as it is related to the oxidation of fuels, redox carriers, and the generation of ATP.

CELL ENERGETICS

Cells manage energy in the form of chemical reactions that change molecules. This often involves activities such as the making or breaking of bonds and the transfer of electrons. Not all cellular reactions are equal with respect to energy. Some release energy, and others require it to proceed. For example, a reaction that proceeds as follows:

$$X + Y \xrightarrow{\text{Enzyme}} Z + \text{Energy}$$

releases energy as it goes forward. This type of reaction is termed **exergonic.*** Energy of this type is considered free—it is available for doing cellular work. Energy transactions such as the following:

$$\text{Energy} + A + B \xrightarrow{\text{Enzyme}} C$$

are called **endergonic,*** because they are driven forward with the addition of energy. Exergonic and endergonic reactions are coupled, so that released energy is immediately put to use.

Summaries of metabolism might make it seem that cells "create" energy from nutrients, but they do not. What they actually do is extract chemical energy already present in nutrient fuels and apply that energy toward useful work in the cell, much like a gasoline engine releases energy as it burns fuel. The engine does not actually produce energy, but it converts the potential energy of the fuel to the free energy of work.

The processes by which cells handle energy are well understood, and most of them are explainable in basic biochemical terms. At the simplest level, cells possess specialized enzyme systems that trap the energy present in the bonds of nutrients as they are progressively broken (figure 8.12). During exergonic reactions, energy released by electrons is stored in various high-energy phosphate molecules such as ATP. As we shall see, the ability of ATP to temporarily store and release the energy of chemical bonds fuels endergonic cell reactions. Before discussing ATP, let us examine the process behind electron transfer: redox reactions.

A CLOSER LOOK AT BIOLOGICAL OXIDATION AND REDUCTION

We stated earlier that biological systems often extract energy through **redox reactions.** Such reactions always occur in pairs, with an electron donor and an electron acceptor, which constitute a *conjugate pair,* or *redox pair.* The reaction can be represented as follows (see also figure 8.13):

Electron donor + Electron acceptor $\longrightarrow$ Electron donor + Electron acceptor

(Reduced)   (Oxidized)   (Oxidized)   (Reduced)

This process salvages electrons along with their inherent energy, and it changes the energy balance, leaving the reduced compound with less energy than the oxidized one. The released energy can be captured to **phosphorylate,** or add an inorganic phosphate, to ADP or to some other compound. This process stores the energy in a high-energy molecule (ATP, for example). In many cases, the cell does not handle electrons as discrete entities but rather as parts of an atom such as hydrogen. For simplicity's sake, we will continue to use the term *electron transfer,* but keep in mind that hydrogens are involved in the transfer process. The removal of hydrogens (a hydrogen atom consists of a single proton and a single electron) from a compound during a redox reaction is called **dehydrogenation.** The job of handling these protons and electrons falls to one or more carriers, which function as short-term repositories for the electrons until they can be transferred (figure 8.13). As we shall see, dehydrogenations are an essential supplier of electrons for the respiratory electron transport system.

Electron Carriers: Molecular Shuttles

Electron carriers resemble shuttles that are alternately loaded and unloaded, repeatedly accepting and releasing electrons and hydrogens to facilitate the transfer of redox energy. Most carriers are coenzymes that transfer both electrons and hydrogens, but some transfer electrons only. The most common carrier is NAD (nicotinamide adenine dinucleotide), which carries hydrogens (and a pair of electrons) from dehydrogenation reactions (figure 8.14). Reduced NAD can be represented in various ways. Because 2 hydrogens are removed, the actual carrier state is $NADH + H^+$, but this is somewhat cumbersome, so we will represent it with the shorter NADH. In catabolic pathways, electrons are extracted and carried through a series of redox reactions until the **final electron acceptor** at the end of a particular pathway is reached (see figure 8.12). In aerobic metabolism, this acceptor is molecular oxygen; in anaerobic metabolism, it is some other inorganic or organic compound. Other common redox carriers are FAD, NADP (NAD phosphate), coenzyme A, and the compounds of the respiratory chain, which are fixed into membranes.

---

* **exergonic** (ex-er-gon′-ik) Gr. *exo,* without, and *ergon,* work. In general, catabolic reactions are exergonic.

* **endergonic** (en-der-gon′-ik) Gr. *endo,* within, and *ergon,* work. Anabolic reactions tend to be endergonic.

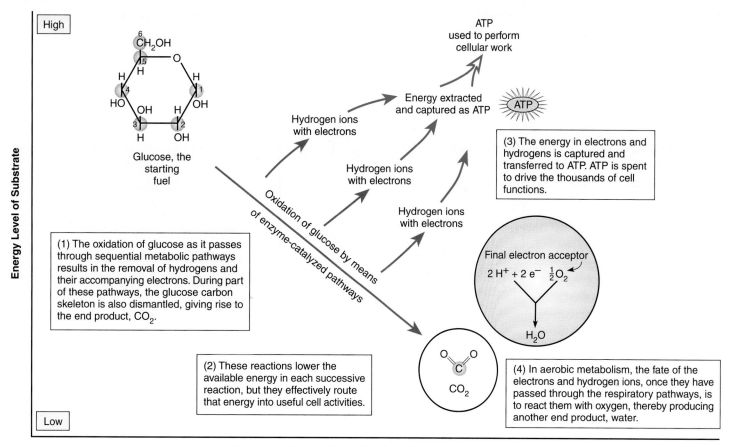

Progress of Energy Extraction Over Time

FIGURE 8.12

A simplified model that summarizes the cell's energy machine. The central events of cell energetics include the release of energy during the systematic dismantling of a fuel such as glucose. This is achieved by the shuttling of hydrogens and electrons to sites in the cell where their energy can be transferred to ATP.

ADENOSINE TRIPHOSPHATE: METABOLIC MONEY

In what ways do cells extract chemical energy from electrons, store it, and then tap the storage sources? To answer these questions, we must look more closely at the powerhouse molecule, adenosine triphosphate. ATP has also been described as metabolic money because it can be earned, banked, saved, spent, and exchanged. As a temporary energy repository, ATP provides a connection between energy-yielding catabolism and all other cellular activities that require energy. Some clues to its energy-storing properties lie in its unique molecular structure (figure 8.15).

The Molecular Structure of ATP

ATP is a three-part molecule consisting of a nitrogen base (adenine) linked to a 5-carbon sugar (ribose), with a chain of three phosphate groups bonded to the ribose (figure 8.15). The type, arrangement, and especially the proximity of atoms in ATP

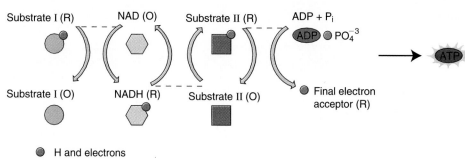

FIGURE 8.13

The role of electron carriers in redox reactions. Electrons and hydrogens are transferred by paired or coupled reactions in which the donor compound is oxidized (loses electrons) and the acceptor is reduced (gains electrons). A molecule such as NAD can serve as a transient carrier of the electrons between substrates. During this series of reactions, some energy given off by electron transfer is used to synthesize ATP. To complete the reaction, the electrons are passed to a final electron acceptor.

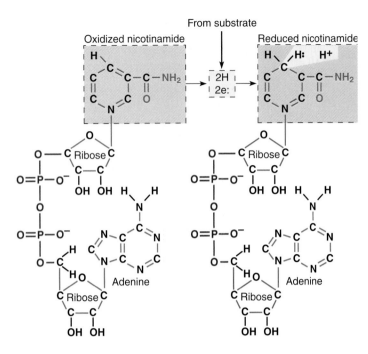

FIGURE 8.14

Details of NAD reduction. This coenzyme contains the vitamin nicotinamide (niacin) and the purine adenine attached to double ribose phosphate molecules (a dinucleotide). The principal site of action is on the nicotinamide (boxed area). Hydrogens and electrons donated by a substrate interact with a carbon on the top of the ring. One hydrogen bonds there, carrying two electrons (H:), and the other hydrogen is carried in solution as H+ (a proton).

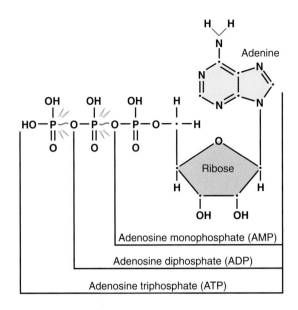

~ Bond that releases energy when broken

FIGURE 8.15

The structure of adenosine triphosphate (ATP) and its partner compounds, ADP and AMP.

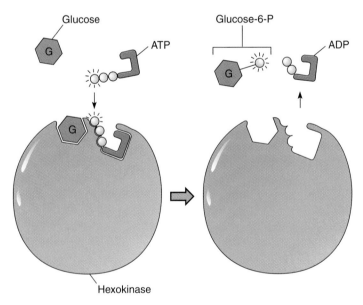

FIGURE 8.16

An example of phosphorylation of glucose by ATP. The first step in catabolizing glucose is the addition of a phosphate from ATP by an enzyme called hexokinase. This use of high-energy phosphate as an activator is a recurring feature of many metabolic pathways.

combine to form a compatible but unstable high-energy molecule. The high energy of ATP originates in the orientation of the phosphate groups, which are relatively bulky and carry negative charges. The proximity of these repelling electrostatic charges imposes a strain that is most acute on the bonds between the last two phosphate groups. The strain on the phosphate bonds accounts for the energetic quality of ATP, because removal of the terminal phosphates releases the bond energy.

Breaking the bonds between two successive phosphates yields adenosine diphosphate (ADP) and adenosine monophosphate (AMP). It is worthwhile noting how economically a cell manages its pool of adenosine nucleotides. AMP derivatives help form the backbone of RNA and are also a major component of certain coenzymes (NAD, FAD, and coenzyme A).

The Metabolic Role of ATP

Cellular reactions involving ATP are necessarily intertwined and balanced. ATP expenditure must inevitably be followed by ATP regeneration, and so on in a never-ending cycle in an active cell. In many instances, the energy released during ATP hydrolysis powers biosynthesis by activating individual subunits before they are enzymatically linked together. ATP is also used to prepare a molecule for catabolism such as the double phosphorylation of a 6-carbon sugar during the early stages of glycolysis (figure 8.16).

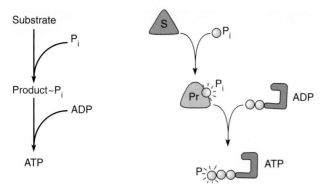

FIGURE 8.17

🔘 **ATP formation at the substrate level shown in outline and illustrated form.** The inorganic phosphate (P_i) and the substrates form a bond with high potential energy. In a reaction catalyzed by ATP synthase, the phosphate is transferred to ADP, thereby producing ATP.

The formation of ATP occurs through a reversal of the process by which it was spent. In heterotrophs, the energy infusion that regenerates a high-energy phosphate comes from certain steps of catabolic pathways, in which nutrients such as carbohydrates are degraded and yield energy. ATP is formed when substrates or electron carriers add a high-energy phosphate bond to ADP. Some ATP molecules are formed through *substrate-level phosphorylation;* that is, energy is released directly from a substrate to ADP (figure 8.17). Other ATPs are formed through *oxidative phosphorylation,* a series of redox reactions occurring during the final phase of the respiratory pathway. Phototrophic organisms have a system of *photophosphorylation,* in which the ATP is formed through a series of sunlight-driven reactions (discussed in more detail in chapter 26).

CHAPTER CHECKPOINTS

All metabolic processes require the constant input and expenditure of some form of usable energy. Chemical energy is the currency that runs the metabolic processes of the cell, but many forms of energy are involved in cell metabolism.

Chemical energy is obtained from the electrons of nutrient molecules through catabolism. It is used to perform the cellular "work" of biosynthesis, movement, membrane transport, and growth.

Energy is extracted from nutrient molecules by redox reactions. A redox pair of substances passes electrons and hydrogens between them. The donor substance loses electrons, becoming oxidized. The acceptor substance gains electrons, becoming reduced.

ATP is the energy molecule of the cell. It donates free energy to anabolic reactions and is continuously regenerated by three phosphorylation processes: substrate phosphorylation, oxidative phosphorylation, and (in certain organisms) photophosphorylation.

Pathways of Bioenergetics

One of the scientific community's greatest achievements was deciphering the biochemical pathways of cells. Initial work with bacteria and yeasts, followed by studies with animal and plant cells,

clearly demonstrated metabolic similarities and strongly supported the concept of the universality of metabolism. The study of the production and use of energy by cells is called **bioenergetics,** including catabolic routes that degrade nutrients and anabolic routes that are involved in cell synthesis. Although these pathways are interconnected and interdependent, anabolic pathways are not simply reversals of catabolic ones. At first glance it might seem more economical to use identical pathways, but having different enzymes and divided pathways allows anabolism and catabolism to proceed simultaneously without interference. For simplicity, we shall focus our discussion on the most common catabolic pathways that will illustrate general principles of other pathways as well.

CATABOLISM: AN OVERVIEW OF NUTRIENT BREAKDOWN AND ENERGY RELEASE

The primary catabolism of fuels that results in energy release in many organisms proceeds through a series of three coupled pathways: (1) **glycolysis,*** also called the Embden-Meyerhof-Parnas (EMP) pathway, (2) the **tricarboxylic acid cycle (TCA),** also known as the citric acid or Krebs cycle,[2] and (3) the **respiratory chain** (electron transport and oxidative phosphorylation). Each segment of the pathway is responsible for a specific set of actions on various products of glucose. The interconnections of these pathways in aerobic respiration are represented in figure 8.18, and their reactions are summarized in table 8.3. In the following sections, we will observe each of these pathways in greater detail.

ENERGY STRATEGIES IN MICROORGANISMS

Nutrient processing is extremely varied, especially in bacteria, yet in most cases it is based on three basic catabolic pathways. In previous discussions, microorganisms were categorized according to their requirement for oxygen gas, and this requirement is related directly to their mechanisms of energy release. As we shall see, **aerobic respiration** is a series of reactions (glycolysis, the TCA cycle, and the respiratory chain) that converts glucose to CO_2 and gives off energy (review figure 8.12). It relies on free oxygen as the final acceptor for electrons and hydrogens and produces a relatively large amount of ATP. Aerobic respiration is characteristic of many bacteria, fungi, protozoa, and animals, and it is the system we will emphasize here. Facultative and aerotolerant anaerobes may use only the glycolysis scheme to incompletely oxidize or **ferment** glucose. In this case, oxygen is not required, organic compounds are the final electron acceptors, and a relatively small amount of ATP is produced. Some strictly anaerobic microorganisms metabolize by means of **anaerobic respiration.** This system involves the same three pathways as aerobic respiration, but it does not use molecular oxygen as the final electron acceptor. Aspects of fermentation and anaerobic respiration are covered in subsequent sections of this chapter.

---

2. The EMP pathway is named for the biochemists who first outlined its steps. TCA refers to the involvement of several organic acids containing three carboxylic acid groups, citric acid being the first tricarboxylic acid formed; Krebs is in honor of Sir Hans Krebs who, with F. A. Lipmann, delineated this pathway, an achievement for which they won the Nobel Prize in 1953.

* glycolysis (gly-kol′-ih-sis) Gr. *glykys,* sweet, and *lysis,* a loosening.

TABLE 8.3

Metabolic Strategies Among Heterotrophic Microorganisms

Scheme	Pathways Involved	Final Electron Acceptor	Products	Chief Microbe Type
Aerobic respiration	Glycolysis, TCA cycle, electron transport	O_2	ATP, CO_2, H_2O	Aerobes; facultative anaerobes
Anaerobic metabolism				
Fermentative	Glycolysis	Organic molecules	ATP, CO_2, ethanol, lactic acid*	Facultative, aerotolerant, strict anaerobes
Respiration	Glycolysis, TCA cycle, electron transport	Various inorganic salts $(NO_3^-, SO_4^{-2}, CO_3^{-3})$	CO_2, ATP, organic acids, H_2S, CH_4, N_2	Anaerobes; some facultatives

*The products of microbial fermentations are extremely varied and include organic acids, alcohols, and gases.

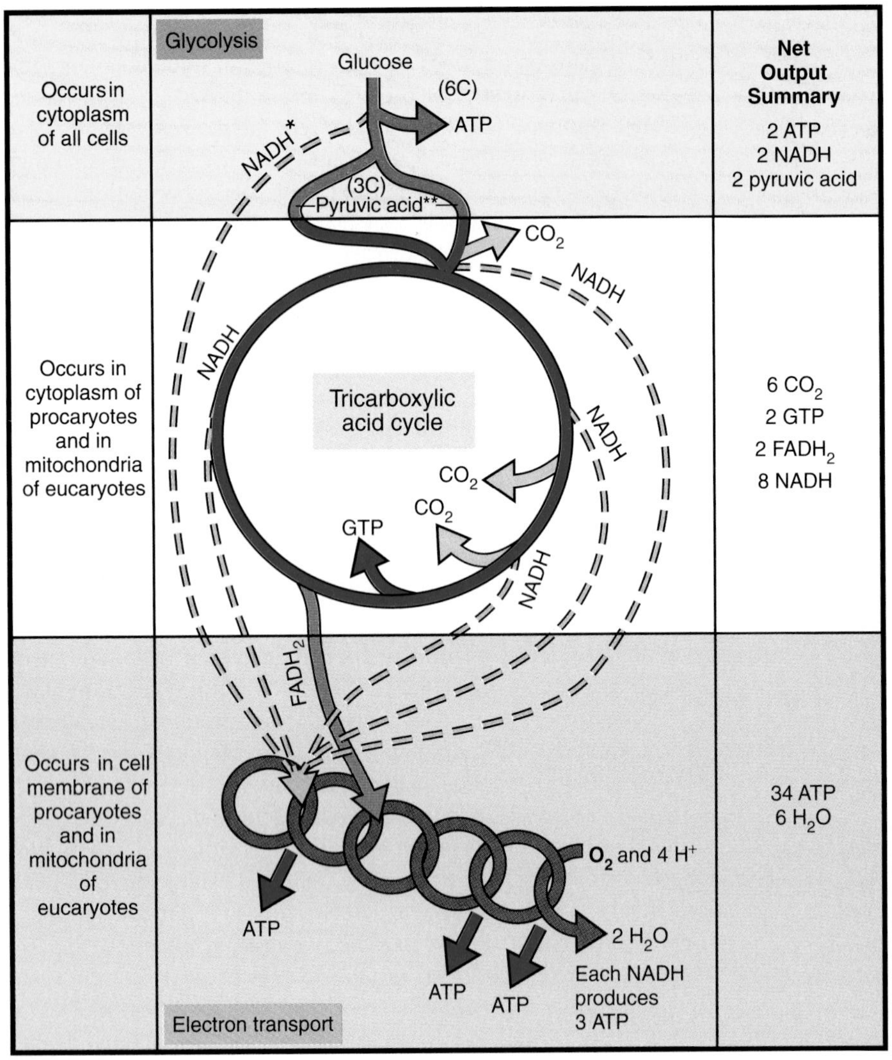

Net Output Summary

2 ATP
2 NADH
2 pyruvic acid

6 CO_2
2 GTP
2 $FADH_2$
8 NADH

34 ATP
6 H_2O

Glycolysis divides the glucose into two 3-carbon fragments called pyruvic acid and produces a small amount of ATP. It does not require oxygen.

The tricarboxylic acid (TCA) cycle receives these 3-carbon pyruvic acid fragments and processes them through redox reactions that extract the electrons and hydrogens. These are shuttled into electron transport to be used in ATP synthesis. CO_2 is an important product of the TCA cycle.

**All reactions in TCA must be multiplied by 2 for summary because each glucose generates 2 pyruvic acids.

Transport of electrons in the third phase generates a large amount of ATP; in the final step, the electrons and hydrogens are received by oxygen, which forms water. Both TCA and electron transport require oxygen to proceed.

FIGURE 8.18

Overview of the flow, location, and products of pathways in aerobic respiration. Glucose is degraded through a gradual stepwise process to carbon dioxide and water, while simultaneously extracting energy.

*Note that the NADH = = = transfers H^+ and e^- from the first 2 pathways to the 3rd.

Aerobic Respiration

Aerobic respiration is a series of enzyme-catalyzed reactions in which electrons are transferred from fuel molecules such as glucose to oxygen as a final electron acceptor. This pathway is the principal energy-yielding scheme for aerobic heterotrophs, and it provides both ATP and metabolic intermediates for many other pathways in the cell, including those of protein, lipid, and carbohydrate synthesis (figure 8.18).

Aerobic respiration in microorganisms can be summarized by an equation:

$$\text{Glucose } (C_6H_{12}O_6) + 6\,O_2 + 38\,\text{ADP} + 38\,P_i \rightarrow$$

$$6\,CO_2 + 6\,H_2O + 38\,\text{ATP}$$

The seeming simplicity of the equation for aerobic respiration conceals its complexity. Fortunately, we do not have to present all of the details to address some important concepts concerning its reactants and products. These ideas are (1) the steps in the oxidation of glucose, (2) the involvement of coenzyme carriers and the final electron acceptor, (3) where and how ATP originates, (4) where carbon dioxide originates, (5) where oxygen is required, and (6) where water originates.

Glucose: The Starting Compound   Carbohydrates such as glucose are good fuels because these compounds are readily oxidized; that is, they are superior hydrogen and electron donors. The enzymatic withdrawal of hydrogen from them also removes electrons that can be used in energy transfers. The end products of the conversion of these carbon compounds are energy-rich ATP and energy-poor carbon dioxide and water. Polysaccharides (starch, glycogen) and disaccharides (maltose, lactose) are stored sources of glucose for the respiratory pathways. Although we use glucose as the main starting compound, other hexoses (fructose, galactose) and fatty acid subunits can enter the pathways of aerobic respiration as well, as we will see in a later section of this chapter.

Glycolysis: The Starting Lineup
An anaerobic process called **glycolysis** enzymatically converts glucose through several steps into **pyruvic acid.** Depending on the organism and the conditions, it may be only the first phase of aerobic respiration, or it may serve as the primary metabolic pathway (fermentation). Glycolysis provides a significant means to synthesize a small amount of ATP anaerobically and also to generate pyruvic acid, an essential intermediary metabolite.

Steps in the Glycolytic Pathway   Glycolysis, or the EMP pathway, proceeds along nine linear steps, starting with glucose and ending with pyruvic acid (figure 8.19). The first portion of the EMP pathway involves activation of the substrate, and the steps following involve oxidation reactions of the glucose fragments, the synthesis of ATP, and the formation of pyruvic acid. *Although each step of metabolism is catalyzed by a specific enzyme, we will not mention it for most reactions.* The following outline lists the principal steps of glycolysis.

1. **Glucose** is phosphorylated by means of an **ATP** acting with the catalyst hexokinase. The product is **glucose-6-phosphate.** (Numbers in chemical names refer to the position of the phosphate on the carbon skeleton.) This is a way of "priming" the system and keeping the glucose inside the cell.
2. Glucose-6-phosphate is converted to its isomer, **fructose-6-phosphate,** by phosphoglucoisomerase.
3. Another **ATP** is spent in phosphorylating the first carbon of fructose-6-phosphate, which yields **fructose-1,6-bisphosphate.**

Up to this point, no energy has been released, no oxidation-reduction has occurred, and, in fact, 2 ATPs have been used. In addition, the molecules remain in the 6-carbon state.

4. Now doubly activated, fructose-1,6-bisphosphate is split into two 3-carbon fragments: **glyceraldehyde-3-phosphate**

(G-3-P) and **dihydroxyacetone phosphate (DHAP).** These molecules are isomers, and DHAP is enzymatically converted to G-3-P, which is the more reactive form for subsequent reactions.

The effect of the splitting of fructose bisphosphate is to double every subsequent reaction, because where there was once a single molecule, there are now two to be fed into the remaining pathways.

5. Each molecule of glyceraldehyde-3-phosphate becomes involved in the single oxidation-reduction reaction of glycolysis, a reaction that sets the scene for ATP synthesis. Two reactions occur simultaneously and are catalyzed by the same enzyme, called glyceraldehyde-3-phosphate dehydrogenase. First, the coenzyme **NAD** picks up hydrogen from G-3-P, forming **NADH.** This step is followed by the addition of an inorganic phosphate (PO_4^{-3}) to form a high-energy bond on the third carbon of the G-3-P substrate. The product of these reactions is **bisphosphoglyceric acid.**

In aerobic organisms, the NADH formed during this step will undergo further reactions in the electron transport system, where the final H acceptor will be oxygen and additional ATPs will be generated (see figure 8.18). In organisms that ferment glucose anaerobically, the NADH will be oxidized back to NAD, and the hydrogen acceptor will be an organic compound.

6. One of the high-energy phosphates of bisphosphoglyceric acid is donated to **ADP,** resulting in a molecule of **ATP.** The product of this reaction is **3-phosphoglyceric acid.**
7,8. During this phase, a substrate for the synthesis of a second ATP is produced in two substeps. First, 3-phosphoglyceric acid is converted to **2-phosphoglyceric acid** through the shift of a phosphate from the third to the second carbon. Then, the removal of a water molecule from 2-phosphoglyceric acid produces **phosphoenolpyruvic acid** and generates another **high-energy phosphate.**
9. In the final reaction of glycolysis, phosphoenolpyruvic acid gives up its high-energy phosphate to form a second **ATP.** This reaction, catalyzed by pyruvate kinase, also produces **pyruvic acid** (pyruvate), a compound with many roles in metabolism.

The 2 ATPs formed during steps 6 and 9 are examples of substrate-level phosphorylation, in that the high-energy phosphate is transferred directly from a substrate to ADP. These reactions are catalyzed by kinases, special types of transferase that can phosphorylate a substrate. Because both molecules that arose in step 4 undergo these reactions, an overall total of 4 ATPs is generated in the partial oxidation of a glucose to 2 pyruvic acids. However, 2 ATPs were expended for steps 1 and 3, so the net number of ATPs available to the cell from these reactions is 2.

PYRUVIC ACID—A CENTRAL METABOLITE

Pyruvic acid occupies an important position in several pathways, and different organisms handle it in different ways (figure 8.20). In strictly aerobic organisms and some anaerobes, pyruvic acid enters the TCA cycle for further processing and energy release. Facultative anaerobes can adopt a fermentative metabolism, in which pyruvic acid is further reduced into acids or other products.

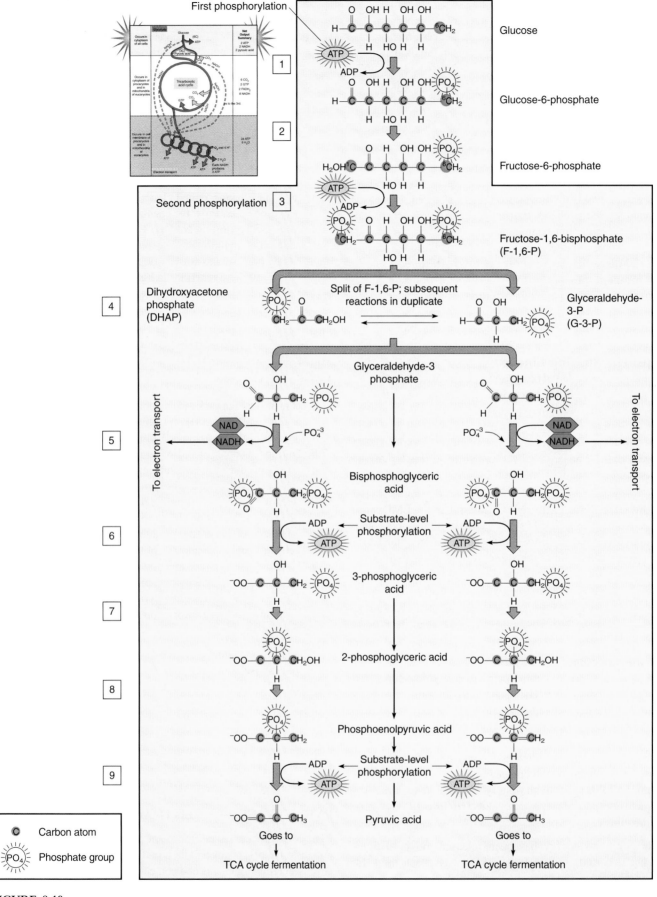

FIGURE 8.19

The reactions of the glycolysis system (Embden-Meyerhof-Parnas pathway). Note that this is an enlarged, more detailed view of the top phase depicted in figure 8.18.

The NADH formed during this reaction will be shuttled into electron transport and used to generate ATP; its formation is one of five dehydrogenations associated with the TCA cycle and accounts for the greater output of ATP in aerobic respiration. Keep in mind that all reactions described actually happen twice for each glucose because of the two pyruvates that are released during glycolysis. An important feature of this pathway is that acetyl groups from the breakdown of certain fats can enter the pathway at this same point (see figure 8.26).

The acetyl groups are subsequently fed into the tricarboxylic acid cycle and combined with a 4-carbon oxaloacetate molecule to form citric acid, a 6-carbon molecule. Although this seems a cumbersome way to dismantle such a small molecule, it is necessary in biological systems for extracting larger amounts of energy from the remaining fragment of acetyl.

Steps in the TCA Cycle

As we have seen, a cyclic pathway is one in which the starting compound is regenerated at the end. The tricarboxylic acid cycle has eight steps, beginning with citric acid formation and ending with the organic acid **oxaloacetic acid**[3] (figure 8.21). As we take a single spin around the TCA cycle, it will be instructive to keep track of (1) the numbers of carbons of each substrate and product (#C), (2) reactions where CO_2 is generated, (3) the involvement of the electron carriers NAD and FAD, and (4) the site of ATP synthesis. The reactions in the TCA cycle are:

1. **Oxaloacetic acid** (oxaloacetate; 4C) reacts with the **acetyl group** (2C) on acetyl CoA, thereby forming **citric acid** (citrate; 6C) and releasing coenzyme A so it can join with another acetyl group.
2. Citric acid is converted to its isomer, **isocitric acid** (isocitrate; 6C), to prepare this substrate for the decarboxylation and dehydrogenation of the next step.
3. Isocritic acid is acted upon by an enzyme complex including NAD or NADP (depending on the organism), in a reaction that generates NADH or NADPH, splits off a carbon dioxide, and leaves **α-ketoglutaric acid** (α-ketoglutarate; 5C).
4. Alpha-ketoglutaric acid serves as a substrate for the last decarboxylation reaction and yet another redox reaction involving coenzyme A and yielding NADH. The product is the high-energy compound **succinyl CoA** (4C).

At this point, the cycle has completed the formation of 3 CO_2s that balance out the original 3-carbon pyruvic acid that began the TCA. The remaining steps are needed not only to regenerate the oxaloacetic acid to start the cycle again but also to extract more energy from the intermediate compounds leading to oxaloacetic acid.

5. Succinyl CoA is the source of the one substrate level phosphorylation in the TCA cycle. In most microbes, it proceeds with the formation of ATP. The product of this reaction is **succinic acid** (succinate; 4C).

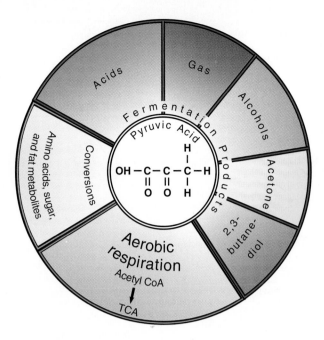

FIGURE 8.20

The fates of pyruvic acid (pyruvate). This metabolite is an important hub in the processing of nutrients by microbes. It may be fermented anaerobically to several end products or oxidized completely to CO_2 and H_2O through the TCA cycle and the electron transport system. It can also serve as a source of raw material for synthesizing amino acids and carbohydrates.

THE TRICARBOXYLIC ACID CYCLE—A CARBON AND ENERGY WHEEL

As we have seen, the anaerobic oxidation of glucose yields a comparatively small amount of energy and gives off pyruvic acid. Pyruvic acid is still energy-rich, containing a number of extractable hydrogens and electrons to power ATP synthesis, but this can be achieved only through the work of the second and third phases of respiration, in which pyruvic acid is converted to CO_2 and H_2O (see figure 8.18). In the following section, we will examine the initial phase of this process, that takes place in the mitochondrial matrix in eucaryotes and in the cytoplasm of bacteria.

The Processing of Pyruvic Acid

To connect the glycolysis pathway to the **tricarboxylic acid** cycle, the pyruvic acid is first converted to a starting compound for that cycle (figure 8.21). This step involves the first oxidation-reduction reaction of this phase of respiration, and it also releases the first carbon dioxide molecule. It involves a cluster of enzymes and **coenzyme A** that participate in the dehydrogenation (oxidation) of pyruvic acid, the reduction of NAD to NADH, and the decarboxylation of pyruvic acid to a 2-carbon **acetyl** group. The acetyl group remains attached to coenzyme A, forming **acetyl coenzyme A** (acetyl CoA) that feeds into the TCA cycle. Pyruvate dehydrogenase, the enzyme complex that makes this reaction possible, is huge. In *E. coli,* it contains four vitamins and has a molecular weight of 6 million—larger than a ribosome!

---

3. In biochemistry, the terms used for organic acids appear as either the acid form (oxaloacetic acid) or its salt (oxaloacetate).

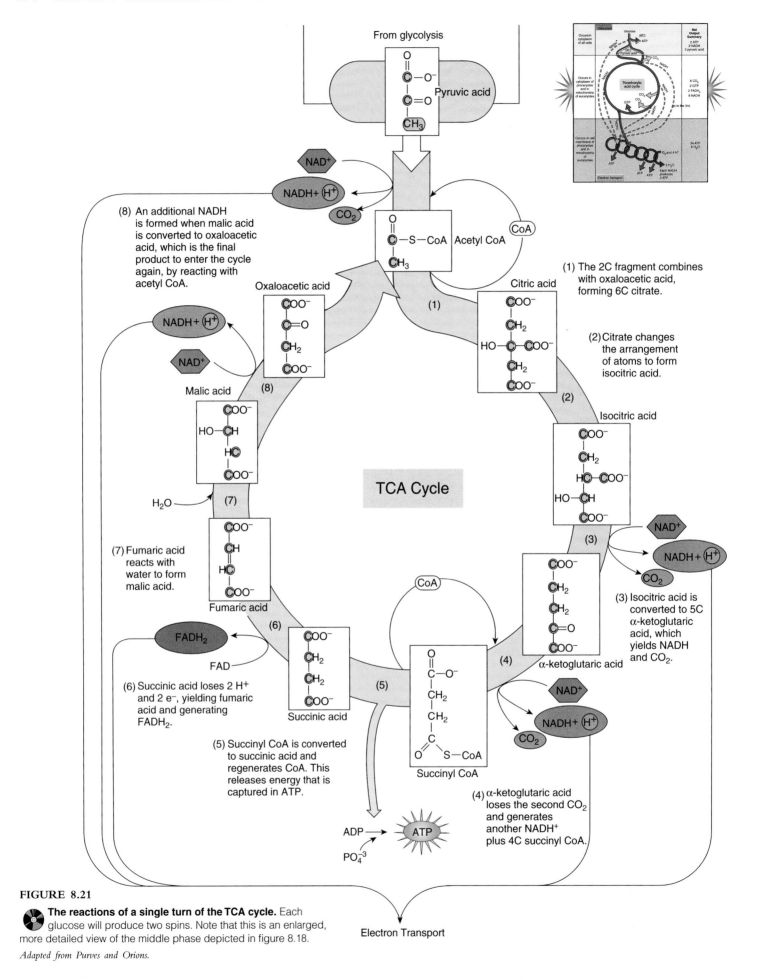

FIGURE 8.21

The reactions of a single turn of the TCA cycle. Each glucose will produce two spins. Note that this is an enlarged, more detailed view of the middle phase depicted in figure 8.18.

Adapted from Purves and Orions.

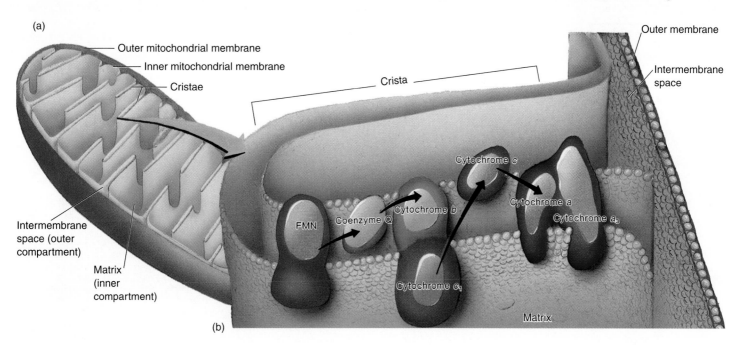

FIGURE 8.22

 The action sites of the mitochondrion. (a) Location of the electron transport scheme with respect to cristae membranes and matrix. **(b)** Exploded view of electron transport chain, showing orientation of electron carriers along the crista membrane.

6. Succinic acid next becomes dehydrogenated, but in this case, the electron and H^+ acceptor is **flavin adenine dinucleotide (FAD)**. The enzyme that catalyzes this reaction, succinyl dehydrogenase, is found in the bacterial cell membrane and mitochondrial crista of eucaryotic cells. As a consequence, the $FADH_2$ directly enters the electron transport system and bypasses the first step shown in figure 8.23. **Fumaric acid** (fumarate; 4C) is the product of this reaction.

7. The addition of water to fumaric acid (called hydration) results in **malic acid** (malate; 4C). This is one of the few reactions in respiration that directly incorporates water.

8. Malic acid is dehydrogenated (with formation of a final NADH), and **oxaloacetic acid** is formed. This step brings the cycle back to its original starting position, where oxaloacetic acid can react with acetyl coenzyme A.

THE RESPIRATORY CHAIN: ELECTRON TRANSPORT AND OXIDATIVE PHOSPHORYLATION

We now come to the energy chain, which is the final "processing mill" for electrons and hydrogen and the major generator of ATP. Overall, the electron transport system (ETS) consists of a chain of special redox carriers that receive electrons from reduced carriers (NADH, $FADH_2$) generated by glycolysis and the TCA cycle and shuttle them in a sequential and orderly fashion (see figure 8.18). The flow of electrons down this chain is highly energetic and gives off ATP at various points. The step that finalizes the transport process is the acceptance of electrons and hydrogen by oxygen, producing water. Some variability exists from one organism to another, but the principal compounds that carry out these complex reactions are NADH dehydrogenase, flavopro-

teins, coenzyme Q *(ubiquinone),** and *cytochromes.** The cytochromes contain a tightly bound metal atom at their center that is actively involved in accepting electrons and donating them to the next carrier in the series. The highly compartmentalized structure of the respiratory chain is an important factor in its function. Note in figure 8.22 that the electron transport carriers and enzymes are embedded in the inner mitochondrial membranes in eucaryotes. The equivalent structure for housing them in bacteria is the cell membrane.

Elements of Electron Transport: The Energy Cascade

The principal questions about the electron transport system are: How are the electrons passed from one carrier to another in the series? How is this progression coupled to ATP synthesis? and, Where and how is oxygen utilized? Although the biochemical details of this process are rather complicated, the basic reactions consist of a number of redox reactions now familiar to us. In general, the seven carrier compounds and their enzymes are arranged in linear sequence and are reduced and oxidized in turn.

The sequence of electron carriers in the respiratory chain of most aerobic organisms is (1) NADH dehydrogenase, which is closely associated in a complex with the adjacent carrier, which is (2) flavin mononucleotide (FMN); (3) coenzyme Q; (4) cytochrome *b;* (5) cytochrome c_1; (6) cytochrome *c;* and

---

* **ubiquinone** (yoo-bik′-wih-nohn) L. *ubique,* everywhere. A type of chemical, similar to vitamin K, that is very common in cells.

* **cytochrome** (sy′-toh-krohm) Gr. *cyto,* cell, and *kroma,* color. Cytochromes are pigmented, iron-containing molecules similar to hemoglobin.

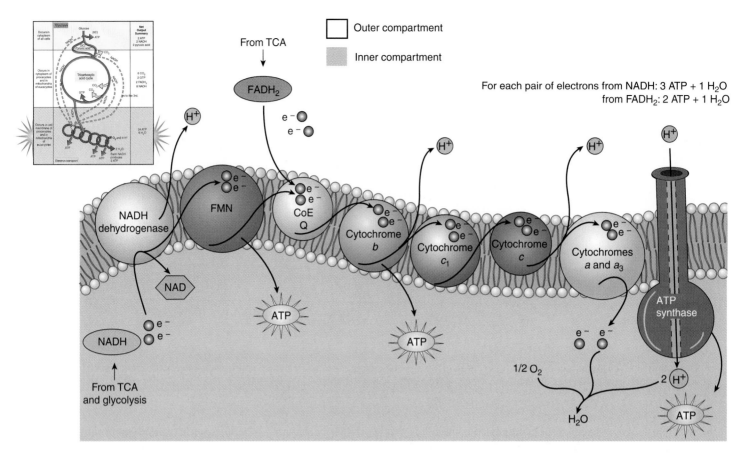

FIGURE 8.23

The electron transport system and oxidative phosphorylation. Starting at NADH dehydrogenase, electrons brought in from the TCA cycle by NADH are passed along the chain of electron transport carriers. Each adjacent pair of transport molecules undergoes a redox reaction. At three sites along this chain, this electron transport is coupled to H^+ movement across the membrane. These actions drive ATP synthesis as shown in figure 8.24. Note that this is an enlarged, more detailed view of the bottom phase depicted in figure 8.18.

(7) cytochromes a and a_3, which are complexed together. Conveyance of the NADHs from glycolysis and the TCA cycle to the first carrier sets in motion the remaining six steps. With each redox exchange, the energy level of the reactants is lessened. The released energy is captured and used by the **ATP synthase** complex, stationed along the cristae in close association with the ETS carriers. Each NADH that enters electron transport gives rise to 3 ATPs (figure 8.23). This coupling of ATP synthesis to electron transport is termed **oxidative phosphorylation.** Since $FADH_2$ from the TCA cycle enters the cycle after the NAD and FMN complex reactions, it has less energy to release, and 2 ATPs result from its processing.

The Formation of ATP and Chemiosmosis

What biochemical processes are involved in coupling electron transport to the production of ATP? We will first look at the system in eucaryotes, which have the components of electron transport embedded in a precise sequence on mitochondrial membranes (figure 8.22). They are stationed between the inner mitochondrial matrix and the outer intermembrane space (figure 8.23). According to a widely accepted concept called the **chemiosmotic hypothesis,** as the electron transport carriers shuttle electrons, they actively pump hydrogen ions (protons) into the outer compartment of the mitochondrion. This process sets up a

concentration gradient of hydrogen ions called the *proton motive force (PMF)*. The PMF generates a difference in charge between the outer membrane compartment ($+$) and the inner membrane compartment ($-$) (figure 8.24a).

Separating the charge has the effect of a battery, which can temporarily store potential energy. This charge will be maintained by the impermeability of the inner cristae membranes to H^+. The only site where H^+ can diffuse into the inner compartment is at the ATP synthase complex, which sets the stage for the final processing of H^+ leading to ATP synthesis.

ATP synthase is a complex enzyme composed of two large units, F_0 and F_1. It is embedded in the membrane but can rotate like a motor and trap chemical energy. As the H^+ ions flow through the F_0 center of the enzyme by diffusion, the F_1 compartments pull in ADP and P_i. Rotation causes a three-dimensional change in the enzyme that bonds these two molecules, thereby releasing ATP into the inner compartment (figure 8.24c). The enzyme is then rotated back to the start position and will continue the process.

Bacterial ATP synthesis occurs by means of this same overall process. However, bacteria have the ETS stationed in the cell membrane, and the direction of the proton movement is from the cytoplasm to the periplasmic space. This difference will affect the amount of ATP produced (discussed in the next section). In both cell types, the chemiosmotic theory has been supported by tests

Intermembrane space
(outer compartment)

Matrix (inner
compartment)

Crista

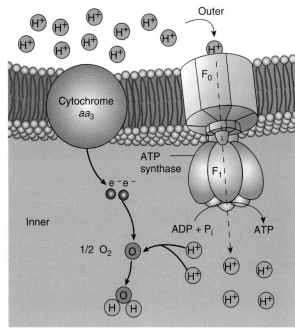

(a) As the carriers in the mitochondrial cristae transport electrons, they also actively pump H^+ ions (protons) to the intermembrane space, producing a chemical and charge gradient between the outer and inner mitochondrial compartments.

(c) The distribution of electric potential across the membrane drives the synthesis of ATP by ATP synthase. The rotation of this enzyme couples diffusion of H^+ to the inner compartment with the bonding of ADP and P_i. The final event of electron transport is the reaction of the electrons with the accumulated H^+ and O_2 to form metabolic H_2O. This step is catalyzed by cytochrome oxidase (cytochrome aa_3)

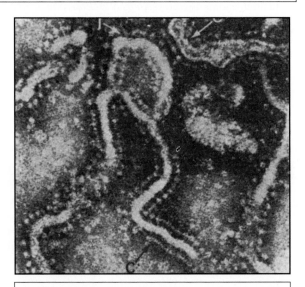

(b) Photomicrograph of isolated cristae captures the ATP synthase complex as hundreds of tiny "lollipops" (arrows) projecting from its surface.

FIGURE 8.24

Chemiosmosis—the force behind ATP synthesis.

(b) Garret and Grisham, Biochemistry *(Saunders, 1995), p. 647.*

showing that oxidative phosphorylation is blocked if the mitochondrial or bacterial cell membranes are disrupted.

Potential Yield of ATPs from Oxidative Phosphorylation

The total of five NADHs (four from the TCA cycle and one from glycolysis) can be used to synthesize:

15 ATPs for ETS ($5 \times 2 \times 3$ per electron pair)

and

$$15 \times 2 = 30 \text{ ATPs per glucose}$$

The single FADH produced during the TCA cycle results in:

2 ATPs per electron pair

and

$$2 \times 2 = 4 \text{ ATPs per glucose}$$

Table 8.4 summarizes the total of ATP and other products for the entire aerobic pathway. These totals are the potential yields possible but may not be fulfilled by many organisms.

SUMMARY OF AEROBIC RESPIRATION

Originally, we presented a summary equation for respiration. We are now in a position to tabulate the input and output of this equation at various points in the pathways and sum up the final

TABLE 8.4

Summary of Aerobic Respiration for One Glucose Molecule

	Glycolysis*	Net Output	TCA Cycle*	Net Output	Respiratory Chain	Net Output	Total Net Output per Glucose
ATP produced	$2 \times 2 =$	4	$1 \times 2 =$	2	$17 \times 2 =$	34	$40 - 2$ (used) $= 38$**
ATP used	2		0		0		
NADH produced	$1 \times 2 =$	2	$4 \times 2 =$	8	0		10
FADH produced	0		$1 \times 2 =$	2	0		2
CO_2 produced	0		$3 \times 2 =$	6	0		6
O_2 used	0		0		$3 \times 2 =$	6	
H_2O produced	2		0		$3 \times 2 =$	6	$8 - 2$ (used) $= 6$
H_2O used	0		2		0		

*Products are multiplied by 2 because the first figure represents the amount for only one trip through the pathway, and two molecules make this trip for each glucose.
**This amount can vary among microbes.

ATP. Close examination of table 8.4 will review several important facets of aerobic respiration:

1. The total yield of ATP is 40: 4 from glycolysis, 2 from the TCA cycle, and 34 from electron transport. However, since 2 ATPs were expended in early glycolysis, this leaves a maximum of **38 ATPs.**

The actual totals may be lower in certain eucaryotic cells because energy is expended in transporting the NADH across the mitochondrial membrane. Certain aerobic bacteria come closest to achieving the full total of 38 because they lack mitochondria and thus do not have to use ATP in transport of NADH across the outer mitochondrial membrane.

2. Six carbon dioxide molecules are generated during the TCA cycle.
3. Six oxygen molecules are consumed during electron transport.
4. Six water molecules are produced in electron transport and 2 in glycolysis, but because 2 are used in the TCA cycle, this leaves a net number of 6.

The Terminal Step

The terminal step, during which oxygen accepts the electrons, is catalyzed by cytochrome aa_3, also called cytochrome oxidase. This large enzyme complex is specifically adapted to receive electrons from cytochrome c, pick up hydrogens from solution, and react with oxygen to form a molecule of water (figure 8.24c). This reaction, though in actuality more complex, is summarized as follows:

$$2 H^+ + 2 e^- + \frac{1}{2} O_2 \rightarrow H_2O$$

Most eucaryotic aerobes have a fully functioning cytochrome system, but bacteria exhibit wide-ranging variations in this part of the system. Some species lack one or more of the redox steps; others have several alternative electron transport schemes. Because many bacteria lack cytochrome c oxidase, this variation can be used to differentiate among certain genera of bacteria. The oxidase test (see figure 18.29) can detect this enzyme. It is used to help identify members of the genera *Neisseria* and *Pseudomonas* and some

species of *Bacillus*. Another variation in the cytochrome system is evident in certain bacteria (*Klebsiella, Enterobacter*) that can grow even in the presence of cyanide because they lack cytochrome oxidase. Cyanide will cause rapid death in humans and other eucaryotes because it blocks cytochrome oxidase, thereby completely terminating aerobic respiration, but it is harmless to these bacteria.

A potential side reaction of the respiratory chain in aerobic organisms is the incomplete reduction of oxygen to superoxide ion (O_2-) and hydrogen peroxide (H_2O_2). Because these toxic oxygen products (previously discussed in chapter 7) can be very damaging to cells, aerobes have various neutralizing enzymes (*superoxide dismutase* and *catalase*). One exception is the genus *Streptococcus*, which can grow well in oxygen yet lacks both cytochromes and catalase. The tolerance of these organisms to oxygen can be explained by the neutralizing effects of a special peroxidase. The lack of cytochromes, catalase, and peroxidases in anaerobes as a rule limits their ability to process free oxygen and contributes to its toxic effects on them.

Alternate Catabolic Pathways

Certain bacteria follow a different pathway in carbohydrate catabolism. The **phosphogluconate pathway** (also called the hexose monophosphate shunt) provides ways to anaerobically oxidize glucose and other hexoses, to release ATP, to produce large amounts of NADPH, and to process pentoses (5-carbon sugars). This pathway, common in heterolactic fermentative bacteria, yields various end products, including lactic acid, ethanol, and carbon dioxide. Furthermore, it is a significant intermediate source of pentoses for nucleic acid synthesis.

ANAEROBIC RESPIRATION

Some bacteria have evolved an anaerobic respiratory system that functions like the aerobic cytochrome system except that it utilizes oxygen-containing salts, rather than free oxygen, as the final electron acceptor. Of these, the nitrate (NO_3^-) and nitrite (NO_2^-) reduction systems are best known. The reaction in species such as *Escherichia coli* is represented as:

Nitrate reductase

$$\downarrow$$

$$NO_3^- + NADH \rightarrow NO_2^- + H_2O + NAD^+$$

The enzyme nitrate reductase catalyzes the removal of oxygen from nitrate, leaving nitrite and water as products. A test for this reaction is one of the physiological tests used in identifying bacteria.

Some species of *Pseudomonas* and *Bacillus* possess enzymes that can further reduce nitrite to nitric oxide (NO), nitrous oxide (N₂O), and even nitrogen gas (N₂). This process, called *denitrification,* is a very important step in recycling nitrogen in the biosphere. Other oxygen-containing nutrients reduced anaerobically by various bacteria are carbonates and sulfates. None of the anaerobic pathways produce as much ATP as aerobic respiration.

CHAPTER CHECKPOINTS

Bioenergetics describes metabolism in terms of production, utilization, and transfer of energy by cells.

Catabolic pathways release energy through three pathways: glycolysis, the tricarboxylic acid cycle, and the respiratory electron transport system.

Cellular respiration is described by the nature of the final electron acceptor. Aerobic respiration implies that O_2 is the final hydrogen acceptor. Anaerobic respiration implies that some other molecule is the final hydrogen acceptor. If the final hydrogen acceptor is an organic molecule, the anaerobic process is considered fermentation.

Carbohydrates are preferred cell energy sources because they are superior hydrogen (electron) donors.

Glycolysis is the catabolic processes by which glucose is oxidized and converted into two molecules of pyruvic acid, with a net gain of 2 ATP. This process is also called substrate phosphorylation.

The tricarboxylic acid cycle processes the 3-carbon pyruvic acid and generates three CO_2 molecules. The electrons it releases are transferred to redox carriers for energy harvesting.

The electron transport chain generates free energy through sequential redox reactions collectively called oxidative phosphorylation. This energy is used to generate up to 36 ATP for each glucose molecule catabolized.

THE IMPORTANCE OF FERMENTATION

Of all the results of pyruvate metabolism, probably the most varied is fermentation. Technically speaking, **fermentation*** is the incomplete oxidation of glucose or other carbohydrates in the absence of oxygen, a process that uses organic compounds as the terminal electron acceptors and yields a small amount of ATP. Over time, the term *fermentation* has acquired several looser connotations. Originally, Pasteur called the microbial action of yeast during wine production *ferments* (Historical Highlights 8.4), and to this day, biochemists use the term in reference to the production of ethyl alcohol by yeasts acting on glucose and other carbohydrates. Fermentation is also what bacteriologists call the formation

of acid, gas, and other products by the action of various bacteria on pyruvic acid. The process is a common metabolic strategy among bacteria. Industrial processes that produce chemicals on a massive scale through the actions of microbes are also called fermentations (see chapter 26). Each of these usages is acceptable for one application or another.

It may seem that fermentation would yield only meager amounts of energy (2 ATPs maximum per glucose). What actually happens, however, is that many bacteria can grow as fast as they would in the presence of oxygen. This rapid growth is made possible by an increase in the rate of glycolysis. From another standpoint, fermentation permits independence from molecular oxygen and allows colonization of anaerobic environments. It also enables microorganisms with a versatile metabolism to adapt to variations in the availability of oxygen. For them, fermentation provides a means to grow even when oxygen levels are too low for aerobic respiration.

Bacteria that digest cellulose in the rumens of cattle (as described in chapter 7) are largely fermentative. After initially hydrolyzing cellulose to glucose, they ferment the glucose to organic acids, which are then absorbed as the bovine's principal energy source. Even human muscle cells can undergo a form of fermentation that permits short periods of activity after the oxygen supply in the muscle has been exhausted. Muscle cells convert pyruvic acid into lactic acid, which allows anaerobic production of ATP to proceed for a time. But this cannot go on indefinitely, and after a few minutes, the accumulated lactic acid causes muscle fatigue.

Products of Fermentation

Alcoholic beverages (wine, beer, whiskey) are perhaps the most prominent among fermentation products; others are solvents (acetone, butanol), organic acids (lactic, acetic), dairy products, and many other foods. Derivatives of proteins, nucleic acids, and other organic compounds are fermented to produce vitamins, antibiotics, and even hormones such as hydrocortisone.

Fermentation products can be grouped into two general categories: alcoholic fermentation products and acidic fermentation products (figure 8.25). **Alcoholic fermentation** occurs in yeast species that have metabolic pathways for converting pyruvic acid to ethanol. This process involves a decarboxylation of pyruvic acid to acetaldehyde, followed by a reduction of the acetaldehyde to ethanol. In oxidizing the NADH formed during glycolysis, this step regenerates NAD, thereby allowing the glycolytic pathway to continue. These processes are crucial in the production of beer and wine, though the actual techniques for arriving at the desired amount of ethanol and the prevention of unwanted side reactions are important tricks of the brewer's trade (see Historical Highlights 8.4). Note that the products of alcoholic fermentation are not only ethanol but also CO_2, a gas that accounts for the bubbles in champagne and beer.

Alcohols other than ethanol can be produced during bacterial fermentation pathways. Certain clostridia produce butanol and isopropanol through a complex series of reactions. Although this process was once an important source of alcohols for industrial use, it has been largely replaced by a nonmicrobial petroleum process.

---

* fermentation (fur-men-tay′-shun) L. *fervere,* to boil, or *fermentatum,* leaven or yeast.

HISTORICAL HIGHLIGHTS 8.4
Pasteur and the Wine-to-Vinegar Connection

The microbiology of alcoholic fermentation was greatly clarified by Louis Pasteur after French wine makers hired him to uncover the causes of periodic spoilage in wines. Especially troublesome was the conversion of wine to vinegar and the resultant sour flavor. Up to that time, wine formation had been considered strictly a chemical process. After extensively studying beer making and wine grapes, Pasteur concluded that wine, both fine and not-so-fine, was the result of microbial action on the juices of the grape and that wine "disease" was caused by contaminating organisms that produced undesirable products such as acid. Although he did not know it at the time, the bacterial contaminants responsible for the acidity of the spoiled wines were likely to be *Acetobacter* or *Gluconobacter* introduced by the grapes, air, or winemaking apparatus. These common gram-negative genera further oxidized ethanol to acetic acid and are presently used in commercial vine-

gar production. The following formula shows how this is accomplished:

$$H-\underset{\underset{H}{|}}{\overset{\overset{H}{|}}{C}}-\underset{\underset{H}{|}}{\overset{\overset{H}{|}}{C}}-OH \longrightarrow H-\underset{\underset{H}{|}}{\overset{\overset{H}{|}}{C}}-\overset{O}{\underset{OH}{C}} + 2H^+$$

Ethanol   Acetic acid

Pasteur's far-reaching solution to the problem is still with us today—mild heating, or *pasteurization,* of the grape juice to destroy the contaminants, followed by inoculation of the juice with a pure yeast culture. It also introduced the concept that a single microbe could be responsible for both a desired product and a disease. The topic of wine making is explored further in chapter 26.

The pathways of **acidic fermentation** are extremely varied (see figure 8.25). Lactic acid bacteria ferment pyruvate in the same way that humans do—by reducing it to lactic acid. If the product of this fermentation is mainly lactic acid, as in certain species of *Streptococcus* and *Lactobacillus,* it is termed *homolactic.* The souring of milk is due largely to the production of this acid by bacteria. When glucose is fermented to a mixture of lactic acid, acetic acid, and carbon dioxide, as is the case with *Leuconostoc* and other species of *Lactobacillus,* the process is termed *heterolactic fermentation.*

Many members of the family Enterobacteriaceae (*Escherichia, Shigella,* and *Salmonella*) possess enzyme systems for converting pyruvic acid to several acids simultaneously (figure 8.26). **Mixed acid fermentation** produces a combination of acetic, lactic, succinic, and formic acids, and it lowers the pH of a medium to about 4.0. *Propionibacterium* produces primarily propionic acid, which gives the characteristic flavor to Swiss cheese while fermentation gas (CO_2) produces the holes. Some members also further decompose formic acid completely to carbon dioxide and hydrogen gases. Because enteric bacteria commonly occupy the intestine, this fermentative activity accounts for the accumulation of some types of gas—primarily CO_2 and H_2—in the intestine. Some bacteria reduce the organic acids and produce the neutral end product 2,3-butanediol (Microbits 8.5).

We have provided only a brief survey of fermentation products, but it is worth noting that microbes can be harnessed to synthesize a variety of other substances by varying the raw materials provided them. In fact, so broad is the meaning of the word *fermentation* that the large-scale industrial syntheses by microorganisms often utilize entirely different mechanisms from those described here, and they even occur aerobically, particularly in antibiotic, hormone, vitamin, and amino acid production (see chapter 26).

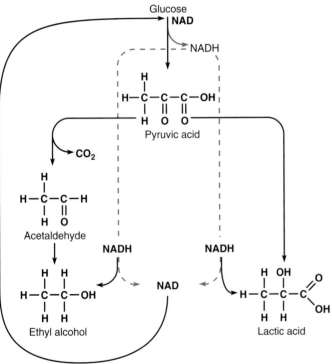

FIGURE 8.25

The chemistry of fermentation systems that produce acid and alcohol. In both cases, the final electron acceptor is an organic compound. In yeasts, pyruvic acid is decarboxylated to acetaldehyde, and the NADH given off in the glycolytic pathway reduces it to ethyl alcohol. In homolactic fermentative bacteria, pyruvic acid is reduced by NADH to lactic acid. Both systems regenerate NAD to feed back into glycolysis or other cycles.

MICROBITS 8.5
Fermentation and Biochemical Testing

The knowledge and understanding of the fermentation products of given species are important not only in industrial production but also in identifying bacteria by biochemical tests. Fermentation patterns in enteric bacteria, for example, are an important identification tool. Specimens are grown in media containing various carbohydrates, and the production of acid or acid and gas is noted (see figure 3.11). For instance, *Escherichia* ferments the milk sugar lactose, whereas *Shigella* and *Proteus* do not. On the basis of its gas production during glucose fermentation,

Escherichia can be further differentiated from *Shigella,* which ferments glucose but does not generate gas. Other enteric bacteria are separated on the basis of whether glucose is fermented to mixed acids or to 2,3-butanediol. *Escherichia coli* produces mixed acids, whereas *Enterobacter* and *Serratia* form primarily 2,3-butanediol. These are part of the IMViC testing described in chapter 20. The M refers to the methyl red test for mixed acids, and the V refers to the Voges-Proskauer test for the butanediol pathway.

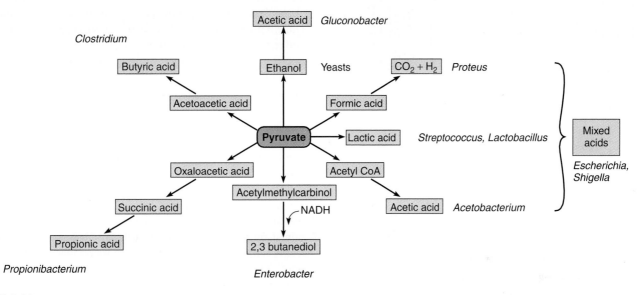

FIGURE 8.26
Miscellaneous products of pyruvate fermentation and the bacteria involved in their production.

Biosynthesis and the Crossing Pathways of Metabolism

Our discussion now turns from catabolism and energy extraction to anabolic functions and biosynthesis. In this section we will overview aspects of intermediary metabolism, including amphibolic pathways, the synthesis of simple molecules, and the synthesis of macromolecules.

THE FRUGALITY OF THE CELL—WASTE NOT, WANT NOT

It must be obvious by now that cells have mechanisms for careful management of carbon compounds. Rather than being dead ends, most catabolic pathways contain strategic molecular intermediates that can be diverted into anabolic pathways. In this way, a given molecule can serve multiple purposes, and the maximum benefit can be derived from all nutrients and metabolites of the cell pool.

The property of a system to function in both catabolism and anabolism is **amphibolism.***

The amphibolic nature of intermediary metabolism can be better appreciated if we look at some examples of the branch points and crossroads of several metabolic pathways (figure 8.27). The pathways of glucose catabolism are an especially rich "metabolic marketplace." The principal sites of amphibolic interaction occur during glycolysis (glyceraldehyde-3-phosphate and pyruvic acid) and the TCA cycle (acetyl coenzyme A and various organic acids).

Amphibolic Sources of Cellular Building Blocks

Glyceraldehyde-3-phosphate can be diverted away from glycolysis and converted into precursors for amino acid, carbohydrate, and triglyceride (fat) synthesis. (A precursor molecule is a compound that is the source of another compound.) Earlier we noted the

* amphibolism (am-fee-bol´-izm) Gr. *amphi,* two-sided, and *bole,* a throw.

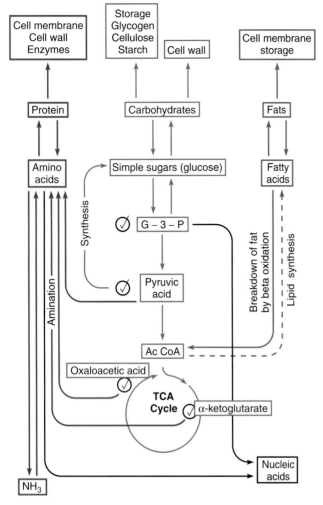

FIGURE 8.27

A summary of metabolic interactions. Many intermediate compounds serve as amphibolic branch points (✓). With comparatively small modifications, these compounds can be converted into other compounds and enter a different pathway. Note that catabolism of glucose (center) furnishes numerous intermediates for amino acid, fat, nucleic acid, and carbohydrate synthesis. Likewise, molecules arising from fat metabolism can be channeled into the TCA cycle.

TCA cycle at acetyl coenzyme A. This aerobic process, called *beta oxidation,* can provide a large amount of energy. Oxidation of a 6-carbon fatty acid yields 50 ATP, compared with 38 for a 6-carbon sugar.

Two metabolites of carbohydrate catabolism that the TCA cycle spins off, oxaloacetic acid and α-ketoglutaric acid, are essential intermediates in the synthesis of certain amino acids. This occurs through **amination,** the addition of an amino group to a carbon skeleton (figure 8.28*a*). A certain core group of amino acids can then be used to synthesize others. Amino acids and carbohydrates can be interchanged through transamination (figure 8.28*b*).

Pathways that synthesize the nitrogen bases (purines, pyrimidines), which are components of DNA and RNA, originate in amino acids and so can be dependent on intermediates from the TCA cycle as well. Because the coenzymes NAD, NADP, FAD, and others contain purines and pyrimidines similar to the nucleic acids, their synthetic pathways are also dependent on amino acids. During times of carbohydrate deprivation, organisms can likewise convert amino acids to intermediates of the TCA cycle by deamination (removal of an amino group) and thereby derive energy from proteins. Deamination results in the formation of nitrogen waste products such as ammonium ions or urea (figure 8.28*c*).

Formation of Macromolecules

Monosaccharides, amino acids, fatty acids, nitrogen bases, and vitamins—the building blocks that make up the various macromolecules and organelles of the cell—come from two possible sources. They can enter the cell from the outside as nutrients, or they can be synthesized through various cellular pathways. The degree to which an organism can synthesize its own building blocks is determined by its genetic makeup, a factor that varies tremendously from group to group. In chapter 7 we found that autotrophs require only CO_2 as a carbon source, a few minerals to synthesize all cell substances, and no organic nutrients. Some heterotrophic organisms (*E. coli,* yeasts) are also very efficient in that they can synthesize all cellular substances from minerals and one organic carbon source such as glucose. Compare this with a strict parasite that has few synthetic abilities of its own and drains most precursor molecules from the host.

Whatever their source, once these building blocks are added to the metabolic pool, they are available for synthesis of polymers by the cell. The details of synthesis vary among the types of macromolecules, but all of them involve the formation of bonds by specialized enzymes and the expenditure of ATP (see figure 8.8*a*).

Amino Acids, Protein Synthesis, and Nucleic Acid Synthesis

Proteins account for a large proportion of a cell's constituents. They are essential components of enzymes, the cell membrane, the cell wall, and cell appendages. As a general rule, 20 amino acids are needed to make these proteins (see chapter 2). Although some organisms (*E. coli,* for example) have pathways that will synthesize all 20 amino acids, others (especially animals) lack some or all of the pathways for amino acid synthesis and must acquire the essential ones from their diets. Protein synthesis itself is a complex process that requires a genetic blueprint and the operation of intricate cellular machinery, as we shall see in chapter 9.

numerous directions that pyruvic acid catabolism can take. In terms of synthesis, pyruvate also plays a pivotal role in providing intermediates for amino acids. In the event of an inadequate glucose supply, it serves as the starting point in glucose synthesis from various metabolic intermediates, a process called *gluconeogenesis.**

The acetyl group that starts the TCA cycle is another extremely versatile metabolite that can be fed into a number of synthetic pathways. This 2-carbon fragment can be converted as a single unit into one of several amino acids, or a number of these fragments can be condensed into hydrocarbon chains that are important bases for fatty acid and lipid synthesis. Note that the reverse is also true—fats can be degraded to acetyl and thereby enter the

---

* gluconeogenesis (gloo″-koh-nee″-oh-gen′-uh-sis) Literally, the formation of "new" glucose.

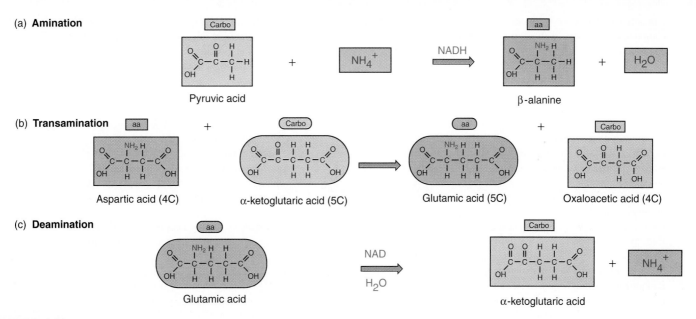

FIGURE 8.28

Reactions that produce and convert amino acids. All of them require energy as ATP or NAD and specialized enzymes. **(a)** Through amination (the addition of an ammonium molecule [amino group]), a carbohydrate can be converted to an amino acid. **(b)** Through transamination (transfer of an amino group from an amino acid to a carbohydrate fragment), metabolic intermediates can be converted to amino acids that are in low supply. **(c)** Through deamination (removal of an amino group), an amino acid can be converted to a useful intermediate of carbohydrate catabolism. This is how proteins are used to derive energy. Ammonium is one waste product.

DNA and RNA are important for the hereditary continuity of cells and the overall direction of protein synthesis. Because nucleic acid synthesis is a major topic of genetics and is closely allied to protein synthesis, it will likewise be covered in chapter 9.

Carbohydrate Biosynthesis

The role of glucose in bioenergetics is so crucial that its biosynthesis is ensured by several alternative pathways. Certain structures in the cell depend on an adequate supply of glucose as well. It is the major component of the cellulose cell walls of some eucaryotes and of certain storage granules (starch, glycogen). Monosaccharides other than glucose are important in the synthesis of bacterial cell walls. Peptidoglycan contains a linked polymer of muramic acid and glucosamine. Carbohydrates (deoxyribose, ribose) are also an essential building block in nucleic acids. Polysaccharides are the predominant components of cell surface structures such as capsules and the glycocalyx, and they are commonly found in slime layers (dextran).

CHAPTER CHECKPOINTS

"Intermediary metabolism" refers to the metabolic pathways that use intermediate compounds and connect anabolic and catabolic reactions.

Amphibolic compounds are the "crossroads compounds" of intermediary metabolism. They not only participate in catabolic pathways but also are precursor molecules to biosynthetic pathways.

Biosynthetic pathways utilize building-block molecules from two sources: the environment and the cell's own catabolic pathways. Microorganisms construct macromolecules from these monomers using ATP and specialized enzymes particular to each species.

Proteins are essential macromolecules in all cells because they function as structural constituents, enzymes, and cell appendages.

Carbohydrates are crucial as energy sources, cell wall constituents, and components of nucleotides.

CHAPTER CAPSULE WITH KEY TERMS

I. Microbial Metabolism
  A. **Metabolism** is the sum of cellular chemical and physical activities; it involves chemical changes to reactants and the release of products using well-established pathways.
  B. Metabolism is a complementary process consisting of **anabolism,** synthetic reactions that convert small molecules into large molecules, and **catabolism,** in which large molecules are degraded. Together, they generate thousands of intermediate molecular states, called **metabolites,** which are regulated at many levels.

II. Enzymes: Metabolic Catalysts
  A. Metabolism is made possible by organic **catalysts,** or **enzymes,** that speed up reactions by lowering the **energy of activation.** Enzymes are not consumed and can be reused. Each enzyme acts specifically upon its assigned metabolite, called the **substrate.**
  B. *Enzyme Structure:* Depending upon its composition, an enzyme is either **conjugated** or **simple.** A conjugated enzyme consists of a protein component called the **apoenzyme** and one or more activators called **cofactors.** Some cofactors are organic molecules called **coenzymes,** and others are inorganic elements,

typically metal ions. To function, a conjugated enzyme must be a complete **holoenzyme** with all its parts. Simple enzymes are composed solely of protein.

C. *Enzyme Specificity:* Substrate attachment occurs in the special pocket called the **active, or catalytic, site.** In order to fit, a substrate must conform to the active site of the enzyme. This three-dimensional state is determined by the amino acid content, sequence, and folding of the apoenzyme. Thus, enzymes are usually **substrate-specific.**

D. *Cofactors:* Metallic cofactors impart greater reactivity to the enzyme-substrate complex. Coenzymes such as **NAD** (nicotinamide adenine dinucleotide) are transfer agents that pass functional groups from one substrate to another. Coenzymes usually contain **vitamins.**

E. *Enzyme Classification:* Enzyme names consist of a prefix derived from the type of reaction or the substrate and the ending -*ase.* By convention, enzymes may also be classified according to their location of action. Thus, an **exoenzyme** is secreted, but an **endoenzyme** is not. Moreover, a **constitutive enzyme** is regularly found in a cell, whereas an **induced enzyme** is synthesized only if its substrate is present.

F. *Types of Enzyme Function:* Metabolic reactions vary. The release of water that comes with formation of new covalent bonds is a **condensation reaction. Hydrolysis reactions** involve addition of water to break bonds. **Functional groups** may be added, removed, or traded in many reactions. Coupled **redox reactions** transfer electrons and protons (H^+) from one substrate to another. Compounds yielding electrons are another. Compounds yielding electrons are **oxidized,** whereas those gaining electrons are **reduced.**

G. *Enzyme Sensitivity:* Enzymes are **labile** (unstable) and function only within narrow operating ranges of temperature and pH, and they are especially vulnerable to **denaturation.** Enzymes are vital metabolic links and thus constitute easy targets for many harmful physical and chemical agents.

III. **Regulation of Enzymatic Activity**

A. Regulatory controls can act on enzymes directly or on the process that gives rise to the enzymes.

1. A substance that resembles the normal substrate and can occupy the same active site is said to exert **competitive inhibition.**

2. In **feedback** (end product) **control,** the concentration of the product at the end of a pathway blocks the action of a key enzyme by **feedback inhibition.** This is found in **allosteric enzymes,** which have a **regulatory site** different from the active site. When a product attaches here, the enzyme's action is blocked.

3. Another mechanism, **feedback repression,** inhibits at the genetic level by controlling the synthesis of key enzymes. Enzyme **induction** conserves cell resources by producing enzymes only when the appropriate substrate is present.

IV. **Major Pathways of Bioenergetics**

A. *The Production and Use of Energy:*

1. **Energy** is the capacity of a system to perform work. It is consumed in **endergonic reactions** and is released in **exergonic reactions.** The freed energy is associated with electrons that can be temporarily captured and transferred to high-energy molecules.

2. Extracting energy requires a series of **electron carriers** arrayed in a downhill **redox chain** between electron donors

and electron acceptors. In **oxidative phosphorylation,** energy is transferred to inorganic phosphate, causing it to form high-energy compounds such as ATP.

B. *Principal Pathways in Oxidation of Glucose:* Carbohydrates, such as glucose, are energy-rich because they can yield a large number of electrons per molecule. Glucose is dismantled in two stages.

1. **Glycolysis** is a pathway that degrades glucose to pyruvic acid in the absence of oxygen.

2. Important intermediates in glycolysis are glucose-6-phosphate, fructose-1,6-bisphosphate, glyceraldehyde-3-phosphate, bisphosphoglyceric acid, phosphoglyceric acids, phosphoenolpyruvate, and pyruvic acid.

C. *Fate of Pyruvic Acid in TCA and Electron Transport:*

1. Pyruvic acid is processed in **aerobic respiration** via the **tricarboxylic acid (TCA) cycle** and its associated electron transport chain.

2. **Acetyl coenzyme A** is a product of pyruvic acid processing. This compound undergoes further oxidation and decarboxylation in the TCA cycle, which generates ATP, CO_2, and H_2O.

3. Important intermediary metabolites are pyruvate, oxaloacetate, citrate, isocitrate, α-ketoglutarate, succinate, fumarate, and malate.

4. The respiratory chain completes energy extraction. Important redox carriers of the electron transport system are NAD, **FAD** (flavin adenine dinucleotide), coenzyme Q, and cytochromes.

5. The **chemiosmotic hypothesis** is a conceptual model that explains the origin and maintenance of electropotential gradients across a membrane that leads to ATP synthesis, by **ATP synthase.**

6. The final electron **acceptor** in aerobic respiration is oxygen. In **anaerobic respiration,** sulfate, nitrate, or nitrite serve this function.

D. **Fermentation** is anaerobic respiration in which both the electron donor and final electron acceptors are organic compounds.

1. Fermentation enables anaerobic and facultative microbes to survive in environments devoid of oxygen. Production of alcohol, vinegar, and certain industrial solvents relies upon fermentation. Fermentation products also play a part in identifying some bacteria.

2. The **phosphogluconate pathway** is an alternative anaerobic pathway for hexose oxidation that also provides for the synthesis of NADPH and pentoses.

E. *Versatility of Glycolysis and TCA Cycle:* Many pathways of metabolism are bidirectional, or **amphibolic,** pathways that can be adapted to serving several functions.

1. Metabolites of these pathways double as building blocks and sources of energy. Intermediates such as pyruvic acid are convertible into amino acids through **amination.** Amino acids can be **deaminated** and used as energy sources. Components for purines and pyrimidines are derived from amino acid pathways.

2. Two-carbon acetyl molecules from pyruvate **decarboxylation** can be used in fatty acid synthesis. Combined with glyceraldehyde-3-phosphate, these fatty acids yield triglyceride, a typical storage fat. Alternately, fats can be broken down and fed into the respiratory pathways by **beta oxidation.**

MULTIPLE-CHOICE QUESTIONS

1. _____ is another term for biosynthesis.
   a. catabolism   c. metabolism
   b. anabolism   d. catalyst

2. Catabolism is a form of metabolism in which _____ molecules are converted into _____ molecules.
   a. large, small   c. amino acid, protein
   b. small, large   d. food, storage

3. An enzyme _____ the activation energy required for a chemical reaction.
   a. increases   c. lowers
   b. converts   d. catalyzes

4. An enzyme
   a. becomes part of the final products
   b. is nonspecific for substrate
   c. is consumed by the reaction
   d. is heat and pH labile

5. An apoenzyme is where the _____ is located.
   a. cofactor   c. redox reaction
   b. coenzyme   d. active site

6. Many coenzymes are
   a. metals   c. proteins
   b. vitamins   d. substrates

7. To digest cellulose in its environment, a fungus produces a/an
   a. endoenzyme   c. catalase
   b. exoenzyme   d. polymerase

8. In negative feedback control of enzymes, a buildup in the amount of _____ decreases the activity in the enzyme.
   a. substrate   c. product
   b. reactant   d. ATP

9. Energy in biological systems is primarily
   a. electrical   c. radiant
   b. chemical   d. mechanical

10. Energy is carried from catabolic to anabolic reactions in the form of _____.
    a. ADP
    b. high-energy ATP bonds
    c. coenzymes
    d. inorganic phosphate

11. Exergonic reactions
    a. release potential energy
    b. consume energy
    c. form bonds
    d. occur only outside the cell

12. A reduced compound is
    a. NAD   c. NADH
    b. FAD   d. ADP

13. Most oxidation reactions in microbial bioenergetics involve the
    a. removal of electrons and hydrogens
    b. addition of electrons and hydrogens
    c. addition of oxygen
    d. removal of oxygen

14. A product or products of glycolysis is/are
    a. ATP   c. CO_2
    b. H_2O d. both a and b

15. Fermentation of a glucose molecule gives off a net number of _____ ATPs.
    a. 4   c. 40
    b. 2   d. 0

16. Complete oxidation of glucose in aerobic respiration yields a net output of _____ ATP.
    a. 40   c. 38
    b. 6   d. 2

17. The compound that enters the TCA cycle from glycolysis is
    a. citric acid   c. pyruvic acid
    b. oxaloacetic acid   d. acetyl coenzyme A

18. The $FADH_2$ formed during the TCA cycle enters the electron transport system at which site?
    a. NADH dehydrogenase
    b. cytochrome
    c. coenzyme Q
    d. ATP synthase

19. ATP synthase complexes can generate _____ ATPs for each NADH that enters electron transport.
    a. 1   c. 3
    b. 2   d. 4

20. **Matching.** Match the term a, b, or c with the metabolic pathway where it occurs.
    a. glycolysis (Embden-Meyerhof-Parnas)
    b. TCA (Krebs) cycle
    c. electron transport/oxidative phosphorylation

    _____ H^+ and e^- are delivered to O_2 as the final acceptor.
    _____ Pyruvic acid is formed.
    _____ GTP is formed.
    _____ H_2O is produced.
    _____ CO_2 is formed.
    _____ Fructose bisphosphate is split into two 3-carbon fragments.
    _____ $NADH^+$ is oxidized.
    _____ ATP synthase is active.

CONCEPT QUESTIONS

1. Show diagrammatically the interaction of holoenzyme and its substrate and general products that can be formed from a reaction.

2. Give the general name of the enzyme that:
   a. synthesizes ATP; digests RNA
   b. phosphorylates glucose
   c. catalyzes the formation of acetyl from pyruvic acid (just before step 1 in the TCA cycle)
   d. reduces pyruvic acid to lactic acid
   e. reduces nitrate to nitrite

3. a. Explain what an allosteric enzyme is and how negative feedback works. Two steps in glycolysis are catalyzed by allosteric enzymes. These are: (1) phosphoglucoisomerase, which catabolizes step 2, and (2) pyruvate kinase, which catabolizes step 9.
   b. Suggest what metabolic products might regulate these enzymes.
   c. How might one place these regulators in figure 8.19?

4. Explain how oxidation of a substrate proceeds without oxygen.

5. In the following redox pairs, which compound is reduced and which is oxidized?
   a. NAD and NADH
   b. $FADH_2$ and FAD
   c. lactic acid and pyruvic acid
   d. NO_3^- and NO_2^-
   e. Ethanol and acetaldehyde

6. Discuss the relationship of:
   a. anabolism to catabolism
   b. ATP to ADP
   c. glycolysis to fermentation
   d. electron transport to oxidative phosphorylation

7. a. What is meant by the concept of the "final electron acceptor"?
   b. What are the final electron acceptors in aerobic, anaerobic, and fermentative metabolism?

8. Name the major ways that substrate-level phosphorylation is different from oxidative phosphorylation.

9. Compare the location of glycolysis, TCA cycle, and electron transport in procaryotic and eucaryotic cells.

10. a. Outline the basic steps in glycolysis, indicating where ATP is used and given off.
    b. Where does NADH originate, and what is its fate in an aerobe?
    c. What is the fate of NADH in a fermentative organism?

11. a. What is the source of ATP in the TCA cycle?
    b. How many ATPs are formed from the original glucose molecule?
    c. How does the total of ATPs generated differ between bacteria and many eucaryotes? What causes this difference?

12. Name the sources of oxygen in bacteria that use anaerobic respiration.

13. a. Summarize the chemiosmotic theory of ATP formation.
    b. What is unique about the actions of ATP synthase?

14. How are aerobic and anaerobic respiration different?

15. Compare the general equation for aerobic metabolism with table 8.4 and verify that all figures balance.

16. Water is one of the end products of aerobic respiration. Where in the metabolic cycles is it formed, and where is it used?

17. Briefly outline the use of certain metabolites of glycolysis and TCA in amphibolic pathways.

CRITICAL-THINKING QUESTIONS

1. Using the concept of fermentation, describe the microbial (biochemical) mechanisms that cause milk to sour.

2. *Trichomonas vaginalis* is a protozoan agent of a sexually transmitted disease in humans. It lives on the vaginal or urinary mucous membranes, feeding off dead cells and glycogen. Astonishingly, this eucaryote completely lacks mitochondria. Predict what sort of metabolism it must have.

3. Explain some of the main functions of vitamins and why they are essential growth factors in human and microbial metabolism.

4. Describe some of the special adaptations of the enzymes found in extremophiles.

5. Beer production requires an early period of rapid aerobic metabolism of glucose by yeast. Given that anaerobic conditions are necessary to produce alcohol, can you explain why this step is necessary?

6. How many ATPs would be formed as a result of aerobic respiration if cytochrome oxidase were missing from the respiratory chain (as is the case with many bacteria)?

7. Draw a model of ATP synthase in three dimensions, showing how it works. Where in the mitochondrion does the ATP supply collect?

8. Microorganisms are being developed to control human-made pollutants and oil spills that are metabolic poisons to animal cells. A promising approach has been to genetically engineer bacteria to degrade these chemicals. What is actually being manipulated in these microbes?

INTERNET SEARCH TOPICS

1. Look up fermentation on the WWW, and outline some of the products made by this process.

2. Find websites that feature three-dimensional views of enzymes and compare the structures and functions of five of the enzymes shown.

Microbial Genetics

The study of modern genetics is really a study of the language of the cell, a special language found in deoxyribonucleic acid—DNA. In this chapter we shall investigate the structure and function of this molecule and explore how it is copied, how its language is interpreted into useful cell products, how it is controlled, how it changes, how it is transferred from one cell to another, and its applications in microbial genetics.

The search for the structure of the genetic material is one of the most compelling stories in biology. See Historical Highlights 9.1 for a short history of this saga.

Chapter Overview

- Genetics is the study of the structure and function of the genomes of biological entities.
- The primary levels of heredity lie in the chromosomes, genes, and the DNA molecule.
- DNA is a very elongate molecule composed of deoxyribose, phosphate, and nitrogenous bases arranged in a double helix.
- The two DNA strands are held together by pair bonding between matched sets of purine and pyrimidine bases: adenine pairs with thymine; cytosine pairs with guanine.
- The arrangement of the base pairs along the length of DNA provides detailed instructions for the formation of proteins, key components of cell structure and metabolism.
- The DNA molecule must be replicated for the distribution of genetic material to offspring.
- Interpretation of the DNA is accomplished by transcription of its code into helper molecules of RNA that cooperate to translate the code into proteins.
- During protein synthesis on the ribosomes, codons of mRNA are sequentially matched by a complementary anticodon on tRNA, which ensures the correct amino acid is added to the protein.
- Viruses contain various forms of DNA and RNA that are translated by the genetic machinery of their host cells to form functioning viral particles.
- The genetic activities of cells are highly regulated by operons, groups of genes that interact as a unit to control the use or synthesis of metabolic substances.

The men who cracked the code of life. Dr. James Watson (left) and Dr. Francis Crick (right) stand next to their model that finally explained the structure of DNA in 1953.

- DNA undergoes mutations, permanent changes in its language, that alter the expression of genes and serve as a force in evolution of organisms.
- Bacteria undergo genetic recombination through transfer of genes in the form of small pieces of DNA, transposons, or circular plasmids.
- Forms of recombination in bacteria include conjugation, transformation, and transduction.

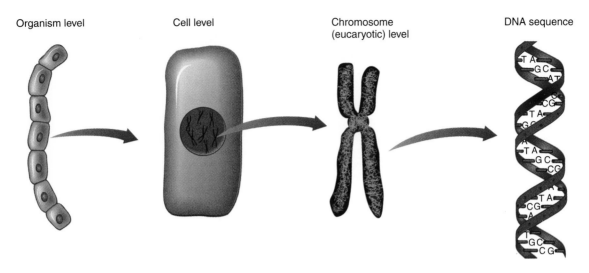

Organism level   Cell level   Chromosome (eucaryotic) level   DNA sequence

FIGURE 9.1

Levels of genetic study. The operations of genetics can be observed at the levels of organism, cell, chromosome, and DNA sequence.

Introduction to Genetics and Genes: Unlocking the Secrets of Heredity

Genetics* is the study of the inheritance, or **heredity,*** of living things. It is a wide-ranging science that explores the transmission of biological properties (traits) from parent to offspring, the expression and variation of those traits, the structure and function of the genetic material, and how this material changes. The study of genetics exists on several levels (figure 9.1). Organism genetics observes the heredity of the whole organism or cell; chromosome genetics examines the characteristics and actions of chromosomes; and molecular genetics deals with the biochemistry of the genes. All of these levels are useful areas of exploration, but in order to understand the expressions of microbial structure, physiology, mutations, and pathogenicity, we need to examine the operation of genes at the cellular and molecular levels. The study of microbial genetics provides a greater understanding of human genetics and an increased appreciation for the astounding advances in genetic engineering we are currently witnessing (chapter 10).

THE NATURE OF THE GENETIC MATERIAL

For a species to survive, it must have the capacity of self-replication. In single-celled microorganisms, reproduction involves the division of the cell by means of binary fission, budding, or mitosis, but these forms of reproduction involve a more significant activity than just simple cleavage of the cell mass. Because the genetic material is responsible for inheritance, it must be accurately duplicated and separated into each daughter cell to ensure its normal function. This genetic material is a long, encoded molecule of DNA that can be studied on several levels. Before we look at how DNA is copied, let us explore the organization of this genetic material, proceeding from the general to the specific.

The Levels of Structure and Function of the Genome

The **genome** is the sum total of genetic material of a cell. Although most of the genome exists in the form of chromosomes, genetic material can appear in nonchromosomal sites as well (figure 9.2). For example, bacteria and some fungi contain tiny extra pieces of DNA (plasmids), and certain organelles of eucaryotes (the mitochondria and chloroplasts) are equipped with their own genetic programs. Genomes of cells are composed exclusively of DNA, but viruses contain either DNA or RNA as the principal genetic material. Although the specific genome of an individual organism is unique, the general pattern of nucleic acid structure and function is similar among all organisms.

In general, a **chromosome** is a discrete cellular structure composed of a neatly packaged elongate DNA molecule. The chromosomes of eucaryotes and bacterial cells differ in several respects. The structure of eucaryotic chromosomes consists of a DNA molecule tightly wound around histone proteins (see Microbits 9.3), whereas a bacterial chromosome (chromatin body) is condensed and secured into a packet by means of histonelike proteins. Eucaryotic chromosomes are located in the nucleus; they vary in number from a few to hundreds; they can occur in pairs (diploid) or singles (haploid); and they appear elongate. In contrast, most bacteria have a single, circular chromosome, although exceptions exist in a few bacteria that have linear or multiple chromosomes.

The chromosomes of all cells are subdivided into basic informational packets called genes. A **gene** can be defined from more than one perspective. In classical genetics, the term refers to the fundamental unit of heredity responsible for a given trait in an organism. In the molecular and biochemical sense, it is a site on the chromosome that provides information for a certain cell function. More specifically still, it is a certain segment of DNA that contains the necessary code to make a protein or RNA molecule. This last definition of a gene will be emphasized in this chapter. For an analogy that clarifies the relationship of the genome, chromosomes, genes, and DNA, see Spotlight on Microbiology 9.2.

---

* **genetics** L. *genesis,* birth, generation.

* **heredity** L. *hereditas,* heirship. All the characteristics genetically inherited by an organism.

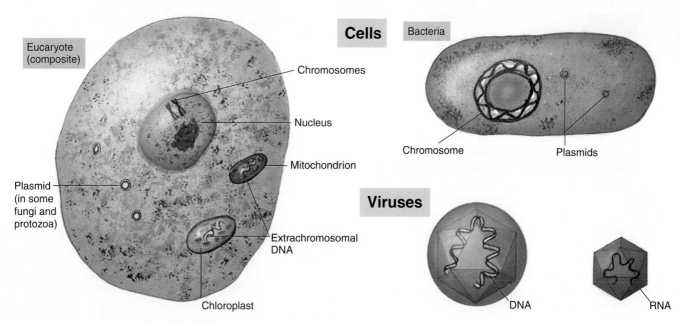

Cells

Eucaryote
(composite)

Chromosomes

Nucleus

Mitochondrion

Plasmid
(in some
fungi and
protozoa)

Extrachromosomal
DNA

Chloroplast

Bacteria

Chromosome

Plasmids

Viruses

DNA

RNA

FIGURE 9.2

The general location and forms of the genome in selected cell types and viruses. (not exactly to scale)

The Size and Packaging of Genomes

Genomes vary greatly in size. The smallest viruses have four or five genes; the bacterium *Escherichia coli* has a single chromosome containing 4,288 genes, and a human cell packs about ten times that many into 46 chromosomes. The exact number is not yet known, but is estimated to be somewhere between 30,000 and 50,000 genes. The total length of DNA relative to cell size is notorious: the chromosome of *E. coli* would measure about 1 mm if unwound and stretched out linearly, and yet this fits within a cell that measures only about 2 μm across, making the DNA 500 times as long as the cell (figure 9.3). Still, the bacterial chromosome takes up only about one-third to one-half of the cell's volume. Likewise, if the sum of all DNA contained in the 46 human chromosomes were unraveled and laid end to end, it would measure about 6 feet. This means that the DNA is about 180,000 times longer than a cell 10 μm wide and a million times longer than the width of the nucleus. How can such elongated genomes fit into the miniscule volume of a cell, and in the case of eucaryotes, into an even smaller compartment, the nucleus? The answer lies in the regular coiling of the DNA chain (Microbits 9.3).

THE DNA CODE: A SIMPLE YET PROFOUND MESSAGE

Examining the function of DNA at the molecular level requires an even closer look at its structure. To do this we will imagine being able to magnify a small piece of a gene about 5 million times. What such fine scrutiny will disclose is one of the great marvels of biology. Our first view of DNA in chapter 2 revealed that it is a gigantic molecule, a type of nucleic acid, with two strands combined into a double helix (see figures 2.23 and 2.25). The general structure is universal, except in some viruses that contain single-stranded DNA. The basic unit of DNA structure is a **nucleotide,** a molecule composed of **phosphate, deoxyribose sugar,** and a **nitrogenous base.** The nucleotides covalently bond to form a sugar-phosphate linkage that becomes the backbone of each strand. Each sugar attaches in a repetitive pattern to two phosphates. One of the bonds is to the number 5′ (read "five prime") carbon on deoxyribose, and the other is to the 3′ carbon, which confers a certain order and direction on each strand (figure 9.4).

The nitrogenous bases, **purines** and **pyrimidines,** attach by covalent bonds at the 1′ position of the sugar. They span the center of the molecule and pair with appropriate complementary bases from the other side of the helix. The paired bases are so aligned as

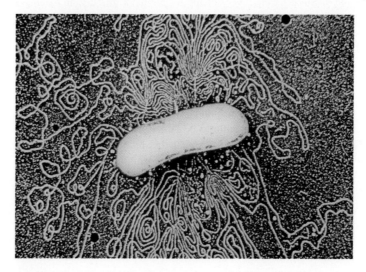

FIGURE 9.3

An *Escherichia coli* cell disrupted to release its DNA molecule. The cell has spewed out its single, uncoiled DNA strand into the surrounding medium.

The search for the primary molecules of heredity was a serious focus throughout the first half of the twentieth century. At first many biologists thought that protein was the genetic material. An important milestone occurred in 1944 when Oswald Avery, Colin MacLeod, and Maclyn McCarty purified DNA and demonstrated at last that it was indeed the blueprint for life. This was followed by an avalanche of research, which continues today.

One area of extreme interest concerned the molecular structure of DNA. In 1951, American biologist James Watson and English physicist Francis Crick collaborated on solving the DNA puzzle. Although they did little of the original research, they were intrigued by several findings from other scientists. It had been determined by Erwin Chargaff that any model of DNA structure would have to contain deoxyribose, phosphate, purines, and pyrimidines arranged in a way that would provide variation and a simple way of copying itself. Watson and Crick spent long hours constructing models with cardboard cutouts and kept alert for any and every bit of information that might give them an edge.

Two English biophysicists, Maurice Wilkins and Rosalind Franklin, had been painstakingly collecting data on X-ray crystallographs of DNA for several years. With this technique, molecules of DNA bombarded by X rays produce a photographic image that can predict the three-dimensional structure of the molecule. After being allowed to view certain X-ray data, Watson and Crick noticed an unmistakable pattern: The molecule appeared to be a double helix. Gradually, the pieces of the puzzle fell into place, and a final model was assembled—a model that explained all of the qualities of DNA, including how it is copied (see chapter opening photo). Although Watson and Crick were rightly hailed for the clarity of their solution, it must be emphasized that their success was due to the considerable efforts of a number of English and American scientists. This historic discovery showed that the tools of physics and chemistry have useful applications in biological systems, and it also spawned ingenious research in all areas of molecular genetics.

Since the discovery of the double helix in 1953, an extensive body of biochemical, microscopic, and crystallographic analysis has left little doubt that the model first proposed by Watson and Crick is correct. Newer techniques using scanning tunneling microscopy produce three-dimensional images of DNA magnified 2 million times. These images verify the helical shape and twists of DNA represented by models.

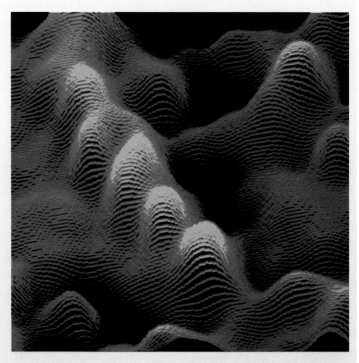

The first direct glimpse at DNA's structure. This false-color scanning tunneling micrograph of calf thymus gland DNA (2,000,000×) brings out the well-defined folds in the helix.

to be joined by hydrogen bonds. Such weak bonds are easily broken, allowing the molecule to be "unzipped" into its complementary strands. Later we will see that this feature is of great importance in gaining access to the information encoded in the nitrogenous base sequence. Pairing of purines and pyrimidines is not random; it is dictated by the formation of hydrogen bonds between certain bases. Thus, in DNA, the purine **adenine** (A) pairs with the pyrimidine **thymine** (T), and the purine **guanine** (G) pairs with the pyrimidine **cytosine** (C). New research also indicates that the bases are attracted to each other in this pattern because each has a complementary three-dimensional shape that matches its pair. Although the base-pairing partners generally do not vary, the sequence of base pairs along the DNA molecule can assume any order, resulting in an infinite number of possible nucleotide sequences.

Other important considerations of DNA structure concern the nature of the double helix itself. The halves are not parallel or oriented in the same direction. One side of the helix runs in the opposite direction of the other, in an *antiparallel arrangement* (figure 9.4b). The order of the bond between the carbon on deoxyribose and the phosphates is used to keep track of the direction of the two sides of the helix. Thus, one helix runs from the 5′ to 3′ direction, and the other runs from the 3′ to 5′ direction. This characteristic is a significant factor in DNA synthesis and translation. As apparently perfect and regular as the DNA molecule may seem, it is not exactly symmetrical. The torsion in the helix and the stepwise stacking of the nitrogen bases produce two different-sized surface features, the major and minor grooves (figure 9.4c).

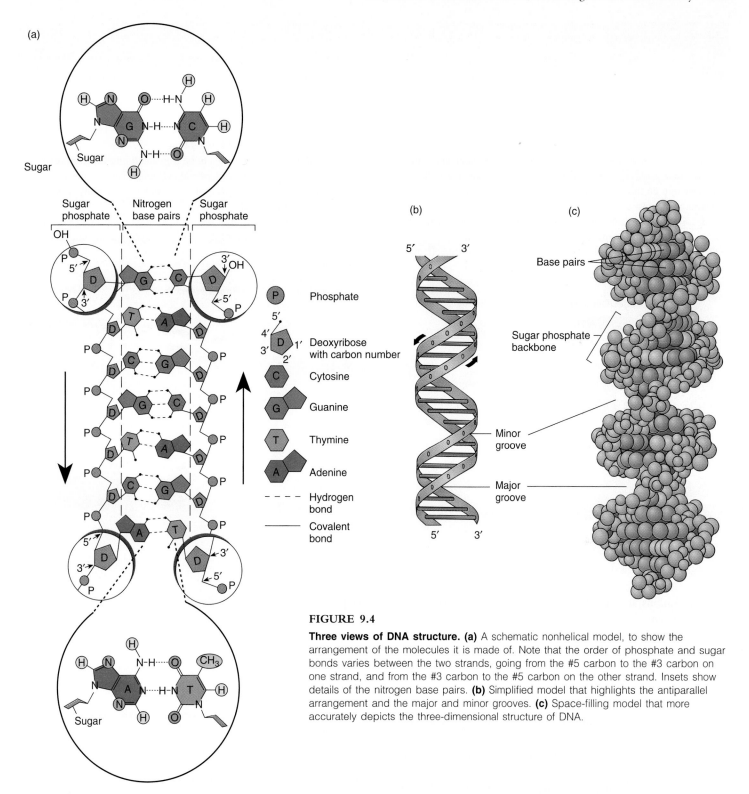

FIGURE 9.4

Three views of DNA structure. (a) A schematic nonhelical model, to show the arrangement of the molecules it is made of. Note that the order of phosphate and sugar bonds varies between the two strands, going from the #5 carbon to the #3 carbon on one strand, and from the #3 carbon to the #5 carbon on the other strand. Insets show details of the nitrogen base pairs. **(b)** Simplified model that highlights the antiparallel arrangement and the major and minor grooves. **(c)** Space-filling model that more accurately depicts the three-dimensional structure of DNA.

THE SIGNIFICANCE OF DNA STRUCTURE

The nitrogen bases influence DNA in two major ways:

1. **Maintenance of the code during reproduction.** The constancy of base-pairing guarantees that the code will be retained during cell growth and division. When the two strands are separated, each one provides a **template** (pattern or model) for the replication (exact copying) of a new molecule (figure 9.5). Because the sequence of one strand automatically gives the sequence of its partner, the code can be duplicated with fidelity.

2. **Providing variety.** The order of bases along the length of the DNA strand constitutes the genetic program, or the language, of the DNA code. Adding to our earlier

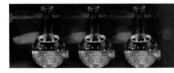

The relationship of the genome, chromosomes, and genes to one another and to DNA is analogous to a collection of videotapes of family events. The whole library of tapes (genome) contains several individual cassettes (chromosomes); the tape on the cassette is divided sequentially into several separate events (genes), which may be selected and played to generate a picture on the television screen (product). This analogy works in several other ways:

1. At all levels (library, cassette, event), the basic informational unit is still the tape itself (the DNA molecule).
2. Like a tape, the DNA molecule can be copied, spliced, and edited.

3. Somewhat like the spools of a cassette, the long DNA of a chromosome is carefully wrapped up so that it is compact, easy to read, and will not get tangled.
4. For the tape (DNA code) to be translated into images (cell product), a tape player (special cell machinery) is necessary.
5. The tape, like the DNA molecule, makes sense only if played in a certain direction.
6. The entire collection of tapes, like the genome, is a store of information. Not all of it is being "played" at any one time.

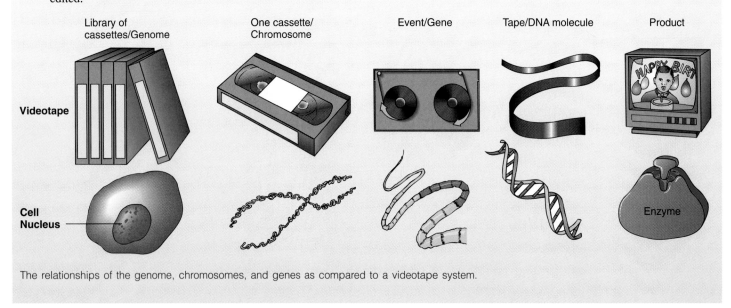

The relationships of the genome, chromosomes, and genes as compared to a videotape system.

definitions, the message present in a gene is a precise arrangement of these bases, and the genome is the collection of all DNA bases that, in an ordered combination, are responsible for the unique qualities of each organism.

It is tempting to ask how such a seemingly simple code can account for the extreme differences among forms as diverse as a virus, *E. coli,* and a human. The English language, based on 26 letters, can create an infinite variety of words, but how can an apparently complex genetic language such as DNA be based on just four nitrogen base "letters"? A mathematical example can explain the possibilities. For a segment of DNA that is 1,000 nucleotides long, there are $4^{1,000}$ different sequences possible. Carried out, this number would approximate 1.5×10^{602}, a number so huge that it literally provides endless degrees of variation.

DNA REPLICATION: PRESERVING THE CODE AND PASSING IT ON

It has been established that the sequence of bases along the length of a gene constitutes the language of DNA. For this language to

be preserved for hundreds of generations, it will be necessary for the genetic program to be duplicated and passed on to each offspring. This process of duplication is called **DNA replication.** In the following example, we will show replication in bacteria, but with some exceptions, it also applies to the process as it works in eucaryotes and some viruses. Early in binary fission, the metabolic machinery of a bacterium responds to a message and initiates the duplication of the chromosome. This DNA replication must be completed during a single generation time (around 20 minutes in *E. coli*).

The Overall Replication Process

What features allow the DNA molecule to be exactly duplicated, and how is its integrity retained? DNA replication requires a careful orchestration of the actions of 30 different enzymes (partial list in table 9.1), which separate the strands of the existing DNA molecule, copy its template, and produce two complete daughter molecules. A simplified version of replication is shown in figure 9.5 and includes the following: (1) uncoiling the parent DNA molecule, (2) unzipping the hydrogen bonds between the base pairs, thus separating the two strands and exposing the nucleotide sequence of the

MICROBITS 9.3
The Packaging of DNA: Winding, Twisting, and Coiling

The analogy of DNA to a cassette tape is imperfect because the DNA molecule is not perfectly wound around a spool. Packing the mass of DNA into the cell involves further levels of DNA structure called super-coils or superhelices. In the simpler system of procaryotes, the circular chromosome is packaged by the action of a special enzyme called a *topoisomerase** (specifically, DNA *gyrase*).* This enzyme coils the chromosome into a tight bundle by introducing a reversible series of twists into the DNA molecule. The system in eucaryotes is more complex, with three or more levels of coiling. First, the DNA molecule of a chromosome, which is linear, is wound twice around the histone proteins, creating a chain of *nucleosomes*.* The nucleosomes fold in a spiral formation upon one another. An even greater supercoiling occurs when this spiral arrange-ment further twists on its radius into a giant spiral with loops radiating from the outside. This extreme degree of compactness is what makes the eucaryotic chromosome visible during mitosis (see figure 5.6). In addition to reducing the volume occupied by DNA, supercoiling solves the problem of keeping the chromosomes from getting tangled during cell division, and it protects the code from massive disruptions due to breakage.

---

* **topoisomerase** (tah″-poh-eye-saw′-mur-ayce) Any enzyme that changes the configuration of DNA.

* **gyrase** (jy′-rayce) L. *gyros,* ring or circle. A bacterial enzyme that produces super-coils.

* **nucleosomes** "Nucleus bodies" arranged like beads on a chain.

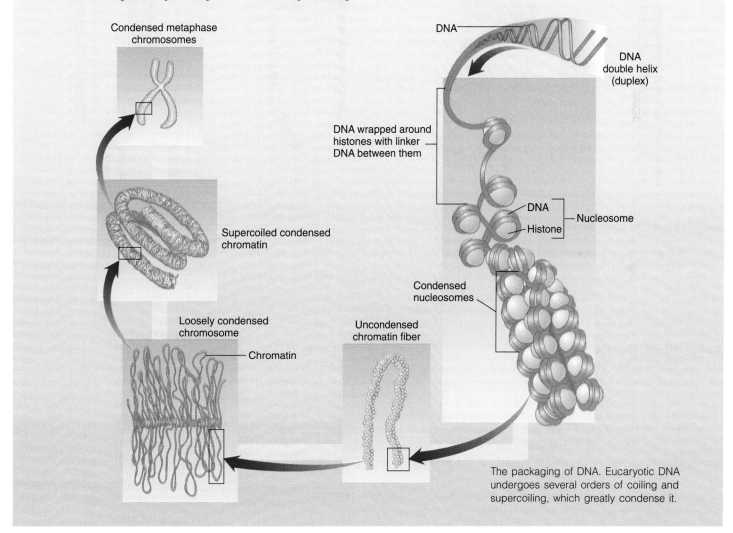

The packaging of DNA. Eucaryotic DNA undergoes several orders of coiling and supercoiling, which greatly condense it.

helix to serve as **templates,** and (3) synthesizing two double strands by attachment of the correct complementary nucleotides to each single-stranded template. It is worth noting that each daughter molecule will be identical to the parent in composition, but neither one is completely new; the strand that serves as a template is an original parental DNA strand. The preservation of the parent molecule in this way, termed *semiconservative replication,* helps explain the reliability and fidelity of replication.

(a) (b) (c) (d)

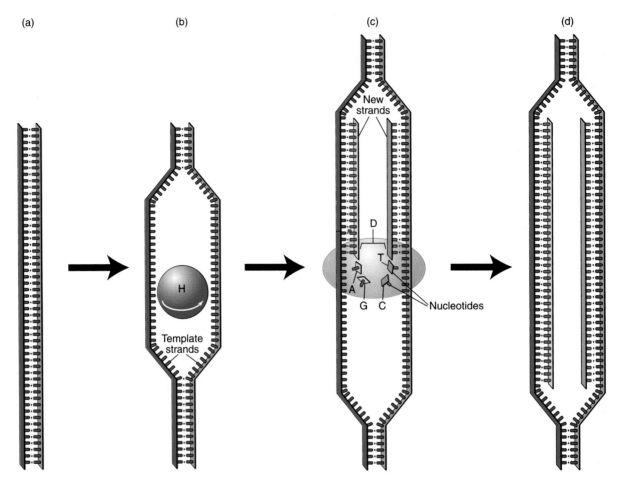

FIGURE 9.5

Simplified steps to show the semiconservative replication of DNA. (a, b) The two strands of the double helix are unwound by a helicase (H), which disrupts the hydrogen bonds and exposes the nitrogen base codes of DNA. Each single strand formed will serve as a template to synthesize a new double strand of DNA. **(c)** A DNA polymerase (D) proceeds along the DNA molecule, attaching the correct nucleotides according to the pattern of the template. An A on the template will pair with a T on the new molecule, and a C will pair with a G. **(d)** The resultant new DNA strands contain one strand of the newly synthesized DNA and the original template strand. The integrity of the code is kept intact because the linear arrangement of the bases is maintained during this process.

Refinements and Details of Replication

The circular bacterial DNA molecule replicates by means of a special configuration called a **replicon.** Replication begins at a precise initiation site containing a certain repeated sequence called a *palindrome,* which forms the origin of replication. In language, a palindrome is a word, phrase, or sentence that reads the same both forward and backward: for example, radar; and madam I'm Adam. A DNA palindrome, also called an *inverted repeat,* might read as follows:

$$\overrightarrow{\text{GCTAGC}}$$

$$\underleftarrow{\text{CGATCG}}$$

Note that the DNA format reads this way using the complementary strands. At these sites, enzymes called *helicases* ("unzipping enzymes") untwist the helix and break its hydrogen bonds, thereby creating two separate strands.

Replication begins when an RNA *primer* is synthesized and enters at the initiation site (figure 9.6a). DNA polymerase III cannot begin synthesis unless it has this short strand of RNA to serve as a starting point for adding nucleotides. Because the bacterial

DNA molecule is circular, opening of the circle forms two **replication forks,** each containing its own set of replication enzymes. The DNA polymerase III is a huge enzyme complex that encircles the replication fork and adds nucleotides in accordance with the tem-

TABLE 9.1

Some Enzymes Involved in DNA Replication and Their Functions

Enzyme	Function
Helicase	Unzipping the DNA helix
Primase	Synthesizing an RNA primer
DNA polymerase III	Adding bases to the new DNA chain; proofreading the chain for mistakes
DNA polymerase I	Removing primer, closing gaps, repairing mismatches
Ligase	Final binding of nicks in DNA during synthesis and repair
Gyrase	Supercoiling

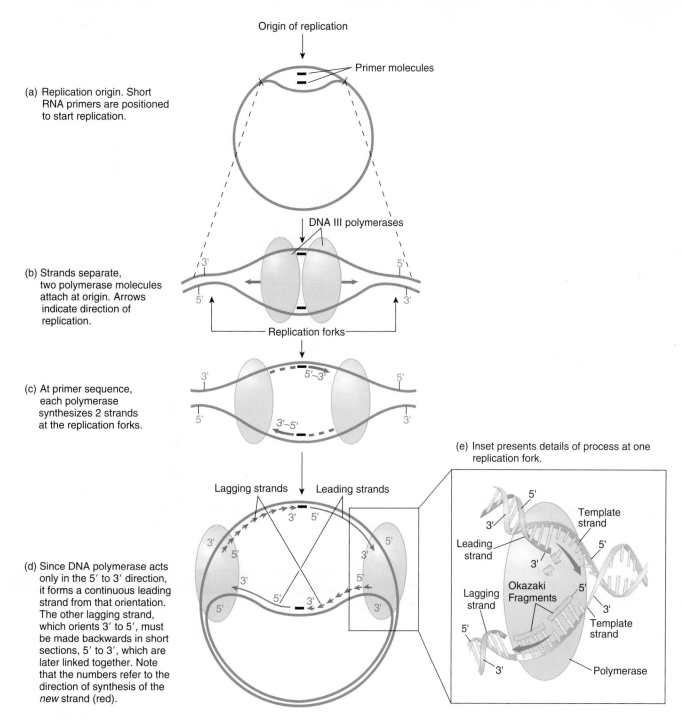

(a) Replication origin. Short RNA primers are positioned to start replication.

(b) Strands separate, two polymerase molecules attach at origin. Arrows indicate direction of replication.

(c) At primer sequence, each polymerase synthesizes 2 strands at the replication forks.

(d) Since DNA polymerase acts only in the 5′ to 3′ direction, it forms a continuous leading strand from that orientation. The other lagging strand, which orients 3′ to 5′, must be made backwards in short sections, 5′ to 3′, which are later linked together. Note that the numbers refer to the direction of synthesis of the *new* strand (red).

(e) Inset presents details of process at one replication fork.

FIGURE 9.6

The bacterial replicon: a model for DNA synthesis. (a) Circular DNA has a special origin site where replication originates. **(b)** When strands are separated, two replication forks form, and a DNA polymerase III enters at each fork. **(c)** Starting at the primer sequence, both polymerases move along the template strands (blue), synthesizing the new strands (red) at each fork. **(d)** DNA polymerase works only in the 5′ to 3′ direction, necessitating a different pattern of replication at each fork. Because the leading strand orients in the 5′ to 3′ direction, it will be synthesized continuously. The lagging strand, which orients in the opposite direction, can only be synthesized in short sections, 5′ to 3′, which are later linked together. **(e)** Inset presents details of process at one replication fork, and shows the Okazaki fragments and the relationship of the template, leading, and lagging strands.

plate pattern. As synthesis proceeds, the forks continually open up to expose the template for replication (figure 9.6).

Because DNA polymerase is correctly oriented for synthesis *only* in the 5′ to 3′ direction of the new molecule (red) strand, only one strand, called the **leading strand,** can be synthesized as a continuous, complete strand. The strand with the opposite orientation

(3′ to 5′) is termed the **lagging strand** (figure 9.6*d,e*). Because it cannot be synthesized continuously, the polymerase adds nucleotides a few at a time in the direction away from the fork (5′ to 3′). As the fork opens up a bit, the next segment is synthesized backwards to the point of the previous segment, a process repeated at both forks until synthesis is complete. In this way, the DNA polymerase is able

(a)

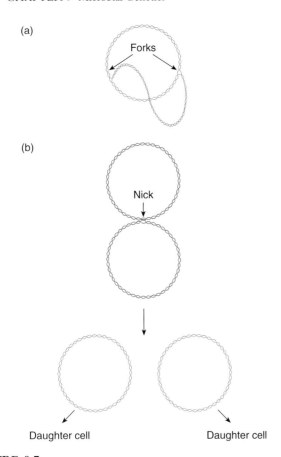

(b)

FIGURE 9.7

Completion of chromosome replication in bacteria. (a) As replication proceeds, one strand loops down. **(b)** Final separation is achieved through repair and the release of two completed molecules. The daughter cells receive these during binary fission.

to synthesize the two new strands simultaneously. This manner of synthesis produces one strand containing short fragments of DNA (100 to 1,000 bases long) called *Okazaki fragments.* These fragments are attached to the growing end of the lagging strand by another enzyme called **DNA ligase.**

Elongation and Termination of the Daughter Molecules The addition of nucleotides proceeds at an astonishing pace, estimated in some bacteria at 750 bases per second at each fork! As replication proceeds, the duplicated strand loops down (figure 9.7*a*). When the forks come full circle and meet, ligases move along the lagging strand to begin the initial linking of the fragments and to complete synthesis and separation of the two circular daughter molecules (figure 9.7*b*). The DNA polymerase I removes the RNA primers used to initiate DNA synthesis and replaces them with DNA.

Like any language, DNA is occasionally "misspelled" when an incorrect base is added to the growing chain. Studies have shown that such mistakes are made once in approximately 10^8 to 10^9 bases, but most of these are corrected. If not corrected, they become mutations and can lead to serious cell dysfunction and even death. Because continued cellular integrity is very dependent on accurate replication, cells have evolved their own proofreading function for DNA. DNA polymerase III, the enzyme that elongates the

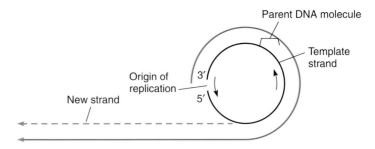

FIGURE 9.8

 **Simplified model of rolling circle DNA replication.** This pattern occurs in some viruses and plasmids. As the parent DNA rotates, a single strand (green) is synthesized on one template strand, and this new strand is rolled off the circle. A complementary strand (blue) is then formed in sections using this new strand, thus creating a double strand.

molecule, can also detect incorrect, unmatching bases, excise them, and replace them with the correct base. DNA polymerase I can also proofread the molecule and repair damaged DNA.

Replication in Other Biological Systems The replication pattern of eucaryotes is similar to that of procaryotes. It also uses replicons and a variety of DNA polymerases, and replication proceeds both directions from the point of origin. The main difference is that the eucaryotic chromosomes are linear and require hundreds to thousands of replicons acting simultaneously in order to complete replication efficiently.

A novel form of DNA synthesis called **rolling circle** occurs in circular genetic material found in plasmids and some bacterial viruses (figure 9.8). In general, it involves the replication of one strand of the parent DNA, forming a single strand that rolls off the circle. This strand is then replicated, thus converting it to a double-stranded duplicate of the original DNA molecule.

CHAPTER CHECKPOINTS

Nucleic acids are molecules that contain the blueprints of life in the form of genes. DNA is the blueprint molecule for all cellular organisms. The blueprints of viruses, however, can be either DNA or RNA.

The total amount of DNA in an organism is termed its genome. The genome of each species contains a unique arrangement of genes that define its appearance (phenotype), metabolic activities, and pattern of reproduction.

The genome of procaryotes is quite small compared with the genomes of eucaryotes. Bacterial DNA consists of a few thousand genes in one circular chromosome. Eucaryotic genomes range from thousands to hundreds of thousands of genes. Their DNA is packaged in tightly wound spirals arranged in discrete chromosomes.

DNA copies itself just before cellular division by the process of semiconservative replication. Semiconservative replication means that each "old" DNA strand is the template upon which each "new" strand is synthesized.

The circular bacterial chromosome is replicated at two forks as directed by DNA polymerase III. At each fork, two new strands are synthesized—one continuously and one in short fragments, and mistakes are proofread and removed.

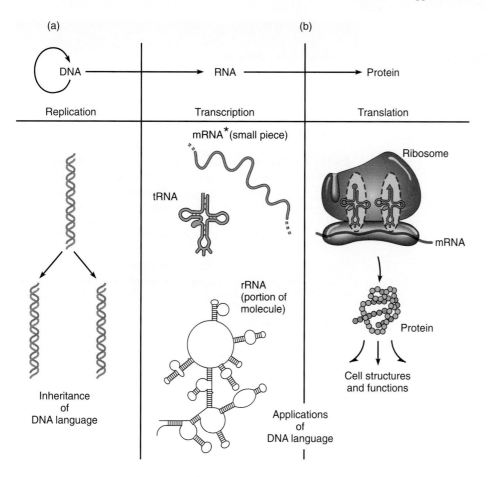

(a)

DNA → RNA → Protein

Replication   Transcription   Translation

mRNA*(small piece)

tRNA

Ribosome

(b)

mRNA

rRNA
(portion of
molecule)

Protein

Cell structures
and functions

Inheritance
of
DNA language

Applications
of
DNA language

*The sizes of RNA are not to scale—tRNA and mRNA are enlarged to show details.

FIGURE 9.9

Summary of the flow of genetic information in cells. DNA is the ultimate storehouse and distributor of genetic information. **(a)** DNA can be replicated and passed to offspring. **(b)** DNA must be deciphered into a usable cell language. It does this by transcribing its code into RNA helper molecules that translate that code into protein.

Applications of the DNA Code: Transcription and Translation

We have explored how the genetic message in the DNA molecule is conserved through replication. Now we must consider the precise role of DNA in the cell. Given that the sequence of bases in DNA is a genetic code, just what is the nature of this code and how is it utilized by the cell? Although the DNA library is full of critical information, the molecule itself does not perform cell processes directly. Its stored information is conveyed to RNA molecules, which carry out instructions. The concept that genetic information flows from DNA to RNA to protein is a central theme of molecular biology (figure 9.9). More precisely, it states that the master code of DNA is first copied onto an RNA molecule called a messenger by means of **transcription,** and the RNA message is decoded by special cell components into proteins during **translation.** The principal exceptions to this pattern are found in RNA viruses, which convert RNA to other RNA, and in retroviruses, which convert RNA to DNA.

THE GENE-PROTEIN CONNECTION

Genes fall into three basic categories: *structural genes* that code for proteins, genes that code for RNA, and *regulatory genes* that control gene expression. The sum of all of these types of genes constitutes an organism's distinctive genetic makeup, or **genotype.*** The expression of the genotype creates traits (certain structures or functions) referred to as the **phenotype.*** Just as a person inherits a combination of genes (genotype) that gives a certain eye color or height (phenotype), a bacterium inherits genes that direct the formation of a flagellum, and a virus contains genes for its capsid structure.

The Triplet Code and the Relationship to Proteins

Several questions invariably arise concerning the relationship between genes and cell function. For instance, how does gene structure lead to the expression of traits in the individual, and what features of gene expression cause one organism to be so distinctly different from another? For answers, we must turn to the correlation between gene and protein structure. We know that each structural gene is a linear sequence of nucleotides that codes for a protein. Because each protein is different, each gene must also differ somehow in its composition. In fact, the language of DNA exists in the order of groups of three consecutive bases called **triplets** on one DNA strand (figure 9.10). Thus, one gene differs from another in its

---

* genotype (jee′-noh-typ) Gr. *gennan,* to produce, and *typos,* type.

* phenotype (fee′-noh-typ) Gr. *phainein,* to show. The physical manifestation of gene expression.

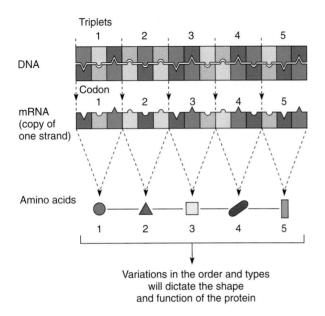

FIGURE 9.10

Simplified view of the DNA-protein relationship. The DNA molecule is a continuous chain of base pairs, but the sequence must be interpreted in groups of three base pairs (a triplet). Each triplet as copied into mRNA codons will translate into one amino acid; consequently, the ratio of base pairs to amino acids is 3:1.

TABLE 9.2

Types of Ribonucleic Acid

RNA Type	Contains Codes For	Function in Cell	Translated
Messenger (mRNA)	Sequence of amino acids in protein	Carries the DNA master code to the ribosome	Yes
Transfer (tRNA)	A cloverleaf tRNA to carry amino acids	Brings amino acids to ribosome	Helps in translation
Ribosomal (rRNA)	Several large structural rRNA molecules	Forms the major part of a ribosome and participates in protein synthesis	No
Primer	An RNA that can begin DNA replication	Primes DNA	No
Ribozymes	RNA enzymes, parts of splicer enzymes	Remove introns from other RNAs in eucaryotes	No

composition of triplets. An equally important part of this concept is that each triplet represents a code for a particular amino acid. When the triplet code is transcribed and translated, it dictates the type and order of amino acids in a polypeptide (protein) chain.

The final key points that connect DNA and protein function are:

1. A protein's primary structure—the order and type of amino acids in the chain—determines its characteristic shape and function.
2. Proteins ultimately determine phenotype, the expression of all aspects of cell function and structure. Put more simply, living things are what their proteins make them.
3. DNA is mainly a blueprint that tells the cell which kinds of proteins to make and how to make them.

THE MAJOR PARTICIPANTS IN TRANSCRIPTION AND TRANSLATION

Transcription, the formation of RNA using the DNA code, and translation, the synthesis of proteins, are highly complex. A number of components participate: most prominently, messenger RNA, transfer RNA, ribosomes, several types of enzymes, and a storehouse of raw materials. After first examining each of these components, we shall see how they come together in the assembly line of the cell.

RNAs: Tools in the Cell's Assembly Line

Ribonucleic acid is an encoded molecule like DNA, but its general structure is different in several ways: (1) It is a **single-stranded molecule;** that is, it is a single helix. In some cases, it does form

secondary and tertiary structures, but these structures still arise from a single strand (figure 9.9). (2) RNA contains **uracil,** instead of thymine, as the complementary base-pairing mate for adenine. This does not change the inherent DNA code in any way because the uracil still follows the pairing rules. (3) Although RNA, like DNA, contains a backbone that consists of alternating sugar and phosphate molecules, the sugar in RNA is **ribose** rather than deoxyribose. The many functional types of RNA range from small regulatory pieces to large structural ones (table 9.2). All types of RNA are formed through transcription of a DNA gene, but only mRNA is further translated into another type of molecule (protein).

Messenger RNA: Carrying DNA's Message

Messenger RNA (**mRNA**) is a transcript (copy) of a structural gene or genes complementary to DNA. It is synthesized by a process similar to DNA replication, and the complementary base-pairing rules ensure that the code will be faithfully copied in the mRNA transcript. The message of this transcribed strand is displayed in a series of triplets called **codons** (figure 9.11a), and the length of the mRNA molecule varies from about 100 nucleotides to several thousands. The details of transcription and the function of mRNA in translation will be covered shortly.

Transfer RNA: The Key to Translation

Transfer RNA (**tRNA**) is also a copy of a specific region of DNA; however, it differs from mRNA. It is uniform in length, being 75–95 nucleotides in length, and it contains sequences of bases that form hydrogen bonds with complementary sections of the same tRNA strand. At these points, the molecule bends back upon itself into several *hairpin loops,* giving the molecule a secondary

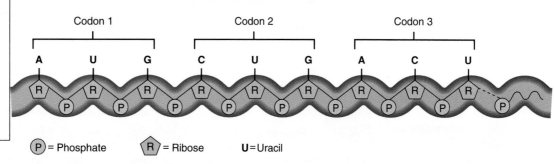

(a) Messenger RNA (mRNA). A short piece of messenger RNA (mRNA) illustrates the general structure of RNA: single strandedness, repeating phosphate-ribose sugar backbone single nitrogen bases with uracil instead of thymine.

Codon 1   Codon 2   Codon 3

A U G   C U G   A C U

P = Phosphate   R = Ribose   U = Uracil

(b) Transfer RNA (tRNA). Transfer RNA (tRNA) can loop back on itself to form intrachain hydrogen bonds. The result of the secondary structure is a cloverleaf structure, shown here in simplified form. At its bottom is an anticodon that specifies the attachment of a particular amino acid at the 3′ end. At right is a three-dimensional view of the tertiary structure of tRNA.

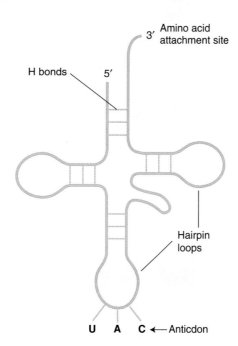

H bonds   5′

3′ Amino acid attachment site

Hairpin loops

U A C ← Anticdon

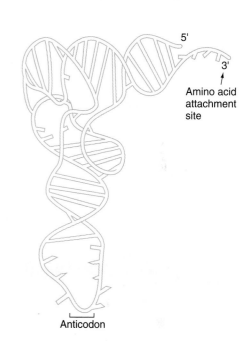

5′

3′ Amino acid attachment site

Anticodon

FIGURE 9.11

Characteristics of messenger and transfer RNA.

cloverleaf structure that folds even further into a complex, three-dimensional helix (figure 9.11*b*). This compact molecule is an adaptor that converts RNA language into protein language. The bottom loop of the cloverleaf exposes a triplet, the **anticodon,** that both designates the specificity of the tRNA and complements mRNA's codons. At the opposite end of the molecule is a binding site for the amino acid that is specific for that tRNA's anticodon. For each of the 20 amino acids (see table 2.5), there is at least one specialized type of tRNA to carry it. The charging of the tRNA takes place in two enzyme-driven steps: First an ATP activates the amino acid, and then this group binds to the acceptor end of the tRNA. Because tRNA is the molecule that will convert the master code on mRNA into a protein, the accuracy of this step is crucial.

The Ribosome: A Mobile Molecular Factory for Translation

The procaryotic (70S) ribosome is a particle composed of tightly packaged ribosomal RNA (rRNA) and protein. The rRNA component of the ribosome is also a long polynucleotide molecule. It

forms complex three-dimensional figures that contribute to the structure and function of ribosomes (see figure 9.9). The interactions of proteins and rRNA create the two subunits of the ribosome that engage in final translation of the genetic code (see figure 9.13). A metabolically active bacterial cell can accommodate up to 20,000 of these miniscule factories—all actively engaged in reading the genetic program, taking in raw materials, and emitting proteins at an impressive rate.

TRANSCRIPTION: THE FIRST STAGE OF GENE EXPRESSION

During transcription, the DNA code is converted to RNA through several stages, directed by a huge and very complex enzyme system, **RNA polymerase** (figure 9.12). Only one strand of the DNA—the **template strand**—contains meaningful instructions for synthesis of a functioning polypeptide. In most cases, the triplet on DNA that signals the start of transcription is TAC, which will transcribe as AUG, the start codon on mRNA. The nontranscribed

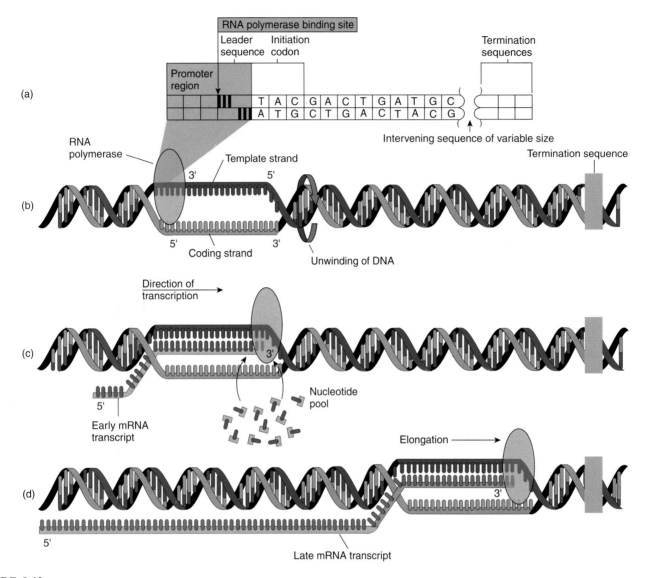

FIGURE 9.12

The major events in mRNA synthesis or transcription. (a) Overall view of a gene. Each gene contains a specific promoter region and a leader sequence for guiding the beginning of transcription. This is followed by the region of the gene that codes for a polypeptide and ends with a series of terminal sequences that stop translation. **(b)** DNA is unwound at the promoter by RNA polymerase. Only one strand of DNA, called the template strand, is copied by the RNA polymerase. This strand runs in the 3′ to 5′ direction. **(c)** As the RNA polymerase moves along the strand, it adds complementary nucleotides as dictated by the DNA template, forming the single-stranded mRNA that reads in the 5′ to 3′ direction. **(d)** The polymerase continues transcribing until it reaches a termination site and the mRNA transcript is released for translation. Note that the section of the DNA that has been transcribed is rewound into its original configuration.

strand is called the *coding strand*. The strand of DNA that serves as a template varies from one gene to another.

Transcription is initiated when RNA polymerase recognizes a segment of the DNA called the *promoter region* that lies near the beginning of the gene segment to be transcribed (figure 9.12). This region is usually some version of a palindrome such as TATATA or TATA. It is here that the polymerase begins building the mRNA chain. The polymerase binds to a starting point and unwinds part of the DNA helix to expose the nitrogen bases on the template strand.

During elongation, which proceeds in the 3′ to 5′ direction, the mRNA is assembled by the addition of nucleotides that are complementary to the DNA template. Be reminded that uracil (U) is placed as adenine's complement. As elongation continues, the part of DNA already transcribed is rewound into its original helical

form. At termination, the polymerases recognize another code that signals the separation and release of the mRNA strand, also called the **transcript.** How long is the mRNA? The very smallest mRNA might consist of 100 bases; an average-sized mRNA might consist of 1,200 bases; and a large one, of several thousand.

TRANSLATION: THE SECOND STAGE OF GENE EXPRESSION

In translation, all of the elements needed to synthesize a protein, from the mRNA to the amino acids, are brought together on the ribosomes (figure 9.13). The entire process proceeds through the stages of initiation, elongation, termination, and protein folding and processing.

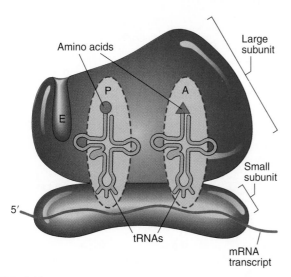

Amino acids

Large subunit

Small subunit

5′

tRNAs

mRNA transcript

FIGURE 9.13

The "players" in translation. A ribosome serves as the stage for protein synthesis. Assembly of the small and large subunits results in specific sites for holding the mRNA and two tRNAs with their amino acids.

Initiation of Translation

The mRNA molecule leaves the DNA transcription site and is transported to ribosomal staging sites in the cytoplasm. Ribosomal subunits are specifically adapted to assembling and forming sites to hold the mRNA and tRNAs. The ribosome thus recognizes these molecules and stabilizes reactions between them. The small subunit binds to the 5′ end of the mRNA, and the large subunit supplies enzymes for making peptide bonds on the protein. This part is like the videotape analogy, with the end of the mRNA being strung through the ribosome somewhat like a tape through the head of a tape player so that it will be read on the proper frame. The mRNA usually has a short leader that does not encode part of the protein (like the first part of a tape before the movie begins). As the ribosome begins to scan the mRNA, it encounters the START codon, which is almost always **AUG** (and rarely, GUG).

With the mRNA message in place on the assembled ribosome, the next step in translation involves entrance of tRNAs with their amino acids (figure 9.13). The pool of cytoplasm around the region contains a complete array of tRNAs, previously charged by having the correct amino acid attached. The step in which the complementary tRNA meets with the mRNA code is guided by the two sites on the large subunit of the ribosome called the **P site** (left) and the **A site** (right).[1] Think of these sites as shallow depressions in the larger subunit of the ribosome, each of which accommodates a tRNA. The ribosome also has an exit or **E site** that signals the release of used tRNAs.

The Master Genetic Code: The Message in Messenger RNA

By convention, the master genetic code is represented by the mRNA codons and the amino acids they specify (figure 9.14). Except in a very few cases, this code is universal, whether for

---

1. P stands for peptide site; A stands for aminoacyl site (a charged amino acid); E stands for exit site.

		Second Position			
	U	**C**	**A**	**G**	
U	UUU UUC } Phenylalanine UUA UUG } Leucine	UCU UCC UCA UCG Serine	UAU UAC } Tyrosine UAA UAG } STOP**	UGU UGC } Cysteine UGA STOP** UGG Tryptophan	U C A G
C	CUU CUC CUA CUG Leucine	CCU CCC CCA CCG Proline	CAU CAC } Histidine CAA CAG } Glutamine	CGU CGC CGA CGG Arginine	U C A G
A	AUU AUC Isoleucine AUA AUG START Methionine*	ACU ACC ACA ACG Threonine	AAU AAC } Asparagine AAA AAG } Lysine	AGU AGC } Serine AGA AGG } Arginine	U C A G
G	GUU GUC GUA GUG Valine	GCU GCC GCA GCG Alanine	GAU GAC } Aspartic acid GAA GAG } Glutamic acid	GGU GGC GGA GGG Glycine	U C A G

First Position (left vertical label); Third Position (right vertical label)

* This codon initiates translation.
**For these codons, which give the orders to stop translation, there are no corresponding tRNAs and no amino acids.

FIGURE 9.14

The Genetic Code: Codons of mRNA That Specify a Given Amino Acid. The master code for translation is found in the mRNA codons.

procaryotes, eucaryotes, or viruses. It is worth noting that once the triplet code on mRNA is known, the original DNA sequence, the complementary tRNA code, and the types of amino acids in the protein are automatically known (figure 9.15). One cannot predict from protein structure what the exact mRNA and DNA codons are because of a factor called **degeneracy,** meaning that more than one codon can translate into a given amino acid.

In figure 9.14, the mRNA codons and their corresponding amino acid specificities are given. Because there are 64 different triplet codes [2] and only 20 different amino acids, it is not surprising that some amino acids are represented by several codons. For example, leucine and serine can each be represented by any of six different triplets, and only tryptophan and methionine are represented by a single codon. In such codons as leucine, only the first two nucleotides are required to encode the correct amino acid, and the third nucleotide does not change its sense. This property, called *wobble,* is thought to permit some variation or mutation without harming the message.

The Beginning of Protein Synthesis

With mRNA serving as the guide, the stage is finally set for actual protein assembly. The correct **tRNA** (labeled 1 on figure 9.16) enters the P site and binds to the **start codon (AUG)** presented by the mRNA. Rules of pairing dictate that the anticodon of this tRNA must be complementary to the mRNA codon AUG, thus the tRNA with anticodon UAC will first occupy site P. It happens that the amino acid carried by the initiator tRNA in bacteria is formyl *methionine* ($_f$Met; see figure 9.14), though in many cases, it may not remain a permanent part of the finished protein.

Continuation and Completion of Protein Synthesis: Elongation and Termination

While reviewing the dynamic process of protein assembly, you will want to remain aware that the ribosome shifts its "reading frame" to the right along the mRNA from one codon to the next. This brings the next codon into place on the ribosome and makes a space for the next tRNA to enter the A position. A peptide bond is formed between the amino acids on the adjacent tRNAs, and the polypeptide grows in length (figure 9.16).

Elongation begins with the filling of the A site by a second **tRNA** (2 on figure 9.16). The identity of this tRNA and its amino acid is dictated by the second mRNA codon.

The entry of tRNA 2 into the A site brings the two adjacent tRNAs in favorable proximity for a peptide bond to form between the amino acids (aa) they carry. The $_f$Met is transferred from the first tRNA to aa 2, resulting in two coupled amino acids called a dipeptide (figure 9.16*b*).

For the next step to proceed, some room must be made on the ribosome, and the next codon in sequence must be brought into position for reading. This process is accomplished by *translocation,* the enzyme-directed shifting of the ribosome to the right along the mRNA strand, which causes the blank tRNA (1) to be discharged from the ribosome (figure 9.16*c*) at the E site. This also shifts the tRNA holding the dipeptide into P position. Site A is temporarily left empty. The tRNA that has been released is now free to drift off

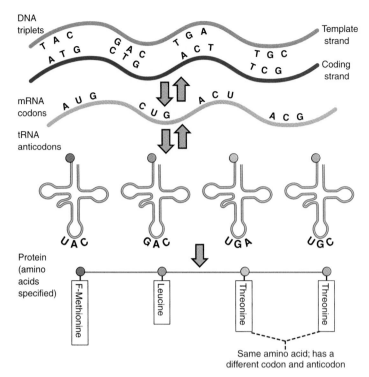

FIGURE 9.15

Interpreting the DNA code. If the DNA code is known, the mRNA codon can be surmised. If a codon is known, the anticodon can be determined, but it may not be possible to determine exact codon sequence from the anticodons because of the degeneracy of the code. It is also not possible to determine the exact codon or anticodon from protein structure.

into the cytoplasm and become recharged with an amino acid for later additions to this or another protein.

The stage is now set for the insertion of tRNA 3 at site A as directed by the third mRNA codon (figure 9.16*d*). This insertion is followed once again by peptide bond formation between the dipeptide and aa 3 (making a tripeptide), splitting of the peptide from tRNA 2, and translocation. This releases tRNA 2, shifts mRNA to the next position, moves tRNA 3 to position P, and opens position A for the next tRNA (4). From this point on, peptide elongation proceeds repetitively by this same series of actions out to the end of the mRNA.

The termination of protein synthesis is not simply a matter of reaching the last codon on mRNA. It is brought about by the presence of at least one special codon occurring just after the codon for the last amino acid. Termination codons—UAA, UAG, and UGA—are codons for which there is no corresponding tRNA. Although they are often called **nonsense codons,** they carry a necessary and useful message: *Stop* here. When this codon is reached, a special enzyme breaks the bond between the final tRNA and the finished polypeptide chain, releasing it from the ribosome.

Before newly made proteins can carry out their structural or enzymatic roles, they often require finishing touches. Even before the peptide chain is released from the ribosome, it begins folding upon itself to achieve its biologically active tertiary conformation. Other alterations, called *posttranslational* modifications, may be necessary. Some proteins must have the starting amino acid (formyl

---

2. $64 = 4^3$ (the 4 different codons in all possible combinations of 3).

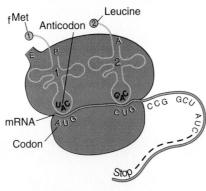

(a) Entrance of tRNAs 1 and 2

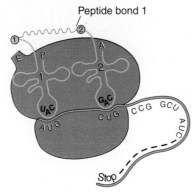

(b) Formation of peptide bond 1

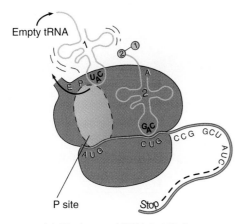

(c) Discharge of tRNA 1 at E site

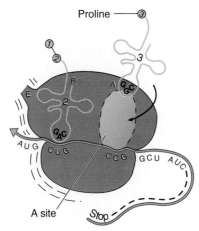

(d) First translocation; enter tRNA 3 by ribosome

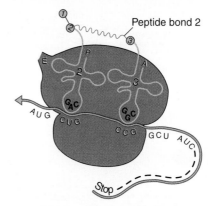

(e) Formation of peptide bond

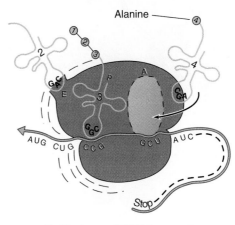

(f) Discharge of tRNA 2; second translocation; enter tRNA 4

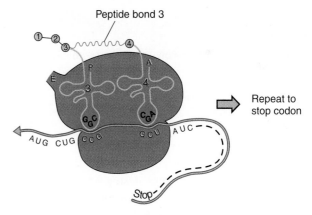

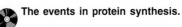

(g) Formation of peptide bond

FIGURE 9.16

The events in protein synthesis.

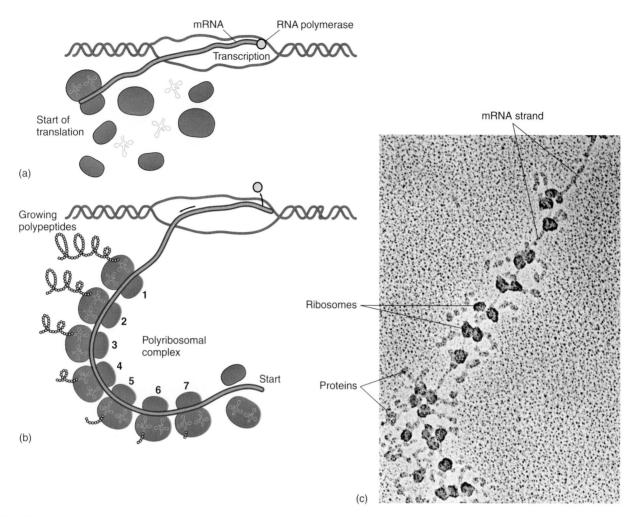

FIGURE 9.17

Speeding up the protein assembly line bacteria. (a) The mRNA transcript encounters ribosomal parts immediately as it leaves the DNA. **(b)** The ribosomal factories assemble along the mRNA in a chain, each ribosome reading the message and translating it into protein. Many products will thus be well along the synthetic pathway before transcription has even terminated. **(c)** Photomicrograph of a polyribosomal complex in action. Note that the protein "tails" vary in length depending on the stage of translation.

methionine) clipped off; proteins destined to become complex enzymes have cofactors added; and some join with other completed proteins to form quaternary levels of structure.

The machinelike operation of transcription and translation is amazing in its precision. Protein synthesis in bacteria is both efficient and rapid. At 37°C, 12–17 amino acids per second are added to a growing peptide chain. An average protein consisting of about 400 amino acids requires less than half a minute for complete synthesis. Further efficiency is gained when the translation of mRNA starts while transcription is still occurring (figure 9.17). A single mRNA is long enough to be fed through more than one ribosome simultaneously. This permits the synthesis of hundreds of protein molecules from the same mRNA transcript arrayed along a chain of ribosomes. This **polyribosomal complex** is indeed an assembly line for mass production of proteins. Protein synthesis consumes an enormous amount of energy. Nearly 1,200 ATPs are required just for synthesis of an average-sized protein.

EUCARYOTIC TRANSCRIPTION AND TRANSLATION: SIMILAR YET DIFFERENT

There are a few differences between procaryotic and eucaryotic gene expression. Eucaryotic DNA lies in the nucleus, so that is where mRNA originates. To be translated, it has to pass through the pores in the nuclear envelope to the ribosomes. Simultaneous transcription and translation as exhibited in procaryotes is not possible.

Eucaryotes and procaryotes share many similarities in protein synthesis. The start codon in eucaryotes is also AUG, but it codes for methionine alone. Another difference is that eucaryotic mRNAs code for just one protein, unlike bacterial mRNAs, which often contain several genes in series.

We have given the simplified definition of a gene that works well for procaryotes, but most eucaryotic genes are not colinear[3]—

3. Colinearity means that the base sequence can be read directly into a series of amino acids.

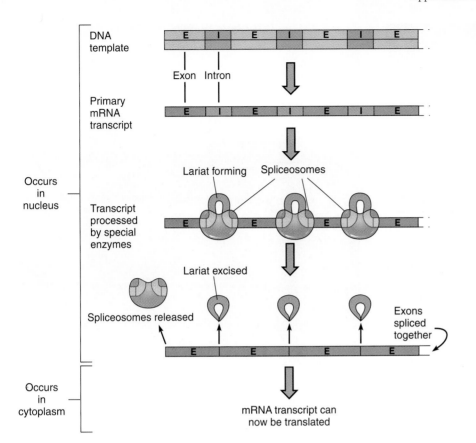

DNA template

Exon Intron

Primary mRNA transcript

Occurs in nucleus

Transcript processed by special enzymes

Lariat forming   Spliceosomes

Lariat excised

Spliceosomes released

Exons spliced together

Occurs in cytoplasm

mRNA transcript can now be translated

FIGURE 9.18

The split gene of eucaryotes. Eucaryotic genes have an additional complicating factor in their translation. Their coding sequences, or exons (E), are interrupted at intervals by segments called introns (I) that are not part of that protein's code. Introns are transcribed but not translated, which necessitates their removal by RNA splicing enzymes before translation.

meaning that they do *not* exist as an uninterrupted series of triplets coding for a protein. A eucaryotic gene contains the code for a protein, but located along the gene are one to several intervening sequences of bases, called **introns,** that do not code for product. Introns are interspersed between coding regions, called **exons,** that will be translated into product (figure 9.18). We can use words as examples. A short section of colinear procaryotic gene might read TOM SAW OUR DOG DIG OUT; a eucaryotic gene that codes for the same portion would read TOM SAW XZKP FPL OUR DOG QZWVP DIG OUT. The recognizable words are the exons, and the nonsense letters represent the introns.

This unusual genetic architecture, sometimes called a **split gene,** requires further processing before translation (figure 9.18). Transcription of the entire gene with both exons and introns occurs first. Next a type of RNA and protein called a spliceosome, recognizes the exon-intron junctions and enzymatically cuts through them. The action of this splicer enzyme loops the introns into lariat-shaped pieces, excises them, and joins the exons end to end. By this means, a strand of mRNA with no intron material is produced. This completed mRNA strand can then proceed to the cytoplasm to be translated.

At first glance, this system seems to be a cumbersome way to make a transcript, and the value of this extra genetic baggage is still the subject of much debate. Several different types of introns have been discovered, some of which do code for cell substances. One particular intron discovered in yeast gives the code for a reverse transcriptase, and other introns can be translated into endonucleases. Some experts hypothesize that introns represent a "sink" for extra bits of genetic material that could be available for splicing into existing genes, thus promoting genetic change and evolution. There is also evidence that introns serve as genetic regulators and in ribosomal assembly.

THE GENETICS OF ANIMAL VIRUSES

The genetic nature of viruses was described in chapter 6. Viruses essentially consist of one or more pieces of DNA or RNA enclosed in a protective coating. Above all, they are genetic parasites that require access to their host cell's genetic and metabolic machinery to be replicated, transcribed, and translated, and they also have the potential for genetically changing the cells. Because they contain only those genes needed for the production of new viruses, the genomes of viruses tend to be very compact and economical. In fact, this very simplicity makes them excellent subjects for the study of gene function.

The genetics of viruses is quite diverse (see chapters 24 and 25). In many viruses, the nucleic acid is linear in form; in others, it is circular. The genome of most viruses exists in a single molecule, though in a few, it is segmented into several smaller molecules. Most viruses contain normal double-stranded (ds) DNA or single-stranded (ss) RNA, but other patterns exist. There are ssDNA viruses, dsRNA viruses, and retroviruses, which work backward by making dsDNA from ssRNA. In some instances, viral genes overlap one another, and in a few DNA viruses, both strands contain a translatable message.

TABLE 9.3

Patterns of Genetic Flow in Viruses

DNA Viruses
  dsDNA → dsDNA (semiconservative)
  ssDNA → dsDNA → ssDNA (only one virus group)

RNA Viruses
  (+) ssRNA → (−) ssRNA → (+) ssRNA
  (−) ssRNA → (+) ssRNA → (−) ssRNA
  ssRNA → ssDNA → dsDNA → ssRNA (retroviruses)
  dsRNA → ssRNA → dsRNA (conservative)

A few generalities can be stated about viral genetics. In all cases, the viral nucleic acid penetrates the cell and is introduced into the host's gene-processing machinery at some point. In successful infection, an invading virus instructs the host's machinery to synthesize large numbers of new virus particles by a mechanism specific to a particular group. Replication of the DNA molecule of DNA animal viruses occurs in the nucleus, where the cell's DNA replication machinery lies (except in the poxviruses); the genome of most RNA viruses is replicated in the cytoplasm (except in the orthomyxoviruses [influenza]). In all viruses, viral mRNA is translated into viral proteins on host cell ribosomes using host tRNA. In the next section, we will briefly observe some major patterns of genetic replication (table 9.3).

Replication, Transcription, and Translation of dsDNA Viruses

Replication of dsDNA viruses is divided into phases (figure 9.19). During the early phase, viral DNA enters the nucleus, where several genes are transcribed into a messenger RNA. This transcript moves into the cytoplasm to be translated into viral proteins (enzymes) needed to replicate the viral DNA; this replication occurs in the nucleus. The host cell's own DNA polymerase is often involved, though some viruses (herpes, for example) have their own. During the late phase, other parts of the viral genome are transcribed and translated into proteins required to form the capsid and other structures. The new viral genomes and capsids are assembled, and the mature viruses are released by budding or cell disintegration.

Double-stranded DNA viruses interact directly with the DNA of their host cell. In some viruses, the viral DNA becomes silently *integrated* into the host's genome by insertion at a particular site on the host genome (figure 9.19). This integration may later lead to the transformation[4] of the host cell into a cancer cell and the production of a tumor. Several DNA viruses, including hepatitis B (HBV), the herpesviruses, and papillomaviruses (warts), are known to be initiators of cancers and are thus termed *oncogenic*.* The mechanisms of transformation and oncogenesis involve special genes called oncogenes that can regulate cellular genomes (see chapter 17).

---

4. The process of genetic change in a cell, leading to malignancy.

  * oncogenic (ahn″-koh-jen′-ik) Gr. *onkos*, mass, and *gennan*, to produce. Refers to any cancer-causing process. Viruses that do this are termed oncoviruses.

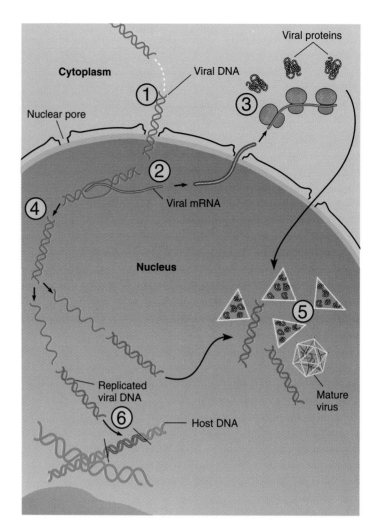

FIGURE 9.19

Genetic stages in the multiplication of double-stranded DNA viruses. The virus penetrates the host cell and releases DNA, which **(1)** enters the nucleus and **(2)** is transcribed. Other events are: **(3)** Viral mRNA is translated into structural proteins; proteins enter the nucleus. **(4)** Viral DNA is replicated repeatedly in the nucleus. **(5)** Viral DNA and proteins are assembled into a mature virus in the nucleus. **(6)** Because it is double-stranded, the viral DNA can insert itself into host DNA (latency).

Replication, Transcription, and Translation of RNA Viruses

RNA viruses exhibit several differences from DNA viruses. Their genomes are smaller and less stable; they enter the host cell already in an RNA form; and the virus cycle occurs entirely in the cytoplasm for most viruses. RNA viruses can have one of the following genetic messages: a positive-sense genome (+) that comes ready to be translated into proteins, a negative-sense genome (−) that must be converted to positive before translation, a positive-sense genome (+) that can be converted to DNA, or a dsRNA genome.

Positive-Sense Single-Stranded RNA Viruses   Positive-sense RNA viruses such as polio and hepatitis A virus must first replicate a negative strand as a master template to produce more positive strands. Shortly after the virus uncoats in the cell, its positive strand is translated into a large protein that is soon cleaved into individual

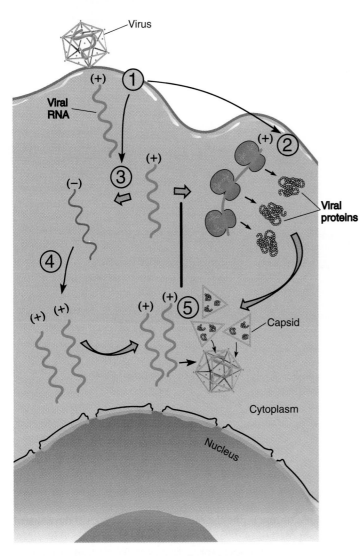

FIGURE 9.20

Replication of positive-sense, single-stranded RNA viruses. In general, these viruses do not enter the nucleus. **(1)** Penetration and uncoating of viral RNA. **(2)** Because it is positive in sense and single-stranded, the RNA can be directly translated on host cell ribosomes into various necessary viral proteins. **(3)** A negative genome is synthesized against the positive template to produce large numbers of positive genomes for final assembly. **(4)** The negative template is then used to synthesize a series of positive replicates. **(5)** RNA strands and proteins assemble into mature viruses.

functional units, one of which is a polymerase that initiates the replication of the viral strand (figure 9.20). Replication of a single-stranded positive-sense strand is done in two steps. First, a negative strand is synthesized using the parental positive strand as a template by the usual base-pairing mechanism. The resultant negative strand becomes a master template against which numerous positive daughter strands are made. Further translation of the viral genome produces large numbers of structural proteins for final assembly and maturation of the virus.

RNA Viruses with Reverse Transcriptase: Retroviruses A most unusual class of viruses has a unique capability to reverse the order of the flow of genetic information. Thus far in our discussion,

all genetic entities have shown the patterns DNA → DNA, DNA → RNA, or RNA → RNA. Retroviruses, including HIV, the cause of AIDS, and HTLV I, a cause of one type of human leukemia, synthesize DNA using their RNA genome as a template (see figure 25.16). They accomplish this by means of an enzyme, **reverse transcriptase,** that comes packaged with each virus particle. This enzyme synthesizes a single-stranded DNA against the viral RNA template and then directs the formation of a complementary strand of this ssDNA, resulting in a double strand of viral DNA. The dsDNA strand enters the nucleus, where it can be integrated into the host genome and transcribed by the usual mechanisms into new viral ssRNA. Translation of the viral RNA yields viral proteins for final virus assembly. The capacity of a retrovirus to become inserted into the host's DNA as a provirus has several possible consequences. In some cases, these viruses are oncogenic and are known to transform cells and produce tumors. It allows the AIDS virus to remain latent in an infected cell until a stimulus activates it to continue a productive cycle.

CHAPTER CHECKPOINTS

Information in DNA genes is converted to actual substances by the process of transcription and translation. The substances produced are structural proteins and enzymes. Enzymes, in turn, control an organism's metabolic activities.

DNA triplets specify the same mRNA codons, regardless of the species in which they occur. The language of DNA and mRNA extends into the noncellular world of viruses as well.

The processes of transcription and translation are similar but not identical for procaryotes and eucaryotes. Eucaryotes transcribe DNA in the nucleus, remove its introns, and translate it in the cytoplasm. Bacteria transcribe and translate simultaneously because there are no nucleus and introns. Bacterial polyribosomes attach themselves directly to the mRNA as it is released by its DNA template.

Viruses replicate by utilizing the transcription and translation processes of cellular organisms. They invade host cells and "force" them to transcribe and translate viral genes into new virus particles. In some cases, viruses connect themselves to the host genome and are passed on to its progeny, some of which will become virus factories many generations later.

The genetic material of viruses can be either DNA or RNA occurring in single, double, positive-, or negative-sense strands. Viral DNA contains just enough information to force a cell to become a virus factory.

Genetic Regulation of Protein Synthesis and Metabolism

In chapter 8 we surveyed the metabolic reactions in cells and the enzymes involved in those reactions. At that time, we mentioned types of metabolic regulation that are genetic in origin. These control mechanisms ensure that genes are active only when their products are required. In this way, enzymes will be produced to appropriately reflect nutritional status and prevent the waste of energy and materials in dead-end synthesis. Genetic function in procaryotes is

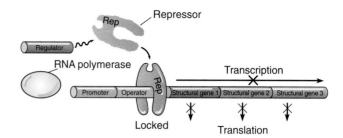

(a) Operon Off. In the absence of lactose, a repressor protein (the product of a regulatory gene) attaches to the operator gene of the operon. This effectively locks the operator and prevents any transcription of structural genes downstream (to its right). Suppression of transcription (and consequently, of translation) prevents the unnecessary synthesis of enzymes for processing lactose.

(b) Operon On. Upon entering the cell, the substrate (lactose) becomes a genetic inducer by attaching to the repressor, which loses its grip and falls away. The operator is now free to initiate transcription, and enzymatic products of translated genes perform the necessary reactions on their lactose substrate.

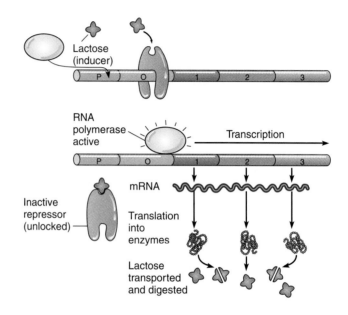

FIGURE 9.21

The lactose operon in bacteria: how inducible genes are controlled by substrate.

regulated by a specific collection of genes called an **operon.*** Operons use a coordinated set of control genes (regulators, promoters, and operators) to govern the operation of related structural genes. Such gene regulation responds to external stimuli and usually occurs at the level of transcription. Operons act in one of two ways: (1) The operon can be turned on *(induced)* by the substrate of the enzyme for which the structural genes code, or (2) the operon can be turned off *(repressed)* by the product its enzymes synthesize. Although operons are the primary regulators in bacteria, comparable devices probably govern the much more complicated eucaryotic metabolism as well.

THE LACTOSE OPERON: A MODEL FOR INDUCIBLE GENE REGULATION IN BACTERIA

The best understood cell system for explaining control through genetic induction is the **lactose *(lac)* operon.** This concept, first postulated in 1961 by François Jacob and Jacques Monod, accounts for the regulation of lactose metabolism in *Escherichia coli.* Many other operons with similar modes of action have since been identified, and together they furnish convincing evidence that the environment of a cell can have great impact on gene expression.

The lactose operon is made up of three segments, or *loci,** of DNA: (1) the **regulator,** composed of the gene that codes for a pro-

tein capable of repressing the operon (a **repressor**); (2) the **control locus,** composed of two genes, the **promoter** (recognized by RNA polymerase) and the **operator,** a sequence where transcription of the structural genes is initiated; and (3) the **structural locus,** made up of three genes, each coding for a different enzyme needed to catabolize lactose (figure 9.21). One of the enzymes, β-galactosidase, hydrolyzes the lactose into its monosaccharides; another, permease, brings lactose across the cell membrane.

In bacteria, structural genes required for the metabolism of a nutrient tend to be arranged in series. This is an efficient strategy that permits genes for a particular metabolic pathway to be induced or repressed in unison. The enzymes of the operon are of the inducible sort mentioned in chapter 8. The promoter, operator, and structural components lie in contiguous order, but the regulator can be at a distant site.

In inductive systems like the *lac* operon, the operon is normally in an *off mode* and does not initiate enzyme synthesis when the appropriate substrate is absent (figure 9.21a). How is the operon maintained in this mode? The key is in the repressor protein that is coded by the regulatory locus. This relatively large molecule has sites for allosteric binding, one for the operator and another for lactose. In the absence of lactose, this repressor binds with the

---

* operon (op´-ur-on) L. *opera,* exertion. A regulatory site for gene expression.

* loci (loh´-sy) sing. locus (loh´-kus) L. *locus,* a place. The site on a chromosome occupied by a gene.

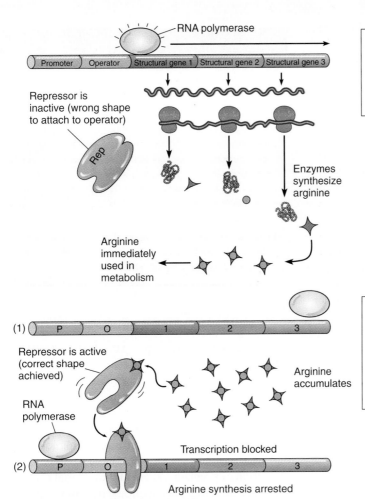

(a) **Operon On.** A repressible operon remains on when its nutrient products (here, arginine) are in great demand by the cell because the repressor remains inactive at low nutrient levels.

(b) **Operon Off.** The operon is repressed when (1) arginine builds up and, serving as a corepressor, activates the repressor. (2) The repressor complex affixes to the operator and blocks the RNA polymerase and further transcription of genes for arginine synthesis.

FIGURE 9.22
Repressible operon: control of a gene through excess nutrient.

operator locus, thereby blocking the transcription of the structural genes lying downstream. Think of the repressor as a lock on the operator, and if the operator is locked, the structural genes cannot be transcribed.

If lactose is added to the cell's environment, it triggers several events that turn the operon *on*. Because lactose is ultimately responsible for stimulating protein synthesis, it is called an *inducer*. The binding of lactose to the repressor protein causes a conformational change in the repressor that dislodges it from the operator segment (figure 9.21*b*). The control segment that was previously inactive is opened up, and RNA polymerase can now bind to the promoter. The structural genes are transcribed in a single unbroken transcript coding for all three enzymes. During translation, however, each gene is used to synthesize a separate protein.

As lactose is depleted, further enzyme synthesis is not necessary, so the order of events reverses. At this point there is no longer sufficient lactose to inhibit the repressor, hence the repressor is again free to attach to the operator. The operator is locked, and transcription of the structural genes and protein synthesis related to lactose both stop.

A fine but important point about the *lac* operon is that it functions only in the absence of glucose. Glucose is the preferred carbon source because it can be used immediately in growth, and does not require induction of an operon. When glucose is present, the necessary enzymes to bring lactose into the cells are not produced.

A REPRESSIBLE OPERON

Bacterial systems for synthesis of amino acids, purines and pyrimidines and many other processes work on a slightly different principle—that of repression. Similar factors such as repressor proteins, operators, and a series of structural genes exist for this operon, but with some important differences. Unlike the *lac* operon, this operon is normally in the *on* mode and will be turned *off* only when this nutrient is no longer required. The nutrient plays an additional role as a **corepressor** needed to block the action of the operon.

A growing cell that needs the amino acid arginine (arg) effectively illustrates the operation of a repressible operon. Under these conditions, the *arg* operon is set to *on,* and arginine is being actively synthesized through the action of its enzymatic products (figure 9.22*a*). In an active cell, the arginine will be used immediately, and the repressor will remain inactive because there is too little free arginine to activate it. As the cell's metabolism begins to slow down, however, the synthesized arginine will no longer be used up and will accumulate. The free arginine is then available to act as a corepressor by attaching to the repressor. This reaction produces a functioning repressor that locks the operator and stops further transcription and arginine synthesis (figure 9.22b).

Analogous gene control mechanisms in eucaryotic cells are not as well understood, but it is known that gene function can be altered by intrinsic regulatory segments similar to operons. Some molecules, called transcription factors, insert on the grooves of the DNA molecule and enhance transcription of specific genes. Examples include zinc "fingers" and leucine "zippers." Evidence of certain kinds of regulation (or its loss) can be detected in cancer cells, where genes called oncogenes abnormally override built-in genetic controls. An oncogene can exert its control by causing the formation of a chemical that permanently activates a part of the genome controlling cell growth. A cell with no constraints on the number of cell divisions can grow out of all normal bounds into tumors and leukemias (see chapter 17).

ANTIBIOTICS THAT AFFECT TRANSCRIPTION AND TRANSLATION

Naturally occurring cell nutrients are not the only agents capable of modifying gene expression. Some infection therapy is based on the concept that certain drugs react with DNA, RNA, and ribosomes and thereby alter genetic expression (see chapter 12). Treatment with such drugs is based on an important premise: that growth of the infectious agent will be inhibited by blocking its protein-synthesizing machinery selectively, without disrupting the cell synthesis of the patient receiving the therapy.

Drugs that inhibit protein synthesis exert their influence on transcription or translation. For example, the rifamycins used in therapy for tuberculosis bind to RNA polymerase, blocking the initiation step of transcription, and are selectively more active against bacterial RNA polymerase than the corresponding eucaryotic enzyme. Actinomycin D binds to DNA and halts mRNA chain elongation, but its mode of action is not selective for bacteria. For this reason, it is very toxic and never used to treat bacterial infections, though it can be applied in tumor treatment.

The ribosome is a frequent target of antibiotics that inhibit ribosomal function and ultimately protein synthesis. The value and safety of these antibiotics again depend upon the differential susceptibility of procaryotic and eucaryotic ribosomes. One problem with drugs that selectively disrupt procaryotic ribosomes is that the mitochondria of humans contain a procaryotic type of ribosome, and these drugs may inhibit the function of the host's mitochondria. One group of antibiotics (including erythromycin and spectinomycin) prevents translation by interfering with the attachment of mRNA to ribosomes. Chloramphenicol, lincomycin, and tetracycline bind to the ribosome in a way that blocks the elongation of the polypeptide, and aminoglycosides (such as streptomycin) inhibit peptide initiation and elongation. It is interesting to note that these drugs have served as important tools to explore genetic events because they can arrest specific stages in these processes.

CHAPTER CHECKPOINTS

Gene expression must be orchestrated to coordinate the organism's needs with nutritional resources. Genes can be turned "on" and "off" by specific molecules, which expose or hide their nucleotide codes for transcribing proteins. Most, but not all, of these proteins are enzymes.

Operons are collections of genes in bacteria that code for products with a coordinated function. They include genes for regulatory, operational, and structural components of the cell. Nutrients can combine with regulator gene products to turn a set of structural genes on (inducible genes) or off (repressible genes). The *lac* (lactose) operon is an example of an inducible operon. The *arg* (arginine) operon is an example of a repressible operon.

The rifamycins, tetracyclines, and aminoglycosides are classes of antibiotics that are effective because they interfere with transcription and translation processes in microorganisms.

Mutations: Changes in the Genetic Code

As precise and predictable as the rules of genetic expression seem, permanent changes do occur in the genetic code. Indeed, genetic change is the driving force of evolution. In microorganisms, such changes may become evident in altered gene expression such as the appearance or disappearance of anatomical or physiological traits. For example, a normally colored bacterium can lose its ability to form pigment, or a strain of the malarial parasite can develop resistance to a drug. Phenotypic changes of this type are ultimately due to changes in the genotype. Any permanent, inheritable change in the genetic information of the cell is a **mutation.** On a strictly molecular level, a mutation is an alteration in the nitrogen base sequence of DNA. It can involve the loss of base pairs, the addition of base pairs, or a rearrangement in the order of base pairs. Do not confuse this with genetic recombination, in which microbes transfer whole segments of genetic information between themselves.

A microorganism that exhibits a natural, nonmutated characteristic is known as a **wild type,** or wild strain. If a microorganism bears a mutation, it is called a **mutant strain.** Mutant strains can show variance in morphology, nutritional characteristics, genetic control mechanisms, resistance to chemicals, temperature preference, and nearly any type of enzymatic function. Mutant strains are very useful for tracking genetic events, unraveling genetic organization, and pinpointing genetic markers. A classic method of detecting mutant strains involves addition of various nutrients to a culture to screen for its use of that nutrient. For example, in a culture of a wild-type bacterium that is lactose-positive (meaning it has the necessary enzymes for fermenting this sugar), a small number of mutant cells have become lactose-negative, having lost the capacity to ferment this sugar. If the culture is plated on a medium containing indicators for fermentation, each colony can be observed for its fermentation reaction, and the negative strain isolated. Another standard method of detecting and isolating microbial mutants is by replica plating (figure 9.23).

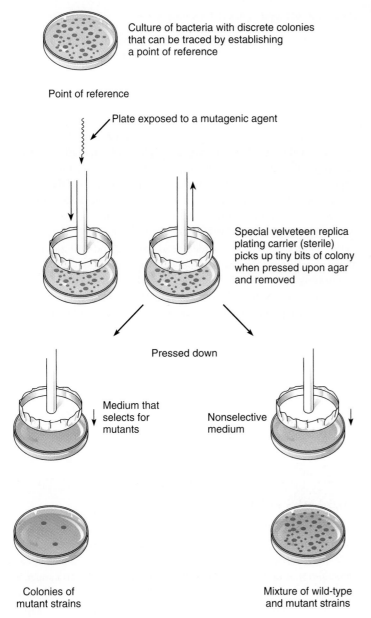

FIGURE 9.23

The general basis of replica plating. This method was developed by Joshua Lederberg for detecting and isolating mutant strains of microorganisms.

Culture of bacteria with discrete colonies that can be traced by establishing a point of reference

Point of reference

Plate exposed to a mutagenic agent

Special velveteen replica plating carrier (sterile) picks up tiny bits of colony when pressed upon agar and removed

Pressed down

Medium that selects for mutants

Nonselective medium

Colonies of mutant strains

Mixture of wild-type and mutant strains

TABLE 9.4

Selected Mutagenic Agents and Their Effects

Agent	Effect
Chemical	
Nitrous acid, bisulfite	Removes an amino group from some bases
Ethidium bromide	Inserts between the paired bases
Acridine dyes	Cause frameshifts due to insertion between base pairs
Nitrogen base analogs	Compete with natural bases for sites on replicating DNA
Radiation	
Ionizing (gamma rays, X rays)	Form free radicals that cause single or double breaks in DNA
Ultraviolet	Causes cross-links between adjacent pyrimidines

with DNA in a disruptive manner (table 9.4). The carefully controlled use of mutagens has proved a useful way to induce mutant strains of microorganisms for study.

Chemical mutagenic agents such as acridine dyes insert completely across the DNA helices between adjacent bases to produce a frameshift mutation and distort the helix (figure 9.24*a*). Analogs[5] of the nitrogen bases (5-bromodeoxyuridine and 2-aminopurine, for example) are chemical mimics of natural bases that are incorporated into DNA during replication. Addition of these abnormal bases leads to mistakes in base-pairing. Many chemical mutagens are also carcinogens, or cancer-causing agents (see the discussion of the Ames test in a later section of this chapter).

Physical agents that alter DNA are primarily types of radiation. High-energy gamma rays and X rays introduce major physical changes into DNA, and it accumulates breaks that may not be repairable. Ultraviolet (UV) radiation induces abnormal bonds between adjacent pyrimidines that prevent normal replication (see figure 11.10). Exposure to large doses of radiation can be fatal, which is why radiation is so effective in microbial control; it can also be carcinogenic in animals.

CATEGORIES OF MUTATIONS

Mutations range from large mutations, in which large genetic sequences are gained or lost (see later section on transposons), to small ones that affect only a single base on a gene. These latter mutations, which involve addition, deletion, or substitution of a few bases, are called **point mutations.** The effects of some point mutations are described in table 9.5.

To understand how a change in DNA influences the cell, remember that the DNA code appears in a particular order of triplets (three bases) that is transcribed into mRNA codons, each of which specifies an amino acid. A permanent alteration in the DNA that is

CAUSES OF MUTATIONS

A mutation is described as spontaneous or induced, depending upon its origin. A **spontaneous mutation** is a random change in the DNA arising from mistakes in replication or the detrimental effects of natural background radiation (cosmic rays) on DNA. The frequency of spontaneous mutations has been measured for a number of organisms. Mutation rates vary tremendously, from one mutation in 10^5 replications (a high rate) to one mutation in 10^{10} replications (a low rate). The rapid rate of bacterial reproduction allows these mutations to be observed more readily in bacteria than in animals.

Induced mutations result from exposure to known **mutagens,** which are primarily physical or chemical agents that interact

---

5. An analog is a chemical structured very similarly to another chemical except for minor differences in functional groups.

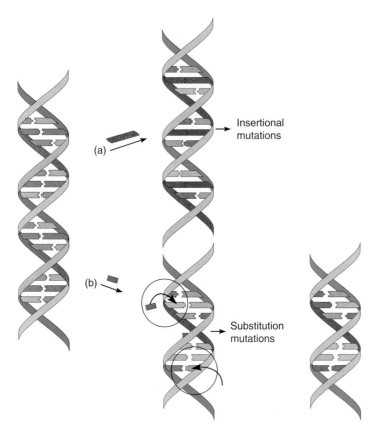

FIGURE 9.24

Two types of point mutation, showing modes of action of chemical mutagenic agents. (a) Some mutagens (ethidium bromide) insert into the helix and interfere with its replication. **(b)** Other chemicals act as imposters. Owing to their structural similarity to a regular nitrogenous base, they can fit in its place, but they cannot function like the regular base.

copied faithfully into mRNA and translated can change the structure of the protein. A change in a protein can likewise change the morphology and physiology of a cell. Most mutations have a harmful effect on the cell, leading to cell dysfunction or death; these are called lethal mutations. Neutral mutations produce neither adverse nor helpful changes. A small number of mutations are beneficial in that they provide the cell with a useful change in structure or physiology.

A change in the code that leads to placement of a different amino acid is called a **missense mutation.** A missense mutation can do one of the following: (1) create a faulty, nonfunctional protein, (2) produce a different but functional protein, or (3) cause no significant alteration in protein function.

A **nonsense mutation,** on the other hand, changes a normal codon into a stop codon that does not code for an amino acid and stops the production of the protein wherever it occurs. A nonsense mutation almost always results in a nonfunctional protein. A **silent mutation** alters a base but does not change the amino acid and thus has no effect. Because of the degeneracy of the code, ACU, ACC, ACG, and ACA all code for threonine, so a mutation that changes only the last base will not alter the sense of the message in any way. A **back-mutation** occurs when a gene that has undergone mutation reverses (mutates back) to its original base composition.

REPAIR OF MUTATIONS

Earlier we indicated that DNA has a proofreading mechanism to repair mistakes in replication that might otherwise become permanent (see page 258). Because mutations are potentially life-threatening, the cell has additional systems for finding and repairing DNA that has been damaged by various mutagenic agents and processes. Most ordinary DNA damage is resolved by enzymatic systems specialized for finding and fixing such defects.

DNA that has been damaged by ultraviolet radiation can be healed by **photoactivation** or **light repair.** This repair mechanism requires visible light and a light-sensitive enzyme, DNA photolyase, which can detect and attach to the damaged areas (sites of abnormal pyrimidine binding). Ultraviolet repair mechanisms are successful only for a relatively small number of UV mutations. Cells cannot repair severe, widespread damage and will die. In humans, the genetic disease *xeroderma pigmentosa* is due to nonfunctioning genes for the enzyme photolyase. Persons suffering from this rare disorder develop severe skin cancers; this relation provides strong evidence for a link between cancer and mutations.

Mutations can be excised by a series of enzymes that remove the incorrect bases and add the correct ones. This process is known as *excision repair.* First, enzymes break the bonds between the bases and the sugar-phosphate strand at the site of the error. A different enzyme subsequently removes the defective bases one at a time, leaving a gap that will be filled in by DNA polymerase I and ligase (figure 9.25). A repair system can also locate **mismatched bases** that were missed during proofreading: for example, C mistakenly paired with A, or G with T. The base must be replaced soon after the mismatch is made, or it will not be recognized by the repair enzymes.

THE AMES TEST

New agricultural, industrial, and medicinal chemicals are constantly being added to the environment, and exposure to them is widespread. The discovery that many such compounds are mutagenic and that up to 83% of these mutagens are linked to cancer is significant. Although animal testing has been a standard method of detecting chemicals with carcinogenic potential, a more rapid screening system called the **Ames test**[6] is also commonly used. In this ingenious test, the experimental subjects are bacteria whose gene expression and mutation rate can be readily observed and monitored. The premise is that any chemical capable of mutating bacterial DNA can similarly mutate mammalian (and thus human) DNA and is therefore potentially hazardous.

One indicator organism in the Ames test is a mutant strain of *Salmonella typhimurium*[7] that has lost the ability to synthesize the amino acid histidine, a defect highly susceptible to back-mutation because the strain also lacks DNA repair mechanisms. Mutations that cause reversion to the wild strain, which is capable of synthesizing histidine, occur spontaneously at a low rate. A test agent is considered a mutagen if it enhances the rate of back-mutation beyond levels that would occur spontaneously. One variation on this testing procedure is outlined in figure 9.26. The Ames test has

---

6. Named for its creator, Bruce Ames.

7. *S. typhimurium* inhabits the intestine of poultry and causes food poisoning in humans. It is used extensively in genetic studies of bacteria.

TABLE 9.5

Classification of Major Types of Mutations

Type of Mutation	Description	Effect of Mutation, Using Language Examples

Categories Based on Alteration of Base Sequence in DNA

Frameshift   Addition or loss of one or two bases in a gene changes reading frame, so that codon triplets no longer read in correct register.

Deletion   Removal of bases; triplet codes are offset by loss.   **Example**

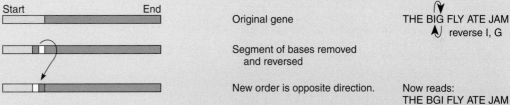

	Original gene	SUE SAW HER HEN SET
		delete E
	A piece is deleted at D.	
	Segment of gene right of deletion shifts left and attaches at point D.	Now reads: SUS AWH ERH ENS ET

Insertion   Addition of bases; triplet codes are offset by gain.

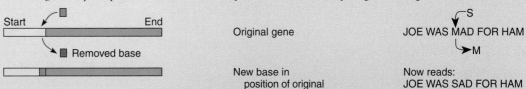

	Original gene	SAM DID HIS SPY ACT
		insert X
	Sequence inserted at P	
	Segment of gene right of P shifts right and attaches at point P.	Now reads: SAM XDI DHI SSP YAC T

Non-Frameshift

Inversion   Adjacent bases exchange positions, which creates an error in base-pairing and a change in one or two codons.

	Original gene	THE BIG FLY ATE JAM
		reverse I, G
	Segment of bases removed and reversed	
	New order is opposite direction.	Now reads: THE BGI FLY ATE JAM

Substitution   A wrong base is put in place of a correct base, which produces an error in base-pairing and a change in codon.

		S
	Original gene	JOE WAS MAD FOR HAM
	Removed base	M
	New base in position of original	Now reads: JOE WAS SAD FOR HAM

Categories Based on Overall Effect of Mutation

Silent	Change in base that causes no alteration in codon	GGT to GGA still gives proline.
Missense	Change in base that causes alteration in one codon with consequences that range from none to severe, depending upon the nature of the amino acid change	Weaker mutation substitutes different amino acid but does not greatly affect protein function: e.g., TAA mutated to GAA substitutes leucine for isoleucine.
		Harmful mutation changes amino acid and protein structure so that function is not competent: e.g., CTT to CAT substitutes valine for glutamic acid, changing the folding pattern of the protein.
Nonsense	Change in base that causes the development of a STOP codon somewhere along the gene	Will always interrupt protein synthesis and cause a nonfunctional, incomplete protein; outcome can be lethal if this is an essential gene with only a few copies.

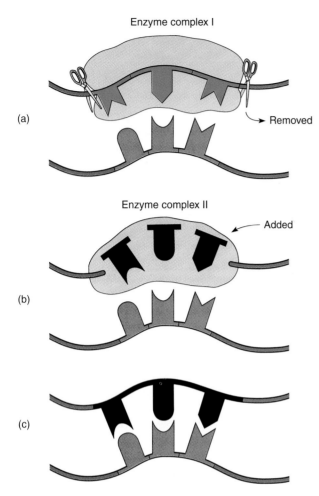

Enzyme complex I

(a)

→ Removed

Enzyme complex II

→ Added

(b)

(c)

FIGURE 9.25

Excision repair of mutation by enzymes. (a) The first enzyme complex recognizes one or several incorrect bases and removes them. **(b)** The second complex (DNA polymerase I and ligase) places correct bases and seals the gaps. **(c)** Repaired DNA.

proved invaluable for screening an assortment of environmental and dietary chemicals for mutagenesis and carcinogenicity without resorting to more expensive and time-consuming animal studies.

POSITIVE AND NEGATIVE EFFECTS OF MUTATIONS

Many mutations are not repaired. How the cell copes with them depends on the nature of the mutation and the strategies available to that organism. Mutations that are permanent and heritable will be passed on to the offspring of organisms and new viruses and will be a long-term part of the gene pool. Some mutations are harmful to organisms; others provide adaptive advantages.

If a mutation leading to a nonfunctional protein occurs in a gene for which there is only a single copy, as in haploid or simple organisms, the cell will probably die. This happens when certain mutant strains of *E. coli* acquire mutations in the genes needed to repair damage by UV radiation. Mutations of the human genome affecting the action of a single protein (mostly enzymes) are responsible for more than 3,500 diseases (see chapter 10).

Although most spontaneous mutations are not beneficial, a small number contribute to the success of the individual and the

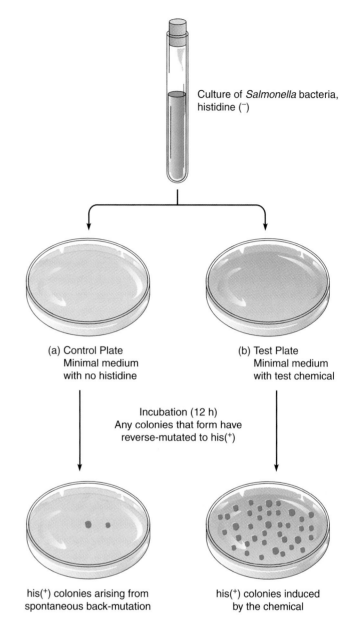

Culture of *Salmonella* bacteria, histidine ($^-$)

(a) Control Plate
Minimal medium
with no histidine

(b) Test Plate
Minimal medium
with test chemical

Incubation (12 h)
Any colonies that form have
reverse-mutated to his($^+$)

his($^+$) colonies arising from
spontaneous back-mutation

his($^+$) colonies induced
by the chemical

FIGURE 9.26

The Ames test. This test is based on a strain of *Salmonella typhimurium* that cannot synthesize histidine (his$^-$). It lacks the enzymes to repair DNA so that mutations show up readily, and has leaky cell walls that permit the ready entrance of chemicals. Many potential carcinogens (benzanthracene and aflatoxin, for example) are mutagenic agents only after being acted on by mammalian liver enzymes, so an extract of these enzymes is added to the test medium. **(a)** In the control setup, bacteria are plated on a histidine-free medium containing liver enzymes but lacking the test agent. **(b)** The experimental plate is prepared the same way except that it contains the test agent. After incubation, plates are observed for colonies. Any colonies developing on the plates are due to a back-mutation in a cell, which has reverted it to a his($^+$) strain. The degree of mutagenicity of the chemical agent can be calculated by comparing the number of colonies growing on the control plate with the number on the test plate. Chemicals that produce an increased incidence of back-mutation are considered carcinogens.

population by creating variant strains with alternate ways of expressing a trait. Microbes are not "aware" of this advantage and do not direct these changes; they simply respond to the environment they encounter. Those organisms with beneficial mutations can

TABLE 9.6

Types of Intermicrobial Exchange

Mode	Factors Involved	Direct or Indirect*	Genes Transferred
Conjugation	Donor cell with pilus Fertility plasmid in donor Both donor and recipient alive Gram-negative and gram-positive bacteria	Direct	Drug resistance; resistance to metals; toxin production; enzymes; adherence molecules; degradation of toxic substances; uptake of iron
Transformation	Free donor DNA (fragment) Live, competent recipient cell	Indirect	Polysaccharide capsule; unlimited with cloning techniques
Transduction	Donor is lysed bacterial cell Defective bacteriophage is carrier of donor DNA Live, competent recipient cell of same species as donor	Indirect	Toxins; enzymes for sugar fermentation; drug resistance

Direct means the donor and recipient are in contact during exchange; indirect means they are not.

more readily adapt, survive, and reproduce. In the long-range view, mutations and the variations they produce are the raw materials for change in the population and, thus, for evolution.

Mutations that create variants occur frequently enough that any population contains mutant strains for a number of characteristics, but as long as the environment is stable, the population will remain stable. When the environment changes, however, it can become hostile for the survival of certain individuals, and only those microbes bearing protective mutations can **adapt** to the new environment and survive. In this way, the environment **naturally selects** certain mutant strains that will reproduce, give rise to subsequent generations, and in time, be the dominant strain in the population. Through these means, any change that confers an advantage during selection pressure will be retained by the population. One of the clearest models for this sort of selection and adaptation is acquired drug resistance in bacteria (see chapter 12). One of the fascinating mechanisms bacteria have developed for increasing their adaptive capacity is genetic exchange (called genetic recombination).

DNA Recombination Events

Genetic recombination, or hybridization through sexual reproduction, is an important foundation of genetic variation in eucaryotes. Although bacteria have no exact equivalent to sexual reproduction, they exhibit a primitive means for sharing or recombining parts of their genome. An event in which one bacterium donates DNA to another bacterium is a type of genetic transfer termed **recombination.** The properties of bacteria that influence recombination are the presence of extrachromosomal DNA and the versatility bacteria display in interchanging genes. The end result of recombination is a new strain different from both the donor and the original recipient strain. Genetic exchanges have tremendous effects on the genetic diversity of bacteria, and unlike spontaneous mutations, are generally beneficial to them. They provide additional genes for resistance to drugs and metabolic poisons, new nutritional and metabolic capabilities, and increased virulence and adaptation to the environment in general.

TRANSMISSION OF GENETIC MATERIAL IN BACTERIA

DNA transmitted between bacteria involves **plasmids,** closed circular molecules of DNA separate from chromosomes, or small chromosomal fragments that have escaped from a lysed cell. Plasmids are not necessary to basic bacterial survival, but they can be genetically useful and confer greater versatility or adaptability on the recipient cell. In addition to being transferrable, donated DNA can be integrated into the chromosome of the recipient and replicated and transmitted to progeny during cell division. Plasmids and integrated genes are transcribed and translated along with the regular chromosome. Genetic recombination occurs at a low rate in natural populations, and it can also be induced in the laboratory. In chapter 10 we will see that bacteria, viruses, and some yeasts are capable of receiving and using the DNA of other organisms that is grafted into a plasmid.

Depending upon the mode of transmission, the means of genetic recombination is called conjugation, transformation, or transduction. **Conjugation*** requires the attachment of two related species and the formation of a bridge that can transport DNA. **Transformation*** entails the transfer of naked DNA and requires no special vehicle. **Transduction*** is DNA transfer mediated through the action of a bacterial virus (table 9.6).

Conjugation: Bacterial Sex

Conjugation is a mode of sexual mating in which a plasmid or other genetic material is transferred by a donor to a recipient cell via a direct connection (figure 9.27). It occurs primarily in gram-negative bacteria, but many gram-positive cells can conjugate. In gram-negative cells, the donor has a plasmid (**fertility,** or **F' factor**) that allows the synthesis of a **pilus,** or **conjugative pilus.** The recipient cell is a related species or genus that has a recognition site on its surface. A cell's role in conjugation is denoted by F^+ for the cell that has the F plasmid and by F^- for the cell that lacks it. Contact is

---

* conjugation (kahn″-jew-gay′-shun) L. *conjugatus,* yoked together.

* transformation (trans-for-may′-shun) L. *trans,* across, and *formatio,* to form. This term is also used to mean the cancerous (malignant) conversion of cells.

* transduction (trans-duk′-shun) L. *transducere,* to lead across.

(a, b) **Process of Conjugation**

(c) **F Factor Transfer**

(d) **Hfr Transfer**

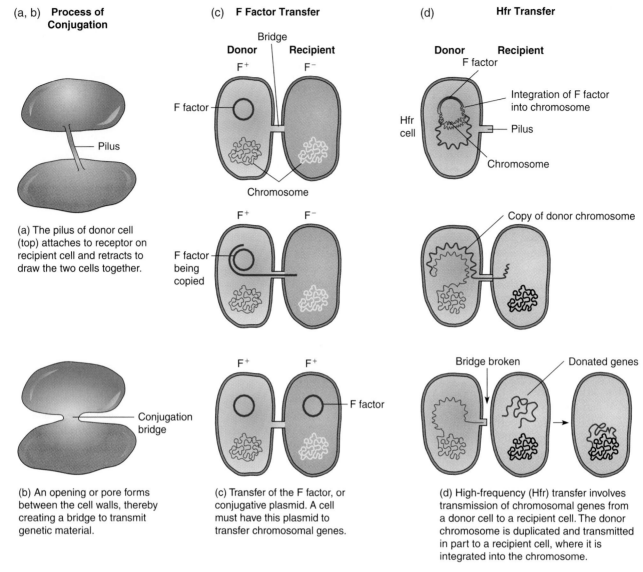

(a) The pilus of donor cell (top) attaches to receptor on recipient cell and retracts to draw the two cells together.

(b) An opening or pore forms between the cell walls, thereby creating a bridge to transmit genetic material.

(c) Transfer of the F factor, or conjugative plasmid. A cell must have this plasmid to transfer chromosomal genes.

(d) High-frequency (Hfr) transfer involves transmission of chromosomal genes from a donor cell to a recipient cell. The donor chromosome is duplicated and transmitted in part to a recipient cell, where it is integrated into the chromosome.

FIGURE 9.27

Conjugation: genetic transmission through direct contact between two cells.

made when a pilus grows out from the F^+ cell, attaches to the surface of the F^- cell, contracts, and draws the two cells together (see figure 4.9). In both gram-positive and gram-negative cells, an opening is created between the connected cells, and the replicated DNA passes across from one cell to the other. Conjugation is a very conservative process, in that the donor bacterium generally retains a copy of the genetic material being transferred.

There are hundreds of conjugative plasmids with some variations in their properties. One of the best understood plasmids is the F factor in *E. coli,* which exhibits these patterns of transfer:

1. The donor (F^+) cell makes a copy of its F factor and transmits this to a recipient (F^-) cell. The F^- cell is thereby changed into an F cell capable of producing a pilus and conjugating with other cells (figure 9.27c). No additional donor genes are transferred at this time.

2. In high-frequency recombination (Hfr) donors, the fertility factor has been integrated into the F^+ donor chromosome.

The term *high-frequency recombination* was adopted to denote that a cell with an integrated F factor transmits its chromosomal genes at a higher frequency than other cells.

The F factor can direct a more comprehensive transfer of part of the donor chromosome to a recipient cell. This transfer occurs through duplication of the DNA by means of the rolling circle mechanism. One strand of DNA is retained by the donor, and the other strand is transported across to the recipient cell (figure 9.27d). The F factor may not be transferred during this process. The transfer of an entire chromosome takes about 100 minutes, but the pilus bridge between cells is ordinarily broken before this time, and rarely is the entire genome of the donor cell transferred.

Conjugation has great biomedical importance. Special **resistance (R) plasmids,** or **factors,** that bear genes for resisting antibiotics and other drugs are commonly shared among bacteria through conjugation. Transfer of R factors can confer multiple resistance to antibiotics such as tetracycline, chloramphenicol, strep-

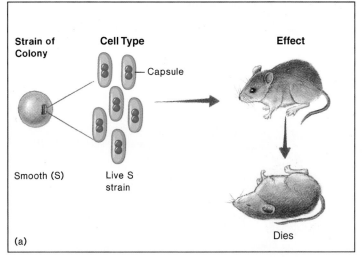

Strain of Colony | Cell Type | Effect

Capsule

Smooth (S)   Live S strain

Dies

(a)

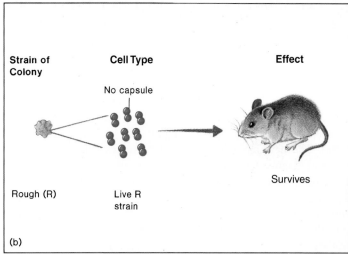

Strain of Colony | Cell Type | Effect

No capsule

Rough (R)   Live R strain

Survives

(b)

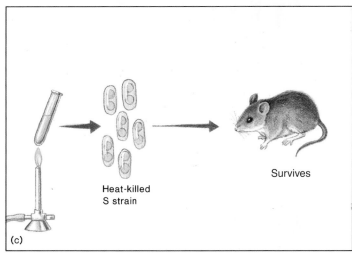

Heat-killed S strain

Survives

(c)

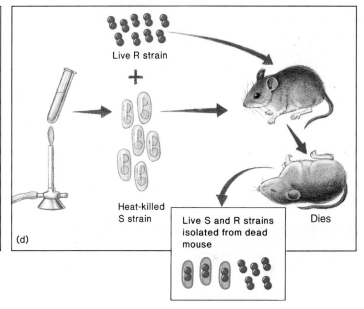

Live R strain

+

Heat-killed S strain

Live S and R strains isolated from dead mouse

Dies

(d)

FIGURE 9.28

 Griffith's classic experiment in transformation. In essence, this experiment proved that DNA released from a killed cell can be acquired by a live cell. The cell receiving this new DNA is genetically transformed—in this case, from a nonvirulent strain to a virulent one.

tomycin, sulfonamides, and penicillin. This phenomenon is discussed further in chapter 12. Other types of R factors carry genetic codes for resistance to heavy metals (nickel and mercury) or for synthesizing virulence factors (toxins, enzymes, and adhesion molecules) that increase the pathogenicity of the bacterial strain. Conjugation studies have also provided an excellent way to map the bacterial chromosome.

Transformation: Capturing DNA from Solution

One of the cornerstone discoveries in microbial genetics was made in the late 1920s by the English biochemist Frederick Griffith working with *Streptococcus pneumoniae* and laboratory mice. The pneumococcus exists in two major strains based on the presence of the capsule, colonial morphology, and pathogenicity. Encapsulated strains bear a smooth (S) colonial appearance and are virulent; strains lacking a capsule have a rough (R) appearance and are nonvirulent (see figure 4.12). (Recall that the capsule protects a bac-

terium from the phagocytic host defenses.) To set the groundwork, Griffith showed that when a mouse was injected with a live, virulent (S) strain, it soon died (figure 9.28*a*). When another mouse was injected with a live, nonvirulent (R) strain, the mouse remained alive and healthy (figure 9.28*b*). Next he tried a variation on this theme. First, he heat-killed an S strain and injected it into a mouse, which remained healthy (figure 9.28*c*). Then came the ultimate test: Griffith injected both dead S cells and live R cells into a mouse, with the result that the mouse died from pneumococcal blood infection (figure 9.28*d*). If killed bacterial cells do not come back to life and the nonvirulent strain was harmless, why did the mouse die? Although he did not know it at the time, Griffith had demonstrated that dead S cells, while passing through the body of the mouse, broke open and released some of their DNA (by chance, that part containing the genes for making a capsule). A few of the live R cells subsequently picked up this loose DNA and were **transformed** by it into virulent, capsule-forming strains.

Later studies supported the concept that a chromosome released by a lysed cell breaks into fragments small enough to be accepted by a recipient cell and that DNA, even from a dead cell, retains it code. This nonspecific acceptance by a bacterial cell of small fragments of soluble DNA from the surrounding environment is termed **transformation.** Transformation is apparently facilitated by special DNA-binding proteins on the cell wall that capture DNA from the surrounding medium. Cells that are capable of accepting genetic material through this means are termed *competent*. The new DNA is processed by the cell membrane and transported into the cytoplasm, where it is inserted into the bacterial chromosome. Transformation is a natural event among both gram-positive streptococci and gram-negative *Haemophilus* and *Neisseria* species. In addition to genes coding for the capsule, bacteria also exchange genes for antibiotic resistance and bacteriocin synthesis in this way.

Because transformation requires no special appendages, and the donor and recipient cells do not have to be in direct contact, the process is useful for certain types of recombinant DNA technology. With this technique, foreign genes from a completely unrelated organism are inserted into a plasmid, which is then introduced into a competent bacterial cell through transformation. These recombinations can be carried out easily in a test tube, and human genes can be experimented upon and even expressed outside the human body by placing them in a microbial cell. This same phenomenon in eucaryotic cells, termed *transfection,* is an essential aspect of genetically engineered yeasts, bird embryos, and mice, and it has been proposed as a future technique for curing genetic diseases in humans. These topics are covered in more detail in chapter 10.

Transduction: The Case of the Piggyback DNA

Bacteriophages (bacterial viruses) have been previously described as destructive bacterial parasites. Infection by a virus does not always kill the host cell, however, and viruses can in fact serve as genetic vectors (an entity that can bring foreign DNA into a cell). The process by which a bacteriophage serves as the carrier of DNA from a donor cell to a recipient cell is **transduction.** Although it occurs naturally in a broad spectrum of bacteria, the participating bacteria in a single transduction event must be the same species because of the specificity of viruses for host cells. The events in transduction are shown in figure 9.29.

There are two versions of transduction. In *generalized transduction*, random fragments of disintegrating host DNA are taken up by the phage during assembly. Virtually any gene from the bacterium can be transmitted through this means. In *specialized transduction,* a highly specific part of the host genome is regularly incorporated into the virus. This specificity is explained by the prior existence of a temperate prophage inserted in a fixed site on the bacterial chromosome. When activated, the prophage DNA separates from the bacterial chromosome, carrying a small segment of host genes with it. During a lytic cycle, these specific viral-host gene combinations are incorporated by the viral particles and carried to another bacterial cell.

Several cases of specialized transduction have biomedical importance. The virulent strains of bacteria such as *Corynebacterium diphtheriae, Clostridium* spp., and *Streptococcus pyogenes* all produce toxins with profound physiological effects, whereas nonvirulent strains do not produce toxins. It turns out that toxicity

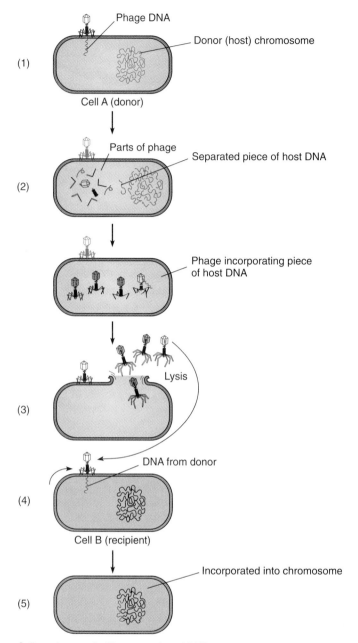

FIGURE 9.29

Generalized transduction: genetic transfer by means of a virus carrier.
(1) A phage infects cell A (the donor cell) by normal means. **(2)** During replication and assembly, a phage particle incorporates a segment of bacterial DNA by mistake. **(3)** Cell A then lyses and releases the mature phages, including the genetically altered one. **(4)** The altered phage adsorbs to and penetrates another host cell (cell B), injecting the DNA from cell A rather than viral nucleic acid. **(5)** Cell B receives this donated DNA, which recombines with its own DNA. Because the virus is defective (biologically inactive as a virus), it is unable to complete a lytic cycle. The transduced cell survives and can use this new genetic material.

arises from the expression of bacteriophage genes that have been introduced by transduction. Only those bacteria infected with a temperate phage are toxin formers. Other instances of transduction are seen in staphylococcal transfer of drug resistance and in the transmission of gene regulators in gram-negative rods *(Escherichia, Salmonella).*

Transposons: "This Gene Is Jumpin'"   One type of genetic transferral of great interest involves **transposons.** These transposable elements have the distinction of shifting from one part of the genome to another and so are termed "jumping genes." When the idea of their existence in corn plants was first postulated by the geneticist Barbara McClintock, it was greeted with some skepticism, because it had long been believed that the locus of a given gene was set and that genes did not or could not move around. Now it is evident that jumping genes are widespread among procaryotic and eucaryotic cells and viruses.

All transposons share the general characteristic of traveling from one location to another on the genome—from one chromosomal site to another, from a chromosome to a plasmid, or from a plasmid to a chromosome (figure 9.30). Because transposons occur in plasmids, they can also be transmitted from one cell to another in bacteria and a few eucaryotes. Some transposons replicate themselves before jumping to the next location, and others simply move without replicating first.

Transposons can be recognized by the palindromic sequences at each end, some of which are hundreds of bases long. These regions permit discovery and removal of the transposon sequence, and they dictate the site of insertion into DNA when the transposon relocates.

The overall effect of transposons—to scramble the genetic language—can be beneficial or adverse, depending upon such variables as where insertion occurs in a chromosome, what kinds of genes are relocated, and the type of cell involved. On the beneficial side, transposons are known to be involved in (1) the creation of different genetic combinations, necessary for the high levels of variation in antibodies and receptor molecules in cells (see chapter 15); (2) changes in traits such as colony morphology, pigmentation, pili, and antigenic characteristics; (3) replacement of damaged DNA; and (4) the intermicrobial transfer of drug resistance (in bacteria). On the negative side, rearrangement of DNA that leads to mutations such as deletions, insertions, translocations, inversions, and chromosome breakage can be disruptive and even lethal. Retroviruses such as the AIDS virus behave as transposons by randomly inserting in the genome in ways that can lead to severe cell dysfunction (see chapters 17 and 25).

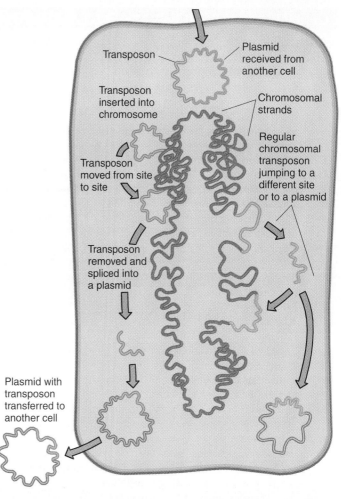

FIGURE 9.30

Transposons: shifting segments of the genome. Potential mechanisms in the movement of transposons in bacterial cells. Some transposons are plasmids that are shifted from site to site; others are regular parts of chromosomes that are moved out of one site and spliced into another.

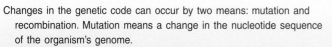

CHAPTER CHECKPOINTS

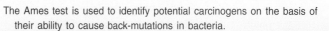

Changes in the genetic code can occur by two means: mutation and recombination. Mutation means a change in the nucleotide sequence of the organism's genome.

Recombination means the addition of genes from an outside source, such as a virus or another cell.

Mutations can be either spontaneous or induced by exposure to some external mutagenic agent.

All cells have enzymes that repair damaged DNA. When the degree of damage exceeds the ability of the enzymes to make repairs, mutations occur.

Mutation-induced changes in DNA nucleotide sequencing range from a single nucleotide to addition or deletion of large sections of genetic material.

The Ames test is used to identify potential carcinogens on the basis of their ability to cause back-mutations in bacteria.

Genetic recombination occurs in eucaryotes through sexual reproduction. In bacteria, recombination occurs through the processes of transformation, conjugation, and transduction and the incorporation of lysogenic viruses into the genome.

Transposons are genes that can relocate from one part of the genome to another, causing rearrangement of genetic material. Such rearrangements have both beneficial and harmful consequences for the organism involved.

CHAPTER CAPSULE WITH KEY TERMS

I. Genes and the Genetic Material

A. **Genetics** is the study of **heredity,** and the **genome** is the sum total of genetic material of a cell.

B. A **chromosome** is composed of DNA in all organisms; **genes** are specific segments of the elongate DNA molecule. Genes code for polypeptides and proteins that become enzymes, antibodies, or structures in the cell.

II. Gene Structure and Replication

A. A **gene** consists of DNA, a double helix formed from linked **nucleotides** composed of a **phosphate, deoxyribose sugar,** and a **nitrogen base—purine** or **pyrimidine.**

B. The backbone of the molecule is formed of **antiparallel** strands of repeating deoxyribose sugar-phosphate units that are linked together by the base-pairing of adenine with thymine and cytosine with guanine. The order of base pairs in DNA constitutes the genetic code. The very long DNA molecule must be highly coiled to fit into the cell.

C. Pairing ensures the accuracy of the copying of DNA synthesis or **replication.**

   1. Replication is **semiconservative** and requires enzymes such as **helicase, DNA polymerase, ligase,** and **gyrase.**

   2. These components in conjunction with the chromosome being duplicated constitute a **replicon.** The unzipped strands of DNA function as **templates.** Synthesis proceeds along two **replication forks,** each with a **leading strand** and a **lagging strand.**

III. Gene Function

A. Processing of genetic information proceeds from DNA to RNA and to protein. RNA synthesis is called **transcription,** and protein synthesis is called **translation.**

B. An organism's genetic makeup, or **genotype,** provides information for the expression of traits, called the **phenotype.** The genetic code is organized into **triplets** of nitrogen bases, and each triplet (codon) corresponds to a particular amino acid in a protein.

C. *Types of RNA:* Unlike DNA, RNA is single-stranded and contains **uracil** instead of thymine and **ribose** instead of deoxyribose.

   1. Chief forms are messenger, transfer, and ribosomal RNA (**mRNA, tRNA,** and **rRNA).**

   2. Triplets of bases on mRNA called **codons** convey genetic information for protein structure. The corresponding **anticodon** on tRNA ensures delivery of the appropriate amino acid. Ribosomes combine rRNA and protein to form small and large subunits that provide staging sites for translation.

D. *Transcription and Translation:*

   1. Transcription begins when **RNA polymerase** recognizes a promoter region on DNA and elongation of the new strand proceeds in the 5′ to 3′ direction. Only one DNA strand, the **template,** is copied. The mRNA strand goes to a ribosome for translation.

   2. Translation begins at the **AUG** or **start codon.** Protein assembly proceeds as tRNAs enter the ribosome with anticodons complementary to the codons of the mRNA. Peptide bonds are formed between amino acids on adjacent tRNAs. This process continues until the **nonsense (stop) codon** on mRNA is reached.

E. *Eucaryotic Gene Expression:* Eucaryotes have **split genes** interrupted by noncoding regions called **introns** that separate coding segments called **exons.** The synthesis of the final mRNA transcript requires splicing to delete stretches that correspond to introns.

IV. The Genetics of Animal Viruses

A. Genomes of viruses can be linear or circular; segmented or not; made of double-stranded (ds) DNA, single-stranded (ss) DNA, ssRNA, or dsRNA.

B. In general, DNA viruses replicate in the nucleus, RNA viruses in the cytoplasm.

C. Retroviruses synthesize dsDNA from ssRNA.

D. The DNA of some viruses can be silently integrated into the host's genome. Integration by **oncogenic** viruses can lead to **transformation** of the host cell into an immortal cancerous cell.

E. RNA viruses have strand polarity (positive- or negative-sense genome) and double-strandedness.

V. Regulation of Genetic Function

A. Protein synthesis and metabolism are regulated by **gene induction** or **repression,** as controlled by an **operon.**

B. An operon is a DNA unit of **regulatory genes** (made up of **regulators, promoters,** and **operators**) that controls the expression of **structural genes** (which code for enzymes and structural peptides).

   1. **Inducible operons** such as the **lactose operon** are normally *off* but are turned *on* by a lactose inducer.

   2. **Repressible operons** govern anabolism and are usually *on,* but can be shut *off* when the end product is no longer needed.

   3. Certain antibiotics affect transcription and translation. Adverse modes of action include interference with RNA polymerase, RNA elongation, and ribosomal activity.

VI. Gene Mutation

A. Genome changes in microbes come from **mutations** and intermicrobial genetic exchanges.

B. The term **wild type,** or strain, denotes the original form; **mutant strain** refers to the altered version.

C. Mutations benefit organisms through an adaptive advantage, but some are lethal.

D. *Types of Mutations:*

   1. Mutations are **spontaneous** if they occur randomly and **induced** if they are due to directed chemical or physical agents called **mutagens.**

   2. **Point mutations** entail addition, removal, or substitution of a few bases.

      a. A **missense mutation** leads to amino acid substitution.

      b. A **nonsense mutation** arrests peptide synthesis without amino acid insertion.

      c. A **silent mutation** causes base substitution without amino acid substitution.

      d. A **back-mutation** is a reversion to the original base composition.

E. *Repair of Mutations:* Enzymes can identify and repair mutations. Some enzymes locate **mismatched** bases and engage in repairs. DNA damaged by ultraviolet light can be corrected by **photoactivation** or light **repair.**

F. The **Ames test** is a mutation-screening method based on the susceptibility of mutant *Salmonella typhimurium* to a back-mutation. It is used to measure the mutagenicity of chemicals.

VII. DNA Recombination

A. Intermicrobial transfer and genetic recombination permit gene sharing between bacteria.

1. In **conjugation,** one bacterium donates a **plasmid (fertility,** or **F factor)** to a compatible recipient via a **conjugative pilus.** Conjugation resulting in transfer of **resistance,** or **R, plasmids** is biomedically important.
2. In bacterial **transformation,** naked DNA is transferred without special carriers. It was first demonstrated in smooth and rough pneumococcal strains.
3. **Transduction** is transfer of host DNA by bacteriophages. Drug resistance and virulence genes can be transferred by this route.
B. **Transposons** are large DNA sequences that regularly depart and reinsert into chromosomal and plasmid sites within the same cell or between cells. These rearrangements in DNA generate mutations and variations in chromosome structure.

MULTIPLE-CHOICE QUESTIONS

1. What is the smallest unit of heredity?
   a. chromosome   c. codon
   b. gene   d. nucleotide

2. A nucleotide contains which of the following?
   a. 5 C sugar
   b. nitrogen base
   c. phosphate
   d. b and c only
   e. all of these

3. The nitrogen bases in DNA are bonded to the
   a. phosphate   c. ribose
   b. deoxyribose   d. hydrogen

4. DNA replication is semiconservative because the ____ strand will become half of the ____ molecule.
   a. RNA, DNA
   b. template, finished
   c. sense, mRNA
   d. codon, anticodon

5. In DNA, adenine is the complementary base for ____, and cytosine is the complement for ____.
   a. guanine, thymine
   b. uracil, guanine
   c. thymine, guanine
   d. thymine, uracil

6. The base pairs are held together primarily by
   a. covalent bonds   c. ionic bonds
   b. hydrogen bonds   d. gyrases

7. Why must the lagging strand of DNA be replicated in short pieces?
   a. because of limited space
   b. otherwise, the helix will become distorted
   c. the DNA polymerase can synthesize in only one direction
   d. to make proofreading of code easier

8. Messenger RNA is formed by ____ of a gene on the DNA template strand.
   a. transcription   c. translation
   b. replication   d. transformation

9. Transfer RNA is the molecule that
   a. contributes to the structure of ribosomes
   b. adapts the genetic code to protein structure
   c. transfers the DNA code to mRNA
   d. provides the master code for amino acids

10. As a general rule, the template strand on DNA will always begin with
    a. TAC   c. ATG
    b. AUG   d. UAC

11. The *lac* operon is usually in the ____ position and is activated by a/an ____ molecule.
    a. on, repressor   c. on, inducer
    b. off, inducer   d. off, repressor

12. The repressible operon is important in regulating ____.
    a. amino acid synthesis
    b. DNA replication
    c. sugar metabolism
    d. ATP synthesis

13. For mutations to have an effect on populations of microbes, they must be
    a. inheritable   d. a and b
    b. permanent   e. all of the above
    c. beneficial

14. Which of the following characteristics is *not* true of a plasmid?
    a. It is a circular piece of DNA.
    b. It is required for normal cell function.
    c. It is found in bacteria.
    d. It can be transferred from cell to cell.

15. Which genes can be transferred by all three methods of intermicrobial transfer?
    a. capsule production
    b. toxin production
    c. F factor
    d. drug resistance

16. Which of the following would occur through specialized transduction?
    a. acquisition of Hfr plasmid
    b. transfer of genes for toxin production
    c. transfer of genes for capsule formation
    d. transfer of a plasmid with genes for degrading pesticides

17. **Multiple Matching.** Fill in the blanks with all the letters of the words at the right that apply.

    ____ genetic transfer that occurs after the donor is dead

    ____ carries the codon

    ____ carries the anticodon

    ____ a process synonymous with mRNA synthesis

    ____ bacteriophages participate in this transfer

    ____ a process requiring an F^+ pilus

    ____ duplication of the DNA molecule

    ____ process in which transcribed DNA code is deciphered into a polypeptide

    ____ involves plasmids

    a. replication   f. mRNA
    b. tRNA   g. transcription
    c. conjugation   h. transformation
    d. ribosome   i. translation
    e. transduction   j. none of these

CONCEPT QUESTIONS

1. Compare the genetic material of eucaryotes, bacteria, and viruses in terms of general structure, size, and mode of replication.

2. Briefly describe how DNA is packaged to fit inside a cell.

3. Describe what is meant by the antiparallel arrangement of DNA.

4. On paper, replicate the following segment of DNA:

   5′ A T C G G C T A C G T T C A C 3′

   3′ T A G C C G A T G C A A G T G 5′

   a. Show the direction of replication of the new strands and explain what the lagging and leading strands are.
   b. Explain how this is semiconservative replication. Are the new strands identical to the original segment of DNA?

5. Name several characteristics of DNA structure that enable it to be replicated with such great fidelity generation after generation.

6. Explain the following relationship: DNA formats RNA, which makes protein.

7. What message does a gene provide? How is the language of the gene expressed?

8. If a protein is 3,300 amino acids long, how many nucleotide pairs long is the gene sequence that codes for it?

9. a. What is a palindrome in DNA and what is one function?
   b. Draw a short sequence of a palindrome.

10. Compare the structure and functions of DNA and RNA.

11. a. Where does transcription begin?
    b. What are the template and coding strands of DNA?
    c. Why is only one strand transcribed, and is the same strand of DNA always transcribed?

12. Compare and contrast the actions of DNA and RNA polymerase.

13. What are the functions of start and nonsense codons? Give examples of them.

14. The following sequence represents triplets on DNA:

    TAC CAG ATA CAC TCC CCT GCG ACT

    a. Give the mRNA codons and tRNA anticodons that correspond with this sequence, and then give the sequence of amino acids in the polypeptide.
    b. Provide another mRNA strand that can be used to synthesize this same protein.
    c. Looking at figure 9.15, give the type and order of the amino acids in the peptide.

15. a. Summarize how bacterial and eucaryotic cells differ in gene structure, transcription, and translation.
    b. Discuss the roles of exons and introns.

16. Compare DNA viruses with RNA viruses in their general methods of nucleic acid synthesis and viral replication.

17. a. Compare and contrast the *lac* operon with a repressible operon system.
    b. How is the *lac* system related to the feedback control of enzymes mentioned in chapter 8?

18. What is the premise of the Ames test?

19. Describe the principal types of mutations. Give an example of a mutation that is beneficial and one that is lethal or harmful.

20. a. Compare conjugation, transformation, and transduction on the basis of general method, nature of donor, and nature of recipient.
    b. List some examples of genes that can be transferred by intermicrobial transfer.

21. By means of a flowchart, show the possible jumps that a transposon can make. Show the involvement of viruses in its movement.

CRITICAL-THINKING QUESTIONS

1. A simple test you can do to demonstrate the coiling of DNA in bacteria is to open a large elastic band, stretch it taut, and twist it. First it will form a loose helix, then a tighter helix, and finally, to relieve stress, it will twist back upon itself. Further twisting will result in a series of knotlike bodies; this is how bacterial DNA is condensed.

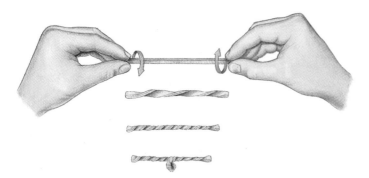

2. Knowing that retroviruses operate on the principle of reversing the direction of transcription from RNA to DNA, propose a drug that might possibly interfere with their replication.

3. Using the piece of DNA in concept question 14, show a deletion, an insertion, a substitution, and an inversion. Which ones are frameshift mutations? Are any of your mutations nonsense? Missense? (Use the universal code to determine this.)

4. Using figure 9.14 and table 9.5, go through the steps in mutation of a codon followed by its transcription and translation that will give the end result in silent, missense, and nonsense mutations.

5. Explain the principle of "wobble" and find four amino acids that are encoded by wobble bases (figure 9.14). Suggest some benefits of this phenomenon to microorganisms.

6. Suggest a reason for having only one strand of DNA serve as a source of useful genetic information. What could be some possible functions of the coding strand?

7. The enzymes required to carry out transcription and translation are themselves produced through these same processes. Speculate which may have come first in evolution—proteins or nucleic acids—and explain your choice.

8. Why can one not reliably predict the sequence of nucleotides on mRNA or DNA by observing the amino acid sequence of proteins?

9. Speculate on the manner in which transposons can be involved in cancer.

10. Explain what is meant by the expression: phenotype = genotype + environment.

INTERNET SEARCH TOPICS

1. Look up variations on DNA structure by entering the terms Z-DNA and B-DNA. Compare these forms of DNA with the one presented in this chapter.

2. Find information on introns. Explain at least five current theories as to their possible functions.

Genetic Engineering:
A Revolution in Molecular Biology

I magine having the power to cure genetic diseases, to identify an infectious agent or diagnose early cancer in a few hours from a single cell, and to witness the entire human genome readout by a computer. Such sensational prospects at one time may have seemed fantastic, but today they are realities. These feats and many others have been placed within our grasp by **genetic engineering*** and **biotechnology.** Scientists from diverse disciplines now have at their disposal a variety of powerful molecular tools to manipulate, alter, and analyze the genetic material of microbes, plants, animals, and viruses. This relatively young science has ushered in a new era in the history of humankind, and it has spawned an explosion of medical, agricultural, and industrial applications.

The genetic revolution has forced every person to become DNA-literate. The media keep us constantly updated on genetic news events ranging from provocative to profound: eating genetically engineered plants to get vaccinated, creating animals that synthesize human hormones, deciphering the ancient DNA of fossils, using viruses to cure cancer and other diseases.

We have been challenged to consider the accuracy of DNA fingerprinting evidence in court and the safety of eating genetically modified seeds, vegetables, and animal products. The ability to change genetic blueprints and create totally new types of organisms is so alarming that a group of scientists and religious leaders have strongly urged the government to legislate against genetic alteration of human gametes and embryos. Some environmental activists have successfully blocked the release of engineered microbes and plants into natural habitats. At the end of this chapter, we will further explore these and other **bioethical*** considerations (Microbits 10.2).

This patch of glowing multicolored bands is a small sample of an automated DNA sequence. The order of bands gives the sequence of bases in the DNA code, and the color of the band tells us which base it is.

Chapter Overview

- The study of DNA has developed into a large-scale industry for manipulating and modifying the genetic material of organisms.
- This science, called biotechnology, aims to develop products, microbes, plants, and animals for use in commercial, agricultural, research, and medical applications.
- DNA is the focus of bioengineering, or genetic engineering technology, which is based on the ease with which DNA can be isolated, handled, and modified.

---

* **genetic engineering** Sometimes called bioengineering—defined as the direct, deliberate modification of an organism's genome. Biotechnology is a more encompassing term that includes the use of DNA, genes, or genetically altered organisms in commercial production.

* **bioethics** A field that relates biological issues to human conduct and moral judgment.

- The DNA double helix readily parts into its two strands; it can be cut with endonucleases, and inserted at exacting sites on other DNA.
- Segments of DNA can be separated by electrical currents, synthesized by machine, and visualized by means of probes, which are some of the bases of identification.
- Technology has developed methods for sequencing DNA that reveal the order of nucleotides it contains; DNA can also be amplified by the polymerase chain reaction DNA.
- Recombinant DNA (rDNA) technology is a branch of genetic engineering that inserts foreign genes into cells, thereby creating modified organisms that can express these donated genes.
- This process of inserting and propagating DNA is called cloning, and it requires special vectors to transport the genes into cloning hosts for maintenance and expression.

- Cloning creates microbes that carry isolated genes from other organisms, giving rise to living gene libraries that can provide a source of known DNA for study and commercial uses.
- Recombinant cloning hosts can be induced to synthesize the products coded for by the gene, thereby providing a source of hormones, drugs, enzymes, immune factors, and other commercial substances.
- Techniques similar to rDNA are used to create genetically modified organisms such as microbes, plants, and animals that are transfected with selected genes to create new strains for medical, agricultural, genetic research, and pharmaceutical purposes.
- Genetic treatments are a method for correcting inherited defects by inserting the natural, normal gene back into the genome through viruses and other techniques.
- Genomes of organisms can be ordered in a map that shows the order of large genetic elements, genes, and even nucleotides that are essential for analyzing the structure and functions of the genes it contains.
- Genetic fingerprinting is a method of arraying an entire genome in a way to show its unique qualities so it can be used in comparisons and identification. It is an essential tool of forensics, pedigree analysis, and microbe identification.
- Bioethics is the field that analyzes the effects of the new DNA technology on society. It will be critically involved in helping develop guidelines for managing the many issues that it creates.

Basic Elements and Applications of Genetic Engineering

Information on genetic engineering and its biotechnological applications is growing at such an expanding rate that some new discovery or product is disclosed almost on a daily basis. To keep this subject somewhat manageable, we will present essential concepts and applications that are both conventional and as up-to-date as possible. As an organizational aid, the topics are divided into the following six major sections, each designated with a roman numeral: I. Tools and Techniques of Genetic Engineering; II. Methods in Recombinant DNA Technology; III. Biochemical Products of Recombinant DNA Technology; IV. Genetically Modified Organisms; V. Genetic Treatments; and VI. Genome Analysis. Figure 10.1 summarizes the main categories to be covered in these topics.

I. Tools and Techniques of Genetic Engineering

Genetic studies for over 30 years have provided a number of insights into certain measurable and predictable activities of nucleic acids. Unraveling this information also had a practical benefit, which was the development of highly sophisticated laboratory tools and techniques. Traditional methods of genetic analysis have involved observing phenotypic inheritance in a test subject. But with these newer techniques, discovery is more likely to involve seeing actual DNA patterns on electrophoresis gels, locating the exact gene on a chromosome, and producing computer-generated gene

sequences. Each tool or technique described in the next several sections does not stand alone but is usually an important part of an overall methodology.

DNA: AN AMAZING MOLECULE

As we learned in chapter 9, the structure of DNA provides a complex code that is the guide for making proteins. But, as a subject for study, it provides numerous intriguing features. Its code, when deciphered, can tell us a great deal about biological structure and function. From the standpoint of the laboratory, it lends itself to several methods of modification and manipulation. The code that makes up its language is like deciphering a puzzle. It can be wound up and untwisted like a rubber band, opened into two strands like a zipper, and diced up into shorter pieces that can be spliced to other DNA.

One useful property of DNA is that it readily **anneals,*** meaning that it changes its binding properties in response to heating and cooling. Exposure to temperatures just below boiling (90°–95°C) causes DNA to become temporarily denatured (figure 10.2). When heat breaks the hydrogen bonds holding the double helix together, it separates longitudinally into two strands. Each strand displays its nucleotide code so that the DNA in this form can be subjected to tests or replicated. When heating is followed by slow cooling, two single DNA strands rejoin (renature) by hydrogen bonds at complementary sites. As we shall see, annealing is a necessary feature of the polymerase chain reaction (PCR) and nucleic acid probes described later.

Enzymes for Dicing, Splicing, and Reversing Nucleic Acids

The polynucleotide strands of DNA can also be clipped crosswise at selected positions by means of enzymes called **restriction endonucleases.*** These enzymes recognize foreign DNA and are capable of digesting or hydrolyzing DNA bonds. In the bacterial cell, this ability protects against the incompatible DNA of bacteriophages or plasmids. In the biotechnologist's lab, the enzymes can be used to cleave DNA at desired sites and are a must for the techniques of recombinant DNA technology.

So far, hundreds of restriction endonucleases have been discovered in bacteria. Each type has a known sequence of 4 to 10 base pairs as its target, so sites of cutting can be finely controlled. These enzymes have the unique property of recognizing and clipping at base sequences called **palindromes** (figure 10.2c). Also see page 256.

Endonucleases are named by combining the first letter of the bacterial genus, the first two letters of the species, and the endonuclease number. Thus, EcoRI* is the first endonuclease found in *Escherichia coli,* and HindIII* is the third endonuclease discovered in *Haemophilus influenzae* Type d (figure 10.2b).

---

* **anneal** (ah-neel') O. E. *anaelan,* to kindle. A process of heating and slowly cooling a substance to change its properties.

* **restriction endonuclease** The meaning of restriction is that the enzymes do not act upon the bacterium's DNA; an endonuclease nicks DNA internally, not at the ends.

* **EcoRI** is pronounced "ee'-koh-arr-one."

* **HindIII** is pronounced "hindy-three."

Endonucleases are used to restrict DNA into smaller pieces for further study as well as to remove and insert it during recombinant DNA techniques, described in a subsequent section. Endonucleases such as HaeIII make straight, blunt cuts on DNA. But more often (EcoRI and HindIII, for example), the enzymes make staggered symmetrical cuts that leave short tails called "sticky ends." Such adhesive tails will base-pair with complementary tails on other DNA fragments or plasmids. This effect makes it possible to splice genes into specific sites. Circularization into a closed circle can occur when the complemen-

tary tails are on opposite ends of the same linear DNA fragment (figure 10.2c).

The pieces of DNA produced by restriction endonucleases are termed *restriction fragments*. Because genomes of members of the same species can vary in the cutting pattern by specific endonucleases, it is possible to detect genetic differences called *restriction fragment length polymorphisms* (RFLPs) that are useful in analyzing genetic relationships. Hundreds of cleavage sites that produce RFLPs are sprinkled throughout genomes. Because RFLPs serve as a type of *genetic marker,* they can help lo-

I. Tools and Techniques of Genetic Engineering

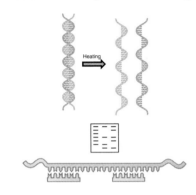

- Annealing: The DNA strand can be separated into two strands with heat.

- Restriction endonucleases cut DNA lengthwise at specific sites into smaller fragments for analysis.

- DNA polymerases, reverse transcriptases, and ligases are other technologically useful enzymes for manipulating DNA.

- Electrophoresis: DNA fragments migrate through a gel in an electrical field and can be sorted by size.

- Gene probes and blots: Probes are small labeled fragments of known DNA content used to identify unknown DNA through hybridization. Blots make points of hybridization visible by means of reporters such as dyes and radioactive isotopes.

- DNA sequencing: The order of bases in DNA can be determined by special techniques.

- Polymerase chain reaction: A rapid technique for amplifying small amounts of DNA so it can be identified, sequenced, or otherwise analyzed.

II. Methods in Recombinant DNA Technology

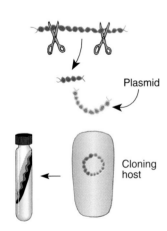

- A gene from a genetic donor is inserted in a cloning vector and used to transform a cloning host to create a genetically engineered organism.

- Gene isolation: Specific genes are isolated from a variety of viruses, bacteria, plants, and animals by means of endonucleases, complementary DNA (cDNA), and the polymerase chain reaction.

- Cloning hosts are bacteria or other microbes that can be transformed to maintain, replicate, and express foreign genes.

- Constructing gene libraries: An isolated gene can be inserted into a cloning host, which is cultured to maintain the gene in a living DNA library.

- Plasmids are small, circular DNA vectors that can carry foreign genes to be cloned.

- Artificial chromosomes are large-capacity vectors for transferring and storing isolated genetic material.

III. Biochemical Products of Recombinant DNA Technology

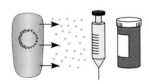

- Recombinant product expression and isolation: The recombinant cloning host is induced to synthesize the protein that is encoded by the donor DNA.

- Protein products include a variety of drugs, enzymes, hormones, pesticides, and vaccines.

FIGURE 10.1

Foundations and applications of genetic engineering and DNA technology. Each roman numeral heading corresponds to a major section covered in the text.

cate specific sites along a DNA strand and are thus useful in preparation of gene maps and DNA profiles (see figures 10.16 and 10.17).

Another enzyme, called a **ligase,** is necessary to seal the sticky ends together by rejoining the phosphate-sugar bonds cut by endonucleases. Its main application is in final splicing of genes into plasmids and chromosomes.

An enzyme called **reverse transcriptase** is best known for its role in the replication of the AIDS virus and other retroviruses. It also provides geneticists with a valuable tool for con-

verting RNA into DNA. Copies called **complementary DNA,** or **cDNA,** can be made from messenger, transfer, ribosomal, and other forms of RNA. The technique provides a valuable means of synthesizing eucaryotic genes from mRNA transcripts. The advantage is that the synthesized gene will be free of the intervening sequences (introns) that can complicate the management of eucaryotic genes in genetic engineering. Complementary DNA can also be used to analyze the nucleotide sequence of RNAs, such as those found in ribosomes and transfer RNAs (see figure 4.29).

IV. Genetically Modified Organisms (GMOs)

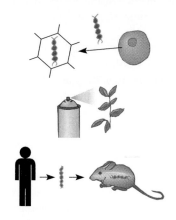

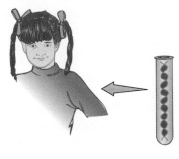

- Viruses have been engineered to produce vaccines, cancer treatments, and vectors.

- Bacteria have been engineered to carry pesticides, bioremediate toxic spills, and prevent frost on plants.

- Transgenic plants are engineered with *Agrobacterium* for pesticide resistance, disease resistance, higher vitamin content, insect resistance, nitrogen fixation, and improved nutrition and storage.

- Transgenic animals are engineered at the embryo stage to create animals with genetic defects, as sources of human proteins, and to improve the growth rate and productivity of livestock.

V. Genetic Treatments

- Genetic material is inserted into human cells to mitigate the effects of genetic diseases such as sickle-cell anemia, cystic fibrosis, and cancer.

VI. Genome Analysis

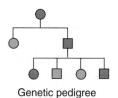

Chromosome Map

- Maps of genomic and mitochondrial DNA have been completed for hundreds of organisms and viruses.

- Completely sequenced maps permit predicting protein structure.

- Maps show relationships between organisms and allow tracing of evolutionary trends.

- DNA can be screened and profiled through a fingerprinting technique involving electrophoresis of restricted fragments and their development with specific markers.

Genetic pedigree

(a) DNA annealing process. DNA responds to heat by denaturing—losing its hydrogen bonding, and thereby separating into its two strands. When cooled, the two strands rejoin at complementary sites. The two strands need not be from the same organisms as long as they have matching sites.

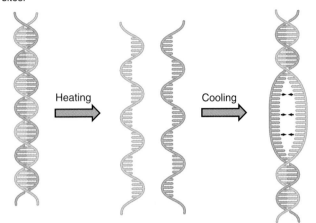

(b) Examples of palindromes and cutting patterns.

Endonuclease	EcoRI	HindIII	HaeIII
Cutting pattern	G A A T T C C T T A A G	A A G C T T T T C G A A	G G C C C C G G

(c) Action of restriction endonucleases. (1) A restriction endonuclease recognizes and cleaves DNA at the site of a specific palindromic sequence. Cleavage can produce staggered tails called sticky ends that accept complementary tails for gene splicing. (2) One example of how the sticky ends can be used. When formed at the ends of a DNA, they bind a linear molecule into a circular one. This is one method of producing a circular plasmid.

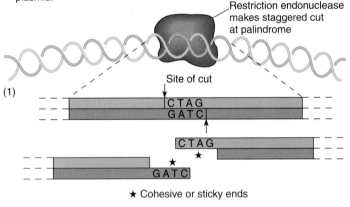

Restriction endonuclease makes staggered cut at palindrome

Site of cut

(1)

CTAG
GATC

CTAG
GATC

★ Cohesive or sticky ends

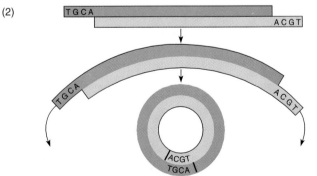

(2)

TGCA
ACGT

TGCA
ACGT

ACGT
TGCA

FIGURE 10.2

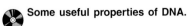
Some useful properties of DNA.

Analysis of DNA

One way to produce a readable pattern of DNA fragments is through **gel electrophoresis.** In this technique, samples are placed in compartments (wells) in a soft agar gel and subjected to an electrical current. Flow of electricity from the negative pole to the positive pole causes the DNA pieces to migrate in the gel substrate toward the positive pole. The rate of movement is based primarily on the size of the fragments. The larger fragments move more slowly and remain nearer the top of the gel, whereas the smaller fragments migrate faster and are positioned farther from the wells. The positions of DNA fragments are determined by staining the gel or developing it with probes and comparing the pattern against a known standard as in a DNA fingerprint (figure 10.3; see figure 10.16). Electrophoresis patterns can be quite distinctive and are very useful in characterizing DNA fragments and comparing the degree of genetic similarities among samples.

Nucleic Acid Hybridization and Probes

Two different nucleic acids can **hybridize** by uniting at their complementary sites. All different combinations are possible: Single-stranded DNA can unite with other single-stranded DNA or RNA, and RNA can hybridize with other RNA. This property has been the inspiration for specially formulated oligonucleotide tracers called **gene probes.** Hybridization probes have practical value because they can detect specific nucleotide sequences in unknown samples. So that areas of hybridization can be visualized, the probes carry reporter molecules such as radioactive labels, which are isotopes that emit radiation, or luminescent labels, which give off visible light. Reactions can be revealed by placing photographic film in contact with the test reaction. Fluorescent probes contain dyes that can be visualized with ultraviolet radiation, and enzyme-linked probes react with substrate to release colored dyes.

When probes hybridize with an unknown sample of DNA or RNA, they tag the precise area and degree of hybridization and help determine the nature of nucleic acid present in a sample. In a method called the *Southern blot,*[1] DNA fragments are first separated by electrophoresis and then denatured and immobilized on a special filter. A DNA probe is then incubated with the sample, and wherever this probe encounters the segment for which it is complementary, it will attach and form a hybrid. Development of the hybridization pattern will show up as one or more bands (figure 10.3). This method is a sensitive and specific way to isolate fragments from a complex mixture and to find specific gene sequences on DNA. Southern blotting is also one of the important first steps for preparing isolated genes.

Probes are commonly used for diagnosing the cause of an infection from a patient's specimen and identifying a culture of an unknown bacterium or virus. The method for doing this was first outlined in chapter 4 (see figure 4.28). A simple and rapid method called a hybridization test does not require electrophoresis. DNA from a test sample is isolated, denatured, placed on an absorbent filter, and combined with a microbe-specific probe (figure 10.4; see figure 24.21). The blot is then developed and observed for areas of

---

1. Named for its developer, E. M. Southern. The "Northern" blot is a similar method used to analyze RNA.

(a)

Restriction endonucleases selectively cleaving sites of DNA

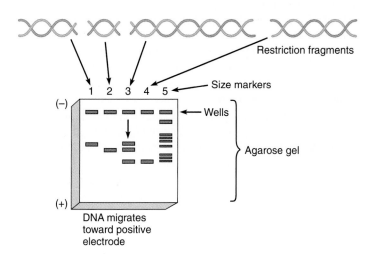

Restriction fragments

Size markers

(−)

Wells

Agarose gel

(+)

DNA migrates toward positive electrode

(b)

Result, following development

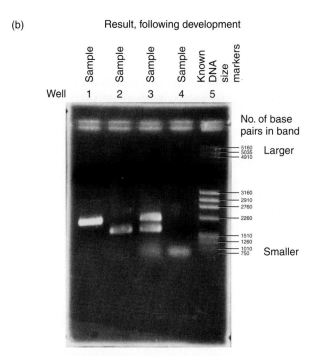

Well 1 2 3 4 5

Sample Sample Sample Sample Known DNA size markers

No. of base pairs in band

Larger

5160
5035
4910

3160
2910
2760

2260

1510
1260

1010
750

Smaller

FIGURE 10.3

Revealing the patterns of DNA with electrophoresis. (a) After cleavage into fragments, DNA is loaded into wells on one end of an agarose gel. When an electrical current is passed through the gel (from the negative pole to the positive pole), the DNA, being negatively charged, migrates toward the positive pole. The larger fragments, measured in numbers of base pairs, migrate more slowly and remain nearer the wells than the smaller (shorter) fragments. **(b)** An actual developed gel (here using radioactive labels) reveals a separation pattern of the fragments of DNA. The size of a given DNA band can be determined by comparing it to a known set of markers (lane 5) called a ladder. This method can be used for general screening of DNA or for genetic fingerprints.

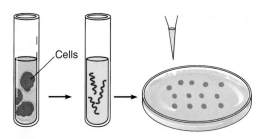

Cells

(a) A sample of the test specimen (cells) is disrupted to release its DNA.

(b) The DNA is blotted on a support matrix, where it binds.

(c) The DNA is denatured to separate into its two strands, which are immobilized on the matrix.

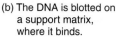

Probes

Does ____ not fit

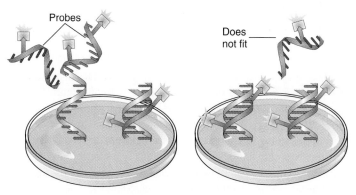

(d) Labeled probes, consisting of a short single strand of radioactive or fluorescent nucleotides, hybridize with any segments on the unknown test strands to which they are complementary. Unreacted probes are rinsed away, leaving only those that have attached.

(e) Results of a blot hybridization test for human papillomavirus, the cause of genital warts and some forms of uterine cancer. Here, a radioactive probe has detected the virus in five patient specimens (dark spots).

FIGURE 10.4

Conducting a DNA probe hybridization test. Cells and DNA are greatly disproportionate to show reaction.

hybridization. Commercially available diagnostic kits are now on the market for identifying intestinal pathogens such as *E. coli, Salmonella, Campylobacter, Shigella, Clostridium difficile*, rotaviruses, and adenoviruses. Other bacterial probes exist for *Mycobacterium, Legionella, Mycoplasma,* and *Chlamydia*; viral probes are available for herpes simplex and zoster, papilloma (genital warts), hepatitis A and B, and AIDS (figure 10.4*e*). DNA probes have also been developed for human genetic markers and some types of cancer.

With another method, called *fluorescent in situ hybridization* (FISH), probes are applied to intact cells and observed microscopically for the presence and location of specific genetic marker sequences on genes. It is a very effective way to locate genes on chromosomes. In situ techniques can also be used to identify unknown bacteria living in natural habitats without having to culture them, and they can be used to detect RNA in cells and tissues.

Methods Used to Size, Synthesize, and Sequence DNA

The relative sizes of nucleic acids are usually denoted by the number of base pairs (bp) or nucleotides they contain. For example, the palindromic sequences recognized by endonucleases are usually 4 to 10 bp in length, an average gene in *E. coli* is approximately 1,300 bp, or 1.3 kilobases (kb), and its entire genome is approximately 4.7 million base pairs (Mb). The DNA of the human mitochondrion contains 16 kb, and the Epstein-Barr virus (a cause of infectious mononucleosis) has 172 kb. Humans have approximately 3.5 billion base pairs (Bb) arrayed along 46 chromosomes.

Large segments of DNA or RNA are difficult to handle and analyze, but this problem can be solved by using far shorter pieces of DNA or RNA called **oligonucleotides.*** Oligonucleotides vary in length from 2 to 200 bp, although the most common ones are about 20 to 30 bp. Although they can be isolated from cells, for most applications oligonucleotides are tailor-made on a DNA synthesizer that adds the desired nucleotides in a controlled series of steps that limits the length to about 200 nucleotides. Other oligonucleotides are synthesized by PCR (discussed later).

DNA Sequencing: Determining the Exact Genetic Code

Analysis of DNA by its size, restriction patterns, and hybridization characteristics is instructive, but the most detailed information comes from determining the actual order and types of bases in DNA. This process, called **DNA sequencing,** provides a high resolution blueprint for all types of DNA, including genomic, cDNA, artificial chromosomes, plasmids, and cloned genes. The two most common sequencing techniques were developed by Frederick Sanger and the team of M. Maxim and W. Gilbert. We will cover the basic techniques involved in the Sanger sequencer, based on the synthesis and analysis of a complementary strand in a test tube (figure 10.5).

Since most DNA being sequenced is very long, it is made more manageable by cutting into a large number of shorter fragments and separated. The test strands, typically several hundred nucleotides long, are then denatured to expose single strands that will serve as templates to synthesize complementary strands. The fragments are divided into four separate tubes that contain primers to set the start point for the synthesis to begin on one of the strands. The nucleotides will attach at the $3'$ end of the primer, using the template strands as a guide, essentially the same as regular DNA synthesis as performed in a cell.

The tubes are incubated with the necessary DNA polymerase and all four of the regular nucleotides needed to carry out the process of elongating the complementary strand. The tracer molecules used in sequencing are a modified type of dedeoxy (dd) nucleotides that lack a hydroxyl group required for the attachment of the next nucleotide, so that they interrupt further synthesis at that site. To keep track of this process, each of the four reaction tubes contains only one version of a given dedeoxy nucleotide, and each dd nucleotide type is labeled with a fluorescent dye of a different color. As the reaction proceeds, strands elongate by adding normal nucleotides, but a small percentage of fragments will randomly incorporate the complementary dd nucleotide and be terminated. Eventually, all possible positions in the sequence will incorporate a terminal fluorescent nucleotide, thus producing a series of strands that reflect the correct sequence.

The reaction products are placed into four wells (G, C, A, T) of a polyacrylamide gel, which is sensitive enough to separate strands that differ by only a single nucleotide in length. Electrophoresis separates the fragments in order according to both size and lane, and only the fragments carrying the fluorescent labels will be readable on the gel. The gel indicates the comparative orientation of the bases, which graduate in size from the smallest fragments that terminate early and migrate farthest (bottom of gel) to successively longer fragments (moving stepwise to the top). Reading the order of first appearance of a given gel fragment in the G, C, A, or T lanes provides the correct sequence of bases of the complementary strand, and it also allows one to infer the sequence of the template strand as well. This method of sequencing is remarkably accurate, with only about one mistake in every 1,000 bases. For managing the giant genomes of humans, mice, and other organisms, it has been necessary to automate this method using high-density arrays that can be rapidly sequenced and displayed by a computer (see chapter opening photo).

Polymerase Chain Reaction: A Molecular Xerox Machine for DNA

Some of the techniques used to analyze DNA and RNA are limited by the small amounts of test nucleic acid or by the complex steps involved. This problem was largely solved by the invention of a simple, versatile way to amplify DNA called the **polymerase chain reaction (PCR).** This technique rapidly increases the amount of DNA in a sample without the need for making cultures or carrying out complex purification techniques. It is so sensitive that it holds the potential to detect cancer from a single cell or to diagnose an infection from a single gene copy. It is comparable to being able to pluck a single DNA "needle" out of a "haystack" of other molecules and make unlimited copies of the DNA. The rapid rate of PCR makes it possible to replicate a target DNA from a few copies to billions of copies in a few hours. It can amplify DNA fragments that consist of a few base pairs to whole genomes containing several million base pairs.

---

* oligonucleotides (aw″-lig-oh-noo′-klee-oh-tydz) Gr. *oligo*, few or scant.

(1) Isolated unknown DNA fragment (one strand will be shown for clarity).

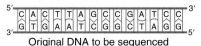

Original DNA to be sequenced

(2) DNA is annealed to produce single template strand.

(3) Strand is labeled with specific primer molecule.

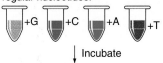

(4) DNA polymerase and regular nucleotide mixture (ATP, CTP, GTP, and TTP) are added; ddG, ddA, ddC, and ddT, each labeled with a different color, are placed in separate reaction tubes with the regular nucleotides.

+G   +C   +A   +T

↓ Incubate

(5) Newly replicated strands are terminated at the point of addition of a dd nucleotide.

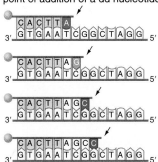

(6) A schematic view of how all possible positions on the fragment are occupied by a labeled nucleotide.

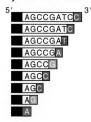

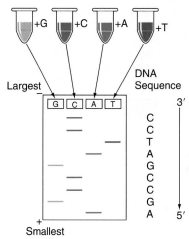

(7) Running the reaction tubes in four separate gel lanes separates them by size and nucleotide type. Reading from bottom to top, one base at a time, provides the correct DNA sequence.

FIGURE 10.5
Steps in a Sanger DNA sequence technique.

To understand the idea behind PCR, it will be instructive to review figure 9.6, which describes synthesis of DNA as it occurs naturally in cells. The PCR method uses essentially the same events, with the opening up of the double strand, using the exposed strands as templates, and the addition of primers.

Initiating the reaction requires a few specialized ingredients (figure 10.6). The **primers** are synthetic oligonucleotides of a known sequence of 15 to 30 bases that serve as landmarks to indicate where DNA amplification will begin. Depending upon the purposes and what is known about the DNA being replicated, the primers can be random, attaching to any sequence they may fit, or they may be highly specific and chosen to amplify a known gene. So that DNA templates do not recombine and interfere with DNA synthesis, processing must be carried out at a relatively high temperature. This necessitates the use of special **DNA polymerases** isolated from thermophilic bacteria. Examples of these unique enzymes are Taq polymerase obtained from *Thermus aquaticus* (see

Microbits 7.4) and Vent polymerase from *Thermococcus litoralis*. Another useful component of PCR is a machine called a thermal recycler that automatically initiates the cyclic temperature changes.

The PCR technique operates by repetitive cycling of three basic steps: denaturation, priming, and extension. The first step involves heating target DNA to 94°C to separate it into two strands. Next, the system is cooled to between 50° and 65°C, depending on the content of G–C and A–T base pairs, if it is known. Recall that G–C pairs have a third hydrogen bond, which will make the helix harder to separate if the DNA contains higher amounts of these bases. Primers are added in a concentration that favors binding to the complementary strand of test DNA. This reaction prepares the two DNA strands, now called **amplicons,** for synthesis. In the third phase, which proceeds at 72°C, DNA polymerase and raw materials in the form of nucleotides are added. Beginning at the free end of the primers on both strands, the polymerases extend the molecule by adding appropriate nucleotides and produce two complete strands of DNA.

(a) In cycle 1, the DNA to be amplified is annealed, primed, and replicated by a polymerase that can function at high temperature. The two resulting strands then serve as templates for a second cycle of annealing, priming, and synthesis.*

Cycle 1

DNA Sample

Heat to 94°C

Denaturation

Strands separate

50° to 65°C

Priming

Oligonucleotide primers attach at ends of strands to promote replication of amplicons

Amplicons   Primer   Primer

Extension

DNA polymerase synthesizes complementary strand

72°C

Polymerase

Cycle 1

Cycle 2

Denaturation

2 copies   New strand

New strand   Original strands

Heat to 94°C

Priming

72°C

Extension

4 copies

Cycle 2

Cycles 3, 4, . . . repeat same steps

(b) A view of the process after 7 cycles, with 64 copies of amplified DNA. Continuing this process for 20 to 40 cycles can produce millions of identical DNA molecules.

*For simplicitys' sake, we have omitted the elongation of the complete original parent strand during the first cycles. Ultimately, templates that correspond only to the smaller fragments dominate and become the primary population of replicated DNA.

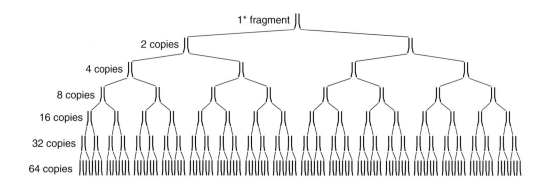

1* fragment

2 copies

4 copies

8 copies

16 copies

32 copies

64 copies

FIGURE 10.6

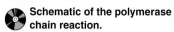

Schematic of the polymerase chain reaction.

It is through repetition of these same steps that DNA becomes amplified. When the DNAs formed in the first cycle are denatured, they become amplicons to be primed and extended in the second cycle. Each subsequent cycle converts the new DNAs to amplicons and doubles the number of copies. The number of cycles required to produce a million molecules is 20, but the process could theoretically be carried out to 40 cycles. One significant advantage of this technique has been its natural adaptability to automation. A PCR machine can perform 20 cycles on nearly 100 samples in 2 or 3 hours.

The amplified DNA comes in a form ready for restriction, electrophoresis, probe identification, and sequencing. PCR can be adapted to analyze RNA by initially converting an RNA sample to DNA with reverse transcriptase. This cDNA can then be amplified by PCR in the usual manner. It is by such means that ribosomal RNA and messenger RNA are readied for sequencing. The polymerase chain reaction has found prominence as a powerful workhorse of molecular biology, medicine, and biotechnology. It often plays an essential role in gene mapping, the study of genetic defects and cancer, forensics, taxonomy, and evolutionary studies.

For all of its advantages, PCR has some problems. A serious concern is the introduction and amplification of nontarget DNA from the surrounding environment. Such contamination can be minimized by using equipment and rooms dedicated for DNA analysis maintained with the utmost degree of cleanliness. Problems with contaminants can also be reduced by using gene-specific primers and treating samples with special enzymes that can degrade the contaminating DNA before it is amplified.

CHAPTER CHECKPOINTS

The genetic revolution has produced a wide variety of industrial technologies that translate and radically alter the blueprints of life. The potential of biotechnology promises not only improved quality of life and enhanced economic opportunity but also serious ethical dilemmas. Increased public understanding is essential for developing appropriate guidelines for responsible use of these revolutionary techniques.

Genetic engineering utilizes a wide range of methods that physically manipulate DNA for purposes of visualization, sequencing, hybridizing, and identifying specific sequences. The tools of genetic engineering include specialized enzymes, gel electrophoresis, DNA sequencing machines, and gene probes. The polymerase chain reaction (PCR) technique amplifies small amounts of DNA into larger quantities for further analysis.

II. Methods in Recombinant DNA Technology: How to Imitate Nature

The primary intent of **recombinant DNA (rDNA) technology** is to deliberately remove genetic material from one organism and combine it with that of a different organism. Its origins can be traced to 1970, when microbiologists first began to duplicate the clever tricks bacteria do naturally with bits of extra DNA such as plasmids, transposons, and proviruses. The discovery that bacteria can readily accept, replicate, and express foreign DNA made them powerful agents for studying the genes of other organisms in isola-

tion. The practical applications of this work were soon realized by biotechnologists. Bacteria could be genetically engineered to mass-produce substances such as hormones, enzymes, and vaccines that were difficult to synthesize by the usual industrial methods.

Figure 10.7 provides an overview of the recombinant DNA procedure. An important objective of this technique is to form genetic **clones.*** Cloning involves the removal of a selected gene from an animal, plant, or microorganism (the genetic donor) followed by its propagation in a different host organism. Cloning requires that the desired donor gene first be selected, excised by restriction endonucleases, and isolated. The gene is next inserted into a **vector** (usually a plasmid or a virus) that will insert the DNA into a **cloning host.** The cloning host is usually a bacterium or a yeast that can replicate the gene and translate it into the protein product for which it codes. In the next section, we examine the elements of gene isolation, vectors, and cloning hosts and show how they participate in a complete rDNA procedure.

TECHNICAL ASPECTS OF RECOMBINANT DNA AND GENE CLONING

The first hurdles in cloning a target gene are to locate its exact site on the genetic donor's chromosome and to isolate it. Some of the first isolated genes came from viruses, which have extremely small and manageable genomes, and later genes were isolated from bacteria and yeasts. At first, the complexity of the human genome made locating specific genes very difficult. However, this process has been greatly improved by tools for excising, mapping, and identifying genes. Among the most common strategies for obtaining genes in an isolated state are:

1. The DNA is removed from cells and separated into fragments by endonucleases. Each fragment is then inserted into a vector and cloned. The cloned fragments are Southern blotted and probed to identify desired sequences. This is a long and tedious process, because each fragment of DNA must be examined for the cloned gene. For example, if the human genome were separated into 20 kb fragments, there would be at least 150,000 clones to test.
2. A gene can be synthesized from isolated mRNA transcripts using reverse transcriptase (cDNA).
3. DNA can be synthesized artificially using the polymerase chain reaction.

Although gene cloning and isolation can be very laborious, a fortunate outcome is that, once isolated, genes can be maintained in a cloning host and vector just like a microbial pure culture. *Genomic libraries,* which are collections of isolated and characterized genes, now exist for millions of genes from hundreds of organisms and viruses.

Characteristics of Cloning Vectors

A good recombinant vector has two indispensable qualities: It must be capable of carrying a significant piece of the donor DNA, and it must be readily accepted by the cloning host. Plasmids are excellent

---

* clone (klohn) Gr. *klon,* young shoot or twig. Do not confuse the molecular biology use of clone with other uses. For example, a clone could be an organism that is genetically identical to its parent.

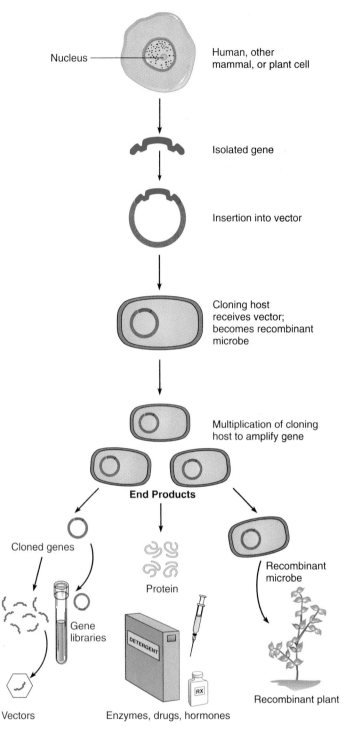

Nucleus

Human, other mammal, or plant cell

Isolated gene

Insertion into vector

Cloning host receives vector; becomes recombinant microbe

Multiplication of cloning host to amplify gene

End Products

Cloned genes

Gene libraries

Protein

Recombinant microbe

DETERGENT

RX

Vectors

Enzymes, drugs, hormones

Recombinant plant

FIGURE 10.7

An overview of recombinant DNA techniques. Although the techniques have slight variations, the general events include isolating a desired gene from an animal, plant, virus, or bacterium; inserting the gene into a vector; cloning the vector with its foreign gene in a cloning host; and isolating an end product. This is the basis of a variety of applications in gene analysis, medicine, industry, and agriculture.

vectors because they are small, well characterized, and easy to manipulate, and they can be transferred into appropriate host cells through transformation. Bacteriophages also serve well because they have the natural ability to inject DNA into bacterial hosts

through transduction. A common vector in early work was an *E. coli* plasmid that carries genetic markers for resistance to antibiotics, although it is restricted by the relatively small amount of foreign DNA it can accept. A modified phage vector, the *Charon*[2] phage, is missing large sections of its genome, so it can carry a fairly large segment of foreign DNA. In addition, hybrid vectors have been developed by splicing two different vectors together. One that combines both a plasmid and a phage, called a *cosmid,* is capable of carrying relatively large genomic sequences. A hybrid *E. coli–*yeast vector can be inserted in both bacterial and yeast cloning hosts.

For a vector to function as intended, it must carry various specialized genes (figure 10.8). Shuttle vectors are chosen expressly for their ability to be accepted by the cloning host. Vectors also contain expression genes for directing the start of replication, transcription, and translation, and established sites for the action of restriction endonucleases. It is also essential to have some way to determine which cells of a culture have accepted the vector. One method involves adding genes for resistance to antibiotics (ampicillin, tetracycline). Cloning hosts can then be screened on selective media containing the antibiotic, and recombinant clones are clearly evident by their ability to grow on these media (see figure 10.10).

One limitation of plasmids and even cosmids is that they are unable to carry the huge genome sequences (especially from eucaryotic cells) that are part of many studies. One answer to this limitation has been the development of **artificial chromosomes,** which are large, genome-sized segments of cloned DNA carried by bacteria (BACs) and yeast (YACs). BACs can carry up to 300 kb of DNA. Yeast plasmids are able to copy even larger stretches of foreign DNA and can enhance the ability to replicate more complex genomes outside of their natural host.

Characteristics of Cloning Hosts

The best cloning hosts possess several key characteristics (table 10.1). The traditional cloning host and the one still used in the majority of experiments is *Escherichia coli.* Because this bacterium was the original recombinant host, the protocols using it are well established, relatively easy, and reliable. Hundreds of specialized cloning vectors have been developed for it. The main disadvantage with this species is its lack of versatility in correctly expressing eucaryotic genes. One alternative host for certain industrial processes and research is the yeast *Saccharomyces cerevisiae,* which, being eucaryotic, already possesses mechanisms for processing and modifying eucaryotic gene products. Certain techniques may also employ different bacteria *(Bacillus subtilis),* animal cell culture, and even live animals and plants to serve as cloning hosts. In our coverage, we present the recombinant process as it is performed in bacteria and yeasts.

CONSTRUCTION OF A CHIMERA, INSERTION INTO A CLONING HOST, AND GENETIC EXPRESSION

This section illustrates the main steps in rDNA technology to produce a drug called alpha-2 interferon (Roferon-A). This form of interferon is used to treat cancers such as hairy-cell leukemia and

---

2. Named for the mythical boatman in Hades who carried souls across the River Styx.

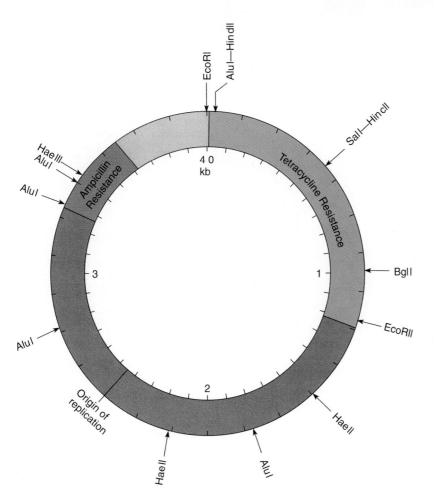

FIGURE 10.8

Partial map of the pBR322 plasmid of *Escherichia coli.* Arrows delineate sites cleaved by various restriction endonucleases. Numbers indicate the size of regions in kilobases (kb). Other important components are genes for ampicillin and tetracycline resistance and the origin of replication.

Kaposi's sarcoma in AIDS patients. The human alpha interferon gene is a linear molecule of approximately 500 bp that codes for a polypeptide of 166 amino acids. It was originally isolated and identified from human blood cells and prepared from processed mRNA transcripts that are free of introns. This is a necessary step because the bacterial cloning host has none of the machinery needed to excise this nontranslated part of a gene.

The first step in cloning is to prepare the isolated interferon gene for splicing into an *E. coli* plasmid (figure 10.9). One way this is accomplished is to attach terminal nucleotides (sticky ends) that can mate with the plasmid after it has been digested with an endonuclease. In the case of the HindIII endonuclease, for example, the gene

will require one sticky end that reads AA and another that reads TT. In the presence of the endonuclease, the plasmid's circular molecule is nicked open, and its sticky ends are exposed (figure 10.9). When the gene and plasmid are placed together, their free ends base-pair, and a ligase makes the final covalent seals. The resultant gene and plasmid combination is called a *chimera.**

Following this procedure, the chimera is introduced by transformation into the cloning host, a special laboratory strain of *E. coli* that lacks any extra plasmids that could complicate the expression of the gene. Because the recombinant plasmid enters only some of the cloning host cells, it is necessary to locate these recombinant clones. Cultures are plated out on medium containing ampicillin, and only those clones that carry the plasmid with ampicillin resistance can form colonies (figure 10.10). These recombinant colonies are selected from the plates and cultured. As the cells multiply, the plasmid is replicated along with the cell's chromosome. In a few hours of growth, there can be billions of cells, each containing the interferon gene.

The bacteria's ability to express the eucaryotic gene is ensured, because the plasmid has been modified with the necessary transcription and translation recognition sequences. As the *E. coli* culture grows in a special medium under optimal conditions, it transcribes and translates the interferon gene, synthesizes the

---

TABLE 10.1

Desirable Features in a Microbial Cloning Host

Rapid overturn, fast growth rate
Can be grown in large quantities using ordinary culture methods
Nonpathogenic
Genome that is well delineated (mapped)
Capable of accepting plasmid or bacteriophage vectors
Maintains foreign genes through multiple generations
Will secrete a high yield of proteins from expressed foreign genes

---

*chimera (ky-meer′-uh) Gr. *chimaira,* a monster with a lion's head, a goat's body, and a serpent's tail. It can also refer to a product formed by fusion of two different organisms.

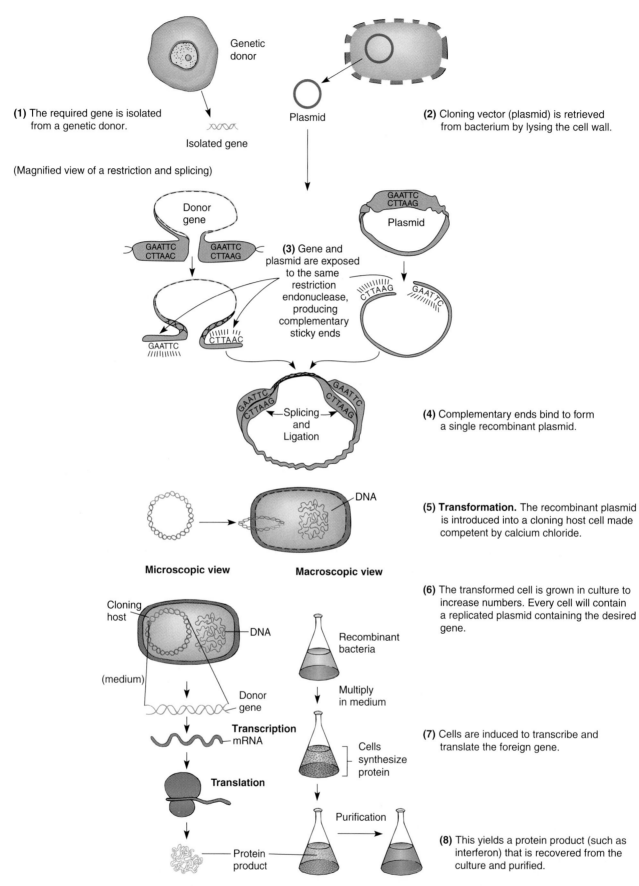

(1) The required gene is isolated from a genetic donor.

Genetic donor

Isolated gene

Plasmid

(2) Cloning vector (plasmid) is retrieved from bacterium by lysing the cell wall.

(Magnified view of a restriction and splicing)

Donor gene

GAATTC
CTTAAC

GAATTC
CTTAAG

GAATTC
CTTAAG

Plasmid

(3) Gene and plasmid are exposed to the same restriction endonuclease, producing complementary sticky ends

CTTAAG

GAATTC

GAATTC

CTTAAC

GAATTC
CTTAAG

GAATTC
CTTAAG

Splicing and Ligation

(4) Complementary ends bind to form a single recombinant plasmid.

DNA

Microscopic view

Macroscopic view

(5) Transformation. The recombinant plasmid is introduced into a cloning host cell made competent by calcium chloride.

(6) The transformed cell is grown in culture to increase numbers. Every cell will contain a replicated plasmid containing the desired gene.

Cloning host

DNA

(medium)

Donor gene

Recombinant bacteria

Multiply in medium

Transcription
mRNA

Cells synthesize protein

(7) Cells are induced to transcribe and translate the foreign gene.

Translation

Purification

Protein product

(8) This yields a protein product (such as interferon) that is recovered from the culture and purified.

FIGURE 10.9

Steps in recombinant DNA, gene cloning, and product retrieval.

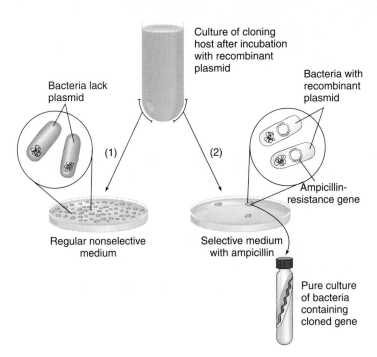

Bacteria lack plasmid

Culture of cloning host after incubation with recombinant plasmid

Bacteria with recombinant plasmid

(1)   (2)

Ampicillin-resistance gene

Regular nonselective medium

Selective medium with ampicillin

Pure culture of bacteria containing cloned gene

FIGURE 10.10

One method for screening clones of bacteria that have been transformed with the donor gene. (1) Plating the culture on nonselective medium will not separate the transformed cells from normal cells, which lack the plasmid. **(2)** Plating on selective medium containing ampicillin will permit only cells containing the plasmid to multiply. Colonies growing on this medium that carry the cloned gene can be used to make a culture for gene libraries, industrial production, and other processes. (Plasmids are shown disproportionate to size of cell.)

peptide, and secretes it into the growth medium. At the end of the process, the cloning cells and other chemical and microbial impurities are removed from the medium. Final processing to excise a terminal amino acid from the peptide yields the interferon product in a relatively pure form (see figure 10.9). The scale of this procedure can range from test tube size to gigantic industrial vats that can manufacture thousands of gallons of product (see figure 26.40).

Although the process we have presented here produces interferon, some variation of it can be used to mass produce a variety of hormones, enzymes, and agricultural products such as pesticides.

CHAPTER CHECKPOINTS

Recombinant DNA techniques combine DNA from different sources to produce microorganism "factories" that produce hormones, enzymes, and vaccines on an industrial scale. Cloning is the process by which genes are removed from the original host and duplicated for transfer into a cloning host by means of cloning vectors.

Plasmids, bacteriophages, and cosmids are types of cloning vectors used to transfer recombinant DNA into a cloning host. Cloning hosts are simply organisms that readily accept recombinant DNA, grow easily, and synthesize large quantities of specific gene product.

III. Biochemical Products of Recombinant DNA Technology

Recombinant DNA technology is used by pharmaceutical companies to manufacture medications that cannot be manufactured by any other means. Diseases such as diabetes and dwarfism, caused by the lack of an essential hormone, are treated by replacing the missing hormone. Porcine and bovine insulin were once the only forms available to treat diabetes, even though such animal products can cause allergic reactions in sensitive individuals. In contrast, dwarfism cannot be treated with animal growth hormones, so originally the only source of human growth hormone (HGH) was the pituitaries of cadavers. At one time, not enough HGH was available to treat the thousands of children in need.

Recombinant technology changed the outcome of these and many other conditions by enabling large-scale manufacture of lifesaving hormones and enzymes of human origin. Recombinant human insulin can now be prescribed for diabetics, and recombinant HGH can now be administered to children with dwarfism, a premature aging disease called progeria, or failure to grow. Other protein-based hormones, enzymes, and vaccines produced through rDNA are summarized in table 10.2. In addition to the obvious medical applications, numerous agricultural and industrial products are being developed at a rapid rate. An especially valuable aspect is that the techniques of genetic engineering can be combined with other areas of biotechnology to manufacture commercial quantities of substances such as enzymes.

Nucleic acid products also have a number of medical applications. A new development in vaccine formulation involves using microbial DNA as a stimulus for the immune system. So far, animal tests using DNA vaccines for AIDS and influenza indicate that this may be a breakthrough in vaccine design. Recombinant DNA could also be used to produce DNA-based drugs for the types of gene and antisense therapy discussed in the genetic treatments section.

IV. Genetically Modified Organisms

The process of artificially introducing foreign genes into organisms is termed *transfection,* and the recombinant organisms produced in this way are called *transgenic* or genetically modified organisms (GMOs). Foreign genes have been inserted into a variety of microbes, plants, and animals through recombinant DNA techniques developed especially for them. Transgenic "designer" organisms are available for a variety of biotechnological applications. Because they are unique life forms that would never have otherwise occurred, they can be patented.

RECOMBINANT MICROBES: MODIFIED BACTERIA AND VIRUSES

One of the first practical applications of rDNA in agriculture was to create a genetically altered strain of the bacterium *Pseudomonas syringae*. The wild strain ordinarily contains an ice nucleation gene that promotes ice or frost formation on moist plant surfaces. Genetic alteration of the frost gene using recombinant plasmids created a different strain that could prevent ice crystals from forming.

TABLE 10.2

Current Protein Products from Recombinant DNA Technology

Immune Treatments

Interferons—peptides used to treat some types of cancer, multiple sclerosis, and viral infections such as hepatitis and genital warts

Interleukins—types of cytokines that regulate the immune function of white blood cells; used in cancer treatment

Orthoclone—an immune suppressant in transplant patients

Macrophage colony stimulating factor (GM-CSF)—used to stimulate bone marrow activity after bone marrow grafts

Tumor necrosis factor (TNF)—used to treat cancer

Hormones

Erythropoietin (EPO)—a peptide that stimulates bone marrow used to treat some forms of anemia

Tissue plasminogen activating factor (tPA)—can dissolve potentially dangerous blood clots

Hemoglobin A—form of artificial blood to be used in place of real blood for transfusions

Factor VIII—needed as replacement blood-clotting factor in type A hemophilia

Relaxin—an aid to childbirth

Proleukin—a drug to treat kidney cancer

Enzymes

rH DNase (pulmozyme)—a treatment that can break down the thick lung secretions of cystic fibrosis

Antitrypsin—replacement therapy to benefit emphysema patients

PEG-SOD—a form of superoxide dismutase that minimizes damage to brain tissue after severe trauma

Vaccines

Vaccines for hepatitis B and *Haemophilus influenzae* Type b meningitis

Experimental malaria and AIDS vaccines based on recombinant surface antigens

Miscellaneous

Bovine growth hormone or bovine somatotropin (BST)—given to cows to increase milk production

Apolipoprotein—to deter the development of fatty deposits in the arteries and to prevent strokes and heart attacks

Spider silk—a light, tough fabric for parachutes and bulletproof vests

A commercial product called Frostban has been successfully applied to stop frost damage in strawberry and potato crops. A strain of *Pseudomonas fluorescens* has been engineered with the gene from a bacterium *(Bacillus thuringiensis)* that codes for an insecticide. These recombinant bacteria are released to colonize plant roots and help destroy invading insects. All releases of recombinant microbes must be approved by the Environmental Protection Agency (EPA) and are closely monitored. So far, extensive studies have shown that such microbes do not proliferate in the environment and probably pose little harm to humans.

Microbiologists have invented an unusual means of assaying drug resistance in the pathogen *Mycobacterium tuberculosis*. They engineered a bacteriophage with the genes for luciferase, an enzyme from fireflies that causes light emission. Live *M. tuberculosis cells* infected with these viruses light up; dead or dying cells do not. The test is conducted by incubating cultures with the usual drugs used in treatment and observing the viability of the culture. Dying cells indicate that the drug is working, and a healthy culture indicates drug resistance. The advantage to this type of test is that results can be obtained in a few days rather than the several weeks usually required.

Viruses can be genetically engineered for transfection purposes. Several viral vectors have been developed for gene therapy and for an experimental AIDS vaccine (see figure 25.20). In Australia, a mousepox virus was developed for release into habitats of mice that have become a serious pest. Unfortunately, the virus was so lethal, that its use was halted for fear it may jump hosts.

Another very significant bioengineering interest has been to create microbes to bioremediate disturbed environments. Biotechnologists have already developed and tested several types of bacteria that clean up oil spills and degrade pesticides and toxic substances (see chapter 26).

TRANSGENIC PLANTS: IMPROVING CROPS AND FOODS

Two unusual species of bacteria in the genus *Agrobacterium* are the original genetic engineers of the plant world. These pathogens live in soil and can invade injured plant tissues. Inside the wounded tissue, the bacteria transfer a discrete DNA fragment called T-DNA into the host cell. The DNA is integrated into the plant cell's chromosome, where it directs the synthesis of nutrients to feed the bacteria and hormones to stimulate plant growth. In the case of *Agrobacterium tumefaciens*, increased growth causes development of a tumor called *crown gall disease*, a mass of undifferentiated tissue on the stem. The other species, *A. rhizogenes*, attacks the roots and transforms them into abnormally overgrown "hairy roots."

The capacity of these bacteria, especially *A. tumefaciens*, to transfect host cells can be attributed to a large plasmid termed **Ti** (tumor-inducing). This plasmid inserts into the genomes of the infected plant cells and transforms them (figure 10.11). Even after the bacteria in a tumor are dead, the plasmid genes remain in the cell nucleus and the tumor continues to grow. This plasmid is a perfect vector for inserting foreign genes into plant genomes. The procedure involves removing the Ti plasmid, inserting a previously isolated gene into it, and returning it to *Agrobacterium*. Infection of the plant by the recombinant bacteria automatically transfers the plasmid with the foreign gene into the plant cells. The insertion of genes with a Ti plasmid works primarily to engineer important dicot plants such as potatoes, tomatoes, cotton, and grapes.

For plants that cannot be transformed with *Agrobacterium*, seed embryos have been successfully implanted with additional genes using gene guns—small "shotguns" that shoot the genes into plant embryos with tiny gold bullets. Agricultural officials had approved more than 100 plants for field testing through the year 2000. This movement toward release of engineered plants into the environment has led to some controversy. Many plant geneticists and ecologists are seriously concerned that transgenic plants will share their genes for herbicide, pesticide, and virus resistance with natural plants, leading to "superweeds" that could flourish and become indestructible. The Department of Agriculture is carefully regulating all releases of transgenic plants.

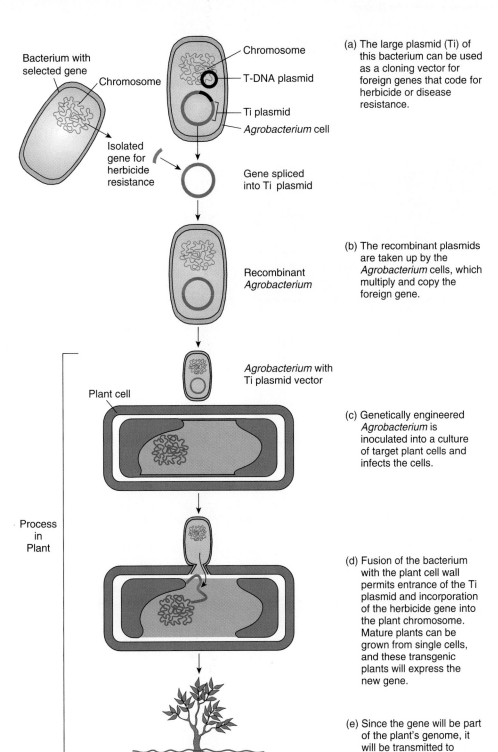

(a) The large plasmid (Ti) of this bacterium can be used as a cloning vector for foreign genes that code for herbicide or disease resistance.

(b) The recombinant plasmids are taken up by the *Agrobacterium* cells, which multiply and copy the foreign gene.

(c) Genetically engineered *Agrobacterium* is inoculated into a culture of target plant cells and infects the cells.

(d) Fusion of the bacterium with the plant cell wall permits entrance of the Ti plasmid and incorporation of the herbicide gene into the plant chromosome. Mature plants can be grown from single cells, and these transgenic plants will express the new gene.

(e) Since the gene will be part of the plant's genome, it will be transmitted to offspring in seeds.

FIGURE 10.11

Bioengineering of plants. Most techniques employ a natural tumor-producing bacterium called *Agrobacterium tumefaciens.*

Another strategy for engineering plants is to insert a strand of antisense RNA that turns off the expression of a target gene (figure 10.12, which illustrates the antisense concept). Researchers at the University of California at Davis transformed a strain of tomatoes with antisense RNA to block the gene that controls fruit ripening. Plants with this gene produce normal fruits that lack certain enzymes necessary for ripening. As a result, the tomato does not ripen too rapidly, and it can be left on the vine to develop its natural fla-vor. It can be induced to ripen using ethylene gas. See table 10.3 for examples of plants that have been engineered.

TRANSGENIC ANIMALS: ENGINEERING EMBRYOS

The inclination to be genetically engineered also exists for animals. In fact, animals are so amenable to transfection that several hundred

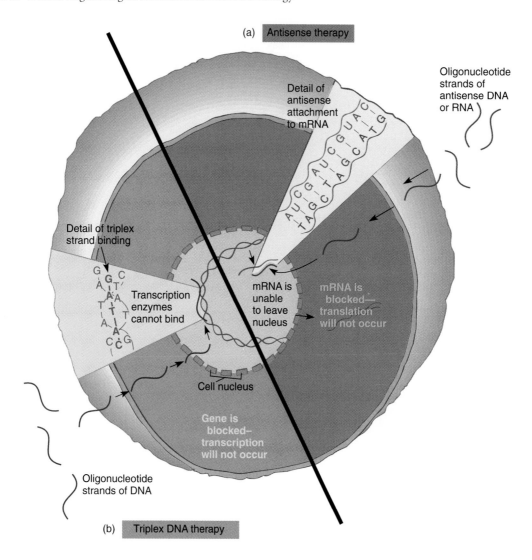

FIGURE 10.12

Mechanisms of antisense DNA and triplex DNA. These genetic medicines are designed to prevent the expression of an undesirable gene or virus. Both use oligonucleotides of known sequence that can bind with genetic targets that are the cause of disease. The site of action is in the nucleus. **(a)** When antisense strands enter the nucleus, they bind to complementary areas on an existing mRNA molecule. Such bound RNA will be unavailable for translation on the cell's ribosomes, and no protein will be made. **(b)** Triplex DNA strands attach by base-pairing to regulatory sites and other sites on the chromosome itself. This action masks the points of attachment for enzymes of transcription, and mRNA cannot be made. Without mRNA, there will be no translation or protein formed.

strains of transgenic animals have been introduced by research and industry. One reason for this movement toward animals is that, unlike bacteria and yeasts, they can express human genes in organs and organ systems that are very similar to those of humans. This advantage has led to the design of animal models to study human genetic diseases and then to use these natural systems to test new genetic therapies before they are used in humans. Animals can also be engineered to become "factories" to manufacture human proteins and other products.

The most effective way to insert genes into animals is by using a virus to transfect the fertilized egg or early embryo (germ line engineering) (figure 10.13). Foreign genes can be delivered into an egg by pulsing the egg with high voltage or by injection. The success rate does not have to be high, because a transgenic animal will usually pass the genes on to its offspring. The dominant animals in the genetic engineer's laboratory are transgenic forms of mice.

Mice are so easy to engineer that in some ways they have become as prominent in genetic engineering as *E. coli*. In some early experiments, fertilized mouse embryos were injected with human genes for growth hormone, producing supermice twice the size of normal mice. Laboratories have developed thousands of genetically modified mice.

The so-called knockout mouse has become a standard way of producing animals with tailor-made genetic defects. The technique involves transfecting a mouse embryo with a defective gene and cross-breeding the progeny through several generations. Eventually, mice will be born with two defective genes and will express the disease. Mouse models exist for cystic/fibrosis, hardening of the arteries, Gaucher's disease (a lysosomal storage disease), Alzheimer disease, and sickle-cell anemia. See table 10.4 for a survey of applications of transgenic animals used in animal husbandry and the drug industry.

TABLE 10.3

Examples of Engineered (Transgenic) Plants

Plant	Trait	Results
Nicotiana tabacum (tobacco)	Herbicide resistance	 Tobacco plants in the upper row have been transformed with a gene that provides protection against Buctril, a systemic herbicide. Plants in the lower row are normal and not transformed. Both groups were sprayed with Buctril and allowed to sit for 6 days. (The control plants at the beginning of each row were sprayed with a blank mixture lacking Buctril.)
Pisum sativum (garden pea)	Pest protection	 Pea plants were engineered with a gene that prevents digestion of the seed starch (see seeds on the left). This gene keeps tiny insects called weevils from feeding on the seeds. Seeds on the right are from plants that were not engineered and are suffering from weevil damage (note holes).
Oryza sativa (rice)	Viral resistance	A new form of rice plant was created with genes for resistance to the rice stripe virus. The virus is a significant pathogen that severely stunts young plants and reduces crop yields.

Several embryos recovered from sacrificed female

Embryos transferred to a depression slide containing culture medium

Culture medium

Oil

As embryo is held in place, DNA is injected into pronucleus.

Holding pipette

Pronucleus

DNA to be injected

Injection pipette

Several injected embryos are placed into oviduct of receptive female.

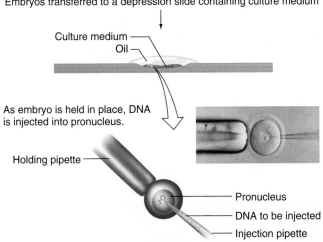

FIGURE 10.13
How transgenic mice are created.

CHAPTER CHECKPOINTS

Bioengineered hormones, enzymes, and vaccines are safer and more effective than similar substances derived from animals. Recombinant DNA drugs and vaccines are useful alternatives to traditional treatments for disease.

Transfection is the process by which foreign genes are introduced into organisms. The transfected organism is termed *transgenic.* Transgenic microorganisms are genetically designed for medical diagnosis, crop improvement, pest reduction, and bioremediation. Transgenic animals are genetically designed to model genetic therapies, improve meat yield, or synthesize specific biological products.

V. Genetic Treatments: Introducing DNA into the Body

GENE THERAPY

One result of the research with transgenic animals is **gene therapy,** a technique for replacing a faulty gene with a normal one in people with fatal or extremely debilitating genetic diseases. The inherent benefit of this therapy is to permanently cure the physiological dysfunction by repairing the genetic defect. There are two strategies for this therapy. In *ex vivo* therapy, the normal gene is cloned in vectors such as retroviruses (mouse leukemia virus) or adenoviruses that are infectious but relatively harmless. Tissues removed from the patient are incubated with these genetically modified viruses to transfect them with the normal gene. The transfected cells are then reintroduced into the patient's body by transfusion (figure 10.14). In contrast, the *in vivo* type of therapy skips the intermediate step of incubating excised patient tissue. Instead, the naked DNA or a virus vector is directly introduced into the patient's tissues.

So far, therapy has been hampered by a number of difficulties. The systems for delivering the gene into target cells are not completely successful, and the rate of productive transfection by the viruses is very low. Some experts fear that the virus could insert the DNA into a site that turns on an oncogene or could activate other dormant viruses. Alternative delivery methods are being tested to deliver the product more precisely. These include gene guns similar to those previously described for plants and liposomes, tiny spheres with an exterior lipid bilayer and aqueous center that can carry genes across cell membranes.

Other problems have to do with incorporation, retention, and expression of the DNA by the cell. In some cells, the DNA appears to be picked up and held free in the nucleus. The transfected cells can even translate the gene for a limited time, but they may not duplicate or pass it on during cell division, and these cells often have a limited lifespan. One solution tailored for diseases of blood cells is to genetically engineer the long-lived stem cells in bone marrow. If enough stem cells take up and retain the replacement gene, it will be inherited by new blood cells for the life of the patient.

The first gene therapy experiment in humans was initiated in 1990 by researchers at the National Institutes of Health. The subject was a 4-year-old girl suffering from a severe immunodeficiency disease caused by the lack of the enzyme adenosine deaminase (ADA). She was transfused with her own blood cells that had been engineered to contain a functional ADA gene. Later, other children were given the same type of therapy. So far, the children have shown remarkable improvement and continue to be healthy, but the treatment is not permanent and must be repeated.

Single-gene defects such as hemophilia and sickle-cell anemia are possible candidates for therapy. Other clinical trials are summarized in table 10.5. Several patients have been treated suc-

TABLE 10.4

Pharmaceutical Production by Some Transgenic Animals

Treatment For :	Selected Animal	Protein Expressed in Milk	Application
Hemophilia	Pig	Factor VIII and IX	Blood-clotting factor
Emphysema	Sheep	Alpha antitrypsin	Supplemental protein—lack of this protein leads to emphysema
Cancer	Rabbit	Human interleukin-2	Stimulates the production of T lymphocytes to fight selected cancers
Septicemia	Cow	Lactoferrin	Iron-binding protein inhibits the growth of bacteria and viral infection
Surgery, trauma, burns	Cow	Human albumin	Return blood volume to its normal level
GHD (growth hormone deficiency)	Goat	HGH (human growth hormone)	Supplemental protein improves bone metabolism, density, and strength
Heart attack, stroke	Goat	Tissue plasmagen activator (tPA)	Dissolves blood clots to reduce heart damage (if administered within 3 hours)

These two mice are genetically the same strain, except that the one on the right has been transfected with the human growth hormone gene (shown in its circular plasma vector).

Little piglets have been genetically engineered to synthesize human Factor VIII, a clotting agent used by type A hemophiliacs. The females will produce it in their milk when mature. It is estimated that only a few hundred pigs would be required to meet the world's demand for this drug.

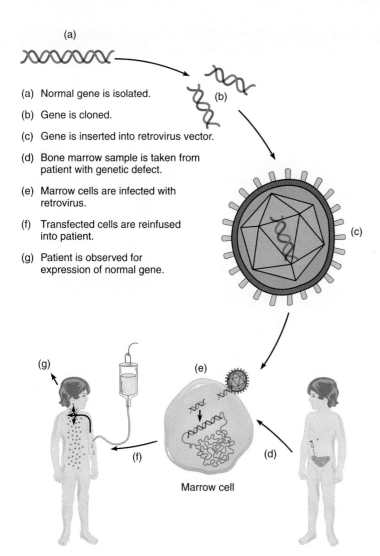

(a) Normal gene is isolated.

(b) Gene is cloned.

(c) Gene is inserted into retrovirus vector.

(d) Bone marrow sample is taken from patient with genetic defect.

(e) Marrow cells are infected with retrovirus.

(f) Transfected cells are reinfused into patient.

(g) Patient is observed for expression of normal gene.

Marrow cell

FIGURE 10.14

Protocol for the *ex vivo* type of gene therapy in humans.

cessfully for another type of severe combined immunodeficiency syndrome called X-1–linked SCID that is due to a missing enzyme required for mature immune cells. This is the first instance of an apparent cure. Full function has been restored to several hemophilic children, using a similar technique. Medical researchers have been able to correct the damage in certain vascular diseases by infusing growth factor genes into their circulatory systems. In trials for cystic fibrosis (CF) therapy, adenoviruses carrying a normal CF gene are delivered directly into the nose and lungs. So far, the results of this therapy have been less successful, probably because of a low acceptance rate for take-up of the gene by respiratory cells.

The ultimate sort of gene therapy is gene line therapy, in which genes are inserted into an egg, sperm, or early embryo. Although tests in animals point to its possible effectiveness, its use on humans has been rather controversial—although a private institute has invested several hundred million dollars to begin this process.

ANTISENSE AND TRIPLEX DNA TECHNOLOGY: GENETIC MEDICINES

Up to now we have considered the use of recombinant hormones and other proteins to replace a missing or dysfunctional protein and gene therapy to restore a missing or defective gene. Biomedical researchers have been exploring yet a third type of therapy with genetic drugs called **antisense** and **triplex agents.** With this approach, oligonucleotides are delivered into cells to block transcription or translation of a specific DNA site, thereby bypassing the problem gene and protein altogether.

Antisense DNA: Targeting Messenger RNA

Note that the term *antisense* as used in molecular genetics does *not* mean "no sense." It is used to describe any nucleic acid strand with a base sequence that is complementary to the "sense" or translatable strand. For example, DNA contains a template strand that is transcribed and a matching strand that is not usually transcribed. We can also apply this terminology to RNA. Messenger RNA is considered the translatable strand, and a strand with its complementary sequence of nucleotides would be the antisense strand. To illustrate: If an mRNA sequence read AUGCGAGAC, then an antisense RNA strand for it would read UACGCUCUG. In practice, most antisense agents are made of DNA because the technology for large-scale RNA synthesis is less feasible. The therapy takes advantage of the readiness with which an antisense DNA can hybridize with RNA. It is worth noting that the concept of antisense nucleic acids did not originate with genetic engineers. It appears to be one mechanism by which organisms naturally control gene expression.

When an adequate dose of antisense DNA is delivered across the cell membrane into the cytoplasm and nucleus, it binds to specific sites on any mRNAs that are the targets of therapy. As a result, the reading of that mRNA transcript on ribosomes will be blocked, and the gene product will not be synthesized (see figure 10.12*a*).

Animal experiments show that antisense drugs are relatively nontoxic and safe to use, and cell culture tests indicate that they can retard synthesis of an undesirable protein. But, as with most new therapies, there are also some stumbling blocks. One is the drugs' high cost and the need to administer them throughout the course of a person's life. Another drawback has been the inability to get a significant dose of the agent into cells. Whether the DNA could become a permanent part of the genome of cells and lead to cancer and other diseases is another concern.

It is also clear that this therapy will not work for many genetic-based diseases such as cystic fibrosis and sickle-cell anemia that are caused by the lack of a functioning gene. However, it promises to be most effective in preventing diseases caused by undesirable gene expression and by viruses. Diseases that might be amenable to this therapy are cancers, autoimmune diseases such as multiple sclerosis, infections such as hepatitis B and AIDS, and Alzheimer disease. Human clinical trials are under way for blocking the replication of human papillomavirus (the cause of genital warts and cervical cancer) and the AIDS virus. Trials of acute and chronic myelogenous leukemias, caused by the abnormal expression of a cancer gene, are also being carried out (table 10.5). The

TABLE 10.5

Sample of Human Genetic Defects Targeted for Gene Therapy

Disease	Nature of Defect	Gene Replacement Strategy
Adenosine deaminase (ADA) deficiency	Severe immunodeficiency due to destruction of lymphocytes by toxic metabolic product.	ADA gene inserted into bone marrow white blood cells using virus vector or liposomes.
X-1–linked SCID	Enzyme needed for maturing B and T cells is missing.	Bone marrow is modified by means of engineered virus.
Cystic fibrosis	Lack of necessary protein for conducting chloride ions across membranes causes buildup of mucus and secretions in lungs, pancreas, other organs.	CFTR gene placed in adenovirus or liposomes and inoculated into nasal cavity or lungs.
Sickle-cell anemia, beta-thalassemia	Lack of correct gene to make the normal B globin protein of hemoglobin.	Hemoglobin A (HbA) gene placed in bone marrow cells by vector.
Hemophilia B	Lack of gene to produce factor VIII, needed for blood clotting.	Normal gene for factor VIII incorporated into marrow cells.
Gaucher's disease	Deficiency in an enzyme that is needed to metabolize glucocerebroside leads to its buildup in lysosomes.	Insertion of normal glucocerebrosidase gene into marrow cells.
Muscular dystrophy	Lack of gene for producing dystrophin necessary for normal muscle development.	Dystrophin gene delivered into muscle tissue by one of several methods.
AIDS	Infection by HIV destroys important immune cells (T cells) and causes severe loss of immune function.	Fibroblasts are infected with gene that stimulates an immune response against HIV; other test uses ribozyme enzyme to block virus activity.
Malignant melanoma	Severest form of skin cancer that does not respond well to traditional therapy.	Insertion of B7 gene into tumor cells to boost the natural white blood cell response against the cancer.
Acute myelogenous leukemia	Blood disease caused by overexpression of a gene.	Genetic modification of cells by replacement of defective gene; also addition of antisense DNA to block the actions of cancer genes.

alteration of plant genes with antisense agents has already proved itself in agricultural applications.

Triplex DNA: Adding a Third Strand

Although the usual structure of DNA is a helix composed of two strands, researchers have known for years that DNA can exist in other naturally and artificially induced formats. One of these unusual variations, **triplex DNA,** is a triple helix formed when a third strand of DNA inserts into the major groove of the molecule. The inserted strand forms hydrogen bonds with the purine bases on one of the adjacent strands (see figure 10.12b). It seems logical that the template of DNA that has an extra strand wedged into its structure can be relatively inaccessible to normal transcription. Indeed, it is theorized that this may be another natural means for cells to control their genetic expression.

The potential therapeutic applications of triplex DNA are still in their infancy. So far, oligonucleotides have been synthesized to form triplex DNA that interacts with regulatory sequences in genes. It can interrupt the action of transcription factors and prevent RNA polymerase from forming mRNA transcripts. Studies performed on cell cultures revealed that the genes coding for oncogenes, viruses, and the receptor for interleukin-2 (an immune stimulant) could be blocked with triplex DNA. In the clinical trials to come, this therapy will no doubt be plagued with some of the same problems mentioned for the antisense approach.

VI. Genome Analysis: Maps, Fingerprints, and Family Trees

DNA technology has opened the way to substantial discoveries in basic genetics. It has provided a means of deriving a complete genetic knowledge of any organism or virus. Such knowledge will in turn lead to understanding of diverse biological phenomena, including

"We finished the genome map, now we can't figure out how to fold it!"

Cartoon by John Chase. Reprinted by permission.

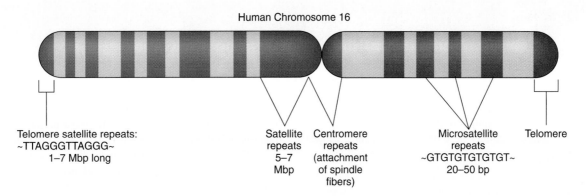

FIGURE 10.15

A physical map of chromosome 16. This map shows areas on the chromosome that are devoted to support DNA, most of which consist of highly repetitive nucleotides.

development, virulence, disease, and evolution. Its influence is being felt in numerous other fields such as forensics (medical law), genetic pedigreeing, anthropology, neurobiology, and archeology.

GENOME MAPPING AND SCREENING: AN ATLAS OF THE GENOME

We have seen a variety of methods for accessing the genomes of organisms, but the most useful information comes from knowing the sequential makeup of the genetic material. Genetic engineers find it very informative to know the *locus,* or exact position of a particular gene on a chromosome, and they also seek information on the types and numbers of *alleles,* which are sites that vary from one individual to another. The process of determining location of loci and other qualities of genomic DNA is called **mapping.** Maps vary in resolution and applications. *Linkage maps* show the relative proximity and order of genes on a chromosome and are relatively low resolution because only a few exact locations are mapped. *Physical maps* are more detailed arrays that depict not only the relative positions of distinct sections of DNA, but give the numerical size in base pairs (figure 10.15). This technology uses restriction fragments and fluorescent hybridization probes to visualize the position of a particular selected site along a segment of DNA.

By far the most detailed maps are *sequence maps,* which are produced by the sequencers we discussed earlier in the chapter. They give an exact order of bases in a plasmid, chromosome, or entire genome. Because this level of resolution is most promising for understanding the nature of the genes, what they code for, and their functions, this form of map dominates the research programs. When the Human Genome Project was launched in 1986 to map the human genome, it was expected to require 15 to 20 years and cost 3 billion dollars. Due to advances in technology, it was mostly completed by the year 2000 (Microbits 10.1).

Other genome projects have also been highly successful, so much so that the genomes of over 1,000 organisms, viruses, and organelles have been fully sequenced. As of 2001, this includes around 600 viruses, 125 procaryotes (bacteria and archaea), and over 300 eucaryotes (including human, mouse, *C. elegans* (a

nematode worm), yeast, fruitfly, *Arabidopsis* (a small flowering plant), and rice. One of the remarkable discoveries in this huge enterprise has been how similar the genomes of relatively unrelated organisms are. Humans share around 80% of their DNA codes with mice, about 60% with rice, and even 30% with the worm *C. elegans.*

Another interesting surprise has been that a very large amount of the human genome contains DNA sequences that do not code for cell products. Ninety-seven percent of it is made up of unusual support DNA that functions in chromosome stabilizing and division, gene regulation, and ribosome assembly.

One subgroup of support DNA is composed of short sections of highly repetitive DNA, meaning that the same nucleotide sequence is repeated 10 to several thousand times (figure 10.15). Three identifiable types are: (1) *Satellite DNA* consisting of 1 to 7 million base pairs located mainly at the ends and centers of chromosomes. It appears to maintain the integrity of chromosomes. (2) *Minisatellite DNA* is a series of 200 to 3,000 base pair repeats scattered throughout the chromosomes. One type, called variable number tandem repeats (VNTRs), is an important type of restriction fragment used in mapping genes and tracing inheritance. (3) *Microsatellites* are even tinier (20–50 bp) repetitive blocks of DNA.

Although sequencing provides the ultimate genetic map, it does not automatically tell what the exact genes and alleles are. Analyzing and storing this massive amount of new data requires specialized computers. Because it is information of both a biological and mathematical content, two whole new disciplines have grown up around managing these data: *genomics* and *bioinformatics.* The job of genomics and bioinformatics will be to analyze and classify genes, determine protein sequences, and ultimately determine the function of the genes. This fact is likely to bring another golden era in biology and microbiology, that of unraveling the complexity of relationships of genes and gene regulation. In time, it will provide a complete understanding of such phenomena as normal cell function, disease, development, aging, and many other issues. In addition, it will allow us to characterize the exact genetic mechanisms behind pathogens, and allow new treatments to be developed against them.

"Achieving this milestone is an exhilarating moment in history, and a credit to the ingenuity and dedication of some of the brightest scientists of the current generation."

Francis Collins, head of the U.S. Genome Project

Imagine trying to develop a detailed map from New York to Los Angeles one inch at a time, with few known landmarks to use as references. This is something akin to what molecular geneticists have done in mapping the human genome. When finally finished, its sequence of around 40,000 genes and 3.2 billion base pairs will fill a million pages of text. The full map will be like a "human gene dictionary" or "encyclopedia of humans" composed of 24 volumes, one for each chromosome type. How was this feat accomplished? If it had continued at the pace of the first sequencing attempts, it would have taken 200 years. Ultimately, it took 12 major U.S. centers and hundreds of smaller public and private sites, and included thousands of individuals in 50 countries working cooperatively over 14 years.

The key to developing an accurate map is analogous to first setting up markers along the DNA that can be used as points of reference (see figure 10.15). Developing these resolution maps involves a process called "walking a chromosome." For this, the investigators clone large segments (about 125 kb) of DNA in BAC or YAC vectors and screen these clones for equally spaced unique sequences that are found only in human DNA. These markers are called sequence tagged sites, or STSs, and expressed sequence tags, or ESTs. Known tagged sequences at opposite ends of a cloned site can be compared with other clones to find areas of overlap. This process required at least 30,000 of these markers to be

laid down in accurate order to serve as landmarks. Over 100,000 of these have been discovered to date. The result is something called a "contig (for contiguous) map" that gives a fairly precise positioning of the cloned fragments of DNA and, eventually, of exact gene locations.

The contig map paved the way for the complete sequencing of the DNA in the fragments to form a structural map. The current sequencing was carried out in large-scale centers using giant sequencing machines that collectively determine the order of millions of base pairs in a few days. DNA sequencing requires specialized computers that can store billions of megabytes of data. During sequencing of the human genome, a supercomputer had to perform around 500 quintillion calculations. In the spring of 2000, Celera Genomics completed a rough draft of the genome, about 90% of all nucleotides. The ordering of the sequence is expected to be completed by 2003. Approximately 50% of individual genes have been located.

The outcome of the Human Genome Project promises to forever change the human condition. With a map of the human genome, the genetic background of the 5,000 genetic diseases could eventually be disclosed, clearing the way for extensive genetic screening. It would not only enable families to know in advance their risks for certain diseases but would also guide decisions on correcting or alleviating them through gene and other forms of therapy. As new genes are discovered on a weekly basis, it becomes increasingly possible to be tested for a wide variety of genetic diseases. This testing goes hand in hand with methods of genetic analysis described in the section on DNA fingerprinting.

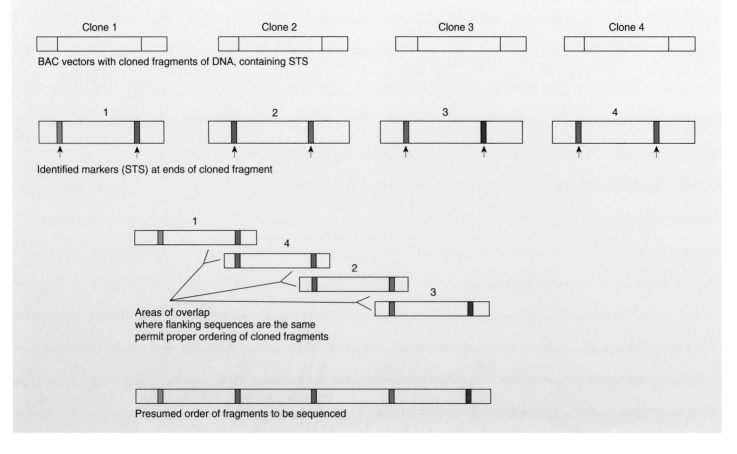

Clone 1 Clone 2 Clone 3 Clone 4

BAC vectors with cloned fragments of DNA, containing STS

1 2 3 4

Identified markers (STS) at ends of cloned fragment

1
4
2
3

Areas of overlap
where flanking sequences are the same
permit proper ordering of cloned fragments

Presumed order of fragments to be sequenced

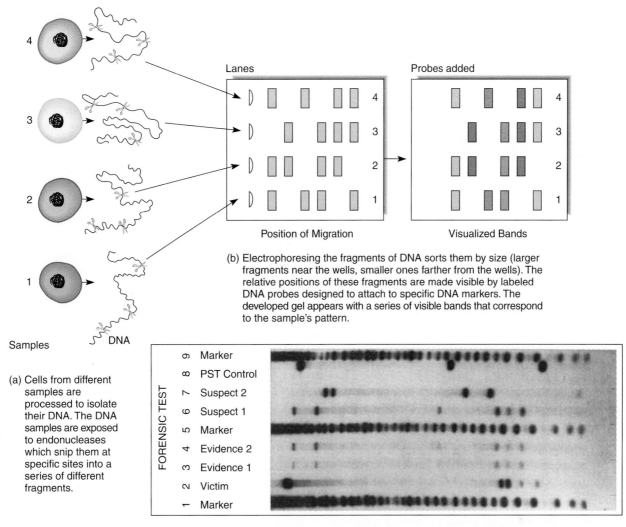

(b) Electrophoresing the fragments of DNA sorts them by size (larger fragments near the wells, smaller ones farther from the wells). The relative positions of these fragments are made visible by labeled DNA probes designed to attach to specific DNA markers. The developed gel appears with a series of visible bands that correspond to the sample's pattern.

(a) Cells from different samples are processed to isolate their DNA. The DNA samples are exposed to endonucleases which snip them at specific sites into a series of different fragments.

(c) An actual DNA fingerprint used in a rape trial. Control lanes with known markers are in lanes 1, 5, 8, and 9. The second lane contains a sample of DNA from the victim's blood. Evidence samples 1 and 2 (lanes 3 and 4) contain semen samples taken from the victim. Suspects 1 and 2 (lanes 6 and 7) were tested. Can you tell by comparing evidence and suspect lanes which individual committed the rape?

FIGURE 10.16
DNA fingerprints: the bar codes of life.

DNA FINGERPRINTING: A UNIQUE PICTURE OF A GENOME

Although DNA is based on a structure of nucleotides, the exact way these nucleotides are combined is unique for each organism. It is now possible to apply DNA technology in a manner that emphasizes these differences and arrays the entire genome in a pattern for comparison. **DNA fingerprinting** (also called DNA typing or profiling) is best known as a tool of forensic science first devised in the mid-1980s by Alex Jeffreys of Great Britain (figure 10.16).

Several of the methods discussed previously in this chapter are involved in the creation of a DNA fingerprint. Techniques such as the use of restriction endonucleases for cutting DNA precisely, PCR amplification for increasing the number of copies of a certain genome, electrophoresis to separate fragments, hybridization probes to locate specific loci and alleles, and Southern blotting for producing a visible record are all employed by biotechnologists. The specificity of the DNA fingerprint is greatly increased by using known markers that can detect the numerous polymorphisms that exist in genomes. Unknown genomes can also be typed for comparative purposes with a variety of technologies such as the RFLP type of genetic marker discussed earlier. Some newer systems for detecting variations in the DNA code are the random amplification of polymorphic DNA (RAPD) and short tandem repeats (STRs). STRs are small blocks of base pairs repeated end to end that vary in number and position in each individual. Even more resolution can be obtained with single nucleotide polymorphisms (SNP), which have over 30,000 variations. Companies are currently adapting these to a miniaturized method of processing with computer chips that can hold thousands of sequences.

Regardless of the method used, the primary function of the fingerprint is to provide a snapshot of the genome, displaying it as a unique picture of an individual organism. To get some idea of

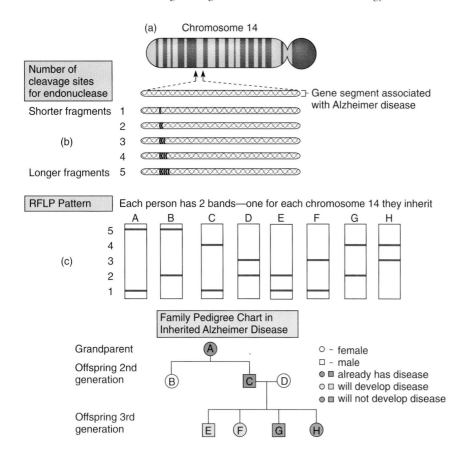

FIGURE 10.17

Pedigree analysis based on genetic screening for familial type Alzheimer disease. (a) The site for this disease is on chromosome 14. **(b)** Gene mapping has determined that this chromosome segment has five variations, depending on the number and size of cleavage sites. **(c)** A Southern blot using the restriction fragments (RFLPs) is performed on a family with three generations. Each person will have two bands that correspond with each of the chromosomes they have inherited. The grandmother (A) with pattern 1,5 and her son (C) with pattern 1,4 both already have the disease. Patterns for his sister (B), his wife (D), and their four children (E, F, G, and H) are also shown. Comparing the band patterns of A and B with those of other family members makes it possible to predict which children will inherit the gene for Alzheimer disease. Because the band shared by A and C is 1, we know that this is the variant associated with Alzheimer disease. Children E (1,2) and F (1,3) will develop the disease, and children G (2,4) and H (3,4) will not. It is also evident that this form of Alzheimer disease is inherited as a dominant trait.

how useful this technology can be for determining genetic relatedness, consider the following example. Only identical twins would have the exact same DNA fingerprint, and close family members will probably share many of the same (but not exact) alleles. The probability of two unrelated persons having an identical pattern using just a single marker (band) is calculated to be about 1 in 3. If the number of markers is increased to 10, the probability that there would be an exact match decreases dramatically—to 1 in 20,000. If you increase this again to 24 markers, the chance that two people would have the same fingerprint pattern is an amazing 1 in 17 billion. The most powerful uses of DNA fingerprinting are in forensics, detecting genetic diseases, determining parentage, analyzing the family trees of humans and animals, identifying microorganisms, and tracing the lineage of ancient organisms.

One of the first uses of the technique of DNA fingerprinting was in forensic medicine. Besides real fingerprints, criminals often leave other evidence at the site of a crime—a hair, a piece of skin or fingernail, semen, blood, or saliva. Because it can be combined with amplifying techniques such as PCR, DNA fingerprinting provides a way to test even specimens that have only minute amounts of DNA, or even old specimens. Although contamination has been a problem in interpretation, new refinements in technique have made this less of a problem than it once was. If enough markers are used, it is possible to give probabilities in the billions and even trillions that the DNA matches evidence at the crime scene.

DNA fingerprinting is so widely accepted that it is now routine to use it as trial evidence. It has been used to convict and exonerate thousands of accused defendants in criminal cases. In fact, the technology is so reliable that many older cases for which samples

are available are being reopened and solved. The demand has been so high that many forensic units are overwhelmed by requests for testing. The FBI is currently setting up a national database of DNA fingerprints, and several states allow DNA fingerprinting of felons in prison (so-called DNA dogtags). In a most amazing application of this knowledge, forensic scientists were able to positively identify the culprits in the World Trade Center bombing and the Unabomber, Ted Kaczynski, by DNA recovered from the back of postage stamps they had licked!

Fingerprinting can also be used to detect genes that increase the risk for hereditary diseases and the patterns of inheritance in conditions such as Huntington disease, cystic fibrosis, and Alzheimer disease (figure 10.17). Test DNA from various family members is reacted with gene-specific probes that will show exactly which restriction sites (genetic markers) the person possesses. The genetic markers are not the actual gene that causes the disease, but they are located nearby and are inherited in exactly the same pattern. The genetic family tree for many diseases can be worked out by comparing fingerprint patterns in a family that is known to carry and express the gene. Futurists predict that a DNA fingerprint will be made for every child at birth and kept on file, much as footprints and regular fingerprints have been used until now.

This same method of analysis can be used for clarifying human paternity and maternity, the pedigree (genetic ancestry) of domestic animals, and the genetic diversity of animals bred in zoos.

The Soviet government tested the remains of what they believed were the family of Tsar Nicholas, executed in 1918. The samples were compared with the DNA of known family members. The analysis confirmed that the bodies were indeed those of the

MICROBITS 10.2
Bioethics Confronts Bioengineering

No one would minimize the real and potential benefits to be derived from genetic engineering. Yet, in our endorsement of this technology, we must also at least consider its possible adverse impact. Some alarmists feel that tampering with the genes of humans and other organisms is tantamount to playing God, and they predict dire sociological, ecological, and medical consequences. Jeremy Rifkin of the Foundation on Economic Trends voiced his group's fear:

> We are embarking on a very potentially troublesome journey, where we begin to reduce all other animals on this planet to genetically engineered products. . . . We will increasingly think of ourselves as just gene codes and blueprints and programs that can be tinkered with.

Other critics argue that creating entirely new organisms and releasing them into the earth's habitats may upset the balance of nature. They point out that the same techniques that are used to engineer plants to resist herbicides and bacteria to make insulin could also create a virus that inserts into the human genome and causes cancer or other diseases.

The Problem with GMOs

One of the most publicized issues has been the use of genetically modified organisms in food crops. A serious controversy has erupted over transgenic corn and soybeans. The corn is engineered to contain an insecticide gene from the bacterium, *Bacillus thuringiensis* (Bt). The soybeans contain genes for herbicide resistance. Over 50% of the soybeans and 20% of the corn planted worldwide are genetically modified. Although they are not intended for use in human food, in several instances, they have wound up in processed products, leading to concerns about safety.

In Europe and America, protesters have marched to demand the end to "frankenfoods." One possible problem is that these foods may give rise to allergies or other sensitivities. Ecologists fear that the insecticide gene will be spread to other plants and destroy the "good" insects. It has already been found lethal to monarch butterflies. In the end, the advantages of these new plants must be weighed against their adverse impact.

Brave New World?

Other consequences of this technology are far-reaching. For instance, is the world heading toward a time when normal children will be given genetically engineered growth hormone to increase size? A controlled study was conducted to assess the effects of recombinant growth hormone on the development of children who are healthy but somewhat small for their age. Data from these trials indicate that hormone therapy does increase height if given early, but the increases are not dramatic. The perception of small size as a "disease" that needs to be treated is very distressing to some medical experts, and the treatment is currently limited to treating children who are deficient in this hormone. Yet supporters feel that the use of engineered hormones is really no different from cosmetic surgery and should be available to people who want it.

Detailed analysis of the genetic blueprint of human beings will not only provide a greater understanding of what causes genetic diseases, but it will also allow screening for any defective genes that could lead to disease. Such a possibility will also give rise to numerous ethical and le-

A demonstrator at a biotechnology conference protests the development and sale of genetically modified foods.

gal questions. Opponents have already objected vehemently to genetic screening as an invasion of privacy, not unlike requiring genetic "social security numbers." Might such tests be demanded as a means of determining one's fitness for a job or insurance coverage? The federal government has announced its opposition to this use of screening, but has yet to offer legislation to control it.

It is far more likely that the ability to profile an individual's genetic makeup will have beneficial effects. Children with serious or life-threatening defects may require prompt therapy to correct or alleviate the effects of the disease. In pinpointing the risk for cancer and other serious diseases, screening will help people make meaningful life-style and medical decisions. In addition, genetic counseling would be greatly improved since parents with a complete genetic profile could determine in advance what potentially harmful genes their children might inherit.

Tests are now available to determine the sex of an eight-cell-stage embryo and whether it carries certain genetic diseases. Will these techniques be used to promote sex selection, or will they influence decisions to terminate pregnancy on the basis of genetic defects? Although it has not yet been attempted in humans, gene line therapy (correcting genetic defects in the zygote or early embryo) will be a distinct possibility in the future. The most vocal critics imagine a day when parents will be able to create superbabies by requesting that certain favorable genes be added to their children's genomes. Others, however, feel that gene line therapy would be the ultimate solution to correcting genetic defects that diminish the quality of life.

continued on next page

The Bottom Line

An inevitable outcome of bioengineering has been its potential for profit. By 2001, over a hundred bioengineered products are expected to earn as much as $30 billion in revenues. With thousands of biotechnology companies worldwide, each actively developing various drugs, research tools, or industrial applications, this industry will continue to expand. Research and development on new drugs and treatments often involves billion-dollar monetary investments. At the present time, pharmaceutical companies are testing and requesting approval for over 400 new biotech medicines. Most of these are for treating cancer, AIDS, vaccines, and gene therapy.

Most biotechnology companies would like to have rights to exclusive marketing of the drugs, procedures, animals, plants, and microbes they have developed. Several hundred patents have been issued for unique and commercially useful organisms and drugs, and over 10,000 have been applied for. Currently, one of the most provocative questions is: Who owns the human genome? Several companies have patented human databases for genes that influence breast cancer, colon cancer, and obesity. Because the DNA was isolated and characterized, it existed in a form different from nature. Such patents have been used to develop test kits to detect defective genes and to assess risk, but later they may be used to develop therapies.

Protecting the public's best interest will require that bioethicists and representatives of many disciplines—science, law, medicine, business, government, and education—continue to discuss the basic moral and economic issues and to develop guidelines for the future. The main problem may well be whether laws can keep up with the rapid advances in the field. There is no doubt that bioengineering will eventually touch all of our lives and even change the ways we regard the world.

Romanovs. The same technique also proved that a woman claiming to be their missing daughter, Anastasia, could not have been related—putting that legend finally to rest.

The U.S. military has begun keeping DNA profiles on their personnel. When the remains of several "unknown soldiers" who had died in the Persian Gulf War were too mutilated and mixed to be properly identified, the DNA content was analyzed. A comparison with known DNA profiles of soldiers missing in action allowed the remains to be identified and separated.

Infectious disease laboratories are developing numerous test procedures to profile bacteria and viruses. Profiling is used to identify the gonococcus, *Chlamydia, syphilis spirochete,* and *M. tuberculosis* (see figure F in Introduction to Identification Techniques in Medical Microbiology, page 541). It is also an essential tool for determining genetic relationships between microbes.

Because even fossilized DNA can remain partially intact, ancient DNA is being studied for comparative and evolutionary purposes. Anthropologists have traced the migrations of ancient people by analyzing the mitochondrial DNA found in bone fragments. Mitochondrial DNA (mtDNA) is less subject to degradation than is chromosomal DNA and can be used as an evolutionary time clock. A group of evolutionary biologists have been able to recover Neanderthal mtDNA and use it to show that this group of ancient primates could not have been an ancestor of humans. Biologists studying canine origins have also been able to show that the wolf is the ancestral form for modern canines.

CHAPTER CHECKPOINTS

Gene therapy is the replacement of faulty host genes with functional genes by use of cloning vectors such as viruses and liposomes. This type of transfection can be used to treat genetic disorders and acquired disease. Antisense DNA and triplex DNA are used to block expression of undesirable host genes in plants and animals as well as those of intracellular parasites.

DNA technology has advanced understanding of basic genetic principles that have significant applications in a wide range of disciplines, particularly medicine, evolution, forensics, and anthropology.

The Human Genome Project has not only identified sites of specific human genes but has also clarified the functions of a special type of DNA (support DNA) that does not code for proteins.

DNA fingerprinting is a technique by which individuals are identified for purposes of medical diagnosis, genetic ancestry, and forensics.

Mitochondrial DNA analysis is being used to trace evolutionary origins in animals and plants.

CHAPTER CAPSULE WITH KEY TERMS

Sciences that manipulate, alter, and analyze the genetic material of microbes, plants, animals, and viruses are **genetic engineering** (bioengineering) and **biotechnology**. Genetic engineering is defined as the direct, deliberate modification of an organism's genome, and biotechnology is the use of DNA or genetically altered organisms in commercial production.

In addition to biology and related fields, the influence of applied genetics is felt in numerous other areas such as forensics (medical law), agriculture, industry, anthropology, and archaeology. These powerful and useful tools promise to change many areas of law, commerce, and medicine (refer to figure 10.1).

I. **Techniques in Genetic Engineering**
   A. Amazing DNA
      1. DNA readily **anneals** (becomes temporarily denatured) when heat breaks the hydrogen bonds holding the double helix together, and it separates into two strands. Slow cooling causes single DNA strands to rejoin (renature) at complementary sites.
      2. DNA can also be clipped crosswise at specific sites by means of bacterial enzymes called **restriction endonucleases**. These enzymes leave short tails called sticky ends that base-pair with complementary tails on linear DNA or plasmids.

Endonucleases allow splicing of genes into specific sites and create fragments that circularize into a plasmid.

3. Fragments produced by restriction endonucleases—*restriction fragment length polymorphisms* (**RFLPs**)—vary in length and are inherited in predictable patterns that make them useful *markers* of unique genetic characteristics.

4. **Ligases** rejoin sticky ends made by endonucleases for final splicing of genes into plasmids and chromosomes.

5. **Reverse transcriptase** can convert RNA into DNA. Complementary, or **cDNA,** of messenger, transfer, and ribosomal RNA provide a valuable means of synthesizing eucaryotic genes from mRNA transcripts.

6. **Gel electrophoresis** can produce a readable pattern of DNA fragments. When samples in gel are subjected to an electrical current, the DNA pieces migrate in the gel substrate toward the positive pole, forming a pattern based on fragment size. The pattern is analyzed by comparing it against known standards to characterize genetic similarities.

7. **Oligonucleotides** are short fragments of DNA or RNA usually made by DNA synthesizers. The exact sequence of the base pairs in DNA is automatically analyzed by sequencers. Sequencing is critical for detailed gene analysis.

B. Nucleic Acid Hybridization and Probes
 1. The ability of single-stranded nucleic acids to **hybridize** or join together at complementary sites makes it possible to use **gene** or **hybridization probes** to detect specific nucleotide sequences.
    a. Hybridization probes can identify unknown samples or isolate precise fragments from a complex mixture. In the *Southern blot* method, probes are used to label a gene or nucleotide sequence on an electrophoretic gel; the blot hybridization is a rapid, direct probe of a sample. Probes are used for diagnosing the cause of an infection from a patient's specimen and identifying a culture of an unknown bacterium or virus with a microbe-specific probe.
    b. *In situ hybridization* uses probes on intact cells to visualize the presence and location of specific nucleic acids such as genes on chromosomes or RNA in cells and tissues.
    c. DNA sequencing machines can produce highly accurate maps of the exact sequence of DNA. The Sanger sequencer synthesizes and analyzes DNA fragments using fluorescent dyes to label bases.
 2. The **polymerase chain reaction (PCR)** is a valuable tool that can amplify the amount of DNA in a sample from a few copies to billions of copies in a few hours.
    a. The PCR technique operates by repetitive cycling of three basic steps: (1) **denaturation**—heating target DNA to separate it into two strands, called **amplicons;** (2) addition of **primers** (synthetic oligonucleotides) to serve as guides for positioning the start of DNA amplification; (3) **extension**—thermostable DNA polymerase synthesizes the complementary strand by adding appropriate nucleotides.
    b. After the first cycle, the DNA strands will be denatured to be primed and extended in the second cycle, thereby doubling the number of copies to four. This doubling can be continued for 20 to 30 cycles.

II. **Methods in Recombinant DNA Technology**
 A. **Technical aspects of recombinant DNA (rDNA) technology** methods deliberately remove genetic material from one organism and combine it with that of a different organism.
  1. An important objective of rDNA is the formation of genetic **clones,** or duplicates. Cloning involves the selection and removal of a foreign gene from an animal, plant, or microorganism (the genetic donor) followed by its propagation in a different host organism. Insertion of the foreign gene is accomplished by means of a **vector** (usually a plasmid or virus) that will carry the DNA into a **cloning host,** usually a bacterium or yeast.
  2. A target gene can be isolated or prepared by (1) analysis of the genetic donor's chromosomes through Southern blotting and screening techniques; (2) synthesis by reverse transcription from an mRNA transcript; and (3) synthesis by machines. Cloned genes are maintained in vectors, producing genomic libraries of donor genes.
 B. Characteristics of Cloning Vectors and Cloning Hosts
  1. The best recombinant vectors are plasmids and bacteriophages that carry a significant piece of donor DNA and can be readily transferred into appropriate host cells.
  2. Examples include *E. coli* plasmids; a modified phage vector, the *Charon* phage; and a hybrid vector formed by merging a plasmid and a phage, called a *cosmid.* Other systems for cloning large pieces of foreign DNA are *yeast artificial chromosomes* (YACs) and *bacterial artificial chromosomes* (BACs).
  3. The best cloning hosts have a rapid growth rate, are nonpathogenic, have a well-delineated, simple genome, can function with plasmids and virus vectors, and will replicate and express foreign genes. Common cloning hosts are special strains of *E. coli* and *Saccharomyces cerevisiae.*
 C. Construction of a Chimera—The Recombination Process
  1. In the first cloning step, the foreign gene is excised with an endonuclease and spliced into a plasmid that has been cut with the same endonuclease so that the terminal nucleotides of the two will properly mate; the bonds are sealed by a ligase.
  2. The recombinant plasmid, or *chimera,* can be introduced by transformation into the proper bacterial cloning host. Recombinant clones are identified by their antibiotic resistance.
  3. Progeny cells containing the chimera gene can express the foreign gene by synthesizing the polypeptide it codes for and secreting it.
  4. After final processing, a relatively pure product is left. This process may be done on an industrial scale to mass-produce a variety of hormones, enzymes, and agricultural products.

III. **Biochemical Products of Recombinant DNA Technology**
Recombinant DNA techniques give rise to biochemical products: genetically recombined microbes, plants, and animals; medicines for gene therapy; and methods for genomic analysis.

Recombinant DNA technology is used to manufacture medications for diseases, such as diabetes and dwarfism, caused by the lack of an essential hormone. Various recombinant hormones are available to treat medical conditions.

IV. **Genetically Modified Organisms (GMOs)**
Artificially introducing foreign genes into organisms is termed *transfection,* and recombinant organisms are called *transgenic.* Foreign genes have been used to engineer unique microbes, plants, and animals for a variety of biotechnological applications.
 A. Recombinant Bacteria and Viruses
  1. Recombinant DNA is used to genetically alter *Pseudomonas syringae* to make a commercial product called Frostban that can prevent ice crystals from forming on plants in the field. A natural insecticide gene was added to *Pseudomonas fluorescens* to colonize plant roots. Biotechnologists have developed bacteria to clean up oil spills and degrade pollutants.

2. Viral vectors have been developed for testing drug resistance in *Mycobacterium tuberculosis,* gene therapy, an experimental AIDS vaccine, and to introduce resistance genes into crops.

B. Transgenic Plants

Agrobacterium bacteria can invade plant cells and integrate their DNA into the genome causing a tumor called *crown gall disease.* This ability makes them valuable for inserting foreign genes into plant genomes through a special **Ti plasmid** using rDNA techniques. Infection of the plant with the recombinant bacteria automatically transfers the foreign gene into its cells. Transfection of dicot plants has led to built-in pesticide and pathogen resistance in many species. Other plants are transfected with gene guns that forcefully impel genes into plant embryos.

C. Transgenic Animals

1. Several hundred strains of transgenic animals have been introduced by research and industry to study genetic diseases, to test new genetic therapies, and to become "factories" to manufacture proteins and other products.

2. Common transgenic animals are mice transfected by inserting genes into the embryo (gene line engineering). Mice have been engineered with human genes for growth hormone and to create animal models for human diseases (refer to table 10.4 for details).

V. **Genetic Treatments**

A. **Gene therapy** has the potential to cure genetic diseases by replacing a faulty gene. The *ex vivo* method uses virus vectors containing the normal gene to transfect human tissues in a test tube. The transfected cells are then reintroduced into the patient's body. In *in vivo* therapy, the body is directly infected with virus vectors.

B. Antisense and Triplex DNA Technology

1. **Antisense** and **triplex agents** are oligonucleotide drugs delivered into cells to block undesirable expression of genes.

Antisense refers to a nucleic acid strand that is complementary to the sense, or translatable, strand. Antisense drugs (usually DNA) are chemically modified agents that bind to a target mRNA and interfere with its reading on ribosomes.

2. **Triplex DNA** is a triple helix formed when a third strand of DNA forms hydrogen bonds with the purine bases on one of the helices. This extra strand can make the DNA template inaccessible to normal transcription.

VI. **Genome Analysis**

A. **Gene mapping** is a way to demarcate the nature of a genome. Location, or physical maps, indicate sites of the genes on chromosomes, and sequence maps provide the order of base pairs in a gene.

1. The Human Genome Project is a long-term attempt to map the 40,000 genes in the human genome.

2. The completed map will present detailed genetic information on inherited diseases and will allow genetic screening, enabling families to know their risks for certain diseases.

B. DNA Fingerprinting

The exact way DNA nucleotides are combined is unique for each individual, and DNA technology can be used to array the genome in patterns for comparison. It involves releasing DNA from cells, isolating it, and exposing it to restriction endonucleases that cleave the DNA into a set of relatively unique fragments. After the fragments have been separated by gel electrophoresis, probes are applied to highlight specific restriction landmarks (RFLPs).

1. The pattern of bands is called a **DNA fingerprint.** This technique can identify hereditary relationships and inheritance patterns of genetic diseases.

2. It is also used to keep genetic records in the military and to analyze ancient DNA for comparative and evolutionary studies. Anthropologists and historians have applied the technology in tracing the possible origins of humans.

MULTIPLE-CHOICE QUESTIONS

1. Which gene is incorporated into plasmids to detect recombinant cells?
   a. restriction endonuclease
   b. virus receptors
   c. a gene for antibiotic resistance
   d. reverse transcriptase

2. Which of the following is *not* essential to carry out the polymerase chain reaction?
   a. primers
   b. DNA polymerase
   c. gel electrophoresis
   d. high temperature

3. Which of the following is usually involved in making a nucleic acid hybrid?
   a. Taq polymerase
   b. a labeled probe
   c. a plasmid
   d. RNA

4. What do we call the synthetic unit of the polymerase chain reaction?
   a. annealer
   b. ligase
   c. amplicon
   d. primer

5. The function of ligase is to
   a. rejoin segments of DNA
   b. make longitudinal cuts in DNA
   c. synthesize cDNA
   d. break down ligaments

6. The pathogen of plant roots that is used as a cloning host is
   a. *Pseudomonas*
   b. *Agrobacterium*
   c. *Escherichia coli*
   d. *Saccharomyces cerevisiae*

7. Which of the following is a site that could be clipped by an endonuclease?
   a. ATCGATCGTAGCTAGC
   b. AAGCTTTTCGAA
   c. ATATATATATATAT
   d. ACCATTGGTA

8. The antisense DNA strand that complements mRNA AUGCGCGAC is
   a. UACGCUCUG
   b. GTCTCGCAT
   c. TACGCTCTG
   d. DNA cannot complement mRNA

9. Which DNA fragment will be closest to the top (negative pole) of an electrophoretic gel?
   a. 450 bp   c. 5 kb
   b. 3,560 bp   d. 1,500 bp

10. Which of the following is a primary participant in cloning an isolated gene?
    a. restriction endonuclease
    b. vector
    c. host organism
    d. all of these

11. For which of the following would a nucleic acid probe *not* be used?
    a. locating a gene on a chromosome
    b. developing a Southern blot
    c. identifying a microorganism
    d. constructing a recombinant plasmid

12. **Single Matching.** Match the term with its description:
    ____ nucleic acid probe
    ____ antisense strand
    ____ template strand
    ____ reverse transcriptase
    ____ Taq polymerase
    ____ triplex DNA
    ____ primer
    ____ restriction endonuclease
    a. enzyme that transcribes RNA into DNA
    b. DNA molecule with an extra strand inserted
    c. the nontranslated strand of DNA or RNA
    d. enzyme that snips DNA at palindromes
    e. oligonucleotide that initiates the PCR
    f. strand of nucleic acid that is transcribed or translated
    g. thermostable enzyme for synthesizing DNA
    h. oligonucleotide used in hybridization

CONCEPT QUESTIONS

1. Define genetic engineering and biotechnology, and summarize the important purposes of these fields. Review the use of the terms *genome, chromosome, gene, DNA,* and *RNA* from chapter 9.

2. a. Describe the processes involved in denaturing and renaturing of DNA.
   b. What is useful about this procedure?
   c. Why is it necessary to denature the DNA in the Southern blot test?
   d. How would the Southern blot be used with PCR?

3. a. Define restriction endonuclease.
   b. Define palindrome and draw three different palindromic sequences.
   c. Using nucleotide letters, show how a piece of DNA can be cut to circularize it.
   d. What are restriction length polymorphisms, and how are they used?

4. Explain how electrophoresis works and the general way that DNA is sized. Estimate the size of the DNA fragment in base pairs in the first lane of the gel in figure 10.3*b*. Define oligonucleotides, explain how they are formed, and give three uses for them.

5. a. Briefly describe the functions of DNA synthesizers and sequencers.
   b. How would you make a copy of DNA from an mRNA transcript?
   c. Show how this process would look, using base notation.
   d. What is this DNA called?
   e. Why would it be an advantage to synthesize eucaryotic genes this way?

6. Go through the basic steps of the Sanger method to sequence DNA, and make a simple drawing to help visualize it.

7. a. Explain the meaning of the shorthand in the next column to represent the polymerase chain reaction.
   b. Describe the effects of temperature change in PCR.
   c. Exactly what are the functions of the primer and Taq polymerase?
   d. Explain why the PCR is unlikely to amplify contaminating bacterial DNA in a sample of human DNA.
   e. Explain how PCR could be used to pick a gene out of a complex genome and amplify it.

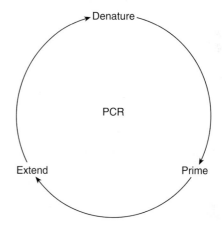

8. a. What characteristics of plasmids and bacteriophages make them good cloning vectors?
   b. Name several types of vectors, and explain what benefits they have.
   c. List the types of genes that they can contain.

9. a. Describe the principles behind recombinant DNA technology.
   b. Outline the main steps in cloning a gene.
   c. Once cloned, how can this gene be used?
   d. Characterize several ways that recombinant DNA technology can be used.

10. a. What characteristics of bacteria make them good cloning hosts?
    b. What is one way to determine whether a bacterial culture has received a recombinant plasmid?

11. a. What is transfection, and what are transgenic organisms?
    b. Explain how *Agrobacterium* is used to transfect plants.
    c. Describe a method for transfecting animals.
    d. Summarize some of the uses for transgenic plants and animals.

12. a. What is the main difference between *ex vivo* and *in vivo* gene therapy?
    b. Does the virus vector used in gene therapy replicate itself in the host cell?
    c. Why would this not be a good idea?
    d. What are some of the main problems with gene therapy?

13. a. Describe the molecular mechanisms by which a DNA antisense molecule could work as a genetic medicine.
    b. Do the same for triplex DNA.
    c. Are the therapies permanent?
    d. Why or why not?

14. a. What is a gene map?
    b. Show by a diagram how chromosomal, physical, and sequence maps are different.
    c. Which organisms are being mapped, and what uses will these maps perform?

    d. Why did the human genome map require 15 years to complete?
    e. What are some possible effects of knowing the genetic map of humans?

15. a. Describe what a DNA fingerprint is and why, and how restriction fragments can be used to form a unique DNA pattern.
    b. Discuss briefly how DNA fingerprinting is being used routinely by biomedicine, the law, the military, and human biology.

CRITICAL-THINKING QUESTIONS

1. a. Give an example of a benefit of genetic engineering to society and a possible adverse outcome.
   b. Give an example of an ecological benefit and a possible adverse side effect.

2. a. In reference to Microbits 10.2, what is your opinion of the dangers associated with genetic engineering?
   b. Most of us would agree to growth hormone therapy for a child with dwarfism, but how do we deal with parents who want to give growth hormones to their 8-year-old son so that he will be "better at sports?"

3. a. If gene probes, fingerprinting, and mapping could make it possible for you to know of future genetic diseases in you or one of your children, would you wish to use this technology to find out?
   b. What if it were used as a screen for employment or insurance?

4. a. Can you think of a reason that bacteria make restriction endonucleases?
   b. What is it about the endonucleases that prevents bacteria from destroying their own DNA?
   c. Look at figure 10.5, and determine the correct DNA sequence for the fragment's template strand. (Be careful of orientation.)
   d. Which suspect was more likely the rapist, according to the fingerprint in figure 10.16c?

5. a. Describe how a virus might be genetically engineered to make it highly virulent.
   b. Can you trace the genetic steps in the development of a tomato plant that has become frost-free from the addition of a flounder's antifreeze genes? (Hint: You need to use *Agrobacterium*.)
   c. What is a different approach to preventing frost on tomatoes?

6. a. Give three different methods to treat or cure cystic fibrosis using the techniques of genetic engineering.
   b. Outline a way that gene therapy could be used to treat a patient with genetic defects due to the presence of extra genes (trisomies as described in chapter 17).

7. You have obtained a blood sample in which only red blood cells are left to analyze.
   a. Can you conduct a DNA analysis of this blood?
   b. If no, explain why.
   c. If yes, explain what you would use to analyze it and how to do it.

8. The way that PCR amplifies DNA is similar to the doubling in a population of growing bacteria; a single DNA strand is used to synthesize 2 DNA strands, which become 4, then 8, then 16, etc. If a complete cycle takes 3 minutes,
   a. how many strands of DNA would theoretically be present after 10 minutes?
   b. after 30 minutes?
   c. after 1 hour?

9. a. Design an antisense DNA drug for this gene: TACGGCTATATTCCGGGC.
   b. Design an antisense RNA drug for the same gene.
   c. How would a triplex drug for this gene work?
   d. Describe how antisense DNA works to block the replication of viruses.

10. a. How would you regard the release of a bioengineered bacterium in your own backyard?
    b. Describe any moral, ethical, or biological problems associated with eating tomatoes from an engineered plant or pork from a transgenic pig.
    c. What are the moral considerations of using transgenic animals to manufacture various human products?

11. You are on a jury to decide whether a person committed a homicide and you have to weigh DNA fingerprinting evidence. Two different sets of fingerprints were done: one that tested 5 markers and one that tested 10. Both sets match the defendant's profile.
    a. Which one is more reliable and why?
    b. How important is the fingerprint if you are told that the fingerprint pattern occurs in one person out of 10,000 in the general population?
    c. Would knowing that the defendant lived in the same apartment building as the victim have any effect on your decision?

12. a. How do you suppose the fish and game department, using DNA evidence, could determine whether certain individuals had poached a deer or if a particular mountain lion had eaten part of a body.
    b. Can you think of some reasons that it would *not* be possible to recreate dinosaurs using the technology we have described in this chapter?

13. a. Look at the pedigree chart for Alzheimer disease. How many genes for this disease must be inherited for it to be expressed?
    b. What kinds of offspring could be produced if a person with the genotype of 1,3 married a person with genotype 1,5?
    c. If cystic fibrosis has the same general profile of inheritance as Alzheimer disease, yet it is recessive (requires two genes to be expressed), what would be the future of the offspring in the pedigree chart?

14. Who actually owns the human genome?
    a. Make cogent arguments on various sides of the question.
    b. Explain the steps required in producing a structural map (base sequence) of DNA.

15. It is possible that by 2010, we can have a detailed understanding of normal and pathologic phenotypes. Discuss the idea that the human genome map (genotype) is now a phenotype.

INTERNET SEARCH TOPICS

1. If you are interested in discovering more about the human genome projects, there are dozens of websites on the Internet; for example, for human genome centers, go to <http://www.ornl.gov/hgmis/CENTERS.HTML>, nhgri.nih.gov, or type Genbank into a search engine such as Alta Vista.

2. Access the website Biotech Basics and look up information on the New Leaf Plus potato. How has it been genetically engineered? What impact will it have on the resources involved in growing these potatoes?

Physical and Chemical Control of Microbes

T he natural condition of humanity is to share surroundings with a large, diverse population of microorganisms. The complete exclusion of microbes from the environment is not only impossible but of questionable value. In many instances, however, our health and comfort can depend on the ability to destroy, inhibit, and remove microbes in the habitats we share. These techniques, also known as antimicrobial control, are very broad in scope. They include routine activities such as cleaning, refrigeration, and cooking. They are also of central importance in the medical, dental, and commercial settings to prevent infection and spoilage, and to ensure the safety of food, water, and other products. Both this chapter and the following one will survey important aspects of microbial control.

Chapter Overview

- The control of microbes in the environment, on the body, and in products is a constant concern of health care and industry.
- Control measures known as antimicrobial techniques use physical and chemical agents to target contaminants and other undesirable microbes.
- Antimicrobial control methods sterilize, disinfect, antisepticize, and sanitize materials as a means of preventing infection, spoilage, and other harmful microbial activities.
- Antimicrobial agents are designed to destroy, inhibit, or remove microbes.
- Microbes exhibit wide degrees of resistance to antimicrobial agents and require a variety of methods to control them.
- Factors that affect the action of antimicrobial methods are time, temperature, type of agent, microbial load, types of microbes, and the nature of the material being treated.
- Microbicidal agents kill microbes by causing irreversible damage to some part of the microbe, thereby permanently preventing it from reproducing.
- Microbistatic agents stop microbes from reproducing, but not permanently.
- Mechanical agents physically remove microbes from materials, but do not necessarily kill or inhibit them.
- Antimicrobial agents damage microbes by disrupting structure (cell wall, cell membrane), altering proteins, stopping synthesis, and inactivating genetic material.

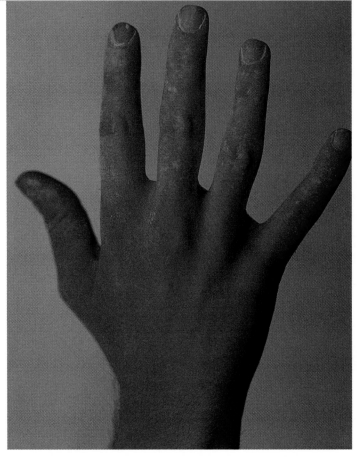

A hand illuminated by UV radiation highlights the regions of skin and nails (lighter glowing areas) that still harbor microbes after a brief wash. The humble handscrub remains our first line of control against infection, but it must be done correctly and last for a long enough time.

- Heat is the most important agent in controlling microorganisms; it can be delivered in either moist or dry form.
- Methods that use heat include steam sterilization, pasteurization, and dry ovens.
- Radiation exposes materials to high-energy waves and particles that can penetrate microbes. Examples are ionizing and ultraviolet radiation.
- Chemical solutions may be used as disinfectants, sterilants, antiseptics, cleaners, and preservatives.
- The major classes of antimicrobial chemicals include halogens (iodine and chlorine compounds), alcohols, phenolics, peroxides, heavy metals, detergents, aldehydes, and gases.

HISTORICAL HIGHLIGHTS 11.1
Microbial Control in Ancient Times

No one knows for sure when humans first applied methods that could control microorganisms, but perhaps the discovery and use of fire in prehistoric times was the starting point. We do know that records describing simple measures to control decay and disease appear from civilizations that existed several thousand years ago. We know, too, that these ancient people had no concept that germs caused disease, but they did have a mixture of religious beliefs, skills in observing natural phenomena, and possibly, a bit of luck. This combination led them to carry out simple and sometimes rather hazardous measures that contributed to the control of microorganisms,

Salting, smoking, pickling, and drying foods and exposing food, clothing, and bedding to sunlight were prevalent practices among early civilizations. The Egyptians showed surprising sophistication and understanding of decomposition by embalming the bodies of their dead with strong salts and pungent oils. They introduced filtration of wine and water as well. The Greeks and Romans burned clothing and corpses during epidemics, and they stored water in copper and silver containers. The armies of Alexander the Great reportedly boiled their drinking water and buried their wastes. Burning sulfur to fumigate houses and applying sulfur as a skin ointment also date from this approximate era.

During the great plague pandemic of the Middle Ages, it was commonplace to bury corpses in mass graves, burn the clothing of plague victims, and ignite aromatic woods in the houses of the sick in the belief that fumes would combat the disease. In a desperate search for some sort of protection, survivors wore peculiar garments and anointed their bodies with herbs, strong perfume, and vinegar. These attempts may sound foolish and antiquated, but it now appears that they may have had some benefits. Burning wood releases formaldehyde, which could have acted as a disinfectant; herbs, perfume, and vinegar contain mild antimicrobial substances. Each of these early methods, although somewhat crude, laid the foundations for microbial control methods that are still in use today.

Illustration of protective clothing used by doctors in the 1700s to avoid exposure to plague victims. The beaklike portion of the hood contained volatile perfumes to protect against foul odors and possibly inhaling "bad air."

Controlling Microorganisms

Much of the time in our daily existence, we take for granted tap water that is drinkable, food that is not spoiled, shelves full of products to eradicate "germs," and drugs to treat infections. Controlling our degree of exposure to potentially harmful microbes is a monumental concern in our lives, and it has a long and eventful history (Historical Highlights 11.1).

GENERAL CONSIDERATIONS
IN MICROBIAL CONTROL

The methods of microbial control belong to the general category of *decontamination* procedures, in that they destroy or remove contaminants. In microbiology, contaminants are microbes present at a given place and time that are undesirable or unwanted. Most decontamination methods employ either **physical agents,** such as heat or *radiation,* or **chemical agents** such as disinfectants and antiseptics.

This separation is convenient, even though the categories overlap in some cases; for instance, radiation can cause damaging chemicals to form, or chemicals can generate heat. A flowchart (figure 11.1) summarizes the major applications and aims in microbial control.

RELATIVE RESISTANCE OF MICROBIAL FORMS

The primary targets of microbial control are microorganisms capable of causing infection or spoilage that are constantly present in the external environment and on the human body. This targeted population is rarely simple or uniform; in fact, it often contains mixtures of microbes with extreme differences in resistance and harmfulness. Contaminants that can have far-reaching effects if not adequately controlled include bacterial vegetative cells and endospores, fungal hyphae and spores, yeasts, protozoan trophozoites and cysts, worms, insects and their eggs, and viruses. The following scheme compares the general resistance these forms have to physical and chemical methods of control:

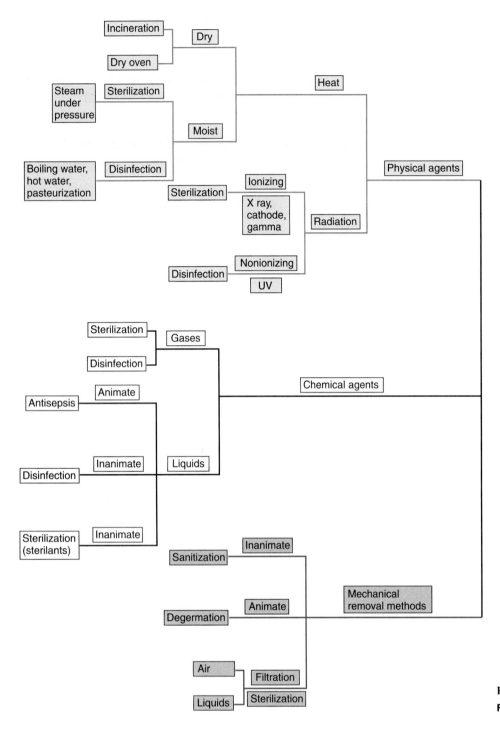

FIGURE 11.1
Flowchart of microbial control measures.

Highest resistance

Bacterial endospores

Moderate resistance

Protozoan cysts; some fungal sexual spores (zygospores); some viruses. In general, naked viruses are more resistant than enveloped forms. Among the most resistant viruses are the hepatitis B virus and the poliovirus. Bacteria with more resistant vegetative cells are *Mycobacterium tuberculosis, Staphylococcus aureus,* and *Pseudomonas* species.

Least resistance

Most bacterial vegetative cells; ordinary fungal spores and hyphae; enveloped viruses; yeasts; and trophozoites

Actual comparative figures on the requirements for destroying various groups of microorganisms are shown in table 11.1. Because bacterial endospores are the most resistant microbial entities, their destruction constitutes the goal of processes that sterilize (see definition in following section), because any process that kills them will invariably kill all less resistant forms. Other methods of control

TABLE 11.1
Relative Resistance of Bacterial Endospores and Vegetative Cells to Control Agents

Method	Spores*	Vegetative Forms*	Relative Resistance**
Heat (moist)	120°C	80°C	1.5×
Radiation (X-ray) dosage	0.4 Mrad	0.1 Mrad	4×
Ultraviolet rays (exposure time)	1.5 h	10 min	9×
Sterilizing gas (ethylene oxide)	1,200 mg/1	700 mg/1	1.7×
Sporicidal liquid (2% glutaraldehyde)	3 h	10 min	18×

*Values are based on methods (concentration, exposure time, intensity) that are required to destroy the most resistant pathogens in each group.
**The greater resistance of spores versus vegetative cells given as an average figure.

(disinfection, antisepsis) act primarily upon microbes that are less hardy than endospores.

TERMINOLOGY AND METHODS OF MICROBIAL CONTROL

Through the years, a growing terminology has emerged for describing and defining measures that control microbes. To complicate matters, the everyday use of some of these terms can at times be vague and inexact. For example, occasionally one may be directed to "sterilize" or "disinfect" a patient's skin, even though this usage does not fit the technical definition of either term. To lay the groundwork for the concepts in microbial control to follow, we present here a series of concepts, definitions, and usages in antimicrobial control.

Sterilization

Sterilization is a process that destroys or removes all viable microorganisms, including viruses. Any material that has been subjected to this process is said to be **sterile.*** These terms should be used only in the strictest sense for methods that have been proved to sterilize. An object cannot be slightly sterile or almost sterile—it is either sterile or not sterile. Control methods that sterilize are generally reserved for inanimate objects, because sterilizing parts of the human body would call for such harsh treatment that it would be highly dangerous and impractical. As we shall see in chapter 13, many internal parts of the body—the brain, muscles, and liver, for example—are naturally free of microbes.

Sterilized products—surgical instruments, syringes, and commercially packaged foods, just to name a few—are essential to human well-being. Although most sterilization is performed with a physical agent such as heat, a few chemicals called *sterilants* can be classified as sterilizing agents because of their ability to destroy spores.

At times, sterilization is neither practicable nor necessary, and only certain groups of microbes need to be controlled. Some antimicrobial agents eliminate only the susceptible vegetative states of microorganisms but do not destroy the more resistant endospore and cyst stages. Keep in mind that the destruction of spores is not always a necessity, because most of the infectious diseases of humans and animals are caused by non-spore-forming microbes.

Microbicidal Agents

The root *-cide,* meaning to kill, can be combined with other terms to define an antimicrobial agent aimed at destroying a certain group of microorganisms. For example, a **bactericide** is a chemical that destroys bacteria except for those in the endospore stage. It may or may not be effective on other microbial groups. A **fungicide** is a chemical that can kill fungal spores, hyphae, and yeasts. A **virucide** is any chemical known to inactivate viruses, especially on living tissue. A **sporicide** is an agent capable of destroying bacterial endospores. A sporicidal agent can also be a sterilant because it can destroy the most resistant of all microbes.

Agents That Cause Microbistasis

The Greek words *stasis* and *static* mean to stand still. They can be used in combination with various prefixes to denote a condition in which microbes are temporarily prevented from multiplying but are not killed outright. Although killing or permanently inactivating microorganisms is the usual goal of microbial control, microbistasis does have meaningful applications. **Bacteriostatic** agents prevent the growth of bacteria on tissues or on objects in the environment, and *fungistatic* chemicals inhibit fungal growth. Materials used to control microorganisms in the body (antiseptics and drugs) have microbistatic effects because many microbicidal compounds can be too toxic to human cells.

Germicides, Disinfection, Antisepsis

A **germicide,*** also called a *microbicide,* is any chemical agent that kills pathogenic microorganisms. A germicide can be used on inanimate (nonliving) materials or on living tissue, but it ordinarily cannot kill resistant microbial cells. Any physical or chemical agent that kills "germs" is said to have **germicidal** properties.

The related term, **disinfection,*** refers to the use of a physical process or a chemical agent (a **disinfectant**) to destroy vegetative pathogens but not bacterial endospores. It is important to note that disinfectants are normally used only on inanimate objects because, in the concentrations required to be effective, they can be toxic to human and other animal tissue. Disinfection processes also remove the harmful products of microorganisms (toxins) from materials. Examples of disinfection include applying a solution of 5% bleach to an examining table, boiling food utensils used by a sick person, and immersing thermometers in an iodine solution between uses.

In modern usage, **sepsis** is defined as the growth of microorganisms in the body or the presence of microbial toxins in blood

---

*sterile (ster′ -ill) Gr. *steria,* barren. (This has another, older meaning that connotes the inability to produce offspring.)

---

*germicide (jer′-mih-syd) L. *germen,* germ, and *caedere,* to kill. Germ is a common term for a pathogenic microbe.

*disinfection (dis″-in-fek′-shun) L. *dis,* apart, and *inficere,* to corrupt.

and other tissues. The term **asepsis*** refers to any practice that prevents the entry of infectious agents into sterile tissues and thus prevents infection. Aseptic techniques commonly practiced in health care range from sterile methods that exclude all microbes to **antisepsis.*** In antisepsis, chemical agents called **antiseptics** are applied directly to exposed body surfaces (skin and mucous membranes), wounds, and surgical incisions to destroy or inhibit vegetative pathogens. Examples of antisepsis include preparing the skin before surgical incisions with iodine compounds, swabbing an open root canal with hydrogen peroxide, and ordinary handwashing with a germicidal soap.

Methods That Reduce the Numbers of Microorganisms

Several applications in commerce and medicine do not require actual sterilization, disinfection, or antisepsis but are based on reducing the levels of microorganisms so that the possibility of infection or spoilage is greatly decreased. Restaurants, dairies, breweries, and other food industries consistently handle large numbers of soiled utensils that could readily become sources of infection and spoilage. These industries must keep microbial levels to a minimum during preparation and processing. **Sanitization*** is any cleansing technique that mechanically removes microorganisms (along with food debris) to reduce the level of contaminants. A **sanitizer** is a compound such as soap or detergent used to perform this task.

Cooking utensils, dishes, bottles, cans, and used clothing that have been washed and dried may not be completely free of microbes, but they are considered safe for normal use (sanitary). Air sanitization with ultraviolet lamps reduces airborne microbes in hospital rooms, veterinary clinics, and laboratory installations. It is important to note that some sanitizing processes (such as dishwashing machines) are rigorous enough to sterilize objects, but this is not true of all sanitization methods.

It is often necessary to reduce the numbers of microbes on the human skin through **degermation.** This process usually involves scrubbing the skin or immersing it in chemicals, or both. It also emulsifies oils that lie on the outer cutaneous layer and mechanically removes potential pathogens on the outer layers of the skin. Examples of degerming procedures are the surgical handscrub, the application of alcohol wipes to the skin, and the cleansing of a wound with germicidal soap and water. The concepts of antisepsis and degermation clearly overlap, since a degerming procedure can simultaneously be antiseptic, and vice versa.

WHAT IS MICROBIAL DEATH?

Death is a phenomenon that involves the permanent termination of an organism's vital processes. Signs of life in complex organisms such as animals are self-evident, and death is made clear by loss of nervous function, respiration, or heartbeat. In contrast, death in microscopic organisms that are composed of just one or a few cells is often hard to detect, because they reveal no conspicuous vital signs

to begin with. Lethal agents (such as radiation and chemicals) do not necessarily alter the overt appearance of microbial cells. Even the loss of movement in a motile microbe cannot be used to indicate death. This fact has made it necessary to develop special qualifications that define and delineate microbial death.

The destructive effects of chemical or physical agents occur at the level of a single cell. As the cell is continuously exposed to an agent such as intense heat or toxic chemicals, various cell structures become dysfunctional, and the entire cell can sustain irreversible damage. At present, the most practical way to detect this damage is to determine if a microbial cell can still reproduce when exposed to a suitable environment. If the microbe is so disturbed metabolically and structurally that it is unable to multiply, then it is no longer viable. *The permanent loss of reproductive capability, even under optimum growth conditions, has become the accepted microbiological definition of death.*

Factors That Affect Death Rate

The ability to define microbial death has tremendous theoretical and practical importance. It allows medicine and industry to test the conditions required to destroy microorganisms, to pinpoint the ways that antimicrobial agents kill cells, and to establish standards of sterilization and disinfection in these fields. Hundreds of testing procedures have been developed for evaluating physical and chemical agents. Appendix B presents a general breakdown of certain types of antimicrobial tests, along with the kinds of variables that must be controlled.

The cells of a culture show marked variations in susceptibility to a given microbicidal agent. Death of the whole population is not instantaneous but occurs in a logarithmic manner and requires a certain time of exposure (figure 11.2*a*). The most susceptible cells (younger, actively growing cells) die immediately, whereas less susceptible cells (older, inactive ones) have greater resistance and require longer exposure times. Eventually, a point is reached at which survival of any cells is highly unlikely; this point is equivalent to sterilization.

The effectiveness of a particular agent is governed by several factors besides time. The following additional factors influence the action of antimicrobial agents:

1. The number of microorganisms (figure 11.2*b*). A higher load of contaminants requires more time to destroy.
2. The nature of the microorganisms in the population (figure 11.2*c*). In most actual circumstances of disinfection and sterilization, the target population is not a single species of microbe but a mixture of bacteria, fungi, spores, and viruses, presenting an even greater spectrum of microbial resistance.
3. The temperature and pH of the environment.
4. The concentration (dosage, intensity) of the agent. For example, UV radiation is most microbicidal at 260 nm; most disinfectants are more active at higher concentrations.
5. The mode of action of the agent (figure 11.2*d*). How does it kill or inhibit the microorganism?
6. The presence of solvents, interfering organic matter, and inhibitors. Large amounts of saliva, blood, and feces can inhibit the actions of disinfectants and even of heat.

The influence of these factors will be discussed in greater detail in subsequent sections.

---

*asepsis (ay-sep′-sis) Gr. *a*, no or none, and *sepsis*, decay.

*antisepsis *anti*, against.

*sanitization (san″-ih-tih-zay′-shun) L. *sanitas*, health.

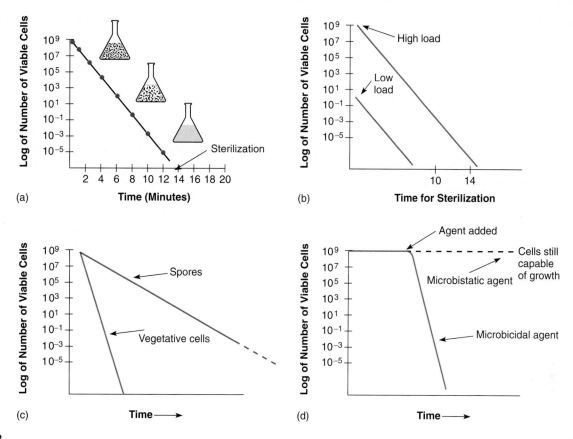

FIGURE 11.2

Factors that influence the rate at which microbes are killed by antimicrobial agents. (a) Length of exposure to the agent. During exposure to a chemical or physical agent, all cells of a microbial population, even a pure culture, do not die simultaneously. Over time, the number of viable organisms remaining in the population decreases logarithmically, giving a straight-line relationship on a graph. The point at which the number of survivors is infinitesimally small is considered sterilization. **(b)** Effect of the microbial load. **(c)** Relative resistance of spores versus vegetative forms. **(d)** Action of the agent, whether destructive or inhibitory.

HOW ANTIMICROBIAL AGENTS WORK: THEIR MODES OF ACTION

An antimicrobial agent's adverse effect on cells is known as its *mode* (or *mechanism*) *of action*. Many antimicrobial agents affect more than one cellular target and may inflict both primary and secondary damages that eventually lead to cell death. Agents can be classified according to the following degrees of selectiveness: (1) agents that are less selective in their scope of destructiveness, inflict severe damage on many cell parts, and are generally very biocidal (heat, radiation, some disinfectants); (2) moderately selective agents with intermediate specificity (certain disinfectants and antiseptics); and (3) more selective agents (drugs) whose target is usually limited to a specific cell structure or function and whose effectiveness is restricted only to certain microbes.

The cellular targets of physical and chemical agents fall into four general categories: (1) the cell wall, (2) the cell membrane, (3) cellular synthetic processes (DNA, RNA), and (4) proteins.

The Effects of Agents on the Cell Wall
The cell wall maintains the structural integrity of bacterial and fungal cells. Several types of chemical agents damage the cell wall by blocking its synthesis, digesting it, or breaking down its surface. A cell deprived of a functioning cell wall becomes fragile and is lysed very easily. Examples of this mode of action include some antimicrobial drugs (penicillins) that interfere with the synthesis of the cell wall in bacteria (see figure 12.3). Detergents and alcohol can also disrupt cell walls, especially in gram-negative bacteria.

How Agents Affect the Cell Membrane
All microorganisms have a cell membrane composed of lipids and proteins, and even some viruses have an outer membranous envelope. As we learned in previous chapters, a cell's membrane provides a two-way system of transport. If this membrane is disrupted, a cell loses its selective permeability and can neither prevent the loss of vital molecules nor bar the entry of damaging chemicals. Loss of those abilities leads to cell death. Detergents called **surfactants*** work as microbicidal agents because they lower the surface tension of cell membranes. Surfactants are polar molecules with hydrophilic and hydrophobic regions that can physically bind to the lipid layer and penetrate the internal hydrophobic region of membranes. In effect, this process "opens up" the once tight interface, leaving leaky spots that allow injurious chemicals to seep into the cell and important ions to seep out (figure 11.3).

---

*surfactant (sir-fak´-tunt) A word derived from **sur**face-**act**ing a**gent**.

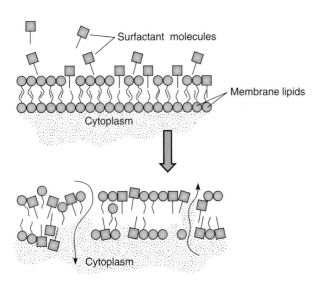

FIGURE 11.3

Mode of action of surfactants on the cell membrane. Surfactants inserting in the lipoidal layers disrupt it and create abnormal channels that alter permeability and cause leakage both into and out of the cell.

Agents That Affect Protein and Nucleic Acid Synthesis

Microbial life depends upon an orderly and continuous supply of proteins to function as enzymes and structural molecules. As we saw in chapter 9, these proteins are synthesized on the ribosomes through a complex process called translation. Any agent that interferes with accurate translation also prevents synthesis of a complete, functioning protein. For instance, the antibiotic chloramphenicol binds to the ribosomes of bacteria in a way that stops peptide bonds from forming. In its presence, many bacterial cells are inhibited from forming proteins required in growth and metabolism and are thus inhibited from multiplying. Most of the agents that block protein synthesis are drugs used in antimicrobial therapy. These drugs will be discussed in greater detail in chapter 12.

The nucleic acids are likewise necessary for the continued functioning of microbes. DNA must be regularly replicated and transcribed in growing cells, and any agent that either impedes these processes or changes the genetic code is potentially antimicrobial. Some agents bind irreversibly to DNA, preventing both transcription (formation of RNA) and translation; others are mutagenic agents. Gamma, ultraviolet, or X radiation causes mutations that result in permanent inactivation of DNA. Chemicals such as formaldehyde and ethylene oxide also interfere with DNA and RNA function.

Agents That Alter Protein Function

A microbial cell contains large quantities of proteins that function properly only if they remain in a normal three-dimensional configuration called the *native state*. The antimicrobial properties of some agents arise from their capacity to disrupt, or **denature,** proteins. In general, denaturation occurs when the bonds that maintain the secondary and tertiary structure of the protein are broken. Breaking these bonds will cause the protein to unfold or create random, irregular loops and coils (figure 11.4). One way that proteins can be denatured is through coagulation by moist heat (the same reaction seen in the irreversible solidification of the white of an egg when

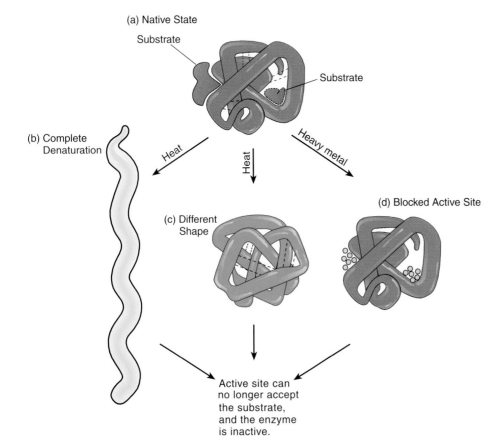

FIGURE 11.4

Modes of action affecting protein function.
(a) The functional native state is maintained by bonds that create active sites to fit the substrate. Some agents denature the protein by breaking all or some secondary and tertiary bonds. Results are **(b)** complete unfolding or **(c)** random bonding and incorrect folding. **(d)** Some agents react with functional groups on the active site and interfere with bonding.

boiled). Chemicals such as strong organic solvents (alcohols, acids) and phenolics also coagulate proteins. Other antimicrobial agents, such as metallic ions, attach to the active site of the protein and prevent it from interacting with its correct substrate. Regardless of the exact mechanism, such losses in normal protein function can promptly arrest metabolism.

Practical Concerns in Microbial Control

Numerous considerations govern the selection of a workable method of microbial control. The following are among the most pressing concerns: (1) Does the application require sterilization, or is disinfection adequate? In other words, must spores be destroyed, or is it necessary to destroy only vegetative pathogens? (2) Is the item to be reused or permanently discarded? If it will be discarded, then the quickest and least expensive method should be chosen. (3) If it will be reused, can the item withstand heat, pressure, radiation, or chemicals? (4) Is the control method suitable for a given application? (For example, ultraviolet radiation is a good sporicidal agent, but it will not penetrate solid materials.) Or, in the case of a chemical, will it leave an undesirable residue? (5) Will the agent penetrate to the necessary extent? (6) Is the method cost- and labor-efficient, and is it safe?

A remarkable variety of substances can require sterilization. They run the gamut from durable solids such as rubber to sensitive liquids such as serum, and from air to tissue grafts. Hundreds of situations requiring sterilization confront the network of persons involved in health care, be it technician, nurse, doctor, or manufacturer, and no universal method works well in every case.

Considerations such as cost, effectiveness, and method of disposal are all important. For example, the disposable plastic items such as catheters and syringes that are used in invasive medical procedures have the potential for infecting the tissues. These must be sterilized during manufacture by a nonheating method (gas or radiation), because heat can damage delicate plastics. After these items have been used, it is often necessary to destroy or decontaminate them before they are discarded because of the potential risk to the handler (from needlesticks). Steam sterilization, which is quick and sure, is a sensible choice at this point, because it does not matter if the plastic is destroyed. Health care workers are held to very high standards of infection prevention (see universal precautions, page 540).

CHAPTER CHECKPOINTS

Microbial control methods involve the use of physical and chemical agents to eliminate or reduce the numbers of microorganisms from a specific environment.

Microbial control methods are used to prevent the spread of infectious agents, retard spoilage, and keep commercial products safe.

The population of microbes that cause spoilage or infection varies widely in species composition, resistance, and harmfulness, so microbial control methods must be adjusted to fit individual situations.

The type of microbial control is indicated by the terminology used. Sterilization agents destroy all viable organisms, including viruses. Antisepsis, disinfection, and sanitization agents reduce the numbers of viable microbes to a specified level.

Antimicrobic agents are described according to their ability to destroy or inhibit microbial growth. Microbicidal agents cause microbial death. They are described by what they are -cidal for: sporocides, bactericides, fungicides, viricides.

An antiseptic agent is applied to living tissue to destroy or inhibit microbial growth.

A disinfectant agent is used on inanimate objects to destroy vegetative pathogens but not bacterial endospores.

Sanitization reduces microbial numbers on inanimate objects to safe levels by physical or chemical means.

Degermation refers to the process of mechanically removing microbes from the skin.

Microbial death is defined as the permanent loss of reproductive capability in microorganisms.

Antimicrobic agents attack specific cell sites to cause microbial death or damage. Any given antimicrobic agent attacks one of four major cell targets: the cell wall, the cell membrane, biosynthesis pathways for DNA or RNA, or protein (enzyme) function.

Methods of Physical Control

Microorganisms have adapted to the tremendous diversity of habitats the earth provides, even severe conditions of temperature, moisture, pressure, and light. For microbes that normally withstand such extreme physical conditions, our attempts at control would probably have little effect. Fortunately, the vast majority of microbes that must be controlled are not adapted to such extremes and are readily controlled by abrupt changes in environment. Most prominent among antimicrobial physical agents is heat. Other less widely used agents include radiation, filtration, ultrasonic waves, and even cold. The following sections will examine some of these methods and explore their practical applications in medicine, commerce, and the home.

HEAT AS AN AGENT OF MICROBIAL CONTROL

A sudden departure from a microbe's temperature of adaptation is likely to have a detrimental effect on it. As a rule, elevated temperatures (exceeding the maximum) are microbicidal, whereas lower temperatures (below the minimum) tend to have inhibitory or microbistatic effects. The two physical states of heat used in microbial control are moist and dry. **Moist heat** occurs in the form of hot water, boiling water, or steam (vaporized water). In practice, the temperature of moist heat usually ranges from 60° to 135°C. As we shall see, the temperature of steam can be regulated by adjusting its pressure in a closed container. The expression **dry heat** denotes air with a low moisture content that has been heated by a flame or electric heating coil. In practice, the temperature of dry heat ranges from 160°C to several thousand degrees Celsius.

Mode of Action and Relative Effectiveness of Heat

In addition to their physical state, moist and dry heat differ in their efficiency. At a given temperature, moist heat works several times faster than dry heat. At the same length of exposure, moist heat kills cells at a lower temperature than dry heat (table 11.2). Both forms of heat disrupt important cell components, but it appears that their

TABLE 11.2

Comparison of Times and Temperatures to Achieve Sterilization with Moist and Dry Heat

	Temperature	Time to Sterilize
Moist Heat	121°C	15 min
	125°C	10 min
	134°C	3 min
Dry Heat	121°C	600 min
	140°C	180 min
	160°C	120 min
	170°C	60 min

TABLE 11.3

Thermal Death Times of Various Endospores

Organism	Temperature	Time of Exposure to Kill Spores
Moist Heat		
Bacillus subtilis	121°C	1 min
B. stearothermophilis	121°C	12 min
Clostridium botulinum	120°C	10 min
Cl. tetani	105°C	10 min
Dry Heat		
Bacillus subtilis	121°C	120 min
B. stearothermophilis	140°C	5 min
Clostridium botulinum	120°C	120 min
Cl. tetani	100°C	60 min

specific modes of action are different. Exposure to moist heat generally coagulates and therefore denatures cell proteins. The greater effectiveness of moist heat has been attributed to this particular mode of action, because protein (enzyme) denaturation occurs more rapidly and at a lower temperature if moisture is present. Components such as the membrane, ribosomes, DNA, and RNA are also damaged by moist heat.

Dry heat also has several effects on microbes. For instance, it can oxidize cells and, at extremely high temperatures, reduce them to ashes. It dehydrates cell components and can denature proteins and DNA, but proteins are more stable in dry heat than in moist heat, and higher temperatures are required to inactivate them.

Heat Resistance and Thermal Death of Spores and Vegetative Cells

Bacterial endospores exhibit the greatest resistance, and vegetative states of bacteria and fungi are the least resistant to both moist and dry heat. Destruction of spores usually requires temperatures above boiling (table 11.3), although resistance varies widely. In boiling water (100°C), the spores of *Bacillus anthracis* (the agent of anthrax) can be destroyed in a few minutes, whereas the spores of some thermophilic and anaerobic species can require several hours.

Vegetative cells also vary in their sensitivity to heat, though not to the same extent as spores (table 11.4). Among bacteria, the death times with moist heat range from 50°C for 3 minutes *(Neisseria gonorrhoeae)* to 60°C for 60 minutes *(Staphylococcus aureus)*. It is worth noting that vegetative cells of sporeformers are just as susceptible as vegetative cells of non-sporeformers and that pathogens are neither more nor less susceptible than nonpathogens. Other microbes, including fungi (yeasts, molds, and some of their spores), protozoa, and worms, are rather similar in their sensitivity to heat. Viruses are surprisingly resistant to heat, with a tolerance range extending from 55°C for 2 to 5 minutes (adenoviruses) to 60°C for 600 minutes (hepatitis virus). For practical purposes, all non-heat-resistant forms of bacteria, yeasts, molds, protozoa, worms, and viruses are destroyed by exposure to 80°C for 20 minutes.

Practical Concerns in the Use of Heat: Thermal Death Measurements

Adequate sterilization requires that both temperature and length of exposure be considered. As a general rule, higher temperatures allow shorter exposure times, and lower temperatures require longer

exposure times. A combination of these two variables constitutes the **thermal death time,** or TDT, defined as the shortest length of time required to kill all microbes at a specified temperature. The TDT has been experimentally determined for the microbial species that are common or important contaminants in various heat-treated materials. Another way to compare the susceptibility of microbes to heat is the thermal death point (TDP), defined as the lowest temperature required to kill all microbes in a sample in 10 minutes.

Many perishable substances are processed with moist heat. Some of these products are intended to remain on the shelf at room temperature for several months or even years. The chosen heat treatment must render the product free of agents of spoilage or disease. At the same time, the quality of the product and the speed and cost of processing must be considered. For example, in the commercial preparation of canned green beans, one of the cannery's

TABLE 11.4

Average Thermal Death Times of Vegetative Stages of Microorganisms

Microbial Type	Temperature	Time (Min)
Non-spore-forming pathogenic bacteria	58°C	28
Non-spore-forming nonpathogenic bacteria	61°C	18
Vegetative stage of spore-forming bacteria	58°C	19
Fungal spores	76°C	22
Yeasts	59°C	19
Heat inactivation of viruses		
Nonenveloped	57°C	29
Enveloped	54°C	22
Protozoan trophozoites	46°C	16
Protozoan cysts	60°C	6
Worm eggs	54°C	3
Worm larvae	60°C	10

greatest concerns is to prevent growth of the agent of botulism. From several possible TDTs for *Clostridium botulinum* spores, the cannery must choose one that kills all spores but does not turn the beans to mush. Out of these many considerations emerges an optimal TDT for a given processing method. Commercial canneries heat low-acid foods at 121°C for 30 minutes, a treatment that sterilizes these foods. Because of such strict controls in canneries, cases of botulism due to commercially canned foods are rare.

Common Methods of Moist Heat Control

The four ways that moist heat is employed to sterilize or disinfect are (1) steam under pressure, (2) live, nonpressurized steam, (3) boiling water, and (4) pasteurization.

Steam Under Pressure   A temperature of 100°C is the highest that steam can reach under normal atmospheric pressure at sea level. This pressure is measured at 15 pounds per square inch (psi), or 1 atmosphere. In order to raise the temperature of steam above this point, it must be pressurized in a closed chamber. This phenomenon is explained by the physical principle that governs the behavior of gases under pressure. When a gas is compressed, its temperature rises in direct relation to the amount of pressure. So, when the pressure is increased to 5 psi above normal atmospheric pressure, the temperature of steam rises to 109°C. When the pressure is increased to 10 psi above normal, its temperature will be 115°C, and at 15 psi (a total of 2 atmospheres), it will be 121°C. It is not the pressure by itself that is killing microbes, but the increased temperature it produces.

Such pressure-temperature combinations can be achieved only with a special device that can subject pure steam to pressures greater than 1 atmosphere. Health and commercial industries use an **autoclave** for this purpose, and a comparable home appliance is the pressure cooker. Autoclaves have a fundamentally similar plan: a cylindrical metal chamber with an airtight door on one end and racks to hold materials (figure 11.5). Its construction includes a complex network of valves, pressure and temperature gauges, and ducts for regulating and measuring pressure and conducting the steam into the chamber. Sterilization is achieved when the steam condenses against the objects in the chamber and gradually raises their temperature.

Experience has shown that the most efficient pressure-temperature combination for achieving sterilization is 15 psi, which yields 121°C. It is possible to use higher pressure to reach higher temperatures (for instance, increasing the pressure to 30 psi raises the temperature 11°C), but doing so will not significantly reduce the exposure time and can harm the items being sterilized. It is important to avoid overpacking or haphazardly loading the chamber, which prevents steam from circulating freely around the contents and impedes the full contact that is necessary. The duration of the process is adjusted according to the bulkiness of the items in the load (thick bundles of material or large flasks of liquid) and how full the chamber is. The range of holding times varies from 10 minutes for light loads to 40 minutes for heavy or bulky ones; the average time is 20 minutes.

The autoclave is a superior choice to sterilize heat-resistant materials such as glassware, cloth (surgical dressings), rubber (gloves), metallic instruments, liquids, paper, some media, and some heat-resistant plastics. If the items are heat-sensitive (plastic Petri plates) but will be discarded, the autoclave is still a good choice. However, the autoclave is ineffective for sterilizing substances that repel moisture (oils, waxes, powders).

Intermittent Sterilization   Selected substances that cannot withstand the high temperature of the autoclave can be subjected to **intermittent sterilization,** also called *tyndallization.*[1] This technique requires a chamber to hold the materials and a reservoir for boiling water. Items in the chamber are exposed to free-flowing steam for 30 to 60 minutes. This temperature is not sufficient to reliably kill spores, so a single exposure will not suffice. On the assumption that surviving spores will germinate into less resistant vegetative cells, the items are incubated at appropriate temperatures for 23 to 24 hours, and then again subjected to steam treatment. This cycle is repeated for 3 days in a row. Because the temperature never gets above 100°C, highly resistant spores that do not germinate may survive even after 3 days of this treatment.

Intermittent sterilization is used most often to process heat-sensitive culture media, such as those containing sera, egg, or carbohydrates (which can break down at higher temperatures) and some canned foods. It is probably not effective in sterilizing items such as instruments and dressings that provide no environment for spore germination, but it certainly can disinfect them.

Pasteurization: Disinfection of Beverages   Fresh beverages such as milk, fruit juices, beer, and wine are easily contaminated during collection and processing. Because microbes have the potential for spoiling these foods or causing illness, heat is frequently used to reduce the microbial load and destroy pathogens. **Pasteurization** is a technique in which heat is applied to liquids to kill potential agents of infection and spoilage, while at the same time retaining the liquid's flavor and food value. Ordinary pasteurization techniques require special heat exchangers that expose the liquid to 71.6°C for 15 seconds (flash method) or to 63°–66°C for 30 minutes (batch method). The first method is preferable because it is less likely to change flavor and nutrient content, and it is more effective against certain resistant pathogens such as *Coxiella* and *Mycobacterium*. Although these treatments inactivate most viruses and destroy the vegetative stages of 97–99% of bacteria and fungi, they do not kill endospores or *thermoduric* species (mostly nonpathogenic lactobacilli, micrococci, and yeasts). Milk is not sterile after regular pasteurization. In fact, it can contain 20,000 microbes per milliter or more, which explains why even an unopened carton of milk will eventually spoil. Newer techniques can also produce *sterile milk* that has a storage life of 3 months. This milk is processed with ultrahigh temperature (UHT)—134°C for 1 to 2 seconds (see chapter 26).

One important aim in pasteurization is to prevent the transmission of milk-borne diseases from infected cows or milk handlers. The primary targets of pasteurization are non-spore-forming pathogens: *Salmonella* species (a common cause of food infection), *Campylobacter jejuni* (acute intestinal infection), *Listeria monocytogenes* (listeriosis), *Brucella* species (undulant fever), *Coxiella*

---

1. Named for the British physicist John Tyndall who did early experiments with sterilizing procedures.

(a)

burnetii (Q fever), *Mycobacterium bovis* and *M. tuberculosis,* and several enteric viruses.

Pasteurization also has the advantage of extending milk storage time, and it can also be used by some wineries and breweries to stop fermentation and destroy contaminants.

Boiling Water: Disinfection   A simple boiling water bath or chamber can quickly decontaminate items in the clinic and home. Because a single processing at 100°C will not kill all resistant cells, this method can be relied on only for disinfection and not for sterilization. Exposing materials to boiling water for 30 minutes will kill most non-spore-forming pathogens, including resistant species such as the tubercle bacillus and staphylococci. Probably the greatest disadvantage with this method is that the items can be easily recontaminated when removed from the water. Boiling is also a recommended method of disinfecting unsafe drinking water. In the home, boiling water is a fairly reliable way to sanitize and disinfect materials for babies, food preparation, and utensils, bedding, and clothing from the sickroom.

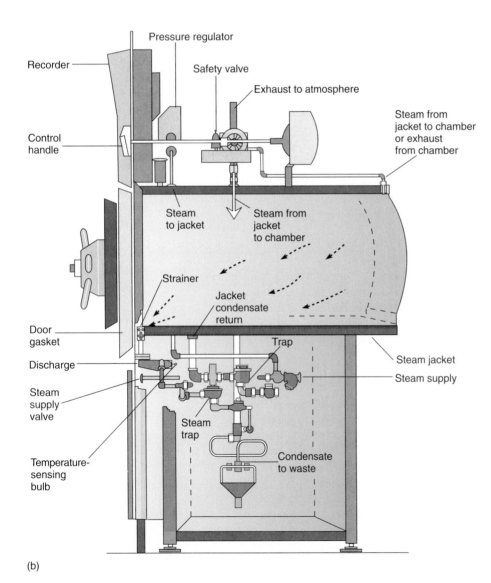

(b)

FIGURE 11.5

Steam sterilization with the autoclave.
(a) A large automatic autoclave used in sterilization by drug companies. **(b)** Cutaway section, showing autoclave components.

(b) From John J. Perkins, Principles and Methods of Sterilization in Health Science, *2nd ed., 1969. Courtesy of Charles C. Thomas, Publisher, Springfield, Illinois.*

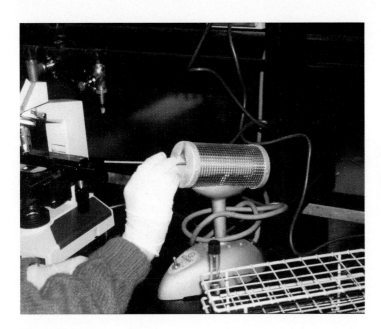

FIGURE 11.6

Dry heat incineration. Infrared incinerator with shield to prevent spattering of microbial samples during flaming.

Dry Heat: Hot Air and Incineration

Dry heat is not as versatile or as widely used as moist heat, but it has several important sterilization applications. The temperatures and times employed in dry heat vary according to the particular method, but in general, they are greater than with moist heat. **Incineration** in a flame or electric heating coil is perhaps the most rigorous of all heat treatments. The flame of a Bunsen burner reaches 1,870°C at its hottest point, and furnaces/incinerators operate at temperatures of 800°–6,500°C. Direct exposure to such intense heat ignites and reduces microbes and other substances to ashes and gas.

Incineration of microbial samples on inoculating loops and needles using a Bunsen burner is a very common practice in the microbiology laboratory. This method is fast and effective, but it is also limited to metals and heat-resistant glass materials. Incinerators (figure 11.6) are regularly employed in hospitals and research labs for complete destruction and disposal of infectious materials such as syringes, needles, cultural materials, dressings, bandages, bedding, animal carcasses, and pathology samples.

The hot-air oven provides another means of dry-heat sterilization. The so-called **dry oven** is usually electric (occasionally gas) and has coils that radiate heat within an enclosed compartment. Heated, circulated air transfers its heat to the materials in the oven. Sterilization requires exposure to 150°–180°C for 2 to 4 hours, which ensures thorough heating of the objects and destruction of spores.

The dry oven is used in laboratories and clinics for heat-resistant items that do not sterilize well with moist heat. Substances appropriate for dry ovens are glassware, metallic instruments, powders, and oils that steam does not penetrate well. This method is not suitable for plastics, cotton, and paper, which may burn at the high temperatures, or for solutions, which will dry out. Another limitation is the time required for it to work.

THE EFFECTS OF COLD AND DESICCATION

The principal benefit of cold treatment is to slow growth of cultures and microbes in food during processing and storage. *It must be emphasized that cold merely retards the activities of most microbes.* Although it is true that some microbes are killed by cold temperatures, most are not adversely affected by gradual cooling, long-term refrigeration, or deep-freezing. In fact, freezing temperatures, ranging from −70°C to −135°C, provide an environment that can preserve cultures of bacteria, viruses, and fungi for long periods. Some psychrophiles grow very slowly even at freezing temperatures and can continue to secrete toxic products. Unawareness of these facts is probably responsible for numerous cases of food poisoning from frozen foods that have been defrosted at room temperature and then inadequately cooked. Pathogens able to survive several months in the refrigerator are *Staphylococcus aureus, Clostridium* species (sporeformers), *Streptococcus* species, and several types of yeasts, molds, and viruses. A recent outbreak of *Salmonella* food infection from homemade ice cream attests to the unreliability of freezing temperatures to kill pathogens.

Vegetative cells directly exposed to normal room air gradually become dehydrated, or *desiccated.** More delicate pathogens such as *Streptococcus pneumoniae,* the spirochete of syphilis, and *Neisseria gonorrhoeae* can die after a few hours of air-drying, but many others are not killed and some are even preserved. Endospores of *Bacillus* and *Clostridium* are viable for millions of years under extremely arid conditions. Staphylococci and streptococci in dried secretions, and the tubercle bacillus surrounded by sputum, can remain viable in air and dust for lengthy periods. Many viruses (especially nonenveloped) and fungal spores can also withstand long periods of desiccation. Desiccation can be a valuable way to preserve foods because it greatly reduces the amount of water available to support microbial growth.

It is interesting to note that a combination of freezing and drying—*lyophilization**—is a common method of preserving microorganisms and other cells in a viable state for many years. Pure cultures are frozen instantaneously and exposed to a vacuum that rapidly removes the water (it goes right from the frozen state into the vapor state). This method avoids the formation of ice crystals that would damage the cells. Although not all cells survive this process, enough of them do to permit future reconstitution of that culture.

As a general rule, chilling, freezing, and desiccation should not be construed as methods of disinfection or sterilization because their antimicrobial effects are erratic and uncertain, and one cannot be sure that pathogens subjected to them have been killed.

RADIATION AS A MICROBIAL CONTROL AGENT

Another type of energy that exerts antimicrobial effects is radiation. For our purposes, **radiation** is defined as energy emitted from atomic activities and dispensed at high velocity through matter or space. Radiation can behave as waves or as particles, depending upon the conditions. Here we will consider the wavelike character of electromagnetic radiation and the particulate character of particle radiation. The wavelengths of **electromagnetic radiation** range

---

*desiccate (des'-ih-kayt) To dry at normal environmental temperatures.

*lyophilization (ly-off″-il-ih-za′-shun) Gr. *lyein,* to dissolve, and *philein,* to love.

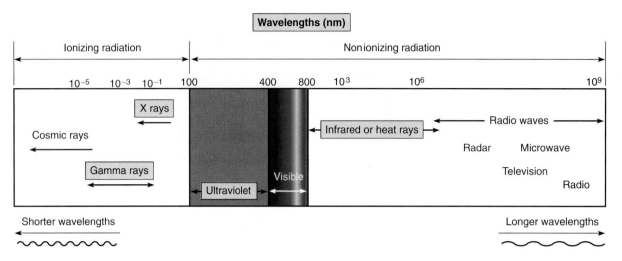

FIGURE 11.7

Electromagnetic radiation used in chemical control. Short wavelengths of gamma and X rays are the most manageable and practical forms of ionizing radiation. Nonionizing radiation, such as ultraviolet rays in the range of 240 to 280 nm and infrared (heat) rays, are fair microbicidal agents. Other waves are not routinely used in microbial control.

from high-energy, short-wavelength gamma rays at one extreme to low-energy, very long radio waves at the other (figure 11.7). For the most part, only electromagnetic radiations at the gamma ray, X-ray (also called roentgen ray), and ultraviolet ray (UV) levels are suitable for microbial control. **Particle radiation** consists of subatomic particles such as electrons, protons, and neutrons that have been freed from the atom. The only particle radiation feasible for antimicrobial applications is the high-speed electron (also called a β particle or cathode ray).

Visible light does exert some antimicrobial effects, but it is not predictable enough in its effects to control all microbes. Infrared rays can be destructive by virtue of the heat they produce, but they have fewer applications than other heating methods. Microwaves also apparently kill microorganisms through heat production, but the antimicrobial effects of long radio waves are minimal.

Modes of Action of Ionizing Versus Nonionizing Radiation

The actual physical effects of radiation on microbes can be understood by visualizing the process of *irradiation,* or bombardment with radiation, at the cellular level (figure 11.8). When a cell is bombarded by certain waves or particles, its molecules absorb some of the available energy, leading to one of two consequences: (1) If the radiation ejects orbital electrons from an atom, it causes ions to form. This is the effect of **ionizing radiation.** One of the most sensitive targets for ionizing radiation is the DNA molecule, which can sustain mutations on a broad scale. Secondary lethal effects appear to be chemical changes in organelles and the production of toxic substances. Gamma rays, X rays, and high-speed electrons are all ionizing in their effects. (2) **Nonionizing radiation,** best exemplified by UV, excites atoms by raising them to a higher energy state, but it does not ionize them. This atomic excitation, in turn, leads to abnormal linkages within molecules such as DNA and is thus a source of mutations (see chapter 9).

Ionizing Radiation: Gamma Rays, X Rays, and Cathode Rays

Over the past several years, ionizing radiation has become safer and more economical to use, and its applications have mushroomed. It is a highly effective alternative for sterilizing materials that are sensitive to heat or chemicals. Because it sterilizes in the absence of heat, irradiation is a type of *cold sterilization.* Devices that emit ionizing rays include gamma-ray machines containing radioactive cobalt, X-ray machines similar to those used in medical diagnosis, and cathode-ray machines that operate like the vacuum tube in a television set. Items are placed in these machines and irradiated for a short time with a carefully chosen dosage. The dosage is measured in rads (**r**adiation **a**bsorbed **d**ose). Depending on the application, exposure ranges from 0.5 to 5 megarads (Mrad; a megarad is equal to 1,000,000 rads). Although all ionizing radiations can penetrate solids and liquids, gamma rays are most penetrating, X rays are intermediate, and cathode rays least penetrating.

Applications of Ionizing Radiation

Even though various agencies have been treating food with ionizing radiation for over 50 years, it remains a controversial method. Consumer organizations are concerned that radiation poses a danger both in potential toxicity and in radioactivity that it can then impart to foods. Two known problems that can occur with irradiated food are (1) changes in flavor and nutrition and (2) the possibility of introducing undesirable chemical reactions. These disadvantages have been reduced by carrying out the process in very cold, oxygen-free chambers with extremely low doses of radiation. Irradiated food cannot be sold to consumers without clear labeling that this method has been used (figure 11.9).

Ionizing radiation is currently approved in the United States for cutting down the microbial load and for disinfecting cured meats, some spices, and seasonings. It is also used on fresh pork to destroy *Trichinella* worms (the cause of trichinosis); on chicken carcasses and beef to control *Salmonella* and *E. coli;* on wheat to

Ionizing Radiation

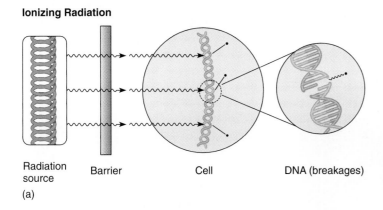

Nonionizing Radiation

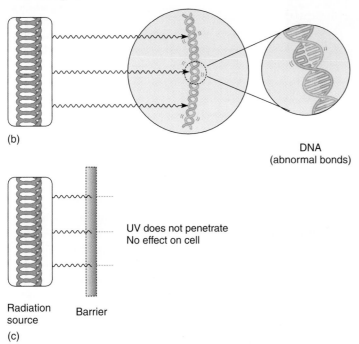

DNA
(abnormal bonds)

UV does not penetrate
No effect on cell

Radiation
source   Barrier

(c)

FIGURE 11.8

Cellular effects of irradiation. (a) Ionizing radiation can penetrate a solid barrier, bombard a cell, enter it, and dislodge electrons from molecules. Breakage of DNA creates massive mutations. **(b)** Nonionizing radiation enters a cell, strikes molecules, and excites them. The effect on DNA is mutation by formation of abnormal bonds. **(c)** A solid barrier cannot be penetrated by nonionizing radiation.

kill insect eggs; and on fresh fruits and vegetables to cut down on surface microbes that can increase the rate of spoilage. See chapter 26 for further discussion on irradiation of food.

Sterilizing medical products with ionizing radiation is a rapidly expanding field. Drugs, vaccines, medical instruments (especially plastics), syringes, surgical gloves, and tissues such as bone, skin, and heart valves for grafting all lend themselves to this mode of sterilization. Its main advantages include speed, high penetrating power (it can sterilize materials through outer packages and wrappings), and the absence of heat. Its main disadvantages are potential dangers to machine operators from exposure to radiation and possible damage to some materials.

Nonionizing Radiation: Ultraviolet Rays

Sunlight is the natural source of ultraviolet rays, which accounts for its microbicidal effects. **Ultraviolet radiation** ranges in wavelength from approximately 100 nm to 400 nm. It is most lethal from 240 nm to 280 nm (with a peak at 260 nm). In everyday practice, the source of UV radiation is the germicidal lamp, which generates radiation at 254 nm. Owing to its lower energy state, UV radiation is not as penetrating as ionizing radiation. Because UV radiation passes readily through air, slightly through liquids, and only poorly through solids, the object to be disinfected must be directly exposed to it for full effect.

As UV radiation passes through a cell, it is initially absorbed by DNA. Specific molecular damage occurs on the pyrimidine bases (thymine and cytosine), which form abnormal linkages with each other called *pyrimidine dimers* (figure 11.10). These bonds occur between adjacent bases on the same DNA strand and interfere with normal DNA replication and transcription. The results are inhibition of growth and cellular death. In addition to altering DNA, UV radiation also disrupts cells by generating toxic photochemical products (free radicals). Ultraviolet rays are a powerful tool for destroying fungal cells and spores, bacterial vegetative cells, protozoa, and viruses. Bacterial spores are about 10 times more resistant to radiation than are vegetative cells, but they can be killed by increasing the time of exposure.

Applications of UV Radiation UV radiation is usually directed at disinfection rather than sterilization. Germicidal lamps can cut down on the concentration of airborne microbes as much as 99%. They are used in hospital rooms, operating rooms, schools, food preparation areas, and dental offices. Ultraviolet disinfection of air has proved effective in reducing postoperative infections, preventing the transmission of infections by respiratory droplets, and curtailing the growth of microbes in food-processing plants and slaughterhouses.

Ultraviolet irradiation of liquids requires special equipment to spread the liquid into a thin, flowing film that is exposed directly to a lamp. This method can be used to treat drinking water (figure 11.11) and to purify other liquids (milk and fruit juices) as an alternative to heat. Ultraviolet treatment has proved effective in freeing vaccine antigens and plasma from contaminants. The surfaces of solid, nonporous materials such as walls and floors, as well as meat, nuts, tissues for grafting, and drugs, have been successfully disinfected with UV.

One major disadvantage of UV is its poor powers of penetration through solid materials such as glass, metal, cloth, plastic, and even paper. Another drawback to UV is the damaging effect of overexposure on human tissues, including sunburn, retinal damage, cancer, and skin wrinkles.

SOUND WAVES IN MICROBIAL CONTROL

High-frequency sound (sonic) waves beyond the sensitivity of the human ear are known to disrupt cells. These frequencies range from 15,000 to more than 200,000 cycles per second (supersonic to ultrasonic). Sonication transmits vibrations through a water-filled chamber (sonicator) to induce pressure changes and create intense points of turbulence that can stress and burst cells in the vicinity. Gram-negative rods are most sensitive to ultrasonic vibrations, and

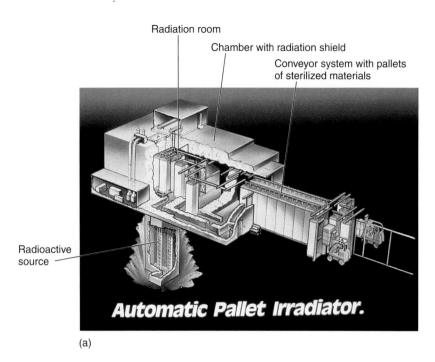

Radiation room
Chamber with radiation shield
Conveyor system with pallets of sterilized materials
Radioactive source
Automatic Pallet Irradiator.

(a)

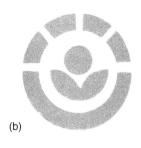

(b)

FIGURE 11.9

Sterilization with ionizing radiation. (a) This irradiation machine uses radioactive cobalt 60 as a gamma radiation source to sterilize fruits, vegetables, meats, fish, and spices. Although this method has stirred some controversy regarding its safety, it is gaining in acceptance because of its ability to increase shelf-life and reduce food-borne infections. **(b)** Regulations dictate that this universal symbol for radioactivity must be affixed to all irradiated materials.

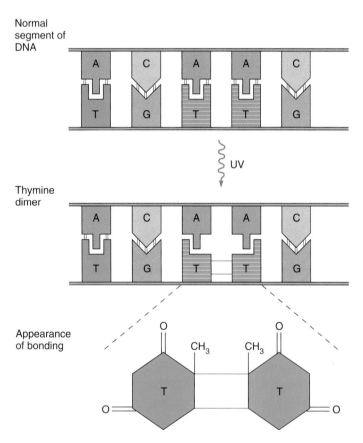

Normal segment of DNA

| A | C | A | A | C |
| T | G | T | T | G |

UV

Thymine dimer

| A | C | A | A | C |
| T | G | T | T | G |

Appearance of bonding

FIGURE 11.10

Formation of pyrimidine dimers by the action of ultraviolet (UV) radiation. This shows what occurs when two adjacent thymine bases on one strand of DNA are induced by UV rays to bond laterally with each other. The result is a thymine dimer shown in greater detail. Dimers can also occur between adjacent cytosines and thymine and cytosine bases. If they are not repaired, dimers can prevent that segment of DNA from being correctly replicated or transcribed. Massive dimerization is lethal to cells.

gram-positive cocci, fungal spores, and bacterial spores are most resistant to them. Sonication also forcefully dislodges foreign matter from objects. Heat generated by sonic waves (up to 80°C) also appears to contribute to an antimicrobial action. Ultrasonic devices are used in dental and some medical offices to clear debris and saliva from instruments before sterilization and to clean dental restorations. However, most sonic machines are not predictable enough to be used in disinfection or sterilization. Other types of ultrasonic devices are available for medical diagnosis and for removing plaque and calculus from teeth.

STERILIZATION BY FILTRATION: TECHNIQUES FOR REMOVING MICROBES

Filtration is an effective method to remove microbes from air and liquids. In practice, a fluid is strained through a filter with openings large enough for the fluid to pass through but too small for microorganisms to pass through (figure 11.12).

Most modern microbiological filters are thin membranes of cellulose acetate, polycarbonate, and a variety of plastic materials (Teflon, nylon) whose pore size can be carefully controlled and standardized. Ordinary substances such as charcoal, diatomaceous earth, or unglazed porcelain are also used in some applications. Viewed microscopically, most filters are perforated by very precise, uniform pores (figure 11.12b). The pore diameters vary from coarse (8 μm) to ultrafine (0.02 μm), permitting selection of the minimum particle size to be trapped. Those with the smallest pore diameters permit true sterilization by removing viruses, and some will even remove large proteins. A sterile liquid filtrate is typically produced by suctioning the liquid through a sterile filter into a presterilized container. These filters are also used to separate mixtures of microorganisms and to enumerate bacteria in water analysis (see chapter 26).

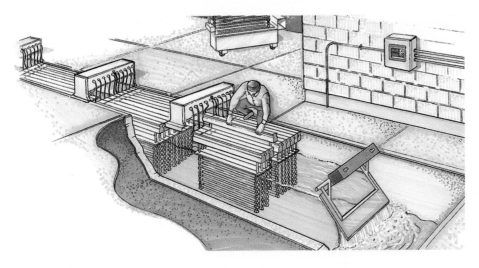

FIGURE 11.11

An ultraviolet (UV) treatment system for disinfection of wastewater. Water flows through racks of UV lamps and is exposed to 254 nm UV radiation. This system has a capacity of several million gallons per day and can be used as an alternative to chlorination.

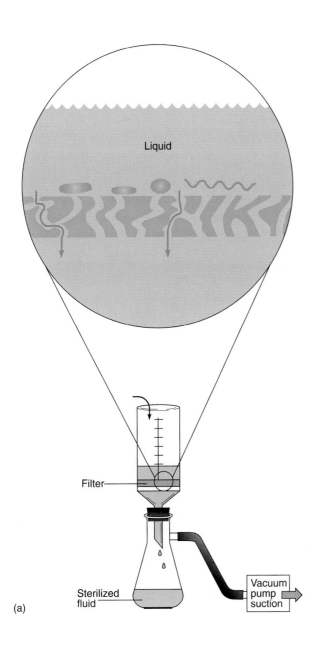

(a)

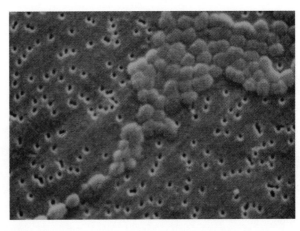

(b)

FIGURE 11.12

Membrane filtration. (a) Vacuum assembly for achieving filtration of liquids through suction. Inset shows filter as seen in cross section, with tiny passageways (pores) too small for the microbial cells to enter but large enough for liquid to pass through. **(b)** Scanning electron micrograph of filter, showing relative size of pores and bacteria trapped on its surface (5,900×).

Applications of Filtration Sterilization   Filtration sterilization is used to prepare liquids that cannot withstand heat, including serum and other blood products, vaccines, drugs, IV fluids, enzymes, and media. Filtration has been employed as an alternative to sterilize milk and beer without altering their flavor. It is also an important step in water purification. Its usage extends to filtering out particulate impurities (crystals, fibers, and so on) that can cause severe reactions in the body. It has the disadvantage of not removing soluble molecules (toxins) that can cause disease. Filtration is also an efficient means of removing airborne contaminants that are a common source of infection and spoilage. High-efficiency particulate air (HEPA) filters are widely used to provide a flow of sterile air to hospital rooms and sterile rooms.

Physical methods of microbial control include heat, cold, radiation, and drying.

Heat is the most widely used method of microbial control. It is used in combination with water (moist heat) or as dry heat (oven, flames).

The thermal death time (TDT) is the shortest length of time required to kill all microbes at a specific temperature. The TDT is longest for spore-forming bacteria and certain viruses.

The thermal death point (TDP) is the lowest temperature at which all microbes are killed in a specified length of time (10 minutes).

Autoclaving, or steam sterilization, is the process by which steam is heated under pressure to sterilize a wide range of materials in a comparatively short time (minutes to hours). It is effective for most materials except water-resistant substances such as oils, waxes, and powders.

Boiling water and pasteurization of beverages disinfect but do not sterilize materials.

Dry heat is microbicidal under specified times and temperatures. Flame heat, or incineration, is microbicidal. It is used when total destruction of microbes and materials is indicated.

Chilling, freezing, and desiccation are microbistatic but not microbicidal. They are *not* considered true methods of disinfection because they are not consistent in their effectiveness.

Ionizing radiation (cold sterilization) by gamma rays and X rays is used to sterilize medical products, meats, and spices. It damages DNA and cell organelles by producing disruptive ions.

Ultraviolet light, or nonionizing radiation, has limited penetrating ability. It is therefore restricted to disinfecting air and certain liquids.

Ultrasound is microbistatic to most microbes, but it is microbicidal to gram-negative bacteria. It is used primarily to reduce microbial load from inanimate objects.

Sterilization by filtration removes microbes from heat-sensitive liquids and circulating air. The pore size of the filter determines what kinds of microbes are removed.

Chemical Agents in Microbial Control

Chemical control of microbes probably emerged as a serious science in the early 1800s, when physicians used chloride of lime and iodine solutions to treat wounds and to wash their hands before surgery. At the present time, approximately 10,000 different antimicrobial chemical agents are manufactured; probably 1,000 of them are used routinely in the allied health sciences and the home. There is a genuine need to avoid infection and spoilage, but the abundance of products available to "kill germs, disinfect, antisepticize, clean and sanitize, deodorize, fight plaque, and purify the air" indicates a preoccupation with eliminating microbes from the environment that, at times, seems excessive (Spotlight on Microbiology 11.2).

Antimicrobial chemicals occur in the liquid, gaseous, or even solid state and vary from disinfectants and antiseptics to sterilants and preservatives (chemicals that inhibit the deterioration of substances). Liquid agents contain water, alcohol, or a mixture of the two, with various dissolved solutes. Solutions containing pure water as the solvent are termed *aqueous,* whereas those with pure alcohol or water-alcohol mixtures are termed *tinctures.*

CHOOSING A MICROBICIDAL CHEMICAL

The choice and appropriate use of antimicrobial chemical agents is of constant concern in medicine and dentistry. Although actual clinical practices of chemical decontamination vary widely, some desirable qualities in a germicide have been identified, including: (1) rapid action even in low concentrations, (2) solubility in water or alcohol and long-term stability, (3) broad-spectrum microbicidal action without being toxic to human and animal tissues, (4) penetration of inanimate surfaces to sustain a cumulative or persistent action, (5) resistance to becoming inactivated by organic matter, (6) noncorrosive or nonstaining properties, (7) sanitizing and deodorizing properties, and (8) inexpensiveness and ready availability. As yet, no chemical can completely fulfill all of those requirements, but glutaraldehyde and hydrogen peroxide approach this ideal. At the same time, we should question the rather overinflated claims made about certain commercial agents such as mouthwashes and disinfectant air sprays.

Germicides are evaluated in terms of their effectiveness in destroying microbes on medical and dental materials. The three levels of chemical decontamination procedures are *high, intermediate,* and *low* (table 11.5). High-level germicides kill endospores, and, if properly used, are sterilants. Materials that necessitate high-level control are medical devices—for example, catheters, heart-lung equipment, and implants—that are not heat-sterilizable and are intended to enter body tissues during medical procedures. Intermediate-level germicides kill fungal (but not bacterial) spores, resistant pathogens such as the tubercle bacillus, and viruses. They are used to disinfect items (respiratory equipment, endoscopes) that come into intimate contact with the mucous membranes but are noninvasive. Low levels of disinfection eliminate only vegetative bacteria, vegetative fungal cells, and some viruses. They are used to clean materials such as electrodes, straps, and furniture that touch the skin surfaces but not the mucous membranes.

FACTORS THAT AFFECT THE GERMICIDAL ACTIVITY OF CHEMICALS

Factors that control the effect of a germicide include the nature of the microorganisms being treated, the nature of the material being treated, the degree of contamination, the time of exposure, and the strength and chemical action of the germicide (table 11.6). Standardized procedures for testing the effectiveness of germicides are summarized in appendix B. The modes of action of most germicides involve the cellular targets discussed in an earlier section of this chapter: proteins, nucleic acids, the cell wall, and the cell membrane.

A chemical's strength or concentration is expressed in various ways, depending upon convention and the method of preparation. The content of many chemical agents can be expressed by more than one notation. In dilutions, a small volume of the liquid chemical (solute) is diluted in a larger volume of solvent to achieve a certain ratio. For example, a common laboratory phenolic disinfectant such as Lysol is usually diluted 1:200; that is, one part of chemical has been added to 200 parts of water by volume. Solutions such as chlorine that are effective in very high dilutions are

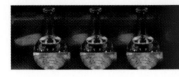

SPOTLIGHT ON MICROBIOLOGY   11.2
Pathogen Paranoia: "The Only Good Microbe Is a Dead Microbe"

The sensational publicity over outbreaks of infections such as influenza, tuberculosis, and microbial food poisoning has monumentally influenced the public view of microorganisms. Thousands of articles have sprinkled the news services over the past five years. On the positive side, this glut of information has improved people's awareness of the importance of microorganisms. And, certainly, such knowledge can be seen as beneficial when it leads to well-reasoned and sensible choices, such as using greater care in handwashing, food handling, and personal hygiene. But sometimes a little knowledge can be dangerous. The trend also seems to have escalated into an obsessive fear of "germs" lurking around every corner and a fixation on eliminating microbes from the environment and the human body.

As might be expected, commercial industries have found a way to capitalize on those fears. Since 1997, over 300 new products that contain antibacterial or germicidal chemicals have been marketed. A widespread array of cleansers and commonplace materials have already had antimicrobic chemicals added. First it was hand soaps and dishwashing detergents, and eventually the list grew to include shampoos, laundry aids, hand lotions, foot pads for shoes, deodorants, sponges and scrub pads, kitty litter, acne medication, cutting boards, garbage bags, toys, and toothpaste.

By far, the prevalent chemical agent routinely added to these products is a phenolic called *triclosan* (Irgasan). This substance is fairly mild and nontoxic and does indeed kill most pathogenic bacteria. However, it does not reliably destroy viruses or fungi and has been linked to cases of skin rashes due to hypersensitivity.

One unfortunate result of the negative news on microbes is how it fosters the feeling that all microbes are harmful, We must not forget that most human beings manage to remain healthy despite the fact that they live in continual intimate contact with microorganisms. We really do not have to be preoccupied with microbes every minute or feel overly concerned that the things we touch, drink, or eat are sterile, as long as they are somewhat clean and free of pathogens. For most of us, resistance to infection is well maintained by our numerous host defenses.

Medical experts are concerned that the widespread overuse of these antibacterial chemicals could favor the survival and growth of resistant strains of bacteria. A study reported in 2000 that many pathogens such as *Mycobacterium tuberculosis* and *Pseudomonas* are naturally resistant to triclosan, and that *E. coli* and *Staphylococcus aureus* have already demonstrated decreased sensitivity to it. The widespread use of this chemical may actually select for "super microbes" that survive ordinary disinfection. Another outcome of overuse of environmental germicides is to reduce the natural contact with microbes that is required to maintain the normal resident flora and stimulate immunities. Constant use of these agents could shift the balance in the normal flora of the body by killing off harmless or beneficial microbes.

Infectious disease specialists urge a happy medium approach. Instead of filling the home with questionable germicidal products, they encourage cleaning with traditional soaps and detergents, reserving more potent products to reduce the spread of infection among household members.

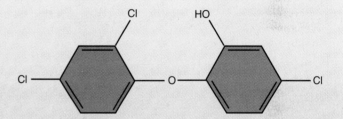

The molecular structure of triclosan, also known as Irgasan and Ster-Zac, a phenol-based chemical that destroys bacteria by disrupting cell walls and membranes.

expressed in parts per million (ppm). In percent solutions, the solute is added to water by weight or volume to achieve a certain percentage in the solution. Alcohol, for instance, is used in percentages ranging from 50% to 95%. In general, solutions of low dilution or high percentage have more of the active chemical (are more concentrated) and tend to be more germicidal, but expense and potential toxicity can necessitate using the minimum strength that is effective.

Another factor that contributes to germicidal effectiveness is the length of exposure. Most compounds require adequate contact time to allow the chemical to penetrate and to act on the microbes present. The composition of the material being treated must also be considered. Smooth, solid objects are more reliably disinfected than are those with pores or pockets that can trap soil. An item contaminated with common biological matter such as serum, blood, saliva, pus, fecal material, or urine presents a problem in disinfection.

Large amounts of organic material can hinder the penetration of a disinfectant and, in some cases, form bonds that reduce its activity. Adequate cleaning of instruments and other reusable materials ensures that the germicide or sterilant will better accomplish the job for which it was chosen.

GERMICIDAL CATEGORIES ACCORDING TO CHEMICAL GROUP

Several general groups of chemical compounds are widely used for antimicrobial purposes in medicine and commerce (see table 11.5). Prominent agents include halogens, heavy metals, alcohols, phenolic compounds, oxidizers, aldehydes, detergents, and gases. These groups will be surveyed in the following section from the standpoint of each agent's specific forms, modes of action, indications for use, and limitations.

TABLE 11.5

Qualities of Chemical Agents Used in Health Care

Agent	Target Microbes	Level of Activity	Toxicity	Comments
Chlorine	Sporicidal (slowly)	Intermediate	Gas is highly toxic; solution irritates skin	Inactivated by organics; unstable in sunlight
Iodine	Sporicidal (slowly)	Intermediate	Can irritate tissue; toxic if ingested	Iodophors★ are milder forms
Phenolics	Some bacteria, viruses, fungi	Intermediate to low	Can be absorbed by skin; can cause CNS damage	Poor solubility; expensive
Alcohols	Most bacteria, viruses, fungi	Intermediate	Toxic if ingested; a mild irritant; dries skin	Inflammable, fast-acting
Hydrogen peroxide,★ stabilized	Sporicidal	High	Toxic to eyes; toxic if ingested	Improved stability; works well in organic matter
Quaternary ammonium compounds	Some bactericidal, virucidal, fungicidal activity	Low	Irritating to mucous membranes; poisonous if taken internally	Weak solutions can support microbial growth; easily inactivated
Soaps	Certain very sensitive species	Very low	Nontoxic; few if any toxic effects	Used for removing soil, oils, debris
Mercurials	Weakly microbistatic	Low	Highly toxic if ingested, inhaled, absorbed	Easily inactivated
Silver nitrate	Bactericidal	Low	Toxic, irritating	Discolors skin
Glutaraldehyde★	Sporicidal	High	Can irritate skin; toxic if absorbed	Not inactivated by organic matter; unstable
Formaldehyde	Sporicidal	High to intermediate	Very irritating; fumes damaging, carcinogenic	Slow rate of action; limited applications
Ethylene oxide gas★	Sporicidal	High	Very dangerous to eyes, lungs; carcinogenic	Explosive in pure state; good penetration; materials must be aerated
Dyes	Weakly bactericidal, fungicidal	Low	Low toxicity	Stain materials, skin
Chlorhexidine★	Most bacteria, some viruses, fungi	Low to intermediate	Low toxicity	Fast-acting, mild, has residual effects

★*These forms approach the ideal by having many of the following characteristics: broad spectrum, low toxicity, fast action, penetrating abilities, residual effects, stability, potency in organic matter, and solubility.*

The Halogen Antimicrobial Chemicals

The **halogens*** are fluorine, bromine, chlorine, and iodine, a group of nonmetallic elements that commonly occur in minerals, seawater, and salts. Although they can exist in either the ionic (halide) or nonionic state, most halogens exert their antimicrobial effect primarily in the nonionic state, not the halide state (chloride, iodide, for example). Because fluorine and bromine are difficult and dangerous to handle, and are no more effective than chlorine and iodine, only the latter two are used routinely in germicidal preparations. These elements are highly effective components of disinfectants and antiseptics because they are microbicidal and not just microbistatic, and they are sporicidal with longer exposure. For these reasons, halogens are the active ingredients in nearly one-third of all antimicrobial chemicals currently marketed.

Chlorine and Its Compounds Chlorine has been used for disinfection and antisepsis for approximately 200 years. The major forms used in microbial control are liquid and gaseous chlorine

---

***halogens** (hay′-loh-jenz) Gr. *halos,* salt, and *gennan,* to produce.

(Cl_2), hypochlorites (OCl), and chloramines (NH_2Cl). In solution, these compounds combine with water and release hypochlorous acid $(HOCl)$, which oxidizes the sulfhydryl $(S—H)$ group on the amino acid cysteine and interferes with disulfide $(S—S)$ bridges on numerous enzymes. The resulting denaturation of the enzymes is permanent and suspends metabolic reactions. Chlorine kills not only bacterial cells and endospores but also fungi and viruses. The major limitations of chlorine compounds are: (1) They are ineffective if used at an alkaline pH; (2) excess organic matter can greatly reduce their activity; and (3) they are relatively unstable, especially if exposed to light.

Chlorine Compounds in Disinfection and Antisepsis Gaseous and liquid chlorine are used almost exclusively for large-scale disinfection of drinking water, sewage, and wastewater from such sources as agriculture and industry. Chlorination to a concentration of 0.6 to 1.0 parts of chlorine per million parts of water will ensure that water is safe to drink. This rids the water of most pathogenic vegetative microorganisms without unduly affecting its taste (some persons may debate this).

TABLE 11.6

Required Concentrations and Times for Chemical Destruction of Selected Microbes

Organism	Concentration	Time
Agent: Aqueous Iodine		
Staphylococcus aureus★	2%	2 min
Escherichia coli★★	2%	1.5 min
Enteric viruses	2%	10 min
Agent: Chlorine		
Mycobacterium tuberculosis★	50 ppm	50 sec
Entamoeba cysts (protozoa)	0.1 ppm	150 min
Hepatitis A virus	3 ppm	30 min
Agent: Phenol		
Staphylococcus aureus	1:85 dil	10 min
Escherichia coli	1:75 dil	10 min
Agent: Ethyl Alcohol		
Staphylococcus aureus	70%	10 min
Escherichia coli	70%	2 min
Poliovirus	70%	10 min
Agent: Hydrogen Peroxide		
Staphylococcus aureus	3%	12.5 sec
Neisseria gonorrhoeae	3%	0.3 sec
Herpes simplex virus	3%	12.8 sec
Agent: Quaternary Ammonium Compound		
Staphylococcus aureus	450 ppm	10 min
Salmonella typhi★★	300 ppm	10 min
Agent: Silver Ions		
Staphylococcus aureus	8 µg/ml	48 h
Escherichia coli	2 mg/ml	48 h
Candida albicans (yeast)	14 mg/ml	48 h
Agent: Glutaraldehyde		
Staphylococcus aureus	2%	<1 min
Mycobacterium tuberculosis	2%	<10 min
Herpes simplex virus	2%	<10 min
Agent: Ethylene Oxide Gas		
Streptococcus faecalis	500 mg/l	2–4 min
Influenza virus	10,000 mg/l	25 h
Agent: Chlorhexidine		
Staphylococcus aureus	1:10 dil	15 sec
Escherichia coli	1:10 dil	30 sec

★*Gram-positive vegetative bacterium.*
★★*Gram-negative vegetative bacterium.*

Hypochlorites are perhaps the most extensively used of all chlorine compounds. The scope of applications is broad, including sanitization and disinfection of food equipment in dairies, restaurants, and canneries and treatment of swimming pools, spas, drinking water, and even fresh foods. Hypochlorites are used in the allied health areas to treat wounds and to disinfect equipment, bedding, and instruments. Common household bleach is a weak solution (5%) of sodium hypochlorite that serves as an all-around disinfectant, deodorizer, and stain remover.

Chloramines (dichloramine, halazone) are being employed more frequently as an alternative to pure chlorine in treating water supplies. Because standard chlorination of water is now believed to produce unsafe levels of cancer-causing substances such as trihalomethanes, some water districts have been directed by federal agencies to adopt chloramine treatment of water supplies. Chloramines also serve as sanitizers and disinfectants and for treating wounds and skin surfaces.

Iodine and Its Compounds   Iodine is a pungent black chemical that forms brown-colored solutions when dissolved in water or alcohol. The two primary iodine preparations are *free iodine* in solution (I_2) and *iodophors*. Iodine rapidly penetrates the cells of microorganisms, where it apparently disturbs a variety of metabolic functions by interfering with the hydrogen and disulfide bonding of proteins (similar to chlorine). All classes of microorganisms are killed by iodine if proper concentrations and exposure times are used. Iodine activity is not as adversely affected by organic matter and pH as chlorine is.

Applications of Iodine Solutions   Aqueous iodine contains 2% iodine and 2.4% sodium iodide; it is used as a topical antiseptic before surgery and occasionally as a treatment for burned and infected skin. A stronger iodine solution (5% iodine and 10% potassium iodide) is used primarily as a disinfectant for plastic items, rubber instruments, cutting blades, and thermometers. **Iodine tincture** is a 2% solution of iodine and sodium iodide in 70% alcohol that can be used in skin antisepsis. Because iodine can be extremely irritating to the skin and toxic when absorbed, strong aqueous solutions and tinctures (5–7%) are no longer considered safe for routine antisepsis. Iodine tablets are available for disinfecting water during emergencies or destroying pathogens in impure water supplies.

Iodophors are complexes of iodine and a neutral polymer such as a polyvinylalcohol. This formulation allows the slow release of free iodine and increases its degree of penetration. These compounds have largely replaced free iodine solutions in medical antisepsis because they are less prone to staining or irritating tissues. Common iodophor products marketed as Betadine, Povidone (PVP), and Isodine contain 2–10% of available iodine. They are used to prepare skin and mucous membranes for surgery and injections, in surgical handscrubs, to treat burns, and to disinfect equipment and surfaces. A recent study showed that Betadine solution is an effective means of preventing eye infections in newborn infants, and it may replace antibiotics and silver nitrate as the method of choice.

Phenol and Its Derivatives

Phenol (carbolic acid) is an acrid, poisonous compound derived from the distillation of coal tar. First adopted by Joseph Lister in 1867 as a surgical germicide, phenol was the major antimicrobial chemical until other phenolics with fewer toxic and irritating effects were developed. Solutions of phenol are now used only in certain limited cases, but it remains one standard against which other phenolic disinfectants are rated. Substances chemically related to phenol are often referred to as phenolics. Hundreds of these chemicals are now available.

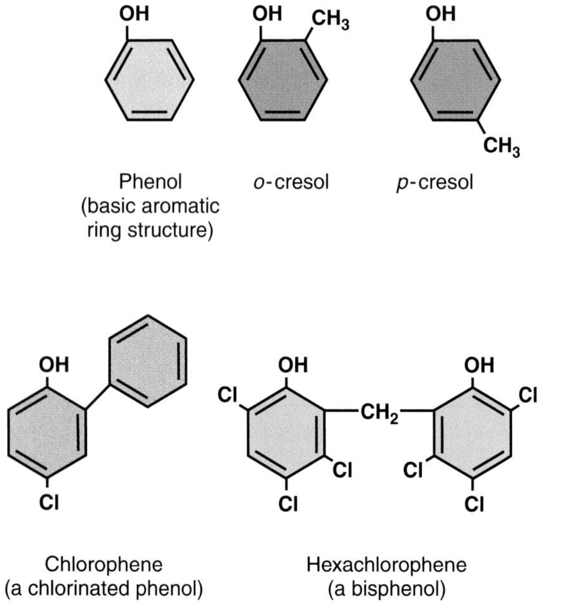

Phenol
(basic aromatic
ring structure)

o-cresol

p-cresol

Chlorophene
(a chlorinated phenol)

Hexachlorophene
(a bisphenol)

FIGURE 11.13

Some phenolics. All contain a basic aromatic ring, but they differ in the types of additional compounds such as Cl and CH_3.

Phenolics consist of one or more aromatic carbon rings with added functional groups (figure 11.13). Among the most important are alkylated phenols (cresols), chlorinated phenols, and bisphenols. In high concentrations, they are cellular poisons, rapidly disrupting cell walls and membranes and precipitating proteins; in lower concentrations, they inactivate certain critical enzyme systems. The phenolics are strongly microbicidal and will destroy vegetative bacteria (including the tubercle bacillus), fungi, and most viruses (not hepatitis B), but they are not reliably sporicidal. Their continued activity in the presence of organic matter and their detergent actions contribute to their usefulness. Unfortunately, the toxicity of many of the phenolics makes them too dangerous to use as antiseptics.

Applications of Phenolics Phenol itself is still used for general disinfection of drains, cesspools, and animal quarters, but it is seldom applied as a medical germicide. The cresols are simple phenolic derivatives that are combined with soap for intermediate or low levels of disinfection in the hospital. Lysol and creolin, in a 1–3% emulsion, are common household versions of this type.

The bisphenols are also widely employed in commerce, clinics, and the home. One type, orthophenyl phenol, is the major ingredient in disinfectant aerosol sprays. This same phenolic is also found in some proprietary compounds (Lysol) often used in hospital and laboratory disinfection. One particular bisphenol, hexachlorophene, was once a common additive of cleansing soaps (pHisoHex) used in the hospital and home. When hexachlorophene was found to be absorbed through the skin and a cause of neurological damage, it was no longer available without a prescription. It is occasionally used to control outbreaks of skin infections.

Perhaps the most widely used phenolic is **triclosan,** chemically known as dichlorophenoxyphenol (see Spotlight on Microbes 11.2). It is the antibacterial compound added to dozens of products, from soaps to kitty litter. It acts as both a disinfectant and antiseptic and is broad-spectrum in its effects.

Chlorhexidine

The compound chlorhexidine (Hibiclens, Hibitane) is a complex organic base containing chlorine and two phenolic rings. It is somewhat related to the cationic detergents in its mode of action, being both a surfactant and a protein denaturant. At moderate to high concentrations, it is bactericidal for both gram-positive and gram-negative bacteria but inactive against spores. Its effects on viruses and fungi vary. It possesses distinct advantages over many other antiseptics because of its mildness, low toxicity, and rapid action, and it is not absorbed into deeper tissues to any extent. Alcoholic or aqueous solutions of chlorhexidine are now commonly used for handscrubbing, preparing skin sites for surgical incisions and injections, and whole body washing. Chlorhexidine solution also serves as an obstetric antiseptic, a neonatal wash, a wound degermer, a mucous membrane irrigant, and a preservative for eye solutions.

Alcohols As Antimicrobial Agents

Alcohols are colorless hydrocarbons with one or more —OH functional groups. Of several alcohols available, only ethyl and isopropyl are suitable for microbial control. Methyl alcohol is not very microbicidal, and more complex alcohols are either poorly soluble in water or too expensive for routine use. Alcohols are employed alone in aqueous solutions or as solvents for tinctures (iodine, for example). Alcohol's mechanism of action depends in part upon its concentration. Concentrations of 50% and higher dissolve membrane lipids, disrupt cell surface tension, and compromise membrane integrity. Alcohol that has entered the protoplasm denatures proteins through coagulation, but only in alcohol-water solutions of 50–95%. Absolute alcohol (100%) dehydrates cells and inhibits their growth but is generally not a protein coagulant.

Although useful in intermediate- to low-level germicidal applications, alcohol does not destroy bacterial spores at room temperature. Alcohol can, however, destroy resistant vegetative forms, including tubercle bacilli and fungal spores, provided the time of exposure is adequate. Alcohol is generally more effective in inactivating enveloped viruses than the more resistant nonenveloped viruses such as poliovirus and hepatitis A virus.

Applications of Alcohols Ethyl alcohol, also called ethanol or grain alcohol, is known for being germicidal, nonirritating, and inexpensive. Solutions of 70% to 95% are routinely used as skin degerming agents because the surfactant action removes skin oil, soil, and some microbes sheltered in deeper skin layers. One limitation to its effectiveness is the rate at which it evaporates. Ethyl alcohol is occasionally used to disinfect electrodes, face masks, and thermometers, which are first cleaned and then soaked in alcohol for 15 to 20 minutes. Isopropyl alcohol, sold as rubbing alcohol, is even more microbicidal and less expensive than ethanol, but these benefits must be weighed against its toxicity. It must be used with caution in disinfection or skin cleansing, because inhalation of its vapors can adversely affect the nervous system.

Hydrogen Peroxide and Related Germicides

Hydrogen peroxide (H_2O_2) is a colorless, caustic liquid that decomposes in the presence of light, metals, or catalase into water and oxygen gas. Peroxide solutions have been used for about 50 years. Early formulations were unstable and inhibited by organic matter, but manufacturing methods now permit synthesis of H_2O_2 so stable that even dilute solutions retain activity through several months of storage.

The germicidal effects of hydrogen peroxide are due to the direct and indirect actions of oxygen. Oxygen forms hydroxyl free radicals (.OH), which, like the superoxide radical (see chapter 7), are highly toxic and reactive to cells. Although most microbial cells produce catalase to inactivate the metabolic hydrogen peroxide, it cannot neutralize the amount of hydrogen peroxide entering the cell during disinfection and antisepsis. Hydrogen peroxide is bactericidal, virucidal, and fungicidal and, in higher concentrations, sporicidal.

Applications of Hydrogen Peroxide   As an antiseptic, 3% hydrogen peroxide serves a variety of needs, including skin and wound cleansing, bedsore care, and mouthwashing. It is especially useful in treating infections by anaerobic bacteria because of the lethal effects of oxygen on these forms. Hydrogen peroxide is also a versatile disinfectant for soft contact lenses, surgical implants, plastic equipment, utensils, bedding, and room interiors. Hydrogen peroxide solutions and vapors are the latest method in flash sterilization of food packaging equipment and other industrial processes.

Another compound with effects similar to those of hydrogen peroxide is ozone (O_3), used to disinfect air, water, and industrial air conditioners and cooling towers.

Chemicals with Surface Action: Detergents

Detergents are complex organic substances that act as surfactants. They carry either an anionic (negative) or cationic (positive) charge. Most anionic (negatively charged) detergents have limited microbicidal power. Soaps belong to this group. Cationic detergents are the most effective, especially the quaternary ammonium compounds (usually shortened to *quats*).

All detergents have a general molecular structure that includes a long-chain hydrocarbon residue and a highly charged, or polar, group (figure 11.14). This configuration makes them soluble in lipid-based structures, and they act as surfactants. These properties can have several effects, but chief among them is the disruption of the cell membrane and the loss of its selective permeability. Quaternary ammonium compounds cause leakage of microbial cytoplasm, precipitate proteins, and inhibit metabolism. Because of their ability to interact with surfaces, detergents make good wetting agents, cleansing agents, and emulsifiers.

The range of activity of detergents is varied. Quaternary ammonium compounds, if used at medium concentrations, are effective against some gram-positive bacteria, viruses, fungi, and algae. In low concentrations, they have only microbistatic effects. The quats are ineffective against the tubercle bacillus, hepatitis virus, *Pseudomonas,* and spores at any concentration. Their activity can be greatly reduced in the presence of organic matter, and they function best in alkaline solutions. As a result of their limitations, quats are rated only for low-level disinfection in the clinical setting.

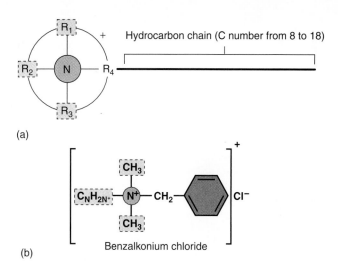

(a)

(b)

FIGURE 11.14

The structure of detergents. (a) In general, detergents are polar molecules with a positively charged head and at least one long, uncharged hydrocarbon chain. The head contains a central nitrogen nucleus with various alkyl (R) groups attached. **(b)** A common quaternary ammonium detergent, benzalkonium chloride.

Applications of Detergents and Soaps   Quaternary ammonium compounds include benzalkonium chloride, Zephiran, and cetylpyridinium chloride (Ceepryn). In dilutions ranging from 1:100 to 1:1,000, quats are mixed with cleaning agents to simultaneously disinfect and sanitize floors, furniture, equipment surfaces, and restrooms. They are used to clean restaurant eating utensils, food-processing equipment, dairy equipment, and clothing. They are common preservatives for ophthalmic solutions and cosmetics. Their level of disinfection is far too low for disinfecting medical instruments.

Soaps are alkaline compounds made by combining the fatty acids in oils with sodium or potassium salts. In usual practice, soaps are only weak microbicides, and they destroy only highly sensitive forms such as the agents of gonorrhea, meningitis, and syphilis. The common hospital pathogen *Pseudomonas* is so resistant to soap that various species grow abundantly in soap dishes. Soaps function primarily as cleansing agents and sanitizers in industry and the home. The superior sudsing and wetting properties of soaps help to mechanically remove large amounts of surface soil, greases, and other debris that contains microorganisms. Soaps gain greater germicidal value when mixed with agents such as chlorhexidine or iodine. They can be used for cleaning instruments before heat sterilization, degerming patients' skin, routine handwashing by medical and dental personnel, and preoperative handscrubbing. Vigorously brushing the hands with germicidal soap over a 15-minute period is an effective way to remove dirt, oil, and surface contaminants as well as some resident microbes, but it will never sterilize the skin (Historical Highlights 11.3 and figure 11.15).

Heavy Metal Compounds

Various forms of the metallic elements mercury, silver, gold, copper, arsenic, and zinc have been applied in microbial control over several

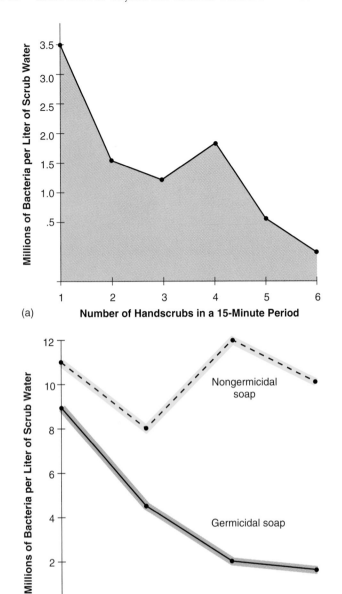

(a)

Number of Handscrubs in a 15-Minute Period

(b)

Days

FIGURE 11.15

Graphs showing effects of handscrubbing. (a) Reduction of microbes on the hands during a single (15-minute) episode of handwashing with nongermicidal soap. The increase in the fourth scrub is due to a high level of resident microbes uncovered in that skin layer. **(b)** Comparison of scrubbing over several days with a nongermicidal soap versus a germicidal soap. Germicidal soap has persistent effects on skin over time, keeping the microbial count low. Without germicide, soap does not show this sustained effect.

centuries. These are often referred to as heavy metals because of their relatively high atomic weight. However, from this list, only preparations containing mercury and silver still have any significance as germicides. Although some metals (zinc, iron) are actually needed in small concentrations as cofactors on enzymes, the higher molecular weight metals (mercury, silver, gold) can be very toxic, even in minute quantities (parts per million). This property of having antimicrobial effects in exceedingly small amounts is called an

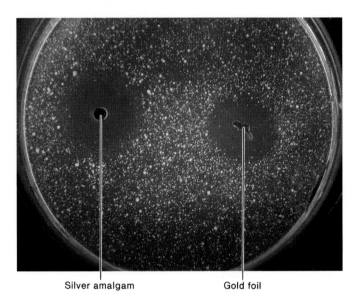

Silver amalgam   Gold foil

FIGURE 11.16

Demonstration of the oligodynamic action of heavy metals. A pour plate inoculated with saliva has small fragments of heavy metals pressed lightly into it. During incubation, clear zones indicating growth inhibition developed around both fragments. The slightly larger zone surrounding the amalgam (used in tooth fillings) probably reflects the synergistic effect of the silver and mercury it contains.

oligodynamic* action (figure 11.16). Heavy metal germicides contain either an inorganic or an organic metallic salt, and they come in the form of aqueous solutions, tinctures, ointments, or soaps.

Mercury, silver, and most other metals exert microbicidal effects by binding onto functional groups, including sulfhydryls, hydroxyls, amines, and phosphates. This binding inactivates proteins and rapidly brings metabolism to a standstill (see figure 11.4*c*). This mode of action can destroy many types of microbes, including vegetative bacteria, fungal cells and spores, algae, protozoa, and viruses (but not endospores).

Unfortunately, there are several drawbacks to using metals in microbial control: (1) Metals are very toxic to humans if ingested, inhaled, or absorbed through the skin, even in small quantities, for the same reasons that they are toxic to microbial cells; (2) they commonly cause allergic reactions; (3) large quantities of biological fluids and wastes neutralize or depress their actions; and (4) microbes can develop resistance to metals. Health and environmental considerations have dramatically reduced the use of metallic antimicrobic compounds in medicine, dentistry, commerce, and agriculture.

Applications of Heavy Metals Weak (0.001% to 0.2%) organic mercury tinctures such as thimerosal (Merthiolate) and nitromersol (Metaphen) are fairly effective antiseptics and infection preventives, but they should never be used on broken skin because they are harmful and can delay healing. The organic mercurials also serve as preservatives in cosmetics and ophthalmic solutions. Mercurochrome, that old staple of the medicine cabinet, is now considered among the poorest of antiseptics.

A silver compound with several applications is silver nitrate

---

*oligodynamic (ol″-ih-goh-dy-nam′-ik) Gr. *oligos,* little, and *dynamis,* power.

More than a hundred years ago, before sterile gloves were a routine part of medical procedures, the hands remained bare during surgery. Realizing the danger from microbes, medical practitioners attempted to sterilize the hands of surgeons and their assistants to prevent surgical infections. Several stringent (and probably very painful) techniques involving strong chemical germicides and vigorous scrubbing were practiced. Here are a few examples.

In Schatz's method, the hands and forearms were first cleansed by brisk scrubbing with liquid soap for 3 to 5 minutes, then soaked in a saturated solution of permanganate at a temperature of 110°F until they turned a deep mahogany brown. Next, the limbs were immersed in saturated oxalic acid until the skin became decolorized. Then, as if this were not enough, the hands and arms were rinsed with sterile limewater and washed in warm bichloride of mercury for 1 minute.

Or, there was Park's method (more like a torture). First, the surfaces of the hands and arms were rubbed completely with a mixture of cornmeal and green soap to remove loose dirt and superficial skin. Next, a paste of water and mustard flour was applied to the skin until it began to sting. This potion was rinsed off in sterile water, and the hands and arms were then soaked in hot bichloride of mercury for a few minutes, during which the solution was rubbed into the skin.

Another method once earnestly suggested for getting rid of microorganisms was to expose the hands to a hot-air cabinet to "sweat the germs" out of skin glands. Pasteur himself advocated a quick flaming of the hands to maintain asepsis.

The old dream of sterilizing the skin was finally reduced to some basic realities: The microbes entrenched in the epidermis and skin glands cannot be completely eradicated even with the most intense efforts, and the skin cannot be sterilized without also seriously damaging it. Because this is true for both medical personnel and their patients, the chance always exists that infectious agents can be introduced during invasive medical procedures. Safe surgery had to wait until 1890, when rubber gloves were first made available for placing a sterile barrier around the hands. Of course, this did not mean that skin cleansing and antiseptic procedures were abandoned or downplayed. A thorough scrubbing of the skin, followed by application of an antiseptic, is still needed to remove the most dangerous source of infections—the superficial contaminants constantly picked up from whatever we touch.

(a)

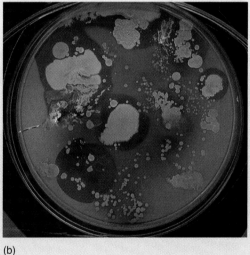

(b)

Microbes on normal unwashed hands. **(a)** A scanning electron micrograph of a piece of skin from a fingertip shows clusters of bacteria perched atop a fingerprint ridge (47,000×). **(b)** Heavy growth of microbial colonies on a plate of blood agar. This culture was prepared by passing an open sterile plate around a classroom of 30 students and having each one touch its surface. After incubation, a mixed population of bacteria and fungi appeared. Clear zones in agar can be indicative of pathogens.

($AgNO_3$) solution. Crede introduced it in the late nineteenth century for preventing gonococcal infections in the eyes of newborn infants who had been exposed to an infected birth canal. This preparation is not used as often now because many pathogens are resistant to it. It has been replaced by antibiotics in most instances. Solutions of silver nitrate (1–2%) can also be used as topical germicides on mouth ulcers and occasionally root canals. Silver sulfadiazine ointment, when added to dressings, effectively prevents in-

fection in second- and third-degree burn patients, and pure silver is now incorporated into catheters to prevent urinary tract infections in the hospital. Colloidal silver preparations are mild germicidal ointments or rinses for the mouth, nose, eyes, and vagina.

Aldehydes As Germicides

Organic substances bearing a —CHO functional group (a strong reducing group) on the terminal carbon are called aldehydes. Several common substances such as sugars and some fats are technically aldehydes. The two aldehydes used most often in microbial control are **glutaraldehyde** and **formaldehyde.**

Glutaraldehyde is a yellow acidic liquid with a mild odor. The molecule's two aldehyde groups favor the formation of polymers. The mechanism of activity involves cross-linking protein molecules on the cell surface. In this process, amino acids are alkylated, meaning that a hydrogen atom on an amino acid is replaced by the glutaraldehyde molecule itself (figure 11.17). It can also irreversibly disrupt the activity of enzymes within the cell. Glutaraldehyde is rapid and broad-spectrum, and is one of the few chemicals officially accepted as a sterilant and high-level disinfectant. It kills spores in 3 hours and fungi and vegetative bacteria (even *Mycobacterium* and *Pseudomonas*) in a few minutes. Viruses, including the most resistant forms, appear to be inactivated after relatively short exposure times. Glutaraldehyde retains its potency even in the presence of organic matter, is noncorrosive, does not damage plastics, and is less toxic or irritating than formaldehyde. Its principal disadvantage is that it is somewhat unstable, especially with increased pH and temperature.

Formaldehyde is a sharp, irritating gas that readily dissolves in water to form an aqueous solution called *formalin.* Full saturation of formaldehyde (37%) produces a solution of 100% formalin. The chemical is microbicidal through its attachment to nucleic acids and functional groups of amino acids. Formalin is an intermediate- to high-level disinfectant, although it acts more slowly than glutaraldehyde. Formaldehyde's extreme toxicity (it is classified as a carcinogen) and irritating effects on the skin and mucous membranes greatly limit its clinical usefulness.

Applications of the Aldehydes Glutaraldehyde is a milder chemical for sterilizing materials that are damaged by heat. Commercial products (Cidex, Glutarol) diluted to 2% are used to sterilize respiratory therapy equipment, hemostats, fiberoptic endoscopes (laparoscopes, arthroscopes), and anesthetic devices. Glutaraldehyde is employed in dental offices for practical disinfection of instruments (usually in combination with autoclaving) because of its ability to destroy hepatitis B and other blood-borne viruses. It can also be used to preserve vaccines, sanitize poultry carcasses, and degerm cow's teats.

Formalin tincture (8%) has limited use as a disinfectant for surgical instruments, and 1% formalin solution is still used to disinfect some reusable kidney dialysis instruments. Any object that comes into intimate contact with the body must be thoroughly rinsed with sterile water to remove the formalin residue. It is, after all, one of the active ingredients in embalming fluid.

Gaseous Sterilants and Disinfectants

Processing inanimate substances with chemical vapors, gases, and aerosols provides a versatile alternative to heat or liquid chemicals. Currently, those vapors and aerosols having the broadest applications are ethylene oxide (ETO), propylene oxide, and betapropiolactone (BPL).

Ethylene oxide is a colorless substance that exists as a gas at room temperature. It is very explosive in air, a feature that can be eliminated by combining it with a high percentage of carbon dioxide or fluorocarbon. Like the aldehydes, ETO is a very strong alkylating

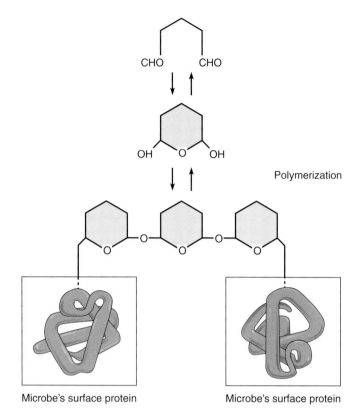

Polymerization

Microbe's surface protein   Microbe's surface protein

FIGURE 11.17

Actions of glutaraldehyde. The molecule forms a closed ring that can polymerize. When these alkylating polymers react with amino acids, they cross-link and inactivate proteins.

agent, and it reacts vigorously with functional groups of DNA and proteins. Through these actions, it blocks both DNA replication and enzymatic actions. Ethylene oxide is the only gas generally accepted for chemical sterilization because, when employed according to strict procedures, it is a sporicide. A specially designed ETO sterilizer called a *chemiclave,* a variation on the autoclave, is equipped with a chamber, gas ports, and temperature, pressure, and humidity controls (figure 11.18). Ethylene oxide is rather penetrating but relatively slow-acting, requiring from 90 minutes to 3 hours. Some items absorb ETO residues and must be aerated for several hours after exposure to ensure dissipation of as much residual gas as possible. For all of its effectiveness, ETO has some unfortunate features. Its explosiveness makes it dangerous to handle; it can damage the lungs, eyes, and mucous membranes if contacted directly; and it is rated as a carcinogen by the government.

Applications of Gases and Aerosols Ethylene oxide (carboxide, cryoxide) is an effective way to sterilize and disinfect plastic materials and delicate instruments in hospitals and industries. It can safely sterilize prepackaged medical devices, surgical supplies, syringes, and disposable Petri plates. Ethylene oxide has been used extensively to disinfect sugar, spices, dried foods, and drugs.

Propylene oxide is a close relative of ETO, with similar physical properties and mode of action, although it is less toxic. Because it breaks down into a relatively harmless substance, it is safer than ETO for sterilization of foods (nuts, powders, starches, spices).

and malachite green are very active against gram-positive species of bacteria and various fungi, they are incorporated into solutions and ointments to treat skin infections (ringworm, for example). The yellow acridine dyes, acriflavine and proflavine, are sometimes utilized for antisepsis and wound treatment in medical and veterinary clinics. For the most part, dyes will continue to have limited applications because they stain and have a narrow spectrum of activity.

Acids and Alkalies

Conditions of very low or high pH can destroy or inhibit microbial cells; but they are limited in applications due to their corrosive, caustic, and hazardous nature. Aqueous solutions of ammonium hydroxide remain a common component of detergents, cleansers, and deodorizers. Organic acids are widely used in food preservation because they prevent spore germination and bacterial and fungal growth and because they are generally regarded as safe to eat. Acetic acid (in the form of vinegar) is a pickling agent that inhibits bacterial growth; propionic acid is commonly incorporated into breads and cakes to retard molds; lactic acid is added to sauerkraut and olives to prevent growth of anaerobic bacteria (especially the clostridia), and benzoic and sorbic acids are added to beverages, syrups, and margarine to inhibit yeasts.

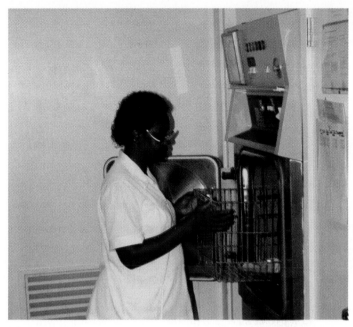

(a)

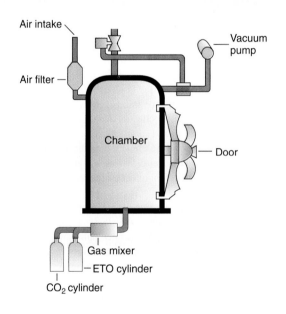

(b)

FIGURE 11.18

Sterilization using gas. (a) A chemiclave, an automatic ethylene oxide sterilizer. **(b)** The machine is equipped with gas canisters containing ethylene oxide (ETO) and carbon dioxide, a chamber to hold items, and mechanisms that evacuate gas and introduce air.

Betapropiolactone (BPL) is a rapidly microbicidal but somewhat toxic solution that is used for disinfecting whole rooms and instruments, sterilizing bone and arterial grafts, and inactivating viruses in vaccines.

Dyes As Antimicrobial Agents

Dyes are important in staining techniques and as selective and differential agents in media; they are also a primary source of certain drugs used in chemotherapy. Because aniline dyes such as crystal violet

·CHAPTER CHECKPOINTS

Chemical agents of microbial control are classified by their physical state and chemical nature.

Chemical agents can be either microbicidal or microbistatic. They are also classified as high-, medium-, or low-level germicides.

Factors that determine the effectiveness of a chemical agent include the type and numbers of microbes involved, the material involved, the strength of the agent, and the exposure time.

Halogens are effective chemical agents at both microbicidal and microbistatic levels. Chlorine compounds disinfect water, food, and industrial equipment. Iodine is used as either free iodine or iodophor to disinfect water and equipment. Iodophors are also used as antiseptic agents.

Phenols are strongly microbicidal agents used in general disinfection. Milder phenol compounds, the bisphenols, are also used as antiseptics.

Alcohols dissolve membrane lipids and destroy cell proteins. Their action depends upon their concentration, but they are generally only microbistatic.

Hydrogen peroxide is a versatile microbicide that can be used as an antiseptic for wounds and a disinfectant for utensils. A high concentration is an effective sporicide.

Surfactants are of two types: detergents and soaps. They reduce cell membrane surface tension, causing membrane rupture. Cationic detergents, or quats, are low-level germicides limited by the amount of organic matter present and the microbial load.

Aldehydes are potent sterilizing agents and high-level disinfectants that irreversibly disrupt microbial enzymes.

Ethylene oxide is the only accepted gaseous sterilant. It is effective for batch sterilizing medical materials, but it is explosive, toxic to tissues, and carcinogenic.

CHAPTER CAPSULE WITH KEY TERMS

I. Physical Methods for Controlling Microorganisms

A. Moist Heat: Use of hot water or steam in **disinfection** to kill vegetative cells or in **sterilization** to kill spores. Mode of action: Denaturation of proteins, destruction of membranes and DNA.

1. *For sterilization:*
   a. **Autoclave** uses steam under pressure (15 psi/121°C/10–40 min). Applied on heat-resistant materials that steam can penetrate. Not recommended for oils, powders, heat-sensitive materials.
   b. **Intermittent sterilization** uses free-flowing, unpressurized steam at 100°C applied for 30 to 60 minutes on 3 successive days. Used for substances that cannot be autoclaved, especially media and certain canned foods.

2. *For disinfection:*
   a. **Pasteurization** is application of heat less than 100°C to liquids; employs the *flash method* to heat liquid to 71°C for 15 seconds. Targets non-spore-forming milk-borne pathogens, such as *Salmonella* and *Listeria,* and reduces spoilage by lowering the overall microbial count; also applied to beer and wine. Only reduces microbial content; many thermodurics survive.
   b. **Boiling** water at 100°C used to destroy pathogens (not spores) on household supplies; materials may become recontaminated.

B. Dry Heat: Use of hot air with low moisture content to sterilize.

1. Mode of action: Depending on the temperature, combusts, dehydrates, or coagulates proteins.

2. **Incineration:**
   a. *Open flame:* Materials are placed for a few seconds in an 800° to 1,800°C flame of Bunsen burner.
   b. *Furnace:* Incinerator chamber equipped with 600°–1,200°C flame that burns materials to ashes. Used for decontamination of hospital and industrial wastes; may add pollutants to atmosphere.

3. **Dry oven:** Closed chamber heated to 150°–180°C for 2 to 4 hours. Used for sterilizing metals and glass, but too vigorous for liquids, plastics, paper, cloth.

C. Cold Temperatures

1. Refrigeration (0°–15°C) or freezing (below 0°C) is not a reliable mode of disinfection. It is only **microbistatic,** meaning it slows the growth rate of most microbes.
2. Refrigeration is used for preserving foods, media, and cultures.

D. Drying/desiccation is the gradual withdrawal of water from cells by exposure to room air, leading to metabolic inhibition. Effective in preservation, but not reliable for pathogens, and not really an effective method for infection control.

E. Radiation/Irradiation

1. Use of energy in the form of waves (**electromagnetic**) and particles that can be transmitted through space. A method of **cold sterilization**.
2. **Ionizing** radiation uses high-energy, short waves and particles that can dislodge electrons from atoms. This causes direct damage to DNA by the formation of breaks and mutations. Types are *gamma rays, X rays, cathode rays* (high-speed electrons) used in **irradiation** and used to sterilize heat-sensitive medical materials in packages; also to sterilize and increase storage time of fresh products; very penetrating. Somewhat more expensive and dangerous than other methods.
3. **Nonionizing** radiation uses moderate energy, medium-length **ultraviolet waves** that excite, but do not ionize, atoms.
   a. Rays act on DNA molecule, forming **dimers** between adjacent **pyrimidines,** and produce toxic products.
   b. UV lamps disinfect air in medicine and industry; treat water, sera, vaccines, drugs; disinfect solid surfaces.
   c. Does not penetrate most solids and can damage human tissues (skin, eyes).

F. Filtration involves the physical removal of microbes from liquids and air by trapping them in fine filters. The fluid passes through tiny pores into a container. Can be used to remove viruses. Filters can sterilize heat-sensitive liquids (vaccines, serum, drugs, media, water). Can remove microbes from air in hospital rooms, isolation units, and "clean rooms."

II. Chemical Control of Microorganisms

A. General uses: **Disinfectants, antiseptics, sterilants,** preservatives, **sanitizers, degermers.**

B. Physical states: Solutions come in **aqueous** form (chemicals dissolved in water) and **tincture** form (chemicals dissolved in alcohol); solutions expressed as dilutions, percents, and parts per million.

C. Halogens

1. **Chlorine:** Forms are Cl_2, hypochlorites, chloramines.
   a. Denaturation of proteins by disrupting disulfide bonds; can be sporicidal with adequate time.
   b. Elemental chlorine (1–2 ppm) disinfects water by destroying pathogenic vegetative pathogens.
   c. Hypochlorites (chlorine bleach) are used extensively for disinfection and sanitization.
   d. Chloramines are clinical disinfectants and antiseptics and alternative water disinfection agents.
   e. Chemical action may be retarded by high levels of organic matter.

2. **Iodine:** Forms are free iodine (I_2) and iodophors (iodine complexed to organic polymers).
   a. Interfere with protein interchain bonds, causing denaturation; can be sporicidal.
   b. Weak solutions are topical antiseptics; iodine tincture is for a high degree of surgical asepsis and disinfection.
   c. Iodine solutions stain and corrode. Stronger solutions are too irritating and toxic to use on tissue.
   d. **Iodophors** are aqueous solutions with 2–10% iodine (Betadine, Povidone); they are milder medical and dental degerming agents, disinfectants, and ointments.

D. Phenolics

1. Phenols act on microbes by disrupting cell membranes and precipitating proteins. They are **bactericidal, fungicidal,** and **virucidal,** but not sporicidal.
2. Phenol is an older toxic disinfectant; 1–3% cresol in soap is a common housekeeping disinfectant and cleaner. Orthophenyl phenol (Lysol) is a milder phenolic used in air sprays and hospital disinfectants. Triclosan is an antibacterial additive to soaps, detergents, and numerous household products.
3. Most phenolics are too toxic for use as antiseptics and are relatively insoluble.

E. Chlorhexidine is available as Hibiclens and Hibitane.

1. Is a surfactant and protein denaturant, with bactericidal, some antiviral and antifungal effects, but not sporicidal.

2. An aqueous or alcohol solution is a skin degerming agent for preoperative scrubs, skin cleaning, and burns.
3. Product is relatively mild, nontoxic, and fast-acting.

F. Alcohols include ethyl and isopropyl, usually in solutions of 50–95%.
   1. Act as a **surfactant** to dissolve membrane lipids and coagulate proteins; work on bacterial vegetative cells and fungi, but are not sporicidal.
   2. Ethyl alcohol (70–95%) is a germicide in the clinic, laboratory; used at home for degerming, low-level disinfection, and sanitization.
   3. Isopropyl alcohol has similar uses, though it is more toxic and less safe.

G. Hydrogen peroxide is available in weak (3%) to strong (25%) solutions that produce highly active hydroxyl-free radicals and damage proteins and DNA molecules; also decomposes to water and O_2 gas, which can be toxic to anaerobes; strong solutions are sporicidal.
   1. 3% hydrogen peroxide is used for skin and wound antisepsis, care of mucous membrane infections; and disinfection of utensils.
   2. 6–25% can be used to sterilize equipment.
   3. Decomposes in presence of light and catalase.

H. Detergents and Soap
   1. Cationic detergents are known as **quaternary ammonium compounds** or **quats.** They are surfactants that alter cell permeability. Their action is limited to some bacteria and fungi; they are not sporicidal.
   2. Benzalkonium and cetylpyridinium chlorides are quats used in environmental disinfection, as cleansers in clinics and the food industry, and as preservatives.
   3. Soaps are not very microbicidal but function in the mechanical removal of grease and soil on skin, utensils, and environmental surfaces.
   4. Soaps do not destroy the tubercle bacillus or hepatitis B virus and are inactivated by large quantities of organic matter.

I. Heavy metal compounds include mercury and silver solutions and tinctures; work in high dilutions (**oligodynamic action**) and precipitate proteins; are germicidal but not sporicidal.
   1. Organic mercurial tinctures (thimerosal, nitromersol) are skin antiseptics and preservatives.
   2. Silver nitrate solutions are used as a topical antiseptic; silver sulfadiazine ointment can prevent burn infections; colloidal silver preparations are used for mouth and eye rinses.
   3. Metal solutions are highly toxic, cause allergies, and are neutralized by organic substances.

J. Aldehydes include **glutaraldehyde** and **formaldehyde** or formalin solutions. They kill cells by alkylation of amino and nucleic acids, with wide microbicidal action.
   1. Glutaraldehyde in 2% solutions (Cidex) is used as a **sterilant** for heat-sensitive instruments in the medical and dental office, and for some types of environmental disinfection.
   2. Formalin in 1–8% is limited to disinfection of some instruments, rooms, and as a preservative.
   3. Glutaraldehyde is somewhat unstable; formaldehyde is toxic and irritating.

K. Gases and aerosols include **ethylene oxide (ETO),** propylene oxide, and betapropiolactone, which inactivate nucleic acids and proteins. All are sporicidal, but only ETO is an approved **sterilant.**
   1. Ethylene oxide is added to a special chamber for sterilizing plastic supplies and to treat spices and dried foods.
   2. Propylene oxide can disinfect food (nuts, starch).
   3. Betapropiolactone is used to disinfect rooms and some biological materials.
   4. ETO is somewhat dangerous because of its explosiveness and toxicity, and it is slow-acting. Betapropiolactone is toxic and has poor penetration.

MULTIPLE-CHOICE QUESTIONS

1. A microbicidal agent has what effect?
   a. sterilizes
   b. inhibits microorganisms
   c. is toxic to human cells
   d. destroys microorganisms

2. Microbial control methods that kill _____ are able to sterilize.
   a. viruses   c. endospores
   b. the tubercle bacillus   d. cysts

3. Any process that destroys the non-spore-forming contaminants on inanimate objects is
   a. antisepsis   c. sterilization
   b. disinfection   d. degermation

4. Sanitization is a process by which
   a. the microbial load on objects is reduced
   b. objects are made sterile with chemicals
   c. utensils are scrubbed
   d. skin is debrided

5. An example of an agent that lowers the surface tension of cells is
   a. phenol   c. alcohol
   b. chlorine   d. formalin

6. High temperatures _____ and low temperatures _____ .
   a. sterilize, disinfect
   b. kill cells, inhibit cell growth
   c. denature proteins, burst cells
   d. speed up metabolism, slow down metabolism

7. The temperature-pressure combination for an autoclave is
   a. 100°C and 4 psi   c. 131°C and 9 psi
   b. 121°C and 15 psi   d. 115°C and 3 psi

8. Microbe(s) that is/are the target(s) of pasteurization include:
   a. *Clostridium botulinum*
   b. *Mycobacterium* species
   c. *Salmonella* species
   d. both b and c

9. Ionizing radiation removes _____ from atoms.
   a. protons   c. electrons
   b. waves   d. ions

10. The primary mode of action of nonionizing radiation is to
    a. produce superoxide ions
    b. make pyrimidine dimers
    c. denature proteins
    d. break disulfide bonds

11. The most versatile method of sterilizing heat-sensitive liquids is
    a. UV radiation
    b. exposure to ozone
    c. beta propiolactone
    d. filtration

12. _____ is the iodine antiseptic of choice for wound treatment.
    a. Eight percent tincture
    b. Five percent aqueous
    c. Iodophor
    d. Potassium iodide solution

13. A chemical with sporicidal properties is
    a. phenol
    b. alcohol
    c. quaternary ammonium compound
    d. glutaraldehyde

14. Silver nitrate is used
    a. in antisepsis of burns
    b. as a mouthwash
    c. to treat genital gonorrhea
    d. to disinfect water

15. Detergents are
    a. high-level germicides
    b. low-level germicides
    c. excellent antiseptics
    d. used in disinfecting surgical instruments

16. Which of the following is an approved sterilant?
    a. chlorhexidine
    b. betadyne
    c. ethylene oxide
    d. ethyl alcohol

CONCEPT QUESTIONS

1. Compare sterilization with disinfection and sanitization. Describe the relationship of the concepts of sepsis, asepsis, and antisepsis.

2. a. Explain the effect of a tuberculocide.
   b. What would be the effect of a pseudomonicide?
   c. What does a virustatic agent do?

3. a. Briefly explain how the type of microorganisms present will influence the effectiveness of exposure to antimicrobial agents.
   b. Explain how the numbers of contaminants can influence the measures used to control them.

4. a. Precisely what is microbial death?
   b. Why does a population of microbes not die instantaneously when exposed to an antimicrobial agent?

5. Why are antimicrobial processes inhibited in the presence of extraneous organic matter?

6. Describe four modes of action of antimicrobial agents, and give a specific example of how each works.

7. a. Summarize the nature, mode of action, and effectiveness of moist and dry heat.
   b. Compare the effects of moist and dry heat on vegetative cells and spores.

8. How can the temperature of steam be raised above 100°C? Explain the relationship involved.

9. a. What do you see as a basic flaw in tyndallization?
   b. In boiling water devices?
   c. In incineration?
   d. In ultrasonic devices?

10. What are several microbial targets of pasteurization?

11. Explain why desiccation and cold are not reliable methods of disinfection.

12. a. What are some advantages of ionizing radiation as a method of control?
    b. Some disadvantages?

13. a. What is the precise mode of action of ultraviolet radiation?
    b. What are some disadvantages to its use?

14. What are the superior characteristics of iodophors over free iodine solutions?

15. a. What does it mean to lower the surface tension?
    b. What cell parts will be most affected by surface active agents?

16. a. Name one chemical for which the general rule that a higher concentration is more effective is *not* true.
    b. What is a sterilant?

17. Name the principal sporicidal chemical agents.

18. Why is hydrogen peroxide solution so effective against anaerobes?

19. Give the uses and disadvantages of the heavy metal chemical agents, glutaraldehyde, and the sterilizing gases.

20. What does it mean to say that a chemical has an oligodynamic action?

21. a. Define cold sterilization.
    b. Name three totally different methods that qualify for this definition.

CRITICAL-THINKING QUESTIONS

1. What is wrong with this statement: "The patient's skin was sterilized with alcohol"? What would be the more correct wording?

2. For each item on the following list, give a reasonable method of sterilization. You cannot use the same method more than three times; the method must sterilize, not just disinfect; and the method must not destroy the item or render it useless. After considering a workable method, think of a method that would not work. Note: Where an object containing something is given, you must sterilize everything (for example, both the jar and the Vaseline in it). Some examples of

methods are autoclave, ethylene oxide gas, dry oven, and ionizing radiation.

room air	inside of a refrigerator
blood in a syringe	wine
serum	a jar of Vaseline
a pot of soil	a child's toy
plastic Petri plates	fruit in plastic bags
heat-sensitive drugs	rubber gloves
cloth dressings	disposable syringes

leather shoes from a thrift shop   talcum powder
a cheese sandwich   milk
human hair (for wigs)   orchid seeds
a flask of nutrient agar   metal instruments
an entire room (walls, floor, etc.)

3. a. Graph the data on tables 11.3 and 11.4, plotting the time on the Y axis and the temperature on the X axis for three different organisms.
   b. What conclusions can you draw regarding the effects of time and temperature on thermal death?
   c. Is there any difference between the graph for a sporeformer and the graph for a non-sporeformer?

4. Can you think of situations in which the same microbe would be considered a serious contaminant in one case and completely harmless in another?

5. A supermarket/drugstore assignment: With a partner, look at the labels of 25 different products used to control microbes, and make a list of their active ingredients, their suggested uses, and information on toxicity and precautions.

6. Devise an experiment that will differentiate between bacteriocidal and bacteriostatic effects.

7. There is quite a bit of concern that chlorine used as a water purification chemical presents serious dangers. What alternative methods covered by this chapter could be used to purify water supplies and yet keep them safe from contamination?

8. The shelf-life and keeping qualities of fruit and other perishable foods are greatly enhanced through irradiation, saving industry and consumers billions of dollars. How would you personally feel about eating a piece of irradiated fruit? How about spices that had been sterilized with ETO?

9. The microbial levels in saliva are astronomical ($<10^6$ cells per ml, on average). What effect do you think a mouthwash can have against this high number? Comment on the claims made for such products.

10. Can you think of some innovations the health care community can use to deal with medical waste and its disposal that prevent infection but are ecologically sound?

INTERNET SEARCH TOPIC

Look up information on an infectious agent called a prion with respect to how it is transmitted and its resistance. What methods of antimicrobial control are required to destroy it?

CHAPTER 12

Drugs, Microbes, Host— The Elements of Chemotherapy

T he subject of this chapter is antimicrobial chemotherapy—the use of chemicals to control or prevent infection. As you will see, the administration of a drug has an impact beyond simply the action of the drug on the microbe. It also includes any effects microbes exert against the drug and, because drugs work inside the body, interactions between the drug and the patient's tissues. The principal areas to be covered here are general concepts of chemotherapy; mechanisms of drug action; drug resistance; antibacterial, antifungal, antihelminthic, and antiviral drugs; adverse human reactions to drugs; methods of drug testing; and factors in drug selection.

The father of modern antibiotics. A Scottish physician, Sir Alexander Fleming, accidently discovered penicillin when his keen eye noticed that colonies of bacteria were being lysed by a fungal contaminant. He isolated the active ingredient and set the scene for the development of the drug ten years later.

Chapter Overview

- Antimicrobial chemotherapy is the treatment and prevention of infectious diseases by means of chemicals called drugs.
- Antimicrobial drugs include antibiotics derived from bacteria and fungi, and synthetic drugs produced by chemical reactions.
- Narrow-spectrum antimicrobial drugs affect a small range of microbes, and broad-spectrum drugs affect a wider range of microbes.
- Chemotherapy involves a complex interaction between the microbe, drug, and host.
- The best antimicrobial drugs have low toxicity to humans and lack other side effects such as drug resistance, allergy, and disruption of natural flora.
- The primary action of antimicrobial drugs is to interfere with some specific component of the microbe's structure, its enzymes, or synthesis of proteins and other molecules.
- Drug resistance is a process by which microbes develop genetic changes that allow them to circumvent the effects of a drug.
- Hundreds of drugs have been developed for treating bacterial, fungal, protozoan, helminthic, and viral infections.
- The predominant antibacterial drug classes are the penicillins, cephalosporins, tetracyclines, sulfa drugs, and fluoroquinolones.
- Selecting a drug for therapy is based upon the microbe's sensitivity to the drug, the drug's toxicity, and the health of the patient.
- Adverse side effects of drugs include damage to skin, liver, kidney, circulatory system, nervous system, and gastrointestinal tract.
- Drugs may cause allergies and disrupt the host's normal flora, leading to other infections.
- Antimicrobial drugs are often overprescribed, ineffective, taken in incorrect doses for too short a time, and broadcast into the environment in livestock feeds.
- People need to become aware of the correct guidelines for drug therapy as a way to protect drug diversity and prevent drug resistance.

Early human cultures relied on various types of primitive medications such as potions, poultices, and mudplasters. Many were concocted from plant, animal, and mineral products that had been found, usually through trial and error or accident, to have some curative effect upon ailments and complaints. In one ancient Chinese folk remedy, a fermented soybean curd was applied to skin infections. The Greeks used wine and plant resins (myrrh and frankincense), rotting wood, and various mineral salts to treat diseases. Some folk medicines were occasionally effective, but most of them were probably witches' brews that either had no effect or were even harmful. It is interesting that the Greek word *pharmakeutikos* originally meant the practice of witchcraft. These ancient remedies were handed down from generation to generation, but it was not until the Middle Ages that a specific disease was first treated with a specific chemical. Dosing syphilitic patients with inorganic arsenic and mercury compounds may have proved the ancient axiom: Graviora quaedum sunt remedia periculus. ("Some remedies are worse than the disease.")

An enormous breakthrough in the science of drug therapy came with the germ theory of infection by Robert Koch (see chapter 1). This allowed disease treatment to focus on a particular microbe, which in turn opened the way for Paul Ehrlich to formulate the first theoretical concepts in chemotherapy in the late 1800s. Ehrlich had observed that certain dyes affixed themselves to specific microorganisms and not to animal tissues. This observation led to the profound idea that if a drug was properly selective in its actions, it would zero in on and destroy a microbial target and leave human cells unaffected. His first discovery was an arsenic-based drug that was very toxic to the spirochete of syphilis but, unfortunately, to humans as well. Ehrlich systematically altered this parent molecule, creating numerous derivatives. Finally, on the 606th try, he arrived at a compound he called *salvarsan*. This drug had some therapeutic merit and was used for a few years, but it eventually had to be discontinued because it was still not selective enough in its toxicity. Ehrlich's work had laid important foundations for many of the developments to come.

Another pathfinder in early drug research was Gerhard Domagk, whose discoveries in the 1930s launched a breakthrough in therapy that marked the true beginning of broad-scale usage of antimicrobic drugs. Domagk showed that the red dye prontosil was chemically changed by the body into a compound with specific activity against bacteria. This substance was sulfonamide—the first *sulfa* drug. In a short time, the structure of this drug was determined, and it became possible to synthesize it on a wide scale and to develop scores of other sulfonamide drugs. Although these drugs had immediate applications in therapy (and still do), still another fortunate discovery was needed before the golden age of antibiotics could really blossom.

The discovery of antibiotics dramatically demonstrates how developments in science and medicine often occur through a combination of accident, persistence, collaboration, and vision. In 1928 in the London laboratory of Sir Alexander Fleming, a plate of *Staphylococcus aureus* became contaminated with the mold *Penicillium notatum*. Observing these plates, Fleming noted that the colonies of *Staphylococcus* were evidently being destroyed by some activity of the nearby *Penicillium* colonies (see chapter opening photo). Struck by this curious phenomenon, he extracted from the fungus a compound he called penicillin and showed that it was responsible for the inhibitory effects.

Although Fleming understood the potential for penicillin, he was unable to develop it. A decade after his discovery, English chemists Howard Florey and Ernst Chain worked out methods for industrial production of penicillin to help in the war effort. Clinical trials conducted in 1941 ultimately proved its effectiveness, and cultures of the mold were brought to the United States for an even larger-scale effort. When penicillin was made available to the world's population, at first it seemed a godsend. But in time, because of extreme overuse and misunderstanding of its capabilities, it also became the model for one of the most serious drug problems—namely, drug resistance. By the 1950s, the pharmaceutical industry had entered an era of drug research and development that soon made penicillin only one of a large assortment of antimicrobic drugs.

Principles of Antimicrobial Therapy

A hundred years ago in the United States, one out of three children was expected to die of an infectious disease before the age of five. Early death or severe lifelong debilitation from scarlet fever, diphtheria, tuberculosis, meningitis, and many other bacterial diseases was a fearsome yet undeniable fact of life to most of the world's population. The introduction of modern drugs to control infections in the 1930s began as a medical revolution that has added significantly to the lifespan and health of humans. It is no wonder that for many years, antibiotics in particular were regarded as the miracle cure-all for infectious diseases. In later discussions, we will evaluate this misconception in light of the shortcomings of chemotherapy. Although antimicrobic drugs have greatly reduced the incidence of certain infections, they have definitely not eradicated infectious disease and probably never will. In fact, in some parts of the world, mortality rates from infectious diseases are as high as before the arrival of antimicrobic drugs. Nevertheless, humans have been taking medicines to try to control diseases for thousands of years (Historical Highlights 12.1).

THE TERMINOLOGY OF CHEMOTHERAPY

Any chemical used in treatment, relief, or **prophylaxis*** of disease is defined as a **chemotherapeutic drug** or agent. When chemotherapeutic drugs are given as a means to control infection, the practice is termed **antimicrobial chemotherapy.*** Antimicrobial drugs

---

* prophylaxis (proh″-fih-lak′-sis) Gr. *prophylassein,* to keep guard before. A process that prevents infection or disease in a person at risk.

* chemotherapy (kee″-moh-ther′-uh-pee) Gr. *chemieia,* chemistry, and *therapeia,* service to the sick. Use of drugs to treat disease.

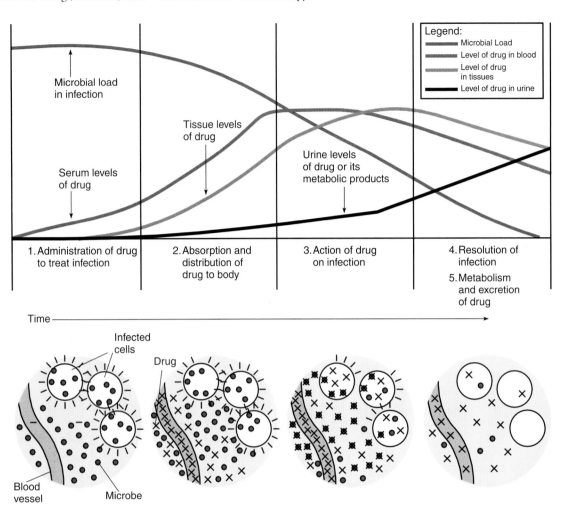

FIGURE 12.1

Anti-infective therapy. The course of events in chemotherapy involves an interaction between patient, drug, and infectious agent. The top part of the figure compares variations in infection intensity and drug levels over time. The lower part represents happenings at the microscopic level in tissues during this same time period.

(also termed anti-infective drugs) are a special class of compounds capable even in high dilutions of destroying or inhibiting microorganisms. The origin of modern antimicrobial drugs is varied. The **antibiotics*** are substances produced by the natural metabolic processes of some microorganisms that can inhibit or destroy other microorganisms. **Synthetic** antimicrobial drugs are derived in the laboratory from dyes or other organic compounds, through chemical reactions. Although separation into these two categories has been traditional, they tend to overlap, because most antibiotics, termed *semisynthetic,* are now chemically altered in the laboratory. The current trend is to use the term **antimicrobic** for all antimicrobial drugs, regardless of origin.

Antimicrobial drugs vary in their scope of activity. The so-called **narrow-spectrum** agents are effective against a limited array of different microbial types. Examples are bacitracin, an antibiotic whose inhibitory effects extend mainly to certain gram-positive bacteria, or griseofulvin, which is used chiefly in fungal skin infections. **Broad-spectrum** or *extended-spectrum* agents are active against a wider range of different microbes. The targets of antibiotics in the tetracycline group, for example, are a variety of gram-positive and gram-negative bacteria, rickettsias, mycoplasmas, and even protozoa. The concept of spectrum of activity has also become blurred to some extent, because an agent can be induced by artificial means to change its spectrum.

DRUG, MICROBE, HOST—SOME BASIC INTERACTIONS

The broad concept of anti-infective therapy revolves around three interacting factors: the drug, the microorganism, and the infected host. To put it simply, the drug should destroy the infectious agent without harming the host's cells. This process can be visualized more tangibly as a series of stages (figure 12.1): (1) The drug is administered to the host via a designated route. Delivery is primarily by oral, circulatory, muscular, and cutaneous routes. (2) The drug is dissolved in body fluids. (3) The drug is delivered to the infected area (extracellular or intracellular). (4) The drug destroys the infectious agent or inhibits its growth. (5) The drug is eventually

---

* **antibiotic** (an-tee´-by-aw″-tik) Gr. *anti,* against, and *bios,* life.

TABLE 12.1
Characteristics of the Ideal Antimicrobial Drug

- Selectively toxic to the microbe but nontoxic to host cells
- Microbicidal rather than microbistatic
- Relatively soluble and functions even when highly diluted in body fluids
- Remains potent long enough to act and is not broken down or excreted prematurely
- Not subject to the development of antimicrobial resistance
- Complements or assists the activities of the host's defenses
- Remains active in tissues and body fluids
- Readily delivered to the site of infection
- Not excessive in cost
- Does not disrupt the host's health by causing allergies or predisposing the host to other infections

excreted or broken down by the host's organs, ideally without harming them. Among the factors that affect the outcome of drug therapy are those that aid or block drug action and those that influence the behavior of microbe or host tissues. Favorable features of antimicrobial drugs in relation to the infectious agents and host are presented in table 12.1.

THE ORIGINS OF ANTIMICROBIAL DRUGS

Nature is undoubtedly the most prolific producer of antimicrobial drugs. Antibiotics, after all, are common metabolic products of aerobic spore-forming bacteria and fungi. By inhibiting the growth of other microorganisms in the same habitat (antagonism), antibiotic producers presumably enjoy less competition for nutrients and space. The greatest numbers of antibiotics are derived from bacteria in the genera *Streptomyces* and *Bacillus* and molds in the genera *Penicillium* and *Cephalosporium*. Chemists have sought to decipher the structure of the more useful antibiotics in order to produce related semisynthetic compounds with specialized features (Microbits 12.2).

CHAPTER CHECKPOINTS

Antimicrobial chemotherapy involves the use of drugs to control infection on or in the body.

Antimicrobial drugs are produced either synthetically or from natural sources. They inhibit or destroy microbial growth in the infected host. Antibiotic drugs are the subset of antimicrobics produced by the natural metabolic processes of microorganisms.

Antimicrobial drugs are classified by their range of effectiveness. Broad-spectrum antimicrobics are effective against many types of microbes. Narrow-spectrum antimicrobics are effective against a limited group of microbes.

Antimicrobial therapy involves three interacting factors: the drug, the microbe, and the infected host.

Spore-forming bacteria and fungi are the primary sources of most antibiotics. The molecular structures of these compounds can be chemically altered to form additional semisynthetic antimicrobics.

Interactions Between Drug and Microbe

The primary effect of antimicrobial drugs is either to disrupt the cell processes or structures of bacteria, fungi, and protozoa or to inhibit virus replication. Most of the drugs used in chemotherapy interfere with the function of enzymes required to synthesize or assemble macromolecules, or they destroy structures already formed in the cell. Above all, drugs should be **selectively toxic,** which means they should kill or inhibit microbial cells without simultaneously damaging host tissues. This concept of selective toxicity is central to chemotherapy, and the best drugs are those that block the actions or synthesis of molecules in microorganisms but not in vertebrate cells. Examples of drugs with selective toxicity are those that block the synthesis of the cell wall in bacteria (penicillins). They have low toxicity and few direct effects on human cells because human cells lack a wall and are thus neutral to this action of the antibiotic. Among the most toxic to human cells are drugs that act upon a common structure such as the cell membrane (amphotericin B, for example). In a later section, we will address the circumstances in which the ideal of selective toxicity is not met.

MECHANISMS OF DRUG ACTION

Antimicrobial drugs injure microbes through several different mechanisms. Microbicidal drugs lyse and kill microorganisms by inflicting direct damage upon specific cellular targets. Microbistatic drugs interfere with cell division, and thus inhibit reproduction. Drugs that only inhibit growth are not considered directly responsible for subsequent microbial death; their primary importance is to keep the microbes from multiplying and prevent their spread, thus allowing the host defenses an opportunity to destroy and remove the infectious agent.

Antimicrobial drugs function specifically in one of the following ways (figure 12.2): (1) They inhibit cell wall synthesis; (2) they inhibit nucleic acid synthesis or function; (3) they inhibit protein synthesis; or (4) they interfere with the function of the cell membrane. These categories are not completely discrete, and some effects can overlap. In chapter 11 we presented an overall picture of antimicrobial mechanisms; here, we include selected examples of antimicrobics that will illustrate the general ways that drugs act upon microbial cells. It is worth noting that much knowledge of cellular structure and function has emerged as an added bonus from research on the effects of drugs.

Antimicrobial Drugs That Affect the Bacterial Cell Wall

The cell walls of most bacteria contain a rigid girdle of peptidoglycan, which protects the cell against rupture from hypotonic environments. Active cells must constantly synthesize new peptidoglycan and transport it to its proper place in the cell envelope. Drugs such as penicillins and cephalosporins react with one or more of the enzymes required to complete this process, causing the cell to develop weak points at growth sites and to become osmotically fragile (figure 12.3). Antibiotics that produce this effect are considered bactericidal, because the weakened cell is subject to lysis. It is essential to note that most of these antibiotics are active only in young, growing cells, because old, inactive, or dormant cells do not

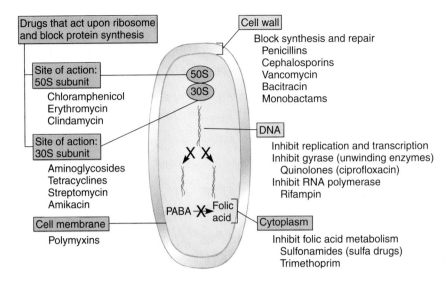

FIGURE 12.2

Primary sites of action of antimicrobic drugs on bacterial cells. See text for more discussion of mechanisms.

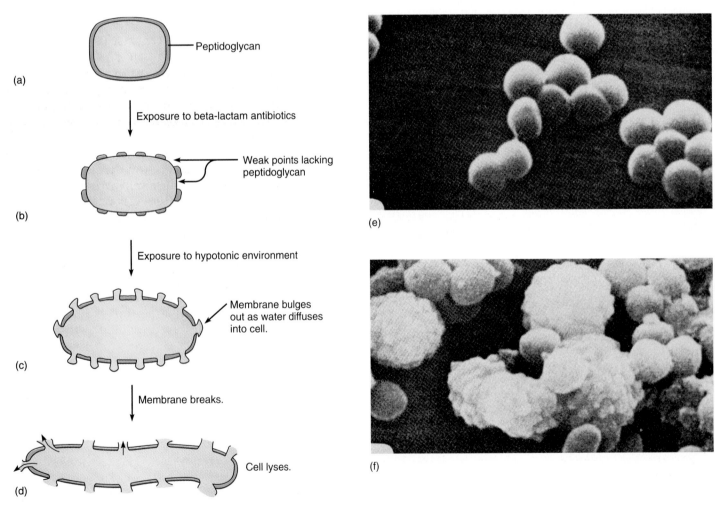

FIGURE 12.3

The consequences of exposing a growing cell to antibiotics that prevent cell wall synthesis. (a) Coccus is exposed to beta-lactam agent (cephalosporin). **(b)** Weak points develop where peptidoglycan is incomplete. **(c)** The weakened cell is exposed to a hypotonic environment. **(d)** The cell lyses. **(e)** Scanning electron micrograph of bacterial cells in their normal state (10,000×). **(f)** Scanning electron micrograph of the same cells in the drug-affected state, showing surface bulges (10,000×).

MICROBITS 12.2
A Modern Quest for Designer Drugs

Once the first significant drug was developed, the world immediately witnessed a scientific scramble to find more antibiotics. This search was advanced on several fronts. Hundreds of investigators began the laborious task of screening samples from soil, dust, muddy lake sediments, rivers, estuaries, oceans, plant surfaces, compost heaps, sewage, skin, and even the hair and skin of animals for antibiotic-producing bacteria and fungi. This intense effort has paid off over the past 50 years, because more than 10,000 antibiotics have been discovered (though surprisingly, only a relatively small number have been applicable to chemotherapy). Finding a new antimicrobial substance is only a first step. The complete pathway of drug development from discovery to therapy takes between 12 and 24 years at a cost of billions of dollars.

Antibiotics are products of fermentation pathways that occur in many spore-forming bacteria and fungi. The role of antibiotics in the lives of these microbes must be important since the genes for antibiotic production are preserved in evolution. Some experts theorize that antibiotic-releasing microorganisms can inhibit or destroy nearby competitors or predators; others propose that antibiotics play a part in spore formation. Whatever benefit the microbes derive, these compounds have been extremely profitable for humans. Every year, the pharmaceutical industry farms vast quantities of microorganisms and harvests their products to treat diseases caused by other microorganisms. Researchers have facilitated the work of nature by selecting mutant species that yield more abundant or useful products, by varying the growth medium, or by altering the procedures for large-scale industrial production (see chapter 26).

Another approach in the drug quest is to chemically manipulate molecules by adding or removing functional groups. Drugs produced in this way are designed to have advantages over other, related drugs. Using the **semisynthetic** method, a natural product of the microorganism is joined with various preselected functional groups. The antibiotic is reduced to its basic molecular framework (called the nucleus), and to this nucleus specially selected side chains (R groups) are added. A case in point is the metamorphosis of the semisynthetic penicillins. The nucleus is an inactive penicillin derivative called aminopenicillanic acid, which has an opening on the number 6 carbon for addition of R groups. A particular carboxylic acid (R group) added to this nucleus can "fine-tune" the penicillin, giving it special characteristics. For instance, some R groups will make the product resistant to penicillinase (methicillin), some confer a broader activity spectrum (ampicillin), and others make the product acid-resistant (penicillin V). Cephalosporins and tetracyclines also exist in several semisynthetic versions. The potential for using bioengineering techniques to design drugs seems almost limitless, and, indeed, several drugs have already been produced by manipulating the genes of antibiotic producers. Some investigators are looking for new types of drugs in plants and animals (see Medical Microfile 1.3).

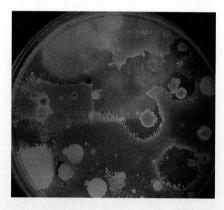

A plate with several discrete colonies of soil bacteria was sprayed with a culture of *Escherichia coli* and incubated. Zones of inhibition (clear areas with no growth) surrounding several colonies indicate species that produce antibiotics.

Synthesizing penicillins. **(a)** The original penicillin G molecule is a fermentation product of *Penicillium chrysogenum* that appears somewhat like a house with a removable patio on the left. This house without the patio is the basic nucleus called aminopenicillanic acid. **(b–d)** Various new fixtures (R groups) can be added to change the properties of the drug. These R groups will produce different penicillins: **(b)** methicillin; **(c)** ampicillin; and **(d)** penicillin V.

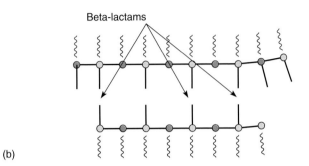

(a)

(b)

FIGURE 12.4

The mode of action of beta-lactam antibiotics on the bacterial cell wall.
(a) Intact peptidoglycan has chains of NAM (N-acetyl muramic acid) and
NAG (N-acetyl glucosamine) glycans cross-linked by peptide bridges.
(b) The beta-lactam antibiotics block the peptidases that link the cross-
bridges between NAMs, thereby greatly weakening the cell wall meshwork.

synthesize peptidoglycan. (One exception is a new class of antibiotics called the penems.)

Cycloserine inhibits the formation of the basic peptidoglycan
subunits, and vancomycin hinders the elongation of the peptidoglycan. The beta-lactams (penicillins and cephalosporins) bind and
block peptidases that cross-link the glycan molecules, thereby

interrupting the completion of the cell wall (figure 12.4). Penicillins
that do not penetrate the outer membrane are less effective against
gram-negative bacteria, but broad-spectrum penicillins and
cephalosporins can cross the cell walls of gram-negative species.

Antimicrobial Drugs That Affect Nucleic Acid Synthesis

As you will recall from chapter 9, the metabolic pathway that generates DNA and RNA is a long, enzyme-catalyzed series of reactions. Like any complicated process, it is subject to breakdown at
many different points along the way, and inhibition at any given
point in the sequence can block subsequent events. Antimicrobial
drugs interfere with nucleic acid synthesis by blocking synthesis of
nucleotides, inhibiting replication, or stopping transcription. Because functioning DNA and RNA are required for proper translation as well, the effects on protein metabolism can be far-reaching.

Sulfonamides One of the more extensively studied modes of action is that of the **sulfonamides** (sulfa drugs). Because these synthetic drugs interfere with an essential metabolic process in bacteria,
they represent a model for **competitive inhibition.** They act as
structural or **metabolic analogs*** that mimic the natural substrate of
an enzyme and vie for its active site. In practice, sulfa drugs are
very similar to the natural metabolic compound PABA (para-
aminobenzoic acid) required by bacteria to synthesize the coenzyme
tetrahydrofolic acid, which participates in the synthesis of purines
and certain amino acids. A sulfonamide molecule has high affinity
for the PABA site on the enzyme and can successfully compete in a
"chemical race" with PABA to occupy those sites (figure 12.5).

---

* analog (an'-uh-log) A compound whose configuration closely resembles another
compound required for cellular reactions.

FIGURE 12.5

🔘 **The mode of action of sulfa drugs.** This
class of drugs acts as a structural analog of
PABA, a molecule that is required for a step in
folic acid synthesis (upper panel). Sulfa drugs
can bind to the same active site on the enzyme
because of its similar shape. Because sulfa is the
incorrect substrate, it cannot be added to the folic
acid complex, and further synthesis stops (lower
panel).

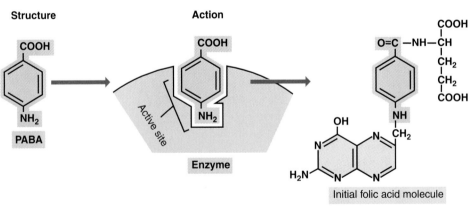

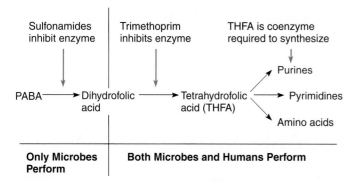

FIGURE 12.6

🔘 **A scheme for the synthesis and function of folic acid and points of inhibition by antimetabolites.** Sulfa drugs (sulfonamides) inhibit the first step, and trimethoprim blocks the second; together, they have a synergistic action that inhibits the entire pathway leading to nitrogen base and amino acid synthesis.

Sulfonamides ultimately cause an inadequate supply of tetrahydrofolic acid for purine production, which invariably halts nucleic acid synthesis and prevents bacterial cells from multiplying.

Sulfonamides are valuable in therapy because they inhibit bacteria and certain fungi, but not mammalian cells. This one-sided inhibition stems from a basic nutritional difference between humans and microorganisms. Although humans require tetrahydrofolic acid for nucleic acid synthesis as much as bacteria do, humans cannot synthesize it because human cells lack this special enzymatic system. Thus, it is an essential nutrient (vitamin) that must come from the diet, and human metabolism cannot be inhibited by sulfa drugs.

Other Nucleic Acid Inhibitors   Trimethoprim is a metabolic analog that competes for a site on another enzyme needed for the process of folic acid synthesis (figure 12.6). Blocking this enzyme also interrupts the synthesis of purines and pyrimidines. Trimethoprim is often given simultaneously with sulfonamides to achieve a **synergistic** effect. In pharmacology, this refers to an additive effect achieved by two drugs working together, thus requiring a lower dose of each.

Other antimicrobics inhibit DNA synthesis. Chloroquine (an antimalarial drug) binds and cross-links the double helix. The newer broad-spectrum quinolones inhibit DNA unwinding enzymes or helicases, thereby stopping DNA transcription. Antiviral drugs that are analogs of purines and pyrimidines (including azidothymidine [AZT] and acyclovir) insert in the viral nucleic acid and block further replication.

Drugs That Block Translation

Most inhibitors of translation, or protein synthesis, react with the ribosome-mRNA complex. Although human cells also have ribosomes, the ribosomes of eucaryotes are different in size and structure from those of procaryotes, so these antimicrobics usually have a selective action against bacteria. One potential therapeutic consequence of drugs that bind to the procaryotic ribosome is the damage they can do to eucaryotic mitochondria, which contain a procaryotic type of ribosome. Two possible targets of ribosomal inhibition are the 30S subunit and the 50S subunit (figure 12.7). Aminoglycosides

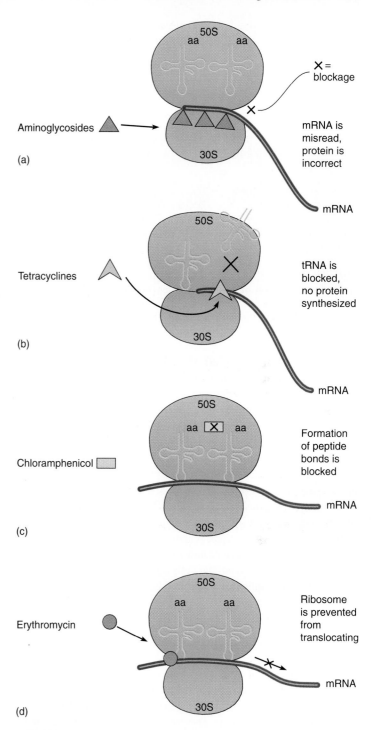

FIGURE 12.7

🔘 **Sites of inhibition on the procaryotic ribosome and major antibiotics that act on these sites.** All have the general effect of blocking protein synthesis. Blockage actions are indicated by X.

(streptomycin, gentamicin, for example) insert on sites on the 30S subunit and cause the misreading of the mRNA, leading to abnormal proteins. Tetracyclines block the attachment of tRNA on the A acceptor site and effectively stop further synthesis. Other antibiotics attach to sites on the 50S subunit in a way that prevents the formation of peptide bonds (chloramphenicol) or inhibits translocation of the subunit during translation (erythromycin).

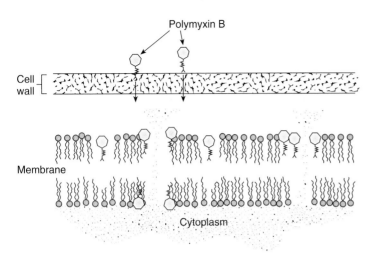

FIGURE 12.8

The detergent action of polymyxin. After passing through the cell wall of gram-negative bacteria, polymyxin binds to the cell membrane and forms abnormal openings that cause the membrane to become leaky.

Drugs That Disrupt Cell Membrane Function

A cell with a damaged membrane invariably dies from disruption in metabolism or lysis and does not even have to be actively dividing to be destroyed. The antibiotic classes that damage cell membranes have specificity for a particular microbial group, based on differences in the types of lipids in their cell membranes.

Polymyxins interact with membrane phospholipids, distort the cell surface, and cause leakage of proteins and nitrogen bases, particularly in gram-negative bacteria (figure 12.8). The polyene antifungal antibiotics (amphotericin B and nystatin) form complexes with the sterols on fungal membranes; these complexes cause abnormal openings and seepage of small ions. Unfortunately, this selectivity is not exact, and the universal presence of membranes in microbial and animal cells alike means that most of these antibiotics can be quite toxic to humans.

THE ACQUISITION OF DRUG RESISTANCE

The wide-scale use of antimicrobics soon led to microbial **drug resistance,** an adaptive response in which microorganisms begin to tolerate an amount of drug that would ordinarily be inhibitory. The development of mechanisms for circumventing or inactivating antimicrobic drugs is due largely to genetic versatility and adaptability of microbial populations. The property of drug resistance can be intrinsic as well as acquired. Intrinsic drug resistance exists naturally and is not acquired through specific genetic changes. In our context, the term *drug resistance* will mean resistance acquired by microbes that had formerly been sensitive to the drug.

How Does Drug Resistance Develop?

The genetic events most often responsible for drug resistance are chromosomal mutations or extrachromosomal DNA that are transferred from a resistant species to a sensitive one. Chromosomal drug resistance usually results from spontaneous random mutations in bacterial populations. The chance that such a mutation will be advantageous is minimal, and the chance that it will confer resistance to a specific drug is lower still. Nevertheless, given the huge

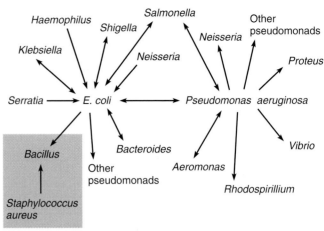

FIGURE 12.9

Spread of resistance factors. This figure traces documented evidence of known cases in which R factors have been transferred among pathogens. Such promiscuous exchange of drug resistance occurs primarily by conjugation and transduction. Most of the bacteria are gram-negative, but *Bacillus* and *Staphylococcus* are unrelated gram-positive genera. This phenomenon is responsible for the rapid spread of drug-resistant microbes.

Source: Data from Young and Mayer, Review of Infectious Diseases, *1:55, 1979.*

numbers of microorganisms in any population and the constant rate of mutation, such mutations do occur. The end result varies from slight changes in microbial sensitivity, which can be overcome by larger doses of the drug, to complete loss of sensitivity.

Resistance associated with intermicrobial transfer originates from plasmids called **resistance (R) factors** that are transferred through conjugation, transformation, or transduction. Studies have shown that plasmids encoded with drug resistance are naturally present in microorganisms before they have been exposed to the drug. Such traits are "lying in wait" for an opportunity to be expressed and to confer adaptability on the species. Many bacteria also maintain transposable drug resistance sequences (transposons) that are duplicated and inserted from one plasmid to another or from a plasmid to the chromosome. Chromosomal genes and plasmids containing codes for drug resistance are faithfully replicated and inherited by all subsequent progeny. This sharing of resistance genes accounts for the rapid proliferation of drug-resistant species (figure 12.9). A growing body of evidence points to the ease and frequency of gene transfers in nature, from totally unrelated bacteria living in the body's normal flora and the environment.

Specific Mechanisms of Drug Resistance

In general, a microorganism loses its sensitivity to a drug by expressing genes that stop the action of the drug. Gene expression takes the form of: (1) synthesis of enzymes that inactivate the drug, (2) decrease in cell permeability and uptake of the drug, (3) change in the number or affinity of the drug receptor sites, or (4) modification of an essential metabolic pathway. Some bacteria can become resistant indirectly by lapsing into dormancy, or, in the case of penicillin, by converting to a cell-wall-deficient form (L form) that penicillin cannot affect.

(a) Drug inactivation

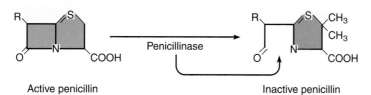

Active penicillin   Inactive penicillin

> Inactivation of a drug like penicillin by penicillinase, an enzyme that cleaves a portion of the molecule and renders it inactive.

(b) Decreased permeability/change in shape of receptor

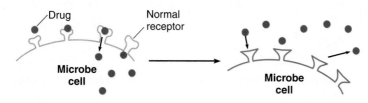

> The receptor that transports the drug is altered, so that the drug cannot enter the cell.

(c) Activation of drug pumps

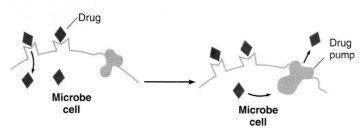

> Specialized membrane proteins are activated and continually pump the drug out of the cell.

(d) Use of alternate metabolic pathway

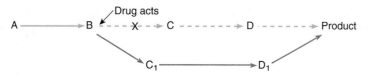

> The drug has blocked the usual metabolic pathway, so the microbe circumvents it by using an alternate, unblocked pathway that achieves the required outcome.

FIGURE 12.10

Examples of mechanisms of acquired drug resistance.

Drug Inactivation Mechanisms   Microbes inactivate drugs by producing enzymes that permanently alter drug structure. One example, bacterial exoenzymes called **beta-lactamases,** hydrolyze the *beta-lactam** ring structure of some penicillins and cephalosporins. Two beta-lactamases—*penicillinase* and *cephalosporinase*—disrupt the structure of certain penicillin or cephalosporin molecules so their activity is lost (figure 12.10a). So many strains of *Staphylococcus aureus* produce penicillinase that regular penicillin is rarely a possible therapeutic choice. Now that some strains of *Neisseria gonorrhoeae,* called **PPNG,**[1] have also acquired penicillinase, alternative drugs are required to treat gonorrhea. A large number of other gram-negative species are inherently resistant to some of the penicillins and cephalosporins because of naturally occurring beta-lactamases.

Decreased Drug Permeability or Increased Drug Transport   The resistance of some bacteria can be due to a mechanism that prevents the drug from entering the cell and acting on its target. For example, the outer membrane of the cell wall of certain gram-negative bacteria is a natural blockade for some of the penicillin drugs. Resistance to the tetracyclines can arise from plasmid-encoded proteins that pump the drug out of the cell. Resistance to the aminoglycoside antibiotics is a special case in which microbial cells have lost the capacity to transport the drug intracellularly (figure 12.10b).

Many bacteria possess **multidrug resistant (MDR) pumps** that actively transport drugs and other chemicals out of cells. These pumps are proteins encoded by plasmids and chromosomes. They are stationed in the cell membrane and expel molecules by a proton-motive force similar to ATP synthesis (figure 12.10c). They confer drug resistance on many gram-positive pathogens *(Staphylococcus, Streptococcus)* and gram-negative pathogens *(Pseudomonas, E. coli).* Because they lack selectivity, one type of pump can expel a broad array of antimicrobial drugs, detergents, and other toxic substances.

---

1. Penicillinase-producing *Neisseria gonorrhoeae.*

* beta-lactam (bay'-tuh-lak'-tam) Molecular structure shown in figure 12.12.

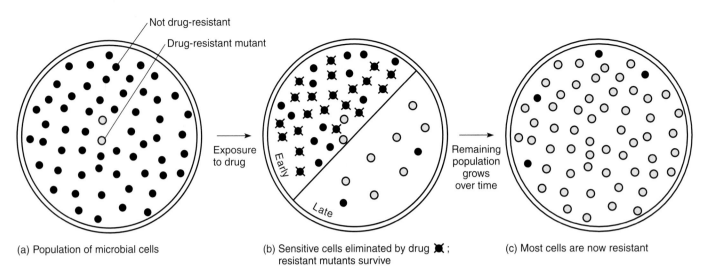

(a) Population of microbial cells

(b) Sensitive cells eliminated by drug ;
resistant mutants survive

(c) Most cells are now resistant

FIGURE 12.11

The events in natural selection for drug resistance. (a) Populations of microbes can harbor some members with a prior mutation that confers drug resistance. **(b)** Environmental pressure (here, the presence of the drug) selects for survival of these mutants. **(c)** They eventually become the dominant members of the population.

Change of Drug Receptors

Because most drugs act on a specific target such as protein, RNA, DNA, or membrane structure, microbes can circumvent drugs by altering the nature of this target. In bacteria resistant to rifampin and streptomycin, the structure of key proteins has been altered so that these antibiotics can no longer bind (figure 12.10*b*). Erythromycin and clindamycin resistance is associated with an alteration on the 50S ribosomal binding site. Penicillin resistance in *Streptococcus pneumoniae* and methicillin resistance in *Staphylococcus aureus* are related to an alteration in the binding proteins in the cell wall. Fungi can become resistant by decreasing their synthesis of ergosterol, the principal receptor for certain antifungal drugs.

Changes in Metabolic Patterns

The action of antimetabolites can be circumvented if a microbe develops an alternative metabolic pathway or enzyme (figure 12.10*d*). Sulfonamide and trimethoprim resistance develops when microbes deviate from the usual patterns of folic acid synthesis. Fungi can acquire resistance to flucytosine by completely shutting off certain metabolic activities.

Natural Selection and Drug Resistance

So far, we have been considering drug resistance at the cellular and molecular levels, but its full impact is felt only if this resistance occurs throughout the cell population. Let us examine how this might happen and its long-term therapeutic consequences. Recall that any large population of microbes is likely to contain a few individual cells that are already drug-resistant because of prior mutations or transfer of plasmids (figure 12.11*a*). As long as the drug is not present in the habitat, the numbers of these resistant forms will remain low because they have no particular growth advantage. But if the population is subsequently exposed to this drug (figure 12.11*b*), sensitive individuals are inhibited or destroyed, and resistant forms survive and proliferate. During subsequent population growth, all offspring of these resistant microbes will inherit this drug resistance. In time, the replacement population will have a preponderance of the drug-resistant forms and can eventually become completely resistant (figure 12.11*c*). In ecological terms, the environmental factor (in this case, the drug) has put selection pressure on the population, allowing the more "fit" microbe (the drug-resistant one) to survive, and the population has evolved to a condition of drug resistance. Natural selection for drug-resistant forms is apparently a common phenomenon. It takes place most frequently in various natural habitats, laboratories, and medical environments, but it occasionally occurs within the bodies of humans and animals during drug therapy (Spotlight on Microbiology 12.3).

CHAPTER CHECKPOINTS

Microorganisms are termed drug-resistant when they are no longer inhibited by an antimicrobic to which they were previously sensitive.

Drug resistance is genetic; microbes develop or acquire genes that code for methods of inactivating or escaping the antimicrobic. Resistance is selected for in environments where antimicrobics are present in high concentrations, such as in hospitals.

Microbial drug resistance develops through the selection of preexisting random mutations and through acquisition of resistance genes from other microorganisms.

Varieties of microbial drug resistance include drug inactivation, decreased drug uptake, decreased drug receptor sites, and modification of metabolic pathways formerly attacked by the drug.

Widespread indiscriminate prescribing of antimicrobics has resulted in an explosion of microorganisms resistant to all common drugs, including the following: *Streptococcus*, *Staphylococcus*, all the Enterobacteriaceae, *Salmonella*, *Shigella*, and *Mycobacterium tuberculosis*.

SPOTLIGHT ON MICROBIOLOGY 12.3
The Rise of Drug Resistance

Evolution of Antimicrobial Therapy in a Nutshell

Year 2000 B.C. "Here, eat this root."

Year A.D. 1000 "That root is heathen.
  Here, say this prayer."

Year 1850 "That prayer is superstitious.
  Here, drink this potion."

Year 1920 "That potion is snake oil.
  Here, swallow this pill."

Year 1945 "That pill is ineffective.
  Here, take this penicillin."

Year 1955 "OOPS, bugs mutated.
  Here, take this tetracycline."

Years 1960–1999 Thirty-nine more OOPS's.
  "Here, take this more powerful antibiotic."

Year 2000 The bugs have won! "Here, eat this root."

Anonymous observation from the *World Health Report on Infectious Diseases, 2000*

There has been an unrealistic tendency to assume that science would come to the rescue and solve the problem of drug resistance. If drug companies just keep making more and better antimicrobics, soon infectious diseases will be vanquished. This unfortunate attitude has vastly underestimated the extreme versatility and adaptability of microorganisms and the complexity of the task. It is a fact of nature that if you expose a large number of microbes to a variety of drugs, there will always be some genetically favored individuals that survive and thrive. The AIDS virus (HIV) is so prone to drug resistance that it can become resistant during the first few weeks of therapy in one individual. Because it mutates so rapidly, this means that in most cases, it will eventually become resistant to all drugs that have been developed so far.

Ironically, thousands of patients die every year in the United States from infections that lack effective drugs, and 60% of hospital infections are caused by drug-resistant microbes. For many years, concerned observers reported the gradual development of drug resistance in staphylococci, *Salmonella,* and gonococci. But during the past decade, the scope of the problem has escalated. It is now a common event to discover microbes that have become resistant to relatively new drugs in a very short time. In fact, many strains of pathogens have multiple drug resistance, and a few are resistant to all drugs. Following is a table of some notable examples:

TABLE 12.A

Organism	Drug	Year and Prevalence of Resistance	
Pneumococci		**1989**	**2000**
	Penicillin	Rare	30%
	Erythromycin	0	15
	Cephalosporin	0	14
		1990	**2000**
Gonococci	Penicillin	10	70%
	Fluoroquinolone	1.4%	10%
Enterococci	Vancomycin	Rare	50%
		1995	**2000**
Campylobacter	Fluoroquinolones	Rare	10%

The Hospital Factor

The clinical setting is a prolific source of drug-resistant strains of bacteria. This environment continually exposes pathogens to a variety of drugs, since drugs are inevitably released during therapy. The hospital also maintains patients with weakened defenses, making them highly susceptible to pathogens. A classic example occurred with *Staphylococcus aureus* and penicillin. In the 1950s, hospital strains began to show resistance to this drug, and because of indiscriminate use, they became nearly 100% resistant in 30 years. In a short time, MRSA (multiply-resistant *S. aureus*) strains appeared, which can tolerate nearly all antibiotics. Up until recently, MRSA has been sensitive to the drug of last resort, vancomycin, but due to transfer of resistant plasmids, it appears to be showing a level of resistance to this as well. To complicate this problem, strains of MRSA are now being spread into the community.

Drugs in Animal Feeds

Another practice that has contributed significantly to growing drug resistance is the addition of antimicrobics to livestock feed, with the idea of decreasing infections and thereby improving animal health. This practice has had serious impact in both the United States and Europe. Enteric bacteria such as *Salmonella, Escherichia coli,* and enterococci that live as normal intestinal flora of these animals readily share resistance plasmids and are

(continued on next page)

Survey of Major Antimicrobial Drug Groups

Scores of antimicrobial drugs are marketed in the United States. Although the medical and pharmaceutical literature contains a wide array of names for antimicrobics, most of them are variants of a small number of drug families. About 260 different antimicrobial drugs currently classified in 20 drug families. Drug reference books may give the impression that there are 10 times that many because various drug companies assign different trade names to the very same generic drug. Ampicillin, for instance, is available under 50 different names. Most antibiotics are useful in controlling bacterial infections, though we shall also consider a number of antifungal, antiviral, and antiprotozoan drugs. Table 12.2 summarizes some major infectious agents, the diseases they cause, and the drugs indicated to treat them (drugs of choice).

constantly selected and amplified by exposure to drugs. These pathogens subsequently "jump" to humans and cause drug-resistant infections, oftentimes at epidemic proportions. In a deadly outbreak of *Salmonella* infection in Denmark, the pathogen was found to be resistant to seven different antimicrobics. In the United States, a strain of fluoroquinolone-resistant *Campylobacter* from chickens caused over 5,000 cases of food infection in the late 1990s. The opportunistic pathogen called VRE (vancomycin-resistant enterococcus) has been traced to the use of a vancomycin-like drug in cattle feed. It is now one of the most tenacious of hospital-acquired infections for which there are few drug choices. To attempt to curb this source of resistance, Europe and the United States have banned the use of human drugs in animal feeds.

Worldwide Drug Resistance

The drug dilemma has become a widespread problem, affecting all countries and socioeconomic groups. In general, the majority of infectious diseases, whether bacterial, fungal, protozoan, or viral, are showing increases in drug resistance. In parts of India, the main drugs used to treat cholera (furazolidone, ampicillin) have gone from being highly effective to essentially useless in 10 years. In Southeast Asia 98% of gonococcus infections are multidrug resistant. Malaria, tuberculosis, and typhoid fever pathogens are gaining in resistance, with few alternate drugs to control them. To add to the problem, global travel and globalization of food products means that drug resistance can be rapidly exported.

In countries with adequate money to pay for antimicrobics, most infections will be treated, but at some expense. In the United States alone, the extra cost for treating the drug-resistant variety is around $10 billion per year. In many developing countries, drugs are mishandled by overuse and underuse, either of which can contribute to drug resistance. Many countries that do not regulate the sale of prescription drugs make them readily available to purchase over the counter. For example, the antituberculosis drug INH (isoniazid) is sometimes used as a "lung vitamin" to improve health, and antibiotics are taken in the wrong dose and wrong time for undiagnosed conditions. This means that these countries serve as breeding grounds for drug resistance that can eventually be carried to other countries.

It is clear that we are in a race with microbes and we are falling behind. If the trend is not contained, the world may indeed return to a time when there are few effective drugs left. We simply cannot develop them as rapidly as microbes can develop resistance. In this light, it is

TABLE 12.B
Strategies to Limit Drug Resistance by Microorganisms

Drug Usage
- Physicians have the responsibility for making an accurate diagnosis and prescribing the correct drug therapy.
- Patients must comply with and carefully follow the physician's guidelines. It is important for the patient to take the correct dosage, by the best route, for the appropriate period. This diminishes the selection for mutants that can resist low drug levels, and ensures elimination of the pathogen.
- Combined therapy: Administering two or more drugs together increases the chances that at least one of the drugs will be effective and that a resistant strain will not be able to persist. The basis for this method lies in the unlikelihood of simultaneous resistance to several drugs.

Drug Research
- Developing shorter-term, higher-dose antimicrobics that are more effective, less expensive, and have fewer side effects.
- Pharmaceutical companies continue to seek new antimicrobic drugs with structures that are not readily inactivated by microbial enzymes or drugs with modes of action that are not readily circumvented.

Long-Term Strategies
Antimicrobics (especially those for bacterial infections) are over-produced, overprescribed, and used inappropriately on a very wide scale.
- Proposals to reduce the abuses range from educational programs for health workers to requiring written justification from the physician on all antibiotics prescribed.
- Especially valuable antimicrobics may be restricted in their use to only one or two types of infections.
- The addition of antimicrobics to animal feeds must be curtailed worldwide.
- Increase government programs which make effective therapy available to low-income populations.
- Use vaccines to provide alternative protection.

essential to fight the battle on more than one front. Table 12.B summarizes the several critical strategies to give us an edge in controlling drug resistance.

ANTIBACTERIAL DRUGS

Penicillin and Its Relatives
The **penicillin** group of antibiotics, named for the parent compound, is a large, diverse group of compounds, most of which end in the suffix *-cillin*. Although penicillins could be completely synthesized in the laboratory from simple raw materials, it is more practical and economical to obtain natural penicillin through microbial fermentation. The natural product can then be used either

in unmodified form or to make semisynthetic derivatives. *Penicillium chrysogenum* is the major source of the drug. All penicillins consist of three parts: a thiazolidine ring, a beta-lactam ring, and a variable side chain that dictates its microbicidal activity (figure 12.12).

Subgroups and Uses of Penicillins The characteristics of certain penicillin drugs are shown in table 12.3. Penicillins G and V

TABLE 12.2

Selected Survey of Chemotherapeutic Agents in Infectious Diseases

Infectious Agent	Typical Infection	Drugs of Choice*
Bacteria		
Gram-Positive Cocci		
Staphylococcus aureus	Abscess, skin, toxic shock	Pencillins, vancomycin, cephalosporin
Streptococcus pyogenes	Strep throat, erysipelas, rheumatic fever	Penicillin, cephalosporin, erythromycin
S. pneumoniae	Pneumonia	Penicillin, if sensitive, cephalosporin, erythromycin
Gram-Positive Rods		
Bacillus	Anthrax	Penicillin, doxycycline
Clostridium	Gas gangrene, tetanus	Penicillin, cephalosporin, clindamycin
Corynebacterium	Diphtheria	Erythromycin, penicillin
Acid-Fast Rods		
Mycobacterium	Tuberculosis	(Isoniazid, rifampin, pyrazinamide)*, ethambutol, streptomycin
M. leprae	Leprosy	(Dapsone, rifampin, clofazimine)**
Nocardia	Nocardiosis	Sulfamethoxazole-Trimethoprim (SxT), cephalosporin
Gram-Negative Cocci		
Neisseria gonorrhoeae	Gonorrhea	Ceftriaxone, ciprofloxacin, spectinomycin
N. meningitidis	Meningitis	Penicillin G, ceftriaxone, ampicillin
Gram-Negative Rods		
Bordetella	Pertussis	Erythromycin
Brucella	Brucellosis	Tetracycline, rifampin, gentamicin
Escherichia coli	Sepsis, diarrhea, urinary tract infection	Cephalosporin, quinolone, SxT**
Haemophilus influenzae	Meningitis	Cefotaxime, cephtriaxone
Legionella	Legionnaires disease	Erythromycin, rifampin
Pseudomonas	Opportunistic lung and burn infections	Ticarcillin, aminoglycoside
Salmonella	Typhoid fever, salmonellosis	Cefalosporin, quinolone
Shigella	Dysentery	Quinolone, SxT, ampicillin
Vibrio cholerae	Cholera	Tetracyclines, SxT
Yersinia pestis	Plague	Streptomycin, gentamicin
Spirochetes		
Borrelia	Lyme disease	Ceftriaxone, doxycycline
Treponema pallidum	Syphilis	Penicillin, tetracyclines
Mycoplasma	Pneumonia, urinary infections	Erythromycin, azithromycin, clarithromycin
Rickettsia	Rocky Mountain spotted fever	Tetracyclines, chloramphenicol
Chlamydia	Urethritis, vaginitis	Azithromycin, doxycycline, erythromycin, quinolones

(continued on next page)

are the most important natural forms. Penicillin is considered the drug of choice for infections by known sensitive, gram-positive cocci (most streptococci) and some gram-negative bacteria (meningococci and the spirochete of syphilis).

Certain semisynthetic penicillins such as ampicillin, carbenicillin, and amoxicillin have broader spectra and thus can be used to treat infections by gram-negative enteric rods. Penicillinase-resistant penicillins such as methicillin, nafcillin, and cloxacillin are useful in treating infections caused by some penicillinase-producing bacteria. Mezlocillin and azlocillin have such an extended spectrum that they can be substituted for combinations of antibiotics. All of the cillin drugs are relatively mild and well tolerated because of their specific mode of action on cell walls (which humans lack). The primary problems in therapy include allergy and resistant strains of pathogens.

The Cephalosporin Group of Drugs

The **cephalosporins** are a comparatively new group of antibiotics that currently account for a majority of all antibiotics administered. The first compounds in this group were isolated in the late 1940s from the mold *Cephalosporium acremonium*. Cephalosporins are similar to penicillins in their beta-lactam structure that can be synthetically altered (figure 12.13) and in their mode of action. The generic names of these compounds are often recognized by the presence of the root *cef, ceph,* or *kef* in their names.

Subgroups and Uses of Cephalosporins The cephalosporins are versatile. They are relatively broad-spectrum, resistant to penicillinases, and cause fewer allergic reactions than penicillins. Although some cephalosporins are given orally, many are

TABLE 12.2

continued

Infectious Agent	Typical Infection	Drugs of Choice*
Fungi		
Systemic Mycoses		
Aspergillus	Aspergillosis	Amphotericin B, azoles, flucytosine
Blastomyces	Blastomycosis	Ketoconazole, amphotericin B
Candida albicans	Candidiasis	Amphotericin B, fluconazole
Coccidioides immitis	Valley fever	Amphotericin B, azoles
Cryptococcus neoformans	Cryptococcosis	Amphotericin B, fluconazole, flucytosine
Pneumocystis carinii	Pneumonia (PCP)	SxT, pentamidine
Sporothrix schenckii	Sporotrichosis	Iodides, itraconazole
Protozoa		
Entamoeba histolytica	Amebiasis	Metronidazole, tetracycline, paromomycin
Giardia lamblia	Giardiasis	Quinacrine, metronidazole
Plasmodium	Malaria	Chloroquine, primaquine, quinine
Toxoplasma gondii	Toxoplasmosis	Pyrimethamine, sulfadiazine
Trichomonas vaginalis	Trichomoniasis	Metronidazole
Trypanosoma cruzi	Chagas disease	Nitrifurimox***
T. brucei	Sleeping sickness	Suramin,*** pentamidine
Helminths		
Ascaris	Ascariosis	Mebendazole/pyrantel, piperazine
Cestodes	Tapeworm	Niclosamide, praziquantel
Schistosoma	Schistosomiasis	Praziquantel, metrifonate
Various fluke infections		Praziquantel, tetrachlorethylene, bithionol
Various roundworm infections		Mebendazole, thiabendazol, piperazine
Viruses		
Herpesvirus	Genital herpes, oral herpes, shingles	Acyclovir, ganciclovir
HIV	AIDS	(AZT, protease inhibitors), ddI, ddC, d4T
Orthomyxovirus	Type A influenza	Amantadine, rimantidine

*More or less in order of preference.
**() Usually given in combination.
***Available in the United States only from the Drug Service of the Centers for Disease Control.

poorly absorbed from the intestine and must be administered **parenterally,*** by injection into a muscle or a vein.

Three generations of cephalosporins exist, based upon their antibacterial activity. First-generation cephalosporins such as cephalothin and cefazolin are most effective against gram-positive cocci and a few gram-negative bacteria. Second-generation forms include cefaclor and cefonacid, which are more effective than the first-generation forms in treating infections by gram-negative bacteria such as *Enterobacter, Proteus,* and *Haemophilus.* Third-generation cephalosporins, such as cephalexin (Keflex) and cefotaxime, are broad-spectrum with especially well-developed activity against enteric bacteria that produce beta-lactamases. Ceftriaxone (rocephin) is a new semisynthetic broad-spectrum drug for treating a wide variety of respiratory, skin, urinary, and nervous system infections.

Other Beta-Lactam Antibiotics
Related antibiotics include imipenem, a broad-spectrum drug for infections with aerobic and anaerobic pathogens. It is active in very small concentrations and can be taken by mouth with few side effects. Aztreonam, isolated from the bacterium *Chromobacterium violaceum,* is a newer narrow-spectrum drug for treating pneumonia, septicemia, and urinary tract infections by gram-negative aerobic bacilli.

The Aminoglycoside Drugs
Antibiotics composed of two or more amino sugars and an aminocyclitol (6-carbon) ring are referred to as **aminoglycosides** (figure 12.14). These complex compounds are exclusively the products of various species of soil *actinomycetes** in the genera *Streptomyces* (figure 12.15) and *Micromonospora.*

---

* parenterally (par-ehn'-tur-ah-lee) Gr. *para,* beyond, and *enteron,* intestine. A route of drug administration other than the gastrointestinal tract.

* actinomycetes (ak"-tin-oh-my-see'-teez) Gr. *actinos,* ray, and *myces,* fungus. A group of filamentous, funguslike bacteria.

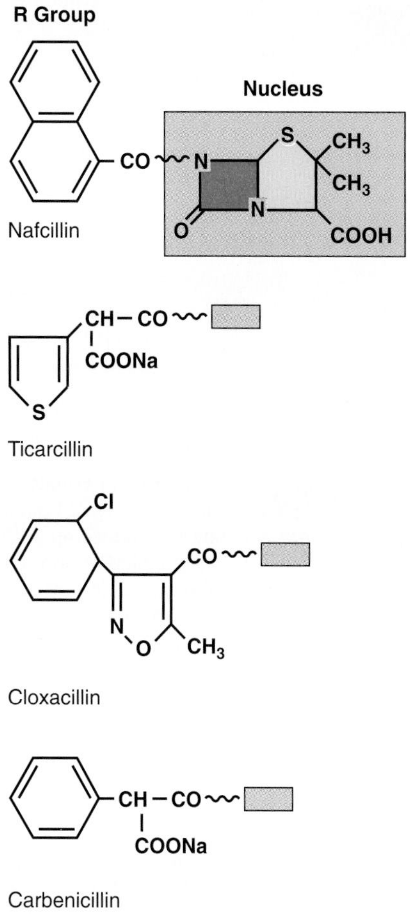

R Group

Nucleus

Nafcillin

Ticarcillin

Cloxacillin

Carbenicillin

FIGURE 12.12

Chemical structure of penicillins. All penicillins contain a thiazolidine ring (yellow) and a beta-lactam ring (red), but each differs in the nature of the side chain (R group), which is also responsible for differences in biological activity.

Subgroups and Uses of Aminoglycosides The aminoglycosides have a relatively broad antimicrobial spectrum because they inhibit protein synthesis. They are especially useful in treating infections caused by aerobic gram-negative rods and certain gram-positive bacteria. Streptomycin is among the oldest of the drugs and has gradually been replaced by newer forms with less mammalian toxicity. It is still the antibiotic of choice for treating bubonic plague and tularemia and is considered a good antituberculosis agent. Gentamicin is less toxic and is widely administered for infections caused by gram-negative rods (*Escherichia, Pseudomonas, Salmonella,* and *Shigella).* Two relatively new aminoglycosides, tobramycin and amikacin, are also used for gram-negative bacillary infections and have largely replaced kanamycin.

Tetracycline Antibiotics

In 1948, a colony of *Streptomyces* isolated from a soil sample gave off a substance, aureomycin, with strong antimicrobic properties. This antibiotic was used to synthesize its relatives terramycin and tetracycline. These natural parent compounds and semisynthetic derivatives are known as the **tetracyclines** (figure 12.16a). Their action of binding to ribosomes and blocking protein synthesis accounts for the broad-spectrum effects in the group.

Subgroups and Uses of Tetracyclines The scope of microorganisms inhibited by tetracyclines includes gram-positive and gram-negative rods and cocci, aerobic and anaerobic bacteria, mycoplasmas, rickettsias, and spirochetes. Tetracycline compounds such as doxycycline and minocycline are administered orally to treat several sexually transmitted diseases, Rocky Mountain spotted fever, typhus, *Mycoplasma* pneumonia, cholera, leptospirosis, acne, and even some protozoan infections. Although generic tetracycline is low in cost and easy to administer, its side effects—namely, gastrointestinal disruption and deposition in hard tissues—can limit its use (see table 12.4).

TABLE 12.3

Characteristics of Selected Penicillin Drugs

Name	Spectrum of Action	Uses, Advantages	Disadvantages
Penicillin G	Narrow	Best drug of choice when bacteria are sensitive; low cost; low toxicity	Can be hydrolyzed by penicillinase; allergies occur; requires injection
Penicillin V	Narrow	Good absorption from intestine; otherwise, similar to penicillin G	Hydrolysis by penicillinase; allergies
Oxacillin, dicloxacillin	Narrow	Not susceptible to penicillinase; good absorption	Allergies; expensive
Methicillin, nafcillin	Narrow	Not usually susceptible to penicillinase	Poor absorption; allergies; growing resistance
Ampicillin	Broad	Works on gram-negative bacilli	Can be hydrolyzed by penicillinase; allergies; only fair absorption
Amoxicillin	Broad	Gram-negative infections; good absorption	Hydrolysis by penicillinase; allergies
Carbenicillin	Broad	Same as ampicillin	Poor absorption; used only parenterally
Azlocillin, mezlocillin Ticarcillin	Very broad	Effective against *Pseudomonas* species; low toxicity compared with aminoglycosides	Allergies, susceptible to many beta-lactamases

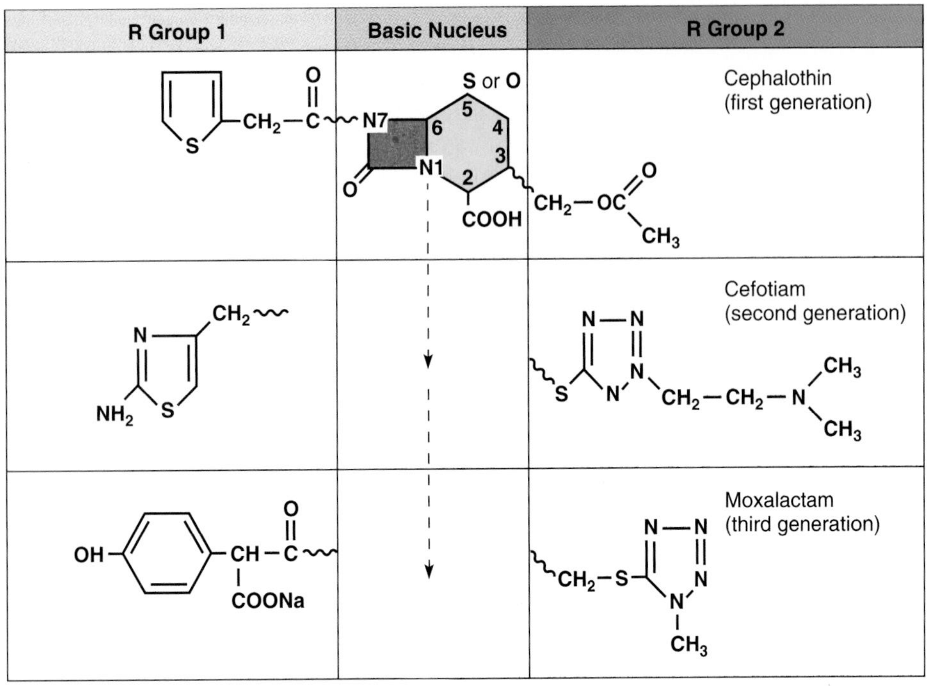

R Group 1	Basic Nucleus	R Group 2
(thiophene) $CH_2-C=O$	$N7$ S or O 5 6 4 $N1$ 3 O 2 $COOH$	Cephalothin (first generation) $CH_2-O-C=O$ CH_3
(thiazole) CH_2 N NH_2 S		Cefotiam (second generation) $S-N-N$ $N-N$ CH_2-CH_2-N CH_3 CH_3
HO (phenol) $CH-C=O$ $COONa$		Moxalactam (third generation) $N-N$ $CH_2-S-N-N$ CH_3

FIGURE 12.13

The structure of cephalosporins. Like penicillin, they have a beta-lactam ring (red), but they have a different main ring (yellow). However, unlike penicillins, they have two sites for placement of R groups (at positions 3 and 7). This makes possible several generations of molecules with greater versatility in function and complexity in structure.

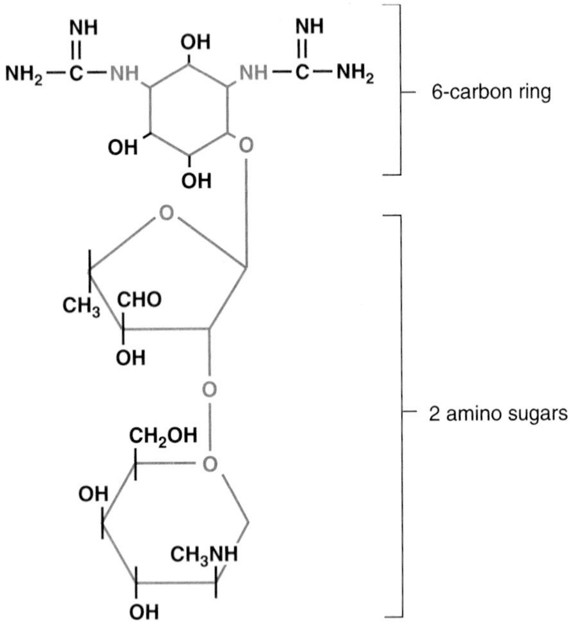

NH_2-C-NH NH OH $NH-C-NH_2$ NH
OH
OH O — 6-carbon ring
CH_3 CHO
OH
O
CH_2OH O
OH
CH_3NH — 2 amino sugars
OH

FIGURE 12.14

The structure of streptomycin. Colored portions of the molecule show the general arrangement of an aminoglycoside.

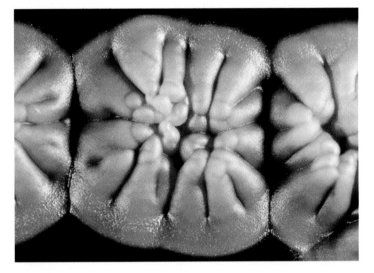

FIGURE 12.15

A colony of Streptomyces, one of nature's most prolific antibiotic producers.

Chloramphenicol

Originally isolated in the late 1940s from *Streptomyces venezuelae*, **chloramphenicol** is a potent broad-spectrum antibiotic with a unique nitrobenzene structure (figure 12.16*b*). Its primary effect on cells is to block peptide bond formation and protein synthesis. It is one type of antibiotic that is no longer derived from the natural source but is entirely synthesized through chemical processes. Although this drug is fully as broad-spectrum as the tetracyclines, it is

so toxic to human cells that its uses are restricted. A small number of people undergoing long-term therapy with this drug incur irreversible damage to the bone marrow that usually results in a fatal form of aplastic anemia.[2] Its administration is now limited to typhoid fever, brain abscesses, and rickettsial and chlamydial infections for which an alternative therapy is not available. Chloramphenicol

2. A failure of the blood-producing tissue that results in very low levels of blood cells.

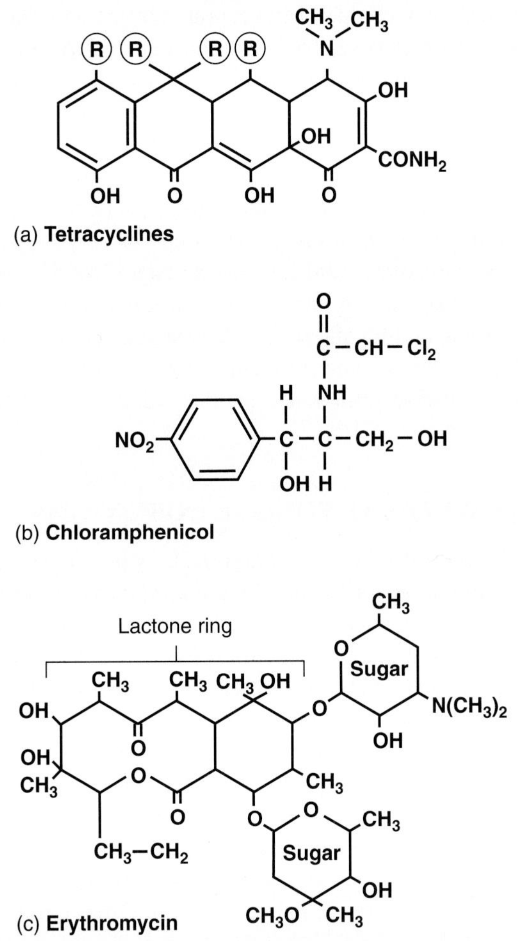

(a) **Tetracyclines**

(b) **Chloramphenicol**

(c) **Erythromycin**

FIGURE 12.16

Structures of miscellaneous broad-spectrum antibiotics.
(a) Tetracyclines. These are named for their regular group of four rings. The several types vary in structure and activity by substitution at the four R groups. **(b)** Chloramphenicol. **(c)** Erythromycin, an example of a macrolide drug. Its central feature is a large lactone ring to which two hexose sugars are attached.

should never be given in large doses repeatedly over a long time period, and the patient's blood must be monitored during therapy.

Erythromycin, Clindamycin, Vancomycin, Rifamycin

Erythromycin is a macrolide antibiotic first isolated in 1952 from a strain of *Streptomyces*. Its structure consists of a large lactone ring with sugars attached (figure 12.16c). This drug is relatively broad-spectrum and of fairly low toxicity. Its mode of action is to block protein synthesis by attaching to the ribosome. It is administered orally as the drug of choice for *Mycoplasma* pneumonia, legionellosis, *Chlamydia* infections, pertussis, and diphtheria and as a prophylactic drug prior to intestinal surgery. It also offers a useful substitute for dealing with penicillin-resistant streptococci and gonococci and for treating syphilis and acne. Newer semisynthetic macrolides include *clarithromycin* and *azithromycin*. Both drugs are useful for middle ear, respiratory, and skin infections and have also been approved for *Mycobacterium* (MAC) infections in AIDS patients. Clarithromycin has additional applications in controlling infectious stomach ulcers.

Clindamycin is a broad-spectrum antibiotic related to lincomycin. The tendency of clindamycin to cause adverse reactions in the gastrointestinal tract limits its applications to (1) serious infections in the large intestine and abdomen due to anaerobic bacteria (*Bacteroides* and *Clostridium*) that are unresponsive to other antibiotics, (2) infections with penicillin-resistant staphylococci, and (3) acne medications applied to the skin.

Vancomycin is a narrow-spectrum antibiotic most effective in treating staphylococcal infections in cases of penicillin and methicillin resistance or in patients with an allergy to penicillins. It has also been chosen to treat *Clostridium* infections in children and endocarditis (infection of the lining of the heart) caused by *Enterococcus faecalis*. Because it is very toxic and hard to administer, vancomycin is usually restricted to the most serious, life-threatening conditions.

Another product of the genus *Streptomyces* is rifamycin, which is altered chemically into **rifampin.** It is somewhat limited in spectrum because the molecule cannot pass through the cell envelope of many gram-negative bacilli. It is mainly used to treat infections by several gram-positive rods and cocci and a few gram-negative bacteria. Rifampin figures most prominently in treating mycobacterial infections, especially tuberculosis and leprosy, but it is usually given in combination with other drugs to prevent development of resistance. Rifampin is also recommended for prophylaxis in *Neisseria meningitidis* carriers and their contacts, and it is occasionally used to treat *Legionella, Brucella,* and *Staphylococcus* infections.

The *Bacillus* Antibiotics: Bacitracin and Polymyxin

Bacitracin is a narrow-spectrum peptide antibiotic produced by a strain of the bacterium *Bacillus subtilis*. Since it was first isolated, its greatest claim to fame has been as a major ingredient in a common drugstore antibiotic ointment (Neosporin) for combating superficial skin infections by streptococci and staphylococci. For this purpose, it is usually combined with neomycin (an aminoglycoside) and polymyxin.

Bacillus polymyxa is the source of the **polymyxins,** narrow-spectrum peptide antibiotics with a unique fatty acid component that contributes to their detergent activity (see figure 12.8). Only two polymyxins—B and E (also known as colistin)—have any routine applications, and even these are limited by their toxicity to the kidney. Either drug can be indicated to treat drug-resistant *Pseudomonas aeruginosa* and severe urinary tract infections caused by other gram-negative rods.

New Classes of Antibiotics

Most new antibiotics are formulated from the traditional drug classes. Two recent additions that are completely new drug types are fosfomycin and synercid. **Fosfomycin** trimethamine is a phosphoric acid agent effective as alternate treatment for urinary tract infections caused by enteric bacteria. **Synercid** is a combined antibiotic from the streptogramin group of drugs. It is effective against *Staphylococcus* and *Enterococcus* species that cause endocarditis and surgical infections, and against resistant strains of *Streptococcus*. It is one of the main choices when other drugs are ineffective due to resistance. Many experts are urging physicians to use these medications only when no other drugs are available to slow the development of drug resistance.

Nucleus   **R Group**

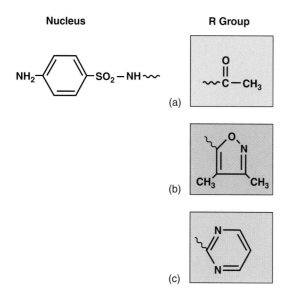

FIGURE 12.17

The structures of some sulfonamides. **(a)** Sulfacetamide, **(b)** sulfadiazine, and **(c)** sulfisoxazole.

SYNTHETIC ANTIBACTERIAL DRUGS

The synthetic antimicrobics as a group do not originate from bacterial or fungal fermentations. Some were developed from aniline dyes, and others were originally isolated from plants. Although they have been largely supplanted by antibiotics, several types are still useful.

The Sulfonamides, Trimethoprim, and Sulfones

The very first modern antimicrobic drugs were the **sulfonamides,** or sulfa drugs, named for sulfanilamide, an early form of the drug (figure 12.17). Although thousands of sulfonamides have been formulated, only a few have gained any importance in chemotherapy. Because of its solubility, sulfisoxazole is the best agent for treating shigellosis, acute urinary tract infections, and certain protozoan infections. Silver sulfadiazine ointment and solution are prescribed for treatment of burns and eye infections. In many cases, sulfamethoxazole is given in combination with **trimethoprim** (Septra, Bactrim) to take advantage of the synergistic effect of the two drugs. This combination is one of the primary treatments for *Pneumocystis carinii* pneumonia (PCP) in AIDS patients.

Sulfones are compounds chemically related to the sulfonamides but lacking their broad-spectrum effects. This lack does not diminish their importance as key drugs in treating leprosy. The most active form is dapsone, usually given in combination with rifampin and clofazamine (an antibacterial dye) over long periods.

Miscellaneous Antibacterial Agents   Isoniazid (INH) is bactericidal to *Mycobacterium tuberclosis,* but only against growing cells. Oral doses are indicated for both active tuberculosis and prophylaxis in cases of a positive TB test. Ethambutol, a closely related compound, is effective in treating the early stages of tuberculosis.

Much excitement has been generated by the new class of synthetic drugs chemically related to quinine called **fluoroquinolones.** These drugs exhibit several ideal traits, including potency and broad spectrum. Even in minimal concentrations, quinolones inhibit a wide variety of gram-positive and gram-negative bacterial species. In addition, they are readily absorbed from the intestine. The principal quinolones, norfloxacin and ciprofloxacin, have been successful in therapy for urinary tract infections, sexually transmitted diseases, gastrointestinal infections, osteomyelitis, respiratory infections, and soft tissue infections. Newer drugs in this category are sparfloxacin and levofloxacin. These agents are especially recommended for pneumonia, bronchitis, and sinusitis. Side effects that limit the use of quinolones include seizures and other brain disturbances.

AGENTS TO TREAT FUNGAL INFECTIONS

Because the cells of fungi are eucaryotic, they present special problems in chemotherapy. For one, the great majority of chemotherapeutic drugs are designed to act on bacteria and are generally ineffective in combating fungal infections. For another, the similarities between fungal and human cells often mean that drugs toxic to fungal cells are also capable of harming human tissues. A few agents with special antifungal properties have been developed for treating systemic and superficial fungal infections. Four main drug groups currently in use are the macrolide polyene antibiotics, griseofulvin, synthetic azoles, and flucytosine (figure 12.18).

Macrolide polyenes, represented by **amphotericin B** (named for its acidic and basic—amphoteric—properties) and **nystatin** (for New York State, where it was discovered), have a structure that mimics the lipids in some cell membranes. Amphotericin B (fungizone) is by far the most versatile and effective of all antifungals. Not only does it work on most fungal infections, including skin and mucous membrane lesions caused by *Candida albicans,* but it is one of the few drugs that can be injected to treat systemic fungal infections such as histoplasmosis and cryptococcus meningitis. Nystatin is used only topically or orally to treat candidiasis of the skin and mucous membranes, but it is not useful for subcutaneous or systemic fungal infections or for ringworm.

Griseofulvin is an antifungal product especially active in certain dermatophyte infections such as athlete's foot. The drug is deposited in the epidermis, nails, and hair, where it inhibits fungal growth. Because complete eradication requires several months and griseofulvin is relatively nephrotoxic, this therapy is given only for the most stubborn cases.

The **azoles** are broad-spectrum antifungal agents with a complex ringed structure. The most effective drugs are ketoconazole, fluconazole, clotrimazole, and miconazole. Ketoconazole is used orally and topically for cutaneous mycoses, vaginal and oral candidiasis, and some systemic mycoses. Fluconazole can be used in selected patients for AIDS-related mycoses such as aspergillosis and cryptococcus meningitis. Clotrimazole and miconazole are used mainly as topical ointments for infections in the skin, mouth, and vagina.

Flucytosine is an analog of cytosine that has antifungal properties. It is rapidly absorbed after oral therapy, and it is readily dissolved in the blood and cerebrospinal fluid. Alone, it can be used to

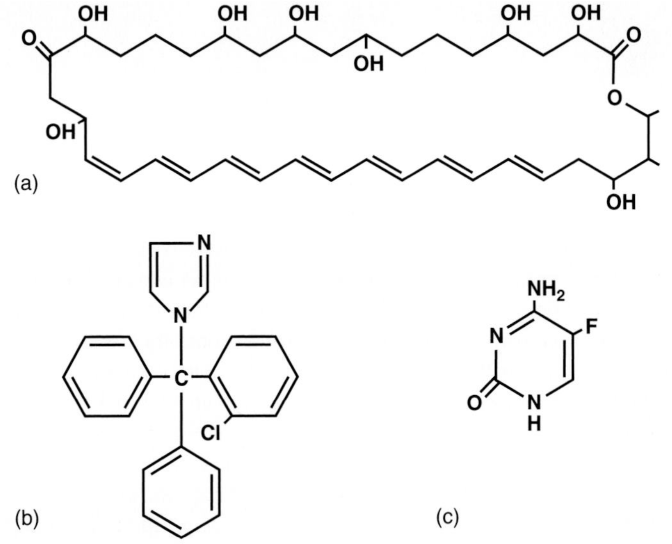

FIGURE 12.18

Some antifungal drug structures. (a) Polyenes. The example shown is amphotericin B (proposed structure), a complex steroidal antibiotic that inserts into fungal cell membranes. **(b)** Clotrimazole, one of the azoles. **(c)** Flucytosine, a structural analog of cystosine that contains fluoride.

treat certain cutaneous mycoses. Now that many fungi are resistant to flucytosine, it must be combined with amphotericin B to effectively treat systemic mycoses.

ANTIPARASITIC CHEMOTHERAPY

The enormous diversity among protozoan and helminth parasites and their corresponding therapies reaches far beyond the scope of this textbook; however, a few of the more common drugs will be surveyed here and again in chapter 23 (see table 23.5). Presently, a small number of approved and experimental drugs are used to treat malaria, leishmaniasis, trypanosomiasis, amebic dysentery, and helminth infections, but the need for new and better drugs has spurred considerable research in this area.

Antimalarial Drugs: Quinine and Its Relatives

Quinine, extracted from the bark of the cinchona tree, was the principal treatment for malaria for hundreds of years, but it has been replaced by the synthesized quinolines, mainly chloroquine and primaquine, which have less toxicity to humans. Because there are several species of *Plasmodium* (the malaria parasite) and many stages in its life cycle, no single drug is universally effective for every species and stage, and each drug is restricted in application. For instance, primaquine eliminates the liver phase of infection, and chloroquine suppresses acute attacks associated with infection of red blood cells. Chloroquine is taken alone for prophylaxis and suppression of acute forms of malaria. Primiquine is administered to patients with relapsing cases of malaria. Although plain quinine was abandoned for a time because of its toxicity, the development of chloroquine-resistant *Plasmodium* in South America and Southeast Asia restored quinine to a role in treating drug-resistant infections.

Chemotherapy for Other Protozoan Infections

A widely used amebicide, metronidazole (Flagyl), is effective in treating mild and severe intestinal infections and hepatic disease caused by *Entamoeba histolytica*. Given orally, it also has applications for infections by *Giardia lamblia* and *Trichomonas vaginalis*. Other drugs with antiprotozoan activities are quinicrine (a quinine-based drug), sulfonamides, and tetracyclines.

Antihelminthic Drug Therapy

Treating helminthic infections has been one of the most difficult and challenging of all chemotherapeutic tasks. Flukes, tapeworms, and roundworms are much larger parasites than other microorganisms and, being animals, have greater similarities to human physiology. Also, the usual strategy of using drugs to block their reproduction is usually not successful in eradicating the adult worms. The most effective drugs immobilize, disintegrate, or inhibit the metabolism of all stages of the life cycle.

Mebendazole and thiabendazole are broad-spectrum antiparasitic drugs used in several roundworm and tapeworm intestinal infestations. These drugs work locally in the intestine to inhibit the function of the microtubules of worms, eggs, and larvae, which interferes with their glucose utilization and disables them. The compounds pyrantel and piperazine paralyze the muscles of intestinal roundworms. Niclosamide destroys the scolex and the adjoining proglottids of tapeworms, thereby loosening the worm's holdfast. In these forms of therapy, the worms are unable to maintain their grip on the intestinal wall and are expelled along with the feces by the normal peristaltic action of the bowel. Two newer antihelminthic drugs are praziquantel, a treatment for various tapeworm and fluke infections, and ivermectin, a veterinary drug now used for strongyloidiasis and oncocercosis in humans.

ANTIVIRAL CHEMOTHERAPEUTIC AGENTS

The treatment of viral infections with chemotherapeutic agents is relatively new. Traditionally, viral infections have been prevented by vaccination, but useful vaccines have not been developed for many viruses. Epidemics of AIDS, genital warts, and influenza, not to mention the common cold, continue to emphasize the need for effective antiviral drugs. The existing antiviral agents have some limitations such as narrow spectrum and acting only intracellularly. Many of them are toxic as well.

Most compounds have their effects on the completion of the virus cycle. Three major modes of action are: (1) barring complete penetration of the virus into the host cell, (2) blocking the transcription and translation of viral molecules, and (3) preventing the maturation of viral particles. Although antiviral drugs protect uninfected cells by keeping viruses from being synthesized and released, most are unable to destroy extracellular viruses or those in a latent state.

Several antiviral agents mimic the structure of nucleotides and compete for sites on replicating DNA. The incorporation of these synthetic nucleotides inhibits further DNA synthesis. **Acyclovir** (Zovirax) and its relatives are synthetic purine compounds that block DNA synthesis in a small group of viruses, particularly the herpesviruses. In the topical form, it is most effective in controlling the primary attack of facial or genital herpes.

MICROBITS 12.4
Household Remedies—From Apples to Zinc

Who would have thought that drinking a glass of apple juice, swallowing a clove of garlic, or eating yogurt could be a reliable therapy for infections? A series of research findings from the past few years seems to point to a possible role for these as medicinal aids. Apple juice, fruit juices, and even tea contain natural antiviral substances, thought to be organic acids, that kill the poliovirus and coxsackievirus. Drinking beverages that contain these substances can help prevent the passage of those viruses into the intestine (their usual site of entry). Could this be a reason that "an apple a day keeps the doctor away"?

Another common home remedy has been drinking cranberry juice to reduce the symptoms of urinary tract infections. New research indicates that tannins in this product are excreted through the kidney and can block the attachment of pili by pathogens such as *Escherichia coli.* Controlled studies now support the benefits of zinc ions to control the common cold. It has been suggested that the zinc attaches to cell receptors and blocks the attachment of cold viruses. Various tablets and lozenges are now sold over the counter as cold deterrents.

The therapeutic benefits of certain foods are often surprising. Yogurt made with live cultures has been shown in controlled studies to inhibit *Staphylococcus* infections. Further examination demonstrated that some compound given off by *Lactobacillus* in yogurt prevents the adhesion of pathogens to tissues. This may explain the benefits of eating yogurt to control infections of the gastrointestinal tract and vagina. Research indicates that compounds in garlic extract inactivate viruses and destroy bacteria. If all else fails, one should not overlook the recuperative powers of chicken soup, sometimes known as "Jewish penicillin." This, too, has been found in controlled scientific tests to shorten the length and relieve the symptoms of colds, though the active ingredients have not been isolated. It appears that a timely trip to the kitchen could be as beneficial as one to the medicine cabinet. Perhaps the saying "feeding a cold and starving a fever" has some basis in fact.

Intravenous or oral acyclovir therapy can reduce the severity of primary and recurrent genital herpes episodes. Some newer relatives (valacyclovir) are more effective and require fewer doses. Famciclovir is used to treat shingles and chickenpox caused by the herpes zoster virus, and gancyclovir is approved to treat cytomegalovirus infections of the eye. An analog of adenine, *vidarabine,* is also effective against the herpesviruses. **Ribavirin** is a guanine analog used in aerosol form to treat life-threatening infections by the respiratory syncytial virus (RSV) in infants and some types of viral hemorrhagic fever.

Azidothymidine (**AZT** or zidovudine) is a thymine analog used exclusively to treat AIDS patients. This drug is specific for HIV by preventing the natural action of the viral reverse transcriptase and blocking further DNA synthesis and viral replication (see figure 25.19). It is indicated for patients with full-blown AIDS and for those in the earlier phases of disease. It is also effective as a preventive measure in pregnant HIV-positive mothers and in people accidently exposed to blood or other fluids. Other approved anti-AIDS drugs that also act as nucleotide analogs are didanosine (ddI), zalcitabine (ddC), and stavudine (d4T). These drugs are used as alternatives to AZT or in combination with it to improve the clinical condition of the patient.

Recent additions to AIDS treatment regimens are **protease inhibitors** such as saquinavir and ritonivir. These drugs have the benefit of preventing the assembly of functioning viral particles. When used in combination with nucleotide analogs, protease inhibitors can greatly slow the progression of AIDS. See chapter 25 for further coverage of this topic.

Amantadine and its relative, rimantidine, are restricted almost exclusively to treating infections by influenza A virus. Relenza and tamiflu medications are effective for influenza B and useful as prophylactics as well. Because the action of these drugs is to inhibit the uncoating of the viral RNA, they must be given rather early in an infection. In addition to these antiviral drugs, dozens of other agents are under investigation. For a novel approach for controlling viruses see Microbits 12.4.

A sensible alternative to artificial drugs has been a human-based substance, **interferon (IFN).** Interferon is a carbohydrate-containing protein produced primarily by fibroblasts and leukocytes in response to various immune stimuli. It has numerous biological activities, including antiviral and anticancer properties. Studies have shown that it is a versatile part of animal host defenses, having a great import in natural immunities. (Its mechanism is discussed in chapter 14.)

The first investigations of interferon's antiviral activity were limited by the extremely minute quantities that could be extracted from human blood. Several types of interferon are currently produced by the recombinant DNA technology techniques outlined in chapter 10. Extensive clinical trials have tested its effectiveness in viral infections and cancer. Some of the known therapeutic benefits of interferon include: (1) reducing the time of healing and some of the complications in certain infections (mainly of herpesviruses), (2) preventing or reducing some symptoms of cold and papillomaviruses (warts), (3) slowing the progress of certain cancers, including bone cancer and cervical cancer, and certain leukemias and lymphomas, (4) treating a rare cancer called hairy-cell leukemia, hepatitis C (a viral liver infection), genital warts, and Kaposi's sarcoma in AIDS patients.

Characteristics of Host–Drug Reactions

Although selective antimicrobial toxicity is the ideal constantly being sought, chemotherapy by its very nature involves contact with foreign chemicals that can harm human tissues. In fact, estimates indicate that at least 5% of all persons taking an antimicrobial drug experience some type of serious adverse reaction to it. The major **side effects** of drugs fall into one of three categories: direct damage to tissues through toxicity, allergic reactions, and disruption in the balance of normal microbial flora. The damage incurred by antimicrobial drugs can be short-term and reversible or permanent, and it ranges in severity from cosmetic to lethal. Table 12.4 summarizes drug groups and their major side effects.

TOXICITY TO ORGANS

Certain drugs adversely affect the following organs: the liver (hepatotoxic), kidneys (nephrotoxic), gastrointestinal tract, cardiovascular system and blood-forming tissue (hemotoxic), nervous system (neurotoxic), respiratory tract, skin, bones, and teeth.

Because the liver is responsible for metabolizing and detoxifying foreign chemicals in the blood, it can be damaged by a drug or its metabolic products. Injury to liver cells can result in enzymatic abnormalities, fatty liver deposits, hepatitis, and liver failure. The kidney is involved in excreting drugs and their metabolites. Some drugs irritate the nephron tubules, creating changes that interfere with their filtration abilities. Drugs such as sulfonamides crystallize in the kidney pelvis and form stones that can obstruct the flow of urine.

The most common complaint associated with oral antimicrobial therapy is diarrhea, which can progress to severe intestinal irritation or colitis. Although some drugs directly irritate the intestinal lining, the usual gastrointestinal complaints are caused by disruption of the intestinal microflora (discussed in a subsequent section).

Many drugs given for parasitic infections are toxic to the heart, causing irregular heartbeats and even cardiac arrest in extreme cases. Chloramphenicol can severely depress blood-forming cells in the bone marrow, resulting in either a reversible or a permanent (fatal) anemia. Some drugs hemolyze the red blood cells, others reduce white blood cell counts, and still others damage platelets or interfere with their formation, thereby inhibiting blood clotting.

Certain antimicrobics act directly on the brain and cause seizures. Others, such as aminoglycosides, damage nerves (very commonly, the 8th cranial nerve), leading to dizziness, deafness, or motor and sensory disturbances. When drugs block the transmission of impulses to the diaphragm, respiratory failure can result.

The skin is a frequent target of drug-induced side effects. The skin response can be a symptom of drug allergy or a direct toxic effect. Some drugs interact with sunlight to cause photodermatitis, a skin inflammation. Tetracyclines are contraindicated (not advisable) for children from birth to 8 years of age because they bind to the enamel of the teeth, creating a permanent gray to brown discoloration (figure 12.19). Pregnant women should avoid tetracyclines because they cross the placenta and can be deposited in the developing fetal bones and teeth.

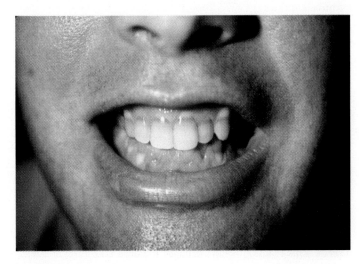

FIGURE 12.19

Drug induced side effect. An adverse effect of tetracycline given to young children is the permanent discoloration of tooth enamel.

TABLE 12.4

Major Adverse Toxic Reactions to Common Drug Groups

Antimicrobic Drug	Primary Tissue Affected	Primary Damage or Abnormality Produced
Antibacterials		
Penicillin G	Skin	Rash
Carbenicillin	Platelets	Abnormal bleeding
Ampicillin	GI tract	Diarrhea and enterocolitis
Cephalosporins	Platelet function	Inhibition of prothrombin synthesis
	White blood cells	Decreased circulation
	Kidney	Nephritis
Tetracyclines	GI tract	Diarrhea and enterocolitis
	Teeth, bones	Discoloration of tooth enamel
	Skin	Reactions to sunlight (photosensitization)
Chloramphenicol	Bone marrow	Injury to red and white blood cell precursors
Aminoglycosides (streptomycin, gentamicin, amikacin)	GI tract, hair cells in cochlea, vestibular cells, neuromuscular, kidney tubules	Diarrhea and enterocolitis; malabsorption; loss of hearing, dizziness, kidney damage
Isoniazid	Liver	Hepatitis
	Brain	Seizures
	Skin	Dermatitis
Sulfonamides	Kidney	Formation of crystals; blockage of urine flow
	Red blood cells	Hemolysis
	Platelets	Reduction in number
Polymyxin	Kidney	Damage to membranes of tubule cells
	Neuromuscular	Weakened muscular responses
Quinolones (ciprofloxacin, norfloxacin)	Nervous system, bones, GI tract	Headache, dizziness, tremors, GI distress
Rifampin	Liver	Damage to hepatic cells
	Skin	Dermatitis
Antifungals		
Amphotericin B	Kidney	Disruption of tubular filtration
Flucytosine	White blood cells	Decreased number
Antiprotozoan Drugs		
Metronidazole	GI tract	Nausea, vomiting
Chloroquine	GI tract	Vomiting
	Brain	Headache
	Skin	Itching
Antihelminthics		
Niclosamide	GI tract	Nausea, abdominal pain
Pyrantel	GI tract	Irritation
	Brain	Headache, dizziness
Antivirals		
Acyclovir	Brain	Seizures, confusion
	Skin	Rash
Amantadine	Brain	Nervousness, light-headedness
	GI tract	Nausea
AZT	Bone marrow	Immunosuppression, anemia

ALLERGIC RESPONSES TO DRUGS

One of the most frequent drug reactions is heightened sensitivity, or *allergy*. This reaction occurs because the drug acts as an antigen (a foreign material capable of stimulating the immune system) and stimulates an allergic response. This response can be provoked by the intact drug molecule or by substances that develop from the body's metabolic alteration of the drug. In the case of penicillin, for instance, it is not the penicillin molecule itself that causes the allergic response but a product, *benzlpenicilloyl*. Allergic reactions have been reported for every major type of antimicrobic drug, but the penicillins account for the greatest number of antimicrobic allergies, followed by the sulfonamides.

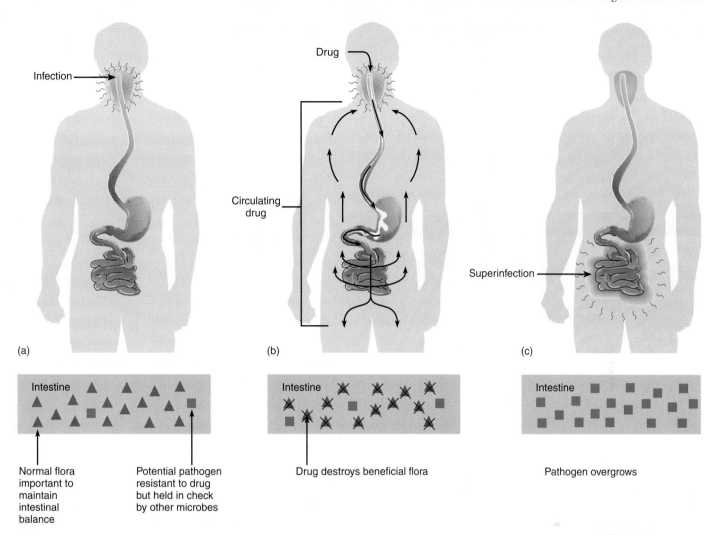

FIGURE 12.20

The role of antimicrobics in disrupting microbial flora and causing superinfections. (a) A primary infection in the throat is treated with an oral antibiotic. **(b)** The drug is carried to the intestine and is absorbed into the circulation. **(c)** The primary infection is cured, but drug-resistant pathogens have survived and create an intestinal superinfection.

People who are allergic to a drug become sensitized to it during the first contact, usually without symptoms. Once the immune system is sensitized, a second exposure to the drug can lead to a reaction such as a skin rash (hives), respiratory inflammation, and, rarely, anaphylaxis, an acute, overwhelming allergic response that develops rapidly and can be fatal. (This topic is discussed in greater detail in chapter 17.)

SUPPRESSION AND ALTERATION OF THE MICROFLORA BY ANTIMICROBICS

Most normal, healthy body surfaces, such as the skin, large intestine, outer openings of the urogenital tract, and oral cavity, provide numerous habitats for a virtual "garden" of microorganisms. These normal colonists or residents, called the **flora*** or microflora, consist mostly of harmless or beneficial bacteria, but a small number can potentially be pathogens. Although we shall defer a more detailed discussion of this topic to chapter 13, here we focus on the general effects of drugs on this population.

If a broad-spectrum antimicrobic is introduced into the host to treat infection, it will destroy microbes regardless of their roles in the balance, affecting not only the targeted infectious agent but also many others in sites far removed from the original infection (figure 12.20). When this therapy destroys beneficial resident species, the pathogens that were once in small numbers begin to overgrow and cause disease. This complication is called a **superinfection.**

Some common examples demonstrate how a disturbance in microbial flora leads to replacement flora and superinfection. A broad-spectrum cephalosporin used to treat a urinary tract infection by *Escherichia coli* will cure the infection, but it will also destroy the lactobacilli in the vagina that normally maintain a protective acidic environment there. The drug has no effect, however, on *Candida albicans,* a yeast that also resides in normal vaginas. Released

---

* **flora** (flor´-ah) Gr. *flora,* the goddess of flowers. The microscopic life present in a particular location.

from the inhibitory pH normally provided by lactobacilli, the yeasts proliferate and cause an infection. *Candida* can cause similar superinfections of the oropharynx (thrush) and the large intestine.

Oral therapy with tetracyclines, clindamycin, and broad-spectrum penicillins and cephalosporins is associated with a serious and potentially fatal condition known as *antibiotic-associated colitis* (pseudomembranous colitis). This condition is due to the overgrowth in the bowel of *Clostridium difficile,* a spore-forming bacterium that is resistant to the antibiotic. It invades the intestinal lining and releases toxins that induce diarrhea, fever, and abdominal pain.

CHAPTER CHECKPOINTS

The three major side effects of antimicrobics are toxicity to organs, allergic reactions, and problems resulting from suppression or alteration of normal flora.

Antimicrobics that destroy most but not all normal flora allow the unaffected normal flora to overgrow, causing a superinfection.

Considerations in Selecting an Antimicrobic Drug

Before actual antimicrobic therapy can begin, it is important that at least three factors be known: (1) the nature of the microorganism causing the infection; (2) the degree of the microorganism's susceptibility (also called sensitivity) to various drugs; and (3) the overall medical condition of the patient.

IDENTIFYING THE AGENT

Identification of infectious agents from body specimens should be attempted as soon as possible. It is especially important that such specimens be taken before the antimicrobic drug is given, just in case the drug eliminates the infectious agent. Direct examination of body fluids, sputum, or stool is a rapid initial method for detecting and perhaps even identifying bacteria or fungi. A doctor often begins the therapy on the basis of such immediate findings. The choice of drug will be based on experience with drugs that are known to be effective against the microbe; this is called the "informed best guess." For instance, if a sore throat appears to be caused by *Streptococcus pyogenes,* the physician might prescribe penicillin, because this species seems to be universally sensitive to it so far. If the infectious agent is not or cannot be isolated, epidemiologic statistics may be required to predict the most likely agent in a given infection. For example, *Streptococcus pneumoniae* accounts for the majority of cases of meningitis in children, followed by *Neisseria meningitidis* and *Haemophilus influenzae.*

TESTING FOR THE DRUG SUSCEPTIBILITY OF MICROORGANISMS

Testing is essential in those groups of bacteria commonly showing resistance, primarily *Staphylococcus* species, *Neisseria gonorrhoeae, Streptococcus pneumoniae,* and *Enterococcus faecalis,* and the aerobic gram-negative enteric bacilli. However, not all infectious agents require antimicrobial sensitivity testing. Drug testing in fungal or protozoan infections is difficult and is often unnecessary. When certain groups, such as group A streptococci and all anaerobes (except *Bacteroides*), are known to be uniformly susceptible to penicillin G, testing may not be necessary unless the patient is allergic to penicillin.

Selection of a proper antimicrobial agent begins by demonstrating the *in vitro* activity of several drugs against the infectious agent by means of standardized methods. In general, these tests involve exposing a pure culture of the bacterium to several different drugs and observing the effects of the drugs on growth.

The *Kirby-Bauer* technique is an agar diffusion test that provides useful data on antimicrobic susceptibility. In this test, the surface of a plate of special medium is seeded with the test bacterium, and small discs containing a premeasured amount of antimicrobic are dispensed onto the bacterial lawn. After incubation, the zone of inhibition surrounding the discs is measured and compared with a standard for each drug (figure 12.21 and table 12.5). The profile of antimicrobic sensitivity, or *antibiogram,* provides data for drug selection. The Kirby-Bauer procedure is less effective for bacteria that are anaerobic, highly fastidious, or slow-growing (*Mycobacterium*). A different diffusion system that provides additional information on drug effectiveness is the E-test (figure 12.22).

More sensitive and quantitative results can be obtained with tube dilution tests. First the antimicrobic is diluted serially in tubes of broth, and then each tube is inoculated with a small uniform sample of pure culture, incubated, and examined for growth (turbidity). The smallest concentration (highest dilution) of drug that visibly inhibits growth is called the **minimum inhibitory concentration,** or MIC. The MIC is useful in determining the smallest effective dosage of a drug and in providing a comparative index against other antimicrobics (figure 12.23 and table 12.6). In many clinical laboratories, these antimicrobic testing procedures are performed in automated machines that can test dozens of drugs simultaneously.

THE MIC AND THERAPEUTIC INDEX

The results of antimicrobic sensitivity tests guide the physician's choice of a suitable drug. If therapy has already commenced, it is imperative to determine if the tests bear out the use of that particular drug. Once therapy has begun, it is important to observe the patient's clinical response, because the *in vitro* activity of the drug is not always correlated with its *in vivo* effect. When antimicrobic treatment fails, the failure is due to (1) the inability of the drug to diffuse into that body compartment (the brain, joints, skin); (2) a few resistant cells in the culture that did not appear in the sensitivity test; or (3) an infection caused by more than one pathogen (mixed), some of which are resistant to the drug. If therapy does fail, a different drug, combined therapy, or a different method of administration must be considered.

Many factors influence the choice of an antimicrobic drug besides microbial sensitivity to it. The nature and spectrum of the drug, its potential adverse effects, and the condition of the patient can be critically important. When several antimicrobic drugs are

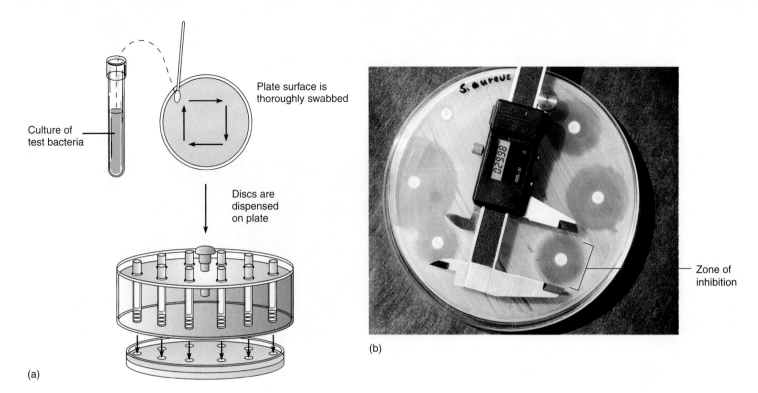

(a)

(b)

Culture of test bacteria

Plate surface is thoroughly swabbed

Discs are dispensed on plate

Zone of inhibition

FIGURE 12.21

Technique for preparation and interpretation of disc diffusion tests. (a) Standardized methods are used to seed a lawn of bacteria over the medium. A dispenser delivers several drugs onto a plate, followed by incubation. **(b)** Interpretation of results. During incubation, antimicrobics become increasingly diluted as they diffuse out of the disc into the medium. If the test bacterium is sensitive to a drug, a zone of inhibition develops around its disc. The larger the size of this zone, the greater is the bacterium's sensitivity to the drug. The diameter of each zone is measured in millimeters and evaluated for susceptibility or resistance by means of a comparative standard (see table 12.5). Drug resistance can be detected by a small or nonexistent zone, or by tiny colonies within the zone of inhibition.

available for treating an infection, final drug selection advances to a new series of considerations. In general, it is better to choose the narrowest-spectrum drug of those that are effective if the causative agent is known. This decreases the potential for superinfections and other adverse reactions.

Because drug toxicity is of concern, it is best to choose the one with high selective toxicity for the infectious agent and low human toxicity. The **therapeutic index (TI)** is defined as the ratio of the dose of the drug that is toxic to humans as compared to its minimum effective (therapeutic) dose. The closer these two figures are

TABLE 12.5

Results of Kirby-Bauer Test

Drug	Zone Sites (mm) Required For:		Actual Result (mm) for *Staphylococcus aureus*	Evaluation
	Susceptibility (S)	Resistance (R)		
Bacitracin	>13	<8	15	S
Chloramphenicol	>18	<12	20	S
Erythromycin	>18	<13	25	S
Gentamicin	>13	<12	16	S
Kanamycin	>18	<13	20	S
Neomycin	>17	<12	12	R
Penicillin G	>29	<20	10	R
Polymyxin B	>12	<8	10	R
Streptomycin	>15	<11	11	R
Vancomycin	>12	<9	15	S
Tetracycline	>19	<14	25	S

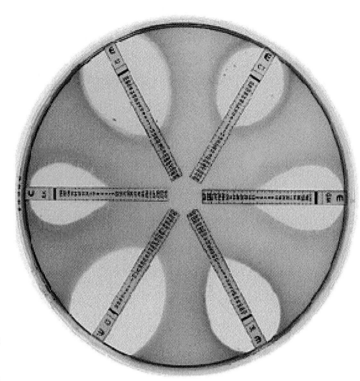

These E-tests have been carried out on an isolate of *Pseudomonas aeruginosa*.

The following MICs (μg) have been determined:

Aztreonam (AT)	1.5
Amikacin (AK)	4
Cefotaxime (C)	4
Gentamicin (GM)	4.5
Ciprofloxacin (CI)	0.125
Piperacillin /Tazobactam (PT)	2

On the basis of these results, treatment with ciprofloxacin would be either commenced (or continued, if the patient is very ill and treatment has already commenced).

FIGURE 12.22

Alternate to the Kirby-Bauer procedure.
Another diffusion test is the E-test, which uses a strip to produce the zone of inhibition. The advantage of the E-test is that the strip contains a gradient of drug calibrated in μg. This way, the MIC can be measured by observing the mark on the strip that corresponds to the edge of the zone of inhibition. This test can also be used to detect and quantify developing resistance.

(the smaller the ratio), the greater is the potential for toxic drug reactions. For example, a drug that has a therapeutic index of:

$$\frac{10 \ \mu g/ml: \text{toxic dose}}{9 \ \mu g/\mu l \ (\text{MIC})} \quad \boxed{\text{TI}=1.1}$$

is a riskier choice than one with a therapeutic index of:

$$\frac{10 \ \mu g/ml}{1 \ \mu g/ml} \quad \boxed{\text{TI}=10}$$

Drug companies recommend dosages that will inhibit the microbes but not adversely affect patient cells. When a series of drugs being considered for therapy have similar MICs, the drug with the highest therapeutic index usually has the widest margin of safety.

The physician must also take a careful history of the patient to discover any preexisting medical conditions that will influence the activity of the drug or the response of the patient. A history of allergy to a certain class of drugs should preclude the administration of that drug. Underlying liver or kidney disease will ordinarily necessitate the modification of drug therapy because these organs play such an important part in metabolizing or excreting the drug. Infants, the elderly, and pregnant women require special precautions. For example, age can diminish gastrointestinal absorption and organ function, and most antimicrobial drugs cross the placenta and could affect fetal development.

The intake of other drugs must be carefully scrutinized, because incompatibilities can result in increased toxicity or failure of one or more of the drugs. For example, the combination of aminoglycosides and cephalosporins increases nephrotoxic effects; antacids reduce the absorption of isoniazid; and the interaction of tetracycline or rifampin with oral contraceptives can abolish the contraceptive's effect. Some drugs (penicillin with certain aminoglycosides, or amphotericin B with flucytosine) act synergistically, so that reduced doses of each can be used in combined therapy. Other concerns in choosing drugs include any genetic or metabolic abnormalities in the patient, the site of infection, the route of administration, and the cost of the drug.

The Art and Science of Choosing an Antimicrobial Drug
Even when all the information is in, the final choice of a drug is not always easy or straightforward. Consider the case of an elderly alcoholic patient with pneumonia caused by *Klebsiella* and complicated by diminished liver and kidney function. All drugs must be given parenterally because of prior damage to the gastrointestinal lining and poor absorption. Drug tests show that the infectious agent is sensitive to third-generation cephalosporins, gentamicin, imipenem and azlocillin. The patient's history shows previous allergy to the penicillins, so these would be ruled out. Drug interactions occur between alcohol and the cephalosporins, which are also associated with serious bleeding in elderly patients, so this may not be a good choice. Aminoglycosides such as gentamicin are nephrotoxic and poorly cleared by damaged kidneys. Imipenem causes intestinal discomfort, but it has less toxicity and would be a viable choice.

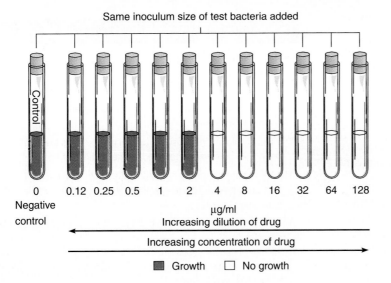

Same inoculum size of test bacteria added

0 0.12 0.25 0.5 1 2 4 8 16 32 64 128
Negative control
μg/ml
Increasing dilution of drug
Increasing concentration of drug

■ Growth □ No growth

FIGURE 12.23

Tube dilution test for determining the minimum inhibitory concentration (MIC). The antibiotic is diluted serially through tubes of liquid nutrient from right to left. All tubes are inoculated with an identical sample of a test bacterium and then incubated. The first tube on the left is a control that lacks the drug and shows maximum growth. The dilution of the first tube in the series that shows no growth (no turbidity) is the MIC.

In the case of a cancer patient with severe systemic *Candida* infection, there will be fewer criteria to weigh. Intravenous amphotericin B or fluconazole are the only possible choices, despite drug toxicity and other possible adverse side effects. In a life-threatening situation, in which a dangerous chemotherapy is perhaps the only chance for survival, the choices are reduced and the priorities are different.

AN ANTIMICROBIC DRUG DILEMMA

We began this chapter with a view of the exciting strides made in chemotherapy during the past few years, but we must end it on a note of qualification and caution. There is now a worldwide problem in the management of antimicrobic drugs, which rank second only to some nervous system drugs in overall usage. The remarkable progress in treating many infectious diseases has spawned a view of antimicrobics as a "cure-all" for infections as diverse as the common cold and acne. And, although it is true that nothing is as dramatic as curing an infectious disease with the correct antimicrobic drug, in many instances, drugs have no effect or can be harmful. The depth of this problem can perhaps be appreciated better with a few statistics:

1. Roughly 200 million prescriptions for antibiotics are written in the United States every year. One study disclosed that the percentage of incorrect prescriptions ranges from 40 to 90%. Many prescriptions are for bacterial infections, even though the infection is viral in origin. Many drugs are also misprescribed as to type, dosage, or length of therapy. Such overuse of antimicrobics is known to increase the development of antimicrobial resistance, harm the patient, and waste billions of dollars.

2. Drugs are often prescribed without benefit of culture or susceptibility testing, even when such testing is clearly warranted.

3. There is a tendency to use a "shotgun" antimicrobial therapy for minor infections, which involves administering a broad-spectrum drug instead of a more specific narrow-spectrum one. This practice can lead to superinfections and other adverse reactions. Tetracyclines and chloramphenicol are still prescribed routinely for infections that would be treated more effectively with narrower spectrum, less toxic drugs.

4. More expensive newer drugs are chosen when a less costly older one would be just as effective. Among the most expensive drugs are cephalosporins and the longer-acting

TABLE 12.6

Comparative MICs (μg/ml) for Common Drugs and Pathogens

Bacterium	Penicillin G	Ampicillin	Sulfamethoxazole	Tetracycline	Cefaclor
Staphylococcus aureus	4.0	0.05	3.0	0.3	4.0
Enterococcus faecalis	3.6	1.6	100.0	0.3	60.0
Neisseria gonorrhoeae	0.5	0.5	5.0	0.8	2.0
Escherichia coli	100.0	12.0	3.0	6–50.0	3.0
Pseudomonas aeruginosa	>500.0	>200.0	NA	>100.0	NA
Salmonella species	12.0	6.0	10.0	1.0	0.8
Clostridium	0.16	NA	NA	3.0	12.0

NA = not available

tetracyclines, yet these are among the most commonly prescribed antibiotics.

5. Tons of excess antimicrobic drugs produced in this country are exported to other countries, where controls are not as strict. Nearly 200 different antibiotics are sold over the counter in Latin America and Asian countries. It is common to self-medicate without understanding the correct medical indication. Drugs used in this way are largely ineffectual but, worse yet, they are known to be responsible for emergence of drug-resistant bacteria that subsequently cause epidemics.

The medical community recognizes that most physicians are motivated by important and prudent concerns, such as the need for immediate therapy to protect a sick patient and for defensive medicine to provide the very best care possible, but many experts feel that more education is needed for both physicians and patients concerning the proper occasions for prescribing antibiotics. In the final analysis, every allied health professional should be critically aware not only of the admirable and utilitarian nature of antimicrobics but also of their limitations.

CHAPTER CHECKPOINTS

The three major considerations necessary to choose an effective antimicrobic are the nature of the infecting microbe, the microbe's sensitivity to available drugs, and the overall medical status of the infected host.

The Kirby-Bauer test identifies antimicrobics that are effective against a specific infectious bacterial isolate.

The MIC (minimum inhibitory concentration) identifies the smallest effective dose of an antimicrobic toxic to the infecting microbe.

The therapeutic index is a ratio of the amount of drug toxic to the infected host and the MIC. The smaller the ratio, the greater the potential for toxic host-drug reactions.

The effectiveness of antimicrobic drugs is being compromised by several alarming trends: inappropriate prescription, use of broad-spectrum instead of narrow-spectrum drugs, use of higher-cost drugs, sale of over-the-counter antimicrobics in other countries, and lack of sufficient testing before prescription.

CHAPTER CAPSULE WITH KEY TERMS

I. **Antimicrobial Chemotherapy**
   A. Chemotherapeutic Drugs: Used in treatment of infections, control of microbes in the body.
      1. Categories of antimicrobics
         a. **Antibiotics** are chemicals derived from bacteria and molds, natural or **semisynthetic** (part natural, part synthetic).
         b. **Synthetic drugs** are derived completely from industrial processes.
         c. Drugs may be **narrow-spectrum** or **broad-spectrum,** microbicidal or microbistatic.
      2. Scope: Types include antibacterial (largest number), antifungal, antiprotozoan, antihelminthic, antiviral.
      3. Ideal qualities of antimicrobics
         a. **Selective toxicity,** meaning high toxicity to microorganisms, low toxicity to animals.
         b. Potency unaltered by dilution.
         c. Stability and solubility in tissue fluids.
         d. Lack of disruption to host's immune system or microflora.
         e. Exempt from drug resistance.
      4. Route of administration: By mouth; **parenteral** (injection into vein, muscle); topical on skin, mucous membranes.
      5. Major adverse results of chemotherapy
         a. Toxicity ranging from slight, short-term damage to permanent debilitation.
         b. Allergic reactions.
         c. Disruption of normal microbial flora of body.
         d. **Superinfections** by resistant species.
         e. **Drug resistance** is a serious outcome that results when pathogens develop tolerance to drugs through mutation or transfer of resistance plasmids (**R factors).**
      6. Special clinical approaches
         a. **Prophylaxis,** administering antimicrobic drugs to prevent infections in highly susceptible persons.
         b. **Combined therapy,** administering two or more antimicrobics simultaneously to circumvent drug

resistance or to achieve **synergism,** the additive or magnified effectiveness of certain drugs working together.
      7. Stages in selection of proper drug
         a. Identification of microbe *in vitro;* testing antimicrobial sensitivity or susceptibility (determining the **MIC**).
         b. Assessing the **therapeutic index.**
         c. Final drug selection weighs potential effectiveness, toxicity, *in vivo* effects, spectrum, and the medical condition of the patient.
      8. Perspectives in antimicrobic abuse: Drugs are overprescribed, overproduced, and used inappropriately on a worldwide basis, with unfortunate medical and economic consequences.

II. **Antibacterial Antibiotics**
   A. **Penicillins**
      1. Types/source: **Beta-lactam**–based drugs; from *Penicillium chrysogenum* mold; natural form is penicillin G; semisynthetic forms vary in spectrum and applications.
      2. Mode of action: Bactericidal; blocks the completion of the cell wall, causes weak points and cell rupture.
      3. Examples are narrow-spectrum penicillin and broad-spectrum ampicillin and methicillin.
      4. Problems in therapy: Penicillins are not highly toxic, but cause of allergic reactions; bacterial resistance to some forms of penicillin occurs through **beta-lactamase** (for example, **penicillinase**).
   B. **Cephalosporins**
      1. Types/source: Natural and semisynthetic forms from *Cephalosporium* mold.
      2. Mode of action: Similar to penicillins; inhibit peptidoglycan synthesis.
      3. Specific uses/spectra
         a. Effective against gram-positive cocci and gram-negative cocci and rods.
         b. Broad-spectrum, especially for gram-negative enteric rods and gram-positive cocci.

    4. Problems in therapy: Adverse blood and kidney reactions; superinfections; allergic reactions; bacterial resistance through **cephalosporinases.**

C. **Aminoglycosides**
1. Types/source/spectrum: Mostly narrow-spectrum, from *Streptomyces.*
2. Mode of action: Interference with the bacterial ribosome, inhibition of protein synthesis.
3. Examples
  a. Streptomycin, narrow-spectrum, for tuberculosis therapy.
  b. Gentamicin, for gram-negative enteric infections.
4. Problems in therapy: Toxic reactions to 8th cranial nerve, kidney damage, intestinal disturbances; drug resistance.

D. **Tetracyclines** and **Chloramphenicol**
1. Source/spectrum: Very broad-spectrum drugs originally from species of *Streptomyces* (now semi- or fully synthetic).
2. Mode of action: Interfere with translation (protein synthesis).
3. Examples
  a. Tetracyclines very broad-spectrum against numerous types of gram-positives, -negatives, rickettsias, and mycoplasmas.
  b. Chloramphenicol (chloromycetin) limited by toxicity; indicated for serious infections where there is no alternative.
4. Problems in therapy: Tetracycline may lead to hepatotoxicity, gastric disturbance, discoloration of tooth enamel in children, superinfections; chloramphenicol may damage bone marrow.

E. **Erythromycin, Clindamycin, Vancomycin, Rifampin**
1. Modes of action: Erythromycin and clindamycin disrupt protein synthesis; vancomycin interferes with early cell wall synthesis; rifampin inhibits RNA synthesis.
2. Specific uses/spectra
  a. Erythromycin is extended spectrum treatment for penicillin-resistant bacteria.
  b. Clindamycin used for intestinal infections by anaerobes.
  c. Vancomycin applied in life-threatening staphylococcal infections.
  d. Rifampin used for tuberculosis and leprosy.
3. Problems in therapy: Vancomycin is neurotoxic; clindamycin and erythromycin can harm the GI tract; rifampin is hepatotoxic; resistant bacteria occur for all.

F. **Bacitracin** and **Polymyxin**
1. Spectrum: Narrow-spectrum.
2. Mode of action
  a. Bacitracin prevents synthesis of the cell wall of gram-positive bacteria.
  b. Polymyxin has detergent action that disrupts cell membrane of gram-negative bacteria.
3. Specific uses
  a. Bacitracin used in ointments with neomycin for skin infections.
  b. Polymyxins used to treat *Pseudomonas* infection or in skin ointments.
4. Problems in therapy: Polymyxin can cause nephrotoxic and neuromuscular reactions; bacitracin is useful only for topical applications.

III. **Synthetic Antibacterial Drugs**
A. **Sulfonamides** (Sulfa Drugs)
1. Types/spectrum: All types have similar basic structure; commonest is sulfisoxazole; relatively broad-spectrum.

2. Mode of action: Acts as an **antimetabolite,** a **metabolic analog** that causes **competitive inhibition,** resulting in blockage in nucleic and amino acid synthesis.
3. Specific uses: Urinary tract infections, nocardiosis, burn and eye infections; often combined with trimethoprim.
4. Problems in therapy: Allergy and formation of crystals in kidney.

B. Miscellaneous
1. **Trimethoprim,** medication for urinary, respiratory, and gastrointestinal infections; a competitive inhibitor in nucleic acid synthesis; can cause bone marrow damage.
2. Dapsone, a major antileprosy drug, combined with rifampin.
3. **Isoniazid** (INH), an antituberculosis drug; blocks synthesis of cell wall of mycobacteria; may damage liver.
4. **Fluoroquinolones** (ciprofloxacin), promising new broad-spectrum drugs.

C. Drugs for Fungal Infections
1. Polyenes: Amphotericin B and nystatin, antibiotics that disrupt cell membrane by detergent action. Amphotericin is a key drug in systemic fungal infections; nystatin is used for skin and membrane candidiasis. Both are commonly nephrotoxic.
2. Azoles: Synthetic drugs that interfere with membrane synthesis; ketoconazole, fluconazole, and miconazole are used for various levels of infection; can cause liver damage.
3. Flucytosine: Synthetic inhibitor of DNA synthesis; used alone or in combination with amphotericin for systemic mycoses; may lower WBC count; fungal resistance.

D. Drugs for Protozoan Infections
1. Quinines: Types are chloroquine, primaquine, and quinine used to manage malaria depending on sensitivity and stage in the parasite's cycle; can cause intestinal symptoms and eye disturbances; resistance and complexity of life cycle are main hurdles.
2. Others: Metronidazole (flagyl) for amebiasis, giardiasis, and trichomonas infections; suramin, melarsoprol, indicated in treatment of African trypanosomiasis; nitrifurimox, for acute South American trypanosomiasis.

E. Drugs for Helminth Infections
1. Mebendazole, thiabendazole, and praziquantel, all-purpose agents in treating intestinal roundworm, tapeworm, and some fluke infestations.
2. Pyrantel and piperazine, primarily for intestinal roundworms.
3. Niclosamide, for tapeworms.
4. Taken orally, cure occurs only upon incapacitation or death of worms and eggs followed by their expulsion in feces.

F. Drugs for Viral Infections
Most antivirals function intracellularly to block some part of virus multiplication; main drawbacks are resistance and toxicity.
1. **Acyclovir** and vidarabine are synthetic nitrogen bases that block synthesis of viral components in herpesviruses.
2. **Amantadine,** restricted to treating influenza A infections.
3. **AZT** and **protease inhibitors** are anti-AIDS drugs.
4. **Interferon,** a naturally occurring protein that can be useful in reducing symptoms of some viral infections and treating a few cancers.

MULTIPLE-CHOICE QUESTIONS

1. A compound synthesized by bacteria or fungi that destroys or inhibits the growth of other microbes is a/an
   a. synthetic drug
   b. antibiotic
   c. antimicrobic drug
   d. competitive inhibitor

2. Which statement is *not* an aim in the use of drugs in antimicrobial chemotherapy? The drug should:
   a. have selective toxicity
   b. be active even in high dilutions
   c. be broken down and excreted rapidly
   d. be microbicidal

3. Drugs that prevent the formation of the bacterial cell wall are
   a. quinolones
   b. beta-lactams
   c. tetracyclines
   d. aminoglycosides

4. Sulfonamide drugs initially disrupt which process?
   a. folic acid synthesis
   b. transcription
   c. PABA synthesis
   d. protein synthesis

5. Microbial resistance to drugs is acquired through
   a. conjugation
   b. transformation
   c. transduction
   d. all of these

6. R factors are _____ that contain a code for _____.
   a. genes, replication
   b. plasmids, drug resistance
   c. transposons, interferon
   d. plasmids, conjugation

7. When a patient's immune system becomes reactive to a drug, this is an example of
   a. superinfection
   b. drug resistance
   c. allergy
   d. toxicity

8. An antibiotic that disrupts the normal flora can cause
   a. the teeth to turn brown
   b. aplastic anemia
   c. a superinfection
   d. hepatotoxicity

9. Most antihelminthic drugs function by
   a. weakening the worms so they can be flushed out by the intestine
   b. inhibiting worm metabolism
   c. blocking the absorption of nutrients
   d. inhibiting egg production

10. An example of an antiviral drug that can prevent a viral nucleic acid from being replicated is
    a. azidothymidine
    b. acyclovir
    c. amantadine
    d. both a and b

11. Which of the following effects do antiviral drugs *not* have?
    a. killing extracellular viruses
    b. stopping virus synthesis
    c. inhibiting virus maturation
    d. blocking virus receptors

12. Which of the following modes of action would be most selectively toxic?
    a. interrupting ribosomal function
    b. dissolving the cell membrane
    c. preventing cell wall synthesis
    d. inhibiting DNA replication

13. The MIC is the _____ of a drug that is required to inhibit growth of a microbe.
    a. largest concentration
    b. standard dose
    c. smallest concentration
    d. lowest dilution

14. An antimicrobic drug with a _____ therapeutic index is a better choice than one with a _____ therapeutic index.
    a. low, high
    b. high, low

CONCEPT QUESTIONS

1. Differentiate between antibiotics and synthetic drugs.

2. a. Differentiate between narrow-spectrum and broad-spectrum antibiotics.
   b. Can you determine why some drugs have narrower spectra than others? (Hint: Look at their mode of action.)
   c. How might one determine whether a particular antimicrobic is broad- or narrow-spectrum?

3. a. What is the major source of antibiotics?
   b. What appears to be the natural function of antibiotics?

4. a. Using the diagram at right as a guide, briefly explain how the three factors in drug therapy interact.
   b. What drug characteristics will make treatment most effective?
   c. What are the major aims of new antimicrobic drugs?
   d. Which of your answers to question *c* do you think is the most important?

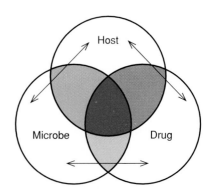

5. a. Explain the major modes of action of antimicrobial drugs, and give an example of each.
   b. What is competitive inhibition?

c. What is the basic reason that a metabolic analog molecule can inhibit metabolism?

d. What are the long-term effects of drugs that block transcription?

e. Why would a drug that blocks translation on the ribosomes of bacteria also affect human cells?

f. Why do drugs that act on bacterial and fungal membranes generally have high toxicity?

6. a. Explain the phenomenon of drug resistance from the standpoint of microbial genetics (include a description of R factors).

b. Multiple drug resistance is becoming increasingly common in microorganisms. Explain how one bacterium can acquire resistance to several drugs.

7. a. Explain four general ways that microbes evade the effects of drugs.

b. What is the effect of beta-lactamase?

8. What causes mutated or plasmid-altered strains of drug-resistant microbes to persist in a population?

9. Construct a chart that summarizes the modes of action and applications of the major groups of antibacterial drugs (antibiotics and synthetics), antifungal drugs, antiparasitic drugs, and antiviral drugs.

10. a. Explain why there are so few antifungal, antiparasitic, and antiviral drugs.

b. What effect do nitrogen-base analogs have upon viruses?

c. Summarize the origins and biological actions of interferon.

11. a. Generally overview the adverse effects of antimicrobic drugs on the host.

b. On what basis can one explain allergy to drugs?

c. Describe the stages in a superinfection.

12. a. Outline the steps in antimicrobic susceptibility testing.

b. Compare the interpretation of the Kirby-Bauer technique and the E test with the MIC technique.

c. What is the therapeutic index, and how is it used?

CRITICAL-THINKING QUESTIONS

1. Describe the events that occur when an oral drug is taken to treat (1) a skin infection or (2) meningitis (an infection of the meninges of the brain).

2. Occasionally, one will hear the expression that a microbe has become "immune" to a drug.

a. What is a better way to explain what is happening?

b. Explain a simple test one could do to determine if drug resistance was developing in a culture.

3. a. Can you think of additional ways that drug resistance can be prevented?

b. What can health care workers do?

c. What can one do on a personal level?

4. Drugs are often given to surgical patients, to dental patients with heart disease, or to healthy family members exposed to contagious infections.

a. What word would you use to describe this use of drugs?

b. What is the purpose of this form of treatment?

c. Explain some potential undesired effects of this form of therapy.

5. a. Your pregnant neighbor has been prescribed a daily dose of oral tetracycline for acne. Do you think this therapy is advisable for her? Why or why not?

b. A woman has been prescribed a broad-spectrum oral cephalosporin for a strep throat. What are some possible consequences in addition to cure of the infected throat?

c. A man has a severe case of gastroenteritis that is negative for bacterial pathogens. A physician prescribes an oral antibacterial drug in treatment. What are your opinions of this therapy?

6. You have been directed to take a sample from a growth-free portion of the zone of inhibition in the Kirby-Bauer test and inoculate it onto a plate of nonselective medium.

a. What does it mean if growth occurs on the new plate?

b. What if there is no growth?

7. In cases in which it is not possible to culture or drug test an infectious agent (such as middle ear infection), how would the appropriate drug be chosen?

8. Using the results in tables 12.5 and 12.6 and reviewing drug characteristics, choose an antimicrobic for each of the following situations (explain your choice):

a. for an adult patient suffering from *Mycoplasma* pneumonia

b. for a child with meningitis (drug must enter into cerebrospinal fluid)

c. for a patient with allergy to erythromycin

d. for a urinary tract infection by *Enterococcus*

e. for gonorrhea

9. What factors can play a part in drug synergism?

10. How would you personally feel about being told by a physician that your infection cannot be cured by an antibiotic, and that the best thing to do is go home, drink a lot of fluids, and take aspirin or other symptom-relieving drugs?

11. Refer to the tube dilution test shown in figure 12.23, and give the MIC of the drug being tested.

12. a. Explain the basis for combined therapy.

b. Give reasons why it could be helpful to use combined therapy in treating HIV infection.

INTERNET SEARCH TOPICS

1. Look up multidrug efflux pumps on a search engine and determine how these molecules confer drug resistance on bacteria.

2. Locate information on new types of antibacterial drugs called linezolid, evevnimycin, daptomycin, and synercid. Determine their mode of action and indications for use.

Microbe-Human Interactions: Infection and Disease

The human body exists in a state of dynamic equilibrium with microorganisms. In the healthy individual, this balance is maintained as a peaceful coexistence and lack of disease. But on occasion, the balance tips in favor of the microorganism, and an infection or disease results. In this chapter, we explore each component of the host-parasite relationship, beginning with the nature and function of normal flora, moving to the stages of infection and disease, and closing with a study of epidemiology and the patterns of disease in populations. These topics will set the scene for chapters 14 and 15, which deal with the ways the host defends itself against assault by microorganisms.

Chapter Overview

- Humans are constantly exposed to microbes in their environment, mostly without harm.
- When microbes colonize the human body, sometimes they become part of the normal flora and other times, they may cause infection and disease.
- Normal flora are bacteria, fungi, and protozoa that reside naturally in the skin, gastrointestinal tract, mucous membranes, and genitourinary tracts, where they provide a stabilizing balance.
- Pathogens invade the body and cause harm to the tissues by means of virulence factors.
- An infection occurs when an adequate dose of pathogens gains access to the body through a certain route, and subsequently adheres, grows, and disrupts tissues.
- Pathogens produce virulence factors such as toxins and enzymes to help them invade the host and damage cells.
- Types of infections include local, systemic, blood, latent, primary, secondary, acute, chronic, and asymptomatic.
- Infections and diseases often follow predictable patterns involving an incubation period, invasion period, and convalescent period.
- The effects of infections and diseases are manifest as symptoms and signs, which may lead to long-term, permanent damage.
- Epidemiology is a field that studies the patterns of disease occurrence in a population.
- Epidemiologists are concerned with monitoring the numbers of cases; the geographic distribution; the sex, age, and ethnicity of affected people; and the mortality rate.
- Disease statistics are reported to various public health agencies to keep track of regional, sporadic, epidemic, and worldwide levels of distribution.

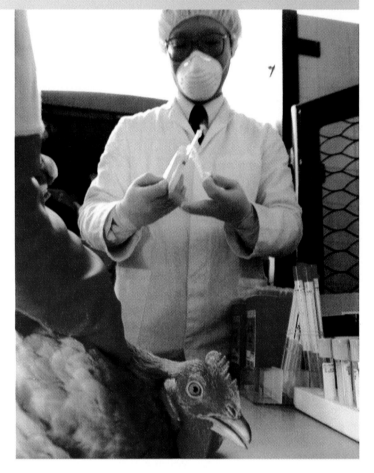

A virologist prepares to take blood from a chicken to screen for infection with a strain of influenza. This deadly virus can cause widespread illness in chickens and is readily transmissible to humans. Public health agencies monitor both domestic and wild animals to control the spread of infectious diseases.

- Pathogens originate from humans, animals, food, water, and the environment. Some are transmissible and some are not.
- Transmissible infectious diseases may be spread by direct or indirect means by overtly infected people, carriers, vectors, and vehicles.
- A significant number of infections are acquired through exposure to the hospital environment.
- The causative agents for diseases are determined by a standard set of scientific trials developed by Koch.

MEDICAL MICROFILE 13.1
Patterns of Microbe-Human Interactions

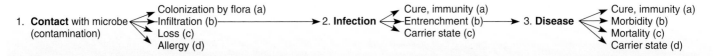

1. **Contact** with microbe (contamination) →
   - Colonization by flora (a)
   - Infiltration (b) →
   - Loss (c)
   - Allergy (d)

→ 2. **Infection** ←
   - Cure, immunity (a)
   - Entrenchment (b) →
   - Carrier state (c)

→ 3. **Disease** ←
   - Cure, immunity (a)
   - Morbidity (b)
   - Mortality (c)
   - Carrier state (d)

1. Microbes are first acquired on the exposed areas of the body through contact with other living things or the environment. **The instances and consequences of microbe-human contact vary with the type of microbe, the condition of the body, and the actions of the host:**
   a. After initial contact, some microbes take up residence as part of the normal flora, a diverse yet relatively stable collection of microbes that are important in maintaining the host equilibrium. (Note that even normal flora may, under certain circumstances, cause infection.)
   b. Some microbes with greater infectious potential may avoid host defenses, infiltrate the body, and cause an infection; that is, they act as **pathogens.** Many factors determine whether an infection takes place, but in general, the greater the **virulence* of the microbe, the more likely it is that infection will take place.**
   c. Some microbes (the transients) stay on the body for only a short time and are destroyed by the host defenses or removed by an action such as cleaning or antisepsis.
   d. Some microbes or microbial products result in hypersensitivity reactions, or allergy.

2. **In the event of infection, three outcomes are possible:**
   a. The immune system may arrest the infection before injury to tissues and organs occurs.
   b. The infectious agent is not arrested and becomes entrenched in tissues, where its **pathologic*** effects cause some degree of damage.
   c. The microbe may be harbored inconspicuously for varying lengths of time (carrier state).

3. **If infection proceeds to disease, several outcomes are possible:**
   a. The immune system may eventually arrest the microbe and stop the disease process.
   b. Damaged tissues and organs may lead to dysfunction or **morbidity.***
   c. Damage may be severe enough to cause death, or mortality.
   d. A carrier state is established.

***virulence** (veer-yoo-lents) L. *virulentia,* virus, poison.

* **pathologic** (path″-uh-loj′-ik) A disease state caused by structural and functional damage to tissues.

* **morbidity** (mor-bih′-dih-tee) L. *morbidis,* sick. A condition of being diseased.

The Human Host

In chapter 7, several of the basic interrelationships between humans and microorganisms were considered. Most of the microbes inhabiting the human body benefit from the nutrients and protective habitat it provides. From the human point of view, these relationships run the gamut from mutualism to commensalism to parasitism and can have beneficial, neutral, or harmful effects (see page 204). A common characteristic of all microbe-human relationships, regardless of where they lead, is that they begin with contact.

CONTACT, INFECTION, DISEASE—A CONTINUUM

The body surfaces are constantly exposed to microbes. Some microbes become implanted there as colonists (normal flora), some are rapidly lost (transients), and others invade the tissues. Such intimate contact with microbes inevitably leads to **infection,** a condition in which pathogenic microorganisms penetrate the host defenses, enter the tissues, and multiply. When the cumulative effects of the infection damage or disrupt tissues and organs, a **disease** results. A disease is defined as any deviation from health. There are hundreds of different diseases caused by such factors as infections, diet, genetics, and aging. In this chapter, however, we will discuss

only **infectious disease**—the disruption of a tissue or organ caused by microbes or their products.

The pattern of the host-parasite relationship can be viewed as a series of stages that begins with contact, progresses to infection, and ends in disease (Medical Microfile 13.1). Because of numerous factors relating to host resistance and degree of pathogenicity, not all contacts lead to infection and not all infections lead to disease. In fact, contamination without infection and infection without disease are the rule. Before we consider further details of infection and disease, let us examine the fascinating relationship between humans and their resident flora.

RESIDENT FLORA: THE HUMAN AS A HABITAT

With its constant source of nourishment and moisture, relatively stable pH and temperature, and extensive surfaces upon which to settle, the human body provides a favorable habitat for an abundance of microorganisms. In fact, it is so favorable that, cell-for-cell, microbes outnumber human cells ten to one! The large and mixed collection of microbes adapted to the body has been variously called the **normal resident flora,** or *indigenous* flora,*

---

* **indigenous** (in-dih′-juh-nus) Belonging or native to.

TABLE 13.1
Sites That Harbor a Normal Flora
Skin and its contiguous mucous membranes
Upper respiratory tract
Gastrointestinal tract (various parts)
Outer opening of urethra
External genitalia
Vagina
External ear canal
External eye (lids, conjunctiva)

TABLE 13.2
Sterile (Microbe-Free) Anatomical Sites and Fluids
All Internal Tissues and Organs
Heart and circulatory system
Liver
Kidneys and bladder
Lungs
Brain and spinal cord
Muscles
Bones
Ovaries/testes
Glands (pancreas, salivary, thyroid)
Sinuses
Middle and inner ear
Internal eye
Fluids Within an Organ or Tissue
Blood
Urine in kidneys, ureters, bladder
Cerebrospinal fluid
Saliva prior to entering the oral cavity
Semen prior to entering the urethra
Amniotic fluid surrounding the embryo and fetus

though some microbiologists prefer to use the terms *microflora* and *commensals*. The normal residents include an array of bacteria, fungi, protozoa, and, to a certain extent, viruses and arthropods.

Acquiring Resident Flora

Like a virgin planet being settled by alien beings, humans begin to acquire microflora from their earliest contacts with their immediate environment. Development of the flora proceeds in a relatively orderly and characteristic way. Of the seemingly limitless influx of microbes, only some are capable of persisting in and on the body. The vast majority are destroyed by host defenses or are removed. Those that remain survive by adapting to a particular microhabitat—one that fills their requirements for food, moisture, presence or lack of oxygen, layers of dead cells, and even other microorganisms. In general, most anatomical surfaces directly exposed to the environment can and do harbor microbes. As indicated in table 13.1, these surfaces are mainly the skin and mucous membranes, parts of the inner surface of the gastrointestinal tract, and openings to the cutaneous surface from the urinary, respiratory, and reproductive tracts. By contrast, organs and fluids inside the body cavity and the central nervous system remain free of normal flora during life because host defense mechanisms maintain their sterility (table 13.2). Although microorganisms can transiently enter these sites, they do not normally become established there.

Interactions between host and indigenous flora have evolved over millions of years into complex and dynamic relationships that directly and indirectly influence many aspects of body function. For the most part, effects of normal flora are protective or nutritional. For instance, the flora can modify the microhabitat by altering pH and oxygen tension; excreting chemicals such as fatty acids, gases, alcohol, and antibiotics; or creating barriers. Such reactions can favor colonization by more beneficial types of microbes. They can also inhibit potential pathogens by restricting their population size and activity through antagonism and competition for nutrients. Other expected functions of the normal flora are discussed in a later section on germ-free animals.

Although relatively stable, the flora can fluctuate to a limited extent with general health, age, variations in diet, hygiene, hormones, and drug therapy. Most of the microorganisms are not serious pathogens and remain harmless as long as they do not penetrate superficial skin and mucosal barriers. However, a few species have the potential to become medically important if the balance of the

flora shifts. This imbalance is seen in the overgrowth of *Clostridium difficile* or *Candida albicans* in patients undergoing antibiotic therapy. After being temporarily disturbed, the normal balance of flora is restored by surviving residents harbored in protected areas and by contact with the environment. Members of the flora can also become infectious if the health of the host is compromised (as by cancer or AIDS), allowing opportunists (such as the agents of pneumococcal and *Pneumocystis* pneumonia) to overcome the weakened host defenses (see table 13.5). It is notable that some humans harbor organisms that can be pathogenic to others. As a consequence, the flora is an important reservoir for bacterial pathogens such as *Staphylococcus aureus*, *Corynebacterium diphtheriae*, and *Neisseria meningitidis*.

Initial Colonization of the Newborn

The uterus and its contents are normally sterile during embryonic and fetal development and remain essentially germ-free until just before birth. The event that first exposes the infant to microbes is the breaking of the fetal membranes, at which time microbes from the mother's vagina can enter the womb. Comprehensive exposure occurs during the birth process itself, when the baby unavoidably comes into intimate contact with the birth canal (figure 13.1). Within 8 to 12 hours after delivery, the newborn typically has been colonized by bacteria such as streptococci, staphylococci, and lactobacilli, acquired primarily from its mother. The nature of the flora initially colonizing the large intestine depends upon whether the baby is bottle- or breast-fed. Bottle-fed infants (receiving milk or a milk-based formula) tend to acquire a mixed population of coliforms, lactobacilli, enteric streptococci, and staphylococci. In contrast, the intestinal flora of breast-fed infants consists primarily of

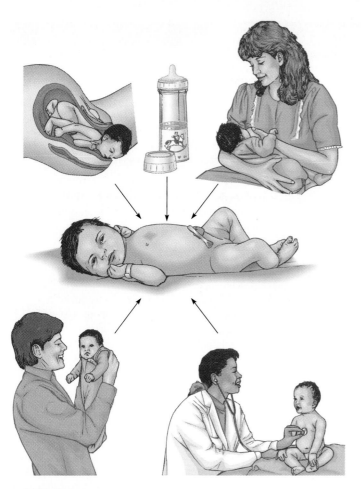

FIGURE 13.1
The origins of flora in newborns.

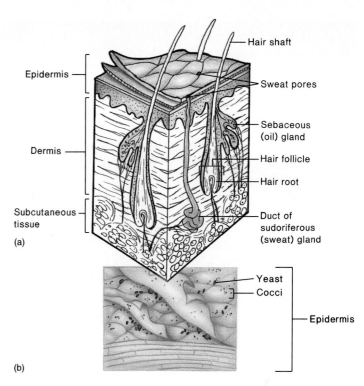

FIGURE 13.2

🔲 **The landscape of the skin.** **(a)** The epidermis, along with associated glands and follicles (colored), provides rich and diverse habitats. Noncolored regions (dermis and subcutaneous layers) are free of microbial flora. **(b)** A highly magnified view of the skin surface reveals beds of rod- and coccus-shaped bacteria and yeasts beneath peeled-back skin flakes.

Bifidobacterium species whose growth is favored by a growth factor from the milk. This bacterium metabolizes sugars into acids that protect the infant from infection by certain intestinal pathogens. The skin, gastrointestinal tract, and portions of the respiratory and genitourinary tract all continue to be colonized as contact continues with family members, hospital personnel, the environment, and food.

Milestones that contribute to development of the adult pattern of flora are eruption of teeth, weaning, and introduction of the first solid food. Although exposure to microbes is unavoidable and even necessary for the maturation of the infant's flora, contact with pathogens is dangerous, because the neonate is not yet protected by a full complement of flora, and owing to its immature immune defenses, is extremely susceptible to infection.

INDIGENOUS FLORA OF SPECIFIC REGIONS

Although we tend to speak of the flora as a single unit, it is a complex mixture of hundreds of species, differing somewhat in quality and quantity from one individual to another. Studies of the flora have shown that most people harbor certain specially adapted bacteria, fungi, and protozoa (table 13.3).

Flora of the Human Skin

The skin is the largest and most accessible of all organs. Its major layers are the epidermis, an outer layer of dead cells continually being sloughed off and replaced, and the dermis, which lies atop the subcutaneous layer of tissue (figure 13.2*a*). Depending on its location, skin also contains hair follicles and several types of glands, and the outermost surface is covered with a protective, waxy cuticle that can help microbes adhere. The normal flora resides only in or on the dead cell layers, and except for in follicles and glands, it does not extend into the dermis or subcutaneous levels. The nature of the population varies according to site. Oily, moist skin supports a more prolific flora than dry skin. Humidity, occupational exposure, and clothing also influence its character. Transition zones where the skin joins with the mucous membranes of the nose, mouth, and external genitalia harbor a particularly rich flora.

Ordinarily, there are two cutaneous populations. The **transient population,** or exposed flora, clings to the skin surface but does not ordinarily grow there. It is acquired by routine contact, and it varies markedly from person to person and over time. The transient flora includes any microbe a person has picked up, often species that do not ordinarily live on the body, and it is greatly influenced by the hygiene of the individual.

TABLE 13.3

Life on Humans: Sites Containing Well-Established Flora and Representative Examples

Anatomic Sites	Common Genera	Remarks
Skin	**Bacteria:** *Staphylococcus, Micrococcus, Corynebacterium, Propionibacterium, Mycobacterium*	Microbes live only in upper dead layers of epidermis, glands, and follicles; dermis and layers below are sterile.
	Fungi: *Malassezia* yeast	Dependent on skin lipids for growth.
	Arthropods: *Demodix* mite	Present in sebaceous glands and hair follicles.
Gastrointestinal Tract Oral cavity	**Bacteria:** *Streptococcus, Neisseria, Veillonella, Staphylococcus, Fusobacterium, Lactobacillus, Bacteroides, Corynebacterium, Actinomyces, Eikenella, Treponema, Haemophilus*	Colonize the epidermal layer of cheeks, gingiva, pharynx; surface of teeth; found in saliva in huge numbers.
	Fungi: *Candida* species	Can cause thrush.
	Protozoa: *Trichomonas tenax, Entamoeba gingivalis*	Frequent the gingiva of persons with poor oral hygiene.
Large intestine and rectum	**Bacteria:** *Bacteroides, Fusobacterium, Eubacterium, Bifidobacterium, Clostridium,* fecal streptococci, *Peptococcus, Lactobacillus,* coliforms (*Escherichia, Enterobacter*)	Sites of lower gastrointestinal tract other than large intestine and rectum have sparse or nonexistent flora. Flora consists predominantly of strict anaerobes; other microbes are aerotolerant or facultative.
	Fungi: *Candida*	
	Protozoa: *Entamoeba coli, Trichomonas hominis*	Feed on waste materials in the large intestine.
Upper Respiratory Tract	Microbial population exists in the nasal passages, throat, and pharynx; owing to proximity, flora is similar to that of oral cavity	Trachea and bronchi have a sparse population; smaller breathing tubes and alveoli have no normal flora and are essentially sterile.
Genital Tract	**Bacteria:** *Lactobacillus, Streptococcus, Corynebacterium, Escherichia, Mycobacterium*	In females, flora occupies the external genitalia and vaginal and cervical surfaces; internal reproductive structures normally remain sterile. Flora responds to hormonal changes during life.
	Fungi: *Candida*	Cause of yeast infections.
Urinary Tract	**Bacteria:** *Staphylococcus, Streptococcus,* coliforms	In females, flora exists only in the first portion of the urethral mucosa; the remainder of the tract is sterile. In males, the entire reproductive and urinary tract is sterile except for a short portion of the anterior urethra.

The **resident population** lives and multiplies in deeper layers of the epidermis and in glands and follicles (figure 13.2*b*). The composition of the resident flora is more stable, predictable, and less influenced by hygiene than is the transient flora. The normal skin residents consist primarily of bacteria (notably *Staphylococcus, Corynebacterium,* and *Propionibacterium*) and yeasts. Moist skin folds, especially between the toes, tend to harbor fungi, whereas lipophilic mycobacteria and staphylococci are prominent in sebaceous[1] secretions of the axilla, external genitalia, and external ear canal. One species, *Mycobacterium smegmatis,* lives in the cheesy secretion, or *smegma,* on the external genitalia of men and women.

Flora of the Gastrointestinal Tract

The gastrointestinal (GI) tract receives, moves, digests, and absorbs food; it also removes waste. It encompasses the oral cavity, esophagus, stomach, small intestine, large intestine, rectum, and anus.

Stating that the GI tract harbors flora may seem to contradict our earlier statement that internal organs are sterile, but it is not an exception to the rule. How can this be true? In reality, the GI tract is a long, hollow tube (with numerous pockets and curves), bounded by the mucous membranes of the oral cavity on one extreme and those of the anus on the other. Because the innermost surface of this tube is exposed to the environment, it is topographically outside the body, so to speak (figure 13.3). This architecture permits ingested materials to be processed before they cross the mucosal epithelium, but at the same time, it creates numerous niches for microbes.

The shifting conditions of pH, oxygen tension, and differences in microscopic anatomy of the GI tract are reflected by the variations in or distribution of the flora (figure 13.4). Some microbes remain attached to the mucous epithelium or its associated structures, and others dwell in the *lumen.** Although the abundance of nutrients invites microbial growth, the only areas that

---

1. From sebum, a lipid material secreted by glands in the hair follicles.

* lumen (loo′-men) The space within a tubular structure.

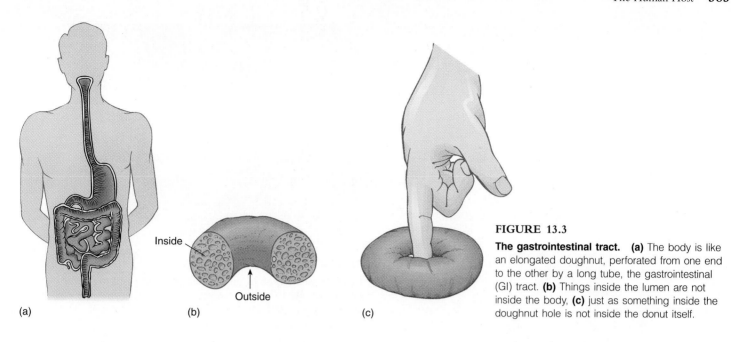

Inside

Outside

(a)   (b)   (c)

FIGURE 13.3

The gastrointestinal tract. **(a)** The body is like an elongated doughnut, perforated from one end to the other by a long tube, the gastrointestinal (GI) tract. **(b)** Things inside the lumen are not inside the body, **(c)** just as something inside the doughnut hole is not inside the donut itself.

harbor appreciable permanent flora are the oral cavity, large intestine, and rectum. The esophagus undergoes wavelike contractions (peristalsis), a process that constantly flushes microorganisms; the stomach acid inhibits most microbes; and peristalsis and digestive enzymes help exclude flora from all but the terminal segment of the small intestine.

Flora of the Mouth   The oral cavity has a unique flora that is among the most diverse and abundant of the body. Microhabitats, including the cheek epithelium, gingiva, tongue, floor of the mouth, and tooth enamel, provide numerous adaptive niches for hundreds of different species to colonize. The most common residents are aerobic *Streptococcus* species—*S. sanguis, S. salivarius, S. mitis*— that colonize the smooth superficial epithelial surfaces. Two species, *S. mutans* and *S. sanguis,* make a major contribution to dental caries by forming sticky dextran slime layers in the presence of simple sugars. The adherence of dextrans to the tooth surface establishes a medium that attracts other bacteria (see figure 21.30).

Once the teeth have erupted, an anaerobic habitat is established in the gingival crevice that favors the colonization of anaerobic bacteria that can be involved in dental caries and periodontal[2] infections. The instant that saliva is secreted from ducts into the oral cavity, it becomes laden with resident and transient flora. Saliva normally has a high bacterial count (up to 5×10^9 cells per milliliter), a fact that tends to make mouthwashes rather ineffective and a human bite very dangerous.

Flora of the Large Intestine   The flora of the intestinal tract has complex and profound interactions with the host. The large intestine (cecum and colon) and the rectum harbor a huge population of microbes (10^8–10^{11} per gram of feces) (figure 13.5). So abundant and prolific are these microbes that they constitute 10–30% of the

fecal volume. Even an individual on a long-term fast passes feces consisting primarily of bacteria.

Because of the state of the intestinal environment, the predominant fecal flora are strictly anaerobic bacteria *(Bacteroides, Bifidobacterium, Fusobacterium,* and *Clostridium)*. **Coliforms*** such as *Escherichia coli, Enterobacter,* and *Citrobacter* are present in smaller numbers. Many species ferment waste materials in the feces, generating vitamins (B_{12}, vitamin K, pyridoxine, riboflavin, and thiamine) and acids (acetic, butyric, and propionic acids) of potential value to the host. Occasionally significant are bacterial digestive enzymes that convert disaccharides to monosaccharides or promote steroid metabolism.

Intestinal bacteria contribute to intestinal odor by producing *skatole,** amines, and gases (CO_2, H_2, CH_4, and H_2S). Intestinal gas is known in polite circles as *flatus,* and the expulsion of it as *flatulence.* Some of the gas arises through the action of bacteria on dietary carbohydrate residues from vegetables such as cabbage, corn, and beans. The bacteria produce an average of 8.5 liters of gas daily, but only a small amount is ejected in flatus. Combustible gases occasionally form an explosive mixture in the presence of oxygen that has reportedly ignited during intestinal surgery and ruptured the colon!

Late in childhood, members of certain ethnic groups lose the ability to secrete the enzyme lactase. When they ingest milk or other lactose-containing dairy products, lactose is acted upon instead by intestinal bacteria, and severe intestinal distress can result. The recommended treatment for this deficiency is avoiding these foods or eating various lactose-free substitutes.

---

2. Situated or occurring around the tooth.

---

* **coliform** (koh´-lih-form) L. *colum,* a sieve. Gram-negative, facultatively anaerobic, and lactose-fermenting microbes.

* **skatole** (skat´-ohl) Gr. *skatos,* dung. One chemical that gives feces its characteristic stench.

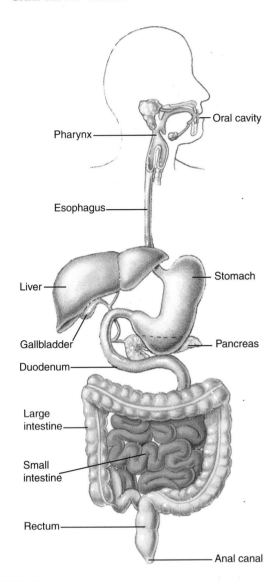

FIGURE 13.4

Distribution of flora. Areas of the gastrointestinal tract that shelter resident flora are highlighted in color. Noncolored areas do not regularly harbor residents.

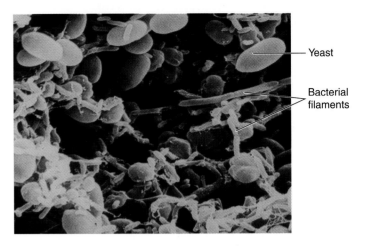

FIGURE 13.5

Electron micrograph of the mucous membrane of the large intestine. Microcolonies of yeasts and long, filamentous bacteria located in pockets are visible.

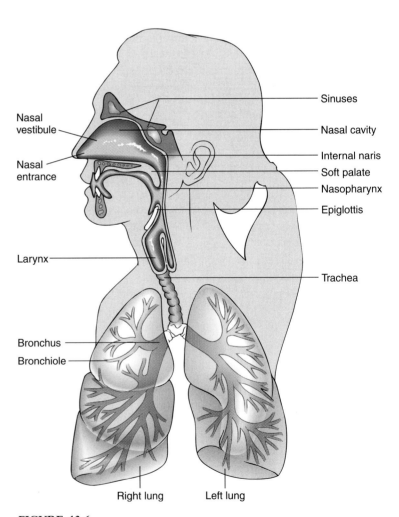

FIGURE 13.6

Colonized regions of the respiratory tract. The moist mucous blanket of the nasopharynx has a well-entrenched flora (pink). Some colonization occurs in the upper trachea, but lower regions of bronchi, bronchioles, and lungs lack resident microbes.

Flora of the Respiratory Tract

The first microorganisms to colonize the upper respiratory tract (nasal passages and pharynx) are predominantly oral streptococci. Inhaled air regularly contains microbes that are filtered out, destroyed, or expelled, although some can adapt to specific regions of this habitat. *Staphylococcus aureus* preferentially resides in the nasal entrance, nasal vestibule, and anterior nasopharynx, and *Neisseria* species take up residence in the mucous membranes of the nasopharynx behind the soft palate (figure 13.6). Lower still are assorted streptococci and species of *Haemophilus* that colonize the tonsils and lower pharynx. Conditions lower in the respiratory tree (bronchi and lungs) are unfavorable habitats for permanent residents.

Flora of the Genitourinary Tract

The regions of the genitourinary tract that harbor microflora are the vagina and outer opening of the urethra in females and the anterior urethra in males (figure 13.7). The internal reproductive organs are

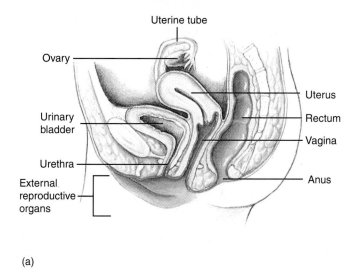

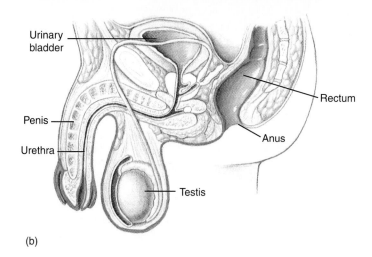

(a)

(b)

FIGURE 13.7

Location of (a) female and (b) male genitourinary flora (indicated by color).

kept sterile through physical barriers such as the cervical plug and other host defenses. The kidney, ureter, bladder, and upper urethra are presumably kept sterile by urine flow and regular bladder emptying. Because the urethra in women is so short (about 3.5 cm long), it can form a passage for bacteria to the bladder and lead to urinary tract infections. The principal residents of the urethra are nonhemolytic streptococci, staphylococci, corynebacteria, and occasionally, coliforms.

The vagina presents a notable example of how changes in physiology can greatly influence the composition of the normal flora. An important factor influencing these changes in women is the hormone estrogen. Estrogen normally stimulates the vaginal mucosa to secrete the carbohydrate glycogen, which certain bacteria (primarily *Lactobacillus* species) ferment, thus lowering the pH to about 4.5. Before puberty, a girl produces little estrogen and little glycogen, and has a vaginal pH of about 7. These conditions favor the establishment of diphtheroids,[3] staphylococci, streptococci, and some coliforms. As hormone levels rise at puberty, the vagina begins to deposit glycogen, and the flora shifts to the acid-producing lactobacilli. It is thought that the acidic pH of the vagina during this time prevents the establishment and invasion of microbes with potential to harm a developing fetus. The estrogen-glycogen effect continues, with minor disruptions, throughout the childbearing years until menopause, when the flora returns to a mixed population similar to that of prepuberty. These transitions are not abrupt, but occur over several months to years.

STUDIES WITH GERM-FREE ANIMALS

For years, questions lingered about how essential the microbial flora is to normal life and what functions various members of the flora might serve. The need for animal models to further investigate these questions led eventually to development of laboratory strains of **germ-free,** or **axenic,** mammals and birds. The techniques and

facilities required for producing and maintaining a germ-free colony are exceptionally rigorous. After the young mammals are taken from the mother aseptically by cesarian section, they are immediately transferred to a sterile isolator or incubator (figure 13.8). The newborns must be fed by hand through gloved ports in the isolator until they can eat on their own, and all materials entering their chamber must be sterile. Rats, mice, rabbits, guinea pigs, monkeys, dogs, hamsters, and cats are some of the mammals raised in the germ-free state.

Experiments with germ-free animals are of two basic varieties: (1) general studies on how the lack of normal microbial flora influences the nutrition, metabolism, and anatomy of the animal, and (2) *gnotobiotic** studies, in which the germ-free subject is inoculated either with a single type of microbe to determine its individual

FIGURE 13.8

Sterile enclosure for rearing and handling germ-free laboratory animals.

---

3. Any nonpathogenic species of *Corynebacterium*.

* **gnotobiotic** (noh″-toh-by-ah′-tik) Gr. *gnotos,* known, and *biota,* the organisms of a region.

TABLE 13.4

Effects and Significance of Experiments with Germ-Free Subjects

Effect in Germ-Free Animal	Significance
Enlargement of the cecum; other degenerative diseases of the intestinal tract of rats, rabbits, chickens	Microbes are needed for normal intestinal development
Vitamin deficiency in rats	Microbes are a significant nutritional source of vitamins
Underdevelopment of immune system in most animals	Microbes are needed to stimulate development of certain host defenses
Absence of dental caries and periodontal disease in dogs, rats, hamsters	Normal flora are essential in caries formation and gum disease
Heightened sensitivity to enteric pathogens (*Shigella, Salmonella, Vibrio cholerae*) and to fungal infections	Normal flora are antagonistic against pathogens
Lessened susceptibility to amebic dysentery	Normal flora facilitate the completion of the life cycle of the ameba in the gut

effect or with several known microbes to determine interrelationships. Results are validated by comparing the germ-free group with a conventional, normal control group. Table 13.4 summarizes some major conclusions arising from studies with germ-free animals.

A dramatic characteristic of germ-free animals is that they live longer and have fewer diseases than normal controls, as long as they remain in a sterile environment. From this standpoint, it is clear that the flora is not needed for survival and may even be the source of infectious agents. At the same time, it is also clear that axenic life is highly impractical. Additional studies have revealed important facts about the effect of the flora on various organs and systems. For example, the flora contributes significantly to the development of the immune system. When germ-free animals are placed in contact with normal control animals, they gradually develop a flora similar to that of the controls. However, germ-free subjects are less tolerant of microorganisms and can die from infections by relatively harmless species. This susceptibility is due to the immature character of the immune system of germ-free animals. These animals have a reduced number of certain types of white blood cells and slower antibody response.

Gnotobiotic experiments have clarified the dynamics of several infectious diseases. Perhaps the most striking discoveries were made in the case of oral diseases. For years, the precise involvement of microbes in dental caries had been ambiguous. Studies with germ-free rats, hamsters, and beagles confirmed that caries development is influenced by heredity, a diet high in sugars, and poor oral hygiene. Even when all these predisposing factors are present, however, germ-free animals still remain free of caries unless they

have been inoculated with specific bacteria. Further discussion on dental diseases is found in chapter 21.

The ability of known pathogens to cause infection can also be influenced by normal flora, sometimes in opposing ways. Studies have indicated that germ-free animals are highly susceptible to experimental infection by the enteric pathogens *Shigella* and *Vibrio*, whereas normal animals are less susceptible, presumably because of their protective flora. In marked contrast, *Entamoeba histolytica* (the agent of amebic dysentery) is more pathogenic in the normal animal than in the germ-free animal. One explanation for this phenomenon is that *E. histolytica* must feed on intestinal bacteria to complete its life cycle.

CHAPTER CHECKPOINTS

Humans are contaminated with microorganisms from the moment of birth onward. An infection is a condition in which contaminating microorganisms overcome host defenses, multiply, and cause disease, damaging tissues and organs.

The resident or normal flora includes bacteria, fungi, and protozoa.

There are two types of cutaneous populations of microbes: the transients, which cling to but do not grow on the superficial layers of the skin, and the more permanent residents, which reside in the deeper layers of the epidermis and its glands.

The flora of the alimentary canal is confined primarily to the mouth, large intestine, and rectum.

The flora of the respiratory tract extends from the nasal cavity to the lower pharynx.

The flora of the genitourinary tract is restricted to the urethral opening in males and to the urethra and vagina in females.

Special germ-free, or axenic, animals have been developed for testing the effects of the normal flora and how they influence diseases.

The Progress of an Infection

When the association between host and parasite tilts in the direction of infection and disease, a particular series of events is set into motion. The microbe enters the body, attaches itself, invades (crosses host barriers), multiplies in a target tissue, and finally is released to the exterior by various pathways (figure 13.9).

The type and severity of an infection depend on numerous factors, most of which are related to the pathogenicity of the microbe and the condition of the host. *Pathogenicity,* we learned earlier, is a broad concept that describes a microorganism's potential to cause an infection or disease. Pathogenic microbes have been traditionally divided into two categories, depending upon the nature of the microbe-host relationship. **True pathogens** (primary pathogens) are capable of causing infection and disease in healthy persons with normal immune defenses. Examples include the influenza virus, plague bacillus, rabies virus, malarial protozoan, and other microorganisms with well-developed qualities of virulence (discussion follows). Primary pathogens are generally associated with a distinct, recognizable disease. The degree of pathogenicity

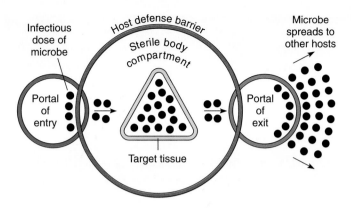

FIGURE 13.9

An overview of the events in infection.   An adequate dose of an infectious agent overcomes a defense barrier and enters the sterile tissues of a host through one of several portals of entry. From here, it moves or is carried to a specific organ or tissue, called the target tissue, where it multiplies and usually causes some degree of damage. The exit of the pathogen through the same or another portal can facilitate its transmission to another host.

TABLE 13.5
Factors That Weaken Host Defenses and Increase Susceptibility to Infection*

- Old age and extreme youth (infancy, prematurity)
- Genetic defects in immunity and acquired defects in immunity (AIDS)
- Surgery and organ transplants
- Organic disease: cancer, liver malfunction, diabetes
- Chemotherapy/immunosuppressive drugs
- Physical and mental stress
- Other infections

*These conditions compromise defense barriers or immune responses.

of true pathogens ranges from weak to potent; for instance, infection with the cold virus is mild and nonfatal, whereas infection with the rabies virus is nearly 100% fatal.

Opportunistic pathogens cause disease when the host's defenses are compromised or when they become established in a part of the body that is not natural to them. Opportunists are not considered pathogenic to the normal, healthy person and do not usually have well-developed virulence properties. It could be argued that most microbes have some capacity for pathogenicity. For example, a severely compromised host may be challenged by a large inoculum in a vulnerable site. Another side of this idea is that anyone who becomes infected may be temporarily compromised to an extent. Factors that greatly **predispose** a person to infections, both primary and opportunistic, are shown in table 13.5.

Recognizing that classifying pathogens as either true or opportunistic may be an oversimplification, the Centers for Disease Control and Prevention has adopted a system of biosafety categories for pathogens based on their degree of pathogenicity and the relative danger in handling them. With this system, presented in Table M.1, page 542, pathogens are assigned a level or class. Microbes not known to cause disease in humans (such as *Micrococcus luteus*) are given class 1 status; moderate risk agents such as *Staphylococcus aureus* are assigned to class 2; readily transmitted and virulent agents such as *Mycobacterium tuberculosis* are in class 3; and the highest risk microbes (deadly pathogens such as rabies and Ebola fever viruses) are in class 4.

Differences in pathogenicity can be accounted for on the basis of virulence. Virulence takes into account the ability of microbes to invade a host and produce toxins (toxigenicity). These properties are called **virulence factors.** Virulence can be due to single or multiple factors. In some microbes, the causes of virulence are clearly established, but in others they are not. Although virulence is sometimes used interchangeably with pathogenicity, it is actually not a synonym. Pathogenicity is a more general term for comparing degrees of microbial infectiousness. In the following

section, we examine the effects of virulence factors while simultaneously outlining the stages in the progress of an infection.

THE PORTAL OF ENTRY: GATEWAY TO INFECTION

To initiate an infection, a microbe enters the tissues of the body by a characteristic route, the **portal of entry,** usually a cutaneous or membranous boundary. The source of the infectious agent can be **exogenous,** originating from a source outside the body (the environment or another person or animal), or **endogenous,** already existing on or in the body (normal flora or latent infection).

For the most part, the portals of entry are the same anatomical regions that also support normal flora: the skin, gastrointestinal tract, respiratory tract, and urogenital tract. The majority of pathogens have adapted to a specific portal of entry, one that provides a habitat for further growth and spread. This adaptation can be so restrictive that if certain pathogens enter the "wrong" portal, they will not be infectious. For instance, inoculation of the nasal mucosa with the influenza virus invariably gives rise to the flu, but if this virus contacts only the skin, no infection will result. Likewise, contact with athlete's foot fungi in small cracks in the toe webs can induce an infection, but inhaling the fungus spores will not infect a healthy individual. Occasionally, an infective agent can enter by more than one portal. For instance, *Mycobacterium tuberculosis* enters through both the respiratory and gastrointestinal tracts, and *Corynebacterium diphtheriae* can infect by way of the throat and the skin. Common pathogens in the genera *Streptococcus* and *Staphylococcus* have adapted to invasion through several portals of entry such as the skin, urogenital tract, and respiratory tract.

Infectious Agents That Enter the Skin

The skin is a very common portal of entry. The actual sites of entry are usually nicks, abrasions, and punctures (many of which are tiny and inapparent) rather than smooth, unbroken skin. *Staphylococcus aureus* (the cause of boils), *Streptococcus pyogenes* (an agent of impetigo), the fungal dermatophytes, and agents of gangrene and tetanus gain access through damaged skin. The viral agent of cold sores (herpes simplex, type 1) enters through the mucous membranes near the lips.

The older term for STD—*venereal*—is derived from Venus, the Latin name of the goddess of love. For centuries, sexually oriented diseases were considered a curse from God, with the greater share of blame attributed to women. During biblical times, so great was the condemnation of women afflicted with sex-associated diseases that many of them were executed. Attempts to treat or control infection during the early twentieth century were squelched by clergymen protesting that, "If God had wanted to eradicate venereal disease, he would not have given it to women in the first place." Posters from World Wars I and II and even into the 1950s flaunt the role of women in these diseases and depict them in the most debasing ways. Just as it is clear that STDs can be stigmatizing, it is clear that men share equally in their transmission. The term *sexually transmitted disease* recognizes this fact.

STDs account for an estimated 4% of infections worldwide, and approximately 15 million new cases occur in the United States every year, mostly among young, sexually active people. The most recent available statistics (1998) for the estimated incidence of common STDs comes from the CDC:

STD	Estimated Number of New Cases
Human papillomavirus	5,500,000
Trichomoniasis	5,000,000
Chlamydiosis	3,000,000
Herpes simplex	1,000,000
Gonorrhea	650,000
Hepatitis B	77,000
Syphilis	70,000
AIDS	40,000

Some infectious agents create their own passageways into the skin using digestive enzymes. For example, certain helminth worms burrow through the skin directly to gain access to the tissues. Other infectious agents enter through bites. The bites of insects, ticks, and other animals offer an avenue to a variety of viruses, rickettsias, and protozoa. An artificial means for breaching the skin barrier is contaminated hypodermic needles by intravenous drug abusers. Users who inject drugs are predisposed to a disturbing list of well-known diseases: hepatitis, AIDS, tetanus, tuberculosis, osteomyelitis, and malaria. A resurgence of some of these infections is directly traceable to drug use. Contaminated needles often contain bacteria from the skin or environment that induce heart disease (endocarditis), lung abscesses, and chronic infections at the injection site.

Although the conjunctiva, the outer protective covering of the eye, is ordinarily a relatively good barrier to infection, bacteria such as *Haemophilus aegyptius* (pinkeye), *Chlamydia trachomatis* (trachoma), and *Neisseria gonorrhoeae* have special affinity for this membrane. Foreign bodies in the eye or minor injuries can serve as a portal for herpes simplex virus and *Acanthamoeba* (a protozoan that can contaminate homemade contact lens solutions).

The Gastrointestinal Tract as Portal

The gastrointestinal tract is the portal of entry for pathogens contained in food, drink, and other ingested substances. They are adapted to survive digestive enzymes and abrupt pH changes. Most enteric pathogens possess specialized mechanisms for entering and localizing in the mucosa of the small or large intestine. The best-known enteric agents of disease are gram-negative rods in the genera *Salmonella, Shigella, Vibrio,* and certain strains of *Escherichia coli* (see Spotlight on Microbiology 20.3). Viruses that enter through the gut are poliovirus, hepatitis A virus, echovirus, and rotavirus. Important enteric protozoans are *Entamoeba histolytica* (amebiasis) and *Giardia lamblia* (giardiasis). Although the anus is not a typical portal of entry, it becomes one in people who practice anal sex.

The Respiratory Portal of Entry

The oral and nasal cavities are also the gateways to the respiratory tract, the portal of entry for the greatest number of pathogens. The extent to which an agent is carried into the respiratory tree is based primarily on its size. In general, small cells and particles are inhaled more deeply than larger ones. Infectious agents with this portal of entry include the bacteria of streptococcal sore throat, meningitis, diphtheria, and whooping cough and the viruses of influenza, measles, mumps, rubella, chickenpox, and the common cold. Pathogens that are inhaled into the lower regions of the respiratory tract (bronchioles and lungs) can cause **pneumonia,** an inflammatory condition of the lung. Bacteria (*Streptococcus pneumoniae, Klebsiella, Mycoplasma*) and fungi (*Cryptococcus* and *Pneumocystis*) are a few of the agents involved in pneumonias. All types of pneumonia are on the increase owing to the greater susceptibility of AIDS patients to them. Other agents causing unique recognizable lung diseases are *Mycobacterium tuberculosis* and fungal pathogens such as *Histoplasma*.

Urogenital Portals of Entry

The urogenital tract is the portal of entry for pathogens that are contracted by sexual means (intercourse or intimate direct contact). The older term, *venereal disease,* has been replaced by the more descriptive term, **sexually transmitted disease** (STD; Microbits 13.2). The microbes of STDs enter the skin or mucosa of the penis, external genitalia, vagina, cervix, and urethra. Some can penetrate an unbroken surface; others require a cut or abrasion. The once predominant sexual diseases syphilis and gonorrhea have been supplanted by a large and growing list of STDs headed by genital warts, chlamydia, and herpes. Evolving sexual practices have increased the incidence of STDs that were once uncommon, and diseases that were not originally considered STDs are now so classified.[4] Other common sexually transmitted agents are HIV (AIDS virus), *Trichomonas* (a protozoan), *Candida albicans* (a yeast), and hepatitis B virus.

---

4. Amebic dysentery, scabies, salmonellosis, and *Strongyloides* worms are examples.

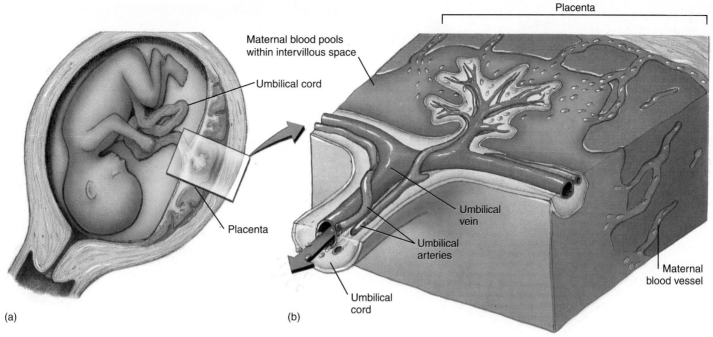

FIGURE 13.10

Transplacental infection of the fetus. **(a)** Fetus in the womb. **(b)** In a closer view, microbes are shown penetrating the maternal blood vessels and entering the blood pool of the placenta. They then invade the fetal circulation by way of the umbilical vein.

Pathogens That Infect During Pregnancy and Birth

The placenta is an exchange organ formed by maternal and fetal tissues that separates the blood of the developing fetus from that of the mother, yet permits diffusion of dissolved nutrients and gases to the fetus. The placenta is ordinarily an effective barrier against microorganisms in the maternal circulation. However, a few microbes such as the syphilis spirochete can cross the placenta, enter the umbilical vein, and spread by the fetal circulation into the fetal tissues (figure 13.10).

Other infections, such as herpes simplex, occur perinatally when the child is contaminated by the birth canal. The common infections of fetus and neonate are grouped together in a unified cluster, known by the acronym **STORCH,** that medical personnel must monitor. STORCH stands for **s**yphilis, **t**oxoplasmosis, **o**ther diseases (hepatitis B, AIDS, and chlamydia), **r**ubella, **c**ytomegalovirus, and **h**erpes simplex virus. The most serious complications of STORCH infections are spontaneous abortion, congenital abnormalities, brain damage, prematurity, and stillbirths.

THE SIZE OF THE INOCULUM

Another factor crucial to the course of an infection is the quantity of microbes in the inoculating dose. For most agents, infection will proceed only if a minimum number, called the *infectious dose* (ID), is present. This number has been determined experimentally for many microbes. In general, microorganisms with smaller infectious doses have greater virulence. The ID varies from the astonishing number of 1 rickettsial cell in Q fever to about 10 infectious cells in tuberculosis, giardiasis, and coccidioidomycosis. The ID is 1,000 cells for gonorrhea and 10,000 cells for typhoid fever in contrast to 1,000,000,000 cells in cholera. Lack of an infectious

dose will generally not result in an infection. But if the quantity is far in excess of the ID, the onset of disease can be extremely rapid. Even weakly pathogenic species can be rendered more virulent with a large inoculum.

MECHANISMS OF INVASION AND ESTABLISHMENT OF THE PATHOGEN

Following entry of the pathogen, the next stage in infection requires that the pathogen (1) bind to the host, (2) penetrate its barriers, and (3) become established in the tissues. How the pathogen achieves these ends greatly depends upon its specific biochemical and structural characteristics.

How Pathogens Attach

Adhesion is a process by which microbes gain a more stable foothold at the portal of entry. Because this often involves a specific interaction between molecules on the microbial surface and receptors on the host cell, adhesion can determine the specificity of a pathogen for its host organism and in some cases the specificity of a pathogen for a particular cell type. Once attached, the pathogen is poised advantageously to invade the sterile body compartments. Bacterial pathogens attach most often by mechanisms such as fimbriae (pili), flagella, and adhesive slimes or capsules; viruses attach by means of specialized receptors (figure 13.11). Protozoa can infiltrate by means of their organelle of locomotion; for example, *Balantidium coli* is said to penetrate by the boring action of its cilia. In addition, parasitic worms are mechanically fastened to the portal of entry by suckers, hooks, and barbs (see chapter 23). Adhesion methods of various microbes and the diseases they lead to are shown in table 13.6.

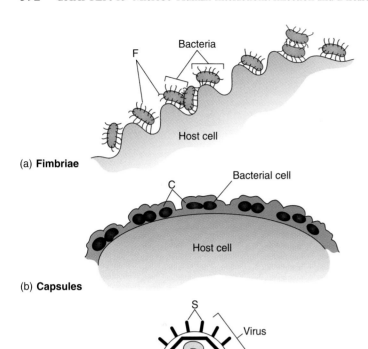

(a) **Fimbriae**

(b) **Capsules**

(c) **Spikes**

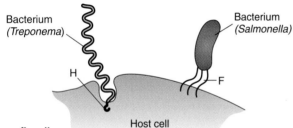

(d) **Hooks or flagella**

FIGURE 13.11

 Mechanisms of adhesion by pathogens. **(a)** Fimbriae (F), minute bristlelike appendages. **(b)** Adherent extracellular capsules (C) made of slime or other sticky substances. **(c)** Viral envelope spikes (S). **(d)** Specialized cell hooks (H) or flagella (F). (See table 13.6 for specific examples.)

TABLE 13.6

Adhesion Properties of Microbes

Microbe	Disease	Adhesion Mechanism
Neisseria gonorrhoeae	Gonorrhea	Fimbriae attach to genital epithelium
Escherichia coli	Diarrhea	Well-developed K antigen capsule
Shigella	Dysentery	Fimbriae can attach to intestinal epithelium
Vibrio	Cholera	Glycocalyx anchors microbe to intestinal epithelium
Treponema	Syphilis	Tapered hook embeds in host cell
Mycoplasma	Pneumonia	Specialized tip at ends of bacteria fuse tightly to lung epithelium
Pseudomonas aeruginosa	Burn, lung infections	Fimbriae and slime layer
Streptococcus pyogenes	Pharyngitis, impetigo	Lipotechoic acid and M-protein anchor cocci to epithelium
Streptococcus mutants, S. sobrinus	Dental caries	Dextran slime layer glues cocci to tooth surface
Influenza virus	Influenza	Viral spikes react with receptor on cell surface
Poliovirus	Polio	Capsid proteins attach to receptors on susceptible cells
HIV	AIDS	Viral spikes adhere to white blood cell receptor
Giardia lamblia (protozoan)	Giardiasis	Small suction disc on underside attaches to intestinal surface

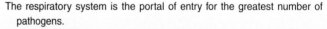

CHAPTER CHECKPOINTS

Microbial infections result when a microorganism penetrates host defenses, multiplies, and damages host tissue. The *pathogenicity* of a microbe refers to its ability to cause infection or disease. The *virulence* of a pathogen refers to the degree of damage it inflicts on the host tissues.

True pathogens cause infectious disease in healthy hosts, whereas *opportunistic pathogens* become infectious only when the host immune system is compromised in some way.

The site at which a microorganism first contacts host tissue is called the *portal of entry*. Most pathogens have one preferred portal of entry, although some have more than one.

The respiratory system is the portal of entry for the greatest number of pathogens.

The *infectious dose,* or ID, refers to the minimum number of microbial cells required to initiate infection in the host. The ID varies widely among microbial species.

Fimbriae, flagella, hooks, and adhesive capsules are types of adherence factors by which pathogens physically attach to host tissues.

How Virulence Factors Contribute to Tissue Damage

Virulence factors from a microbe's perspective are simply adaptations it uses to invade and establish itself in the host. These same factors determine the degree of tissue damage that occurs. Effects vary from pathogens that invade the tissues but multiply with relatively minor effects, such as cold viruses, to very harmful pathogens (the tetanus bacillus and AIDS virus, for instance) that severely damage and even kill the host. For convenience, we will divide the virulence factors into exoenzymes, toxins, and antiphagocytic factors (figure 13.12). Although this distinction is useful, there is often a very fine line between enzymes and toxins because many substances called toxins actually function as enzymes.

Extracellular Enzymes   Many pathogenic bacteria, fungi, protozoa, and worms secrete **exoenzymes** that break down and inflict damage on tissues. Other enzymes dissolve the host's defense barriers and promote the spread of microbes to deeper tissues.

Examples of enzymes are: (1) mucinase, which digests the protective coating on mucous membranes and is a factor in amebic dysentery; (2) keratinase, which digests the principal component of skin and hair, and is secreted by fungi that cause ringworm; (3) collagenase, which digests the principal fiber of connective tissue and is an invasive factor of *Clostridium* species and certain worms; and (4) hyaluronidase, which digests hyaluronic acid, the ground substance that cements animal cells together. This enzyme is an important virulence factor in staphylococci, clostridia, streptococci, and pneumococci.

Some enzymes react with components of the blood. Coagulase, an enzyme produced by pathogenic staphylococci, causes clotting of blood or plasma. By contrast, the bacterial kinases (streptokinase, staphylokinase) do just the opposite, dissolving fibrin clots and expediting the invasion of damaged tissues. In fact, one form of streptokinase (streptase) is marketed as a therapy to dissolve blood clots in patients with problems with thrombi and emboli.[5]

Bacterial Toxins: A Potent Source of Cellular Damage   A **toxin** is a specific chemical product of microbes, plants, and some animals that is poisonous to other organisms. **Toxigenicity,** the power to produce toxins, is a genetically controlled characteristic of many species and is responsible for the adverse effects of a variety of diseases generally called **toxinoses.** A type of toxinosis in which the toxin is spread by the blood from the site of infection is a **toxemia** (tetanus and diphtheria, for example), whereas one caused by ingestion of toxins is an **intoxication** (botulism). A toxin is named according to its specific target of action: Neurotoxins act on the nervous system; enterotoxins act on the intestine; hemotoxins lyse red blood cells; and nephrotoxins damage the kidneys.

A more traditional scheme classifies toxins according to their origins (figure 13.13). An unbound toxin molecule secreted by a living bacterial cell into the infected tissues is an **exotoxin.** A toxin that is not secreted but is released only after the cell is damaged or lysed is an **endotoxin.** Other important differences between the two

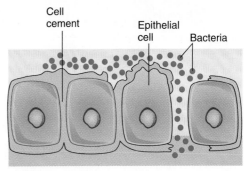

(a) **Exoenzymes**

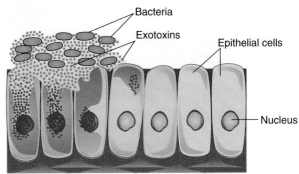

(b) **Toxins**

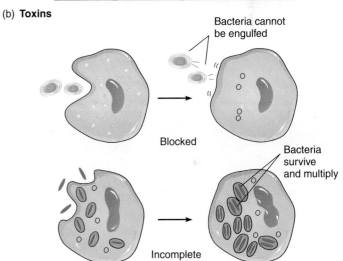

(c) **Phagocytosis**

FIGURE 13.12

The function of exoenzymes, toxins, and phagocyte blockers in invasiveness.   **(a)** Exoenzymes. Bacteria produce extracellular enzymes that dissolve extracellular barriers and penetrate through or between cells to underlying tissues. **(b)** Toxins (primarily exotoxins) secreted by bacteria diffuse to target cells, which are poisoned and disrupted. **(c)** Blocked (top) or incomplete (bottom) phagocytosis. Phagocytosis is not effective for several reasons. For example, some pathogens have a protective coating that makes them hard to engulf, or the phagocyte ingests them but the microbes can still multiply.

groups, summarized in table 13.7, are generally chemical and medical in nature.

Exotoxins are proteins with a strong specificity for a target cell and extremely powerful, sometimes deadly, effects. They generally affect cells by damaging the cell membrane and initiating

---

5. These conditions are intravascular blood clots that can cause circulatory obstructions.

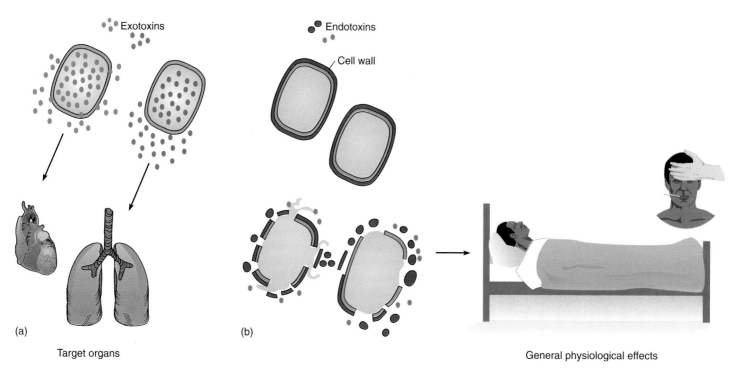

(a) Target organs

(b) General physiological effects

FIGURE 13.13

The origins and effects of circulating exotoxins and endotoxins. **(a)** Exotoxins, given off by live cells, have highly specific targets and physiological effects. **(b)** Endotoxins, given off when the cell wall of gram-negative bacteria disintegrates, have more generalized physiological effects.

TABLE 13.7		
Differential Characteristics of Bacterial Exotoxins and Endotoxins		
Characteristic	**Exotoxins**	**Endotoxins**
Toxicity	Toxic in minute amounts	Toxic in high doses
Effects on the Body	Specific to a cell type (blood, liver, nerve)	Systemic: fever, inflammation
Chemical Composition	Polypeptides	Lipopolysaccharide of cell wall
Heat Denaturation at 60°C	Unstable	Stable
Toxoid Formation	Convert to toxoid★	Do not convert to toxoid
Immune Response	Stimulate antitoxins★★	Do not stimulate antitoxins
Fever Stimulation	Usually not	Yes
Manner of Release	Secreted from live cell	Released by cell during lysis
Typical Sources	A few gram-positive and gram-negative	All gram-negative bacteria

★*A toxoid is an inactivated toxin used in vaccines.*
★★*An antitoxin is an antibody that reacts specifically with a toxin.*

lysis or by disrupting intracellular function. **Hemolysins**★ are a class of bacterial exotoxin that disrupts the cell membrane of red blood cells (and some other cells too). This damage causes the red blood cells to **hemolyze**—to burst and release hemoglobin pigment. Hemolysins that increase pathogenicity include the streptolysins of *Streptococcus pyogenes* and the alpha (α) and beta (β) toxins of *Staphylococcus aureus*. When colonies of bacteria growing on blood agar produce hemolysin, distinct zones appear around the colony. The pattern of hemolysis is often used to identify bacteria and determine their degree of pathogenicity (see chapter 18).

The toxins of diphtheria, tetanus, and botulism, among others, attach to a particular target cell, become internalized, and interrupt an essential cell pathway. The consequences of cell disruption depend upon the target. One toxin of *Clostridium tetani* blocks the action of certain spinal neurons; the toxin of *Clostridium botulinum* prevents the transmission of nerve-muscle stimuli; pertussis toxin inactivates the respiratory cilia; and cholera toxin provokes profuse salt and water loss from intestinal cells. More details of the pathology of exotoxins are found in later chapters on specific diseases.

Endotoxins belong to a class of chemicals called lipopolysaccharides (LPS), which are part of the outer membrane of gram-negative cell walls (see Medical Microfile 20.1). When gram-negative bacteria cause infections, some of them eventually lyse and release these LPS molecules into the infection site or into the circulation. Endotoxins differ from exotoxins in having a variety of systemic

★ **hemolysin** (hee-mahl′-uh-sin) Gr. *haima*, blood, and *lysis*, dissolution. A substance that causes hemolysis.

MEDICAL MICROFILE   13.3
The Classic Stages of Clinical Infections

As the body of the host responds to the invasive and toxigenic activities of a parasite, it passes through four distinct phases of infection and disease: the incubation period, the prodromium, the period of invasion, and the convalescent period.

The **incubation period** is the time from initial contact with the infectious agent (at the portal of entry) to the appearance of the first symptoms. During the incubation period, the agent is multiplying at the portal of entry but has not yet caused enough damage to elicit symptoms. Although this period is relatively well defined and predictable for each microorganism, it does vary according to host resistance, degree of virulence, and distance between the target organ and the portal of entry (the farther apart, the longer the incubation period). Overall, an incubation period can range from several hours in pneumonic plague to several years in leprosy. The majority of infections, however, have incubation periods ranging between 2 and 30 days.

The earliest notable symptoms of infection appear as a vague feeling of discomfort, such as head and muscle aches, fatigue, upset stomach, and general malaise. This short period (1–2 days) is known as the **prodromal stage.** The infectious agent next enters a **period of invasion,** during which it multiplies at high levels, exhibits its greatest toxicity, and becomes well established in its target tissue. This period is often marked by fever and other prominent and more specific signs and symptoms, which can include cough, rashes, diarrhea, loss of muscle control, swelling, jaundice, discharge of exudates, or severe pain, depending on the particular infection. The length of this period is extremely variable.

As the patient begins to respond to the infection, the symptoms decline—sometimes dramatically, other times slowly. During the recovery that follows, called the **convalescent period,** the patient's strength and health gradually return owing to the healing nature of the immune response. An infection that results in death is called **terminal.** This term is particularly applicable if the infection is the immediate cause of death in a patient already suffering from a degenerative disease or cancer. Thus, an alcoholic might succumb to terminal pneumonia, or an AIDS patient will actually die from a secondary infection.

The transmissibility of the microbe during these four stages must be considered on an individual basis. A few agents are released mostly during incubation (measles, for example); many are released during the invasive period *(Shigella)*; and others can be transmitted during all of these periods (hepatitis B).

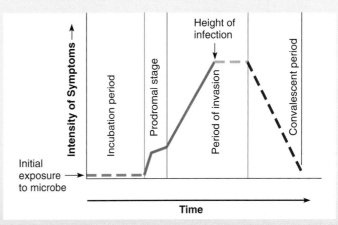

Stages in the course of infection and disease. Dashed lines represent periods with a variable length.

effects on tissues and organs. Depending upon the amounts present, endotoxins can cause fever, inflammation, hemorrhage, and diarrhea. Blood infection by gram-negative bacteria such as *Salmonella, Shigella, Neisseria meningitidis,* and *Escherichia coli* are particularly dangerous, in that it can lead to fatal endotoxic shock.

How Microbes Escape Phagocytosis   *Antiphagocytic factors* are another type of virulence factor used by some pathogens to avoid certain white blood cells called phagocytes. These cells would ordinarily engulf and destroy pathogens by means of enzymes and other antibacterial chemicals (see chapter 14). The antiphagocytic factors of resistant microorganisms help them to circumvent some part of the phagocytic process (see figure 13.12*c*). The most aggressive strategy involves bacteria that kill phagocytes outright. Species of both *Streptococcus* and *Staphylococcus* produce **leukocidins,** substances that are toxic to white blood cells. Some microorganisms secrete an extracellular surface layer (slime or capsule) that makes it physically difficult for the phagocyte to engulf them. *Streptococcus pneumoniae, Salmonella typhi, Neisseria meningitidis,* and *Cryptococcus neoformans* are notable examples. Some

bacteria are well adapted to survival inside phagocytes after ingestion. For instance, pathogenic species of *Legionella, Mycobacterium,* and many rickettsias are readily engulfed but are capable of avoiding further destruction. The ability to survive intracellularly in phagocytes has special significance because it provides a place for the microbes to hide, grow, and be spread throughout the body.

Establishment, Spread, and Pathologic Effects

Aided by virulence factors, microbes eventually settle in a particular target organ and continue to cause damage at the site. The type and scope of injuries inflicted during this process account for the typical stages of an infection (Medical Microfile 13.3), the patterns of the infectious disease, and its manifestations in the body.

In addition to the adverse effects of enzymes, toxins, and other factors, multiplication by a pathogen frequently weakens host tissues. Pathogens can obstruct tubular structures such as blood vessels, lymphatic channels, fallopian tubes, and bile ducts. Accumulated damage can lead to cell and tissue death, a condition called **necrosis.** Although viruses do not produce toxins or destructive enzymes, they destroy cells by multiplying in and lysing them. Many

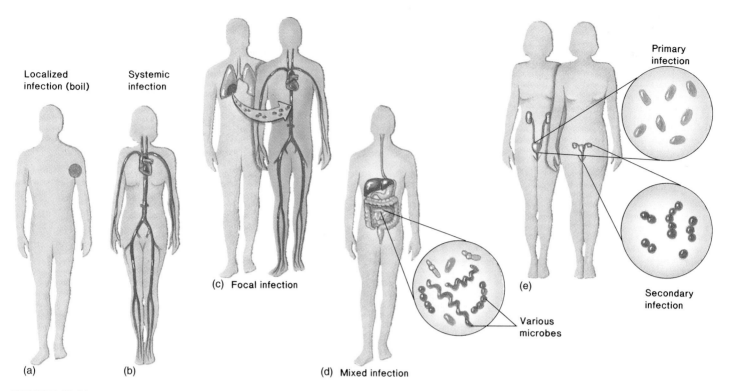

FIGURE 13.14

The occurrence of infections with regard to location, type of microbe, and length of time. **(a)** A localized infection, in which the pathogen is restricted to one specific site. **(b)** Systemic infection, in which the pathogen spreads through circulation to many sites. **(c)** A focal infection occurs initially as a local infection, but circumstances cause the microbe to be carried to other sites systemically. **(d)** A mixed infection, in which the same site is infected with several microbes at the same time. **(e)** In a primary-secondary infection, an initial infection is complicated by a second one in a different location (usually) and caused by a different microbe.

of the cytopathic effects of viral infection arise from the impaired metabolism and death of cells (see chapter 6).

Patterns of Infection   Patterns of infection are many and varied (figure 13.14). In the simplest situation, a **localized infection,** the microbe enters the body and remains confined to a specific tissue (figure 13.14*a*). Examples of localized infections are boils, fungal skin infections, and warts.

Many infectious agents do not remain localized but spread from the initial site of entry to other tissues. In fact, spreading is necessary for pathogens, such as rabies and hepatitis A virus, whose target tissue is some distance from the site of entry. The rabies virus travels from a bite wound along nerve tracts to its target in the brain, and the hepatitis A virus moves from the intestine to the liver via the circulatory system. When an infection spreads to several sites and tissue fluids, usually in the bloodstream, it is called a **systemic infection** (figure 13.14*b*). Examples of systemic infections are viral diseases (measles, rubella, chickenpox, and AIDS); bacterial diseases (brucellosis, anthrax, typhoid fever, and syphilis); and fungal diseases (histoplasmosis and cryptococcosis). Infectious agents can also travel to their targets by means of nerves (as in rabies) or cerebrospinal fluid (as in meningitis).

A **focal infection** is said to exist when the infectious agent breaks loose from a local infection and is carried into other tissues (figure 13.14*c*). This pattern is exhibited by tuberculosis or by streptococcal pharyngitis, which gives rise to scarlet fever. In the condition called **toxemia,**[6] the infection itself remains localized at the portal of entry, but the toxins produced by the pathogens are carried by the blood to the actual target tissue. In this way, the target of the bacterial cells can be different from the target of their toxin (see discussions of tetanus and diphtheria in chapter 19).

An infection is not always caused by a single microbe. In a **mixed infection,** several agents establish themselves simultaneously at the infection site (figure 13.14*d*). In some mixed or synergistic infections, the microbes cooperate in breaking down a tissue. In other mixed infections, one microbe creates an environment that enables another microbe to invade. Gas gangrene, wound infections, dental caries, and human bite infections tend to be mixed.

Some diseases are described according to a sequence of infection. When an initial, or **primary, infection** is complicated by another infection caused by a different microbe, the second infection is termed a **secondary infection** (figure 13.14*e*). This pattern often occurs in a child with chickenpox (primary infection) who may scratch his pox and infect them with *Staphylococcus aureus* (secondary infection). The secondary infection need not be in the same site as the primary infection, and it usually indicates altered host defenses.

---

6. Not to be confused with toxemia of pregnancy, which is a metabolic disturbance and not an infection.

Words in medicine have great power and economy. A single technical term can often replace a whole phrase or sentence, thereby saving time and space in patient charting. The beginning student may feel overwhelmed by what seems like a mountain of new words. However, having a grasp of a few root words and a fair amount of anatomy can help you learn many of these words and even deduce the meaning of unfamiliar ones. Some examples of medical shorthand follow.

The suffix -**itis** means an inflammation and, when affixed to the end of an anatomical term, indicates an inflammatory condition in that location. Thus, meningitis is an inflammation of the meninges surrounding the brain; encephalitis is an inflammation of the brain itself; hepatitis involves the liver; vaginitis, the vagina; gastroenteritis, the intestine; and otitis media, the middle ear. Although not all inflammatory conditions are caused by infections, many infectious diseases inflame their target organs.

The suffix -**emia** is derived from the Greek word *haeima,* meaning blood. When added to a word, it means "associated with the blood." Thus, septicemia means sepsis (infection) of the blood; bacteremia, bacteria in the blood; viremia, viruses in the blood; and fungemia, fungi in the blood. It is also applicable to specific conditions such as toxemia, gonococcemia, and spirochetemia.

The suffix -**osis** means "a disease or morbid process." It is frequently added to the names of pathogens to indicate the disease they cause: for example, listeriosis, histoplasmosis, toxoplasmosis, shigellosis, salmonellosis, and borreliosis. A variation of this suffix is -*iasis,* as in trichomoniasis and candidiasis.

The suffix -**oma** comes from the Greek word *onkomas* (swelling) and means tumor. Although the root is often used to describe cancers (sarcoma, melanoma), it is also applied in some infectious diseases that cause masses or swellings (tuberculoma, leproma).

Infections that come on rapidly, with severe but short-lived effects, are called **acute infections.** Infections that progress and persist over a long period of time are **chronic infections. Subacute infections** do not come on as rapidly as acute infections or persist as long as chronic ones. Microbits 13.4 illustrates other common terminology used to describe infectious diseases.

SIGNS AND SYMPTOMS: WARNING SIGNALS OF DISEASE

When an infection causes **pathologic** changes leading to disease, it is often accompanied by a variety of signs and symptoms. A **sign** is any objective evidence of disease as noted by an observer; a **symptom** is the subjective evidence of disease as sensed by the patient. In general, signs are more precise than symptoms, though both can have the same underlying cause. For example, an infection of the brain might present with the sign of bacteria in the spinal fluid and symptom of headache. Or a streptococcal infection might produce a sore throat (symptom) and inflamed pharynx (sign). Disease indicators that can be sensed and observed can qualify as either a sign or a symptom. When a disease can be identified or defined by a certain complex of signs and symptoms, it is termed a **syndrome.** Signs and symptoms with considerable importance in diagnosing infectious diseases are shown in table 13.8.

Signs and Symptoms of Inflammation

The earliest symptoms of disease result from the activation of the body defense process called **inflammation.*** The inflammatory response includes cells and chemicals that respond nonspecifically

to disruptions in the tissue. This subject is discussed in greater detail in chapter 14, but it is worth noting here that many signs and symptoms of infection are caused by the mobilization of this system. Some common symptoms of inflammation include fever, pain, soreness, and swelling. Signs of inflammation include **edema,*** the accumulation of fluid in an afflicted tissue; *granulomas* and *abscesses,* walled-off collections of inflammatory cells and microbes in the tissues; and *lymphadenitis,* swollen lymph nodes.

Rashes and other skin eruptions are common symptoms and signs in many diseases, and because they tend to mimic each other, it can be difficult to differentiate among diseases on this basis alone. The general term for the site of infection or disease is **lesion.*** Skin lesions can be restricted to the epidermis and its glands and follicles, or they can extend into the dermis and subcutaneous regions. The lesions of some infections undergo characteristic changes in appearance during the course of disease and thus fit more than one category (see Medical Microfile, page 536, and figure 24.1).

Signs of Infection in the Blood

Changes in the number of circulating white blood cells, as determined by special counts, are considered to be signs of possible infection. *Leukocytosis*** is an increase in the level of white blood cells, whereas *leukopenia*** is a decrease. Other signs of infection revolve around the occurrence of a microbe or its products in the

---

* inflammation (in-flam′-uh-tor″-ee) L. *inflammatio,* to set on fire.

* edema (uh-dee′-muh) Gr. *oidema,* swelling.

* lesion (lee′-zhun) L. *laesio,* to hurt.

* leukocytosis (loo″-koh′-sy-toh′-sis) From *leukocyte,* a white blood cell, and the suffix -*osis.*

* leukopenia (loo″-koh-pee′-nee-uh) From *leukocyte* and *penia,* a loss or lack of.

TABLE 13.8

Common Signs and Symptoms of Infectious Diseases

Signs	Symptoms
Fever	Chills
Septicemia	Pain, ache, soreness, irritation
Microbes in tissue fluids	Nausea
Chest sounds	Malaise, fatigue
Skin eruptions	Chest tightness
Leukocytosis	Itching
Leukopenia	Headache
Swollen lymph nodes	Nausea
Abscesses	Abdominal cramps
Tachycardia (increased heart rate)	Anorexia (lack of appetite)
Antibodies in serum	Sore throat

blood. The clinical term for blood infection, **septicemia,** refers to a general state in which microorganisms are multiplying in the blood and are present in large numbers. When small numbers of bacteria or viruses are found in the blood, the correct terminology is **bacteremia** or **viremia,** which means that these microbes are present in the blood but are not necessarily multiplying.

During infection, a normal host will invariably show signs of an immune response in the form of antibodies in the serum or some type of sensitivity to the microbe. This fact is the basis for several serological tests used in diagnosing infectious diseases such as AIDS or syphilis. Such specific immune reactions indicate the body's attempt to develop specific immunities against pathogens. We will concentrate on this role of the host defenses in chapters 14 and 15.

Infections That Go Unnoticed

It is rather common for an infection to produce no noticeable symptoms, even though the microbe is active in the host tissue. In other words, although infected, the host does not manifest the disease. Infections of this nature are known as **asymptomatic,** *subclinical,* or *inapparent* because the patient experiences no symptoms or disease and does not seek medical attention. However, it is important to note that most infections are attended by some sort of sign. In the section on epidemiology, we further address the significance of subclinical infections in the transmission of infectious agents.

THE PORTAL OF EXIT: VACATING THE HOST

Earlier, we introduced the idea that a parasite is considered *unsuccessful* if it kills its host. A parasite is equally unsuccessful if it does not have a provision for leaving its host and moving to other susceptible hosts. With few exceptions, pathogens depart by a specific avenue called the **portal of exit** (figure 13.15). In most cases, the pathogen is shed or released from the body through secretion, excretion, discharge, or sloughed tissue. The usually very high num-

ber of infectious agents in these materials increases both virulence and the likelihood that the pathogen will reach other hosts. In many cases, the portal of exit is the same as the portal of entry, but a few pathogens use a different route. As we see in the next section, the portal of exit concerns epidemiologists because it greatly influences the dissemination of infection in a population.

Respiratory and Salivary Portals

Mucus, sputum, nasal drainage, and other moist secretions are the media of escape for the pathogens that infect the lower or upper respiratory tract. The most effective means of releasing these secretions are coughing and sneezing (see figure 13.22), although they can also be released during talking and laughing. Tiny particles of liquid released into the air form aerosols or droplets that can spread the infectious agent to other people. The agents of tuberculosis, influenza, measles, and chickenpox most often leave the host through airborne droplets. Droplets of saliva are the exit route for several viruses, including those of mumps, rabies, and infectious mononucleosis.

Skin Scales

The outer layer of the skin and scalp are constantly being shed into the environment. A large proportion of household dust is actually composed of skin scales. A single person can shed several billion skin cells a day, and some persons, called shedders, disseminate massive numbers of bacteria into their immediate surroundings. Skin lesions and their exudates can serve as portals of exit in warts, fungal infections, boils, herpes simplex, smallpox, and syphilis.

Fecal Exit

Feces is a very common portal of exit. Some intestinal pathogens grow in the intestinal mucosa and create an inflammation that increases the motility of the bowel. This increased motility speeds up peristalsis, resulting in diarrhea, and the more fluid stool provides a rapid exit for the pathogen. A number of helminth worms release cysts and eggs through the feces. Feces containing pathogens are a public health problem when allowed to contaminate drinking water or when used to fertilize crops (see chapter 23).

Urogenital Tract

A number of agents involved in sexually transmitted infections leave the host in vaginal discharge or semen. This is also the source of neonatal infections such as herpes simplex, *Chlamydia,* and *Candida albicans,* which infect the infant as it passes through the birth canal. Less commonly, certain pathogens that infect the kidney are discharged in the urine: for instance, the agents of leptospirosis, typhoid fever, tuberculosis, and schistosomiasis.

Removal of Blood or Bleeding

Although the blood does not have a direct route to the outside, it can serve as a portal of exit when it is removed or released through a vascular puncture made by natural or artificial means. Blood-feeding animals such as ticks and fleas are common transmitters of

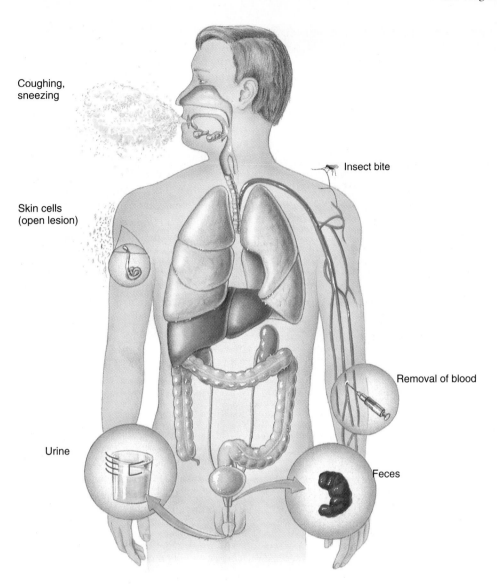

Coughing, sneezing

Skin cells (open lesion)

Insect bite

Removal of blood

Urine

Feces

FIGURE 13.15

Major portals of exit of infectious diseases.

pathogens (see Microbits 21.5). The AIDS and hepatitis viruses are transmitted by shared needles or through small gashes in a mucous membrane caused by sexual intercourse. Blood donation is also a means for certain microbes to leave the host, though this means of exit is now unusual because of close monitoring of the donor population and blood used for transfusions.

THE PERSISTENCE OF MICROBES AND PATHOLOGIC CONDITIONS

The apparent recovery of the host does not always mean that the microbe has been completely removed or destroyed by the host defenses. After the initial symptoms in certain chronic infectious diseases, the infectious agent retreats into a dormant state called **latency.** Throughout this latent state, the microbe can periodically

become active and produce a recurrent disease. The viral agents of herpes simplex, herpes zoster, hepatitis B, AIDS, and Epstein-Barr can persist in the host for long periods. The agents of syphilis, typhoid fever, tuberculosis, and malaria also enter into latent stages. The person harboring a persistent infectious agent may or may not shed it during the latent stage. If it is shed, such persons are chronic carriers who serve as sources of infection for the rest of the population.

Some diseases leave **sequelae*** in the form of long-term or permanent damage to tissues or organs. For example, meningitis can result in deafness, a strep throat can lead to rheumatic heart disease, Lyme disease can cause arthritis, and polio can produce paralysis.

---

* sequelae (suh-kwee′-lee) L. *sequi,* to follow.

CHAPTER CHECKPOINTS

Exoenzymes, toxins, and antiphagocytic factors are the three main types of *virulence factors* pathogens utilize to combat host defenses and damage host tissue.

Exotoxins and endotoxins differ in their chemical composition and tissue specificity.

Antiphagocytic factors produced by microorganisms include leukocidins, capsules, and factors that resist digestion by white blood cells.

Patterns of infection vary with the pathogen or pathogens involved. They range from local and focal to systemic.

A mixed infection is caused by two or more microorganisms simultaneously.

Infections can be characterized by their sequence as primary or secondary and by their duration as either acute or chronic.

An infectious disease is characterized by both objective signs and subjective symptoms.

Infectious diseases that are asymptomatic or subclinical nevertheless produce clinical signs.

The portal of exit by which a pathogen leaves its host is usually but not always the same as the portal of entry.

The portals of exit and entry determine how pathogens spread in a population.

Some pathogens persist in the body in a latent state; others cause long-term diseases called *sequelae*.

Epidemiology: The Study of Disease in Populations

So far, our discussion has revolved primarily around the impact of an infectious disease in a single individual. Let us now turn our attention to the effects of diseases on the community—the realm of **epidemiology.*** By definition, this term involves the study of the frequency and distribution of disease and other health-related factors in defined human populations. It involves many disciplines—not only microbiology, but anatomy, physiology, immunology, medicine, psychology, sociology, ecology, and statistics—and it considers many diseases other than infectious ones, including heart disease, cancer, drug addiction, and mental illness. The epidemiologist is a medical sleuth who collects clues on the causative agent, pathology, sources and modes of transmission, and tracks the numbers and distribution of cases of disease in the community. In fulfilling these demands, the epidemiologist asks who, when, where, how, why, and what? about diseases. The outcome of these studies helps public health departments develop prevention and treatment programs and establish a basis for predictions.

WHO, WHEN, AND WHERE? TRACKING DISEASE IN THE POPULATION

Epidemiologists are concerned with all of the factors covered earlier in this chapter: virulence, portals of entry and exit, and the course

---

* epidemiology (ep″-ih-dee-mee-ahl′-uh-gee) Gr. *epidemios,* prevalent.

of disease. But they are also interested in **surveillance**—that is, collecting, analyzing, and reporting data on the rates of occurrence, mortality, morbidity, and transmission of infections. Surveillance involves keeping data for a large number of diseases seen by the medical community and reported to public health authorities. By law, certain **reportable,** or notifiable, diseases must be reported to authorities; others are reported on a voluntary basis.

A well-developed network of individuals and agencies at the local, district, state, national, and international levels keeps track of infectious diseases. Physicians and hospitals report all notifiable diseases that are brought to their attention. Case reporting can focus on a single individual or collectively on group data.

Local public health agencies first receive the case data and determine how they will be handled. In most cases, health officers investigate the history and movements of patients to trace their prior contacts and to control the further spread of the infection as soon as possible through drug therapy, immunization, and education. In sexually transmitted diseases, patients are asked to name their partners so that these persons can be notified, examined, and treated. It is very important to maintain the confidentiality of the persons in these reports. The principal government agency responsible for keeping track of infectious diseases nationwide is the Centers for Disease Control and Prevention (CDC) in Atlanta, Georgia, which is a part of the United States Public Health Service. The CDC publishes a weekly notice of diseases (the *Morbidity and Mortality Report*) that provides weekly and cumulative summaries of the case rates and deaths for about 50 notifiable diseases, highlights important and unusual diseases, and presents data concerning disease occurrence in the major regions of the United States. Ultimately, the CDC shares its statistics on disease with the World Health Organization (WHO) for worldwide tabulation and control.

Epidemiologic Statistics: Frequency of Cases

The **prevalence** of a disease is the total number of existing cases with respect to the entire population. It is a cumulative statistic, usually represented as the percentage of the population having a particular disease at any given time. Disease **incidence** measures the number of new cases over a certain time period, as compared with the general healthy population. This statistic, also called the case, or morbidity, rate, indicates both the rate and the risk of infection. The equations used to figure these rates are:

$$\text{Prevalence} = \frac{\text{Total number of cases in population}}{\text{Total number of persons in population}} \times 100 = \%$$

$$\text{Incidence} = \frac{\text{Number of new cases}}{\text{Number of healthy persons}} = \text{Ratio}$$

As an example, let us use a classroom of 50 students exposed to a new strain of influenza. Before exposure, the prevalence and incidence in this population are both zero (0/50). If in one week, 5 out of the 50 people contract the disease, the prevalence is 5/50 = 10%, and the incidence is 1 in 9 (5 cases compared with 45 healthy persons). If after 2 weeks, 5 more students contract the flu, the prevalence becomes 5 + 5 = 10/50 = 20%, and the incidence becomes 1 in 8 (5/40). When dealing with large populations, the incidence is usually given in numbers of cases per 1,000 or 100,000 population.

Pertussis (whooping cough)—Reported cases by year, United States, 1960–2000

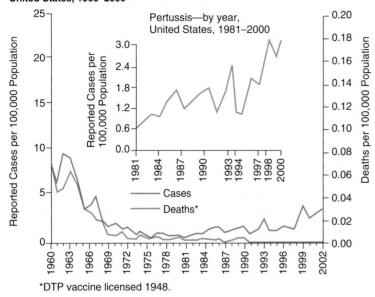

*DTP vaccine licensed 1948.

(a) Case rates (incidence) of pertussis, per 100,000 population. The inset magnifies the recent incidence of epidemics (peaks on the graph) that appear to coincide with lapses in population immunity.

Pertussis (whooping cough)—Reported cases by age group, United States, 1998

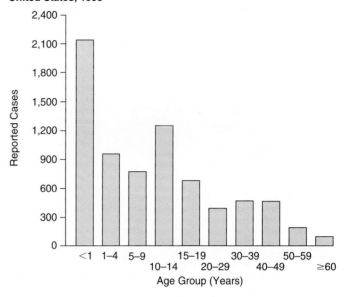

(b) A histogram graphs data of the total case rate of pertussis according to the age group affected. Notice that the majority of cases occur in children from birth to 4 years of age.

Gonorrhea—Reported cases per 100,000 population, United States, 2000

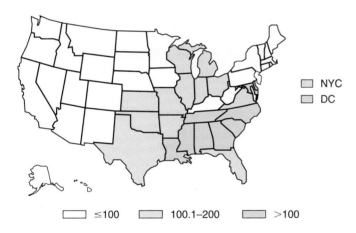

(c) Reported cases of gonorrhea by a statewide analysis. States with the highest prevalence are in the Midwest and South. The average overall rate of infection is approximately 140 cases/100,000 population (currently on the rise—see figure 18.22).

Hepatitis B—Reported cases per 100,000 population, United States and territories, 2000

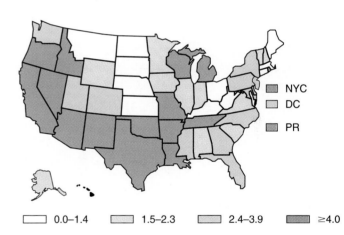

(d) Reported cases of hepatitis B show a higher case rate in the West and Southwest. The CDC estimates that 5% of U.S. residents are infected with the virus (see figure 24.16).

FIGURE 13.16

Methods for analyzing epidemiologic data.

Statistics from U.S. Department of Health and Human Services, Centers for Disease Control and Prevention, Atlanta, GA.

The changes in incidence and prevalence are usually followed over a seasonal, yearly, and long-term basis and are helpful in predicting trends (figure 13.16). Statistics of concern to the epidemiologist are the rates of disease with regard to sex, race, or geographic region (figure 13.16c, d). Also of importance is the **mortality rate,** which measures the total number of deaths in a population due to a certain disease. Over the past century, the overall death rate from infectious diseases has dropped, although the number of persons afflicted with infectious diseases (the **morbidity rate**) has remained relatively high.

Monitoring statistics also makes it possible to define the frequency of a disease in the population (figure 13.17). An infectious

(a) **Endemic Occurrence**   • Outbreaks

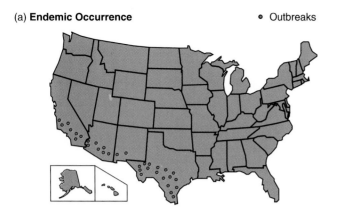

(b) **Sporadic Occurrence**

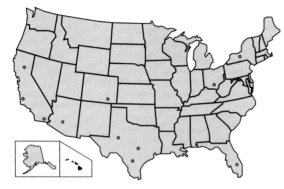

(c) **Epidemic Occurrence**

(d) **Pandemic Occurrence**

FIGURE 13.17

Patterns of infectious disease occurrence. **(a)** In endemic occurrence, cases are concentrated in one area at a relatively stable rate. **(b)** In sporadic occurrence, a few cases occur randomly over a wide area. **(c)** An epidemic is an increased number of cases that often appear in geographic clusters. **(d)** Pandemic occurrence means that an epidemic ranges over more that one continent.

disease that exhibits a relatively steady frequency over a long time period in a particular geographic locale is **endemic** (figure 13.17*a*). For example, Lyme disease is endemic to certain areas of the United States where the tick vector is found. A certain number of new cases are expected in these areas every year. When a disease is **sporadic,** occasional cases are reported at irregular intervals in unpredictable locales (figure 13.17*b*). Tetanus and diphtheria are reported sporadically in the United States (fewer than 50 cases a year). When statistics indicate that the prevalence of an endemic or sporadic disease is increasing beyond what is expected for that population, the pattern is described as an **epidemic** (figure 13.17*c*). (See figure 13.16 for an idea of how epidemics look on graphs.) An epidemic exists when an increasing trend is observed in a particular population. The time period is not defined—it can range from hours in food poisoning to years in syphilis—nor is an exact percentage of increase needed before an outbreak can qualify as an epidemic. Several epidemics occur every year in the United States, most recently among STDs such as chlamydia and gonorrhea. The spread of an epidemic across continents is a **pandemic,** as exemplified by AIDS and influenza (figure 13.17*d*).

One important epidemiologic truism might be called the "iceberg effect." Regardless of case reporting and public health screening, a large number of cases of infection in the community go undiagnosed and unreported. (For a list of reportable diseases in the United States, see table 13.9.) In the instance of salmonellosis,

TABLE 13.9

Notifiable Diseases in the United States*

AIDS	Lyme disease
Amebiasis	Lymphogranuloma venereum
Anthrax	Malaria
Arboviral infections	Measles (rubeola)
Aseptic meningitis	Meningococcal infections
Botulism	Mumps
Brucellosis	Pertussis
Chancroid	Plague
Chickenpox	Poliomyelitis
Cholera	Psittacosis
Chlamydiosis	Rabies
Cryptosporidiosis	Rheumatic fever
Diphtheria	Rocky Mountain spotted fever
Encephalitis	Rubella
Enterovirus	Salmonellosis
Escherichia coli 0157:H7	Shigellosis
Gonorrhea	Syphilis
Hansen's disease (leprosy)	Tetanus
Haemophilus influenzae meningitis	Toxic shock syndrome
Hepatitis A	Trichinosis
Hepatitis B	Tuberculosis
Hepatitis C	Tularemia
Hepatitis, other	Typhoid fever
Influenza	Typhus
Legionellosis	Yellow fever
Leptospirosis	

Depending on the state, some of these diseases are reported only if they occur at epidemic levels; others must be reported on a case-by-case basis. You can request a list of reportable infectious diseases in your state by calling the State Department of Health.

approximately 40,000 cases are reported each year. Epidemiologists estimate that the actual number is more likely somewhere between 400,000 and 4,000,000 (see figure 20.15). The iceberg effect can be even more lopsided for sexually transmitted diseases or for infections that are not brought to the attention of reporting agencies.

Investigative Strategies of the Epidemiologist

Initial evidence of a new disease or an epidemic in the community is fragmentary. A few sporadic cases are seen by physicians and are eventually reported to authorities. However, it can take several reports before any alarm is registered. Epidemiologists and public health departments must piece together odds and ends of data from a series of apparently unrelated cases and work backward to reconstruct the epidemic pattern. A completely new disease requires even greater preliminary investigation, because the infectious agent must be isolated and linked directly to the disease (see the discussion of Koch's postulates in a later section of this chapter). All factors possibly impinging on the disease are scrutinized. Investigators search for *clusters* of cases indicating spread between persons or a public (common) source of infection; they also look at possible contact with animals, contaminated food, water, and public facilities, and at human interrelationships or changes in community structure. Out of this maze of case information, the investigators hope to recognize a pattern that indicates the source of infection so that they can quickly move to control it.

RESERVOIRS: WHERE PATHOGENS PERSIST

In order for an infectious agent to continue to exist and be spread, it must have a permanent place to reside. The **reservoir** is the primary habitat in the natural world from which a pathogen originates. Often it is a human or animal carrier, although soil, water, and plants are also reservoirs. The reservoir can be distinguished from the infection **source,** which is the individual or object from which an infection is actually acquired. In diseases such as syphilis, the reservoir and the source are the same (the human body). In the case of hepatitis A, the reservoir (a human carrier) is usually different from the source of infection (contaminated food).

Living Reservoirs

Many pathogens continue to exist and spread because they are harbored by members of a host population. Persons or animals with frank symptomatic infection are obvious sources of infection, but a **carrier** is, by definition, an individual who *inconspicuously* shelters a pathogen and spreads it to others without any notice. Although human carriers are occasionally detected through routine screening (blood tests, cultures) and other epidemiologic devices, they are unfortunately very difficult to discover and control. As long as a pathogenic reservoir is maintained by the carrier state, the disease will continue to exist in that population, and the potential for epidemics will be a constant threat. The duration of the carrier state can be short- or long-term, and actual infection of the carrier may or may not be involved.

Several situations can produce the carrier state. **Asymptomatic** (apparently healthy) **carriers** are indeed infected, but as previously indicated, they show no symptoms (figure 13.18*a*). A few asymptomatic infections (gonorrhea and genital warts, for instance)

can carry out their entire course without overt manifestations. Other asymptomatic carriers, called *incubation carriers,* spread the infectious agent during the incubation period. For example, AIDS patients can harbor and spread the virus for months and years before their first symptoms appear. Recuperating patients without symptoms are considered *convalescent carriers* when they continue to shed viable microbes and convey the infection to others. Diphtheria patients, for example, spread the microbe for up to 30 days after the disease has subsided.

An individual who shelters the infectious agent for a long period after recovery because of the latency of the infectious agent is a *chronic carrier.* Patients who have recovered from tuberculosis, hepatitis, and herpes infections frequently carry the agent chronically. About one in 20 victims of typhoid fever continues to harbor *Salmonella typhi* in the gallbladder for several years, and sometimes for life. The most infamous of these was "Typhoid Mary," a cook who spread the infection to hundreds of victims in the early 1900s.

The **passive carrier** state is of great concern during patient care (see a later section on nosocomial infections). Medical and dental personnel who must constantly handle materials that are heavily contaminated with patient secretions and blood are at risk for picking up pathogens mechanically and accidently transferring them to other patients (figure 13.18*b*). Proper handwashing, handling of contaminated materials, and aseptic techniques greatly reduce this likelihood.

Animals As Reservoirs and Sources   Up to now, we have lumped animals with humans in discussing living reservoirs or carriers, but animals deserve special consideration as vectors of infections. The word **vector** is used by epidemiologists to indicate a live animal that transmits an infectious agent from one host to another. (The term is sometimes misused to include any object that spreads disease.) The majority of vectors are arthropods such as fleas, mosquitoes, flies, and ticks, although larger animals can also spread infection—for example, mammals (rabies), birds (psittacosis), or lower vertebrates (salmonellosis).

By tradition, vectors are placed into one of two categories, depending upon the animal's relationship with the microbe (figure 13.19). A **biological vector** actively participates in a pathogen's life cycle, serving as a site in which it can multiply or complete its life cycle. A biological vector communicates the infectious agent to the human host by biting, aerosol formation, or touch. In the case of biting vectors, the animal can (1) inject infected saliva into the blood (the mosquito), (2) defecate around the bite wound (the flea) (figure 13.19*a*), or (3) regurgitate blood into the wound (the tsetse fly). More detailed discussions of the roles of biological vectors are found in chapters 20, 21, 23, and 25.

Mechanical vectors are not necessary to the life cycle of an infectious agent and merely transport it without being infected. The external body parts of these animals become contaminated when they come into physical contact with a source of pathogens. The agent is subsequently transferred to humans indirectly by an intermediate such as food or, occasionally, by direct contact (as in certain eye infections). Houseflies (figure 13.19*b*) are noxious mechanical vectors. They feed on decaying garbage and feces, and while they are feeding, their feet and mouthparts easily become contaminated. They also regurgitate juices onto food to soften and

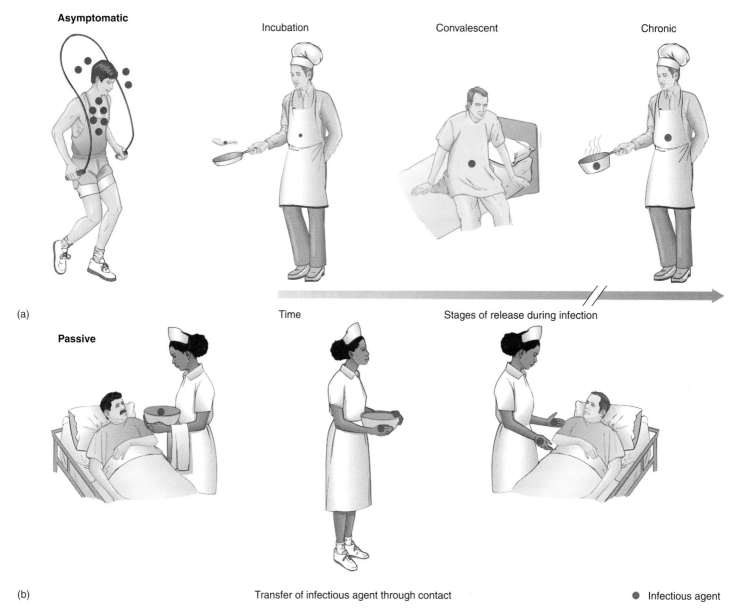

(a)

Asymptomatic Incubation Convalescent Chronic

Time Stages of release during infection

(b) Passive

Transfer of infectious agent through contact ● Infectious agent

FIGURE 13.18

Types of carriers. **(a)** An asymptomatic carrier is infected without symptoms. Incubation carriers are in the early stages of infection; convalescent carriers are in the late stages of recovery; chronic carriers sequester the microbe for long periods after the infection is over. **(b)** A passive carrier is contaminated but not infected.

digest it. Flies spread more than 20 bacterial, viral, protozoan, and worm infections. Other nonbiting flies transmit tropical ulcers, yaws, and trachoma (see chapter 21). Cockroaches, which have similar unsavory habits, play a role in the mechanical transmission of fecal pathogens as well as contributing to allergy attacks in asthmatic children.

Many vectors and animal reservoirs spread their own infections to humans. An infection indigenous to animals but naturally transmissible to humans is a **zoonosis.*** In these types of infections, the human is essentially a dead-end host and does not contribute to the

natural persistence of the microbe. Some zoonotic infections (rabies, for instance) can have multihost involvement, and others can have very complex cycles in the wild (see plague in chapter 20). Zoonotic spread of disease is promoted by close associations of humans with animals, and people in animal-oriented or outdoor professions are at greatest risk. At least 150 zoonoses exist worldwide; the most common ones are listed in table 13.10. Zoonoses make up a full 70% of all new emerging diseases worldwide. It is worth noting that zoonotic infections are impossible to completely eradicate without also eradicating the animal reservoirs. Attempts have been made to eradicate mosquitoes and certain rodents, but it is inconceivable that such extreme measures would ever be tried on wild birds or mammals.

---

* zoonosis (zoh″-uh-noh′-sis) Gr. *zoion,* animal, and *nosos,* disease.

(a) Biological vectors are infected.

Relative true sizes:

(b) Mechanical vectors are not infected.

FIGURE 13.19

Two types of vectors. **(a)** Biological vectors serve as hosts during pathogen development. They include the flea, a carrier of bubonic plague and murine typhus. **(b)** Mechanical vectors such as the housefly ingest filth and transport pathogens on their feet and mouthparts.

TABLE 13.10

Animal Hosts and Their Common Zoonotic Infections

	Domestic Animals					Wild Animals			
Disease	**Cats**	**Dogs**	**Cattle**	**Horses**	**Poultry**	**Arthropods**	**Birds**	**Rodents**	**Primates**
Viruses									
Rabies	+	+	+	−	−	−	−	+	+
Yellow fever	−	−	−	−	−	+	−	+	+
Viral fevers	−	−	−	−	−	+	−	+	−
Hantavirus	−	−	−	−	−	−	−	+	−
Influenza	−	−	−	−	+	−	+	−	−
Bacteria									
Q fever	−	−	+	−	−	−	+	+	−
Rocky Mountain spotted fever	−	+	−	−	−	+	−	−	−
Psittacosis	−	−	−	−	+	−	+	−	−
Leptospirosis	+	+	+	+	−	−	−	+	+
Anthrax	+	+	+	+	−	−	−	+	−
Brucellosis	−	+	+	−	−	−	−	−	−
Listeriosis	−	+	+	+	−	−	+	+	−
Plague	+	+	−	−	−	+	−	+	−
Salmonellosis	+	+	+	+	+	+	+	+	+
Tularemia	+	+	−	+	−	+	+	+	−
Miscellaneous									
Ringworm	+	+	+	+	−	−	−	+	+
Toxoplasmosis	+	−	+	−	−	−	+	+	−
Trypanosomiasis	+	+	+	−	−	+	−	+	+
Larval migrans	+	+	−	−	−	−	−	−	−
Trichinosis	−	−	−	+	−	−	−	−	−
Tapeworm	−	−	+	−	−	−	−	−	−
Scabies (mange)	+	+	+	+	+	−	−	−	−

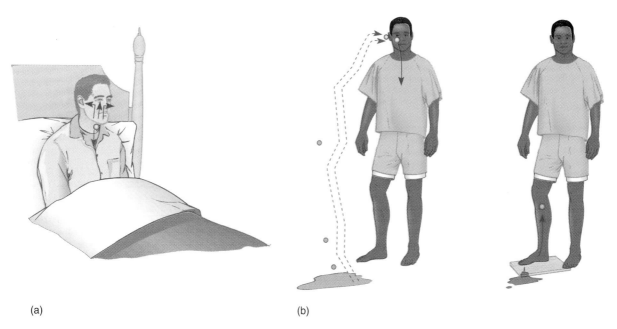

(a)   (b)

FIGURE 13.20

How non-communicable infections are acquired.   **(a)** From normal flora invading tissues and **(b)** from contact with microbes that live in the soil and water. Note that both infections are not acquired from another infected person, either directly or indirectly.

Nonliving Reservoirs

It is clear that microorganisms have adapted to nearly every habitat in the biosphere. They thrive in soil and water and often find their way into the air. Although most of these microbes are saprobic and cause little harm and considerable benefit to humans, some are opportunists and a few are regular pathogens. Because human hosts are in regular contact with these environmental sources, acquisition of pathogens from natural habitats is of diagnostic and epidemiologic importance.

Soil harbors the vegetative forms of bacteria, protozoa, helminths, and fungi, as well as their resistant or developmental stages such as spores, cysts, ova, and larvae. Regular bacterial pathogens include the anthrax bacillus and species of *Clostridium* that are responsible for gas gangrene, botulism, and tetanus. Pathogenic fungi in the genera *Coccidioides* and *Blastomyces* are spread by spores in the soil and dust. The invasive stages of the hookworm *Necator* occur in the soil. Natural bodies of water carry fewer nutrients than soil does but still support a number of pathogenic species such as *Legionella, Cryptosporidium,* and *Giardia.*

HOW AND WHY? THE ACQUISITION AND TRANSMISSION OF INFECTIOUS AGENTS

Infectious diseases can be categorized on the basis of how they are acquired. A disease is **communicable** when an infected host can transmit the infectious agent to another host and establish infection in that host. (Although this is standard terminology, one must realize that it is not the disease that is communicated, but the microbe. Also be aware that the word *infectious* is sometimes used interchangeably with the word *communicable,* but this is not precise us-

age.) The transmission of the agent can be direct or indirect, and the ease with which the disease is transmitted varies considerably from one agent to another. If the agent is highly transmissible, especially through direct contact, the disease is **contagious.** Influenza and measles move readily from host to host and thus are contagious, whereas leprosy is only weakly communicable. Because they can be spread through the population, communicable diseases will be our main focus in the following sections.

In contrast, a **non-communicable** infectious disease does *not* arise through transmission of the infectious agent from host to host. The infection and disease are acquired through some other, special circumstance. Non-communicable infections occur primarily when a compromised person is invaded by his own microflora (as with certain pneumonias, for example) or when an individual has accidental contact with a facultative parasite that exists in a nonliving reservoir such as soil (figure 13.20). Some examples are certain mycoses, acquired through inhalation of fungal spores, and tetanus, in which *Clostridium tetani* spores from a soiled object enter a cut or wound. Persons thus infected do not become a source of disease to others.

Patterns of Transmission in Communicable Diseases

The routes or patterns of disease transmission are many and varied. The spread of diseases is by direct or indirect contact with animate or inanimate objects and can be horizontal or vertical. The term *horizontal* means the disease is spread through a population from one infected individual to another; *vertical* signifies transmission from parent to offspring via the ovum, sperm, placenta, or milk. The extreme complexity of transmission patterns among microorganisms makes it very difficult to generalize. However, for easier

through their activities (figure 13.21). The transmitter of the infectious agent can be either openly infected or a carrier.

Indirect Spread by Vehicles: Contaminated Materials The term **vehicle** specifies any inanimate material commonly used by humans that can transmit infectious agents. A *common vehicle* is a single material that serves as the source of infection for many individuals. Some specific types of vehicles are food, water, various biological products (such as blood, serum, and tissue), and fomites. A **fomite*** is an inanimate object that harbors and transmits pathogens. The list of possible fomites is as long as your imagination allows. Probably highest on the list would be objects commonly in contact with the public such as doorknobs, telephones, push buttons, and faucet handles that are readily contaminated by touching. Shared bed linens, handkerchiefs, toilet seats, toys, eating utensils, clothing, personal articles, and syringes are other examples. Although paper money is impregnated with a disinfectant to inhibit microbes, pathogens are still isolated from bills as well as coins.

Outbreaks of food poisoning often result from the role of food as a common vehicle. The source of the agent can be soil, the handler, or a mechanical vector. In the type of transmission termed the *oral-fecal route,* a fecal carrier with inadequate personal hygiene contaminates food during handling, and an unsuspecting person ingests it. Hepatitis A, amebic dysentery, shigellosis, and typhoid fever are often transmitted this way. Because milk provides a rich growth medium for microbes, it is a significant means of transmitting pathogens from diseased animals, infected milk handlers, and environmental sources of contamination. The agents of brucellosis, tuberculosis, Q fever, salmonellosis, and listeriosis are transmitted by contaminated milk. Water that has been contaminated by feces or urine can carry *Salmonella, Vibrio* (cholera) viruses (hepatitis A, polio), and pathogenic protozoans *(Giardia, Cryptosporidium).*

Indirect Spread by Airborne Route: Droplet Nuclei and Aerosols Unlike soil and water, outdoor air cannot provide nutritional support for microbial growth and seldom transmits airborne pathogens. On the other hand, indoor air (especially in a closed space) can serve as an important medium for the suspension and dispersal of certain respiratory pathogens via droplet nuclei and aerosols. **Droplet nuclei** are dried microscopic residues created when microscopic pellets of mucus and saliva are ejected from the mouth and nose. They are generated forcefully in an unstifled sneeze or cough (figure 13.22) or mildly during other vocalizations. Although the larger beads of moisture settle rapidly, smaller particles evaporate and remain suspended for longer periods. Droplet nuclei are implicated in the spread of hardier pathogens such as the tubercle bacillus and the influenza virus. **Aerosols** are suspensions of fine dust or moisture particles in the air that contain live pathogens. Q fever is spread by dust from animal quarters, and psittacosis by aerosols from infected birds. An unusual outbreak of coccidioidomycosis (a lung infection) occurred during the 1994 southern California earthquake. Epidemiologists speculate that disturbed hillsides and soil gave off clouds of dust containing the spores of *Coccidioides.*

---

* fomite (foh′-myt) L. *fomes,* tinder.

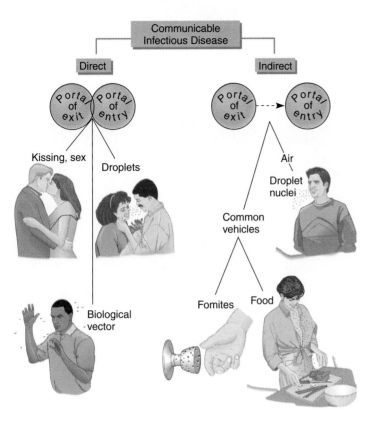

FIGURE 13.21

Summary of how communicable infectious diseases are acquired. Direct mechanisms of transmission involve physical contact between two hosts. Indirect modes of transmission require some object or material to transfer the infectious agent between hosts.

organization, we will divide microorganisms into two major groups, as shown in figure 13.21: transmission by direct contact or transmission by indirect routes (with the latter category divided into vehicle and airborne transmission).

Modes of Direct Transmission In order for microbes to be directly transferred, some type of contact must occur between the skin or mucous membranes of the infected person and that of the infectee. It may help to think of this route as the portal of exit meeting the portal of entry without the involvement of an intermediate object or substance (figure 13.21). Included in this category are fine droplets sprayed directly upon a person during sneezing or coughing (as distinguished from droplet nuclei that are transmitted some distance by air). Most sexually transmitted diseases are spread directly. In addition, infections that result from kissing, nursing, placental transfer, or bites by biological vectors are direct. Most obligate parasites are far too sensitive to survive for long outside the host and can be transmitted only through direct contact.

Routes of Indirect Transmission For microbes to be considered indirectly transmitted, the infectious agent must pass from an infected host to an intermediate conveyor and from there to another host. This form of communication is especially pronounced when the infected individuals contaminate inanimate objects, food, or air

FIGURE 13.22

The explosiveness of a sneeze. Special photography dramatically captures droplet formation in an unstifled sneeze. Even the merest attempt to cover a sneeze with one's hand will reduce this effect considerably. When such droplets dry and remain suspended in air, they are droplet nuclei.

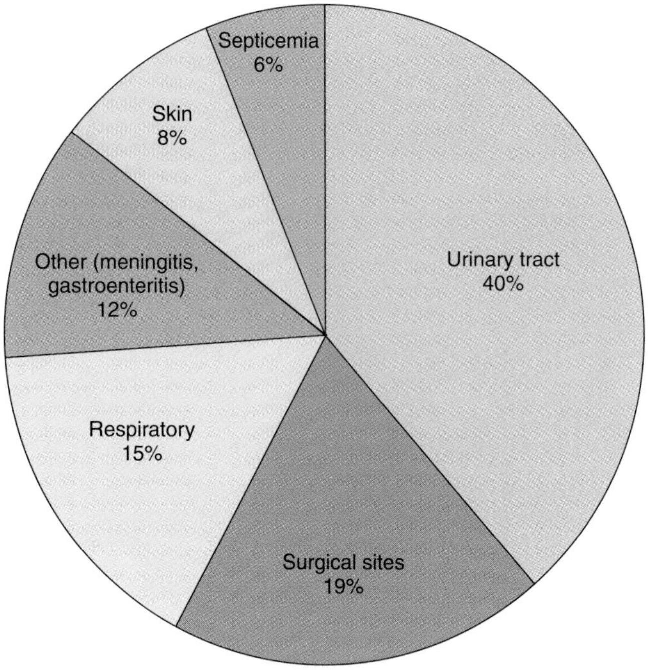

FIGURE 13.23

Most common nosocomial infections. Relative frequency by body site.

NOSOCOMIAL INFECTIONS: THE HOSPITAL AS A SOURCE OF DISEASE

Infectious diseases acquired as a result of a hospital stay are known as **nosocomial*** infections. This concept seems incongruous at first thought, because a hospital is regarded as a place to get treatment for a disease, not a place to acquire a disease. Yet it is not uncommon for a surgical patient's incision to become infected or a burn patient to develop a case of pneumonia in the clinical setting. The rate of nosocomial infections can be as low as 0.1% or as high as 20% of all admitted patients depending on the clinical setting, with an average of about 5%. In light of the number of admissions, this amounts to from 2 to 4 million cases a year, which result in 20,000 to 40,000 deaths. Nosocomial infections cost time and money. By one estimate, they amount to 8 million days of hospitalization a year at an additional cost of $5 to $10 billion.

So many factors unique to the hospital environment are tied to nosocomial infections that a certain number of infections are virtually unavoidable. After all, the hospital both attracts and creates compromised patients, and it serves as a collection point for pathogens. Some patients become infected when surgical procedures or lowered defenses permit resident flora to invade their bodies. Other patients acquire infections directly or indirectly from fomites, medical equipment, other patients, medical personnel, visitors, air, and water.

The health care process itself increases the likelihood that infectious agents will be transferred from one patient to another.

Treatments using reusable instruments such as respirators and thermometers constitute a possible source of infectious agents. Indwelling devices such as catheters, prosthetic heart valves, grafts, drainage tubes, and tracheostomy tubes form a ready portal of entry and habitat for infectious agents. An additional problem is the tendency for drug-resistant strains of microorganisms to develop in hospitals, thereby further complicating treatment.

The most common nosocomial infections involve the urinary tract, the respiratory tract, and surgical incisions (figure 13.23). Gram-negative intestinal flora *(Escherichia coli, Klebsiella, Pseudomonas)* are cultured in more than half of patients with nosocomial infections (see figure 20.7). Gram-positive bacteria (staphylococci and streptococci) and yeasts make up most of the remainder. True pathogens such as *Mycobacterium tuberculosis, Salmonella,* hepatitis B, and influenza virus can be transmitted in the clinical setting as well.

The potential seriousness and impact of nosocomial infections have required hospitals to develop committees that monitor infectious outbreaks and develop guidelines for infection control and aseptic procedures.

Surgical asepsis involves a high level of disinfection, antisepsis, and sterilization as would be required to maintain a microbe-free surgery. Instruments, dressings, sponges, and all other supplies coming into contact with the patient are sterilized. Personnel must be fully covered in sterile garments, and room surfaces and air must be thoroughly disinfected.

Medical asepsis includes any practice that lowers the load of infectious microbes in patients, personnel, and the hospital environment. Included are handwashing, decontamination procedures, and *isolation* of patients. In isolation, various barriers (gloves,

---

* nosocomial (nohz″-oh-koh′-mee-al) Gr. *nosos,* disease, and *komeion,* to take care of. Originating from a hospital or infirmary.

mask, gown) and other techniques are used in the patient's room to prevent the entry or exit of infectious agents on health care workers, visitors, and into the surroundings. The aim of most isolation is to contain the spread of infectious agents from infected patients or carriers, and one category (reverse isolation) prevents highly compromised patients from coming into contact with pathogens from the outside environment. Table 13.11 summarizes guidelines for the major types of isolation.

An essential member of the infection control team is the *infection control officer*. This person oversees all hospital personnel and procedures to minimize the spread of infection. The infection control officer's responsibilities include keeping track of infections in the wards, determining possible epidemics, relaying this information to the rest of the team, seeking out breaches in asepsis in nursing care or surgery, and training others in aseptic techniques.

Control procedures critical to reducing nosocomial infections are proper handwashing and surgical scrub techniques; proper disposal procedures for contaminated substances and blood; patient isolation; antibiotic prophylaxis; strict sterilization, disinfection, and sanitization procedures; restricting infected personnel; and immunizing personnel.

Another high-risk group for infections in the hospital are patient caregivers. The very nature of their work exposes them to needlesticks (a type of inoculation infection), infectious secretions and blood, and physical contact with the patient. The same practices that interrupt the routes of infection in the patient can also protect the health worker. It is for this reason that most hospitals have adopted universal precautions that recognize that all secretions from all persons in the clinical setting are potentially infectious and that transmission can occur in either direction. See the section "Introduction to Identification Techniques", that follows chapter 17, for a full description of these precautions.

WHICH AGENT IS THE CAUSE? USING KOCH'S POSTULATES TO DETERMINE ETIOLOGY

An essential aim in the study of infection and disease is determining the precise **etiologic,** or causative, agent. In our modern technological age, we take for granted that a certain infection is caused by a certain microbe, but such has not always been the case. More than a century ago, Robert Koch realized that in order to prove the germ theory of disease he would have to develop a standard for determining causation that would stand the test of scientific scrutiny. Out of his experimental observations on the transmission of anthrax in cows came a series of proofs, called **Koch's postulates,** that established the principal criteria for etiologic studies (figure 13.24). These postulates direct an investigator to (1) find evidence of a particular microbe in every case of a disease, (2) isolate that microbe from an infected subject and cultivate it artificially in the laboratory, (3) inoculate a susceptible healthy subject with the laboratory isolate and observe the resultant disease, and (4) reisolate the agent from this subject.

Valid application of Koch's postulates requires attention to several critical details. Each isolated culture must be pure, observed microscopically, and identified by means of characteristic tests; the first and second isolate must be identical; and the pathologic

TABLE 13.11

Levels of Isolation Used in Clinical Settings

Type of Isolation*	Protective Measures**	To Prevent Spread of
Enteric Precautions	Gowns and gloves must be worn by all persons having direct contact with patient; masks not required; special precautions taken for disposing of feces and urine	Diarrheal diseases; *Shigella, Salmonella,* and *Escherichia coli* gastroenteritis; cholera; hepatitis A; rotavirus; and giardiasis
Respiratory Precautions	Private room with closed door is necessary; gowns and gloves not required; masks usually indicated; items contaminated with secretions must be disinfected	Tuberculosis, measles, mumps, meningitis, pertussis, rubella, chickenpox
Drainage and Secretion Precautions	Gowns and gloves required for all persons; masks not needed; contaminated instruments and dressings require special precautions	Staphylococcal and streptococcal infections; gas gangrene; herpes zoster; burn infections
Strict Isolation	Private room with closed door required; gowns, masks, and gloves must be worn by all persons; contaminated items must be wrapped and sent to central supply for decontamination	Mostly highly virulent or contagious microbes; includes diphtheria, some types of pneumonia, extensive skin and burn infections, disseminated herpes simplex and zoster
Reverse Isolation (Also Called Protective Isolation)	Same guidelines as for strict isolation; room may be ventilated by unidirectional or laminar airflow filtered through a high-efficiency particulate air (HEPA) filter that removes most airborne pathogens; infected persons must be barred	Used to protect patients extremely immunocompromised by cancer therapy, surgery, genetic defects, burns, prematurity, or AIDS and therefore vulnerable to opportunistic pathogens

*Precautions are based upon the primary portal of entry and communicability of the pathogen.
**In all cases, visitors to the patient's room must report to the nurses' station before entering the room; all visitors and personnel must wash their hands upon entering and leaving the room.

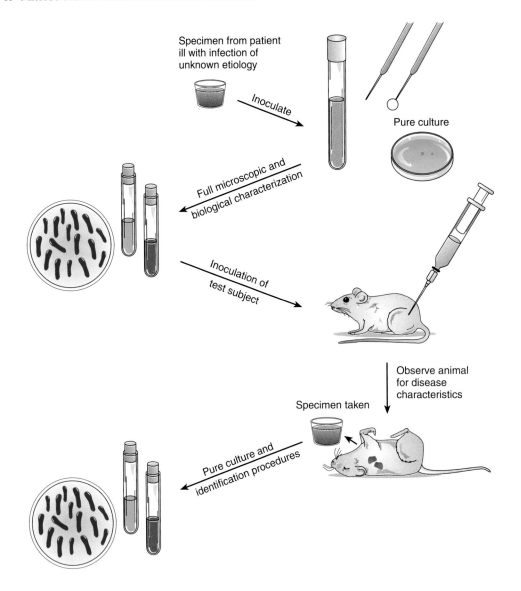

FIGURE 13.24

Koch's postulates: Is this the etiologic agent? The microbe in the initial and second isolations and the disease in the patient and experimental animal must be identical for the postulates to be satisfied.

effects, signs, and symptoms of the disease in the first and second subject must be the same. Once established, these postulates were rapidly put to the test, and within a short time, they had helped determine the causative agents of tuberculosis, diphtheria, and plague. Today, most infectious diseases have been directly linked to a known infectious agent.

Koch's postulates continue to play an essential role in modern epidemiology. Every decade, new diseases challenge the scientific community and require application of the postulates. Prominent examples are toxic shock syndrome, AIDS, Lyme disease, and Legionnaires disease (named for the American Legion members who first contracted a mysterious lung infection in Philadelphia).

Koch's postulates are reliable for most infectious diseases, but they cannot be completely fulfilled in certain situations. For ex-

ample, some infectious agents are not readily isolated or grown in the laboratory. If one cannot elicit a similar infection by inoculating it into an animal, it is very difficult to prove the etiology. In the past, scientists have attempted to circumvent this problem by using human subjects (Historical Highlights 13.5).

A small but vocal group of critics has claimed that the postulates have not been adequately carried out for AIDS and thus, that HIV cannot be claimed as the causative agent, despite overwhelming evidence from observing infected humans and primates that it is. Cases of accidental infection through exposure to blood have yielded much proof. In all cases, subjects developed an early virus syndrome, carried high virus levels, and later developed AIDS. One study with baboons revealed that they do indeed develop symptoms of AIDS when infected with a variant of HIV.

In this day and age, human beings are not used as subjects for determining the cause of infectious disease, but in earlier times they were. A long tradition of human experimentation dates well back into the eighteenth century, with the subject frequently the experimenter himself. In some studies, mycologists inoculated their own skin and even that of family members with scrapings from fungal lesions to demonstrate that the disease was transmissible. In the early days of parasitology, it was not uncommon for a brave researcher to swallow worm eggs in order to study the course of his infection and the life cycle of the worm.

In a sort of reverse test, a German colleague of Koch's named Max von Petenkofer believed so strongly that cholera was *not* caused by a bacterium that he and his assistant swallowed cultures of the vibrio. Fortunately for them, they did not acquire serious infection. Many self-experimenters have not been so fortunate.

One of the most famous cases is that of Jesse Lazear, a Cuban physician who worked with Walter Reed on the etiology of yellow fever in 1900. Dr. Lazear was convinced that mosquitoes were directly involved in the spread of yellow fever, and by way of proof, he allowed himself and two volunteers to be bitten by mosquitoes infected with the blood of yellow fever patients. Although all three became ill as a result of this exposure, Dr. Lazear's sacrifice was the ultimate one—he died of yellow fever. Years later, paid volunteers were used to completely fulfill the postulates.

Painting of Dr. Jesse Lazear exposing the arm of James Carroll to a mosquito infected with the yellow fever virus. Human volunteers contributed significantly to our understanding of the transmission of yellow fever and many other diseases.

Dr. Lazear was not the first martyr in this type of cause. Fifteen years previously, a young Peruvian medical student, Daniel Carrion, attempted to prove that a severe blood infection, Oroya fever, had the same etiology as *verucca peruana,* an ancient disfiguring skin disease. After inoculating himself with fluid from a skin lesion, he developed the severe form and died, becoming a national hero. In his honor, the disease now identified as bartonellosis is also sometimes called Carrion's disease. Eventually, the microbe *(Bartonella bacilliformis)* was isolated, a monkey model was developed, and the sand fly was shown to be the vector for this disease.

For many years, syphilis and gonorrhea were thought to be different stages of the same disease because of an unfortunate experiment. In an ironic twist of fate, Fritz Schaudinn eventually proved the causation of syphilis, only to die later from amebic dysentery—which he had acquired from self-experimentation.

The incentive for self-experimentation has continued, but present-day researchers are more likely to test experimental vaccines than dangerous microbes. One exception to this was Dr. J. Robert Warren and Barry Marshall who tested the effects of *Helicobacter pylori* by swallowing a culture. Although they didn't get ulcers, they helped to establish the pathogenicity of the microbe. Even more recently, a parasitologist fed several of his patients live worm eggs to treat their inflammatory bowel disease, and it actually improved their condition significantly.

CHAPTER CHECKPOINTS

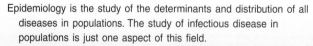 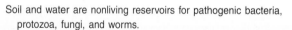

Epidemiology is the study of the determinants and distribution of all diseases in populations. The study of infectious disease in populations is just one aspect of this field.

Data on specific, reportable diseases is collected by local, national, and worldwide agencies.

The *prevalence* of a disease is the percentage of existing cases in a given population. The disease *incidence,* or *morbidity rate,* is the ratio of newly infected to uninfected members of a population.

The disease frequency is described as sporadic, epidemic, pandemic, or endemic.

The primary habitat of a pathogen is called its reservoir. A human reservoir is also called a carrier.

Animals can be either reservoirs or vectors of pathogens. An infected animal is a biological vector. Uninfected animals, especially insects, that transmit pathogens mechanically are called mechanical vectors.

Soil and water are nonliving reservoirs for pathogenic bacteria, protozoa, fungi, and worms.

A communicable disease can be transmitted from an infected host to others, but not all infectious diseases are communicable.

The spread of infectious disease from person to person is called horizontal transmission. The spread of infectious disease from parent to offspring is called vertical transmission.

Infectious diseases are spread by either direct or indirect routes of transmission. Vehicles of indirect transmission include soil, water, food, droplet nuclei, and fomites (inanimate objects).

Nosocomial infections are acquired in a hospital from surgical procedures, equipment, personnel, and exposure to drug-resistant microorganisms.

Causative agents of infectious disease must be isolated and identified according to Koch's postulates.

CHAPTER CAPSULE WITH KEY TERMS

I. **Contact-Infection-Disease: The Host-Parasite Relationship**
The human body is constantly in contact (contaminated) with microbes. Some are pathogens that may cause an **infection** by circumventing the host defense system, entering normally sterile tissues, and multiplying there. When infections lead to a disruption in tissues, **infectious disease** results. The outcome is highly variable, but most contacts do not result in infection, and most infections do not lead to disease.
   A. The Body As a Habitat
      1. The **resident flora** or **microflora** is a huge and rich mixed population of microorganisms residing on body surfaces exposed to the environment, including the skin, mucous membranes, parts of the gastrointestinal tract, urinary tract, reproductive tract, and upper respiratory tract.
      2. Anatomical sites lying within the body cavity (organs) and fluids (blood, urine) in those sites do not harbor flora.
      3. *Colonization:* Begins just prior to birth and continues over an individual's life; variations occur in response to individual differences in age, diet, hygiene, and health.
      4. *Role of Flora*
         a. Bacteria may maintain a balance in the normal conditions.
         b. Studies with **axenic** animals (free of any normal flora) show that flora contribute to the development of the immune and gastrointestinal systems and also to some diseases (dental caries).
         c. Normal flora are sometimes agents of infection.
   B. Factors Affecting the Course of Infection and Disease
      1. *Pathogenicity and Virulence*
         a. **Pathogenicity** is the property of microorganisms to cause infection and disease.
         b. **Virulence** is the precise factors used by the microbe to invade and damage host tissues; it helps define the degree of pathogenicity.
         c. Pathogenicity varies with a microbe's ability to invade or harm host tissues and with the condition of host defenses.
         d. A **true pathogen** produces **virulence factors** that allow it to readily evade host defenses and to harm host tissues. True pathogens can infect normal, healthy hosts with intact defenses.
         e. An **opportunistic pathogen** is not highly virulent but can cause disease in persons whose host defenses are compromised by *predisposing conditions* such as age, genetic defects, medical procedures, and underlying organic disease.
      2. *Mechanisms of Infection and Disease*
         a. The **portal of entry** is the route by which microbes enter the tissues, primarily via skin, alimentary tract, respiratory tract **(pneumonia),** urogenital tract **(sexually transmitted diseases),** or placenta.
         b. Pathogens that come from outside the body are **exogenous;** those that originate from normal flora are **endogenous.**
         c. The size of the *infectious dose* is of great importance.
         d. In the process of **adhesion,** a microbe attaches to the host cell by means of fimbriae, flagella, capsules, or receptors that position it for invasion.
      3. *Virulence Factors*
         a. **Exoenzymes** digest epithelial tissues, disrupt tissues, and permit invasion.

   b. **Toxigenicity** is a microbe's capacity to produce toxins at site of multiplication which affect cellular targets.
      (1) **Toxinoses** are diseases caused by toxins that damage structure or function of host cells.
      (2) **Toxemia** refers to toxins absorbed into the blood.
      (3) **Intoxication** means ingestion of toxins.
      (4) An **exotoxin** is a protein secreted by living bacteria with powerful effects on a specific organ. Examples are **hemolysins** and tetanus and diphtheria toxins.
      (5) An **endotoxin** is the lipopolysaccharide portion of a gram-negative cell wall released when a bacterial cell dies; causes generalized symptoms such as fever.
   c. **Antiphagocytic factors** include leukocidins (white blood cell poisons) and capsules.
   C. Effects on Target Organ/Spread of Infection
      1. *Patterns of Infection:* Stages in Infection/Disease
         a. **Incubation period,** the period from contact with infectious agent until appearance of first symptoms.
         b. **Prodromium,** a short period of initial, vague symptoms.
         c. **Period of invasion,** a variable period during which microbe multiplies in high numbers and causes severest symptoms.
         d. **Convalescent period,** a period of recovery, with decline of symptoms.
      2. *Type of Infections/Diseases*
         a. **Localized infection,** microbe remains in isolated site.
         b. **Systemic infection,** microbe is spread through the tissues by circulation.
         c. **Focal infection,** microbe spreads from local site to entire body (systemic).
         d. **Mixed infection,** several microbes cause one type of infection simultaneously.
         e. **Primary infection,** the initial infection in a series.
         f. **Secondary infection,** a second infection that complicates a primary infection.
         g. **Septicemia** and **bacteremia** refer to microbes in the blood.
         h. **Acute infection** appears suddenly, has a short course, and is relatively severe.
         i. **Chronic infection** persists over a long period of time.
         j. **Subacute infection** has a pattern between acute and chronic.
      3. *Signs and Symptoms:* Manifestations of disease, indicators of **pathologic** effects on target organs.
         a. A **sign** is objective, measurable evidence noted by an observer. Examples include septicemia, change in number of white blood cells; skin **lesions; inflammation; necrosis,** lysis or death of tissue.
         b. A **symptom** is a subjective effect of disease as sensed by patient. Examples are pain, fatigue, and nausea.
         c. A **syndrome** is a disease that manifests as a predictable complex of symptoms; infections that do not show symptoms are called *asymptomatic, subclinical,* or *inapparent.*
         d. Through the **portal of exit,** microbe is released with bodily secretions and discharges to have access to new host; portals include respiratory droplets from sneezing, coughing, saliva, skin, feces, urogenital tract (urine, mucus, semen), and blood.
         e. A microbe may become dormant **(latent)** and cause recurrent infections. Damaging effects that remain in organs and tissues after infection are **sequelae.**

II. **Epidemiology**

   A. **Epidemiology** is a science that determines the factors influencing causation, frequency, and distribution of disease in a community.

      1. Epidemiologists are involved in **surveillance** of reportable diseases in populations and consider measures to protect the public health.

      2. They are concerned with disease statistics such as **prevalence** (the total number of cases), **incidence** (the number of new cases), **morbidity** (general health of the population), and **mortality** (death).

   B. Frequency of Disease

      1. **Endemic,** a disease constantly present in a certain geographic area.

      2. **Sporadic,** a disease that occurs occasionally with no predictable pattern.

      3. **Epidemic,** sudden outbreak of disease in which numbers increase beyond expected trends.

      4. **Pandemic,** worldwide epidemic.

   C. Origin of Pathogens

      1. The **reservoir** is a place where the pathogen ultimately originates (its habitat).

      2. **Source** of infection refers to the immediate origin of an infectious agent.

      3. **Carrier** is an individual that inconspicuously shelters a pathogen and spreads it to others.

         a. **Asymptomatic carrier** is infected without symptoms.

         b. **Incubation carriers** carry early in disease.

         c. **Convalescent carriers** carry in last phases of recovery.

         d. **Chronic carriers** carry for long periods after recovery.

         e. **Passive carriers** are uninfected but convey infectious agents from infected persons to uninfected ones by hand and instrument contact.

   D. Vectors/Zoonoses

      1. A **vector** is an animal that transmits pathogens.

      2. A **biological vector** is an alternate animal host (mosquito, flea) that assists in completion of life cycle of microbe.

      3. A **mechanical vector** is an animal that does not host microbial life cycle, but is a short-term transmitter (housefly).

      4. A **zoonosis** is an infection for which animals are the natural reservoir and host that can be transmitted to humans.

   E. Acquisition of Infection

      1. **Communicable infectious disease** occurs when pathogen is transmitted from host to host directly or indirectly; **contagious diseases** are readily transmissible through direct contact.

      2. **Non-communicable diseases** are not spread from host to host; acquired from one's own flora (pneumonia) or from a nonliving environmental reservoir (tetanus).

      3. *Direct Transmission*

      Infectious agent is spread through direct contact of portal of exit with portal of entry (STDs, herpes simplex).

      4. *Indirect Transmission*

         a. A material (**vehicle**) contaminated with pathogens serves as intermediate source of infections.

         b. A **fomite** is an inanimate object contaminated with pathogens (public facilities, personal items).

         c. Food serves as a vehicle.

         d. **Droplet nuclei** are airborne dried particles containing infectious agents, formed by sneezing and coughing.

      5. *Nosocomial infections* are infectious diseases that originate in the hospital or clinical setting.

         a. They commonly occur among surgical and chronically ill patients.

         b. Hospitals monitor various asepsis procedures to help reduce the number of infections.

         c. Isolation of patients and other universal precautions are necessary controls.

      6. *Koch's postulates* defines a series of criteria that must be followed to determine the **etiologic** (causative) agent of disease.

MULTIPLE-CHOICE QUESTIONS

1. The best descriptive term for the resident flora is
  a. commensals   c. pathogens
  b. parasites   d. mutualists

2. Resident flora is commonly found in the
  a. stomach   c. salivary glands
  b. kidney   d. urethra

3. Resident flora is absent from the
  a. pharynx   c. intestine
  b. lungs   d. hair follicles

4. Virulence factors include
  a. toxins   c. capsules
  b. enzymes   d. all of these

5. The specific action of hemolysins is to
  a. damage white blood cells   c. damage red blood cells
  b. cause fever   d. cause leukocytosis

6. The _____ is the time that lapses between encounter with a pathogen and the first symptoms.
  a. prodromium   c. period of convalescence
  b. period of invasion   d. period of incubation

7. A short period early in a disease that manifests with general malaise and achiness is the
  a. period of incubation   c. sequela
  b. prodromium   d. period of invasion

8. The presence of a few bacteria in the blood is termed
  a. septicemia   c. bacteremia
  b. toxemia   d. a secondary infection

9. A _____ infection is acquired in a hospital.
  a. subclinical   c. nosocomial
  b. focal   d. zoonosis

10. A/an _____ is a passive animal transporter of pathogens.
  a. zoonosis   c. mechanical vector
  b. biological vector d. asymptomatic carrier

11. An example of a non-communicable infection is:
  a. measles   c. tuberculosis
  b. leprosy   d. tetanus

12. A general term that refers to an increased white blood cell count is
  a. leukopenia   c. leukocytosis
  b. inflammation d. leukemia

13. A positive antibody test for HIV would be a _____ of infection.
    a. sign
    c. syndrome
    b. symptom
    d. sequela

14. Which of the following would *not* be a portal of entry?
    a. the meninges
    c. skin
    b. the placenta
    d. small intestine

15. Which of the following is *not* a condition of Koch's postulates?
    a. isolate the causative agent of a disease
    b. cultivate the microbe in a lab
    c. inoculate a test animal to observe the disease
    d. test the effects of a pathogen on humans

CONCEPT QUESTIONS

1. Differentiate between contamination, infection, and disease. What are the possible outcomes in each?

2. How are infectious diseases different from other diseases?

3. Name the general body areas that are sterile. Why is the inside of the intestine not sterile like many other organs?

4. What causes variations in the flora of the newborn intestine?

5. What factors influence how the flora of the vagina develops?

6. Why must axenic young be delivered by cesarian section?

7. Explain several ways that true pathogens differ from opportunistic pathogens.

8. a. Distinguish between pathogenicity and virulence.
   b. Define virulence factors, and give examples of them in gram-positive and gram-negative bacteria, viruses, and parasites.

9. Describe the course of infection from contact with the pathogen to its exit from the host.

10. a. Explain why most microbes are limited to a single portal of entry.
    b. For each portal of entry, give a vehicle that carries the pathogen and the course it must travel to invade the tissues.
    c. Explain how the portal of entry could differ from the site of infection.

11. Differentiate between exogenous and endogenous infections.

12. a. What factors possibly affect the size of the infectious dose?
    b. Name five factors involved in microbial adhesion.

13. Which body cells or tissues are affected by hemolysins, leukocidins, hyaluronidase, kinases, tetanus toxin, pertussis toxin, and enterotoxin?

14. Compare and contrast: systemic versus local infections; primary versus secondary infections; infection versus intoxication.

15. What are the differences between signs and symptoms? (First put yourself in the place of a patient with an infection and then in the place of a physician examining you. Describe what you would feel and what the physician would detect upon examining the affected area.)

16. a. What are some important considerations about the portal of exit?
    b. Name some examples of infections and their portals of exit.

17. Complete the table:

	Exotoxins	Endotoxins
Chemical makeup		
General source		
Degree of toxicity		
Effects on cells		
Symptoms in disease		
Examples		

18. a. Explain what it means to be a carrier of infectious disease.
    b. Describe four ways that humans can be carriers.
    c. What is epidemiologically and medically important about carriers in the population?

19. a. Outline the science of epidemiology and the work of an epidemiologist.
    b. Using the following statistics, based on number of reported cases, determine which show endemic, sporadic, or epidemic patterns. Explain how you can determine each type.

United States Region	Disease Statistics	
Chlamydiosis	**1998**	**2000**
Northeast	20,090	22,500
East Coast	61,510	61,400
Central	137,400	149,700
Southeast	246,200	259,900
West	132,100	136,100
Total cases	597,300	629,600
Lyme disease		
Northeast	5,056	4,500
East Coast	9,311	8,250
Central	1,091	672
Southeast	1,160	1,420
West	183	193
Total cases	16,801	15,035
Rubella		
Northeast	38	6
East Coast	150	22
Central	43	118
Southeast	114	59
West	19	24
Total cases	364	229

    c. Explain what would have to occur for these diseases to have a pandemic distribution.

20. Distinguish between mechanical and biological vectors, giving one example of each.

21. a. Explain the precise difference between communicable and non-communicable infectious diseases.
    b. Between direct and indirect modes of transmission.
    c. Between vectors and vehicles as modes of transmission.

22. a. Nosocomial infections can arise from what two general sources?
    b. From this chapter and figure 20.7, outline the major agents involved in nosocomial infections.
    c. Do they tend to be true pathogens or opportunists?
    d. Outline the two types of hospital asepsis and define isolation.
    e. What is the work of an infection control officer?

23. a. List the main features of Koch's postulates.
    b. Why is it so difficult to prove them for some diseases?

CRITICAL-THINKING QUESTIONS

1. a. Discuss the relationship between the vaginal residents and the colonization of the newborn.
   b. Can you think of some serious medical consequences of this relationship?
   c. Why would normal flora cause some infections to be more severe and other infections to be less severe?

2. If the following patient specimens produced positive cultures when inoculated and grown on appropriate media, indicate whether this result indicates a disease state and why or why not:

Urine	Throat
Lung biopsy	Feces
Saliva	Blood
Cerebrospinal fluid	Urine from bladder
Liver biopsy	Semen

   What are the important clinical implications of positive blood or cerebrospinal fluid?

3. Pretend that you have been given the job of developing a colony of germ-free cockroaches.
   a. What will be the main steps in this process?
   b. What possible experiments can you do with these animals?

4. If healthy persons are resistant to infection with opportunists or weak pathogens, what is the expected result if a compromised person is exposed to a true pathogen?

5. Explain how the endotoxin gets into the bloodstream of a patient with endotoxic shock.

6. You are a physician following the course of an infection in a small child who has been exposed to scarlet fever. Describe the main events that occur, what is happening during each stage, and the causes of each sign and symptom.

7. Describe each of the following infections using correct technical terminology. (Descriptions may fit more than one category.) Use terms such as primary, secondary, nosocomial, STD, mixed, latent, toxemia, chronic, zoonotic, asymptomatic, local, systemic, -itis, -emia.

   Caused by needlestick in dental office
   Pneumocystis pneumonia in AIDS patient
   Bubonic plague from rat flea bite
   Diphtheria
   Undiagnosed chlamydiosis
   Acute necrotizing gingivitis
   Syphilis of long duration
   Large numbers of gram-negative rods in the blood
   A boil on the back of the neck
   An inflammation of the meninges
   Scarlet fever

8. a. Using statistics in the endpapers of this book, find several reportable diseases that show an increase in numbers of cases that could indicate an outbreak or epidemic.
   b. Provide some possible explanations for regional differences in the distribution of gonorrhea and hepatitis (figure 13.16).

9. Name 10 fomites that you came into contact with today.

10. Describe what parts of Koch's postulates were unfulfilled by Dr. Lazear's experiment described in Historical Highlights 13.5.

11. a. Suggest several reasons why urinary tract, respiratory tract, and surgical infections are the most common nosocomial infections.
    b. Name several measures that health care providers must exercise at all times to prevent or reduce nosocomial infections.

INTERNET SEARCH TOPICS

Use an Internet search engine to look up the following topics. Write a short summary of what you discover on your searches.

Sentinel chickens
Tuskegee syphilis experiment
Koch's postulates verified for AIDS and HIV

CHAPTER 14

The Nature of Host Defenses

The survival of the host depends upon an elaborate network of defenses that keeps harmful microbes and other foreign materials from penetrating the body. Should they penetrate, additional host defenses are summoned to prevent them from becoming established in tissues. Defenses involve barriers, cells, and chemicals, and they range from nonspecific to specific and from inborn to acquired. This chapter introduces the main lines of defense intrinsic to all humans. Topics included in this survey are the anatomical and physiological systems that detect, recognize, and destroy foreign substances and the general adaptive responses that account for an individual's long-term immunity or resistance to infection and disease.

Chapter Overview

- The body has a complex overlapping series of defenses that protect it against invasion by harmful microbes and other foreign matter.
- Defenses exist at several levels of development and specificity.
- First line defenses are inborn physical barriers such as skin, and second line defenses are non-specific, protective reactions in the fluid compartments such as phagocytosis.
- Third line defenses are aimed at a specific pathogen and give a long-term form of protection that will come into play if that pathogen is ever encountered again.
- The immune system relies on a vast network of cells that circulate through the tissues to search for, detect, and destroy foreign invaders such as bacteria and viruses.
- The systems of the body that are most involved in immune function include the blood, lymphoid organs and tissues, and the reticuloendothelial system. These systems freely communicate among one another.
- White blood cells, or leukocytes, are formed in the red bone marrow and released into circulation. They migrate out of the blood into the tissues to carry out some of their complex functions relating to phagocytosis, inflammation, antibody production, and pathogen killing.
- The lymphoid organs such as the spleen and lymph nodes receive and transport tissue fluids and white blood cells, and are important sites of surveillance and immune reactions.
- Inflammation is a sequential protective response to injury that stimulates the influx of beneficial cells and chemicals that can protect against infection and further damage at that site.
- Several types of chemical substances, called mediators, are released by cells during inflammation and other immune responses. These

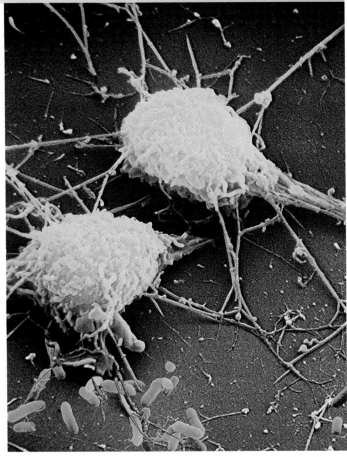

Like an octopus, a macrophage sends out long, tentacled pseudopods to capture its *Escherichia coli* prey (blue cells). False-color scanning electron micrograph (22,000×).

include chemicals that change the size of blood vessels, stimulate the migration of white blood cells, and initiate fever.
- Interferon is a nonspecific immune mediator that inhibits the replication of viruses and regulates a variety of immune responses.
- The complement system is an organized chain reaction of chemicals acting sequentially to lyse cells and viruses.
- Phagocytes are specialized cells that function in engulfment and clearance of foreign molecules, cells, viruses, and particles. They contain numerous enzymes and toxic chemicals to carry out this function.
- Acquired specific immunities provided by B and T lymphocytes provide specific protection against infection and are essential to survival.

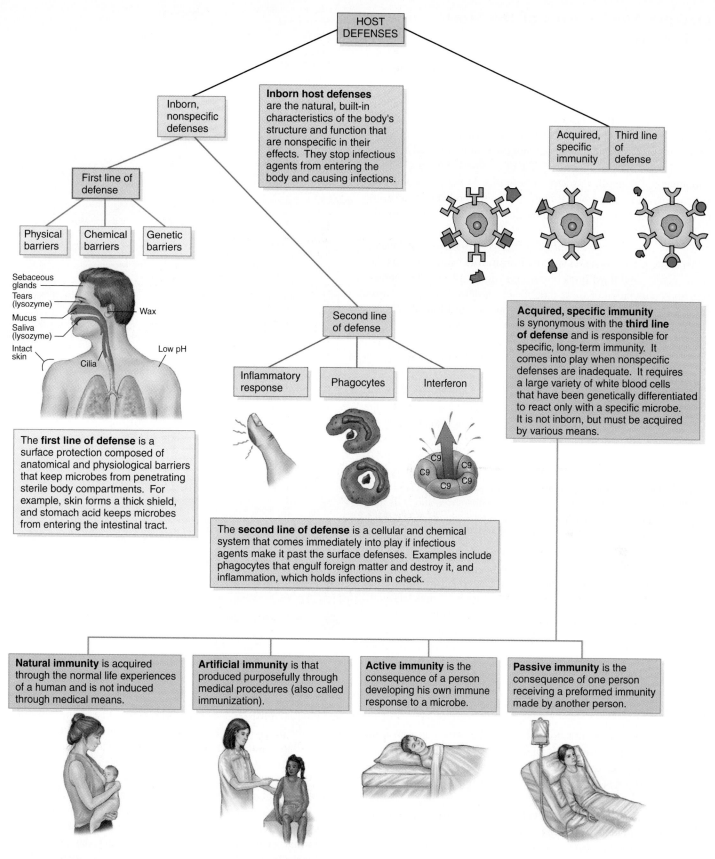

FIGURE 14.1

Flowchart summarizing the major components of the host defenses.   Defenses are classified into one of two general categories: either inborn and nonspecific or acquired and specific. These can be further subdivided into the first, second, and third lines of defense, each being characterized by a different level and type of protection. The third line of defense is the most complex and is responsible for immunity.

Defense Mechanisms of the Host in Perspective

In chapter 13 we explored the host-parasite relationship, with emphasis on the role of microorganisms in disease. In this chapter we examine the other side of the relationship—that of the host defending itself against microorganisms. As previously stated, in light of the unrelenting contamination and colonization of humans, it is something of a miracle that we are not constantly infected and diseased. This does *not* happen because of a remarkable, fascinating, and amazingly complex system of defenses. In the war against all sorts of invaders, microbial and otherwise, the body erects a series of barriers, sends in an army of cells, and emits a flood of chemicals to protect tissues from harm.

The host defenses embrace a multilevel network of innate, nonspecific protections and specific **immunities*** referred to as the first, second, and third lines of defense (figure 14.1). The interaction and cooperation of these three levels of defense normally provide complete protection against infection. The **first line of defense** includes any barrier that blocks invasion at the portal of entry. This mostly nonspecific line of defense limits access to the internal tissues of the body. However, it is not considered a true immune response because it does not involve recognition of a specific foreign substance but is very general in action. The **second line of defense** is a slightly more internalized system of protective cells and fluids that includes inflammation and phagocytosis. It acts rapidly at both the local and systemic levels once the first line of defense has been circumvented. The highly specific **third line of defense** is acquired on an individual basis as each foreign substance is encountered by white blood cells called lymphocytes. The reaction with each different microbe produces unique protective substances and cells that can come into play if that microbe is encountered again. The third line of defense provides long-term immunity.

The human systems are armed with various levels of defense that do not operate in a completely separate fashion; most defenses overlap and are even redundant in some of their effects. This "immunological overkill" literally bombards microbial invaders with an entire assault force, making their survival unlikely. Because of the interwoven nature of host defenses, we will introduce basic concepts of structure and function that will prepare you for later information on specific reactions of the immune system (see chapter 15).

BARRIERS AT THE PORTAL OF ENTRY: A FIRST LINE OF DEFENSE

A number of defenses are a normal part of the body's anatomy and physiology. These natural, inborn, nonspecific defenses can be divided into physical, chemical, and genetic barriers that impede the entry of microbes at the site of first contact (figure 14.2).

Physical or Anatomical Barriers at the Body's Surface

The skin and mucous membranes of the respiratory and digestive tracts have several built-in defenses. The outermost layer (stratum corneum) of the skin is composed of epithelial cells that have become compacted, cemented together, and impregnated with an in-

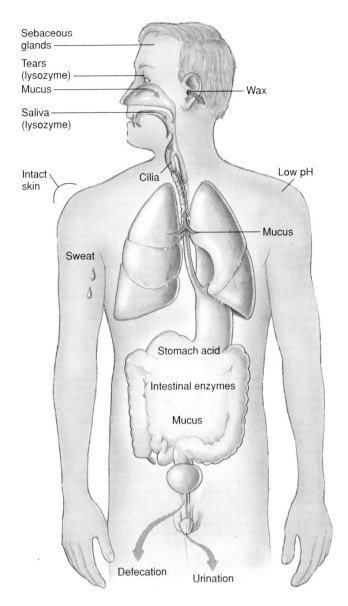

FIGURE 14.2

The primary physical and chemical defense barriers.

soluble protein, keratin. The result is a thick, tough layer that is highly impervious and waterproof. Few pathogens can penetrate this unbroken barrier, especially in regions such as the soles of the feet or the palms of the hands, where the stratum corneum is much thicker than on other parts of the body. Other cutaneous barriers include hair follicles and skin glands. The hair shaft is periodically extruded, and the follicle cells are *desquamated*.* The flushing effect of sweat glands also helps remove microbes.

The mucocutaneous membranes of the digestive, urinary, and respiratory tracts and of the eye are moist and permeable. Despite the normal wear and tear upon these epithelia, damaged cells are rapidly replaced. The mucous coat on the free surface of some

---

* immunity (im-yoo′-nih-tee) Gr. *immunis*, free, exempt. A state of resistance to infection.

* desquamate (des′-kwuh-mayt) L. *desquamo*, to scale off. The casting off of epidermal scales.

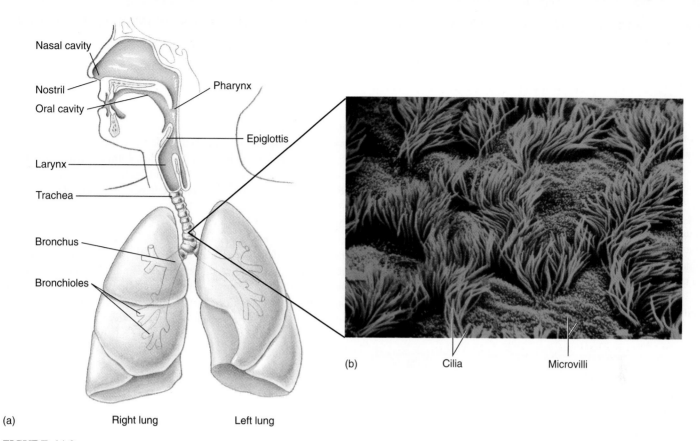

Nasal cavity

Nostril

Oral cavity

Pharynx

Epiglottis

Larynx

Trachea

Bronchus

Bronchioles

(a)   Right lung   Left lung

(b)   Cilia   Microvilli

FIGURE 14.3

The ciliary defense of the respiratory tree.   **(a)** The epithelial lining of the airways contains a brush border of cilia to entrap and propel particles upward toward the pharynx. **(b)** Tracheal mucosa (5,000×).

membranes impedes the entry of bacteria. Blinking and *lacrimation** flush the eye's surface with tears and rid it of irritants. The constant flow of saliva helps carry microbes into the harsh conditions of the stomach. Vomiting and defecation also evacuate noxious substances or microorganisms from the body.

The respiratory tract is constantly guarded from infection by elaborate and highly effective adaptations. Nasal hair traps larger particles. Rhinitis, the copious flow of mucus and fluids that occurs in allergy and colds, exerts a flushing action. In the respiratory tree (primarily the trachea and bronchi), a ciliated epithelium (called the ciliary escalator) conveys foreign particles entrapped in mucus toward the pharynx to be removed (figure 14.3). Irritation of the nasal passage reflexly initiates a sneeze, which expels a large volume of air at high velocity. Similarly, the acute sensitivity of the bronchi, trachea, and larynx to foreign matter triggers coughing, which ejects irritants.

The genitourinary tract derives partial protection from the continuous trickle of urine through the ureters and from periodic bladder emptying that flushes the urethra.

The composition and protective effect exerted by flora were discussed in chapter 13. Even though the resident flora does not

constitute an anatomical barrier, its presence can block the access by pathogens to epithelial surfaces and can create an unfavorable environment for pathogens.

Nonspecific Chemical Defenses

The skin and mucous membranes offer a variety of chemical defenses. Sebaceous secretions exert an antimicrobial effect, and specialized glands such as the meibomian glands of the eyelids lubricate the conjunctiva with an antimicrobial secretion. An additional defense in tears and saliva is **lysozyme,** an enzyme that hydrolyzes the peptidoglycan in the cell wall of bacteria. The high lactic acid and electrolyte concentrations of sweat and the skin's acidic pH and fatty acid content are also inhibitory to many microbes. Likewise, the hydrochloric acid in the stomach renders protection against many pathogens that are swallowed, and the intestine's digestive juices and bile are potentially destructive to microbes. Even semen contains an antimicrobial chemical that inhibits bacteria, and the vagina has a protective acidic pH maintained by normal flora.

Genetic Defenses

Some hosts are genetically immune to the diseases of other hosts. One explanation for this phenomenon is that some pathogens have such great specificity for one host species that they are incapable of infecting other species. One way of putting it is: "Humans can't

---

* **lacrimation** (lak″-rih-may′-shun) L. *lacrimatio,* tears.

acquire distemper from cats, and cats can't get mumps from humans." This specificity is particularly true of viruses, which can invade only by attaching to a specific host receptor. But it does not hold true for zoonotic infectious agents that attack a broad spectrum of animals. Genetic differences in susceptibility can also exist within members of one species. Humans carrying a gene or genes for sickle-cell anemia are resistant to malaria. Genetic differences also exist in susceptibility to tuberculosis, leprosy, and certain systemic fungal infections.

The vital contribution of barriers is clearly demonstrated in people who have lost them or never had them. Patients with severe skin damage due to burns are extremely susceptible to infections; those with blockages in the salivary glands, tear ducts, intestine, and urinary tract are also at greater risk for infection. But as important as it is, the first line of defense alone is not sufficient to protect against infection. Because many pathogens find a way to circumvent the barriers by using their virulence factors (discussed in chapter 13), a whole new set of defenses—inflammation, phagocytosis, specific immune responses—are brought into play.

CHAPTER CHECKPOINTS

The multilevel, interconnecting network of host protection against microbial invasion is organized into three lines of defense.

- The first line consists of physical and chemical barricades provided by the skin and mucous membranes.

- The second line encompasses all the nonspecific cells and chemicals found in the tissues and blood.

- The third line, the specific immune response, is customized to react to specific antigens of a microbial invader. This response immobilizes and destroys the invader every time it appears in the host.

Introducing the Immune System

Immunology is the study of all biological, chemical, and physical events surrounding the function of the immune system. This field has mushroomed into an exciting, precedent-setting area of molecular biology that dominates the progress of many areas of biology and medicine. Several recent Nobel prizes in medicine were presented for work in immunology. Breakthroughs in the areas of cancer, AIDS, or therapy emerge from the immunologic research community on a constant basis.

A study of immunities necessitates examining a number of interrelated concepts. For example, because immune reactions actually happen at the molecular level, we must understand the concepts of foreignness, surveillance, recognition, and specificity. In order to discuss phagocytosis, we must know something about blood cells; in turn, in order to discuss inflammation, we must understand the roles of body fluids and chemical mediators; and in order to understand specific immunities, we must know something about all of these concepts and several related ones.

THE MOLECULAR BASIS OF IMMUNE RESPONSES

Whenever foreign material such as an infectious agent enters the tissues, the cells of the immune system are rapidly enlisted in a formidable molecular interchange. To understand this reaction, we must first define what constitutes a foreign substance. In the most general sense, a **foreign material** is something that can be recognized or distinguished as not being a natural part of an organism's body. The body has a fleet of white blood cells strategically located to explore continually through its tissues, compartments, and fluids like billions of tiny fingers, feeling and evaluating what is there and searching for any molecules that are new or different. This process of scouting the tissues for foreign molecules and other possibly threatening particles is **surveillance** (figure 14.4).

At the same time that white blood cells survey the tissues, they also evaluate and differentiate the molecules they detect by a process of **recognition.** Inherent in recognition is the capacity of the probing cells to sort out the **natural markers**[1] **of the body** (also known as **self**) from the **foreign markers (nonself).** This is an essential step, because the immune system is programmed to react as if nonself is potentially harmful and should be earmarked for destruction, whereas self is not. In this hunt for things that are not part of self, the immune system has a highly refined "sense of touch." When nonself is discovered and recognized, a whole battalion of responses is brought into play to **entrap and destroy** (kill, neutralize) the invading offender.

The main players in these reactions are molecules called **markers** that protrude from the cell surface like minuscule signposts announcing that cell or molecule's identity. The identity of a given marker is based on its **specificity,** a unique configuration that dictates the kinds of immune responses it can elicit. Markers can also be **receptors,** molecules that bind specifically with complementary molecules in ways that signal, communicate, and trigger reactions inside the cell.

An even more specific term for a foreign, nonself molecule is **antigen.*** An antigen is a substance, often a surface marker, that evokes a specific immune response. As we shall see later in this chapter and in the next, antigens are responsible for eliciting specific reactions that occur during infection, vaccination, and allergies.

CHAPTER CHECKPOINTS

The immune system operates first as a surveillance system that discriminates between the host's self identity markers and the nonself identity markers of foreign cells. When it recognizes that a marker or antigen is foreign, or nonself, the immune system tailors its response specifically to each different antigen. As far as the immune system is concerned, if an antigen is not self, it is foreign, does not belong, and must be destroyed.

---

1. The term *marker* is also employed in genetics in a different sense—that is, to denote a detectable characteristic of a particular genetic mutant. A genetic marker may or may not be a surface marker.

* **antigen** (an'-tih-jen) From *antibody* + *gen,* to produce. Originally, any substance that elicits the production of antibodies, but the term is now used to include other kinds of reactions.

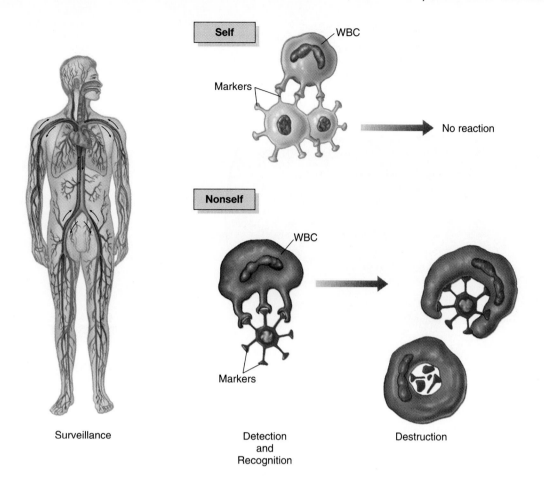

FIGURE 14.4

Search, recognize, and destroy is the mandate of the immune system. White blood cells are equipped with a very sensitive sense of "touch." As they sort through the tissues, they feel surface markers that help them determine what is self and what is not. When self markers are recognized, no response occurs. However, when nonself is detected, a reaction to destroy it is mounted.

Systems Involved in Immune Defenses

Unlike many systems, the **immune system** does not exist in a single, well-defined site; rather, it encompasses a large, complex, and diffuse network of cells and fluids that permeate every organ and tissue. It is this very arrangement that promotes the surveillance and recognition processes that help screen the body for harmful substances.

The body is partitioned into several fluid-filled spaces called the intracellular, extracellular, lymphatic, cerebrospinal, and circulatory compartments. Although these compartments are physically separated, they have numerous connections. Their structure and position permit extensive interchange and communication (figure 14.5). Among the body compartments that participate in immune function are (1) the *reticuloendothelial* system (RES),* (2) the spaces surrounding tissue cells that contain *extracellular fluid (ECF),* (3) the *bloodstream,* and (4) the *lymphatic system.* In the following section, we consider the anatomy of these

main compartments and how they interact in the second and third lines of defense.

THE COMMUNICATING BODY COMPARTMENTS

For effective immune responsiveness, the activities in one fluid compartment must be conveyed to other compartments. Let us see how this occurs by viewing tissue at the microscopic level (figure 14.5*a*). At this level, clusters of tissue cells are in direct contact with the reticuloendothelial system (RES) and the extracellular fluid (ECF). Other compartments (vessels) that penetrate at this level are blood and lymphatic capillaries. This means that cells and chemicals that originate in the RES and ECF can diffuse or migrate into the blood and lymphatics; any products of a lymphatic reaction can be transmitted directly into the blood through the connection between these two systems; and certain cells and chemicals originating in the blood can move through the vessel walls into the extracellular spaces and migrate into the lymphatic system.

The flow of events among these systems depends on where an infectious agent or foreign substance first intrudes. A typical progression might begin in the extracellular spaces and RES, move to

---

* reticuloendothelial (reh-tik″-yoo-loh-en″-doh-thee′-lee-al) L. *reticulum,* a small net, and *endothelium,* lining of the blood vessel.

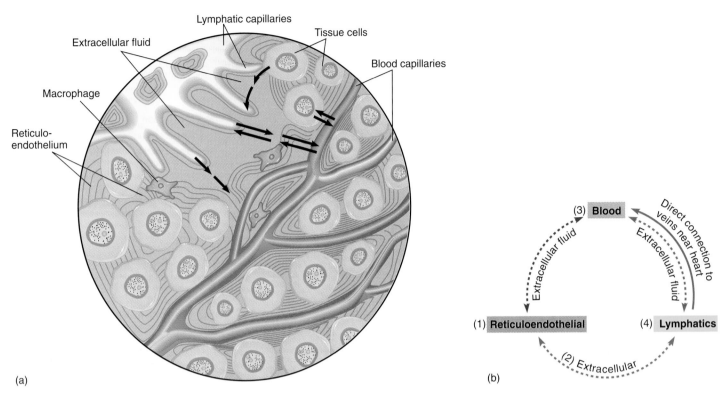

FIGURE 14.5

Connections between the body compartments. **(a)** The meeting of the major fluid compartments at the microscopic level.
(b) Schematic view of the main fluid compartments and how they form a continuous cyclic system of exchange. Reactions in one section are rapidly broadcast to the others.

the lymphatic circulation, and ultimately end up in the bloodstream. Regardless of which compartment is first exposed, an immune reaction in any one of them will eventually be communicated to the others at the microscopic level. An obvious benefit of such an integrated system is that no cell of the body is far removed from competent protection, no matter how isolated. Let us take a closer look at each of these compartments.

Immune Functions of the Reticuloendothelial System

The tissues of the body are permeated by a support network of connective tissue fibers, or a *reticulum,* that originates in the cellular basal lamina, interconnects nearby cells, and meshes with the massive connective tissue network surrounding all organs. This network, called the **reticuloendothelial system** (figure 14.6) is intrinsic to the immune function because it provides a passageway within and between tissues and organs. It also coexists with and helps form a niche for a collection of phagocytic cells termed the **mononuclear phagocyte system.** The RES is heavily endowed with white blood cells called macrophages waiting to attack passing foreign intruders as they arrive in the skin, lungs, liver, lymph nodes, spleen, and bone marrow.

Origin, Composition, and Functions of the Blood

The circulatory system consists of the circulatory system proper, which includes the heart, arteries, veins, and capillaries that circulate the blood, and the lymphatic system, which includes lymphatic

vessels and lymphatic organs (lymph nodes) that circulate lymph (see figure 14.14). As we shall see, these two circulations parallel, interconnect with, and complement one another.

The substance that courses through the arteries, veins, and capillaries is **whole blood,** a liquid connective tissue consisting of **blood cells** (formed elements) suspended in **plasma** (ground substance). One can visualize these two components with the naked eye when a tube of *unclotted* blood is allowed to sit or is spun in a centrifuge. The cells' density causes them to settle into an opaque layer at the bottom of the tube, leaving the plasma, a clear, yellowish fluid, on top (figure 14.7). The terms *plasma* and *serum* are sometimes mistakenly interchanged, and though these substances have the same basic origin, they differ in one important way. Because **serum** is the fluid extruded from *clotted* blood and the clotting proteins enter into the formation of the clot, serum does not contain the clotting proteins that plasma does. An easy way to remember this is: "Plasma can clot and serum cannot." Their composition is otherwise very similar.

Fundamental Characteristics of Plasma Plasma contains hundreds of different chemicals produced by the liver, white blood cells, endocrine glands, and nervous system and absorbed from the digestive tract. The main component of this fluid is water (92%), and the remainder consists of proteins such as albumin and globulins (including antibodies), other immunochemicals, fibrinogen and other clotting factors, hormones, nutrients (glucose, amino

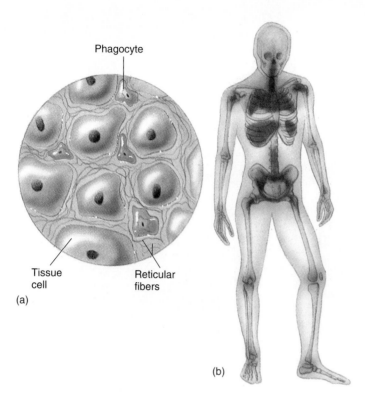

Phagocyte

Tissue
cell

Reticular
fibers

(a)

(b)

FIGURE 14.6

🖝 **The reticuloendothelial system occurs as a pervasive, continuous connective tissue framework throughout the body.** **(a)** This system begins at the microscopic level with a fibrous support network (reticular fibers) enmeshing each cell. This web connects one cell to another within a tissue or organ and provides a niche for phagocytes, which can crawl within and between tissues. **(b)** The degrees of shading in the body indicate variations in phagocyte concentration (darker = greater).

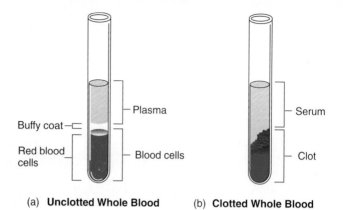

Buffy coat

Plasma

Red blood
cells

Blood cells

Serum

Clot

(a) **Unclotted Whole Blood**   (b) **Clotted Whole Blood**

FIGURE 14.7

The macroscopic composition of whole blood. **(a)** When blood containing anticoagulants is allowed to sit for a period, it stratifies into a clear layer of plasma, a thin layer of off-white material called the buffy coat (which contains the white blood cells), and a layer of red blood cells in the bottom, thicker layer. **(b)** When blood is drawn and allowed to clot, a clear layer of serum is formed above the clot.

acids, fatty acids), ions (sodium, potassium, calcium, magnesium, chloride, phosphate, bicarbonate), dissolved gases (O_2 and CO_2), and waste products (urea). These substances support the normal physiological functions of nutrition, development, protection, homeostasis, and immunity. We return to the subject of plasma and its function in immune interactions later in this chapter and in chapter 15.

A Survey of Blood Cells The production of blood cells, or **hemopoiesis,*** begins early in embryonic development in the yolk sac (an embryonic membrane). Later it is taken over by the liver and lymphatic organs, and it is finally assumed entirely and permanently by the red bone marrow (figure 14.8). Although much of a newborn's red marrow is devoted to hemopoietic function, the active marrow sites gradually recede, and by the age of 4 years, only the ribs, sternum, pelvic girdle, flat bones of the skull and spinal column, and proximal portions of the humerus and femur are devoted to blood cell production.

The relatively short life of blood cells demands a rapid turnover that is continuous throughout a human lifespan. The pri-

mary precursor of new blood cells is a pool of undifferentiated cells called pluripotential **stem cells**[2] maintained in the marrow. During development, these stem cells proliferate and *differentiate*—meaning that immature or unspecialized cells develop the specialized form and function of mature cells. The primary lines of cells that arise from this process produce red blood cells (RBCs, or erythrocytes), white blood cells (WBCs, or leukocytes), and platelets (thrombocytes). The white blood cell lines are programmed to develop into several secondary lines of cells during the final process of differentiation (figure 14.9). These committed lines of WBCs are largely responsible for immune function.

The **white blood cells,** or **leukocytes,*** are traditionally evaluated by their reactions with hematologic stains that contain a mixture of dyes and can differentiate cells by color and morphology. When this stain used on blood smears is evaluated by the light microscope, the leukocytes appear either with or without noticeable colored granules in the cytoplasm and, on that basis, are divided into two groups: **granulocytes** and **agranulocytes.** Greater magnification reveals that even the agranulocytes have tiny granules in their cytoplasm, so some hematologists also use the appearance of the nucleus to distinguish them. Granulocytes have a lobed nucleus, and agranulocytes have an unlobed, rounded nucleus. Note both of these characteristics in circulating leukocytes (see figures 14.9 and 14.10).

Granulocytes The types of granular leukocytes present in the bloodstream are neutrophils, eosinophils, and basophils. All three are known for prominent cytoplasmic granules that stain with some combination of acidic dye (eosin) or basic dye (methylene blue). Although these granules are useful diagnostically, they also function in numerous physiological events.

---

* **hemopoiesis** (hee″-moh-poy-ee′-sis) Gr. *haima,* blood, and *poiesis,* a making. Also called hematopoiesis.

2. Pluripotential stem cells can develop into several different types of blood cells; unipotential cells have already committed to a specific line of development.

* **leukocyte** (loo′-koh-syte) Gr. *leukos,* white, and *kytos,* cell. The whiteness of unstained WBCs is best seen in the white layer, or buffy coat, of sedimented blood.

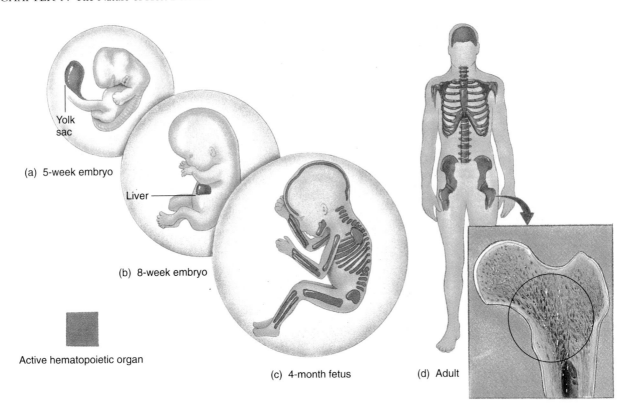

FIGURE 14.8

Stages in hemopoiesis. The sites of blood cell production change as development progresses from **(a, b)** yolk sac and liver in the embryo, to **(c)** extensive bone marrow sites in the fetus and **(d)** selected bone marrow sites in the child and adult. **(Inset)** Red marrow occupies the spongy bone (circle).

Neutrophils* are distinguished from other leukocytes by their conspicuous, lobed nuclei and by their fine, pale lavender granules. In cells newly released from the bone marrow, the nuclei are horseshoe-shaped, but as they age, they form multiple lobes (up to five). These cells, also called **polymorphonuclear* neutrophils (PMNs),** make up 55% to 90% of the circulating leukocytes—about 25 billion cells in the circulation at any given moment. The main work of the neutrophils is in phagocytosis.[3] Their high numbers in both the blood and tissues suggest a constant challenge from resident microflora and environmental sources. Most of the cytoplasmic granules carry digestive enzymes and other chemicals that degrade the phagocytosed materials (see the discussion of phagocytosis later in this chapter). The average neutrophil lives only about 8 days, spending much of this time in the tissues and only about 6 to 12 hours in circulation.

Eosinophils* are readily distinguished in a stain preparation by their larger, orange to red (eosinophilic) granules and bilobed nucleus. They are much more numerous in the bone marrow and the spleen than in the circulation, contributing only 1% to 3% of the total WBC count. The role of the eosinophil in the immune system is not fully defined, though several functions have been suggested. Their granules contain peroxidase, lysozyme, and other digestive enzymes. Eosinophils appear to be weakly phagocytic for bacteria, foreign particles, and antigen-antibody complexes. High levels of eosinophils are observed in infections by helminth parasites and fungi, and it is now an accepted fact that they play a major role in destroying these large eucaryotic parasites. Much evidence is accumulating on the role of eosinophils as mediators in immediate allergies such as asthma and anaphylaxis (see chapter 17).

Basophils* are characterized by pale-stained, constricted nuclei and very prominent dark blue to black granules. They are the scarcest type of leukocyte, making up less than 0.5% of the total circulating WBCs in a normal individual. Basophils share some morphological and functional similarities with widely distributed tissue cells called **mast* cells.** Although these two cell types were once regarded as identical, mast cells are nonmotile elements bound to connective tissue around blood vessels, nerves, and epithelia, and basophils are motile elements derived from bone marrow. With

---

3. The neutrophil is sometimes called a microphage, or "small eater."

* neutrophil (noo′-troh-fil) L. *neuter,* neither, and Gr. *philos,* to love. The granules are neutral and do not react markedly with either acidic or basic dyes. In clinical reports, they are often called "polys" or PMNs for short.

* polymorphonuclear Gr. *poly,* many; *morph,* shape; and *nuclear,* nucleus.

* eosinophil (ee″-oh-sin′-oh-fil) Gr. *eos,* dawn, rosy, and *philos,* to love. Eosin is a red, acidic dye attracted to the granules.

---

* basophil (bay′-soh-fil) Gr. *basis,* foundation. The granules attract to basic dyes.

* mast From Ger. *mast,* food. Early cytologists thought these cells were filled with food vacuoles.

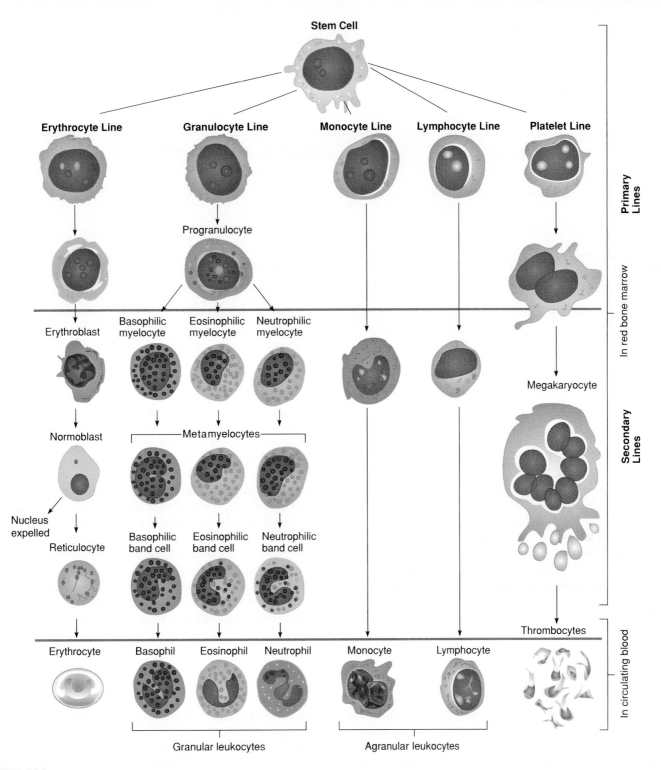

Stem Cell

Erythrocyte Line   Granulocyte Line   Monocyte Line   Lymphocyte Line   Platelet Line

Progranulocyte

Primary Lines

In red bone marrow

Erythroblast   Basophilic myelocyte   Eosinophilic myelocyte   Neutrophilic myelocyte

Normoblast   Metamyelocytes

Megakaryocyte

Secondary Lines

Nucleus expelled

Reticulocyte   Basophilic band cell   Eosinophilic band cell   Neutrophilic band cell

Thrombocytes

In circulating blood

Erythrocyte   Basophil   Eosinophil   Neutrophil   Monocyte   Lymphocyte

Granular leukocytes   Agranular leukocytes

FIGURE 14.9

The development of blood cells and platelets.   Each cell type in circulating blood (bottom row) is ultimately derived from an undifferentiated stem cell in the red marrow. During differentiation, the stem cell gives rise to several cell lines that become more and more specialized. Mature cells are released into the circulatory system.

regard to immune function, both types act primarily in immediate allergy and inflammation. Their granules release chemical mediators such as histamine, serotonin, heparin, and several enzymes with pronounced physiological effects. As we shall see in chapter 17, many allergy symptoms are directly attributable to the effects of these chemicals on tissues and organs.

Agranulocytes   Agranular leukocytes have globular, nonlobed nuclei and lack prominent cytoplasmic granules when viewed with the light microscope. The two general types are monocytes and lymphocytes.

Lymphocytes are the second most common WBC in the blood, comprising 20% to 35% of the total circulating leukocytes.

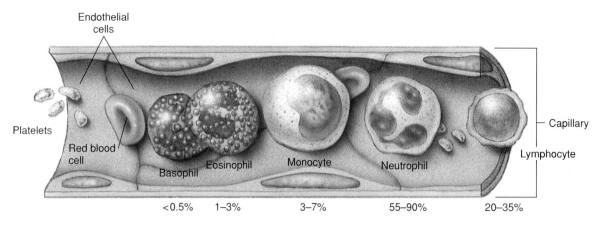

FIGURE 14.10

The microanatomy and circulating cells of the bloodstream. A cutaway view of a capillary reveals a histological picture of thin, tile-like endothelial cells that form a thin, one-cell-thick tube. The cell types and relative proportions of circulating white blood cells are also shown.

The fact that their overall number throughout the body is among the highest of all cells indicates how important they are to immunity. One estimate suggests that about one-tenth of all adult body cells are lymphocytes, exceeded only by erythrocytes and fibroblasts. In a stained blood smear, most lymphocytes are small, spherical cells with a uniformly dark blue, rounded nucleus surrounded by a thin fringe of clear, pale blue cytoplasm, although in tissues, they can become much larger and can even mimic monocytes in appearance (see figure 14.9). Lymphocytes exist as two functional types—the bursal-equivalent, or **B lymphocytes** (**B cells,** for short), and the thymus-derived, or **T lymphocytes** (**T cells,** for short). B cells were first demonstrated in and named for a special lymphatic gland of chickens called the *bursa of Fabricius,* the site for their maturation in birds. In humans, B cells mature in special bone marrow sites, but humans do not have a bursa. T cells mature in the thymus gland in all birds and mammals. Both populations of cells are transported by the bloodstream and lymph and move about freely between lymphoid organs and connective tissue.

Lymphocytes are the key cells of the third line of defense and the specific immune response (figure 14.11). When stimulated by foreign substances (antigens), lymphocytes are transformed into activated cells that neutralize and destroy that foreign substance. The contribution of B cells is mainly in **humoral immunity,**[4] defined as protective molecules carried in the fluids of the body. When activated B cells divide, they form specialized *plasma cells,* which produce **antibodies,** large protein molecules that interlock with an antigen and participate in its destruction. Activated T cells engage in a spectrum of immune functions characterized as **cell-mediated immunity (CMI)** in which T cells help, suppress, and modulate immune functions and kill foreign cells. The action of both classes of lymphocytes accounts for the recognition and memory typical of immunity. So important are lymphocytes to the defense of the body that a large portion of chapter 15 is devoted to their reactions.

Monocytes are generally the largest of all white blood cells and the third most common in the circulation (3–7%). As they mature, the nucleus becomes oval- or kidney-shaped—indented on one side, off-center, and often contorted with fine wrinkles. The pale blue cytoplasm holds many fine vacuoles containing digestive enzymes. Monocytes are discharged by the bone marrow into the bloodstream, where they live as phagocytes for a few days. Later they leave the circulation to undergo final differentiation into **macrophages** (see figure 14.20). Unlike many other WBCs, the monocyte-macrophage series is relatively long-lived and retains an ability to multiply. Macrophages are among the most versatile and important of cells. Most immune reactions involve them either directly or indirectly. In general, they are responsible for (1) many types of specific and nonspecific phagocytic and killing functions (they assume the job of cellular housekeepers, "mopping up the messes" created by infection and inflammation); (2) processing foreign molecules and presenting them to lymphocytes; and (3) secreting biologically active compounds that assist, mediate, attract, and inhibit immune cells and reactions. We touch upon these functions in several ensuing sections.

Erythrocyte and Platelet Lines The **erythrocyte** line of cells goes through a process in which the nucleus is finally extruded. The resultant red blood cells are simple, biconcave sacs of hemoglobin that transport oxygen and carbon dioxide to and from the tissues (see figure 14.10). These are the most numerous of circulating blood cells, appearing in stains as small pink circles. Red blood cells do not ordinarily have immune functions, though they can be the target of immune reactions (see chapter 17).

Platelets, or *thrombocytes,* are formed elements in circulating blood that are *not* whole cells. They originate in the bone

---

4. In reference to the humors, the liquids of the body. Humoral immunity includes antibodies, complement, and interferon.

* **antibody** (an'-tih-bahd"-ee) Gr. *anti*, against, and O.E., *bodig*, body.

* **monocyte** (mon'-oh-syte) From *mono*, one, and *cytos*, cell.

* **macrophage** (mak'-roh-fayj) Gr. *macro*, large, and *phagein*, to eat. They are the "large eaters" of the tissues.

* **erythrocyte** (eh-rith'-roh-syte) Gr. *erythros*, red. The red color comes from hemoglobin.

* **platelet** (playt'-let) Gr. *platos*, flat, and *let*, small.

* **thrombocyte** (throm'-boh-syte) Gr. *thrombos*, clot.

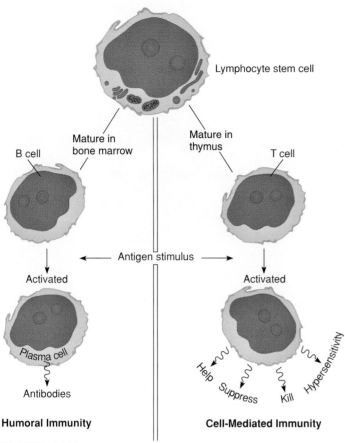

FIGURE 14.11

Summary of the general development and functions of lymphocytes, which are the cornerstone of specific immune reactions. B cells and T cells arise from the same stem cell but later diverge into two cell lines. Their appearances are similar, and one cannot differentiate them on the basis of staining. Note the relatively large nucleus—cytoplasm ratio and the lack of granules.

marrow when a giant multinucleate cell called a *megakaryocyte* disintegrates into numerous tiny, irregular-shaped pieces, each containing bits of the cytoplasm and nucleus (see figure 14.9). In stains, platelets are blue-gray with fine red granules and are readily distinguished from cells by their small size. Platelets function primarily in hemostasis (plugging broken blood vessels to stop bleeding) and in releasing chemicals that act in blood clotting and inflammation.

Unique Dynamic Characteristics of White Blood Cells

Many lymphocytes and phagocytes make regular journeys from the blood and lymphatics to the tissues and back again to the circulation as part of the constant surveillance of the compartments. In order for these WBCs to complete this circuit, they adhere to the inner walls of the smaller blood vessels. From this position, they are poised to migrate out of the blood into the tissue spaces by a process called **diapedesis.***

Diapedesis, also known as transmigration, is aided by several related characteristics of WBCs. For example, they are actively motile and readily change shape. This phenomenon is also assisted by the nature of the endothelial cells lining the venules. They con-

tain complex adhesive receptors that capture the WBCs and participate in their transport from the venules into the extracellular spaces (figure 14.12).

Another factor in the migratory habits of these WBCs is **chemotaxis.*** This is defined as the tendency of cells to migrate in response to a specific chemical stimulus given off at a site of injury or infection (see inflammation and phagocytosis later in this chapter). Through this means, cells swarm from many compartments to the site of infection and remain there to perform general and specific immune functions. These basic properties are absolutely essential for the sort of intercommunication and deployment of cells required for most immune reactions (figure 14.12).

Components and Functions of the Lymphatic System

The lymphatic system is a compartmentalized network of vessels, cells, and specialized accessory organs (figure 14.13). It begins in the farthest reaches of the tissues as tiny capillaries that transport a special fluid (lymph) through an increasingly larger tributary system of vessels and filters (lymph nodes), and it leads to major vessels that drain back into the regular circulatory system. Some major functions of the lymphatic system are (1) to provide an auxiliary route for the return of extracellular fluid to the circulatory system proper,[5] (2) to act as a "drain-off" system for the inflammatory response, and (3) to render surveillance, recognition, and protection against foreign materials through a system of lymphocytes, phagocytes, and antibodies.

Lymphatic Fluid   Lymph is a plasmalike liquid carried by the lymphatic circulation. It is formed when certain blood components move out of the blood vessels into the extracellular spaces and diffuse or migrate into the lymphatic capillaries. Thus, the composition of lymph parallels that of serum in many ways. It is made up of water, dissolved salts, and 2% to 5% protein (especially antibodies and albumin). Like blood, it also transports numerous white blood cells (especially lymphocytes) and miscellaneous materials such as fats, cellular debris, and infectious agents that have gained access to the tissue spaces. Unlike blood, red blood cells are not normally found in lymph.

Lymphatic Vessels   The system of vessels that transports lymph is constructed along the lines of regular blood vessels. The tiniest ones, lymphatic capillaries, accompany the blood capillaries and permeate all parts of the body except the central nervous system and certain organs such as bone, placenta, or thymus. The thin capillary walls make them permeable to extracellular fluid. The density of lymphatic vessels is particularly high in the hands, feet, and around the areolae of the breasts. Unlike the bloodstream, the lymphatic system does not circulate cyclically, but in one direction—from the extremities toward the heart (figure 14.14). The lymphatic capillary network feeds into a series of fewer but larger vessels that, in turn, drain finally into two main ducts (the thoracic duct and the right lymphatic duct), which empty the lymph into the main circulation at the large veins at the base of the neck.

---

5. The importance of this function is most evident in cases of impaired lymphatic drainage as seen in patients with filariasis, a roundworm infection (see figure 23.20).

* diapedesis (dye″-ah-puh-dee′-sis) Gr. *dia*, through, and *pedan*, to leap.

* chemotaxis (kee-moh-tak′-sis) NL *chemo*, chemical, and *taxis*, arrangement.

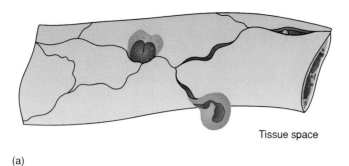

(a)

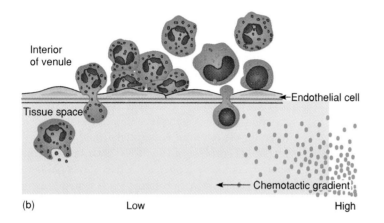

Interior of venule

Tissue space

←Endothelial cell

←Chemotactic gradient

(b)   Low   High

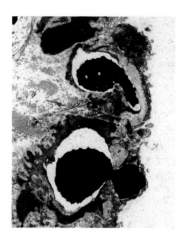

(c)

FIGURE 14.12

Diapedesis and chemotaxis of leukocytes.
(a) View of a venule depicts white blood cells squeezing themselves between spaces in the blood vessel wall through diapedesis. **(b)** This process, shown in cross section, indicates how the pool of leukocytes adheres to the endothelial wall. From this site, they are poised to migrate out of the vessel into the tissue space. **(c)** This photograph captures neutrophils in the process of diapedesis.

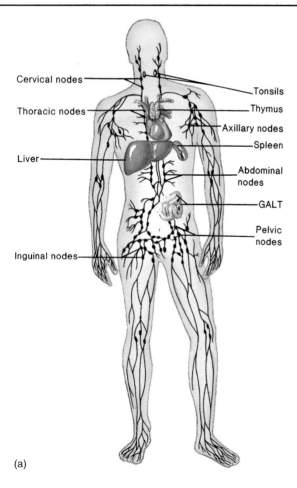

Cervical nodes
Thoracic nodes
Liver
Inguinal nodes
Tonsils
Thymus
Axillary nodes
Spleen
Abdominal nodes
GALT
Pelvic nodes

(a)

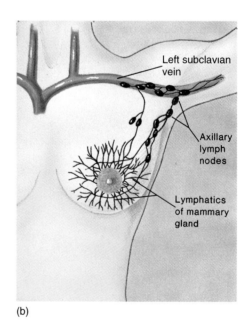

Left subclavian vein
Axillary lymph nodes
Lymphatics of mammary gland

(b)

FIGURE 14.13

General components of the lymphatic system. **(a)** This system consists of an anastomosing, branching network of vessels that permeate most tissues of the body. Note the especially high concentration of this network in the hands, feet, and breasts. **(b)** Organs that form part of this system are lymph nodes, clustered at major drainage points (armpit, groin, intestine). Other lymphatic structures are the spleen, part of the small intestine (gut-associated lymphoid tissue, GALT), the thymus, and the tonsils.

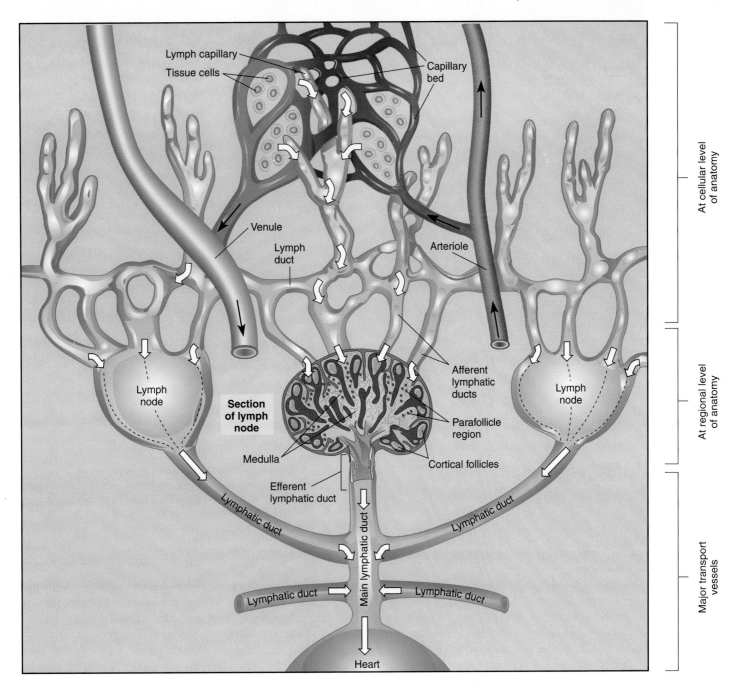

FIGURE 14.14

The circulatory scheme of the lymphatic vessels.   The lymphatic capillaries originate at the finest level of the tissues as tiny, fingerlike projections that absorb excess tissue fluid and transport it into a network of fewer and larger vessels (ducts). These ducts drain their fluid (lymph) into the lymph nodes. The center insert shows the anatomy of a lymph node. Its afferent lymphatic ducts penetrate into the sinuses. As the lymph percolates through these sinuses, it is filtered and exposed to populations of B and T lymphocytes that are segregated into specific sites. After filtration, the lymph leaves the node by a single afferent duct from which it is transported into a larger series of drainage vessels. These vessels empty their load of lymph into large main ducts near the heart. By this means, lymphatic chemicals and cells are returned to the blood.

Lymphoid Organs and Tissues   Other organs and tissues that perform lymphoid functions are the lymph nodes (glands), thymus, spleen, and clusters of tissues in the gastrointestinal tract (gut-associated lymphoid tissue; GALT) and the pharynx (the tonsils, for example). A trait common to these organs is a loose connective tissue framework that houses aggregations of lymphocytes, the important class of white blood cells mentioned previously.

Lymph Nodes   Lymph nodes are small, encapsulated, bean-shaped organs stationed, usually in clusters, along lymphatic channels and large blood vessels of the thoracic and abdominal cavities (see figure 14.13). Major aggregations of nodes occur in the loose connective tissue of the armpit (axillary nodes), groin (inguinal nodes), and neck (cervical nodes). Both the location and architecture of these nodes clearly specialize them for filtering out materials

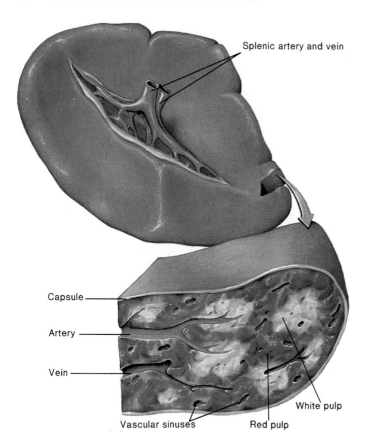

FIGURE 14.15

The anatomy of the spleen. Enlarged section depicts the major sites of white blood cell activity (white pulp) and red blood cell storage (red pulp).

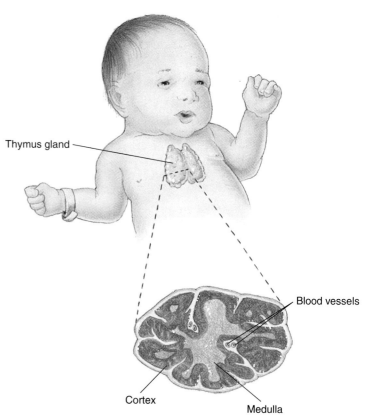

FIGURE 14.16

The thymus gland. Immediately after birth, the thymus is a large organ that nearly fills the region over the midline of the upper thoracic region. In the adult, however, it is proportionately smaller (to compare, see figure 14.13a). Section shows the main anatomical regions of the thymus. Immature T cells enter through the cortex and migrate into the medulla as they mature.

that have entered the lymph and providing appropriate cells and niches for immune reactions.

A view of a single sectioned lymph node reveals its filtering and cellular response systems (figure 14.14). Incoming lymphatic vessels (afferent lymphatic ducts) transport the lymph into sinuses in the node that contain segregated populations of lymphocytes. The central zone, or *medulla,** is an accumulation point for lymph and cells passing through the node. The surrounding germinal centers in the *cortex** are packed with T and B cells. This system of sinuses and discrete lymphocyte zones filters out particulate materials (microbes, for instance) and contributes WBCs to the lymph as it passes through. Many of the initial encounters between lymphocytes and microbes that result in specific immune responses occur in the lymph nodes.

Spleen The spleen is a lymphoid organ in the upper left portion of the abdominal cavity. It is somewhat similar to a lymph node in its basic structure and function, except that the spleen circulates blood instead of lymph (figure 14.15). It consists of a connective tissue network of vascular sinuses called the *red pulp,* which is rich in macrophages, neutrophils, and erythrocytes. Nestled within the

red pulp are compact regions of lymphocytes called the *white pulp.* B cells and T cells occupy separate predetermined areas of the white pulp. The spleen serves as an important station for phagocytosis of foreign matter and immune reactions against bacteria, and it also removes and breaks down worn erythrocytes. Although adults whose spleens have been surgically removed can live a relatively normal life, asplenic children are extremely immunocompromised.

The Thymus: Site of T-Cell Maturation The **thymus*** originates in the embryo as two lobes in the pharyngeal region that fuse into a triangular structure. The size of the thymus is greatest proportionately at birth (figure 14.16), and it continues to exhibit high rates of activity and growth until puberty, after which it begins to shrink gradually through adulthood. During the last few decades of life, its function is greatly diminished, because its primary work has been completed. The thymus is sectioned into a medulla, composed of special epithelial cells, and a cortex, containing undifferentiated lymphocytes called thymocytes. Under the influence of thymic hormones, thymocytes develop specificity and are released into the circulation as mature T cells. The T cells subsequently migrate to and

---

* medulla (meh-dul´-ah) L. *medius,* middle or marrow.

* cortex (kor´-teks) L. *cortex,* bark, rind, shell.

---

* thymus (thigh´-mus) Gr. *thymos,* soul, mind.

settle in other lymphoid organs (for example, the lymph nodes and spleen), where they occupy the specific sites described previously.

The thymus gland was once thought to have no important function. Medical science was so mistaken about its significance that children's necks were sometimes irradiated to "cure" a condition called "enlarged thymus." Experiments in the early 1960s finally clarified its link to lymphocyte development. If the thymus is surgically removed in neonates, they become severely immuno-deficient and fail to thrive, and babies born without a functional thymus are vulnerable to a disease called DiGeorge syndrome (see page 523). Adults have developed enough mature T cells that removal of the thymus or reduction in its function has milder effects. Do not confuse the thymus with the thyroid gland, which is located nearby but has an entirely different function.

Miscellaneous Lymphoid Tissue   At many sites on or just beneath the mucosa of the gastrointestinal and respiratory tracts lie discrete bundles of lymphocytes. The positioning of this diffuse system provides an effective first-strike potential against the constant influx of microbes and other foreign materials in food and air. In the pharynx, a ring of tissues called the **tonsils** provides an active source of lymphocytes. The breasts of pregnant and lactating women also become temporary sites of antibody-producing lymphoid tissues (see colostrum, Medical Microfile 15.4). The intestinal tract houses the best-developed collection of lymphoid tissue, called **gut-associated lymphoid tissue,** or **GALT.** Examples of GALT include the appendix, the lacteals (special lymphatic vessels stationed in each intestinal villus), and *Peyer's patches,* compact aggregations of lymphocytes in the ileum of the small intestine. GALT provides immune functions against intestinal pathogens and is a significant source of some types of antibodies.

CHAPTER CHECKPOINTS

The immune system is a complex collection of fluids and cells that penetrate every organ, tissue space, fluid compartment, and vascular network of the body. The four major subdivisions of this system are the RES, the ECF, the blood vascular system, and the lymphatic system.

The RES, or reticuloendothelial system, is a network of connective tissue fibers inhabited by macrophages ready to attack and ingest microbes invading the first and second lines of defense.

The ECF, or extracellular fluid, compartment surrounds all tissue cells and is penetrated by both blood and lymph vessels, which bring all components of the second and third line of defense to attack infectious microbes.

The blood contains both specific and nonspecific defenses. Nonspecific cellular defenses include the granulocytes and macrophages. The two components of the specific immune response are the T lymphocytes, which provide specific cell-mediated immunity, and the B lymphocytes, which produce specific antibody or humoral immunity.

The lymphatic system has three functions: (1) It returns tissue fluid to general circulation; (2) it carries away excess fluid in inflamed tissues; (3) it concentrates and processes foreign invaders and initiates the specific immune response. Important sites of lymphoid tissues are lymph nodes, spleen, thymus, tonsils, and GALT.

Nonspecific Immune Reactions of the Body's Compartments

Now that we have introduced the principal anatomical and physiological framework of the immune system, let us address some mechanisms that play important roles in host defenses: inflammation, phagocytosis, interferon, and complement. Because of the generalized nature of these defenses, they are primarily nonspecific in their effects, but they also support and interact with the specific immune responses described in chapter 15.

THE INFLAMMATORY RESPONSE: A COMPLEX CONCERT OF REACTIONS TO INJURY

At its most general level, the inflammatory response is a reaction to any traumatic event in the tissues. It is so very commonplace that most of us manifest inflammation in some way every day. It appears in the nasty flare of a cat scratch, the blistering of a burn, the painful lesion of an infection, and the symptoms of allergy. It is readily identifiable by a classic series of signs and symptoms characterized succinctly by four Latin terms: *rubor, calor, tumor,* and *dolor.* Rubor (redness) is caused by increased circulation and vasodilation in the injured tissues; calor (warmth) is the heat given off by the increased flow of blood; tumor (swelling) is caused by increased fluid escaping into the tissues; and dolor (pain) is caused by the stimulation of nerve endings (figure 14.17). Although these manifestations can be unpleasant, they serve an important warning that injury has taken place and set in motion responses that save the body from further injury.

Factors that can elicit inflammation include trauma from infection (the primary emphasis here), tissue injury or necrosis due to physical or chemical agents, and specific immune reactions. Although the details of inflammation are very complex, its chief functions can be summarized as follows: (1) to mobilize and attract immune components to the site of the injury, (2) to set in motion mechanisms to repair tissue damage and localize and clear away harmful substances, and (3) to destroy microbes and block their further invasion (figure 14.18). The inflammatory response is

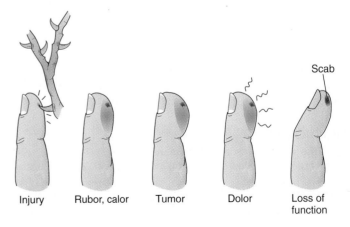

FIGURE 14.17

The response to injury.   This classic checklist encapsulates the reactions of the tissues to an assault. Each of the events is an indicator of one of the mechanisms of inflammation described in this chapter.

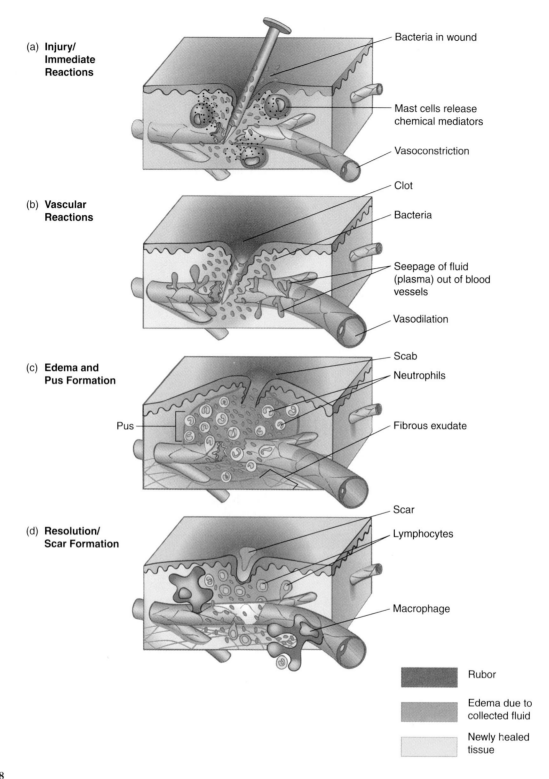

(a) **Injury/ Immediate Reactions**

Bacteria in wound

Mast cells release chemical mediators

Vasoconstriction

(b) **Vascular Reactions**

Clot

Bacteria

Seepage of fluid (plasma) out of blood vessels

Vasodilation

(c) **Edema and Pus Formation**

Scab

Neutrophils

Pus

Fibrous exudate

(d) **Resolution/ Scar Formation**

Scar

Lymphocytes

Macrophage

Rubor

Edema due to collected fluid

Newly healed tissue

FIGURE 14.18

The major events in inflammation. **(a)** Injury → Reflex narrowing of the blood vessels (vasoconstriction) lasting for a short time → Release of chemical mediators into area. **(b)** Increased diameter of blood vessels (vasodilation) → Increased blood flow → Increased vascular permeability → Leakage of fluid (plasma) from blood vessels into tissues (exudate formation). **(c)** Edema → Infiltration of site by neutrophils and accumulation of pus. **(d)** Macrophages and lymphocytes → Repair, either by complete resolution and return of tissue to normal state or by formation of scar tissue.

Not every aspect of inflammation is protective or results in the proficient resolution of tissue damage. As one looks over a list of diseases, it is rather striking how many of them are due in part or even completely to an overreactive or dysfunctional inflammatory response.

Some "itis" reactions mentioned in chapter 13 are a case in point (see Microbits 13.4). Inflammatory exudates that build up in the brain in African trypanosomiasis, cryptococcosis, and other brain infections can be so injurious to the nervous system that impairment is permanent. Frequently, an inflammatory reaction that walls off the pathogen leads to an abscess, a swollen mass of neutrophils and dead, liquefied tissue that can harbor live pathogens in the center. Abscesses are a prominent feature of staphylococcal, amebic, and enteric infections.

Other pathologic manifestations of chronic diseases—for example, the tubercles of tuberculosis, the lesions of late syphilis, the disfiguring nodules of leprosy, and the cutaneous ulcers of leishmaniasis—are due to an aberrant tissue response called *granuloma formation* (see figure 19.16). Granulomas develop not only in response to microbes but also in response to inanimate foreign bodies (sutures and mineral grains that are

difficult to break down). This condition is initiated when neutrophils ineffectively and incompletely phagocytose the pathogens or materials involved in an inflammatory reaction. The macrophages then enter to clean up and attempt to phagocytose the dead neutrophils and foreign substances, but they fail to completely manage them. They respond by storing these ingested materials in vacuoles and becoming inactive. Over a given time period, large numbers of adjacent macrophages fuse into giant, inactive multinucleate cells called foreign body giant cells. These sites are further infiltrated with lymphocytes. The resultant collections make the tissue appear granular—hence, the name. A granuloma can exist in the tissue for months, years, or even a lifetime.

Medical science is rapidly searching for new applications for the massive amount of new information on inflammatory mediators. One highly promising area appears to be the use of chemokine inhibitors that could reduce chemotaxis and the massive, destructive influx of leukocytes. Such therapy could ultimately be used for certain cancers, hardening of arteries, and Alzheimer disease.

a powerful defensive reaction, a means for the body to maintain stability and restore itself after an injury. But when it is chronic, it has the potential to actually *cause* tissue injury, destruction, and disease (Medical Microfile 14.1).

THE STAGES OF INFLAMMATION

The process leading to inflammation is a dynamic, predictable sequence of events that can be acute, lasting from a few minutes or hours, to chronic, lasting for days, weeks, or years. Once the initial injury has occurred, a chain reaction takes place at the site of damaged tissue, summoning beneficial cells and fluids into the injured area. As an example, we will look at an injury at the microscopic level and observe the flow of major events (figure 14.18).

Vascular Changes: Early Inflammatory Events

Following an injury, some of the earliest changes occur in the vasculature (arterioles, capillaries, venules) in the vicinity of the damaged tissue. These changes are controlled by nervous stimulation, **chemical mediators,** and **cytokines*** released by blood cells, tissue cells, and platelets in the injured area. Some mediators are *vasoactive*—that is, they affect the endothelial cells and smooth muscle cells of blood vessels, and others are **chemotactic factors,** also called **chemokines,** that affect white blood cells. Inflammatory mediators cause fever, stimulate lymphocytes, prevent virus spread, and cause allergic symptoms (Medical Microfile 14.2 and figure 14.19). Although the constriction of arterioles is stimulated

first, it lasts for only a few seconds or minutes and is followed in quick succession by the opposite reaction, vasodilation. The overall effect of vasodilation is to increase the flow of blood into the area, which facilitates the influx of immune components and also causes redness and warmth.

Edema: Leakage of Vascular Fluid into Tissues

Some vasoactive substances cause the endothelial cells surrounding postcapillary venules to contract and form gaps through which blood-borne components exude into the extracellular spaces. The fluid part that escapes is called the *exudate*. Accumulation of this fluid in the tissues gives rise to local swelling and hardness called **edema.** The edematous exudate contains varying amounts of plasma proteins, such as globulins, albumin, the clotting protein fibrinogen, blood cells, and cellular debris. Depending upon its content, the exudate varies from serous (clear) to serosanguinous (containing red blood cells) to purulent (containing pus). In some types of edema, the fibrinogen is converted to fibrin threads that enmesh the injury site. Within an hour, multitudes of neutrophils responding chemotactically to special signaling molecules converge on the injured site (see figure 14.18c).

The Benefits of Edema and Chemotaxis   Both the formation of edematous exudate and the infiltration of neutrophils are physiologically beneficial activities. The influx of fluid dilutes toxic substances, and the fibrin clot can effectively trap microbes and prevent their further spread. The neutrophils that aggregate in the inflamed site are immediately involved in phagocytosing and destroying bacteria, dead tissues, and particulate matter (by mechanisms discussed in a later section on phagocytosis). In some types

---

* **cytokine** (sy'-toh-kyne) Gr. *cytos,* cell, and *kinein,* to move. A protein or polypeptide produced by WBCs that regulates host defenses.

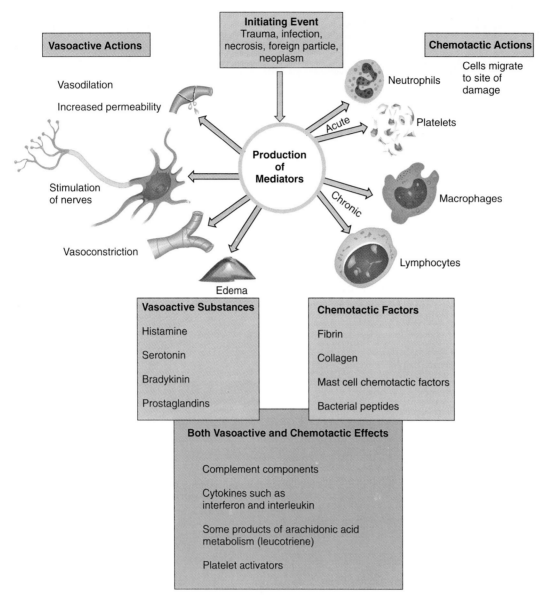

FIGURE 14.19

Chemical mediators of the inflammatory response and their effects.

of inflammation, accumulated phagocytes contribute to **pus,** a whitish mass of cells, liquefied cellular debris, and bacteria. Certain bacteria (streptococci, staphylococci, gonococci, and meningococci) are especially powerful attractants for neutrophils and are thus termed **pyogenic,** or pus-forming, bacteria.

Late Reactions of Inflammation   Sometimes a mild inflammation can be resolved by edema and phagocytosis. Inflammatory reactions that are more long-lived attract a collection of monocytes, lymphocytes, and macrophages to the reaction site. Clearance of pus, cellular debris, dead neutrophils, and damaged tissue is performed by macrophages, the only cells that can engulf and dispose of such large masses. At the same time, B lymphocytes react with foreign molecules and cells by producing specific antimicrobial proteins (antibodies), and T lymphocytes kill intruders directly. Late in

the process the tissue is completely repaired, if possible, or replaced by connective tissue in the form of a scar (see figure 14.18*d*). If the inflammation cannot be relieved or resolved in this way, it can become chronic and create a long-term pathologic condition.

Fever: An Adjunct to Inflammation

An important systemic component of inflammation is **fever,** defined as an abnormally elevated body temperature. Although fever is a nearly universal symptom of infection, it is also associated with certain allergies, cancers, and other organic illnesses. Fevers whose causes are unknown are called fevers of unknown origin, or FUO.

The body temperature is normally maintained by a control center in the hypothalamus. This thermostat regulates the body's heat production and heat loss and sets the core temperature at around 37°C (98.6°F), with slight fluctuations (1°F) during a daily

Just as the nervous system is coordinated by a complex communications network, so too is the immune system. Hundreds of small, active molecules are constantly being secreted to regulate, stimulate, suppress, and otherwise control the many aspects of cell development, inflammation, and immunity. These substances are the products of several types of cells, including monocytes, macrophages, lymphocytes, fibroblasts, mast cells, platelets, and endothelial cells of blood vessels. Their effects may be local or systemic, short-term or long-lasting, nonspecific or specific, protective or pathologic.

In recent times, the field of cytokines has become so increasingly complex that we can include here only an overview of the major groups of important cytokines and other mediators. The major functional types can be categorized into (1) cytokines that mediate nonspecific immune reactions such as inflammation and phagocytosis, (2) cytokines that regulate the growth and activation of lymphocytes, (3) cytokines that activate immune reactions during inflammation, (4) hemopoiesis factors for white blood cells, (5) vasoactive mediators, and (6) miscellaneous inflammatory mediators.

Nonspecific Mediators of Inflammation and Immunity

- *Tumor necrosis factor (TNF)*, a substance from macrophages that increases chemotaxis and phagocytosis and stimulates other cells to secrete inflammatory cytokines. It also serves as an endogenous pyrogen that induces fever, increases blood coagulation, suppresses bone marrow, and causes wasting of the body called cachexia.
- *Interferon (IFN), alpha and beta,* produced by leukocytes and fibroblasts, inhibits virus replication and cell division and increases the action of certain lymphocytes that kill other cells.
- *Interleukin* (IL) 1*, a product of macrophages and epithelial cells that has many of the same biological activities as TNF, such as inducing fever and activation of certain white blood cells.
- *Interleukin-6,* secreted by macrophages, lymphocytes, and fibroblasts. Its primary effects are to stimulate the growth of B cells and to increase the synthesis of liver proteins.
- *Various chemokines.* By definition, chemokines are cytokines that stimulate the movement and migration of white blood cells (chemotactic factors). Included among these are complement C5A, interleukin 8, and platelet factor.

Cytokines That Regulate Lymphocyte Growth in Activation

- *Interleukin-2,* the primary growth factor from T cells. Interestingly, it acts on the same cells that secrete it. It stimulates mitosis and secretion of other cytokines. In B cells, it is a growth factor and stimulus for antibody synthesis.
- *Interleukin-4,* a stimulus for the production of allergy antibodies; inhibits macrophage actions; favors development of T cells.

Cytokines That Activate Specific Immune Reactions

- *Gamma interferon,* a T-cell derived mediator whose primary function is to activate macrophages. It also promotes the differentiation of T and B cells, activates neutrophils, and stimulates diapedesis.
- *Interleukin-5* activates eosinophils and B cells; *interleukin-10* inhibits macrophages and B cells; and *interleukin-12* activates T cells and killer cells.

Vasoactive Mediators

- *Histamine,* a vasoactive mediator produced by mast cells and basophils that causes vasodilation, increased vascular permeability, and mucus production. It functions primarily in inflammation and allergy.
- *Serotonin,* a mediator produced by platelets and intestinal cells that causes smooth muscle contraction, inhibits gastric secretion, and acts as a neurotransmitter.
- *Bradykinin,* a vasoactive amine from the blood or tissues that stimulates smooth muscle contraction and increases vascular permeability, mucus production, and pain. It is particularly active in allergic reactions.

Miscellaneous Inflammatory Mediators

- *Prostaglandins,* produced by most body cells; complex chemical mediators that can have opposing effects (for example, dilation or constriction of blood vessels) and are powerful stimulants of inflammation and pain.
- *Leukotrienes* stimulate the contraction of smooth muscle and enhance vascular permeability. They are implicated in the more severe manifestations of immediate allergies (constriction of airways).
- *Platelet-activating factor,* a substance released from basophils, causes the aggregation of platelets and the release of other chemical mediators during immediate allergic reactions.

---

* **Interleukin** is a term that refers to a group of small peptides originally isolated from leukocytes. There are currently 20 known interleukins. We now know that other cells besides leukocytes can synthesize them and that they have a variety of biological activities. Functions of some selected examples will be presented in chapter 15.

MEDICAL MICROFILE 14.3
Some Facts About Fever

Fever is such a prevalent reaction that it is a prominent symptom of hundreds of diseases. For thousands of years, people believed fever was part of an innate protective response. Hippocrates offered the idea that it was the body's attempt to burn off a noxious agent. Sir Thomas Sydenham wrote in the seventeenth century: "Why, fever itself is Nature's instrument!" So widely held was the view that fever could be therapeutic that pyretotherapy (treating disease by inducing an intermittent fever) was once used to treat syphilis, gonorrhea, leishmaniasis (a protozoan infection), and cancer. This attitude fell out of favor when drugs for relieving fever (aspirin) first came into use in the early 1900s, and an adverse view of fever began to dominate.

Changing Views of Fever

In recent times, the medical community has returned to the original concept of fever as more healthful than harmful. Experiments with vertebrates indicate that fever is a universal reaction, even in cold-blooded animals such as lizards and fish. A study with *febrile** mice and frogs

---

*febrile (fee'-bril) L. *febris*, fever. Feverish.

indicated that fever increases the rate of antibody synthesis. Work with tissue cultures showed that increased temperatures stimulate the activities of T cells and increase the effectiveness of interferon. Artificially infected rabbits and pigs allowed to remain febrile survive at a higher rate than those given suppressant drugs. Fever appears to enhance phagocytosis of staphylococci by neutrophils in guinea pigs and humans.

Hot and Cold: Why Do Chills Accompany Fever?

Fever almost never occurs as a single response; it is usually accompanied by chills. What causes this oddity—that a person flushed with fever periodically feels cold and trembles uncontrollably? The explanation lies in the natural physiological interaction between the thermostat in the hypothalamus and the temperature of the blood. For example, if the thermostat has been set (by pyrogen) at 102°F but the blood temperature is 99°F, the muscles are stimulated to contract involuntarily (shivering) as a means of producing heat. In addition, the vessels in the skin constrict, creating a sensation of cold, and the piloerector muscles in the skin cause "goose bumps" to form.

---

cycle. Fever is initiated when a circulating substance called **pyrogen*** resets the hypothalamic thermostat to a higher setting. This change signals the musculature to increase heat production and peripheral arterioles to decrease heat loss through vasoconstriction (Medical Microfile 14.3). Fevers range in severity from low-grade (37.7°–38.3°C or 100°–101°F) to moderate (38.8°–39.4°C, or 102°–103°F) to high (40.0°–41.1°C, or 104°–106°F). Pyrogens are described as *exogenous* (coming from outside the body) or *endogenous* (originating internally). Exogenous pyrogens are products of infectious agents such as viruses, bacteria, protozoans, and fungi. One well-characterized exogenous pyrogen is endotoxin, the lipopolysaccharide found in the cell walls of gram-negative bacteria. Blood, blood products, vaccines, or injectable solutions can also contain exogenous pyrogens. Endogenous pyrogens are liberated by monocytes, neutrophils, and macrophages during the process of phagocytosis and appear to be a natural part of the immune response. Two potent pyrogens released by macrophages are interleukin-1 and tumor necrosis factor.

Benefits of Fever The association of fever with infection strongly suggests that it serves a beneficial role, a view still being debated but gaining acceptance. Aside from its practical and medical importance as a sign of a physiological disruption, increased body temperature has additional benefits:

- Fever inhibits multiplication of temperature-sensitive microorganisms such as the poliovirus, cold viruses, herpes zoster virus, systemic and subcutaneous fungal pathogens, *Mycobacterium* species, and the syphilis spirochete.
- Fever impedes the nutrition of bacteria by reducing the availability of iron. It has been demonstrated that during fever, the macrophages stop releasing their iron stores, which could retard several enzymatic reactions needed for bacterial growth.
- Fever increases metabolism and stimulates immune reactions and naturally protective physiological processes. It speeds up hematopoiesis, phagocytosis, and specific immune reactions.

Treatment of Fever With this revised perspective on fever, whether to suppress it or not can be a difficult decision. Some advocates feel that a slight to moderate fever in an otherwise healthy person should be allowed to run its course, in light of its potential benefits and minimal side effects. All medical experts do agree that high and prolonged fevers, or fevers in patients with cardiovascular disease, seizures, and respiratory ailments, are risky and must be treated immediately with suppressant drugs. The classic therapy for fever is an *antipyretic** drug such as aspirin or acetaminophen (Tylenol) that lowers the setting of the hypothalamic center

---

* **pyrogen** (py'-roh-jen) Gr. *pyr*, fire, and *gennan*, produce. As in funeral pyre and pyromaniac.

* **antipyretic** (an"-tih-py-reh'-tik) L. *anti*, against, and *pyretos*, fever. An agent that relieves fever.

and restores normal temperature. Any physical technique that increases heat loss (tepid baths, for example) can also help reduce the core temperature.

PHAGOCYTES: THE EVER–PRESENT BUSYBODIES OF INFLAMMATION AND SPECIFIC IMMUNITY

By any standard, a phagocyte represents an impressive piece of living machinery, meandering through the tissues to seek, capture, and destroy a target. The general activities of phagocytes are (1) to survey the tissue compartments and discover microbes, particulate matter (dust, carbon particles, antigen-antibody complexes), and injured or dead cells; (2) to ingest and eliminate these materials; and (3) to extract immunogenic information (antigens) from foreign matter. It is generally accepted that all cells have some capacity to engulf materials, but *professional phagocytes* do it for a living. The three main types of phagocytes are neutrophils, monocytes, and macrophages.

Granulocytic Phagocytes: Neutrophils and Eosinophils

As previously stated, neutrophils are general-purpose phagocytes that react early in the inflammatory response to bacteria and other foreign materials and to damaged tissue (see figure 14.18). A common sign of bacterial infection is a high neutrophil count in the blood (neutrophilia), and neutrophils are also a primary component of pus. Eosinophils are attracted to sites of parasitic infections and antigen-antibody reactions, though they play only a minor phagocytic role.

Macrophage: King of the Phagocytes

After emigrating out of the bloodstream into the tissues, monocytes are transformed by various inflammatory mediators into macrophages. This process is marked by an increase in size and by enhanced development of lysosomes and other organelles (figure 14.20). At one time, macrophages were classified as either fixed (adherent to tissue) or wandering, but this terminology can be misleading. All macrophages retain the capacity to move about. Whether they reside in a specific organ or wander depends upon their stage of development and the immune stimuli they receive. Specialized macrophages called *histiocytes* migrate to a certain tissue and remain there during their lifespan. Examples are alveolar (lung) macrophages, the Kupffer cells in the liver, Langerhans cells in the skin (figure 14.21), and macrophages in the spleen, lymph nodes, bone marrow, kidney, bone, and brain. Other macrophages do not reside permanently in a particular tissue and drift nomadically throughout the RES. Not only are macrophages dynamic scavengers, but they also process foreign substances and prepare them for reactions with B and T lymphocytes (see chapter 15).

Mechanisms of Phagocytic Discovery, Engulfment, and Killing

Although the term *phagocytosis* literally means the engulfment of particles by cells, phagocytes actually endocytose both particulate and liquid substances. But phagocytosis is more than just the physical process of engulfment, because phagocytes also actively attack and

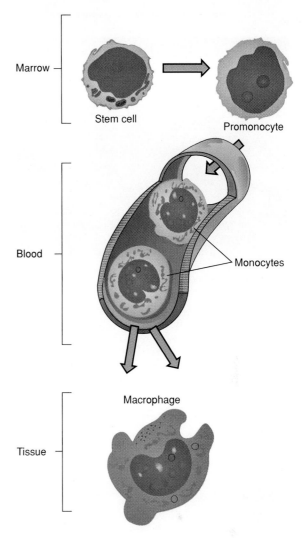

FIGURE 14.20

The developmental stages of monocytes and macrophages. The cells progress through maturational stages in the bone marrow and peripheral blood. Once in the tissues, a macrophage can remain nomadic or take up residence in a specific organ.

dismantle foreign cells with a wide array of antimicrobial substances. The events in phagocytosis include chemotaxis, ingestion, phagolysosome formation, destruction, and excretion (figure 14.22).

Chemotaxis and Ingestion   Phagocytes migrate into a region of inflammation with a deliberate sense of direction, attracted by a gradient of stimulant products from the parasite and host tissue at the site of injury. On the scene of an inflammatory reaction, they often trap cells or debris against the fibrous network of connective tissue or the wall of blood and lymphatic vessels. Phagocytosis is often accompanied by *opsonization* (discussed again in chapter 15). This is a process that coats the surface of microorganisms with antibodies or complement, thereby facilitating recognition and engulfment. Once the phagocyte has made contact with its prey, it extends pseudopods that enclose the cells or particles in a pocket and internalize them in a vacuole called a *phagosome*.

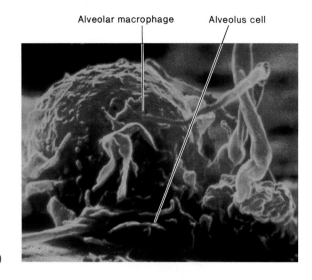

Alveolar macrophage   Alveolus cell

(a)

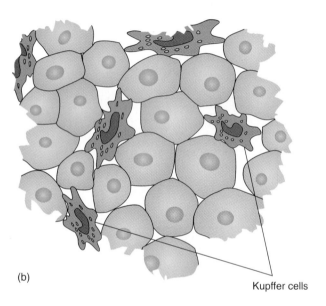

(b)

Kupffer cells

Langerhans cells

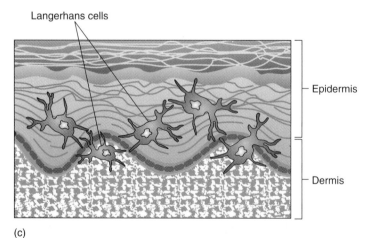

Epidermis

Dermis

(c)

FIGURE 14.21

Sites containing macrophages. **(a)** Scanning electron micrograph view of a lung with an alveolar macrophage. **(b)** Liver tissue with Kupffer cells. **(c)** Langerhans cells deep in the epidermis.

Phagolysosome Formation and Killing In a short time, *lysosomes* migrate to the scene of the phagosome and fuse with it to form a *phagolysosome*. Other granules containing antimicrobial chemicals are released into the phagolysosome, forming a potent brew designed to poison and then dismantle the ingested material. The destructiveness of phagocytosis is evident by the death of bacteria within 30 minutes after contacting this battery of antimicrobial substances.

Destruction and Elimination Systems Two separate systems of destructive chemicals await the microbes in the phagolysosome. The oxygen-dependent system elaborates several substances that were described in chapters 7 and 11. Myeloperoxidase, an enzyme found in granulocytes, forms halogen ions (OCl^-) that are strong oxidizing agents. Other products of oxygen metabolism such as hydrogen peroxide, the superoxide anion (O_2^-), activated or so-called singlet oxygen ($^1O_2$), and the hydroxyl free radical (.OH) separately and together have formidable killing power. Other mechanisms that come into play are the liberation of lactic acid, lysozyme, and *nitric oxide* (NO), a powerful mediator that kills bacteria and inhibits viral replication. Cationic proteins that injure bacterial cell membranes and a number of proteolytic and other hydrolytic enzymes complete the job. The small bits of undigestible debris are exocytosed and released.

CONTRIBUTORS TO THE BODY'S CHEMICAL IMMUNITY

Interferon: Antiviral Cytokines and Immune Stimulants

Interferon (IFN) was described in chapter 12 as a small protein produced naturally by certain white blood and tissue cells that is used in therapy against certain viral infections and cancer. Although the interferon system was originally thought to be directed exclusively against viruses, it is now known to be involved also in defenses against other microbes and in immune regulation and intercommunication. Three major types are *alpha interferon,* a product of lymphocytes and macrophages; *beta interferon,* a product of fibroblasts and epithelial cells; and *gamma interferon,* a product of T cells.

All three classes of interferon are produced in response to viruses, RNA, immune products, and various antigens. Their biological activities are extensive. In all cases, they bind to cell surfaces and induce changes in genetic expression, but the exact results vary. In addition to antiviral effects discussed in the next section, all three IFNs can inhibit the expression of cancer genes and have tumor suppressor effects. Alpha and beta IFN stimulate phagocytes, and gamma IFN is an immune regulator of macrophages and T and B cells.

Characteristics of Antiviral Interferon The binding of a virus to the receptors of an infected cell sends a signal into the cell nucleus that activates the genes coding for interferon (figure 14.23). As interferon is synthesized, it is rapidly secreted by the cell into the extracellular spaces. The action of antiviral interferon is indirect; it does not kill or inhibit the virus directly. After diffusing to nearby, uninfected cells and entering them, IFN activates a gene complex that codes for another protein. This second protein, not

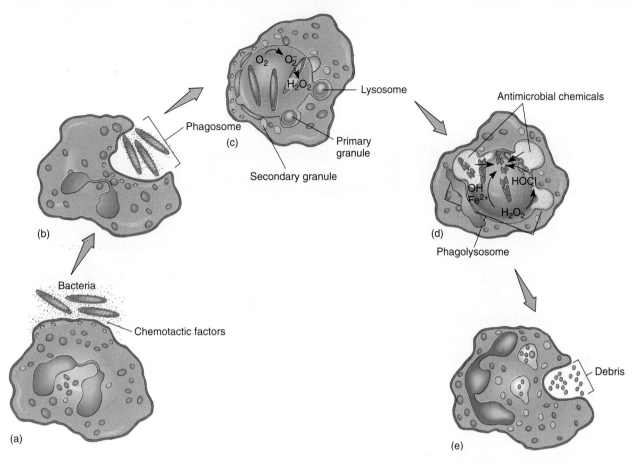

FIGURE 14.22

The phases in phagocytosis. **(a)** Chemotaxis. **(b)** Contact and ingestion (forming a phagosome). **(c)** Formation of phagolysosome (granules fuse with phagosome). **(d)** Killing, digestion of the microbe. **(e)** Release of debris.

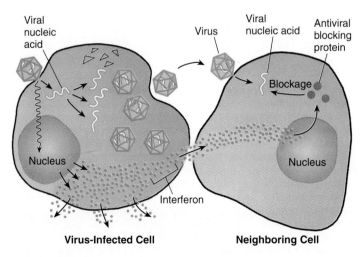

FIGURE 14.23

The antiviral activity of interferon. When a cell is infected, its nucleus is triggered to transcribe and translate the interferon (IFN) gene. The interferon diffuses out of the infected cell into nearby (uninfected) cells, where it enters the nucleus. Here IFN activates a gene for synthesizing a peptide that blocks viral replication. Note that the original cell is not protected by IFN and that IFN does not prevent viruses from invading the protected cells.

interferon itself, interferes with the multiplication of viruses. Interferon is not virus-specific, so its synthesis in response to one type of virus will also protect against other types. Because this inhibitory protein is the direct inhibitor of viruses, it has been a valuable treatment for a number of virus infections.

Other Roles of Interferon Interferons are also important immune regulatory cytokines that activate or instruct the development of white blood cells. For example, alpha interferon produced by T lymphocytes activates a subset of cells called natural killer (NK) cells. In addition, one type of beta interferon plays a role in the maturation of B and T lymphocytes and in inflammation. Gamma interferon inhibits cancer cells, stimulates B lymphocytes, activates macrophages, and enhances the effectiveness of phagocytosis.

Complement: A Versatile Backup System

Among its many overlapping functions, the immune system has another complex and multiple-duty system called **complement (C factor)** that, like inflammation and phagocytosis, is brought into play at several levels. The complement system, named for its property of "completing" immune reactions, consists of 20 blood proteins that work in concert to destroy bacteria and certain viruses. The sources of complement factors are liver hepatocytes, lymphocytes,

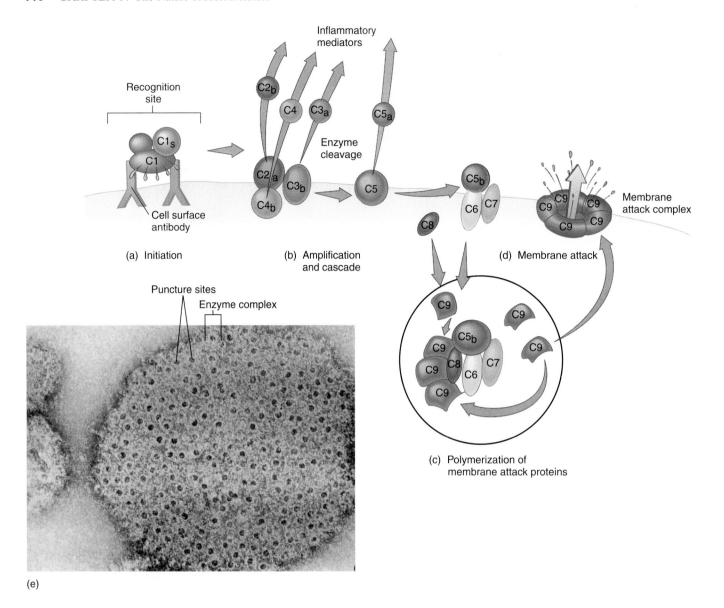

(a) Initiation

(b) Amplification and cascade

(c) Polymerization of membrane attack proteins

(d) Membrane attack

(e)

and monocytes. Some knowledge of this important system will help in your understanding of topics in chapters 15 and 16.

The concept of a cascade reaction is helpful in understanding how complement functions. A cascade reaction is a sequential physiological response like that of blood clotting, in which the first substance in a chemical series activates the next substance, which activates the next, and so on, until a desired end product is reached. Complement exhibits two schemes, the *classical pathway* and the *alternative pathway,* which differ in how they are activated and in speed and efficiency. The two pathways merge at the final stages into a common pathway with a similar end result (figure 14.24). Since the complement numbers (C1–C9) are based on the order of their discovery, be aware that factors C1–C4 do not appear in numerical order during activation (table 14.1).

Overall Stages in the Complement Cascade In general, the complement cascade includes the three stages of *initiation, amplification and cascade,* and *membrane attack.* At the outset, an initiator

FIGURE 14.24

Steps in the classical complement pathway at a single site.
(a) Initiation. The classical pathway begins when C1 components bind to a recognition site on antibodies attached to a foreign cell.
(b) Amplification and cascade. The C1 complex is an enzyme that activates a second series of components, C4 and C2. When these have been enzymatically cleaved into separate molecules, they become a second enzyme complex that activates C3. At this same site, C3 binds to C5 and cleaves it to form a product that is tightly bound to the membrane.
(c) Polymerization. C5 is a reactive site for the final assembly of an attack complex. In series, C6, C7, and C8 aggregate with C5 and become integrated into the membrane. They form a substrate upon which the final component, C9, can bind. Up to 15 of these C9 units ring the central core of the complex. **(d) Membrane attack.** The final product of these reactions is a large, donut-shaped enzyme that punctures small pores through the membrane, leading to cell lysis. **(e)** An electron micrograph (187,000×) of a cell reveals multiple puncture sites over its surface. The lighter, ringlike structures are the actual enzyme complex.

TABLE 14.1
Complement Proteins

Pathway	Component
Classical **(Initial Portion)** Rapid and efficient	C1q C1r C1s C4 C2 C3
Lectin **(Proteins That Bind Mannans)**	**Membrane Attack Components (Common to Both Pathways)** C5 C6 C7 C8 C9
Alternative or Properdin (Initial Portion) Slower and less efficient	Properdin Factor B Factor D Factor C3b and Mg

TABLE 14.2
Substances That Activate the Complement Pathways

Activators in the Classical Pathways	Activators in the Alternative Pathway
Complement-fixing antibodies: IgG, IgM	Cell wall components; e.g., yeast and bacteria
Bacterial lipopolysaccharide	Viruses; e.g., influenza virus
Pneumococcal C-reactive protein	Parasites; e.g., *Schistosoma*
Retroviruses	Fungi; e.g., *Cryptococcus*
Polynucleotides	Some tumor cells
Mitochondrial membranes	X-ray opaque media, dialysis membranes

(such as microbes, cytokines, and antibodies; table 14.2) reacts with the first complement chemical, which propels the reaction on its course. This process develops a recognition site on the surface of the target cell where the initial C components will bind. Through a stepwise series, each component reacts with another on or near the recognition site. In the C2–C5 series, enzymatic cleavage produces several inflammatory cytokines. Other details of the pathways differ, but whether classical or alternative, the functioning end product is a large ring-shaped protein termed the *membrane attack complex.* This complex can digest holes in the cell membranes of bacteria, cells, and enveloped viruses, thereby destroying them (figure 14.24a–e). The two pathways are described in more detail in Microbits 14.4.

Specific Immunities: The Third and Final Line of Defense

When host barriers and nonspecific defenses fail to control an infectious agent, a person with a normal functioning immune system has an extremely substantial mechanism to resist the pathogen—the third, specific line of immunity. This aspect of immunity is the resistance developed after contracting childhood ailments such as chickenpox or measles that provides long-term protection against future attacks. This sort of immunity is not innate, but adaptive; it is acquired only after an immunizing event such as an infection. The absolute need for acquired or adaptive immunity is impressively documented in children who have genetic defects in this system or in AIDS patients who have lost it. Even with heroic measures to isolate the patient, combat infection, or restore lymphoid tissue, the victim is constantly vulnerable to life-threatening infections.

Acquired specific immunity is the product of a dual system that we have previously mentioned—the B and T lymphocytes. During fetal development, these lymphocytes undergo a selective process that specializes them for reacting only to one specific antigen. During this time, **immunocompetence,** the ability of the body to react with myriad foreign substances, develops. An infant is born with the theoretical potential to acquire millions of different immunities.

Two features that most characterize this third line of defense are **specificity** and **memory.** Unlike mechanisms such as anatomical barriers or phagocytosis, acquired immunity is highly selective. For example, the antibodies produced during an infection against the chickenpox virus will function against that virus and not against the measles virus (figure 14.25a). The property of memory pertains to the rapid mobilization of lymphocytes that have been programmed to "recall" their first engagement with the invader and rush to the attack once again (figure 14.25b). This complex and fascinating response will be covered more extensively in chapter 15.

CHAPTER CHECKPOINTS

Nonspecific immune reactions are generalized responses to invasion, regardless of the type. These include inflammation, phagocytosis, interferon, and complement.

The four symptoms of inflammation are rubor (redness), calor (heat), tumor (edema), and dolor (pain).

Fever is another component of nonspecific immunity. It is caused by both endogenous and exogenous pyrogens. Fever increases the rapidity of the host immune responses and reduces the viability of many microbial invaders.

Macrophages are activated monocytes. Along with neutrophils (PMNs), they are the key phagocytic agents of nonspecific response to disease.

The plasma contains complement, a nonspecific group of chemicals that works with the third line of defense to attack foreign cells.

CHAPTER CHECKPOINTS

B and T lymphocytes are the agents of the specific immune response. Unlike the nonspecific responses, it develops after exposure to a specific antigen and is tailor-made to it. In addition, the B and T cells "remember" the antigen and respond much more rapidly on subsequent exposures.

MICROBITS 14.4
How Complement Works

The **classical pathway** is a part of the specific immune response covered in chapter 15. It is initiated when antibody, called complement-fixing antibody, attaches to antigen on the surface of a membrane. The first chemical, C1, is a large complex of three molecules, C1q, C1r, and C1s (see table 14.1). When the C1q subunit has recognized and bound two or more antibodies, the C1r subunit cleaves the C1s proenzyme, and an activated enzyme emerges. During amplification, the C1s enzyme has as its primary targets proenzymes C4 and C2. Through the enzyme's action, C4 is converted into C4a and C4b, and C2 is converted into C2a and C2b. C4b and C2a fragments remain attached as an enzyme, C3 convertase, whose substrate is factor C3. The cleaving of C3 yields subunits C3a and C3b. C3b has the property of binding strongly with the cell membrane in close association with the component C5, and it also forms an enzyme complex with C4b–C2b that converts C5 into two fragments, C5a and C5b. C5b will form the nucleus for the membrane attack complex. This is the point at which the two pathways merge. From this point on, C5b reacts with C6 and C7 to form a stable complex inserted in the membrane. Addition of C8 to the complex causes the polymerization of several C9 molecules into a giant ring-shaped membrane attack complex that bores ring-shaped holes in the membrane, which lyse the target cell and cause it to disintegrate or destroy the virus (figure 14.24d,e).

The **alternative pathway,** sometimes called the **properdin* pathway,** is not specific to a particular microbe. It can be initiated by a wide variety of microbes, tumors, and cell walls. It requires a different group of serum proteins—factors B, D, and P (properdin), C3b, and magnesium in the initiation and amplification phases rather than C1, C2, or C4 components (see table 14.1). The remainder of the steps occur as in the classical pathway. The principal function of the alternative system is to provide a slower but less specific means of lysing foreign cells (especially gram-negative bacteria) and viruses.

You will notice that at many of the steps of the classical pathway, two molecules are given off. One of these continues in the formation of the membrane attack complex, and the other (C2a, C4a, C3a, or C5a) goes on to become a cytokine or stimulant of inflammation and other immune reactions. Complement components also behave as one type of opsonin that promotes phagocytosis. Complement can participate in inflammation and allergy by causing liberation of vasoactive substances from mast cells and basophils. C3a and C5a are so potent in this response that injecting only one quadrillionth of a gram elicits an immediate flare-up at the site.

*properdin (proh'-pur-din) L. *pro,* before, and *perdere,* to destroy.

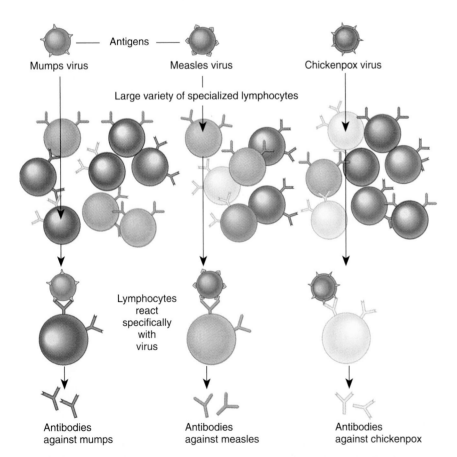

(a) Specificity: Viruses and other infectious agents contain antigen molecules that are specific to a single type of lymphocyte. One result of binding will be the production of virus-specific antibodies.

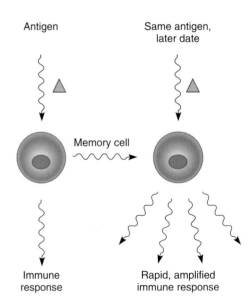

(b) Memory: First contact with antigen creates a memory and quick recall upon second and other future contacts with that antigen.

FIGURE 14.25

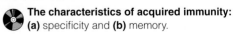 The characteristics of acquired immunity: **(a)** specificity and **(b)** memory.

CHAPTER CAPSULE WITH KEY TERMS

I. **The Three Levels of Host Defenses**
   A. The **first line of defense** is composed of barriers that block the pathogen at the portal of entry; the **second line of defense** includes generalized protective cells and fluids in tissues; and the **third line of defense** includes specific immune reactions with microbes that are required for survival.
   B. *First Line Defenses:* Physical barriers are anatomical structures such as skin, mucous membranes, and cilia; chemical barriers include lysozyme in tears and saliva, and gastric acidity; genetic barriers can be a lack of susceptibility to an infectious agent due to the specialization of a microbe for an exact host.

II. **Immunity/Immunology**
   A. **Immunity** is the development of specific resistance to an infectious agent. Functions of the immune system are: **surveillance** by specialized **white blood cells (WBCs)** for **markers** or receptors (molecules on surface of cells) in tissues; **recognition** by WBCs that detect and differentiate normal markers **(self)** from **foreign** markers called **antigens (nonself).** WBCs are programmed to destroy and remove any foreign material, while self is usually unaffected.
   B. *Components of Immunity:* The immune system is a diffuse network of cells, fibers, chemicals, fluids, tissues, and organs that permeates the body. The structure at the cellular level includes separate but interconnected compartments: the **reticuloendothelial** and **mononuclear phagocyte system,** a continuous network of fibers and phagocytes that surrounds the tissues and organs; the **extracellular fluid (ECF),** a liquid environment in which all cells are bathed; the **lymphatic system,** a series of vessels and organs that carry *lymph* from tissues; and the **bloodstream,** which circulates blood to all organs. The constant communication among the compartments ensures that a reaction in one will be transmitted to another.

III. **Circulatory System: Blood and Lymphatics**
   A. *Composition of Whole Blood:* **Plasma** is a clear, complex liquid that contains nutrients, ions, gases, hormones, antibodies, albumin, and waste products dissolved in water. **Serum** is plasma minus the clotting factors. **Blood cells** are formed by **hemopoiesis** in particular bone marrow sites. From **stem cells,** three main lines of cells are differentiated—white blood cells **(leukocytes),** red blood cells **(erythrocytes),** and megakaryocytes that give rise to **platelets.**
   B. *Functions of Leukocytes:* Leukocytes, the primary cells of host defenses and immunity, are either **granulocytes** or **agranulocytes.** Granulocytes contain distinct granules in cytoplasm and include **neutrophils,** which function as phagocytes; **eosinophils,** which function in worm and fungal infections; **basophils,** which are involved in allergic responses, along with tissue **mast cells.** Agranulocytes lack noticeable granules and include **monocytes** and **lymphocytes.** Monocytes function as blood phagocytes and give rise to **macrophages** in tissues. Two main types of lymphocytes are **B cells,** which produce **antibodies** as part of **humoral immunities,** and **T cells,** which participate in **cell-mediated immunities (CMIs).** Leukocytes are motile by ameboid motion, can be carried between the endothelial cells of small blood vessels **(diapedesis)** and enter surrounding tissues, and can respond to tissue injury or infection by migrating toward chemical signals **(chemotaxis).**

IV. **Lymphatic System**
   The lymphatic system begins as fine capillaries in tissues that gradually join together into larger vessels that eventually drain into the blood circulation. The vessels transport **lymph,** a fluid that contains serum components and white blood cells. Lymphatic organs include **lymph nodes,** compact filters where lymphocytes aggregate and where immune challenges occur; the **spleen,** a blood filter and repository of immune cells; and the **thymus,** where T cells mature. Other lymphatic tissue includes tonsils, gut-associated lymphoid tissue **(GALT),** and Peyer's patches.

V. **Generalized Immune Reactions**
   A. *Inflammatory Response:* This complex system responds to tissue injury (infection, burn, allergy) by mobilizing the immune system against pathogens, repairing damage, and clearing infection. Its common signs and symptoms are redness, heat, swelling, and pain.
   B. *Stages of Inflammatory Response:* Blood vessels narrow and then dilate in response to **chemical mediators** and **cytokines** released by injured tissues and white blood cells. Next, the buildup of fluid from **edema** swells the tissues and keeps infection from spreading, and attracts neutrophils to engulf debris and microbes. WBCs, microbes, debris, and fluid collect to form **pus;** macrophages clean up the residue of inflammation; lymphocytes carry out immune reactions such as antibody formation; and healing occurs. Long-term inflammation can result in injury and disease (granuloma). **Fever,** an increase in body temperature above normal, is due to **pyrogens,** substances that alter the temperature setting in the brain. Fever can slow microbial multiplication and stimulate the immune response.
   C. **Phagocytosis** is a process whereby foreign materials are engulfed and destroyed. Neutrophils engulf small particles, microbes, molecules; macrophages are larger cells that scavenge large packets of cellular debris and extract antigenic information. Macrophages live in a specific tissue or organ (liver, lung, skin) or are free and wandering. After materials are engulfed by the cell into a **phagosome** vacuole, **lysosomes** containing powerful chemicals unite with the phagosome and destroy its contents.

VI. **Important Chemical Defenses**
   A. **Interferon (IFN)** is a family of proteins produced by leukocytes and fibroblasts in response to infection, cancer, or various immune signals. *Alpha* and *beta* interferon are natural infection-fighting substances, and *gamma* interferon is an immune activator and regulator. All three also function as antiviral and anticancer cytokines. IFN works by inhibiting the synthesis of viral proteins and by increasing the cellular immune defenses.
   B. **Complement** is a complex chemical defense system that destroys certain pathogens and produces chemical mediators. It involves chemicals called complement **(C factor)** that act in cascade fashion: One component activates the next in line, which activates the next, and so on. Two major pathways are: (1) classical, which involves activation of complement by specific antibody; and (2) alternative, which is a nonspecific reaction to infections. The result of complement activation is a huge structural protein, the **membrane attack complex,** that can kill cells and inactivate viruses by digesting holes in their membranes.

VII. **Characteristics of Acquired Immunities**
   Immunocompetent individuals possess a third line of defense that is acquired only after direct exposure to an infectious agent. It is made possible by a large repertoire of lymphocytes present from birth that mounts an individualized response to each different infectious agent. These responses are extremely **specific** in their effects and give rise to **immunologic memory,** which provides protection in the case of reexposure to the same pathogen.

MULTIPLE-CHOICE QUESTIONS

1. An example of a nonspecific chemical barrier to infection is
   a. unbroken skin
   b. lysozyme in saliva
   c. cilia in respiratory tract
   d. all of these

2. Which nonspecific host defense is associated with the trachea?
   a. lacrimation
   b. ciliary lining
   c. desquamation
   d. lactic acid

3. Which of the following blood cells function primarily as phagocytes?
   a. eosinophils
   b. basophils
   c. lymphocytes
   d. neutrophils

4. Which of the following is not a lymphoid tissue?
   a. spleen
   b. thyroid gland
   c. lymph nodes
   d. GALT

5. What is included in GALT?
   a. thymus
   b. Peyer's patches
   c. tonsils
   d. breast lymph nodes

6. Monocytes are ____ leukocytes that develop into ____.
   a. granular, phagocytes
   b. agranular, mast cells
   c. agranular, macrophages
   d. granular, T cells

7. Which of the following inflammatory signs specifies pain?
   a. tumor
   b. dolor
   c. calor
   d. rubor

8. An example of an inflammatory mediator that stimulates vasodilation is
   a. histamine
   b. collagen
   c. complement C5a
   d. interferon

9. ____ is an example of an inflammatory mediator that stimulates chemotaxis.
   a. Endotoxin
   b. Serotonin
   c. Fibrin clot
   d. Interleukin-2

10. An example of an exogenous pyrogen is
    a. interleukin-1
    b. complement
    c. interferon
    d. endotoxin

11. ____ interferon is secreted by ____ and is involved in destroying viruses.
    a. Gamma, fibroblasts
    b. Beta, lymphocytes
    c. Alpha, natural killer cells
    d. Beta, fibroblasts

12. Which of the following substances is *not* produced by phagocytes to destroy engulfed microorganisms?
    a. hydroxyl radicals
    b. superoxide anion
    c. hydrogen peroxide
    d. bradykinin

13. Which of the following is the end product of the complement system?
    a. properdin
    b. cascade reaction
    c. membrane attack complex
    d. complement factor C9

14. Immunologic memory refers to the ability of the immune system to
    a. recognize millions of different antigens
    b. react with millions of different antigens
    c. migrate from the blood vessels into the tissues
    d. recall a previous immune response

15. Which subset of WBCs accounts for acquired, specific immunity?
    a. monocytes
    b. B cells
    c. T cells
    d. both b and c

CONCEPT QUESTIONS

1. a. Explain the functions of the three lines of defense.
   b. Which is the most essential to survival?
   c. What is the difference between nonspecific host defenses and immune responses?

2. a. Describe the main elements of the process through which the immune system distinguishes self from nonself.
   b. How is surveillance of the tissues carried out?
   c. What is responsible for it?
   d. What does the term *foreign* mean in reference to the immune system?
   e. Define the term *antigen*.

3. a. Trace the complete cycle of a bacterium through the immune compartments, starting with the blood, the RES, the lymphatics, and the ECF. (You will start and end at the same point.)
   b. What is the direct connection between the bloodstream and the lymphatic circulation?

4. a. What are the main components of the reticuloendothelial system?
   b. Why is it also called the mononuclear phagocyte system?
   c. How does it communicate with tissues and the vascular system?

5. a. Prepare a simplified outline of the cell lines of hemopoiesis.
   b. What is a stem cell?
   c. Review the anatomical locations of hemopoiesis at various stages of development.

6. a. Differentiate between granulocytes and agranulocytes.
   b. Describe the main cell types in each group, their functions, and their incidence in the circulation.

7. a. What is the principal function of lymphocytes?
   b. Differentiate between the two lymphocyte types and between humoral and cell-mediated immunity.

8. a. Why are platelets called formed elements?
   b. What are their functions?
   c. How are they associated with inflammation?

9. a. Explain the processes of diapedesis and chemotaxis, and show how they interrelate.
   b. Referring to question 3, explain how these processes can account for the movements of leukocytes through the fluid compartments.

10. a. What is lymph, and how is it formed?
    b. Why are white cells but not red cells normally found in it?
    c. What are the functions of the lymphatic system?
    d. Explain the filtering action of a lymph node.
    e. What is GALT, and what are its functions?

11. Differentiate between the terms *pyogenic* and *pyrogenic*. Use examples to explain the impact they have on the immune response.

12. Briefly account for the origins and actions of the major types of inflammatory mediators and cytokines.

13. a. Describe the events that give rise to macrophages.
    b. What types of macrophages are there, and what are their principal functions?

14. a. Outline the major phases of phagocytosis.
    b. In what ways is a phagocyte a tiny container of disinfectants?

15. a. Briefly describe the three major types of interferon, their sources, and their biological effects.
    b. Describe the mechanism by which interferon acts as an antiviral compound.

16. a. Describe the general complement reaction in terms of a cascade.
    b. What is the end result of complement activation?
    c. What are some other functions of complement components?

17. What are immunocompetence, immunologic specificity, and immunologic memory?

CRITICAL-THINKING QUESTIONS

1. Suggest some reasons that there is so much redundancy of action and there are so many interacting aspects of immune responses.

2. a. What is probably missing in children born without a functioning lymphocyte system?
    b. What is the most important component extracted in bone marrow transplants?

3. a. What is the likelihood that plants have some sort of immune protection?
    b. Explain your reasoning.

4. a. What substances contained in plasma are absent from serum? (Which of the two would be used to perform a clotting test?)
    b. If one wishes to give antibodies to treat infections, which is better to use, plasma or serum, or does it matter?

5. A patient's chart shows an increase in eosinophil levels.
    a. What does this cause you to suspect?
    b. What if the basophil levels are very high?

6. How can adults continue to function relatively normally after surgery to remove the thymus, tonsils, spleen, or lymph nodes?

7. a. What actions of the inflammatory and immune defenses account for swollen lymph nodes and leukocytosis?
    b. What is pus, and what does it indicate?
    c. In what ways can edema be beneficial?
    d. In what ways is it harmful?

8. An obsolete treatment for syphilis involved inducing fever by deliberately infecting patients with the agent of relapsing fever. A recent but failed AIDS treatment involved applying heat to the body to induce hyperthermia. Can you provide some possible explanations behind these peculiar forms of treatment?

9. Patients with a history of tuberculosis often show scars in the lungs and experience recurrent infection. Account for these effects on the basis of the inflammatory response.

10. Explain why the alternative pathway in complement action is nonspecific and the classical pathway is more specific in its effects, even though they arrive at the same end product.

11. *Shigella, Mycobacterium,* and numerous other pathogens have developed mechanisms that prevent them from being killed by phagocytes.
    a. Suggest two or three factors that help them avoid destruction by the powerful antiseptics in macrophages.
    b. In addition, suggest the potential implications that these infected macrophages can have in the development of disease.

12. Macrophages perform the final job of removing tissue debris and other products of infection. Indicate some of the possible effects when these scavengers cannot successfully complete the work of phagocytosis.

13. Account for the several inflammatory symptoms that occur in the injection site when one has been vaccinated against influenza and tetanus.

14. a. Knowing that fever is potentially both harmful and beneficial, what are some possible guidelines for deciding whether to suppress it or not?
    b. What is the specific target organ of a fever-suppressing drug?

INTERNET SEARCH TOPICS

1. Type in the term *cytokine* on a search engine and look for a cytokine Web page. Find information relating to the number of currently known substances, and describe two medical applications of cytokine therapy.

2. Find a website that deals with the uses of interferon in therapy. Name viral infections and cancers it is used to treat. Are there any adverse side effects?

The Acquisition of Specific Immunity and Its Applications

T he primary focus of this chapter is the remarkable system of lymphocytes that are responsible for specific, acquired immunities. In our more detailed examination, we will discover the complex adaptations for defense against microbes, cancer, and toxic substances. This background will prepare you for coverage of vaccination, immune testing, allergy, and immune deficiency in chapters 16 and 17.

Chapter Overview

- The most specific host defenses are derived from a dual system of lymphocytes that are genetically programmed to react with foreign substances (antigens) found in microbes and other organisms.
- These cells carry glycoprotein receptors that dictate their specificity and reactivity.
- B lymphocytes have antibody receptors, T lymphocytes have T-cell receptors, and macrophages have histocompatibility receptors such as MHC and HLA.
- B cells and T cells arise in the bone marrow, where they proliferate and develop extreme variations in the expression of receptor genes.
- Differentiation of lymphocytes creates billions of genetically different clones that each have a unique specificity for antigen.
- The B cells reach final maturity in special bone marrow sites, and the T cells reach final maturity in the thymus gland.
- Both types of lymphocytes home (migrate) to separate sites in lymphoid tissue where they serve as a constant source of immune cells primed to respond to their correct antigen.
- Antigens are foreign cells, viruses, and molecules that meet a required size and complexity. They are capable of triggering immune reactions by lymphocytes.
- The B and T cells react with antigens through a complex series of cooperative events that involve the processing of antigens by macrophages and the assistance of helper T cells and cytokine stimulants (interleukins).
- B cells activated by antigen enter the cell cycle and mitosis, giving rise to plasma cells that secrete antibodies (humoral immunity) and long-lived memory cells.
- Antibodies have binding sites that affix tightly to an antigen and hold it in place for agglutination, opsonization, complement fixation, and neutralization.

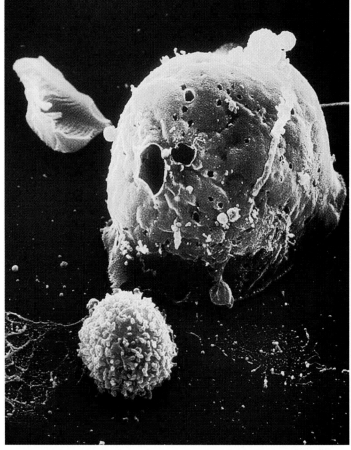

A cytotoxic T cell (lower blue cell) has mounted a successful attack on a tumor cell (larger yellow cell). These small killer cells perforate their cellular targets with holes that lead to lysis and death.

- The amount of antibodies increases during the initial contact with antigen and rises rapidly during subsequent exposures due to memory cells ready for immediate reactions.
- T cells have various receptors that signal their ability to respond as helper cells, suppressor cells, and cytotoxic cells that kill complex pathogens and cancer cells.
- Acquired immunities fall into the categories of natural, artificial, active, and passive.

Further Explorations into the Immune System

In chapter 14 we described the capacity of the immune system to survey, recognize, and react to foreign cells and molecules, and we overviewed the characteristics of nonspecific host defenses, blood cells, phagocytosis, inflammation, and complement. In addition, we introduced the concepts of acquired immunity and specificity. In this chapter we take a closer look at those topics.

The elegance and complexity of immune function are largely due to lymphocytes working closely together with macrophages. To simplify and clarify the network of immunologic development and interaction, we present it here as a series of five stages, with each stage covered in a separate section (figure 15.1). The principal stages include the following: I. lymphocyte development and differentiation; II. the processing of antigens; III. the challenge of B and T lymphocytes by antigens; IV. B lymphocytes and the production and activities of antibodies; and V. T-lymphocyte responses. Notice that as each stage is covered, we will simultaneously be following the sequence of an immune response.

The Dual Nature of Specific Immune Responses

In the following overview, color coded Roman numerals correspond with the subsequent text sections that cover these topics in more detail.

I. DEVELOPMENT OF THE DUAL LYMPHOCYTE SYSTEM

Lymphocytes are central to immune responsiveness. They undergo a sequential development that begins in the embryonic yolk sac and shifts to the liver and bone marrow. Although all lymphocytes arise from the same basic stem cell type, at some point in development, they diverge into two distinct types. Final maturation of B cells occurs in specialized bone marrow sites and that of T cells occurs in the thymus. This process commits each individual B cell or T cell to one specificity. Both cell types subsequently migrate to precise, separate areas in the lymphoid organs (for instance, nodes and spleen, as described in chapter 14) to serve as a lifelong continuous source of immune responsiveness.

II. ENTRANCE AND PROCESSING OF ANTIGENS AND CLONAL SELECTION

When foreign cells, or antigens, enter a fluid compartment of the body, they are immediately swept up in an interconnected network of the lymphatics, blood, and the reticuloendothelial system (RES). In these sites the antigenic materials are met by a battery of cells that work together to screen, entrap, and eliminate them. Certain specialized macrophages are usually the first cells to recognize and react. They ingest and process antigens and present them to the lymphocytes that are specific for that antigen. This selection process is the trigger that activates the lymphocytes. In most cases, the response of B cells also requires the additional assistance of special classes of T cells.

III.B AND III.T ACTIVATION OF LYMPHOCYTES AND CLONAL EXPANSION

When challenged by antigen, both B cells and T cells further differentiate and proliferate. The multiplication of a particular lymphocyte creates a clone, or group of genetically identical cells, some of which are memory cells that will ensure future reactiveness against that antigen. Because the B-cell and T-cell responses depart notably from this point in the sequence, they will be summarized separately.

IV. PRODUCTS OF B LYMPHOCYTES: ANTIBODY STRUCTURE AND FUNCTIONS

The active progeny of a dividing B-cell clone are called plasma cells. These cells are programmed to synthesize and secrete antibodies into the tissue fluid. When these antibodies attach to the antigen for which they are specific, the antigen is marked for destruction or neutralization. Because secreted antibody molecules circulate freely in the tissue fluids, lymph, and blood, the immunity they provide is humoral.

V. HOW T CELLS RESPOND TO ANTIGEN: CELL-MEDIATED IMMUNITY (CMI)

T-cell types and responses are extremely varied. When activated (sensitized) by antigen, a T cell gives rise to one of several types of progeny, each involved in a cell-mediated immune function. The four main functional types of T cells are: (1) helper cells that assist in immune reactions, (2) suppressor cells that suppress and regulate immune reactions, (3) cytotoxic, or killer, cells that destroy specific target cells, and (4) delayed hypersensitivity cells that function in certain allergic reactions. Although T cells secrete cytokines that help destroy antigen or regulate immune responses, they do not produce antibodies.

CHAPTER CHECKPOINTS

Acquired specific immunity is an elegant but complex matrix of interrelationships between lymphocytes and macrophages consisting of several stages.

Stage I. Lymphocytes originate in hemopoietic tissue but go on to diverge into two distinct types: B cells, which produce antibody, and T cells, which produce cytokines that initiate and coordinate the entire immune response.

Stage II. Macrophages detect invading foreign antigens and present them to lymphocytes, which recognize the antigen and initiate the specific immune response.

Stage III. Lymphocytes proliferate, producing clones of progeny that include groups of responder cells and memory cells.

Stage IV. Activated B lymphocytes become plasma cells that produce and secrete large quantities of antibodies.

Stage V. Activated T lymphocytes differentiate into one of four subtypes, which regulate and participate directly in the specific immune responses.

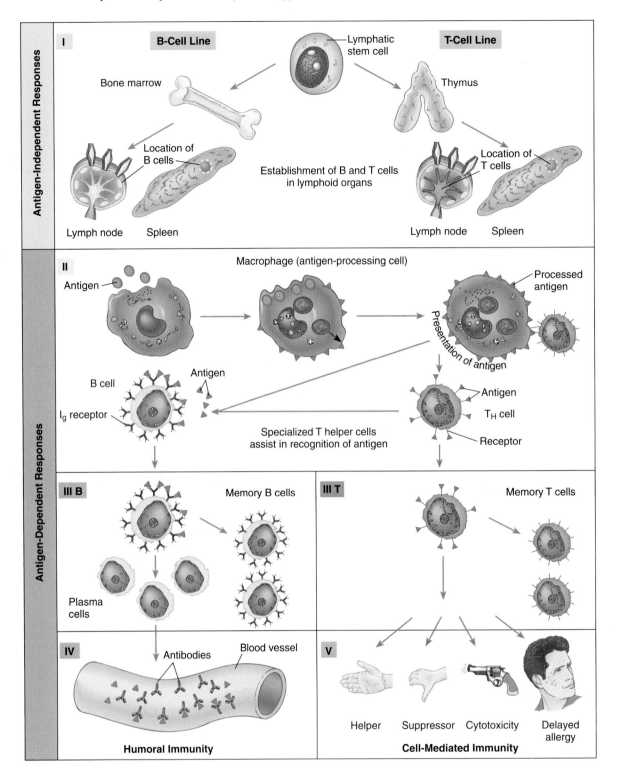

FIGURE 15.1

Overview of the stages of lymphocyte development and function. **I.** Development of B- and T-lymphocyte specificity and migration to lymphoid organs. **II.** Antigen processing by macrophage and presentation to lymphocytes; assistance to B cells by T cells. **III B** and **III T.** Lymphocyte activation, clonal expansion, and formation of memory B and T cells. **IV.** Humoral immunity. B-cell line produces antibodies to react with the original antigen. **V.** Cell-mediated immunity. Activated T cells perform various functions, depending on the signal and type of antigen.

Essential Preliminary Concepts for Understanding Immune Reactions of Sections I–V

Before we examine lymphocyte development and function in greater detail, we must initially review concepts such as the unique structure of molecules (especially proteins), the characteristics of cell surfaces (membranes and envelopes), the ways that genes are expressed, and immune recognition and identification of self and nonself. Ultimately, the shape and function of protein receptors and markers protruding from the surfaces of cells are the result of genetic expression, and these molecules are responsible for specific immune recognition and, thus, immune reactions.

MARKERS ON CELL SURFACES INVOLVED IN RECOGNITION OF SELF AND NONSELF

Chapter 14 touched on the fundamental idea that cell markers or receptors confer specificity and identity. A given cell can express several different receptors, each type playing a distinct and significant role in detection, recognition, and cell communication. Major functions of receptors are: (1) to perceive and attach to nonself or foreign molecules (antigens), (2) to promote the recognition of self molecules, (3) to receive and transmit chemical messages among other cells of the system, and (4) to aid in cellular development. Because of their importance in the immune response, we will concentrate here on the major receptors of lymphocytes and macrophages.

How Are Receptors Formed?

The nature of cell receptors is dictated by which specific elements of the genome are active during development. As a cell matures— be it liver or brain cell, lymphocyte or macrophage—certain genes that code for the cell receptors will be transcribed and translated into protein products with a distinctive shape, specificity, and function. This receptor is modified and packaged by the endoplasmic reticulum and Golgi complex. It is ultimately inserted into the cell membrane so as to be accessible to antigens, other cells, and chemical mediators (figure 15.2). Note that the receptors of many cells, including lymphocytes and macrophages, are glycoproteins with additional carbohydrate fragments added.

Major Histocompatibility Complex

One set of genes that codes for human cell receptors is the **major histocompatibility complex (MHC).** This gene complex gives rise to a series of glycoproteins (called MHC antigens) found on all cells except red blood cells. Because these markers were first identified in humans on the surface of white blood cells, the MHC is also known as the **human leukocyte antigen (HLA)** system. This receptor complex plays a vital role in recognition of self by the immune system and in rejection of foreign tissue. Genes that regulate and code for the MHC of humans are located on the sixth chromosome, clustered in a multigene complex of three subgroups called class I, class II, and class III (figure 15.3).

The functions of the three MHC groups have been identified. Class I genes code for markers that display unique characteristics of self and allow for the recognition of self molecules and the regula-

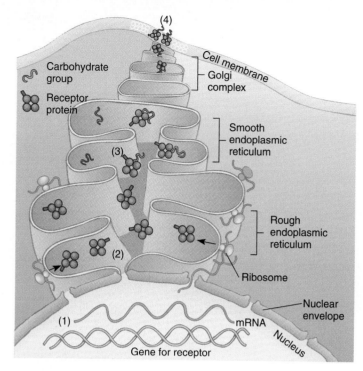

FIGURE 15.2

Receptor formation in a developing cell. **(1)** Gene coding for the receptor is transcribed. **(2)** In the endoplasmic reticulum, mRNA is translated into the protein portion of the receptor. **(3)** A carbohydrate side chain is added, forming glycoprotein. **(4)** The finished receptor is transported to and inserted into the cell membrane.

tion of immune reactions. This class is required by certain T cells. The system is rather complicated in its details, but in general, each human being inherits a particular combination of class I MHC (HLA) genes in a relatively predictable fashion. Although millions of different combinations and variations of these genes are possible among humans, the closer the relationship, the greater the probability for similarity in MHC profile (see figure 17.19). Individual differences in the exact inheritance of MHC genes, however, make it highly unlikely that even closely related persons will express an identical MHC profile. This fact introduces an important recurring theme: Although humans are genetically the same species, the cells of each individual express molecules that are foreign (antigenic) to other humans, which is how the term *histo-* (tissue) compatibility (acceptance) originated. This fact necessitates testing for HLA and other antigens when blood is transfused and organs are transplanted (see graft rejection, chapter 17).

Class II MHC genes also code for immune regulatory receptors, but in this case, for receptors that recognize and react with foreign antigens. This system is located primarily on macrophages and B cells, and it is involved in presenting antigens to T cells during cooperative immune reactions.

Unlike the other two classes, which code for molecules that are inserted into cell surfaces, class III MHC genes code for certain secreted complement components such as C2 and C4 (described in chapter 14).

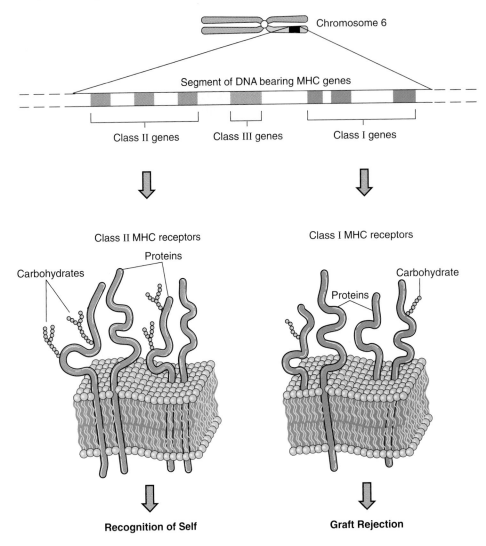

FIGURE 15.3

Glycoprotein receptors of the human major histocompatibility (human leukocyte antigen) gene complex (MHC).

Lymphocyte Receptors and Specificity to Antigen

The part lymphocytes play in immune surveillance and recognition emphasizes the essential role of their receptors. Although they possess MHC antigens for recognizing self, they also carry receptors on their membranes that recognize foreign antigens. Antigen molecules exist in great diversity; there are potentially millions and even billions of unique types. The many sources of antigens include microorganisms as well as an awesome array of chemical compounds in the environment. One of the most fascinating questions in immunology is: How can the lymphocyte receptors be varied to react with such a large number of different antigens? After all, it is generally accepted that there will have to be a different lymphocyte receptor for each unique antigen. Some questions that naturally follow are: How can a cell accommodate enough genetic information to respond to millions or even billions of antigens? When, where, and how does the capacity to distinguish native from foreign tissue arise? To answer these questions, we must first introduce a central theory of immunity.

THE ORIGIN OF DIVERSITY AND SPECIFICITY IN THE IMMUNE RESPONSE

The Clonal Selection Theory and Lymphocyte Development

Research findings have shown that lymphocytes use only about 500 genes to produce a tremendous variety of specific receptors. To understand how this occurs, we turn to a widely accepted concept called the **clonal selection theory.** According to this theory, undifferentiated lymphocytes undergo genetic mutations and recombinations while they proliferate in the embryo (figure 15.4). This process leads to extreme variations in the expression of their receptor specificity. The mechanism might be pictured as follows: As an immature, undifferentiated lymphocyte (A) divides, its receptor genes recombine randomly to produce differing genetic codes in the two daughter cells. Because of slight differences in how the genes are spliced together, daughter cell A1 receives a combination of genes and a specificity that are different from those

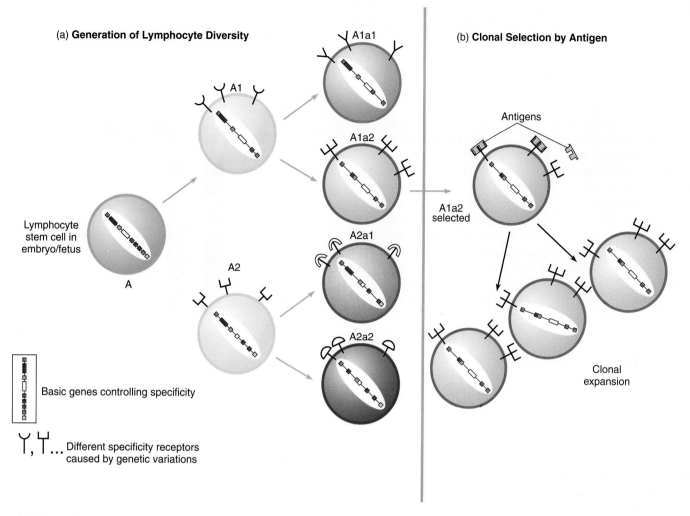

(a) Generation of Lymphocyte Diversity

A1a1

A1

A1a2

Lymphocyte
stem cell in
embryo/fetus

A

A2a1

A2

A2a2

Basic genes controlling specificity

Y, ⅄ ... Different specificity receptors
caused by genetic variations

(b) Clonal Selection by Antigen

Antigens

A1a2
selected

Clonal
expansion

FIGURE 15.4

Clonal selection theory, the origins of lymphocyte specificity and proliferation.   **(a)** During the development of stem cells, the genes that program for specificity to antigen undergo variations so that lymphocyte daughter cells inherit differing genetic programs. These alterations are expressed in the nature of the protein and the shape of the receptor. The potential for genetic and receptor variation is so great that hundreds of millions of genetically different clones can be formed. The receptor is important for providing a specific recognition and reaction site for an antigen, but this part of development does not require the actual presence of the antigen. **(b)** Clonal selection and expansion when antigen is present. The complete repertoire of lymphocytes has developed by the time a child is born, so that any antigen entering the system will automatically encounter and select its prespecified lymphocyte and stimulate it to expand into a genetically identical group or clone, each of which can respond to the same antigen. Note that all receptors on a single lymphocyte are identical.

of daughter cell A2. By the same process, the progeny of A1 and A2 will likewise receive a different genetic program, and so on, for millions of different lymphocytes. The ultimate impact of these genetic variations is that the amino acid composition of the receptor, and thus its shape and specificity, will be different for each lymphocyte type. Each genetically distinct group of lymphocytes that possesses the same specificity is called a **clone.** Although B cells and T cells have different kinds of receptors on their surfaces, the general principles of genetics and acquisition of specificity are the same for both.

As a result of gene reassortment on such a massive scale, the lymphoid system eventually can develop (in theory) at least a billion different clones of lymphocytes. So, by the time of birth, the

ability to react with a tremendous variety of antigens is already in place in the lymphoid tissue. Two important generalities of immune responsiveness one can derive from the clonal selection theory are that: (1) lymphocyte specificity is preprogrammed, existing in the genetic makeup before an antigen has ever entered the system, and (2) each genetically different type of lymphocyte expresses only a single specificity.

According to another part of the clonal selection theory, any lymphocyte that could possibly mount a harmful response against *self* molecules is eliminated or suppressed. Immunologists call this feature **tolerance to self.** Some immune diseases (called autoimmune) are believed to arise when the immune system loses this tolerance and attacks self (see chapter 17).

The final part of the clonal selection theory proposes that the first introduction of each distinct type of antigen into the immune system selects a genetically distinct lymphocyte and causes it to expand into a clone of cells that can react to that antigen.

The Specific B-Cell Receptor: An Immunoglobulin Molecule

In the case of B lymphocytes, the receptor genes that undergo the recombination described are those governing **immunoglobulin*** **(Ig)** synthesis. Immunoglobulins are large glycoprotein molecules that serve as the specific receptors of B cells and as antibodies. The basic immunoglobulin molecule is a composite of four polypeptide chains: a pair of identical heavy (H) chains and a pair of identical light (L) chains (figure 15.5). One light chain is bonded to one heavy chain, and the two heavy chains are bonded to one another with disulfide bonds, creating a symmetrical, Y-shaped arrangement. The ends of the forks formed by the light and heavy chains contain pockets, called the **antigen binding sites.** It is these sites that can be highly variable in shape to fit a wide range of antigens. This extreme versatility is due to **variable regions (V)** where amino acid composition is highly varied from one clone of B lymphocytes to another. The remainder of the light chains and heavy chains consist of **constant regions (C)** whose amino acid content does not vary greatly from one antibody to another. Although we will subsequently discuss immunoglobulins and their function as antibodies, for now, we will concentrate on the genetics that explain the origins of lymphocyte specificity.

Development of the Receptors During Lymphocyte Maturation

The genes that code for immunoglobulins lie on three different chromosomes. An undifferentiated lymphocyte has about 150 different genes that code for the variable region of light chains and a total of about 250 genes for the variable and **diversity regions (D)** of the heavy chains. It has only a few genes for coding for the constant regions and a few for the **joining regions (J)** that join segments of the molecule together. Owing to genetic recombination during development, only the selected (V and D) receptor genes are active in the mature cell, and all the other V and D genes are inactive (figure 15.6). This is how the singular specificity of lymphocytes arises.

It is helpful to think of the immunoglobulin genes as blocks of information that code for a polypeptide lying in sequence along a chromosome. During development of each lymphocyte the genetic blocks are independently segregated, randomly selected, and assembled as follows:

- For a heavy chain, a variable region gene and diversity region gene are selected from among the hundreds available and spliced to one joining region gene and one constant region gene.
- For a light chain, one variable, one joining, and one constant gene are spliced together.

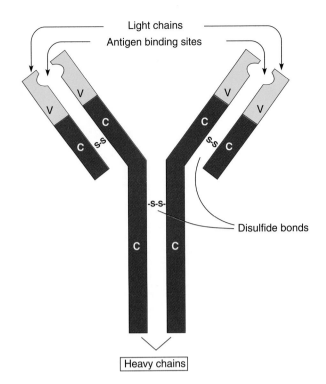

FIGURE 15.5

Simplified structure of an immunoglobulin molecule. The main components are four polypeptide chains—two identical light chains and two identical heavy chains bound by disulfide bonds as shown. Each chain consists of a variable region (V) and a constant region (C). The variable regions of light and heavy chains form a binding site for antigen.

- After transcription and translation of each gene complex into a polypeptide, a heavy chain combines with a light chain to form half an immunoglobulin; two of these combine to form a completed monomer (figure 15.6).

Once synthesized, the immunoglobulin product is transported to the cell membrane and inserted there to act as a receptor that expresses the specificity of that cell and to react with an antigen as shown in figure 15.2. The first receptor on most B cells is a small form of IgM, and mature B cells carry IgD receptors (see table 15.2, page 463). It is notable that for each lymphocyte, the genes that were selected for the variable region, and thus for its specificity, will be locked in for the rest of the life of that lymphocyte and its progeny. We will discuss the different classes of Ig molecules further in section IV.

T-Cell Receptors

The T-cell receptor for antigen belongs to the same protein family as the B-cell receptor. It is similar to B cells in being formed by genetic modification, having variable and constant regions, being inserted into the membrane, and having an antigen binding site formed from two parallel polypeptide chains (figure 15.7). Unlike the immunoglobulins, the T-cell receptor is relatively small and does not appear to have a humoral function. The several other receptors of T cells are described in a later section.

---

* immunoglobulin (im″-yoo-noh-glahb′-yoo-lin) The technical name for an antibody.

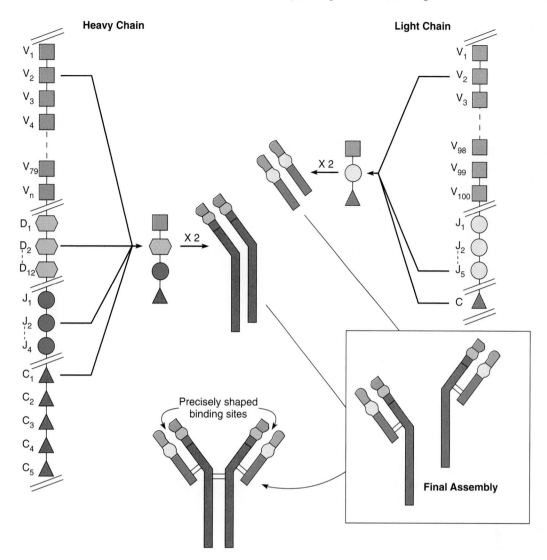

FIGURE 15.6

A simplified look at immunoglobulin genetics.   The phenomenon of B-cell differentiation could be compared to a "cutting and pasting" process, in which the final gene that codes for a heavy or light chain is assembled by splicing blocks of genetic material from several regions. *(Left)* The heavy-chain gene is composed of genes from four separate segments (V, D, J, and C) that are transcribed and translated to form a single polypeptide chain. *(Right)* The light-chain genes are put together like heavy ones, except that the final gene is spliced from three gene groups (V, J, and C). During final assembly, first the heavy and light chains are bound, and then the heavy-light combinations are connected to form the immunoglobulin molecule.

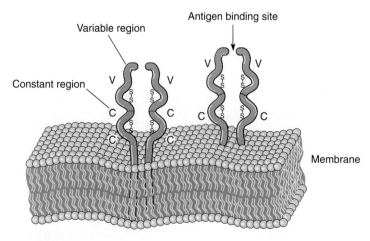

FIGURE 15.7

Proposed structure of the T-cell receptor for antigen.

CHAPTER CHECKPOINTS

The surfaces of all cell membranes contain identity markers known as protein receptors. These cell markers function in identification, communication, and cell development. They are also self identity markers.

Human cell markers are genetically determined by MHCs. MHC I codes for identity markers on all host cells. MHC II codes for immune receptors on macrophages and B cells. MHC III codes for secretion of complement C2 and C4.

The clonal selection theory explains that during prenatal development, both B and T cells develop millions of genetically different clones through independent segregation, random reassortment, and mutation. Together these clones possess enough genetic variability to respond to many millions of different antigens. Each clone, however, can respond to only one specific antigen.

The Lymphocyte Response System in Depth

Now that you have a working knowledge of some factors in the development of immune specificity, let us look at each stage of an immune response as originally outlined in figure 15.1.

I. DEVELOPMENT OF THE DUAL LYMPHOCYTE SYSTEM: THE STAGES IN ORIGIN, DIFFERENTIATION, AND MATURATION

Development generally follows a similar general pattern in both types of lymphocytes (figure 15.8). Starting in embryonic and fetal stages, stem cells in the yolk sac, liver, and bone marrow give rise to immature lymphocytes that are released into the circulation. Because these undifferentiated cells cannot yet react with antigens, they must first undergo developmental changes at some specific anatomical location (table 15.1). This maturation occurs along two separate lines that will characterize all future responses. The fully mature B and T lymphocytes are released and ultimately take up residence in various lymphoid organs. Lymphocyte differentiation and immunocompetence are basically complete by the late fetal or early neonatal period.

Specific Happenings in B-Cell Maturation

The site of B-cell maturation was first discovered in birds, which have an organ in the intestine called the bursa. For some time, the human bursal equivalent was not established. Now it is known to be certain bone marrow sites that harbor *stromal cells*. These huge cells nurture the lymphocyte stem cells and provide hormonal signals that initiate B-cell development. As a result of gene modification and selection, hundreds of millions of distinct B cells develop. These naive lymphocytes "home" to specific sites in the lymph nodes, spleen, and gut-associated lymphoid tissue (GALT), where they adhere to specific binding molecules. Here they will come into contact with antigens throughout life. In addition to having immunoglobulins as surface receptors, a fully differentiated B cell has a distinctively rough appearance under high magnification, with numerous microvillus projections.

Specific Happenings in T-Cell Maturation

The maturation of T cells and the development of their specific receptors are directed by the thymus gland and its hormones. Due to the complexity of T-cell function, there are at least seven classes of T-cell receptors or markers, termed the CD cluster (abbreviated from cluster of differentiation—CD1, CD2, etc.). Some receptors (see figure 15.7) recognize antigen molecules bound to cells on MHC receptors; others recognize receptors on B cells, other T cells, and macrophages. Like B cells, T cells also migrate to lymphoid organs and occupy specific sites. Additional characteristics typical of mature T cells are smaller and fewer microvilli under high magnification and the property of rosette formation when mixed with normal sheep red blood cells (see figure 16.13). It has been estimated that 25×10^9 T cells pass between the lymphatic and general circulation per day.

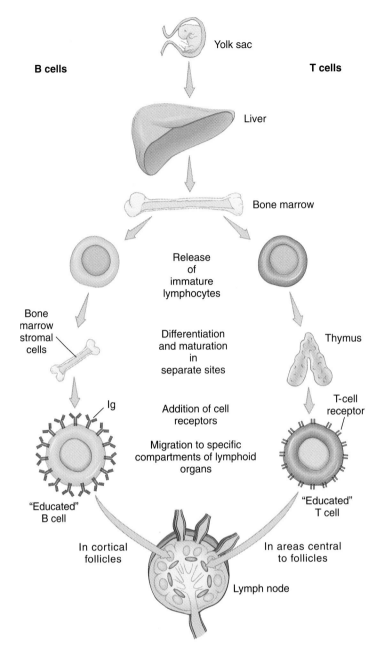

FIGURE 15.8

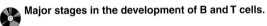 Major stages in the development of B and T cells.

II. ENTRANCE AND PROCESSING OF ANTIGENS AND CLONAL SELECTION

Having reviewed the characteristics of lymphocytes, let us now examine the properties of antigens, the substances that cause them to react. As we reported in chapter 14, an **antigen (Ag)**[1] is a substance that provokes an immune response in specific lymphocytes. The property of behaving as an antigen is called **antigenicity.** The term *immunogen* is another term of reference for a substance that can elicit an immune response. To be perceived as an antigen or

---

1. Originally, **anti**body **gen**erator.

TABLE 15.1

Contrasting Properties of B-Cell and T-Cell Lines

	B Cells	T Cells
Site of Maturation	Bone marrow	Thymus
Nature of Surface Markers	Immunoglobulin	Several CD receptors
Texture of Surface	Rough	Smoother
Circulation in Blood	Low numbers	High numbers
Receptors for Antigen	Immunoglobulin	T-cell receptor
Distribution in Lymphatic Organs	Cortex (in follicles)	Paracortical sites (interior to the follicles)
Product of Antigenic Stimulation	Plasma cells and memory cells	Several types of sensitized T cells and memory cells
General Functions	Production of antibodies	Cells function in helping, suppressing, killing, delayed hypersensitivity; synthesize cytokines

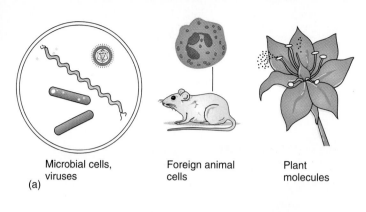

Microbial cells, viruses   Foreign animal cells   Plant molecules

(a)

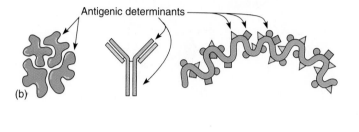

Antigenic determinants

(b)

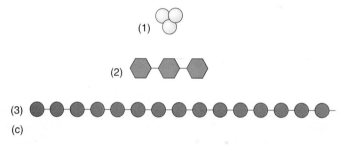

(1)

(2)

(3)

(c)

FIGURE 15.9

Characteristics of antigens. **(a)** Whole cells and viruses make good immunogens. **(b)** Complex molecules with several antigenic determinants make good immunogens. **(c)** Poor immunogens include small molecules not attached to a carrier molecule *(1)*, simple molecules *(2)*, and large but repetitive molecules *(3)*.

immunogen, a substance must meet certain requirements in foreignness, shape, size, and accessibility.

Characteristics of Antigens

One important characteristic of an antigen is that it be perceived as **foreign,** meaning that it is not a normal constituent of the body. Whole microbes or their parts, cells, or substances that arise from other humans, animals, plants, and various molecules all possess this quality of foreignness and thus are potentially antigenic to the immune system of an individual (figure 15.9). Molecules of complex composition such as proteins and protein-containing compounds prove to be more immunogenic than repetitious polymers composed of a single type of unit. Most materials that serve as antigens fall into the following chemical categories:

- Proteins and polypeptides (enzymes, albumin, antibodies, hormones, exotoxins)
- Lipoproteins (cell membranes)
- Glycoproteins (blood cell markers)
- Nucleoproteins (DNA complexed to proteins, but not pure DNA)
- Polysaccharides (certain bacterial capsules) and lipopolysaccharides

Effects of Molecular Shape and Size   To initiate an immune response, a substance must also be large enough to "catch the attention" of the surveillance cells. Molecules with a molecular weight (MW) of less than 1,000 are seldom complete antigens, and those between 1,000 MW and 10,000 MW are weakly so. Complex macromolecules approaching 100,000 MW are the most immunogenic, a category also dominated by large proteins. Note that large size alone is not sufficient for antigenicity; glycogen, a polymer of glucose with a highly repetitious structure, has an MW over 100,000 and is not normally antigenic, whereas insulin, a protein with an MW of 6,000, can be antigenic.

A lymphocyte's capacity to discriminate differences in molecular shape is so fine that it recognizes and responds to only a portion of the antigen molecule. This molecular fragment, called the **antigenic determinant,** is the primary signal that the molecule is foreign (figure 15.10). The particular tertiary structure and shape of this determinant must conform like a key to the receptor "lock" of the lymphocyte, which then responds to it. Certain amino acids accessible at the surface of proteins or protruding carbohydrate side chains are typical examples. Many foreign cells and molecules are very complex antigenically, with numerous determinants, each of which will elicit a separate and different lymphocyte response.

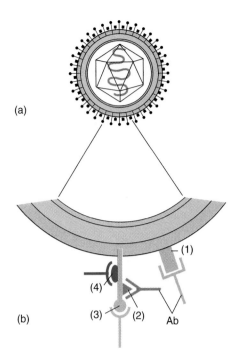

FIGURE 15.10

Mosaic antigens. **(a)** Microbes such as viruses present various sites that serve as separate antigenic determinants. **(b)** Inset indicates that each determinant *(1, 2, 3, 4)* will stimulate a different lymphocyte and antibody response.

Examples of these multiple, or *mosaic,* antigens include bacterial cells containing cell wall, membrane, flagellar, capsular, and toxin antigens; and viruses, which express various surface and core antigens (figure 15.10).

Small foreign molecules that consist only of a determinant group and are too small by themselves to elicit an immune response are termed **haptens.** However, if such an incomplete antigen is linked to a larger carrier molecule, the combination develops immunogenicity (figure 15.11). The carrier group contributes to the size of the complex and enhances the proper spatial orientation of the determinative group, while the hapten serves as the antigenic determinant. Haptens include such molecules as drugs, metals, and

ordinarily innocuous household, industrial, and environmental chemicals. Many haptens develop antigenicity in the body by combining with large carrier molecules such as serum proteins (see allergy in chapter 17).

Special Types of Antigens So far, we have emphasized the role of microbial antigens, but tissues and proteins (including enzymes and antibodies) from other humans and animals are also antigenic. On occasion, even a part of the body can take on the character of an antigen. During lymphocyte differentiation, immune tolerance to self tissue occurs, but a few anatomical sites can contain sequestered (hidden) molecules that escape this assessment. Such molecules, called **autoantigens,** can occur in tissues (eye, thyroid gland, for example) that are walled off early in embryonic development before the surveillance system is in complete working order. Because tolerance to these substances has not yet been established, they can subsequently be mistaken as foreign; this mechanism appears to account for some types of autoimmune diseases such as rheumatoid arthritis (chapter 17).

Because each human being is genetically and biochemically unique (except for identical twins), the proteins and other molecules of one person can be antigenic to another. **Alloantigens**[*] are cell surface markers and molecules that occur in some members of the same species but not in others. Alloantigens are the basis for an individual's blood group (see chapter 17) and major histocompatibility profile, and they are responsible for incompatibilities that can occur in blood transfusion or organ grafting.

On the other hand, different organisms can possess some molecules, called **heterophilic,**[*] or heterogenetic, antigens, with the same or a similar determinant group. These antigens stimulate a response from the same lymphocyte clone even when they are from totally different sources. Antibodies raised against a heterophilic antigen from one organism will **cross-react** with a similar or identical antigen from another source. Examples of antigens of different origins with heterophilic determinants are carbohydrate residues on the surfaces of bacteria and red blood cells, the antigens of Group A

---

[*] alloantigen (al′-oh) Gr. *allos,* other.

[*] heterophile (het′-ur-oh-fyl) Gr. *hetero,* other, and *phile,* to love.

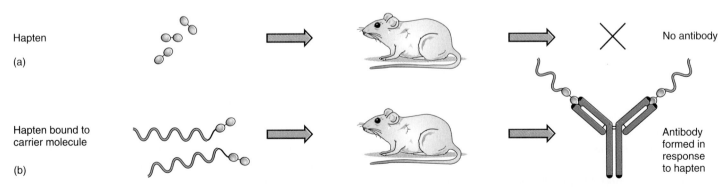

Hapten

(a)

Hapten bound to carrier molecule

(b)

No antibody

Antibody formed in response to hapten

FIGURE 15.11

The hapten-carrier phenomenon. **(a)** Haptens are too small to be discovered by an animal's immune system; no response. **(b)** A hapten bound to a large molecule will serve as an antigenic determinant and stimulate a response and an antibody that is specific for it.

MICROBITS 15.1
An Immunologic Bargain Basement

Realizing the great variety and quantity of antigens that find their way into the body, one might wonder how this microscopic flotsam and jetsam can ever meet up with the right lymphocyte. After all, the body's fluid compartments are relatively large, and antigens are constantly entering through some portal or another. The events of this awesome sorting process could be likened to a rummage sale in a microscopic department store. The store (lymph node) is crowded with large numbers of all types of customers (lymphocytes), and it contains all sorts of merchandise (antigens). Picture the mounds of antigenic merchandise arrayed on tables and thousands of lymphocyte customers milling around, pawing through the piles of materials with their receptor hands to find just the right specificity (size, fit, color, shape, and brand). The sales personnel (macrophages and T helper cells) display the merchandise, help in the selection, and expedite the sale (contact of the antigen with its matching lymphocyte). Like customers, some lymphocytes become "experienced shoppers." On subsequent shopping sprees, they retain a memory of previous sales and are even more efficient.

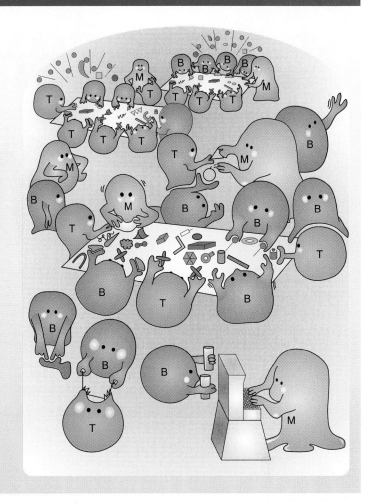

streptococci and human heart tissue, and cardiolipin, a phospholipid present in a wide assortment of living things. Heterophilic antigens may play a part in diseases such as rheumatic fever and in false positive diagnostic tests (as occur in syphilis).

Some bacterial toxins, called *superantigens,* are potent stimuli for T cells. Their presence in an infection activates T cells at a rate 100 times greater than ordinary antigens. The result can be an overwhelming release of cytokines and cell death. Such diseases as toxic shock syndrome (see chapter 18) and certain autoimmune diseases (see chapter 17) are associated with this class of antigens.

Antigens that evoke allergic reactions, called **allergens,** will be characterized in detail in chapter 17.

Host Response to Antigens: A Cooperative Affair
The basis for most immune responses is the encounter between antigens and white blood cells. Microbes and other foreign substances enter most often through the respiratory or gastrointestinal mucosa and less frequently through other mucous membranes, the skin, or across the placenta. Antigens introduced intravenously be-

come localized in the liver, spleen, bone marrow, kidney, and lung. If introduced by some other route, antigens are carried in lymphatic fluid and concentrated by the lymph nodes. The lymph nodes and spleen are important in concentrating the antigens and circulating them thoroughly through all areas populated by lymphocytes so that they come into contact with the proper clone. To help imagine this process, see the analogy in Microbits 15.1.

The Role of Macrophages: Antigen Processing and Presentation
In most immune reactions, the antigen must be further acted upon and formally presented to lymphocytes by special macrophages or other cells called **antigen-processing cells (APCs).** One of the most prominent APCs is a large **dendritic*** cell that engulfs the antigen and modifies it so that it will be more immunogenic and recognizable to lymphocytes. After processing is complete, the

* **dendritic** (den′-drih-tik) Gr. *dendron,* tree. In reference to the long, branchlike extensions of the cell membrane.

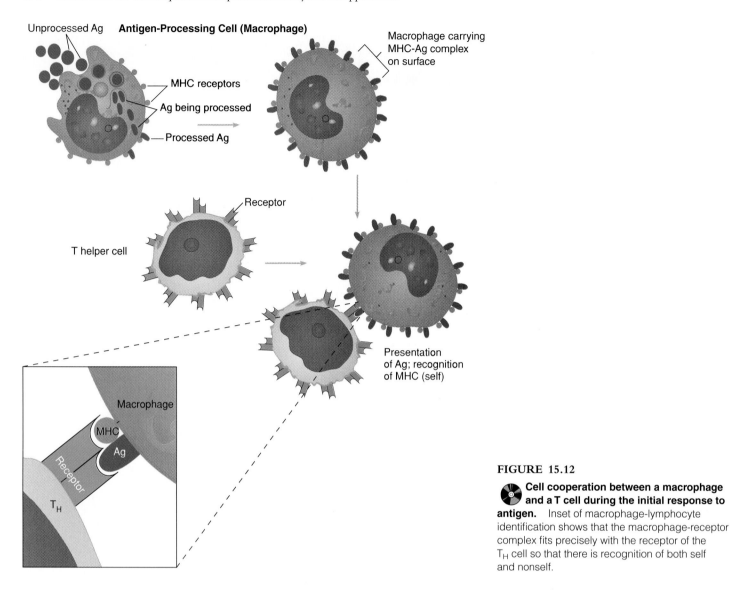

Unprocessed Ag

Antigen-Processing Cell (Macrophage)

MHC receptors

Ag being processed

Processed Ag

Macrophage carrying
MHC-Ag complex
on surface

Receptor

T helper cell

Presentation
of Ag; recognition
of MHC (self)

Macrophage

MHC

Receptor

Ag

T$_H$

FIGURE 15.12

Cell cooperation between a macrophage and a T cell during the initial response to antigen. Inset of macrophage-lymphocyte identification shows that the macrophage-receptor complex fits precisely with the receptor of the T$_H$ cell so that there is recognition of both self and nonself.

antigen is moved to the surface of the APC and bound to the MHC receptor so that it will be readily accessible to the lymphocytes during presentation (figure 15.12).

Presentation of Antigen to the Lymphocytes and Its Early Consequences

For lymphocytes to respond to the APC-bound antigen, certain conditions must be met. **T-cell-dependent antigens,** usually protein-based, require recognition steps between the macrophage, antigen, and lymphocytes. The first cells on the scene to assist the macrophage in activating B cells and other T cells are a special class of **helper T cells (T$_H$).** This class of T cell bears a receptor that binds simultaneously with the class II MHC receptor on the macrophage and with one site on the antigen (figure 15.12). Once identification has occurred, a cytokine, **interleukin*-1 (IL-1),** produced by the macrophage, activates this T helper cell. The T$_H$ cell, in turn, produces a different cytokine, **interleukin-2 (IL-2),** that stimulates a general increase in activity of committed B and T cells

(see Spotlight on Microbiology 15.3, page 467). The manner in which B and T cells subsequently become activated by the macrophage–T helper cell complex and their individual responses to antigen will be addressed separately in sections III B and IV, and III T and V.

A few antigens can trigger a response from B lymphocytes without the cooperation of macrophages or T helper cells. These **T-cell-independent** antigens are usually simple molecules such as carbohydrates with many repeating and invariable determinant groups. Examples include lipopolysaccharide from the cell wall of *Escherichia coli* and polysaccharide from the capsule of *Streptococcus pneumoniae.* Because so few antigens are of this type, most B-cell reactions require helper T cells.

III. B ACTIVATION OF B LYMPHOCYTES: CLONAL EXPANSION AND ANTIBODY PRODUCTION

The immunologic activation of most B cells requires a series of events (figure 15.13):

1. **Clonal selection and binding of antigen.** In this case, a pre-committed B cell of a particular clonal specificity picks up the

---

* **interleukin** (in″-tur-loo′-kin) A peptide that carries signals between white blood cells.

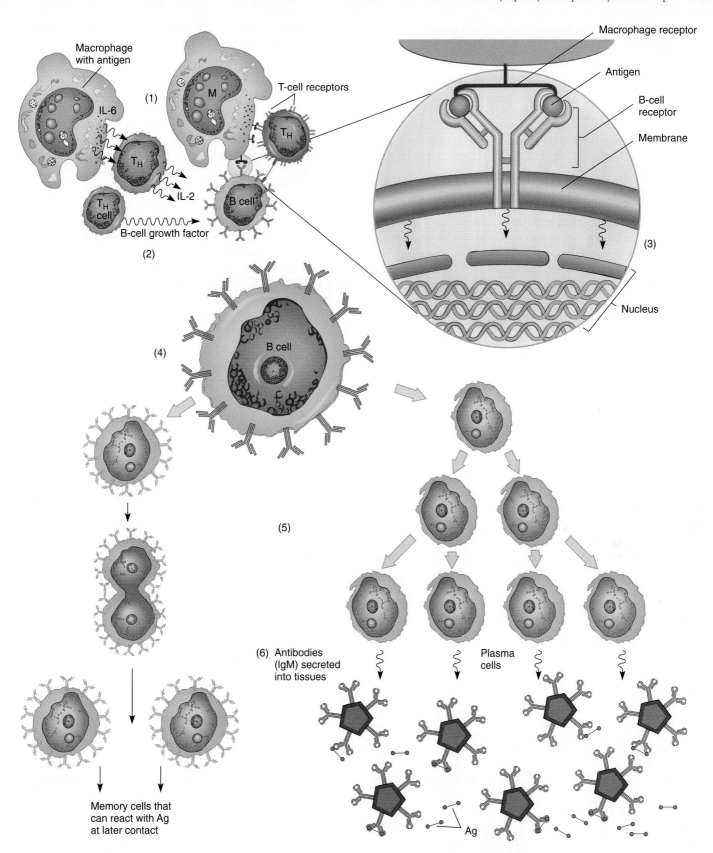

FIGURE 15.13

Events in B-cell activation.   **(1)** Clonal selection. Antigen is presented to the B cell by a macrophage; activated T cells facilitate recognition by binding to both the macrophage and the B cell and **(2)** produce growth factors that stimulate the B cell. **(3, inset)** The binding of antigen on the immunoglobulin receptors of the B cell transmits a signal to the cell nucleus to begin activation of the B cell. **(4)** The B cell is transformed into an activated cell. **(5)** The activated cell undergoes mitotic divisions and clonal expansion into memory and plasma cells. **(6)** Plasma cells secrete antibodies that can react with antigens.

antigen itself or is presented with the antigen by the macrophage complex so that the antigen is bound to the B-cell receptors. At the same time, the MHC/Ag receptor on the B cell is bound to the T_H cell, thereby ensuring self recognition.

2. **Instruction by chemical mediators.** The B cell receives developmental signals from macrophages and T cells (interleukin-2 and -6) and various other growth factors, such as IL-4 and -5.

3. The combination of these stimuli on the membrane receptors causes a signal to be transmitted internally to the B-cell nucleus.

4. This event triggers **B-cell activation.** An activated B cell undergoes an increase in DNA synthesis, organelle bulk, and size in preparation for entering the cell cycle and mitosis.

5. **Clonal expansion.** A stimulated B cell multiplies through successive mitotic divisions and produces a large population of genetically identical daughter cells. Some cells that stop short of becoming fully differentiated are **memory cells,** which remain for long periods to react with that same antigen at a later time. This reaction also expands the clone size, so that subsequent exposure to that antigen provides more cells with that specificity. This expansion of the clone size accounts for the increased memory response. By far the most numerous progeny are large, specialized, terminally differentiated B cells called **plasma cells.**

6. **Antibody production and secretion.** The primary action of plasma cells is to secrete into the surrounding tissues copious amounts of antibodies with the same specificity as the original receptor (figure 15.13). Although an individual plasma cell can produce around 2,000 antibodies per second, production does not continue indefinitely because of regulation from the T suppressor (T_S) class of cells. The plasma cells do not survive for long and deteriorate after they have synthesized antibodies.

CHAPTER CHECKPOINTS

Immature lymphocytes released from hemopoietic tissue migrate (home) to one of two sites for further development. B cells mature in the stromal cells of the bone marrow. T cells mature in the thymus.

Antigens or immunogens are proteins or other complex molecules of high molecular weight that trigger the immune response in the host.

Lymphocytes respond to a specific portion of an antigen called the antigenic determinant. A given microorganism has many such determinants, all of which stimulate individual specific immune responses.

Haptens are molecules that are too small to trigger an immune response alone but can be immunogenic when they attach to a larger substance, such as host serum protein.

Autoantigens and allergens are types of antigens that cause damage to host tissue as a consequence of the immune response.

Macrophages or other antigen-processing cells (APCs) bind foreign antigen to their cell surfaces for presentation to lymphocytes. Physical contact between the APC, T cells, and B cells activates these lymphocytes to proceed with their respective immune responses.

IV. PRODUCTS OF B LYMPHOCYTES: ANTIBODY STRUCTURE AND FUNCTIONS

The Structure of Immunoglobulins

Earlier we saw that a basic immunoglobulin (Ig) molecule contains four polypeptide chains connected by disulfide bonds. Let us view this structure once again, using an **IgG molecule** as a model. Two functionally distinct segments called *fragments* can be differentiated. The two "arms" that bind antigen are termed **antigen binding fragments (Fabs),** and the rest of the molecule is the **crystallizable fragment (Fc),** so called because it has been crystallized in pure form. The distal end of each Fab fragment (consisting of the variable regions of the heavy and light chains) folds into a groove that will accommodate one antigenic determinant. The presence of a special *hinge* region at the site of attachment between the Fab and Fc fragments allows swiveling of the Fab fragments. In this way, they can change their angle to accommodate nearby antigen sites that vary slightly in distance and position. The Fc fragment is involved in binding to various cells and molecules of the immune system itself. Figure 15.14 shows three views of antibody structure.

Antibody–Antigen Interactions and the Function of the Fab

The site on the antibody where the antigenic determinant inserts is composed of a *hypervariable region* whose amino acid content can be extremely varied. Antibodies differ somewhat in the exactness of this groove for antigen, but a certain complementary fit is necessary for the antigen to be held effectively (figure 15.15). The specificity of antigen binding sites for antigens is very similar to enzymes and substrates (in fact, some antibodies are used as enzymes; see Microbits 8.2). So specific are some immunoglobulins for antigen that they can distinguish between a single functional group of a few atoms. Because the specificity of the Fab sites is identical, an Ig molecule can bind antigenic determinants on the same cell or on two separate cells and thereby link them.

The principal activity of an antibody is to unite with, immobilize, call attention to, or neutralize the antigen for which it was formed (figure 15.16). Antibodies called **opsonins** stimulate **opsonization,*** a process in which microorganisms or other particles are coated with specific antibodies so that they will be more readily recognized by phagocytes, which dispose of them. Opsonization has been likened to putting handles on a slippery object to provide phagocytes a better grip. The capacity for antibodies to aggregate, or *agglutinate,* antigens is the consequence of their cross-linking cells or particles into large clumps. This is a principle behind certain immune tests discussed in chapter 16. The interaction of an antibody with complement can result in the specific rupturing of cells and some viruses. In **neutralization** reactions, antibodies fill the surface receptors on a virus or the active site on a molecule to prevent it from functioning normally. **Antitoxins** are a special type of antibody that neutralize bacterial exotoxins. It should be noted that not all antibodies are protective; some neither benefit nor harm, and a few actually cause diseases (see autoimmunity in chapter 17).

---

* opsonization (ahp″-son-uh-zay′-shun) Gr. *opsonein,* to prepare food.

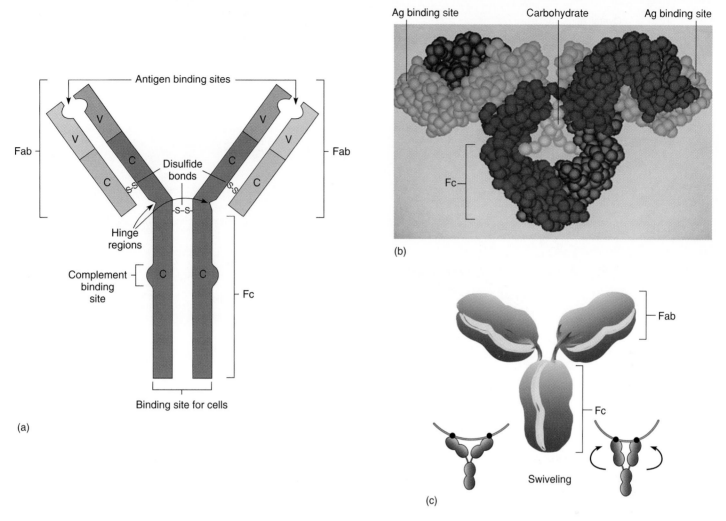

FIGURE 15.14

Working models of antibody structure.   **(a)** Diagrammatic view of IgG depicts the principal functional areas (Fabs and Fc) of the molecule. **(b)** Realistic model of immunoglobulin shows the tertiary and quaternary structure achieved by additional intrachain and interchain bonds and the position of the carbohydrate component. **(c)** The "peanut" model of IgG helps illustrate swiveling of Fabs relative to one another and to Fc.

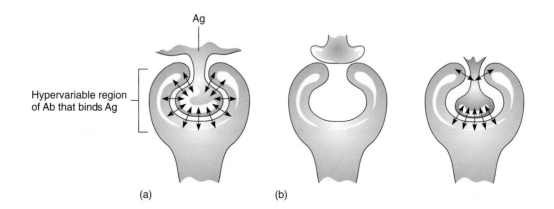

FIGURE 15.15

Antigen-antibody binding.   The union of antibody (Ab) and antigen (Ag) is characterized by a certain degree of fit and is supported by weak linkages such as hydrogen bonds and electrostatic attraction. **(a)** In a snug fit such as that shown here, there is great opportunity for attraction and strong attachment. The strength of this union confers high affinity. **(b)** Examples of the relationship of other antigens with this same antibody. The first Ag clearly cannot be accommodated. The second (purple) antigen is not a perfect fit, but it can bind to the antibody.

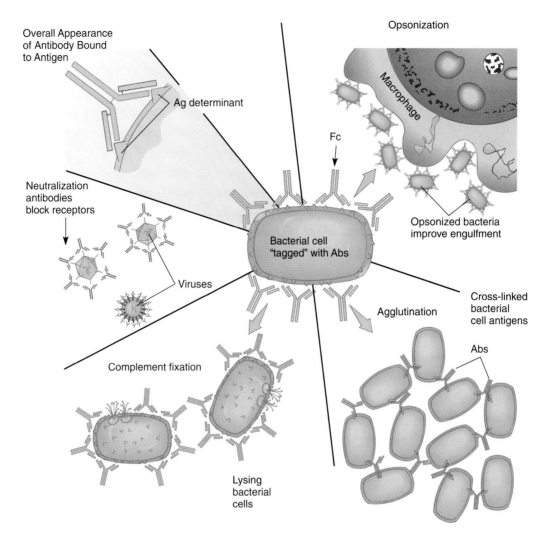

FIGURE 15.16

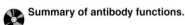 **Summary of antibody functions.**

Functions of the Crystallizable Fragment: Interactions with Self

Although the Fab fragments bind antigen, the Fc fragment has a different binding function. In most classes of immunoglobulin, the proximal end of Fc contains an effector molecule that can bind to certain receptors on the membrane of cells, such as macrophages, neutrophils, eosinophils, mast cells, basophils, and lymphocytes. The effect on an antibody's Fc fragment binding to a cell receptor depends upon that cell's role. In the case of opsonization, the attachment of antibody to foreign cells and viruses exposes the Fc fragments to phagocytes. Certain antibodies have receptors on the Fc portion for fixing complement, and in some immune reactions, the binding of Fc causes the release of cytokines. For example, the antibody of allergy (IgE) binds to basophils and mast cells, which causes the release of allergic mediators such as histamine (see chapter 17). The size and amino acid composition of Fc also determine an antibody's permeability, its distribution in the body, and its class.

Accessory Molecules on Immunoglobulins

All antibodies contain molecules in addition to the basic polypeptides. Varying amounts of carbohydrates are affixed to the constant regions in most instances (table 15.2). Two additional accessory molecules are the *J chain* that joins the monomers of IgA and IgM, and the *secretory component,* which helps move Ig across mucous membranes. These proteins occur only in certain immunoglobulin classes.

The Classes of Immunoglobulins

Immunoglobulins exist as structural and functional classes called *isotypes* (compared and contrasted in table 15.2). The differences in these classes are due primarily to variations in the Fc fragment and its accessory molecules. The classes are differentiated with shorthand names (Ig, followed by a letter: IgG, IgA, IgM, IgD, IgE).

The structure of **IgG** has already been presented. It is a monomer produced by memory cells responding the second time to a given antigenic stimulus. It is by far the most prevalent antibody circulating throughout the tissue fluids and blood. It has numerous

TABLE 15.2

Characteristics of the Immunoglobulin (Ig) Classes

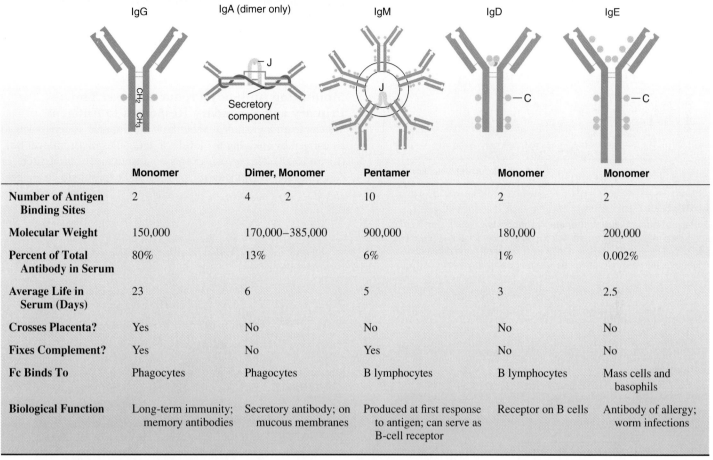

	IgG	IgA (dimer only)	IgM	IgD	IgE
	Monomer	**Dimer, Monomer**	**Pentamer**	**Monomer**	**Monomer**
Number of Antigen Binding Sites	2	4 2	10	2	2
Molecular Weight	150,000	170,000–385,000	900,000	180,000	200,000
Percent of Total Antibody in Serum	80%	13%	6%	1%	0.002%
Average Life in Serum (Days)	23	6	5	3	2.5
Crosses Placenta?	Yes	No	No	No	No
Fixes Complement?	Yes	No	Yes	No	No
Fc Binds To	Phagocytes	Phagocytes	B lymphocytes	B lymphocytes	Mass cells and basophils
Biological Function	Long-term immunity; memory antibodies	Secretory antibody; on mucous membranes	Produced at first response to antigen; can serve as B-cell receptor	Receptor on B cells	Antibody of allergy; worm infections

C = carbohydrate.
J = J chain.

functions: It neutralizes toxins, opsonizes, and fixes complement, and it is the only antibody capable of crossing the placenta.

The two forms of **IgA** are: (1) a monomer that circulates in small amounts in the blood and (2) a dimer that is a significant component of the mucous and serous secretions of the salivary glands, intestine, nasal membrane, breast, lung, and genitourinary tract. The dimer, called **secretory IgA,** is formed in a plasma cell by two monomers attached by a J piece. To facilitate the transport of IgA across membranes, a secretory piece is later added by the gland cells themselves. IgA coats the surface of these membranes and appears free in saliva, tears, colostrum, and mucus. It confers the most important specific local immunity to enteric, respiratory, and genitourinary pathogens. Its contribution in protecting newborns who derive it passively from nursing is mentioned in Medical Microfile 15.4, page 471.

IgM (M for *macro*) is a huge molecule composed of five monomers (making it a pentamer) attached by the Fc receptors to a central J chain. With its 10 binding sites, this molecule has tremendous avidity for antigen (*avidity* means the capacity to bind antigens). It is the first class synthesized by a plasma cell following its first encounter with antigen. Its complement-fixing and opsonizing qualities make it an important antibody in many immune reactions. It circulates mainly in the blood and is far too large to cross the placental barrier.

IgD is a monomer found in miniscule amounts in the serum, and it does not fix complement, opsonize, or cross the placenta. Its main function is to serve as a receptor for antigen on B cells, usually along with IgM. It seems to be the triggering molecule for B-cell activation, and it can also play a role in immune suppression.

IgE is also an uncommon blood component unless one is allergic or has a parasitic worm infection. Its Fc region interacts with receptors on mast cells and basophils. Its biological significance is to stimulate an inflammatory response through the release of potent physiological substances by the basophils and mast cells. Because inflammation would enlist blood cells such as eosinophils and lymphocytes to the site of infection, it would certainly be one defense against parasites. Unfortunately, IgE has another, more insidious effect—that of mediating anaphylaxis, asthma, and certain other allergies (see chapter 17).

Evidence of Antibodies in Serum

Regardless of the site where antibodies are first secreted, a large quantity eventually ends up in the blood by way of the body's communicating networks. If one submits a sample of **antiserum** (serum containing specific antibodies) to electrophoresis, the major groups of proteins migrate in a pattern consistent with their mobility and

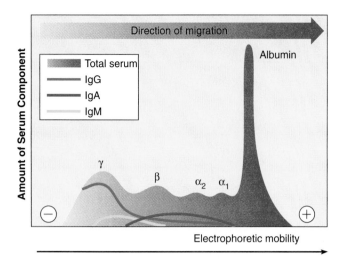

FIGURE 15.17

Pattern of human serum after electrophoresis. When antiserum is subjected to electrical current, the various proteinaceous components are separated into bands. This is a means of separating the different antibodies and serum proteins as well as quantifying them.

size (figure 15.17). The albumins show up in one band, and the globulins in four bands called alpha-1 (α_1), alpha-2 (α_2), beta (β), and gamma (γ) globulins. Most of the globulins represent antibodies, which explains how the term *immunoglobulin* was derived. **Gamma globulin** is composed primarily of IgG, whereas β and α_2 globulins are a mixture of IgG, IgA, and IgM. As we will see in chapter 16, the gamma globulin fraction of serum is important in immune therapies.

Monitoring Antibody Production over Time: Primary and Secondary Responses to Antigens

We can learn a great deal about how the immune system reacts to an antigen by studying the levels of antibodies in serum over time (figure 15.18). This level is expressed quantitatively as the **titer,**[*] or concentration of antibodies. Upon the first exposure to an antigen,

---

[*] titer (ty'-tur) Fr. *titre*, standard. One method for determining titer is shown in figure 16.5.

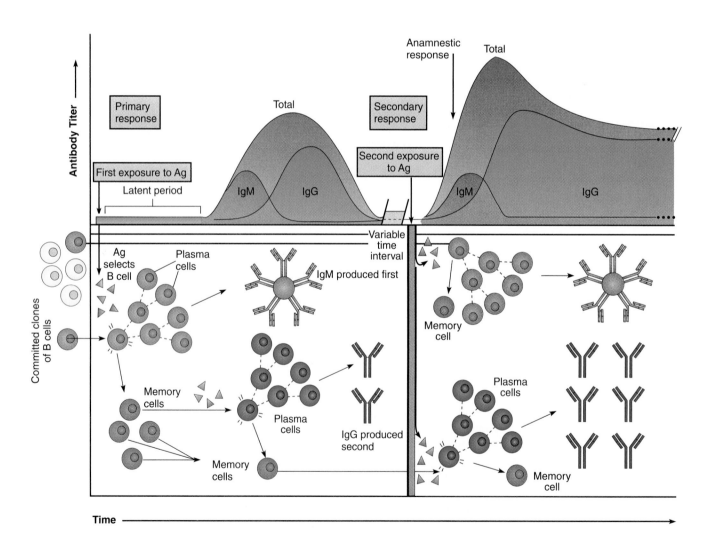

FIGURE 15.18

Primary and secondary responses to antigens. *(Top)* The pattern of antibody titer and subclasses as monitored during initial and subsequent exposure to the same antigen. *(Bottom)* A view of the B-cell responses that account for the pattern. Depicted are clonal selection, clonal expansion, production of memory cells and plasma cells, and the predominant antibody class occurring at first and second contact with antigen (Ag).

the system undergoes a **primary response.** The earliest part of this response, the *latent period,* is marked by a lack of antibodies for that antigen, but much activity is occurring. During this time, the antigen is being concentrated in lymphoid tissue, and is being processed by the correct clones of B lymphocytes. As plasma cells synthesize antibodies, the serum titer increases to a certain plateau and then tapers off to a low level over a few weeks or months. When the class of antibodies produced during this response is tested, an important characteristic of the response is uncovered. It turns out that, early in the primary response, most of the antibodies are the IgM type, which is the first class to be secreted by naive plasma cells. Later, the class of the antibodies (but not their specificity) is switched to IgG or some other class (IgA or IgE).

When the immune system is exposed again to the same immunogen within weeks, months, or even years, a **secondary response** occurs. The rate of antibody synthesis, the peak titer, and the length of antibody persistence are greatly increased over the primary response. The rapidity and amplification seen in this response are attributable to the memory B cells that were formed during the primary response. Because of its association with recall, the secondary response is also called the **anamnestic* response.** The advantage of this response is evident: It provides a quick and potent strike against subsequent exposures to infectious agents. This memory effect forms the basis for giving **boosters**—additional doses of vaccine—to increase the serum titer.

Monoclonal Antibodies: Useful Products from Cancer Cells

The value of antibodies as tools for locating or identifying antigens is well established. For many years, antiserum extracted from human or animal blood was the main source of antibodies for tests and therapy, but most antiserum has a basic problem. It contains **polyclonal antibodies,** meaning that it is a mixture of different antibodies because it reflects dozens of immune reactions from a wide variety of B-cell clones. This characteristic is to be expected, because several immune reactions may be occurring simultaneously, and even a single species of microbe can stimulate several different types of antibodies. Certain applications in immunology require a pure preparation of **monoclonal antibodies (MABs)** that originate from a single clone and have a single specificity for antigen.

The technology for producing monoclonal antibodies is possible by hybridizing cancer cells and plasma cells *in vitro* (figure 15.19). This technique began with the discovery that tumors isolated from multiple *myelomas** in mice consist of identical plasma cells. These monoclonal plasma cells secrete a strikingly pure form of antibodies with a single specificity and continue to divide indefinitely. Immunologists recognized the potential in these plasma cells and devised a **hybridoma** approach to creating MABs. The basic idea behind this approach is to hybridize or fuse a myeloma cell with a normal plasma cell from a mouse spleen to create an immortal cell that secretes a supply of functional antibodies with a single specificity.

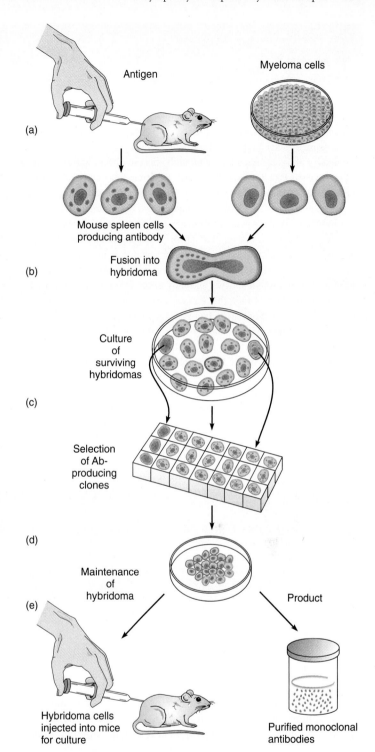

FIGURE 15.19

Summary of the technique for producing monoclonal antibodies by hybridizing myeloma tumor cells with normal plasma cells.   (a) A normal mouse is inoculated with an antigen having the desired specificity, and plasma cells are isolated from its spleen. A special strain of mouse provides the myeloma cells. **(b)** The two cell populations are mixed with polyethylene glycol, which causes some cells in the mixture to fuse and form hybridomas. **(c)** Surviving cells are cultured and separated into individual wells. **(d)** Tests are performed on each hybridoma to determine the specificity of the antibody (Ab) it secretes. **(e)** A hybridoma with the desired specificity is grown in tissue culture; antibody product is then isolated and purified. The hybridoma is maintained in a susceptible mouse for future use.

---

* anamnestic (an-am-ness'-tik) Gr. *anamnesis,* a recalling.

* myeloma (my-uh-loh'-muh) Gr. *myelos,* marrow, and *oma,* tumor. A malignancy of the bone marrow.

MEDICAL MICROFILE   15.2
Monoclonal Antibodies: Variety Without Limit

Imagine releasing millions of tiny homing pigeons into a molecular forest and having them navigate directly to their proper roost, and you have some sense of what monoclonal antibodies can do. Laboratories use them to identify antigens, receptors, and antibodies; to differentiate cell types (T cells versus B cells) and cell subtypes (different sets of T cells); to diagnose diseases such as cancer and AIDS; and to identify bacteria and viruses.

A number of promising techniques have been directed toward using monoclonals as drugs. When genetic engineering is combined with hybridoma technology, the potential for producing antibodies of almost any desired specificity and makeup is possible. For instance, *chimeric* MABs (monoclonal antibodies) that are part human and part mouse have been produced through splicing antibody genes. It is even possible to design an antibody molecule that has antigen binding sites with different specificities. Monoclonals can be hybridized with plant or bacterial toxins to form **immunotoxin** complexes that attach to a target cell and poison it. The most exciting prospect of this therapy is that it can destroy a specified cancer cell and not harm normal cells. Monoclonals are currently employed in treatment of cancers such as breast and colon cancers. Drugs such as Rituxan and herceptin bind to cancer cells and trigger their death.

Such antibodies could also be used to suppress allergies, autoimmunities, and graft rejection. Drugs such as OKT3 and Orthoclone are currently used to prevent the rejection of organ transplants by incapacitating cytotoxic T cells. Now that researchers have genetically engineered plants and mice to produce human MABs, therapeutic uses will continue to expand. MABs are undergoing clinical trials to treat heart, lung, and inflammatory diseases.

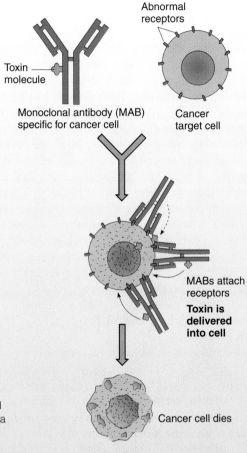

The mechanism of an immunotoxin. A potential therapy for cancer and immune dysfunctions is the use of a monoclonal antibody specific for a certain tumor that carries a potent toxin molecule. The antibody would circulate to cancer cells and deliver the toxin to them. Normal body cells would be unharmed.

The introduction of this technology has the potential for numerous biomedical applications. Monoclonal antibodies have provided immunologists with excellent standardized tools for studying the immune system and for expanding disease diagnosis and treatment. Most of the successful applications thus far use MABs in *in vitro* diagnostic testing and research. Although injecting monoclonal antibodies to treat human disease is an exciting prospect, so far this therapy has been stymied because most MABs are of mouse origin, and many humans will develop hypersensitivity to them. The development of human MABs and other novel approaches using genetic engineering is currently under way (Medical Microfile 15.2).

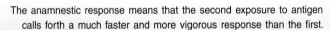

CHAPTER CHECKPOINTS

B cells produce five classes of antibody: IgM, IgG, IgA, IgD, and IgE. IgM and IgG predominate in plasma. IgA predominates in body secretions. IgD binds to B cells as an antigen receptor. IgE binds to tissue cells, promoting inflammation.

Antibodies bind physically to the specific antigen that stimulates their production, thereby immobilizing the antigen and enabling it to be destroyed by other components of the immune system.

The anamnestic response means that the second exposure to antigen calls forth a much faster and more vigorous response than the first.

Monoclonal, or pure, antibodies can be produced commercially by fusing a plasma cell with a myeloma cell to produce an immortal hybridoma.

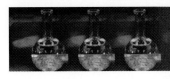

Although the immunities of T cells are usually thought of as cell-mediated, one must not overlook the fact that T cells are also prolific chemical factories. The products of sensitized T cells that communicate with and act upon other cells are a type of cytokine called **lymphokines.** Lymphokines have diverse effects. Some (interferon and interleukin) regulate immune reactions; some mediate inflammation (chemotactic factors); and others (lymphotoxins) kill whole cells. In the previous chapter, we mentioned the role of **gamma interferon** in regulating B cells and T cells. It also activates natural killer cells (NK) and stimulates macrophages. **Interleukin-2** was discussed in conjunction with T helper cells and its role as a growth promoter and general stimulus for lympho-

cyte activity. **Interleukin-3** is a powerful stimulus for stem cell development in the bone marrow. Several other interleukins (IL-4, -5, -6, -7) promote proliferation, differentiation, and secretion of B and T cells.

The mechanisms by which cytotoxic lymphocytes destroy their whole cell targets is just being understood. It appears that they seek out the foreign cell's membrane and secrete various metabolic products into it. The compounds that damage the target cell are termed **lymphotoxins.** These molecular poisons can disrupt the cell membrane by means of specialized proteins called perforins. They also precipitate the lysis of the cell nucleus and initiate cell death (see figure 15.21).

III. T ACTIVATION OF T LYMPHOCYTES AND
V. HOW T CELLS RESPOND TO ANTIGEN: CELL-MEDIATED IMMUNITY (CMI)

During the time that B cells have been actively responding to antigens, the T-cell limb of the system has been similarly engaged. The responses of T cells, however, are **cell-mediated** immunities, which require the direct involvement of T lymphocytes throughout the course of the reaction. These reactions are among the most complex and diverse in the immune system and involve several subsets of T cells whose particular actions are dictated by CD receptors. All mature T cells have CD2 (the cause of rosetting), but **CD4** and **CD8** are found only on certain classes (table 15.3). T cells are restricted; that is, they require some type of MHC (self) recognition before they can be activated, and all produce cytokines with a spectrum of biological effects (Spotlight on Microbiology 15.3).

T cells have notable differences in function from B cells. Rather than making antibodies to control foreign antigens, the whole T cell acts directly in contact with the antigen. They also stimulate other T cells, B cells, and phagocytes.

The Activation of T Cells and Their Differentiation into Subsets

The mature T cells in lymphoid organs are primed to react with antigens that have been processed and presented to them by macrophages. A T cell is initially **sensitized** when antigen is bound to its receptor. By mechanisms not yet fully characterized, sensitization leads to the final differentiation of the cell into one of four functionally specialized subsets: helper, suppressor, cytotoxic, or delayed hypersensitivity T cells (table 15.3 and figure 15.20). As with B cells, activated T cells transform into lymphoblasts in preparation for mitotic divisions, and they divide into one of the subsets of effector cells and memory cells that can interact with the antigen upon subsequent contact. Memory T cells are some of the longest-lived blood cells known (70 years in one well-documented case).

T Helper (T_H) Cells   Helper cells play a central role in assisting with immune reactions to antigens, including those of B cells and other T cells. They do this directly by receptor contact and indirectly by releasing cytokines such as interleukin-2, which stimulates the primary growth and activation of B and T cells, and interleukins-4, -5, and -6, which stimulate various activities of B cells. T helper cells are the most prevalent type of T cell in the blood and lymphoid organs, making up about 65% of this population. The severe depression of this class of T cells (with CD4 receptors) by HIV is what largely accounts for the immunopathology of AIDS.

TABLE 15.3
Characteristics of Subsets of T Cells

Subset	Shorthand Designations	Functions/ Important Features
T helper cells	T_H, T_4	Assist B cells in recognition of antigen; assist other subsets of T cells in recognition and reaction to antigen; identified by CD4 receptors
T suppressor cells	T_S, T_8	Regulate immune reactions; cells limit the extent of antibody production; block some T-cell activity; carry CD5 and CD8 receptors
Cytotoxic (killer) cells	T_C, T_K	Destroy a target foreign cell by lysis; important in destruction of complex microbes, cancer cells, virus-infected cells; graft rejection; allergy; also have CD5 and CD8 receptors
Delayed hypersensitivity cells	T_D, T_{DTH}	Responsible for allergies occurring several hours or days after contact; skin reactions as in tuberculin test

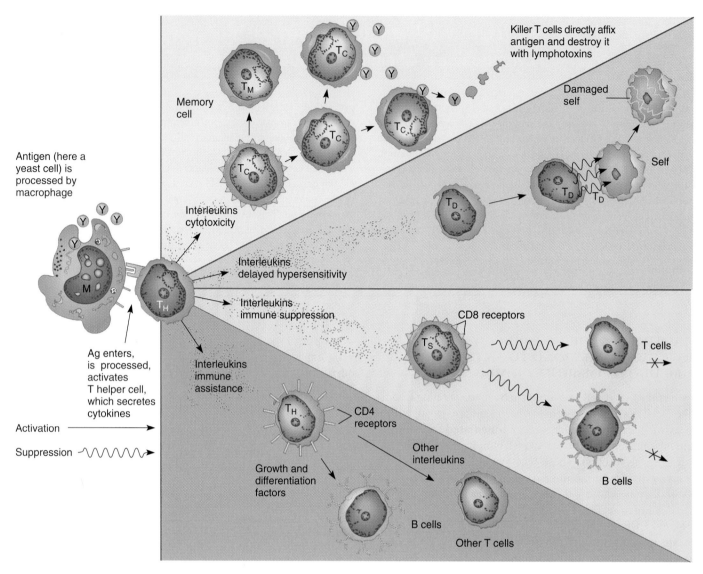

FIGURE 15.20

Overall scheme of T-cell activation and differentiation into different types of T cells. T_C cells destroy certain microbes and foreign cells; T_D cells react with tissues and cause a type of hypersensitivity that damages self; T_S cells inhibit the actions of B and T cells; T_H cells assist in the actions of B and T cells.

T Suppressor (T_S) Cells   Not all T cells are involved in activating immune responses. The T suppressor (T_S) cells actually do just the opposite—they inhibit them. These T cells appear to be a special class of lymphocytes that carry the CD8 receptor and are considered to be primarily regulatory in function. Although the precise mechanisms are not yet understood, T_S cells are known to inhibit the functions of B cells and other T cells. They do this by producing various specific protein inhibitors that keep antigen processing cells and lymphocytes from reacting to antigens. The benefit of this suppressor function is to restrict random immune responses that could become destructive or uncontrolled. T_S cells also appear to play a role in development of tolerance to self and foreign antigens. It seems likely that abnormalities in the suppressor cells are an underlying basis for certain autoimmune diseases and cancers.

Cytotoxic T (T_C) Cells: Cells That Kill Other Cells   Cytotoxicity is the capacity of certain T cells to kill a specific target cell. It is a fascinating and powerful property that accounts for much of our immunity to foreign cells and cancer, and yet, under some circumstances, it can lead to disease. For a *killer T cell* to become activated, it must recognize foreign receptors presented to it, and mount a direct attack upon the target cell. After activation, the T_C cell delivers a dose of several cytokines that severely injures the target cell (figure 15.21). This process involves the formation of pore-forming proteins, or **perforins**,[*] that attack the target cell membrane and create pores through which ions can leak. The loss of selective permeability is followed by target cell death through a process called *apoptosis* (ah-poh-toh´-sis). The apoptosis

---

[*] **perforin** From the term *perforate* or to penetrate with holes.

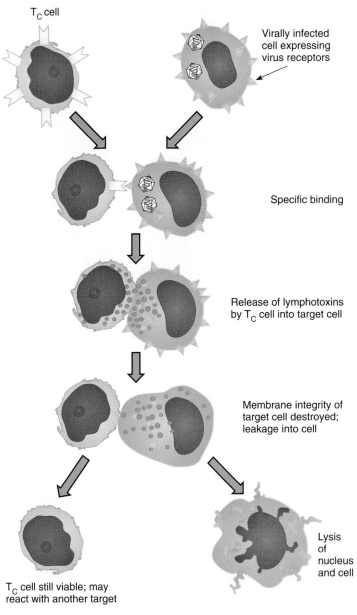

FIGURE 15.21

Stages of cell-mediated cytotoxicity and the action of lymphotoxins on virus-infected target cells.

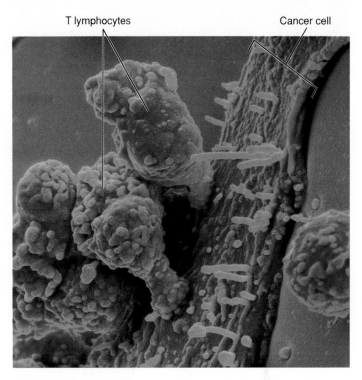

FIGURE 15.22

Cytotoxicity.   A large cancer cell is attacked by killer T cells, which release lymphotoxins directly into the cancer and eventually leave behind a lifeless shell.

is genetically programmed and results in destruction of the nucleus and complete cell lysis.

Target Cells That T_C Cells Can Destroy Include the Following:

- Fungi, protozoans, and complex bacteria (mycobacteria).
- Virally infected cells (figure 15.21). Cytotoxic cells recognize these because of telltale virus receptors expressed on their surface. Cytotoxic defenses are an essential protection against viruses.
- Cancer cells. T cells constantly survey the tissues and immediately attack any abnormal cells they encounter (figure 15.22). The importance of this function is clearly demonstrated in the susceptibility of T-cell-deficient people to cancer (chapter 17).

- Cells from other animals and humans. Cytotoxic CMI is the most important factor in **graft rejection.** In this instance, the T_C cells attack the foreign tissues that have been implanted into a recipient's body.

Other Types of Killer Cells   **Natural killer (NK)** cells are a type of lymphocyte related to T cells that lack specificity for antigens. They circulate through the spleen, blood, and lungs and are probably the first killer cells to attack cancer cells and virus-infected cells. They destroy such cells by similar mechanisms as T cells. Their activities are acutely sensitive to cytokines such as interleukin-12 and α and β interferon.

Delayed Hypersensitivity T (T_D) Cells   Although immediate allergies such as hay fever and anaphylaxis are mediated by antibodies, certain delayed responses to allergens (the tuberculin reactions, for example) are initiated by special T cells. These reactions are discussed in chapter 17.

A PRACTICAL SCHEME FOR CLASSIFYING SPECIFIC IMMUNITIES

The means by which humans acquire immunities can be conveniently encapsulated within four interrelated categories: active, passive, natural, and artificial.

> **Active immunity** occurs when an individual receives an immune stimulus (antigen) that activates the B and T cells, causing the body to produce immune substances

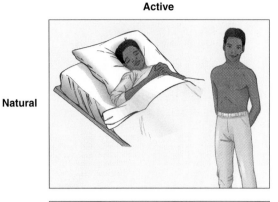
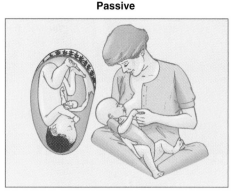

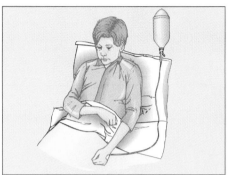

Active **Passive**

Natural

Artificial

FIGURE 15.23

Categories of acquired immunities.   Natural immunities, which occur during the normal course of life, are either active (acquired from an infection and then recovering) or passive (antibodies donated by the mother to her child). Artificial immunities are acquired through medical practices and can be active (vaccinations with antigen, to stimulate an immune response) or passive (immune therapy with a serum containing antibodies).

such as antibodies. Active immunity is marked by several characteristics: (1) It is an essential attribute of an immunocompetent individual; (2) it creates a memory that renders the person ready for quick action upon reexposure to that same antigen; (3) it requires several days to develop; and (4) it lasts for a relatively long time, sometimes for life. Active immunity can be stimulated by natural or artificial means.

Passive immunity occurs when an individual receives immune substances (antibodies) that were produced actively in the body of another human or animal donor. The recipient is protected for a time even though he or she has not had prior exposure to the antigen. It is characterized by: (1) lack of memory for the original antigen, (2) lack of production of new antibodies against that disease, (3) immediate onset of protection, and (4) short-term effectiveness, because antibodies have a limited period of function, and ultimately, the recipient's body disposes of them. Passive immunity can also be natural or artificial in origin.

Natural immunity encompasses any immunity that is acquired during the normal biological experiences of an individual rather than through medical intervention.

Artificial immunity is protection from infection obtained through medical procedures. This type of immunity is induced by immunization with vaccines and immune serum.

Figure 15.23 illustrates the various possible combinations of acquired immunities.

Natural Active Immunities: Getting the Infection

After recovering from infectious disease, a person may be actively resistant to reinfection for a period that varies according to the disease. In the case of childhood viral infections such as measles, mumps, and rubella, this natural active stimulus provides lifelong immunity. Other diseases result in a less extended immunity of a few months to years (such as pneumococcal pneumonia and shigellosis), and reinfection is possible. Even a subclinical infection can stimulate natural active immunity. This probably accounts for the fact that some people are immune to an infectious agent without ever having been noticeably infected with or vaccinated for it.

Natural Passive Immunity: Mother to Child

Natural, passively acquired immunity occurs only as a result of the prenatal and postnatal, mother-child relationship. During fetal life, IgG antibodies circulating in the maternal bloodstream are small enough to pass or be actively transported across the placenta. Antibodies against tetanus, diphtheria, pertussis, and several viruses regularly cross the placenta. This natural mechanism provides an infant with a mixture of many maternal antibodies that can protect it for the first few critical months outside the womb, while its own immune system is gradually developing active immunities. Depending upon the microbe, passive protection lasts anywhere from a few months to a year. But eventually, the

MEDICAL MICROFILE 15.4
Breast Feeding: The Gift of Antibodies

An advertising slogan from the past claims that cow's milk is "nature's most nearly perfect food." One could go a step further and assert that human milk is nature's *perfect* food for young humans. Clearly, it is loaded with essential nutrients, not to mention being available on demand from a readily portable, hygienic container that does not require refrigeration or warming. But there is another and perhaps even greater benefit. During lactation, the breast becomes a site for the proliferation of lymphocytes that produce IgA, a special class of antibody that protects the mucosal surfaces from local invasion by microbes. The very earliest secretion of the breast, a thin, yellow milk called *colostrum,* is very high in IgA. These antibodies form a protective coating in the gastrointestinal tract of a nursing infant that guards against infection by a number of enteric pathogens (*Escherichia coli, Salmonella,* poliovirus, rotavirus). Protection at this level is especially critical because an infant's own IgA and natural intestinal barriers are not yet developed. As with immunity

in utero, the necessary antibodies will be donated only if the mother herself has active immunity to the microbe through a prior infection or vaccination.

In recent times, the ready availability of artificial formulas and the changing life-styles of women have reduced the incidence of breast feeding. Where adequate hygiene and medical care prevail, bottle-fed infants get through the critical period with few problems, because the foods given them are relatively sterile and they have received protection against some childhood infections in utero. Mothers in developing countries with untreated water supplies or poor medical services are strongly discouraged from using prepared formulas, because they can actually inoculate the baby's intestine with pathogens from the formula. Millions of neonates suffer from severe and life-threatening diarrhea that could have been prevented by the hygienic practice of nursing.

infant's body clears the antibody. Most childhood vaccinations are timed so that there is no lapse in protection against common childhood infections.

Another source of natural passive immunity comes to the baby by way of mother's milk (Medical Microfile 15.4). Although the human infant acquires 99% of natural passive immunity in utero and only about 1% through nursing, the milk-borne antibodies provide a special type of intestinal protection that is not forthcoming from transplacental antibodies.

Artificial Immunity: Immunization

Immunization is any clinical process that produces immunity in a subject. Because it is often used to give advance protection against infection, it is also called *immunoprophylaxis.* The use of these terms is sometimes imprecise, thus it should be stressed that active immunization, in which a person is administered antigen, is synonymous with vaccination, and that passive immunization, in which a person is given antibodies, is a type of immune therapy.

Vaccination: Artificial Active Immunization

The term **vaccination** originated from the Latin word *vacca* (cow), because the cowpox virus was used in the first preparation for active immunization against smallpox (see Historical Highlights 16.1). Vaccination exposes a person to a specially prepared microbial (antigenic) stimulus, which then triggers the immune system to produce antibodies and lymphocytes to protect the person upon future exposure to that microbe. As with natural active immunity, the degree and length of protection vary. Commercial vaccines are currently available for about 26 diseases. Methods of vaccine antigen selection and modes of vaccination are discussed more fully in chapter 16.

Immunotherapy: Artificial Passive Immunization

In **immunotherapy,** a patient at risk for acquiring a particular infection is administered a preparation that contains specific antibodies against that infectious agent. In the past, these therapeutic substances were obtained by vaccinating animals (horses in particular), then taking blood and extracting the serum. However, horse serum is now used only in limited situations because of the potential for hypersensitivity to it. Pooled human serum from donor blood (gamma globulin) and **immune serum globulins** containing high quantities of antibodies are more frequently used. Immune serum globulins are used to protect people who have been exposed to hepatitis, measles, and rubella. More specific immune serum, obtained from patients recovering from a recent infection, is useful in preventing and treating hepatitis B, rabies, pertussis, and tetanus.

A final outline summarizing the system of host defenses covered in chapters 14 and 15 was presented in figure 14.1. You will want to use this resource to review major aspects of immunity and to guide you in answering certain questions (see concept question 17).

CHAPTER CHECKPOINTS

T cells do not produce antibodies. Instead, they produce different cytokines that play diverse roles in the immune response. Each subset of T cell produces a particular cytokine that stimulates lymphocytes or destroys foreign cells.

Active immunity means that your body produces antibodies to a disease agent. If you contract the disease, you can develop natural active immunity. If you are vaccinated, your body will produce artificial active immunity.

In passive immunity, you receive antibodies from another person. Natural passive immunity comes from the mother. Artificial passive immunity is administered medically.

CHAPTER CAPSULE WITH KEY TERMS

Specific Immunity[*]

I. **Development of Lymphocyte Specificity/Receptors**
   A. **Acquired immunity** involves the reactions of B and T lymphocytes to foreign molecules, or **antigens.** Before they can react, each lymphocyte must undergo differentiation into its final functional type by developing protein receptors for antigen, the specificity of which is genetically controlled and unique for each type of lymphocyte.
   B. The **clonal selection theory** explains this process. Genetic recombination and mutation during embryonic and fetal development produce billions of different lymphocyte clones, each bearing a different receptor. This provides a huge lymphocyte repertoire required to react with antigens.
   C. **Tolerance to self,** the elimination of any lymphocyte clones that can attack self, occurs during this time.
   D. The receptors on B cells are **immunoglobulin (Ig)** molecules, and receptors on T cells are smaller glycoprotein molecules.
   E. Other receptors needed in recognition are governed by the **major histocompatibility (MHC)** gene complex, which is also referred to as the **human leukocyte antigen (HLA)** complex.
   F. Expression of these genes gives rise to receptors on most cells that govern cell communication and recognition of self and antigens.
   G. *B-Cell Maturation:* Immature B stem cells originate in the yolk sac, liver, and bone marrow and differentiate into mature cells under the influence of special stromal cells in the bone marrow. Mature cells acquire Ig receptors and home to predetermined sites in lymphoid organs, ready to react with antigen.
   H. *T-Cell Maturation:* Immature T stem cells originate in the same areas of the embryo and fetus but mature under the influence of the thymus gland. Specificity is acquired through addition of CD receptors, and mature cells home to different sites in lymphoid organs.

II. **Introduction of Antigens/Immunogens**
   A. An antigen (Ag) is any substance that stimulates an immune response.
      1. Requirements for **antigenicity** include foreignness (recognition as nonself), large size, and complexity of cell or molecule.
      2. Foreign cells and large complex molecules (over 10,000 MW) are most antigenic.
      3. Foreign molecules less than 1,000 MW (**haptens**) are not antigenic unless attached to a larger carrier molecule.
      4. The **antigenic determinant** is the small molecular group of the foreign substance that is recognized by lymphocytes. Cells, viruses, and large molecules can have numerous antigenic determinants.
   B. Special categories of antigens include:
      1. **Autoantigens,** molecules on self tissues for which tolerance is inadequate.
      2. **Alloantigens,** cell surface markers of one individual that are antigens to another of that same species.
      3. **Heterophilic antigens,** molecules from unrelated species that bear similar antigenic determinants.
      4. *Superantigens,* complex bacterial toxins.
      5. **Allergen,** the antigen that provokes allergy.

III. **Cooperation in Immune Reactions to Antigen**
   A. **T-cell-dependent antigens** must be processed by special macrophages, the **antigen-processing cells (APCs).** An APC alters the antigen and attaches it to its MHC receptor for presentation to lymphocytes.
   B. Antigen presentation involves a direct collaboration among the macrophage, a T helper (T_H) cell, and an antigen-specific B or T cell.
   C. Cytokines involved are **Interleukin-1** from the APC which activates the T_H cells, and **interleukin-2** produced by the T_H cell, which activates B and other T cells.

IV. **B-Cell Activation and Antibody Production**
   A. When B cells receive the antigen and are stimulated by B-cell growth and differentiation factors, they enter the cell cycle in preparation for mitosis and **clonal expansion.**
   B. Divisions give rise to **plasma cells** that secrete antibodies and **memory cells** that can react to that same antigen later.
      1. *Nature of Antibodies* (Immunoglobulins): A single immunoglobulin molecule (monomer) is a large Y-shaped protein molecule consisting of four polypeptide chains. It contains two identical fragments (**Fab**) with ends that form the active site that binds with a unique specificity to an antigen. The single fragment of the antibody (**Fc**) binds to self.
      2. *Antigen-Antibody (Ag-Ab) Reactions* include **opsonization, neutralization** by **antitoxin, agglutination,** and complement fixation.
         a. The Fc portion can bind to various body cells and mediate inflammation and allergy.
      3. The five **antibody classes,** which differ in size and function, are **IgG, IgA** (secretory Ab), **IgM, IgD,** and **IgE.**
      4. *Antibodies in Serum (Antiserum):*
         a. Serum antibodies can be identified through electrophoresis and quantified by testing the **titer** (levels of antibodies) over time.
         b. The first introduction of an Ag to the immune system produces a **primary response,** with a gradual increase in Ab titer.
         c. The second contact with the same Ag produces a **secondary,** or **anamnestic, response,** due to memory cells produced during initial response.
      5. **Monoclonal antibodies (MAB)** are single specificity antibodies formed by fusing a mouse B cell with a cancer cell. MABs are used in diagnosis of disease, identification of microbes, and therapy.

V. **T Cells and Cell-Mediated Immunity (CMI)**
   A. T cells act directly against antigens and foreign cells. T cells secrete cytokines (**interleukin, interferon, lymphotoxins**) that act on other cells. *Sensitized T cells* proliferate into long-lasting memory T cells and one of the following depending upon its **CD** receptor:
      1. **T helper cells** (T_H) assist other T cells and B cells.
      2. **T suppressor cells** (T_S) limit the actions of other T cells and B cells.
      3. **Cytotoxic,** or **killer, T cells** (T_C) destroy complex foreign or abnormal cells by forming **perforins** that lyse the cell.
      4. **Delayed hypersensitivity cells** (T_D) cause a form of hypersensitivity.

---

[*] Also review figure 15.1.

VI. **Classification of Acquired Immunity**
   A. Immunities acquired through B and T lymphocytes can be classified by a simple system.
      1. **Natural immunity** is acquired as part of normal life experiences.
      2. **Artificial immunity** is acquired through medical procedures such as **immunization.**
      3. **Active immunity** results when a person is challenged with antigen that stimulates production of antibodies. It creates memory, takes time, and is lasting.
      4. In **passive immunity,** preformed antibodies are donated to an individual. It does not create memory, acts immediately, and is short term.
   B. Combinations of acquired immunity are:
      1. **Natural active,** acquired upon infection and recovery, and **natural passive,** acquired by a child through placenta and breast milk.
      2. **Artificial active (vaccination)**, acquired through inoculation with a selected antigen, and **artificial passive,** administration of **immune serum** or globulin.

MULTIPLE-CHOICE QUESTIONS

1. The primary B-cell receptor is
   a. IgD
   b. IgA
   c. IgE
   d. IgG

2. In humans, B cells mature in the ____, and T cells mature in the ____.
   a. GALT, liver
   b. bursa, thymus
   c. bone marrow, thymus
   d. lymph nodes, spleen

3. Small, simple molecules are ____ antigens.
   a. poor
   b. never
   c. good
   d. heterophilic

4. An example of a mosaic antigen is
   a. albumin
   b. a lipopolysaccharide molecule
   c. a virus
   d. a hapten

5. Which type of cell actually secretes antibodies?
   a. T cells
   b. macrophages
   c. plasma cells
   d. monocytes

6. The cross-linkage of antigens by antibodies is known as
   a. opsonization
   b. a cross-reaction
   c. agglutination
   d. complement fixation

7. The greatest concentration of antibodies is found in the ____ fraction of the serum.
   a. gamma globulin
   b. albumin
   c. beta globulin
   d. alpha globulin

8. ____ is a measurement of the relative amount of immunoglobulin in the serum.
   a. Gamma globulin
   b. Secretory antibody
   c. Antibody titer
   d. Complement fixation

9. A cytokine that stimulates the activity of B and T cells is
   a. interleukin-2
   b. opsonin
   c. lymphotoxin
   d. interleukin-1

10. ____ T cells assist in the functions of certain B cells and other T cells.
   a. Sensitized
   b. Cytotoxic
   c. Helper
   d. Natural killer

11. T_C cells are important in controlling
   a. virus infections
   b. allergy
   c. autoimmunity
   d. all of these

12. Vaccination is synonymous with ____ immunity.
   a. natural active
   b. artificial passive
   c. artificial active
   d. natural passive

13. Fusion between a plasma cell and a tumor cell creates a
   a. lymphoblast
   b. hybridoma
   c. natural killer cell
   d. myeloma

14. T cells are the source of which cytokines?
   a. interleukin
   b. interferon
   c. lymphotoxin
   d. a and b
   e. all of the choices

15. **Multiple matching.** Place all possible matches in the space at the left.

   ____ IgG   a. Found in mucous secretions
   ____ IgA   b. A monomer
   ____ IgD   c. A dimer
   ____ IgE   d. Has greatest number of Fabs
   ____ IgM   e. Is primarily a surface receptor for B cells
                     f. Major Ig of primary response to Ag
                     g. Major Ig of secondary response to Ag
                     h. Crosses the placenta
                     i. Fixes complement
                     j. Involved in allergic reactions

CONCEPT QUESTIONS

1. a. What function do receptors play in specific immune responses?
   b. How can receptors be made to vary so widely?

2. Describe the major histocompatibility complex, and explain how it participates in immune reactions.

3. a. Evaluate the following statement: Each different lymphocyte type must have a unique receptor to react with antigen.
   b. How many different Ags might one be expected to meet up with during life?

4. a. What constitutes a clone of lymphocytes?
   b. Explain the clonal selection theory of antibody specificity and diversity.
   c. During development, when is antigen not needed, and why is it not needed?
   d. When is antigen needed?
   e. Why must the body develop tolerance to self?

5. a. Trace the development of the B-cell receptor from gene to cell surface.
   b. What is the structure of the receptor?
   c. What is the function of the variable regions?

6. a. Trace the origin and development of B lymphocytes; of T lymphocytes.
   b. What is happening during lymphocyte maturation?

7. Describe three ways that B cells and T cells are similar and at least five major ways in which they are different.

8. a. What is an antigen or immunogen?
   b. What is the antigenic determinant?
   c. How do foreignness, size, and complexity contribute to antigenicity?
   d. What is a mosaic antigen?
   e. Why are haptens by themselves not antigenic, even though foreign?
   f. How can they be made to behave as antigens?

9. a. Differentiate among autoantigens, alloantigens, and heterophile antigens.
   b. Explain briefly what importance each has in immune reactions.

10. a. Describe the actions of an antigen-processing cell.
    b. What is the difference between a T-cell-dependent and T-cell-independent response?

11. a. Trace the immune response system, beginning with the entry of a T-cell-dependent antigen, antigen processing, presentation, the cooperative response among the macrophage and lymphocytes, and the reactions of activated B and T cells.
    b. What are the actions of interleukins-1 and -2?

12. a. On what basis is a particular B-cell clone selected?
    b. How are B cells activated, and what events are involved in this process?
    c. What happens after B cells are activated?
    d. What are the functions of plasma cells, clonal expansion, and memory cells?

13. a. Describe the structure of immunoglobulin.
    b. What are the functions of the Fab and Fc portions?
    c. Describe four or five ways that antibodies function in immunity.
    d. Describe the attachment of Abs to Ags. (What eventually happens to the Ags?)

14. a. Contrast the primary and secondary response to Ag.
    b. Explain the type, order of appearance, and amount of immunoglobulin in each response and the reasons for them.
    c. What causes the latent period? The anamnestic response?
    d. Explain how monoclonal and polyclonal antibodies are different.
    e. Outline the basic steps in production of monoclonal antibodies.
    f. Describe several possible applications of monoclonals in medicine.

15. a. Why are the immunities involving T cells called cell-mediated?
    b. How do T cells become sensitized?
    c. Summarize the function of each category of T cell and the types of receptors with which they are associated. Define cytokines, and provide some examples of them.
    d. How do cytotoxic cells kill their target?
    e. Why would the immune system naturally require suppression?
    f. What is a natural killer cell, and what are its functions?

16. a. Contrast active and passive immunity in terms of how each is acquired, how long it lasts, whether memory is triggered, how soon it becomes effective, and what immune cells and substances are involved.
    b. Name at least two major ways that natural and artificial immunities are different.

17. **Multiple matching.** (Summarizes information from chapters 14 and 15 [see figure 14.1].) In the blanks on the left, place the letters of all of the host defenses and immune responses in the right column that can fit the description.

____ vaccination for tetanus	a. active
____ lysozyme in tears	b. passive
____ immunization with horse serum	c. natural
____ in utero transfer of antibodies	d. artificial
____ booster injection for diphtheria	e. acquired
____ recovery from a case of mumps	f. innate, inborn
____ colostrum	g. chemical barrier
____ interferon	h. mechanical barrier
____ action of neutrophils	i. genetic barrier
____ injection of gamma globulin	j. specific
____ recovery from a case of mumps	k. nonspecific
____ edema	l. inflammatory response
____ humans having protection from canine distemper virus	m. second line of defense
____ stomach acid	n. none of these
____ cilia in trachea	
____ asymptomatic chickenpox	
____ complement	

CRITICAL-THINKING QUESTIONS

1. What is the advantage of having lymphatic organs screen the body fluids, directly and indirectly?

2. Cells contain built-in suicide genes to self-destruct by apoptosis under certain conditions. Can you explain why development of the immune system might depend in part on this sort of adaptation?

3. Double-stranded DNA is a large, complex molecule, but it is not generally immunogenic unless it is associated with proteins or carbohydrates. Can you think why this might be so? (Hint: How universal is DNA?)

4. How would you go about producing monoclonal antibodies that would participate in the destruction of cancer cells but would not kill normal human cells?

5. Explain how it is possible for people to give a false positive reaction in blood tests for syphilis, AIDS, and infectious mononucleosis.

6. Describe the cellular/microscopic pathology in the immune system of AIDS patients that results in opportunistic infections and cancers.

7. Explain why most immune reactions result in a polyclonal collection of antibodies.

8. a. Why does rising titer of Abs indicate infection?
   b. Why do some vaccinations require three or four boosters?

9. a. What event must occur for passive immunity to exist at all?
   b. Armed with this knowledge, suggest an effective way to immunize a fetus.
   c. Is this method uniformly safe?
   d. What would be an even safer way to ensure that fetuses get necessary antibodies?

10. a. Combine information on the functions of different classes of Ig to explain the exact mechanisms of natural passive immunity (both transplacental and colostrum-induced).
    b. Why are these sorts of immunity short-term?

11. Using the pattern along the lines of Microbits 15.1, develop an analogy that compares the clonal selection theory of antibody diversity and specificity to buying clothes right off the rack versus having them tailor-made.

12. Using words and arrows, complete a flow outline of an immune response, beginning with entrance of antigen; include processing, cell interaction, involvement of cytokines, and the end results for B and T cells.

INTERNET SEARCH TOPICS

1. Access information on lymphokine-activated killer cells (LAK) and tumor-infiltrating leukocytes (TIL) as clinical therapy approaches for cancer. Describe the therapy and its effectiveness for various cancers.

2. Look up apoptosis; find three or four papers that feature research on this topic. What are some potential applications of the concept of programmed cell death? What causes the cells to die?

CHAPTER 1

Immunization and Immune Assays

An expanded knowledge of immune function has yielded significant breakthroughs in medical technology for manipulating and monitoring the immune system. One of the most practical benefits of immunology is administering vaccines and other immune treatments against common infectious diseases such as hepatitis and tetanus that once caused untold sickness and death. Another valuable application of this technology is testing the blood for signs of infection or disease. It is routine medical practice to diagnose such infections as HIV, syphilis, hepatitis B, and rubella by means of immunologic analysis. Both separately and in combination, these methods have made sweeping contributions to individual and community health by improving diagnosis and treatment, and controlling the spread of disease. Many hopes for the survival of humankind lie in our ability to harness the amazing workings of the immune system.

Chapter Overview

- Discoveries in basic immune function have created powerful medical tools to artificially induce protective immunities and to determine the status of the immune system.
- Immunization may be administered by means of passive and active methods.
- Passive immunity is acquired by infusing antiserum taken from other patients' blood that contains high levels of protective antibodies. This form of immunotherapy is short-lived.
- Active methods involve administering a vaccine against an infectious agent that may be encountered in the future. This form of protection provides a longer-lived immunity.
- Vaccines contain some form of microbial antigen that has been altered so that it stimulates a protective immune response without causing the disease.
- Vaccines can be made from whole dead or live cells and viruses, parts of cells or viruses, or by recombinant DNA techniques.
- Reactions between antibodies and antigens provide specific and sensitive tests that can be used in diagnosis of disease and identification of pathogens.
- Serology involves the testing of a patient's blood serum for antibodies that can indicate a current or past infection and the degree of immunity.
- Tests that produce visible interactions of antibodies and antigens include agglutination, precipitation, and complement fixation.

Detail from "The Cowpock," an 1808 etching that caricatured the worst fears of the English public concerning Edward Jenner's smallpox vaccine.

- Assays can be used to separate antigens and antibodies and visualize them with radioactivity or fluorescence. These include immunoelectrophoresis, the Western blot, and direct and indirect immunoassays.

Practical Applications of Immunologic Function

A knowledge of the immune system and its responses to antigens has provided extremely valuable biomedical applications in two major areas: (1) use of antiserum and vaccination to provide artificial protection against disease and (2) diagnosis of disease through immunologic testing.

The basic notion of immunization has existed for thousands of years. It probably stemmed from the observation that persons who had recovered from certain communicable diseases rarely if ever got a second case. Undoubtedly, the earliest crude attempts involved bringing a susceptible person into contact with a diseased person or animal. The first recorded attempt at immunization occurred in sixth century China. It consisted of drying and grinding up smallpox scabs and blowing them with a straw into the nostrils of vulnerable family members. By the tenth century, this practice had changed to the deliberate inoculation of dried pus from the smallpox pustules of one patient into the arm of a healthy person, a technique later called **variolation** (variola is the smallpox virus). This method was used in parts of the Far East for centuries before Lady Mary Montagu brought it to England in 1721. Although the principles of the technique had some merit, unfortunately, many recipients and their contacts died of smallpox. This outcome vividly demonstrates a cardinal rule for a workable vaccine: It must contain an antigen that will provide protection but not cause the disease. Variolation was so controversial that any English practitioner caught doing it was charged with a felony.

Eventually, this human experimentation paved the way for the first really effective vaccine, developed by the English physician Edward Jenner in 1796 (see chapter 24). Jenner conducted the first scientifically controlled study, one that had a tremendous impact on the advance of medicine. His work gave rise to the words **vaccine** and **vaccination** (from L., *vacca*, cow), which now apply to any immunity obtained by inoculation with selected antigens. Jenner was inspired by the case of a dairymaid who had been infected by a pustular infection called cowpox. This is a related virus that afflicts cattle but causes a milder condition in humans. She explained that she and other milkmaids had remained free of smallpox. Other residents of the region expressed a similar confidence in the cross-protection of cowpox. To test the effectiveness of this new vaccine, Jenner prepared material from human cowpox lesions and inoculated a young boy. When challenged 2 months later with an injection of crusts from a smallpox patient, the boy proved immune.

Jenner's discovery—that a less pathogenic agent could confer protection against a more pathogenic one—is especially remarkable in view of the fact that microscopy was still in its infancy and the nature of viruses was unknown. At first, the use of the vaccine was regarded with some fear and skepticism (see chapter opening illustration). When Jenner's method proved successful and word of its significance spread, it was eventually adopted in many other countries. Eventually, the original virus mutated into a unique strain (*vaccina* virus) that was the modern basis of the vaccine. Now that smallpox is no longer a threat, smallpox vaccination has been essentially discontinued in most areas.

Other historical developments in vaccination included using heat-killed bacteria in vaccines for typhoid fever, cholera, and plague and techniques for using neutralized toxins for diphtheria and tetanus. Throughout the history of vaccination, there have been vocal opponents and minimizers, but numbers do not lie. Whenever a vaccine has been introduced, the prevalence of that disease has declined.

IMMUNIZATION: METHODS OF MANIPULATING IMMUNITY FOR THERAPEUTIC PURPOSES

The concept of artificially induced immunity was introduced in chapters 14 and 15. Methods that actively or passively immunize people are widely used in disease prevention and treatment. In the case of passive immunization, a patient is given preformed antibodies, which is actually a form of **immunotherapy.** In the case of active immunization, a patient is vaccinated with a microbe or its antigens, providing a form of advance protection.

Immunotherapy: Artificial Passive Immunity

The first attempts at passive immunization involved the transfusion of horse serum containing antitoxins to prevent tetanus and to treat patients exposed to diphtheria. Since then, antisera from animals have been replaced with products of human origin that function with various degrees of specificity. **Immune serum globulin (ISG),** sometimes called *gamma globulin,* contains immunoglobulin extracted from the pooled blood of at least 1,000 human donors. The method of processing ISG concentrates the antibodies to increase potency and eliminates potential pathogens (such as the hepatitis B and HIV viruses). It is a treatment of choice in preventing measles and hepatitis A and in replacing antibodies in immunodeficient patients. Most forms of ISG are injected intramuscularly to minimize adverse reactions, and the protection it provides lasts 2 to 3 months.

A preparation called **specific immune globulin (SIG)** is derived from a more defined group of donors. Companies that prepare SIG obtain serum from patients who are convalescing and in a hyperimmune state after such infections as pertussis, rabies, tetanus, chickenpox, and hepatitis B. These globulins are preferable to ISG because they contain higher titers of specific antibodies obtained from a smaller pool of patients. Although useful for prophylaxis in persons who have been exposed or may be exposed to infectious agents, these sera are often limited in availability.

When a human immune globulin is not available, antisera and antitoxins of animal origin can be used. Sera produced in horses are available for diphtheria, botulism, and spider and snake bites. Unfortunately, the presence of horse antigens can stimulate allergies such as serum sickness or anaphylaxis (see chapter 17). Although donated immunities only last a relatively short time, they act immediately and can protect patients for whom no other useful medication or vaccine exists.

Artificial Active Immunity: Vaccination

Active immunity can be conferred artificially by **vaccination**—exposing a person to material that is antigenic but not pathogenic. The discovery of vaccination was one of the farthest reaching and most important developments in medical science (Historical Highlights 16.1). The basic principle behind vaccination is to stimulate a primary and secondary anamnestic response (see figure 15.18) that primes the immune system for future exposure to a virulent pathogen. If this pathogen enters the body, the immune response will be immediate, powerful, and sustained.

Vaccines have profoundly reduced the prevalence and impact of many infectious diseases that were once common and often deadly. In this section, we survey the principles of vaccine preparation and important considerations surrounding vaccine indication and safety. (Vaccines are also given specific consideration in later chapters on bacterial and viral diseases.)

Principles of Vaccine Preparation A vaccine must be considered from the standpoints of antigen selection, effectiveness, ease in administration, safety, and cost. In natural immunity, an infectious agent stimulates appropriate B and T lymphocytes and creates memory clones. In artificial active immunity, the objective is to obtain this same response with a modified version of the microbe or its components. A safe and effective vaccine should mimic the natural protective response, not cause a serious infection or other disease, have long-lasting effects in a few doses, and be easy to administer. Most vaccine preparations contain one of the following antigenic stimulants (figure 16.1): (1) killed whole cells or inactivated viruses, (2) live, attenuated cells or viruses, (3) antigenic components of cells or viruses, or (4) genetically engineered microbes or microbial antigens. A survey of the major licensed vaccines and their indications is presented in table 16.1.

Large, complex antigens such as whole cells or viruses are very effective immunogens. Depending on the vaccine, these are either killed or attenuated. **Killed** or **inactivated vaccines** are prepared by cultivating the desired strain or strains of a bacterium or virus and treating them with formalin, radiation, heat, or some other agent that does not destroy antigenicity. One type of vaccine for the bacterial disease typhoid fever is of this type (see chapter 20). Salk polio vaccine and influenza vaccine contain inactivated viruses. Because the microbe does not multiply, killed vaccines often require a larger dose and more boosters to be effective.

A number of vaccines are prepared from **live, attenuated*** microbes. **Attenuation** is any process that substantially lessens or negates the virulence of viruses or bacteria. It is usually achieved by modifying the growth conditions or manipulating microbial genes in a way that eliminates virulence factors. Attenuation methods include long-term cultivation, selection of mutant strains that grow at colder temperatures (cold mutants), passage of the microbe through unnatural hosts or tissue culture, and removal of virulence genes. The vaccine for tuberculosis (BCG) was obtained after 13 years of subculturing the agent of bovine tuberculosis (see chapter 19). Vaccines for measles, mumps, polio (Sabin), and rubella contain live, nonvirulent viruses. The advantages that favor live preparations are: (1) Viable microorganisms can multiply and produce infection (but not disease) like the natural organism; (2) they confer long-lasting protection; and (3) they usually require fewer doses and boosters than other types of vaccines. Disadvantages of using

(a) Whole-Cell Vaccines

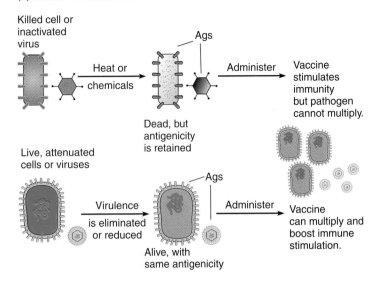

Killed cell or inactivated virus

Heat or chemicals → Dead, but antigenicity is retained — Ags

Administer → Vaccine stimulates immunity but pathogen cannot multiply.

Live, attenuated cells or viruses

Virulence is eliminated or reduced → Alive, with same antigenicity — Ags

Administer → Vaccine can multiply and boost immune stimulation.

(b) Acellular or Subunit Vaccine

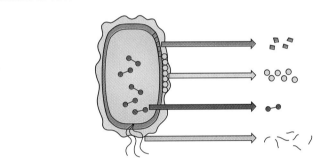

(c) Recombinant Vaccine

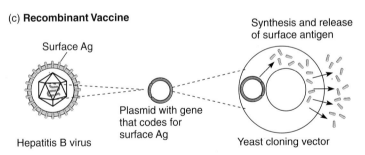

Surface Ag

Hepatitis B virus

Plasmid with gene that codes for surface Ag

Yeast cloning vector

Synthesis and release of surface antigen

FIGURE 16.1

Strategies in vaccine design. **(a)** Whole cells or viruses, killed or attenuated. **(b)** Acellular or subunit vaccines are made by disrupting the microbe to release various molecules or cell parts that can be isolated and purified. **(c)** Recombinant vaccines are made by isolating a gene for antigenicity from the pathogen (here a hepatitis virus) and splicing it into a plasmid. Insertion of the recombinant plasmid into a cloning host (yeast) results in the production of large amounts of viral surface antigen to use in vaccine preparation.

live microbes in vaccines are that they require special storage facilities, can be transmitted to other people, and can mutate back to a virulent strain (see polio, chapter 25).

If the exact antigenic determinants that stimulate immunity are known, it is possible to produce a vaccine based on a selected component of a microorganism. These vaccines for bacteria are called **acellular** or **subcellular vaccines.** For viruses, they are

TABLE 16.1

Currently Approved Vaccines

Disease/Preparation	Route of Administration	Recommended Usage/Comments
Contain Killed Whole Bacteria		
Cholera	Subcutaneous (SQ) injection	For travelers; effect not long-term
Typhoid	Intramuscular (IM)	For travelers only; efficacy variable
Plague	SQ	For exposed individuals and animal workers; variable protection
Contain Live, Attenuated Bacteria		
Tuberculosis (BCG)	Intradermal (ID) injection	For high-risk occupations only; protection variable
Acellular Vaccines (Capsular Polysaccharides)		
Meningitis (meningococcal)	SQ	For protection in high-risk infants, military recruits; short duration
Meningitis (*Haemophilus influenzae*)	IM	For infants and children; may be administered with DTaP
Pneumococcal pneumonia	IM or SQ	Important for people at high risk: the young, elderly, and immunocompromised; moderate protection
Pertussis (aP)	IM	For newborns and children; contains recombinant protein antigens
Toxoids (Formaldehyde-Inactivated Bacterial Exotoxins)		
Diphtheria	IM	A routine childhood vaccination; highly effective in systemic protection
Tetanus	IM	A routine childhood vaccination; highly effective
Botulism	IM	Only for exposed individuals such as laboratory personnel
Contain Inactivated Whole Viruses		
Poliomyelitis (Salk)	IM	Routine childhood vaccine; now used as first choice
Rabies	IM	For victims of animal bites or otherwise exposed; effective
Influenza	IM	For high-risk populations; requires constant updating for new strains; immunity not durable
Hepatitis A	IM	Protection for travelers, institutionalized people
Contain Live, Attenuated Viruses		
Adenovirus infection	Oral	For immunizing military recruits
Measles (rubeola)	SQ	Routine childhood vaccine; very effective
Mumps (parotitis)	SQ	Routine childhood vaccine; very effective
Poliomyelitis	Oral	Routine childhood vaccine; very effective, but can cause polio
Rubella	SQ	Routine childhood vaccine; very effective
Chickenpox (varicella)	SQ	Routine childhood vaccine; immunity can diminish over time
Yellow fever	SQ	Travelers, military personnel in endemic areas
Subunit Viral Vaccines		
Hepatitis B	IM	Recommended for all children, starting at birth; also for health workers and others at risk
Influenza	IM	See influenza above
Recombinant Vaccines		
Hepatitis B	IM	Used more often than subunit, but for same groups
Pertussis	IM	See acellular above
Lyme Disease (LymeRIX)	IM	Made from the surface protein; used for persons with increased risk of exposure to ticks

called **subunit vaccines.** The antigen used in these vaccines may be taken from cultures of the microbes, produced by rDNA technology, or synthesized chemically.

Examples of component antigens currently in use are the capsules of the pneumococcus and meningococcus, the protein surface antigen of anthrax, and the surface proteins of hepatitis B virus. A special type of vaccine is the **toxoid,*** which consists of a purified bacterial exotoxin that has been chemically denatured. By

eliciting the production of antitoxins that can neutralize the natural toxin, toxoid vaccines provide protection against toxinoses such as diphtheria and tetanus.

New Vaccine Strategies

Despite considerable successes, dozens of bacterial, viral, protozoan, and fungal diseases still remain without a functional vaccine. Of all of the challenges facing vaccine specialists, probably the most difficult has been choosing a vaccine antigen that is safe and that properly stimulates immunity. Currently, much attention is

---

***toxoid** (tawks´-oyd) Toxinlike.

being focused on newer strategies for vaccine preparation that employ antigen synthesis, recombinant DNA, and gene cloning technology.

When the exact composition of an antigenic determinant is known, it is possible to synthesize it. This ability permits preservation of antigenicity while greatly increasing antigen purity and concentration. The malaria vaccine currently being used in areas of South America and Africa is composed of three synthetic peptides from the parasite. Several biotechnology companies are exploring the possibility of using plants to synthesize microbial proteins, potentially leading to mass production of edible vaccine antigens.

Some of the genetic engineering concepts introduced in chapter 10 offer novel approaches to vaccine development. These methods are particularly effective in designing vaccines for obligate parasites that are difficult or expensive to culture, such as the syphilis spirochete or the malaria parasite. This technology provides a means of isolating the genes that encode various microbial antigens, inserting them into plasmid vectors, and cloning them in appropriate hosts. The outcome of recombination can be varied as desired. For instance, the cloning host can be stimulated to synthesize and secrete a protein product (antigen), which is then harvested and purified (figure 16.1c). This is how certain vaccines for hepatitis B rotavirus and Lyme disease are prepared. Antigens from the agents of syphilis, *Schistosoma,* and influenza have been similarly isolated and cloned and are currently being considered as potential vaccine material.

Another ingenious technique using genetic recombination has been nicknamed the *Trojan horse* vaccine. The term derives from an ancient legend in which the Greeks sneaked soldiers into the fortress of their Trojan enemies by hiding them inside a large, mobile wooden horse. In the microbial equivalent, genetic material from a selected infectious agent is inserted into a live carrier microbe that is nonpathogenic. In theory, the recombinant microbe will multiply and express the foreign genes, and the vaccine recipient will be immunized against the microbial antigens. Vaccinia, the virus originally used to vaccinate for smallpox, and adenoviruses have proved practical agents for this technique. Vaccinia is used as the carrier in one of the experimental vaccines for AIDS (see figure 25.19), herpes simplex 2, leprosy, and tuberculosis.

DNA vaccines are being hailed as the most promising of all of the newer approaches to immunization. The technique in these formulations is very similar to gene therapy as described in figure 10.13, except in this case, microbial (not human) DNA is inserted into a plasmid vector and inoculated into a recipient (figure 16.2a). The expectation is that the human cells will take up some of the plasmids and express the microbial DNA in the form of proteins. Because these proteins are foreign, they will be recognized during immune surveillance and cause B and T cells to be sensitized and form memory cells.

Experiments with animals have shown that these vaccines are very safe and that only a small amount of the foreign antigen need be expressed to produce effective immunity. Another advantage to this method is that any number of potential microbial proteins can be expressed, making the antigenic stimulus more complex and improving the likelihood that it will stimulate both antibody and cell-mediated immunity. At the present time, over 30 DNA-based vaccines are being tested in animals. Vaccines for Lyme disease, hepatitis C, herpes simplex, influenza, tuberculosis, papillomavirus, and malaria are undergoing animal trials, most with encouraging results.

Vaccine effectiveness relies, in part, on the production of antibodies that closely fit the natural antigen. Realizing that such reactions are very much like a molecular jigsaw puzzle, researchers have proposed an entirely new concept in vaccines. The *anti-idiotype vaccine* is based on the principle that the antigen binding (variable) region, or *idiotype,** of a given antibody (A) can be antigenic to a genetically different recipient and can cause that recipient's immune system to produce antibodies (B; also called anti-idiotypic antibodies) specific for the variable region on antibody A (figure 16.2b). The purpose for making anti-idiotypic antibodies is that they will display an identical configuration as the desired antigen and can be used in vaccines. This method avoids administering a microbial antigen, thus reducing the potential for dangerous side effects. Using monoclonal antibodies, this approach has been used to mimic the surface antigen of hepatitis B virus and *Trypanosoma* with some success.

Route of Administration and Side Effects of Vaccines

Most vaccines are injected by subcutaneous, intramuscular, or intradermal routes. Oral vaccines are available for only two diseases (table 16.1), but they have some distinct advantages. An oral dose of a vaccine can stimulate protection (IgA) on the mucous membrane of the portal of entry. Oral vaccines are also easier to give, more readily accepted, and well tolerated. Some vaccines require the addition of a special binding substance, or *adjuvant.** An adjuvant is any compound that enhances immunogenicity and prolongs antigen retention at the injection site. The adjuvant precipitates the antigen and holds it in the tissues so that it will be released gradually. Its gradual release, presumably, facilitates contact with macrophages and lymphocytes. Common adjuvants are alum (aluminum hydroxide salts), Freund's adjuvant (emulsion of mineral oil, water, and extracts of mycobacteria), and beeswax.

Vaccines must go through many years of trials in experimental animals and humans before they are licensed for general use. Even after they have been approved, like all therapeutic products, they are not without complications. The most common of these are local reactions at the injection site, fever, allergies, and other adverse reactions. Relatively rare reactions (about 1 case out of 220,000 vaccinations) are panencephalitis (from measles vaccine), back-mutation to a virulent strain (from polio vaccine), disease due to contamination with dangerous viruses or chemicals, and neurological effects of unknown cause (from pertussis and swine flu vaccines). Some patients experience allergic reactions to the medium (eggs or tissue culture) rather than to vaccine antigens. Some recent studies have attempted to link childhood vaccinations to later development of diabetes, asthma, and autism. After thorough examination of records, epidemiologists have found no convincing evidence for a connection to these diseases.

---

*idiotype (id´-ee-oh-type) Gr. *idios,* own, peculiar. Another term for the antigen binding site.

*adjuvant (ad´-joo-vunt) L. *adjuvare,* to help.

(a) **Technology for DNA vaccines**

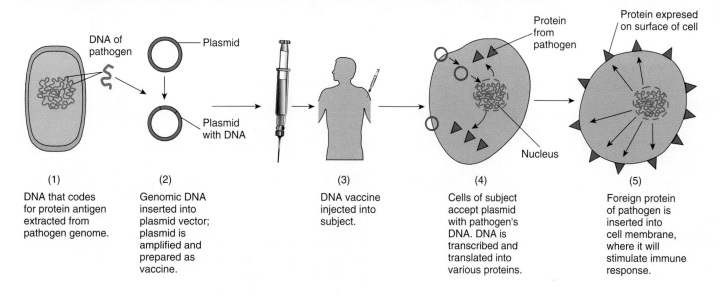

(1)	(2)	(3)	(4)	(5)
DNA that codes for protein antigen extracted from pathogen genome.	Genomic DNA inserted into plasmid vector; plasmid is amplified and prepared as vaccine.	DNA vaccine injected into subject.	Cells of subject accept plasmid with pathogen's DNA. DNA is transcribed and translated into various proteins.	Foreign protein of pathogen is inserted into cell membrane, where it will stimulate immune response.

(b) **Technology for anti-idiotypic vaccines**

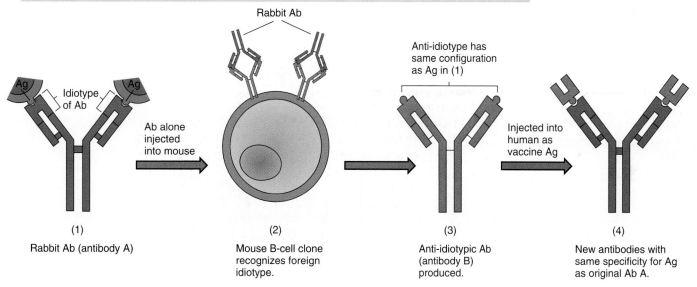

(1)	(2)	(3)	(4)
Rabbit Ab (antibody A)	Mouse B-cell clone recognizes foreign idiotype.	Anti-idiotypic Ab (antibody B) produced.	New antibodies with same specificity for Ag as original Ab A.

FIGURE 16.2

Other technologies useful in preparing vaccines.   **(a)** DNA vaccines contain all or part of the pathogen's DNA, which is used to "infect" recipient's cells. Processing of the DNA leads to production of an antigen protein that can stimulate a specific response against that pathogen. **(b)** Anti-idiotypic vaccines use antibodies as foreign proteins to mimic the structure of the natural antigen. See text for details.

When known or suspected adverse effects have been detected, vaccines are altered or withdrawn. Most recently, the whole-cell pertussis vaccine was replaced by the acellular capsule (aP) form when it was associated with adverse neurological effects. The live oral rotavirus vaccine had to be withdrawn when children experienced intestinal blockage. Polio vaccine was switched from live, oral to inactivated when too many cases of paralytic disease occurred from back-mutated vaccine stocks. Vaccine companies are also phasing out certain preservatives (thimerosal) that are thought to cause allergies and other potential side effects.

Professionals involved in giving vaccinations must understand their inherent risks but also realize that the risks from the infectious disease almost always outweigh the chance of an adverse vaccine reaction. The greatest caution must be exercised in giving live vaccines to immunocompromised or pregnant patients, the latter because of possible risk to the fetus.

To Vaccinate: Why, Whom, and When?

Vaccination confers long-lasting, sometimes lifetime, protection in the individual, but an equally important effect is to protect the public health. Vaccination is an effective method of establishing

TABLE 16.2

Recommended Regimen and Indications for Routine Vaccinations

Shaded bars indicate that a dose can be given once during this time frame

Routine Schedules for Children and Adults

Vaccine	Birth	2 Months	4 Months	6 Months	12 Months	15 Months	18 Months	4–6 Years	11–12 Years	Adults	Comments
Mixed vaccines											
Diphtheria, Tetanus, Pertussis[1] (DTaP)		DTaP	DTaP	DTaP	DTaP one dose			DTaP	Td or T		Td is tetanus/diphtheria; T is tetanus alone; either one should be given as booster every 10 years.
Measles,[2] Mumps, Rubella (MMR)					MMR			MMR or MMR			First dose varies with disease incidence; booster given at either 4–6 or 11–12 years.
Haemophilus influenzae type b (Hib)		Hib	Hib	Hib	Hib						Schedule depends upon source of vaccine; given with DTaP as TriHIBit
Single vaccines											
Poliovirus (IPV)[3]		IPV	IPV	IPV	one dose			IPV			Similar schedule to DTaP; injected vaccine
Hepatitis B (HB)	HB option 1		HB option 2	HB option 3							Option depends upon condition of infant; 3 doses given
Chickenpox (CPV)					CPV one dose				CPV two doses		Cannot be given to children <1 year old
Pneumococcus Vaccine (PV)		PV	PV	PV	PV						Used to protect children against otitis media

[1]DTaP. The diphtheria–tetanus–acellular pertussis vaccine has replaced the DTP.

[2]Measles vaccine (Attenuvax) can be given alone to children during epidemics or to adults immunized before 1970.

[3]IPV—inactivated polio vaccine—is now indicated as a safer alternative to the oral polio vaccine.

Used in Cases of Specific Risk Due to Occupational or Other Exposure

Vaccine	Group Targeted
Hepatitis B (Recombivax)	Health care personnel; people exposed through life-style
Hepatitis A (Havrix)	Children 2–14 years who live in areas of high prevalence
Pneumococcus (Pneumovax)	Elderly patients, children with sickle-cell anemia
Influenza, polio, tuberculosis (BCG)	Hospital, laboratory, health care workers
Rabies, plague, Lyme disease	People whose jobs involve contact with animals (veterinarians, forest rangers); known or suspected exposure to rabid animal; living in areas of high incidence
Cholera, hepatitis B, hepatitis A, measles, yellow fever, meningococcal meningitis, polio, rabies, typhoid, plague	Travelers to endemic regions, including military recruits (varies with geographic destination)

herd immunity in the population. According to this concept, individuals immune to a communicable infectious disease will not be carriers, and therefore the occurrence of that microbe will be reduced. With a larger number of immune individuals in a population (herd), it will be less likely that an unimmunized member of the population will encounter the agent. In effect, collective immunity through mass immunization confers indirect protection on the nonimmune (such as children). Herd immunity maintained through immunization is an important force in averting epidemics.

Vaccination is recommended for all typical childhood diseases for which a vaccine is available and for people in certain special circumstances (health workers, travelers, military personnel). Table 16.2 outlines a general schedule for childhood immunization and the indication for special vaccines. Not only are some vaccines mixtures of antigens, but in certain cases several vaccines are also administered simultaneously. Examples are military recruits who receive as many as 15 injections within a few minutes and children who receive boosters for DTaP and polio at the same time they receive the MMR vaccine. Experts doubt that immune interference (inhibition of one immune response by another) is a significant problem in these instances, and the mixed vaccines are carefully balanced to prevent this eventuality. The main problem with simultaneous administration is that side effects can be amplified.

Knowledge of the specific immune response has two practical
applications: (1) commercial production of antisera and vaccines and
(2) development of rapid, sensitive methods of disease diagnosis.

Artificial passive immunity usually involves administration of antiserum,
and occasionally B and T cells. Antibodies collected from donors
(human or otherwise) are injected into people who need protection
immediately. Examples include ISG (immune serum globulin) and SIG
(specific immune globulin).

Artificial active agents are vaccines that provoke a protective immune
response in the recipient but do not cause the actual disease.
Vaccination is the process of challenging the immune system with a
specially selected antigen. Examples are (1) killed or inactivated
microbes, (2) live, attenuated microbes, (3) subunits of microbes, and
(4) genetically engineered microbes or microbial parts.

Vaccination programs seek to protect the individual directly through
raising the antibody titer and indirectly through the development of
herd immunity.

SEROLOGICAL AND IMMUNE TESTS: MEASURING THE IMMUNE RESPONSE *IN VITRO*

The antibodies formed during an immune reaction are important in
combating infection, but they hold additional practical value. Characteristics of antibodies (such as their quantity or specificity) can
reveal the history of a patient's contact with microorganisms or
other antigens. This is the underlying basis of **serological testing.**
Serology is the branch of immunology that traditionally deals with
in vitro diagnostic testing of serum. Serological testing is based on
the familiar concept that antibodies have extreme specificity for
antigens, so when a particular antigen is exposed to its specific antibody, it will fit like a hand in a glove. The ability to visualize this
interaction by some means provides a powerful tool for detecting,
identifying, and quantifying antibodies—or for that matter, antigens. The scheme works both ways, depending on the situation.
One can detect or identify an unknown antibody using a known
antigen, or one can use an antibody of known specificity to help detect or identify an unknown antigen (figure 16.3). Modern serological testing has grown into a field that tests more than just serum.
Urine, cerebrospinal fluid, whole tissues, and saliva can also be
used to determine the immunologic status of patients. These and
other immune tests are helpful in confirming a suspected diagnosis
or in screening a certain population for disease.

General Features of Immune Testing

The strategies of immunologic tests are diverse, and they underline
some of the brilliant and imaginative ways that antibodies and antigens can be used as tools. We will summarize them under the head-
ings of agglutination, precipitation, immunodiffusion, complement
fixation, fluorescent antibody tests, and immunoassay tests. First
we will overview the general characteristics of immune testing, and
we will then look at each type separately.

The most effective serological tests have a high degree of
specificity and sensitivity (figure 16.4). **Specificity** is the property
of a test to focus upon only a certain antibody or antigen and not to

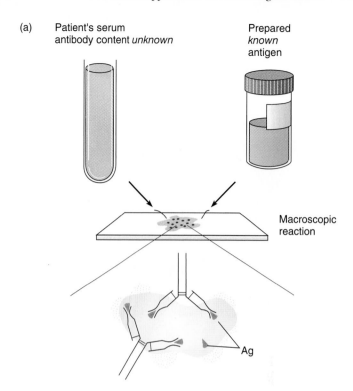

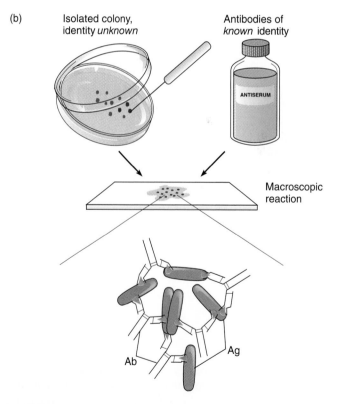

FIGURE 16.3

Basic principles of testing using antibodies and antigens. **(a)** In
serological diagnosis of disease, a blood sample is scanned for the
presence of antibody using an antigen of known specificity. A positive
reaction is usually evident as some visible sign, such as color change
or clumping, that indicates a specific interaction between antibody and
antigen. (The reaction at the molecular level is rarely observed.) **(b)** An
unknown microbe is mixed with serum containing antibodies of known
specificity, a procedure known as serotyping. Microscopically or
macroscopically observable reactions indicate a correct match between
antibody and antigen and permit identification of the microbe.

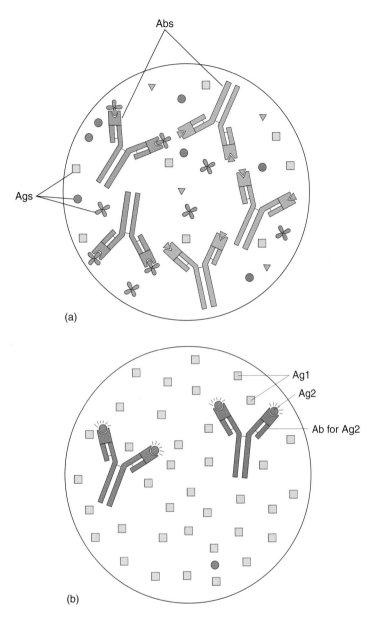

FIGURE 16.4

Specificity and sensitivity in immune testing.   **(a)** This test shows specificity in which an antibody (Ab) attaches with great exactness with only one type of antigen (Ag). **(b)** Sensitivity is demonstrated by the fact that Ab can locate Ag, even when it is greatly diluted.

react with unrelated or distantly related ones. **Sensitivity** means that the test can detect even very small amounts of antibodies or antigens that are the targets of the test. New systems using monoclonal antibodies have greatly improved specificity, and those using radioactivity, enzymes, and electronics have improved sensitivity.

Visualizing Antigen-Antibody Interactions   The primary basis of most tests is the binding of an antibody (Ab) to a specific molecular site on an antigen (Ag). Because this reaction cannot be readily seen without an electron microscope, tests involve some type of endpoint reaction visible to the naked eye or with regular magnification that tells whether the result is positive or negative. In

the case of large antigens such as cells, Ab binds to Ag and creates large clumps or aggregates that are visible macroscopically or microscopically (figure 16.5*a*). Smaller Ag-Ab complexes that do not result in readily observable changes will require special indicators in order to be visualized. Endpoints are often revealed by dyes or fluorescent reagents that can tag molecules of interest. Similarly, radioactive isotopes incorporated into antigens or antibodies constitute sensitive tracers that are detectable with photographic film.

An antigen-antibody reaction can be used to read a **titer,** or the quantity of antibodies in the serum. Titer is determined by serially diluting a sample in tubes or in a multiple-welled microtiter plate and mixing it with antigen (figure 16.5*b*). It is expressed as the highest dilution of serum that produces a visible reaction with an antigen. The more a sample can be diluted and yet still react with antigen, the greater is the concentration of antibodies in that sample and the higher is its titer. Interpretation of testing results is discussed in Medical Microfile 16.2.

Agglutination and Precipitation Reactions

The essential differences between agglutination and precipitation are in size, solubility, and location of the antigen. In agglutination, the antigens are whole cells such as red blood cells or bacteria with determinant groups on the surface. In precipitation, the antigen is a soluble molecule. In both instances, when Ag and Ab are optimally combined so that neither is in excess, one antigen is interlinked by several antibodies to form an insoluble, three-dimensional aggregate so large that it cannot remain suspended and it settles out (figure 16.5*a*).

Agglutination Testing   Agglutination is discernible because the antibodies cross-link the antigens to form visible clumps. Agglutination tests are performed routinely by blood banks to determine ABO and Rh (Rhesus) blood types in preparation for transfusions. In this type of test, antisera containing antibodies against the blood group antigens on red blood cells are mixed with a small sample of blood and read for the presence or absence of clumping (see figure 17.10). The *Widal test* is an example of a tube agglutination test for diagnosing salmonelloses and undulant fever. In addition to detecting specific antibody, it also gives the serum titer (figure 16.5*b*).

Numerous variations of agglutination testing exist. The Rapid Plasma Reagin (RPR) test is one of several tests commonly used to test for antibodies to syphilis. The cold agglutinin test, named for antibodies that react only at lower temperatures (4°–20°C), was developed to diagnose *Mycoplasma* pneumonia. The *Weil-Felix reaction* is an agglutination test sometimes used in diagnosing rickettsial infections.

In some tests, special agglutinogens have been prepared by affixing antigen to the surface of an inert particle. In *latex agglutination* tests, the inert particles are tiny latex beads. Kits using latex beads are available for assaying pregnancy hormone in the urine, identifying *Candida* yeasts and bacteria (staphylococci, streptococci, and gonococci), and diagnosing rheumatoid arthritis.

In *viral hemagglutination* testing, agglutinogen is a red blood cell that reacts naturally with certain viral antigens. The RBCs are mixed with a known virus and a patient's serum of unknown content (figure 16.6). The test is interpreted differently than other agglutination tests because it is based on a competition between the

What if a patient's serum gives a positive reaction—is **seropositive**—in a serological test? In most situations, it means that antibodies specific for a particular microbe have been detected in the sample. But one must be cautious in proceeding to the next level of interpretation. The mere presence of antibodies does not necessarily indicate that the patient has a disease but only that he or she has possibly had contact with a microbe or its antigens through infection or vaccination. In screening tests for determining a patient's history (rubella, for instance), knowing that a certain titer of antibodies is present can be significant, because it shows that the person has some protection. However, when the test is being used to diagnose current disease, a series of tests to show a rising titer of antibodies is necessary. The accompanying figure indicates how such a test can be used to diagnose patients who have nonspecific symptoms that could fit several diseases. Lyme disease, for instance, can be mistaken for arthritis or viral infections. In the first group, note that the antibody titer against *Borrelia burgdorferi* increased steadily over a 6-week period. A control group that shared similar symptoms did not exhibit a rise in titer for this microbe.

Another important consideration in testing is the occasional appearance of biological **false positives.** These are results in which a patient's serum shows a positive reaction, even though, in reality, he is not or has not been infected by the microbe. False positives such as those

in syphilis and AIDS testing arise when antibodies or other substances present in the serum **cross-react** with the test reagents, producing a positive result. Such false results may require retesting by a method that greatly minimizes cross-reactions.

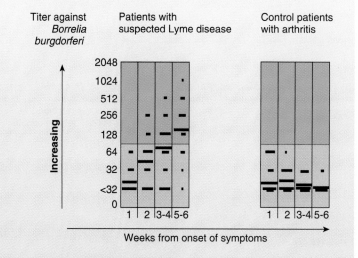

(a) **Agglutination**

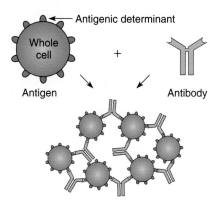

Microscopic appearance of clumps

Precipitation

Cell-free molecule in solution

Microscopic appearance of precipitate

(b) The Tube Agglutination Test

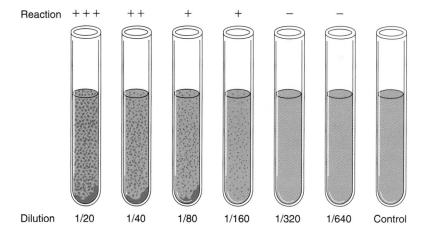

A sample of patient's serum is serially diluted with saline. The dilution is made in a way that halves the number of antibodies in each subsequent tube. An equal amount of the antigen (here, blue bacterial cells) is added to each tube. The control tube has antigen, but no serum. After incubation and centrifugation, each tube is examined for agglutination clumps as compared with the control, which will be cloudy and clump-free. The titer is defined as the dilution of the last tube in the series that shows agglutination.

FIGURE 16.5

Cellular/molecular view of agglutination and precipitation reactions that produce visible antigen-antibody complexes. Although IgG is shown as the Ab, IgM is also involved in these reactions.

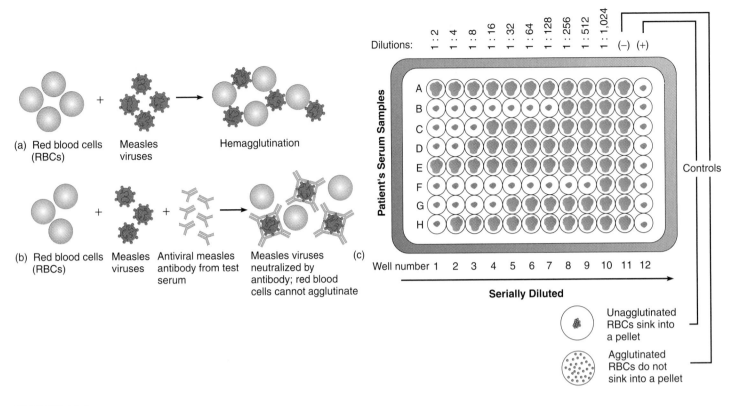

FIGURE 16.6

🔴 **Theory and interpretation of viral hemagglutination.** **(a)** In an antibody-free system, the natural tendency of measles virus to combine and interlink red blood cells causes an agglutination reaction. **(b)** In a test system using antibodies specific to the measles virus, these Abs will react with the viruses and prevent them from binding to red blood cells, thereby blocking agglutination. Thus, no agglutination means a positive reaction for antibody. **(c)** In an actual test, a multiwell microtiter plate is set up to run dilutions of several patients' sera. In a negative test, the agglutinated RBCs fill up most of the well; in a positive test, the unagglutinated RBCs fall into a small pile (pellet) in the well's bottom. (This test allows the reading of titer as well.)

RBCs and the antibodies for the virus antigens. If the patient's serum does *not* contain antibodies specific to the virus, the virus reacts with the RBCs instead and agglutinates them. Thus, agglutination here indicates no antibodies and a seronegative result. On the other hand, if the specific antibodies for the virus are present, they attach to the virus particles, the RBCs remain free, and there is no agglutination. As performed in the wells of microtiter plates, the difference is apparent to the naked eye (figure 16.6). Several viral diseases (measles, rubella, mumps, mononucleosis, and influenza) can be diagnosed with this test.

Precipitation Tests In precipitation reactions, the soluble antigen is precipitated (made insoluble) by an antibody. This reaction is observable in a test tube in which antiserum has been carefully laid over an antigen solution (figure 16.7a). At the point of contact, a cloudy or opaque zone forms.

One example of this technique is the VDRL (Veneral Disease Research Lab) test that also detects antibodies to syphilis. Although it is a good screening test, it contains a heterophilic antigen (cardiolipin) that may give rise to false positive results. Although precipitation is a useful detection tool, the precipitates are so easily disrupted in liquid media that most precipitation reactions are carried out in agar gels. These substrates are sufficiently soft to allow the reactants (Ab and Ag) to freely diffuse, yet firm enough to hold

the Ag-Ab precipitate in place. One technique with applications in microbial identification and diagnosis of disease is the **double diffusion** (Ouchterlony) method. It is called double diffusion because it involves diffusion of both antigens and antibodies. The test is performed by punching a pattern of small wells into an agar medium and filling them with test antigens and antibodies. A band forming between two wells indicates that antibodies from one well have met and reacted with antigens from the other well. Variations on this technique provide several results (figure 16.7).

Immunoelectrophoresis constitutes yet another refinement of diffusion and precipitation in agar. With this method, a serum sample is first electrophoresed to separate the serum proteins as previously shown (see figure 15.17). Antibodies that react with specific serum proteins are placed in a trough parallel to the direction of migration, forming reaction arcs specific for each protein (figure 16.8). This test is widely used to detect disorders in the production of antibodies. *Counterimmunoelectrophoresis,* which uses an electrical current to speed up the migration of antibody and antigen, is a newer technique for identifying bacterial and viral antigens in blood.

The Western Blot for Detecting Proteins

The **Western blot** test is somewhat similar to the previous tests because it involves the electrophoretic separation of proteins, followed by an immunoassay to detect these proteins. This test is a

(a)

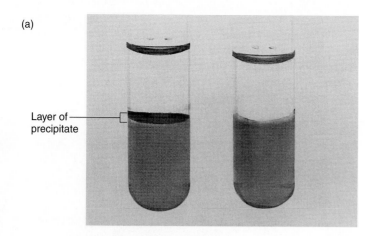

Layer of precipitate

(b) **I.**

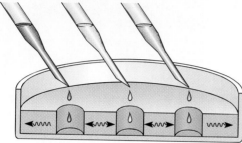

Side view

I. In one method of setting up a double diffusion test, wells are punctured in soft agar, and antibodies (Ab) and antigens (Ag) are added in a pattern. As the contents of the wells diffuse toward each other, a number of reactions can result.

II. Identity between antigens

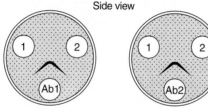

II. When a well containing Ab is placed near two Ag wells, lines of precipitate that form between the Ab-Ag wells are indicative of various characteristics of the test antigens. A continuous line indicates a distinct Ag-Ab reaction. In reaction II, the smooth Vs between wells 1 and 2 and the Ab show that they are the same antigen and antibody.

III. Nonidentity between antigens

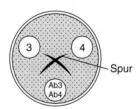

Spur

III. When two lines cross each other to form a spur, this indicates a lack of identity between the Ags, and in this case, they lack any antigenic similarities.

IV. Partial identity between antigens

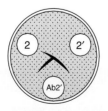

IV. The formation of a spur on one side only demonstrates that the two antigens share one antigenic determinant but differ in other antigenic characteristics.

1,2,3,4 = Antigens
2′ = Antigen 2 with added determinant
Ab 1,2,3,4 = Antibodies directed against Ags 1,2,3,4

FIGURE 16.7

Precipitation reactions. **(a)** A tube precipitation test for streptococcal group antigens. Specific antiserum has been placed in the bottom of the tubes and antigen solution carefully overlaid to form a zone of contact. The left-hand tube has developed a heavy band of precipitate indicative of a positive reaction between antibody and antigen. The right-hand tube is negative. **(b)** Double-diffusion (Ouchterlony) tests in a semisolid matrix.

(a) From Gillies and Dodds, Bacteriology Illustrated, 5th ed., fig. 14. Reprinted by permission of Longman Group Ltd.

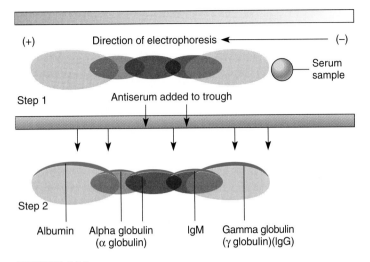

Step 1

Step 2

Albumin | Alpha globulin (α globulin) | IgM | Gamma globulin (γ globulin)(IgG)

FIGURE 16.8

Immunoelectrophoresis of normal human serum. *Step 1.* Proteins are separated by electrophoresis on a gel. *Step 2.* To identify the bands and increase visibility, antiserum containing antibodies specific for serum proteins is placed in a trough and allowed to diffuse toward the bands. This diffusion produces a pattern of numerous arcs representing major serum components.

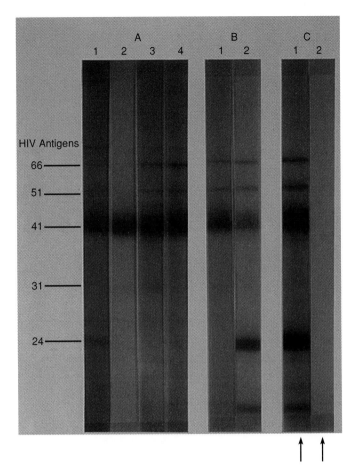

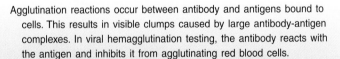

FIGURE 16.9

The Western blot procedure. The example shown here tests for antibodies to specific HIV antigens. The test strips are prepared by electrophoresing several of the major HIV surface and core antigens and then blotting them onto special filters. The test strips are incubated with a patient's serum and developed with a radioactive or colorimetric label. Sites where HIV antigens have bound antibodies show up as bands. Patients' sera are then compared with a positive control strip (C1) containing antibodies for all HIV antigens. In this series, A strips are from patients with fully developed AIDS, B strips are from patients in the early phase of HIV disease, and strip C2 is HIV negative (no antibodies to HIV).

Courtesy of Philip D. Markham, Advanced BioScience Laboratories, Inc.

counterpart of the Southern blot test for identifying DNA (see figure 10.4). It is a highly specific and sensitive way to identify or verify a particular protein (antibody or antigen) in a sample (figure 16.9). First, the test material is electrophoresed in a gel to separate out particular bands. The gel is then transferred to a special blotter that binds the reactants in place. The blot is developed by incubating it with a solution of antibody or antigen that has been labeled with radioactive, fluorescent, or luminescent labels. Sites of specific binding will appear as a pattern of bands that can be compared with known positive and negative samples. This is currently the verification test for people who are antibody-positive for HIV in the ELISA test (described in a later section), because it tests more types of antibodies and is less subject to misinterpretation than are other antibody tests. The technique has significant applications for detecting microbes and their antigens in specimens.

CHAPTER CHECKPOINTS

Serological tests can test for either antigens or antibodies. Most are *in vitro* assessments of antigen-antibody reactivity from a variety of body fluids. The basis of these tests is an antigen-antibody reaction made visible through the processes of agglutination, precipitation, immunodiffusion, complement fixation, fluorescent antibody, and immunoassay techniques.

One measurement is the *titer,* described as the concentration of antibody in serum. It is the highest dilution of serum that gives a visible antigen-antibody reaction. The higher the titer, the greater the level of antibody present.

Agglutination reactions occur between antibody and antigens bound to cells. This results in visible clumps caused by large antibody-antigen complexes. In viral hemagglutination testing, the antibody reacts with the antigen and inhibits it from agglutinating red blood cells.

In precipitation reactions, soluble antigen and antibody react to form insoluble, visible precipitates. Precipitation reactions can also be visualized by adding radioactive or enzyme markers to the antigen-antibody complex.

In immunoelectrophoresis techniques such as the Western blot, proteins that have been separated by electrical current are identified by labeled antibodies. HIV infections are verified with this method.

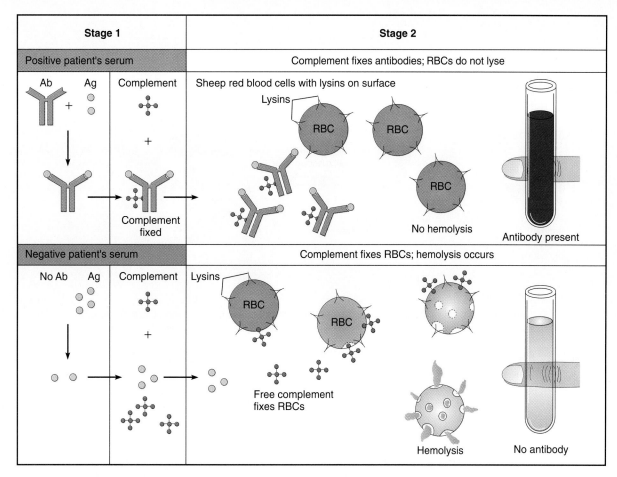

Stage 1	Stage 2		
Positive patient's serum	**Complement fixes antibodies; RBCs do not lyse**		

Ab   Ag   Complement   Sheep red blood cells with lysins on surface

Lysins

RBC   RBC

RBC

Complement fixed

No hemolysis   Antibody present

| **Negative patient's serum** | **Complement fixes RBCs; hemolysis occurs** | | |

No Ab   Ag   Complement   Lysins

RBC

RBC

Free complement fixes RBCs

Hemolysis   No antibody

FIGURE 16.10

Complement fixation test.   In this example, a patient's serum is being tested for antibodies to a certain infectious agent. In reading this test one observes the cloudiness of the tube. If it is cloudy, the RBCs are not hemolyzed and the test is positive. If it is clear and pink, the RBCs are hemolyzed and the test is negative.

Complement Fixation

An antibody that requires (fixes) complement to complete the lysis of its antigenic target cell is termed a **lysin** or **cytolysin.** When lysins act in conjunction with the intrinsic complement system on red blood cells, the cells hemolyze (lyse and release their hemoglobin). This lysin-mediated hemolysis is the basis of a group of tests called **complement fixation,** or CF (figure 16.10).

Complement fixation testing uses four components—antibody, antigen, complement, and sensitized sheep red blood cells—and it is conducted in two stages. In the first stage, the test antigen is allowed to react with the test antibody (at least one must be of known identity) in the absence of complement. If the Ab-Ag are specific for each other, they form complexes. To this mixture, purified complement proteins from guinea pig blood are added. If antibody and antigen have complexed during the previous step, they attach, or **fix,** the complement to them, thus preventing it from participating in further reactions. The extent of this complement fixation is determined in the second stage; sheep RBCs are mixed with sheep-specific lysins. This mixing will produce an indicator complex that can also fix complement. Contents of the stage 1 tube are

mixed with the stage 2 tube and observed for hemolysis, which can be observed with the naked eye as a clearing of the solution. If hemolysis *does not* occur, it means that the complement was used up by the first stage Ab-Ag complex and that the unknown antigen or antibody was indeed present. This result is considered positive. If hemolysis *does* occur, it means that unfixed complement from tube 1 reacted with the RBC complex instead, thereby causing lysis of the sheep RBCs. This result is negative for the antigen or antibody that was the target of the test. Complement fixation tests are somewhat complicated, yet they are invaluable in diagnosing certain viral, rickettsial, fungal, and parasitic infections.

The antistreptolysin O (ASO) titer test measures the levels of antibody against the streptolysin toxin, an important hemolysin of group A streptococci. It employs a technique related to complement fixation. A serum sample is exposed to known suspensions of streptolysin and then allowed to incubate with RBCs. Lack of hemolysis indicates antistreptolysin antibodies in the patient's serum that have neutralized the streptolysin and prevented hemolysis. This is an important verification procedure for scarlet fever, rheumatic fever, and other related streptococcal syndromes (see chapter 18).

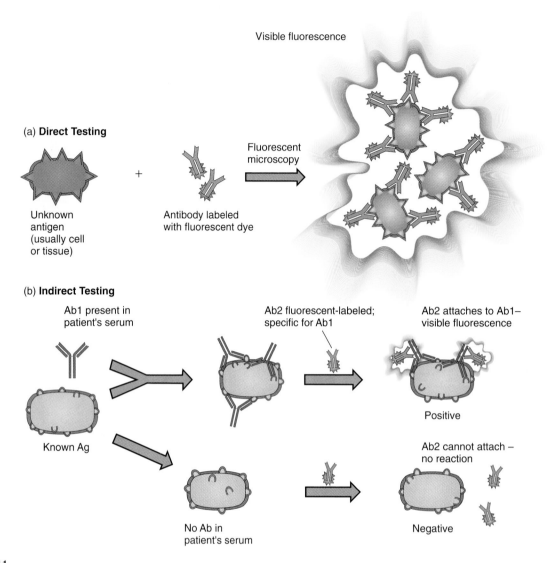

Visible fluorescence

(a) Direct Testing

Unknown
antigen
(usually cell
or tissue)

+

Antibody labeled
with fluorescent dye

Fluorescent
microscopy

(b) Indirect Testing

Ab1 present in
patient's serum

Ab2 fluorescent-labeled;
specific for Ab1

Ab2 attaches to Ab1—
visible fluorescence

Positive

Known Ag

No Ab in
patient's serum

Ab2 cannot attach –
no reaction

Negative

FIGURE 16.11

Immunofluorescence testing. The success of these techniques is contingent upon the accumulation of sufficient labeled antibody (Ab) in one site that it will show up as fluorescing cells or masses. **(a)** Direct: Unidentified antigen (Ag) is directly tagged with fluorescent Ab. **(b)** Indirect: Ag of known identity is used to assay unknown Ab; a positive reaction occurs when the second Ab (with fluorescent dye) affixes to the first Ab.

Miscellaneous Serological Tests

A test that relies on changes in cellular activity as seen microscopically is the *Treponema pallidum immobilization* (TPI) test for syphilis. The impairment or loss of motility of the *Treponema* spirochete in the presence of test serum and complement indicates that the serum contains anti-*Treponema pallidum* antibodies (see chapter 21). In *toxin neutralization* tests, a test serum is incubated with the microbe that produces the toxin. If the serum inhibits the growth of the microbe, one can conclude that antitoxins are present.

Serotyping is an antigen-antibody technique for identifying, classifying, and subgrouping certain bacteria into categories called *serotypes,* using antisera for cell antigens such as the capsule, flagellum, and cell wall. It is widely used in typing *Salmonella* species and strains and is the basis for the numerous serotypes of streptococci. The Quellung test, which identifies serotypes of the pneumococcus involves a precipitation reaction in which antibodies react with the capsular polysaccharide. Although the reaction makes the capsule seem to swell, it is actually creating a zone of Ab-Ag complex on the cell's surface (see figure 18.20).

Fluorescent Antibodies and Immunofluorescence Testing

The property of dyes such as fluorescein and rhodamine to emit visible light in response to ultraviolet radiation was discussed in chapter 3. This property of fluorescence has found numerous applications in diagnostic immunology. The fundamental tool in immunofluorescence testing is a fluorescent antibody—a monoclonal antibody labeled by a fluorescent dye (fluorochrome).

The two ways that fluorescent antibodies (FABs) can be used are shown in figure 16.11. In *direct testing,* an unknown test specimen or antigen is fixed to a slide and exposed to a fluorescent antibody solution of known composition. If the antibodies are complementary to antigens in the material, they will bind to it. After the slide is rinsed to remove unattached antibodies, it is observed with the fluorescent microscope. Fluorescing cells or specks indicate the presence of Ab-Ag complexes and a positive result. These tests are valuable for identifying and locating antigens on the surfaces of cells or in tissues and in identifying the disease agents of syphilis (see figure D, page 539), gonorrhea, chlamydiosis, whooping cough, Legionnaires' disease, plague, trichomoniasis, meningitis, and listeriosis.

In *indirect testing* methods, the fluorescent antibodies are *anti-isotypic* antibodies made to react with the Fc region of another antibody (remember that antibodies can be antigenic). In this scheme, an antigen of known character (a bacterial cell, for example) is combined with a test serum of unknown antibody content. The fluorescent antibody solution that can react with the unknown antibody is applied and rinsed off to visualize whether the serum contains antibodies that have affixed to the antigen. A positive test shows fluorescing aggregates or cells, indicating that the fluorescent antibodies have combined with the unlabeled antibodies. In a negative test, no fluorescent complexes will appear. This technique is frequently used to diagnose syphilis (FTA-ABS) and various viral infections discussed in chapters 24 and 25.

Immunoassays: Tests of Great Sensitivity

The elegant tools of the microbiologist and immunologist are being used increasingly in athletics, criminology, government, and business to test for trace amounts of substances such as hormones, metabolites, and drugs. But traditional techniques in serology are not refined enough to detect a few molecules of these chemicals. Extremely sensitive alternative methods that permit rapid and accurate measurement of trace antigen or antibody are called **immunoassays.** Examples of the technology for detecting an antigen or antibody in minute quantities include radioactive isotope labels, enzyme labels, and sensitive electronic sensors. Many of these tests are based on specifically formulated monoclonal antibodies.

Radioimmunoassay (RIA)   Antibodies or antigens labeled with a radioactive isotope can be used to pinpoint minute amounts of a corresponding antigen or antibody. Although very complex in practice, these assays compare the amount of radioactivity present in a sample before and after incubation with a known, labeled antigen or antibody. The labeled substance competes with its natural, nonlabeled partner for a reaction site. Large amounts of a bound radioactive component indicate that the unknown test substance was not present. The amount of radioactivity is measured with an isotope counter or a photographic emulsion (autoradiograph). Radioimmunoassay has been employed to measure the levels of insulin and other hormones and to diagnose allergies, chiefly by the radioimmunosorbent test (RIST) for measurement of IgE in allergic patients and the radioallergosorbent test (RAST) to standardize allergenic extracts (chapter 17).

Enzyme-Linked Immunosorbent Assay (ELISA)   The **ELISA test,** also known as enzyme immunoassay (EIA), contains an enzyme-antibody complex that can be used as a color tracer for antigen-antibody reactions. The enzymes used most often are horseradish peroxidase and alkaline phosphatase, both of which release a dye (chromogen) when exposed to their substrate. This technique also relies on a solid support such as a plastic microtiter plate that can *adsorb* (attract on its surface) the reactants (figure 16.12).

As with immunofluorescence, this technique is applicable for both direct and indirect assays. In *direct ELISA,* or sandwich, tests, a known antibody is absorbed to the bottom of a well and incubated with a solution containing unknown antigen (figure 16.12*a*). After excess unbound components have been rinsed off, an enzyme-antibody indicator that can react with the antigen is added. If antigen is present, it will attract the indicator-antibody and hold it in place. Next, the substrate to the enzyme is placed in the wells and incubated. Enzymes affixed to the antigen will hydrolyze the substrate and release a colored dye. Thus, any color developing in the wells is a positive result. Lack of color means that the antigen was not present and that the subsequent rinsing removed the enzyme-antibody complex.

The *indirect ELISA* test can detect antibodies in a serum sample. As with other indirect tests, the final positive reaction is achieved by means of an antibody-antibody reaction. Like the direct ELISA, the indicator antibody is complexed to an enzyme that produces a color change with positive serum samples (figure 16.12*b*). The starting reactant is a known antigen that is adsorbed to the surface of a well. To this, an unknown serum is added. After rinsing, an enzyme-Ab reagent that can react with the test antibody is placed in the well. The substrate to the enzyme is then added, and the wells are scanned for color changes. Color development indicates that all the components reacted and that the antibody was present in the patient's serum. This is the common screening test for the antibodies to HIV (AIDS virus), various rickettsial species, *Salmonella,* the cholera vibrio, and *Helicobacter,* a cause of gastric ulcers. Because false positives can occur, a verification test may be necessary (such as Western blot for HIV).

A newer technology uses electronic monitors that directly read out antibody-antigen reactions. Without belaboring the technical aspects, these systems contain computer chips that sense the minute changes in electrical current given off when an antibody binds to antigen. The potential for sensitivity is extreme; it is thought that amounts as small as 12 molecules of a substance can be detected in a sample. In another procedure, antibody-substrate molecules are incubated with sample and then exposed to the enzyme alkaline phosphatase. If the antibody is bound, the enzyme reacts with the substrate and causes visible light to be emitted. The light can be detected by machines or photographic films.

Tests That Differentiate T Cells and B Cells

So far we have concentrated on tests that identify antigens and antibodies in samples, but techniques also exist that differentiate between B cells and T cells and can quantify subsets of each. Information on the types and numbers of lymphocytes in blood and other samples is a common way to evaluate immune dysfunctions such as those in AIDS, immunodeficiencies, and cancer. A simple method for identifying T cells is to mix them with untreated sheep red blood cells. Receptors on the T cells bind the RBCs into a flowerlike cluster called a *rosette formation* (figure 16.13*a*). Rosetting can also occur in B cells if one uses Ig-coated bovine RBCs or mouse erythrocytes.

(a) **Direct Antibody Sandwich Method for Unknown Antigen**

(b) **Indirect Immunosorbent Assay for Unknown Antibody**

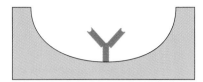

Antibody is adsorbed to well.

Antigen is adsorbed to well.

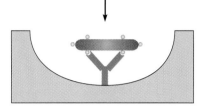

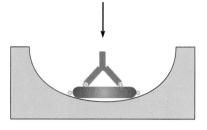

Test antigen is added; if complementary, antigen binds to antibody.

Test antiserum is added; if antibody is complementary, it binds to antigen.

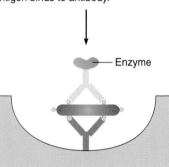

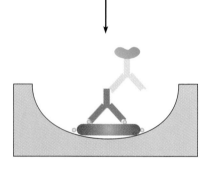

Enzyme

Enzyme-linked antibody specific for test antigen then binds to antigen, forming sandwich.

Enzyme-linked antibody specific for test antibody binds to it.

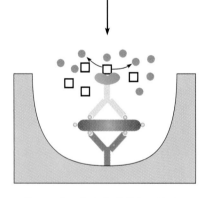

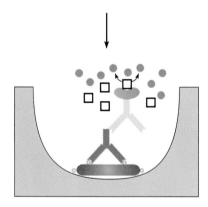

Enzyme's substrate (□) is added, and reaction produces a visible color change (●).

Enzyme's substrate (□) is added, and reaction produces a visible color change (●).

FIGURE 16.12

Methods of ELISA testing. The success of these tests depends on adequate rinsing between steps to remove unreacted or nonspecific components. **(a)** Direct: Antibody sandwich method. **(b)** Indirect: Immunosorbent assay. One type of screening test for HIV infection is performed as in **(b).**

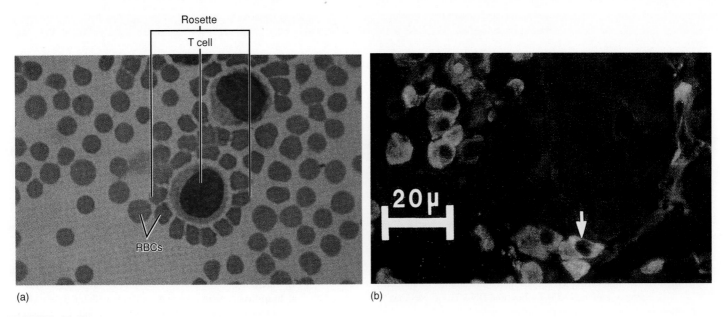

FIGURE 16.13

Tests for characterizing T cells and B cells.   **(a)** Photomicrograph of rosette formation that identifies T cells. **(b)** Plasma cells (arrow) highlighted by fluorescent antibodies.

Indirect fluorescent techniques have been developed to differentiate between T cells and B cells as well as to subgroup them (figure 16.13b). These subgroup tests utilize monoclonal antibodies produced in response to specific cell markers. B-cell tests categorize different stages in B-cell development and are very useful in characterizing B-cell cancers. Tests that can help differentiate the CD4, CD8, and other T-cell subsets are important in monitoring AIDS and other immunodeficiency diseases.

In Vivo Testing

Probably the first immunologic tests were performed not in a test tube but on the body itself. A classic example of one such technique is the *tuberculin test,* which uses a small amount of purified protein derivative (PPD) from *Mycobacterium tuberculosis* injected into the skin. The appearance of a red, raised lesion can indicate previous exposure to tuberculosis (see figure 17.18a). In practice, *in vivo* tests employ principles similar to serological tests, except in this case an antigen or an antibody is introduced into a patient to elicit some sort of visible reaction. Like the tuberculin test, some of these diagnostic skin tests are useful for evaluating infections due to fungi (coccidioidin and histoplasmin tests, for example) or allergens (see skin testing in chapter 17).

CHAPTER CHECKPOINTS

Complement fixation involves a two-part procedure in which complement fixes to a specific antibody if present, or to red blood cell antigens, if antibody is absent. Lack of RBC hemolysis is indicative of a positive test.

Serological tests can measure the degree to which host antibody binds directly to disease agents or toxins. This is the principle behind tests for syphilis and rheumatic fever.

Direct fluorescent antibody tests indicate presence of an antigen and are useful in identifying infectious agents. Indirect fluorescent tests indicate the presence of a particular antibody and can diagnose infection.

Immunoassays can detect very small quantities of antigen, antibody, or other substances. Radioimmunoassay uses radioisotopes to detect trace amounts of biological substances.

The ELISA test uses enzymes and dyes to detect antigen-antibody complexes. It is widely used to detect viruses, bacteria, and antibodies in HIV infection.

Technicians use precise assays to differentiate between B and T cells and to identify subgroups of these cells for disease diagnosis.

In vivo serological testing, such as the tuberculin test, involves subcutaneous injection of antigen to elicit a visible antigen-antibody response in the host.

CHAPTER CAPSULE WITH KEY TERMS

Biomedical Applications of Specific Immunity

I. *Immunization:* **Producing immunity by medical intervention**

A. Passive immunotherapy includes administering immune serum globulin and specific immune globulins pooled from donated serum to prevent infection and disease in those at risk; antisera and antitoxins from animals are occasionally used.

B. Active immunization is synonymous with **vaccination**; provides an antigenic stimulus that does not cause disease but can produce long-lasting, protective immunity. **Vaccines** are made with:

1. Killed whole cells or inactivated viruses that do not reproduce but are antigenic.
2. Live, attenuated cells or viruses that are able to reproduce but have lost virulence.
3. Acellular or subunit components of microbes such as surface antigen or neutralized toxins (toxoids).
4. Genetic engineering techniques, including cloning of antigens, recombinant attenuated microbes, and DNA-based vaccines.

C. Boosters (additional doses) are often required.

D. Vaccination increases herd immunity, protection provided by mass immunity in a population.

II. *Serological/Immune Testing:* **Serology** is a science that attempts to detect signs of infection in a patient's serum such as antibodies specific for a microbe.

A. The basis of serological tests is that Abs specifically bind to Ag *in vitro.* An Ag of known identity will react with antibodies in an unknown serum sample. The reverse is also true; known antibodies can be used to detect and type antigens.

B. These Ag-Ab reactions are visible in the form of obvious clumps and precipitates, color changes, or the release of radioactivity. Test results are read as positive or negative.

C. Desirable properties of tests are **specificity** and **sensitivity.**

D. Types of Tests

1. In **agglutination** tests, antibody cross-links whole-cell antigens, forming complexes that settle out and form visible clumps in the test chamber; examples are tests for blood type, some bacterial diseases, and viral diseases.

2. **Double diffusion precipitation** tests involve the diffusion of Ags and Abs in a soft agar gel, forming zones of **precipitation** where they meet.

3. **In immunoelectrophoresis,** migration of serum proteins in gel is combined with precipitation by antibodies.

4. The **Western blot** test separates antigen into bands. After the gel is affixed to a blotter, it is reacted with a test specimen and developed by radioactivity or with dyes.

5. **Complement fixation** tests detect **lysins**—antibodies that fix complement and can lyse target cells. It involves first mixing test Ag and Ab with complement and then with sensitized sheep RBCs. If the complement is fixed by the Ag-Ab, the RBCs remain intact, and the test is positive. If RBCs are hemolyzed, specific antibodies are lacking.

6. In **direct assays**, known marked Ab is used to detect unknown Ag (microbe).
   a. In **indirect testing,** known Ag reacts with unknown Ab, and the reaction is made visible by a second Ab that can affix to and identify the unknown Ab.
   b. **Immunofluorescence testing** uses **fluorescent antibodies** (FABs tagged with fluorescent dye) either directly or indirectly to visualize cells or cell aggregates that have reacted with the FABs.

7. **Immunoassays** are highly sensitive tests for Ag and Ab.
   a. In **radioimmunoassay**, Ags or Abs are labeled with radioactive isotopes and traced.
   b. The **enzyme-linked immunosorbent assay (ELISA)** can detect unknown Ag or Ab by direct or indirect means. A positive result is visualized when a colored product is released by an enzyme-substrate reaction.
   c. Tests are also available to differentiate B cells from T cells and their subtypes.

8. With *in vivo* testing, Ags are introduced into the body directly to determine the patient's immunologic history.

MULTIPLE-CHOICE QUESTIONS

1. A living microbe with reduced virulence that is used for vaccination is considered
   - a. a toxoid
   - b. attenuated
   - c. denatured
   - d. an adjuvant

2. A vaccine that contains parts of viruses is called
   - a. acellular
   - b. recombinant
   - c. subunit
   - d. attenuated

3. Widespread immunity that protects the population from the spread of disease is called
   - a. seropositivity
   - b. cross-reactivity
   - c. epidemic prophylaxis
   - d. herd immunity

4. DNA vaccines contain _____ DNA that stimulates cells to make _____ antigens.
   - a. human, RNA
   - b. microbial, protein
   - c. human, protein
   - d. microbial, polysaccharide

5. Administration of immune serum globulin is a form of _____ immunization that _____.
   - a. active, prevents infection
   - b. passive, provides long-term immunity
   - c. therapeutic, prevents disease
   - d. prophylactic, stimulates the immune system

6. What is the purpose of an adjuvant?
   - a. to kill the microbe
   - b. to stop allergic reactions
   - c. to improve the contact between the antigen and lymphocytes
   - d. to make the antigen more soluble in the tissues

7. An example of a recombinant DNA vaccine is
   - a. tetanus
   - b. MMR
   - c. polio
   - d. hepatitis B

8. In agglutination reactions, the antigen is a _____; in precipitation reactions, it is a _____.
   a. soluble molecule, whole cell
   b. whole cell, soluble molecule
   c. bacterium, virus
   d. protein, carbohydrate

9. Which reaction requires complement?
   a. hemagglutination
   b. precipitation
   c. hemolysis
   d. toxin neutralization

10. A patient with a _____ titer of antibodies to an infectious agent has greater protection than a patient with a _____ titer.
    a. high, low   c. negative, positive
    b. low, high   d. old, new

11. Direct immunofluorescence tests use a labeled antibody to identify _____.
    a. an unknown microbe
    b. an unknown antibody
    c. fixed complement
    d. agglutinated antigens

12. The Western blot test can identify
    a. unknown antibodies
    b. unknown antigens
    c. specific DNA
    d. both a and b

13. An example of an *in vivo* serological test is
    a. indirect immunofluorescence
    b. radioimmunoassay
    c. tuberculin test
    d. complement fixation

14. **Multiple Matching.** For the following list of vaccines, first look up the disease for which the vaccine is intended (if not apparent). Then list the letters of all the descriptions that fit. (There may be more than one possible answer.)

    _____ BCG
    _____ Measles
    _____ Pertussis
    _____ Tetanus
    _____ Mumps
    _____ Diphtheria
    _____ Hepatitis B
    _____ Hib
    _____ Polio
    _____ Rubella
    _____ Pneumococcus

    a. Uses live, attenuated microbes
    b. Based on a toxoid
    c. Subunit vaccine
    d. Used in mixed vaccine
    e. Does not require boosters
    f. Uses killed, whole cells
    g. Uses killed, whole viruses
    h. Routinely given in childhood
    i. Mainly for people at high risk
    j. Made by genetic engineering

CONCEPT QUESTIONS

1. a. Name three products used in passive artificial immunization.
   b. What are the primary reasons for using these substances?
   c. What are some disadvantages of this form of immunization?
   d. What is the difference between immunization used for prophylaxis and that used for treatment?

2. a. Outline the strategies for developing vaccines, and give specific examples for each method.
   b. By what means are microorganisms attenuated?
   c. What is the purpose of an adjuvant?
   d. Describe the way that a Trojan horse vaccine works.

3. a. What are the advantages and disadvantages of a killed vaccine; a live, attenuated vaccine; a subunit vaccine; a recombinant vaccine; and a DNA vaccine?
   b. Use an outline to explain how an inoculation with tetanus toxoid will protect a person the next time he or she steps on a dirty piece of glass.

4. a. Describe the concept of herd immunity.
   b. How does vaccination contribute to its development in a community?
   c. Give some possible explanations for recent epidemics of diphtheria and whooping cough.

5. a. What is the basis of serology and serological testing?
   b. Differentiate between specificity and sensitivity.
   c. Describe several general ways that Ag-Ab reactions are detected.

6. a. What does seropositivity mean?
   b. In what ways is the antibody titer of serum important?
   c. What causes false positive tests?

7. a. Explain how agglutination and precipitation reactions are alike.
   b. In what ways are they different?
   c. Make a drawing of the manner in which antibodies cross-link the antigens in agglutination and precipitation reactions.
   d. Give examples of several tests that employ the two reactions.

8. a. What is meant by complement fixation? What are cytolysins?
   b. What is the purpose of using sheep red blood cells in this test?

9. a. Explain the differences between direct and indirect procedures in serological or immunoassay tests.
   b. How is fluorescence detected?
   c. How is the reaction in a radioimmunoassay detected?
   d. How does a positive reaction in an ELISA test appear?

10. a. Briefly describe the principles and give an example of the use of a specific test using immunoelectrophoresis, Western blot, complement fixation, fluorescence testing (direct and indirect), and immunoassays (direct and indirect ELISA).
    b. Explain a rapid microscopic method for differentiating T cells from B cells.

CRITICAL-THINKING QUESTIONS

1. Describe the relationship between an antitoxin, a toxin, and a toxoid.

2. It is often said that a vaccine does not prevent infection; rather, it primes the immune system to undergo an immediate response to prevent an infection from spreading. Explain what is meant by this statement, and outline what is happening at the cellular/molecular level from the time of vaccination until infection with the infectious agent actually occurs.

3. At least three boosters are given for DTaP vaccines.
   a. Explain what each subsequent booster does, and why more than one is needed.
   b. Which features of the immune system allow it to react efficiently with 10 to 15 different vaccine antigens simultaneously?

4. Explain how to design a vaccine that could:
   a. induce protective IgA in the intestine
   b. give immunity to dental caries
   c. protect against the liver phase of the malaria parasite
   d. be derived from a single microbe and immunize against two different infectious diseases

5. a. Suggest several reasons that it could be risky to administer a vaccine containing a live, attenuated DNA virus.
   b. Explain what is involved in making a DNA vaccine.
   c. Explain why measles vaccine does not reliably protect infants if it is given before 12 months of age.

6. a. Determine the vaccines you have been given and those for which you will require periodic boosters.
   b. Suggest vaccines you may need in the future.

7. When traders and missionaries first went to the Hawaiian Islands, the natives there experienced severe disease and high mortality rates from smallpox, measles, and certain STDs.
   a. Explain what factors are involved in the sudden prevalence of disease in previously unexposed populations.
   b. Explain the ways in which vaccination has been responsible for the worldwide eradication of diseases such as smallpox and polio.

8. Why do some tests for antibody in serum (such as for HIV and syphilis) require backup verification with additional tests at a later date?

9. a. Look at figure 16.5b. What is the titer as shown?
   b. If the titer had been 1:40, what interpretation would be made as to the immune status of the patient?
   c. What would it mean if a test 2 weeks later revealed a titer of 1:1280?
   d. What would it mean if no agglutination had occurred in any tube?
   e. From figure 16.6c, can you tell which patients have measles antibody and which do not?
   f. What are the titers of the Ab-positive patients?

10. Why do we interpret positive hemolysis in the complement fixation test to mean negative for the test substance?

11. Explain how an immunoassay method could use monoclonal antibodies to differentiate between B and T cells and between different subsets of T cells.

INTERNET SEARCH TOPICS

1. Find information on vaccines for herpes simplex, human papillomavirus, group B *Streptococcus*, cancer, and dental diseases.

2. Find descriptions of current tests for HIV infection such as ELISA, Western blot, and antigen tests; determine the indications for use and their relative sensitivity and specificity.

Disorders in Immunity

H umans possess a powerful and intricate system of defense, which by its very nature also carries the potential to cause injury and disease. In most instances, a defect in immune function is expressed in commonplace, but miserable, symptoms such as those of hay fever and dermatitis. But abnormal or undesirable immune functions are also actively involved in debilitating or life-threatening diseases such as asthma, anaphylaxis, rheumatoid arthritis, graft rejection, and cancer.

Chapter Overview

- The immune system is subject to several types of dysfunctions termed immunopathologies.
- Some dysfunctions are due to abnormally heightened responses to antigens as manifested in allergies, hypersensitivities, and autoimmunities.
- Some dysfunctions are due to the reduction or loss in protective immune reactions due to genetic or environmental causes, as exemplified by immunodeficiencies and cancer.
- Some immune damage is caused by normal actions that are directed at foreign tissues placed in the body for therapy, such as transfusions and transplants.
- Hypersensitivities are divided into immediate, antibody-mediated, immune complex, and delayed allergies.
- Allergens are the foreign molecules from the environment, other organisms, or even from the body that cause a hypersensitive or allergic response.
- Most hypersensitivities require an initial sensitizing event, followed by a later contact that causes symptoms.
- The immediate type of allergy is mediated by special types of B cells that produce an antibody called IgE. IgE can cause mast cells to release allergic chemicals such as histamine that stimulate symptoms.
- Examples of immediate allergies are atopy, asthma, food allergies, and anaphylaxis.
- Another type of hypersensitivity arises from the action of other antibodies (IgG and IgM) that can fix complement and lyse foreign cells. An example is the reaction due to incompatible blood transfusions or placental transfer.
- Immune complex reactions are caused by large amounts of circulating antibodies against foreign molecules accumulating in tissues and organs.
- Autoimmune diseases are due to the production of B and T cells that are abnormally sensitized to react with the body's natural molecules and

This delicate poison ivy plant, along with its close relatives poison oak and sumac, is one of the most common causes of allergy in the United States. Its leaves contain an oil that many people are sensitive to. The nature of its effects are covered in this chapter.

thus can damage cells and tissues. Some examples of these diseases include rheumatoid arthritis, systemic lupus erythematosus, myasthenia gravis, and multiple sclerosis.

- T cells are involved in delayed-type hypersensitivities wherein allergens cause damage to cells and graft rejection.
- Immunodeficiencies are pathologies in which B and T cells and other immune cells are missing or destroyed. They may be inborn and genetic or acquired.
- The primary outcome of immunodeficiencies is manifest in recurrent infections and lack of immune competence.
- Cancer is an abnormal overgrowth of cells due to a genetic defect and the lack of effective immune surveillance.

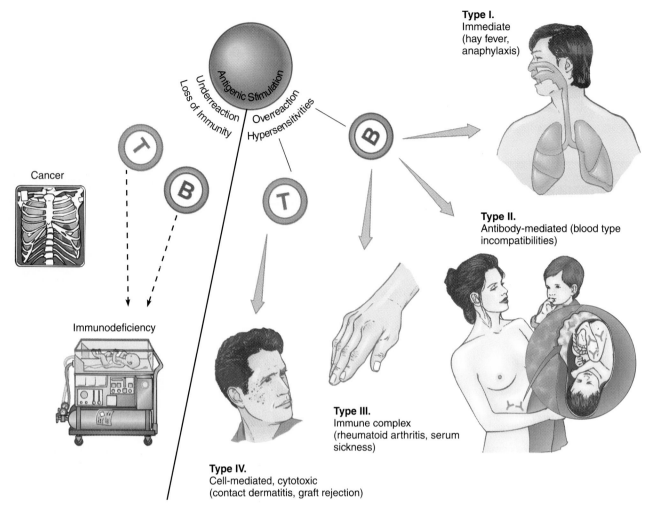

Type I.
Immediate
(hay fever,
anaphylaxis)

Type II.
Antibody-mediated (blood type
incompatibilities)

Type III.
Immune complex
(rheumatoid arthritis, serum
sickness)

Type IV.
Cell-mediated, cytotoxic
(contact dermatitis, graft rejection)

Antigenic Stimulation

Underreaction
Loss of Immunity

Overreaction
Hypersensitivities

Cancer

Immunodeficiency

FIGURE 17.1
Overview of diseases of the immune system. Just as the system of T cells and B cells provides necessary protection against infection and disease, the same system can cause serious and debilitating conditions by overreacting or underreacting to immune stimuli.

The Immune Response: A Two-Sided Coin

With few exceptions, our previous discussions of the immune response have centered around its numerous beneficial effects. The precisely coordinated system that seeks out, recognizes, and destroys an unending array of foreign materials is clearly protective, but it also presents another side—a side that promotes rather than prevents disease. In this chapter, we will survey **immunopathology,** the study of disease states associated with overreactivity or underreactivity of the immune response (figure 17.1). In the cases of **allergies** and **autoimmunity,** the tissues are innocent bystanders attacked by excessive immunologic functions. In **grafts** and **transfusions,** a recipient reacts to the foreign tissues and cells of another individual. In **immunodeficiency diseases,** immune function is incompletely developed, suppressed, or destroyed. **Cancer** falls into a special category, because it is both a cause and an effect of immune dysfunction. As we shall see, one fascinating by-product of studies of immune disorders has been our increased understanding of the basic workings of the immune system.

Overreactions to Antigens: Allergy/Hypersensitivity

The term **allergy*** means a condition of altered reactivity or exaggerated immune response that is manifested by inflammation. Although it is sometimes used interchangeably with **hypersensitivity,** some experts refer to immediate reactions such as hay fever as allergies and to delayed reactions as hypersensitivities. Allergic individuals are acutely sensitive to repeated contact with antigens, called **allergens,** that do not noticeably affect nonallergic individuals. Although the general effects of hyperactivity are detrimental, we must be aware that it involves the very same types of immune reactions as those at work in protective immunities. These include humoral and cell-mediated actions, the inflammatory response, phagocytosis, and complement. Such an association means that all humans have the potential to develop hypersensitivity under particular circumstances.

---

* allergy (al′-er-jee) Gr. *allos,* other, and *ergon,* work.

TABLE 17.1

Hypersensitivity States

Type		Systems and Mechanisms Involved	Examples
I.	Immediate hypersensitivity	IgE-mediated; involves mast cells, basophils, and allergic mediators	Anaphylaxis, atopic allergies such as hay fever, asthma
II.	Antibody-mediated	IgG, IgM antibodies act upon cells with complement and cause cell lysis; includes some autoimmune diseases	Blood group incompatibility, pernicious anemia; myasthenia gravis
III.	Immune complex–mediated	Antibody-mediated inflammation; circulating IgG complexes deposited in basement membranes of target organs; includes some autoimmune diseases	Systemic lupus erythematosus; rheumatoid arthritis; serum sickness; rheumatic fever
IV.	T cell–mediated	Delayed hypersensitivity and cytotoxic reactions in tissues	Infection reactions; contact dermatitis; graft rejection; some types of autoimmunity such as diabetes

Originally, allergies were defined as either immediate or delayed, depending upon the time lapse between contact with the allergen and onset of symptoms. Subsequently, they were differentiated as humoral versus cell-mediated. But as information on the nature of the allergic immune response accumulated, it became evident that, although useful, these schemes oversimplified what is really a very complex spectrum of reactions. The most widely accepted classification, first introduced by immunologists P. Gell and R. Coombs, includes four major categories: type I (atopy and anaphylaxis), type II (IgG- and IgM-mediated cell damage), type III (immune complex), and type IV (delayed hypersensitivity) (table 17.1). In general, types I, II, and III involve a B-cell–immunoglobulin response, and type IV involves a T-cell response (figure 17.1). The antigens that elicit these reactions can be exogenous, originating from outside the body (microbes, pollen grains, and foreign cells and proteins), or endogenous, arising from self tissue (autoimmunities).

One of the reasons allergies are easily mistaken for infections is that both involve damage to the tissues and thus trigger the inflammatory response (see figure 14.18). Many symptoms and signs of inflammation (redness, heat, skin eruptions, edema, and granuloma) are prominent features of allergies.

CHAPTER CHECKPOINTS

Immunopathology is the study of diseases associated with excesses and deficiencies of the immune response. Such diseases include allergies, autoimmunity, grafts, transfusions, immunodeficiency disease, and cancer.

An allergy or hypersensitivity is an exaggerated immune response that injures or inflames tissues.

There are four categories of hypersensitivity reactions: type I (atopy and anaphylaxis), type II (transfusion reactions), type III (immune complex reactions), and type IV (delayed hypersensitivity reactions).

Antigens that trigger hypersensitivity reactions are allergens. They can be either exogenous (originate outside the host) or endogenous (involve the host's own tissue).

Type I Allergic Reactions: Atopy and Anaphylaxis

All type I allergies share a similar physiological mechanism, are immediate in onset, and are associated with exposure to specific antigens. However, it is convenient to recognize two subtypes: **Atopy*** is any chronic local allergy such as hay fever or asthma; **anaphylaxis*** is a systemic, often explosive reaction that involves airway obstruction and circulatory collapse. In the following sections, we will consider the epidemiology of type I allergies, allergens and routes of inoculation, mechanisms of disease, and specific syndromes.

EPIDEMIOLOGY AND MODES OF CONTACT WITH ALLERGENS

Allergies exert profound medical and economic impact. Allergists (physicians who specialize in treating allergies) estimate that about 10% to 30% of the population is prone to atopic allergy. It is generally acknowledged that self-treatment with over-the-counter medicines accounts for significant underreporting of cases. The 35 million people afflicted by hay fever (15–20% of the population) spend about half a billion dollars annually for medical treatment. The monetary loss due to employee debilitation and absenteeism is immeasurable. The majority of type I allergies are relatively mild, but certain forms such as asthma and anaphylaxis may require hospitalization and cause death. About 2 million people in the United States suffer from asthma.

The predisposition for type I allergies is inherited. Be aware that what is hereditary is a generalized *susceptibility,* not the allergy to a specific substance. For example, a parent who is allergic to ragweed pollen can have a child who is allergic to cat hair. The prospect of a child's developing atopic allergy is at least 25% if one parent is atopic, increasing up to 50% if grandparents or siblings are also afflicted. The actual basis for atopy appears to be a genetic

---

* atopy (at′-oh-pee) Gr. *atop,* out of place.

* anaphylaxis (an″-uh-fih-lax′-us) Gr. *ana,* excessive, and *phylaxis,* protection.

program that favors allergic antibody (IgE) production, increased reactivity of mast cells, and increased susceptibility of target tissue to allergic mediators. Allergic persons often exhibit a combination of syndromes, such as hay fever, eczema, and asthma.

Other factors that affect the presence of allergy are age, infection, and geographic locale. New allergies tend to crop up throughout an allergic person's life, especially as new exposures occur after moving or changing life-style. In some persons, atopic allergies last for a lifetime; others "outgrow" them, and still others suddenly develop them later in life. Some features of allergy are not yet completely explained.

THE NATURE OF ALLERGENS AND THEIR PORTALS OF ENTRY

As with other antigens, allergens have certain immunogenic characteristics. Not unexpectedly, proteins are more allergenic than carbohydrates, fats, or nucleic acids. Some allergens are haptens, non-proteinaceous substances with a molecular weight of less than 1,000 that can form complexes with carrier molecules in the body (see figure 15.11). Organic and inorganic chemicals found in industrial and household products, cosmetics, food, and drugs are commonly of this type. Table 17.2 lists a number of common allergenic substances.

Allergens typically enter through epithelial portals in the respiratory tract, gastrointestinal tract, and skin. The mucosal surfaces of the gut and respiratory system present a thin, moist surface that is normally quite penetrable. The dry, tough keratin coating of skin is less permeable, but access still occurs through tiny breaks, glands, and hair follicles. It is worth noting that the organ of allergic expression may or may not be the same as the portal of entry.

Airborne environmental allergens such as pollen, house dust, dander (shed skin scales), or fungal spores are termed *inhalants.* Each geographic region harbors a particular combination of airborne substances that varies with the season and humidity (figure 17.2*a*). Pollen, the most common offender, is given off seasonally by the reproductive structures of pines and flowering plants (weeds, trees, and grasses). Unlike pollen, mold spores are released throughout the year and are especially profuse in moist areas of the home and garden. Airborne animal hair and dander, feathers, and the saliva of dogs and cats are common sources of allergens. The component of house dust that appears to account for most dust al-

lergies is not soil or other debris, but the decomposed bodies of tiny mites that commonly live in this dust (figure 17.2*b*). Some people are allergic to their work, in the sense that they are exposed to allergens on the job. Examples include florists, beauty operators, woodworkers, farmers, drug processors, welders, and plastics manufacturers whose work can aggravate inhalant allergies.

Allergens that enter by mouth, called *ingestants,* often cause food allergies: *Injectant* allergies are an important adverse side effect of drugs or other substances used in diagnosing, treating, or preventing disease. A natural source of injectants is venom from stings by hymenopterans, a family of insects that includes honeybees and wasps. *Contactants* are allergens that enter through the skin. Many contact allergies are of the type IV, delayed variety discussed later in this chapter.

MECHANISMS OF TYPE I ALLERGY: SENSITIZATION AND PROVOCATION

What causes some people to sneeze and wheeze every time they step out into the spring air, while others suffer no ill effects? In order to answer this question, we must examine what occurs in the tissues of the allergic individual that does not occur in the normal person. In general, type I allergies develop in stages (figure 17.3). The initial encounter with an allergen provides a **sensitizing dose** that primes the immune system for a subsequent encounter with that allergen but generally elicits no signs or symptoms. The memory cells and immunoglobulin are then ready to react with a subsequent **provocative dose** of the same allergen. It is this dose that precipitates the signs and symptoms of allergy. Despite numerous anecdotal reports of people showing an allergy upon first contact with an allergen, it is generally believed that these individuals unknowingly had contact at some previous time. Fetal exposure to allergens from the mother's bloodstream is one possibility, and foods can be a prime source of "hidden" allergens such as penicillin.

The Physiology of IgE-Mediated Allergies

During primary contact and sensitization, the allergen penetrates the portal of entry (figure 17.3*a*). When large particles such as pollen grains, hair, and spores encounter a moist membrane, they release molecules of allergen that pass into the tissue fluids and lymphatics. The lymphatics then carry the allergen to the lymph nodes, where specific clones of B cells recognize it, are activated, and proliferate into plasma cells. These plasma cells produce **immunoglobulin E (IgE),** the antibody of allergy. IgE is different from other immunoglobulins in having an Fc receptor region with great affinity for mast cells and basophils. The binding of IgE to these cells in the tissues sets the scene for the reactions that occur upon repeated exposure to the same allergen (see figures 17.3*b* and 17.6).

The Role of Mast Cells and Basophils

The most important characteristics of mast cells and basophils relating to their roles in allergy are:

1. Their ubiquitous location in tissues. Mast cells are located in the connective tissue of virtually all organs, but particularly high concentrations exist in the lungs, skin, gastrointestinal tract, and genitourinary tract. Basophils circulate in the blood but migrate readily into tissues.

TABLE 17.2			
Common Allergens, Classified by Portal of Entry			
Inhalants	**Ingestants**	**Injectants**	**Contactants**
Pollen	Food	Hymenopteran	Drugs
Dust	(chocolate,	venom (bee,	Cosmetics
Mold spores	wheat, eggs,	wasp)	Heavy metals
Dander	milk, nuts,	Drugs	Detergents
Animal hair	strawberries;	Vaccines	Formalin
Insect parts	fish)	Serum	Rubber
Formalin	Food additives	Enzymes	Glue
Drugs	Drugs (aspirin,	Hormones	Solvents
Enzymes	penicillin)		Dyes

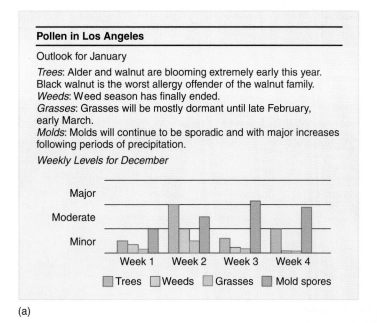

Pollen in Los Angeles

Outlook for January

Trees: Alder and walnut are blooming extremely early this year. Black walnut is the worst allergy offender of the walnut family.
Weeds: Weed season has finally ended.
Grasses: Grasses will be mostly dormant until late February, early March.
Molds: Molds will continue to be sporadic and with major increases following periods of precipitation.

Weekly Levels for December

(a)

(b)

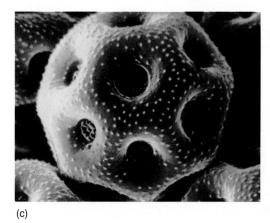

(c)

FIGURE 17.2

Monitoring airborne allergens. **(a)** The air in heavily vegetated places with a mild climate is especially laden with allergens such as pollen and mold spores. In just one month in southern California, weed pollen subsided to near zero and mold spore levels doubled. **(b)** Because the dust mite *Dermatophagoides* feeds primarily on human skin cells in house dust, these mites are found in abundance in bedding and carpets. Airborne mite feces and particles from their bodies are an important source of allergies. **(c)** Scanning electron micrograph of a single pollen grain from a rose (6,000×). Millions of these are released from a single flower.

(a) Source: Data from the Asthma and Allergy Foundation of America, Los Angeles, CA.

2. Their capacity to bind IgE during sensitization (figure 17.3). Each cell carries 30,000 to 100,000 cell receptors that attract 10,000 to 40,000 IgE antibodies.
3. Their cytoplasmic granules (secretory vesicles), which contain physiologically active cytokines (histamine, serotonin—introduced in chapter 14).
4. Their tendency to **degranulate** (figures 17.3*b* and 17.4), or release the contents of the granules into the tissues when properly stimulated by allergen.

Let us now see what occurs when sensitized cells are challenged with allergen a second time.

The Second Contact with Allergen

After sensitization, the IgE-primed mast cells can remain in the tissues for years. Even after long periods without contact, a person can retain the capacity to react immediately upon reexposure. The next time allergen molecules contact these sensitized cells, they bind across adjacent receptors and stimulate degranulation. As chemical mediators are released, they diffuse into the tissues and bloodstream. Cytokines give rise to numerous local and systemic reactions, many of which appear quite rapidly (figure 17.3*b*). The symptoms of allergy are not caused by the direct action of allergen on tissues but by the physiological effects of mast cell mediators on target organs.

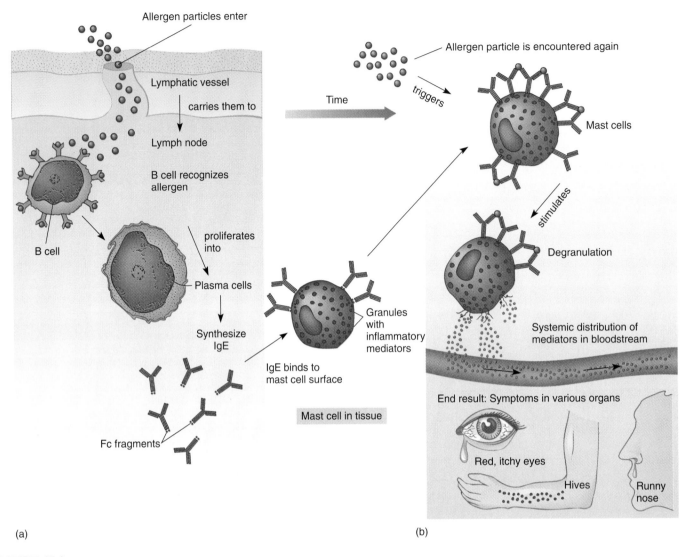

Sensitization/IgE Production

Allergen particles enter

Lymphatic vessel

carries them to

Lymph node

B cell recognizes allergen

B cell

proliferates into

Plasma cells

Synthesize IgE

IgE binds to mast cell surface

Granules with inflammatory mediators

Mast cell in tissue

Fc fragments

Time

Subsequent Exposure to Allergen

Allergen particle is encountered again

triggers

Mast cells

stimulates

Degranulation

Systemic distribution of mediators in bloodstream

End result: Symptoms in various organs

Red, itchy eyes

Hives

Runny nose

(a)

(b)

FIGURE 17.3

A schematic view of cellular reactions during the type I allergic response.   **(a)** Sensitization (initial contact with sensitizing dose).
(b) Provocation (later contacts with provocative dose).

CYTOKINES, TARGET ORGANS, AND ALLERGIC SYMPTOMS

Numerous substances involved in mediating allergy (and inflammation) have been identified. The principal chemical mediators produced by mast cells and basophils are histamine, serotonin, leukotriene, platelet-activating factor, prostaglandins, and bradykinin (figure 17.4). These chemicals, acting alone or in combination, account for the tremendous scope of allergic symptoms. For some theories pertaining to this function of the allergic response, see Medical Microfile 17.1. Targets of these mediators include the skin, upper respiratory tract, gastrointestinal tract, and conjunctiva. The general responses of these organs include rashes, itching, redness, rhinitis, sneezing, diarrhea, and shedding of tears. Systemic targets

include smooth muscle, mucous glands, and nervous tissue. Because smooth muscle is responsible for regulating the size of blood vessels and respiratory passageways, changes in its activity can profoundly alter blood flow, blood pressure, and respiration. Pain, anxiety, agitation, and lethargy are also attributable to the effects of mediators on the nervous system.

Histamine* is the most profuse and fastest-acting allergic mediator. It is a potent stimulator of smooth muscle, glands, and eosinophils. Histamine's actions on smooth muscle vary with location. It *constricts* the smooth muscle layers of the small bronchi and intestine, thereby causing labored breathing and increased

---

* histamine (his´-tah-meen) Gr. *histio*, tissue, and amine.

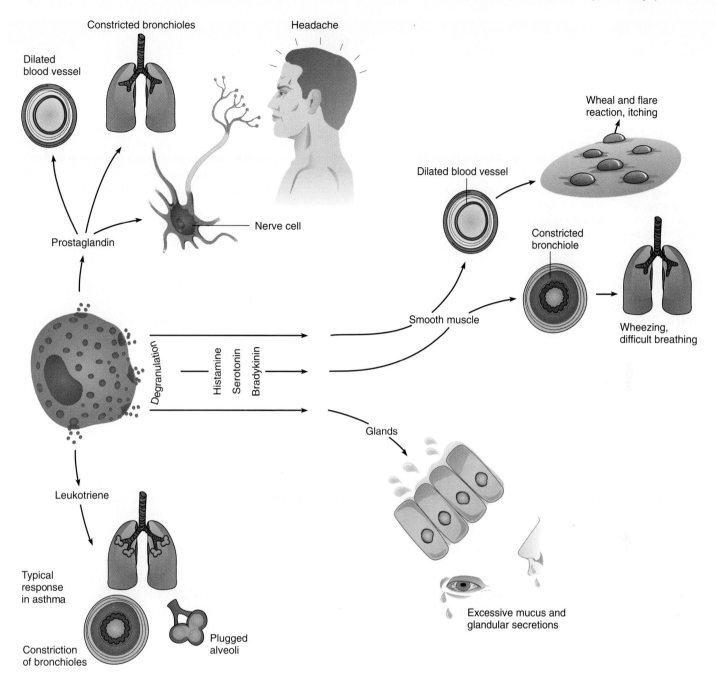

FIGURE 17.4

 The spectrum of reactions to inflammatory cytokines and the common symptoms they elicit in target tissues and organs. Note the extensive overlapping effects.

intestinal motility. In contrast, histamine *relaxes* vascular smooth muscle and dilates arterioles and venules. It is responsible for the *wheal* and flare* reaction in the skin (see figure 17.6*a*), pruritis (itching), and headache. More severe reactions (such as anaphylaxis) can be accompanied by edema and vascular dilation, which lead to hypotension, tachycardia, circulatory failure, and, fre-

quently, shock. Salivary, lacrimal, mucous, and gastric glands are also histamine targets.

Although the role of **serotonin*** in human allergy is uncertain, its effects appear to complement those of histamine. In experimental animals, serotonin increases vascular permeability, capillary dilation, smooth muscle contraction, intestinal peristalsis, and respiratory rate, but it diminishes central nervous system activity.

---

* **wheal** (weel) A smooth, slightly elevated, temporary welt that is surrounded by a flushed patch of skin (flare).

---

* **serotonin** (ser″-oh-toh′-nin) L. *serum,* whey, and *tonin,* tone.

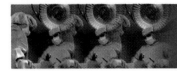

MEDICAL MICROFILE 17.1
Of What Value Is Allergy?

Why would humans and other mammals evolve an allergic response that is capable of doing so much harm and even causing death? It is unlikely that this limb of immunity exists merely to make people miserable; it must have a role in protection and survival. What are the underlying biological functions of IgE, mast cells, and the array of potent cytokines? Analysis has revealed that, although allergic persons have high levels of IgE, trace quantities are present even in the sera of nonallergic individuals, just as mast cells and inflammatory chemicals are also part of normal human physiology. It is generally believed that one important function of this system is to defend against helminth worms that are ubiquitous human parasites. In chapter 14, we learned that inflammatory mediators serve valuable functions, such as increasing blood flow and vascular permeability to summon essential immune components to an injured site. They are also responsible for increased mucous secretion, gastric motility, sneezing, and coughing, which help expel noxious agents. The difference is that, in allergic persons, the quantity and quality of these reactions are excessive and uncontrolled.

Before the specific types were identified, **leukotriene*** was known as the "slow-reacting substance of anaphylaxis" for its property of inducing gradual contraction of smooth muscle. This type of leukotriene is responsible for the prolonged bronchospasm, vascular permeability, and mucous secretion of the asthmatic individual. Other leukotrienes stimulate the activities of polymorphonuclear leukocytes.

Platelet-activating factor is a lipid released by basophils, neutrophils, monocytes, and macrophages that causes platelet aggregation and lysis. The physiological response to stimulation by this factor is similar to that of histamine, including increased vascular permeability, pulmonary smooth muscle contraction, pulmonary edema, hypotension, and a wheal and flare response in the skin.

Prostaglandins* are a group of powerful inflammatory agents. Normally, these substances regulate smooth muscle contraction (for example, they stimulate uterine contractions during delivery). In allergic reactions, they are responsible for vasodilation, increased vascular permeability, increased sensitivity to pain, and bronchoconstriction. Certain anti-inflammatory drugs work by preventing the actions of prostaglandins.

Bradykinin* is related to a group of plasma and tissue peptides known as kinins that participate in blood clotting and chemotaxis. In allergy, it causes prolonged smooth muscle contraction of the bronchioles, dilatation of peripheral arterioles, increased capillary permeability, and increased mucous secretion.

SPECIFIC DISEASES ASSOCIATED WITH IgE- AND MAST CELL–MEDIATED ALLERGY

The mechanisms just described are basic to hay fever, allergic asthma, food allergy, drug allergy, eczema, and anaphylaxis. In this section, we cover the main characteristics of these conditions, followed by methods of detection and treatment.

Atopic Diseases

Hay fever is a generic term for **allergic rhinitis,*** a seasonal reaction to inhaled plant pollen or molds, or a chronic, year-round reaction to a wide spectrum of airborne allergens or inhalants (see table 17.2). The targets are typically respiratory membranes, and the symptoms include nasal congestion; sneezing; coughing; profuse mucous secretion; itchy, red, and teary eyes; and mild bronchoconstriction.

Asthma* is a respiratory disease characterized by episodes of impaired breathing due to severe bronchoconstriction. The airways of asthmatic people are exquisitely responsive to minute amounts of inhalant allergens, food, or other stimuli, such as infectious agents. The symptoms of asthma range from occasional, annoying bouts of difficult breathing to fatal suffocation. Labored breathing, shortness of breath, wheezing, cough, and ventilatory *rales** are present to one degree or another. The respiratory tract of an asthmatic person is chronically inflamed and severely overreactive to allergy chemicals, especially leukotrienes and serotonin from pulmonary mast cells. Other pathologic components are thick mucous plugs in the air sacs and lung damage that can result in long-term respiratory compromise. An imbalance in the nervous control of the respiratory smooth muscles is apparently involved in asthma, and the episodes are influenced by the psychological state of the person, which strongly supports a neurological connection.

The number of asthma sufferers in the United States is estimated at 10 million, with nearly one-third of them children. For reasons that are not completely understood, asthma is on the increase, and deaths from it have doubled since 1982, even though effective agents to control it are more available now than they have ever been before. A recent study of inner-city children has correlated high levels of asthma to contact with cockroach antigens in their living quarters. Nearly 40% of children age 10 or younger showed extreme sensitivity to the droppings and remains of these insects.

---

* leukotriene (loo″-koh-try′-een) Gr. *leukos,* white blood cell, and *triene,* a chemical suffix.

* prostaglandin (pross″-tah-glan′-din) From prostate gland. The substance was originally isolated from semen.

* bradykinin (brad″-ee-kye′-nin) Gr. *bradys,* slow, and *kinein,* to move.

---

* rhinitis (rye-nye′-tis) Gr. *rhis,* nose, and *itis,* inflammation.

* asthma (az′-muh) The Greek word for gasping.

* rales (rails) Abnormal breathing sounds.

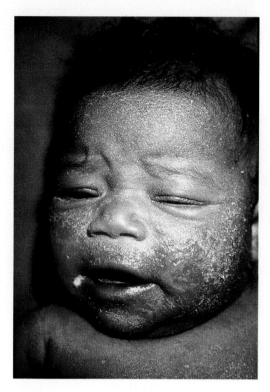

FIGURE 17.5

Atopic dermatitis, or eczema. Vesicular, encrusted lesions are typical in afflicted infants. This condition is prevalent enough to account for 1% of pediatric care.

Atopic dermatitis is an intensely itchy inflammatory condition of the skin, sometimes also called **eczema.*** Sensitization occurs through ingestion, inhalation, and, occasionally, skin contact with allergens. It usually begins in infancy with reddened, vesicular, weeping, encrusted skin lesions. It then progresses in childhood and adulthood to a dry, scaly, thickened skin condition (figure 17.5). Lesions can occur on the face, scalp, neck, and inner surfaces of the limbs and trunk. The itchy, painful lesions cause considerable discomfort, and they are often predisposed to secondary bacterial infections. An anonymous writer once aptly described eczema as "the itch that rashes" or "one scratch is too many but one thousand is not enough."

Food Allergy

The ordinary diet contains a vast variety of compounds that are potentially allergenic. It is generally believed that food allergies are due to a digestive product of the food or to an additive (preservative or flavoring). Although the mode of entry is intestinal, food allergies can also affect the skin and respiratory tract. Gastrointestinal symptoms include vomiting, diarrhea, and abdominal pain. In severe cases, nutrients are poorly absorbed, leading to growth retardation and failure to thrive in young children. Other manifestations of food allergies include eczema, hives, rhinitis, asthma, and occasionally, anaphylaxis. Classic food hypersensitivity involves IgE and degranulation of mast cells, but not all reactions involve this mechanism. The most common food allergens come from peanuts, fish, cow's milk, eggs, shellfish, and soybeans.

Drug Allergy

Modern chemotherapy has been responsible for many medical advances. Unfortunately, it has also been hampered by the fact that drugs are foreign compounds capable of stimulating allergic reactions. In fact, allergy to drugs is one of the most common side effects of treatment (present in 5–10% of hospitalized patients). Depending upon the allergen, route of entry, and individual sensitivities, virtually any tissue of the body can be affected, and reactions range from mild atopy to fatal anaphylaxis. Compounds implicated most often are antibiotics (penicillin is number one in prevalence), synthetic antimicrobics (sulfa drugs), aspirin, opiates, and anaesthetics. The actual allergen is not the intact drug itself but a hapten given off when the liver processes the drug. Some forms of penicillin sensitivity are due to the presence of small amounts of the drug in meat, milk, and other foods and to exposure to *Penicillium* mold in the environment.

ANAPHYLAXIS: AN OVERPOWERING SYSTEMIC REACTION

The term **anaphylaxis,** or **anaphylactic shock,** was first used to denote a reaction of animals injected with a foreign protein. Although the animals showed no response during the first contact, upon reinoculation with the same protein at a later time, they exhibited acute symptoms—itching, sneezing, difficult breathing, prostration, and convulsions—and many died in a few minutes. Two clinical types of anaphylaxis are distinguished in humans. *Cutaneous anaphylaxis* is the wheal and flare inflammatory reaction to the local injection of allergen. *Systemic anaphylaxis,* on the other hand, is characterized by sudden respiratory and circulatory disruption that can be fatal in a few minutes. In humans, the allergen and route of entry are variable, though bee stings and injections of antibiotics or serum are implicated most often. Bee venom is a complex material containing several allergens and enzymes that can create a sensitivity that can last for decades after exposure.

The underlying physiological events in systemic anaphylaxis parallel those of atopy, but the concentration of chemical mediators and the strength of the response are greatly amplified. The immune system of a sensitized person exposed to a provocative dose of allergen responds with a sudden, massive release of chemicals into the tissues and blood, which act rapidly on the target organs. Anaphylactic persons have been known to die in 15 minutes from complete airway blockage.

DIAGNOSIS OF ALLERGY

Because allergy mimics infection and other conditions, it is important to determine if a person is actually allergic. If possible or necessary, it is also helpful to identify the specific allergen or allergens. Allergy diagnosis involves several levels of tests, including nonspecific, specific, *in vitro,* and *in vivo* methods.

A new test that can distinguish whether a patient has experienced an allergic attack measures elevated blood levels of tryptase, an enzyme released by mast cells that increases during an allergic response. Several types of specific *in vitro* tests can determine

---

* eczema (eks′-uh-mah; also ek-zeem′-uh) Gr. *ekzeo,* to boil over.

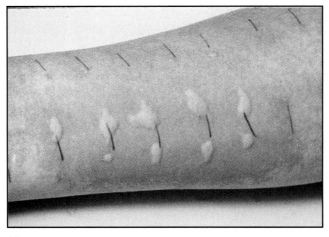

(a)

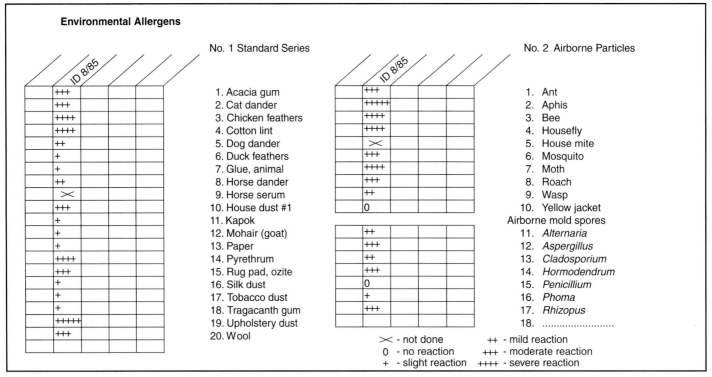

Environmental Allergens

No. 1 Standard Series

ID 8/85

| +++ |
| +++ |
| ++++ |
| ++++ |
| ++ |
| + |
| + |
| ++ |
| ⨯ |
| +++ |
| + |
| + |
| + |
| ++++ |
| +++ |
| + |
| + |
| + |
| +++++ |
| +++ |
| |

1. Acacia gum
2. Cat dander
3. Chicken feathers
4. Cotton lint
5. Dog dander
6. Duck feathers
7. Glue, animal
8. Horse dander
9. Horse serum
10. House dust #1
11. Kapok
12. Mohair (goat)
13. Paper
14. Pyrethrum
15. Rug pad, ozite
16. Silk dust
17. Tobacco dust
18. Tragacanth gum
19. Upholstery dust
20. Wool

No. 2 Airborne Particles

ID 8/85

| +++ |
| +++++ |
| ++++ |
| ++++ |
| ⨯ |
| +++ |
| ++++ |
| +++ |
| ++ |
| 0 |

1. Ant
2. Aphis
3. Bee
4. Housefly
5. House mite
6. Mosquito
7. Moth
8. Roach
9. Wasp
10. Yellow jacket

Airborne mold spores

| ++ |
| +++ |
| ++ |
| +++ |
| 0 |
| + |
| +++ |

11. *Alternaria*
12. *Aspergillus*
13. *Cladosporium*
14. *Hormodendrum*
15. *Penicillium*
16. *Phoma*
17. *Rhizopus*
18. ....

⨯ - not done   ++ - mild reaction
0 - no reaction   +++ - moderate reaction
+ - slight reaction   ++++ - severe reaction

(b)

FIGURE 17.6

A method for conducting an allergy skin test. The forearm (or back) is mapped and then injected with a selection of allergen extracts. The allergist must be very aware of potential anaphylaxis attacks triggered by these injections. **(a)** Close-up of skin wheals showing a number of positive reactions (dark lines are measurer's marks). **(b)** An actual skin test record for some common environmental allergens with a legend for assessing them.

the allergic potential of a patient's blood sample. The leukocyte histamine-release test measures the amount of histamine released from the patient's basophils when exposed to a specific allergen. Serological tests that use radioimmune assays (see chapter 16) to reveal the quantity and quality of IgE are also clinically helpful.

Skin Testing

A useful *in vivo* method to detect precise atopic or anaphylactic sensitivities is skin testing. With this technique, a patient's skin is injected, scratched, or pricked with a small amount of a pure al-

lergen extract. Hundreds of these allergen extracts contain common airborne allergens (plant and mold pollen) and more unusual allergens (mule dander, theater dust, bird feathers). Unfortunately, skin tests for food allergies using food extracts are unreliable in most cases. In patients with numerous allergies, the allergist maps the skin on the inner aspect of the forearms or back and injects the allergens intradermally according to this predetermined pattern (figure 17.6a). Approximately 20 minutes after antigenic challenge, each site is appraised for a wheal response indicative of histamine release. The diameter of the wheal is

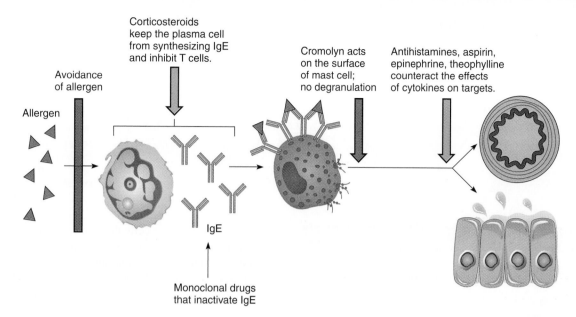

Corticosteroids keep the plasma cell from synthesizing IgE and inhibit T cells.

Avoidance of allergen

Allergen

Cromolyn acts on the surface of mast cell; no degranulation

Antihistamines, aspirin, epinephrine, theophylline counteract the effects of cytokines on targets.

IgE

Monoclonal drugs that inactivate IgE

FIGURE 17.7
Strategies for circumventing allergic attacks.

measured and rated on a scale of 0 (no reaction) to 4+ (greater than 15 mm). Figure 17.6*b* shows skin test results for a person with extreme inhalant allergies.

TREATMENT AND PREVENTION OF ALLERGY

In general, the methods of treating and preventing type I allergy involve (1) avoiding the allergen, though this may be very difficult in many instances; (2) taking drugs that block the action of lymphocytes, mast cells, or chemical mediators; and (3) undergoing desensitization therapy.

It is not possible to completely prevent initial sensitization, since there is no way to tell in advance if a person will develop an allergy to a particular substance. The practice of delaying the introduction of solid foods apparently has some merit in preventing food allergies in children, though even breast milk can contain allergens ingested by the mother. Although rigorous cleaning and air conditioning can reduce contact with airborne allergens, it is not feasible to isolate a person from all allergens, which is the reason drugs are so important in control.

Therapy to Counteract Allergies

The aim of antiallergy medication is to block the progress of the allergic response somewhere along the route between IgE production and the appearance of symptoms (figure 17.7). Oral anti-inflammatory drugs such as corticosteroids inhibit the activity of lymphocytes and thereby reduce the production of IgE, but they also have dangerous side effects and should not be taken for prolonged periods. Some drugs block the degranulation of mast cells and reduce the levels of inflammatory cytokines. The most effective of these are diethylcarbamazine and cromolyn. Asthma and rhinitis sufferers can find relief with a new drug that blocks synthesis of leukotriene and a monoclonal antibody that inactivates IgE (Xolair).

Widely used medications for preventing symptoms of atopic allergy are *antihistamines,* the active ingredients in most over-the-counter allergy-control drugs. Antihistamines interfere with histamine activity by binding to histamine receptors on target organs. Most of them have major side effects, however, such as drowsiness. Newer antihistamines lack this side effect because they do not cross the blood-brain barrier. Other drugs that relieve inflammatory symptoms are aspirin and acetaminophen, which reduce pain by interfering with prostaglandin, and theophylline, a bronchodilator that reverses spasms in the respiratory smooth muscles. Persons who suffer from anaphylactic attacks are urged to carry at all times injectable epinephrine (adrenaline) and an identification tag indicating their sensitivity. An aerosol inhaler containing epinephrine can also provide rapid relief. Epinephrine reverses constriction of the airways and slows the release of allergic mediators.

Approximately 70% of allergic patients benefit from controlled injections of specific allergens as determined by skin tests. This technique, called **desensitization** or **hyposensitization,** is a therapeutic way to prevent reactions between allergen, IgE, and mast cells. The allergen preparations contain pure, preserved suspensions of plant antigens, venoms, dust mites, dander, and molds (but so far, hyposensitization for foods has not proved very effective). The immunologic basis of this treatment is open to differences in interpretation. One theory suggests that injected allergens stimulate the formation of high levels of allergen-specific IgG (figure 17.8). It has been proposed that these IgG *blocking antibodies* remove allergen from the system before it can bind to IgE, thus preventing the degranulation of mast cells. It is also possible that allergen delivered in this fashion combines with the IgE itself and prevents it from reacting with the mast cells. Some experts suggest that the therapy induces specific clones of suppressor T cells to block the production of IgE by B cells.

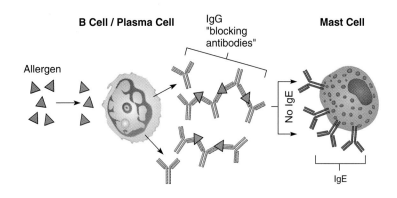

B Cell / Plasma Cell

IgG "blocking antibodies"

Mast Cell

Allergen

No IgE

IgE

FIGURE 17.8

 **The blocking antibody theory for allergic desensitization.** An injection of allergen causes IgG antibodies to be formed instead of IgE; these blocking antibodies cross-link and effectively remove the allergen before it can react with the IgE in the mast cell.

CHAPTER CHECKPOINTS

Type I hypersensitivity reactions result from excessive IgE production in response to an exogenous antigen.

The two kinds of type I hypersensitivities are atopy, a chronic, local allergy, and anaphylaxis, a systemic, potentially fatal allergic response.

The predisposition to type I hypersensitivities is inherited, but age, geographic locale, and infection also influence allergic response.

Type I allergens include inhalants, ingestants, injectants, and contactants.

The portals of entry for type I antigens are the skin, respiratory tract, gastrointestinal tract, and genitourinary tract.

Type I hypersensitivities are set up by a sensitizing dose of allergen and expressed when a second provocative dose triggers the allergic response. The time interval between the two can be many years.

The primary participants in type I hypersensitivities are IgE, basophils, mast cells, and agents of the inflammatory response.

Allergies are diagnosed by a variety of *in vitro* and *in vivo* tests that assay specific cells, IgE, and local reactions.

Allergies are treated by medications that interrupt the allergic response at certain points. Allergic reactions can often be prevented by desensitization therapy.

Type II Hypersensitivities: Reactions That Lyse Foreign Cells

The diseases termed type II hypersensitivities are a complex group of syndromes that involve complement-assisted destruction (lysis) of cells by antibodies (IgG and IgM) directed against those cells' surface antigens. This category includes transfusion reactions and some types of autoimmunities (discussed in a later section). The cells targeted for destruction are often red blood cells, but other cells can be involved.

HUMAN BLOOD TYPES

Chapters 14 and 15 described the functions of unique surface receptors or markers on cell membranes. Ordinarily, these receptors play essential roles in transport, recognition, and development, but they become medically important when the tissues of one person are placed into the body of another person. Blood transfusions and organ donations introduce alloantigens (molecules that differ in the

same species) on donor cells that are recognized by the lymphocytes of the recipient. These reactions are not really immune dysfunctions as allergy and autoimmunity are. The immune system is in fact working normally, but it is not equipped to distinguish between the desirable foreign cells of a transplanted tissue and the undesirable ones of a microbe.

THE BASIS OF HUMAN ABO ANTIGENS AND BLOOD TYPES

The existence of human blood types was first demonstrated by an Austrian pathologist, Karl Landsteiner, in 1904. While studying incompatibilities in blood transfusions, he found that the serum of one person could clump the red blood cells of another. Landsteiner identified four distinct types, subsequently called the ABO blood groups.

Like the MHC antigens on white blood cells, the ABO antigen markers on red blood cells are genetically determined and composed of glycoproteins. These ABO antigens are inherited as two (one from each parent) of three alternative *alleles:** A, B, or O. A and B alleles are dominant over O and codominant with one another. As table 17.3 indicates, this mode of inheritance gives rise to four blood types (phenotypes), depending on the particular combination of genes. Thus, a person with an *AA* or *AO* genotype has **type A** blood; genotype *BB* or *BO* gives **type B;** genotype *AB*

---

* allele (ah-leel′) Gr. *allelon*, of one another. An alternate form of a gene for a given trait.

TABLE 17.3

Characteristics of ABO Blood Groups

Genotype	Phenotype A or B* RBC Antigen	Prevalence in Population**	Serum Content of Antibodies
OO	Neither	Most common	Both anti-a and anti-b
AA, AO	A	Second most common	Anti-b
BB, BO	B	Third most common	Anti-a
AB	AB	Least common	Neither antibody

*Capital letters generally denote antigen; lowercase denotes antibody.
**True of most large populations of mixed racial and ethnic groups.

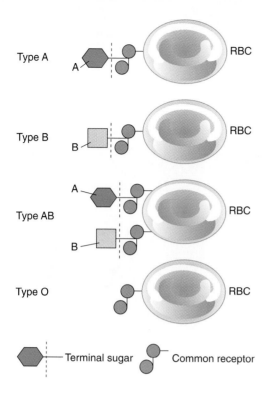

FIGURE 17.9

The genetic/molecular basis for the A and B antigens (receptors) on red blood cells. In general, persons with blood types A, B, and AB inherit a gene for the enzyme that adds a certain terminal sugar to the basic RBC receptor. Type O persons do not have such an enzyme and lack the terminal sugar.

produces **type AB;** and genotype *OO* produces **type O.** Some important points about the blood types are: (1) They are named for the dominant antigen(s); (2) the RBCs of type O persons have antigens, but not A and B antigens; and (3) tissues other than RBCs carry A and B antigens.

The actual origin of the AB antigens and blood types are shown in figure 17.9. The A and B genes each code for an enzyme that adds a terminal carbohydrate to RBC receptors during maturation. RBCs of type A contain an enzyme that adds N-acetylgalactosamine to the receptor; RBCs of type B have an enzyme that adds D-galactose; RBCs of type AB contain both enzymes that add both carbohydrates; and RBCs of type O lack the genes and enzymes to add a terminal molecule.

ANTIBODIES AGAINST A AND B ANTIGENS

Although an individual does not normally produce antibodies in response to his or her own RBC antigens, the serum can contain antibodies that react with blood of another antigenic type, even though contact with this other blood type has *never* occurred. These performed antibodies account for the immediate and intense quality of transfusion reactions. As a rule, type A blood contains antibodies (anti-b) that react against the B antigens on type B and AB red blood cells. Type B blood contains antibodies (anti-a) that react with A antigen on type A and AB red blood

cells. Type O blood contains antibodies against both A and B antigens. Type AB blood does not contain antibodies against either A or B antigens[1] (table 17.3). What is the source of these anti-a and anti-b antibodies? It appears that they develop in early infancy because of exposure to certain heterophile antigens that are widely distributed in nature. These antigens are surface molecules on bacteria and plant cells that mimic the structure of A and B antigens. Exposure to these sources stimulates the production of corresponding antibodies.[2]

Clinical Concerns in Transfusions

The presence of ABO antigens and a, b antibodies underlie several clinical concerns in giving blood transfusions. First, the individual blood types of donor and recipient must be determined. By use of a standard technique, drops of blood are mixed with antisera that contain antibodies against the A and B antigens and are then observed for the evidence of agglutination (figure 17.10).

Knowing the blood types involved makes it possible to determine which transfusions are safe to do. The general rule of compatibility is that the RBC antigens of the donor must not be agglutinated by antibodies in the recipient's blood (figure 17.11). The ideal practice is to transfuse blood that is a perfect match (A to A, B to B). But even in this event, blood samples must be cross-matched before the transfusion because other blood group incompatibilities can exist. This test involves mixing the blood of the donor with the serum of the recipient to check for agglutination.

Under certain circumstances (emergencies, the battlefield), the concept of universal transfusions can be used. To appreciate how this works, we must apply the rule stated in the previous paragraph. Type O blood lacks A and B antigens and will not be agglutinated by other blood types, so it could theoretically be used in any transfusion. Hence, a person with this blood type is called a *universal donor*. Because type AB blood lacks agglutinating antibodies, an individual with this blood could conceivably receive any type of blood. Type AB persons are consequently called *universal recipients*. Although both types of transfusions involve antigen-antibody incompatibilities, these are of less concern because of the dilution of the donor's blood in the body of the recipient. Additional RBC markers that can be significant in transfusions are the Rh, MN, and Kell antigens (see next sections).

Transfusion of the wrong blood type causes various degrees of adverse reaction. The severest reaction is massive hemolysis when the donated red blood cells react with recipient antibody and trigger the complement cascade (figure 17.11). The resultant destruction of red cells leads to systemic shock and kidney failure brought on by the blockage of glomeruli (blood-filtering apparatus) by cell debris. Death is a common outcome. Other reactions caused by RBC destruction are fever, anemia, and jaundice. A transfusion reaction is managed by immediately halting the transfusion, administering drugs to remove hemoglobin from the blood, and beginning another transfusion with red blood cells of the correct type.

---

1. Why would this be true? The answer lies in the first sentence of the paragraph.

2. Evidence comes from germ-free chickens, which do not have antibodies against the antigens of blood types, whereas normal chickens possess these antibodies.

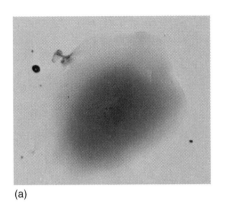

(a)

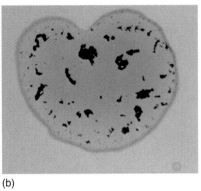

(b)

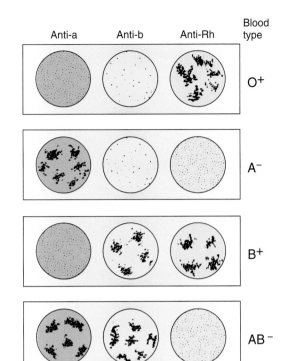

Anti-a	Anti-b	Anti-Rh	Blood type
			O⁺
			A⁻
			B⁺
			AB⁻

(c)

FIGURE 17.10

Interpretation of blood typing. In this test, a drop of blood is mixed with a specially prepared antiserum known to contain antibodies against the A, B, or Rh antigens. **(a)** If that particular antigen is not present, the red blood cells in that droplet do not agglutinate and form an even suspension. **(b)** If that antigen is present, agglutination occurs and the RBCs form visible clumps. **(c)** Several patterns and their interpretations. Anti-a, anti-b, and anti-Rh are shorthand for the antiserum applied to the drops. (In general, O^+ is the most common blood type, and AB^- is the rarest.)

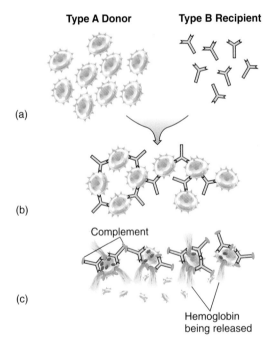

Type A Donor   **Type B Recipient**

(a)

(b)

Complement

(c)

Hemoglobin being released

FIGURE 17.11

⊙ **Microscopic view of a transfusion reaction.** **(a)** Incompatible blood. The red blood cells of the type A donor contain antigen A, while the serum of the type B recipient contains anti-a antibodies that can agglutinate donor cells. **(b)** Agglutination particles can block the circulation in vital organs. **(c)** Activation of the complement by antibody on the RBCs can cause hemolysis and anemia. This sort of incorrect transfusion is very rare because of the great care taken by blood banks to ensure a correct match.

THE RH FACTOR AND ITS CLINICAL IMPORTANCE

Another RBC antigen of major clinical concern is the **Rh factor** (or D antigen). This factor was first discovered in experiments exploring the genetic relationships among animals. Rabbits inoculated with the RBCs of rhesus monkeys produced an antibody that also reacted with human RBCs. Further tests showed that this monkey antigen (termed Rh for rhesus) was present in about 85% of humans and absent in the other 15%. The details of Rh inheritance are more complicated than those of ABO, but in simplest terms, a person's Rh type results from a combination of two possible alleles—a dominant one that codes for the factor and a recessive one that does not. A person inheriting at least one Rh gene will be Rh^+; only those persons inheriting two recessive genes are Rh^-. This factor is denoted by a symbol above the blood type, as in O^+ or AB^- (see figure 17.10c). However, unlike the ABO antigens, exposure to normal flora does not sensitize Rh^- persons to the Rh factor. The only ways one can develop antibodies against this factor are through placental sensitization or transfusion.

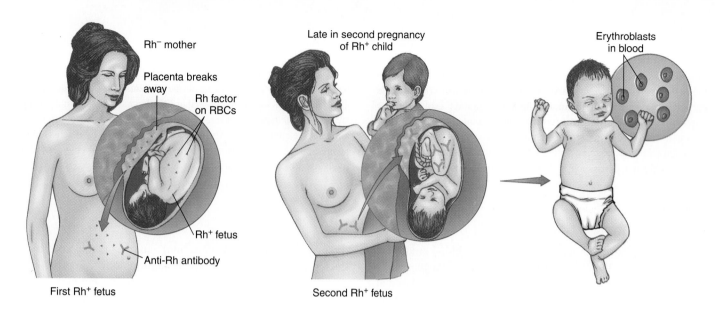

(a) **The development and aftermath of Rh sensitization.**

Rh⁻ mother

Placenta breaks away

Rh factor on RBCs

Rh⁺ fetus

Anti-Rh antibody

First Rh⁺ fetus

Late in second pregnancy of Rh⁺ child

Second Rh⁺ fetus

Erythroblasts in blood

(b) **Prevention of erythroblastosis fetalis with anti-Rh immune globulin (RhoGAM)**

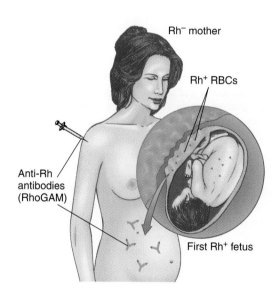

Rh⁻ mother

Rh⁺ RBCs

Anti-Rh antibodies (RhoGAM)

First Rh⁺ fetus

FIGURE 17.12

Development and control of Rh incompatibility.   **(a)** Initial sensitization of the maternal immune system to fetal Rh factor occurs when fetal cells leak into the mother's circulation late in pregnancy, or during delivery, when the placenta tears away. The child will escape hemolytic disease in most instances, but the mother, now sensitized, will be capable of an immediate reaction to a second Rh⁺ fetus and its Rh-factor antigen. At that time, the mother's anti-Rh antibodies pass into the fetal circulation and elicit severe hemolysis in the fetus and neonate. **(b)** Prevention of erythroblastosis fetalis with anti-Rh immune globulin (RhoGAM). Injecting a mother who is at risk with RhoGAM during her first Rh⁺ pregnancy helps to inactivate and remove the fetal Rh-positive cells before her immune system can react and develop sensitivity.

Hemolytic Disease of the Newborn and Rh Incompatibility

The potential for placental sensitization occurs when a mother is Rh⁻ and her unborn child is Rh⁺. The obvious intimacy between mother and fetus makes it possible for fetal RBCs to leak into the mother's circulation during childbirth, when the detachment of the placenta creates avenues for fetal blood to enter the maternal circulation. The mother's immune system detects the foreign Rh factors on the fetal RBCs and is sensitized to them by producing antibodies and memory B cells. The first Rh⁺ child is usually not affected because the process begins so late in pregnancy that the child is born before maternal sensitization is completed. However, the mother's immune system has been strongly primed for a second contact with this factor in a subsequent pregnancy (figure 17.12*a*).

In the next pregnancy with an Rh⁺ fetus, fetal blood cells escape into the maternal circulation late in pregnancy and elicit a memory response. The fetus is at risk when the maternal anti-Rh antibodies cross the placenta into the fetal circulation, where they affix to fetal RBCs and cause complement-mediated lysis. The outcome is a potentially fatal **hemolytic disease of the newborn (HDN)** called *erythroblastosis fetalis* (eh-rith″-roh-blas-toh′-sis fee-tal′-is). This term is derived from the presence of immature nucleated RBCs called erythroblasts in the blood. They are released into the infant's circulation to compensate for the massive destruction of RBCs by maternal antibodies. Additional symptoms are severe anemia, jaundice, and enlarged spleen and liver.

Maternal-fetal incompatibilities are also possible in the ABO blood group, but adverse reactions occur less frequently than with

MEDICAL MICROFILE 17.2

Why Doesn't a Mother Reject Her Fetus?

Think of it: Even though mother and child are genetically related, the father's genetic contribution guarantees that the fetus will contain molecules that are antigenic to the mother. In fact, with the recent practice of implanting one woman with the fertilized egg of another woman, the surrogate mother is carrying a fetus that has no genetic relationship to her. Yet, even with this essentially foreign body inside the mother, dangerous immunologic reactions such as Rh incompatibility are rather rare. In what ways do fetuses avoid the surveillance of the mother's immune system? The answer appears to lie in the placenta and embryonic tissues. The fetal components that contribute to these tissues are not strongly antigenic, and they form a barrier that keeps the fetus isolated in its own antigen-free environment. The placenta is surrounded by a dense, many-layered envelope that prevents the passage of maternal cells, and it actively absorbs, removes, and inactivates circulating antigens.

Rh sensitization because the antibodies to these blood group antigens are IgM rather than IgG and are unable to cross the placenta in large numbers. In fact, the maternal-fetal relationship is a fascinating instance of foreign tissue not being rejected, despite the extensive potential for contact (Medical Microfile 17.2).

Preventing Hemolytic Disease of the Newborn

Once sensitization of the mother to Rh factor has occurred, all other Rh^+ fetuses will be at risk for hemolytic disease of the newborn. Prevention requires a careful family history of an Rh^- pregnant woman. It can predict the likelihood that she is already sensitized or is carrying an Rh^+ fetus. It must take into account other children she has had, their Rh types, and the Rh status of the father. If the father is also Rh^-, the child will be Rh^- and free of risk, but if the father is Rh^+, the probability that the child will be Rh^+ is 50% or 100%, depending on the exact genetic makeup of the father. If there is any possibility that the fetus is Rh^+, the mother must be passively immunized with antiserum containing antibodies against the Rh factor (Rh_o [D] *immune globulin*, or RhoGAM*). This antiserum, injected at 28 to 32 weeks and again immediately after delivery, reacts with any fetal RBCs that have escaped into the maternal circulation, thereby preventing the sensitization of the mother's immune system to Rh factor (figure 17.12b). Anti-Rh antibody must be given with each pregnancy that involves an Rh^+ fetus. It is ineffective if the mother has already been sensitized by a prior Rh^+ fetus or an incorrect blood transfusion, which can be determined by a serological test. As in ABO blood types, the Rh factor should be matched for a transfusion, although it is acceptable to transfuse Rh^- blood if the Rh type is not known.

OTHER RBC ANTIGENS

Although the ABO and Rh systems are of greatest medical significance, about 20 other red blood cell antigen groups have been discovered. Examples are the *MN, Ss, Kell,* and *P* blood groups. Because of incompatibilities that these blood groups present, transfused blood is screened to prevent possible cross-reactions. The study of these blood antigens (as well as ABO and Rh) has given rise to other useful applications. For example, they can be useful in

forensic medicine (crime detection), studying ethnic ancestry, and tracing prehistoric migrations in anthropology. Many blood cell antigens are remarkably hardy and can be detected in dried blood stains, semen, and saliva. Even the 2,000-year-old mummy of King Tutankhamen has been typed A_2MN!

CHAPTER CHECKPOINTS

Type II hypersensitivity reactions occur when preformed antibodies react with foreign cell–bound antigens. The most common type II reactions occur when transfused blood is mismatched to the recipient's ABO type. IgG or IgM antibodies attach to the foreign cells, resulting in complement fixation. The resultant formation of membrane attack complexes lyses the donor cells.

Type II hypersensitivities are stimulated by antibodies formed against red blood cell (RBC) antigens or against other cell-bound antigens following prior exposure.

Complement, IgG, and IgM antibodies are the primary mediators of type II hypersensitivities.

The concepts of universal donor (type O) and universal recipient (type AB) apply only under emergency circumstances. Cross-matching donor and recipient blood is necessary to determine which transfusions are safe to perform.

Type II hypersensitivities can also occur when Rh^- mothers are sensitized to Rh^+ RBCs of their unborn babies and the mother's anti-Rh antibodies cross the placenta, causing hemolysis of the newborn's RBCs. This is called hemolytic disease of the newborn, or erythroblastosis fetalis.

Type III Hypersensitivities: Immune Complex Reactions

Type III hypersensitivity involves the reaction of soluble antigen with antibody and the deposition of the resulting complexes in basement membranes of epithelial tissue. It is similar to type II, because it involves the production of IgG and IgM antibodies after repeated exposure to antigens and the activation of complement. Type III differs from type II because its antigens are not attached to the surface of a cell. The interaction of these antigens with antibodies

---

* RhoGAM Immunoglobulin fraction of human anti-Rh serum, prepared from pooled human sera.

produces free-floating complexes that can be deposited in the tissues, causing an **immune complex reaction** or disease. This category includes therapy-related disorders (serum sickness and the Arthus reaction) and a number of autoimmune diseases (such as glomerulonephritis and lupus erythematosus).

MECHANISMS OF IMMUNE COMPLEX DISEASE

After initial exposure to a profuse amount of antigen, the immune system produces large quantities of antibodies that circulate in the fluid compartments. When this antigen enters the system a second time, it reacts with the antibodies to form antigen-antibody complexes (figure 17.13). These complexes summon various inflammatory components such as complement and neutrophils, which would ordinarily eliminate Ag-Ab complexes as part of the normal immune response. In an immune complex disease, however, these complexes are so abundant that they deposit in the *basement membranes** of epithelial tissues and become inaccessible. In response to these events, neutrophils release lysosomal granules that digest tissues and cause a destructive inflammatory condition. The symptoms of type III hypersensitivities are due in great measure to this pathologic state.

TYPES OF IMMUNE COMPLEX DISEASE

During the early tests of immunotherapy using animals, hypersensitivity reactions to serum and vaccines were common. In addition to anaphylaxis, two syndromes, the **Arthus reaction**[3] and **serum sickness,** were identified. These syndromes are associated with certain types of passive immunization (especially with animal serum).

Serum sickness and the Arthus reaction are like anaphylaxis in requiring sensitization and preformed antibodies. Characteristics that set them apart are: (1) They depend upon IgG, IgM, or IgA (precipitating antibodies) rather than IgE; (2) they require large doses of antigen (not a miniscule dose as in anaphylaxis); and (3) they have delayed symptoms (a few hours to days). The Arthus reaction and serum sickness differ from each other in some important ways. The Arthus reaction is a *localized* dermal injury due to inflamed blood vessels in the vicinity of any injected antigen. Serum sickness is a *systemic* injury initiated by antigen-antibody complexes that circulate in the blood and settle into membranes at various sites.

The Arthus Reaction

The Arthus reaction is usually an acute response to a second injection of vaccines (boosters) or drugs at the same site as the first injection. In a few hours, the area becomes red, hot to the touch, swollen, and very painful. These symptoms are mainly due to the destruction of tissues in and around the blood vessels and the release of histamine from mast cells and basophils. Although the reaction is usually self-limiting and rapidly cleared, intravascular blood clotting can occasionally cause necrosis and loss of tissue.

---

3. Named after Maurice Arthus, the physiologist who first identified this localized inflammatory response.

* **Basement membranes** are basal partitions of epithelia that normally filter out circulating antigen-antibody complexes.

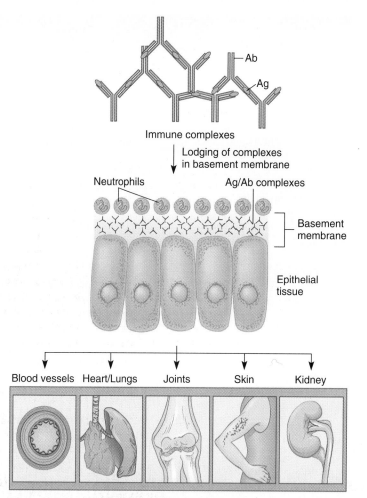

Major organs that can be targets of immune complex deposition

FIGURE 17.13

The background of immune complex disease. In general, circulating immune complexes become lodged in the basement membrane of the epithelia and cause vascular damage and organ malfunction.

Serum Sickness

Serum sickness was named for a condition that appeared in soldiers after repeated injections of horse serum to treat tetanus. It can also be caused by injections of animal hormones and drugs. The immune complexes enter the circulation, are carried throughout the body, and are eventually deposited in blood vessels of the kidney, heart, skin, and joints (figure 17.13). The condition can become chronic, causing symptoms such as enlarged lymph nodes, rashes, painful joints, swelling, fever, and renal dysfunction.

AN INAPPROPRIATE RESPONSE AGAINST SELF, OR AUTOIMMUNITY

The immune diseases we have covered so far are all caused by foreign antigens. In the case of **autoimmunity,** an individual actually develops hypersensitivity to himself. This pathologic process accounts for **autoimmune diseases,** in which **autoantibodies** and, in certain cases, T cells mount an abnormal attack against self antigens. The scope of autoimmune diseases is extremely varied. In

TABLE 17.4

Selected Autoimmune Diseases

Disease	Target	Type of Hypersensitivity	Characteristics
Systemic lupus erythematosus (SLE)	Systemic	III	Inflammation of many organs; antibodies against red and white blood cells, platelets, clotting factors, nucleus
Rheumatoid arthritis and ankylosing spondylitis	Systemic	III	Vasculitis; frequent target is joint lining; antibodies against other antibodies (rheumatoid factor)
Scleroderma	Systemic	II	Excess collagen deposition in organs; antibodies formed against many intracellular organelles
Hashimoto's thyroiditis	Thyroid	II	Destruction of the thyroid follicles
Graves disease	Thyroid	II	Antibodies against thyroid-stimulating hormone receptors
Pernicious anemia	Stomach lining	II	Antibodies against receptors prevent transport of vitamin B_{12}
Myasthenia gravis	Muscle	II	Antibodies against the acetylcholine receptors on the nerve-muscle junction alter function
Type I diabetes	Pancreas	II	Antibodies stimulate destruction of insulin-secreting cells
Type II diabetes	Insulin receptor	II	Antibodies block attachment of insulin
Multiple sclerosis	Myelin	II	T cells and antibodies sensitized to myelin sheath destroy neurons
Goodpasture syndrome (glomerulonephritis)	Kidney	II	Antibodies to basement membrane of the glomerulus damage kidneys
Rheumatic fever	Heart	II	Antibodies to group A *Streptococcus* cross-react with heart tissue

general, they can be differentiated as *systemic,* involving several major organs, or *organ-specific,* involving only one organ or tissue. They usually fall into the categories of type II or type III hypersensitivity, depending upon how the autoantibodies bring about injury. Some major autoimmune diseases, their targets, and basic pathology are presented in table 17.4.

Genetic and Gender Correlation in Autoimmune Disease

In most cases, the precipitating cause of autoimmune disease remains obscure, but we do know that susceptibility is determined by genetics and influenced by gender. Cases cluster in families, and even unaffected members tend to develop the autoantibodies for that disease. More direct evidence comes from studies of the major histocompatibility gene complex. Particular genes in the class I and II major histocompatibility complex (see figure 15.3) coincide with certain autoimmune diseases. For example, autoimmune joint diseases such as rheumatoid arthritis and ankylosing spondylitis are more common in persons with the B-27 HLA type; systemic lupus erythematosus, Graves disease, and myasthenia gravis are associated with the B-8 HLA antigen. Why autoimmune diseases (except ankylosing spondylitis) afflict more females than males also remains a mystery. Females are more susceptible during childbearing years than before puberty or after menopause, suggesting a possible hormonal relationship.

The Origins of Autoimmune Disease

Very low titers of autoantibodies in otherwise healthy individuals suggest some normal function for them. A moderate, regulated amount of autoimmunity is probably required to dispose of old cells and cellular debris. Disease apparently arises when this regulatory or recognition apparatus goes awry. Attempts to explain the origin of autoimmunity include the following theories.

The *sequestered antigen theory* explains that during embryonic growth, some tissues are immunologically privileged; that is, they are sequestered behind anatomical barriers and cannot be scanned by the immune system (figure 17.14a). Examples of these sites are regions of the central nervous system, which are shielded by the meninges and blood-brain barrier; the lens of the eye, which is enclosed by a thick sheath; and antigens in the thyroid and testes, which are sequestered behind an epithelial barrier. Eventually the antigen becomes exposed by means of infection, trauma, or deterioration, and is perceived by the immune system as a foreign substance.

According to the *clonal selection theory,* the immune system of a fetus develops tolerance by eradicating all self-reacting lymphocyte clones, called *forbidden clones,* while retaining only those clones that react to foreign antigens. Some of these clones may survive, and since they have not been subjected to this tolerance process, they can attack tissues with self antigens.

The *theory of immune deficiency* proposes that mutations in the receptor genes of some lymphocytes render them reactive to self or that a general breakdown in the normal T-suppressor function sets the scene for inappropriate immune responses.

Some autoimmune diseases appear to be caused by *molecular mimicry,* in which microbial antigens bear molecular determinants similar to normal human cells. An infection could cause formation of antibodies that can cross-react with tissues. This is one purported explanation for the pathology of rheumatic fever. Autoimmune disorders such as type I diabetes and multiple sclerosis

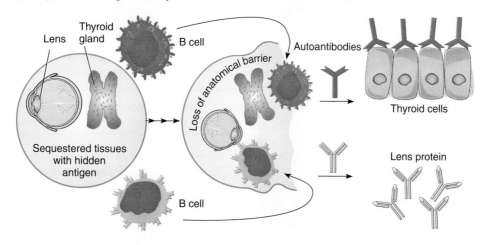

(a) **Sequestered Antigen Theory**

Lens

Thyroid gland

B cell

Loss of anatomical barrier

Autoantibodies

Thyroid cells

Sequestered tissues with hidden antigen

B cell

Lens protein

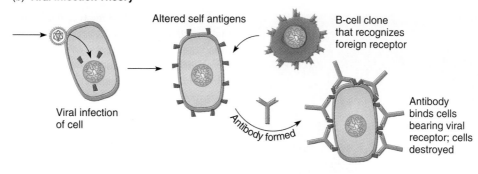

(b) **Viral Infection Theory**

Altered self antigens

B-cell clone that recognizes foreign receptor

Viral infection of cell

Antibody formed

Antibody binds cells bearing viral receptor; cells destroyed

FIGURE 17.14

Possible explanations for autoimmunity. **(a)** Self antigens are sequestered and later incorrectly identified as a foreign antigen by B lymphocytes. **(b)** Self antigens are altered by viral infection, which causes an immune response against the perceived foreign antigens.

are likely triggered by *viral infection*. Viruses can noticeably alter cell receptors, thereby causing immune cells to attack the tissues bearing viral receptors (figure 17.14*b*).

Examples of Autoimmune Disease

Systemic Autoimmunities One of the severest chronic autoimmune diseases is **systemic lupus erythematosus*** (SLE, or lupus). This name originated from the characteristic rash that spreads across the nose and cheeks in a pattern suggesting the appearance of a wolf (figure 17.15*a*). Although the manifestations of the disease vary considerably, all patients produce autoantibodies against a great variety of organs and tissues. The organs most involved are the kidneys, bone marrow, skin, nervous system, joints, muscles, heart, and GI tract. Antibodies to intracellular materials such as the nucleoprotein of the nucleus and mitochondria are also common.

In SLE, autoantibody-autoantigen complexes appear to be deposited in the basement membranes of various organs. Kidney failure, blood abnormalities, lung inflammation, myocarditis, and skin lesions are the predominant symptoms. One form of chronic

lupus (called discoid) is influenced by exposure to the sun and primarily afflicts the skin. The etiology of lupus is still a puzzle. It is not known how such a generalized loss of self-tolerance arises, though viral infection or loss of T-cell suppressor function are suspected. The fact that women of childbearing years account for 90% of cases indicates that hormones may be involved. The diagnosis of SLE can usually be made with blood tests. Antibodies against the nucleus (ANA) and various tissues (detected by indirect fluorescent antibody or radioimmune assay techniques) are common, and a positive test for the lupus factor (an antinuclear factor) is also very indicative of the disease.

Rheumatoid arthritis,* another systemic autoimmune disease, incurs progressive, debilitating damage to the joints. In some patients, the lung, eye, skin, and nervous system are also involved. In the joint form of the disease, autoantibodies form immune complexes that bind to the synovial membrane of the joints and activate phagocytes and stimulate release of cytokines. Chronic inflammation leads to scar tissue and joint destruction. The joints in the hands and feet are affected first, followed by the knee and hip joints (figure 17.15*b*). The precipitating cause in rheumatoid arthritis is

---

* systemic lupus erythematosus (sis-tem′-ik loo′-pis air″-uh-theem-uh-toh′-sis) L. *lupus,* wolf, and *erythema,* redness.

* rheumatoid arthritis (roo′-muh-toyd ar-thry′-tis) Gr. *rheuma,* a moist discharge, and *arthron,* joint.

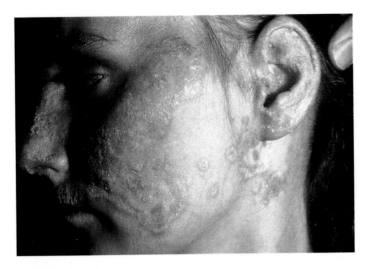

(a)

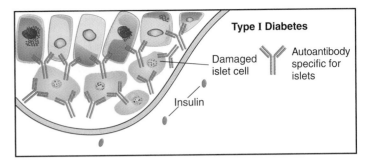

FIGURE 17.16

The autoimmune component in diabetes mellitus, type I.
Autoantibodies produced against the beta cells of the islets of
Langerhans destroy the cells and greatly reduce insulin synthesis.

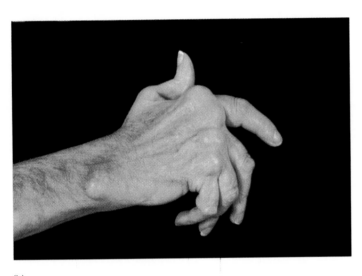

(b)

FIGURE 17.15

Common autoimmune diseases. **(a)** Systemic lupus erythematosus.
One symptom is a prominent rash across the bridge of the nose and on
the cheeks. These papules and blotches can also occur on the chest
and limbs. **(b)** Rheumatoid arthritis commonly targets the synovial
membrane of joints. Over time, chronic inflammation causes thickening
of this membrane, erosion of the articular cartilage, and fusion of the
joint. These effects severely limit motion and can eventually swell and
distort the joints.

not known, though infectious agents such as Epstein-Barr virus
have been suspected. The most common feature of the disease is the
presence of an IgM antibody, called rheumatoid factor (RF), directed against other antibodies. This does not cause the disease but
is used mainly in diagnosis. Some relief can be achieved with anti-
inflammatory agents, immunosuppressive drugs, and gold salt injections in some individuals.

Autoimmunities of the Endocrine Glands   On occasion, the
thyroid gland is the target of autoimmunity. The underlying cause
of **Graves disease** is the attachment of autoantibodies to receptors

on the follicle cells that secrete the hormone thyroxin. The abnormal stimulation of these cells causes the overproduction of this hormone and the symptoms of hyperthyroidism. In **Hashimoto
thyroiditis,** both autoantibodies and T cells are reactive to the thyroid gland, but in this instance, they reduce the levels of thyroxin by
destroying follicle cells and by inactivating the hormone. As a result of these reactions, the patient suffers from hypothyroidism.

The pancreas and its hormone, insulin, are other autoimmune targets. Insulin, secreted by the beta cells in the pancreas,
regulates and is essential to the utilization of glucose by cells.
Diabetes mellitus is caused by a dysfunction in insulin production
or utilization (figure 17.16). Type I diabetes (also termed insulin-
dependent diabetes) is associated with autoantibodies and sensitized T cells that damage the beta cells. A complex inflammatory
reaction leading to lysis of these cells greatly reduces the amount
of insulin secreted.

Neuromuscular Autoimmunities Myasthenia gravis* is named
for the pronounced muscle weakness that is its principal symptom.
Although the disease afflicts all skeletal muscle, the first effects are
usually felt in the muscles of the eyes and throat. Eventually, it can
progress to complete loss of muscle function and death. The classic
syndrome is caused by autoantibodies binding to the receptors for
acetylcholine, a chemical required to transmit a nerve impulse
across the synaptic junction to a muscle (figure 17.17). The immune
attack so severely damages the muscle cell membrane that transmission is blocked and paralysis ensues. Current treatment usually
includes immunosuppressive drugs and therapy to remove the autoantibodies from the circulation. Experimental therapy using im-
munotoxins to destroy lymphocytes that produce autoantibodies
shows some promise.

Multiple sclerosis* **(MS)** is a paralyzing neuromuscular disease associated with lesions in the insulating myelin sheath that
surrounds neurons in the white matter of the central nervous system. The underlying pathology involves damage to the sheath by

---

* myasthenia gravis (my″-us-thee′-nee-uh grah′-vis) Gr. *myo*, muscle, *astheneia*,
  weakness, and *gravida*, heavy.

* sclerosis (skleh-roh′-sis) Gr. *sklerosis*, hardness.

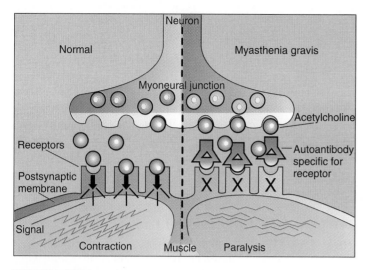

FIGURE 17.17

 **Proposed mechanisms for involvement of autoantibodies in myasthenia gravis.** Antibodies developed against receptors on the postsynaptic membrane block them so that acetylcholine cannot bind and muscle contraction is inhibited.

both T cells and autoantibodies that severely compromises the capacity of neurons to send impulses. The principal motor and sensory symptoms are muscular weakness and tremors, difficulties in speech and vision, and some degree of paralysis. Most MS patients first experience symptoms as young adults, and they tend to experience remissions (periods of relief) alternating with recurrences of disease throughout their lives. Convincing evidence from studies of the brain tissue of MS patients points to a strong connection between the disease and infection with human herpesvirus 6 (see chapter 24). The disease can be treated passively with monoclonal antibodies that target T cells, and a vaccine containing the myelin protein has shown beneficial effects. Immunosuppressants such as cortisone and interferon B may also alleviate symptoms.

CHAPTER CHECKPOINTS

Type III hypersensitivities are induced when a profuse amount of antigen enters the system and results in large quantities of antibody formation.

Type III hypersensitivity reactions occur when large quantities of antigen react with host antibody to form small, soluble immune complexes that settle in tissue cell membranes, causing chronic destructive inflammation. The reactions appear hours or days after the antigen challenge.

The mediators of type III hypersensitivity reactions include soluble IgA, IgG, or IgM, and agents of the inflammatory response.

Two kinds of type III hypersensitivities are localized (Arthus) reactions and systemic (serum sickness). Arthus reactions occur at the site of injected drugs or booster immunizations. Systemic reactions occur when repeated antigen challenges cause systemic distribution of the immune complexes and subsequent inflammation of joints, lymph nodes, and kidney tubules.

Autoimmune hypersensitivity reactions occur when autoantibodies or host T cells mount an abnormal attack against self antigens. Autoimmune antibody responses can be either local or systemic type II or type III

hypersensitivity reactions. Autoimmune T-cell responses are type IV hypersensitivity reactions.

Susceptibility to autoimmune disease appears to be influenced by gender and by genes in the MHC complex.

Autoimmune disease may be an excessive response of a normal immune function, the appearance of sequestered antigens, "forbidden" clones of lymphocytes that react to self antigens, or the result of alterations in the immune response caused by infectious agents, particularly viruses.

Examples of autoimmune diseases include systemic lupus erythematosus, rheumatoid arthritis, diabetes mellitus, myasthenia gravis, and multiple sclerosis.

Type IV Hypersensitivities: Cell-Mediated (Delayed) Reactions

The adverse immune responses we have covered so far are explained primarily by B-cell involvement and antibodies. But type IV hypersensitivity involves primarily the T-cell branch of the immune system. Type IV immune dysfunction has traditionally been known as delayed hypersensitivity because the symptoms arise one to several days following the second contact with an antigen. In general, type IV diseases result when T cells respond to self tissues or transplanted foreign cells. Examples of type IV hypersensitivity include delayed allergic reactions to infectious agents, contact dermatitis, and graft rejection.

DELAYED-TYPE HYPERSENSITIVITY

Infectious Allergy
A classic example of a delayed-type hypersensitivity occurs when a person sensitized by tuberculosis infection is injected with an extract (tuberculin) of the bacterium *Mycobacterium tuberculosis*. The so-called **tuberculin reaction** is an acute skin inflammation at the injection site appearing within 24 to 48 hours. So useful and diagnostic is this technique for detecting present or prior tuberculosis that it is the chosen screening device (figure 17.18*a*). Other infections that use similar skin testing are leprosy, syphilis, histoplasmosis, toxoplasmosis, and candidiasis. This form of hypersensitivity arises from time-consuming cellular events involving the T_D class of cells. After these cells receive processed microbial antigens from macrophages, they release broad-spectrum cytokines that attract inflammatory cells to the site—particularly mononuclear cells, fibroblasts, and other lymphocytes. In a chronic infection (tertiary syphilis, for example), extensive damage to organs can occur through granuloma formation.

Contact Dermatitis
The most common delayed allergic reaction, **contact dermatitis,** is caused by exposure to resins in poison ivy or poison oak (Spotlight on Microbiology 17.3), to simple haptens in household and personal articles (jewelry, cosmetics, elasticized undergarments), and to certain drugs. Like immediate atopic dermatitis, the reaction to these allergens requires a sensitizing and a provocative dose. The

(a)

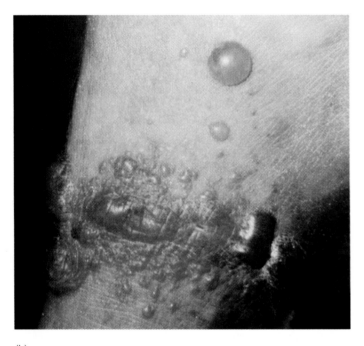

(b)

FIGURE 17.18

🔵 **Common delayed-type reactions.** **(a)** Positive tuberculin test. Intradermal injection of tuberculin extract in a person sensitized to tuberculosis yields a slightly raised red bump greater than 10 mm in diameter. **(b)** Contact dermatitis from poison oak, showing various stages of involvement: blisters, scales, and thickened patches.

allergen first penetrates the outer skin layers, is processed by Langerhans cells (skin macrophages), and is presented to T cells. When subsequent exposures attract lymphocytes and macrophages to this area, these cells give off enzymes and inflammatory cytokines that severely damage the epidermis in the immediate vicinity. This response accounts for the intensely itchy papules and blisters that are the early symptoms (figure 17.18b). As healing progresses, the epidermis is replaced by a thick, horny layer. Depending upon the dose and the sensitivity of the individual, the time from initial contact to healing can be a week to 10 days.

T CELLS AND THEIR ROLE IN ORGAN TRANSPLANTATION

Transplantation or grafting of organs and tissues is a common medical procedure. Although it is life-giving, this technique is plagued by the natural tendency of lymphocytes to seek out foreign antigens and mount a campaign to destroy them. The bulk of the damage that occurs in graft rejections can be attributed to expression of cytotoxic T cells and other killer cells. This section will cover the mechanisms involved in graft rejection, tests for transplant compatibility, reactions against grafts, prevention of graft rejection, and types of grafts.

The Genetic and Biochemical Basis for Graft Rejection

In chapter 15, we discussed the role of major histocompatibility (MHC or HLA) genes and receptors in immune function. In general, the genes and receptors in MHC classes I and II are extremely important in recognizing self and in regulating the immune response. These receptors also set the events of graft rejection in motion. The MHC genes of humans are inherited from among a large pool of genes, so the cells of each person can exhibit variability in the pattern of cell surface molecules (figure 17.19). The pattern is identical in different cells of the same person and can be similar in related siblings and parents, but the more distant the relationship, the less likely that the MHC genes and receptors will be similar. When donor tissue (a graft) displays surface receptors of a different MHC class, the T cells of the recipient (called the host) will recognize its foreignness and react against it.

T Cell-Mediated Recognition of Foreign MHC Receptors

Host Rejection of Graft When the cytotoxic T cells of a host recognize foreign class I MHC receptors on the surface of grafted cells, they release interleukin-2 as part of a general immune mobilization. Receipt of this stimulus amplifies helper and cytotoxic T cells specific to the foreign antigens on the donated cells. The cytotoxic cells bind to the grafted tissue and secrete lymphokines that begin the rejection process within 2 weeks of transplantation (figure 17.20a). Late in this process, antibodies formed against the graft tissue contribute to immune damage. A final blow is the destruction of the vascular supply, promoting death of the grafted tissue.

Graft Rejection of Host In certain severe immunodeficiencies, the host cannot or does not reject a graft. But this failure may not protect the host from serious damage, because graft incompatibility is a two-way phenomenon. Some grafted tissues (especially bone marrow) contain an indigenous population called passenger lymphocytes. This makes it quite possible for the graft to reject the host, causing **graft versus host disease (GVHD)** (figure 17.20b). Since any host tissue bearing MHC receptors foreign to the graft can be attacked, the effects of GVHD are widely systemic and toxic. A papular, peeling skin rash is the most common symptom. Other organs affected are the liver, intestine, muscles, and mucous membranes. GVHD occurs in approximately 30% of bone marrow transplants within 100 to 300 days of the graft. A relatively high percentage of recipients die from its effects.

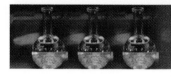

SPOTLIGHT ON MICROBIOLOGY 17.3
Pretty, Pesky, Poisonous Plants

As a cause of allergic contact dermatitis (affecting about 10 million people a year), nothing can compare with a single family of plants belonging to the genus *Toxicodendron*. At least one of these plants—either poison ivy, poison oak, or poison sumac—flourishes in the forests, woodlands, or along the trails of most regions of America. The allergen in these plants, an oil called urushiol, has such extreme potency that a pinhead-sized amount could spur symptoms in 500 people, and it is so long-lasting that botanists must be careful when handling 100-year-old plant specimens. Although degrees of sensitivity vary among individuals, it is estimated that 85% of all Americans are potentially hypersensitive to this compound. Some people are so acutely sensitive that even the most miniscule contact, such as handling pets or clothes that have touched the plant or breathing vaporized urushiol, can trigger an attack.

Humans first become sensitized by contact during childhood. Individuals at great risk (firefighters, hikers) are advised to determine their degree of sensitivity using a skin test, so that they can be adequately cautious and prepared. Some odd remedies include skin potions containing bleach, buttermilk, ammonia, hair spray, and meat tenderizer. Commercial products are available for blocking or washing away the urushiol. Allergy researchers are currently testing oral vaccines containing a form of urushiol, which seem to desensitize experimental animals. An effective method using poison ivy desensitization injection is currently available to people with extreme sensitivity.

Learning to identify these common plants can prevent exposure and sensitivity. One old saying that might help warns, "Leaves of three, let it be; berries white, run with fright." (See chapter opening photo of poison ivy.)

Poison oak

Poison sumac

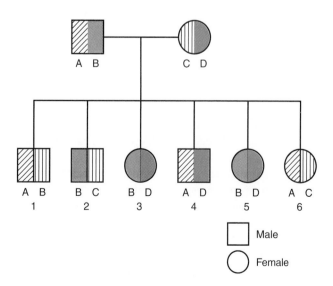

FIGURE 17.19

The pattern of inheritance of MHC (HLA) genes. A simplified version of the human leukocyte antigen (HLA) complex in a family. In this example, there are two genes in the complex, and each parent has a different set of genes (A/B and C/D). A child can inherit one of four different combinations. Out of six children, two sets (1 and 6, 3 and 5) have identical HLA genes and are good candidates for exchange grafts. Children sharing one gene (for example, 1, 4, and 6 share antigen A) are close matches, but two pairs of children (2 and 4, 3 and 6) do not match at all.

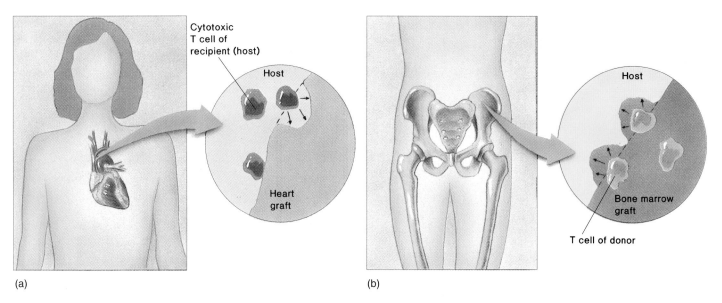

FIGURE 17.20

Potential reactions in transplantation. **(a)** The host's immune system (primarily cytotoxic T cells) encounters the cells of the donated organ (heart) and rejects the organ by secreting cytokines. **(b)** Grafted tissue (bone marrow) contains endogenous T cells that recognize the host's tissues as foreign and mount a cytokine attack. The recipient will develop symptoms of graft versus host disease.

Classes of Grafts

Grafts are generally classified according to the genetic relationship between the donor and the recipient. Tissue transplanted from one site on an individual's body to another site on his body is known as an **autograft.** Typical examples are skin replacement in burn repair and the use of a vein to fashion a coronary artery bypass. In an **isograft,** tissue from an identical twin is used. Because isografts do not contain foreign antigens, they are not rejected, but this type of grafting has obvious limitations. **Allografts,** the most common type of grafts, are exchanges between genetically different individuals belonging to the same species (two humans). A close genetic correlation is sought for most allograft transplants (see next section). A **xenograft** is a tissue exchange between individuals of different species. Until rejection can be better controlled, most xenografts are experimental or for temporary therapy only.

Avoiding and Controlling Graft Incompatibility

Graft rejection can be averted or lessened by directly comparing the tissue of the recipient with that of potential donors. Several **tissue matching** procedures are used. In the *mixed lymphocyte reaction (MLR),* lymphocytes of the two individuals are mixed and incubated. If an incompatibility exists, some of the cells will become activated and proliferate. *Tissue typing* is similar to blood typing, except that specific antisera are used to disclose the HLA antigens on the surface of lymphocytes. In most grafts (one exception is bone marrow transplants), the ABO blood type must also be matched. Although a small amount of incompatibility is tolerable in certain grafts (liver, heart, kidney), a closer match is more likely to be successful, so the closest match possible is sought.

Drugs That Suppress Allograft Rejection Despite an international computerized hotline for matching recipients with donors and a greater availability of viable organs than in the past, an ideal match between donor and recipient is still the exception rather than the rule, and some sort of immunosuppressive therapy to overcome

rejection is usually required. Rejection can be controlled with agents such as cyclosporin A, methotrexate, prednisone, and a monoclonal antibody OKT3. Except for cyclosporin A, intervention with drugs can be complicated by general suppression of the immune system (especially T cells) and frequent opportunistic infections.

Cyclosporin A is a polypeptide isolated from a fungus. It has dramatically improved the survival rate of allograft patients (kidney, heart, liver, and bone marrow) and has reduced the incidence of fatal infections. Although its action is not entirely understood, cyclosporin appears to block the activation of T helper cells and interfere with the release of interleukin-2. What makes this drug so valuable is that it does not inhibit important lymphoid cells and phagocytes, and the body is better able to ward off infections. Its adverse effects of kidney toxicity and increased blood pressure can be reduced by adjusting the dose and monitoring blood levels of the drug. Because of its ability to inhibit undesirable T-cell activity, cyclosporin is also being used to treat autoimmune diseases such as type I diabetes and rheumatoid arthritis. Newer drugs aim to block the binding of IL-2 on T cells.

Types of Transplants

Today, transplantation is a recognized medical procedure whose benefit is reflected in several thousand transplants each year. It has been performed on every major organ, including parts of the brain. The most frequent transplant operations involve skin, heart, kidney, coronary artery, cornea, and bone marrow. The sources of organs and tissues are live donors (kidney, skin, bone marrow, liver), cadavers (heart, kidney, cornea), and fetal tissues. In the past decade, we have witnessed some unusual types of grafts. For instance, the fetal pancreas has been implanted as a potential treatment for diabetes, and fetal brain tissues for Parkinson disease. Part of a liver has been transplanted from a live parent to a child, and parents have donated a lobe from their lungs to help restore function in their children with severe cystic fibrosis.

MEDICAL MICROFILE °17.4
The Mechanics of Bone Marrow Transplantation

In some ways, bone marrow is the most exceptional form of transplantation. It does not involve invasive surgery in either the donor or recipient, and it permits the removal of tissue from a living donor that is fully replaceable. While the donor is sedated, a bone marrow/blood sample is aspirated by inserting a special needle into an accessible marrow cavity. The most favorable sites are the crest and spine of the ilium (major bone of the pelvis). During this procedure, which lasts 1 to 2 hours, 3% to 5% of the donor's marrow is withdrawn in 20 to 30 separate extractions. Between 500 and 800 ml of marrow is removed. The donor may experience some pain and soreness, but there are rarely any serious complications. In a few weeks, the depleted marrow will naturally replace itself. Implanting the harvested bone marrow is rather convenient, because it is not necessary to place it directly into the marrow cavities of the recipient. Instead, it is dripped intravenously into the circulation, and the new marrow cells automatically settle in the appropriate bone marrow regions. The survival and permanent establishment of the marrow cells are increased by administering various growth factors and stem cell stimulants to the patient.

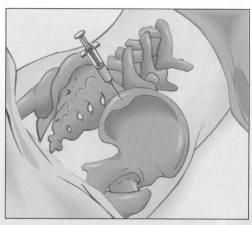

Removal of a bone marrow sample for transplantation. Samples are removed by inserting a needle into the spine or crest of the ilium. (The ilium is a prolific source of bone marrow.)

Bone marrow transplantation is a rapidly growing medical procedure for patients with immune deficiencies, aplastic anemia, leukemia and other cancers, and radiation damage. This procedure is extremely expensive, costing up to $200,000 per patient. Before bone marrow from a closely matched donor can be infused (Medical Microfile 17.4), the patient is pretreated with chemotherapy and whole-body irradiation, a procedure designed to destroy his own blood stem cells and thus prevent rejection of the new marrow cells. Within 2 weeks to a month after infusion, the grafted cells are established in the host. Because donor lymphoid cells can still cause GVHD, anti-rejection drugs may be necessary. An amazing consequence of bone marrow transplantation is that a recipient's blood type may change to the blood type of the donor.

CHAPTER CHECKPOINTS

Type IV hypersensitivity reactions occur when cytotoxic T cells attack either self tissue or transplanted foreign cells. Type IV reactions are also termed delayed hypersensitivity reactions because they occur hours to days after the antigenic challenge.

Type IV hypersensitivity reactions are mediated by T lymphocytes and are carried out against foreign cells that show both a foreign MHC and a nonself receptor site.

Examples of type IV reactions include the tuberculin reaction, contact dermatitis, and mismatched organ transplants (host rejection and GVHD reactions).

The four classes of transplants or grafts are determined by the degree of MHC similarity between graft and host. From most to least similar, these are: autografts, isografts, allografts, and xenografts.

Graft rejection can be minimized by tissue matching procedures, immunosuppressive drugs, and use of tissues that do not provoke a type IV response.

Immunodeficiency Diseases: Hyposensitivity of the Immune System

It is a marvel that development and function of the immune system proceed as normally as they do. On occasion, however, an error occurs and a person is born with or develops weakened immune responses. In many cases, these very "experiments" of nature have provided penetrating insights into the exact functions of certain cells, tissues, and organs because of the specific signs and symptoms shown by the immunodeficient individuals. The predominant consequences of immunodeficiencies are recurrent, overwhelming infections, often with opportunistic microbes. Immunodeficiencies fall into two general categories: *primary diseases,* present at birth (congenital) and usually stemming from genetic errors, and *secondary diseases,* acquired after birth and caused by natural or artificial agents (table 17.5).

PRIMARY IMMUNODEFICIENCY DISEASES

Deficiencies affect both specific immunities such as antibody production and less-specific ones such as phagocytosis. Consult figure 17.21 to survey the places in the normal sequential development of lymphocytes where defects can occur and the possible consequences. In many cases, the deficiency is due to an inherited abnormality, though the exact nature of the abnormality is not known for a number of diseases. Because the development of B cells and T cells departs at some point, an individual can lack one or both cell lines. It must be emphasized, however, that some deficiencies affect other cell functions. For example a T-cell deficiency can affect B-cell function because of the role of T helper cells. In some deficiencies, the lymphocyte in question is completely absent or is present at very low levels, whereas in others, lymphocytes are present but do not function normally.

TABLE 17.5

General Categories of Immunodeficiency Diseases

Primary Immune Deficiencies (Genetic)	Secondary Immune Deficiencies (Acquired)
B-Cell Defects (Low Levels of B Cells and Antibodies) Agammaglobulinemia (X-linked, non-sex-linked) Hypogammaglobulinemia Selective immunoglobulin deficiencies	**From Natural Causes** Infection: AIDS, leprosy, tuberculosis, measles Other disease: cancer, diabetes Nutrition deficiencies Stress Pregnancy Aging
T-Cell Defects (Lack of All Classes of T Cells) Thymic aplasia (DiGeorge syndrome) Chronic mucocutaneous candidiasis	
Combined B-Cell and T-Cell Defects (Usually Caused by Lack or Abnormality of Lymphoid Stem Cell) Severe combined immunodeficiency disease (SCID) Adenosine deaminase (ADA) deficiency Wiskott-Aldrich syndrome Ataxia-telangiectasia	**From Immunosuppressive Agents** Irradiation Severe burns Steroids (cortisones) Drugs to treat graft rejection and cancer Removal of spleen
Phagocyte Defects Chédiak-Higashi syndrome Chronic granulomatous disease of children Lack of surface adhesion molecules	
Complement Defects Lacking one of C components Hereditary angioedema Associated with rheumatoid diseases	

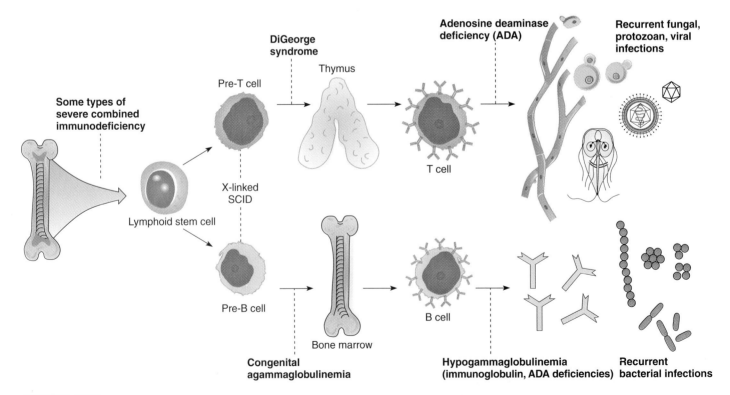

FIGURE 17.21

 The stages of development and the functions of B cells and T cells, whose failure causes immunodeficiencies. Dotted lines represent the phases in development where breakdown can occur.

Clinical Deficiencies in B-Cell Development or Expression

Genetic deficiencies in B cells usually appear as an abnormality in immunoglobulin expression. In some instances, only certain immunoglobulin classes are absent; in others, the levels of all types of immunoglobulins (Ig) are reduced. A significant number of B-cell deficiencies are X-linked (also called sex-linked) recessive traits, meaning that the gene occurs on the X chromosome and the disease appears primarily in male children.

The term **agammaglobulinemia** literally means the absence of gamma globulin, the primary fraction of serum that contains immunoglobulins. Because it is very rare for Ig to be completely absent, some physicians prefer the term **hypogammaglobuline-mia.*** Both sex-linked and *autosomal*[4] recessive types of this rare syndrome occur. In both cases, mature B cells are absent, the level of antibodies is greatly reduced, and lymphoid organs are incompletely developed. T-cell function in these patients is usually normal. The symptoms of recurrent, serious bacterial infections usually appear about 6 months after birth. The bacteria most often implicated are pyogenic cocci, *Pseudomonas,* and *Haemophilus influenzae,* and the most common infection sites are the lungs, sinuses, meninges, and blood. Many Ig-deficient patients can have recurrent infections with viruses and protozoa, as well. Patients often manifest a wasting syndrome and have a reduced lifespan, but modern therapy has improved their prognosis. The current treatment for this condition is passive immunotherapy with immune serum globulin and continuous antibiotic therapy.

The lack of a particular class of immunoglobulin is a relatively common condition. Although genetically controlled, its underlying mechanisms are not yet clear. **IgA deficiency** is the most prevalent, occurring in about one person in 600. Such persons have normal quantities of B cells and other immunoglobulins, but they are unable to synthesize IgA. Consequently, they lack protection against local microbial invasion of the mucous membranes and suffer recurrent respiratory and gastrointestinal infections. The usual treatment using Ig replacement does not work, because conventional preparations are high in IgG, not IgA.

Clinical Deficiencies in T-Cell Development or Expression

Due to their critical role in immune defenses, a genetic defect in T cells results in a broad spectrum of disease, including severe opportunistic infections, wasting, and cancer. In fact, a dysfunctional T-cell line is usually more devastating than a defective B-cell line because T helper cells are required to assist in most specific immune reactions. The deficiency can occur anywhere along the developmental spectrum, from thymus to mature, circulating T cells.

Abnormal Development of the Thymus   The most severe of the T-cell deficiencies involve the congenital absence or immaturity of the thymus gland. Thymic aplasia, or **DiGeorge syndrome,** results when the embryonic third and fourth pharyngeal pouches fail to develop. Some cases are associated with a deletion in chromo-

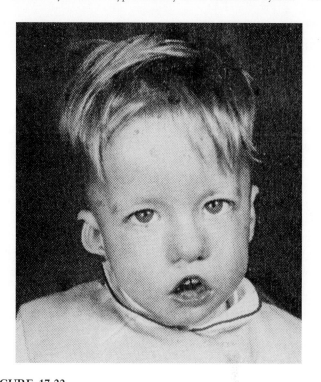

FIGURE 17.22

Facial characteristics of a child with DiGeorge syndrome.   Typical defects include low-set, deformed earlobes; wide-set, slanted eyes; a small, bowlike mouth; and the absence of a philtrum (the vertical furrow between the nose and upper lip).

some 22 (figure 17.22). The accompanying lack of cell-mediated immunity makes children highly susceptible to persistent infections by fungi, protozoa, and viruses. Common, usually benign, childhood infections such as chickenpox, measles, or mumps can be overwhelming and fatal in these children. Even vaccinations using attenuated microbes pose a danger. Other symptoms of thymic failure are reduced growth, wasting of the body, unusual facial characteristics, and an increased incidence of lymphatic cancer. These children can have reduced antibody levels, and they are unable to reject transplants. The major therapy for them is a transplant of thymus tissue.

Severe Combined Immunodeficiencies: Dysfunction in B and T Cells

Severe combined immunodeficiencies (SCIDs) are the most dire and potentially lethal of the immunodeficiency diseases because they involve dysfunction in both lymphocyte systems. Some SCIDs are due to the complete absence of the lymphocyte stem cell in the marrow; others are attributable to the dysfunction of B cells and T cells later in development. Infants with SCID usually manifest the T-cell deficiencies within days after birth by developing candidiasis, sepsis, pneumonia, or systemic viral infections. This debilitating condition appears to have several forms. In the two most common forms, Swiss-type agammaglobulinemia and thymic alymphoplasia, the numbers of all types of lymphocytes are extremely low, the blood antibody content is greatly diminished, and the thymus and cell-mediated immunity are poorly developed. Both diseases are due to a genetic defect in the development of the lymphoid cell line.

---

4. Meaning that the gene is located on a chromosome other than a sex chromosome.

* hypogammaglobulinemia (hy′-poh-gem-ah-glob-yoo-lin-ee′-mee-ah) Gr. *Hypo,* denoting a lowered level of Ig.

SPOTLIGHT ON MICROBIOLOGY 17.5
An Answer to the Bubble Boy Mystery

David Vetter, the most famous SCID child, lived all but the last 2 weeks of his life in a sterile environment to isolate him from the microorganisms that could have quickly ended his life. When medical tests performed before birth had indicated that David might inherit this disease, he was delivered by cesarian section and immediately placed in a sterile isolette. From that time, he lived in various plastic chambers—ranging from room-size to a special suit that allowed him to walk outside. Remarkably, he developed into a well-adjusted child, even though his only physical contact with others was through special rubber gloves. When he was 12, his doctors decided to attempt a bone marrow transplant that might allow him to live free of his bubble prison. David was transplanted with his sister's bone marrow, but the marrow harbored a common herpesvirus called Epstein-Barr virus. Because he lacked any form of protective immunities against this oncogenic virus, a cancer spread rapidly through his body. Despite the finest medical care available, David died a short time later from the metastatic cancer.

In 1993, after several years of study, researchers discovered the basis for David's immunodeficiency. He had inherited an X-linked form (called XSCID) that arises from a defective genetic code in the receptors for interleukin-2, interleukin-4, and interleukin-7. The defect prevents the receptors on T cells and B cells from receiving the appropriate inter-

leukin signals for growth, development, and reactivity. The end result is that both cytotoxic immunities and antibody-producing systems are shut down, leaving the body defenseless against infections and cancer.

David Vetter, the boy in the plastic bubble.

A rarer form of SCID is **adenosine deaminase (ADA) deficiency,** which is caused by an autosomal recessive defect in the metabolism of adenosine. In this case, lymphocytes develop, but a metabolic product builds up abnormally and selectively destroys them. Infants with ADA deficiency are subject to recurrent infections and severe wasting typical of severe deficiencies. A small number of SCID cases are due to a developmental defect in receptors for B and T cells. An X-linked deficiency in interleukin receptors was responsible for the disease of David, the child in the "plastic bubble" (Spotlight on Microbiology 17.5). Another newly identified condition, *bare lymphocyte syndrome,* is caused by the lack of genes that code for class II MHC receptors.

Because of their profound lack of specific adaptive immunities, SCID children require the most rigorous kinds of aseptic techniques to protect them from opportunistic infections. Aside from life in a sterile plastic bubble, the only serious option for their long-time survival is total replacement or correction of dysfunctional lymphoid cells. Some infants can benefit from fetal liver or bone marrow grafts. Although transplanting compatible bone marrow has been about 50% successful in curing the disease, it is complicated by graft versus host disease. The condition of some ADA-deficient patients has been partly corrected by periodic transfusions of blood containing large amounts of the normal enzyme. A more lasting treatment for both X-linked and ADA types of SCID has been found in gene therapy—insertion of a gene to replace the defective gene (see figure 10.14). Several children have apparently been cured by transfecting their bone marrow stem cells with the normal gene for ADA.

SECONDARY IMMUNODEFICIENCY DISEASES

Secondary acquired deficiencies in B cells and T cells are caused by one of four general agents: (1) infection, (2) organic disease, (3) chemotherapy, or (4) radiation.

The most recognized infection-induced immunodeficiency is **AIDS.** This syndrome is caused when several types of immune cells, including T helper cells, monocytes, macrophages, and antigen presenting cells, are infected by the human immunodeficiency virus (HIV). It is generally thought that the depletion of T helper cells and functional impairment of immune responses ultimately account for the cancers and opportunistic protozoan, fungal, and viral infections associated with this disease. See chapter 25 for an extensive discussion of AIDS. Other infections that can deplete immunities are measles, leprosy, and malaria.

Cancers that target the bone marrow or lymphoid organs can be responsible for extreme malfunction of both humoral and cellular immunity (see next section). In leukemia, the massive number of cancer cells compete for space and literally displace the normal cells of the bone marrow and blood. Plasma cell tumors produce large amounts of nonfunctional antibodies, and thymus gland tumors cause severe T-cell deficiencies.

An ironic outcome of life-saving medical procedures is the possible suppression of a patient's immune system. For instance, some immunosuppressive drugs that prevent graft rejection by T cells can likewise suppress beneficial immune responses. Although radiation and anticancer drugs are the first line of therapy for many types of cancer, both agents are extremely damaging to the bone marrow and other body cells.

Cancer: Cells Out of Control

The term **cancer** comes from the Latin word for crab and presumably refers to the appendage-like projections that a spreading tumor develops. A cancer is defined as the new growth of abnormal cells. The disease is also known by the synonym *neoplasm,** or by more specific terms that usually end in the suffix *-oma*. **Oncology,** the field of medicine that specializes in cancer, is growing with such great rapidity that we can only survey major concepts as they relate to immunology, including the basic characteristics, classification, and origins of cancer.

CHARACTERISTICS AND CLASSIFICATION OF TUMORS AND CANCERS

An abnormal growth or tumor is generally characterized as benign or malignant. A **benign tumor** is a self-contained mass within an organ that does not spread into adjacent tissues. The mass is usually slow-growing, rounded, and not greatly different from its tissue of origin. Benign tumors do not ordinarily cause death unless they grow into a critical space such as the heart valves or brain ventricles. The feature that most distinguishes a **malignant tumor** (cancer) is uncontrolled growth of abnormal cells within normal tissue. As a general rule, cancer originates in cells such as skin and bone marrow that have retained the capacity to divide, while mature cells that have lost this power (neurons, for instance) do not commonly become cancerous.

Most cancerous growths show other characteristics, including (1) disorganized behavior and independence from surrounding normal tissues, (2) permanent loss of cell differentiation, and (3) expression of special markers on their surface. The cells of a malignant tumor also do not remain encapsulated but tend to spread both locally and distantly to other organs. As the initial (primary) tumor grows, it invades nearby tissues, enters lymphatic and blood vessels, and establishes secondary tumors in remote sites. This property of spreading is called **metastasis,** and the cancer is said to *metastasize*. Malignant tumors range in effects from those seen in pancreatic cancer, which spread rapidly and are almost always fatal, to others as occur in basal cell carcinoma of the skin, which are much less aggressive, more treatable, and seldom fatal.

Another tumor-classifying scheme relies on the tissue or cell of primary origin. Although cancers are commonly referred to sim-

ply as bladder cancer, breast cancer, or liver cancer, their technical names more clearly define the tumor origin. In general, cancers originating from epithelial tissues are called **carcinomas,** and those originating from mesenchyme (embryonic connective tissue) are **sarcomas.** Combining these terms with the tissue of origin supplies the full descriptive name. For example, adenocarcinoma develops in glandular tissue such as the pancreas or thyroid; squamous cell carcinoma arises in the skin epidermis; retinoblastoma occurs in the retina; lymphosarcoma in the lymph nodes; hepatoma (hepatic sarcoma) in the liver; and melanoma in the skin melanocytes. A special name, leukemia, denotes cancer of the blood-forming tissues. Considering the large number of cell types in a given organ, the great diversity of tumors (more than 100 clinical types) is to be expected.

Epidemiology of Cancer

One-third of all U.S. citizens will develop cancer some time during their lifetime. Cancer is the second leading cause of death in humans (following heart and circulatory disease). Annually, approximately 850,000 new cases are diagnosed, and 500,000 people die from various cancers. Cancer rates also vary according to gender, race, ethnic group, age, geographic region, occupational exposure to carcinogens, diet, and life-style. Figure 17.23 compiles the most recent statistics on cancer incidence in the United States for mixed populations. Cancers currently on the increase are melanoma (presumably due to increased levels of UV radiation entering the atmosphere), Kaposi's sarcoma (in AIDS patients), thyroid cancer, esophageal cancer, and prostate cancer (because of an improved method of detection). Lung cancer accounts for the largest number of deaths in both sexes. The second most common cause of cancer deaths is prostate cancer in men and breast cancer in women.

It is increasingly evident that susceptibility to certain cancers is inherited. At least 30 neoplastic syndromes that run in families have been described. A particularly stunning discovery is that some cancers in animals are actually due to endogenous viral genomes passed from parent to offspring in the gametes. Many cancers reflect an interplay of both hereditary and environmental factors. An example is lung cancer, which is associated mainly with tobacco smoking. But not all smokers develop lung cancer, and not all lung cancer victims smoke tobacco.

Proposed Mechanisms of Cancer

Cancer is clearly a complex disease associated with numerous physical, chemical, and biological agents (table 17.6). For years, there was no clear unifying explanation for how so many different factors could cause normal cells to become cancerous. But recent research findings link all cancers to genetic damage or inherited genetic predispositions that alter the function of certain genes found normally in all cells. The evidence for the interrelationship between genes and cancer is derived from the following observations: (1) Cancer cells often have damaged chromosomes; (2) a specific alteration in a gene can lead to cancer; (3) the predisposition for some cancers is inherited; (4) rates of cancer are highest in

---

* **neoplasm** (nee'-oh-plazm) Gr. *neo,* new, and *plasm,* formation.

* **carcinoma** (kar″-sih-noh'-mah) Gr. *karkinos,* crab.

* **sarcoma** (sar-koh'-mah) Gr. *sarcos,* flesh, and *oma,* tumor. Sometimes referred to as soft-tissue tumors.

2000 Cancer Incidence
by Site in Men

2000 Cancer Incidence
by Site in Women

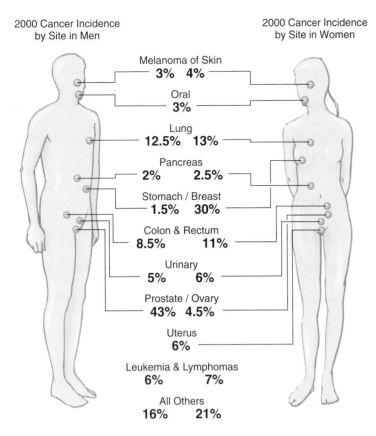

Melanoma of Skin
3%   4%

Oral
3%

Lung
12.5%   13%

Pancreas
2%   2.5%

Stomach / Breast
1.5%   30%

Colon & Rectum
8.5%   11%

Urinary
5%   6%

Prostate / Ovary
43%   4.5%

Uterus
6%

Leukemia & Lymphomas
6%   7%

All Others
16%   21%

FIGURE 17.23

Incidence of cancer in men and women as of 2000. Note that breast cancer is the most common type in women, and prostate cancer predominates in men. Data (unofficial) are from the American Cancer Society.

TABLE 17.6

Selected Cancer-Triggering Agents and Their Diseases

Agent	Type of Exposure	Cancer Type
Physical Agents		
Ionizing radiation	Occupational exposure, therapy	Leukemia, many others
UV radiation	Sunlight, suntanning lamps	Skin
Asbestos	Industrial, building insulation	Lung
Chemical Agents		
Tobacco	Smoking, chewing	Oral, lung
Alcohol	Ingestion	Oral, esophageal, liver
Arsenic	Manufacturing, mining	Lung, skin, liver, breast
Benzene	Various industrial processes	Leukemia
Polycyclic hydrocarbons	Medicines, industrial use (coal tar derivatives)	Lung, skin
Vinyl chloride	Manufacture of plastic	Liver
Aflatoxin	Fungus-contaminated foods	Liver
Aromatic amines	Manufacturing processes	Bladder
Heavy metal dust	Industry	Lung, nasal
Anabolic steroids	Medication	Liver
Estrogens	Medication	Vaginal, liver
Radon gas	Mines, homes	Lung
Biological Agents (Acquired Through Infection)		
Retrovirus	T-cell leukemia	
Papillomaviruses	Cervical cancer	
Polyoma viruses	Various tumors in mice and hamsters	
Epstein-Barr virus	Burkitt lymphoma; nasopharyngeal carcinoma	
Hepatitis B virus	Liver cancer	

individuals who cannot repair damaged DNA; (5) mutagenic agents cause cancer; (6) cells contain genes that can be transformed to cancer-causing oncogenes; and (7) tumor-suppressor genes (anti-oncogenes) exist in the normal genome. These discoveries and other advances in cancer research have provided the seeds for a general unifying theory for cancer.

Oncogene: A Normal Gene Gone Haywire? Cancer is another experiment of nature that continues to inform us about normal cell function. In the section on gene regulation in chapter 9, we learned that normal cell operations are strictly controlled by regulatory genes. Extracellular chemical signals are received by receptors, which transmit them to a regulatory complex in the nucleus that switches the appropriate genes on or off. During the rapid growth of a young individual, cells are extremely active, and the switch that initiates cell division is operating much of the time. By contrast, most adult cells exist in a nondividing state, except for divisions involved in repair and maintenance. This state of balance ensures that cells divide at a rate compatible with normal development and function.

Cancer results when the system that organizes cell division is short-circuited. Current theories explain that the gene complex controlling cell division contains a gene termed a **proto-oncogene** that regulates the onset of mitosis. The proto-oncogene is, in turn, regulated by another gene called a **tumor-suppressor gene,** or **anti-**oncogene, that prevents the proto-oncogene from acting continuously (figure 17.24*a*) and thereby keeps the cell division cycle operating normally. It is now clear that genetic disruptions (mutation, chromosomal alterations) can affect the normal actions of the proto-oncogenes and tumor-suppressor genes. If a genetic error keeps the proto-oncogene switched on, it converts to an **oncogene.** The oncogene is then programmed to override the normal mitotic controls and cause the cell to divide continuously (figure 17.24*b*). This process of cell immortalization is termed **transformation.** A number of happenings might explain how an oncogene goes permanently out of control and cannot be turned off. The antioncogene system may be missing or may have been inactivated by mutations, or new genetic material may have been added to it. Another possibility is that the oncogene products have been altered by mutation and no longer respond to regulation (remember that genes code for protein products). Some genes express their oncogenic

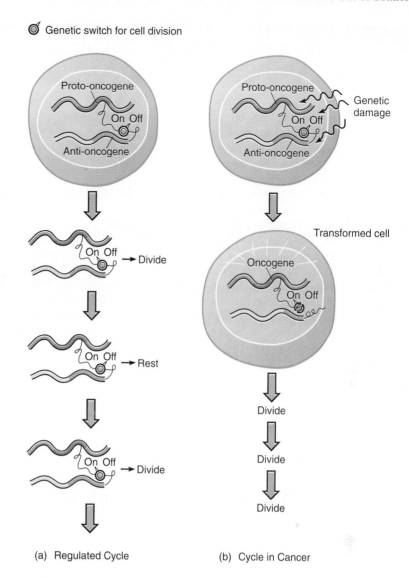

Genetic switch for cell division

FIGURE 17.24

Common pathway for neoplasias. **(a)** Normal cell cycle and genetic controls. **(b)** Event that stimulates transformation into a cancer cell. Any agent that can alter the genetic control mechanisms can cause a cell to divide at incorrect times and in incorrect places.

potential after jumping from one site on a chromosome to another. New oncogenes are being discovered at a rapid pace, and their modes of action are gradually being deciphered (Spotlight on Microbiology 17.6).

The Role of Viruses in Cancer   One documented oncogenic change occurs when a cell is infected by a retrovirus (figure 17.25; see chapter 25). Once inside a host cell, these RNA viruses utilize an enzyme called reverse transcriptase to synthesize viral DNA, which is subsequently inserted at a particular site on a host chromosome. Some retroviruses, such as the Rous sarcoma virus of chickens, carry viral oncogenes whose products cause transformation of host cells into cancer cells. Other retroviral genomes insert on host chromosomes at regulatory sites; their insertion at those sites can lead to activation of the cell's own proto-oncogenes. One type of cancer in humans, T-cell leukemia, is attributable to insertional mutagenesis by a retrovirus. DNA viruses are also involved in cancer. Human papillomavirus (HPV), which causes genital warts and is strongly implicated in cervical carcinoma, inserts its genome directly into a chromosome in a way that overrides the usual cellular growth controls. Epstein-Barr virus (EBV), a cause of infectious mononucleosis and Burkitt lymphoma, induces a chromosomal alteration that can interfere with cell death.

Chromosomal Damage and Cancer   A number of cancers have been associated with gross chromosomal changes or other aberrations that occur during mitotic divisions (figure 17.26). In the form of chromosomal disruption called *translocation* (figure 17.26*a*), a segment of one chromosome is transferred to another chromosome that is not its homologue. The abnormal placement of the translocated genes is believed to act as a strong oncogenic stimulus. Burkitt lymphoma (see figure 24.13), a malignant tumor of lymphocytes, is associated with the translocation of part of chromosome 8 to chromosome 14. Most cases of myelogenous leukemia are due to the translocation of a proto-oncogene on chromosome 9 to a different site on chromosome 22.

Certain cancers appear to result from a mistake in mitosis that produces cells bearing three rather than the usual pair of a particular chromosome. Such *trisomies* (figure 17.26*b*) occurring with chromosome 8 and 12 can lead to leukemia, and an extra copy of

The outpouring of genetic discovery originating from the Human Genome Project has had a monumental impact on identification of genes involved in cancers. Remarkable progress is being made in identifying cancer genes, mapping them to specific areas of chromosomes, and developing special markers for these genes.

What has followed naturally from this new knowledge on exact locations of genes has been a tremendous activity in disclosing the ways these genes cause cancer. It is becoming clear that cancer development involves many complex interactions between genes, their products, and external signals received by the cell. There is no single mechanism. Consider the fact that at least eight different genetic defects lead to colon cancer and nearly that many are implicated in breast cancers. The one unifying pathway for all of these cancers, however, is a genetic alteration that disrupts the normal cell division cycle and transforms a normal cell into a cancer cell. In a large number of cases, cancer cells appear when the cell cycle clock in the nucleus can no longer respond appropriately to growth signals.

The Role of Oncogenes in the Cell Cycle

Cell chemicals called cyclins that control the phases in mitosis and cell growth can explain how some types of cancer arise. They act as chemical signals to promote cell division in a timely and orderly fashion. One of them, cyclin D1, is important for the early growth phase of cells. This substance turns on a set of enzymes called kinases that stimulate the replication of DNA. It turns out that the gene that codes for cyclin D1 (also called bc11) is a type of oncogene. When a defect in the gene causes the cyclin to be overproduced or produced at the wrong time, the cell goes into a nonstop division mode. Several other oncogenes apparently act at the level of the cyclins.

The Role of Tumor-Suppressor Genes

Another important model for cancer formation involves defective tumor-suppressor genes (TSGs), of which there are presently about a dozen known. One example is the human retinoblastoma (Rb) gene on chromosome 13, which prevents retinal cancer and other cancers. Individuals who have inherited defective genes, or whose normal genes have been mutated by carcinogens, can develop retinal tumors, osteosarcoma (a form of bone cancer), breast cancer, and a form of lung cancer. Molecular biologists determined that when the normal protein product of the Rb gene binds to a specific site on DNA, it blocks cellular growth and prevents the cells from growing out of control. However, when these genes are mutated or damaged, there is no suppression. Another fascinating tumor-suppressor gene, p53, becomes altered by simple missense mutations and has been implicated in 51 different tumors. p53 is a master regulator switch for the cell cycle that blocks the expression of the cyclin genes. When it is defective and the production of cyclin is altered, the cycle becomes unregulated and out of control. A second master regulator gene, P-TEN, codes for a protein to slow cell division. Disruption of this gene through mutation creates an aggressive cancer cell that is responsible for several rapidly growing tumors.

Programmed Cell Suicide

One normal developmental activity of cells is to undergo natural, programmed death, or *apoptosis*. This is one way that the body gets rid of abnormal, diseased, or old cells, and it accounts for various aspects of aging. Apoptosis is genetically coded by special "suicide genes" that are part of the genome of all cells. Now it appears that a disruption in this process may figure in some types of cancer. Because a cancer cell is an abnormal cell, it would ordinarily be programmed to die by apoptosis. But if genetic defects keep the suicide gene from being switched on, the cells never receive a death signal and thus become immortalized. Recent research on the role of viruses and cell death has shown that some viruses, such as the Epstein-Barr virus, contain a gene called C-MYC that interferes with the onset of apoptosis and thereby allows infected cells to survive. The EB virus is a known cause of Burkitt lymphoma. Other viruses that block cell death are human adenoviruses and papillomaviruses (warts).

Inability to Repair Mutations

Nearly one in every 200 people in the Western Hemisphere carries a defective gene that can lead to colon cancer, making this one of the most common genetic defects. If expressed, the defective gene leads to a condition called hereditary nonpolyposis colon cancer (HNPCC). The cause of this cancer is yet another variation on the theme of genetic defects. In this case, the defect occurs in the repair mechanisms for DNA. The genes that ordinarily regulate the repair of mismatched nucleotides are defective and can no longer detect and repair mutations. This condition leads to accumulated mutations and increases the likelihood of cancer.

Many discoveries in the basics of cancer genetics are already finding applications. In the future, it may be possible to use gene therapy to cure some types of cancer. Biotechnology and drug companies are already in the process of developing specialized therapies and cancer drugs based on molecular genetics (see chapter 10). A drug based on antisense technology has been successfully used to inactivate abnormal genes in melanoma cancers. Now that the exact locations and DNA sequences of oncogenes and tumor-suppressor genes are known, genetic probes and other markers are being developed at a rapid rate. The time is rapidly approaching when routine screening for susceptibility to many cancers will be as commonplace as routine medical tests. Other important developments are rapidly occurring in immunotherapy using monoclonal antibodies that inactivate receptors on cancer cells. This type of therapy has been used successfully on breast cancer.

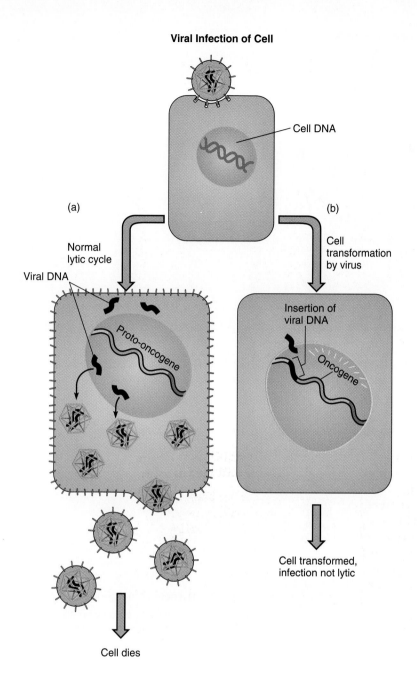

Viral Infection of Cell

Cell DNA

(a)

(b)

Normal
lytic cycle

Viral DNA

Proto-oncogene

Cell
transformation
by virus

Insertion of
viral DNA

Oncogene

Cell transformed,
infection not lytic

Cell dies

FIGURE 17.25

Outcomes of viral infection. **(a)** A virus can go through the normal lytic cycle and destroy its host cell. **(b)** A possible mechanism for viral induction of cancer. DNA viruses and retroviruses can insert DNA into the host's genome. If it inserts into a sensitive site (near a proto-oncogene), the cell can be converted into a tumor cell.

chromosome 7 is associated with melanoma. Several types of leukemia can be traced to the *loss* of a whole chromosome (figure 17.26*c*). The loss of part of a chromosome during cell division (chromosomal deletion) is also an important factor in some cancers. A large number of colorectal carcinomas manifest a deletion of a small part of chromosome 17 (figure 17.26*d*), and retinoblastoma occurs through deletions on chromosome 13 (see Spotlight on Microbiology 17.6).

Sufficient evidence indicates that some cells become cancerous because of *gene amplification* (figure 17.26*e*), a process whereby numerous extra copies of a gene are produced during DNA replication. When the copied gene is a proto-oncogene, the likelihood of oncogenic transformation is greatly increased. Certain tumors arising in squamous cell tissue or tissue of the lung, brain, and breast are linked to gene amplification.

The Role of Carcinogens   What is the true etiologic role of chemical and physical carcinogenic agents (see table 17.6)? It appears that carcinogens genetically alter or damage DNA in a way that activates or deregulates a proto-oncogene and transforms it into an oncogene. Although physical agents such as X rays and UV radiation have well-developed powers to mutate or damage DNA (see figure 11.8), most chemical carcinogens operate in a stepwise fashion that requires at least two different stimuli. The first chemical, the *initiating stimulus,* is metabolized by the liver into a more reactive chemical that permanently alters the DNA of a particular target cell. This is the carcinogenic chemical targeted by the Ames test (see chapter 9). Before the altered cell becomes neoplastic, it must be acted upon by a second chemical stimulus, called a *promoter* or *co-carcinogen,* that completes the transformation to cancer.

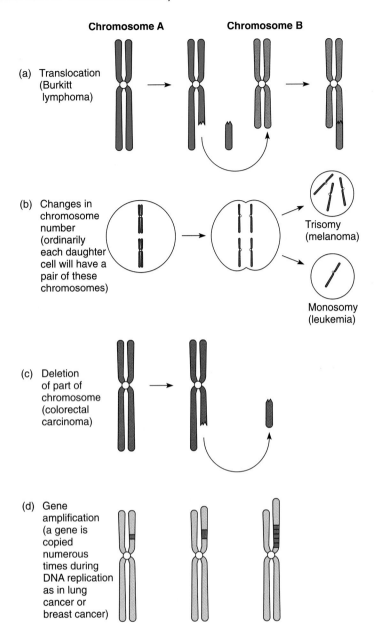

Chromosome A   **Chromosome B**

(a) Translocation (Burkitt lymphoma)

(b) Changes in chromosome number (ordinarily each daughter cell will have a pair of these chromosomes)

Trisomy (melanoma)

Monosomy (leukemia)

(c) Deletion of part of chromosome (colorectal carcinoma)

(d) Gene amplification (a gene is copied numerous times during DNA replication as in lung cancer or breast cancer)

FIGURE 17.26

Alterations in chromosomes that can be implicated in some cancers (a–d).

Progression of Cancer

Transformation is marked by the appearance of aberrant histology and physiology in the malignant cell and its progeny. Cancer cells are frequently pleomorphic. They may appear as giant cells with bizarre shapes; show an abnormal, vacuolated cytoplasm; or possess enlarged, multiple nuclei. They express surface markers that may or may not be normal. Some markers are surface receptors for receiving stimuli from *growth factors*,[5] small polypeptides involved in cellular development that can favor cancerous growth. Some tumors secrete chemical factors that stimulate growth of additional circulatory vessels, an effect not unlike "self-feeding." An important feature of malignant cells is the loss of cohesiveness that keeps normal cells aggregated, so that they readily become dislodged from a tumor, metastasize into the circulation, and are car-

ried to distant sites. Such invasive tumor cells also have the property of binding to and disrupting connective tissue barriers surrounding tissues and organs.

THE FUNCTION OF THE IMMUNE SYSTEM IN CANCER

What role does the immune system play in controlling cancer? A concept that readily explains the detection and elimination of cancer cells is **immune surveillance.** Many experts postulate that cells with cancer-causing potential arise constantly in the body but that the immune system ordinarily discovers and destroys these cells, thus keeping cancer in check. Experiments with mammals have amply demonstrated that components of cell-mediated immunity interact with tumors and their antigens. The primary types of cells that operate in surveillance and destruction of tumor cells are

_____

5. It is felt that certain oncogenes are the source of these growth factors.

cytotoxic T cells, natural killer (NK) cells, and macrophages. It appears that these cells recognize abnormal or foreign surface markers on the tumor cells and destroy them by mechanisms shown in figure 15.21. Antibodies help destroy tumors by interacting with macrophages and natural killer cells. This involvement of the immune system in destroying cancerous cells has inspired specific therapies (see Spotlight on Microbiology 17.6).

How do we account for the commonness of cancer in light of this powerful range of defenses against tumors? To an extent, the answer is simply that, as with infection, the immune system can and does fail. In some cases, the cancer may not be immunogenic enough; it may retain self markers and not be targeted by the surveillance system. In other cases, the tumor antigens may have mutated to escape detection. As we saw earlier, patients with immunodeficiencies such as AIDS and SCID are more susceptible to various cancers because they lack essential T-cell or cytotoxic functions. It may turn out that most cancers are associated with some sort of immunodeficiency, even a slight or transient one.

CHAPTER CHECKPOINTS

Cancer is caused by genetic transformation of normal host cells into malignant cells. These transformed cells perform no useful function but, instead, grow unchecked and interfere with normal tissue function.

Benign tumors are self-contained and slow-growing and do not differ greatly from their tissues of origin.

Malignant tumors are invasive, fast-growing, and very different from their tissues of origin. They metastasize into organs remote from the site of origin.

Possible causes of cancerous transformation include irregularities in mitosis, genetic damage, activation of oncogenes, and infection by retroviruses.

Cytotoxic T cells, NK cells, and macrophages Identify and destroy transformed cancerous cells by recognizing and attaching to foreign surface markers on the transformed cells. Cancerous cells survive when these mechanisms fail.

Immunotherapy offers the most promise in treating cancers, compared with surgery, radiation, or conventional chemotherapy.

CHAPTER CAPSULE WITH KEY TERMS

I. **Immunopathology**

The study of disease states involving the malfunction of the immune system is called **immunopathology.**

  A. **Allergy,** or **hypersensitivity,** is an exaggerated, adverse expression of certain immune responses.

  B. Abnormal responses to foreign antigens are characteristic of **immune complex disease;** undesirable reactions to foreign tissues are graft rejections.

  C. **Autoimmunity** involves abnormal responses to self antigens. A deficiency or loss in immune function is called **immunodeficiency.**

II. **Allergy/Hypersensitivity**

  A. An allergic reaction is a heightened immune response to antigens involving misdirected but natural mechanisms of humoral immunity, cellular immunity, and inflammation.

  B. The four types of immune reactions are classified according to the type of lymphocyte involved, type of antigen, and nature of damage.

III. **Type I Hypersensitivities**

  A. Immediate-onset allergies involve contact with **allergens,** antigens that affect certain people; susceptibility is inherited; allergens enter through four portals: Inhalants are breathed in (pollen, dust); ingestants are swallowed (food, drugs); injectants are inoculated (drugs, bee stings); contactants react on skin surface (cosmetics, glue).

  B. *Mechanism:* On first contact with allergen, specific B cells react with allergen and form a special antibody class called **IgE,** which affixes by its Fc receptor to **mast cells** and **basophils.**

    1. This **sensitizing dose** primes the allergic response system.

    2. Upon subsequent exposure with a **provocative dose,** the same allergen binds to the IgE-mast cell complex.

    3. This causes **degranulation,** release of intracellular granules containing mediators (**histamine,** serotonin, leukotriene, prostaglandin), with physiological effects such as vasodilation and bronchoconstriction.

    4. Symptoms are rash, itching, redness, increased mucous discharge, pain, swelling, and difficulty in breathing.

  C. *Specific Diseases*

    1. An **atopic allergy** is a local reaction to an allergen.

    2. **Allergic rhinitis (hay fever)** is a seasonal respiratory allergy.

    3. **Asthma** is a chronic respiratory condition.

    4. **Atopic dermatitis (eczema)** is characterized by an itchy skin rash.

    5. **Food allergy** involves respiratory, cutaneous, and skin reactions to common foodstuffs.

  D. **Systemic anaphylaxis** is an acute, extreme reaction to allergens that results in severe respiratory and circulatory symptoms. Death may occur through compromised respiration and circulatory collapse.

  E. *Diagnosis of allergy* can be made by a histamine release test on basophils; serological assays for IgE; and skin testing, which injects allergen into the skin and mirrors the degree of reaction.

  F. *Control of allergy* involves drugs to interfere with the action of histamine, inflammation, and release of cytokines from mast cells. Desensitization therapy involves the administration of purified allergens.

IV. **Type II Hypersensitivities**

  A. Type II reactions involve the interaction of antibodies, foreign cells, and complement, leading to lysis of the foreign cells. In transfusion reactions, humans may become sensitized to special antigens on the surface of the red blood cells of other humans.

  B. The **ABO blood groups** are genetically controlled: Type A blood has A antigens on the RBCs; type B has B antigens; type AB has both A and B antigens; and type O has neither antigen. People produce antibodies against A or B antigens if they lack these antigens. Antibodies can react with antigens if the wrong blood type is transfused.

  C. **Rh factor** is another RBC antigen that becomes a problem if an Rh^- mother is sensitized by an Rh^+ fetus. A second fetus can receive antibodies she has made against the factor and develop **hemolytic disease of the newborn.** Prevention involves therapy with **Rh immune globulin.**

V. **Type III Immune Complex Reactions**

A. Exposure to a large quantity of soluble foreign antigens (serum, drugs) stimulates antibodies that produce small, soluble Ag-Ab complexes.

1. These **immune complexes** are trapped in various organs and tissues, which incites a damaging inflammatory response.

2. **Arthus reaction** is a local reaction to a series of injected antigens in the same body site; it may lead to tissue destruction.

3. **Serum sickness** is a systemic disease resulting from repeated injections of foreign proteins; high levels of circulating immune complexes are deposited in various organs and cause tissue damage.

VI. **Autoimmunity**

A. In certain type II and III hypersensitivities, the immune system has lost tolerance to self molecules (autoantigens) and forms autoantibodies and sensitized T cells against them. Disruption of function can be systemic or organ specific.

B. Autoimmune diseases are genetically determined and more common in females.

1. **Systemic lupus erythematosus (SLE)** is a chronic, systemic disease in which antibodies are deposited in the kidney, skin, lungs, and heart.

2. **Rheumatoid arthritis** is a chronic systemic autoimmunity in the joints; appears to be associated with immune complexes that cause chronic inflammation and scar tissue.

3. **Endocrine autoimmunities** include Grave disease, Hashimoto thyroiditis, and types I and II **diabetes mellitus.**

4. **Myasthenia gravis** is an immune attack upon the myoneural junction, with muscle paralysis.

5. **In multiple sclerosis,** T cells and antibodies damage the myelin sheath of nerve cells; accompanied by motor and sensory loss.

VII. **Type IV Cell-Mediated Hypersensitivity**

A delayed response to antigen involving the activation of and damage by T cells.

A. *Delayed allergic response:* Skin response to allergens, including infectious agents. Example is **tuberculin reaction;** contact dermatitis is caused by exposure to plants (ivy, oak) and simple environmental molecules (metals, cosmetics); cytotoxic T cells acting on allergen elicit a skin reaction.

B. *Graft rejection:* Reaction of cytotoxic T cells directed against foreign cells of a grafted tissue; involves recognition of foreign HLA by T cells and rejection of tissue.

1. Host may reject graft; graft may reject host.

2. Types of grafts include: **autograft,** from one part of body to another; **isograft,** grafting between identical twins; **allograft,** between two members of same species; **xenograft,** between two different species.

3. All major organs may be successfully transplanted.

4. Allografts require **tissue match** (HLA antigens must correspond); rejection is controlled with drugs.

VIII. **Immunodeficiency Diseases**

Components of the immune response system are absent. Deficiencies involve B and T cells, phagocytes, and complement.

A. **Primary immunodeficiency** is genetically based, congenital; defect in inheritance leads to lack of B-cell activity, T-cell activity, or both.

B. **B-cell defect** is called **agammaglobulinemia;** patient lacks antibodies; serious recurrent bacterial infections result. In Ig deficiency, one of the classes of antibodies is missing or deficient.

C. In **T-cell defects,** the thymus is missing or abnormal. In **DiGeorge syndrome,** the thymus fails to develop; afflicted children experience recurrent infections with eucaryotic pathogens and viruses; immune response is generally underdeveloped.

D. In **severe combined immunodeficiency (SCID),** both limbs of the lymphocyte system are missing or defective; no adaptive immune response exists; fatal without replacement of bone marrow or other therapies.

E. **Secondary (acquired) immunodeficiency** is due to damage after birth (infections, drugs, radiation). AIDS is the most common of these; T helper cells are main target; deficiency manifests in numerous opportunistic infections and cancers.

IX. **Cancer and the Immune System**

A. **Cancer** is characterized by overgrowth of abnormal tissue, also known as a *neoplasm.* It appears to arise from malfunction of immune surveillance.

1. **Tumors** can be **benign** (a nonspreading local mass of tissue) or **malignant** (a cancer) that spreads (metastasizes) from the tissue of origin to other particular sites by means of the circulation.

2. Malignant tumors may be **carcinomas,** originating from epithelial tissue, or **sarcomas,** originating from embryonic connective tissue.

3. Cancers occur in nearly every cell type (except mature, nondividing cells).

B. Cancer cells appear to share a common basic mechanism involving some type of gene alteration that turns a normal gene **(proto-oncogene)** into an **oncogene.**

1. The oncogene **transforms** the cell and triggers uncontrolled growth and abnormal structure and function.

2. Gene alterations may be due to chromosome disruptions, viral infection, or the actions of chemical and physical **carcinogens.**

3. Malignant cancer cells display special markers, respond to growth factors, and lose their cohesiveness.

4. Treatment is with surgery, chemicals, radiation, and immunotherapy.

MULTIPLE-CHOICE QUESTIONS

1. Pollen is which type of allergen?
   a. contactant
   b. ingestant
   c. injectant
   d. inhalant

2. B cells are responsible for which allergies?
   a. asthma
   b. anaphylaxis
   c. tuberculin reactions
   d. both a and b

3. Which allergies are T cell–mediated?
   a. type I
   b. type II
   c. type III
   d. type IV

4. The contact with allergen that results in symptoms is called the
   a. sensitizing dose
   b. degranulation dose
   c. provocative dose
   d. desensitizing dose

5. Production of IgE and degranulation of mast cells are involved in
   a. contact dermatitis
   b. anaphylaxis
   c. Arthus reaction
   d. both a and b

6. The direct, immediate cause of allergic symptoms is the action of
    a. the allergen directly on smooth muscle
    b. the allergen on B lymphocytes
    c. allergic mediators released from mast cells and basophils
    d. IgE on smooth muscle

7. Theoretically, type ___ blood can be donated to all persons because it lacks ___.
    a. AB, antibodies   c. AB, antigens
    b. O, antigens   d. O, antibodies

8. An example of a type III immune complex disease is
    a. serum sickness   c. graft rejection
    b. contact dermatitis   d. atopy

9. Type II hypersensitivities are due to
    a. IgE reacting with mast cells
    b. activation of cytotoxic T cells
    c. IgG-allergen complexes that clog epithelial tissues
    d. complement-induced lysis of cells in the presence of antibodies

10. Production of autoantibodies may be due to
    a. emergence of forbidden clones of B cells
    b. production of antibodies against sequestered tissues
    c. infection-induced change in receptors
    d. all of these are possible

11. Rheumatoid arthritis is an ___ that affects the ___.
    a. immunodeficiency disease, muscles
    b. autoimmune disease, nerves
    c. allergy, cartilage
    d. autoimmune disease, joints

12. A positive tuberculin skin test is an example of
    a. a delayed-type allergy
    b. acute contact dermatitis
    c. autoimmunity
    d. eczema

13. Which disease would be most similar to AIDS in its pathology?
    a. X-linked agammaglobulinemia
    b. SCID
    c. ADA deficiency
    d. DiGeorge syndrome

14. A general feature common to all cancers is a ___ that leads to ___.
    a. carcinogenic agent, carcinoma
    b. mutation, gene amplification
    c. genetic defect, uncontrolled cell growth
    d. proto-oncogene, anti-oncogene

15. The property whereby cancer cells display abnormal anatomy and growth patterns is called
    a. metastasis
    b. oncogenic
    c. carcinogenic
    d. malignant

16. A cancer associated with viral infections includes
    a. cervical carcinoma
    b. Burkitt lymphoma
    c. T-cell leukemia
    d. all of these

CONCEPT QUESTIONS

1. a. Define allergy and hypersensitivity.
   b. What accounts for the reactions that occur in these conditions?
   c. What does it mean when a reaction is immediate or delayed?
   d. Give examples of each type.

2. Describe several factors that influence types and severity of allergic responses.

3. a. How are atopic allergies similar to anaphylaxis?
   b. How are they different?

4. a. How do allergens gain access to the body?
   b. What are some examples of allergens that enter by these portals?

5. a. Trace the course of a pollen grain through sensitization and provocation in type I allergies.
   b. Include in the discussion the role of mast cells, basophils, IgE, and allergic mediators.
   c. Outline the target organs and symptoms of the principal atopic diseases and their diagnosis and treatment.

6. a. Describe the allergic response that leads to anaphylaxis. Include its usual causes, how it is diagnosed and treated, and two effective physiological targets for treatment.
   b. Explain how hyposensitization is achieved and suggest two mechanisms by which it might work.

7. a. What is the mechanism of type II hypersensitivity?
   b. Why are the tissues of some people antigenic to others?
   c. Would we be concerned about this problem if it were not for transfusions?
   d. What is the actual basis of the four ABO and Rh blood groups?
   e. Where do we derive our natural hypersensitivities to the A or B antigens that we do not possess?
   f. How does a person become sensitized to Rh factor? List consequences.

8. Explain the rules of transfusion. Illustrate what will happen if type A blood is accidently transfused into a type B person.

9. a. Contrast type II and type III hypersensitivities with respect to type of antigen, antibody, and manifestations of disease.
   b. What is immune complex disease?
   c. Differentiate between the Arthus reaction and serum sickness.

10. a. Explain the pathologic process in autoimmunity.
    b. Draw diagrams that explain five possible mechanisms for the development of autoimmunity.
    c. Describe four major types of autoimmunity, comparing target organs and symptoms.

11. Compare and contrast type I (atopic) and type IV (delayed) hypersensitivity as to mechanism, symptoms, eliciting factors, and allergens.

12. a. What is the molecular/cellular basis for a host rejecting the graft tissue?
    b. Account for a graft rejecting the host.
    c. Compare the four types of grafts.
    d. What does it mean to say that two tissues constitute a close match?
    e. Describe the procedure involved in a bone marrow transplant.

13. a. In general, what causes primary immunodeficiencies?
    b. Acquired immunodeficiencies?
    c. Why can T-cell deficiencies have greater impact than B-cell deficiencies?
    d. What kinds of symptoms accompany a B-cell defect?
    e. A T-cell defect?
    f. Combined defects?
    g. Give examples of specific diseases that involve each type of defect.

14. a. Name several medical conditions that require immunosuppressive drugs.
    b. Name some immunosuppressive drugs, and explain what they do.

15. a. Define cancer, and give some synonyms.
    b. Differentiate between a benign tumor and a malignant tumor, and give examples.
    c. How are cancers technically differentiated?

16. a. Describe a possible mechanism to account for transformation of a normal cell to a cancer cell.
    b. Describe the possible genetic and environmental factors that increase oncogenesis.
    c. How do infectious agents such as viruses cause cancer?

17. Relate how the immune system is involved in cancer.

18. **Multiple Matching.** Choose the description that best fits the cytokines.

    ____ histamine   ____ leukotriene
    ____ prostaglandin   ____ platelet factor
    ____ bradykinin

    a. a peptide involved in blood clotting and chemotaxis
    b. a lipid that aggregates and lyses thrombocytes
    c. causes prolonged bronchospasm and mucous secretion in the lungs
    d. increases inflammation and sensitivity to pain
    e. a potent stimulus for smooth muscle and glandular secretion

CRITICAL-THINKING QUESTIONS

1. a. Discuss the reasons that the immune system is sometimes called a double-edged sword.
   b. Suggest a possible function of allergy.

2. A 3-week-old neonate develops severe eczema after being given penicillin therapy for the first time. Can you explain what has happened?

3. Can you explain why a person is allergic to strawberries when he eats them, but shows a negative skin test to them?

4. a. Where in the course of type I allergies do antihistamine drugs work?
   b. Cortisone?
   c. Desensitization?

5. a. Although we call persons with type O blood universal donors and those with type AB blood universal recipients, what problems might accompany transfusions involving O donors or AB recipients?
   b. What is the truly universal donor and recipient blood type? (Include Rh type.)
   c. Can you explain how a person could have a genotype for type A or B blood but a phenotype for type O?

6. Why would it be necessary for an Rh⁻ woman who has had an abortion, miscarriage, or an ectopic pregnancy to be immunized against the Rh factor?

7. a. Describe three circumstances that might cause antibodies to develop against self tissues.
   b. Can you explain how people with autoimmunity could develop antibodies against intracellular components (nucleus, mitochondria, and DNA)?

8. Would a person show allergy to poison oak upon first contact? Explain.

9. a. Looking at figure 17.19, predict which family members would be good allograft pairs and which would be incompatible.
   b. Why might a graft be rejected even between closely matched siblings?
   c. Transplanting a baboon heart into a baby is what kind of graft?

10. Why are primary immunodeficiencies considered experiments of nature?

11. Why are defective genes found on the X chromosome expressed most often in males?

12. a. Explain why babies with agammaglobulinemia do not develop opportunistic infections until about 6 months after birth.
    b. Explain why people with B-cell deficiencies can benefit from artificial passive immunotherapy. Explain whether vaccination would work for them.

13. a. Why are SCIDS children unable to reject grafts?
    b. What would be the major problem in most bone marrow transplantations for these children?
    c. Draw a general outline of the events that cause SCIDS in the child called David.
    d. Why did David acquire cancer?

14. In what ways can cancer be both a cause and a symptom of immunodeficiency?

15. What features of cancer cells account for metastasis?

16. In what ways could cancer be inherited?

17. If a chemical is found to be a mutagen by the Ames test, what other factors will be required for it to induce cancer?

18. Most cancer therapy removes or destroys the metastatic cells. Describe an alternate technology to prevent cancer. (Hint: You may have to refer to chapter 10.)

INTERNET SEARCH TOPICS

Go to a search engine and locate websites involving these areas of interest:

1. New allergy drugs that block the action of IgE. What is their mechanism?

2. Poison ivy or oak hypersensitivity. How can you identify and treat it?

3. Bone marrow transplants. Locate information on registering to donate marrow and the process surrounding testing and transplantation.

Introduction to Identification
Techniques in Medical Microbiology

Preface to the Survey of Microbial Diseases

The next eight chapters cover the most clinically significant bacterial, fungal, parasitic, and viral diseases in the world, with an emphasis on those most prevalent in North America. This survey will encompass the morphology, physiology, and virulence of the microbes and the epidemiology, pathology, treatment, and prevention of the diseases they cause. The bacteria are divided into the gram-positive and gram-negative cocci (chapter 18); the gram-positive bacilli (chapter 19); the gram-negative bacilli (chapter 20); miscellaneous bacteria and dental infections (chapter 21); fungi (chapter 22); protozoan and helminth parasites (chapter 23); DNA viruses (chapter 24); and RNA viruses (chapter 25).

Several explanatory concepts can be found in previous chapters. Information on general methods of specimen processing is in figures 3.1 and 6.23 and in chapters 4 and 10. Other references include media (chapter 3), chemotherapy (chapter 12), and host-parasite relations and epidemiology (chapter 13). Particular attention is called to Microbits 13.4 on the terminology of disease, Medical Microfile M.1 on skin lesions, and the discussion of vaccines and serological tests (chapter 16). An extensive reference that lists organisms and diseases by site of infection is located in appendix D.

The investigation and diagnosis of these diseases depend to a considerable degree on a series of basic laboratory techniques. This section summarizes the methods used by clinical microbiologists and other medical personnel to collect, isolate, and identify infectious agents.

On the Track of the Infectious Agent: Specimen Collection

Regardless of the method of diagnosis, specimen collection is the common point that guides the health care decisions of every member of a clinical team. Indeed, the success of identification and treatment depends on how specimens are collected, handled, and stored. Specimens can be taken by a medical technologist, nurse, physician, or even by the patient himself. However, it is imperative that general aseptic procedures be used, including sterile sample containers and other tools to prevent contamination from the environment or the patient. Figure A delineates the most common sampling sites and procedures.

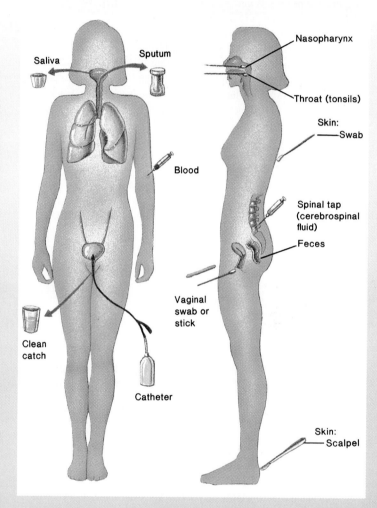

FIGURE A

Sampling sites and methods of collection for clinical laboratories.

In sites that normally contain resident microflora, care should be taken to sample only the infected site and not surrounding areas. For example, throat and nasopharyngeal swabs should not touch the tongue, cheeks, or saliva. Saliva is an especially undesirable contaminant because it contains millions of bacteria per milliliter, most of which are normal flora. Saliva samples are occasionally taken for dental diagnosis by having the patient expectorate into a container. Depending on the nature of the lesion, skin can be

MEDICAL MICROFILE   M.1
Rashes, Rushes, and Flushes

Diseases are often manifested by the presence of skin **lesions*** of a particular characteristic appearance. A small, flat-colored skin lesion is a *macule.** The color may be red, tan, or white. Crops of these spots are early signs of measles and rubella. A raised lesion composed of solid tissue is a *papule;** it varies in texture from smooth to rough and can be any number of colors. Often, papules are a later stage in the development of a macule, producing the so-called maculopapular rash. A thicker papule that penetrates into the dermis is a *nodule.** Leprosy and several fungus infections produce nodular lesions.

Many lesions contain some sort of liquid or semiliquid material called an exudate. A *vesicle** is a small blister that contains clear, yellowish fluid, and a *bulla** is a large, fluid-filled blister. Vesicles are characteristic of herpes simplex, impetigo, and chickenpox (a pock is one type of vesicle); bullae are seen in some types of staphylococcal skin infections. If the space in the lesion is filled with pus (a mixture of fluid, white blood cells, tissue debris, and bacteria), it is a *pustule.** Acne lesions are types of pustules. Vesicles, bullae, and pustules have a fragile surface that can burst and release the exudate, which dries and forms a *crust.*

When a lesion results from skin being sloughed, the process is known as *erosion* or *ulceration.* Erosion is a superficial loss of the epidermis, whereas an ulcer is a deeply penetrating open sore. The chancre of syphilis and the eschar of anthrax are erosive types of lesions; large, draining ulcers appear in leishmaniasis and some worm infections.

---

* lesion (lee'-zhun) L. *laesio,* to hurt.

* macule (mak'-yool) L. *maculatus,* spotted.

* papule (pap'-yool) L. *papula,* pimple.

* nodule (nawd'-yool) L. *nodulus,* little knot.

* vesicle (ves'-ik-ul) L. *vesicula,* small bladder.

* bulla (byoo'-lah) L. *bulla,* bubble.

* pustule (pust'-yool) L. *pustula,* blister or pimple.

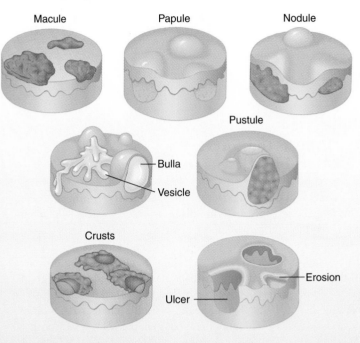

Cutaway views of the skin, comparing various lesions.

Some skin manifestations are due to alterations in circulation. *Erythema** is a confluent reddening of the skin due to increased blood flow. The rash of scarlet fever and the peculiar migrating skin eruption of Lyme disease are types of erythema. Hemorrhaging into the skin produces a brown to purple discoloration called *purpura,** as seen in bubonic plague and meningococcal meningitis.

---

* erythema (air-ih-thee'-mah) Gr. *erythema,* flush upon the skin.

* purpura (pur'-pur-ah) L. *purpura,* purple.

swabbed or scraped with a scalpel to expose deeper layers. The mucous lining of the vagina, cervix, or urethra can be sampled with a swab or applicator stick.

Urine is taken aseptically from the bladder with a thin tube called a catheter. Another method, called a "clean catch," is taken by washing the external urethra and collecting the urine midstream. The latter method inevitably incorporates a few normal flora into the sample, but these can usually be differentiated from pathogens in an actual infection. Sputum, the mucous secretion that coats the lower respiratory surfaces, especially the lungs, is discharged by coughing or taken by catheterization to avoid contamination with saliva. Sterile materials such as blood, cerebrospinal fluid, and tissue fluids must be taken by sterile needle aspiration. Antisepsis of the puncture site is extremely important in these cases. Additional sources of specimens are the vagina, eye, ear canal, nasal cavity (all by swab), and diseased tissue that has been surgically removed (biopsied).

After proper collection, the specimen is promptly transported to a lab and stored appropriately (usually refrigerated) if it must be held for a time. Nonsterile samples in particular, such as urine, feces, and sputum, are especially prone to deterioration at room temperature. Special swab and transport systems are designed to collect the specimen and maintain it in stable condition for several hours. These devices contain nonnutritive maintenance media (so the microbes do not grow), a buffering system, and an anaerobic environment to prevent possible destruction of oxygen-sensitive bacteria.

OVERVIEW OF LABORATORY TECHNIQUES

The routes taken in specimen analysis are the following: (1) direct tests using microscopic, immunologic, or other specific methods that provide immediate clues as to the identity of the microbe or microbes in the sample and (2) cultivation, isolation, and identification of pathogens using a wide variety of general and specific

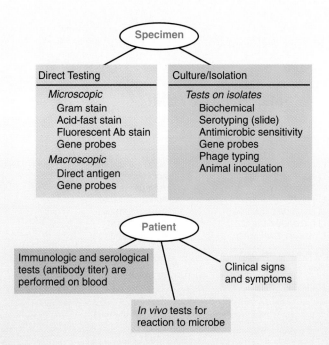

FIGURE B

A scheme of specimen isolation and identification.

tests (figure B). Most test results fall into two categories: **presumptive data,** which place the isolated microbe (isolate) in a preliminary category such as a genus, and more specific, **confirmatory data,** which provide more definitive evidence of a species. Some tests are more important for some groups of bacteria than for others. The total time required for analysis ranges from a few minutes in a streptococcal sore throat to several weeks in tuberculosis.

Results of specimen analysis are entered in a summary patient chart (figure C) that can be used in assessment and treatment regimens. The type of antimicrobial drugs chosen for testing varies with the type of bacterium isolated.

Some diseases are diagnosed without the need to identify microbes from specimens. Serological tests on a patient's serum can detect signs of an antibody response. One method that clarifies whether a positive test indicates current or prior infection is to take two samples several days apart to see if the antibody titer is rising. Skin testing can pinpoint a delayed allergic reaction to a microorganism (see chapters 16 and 17). These tests are also important in screening the general population for exposure to an infectious agent such as rubella or tuberculosis.

Because diagnosis is both a science and an art, the ability of the practitioner to interpret signs and symptoms of disease can be very important. AIDS, for example, is usually diagnosed by serological tests and a complex of signs and symptoms without ever isolating the virus. Some diseases (athlete's foot, for example) are diagnosed purely by the typical presenting symptoms and may require no lab tests at all.

Immediate Direct Examination of Specimen

Direct microscopic observation of a fresh or stained specimen is one of the most rapid methods of determining presumptive and sometimes confirmatory characteristics (see figures 18.26 and 22.25). Stains most often employed for bacteria are the Gram stain (see Microbits 4.2) and the acid-fast stain (see figure 19.14). For many species these ordinary stains are useful, but they do not work with certain organisms. Direct fluorescence antibody (DFA) tests can highlight the presence of the microbe in patient specimens by means of labeled antibodies (figure D). DFA tests are particularly useful for bacteria, such as the syphilis spirochete, that are not readily cultivated in the laboratory or if rapid diagnosis is essential for the survival of the patient.

Another way that specimens can be analyzed is through *direct antigen testing,* a technique similar to direct fluorescence in that known antibodies are used to identify antigens on the surface of bacterial isolates. But in direct antigen testing, the reactions can be seen with the naked eye. Quick test kits that greatly speed clinical diagnosis are available for *Staphylococcus aureus, Streptococcus pyogenes* (see figure 18.14), *Neisseria gonorrhoeae, Haemophilus influenzae,* and *Neisseria meningitidis.* However, when the microbe is very sparse in the specimen, direct testing is like looking for a needle in a haystack, and more sensitive methods are necessary.

Cultivation of Specimen

Isolation Media   Such a wide variety of media exist for microbial isolation that a certain amount of preselection must occur, based on the nature of the specimen. In cases where the suspected pathogen is present in small numbers or is easily overgrown, the specimen can be initially enriched with specialized media. In specimens such as urine and feces that have high bacterial counts and a diversity of species, selective media are used (see figure 20.9). In most cases, specimens are also inoculated into differential media that define such characteristics as reactions in blood (blood agar) and fermentation patterns (mannitol salt and MacConkey agar). A patient's blood is usually cultured in a special bottle of broth that can be periodically sampled for growth. Numerous other examples of isolation, differential, and biochemical media were presented in chapter 3 and appear in various sections of chapters 18, 19, 20, and 21. So that subsequent steps in identification will be as accurate as possible, all work must be done from isolated colonies or pure cultures, because working with a mixed or contaminated culture gives misleading and inaccurate results. From such isolates, clinical microbiologists obtain information about a pathogen's microscopic morphology and staining reactions, cultural appearance, motility, oxygen requirements, and biochemical characteristics.

Biochemical Testing   The physiological reactions of bacteria to nutrients and other substrates provide excellent indirect evidence of the types of enzyme systems present in a particular species. Many of these tests are based on the following scheme:

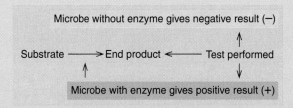

MICROBIOLOGY UNIT

DATE, TIME & PERSON COLLECTING	SPECIMEN NUMBER	ANTIBIOTIC THERAPY	TENTATIVE DIAGNOSIS

SOURCE OF SPECIMEN		TEST REQUEST	

SOURCE OF SPECIMEN

- ☐ THROAT
- ☐ SPUTUM
- ☐ STOOL
- ☐ CERVIX
- ☐ AEROSOL INDUCED SPUTUM
- ☐ WOUND - SPECIFY SITE _____
- ☐ OTHER - SPECIFY _____

- ☐ BLOOD
- ☐ URINE - CLEAN CATCH
- ☐ URINE - CATH
- ☐ BRONCHIAL WASHING

TEST REQUEST

- ☐ GRAM STAIN
- ☐ ROUTINE CULTURE
- ☐ SENSITIVITY
- ☐ MIC
- ☐ ANAEROBIC CULTURE
- ☐ R/O GROUP A STREP
- ☐ WRIGHT STAIN (WBC)

- ☐ ACID FAST SMEAR
- ☐ ACID FAST CULTURE
- ☐ FUNGUS WET MOUNT
- ☐ FUNGUS CULTURE
- ☐ PARASITE STUDIES
- ☐ OCCULT BLOOD
- ☐ PCR ANALYSIS

DO NOT WRITE BELOW THIS LINE –# FOR LAB USE ONLY

GRAM STAIN (4+ NUMEROUS; 3+ MANY; 2+ MODERATE; 1+FEW; 0 NONE SEEN)

COCCI: GRAM POS._____ GRAM NEG._____ W B C._____

BACILLI: GRAM POS._____ GRAM NEG._____ EPITHELIAL CELLS_____

INTRACELLULAR & EXTRACELLULAR GRAM-NEGATIVE DIPLOCOCCI_____

YEAST_____   ☐ No organisms seen.

FUNGUS: WET MOUNT ☐ No mycotic elements or budding structures seen.

☐ _____

CULTURE ☐ _____

AFB: SMEAR ☐ No acid fast bacilli seen.

☐ _____

CULTURE ☐ _____

PARASITE STUDIES: DIRECT:_____

CONCENTRATE:_____

PERMANENT:_____

OCCULT BLOOD:

APPEARANCE OF STOOL:_____

OCCULT BLOOD:_____

COLONY COUNT: Urine organisms/ml. ☐ _____ ☐ > 100,000

MISCELLANEOUS RESULTS:

- ☐ NO GROWTH IN: ☐ 2 DAYS ☐ 3 DAYS ☐ 5 DAYS ☐ 7 DAYS
- ☐ NORMAL FLORA ISOLATED
- ☐ NO ENTEROPATHOGENS ISOLATED
- ☐ SPUTUM UNACCEPTABLE FOR CULTURE — REPRESENTS SALIVA — NEW SPECIMEN REQUESTED
- ☐ URINE > 2 COLONY TYPES PRESENT REPRESENT CONTAMINATION — NEW SPECIMEN REQUESTED

CULTURE RESULTS

1+ FEW   3+ MANY

2+ MODERATE   4+ NUMEROUS

ANAEROBES	☐ BACTEROIDES
	☐ CLOSTRIDIUM
	☐ PEPTOSTREPTOCOCCUS
	☐
	☐
ENTERICS	☐ ESCHERICHIA COLI
	☐ ENTEROBACTER
	☐ KLEBSIELLA
	☐ PROTEUS
	☐
STAPH-YLOCOCCUS	☐ AUREUS
	☐ EPIDERMIDIS
	☐ SAPROPHYTICUS
STREP-TOCOCCUS	☐ GROUP A
	☐ GROUP B
	☐ GROUP D ENTEROCOCCI
	☐ GROUP D NON ENTEROCOCCI
	☐ PNEUMONIAE
	☐ VIRIDANS
	☐
YEAST	☐ CANDIDA
	☐
OTHER ISOLATES	☐ PSEUDOMONAS
	☐ HAEMOPHILUS
	☐ GARDNERELLA VAGINALIS
	☐ NEISSERIA
	☐ CL. DIFFICILE
	☐

SENSITIVITY TESTS

NOTE: Bacteria with intermediate susceptibility may not respond satisfactorily to therapy.	AMIKACIN	AMPICILLIN	BETA LACTAMASE PRODUCTION	CARBENICILLIN	CEFAZOLIN	CEFOTAXIME	CEFOXITIN	CEFUROXIME	CHLORAMPHENICOL	CLINDAMYCIN	ERYTHROMYCIN	GENTAMICIN	METHICILLIN	METRONIDAZOLE	NITROFURANTOIN	PENICILLIN	IMMUNOLOGY	TETRACYCLINE	TOBRAMYCIN	TRIMETHO PRIM- SULFAME- THOXAZOLE	VANCOMYCIN	CIPROFLOX
A																						
B																						
C																						

☐ COMMENTS: _____

DATE _____ TECHNOLOGIST _____

706-30A

FIGURE C

Example of a clinical form used to report data on patients' specimens.

The microbe is cultured in a medium with a special substrate and then tested for a particular end product. The presence of the end product indicates that the enzyme is expressed in that species; its absence means it lacks the enzyme for utilizing the substrate in that particular way. These types of reactions are particularly meaningful in bacteria, which are haploid and generally express their genes for utilizing a given nutrient.

Among the prominent biochemical tests are carbohydrate fermentation (acid and/or gas); hydrolysis of gelatin, starch, and other polymers; enzyme actions such as catalase, oxidase, and coagulase;

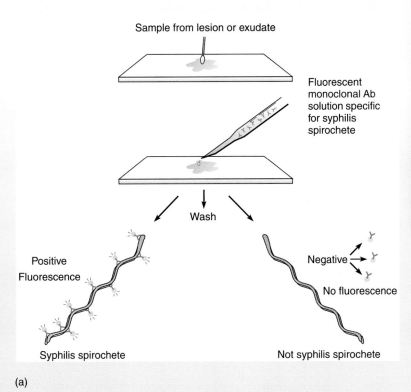

(a)

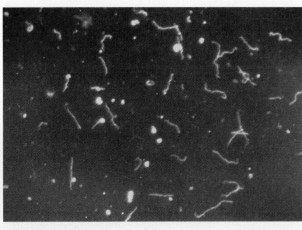

(b)

FIGURE D

(a) Direct fluorescent antigen test results for *Treponema pallidum,* the syphilis spirochete, and an unrelated spirochete. **(b)** Photomicrograph of this technique used on a blood sample from a syphilitic patient.

and various by-products of metabolism. Examples of these tests are presented in chapters 18, 19, and 20. Many are presently performed with rapid, miniaturized systems that can simultaneously determine up to 23 characteristics in small individual cups or spaces (see figures 18.7 and 20.10). An important plus, given the complexity of biochemical profiles, is that such systems are readily adapted to computerized analysis.

Common schemes for identifying bacteria are somewhat artificial but convenient. They are based on easily recognizable characteristics such as motility, oxygen requirements, Gram stain reactions, shape, spore formation, and various biochemical reactions. Schemes can be set up as flowcharts (figure E) or keys that trace a route of identification by offering pairs of opposing characteristics (positive versus negative, for example) from which to select. Eventually, an endpoint is reached, and the name of a genus or species that fits that particular combination of characteristics appears. Diagnostic tables that provide more complete information are preferred by many laboratories because variations from the general characteristics used on the flowchart can be misleading. Both systems are used in this text.

Miscellaneous Tests   When morphological and biochemical tests are insufficient to complete identification, other tests come into play. Commercial **serotyping** kits are important for identifying isolates in the genera *Salmonella, Shigella, Streptococcus,* and *Staphylococcus.* In general, a bit of culture is mixed with specific, animal-derived antisera or monoclonal antibodies that agglutinate the cells if they are complementary (technique shown in figure 16.3).

Bacteria host viruses called bacteriophages that are very species- and strain-specific. Such selection by a virus for its host is useful in typing some bacteria, primarily *Staphylococcus* and *Salmonella.* The technique of **phage typing** involves inoculating a lawn of cells onto a Petri dish, mapping it off into blocks, and applying a different phage to each block. Cleared areas corresponding to lysed cells indicate sensitivity to that phage. Phage typing is chiefly used for tracing strains of bacteria in epidemics.

Animals must be inoculated to cultivate bacteria such as *Mycobacterium leprae* and *Treponema pallidum,* whereas avian embryos and cell cultures are used to grow rickettsias, chlamydias, and viruses. Animal inoculation is also occasionally used to test bacterial or fungal virulence.

Modern molecular biology has provided some extremely specific and sensitive systems for identifying microbes. The value of such molecular testing is that it measures similarities or differences in the actual molecular and genetic structure of microbes, not just in morphological appearance or physiological reactions.

The uses of these methods are expanding rapidly. Some of them have been discussed and illustrated in prior chapters, and others will be covered with individual infectious agents. **Gene probes** and hybridization techniques based on unique arrangements of nucleotides in DNA are gaining in importance (see figure 4.28 and figure F). Probes are small segments of single-stranded DNA or RNA with a known sequence that are mixed with nucleic acid from an unknown microbe. Hybridization of the known strand with the unknown strand is highly suggestive of a matching identity. Gene probes are currently used to identify *Escherichia coli* and other gram-negative bacteria, *Mycobacterium, Mycoplasma, Legionella, Chlamydia, Treponema,* and numerous viruses.

Many nucleic acid assays use the polymerase chain reaction (PCR). This method can amplify DNA present in samples even in

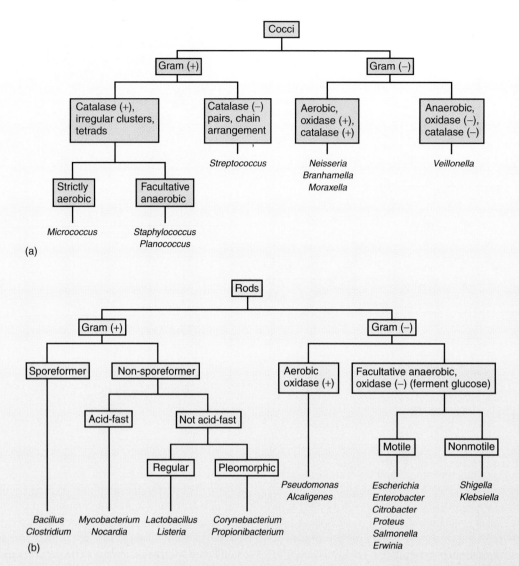

FIGURE E

Flowchart to separate primary genera of gram-positive and gram-negative bacteria. (a) Cocci and (b) rods involved in human diseases.

tiny amounts, which greatly improves the sensitivity of the test (see figure 10.6). PCR tests are being used or developed for a wide variety of bacteria, viruses, protozoa, and fungi.

Antimicrobic sensitivity tests are not only important in determining the drugs to be used in treatment (see figure 12.21), but the patterns of sensitivity can also be used in presumptive identification of some species of *Streptococcus* (see figure 18.14*a*), *Pseudomonas,* and *Clostridium.* Antimicrobics are also used as selective agents in many media.

Determining Clinical Significance of Cultures Questions that can be difficult but necessary to answer in this era of debilitated patients and opportunists are: Is an isolate clinically important, and how do you decide whether it is a contaminant or just part of the normal flora? The number of microbes in a sample is one useful criterion. For example, a few colonies of *Escherichia coli* in a urine sample can simply indicate normal flora, whereas several hundred can mean active infection. In contrast, the presence of a single colony of a true pathogen such as *Mycobacterium tuberculosis* in sputum or an opportunist in sterile sites such as cerebrospinal fluid or blood is highly suggestive of its role in disease. Furthermore, the

repeated isolation of a relatively pure culture of any microorganism can mean it is an agent of disease, though care must be taken in this diagnosis. Another problem facing the laboratory technician is that of differentiating a pathogen from species in the normal flora that are similar in morphology from their more virulent relatives.

Universal Blood and Body Fluid Precautions

Medical and dental settings require stringent measures to prevent the spread of nosocomial (clinically acquired) infections from patient to patient, from patient to worker, and from worker to patient. But even with precautions, the rate of such infections is rather high. Recent evidence indicates that more than one-third of nosocomial infections could be prevented by consistent and rigorous infection control methods.

Previously, control guidelines were disease-specific, and clearly identified infections were managed with particular restrictions and techniques. With this arrangement, personnel tended to handle materials labeled *infectious* with much greater care than

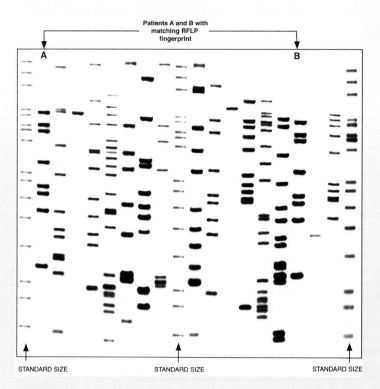

Patients A and B with matching RFLP fingerprint

A · · · · · B · · ·

STANDARD SIZE STANDARD SIZE STANDARD SIZE

FIGURE F

DNA typing of restriction fragment length polymorphisms (RFLP) for *Mycobacterium tuberculosis*. The pattern shows results for various strains isolated from 17 patients. Bands were developed by DNA hybridization using probes specific to genes from several *M. tuberculosis* strains. Lanes 1, 10, and 20 provide size markers for reference. Patients A and B are infected with the same common strain of the pathogen.

those that were not so labeled. The AIDS epidemic spurred a reexamination of that policy. Because of the potential for increased numbers of undiagnosed HIV-infected patients, the Centers for Disease Control and Prevention laid down more-stringent guidelines for handling patients and body substances. These guidelines have been termed **universal precautions (UPs),** because they are based on the assumption that all patient specimens could harbor infectious agents and so must be treated with the same degree of care. They also include body substance isolation (BSI) techniques to be used in known cases of infection.

It is worth mentioning that these precautions are designed to protect all individuals in the clinical setting—patients, workers, and the public alike. In general, they include techniques designed to prevent contact with pathogens and contamination and, if prevention is not possible, to take purposeful measures to decontaminate potentially infectious materials.

The universal precautions recommended for all health care settings are:

1. Barrier precautions, including masks and gloves, should be taken to prevent contact of skin and mucous membranes with patients' blood or other body fluids. Because gloves can develop small invisible tears, double gloving decreases the risk further. For protection during surgery, venipuncture, or emergency procedures, gowns, aprons, and other body coverings should be worn. Dental workers should wear eyewear and face shields to protect against splattered blood and saliva.

2. More than 10% of health care personnel are pierced each year by sharp (and usually contaminated) instruments. These accidents carry risks not only for AIDS but also for hepatitis B, hepatitis C, and other diseases. Preventing inoculation infection requires vigilant observance of proper techniques. All disposable needles, scalpels, or sharp devices from invasive procedures must immediately be placed in puncture-proof containers for sterilization and final discard. Under no circumstances should a worker attempt to recap a syringe, remove a needle from a syringe, or leave unprotected used syringes where they pose a risk to others. Reusable needles or other sharp devices must be heat-sterilized in a puncture-proof holder before they are handled. If a needlestick or other injury occurs, immediate attention to the wound, such as thorough degermation and application of strong antiseptics, can prevent infection.

3. Dental handpieces should be sterilized between patients, but if this is not possible, they should be thoroughly disinfected with a high-level disinfectant (peroxide, hypochlorite). Blood and saliva should be removed completely from all contaminated dental instruments and intraoral devices prior to sterilization.

4. Hands and other skin surfaces that have been accidently contaminated with blood or other fluids should be scrubbed immediately with a germicidal soap. Hands should likewise be washed after removing rubber gloves, masks, or other barrier devices.

5. Because saliva can be a source of some types of infections, barriers should be used in all mouth-to-mouth resuscitations.

6. Health care workers with active, draining skin or mucous membrane lesions must refrain from handling patients or equipment that will come into contact with other patients. Pregnant health care workers risk infecting their fetuses and must pay special attention to these guidelines. Personnel should be protected by vaccination whenever possible.

Isolation procedures for known or suspected infections should still be instituted on a case-by-case basis.

LABORATORY BIOSAFETY LEVELS AND CLASSES OF PATHOGENS

Personnel handling infectious agents in the laboratory must be protected from possible infection through special risk management or containment procedures. These involve the following: (1) carefully observing standard laboratory aseptic and sterile procedures while handling cultures and infectious samples; (2) using large-scale sterilization and disinfection procedures; (3) restricting eating, drinking, and smoking; and (4) wearing personal protective items such as gloves, masks, safety glasses, laboratory coats, boots, and headgear. Some circumstances also require additional protective equipment such as biological safety cabinets for inoculations and specially engineered facilities to control materials entering and leaving the laboratory in the air and on personnel. Table M.1 summarizes the primary biosafety levels and agents of disease as characterized by the Centers for Disease Control and Prevention.

TABLE M.1

Primary Biosafety Levels and Agents of Disease

Biosafety Level	Facilities and Practices	Risk of Infection and Class of Pathogens
1	Standard, open bench, no special facilities needed; typical of most microbiology teaching labs; access may be restricted.	Low infection hazard; microbes not generally considered pathogens and will not colonize the bodies of healthy persons; *Micrococcus luteus, Bacillus megaterium, Lactobacillus, Saccharomyces.*
2	At least Level 1 facilities and practices, plus personnel must be trained in handling pathogens; lab coats and gloves required; safety cabinets may be needed; biohazard signs posted; access restricted.	Agents with moderate potential to infect; class 2 pathogens can cause disease in healthy people but can be contained with proper facilities; most pathogens belong to class 2; includes *Staphylococcus aureus, Escherichia coli, Salmonella* spp., *Corynebacterium diphtheriae;* pathogenic helminths; hepatitis A, B, and rabies viruses; *Cryptococcus* and *Blastomyces.*
3	Minimum of Level 2 facilities and practices; all manipulation performed in safety cabinets; lab designed with special containment features; only personnel with special clothing can enter; no unsterilized materials can leave the lab; personnel warned, monitored, and vaccinated against infection dangers.	Agents can cause severe or lethal disease especially when inhaled; class 3 microbes include *Mycobacterium tuberculosis, Francisella tularensis, Yersinia pestis, Brucella* spp., *Coxiella burnetii, Coccidioides immitis,* and yellow fever, WEE, and AIDS viruses.
4	Minimum of Level 3 facilities and practices; facilities must be isolated with very controlled access; clothing changes and showers required for all people entering and leaving; materials must be autoclaved or fumigated prior to entering and leaving lab.	Agents being handled are highly virulent microbes that pose extreme risk for morbidity and mortality when inhaled in droplet or aerosol form; most are exotic flaviviruses; arenaviruses, including Lassa fever virus; or filoviruses, including Ebola and Marburg viruses.

REVIEW OF CLINICAL LABORATORY TECHNOLOGY WITH KEY TERMS

Concerns of the Clinical Laboratory

The lab provides technical support in determining the causative agents of disease and the antimicrobics to be used for treatment. The **method of specimen collection** is crucial; must be done aseptically; swabs used for nasopharynx, vagina, skin; urine collected as free flow; sterile urine requires a catheter; sputum is coughed up or aspirated from lung; blood and spinal fluid taken by puncture of sterile site with needle.

I. Handling

Samples are transported to lab; those containing fragile pathogens must be placed in special environment, processed rapidly, or refrigerated for a short time.

II. Laboratory Protocols

Identification of microbe and diagnosis of disease require a wide array of procedures, including:
a. direct specimen testing;
b. cultivation and isolation;
c. examination of biochemistry, antigenicity, and genetics of microbe; and
d. examination of patient.
Collected data fall into categories of presumptive evidence or confirmatory evidence of a species.

III. Testing

Direct tests include microscopic examination of stained specimen, direct fluorescent antibody tests, and macroscopic antigen tests, all of which provide rapid clinical data. **Cultivation** is usually required for confirmation.

A. Initial isolation of pathogen is performed on some types of selective, enrichment, or differential media, chosen according to the expected nature of the specimen and pathogen.
1. Cultures are examined for number and types of colonies.
2. Decision is made as to whether the isolates are normal flora, contaminants of sampling, or actual cause of infection.
B. Isolated colonies are examined for microscopic and macroscopic morphology.
C. Further tests include motility, special stains, and the presence of enzyme systems for processing nutrients; hundreds of biochemical tests are available to determine these characteristics; the most important ones are selected in accordance with evidence of a given group or genus.
1. Miniaturized methods are in common use.
2. Computerized analysis of results is helpful.
3. Collected data are processed through flowcharts or tables to arrive at the species that most closely fits the unknown isolate.
D. Other tests used in determining species are **serotyping, antimicrobic sensitivity, genetic analysis;** use of animals may be necessary to isolate and confirm some microbes.
E. **Immunologic tests** are performed on the patient's serum or body: Serological tests give evidence of antibodies in serum; skin testing (such as tuberculin) shows reactions of body to proteins derived from a pathogen; both are helpful in screening for past or current infection; diagnosis also requires observation of clinical signs and symptoms.

REVIEW QUESTIONS

1. Why do specimens need to be taken aseptically even when nonsterile sites are being sampled and selective media are to be used?

2. Explain the general principles in specimen collection.

3. a. What is involved in direct specimen testing?
   b. In presumptive and confirmatory tests?
   c. In cultivating and isolating the pathogen?
   d. In biochemical testing?
   e. In gene probes?
   f. Explain which techniques are more sensitive and specific, in general, and why this would be the case.

4. Differentiate between the serological tests used to identify isolated cultures of pathogens and those used to diagnose disease from patients' serum.

5. Why is it important to prevent microbes from growing in specimens?

6. Why is speed so important in the clinical laboratory?

7. Summarize the important points in determining if a clinical isolate is involved in infection.

8. a. What is the primary, underlying principle behind use of universal precautions?
   b. Describe 5 different medical and dental situations that require special methods of protection, and delineate what these are.

9. Explain the four biosafety levels and what facilities are required at each level.

10. Differentiate between macules, papules, pustules, vesicles, nodules, and erythema.

The Cocci of Medical Importance

G ram-positive and gram-negative cocci are among the most significant infectious agents of humans. Because these bacteria tend to stimulate pus formation, they are often referred to collectively as the **pyogenic cocci.** The most common infectious species in this group belong to four genera: *Staphylococcus, Streptococcus, Enterococcus,* and *Neisseria.*

Chapter Overview

- Gram-positive and gram-negative cocci are significant causes of infections and disease in the community and clinical settings.
- Members of the genus *Staphylococcus* are gram-positive cocci that commonly inhabit the human skin and mucous membranes yet are resistant enough to survive drying, heat, and other harsh environmental conditions.
- *Staphylococcus aureus,* the most serious pathogen in the group, may be identified by its production of coagulase, hemolysis, and several types of virulence factors that aid its invasion and establishment in tissues.
- Among the most notable staphylococcal skin infections are furuncles, carbuncles, and impetigo; important systemic infections are toxic shock syndrome, osteomyelitis, and scalded skin syndrome.
- *Streptococcus* bacteria are gram-positive cocci in chains that are widespread in humans, animals, and the environment. Human species are commensals and pathogens of the skin, mucous membranes, and large intestine.
- The primary pathogen is *Streptococcus pyogenes* (group A), which has a variety of virulence factors including surface antigens, toxins, and enzymes.
- The primary diseases of group A include skin infections (pyoderma, erysipelas, and pharyngitis), and systemic infections (scarlet fever, rheumatic fever, and glomerulonephritis).
- Other streptococcal diseases include subacute endocarditis, tooth abscesses, dental caries, streptococcal pneumonia, and otitis media (*S. pneumoniae*).
- The *Neisseria* are gram-negative diplococci that reside in the mucous membranes of humans and animals.
- Two major infectious agents are *N. gonorrhoeae,* the cause of gonorrhea, and *N. meningitidis,* the cause of one form of meningitis.
- The pathogenicity of *Neisseria* comes from capsules, pili, and endotoxins.

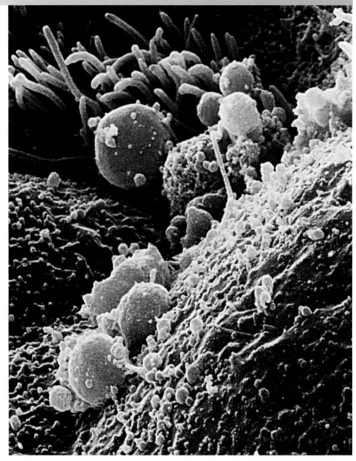

Several tiny spherical cells of the pneumococcus are firmly attached to epithelial cells of the respiratory tract. Early toxic damage to the cells and loss of cilia are evident. This pathogen is a frequent resident of the human pharynx, but it is also the major cause of bacterial pneumonia (3,750×).

- Gonorrhea is a common STD that presents symptoms of urethritis, pelvic inflammatory disease, and ophthalmic inflammation.
- Meningococcal meningitis invades the nasopharynx and spreads to the meninges, brain, spinal cord, and general circulation.
- Control of infections by all cocci requires use of universal precautions, care in testing and selection of appropriate antimicrobial drugs, and in a few cases, vaccination.

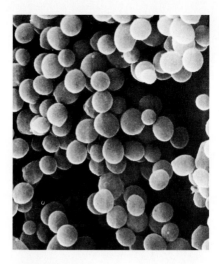

FIGURE 18.1

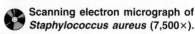 **Scanning electron micrograph of Staphylococcus aureus (7,500×).** This genus owes its name to the grapelike (Gr. *staphyle*) appearance of the clusters.

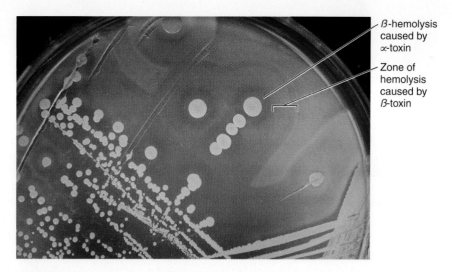

β-hemolysis caused by α-toxin

Zone of hemolysis caused by β-toxin

FIGURE 18.2

Blood agar plate growing Staphylococcus aureus. Some strains show two zones of hemolysis. The relatively clearer inner zone is caused by α-toxin, whereas the outer zone is fuzzy and appears only if the plate has been refrigerated. This outer zone is the β, or "hot-cold," hemolysin that shows up when the plate is refrigerated.

General Characteristics of the Staphylococci

The genus **Staphylococcus** is a common inhabitant of the skin and mucous membranes, and it accounts for a considerable proportion of human infections (often called "staph" infections). Its spherical cells are arranged primarily in irregular clusters and occasionally in short chains and pairs (figure 18.1). Although typically gram-positive, stains from older cultures and clinical specimens sometimes do not stain true. As a group, the staphylococci lack spores and flagella and may be encapsulated.

Currently 31 species have been placed in the genus *Staphylococcus*, but the most important human pathogens are (1) *S. aureus*,[1] (2) *S. epidermidis, S. capitis,* and *S. hominis,* close relatives that are common skin flora, and (3) *S. saprophyticus.* Of these, *S. aureus* is considered the most serious pathogen, although the other species have become increasingly associated with opportunistic infections and can no longer be regarded as harmless commensals.

As a group, the staphylococci are responsible for about 13% of the 2 million hospital infections reported each year. They are implicated in nearly 80,000 deaths nationwide. The following sections focus on the significant characteristics of each species separately.

GROWTH AND PHYSIOLOGICAL CHARACTERISTICS OF *STAPHYLOCOCCUS AUREUS*

Staphylococcus aureus grows in large, round, opaque colonies (figure 18.2) at an optimum of 37°C, though it can grow anywhere between 10°C and 46°C. The species is a facultative anaerobe whose growth is enhanced in the presence of O_2 and CO_2. Its nutrient re-

quirements can be satisfied by routine laboratory media, and most strains are metabolically versatile; that is, they can digest proteins and lipids and ferment a variety of sugars. This species is considered the most resistant of all non-spore-forming pathogens, with well-developed capacities to withstand high salt (7.5–10%), extremes in pH, and high temperatures (up to 60°C for 60 minutes). It also remains viable after months of air drying and resists the effects of many disinfectants and antibiotics. These properties contribute to the reputation of *S. aureus* as a troublesome hospital pathogen.

The following sections deal with the major determinants of pathogenicity that fall into the categories of enzymes and toxins. Perhaps no other bacterial pathogen produces as many virulence factors as does *S. aureus*. Because known virulent strains lack one or more of the known virulence factors, no single factor accounts totally for virulence, leading some experts to theorize that overall staphylococcal pathogenicity is due to a combination of factors.

The Enzymes of *S. aureus*

Pathogenic *S. aureus* typically produces **coagulase,** an enzyme that coagulates plasma and blood (see figure 18.6*b*). The precise importance of coagulase to the disease process remains uncertain. It may be that coagulase causes fibrin to be deposited around staph cells. Fibrin can stop the action of host defenses such as phagocytosis, or it may promote adherence to tissues. Because 97% of all human isolates of *S. aureus* produce this enzyme, its presence is considered the most diagnostic species characteristic.

An enzyme that appears to promote invasion is hyaluronidase, or the "spreading factor," which digests the intercellular "glue" (hyaluronic acid) that binds connective tissue in host tissues. Other enzymes associated with *S. aureus* include staphylokinase, which digests blood clots; a nuclease that digests DNA (DNase); and lipases that help bacteria colonize oily skin surfaces. Enzymes that inactivate penicillin (penicillinase) and other antimicrobial drugs are produced by a majority of strains, giving them multiple resistance.

---

1. From L. *aurum,* gold. Named for the tendency of some strains to produce a golden-yellow pigment, though this characteristic is not common enough to be used as a criterion for identification.

Toxic shock syndrome (TSS) was first identified as a discrete clinical entity in 1978 in young women using vaginal tampons, even though the disease has probably existed for much longer. At first, the precise link between TSS and the use of tampons was a mystery. Then Harvard Medical School researchers discovered that ultra-absorbent brands of tampons bind magnesium ions, thereby creating a habitat for increased colonization and growth of vaginal *S. aureus* and increased TSS toxin production. The toxin enters the bloodstream and causes a series of reactions, including fever, vomiting, rash, and renal, liver, blood, and muscle involve-ment, which are sometimes fatal. A small percentage of infections occurs in women using diaphragm or sponge-type contraceptives. Cases of TSS have also been reported in children, men, and nonmenstruating women and in patients recovering from nasal surgery. The evidence implicating ultra-absorbent tampons as a major factor in the disease was so compelling that they were taken off the market. From the highest yearly count of 880 TSS cases in 1981, the number gradually decreased to about 100 cases in 2000.

The Toxins of *S. aureus*

Most of the toxic products of this species are classified as blood cell toxins (hemolysins and leukocidins), intestinal toxins, and epithelial toxins. The **hemolysins,** for example, lyse red blood cells by disrupting their membranes. All of the staphylococcal hemolysins produce a zone of hemolysis in blood agar (see figure 18.2). The most far-reaching in its biological effects is a form of hemolysin called α (alpha)-toxin. This powerful substance lyses the red blood cells of various mammals and damages leukocytes, skeletal muscle, heart muscle, and renal tissue as well. It is regarded as a significant contributor to the pathologic process. Other powerful hemolysins isolated from human strains include β (beta)-toxin; δ (delta)-toxin, and γ (gamma)-toxin. (Note that the Greek letters used for the staphylococcal hemolysins do not correspond to the Greek letters describing general patterns of hemolysis in chapter 13. For example, α-toxin produces β-hemolysis.)

Other staphylococcal exotoxins include **leukocidin,** which damages cell membranes of neutrophils and macrophages, causing them to lyse. This toxin probably helps incapacitate the phagocytic line of defense. Some strains produce exotoxins called **enterotoxins** that act upon the gastrointestinal tract of humans. A few strains produce an **exfoliative*** toxin that separates the epidermal layer from the dermis and causes the skin to peel away. This toxin is responsible for staphylococcal scalded skin syndrome, in which the skin looks burned (see figure 18.5). The most recent toxin brought to light is **toxic shock syndrome toxin (TSST)** (Medical Microfile 18.1). The presence of this toxin in victims of toxic shock syndrome indicates its probable role in the development of this dangerous condition. The contributions of these toxins and enzymes to disease are discussed in the section on pathology.

Epidemiology and Pathogenesis of *S. aureus*

It is surprising that a bacterium with such great potential for virulence as *Staphylococcus aureus* is a common, intimate human associate. The microbe is present in most environments frequented by humans and is readily isolated from fomites. Colonization of the infant begins within hours after birth and continues throughout life.

The carriage rate for normal healthy adults varies anywhere from 20% to 60%, and the pathogen tends to be harbored intermittently rather than chronically. Carriage occurs mostly in the anterior nares (nostrils) and, to a lesser extent, in the skin, nasopharynx, and intestine. Usually this colonization is not associated with symptoms, nor does it ordinarily lead to disease in carriers or their contacts. Circumstances that predispose an individual to infection include poor hygiene and nutrition, tissue injury, preexisting primary infections, diabetes mellitus, and immunodeficiency states. Staph infections in the newborn nursery and surgical wards are the third most common nosocomial infection. The so-called "hospital strains" can readily spread in an epidemic pattern within and outside the hospital.

THE SCOPE OF CLINICAL STAPHYLOCOCCAL DISEASE

Depending on the degree of invasion or toxin production by *S. aureus,* disease ranges from localized to systemic. A local staphylococcal infection often presents as an inflamed, fibrous lesion enclosing a core of pus called an **abscess** (figure 18.3). Toxigenic disease can present itself as a toxemia due to production of toxins in the body or as food intoxication, the ingestion of preformed toxin in food.

Localized Cutaneous Infections

Staphylococcus usually invades the skin through wounds, follicles, or skin glands. The most common infection is a mild, superficial inflammation of hair follicles (**folliculitis**) or glands (hidradenitis). Although these lesions are usually resolved with no complications, they can lead to infections of subcutaneous tissues. A **furuncle*** (boil) results when the inflammation of a single hair follicle or sebaceous gland progresses into a large, red, and extremely tender abscess or pustule (figure 18.3b). Furuncles often occur in clusters (furunculosis) on parts of the body such as the buttocks, axillae, and back of the neck, where skin rubs against other skin or clothing. A **carbuncle*** is a larger (sometimes as big as a baseball) and deeper

---

* exfoliative (eks-foh′-lee-ay″-tiv) L. *exfoliatio,* a falling off in layers.

* furuncle (fur′-unkl) L. *furunculus,* little thief.

* carbuncle (car′-bunkl) L. *carbunculus,* little coal.

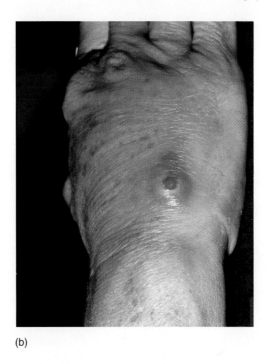

(b)

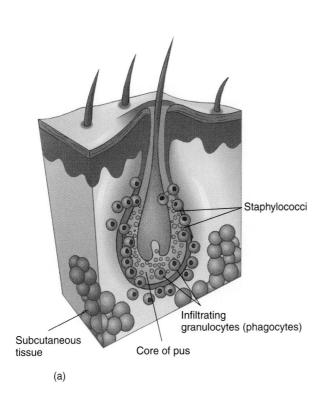

(a)

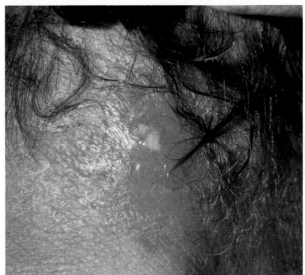

(c)

FIGURE 18.3

Cutaneous lesions of *Staphylococcus aureus.* Fundamentally, all are skin abscesses that vary in size, depth, and degree of tissue involvement. **(a)** Sectional view of a boil or furuncle, a single pustule that develops in a hair follicle or gland and is the classic lesion of the species. The inflamed infection site becomes abscessed when masses of phagocytes, bacteria, and fluid are walled off by fibrin. **(b)** A furuncle on the back of the hand. **(c)** A carbuncle on the back of the neck. Carbuncles are massive deep lesions that result from multiple, interconnecting furuncles. Swelling and rupture into the surrounding tissues can be marked.

lesion created by aggregation and interconnection of a cluster of furuncles. It is usually found in areas of thick, tough skin such as on the back of the neck (figure 18.3c). Carbuncles are extremely painful and can even be fatal in elderly patients when they give rise to systemic disease. One staphylococcal skin infection not confined to follicles and skin glands is the *bullous* type of **impetigo.*** It is characterized by bubblelike epidermal swellings that can break and peel away like a localized form of scalded skin syndrome (see figure 18.5a). This form of impetigo is most common in newborns.

---

* impetigo (im-puh-ty'-goh) L. *impetus*, to attack. Another type is caused by group A streptococci.

Miscellaneous Systemic Infections

Most systemic staphylococcal infections have a focal pattern, spreading from a local cutaneous infection to other sites, a common one being bone (figure 18.4). In **osteomyelitis,** the pathogen is established in the highly vascular metaphyses of a variety of bones (often the femur, tibia, ankle, or wrist). Abscess formation in the affected area results in an elevated, tender lump and necrosis of the bony tissue (figure 18.4b). Symptoms of osteomyelitis include fever, chills, pain, and muscle spasm. This form of osteomyelitis is seen most frequently in growing children, adolescents, and intravenous drug abusers. Another type, secondary osteomyelitis, develops after traumatic injury (compound fracture) or surgery in cancer and diabetes patients.

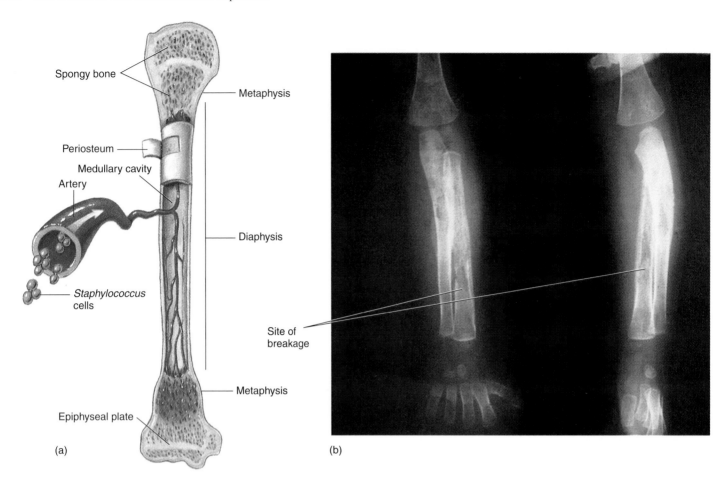

FIGURE 18.4

Staphylococcal osteomyelitis in a long bone. **(a)** In the most common form, the bacteria spread in the circulation from some other infection site, enter the artery, and lodge in the small vessels in bony pockets of the metaphysis or diaphysis. Growth of the cells causes inflammation and damage that are manifest as swelling and necrosis. **(b)** Two X-ray views of a ruptured ulna caused by osteomyelitis.

Systemic staphylococcal infections affect a variety of organs. Because these bacteria inhabit the nasopharynx, they can be aspirated into the lungs and cause a form of pneumonia involving multiple lung abscesses and symptoms of fever, chest pain, and bloody sputum. Although staphylococci account for only a small proportion of such pneumonia cases, the fatality rate is 50%. Most cases occur in infants and children suffering from cystic fibrosis and measles; this form of pneumonia is also one of the most serious complications of influenza in elderly patients.

Staphylococcal bacteremia causes a high mortality rate among hospitalized patients with chronic disease. Its primary origin is bacteria that have been released from an infection site or from colonized medical devices (catheters, shunts). Circulating bacteria transported to the kidneys, liver, and spleen often form abscesses and give off toxins into the circulation. One consequence of staphylococcal bacteremia is a fatal form of endocarditis associated with the colonization of the heart's lining, cardiac abnormalities, and rapid destruction of the valves. Infection of the joints can produce a deforming arthritis (pyoarthritis). A severe form of meningitis (accounting for about 15% of cases) occurs when *S. aureus* invades the cranial vault.

Toxigenic Staphylococcal Disease

Disorders due strictly to the toxin production of *S. aureus* are food intoxication, scalded skin syndrome, and toxic shock syndrome (see Medical Microfile 18.1). Enterotoxins produced by certain strains are responsible for the most common type of food poisoning in the United States (see chapter 26). This illness is associated with eating foods such as custards, sauces, cream pastries, processed meats, chicken salad, or ham that have been contaminated by handling and then left unrefrigerated for a few hours. Because of the high salt tolerance of *S. aureus*, even foods containing salt as a preservative are not exempt. The toxins produced by the multiplying bacteria do not noticeably alter the food's taste or smell. Enterotoxins are heat-stable (inactivation requires 100°C for at least 30 minutes), so heating the food after toxin production may not prevent disease. The ingested toxin acts upon the gastrointestinal epithelium and stimulates nerves, with acute symptoms of cramping, nausea, vomiting, and diarrhea that appear in 2 to 6 hours. Recovery is rapid, usually within 24 hours.

Children with infection of the umbilical stump or eyes are susceptible to a toxemia called **staphylococcal scalded skin syndrome (SSSS).** Upon reaching the skin, this toxin induces a painful, bright red flush over the entire body that first blisters and

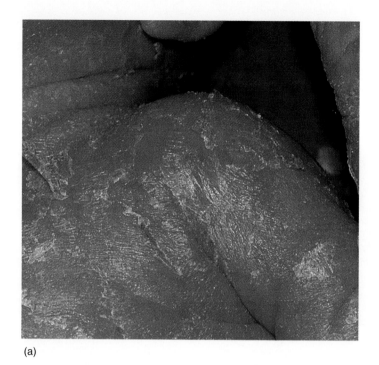

(a)

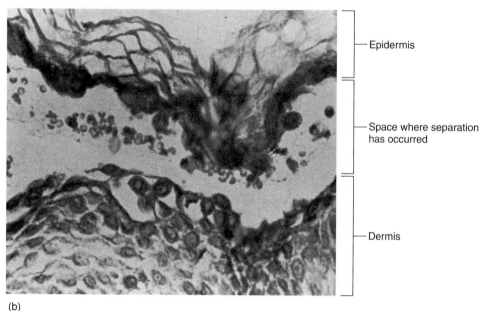

Epidermis

Space where separation
has occurred

Dermis

(b)

FIGURE 18.5

Staphylococcal scalded skin syndrome (SSSS) in a newborn child.
(a) Exfoliative toxin produced in local infections causes blistering and peeling away of the outer layer of skin. **(b)** Photomicrograph of a segment of skin affected with SSSS. The point of epidermal shedding, or desquamation, is at the dermis. The lesions will heal well because the level of separation is so superficial.

then causes desquamation of the epidermis (figure 18.5). The vast majority of SSSS cases have been described in infants and children under the age of 4. This same exfoliative toxin causes the local reaction in bullous impetigo, which can afflict people of all ages.

HOST DEFENSES AGAINST *S. AUREUS*

Despite regular close contact throughout life, humans have a well-developed resistance to staphylococcal infections. Studies performed on human volunteers established that an injection of several hundred thousand staphylococcal cells into unbroken skin was not sufficient to cause abscess formation. Yet, when sutures containing only a few hundred cells were sewn into the skin, a classic lesion rapidly formed, indicating the aggravating effect of a foreign body in infection. Specific antibodies are produced against most of the staphylococcal antigens and intracellular substances, though none of these antibodies appears to effectively immunize a person against reinfection for long periods (except for the antitoxin to SSSS toxin). The most powerful defense lies in the phagocytic response by neutrophils and macrophages. When aided by the opsonic action of complement, phagocytosis effectively disposes of staphylococci that have gained access to tissues. Inflammation and cell-mediated immunity that stimulate abscess formation also help contain staphylococci and prevent their further spread.

THE OTHER STAPHYLOCOCCI

Because they lack the enzyme coagulase, other species in the genus *Staphylococcus* are called the **coagulase-negative staphylococci (CNS).** Several species are included in this group, some of human

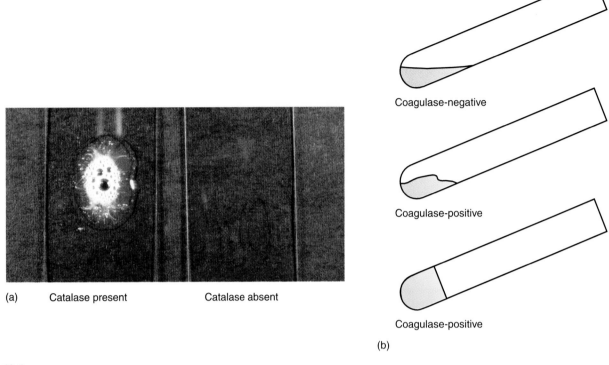

(a)   Catalase present   Catalase absent

Coagulase-negative

Coagulase-positive

Coagulase-positive

(b)

FIGURE 18.6

Tests for differentiating the genus *Staphylococcus* from *Streptococcus* and for identifying *Staphylococcus aureus*. **(a)** In the *slide catalase test,* adding 3% solution of hydrogen peroxide to a small sample of isolate will cause the sample to bubble vigorously if catalase is present and remain undisturbed if catalase is lacking. **(b)** Staphylococcal coagulase is an enzyme that reacts with factors in plasma to initiate clot formation. In the coagulase test, a tube of plasma is inoculated with an isolate. If it remains liquid and flows, the test is negative. If the plasma develops a lump or becomes completely clotted, the test is positive.

origin and others of nonhuman, mammalian origin. Although this group was once considered clinically nonsignificant, its importance has greatly increased over the past 20 years. Coagulase-negative staphylococci currently account for a large proportion of nosocomial and opportunistic infections in immunocompromised patients.

The normal habitat of *Staphylococcus epidermidis* is the skin and mucous membranes. *Staphylococcus hominis* lives in skin areas high in apocrine glands, and *S. capitis* populates the scalp, face, and external ear. From these sites, the bacteria are positioned to enter breaks in the protective skin barrier. Infections usually occur after surgical procedures, such as insertion of shunts, catheters, and various prosthetic devices that require incisions through the skin. It appears that any foreign object supports colonization and proliferation of these bacteria, which develop thick adherent capsules. It is particularly alarming that the rate of these infections has increased owing to technological advances in maintaining patients. Although *S. epidermidis* is not as invasive or toxic as *S. aureus,* it can also cause endocarditis, bacteremia, and urinary tract infections.

Less is known about the distribution and epidemiology of *S. saprophyticus.* It is an infrequent resident of the skin, lower intestinal tract, and vagina, but its primary importance is in urinary tract infections. For unknown reasons, urinary *S. saprophyticus* infection is found almost exclusively in sexually active adolescent women and is the second most common cause of urinary infections in this group.

IDENTIFICATION OF *STAPHYLOCOCCUS* IN CLINICAL SAMPLES

Staphylococci are frequently isolated from pus, tissue exudates, sputum, urine, and blood. To prevent inadvertently culturing staphylococci from the environment or from normal residents, great care in collection must be exercised. Primary isolation is achieved by inoculation on sheep or rabbit blood agar, or in heavily contaminated specimens, selective media such as mannitol salt agar are used. In addition to culturing, the specimen can be Gram stained and observed for irregular clusters of gram-positive cocci. Because differentiating among gram-positive cocci is not possible using colonial and morphological characteristics alone, other presumptive tests are required (see figure E in "Introduction to Identification Techniques in Medical Microbiology"). The production of catalase, an enzyme that breaks down hydrogen peroxide accumulated during oxidative metabolism, can be used to differentiate the staphylococci, which produce it, from the streptococci, which do not (figure 18.6a). The property of *Staphylococcus* to grow anaerobically and to ferment sugars separates it from *Micrococcus,* a nonpathogenic genus that is a common specimen contaminant.

One key technique for separating *S. aureus* from other species of *Staphylococcus* is the coagulase test (figure 18.6b). By definition, any isolate that coagulates plasma is *S. aureus,* and all others are coagulase-negative. Rapid multitest systems are used routinely to collect other physiological information (figure 18.7).

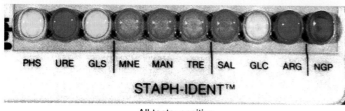

All tests: positive

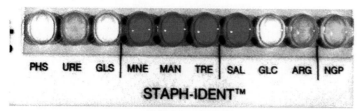

Inoculated: all negative

FIGURE 18.7

Miniaturized test system used in further identification of _Staphylococcus_ isolates. Microbiologists can separate 19 species with this system. The cupules contain substrates that detect phosphatase production (PHS), urea hydrolysis (URE), glucosidase production (GLS), mannose fermentation (MNE), mannitol fermentation (MAN), trehalose fermentation (TRE), salicin fermentation (SAL), glucuronidase production (GLC), arginine hydrolysis (ARG), and galactosidase production (NGP).

Table 18.1 lists some of the characteristics that further separate the five species most often involved in human infections. An important confirming identification of _S. aureus_ can be made with a latex bead agglutination test. It is based on the surface protein (A) that can bind to IgG antibodies.

Differentiating the common coagulase-negative staphylococci from one another relies primarily upon the novobiocin resistance of _S. saprophyticus,_ mannose fermentation in _S. epidermidis,_ the lack of urease production in _S. capitis,_ and lack of anaerobic growth in _S. hominis._

CLINICAL CONCERNS IN STAPHYLOCOCCAL INFECTIONS

The staphylococci are notorious in their acquisition of resistance to new drugs, and they continue to defy attempts at medical control. Because resistance to common drugs is likely, antimicrobic susceptibility tests are essential in selecting a correct therapeutic agent (see chapter 12).

Ninety-five percent of strains of _S. aureus_ have acquired genes for penicillinase, which makes them resistant to the traditional drugs—penicillin and ampicillin. Even more problematic are strains of MRSA (**m**ethicillin-**r**esistant _Staphylococcus aureus_) that carry multiple resistance to a wide range of antimicrobics, including methicillin, gentamicin, cephalosporins, tetracycline, erythromycin, and even quinolones. A few strains have acquired resistance to all major drug groups except vancomycin. Most of these isolates are found in hospitals, where they account for 40% of staph infections. Now an even greater cause for concern are reports of vancomycin resistance. Infectious disease specialists are monitoring this new trend and introducing drug replacements such as synercid (see chapter 12).

Treatment of Staph Infections

Clinical experience has shown that abscesses will not respond to therapy unless they are surgically perforated and cleared of pus and foreign bodies. Severe systemic conditions such as endocarditis, septicemia, osteomyelitis, pneumonia, and toxemia respond slowly and require intensive, lengthy therapy by oral or injected drugs or some combination of the two. Two vaccines are now in development after successful clinical trials. These may be available within the next two or three years for high-risk patients.

Prevention of Staph Infections

The principal reservoir of the pathogenic staphylococci can never be eliminated. As long as there are humans, there will be carriers and infections. It is difficult to block the colonization of the human body, but one can minimize the opportunity for nosocomial

---

TABLE 18.1

Separation of Clinically Important Species of _Staphylococcus_

Test	_S. aureus_	_S. epidermidis_	_S. saprophyticus_	_S. capitis_	_S. hominis_
Coagulase	+★	−	−	−	−
Anaerobic growth	+	+	+	+	−
Urease production	+	+	+	−	+
Mannitol fermentation (aerobic)	+★	−	W	+	−
Mannose fermentation	+	+	−	+	−
Trehalose fermentation	+	−	+	−	+/−
β-hemolysis by α-toxin	+	−	−	−	−
Produces DNase and RNase	+	−	−	−	−
Sensitive to lysostaphin	+	−	−	−	−
Susceptible to novobiocin	+	+	−	+	+
Pathogenicity	Primary	Opportunistic	Opportunistic	Opportunistic	Opportunistic

★_A few strains test negative for this._
W = May occur, but weakly.

infection by careful hygiene and adequate cleansing of surgical incisions and burns. Controlling the increasing incidence of nosocomial outbreaks, especially among people at greatest risk (surgical patients and neonates) is a real challenge for hospital staff. Patients or hospital personnel who are asymptomatic carriers or have frank infections constitute the most common source. The burden of effective deterrence again falls to meticulous handwashing, proper disposal of infectious dressings and discharges, isolation of people with open lesions, and attention to indwelling catheters and needles.

When possible, other preventive measures must be instituted. For example, hospital workers who carry *S. aureus* nasally may be barred from nurseries, operating rooms, and delivery rooms. Occasionally, carriers are treated for several months with a combination of antimicrobic drugs (rifampin and dicloxacillin, for example).

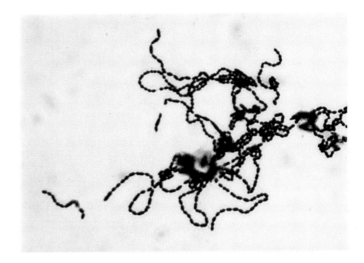

FIGURE 18.8

 One of the truly great sights in microbiology—a freshly isolated *Streptococcus* with its long, intertwining chains. These occur only in liquid media and are especially well-developed if nutrients are in limited supply (1,000×).

CHAPTER CHECKPOINTS

The genus *Staphylococcus* contains 31 species. Most of these are human commensals, but *S. aureus*, *S. epidermidis*, and *S. saprophyticus* and others can be pathogenic.

All species in the genus *Staphylococcus* are gram-positive facultative anaerobes that tolerate extremes of temperature, osmotic pressure, and drying. All are catalase-positive.

Coagulase-positive staphylococci are by definition *S. aureus*, the most resistant of all non-spore-forming pathogenic bacteria.

S. aureus produces an impressive array of virulence factors that enable it to resist phagocytosis, destroy host tissue, and invade the blood. These include coagulase, hemolysins, leukocidins, enzymes that digest host tissue, and toxins that attack the epidermis, vascular smooth muscle, and the intestinal epithelium. These virulence factors enable *S. aureus* to cause both local and systemic infections.

Abscesses are localized staphylococcal infections. They occur predominantly in hair follicles and glands.

S. aureus causes disease mainly when the host is weakened by poor health, injury, disease, surgery, or immunodeficiency.

S. epidermidis and *S. saprophyticus* are coagulase-negative staphylococci that cause opportunistic infections. *S. epidermidis* causes nosocomial skin incision infections, and *S. saprophyticus* causes urinary tract infections.

Overuse and inappropriate prescription of antimicrobial drugs have selected for multiple drug-resistant strains of *S. aureus* in hospitals, causing serious problems in treatment. All *S. aureus* isolates should therefore be checked for antimicrobic sensitivity so that an effective therapeutic can be prescribed.

Consistent practice of universal precautions by *all* hospital staff is necessary to prevent nosocomial staphylococcal infections.

General Characteristics of the Streptococci and Related Genera

The genus ***Streptococcus**** includes a large and varied group of bacteria. Some are normal residents or agents of disease in humans and animals; others are free-living in the environment. Members of this group are known for the arrangement of cocci in long, beadlike chains. The length of these chains varies, and it is common to find them in pairs (figure 18.8). The general shape of the cells is spherical, but they can also appear ovoid or rodlike, especially in actively dividing young cultures.

Streptococci are non-spore-forming and nonmotile (except for an occasional flagellated strain), and they can form capsules and slime layers. They are facultative anaerobes that ferment a variety of sugars, usually with the production of lactic acid (homofermentative). Streptococci do not form catalase, but they do have a peroxidase system for inactivating hydrogen peroxide, which allows their survival even in the presence of oxygen. Most parasitic forms are fastidious in nutrition and require enriched media for cultivation. Colonies are usually small, nonpigmented, and glistening. Most members of the genus are quite sensitive to drying, heat, and disinfectants and seldom develop drug resistance, though pneumococci and enterococci are notable exceptions.

Species of *Streptococcus* have traditionally been classified according to a system developed by Rebecca Lancefield in the 1930s. She discovered that the cell wall carbohydrates (antigens) of various cultures stimulated formation of antibodies with differing specificities. She characterized 14 different groups using an alphabetic system (A, B, C). An alternative division method of grouping is based on their reaction in blood agar (figure 18.9). Those producing a zone of β-hemolysis on sheep blood agar are members of Lancefield groups A, B, C, G, and certain strains of D, and those producing α-hemolysis are *S. pneumoniae* and the *viridans**** streptococci. Several species of nonhemolytic cocci exist, but most of them are of less clinical significance. Table 18.2 summarizes the streptococcal groups, their habitats, and their pathogenicity.

---

* *Streptococcus* Gr. *streptos*, winding, twisted. The chain arrangement is the result of division in only one plane.

* *viridans* (vih′-rih-denz) L. *viridis*, green. This term comes from the green color that develops around colonies that produce α-hemolysis.

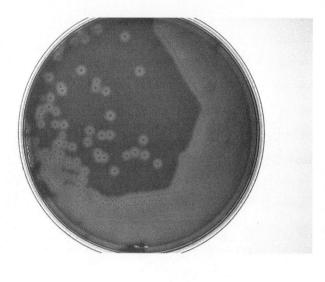

(a)

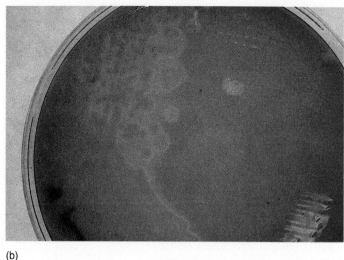

(b)

FIGURE 18.9

Hemolysis patterns on blood agar may be used to separate streptococci into major subgroups. **(a)** Colonies of *Streptococcus pyogenes* showing β-hemolysis, which forms clear zones with completely lysed red blood cells. **(b)** Colonies of *S. pneumoniae* demonstrate the less-discrete greenish zones typical of α-hemolysis.

TABLE 18.2

Major Species of *Streptococcus* and Related Genera

Species	Lancefield Group	Hemolysis Type	Habitat	Pathogenicity to Humans
S. pyogenes	**A**	Beta (β)	**Human throat**	**Skin, throat infections, scarlet fever**
S. agalactiae	**B**	β	**Human vagina, cow udder**	**Neonatal, wound infections**
S. equisimilis	C	β	Swine, cows, horses	Pharyngitis, endocarditis
S. equi, S. zooepidemicus	C	β	Various mammals	Rare, in abscesses
S. dysgalactiae	C	β	Cattle	Rare
Enterococcus faecalis	**D**	α, β, N	**Human, animal intestine**	**Endocarditis, UTI★**
E. faecium, E. durans	D	Alpha (α)	Human, animal intestine	Similar to *E. faecalis*
S. bovis	D	N	Cattle	Subacute endocarditis, bacteremia
S. anginosus	F, G, L	β	Humans, dogs	Endocarditis, URT★★ infections
S. sanguis	**H**	α	**Human oral cavity**	**Endocarditis, dental caries**
S. salivarius	K	N	Human saliva	Endocarditis
Lactococcus lactis	N	V	Dairy products	Very rare
S. mutans	**NI★★★**	N	**Human oral cavity**	**Dental caries**
S. uberis, S. acidominimus	NI	V	Domestic mammals	Rare
S. mitior	O, M	α	Human oral cavity	Tooth abscess, endocarditis
S. milleri	F	N	URT	Endocarditis, organ abscess
S. pneumoniae	**NI**	α	**Human RT**	**Bacterial pneumonia**

Note: Species in bold type are the most significant sources of human infection and disease.
N = none   V = varies
★Urinary tract infection.
★★Upper respiratory tract.
★★★No group C-carbohydrate identified.

Despite the large number of streptococcal species, human disease is most often associated with **S. pyogenes, S. agalactiae, viridans streptococci, S. pneumoniae,** and **Enterococcus faecalis.** The latter species was formerly called *Streptococcus faecalis* until DNA hybridization tests indicated that it is a distinct genus. Several members of groups C and G are commensals of domestic animals that can also colonize and infect humans. In fact, streptococci are notorious for being shared between humans and their pets.

β-HEMOLYTIC STREPTOCOCCI: STREPTOCOCCUS PYOGENES

By far the most serious streptococcal pathogen of humans is *Streptococcus pyogenes*, the main representative of group A. It is a relatively strict parasite, inhabiting the throat, nasopharynx, and occasionally the skin of humans. The involvement of this species in severe disease is partly due to the substantial array of

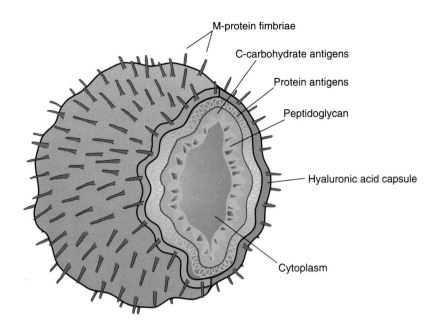

M-protein fimbriae

C-carbohydrate antigens

Protein antigens

Peptidoglycan

Hyaluronic acid capsule

Cytoplasm

FIGURE 18.10

Cutaway view of group A *Streptococcus*. The outermost fringe consists of fimbriae composed in part of M-protein, group-specific substance. Other layers making up the cell envelope are the capsule, protein antigens, and C-carbohydrate antigens.

surface antigens, toxins, and enzymes it generates, although even highly virulent strains do not produce all of the toxins and enzymes.

Cell Surface Antigens and Virulence Factors

Streptococci display numerous surface antigens (figure 18.10). C-carbohydrates, specialized polysaccharides or teichoic acids found on the surface of the cell wall, are the basis for Lancefield groups. Their apparent contribution to pathogenesis is to protect the bacterium from being dissolved by the lysozyme defense of the host. Lipoteichoic acid accounts for the adherence of *S. pyogenes* to epithelial cells in the skin or pharynx. Another type-specific molecule is the *M-protein,* of which about 80 different subtypes exist. This substance is the main component of fimbriae, the spiky surface projections that contribute to virulence by resisting phagocytosis and improving adherence. A capsule is formed by most *S. pyogenes* strains, but it remains attached to the cell surface only during the rapid growth phase of a culture and is probably not as important to virulence as are other factors.

Major Extracellular Toxins Group A streptococci owe several pathologic properties to the effects of hemolysins called **streptolysins.** The two types are streptolysin O (SLO) and streptolysin S (SLS).[2] Although both types cause β-hemolysis of sheep blood agar, SLS produces the major form of surface hemolysis, whereas SLO produces hemolysis only in deep (anaerobic) colonies. Both hemolysins rapidly injure many cells and tissues, including leukocytes, liver, and heart muscle.

A key toxin in the development of scarlet fever (discussed in a later section of this chapter) is (**erythrogenic***) **pyrogenic toxin.** This toxin is responsible for the bright red rash typical of this disease, and it also induces fever by acting upon the temperature reg-

ulatory center. Only lysogenic strains of *S. pyogenes* that contain genes from a temperate bacteriophage can synthesize this toxin.

Some of the streptococcal toxins (pyrogenic and streptolysin O) contribute to increased tissue injury by acting as *superantigens.* These toxins are strong stimulants of monocytes and T lymphocytes. When activated, these cells proliferate and produce *tumor necrosis factor,* which in high quantities leads to vascular injury. This is the likely mechanism for the severe pathology of toxic shock syndrome and necrotizing fasciitis.

Major Extracellular Enzymes Several enzymes for digesting macromolecules are given off by the group A streptococci, though their role in pathogenesis is somewhat obscure. Streptokinase, similar to staphylokinase, activates a pathway leading to the digestion of fibrin clots and may play a role in invasion. Hyaluronidase breaks down the binding substance in connective tissue and promotes spreading of the pathogen into the tissues, while streptodornase (DNase) liquefies purulent discharges by hydrolyzing DNA.

Epidemiology and Pathogenesis of *S. pyogenes*

Streptococcus pyogenes has traditionally been linked with a diverse spectrum of infection and disease. Before the era of antibiotics, it accounted for a major portion of serious human infections and deaths from such diseases as rheumatic fever and puerperal sepsis. Although its importance has diminished in the past 50 years, recent epidemics have reminded us of its potential for sudden and serious illness (Spotlight on Microbiology 18.2).

Humans are the only significant reservoir for *S. pyogenes.* Healthy or subclinical carriers of virulent strains constitute about 5–15% of the population. Infection is generally transmitted through direct contact, droplets, and, occasionally, food or fomites. The bacteria invade at periods of lowered host resistance, primarily through the skin and pharynx. The incidence and types of infections are altered by climate, season, and living conditions. Skin infections occur more frequently during the warm temperatures of summer and fall, and pharyngeal infections increase in the winter

---

2. In SLO, O stands for oxygen because the substance is inactivated by oxygen. SLO is produced by most strains of *S. pyogenes.* In SLS, S stands for serum because the substance has an affinity for serum proteins. SLS is oxygen-stable.

* **erythrogenic** (eh-rith″-roh-jen′-ik) Gr. *erythros,* red, and *gennan,* to produce.

Strep infections are "occupational diseases of childhood" that usually follow a routine and uncomplicated course. The greatest cause for concern are those few occasions when such infections erupt into far more serious ailments. One dramatic example is **necrotizing fasciitis,** a complication of *S. pyogenes* infection that has been known for hundreds of years. Recent small outbreaks of the disease in England and the United States have received heavy publicity as the "flesh-eating disease" caused by "killer bacteria." It should be emphasized that cases of this disease are rather rare, but its potential for harm is high. It can begin with an innocuous cut in the skin and spread rapidly into nearby tissue, causing severe disfigurement and even death.

There is really no mystery to the pathogenesis of necrotizing fasciitis. It begins very much like impetigo and other skin infections: Streptococci on the skin are readily introduced into small abrasions or cuts, where they begin to grow rapidly. The term "flesh-eating" for this effect is really a misnomer. These strains of group A streptococci have great toxigenicity and invasiveness by releasing special enzymes and toxins. Their enzymes digest the connective tissue in skin, and their toxins poison the epidermal and dermal tissue. As the flesh is killed, it separates and sloughs off, forming a pathway for the bacteria to spread into deeper tissues such as muscle. More dangerous infections involve a mixed infection with anaerobic bacteria and systemic spread of the toxin to other organs. It is true that some patients have lost parts of their limbs and faces and others have suffered amputation, but early diagnosis and treatment can prevent these complications. Fortunately, even virulent strains of *Streptococcus pyogenes* are not highly drug resistant.

More mysterious, however, is the background epidemiology involved in these virulent new outbreaks. The incidence of aggressive infections had been gradually declining throughout the antibiotic era, but then in the latter 1980s, clusters began cropping up again. Bacteria isolated from cases in the Rocky Mountains and the East appeared to be

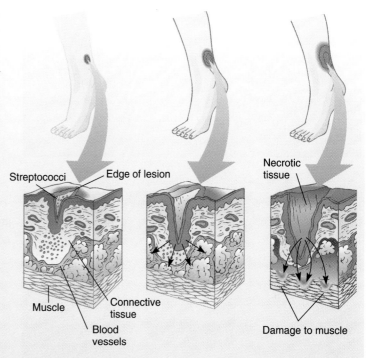

The phases of *Streptococcus pyogenes*–induced necrotizing fasciitis.

new, more virulent strains. Infectious disease experts theorize that these new strains of streptococci are probably mutants that have acquired toxin genes from infecting viruses.

---

* **necrotizing fasciitis** (nee′-kroh-ty″-zing fass″-ee-eye′-tis) Gr. *nekrosis*, deadness, and L. *fascia*, the connective tissue sheath around muscles and other organs.

---

months. Children 5 to 15 years of age are the predominant group affected by both types of infections. In addition to local cutaneous and throat infections, *S. pyogenes* can give rise to a variety of systemic infections and progressive sequelae if not properly treated.

Skin Infections   When virulent streptococci invade a nick in the skin or the mucous membranes of the throat, an inflammatory primary lesion is produced. Pyogenic infections appearing after local invasion of the skin are pyoderma or erysipelas; those developing in the throat are pharyngitis or tonsillitis.

Pyoderma, or streptococcal impetigo, is marked by burning, itching papules that break and form a highly contagious yellow crust (figure 18.11*a*). Impetigo often occurs in epidemics among school children, and it is also associated with insect bites, poor hygiene, and crowded living conditions. A slightly more invasive form of skin infection is **erysipelas.** The pathogen usually enters through a small wound or incision on the face or extremities and eventually spreads to the dermis and subcutaneous tissues. Early

symptoms are edema and redness of the skin near the portal of entry, fever, and chills. The lesion begins to spread outward, producing a slightly elevated edge that is noticeably red, hot, and often vesicular (figure 18.11*b*). Depending on the depth of the lesion and how the infection progresses, cutaneous lesions can remain superficial or produce long-term systemic complications. Severe cases involving large areas of skin are occasionally fatal.

Most people associate streptococci with the condition called strep throat or, more technically, **streptococcal pharyngitis** (tonsillitis). Estimates indicate that most humans will acquire this particular infection at some time during their lifetime. The organism multiplies in the tonsils or pharyngeal mucous membranes, causing redness, edema, enlargement, and extreme tenderness that make swallowing difficult and painful (figure 18.12). These symptoms can be accompanied by fever, headache, nausea, and abdominal pain. Other signs include a *purulent* exudate over the tonsils, swollen lymph nodes, and occasionally white, pus-filled nodules on the tonsils.

---

* **erysipelas** (er″-ih-sip′-eh-las) Gr. *erythros*, red, and *pella*, skin.

* **purulent** (puh′-roo-lent) L. *purulentus*, inflammation. In reference to pus.

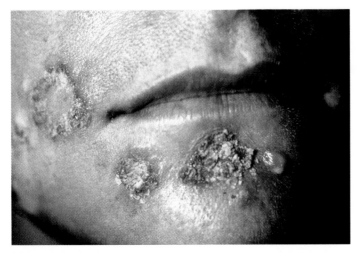

(a)

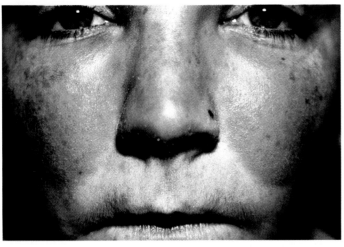

(b)

FIGURE 18.11

⊙ **Streptococcal skin infections.** **(a)** Impetigo lesions on the face. **(b)** Erysipelas of the face. Although it is a superficial infection, the inflammatory reaction spreads horizontally from the initial point of entry. Tender, red, puffy lesions have a sharp border that can burst and release fluid.

Systemic Infections Throat infection can lead to **scarlet fever** (scarlatina) when it involves a strain of *S. pyogenes* carrying a prophage that codes for pyrogenic toxin. Systemic spread of this toxin results in high fever and a bright red, diffuse rash over the face, trunk, inner arms and legs, and even the tongue. Within 10 days, the rash and fever usually disappear, often accompanied by desquamation (sloughing) of the epidermis. Many cases of pharyngitis and even scarlet fever are mild and uncomplicated, but on occasion they elicit an inflammatory reaction that leads to severe sequelae. The dissemination of streptococci into the lymphatics and the blood can give rise to septicemia, an infrequent complication limited to people who are already weakened by underlying disease. Another relatively rare infection is *S. pyogenes* pneumonia, which accounts for less than 5% of bacterial pneumonias and is usually secondary to influenza or some other pulmonary disorder. A recent

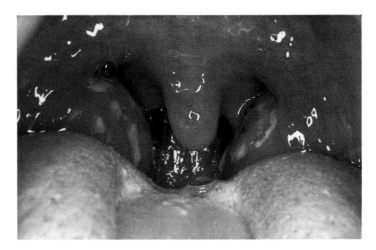

FIGURE 18.12

The appearance of the throat in pharyngitis and tonsillitis. The pharynx and tonsils become bright red and suppurative. Whitish pus nodules may also appear on the tonsils.

invasive infection, called *streptococcal toxic shock syndrome*, has appeared in the United States and Europe. It begins as a profound bacteremia and deep tissue infection and rapidly progresses to multiple organ failure. Even with treatment, around 30% of patients die.

Long-Term Complications of Group A Infections
The principal sequelae that appear within a few weeks after group A streptococcal infections are (1) **rheumatic* fever (RF),** a delayed inflammatory condition of the joints, heart, and subcutaneous tissues, and (2) **acute glomerulonephritis (AGN),** a disease of the kidney glomerulus and tubular epithelia. The actual mechanisms for these streptococcal sequelae have yet to be completely explained, though several theories have been proposed. It is possible that tissue injury in both diseases is tied directly to streptococcal invasion of tissues or to the action of toxins such as SLO and SLS. According to the autoimmunity theory, antibodies formed against the streptococcal cell wall and membranes cross-react with molecules in the host that are similar to these bacterial antigens. In the heart, these antibody-antigen reactions trigger inflammation and injury of cardiac tissues, and in the kidney, immune complexes deposited in the epithelia inflame and damage the filtering apparatus (see chapter 17).

Rheumatic fever usually follows an overt or subclinical case of streptococcal pharyngitis or tonsillitis in children. Its major clinical features are carditis, an abnormal electrocardiogram, painful arthritis, chorea, nodules under the skin, and fever.[3] The course of the syndrome extends from 3 to 6 months, usually without lasting damage. In patients with severe carditis, however, extensive damage to the heart valves and muscle can occur (figure 18.13). Although the degree of permanent damage does not usually reveal itself until middle age, it is often extensive enough to require the re-

---

3. Carditis is inflammation of heart tissues; chorea is a nervous disorder characterized by involuntary, jerky movements.

* rheumatic (roo-mat′-ik) Gr. *rheuma,* flux. Involving inflammation of joints, muscles, and connective tissues.

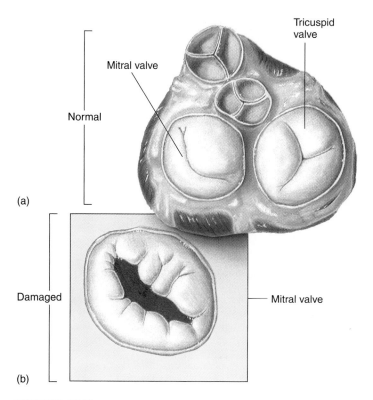

Normal

Mitral valve

Tricuspid valve

(a)

Damaged

Mitral valve

(b)

FIGURE 18.13

The cardiac complications of rheumatic fever. Pathologic processes of group A streptococcal infection can extend to the heart. In this example, it is believed that cross-reactions between streptococcal-induced antibodies and heart proteins have a gradual destructive effect on the atrioventricular valves (especially the mitral valve) or semilunar valves. Scarring and deformation change the capacity of the valves to close and shunt the blood properly. **(a)** Normal valves, viewed from above. **(b)** Inset reveals scar tissue on a damaged mitral valve.

placement of damaged valves. Heart disease due to RF is on the wane because of medical advances in diagnosing and treating streptococcal infections, but it is still responsible for thousands of deaths each year in the United States and many times that number in the rest of the world.

In AGN, the kidney cells are so damaged that they cannot adequately filter blood. The first symptoms are nephritis (appearing as swelling in the hands and feet and low urine output), increased blood pressure, and occasionally heart failure. Urine samples are extremely abnormal, with high levels of red and white blood cells and protein. AGN can clear up spontaneously, can become chronic and lead to kidney failure later, or can be immediately fatal.

Two antibodies can give long-term protection against group A infections. One is the type-specific antibody produced in response to the M-protein, though it will not prevent infections with a different strain (which explains why a person can have recurring cases of strep throat). The other is a neutralizing antitoxin against the erythrogenic toxin that prevents the fever and rash of scarlet fever. Serological tests for anti-streptococcal antibodies are used primarily to detect continuing or recent streptococcal infection and to judge a patient's susceptibility to rheumatic fever and glomerulonephritis.

GROUP B: *STREPTOCOCCUS AGALACTIAE*

Several other species of β-hemolytic *Streptococcus* in groups B, C, and D live among the normal flora of humans and other mammals and can be isolated in clinical specimens from diseased human tissue. The group B streptococci (GBS), represented by the species *S. agalactiae,* demonstrate clearly how the distribution of a parasite can change in a relatively short time. It regularly resides in the human vagina, pharynx, and large intestine. A strain found in cattle is a frequent cause of bovine mastitis.[4] The major result of human colonization has been a sudden increase in serious infections in newborns and compromised people.

Streptococcus agalactiae has been chiefly implicated in neonatal, wound, and skin infections and in endocarditis. People suffering from diabetes and vascular disease are particularly susceptible to wound infections. Because of its location in the vagina, GBS can be transferred to the infant during delivery, sometimes with dire consequences. An early-onset infection develops a few days after birth and is accompanied by sepsis, pneumonia, and high mortality. This bacterial pathogen is the most prevalent cause of neonatal pneumonia, sepsis, and meningitis in the United States and Europe. Approximately 15,000 babies a year acquire infection, with 5,000 deaths in the United States alone. A later complication comes on in 2 to 6 weeks, with symptoms of meningitis—fever, vomiting, and seizures. About 20% of children have long-term neurological damage. Because most cases occur in the hospital, personnel must be aware of the risk of passively transmitting this pathogen, especially in the neonatal and surgical units. Pregnant women should be screened for colonization in the third trimester and immunized with globulin and treated with a course of antibiotics if infection is found. A GBS vaccine based on the capsular polysaccharide will be available for vaccinating women in the near future.

GROUP D ENTEROCOCCI AND GROUPS C AND G STREPTOCOCCI

Enterococcus faecalis, E. faecium, and *E. durans* are called the enterococci because they are normal colonists of the human large intestine. Two other members of Group D, *Streptococcus bovis* and *S. equinus,* are nonenterococci that colonize other animals and occasionally humans. Infections caused by *E. faecalis* arise most often in elderly patients undergoing surgery and affect the urinary tract, wounds, blood, the endocardium, the appendix, and other intestinal structures. Enterococci are emerging as serious nosocomial opportunists, primarily because of the rising incidence of multi-drug-resistant strains and the ease with which they are transferred from person to person.

Groups C and G are common flora of domestic animals but are frequently isolated from the human upper respiratory tract. Occasionally, they imitate group A strep in causing pharyngitis and glomerulonephritis. More often, they cause bacteremia and disseminated deep-seated infections in severely compromised patients.

---

4. Inflammation of the mammary gland.

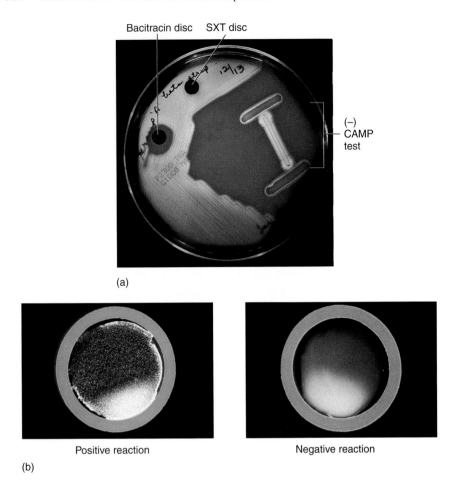

(a)

Positive reaction   Negative reaction

(b)

FIGURE 18.14

Streptococcal tests. **(a)** Bacitracin disc test. With very few exceptions, only *Streptococcus pyogenes* is sensitive to a minute concentration (0.02 μg) of bacitracin. Any zone of inhibition around the B disc is interpreted as a presumptive indication of this species. (Note: Group A streptococci are negative for SXT sensitivity and the CAMP test.) **(b)** A rapid, direct test kit for diagnosis of group A infections. With this method, a patient's throat swab is introduced into a system composed of latex beads and monoclonal antibodies. *(Left)* In a positive reaction, the C-carbohydrate on group A streptococci produces visible clumps. *(Right)* A smooth, milky reaction is negative.

LABORATORY IDENTIFICATION TECHNIQUES

Because streptococcal disease can come on rapidly, it is beneficial to routinely culture throat and skin infections in children. The failure to recognize group A streptococcal infections (even very mild ones) can have devastating effects. Rapid cultivation and diagnostic techniques that will ensure proper treatment and prevention measures are essential. Several companies have developed rapid diagnostic test kits to be used in clinics or offices to detect group A streptococci from pharyngeal swab samples. These tests, based on monoclonal antibodies that react with the C-carbohydrates of group A, are highly specific and sensitive (figure 18. 14*b*). They are primarily useful as a method for screening patients who test positive.

Complete identification usually necessitates cultivating a specimen on sheep blood agar plates and occasionally enrichment media. Group A streptococci are by far the most common β-hemolytic isolates in human lesions, but lately an increased number of infections by group B streptococci and the existence of β-hemolytic enterococci have made it important to use differentiation tests. (Groups C and G are also β-hemolytic, but they are most often associated with infections in other mammals and are only infrequently found in humans.) A positive bacitracin disc test (figure 18. 14*a*) provides important evidence of group A. Other important characteristics that separate the various β-hemolytic streptococci are presented in table 18.3.

Group B streptococci are differentiated from groups A and D by the CAMP reaction (figure. 18.15) and hippurate hydrolysis,

both of which are positive for group B and negative for groups A and D. *Enterococcus faecalis* (the predominant enterococcus) differs from the streptococci in its resistance to heat, 6.5% salt, and low concentrations of penicillin G. For these reasons, certain strains can be confused with *Staphylococcus aureus,* but a Gram stain and catalase test help differentiate these two. Because *E. faecalis* is sometimes found in the pharynx, strains can also be mistaken for *Streptococcus pyogenes,* but the ability of *E. faecalis* to hydrolyze esculin in 40% bile salt, and its resistance to bacitracin help distinguish it. All groups can also be differentiated with specific agglutination tests, available in commercial kits.

TREATMENT AND PREVENTION OF STREPTOCOCCAL INFECTIONS

Antimicrobic therapy is aimed at curing infection and preventing complications. All strains of *S. pyogenes* continue to be very sensitive to penicillin or one of its derivatives. In cases of pharyngitis, small children receive an intramuscular injection of 600,000 units of benzathine penicillin to achieve the necessary circulating levels for a period of 10 days; larger children receive 900,000 units, and adults 1.2 million units. An alternative therapy often used for impetigo is oral penicillin V taken for 10 days. In cases of penicillin allergy, another drug such as erythromycin or a cephalosporin is prescribed.

The only certain way to arrest rheumatic fever or acute glomerulonephritis is to treat the preceding infection, because once

TABLE 18.3

Scheme for Differentiating β-Hemolytic Streptococci

Characteristic	Group A (S. pyogenes)	Group B (S. agalactiae)	Groups C/G (S. equisimilis)	Group D (Enterococcus faecalis)
Bacitracin sensitivity	+	−	−	−
CAMP factor*	−	+	−	−
Esculin** hydrolysis in 40% bile	−	−	−	+
SxT sensitivity***	−	−	+	−
Growth at 45°C	−	−	−	+
Growth in 6.5% salt	−	+	−	+
Hippurate hydrolysis	−	+	−	−
PYR test****	+	−	−	+

*Name is derived from the first letters of its discoverers. CAMP is a diffusable substance of group B, which lyses sheep red blood cells in the presence of staphylococcal hemolysin.
**A sugar that can be split into glucose and esculetin by the group D streptococci.
***Sulfa and trimethoprim. The test is performed (like bacitracin) with discs containing this combination drug.
****PYR for L pyrrolidonyl-β-naphthylamide. This tests for an enzyme that is found mainly in group A and enterococci.

these two pathologic states have developed, there are no specific treatments. A vaccine for group A streptococci that can be delivered through a nasal spray is currently undergoing clinical trials.

Some physicians recommend that people with a history of rheumatic fever or recurring strep throat receive continuous, long-term penicillin prophylaxis. Mass prophylaxis has also been indicated for young people exposed to epidemics in boarding schools, military camps, and other institutions and for known carriers. Tonsillectomy (removal of the tonsils) has been used extensively in the past, but with questionable effectiveness, because these bacteria can infect other parts of the throat. In hospitals, carriers of *S. pyogenes* should not be allowed to work with surgical, obstetric, or immunocompromised patients. Patients with group A infections must be isolated, and high-level precautions must be practiced in handling infectious secretions.

The antibiotic treatment of choice for group B streptococcal infection is penicillin G; alternatives are vancomycin and cephalosporins. Some physicians advocate routine penicillin prophylaxis in colonized mothers and infants, but others fear that this practice can increase the rate of allergies and infections by resistant strains. Artificial passive immunization with human immunoglobulins is currently being considered for treatment and prevention in mothers and infants at high risk. Treating enterococcal infection usually requires combined therapy with ampicillin and aminoglycoside (gentamicin) to take advantage of the synergism of these drugs and to overcome antimicrobic resistance.

α-HEMOLYTIC STREPTOCOCCI: THE VIRIDANS GROUP

The viridans category encompasses a large and complex group of human streptococci not entirely groupable by Lancefield serology. They are the most numerous and widespread residents of the oral cavity (gingiva, cheeks, tongue, saliva) and are also found in the nasopharynx, genital tract, and skin. These species can cause serious systemic infections, although most of them are opportunists and lack the full complement of toxins and enzymes that occur in group A. The one characteristic shared by all species (*Streptococcus mitis, S. mutans, S. milleri, S. salivarius, S. sanguis*) is α-hemolysis. Other characteristics used to distinguish them are too numerous to include here.

Because viridans streptococci are not highly invasive, their entrance into tissues usually occurs through dental or surgical instrumentation and manipulation. These organisms are constant

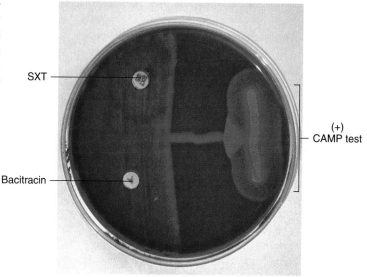

FIGURE 18.15

Tests for characterizing other β-hemolytic streptococci. In the CAMP test, a special strain of β-hemolytic *Streptococcus aureus* that reacts synergistically with group B *Streptococcus* (GBS) is streaked between two smears of the sample. If the isolate is GBS, increased areas of hemolysis appear where the smears meet. (Note that group B streptococci are also negative for bacitracin and SXT tests.)

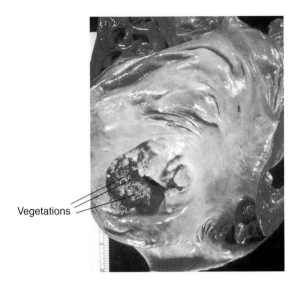

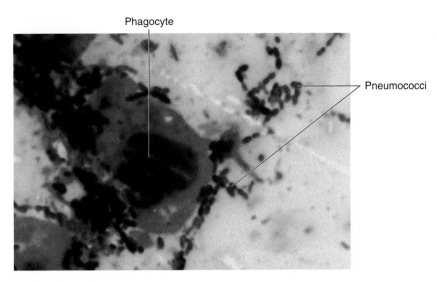

FIGURE 18.16

A view of the heart in subacute bacterial endocarditis. The tricuspid valve has nodular vegetations and distortions on the surface of the leaflets, which constantly release bacteria into the circulation.

FIGURE 18.17

Diagnosing *Streptococcus pneumoniae*. This Gram stain of sputum from a pneumonia patient indicates the morphology of *Streptococcus pneumoniae* as small, pointed diplococci. Large cells interspersed among the pneumococci are phagocytes (600×).

inhabitants of the gums and teeth, and even chewing hard candy or brushing the teeth can provide a portal of entry for them. Dental procedures can lead to bacteremia, meningitis, abdominal infection, and tooth abscesses. But the most important complication of all viridans streptococcal infections is **subacute endocarditis.** In this condition, blood-borne bacteria settle on areas of the heart lining or valves that have previously been injured by rheumatic fever, valve surgery, or the like. Colonization of these surfaces leads to thick layers of bacteria and fibrin called *vegetations* (figure 18.16). As the disease progresses, vegetations increase in size, constantly releasing masses of bacteria into the circulation. These masses, or emboli, can travel to the lungs and brain, creating a crisis in circulation and damage to those organs. Because of its insidious and somewhat concealed course, endocarditis is considered subacute rather than acute or chronic. Symptoms and signs range from fever, heart murmur, and emboli to weight loss and anemia.

Endocarditis is diagnosed almost exclusively by blood culture, and repeated blood samples positive for bacteremia are highly suggestive of it. The goal in treatment is to completely destroy the microbes in the vegetations, which usually can be accomplished by long-term therapy with penicillin G. Because persons with preexisting heart conditions are at high risk for this disease, they usually receive prophylactic antibiotics prior to dental and surgical procedures.

Another very common dental disease involving viridans streptococci is dental caries. In the presence of sugar, *S. mutans* and *S. sanguis* produce slime layers made of glucose polymers that adhere tightly to tooth surfaces. These sticky polysaccharides are the basis for plaque, the adhesive white material on teeth that becomes coinfected with other bacteria and fosters dental disease (see chapter 21).

STREPTOCOCCUS PNEUMONIAE: THE PNEUMOCOCCUS

High on the list of significant human pathogens is ***Streptococcus pneumoniae,*** a unique species that was formerly called *Diplococcus* until its genetic similarity to the streptococci was demonstrated. Because it causes 60–70% of all bacterial pneumonias, *S. pneumoniae* is also referred to as the **pneumococcus.** Gram stains of sputum specimens from pneumonia patients reveal small, lancet-shaped cells arranged in pairs and short chains (figure 18.17). Cultures require complex media such as blood or chocolate agar, and they produce smooth or mucoid colonies and α-hemolysis (see figures 4.12 and 18.9*b*). Growth is improved by the presence of 5–10% CO_2, and cultures die in the presence of oxygen because they lack catalase and peroxidases.

All pathogenic (smooth) strains form rather large capsules, this being their major virulence factor. Rough strains lack a capsule and are nonvirulent. Capsules help the streptococci escape phagocytosis, which is the major host defense in pyogenic infections. They contain a polysaccharide antigen called the specific soluble substance (SSS) that varies chemically among the pneumococcal types and stimulates antibodies of varying specificity. So far, 84 different capsular types (specified by numerals 1, 2, 3, . . .) have been identified, using a technique called the *Quellung* test or capsular swelling reaction (see figure 18.20).

Epidemiology and Pathology of the Pneumococcus

From 5% to 50% of all people carry *S. pneumoniae* as part of the normal flora in the nasopharynx. Although infection is often acquired endogenously from one's own flora, it occasionally occurs after direct contact with respiratory secretions or droplets from

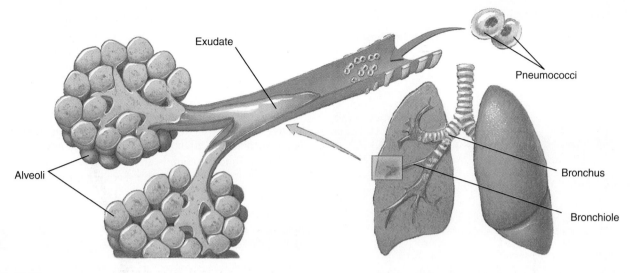

Exudate

Pneumococci

Alveoli

Bronchus

Bronchiole

FIGURE 18.18

The course of bacterial pneumonia. As the pneumococcus traces a pathway down the respiratory tree, it provokes intense inflammation and exudate formation. The blocking of the bronchioles and alveoli by consolidation of inflammatory cells and products is evident.

carriers. *Streptococcus pneumoniae* is very delicate and does not survive long out of its habitat. Factors that favor development of pneumonia are old age, the season (rate of infection is highest in the winter), other diseases (people with underlying lung disease or viral infections have weakened defenses), and living in institutions, which increases the chance of contact with infected people.

The Pathology of Pneumonia Healthy people commonly inhale microorganisms into the respiratory tract without serious consequences because of the host defenses present there. Pneumonia is likely to occur when mucus containing a load of bacterial cells is aspirated from the pharynx into the lungs of susceptible individuals who have lost their defenses. Passing into the bronchioles and alveoli, the pneumococci multiply and induce an overwhelming inflammatory response. This is marked by the release of a torrent of edematous fluids into the lungs. In a form of pneumococcal pneumonia termed **lobar pneumonia,** this fluid accumulates in the alveoli along with red and white blood cells. As the infection and inflammation spread rapidly through the lung, the patient can actually "drown" in his own secretions. If this mixture of exudate, cells, and bacteria solidifies in the air spaces, a condition known as *consolidation** occurs (figure 18.18). In infants and the elderly, the areas of infection are usually spottier and centered more in the bronchi than in the alveoli (bronchial pneumonia).

Symptoms of pneumococcal pneumonia are chills, shaking, rapid breathing, and fever. The patient can experience severe pain in the chest wall, cyanosis (due to compromised oxygen intake), a cough that produces rusty-colored (bloody) sputum, and abnormal breathing sounds. Systemic complications of pneumonia are pleuritis and endocarditis, but pneumococcal bacteremia and meningitis are the greatest danger to the patient.

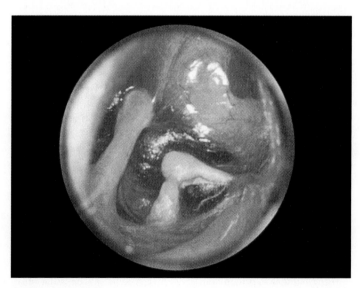

FIGURE 18.19

Pneumococcal otitis media. The acute inflammation and internal pressure of a middle ear infection are evident in the redness and bulging of the tympanum as seen during otoscope examination. Occasionally, the eardrum breaks as shown here. (White objects are the auditory bones.)

In young children, *S. pneumoniae* is a common agent of upper respiratory tract infections that can spread to the meninges and cause meningitis. It is even more common for this agent to gain access to the chamber of the middle ear by way of the eustachian tube and cause a middle ear infection called **otitis media.** This occurs readily in children under 2 years because of their relatively short eustachian tubes. Otitis media is the third most common childhood disease in the United States, and the pneumococcus accounts for the majority of cases. The severe inflammation in the small space of the middle ear induces acutely painful earaches and sometimes even temporary deafness (figure 18.19).

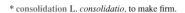
* consolidation L. *consolidatio,* to make firm.

"Swollen" capsule   Cell body

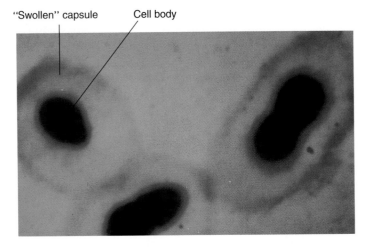

FIGURE 18.20

A positive Quellung, or capsular swelling, test with anticapsular precipitins is confirmatory for *Streptococcus pneumoniae*. It can also be used to identify the precise capsular serotype. The reaction of antibodies with the capsular polysaccharide intensifies the capsule.

Healthy people have high natural resistance to the pneumococcus. The natural mucous and ciliary responses of the respiratory tract help flush out transient organisms. Respiratory phagocytes are also essential in the eradication process, but this system works only if the capsule is coated by opsonins in the presence of complement. After recovery, the individual will be immune to future infections by that particular pneumococcal type. This immunity is the basis for a successful vaccine (see subsequent section on prevention).

Laboratory Cultivation and Diagnosis   A specimen is usually necessary before diagnosing pneumococcal infections because a number of other microbes can cause pneumonia, sepsis, and meningitis. Blood cultures, sputum, pleural fluid, and spinal fluid are common specimens. A Gram-stained specimen is very instructive (see figure 18.17), and presumptive identification can often be made by this method alone. Also highly definitive is the Quellung reaction, a serological test in which sputum is mixed with anti-capsular antisera and observed microscopically. Precipitation of antibodies on the surface of the capsule gives the appearance of swelling (figure 18.20). Alpha-hemolysis helps differentiate the pneumococcus from non-viridans streptococci. Diagnosis can then be confirmed by testing for the agent's sensitivity to the drug optochin, a positive bile solubility, and positive insulin fermentation tests.

Treatment and Prevention of Pneumococcal Infections
The treatment of choice for pneumococcal infections has traditionally been a large daily dose of penicillin G or penicillin V. In recent times, however, clinics have reported increased cases of drug-resistant strains. Drug sensitivity testing is therefore essential. Alternative drugs are cephalosporins, vancomycin, quinolones, and erythromycin. Daily penicillin prophylaxis has recently been suggested to protect children with sickle-cell anemia against recurrent pneumococcal infections. Untreated cases in these children have a mortality rate up to 30%.

Although antibiotics arrest the course of pneumonia, active immunity is important in preventing recurrences. This is the one streptococcal disease for which effective vaccination is available. Polyvalent vaccines (Pneumovax and Pnu-immune) that contain capsular antigens of 23 of the most frequently encountered serotypes are indicated for patients at particularly high risk, including those with sickle-cell anemia, lack of a spleen, congestive heart failure, lung disease, diabetes, kidney disease, and advanced age. These vaccines are effective for about 5 years in 60–70% of those vaccinated. Immunization is also suggested in normal, healthy infants as an effective preventative therapy for otitis media.

CHAPTER CHECKPOINTS

Streptococci are gram-positive cocci arranged in chains and pairs. The genus includes both harmless commensals and formidable pathogens. All *Streptococcus* species are catalase-negative, which distinguishes them from the staphylococci.

Streptococci are classified immunologically by their cell wall antigens (Lancefield groups) and also by their hemolysis of red blood cells. Clinical identification of *Streptococcus* species is based on the sensitivity to certain drugs, presence of the CAMP reaction, and other biochemical tests.

Streptococci that use hemolysins to completely hemolyze red blood cells are termed β-hemolytic. Lancefield groups A, B, C, G, and some members of group D are all β-hemolytic. α-hemolytic streptococci are found in several Lancefield groups. α-hemolysis is caused by a hemolysin that partially clears red blood cells in blood agar cultures.

S. pyogenes is a group A β-hemolytic pathogen. It is extremely pathogenic because of its many virulence factors such as M-protein, tissue-digesting enzymes, and streptolysins S and O, which attack leukocytes, kidney, and heart muscle. Long-term sequelae of *S. pyogenes* infections are rheumatic fever and acute glomerulonephritis (AGN).

Local infections caused by *S. pyogenes* include the skin infections impetigo and erysipelas, as well as pharyngitis ("strep throat").

The group B pathogen *S. agalactiae* is a common opportunist agent of wound, skin, and neonatal infections.

The group D species *Enterococcus faecalis* causes opportunistic infections of wounds, blood, endocardium, and the urinary and gastrointestinal tracts.

Streptococci from groups C and G, usually animal flora, are increasingly isolated in pharyngitis, AGN, and bacteremias in immunocompromised patients.

Alpha-hemolytic streptococci of the viridans group cause dental caries as well as subacute bacterial endocarditis following dental surgery or injury to the oral mucosa.

Alpha-hemolytic *S. pneumoniae* causes 60–70% of all bacterial pneumonias. Young children, the elderly, and immunocompromised individuals are particularly susceptible. However, immunization against *S. pneumoniae* gives effective protection to these groups.

Most streptococcal infections can be cured with penicillin G, except for the enterococci and pneumococci. They require combination drug therapy because of developing resistance to antimicrobial drugs.

The Family Neisseriaceae: Gram-Negative Cocci

Members of the Family Neisseriaceae are residents of the mucous membranes of warm-blooded animals. Most species are relatively innocuous commensals, but two are primary human pathogens with far-reaching medical impact. The genera contained in this group are *Neisseria, Moraxella,* and *Acinetobacter.* Of these, *Neisseria* has the greatest clinical significance.

A distinguishing feature of the **Neisseria*** is their cellular morphology. Rather than being perfectly spherical, the cells are bean-shaped and paired, with their flat sides touching (figure 18.21). None develop flagella or spores, but capsules can be found on the pathogens. The cells are typically gram-negative, possessing an outer membrane in the cell wall and, in many cases, pili.

Most *Neisseria* are strict parasites that do not survive long outside the host, particularly where hostile conditions of drying, cold, acidity, or light prevail. *Neisseria* species are aerobic or microaerophilic and have an oxidative form of metabolism. They produce catalase, enzymes for fermenting various carbohydrates, and the enzyme cytochrome oxidase that can be used at a general level in identification. The pathogenic species, *N. gonorrhoeae* and *N. meningitidis,* require complex enriched media and grow best in an atmosphere containing additional CO_2. We shall concentrate on other features of the pathogenic *Neisseria* in the following sections.

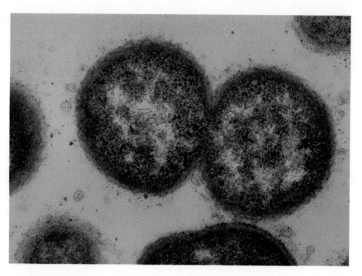

FIGURE 18.21

This transmission electron micrograph of *Neisseria* (52,000×) clearly indicates how the diplococci form.

NEISSERIA GONORRHOEAE: THE GONOCOCCUS

Gonorrhea* has been known as a sexually transmitted disease since ancient times. Its name originated with the Greek physician Claudius Galen, who thought that it was caused by an excess flow of semen. For a fairly long period in history, gonorrhea was confused with syphilis. Later, microbiologists went on to cultivate *N. gonorrhoeae,* also known as the **gonococcus,** and to prove conclusively that it alone was the etiologic agent of gonorrhea.

Factors Contributing to Gonococcal Pathogenicity

The virulence of the gonococcus is due chiefly to the presence of pili and other surface molecules that promote mutual attachment of cocci to each other and invasion and infection of epithelial tissue. In addition to their role in adherence, pili also seem to slow phagocytosis by macrophages and neutrophils. Another contributing factor in pathogenicity is a protease that cleaves the secretory antibody (IgA) on mucosal surfaces and keeps it from working.

Epidemiology and Pathology of Gonorrhea

Gonorrhea is a strictly human infection that occurs worldwide and ranks number five in prevalence among the top five sexually transmitted diseases. Although about 500,000 cases are reported in the United States each year, it is estimated that the actual incidence is much higher—in the millions if one counts asymptomatic infections. Most cases occur in young people (18–24 years old) with

multiple sex partners. Figures on the prevalence of gonorrhea and syphilis over the past 60 years show a fluctuating pattern, apparently corresponding to periods of social and political upheaval, when promiscuity tends to increase (figure 18.22). One interesting effect occurred during the sexual revolution of the 1960s, when reliance on oral contraceptives rather than condoms to prevent pregnancy increased the transmission of the gonococcus. Other trends in gonorrhea are featured in figure 13.16.

Studies done with male volunteers revealed that an infectious dose can range from 100 to 1,000 colony-forming units. *Neisseria gonorrhoeae* does not survive more than one or two hours on fomites and is most infectious when transferred directly to a suitable mucous membrane. Except for neonatal infections, the gonococcus spreads through some form of sexual contact. The pathogen comes in contact with an appropriate portal of entry that is genital or extragenital (rectum, eye, or throat). After attaching to the epithelial surface by pili, the bacteria invade the underlying connective tissue. In 2 to 6 days, this process results in an inflammatory reaction that may or may not produce noticeable symptoms. The infection is asymptomatic in approximately 10% of males and 50% of females. It is this reservoir that is most important in the persistence and spread of the pathogen. The following sections survey the several categories of gonorrhea.

Genital Gonorrhea in the Male   Infection of the urethra elicits urethritis, painful urination, and a yellowish discharge, though a relatively large number of cases are asymptomatic. In most cases, infection is limited to the distal urogenital tract, but it can occasionally spread from the urethra to the prostate gland and epididymis (figure 18.23). Scar tissue formed in the spermatic ducts during healing of an invasive infection can render the individual infertile. This outcome is becoming increasingly rarer with improved diagnosis and treatment regimens.

---

* *Neisseria* (ny-serr'-ee-uh) After the German physician Albert Neisser who first observed the agent of gonorrhea in 1879.

* gonorrhea Gr. *gonos,* seed, and *rhein,* to flow.

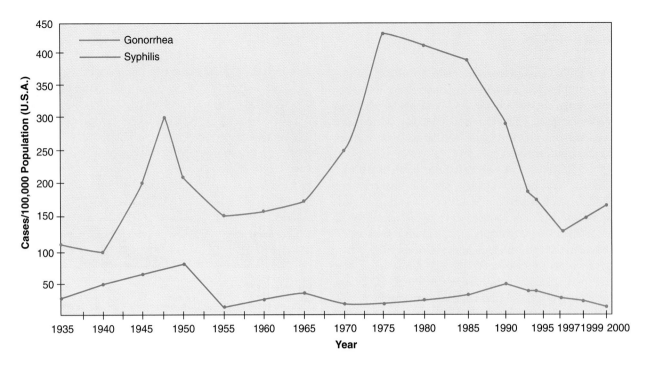

FIGURE 18.22

Comparative graph of two reportable infectious STDs. Gonorrhea was once the most common reportable STD in the United States. Notice the rise in cases (epidemics) corresponding with two periods: 1940–1946 (World War II) and 1965–1975 (the Vietnam War). This pattern is correlated with times of social upheaval and changing sexual mores. Epidemiologists also theorize that increased use of oral contraceptives contributed to the second epidemic period.

Genitourinary Gonorrhea in the Female The proximity of the genital and urinary tract openings increases the likelihood that both systems can be infected during sexual intercourse. A mucopurulent or bloody vaginal discharge occurs in about half the cases, along with painful urination if the urethra is affected. Major complications occur when the infection ascends from the vagina and cervix to higher reproductive structures such as the uterus and fallopian tubes (figure 18.24). One disease resulting from this progression is **salpingitis,*** also known as **PID** (pelvic inflammatory disease), a condition characterized by fever, abdominal pain, and tenderness. It is not unusual for the microbe to become involved in mixed infections with anaerobic bacteria. The buildup of scar tissue from these infections can block the fallopian tubes, causing sterility and ectopic pregnancies.

Extragenital Gonococcal Infections in Adults Extragenital sexual transmission and carriage of the gonococcus are not uncommon. Anal intercourse can lead to proctitis, and oral copulation can result in pharyngitis and gingivitis. These complications appear most often in homosexual males. Careless personal hygiene can account for self-inoculation of the eyes and a serious form of conjunctivitis. In a small number of cases, the gonococcus enters the bloodstream and is disseminated to the joints and skin. Involvement of the wrist and ankle can lead to chronic arthritis and a painful, sporadic, papular rash on the limbs. Rare complications of gonococcal bacteremia are meningitis and endocarditis.

Gonococcal Infections in Children Infants born to gonococcus carriers are in danger of being infected as they pass through the birth canal. Because of the potential harm to the fetus, physicians usually screen pregnant mothers for its presence. Gonococcal eye infections are very serious and often manifest sequelae such as keratitis, ophthalmia neonatorum, and even blindness (figure 18.25). A universal precaution to prevent those complications is the instillation of antibiotics, silver nitrate, or other antiseptics into the conjunctival sac of newborn babies. The pathogen may also infect the pharynx and respiratory tract of neonates. Finding gonorrhea in children other than babies is strong evidence of sexual abuse by infected adults, and it mandates child welfare consultation along with thorough bacteriologic analysis.

Clinical Diagnosis and Control of Gonococcal Infections

Diagnosing gonorrhea is relatively straightforward. The presence of gram-negative diplococci in neutrophils from urethral, vaginal, cervical, or eye exudates is especially diagnostic because gonococci tend to be engulfed and remain viable within phagocytes (figure 18.26). This simple procedure can provide at least presumptive evidence of gonorrhea. It is most successful in males and least successful in asymptomatic infection. Other tests used to identify *Neisseria gonorrhoeae* and differentiate it from related species are discussed in a later section.

Several hundred thousand new cases of gonorrhea occur every year in the United States. Of these cases 10–20% are caused by drug-resistant strains called penicillinase-producing *N. gonorrhoeae* (**PPNG**) or tetracycline-resistant (TRNG). A fairly large proportion

---

* **salpingitis** (sal″-pin-jy′-tis) Gr. *salpinx*, tube, and *itis*, inflammation. An inflammation of the fallopian tubes.

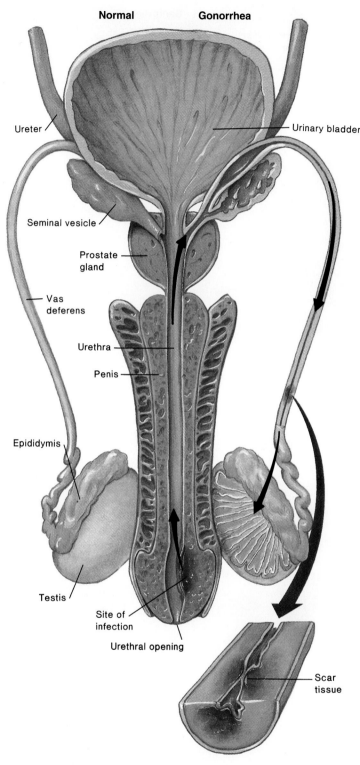

Normal   **Gonorrhea**

Ureter

Urinary bladder

Seminal vesicle

Prostate gland

Vas deferens

Urethra

Penis

Epididymis

Testis

Site of infection

Urethral opening

Scar tissue

FIGURE 18.23

Gonorrheal damage to the male reproductive tract. *(Left)* Frontal view of the male reproductive tract. *(Left side)* The normal, uninfected state. The sperm-carrying ducts are continuous from the testis to the urethral opening. *(Right side)* The route of ascending gonorrhea complications. Infection begins at the tip of the urethra, ascends the urethra through the penis, and passes into the vas deferens. Occasionally, it can even enter the epididymides and testes. *(Inset)* Damage to the ducts carrying sperm can create scar tissue and blockage, which reduce sperm passage and can lead to sterility.

of cases are complicated by a concurrent STD such as chlamydiosis. One strategy to prevent drug resistance and combat coinfections is combined therapy with a broad-spectrum cephalosporin (cephtriaxone) plus tetracycline (doxycycline). An alternative therapy combines a quinoline (ciprofloxacin) with tetracycline. Other drugs with some effectiveness are spectinomycin and azithromycin.

Although gonococcal infections stimulate local production of antibodies and activate the complement system, these responses do not produce lasting immunity, and some people experience recurrent infections.

Gonorrhea is a reportable infectious disease, which means that any physician diagnosing it must forward the information to a public health department. The follow-up to this finding involves tracing sexual partners to offer prophylactic antibiotic therapy. There is a pressing need to seek out and treat asymptomatic carriers and their sexual contacts, but complete control of this group is nearly impossible. Other control measures include education programs that emphasize the effects of all STDs and promote safer sexual practices such as the use of condoms.

NEISSERIA MENINGITIDIS: THE MENINGOCOCCUS

Another serious human pathogen is *Neisseria meningitidis,* a bacterium known commonly as the **meningococcus** and usually associated with epidemic cerebrospinal meningitis. Important factors in meningococcal invasiveness are a polysaccharide capsule, pili, and IgA protease. Although 12 different strains of capsular antigens exist, serotypes A, B, and C are responsible for most cases of infection. Another virulence factor with potent pathologic effects is the lipopolysaccharide (endotoxin) released from the cell wall when the microbe lyses.

Epidemiology and Pathogenesis of Meningococcal Disease

The diseases of *N. meningitidis* have a sporadic or epidemic incidence in late winter or early spring. The continuing reservoir of infection is humans who harbor the pathogen in the nasopharynx. The carriage state, which can last from a few days to several months, exists in 3% to 30% of the adult population and can exceed 50% in institutional settings. The scene is set for transmission when carriers live in close quarters with nonimmune individuals, as might be expected in families, day care facilities, and military barracks. The highest risk groups are young children (6–36 months old) and older children and young adults (10–20 years old). *Neisseria meningitidis* is the most frequent cause of gram-negative **meningitis** in all age groups except older patients, who are more frequently infected by *Haemophilus influenzae* (see figure 20.22). At the present time reports of meningococcal meningitis are on the rise.

Because this bacterium does not survive long in the environment, meningococci are usually acquired through close contact with secretions or droplets. Upon reaching its portal of entry in the nasopharynx, the meningococcus attaches there with pili. In many people, this can result in simple asymptomatic colonization. In the more vulnerable individual, however, the meningococci are engulfed by epithelial cells of the mucosa and penetrate into the nearby blood vessels, thereby damaging the epithelium and causing pharyngitis.

Normal   **Gonorrhea**

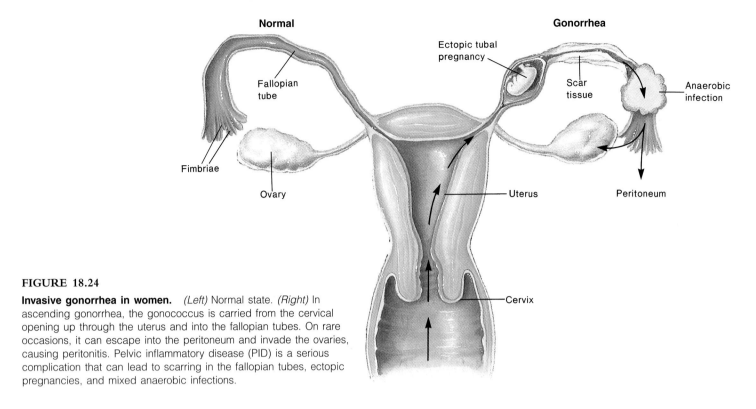

FIGURE 18.24

Invasive gonorrhea in women. *(Left)* Normal state. *(Right)* In ascending gonorrhea, the gonococcus is carried from the cervical opening up through the uterus and into the fallopian tubes. On rare occasions, it can escape into the peritoneum and invade the ovaries, causing peritonitis. Pelvic inflammatory disease (PID) is a serious complication that can lead to scarring in the fallopian tubes, ectopic pregnancies, and mixed anaerobic infections.

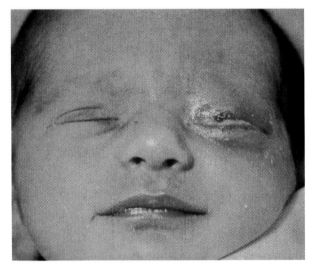

FIGURE 18.25

Gonococcal ophthalmia neonatorum in a week-old infant. The infection is marked by intense inflammation and edema; if allowed to progress, it causes damage that can lead to blindness. Fortunately, this infection is completely preventable and treatable.

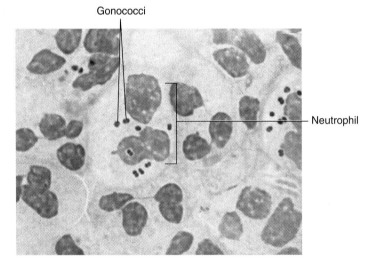

FIGURE 18.26

Gram stain of urethral pus from a patient with gonorrhea (1,000×). Note the intracellular (phagocytosed) gram-negative diplococci in polymorphonuclear leukocytes (neutrophils).

Bacteria entering the blood vessels rapidly permeate the meninges and produce symptoms of meningitis, the most common complication in children. It is marked by fever, sore throat, headache, stiff neck, convulsions, and vomiting. The most serious complications of meningococcal pharyngitis are due to meningococcemia (figure 18.27). The pathogen sheds endotoxin into the generalized circulation, which is a potent stimulus for certain white blood cells. Damage to the blood vessels caused by cytokines leads to vascular collapse, hemorrhage, and crops of lesions called *petechiae** on the trunk and appendages. In a small number of cases, meningococcemia becomes a fulminant disease with a high mortality rate. It has a sudden onset, marked by fever (higher than 40°C), chills, delirium, severe widespread *ecchymoses,** shock, and coma (figure 18.28). Generalized intravascular clotting, cardiac failure, damage to the adrenal glands, and death can occur within a few hours.

---

* petechiae (pee-tee′-kee-ee) Small, nonraised, round purple spots caused by hemorrhage into the skin. Larger spots are called ecchymoses (ek″-ih-moh′-seez).

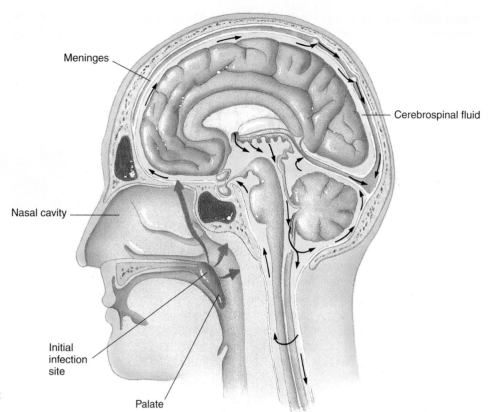

FIGURE 18.27

 Dissemination of the meningococcus from a nasopharyngeal infection.
Bacteria spread to the roof of the nasal cavity, which borders a highly vascular area at the base of the brain. From this location, they can enter the blood and escape into the cerebrospinal fluid. Infection of the meninges leads to meningitis and an inflammatory purulent exudate over the brain surface.

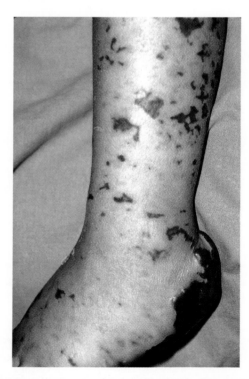

FIGURE 18.28

The appearance of meningococcemia. The blotches on the leg and foot are due to subcutaneous hemorrhages. This condition can occur anywhere on the body, including the mucous membranes and conjunctiva. Endotoxins released during blood infection are thought to be largely responsible for this pathologic state.

Clinical Diagnosis of Meningococcal Disease

Suspicion of bacterial meningitis constitutes a medical emergency, and differential diagnosis must be done with great haste and accuracy, since complications can come on so rapidly and with such lethal consequences. Cerebrospinal fluid, blood, or nasopharyngeal samples are stained and observed directly for the typical gram-negative diplococci. Cultivation may be necessary to differentiate the bacterium from other species (see subsequent section on laboratory diagnosis). Specific rapid tests are also available for detecting the capsular polysaccharide or the cells directly from specimens without culturing.

Immunity, Treatment, and Prevention of Meningococcal Infection

The infection rate in most populations is about 1%, so well-developed natural immunity to the meningococcus appears to be the rule. This is due to a sort of natural immunization that occurs during the early years of life as one is exposed to the meningococcus and its close relatives. Resistance is due to opsonizing antibodies that develop against the capsular polysaccharides in groups A and C and against membrane antigens in group B. Because even treated meningococcal disease has a mortality rate of up to 15%, it is vital that chemotherapy begin as soon as possible with one or more drugs. It can even be given while tests for the causative agent are under way. Penicillin G is the most potent of the drugs available for meningococcal infections; it is generally given in high doses intravenously. If the patient cannot tolerate penicillin, a third-generation cephalosporin is another choice. Patients may also require treatment for shock and intravascular clotting.

When family members, medical personnel, or children in day care have come in close contact with infected people, preventive therapy with rifampin or tetracycline may be warranted. Meningococcal vaccines that contain specific purified capsular antigens are available to protect high-risk groups, especially during epidemics. Group A vaccine protects all ages, but group C vaccine is useful only for individuals over 2 years of age, and a group B vaccine is not yet available.

DIFFERENTIATING PATHOGENIC FROM NONPATHOGENIC *NEISSERIA*

It is usually necessary to differentiate true pathogens from normal *Neisseria* that also live in the human body and can be present in infectious fluids. Immediately after collection, specimens are streaked on Modified Thayer-Martin medium (MTM) or chocolate agar and incubated in a high CO_2 atmosphere. Presumptive identification of the genus is obtained by a Gram stain and oxidase testing on isolated colonies (figure 18.29). Further testing may be necessary to differentiate the two pathogenic species from one another, from other oxidase-positive infectious species, and from normal flora of the oropharynx and genitourinary tract that can be confused with the pathogens. Sugar fermentation, growth patterns, nitrate reduction, and pigment production are useful differentiation tests at this point (table 18.4). Several rapid method identification kits have been developed for this purpose. Many laboratories now rely on a rapid PCR test that can detect gonococcal DNA directly from a urine sample.

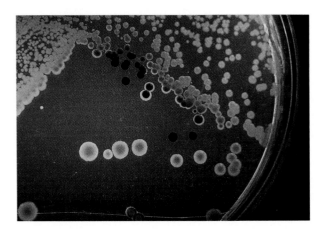

FIGURE 18.29

The oxidase test. A drop of oxidase reagent is placed on a suspected *Neisseria* or *Branhamella* colony. If the colony reacts with the chemical to produce a purple to black color, it is oxidase-positive; those that remain white to tan are oxidase-negative. Because several species of gram-negative rods are also oxidase-positive, this test is presumptive for these two genera only if a Gram stain has verified the presence of gram-negative cocci.

TABLE 18.4

Scheme for Differentiating Gram-Negative Cocci and Coccobacilli

Characteristic or Test	Neisseria gonorrhoeae	N. meningitidis	N. lactamica*	N. sicca	Branhamella catarrhalis	Moraxella spp.	Acinetobacter spp.
Oxidase	+	+	+	+	+	+	−
Growth on MTM	+	+	+	−	+/−	−	−
Growth on nutrient agar	−	−	−	+	+	+	+
Growth on blood agar at 22°C	−	−	+/−	+/−	+	+	+
Yellow pigment	−	−	+	+/−	−	−	−
Require increased CO_2 during isolation	+	−	−	−	−	−	−
Sugar fermentation							
Maltose	−	+	+	+	−	−	+/−
Sucrose	−	−	−	+	−	−	+/−
Lactose	−	−	+	−	−	−	+/−
Nitrate reduction	−	−	−	−	+	+/−	−
Nitrite reduction	−	+/−	++	+	+	−	−
Capsule	+	+	−	+/−	+	+/−	+/−
Human pathogenicity	+	+	+	UDC**	Opportunistic	UDC	Opportunistic

*A weak pathogen, found in the nasopharynx of children and easily mistaken for N. meningitidis.
**Stands for Usually Dismissed as Contaminants, because pathogenicity is very rare.

CHAPTER CHECKPOINTS

Neisseria are fastidious, nonmotile, gram-negative, coffee bean–shaped cocci that are facultative anaerobes and live as commensals in the mucous membranes of mammals. The two significant pathogens in this group are *Neisseria gonorrhoeae* and *Neisseria meningitidis*.

N. gonorrhoeae is the causative agent of gonorrhea, a sexually transmitted disease that causes infections of the male and female reproductive tracts. Genital gonorrhea in males causes urethritis and painful urination. Genital gonorrhea in females can lead to PID and ectopic pregnancies. It also causes sterility, proctitis, and pharyngitis in the sexually active, as well as gonococcal eye infections of the newborn.

Gonorrhea is a reportable STD because it is one of the top five STDs worldwide. It is spread by direct contact with infected carriers, who are often asymptomatic. Humans are the only reservoirs. It is especially prevalent in 18- to 24-year-olds. Gonorrhea is treated by combined drug therapy because of widespread resistance to penicillin and frequent concurrent infection by other STD agents, particularly chlamydia.

Virulence factors of *N. gonorrhoeae* include pili, a capsule, and a protease that inactivates IgA.

N. meningitidis is the causative agent of epidemic cerebrospinal meningitis. It is transmitted by respiratory secretions or droplets from infected carriers.

Meningococcus virulence factors include lipopolysaccharide (endotoxin), IgA protease, pili, and a capsule. Toxins cause cardiac failure, vascular collapse, and clotting disorders. Immunization is available to protect at-risk groups.

Other Gram-Negative Cocci and Coccobacilli

The genera *Branhamella, Moraxella,* and *Acinetobacter* are included in the same family as *Neisseria* because of morphological and biochemical similarities. Most species are either relatively harmless commensals of humans and other mammals or are saprobes living in soil and water. In the past few years, however, one species in particular has emerged as a significant opportunist in hosts with disturbed immune functions. This species, *Branhamella (Moraxella) catarrhalis,** is found in the normal human nasopharynx and can cause purulent disease. It is associated with several clinical syndromes such as meningitis, endocarditis, otitis media, bronchopulmonary infections, and neonatal conjunctivitis. Adult patients with leukemia, alcoholism, malignancy, diabetes, or rheumatoid disease are the most susceptible to it.

Because *Branhamella* is morphologically similar to the gonococcus and the meningococcus, biochemical methods are needed to discriminate it. Key characteristics are the complete lack of carbohydrate fermentation and positive nitrate reduction in *B. catarrhalis.* Treatment of infection by this organism is with erythromycin or cephalosporins, because so many strains of this species produce penicillinase.

*Moraxella** species are short, plump rods rather than cocci, and some exhibit a twitching motility. These bacteria are widely distributed on the mucous membranes of domestic mammals and humans and are generally regarded as weakly pathogenic or nonpathogenic. Some species are rarely implicated in ear infections and conjunctivitis in humans.

The genus *Acinetobacter** is similar morphologically to *Moraxella.* It is a small, paired, gram-negative cell that varies in shape from a true rod to a coccus. However, it is quite different from the members of this family in several ways. For example, its habitat is soil, water, and sewage; it is nonfastidious; and it is oxidase-negative. *Acinetobacter* is a common contaminant in clinical specimens and is ordinarily not considered clinically significant. In rare cases, however, it is clearly an agent of disease. Most infections are nosocomial in origin, affecting traumatized or debilitated people with indwelling catheters or other instrumentation. Septicemia, meningitis, endocarditis, pneumonia, and urinary tract infections have been reported.

CHAPTER CHECKPOINTS

Other Neisseriaceae implicated in infectious disease are *Branhamella* and *Acinetobacter.* Most members, such as *Moraxella* species, are harmless commensals or saprobes, but *Branhamella catarrhalis* causes opportunistic infections in immunocompromised individuals. *Acinetobacter* is a common agent in soil and water and is increasingly isolated in nosocomial infections.

---

* *Branhamella catarrhalis* (bran″-hah-mel′-ah cah-tahr-al′-is) After Sarah Branham, a bacteriologist working on the genus *Neisseria,* and L. *catarrhus,* to flow. It is also known as *Moraxella.*

* *Moraxella* (moh-rak-sel′-uh) From Victor Morax, a Swiss ophthalmologist who worked with the genus.

* *Acinetobacter* (ass″-ih-nee-toh-bak′-tur) Gr. *a,* none, *kinetos,* movement, and *bacterion,* rod.

CHAPTER CAPSULE WITH KEY TERMS

I. **Genus *Staphylococcus***
Nonmotile, non-spore-forming cocci arranged in irregular clusters; facultative anaerobes; fermentative; salt-tolerant, catalase-positive *Staphylococcus aureus* produces a number of virulence factors. Enzymes include coagulase (a confirmatory characteristic), hyaluronidase, staphylokinase, nuclease, and penicillinase. Toxins are β-hemolysins, leukocidin, enterotoxin, exfoliative toxin, and toxic shock syndrome toxin. Microbe carried in nasopharynx and skin.
   A. *Infections:* Target is skin; local **abscess** occurs at site of invasion of hair follicle, gland; manifestations are **folliculitis, furuncle, carbuncle,** and **bullous impetigo.** Other common infections are

osteomyelitis, a focal infection of bone; bacteremia, leading to endocarditis; and pneumonia.
   B. *Toxic Disease:* Food intoxication due to enterotoxin; **staphylococcal scalded skin syndrome (SSSS),** a skin condition that causes desquamation; **toxic shock syndrome,** toxemia in women due to infection of vagina, associated with wearing tampons.
   C. *Principal Coagulase-Negative Staphylococci: S. epidermidis,* a normal resident of skin and follicles; an opportunist and one of the most common causes of nosocomial infections, chiefly in surgical patients with indwelling medical devices or implants. *S. saprophyticus* is a urinary pathogen.

D. *Treatment: S. aureus* has multiple resistance to antibiotics, especially penicillin, ampicillin, and methicillin; drug selection requires sensitivity testing; cephalosporins often used; abscesses require debridement and removal of pus; extreme resistance of staphylococci to harsh environmental conditions makes control difficult, requiring a high level of disinfection and antisepsis.

II. **Streptococci**

A large, varied group of bacteria (about 25 species), containing the genera *Streptococcus* and *Enterococcus*. Cocci are in chains of various lengths; nonmotile, non-spore-forming; often encapsulated; fermentative; catalase-negative; most pathogens fastidious and sensitive to environmental exposure. Classified into Lancefield groups (A–R) according to the type of serological reactions of the cell wall carbohydrate; also characterized by type of hemolysis. The most important sources of human disease are β-hemolytic *S. pyogenes* (group A), *S. agalactiae* (group B), *Enterococcus faecalis* (group D), and α-hemolytic *S. pneumoniae* and the *viridans streptococci*.

A. *β-Hemolytic Streptococci: S. pyogenes* is the most serious pathogen of family; produces several virulence factors, including C-carbohydrates, M-protein (fimbriae), streptokinase, hyaluronidase, DNase, **hemolysins** (SLO, SLS), **pyogenic toxin.**

   1. Microbe resides in nasopharynx of carriers; transmitted through close contact; invades skin and mucous membranes.
   2. **Skin infections** include **pyoderma** (strep **impetigo**) and **erysipelas** (deeper, spreading skin infection).
   3. Systemic conditions include **strep throat** or **pharyngitis,** severe inflammation of throat membranes; may lead to toxemia, called **scarlet fever**—generalized flushing of skin and high fever due to erythrogenic toxin; also causes pneumonia.
   4. Sequelae caused by immune response to streptococcal toxins include **rheumatic fever,** a delayed allergy that damages heart valves and joints, and **glomerulonephritis,** an inflammation that leads to malfunction or destruction of kidney tubules.

B. *Streptococcus agalactiae* (group B), a pathogen increasingly found in the human vagina; causes neonatal, wound, and skin infections, particularly in debilitated persons.

   1. *Enterococcus faecalis* and other enteric group D species are normal flora of intestine; cause opportunistic urinary, wound, and surgical infections.
   2. Group A and group B are treated primarily with some type of pencillin (G is very effective); sensitivity testing may be necessary for enterococci; no vaccines available.

C. *α-Hemolytic Streptococci:* The **viridans streps** *S. mitis, S. salivarius, S. mutans,* and *S. sanguis* constitute oral flora in saliva.

   1. Principal infections are **subacute endocarditis,** mass colonization of heart valves following dental procedures.
   2. *Mutans* and *sanguis* species are the main contributors to plaque and dental disease.

D. *S. pneumoniae,* the **pneumococcus,** has heavily encapsulated, lancet-shaped diplococci; capsule is an important virulence factor—84 types; reservoir is nasopharynx of normal healthy carriers.

E. The pneumococcus is the most common cause of bacterial pneumonia; attacks patients with weakened respiratory defenses; entrance of bacteria into lungs initiates acute, massive inflammatory response that fills lungs (and bronchioles) with fluid; **consolidation** of fluid leads to **lobar pneumonia** and compromised respiration. Other symptoms include fever/chills, cyanosis, cough; **otitis media,** inflammation of middle ear, is common in children; vaccination available for patients at risk.

III. **Gram-Negative Cocci**

Primary pathogens are in the genus *Neisseria,* common residents of mucous membranes. *Neisseria* species are bean-shaped diplococci that have capsules and pili; they are oxidase-positive, non-spore-forming, nonmotile; pathogens are fastidious and do not survive long in the environment.

A. *Neisseria gonorrhoeae:* The **gonococcus,** cause of **gonorrhea;** microbe invades mucous membranes by attaching with **pili;** tends to be located intracellularly in pus cells; among the top five most common STDs; may be transmitted from mother to newborn; asymptomatic carriage is common in both sexes.

   1. Symptoms of **gonorrhea in males** are urethritis, discharge; infection of deeper reproductive structures may cause scarring and infertility. Symptoms of **gonorrhea in females** include vaginitis, urethritis. Ascending infection may lead to **salpingitis (PID),** mixed anaerobic infection of abdomen; common cause of sterility and ectopic tubal pregnancies due to scarred fallopian tubes. **Extragenital infections** may be anal, pharyngeal, conjunctivitis, septicemia, arthritis.
   2. **Infections in newborns** cause eye inflammation, occasionally infection of deeper tissues and blindness; can be prevented by prophylaxis immediately after birth.
   3. Preferred treatment is a combination of cephalosporin and tetracycline due to increasing **PPNG** (penicillin-resistant) strains; no vaccine exists; safe sex practices, lack of promiscuity are important controls.

B. *Neisseria meningitidis:* The **meningococcus** is a prevalent cause of **meningitis.** Agent can be carried in the nasopharynx; invades when resistance is lowered; is spread by close contact; bacterium adheres by capsule and pili. Disease begins when bacteria enter bloodstream, pass into the cranial circulation, multiply in meninges; very rapid onset; initial symptoms are neurologic; **endotoxin** released by pathogen causes hemorrhage and shock; can be fatal; treated with penicillin and/or chloramphenicol; vaccines exist for groups A and C.

C. *Other Gram-Negative Cocci and Coccobacilli:* **Branhamella catarrhalis,** common member of throat flora; is an opportunist in cancer, diabetes, alcoholism. **Moraxella,** short rods that colonize mammalian mucous membranes. **Acinetobacter,** gram-negative coccobacilli that occasionally cause nosocomial infections.

MULTIPLE-CHOICE QUESTIONS

1. Which of the following are pyogenic cocci?
   a. *Streptococcus*
   b. *Staphylococcus*
   c. *Neisseria*
   d. all of these

2. The coagulase test is used to differentiate *Staphylococcus aureus* from
   a. other staphylococci
   b. streptococci
   c. micrococci
   d. enterococci

3. The symptoms in scarlet fever are due to
   a. streptolysin
   b. coagulase
   c. pyrogenic toxin
   d. alpha toxin

4. Penicillin-resistant *Staphylococcus* can be treated with _____ without sensitivity testing.
   a. ampicillin
   b. erythromycin
   c. methicillin
   d. no antibiotic

5. The most severe streptococcal diseases are caused by
   a. group B streptococci
   b. group A streptococci
   c. pneumococci
   d. enterococci

6. Rheumatic fever damages the _____, and glomerulonephritis damages the _____ .
   a. skin, heart
   b. joints, bone marrow
   c. heart valves, kidney
   d. brain, kidney

7. _____ hemolysis is the partial lysis of red blood cells due to bacterial hemolysins.
   a. Gamma
   b. Alpha
   c. Beta
   d. Delta

8. An effective vaccine exists to prevent infections from
   a. *S. aureus*
   b. *S. pyogenes*
   c. *N. gonorrhoeae*
   d. *S. pneumoniae*

9. Viridans streptococci commonly cause
   a. pneumonia
   b. meningitis
   c. subacute endocarditis
   d. otitis media

10. Which genus of bacteria has pathogens that can cause blindness and deafness?
    a. *Streptococcus*
    b. *Staphylococcus*
    c. *Neisseria*
    d. *Branhamella*

11. An important test for identifying *Neisseria* is
    a. production of oxidase
    b. production of catalase
    c. sugar fermentation
    d. beta-hemolysis

12. A complication of genital gonorrhea in both men and women is
    a. infertility
    b. pelvic inflammatory disease
    c. arthritis
    d. blindness

13. The skin blotches in meningitis are due to
    a. skin invasion by *N. meningitidis*
    b. blood clots
    c. erysipelas
    d. endotoxins in the blood

14. Which infectious agent of those covered in the chapter would most likely be acquired from a contaminated doorknob?
    a. *Staphylococcus aureus*
    b. *Streptococcus pyogenes*
    c. *Neisseria meningitidis*
    d. *Streptococcus pneumoniae*

CONCEPT QUESTIONS

1. Differentiate between the pathologies in staphylococcal infections and toxinoses.

2. Explain how the actions of each of the following make all of them virulence factors:
   a. hemolysins   c. kinases
   b. leukocidin   d. hyaluronidase

3. a. Describe the normal habitat of *Staphylococcus aureus*.
   b. Distinguish between the four main "staph" skin infections.
   c. Define the term *abscess*, and describe at least two kinds that are caused by staph species.

4. What does it mean to say osteomyelitis is a focal infection?

5. What conditions favor staph food poisoning?

6. What conditions favor toxic shock syndrome?

7. a. Describe the principal role of the coagulase-negative staphylococci in disease.
   b. Describe the usual habitats of these staphylococci.

8. Compare the symptoms of streptococcal and staphylococcal impetigo.

9. a. Describe the major group A streptococcal infections.
   b. Why is a "strep throat" a cause for concern?

10. Discuss the apparent pathology at work in rheumatic fever and acute glomerulonephritis.

11. a. How are group B streptococci important?
    b. How is the genus *Enterococcus* different from the genus *Streptococcus?*
    c. What is the medical significance of enterococci and group C and G streptococci?

12. a. How does lobar pneumonia arise?
    b. How is it different from bronchial pneumonia?
    c. What causes the patient to get a blue tinge to his mucous membranes?
    d. Explain how children acquire otitis media.

13. a. Compare and contrast the characteristics of male and female genital gonorrhea.
    b. Describe at least two extragenital complications of gonorrhea.
    c. Explain how neonates become infected with *N. gonorrhoeae* and describe the disease symptoms.

14. a. Describe the pathway that *N. meningitidis* takes from infection in the nasopharynx to the brain.
    b. Describe the specific virulence factors that cause symptoms of meningitis.

15. **Single Matching.** Only one description in the right-hand column fits a word in the left-hand column.

____ furuncle	a. complete red blood cell lysis
____ osteomyelitis	b. substance involved in heart valve damage
____ coagulase	
____ pyrogenic toxin	c. dissolves blood clots
____ rheumatic fever	d. enzyme of pathogenic *S. aureus*
____ β-hemolysis	
____ consolidation	e. cutaneous infection of group A streps
____ viridans streptococci	
____ erysipelas	f. solidification of pockets in lung
____ endocarditis	g. unique pathologic feature of *N. gonorrhoeae*
____ streptolysin	
____ streptokinase	h. a boil
	i. cause of tooth abscesses
	j. focal infection of long bones
	k. heart colonization by viridans streps
	l. cause of scarlet fever
	m. long-term sequelae of strep throat

16. **Multiple Matching.** Match the bacterium in the left-hand column with its characteristic. More than one characteristic may fit each bacterium.

____ *Staphylococcus aureus*	a. bacitracin sensitivity
____ *Streptococcus pyogenes*	b. diplococcus
____ *Streptococcus pneumoniae*	c. causes pneumonia
____ *Neisseria gonorrhoeae*	d. complication is glomerulonephritis
____ *Neisseria meningitidis*	
____ *Enterococcus faecalis*	e. infects female reproductive tract
	f. resident of nasopharynx
	g. is in clusters
	h. is in chains
	i. has a capsule
	j. catalase production

CRITICAL-THINKING QUESTIONS

1. a. What is the probable significance of isolating large numbers of *Streptococcus mitis* from a throat swab sample?
   b. What is the significance of isolating five colonies of *Streptococcus pyogenes* in a throat culture?
   c. What is the significance of isolating three *Staphylococcus epidermidis* colonies from a swab culture of an open wound?
   d. What is the significance of isolating 100 colonies of *S. epidermidis* from a swab from an indwelling catheter?
   e. How important is the isolation of 10 colonies of *N. meningitidis* from the nasopharynx?
   f. What if it is isolated from the spinal fluid? Give one possible explanation for finding *Enterococcus faecalis* in the blood.

2. Why is it so important to differentiate *S. aureus* from coagulase-negative staphylococci?

3. You have been handed the problem of diagnosing gonorrhea from a single test. Which one will you choose and why?

4. How do the gram-positive and gram-negative diplococci differ in exact morphology?

5. Why do you suppose there are no useful vaccines for most of the pyogenic cocci?

6. You have been called upon to prevent outbreaks of SSSS in the nursery of a hospital where this strain of *S. aureus* has been isolated.
   a. What will be your main concerns?
   b. What procedures will you establish to address these concerns?
   c. Suppose that a pediatrician on the staff is found to be a nasal carrier of MRSA. How will you deal with this discovery?

7. Explain why pneumonia occurs most often in the elderly and the immunosuppressed.

8. How would an adult get gonococcal conjunctivitis?

9. You have been given the assignment of obtaining some type of culture that could help identify the cause of otitis media in a small child.
   a. What area might you sample?
   b. Why is this disease so common in children?

10. a. Name the three species of pyogenic cocci most commonly implicated in neonatal disease.
    b. Explain how infants might acquire such diseases.

11. Comment on the practice of kissing pets on the face or allowing them to lick or kiss one's face. What are some possible consequences of this?

12. a. For which pyogenic cocci is penicillin *not* a good choice for treatment?
    b. Itemize alternative drugs for pyogenic cocci and the indications for their use.

13. Explain this statement: The prevalence of gonorrhea is higher in women, but the incidence of gonorrhea is higher in men.

14. a. Explain why it is so important to take a detailed cardiovascular and infection history of dental patients.
    b. What is the policy on preventing complications of dental work in patients with heart disease?

15. Case study 1. An elderly man with influenza acquires a case of pneumonia. Gram-positive cocci isolated from his sputum give β-hemolysis on blood agar; the infection is very difficult to treat. Later it is shown that the man shared a room with a patient with bone infection. Isolates from both infections were the same.
    a. What is the probable species?
    b. Justify your answer.
    c. What treatment will be necessary?

16. Case study 2. A person with an inflamed cut on his head went to a physician, who then discovered that microbes had invaded the clot and spread into underlying tissues. Patchy areas developed around the lesion, leading to redness and edema. After a brief hospitalization, the patient died.
    a. What was the probable condition, and which species and virulence factors were involved?
    b. What should the treatment have been?

17. Case study 3. A child is brought to the emergency room in a semiconscious state with a high fever. Earlier he had complained of a stiff neck and headache. A tap of spinal fluid is performed and tested. A Gram stain reveals numerous gram-negative cocci or coccobacilli and WBCs, and a diagnosis of meningitis is made. Owing to the patient's serious condition, immediate drug therapy is necessary. Two species could cause this condition (see table A.8, Appendix D).
    a. Which drug would work on both infections?
    b. What are some rapid tests used to differentiate the two bacteria?

INTERNET SEARCH TOPICS

1. Search for information on new vaccines for staphylococci. Explain what you find about their mechanism and effectiveness.

2. Locate information on MRSA, VISA, VRE, and PPNG. What do the acronyms stand for and what is the current medical status of these drug-resistant cocci?

The Gram-Positive Bacilli of Medical Importance

In terms of sheer numbers, rod-shaped bacteria dominate the field of pathogenic bacteriology. Included among the diseases caused by bacilli are ancient, deadly, and fascinating ones such as bubonic plague, typhoid fever, tetanus, and leprosy, as well as newly emerging ones such as listeriosis and Legionnaires disease. In chapters 19 and 20, we will characterize and differentiate genera and medically important species using a simple modification of a system presented in *Bergey's Manual of Systematic Bacteriology*. This system divides the bacilli by Gram reaction and subgroups them by morphological characteristics and oxygen utilization (see figure E, page 540). Each section covers unique features of the pathogens and the epidemiology, pathology, and control measures of the principal diseases.

Chapter Overview

- Gram-positive bacilli account for a number of significant infectious diseases. General separation into groups can be done according to presence of spores, acid fastness, and cell morphology.
- The major genera of endospore-forming bacilli are *Bacillus* and *Clostridium*.
- Members of the genus *Bacillus* are primarily aerobic soil inhabitants that are noninfectious. The most important pathogen is *B. anthracis*, the cause of anthrax, a zoonotic disease of livestock.
- Members of the genus *Clostridium* are anaerobic inhabitants of soil, vegetation, and occasionally normal flora. Their pathogenesis is due to resistant spores and powerful exotoxins and enzymes.
- Clostridial diseases arising from wounds include tetanus, a neuromuscular disease, and gas gangrene, a soft-tissue infection that damages muscles.
- Clostridia are also implicated in food-borne illness, notably botulism, a food intoxication.
- *Listeria monocytogenes* and *Erysipelothrix rhusiopathiae* are non-spore-forming rods that can infect humans.
- *Listeria* is widespread throughout natural habitats and may be ingested in contaminated food. It causes a systemic infection, listeriosis, which can be very serious in babies and the elderly.
- *Erysipelothrix* causes a zoonosis in pigs and erysipeloid, a skin infection in humans.
- *Corynebacterium* species are gram-positive pleomorphic rods with palisades arrangement.
- The primary pathogen in the group is *Corynebacterium diphtheriae*, the agent of diphtheria, a toxin-induced disease of the throat, heart, and nervous system.

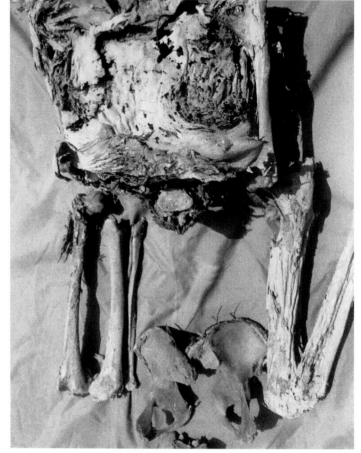

Ancient tuberculosis. A 900-year-old Peruvian mummy was tested for DNA from the tubercle bacillus by means of a highly sensitive PCR technique. Discovery of TB DNA laid to rest the theory that Columbus brought the disease to the Americas from Europe.

- Members of *Mycobacterium* are acid-fast bacilli with complex lipids in their cell walls.
- The most prominent pathogen worldwide is *Mycobacterium tuberculosis*, the cause of tuberculosis, a disease spread by respiratory droplets.
- Tuberculosis begins with a lung infection but can progress chronically to other organs. It is diagnosed by skin testing, X ray, and acid-fast staining.
- *Mycobacterium leprae* is an obligate parasite of humans that is weakly communicable through very close contact.
- The disease leprosy is a slow, progressive infection of skin and nerves that leads to nerve damage, deformities, and thickened nodules in the skin of the face and extremities.
- The gram-positive bacillary diseases for which there is effective vaccination are anthrax, tetanus, diphtheria, and tuberculosis.

Medically Important Gram-Positive Bacilli

The gram-positive bacilli can be subdivided into three general groups based on the presence or absence of endospores and the characteristic of acid-fastness. Further levels of separation correspond to oxygen requirements and cell morphology. This scheme can be organized as follows:

> **Endospore-forming bacilli**
> > Aerobic: *Bacillus*
> > Anaerobic: *Clostridium*
>
> **Non-endospore-forming bacilli**
> > Regular in morphology
> > > *Listeria*
> > > *Erysipelothrix*
> > Irregular in morphology
> > > Aerobic: *Corynebacterium*
> > > Anaerobic: *Propionibacterium*
>
> **Acid-fast bacilli**
> > *Mycobacterium*
> > *Nocardia*
>
> **Non-acid-fast branching filamentous bacilli**
> > Actinomyces

Gram-Positive Spore-Forming Bacilli

Most endospore-forming bacteria are gram-positive, motile, rod-shaped forms in the genera *Bacillus, Clostridium,* and *Sporolactobacillus.* An endospore is a dense survival unit that develops in a vegetative cell in response to nutrient deprivation (figure 19.1; see figure 4.21). The extreme resistance to heat, drying, radiation, and chemicals accounts for the survival, longevity, and ecological niche of sporeformers, and it is also relevant to their pathogenicity.

GENERAL CHARACTERISTICS OF THE GENUS *BACILLUS*

The genus *Bacillus* includes a large assembly of mostly saprobic bacteria widely distributed in the earth's habitats. *Bacillus* species are aerobic and catalase-positive, and, though they have varied nutritional requirements, none is fastidious. The group is noted for its versatility in degrading complex macromolecules, and it is also a common source of antibiotics. Because the primary habitat of many species is the soil, spores are continuously dispersed by means of dust into water and onto the bodies of plants and animals. Despite their ubiquity, the two species with primary medical importance are *B. anthracis,* the cause of anthrax, and *B. cereus,* the cause of one type of food poisoning.

Bacillus anthracis and Anthrax

Bacillus anthracis is among the largest of all bacterial pathogens, composed of block-shaped, angular nonmotile rods 3–5 μm long and 1–1.2 μm wide and central spores that develop under all

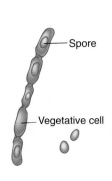
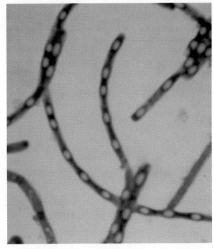

(a)

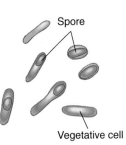
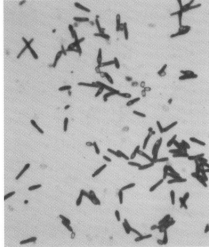

(b)

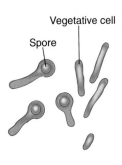
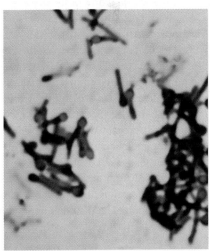

(c)

FIGURE 19.1

Examples of endospore-forming pathogens. **(a)** The morphological appearance of *Bacillus anthracis,* showing centrally placed endospores and a streptobacillus arrangement (600×). **(b)** A stain of *Clostridium perfringens* with central to terminal spores (410×). **(c)** *Cl. tetani* (710×). Its typical tennis racquet morphology is created by terminal spores that swell the sporangium.

growth conditions except in the living body of the host (figure 19.1*a*). Its virulence factors include a polypeptide capsule and exotoxins that in varying combinations produce edema and cell death. For centuries, **anthrax*** has been known as a zoonotic disease of herbivorous livestock (sheep, cattle, goats). It has an important place in the history of medical microbiology because it was Robert Koch's model for developing his postulates in 1877, and later, Louis Pasteur used the disease to prove the usefulness of vaccination.

The anthrax bacillus is a facultative parasite that undergoes its cycle of vegetative growth and sporulation in the soil. Animals become infected while grazing on grass contaminated with spores. When the pathogen is returned to the soil in animal excrement or carcasses, it can sporulate and become a long-term reservoir of infection for the animal population. The majority of anthrax cases are reported in livestock from Africa, Asia, and the Middle East. Most recent cases in the United States have occurred in textile workers handling imported animal hair or hide or products made from them. Because of effective control procedures, the number of cases in the United States is extremely low (fewer than 10 a year).

The circumstances of human infection depend upon the portal of entry. The most common and least dangerous of all forms is **cutaneous anthrax,** caused by spores entering the skin through small cuts and abrasions. Germination and growth of the pathogen in the skin are marked by the production of a papule that becomes increasingly necrotic and later ruptures to form a painless, black **eschar*** (figure 19.2). Handlers of raw wool or hides are at risk of acquiring **pulmonary anthrax** (woolsorter's disease) by inhaling airborne spores. Bacilli grow in the lungs and release exotoxins that produce toxemia with wide-ranging pathologic effects, including capillary thrombosis and cardiovascular shock. The development of septicemia can cause death in a few hours. So fatal is this form of infection that anthrax spores have been a serious choice for biological warfare weapons (Spotlight on Microbiology 19.1). Gastrointestinal anthrax acquired from contaminated meat is another rare but dangerous form of the disease.

Methods of Anthrax Control

Active cases of anthrax are treated with penicillin or tetracycline, but therapy does not lessen the effects of toxemia, so people can still die. A vaccine containing live spores and a toxoid prepared from a special strain of *B. anthracis* are used to protect livestock in areas of the world where anthrax is endemic. Humans should be vaccinated with the purified toxoid if they have occupational contact with livestock or products such as hides and bone or are members of the military. Effective vaccination requires six inoculations given over 1½ years, with yearly boosters. Animals that have died from anthrax must be burned or chemically decontaminated before burial to prevent establishing the microbe in the soil, and imported items containing animal hides, hair, and bone should be gas-sterilized.

Other *Bacillus* Species Involved in Human Disease

Bacillus cereus is a common airborne and dust-borne contaminant that multiplies very readily in cooked foods such as rice, potato, and meat dishes. The spores survive short periods of cooking and

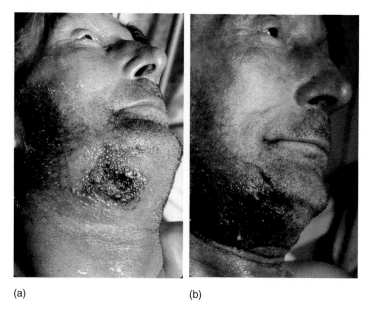

(a)   (b)

FIGURE 19.2

Cutaneous anthrax. **(a)** In early stages, the tissue around the site of invasion is inflamed and edematous. **(b)** A later stage reveals a thick, necrotic lesion called the eschar. It usually separates and heals spontaneously.

reheating; when the food is stored at room temperature, the spores germinate and release enterotoxins. Ingestion of toxin-containing food causes nausea, vomiting, abdominal cramps, and diarrhea. There is no specific treatment, and the symptoms usually disappear within 24 hours.

For many years, most common airborne *Bacillus* species were dismissed as harmless contaminants with weak to nonexistent pathogenicity. However, infections by these species are increasingly reported in immunosuppressed and intubated patients and in drug addicts who do not use sterile needles and syringes. An important contributing factor is that spores are abundant in the environment and the usual methods of disinfection and antisepsis are powerless to control them.

THE GENUS *CLOSTRIDIUM*

Another genus of gram-positive, spore-forming rods that is widely distributed in nature is ***Clostridium.**** It is differentiated from *Bacillus* on the basis of being anaerobic and catalase-negative. The large genus (over 120 species) is extremely varied in its habitats. Saprobic members reside in soil, sewage, vegetation, and organic debris, and commensals inhabit the bodies of humans and other animals. Infections caused by pathogenic species are not normally communicable but occur when spores are introduced into injured skin.

Clostridia cells produce oval or spherical spores that often swell the vegetative cell (see figure 19.1*b,c*). Spores are produced only under anaerobic conditions. Their nutrient requirements are complex, and they can decompose a variety of substrates. They can also synthesize organic acids, alcohols, and other solvents through fermentation. This capacity makes some clostridial species essential tools of the biotechnology industry (see chapter 26). Other

---

* anthrax (an′-thraks) Gr. *anthrax,* carbuncle.

* eschar (ess′-kar) Gr. *eschara,* scab.

* *Clostridium* (klaw-strid′-ee-um) Gr. *closter,* spindle.

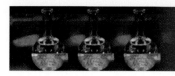

SPOTLIGHT ON MICROBIOLOGY   19.1
A Bacillus That Could Kill Us

Mechanisms of biological warfare are frequently based on the pathogenic potential of microorganisms. Because spore-forming bacteria such as *Bacillus* are so hardy, they have held a particular attraction to scientists as research subjects and, in one case, as the basis of a deadly bomb.

Late in World War II, the mounting outrage over German bombings of London and the pressure for retaliation forced the British and American governments to collaborate on a powerful means of striking back. Given the knowledge that pneumonic anthrax was 100% fatal in a short time period, a highly secret council designed a bomb containing the spores of *Bacillus anthracis.* In the original plan, the detonation of these deadly projectiles over six German cities would have caused a shower of spores to be inhaled by humans and animals. Some estimates predicted that this could have wiped out the population over hundreds of square miles. Late in the war, a plant in Indiana actually began to manufacture the bombs, but by this time, the war had started to wind down. How close the Allies came to dropping these bombs was not disclosed.

During the Cold War, the U.S. Army's Special Operations Division conducted a chilling experiment using *Bacillus subtilis.* The plan was to simulate a possible method of germ warfare by releasing spore aerosols into the air of a crowded Washington, D.C. airport and bus terminal. Agents carried suitcases to spray the air while hundreds of unsuspecting people inhaled "large and acceptably uniform" doses, presumably with no ill effects.

This study and several others like it were based on the disturbing assumption that *B. subtilis* is a harmless air contaminant. It should be clear even to beginning students that many species of bacteria are opportunists and are capable of causing disease, especially if present in large numbers. In fact, *B. subtilis* can cause lung and blood infections in people with weakened immunity. Although flawed, the test did fulfill its original purpose. It was determined that if smallpox virus had been used instead of *B. subtilis,* 300 people in 93 cities would have become infected and that similar exposures at terminals in several other large cities would have spread the disease rapidly to thousands of others. It was also shown that this contamination procedure could be accomplished without detection.

The potential threat from biological warfare is not only a concern of the past. Recent reports state that the governments of Iraq and North Korea have stockpiled huge arsenals of germ agents, including anthrax spores and botulinum toxin (another product of a gram-positive spore-former). There is a fear that these agents could be incorporated into bombs or missiles to be spread over long distances. International agencies are attempting to monitor and force elimination of the deadly agents. All U.S. military personnel are vaccinated to protect them against the use of biowarfare agents.

---

extracellular products, primarily exotoxins, play an important role in various clostridial diseases such as botulism and tetanus.

The Role of Clostridia in Infection and Disease

Clostridial disease can be divided into (1) wound and tissue infections, including myonecrosis, antibiotic-associated colitis, and tetanus, and (2) food intoxication of the perfringens and botulism varieties. Most of these diseases are caused by the release of potent exotoxins that act on specific cellular targets (Medical Microfile 19.2).

Gas Gangrene

The majority of clostridial soft tissue and wound infections are caused by ***Clostridium perfringens,*** *Cl. novyi,* and *Cl. septicum.* The spores of these species can be found in soil, on human skin, and in the human intestine and vagina. The disease they cause has the common name **gas gangrene*** in reference to the gas produced by the bacteria growing in the tissue. It is technically termed anaerobic cellulitis or *myonecrosis.** The conditions that predispose a person to gangrene are surgical incisions, compound fractures, diabetic ulcers, septic abortions, puncture and gunshot wounds, and crushing injuries contaminated by spores from the body or the environment.

Because clostridia are not highly invasive, infection requires damaged or dead tissue that supplies growth factors and an anaerobic environment. The low oxygen tension results from an interrupted blood supply and the presence of aerobic bacteria that deplete oxygen. Such conditions stimulate spore germination, rapid vegetative growth in the dead tissue, and release of exotoxins. *Clostridium perfringens* produces several physiologically active toxins; the most potent one, *alpha toxin* (lecithinase c), causes red blood cell rupture, edema, and tissue destruction (figure 19.3). Additional virulence factors that enhance tissue destruction are collagenase, hyaluronidase, and DNase. The gas formed in tissues, due to fermentation of muscle carbohydrates, can also destroy muscle structure.

Extent and Symptoms of Infection   Two forms of gas gangrene have been identified. In **anaerobic cellulitis,** the bacteria spread within damaged necrotic muscle tissue, producing toxin and gas, but the infection remains localized and does not spread into healthy tissue. The pathology of true **myonecrosis** is more destructive. Toxins produced in large muscles, such as the thigh, shoulder, and buttocks, diffuse into nearby healthy tissue and cause local necrosis there. This damaged tissue then serves as a focus for continued clostridial growth, toxin formation, and gas production. The disease can progress through an entire limb or body area, destroying tissues as it goes (figure 19.4). Initial symptoms of pain, edema, and a bloody exudate in the lesion are followed by fever,

---

* gas gangrene (gang′-green) Gr. *gangraina,* an eating sore. A necrotic condition associated with the release of gases.

* myonecrosis (my″-oh-neh-kro′sis) Gr. *myo,* muscle, and *necros,* dead.

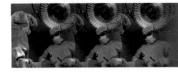

MEDICAL MICROFILE 19.2
Toxicity in the Extreme

Imagine a substance so poisonous that a single microgram of it is a mass lethal dose for 200,000 mice, and a cup of it in concentrated form could quite possibly kill all the humans on the planet. This toxin is 100,000 times more powerful than rattlesnake venom and a million times more potent than strychnine. Although this formidable chemical might sound like the science fiction creation of a mad scientist, it actually comes from the common soil bacterium *Clostridium botulinum.* Curiously, the genus *Clostridium* is the source of two of the most deadly toxins known, botulin and tetanospasmin (from *Cl. tetani*). These two toxins share several other characteristics. For instance, both are polypeptides coded by plasmids and acquired through transduction by a virus. Both are neurotoxins that act on nerve cell processes involved in muscular activity, and both cause loss of voluntary muscle control (paralysis).

But here the similarities end. The diseases these two toxins cause are acquired through generally different patterns. In tetanus, the toxin is formed in the tissues, and in the prominent form of botulism, the toxin is formed in food and then ingested. The mechanism by which the two toxins cause paralysis is also different. Tetanospasmin blocks the activity of inhibitory neurons in the central nervous system, and botulin acts on mononeurons in the peripheral nervous system. Tetanospasmin causes spastic (rigid) paralysis due to hyperactive muscles and sustained contraction (see figure 19.7). In contrast, botulin causes flaccid paralysis by interfering with the transmission of impulses to muscles and blocking contraction (see figure 19.9).

Despite the extreme toxicity of botulin, a form of it called botox has been approved as a drug for treating a variety of muscle ailments. It can be used in extreme dilutions to relax muscles that are overactive. It has been used to successfully treat cross- or wall-eye, blepharospasm, wry neck, spastic vocal cords, and various types of tremors. It minimizes the spasms and pain and improves normal muscle function when administered every 3 to 4 months. Botox is even being injected into facial muscles to control frown creases and wrinkles on the forehead.

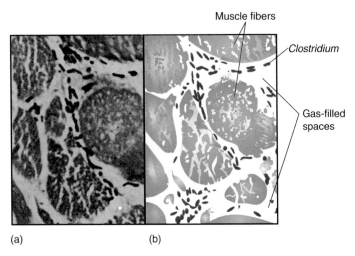

(a)   (b)

FIGURE 19.3

Growth of *Clostridium perfringens* (plump rods), causing gas formation and separation of the fibers.   **(a)** A microscopic analysis of clostridial myonecrosis, showing a histological section of gangrenous skeletal muscle. **(b)** A schematic drawing of the same section.

(a) *From N.A. Boyd et al.,* Journal of Medical Microbiology, *5:459, 1972. Reprinted by permission of Longman Group, Ltd.*

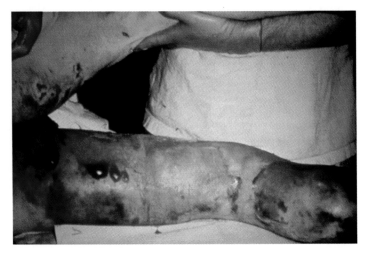

FIGURE 19.4

The clinical appearance of myonecrosis in a compound fracture of the leg.   Necrosis has traveled from the main site of the break to other areas of the leg. Note the severe degree of involvement, with blackening, general tissue destruction, and bubbles on the skin caused by gas formation in underlying tissue.

tachycardia, and blackened necrotic tissue filled with bubbles of gas. Gangrenous infections of the uterus due to septic abortions and clostridial septicemia are particularly serious complications. If treatment is not indicated early, the disease is invariably fatal.

Treatment and Prevention of Gangrene   One of the most effective ways to prevent clostridial wound infections is immediate and rigorous cleansing and surgical repair of deep wounds, decubi-

tus ulcers (bedsores), compound fractures, and infected incisions. *Debridement** of diseased tissue eliminates the conditions that promote the spread of gangrenous infection. This is most difficult in the intestine or body cavity, where only limited amounts of tissue can be removed. Surgery is supplemented by large doses of a broad-spectrum cephalosporin (cefoxitin) or penicillin to control

---

* debridement (dih-breed′-ment) Surgical removal of dead or damaged tissue.

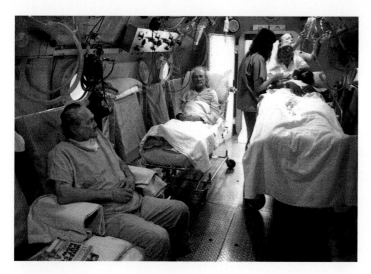

FIGURE 19.5

Decompression treatment for wound infections and diabetic ulcers. This hyperbaric therapy involves holding patients in a chamber that exposes them to higher levels of oxygen to inhibit anaerobic infection and to promote healing.

Los Angeles Times, *December 6, 1993, Metro Section, p. 1. Photographer Rolando Otero.*

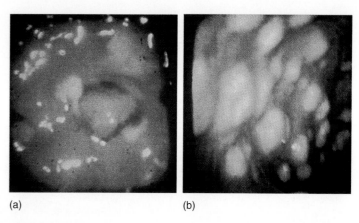

(a)   (b)

FIGURE 19.6

Antibiotic-associated colitis. (a) A mild form with diffuse, inflammatory patches. **(b)** Heavy yellow plaques, or pseudomembranes, typical of more severe cases. Photographs were made by a sigmoidoscope, an instrument capable of photographing the interior of the colon.

infection. Hyperbaric oxygen therapy, in which the affected part is exposed to an increased oxygen gas tension in a pressurized chamber, can also lessen the severity of infection (figure 19.5). The increased oxygen content of the tissues presumably blocks further bacterial multiplication and toxin production. Extensive myonecrosis of a limb may call for surgical removal, or amputation. Because there are so many different antigen subtypes in this group, active immunization is not possible.

Antibiotic-Associated Colitis

A nosocomial infection called **antibiotic-associated** (or pseudomembranous) **colitis** is the second most common intestinal infection after salmonellosis in industrialized countries. The disease is caused by *Clostridium difficile,* a minor but normal resident of the intestine that was once considered relatively harmless. In most instances, this infection is traced to therapy with broad-spectrum antibiotics such as ampicillin, clindamycin, and cephalosporins, and it is a major cause of diarrhea in hospitals. Although *Cl. difficile* is relatively noninvasive, it is able to superinfect the large intestine when the drugs have disrupted the normal flora. It produces enterotoxins that cause areas of necrosis in the wall of the intestine. The predominant symptom is diarrhea commencing late in therapy or even after therapy has stopped. More severe cases exhibit abdominal cramps, fever, and leukocytosis. The colon is inflamed and gradually sloughs off loose, membranelike patches called pseudomembranes consisting of fibrin and cells (figure 19.6). If the condition is not arrested, cecal perforation and death can result.

Mild, uncomplicated cases respond to withdrawal of antibiotics and replacement therapy for lost fluids and electrolytes. More severe infections are treated with oral vancomycin or metronidazole for several weeks until the intestinal flora returns to normal. Because infected persons often shed large numbers of spores in their stools, increased precautions are necessary to prevent spread of the agent to other patients who may be on antimicrobic therapy. Some new techniques on the horizon are vaccination with *Cl. difficile* toxoid and restoration of normal flora by means of a mixed culture of lactobacilli and yeasts.

Tetanus, or Lockjaw

Tetanus* is a neuromuscular disease whose alternate name, **lockjaw,** refers to an early effect of the disease on the jaw muscle. The etiologic agent, *Clostridium tetani,* is a common resident of cultivated soil and the gastrointestinal tracts of animals. Spores usually enter the body through accidental puncture wounds, burns, umbilical stumps, frostbite, and crushed body parts.

The incidence of tetanus is low in North America. Most cases occur among geriatric patients and intravenous drug abusers. The incidence of neonatal tetanus—predominantly the result of an infected umbilical stump or circumcision—is higher in cultures that apply dung, ashes, and mud to these sites to arrest bleeding or as a customary ritual. The disease accounts for several hundred thousand infant deaths a year.

The Course of Infection and Disease The risk for tetanus occurs when spores of *Clostridium tetani* are forced into injured tissue. But the mere presence of spores in a wound is not sufficient to initiate infection because the bacterium is unable to invade damaged tissues readily. It is also a strict anaerobe, and the spores cannot become established unless tissues at the site of the wound are necrotic and poorly supplied with blood, conditions that favor germination.

As the vegetative cells grow, various metabolic products are released into the infection site. Of these, the most serious is **tetanospasmin,** a potent neurotoxin that accounts for the major symptoms of tetanus. The toxin spreads to nearby motor nerve

---

* tetanus (tet′-ah-nus) Gr. *tetanos,* to stretch.

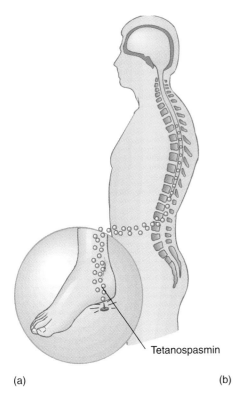

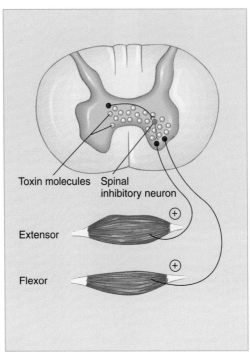

(a)   (b)   (c)

FIGURE 19.7

The events in tetanus.   **(a)** After traumatic injury, bacilli infecting the local tissues secrete tetanospasmin, which is absorbed by the peripheral axons and is carried to the target neurons in the spinal column. **(b)** In the spinal cord, the toxin attaches to the junctions of regulatory neurons that inhibit inappropriate contraction. Released from inhibition, the muscles, even opposing members of a muscle group, receive constant stimuli and contract uncontrollably. **(c)** Muscles contract spasmodically, without regard to regulatory mechanisms or conscious control. Note the clenched jaw typical of risus sardonicus.

endings in the injured tissue, binds to them, and travels by axons to the ventral horns of the spinal cord (figure 19.7). In the spinal column, the toxin binds to specific target sites on the spinal neurons that are responsible for inhibiting skeletal muscle contraction. The toxin inhibits the release of neurotransmitter, and only a small amount is required to initiate the symptoms. The incubation period varies from 4 to 10 days, and shorter incubation periods signify a more serious condition.

Tetanospasmin alters the usual regulation mechanisms for muscle contraction. As a result, the muscles are released from normal inhibition and begin to contract uncontrollably. Powerful muscle groups are most affected, and the first symptoms are clenching of the jaw, followed in succession by extreme arching of the back, flexion of the arms, and extension of the legs (figure 19.8). Lockjaw confers the bizarre appearance of *risus sardonicus* (sarcastic grin) that looks eerily as though the person is smiling (see figure 19.7c). These contractions are intermittent and extremely painful and may be forceful enough to break bones, especially the vertebrae. Death is most often due to paralysis of the respiratory muscles and respiratory collapse. The fatality rate, ranging from 10% to 70%, is highest in cases involving delayed medical attention, a short incubation time, or head wounds. Full recovery requires a few weeks, and other than transient stiffness, no permanent damage to the muscles usually remains.

Treatment and Prevention of Tetanus   Tetanus treatment is aimed at deterring the degree of toxemia and infection and maintaining patient homeostasis. A patient with a clinical appearance suggestive of tetanus should immediately receive antitoxin therapy with human tetanus immune globulin (TIG). Tetanus antitoxin (TAT) from horses may be used, but it is less acceptable because of possible allergic reactions. Although the antitoxin inactivates circulating toxin, it will not counteract the effects of toxin already bound to neurons. Other methods include thoroughly cleansing and removing the afflicted tissue, controlling infection with penicillin or tetracycline, and administering muscle relaxants. The patient may require the assistance of a respirator, and a tracheostomy[1] is sometimes performed to prevent respiratory complications such as aspiration pneumonia or lung collapse.

Tetanus is one of the world's most preventable diseases, chiefly because of an effective vaccine containing tetanus toxoid. During World War II, only 12 cases of tetanus occurred among 2,750,000 wounded soldiers who had been previously vaccinated. The recommended vaccination series for 1- to 3-month-old babies consists of three injections given 2 months apart, followed by booster doses about 1 and 4 years later. Children thus immunized

---

1. The surgical formation of an air passage by perforation of the trachea.

FIGURE 19.8

🔵 **Neonatal tetanus.** Baby with neonatal tetanus, showing spastic paralysis of the paravertebral muscles, which locks the back into a rigid, arched position. Also note the abnormal flexion of the arms and legs.

probably have protection for 10 years. Additional protection against neonatal tetanus may be achieved by vaccinating pregnant women, whose antibodies will be passed to the fetus. Toxoid should also be given to injured persons who have never been immunized, have not completed the series, or whose last booster was received more than 10 years previously. The vaccine can be given simultaneously with passive TIG immunization to achieve immediate and long-term protection.

Clostridial Food Poisoning

Two *Clostridium* species are involved in food poisoning. ***Clostridium perfringens,*** type A, accounts for a mild illness that is the second most common form of food poisoning worldwide, whereas ***Clostridium botulinum*** produces a rarer but more serious intoxication.

Clostridial Gastroenteritis *Clostridium perfringens* spores contaminate many kinds of food, but those most frequently involved in disease are animal flesh (meat, fish) and vegetables (beans) that have not been cooked thoroughly enough to destroy the spores. When these foods are cooled, spores germinate, and the germinated cells multiply, especially if the food is left unrefrigerated. If the food is eaten without adequate reheating, live *Cl. perfringens* cells enter the small intestine and release enterotoxin. The toxin, acting upon epithelial cells, initiates acute abdominal pain, diarrhea, and nausea in 8 to 16 hours. Recovery is rapid, and deaths are extremely rare. *Clostridium perfringens* also causes an enterocolitis infection similar to that caused by *Cl. difficile.* This infectious type of diarrhea is acquired from contaminated food, or it may be transmissible by inanimate objects.

Botulinum Food Poisoning **Botulism*** is an intoxication associated with eating poorly preserved foods, though it can also occur

---

* **botulism** (boch'-oo-lizm) L. *botulis*, sausage. The disease was originally linked to spoiled sausage.

as an infection. Until recent times, it was relatively prevalent and commonly fatal, but modern techniques of food preservation and medical treatment have reduced both its incidence and its fatality rate. However, botulism is a common cause of death in livestock that have grazed on contaminated food and in aquatic birds that have eaten decayed vegetation.

Clostridium botulinum is a spore-forming anaerobe that commonly inhabits soil and water and, occasionally, the intestinal tract of animals. It is distributed worldwide, but occurs most often in the Northern Hemisphere. The species has eight distinctly different types (designated A, B, C_α and C_β, D, E, F, and G), which vary in distribution among animals, regions of the world, and type of exotoxin. Human disease is usually associated with types A, B, E, and F, and animal disease with types A, B, C, D, and E.

The Route of Pathogenesis There is a high correlation between cultural dietary preferences and food-borne botulism. In the United States, the disease is often associated with low-acid vegetables (green beans, corn), fruits, and occasionally meats, fish, and dairy products. Most botulism outbreaks occur in home-processed foods, including canned vegetables, smoked meats, and cheese spreads. The demand for prepackaged convenience foods such as vacuum-packed cooked vegetables and meats has created a new source of risk, but most commercially canned foods are held to very high standards of preservation and are only rarely a source of botulism.

The factors in food processing that lead to botulism are dependent upon several circumstances. Spores are present on the vegetables or meat at the time of gathering and are difficult to remove completely. When contaminated food is bottled and steamed in a pressure cooker that does not reach reliable pressure and temperature, some spores survive (botulinum spores are highly heat-resistant). At the same time, the pressure is sufficient to evacuate the air and create anaerobic conditions. Storage of the bottles at room temperature favors spore germination and vegetative growth. One of the products of metabolism is **botulin,** the most potent microbial toxin known (see Medical Microfile 19.2).

Bacterial growth may not be evident in the appearance of the bottle or can or in the food's taste or texture, and only minute amounts of toxin may be present. Swallowed toxin enters the small intestine and is absorbed into the lymphatics and circulation. From there, it travels to its principal site of action, the neuromuscular junctions of skeletal muscles (figure 19.9). The effect of botulin is to prevent the release of the neurotransmitter substance, acetylcholine, that initiates the signal for muscle contraction. The usual time before onset of symptoms is, 12 to 72 hours, depending on the size of the dose. Neuromuscular symptoms first affect the muscles of the head and include double vision, difficulty in swallowing, and dizziness, but there is no sensory or mental lapse. Although nausea and vomiting can occur at an early stage, they are not common. Later symptoms are descending muscular paralysis and respiratory compromise. In the past, death resulted from stoppage of respiration, but mechanical respirators have reduced the fatality rate to about 10%.

Infant and Wound Botulism In rare instances, *Cl. botulinum* causes infection and toxemia. In these special circumstances, called infant and wound botulism, the spores germinate in the body and produce an infection.

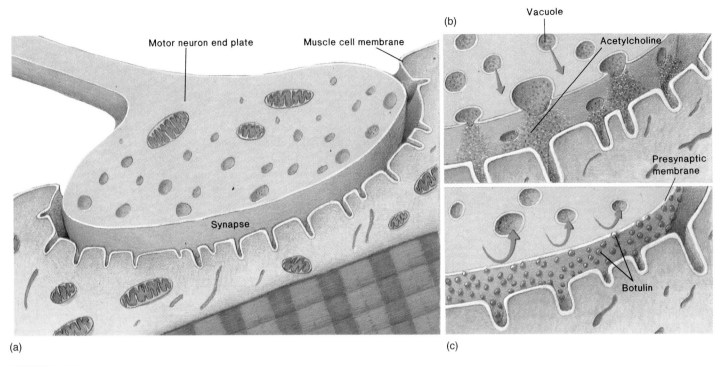

(a)

(b)

(c)

FIGURE 19.9

🔊 **The physiological effects of botulism toxin (botulin).** **(a)** The relationship between the motor neuron and the muscle at the neuromuscular junction. **(b)** In the normal state, acetylcholine released at the synapse crosses to the muscle and creates an impulse that stimulates muscle contraction. **(c)** In botulism, the toxin enters the motor end plate and attaches to the presynaptic membrane, where it blocks release of the transmitter, prevents impulse transmission, and keeps the muscle from contracting.

Infant botulism was first described in the late 1970s in children between the ages of 2 weeks and 6 months who had ingested spores. It is currently the most common type of botulism in the United States, with approximately 80 to 100 cases reported annually. The exact food source is not always known, although raw honey has been implicated in some cases, and the spores are common in dust and soil. Apparently, the immature state of the neonatal intestine and microbial flora allows the spores to gain a foothold, germinate, and give off neurotoxin. As in adults, babies exhibit flaccid paralysis, usually with a weak sucking response, generalized loss of tone (the "floppy baby syndrome"), and respiratory complications. Although adults can also ingest botulinum spores in contaminated vegetables and other foods, the adult intestinal tract normally inhibits this sort of infection.

In **wound botulism,** the spores enter a wound or puncture much as in tetanus, but the symptoms are similar to those of foodborne botulism. Increased cases of this form of botulism are being reported in injecting drug users. The rate of infection is highest in people who inject black tar heroin into the skin.

Treatment and Prevention of Botulism Differentiating botulism from other neuromuscular conditions requires testing the food samples, sampling the patient's blood, and pumping the gastrointestinal tract. The Centers for Disease Control and Prevention provides a source of type A, B, and E trivalent horse antitoxins that must be administered early for greatest effectiveness. Patients are also managed with respiratory and cardiac support systems. Infectious botulism is treated with penicillin to control the microbe's growth and toxin production.

Botulism will always be a potential threat for people who consume home-preserved foods. Preventing it depends on educating the public about the proper methods of preserving and handling canned foods. Pressure cookers should be tested for accuracy in sterilizing, and home canners should be aware of the types of food and conditions likely to cause botulism. Although acidic foods (tomatoes, fruits) are traditionally thought to inhibit the microbe, recent findings indicate that acid content alone may not be sufficient to prevent bacterial growth and toxin production. Other effective preventives include addition of preservatives such as sodium nitrite, salt, or vinegar. Bulging cans or bottles that look or smell spoiled should be discarded, and all home-bottled foods should be boiled for 10 minutes before eating, because the toxin is heat-sensitive and is rapidly inactivated at 100°C. Military researchers are in the process of testing a genetically engineered toxoid that could protect against all strains of the pathogen. It is likely that this will be included in the routine immunizations given to armed service personnel.

Differential Diagnosis of Clostridial Species

Although clostridia are common isolates, their clinical significance is not always immediately evident. Diagnosis frequently depends on the microbial load, the persistence of the isolate on resampling, and the condition of the patient. Laboratory differentiation relies on testing morphological and cultural characteristics, exoenzymes, carbohydrate fermentation, reaction in milk, and toxin production and pathogenicity. Some laboratories use a sophisticated method of gas chromatography that analyzes the chemical differences among species. Other valuable procedures are direct ELISA testing of isolates, toxicity testing in mice or guinea pigs, and serotyping with antitoxin neutralization tests.

Gram-positive bacilli that form endospores are common in soil and water environments. The genera *Bacillus* and *Clostridium* include highly toxigenic and virulent pathogens. *Cl. tetani* and *Cl. botulinum* produce the most deadly toxins known. *B. anthracis* toxin is also deadly in systemic infections. Microbial control of these pathogens must include methods that destroy their highly resistant spores.

Species in the genus *Bacillus* are all aerobic, although a few are facultative anaerobes. Most are also motile. Significant pathogens are *B. anthracis*, the agent of anthrax, a zoonosis of domestic animals, and *B. cereus*, the agent of food poisoning. Important commercial species of *Bacillus* produce antibiotics.

Species in the genus *Clostridium* are strict anaerobes. Significant pathogens are *Cl. tetani*, the agent of tetanus, *Cl. botulinum*, the agent of botulism, *Cl. perfringens* and other species, the cause of gas gangrene and myonecrosis, and *Cl. difficile*, an opportunist that infects the intestine after antibiotic therapy. Commercially important clostridia are industrial producers of organic acids and alcohols.

Gram-Positive Regular Non-Spore-Forming Bacilli

The non-spore-forming gram-positive bacilli are a mixed group of genera subdivided on the basis of morphology and staining characteristics. One loose aggregate of seven genera is characterized as **regular** because they stain uniformly and do not assume pleomorphic shapes. Regular genera include *Lactobacillus, Listeria, Erysipelothrix, Kurthia, Caryophanon, Bronchothrix,* and *Renibacterium. Lactobacillus* is widely distributed in the environment and is a common resident of the intestinal tract and vagina of humans. Some species are also important in processing dairy products. Only rarely are they pathogens. The most significant pathogens in this group are *Listeria monocytogenes* and *Erysipelothrix rhusiopathiae*.

AN EMERGING FOOD-BORNE PATHOGEN: *LISTERIA MONOCYTOGENES*

Listeria monocytogenes* ranges in morphology from coccobacilli to long filaments in palisades formation (table 19.1). Cells show tumbling with one to four flagella and do not produce capsules or spores. *Listeria* is not fastidious and is resistant to cold, heat, salt, pH extremes, and bile.

Epidemiology and Pathology of Listeriosis

The distribution of *L. monocytogenes* is so broad that its reservoir has been difficult to determine. It has been isolated all over the world from water, soil, plant materials, and the intestines of healthy mammals (including humans), birds, fish, and invertebrates. Apparently, the primary reservoir is soil and water, while animals, plants, and food are secondary sources of infection. Most cases of **listeriosis** are associated with ingesting contaminated dairy products, poultry, and meat. Recent epidemics have spurred an in-depth investigation into the prevalence of *L. monocytogenes* in these sources. The pathogen has been isolated in 12% of ground beef and in 25–30% of chicken and turkey carcasses. It was also present in 6% of luncheon meats, hot dogs, and cheeses. Aged cheeses made from raw milk are of special concern because *Listeria* readily survives through long storage and can grow during refrigeration.

Except in cases of pregnancy, human-to-human transmission is probably not a significant factor. A predisposing factor in listeriosis seems to be the weakened condition of host defenses in the intestinal mucosa, since studies have shown that immunocompetent individuals are rather resistant to infection.

---

* *Listeria monocytogenes* (lis-ter′-ee-ah) For Joseph Lister, the English surgeon who pioneered antiseptic surgery; (mah″-noh-sy-toj′-uh-neez) For its effect on monocytes.

TABLE 19.1

Characteristics for Differentiating *L. monocytogenes* from Similar Gram-Positive Bacteria

	Shape	Arrangement	Motility (20°–25°C)	Catalase	Morphology
Listeria monocytogenes	Rods, coccobacilli	Single, in chains	+	+	
Listeria spp.	Rods, coccobacilli	Single, in chains	+	+	
Streptococcus (esp. group D)	Cocci (spheres, ovals)	Pairs, in chains	−	−	
Corynebacterium	Pleomorphic rods	Palisades, single	−	+	
Lactobacillus	Straight rods	Single, in chains	−	−	
Erysipelothrix	Long, slender rods	Filaments	−	−	

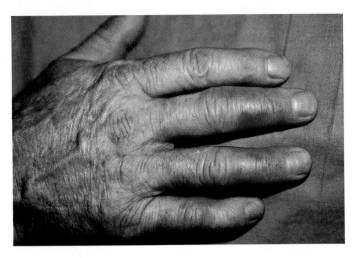

FIGURE 19.10
Erysipeloid on the hand of an animal handler.

Listeriosis in normal adults is often a mild or subclinical infection with nonspecific symptoms of fever, diarrhea, and sore throat. However, listeriosis in immunocompromised patients, fetuses, and neonates usually affects the brain and meninges and results in septicemia. The death rate is around 20% due to the pathogen's high virulence. Pregnant women are especially susceptible to infection, which is transmitted to the infant prenatally when the microbe crosses the placenta or postnatally through the birth canal. Intrauterine infections are widely systemic and usually result in premature abortion and fetal death. Neonatal infections that localize in the meninges cause extensive damage to the nervous system if not treated at an early stage.

Diagnosis and Control of Listeriosis

Diagnosing listeriosis is hampered by the difficulty in isolating it. However, the chances of isolation can be improved by using a procedure called cold enrichment, in which the specimen is held at 4°C and periodically plated onto media, but this procedure can take 4 weeks. *Listeria monocytogenes* can be differentiated from the nonpathogenic *Listeria* and from other bacteria to which it bears a superficial resemblance by characteristics listed in table 19.1. Rapid diagnostic kits using ELISA, immunofluorescence, and gene probe technology are now available for direct testing of dairy products and cultures. Antibiotic therapy should be started as soon as listeriosis is suspected. Ampicillin and trimethoprimsulfamethoxazole are the first choices, followed by erythromycin. Prevention can be improved by adequate pasteurization temperatures and by cooking foods that are suspected of being contaminated with animal manure or sewage. The U.S. government is proposing that all food processors be required to test for the pathogen on their equipment by 2002.

ERYSIPELOTHRIX RHUSIOPATHIAE: A ZOONOTIC PATHOGEN

Epidemiology, Pathogenesis, and Control

Erysipelothrix rhusiopathiae* is a gram-positive rod widely distributed in animals and the environment. Its primary reservoir appears to be the tonsils of healthy pigs. It is also a normal flora of other vertebrates and is commonly isolated from sheep, chickens, and fish. It can persist for long periods in sewage, seawater, soils, and foods. The pathogen causes epidemics of swine erysipelas and sporadic infections in other domestic and wild animals. Humans at greatest risk for infection are those who handle animals, carcasses, and meats, such as slaughterhouse workers, butchers, veterinarians, farmers, and fishermen.

The common portal of entry in human infections is a scratch or abrasion on the hand or arm. The microbe multiplies at the invasion site to produce a disease known as *erysipeloid*, characterized by swollen, inflamed, dark red lesions that burn and itch (figure 19.10). Although the lesions usually heal without complications, rare cases of septicemia and endocarditis do arise. Inflamed red sores on the hands of people in high-risk occupations suggest erysipeloid, but the lesions must be cultured for confirmatory diagnosis. The condition is treated with penicillin or erythromycin. Swine erysipelas can be prevented by vaccinating pigs, but the vaccine does not protect humans. Animal handlers can lower their risk by wearing protective gloves.

CHAPTER CHECKPOINTS

Regular, non-spore-forming gram-positive bacilli are so named because of their consistent shape and staining properties. The seven genera of this group are otherwise diverse in habitat, biochemical properties, and size. Significant pathogens in this group are *Listeria monocytogenes*, the agent of listeriosis, and *Erysipelothrix rhusiopathiae*, the agent of erysipeloid in both humans and animals. *Lactobacillus* species are important commercial producers of many dairy products.

Gram-Positive Irregular Non-Spore-Forming Bacilli

The **irregular,** non-spore-forming bacilli tend to be pleomorphic and to stain unevenly. Of the 20 genera in this category, *Corynebacterium, Mycobacterium,* and *Nocardia* have the greatest clinical significance. These three genera are grouped together because of similar morphological, genetic, and biochemical traits. They produce catalase and possess mycolic acids and a unique type of peptidoglycan in the cell wall. The following sections discuss the primary diseases associated with *Corynebacterium,* a similar genus called *Propionibacterium, Mycobacterium, Actinomyces,* and *Nocardia.*

---

* *Erysipelothrix rhusiopathiae* (er″-ih-sip′-eh-loh-thriks) Gr. *erythros,* red, *pella,* skin, and *thrix,* filament. (ruz″-ee-oh-path′-ee-eye) Gr. *rhusios,* red, and *pathos,* disease.

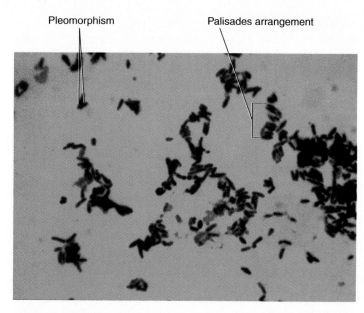

Pleomorphism   Palisades arrangement

FIGURE 19.11

Corynebacterium diphtheriae. A photomicrograph of *Corynebacterium diphtheriae,* showing pleomorphism (especially club forms), metachromatic granules, and palisades arrangement (600×).

occasionally with fomites or contaminated milk. The clinical disease proceeds in two stages: (1) local infection by *Corynebacterium diphtheriae* and (2) toxin production and toxemia. The most common location of primary infection is in the upper respiratory tract (tonsils, pharynx, larynx, and trachea). *Cutaneous diphtheria* is usually a secondary infection manifesting as deep, erosive ulcers that are slow to heal. The bacterium becomes established by means of virulence factors that assist in its attachment and growth. The cells are not ordinarily invasive and usually remain localized at the portal of entry. This form of the disease is on the rise in the United States.

Diphtherotoxin and Toxemia Although infection is necessary for disease, the cardinal determinant of pathogenicity is the production of **diphtherotoxin.** According to studies, this exotoxin is produced only by toxigenic strains of *C. diphtheriae* that carry the structural gene for toxin production acquired from bacteriophages during transduction (see chapter 9). This cytotoxin consists of two polypeptide fragments. Fragment B binds to and is endocytosed by mammalian target cells in the heart and nervous system. Fragment A interacts metabolically with factors in the cytoplasm and arrests protein synthesis.

CORYNEBACTERIUM DIPHTHERIAE

Although several species of *Corynebacterium** are important, most human disease is associated with **C. diphtheriae.** In general morphology, this bacterium is a straight or somewhat curved rod that tapers at the ends, but thin spots in the cell wall cause it to develop pleomorphic club, filamentous, and swollen shapes. Older cells are filled with metachromatic (polyphosphate) granules and can occur in side-by-side palisades arrangement (figure 19.11).

Epidemiology of Diphtheria

For hundreds of years, **diphtheria** was a significant cause of morbidity and mortality, but in the last 50 years, both the number of cases and the fatality rate have steadily declined throughout the world. The current rate for the entire United States is 0.01 cases per million population (figure 19.12). Recent outbreaks of the cutaneous form of the disease have occurred among Native American populations and the homeless. Because many populations harbor a reservoir of healthy carriers, the potential for diphtheria is constantly present. In the mid-1990s, an epidemic occurred in areas of the former Soviet Union, where reduced vaccination programs have put the community at risk. Most cases occur in nonimmunized children from 1 to 10 years of age living in crowded, unsanitary situations.

Pathology of Diphtheria

Exposure to the diphtheria bacillus usually results from close contact with the droplets from human carriers or active infections and

---

* *Corynebacterium* (kor-eye″-nee-bak-ter′-ee-um) Gr. *koryne,* club, and *bakterion,* little rod.

* diphtheria (dif-thee′-ree-ah) Gr. *diphthera,* membrane.

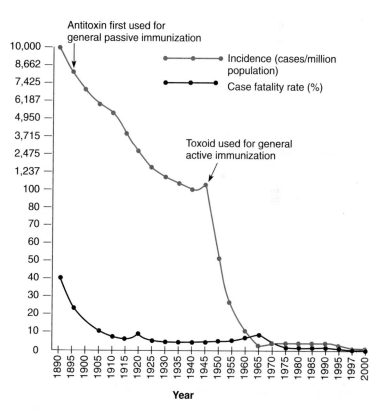

FIGURE 19.12

The incidence and case fatality rates for diphtheria in the United States during the last 100 years. This graph clearly documents the steady decline in cases, starting with the routine use of antitoxin in 1895 and the widespread use of the toxoid vaccine that began in 1945. The percentage of fatalities also dropped dramatically, long before antimicrobic drugs were available. The current residual level of cases is due to a small number of unimmunized carriers in the population.

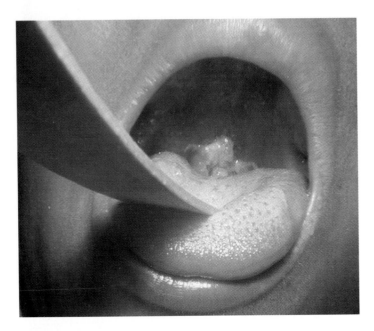

FIGURE 19.13

Diagnosing diphtheria. The clinical appearance in diphtheria infection includes gross inflammation of the pharynx and tonsils marked by grayish patches (a pseudomembrane) and swelling over the entire area.

The toxin affects the body on two levels. Locally, it produces an inflammatory reaction, low-grade fever, sore throat, nausea, vomiting, enlarged cervical lymph nodes, and severe swelling in the neck. One life-threatening complication is the **pseudomembrane,** a greenish-gray film that develops in the pharynx from the solidification of fluid expressed during inflammation (figure 19.13). The pseudomembrane is so leathery and tenacious that attempts to pull it away result in bleeding, and if it forms in the airways, it can cause asphyxiation.

The most dangerous systemic complication is **toxemia,** which occurs when the toxin is absorbed from the throat and carried by the blood to certain target organs, primarily the heart and nerves. The action of the toxin on the heart causes myocarditis and abnormal EKG patterns. Cranial and peripheral nerve involvement can cause muscle weakness and paralysis. Although toxic effects are usually reversible, patients with inadequate treatment often die from asphyxiation, respiratory complications, or heart damage.

Diagnostic Methods for the Corynebacteria

Diphtheria has such great potential for harm that often the physician must make a presumptive diagnosis and begin treatment before the bacteriological analysis is complete. A gray membrane and swelling in the throat are somewhat indicative of diphtheria, although several diseases present a similar appearance. A Gram stain of a membrane or throat specimen can help rule out diphtheria. Epidemiologic factors such as living conditions, travel history, and immunologic history (a positive Schick test) can also aid in initial diagnosis.

A simple stain of *C. diphtheriae* isolates with alkaline methylene blue reveals cells with marked pleomorphism and granulation. Other methods include tests for toxicity and a PCR analysis of the culture. It is important to differentiate *C. diphtheriae* from

"diphtheroids"—similar species often present in clinical materials that are not primary pathogens. *Corynebacterium xerosis** normally lives in the eye, skin, and mucous membranes and is an occasional opportunist in eye and postoperative infections. *Corynebacterium pseudodiphtheriticum,** a normal inhabitant of the human nasopharynx, can colonize natural and artificial heart valves.

Treatment and Prevention of Diphtheria

The adverse effects of toxemia are treated with diphtheria antitoxin (DAT) derived from horses. Prior to injection, the patient must be tested for allergy to horse serum and be desensitized if necessary. The infection is treated with antibiotics from the penicillin or erythromycin family. Bed rest, heart medication, and tracheostomy or bronchoscopy to remove the pseudomembrane may be indicated. Diphtheria can be easily prevented by a series of vaccinations with toxoid, usually given as part of a mixed vaccine against tetanus and pertussis called the DTaP. Currently recommended are three vaccinations, starting at 6 to 8 weeks of age, followed by a booster at 15 months and again at school age. Older children and adults who are not immune to the toxin can be immunized with two doses of diphtheria-tetanus (Td) vaccine.

THE GENUS *PROPIONIBACTERIUM*

*Propionibacterium** resembles *Corynebacterium* in morphology and arrangement, but it differs by being aerotolerant or anaerobic and nontoxigenic. The most prominent species is **P. acnes,** a common resident of the *pilosebaceous** glands of human skin and occasionally the upper respiratory tract. The primary importance of this bacterium is its relationship with the familiar **acne vulgaris** lesions of adolescence. Acne is a complex syndrome influenced by genetic and hormonal factors as well as by the structure of the epidermis, but it is also an infection (Medical Microfile 19.3). *Propionibacterium* is occasionally involved in infections of the eye and artificial joints.

---

* *xerosis* (zee-roh′-sis) Gr. *xerosis,* parched skin.

* *pseudodiphtheriticum* (soo″-doh-dif-ther-it′-ih-kum) Gr. *pseudes,* false, and *diphtheriticus,* of diphtheria.

* *Propionibacterium* (pro″-pee-on″-ee-bak-tee′-ree-um) Gr. *pro,* before, *prion,* fat, and *backterion,* little rod. Named for its ability to produce propionic acid.

* *acnes* (ak′-neez) Gr. *akme,* a point. (The original translators incorrectly changed an *m* to an *n*.)

* *pilosebaceous* (py″-loh-see-bay′-shus) L. *pilus,* hair, and *sebaceous,* tallow.

CHAPTER CHECKPOINTS ✔✔✔

Irregular, non-spore-forming gram-positive bacilli are so named because of irregularities in both morphology and staining properties. They inhabit soil, water, and animal bodies. Three genera contain important pathogens: *Corynebacterium, Propionibacterium,* and *Mycobacterium.*

C. diphtheriae is the agent of diphtheria, an upper respiratory infection. The primary virulence factor is diphtherotoxin, which is toxic to upper respiratory tissue, the peripheral nervous system, and the heart. Immunization provides effective protection.

Propionibacterium acnes is the agent of acne vulgaris, an infection of pilosebaceous glands exacerbated by biochemical changes of adolescence.

MEDICAL MICROFILE 19.3
Acne: The Microbial Connection

Normally, each pilosebaceous gland is a self-contained system for protecting, softening, and lubricating the skin *(1)*. It contains one or more sebaceous glands that continuously release an oily secretion called sebum into the hair follicle. As hair and skin grow, the dead epidermal cells and sebum work their way upward and are discharged from the pore onto the skin surface.

Skin prone to pimples and acne has a structure that traps the mass of sebum and dead cells and clogs the pores. If the skin at the surface swells over the pore's entrance, a closed comedo, or whitehead, results *(2)*; if the pore remains open to the surface but is blocked with a plug of sebum, it is called an open comedo, or blackhead *(3)*. The dark tinge of the blackhead is caused by the accumulation of the pigment melanin, not to uncleanliness. An added factor is overproduction of sebum when the sebaceous gland is stimulated by hormones (especially male). *Propionibacterium* bacteria growing in the follicle release lipases to digest this surplus of oil. The combination of digestive products (fatty acids) and bacterial antigens stimulates an intense local inflammation that bursts the follicle *(4)*. In time, the lesion erupts on the surface as a papule or pustule *(5)*. In some cases, secondary bacterial infections cause deeply scarred lesions. Because acne depends in part on an infection, it can be suppressed with topical and oral antibiotics such as clindamycin or tetracycline. Now after 30 years of widespread use of antimicrobic drugs for treating acne, the bacteria have developed resistance that complicates therapy. Other types of therapy involve chemicals that enhance skin removal (benzoyl peroxide) and slow the production of sebum (Retin A and Accutane).

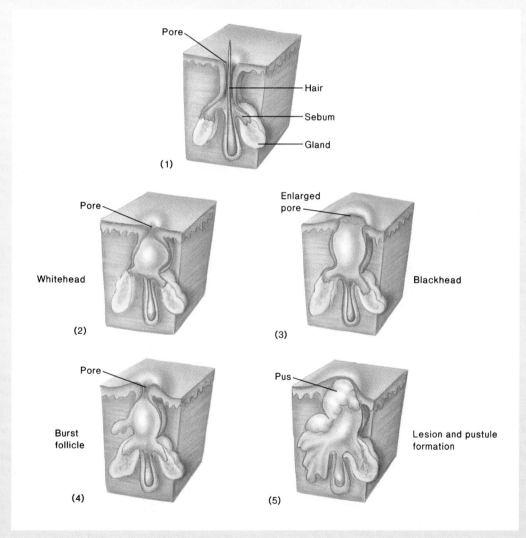

The stages in the formation of an acne lesion.

Mycobacteria: Acid-Fast Bacilli

The genus **Mycobacterium*** is distinguished by its complex layered structure composed of high-molecular-weight mycolic acids and waxes. This high lipid content imparts the characteristic of **acid-fastness** and is responsible for the resistance of the group to drying, acids, and various germicides. The cells of mycobacteria are long, slender, straight, or curved rods with a slight tendency to be filamentous or branching (figure 19.14). Although they usually contain granules and vacuoles, they do not form capsules, flagella, or spores.

Most mycobacteria are strict aerobes that grow well on simple nutrients and media. Compared with other bacteria, the growth rate is generally slow, with generation times ranging from 2 hours to several days. Some members of the genus exhibit colonies containing yellow, orange, or pink carotenoid pigments that require light for development; others are nonpigmented. Many of the 50 mycobacterial species are saprobes living free in soil and water, and several are highly significant human pathogens (table 19.2). Worldwide, millions of people are afflicted with tuberculosis and leprosy. Certain opportunistic species loosely grouped into a category called NTM (**n**on-**t**uberculous **m**ycobacteria) have become an increasing problem in immunosuppressed patients.

MYCOBACTERIUM TUBERCULOSIS: THE TUBERCLE BACILLUS

The *tubercle** bacillus is a long, thin rod that grows in sinuous masses or strands called cords. Unlike many bacteria, it produces no exotoxins or enzymes that contribute to infectiousness. Most strains contain complex waxes and a cord factor (figure 19.15) that contribute to virulence by preventing the mycobacteria from being destroyed by the lysosomes or macrophages. Their survival contributes to further invasion and persistence as intracellular parasites.

Epidemiology and Transmission of Tuberculosis

Mummies from the Stone Age, ancient Egypt, and Peru provide unmistakable evidence that tuberculosis (TB) is an ancient human disease (see chapter opening figure). In fact, it was such a prevalent cause of death that it was called "Captain of the Men of Death" and "White Plague." Its epidemiologic patterns vary with the living conditions in a community or an area of the world. Factors that significantly affect a person's susceptibility to tuberculosis are inadequate nutrition, debilitation of the immune system, poor access to medical care, lung damage, and genetics. People in developing countries are often infected as infants and harbor the microbe for many years until the disease is reactivated in young adulthood. Estimates indicate that possibly one-third of the world's population and 15 million people in the United States carry the TB bacillus.

Cases in the United States show a strong correlation with the age, sex, and recent immigration history of the patient. The highest case rates occur in nonwhite males over 30 years of age and nonwhite females over 60. The highest case rates occur in new immi-

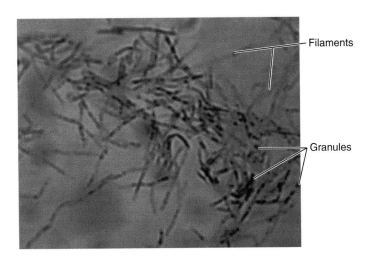

FIGURE 19.14

The microscopic morphology of mycobacteria. Shown here is *Mycobacterium tuberculosis* in an acid-fast stain of sputum from a tubercular patient. Note the irregular morphology, granules, and the filamentous forms (700×).

grants from certain areas of Indochina, South America, and Africa and in AIDS patients. For recent trends in the epidemiology of tuberculosis, see Spotlight on Microbiology 19.4.

The agent of tuberculosis is transmitted almost exclusively by fine droplets of respiratory mucus suspended in the air. The tubercle bacillus is very resistant and can survive for 8 months in fine aerosol particles. Although larger particles become trapped in the mucus and expelled, tinier ones can be inhaled into the bronchioles and alveoli. This effect is especially pronounced among people sharing closed, small rooms with limited access to sunlight and fresh air.

The Course of Infection and Disease

A clear-cut distinction can be made between infection with the tubercle bacillus and the disease it causes. In general, humans are rather easily infected with the bacillus but are resistant to the disease. Estimates project that only about 5% of infected people actually develop a clinical case of tuberculosis. Untreated tuberculosis progresses slowly and is capable of lasting a lifetime, with periods of health alternating with episodes of morbidity. The majority (85%) of TB cases are contained in the lungs, even though disseminated tubercle bacilli can give rise to tuberculosis in any organ of the body. Clinical tuberculosis is divided into primary tuberculosis, secondary (reactivation or reinfection) tuberculosis, and disseminated tuberculosis.

Primary Tuberculosis The minimum infectious dose for lung infection is around 10 cells. The bacilli are phagocytosed by alveolar macrophages and multiply intracellularly. This period of hidden infection is asymptomatic or accompanied by mild fever, but some cells escape from the lungs into the blood and lymphatics. After 3 to 4 weeks, the immune system mounts a complex, cell-mediated assault against the bacilli. The large influx of mononuclear cells into the lungs plays a part in the formation of specific infection sites called **tubercles.** Tubercles are granulomas that consist of a central

---

* *Mycobacterium* (my″-koh-bak-tee′-ree-um) Gr. *myces*, fungus, and *bakterion*, a small rod.

* tubercle (too′-ber-kul) L. *tuberculum*, a swelling or knob.

TABLE 19.2

Differentiation of Important Mycobacterium Species

Species	Primary Habitat	Disease in Humans	Treatment	Rate of Growth*	Pigmentation**
M. tuberculosis	Humans	Tuberculosis (TB)	Combined drugs	S	NP
M. bovis	Cattle	Tuberculosis	Same as TB	S	NP
M. ulcerans	Humans	Skin ulcers	Surgery, grafts	S	NP
M. kansasii	Not clear	Opportunistic lung infection	Difficult, similar to TB	S	PP
M. marinum	Water, fish	Swimming pool granuloma	Tetracycline, rifampin	S	PP
M. scrofulaceum	Soil, water	Scrofula	Removal of lymph nodes	S	PS
M. avium- M. intracellulare complex	Birds	Opportunistic AIDS infection; lung infection like TB	Combined drugs	S	NP
M. fortuitum- M. chelonae complex	Soil, water, animals	Wound abscess; postsurgical infection	4–6-drug regimen; surgery	R	NP
M. phlei	Sputum, soil	Not pathogenic	None	R	PS
M. smegmatis	Smegma, soil	Not pathogenic	None	R	Usually NP
M. leprae	Strict parasite of humans	Leprosy	See text	S	Cannot be grown in artificial media

The mycobacteria are grouped into major categories by their growth rate and their pigment production.
*Growth rate is rapid (R), occurring in less than 7 days, or slow (S), occurring in more than 7 days.
**Photochromogens (PP) develop yellow to dark orange pigment in the presence of light; scotochromogens (PS) synthesize pigment in darkness; and nonpigmented forms (NP) have no color.

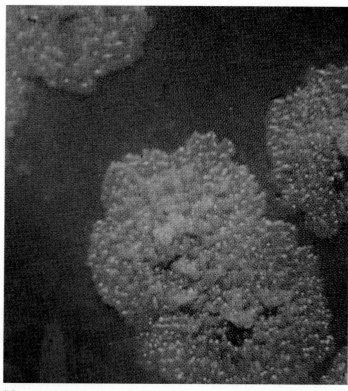

(a)

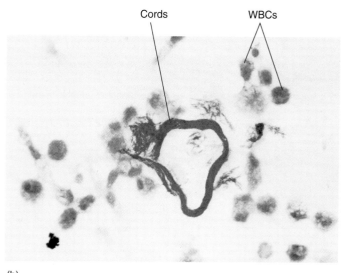

(b)

FIGURE 19.15

Diagnosing tuberculosis.   **(a)** The cultural appearance of *Mycobacterium tuberculosis*. Colonies with a typical granular, waxy pattern of growth. **(b)** Cord formation in infected tissue. Red-stained strands are elongate, massed filaments of TB bacilli.

(a) *From Gillies and Dodds,* Bacteriology Illustrated, *5th ed., Fig. 45, p. 58. Reprinted by permission of Churchill Livingstone.*

SPOTLIGHT ON MICROBIOLOGY 19.4
The Threat from Tuberculosis

After 80 years of decline, a tiny wax-coated bacillus that hides in human lungs has reemerged as a serious threat to the world's health. The hope for eradicating tuberculosis is becoming an unlikely possibility. In many regions of the world, the rates of TB are so high that WHO has requested emergency aid. Nearly eight million new cases occur in parts of Africa, Southeast Asia, and Latin America. With two million deaths per year, TB is second only to AIDS in causing mortality.

North America has also experienced some changes in its TB patterns, especially in the rise of drug resistance. Over the past 10 years, nearly 10% of isolates are resistant to at least one drug, and around 2% are multidrug-resistant (MRTB), meaning that they are not sensitive to two or more of the major TB drugs. Patients are actually dying from TB even while undergoing multidrug therapy! Between 1992 and 1999, these drug-resistant strains had appeared in the majority of states.

Major factors that account for the changing epidemiologic picture are homelessness, drug addiction, the HIV epidemic, reduced government support of TB programs, and more cases occurring in new immigrants. Homelessness and drug addiction have the similar effect of increasing host susceptibility and creating conditions that favor the spread of the TB bacillus. These populations are less likely to have access to good medical care and to comply with the necessarily long course of therapy. Fully one-fifth of cases occur in AIDS patients. The mortality rate in this group is very high: Up to 80% die within 16 weeks of diagnosis.

The lack of well-funded TB control programs has contributed to the epidemic by discontinuing low-cost treatments, by limiting surveillance, and by not following up on treatment regimens. Experts at the CDC believe that multidrug-resistant strains of TB emerged in those patients with active TB who did not follow appropriate drug therapy. The cost of cutting back and "saving money" on TB prevention for the past 15 years has been staggering both in dollars and human lives. Treating a patient in a hospital with active TB costs $25,000 per episode, compared with a few hundred dollars for preventive medicine.

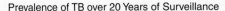

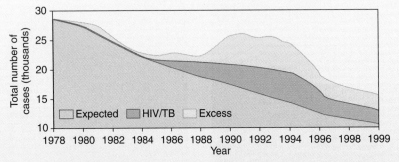

Tuberculosis rates in the United States have declined, but are still higher than originally predicted. This is largely due to high rates of coinfections in AIDS patients, drug resistance, and reduced funding for treatment.

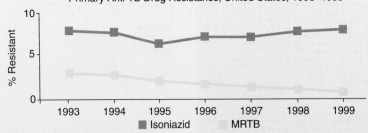

One factor in the prevalence of TB is the existence of numerous strains resistant to anti-tuberculosis drugs such as isoniazid.
Source: Data from Morbidity and Mortality Weekly Report, *U.S. Department of Health and Human Services, Centers for Disease Control and Prevention, Atlanta, GA.*

core containing TB bacilli and enlarged macrophages and an outer wall made of fibroblasts, lymphocytes, and neutrophils (figure 19.16). Although this response further checks spread of infection and helps prevent the disease, it also carries a potential for damage. Frequently, the centers of tubercles break down into necrotic, *caseous** lesions that gradually heal by calcification when normal lung tissue is replaced by calcium deposits. The response of T cells to *M. tuberculosis* proteins also causes a cell-mediated immune

---

* caseous (kay'-see-us) L. *caseus,* cheese. The material formed resembles cheese or curd.

Granuloma cells

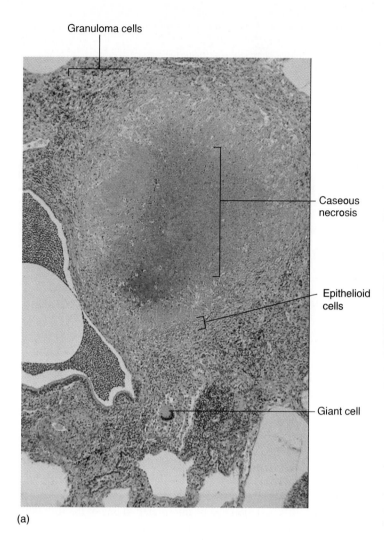

Caseous
necrosis

Epithelioid
cells

Giant cell

(a)

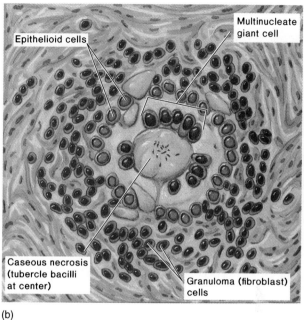

Epithelioid cells

Multinucleate
giant cell

Caseous necrosis
(tubercle bacilli
at center)

Granuloma (fibroblast)
cells

(b)

FIGURE 19.16

Tubercle formation. **(a)** Photomicrograph of a tubercle (16×). **(b)** In this enlarged interpretation, the massive granuloma infiltrate has obliterated the alveoli and set up a dense collar of fibroblasts, lymphocytes (granuloma cells), and epithelioid cells. The core of this tubercle is a caseous (cheesy) material containing the bacilli.

response evident in the **tuberculin reaction,** a valuable diagnostic and epidemiologic tool (see figure 17.18a).

Secondary Reactivation Tuberculosis Although the majority of TB patients recover more or less completely from the primary episode of infection, live bacilli can remain dormant and become reactivated weeks, months, or years later, especially in people with weakened immunity. In chronic tuberculosis, tubercles filled with masses of bacilli expand and drain into the bronchial tubes and upper respiratory tract. Gradually, the patient experiences more severe symptoms, including violent coughing, greenish or bloody sputum, low-grade fever, anorexia, weight loss, extreme fatigue, night sweats, and chest pain. It is the gradual wasting of the body that accounts for an older name for tuberculosis—*consumption.* Untreated secondary disease has nearly a 60% mortality rate.

Extrapulmonary Tuberculosis During the course of secondary TB, the bacilli disseminate rapidly to sites other than the lungs. Organs most commonly involved in **extrapulmonary TB** are the regional lymph nodes, kidneys, long bones, genital tract, brain, and meninges. Because of the debilitation of the patient and the high load of tubercle bacilli, these complications are usually grave.

Renal tuberculosis results in necrosis and scarring of the renal medulla and the pelvis, ureters, and bladder. This damage is accompanied by painful urination, fever, and the presence of blood and the TB bacillus in urine. Genital tuberculosis in males damages the prostate gland, epididymis, seminal vesicle, and testi; and in females, the fallopian tubes, ovaries, and uterus. It often affects reproductive function in both sexes.

Tuberculosis of the bone and joints is a common complication. The spine is a frequent site of infection, though the hip, knee, wrist, and elbow can also be involved. Advanced infiltration of the vertebral column produces degenerative changes that collapse the vertebrae (Pott disease; figure 19.17), resulting in abnormal curvature of the thoracic region (humpback or kyphosis) or of the lumbar region (swayback or lordosis). Neurological damage stemming from compression on nerves can cause extensive paralysis and sensory loss.

Tubercular meningitis is the result of an active brain lesion seeding bacilli into the meninges. Over a period of several weeks, the infection of the cranial compartments can create mental deterioration, permanent retardation, blindness, and deafness. Untreated tubercular meningitis is invariably fatal, and even treated cases can have a 30% to 50% mortality rate.

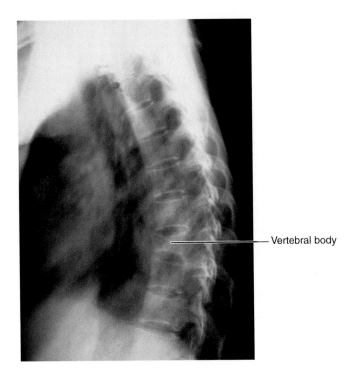

FIGURE 19.17
Tuberculosis of the spinal column (Pott disease) as seen by an X ray.

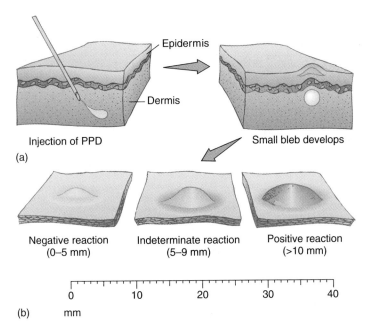

FIGURE 19.18

Skin testing for tuberculosis. (a, b) The Mantoux test. Tuberculin is injected into the dermis. A small bleb from the injected fluid develops but will be absorbed in a short time. After 48 to 72 hours, the skin reaction is rated by the degree (or size) of the reaction. (Views are magnified 3×.) See also figure 17.18a.

Clinical Methods of Detecting Tuberculosis

Clinical diagnosis of tuberculosis traditionally relies upon four techniques: (1) *in vivo* or tuberculin testing, (2) roentgenography (X rays), (3) direct identification of acid-fast bacilli (AFB) in sputum or some other specimen, and (4) cultural isolation and biochemical testing.

Tuberculin Sensitivity and Testing   Because hypersensitivity to tuberculoproteins can persist throughout life, testing for it is an effective way to screen school children, public employees, and health care workers. A positive reaction to these proteins in an unimmunized individual is fairly reliable evidence of recent or past infection or disease. Because vaccination for TB also stimulates delayed hypersensitivity, tuberculin screening is most useful in countries such as the United States where widespread vaccination is not carried out. Physicians should determine a patient's history before giving the tuberculin test.

The test itself involves exposure to a small amount of purified protein derivative (PPD), a standardized solution obtained from culture filtrates of *M. tuberculosis* (figure 19.18a). In the **Mantoux test,** 0.1 ml of PPD (equivalent to 5 tuberculin units) is injected intradermally into the forearm to produce an immediate small bleb. The skin is observed and measured for a red wheal after 48 hours and 72 hours. The three reaction levels include: (1) negative reactions, 5 mm or less in diameter; (2) intermediate reactions, of 5 mm to 9 mm diameter, which indicates doubtful sensitivity and requires retesting; and (3) positive reactions, with a lesion of 10 mm or more (figure 19.18b). Because tuberculin can cause a systemic reaction in sensitive individuals, it should not be injected into known

tuberculin reactors, as these persons often experience severe ulceration and necrosis at the test site.

A positive reaction can be interpreted in one of the following ways: It may be due to recent contact and a new infection, or it may be reactivation of a prior infection. In these cases, the degree of reaction tends to be more pronounced. False positive reactions are caused by a recent BCG vaccination or an infection with a NTM that cross-reacts with the TB bacillus.

A negative skin test also has more than one possible interpretation. (1) In most cases, it means no contact or infection has occurred. (2) It may be too early in the infection for sensitization to appear, indicating a need for retesting after 3 months. (3) Up to 20% of infected subjects are nonreactive and unable to mount a tuberculin reaction. This condition is associated with severe immunocompromise as seen with HIV infection, age, alcoholism, and chronic disease. Skin testing is not a reliable diagnostic indicator in these people.

Roentgenography and Tuberculosis   Chest X rays, or roentgenographs, can help verify TB when other tests have given indeterminate results. X-ray films reveal abnormal radiopaque patches whose appearance and location can be very indicative. Primary tubercular infection presents the appearance of fine areas of infiltration and enlarged lymph nodes in the lower and central areas of the lungs. Secondary tuberculosis films show more extensive infiltration in the upper lungs and bronchi and marked tubercles (figure 19.19). Scars from older infections often show up on X rays and can furnish a basis for comparison with which to identify newly active disease.

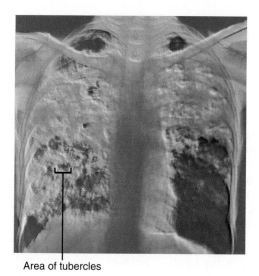

Area of tubercles

FIGURE 19.19

Colorized X ray showing a secondary tubercular infection.

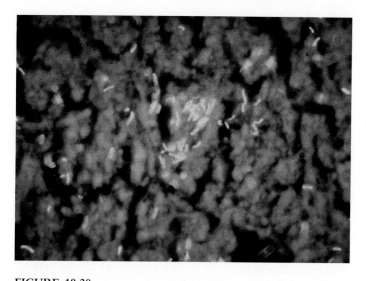

FIGURE 19.20

A fluorescent acid-fast stain of *Mycobacterium tuberculosis* from sputum. Smears are evaluated in terms of the number of AFB seen per field. This quantity is then applied to a scale ranging from 0 to 4+, 0 being no AFB observed and 4+ being more than 9 AFB per field.

Acid-Fast Staining The diagnosis of tuberculosis in people with positive skin tests or X rays can be backed up by acid-fast staining of sputum or other specimens. Several variations on the acid-fast stain are currently in use. The Ziehl-Neelsen stain produces bright red acid-fast bacilli (AFB) against a blue background. Fluorescence staining shows luminescent yellow-green bacilli against a dark background (figure 19.20). The fluorescent acid-fast stain is becoming the method of choice because it is easier to read and provides a more striking contrast.

Laboratory Cultivation and Diagnosis *Mycobacterium tuberculosis* infection is most accurately diagnosed by isolating and identifying the causative agent in pure culture. Because of the specialized expertise and technology required, this is not done by most clinical laboratories as a general rule. In the United States, the handling of specimens and cultures suspected of containing pathogen is strictly regulated by federal laws.

Diagnosis that differentiates between *M. tuberculosis* and other mycobacteria must be accomplished as rapidly as possible so that appropriate treatment and isolation precautions can be instituted. Because specimens are often contaminated with rapid-growing bacteria that will interfere with the isolation of *M. tuberculosis,* they are pretreated with chemicals to remove contaminants and are plated onto selective egg-potato base media (such as Middlebrook 7H11 or Lowenstein-Jensen media). Cultures are incubated under varying temperature and lighting conditions to clarify thermal and pigmentation characteristics and are then observed for signs of growth. Several newer cultivation schemes have shortened the time to several days instead of the 6 to 8 weeks once necessary. Several identification techniques use probes to detect specific genes and can confirm positive specimens early in the infection (see figure F on page 541). Rapid diagnosis is particularly important for public health and treatment considerations.

Management of Tuberculosis

Treatment of TB involves administering drugs for a sufficient period of time to kill the bacilli in the lungs, organs, and macrophages, usually 6 to 24 months. Drug resistance is avoided by a therapeutic regimen that involves combined therapy with at least two drugs selected from a list of eleven, including isoniazid (INH), rifampin, ethambutol, streptomycin, pyrazinamide, thioacetazone, or paraaminosalicylic acid (PAS). The choice of drugs depends upon such considerations as effectiveness, adverse side effects, cost, and special medical problems of the patient. A one-pill regimen called *Rifater* (INH, rifampin, and pyrazinamide) is considered the best combination to effect cure and prevent resistance. If one combination is not working well because of toxicity, drug resistance, or hypersensitivity, a reasonably effective replacement is usually available. The presence of a negative culture or a gradual decrease in the number of AFB on a smear indicates success. However, because permanent cure will not occur if the patient does not comply with drug protocols, there are many relapses. Several programs in the United States have begun using directly observed therapy (DOT) to improve the success of treatment, especially among homeless or isolated patients.

Prevention and Control of Tuberculosis

Although it is essential to identify and treat people with active TB, it is equally important to seek out and treat those in the early stages of infection or at high risk of becoming infected. Treatment groups are divided into tuberculin-positive "converters," who appear to have a reactivated infection, and tuberculin-negative people in high-risk groups such as the contacts of tubercular patients. The standard prophylactic treatment is a daily dose of isoniazid for a year. In the hospital, the use of UV lamps in air-conditioning systems and negative pressure rooms to isolate TB patients can effectively control the spread of infection.

A vaccine based on the attenuated "bacille Calmet-Guerin" (BCG) strain of *M. bovis* is often given to children in countries that have high rates of tuberculosis. Studies have shown that the success rate of vaccinations is around 80% in children and less than that (20–50%) in adults. The length of protection varies from 5 to 15 years. Because the United States does not have as high an incidence as other countries, BCG vaccination is not generally recommended except among certain health professionals and military personnel who may be exposed to TB carriers.

MYCOBACTERIUM LEPRAE: THE LEPROSY BACILLUS

Mycobacterium leprae, the cause of leprosy, was first detected by a Norwegian physician named Gerhard Hansen, and it is sometimes called Hansen's bacillus in his honor. The general morphology and staining characteristics of the leprosy bacillus are similar to those of other mycobacteria, but it is exceptional in two ways: (1) It is a strict parasite that has not been grown in artificial media or human tissue cultures, and (2) it is the slowest growing of all the species. *Mycobacterium leprae* multiplies within host cells in large packets called globi at an optimum temperature of 30°C.

Leprosy* is a chronic, progressive disease of the skin and nerves known for its extensive medical and cultural ramifications. From ancient times, leprosy victims were stigmatized because of the severe disfigurement of the disease and the belief that it was a divine curse. As if the torture of the disease were not enough, leprosy patients once suffered terrible brutalities, including imprisonment under the most gruesome conditions. The modern view of leprosy is more enlightened. We know that it is not readily communicated and that it should not be accompanied by social banishment. Because of the unfortunate connotations associated with the term *leper* (a person who is shunned or ostracized), more acceptable terms such as leprosy patient, Hansen's patient, or leprotic, are preferred.

Epidemiology and Transmission of Leprosy

The incidence of leprosy has recently been declining due to a worldwide control effort. WHO estimates range from 500,000 to 1,000,000 cases, mostly in areas of Asia, Africa, Central and South America, and the Pacific islands where it is endemic. Leprosy is not restricted to warm climates, however, since it also occurs in Siberia, Korea, and northern China. The disease is also reported in a few limited locales in the United States, including parts of Hawaii, Texas, Louisiana, Florida, and California. The total number of new cases reported nationwide is from 300 to 500 per year, many of which are associated with recent immigrants.

The mechanism of transmission among humans is yet to be fully verified. Theories propose that the bacillus is directly inoculated into the skin through contact with a leprotic, that mechanical vectors are involved, or that inhalation of droplet nuclei is a factor.

Although the human body was long considered the sole host and reservoir of the leprosy bacillus, it is now clear that armadillos harbor a mycobacterial species genetically identical to *M. leprae* and may develop a granulomatous disease similar to leprosy.

Whether humans can acquire leprosy from armadillos is not yet known, but it is tempting to speculate that the infection is actually a zoonosis.

Because the leprosy bacillus is not highly virulent, most people who come into contact with it do not develop clinical disease. As with tuberculosis, it appears that health and living conditions influence susceptibility and the course of the disease. An apparent predisposing factor is some defect in the regulation of T cells. Mounting evidence also indicates that some forms of leprosy are associated with a specific genetic marker. Long-term household contact with leprotics, poor nutrition, and crowded conditions increase the risks of infection. Many people become infected as children and harbor the microbe through adulthood.

The Course of Infection and Disease

Once *M. leprae* has entered a portal of entry, macrophages successfully destroy the bacilli, and there are no initial manifestations of infection. But in a small percentage of cases, a weakened or slow macrophage and T-cell response leads to intracellular survival of the pathogen. The usual incubation period varies from 2 to 5 years, with extremes of 3 months to 40 years. The earliest signs of leprosy appear on the skin of the trunk and extremities as small, spotty lesions colored differently from the surrounding skin (figure 19.21). In untreated cases, the bacilli grow slowly in the skin macrophages and Schwann cells of peripheral nerves, and the disease progresses to one of several outcomes.

A useful rating system for leprosy was developed by D.S. Ridley and W. H. Jopling. At the two extremes are tuberculoid leprosy (TL) and lepromatous leprosy (LL) (table 19.3), and in between are borderline tuberculoid (BT), borderline (BB), and borderline lepromatous (BL). Patients can have more than one form of leprosy simultaneously, and one type can progress to another.

Tuberculoid leprosy, the most superficial form, is characterized by asymmetrical, shallow skin lesions containing very few bacilli (figure 19.21). Microscopically, the lesions appear as thin granulomas and enlarged dermal nerves. Damage to these nerves usually results in local loss of pain reception and feeling. This form has fewer complications and is more easily treated than other types of leprosy.

Lepromatous leprosy is responsible for the disfigurations commonly associated with the disease. It is marked by chronicity and severe complications due to widespread dissemination of the bacteria. Leprosy bacilli grow primarily in macrophages in cooler regions of the body, including the nose, ears, eyebrows, chin, and testes. As growth proceeds, the face of the afflicted person develops folds and granulomous thickenings, called **lepromas,** which are caused by massive intracellular overgrowth of *M. leprae* (figure 19.22). Advanced LL causes a loss of sensitivity that predisposes the patient to trauma and mutilation, secondary infections, blindness, and kidney or respiratory failure.

The condition of **borderline leprosy** patients can progress in either direction along the scale, depending upon their treatment and immunologic competence. The most severe effect of intermediate forms of leprosy is early damage to nerves that control the muscles of the hands and feet. The subsequent wasting of the muscles and loss of control produces drop foot and claw hands (figure 19.23). Sensory nerve damage can lead to trauma and loss of fingers and toes.

---

* leprosy (lep'-roh-see) Gr. *lepros,* scaly or rough. Translated from a Hebrew word that refers to uncleanliness.

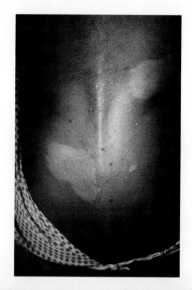

(a)

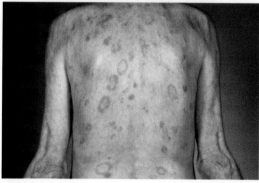

(b)

FIGURE 19.21

Leprosy lesions. Both views are of the tuberculoid form, with shallow, painless lesions. **(a)** Infection in dark-skinned persons manifests as hypopigmented patches or macules. **(b)** In light-skinned individuals, it appears as reddish patches or papules.

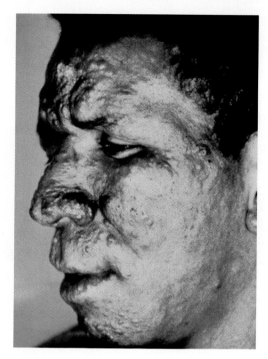

FIGURE 19.22

A clinical picture of lepromatous leprosy (LL). Infection of the nose, lips, chin, and brows produces moderate facial deformation, typical of lepromas.

Diagnosing Leprosy

Leprosy is diagnosed by a combination of symptomology, microscopic examination of lesions, and patient history. A simple yet effective field test in endemic populations is the feather test (figure 19.24). A skin area that has lost sensation and does not itch may be an early symptom. Numbness in the hands and feet, loss of heat and cold sensitivity, muscle weakness, thickened earlobes, and chronic stuffy nose are additional evidence. Laboratory diagnosis relies upon the detection of acid-fast bacilli in smears of skin lesions, nasal discharges, and tissue samples. Knowledge of a patient's prior contacts with leprotics also supports diagnosis. Laboratory isolation of the leprosy bacillus is difficult and not ordinarily attempted.

Treatment and Prevention of Leprosy

Leprosy infection can be controlled by drugs, but therapy is most effective when started before permanent damage to nerves and other tissues has occurred. Because of an increase in resistant strains, multidrug therapy is necessary. Tuberculoid leprosy can be managed with rifampin and dapsone for 6 months. Lepromatous leprosy requires a combination of rifampin, dapsone, and clofazimine until the number of AFB in skin lesions has been substantially reduced (requiring up to 2 years). Then dapsone can be taken alone for an indeterminate period (up to 10 years or more).

Preventing leprosy requires constant surveillance of high-risk populations to discover early cases, chemoprophylaxis of healthy persons in close contact with leprotics, and isolation of leprosy patients. The WHO is currently sponsoring a trial of a vaccine containing killed leprosy bacilli that shows promise in clinical trials.

TABLE 19.3

The Two Major Clinical Forms of Leprosy

Tuberculoid Leprosy	Lepromatous Leprosy
Few bacilli in lesions	Many bacilli in lesions
Few shallow skin lesions in many areas	Numerous deeper lesions concentrated in cooler areas of body
Loss of pain sensation in lesions	Sensory loss more generalized; occurs late in disease
No skin nodules	Gross skin nodules
Occasional mutilation of extremities	Mutilation of extremities common
Reactive to lepromin★	Not reactive to lepromin
Lymph nodes not infiltrated by bacilli	Lymph nodes massively infiltrated by bacilli
Well-developed cell-mediated (T-cell) response	Poorly developed T-cell response

★*Lepromin is an extract of the leprosy bacillus injected intradermally, like tuberculin, to detect delayed allergy to leprosy.*

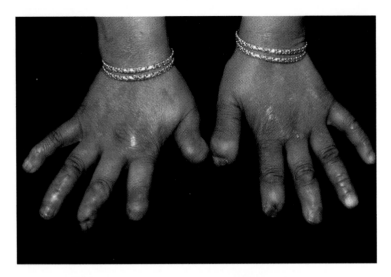

FIGURE 19.23

Deformation of the hands caused by borderline leprosy. The clawing and wasting are chiefly due to nerve damage that interferes with musculoskeletal activity. Individuals in a later phase of the disease can lose their fingers.

INFECTIONS BY NON-TUBERCULOUS MYCOBACTERIA (NTM)

For many years, most mycobacteria were thought to have low pathogenicity for humans (see table 19.2). Saprobic and commensal species are isolated so frequently in soil, drinking water, swimming pools, dust, air, raw milk, and even the human body that both contact and asymptomatic infection appear to be widespread. However, the recent rise in opportunistic and nosocomial mycobacterial infections has demonstrated that many species are far from harmless.

Disseminated Mycobacterial Infection in AIDS Bacilli of the *Mycobacterium avium* complex (MAC) frequently cause secondary infections in AIDS patients with low T-cell counts. In fact, MAC is the third most common cause of death in these patients, after pneumocystis pneumonia and cytomegalovirus infection. These common soil bacteria usually enter through the respiratory tract, multiply, and rapidly disseminate. In the absence of an effective immune counterattack, the bacilli flood the body systems, especially the blood, bone marrow, bronchi, intestine, kidney, and liver. Treatment requires two or more combined drugs such as rifabutin, azithromycin, ethambutol, and rifampin. This therapy usually has to be given for several months or even years.

Non-tuberculous Lung Disease Pulmonary infections caused by commensal mycobacteria have symptoms like a milder form of tuberculosis, but they are not communicable. *Mycobacterium kansaii* infection is endemic to urban areas in the midwestern and southwestern United States and parts of England. It occurs most often in adult white males who already have emphysema or bronchitis. *Mycobacterium fortuitum* complex causes postsurgical skin and soft-tissue infection and pulmonary complications in immunosuppressed patients.

Skin, Lymph Node, and Wound Infections An infection by *M. marinum* has been labeled swimming pool granuloma, because it is a hazard of scraping against the rough concrete surfaces lining

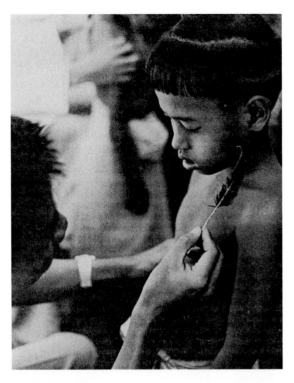

FIGURE 19.24

The feather test for leprosy. The Burmese child is being tickled over her face and trunk to determine loss of fine sensitivity to touch. This simple technique can provide evidence of early infection.

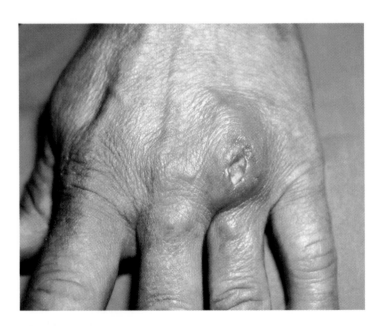

FIGURE 19.25

A chronic swimming pool granuloma of the hand.

swimming pools. The disease starts as a localized nodule, usually on the elbows, knees, toes, or fingers, which then enlarges, ulcerates, and drains (figure 19.25). The granuloma can clear up spontaneously, but it can also persist and require long-term treatment. *Mycobacterium scrofulaceum* causes an infection of the cervical lymph nodes in children living in the Great Lakes region, Canada, and Japan. The bacterium apparently infects the oral cavity and invades the lymph nodes when ingested with food or milk. In most cases, the infection is without complications, but certain children develop *scrofula,* in which the affected lymph nodes ulcerate and drain.

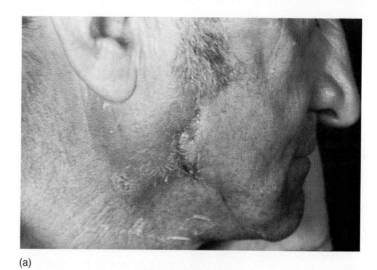
(a)

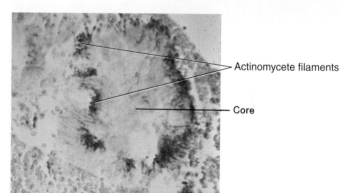

Actinomycete filaments

Core

(b)

FIGURE 19.26

Clinical symptoms and microscopic signs of actinomycosis. **(a)** Early periodontal lesions work their way to the surface in cervicofacial disease. The lesions eventually break out and drain. **(b)** A section of crushed sulfur granule shows filamentous cells surrounding a sulfur-yellow core.

CHAPTER CHECKPOINTS

The genus *Mycobacterium* is distinguished by its acid-fast cell wall composed of mycolic acids and its slow but persistent rate of growth. Significant pathogens are *M. tuberculosis* and *M. leprae.*

M. tuberculosis is an ancient pathogen that annually kills millions of individuals worldwide. It infects the lungs (pulmonary tuberculosis) and other organs of the body (extrapulmonary tuberculosis). Factors in its pathogenicity include high infectivity, cord factor, ability to escape phagocytosis, and the recent development of multiple drug resistance. Reduction of tuberculosis control programs has also contributed to its status as a major pathogen worldwide.

Tuberculosis is diagnosed by the tuberculin skin test, chest X rays, DNA probes, and direct microscopic identification in clinical specimens. Immunization with BCG vaccine offers 20% to 80% protection in areas with high incidence. Multiple drug therapy is effective in most cases.

M. leprae is the causative agent of leprosy (Hansen disease). Its infectivity appears to be either direct contact or droplet transmission. Genetic susceptibility and living conditions of the host may also be factors. There are two forms of leprosy, both of which destroy epithelial tissue and peripheral nerves. Lepromatous leprosy is the more invasive disfiguring form. Tuberculoid leprosy is the milder, superficial form. Treatment with combination drug therapy is effective if begun early. Research for an effective vaccine is under way but has not yet been successful.

Mycobacterium avium complex (MAC), an opportunist responsible for 80% of AIDS-related deaths, is just one of many mycobacterial species increasingly identified as opportunistic and nosocomial agents of infection.

Actinomycetes: Filamentous Bacilli

A group of bacilli closely related to *Mycobacterium* are the pathogenic **actinomycetes.*** These are nonmotile filamentous rods that may be acid-fast. Certain members produce a mycelium-like growth and spores reminiscent of fungi (figure 19.26*b*). They also produce chronic granulomatous diseases. The main genera of actinomycetes involved in human disease are *Actinomyces* and *Nocardia.*

ACTINOMYCOSIS

Actinomycosis is an endogenous infection of the cervicofacial, thoracic, or abdominal regions by species of *Actinomyces* living normally in the human oral cavity, tonsils, and intestine. The cervicofacial form of disease can be a common complication of tooth extraction, poor oral hygiene, and rampant dental caries. The lungs, abdomen, and uterus are also sites of infection.

---

* **actinomycetes** (ak″-tih-noh-my′-seets) Gr. *actinos,* a ray, and *myces,* fungi. Members of the Order Actinomycetales.

Cervicofacial disease is caused by *A. israelii,* which enters a damaged area of the oral mucous membrane and begins to multiply there. Diagnostic signs are swollen, tender nodules in the neck or jaw that give off a discharge containing macroscopic (1–2 mm) sulfur granules (figure 19.26). In most cases, infection remains localized, but bone invasion and systemic spread can occur in people with poor health. Thoracic actinomycosis is a necrotizing lung disorder that can project through the chest wall and ribs. Abdominal actinomycosis is a complication of burst appendices, gunshot wounds, ulcers, or intestinal damage. Uterine actinomycosis has been increasingly reported in women using intrauterine contraceptive devices. Infections are treated with surgical drainage and antibacterial antibiotics (penicillin, erythromycin, tetracycline, and sulfonamides).

Compelling evidence indicates that oral actinomyces play a strategic role in the development of plaque and dental caries. Studies of the oral environment show that *A. viscosus* and certain oral streptococci are the first microbial colonists of the tooth surface. Both groups have specific tooth-binding powers and can adhere to each other and to other species of bacteria (see chapter 21).

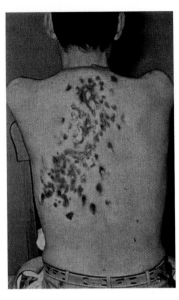

FIGURE 19.27

Nocardiosis. This case of pulmonary disease has extended through the chest wall and ribs to the cutaneous surface.

NOCARDIOSIS

*Nocardia** is a genus of bacilli widely distributed in the soil. Most species are not infectious, but *N. brasiliensis* is a primary pulmonary pathogen, and *N. asteroides* and *N. caviae* are opportunists.

---

* *Nocardia* (noh-kar'-dee-ah) After Edmund Nocard, the French veterinarian who first described the genus.

Nocardioses fall into the categories of pulmonary, cutaneous, or subcutaneous infection. Most cases in the United States are reported in patients with deficient immunity, but a few occur in normal individuals.

Pulmonary nocardiosis is a form of bacterial pneumonia with pathology and symptoms similar to tuberculosis. The lung develops abscesses and nodules and can consolidate. Often the lesions extend to the pleura and chest wall and disseminate to the brain, kidneys, and skin (figure 19.27). *Nocardia brasiliensis* is also one of the chief etiologic agents in mycetoma, discussed in chapter 22.

CHAPTER CHECKPOINTS

Actinomycetes are filamentous, gram-positive bacilli that form a branching mycelium. *Actinomyces* and *Nocardia* are the two pathogenic genera in this group.

Actinomycetes israelii is the causative agent of actinomycosis. This opportunist bacterium infects the mouth, tonsils, and intestine, resulting in nodules that discharge material containing macroscopic sulfur granules. Other actinomycetes are also implicated in dental caries.

Nocardia is a genus of soil bacteria implicated in pulmonary, cutaneous, and subcutaneous infections of immunosuppressed hosts.

CHAPTER CAPSULE WITH KEY TERMS

I. **Gram-Positive Rods**
   A. *Endosporeformers:* Highly resistant spores are produced.
      1. The aerobic genus *Bacillus anthracis* causes **anthrax,** a zoonosis of herbivorous animals; the bacterium inhabits the soil, where it is picked up by grazing animals; humans are infected by contact with animals or their other products; the cutaneous form is manifest by **eschar,** a black ulcer at the site of entrance; the pulmonic form, acquired by inhaling spores, is highly fatal. Anthrax is controlled by antibiotics and animal vaccination.
      2. The anaerobic genus *Clostridium* is common in soil and spread throughout most habitats, even the human body; wound and tissue diseases are due to introduction of the bacterium into damaged tissues; food intoxications arise when pathogenic clostridia grow in food; the basis for disease is powerful **exotoxins** that act upon specific target tissues.
         a. *Cl. perfringens* causes **gas gangrene,** a soft-tissue and muscle infection (**myonecrosis**) in areas of damage caused by injuries or diabetes; spores germinate in anaerobic tissues and produce gas bubbles; may invade nearby healthy tissue; toxin necrotizes muscle and other tissue; treatment is by careful tissue **debridement** and cleansing, hyperbaric oxygen, and antibiotics.
         b. *Cl. difficile* causes **antibiotic-associated colitis,** an acute diarrheal disease associated with disruptions in the GI tract flora due to antimicrobic therapy.
         c. *Cl. tetani* is the cause of **tetanus** or **lockjaw,** a disease associated with dirty puncture wounds or lesions; spores in

anaerobic areas of the wound grow and produce a toxin called **tetanospasmin,** which travels to target cells in the CNS; causes severe uncontrollable spasms of large muscles; treatment is by antitoxin; control is through vaccination with tetanus toxoid.
         d. **Clostridial food poisoning** occurs when *Cl. perfringens* is eaten with inadequately cooked meats and fish, producing toxin in the intestine; marked by severe diarrhea, vomiting; not usually fatal.
         e. **Botulism,** caused by *Cl. botulinum,* is associated with improperly home-canned foods; spores withstand food processing, grow in stored food, and produce **botulin.** Botulism is a true food intoxication; ingested toxin enters circulation and acts on myoneural junctions to block muscle contraction, leading to flaccid paralysis; treated by antitoxin; control by proper canning, heating of foods prior to eating.
         f. **Infant** and **wound botulism** are unusual types of infectious botulism contracted like tetanus.
   B. *Non-spore-forming Rods:* Straight, regular rods and irregular rods.
      1. Straight, nonpleomorphic rods stain evenly. Genera that are regular in morphology include *Listeria monocytogenes* and *Erysipelothrix rhusiopathiae.*
         a. *L. monocytogenes* are widely distributed resistant bacteria that cause **listeriosis;** primary habitat appears to be water, soil, and the intestines; most cases are food infections associated with contaminated dairy products and meats;

disease is mild, self-limited in young adults, but may be severe and complicated in fetuses, neonates, the elderly, or the immunocompromised; treated with penicillin; prevention requires proper pasteurization of milk and dairy products.

    b. *E. rhusiopathiae* is carried by animals (swine) and widely dispersed into the environment; cause of **erysipeloid,** a disease common among occupations that handle animals or animal products; symptom is inflamed red sores on fingers at portal of entry.

2. Irregular, non-spore-forming rods include *Corynebacterium* and *Propionibacterium.*

    a. *C. diphtheriae* is a highly pleomorphic aerobic rod with metachromatic granules and palisades arrangement; agent of *diphtheria,* a disease whose reservoir is healthy human carriers; spread by droplets; occurs first as a localized infection of the throat with tenacious **pseudomembrane** that may cause suffocation; later symptoms are due to a classic **toxemia;** toxin spreads in blood to targets; may damage heart and nervous system; treated by antitoxin, penicillin (erythromycin); controlled by vaccination (DTaP).

    b. *Propionibacterium acnes* is an anaerobic rod found in the skin of humans; is regularly associated with **acne vulgaris** lesions; acts on skin oils to create an inflammatory condition that erupts into pimples.

3. *Acid-Fast Bacilli (AFB):* Contain large amounts of mycolic acids and waxes; are strict aerobes; slender filamentous rods; widely distributed; tend to be resistant to environmental conditions.

    a. *Mycobacterium tuberculosis* causes **tuberculosis (TB),** a common lung infection; tubercle bacillus is spread by droplets in close quarters; susceptible persons have weakened immunities and poor nutrition and living conditions. The disease is divided into (1) **primary pulmonary** infection, often marked by formation of granulomas in the lungs called **tubercles;** (2) **secondary TB,** a severe lung complication that occurs in people with reactivated primary disease; and (3) **disseminated extrapulmonary** infections due to blood-borne TB bacilli carried to bone, kidneys, lymph nodes, or brain; TB diagnosed by tuberculin testing, chest X rays, acid-fast stain for bacilli in sputum, and laboratory cultivation and identification; managed by combined drug therapy, BCG vaccine in some countries.

    b. *M. leprae* causes **leprosy,** a chronic disease that begins in skin and mucous membranes and progresses into nerves; prevalent in endemic regions throughout the world; spread through direct inoculation from **leprotics;** two forms are (1) **tuberculoid,** a superficial infection without skin disfigurement which damages nerves and causes loss of pain perception, and (2) **lepromatous,** a deeply nodular infection that causes severe disfigurement of the face and extremities. In both forms, mutilation of parts may occur due to loss of pain receptors; treatment is by long-term combined therapy.

    c. **NTM** (non-tuberculous mycobacteria) are common environmental species that may be involved in disease. For example, *M. avium* complex (MAC) causes a disseminated disease of AIDS patients, and *M. marinum* causes swimming pool granuloma.

II. **Diseases Caused by Actinomycetes**

The genera *Actinomyces* and *Nocardia* are filamentous bacteria related to mycobacteria that cause chronic infections of the skin and soft tissues.

A. *Actinomycosis: A. israelii* occurs in the human oral cavity, tonsils, intestine; oral disease or surgery may result in cervicofacial infection, also abdominal, thoracic, uterine complications.

B. *Nocardiosis: N. brasiliensis* causes pulmonary disease similar to TB, with abscesses and nodules that may spread to chest wall; one agent of mycetoma.

MULTIPLE-CHOICE QUESTIONS

1. What is the usual habitat of endospore-forming bacteria that are agents of disease?
   a. the intestine of animals
   b. dust and soil
   c. water
   d. foods

2. Most *Bacillus* species are
   a. true pathogens
   b. opportunistic pathogens
   c. nonpathogens
   d. commensals

3. Many clostridial diseases require a/an _____ environment for their development.
   a. living tissue   c. aerobic
   b. anaerobic   d. low-pH

4. *Clostridium perfringens* causes
   a. myonecrosis
   b. food poisoning
   c. antibiotic-induced colitis
   d. both a and b

5. The action of tetanus exotoxin is on the
   a. neuromuscular junction
   b. sensory neurons
   c. spinal interneurons
   d. cerebral cortex

6. The probable habitat of *Listeria* is
   a. the human intestine
   b. animals
   c. soil and water
   d. plants

7. An infection peculiar to swine causes _____ when transmitted to humans.
   a. anthrax   c. tuberculosis
   b. diphtheria   d. erysipeloid

8. TB is spread by
   a. contaminated fomites
   b. food
   c. respiratory droplets
   d. vectors

9. Diagnosis of tuberculosis is by
   a. X ray
   b. Mantoux test
   c. acid-fast stain
   d. all of these

10. The form of leprosy associated with severe disfigurement of the face is
    a. tuberculoid
    b. lepromatous
    c. borderline
    d. papular

11. Soil mycobacteria can be the cause of
    a. tuberculosis
    b. leprosy
    c. swimming pool granuloma

12. Caseous lesions containing inflammatory white blood cells are
    a. lepromas
    b. pseudomembranes
    c. eschars
    d. tubercles

13. Which infectious agent is an obligate parasite?
    a. *Mycobacterium tuberculosis*
    b. *Corynebacterium diphtheriae*
    c. *Mycobacterium leprae*
    d. *Clostridium difficile*

14. Which infectious agents are facultative pathogens?
    a. *Bacillus anthracis*
    b. *Clostridium tetani*
    c. *Clostridium perfringens*
    d. all of these

15. Which infection(s) would be categorized as a zoonosis?
    a. anthrax
    b. gas gangrene
    c. diphtheria
    d. both a and b

16. Actinomycetes are _____ that produce chronic granulomatous diseases.
    a. filamentous bacteria
    b. yeasts
    c. molds
    d. cocci

CONCEPT QUESTIONS

1. a. What is the role of spores in infections?
   b. Describe the general distribution of spore-forming bacteria.

2. a. Briefly outline the epidemiology of anthrax; describe the stages in the development of cutaneous anthrax.
   b. Why is pneumonic anthrax so deadly?
   c. What other *Bacillus* species can be involved in infections or diseases?

3. a. What characteristics of *Clostridium* contribute to its pathogenicity?
   b. Compare the toxigenicity of tetanospasmin and botulin.
   c. What predisposes a patient to clostridial infection?

4. a. Outline the epidemiology of the major wound infections and food intoxications of *Clostridium*.
   b. What is the origin of the gas in gas gangrene?
   c. How does hyperbaric oxygen treatment work?
   d. Why is amputation necessary in some cases?
   e. What is debridement, and how does it prevent some clostridial infections?

5. What is the mechanism of antibiotic-associated colitis?

6. a. What causes the jaw to "lock" in lockjaw?
   b. How would it be possible for patients with no noticeable infection to present with symptoms of tetanus?

7. a. Compare the symptomology of botulism and tetanus.
   b. How are the two conditions alike and different?
   c. What is the difference between food, infant, and wound botulism?

8. a. Describe the epidemiology and route of infection in listeriosis.
   b. Why is listeriosis a serious problem even with refrigerated foods?
   c. Which groups are most at risk for serious complications?

9. Why is erysipeloid an occupation-associated infection?

10. a. What are the distinctive morphological traits of *Corynebacterium*?
    b. Differentiate between diphtheria infection and toxemia.
    c. How can the pseudomembrane be life-threatening?
    d. What is the ultimate origin of diphtherotoxin?

11. a. Outline the unique characteristics of *Mycobacterium*.
    b. What is the epidemiology of TB?
    c. Differentiate between TB infection and TB disease.
    d. What are tubercles?
    e. What is the course of disseminated disease?
    f. Outline the principles of tuberculin testing, chest X rays, and acid-fast staining.
    g. Why does tuberculosis require combined therapy?

12. a. What characteristics make *M. leprae* different from other mycobacteria?
    b. Differentiate between tuberculoid and lepromatous leprosy.
    c. What causes the deformations?
    d. What causes the mutilation of extremities?

13. a. What is the importance of NTM?
    b. Describe the effects of *Mycobacterium avium* complex in AIDS patients.

14. a. Describe the bacteria in the actinomycete group, and explain what characteristics make them similar to fungi.
    b. Briefly describe two of the common diseases caused by this group.

15. **Multiple Matching.** Choose all descriptions that fit.
    _____ *Bacillus*
    _____ *Clostridium*
    _____ *Mycobacterium*
    _____ *Corynebacterium*
    _____ *Listeria*

    1. is acid fast
    2. cells irregular, pleomorphic
    3. can form metachromatic granules
    4. regular-shaped rods
    5. forms endospores
    6. is psychrophilic
    7. is primarily anaerobic
    8. is aerobic
    9. is associated with dairy products
    10. cells can be elongate filaments
    11. shows palisades arrangement

16. **Matching.** Match the disease with the principal portal of entry.

    ____ 1. anthrax
    ____ 2. botulism
    ____ 3. gas gangrene
    ____ 4. antibiotic colitis
    ____ 5. tetanus
    ____ 6. diphtheria
    ____ 7. listeriosis
    ____ 8. tuberculosis
    ____ 9. leprosy

a. skin
b. gastrointestinal tract
c. traumatized tissue
d. respiratory tract
e. urogenital tract
f. placenta
g. none of these

17. **Matching.** Using the same list of diseases as in question 16, match with symptoms and signs from the following list.

    ____ a. loss of cutaneous sensation
    ____ b. double vision
    ____ c. necrosis of muscle tissue
    ____ d. fever
    ____ e. black, cutaneous ulcer
    ____ f. severe diarrhea
    ____ g. difficulty in breathing
    ____ h. heart failure
    ____ i. flaccid paralysis of muscles
    ____ j. spastic paralysis of muscles
    ____ k. severe coughing
    ____ l. pseudomembrane
    ____ m. skin nodules

CRITICAL-THINKING QUESTIONS

1. a. What is the main clinical strategy in preventing gas gangrene?
   b. Why does it work?

2. a. Why is it unlikely that diseases such as tetanus and botulism will ever be completely eradicated?
   b. Name some bacterial diseases in this chapter that could be completely eradicated and explain how.

3. Why is the cause of death similar in tetanus and botulism?

4. a. Why does botulin not affect the senses?
   b. Why does botulism not commonly cause intestinal symptoms?

5. Account for the fact that boiling does not destroy botulism spores but does inactivate botulin.

6. Adequate cooking is the usual way to prevent food poisoning. Why doesn't it work for *Clostridium perfringens* and *Bacillus* food poisoning?

7. a. Why do patients who survive tetanus and botulism often have no sequelae?
   b. How has modern medicine improved the survival rate for these two diseases?

8. What would be the likely consequence of diphtheria infection alone without toxemia?

9. How can one tell that acne involves an infection?

10. a. Do you think the spittoons of the last century were effective in controlling tuberculosis?
    b. Why or why not?

11. a. Provide an explanation for the statement that TB and leprosy are "family diseases."
    b. What, if anything, can be done about multidrug-resistant tuberculosis?

12. Which diseases in this chapter have no real portal of exit?

13. Case history 1. A farm worker sustained a crushing injury to his hand and was taken to a hospital, where he received treatment and tetanus toxoid. During successive weeks, he had several operations to repair the damaged hand and was given antibiotics. After a time, he lost muscle tone and had difficulty talking. Finally, his hand required amputation, and he was given antitoxin.
    a. What do you think the disease was?
    b. Why did the earlier treatments not work?
    c. To what other disease is this similar?

14. Case history 2. Eighty-six people at a St. Patrick's Day celebration developed diarrhea, cramps, and vomiting after eating a traditional dinner of corned beef, cabbage, potatoes, and ice cream. Symptoms appeared within 10 hours of eating, on average. Of those afflicted, 85 had eaten corned beef, one had not. The corned beef had been cooked in an oven the day before the event, stored in the refrigerator, and sliced and placed under heat lamps 1½ hours before serving.
    a. Use this information to diagnose the food poisoning agent.
    b. What sorts of lab tests could one do to verify it as this bacterium?
    c. What would you have done to prevent this incident?

15. Case history 3. An outbreak of gastrointestinal illness was reported in a daycare center. Of 67 people, 14 came down with symptoms of nausea, cramps, and diarrhea within 2 to 3 hours of eating. Food items included fried rice, peas, and apple rings. Symptoms occurred in 14 of the 48 who had eaten the fried rice and in none of those who had not eaten it. The rice had been cooked the night before and refrigerated. The next morning it had been fried with leftover chicken and held without refrigeration until lunchtime.
    a. On the basis of symptoms and food types, what do you expect the etiologic agent to be?
    b. What would you expect to culture from the food?
    c. To what other disease is this one similar?
    d. How can it be differentiated from the agent and disease in case history 2?

INTERNET SEARCH TOPICS

1. Use a search engine to locate websites that cover biological and germ warfare. Look for information on how weapons based on microbes are developed and regulated.

2. Search under tuberculin testing to find information on medical guidelines and interpretations for various test outcomes.

The Gram-Negative Bacilli of Medical Importance

T he gram-negative bacilli comprise a large group of non-spore-forming bacteria adapted to a wide range of habitats and niches. The genera in this category are highly diverse in metabolism and pathogenicity. Because the group is so large and complex, and many of its members are not medically important, we have included only those that are human pathogens.

A large number of representative genera are inhabitants of the large intestine (enteric); some are zoonotic; some are adapted to the human respiratory tract; and still others live in soil and water. It is useful to distinguish between the frank (true) pathogens (*Salmonella, Yersinia pestis, Bordetella,* and *Brucella,* for example), which are infectious to the general population, and the opportunists (*Pseudomonas* and coliforms), which are resident flora that cause infection in people with weakened host defenses.

Chapter Overview

- Gram-negative bacilli are a large, diverse group of bacteria, widely dispersed in the environment and on the bodies of humans and animals. A number of them are agents of human and animal diseases.
- The source of their pathogenesis is derived from the outer membrane of the cell wall, which contains endotoxin and an assortment of enzymes and other virulence factors.
- *Pseudomonas* species are prominent residents of soil and water with serious ecological and medical impact. *Ps. aeruginosa* is an opportunistic pathogen in lung, skin, and surgical infections.
- Respiratory and systemic infections are caused by *Bordetella pertussis,* the cause of whooping cough, and *Legionella pneumophila,* the cause of Legionnaires disease.
- Zoonotic infectious agents in this category are *Brucella,* the cause of brucellosis, which is spread to humans from cattle and pigs; and *Francisella tularensis,* the cause of tularemia, an infection of wild rabbits, rodents, carnivores, and arthropod vectors.
- The largest subgroup represented is the Family Enterobacteriaceae, also called the enterics, which are residents of the large intestine of humans and animals.
- Enterics may infect as either true pathogens or opportunists. Infections range from severe diarrheal and toxigenic illness to urinary tract and wound infections. The group accounts for over 40% of nosocomial infections.
- *Escherichia coli* is the most important coliform, acting as both a true pathogen and an opportunist. It causes severe invasive and toxic diseases such as infant diarrhea and toxemia, as well as urinary tract and surgical infections.

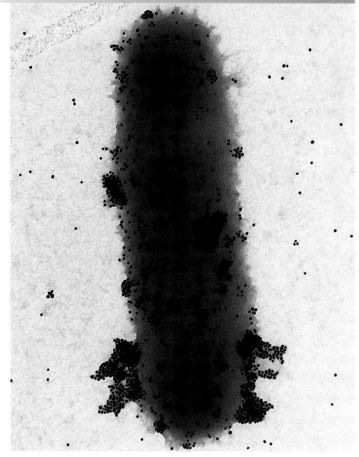

A gram-negative bacterium undergoing lysis following exposure to an antibiotic releases spherical fragments of lipopolysaccharide. This component of the cell wall acts as an endotoxin that seriously disrupts several systems of the body.

- A prominent enteric pathogen is the genus *Salmonella,* the cause of typhoid fever, enteric fever, and food poisoning. Typhoid fever is strictly a human disease that is spread by contaminated food and water. Other salmonelloses are zoonotic in origin, transmitted by food products obtained from infected poultry and cattle.
- Members of the genus *Shigella* originate from humans and are highly transmissible in food that has been contaminated by human feces. *Shigella* toxins damage the large intestine lining and give rise to shigellosis or bacillary dysentery.
- *Yersinia pestis* causes the plague, a zoonosis carried by rodents and acquired from fleas and inhaled bacteria. Bubonic plague is a lymphatic form, and pneumonic plague affects the lungs.

TABLE 20.1

Survey of Gram-Negative Pathogens

Oxygen Requirements	Genus	Species	Disease/Commentary
Aerobes			
	Pseudomonas	*aeruginosa, fluorescens*	Opportunistic pathogens; invade surgical wounds, burns, lungs
	Brucella	*abortus, melitensis*	Brucellosis (undulant fever)
	Francisella	*tularensis*	Tularemia (rabbit fever)
	Bordetella	*pertussis*	Pertussis (whooping cough)
	Legionella	*pneumophila*	Legionellosis; spread by environmental aerosols
Facultative Anaerobes			
The Family Enterobacteriaceae (Oxidase-Negative)			
	Escherichia	*coli*	Numerous strains; pathogens cause diarrhea; opportunists cause urinary tract infections
	Edwardsiella	*tarda*	Gastroenteritis; opportunistic infections of the lungs, surgical incisions, urinary tract
	Citrobacter	Various	⎫
	Klebsiella	species	
	Enterobacter		
	Hafnia		Pathogenic to immunocompromised patients, causing
	Serratia		opportunistic or secondary infections
	Proteus		
	Providencia		
	Morganella		
	Salmonella	↓	Salmonellosis ⎭
		typhi	Typhoid fever
		cholerae-suis	Septicemia, enteric fever
		enteritidis	Gastroenteritis
	Shigella	*dysenteriae*	Bacillary dysentery (shigellosis)
		flexneri	Bacillary dysentery
		boydii	Bacillary dysentery
		sonnei	Bacillary dysentery
	Yersinia	*pestis*	Plague
		pseudotuberculosis	Gastroenteritis; opportunistic
		enterocolitica	Gastroenteritis; opportunistic
The Family Pasteurellaceae (Oxidase-Positive)			
	Pasteurella	*multocida*	Skin abscess; septicemia
	Haemophilus	*influenzae*	Meningitis, epiglottitis
Genera Not Associated with a Family			
	Gardnerella	*vaginalis*	Nonspecific vaginitis
	Eikenella	*corrodens*	Gingival infections
	Streptobacillus	*moniliformis*	Rat bite fever
	Calymmatobacterium	*granulomatis*	Donovanosis or granuloma inguinale
Obligate Anaerobes			
	Bacteroides	*fragilis*	Dominant species in intestine; cause of abdominal infections
		melaninogenicus	Oral soft-tissue infections

- *Haemophilus influenzae* is another cause of meningitis, an invasive infection of the central nervous system that may lead to serious complications.
- Gram-negative bacillary infections for which there is useful vaccination are pertussis, brucellosis, tularemia, typhoid fever, plague, and *Haemophilus* meningitis.

Aerobic Gram-Negative Nonenteric Bacilli

The aerobic gram-negative bacteria of medical importance are a loose assortment of genera including *Pseudomonas,* an opportunistic pathogen; the zoonotic pathogens *Brucella* and *Francisella;* and *Bordetella* and *Legionella,* which are mainly human pathogens. The gram-negative rods can be organized in several ways. To simplify coverage here, they are grouped into three major categories based on oxygen requirements (table 20.1).

One of the most serious consequences of infection with gram-negative bacteria, even if the species involved is not a frank or virulent pathogen, is blood infection. Nationwide, over 100,000 patients a year die from a condition called **septic shock.** Shock is a pathologic state of low blood pressure accompanied by a reduced amount of blood circulating to vital organs, particularly the brain, heart, lungs, and kidneys. Its major symptoms and signs are nausea, tachycardia, cold, clammy skin, and weak pulse. Damage to the organs can elicit respiratory failure, coma, heart failure, and death in a few hours. Although the endotoxins of all gram-negative bacteria can cause shock, the most common clinical cases are due to gram-negative enteric rods.

The adverse factor in gram-negative sepsis is the presence of an endotoxin called lipopolysaccharide (LPS) in the outer membrane of the gram-negative cell wall. Lipopolysaccharide is a potent immune stimulator. The component that accounts for most of the adverse effects is *lipid A* embedded in the external layer of the membrane. This lipid is active only after it is liberated by bacteria that are growing or lysed by host defenses and other factors. In general, lipid A triggers the secretion of interleukins, tumor necrosis factor, and other cytokines by macrophages. In a local infection with only small amounts of endotoxin, the effects of the cytokines are protective. But in massive sepsis, legions of macrophages release profuse amounts into the bloodstream. Some of the cytokines are pyrogenic and induce fever, some activate the complement cascade, and others promote intravascular blood coagulation. The end results are circulation failure, tissue damage, hypotension, and shock.

Ironically, traditional treatments with antibiotics can compound the problem of septic effects because these drugs can actually lyse bacteria (see chapter opening photo). Drug companies are actively developing new drugs that will block the effects of endotoxin. One clinical trial is testing an endotoxin antagonist that would work on all forms of lipid A. Two other possible treatments contain monoclonal antibodies against endotoxin. Although these drugs are not yet available, they show some promise in controlling the more severe effects of endotoxemia.

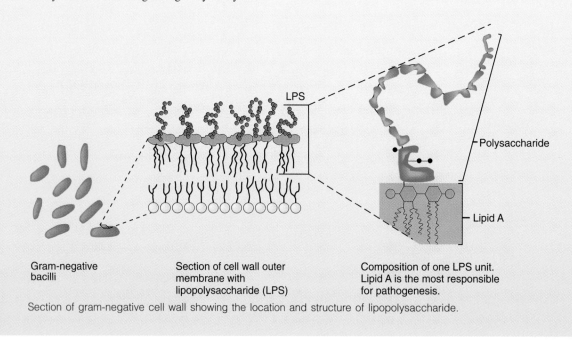

| Gram-negative bacilli | Section of cell wall outer membrane with lipopolysaccharide (LPS) | Composition of one LPS unit. Lipid A is the most responsible for pathogenesis. |

Section of gram-negative cell wall showing the location and structure of lipopolysaccharide.

One notable component of all gram-negative rods is a lipopolysaccharide in the outer membrane of the cell wall that acts as an **endotoxin** (see chapter 13). Because endotoxin in the blood can have severe and far-reaching pathophysiologic effects, gram-negative septicemia, a common *iatrogenic** or nosocomial infection, is a cause for great concern (Medical Microfile 20.1).

PSEUDOMONAS: THE PSEUDOMONADS

The pseudomonads are a large group of free-living bacteria that live primarily in soil, seawater, and fresh water. They also colonize plants and animals and are frequent contaminants in homes and clinical settings. These small, gram-negative rods have a single polar flagellum (figure 20.1), produce oxidase and catalase, and do not ferment carbohydrates. Although they ordinarily obtain energy aerobically through oxidative metabolism, some species can grow anaerobically if provided with a salt such as nitrate. Many species produce green, brown, red, or yellow pigments that diffuse into the medium and change its color.

*Pseudomonas** species are highly versatile. They can adapt to a wide range of habitats and manage to extract needed energy from minuscule amounts of dissolved nutrients. Most members can

* iatrogenic (eye-at″-troh-jen′-ik) Gr. *iatros,* doctor. An infection occurring as a result of medical treatment.

* *Pseudomonas* (soo″-doh-moh′-nas) Gr. *pseudes,* false, and *monas,* a unit.

grow in a medium containing minerals and one simple organic compound. This adaptability accounts for their constant presence in the environment and their capacity to thrive on hosts. These bacteria are also metabolically versatile. They degrade numerous extracellular substances by means of protease, amylase, pectinase, cellulase, and numerous other enzymes.

Pseudomonads have extensive impact on ecology, agriculture, and commerce. They are important decomposers and bioremediators, capable of degrading hundreds of natural substrates. These characteristics make them useful in cleaning up oil spills and clearing pesticides (see chapter 26). Various species have roles as food spoilage agents, plant pathogens, and genetic engineering hosts. Some species produce antibiotics termed pseudomycins that are being tested for their effectiveness in treating fungal infections.

Pseudomonas aeruginosa* is a common inhabitant of soil and water and an intestinal resident in about 10% of normal people. On occasion, it can be isolated from saliva or even the skin. Because the species is resistant to soaps, dyes, quaternary ammonium disinfectants, drugs, drying, and temperature extremes, it is a frequent contaminant of ventilators, intravenous solutions, and anesthesia equipment. Even disinfected instruments, utensils, bathroom fixtures, and mops have been incriminated in hospital outbreaks.

In a pattern similar to that of the enteric bacteria discussed in a later section, *Ps. aeruginosa* is a typical opportunist. It is unlikely to cross healthy, intact anatomical barriers, thus its infectiousness results from invasive medical procedures or weakened host defenses. The conditions that most predispose a person to infection are debilitating illness, immunosuppressant medication, or intravenous injections. Once in the tissues, *Ps. aeruginosa* expresses virulence factors including exotoxins, a phagocytosis-resistant slime layer, and various enzymes that degrade host tissues. It also causes endotoxic shock.

The most common nosocomial *Pseudomonas* infections occur in compromised hosts with severe burns, neoplastic disease, and cystic fibrosis. Complications include pneumonia, urinary tract infections, abscesses, otitis, and corneal disease. *Pseudomonas* septicemia can give rise to diverse and grave conditions such as endocarditis, meningitis, and bronchopneumonia. Infection is especially virulent in premature infants and neonates. Healthy people are subject to outbreaks of skin rashes, urinary tract infections, and external ear infections from community hot tubs and swimming pools. Sponges and washcloths serve as a common reservoir for this species, which when rubbed into the skin can cause a rash (figure 20.2). Wearers of contact lenses are also vulnerable to corneal ulcers from contaminated storage and disinfection solutions.

Unusual characteristics of *Ps. aeruginosa* infections are a grapelike odor and the noticeable color that appears in tissue exudates ("blue pus"). The color is due to the bacterium's production of a blue-green or greenish-yellow pigment (pyocyanin) that can fluoresce (figure 20.3). Lab identification relies on a battery of tests similar to those given for the Enterobacteriaceae (see figure 20.8). The notorious multidrug resistance of *Ps. aeruginosa* makes testing for the drug sensitivity of most isolates imperative. Drugs found effective in controlling infections are the third-generation cephalosporins, aminoglycosides, carbenicillin, polymyxin, quino-

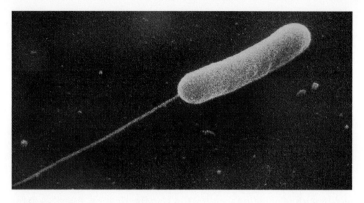

FIGURE 20.1

Electron micrograph of *Pseudomonas aeruginosa* emphasizing its single, polar flagellum.

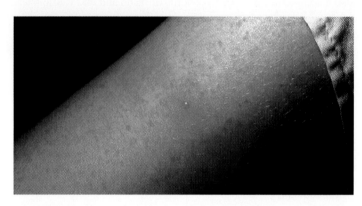

FIGURE 20.2

Skin rashes may be picked up from common sources.   Papulopustular lesion on forearm of patient, which developed subsequent to use of a loofah sponge overgrown with *Pseudomonas aeruginosa*.

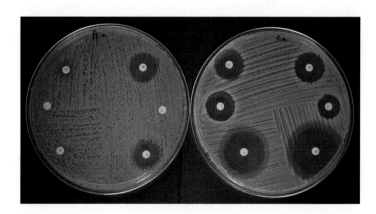

FIGURE 20.3

An antimicrobic sensitivity test of *Pseudomonas aeruginosa* reveals traits typical of the species.   The plate on the left shows its multiple resistance to drugs and the blue-green pigment that diffuses into the medium. A plate of *Staphylococcus aureus* is shown on the right for comparison.

lones, and monobactams. A Swiss company has been testing a vaccine to be used in cystic fibrosis patients. So far, the results have been very positive, providing protection in 72% of the vaccine group.

---

* *aeruginosa* (uh-roo″-jih-noh′-suh) L. *aeruginosa,* full of blue-green copper rust.

BRUCELLA AND BRUCELLOSIS

Malta fever, undulant fever, and **Bang disease**[1] are synonyms for **brucellosis,** a zoonosis transmitted to humans from infected animals or contaminated animal products harboring **Brucella.*** Two species of these tiny, gram-negative coccobacilli are *B. abortus* (from cattle) and *B. suis* (from pigs). Animal brucellosis is an infection of the placenta and fetus that can cause abortion, as typified by Bang disease in cattle. Humans infected with either of these two agents experience a severe febrile illness but not abortion.

Brucellosis occurs worldwide, with concentrations in Europe, Africa, India, and Latin America. It is associated predominantly with occupational contact in slaughterhouses, livestock handling, and the veterinary trade. Infection takes place through contact with blood, urine, placentas, and consumption of raw milk and cheese. Human-to-human transmission is rare. Brucellosis is also a common disease of wild herds of bison and elk. Cattle sharing grazing land with these wild herds often suffer severe outbreaks of Bang disease.

Brucella enters through damaged skin or mucous membranes of the digestive tract, conjunctiva, and respiratory tract. Infected phagocytes carry bacteria into the bloodstream, creating focal lesions in the liver, spleen, bone marrow, and kidney. The cardinal manifestation of human brucellosis is a fluctuating pattern of fever, which is the origin of the common name, undulant fever (figure 20.4). It is also accompanied by chills, profuse sweating, headache, muscle pain and weakness, and weight loss. Fatalities are not common, though the syndrome can last for a few weeks to a year, even with treatment.

The patient's history can be very helpful in diagnosis, as are serological tests of the patient's blood and blood culture of the pathogen. Newer genetic tests are being developed but are not yet widely used. A combination of tetracycline and rifampin or streptomycin taken for 3 to 6 weeks is usually effective in controlling infection. Prevention is effectively achieved by testing and elimination of infected animals, quarantine of imported animals, and pasteurization of milk. Though several types of animal vaccines are available, those developed for humans are either ineffective or unsafe.

FRANCISELLA TULARENSIS AND TULAREMIA

*Francisella tularensis** is the causative agent of **tularemia,** a zoonotic disease of assorted mammals endemic to the northern hemisphere. In several characteristics, it is similar to *Yersinia pestis,* the cause of plague (discussed later in this chapter), and the two species were once included in the same genus, *Pasteurella.* Because the disease has been associated with outbreaks of disease in wild rabbits, it is sometimes called rabbit fever.

Tularemia is abundantly distributed through numerous animal reservoirs and vectors in northern Europe, Asia, and North America, but not in the tropics. This disease is noteworthy for its complex epidemiology and spectrum of symptoms. Although rabbits and rodents (muskrats and ground squirrels) are the chief

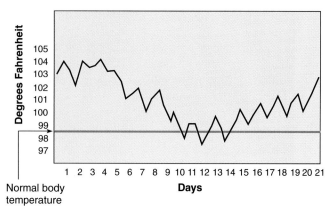

FIGURE 20.4

The temperature cycle in classic brucellosis. Body temperature undulates between day and night and between fever, normal, and subnormal.

Source: Data from A. Smith, Principles of Microbiology, 10th ed., 1985.

reservoirs, other wild animals (skunks, beavers, foxes, opossums) and some domestic animals are implicated as well. Of the more than 50 bloodsucking arthropod vectors known to carry *F. tularensis,* about half also bite humans. Ticks are most often involved, followed by biting flies, mites, and mosquitoes.

Tularemia is strikingly varied in its portals of entry and disease manifestations. In a majority of cases, infection results when the skin or eye is inoculated through contact with infected animals, animal products, contaminated water, and dust. Bites by vectors are also a frequent source of infection, but the disease is not communicated from human to human. With an estimated infective dose of between 10 and 50 organisms, *F. tularensis* is often considered one of the most infectious of all bacteria.

After an incubation period ranging from a few days to three weeks, acute symptoms of headache, backache, fever, chills, malaise, and weakness appear. Clinical manifestations are tied to the portal of entry. They include ulcerative skin lesions, swollen lymph glands, conjunctival inflammation, sore throat, intestinal disruption, and pulmonary involvement. The death rate in systemic and pulmonic forms is 10%, but proper treatment with gentamicin or tetracycline reduces it to almost zero. Because the intracellular persistence of *F. tularensis* can lead to relapses, antimicrobial therapy must not be discontinued prematurely. Protection is afforded by live, attenuated vaccines and protective gloves, masks, and eyewear for laboratory workers and other occupationally exposed personnel.

BORDETELLA PERTUSSIS AND WHOOPING COUGH

*Bordetella pertussis** is a minute, encapsulated coccobacillus responsible for **pertussis,** or **whooping cough,** a communicable childhood affliction that causes an acute respiratory syndrome.

---

1. After B. L. Bang, a Danish physician.

* *Brucella* (broo-sel′-uh) After David Bruce, who isolated the bacterium.

* *Francisella tularensis* (fran-sih-sel′-uh too-luh-ren′-sis) After Edward Francis, one of its discoverers, and Tulare County, California, where the agent was first identified.

* *Bordetella pertussis* (bor-duh-tel′-uh pur-tus′-is) After Jules Bordet, a discoverer of the agent, and L. *per,* severe, and *tussis,* cough.

Contrary to the common impression that the disease is mild and self-limiting, it often causes severe, life-threatening complications in babies. *Bordetella parapertussis* is a closely related species that causes a milder form of the infection.

The primary source of infection is usually other children afflicted with clinical disease. The reservoir appears to be in apparently healthy carriers. Transmission is by direct contact with droplets or inhalation of infectious aerosols. About half of pertussis cases occur in children between birth and 4 years of age. A recent study found evidence that whooping cough is relatively common among older children and adults. In these patients, it is generally a milder disease, presenting as a chronic cough, and it is often misdiagnosed as a cold or the flu.

Although the introduction of an effective vaccine has decreased the prevalence of pertussis in many countries, it is far from an obsolete disease. In developing countries that cannot routinely vaccinate and in industrialized countries with lax vaccination, pertussis is still a major cause of childhood sickness and death. Even in the United States, the case rate has gradually increased between 1981 and 2000 to a total as high as it was in the 1960s. Epidemiologists attribute this upsurge to fewer vaccinations amidst concern over its publicized side effects.

The primary virulence factors of *B. pertussis* are (1) receptors that specifically recognize and bind to ciliated respiratory epithelial cells and (2) toxins that destroy and dislodge ciliated cells (a primary host defense). The loss of the ciliary mechanism leads to buildup of mucus and blockage of the airways. The initial phase of pertussis is the *catarrhal stage,* marked by nasal drainage and congestion, sneezing, and occasional coughing. As symptoms deteriorate into the *paroxysmal stage,* the child experiences recurrent, persistent coughing. Fits of several abrupt, hacking coughs are followed by a deep inhalation that pulls air through the congested larynx and produces a "whoop." Many of the complications of pertussis are due to compromised respiration.

The standard control measure is early vaccination with pertussis vaccine, usually combined with the diphtheria and tetanus vaccines. The CDC has recently revised its recommendations for the type of vaccine to be used. An acellular vaccine (aP) is available for all routine immunizations, starting at 6 weeks of age. This vaccine contains toxoid and other antigens and is well tolerated when given with DT vaccine (DTaP). The vaccine does not give long-term immunity, which is why adults and older children can have recurrences. Other measures to protect susceptible people during epidemics include erythromycin therapy and isolation of cases.

LEGIONELLA AND LEGIONELLOSIS

Legionella is a novel bacterium unrelated to other strictly aerobic gram-negative genera. Although the organisms were originally described in the late 1940s, they were not clearly associated with human disease until 1976. The incident that brought them to the attention of medical microbiologists was a sudden and mysterious epidemic of pneumonia that afflicted 200 American Legion members attending a convention in Philadelphia and killed 29 of them. After 6 months of painstaking analysis, epidemiologists isolated the pathogen and traced its source to contaminated air-conditioning vents in the Legionnaires' hotel. The news media latched onto the

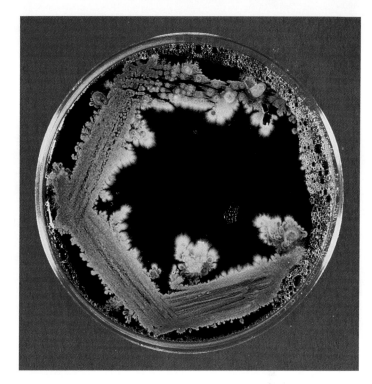

FIGURE 20.5

Legionella pneumophila **colonies on selective charcoal yeast extract medium.**

name **Legionnaires disease,** and the experts named their new isolate *Legionella* (lee″-jun-ell′-uh).

Legionellas are weakly gram-negative motile rods that range in morphology from cocci to filaments. They have fastidious nutrient requirements and can be cultivated only in special media (figure 20.5) and cell cultures. Several species or subtypes have been characterized, but *L. pneumophila* (lung-loving) is the one most frequently isolated in infections.

Legionella's ability to survive and persist in natural habitats has been something of a mystery, yet it appears to be widely distributed in aqueous habitats as diverse as tap water, cooling towers, spas, ponds, and other fresh waters. Recently, researchers determined that the bacteria live in close associations with free-living amebas (figure 20.6). It is released during aerosol formation and can be carried for long distances. Cases have been traced to supermarket vegetable sprayers and the fallout from the Mount Saint Helens volcano.

Studies since 1976 have determined that Legionnaires' disease is not a new disease and that earlier outbreaks occurred sporadically in conjunction with accidental exposure to a common environmental source of the agent. The disease is now known to be worldwide and is prevalent in males over 50 years of age. Nosocomial infections occur most often in elderly patients hospitalized with diabetes, malignant disease, transplants, alcoholism, and lung disease. However, the disease is not communicable from person to person.

Two major clinical forms of the disease are *Legionnaires pneumonia* and *Pontiac fever.* The symptoms of both are a rising

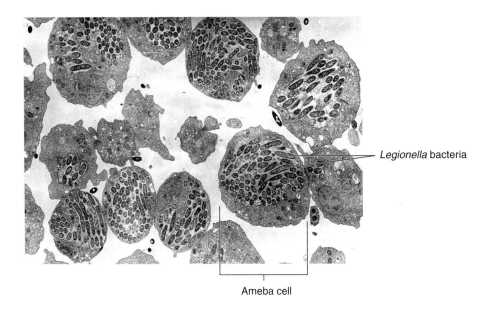

Legionella bacteria

Ameba cell

FIGURE 20.6

***Legionella* living intracellularly in the ameba *Hartmanella*.** Amebas inhabiting natural waters appear to be the reservoir for this pathogen and a means for it to survive in rather hostile environments. The pathogenesis of *Legionella* in humans is likewise dependent on its uptake by and survival in phagocytes.

fever (up to 41°C), cough, diarrhea, and abdominal pain. Legionnaires pneumonia is the more severe disease, progressing to lung consolidation and impaired respiration and organ function. It has a fatality rate of 3–30%. In contrast, Pontiac fever does not lead to pneumonia and rarely causes death. Legionellosis is diagnosed by symptomology and the patient's history. Laboratory analyses include fluorescent antibody staining of specimens, cultivation on charcoal yeast extract (CYE) agar, and DNA probes. The disease is treated with erythromycin alone or in combination with rifampin. Alternative drugs include azithromycin or quinolones. *Legionella*'s wide dispersal in aquatic habitats makes control difficult, though chlorination and regular cleaning of artificial habitats are of some benefit.

CHAPTER CHECKPOINTS

Medically significant gram-negative bacteria are divided into groups based on their oxygen requirements. One common virulence factor in all gram-negative pathogens is endotoxin, the lipopolysaccharide (LPS) component of the cell wall. Endotoxin stimulates excessive release of host defense cytokines, which, in turn, result in high fever, tissue damage, hypotension, and shock.

Medically important gram-negative aerobic bacilli include the opportunistic pathogen *Pseudomonas,* the zoonotic pathogens *Brucella* and *Francisella,* as well as the human pathogens *Bordetella* and *Legionella.*

Pseudomonads normally inhabit soil and water. They are metabolically quite versatile and are able to degrade many substances and survive under minimal conditions.

Pseudomonas aeruginosa is a common nosocomial opportunist, infecting human hosts that are either debilitated or have had invasive medical procedures. Infections occur primarily in patients with severe burns, neoplastic disease, or cystic fibrosis. Healthy individuals acquire skin infections from swimming pools, hot tubs, or contaminated sponges or contact lens solutions.

Brucella infections occur primarily through exposure to infected cattle or pigs but also through drinking unpasteurized milk. Brucellosis is a

systemic infection characterized by alternating periods of fever, sweating, and chills. The infection is carried by neutrophils to many body organs. Combination drug therapy is effective. Prevention includes animal vaccination and quarantine.

Francisella tularensis is the causative agent of tularemia, an arthropod-borne zoonosis infecting rabbits, rodents, and other mammals. Infection occurs through skin, gastrointestinal tract, and respiratory tract. Tularemia is highly transmissible through infected animals, contaminated water, droplets, dust, or through the bites of infected arthropods. Gentamicin is effective in most cases. Immunization provides protection for occupational risk groups.

Bordetella pertussis is the causative agent of whooping cough, or pertussis, a potentially fatal respiratory infection associated with uncontrollable coughing and transmitted through droplets. Childhood immunization is by far the best protection.

Legionella is the causative agent of legionellosis (Legionnaires disease) and the milder Pontiac fever. Both are non-communicable lung infections transmitted by aerosols from air conditioners, cooling towers, and other aquatic habitats. Legionellosis is a potentially fatal disease if untreated, but it responds to antibiotic therapy.

MEDICAL MICROFILE 20.2
Diarrheal Disease

Diarrhea is an acute syndrome of the intestinal tract in which the volume, fluid content, and frequency of bowel movements increase. It is usually a symptom of **gastroenteritis** (inflammation of the lining of the stomach and intestine) and can be accompanied by severe abdominal pain. Diarrhea is generated by several pathologic states—most commonly, infection, intestinal disorders, and food poisoning. Although the human large intestine ordinarily harbors a huge microbial population, most bacterial, protozoan, and viral agents of diarrhea are not members of this normal gut flora but are acquired through contaminated food or water.

Infectious diarrhea has two basic mechanisms. In the toxigenic type of disease, bacteria release **enterotoxins** that bind surface receptors of the small intestine. These toxins disrupt the physiology of epithelial cells and cause increased secretion of electrolytes and water loss, a condition called **secretory diarrhea.** The organism itself does not invade the tissues. Secretory diarrhea is characterized by its large volume, and there is little blood in the stool. This is the mechanism of cholera (see chapter 21) and some types of *Escherichia coli* and *Shigella* infectious damage.

In a more invasive diarrheal disease, the microbe invades the wall of the small or large intestine and does injury to the mucosa and underlying tissue layers. Signs and symptoms include pain in the rectum, blood in the stool, and ulceration of the inner lining. *Salmonella,* other strains of *Shigella* and *E. coli, Campylobacter,* and *Entamoeba histolytica* are responsible for this type of intestinal disease. Regardless of the cause, the loss of fluid that accompanies diarrhea results in severe dehydration and sometimes death. It accounts for 40% of the world's infectious diseases and 18% of deaths worldwide. Infants are especially vulnerable to diarrheal illness because of their smaller fluid reserves and undeveloped immunity.

See Spotlight on Microbiology 20.3 to review a successful support treatment that offsets some of the complications of diarrhea.

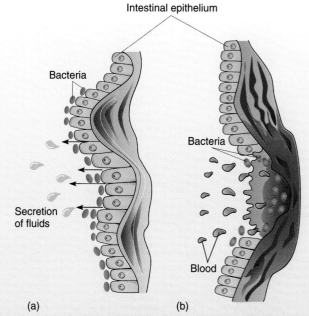

Mechanisms of infectious diarrhea. **(a)** In a toxigenic infection, the microbe remains on the surface of epithelial cells and secretes toxin into the cells. **(b)** In an invasive infection, the microbe breaks down epithelial cells and forms ulcerations, resulting in loss of the intestinal lining and bleeding.

Identification and Differential Characteristics of the Enterobacteriaceae

One large family of gram-negative bacteria that exhibits a considerable degree of relatedness is the Enterobacteriaceae. Although many members of this group inhabit soil, water, and decaying matter, they are also common occupants of the large bowel of humans and animals. All members are small (the average is 1 μm by 2–3 μm) non-spore-forming rods. The bacteria in this group grow best in the presence of air, but they are facultative and can ferment carbohydrates by an anaerobic pathway. This group is probably the most common one isolated in clinical specimens—both as normal flora and as agents of disease.

Enteric pathogens are the most frequent cause of **diarrheal illnesses** (Medical Microfile 20.2), which account for an annual mortality rate of 3 million people and an estimated 4 billion infections worldwide. Counting the total number of cases that receive medical attention, along with estimates of unreported, subclinical cases, enteric illness is probably responsible for more morbidity than any other disease. Prominent pathogenic enterics include *Salmonella, Shigella,* and strains of *Escherichia,* many of which are harbored by human carriers.

Enterics (along with *Pseudomonas* species) also account for more than 50% of all isolates in nosocomial infections (figure 20.7). Their widespread involvement in hospital-acquired infections can be attributed to their constant presence in the hospital environment and their survival capabilities. Enterics are commonly isolated from soap dishes, sinks, and invasive devices such as tracheal cannulas and indwelling catheters. The most important enteric opportunists are *E. coli, Klebsiella, Proteus, Enterobacter, Serratia,* and *Citrobacter.* Interestingly, the diseases caused by these agents usually involve systems other than the gastrointestinal tract, such as the lungs and the urinary tract.

The genera in this family share several key properties: They ferment glucose, reduce nitrates to nitrites, and are oxidase-negative. The family is traditionally divided into two subcategories.

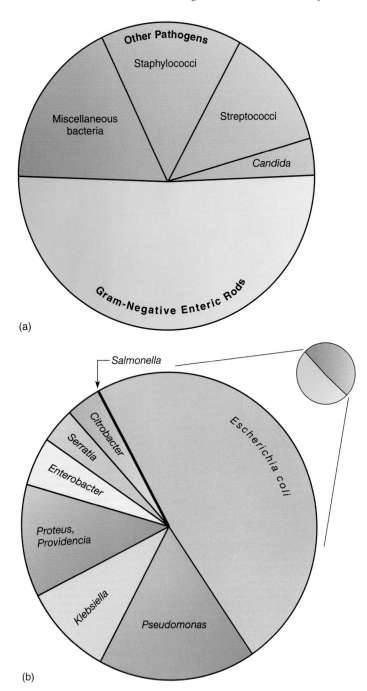

(a)

(b)

FIGURE 20.7

Bacteria that account for the majority of nosocomial infections.
(a) The overall contribution of gram-negative rods versus other agents in hospital infections. **(b)** Comparison of the relative proportion of each genus or species isolated from infections by the gram-negative enteric group.

The **coliforms** include *Escherichia coli* and other gram-negative normal enteric flora that ferment lactose rapidly (within 48 hours). **Noncoliforms** are generally non-lactose-fermenting or slow lactose-fermenting bacteria that are either normal flora or regular pathogens. Although minor exceptions occur, this distinction has considerable practical merit in laboratory, clinical, and epidemio-

logic terms. The following outline provides a framework for organizing the discussion of the Enterobacteriaceae:

> **Coliforms:** Rapid lactose-fermenting enteric bacteria that are normal flora and opportunistic (some strains of *E. coli* are true pathogens). Examples include:
>
> > *Escherichia coli*
> > *Klebsiella*
> > *Enterobacter*
> > *Hafnia*
> > *Serratia*
> > *Citrobacter*
>
> **Noncoliforms in Normal Flora:** Lactose-negative bacteria that are opportunistic, normal gut flora:
>
> > *Proteus*
> > *Morganella*
> > *Providencia*
> > *Edwardsiella*
>
> **True Pathogenic Enterics**
>
> > *Salmonella typhi, S. cholerae-suis, S. enteritidis,*
> > *Shigella dysenteriae, Sh. flexneri, Sh. boydii, Sh. sonnei*
> > *Yersinia enterocolitica, Y. pseudotuberculosis*
>
> **True Pathogenic Nonenteric**
>
> > *Yersinia pestis*

Morphology and staining characteristics are insufficient in themselves to identify the enteric bacilli. Characteristics they share include being non-spore-forming, straight rods that are often motile (exceptions are *Shigella* and *Klebsiella*), catalase-positive, and oxidase-negative. A battery of biochemical tests is commonly used to identify the principal genera and species. Although the details of this process are beyond the scope of this text, a summary will overview the major steps (figure 20.8).

Because fecal specimens contain such a high number of normal flora, they can be inoculated first into enrichment media (selenite or GN—gram-negative—broth) that inhibit the normal flora and favor the growth of pathogens. Enrichment and nonfecal specimens are streaked onto selective, differential media such as MacConkey, EMB, or Hektoen enteric agar and incubated for 24 to 48 hours. Colonies developing on these media will provide the initial separation into lactose-fermenters and non-lactose-fermenters (figure 20.9; see figure 3.9).

After the reactions of isolated colonies are noted, single colonies are subcultured on triple-sugar iron (TSI) agar slants for further analysis. As is evident in figure 20.8, several other tests are required to identify to the level of genus (table 20.2). One series of reactions, the **IMViC** (indole, methyl red, Voges-Proskauer, *i* for pronunciation, and citrate), is a traditional panel that can be used to differentiate among several genera (table 20.3). Commercial identification systems such as the Enterotube (figure 20.10) provide a rapid and compact means of acquiring data required for species identification. In general, whether an isolate is clinically significant depends upon the specimen's origin. For example, normal flora isolated from stool specimens do not usually signify infection in that site, but the same organisms isolated from extraintestinal sites in

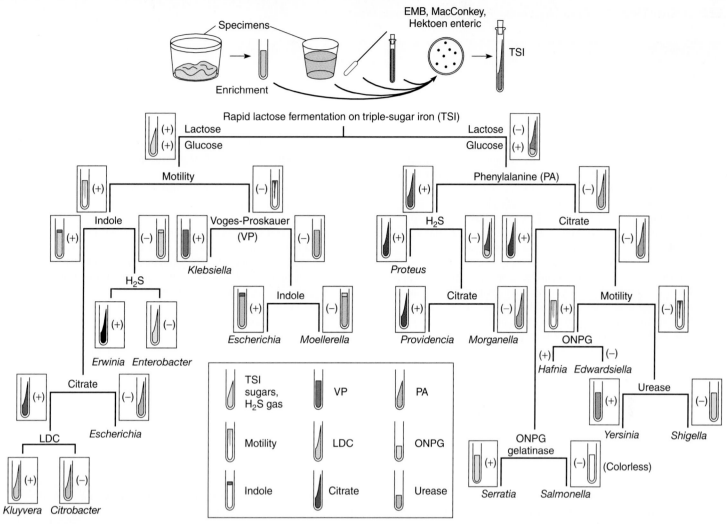

FIGURE 20.8

Procedures for isolating and identifying selected enteric genera.

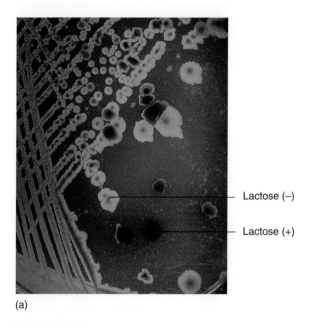

(a)

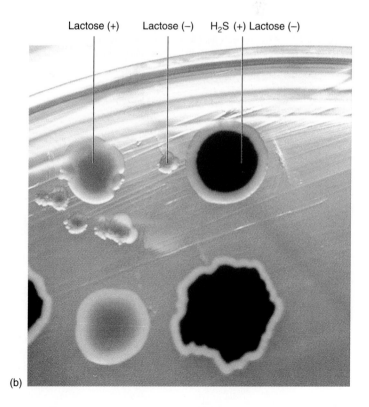

(b)

FIGURE 20.9

Isolation media for enterics, showing differentiating reactions.
(a) Levine's eosin methylene blue (EMB) agar. **(b)** Hektoen enteric agar.
(See table 20.2.)

TABLE 20.2

Protocol in Isolating and Identifying Selected Enteric Genera*

- **MacConkey agar** contains bile salts and crystal violet to inhibit gram-positive bacteria. It contains lactose and neutral red dye to indicate pH changes and differentiate the rapid lactose-fermenting bacteria from lactose nonfermenters or slow fermenters. Lactose (+) colonies are red to pink from the accumulation of acid acting on the neutral red indicator. Lactose (−) colonies are not colored in this way (see figure 3.9).

- **Levine's EMB agar** also contains bile salts, plus eosin and methylene blue dyes that cause lactose fermenters to develop a dark nucleus and sometimes a metallic sheen over the surface. Nonfermenters are pale lavender and non-nucleated (see figure 20.9*a*).

- **Hektoen enteric agar** also contains bile salts and is especially good for isolating pathogens. The bromothymol blue and acid fuchsin indicators differentiate the lactose fermenters (salmon pink to orange) from the nonfermenters (green, blue-green) and also detect the production of H_2S gas, which turns the center of a colony black (see figure 20.9*b*).

- **Utilization of lactose,** a disaccharide, requires two enzymes—a permease to bring lactose into the cell and β-galactosidase to split it into the monosaccharides glucose and galactose. Rapid fermenters on the initial isolation plates and TSI have both enzymes. Slow fermenters have only galactosidase. **ONPG** is chemical shorthand for a test that detects these slow lactose fermenters, which turn a special ONPG medium yellow. Nonfermenters do not react in either test.

- **TSI** is a nonselective medium that indicates some combination of fermentation reactions (primarily lactose and glucose, and with results from isolation medium, possibly sucrose) by means of a phenol red indicator dye. It also reveals gas and hydrogen sulfide (H_2S) production (see figure 3.5). Hydrogen sulfide is a metabolic product of the reduction of an inorganic or organic sulfur source. H_2S is indicated by a reaction with iron salts to form a black precipitate of ferric sulfide.

- The **indole test** indicates the capacity of an isolate to cleave a compound called indole off the amino acid tryptophan. If cleaving occurs, Kovac's reagent reacts with indole to form a bright red ring at the surface of the tube; if it is negative, the Kovac's remains yellow.

- The **methyl red (MR) test** indicates a form of glucose fermentation in which large amounts of mixed acids accumulate in the medium. The pH is lowered to around 4.2, so that methyl red dye remains red when added to the tube. MR-negative bacteria do not lower the pH to this degree, and the tube is yellow to orange in the presence of the dye.

- The **Voges-Proskauer (VP) test** determines whether the product of glucose fermentation is a neutral metabolite called acetyl-methylcarbinol (acetoin). This substance reacts with Barritt's reagent to form a pink to rosy-red tinge in the medium. With negative results, the tube remains brown to yellow.

- **Citrate media** contain citrate as the only usable carbon source and a pH indicator, bromothymol blue. If an isolate can utilize this carbon source, it grows and produces alkaline by-products, resulting in a conversion of the medium's color from green (neutral) to blue (alkaline).

- **Lysine** and **ornithine decarboxylase (LDC, ODC)** detect the presence of enzymes that remove carbon dioxide from amino acids. The end products are alkaline amines that raise the pH and convert the bromocresol purple to a bright lavender color. Negative results are yellow.

- **Urea media** contain 2% urea, a nitrogenous waste product of mammals, and phenol red pH indicator. Some bacteria produce the enzyme urease, which hydrolyses urea into two molecules of ammonium. Ammonium raises the pH of the medium, and the indicator turns bright pink.

- **Phenylalanine (PA) deaminase,** an enzyme produced primarily by the *Proteus* group, removes an amino group from the amino acid phenylalanine to produce phenylpyruvic acid. This reaction is visualized by adding ferric chloride to the tube, which turns olive green when it reacts with the phenylpyruvic acid.

- **Motility** is tested by a hanging drop slide or motility test medium (see figure 3.5).

The metabolic basis for some tests is discussed in chapter 8.

TABLE 20.3

The IMViC Tests for Differentiating Common Opportunistic Enterics

Genus	Indole	Methyl Red	Voges-Proskauer	Citrate
Escherichia	+	+	−	−
Citrobacter	+	+	−	−
Klebsiella/ Enterobacter	−	−	V*	+
Serratia	−	V*	+	+
Proteus	+	−	−	+
Providencia	+	+	−	+
Pseudomonas	−	−	−	V*

V = Species variable.

sputum, blood, urine, and spinal fluid are considered likely agents of opportunistic infections.

ANTIGENIC STRUCTURES AND VIRULENCE FACTORS

The gram-negative enterics have complex surface antigens that are important in pathogenicity and are also the basis of immune responses (figure 20.11). By convention, they are designated **H,** the **flagellar antigen; K,** the **capsule** and/or **fimbrial antigen;** and **O,** the **somatic** or **cell wall antigen.**[2] Not all species carry the H and K antigens, but all have O, the lipopolysaccharide implicated in endotoxic shock. Most species of gram-negative enterics exhibit a

---

2. H comes from the German word *hauch* (breath) in reference to the spreading pattern of motile bacteria grown on moist solid media. K is derived from the German term *kapsel.* O is for *ohne-hauch,* or nonmotile.

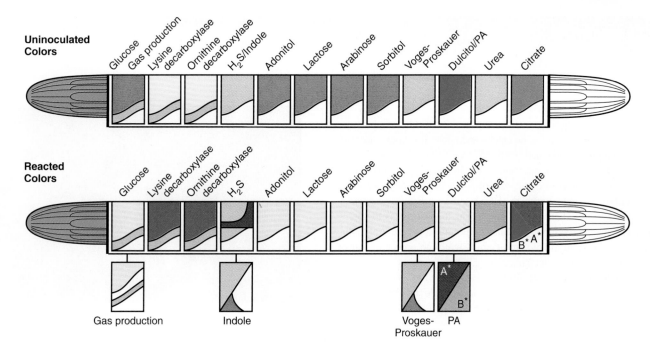

Uninoculated Colors

Glucose · Gas production · Lysine decarboxylase · Ornithine decarboxylase · H₂S/Indole · Adonitol · Lactose · Arabinose · Sorbitol · Voges-Proskauer · Dulcitol/PA · Urea · Citrate

Reacted Colors

Glucose · Lysine decarboxylase · Ornithine decarboxylase · H₂S · Adonitol · Lactose · Arabinose · Sorbitol · Voges-Proskauer · Dulcitol/PA · Urea · Citrate

B* A*

Gas production   Indole   Voges-Proskauer   PA

A*   B*

FIGURE 20.10

BBL Enterotube™ II (Roche Diagnostics), a miniaturized, multichambered tube used for rapid biochemical testing of enterics. An inoculating rod pulled through the length of the tube carries an inoculum to all chambers. Here, an uninoculated tube is compared with one that gives all positive results. For those chambers that give more than one result, the first test is shown on the top (e.g., H₂S) and the second on the bottom (indole).

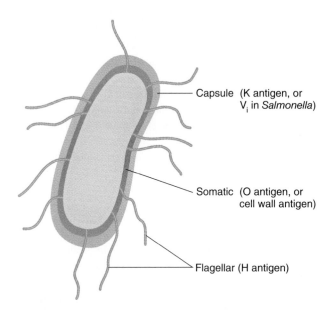

Capsule (K antigen, or Vᵢ in *Salmonella*)

Somatic (O antigen, or cell wall antigen)

Flagellar (H antigen)

FIGURE 20.11

Antigenic structures in gram-negative enteric rods. Variations in the composition of these antigens provide the basis for the serological types found in most genera.

variety of subspecies or serotypes caused by slight variations in the chemical structure of the HKO antigens. These are usually identified by serotyping, a technique in which specific antibodies are mixed with a culture to measure the degree of agglutination or precipitation (see chapter 16). Not only is serotyping helpful in classifying some species, but it also helps pinpoint the source of

outbreaks by a particular serotype, making it an invaluable epidemiologic tool.

The pathogenesis of enterics can be due to endotoxins (see Medical Microfile 20.1), exotoxins, and to the pathogen's capacity to overcome host defenses and multiply in the tissues and blood. Although most enterics are commensals, most also have a latent potential to rapidly alter their pathogenicity. As we saw in chapters 9 and 12, gram-negative bacteria freely transfer chromosomal or plasmid-borne genes that can mediate drug resistance, toxigenicity, and other adaptive traits. When this transference occurs in mixed populations of fecal residents (for example, in the intestines of carriers), it is possible for nonpathogenic or less pathogenic residents to acquire greater virulence (see *E. coli* in the next section). The most common virulence genes transferred contain the codes for enterotoxins, capsules, hemolysins, and fimbriae that promote colonization of the intestinal or urinary epithelia.

Coliform Organisms and Diseases

ESCHERICHIA COLI: THE MOST PREVALENT ENTERIC BACILLUS

Escherichia* coli is the best-known coliform, largely because of its use as a subject for laboratory studies. Although it is called the colon bacillus and sometimes regarded as the predominant species in the intestine of humans, *E. coli* is actually outnumbered 9 to 1 by the strictly anaerobic bacteria of the gut (*Bacteroides* and *Bifidobacterium*). Its prevalence in clinical specimens and infections is

---

* *Escherichia* (ess-shur-eek′-ee-uh) After the German physician Theodor Escherich.

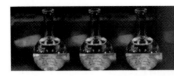

SPOTLIGHT ON MICROBIOLOGY 20.3
Escherichia coli 0157:H7: A Pathogen Known by Its Numbers

The newest pathogenic strain of *Escherichia coli* has been variously associated with fast-food restaurants and supermarket hamburger, but, unlike most new pathogens, it has no catchy nickname. It is known instead by its antigen profile—0 (somatic) type 157 and H (flagellar) type 7. This bacterium is the agent of a spectrum of conditions, ranging from mild gastroenteritis with fever to bloody diarrhea. About 10% of patients develop hemolytic uremic syndrome (HUS), a severe hemolytic anemia that can cause kidney damage and failure. The highest risk is to young children and the elderly and to immunocompromised people. Antibiotic treatments do not seem to help, and there is really no prevention other than avoiding ingestion of the pathogen.

The first human cases of disease were reported in 1982 and since then have continued to increase, probably due to improved diagnosis. Approximately 4,000 cases a year are reported in the United States, although the actual number is possibly as high as 75,000 cases. This new strain has acquired a virulence plasmid for production of Shiga (*Shigella*) toxin that accounts for the more severe disease symptoms.

The pathogen originates from the intestine of dairy cattle, and it is transmitted mainly in contaminated beef, water, and fresh vegetables. Hamburger is a common vehicle because meat processing plants tend to grind meats from several sources together, thereby contaminating large amounts of hamburger with meat from one animal carrier. Millions of pounds of hamburger have been recalled from supermarkets because of common source outbreaks. The largest epidemic in the United States occurred late in 1999, when over 600 people became infected from drinking well water at a fair in New York state.

This emerging infection has been an impetus for several changes from the slaughterhouse to the lab to the kitchen. Critics have proposed that the older visual inspection of meat alone is inadequate to detect bacteria and must be supplemented. The primary way to identify 0157:H7 is through culture on selective media and other special laboratory tests. The extra expense and time involved in such tests would greatly complicate the inspection process and require prolonged storage times. A panel of

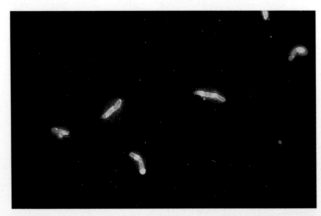

Escherichia coli cells glow in a sample of raw ground beef mixed with fluorescent antibodies specific for the pathogen.

federal officials has recommended that a rapid chemical test that can detect the microbial load be used on all meat and poultry. A group of Montana researchers has developed a 4-hour test that verifies *E. coli* in samples by means of fluorescent antibodies. It is very sensitive and could be supplied in test kits for slaughterhouses, supermarkets, produce suppliers, or even water companies (see figure).

Another result of this newly recognized food-borne pathogen are those small warning labels that now appear on all fresh meat and poultry products to inform the consumer of safe handling instructions (see Microbits 20.4). One guideline is to bring the meat up to a temperature of at least 155°F (68°C) for a few minutes. That would kill most food-borne pathogens. Certainly, ground beef must be cooked completely through and not consumed when it is still rare or pink. Because of its threat, 0157:H7 was listed as an emerging infectious agent and added to the list of reportable diseases.

---

due to its being the most common aerobic and non-fastidious bacterium in the gut. Another mistaken view is that *E. coli* is a harmless commensal. Although many of the 150 strains are not infectious, some have developed greater virulence through plasmid transfer, and others are opportunists.

Examples of Pathogenic Strains of *E. coli*

Enterotoxigenic *E. coli* causes a severe diarrheal illness brought on by two exotoxins, termed heat-labile toxin (LT) and heat-stable toxin (ST), that stimulate heightened secretion and fluid loss. In this way it mimics the pathogenesis of cholera (see figure 21.14). This strain of bacteria also has fimbriae that provide adhesion to the small intestine. **Enteroinvasive** *E. coli* causes an inflammatory disease similar to *Shigella* **dysentery*** that involves invasion and ulceration of the mucosa of the large intestine. **Enteropathogenic**

strains of *E. coli* are linked to a wasting form of infantile diarrhea whose pathogenesis is not well understood. The newest strain to emerge, *E. coli* 0157:H7, causes a hemorrhagic syndrome that can cause permanent damage to the kidney (Spotlight on Microbiology 20.3).

Clinical Diseases of *E. coli*

Most clinical diseases of *E. coli* are transmitted exclusively among humans. Pathogenic strains of *E. coli* are frequent agents of **infantile diarrhea,** the greatest single cause of mortality among babies. In some areas of the world, about 15% to 25% of children 5 years or younger die of diarrhea as either the primary disease or a complication of some other illness. The rate of infection is higher in crowded tropical regions where sanitary facilities are poor and water supplies are contaminated and where adults carry pathogenic strains to which they have developed immunity. The immature, nonimmune neonatal intestine has no protection against these

---

* dysentery (dis′-en-ter″-ee) Gr. *dys,* bad, and *enteria,* bowels.

pathogens entering in unsanitary food or water. The practice of preparing formula from dried powder mixed with contaminated water can be tantamount to inoculating the infant with a pathogen. Often the babies are already malnourished, and further loss of body fluids and electrolytes is rapidly fatal. Mothers in developing countries especially are being urged to nurse their infants so that the children get better and safer nutrition as well as gastrointestinal immunity (see Medical Microfile 15.4).

Despite the popular belief that "Montezuma's revenge," "Delhi belly," and other travel-associated gastrointestinal diseases are caused by more exotic pathogens, up to 70% of **traveler's diarrhea** cases are due to an enterotoxigenic strain of *E. coli*. Travelers pick up more virulent strains while eating or drinking contaminated foods. Within 5 to 7 days, victims are stricken with profuse, watery diarrhea, low-grade fever, nausea, and vomiting. Some travelers develop more chronic enteroinvasive dysentery that resembles *Shigella* or *Salmonella* infection. Oral antimicrobics may be effective in the early phases of infection. Over-the-counter preparations that slow gut motility (kaolin or *Lomotil*) afford symptomatic relief, but they can also serve to retain the pathogen longer and are probably not as effective as bismuth salicylate mixture (Pepto-Bismol), which counteracts the enterotoxin.

Escherichia coli often invades sites other than the intestine. For instance, it causes 50% to 80% of **urinary tract infections (UTI)** in healthy people. Urinary tract infections usually result when the urethra is invaded by its own endogenous bacterial colonists. The infection is more common in women because their relatively short urethras promote ascending infection to the bladder (cystitis) and occasionally the kidneys. Patients with bladder catheters are also at risk for *E. coli* urethritis. Other extraintestinal infections in which *E. coli* is involved are neonatal meningitis, pneumonia, septicemia, and wound infections. These often complicate surgery, endoscopy, tracheostomy, catheterization, renal dialysis, and immunosuppressant therapy.

E. coli and the Coliform Count

Because of its prominence as a normal intestinal bacterium in most humans, *E. coli* is currently one of the indicator bacteria to monitor fecal contamination in water, food, and dairy products. According to this rationale, if *E. coli* is present in a water sample, fecal pathogens such as *Salmonella*, viruses, or even pathogenic protozoa may also be present. Coliforms such as *E. coli* are used because they are present in larger numbers, can survive in the environment, and are easier and faster to detect than true pathogens. If a certain number of coliforms are detected in a sample, the water is judged unsafe to drink. This topic is covered in more detail in chapter 26.

OTHER COLIFORMS

Other coliforms of clinical importance mostly as opportunists are *Klebsiella, Enterobacter, Serratia,* and *Citrobacter* (see figure 20.7). Although these bacteria are enterics, most are infectious when introduced into sites other than the intestine. They cause infection primarily when the natural host defenses fail, often in conjunction with medical procedures. Various members of this group can infect nearly any organ, but the most serious complications involve septicemia and endotoxemia. Because of their widespread occurrence, their drug resistance, and the poor medical state of the

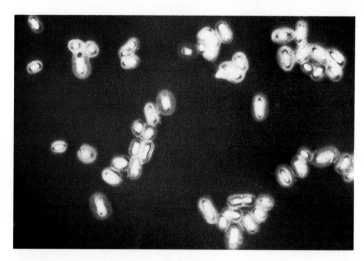

FIGURE 20.12
A capsule stain of *Klebsiella pneumoniae* (1,500×).

patients they infect, complete control will probably never be possible.

In addition to inhabiting the intestines of humans and animals, ***Klebsiella**** are also found in the respiratory tracts of normal individuals. This colonization leads to the chronic lung infections with which the species is associated. Infection by this organism is promoted by the large capsule, which prevents phagocytosis (figure 20.12). Some strains also produce toxins. The most important species in this group is ***Klebsiella pneumoniae,*** a frequent secondary invader and cause of nosocomial pneumonia, meningitis, bacteremia, wound infections, and UTIs.

Enterobacter and a closely related genus, *Hafnia,* commonly inhabit soil, sewage, and dairy products. Urinary tract infections account for the majority of clinical diseases associated with *Enterobacter.* The bacterium is also isolated from surgical wounds, spinal fluid, sputum, and blood. Although sometimes regarded as a minor pathogen, this organism is lethal when introduced into the bloodstream. In one outbreak caused by contaminated intravenous fluids, 150 people were infected and 9 died.

Citrobacter, a regular inhabitant of soil, water, and human feces, is an occasional opportunist in urinary tract infections and bacteremia in debilitated persons.

Serratia occurs naturally in soil and water as well as in the intestine. One species, *Serratia marcescens,**** produces an intense red pigment when grown at room temperature (figure 20.13). This species was once considered so benign that microbiologists used it to trace the movements of air currents in hospitals and over cities and even to demonstrate transient bacteremia after dental extraction. Unfortunately, it is fully capable of invading a compromised host. *Serratia* pneumonia, particularly prevalent in alcoholics, is transmitted by contaminated respiratory-care equipment and infusions. The organism is often implicated in burn and wound infections, as well as in fatal septicemia and meningitis in immunosuppressed patients. To make matters worse, most strains have resistance to several classes of antibiotics.

---

* *Klebsiella* (kleb-see-el'-uh) After Theodor Klebs, a German bacteriologist.

* *Serratia marcescens* (sur-at'-ee-uh) In memory of the great physicist S. Serrati and *marcescens* (mar-sess'-uns) for "fading away."

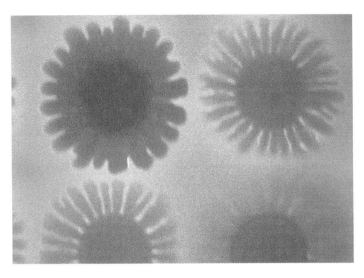

FIGURE 20.13

The brilliant red pigment (prodigiosin) formed by *Serratia marcescens* is striking and beautiful. Imagine how startled eighteenth-century Italians were when bright red dots appeared on the surfaces of cornmeal and communion wafers. In their haste to explain the phenomenon, they proclaimed that the spots were blood and the event was a miracle. When local scientist B. Bizio was called in to solve the mystery, he observed tiny red "fungi" under the microscope.

FIGURE 20.14

The wavelike swarming pattern of *Proteus mirabilis* is typical of this genus. Although this pattern is one of the most striking sights in the microbiology lab, it makes isolation of colonies in mixtures containing *Proteus* very difficult.

Noncoliform Lactose-Negative Enterics

OPPORTUNISTS: *PROTEUS* AND ITS RELATIVES

The three closely related genera *Proteus**, *Morganella**, and *Providencia** are saprobes in soil, manure, sewage, and polluted water and commensals of humans and other animals. Despite widespread distribution, they are ordinarily harmless to a healthy individual. *Proteus* species swarm on the surface of moist agar in an unmistakable concentric pattern (figure 20.14). They are commonly involved in urinary tract infections, wound infections, pneumonia, septicemia, and occasionally infant diarrhea. *Proteus* urinary infections appear to stimulate renal stones and damage due to the urease they produce and the rise in urine pH this causes. *Morganella morganii* and *Providencia* species are involved in similar infections. One opportunist, *P. stuartii*, is a frequent invader of burns. Antibiotic therapy can be hindered by resistance to several antimicrobic agents.

PATHOGENS: *SALMONELLA* AND *SHIGELLA*

The salmonellae and shigellae are distinguished from the coliforms and the *Proteus* group by having well-developed virulence factors, being primary pathogens, and not being normal flora of humans. The illnesses they cause—called **salmonelloses** and **shigelloses**—show some gastrointestinal involvement and diarrhea but often affect other systems as well.

Typhoid Fever and Other Salmonelloses

The most serious pathogen in the genus *Salmonella** is *S. typhi*, the cause of typhoid fever. The other members of the genus are divided into two subgroups: *S. cholerae-suis*, a zoonosis of swine, and *S. enteritidis*, a large superspecies that includes around 1,700 different serotypes, based on variations on the major O, H, and V$_i$ antigens (V$_i$ being the **vi**rulence or capsule antigen in *Salmonella*). Each serotype has a species designation that reflects its source—for example, *S. typhimurium* (found in rats and mice) and *S. newport* (the city of isolation). The genus is so complex that most species determinations are performed by specialists at the Centers for Disease Control and Prevention.

Salmonellae are motile; they ferment glucose with acid and sometimes gas; and most of them produce H$_2$S, but not urease. They are not fastidious, grow readily on most laboratory media, and can survive outside the host in inhospitable environments such as fresh water and freezing temperatures. These pathogens are resistant to chemicals such as bile and dyes (the basis for isolation on selective media) and do not lose virulence after long-term artificial cultivation.

Typhoid fever is so named because it bears a superficial resemblance to typhus, a rickettsial disease, even though the two diseases are otherwise very different. In the United States, the incidence of typhoid fever has remained at a steady rate for the last 30 years, appearing sporadically (figure 20.15). Of the 50 to 100 cases reported annually, roughly half are imported from endemic regions. In other parts of the world, typhoid fever is still a serious health problem responsible for 25,000 deaths each year and probably millions of cases.

The typhoid bacillus usually enters the alimentary canal along with water or food contaminated by feces, although it is occasionally spread by close personal contact. Because humans are

---

* *Proteus* (pro´-tee-us) After Proteus, a Greek sea god who could assume many forms; *Morganella* (mor-gan-ell´-uh) after T. H. Morgan; and *Providencia* (prah-vih-den´-see-uh) after Providence, R.I.

* *Salmonella* (sal-moh-nel´-uh) After Daniel Salmon, an American pathologist.

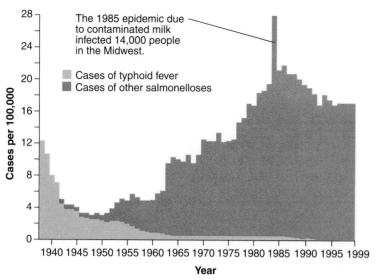

FIGURE 20.15

Data on the prevalence of typhoid fever and other salmonelloses from 1940 to 1999. Nontyphoidal salmonelloses did occur before 1940, but the statistics are not available.

Source: Data from Morbidity and Mortality Weekly Report, *January 9, 1998, Vol. 46. Centers for Disease Control and Prevention, Atlanta, GA.*

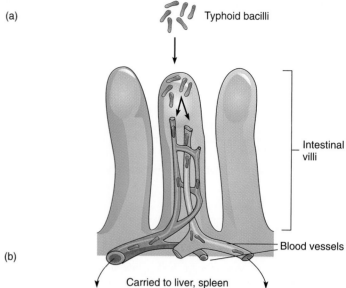

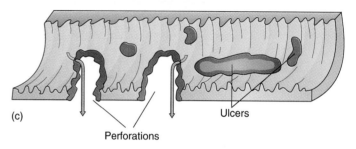

FIGURE 20.16

The phases of typhoid fever. **(a)** Ingested cells invade the small intestinal lining. **(b)** From this site, they enter the bloodstream, causing septicemia and endotoxemia. **(c)** Infection of the lymphatic tissue of the small intestine can produce varying degrees of ulceration and perforation of the intestinal wall.

the exclusive hosts for *S. typhi,* asymptomatic carriers are important in perpetuating and spreading typhoid. Even 6 weeks after convalescence, the bacillus is still shed by about half of recovered patients. A small number of people chronically carry the bacilli for longer periods in the gallbladder; from this site, the bacilli are constantly released into the intestine and feces.

The size of the *S. typhi* dose that must be swallowed to initiate infection is between 1,000 and 10,000 bacilli. At the mucosa of the small intestine, the bacilli adhere and initiate a progressive, invasive infection that leads eventually to septicemia. Symptoms are fever, diarrhea, and abdominal pain. The typhoid bacillus infiltrates the mesenteric lymph nodes and the phagocytes of the liver and spleen. In some people, the small intestine develops areas of ulceration that are vulnerable to hemorrhage, perforation, and peritonitis (figure 20.16). Its presence in the general circulation may lead to nodules or abscesses in the liver or urinary tract. Prompt treatment greatly reduces the possibility of death.

A preliminary diagnosis of typhoid fever can be based upon the patient's history and presenting symptoms, supported by a rising antibody titer (see chapter 16). However, definitive diagnosis requires isolation of the typhoid bacillus. Either chloramphenicol or sulfa-trimethoprim is the drug of choice, though some resistant strains occur. Antimicrobial drugs are usually effective in treating the chronic carrier, but surgical removal of the gallbladder may be necessary in individuals with chronic gallbladder inflammation. Drug companies have produced two new vaccines; one is a live attenuated oral vaccine, and the other is based on the capsular polysaccharide. Both provide temporary protection for travelers and military personnel.

Animal Salmonelloses

Salmonelloses other than typhoid fever are termed enteric fevers, *Salmonella* food poisoning, and gastroenteritis. These diseases are usually less severe than typhoid fever and are ascribed to one of the many serotypes of *Salmonella enteritidis*—most frequently, *S. paratyphi A, S. schottmulleri (paratyphi B), S. hirschfeldii (paratyphi C),* and *S. typhimurium.* A close relative, *Arizona[3] hinshawii,* is a pathogen found in the intestines of reptiles that parallels the salmonellae in its characteristics and diseases.

Nontyphoidal salmonelloses are more prevalent than typhoid fever and are currently holding steady at 40,000 to 50,000 cases a year. The CDC estimates that the true incidence is actually much higher, around 1.5 million cases per year (see figure 20.15). Unlike typhoid, all strains are zoonotic in origin, though humans may become carriers under certain circumstances. *Salmonella* are normal intestinal flora in cattle, poultry, rodents, and reptiles. Animal products such as meat and milk can be readily contaminated during slaughter, collection, and processing. There are inherent risks in eating poorly cooked beef or unpasteurized fresh or dried milk, ice

---

3. Named for the state.

Most food infections or other enteric illnesses caused by *Salmonella, Shigella, Vibrio, Escherichia coli,* and other enteric pathogens are transmitted by the **4 F's—food, fingers, feces,** and **flies**—as well as by water. Individuals can protect themselves from infection by taking precautions when preparing food and traveling, and communities can protect their members by following certain public health measures.

In the Kitchen

A little knowledge and good common sense go a long way. Food-related illnesses can be prevented by:

1. Use of good sanitation methods while preparing food. This includes liberal handwashing, disinfecting surfaces, sanitizing utensils, and preventing contamination by mechanical vectors such as flies and cockroaches.
2. Use of temperature. Sufficient cooking of meats, eggs, and seafood is essential. Most microbes are not heat-resistant and will be killed in 10 minutes at 100°C. Refrigeration and freezing prevent multiplication of bacteria in fresh food that is not to be cooked (but remember, it does not kill them).
3. Do not taste uncooked batter made with raw eggs or ground meats before cooking.

While Traveling

1. Choose food such as sealed or packaged items that are less likely to be contaminated with enteric pathogens.

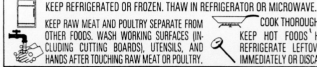

SAFE HANDLING INSTRUCTIONS
THIS PRODUCT WAS PREPARED FROM INSPECTED AND PASSED MEAT AND/OR POULTRY. SOME FOOD PRODUCTS MAY CONTAIN BACTERIA THAT COULD CAUSE ILLNESS IF THE PRODUCT IS MISHANDLED OR COOKED IMPROPERLY. FOR YOUR PROTECTION, FOLLOW THESE SAFE HANDLING INSTRUCTIONS.

KEEP REFRIGERATED OR FROZEN. THAW IN REFRIGERATOR OR MICROWAVE.

KEEP RAW MEAT AND POULTRY SEPARATE FROM OTHER FOODS. WASH WORKING SURFACES (INCLUDING CUTTING BOARDS), UTENSILS, AND HANDS AFTER TOUCHING RAW MEAT OR POULTRY.

COOK THOROUGHLY. KEEP HOT FOODS HOT. REFRIGERATE LEFTOVERS IMMEDIATELY OR DISCARD.

Supermarket label emphasizes precautions in handling, storing, and cooking meat and poultry products, as a way to avoid food-borne disease.

2. Drink bottled water whenever possible.
3. Do not brush your teeth or wash fruit or vegetables in tap water.
4. Pack antidiarrheal medicines and water disinfection tablets.

Community Measures

1. Water purification. Most enteric pathogens are destroyed by ordinary chlorination.
2. Pasteurization of milk.
3. Detection and treatment of carriers.
4. Restriction of carriers from food handling.
5. Constant vigilance during floods and other disasters that result in sewage spillage and contamination of food and drink.

cream, and cheese. A particular concern is the contamination of foods by rodent feces. Several outbreaks of infection have been traced to unclean food storage or to food-processing plants infested with rats and mice.

It is estimated that one out of every three chickens is contaminated with *Salmonella,* and other poultry such as ducks and turkeys are also affected. Eggs are a particular problem because the bacteria may actually enter the egg while the shell is being formed in the chicken. The poultry industry is testing Preempt, a product containing a mixture of normal flora sprayed on newly-hatched chicks to colonize them. This inoculation appears to curb the acquisition of *Salmonella* in a high percentage of treated animals. In general, one should always assume that poultry and poultry products harbor *Salmonella* and handle them accordingly, with clean techniques and adequate cooking (Microbits 20.4). Drug resistance of the salmonellas is on the rise, some of which can be traced to the practice of adding antibiotics to animal feeds.

Most cases are traceable to a common food source such as milk or eggs. Outbreaks have occurred in association with milk, homemade ice cream made with raw eggs, and Caesar salad. Some cases may be due to poor sanitation. In one outbreak, about

60 people became infected after visiting the Komodo dragon (a large lizard) exhibit at the Denver zoo. They apparently neglected to wash their hands after handling the rails and fence of the dragon's cage.

The symptoms, virulence, and prevalence of *Salmonella* infections can be described by means of a pyramid, with typhoid fever at the peak, enteric fevers and gastroenteritis at intermediate levels, and asymptomatic infection at the base. In typhoid and enteric fevers, elevated body temperature and septicemia are much more prominent than gastrointestinal disturbances. Gastroenteritis is manifested by symptoms of vomiting, diarrhea, fluid loss, and mucosal lesions. In otherwise healthy adults, symptoms spontaneously subside after 2 to 5 days; death is infrequent except in debilitated persons.

Serotyping is expensive, complicated, and not necessary before administering treatment. It is attempted primarily when the source of infection must be traced, as in the case of outbreaks and epidemics. A new probe technology that uses PCR can assay and identify *Salmonella* in less than 24 hours. Treatment for complicated cases of gastroenteritis is similar to that for typhoid fever; uncomplicated cases are handled by fluid and electrolyte replacement.

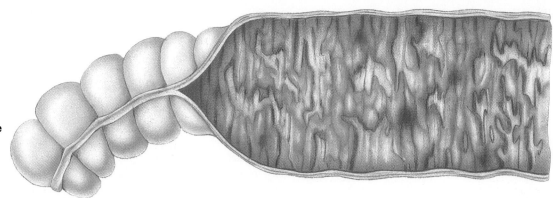

FIGURE 20.17

The appearance of the large intestinal mucosa in *Shigella* (bacillary) dysentery. Note the patches of blood and mucus, the erosion of the lining, and the absence of perforation.

Shigella and Bacillary Dysentery

*Shigella** causes a common but often incapacitating dysentery called shigellosis, which is marked by crippling abdominal cramps and frequent defecation of watery stool filled with mucus and blood. The etiologic agents *(Shigella dysenteriae, Sh. sonnei, Sh. flexneri,* and *Sh. boydii)* are primarily human parasites, though they can infect apes. All produce a similar disease that can vary in intensity. They are nonmotile, nonencapsulated, and not fastidious and do not produce H_2S or urease. The shigellae resemble some types of pathogenic *E. coli* so closely that they are placed in the same subgroup.

Although *Sh. dysenteriae* causes the severest form of dysentery, it is uncommon in the United States and occurs primarily in the Eastern Hemisphere. In the past decade, the prevalent agents in the United States have been *Sh. sonnei* and *Sh. flexneri,* which cause approximately 20,000 to 25,000 cases each year, half of them in children between 1 and 10 years of age. In addition to the usual oral route, shigellosis is also acquired through direct person-to-person contact, largely because of the small infectious dose required (200 cells). The disease is mostly associated with lax sanitation, malnutrition, and crowding and is spread epidemically in daycare centers, prisons, mental institutions, nursing homes, and military camps. As in other enteric infections, *Shigella* can establish a chronic carrier condition in some people that lasts several months.

Shigellosis is different from salmonellosis in that *Shigella* invades the villus cells of the large intestine, rather than the small intestine. In addition, it is not as invasive as *Salmonella* and does not perforate the intestine or invade the blood. It enters the intestinal mucosa by means of lymphoid cells in Peyer's patches. Once in the mucosa, *Shigella* instigates an inflammatory response that causes extensive tissue destruction. The release of endotoxin causes fever, and enterotoxin damages the mucosa and villi. Local areas of erosion give rise to bleeding and heavy mucous secretion (figure 20.17). *Shigella dysenteriae* produces a heat-labile exotoxin (shiga toxin) that has a number of effects, including injury to nerve cells and nerves and damage to the intestine.

Diagnosis is complicated by the coexistence of several alternative candidates for bloody diarrhea such as *E. coli* and the protozoans *Entamoeba histolytica* and *Giardia lamblia.* Isolation and identification follow the usual protocols for enterics. Infection is treated by fluid replacement and oral drugs such as ciprofloxacin and sulfa-trimethoprim (SxT) unless drug resistance is detected, in which case ampicillin or cephalosporins are prescribed. Prevention follows the same steps as for salmonellosis (see Microbits 20.4), and there is no vaccine yet available.

CHAPTER CHECKPOINTS

Enterobacteriaceae are a large family of gram-negative rods. Some are free-living, and others live as enterics, the normal flora of vertebrate intestinal tracts. Enteric species are grouped as coliforms or noncoliforms on the basis of their ability to ferment lactose. Enteric pathogens are identified by sugar fermentations, motility, carbon sources, enzymes, and gas production. Virulence factors of pathogenic species include enterotoxins, exotoxins, capsule, fimbriae, hemolysins, and endotoxin.

Pathogenic *Escherichia coli* diseases are usually limited to the intestine. An exception is *E. coli* 0157:H7, which causes hemolytic anemia and kidney damage. Virulence factors are exotoxins, which cause secretory diarrhea, and fimbriae that enable it to adhere to the intestinal mucosa. *E. coli* is an important opportunist in urinary and nosocomial infections. The presence of *E. coli* in food, water, or dairy products indicates fecal contamination and, therefore, the possible presence of enteric pathogens.

Enteric species *Klebsiella, Enterobacter, Serratia,* and *Citrobacter* are not usually harmful to the healthy host, but they can cause serious, even fatal, nosocomial infections when introduced into nonintestinal sites. *Klebsiella* is the most important secondary invader in this group.

The noncoliform enterics *Proteus, Morganella,* and *Providencia* are also opportunistic pathogens that are often involved in septicemia, wound infection, urinary tract infections, and pneumonia.

Salmonella and *Shigella* are noncoliform pathogens that cause intestinal infections with a wide range of severity, depending on the species.

Salmonella is transmitted directly and also indirectly through contaminated food and water. *Salmonella typhi* causes the most serious disease (typhoid fever). Other species cause salmonellosis food poisoning, gastroenteritis, and enteric fevers. Humans are the only host for *S. typhi,* but other *Salmonella* pathogens are transmitted through poultry and dairy products, human feces, and fomites.

Shigellosis, or bacillary dysentery, is a disease of worldwide distribution. It is caused by several species of *Shigella.* Infections are easily initiated because of the small infective dose. Humans are the only reservoir. Virulence factors are endotoxin and enterotoxins that attack the colon.

---

* *Shigella* (shih-gel'-uh) After K. Shiga, a Japanese physician.

The Enteric *Yersinia* Pathogens

Although formerly classified in a separate category, the genus *Yersinia** has been placed in the Family Enterobacteriaceae on the basis of cultural, biochemical, and serological characteristics. All three species cause zoonotic infections called *yersinioses*. *Yersinia enterocolitica* and *Y. pseudotuberculosis* are intestinal inhabitants of wild and domestic animals that cause enteric infections in humans, and *Y. pestis* is the nonenteric agent of bubonic plague.

Yersinia enterocolitica has been isolated from healthy and sick farm animals, pets, wild animals, and fish, as well as fruits, vegetables, and drinking water. During the incubation period of about 4 to 10 days, the bacteria invade the small intestinal mucosa, and some cells enter the lymphatics and are harbored intracellularly in phagocytes. Inflammation of the ileum and mesenteric lymph nodes gives rise to severe abdominal pain that mimics appendicitis. *Yersinia pseudotuberculosis* shares many characteristics with *Y. enterocolitica*, though infections by the former are more benign and center upon lymph node inflammation rather than mucosal involvement.

NONENTERIC *YERSINIA PESTIS* AND PLAGUE

The word **plague**[4] conjures up visions of death and morbidity unlike any other infectious disease. Although pandemics of plague have probably occurred since antiquity, the first one that was reliably chronicled killed an estimated 100 million people in the sixth century. The last great pandemic occurred in the late 1800s and was transmitted around the world, primarily by rat-infested ships. The disease was brought to the United States through the port of San Francisco around 1906. Eventually, infected rats mingled with native populations of rodents and gradually spread the populations throughout the West and Midwest. The cause of this dread disease is a tiny, harmless-looking gram-negative rod called **Yersinia pestis**, formerly *Pasteurella pestis*, with unusual bipolar staining and capsules (figure 20.18).

Virulence Factors

Strains of the plague bacillus are just as virulent today as in the Middle Ages. Some of the virulence factors that coincide with its high morbidity and mortality are capsular and envelope proteins, which protect against phagocytosis and foster intracellular growth. The bacillus also produces coagulase, which clots blood and is involved in clogging the esophagus in fleas and obstructing blood vessels in humans. Other factors that contribute to pathogenicity are endotoxin and a highly potent murine toxin.

The Complex Epidemiology and Life Cycle of Plague

The plague bacillus exists naturally in many animal hosts, and its distribution is extensive, though the incidence of disease has been reduced in most areas. Plague still exists endemically in large areas of Africa, South America, the Mideast, Asia, and the former USSR, where it sometimes erupts into epidemics such as a recent outbreak in India that infected hundreds of residents. This new surge in cases was attributed to increased populations of rats following the mon-

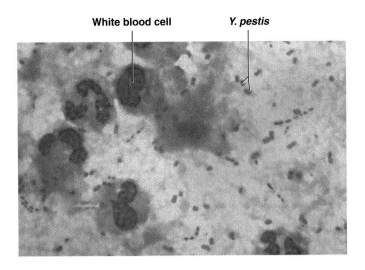

White blood cell   Y. pestis

FIGURE 20.18

A Gram-stained preparation of *Yersinia pestis* in the blood of an infected mouse.   The cells exhibit a distinctive bipolar morphology reminiscent of a safety pin.

soon floods. In the United States, sporadic cases (usually less than 10) occur as a result of contact with wild and domestic animals. No cases of human-to-human transmission have been recorded since 1924. Persons most at risk for developing plague are veterinarians and people living and working near woodlands and forests.

The epidemiology of plague is among the most complex of all diseases. It involves several different types of vertebrate hosts and flea vectors, and its exact cycle varies from one region to another. A general scheme of the cycle is presented in figure 20.19. Humans can develop plague through contact with wild animals **(sylvatic plague),** domestic or semidomestic animals **(urban plague),** or infected humans.

The Animal Reservoirs   The plague bacillus occurs in 200 different species of mammals. The primary long-term *endemic reservoirs* are various rodents such as mice and voles that harbor the organism but do not develop the disease. These hosts spread the disease to other mammals called *amplifying hosts* that become infected with the bacillus and experience massive die-offs during epidemics. These hosts, including rats, ground squirrels, chipmunks, and rabbits, are the usual sources of human plague. The particular mammal that is most important in this process depends on the area of the world. Other mammals (camels, sheep, coyotes, deer, dogs, and cats) can also be involved in the transmission cycle.

Flea Vectors   The principal agents in the transmission of the plague bacillus from reservoir hosts to amplifying hosts to humans are fleas. These tiny, bloodsucking insects (see figure 13.19 and Microbits 21.4) have a special relationship with the bacillus. After a flea ingests a blood meal from an infected animal, the bacilli multiply in its gut. In fleas that effectively transmit the bacillus, the esophagus becomes blocked due to coagulation factors produced by the pathogen. Being unable to feed properly, the ravenous flea jumps from animal to animal in a futile attempt to get nourishment. During this process, regurgitated infectious material is inoculated into the bite wound.

---

4. From the Latin *plaga*, meaning to strike, infest, or afflict with disease, calamity, or some other evil.

* *Yersinia* (yur-sin´-ee-uh) After Alexandre Yersin, a French bacteriologist.

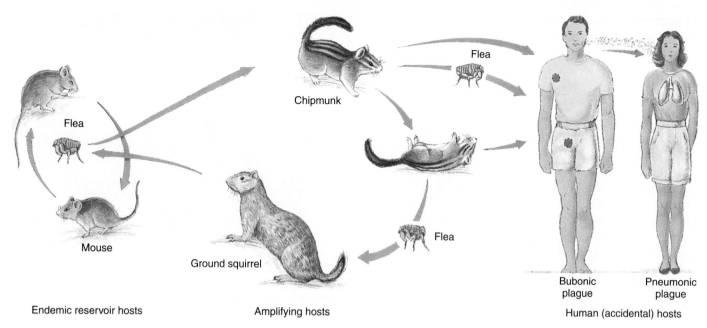

FIGURE 20.19

The infection cycle of *Yersinia pestis* simplified for clarity.

Normally the flea transmits the bacillus within that population, but many fleas are not host-specific and will attempt to feed on other species, even humans. Fleas of rodents such as rats and squirrels are most often vectors in human plague, though occasionally, the human flea is involved. Humans can also be infected by handling infected animals, animal skins, or meat and by inhaling droplets.

Pathology of Plague

The number of bacilli required to initiate a plague infection is small—perhaps 3 to 50 cells. The manifestations of infection lead to bubonic, septicemic, or pneumonic plague. In **bubonic plague,** the plague bacillus multiplies in the flea bite, enters the lymph, and is filtered by the local lymph nodes. Infection causes necrosis and swelling of the node, called a **bubo,*** typically in the groin and less often in the axilla (figure 20.20). The incubation period lasts 2 to 8 days, ending abruptly with the onset of fever, chills, headache, nausea, weakness, and tenderness of the bubo.

Cases of bubonic plague often progress to massive bacterial growth in the blood termed **septicemic plague.** The release of virulence factors causes disseminated intravascular coagulation, subcutaneous hemorrhage, and purpura that may degenerate into necrosis and gangrene. Because of the visible darkening of the skin, the plague has often been called the "black death." In **pneumonic plague,** infection is localized to the lungs and is highly contagious through sputum and aerosols. Without proper treatment, it is invariable fatal.

Diagnosis, Treatment, and Prevention   Because death can ensue as quickly as 2 to 4 days after the appearance of symptoms, prompt diagnosis and treatment of plague are imperative. The pa-

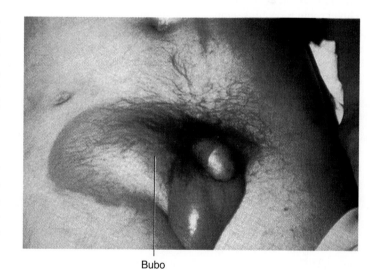

Bubo

FIGURE 20.20

A classic inguinal bubo of bubonic plague.   This hard nodule is very painful and can rupture onto the surface. (The brown discoloration is from iodine.)

tient's history, including recent travel to endemic regions, symptoms, and laboratory findings from bubo aspirates, helps establish a diagnosis. Streptomycin, tetracycline, and chloramphenicol are satisfactory treatments. Treated plague has a 90–95% survival rate.

The menace of plague is proclaimed by its status as one of the internationally quarantinable diseases (the others are cholera and yellow fever). In addition to quarantine during epidemics, plague is controlled by trapping and poisoning rodents near urban and suburban communities and by dusting rodent burrows with insecticide to kill fleas. These methods, however, cannot begin to control the reservoir hosts, so the potential for plague will always be present in

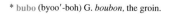

* bubo (byoo′-boh) G. *boubon*, the groin.

endemic areas, especially as humans encroach into rodent habitats. A killed or attenuated vaccine that protects against the disease for a few months is given to military personnel, veterinarians, and laboratory workers.

OXIDASE-POSITIVE NONENTERIC PATHOGENS

Pasteurella multocida

Pasteurella is a zoonotic genus that occurs as normal flora in animals and is mainly of concern to the veterinarian. Of the six recognized species, **Pasteurella multocida*** is responsible for the broadest spectrum of opportunistic infections. For example, poultry and wild fowl are susceptible to cholera-like outbreaks, and cattle are especially prone to epidemic outbreaks of hemorrhagic septicemia or pneumonia known as "shipping fever." The species has also adapted to the nasopharynx of the household cat and is normal flora in the tonsils of dogs. Because many hosts are domesticated animals with relatively close human contact, zoonotic infections are an inevitable and serious complication.

Animal bites or scratches, usually from cats and dogs, cause a local abscess that can spread to the joints, bones, and lymph nodes. Patients with weakened immune function due to liver cirrhosis or rheumatoid arthritis are at great risk for septicemic complications involving the central nervous system and heart. Patients with chronic bronchitis, emphysema, pneumonia, or other respiratory diseases are vulnerable to pulmonary failure. Contrary to the antibiotic resistance of many gram-negative rods, *P. multocida* and other related species are susceptible to penicillin, and tetracycline is an effective alternative.

HAEMOPHILUS: THE BLOOD-LOVING BACILLI

Haemophilus[5] cells are tiny (0.5 × 0.8 mm) gram-negative pleomorphic rods sometimes confused with the genus *Neisseria* in clinical samples. The members of this group tend to be fastidious and sensitive to drying, temperature extremes, and disinfectants. Even though their name means blood-loving, none of these organisms can grow on blood agar alone without special techniques. Some *Haemophilus* species are normal colonists of the upper respiratory tract or vagina, and others (primarily *H. aegyptius, H. parainfluenzae,* and *H. ducreyi*) are virulent species responsible for conjunctivitis, childhood meningitis, and chancroid.

The hemophili have a fastidious requirement for certain factors from blood to complete their metabolic syntheses (figure 20.21). Factor X, hemin, is a necessary component of cytochromes, catalase, and peroxidase. Factor V, nicotinamide adenine dinucleotide (NAD or NADP), is an important coenzyme. These factors are made available to the hemophili in media such as chocolate agar (a form of cooked blood agar; see chapter 3) and Fildes medium.

Haemophilus influenzae was originally named after it was isolated from patients with "flu" about 100 years ago. For over 40 years, it was erroneously proclaimed the causative agent until the real agent, the influenza virus, was discovered. The potential of this

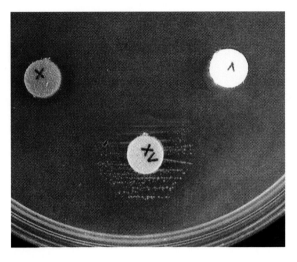

FIGURE 20.21

Discs containing separate or combined factor X and factor V are used to identify or type *Haemophilus* species. *Haemophilus influenzae* will grow only around the disc that has both factors. Other species (not shown here) require factor X or factor V, but not both.

From Gillies and Dodds, Bacteriology Illustrated, *5th ed. Fig. 78 (right), p. 104. Reprinted by permission of Churchill Livingstone.*

species to act as a pathogen remained in question until it was clearly shown to be the agent of **acute bacterial meningitis** in humans. This severe form of meningitis, caused primarily by the b serotype, was once most common in children between 3 months and 5 years of age. The case rates have declined over the past years in this age group likely because of an increased emphasis on vaccination programs. Presently more cases of *H. influenzae* meningitis are reported in the elderly, and meningococcal infection predominates in young infants and children (figure 20.22). In contrast to *Neisseria* meningitis, *Haemophilus* meningitis is not associated with epidemics in the general population but tends to occur as sporadic outbreaks in daycare and family settings. It is transmitted by close contact and nose and throat discharges. Healthy adult carriers are the usual reservoirs of the bacillus.

Haemophilus meningitis is very similar to meningococcal meningitis (see chapter 18), with symptoms of fever, vomiting, stiff neck, and neurological impairment. Untreated cases have a fatality rate of nearly 90%, but even with prompt diagnosis and aggressive treatment, 33% of children sustain residual disability. Other important diseases caused by *H. influenzae* are an intense form of epiglottitis common in older children and young adults that may require immediate intubation or tracheostomy to relieve airway obstruction. This species is also an agent of otitis media, sinusitis, pneumonia, and bronchitis.

Haemophilus infections are usually treated with a combination of chloramphenicol and ampicillin. Outbreaks of disease in families and daycare centers may necessitate rifampin prophylaxis for all contacts. Routine vaccination with a subunit vaccine (Hib) based on type b polysaccharide is recommended for all children, beginning at age 2 months, with three follow-up boosters. It is available in combination with DTaP as TriHiBit.™

---

5. Also spelled *Hemophilus.*

* *Pasteurella multocida* (pas″-teh-rel′-uh mul-toh-see′-duh) From Louis Pasteur, plus L. *multi,* many, and *cidere,* to kill.

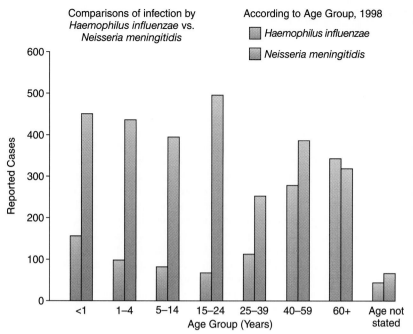

Comparisons of infection by *Haemophilus influenzae* vs. *Neisseria meningitidis*

According to Age Group, 1998

■ *Haemophilus influenzae*
■ *Neisseria meningitidis*

FIGURE 20.22

Comparisons of meningitis incidence in *Haemophilus influenzae* versus *Neisseria meningitidis*. These are the latest statistics available at publication time.

Source: Data from Morbidity and Mortality Weekly Report; Summary of Notifiable Diseases, U.S., 1998, Massachusetts Medical Society for the Centers of Disease Control and Prevention, October 31, 1999.

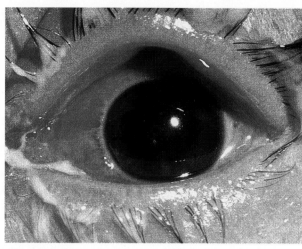

FIGURE 20.23

Acute conjunctivitis, or pinkeye, caused by *Haemophilus aegyptius*.

Haemophilus aegyptius (**Koch-Weeks bacillus**) is an agent of acute communicable **conjunctivitis,** sometimes called **pinkeye.** The subconjunctival hemorrhage that accompanies infection imparts a bright pink tinge to the sclera (figure 20.23). The disease occurs primarily in children, is distributed worldwide, and is spread through contaminated fingers and shared personal items as well as mechanically by gnats and flies. It is treated with antibiotic eyedrops.

Haemophilus ducreyi is the agent of **chancroid** (soft chancre), a sexually transmitted disease prevalent in the tropics and subtropics that afflicts mostly males. It is transmitted by direct contact with infected lesions and is favored by sexual promiscuity and unclean personal habits. After an incubation period lasting 2 to 14 days, lesions develop in the genital or perianal area. First to appear is an inflammatory macule that evolves into a painful, necrotic ulcer similar to those found in lymphogranuloma venereum and syphilis (see chapter 21). Often the regional lymph nodes develop into bubolike swellings that burst open. Although cotrimoxazole and other antimicrobics are effective, infection often recurs.

Haemophilus parainfluenzae and ***H. aphrophilus,*** members of normal oral and nasopharyngeal flora, are involved in infective endocarditis in adults who have underlying congenital or rheumatic heart disease. Such infections typically stem from routine dental procedures, periodontal disease, or some other oral injury.

CHAPTER CHECKPOINTS

Yersinia species are agents of zoonotic infections. *Y. enterocolitica* and *Y. pseudotuberculosis* cause enteric infections, primarily in children. They are transmitted through contaminated food and water. *Y. pestis* causes plague in three forms: bubonic, pneumonic, and septicemic. It is usually transmitted to humans through flea bites. Mice and rodents are the primary reservoirs of this bacillus, but infected fleas spread it to other mammals (amplifying hosts) and from there to humans. Infections respond to antibiotic therapy if treated early, but untreated infections have high mortality rates.

Pasteurella is another genus that causes a wide range of opportunistic zoonoses. *P. multocida* occasionally causes infection in humans through animal bites or scratches.

The genus *Haemophilus* contains both commensals and pathogenic species. *H. influenzae* (Hib) causes acute bacterial meningitis, a serious infection of the brain and its membranes. Vaccination for Hib is routinely given to children.

Other pathogens include *Haemophilus aegyptius,* the causative agent of conjunctivitis; *H. ducreyi,* the agent of soft chancre, an STD; and *H. parainfluenzae* and *H. aphrophilus,* agents of bacterial endocarditis. *Haemophilus* infections respond to combination drug therapy.

CHAPTER CAPSULE WITH KEY TERMS

The **gram-negative rods** are a large group of loosely affiliated families and genera, all of which are non-spore-forming; cell walls of most species contain lipopolysaccharide with **endotoxin,** which causes fever, cardiovascular disruptions, and shock; gram-negative septicemia is a dangerous complication of infection.

I. **Aerobic Rods**

A. *Pseudomonas* species are among the most widely distributed bacteria; thrive even in hostile conditions or where nutrients are scarce; medically, *Ps. aeruginosa* is a common opportunist of medical treatments when normal defenses are compromised; may attack lungs, skin, urinary tract, eyes, ears, hosts with severe burns; may also infect healthy persons; drug resistance limits treatment choices.

B. *Brucella* is a zoonotic genus that causes abortion in cattle, pigs, and goats, and **brucellosis,** or **undulant fever,** in humans.
1. Bacterium is transmitted through direct contact with infected animals or contaminated animal products and ingestion of raw milk; infection occurs in several systems.
2. Treated with streptomycin and rifampin; controlled through animal vaccination, pasteurization.

C. *Francisella tularensis* causes **tularemia,** or "rabbit fever," a zoonosis of rabbits, rodents, and other wild mammals that spreads to humans through direct contact with animals, bites by vectors (ticks), ingestion of contaminated food or water, or by inhalation.
1. Symptoms occur in skin, lymph nodes, lungs, intestine.
2. Treated with gentamicin; vaccine available for risk groups.

D. *Bordetella pertussis* causes a strictly human disease called **pertussis,** or **whooping cough;** contagious and prevalent in children under 6 months; infectious by droplets; attachment of pathogen and toxin production destroy cilial defense and produce a series of coughs that occur in bursts followed by a sudden inspiration; vaccine (acellular pertussis) is a very effective preventive.

E. *Legionella pneumophila* causes **legionellosis,** commonly called **Legionnaires disease;** usual agent is a wide-ranging inhabitant of natural water that survives for months in artificial aquatic environments (cooling towers, air conditioners, taps) and can cause serious lung disease when inhaled.
1. Community infections occur from common environmental sources.
2. Hospital epidemics occur due to contaminated air and water supplies.

II. **Facultative Anaerobic Rods**

The Family **Enterobacteriaceae** is the largest group of gram-negative **enteric** bacteria; small, often motile, fermentative rods occurring in many habitats but often found in animal intestines; informal division of family is made between **coliforms,** normal flora that are rapid lactose fermenters, and **noncoliforms,** genera that do not or only weakly ferment lactose; enterics are predominant bacteria in clinical specimens; some species cause **diarrheal disease** due to **enterotoxins** acting on the intestinal mucosa or to invasion and disruption of the mucosa; identification of group includes series of biochemical and serological tests; primary **antigens** are flagellar (H), cell wall (O), and capsular (K or V$_i$); drug resistance is well entrenched through plasmids; treatment with drugs requires sensitivity testing.

A. *Escherichia coli* exists in several forms; **pathogenic strains** have acquired virulence factors for invasiveness and toxigenicity; cause of **infantile diarrhea,** a complication of malnourished babies fed unsanitary food or water; **traveler's diarrhea** occurs in people who pick up a toxigenic strain from water and food in other countries. *E. coli* is the usual cause of **urinary tract infections** (from normal flora) and nosocomial pneumonia and septicemia.

B. *E. coli* **0157:H7** is an emerging pathogen of cattle and humans; acquired primarily through beef, produce, and water; causes hemolytic uremic syndrome due to kidney damage.

C. Other coliforms are ubiquitous in the hospital environment. *Klebsiella, Enterobacter, Serratia,* and *Citrobacter* account for nosocomial infections associated with tracheostomies, respiratory care equipment, endoscopes, and catheters; other common infections are pneumonia and burn and incision infections.

D. Noncoliform infections are caused by opportunists, pathogenic enterics, and pathogenic nonenterics. **Opportunists** include *Proteus, Morganella,* and *Providencia,* which cause nosocomial infections such as urinary tract infections, wound infections, pneumonia, sepsis, and diarrhea.

III. **True enteric pathogens include *Salmonella* and *Shigella.***

A. *Salmonella* causes **salmonelloses;** most severe disease is **typhoid fever,** caused by *S. typhi;* agent is spread only by humans via unclean food or water.
1. The bacillus crosses wall of small intestine into circulation, is carried to organs, where it may form abscesses.
2. Symptoms include fever, diarrhea, septicemia, and ulceration of small intestine; perforation is a complication.
3. Treatment with chloramphenicol; vaccine available.
4. Other species belong to a serotype of *S. enteritidis,* such as *S. typhimurium* or *S. paratyphi A;* bacilli are common flora of animals; contaminate meat, milk, and eggs; disease somewhat milder but more prevalent than typhoid fever; main infections are enteric fever or gastroenteritis, depending on severity of symptoms; treatment with antibiotics, oral rehydration.

B. *Shigella* causes **shigellosis,** a **bacillary dysentery** characterized by acute painful diarrhea with bloody, mucus-filled stools; species are *Sh. dysenteriae, Sh. sonnei,* and *Sh. flexneri;* all are primarily human parasites; spread by fingers, feces, food, and flies; bacteria remain local to large intestine; damage to intestinal villi causes symptoms; treated with oral antimicrobics and rehydration therapy. Enteric disease is preventable through cleanliness in processing food, adequate cooking and refrigeration, awareness of animal carriers, proper toilet habits, control of water and sewage, monitoring carriers, and avoiding questionable food or water.

IV. **True pathogens in *Yersinia, Pasteurella,* and *Haemophilus* genera.**

A. *Yersinia* causes *yersinioses,* zoonoses spread from mammals to humans; *Y. enterocolitica* and *Y. pseudotuberculosis* cause food infection with appendicitis-like symptoms. *Yersinia pestis* is a nonenteric agent of the **plague,** an ancient virulent disease with complex epidemiology; agent is maintained by relationship between **endemic hosts** (mice) and **amplifying hosts** (rats, squirrels) and **flea vectors** that carry the bacillus between them; humans enter this cycle through flea bite or contact with infected animal; infected humans may pass the agent to other humans. Forms are:
1. **Bubonic plague,** in which multiplication at site of bite creates regional lymphatic swelling, or **bubo.**
2. **Septicemic plague,** a deadly complication marked by hemorrhage.
3. **Pneumonic plague,** a lung infection spread by aerosols. Treatment includes streptomycin, tetracycline; vector and reservoir control and a vaccine can be preventive.

B. Nonenteric pathogens include:
  1. *Pasteurella multocida,* a zoonosis of cattle, poultry, cats, dogs; spread by bites, scratches, other contact; usual disease is abscess and lymph node swelling.
  2. *Haemophilus influenzae* is the most frequent cause of **acute bacterial meningitis** in children between 3 months and 5 years; sporadic infection occurs primarily in daycare and similar settings; passed through respiratory discharges; disease is marked by acute neurological complications, high morbidity, and sequelae; treated with chloramphenicol and ampicillin; vaccination with Hib recommended for children at risk.
  3. *H. aegyptius* causes a form of conjunctivitis called pinkeye.
  4. *H. ducreyi* causes the STD known as **chancroid.**

MULTIPLE-CHOICE QUESTIONS

1. A unique characteristic of many isolates of *Pseudomonas* useful in identification is
   a. fecal odor
   b. fluorescent green pigment
   c. drug resistance
   d. motility

2. Human brucellosis is also known as
   a. Bang disease
   b. undulant fever
   c. rabbit fever
   d. Malta fever

3. *Francisella tularensis* has which portal of entry?
   a. tick bite
   b. intestinal
   c. respiratory
   d. all of these

4. A classic symptom of pertussis is
   a. labored breathing
   b. paroxysmal coughing
   c. convulsions
   d. headache

5. The severe symptoms of pertussis are due to what effect?
   a. irritation of the glottis by the microbe
   b. pneumonia
   c. the destruction of the respiratory epithelium
   d. blocked airways

6. *Escherichia coli* displays which antigens?
   a. capsular
   b. somatic
   c. flagellar
   d. all of these

7. Which of the following is *not* an opportunistic enteric bacterium?
   a. *E. coli*
   b. *Klebsiella*
   c. *Proteus*
   d. *Shigella*

8. Which of the following represents a major difference between *Salmonella* and *Shigella* infections?
   a. mode of transmission
   b. likelihood of septicemia
   c. the portal of entry
   d. presence/absence of fever and diarrhea

9. Complications of typhoid fever are
   a. neurological damage
   b. intestinal perforation
   c. liver abscesses
   d. b and c

10. *Shigella* is transmitted by
    a. food
    b. flies
    c. feces
    d. all of these

11. The bubo of bubonic plague is a/an
    a. ulcer where the flea bite occurred
    b. granuloma in the skin
    c. enlarged lymph node
    d. infected sebaceous gland

12. *Haemophilus influenzae* requires ____ for growth.
    a. hemin
    b. NAD
    c. blood
    d. a and b

13. Circle all diseases for which human vaccines are available:
    a. typhoid fever
    b. shigellosis
    c. *Haemophilus* meningitis
    d. whooping cough
    e. bubonic plague
    f. Legionnaires disease
    g. *E. coli* infantile diarrhea
    h. brucellosis

14. Which of the following are primarily zoonoses?
    a. tularemia
    b. salmonellosis
    c. shigellosis
    d. brucellosis
    e. pasteurellosis
    f. bubonic plague

15. **Single Matching.** Match the infectious agent with its disease.
    ____ *Francisella tularensis*
    ____ *Yersinia pestis*
    ____ *Escherichia coli* 0157:H7
    ____ *Shigella* species
    ____ *Salmonella enteritidis*
    ____ *Salmonella typhi*
    ____ *Pseudomonas aeruginosa*
    ____ *Bordetella pertussis*
    ____ *Legionella pneumophila*
    ____ *Haemophilus aegyptius*
    ____ *Haemophilus influenzae*
    ____ *Haemophilus ducreyi*
    ____ *Pasteurella multocida*

    a. dysentery
    b. local abscess
    c. chancroid
    d. enteric fever
    e. whooping cough
    f. meningitis
    g. typhoid fever
    h. hemolytic uremic syndrome
    i. bubonic plague
    j. Pontiac fever
    k. folliculitis
    l. rabbit fever
    m. pinkeye

CONCEPT QUESTIONS

1. Why are bacteria such as *Pseudomonas* and coliforms so often involved in nosocomial infections?

2. a. Briefly describe the human infections caused by *Pseudomonas, Brucella,* and *Francisella*.
   b. How are they similar?
   c. How are they different?

3. a. What is the pathologic effect of whooping cough?
   b. What factors cause it to predominate in newborn infants?

4. What is unusual about *Legionella*? What is the epidemiologic pattern of the disease?

5. a. Describe the chain of events that result in endotoxic shock.
   b. Describe the key symptoms of endotoxemia.
   c. Differentiate between toxigenic diarrhea and infectious diarrhea.

6. a. Define each of the following: an enteric bacterium, a coliform, and a noncoliform.
   b. Which bacteria in the Family Enterobacteriaceae are true enteric pathogens?
   c. What are opportunists?

7. a. Briefly describe the methods used to isolate and identify enterics.
   b. What is the basis of serological tests, and what is their main use for enterics?

8. a. Explain how *E. coli* can develop increased pathogenicity, using an actual example.
   b. Describe the kinds of infections for which *E. coli* is primarily responsible.
   c. Describe the roles of other coliforms in infections.

9. a. What is salmonellosis?
   b. What is the pattern of typhoid fever?
   c. How does the carrier state occur?
   d. What is the main source of the other salmonelloses?
   e. What kinds of infections do salmonellas cause?

10. What causes the blood and mucus in dysentery?

11. Explain several practices an individual can use to avoid enteric infection and disease at home and when traveling.

12. a. Trace the epidemiologic cycle of plague.
    b. Compare the portal of entry of bubonic plague with that of pneumonic plague.
    c. Why is plague called the black death?

13. Describe the epidemiology and pathology of *Haemophilus influenzae* meningitis.

14. Compare the types of food-related illness discussed in this chapter according to
    a. the Gram reaction of the agent
    b. whether it is a food infection or intoxication
    c. the kinds of food involved

15. a. List the bacteria from this chapter for which general, routine vaccines are given.
    b. For which special groups of bacteria are there vaccines?
    c. For which bacteria are there none?
    d. Why are there no vaccines for these?

16. Briefly outline the zoonotic infections in this chapter, and describe how they are spread to humans.

17. a. Give the portal of entry and target tissues for plague, pertussis, legionellosis, and shigellosis.
    b. Which are primary pulmonary pathogens?

18. Compare and contrast the pathology, diagnosis, and treatment of meningococcal meningitis and *Haemophilus influenzae* meningitis.

CRITICAL-THINKING QUESTIONS

1. What is the logic behind testing for *E. coli* to detect fecal contamination of water?

2. Identify the genera with the following characteristics from figure 20.8:
   a. Lactose ($-$), phenylalanine and urease ($-$), citrate ($+$), ONPG ($-$)
   b. Lactose ($+$), motility ($-$), VP ($-$), Indole ($+$)
   c. Lactose ($+$), motility, indole ($-$), H_2S ($+$)
   d. Lactose ($-$), phenylalanine and urease ($+$), H_2S ($-$), citrate ($+$)

3. Given that so many infections are caused by gram-negative opportunists, what would you predict in the future as the number of compromised patients increases, and why do you make these predictions?

4. An infectious dose of several million cells in enteric infections seems like a lot.
   a. In terms of the size and abundance of microbes, could you see a cluster containing that many cells with the naked eye?
   b. Refer to chapter 7 on microbial growth cycles. About how long would it take an average bacterial species to reach the infectious dose of a million cells starting from a single cell?
   c. Why do enteric diseases require a relatively higher infectious dose than nonenteric diseases?

5. Students in our classes sometimes ask how it is possible for a single enteric carrier to infect 1,000 people at a buffet or for a box turtle to expose someone to food infection. We always suggest that they use their imagination. Provide a detailed course of events that could result from these types of outbreaks.

6. Case study 1. A woman living near a wooded area in a western state discovered her cat carrying a sick mouse. She discarded the mouse, but a neighbor's dog found it and carried it home. In a few days, one child in the neighbor's family got a case of febrile illness that responded to antibiotics. The cat died from an open sore on its neck. Later, the veterinarian who treated the cat developed a fatal pneumonia.
   a. What disease is possible here?
   b. What two possible modes of transmission were demonstrated here?

7. Case study 2. Several persons working in an exercise gym acquired an acute disease characterized by fever, cough, pneumonia, and headache. Treatment with erythromycin cleared it up. The source was never found, but an environmental focus was suspected.
   a. What do you think might have caused the disease?
   b. People in a different gym got skin lesions after sitting in a redwood hot tub. Which pathogen could have caused that?

8. Case study 3. A 3-year-old severely ill child was admitted to a hospital with symptoms of diarrhea, fever, and malaise. Laboratory testing showed abnormal renal and liver values and anemia. She had no history of previous illness, and her food history was a recent meal of teriyaki beef consumed at a local restaurant. She responded to antibiotics.
   a. What was the probable pathogen?
   b. What was the likely source?
   c. What is the pathologic effect of the pathogen?

9. a. Name five bacteria from chapters 19 and 20 that could be used in biological warfare.
   b. What are some possible ways they could be used in warfare?
   c. What is your personal opinion of using microorganisms to gain advantage during war?

10. a. Give your opinion of the U.S. Department of Agriculture's "visual inspection" program to detect pathogenic bacteria in meat.
    b. What is the weakness in this approach?
    c. What are some alternatives to this method of detecting contaminated meat?

11. Looking at the endpapers, determine which of the diseases in this chapter are reportable, which ones are currently increasing in incidence, and which ones are decreasing.

INTERNET SEARCH TOPICS

1. Use the Internet to survey the current prevalence of *E. coli* 0157:H7, *Salmonella, Haemophilus* meningitis, and pertussis in the United States. Search for the latest prevention and control measures used for these diseases/bacteria.

2. Explore Internet sites to find information on new methods of inspecting meat and poultry and of improving food safety (such as pasteurizing eggs and Preempt).

CHAPTER 2

Miscellaneous Bacterial Agents of Disease

A number of agents of bacterial infections do not fit the usual categories of gram-positive or gram-negative rods or cocci. This group includes spirochetes and curviform bacteria, obligate intracellular parasites such as rickettsias and chlamydias, and mycoplasmas. This chapter covers not only those agents but also the vectors that carry them—ticks, lice, and other arthropods. The last section of the chapter surveys oral ecology, dental diseases, and the mixed bacterial infections that are responsible for the diseases of the oral cavity.

Chapter Overview

- Unusual groups of pathogenic bacteria include spirochetes, vibrios, obligate intracellular parasites, and cell-wall-deficient species.
- *Treponema* are thin, flexible, spiral-shaped cells. Syphilis, an STD caused by *Treponema pallidum*, is a complex and chronic disease that progresses through three major phases and a latent period that may extend for many years.
- *Leptospira* is a zoonotic pathogen spread to humans by contact with animal excrement. *Borrelia* spirochetes are arthropod-borne pathogens that cause two forms of relapsing fever, a recurrent febrile infection.
- Lyme disease is a severe systemic syndrome caused by *Borrelia burgdorferi* and transmitted by tick bites. It is marked by a constellation of skin, joint, cardiac, and neurological complications.
- Vibrios are curved motile rods that are widely distributed throughout aquatic environments and animals. The most prominent genus is *Vibrio*, which causes cholera, and several other types of food infection.
- *Vibrio cholerae* is the agent of cholera, a toxin-induced diarrhea that greatly disrupts the fluid/salt balance of the body.
- Other vibrios with wide medical impact are *Campylobacter*, an animal-associated gastrointestinal pathogen, and *Helicobacter*, a resident of the stomach involved in a variety of gastric illnesses.
- Rickettsias are a large group of tiny obligate intracellular bacteria that live in the bodies of ticks and lice. The genus *Rickettsia* causes typhus and Rocky Mountain spotted fever, the most common rickettsial disease in North America.
- Fleas, lice, and ticks are tiny arthropods that feed from humans and other animals. They serve as vectors for a number of rickettsial and borrelial diseases.
- Chlamydias are obligate parasites with an unusual life cycle that varies between a sturdy infectious phase and a destructive intracellular phase.
- *Chlamydia trachomatis* causes a prevalent STD that can damage the reproductive tract and can also infect the eyes and eye membranes.

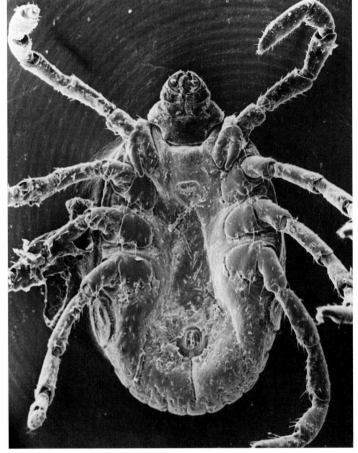

Close-up of the brown dog tick, *Rhiphicephalus sanguineous,* the carrier of a type of bacteria called rickettsias that cause several diseases in humans and are intimately associated with the life cycles of blood-sucking arthropods.

- *C. pneumoniae* infection gives rise to a form of pneumonia, and *C. psittaci* causes a zoonotic infection termed ornithosis.
- Mycoplasmas are the smallest infectious bacteria, known for their lack of a cell wall and their reliance on host membranes for survival. *Mycoplasma pneumoniae* causes a form of chronic pneumonia that lacks typical symptoms.
- Dental infections are due to a mixture of normal oral bacteria that infect when host defenses or oral hygiene have been reduced. Dental infections are associated with the buildup of a complex network of bacteria that produce plaque.
- Plaque buildup initiates the damage to enamel that is characteristic of dental caries; calcified plaque damages the gingiva and leads to gingivitis, periodontitis, and tooth loss.

The Spirochetes

Bacteria called spirochetes have a helical form and a mode of locomotion that appear especially striking in live, unstained preparations using the dark-field or phase-contrast microscope (see figure 21.7). Other traits include a typical gram-negative cell wall and a well-developed periplasmic space that encloses the flagella (called endoflagella or periplasmic flagella) (figure 21.1*a*). Although internal flagella are constrained somewhat like limbs in a sleeping bag, their flexing propels the cell by rotation and even crawling motions. The spirochetes are classified in the Order Spirochaetales, which contains two families and five genera. The majority of spirochetes are free-living saprobes or commensals of animals and are not primary pathogens. But three genera contain major human pathogens: *Treponema, Leptospira,* and *Borrelia* (figure 21.1*b, c, d,* respectively).

TREPONEMES: MEMBERS OF THE GENUS *TREPONEMA*

Treponemes are thin, somewhat regular, coiled cells that live in the oral cavity, intestinal tract, and perigenital regions of humans and animals. The pathogens are strict parasites with complex growth requirements that necessitate cultivating them in live cells. Diseases caused by *Treponema* are called **treponematoses.** The subspecies *Treponema pallidum pallidum** is responsible for venereal and congenital syphilis; the subspecies *T. p. endemicum* causes nonvenereal endemic syphilis, or bejel; and *T. p. pertenue* causes yaws. *Treponema carateum* is the cause of pinta. Infection begins in the skin, progresses to other tissues in gradual stages, and is often marked by periods of healing interspersed with relapses. The major portion of this discussion will center on syphilis, and any mention of *T. pallidum* refers to the subspecies *T. p. pallidum.* Other treponemes of importance are involved in infections of the gingiva (see oral diseases at the end of the chapter).

Treponema pallidum: The Spirochete of Syphilis

The origin of **syphilis*** is an obscure yet intriguing topic of speculation. The disease was first recognized at the close of the fifteenth century in Europe, a period coinciding with the return of Columbus from the West Indies, which led some medical scholars to conclude that syphilis was introduced to Europe from the New World. However, a more probable explanation contends that the spirochete evolved from a related subspecies, perhaps an endemic treponeme already present in the Mediterranean basin. The combination of the immunologically naive population of Europe, the European wars, and sexual promiscuity set the stage for worldwide transmission of syphilis that continues to this day.

---

* *Treponema pallidum* (trep″-oh-nee′-mah pal′-ih-dum) Gr. *trepo,* turn, and *nema,* thread; L. *pallidum,* pale. The spirochete does not stain with the usual bacteriological methods.

* syphilis The term *syphilis* first appeared in a poem entitled "Syphilis sive Morbus Gallicus" by Fracastorius (1530) about a mythical shepherd whose name eventually became synonymous with the disease from which he suffered.

Epidemiology and Virulence Factors of Syphilis

Although infection can be provoked in laboratory animals, the human is evidently the sole natural host and source of *T. pallidum.* It is an extremely fastidious and sensitive bacterium that cannot survive for long outside the host, being rapidly destroyed by heat, drying, disinfectants, soap, high oxygen tension, and pH changes. It survives a few minutes to hours when protected by body secretions

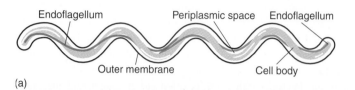

(a)

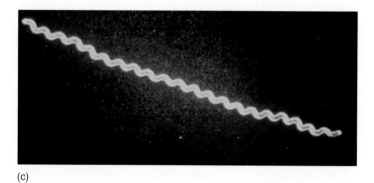

(b)

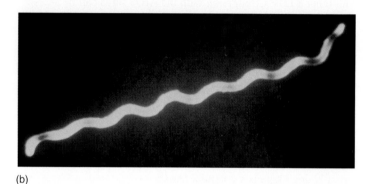

(c)

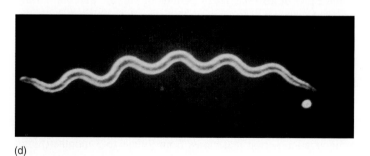

(d)

FIGURE 21.1

Typical spirochete. **(a)** Representation of general spirochete morphology with a pair of endoflagella inserted in the opposite poles, lying beneath the outer membrane and within the periplasmic space. In some spirochetes, the free ends overlap, as shown here. **(b–d)** Variations of the basic helical form. **(b)** *Treponema* has 8–20 evenly spaced coils. **(c)** *Leptospira* has numerous fine, regular coils and one or both ends curved. **(d)** *Borrelia* has 3–10 loose, irregular coils.

and about 36 hours in stored blood. Research with human subjects has demonstrated that the risk of infection from an infected sexual partner is 12% to 30%. Less common modes of transmission are passage to the fetus in utero and laboratory or medical accidents. Syphilitic infection through blood transfusion or exposure to fomites is rare.

Syphilis, like other STDs, has experienced periodic increases during times of social disruption (see figure 18.22). Currently, the number of reported cases is decreasing to the levels of the early 1960s. Because many cases go unreported, the actual incidence is likely to be several times higher than these reports show. Most cases tend to be concentrated in larger metropolitan areas among prostitutes, their contacts, and intravenous drug abusers. Syphilis continues to be a serious problem worldwide, especially in Africa and Asia. Persons with syphilis often suffer concurrent infection with other STDs. Coinfection with the AIDS virus can be an especially deadly combination with a rapidly fatal course.

Pathogenesis and Host Response

Brought into direct contact with mucous membranes or abraded skin, *T. pallidum* binds avidly by its hooked tip to the epithelium (figure 21.2). The number of cells required for infection using human volunteers was established at 57 organisms. At the binding site, the spirochete multiplies and penetrates the capillaries. Within a short time, it moves into the circulation, and the body is literally transformed into a large receptacle for incubating the pathogen—virtually any tissue is a potential target.

Specific factors that account for the virulence of the syphilis spirochete appear to be outer membrane proteins. It produces no toxins and does not appear to kill cells directly. Studies have shown that, although phagocytes seem to act against it and several types of antitreponemal antibodies are formed, cell-mediated immune responses are unable to contain it. The primary lesion occurs when the spirochetes invade the spaces around arteries and stimulate an inflammatory response. Organs are damaged when granulomas form at these sites and block circulation.

Clinical Manifestations Untreated syphilis is marked by distinct clinical stages designated as primary, secondary, and tertiary syphilis (table 21.1). It also has latent periods of varying duration during which the disease is quiescent. The spirochete appears in the lesions and blood during the primary and secondary stages, and thus is communicable at these times. It is largely non-communicable during the tertiary stage, though syphilis can be transmitted during early latency.

Primary Syphilis The earliest indication of syphilis infection is the appearance of a hard **chancre*** at the site of inoculation, after an incubation period that varies from 9 days to 3 months (figure 21.3). The chancre begins as a small, red, hard bump that enlarges and breaks down, leaving a shallow crater with firm margins. The base of the chancre beneath the encrusted surface swarms with spirochetes. Most chancres appear on the internal and external genitalia, but about 20% occur on the lips, oral cavity, nipples, fingers, or around the rectum. Because genital lesions tend to be painless,

---
* chancre (shang'-ker) Fr. for canker; from L. *cancer*, crab. An injurious sore.

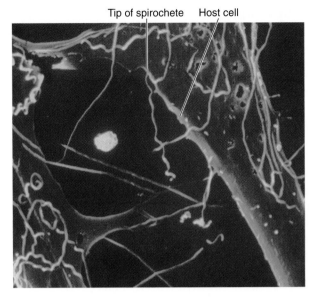

Tip of spirochete   Host cell

FIGURE 21.2

Electron micrograph of the syphilis spirochete attached to cells.
Notice the hooklike nature of the specialized tip of the spirochete.

they may escape notice in some cases. Lymph nodes draining the affected region become enlarged and firm, but systemic symptoms are usually absent. The chancre heals spontaneously without scarring in 3 to 6 weeks, but this healing is deceptive, because the spirochete has escaped into the circulation and is entering a period of tremendous activity.

Secondary Syphilis About 3 weeks to 6 months (average is 6 weeks) after the chancre heals, the secondary stage appears. By then, many systems of the body have been invaded, and the signs and symptoms are more profuse and intense. Initially there is fever, headache, and sore throat, followed by lymphadenopathy and a peculiar red or brown rash that breaks out on all skin surfaces, including the palms and the soles (figure 21.4). Like the chancre, the lesions contain viable spirochetes and disappear spontaneously in a few weeks. The major complications, occurring in the bones, hair follicles, joints, liver, eyes, and brain, can linger for months and years.

Latency and Tertiary Syphilis After resolution of secondary syphilis, about 30% of infections enter a highly varied latent period that can last for 20 years or longer. Latency is divisible into early and late phases, and though antitreponeme antibodies are readily detected, the parasite itself is not. The final stage of disease, late, or **tertiary, syphilis,** is quite rare today because of widespread use of antibiotics to treat other infections. By the time a patient reaches this phase, numerous pathologic complications occur in susceptible tissues and organs. Cardiovascular syphilis results from damage to the small arteries in the aortic wall. As the fibers in the wall weaken, the aorta is subject to distension and fatal rupture. The same pathologic process can damage the aortic valves, resulting in insufficiency and heart failure.

TABLE 21.1

Syphilis: Stages, Symptoms, Diagnosis, and Control

Stage	Average Duration	Clinical Setting	Diagnosis	Treatment
Incubation	3 weeks	No lesion; treponemes adhere and penetrate the epithelium; after multiplying, they disseminate	Asymptomatic phase	Not applicable
Primary	2–6 weeks	Initial appearance of chancre at inoculation site; intense treponemal activity in body; chancre later disappears	Dark-field microscopy; VDRL, FTA-ABS, MHA-TP testing	Benzathine penicillin G, 2×10^6 units; aqueous benzyl or procaine penicillin G, 4.8×10^6 units
Primary latency	2–8 weeks	Healed chancre; little scarring; treponemes in blood; few if any symptoms	Serological tests (+)	As above
Secondary	2–6 weeks after chancre leaves	Skin, mucous membrane lesions; hair loss; patient highly infectious; fever, lymphadenopathy; symptoms can persist for months	Dark-field testing of lesions; serological tests	Double doses of penicillins listed above
Latency	6 months to 8 or more years	Treponemes quiescent unless relapse occurs; lesions can reappear	Seropositive blood test	As above
Tertiary	Variable, up to 20 years	Neural, cardiovascular symptoms; gummas develop in organs; seropositivity	Treponeme may be demonstrated by DNA analysis of tissue	As above

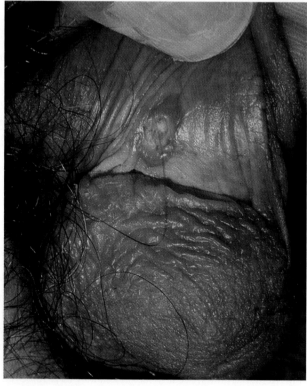

FIGURE 21.3

 A chancre (the lesion of primary syphilis) on the underside of a scrotum. Chancres can appear as solitary lesions or in multiple clusters.

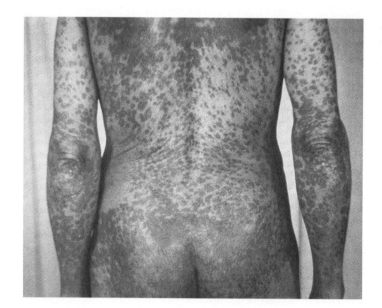

FIGURE 21.4

 Symptom of secondary syphilis. The skin rash in secondary syphilis can form on the trunk, arms, and even palms and soles (this latter location is particularly diagnostic). The rash does not hurt or itch and can persist for months.

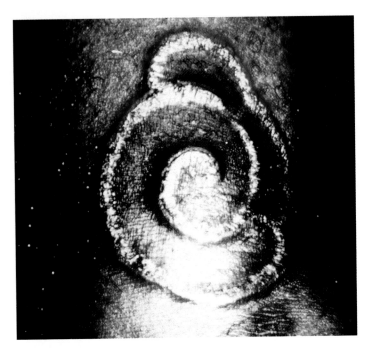

FIGURE 21.5

🔊 **The pathology of late, or tertiary, syphilis.**   A ring-shaped erosive gumma appears on the arm of this patient. Other gummas can be internal.

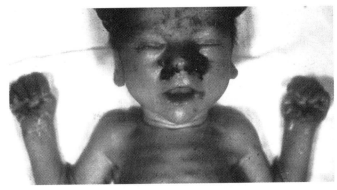

(a)

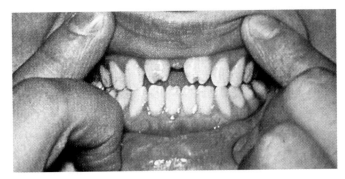

(b)

FIGURE 21.6

🔊 **Congenital syphilis.**   **(a)** An early sign is snuffles, a profuse nasal discharge that obstructs breathing. **(b)** A common characteristic of late congenital syphilis is notched, barrel-shaped incisors (Hutchinson's teeth).

In one form of tertiary syphilis, painful swollen syphilitic tumors called **gummas*** develop in tissues such as the liver, skin, bone, and cartilage (figure 21.5). Gummas are usually benign and only occasionally lead to death, but they can impair function. **Neurosyphilis** can involve any part of the nervous system, but it shows particular affinity for the blood vessels in the brain, cranial nerves, and dorsal roots of the spinal cord. The diverse reactions include severe headaches, convulsions, mental derangement, the Argyll Robertson pupil,[1] atrophy of the optic nerve, and blindness. Destruction of parts of the spinal cord can lead to muscle wasting and loss of activity and coordination.

Congenital Syphilis   *Treponema pallidum* can pass from a pregnant woman's circulation into the placenta and can be carried throughout the fetal tissues. An infection leading to **congenital syphilis** can occur in any of the three trimesters, though it is most common in the second and third. The pathogen inhibits fetal growth and disrupts critical periods of development with varied consequences, ranging from mild to the extremes of spontaneous miscarriage or stillbirth. Early congenital syphilis encompasses the period from birth to 2 years of age and is usually first detected 3 to 8 weeks after birth. Infants often demonstrate such signs as nasal discharge (figure 21.6a), skin eruptions and loss, bone deformation, and nervous system abnormalities. The late form gives rise to an unusual assortment of stigmata in the bones, eyes, inner ear, and joints and causes the formation of Hutchinson's teeth (figure 21.6b). The

number of congenital syphilis cases is closely tied to the incidence in adults, and, because it is sometimes not diagnosed, some children will sustain lifelong disfiguring disease.

Clinical and Laboratory Diagnosis   The pattern of syphilis imposes many complications on diagnosis. Not only do the stages mimic other diseases, but their appearance can be so separated in time as to seem unrelated. The chancre and secondary lesions must be differentiated from bacterial, fungal, and parasitic infections, tumors, and even allergic reactions. Overlapping symptoms of concurrent, sexually transmitted infections such as gonorrhea or chlamydiosis can further complicate diagnosis. The clinician must weigh presenting symptoms, patient history, and microscopic and serological tests in rendering a definitive diagnosis.

One rapid, direct method to diagnose primary, early congenital, and, to a lesser extent, secondary syphilis, is dark-field microscopy of a suspected lesion (figure 21.7). The lesions are gently squeezed or scraped to extract clear serous fluid. A wet mount prepared from the exudate is then observed for the characteristic size, shape, and motility of *T. pallidum*. A single negative test is insufficient to exclude syphilis, because the patient may have removed the organism by washing, so follow-up tests are recommended. Another microscopic test for discerning the spirochete directly in samples is direct immunofluorescence staining with monoclonal antibodies (see figure D, page 539). Patient samples can also be tested

---

1. Perhaps the most common sign still seen today, this condition is caused by adhesions along the inner edge of the iris that fix the pupil's position into a small, irregular circle.

* gumma (goo´-mah) L. *gummi*, gum. A soft tumorous mass containing granuloma tissue.

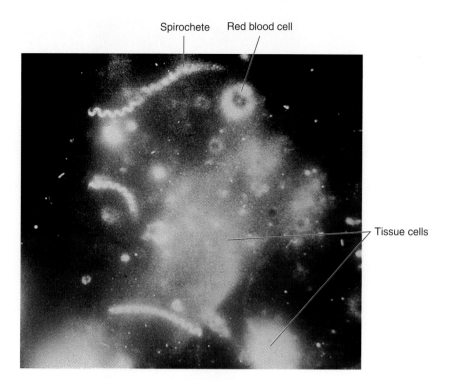

Spirochete   Red blood cell

Tissue cells

FIGURE 21.7

Treponema pallidum **from a syphilitic chancre, viewed with dark-field illumination.** Spirochetes contrast sharply with the red blood cells and tissue cells.

with a DNA probe specific to various spirochete gene sequences (figure 4.28*d*).

Testing Blood for Syphilis   If dark-field or direct antigen tests are negative, serological tests provide valuable diagnostic support. These tests are based upon detection of antibody formed in response to *T. pallidum* infection (see table 21.1). Several of the tests (Rapid Plasma Reagin [RPR], VDRL, Kolmer) are variations on the original test developed by Wasserman using cardiolipin, a natural constituent of many cells, as the antigen. Although anticardiolipin antibodies are not specific for syphilis, the test is an effective way to screen the population for people who may be infected.

Premarital blood tests have traditionally been used as a screening test for syphilis. Blood tests are also suggested for high-risk groups such as homosexuals, male and female prostitutes, people with other STDs, and pregnant women. In the case of a positive result, it is important that a series of serological tests be carried out to detect an elevated antibody titer indicative of active infection, to rule out residual antibodies from a prior cured infection. Because the most common screening tests (RPR and VDRL) are based on reactions to a substance found normally in human tissue, biological false positives can occur, especially in patients with autoimmune diseases or impaired immunity. A more specific test is needed for those who are suspected of having a false-positive result.

Typical of these specific tests is the *T. pallidum* microhemagglutination assay (MHA-TP), which employs red blood cells that have been coated with treponemal antigen. Agglutination of the cells by serum indicates antitreponemal antibodies and infection. Another standard test is an indirect immunofluorescent method called the FTA-ABS (Fluorescent Treponemal Antibody Absorbance) test. The test serum is first absorbed with treponemal cells and reacted with antihuman globulin antibody labeled with fluorescent dyes. If antibodies to the treponeme are present, the

fluorescence on the outside of these cells is highly visible with a fluorescent microscope (see figure 16.11*b*). The most sensitive and specific of all is the *T. pallidum* immobilization (TPI) test, in which live syphilis spirochetes are mixed with the test serum and observed microscopically for loss of motility.

Treatment and Prevention   Penicillin G retains its status as a wonder drug in the treatment of all stages and forms of syphilis. It is given parenterally in large doses with benzathine or procaine to maintain a blood level lethal to the treponeme for at least 7 days (see table 21.1). Alternative drugs (tetracycline and erythromycin) are less effective and are indicated only if penicillin allergy has been documented. It is important that all patients be monitored for compliance or possible treatment failure.

The core of an effective prevention program depends upon detection and treatment of the sexual contacts of syphilitic patients. Public health departments and physicians are charged with the task of questioning the patients and tracing their contacts. All individuals identified as being at risk, even if they show no signs of infection, are given immediate prophylactic penicillin in a single, long-acting dose. The barrier effect of a condom provides superior protection, but washing after sexual intercourse alone is not an adequate protective measure. Protective immunity apparently does arise in humans and in experimentally infected rabbits, which raises the prospect of an effective immunization program in the future. The recent cloning of treponemal surface antigens using recombinant DNA technology will no doubt speed development of vaccines and new diagnostic testing methods.

Nonsyphilitic Treponematoses

The other treponematoses are ancient diseases that closely resemble syphilis in their effects, though they are rarely transmitted sexually or congenitally. These infections, known as bejel, yaws, and pinta, are endemic to certain tropical and subtropical regions of the world, especially rural areas with unsanitary living conditions. The treponemes that cause these infections are nearly indistinguishable from those of syphilis in morphology and behavior. The diseases are slow and progressive and involve primary, secondary, and tertiary stages. They begin with local invasion by the treponeme into the skin or mucous membranes and its subsequent spread to subcutaneous tissues, bones, and joints. Drug therapy with penicillin, erythromycin, or tetracycline remains the treatment of choice for these treponematoses.

Bejel Bejel is also known as endemic syphilis and nonvenereal childhood syphilis. The pathogen, the subspecies *T. pallidum endemicum,* is harbored by a small reservoir of nomadic and seminomadic people in arid areas of the Middle East and North Africa. It is a chronic, inflammatory childhood disease transmitted by direct contact or shared household utensils and other fomites and is facilitated by minor abrasions or cracks in the skin or a mucous membrane. Often the infection begins as small, moist patches in the oral cavity (figure 21.8*a*) and spreads to the skin folds of the body and to the palms.

Yaws Yaws is a West Indian name for a chronic disease known by the regional names bouba, frambesia tropica, and patek. It is endemic to warm, humid, tropical regions of Africa, Asia, and South America. The microbe, subspecies *T. pallidum pertenue,* is readily spread by direct contact with skin lesions or fomites. Crowded living conditions and poor community or personal hygiene are contributing factors. The earliest sign is a large, abscessed papule called the "mother yaw," usually on the legs or lower trunk. After the initial lesion has healed, a secondary crop of moist nodular tumors appears. These tumors erode the skin, periosteum, and bones but do not penetrate to the viscera (figure 21.8*b*). Yaws can be prevented by correcting predisposing conditions, shielding minor skin injuries from mechanical insect vectors, mass treatment, and surveillance for new cases.

Pinta The names *mal del pinto* and *carate* are regional synonyms for pinta, a chronic skin infection caused by *T. carateum.** Transmission evidently requires several years of close personal contact accompanied by poor hygiene and inadequate health facilities. Even though pinta is not currently widespread, the disease is still found in isolated populations inhabiting the tropical forest and valley regions of Latin America. Infection begins in the skin with a dry, scaly papule reminiscent of psoriasis or leprosy. In time, pigmented secondary macules and blanched tertiary lesions appear. Pinta is not life-threatening, but it often creates scars on the afflicted area.

LEPTOSPIRA AND LEPTOSPIROSIS

Leptospires are typical spirochetes marked by tight, regular, individual coils with a bend or hook at one or both ends (figure 21.9). There are only two species in the genus: *Leptospira interrogans,**

---

* *carateum* (kar-uh′-tee-um) From *carate,* the South American name for pinta.

* *Leptospira interrogans* (lep″-toh-spy′-rah in-terr′-oh-ganz) Gr. *leptos,* slender or delicate, and *speira,* a coil; *interrogans* because its appearance suggests a question mark at one end.

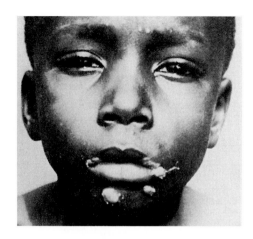

(a)

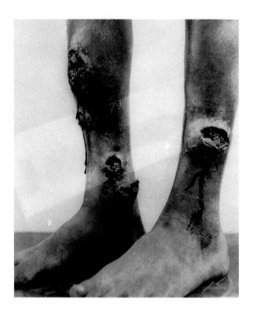

(b)

FIGURE 21.8

Endemic treponematoses. **(a)** Skin and membrane nodules in a young boy with endemic syphilis (bejel). **(b)** The clinical appearance of yaws. The large, draining sores are "mother yaws" that give rise to new lesions.

which causes **leptospirosis** in humans and animals, and *L. biflexa,* a harmless, free-living saprobe. The two species are serologically, genetically, and physiologically distinct. *Leptospira interrogans* demonstrates nearly 200 serotypes distributed among various animal groups, which accounts for the extreme variations in leptospirosis among humans.

Epidemiology and Transmission of Leptospirosis

Leptospirosis is a zoonosis associated with wild animals such as rodents, skunks, raccoons, foxes, and some domesticated animals, particularly horses, dogs, cattle, and pigs. Although these reservoirs are distributed throughout the world, the disease is concentrated mainly in the tropics. Leptospires shed in the urine of an infected animal can survive for several months in neutral or alkaline soil or water. Infection occurs almost entirely through contact of skin abrasions or mucous membranes with animal urine or some environmental source containing urine. It is not associated with animal

Hook

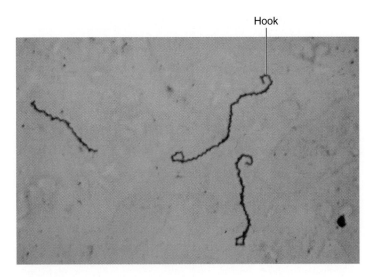

FIGURE 21.9

🔘 *Leptospira interrogans,* **the agent of leptospirosis.**   Note the curved hook at one end of the spirochete. Its relative, *L. biflexa,* has flexible hooks at both ends.

bites, inhalation, or human contact. In the United States, a small number of cases (50–60) each year are reported among older children and young adults exposed to polluted water, and soldiers involved in jungle training.

Pathology of Leptospirosis and Host Response

Leptospirosis proceeds in two phases, and its principal targets are the kidneys, liver, brain, and eyes. During the early, or leptospiremic, phase, the pathogen appears in the blood and cerebrospinal fluid. Symptoms are sudden high fever, chills, headache, muscle aches, conjunctivitis, and vomiting. During the second, or immune, phase, the blood infection is cleared by natural defenses. This period is marked by milder fever, headache due to leptospiral meningitis, and *Weil's syndrome,* a cluster of symptoms characterized by kidney invasion, hepatic disease, jaundice, anemia, and neurological disturbances. Long-term disability and even death can result from injury to the kidneys and liver, but they occur primarily with virulent strains and in elderly patients.

Diagnosis, Treatment, and Prevention

A history of environmental exposure, along with presenting symptoms, can support initial diagnosis of leptospirosis, but definitive diagnosis relies on dark-field microscopy of specimens, *Leptospira* culture, and serological tests. Because leptospiral infection stimulates a strong humoral response, it is possible to test the patient's serum for its antibody titer. A fast, specific, and effective test called the macroscopic slide agglutination test is most often employed for routine screening. Live or formalinized *L. interrogans* is mixed with the patient's serum and observed for agglutination or lysis with a dark-field microscope.

Early treatment with penicillin or tetracycline rapidly reduces symptoms and shortens the course of disease, but delayed therapy is less effective. Strain-specific vaccines made from killed cells are available for humans, dogs, and cattle, but these can confer protection only to a specific endemic strain. Vaccination is aimed at those

with greatest risk such as combat troops training in jungle regions and animal care and livestock workers. The best controls are to wear protective footwear and clothing and to avoid swimming or wading in livestock watering ponds.

BORRELIA: ARTHROPOD-BORNE SPIROCHETES

Members of the genus *Borrelia** are morphologically distinct from other pathogenic spirochetes. They are comparatively larger, ranging from 0.2 to 0.5 μm in width and from 10 to 20 μm in length, and they contain 3 to 10 irregularly spaced and loose coils (see figure 21.1*d*) with an abundance (30–40) of periplasmic flagella. The nutritional requirements of *Borrelia* are so complex that the bacterium can be grown in artificial media only with difficulty.

Human infections with *Borrelia,* termed **borrelioses,** are all transmitted by some type of arthropod vector, usually ticks or lice. The two most important human diseases are relapsing fever and Lyme disease.

Epidemiology of Relapsing Fever

Borrelia hermsii, the cause of tick-borne relapsing fever, is carried by soft ticks (see Microbits 21.4) of the genus *Ornithodoros.* The mammalian reservoirs of this zoonosis are squirrels, chipmunks, and other wild rodents, and the human is generally an accidental host. The spirochetes mature and persist in the salivary glands and intestines of the tick, and both the bite itself and the subsequent scratching initiate infection. Tick-borne relapsing fever occurs sporadically in the United States, usually in campers, backpackers, and forestry personnel who frequent the higher elevations of western states. The incidence of infection is higher in endemic areas of the tropics, especially where rodents have easy access to dwellings.

Whenever famine, war, or natural disasters are coupled with poor hygiene, crowding, and inadequate medical attention, epidemics of louse-borne relapsing fever occur. Such conditions favor the survival and spread of the louse vector *Pediculus humanus* (see Microbits 21.4), which harbors the spirochete *B. recurrentis* in its body cavity. A host is infected when lice are smashed and accidently scratched into a wound or the skin. Louse-borne fever is most common in parts of China, Afghanistan, and Africa.

Pathogenesis and the Nature of Relapses   The pathologic manifestations are similar in tick- and louse-borne relapsing fever. After a 2- to 15-day incubation period, patients experience the abrupt onset of high fever, shaking chills, headache, and fatigue. Later features of the disease include nausea, vomiting, muscle aches, and abdominal pain. Extensive damage to the liver, spleen, heart, kidneys, and cranial nerves occurs in many cases. Half of the patients hemorrhage profusely into organs, and some develop a rash on the shoulders, trunk, and legs. Untreated cases are often lengthy and debilitating and may have a 40% mortality rate.

As the name *relapsing fever* indicates, the fever follows a fluctuating course that is explained by changes in the parasite and the attempts of the immune system to control it (figure 21.10). *Borrelia* have adopted a remarkable strategy for evading the immune system and avoiding destruction. They change surface antigens during growth, so that, in time, the initial antibodies

---

* *Borrelia* (boh-ree′-lee-ah) Named after Amédé Borrel, a French bacteriologist.

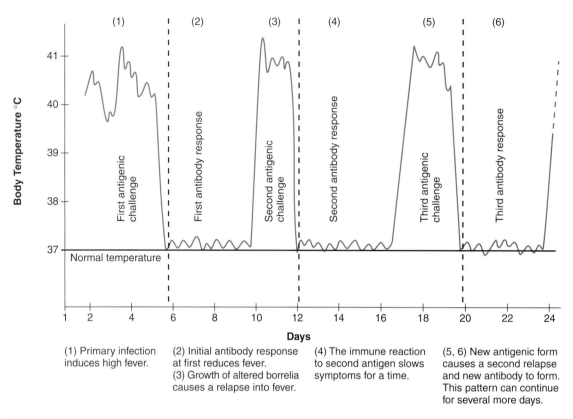

(1) Primary infection induces high fever.

(2) Initial antibody response at first reduces fever.
(3) Growth of altered borrelia causes a relapse into fever.

(4) The immune reaction to second antigen slows symptoms for a time.

(5, 6) New antigenic form causes a second relapse and new antibody to form. This pattern can continue for several more days.

FIGURE 21.10

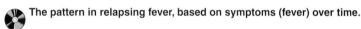
The pattern in relapsing fever, based on symptoms (fever) over time.

become useless. These antigenically altered cells survive, multiply, and cause a second wave of symptoms. Eventually, the immune system forms new antibodies, but it is soon faced with yet another antigenic form. A single strain has been known to generate 24 distinct serological types. Eventually, cumulative immunity against the variety of antigens develops, and complete recovery can occur.

Diagnosis, Treatment, and Prevention   A patient's history of exposure, clinical symptoms, and the presence of *Borrelia* in blood smears are very definitive evidence of borreliosis. Except for pregnant women and children under 7 years old, tetracycline is the treatment of choice. Chloramphenicol, erythromycin, and doxycycline are also effective antimicrobial agents. Because vaccines are not available, prevention of relapsing fever is dependent upon controlling rodents and avoiding tick bites. Louse-borne relapsing fever is effectively arrested by improving hygiene.

Borrelia burgdorferi and Lyme Disease

Lyme disease is the most prominent borreliosis in the United States (Medical Microfile 21.1). Its spirochetal agent, ***Borrelia burgdorferi,**** is transmitted primarily by hard ticks of the genus *Ixodes*. In the northeastern part of the United States, *Ixodes scapularis* (the black-legged deer tick) passes through a complex 2-year cycle that involves two principal hosts (figure 21.11). As a larva or nymph, it feeds on the white-footed mouse, where it picks up the infectious agent. The nymph is relatively nonspecific and will try to

feed on nearly any type of vertebrate, thus it is the form most likely to bite humans. The adult tick reproductive phase of the cycle is completed on deer. In California, the transmission cycle involves *Ixodes pacificus* and the dusky-footed woodrat reservoir.

The incidence of Lyme disease is showing a gradual upward trend from about 10,000 cases per year in 1991 to 14,000 in 2000. This increase may be partly due to improved diagnosis, but it also reflects changes in the numbers of hosts and vectors. The greatest concentrations of Lyme disease are in areas having high mouse and deer populations. Most of the cases have occurred in New York, Pennsylvania, Connecticut, New Jersey, Rhode Island, and Maryland, though the number in the Midwest and West is growing. Highest risk groups include hikers, backpackers, and people living in newly developed communities near woodlands and forests. Peak seasons are the summer and early fall.

Lyme disease is nonfatal but often evolves into a slowly progressive syndrome that mimics neuromuscular and rheumatoid conditions. An early symptom in 70% of cases is a rash at the site of a larval tick bite. The lesion, called *erythema migrans,* looks something like a bull's-eye, with a raised erythematous ring that gradually spreads outward and a pale central region (figure 21.12). Other early symptoms are fever, headache, stiff neck, and dizziness. If not treated or if treated too late, the disease can advance to the second stage, during which cardiac and neurological symptoms (facial palsy) develop. After several weeks or months, a crippling polyarthritis can attack joints, especially in the European strain of the agent. Some people acquire chronic neurological complications that are severely disabling.

---

* *burgdorferi* (berg-dor'-fer-eye) Named for its discoverer, Dr. Willy Burgdorfer.

MEDICAL MICROFILE 21.1
The Disease Named for a Town

In the 1970s, an enigmatic cluster of arthritis cases appeared in the town and surrounding suburbs of Old Lyme, Connecticut. This phenomenon caught the attention of nonprofessionals and professionals alike, whose persistence and detective work ultimately disclosed the unusual nature and epidemiology of Lyme disease. The process of discovery began in the home of Polly Murray, who, along with her family, was beset for years by recurrent bouts of stiff neck, swollen joints, malaise, and fatigue that seemed vaguely to follow a rash from tick bites. When Mrs. Murray's son was diagnosed as having juvenile rheumatoid arthritis, she became skeptical. Conducting her own literature research, she began to discover inconsistencies. Rheumatoid arthritis was described as a rare, noninfectious disease, yet over an 8-year period, she found that 30 of her neighbors had experienced similar illnesses. Eventually this cluster of cases and several others were reported to state health authorities.

The reports caught the attention of Dr. Allen Steere, a rheumatologist with a CDC background. He was able to forge the vital link between the case histories, the disease symptoms, and the presence of unique spirochetes in ticks preserved by some of the patients. These same spirochetes had been previously characterized in 1981 by Dr. Willy Burgdorfer, though he did not realize their importance at the time.

In the years since Lyme disease was formally characterized, retrospective studies showed that this is not really a new disease. The unique bull's-eye rash was reported in Europe at the turn of the century. Recent PCR analysis of tick museum specimens from 50 years ago documents the presence of *Borrelia burgdorferi*. It is now thought that Lyme disease has been present in North America for centuries.

The modern epidemic appears to be the result of an increase in the reservoirs, such as deer and mice, which would naturally amplify tick and spirochete numbers. This effect has been compounded by greater mingling of humans and animals in woodland habitats.

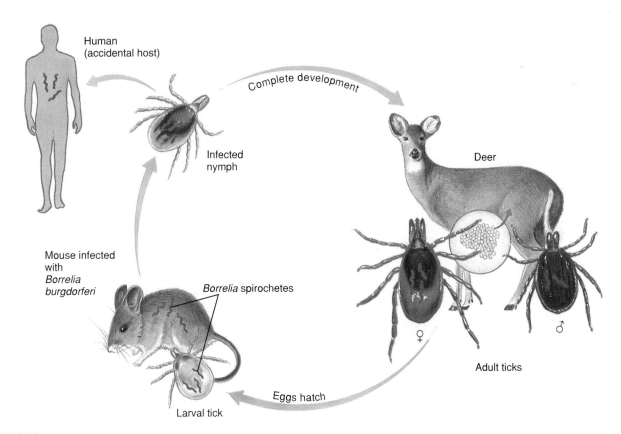

FIGURE 21.11

 The cycle of Lyme disease in the northeastern United States. The exact reservoir hosts vary from region to region in the United States and worldwide.

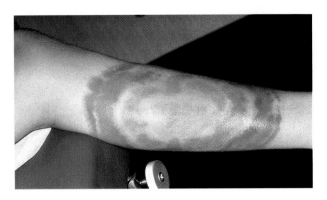

FIGURE 21.12

Lesions of Lyme disease on the lower leg. Note the flat, reddened rings in the form of a bull's-eye. Primary lesions often give rise to large numbers of secondary lesions in other locations.

Diagnosis of Lyme disease can be difficult because of the range of symptoms it presents. Most suggestive are the ring-shaped lesions, isolation of spirochetes from the patient, and serological testing with an ELISA method that tracks a rising antibody titer (see Microbits 16.2). Tests for spirochetal DNA in specimens is especially helpful for late-stage diagnosis. Early treatment with tetracycline and amoxicillin is effective, and other antibiotics such as ceftriaxone and azithromycin are used in late Lyme disease therapy.

Because dogs can also acquire the disease, a vaccine has been marketed to protect them, and a human vaccine for high-risk populations is now offered for people living in areas of high incidence. Anyone involved in outdoor activities should wear protective clothing, boots, leggings, and insect repellent containing DEET.* Individuals exposed to heavy infestation should routinely inspect their bodies for ticks and remove ticks gently without crushing, preferably with forceps or fingers protected with gloves, because it is possible to become infected by tick feces or body fluids.

---

* N,N-Diethy-M-toluamide The active ingredient in OFF! and Cutter repellents.

CHAPTER CHECKPOINTS

The Order Spirochaetales is a group of spirochetes that are mostly harmless saprobes except for pathogens in the genera *Treponema*, *Leptospira*, and *Borrelia*.

Many treponemes are part of the normal flora. Significant pathogens in this genus are *T. p. pallidum*, the agent of epidemic syphilis, *T. p. pertenue*, the agent of yaws, and *T. p. endemicum*, the agent of endemic syphilis. Other species are agents of oral diseases.

T. p. pallidum, a fastidious, obligate parasite, is the agent of syphilis. It is limited to humans. Primary syphilis appears as an infective but painless chancre that usually heals spontaneously. Untreated infection progresses to a systemic condition called secondary syphilis, which is manifest in skin rashes. The spirochetes remain in body organs and can initiate tertiary syphilis after a latency of many years. During this stage, the spirochete can invade the heart, blood vessels, and brain and can form gummas.

Syphilis is difficult to diagnose because of its nonspecific symptoms and the concurrent presence of other STDs. Methods of detection include direct microscopic identification from clinical specimens and serological tests for host antibody (such as the TPI test). Treatment with penicillin G is still effective. Control depends on identification of all sexual contacts and use of barrier contraceptives.

Subspecies of *T. pallidum* are agents of bejel, yaws, and pinta, diseases endemic to specific tropical regions. Their symptoms mimic the stages of syphilis, but all are transmitted nonsexually.

Leptospira interrogans is a hooked spirochete that causes leptospirosis, a tropical zoonosis transmitted through direct contact with the urine of infected animals. It can be diagnosed by serological testing and treated with drugs early in infection.

Spirochetes in the genus *Borrelia* are larger and more loosely coiled than other genera. Pathogens in this group are agents of relapsing fever and Lyme disease. Both are transmitted by arthropods.

The two agents of relapsing fever are *B. hermsii* and *B. recurrentis*. *B. hermsii* is the agent of tick-borne relapsing fever. It is transmitted by soft ticks to humans from rodents. *B. recurrentis* is transmitted among humans by body lice. Relapsing fever is known by the recurrence of symptoms and the tendency of *Borrelia* to change their surface antigens. The disease can be prevented by control of rodent populations, prevention of tick bites, and good personal hygiene.

Borrelia burgdorferi is the agent of Lyme disease, a slowly progressive systemic infection. It is transmitted to humans by hard ticks. The natural hosts are deer and rodents. The most prominent symptom is a bull's-eye rash radiating outward from the bite site. Tetracycline and amoxicillin are effective if administered early.

Other Curviform Gram-Negative Bacteria of Medical Importance

Two genera of curved or short spiral rods prominent in medical bacteriology are *Vibrio** and *Campylobacter.** Vibrios are comma-shaped rods with a single polar flagellum. Campylobacters are short spirals that have one or more flagella and a corkscrew type of

motility. *Vibrio* is a member of the Family Vibrionaceae, and *Campylobacter* is in the Family Spirillaceae.

THE BIOLOGY OF *VIBRIO CHOLERAE*

A freshly isolated specimen of **Vibrio cholerae*** reveals quick, darting cells slightly resembling a wiener or a comma (figure 21.13a). *Vibrio* shares many cultural and physiological characteris-

---

* *Vibrio* (vib′-ree-oh) L. *vibrare*, to shake.

* *Campylobacter* (kam″-pih-loh-bak′-ter) Gr. *campylo*, curved, and *bacter*, rod.

* *cholerae* (kol′-ur-ee) Gr. *chole*, bile. The bacterium was once named *V. comma* for its comma-shaped morphology.

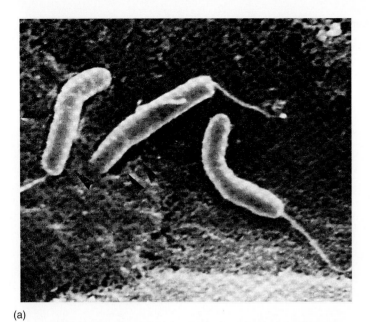

(a)

Vibrios   Villus surface

(b)

FIGURE 21.13

The agent of cholera. **(a)** *Vibrio cholerae,* showing its characteristic curved shape and single polar flagellum. **(b)** Infectious vibrios burrowing into the surface of intestinal villi.

tics with members of the Enterobacteriaceae, a closely related family. They are fermentative and grow on ordinary or selective media containing bile at 37°C. They possess unique O (somatic) antigens, H (flagella) antigens, and membrane receptor antigens that provide some basis for classifying members of the family.

Epidemiology of Cholera

Epidemic cholera, or Asiatic cholera, has been a devastating disease for centuries. Although the human intestinal tract was once thought to be the primary reservoir, it is now known that the parasite is free-living in certain endemic regions.

The pattern of cholera transmission and the onset of epidemics are greatly influenced by the season of the year and the climate. Cold, acidic, dry environments inhibit the migration and survival of *Vibrio,* whereas warm, monsoon, alkaline, and saline conditions favor them. The disease has persisted in a pandemic pattern since 1961, when the *El Tor* biotype began to prevail worldwide. This strain survives longer in the environment, infects a higher number of people, and is more likely to be chronically carried than any other strain. Recent outbreaks in several parts of the world have been traced to giant cargo ships that pick up water in one port and empty it in ports all across the world. It ranks among the top seven causes of morbidity and mortality, affecting several million people in endemic regions of Asia and Africa.

In nonendemic areas such as the United States, the microbe is spread by water and food contaminated by asymptomatic carriers, but it is relatively uncommon. Sporadic outbreaks occasionally occur along the Gulf of Mexico, and the cholera vibrio is sometimes isolated from shellfish in this region.

Pathogenesis of Cholera

After being ingested with food or water, *V. cholerae* encounters the potentially destructive acidity of the stomach. This hostile environment influences the size of the infectious dose (10^8 cells), though certain types of food also shelter the pathogen more readily than others. At the mucosa of the duodenum and jejunum, the vibrios penetrate the mucous barrier using their flagella, adhere to the microvilli of the epithelial cells, and multiply there (figure 21.13*b*). The cells are strictly epipathogens that do not enter the cells or invade the mucosa. The virulence of *V. cholerae* is due entirely to an enterotoxin called *cholera toxin* (CT) that disrupts the normal physiology of intestinal cells. When this toxin binds to specific intestinal receptors, a secondary signaling system is activated. Under the influence of this system, the cells shed large amounts of electrolytes into the intestine, an event that is accompanied by profuse water loss (figure 21.14). Most cases of cholera are mild or self-limited, but in children and weakened individuals, the disease can strike rapidly and violently.

After an incubation period of a few hours to a few days, symptoms begin abruptly with vomiting, followed by copious watery feces called **secretory diarrhea.** This voided fluid contains flecks of mucus, hence the description "rice-water stool." Fluid losses of nearly one liter per hour have been reported in severe cases, and an untreated patient can lose up to 50% of body weight during the course of the disease. The diarrhea causes loss of blood volume, acidosis from bicarbonate loss, and potassium depletion that predispose the patient to muscle cramps, severe thirst, flaccid skin, and sunken eyes, and, in young children, coma and convulsions. Secondary circulatory consequences can include hypotension, tachycardia, cyanosis, and collapse from shock within 18 to 24 hours. If cholera is left untreated, death can occur in less than 48 hours, and the mortality rate approaches 55%.

Diagnosis and Remedial Measures

During epidemics, clinical evidence is usually sufficient to diagnose cholera. But confirmation of the disease is often required for epidemiologic studies and detection of sporadic cases. *Vibrio cholerae* can be readily isolated and identified in the laboratory from stool samples. Direct dark-field microscopic observation reveals

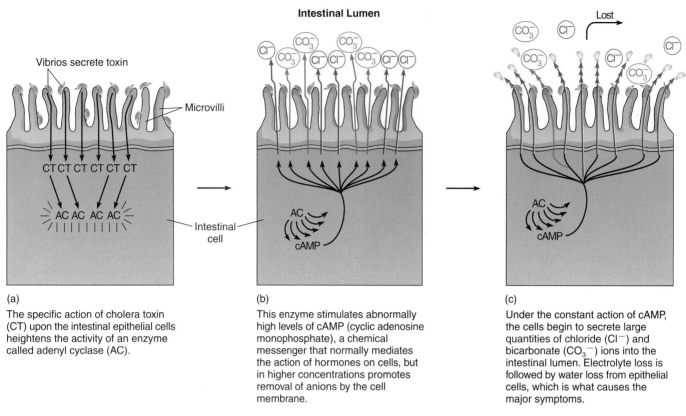

Intestinal Lumen

(a) The specific action of cholera toxin (CT) upon the intestinal epithelial cells heightens the activity of an enzyme called adenyl cyclase (AC).

(b) This enzyme stimulates abnormally high levels of cAMP (cyclic adenosine monophosphate), a chemical messenger that normally mediates the action of hormones on cells, but in higher concentrations promotes removal of anions by the cell membrane.

(c) Under the constant action of cAMP, the cells begin to secrete large quantities of chloride (Cl^-) and bicarbonate (CO_3^-) ions into the intestinal lumen. Electrolyte loss is followed by water loss from epithelial cells, which is what causes the major symptoms.

FIGURE 21.14
Pathogenesis of cholera.

characteristic curved cells with brisk, darting motility as confirmatory evidence. Immobilization or fluorescent staining of feces with group-specific antisera is supportive as well. Difficult or elusive cases can be traced by detecting a rising antitoxin titer in the serum.

The key to cholera therapy is prompt replacement of water and electrolytes, since their loss accounts for the severe morbidity and mortality. This can be accomplished by various rehydration techniques that replace the lost fluid and electrolytes (oral rehydration therapy [ORT]; Medical Microfile 21.2).

Cases in which the patient is unconscious or has complications from severe dehydration require intravenous replenishment as well. Oral antibiotics such as tetracycline and drugs such as trimethoprim-sulfa can terminate the diarrhea in 48 hours, and they also promote recovery and diminish the period of vibrio excretion.

Effective prevention is contingent upon proper sewage disposal and water purification. Detecting and treating carriers with mild or asymptomatic cholera is a serious goal but one that is frequently difficult to attain because of inadequate medical provisions in those countries where cholera is endemic. Vaccines are available for travelers and people living in endemic regions. One contains killed cholera vibrios, but it protects for only 6 months or less. An oral vaccine containing live, attenuated vibrios is an effective alternative.

VIBRIO PARAHAEMOLYTICUS AND VIBRIO VULNIFICUS: PATHOGENS CARRIED BY SEAFOOD

Two additional relatives of *V. cholerae* share its morphology, physiology, and ecological adaptation. *Vibrio parahaemolyticus* and *V. vulnificus* are both salt-tolerant inhabitants of coastal waters and associate with marine invertebrates. In temperate zones, vibrios survive over the winter by settling into the ocean sediment, and, when resuspended by upwelling during the warmer seasons, they become incorporated into the food web, eventually growing on fish, shellfish, and other edible seafood.

Features of *V. parahaemolyticus* Gastroenteritis

Vibrio parahaemolyticus food infection, an acute form of gastroenteritis, was first described in Japan more than 30 years ago. The great majority of cases appear in individuals who have eaten raw, partially cooked, or poorly stored seafood. The vehicles most often implicated are squid, mackerel, sardines, crabs, tuna, shrimp, oysters, and clams. Outbreaks tend to be concentrated along coastal regions during the summer and early fall. The incubation period of nearly 24 hours is followed by explosive, watery diarrhea accompanied by nausea, vomiting, abdominal cramps, and sometimes fever. *Vibrio* toxins cause symptoms that last about 72 hours but can persist for 10 days.

Food infection caused by *V. vulnificus* is very similar in symptoms to that caused by *V. parahaemolyticus*. It is most often associated with ingesting raw oysters and can have a much more severe outcome in critically ill patients with diabetes or liver disease. It is the leading cause of death from food-borne illness in some areas.

Treatment of severe gastroenteritis can require fluid and electrolyte replacement, and occasionally antibiotics, and hospitalization

MEDICAL MICROFILE 21.2
Oral Rehydration Therapy

Drinking a simple sugar-salt drink appears to be a miracle cure for cholera. This relatively simple solution, developed by the World Health Organization, consists of a mixture of the electrolytes sodium chloride, sodium bicarbonate, potassium chloride, and glucose or sucrose dissolved in water (table 21A). When administered early in amounts ranging from 100 to 400 ml/hour, the solution can restore patients in 4 hours, often bringing them literally back from the brink of death. It works even if the individual has diarrhea, because the particular combination of ingredients is well tolerated and rapidly absorbed. Infants and small children who once would have died now survive so often that the mortality rate for treated cases of cholera is near zero. This therapy has several advantages,

especially for countries with few resources. It does not require medical facilities, high-technology equipment, or complex medication protocols. It is also inexpensive, noninvasive, fast-acting, and useful for a number of diarrheal diseases besides cholera.

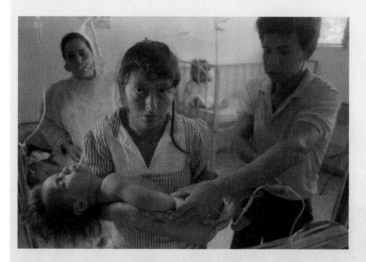

Note the sunken eyes in this child suffering from severe dehydration.

TABLE 21.A
Recommendations for Oral Rehydration Therapy (ORT)

Standard Composition of Oral Rehydration Solution (ORS)	Grams per liter (g/l)	Concentration in Millimoles
NaCl	3.5 g/l	90 (Na) Cl (80) total
$NaHCO_3$	2.5 g/l	30 (HCO_3)
KCl	1.5 g/l	20 (K)
Glucose★	20 g/l	111

Guidelines for Preparation and Treatment
- Components are dissolved in 1 liter of water that has been boiled to disinfect it. It is thoroughly mixed and tasted for saltiness. It should be only slightly salty. This product is similar to a commercial product called Pedialyte or Oralyte.
- Moderately dehydrated children and adults should receive frequent small amounts of ORS at the rate of 4 to 8 ounces per hour over 6 hours. Plain water can be administered in between ORS administrations.
- To check the patient for the return of normal hydration, test the skin turgor (pinch it to see if a fold remains), and observe whether the eyes are still sunken.

★*Sucrose or rice powder may be substituted if glucose is not available.*

may be indicated. Control measures aim to keep the bacterial count in all seafood below the infective dose by continuous refrigeration during transport and storage, sufficient cooking temperatures, and prompt serving. Consumers of raw oysters (or other shellfish) should understand the possible risk of eating raw seafood. In some regions, the USDA mandates that health warnings be posted about this danger in markets and restaurants.

DISEASES OF THE *CAMPYLOBACTER* VIBRIOS

Campylobacters are slender, curved or spiral bacilli propelled by polar flagella at one or both poles, often appearing in S-shaped or gull-winged pairs (figure 21.15). These bacteria tend to be microaerophilic inhabitants of the intestinal tract, genitourinary tract, and oral cavity of humans and animals. Species of *Campylobacter* most significant in medical and veterinary practice are *C. jejuni* and *C. fetus*. A close relative, *Helicobacter* (formerly *Campylobacter*) *pylori,* is an unusual vibrio implicated in certain types of stomach ulcers (Microbits 21.3).

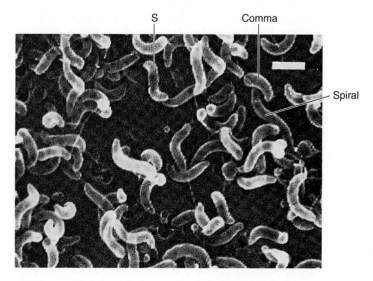

FIGURE 21.15

Scanning micrograph of *Campylobacter jejuni,* showing comma, S, and spiral forms.

MICROBITS 21.3

Helicobacter pylori: It's Enough to Give You Ulcers

Although the human stomach is usually regarded as a hostile habitat, an unusual vibrio, *Helicobacter pylori,* has found its own special niche there. Not only does it thrive in the acidic environment, but evidence has clearly linked it to a variety of gastrointestinal ailments. It is known to cause an inflammatory condition of the stomach lining called gastritis and is implicated in 90% of stomach and duodenal ulcers. It is also an apparent cofactor in the development of a common type of stomach cancer called adenocarcinoma.

The curved cells were first detected by J. Robin Warren in 1979 in stomach biopsies from ulcer patients. He and an assistant, Barry J. Marshall, isolated the microbe in culture and even served as guinea pigs by swallowing a good-sized inoculum to test its effects. Both developed transient gastritis.

Studies have revealed that the pathogen is present in a large proportion of people. It occurs in the stomachs of 25% of healthy middle-aged adults and in more than 60% of adults over 60 years of age. *Helicobacter pylori* is probably transmitted from person to person by the oral-oral or oral-fecal route. It seems to be acquired early in life and carried asymptomatically until its activities begin to damage the digestive mucosa. Because other animals are also susceptible to *H. pylori* and even develop chronic gastritis, it has been proposed that the disease is a zoonosis transmitted from an animal reservoir. Reports on its high frequency in cats suggest that they may be a significant reservoir source. It can also be spread by houseflies acting as mechanical vectors.

Other studies have helped to explain how the vibrio takes up residence in the gastrointestinal tract. First, it bores through the outermost mucus that lines the epithelial tissue. Then it attaches to specific binding sites on the cells and entrenches itself. It turns out that one receptor specific for *Helicobacter* is the same receptor as type O blood, which accounts for the higher rate of ulcers in people with this blood type (1.5–2 times). Another protective adaptation is the formation of urease, an enzyme that converts urea into ammonium and bicarbonate, both alkaline compounds that can neutralize stomach acid. As the immune system recognizes and attacks the pathogen, infiltrating white blood cells damage the epithelium to some degree, leading to chronic active gastritis. In some people, these lesions lead to deeper erosions and ulcers and, eventually, can lay the groundwork for a cancer to develop.

Helicobacter pylori is isolated primarily from biopsy specimens. Because it gives off large amounts of urease, it can be initially identified by inoculating a urease test (see table 20.2). A less invasive technique is the urea breath test. This test involves swallowing labeled urea that gives off radioactive CO_2 in exhaled air if *H. pylori* is present. A PCR test for microbial DNA in biopsies and feces is also being marketed.

Understanding the underlying pathology has useful applications in treatment. Gastritis and ulcers have traditionally been treated with drugs (Tagamet, Zantac) that suppress symptoms by slowing the secretion of acid in the stomach and must be taken continuously for indefinite periods, and relapses are common. The newest recommended therapy is 2 to 4 weeks of clarithromycin to eliminate the bacterial infection, and Zantac, which inhibits the formation of stomach acid. This regimen can actually cure the infection and eliminate symptoms.

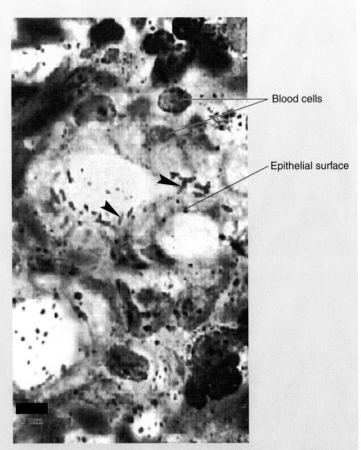

Histological section of stomach showing *Helicobacter pylori* attached to the epithelial surface (arrows). Note infiltration by blood cells indicative of inflammation.

Campylobacter jejuni Enteritis

*Campylobacter jejuni** has recently emerged as a pathogen of such imposing proportions that it is considered one of the most important causes of bacterial gastroenteritis worldwide. Early results of epidemiologic and pathologic studies have shown that this species is a primary pathogen transmitted through contaminated beverages and food, especially water, milk, meat, and chicken.

When ingested, *C. jejuni* cells reach the mucosa at the last segment of the small intestine (ileum) near its junction with the colon; they adhere, burrow through the mucus, and multiply. Symptoms of headache, fever, abdominal pain, and bloody or

---

* *jejuni* (jee-joo′-nye) L. *jejunum.* The small section of intestine between the duodenum and the ileum.

watery diarrhea commence after an incubation period of 1 to 7 days. The mechanisms of pathology appear to involve a heat-labile enterotoxin called CJT that stimulates a secretory diarrhea like that of cholera.

Diagnosis of *C. jejuni* enteritis requires isolation from fecal samples and occasionally blood samples. More rapid presumptive diagnosis can be obtained from direct examination of feces with a dark-field microscope, which accentuates the characteristic curved rods and darting motility. Resolution of infection occurs in most instances with simple, nonspecific rehydration and electrolyte balance therapy. In more severely affected patients, it may be necessary to administer erythromycin, tetracycline, aminoglycosides, or quinolones. Because vaccines are yet to be developed, prevention depends upon rigid sanitary control of water and milk supplies and care in food preparation.

Traditionally of interest to the veterinarian, *C. fetus* (subspecies *venerealis*) causes a sexually transmitted disease of sheep, cattle, and goats. Its role as an agent of abortion in these animals has considerable economic impact on the livestock industry. About 40 years ago, the significance of *C. fetus* as a human pathogen was first uncovered, though its exact mode of transmission in humans is yet to be clarified. This bacterium appears to be an opportunistic pathogen that attacks debilitated persons or women late in pregnancy. Diseases to which *C. fetus* has been linked are meningitis, pneumonia, arthritis, fatal septicemic infection in the newborn, and occasionally, sexually transmitted proctitis in adults.

CHAPTER CHECKPOINTS

The genera *Vibrio* and *Campylobacter* are curved or short spiral-shaped bacteria that are actively motile.

Vibrio cholerae is the agent of epidemic, or Asiatic, cholera. It is spread through contaminated water. Its virulence factor is an enterotoxin that causes profuse diarrhea and dehydration. Diagnosis is by microscopic identification of the bacteria in stool specimens or serological tests. Therapy includes prompt replacement of water and electrolytes followed by antibiotics. Preventive measures focus on effective water purification.

V. parahaemolyticus and *V. vulnificus* are agents of seafood gastroenteritis. They are most often acquired by eating raw shellfish that have been exposed to contaminated wastewater.

Campylobacter jejunum is the agent of campylobacter enteritis. This vibrio is transmitted to humans through contaminated water and animal products. Its enterotoxin, CJT, stimulates a diarrhea similar to cholera. Treatment in severe cases is similar to that for cholera.

Medically Important Bacteria of Unique Morphology and Biology

Bacterial groups that exhibit atypical morphology, physiology, and behavior are the following: (1) rickettsias and chlamydias, obligately parasitic gram-negative coccobacilli, and (2) mycoplasmas, highly pleomorphic bacteria that lack cell walls.

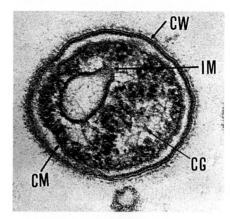

(a)

(b)

FIGURE 21.16

The morphology of *Rickettsia*. **(a)** Several features, including the cell wall (CW), cell membrane (CM), chromatin granules (CG), and mesosome (IM), identify these as tiny, pleomorphic, gram-negative bacteria (185,000×). **(b)** View of rickettsias adhering to the surface of a mouse tissue culture cell.

FAMILY RICKETTSIACEAE

The Family Rickettsiaceae contains about 23 species of pathogens, mostly in the genus *Rickettsia*.* Other members include *Ehrlichia** and the recently renamed *Orientia* (formerly *Rickettsia*). These organisms are known commonly as **rickettsiae,** or **rickettsias,** and the diseases they cause are called **rickettsioses.**

The rickettsias are all obligate to their host cells and require live cells for cultivation; they also spend part of their life cycle in the bodies of arthropods, which serve as vectors. Rickettsioses are among the most important emerging diseases. Six of the 14 recognized diseases have been identified in the last 12 years.

Morphological and Physiological Distinctions of Rickettsias

Rickettsias possess a gram-negative cell wall, binary fission, metabolic pathways for synthesis and growth, and both DNA and RNA. They are among the smallest cells, ranging from 0.3 to 0.6 μm wide and from 0.8 to 2.0 μm long. They are nonmotile pleomorphic rods or coccobacilli (figure 21.16).

---

* *Rickettsia* (rik′-ett′-see-ah) After Howard Ricketts, an American bacteriologist who worked extensively with this group.

* *Ehrlichia* (ur-lik′-ee-ah) After Paul Ehrlich, a German immunologist.

The precise nutritional requirements of the rickettsias have been difficult to demonstrate because of a close association with host cell metabolism. Their obligate parasitism originates from an inability to metabolize AMP, an important precursor to ADP and ATP, which they must obtain from the host. Rickettsias are generally sensitive to environmental exposure, although *R. typhi* can survive several years in dried flea droppings.

Distribution and Ecology of Rickettsial Diseases

The Role of Arthropod Vectors The rickettsial life cycle depends upon a complex exchange between blood-sucking arthropod[2] hosts and vertebrate hosts (Microbits 21.4). Eight tick genera, two fleas, and one louse are involved in the spread of rickettsias to humans. Humans accidently enter the zoonotic life cycles through occupational contact with the animals except in the cases of louse-borne typhus and trench fever. Most vectors apparently harbor rickettsias with no ill effect, but others, like the human body louse, die from typhus infection and do not continuously harbor the pathogen. In certain vectors (Rocky Mountain spotted fever ticks), rickettsias are transferred by eggs of an infected female. Such continuous inheritance of the microbe through multiple generations of ticks creates a long-standing reservoir.

These arthropods feed on the blood or tissue fluids of their mammalian hosts, but not all of them transmit the rickettsial pathogen through direct inoculation with saliva. Ticks directly inoculate the skin lesion from their mouths, but fleas and lice harbor the infectious agents in their intestinal tracts. During their stay on the host's body, these latter insects defecate or are smashed, thereby releasing the rickettsias onto the skin or into a wound. Ironically, scratching the bite helps the pathogen invade deeper tissues.

General Factors in Rickettsial Pathology and Isolation

A common target in rickettsial infections is the endothelial lining of the small blood vessels. The bacteria recognize, enter, and multiply within endothelial cells, causing necrosis of the vascular lining. Among the immediate pathologic consequences are vasculitis, perivascular infiltration by inflammatory cells, vascular leakage, and thrombosis. These pathologic effects are manifested by skin rash, edema, hypotension, and gangrene. Intravascular clotting in the brain accounts for the stuporous mental changes and other neurological symptoms that sometimes occur.

Isolation of most rickettsias from clinical specimens requires a suitable live medium and specialized laboratory facilities, including controlled access and safety cabinets. The usual choices for routine growth and maintenance are the yolk sacs of embryonated chicken eggs, chick embryo cell cultures, and, to a lesser extent, mice and guinea pigs.

SPECIFIC RICKETTSIOSES

Rickettsioses can be differentiated on the basis of their clinical features and epidemiology as: (1) the typhus group; (2) the spotted fever group; (3) scrub typhus, and (4) ehrlichiosis (table 21.2).

Epidemic Typhus and *Rickettsia prowazekii*

Epidemic, or louse-borne, typhus* has been a constant accompaniment to war, poverty, and famine. The extensive investigations of Dr. Howard Ricketts and Stanislas von Prowazek in the early 1900s led to the discovery of the vector and the rickettsial agent, but not without mortal peril; both men died of the very disease they investigated. *Rickettsia prowazekii* was named in honor of their pioneering efforts.

Epidemiology of Epidemic Typhus Humans are the sole hosts of human body lice and the only reservoirs of *R. prowazekii*. The louse spreads infection by defecating into its bite wound or other breaks in the skin. Infection of the eye or respiratory tract can take place by direct contact or inhalation of dust containing dried louse feces, but this is a rarer mode of transmission.

The transfer of lice is increased by crowding, infrequently changing clothing, and sharing clothing. The overall incidence of epidemic typhus in the United States is very low, with no epidemics since 1922. Although no longer common in regions of the world with improved standards of living, epidemic typhus presently persists in regions of Africa, Central America, and South America.

Disease Manifestations and Immune Response in Typhus After entering the circulation, rickettsias pass through an intracellular incubation period of 10 to 14 days. The first clinical manifestations are sustained high fever, chills, frontal headache, and muscular pain. Within 7 days, a generalized rash appears, initially on the trunk, and then spreads to the extremities. Personality changes, oliguria (low urine output), hypotension, and gangrene complicate the more severe cases. Mortality is lowest in children, and as high as 40–60% in patients over 50 years of age.

Recovery usually confers resistance to typhus, but in some cases, the rickettsias are not completely eradicated by the immune response and enter into latency. After several years, a milder recurring form of the disease, known as *Brill-Zinsser disease,* can appear. This disease is seen most often in people who have immigrated from endemic areas and is of concern mainly because they provide a continuous reservoir of the etiologic agent.

Treatment and Prevention of Typhus The standard chemotherapy for typhus is tetracycline or chloramphenicol. Despite antibiotic therapy, however, the prognosis can be poor in patients with advanced circulatory or renal complications. Eradication of epidemic typhus is theoretically possible by exterminating the vector. Widespread dusting of human living quarters with insecticides has provided some environmental control, and individual treatment with an antilouse shampoo or ointment is also effective. Vaccination against the rickettsial agent provides yet another effective method of control.

Epidemiology and Clinical Features of Endemic Typhus

The agent for **endemic typhus** is *Rickettsia typhi (R. mooseri)*, which shares many characteristics with *R. prowazekii* except for pronounced virulence. Synonyms for this rickettsiosis are endemic

---

2. Ticks are in the Class Arachnida, and lice and fleas are in the Class Insecta, both in the Phylum Arthropoda.

* typhus (ty′-fus) Gr. *typhos,* smoky or hazy, underlining the mental deterioration seen in this disease. Typhus is commonly confused with typhoid fever, an unrelated enteric illness caused by *Salmonella typhi.*

Many bacterial pathogens have evolved with and made complex adaptations to the bodies of arthropods. These animals are arachnids (ticks) and insects (fleas and lice) that happen also to be natural ectoparasites of humans, feeding on blood or tissue fluids. In this role as biological vectors, they are an important source of zoonotic infections in humans.

Ticks

Distributed the world over are 810 species of ticks, all of them ectoparasites and about 100 of them vectors of infectious disease.

Soft (argasid) ticks seek specific, sheltered habitats in caves, barns, burrows, or even birdcages, and they parasitize wildlife during nesting seasons and domesticated animals throughout the year. Endemic relapsing fever is transmitted by this type of tick.

Hard (ixodid) ticks have adapted to a wide-ranging life-style, hitchhiking along as their hosts wander through forest, savanna, or desert regions. Depending upon the species, ticks feed during larval, nymph, and adult metamorphic stages. The longevity of ticks is formidable; metamorphosis can extend for 2 years, and adults can survive for 4 years away from a host without feeding. The tiny, unengorged ticks humans pick up from vegetation crawl on the body, embed their mouthparts, and fill with blood, expanding to hundreds of times in size. Ixodid ticks are implicated in Rocky Mountain spotted and Q fevers and human ehrlichosis. Ticks also play host to *Borrelia* (Lyme disease) and arboviruses.

Fleas

Fleas are laterally flattened, wingless insects with well-developed jumping legs and prominent probosci for piercing the skin of warm-blooded animals. They are known for their extreme longevity and resistance, and many are notorious in their nonspecificity, passing with ease from wild or domesticated mammals to humans. In response to mechanical stimulation and warmth, fleas jump onto their targets and crawl about, feeding as they go. A well-known vector is the oriental rat flea, *Xenopsylla cheopis,* that transmits *Rickettsia typhi,* the cause of murine typhus. The flea harbors the pathogen in its gut and periodically contaminates the environment with virulent rickettsias by defecating. This same flea is involved in the transmission of plague (see figure 20.19).

Lice

Lice are small, flat insects equipped with biting or sucking mouthparts. The lice of humans usually occupy head and body hair *(Pediculus humanus)* or pubic, chest, and axillary hair *(Phthirus pubis)*. They feed by gently piercing the skin and sucking blood and tissue fluid. Infection develops when the louse or its feces is inadvertently crushed and rubbed into wounds, skin, eyes, or mucous membranes. Lice are involved in transmitting *R. prowazekii* (epidemic typhus) and *Borrelia recurrentis* (relapsing fever).

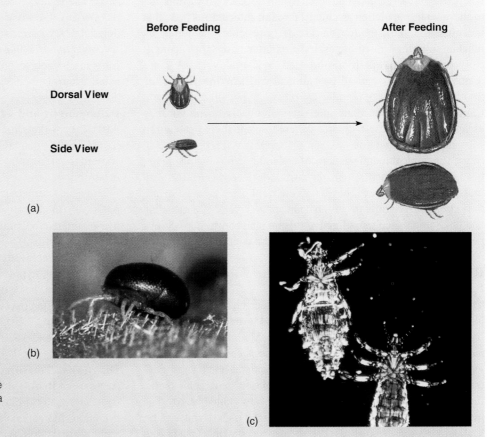

Arthropod vectors. **(a)** Hard tick (*Dermacentor andersoni,* a vector of Rocky Mountain spotted fever) before and after feeding. Because the head of the engorged tick can become embedded in a victim's skin, extreme care must be taken in removing. **(b)** An engorged soft tick, represented here by *Ornithodoros* (the vector of relapsing fever). **(c)** The body louse, a human parasite that transmits epidemic typhus and one form of relapsing fever.

Before Feeding · After Feeding · Dorsal View · Side View · (a) · (b) · (c)

TABLE 21.2

Characteristics of Major Rickettsias Involved in Human Disease

Disease Group	Species	Disease	Vector	Primary Reservoir	Mode of Transmission to Humans	Where Found
Typhus	*Rickettsia prowazekii*	Epidemic typhus	Body louse	Humans	Louse feces rubbed into bite; inhalation	Worldwide
	R. typhi (mooseri)	Murine typhus	Flea	Rodents	Flea feces rubbed into skin; inhalation	Worldwide
Spotted Fever	*R. rickettsii*	Rocky Mountain spotted fever	Tick	Small mammals	Tick bite; aerosols	North and South America
	R. akari	Rickettsialpox	Mite	Mice	Mite bite	Worldwide
Scrub Typhus	*Orientia tsutsugamushi*	—	Immature mite	Rodents	Bite	Asia, Australia, Pacific Islands
Human Ehrlichiosis	*Ehrlichia chaffeensis*	Human monocytic ehrlichiosis	Tick	—	Tick bite	Similar to Rocky Mountain spotted fever
	Ehrlichia species	Human granulocytic ehrlichiosis	Tick	Deer, rodents	Tick bite	Unknown

typhus, murine (mouse) typhus, and flea-borne typhus. The disease occurs in certain parts of Central and South America and in the Southeast, Gulf Coast, and Southwest regions of the United States, where *R. typhi* is regularly harbored by mice and rats. Transmission to the human population is chiefly through infected rat fleas that inoculate the skin, though disease is occasionally acquired by inhalation. In the United States, most reported cases arise sporadically among workers in rat-infested industrial sites. A close relative, *R. felis,* is carried by cats and cat fleas in parts of the Southwest. It causes sporadic cases of a typhuslike disease in California and Texas. Many more cases probably occur, but they escape notice because they are scattered or not severe enough to warrant medical attention.

The clinical manifestations of endemic typhus include fever, headache, muscle aches, and malaise. After 5 days, a skin rash, transient in milder cases, begins on the trunk and radiates toward the extremities. Symptoms dissipate in about 2 weeks. Tetracycline and chloramphenicol are effective therapeutic agents, and various pesticides are available for vector and rodent control.

Rocky Mountain Spotted Fever: Epidemiology and Pathology

The rickettsial disease with greatest impact on people living in North America is **Rocky Mountain spotted fever (RMSF),** named for the place it was first seen—the Rocky Mountains of Montana and Idaho. Ricketts identified the etiologic agent *Rickettsia rickettsii* in smears from infected animals and patients and later discovered that it was transmitted by ticks. Despite its geographic name, this disease occurs infrequently in the western United States. The majority of cases are concentrated in the Southeast and eastern seaboard regions (figure 21.17). It also occurs in Canada and Central and South America. Infections occur most frequently in the

spring and summer, when the tick vector is most active. The yearly rate of RMSF is 20 to 40 cases per 10,000 population, with fluctuations coinciding with weather and tick infestations.

The principal reservoirs and vectors of *R. rickettsii* are hard ticks such as the wood tick *(Dermacentor andersoni),* the American dog tick (*D. variabilis,* among others), and the Lone Star tick (*Amblyomma americanum*). The dog tick is probably most responsible for transmission to humans because it is the major vector in the southeastern United States (figure 21.18).

Pathogenesis and Clinical Manifestations of Spotted Fever
After 2 to 4 days incubation, the first symptoms are sustained fever, chills, headache, and muscular pain. The distinctive spotted rash usually comes on within 2 to 4 days after the prodromium (figure 21.19). Early lesions are slightly mottled like measles, but later ones are macular, maculopapular, and even petechial. In the severest untreated cases, the enlarged lesions merge and can become necrotic, predisposing to gangrene of the toes or fingertips.

Grave manifestations of disease are cardiovascular disruption, including hypotension, thrombosis, and hemorrhage. Conditions of restlessness, delirium, convulsions, tremor, and coma are signs of the often overwhelming effects on the central nervous system (see figure 21.18). Fatalities occur in an average of 20% of untreated cases and 5–10% of treated cases.

Diagnosis, Treatment, and Prevention of Spotted Fever
Any case of Rocky Mountain spotted fever is a cause for great concern and requires immediate treatment, even before laboratory confirmation. Indications sufficiently suggestive to start antimicrobic therapy are the following: (1) a cluster of symptoms, including sudden fever, headache, and rash; (2) recent contact with ticks or dogs;

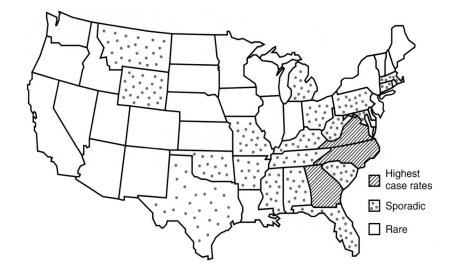

FIGURE 21.17

🔘 **The prevalence of Rocky Mountain spotted fever during the last 5 years.** Most cases are reported in states on the eastern seaboard and southeastern United States.

Highest case rates

Sporadic

Rare

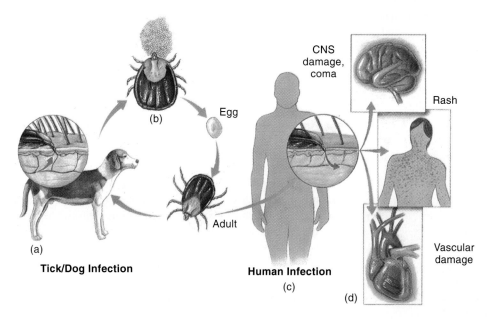

FIGURE 21.18

🔘 **The transmission cycle in Rocky Mountain spotted fever.** **(a)** Dog and wood ticks are the principal vectors. Ticks are infected from a mammalian reservoir during a blood meal. **(b)** Transovarial passage of *Rickettsia rickettsii* to tick eggs serves as a continual source of infection within the tick population. Infected eggs produce infected adults. **(c)** A tick attaches to a human, embeds its head in the skin, feeds, and sheds rickettsias into the bite. **(d)** Systemic involvement includes severe headache, fever, rash, coma, and vascular damage such as blood clots and hemorrhage.

and (3) possible occupational or recreational exposure in the spring or summer.

A recent boon to early diagnosis is a method for staining rickettsias directly in a tissue biopsy using fluorescent antibodies. Isolating rickettsias from the patient's blood or tissues is desirable, but it is expensive and requires specially qualified personnel and laboratory facilities. Specimens taken from the rash lesions are suitable for PCR assay, which is very specific and sensitive and can circumvent the need for culture. Because antibodies appear relatively soon after infection, a change in serum titer detected through an ELISA test can confirm a presumptive diagnosis.

The drug of choice for suspected and known cases is tetracycline (doxycycline) administered for one week. Chloramphenicol is an alternate choice when diagnosis is unclear. A new vaccine, made with rickettsias grown in chick embryo cell cultures, shows some promise in controlling the disease in humans who are at high risk. Other preventive measures follow the pattern for Lyme disease and other tick-borne disease: wearing protective clothing, using insect sprays, and fastidiously removing ticks.

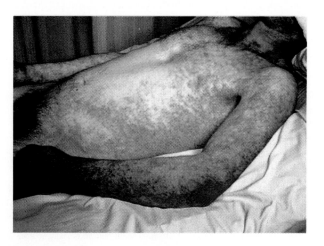

FIGURE 21.19

🔘 **Late generalized rash of Rocky Mountain spotted fever.** In some cases, lesions become hemorrhagic and predispose to gangrene of the extremities.

EMERGING RICKETTSIOSES

Members of *Ehrlichia* share a number of characteristics with *Rickettsia,* including the strict parasitic existence in host cells and the association with ticks. Although *Ehrlichia* have been identified for years as parasites of horses, dogs, and other mammals, human disease has been known only since 1986. Two emergent pathogens are *E. chaffeensis,* the cause of human monocytic erhlichiosis (HME), and an unnamed species that causes human granulocytic erhlichiosis (HGE). Both pathogens have a predisposition for growing in white blood cells, hence the names.

Since 1986, approximately 600 cases of HME have been diagnosed and traced to contact with the Lone Star tick *(Ambylomma).* The incidence of HGE since 1993 has been about 50 cases a year. This disease is carried by *Ixodes scapularis* ticks.

The signs and symptoms of the two diseases are similar. They present with an acute febrile state manifesting headache, muscle pain, and rigors. Most patients recover rapidly with no lasting effects, but around 5% of older chronically ill patients die from disseminated infection. Rapid diagnosis is enabled by PCR tests and indirect fluorescent antibody tests. It can be critical to differentiate or detect coinfection with the Lyme disease *Borrelia,* which is carried by the same tick. Doxycycline will clear up most infections within 7 to 10 days.

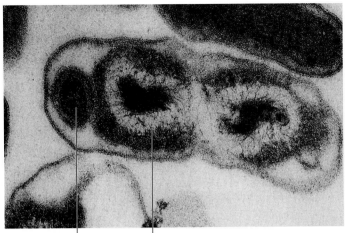

Endospore　　Vegetative cell

FIGURE 21.20

The agent of Q fever. The vegetative cells of *Coxiella burnetii* produce unique endospores that are released when the cell disintegrates. Free spores survive outside the host and are important in transmission.

CHAPTER CHECKPOINTS

The rickettsias and chlamydias are gram-negative obligate intracellular parasites that lack key metabolic enzymes.

The rickettsias are extremely small, pleomorphic rods or coccobacilli that cannot synthesize their own AMP. Most have a complex life-style that cycles between arthropod vectors and vertebrate hosts. Rickettsias are usually transmitted by tick bite or tick feces.

Rickettsial diseases are classified into four groups based on the vector and specific clinical characteristics: the typhus group, the spotted fever group, and ehrlichosis.

R. prowazekii is the agent of epidemic typhus, a systemic infection transmitted by body lice. Invasion of the vascular endothelium can cause necrosis and hypotension. This disease responds well to antibiotic therapy, and recovery usually provides immunity, except for a mild recurrence, Brill-Zinsser disease. Epidemic typhus is associated with overcrowded living conditions.

R. typhi is the agent of endemic, or murine, typhus, which is less virulent than louse-borne typhus. It is transmitted by the feces of infected rat fleas and enters the host when bites are scratched.

R. rickettsii is the agent of Rocky Mountain spotted fever (RMSF), a potentially fatal disease endemic to North America. It is transmitted by the bites of wood ticks, dog ticks, and the Lone Star tick. RMSF produces an infection that if untreated can cause tissue necrosis and cardiovascular and clotting disorders. This disease usually responds to tetracycline if given promptly.

Two new rickettsias in the genus *Ehrlichia* cause human monocytic and granulocytic ehrlichioses, diseases spread by ticks. Both involve infection of white blood cells, fever, and muscle pains.

DISEASES RELATED TO THE RICKETTSIOSES

Q fever[3] was first described in Queensland, Australia. Its origin was mysterious for a time, until Harold Cox working in Montana and Frank Burnet in Australia discovered the agent later named ***Coxiella burnetii.*** This bacterium is similar to rickettsias in being an intracellular parasite, but it is much more resistant because it produces an unusual type of spore (figure 21.20). It is apparently harbored by a wide assortment of vertebrates and arthropods, especially ticks, which play an essential role in transmission between wild and domestic animals. Humans acquire infection largely by means of environmental contamination and airborne spread. Sources of infectious material include urine, feces, milk, and airborne particles from infected animals. The primary portals of entry are the lungs, skin, conjunctiva, and gastrointestinal tract.

Coxiella burnetii has been isolated from most regions of the world. California and Texas have the highest case rates in the United States, though most cases probably go undetected. People at highest risk are farm workers, meat cutters, veterinarians, laboratory technicians, and consumers of raw milk. The clinical manifestations typical of *Coxiella* infection are abrupt onset of fever, chills, head and muscle ache, and occasionally, a rash. Disease is sometimes complicated by pneumonitis, hepatitis, and delayed endocarditis, which are occasionally fatal. Mild or subclinical cases resolve spontaneously, and more severe cases respond to tetracycline therapy. A vaccine is available in many parts of the world and is used on military personnel in the United States. People in contact with livestock and their excrement and secretions should protect themselves.

---

3. For "query," meaning to question, or of unknown origin.

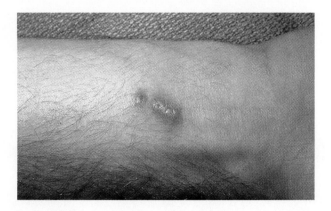

FIGURE 21.21

Cat-scratch disease. A primary nodule appears at the site of the scratch in about 21 days. In time, large quantities of pus collect, and the regional lymph nodes swell.

The Family Bartonellaceae contains the genus *Bartonella,** which now includes species formerly known as *Rochalimaea.* These small gram-negative rods are fastidious but not obligate intracellular parasites, and they can be cultured on blood agar. *Bartonella* species are currently considered a group of emerging pathogens. A disease formerly associated with war is **trench fever.** The causative agent is *Bartonella quintana,* a species that, like epidemic typhus, cycles between humans and lice. Most cases occur in endemic regions of Europe, Africa, and Asia. Highly variable symptoms can include a 5- to 6-day fever (hence 5-day, or quintana, fever); leg pains, especially in the tibial region (shinbone fever); headache; chills; and muscle aches. A macular rash can also occur. The microbe can persist in the blood long after convalescence and is responsible for later relapses.

Bartonella henselae, is the most common agent of **cat-scratch disease (CSD),** an infection connected with a cat scratch or bite. The pathogen can be isolated in over 40% of cats, especially kittens. There are approximately 25,000 cases per year in the United States, 80% of them in children 2 to 14 years old. The symptoms start after 1 to 2 weeks, with a cluster of small papules at the site of inoculation. In a few weeks, the lymph nodes along the lymphatic drainage swell and can become pus-filled (figure 21.21). Most infections remain localized and resolve in a few weeks, but drugs such as tetracycline, erythromycin, and rifampin can be effective therapies. The disease can be prevented by thorough degerming of a cat bite or scratch.

Bartonella is currently an important new pathogen in AIDS patients. It is the cause of **bacillary angiomatosus,** a severe cutaneous and systemic infection. The cutaneous lesions arise as reddish nodules or crusts that can be mistaken for Kaposi's sarcoma. Systems most affected are the liver and spleen, and symptoms are fever, weight loss, and night sweats. Treatment is similar to that for CSD.

OTHER OBLIGATE PARASITIC BACTERIA: THE CHLAMYDIACEAE

Like rickettsias, the chlamydias are obligate parasites that depend on certain metabolic constituents of host cells for growth and maintenance. They show further resemblance to the rickettsias with their small size, gram-negative cell wall, and pleomorphic morphology, but they are markedly different in several aspects of their life cycle. The species of greatest medical significance are *Chlamydia trachomatis,** a very common pathogen involved in sexually transmitted, neonatal, and ocular disease (trachoma); *C. pneumoniae,* the cause of one type of atypical pneumonia; and *C. psittaci,** a zoonosis of birds and mammals that causes ornithosis in humans.

The Biology of *Chlamydia*

Chlamydias alternate between two distinct stages: (1) a small, metabolically inactive, infectious form called the **elementary body** that is released by the infected host cell and (2) a larger, noninfectious, actively dividing form called the **reticulate body** that grows within the host cell vacuoles (figure 21.22). Elementary bodies are tiny, dense spheres shielded by a rigid, impervious envelope that ensures survival outside the eucaryotic host cell. Reticulate bodies are finely granulated and have thin cell walls. Studies of the reticulate bodies indicate that they are energy parasites, entirely lacking enzyme systems for catabolizing glucose and other substrates and for synthesizing ATP, though they do possess ribosomes and mechanisms for synthesizing proteins, DNA, and RNA. Reticulate bodies ultimately differentiate into elementary bodies.

Diseases of *Chlamydia trachomatis*

The reservoir of pathogenic strains of *Chlamydia trachomatis* is the human body. The microbe shows astoundingly broad distribution within the population, often being carried with no symptoms. Elementary bodies are transmitted in infectious secretions, and although infection can occur in all age groups, disease is most severe in infants and children. The two human strains are the **trachoma** strain, which attacks the squamous or columnar cells of mucous membranes in the eyes, genitourinary tract, and lungs, and the **lymphogranuloma venereum (LGV)** strain, which invades the lymphatic tissues of the genitalia.

Chlamydial Diseases of the Eye The two forms of chlamydial eye disease, ocular trachoma and inclusion conjunctivitis, differ in their patterns of transmission and ecology. **Ocular trachoma,** an infection of the epithelial cells of the eye, is an ancient disease and a major cause of blindness in certain parts of the world. Although a few cases occur yearly in the United States, several million cases occur endemically in parts of Africa and Asia. Transmission is favored by contaminated fingers, fomites, flies, and a hot, dry climate.

---

* *Bartonella* (barr″-tun-el′-ah) After A. L. Barton, a Peruvian physician who first described the genus.

* *Chlamydia trachomatis* (klah-mid′-ee-ah trah-koh′-mah-tis) Gr. *chlamys,* a cloak, and *trachoma,* roughness.

* *psittaci* (sih-tah′-see) Gr. *psittacus,* a parrot.

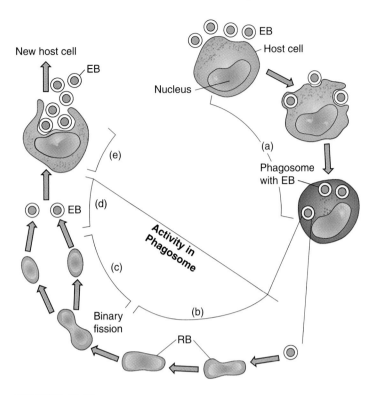

FIGURE 21.22

🔵 **The life cycle of *Chlamydia*.** **(a)** The infectious stage, or elementary body (EB), is taken into phagocytic vesicles by the host cell. **(b)** In the phagosome, each elementary body develops into a reticulate body (RB). **(c)** Reticulate bodies multiply by regular binary fission. **(d)** Mature RBs become reorganized into EBs. **(e)** Completed EBs are released from the host cell.

(a)

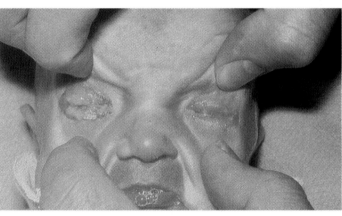

(b)

FIGURE 21.23

🔵 **The pathology of primary ocular chlamydial infection.** **(a)** Ocular trachoma, an early, pebblelike inflammation of the conjunctiva and inner lid in a child. (Note: The eyelid has been retracted to make the lesion more visible.) **(b)** Inclusion conjunctivitis in a newborn. Within 5 to 6 days, an abundant, watery exudate collects around the conjunctival sac. This is currently the most common cause of ophthalmia neonatorum.

The first signs of infection are a mild conjunctival exudate and slight inflammation of the conjunctiva. These are followed by marked infiltration of lymphocytes and macrophages into the infected area. As these cells build up, they impart a pebbled (rough) appearance to the inner aspect of the upper eyelid (figure 21.23*a*). In time, a vascular pseudomembrane of exudate and inflammatory leukocytes forms over the cornea, a condition called *pannus* that lasts a few weeks. Chronic and secondary infections can lead to corneal damage and impaired vision. Early treatment of this disease with tetracycline or sulfa drugs is highly effective and prevents all of the complications. It is a tragedy that in this day of preventive medicine, millions of children worldwide will develop blindness for lack of a few dollars' worth of antibiotics.

Inclusion conjunctivitis is usually acquired through contact with secretions of an infected genitourinary tract. Infantile conjunctivitis develops 5 to 12 days after a baby has passed through the birth canal of its infected mother and is the most prevalent form of conjunctivitis in the United States (100,000 cases per year). The initial signs are conjunctival irritation, a profuse adherent exudate, redness, and swelling (see figure 21.23*b*). Although the disease usually heals spontaneously, trachoma-like scarring occurs often enough to warrant routine prophylaxis of all newborns (as for gonococcal infection) with antibiotics such as erythromycin and tetracycline.

Sexually Transmitted Chlamydial Diseases It has been estimated that *C. trachomatis* is carried in the reproductive tract of up to 10% of all people, with even higher rates among the promiscuous. About 70% of infected women harbor it asymptomatically on the cervix, while 10% of infected males show no signs or symptoms. The potential for this disease to cause long-term reproductive damage has initiated its listing as a reportable disease since 1995. Statistics now show that chlamydiosis is the second most prevalent sexually transmitted disease. The official number of cases reported to the CDC is over 600,000 cases per year, but it probably exceeds that level by 10 times. The prevalence of this infection in young, sexually active teenagers is increasing by 8–10% per year. Medically and socioeconomically, its clinical significance now eclipses gonorrhea, herpes simplex 2, and syphilis.

A syndrome appearing among males with chlamydial infections is an inflammation of the urethra called **nongonococcal urethritis (NGU).** This diagnosis is derived from the symptoms that mimic gonorrhea yet do not involve gonococci. Women with symptomatic chlamydial infection have cervicitis accompanied by a white

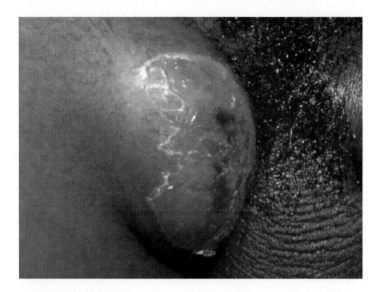

FIGURE 21.24

🔊 **The clinical appearance of advanced lymphogranuloma venereum in a man.** A chronic local inflammation blocks the lymph channels, causing swelling and distortion of the external genitalia.

drainage, endometritis, and salpingitis (pelvic inflammatory disease; PID). As is often the rule with sexually transmitted diseases, chlamydia frequently appears in mixed infections with the gonococcus and other genitourinary pathogens, thereby greatly complicating treatment.

When a particularly virulent strain of *Chlamydia* chronically infects the genitourinary tract, the result is a severe, often disfiguring disease called **lymphogranuloma venereum.**[4] The disease is endemic to regions of South America, Africa, and Asia but occasionally occurs in other parts of the world. Its incidence in the United States is about 500 cases per year. Chlamydias enter through tiny nicks or breaks in the perigenital skin or mucous membranes and form a small, painless vesicular lesion that often escapes notice. Other acute symptoms are headache, fever, and muscle aches. As the lymph nodes near the lesion begin to fill with granuloma cells, they enlarge and become firm and tender (figure 21.24). These nodes, or buboes, can cause long-term lymphatic obstruction that leads to chronic, deforming edema of the genitalia and anus.

Identification, Treatment, and Prevention of Chlamydiosis
Because chlamydias reside intracellularly, specimen sampling requires enough force to dislodge some of the cells from the mucosal surface. Genital samples are taken with a swab inserted a few centimeters into the urethra or cervix, rotated, and removed. Although the most reliable diagnosis comes from culture in chicken embryos, mice, or cell lines, this procedure is too costly and time-consuming to be used routinely in STD clinics; however, it is an essential part of diagnosing neonatal infections. The most sensitive and specific tests currently available are a direct assay of specimens using immunofluorescence (figure 21.25*a*) and a PCR-based probe. Methods useful in diagnosing inclusion conjunctivitis are Giemsa or iodine stains (figure 21.25*b*), but they are not recommended for urogenital specimens because of low sensitivity and the possibility of obtaining false-negative results in asymptomatic patients.

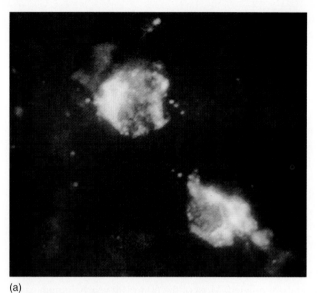

(a)

Inclusion body

Nucleus

(b)

FIGURE 21.25

🔊 **Direct diagnosis of chlamydial infection.** **(a)** A specimen stained with monoclonal antibodies bearing a fluorescent dye. Infected cells glow a bright apple green. **(b)** Direct Giemsa stain of the eyelid scraping of a patient suffering from inclusion conjunctivitis. Large inclusion body represents a phagosome packed with chlamydia in various stages of development.

Urogenital chlamydial infections are most effectively treated with drugs that act intracellularly, such as tetracyclines and azithromycin. Penicillin and aminoglycosides are not effective and must not be used. Because of the high carrier rate and the difficulty in detection, prevention of chlamydial infections is a public health priority. As a general rule, sexual partners of infected people should be treated with drug therapy to prevent infection, and sexually active people can achieve some protection with a condom.

Chlamydia pneumoniae
A strict human pathogen, *Chlamydia pneumoniae,* is distinctly different and not closely related to the other chlamydias. It has been linked to a type of respiratory illness that includes pharyngitis, bronchitis, and pneumonitis. It is usually a mild illness in young

---

4. Also called tropical bubo or lymphogranuloma inguinale.

adults, though it can cause a severe reaction in asthmatic patients that is responsible for increased rates of death in this group.

Chlamydia psittaci and Ornithosis

The term *psittacosis* was adopted to describe a pneumonia-like illness contracted by people working with imported parrots and other psittacine birds in the last century. As outbreaks of this disease appeared in areas of the world having no parrots, it became evident that other birds could carry and transmit the microbe to humans and other animals. In light of this evidence, the generic term **ornithosis*** has been suggested as a replacement.

Ornithosis is a worldwide zoonosis that is carried in a latent state in wild and domesticated birds but becomes active under stressful conditions such as overcrowding. In the United States, poultry have been subject to extensive epidemics that killed as many as 30% of flocks. Infection is communicated to other birds, mammals, and humans by contaminated feces and other discharges that become airborne and are inhaled. Most of the 50 to 100 yearly human cases in the United States occur among poultry and pigeon handlers.

The symptoms of human ornithosis mimic those of influenza and pneumococcal pneumonia. Early manifestations are fever, chills, frontal headache, and muscle aches, and later ones are coughing and lung consolidation. Unchecked, infection can lead to systemic complications involving the meninges, brain, heart, or liver. Although most patients respond well to tetracycline or erythromycin therapy, recovery is often slow and fraught with relapses. Control of the disease is usually attempted by quarantining imported birds and by taking precautions in handling birds, feathers, and droppings.

---

* ornithosis (or″-nih-thoh′-sis) Gr. *ornis*, bird. More than 90 species of birds harbor *C. psittaci.*

CHAPTER CHECKPOINTS

Coxiella burnetii, the cause of Q fever, is a rickettsia-like agent transmitted to humans primarily by milk, meat, and airborne contamination. It infects a broad range of vertebrates and arthropods. It also forms resistant spores. Prevention includes pasteurization of milk.

Bartonella quintana is the agent of trench fever, a systemic infection lasting 5 to 6 days. It is transmitted to humans by body lice and can be recurrent.

Bartonella henselae is the agent of cat-scratch disease, a systemic infection that travels from the intial site along the lymph vessels. This disease responds to antibiotic therapy if not resolved spontaneously.

The chlamydias are small, gram-negative, pleomorphic, intracellular parasites that have no catabolic pathways. They exist in two forms: the elementary body, which is the form transmitted between human hosts through direct contact and body secretions; and the reticulate body, which multiplies intracellularly.

Chlamydia trachomatis is the agent of several STDs: NGU (nongonococcal urethritis), pelvic inflammatory disease, and lymphogranuloma venereum. It also causes ocular trachoma, a serious eye infection.

Chlamydia pneumoniae is the agent of respiratory infections in young adults and asthmatics.

Chlamydia psittaci is the agent of ornithosis, an influenza-like disease carried by birds that has serious systemic complications if untreated. Antibiotic therapy is usually successful.

Mollicutes and Other Cell-Wall-Deficient Bacteria

Bacteria in the Class Mollicutes, also called the **mycoplasmas,** are the smallest self-replicating microorganisms. All of them naturally lack a cell wall (figure 21.26*a*), and except for one genus, all species are parasites of animals and plants. The two most clinically important genera are *Mycoplasma* and *Ureaplasma.* Disease of the respiratory tract has been primarily associated with *Mycoplasma pneumoniae; M. hominis* and *Ureaplasma urealyticum* are implicated in urogenital tract infections.

BIOLOGICAL CHARACTERISTICS OF THE MYCOPLASMAS

Without a rigid cell wall to delimit their shape, mycoplasmas are exceedingly pleomorphic. The small (0.3–0.8 μm), flexible cells assume a spectrum of shapes, ranging from cocci and filaments to doughnuts, clubs, and helices (see figure 4.33). Mycoplasmas are not strict parasites, and they can grow in cell-free media, generate metabolic energy, and synthesize proteins with their own enzymes. However, most are fastidious and require complex media containing sterols, fatty acids, and preformed purines and pyrimidines. Animal species of mycoplasmas are sometimes referred to as membrane parasites because they form intimate associations with membranes of the respiratory and urogenital tracts (figure 21.26*b*). Because mycoplasmas bind to specific receptor sites on cells and adhere so tenaciously that they are not easily removed by usual defense mechanisms, infections are chronic and difficult to eliminate.

Mycoplasma pneumoniae and Atypical Pneumonia

Mycoplasma pneumoniae is a human parasite that is the most common agent of **primary atypical pneumonia (PAP).** This syndrome is atypical in that its symptoms do not resemble those of pneumococcal pneumonia. Primary atypical pneumonia can also be caused by rickettsias, chlamydias, respiratory syncytial viruses, and adenoviruses. Mycoplasmal pneumonia is transmitted by aerosol droplets among people confined in close living quarters, especially families, students, and the military. Community resistance to this pneumonia is high; only 3–10% of those exposed become infected, and fatalities are rare.

Mycoplasma pneumoniae selectively binds to specific receptors of the respiratory epithelium and inhibits ciliary action. Gradual spread of the bacteria over the next 2 to 3 weeks disrupts the cilia and damages the respiratory epithelium. The first symptoms—fever, malaise, sore throat, and headache—are not suggestive of pneumonia. A cough is not a prominent early symptom, and when it does appear, it is mostly unproductive. As the disease progresses, nasal symptoms, chest pain, and earache can develop. The lack of acute illness in most patients has given rise to the nickname "walking pneumonia."

Diagnosis Because a culture can take 2 or 3 weeks, early diagnosis of mycoplasma pneumonia is difficult and relies chiefly on close clinical observation to rule out other bacterial or viral agents. Stains of sputum appear devoid of bacterial cells, leukocyte counts are within normal limits, and X-ray findings are nonspecific. Serological tests based on complement fixation, immunofluorescence, and indirect hemagglutination are useful later in the disease.

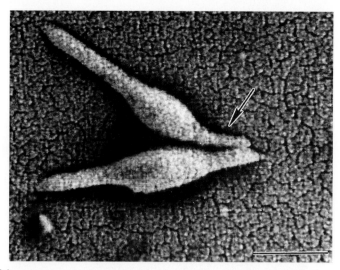

(a)

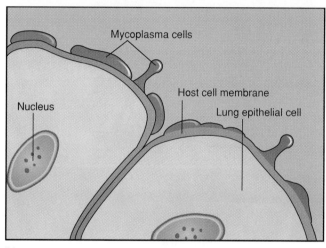

(b)

FIGURE 21.26

The morphology of mycoplasmas. **(a)** A scanning electron micrograph of *Mycoplasma pneumoniae* (bar = 0.5 μm). Note pleomorphic shape and elongate attachment tip (arrow). The cells use this to anchor themselves to host cells. **(b)** Figure depicts how *M. pneumoniae* becomes a membrane parasite that adheres tightly and fuses with the host cell surface. This fusing makes destruction and removal of the pathogen very difficult.

Tetracycline and erythromycin inhibit mycoplasmal growth and help to rapidly diminish symptoms, but they do not stop the shedding of viable mycoplasmas. Patients frequently experience relapses if treatment is not continued for 14 to 21 days. Preventive measures include controlling contamination of fomites, avoiding contact with droplet nuclei, and reducing aerosol dispersion.

Other Mycoplasmas

Mycoplasma hominis and *Ureaplasma urealyticum* are regarded as weak, sexually transmitted pathogens. They are frequently encountered in samples from the urethra, vagina, and cervix of newborns and adults. These species initially colonize an infant at birth and subsequently diminish through early and late childhood. A second period of colonization and persistence is initiated by the onset of sexual intercourse. Evidence linking genital mycoplasmas to human disease is substantial and growing every year. *Ureaplasma urealyticum* is implicated in some types of nonspecific or nongonococcal urethritis and prostatitis. There is increasing evidence that this mycoplasma plays a role in opportunistic infections of the fetus and fetal membranes. It appears to cause some cases of miscarriage, stillbirth, premature birth, and respiratory infections of newborns. *Mycoplasma hominis* is more often associated with vaginitis, pelvic inflammatory disease, and kidney inflammation.

BACTERIA THAT HAVE LOST THEIR CELL WALLS

Exposure of typical walled bacteria to certain drugs (penicillin) or enzymes (lysozyme) can result in wall-deficient bacteria called L forms or L-phase variants (see figure 4.17). L forms are induced or occur spontaneously in numerous species and can even become stable and reproduce themselves, but they are not naturally related to mycoplasmas.

L Forms and Disease

The role of certain L forms in human and animal disease is a distinct possibility, but proving etiology has been complicated because infection is difficult to verify with Koch's postulates. One theory proposes that antimicrobic therapy with cell-wall-active agents induces certain infectious agents to become L forms. In this wall-free state, they resist further treatment with these drugs and remain latent until the therapy ends, at which time they reacquire walls and resume their pathogenic behavior.

Infections with L-phase variants of group A streptococci, *Proteus,* and *Corynebacterium* have been reported, though they are uncommon. In a number of chronic pyelonephritis and endocarditis cases, cell-wall-deficient bacteria have been the only isolates. Research on people with a chronic intestinal syndrome called Crohn disease has uncovered a strong association with a wall-deficient form of *Mycobacterium paratuberculosis,* a close relative of the TB bacillus. When a PCR technique was used to analyze the DNA of colon specimens, it was found that 65% of Crohn patients tested positive for *M. paratuberculosis.* Further studies have shown that this bacterium is found in cow's milk, which may well be the source of infection. Treatment with a drug "cocktail" is gaining support as a means of controlling the disease.

CHAPTER CHECKPOINTS

Mycoplasmas are tiny pleomorphic bacteria that lack a cell wall. Although most species are parasitic, mycoplasmas can be cultured on complex artificial media. They are considered membrane parasites because they bind tightly to epithelial linings of the respiratory and urogenital tracts.

Mycoplasma pneumoniae is the agent of primary atypical pneumonia (walking pneumonia).

M. hominis and *Ureoplasma urealyticum* are agents of sexually transmitted infections of the reproductive tract and kidneys, and more recently, of fetal infections.

M. incognitus is a recently identified pathogen that causes systemic infection by suppressing the immune system.

L forms are wall-deficient variants of walled bacteria such as group A streptococci, *Proteus, Mycobacterium,* and *Corynebacterium* that are occasionally involved in diseases.

Bacteria in Dental Disease

The relationship between humans and their oral microflora is a complex, dynamic microecosystem. The mouth contains a diversity of surfaces for colonization, including the tongue, teeth, gingiva, palate, and cheeks, and it provides numerous aerobic, anaerobic, and microaerophilic microhabitats for the estimated 300 different oral species with which humans coexist. The habitat of the oral cavity is warm, moist, and greatly enriched by the periodic infusion of food. In most humans, this association remains in balance with little adverse effect, but in people with poor oral hygiene, it teeters constantly on the brink of disease.

THE STRUCTURE OF TEETH AND ASSOCIATED TISSUES

Dental diseases can affect nearly any part of the oral cavity, but most of them involve the **dentition** (teeth) and the surrounding supportive structures, collectively known as the **periodontium*** (gingiva, ligaments, membrane, bone) (figure 21.27). A tooth is composed of a flared **crown** that protrudes above the gum and a **root** that is inserted into a bony socket. The outer surface of the crown is protected by a dense coating of **enamel,** an extremely hard, noncellular material composed of tightly packed rods of calcium phosphate ($CaPO_4$) that cannot be replaced once the tooth has matured. The root is surrounded by a layer of cementum, which is anchored by ligaments to the periodontal membrane that lines the socket. The major portion of the tooth inside the crown and root is composed of a highly regular calcified material called dentin, and the core contains a pulp cavity that supplies the living tissues with blood vessels and nerves. The root canal is the portion of the pulp that extends into the roots. The space surrounding the teeth is protected by the **gingiva** (gum), a soft covering composed of connective tissue and a mucous membrane. The primary sites for initial dental infections are the enamel, especially the cusps, and the crevice, or sulcus, formed where the gingiva meets the tooth.

Dental pathology generally affects both hard and soft tissues (figure 21.28). Although both categories of disease are initiated when microbes adhere to the tooth surface and produce dental plaque, their outcomes vary. In the case of dental caries, the gradual breakdown of the enamel leads to invasive disease of the tooth itself, whereas in soft-tissue disease, calcified plaque damages the soft gingival tissues and predisposes them to bacterial invasion. Both diseases are responsible for the loss of teeth, though dental caries are usually implicated in children, and periodontal infections in adults.

HARD-TISSUE DISEASE: DENTAL CARIES

Dental caries* is the most common human disease. It is a complex mixed infection of the dentition that gradually destroys the enamel and often lays the groundwork for the destruction of deeper tissues. It occurs most often on tooth surfaces that are less accessible and harder to clean and on those that provide pockets or crevices where

---

* **periodontium** (per″-ee-oh-don′-shee-um) Gr. *peri,* around, and *odous,* tooth. Gums, bones, and cementum.

* **caries** (kar′-eez) L., rottenness.

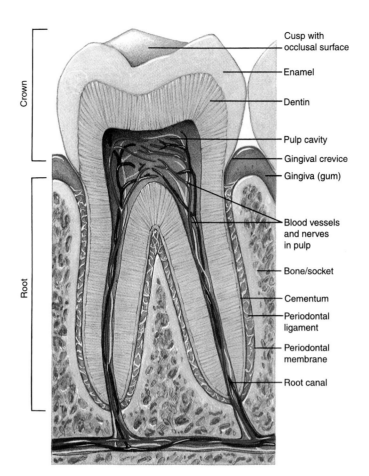

FIGURE 21.27
The anatomy of a tooth.

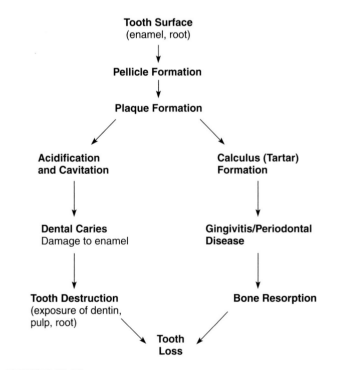

FIGURE 21.28
Summary of the events leading to dental caries, periodontal disease, and bone and tooth loss.

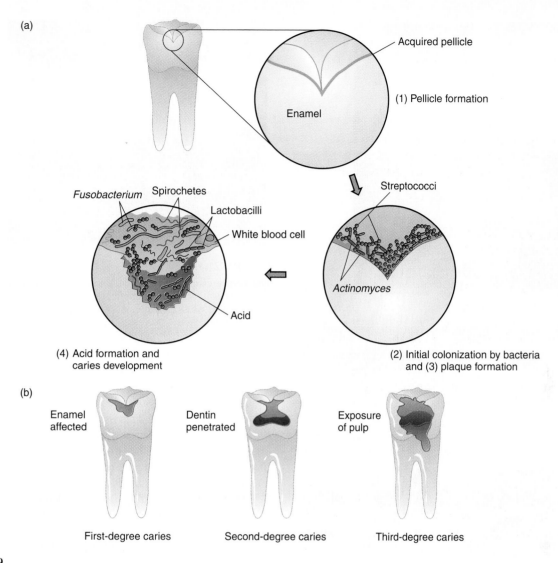

(a)

Acquired pellicle

(1) Pellicle formation

Enamel

Fusobacterium   Spirochetes

Lactobacilli

White blood cell

Streptococci

Actinomyces

Acid

(4) Acid formation and
caries development

(2) Initial colonization by bacteria
and (3) plaque formation

(b)

Enamel
affected

Dentin
penetrated

Exposure
of pulp

First-degree caries

Second-degree caries

Third-degree caries

FIGURE 21.29

Stages in plaque development and cariogenesis.   **(a)** A microscopic view of pellicle and plaque formation, acidification, and destruction of tooth enamel. **(b)** Progress and degrees of cariogenesis.

bacteria can cling. Caries commonly develop on enamel pits and fissures, especially those of the occlusal (grinding) surfaces, though they can also occur on the smoother crown surfaces and subgingivally on the roots.

Over the years, several views have been put forth to explain how dental caries originate. At various times it has been believed that sugar, microbes, or acid cause teeth to rot; however, studies with germ-free animals have now shown that no single factor can account for caries (see figure 1.11 and chapter 13). Caries development occurs in many phases and requires multiple interactions involving the anatomy, physiology, diet, and bacterial flora of the host. The principal stages in the formation of dental caries are pellicle formation, plaque formation, acid production and localization, and enamel etching (figure 21.29).

Plaque Formation

A freshly cleaned tooth is a perfect landscape for colonization by microbes. Within a few moments, it develops a thin, mucous coating called the **acquired pellicle,** which is made up of adhesive sali-

vary proteins. This structure presents a potential substrate upon which certain bacteria first gain a foothold. The process of colonization of the tooth follows a classic pattern of biofilm formation. The most prominent pioneering colonists belong to the genus *Streptococcus.* These gram-positive cocci have adhesive receptors such as fimbriae and slime layers that allow them to cling to the tooth surfaces and to each other, forming a foundation for the mature biofilm aggregate known as **plaque.*** Fed by a diet high in sucrose, glucose, and certain complex carbohydrates, *Streptococcus mutans* and *S. sobrinus* produce sticky polymers of glucose called fructans and glucans. These adhesives bind them to smooth enamel surfaces and form the matrix and bulk of the biofilm. As these primary invaders continue to grow over the tooth surface, they are joined by thin, branching *Actinomyces,* gram-positive bacteria that add to the complex network of early plaque. If plaque is allowed to remain, it is subsequently invaded by other oral bacteria. Among secondary invaders are species of *Lactobacillus, Bacteroides,*

---

* **plaque** (plak) Fr., a patch.

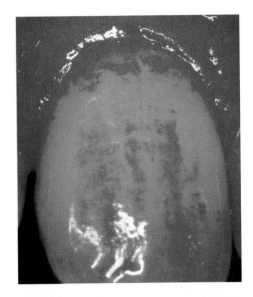

(a)

(b)

FIGURE 21.30

The macroscopic and microscopic appearance of plaque.
(a) Disclosing tablets containing vegetable dye stain heavy plaque accumulations at the junction of the tooth and gingiva. **(b)** Scanning electron micrograph of plaque with long filamentous forms and "corn cobs" that are mixed bacterial aggregates.

Fusobacterium, Porphyromonas, and *Treponema.* Plaque can be highlighted by staining the teeth with disclosing tablets (figure 21.30*a*), and a microscopic view reveals a rich and varied network of bacteria and their products along with epithelial cells and fluids (figure 21.30*b*).

Acid Formation and Localization and Etching of the Enamel

If mature plaque is not removed from sites that readily trap food, it usually evolves into a caries lesion (see figure 21.29). The role of plaque in caries development is related directly to streptococci and lactobacilli, which produce acid as they ferment dietary carbohydrates. If this acid is immediately flushed from the plaque and diluted in the mouth, it has little effect. However, in the denser re-

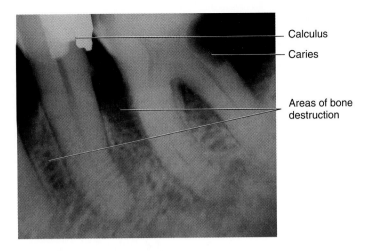

Calculus
Caries
Areas of bone destruction

FIGURE 21.31

The nature of calculus. Radiograph of mandibular premolar and molar, showing calculus on the top and a caries lesion on the right. Bony defects caused by periodontitis affect both teeth.

gions of plaque, the acid can accumulate in direct contact with the enamel surface and lower the pH to below 5, which is acidic enough to begin to dissolve (decalcify) the calcium phosphate of the enamel in that spot. This initial lesion can remain localized in the enamel (first-degree caries) and can be repaired with various inert materials (fillings). Once the deterioration has reached the level of the dentin (second-degree caries), tooth destruction speeds up, and the tooth can be rapidly destroyed. Exposure of the pulp (third-degree caries) is attended by severe tenderness and toothache, and the chance of saving the tooth is diminished.

SOFT-TISSUE (PERIODONTAL) DISEASE

Periodontal disease is so common that 97–100% of the population has some manifestation of it by age 45. Most kinds are due to bacterial colonization and varying degrees of inflammation that occur in response to gingival damage. The most common predisposing condition occurs when the plaque becomes mineralized (calcified) with calcium and phosphate crystals. This process produces a hard, porous substance called **calculus** above and below the gingiva that can induce varying degrees of periodontal damage (figure 21.31).

Calculus and plaque accumulating in the gingival sulcus cause abrasions in the delicate gingival membrane, and the chronic trauma causes a pronounced inflammatory reaction. The damaged tissues become a portal of entry for a variety of bacterial residents. These include anaerobic and gram-negative genera such as *Actinobacillus, Porphyromonas, Bacteroides,* and *Fusobacterium* (spindle-shaped rods), and numerous spirochetes. In response to the mixed infection, the damaged area becomes infiltrated by neutrophils and macrophages, and later by lymphocytes, which cause further inflammation and tissue damage (figure 21.32). The initial signs of **gingivitis** are swelling, loss of normal contour, patches of redness, and increased bleeding of the gingiva. Spaces or pockets of varying depth also develop between the tooth and the gingiva. If this condition persists, a more serious disease called **periodontitis** results. This is the natural extension of the disease into the

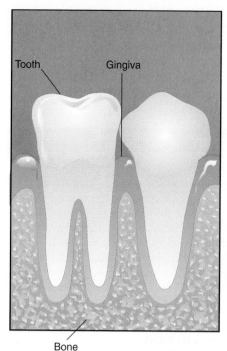

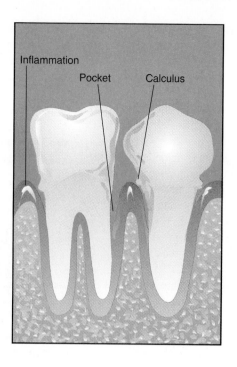

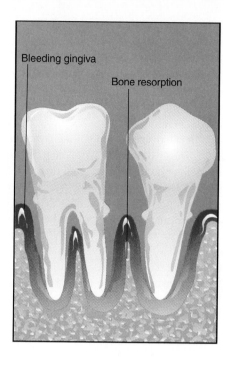

(a) Normal, nondiseased state of tooth, gingiva, and bone.

(b) Calculus buildup and early gingivitis.

(c) Late-stage periodontitis, with tissue destruction, deep pocket formation, loosening of teeth, and bone loss.

FIGURE 21.32

Stages in soft-tissue infection, gingivitis, and periodontitis.

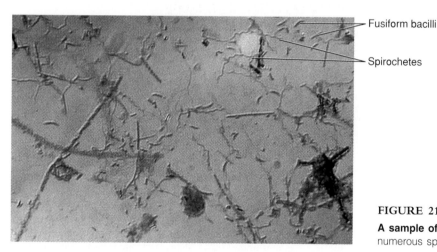

FIGURE 21.33

A sample of exudate from a gingival pocket (560×).   Note the numerous spirochetes and fusiform bacilli.

periodontal membrane and cementum. The deeper involvement increases the size of the pockets and can cause bone resorption severe enough to loosen the tooth in its socket. If the condition is allowed to progress, the tooth can be lost.

The most destructive periodontal disease is **acute necrotizing ulcerative gingivitis (ANUG),** formerly called trench mouth or Vincent's disease. This disease is a synergistic infection involving *Treponema vincentii, Bacteroides gingivalis,* and fusobacteria. These pathogens together produce several invasive factors that cause rapid advancement into the periodontal tissues (figure 21.33). The condition is associated with severe pain, bleeding, pseudo-

membrane formation, and necrosis. ANUG usually results from poor oral hygiene, altered host defenses, or prior gum disease, rather than being communicable. However, it responds well to broad-spectrum antibiotics.

FACTORS IN DENTAL DISEASE

Nutrition and eating patterns have a profound effect on oral diseases. People whose diet is high in refined sugar (sucrose, glucose, and fructose) tend to have more caries, especially if these foods are eaten constantly throughout the day without brushing the teeth. The practice of putting a baby down to nap with a bottle of fruit juice or formula can lead to rampant dental caries ("nursing bottle caries"). In addition to diet, numerous anatomical, physiological, and hereditary factors influence oral diseases. The structure of the tooth enamel can be influenced by genetics and by environmental factors such as fluoride, which strengthens the enamel bonds. Inhibitory factors in saliva such as antibodies and lysozyme can also help prevent dental disease by inhibiting bacterial growth.

The greatest control of dental diseases is preventive dentistry, including regular brushing and flossing to remove plaque, because stopping plaque buildup automatically reduces caries and calculus production. Mouthwashes are relatively ineffective in controlling plaque formation because of the high bacterial content of saliva and the relatively short acting time of the mouthwash. The federal government has blocked advertisements for certain mouthwashes that claim to prevent plaque formation.

Once calculus has formed on teeth, it cannot be removed by brushing but can be dislodged only by special mechanical proce-

dures (scaling) in the dental office. One of the most exciting prospects is the possibility of a vaccine to protect against the primary colonization of the teeth. Some success in inhibiting plaque formation has been achieved in experimental animals with vaccines raised against the whole cells of *Streptococcus mutans* and the fimbriae of *Actinomyces viscosus*.

CHAPTER CHECKPOINTS

The oral cavity is an ecosystem that provides a broad spectrum of niches for a wide variety of microorganisms. A high-sugar diet and poor oral hygiene enhance the growth of bacteria, which cause oral diseases. *Streptococcus*, *Lactobacillus*, and *Actinomyces* are the main agents of dental caries. Invasion of the gums or gingiva by a mixture of spirochetes and gram-negative bacteria causes many gum diseases. ANUG, or trench mouth, is a severe form of gingivitis that can lead to destruction of connective tissue and bone.

CHAPTER CAPSULE WITH KEY TERMS

I. **Miscellaneous Bacterial Infections**
   A. Spirochetes
   Spirochetes are helical, flexible bacteria that move by periplasmic flagella. Several **Treponema** species are obligate parasitic spirochetes with 8 to 12 regular spirals; best observed under dark-field microscope; cause **treponematoses.**
   1. Direct observation of treponeme in tissues and blood tests important in diagnosis.
   2. Cannot be cultivated in artificial media; treatment by large doses of penicillin or tetracycline.
   B. Major Treponematoses
   1. *Syphilis: Treponema pallidum* causes complex progressive disease in adults and children.
      a. **Sexually transmitted syphilis** is acquired through close contact with a lesion; untreated sexual disease occurs in stages over long periods.
         (1) At site of entrance, multiplying treponemes produce a **primary lesion,** a hard ulcer or **chancre,** which disappears as the microbe becomes systemic.
         (2) **Secondary syphilis** occurs when spirochete infects many organs; marked by skin rash, fever, damage to mucous membranes. Both primary and secondary syphilis are communicable. Latent period establishes pathogen in tissues.
         (3) Final, non-communicable **tertiary stage** is marked by tumors called **gummas** and life-threatening cardiovascular and neurological effects.
      b. **Congenital syphilis** is acquired transplacentally; disrupts embryonic and fetal development; survivors may have respiratory, skin, bone, teeth, eye, and joint abnormalities if not treated.
   2. *Nonsyphilitic treponematosis:* Slow progressive cutaneous and bone diseases endemic to specific regions of tropics and subtropics; usually transmitted under unhygienic conditions.
      a. **Bejel** is a deforming childhood infection of the mouth, nasal cavity, body, and hands.
      b. **Yaws** occurs from invasion of skin cut, causing a primary ulcer that seeds a second crop of lesions.

      c. **Pinta** is superficial skin lesion that depigments and scars the skin.
   C. *Leptospira interrogans*
   L. interrogans has very regular coils and a prominent hook; causes **leptospirosis,** a worldwide zoonosis acquired through contact with urine of wild and domestic animal reservoirs. Spirochete enters cut, multiplies in blood and spinal fluid.
   1. Causes muscle aches, headache.
   2. Second phase is marked by Weil's syndrome, which involves kidneys and liver.
   3. Requires early treatment with penicillin or tetracycline.
   D. *Borrelia*
   Borrelia species are loose, irregular spirochetes that cause **borreliosis;** infections are vector-borne (mostly by ticks). Borrelias cause recurrent fever and other symptoms because the spirochete repeatedly changes antigenically and forces the immune system to keep adapting.
   1. *B. hermsii* is zoonotic (in wild rodents) and carried by soft ticks; *B. recurrentis* has a strictly human reservoir and is carried by lice; both species cause relapsing fever.
   2. *Borrelia burgdorferi:* A zoonosis carried by mice and spread by a hard tick *(Ixodes)* that lives on deer and mice; causes **Lyme disease,** a syndrome that occurs endemically in several regions of the United States. Tick bite leads to fever and a prominent ring-shaped rash; if allowed to progress, may cause cardiac, neurological, and arthritic symptoms; can be controlled by antibiotics and by avoiding tick contact.
   E. Curved Bacteria (Vibrios)
   Vibrios are short spirals or sausage-shaped cells with polar flagella.
   1. *Vibrio cholerae:* Causes **epidemic cholera,** a human disease that originated in Asia but is now distributed worldwide in natural waters. Organism is ingested in contaminated food and water; microbe infects the surface of epithelial cells in small intestine, is not invasive; severity of cholera due to potent **cholera toxin** that causes electrolyte and water loss through **secretory diarrhea;** resulting dehydration leads to muscle, circulatory, and neurological symptoms; treatment

with oral rehydration (electrolyte and fluid replacement) completely restores stability; vaccine available.

2. *Vibrio parahaemolyticus:* Causes food infection associated with seawater and seafood; symptoms similar to mild cholera; for prevention, food must be well cooked and refrigerated during storage.

3. *Campylobacter* species: *C. jejuni* is a common cause of severe **gastroenteritis** worldwide; acquired through food or water contaminated by animal feces; enteritis is due to enterotoxin; *C. fetus* causes diseases in pregnant women and fatal septicemia in neonates; *Helicobacter pylori* may be etiologically involved in diseases of the stomach lining such as gastritis and ulcers.

F. Obligate Intracellular Parasitic Bacteria

1. **Rickettsias** and **chlamydias** are tiny, gram-negative rods or cocci that are metabolic and intracellular parasites; diseases treatable with tetracycline and chloramphenicol. *Rickettsia* causes **rickettsioses;** most are zoonoses spread by arthropod vectors.

    a. *Epidemic typhus:* Caused by *Rickettsia prowazekii;* carried by lice; associated with human overcrowding; disease starts with high fever, chills, headache; rash occurs; Brill-Zinsser disease is a chronic, recurrent form.

    b. *Endemic (murine) typhus:* Zoonosis caused by *R. typhi,* harbored by mice and rats; occurs sporadically in areas of high flea infestation; symptoms like epidemic form, but milder.

    c. *Rocky Mountain spotted fever:* Etiologic agent is *R. rickettsii;* zoonosis carried by dog and wood ticks (*Dermacentor*); most cases on eastern seaboard; infection causes distinct spotted, migratory rash; acute reactions include heart damage, CNS damage; disease prevented by drug therapy and avoidance of ticks.

    d. The genus *Ehrlichia* contains two species of rickettsias that have recently been discovered in humans. These tick-borne bacteria cause human monocytic and granulocytic ehrlichiosis.

2. *Characteristics of ectoparasitic, blood-sucking arthropod vectors:* Transmitters of parasites between various vertebrate reservoirs; vectors may pass pathogen to offspring transovarially or to hosts through bite, feces, or mechanical injury and scratching.

    a. **Ticks** are arachnids; hard ticks have a hard shield and soft ticks lack it. They cling to host, feed, and inoculate the host with saliva containing the pathogen; carry Rocky Mountain spotted fever, Q fever, borrelioses, viral fevers.

    b. **Fleas** are flattened insects with jumping legs and a blood-probing mouth; extremely resistant and nonspecific to host; carry murine typhus, bubonic plague.

    c. **Lice** are insects that cling to body hair and gently pierce skin; infection occurs when louse is crushed by scratching and rubbed into skin; carry epidemic typhus, trench fever, and relapsing fever.

G. Diseases Related to the Rickettsioses

1. *Q fever:* Caused by *Coxiella burnetii;* agent has a resistant spore form that can survive out of host; a zoonosis of domestic animals; transmitted by air, dust, unpasteurized milk, ticks; usually inhaled, causing pneumonitis, fever, hepatitis.

2. *Bartonella:* A closely related genus; is the cause of trench fever, spread by lice, and cat-scratch disease, a lymphatic infection associated with a clawing injury by cats.

H. *Chlamydia*

Organisms in the genus *Chlamydia* pass through a transmission phase involving a hardy **elementary body** and an intracellular **reticulate body** that has pathologic effects.

1. *Chlamydia trachomatis:* A strict human pathogen that causes eye diseases and STDs. **Ocular trachoma** is a severe infection that deforms the eyelid and cornea and may cause blindness. **Conjunctivitis** occurs in babies following contact with birth canal; prevented by ocular prophylaxis after birth.

2. *C. trachomatis* causes very common bacterial STDs: nongonococcal urethritis in males, and cervicitis, salpingitis (PID), infertility, and scarring in females; also **lymphogranuloma venereum,** a disfiguring disease of the external genitalia and pelvic lymphatics.

3. *C. pneumoniae* causes an atypical pneumonia that is a serious complication in asthma patients.

4. *Chlamydia psittaci:* Causes **ornithosis,** a zoonosis transmitted to humans from bird vectors; highly communicable among all birds; pneumonia or flulike infection with fever, lung congestion.

I. Mollicutes/Mycoplasmas

Mycoplasmas naturally lack cell walls and are thus highly pleomorphic; not obligate parasites but require special lipids; fuse tightly to host membranes during infection; diseases treated with tetracycline, erythromycin.

1. *Mycoplasma pneumoniae:* Causes **primary atypical pneumonia;** pathogen slowly spreads over interior respiratory surfaces, causing fever, chest pain, and sore throat; *M. hominis* and *Ureaplasma urealyticum* are normal colonists of most persons; may cause urethritis, PID, and other reproductive tract diseases.

2. *L forms:* Bacteria that normally have cell walls but have transiently lost them through drug therapy. They may be involved in certain chronic diseases.

II. **The Role of Mixed Infections in Dental Disease**

The oral cavity contains hundreds of microbial species that participate in interactions between themselves, the human host, and the nutritional role of the mouth; it is continuously vulnerable to infection and disease.

A. Hard-Tissue Disease

Dental caries is a slow, progressive infection of irregular areas of enamel surface; begins with colonization of tooth by slime-forming species of *Streptococcus* and cross adherence with *Actinomyces;* process forms layer of thick, adherent material called **plaque;** this complex layer harbors dense masses of bacteria and extracellular substances.

1. Acid formed by agents in plaque dissolves the inorganic salts in enamel.

2. If unabated, this produces a caries lesion, which may invade dentin and root canal and destroy tooth.

B. Soft-Tissue Disease

Periodontal disease involves **periodontium (gingiva** and surrounding tissues); starts when plaque forms on root of tooth and is mineralized to a hard concretion called **calculus;** this irritates tender gingiva; inflammatory reaction and swelling create **gingivitis** and pockets between tooth and gingiva invaded by bacteria (spirochetes and gram-negative bacilli); tooth socket may be involved **(periodontitis),** and the tooth may be lost.

MULTIPLE-CHOICE QUESTIONS

1. *Treponema pallidum* is cultured in/on
   - a. blood agar
   - b. animal tissues
   - c. serum broth
   - d. eggs

2. A gumma is
   - a. the primary lesion of syphilis
   - b. a syphilitic tumor
   - c. the result of congenital syphilis
   - d. a damaged aorta

3. The treatment of choice for syphilis is
   - a. tetracycline
   - b. antiserum
   - c. penicillin
   - d. sulfa drugs

4. Which of the treponematoses are *not* STDs?
   - a. yaws
   - b. pinta
   - c. syphilis
   - d. both a and b

5. Lyme disease is caused by _____ and spread by _____.
   - a. *Borrelia recurrentis,* lice
   - b. *Borrelia hermsii,* ticks
   - c. *Borrelia burgdorferi,* fleas
   - d. *Borrelia burgdorferi,* ticks

6. Which of the following manifestations occur in Lyme disease?
   - a. arthritis
   - b. rash
   - c. heart disorder
   - d. a and b
   - e. all of these

7. Relapsing fever is spread by
   - a. lice
   - b. ticks
   - c. animal urine
   - d. a and b

8. The primary habitat of *Vibrio cholerae* is
   - a. intestine of humans
   - b. intestine of animals
   - c. natural waters
   - d. exoskeletons of crustaceans

9. The best therapy for cholera is
   - a. oral tetracycline
   - b. oral rehydration therapy
   - c. antiserum injection
   - d. oral vaccine

10. Rickettsias and chlamydias are similar in being
    - a. free of a cell wall
    - b. the cause of eye infections
    - c. carried by arthropod vectors
    - d. obligate intracellular bacteria

11. Which of the following is *not* an arthropod vector of rickettsioses?
    - a. mosquito
    - b. louse
    - c. tick
    - d. flea

12. Chlamydiosis caused by *C. trachomatis* attacks which structure(s)?
    - a. eye
    - b. urethra
    - c. fallopian tubes
    - d. all of these

13. Ornithosis is a _____ infection associated with _____.
    - a. rickettsial, parrots
    - b. chlamydial, mice
    - c. chlamydial, birds
    - d. rickettsial, flies

14. Mycoplasmas attack the _____ of host cells.
    - a. nucleus
    - b. cell walls
    - c. ribosomes
    - d. cell membranes

15. The earliest process that is at the basis of most dental disease is
    - a. acquired pellicle
    - b. acid release
    - c. enamel destruction
    - d. plaque accumulation

16. Dental caries are directly due to
    - a. microbial acid etching away tooth structures
    - b. buildup of calculus
    - c. death of tooth by root infection
    - d. the acquired pellicle

17. Acute necrotizing ulcerative gingivitis is a _____ infection.
    - a. contagious
    - b. mixed
    - c. spirochete
    - d. systemic

CONCEPT QUESTIONS

1. a. Describe the characteristics of *Treponema pallidum* that are related to its transmission.
   b. Name some factors responsible for the current epidemic of syphilis.

2. a. Describe the stages of untreated syphilis infection. Where does the chancre occur, and what is in it?
   b. What is happening as the chancre disappears?
   c. Which stages are symptomatic, and which are communicable?
   d. What is syphilis latency?
   e. For which tissues does the spirochete have an affinity?

3. Describe the conditions leading to congenital syphilis and the long-term effects of the disease.

4. a. Describe the nonspecific and specific tests for syphilis. What do they test for?
   b. Why are they so important?

5. Outline the general characteristics of bejel, yaws, and pinta.

6. Describe the epidemiology and pathology of leptospirosis.

7. a. Describe the three major borrelioses.
   b. How are arthropods involved?
   c. Why are there relapses in relapsing fever?
   d. What does it mean for a patient to be borrelemic or rickettsemic?
   e. How are these conditions related to the role of vectors in the spread of disease?

8. a. Trace the route of the infectious agent from a tick bite to infection.
   b. Do the same for lice.

9. a. Overview the natural history of Lyme disease and its symptoms.
   b. Explain why it may be mistaken for arthritis, allergy, and neurological diseases.

10. a. Briefly, what is the natural history of cholera?
    b. What is its principal pathologic feature?

11. a. What is secretory diarrhea?
    b. How does oral rehydration therapy work?

12. a. Briefly describe the nature of food infection in species of *Vibrio* and the diseases of *Campylobacter.*
    b. What diseases are *Helicobacter pylori* involved in?
    c. Describe its method of invasion and pathogenesis.

13. a. What do rickettsias and chlamydias derive from the host?
    b. How do antibiotics work to control them?

14. a. Outline the life cycles of *Rickettsia prowazekii, Rickettsia rickettsii,* and *Coxiella burnetii.*
    b. What are the predisposing factors for the types of disease caused by each species?
    c. What makes *Coxiella* unique among the bacteria?
    d. How are dogs involved in transmission of RMSF?
    e. What are the general symptoms of rickettsial infections?

15. a. How are chlamydias transmitted?
    b. Describe the major complications of eye infections and STDs.

16. a. What are the pathologic effects of *Mycoplasma pneumoniae?*
    b. Why are the symptoms of mycoplasma pneumonia so mild?
    c. Why doesn't penicillin work on mycoplasma infection?

17. a. In what ways are dental diseases mixed infections?
    b. Discuss the major factors in the development of dental caries and periodontal infections.

18. a. Which diseases in this chapter are zoonoses?
    b. Name them and the major vector involved.

19. a. Find all of the different agents of STDs in this chapter.
    b. Find all of the different agents of pneumonia.
    c. Find all of the agents of gastroenteritis.

20. **Single Matching.** Match each disease in the left column with its vector (or vectors) in the right column.

____ leptospirosis	a. wild animals
____ Lyme disease	b. flea
____ murine typhus	c. tick
____ ornithosis	d. birds
____ relapsing fever	e. louse
____ lymphogranuloma venereum	f. domestic animals
____ cat-scratch disease	g. none of these
____ epidemic typhus	
____ Rocky Mountain spotted fever	
____ Q fever	
____ cholera	

21. **Single Matching.** Match each disease in the left column with its portal of entry in the right column.

____ Q fever	a. skin
____ ornithosis	b. mucous membrane
____ dental caries	c. respiratory tract
____ ANUG	d. urogenital tract
____ mycoplasma	e. eye
____ syphilis	f. oral cavity
____ leptospirosis	g. gastrointestinal tract
____ lymphogranuloma venereum	
____ cholera	
____ Lyme disease	
____ trachoma	
____ *Campylobacter* infection	
____ gastric ulcers	

CRITICAL-THINKING QUESTIONS

1. Why is it so difficult to trace the historical origin of disease, as in syphilis?

2. a. Why does syphilis have such profound effects on the human body?
   b. Why is long-term immunity to syphilis so difficult to achieve?

3. How can congenital syphilis be prevented?

4. How are the nonsyphilitic treponematoses similar to syphilis?

5. a. In view of the fact that cholera causes the secretion of electrolytes into the intestine, explain what causes the loss of water.
   b. What are the principles of osmosis behind this phenomenon?

6. What would be the best type of vaccine for cholera?

7. a. Explain the general relationships of the vector, the reservoir, and the agent of infection.
   b. Can you think of an explanation for Lyme disease having such a low incidence in the southern United States? (Hint: In this region, the larval stages of the tick feed on lizards, not on mice.)

8. Humans are accidental hosts in many vector-borne diseases. What does this indicate about the relationship between the vector and the microbial agent?

9. a. Why can a louse cause infection only once?
   b. Why must it be crushed in order to cause infection?
   c. What kind of vector can infect many individuals?
   d. How do these vectors infect their hosts?

10. a. Why is arthropod vector control so difficult?
    b. Summarize the methods of preventing arthropod-borne disease.

11. a. Which bacteria presented in this chapter can be cultivated on artificial media?
    b. Which require embryos or cell culture?

12. Name four bacterial diseases for which the dark-field microscope is an effective diagnostic tool.

13. Explain how L forms could be involved in disease.

14. Which two infectious agents covered in this chapter would be the most resistant to the environment and why are they so resistant?

15. a. In what way is the oral cavity an ecological system?
    b. What causes an imbalance?
    c. What are some logical ways to prevent dental disease besides removing plaque?

16. Case study 1. A journalist returning from a trip experienced severe fever, vomiting, chills, and muscle aches, followed by symptoms of meningitis and kidney failure. Early tests were negative for septicemia; throat cultures were negative; and penicillin was an effective treatment. Doctors believed the patient's work in the jungles of South America was a possible clue to his disease. What do you think might have been the cause?

17. Case study 2. A man went for a hike in the mountains of New York State and later developed fever and a rash. What two totally different diseases might he have contracted, and what could have been the circumstances of infection?

18. Case study 3. A woman experienced a bout of fever, diarrhea, cramping, and general malaise that lasted for 3 days. She was treated with rehydration therapy but not antibiotics. Cultures and serological tests were negative for any microorganisms tested. The only unusual event that could possibly be linked to her disease was that she had been camping in Michigan and had consumed water from a mountain lake. Name the likely disease.

INTERNET SEARCH TOPICS

1. Go to the World Wide Web to find information on human monocytic and granulocytic ehrlichiosis and bartonellosis. Briefly summarize their epidemiology, including the number of new cases, and suggest what factors may be involved in their emergence.

2. Look up the latest information on pathogenesis of *Helicobacter pylori*. Describe the primary strategies in treating this disease.

The Fungi of Medical Importance

CHAPTER 22

The eucaryotic microbes collectively called fungi were introduced in chapter 5. The profound importance of fungi stems primarily from their role in the earth's ecological balance and their impact on agriculture. To a lesser—but not minor—extent, the fungi are also medically significant, as agents in human disease, allergies, and mycotoxicoses (intoxications due to ingesting fungal toxins). Diseases resulting from fungal infections, primarily by yeasts and molds, are termed mycoses. In this chapter, we will survey the most prevalent mycotic infections, including systemic, cutaneous, and subcutaneous forms. Other topics to be covered include common respiratory allergies and diseases associated with fungal toxins.

Chapter Overview

- Fungi are widespread eucaryotic microorganisms that are the agents of mycotic diseases.
- Most fungal infections are acquired through contact with the environment, and only a few are transmissible from other infected humans.
- The majority of mycotic infections are caused by molds and yeasts.
- Most fungi are not invasive, and most are not obligate parasites.
- Mycotic infections occur in the skin, mucous membranes, and many internal organs and systems.
- True fungal pathogens cause primary pulmonary infections, and they display thermal dimorphism when they enter the tissues.
- Opportunistic fungi invade only when host defenses have been weakened.
- The primary fungal pathogens are endemic to specific regions, whereas the opportunists are distributed worldwide.
- The main systemic mycoses are histoplasmosis, coccidioidomycosis, blastomycosis, and paracoccidioidomycosis.
- The primary subcutaneous infections are sporotrichosis, chromoblastomycosis, and mycetoma.
- The cutaneous mycoses, caused by dermatophytes, include ringworm of the body, scalp, foot, and hand.
- The most common opportunistic mycoses are candidiasis, cryptococcosis, and pneumocystis pneumonia.
- Fungi are important agents of toxic diseases and allergies.
- Fungal infections may be treated with a variety of antimicrobic drugs.

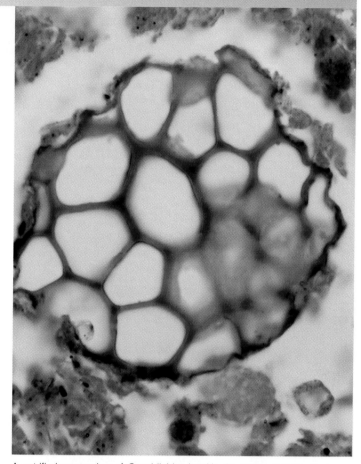

A petrified sporangium of *Coccidioides immitis*, the cause of Valley fever, was unearthed in the lungs of a 600–1,000-year-old American Indian skeleton. This is definitive proof of the ancient origins of this disease.

Fungi As Infectious Agents

Molds and yeasts are so widely distributed in air, dust, fomites, and even among the normal flora that humans are incessantly exposed to them. The fact that the planet's surface is totally dusted with spores led noted California mycologist W. B. Cooke to christen it "our moldy earth." Fortunately, because of the relative resistance of humans and the comparatively nonpathogenic nature of fungi, most exposures do not lead to overt infection. Of the estimated 100,000 fungal species, only about 300 have been linked to disease in animals, though among plants, fungi are the most common and

663

TABLE 22.1

Representative Fungal Pathogens, Degree of Pathogenicity, and Habitat

Microbe	Disease/Infection*	Primary Habitat and Distribution
I. Primary Pathogens		
Histoplasma capsulatum	Histoplasmosis	Soils high in bird guano; Ohio and Mississippi valleys of U.S.; Central and South America; Africa
Blastomyces dermatitidis	Blastomycosis	Presumably soils, but isolation has been difficult; southern Canada; Midwest, Southeast, Appalachia in U.S.; along drainage of major rivers
Coccidioides immitis	Coccidioidomycosis	Highly restricted to alkaline desert soils in southwestern U.S. (California, Arizona, Texas, and New Mexico)
Paracoccidioides brasiliensis	Paracoccidioidomycosis	Soils of rain forests in South America (Brazil, Colombia, Venezuela)
II. Pathogens with Intermediate Virulence		
Sporothrix schenckii	Sporotrichosis	In soil and decaying plant matter; widely distributed
Genera of dermatophytes (*Microsporum, Trichophyton, Epidermophyton*)	Dermatophytosis (various ringworms or tineas)	Human skin, animal hair, soil throughout the world
III. Secondary Pathogens		
Cryptococcus neoformans	Cryptococcosis	Pigeon roosts and other nesting sites (buildings, barns, trees); worldwide distribution
Candida albicans	Candidiasis	Normal flora of human mouth, throat, intestine, vagina; also normal in other mammals, birds; ubiquitous
Aspergillus spp.	Aspergillosis	Soil, decaying vegetation, grains; common airborne contaminants; extremely pervasive in environment
Pneumocystis carinii	Pneumocystis pneumonia (PCP)	Upper respiratory tract of humans, animals
Genera in Mucorales (*Rhizopus, Absidia, Mucor*)	Mucormycosis	Soil, dust; very widespread in human habitation

* *Specific mycotic infections are usually named by adding -mycosis, –iasis, or -osis to the generic name of the pathogen.*

destructive of all pathogens. Human mycotic disease, or **mycosis,*** is associated with true fungal pathogens that exhibit some degree of virulence or with opportunistic pathogens that take advantage of defective resistance (tables 22.1 and 22.2).

TRUE VERSUS OPPORTUNISTIC PATHOGENS

A **true fungal pathogen** is a species that can invade and grow in a healthy, noncompromised animal host. This behavior is contrary to the metabolism and adaptation of fungi, most of which are inhibited by the relatively high temperature and low oxygen tensions of a warm-blooded animal's body. But a small number of fungi have acquired the morphological and physiological adaptations required to survive and grow in this habitat. By far their most striking adaptation is a switch from hyphal cells typical of the mycelial or mold phase to yeast cells typical of the parasitic phase (figure 22.1). This biphasic characteristic of the life cycle is termed *thermal dimorphism** because it is initiated by changing temperature. In general, these organisms grow as molds at 30°C and as yeasts at 37°C.

An **opportunistic fungal pathogen** is different from a true pathogen in several ways (table 22.2). An opportunist has weak to nonexistent invasiveness or virulence, and the host's defenses must

TABLE 22.2

Comparison of True and Opportunistic Fungal Infections

	True Pathogenic Infections	Opportunistic Infections
Degree of Virulence	Well developed	Limited
Condition of Host	Resistance high or low	Resistance low
Primary Portal of Entry	Respiratory	Respiratory, mucocutaneous
Nature of Infection	Usually primary, pulmonary, and systemic; usually asymptomatic	Varies from superficial skin to pulmonary and systemic; usually symptomatic
Nature of Immunity	Well-developed, specific	Weak, short-lived
Infecting Form	Primarily conidial	Conidial or mycelial
Shows Thermal Dimorphism	Strongly	Not usually
Habitat of Fungus	Soil	Varies from soil to flora of humans and animals
Geographic Location	Restricted to endemic regions	Distributed worldwide

---

* mycosis (my-koh´-sis) pl. mycoses; Gr. *mykos,* fungi, and *osis,* a disease process.

* dimorphism (dy-mor´-fizm) Gr. *dimorphos,* having two forms; the property of existing in two distinct cell types.

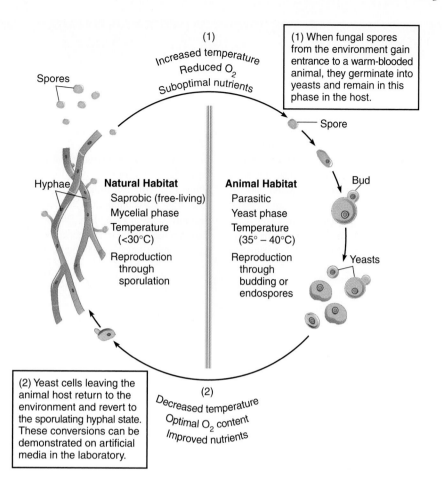

(1)
Increased temperature
Reduced O_2
Suboptimal nutrients

Spores

Hyphae

Natural Habitat
Saprobic (free-living)
Mycelial phase
Temperature
(<30°C)

Reproduction
through
sporulation

(1) When fungal spores
from the environment gain
entrance to a warm-blooded
animal, they germinate into
yeasts and remain in this
phase in the host.

Spore

Bud

Animal Habitat
Parasitic
Yeast phase
Temperature
(35° – 40°C)

Reproduction
through
budding or
endospores

Yeasts

(2) Yeast cells leaving the
animal host return to the
environment and revert to
the sporulating hyphal state.
These conversions can be
demonstrated on artificial
media in the laboratory.

(2)
Decreased temperature
Optimal O_2 content
Improved nutrients

FIGURE 22.1

 The general changes associated with thermal dimorphism, using generic hyphae, spores, and yeasts as examples.

be impaired to some degree for the microbe to gain a foothold. Although some species show both mycelial and yeast stages in their life cycles, they usually do not show thermal dimorphism. Opportunistic pathogens vary in their manifestations from superficial and benign colonizations to deep, chronic systemic disease that is rapidly fatal. Mycoses due to opportunists are an increasingly serious problem (Medical Microfile 22.1).

Some fungal pathogens exist in a category between true pathogens and opportunists. These species are not inherently invasive but can grow when inoculated into the skin wounds or abrasions of healthy people. Examples are *Sporothrix,* the agent of a subcutaneous infection, and the **dermatophytes,*** which cause ringworm and athlete's foot. Some mycologists believe these fungi are undergoing a gradual transformation into true pathogens.

EPIDEMIOLOGY OF THE MYCOSES

Most fungal pathogens do not require a host to complete their life cycles, and the infections they cause are not communicable. Notable exceptions are some dermatophyte and *Candida* species that naturally inhabit the human body and are transmissible. For the remainder of the pathogens, exposure and infection occur when a

human happens upon fungal spores in the environment (usually air, dust, or soil). Unlike the opportunists, true fungal pathogens are distributed in a predictable pattern that coincides with the pathogen's adaptation to the specific climate, soil, or other factors of a relatively restricted geographic region (figure 22.2).

The incidence of all fungal infections has traditionally been difficult to measure because none of the reportable infections monitored by the CDC are caused by fungi. Dermatophytoses are probably the most prevalent, and it is thought that at least 90% of all humans will acquire ringworm or athlete's foot at least once during their lifetimes. Estimates provided by routine skin testing indicate that millions of people have experienced true mycoses, though most cases probably go undiagnosed or misdiagnosed.

Epidemics of mycotic infections can occur after mass exposure to a common source. Memorable incidents include an outbreak of histoplasmosis in people exploring bat caves, and another in South African mine workers who rubbed against contaminated wood planks and came down with sporotrichosis. Coccidioidomycosis is a particular hazard of construction workers and people living in the paths of windstorms. Transmissible fungal infections such as dermatophytoses are readily passed by means of shared personal articles, public facilities, swimming pools, gymnasiums, and contact with infected animals. Candidiasis can be communicated during sexual contact and from mother to child at birth.

---

* **dermatophyte** (der-mah´-toh-fyte˝) Gr. *dermos,* skin, and *phyte,* plant. Fungi were once included in the plant kingdom.

MEDICAL MICROFILE 22.1
Opportunistic Mycoses As Diseases of Medical Progress

At one time, opportunistic fungal infections in hospitalized patients were rather unusual. Common fungi such as *Rhizopus, Geotrichum,* and *Torulopsis* were rarely isolated as etiologic agents 25 years ago. Textbooks from the past describe these agents as common air contaminants that cause infections only under remarkable circumstances. These mycoses were rare primarily because immunodeficient and debilitated patients usually died from their afflictions long before fungal infections could take hold. With the advent of innovative surgeries, drugs, and other therapies, the survival rates and numbers of debilitated patients have significantly increased. Some of the conditions most likely to predispose patients to fungal infections are cancer, AIDS, tuberculosis, immunosuppressive drugs and therapy, chronic organ disease, malnutrition, and alcoholism.

One clinical dilemma that cannot be completely eliminated, even with rigorous disinfection, is the exposure of such patients to potential fungal opportunists from the air, fomites, and even normal flora. Up to 5% of all nosocomial infections are caused by opportunistic fungi. Fungal infections in such high-risk patients progress rapidly and are difficult to diagnose and treat. In one study, 40% of the deaths from clinically acquired infections were caused by fungi.

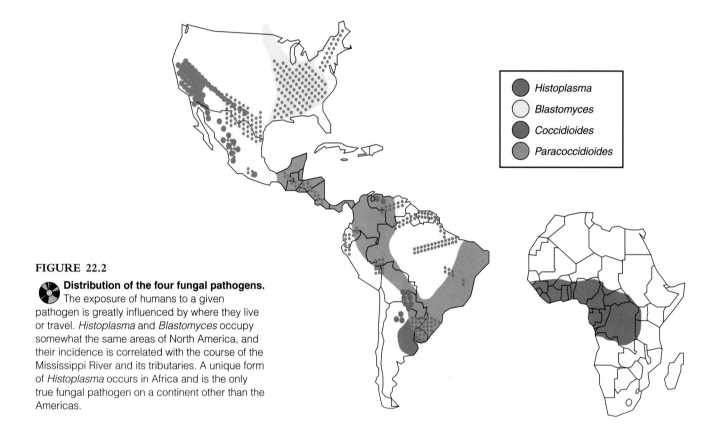

FIGURE 22.2

Distribution of the four fungal pathogens. The exposure of humans to a given pathogen is greatly influenced by where they live or travel. *Histoplasma* and *Blastomyces* occupy somewhat the same areas of North America, and their incidence is correlated with the course of the Mississippi River and its tributaries. A unique form of *Histoplasma* occurs in Africa and is the only true fungal pathogen on a continent other than the Americas.

Legend:
- *Histoplasma*
- *Blastomyces*
- *Coccidioides*
- *Paracoccidioides*

PATHOGENESIS OF THE FUNGI

Mycoses involve complex interactions among the portal of entry, the nature of the infectious dose, the virulence of the fungus, and host resistance. Fungi enter the body mainly via respiratory, mucous, and cutaneous routes. In general, the agents of primary mycoses have a respiratory portal (spores inhaled from the air); subcutaneous agents enter through inoculated skin (trauma); and cutaneous and superficial agents enter through contamination of the skin surface. Spores, hyphal elements, and yeasts can all be infectious, but spores are most often involved because of their durability and abundance.

Thermal dimorphism greatly increases virulence by enabling fungi to tolerate the relatively high temperatures and low O_2 tensions of the body. Fungi in the yeast form can be more invasive than hyphal forms because they grow more rapidly and spread through tissues and blood, while those producing hyphae tend to localize along the course of blood vessels and lymphatics. Specific factors contributing to fungal virulence are the subject of much research. Toxinlike substances have been isolated from several species, but their method of damaging the tissues remains unclear. Fungi also produce various adhesion factors and capsules, hydrolytic enzymes, inflammatory stimulants, and allergens, all of which generate strong host responses.

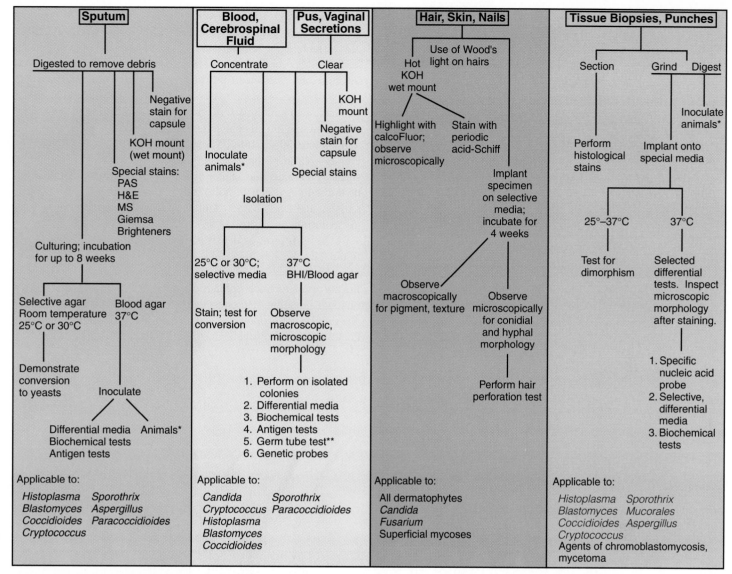

FIGURE 22.3

Methods of processing specimens in fungal disease.

*Animal inoculation is performed only to help diagnose systemic mycoses when other methods are unavailable or indeterminant.
**Some yeasts, when incubated in serum for 2 to 4 hours, sprout tiny hyphal tubes called germ tubes. *Candida albicans* is identified by this characteristic.

The human body is extremely resistant to establishment of fungi. Among its numerous antifungal defenses are the normal integrity of the skin, mucous membranes, and respiratory cilia, but the most important defenses are cell-mediated immunity, phagocytosis, and the inflammatory reaction. Long-term protective immunity can develop for some of the true pathogens, but for the rest, reinfection is a distinct possibility.

DIAGNOSIS OF MYCOTIC INFECTIONS

Satisfactory diagnosis of fungal infections depends mainly upon isolating and identifying the pathogen in the laboratory. Accurate and speedy diagnosis is especially critical to the immunocompromised patient, who must have prompt antifungal chemotherapy. For example, a patient with systemic *Candida* infection can die if the infection is not detected and treated within 5 to 7 days. Because

therapy can vary among the pathogenic fungi, identification to the species level is often necessary.

A suitable specimen can be obtained from sputum, skin scrapings, skin biopsies, cerebrospinal fluid, blood, tissue exudates, urine, or vaginal samples, as determined by the patient's symptoms. Routine laboratory procedures include isolation, microscopic and macroscopic examination, histological stains, serology, and animal inoculation (figure 22.3).

Because culture of the sample can require several days, the immediate direct examination of fresh samples is recommended. Wet mounts can be prepared by mixing a small portion of the sample on a slide with saline, water, or potassium hydroxide to clear the specimen of background debris. The relatively large size and unique appearance of many fungi help them stand out. Large round or oval budding cells are evidence of yeasts, whereas thick,

FIGURE 22.4

🔅 **Special tissue stains called brighteners can amplify the presence of fungal elements in specimens.** They bind tightly to the carbohydrates of the fungal surface. Presented here is a fluorescent view of *Aspergillus fumigatus* taken from a thoracic specimen.

branching strands suggest hyphae. A variety of stains or brighteners are valuable for highlighting fungi in tissue samples (figure 22.4).

The fungal pathogen can be isolated with various solid media such as Sabouraud's dextrose agar, mycosel agar, inhibitory mold medium, and brain-heart-infusion agar. These media can be made more enriched or selective by adding blood, chloramphenicol and gentamicin to inhibit bacteria, or cyclohexamide to slow the growth of undesirable fungal contaminants. Cultures are incubated at room temperature, 30°C, and 37°C over a few days to weeks. Gross colonial morphology, such as the color and texture of the colony's surface and underside, can be very distinctive. Initial identification of the pathogen can be followed by confirmatory physiological, antigen, and DNA-based testing. Examples of certain tests will be given during discussions of specific diseases later in this chapter.

In vitro tests to detect antifungal antibodies in serum (see figure 22.10) can be a useful diagnostic tool in some infections. *In vivo* skin testing for delayed hypersensitivity to fungal antigens is mostly used to trace epidemiologic patterns. It does not verify ongoing infection, and considerable cross-reactivity exists among *Histoplasma, Blastomyces,* and *Coccidioides.* A negative test in a healthy person, however, can rule out infection by these fungi.

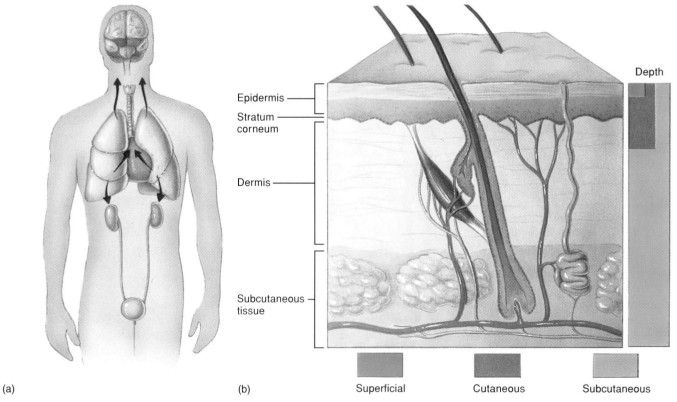

(a) (b)

FIGURE 22.5

Levels of invasion by fungal pathogens. Some species can invade more than one level. **(a)** In systemic (deep) mycoses, the fungus disseminates from the lungs or other sites into the circulation. Fungemia leads to infection of the brain, kidneys, and other organs. **(b)** The skin and its attendant structures provide many potential sites for invasion, including the scalp, smooth skin, hair, and mucous membranes. Differing depths of involvement are: superficial, consisting of extremely shallow epidermal colonizations; cutaneous, involving the stratum corneum and occasionally the upper dermis; and subcutaneous, occurring after a puncture wound has introduced the fungus deeper into the subcutaneous tissues.

CONTROL OF MYCOTIC INFECTIONS

Treatment for fungal infections is based primarily on antifungal drugs, previously covered in chapter 12 and on a disease-by-disease basis in this chapter. Immunization is not usually effective against fungal infections, but work is proceeding on vaccines for coccidioidomycosis and histoplasmosis. Prevention measures are limited to masks and protective clothing to reduce contact with spores, and in some cases, surgical removal of damaged tissues.

ORGANIZATION OF FUNGAL DISEASES

Fungal infections can be presented in several schemes, none of which is completely satisfactory. Traditional methods are based on taxonomic group, location of infection, and type of pathogen. Chapter 5 presents a taxonomic breakdown of the fungi, and this chapter treats them in the following categories according to the type and level of infection they cause and their degree of pathogenicity: (1) **systemic, subcutaneous, cutaneous,** and **superficial** mycoses (the four levels of infection depicted in figure 22.5), and (2) opportunistic mycoses (see table 22.2).

CHAPTER CHECKPOINTS

Fungi are widely distributed in many habitats. Most are harmless saprobes, but approximately 300 species cause mycotic infections in humans. The ubiquitous distribution of fungi ensures that most humans will experience one or more mycoses during their lifetime.

Mycotic infectious agents can be either true pathogens or opportunists. Fungi can cause allergies, toxicoses, and specific infections that are seasonally related. Virulence factors include resistant spores, thermal dimorphism, toxin production, and invasive factors. Infective spores are widely distributed through dust and air.

Fungal infections are categorized by their pathogenicity and the level or type of infection they cause. They can be superficial, cutaneous, subcutaneous, and systemic. True pathogens can initiate infection in a healthy host and exhibit thermal dimorphism, with the yeast form at body temperatures and the mold form at lower temperatures. True pathogens are *Histoplasma, Blastomyces, Coccidioides,* and *Paracoccidioides.*

Opportunists have few if any virulence factors, may or may not have thermal dimorphism, and usually initiate infection in immunodeficient hosts. Examples are *Cryptococcus, Candida,* and *Aspergillus.* Some fungi are not inherently pathogenic but, if introduced subcutaneously, have the potential to cause serious infection. Examples are the dermatophytes and *Sporothrix* species.

Mycoses are not usually communicable, with certain exceptions. The four pathogenic species are endemic to specific ecological regions and may cause epidemics during mass exposure to a common source.

All true pathogens initiate infection through inhalation of spores. The yeast phase develops in the tissues. Host defenses to fungal infection depend on the integrity of all epithelial barriers, a functional inflammatory response, and cell-mediated immunity. Effective treatment of fungal infections requires rapid and accurate diagnosis because drug therapy can vary with the species.

Accurate diagnosis of mycotic agents requires direct microscopic examination of fresh specimens followed by confirmatory isolation on solid media, serological tests for host antibody, and genetic analysis.

Systemic Infections by True Pathogens

The infections of primary fungal pathogens can all be described by the same general model. They are restricted to certain endemic regions of the world. Infection occurs when soil or other matter containing the fungal conidia is disturbed and the spores are inhaled into the lower respiratory tract. The spores germinate in the lungs into yeasts or yeastlike cells and produce an asymptomatic or mild **primary pulmonary infection (PPI)** that parallels tuberculosis. In a small number of hosts, this infection becomes systemic and creates severe, chronic lesions. In a few cases, spores are inoculated into the skin, where they form localized granulomatous lesions. All diseases result in immunity that can be long-term and that manifests clinically as an allergic reaction to fungal antigens.

HISTOPLASMOSIS: OHIO VALLEY FEVER

The most common true pathogen is ***Histoplasma capsulatum,****** the cause of histoplasmosis. This disease has probably afflicted humans since antiquity, though it was not described until 1905 by Dr. Samuel Darling. Through the years, it has been known by various synonyms—Darling's disease, Ohio Valley fever, and reticuloendotheliosis. Certain aspects of its current distribution and epidemiology suggest that it has been an important disease as long as humans have practiced agriculture.

Biology and Epidemiology of *Histoplasma capsulatum*

Histoplasma capsulatum is typically dimorphic. Growth on media below 35°C is characterized by a white or brown, hairlike mycelium, and growth at 37°C on blood agar produces a creamy white, textured colony (figure 22.6).

Histoplasma capsulatum is endemically distributed on all continents except Australia. Its highest rates of incidence occur in the eastern and central regions of the United States (the Ohio Valley). This fungus appears to grow most abundantly in moist soils high in nitrogen content, especially those supplemented by bird and bat *guano.**

A useful tool for determining the distribution of *H. capsulatum* is to inject a fungal extract called *histoplasmin* into the skin and monitor for allergic reactions. Application of this test has verified the extremely widespread distribution of the fungus. In high-prevalence areas such as southern Ohio, Illinois, Missouri, Kentucky, Tennessee, Michigan, Georgia, and Arkansas, 80% to 90% of the population show signs of prior infection. Histoplasmosis prevalence in the United States is estimated at about 500,000 cases per year, with several thousand of them requiring hospitalization and a small number resulting in death.

The spores of the fungus are probably dispersed by the wind and, to a lesser extent, animals. The most striking outbreaks of histoplasmosis occur when concentrations of spores have been dislodged by humans working in parks, bird roosting areas, and old buildings. People of both sexes and all ages incur infection, but adult males experience the majority of cases. The oldest and youngest members of a population are most likely to develop serious disease.

---

* *Histoplasma capsulatum* (his″-toh-plaz′-mah kap″-soo-lay′-tum) Gr. *hist,* tissue, and *plasm,* shape; L. *capsula,* small box.

* **guano** (gwan′-oh) Sp. *huanu,* dung. An accumulation of animal manure.

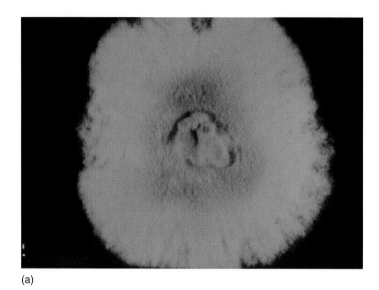

(a)

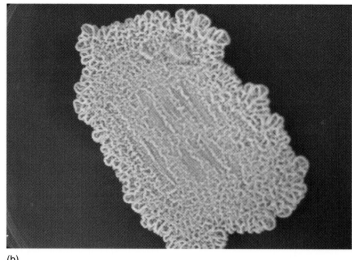

(b)

FIGURE 22.6

Cellular and cultural characteristics of *Histoplasma capsulatum*. (a) A colony at 25°C produces a fuzzy mycelium. (The spores are shown in figure 22.7.) **(b)** A yeast colony (37°C) is dense and waxy.

Infection and Pathogenesis of *Histoplasma*

Histoplasmosis presents a formidable array of manifestations. It can be benign or severe, acute or chronic, and it can show pulmonary, systemic, or cutaneous lesions. Inhaling a small dose of microconidia into the deep recesses of the lung establishes a primary pulmonary infection that is usually asymptomatic. Its primary location of growth is in the cytoplasm of phagocytes such as macrophages. Within these cells, it flourishes and is carried to other sites (figure 22.7). Some people experience mild symptoms such as aches, pains, and coughing, but a few develop more severe symptoms, including fever, night sweats, and weight loss. Primary cutaneous histoplasmosis, in which the agent enters via the skin, is very rare.

The most serious systemic forms of histoplasmosis occur in patients with defective cell-mediated immunity such as AIDS patients. In children, this can lead to liver and spleen enlargement, anemia, circulatory collapse, and death. Adults with systemic disease can acquire lesions in the brain, intestine, adrenal gland, heart, liver spleen, lymph nodes, bone marrow, and skin. Persistent colonization of patients with emphysema and bronchitis causes *chronic pulmonary histoplasmosis,* a complication that has signs and symptoms similar to those of tuberculosis.

Diagnosis and Control of Histoplasmosis

Discovering *Histoplasma* in clinical specimens is a substantial diagnostic indicator. Usually it appears as spherical, "fish-eye" yeasts intracellularly in macrophages and occasionally as free yeasts in samples of sputum and cerebrospinal fluid. Isolating the agent and demonstrating dimorphism help to confirm infection, but this step can require up to 12 weeks. Complement fixation and immunodiffusion serological tests can support a diagnosis by showing a rising antibody titer. Because a positive histoplasmin test does not indicate a new infection, it is not useful in diagnosis.

Undetected or mild cases of histoplasmosis resolve without medical management, but chronic or disseminated disease calls for systemic chemotherapy. The principal drug, amphotericin B, is administered in daily intravenous doses for a few days to a few weeks. Under some circumstances, ketoconazole or other azoles are the drugs of choice. Surgery to remove affected masses in the lungs or other organs can also be effective.

COCCIDIOIDOMYCOSIS: VALLEY FEVER, SAN JOAQUIN VALLEY FEVER, CALIFORNIA DISEASE

Coccidioides immitis* is the etiologic agent of **coccidioidomycosis.** Although the fungus has probably lived in soil for millions of years, human encounters with it are relatively recent and coincide with the encroachment of humans into its habitat (see chapter opening photograph). This unique and fascinating fungus demonstrates the greatest virulence of all mycotic pathogens.

Biology and Epidemiology of *Coccidioides*

The morphology of *C. immitis* is very distinctive. At 25°C, it forms a moist, white to brown colony with abundant, branching, septate hyphae. These hyphae fragment into thick-walled, blocklike **arthroconidia** (arthrospores) at maturity (inset photo, figure 22.8). On special media incubated at 37°–40°C, an arthrospore germinates into the parasitic phase, a small, spherical cell **(spherule).** This structure swells into a giant sporangium that cleaves internally to form numerous endospores that look like bacterial endospores but lack their resistance traits (figure 22.8c).

Coccidioides immitis occurs endemically in various natural reservoirs and casually in areas where it has been carried by wind and animals. Conditions favoring its settlement in a given habitat include high carbon and salt content and a semiarid, relatively hot

---

* *Coccidioides immitis* (kok-sid″-ee-oy′-deez ih′-mih-tis) From *coccidia,* a sporozoan, and L. *immitis,* fierce. The original discoverers thought the microbe looked like a protozoan.

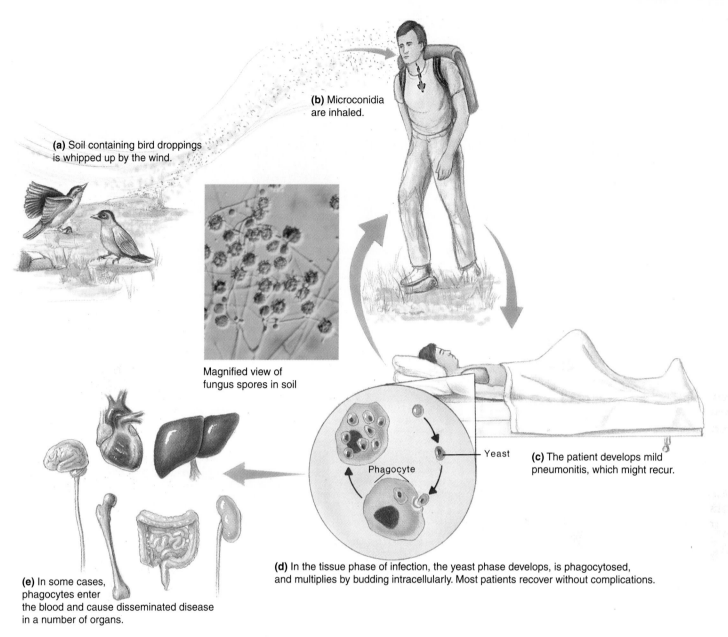

(b) Microconidia are inhaled.

(a) Soil containing bird droppings is whipped up by the wind.

Magnified view of fungus spores in soil

Yeast

Phagocyte

(c) The patient develops mild pneumonitis, which might recur.

(d) In the tissue phase of infection, the yeast phase develops, is phagocytosed, and multiplies by budding intracellularly. Most patients recover without complications.

(e) In some cases, phagocytes enter the blood and cause disseminated disease in a number of organs.

FIGURE 22.7

Events in *Histoplasma* infection and histoplasmosis.

climate. The fungus has been isolated from soils, plants, and a large number of vertebrates. The natural history of *C. immitis* follows a cyclic pattern—a period of dormancy in winter and spring, followed by growth in summer and fall. Growth and spread are greatly increased by cycles of drought and heavy rains, followed by windstorms.

Skin testing has disclosed that the highest incidence of coccidioidomycosis (approximately 100,000 cases per year) is in the southwestern United States, though it also occurs in Mexico and parts of Central and South America. Especially concentrated reservoirs exist in the San Joaquin Valley of California and in southern Arizona. Outbreaks are usually associated with farming activity, archeological digs, and mining.

A highly unusual outbreak of coccidioidomycosis was traced to the Northridge, California earthquake. Clouds of dust bearing loosened spores were given off by landslides, and local winds then carried the dust into outlying residential areas.

Infection and Pathogenesis of Coccidioidomycosis

The arthrospores of *C. immitis* are lightweight and readily inhaled. In the lung they convert to spherules, which swell, sporulate, burst, and release spores that continue the cycle. In 60% of patients, this primary pulmonary infection is inapparent; in the other 40%, it is accompanied by coldlike symptoms such as fever, chest pain, cough, headaches, and malaise. In uncomplicated cases, complete recovery and lifelong immunity are the rule.

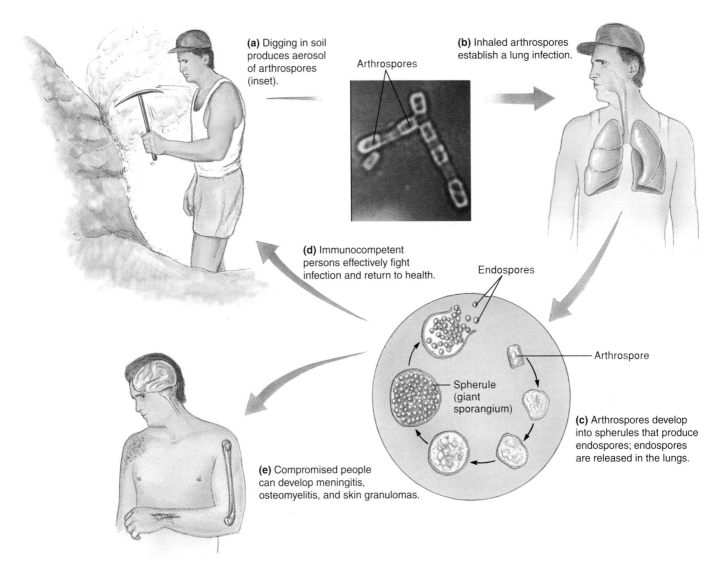

(a) Digging in soil produces aerosol of arthrospores (inset).

Arthrospores

(b) Inhaled arthrospores establish a lung infection.

(d) Immunocompetent persons effectively fight infection and return to health.

Endospores

Arthrospore

Spherule (giant sporangium)

(c) Arthrospores develop into spherules that produce endospores; endospores are released in the lungs.

(e) Compromised people can develop meningitis, osteomyelitis, and skin granulomas.

FIGURE 22.8
Events in *Coccidioides* infection and coccidioidomycosis.

All persons inhaling the arthrospores probably develop some degree of infection, but certain groups have a genetic susceptibility that gives rise to more serious disease. In about five out of a thousand cases, the primary infection does not resolve and progresses with varied consequences. Chronic progressive pulmonary disease is manifested by nodular growths called *fungomas** and cavity formation in the lungs that compromise respiration. Dissemination of the endospores into major organs occasionally takes place in people with impaired cell-mediated immunity, with severe and sometimes fatal results (figure 22.9).

Diagnosis and Control of Coccidioidomycosis

Diagnosis of coccidioidomycosis is straightforward when the highly distinctive spherules are found in sputum, spinal fluid, and biopsies. This finding is further supported by isolation of typical mycelia and spores on Sabouraud's agar and induction of spherules.

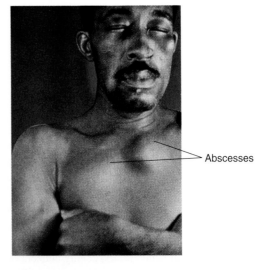

Abscesses

FIGURE 22.9
Disseminated coccidioidomycosis manifested by subcutaneous abscesses in the chest.

_____

* fungoma (fun-joh´-mah) A fungus tumor or growth.

Lines of precipitation   (1)   Antigen (coccidioidin)

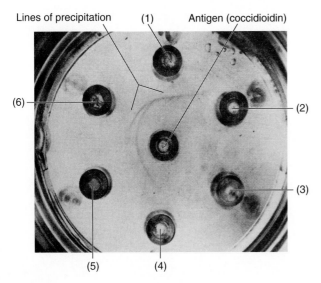

(6)

(2)

(3)

(5)   (4)

FIGURE 22.10

Immunodiffusion testing for coccidioidomycosis.   Small wells are punched into agar, and patients' sera are placed in numbered wells; the center well contains a known fungal antigen such as coccidioidin. Lines of precipitation forming between the outer well and the inner well indicate a reaction between antibodies in the serum and antigen diffusing from the center well. In this test, reactions occurring for sera 1, 5, and 6 are positive evidence of infection. Wells 2, 3, and 4 indicate the absence of antibodies and infection.

Newer specific antigen tests have been effective tools to identify and differentiate *Coccidioides* from other fungi. All cultures must be grown in closed tubes or bottles and opened in a biological containment hood to prevent laboratory infections. Immunodiffusion (figure 22.10) and latex agglutination tests on serum samples are excellent screens for detecting early infection. Skin tests using an extract of the fungus (coccidioidin or spherulin) are of primary importance in epidemiologic studies.

The majority of patients do not require treatment. However, in people with disseminated disease, amphotericin B is administered intravenously. The azole drugs can have some benefit, but they may require a longer term of therapy. Minimizing contact with the fungus in its natural habitat has been of some value. For example, oiling roads and planting vegetation help reduce spore aerosols, and using dust masks while excavating soil prevents workers from inhaling spores.

BIOLOGY OF *BLASTOMYCES DERMATITIDIS:* NORTH AMERICAN BLASTOMYCOSIS

*Blastomyces dermatitidis,** the cause of **blastomycosis,** is another fungal pathogen endemic to the United States. Alternative names for this disease are Gilchrist's disease, Chicago disease, and North American blastomycosis. The dimorphic morphology of *Blastomyces* follows the pattern of other pathogens. Colonies of the saprobic phase are uniformly white to tan, with a thin, septate

---

* *Blastomyces dermatitidis* (blas″-toh-my′-seez der″-mah-tit′-ih-dis) Gr. *blastos,* germ, *myces,* fungus, *dermato,* skin, and *itis,* inflammation.

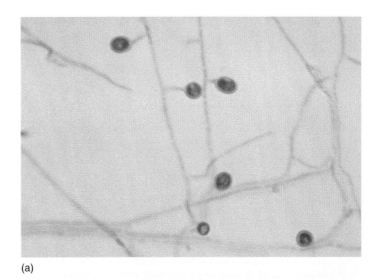

(a)

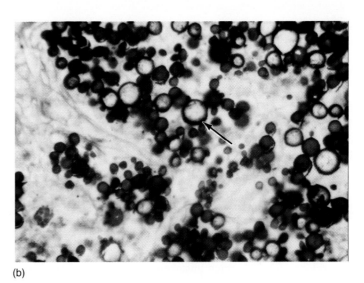

(b)

FIGURE 22.11

The dimorphic nature of *Blastomyces dermatitidis.*   **(a)** Hyphal filaments bear conidia that resemble tiny lollipops. **(b)** The tissue phase as seen in a sputum sample. The arrow points out the very thick cell wall typical of this species.

mycelium and simple ovoid conidia (figure 22.11*a*). Temperature-induced conversion results in a wrinkled, creamy-white colony that yields large, heavy-walled yeasts with buds nearly as large as the mother cell.

Studies indicate that *B. dermatitidis* inhabits areas high in organic matter such as forest soil, decaying wood, animal manure, and abandoned buildings. Its life cycle features dormancy during the warmer, dryer times of the year and growth and sporulation during the colder, wetter seasons. In general, blastomycosis occurs from southern Canada to southern Louisiana and from Minnesota to Georgia. Cases have also been reported in Central America, South America, Africa, and the Middle East. Humans, dogs, cats, and horses are the chief targets of infection, which is usually acquired by inhaling conidia-laden dust from living quarters, farm buildings, or forest litter.

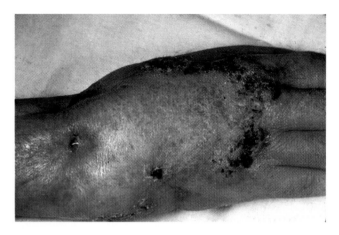

FIGURE 22.12

⊙ **Cutaneous blastomycosis in the hand and wrist as a complication of disseminated infection.** Note the darkly colored, tumorlike vegetations and scar tissue on the hand.

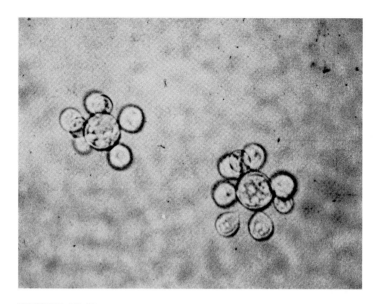

FIGURE 22.13

The morphology of *Paracoccidioides*. A Gridley stain from a skin lesion (1,200×) reveals the central round mother cell with a series of narrow-necked buds that look like the spokes of a wheel.

Infection and Pathology in Blastomycosis

The primary portal of entry of *B. dermatitidis* is the respiratory tract, though it also may enter through accidental inoculation. Inhaling only 10 to 100 conidia is enough to initiate infection. As conidia convert to yeasts and multiply, they encounter macrophages in the lung. A large proportion of primary pulmonary infections are probably symptomatic to some degree.

Mild disease is accompanied by cough, chest pain, hoarseness, and fever. More severe, chronic blastomycosis can progress to the lungs, skin, and numerous other organs. Lung abscesses and tumorlike nodules are often mistaken for cancer. Compared with other fungal diseases, the chronic cutaneous form of blastomycosis is rather common. It frequently begins on the face, hand, wrist, or leg as a subcutaneous nodule that erupts to the skin surface (figure 22.12). Yeasts disseminating into the bone induce symptoms of arthritis and osteomyelitis. Involvement of the central nervous system can bring on headache, convulsions, coma, and mental confusion. Chronic systemic blastomycosis of the spleen, liver, and urogenital tract can last for weeks to years and eventually destroy the host defenses.

Laboratory Diagnosis and Therapy

Microscopic smears of specimens showing large, ovoid yeasts with broad-based buds can provide the most reliable diagnostic evidence of blastomycosis (see figure 22.11*b*). Dimorphic cultures are highly desirable, although they can require several weeks to develop. Diagnosis can be verified by complement fixation and ELISA tests for antibodies. Most skin tests for *Blastomyces* are not useful for general testing because of the frequency of false-positive and false-negative results. Although disseminated infections were once nearly 100% fatal, modern drugs have greatly improved the prognosis. The drug of choice for disseminated and cutaneous disease is amphotericin B, and milder cases typically respond to several months of dihydroxystilbamidine therapy.

PARACOCCIDIOIDOMYCOSIS

The remaining dimorphic fungal pathogen would be a front-runner in a competition for the most tongue-twisting scientific term. ***Paracoccidioides brasiliensis*** * causes **paracoccidioidomycosis,** also known as paracoccidioidal granuloma and South American blastomycosis. Because of its relatively restricted distribution, this disease is the least common of the primary mycoses. *Paracoccidioides* forms a small, nondescript colony with scanty, undistinctive spores at room temperature, but it develops an unusual yeast form at 37°C. The large mother cells sprout small, narrow-necked buds that radiate around the periphery (figure 22.13).

Paracoccidioides has been isolated from the cool, humid soils of tropical and semitropical regions of South and Central America, particularly Brazil, Colombia, Venezuela, Argentina, and Paraguay. People most often afflicted with paracoccidioidomycosis are rural agricultural workers and plant harvesters. Factors that can influence the progress of infection are altered physiology, poor nutrition, and impaired host resistance. Most infections occur in the lungs or the skin, and most of the time they are benign, self-limited events that go completely unnoticed. In a small minority of patients, the pathogen invades the lungs, skin and mucous membranes (especially of the head), and lymphatic organs.

Paracoccidioidomycosis is diagnosed by standard procedures previously described for the other mycoses in this chapter. Closely scrutinizing fresh or stained clinical specimens for the distinctive yeast phase is important, and a cultural follow-up with conversion media is essential. Serological testing can be instrumental in diagnosing and monitoring the course of infection. Principal treatment drugs for disseminated disease, in order of choice, are ketoconazole, amphotericin B, and sulfa drugs.

---

* *Paracoccidioides brasiliensis* (pair″-ah-kok-sid″-ee-oy´-deez brah-sil″-ee-en´-sis) Named for its superficial resemblance to *Coccidioides* and its prevalence in Brazil.

CHAPTER CHECKPOINTS

Primary fungal pathogens have several common characteristics: (1) Each is endemic to a specific region of the world. (2) All cause primary pulmonary infections by inhalation of spores. (3) In most cases, the infection is not life-threatening. (4) Recovery from infection confers lifelong immunity. True fungal pathogens cause systemic infection in certain susceptible groups of people. The spores of some species can also infect the skin.

Histoplasma capsulatum is the causative agent of histoplasmosis, or Ohio Valley fever. It is endemic to the Ohio Valley in the United States and sporadic in much of the world. Airborne spores cause pulmonary, systemic, or cutaneous lesions. The severity of the infection depends on the number of spores inhaled and the immunocompetence of the infected host. *H. capsulatum* is identified by "fish-eye" yeast cells in host macrophages.

Coccidioides immitis is the causative agent of coccidioidomycosis, or Valley fever. It is endemic to salty soils in arid regions of the western United States. Airborne spores cause a primary pulmonary infection that is self-limiting in most cases and results in a lifelong immunity. Rare cases progress to chronic pulmonary disease or systemic infections. *C. immitis* is identified by highly distinctive spherules in fresh tissue specimens.

Blastomyces dermatitidis is the causative agent of blastomycosis, or Gilchrist's disease. It is endemic to North America, Africa, and the Middle East. Like other true pathogens, airborne spores cause a primary pulmonary infection, but the skin is also a site of infection. *B. dermatitidis* is identified by its microscopic appearance in fresh specimens and by serological tests for antibody.

Paracoccidioides brasiliensis causes paracoccidioidomycosis, or South American blastomycosis, endemic to regions of Central and South America. In most cases, infections of the lungs and skin are self-limiting, but persistent infections can be debilitating. *P. brasiliensis* is identified by its unusual budding pattern and by serological tests for antibody.

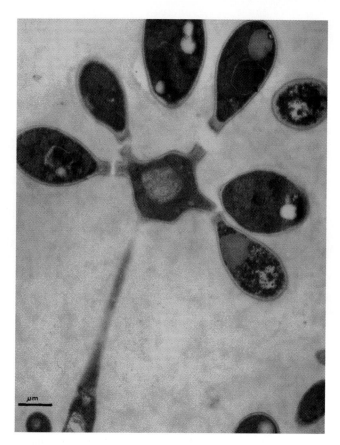

FIGURE 22.14

The microscopic morphology of *Sporothrix schenckii*. Spores develop as floral clusters borne on conidiophores or as single spores pinched off the edge of a hypha.

Subcutaneous Mycoses

When certain fungi are transferred from soil or plants directly into traumatized skin, they can invade the damaged site. Such infections are termed subcutaneous because they involve tissues within and just below the skin. Most species in this group are greatly inhibited by the higher temperatures of the blood and viscera, and only rarely do they disseminate. Nevertheless, these diseases are progressive and can destroy the skin and associated structures. Mycoses in this category are sporotrichosis, chromoblastomycosis, phaeohyphomycosis, and mycetoma.

THE NATURAL HISTORY OF SPOROTRICHOSIS: ROSE-GARDENER'S DISEASE

The cause of **sporotrichosis, *Sporothrix schenckii*,*** is a very common saprobic fungus that decomposes plant matter in soil and humus and exhibits both mycelial and yeast phases (figure 22.14). *Sporothrix* resides in warm, temperate, and moist areas of the trop-

ics and semitropics, though the incidence of sporotrichosis is highest in Africa, Australia, Mexico, and Latin America. A prick by a rose thorn is the classic origin of the cutaneous form of infection and has given rise to its common name. Most episodes of human infection follow contact with thorns, wood, sphagnum moss, bare roots, bark, and other vegetation. Horticulturists, gardeners, farmers, and basket weavers incur the majority of cases. Overt disease is uncommon unless a person has inadequate immune defenses or is exposed to a large inoculum or a virulent strain. Other mammals, notably horses, dogs, cats, and mules, are also susceptible to sporotrichosis.

Pathology, Diagnosis, and Control of Sporotrichosis

In **lymphocutaneous sporotrichosis,** the fungus grows at the site of penetration and develops a small, hard, nontender nodule within a few days to months. This subsequently enlarges, becomes necrotic, breaks through to the skin surface, and drains (figure 22.15). Infection often progresses along the regional lymphatic channels, leaving a chain of lesions at various stages. This condition can persist for several years, though it does not spread in the circulation. When *Sporothrix* conidia are drawn into the lungs, primary pulmonary sporotrichosis can result. Although this infection was once thought to be rare, it is increasingly reported among chronic alcoholics.

---

* *Sporothrix schenckii* (spoh´-roh-thriks shenk´-ee-ee) Gr. *sporos,* seed, and *thrix,* hair. Named for B. R. Schenck, who first isolated it.

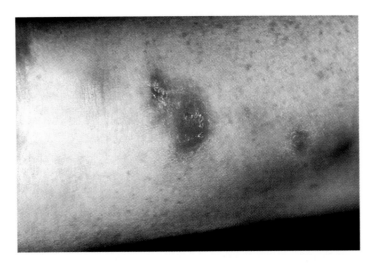

FIGURE 22.15

The clinical appearance of lymphocutaneous sporotrichosis.
A primary sore is accompanied by a series of nodules running along the lymphatic channels of the arm.

Discovery of the agent in tissue exudates, pus, and sputum is difficult because there are usually so few cells in the infection. Special stains can be used to highlight the tiny, cigar-shaped yeasts. Demonstration of typical morphology of isolated cultures and serological tests can confirm the diagnosis. The sporotrichin skin test can be used to determine prior infection. An older but effective drug for sporotrichosis is potassium iodide, given orally in milk or applied topically to open lesions. Amphotericin B and flucytosine can assist in unresponsive cases. The fungus cannot withstand heat, thus local applications of heat packs to lesions have helped resolve some infections. People with occupational exposure should cover their bare limbs as a preventive measure.

CHROMOBLASTOMYCOSIS AND PHAEOHYPHOMYCOSIS: DISEASES OF PIGMENTED FUNGI

Chromoblastomycosis* is a progressive subcutaneous mycosis characterized by highly visible *verrucous** lesions. The principal etiologic agents are a collection of widespread soil saprobes with dark-pigmented mycelia and spores. *Fonsecaea pedrosoi, Phialophora verrucosa,* and *Cladosporium carrionii* are among the most common causative species. **Phaeohyphomycosis,*** a closely related infection, differs primarily in the causative species and the appearance of the infectious agent in tissue. Chromoblastomycotic agents produce very large, thick, yeastlike bodies called sclerotic cells, whereas the agents of phaeohyphomycosis remain typically hyphal.

---

* chromoblastomycosis (kroh″-moh-blas″-toh-my-koh′-sis) Gr. *chroma,* color, and *blasto,* germ. The fungal body is highly colored *in vitro.*

* verrucous (ver-oo′-kus) Tough, warty.

* phaeohyphomycosis (fy″-oh-hy″-foh-my-coh′-sis) Gr. *phaeo,* brown, and *hypho,* thread.

The fungi associated with these infections are of low inherent virulence, and none exhibits thermal dimorphism. Infection occurs when body surfaces, especially the legs and feet, are penetrated by soiled vegetation or inanimate objects. Chromoblastomycosis occurs throughout the world, with peak incidence in the American subtropics and tropics. Men with rural occupations who go barefoot are most vulnerable. After a very long (2- to 3-year) incubation period, a small, colored, warty plaque, ulcer, or papule appears. Because this lesion is generally not painful, many patients do not seek treatment and can go for years watching it progress to a more advanced state. Unfortunately, the patient can aggravate the spread of infection on the body and provoke secondary bacterial infections by scratching the nodules.

Chromoblastomycosis is frequently confused with cancer, syphilis, yaws, and blastomycosis. Diagnosis is based on the clinical appearance of the lesions, microscopic examination of biopsied lesions, and the results of culture. Therapy for the condition includes heat, drugs, surgical removal of early nodules, and amputation in advanced cases. Combined topical amphotericin B and thiabendazole or systemic flucytosine have shown some success in arresting the disease.

The infectious diseases termed phaeohyphomycoses are among the strangest oddities of medical science. The etiologic agents are soil fungi or plant pathogens with brown-pigmented mycelia. Victims of the disease are usually highly compromised patients who have become inoculated with these fungi, which are so widespread in household and medical environments that contact is inevitable. The disease process can be extremely deforming, and the microbe is sometimes hard to identify. Some genera of fungi commonly involved in these mycoses are *Alternaria, Aureobasidium, Curvularia, Dreschlera, Exophiala, Phialophora,* and *Wangiella.* The fungi grow slowly from the portal of entry through the dermis and create enlarged, subcutaneous cysts. In patients with underlying conditions such as endocarditis, diabetes, and leukemia, the fungi can spread into the bones, brain, and lungs.

MYCETOMA: A COMPLEX DISFIGURING SYNDROME

Another disease elicited when soil microbes are accidently implanted into the skin is **mycetoma,** a mycosis usually of the foot or hand that looks superficially like a tumor. It is also called *madura foot* for the Indian region where it was first described. About half of all mycetomas are caused by fungi in the genera *Pseudallescheria* or *Madurella,* though filamentous bacteria called actinomycetes (*Nocardia,* discussed in chapter 19) are often implicated. Mycetomas are endemic to equatorial Africa, Mexico, Latin America, and the Mediterranean, and cases occur sporadically in the United States. Infection begins when bare skin is pierced by a thorn, sliver, leaf, or other type of sharp plant debris. First to appear is a localized abscess in the subcutaneous tissues, which gradually swells and drains (figure 22.16). Untreated cases that spread to the muscles and bones can cause pain and loss of function in the affected body part. Mycetoma has an insidious, lengthy course and is exceedingly difficult to treat.

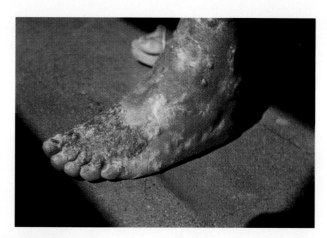

FIGURE 22.16

Early stages of mycetoma of the foot, caused by *Madurella*. Ulceration, swelling, and scarring are visible.

TABLE 22.3

The Dermatophyte Genera and Diseases

Genus	Name of Disease	Principal Targets	How Transmitted
Trichophyton	Ringworm of the scalp, body, beard, and nail Athlete's foot	Hair, skin, nails	Human to human, animal to human
Microsporum	Ringworm of scalp	Scalp hair	Animal to human, soil to human, human to human
	Ringworm of skin	Skin; not nails	
Epidermophyton	Ringworm of the groin and nail	Skin, nails; not hair	Strictly human to human

CHAPTER CHECKPOINTS

Fungal infections that invade traumatized skin are called subcutaneous mycoses. These localized infections rarely become systemic, but they can be very destructive to the skin and its associated components. Sporotrichosis, chromoblastosis, phaeohyphomycosis, and mycetoma are examples.

Sporothrix schenckii is the causative agent of sporotrichosis, or rose-gardener's disease. When introduced subcutaneously, it produces local lesions with the potential to invade lymphatics.

Chromoblastomycosis and phaeohyphomycosis are both caused by certain pigmented soil saprobes that produce characteristic slow-growing skin lesions of low virulence.

Mycetoma, or madura foot, is caused by filamentous fungi that invade traumatized skin. Lesions appear as abscesses and nodules. The systemic form spreads to bones and muscles and is very difficult to treat.

Cutaneous Mycoses

Fungal infections strictly confined to the nonliving epidermal tissues (stratum corneum) and its derivatives (hair and nails) are termed **dermatophytoses.** Common terms used in reference to these diseases are **ringworm,** because they tend to develop in circular, scaly patches (see figure 22.18), and **tinea.*** About 39 species in the genera *Trichophyton, Microsporum,* and *Epidermophyton* are involved in various dermatophytoses. The causative agent of a given type of ringworm differs from person to person and from place to place and is not restricted to a particular genus or species (table 22.3).

** **tinea** (tin´-ee-ah) L., a larva or worm. Early observers thought they were caused by worms.*

CHARACTERISTICS OF DERMATOPHYTES

The dermatophytes are so closely related and morphologically similar that they can be difficult to differentiate. Various species exhibit unique macroconidia, microconidia, and unusual types of hyphae. In general, *Trichophyton* produces thin-walled, smooth macroconidia and numerous microconidia; *Microsporum* produces thick-walled, rough macroconidia and sparser microconidia; and *Epidermophyton* has ovoid, smooth, clustered macroconidia and no microconidia (figure 22.17).

Epidemiology and Pathology of Dermatophytoses

The natural reservoirs of dermatophytes are other humans, animals, and the soil (Medical Microfile 22.2). Important factors that promote infection are the hardiness of the dermatophyte spores (they can last for years on fomites); presence of abraded skin; and intimate contact. Most infections exhibit a long incubation period (months), followed by localized inflammation and allergic reactions to the fungal proteins. As a general rule, infections acquired from animals and soil cause more severe reactions than do infections acquired from other humans, and infections eliciting stronger immune reactions are resolved faster.

Dermatophytic fungi exist throughout the world. Although some species are endemic, they tend to spread rapidly to nonendemic regions through travel. Dermatophytoses endure today as a serious health concern, not because they are life-threatening, but because of the extreme discomfort, stress, pain, and unsightliness they cause.

An Atlas of Dermatophytoses

In the following section, the dermatophytoses are organized according to the area of the body they affect, their mode of acquisition, and their pathologic appearance. Both the common English name and body site and its equivalent Latin name (tinea) are given.

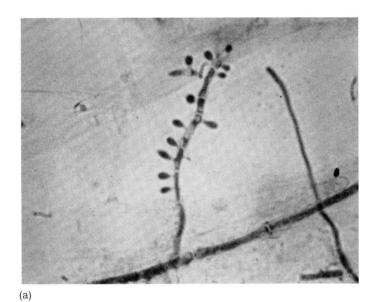

(a)

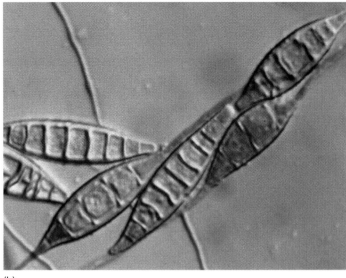

(b)

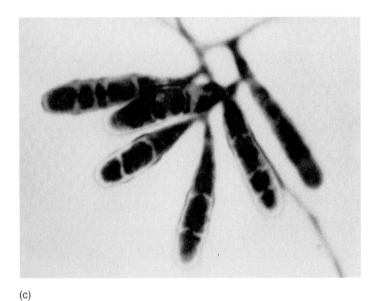

(c)

FIGURE 22.17

Examples of dermatophyte spores. **(a)** Regular, numerous microconidia of *Trichophyton*. **(b)** Macroconidia of *Microsporum canis*, a cause of ringworm in cats, dogs, and humans. **(c)** Smooth-surfaced macroconidia in clusters characteristic of *Epidermophyton*.

Ringworm of the Scalp (Tinea Capitis) This mycosis results from the fungal invasion of the scalp and the hair of the head, eyebrows, and eyelashes (figure 22.18*a*). Very common in children, tinea capitis is acquired from other children and adults or from domestic animals. Manifestations range from small, scaly patches (gray patch), to a severe inflammatory reaction (kerion), to destruction of the hair follicle and permanent hair loss.

Ringworm of the Beard (Tinea Barbae) This tinea, also called "barber's itch," afflicts the chin and beard of adult males. Although once a common aftereffect of unhygienic barbering, it is now contracted mainly from animals.

Ringworm of the Body (Tinea Corporis) This extremely prevalent infection of humans can appear nearly anywhere on the body's glabrous (smooth and bare) skin. The principal sources are other humans, animals, and soil, and it is transmitted primarily by direct contact and fomites (clothing, bedding). The infection usually appears as one or more scaly reddish rings on the trunk, hip, arm, neck, or face (figure 22.18*b*). The ringed pattern is formed when the infection radiates from the original site of invasion into the surrounding skin. Depending on the causal species and the health and hygiene of the patient, lesions vary from mild and diffuse to florid and pustular.

Ringworm of the Groin (Tinea Cruris) Sometimes known as "jock itch," crural ringworm occurs mainly in males on the groin, perianal skin, scrotum, and, occasionally, the penis. The fungus thrives under conditions of moisture and humidity created by profuse sweating or tropical climates. It is transmitted primarily from human to human and is pervasive among athletes and persons living in close situations (ships, military quarters).

MEDICAL MICROFILE 22.2
The Keratin Lovers

The dermatophytic fungi are especially well adapted to breaking down keratin, the primary protein of the epidermal tissues of vertebrates (skin, nails, hair, feathers, and horns). Their affinity for this compound gives them the name *keratophiles*. Examination of infected hairs indicates that these fungi attach to the hair surface and penetrate into its cortex. In time, it grows along the hair's length to the follicle, where it initiates a skin infection. A study of dermatophyte ecology reveals a gradual evolutionary trend from saprobic soil forms that digest keratin but do not parasitize animals, to soil forms that occasionally parasitize animals, to species that are dependent on live animals. Some species can infect a broad spectrum of animals, and others are specific to one particular animal species or region of the body.

One adaptive challenge faced by relatively new fungal parasites is that they are likely to cause severe reactions in the host's skin and to be attacked by the host defenses and eliminated. Thus, the more successful fungi equilibrate with the host by reducing their activity (growth rate, sporulation) and lessen the inflammatory response. Eventually, these dermatophytes become such "good parasites" that they colonize the host for life. A striking example is *Trichophyton rubrum,* a fungus that causes a form of athlete's foot. It has such a tenacious hold and is so hard to cure that its carriers have been called the *"T. rubrum people."*

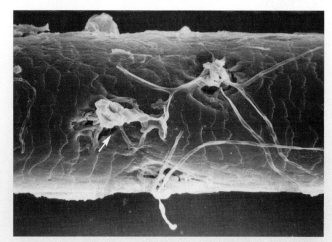

This microscopic view of a human hair shows dermatophyte hyphae growing along the hair and penetrating into its center (arrow).

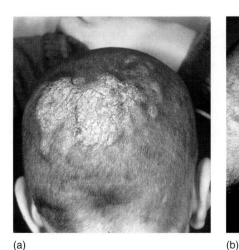

(a)

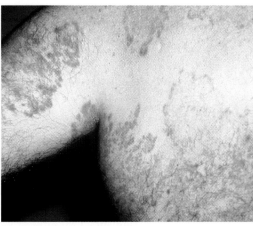

(b)

FIGURE 22.18

Ringworm lesions on the scalp and body vary in appearance. **(a)** Kerion, with deep exudative involvement and complete hair loss in the affected region. **(b)** Widespread lesions over the arm and shoulder have a dramatic ringed appearance that results from the gradual spread of inflammation from the center to the newest area of invasion in a circumferential pattern.

Ringworm of the Foot (Tinea Pedis) Tinea pedis is known by a variety of synonyms, including athlete's foot and jungle rot. The disease is clearly connected to wearing shoes, because it is uncommon in cultures where the people customarily go barefoot (but as you have seen, bare feet have other fungal risks!). Conditions that encase the feet in a closed, warm, moist environment increase the possibility of infection. Tinea pedis is a known hazard in shared facilities such as shower stalls, public floors, and locker rooms. Infections begin with small blisters between the toes that burst, crust over, and can spread to the rest of the foot and nails (figure 22.19*a*).

Ringworm of the Hand (Tinea Manuum) Infection of the hand by dermatophytes is nearly always associated with concurrent infection of the foot. Lesions usually occur on the fingers and palms of one hand, and they vary from white and patchy to deep and fissured.

Ringworm of the Nail (Tinea Unguium) Fingernails and toenails, being masses of keratin, are often sites for persistent fungus colonization. The first symptoms are usually superficial white patches in the nail bed. A more invasive form causes thickening, distortion, and darkening of the nail (figure 22.19*b*). Nail problems

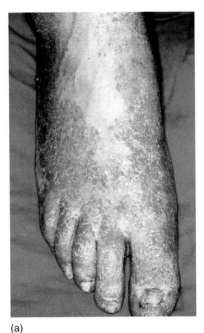

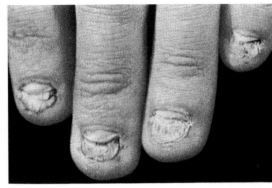

FIGURE 22.19

Ringworm of the extremities. **(a)** *Trichophyton* infection spreading over the foot in a "moccasin" pattern. The chronicity of tinea pedis is attributed to the lack of fatty-acid-forming glands in the feet. **(b)** Ringworm of the nails. Invasion of the nail bed causes some degree of thickening, accumulation of debris, cracking, and discoloration; nails can be separated from underlying structures as shown here.

(a)   (b)

caused by dermatophytes are on the rise as more women wear artificial fingernails, which can provide a portal of entry into the nail bed.

Diagnosis of Ringworm

Dermatologists are most often called upon to diagnose dermatophytoses. Occasionally, the presenting symptoms are so dramatic and suggestive that no further testing is necessary, but in most cases, direct microscopic examination and culturing are needed. Diagnosis of tinea of the scalp caused by some species of *Microsporum* is aided by use of a long-wave ultraviolet lamp that causes infected hairs to fluoresce. Samples of hair, skin scrapings, and nail debris treated with heated KOH show a thin, branching fungal mycelium if infection is present. Culturing specimens on selective media and identifying the species can be important diagnostic aids.

Treatment of the Dermatophytoses

Ringworm therapy is based on the knowledge that the dermatophyte is feeding on dead epidermal tissues. These regions undergo constant replacement from living cells deep in the epidermis, so if multiplication of the fungus can be blocked, the fungus will eventually be sloughed off along with the skin or nail. Unfortunately, this takes time. By far the most satisfactory choice for therapy is a topical antifungal agent. Ointments containing tolnaftate, miconazole, or thiabendazine are applied regularly for several weeks. Some drugs work by speeding up loss of the outer skin layer. Intractable infections can be treated with griseofulvin, but placing a patient on this hepatotoxic and nephrotoxic drug for up to 2 years is probably too risky in most cases. Gentle debridement of skin and ultraviolet light treatments can have some benefit.

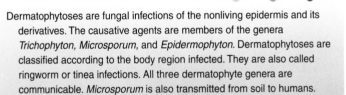

CHAPTER CHECKPOINTS

Dermatophytoses are fungal infections of the nonliving epidermis and its derivatives. The causative agents are members of the genera *Trichophyton, Microsporum,* and *Epidermophyton.* Dermatophytoses are classified according to the body region infected. They are also called ringworm or tinea infections. All three dermatophyte genera are communicable. *Microsporum* is also transmitted from soil to humans.

Superficial Mycoses

Agents of **superficial mycoses** involve the outer epidermal surface and are ordinarily innocuous infections with cosmetic rather than inflammatory effects. **Tinea versicolor**[1] is caused by the yeast *Malassezia furfur,* a normal inhabitant of human skin that feeds on the high oil content of the skin glands. Even though this yeast is very common (carried by nearly 100% of humans tested), in some people its growth elicits mild, chronic scaling and interferes with production of pigment by melanocytes. The trunk, face, and limbs take on a mottled appearance (figure 22.20). The disease is most pronounced in young people who are frequently exposed to the sun. Other skin conditions in which *M. furfur* is implicated are folliculitis, psoriasis, and seborrheic dermatitis. It is also occasionally associated with systemic infections and catheter-associated sepsis in compromised patients.

---

1. Versicolor refers to the color variations this mycosis produces in the skin. Also called pityriasis.

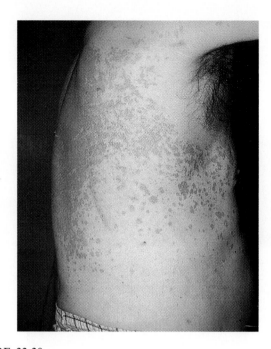

FIGURE 22.20

Tinea versicolor. Mottled, discolored skin pigmentation is characteristic of superficial skin infection by *Malassezia furfur.*

*Piedras** are marked by a tenacious, colored concretion forming on the outside surface of hair shafts (figure 22.21). In **white piedra,** caused by *Trichosporon beigelii,* a white to yellow adherent mass develops on the shaft of scalp, pubic, or axillary hair. At times, the mass is invaded secondarily by brightly colored bacterial contaminants, with startling results. **Black piedra,** caused by *Piedraia hortae,* is characterized by dark-brown to black gritty nodules, mainly on scalp hairs. Neither piedra is common in the United States.

CHAPTER CHECKPOINTS

Superficial mycoses are noninvasive, noninflammatory fungal infections restricted to the hair and outer layer of the epidermis. Superficial mycoses of the hair, or piedra, appear as white or black masses on individual hair shafts. Epidermal infections include tinea versicolor, folliculitis, psoriasis, and seborrheic dermatitis.

Opportunistic Mycoses

Earlier in this chapter, we introduced the concept of opportunistic fungal infections and their predisposing factors (see Medical Microfile 22.1 and table 22.2). The prevailing opportunistic pathogens of humans are the yeasts *Candida* and *Cryptococcus, Pneumocystis,* and a small number of filamentous fungi, primarily *Aspergillus* and certain zygomycetes.

---

* piedra (pee-ay´-drah) Sp., stone. Hard nodules of fungus formed on the hair.

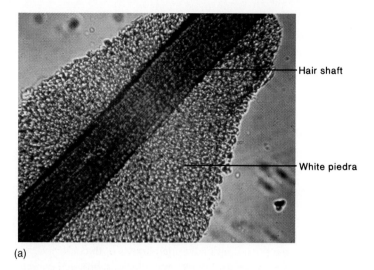

(a)

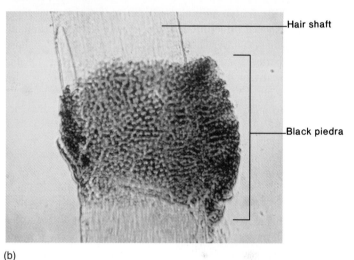

(b)

FIGURE 22.21

Examples of superficial mycoses. **(a)** Light-colored mass on a hair shaft with white piedra. **(b)** Dark, hard concretion enveloping the hair shaft in black piedra (200×).

INFECTIONS BY *CANDIDA:* CANDIDIASIS

*Candida albicans,** an extremely widespread yeast, is the major cause of candidiasis (also called candidosis or moniliasis). Manifestations of infection run the gamut from short-lived, superficial skin irritations to overwhelming, fatal systemic diseases. Microscopically, *C. albicans* has budding cells of varying size that may form both elongate pseudohyphae and true hyphae (see figure 22.23a); macroscopically, it forms off-white, pasty colony with a yeasty odor.

Epidemiology of Candidiasis

Candida albicans occurs as normal flora in the oral cavity, genitalia, large intestine, or skin of 20% of humans. A healthy person with an unimpaired immune system can hold it in check, but the

---

* *Candida albicans* (kan´-dih-dah al´-bih-kanz) L. *candidus,* glowing white, and *albus,* white.

risk of invasion increases with extreme youth, pregnancy, drug therapy, immunodeficiency, and trauma. Any situation that maintains the yeast in contact with moist skin provides an avenue for infection. Although candidiasis is usually endogenous and not contagious, it can be spread in nurseries or through surgery, childbirth, and sexual contact. *Candida albicans* and its close relatives account for nearly 80% of nosocomial fungal infections and 30% of deaths from nosocomial infections in general.

Diseases of *Candida albicans*

Candida albicans causes local infections of the mouth, pharynx, vagina, skin, alimentary canal, and lungs, and it can also disseminate to internal organs. The mucous membranes most frequently involved are the oral cavity and vagina. **Thrush** is a white adherent, patchy infection affecting the membranes of the oral cavity or throat, usually in newborn infants and elderly, debilitated patients (figure 22.22*a*). **Vulvovaginal candidiasis (VC),** known more commonly as **yeast infection,** has widespread occurrence in adult women, especially those who are taking oral antibiotics or those who are diabetic or pregnant, all conditions that can disrupt the normal vaginal flora. Candidal vaginitis also poses a risk for neonates, which can be infected during childbirth, and it can be transmitted to male partners during sexual intercourse. The chief symptoms of VC are a yellow to white discharge, inflammation, painful ulcerations, and itching. The most severe cases spread from the vagina and vulva to the perineum and thighs.

Of all the areas of the gastrointestinal tract, *Candida* most often infects the esophagus and the anus. Esophageal candidiasis, which afflicts 70% of AIDS patients, causes painful, bleeding ulcerations, nausea, and vomiting.

Candidal attack of keratinized structures such as skin and nails, called **onychomycosis,*** is often brought on by predisposing occupational and anatomical factors. People whose occupations require their hands or feet to be constantly immersed in water are at risk for finger and nail invasion. *Intertriginous** infection occurs in moist areas of the body where skin rubs against skin, as beneath the breasts, in the armpit, and between folds of the groin. **Cutaneous candidiasis** can also complicate burns and produce a scaldlike rash on the skin of neonates (figure 22.22*b*).

Candidal blood infection in patients chronically weakened by surgery, bone marrow transplants, advanced cancer, or intravenous drug addiction is usually followed by systemic invasion. The presence of *C. albicans* in the blood is such a serious assault that it causes more human mortalities than any other fungal pathogen. Principal targets of systemic infections are the urinary tract, endocardium, and brain. Patients with valvular disease of the heart or indwelling prosthetic devices are vulnerable to candidal endocarditis, usually caused by other species (*C. tropicalis* and *C. parapsilosis*). *Candida krusei* has reportedly caused terminal infections in bone marrow transplant patients and in recipients of anticancer therapy.

---

* onychomycosis (ahn″-ih-koh-my-koh′-sis) Gr. *onyx,* nail. Infection of the nails by *Candida* or *Aspergillus.*

* intertriginous (in″-ter-trij′-ih-nus) L. *inter,* between, and *trigo,* to rub.

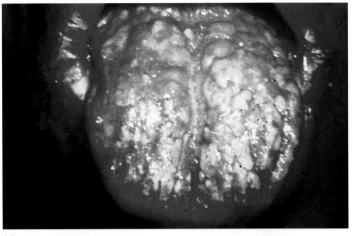

(a)

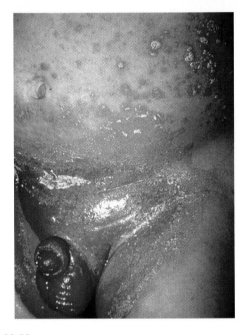

(b)

FIGURE 22.22

Common infections by *Candida albicans.* **(a)** Chronic thrush of the tongue on an adult male with diabetes. **(b)** Severe candidal diaper rash in an infant.

Laboratory Techniques in the Diagnosis of Candidiasis

A presumptive diagnosis of *Candida* infection is made if budding yeast cells and pseudohyphae are found in specimens from localized infections (figure 22.23*a*). Specimens are cultured on standard fungal media incubated at 30°C. Identification is complicated by the numerous species of *Candida* and other look-alike yeasts. Growth on a selective, differential medium containing trypan blue can easily differentiate *Candida* species from the yeast *Cryptococcus* (figure 22.23*b*). Confirmatory evidence of *C. albicans* can also be obtained by the germ tube test, the presence of chlamydospores, and multiple-panel systems that test for biochemical characteristics (figure 22.23*c*). A sensitive DNA amplification technique has been developed for identifying this species directly from clinical specimens.

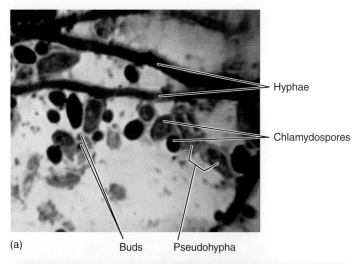

(a)

Hyphae

Chlamydospores

Buds   Pseudohypha

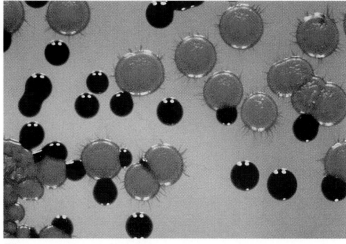

(b)

(c)

FIGURE 22.23

Detection of *Candida albicans.* **(a)** Gram stain of *Candida albicans* in a vaginal smear reveals gram-positive chlamydospores, pseudohyphae, and hyphae. In many cases of vaginal candidiasis, infection is detected during a routine Pap smear. **(b)** Colonies of *Candida albicans* appear pale blue on trypan medium, whereas *Cryptococcus* are dark blue. **(c)** Rapid yeast identification system using biochemical reactions to 12 test substances.

Treatment of Candidiasis

Because candidiasis is almost always opportunistic, it will return in many cases if the underlying disease is not also treated. Therapy for superficial mucocutaneous infection consists of topical antifungal agents (azoles and polyenes). A new class of antifungal drugs, terbinafine, has been approved for treating onychomycosis. Amphotericin B and fluconazole are usually effective in systemic infections. Recurrent bouts of vulvovaginitis are managed by topical azole drug ointments, now available as over-the-counter drugs. Vaginal candidiasis is sexually transmissible, thus it is very important to treat both sex partners to avoid reinfection.

CRYPTOCOCCOSIS* AND *CRYPTOCOCCUS NEOFORMANS*

Another widespread resident of human habitats is the fungus ***Cryptococcus neoformans.**** This yeast has a spherical to ovoid shape, with small, constricted buds and a large capsule that is important in

---

* **cryptococcosis** (krip″-toh-kok-oh´-sis) Older names are torulosis and European blastomycosis.

* *Cryptococcus neoformans* (krip″-toh-kok´-us nee″-oh-for´-manz) Gr. *kryptos*, hidden, *kokkos*, berry, *neo*, new, and *forma*, shape.

its pathogenesis (figure 22.24). Its role as an opportunist is supported by evidence that healthy humans have strong resistance to it and that frank infection occurs primarily in debilitated patients. Most cryptococcal infections *(cryptococcoses)* center around the respiratory, central nervous, and mucocutaneous systems.

Epidemiology of *C. neoformans*

The primary ecological niche of *C. neoformans* is associated with birds. It is prevalent in urban areas where pigeons congregate, and it proliferates in the high nitrogen content of droppings that accumulate on pigeon roosts. Masses of dried yeast cells are readily scattered into the air and dust. The species is sporadically isolated from dairy products, fruits, and the healthy human body (tonsils and skin), where it is considered a contaminant. The highest rates of cryptococcosis occur among patients with AIDS. It causes a severe form of meningitis that is frequently fatal. Other conditions that predispose to infection are steroid treatment, diabetes, and cancer. The disease exists throughout the world, with highest prevalence in the United States. Outbreaks have been reported in construction workers exposed to pigeon roosts, but it is not considered communicable among humans.

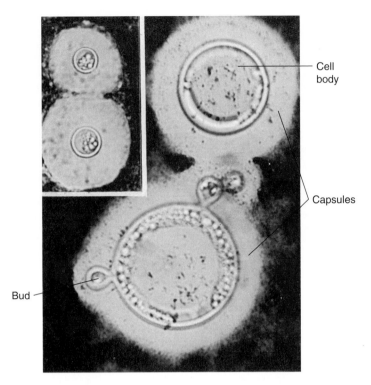

FIGURE 22.24

***Cryptococcus neoformans* from infected spinal fluid stained negatively with India ink.**   Halos around the large spherical yeast cells are thick capsules. Also note the buds forming on one cell. Encapsulation is a useful diagnostic sign for cryptococcosis, although the capsule is fragile and may not show up in some preparations (150×).

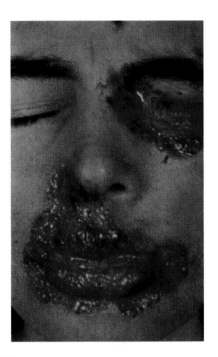

FIGURE 22.25

Cryptococcosis.   A late disseminated case of cutaneous cryptococcosis in which fungal growth produces a gelatinous exudate. The texture is due to the capsules surrounding the yeast cells.

Pathogenesis of Cryptococcosis

The primary portal of entry for *C. neoformans* is respiratory, and most lung infections are subclinical and rapidly resolved. A few patients with pulmonary cryptococcosis develop fever, cough, and nodules in their lungs. The escape of the yeasts into the blood is intensified by weakened host defenses and is attended by severe complications. Sites for which *Cryptococcus* shows an extreme affinity are the brain and meninges. The tumorlike masses formed in these locations can cause headache, mental changes, coma, paralysis, eye disturbances, and seizures. In some cases the infection disseminates into the skin, bones, and viscera (figure 22.25).

Diagnosis and Treatment of Cryptococcosis

The first step in diagnosis of cryptococcosis is negative staining of specimens to detect encapsulated budding yeast cells that do not occur as pseudohyphae. Isolated colonies can be used to perform screening tests that presumptively differentiate *C. neoformans* from the seven other cryptococcal species. Confirmatory results include a negative nitrate assimilation, pigmentation on birdseed agar, and fluorescent antibody tests. Cryptococcal antigen can be detected in a specimen by means of serological tests, and DNA probes can make a positive genetic identification. Systemic cryptococcosis requires immediate treatment with amphotericin B and fluconazole over a period of weeks or months.

PNEUMOCYSTIS CARINII AND PNEUMOCYSTIS PNEUMONIA

Although *Pneumocystis carinii* was discovered in 1909, it remained relatively obscure until it was suddenly propelled into clinical prominence as the agent of ***Pneumocystis carinii* pneumonia (PCP).** PCP is the most frequent opportunistic infection in AIDS patients, most of whom will develop one or more episodes during their lifetimes.

Because this unicellular microbe has characteristics of both protozoa and fungi, its taxonomic status has been somewhat uncertain. The sequence of its ribosomal RNA indicates that it is very similar genetically to the yeast *Saccharomyces*. It differs from most other fungi because it lacks ergosterol, has a weak cell wall, and is an obligate parasite.

Unlike most of the human fungal pathogens, little is known about the life cycle or epidemiology of *Pneumocystis*. Evidently the parasite is a common, relatively harmless resident of the upper respiratory tract. Contact with the agent is so widespread that in some populations a majority of people show serological evidence of infection. Until the AIDS epidemic, symptomatic infections by this organism were very rare, occurring only among the elderly or premature patients that were severely debilitated or malnourished.

Pneumocystis is probably spread in droplet form between humans. In people with intact immune defenses, it is usually held in check by lung phagocytes and lymphocytes, but in those with deficient immune systems, it multiplies intracellularly and extracellularly. The massive numbers of parasites adhere tenaciously to the lung pneumocytes and cause an inflammatory condition. The lung epithelial cells slough off, and a foamy exudate builds up. Symptoms are nonspecific and include cough, fever, shallow respiration, and cyanosis. Improved therapy for AIDS patients (see

Cyst, containing several bodies

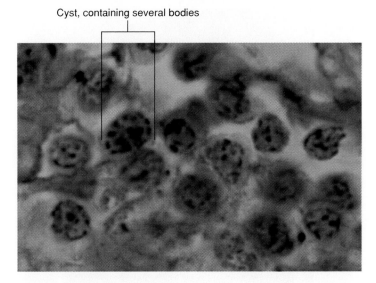

FIGURE 22.26

Pneumocystis carinii **cysts in the lung tissue (stained with methamine silver) of an AIDS patient (500×).**

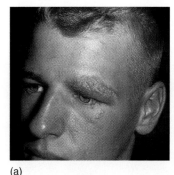
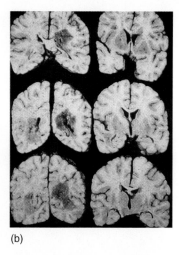

(a)   (b)

FIGURE 22.27

Clinical aspects of aspergillosis.   **(a)** Generalized conjunctival infection accompanied by intense and painful inflammation. (Long-term contact lens wearers must guard against this infection.) **(b)** Brain abscesses (darkened areas) consistent with disseminated disease.

chapter 25) has reduced the incidence of PCP and its mortality rate.

Because human strains of the species have not yet been successfully cultured, PCP must be diagnosed by symptoms and direct examination of lung secretions and tissue (figure 22.26). A DNA amplification probe can provide rapid confirmation of infection. Traditional serological tests for antigens and antibodies are not useful diagnostic tools for *Pneumocystis*.

Traditional antifungal drugs are ineffective against *Pneumocystis* pneumonia. The primary treatments are pentamidine and cotrimoxazole (sulfamethoxazole and trimethoprim) given over a 10-day period. This therapy should be applied even if disease appears mild or is only suspected. Pentamidine in an aerosol form is also given as a prophylactic measure in patients with a low T-cell count. The airways of patients in the active stage of infection often must be suctioned to reduce the symptoms, and additional oxygen is sometimes given.

ASPERGILLOSIS: DISEASES OF THE GENUS *ASPERGILLUS*

Molds of the genus *Aspergillus**** are possibly the most pervasive of all fungi. The estimated 600 species are widely dispersed in dust and air and can be isolated from vegetation, food, and compost heaps. Of the eight species involved in human diseases, the thermophilic fungus *A. fumigatus* accounts for the most cases. Aspergillosis is almost always an opportunistic infection, lately posing a serious threat to AIDS, leukemia, and organ transplant patients. It is also involved in allergies and toxicoses (Medical Microfile 22.3).

Aspergillus infections usually occur in the lungs. People breathing clouds of conidia from the air of granaries, barns, and silos are at greatest risk. In healthy people with extensive exposure, the spores simply germinate in the lungs and form *fungus balls*. Similar benign, noninvasive infections occur in colonization of the sinuses, ear canals, eyelids, and conjunctiva (figure 22.27*a*). In the

more invasive form of aspergillosis, the fungus produces a necrotic pneumonia and disseminates to the brain (figure 22.27*b*), heart, skin, and a wide range of other organs. Systemic aspergillosis usually occurs in very ill hospitalized patients with a poor prognosis.

An obvious branching, septate mycelium with characteristic conidial heads in a specimen is presumptive evidence of aspergillosis (figure 22.28). A culture of the specimen that produces abundant colonies and growth at 37°C will rule out its being only an air contaminant. A radioimmunoassay test for rapidly detecting antigen has been effective in diagnosing some cases. Noninvasive disease can be treated by surgical removal of the fungus tumor or by local drug therapy of the lung using amphotericin or nystatin. Combined therapy of amphotericin B and other antifungal agents is the only effective treatment for systemic disease.

MUCORMYCOSIS: ZYGOMYCETE INFECTION

The zygomycetes are extremely abundant saprobic fungi found in soil, water, organic debris, and food. Their large, prolific sporangia release multitudes of lightweight spores that literally powder the living quarters of humans and usually do little harm besides spoiling foods and rotting fruits and vegetables. But an increasing number of critically ill patients are contracting a disease called *mucormycosis.**** The genera most often involved in mucormycoses are *Rhizopus*, *Absidia*, and *Mucor*.

Underlying debilities that seem to increase the risk for mucormycosis are leukemia, burns, and malnutrition. Uncontrolled diabetes is a special risk because it can lead to reduced pH in the blood and tissue fluids that enhances invasion of the fungus. The fungus enters by way of the nose, lungs, skin, or oral cavity. In rhinocerebral mucormycosis, infection begins in the nasal passages or pharynx and then grows in the blood vessels and up through the palate into the eyes and brain (figure 22.29). Invasion of the lungs occurs most often in leukemia patients, and gastrointestinal

---

* *Aspergillus* (as″-per-jil′-us) L. *aspergere*, to scatter.

* **mucormycosis** (mew″-kor-my-koh′-sis) Any infection caused by members of the Order Mucorales. Also called phycomycosis and zygomycosis.

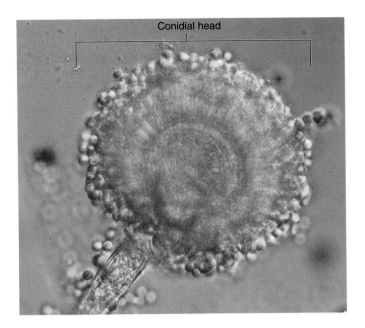

Conidial head

FIGURE 22.28

🔘 **The microscopic appearance of *Aspergillus* in specimens (1,000×).** A large, branched septate mycelium is the most common finding in tissues, but the characteristic conidial heads are also present in some specimens. These molds are not dimorphic and exist only in the hyphal stage.

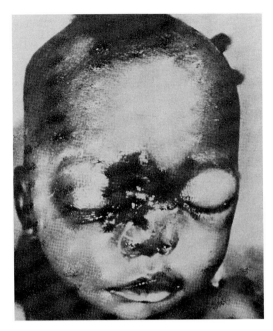

FIGURE 22.29

🔘 **Mucormycosis in a malnourished child.** This infection occurs in individuals so weakened that they usually have little chance for survival.

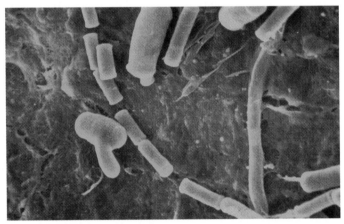

(a)

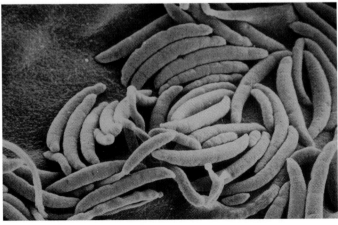

(b)

FIGURE 22.30

Common fungi that can cause uncommon infections. (a) *Geotrichum candidum* with arthrospores (2,000×). (b) *Fusarium* with crescent-shaped conidia (1,750×).

mucormycosis is a complication in children suffering from protein malnutrition (kwashiorkor). Unfortunately, the outlook for most of these patients is not hopeful.

Diagnosis of mucormycosis rests on clinical grounds and biopsies or smears. Very large, thick-walled, nonseptate hyphae scattered through the tissues of a gravely ill diabetic or leukemic patient are very suggestive of mucormycosis. Attempts to isolate the molds from specimens are frequently met with failure, and confusion of the isolates with air contaminants is also a problem. Treatment, which is effective only if commenced early in the infection, consists of surgical removal of infected areas accompanied by high doses of amphotericin B.

MISCELLANEOUS OPPORTUNISTS

It is increasingly clear that nearly any fungus can be implicated in infections when the host's immune defenses are severely compromised. Although it is still relatively unusual to become infected by common airborne fungi, these infections are currently on the rise.

Geotrichosis is a rare mycosis caused by *Geotrichum candidum* (figure 22.30*a*), a mold commonly found in soil, in dairy products, and on the human body. The mold is not highly aggressive and is primarily involved in secondary infection in the lungs of tubercular or highly immunosuppressed patients. Species of *Fusarium* (figure 22.30*b*), a common soil inhabitant and plant pathogen, occasionally infect the eyes, toenails, and burned skin. Mechanical introduction of the fungus into the eye (for example, by contaminated contact lenses) can induce a severe mycotic ulcer. It can also cause a patchy infection of the nail bed somewhat like tinea unguium. *Fusarium* colonization of patients with widespread burns has occasionally led to mortalities. Unusual opportunistic infections have been reported for other commonplace fungi such as

Penicillium, Malassezia, and the red yeast *Rhodotorula.* Even a rare case of systemic infection by the baker's yeast *Saccharomyces* was recently observed.

Fungal Allergies and Intoxications

It should be amply evident by now that fungi are constantly present in the air we breathe. In some people, inhaled fungal spores are totally without effect; in others, they instigate infection; and in still others, they stimulate hypersensitivity. So significant is the impact of fungal allergens that airborne spore counts are performed by many public health departments as a measure of potential risk (see figure 17.2). Among the significant fungal allergies are the following: (1) asthma, often occurring in seasonal episodes; (2) farmer's lung, a chronic and sometimes fatal allergy of agricultural workers exposed to moldy grasses; (3) teapicker's lung; (4) bagassosis, a condition caused by inhaling moldy dust from processed sugarcane debris; and (5) bark stripper's disease, caused by inhaling spores from logs. The role of fungi as toxic agents is discussed in an earlier chapter (see Spotlight on Microbiology 5.2) and in Medical Microfile 22.3.

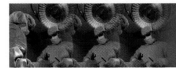

MEDICAL MICROFILE 22.3
Fungi, Food, and Toxins

Ingesting fungi (mycophagy) is a common and often pleasant experience. In some cases, one eats them by choice, as with gourmet mushrooms, and quite frequently, they are unintentionally consumed in dairy products such as yogurt and cottage cheese, fruits, vegetables, and many other foods.

Although most of the fungi we eat accidently are harmless, poisonous ones cause toxic illness. A number of mushrooms and molds contain mycotoxins that act on the central nervous system, liver, gastrointestinal tract, bone marrow, or kidney. The primary cause of **mycotoxicosis** is the consumption of wild mushrooms, a centuries-old practice that has taken the life of many an avid mushroom hunter. The deadliest of all mushrooms is *Amanita phalloides,* the so-called death-angel or destroying angel, whose cap contains enough toxin to kill an adult, though this happens rarely. Clearly, collecting and eating wild mushrooms is recommended only for the most expert mycophagist, and the rest of us should be content to gather mushrooms in the grocery store.

Certain poisonous fungi have been significant in history, and others are tied to ancient religious practices. Rye plants infested by *Claviceps purpurea* develop black masses, known as ergot, on the grains. When such rye grains are inadvertently incorporated into food and ingested, absorption of the ergot alkaloids causes *ergotism.* People thus intoxicated exhibit convulsions, delusions, a burning sensation in the hands (St. Anthony's fire), and symptoms similar to LSD (lysergic acid diethylamide) poisoning. This latter symptom occurs because the alkaloids in *Claviceps* and other mind-altering fungi contain the active ingredient lysergic acid. In the Middle Ages, millions of people died from ergotism, but now the disease is very rare.

Mycotoxicosis is also caused by food contaminated by fungal toxins. One of the most potent of these toxins is *aflatoxin,** a product of *Aspergillus flavus* growing on grains, corn, and peanuts. Although the mold itself is relatively innocuous, the toxin is carcinogenic and hepatotoxic. It is especially lethal to turkeys, ducklings, pigs, and calves that have eaten contaminated feed. In humans, it appears to be a cofactor in liver cancer. The potential hazard aflatoxin poses for domesticated animals and humans necessitates constant monitoring of peanuts, beer, grains, nuts, vegetable oils, animal feed, and even milk. A quick exposure of seeds to ultraviolet radiation is a useful screening test because they glow when contaminated by the toxin.

Toxic fungi are a growing source of environmental acquired illnesses associated with contaminated buildings. *Stachybotrys* spores and toxins inhaled from air are a known cause of serious neurologic disease. Other airborne fungi cause asthma and related respiratory allergies.

---

* aflatoxin (af″-lah-toks′-in) Acronym for *Aspergillus flavus* toxin.

CHAPTER CHECKPOINTS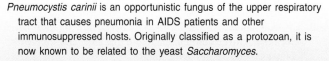

Candida albicans is the causative agent of a wide variety of opportunistic infections. These range from minor skin mycoses to fatal systemic diseases. *Candida* infections are most likely to occur in hosts with lowered resistance, trauma, or invasive devices or those receiving prolonged antibiotic therapy for bacterial infections.

Cryptococcus neoformans is the causative agent of cryptococcosis, a non-communicable mycosis that infects the respiratory, mucocutaneous, and nervous systems. Cryptococcus is commonly found in bird droppings and is spread through air and dust. Those at greatest risk for serious disease are AIDS patients and other immunocompromised hosts.

Aspergillosis is a fungal infection caused by the spores of *Aspergillus* species. Particularly high concentrations of spores occur in dust from granaries, barns, and silos. Inhaled spores germinate in the lung to form localized "fungus balls." Susceptible hosts can develop systemic infections that have a very poor prognosis.

Pneumocystis carinii is an opportunistic fungus of the upper respiratory tract that causes pneumonia in AIDS patients and other immunosuppressed hosts. Originally classified as a protozoan, it is now known to be related to the yeast *Saccharomyces*.

Mucormycosis is caused by saprobic fungi that colonize debilitated hosts with acidosis. Mucormycosis invades through several portals of entry and progresses rapidly to a systemic infection if not treated early.

The genus *Geotrichum* is involved in rare infections of the lungs, eyes, nail bed, and burned skin.

In addition to sometimes being agents of disease, fungi act as allergens and can produce toxins. Many of the allergies are related to agricultural occupations, but some are also common among the general population. *Claviceps* and *Aspergillus* species produce toxins that are hallucinogenic, hepatotoxic, or carcinogenic.

CHAPTER CAPSULE WITH KEY TERMS

I. Fungal Infections: General Principles

A. Microscopic fungi (molds and yeasts) are widespread in nature. A few members cause fungal diseases or **mycoses.**

B. Infectious fungi occur in groups based upon the virulence of the pathogen and the level of involvement, whether **systemic, subcutaneous, cutaneous,** or **superficial.**

C. **True** or **primary pathogens** have virulence factors that allow them to invade and grow in a healthy host.

   1. They are also **thermally dimorphic,** occurring as hyphae in their natural habitat and converting to yeasts while growing as parasites at body temperature (37°C).

   2. Inhaled spores initiate **primary pulmonary infection (PPI),** which can spread to the skin and other systems.

   3. Diseases are not transmissible, and cause a long-term allergic reaction to fungal proteins.

   4. True pathogens are endemic to certain geographic areas.

D. **Opportunistic fungal pathogens** are normal flora or common inhabitants of the environment that invade mainly patients whose host defenses are compromised.

   1. They are only weakly virulent and lack thermal dimorphism.

   2. Infections may be local, cutaneous, or systemic, and are currently on the rise, due to more weakened patients.

E. Some pathogens such as the **dermatophytes** (skin fungi) can infect healthy people, are not highly invasive, and are transmissible.

F. Immunity to fungal infections consists primarily of nonspecific barriers, inflammation, and cell-mediated defenses. Antibodies may be used to detect disease in some cases.

G. Diagnosis and identification require microscopic examination of stained specimens, culturing of pathogen in selective and enriched media, and specific biochemical and *in vitro* serological tests. Skin testing for true and opportunistic pathogens can determine prior disease but is not useful in diagnosis.

H. Control of fungal infections involves drugs such as intravenous amphotericin B, flucytosine, azoles (fluconazole), and nystatin. No useful vaccines exist.

II. Systemic Mycoses Caused by True Pathogens

A. Histoplasmosis (Ohio Valley fever): The agent is *Histoplasma capsulatum,* distributed worldwide, but most prevalent in eastern and central regions of the United States.

   1. Infection is related to the disturbance of soil and animal excreta.

   2. Inhaled conidia produce primary pulmonary infection that may progress to systemic involvement of a variety of organs and chronic lung disease.

B. Coccidioidomycosis (Valley fever): *Coccidioides immitis* has blocklike **arthroconidia** in the free-living stage and **spherules** containing endospores in the lungs.

   1. The agent lives in alkaline soils in semiarid, hot climates, and is endemic to the southwestern United States.

   2. Infection occurs when arthrospores are inhaled from dust. Growth in the lungs creates spherules and nodules; disseminated infection is debilitating.

C. Blastomycosis (Chicago disease): The agent is *Blastomyces dermatitidis,* a free-living species distributed in soil of a large section of the midwestern and southeastern United States.

   1. Inhaled conidia convert to yeasts and multiply in lungs.

   2. The inflammatory response produces cough and fever.

Disease may progress to chronic cutaneous, bone, and nervous system complications.

D. Paracoccidioidomycosis (South American blastomycosis): *Paracoccidioides brasiliensis* is distributed in various regions of Central and South America. Lung infection occurs through inhalation or inoculation of spores, and systemic disease is not common.

III. Subcutaneous Mycoses: Diseases of Tissues in or Below the Skin

A. Sporotrichosis (rose-gardener's disease) is caused by *Sporothrix schenckii,* a free-living fungus that accidently infects the appendages and lungs. The **lymphocutaneous** variety occurs when contaminated plant matter penetrates the skin and the pathogen forms a local nodule, then spreads to nearby lymph nodes.

B. Chromoblastomycosis is a deforming, deep infection of the tissues of the legs and feet by soil fungi that enter a traumatic injury.

C. Mycetoma is a progressive, tumorlike disease of the hand or foot due to chronic fungal infection; may lead to loss of the body part.

IV. Cutaneous Mycoses: Infections Confined to the Skin

A. Dermatophytoses are known as **tinea,** ringworm, and athlete's foot. They are caused by species in the genera *Trichophyton, Microsporum,* and *Epidermophyton,* all of which are adapted to keratinized epidermis (skin, hair, nails).

B. Infections are communicable among humans, animals, and soil; infection is facilitated by moist, chafed skin; inflammatory reactions to fungi cause itching and pain.

   1. **Ringworm of scalp** (tinea capitis) affects scalp and hair-bearing regions of head; hair may be lost. **Ringworm of body** (tinea corporis) occurs as inflamed, red ring lesions anywhere on smooth skin. **Ringworm of groin** (tinea cruris) affects groin and scrotal regions.

   2. **Ringworm of foot and hand** (tinea pedis and tinea manuum) is spread by exposure to public surfaces; occurs between digits and on soles. **Ringworm of nails** (tinea unguium) is a persistent colonization of the nails of the hands and feet that distorts the nail bed.

C. Superficial mycoses affect the outermost epidermal structures. **Tinea versicolor** causes mild scaling, mottling of skin; **white piedra** is whitish or colored masses on the long hairs of the body; and **black piedra** causes dark, hard concretions on scalp hairs.

V. Opportunistic Mycoses

A. Candidiasis is caused by *Candida albicans* and other *Candida* species, common yeasts that normally reside in the mouth, vagina, intestine, and skin.

B. Infection predominates in cases of lowered resistance (babies, pregnancy, drug therapy, AIDS) and arises from normal flora or is transmissible through intimate contact.

   1. **Thrush** occurs as a thick, white, adherent growth on the mucous membranes of mouth and throat.

   2. **Vulvovaginal yeast infection** is a painful inflammatory condition of the female genital region that causes ulceration and whitish discharge.

   3. **Cutaneous candidiasis** occurs in chronically moist areas of skin and in burn patients.

C. Cryptococcosis is caused by *Cryptococcus neoformans,* a widespread encapsulated yeast that inhabits soils around pigeon roosts; common infection of AIDS, cancer, or diabetes patients.

1. Infection of lungs leads to cough, fever, and lung nodules.
2. Dissemination to meninges and brain can cause severe neurological disturbance and death.

D. *Pneumocystis carinii* is a small, unicellular fungus that causes PCP, the most prominent opportunistic infection in AIDS patients. This form of pneumonia forms secretions in the lungs that block breathing and can be rapidly fatal if not controlled with medication.

E. **Aspergillosis** is caused by *Aspergillus*, a very common airborne soil fungus. Inhalation of clouds of spores causes fungus balls in lungs, and it can also cause invasive disease in the eyes, heart, and brain.

F. **Mucormycosis** is caused by molds in genus *Rhizopus* and *Mucor*. These usually harmless air contaminants invade the membranes of the nose, eyes, and brain of people with acidotic (diabetes, malnutrition), with severe consequences.

VI. **Fungal Allergies and Mycotoxicoses**

Airborne fungal spores are common sources of atopic allergies. Contact with fungal toxins leads to **mycotoxicoses.** Most cases are caused by eating poisonous or hallucinogenic mushrooms or food.

MULTIPLE-CHOICE QUESTIONS

1. The ability of a fungus to alternate between hyphal and yeast phases in response to temperature is called
   a. sporulation
   b. conversion
   c. dimorphism
   d. binary fission

2. Which phase of a fungal life cycle is best adapted to growing in a host's body?
   a. yeast
   b. hyphal
   c. conidial
   d. filamentous

3. Primary pathogenic fungi differ from opportunistic fungi in being
   a. more virulent
   b. less virulent
   c. more prevalent
   d. more transmissible

4. True pathogenic fungi
   a. are transmissible from person to person
   b. cause primary pulmonary infections
   c. are endemic to specific geographic areas
   d. both b and c

5. Example(s) of fungal infections that is/are communicable include
   a. dermatomycosis
   b. candidiasis
   c. sporotrichosis
   d. both a and b

6. Histoplasmosis has the greatest endemic occurrence in which region?
   a. Midwest U.S.
   b. Southwest U.S.
   c. Central and South America
   d. North Africa

7. Coccidioidomycosis is endemic to which geographic region in the U.S.?
   a. Southwestern
   b. Northeastern
   c. Pacific Northwest
   d. Southeastern

8. Skin testing with antigen is a useful diagnostic procedure for
   a. histoplasmosis
   b. candidiasis
   c. coccidioidomycosis
   d. blastomycosis

9. A mycetoma is a
   a. fungal tumor of the lung
   b. collection of pus in the brain caused by *Cryptococcus*
   c. deeply invasive fungal infection of the foot or hand
   d. type of ringworm of the scalp

10. Dermatophytic fungi attack the _____ in the _____.
    a. epithelial cells, lungs
    b. melanin, stratum corneum
    c. keratin; skin, nails, hair
    d. bone, foot

11. *Candida albicans* is the cause of _____ , an infection of the _____.
    a. onychomycosis, root canal
    b. moniliasis, lungs
    c. thrush, mouth
    d. salpingitis, uterus

12. Cryptococcosis is associated with
    a. blowing winds
    b. plant debris
    c. pigeon droppings
    d. excavation of ancient dwellings

13. Which fungus does *not* commonly cause systemic infection?
    a. *Blastomyces*
    b. *Histoplasma*
    c. *Cryptococcus*
    d. *Malassezia*

14. Which of the following is the cause of mycotoxicosis?
    a. *Mucor*
    b. *Aspergillus*
    c. *Pneumocystis*
    d. *Cryptococcus*

15. **Single Matching.** Match the disease with its common name:

    _____ coccidioidomycosis
    _____ histoplasmosis
    _____ blastomycosis
    _____ sporotrichosis
    _____ paracoccidioidomycosis
    _____ dermatophytosis
    _____ candidiasis

    a. tinea
    b. San Joaquin Valley fever
    c. South American blastomycosis
    d. Chicago disease
    e. Rose-gardener's disease
    f. yeast infection
    g. Ohio Valley fever

CONCEPT QUESTIONS

1. a. Explain how true pathogens differ from opportunists in physiology, virulence, types of infection, and distribution.
   b. Give examples of each.
   c. Explain the adaptation and importance of thermal dimorphism.

2. a. Why are most fungi considered facultative parasites and fungal infections considered non-communicable?
   b. Give examples of two communicable mycoses.

3. a. What factors are involved in the pathogenesis of fungi?
   b. What parts of fungi are infectious?
   c. What are the primary defenses of humans against fungi?

4. a. What is the role of skin testing in tracing fungal infections?
   b. What causes cross-reactions?
   c. For which groups is it most useful?

5. a. Give an overview of the major steps in fungal identification.
   b. What is one advantage over bacterial identification?
   c. How are *in vitro* and *in vivo* tests different?
   d. Describe an immunodiffusion test; the histoplasmin test.
   e. What biochemical and genetic tests can be conducted in identification?

6. a. What are the general principles involved in treating fungal diseases with drugs?
   b. Why is amphotericin B so toxic?
   c. Name some topical medications and their uses.

7. a. Differentiate between systemic, subcutaneous, cutaneous, and superficial infections.
   b. What is the initial infection site in true pathogens?
   c. What does it mean if an infection disseminates?

8. a. Briefly outline the life cycles of *Histoplasma capsulatum, Coccidioides immitis,* and *Blastomyces dermatitidis.*
   b. In general, what is the source of each pathogen?
   c. Is disease usually overt or subclinical?
   d. To which organs do these pathogens disseminate?
   e. What causes cutaneous disease? How are cutaneous diseases treated?

9. a. What are the most common subcutaneous infections?
   b. Why are these organisms not more invasive?
   c. Describe the life cycle of *Sporothrix* and the disease it causes.
   d. How could one avoid it? Differentiate between chromoblastomycosis and phaeohyphomycosis.

10. a. What are the dermatophytoses?
    b. What is meant by the term *keratophile*?
    c. What do the fungi actually feed upon?
    d. To what do the terms *ringworm* and *tinea* refer?
    e. What are the three main genera of dermatophytes, and what are their reservoirs?

11. a. What are the five major types of ringworm, their common names, symptoms, and epidemiology?
    b. What conditions predispose an individual to these infections?

12. a. What is the difficulty in ridding the epidermal tissues of certain dermatophytes?
    b. What is necessary in treatment?

13. Briefly describe three superficial mycoses.

14. a. What is the relationship of *Candida albicans* to humans?
    b. What medical or other conditions can predispose a person to candidiasis?
    c. Describe the principal infections of women, adults of both sexes, and neonates.
    d. Why should women be screened for candidiasis?
    e. What are the treatments for localized and systemic candidiasis?
    f. If a woman tests positive for infection, besides treating her, what else should be done?

15. a. Describe the pathology of cryptococcosis.
    b. What specimens could be used to diagnose it?
    c. Which group of people is currently most vulnerable to this disease?
    d. What makes the fungus somewhat unique?

16. a. Briefly list the major diseases caused by *Aspergillus* and zygomycetes.
    b. Why are fungus balls not an infection?

17. Briefly outline the epidemiology and pathology of *Pneumocystis carinii.*

CRITICAL-THINKING QUESTIONS

1. Give reasons most humans are not constantly battling fungal infections, considering how prevalent fungi are in the environment.

2. a. Name several medical conditions that compromise the immunities of patients.
   b. Can you think of some solutions to the problems of opportunistic infections in extremely compromised patients?
   c. Is there any such thing as a harmless contaminant with these patients?
   d. Can medical personnel do anything to lessen their patients' exposure to fungi?

3. Describe the consequences if a virulent fungal pathogen infects a severely compromised patient.

4. Using the general guidelines presented in the "Introduction to Identification Techniques in Medical Microbiology," page 535, how can you tell for sure if a fungal isolate is the actual causative agent or merely a contaminant?

5. a. Observe the reactions in figure 22.10 and describe what is happening at the molecular level to make the lines of precipitation in the immunodiffusion tests.
   b. Make a drawing to show this.
   c. What is happening in the wells without the lines?

6. a. Explain why some dermatophytes are considered good parasites.
   b. Why would infections from soil and animal species be more virulent than ones from fungi adapted to humans?

7. What conditions predispose diabetics to fungal infections?

8. a. What is the function of mushroom and other fungal toxins?
   b. How are these toxins harmful, and how are they used to our advantage?

9. a. Explain how *Pneumocystis carinii,* once classified with the protozoan parasites, came to be considered a fungus.
   b. Describe the characteristics that make it protozoan-like and those that are funguslike.
   c. What was the definitive reason for changing its taxonomic position?

10. Case study 1. Mr. Jones is admitted to a hospital with severe chest pain, cough, and fever. The lab tests show tiny cocci-like cells in sputum, and he is treated for bacterial pneumonia and released. However, Mr. Jones does not respond to penicillin or cephalosporin, and the lung condition worsens. After a few days, skin bumps begin to appear on the face and chest. A lung biopsy indicates large sporangia filled with tiny spores, and the patient's history reveals a recent trip to Arizona.
   a. On this basis, how would you diagnose the infection?
   b. What other fungal disease might it suggest?
   c. Which drug should have been given?

11. Case study 2. A male researcher wanting to determine the pathogenesis of a certain fungus that often attacks women did the following experiment on himself. He taped a piece of gauze heavily inoculated with the fungus to the inside of his thigh and left it there for several days, periodically moistening it with sterile saline. After a time, he experienced a severe skin reaction that resembled a burn and the skin peeled off.
   a. Suggest a species of fungus that can cause this effect.
   b. What does this experiment tell you about the factors in pathogenesis related to environment and gender?

12. Case study 3. A worker involved in demolishing an older downtown building is the victim of a severe lung infection.
   a. Without any lab data, what two fungal infections do you suspect?
   b. What is your reasoning?

INTERNET SEARCH TOPICS

1. Look up newer antifungal drugs such as fluconazole and terbinafine, and determine their modes of action and therapeutic uses.

2. Find websites that give information on common fungal infections of AIDS patients and the effects they have on health and survival.

The Parasites of Medical Importance

Parasitology is traditionally the study of eucaryotic parasites. It encompasses protozoa (unicellular animals) and helminths (worms), which are sometimes collectively called the macroparasites, as opposed to the microparasites (bacteria and viruses). This inherently fascinating and practical subject will be presented as a compact survey of the major clinically important protozoa and helminths. This chapter covers elements of their morphology, life cycles, epidemiology, pathogenic mechanisms, and methods of control. Arthropods, which are often included in the field of parasitology, have been previously discussed as vectors for certain viral and bacterial infections in chapters 13, 20, and 21. Other chapters containing information on parasites are chapter 5 (eucaryotes), chapter 7 (nutrition), and chapter 12 (chemotherapy).

Chapter Overview

- Parasitology is the study of infections and diseases caused by protozoans and helminth worms.
- Protozoa are small, usually single-celled amebas, ciliates, flagellates, or apicomplexans that may have complex life cycles with several morphological forms.
- Protozoan diseases are spread by food, by contact with other humans and animals, or by arthropod vectors.
- Protozoan diseases can be managed by drugs, proper hygiene, and vector control.
- Diseases caused by infectious amebas include amebic dysentery, and encephalitis. Ciliates are responsible for balantidiosis.
- Flagellate diseases include trichomoniasis, giardiasis, trypanosomiasis, and leishmaniasis.
- The major apicomplexan diseases are malaria, toxoplasmosis, and cryptosporidiosis.
- Helminth parasites are multicellular worms with specialized adaptations in structure and life cycle for feeding, reproduction, and host transmission.
- Diseases are spread by eggs and larvae that are eaten or burrow into the skin. Some involve transmission by vector hosts.
- Helminth worms include nematodes (roundworms), trematodes (flukes), and cestodes (tapeworms).
- Intestinal nematode diseases include ascariasis, hookworm, strongyloidiasis, and trichinosis.
- Tissue nematode infections are vector-borne filariasis and onchocerciasis.
- Flukes are agents of schistosomiasis and zoonotic liver diseases.

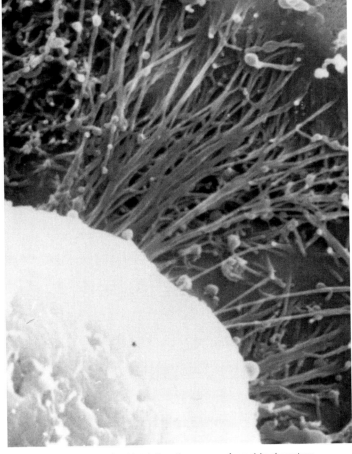

Trophozoite of *Entamoeba histolytica*, the cause of amebic dysentery, displays a fringe of very fine pseudopods it uses to invade and feed on tissue.

- Human tapeworms are acquired from eating poorly cooked beef and pork.
- Control of helminth infections is achieved by administering drugs that paralyze or kill the worms and by interrupting their life cycles.

The Parasites of Humans

Because of their larger size and greater visibility, parasites and their diseases have been known since ancient times. Even so, the science of parasitology is relatively young and still in the early stages of definition at the molecular level. One cannot underestimate the impact of parasites on the everyday lives of humans. The World Health Organization estimates that parasites cause nearly 20 percent of all infectious diseases. Although they are less prevalent in industrialized countries, they affect most humans in both hidden and overt forms at some time in their lives. The distribution of parasitic diseases is influenced by many factors, including rapid travel, immigration, and the increased number of immunocompromised patients. Because parasites can literally travel around the world, it is not uncommon to discover exotic parasitic diseases even in the United States.

Humans play host to dozens of different parasites. A few are indigenous to humans alone and are spread to other humans through direct contact or contaminated food and water. A considerable number are transmitted among human hosts by arthropod vectors. The remainder are zoonoses that humans acquire from contact with pets or other animals.

The current increase in patients with severe immune suppression is changing the epidemiologic pattern of parasitic diseases. AIDS patients in particular play host to a variety of opportunists that dominate their clinical picture. More and more, protozoa that were once rare and obscure pathogens are emerging as the agents of life-threatening diseases.

Typical Protozoan Pathogens

When we first met the protozoa in chapter 5, they were described as single-celled, animal-like microbes usually having some form of motility. The four commonly recognized groups are sarcodinians (amebas), flagellates, ciliates, and apicomplexans. Although the life cycles of pathogenic protozoa can vary a great deal, most of them propagate by simple asexual cell division of the active feeding cell, called a **trophozoite.** Many species undergo formation of a **cyst** that can survive for periods outside the host and can also function as an infective body (see figure 5.27). Others have a more complex life cycle that includes asexual and sexual phases (see later discussion of apicomplexans). The protozoan parasites and diseases to be surveyed here are listed in table 23.1.

INFECTIVE AMEBAS

Entamoeba histolytica and Amebiasis

Amebas are widely distributed in aqueous habitats and are frequent parasites of animals, but only a small number of them have the necessary virulence to invade tissues and cause serious pathology. The most significant pathogenic ameba is ***Entamoeba histolytica.**** The relatively simple life cycle of this parasite alternates between a large trophozoite that is motile by means of pseudopods and a smaller, compact, nonmotile cyst (figure 23.1). The trophozoite lacks most of the organelles of other eucaryotes, and it has a large single nucleus that contains a prominent nucleolus called a *karyosome.* Amebas from fresh specimens are often packed with food vacuoles containing host cells and bacteria. The mature cyst is encased in a thin, yet tough wall and contains four nuclei as well as

---

* *Entamoeba histolytica* (en″-tah-mee′-bah his″-toh-lit′-ih-kuh) Gr. *ento,* within, *histos,* tissue, and *lysis,* a dissolution.

TABLE 23.1	
Major Pathogenic Protozoa, Infections, and Primary Sources	
Protozoan/Disease	**Reservoir/Source**
Ameboid Protozoa	
Amebiasis: *Entamoeba histolytica*	Human/water and food
Brain infection: *Naegleria, Acanthamoeba*	Free-living in water
Ciliated Protozoa	
Balantidiosis: *Balantidium coli*	Zoonotic in pigs
Flagellated Protozoa	
Giardiasis: *Giardia lamblia*	Zoonotic/water and food
Trichomoniasis: *Trichomonas tenax, T. hominis, T. vaginalis*	Human
Hemoflagellates	
Trypanosomiasis: *Trypanosoma brucei, T. cruzi*	Zoonotic/vector-borne
Leishmaniasis: *Leishmania donovani, L. tropica, L. brasiliensis*	Zoonotic/vector-borne
Apicomplexan Protozoa	
Malaria: *Plasmodium vivax, P. falciparum, P. malariae*	Human/vector-borne
Toxoplasmosis: *Toxoplasma gondii*	Zoonotic/vector-borne
Cryptosporidiosis: *Cryptosporidium*	Free-living/water, food
Isosporosis: *Isospora belli*	Dogs, other mammals
Cyclosporiasis: *Cyclospora cayetanensis*	Water/fresh produce

distinctive bodies called *chromatoidals,* which are actually dense clusters of ribosomes.

Epidemiology of Amebiasis   Humans are the primary hosts of *E. histolytica.* Infection is usually acquired by ingesting food or drink contaminated with cysts released by an asymptomatic carrier. The ameba is thought to be carried in the intestine of one-tenth of the world's population, and it kills up to 100,000 people a year. Its geographic distribution is partly due to local sewage disposal and fertilization practices. Occurrence is highest in tropical regions (Africa, Asia, and Latin America), where night soil[1] or untreated sewage is used to fertilize crops and sanitation of water and food can be substandard. Although the prevalence of the disease is lower in the United States, as many as 10 million people could harbor the agent.

Epidemics of amebiasis are infrequent but have been documented in prisons, hospitals, juvenile care institutions, and communities where water supplies are polluted. Amebic infections can also be transmitted by oral-anal sexual contact among homosexual men.

Life Cycle, Pathogenesis, and Control of *E. histolytica*   Amebiasis begins when viable cysts are swallowed and arrive in the small intestine (see figure 5.33) where the alkaline pH and digestive juices of this environment stimulate excystment. The cyst releases

---

1. Human excrement used as fertilizer.

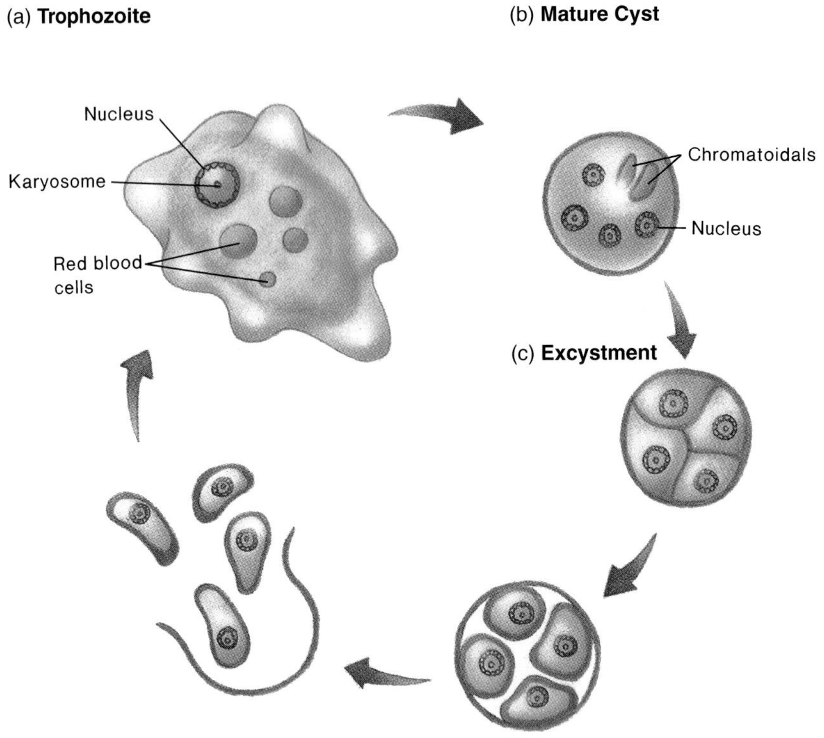

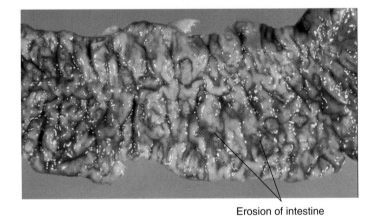

(a) **Trophozoite**

Nucleus

Karyosome

Red blood cells

(b) **Mature Cyst**

Chromatoidals

Nucleus

(c) **Excystment**

FIGURE 23.1

Cellular forms of *Entamoeba histolytica.* **(a)** A trophozoite containing a single nucleus, a karyosome, and red blood cells. **(b)** A mature cyst with four nuclei and two blocky chromatoidals. **(c)** Stages in excystment. Divisions in the cyst create four separate cells, or metacysts, that differentiate into trophozoites and are released.

four trophozoites (figure 23.1*c*), which are swept into the *cecum** and large intestine. There, the trophozoites attach by fine pseudopods (see chapter opening photo), multiply, actively move about, and feed. In about 90% of patients, infection is asymptomatic or very mild, and the trophozoites do not invade beyond the most superficial layer. The severity of the infection can vary with the strain of the parasite, inoculation size, diet, and host resistance.

As hinted by its species name, tissue damage is one of the formidable characteristics of untreated *E. histolytica* infection. Clinical amebiasis exists in intestinal and extraintestinal forms. The initial targets of intestinal amebiasis are the cecum, appendix, colon, and rectum. The ameba secretes enzymes that dissolve tissues, and it actively penetrates deeper layers of the mucosa, leaving erosive ulcerations (figure 23.2). This phase is marked by dysentery (bloody, mucus-filled stools), abdominal pain, fever, diarrhea, and weight loss. The most life-threatening manifestations of intestinal infection are hemorrhage, perforation, appendicitis, and tumorlike growths called amebomas.

Extraintestinal infection occurs when amebas invade the viscera of the peritoneal cavity. The most common site of invasion is the liver. Here, abscesses containing necrotic tissue and trophozoites develop and cause amebic hepatitis. Another regular complication is pulmonary amebiasis. Less frequent targets of infection are the spleen, adrenals, kidney, skin, and brain. Severe forms of the disease result in about a 10% fatality rate.

Entamoeba is harbored by chronic carriers whose intestines favor the encystment stage of the life cycle. Cyst formation cannot occur in active dysentery because the feces are so rapidly flushed from the body, but after recuperation, cysts are continuously shed in feces.

Erosion of intestine

FIGURE 23.2

Intestinal amebiasis and dysentery of the cecum. Red patches are sites of amebic damage to the intestinal mucosa.

Amebic dysentery has symptoms and signs common to other forms of dysentery, especially bacillary dysentery. Initial diagnosis involves examination of a fecal smear for trophozoites or cysts under high magnification and differentiating them from nonpathogenic intestinal amebas that are part of the normal flora (figure 23.3). More definitive identification can be obtained with a stained smear or with a rapid ELISA test using specific monoclonal antibodies. It may be necessary to supplement laboratory findings with clinical data, including symptoms, X rays, and the patient's history, which might indicate possible geographic, occupational, or sexual exposure.

Effective treatment for amebic dysentery usually involves the use of drugs such as iodoquinol that act in the feces, and metronidazole, dehydroemetine, or chloroquine, which work in the tissues. Dehydroemetine is used to control symptoms but will not cure the

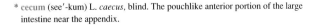

* cecum (see′-kum) L. *caecus,* blind. The pouchlike anterior portion of the large intestine near the appendix.

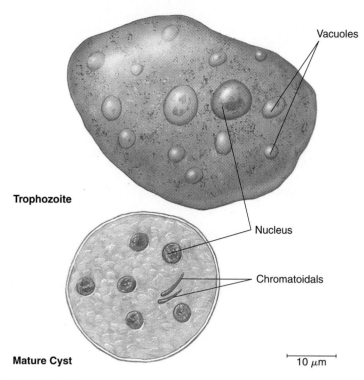

Vacuoles

Trophozoite

Nucleus

Chromatoidals

Mature Cyst

10 μm

FIGURE 23.3

Entamoeba coli. This nonpathogenic intestinal ameba may be mistaken for *E. histolytica*, but it is larger and less motile, with short, blunt pseudopodia. Its cyst has spiny chromatoidals and from one to eight nuclei. Another close relative, *E. dispar*, is also commonly confused with the pathogen.

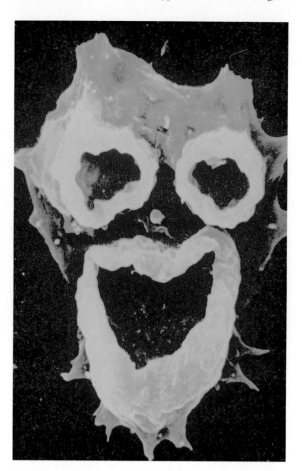

FIGURE 23.4

Scanning electron micrograph of *Naegleria fowleri*. The "eyes" and "mouth" of its face-like appearance are its attachment and feeding structures.

disease. Other drugs are given to relieve diarrhea and cramps, while lost fluid and electrolytes are replaced by oral or intravenous therapy. Infection with *E. histolytica* provokes antibody formation against several antigens, but permanent immunity is unlikely and reinfection can occur. Prevention depends upon concerted efforts similar to those for other enteric diseases (see Microbits 20.4). Because regular chlorination of water supplies does not kill cysts, more rigorous methods such as boiling or iodine are required. Several candidates for an amebiasis vaccine are in development. One is based on a combination of three antigens made by recombinant technology and delivered by the oral route.

Amebic Infections of the Brain

Two common free-living protozoans that cause a rare and usually fatal infection of the brain are ***Naegleria fowleri**** and ***Acanthamoeba.**** Both genera are accidental parasites that invade the body only under unusual circumstances. The trophozoite of *Naegleria* is a small, flask-shaped ameba that moves by means of a single, broad pseudopod and has prominent feeding structures, or amebostomes, that lend it a facelike appearance in scanning electron micrographs (figure 23.4). It forms a rounded, thick-walled, uninucleate cyst that is resistant to temperature extremes and mild chlorination. *Acanthamoeba* has a large, ameboid trophozoite with spiny pseudopods and a double-walled cyst.

Both *N. fowleri* and *Acanthamoeba* ordinarily inhabit standing fresh or brackish water, lakes, puddles, ponds, hot springs, moist soil, and even swimming pools and hot tubs. They are especially abundant in warm water with a high bacterial count.

Most cases of *Naegleria* infection reported worldwide occur in people who have been swimming in warm, natural bodies of fresh water. One epidemic in Australia was due to contamination of a public water supply, and in Belgium, an outbreak was reported among people who had bathed in polluted canal water. Infection can begin when amebas are forced into human nasal passages as a result of swimming, diving, or other aquatic activities. Once the ameba is inoculated into the favorable habitat of the nasal mucosa, it burrows in, multiplies, and subsequently migrates into the brain and surrounding structures. The result is **primary acute meningoencephalitis,** a rapid, massive destruction of brain and spinal tissue that causes hemorrhage and coma and invariably ends in death within a week or so.

Unfortunately, *Naegleria* meningoencephalitis advances so rapidly that treatment usually proves futile. Studies have indicated that early therapy with amphotericin B, sulfadiazine, or tetracycline in some combination can be of some benefit. Because of the wide distribution of the ameba and its hardiness, no general means of control exists. Public swimming pools and baths must be adequately chlorinated and checked periodically for the ameba.

---

* *Naegleria fowleri* (nay-glee′-ree-uh fow′-ler-eye) After F. Nagler and N. Fowler.

* *Acanthamoeba* (ah-kan″-thah-mee′-bah) Gr. *acanthos*, thorn.

Acanthamoeba differs from *Naegleria* in its portal of entry, invading broken skin, the conjunctiva, and occasionally the lungs and urogenital epithelia. Although it causes a meningoencephalitis somewhat similar to that of *Naegleria,* the course of infection is lengthier. At special risk for infection are people with traumatic eye injuries, contact lens wearers, and AIDS patients exposed to contaminated water. Ocular infections can be avoided by carefully tending to injured eyes and using sterile solutions to store and clean contact lenses. Cutaneous and central nervous system infections are complications in AIDS.

AN INTESTINAL CILIATE: *BALANTIDIUM COLI*

Most ciliates are free-living in various aquatic habitats, where they assume a role in the food web as scavengers or as predators upon other microbes. Several species are pathogenic to animals, but only one, *Balantidium* coli, occasionally establishes infection in humans. This giant ciliate (about the size of a period on this page) exists in both trophozoite and cyst states (figure 23.5). Its surface is covered by orderly, oblique rows of cilia, and a shallow depression, the cytostome (cell mouth), is at one end. These parasites will grow only under anaerobic conditions in the presence of the host's bacterial flora.

The large intestine of pigs and, to a lesser extent, sheep, cattle, horses, and primates constitutes the natural habitat of *B. coli,* which are shed as cysts in the feces of these animals. Although pigs are the most common source of *balantidiosis,* it can also be spread by contaminated water and from person to person in institutional settings. Healthy humans and animals are resistant to infection, but compromised hosts are vulnerable to attack.

Like cysts of *E. histolytica,* those of *B. coli* resist digestion in the stomach and small intestine and liberate trophozoites that immediately burrow into the epithelium with their cilia. The resultant erosion of the intestinal mucosa produces varying degrees of irritation and injury, leading to nausea, vomiting, diarrhea, dysentery, and abdominal colic. The protozoan rarely penetrates the intestine or enters the blood. The treatment of choice is oral tetracycline, but if this therapy fails, iodoquinol, nitrimidazine, or metronidazole can be given. Preventive measures are similar to those for controlling amebiasis, with additional precautions to prevent food and drinking water from being contaminated with pig manure.

THE FLAGELLATES (MASTIGOPHORANS)

The natural history of flagellated protozoans spans the ecological spectrum from free-living to commensalistic to parasitic. Pathogenic flagellates play an extensive role in human diseases, ranging from milder or self-limited infections (trichomoniasis and giardiasis) to serious and debilitating vector-borne diseases (trypanosomiasis and leishmaniasis). The major flagellate genera and species and their diseases are given in table 23.2.[2]

---

2. Many flagellates are named in honor of one of their discoverers—for example, *T. cruzi,* O. Cruz; *Leishmania,* W. Leishman; *donovani,* C. Donovan; *Giardia,* A. Giard; *lamblia,* V. Lambl; and *brucei,* D. Bruce. Geographic location (*gambiense, rhodesiense, tropica, brasiliensis, mexicana*) is another basis for naming. Names are also derived from a certain characteristic of the organism—for example, *tenax,* tenacious; *hominis* (human); and *vaginalis,* vaginal habitat.

* *Balantidium* (bal″-an-tid′-ee-um) Gr., A small sac.

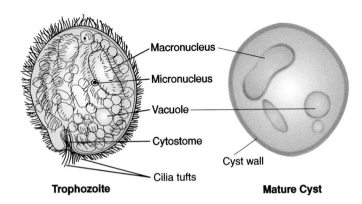

Trophozoite — Macronucleus, Micronucleus, Vacuole, Cytostome, Cilia tufts

Mature Cyst — Cyst wall

FIGURE 23.5

The morphological anatomy of a trophozoite and a mature cyst of *Balantidium coli.*

TABLE 23.2

Flagellate Diseases and Target Organs

Disease	Major Lesion Site	Flagellate
Trichomoniasis		***Trichomonas* species**
Vaginitis, urethritis	Vagina, urethra	*T. vaginalis*
Gingivitis	Opportunist in gums	*T. tenax*
Giardiasis		***Giardia***
Intestinal disease	Duodenum, jejunum	*G. lamblia*
Trypanosomiasis		***Trypanosoma* species**
African sleeping sickness	Skin, viscera, brain	*T. brucei*
Chagas disease	Skin, lymphatics, heart	*T. cruzi*
Leishmaniasis		***Leishmania* species**
Oriental sore	Skin	*L. tropica*
Kala-azar	Viscera	*L. donovani*
Espundia	Skin, mucous membranes	*L. brasiliensis, L. mexicana*

Trichomonads: *Trichomonas* Species

Trichomonads are small, pear-shaped protozoa with four anterior flagella and an undulating membrane, which together produce a characteristic twitching motility in live preparations. They exist only in the trophozoite form and do not produce cysts. Three trichomonads intimately adapted to the human body are *Trichomonas* vaginalis, T. tenax,* and *T. hominis* (figure 23.6).

The most important pathogen, ***Trichomonas vaginalis,*** causes a sexually transmitted disease called **trichomoniasis.** The reservoir for this protozoan is the human urogenital tract, and about 50% of infected people are asymptomatic carriers. Because *T. vaginalis* has no protective cysts, it is a relatively strict parasite that does not survive for long out of the host. Transmission is primarily through contact between genital membranes, although it can on rare occasions be transmitted through communal bathing, public

---

* *Trichomonas* (trik″-oh-moh′-nus) Gr. *thrix,* hair, and *monas,* unit.

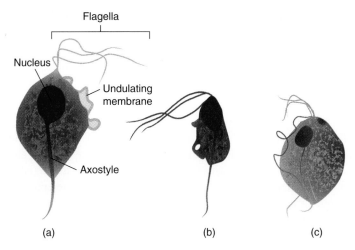

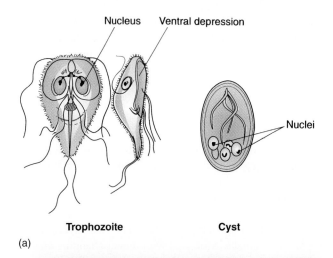

FIGURE 23.6

🔘 **The trichomonads of humans.** **(a)** *Trichomonas vaginalis*, a urogenital pathogen. **(b)** *T. tenax*, a gingival form (infection is rare). **(c)** *T. hominis*, an intestinal species (infection is rare).

facilities, and from mother to child. The incidence of infection is highest among promiscuous young women already infected with other sexually transmitted diseases such as gonorrhea or chlamydia. Approximately 3 million new cases occur every year in the United States, making it the second most prevalent STD.

Symptoms and signs of trichomoniasis in the female include a foul-smelling, green-to-yellow vaginal discharge; vulvitis; cervicitis; and urinary frequency and pain. Severe inflammation of the infection site causes tenderness, edema, chafing, and itching. Males with overt clinical infections experience persistent or recurring urethritis, with a thin, milky discharge and occasionally prostate infection.

Infection is diagnosed by demonstrating the active swimming trichomonads in a wet film of exudate. Alternative methods to detect asymptomatic infection are a Pap smear or culture of the parasite. The infection has been successfully treated with oral and vaginal metronidazole (Flagyl) for one week, and it is essential to treat both sex partners to prevent "ping-pong" reinfection.

Trichomonas tenax is a small trichomonad that ordinarily resides in the oral cavity of 5–10% of humans. Harborage is most common in people with poor oral hygiene or dental disease. *Trichomonas tenax* is the only flagellate in the oral cavity, where it frequents the gingival crevices, feeding on food debris and sloughed cells. Most dental experts agree that it is not a true pathogen but more likely an opportunist in lesions of gingivitis and periodontal pockets. *Trichomonas hominis* (also *Centatrichomonas*) is a resident of the cecum of a small percentage of humans and great apes, but it is believed to be a harmless commensal and is not associated with disease.

Giardia lamblia and Giardiasis

Giardia lamblia* is a pathogenic flagellate first observed by Antonie van Leeuwenhoek in his own feces. For 200 years, it was considered a harmless or weak intestinal pathogen, and only since the 1950s has its prominence as a cause of diarrhea been recog-

---

* *Giardia lamblia* (jee′-ard-ee-uh lam′-blee-uh) Also called *G. intestinalis*.

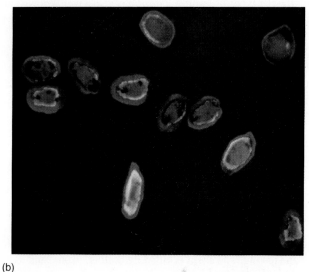

(b)

FIGURE 23.7

🔘 **Trophozoites and cysts of *Giardia lamblia*.** **(a)** The face of a trophozoite with eyes (paired nuclei). Side view shows the ventral depression and origin of the flagella. **(b)** A special DNA probe test uses FISH (fluorescent in situ hybridization) technology to differentiate *Giardia* species using colored dyes. *G. lamblia* glows blue, whereas *G. muris* glows green. In both forms the cell wall antigens are shown in red.

nized. In fact, it is the most common flagellate isolated in clinical specimens. Observed straight on, the trophozoite has a unique, symmetrical heart shape with organelles positioned in such a way that it resembles a face (figure 23.7*a*). Four pairs of flagella emerge from the ventral surface, which is concave and acts like a suction cup for attachment to a substrate. *Giardia* cysts are small, compact, and multinucleate (figure 23.7*a*).

Giardiasis has a complex epidemiologic pattern. The protozoan has been isolated from the intestines of beavers, cattle, coyotes, cats, and human carriers, but the precise reservoir is unclear at this time. Although both trophozoites and cysts escape in the stool, the cysts play a greater role in transmission. Unlike other pathogenic flagellates, *Giardia* cysts can survive for 2 months in the environment. Cysts are usually ingested with water and food or swallowed after close contact with infected people or contaminated objects. Infection can occur with a dose of only 10 to 100 cysts.

Outbreaks of giardiasis point to a spectrum of possible modes of transmission. Community water supplies in areas throughout the United States have been implicated as common vehicles of infection. *Giardia* epidemics have been traced to water from fresh mountain streams as well as chlorinated municipal water supplies in several states. Infections are not uncommon in hikers and campers who used what they thought was clean water from ponds, lakes, and streams in remote mountain areas. Because wild mammals such as muskrats and beavers are intestinal carriers, they could account for cases associated with drinking water from these sources. Obviously, checking water for purity by its appearance is unreliable, because the cysts are too small to be detected.

Cases of food-borne illness have been traced to carriers who contaminate food through unclean personal habits. Daycare centers have reported outbreaks associated with diaper changing and contact with other types of fomites. *Giardia* is also a well-known cause of traveler's diarrhea, which sometimes does not appear until the person has returned home.

Ingested *Giardia* cysts enter the duodenum, germinate, and travel to the jejunum to feed and multiply. Some trophozoites remain on the surface, while others invade the glandular crypts to varying degrees. Superficial invasion by trophozoites causes damage to the epithelial cells, edema, and infiltration by white blood cells, but these effects are reversible. Typical symptoms include diarrhea, abdominal pain, and flatulence. Diagnosis of giardiasis can be difficult because the organism is shed in feces intermittently. A promising DNA test is now available for detection of the cysts in samples (figure 23.7b). The infection can be eradicated with quinacrine or metronidazole.

Now that this parasite is apparently on the increase, many water agencies have had to rethink their policies on water maintenance and testing. The agent is killed by boiling, ozone, and iodine, but, unfortunately, the amount of chlorine used in municipal water supplies does not destroy the cysts. People who must use water from remote sources should assume that it is contaminated and boil or filter it.

HEMOFLAGELLATES: VECTOR-BORNE BLOOD PARASITES

Other parasitic flagellates are called hemoflagellates because of their propensity to live in the blood and tissues of the human host. Members of this group, which includes species of ***Trypanosoma****** and ***Leishmania,*** have several distinctive characteristics in common. They are obligate parasites that cause life-threatening and debilitating zoonoses; they are spread by blood-sucking insects that serve as intermediate hosts; and they are acquired in specific tropical regions. Hemoflagellates have complicated life cycles and undergo morphological changes as they pass between vector and human host. To keep track of these developmental stages, parasitologists have adopted the following set of terms:

amastigote (ay-mas′-tih-goht) Gr. *a-,* without, and *mastix,* whip. The form lacking a free flagellum.

promastigote (proh-mas′-tih-goht) Gr. *pro-,* before. The stage bearing a single, free, anterior flagellum.

---

* *Trypanosoma* (try-pan′-oh-soh-mah) Gr. *trypanon,* borer, and *some,* body.

epimastigote (ep″-ih-mas′-tih-goht) Gr. *epi-,* upon. The flagellate stage, which has a free anterior flagellum and an undulating membrane.

trypomastigote (trih″-poh-mas′-tih-goht) Gr. *trypanon,* auger. The large, fully formed stage characteristic of *Trypanosoma.*

These stages suggest a metamorphic or evolutionary transition from the simple, cystlike amastigote form to the highly complex trypomastigote form. All four stages are present in *T. cruzi,* but *T. brucei* lacks the amastigote and promastigote forms, and *Leishmania* has only the amastigote and promastigote stages (table 23.3). In general, the vector host serves as the site of differentiation for one or more of the stages, and the parasite is infectious to humans only in its fully developed stage.

Trypanosoma Species and Trypanosomiasis

Trypanosoma species are distinguished by their infective stage, the trypomastigote, an elongate, spindle-shaped cell with tapered ends that is capable of eel-like, sinuous motility. Two types of **trypanosomiasis** are distinguishable on a geographical basis. *Trypanosoma brucei* is the agent of **African sleeping sickness,** and *T. cruzi* is the cause of **Chagas disease,** which is endemic to Central and South America. Both exhibit a biphasic life-style alternating between a vertebrate and an insect host. Because the life cycle of *T. cruzi* was covered in chapter 5, only that of sleeping sickness will be discussed in depth.

***Trypanosoma brucei* and Sleeping Sickness** Trypanosomiasis has greatly affected the living conditions of Africans since ancient times. Today, at least 50 million people are at risk, and 30 to 40 thousand new cases occur each year. It imposes an additional hardship when it attacks domestic and wild mammals. The two variants of sleeping sickness are the Gambian (West African) strain, caused by the subspecies *T. b. gambiense,* and the Rhodesian (East African) strain, caused by *T. b. rhodesiense* (figure 23.8a). These geographically isolated types are associated with different ecological niches of the principal tsetse fly vectors. In the West African form, the fly inhabits the dense vegetation along rivers and forests typical of that region, whereas the East African form is adapted to savanna woodlands and lakefront thickets.

The cycle begins when a tsetse fly becomes infected after feeding on an infected reservoir host, such as a wild animal (antelope, pig, lion, hyena), domestic animal (cow, goat), or human (figure 23.8c). In the fly's gut, the trypanosome multiplies, migrates to the salivary glands, and develops into the infectious stage. When the fly bites its host, it releases trypomastigotes into the wound. At this site, the trypanosome multiplies and produces a sore called the primary chancre. From there, the pathogen moves into the lymphatics and the blood (figure 23.8b). Although an immune response occurs, it is counteracted by an unusual adaptation of the trypanosome (Medical Microfile 23.1). The Rhodesian form is acute and advances to brain involvement in 3 to 4 weeks and to death in a few months. The Gambian form, on the other hand, has a longer incubation period, is usually chronic, and may not affect the brain for several years.

Trypanosomiasis affects the lymphatics and areas surrounding blood vessels. Symptoms usually include intermittent fever,

TABLE 23.3

Cellular and Infective Stages of the Hemoflagellates

Genus/Species	Amastigote	Promastigote	Epimastigote	Trypomastigote
Leishmania	Intracellular in human macrophages	Found in sand fly gut; **infective to humans**	Does not occur	Does not occur
Trypanosoma brucei	Does not occur	Does not occur	Present in salivary gland of tsetse fly	In biting mouthparts of tsetse fly; **infective to humans**
Trypanosoma cruzi	Intracellular in human macrophages, liver, heart, spleen	Occurs	Present in gut of reduviid (kissing) bug	In feces of reduviid bug; **transferred to humans**

enlarged spleen, swollen lymph nodes, and joint pain. In both forms, the central nervous system is affected, the initial signs being personality and behavioral changes that progress to lassitude and sleep disturbances. The disease is commonly called sleeping sickness, but in fact, uncontrollable sleepiness occurs primarily in the day and is followed by sleeplessness at night. Signs of advancing neurological deterioration are muscular tremors, shuffling gait, slurred speech, epileptic-like seizures, and local paralysis. Death results from coma, secondary infections, or heart damage.

Sleeping sickness may be suspected if the patient has been bitten by a distinctive fly while living or traveling in an endemic area. Trypanosomes are readily demonstrated in the blood, spinal fluid, or lymph nodes. Chemotherapy is most successful if administered prior to nervous system involvement. Brain infection must be treated with a highly toxic arsenic-based drug called melarsoprol, or a less toxic one, difluoromethylornithine (DFMO). Both drugs are expensive and may be beyond the medical resources of many countries.

Control of trypanosomiasis in western Africa, where humans are the main reservoir hosts, involves eliminating tsetse flies by applying insecticides, trapping flies, or destroying shelter and breeding sites. In eastern regions, where cattle herds and large wildlife populations are reservoir hosts, control is less practical because large mammals are the hosts, and flies are less concentrated in specific sites. In some parts of equatorial Africa, *T. b. gambiense* has recently experienced a resurgence. Epidemics are reported in areas of the Sudan and Zaire, where 20–40% of the population is infected. These latest outbreaks are attributed to a civil war and its disruption in medical services.

FIGURE 23.8

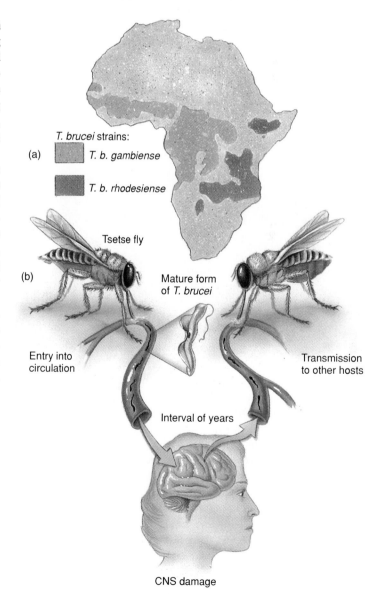

The generalized cycle between humans and the tsetse fly vector. **(a)** The distribution of African trypanosomiasis. **(b)** The saliva of a fly infected with *T. brucei* inoculates the human bloodstream. The parasite matures and invades various organs. In time, its cumulative effects cause central nervous system (CNS) damage. The trypanosome is spread to other hosts through another fly in whose alimentary tract the parasite completes a series of developmental stages.

MEDICAL MICROFILE 23.1
The Trypanosome's Bag of Tricks

The immune system first responds to trypanosome surface antigens by producing immunoglobulins of the IgM type. These antibodies carry out their usual function of tagging the parasite to make it easier to destroy. If this were the end of the matter, sleeping sickness would be a mild infection with few fatalities, but such is not the case. Instead, the trypanosome strikes back with a fascinating genetic strategy in which surviving cells change the structure of their surface glycoprotein antigens. This change in specificity renders the existing IgM ineffective so that the parasite eludes control and multiplies in the blood. When the host fights back by producing IgM against this new antigen, the parasite alters its surface again, the body produces corresponding different IgM, the surface changes again, and so on, in a repetitive cycle that can go on for months or even years. Eventually, the host becomes exhausted and overwhelmed by repeated efforts to catch up with this trypanosome masquerade. This cycle has tremendous impact on the pathology and control of the disease. The presence of the trypanosome in the blood and the severity of symptoms follow a wavelike pattern reminiscent of borreliosis. Development of a vaccine is hindered by the requirement to immunize against at least 100 different antigenic variations.

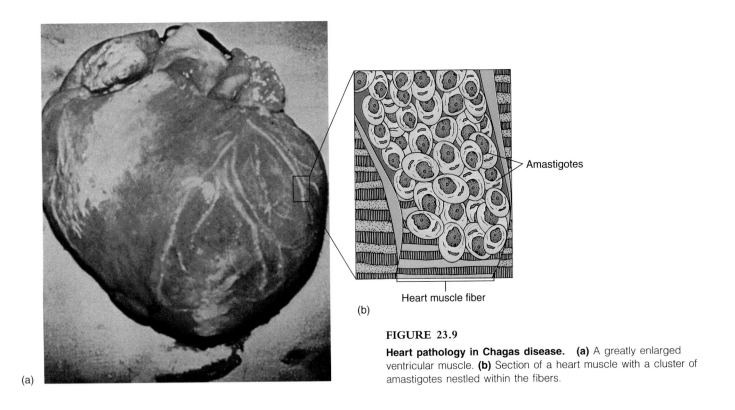

FIGURE 23.9

Heart pathology in Chagas disease. **(a)** A greatly enlarged ventricular muscle. **(b)** Section of a heart muscle with a cluster of amastigotes nestled within the fibers.

***Trypanosoma cruzi* and Chagas Disease** Chagas disease accounts for millions of cases, thousands of deaths, and untold economic loss and human suffering in Latin America. The life cycle of its etiologic agent, *Trypanosoma cruzi* (see chapter 5), parallels that of *T. brucei,* except that the insect hosts are "kissing," or reduviid, bugs that harbor the trypanosome in the hind gut and discharge it in feces. These bugs live throughout Central and South America and often share the living quarters of human and mammalian hosts. Infection occurs when the bug defecates near the site of its own bite, and the human inoculates this wound by accidently rubbing the bug feces into it. Infection is also possible when bug feces fall onto the mucous membranes or conjunctiva and through contact with blood. The course of the illness is somewhat similar to African trypanoso-miasis, marked by a local lesion, fever, and swelling of the lymph nodes, spleen, and liver. Particular targets are the heart muscle and large intestine, both of which harbor masses of amastigotes. The infection of these organs causes enlargement and severe disruption in function (figure 23.9). Chronically infected people with cardiac complications usually die within 2 years.

Diagnosis of Chagas disease is made through microscopic examination of blood and serological testing. Although treatment is sometimes attempted with two drugs, nifurtimox and benzonidazole, they are not effective if the disease has progressed too far, and they have damaging side effects. It has been shown that the prevalence of Chagas disease can be reduced by vector control. Insecticides are of some benefit, but the bugs are very tough and capable

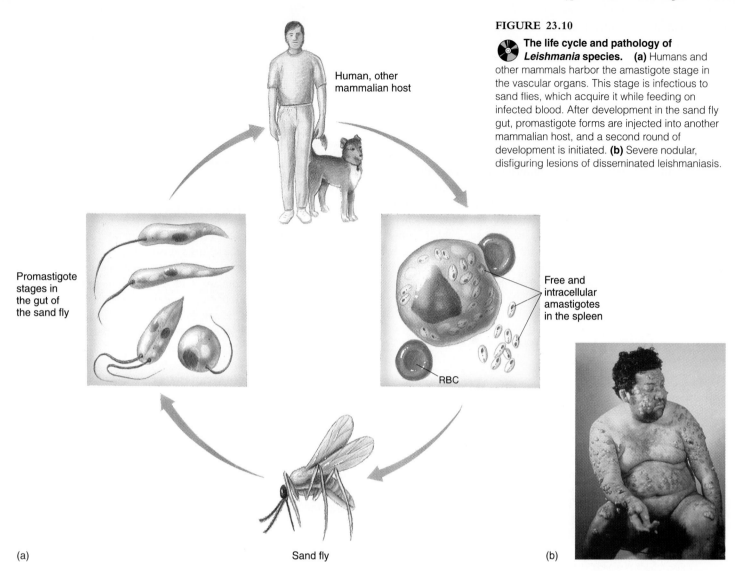

Human, other
mammalian host

Promastigote
stages in
the gut of
the sand fly

Free and
intracellular
amastigotes
in the spleen

RBC

(a)

Sand fly

(b)

FIGURE 23.10

The life cycle and pathology of
***Leishmania* species.** **(a)** Humans and
other mammals harbor the amastigote stage in
the vascular organs. This stage is infectious to
sand flies, which acquire it while feeding on
infected blood. After development in the sand fly
gut, promastigote forms are injected into another
mammalian host, and a second round of
development is initiated. **(b)** Severe nodular,
disfiguring lesions of disseminated leishmaniasis.

of resisting doses that readily kill flies or mosquitoes. Mass collections of reduviid bugs and housing improvements to keep them out of human dwellings have also been effective.

Leishmania Species and Leishmaniasis

Leishmaniasis is a zoonosis transmitted among various mammalian hosts by female **phlebotomine*** flies (sand flies) that require a blood meal to produce mature eggs. *Leishmania* species have adapted to the insect host by timing their own development with periods of blood-taking (figure 23.10). The protozoa enter the fly's gut and then multiply and differentiate into promastigotes that migrate to the proboscis of the fly.

Leishmaniasis is endemic to equatorial regions that provide favorable conditions for sand flies to complete their life cycle. The disease can be transmitted by over 50 species of sand fly, and numerous wild and domesticated animals, especially dogs, serve as other reservoirs. Although humans are usually accidental hosts in the epidemiologic cycle, the fly freely feeds on humans. At particu-

lar risk are travelers or immigrants who have never had contact with the parasite and lack specific immunities.

Leishmania infection begins when an infected fly injects promastigotes while feeding. After being engulfed by macrophages, the parasite converts to an amastigote, and multiplies in the macrophage. The manifestations of the disease vary with the fate of the macrophage. If they remain fixed, the infection stays localized in the skin or mucous membranes, but if the infected macrophages migrate, systemic disease occurs.

Cutaneous leishmaniasis (oriental sore, Baghdad boil) is a localized infection of the capillaries of the skin caused by *L. tropica* in certain Mediterranean, African, and Indian regions and by *L. mexicana* in Latin America. A single small, red papule occurs at the site of the bite and spreads laterally into a large ulcer. A form of mucocutaneous leishmaniasis called espundia, caused by *L. brasiliensis* and endemic to parts of Central and South America, affects both the skin and mucous membranes of the head. After the skin ulcer heals, the protozoa can settle into a chronic infection of the nose, lips, palate, gingiva, and pharynx that is very disfiguring (figure 23.10*b*). Systemic (visceral) leishmaniasis that occurs in parts of Africa, Latin America, and China appears gradually after 2 to 18

---

* **phlebotomine** (flee-bot′-oh-meen) Gr. *phlebo,* vein, and *tomos,* cutting. So called because they feed on blood.

weeks, with a high, intermittent fever; weight loss; and enlargement of internal organs, especially the spleen, liver, and lymph nodes. The most deadly form, *kala azar*, is 75% to 95% fatal in untreated cases. Death is due to complications of infection, including the destruction of blood-forming tissues, anemia, secondary infections, and hemorrhage.

Diagnosis is complicated by the similarity of leishmaniasis to several granulomatous bacterial and fungal infections. The organism can usually be detected in the cutaneous lesion or in the bone marrow during visceral disease. Additional tests include cultivating the parasite, serological tests, and leishmanin skin testing. Chemotherapy with injected pentavalent antimony is usually effective, though pentamidine or amphotericin B may be necessary in resistant cases. Some disease prevention has been achieved by applying insecticides to and eliminating sand fly habitats, as well as by destroying infected dogs (an important reservoir of infection in some areas). A vaccine based on recombinant DNA technology is still under development.

CHAPTER CHECKPOINTS

Parasitic diseases have existed since ancient times but are still a major public health problem today. International travel of people and materials has facilitated the spread of parasites and their vectors worldwide. The increase in immunosuppressed people, particularly those with AIDS, has caused an increase in formerly obscure protozoan infections.

Protozoa are categorized by the type of motility, if any, and the complexity of the infectious agent's life cycle. All protozoans propagate via their trophozoite, or active feeding stage. The resting form, or cyst stage, enables them to exist outside their host. Major human parasites are found in the ameba, ciliate, flagellate, and apicomplexan groups.

The most significant pathogenic ameba is *Entamoeba histolytica*, causative agent of amebic dysentery. It is transmitted through cyst-infected food and water. Humans are its only host. The trophozoites can cause ulcerations in the bowel, which can lead to perforation and infection of other peritoneal organs.

Naegleria fowleri and *Acanthamoeba* are free-living amebas that infect humans exposed to contaminated water. Both cause CNS infections that progress rapidly and are usually fatal.

Balantidium coli is the largest protozoan pathogen and the only ciliate pathogenic to humans. Cysts enter the host through contact with pigs or contaminated water. Trophozoites colonize the intestinal mucosa and cause moderate diarrhea.

The flagellated protozoans contain several major pathogens: *Trichomonas*, *Giardia*, *Trypanosoma*, and *Leishmania*. The most important trichomonad causes an STD called trichomoniasis. *Giardia* flagellates cause a noninvasive infection of the intestinal mucosa, which can proceed to a severe, chronic form in some cases.

The hemoflagellates are systemic invaders that cause debilitating, life-threatening infections of lymphatics, blood vessels, and specific organs. All pathogens in this group are transmitted by arthropod vectors. All have several morphologically distinctive stages to their life cycles.

African sleeping sickness is caused by *Trypanosoma brucei*, which invades the blood, the lymphatics, and the brain. Tsetse flies are the vector and secondary host. *T. cruzi* is the causative agent of Chagas disease. The reduviid bug infects the host through its feces deposited at the bite region. Chagas disease damages the heart and intestine in particular.

Leishmania is the causative agent of leishmaniasis, a zoonosis transmitted by female sand flies that are also the secondary host. Macrophages transmit the infection throughout the human host. The migration of the infected macrophage determines whether the infection is cutaneous or systemic.

APICOMPLEXAN PARASITES

One group of unique protozoa are the apicomplexans, sometimes called the sporozoans. These tiny parasites live on a variety of animal hosts. They are distinct from other protozoans in lacking locomotor organelles in the mature state, though some forms do have a flexing, gliding form of locomotion, and others have flagellated sex cells. Characteristics of the apicomplexan life cycle that add greatly to their complexity are alternations between sexual and asexual phases and between different animal hosts. Most members of the group also form specialized infective bodies that are transmitted by arthropod vectors, food, water, or other means, depending on the parasite. The most important human pathogens belong to *Plasmodium* (malaria), *Toxoplasma* (toxoplasmosis), and *Cryptosporidium* (cryptosporidiosis).

Plasmodium: **The Agent of Malaria**
Throughout human history, including prehistoric times, malaria has been one of the greatest afflictions, in the same ranks as bubonic plague, influenza, and tuberculosis. Even now, as the dominant protozoan disease, it threatens one-third of the world's population every year. The origin of the name is from the Italian words *mal*, bad, and *aria*, air. The superstitions of the Middle Ages explained that evil spirits or mists and vapors arising from swamps caused malaria, because many victims came down with the disease after this sort of exposure. Of course, we now know that a swamp was mainly involved as a habitat for the mosquito vector.

The agent of malaria is an obligate intracellular sporozoan in the genus *Plasmodium*, which contains four species: *P. malariae, P. vivax, P. falciparum,* and *P. ovale*. The human and some primates are the primary vertebrate hosts for these species, which are geographically separate and show variations in the pattern and severity of disease. All forms are spread primarily by the female *Anopheles* mosquito and occasionally by shared needles, blood transfusions, and from mother to fetus. The detailed study of malaria is a large and engrossing field that is far too complicated for this text. We cover only the general aspects of its epidemiology, life cycle, and pathology.

Epidemiology, Life Cycle, and Stages of *Plasmodium* Infection   Although malaria was once distributed throughout most of the world, the control of mosquitoes in temperate areas has successfully restricted it mostly to a belt extending around the equator. Despite this achievement, approximately 300 million to 500 million new cases are still reported each year, about two-thirds of them in Africa. The most frequent victims are children and young adults,

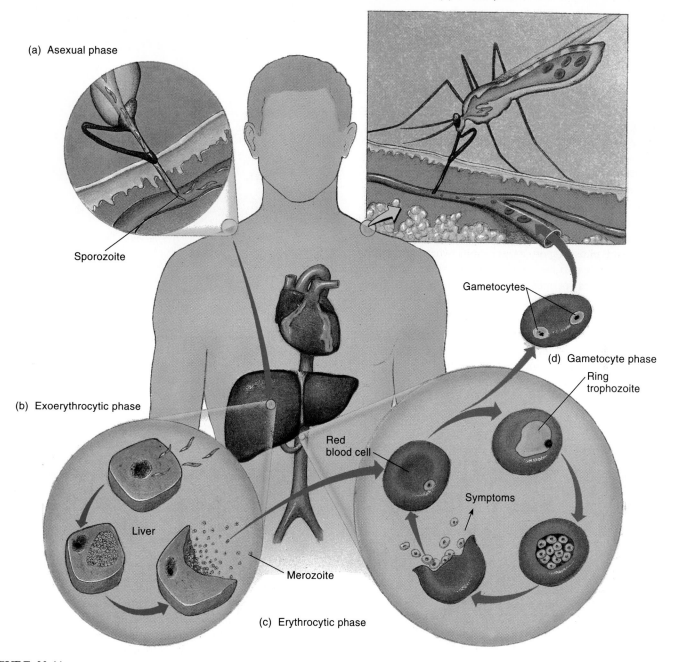

(e) Sexual phase

(a) Asexual phase

Sporozoite

(b) Exoerythrocytic phase

Liver

Merozoite

(c) Erythrocytic phase

Gametocytes

(d) Gametocyte phase

Ring trophozoite

Red blood cell

Symptoms

FIGURE 23.11

The life and transmission cycle of *Plasmodium*, the cause of malaria. **(a)** In the asexual phase in humans, sporozoites enter a capillary through the saliva of a feeding mosquito. **(b)** Exoerythrocytic (liver) phase. Sporozoites invade the liver cells and develop into large numbers of merozoites. **(c)** Erythrocytic phase. Merozoites released into the circulation enter red blood cells. Initial infection is marked by a ring trophozoite; schizogony of the ringed form produces additional merozoites that burst out and infect other red blood cells. **(d)** Gametocytes that develop in certain infected red blood cells are ingested by another mosquito. **(e)** The sexual phase of fertilization and sporozoite formation occurs in the mosquito.

of whom at least 2 million die annually. The total case rate in the United States is about 1,000 to 2,000 new cases a year, most of which occur in immigrants.

Development of the malarial parasite is divided into two distinct phases: the asexual phase, carried out in the human, and the sexual phase, carried out in the mosquito (figure 23.11). The **asexual phase** and infection begin when an infected female *Anopheles* mos-

quito injects saliva containing anticoagulant into the capillary in preparation for taking a blood meal. In the process, she inoculates the blood with motile, spindle-shaped, asexual cells called **sporozoites** (Gr. *sporo,* seed, and *zoon,* animal). The sporozoites circulate through the body and gravitate to the liver in a short time. Within liver cells, the sporozoites undergo asexual division called *schizogony* (Gr. *schizo,* to divide, and *gone,* seed), which generates numerous daughter parasites, or *merozoites.* This phase of *exoery-throcytic development* lasts from 5 to 16 days, depending upon the species of parasite. Its end is marked by eruption of the liver cell, which releases from 2,000 to 40,000 mature merozoites into the circulation.

During the *erythrocytic phase,* merozoites attach to special receptors on red blood cells and invade them, converting in a short time to circular (ring) trophozoites (figure 23.12). This stage feeds upon hemoglobin, grows, and undergoes multiple divisions to produce a cell called a *schizont,* which is filled with more merozoites. Bursting red cells liberate merozoites to infect more red cells. Eventually, certain merozoites differentiate into two types of specialized gametes called *macrogametocytes* (female) and *microgametocytes* (male). Because the human does not provide a suitable environment for the next phase of development, this is the end of the cycle in humans.

The **sexual phase** (sporogony) results when a mosquito draws infected RBCs into her stomach. In the stomach, the microgametocyte releases spermlike gametes that fertilize the larger macrogametes. The resultant diploid cell (oocyst) implants into the stomach wall of the mosquito and undergoes multiple meiotic divisions, releasing sporozoites that migrate to the salivary glands and lodge there. This event completes the sexual cycle and makes the sporozoites available for infecting the next victim.

After a 10- to 16-day incubation period, the first symptoms are malaise, fatigue, vague aches, and nausea followed by bouts of chills, fever, and sweating. These symptoms occur at 48- or 72-hour intervals, as a result of the synchronous rupturing of RBCs. Their interval, length, and regularity reflect the type of malaria. Patients with falciparum malaria, the most malignant type, often manifest persistent fever, rapid pulse, cough, and weakness for weeks without relief. Complications of malaria are hemolytic anemia from lysed blood cells and organ enlargement and rupture due to cellular debris that accumulates in the spleen, liver, and kidneys. Patients with milder, chronic forms of malaria can achieve spontaneous cure after 3 to 5 years, but falciparum malaria has a high death rate in the acute phase, especially in children. Certain kinds of malaria are subject to relapses because some infected liver cells harbor dormant infective cells for up to 5 years.

Some populations are naturally resistant to *Plasmodium* infection. One dramatic example occurs in Africans who carry one gene that codes for sickle-cell hemoglobin. Although carriers are relatively healthy, they produce some abnormal RBCs that prevent the parasite from deriving sufficient nutrition to multiply. The continued existence of sickle-cell anemia in these geographic areas can be directly attributed to the increased survival of genetic carriers. Individuals with sickle-cell anemia have even greater protection from malaria; however, the debilitation of sickle-cell disease outweighs its protective effect.

Diagnosis and Control of Malaria Malaria can be diagnosed definitively by the discovery of a typical stage of *Plasmodium* in stained blood smears (figure 23.12). Other indications are knowledge of the patient's residence or travel in endemic areas and such symptoms as recurring chills, fever, and sweating.

As recently as a decade ago, eradicating malaria seemed possible, but since then, morbidity and mortality have increased in

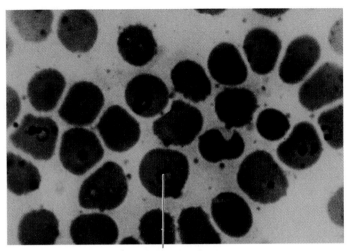

Ring trophozoite

FIGURE 23.12

The ring trophozoite stage in a *Plasmodium falciparum* infection. A smear of peripheral blood shows several ring forms in some red blood cells.

several regions. The two main reasons involve the sheer adaptive and survival capacity of both the parasite and the vector. Standard chemotherapy with antimalarial drugs has selected for drug-resistant strains of the malarial parasite, and mosquitoes have developed resistance to some common insecticides used to control them. The current treatments for malaria include some form of quinine. Chloroquine, the least toxic type, is used in nonresistant forms of the disease. In areas of the world where resistant strains of *P. falciparum* and *P. vivax* predominate, a course of mefloquine or quinine is indicated. Eliminating the parasite from the liver and preventing relapses can be managed with long-term therapy with primaquine or proguenil.

Malaria prevention is attempted through long-term mosquito abatement and human prophylaxis. This includes elimination of water that could serve as a breeding site, and broadcasting of insecticides to reduce populations of adult mosquitoes, especially near human dwellings. Humans can reduce their risk of infection considerably by using netting, screens, and repellents; remaining indoors at night; and taking weekly doses of prophylactic drugs. People with a recent history of malaria must be excluded from blood donation. Even with massive efforts undertaken by the WHO, the prevalence of malaria in endemic areas is still high.

Malaria and Its Elusive Vaccine Several animal and human models have proved that effective vaccination for malaria is possible, but a practical vaccine for broad-scale use has proved more difficult to develop. In fact, malaria is a perfect model to demonstrate the problems of vaccine development. A successful malaria vaccine must be capable of striking a diverse and rapidly changing target. Not only are there four different species, each having different sporozoite, merozoite, and gametocyte stages, but each species can also have different antigenic types of sporozoites and merozoites. A sporozoite vaccine would prevent liver infection, whereas a merozoite vaccine would prevent infection of red blood cells and diminish the pathology that is most responsible for morbidity and mortality. The best vaccines would provide long-term protection against all the phases and variants so that if one immunity were to fail, another could come into play.

The cloning of major plasmodial surface and core antigens and mapping of the parasite's genome have provided new molecular tools for vaccines. It is believed that challenging the immune system with a spectrum of sporozoite and merozoite antigens will increase the chances of attacking most of the life stages of the parasite. Two of the most promising vaccines currently under development are a recombinant DNA version of the sporozoite surface antigens and one that contains the parasite's DNA.

Coccidian Parasites

The Order Coccidiorida of the Apicomplexa contains a number of members that are zoonotic in domestic mammals and birds and can cause severe disease in humans. These unusual protozoans exist in several stages, including dormant oocysts and pseudocysts.

Toxoplasma gondii **and Toxoplasmosis** *Toxoplasma gondii** is an obligate apicomplexan parasite with such extensive cosmopolitan distribution that some experts estimate it affects the majority of the world's population at some time in their lives. In most of these cases, toxoplasmosis goes unnoticed, but disease in the fetus and in immunodeficient people, especially those with AIDS, is severe and often terminal. *Toxoplasma gondii* is a very successful parasite with so little host specificity that it can attack at least 200 species of birds and mammals. However, its primary reservoir and hosts are members of the feline family, both domestic and wild.

To follow the transmission of toxoplasmosis, we must first look at the general stages of *Toxoplasma*'s life cycle in the cat (figure 23.13*a*). The parasite undergoes a sexual phase in the intestine and is then released in feces, where it becomes an infective *oocyst* that survives in moist soil for several months. Ingested oocysts release an invasive asexual tissue phase called a tachyzoite that infects many different tissues and often causes disease in the cat. Eventually, these forms enter an asexual cyst state in tissues, called a pseudocyst. Most of the time, the parasite does not cycle in cats alone and is spread by oocysts to intermediate hosts, usually rodents and birds. The cycle returns to cats when they eat these infected prey animals.

Other vertebrates become a part of this transmission cycle (figure 23.13*b*). Herbivorous animals such as cattle and sheep ingest oocysts that persist in the soil of grazing areas and then develop pseudocysts in their muscles and other organs. Carnivores such as canines are infected by eating pseudocysts in the tissues of carrier animals.

Humans appear to be constantly exposed to the pathogen. The rate of prior infection, as detected through serological tests, can be as high as 90% in some populations. Many cases are caused by ingesting pseudocysts in contaminated meats. A common source is raw or undercooked meat, which is customary fare for certain ethnic groups, such as raw, spiced beef, raw pork, and ground lamb. The grooming habits of cats spread fecal oocysts on their body surfaces, and unhygienic handling of them presents an opportunity to

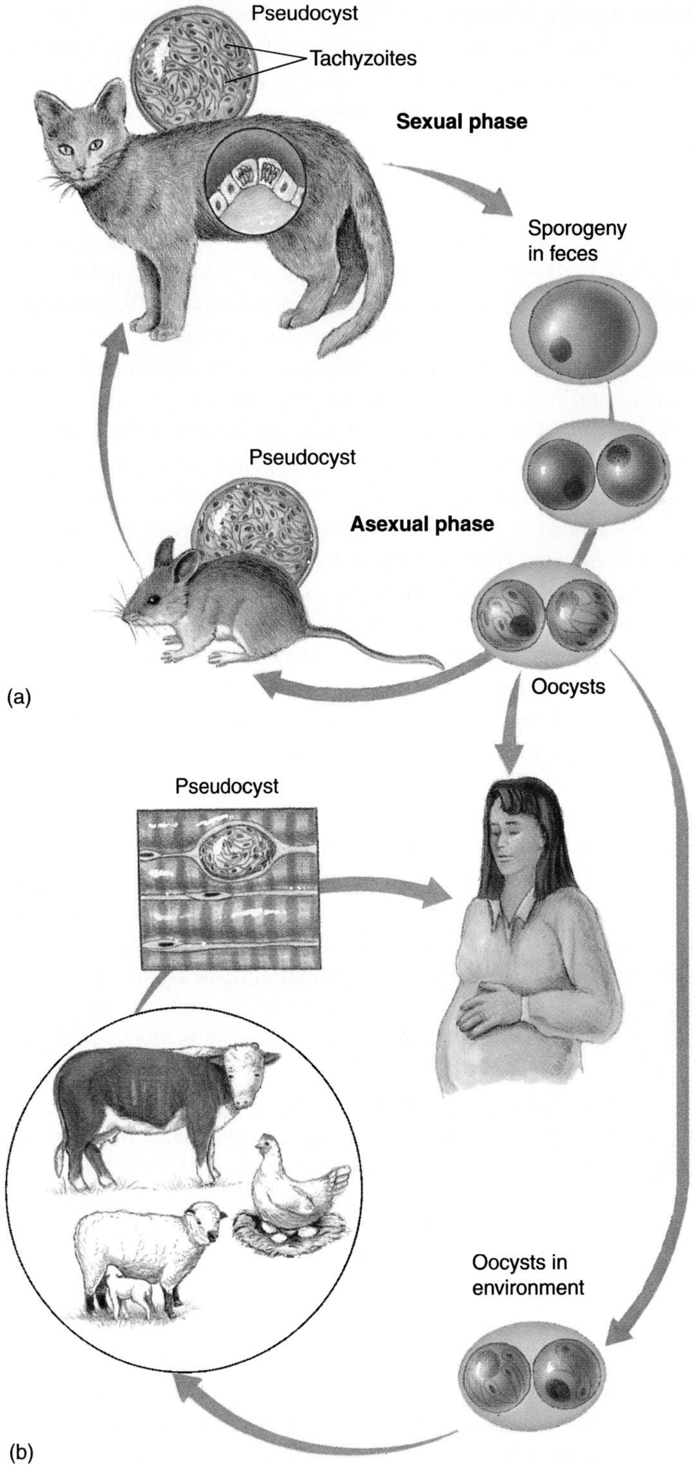

(a)

(b)

FIGURE 23.13

The life cycle and morphological forms of *Toxoplasma gondii*. **(a)** The cycle in cats and their prey. **(b)** The cycle in other animal hosts. The zoonosis has a large animal reservoir (domestic and wild) that becomes infected through contact with oocysts in the soil. Humans can be infected through contact with cats or ingestion of pseudocysts in animal flesh. Infection in pregnant women is a serious complication because of the potential damage to the fetus.

---

* *Toxoplasma gondii* (toks″-oh-plaz′-mah gawn′-dee-eye) L. *toxicum*, poison, and Gr. *plasma*, molded and *gundi*, a small rodent.

ingest oocysts. Infection can also occur when oocysts are inhaled in air or dust contaminated with cat droppings and when tachyzoites cross the placenta to the fetus.

Most cases of toxoplasmosis are asymptomatic or marked by mild symptoms such as sore throat, lymph node enlargement, and low-grade fever. In patients whose immunity is suppressed by infection, cancer, or drugs, the outlook is grim. Chronic, persistent *Toxoplasma* infection can produce extensive brain lesions and can create fatal disruptions of the heart and lungs. A pregnant woman with toxoplasmosis has a 33% chance of transmitting the infection to her fetus. Congenital infection occurring in the first or second trimester is associated with stillbirth and severe abnormalities such as liver and spleen enlargement, liver failure, hydrocephalus, convulsions, and damage to the retina that can result in blindness.

Toxoplasmosis in AIDS and other immunocompromised patients can result from a primary acute or reactivated latent infection. The acute form is a disseminated, rapidly fatal disease with massive brain involvement that is a common complication in AIDS patients. Trophozoites and cysts clustered in intracerebral lesions cause seizures, altered mental state, and coma (figure 23.14) .

Toxoplasmosis shows signs and symptoms that mimic infectious mononucleosis or cytomegalovirus infections. It can be differentiated by means of serological tests that detect antitoxoplasma antibodies, especially those for IgM, which appears early in infection. Disease can also be diagnosed by culture and histological analysis.

The most effective drugs are pyrimethamine and sulfadiazine alone or in combination. Because they do not destroy the cyst stage, they must be given for long periods to prevent recurrent infection.

In view of the fact that the oocysts are so widespread and resistant, hygiene is of paramount importance in controlling toxoplasmosis. There is no such thing as a safe form of raw meat, even salted or spiced. Adequate cooking or freezing below −20°C destroys both oocysts and tissue cysts. Oocysts can also be avoided by washing the hands after handling cats or soil that is possibly contaminated with cat feces, especially sandboxes and litter boxes. Pregnant women should be especially attentive to these rules and should never clean out the cat's litter box.

Sarcocystis and Sarcocystosis Although *Sarcocystis* are parasites of cattle, swine, and sheep, some species have an interesting transmission cycle that involves passage through humans as the final host. Domestic animals are intermediate hosts that pick up infective cysts while grazing on grass contaminated by human excrement. Humans are subsequently infected when they eat beef, pork, or lamb containing tissue cysts, but human-to-human transmission does not occur. Sarcocystosis is an uncommon infection mostly caused by ingesting uncooked meats. Initial symptoms of intestinal sarcocystosis are diarrhea, nausea, and abdominal pain that peak in a few hours after ingestion but can persist for about 2 weeks. There is no specific treatment at the present time.

Cryptosporidium: A Newly Recognized Intestinal Pathogen *Cryptosporidium** is an intestinal apicomplexan that infects a variety of mammals, birds, and reptiles. For many years, **cryp-**

** **Cryptosporidium** (krip″-toh-spo-rid′-ee-um). Gr. *cryptos*, hidden, and *sporos*, seed.*

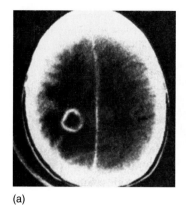
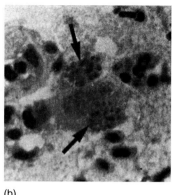

(a)   (b)

FIGURE 23.14

Toxoplasmosis in the brain of an AIDS patient. **(a)** A CT scan of the brain shows a contrast-enhanced image of a cerebral lesion. **(b)** Biopsy tissue stained for *Toxoplasma gondii* contains clusters of intracellular tachyzoites (arrows).

Courtesy of Dr. Henry W. Murray. From Harrison's Principles of Internal Medicine, *12th ed., McGraw-Hill, Inc. Reproduced with permission of the publisher.*

tosporidiosis was considered an intestinal ailment exclusive to calves, pigs, chickens, and other poultry, but it is clearly a zoonosis as well. The organism exhibits a life cycle similar to that of *Toxoplasma*, with a hardy intestinal oocyst and a tissue phase. In transmission pattern it is very similar to *Giardia*.

Cryptosporidiosis has cosmopolitan distribution. Its highest prevalence is in areas with unreliable water and food sanitation. The carrier state occurs in 3% to 30% of the population in developing countries. The susceptibility of the general public to this pathogen has been amply demonstrated by several large-scale epidemics. In the 1990s, 370,000 people developed *Cryptosporidium* gastroenteritis from the municipal water supply in Milwaukee, Wisconsin! Other mass outbreaks of this sort have been traced to contamination of the local water reservoir by livestock wastes. Other studies revealed that at least one-third of all fresh surface waters harbor this parasite. Chlorination is not entirely successful in eradicating the cysts, so most treatment plants use filtration to remove them, but even this method can fail.

Infection is initiated when oocysts are ingested and give rise to sporozoites that penetrate the intestinal cells and live intracellularly in them. The prominent symptoms in patients with clinical disease mimic other types of gastroenteritis, with headache, sweating, vomiting, severe abdominal cramps, and diarrhea. AIDS patients may experience chronic persistent cryptosporidial diarrhea that is one criterion used to diagnose AIDS. The agent can be detected in fecal samples with indirect immunofluorescence and by acid-fast staining of biopsy tissues (figure 23.15*a*). Treatment is presently hampered by the lack of effective drugs and the tendency of immunodeficient patients to suffer relapses.

Isospora belli and Coccidiosis *Isospora* is an intracellular intestinal parasite with a distinctive oocyst stage (figure 23.15*b*). Of several species that colonize mammals, apparently only *Isospora belli** infects humans. The parasite is transmitted in fecally con-

---

** **Isospora belli** (eye-saus′-poh-rah bel′-eye) Gr. *iso*, equal, and *sporos*, seed; L. *bellum*, war.*

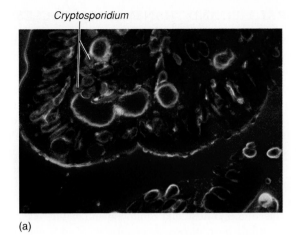

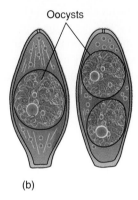

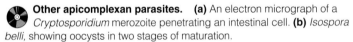

FIGURE 23.15

Other apicomplexan parasites. **(a)** An electron micrograph of a *Cryptosporidium* merozoite penetrating an intestinal cell. **(b)** *Isospora belli,* showing oocysts in two stages of maturation.

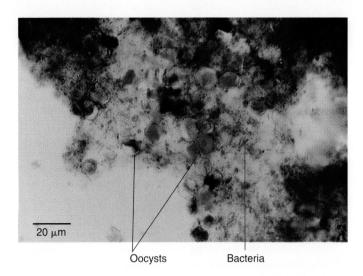

FIGURE 23.16

An acid-fast stain of *Cyclospora* in a human fecal sample. The large (8–10 μm) cysts stain pink to red and have a wrinkled outer wall. Bacteria stain blue.

taminated food or drink and through laboratory contact. Not much is known about the incidence of infection because it is usually asymptomatic or self-limited. The more prominent symptoms are malaise, nausea and vomiting, diarrhea, fatty stools, abdominal colic, and weight loss. In cases requiring treatment, combined therapy with sulfadiazine and pyrimethamine is usually effective.

Cyclospora cayetanensis and Cyclosporiasis *Cyclospora cayetanensis* is an emerging protozoan pathogen related to *Isospora.* Since its first occurrence in 1979, hundreds of outbreaks have been reported in the United States and Canada. Its mode of transmission is oral-fecal, and most cases have been associated with consumption of fresh produce and water, presumably contaminated with feces. Cyclosporiasis occurs worldwide, and although primarily of human origin, it is not spread directly from person to person. Outbreaks have been traced to imported raspberries, salad made with fresh greens, and drinking water. The parasite has also been identified as a significant cause of diarrhea in travelers.

The disease is acquired when oocysts enter the small intestine and invade the mucosa. After an incubation period of about one week, symptoms of watery diarrhea, stomach cramps, bloating, fever, and muscle aches appear. Patients with prolonged diarrheal illness experience anorexia and weight loss.

Diagnosis can be complicated by the lack of recognizable oocysts in the feces. Techniques that improve identification of the parasite are examination of fresh preparations under a fluorescent microscope and an acid-fast stain of a processed stool specimen (figure 23.16). A PCR-based DNA test can also be used to identify *Cyclospora* and differentiate it from other coccidian parasites. This form of analysis is more sensitive and can detect protozoan genetic material even in the absence of cysts. Most cases of infection have been effectively controlled with a combination of trimethoprim and sulfamethoxazole (Bactrim, Septra) for a week. Traditional antiprotozoan drugs are not effective. Some cases of disease may be prevented by cooking or freezing food to kill the oocysts.

Babesia Species and Babesiosis Babesiosis, or piroplasmosis, is historically important for two firsts in microbiology. **Babesia*** was the first protozoan found to be responsible for a disease—specifically, redwater fever.[3] This is an acute febrile hemolytic disease of cattle characterized by the destruction of erythrocytes and the presence of hemoglobin in the urine, hence its name. The disease was also the first one associated with an arthropod vector—a tick. Although the morphology and life cycle of *Babesia* are not commonly understood, they are probably like those of the plasmodia.

Human babesiosis is a relatively rare zoonosis caused by a species of *Babesia* from wild rodents that has been accidently introduced by a hard tick's bite. This infection has been essentially eradicated in the United States, but it remains a problem in various other regions. Most cases occurred in the Northeast, where the *Ixodes* tick vector is located. The infection closely resembles malaria in pathology and symptomology, in that red blood cells burst after they are invaded, and it is diagnosed and treated in a similar fashion.

---

3. Also known as bovine babesiosis, Texas fever, or tick fever.

* *Babesia* (bab-ee′-see″-uh) Named for its discoverer, Victor Babes.

The apicomplexans are obligate intracellular protozoa that lack motility in the mature state. *Plasmodium* and *Babesia* cause serious systemic infections transmitted by mosquitoes and ticks. *Toxoplasma* causes mild, systemic infections in healthy hosts. *Cryptosporidium* causes intestinal infections.

Plasmodium species are the causative agents of malaria, the major protozoan disease worldwide. The *Plasmodium* life cycle is timed to coincide with the reproductive cycle of female *Anopheles* mosquitos, which require human blood for the development of their eggs.

Plasmodium sporozoites, which complete their development in the liver, invade red blood cells as merozoites, and then differentiate into reproductive gametes. Sexual reproduction is completed in the mosquito. Symptoms are caused by the damage to red blood cells. Prevention consists of mosquito abatement procedures and human prophylaxis. Quinine derivatives are the treatment of choice, but resistant strains are increasing.

Toxoplasma gondii causes a mild, systemic infection of healthy hosts, but fatal infections can occur transplacentally and in the immunosuppressed. Humans are accidental hosts; cats are the primary reservoir. Infection is associated with ingestion of cysts.

Cryptosporidium species cause an intestinal infection in both domestic animals and humans. Infection occurs through contact with infected animal feces, primarily through water supplies contaminated with cysts. Healthy hosts recover spontaneously, but the infection is potentially fatal to immunosuppressed hosts.

A Survey of Helminth Parasites

So far our survey has been limited to relatively small infectious agents, but *helminths,** particularly the adults, are comparatively large, multicellular animals with specialized tissues and organs similar to those of the hosts they parasitize. The behavior of these parasites and their adaptations to humans are at once fascinating and grotesque. Imagine having long, squirming white worms suddenly expelled from your nose, feeling and seeing a worm creeping beneath your skin, or having a worm slither across the front of your eye, and you have some idea of the conspicuous nature of these parasites.

Worms that parasitize humans are amazingly diverse, ranging from barely visible roundworms (0.3 mm) to huge tapeworms 25 meters long; and varying from oblong- and crescent-shaped to whip-like or elongate flat ribbons. In the introduction to these organisms in chapter 5, we grouped them into three categories: nematodes (roundworms), trematodes (flukes), and cestodes (tapeworms) and discussed basic characteristics of each group. In this section, we present major facets of helminth epidemiology, life cycles, pathology, and control, followed by an accounting of the major worm diseases. Table 23.4 presents a selected list of medically significant helminths.

---

* helminth Gr. *helmins,* worm.

GENERAL LIFE AND TRANSMISSION CYCLES

The complete life cycle of helminths includes the fertilized egg (embryo), larval, and adult stages. In the majority of helminths, adults derive nutrients and reproduce sexually in a host's body. In nematodes, the sexes are separate and usually different in appearance; in trematodes, the sexes can be either separate or **hermaphroditic,** meaning that male and female sex organs are in the same worm; cestodes are generally hermaphroditic. For a parasite's continued survival as a species, it must complete the life cycle by transmitting an infective form, usually an egg or larva, to the body of another host, either of the same or a different species. By convention, the host in which larval development occurs is the **intermediate (secondary) host,** and adulthood and mating occur in the **definitive (final) host.** A **transport host** is an intermediate host that experiences no parasitic development but is an essential link in the completion of the cycle.

Although individual variations occur in life and transmission cycles, most can be described by one of five basic patterns (figure 23.17). In general, sources for human infection are contaminated food, soil, and water or infected animals, and routes of infection are by oral intake or penetration of unbroken skin. Humans are the definitive hosts for many of the parasites listed in table 23.4, and in about half the diseases, they are also the sole biological reservoir. In other cases, animals or insect vectors serve as reservoirs or are required to complete worm development. In the majority of helminth infections, the worms must leave their host to complete the entire life cycle. Unlike microparasites, adult helminths do not continue to multiply in the host, so a more accurate term for their presence in the host is **infestation.**

GENERAL EPIDEMIOLOGY OF HELMINTH DISEASES

The helminths have a far-reaching impact on human health, economics, and culture. It has been said that there are probably more worm infections in humans than there are humans (due to multiple infections in one individual). Although individuals from all societies and regions play host to worms at some time in their lives, the highest case rates occur among children in rural areas of the tropics and subtropics. The most prevalent diseases and those creating the highest mortality and morbidity are schistosomiasis, filariasis, hookworm disease, ascariasis, and onchocerciasis. In most developed countries with temperate climates, the major parasitic infections are pinworm and trichinosis, though exotic parasitic worms are often encountered in immigrants and travelers. The continuing cycle between parasites and humans is often the result of practices such as using human excrement for fertilizer, exposing bare feet to soil or standing water, eating undercooked or raw meat and fish, and defecating in open soil. In some areas of the world, parasites have been an accepted part of life for thousands of years.

PATHOLOGY OF HELMINTH INFESTATION

Worms exhibit numerous adaptations to the parasitic habit. They have specialized mouthparts for attaching to tissues and for feeding, enzymes with which they liquefy and penetrate tissues, and a cuticle or other covering to protect them from the host

A Survey of Helminth Parasites   **709**

TABLE 23.4

Major Helminths of Humans and Their Modes of Transmission

Classification	Common Name of Disease or Worm	Life Cycle Requirement	Spread to Humans By
Nematodes (Roundworms)			
Intestinal Nematodes			Ingestion
Infective in egg (embryo) stage			
Trichuris trichiura	Whipworm	Humans	Fecal pollution of soil with eggs
Ascaris lumbricoides	Ascariasis	Humans	Fecal pollution of soil with eggs
Enterobius vermicularis	Pinworm	Humans	Close contact
Infective in larval stage			
Necator americanus	New World hookworm	Humans	Fecal pollution of soil with eggs
Ancylostoma duodenale	Old World hookworm	Humans	Fecal pollution of soil with eggs
Strongyloides stercoralis	Threadworm	Humans; may live free	Fecal pollution of soil with eggs
Trichinella spiralis	Trichina worm	Pigs, wild mammals	Consumption of meat containing larvae
Tissue Nematodes			Burrowing of larva into tissue
Wuchereria bancrofti	Filariasis	Humans, mosquitoes	Mosquito bite
Onchocerca volvulus	River blindness	Humans, black flies	Fly bite
Loa loa	Eye worm	Humans, mangrove flies, or deer flies	Fly bite
Dracunculus medinensis	Guinea worm	Humans and *Cyclops* (an aquatic invertebrate)	Ingestion of water containing *Cyclops*
Trematodes			
Schistosoma japonicum	Blood fluke	Humans and snails	Ingestion of fresh water containing larval stage
S. mansoni	Blood fluke	Humans and snails	Ingestion of fresh water containing larval stage
S. haematobium	Blood fluke	Humans and snails	Ingestion of fresh water containing larval stage
Opisthorchis sinensis	Chinese liver fluke	Humans, snails, fish	Consumption of fish
Fasciola hepatica	Sheep liver fluke	Herbivores (sheep, cattle)	Consumption of water and water plants
Paragonimus westermani	Oriental lung fluke	Humans, mammals, snails, freshwater crabs	Consumption of crabs
Cestodes			
Taenia saginata	Beef tapeworm	Humans, cattle	Consumption of undercooked or raw beef
T. solium	Pork tapeworm	Humans, swine	Consumption of undercooked or raw pork
Diphyllobothrium latum	Fish tapeworm	Humans, fish	Consumption of undercooked or raw fish
Hymenolepis nana	Dwarf tapeworm	Humans	Oral-fecal; close contact

defenses. In addition, their organ systems are usually reduced to the essentials: getting food and processing it, moving, and reproducing. Helminths entering by mouth undergo a period of development in the intestine that is often followed by spread to other organs. Helminths that burrow directly into the skin or are inserted by an insect bite quickly penetrate the circulation and are carried throughout the body. In both cases, most worms do not remain in a single location, and they migrate by various mechanisms through major systems, inflicting injury as they go.

Adult helminths eventually take up final residence in sites such as the intestinal mucosa, blood vessels, lymphatics, subcutaneous tissue, skin, liver, lungs, muscles, brain, or even the eyes. Traumatic damage occurring in the course of migration and organ invasion is primarily due to tissue feeding, toxic secretions, and blockages within organs due to high worm load. Pathological effects can include enlargement and swelling of organs, hemorrhage, weight loss, and anemia caused by the parasites' feeding on blood. Despite the popular misconception that intestinal worms cause weight loss by competing for food, weight loss is more likely caused by general malnutrition, malabsorption of food and vitamins due to intestinal damage, lack of appetite, vomiting, and diarrhea.

One common antihelminthic defense is an increase in granular leukocytes called eosinophils that have a specialized capacity to destroy worms (see chapter 14). Although antibodies and sensitized T cells are also formed against worms, their effects are rarely long-term, and reinfection is always a distinct possibility. The inability

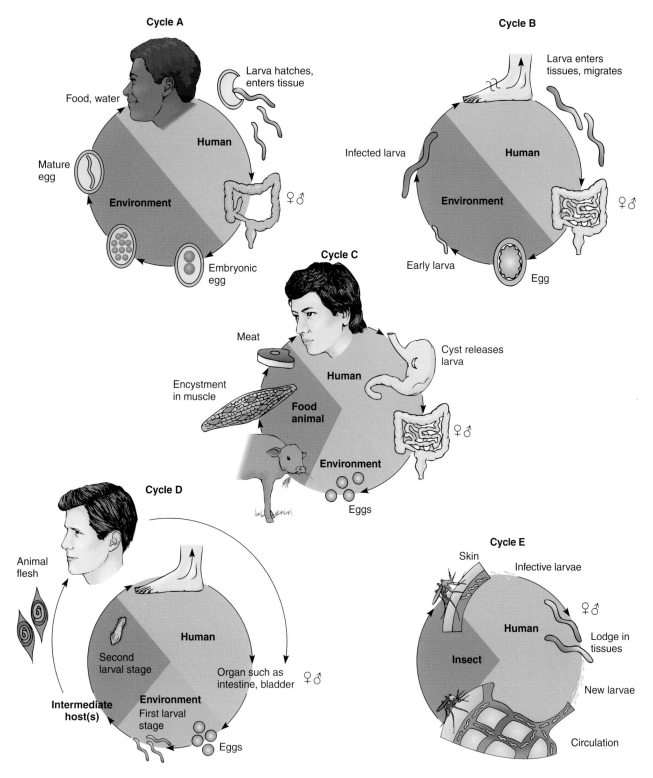

FIGURE 23.17

Five basic helminth life and transmission cycles. In cycle A, the worm develops in intestine; egg is released with feces into environment; eggs are ingested by new host and hatch in intestine (examples: *Ascaris, Trichuris*). In cycle B, the worm matures in intestine; eggs are released with feces; larvae hatch and develop in environment; infection occurs through skin penetration by larvae (example: hookworms). In cycle C, the adult matures in human intestine; eggs are released into environment; eggs are eaten by grazing animals; larval forms encyst in tissue; humans eating animal flesh are infected (example: *Taenia*). In cycle D, eggs are released from human; humans are infected through ingestion or direct penetration by larval phase (examples: *Opisthorchis* and *Schistosoma*). In cycle E, the human is definitive host and carries larval form in blood; insect vector is intermediate host that picks up and transmits larvae through bite (examples: *Wuchereria* and *Onchocerca*).

of the immune system to completely eliminate worm infections appears to be due to their relatively large size, migratory habits, and inaccessibility. The incompleteness of immunity has hampered the development of vaccines for parasitic diseases and has emphasized the use of chemotherapeutic agents instead.

ELEMENTS OF DIAGNOSIS AND CONTROL

Diagnosis of helminthic infestation may require several steps. A differential blood count showing eosinophilia and serological tests indicating sensitivity to helminths both provide indirect evidence of worm infestation. A history of travel to the tropics or immigration from those regions is also helpful, even if it occurred years ago, because some flukes and nematodes persist for decades. The most definitive evidence, however, is the discovery of eggs, larvae, or adult worms in stools, sputum, urine, blood, or tissue biopsies. The worms are sufficiently distinct in morphology that positive identification can be based on any stage, including eggs.

Although several useful antihelminthic medications exist, the cellular physiology of the eucaryotic parasites resembles that of humans, and drugs toxic to them can also be toxic to humans. Some antihelminthic drugs act to suppress a metabolic process that is more important to the worm than to the human. Others inhibit the worm's movement and prevent it from maintaining its position in a certain organ. Therapy is also based on a drug's greater toxicity to the more vulnerable helminths or on the local effects of oral drugs in the intestine. Antihelminthic drugs of choice and their effects are given in table 23.5. In some cases, surgery may be necessary to remove worms or larvae, though this can be difficult if the parasite load is high or not confined.

Preventive measures are aimed at minimizing human contact with the parasite or interrupting its life cycle. In areas where the worm is transmitted by fecally contaminated soil and water, disease rates are significantly reduced through proper sewage disposal, using sanitary latrines, avoiding human feces as fertilizer, and disinfection of the water supply. If the larvae are aquatic and invade through the skin, people should avoid direct contact with infested water. Food-borne disease can be avoided by thoroughly washing and cooking vegetables and meats. Also, because adult worms, larvae, and eggs are sensitive to cold, freezing foods is a highly satisfactory preventive measure. These methods work best if humans are the sole host of the parasite; if they are not, control of reservoirs or vector populations may be necessary.

NEMATODE (ROUNDWORM) INFESTATIONS

Nematodes* are elongate, cylindrical worms that biologists consider one of the most abundant animal groups (a cup of soil or mud can easily contain millions of them). The vast majority are free-living soil and freshwater worms, but around 200 are parasitic, including 50 species that affect humans. Nematodes bear a smooth, protective outer covering, or cuticle, that is periodically shed as the worm grows. They are essentially headless, tapering to a fine point anteriorly. The sexes are distinct, and the male is often smaller than the female, having a hooked posterior containing spicules for coupling with the female (figure 23.18*a*).

---

* nematode (nem´-ah-tohd) Gr. *nema*, thread.

TABLE 23.5

Antihelminthic Therapeutic Agents and Indications for Their Use

Drug	Effect	Used For
Piperazine	Paralyzes worm so it can be expelled in feces	Ascariasis, pinworm, hookworm
Pyrantel	Paralyzes worm so it can be expelled in feces	Ascariasis, pinworm, hookworm, trichinosis
Mebendazole	Blocks key step in worm metabolism	Trichuriasis (whipworm), ascariasis, hookworm, trichinosis
Thiabendazole	Blocks key step in worm metabolism	Strongyloidiasis, guinea worm, hookworm, trichinosis
Albendazole	Blocks key step in worm metabolism	Echinococciasis
Diethylcarbamazine	Unknown, but kills microfilariae; does not kill adult worm	Filariasis, loiasis
Niridazole	Alters worm metabolism	Schistosomiasis
Praziquantel	Interferes with worm metabolism	Schistosomiasis, other flukes, tapeworm
Metrifonate	Paralyzes worm	Schistosomiasis
Niclosamide	Inhibits ATP formation in worm; destroys proglottids but not eggs	Tapeworm
Ivermectin	Blocks nerve transmission	Onchocerciasis

The human parasites are divided into intestinal nematodes, which develop to some degree in the intestine (for example, *Ascaris, Strongyloides,* and hookworms), and tissue nematodes, which spend their larval and adult phases in various soft tissues other than the intestine (for example, various filarial worms).

Intestinal Nematodes (Cycle A)

***Ascaris lumbricoides* and Ascariasis** *Ascaris lumbricoides** is a giant (up to 300 mm long) intestinal roundworm that probably accounts for the greatest number of worm infections (estimated at one billion cases worldwide). Most reported cases in the United States occur in the southeastern states. *Ascaris* spends its larval and adult stages in humans and releases embryonic eggs in feces, which are

---

* *Ascaris lumbricoides* (as´-kah-ris lum″-brih-koy´-dees) Gr. *askaris*, an intestinal worm; L. *lumbricus*, earthworm, and Gr. *eidos*, resemblance.

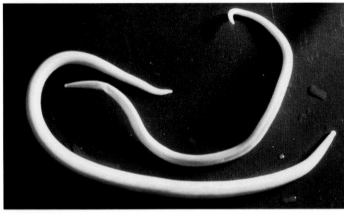

(a)

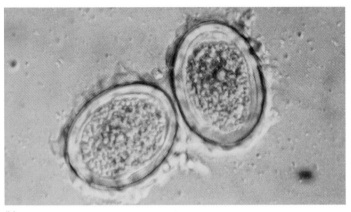

(b)

FIGURE 23.18

Ascaris lumbricoides. **(a)** Adult worms. The female is larger; the male has a hooked end. **(b)** A microscopic view of *A. lumbricoides* ova from a fecal sample.

then spread to other humans through food, drink, or soiled objects placed in the mouth (figure 23.18). The eggs thrive in warm, moist soils and resist cold and chemical disinfectants; they are also sensitive to sunlight, high temperatures, and drying. After ingested eggs hatch in the human intestine, the larvae embark upon an odyssey in the tissues. First they penetrate the intestinal wall and enter its lymphatic and circulatory drainage. From there, they are swept into the right ventricle of the heart, and eventually arrive at the capillaries of the lungs. From this point, the larvae migrate up the respiratory tree to the glottis. Worms entering the throat are swallowed and returned to the small intestine, where they reach adulthood and reproduce, producing up to 200,000 fertilized eggs a day.

Even as adults, male and female worms are not attached to the intestine and retain some of their exploratory ways. They are known to invade the biliary channels of the liver and gallbladder, and on occasion the worms emerge from the mouth and nares. Severe inflammatory reactions mark the migratory route, and allergic reactions such as bronchospasm, asthma, or skin rash can occur. Heavy worm loads can retard the physical and mental development of children. One possibility with intestinal worm infestations is self-reinoculation due to poor personal hygiene.

Trichuris trichiura **and Whipworm Infection** The common name for *Trichuris** *trichiura*—whipworm—refers to its likeness to a miniature buggy whip. Although the worm is morphologically different from *Ascaris,* it has a similar transmission cycle, with humans as the sole host. Trichuriasis has its highest incidence in areas of the tropics and subtropics that have poor sanitation. Embryonic eggs deposited in the soil are not immediately infective and continue development for 3 to 6 weeks in this habitat. Ingested eggs hatch in the small intestine, where the larvae attach, penetrate the outer wall, and go through several molts. The mature adults move to the large intestine and gain a hold with their long, thin heads, while the thicker tail dangles free in the lumen. Following sexual maturation and fertilization, the females eventually lay 3,000 to 5,000 eggs daily into the bowel. The entire cycle requires about 90 days, and untreated infestation can last up to 2 years.

Worms burrowing into the intestinal mucosa can pierce capillaries, cause localized hemorrhage, and provide a portal of entry for secondary bacterial infection. Heavier infestations can cause dysentery, loss of muscle tone, and rectal prolapse, which can prove fatal in children.

Enterobius vermicularis **and Pinworm Infection** The pinworm or seatworm, *Enterobius vermicularis,** was first discussed in chapter 5 (see figure 5.35). Enterobiasis is the most common worm disease of children in temperate zones. The transmission of the pinworm follows a course similar to that of the other enteric parasites. Freshly deposited eggs have a sticky coating that causes them to lodge beneath the fingernails and to adhere to fomites. Upon drying, the eggs become airborne and settle in house dust. Worms are ingested from contaminated food and drink, and self-inoculation from one's own fingers is common. Eggs hatching in the small intestine release larvae, which migrate to the large intestine, mature, and then mate. Itching is brought on when the mature female emerges from the anus and lays eggs. Although infestation is not fatal and most cases are asymptomatic, the afflicted child can suffer from disrupted sleep and sometimes nausea, abdominal discomfort, and diarrhea. A simple, rapid test can be done by pressing a piece of transparent adhesive tape against the anal skin and then applying it to a slide for microscopic evaluation.

Intestinal Helminths (Cycle B)

The Hookworms Two major agents of human hookworm infestation are *Necator americanus,** endemic to the New World, and *Ancylostoma duodenale,** endemic to the Old World, though the two species overlap in parts of Latin America. Otherwise, with respect to transmission, life cycle, and pathology, they are usually lumped together. Humans are sometimes invaded by cat or dog hookworm larvae, sustaining a dermatitis called creeping eruption

---

* *Trichuris* (trik-ur′-is) Gr. *thrix*, hair, and *oura*, tail.

* *Enterobius vermicularis* (en″-ter-oh′-bee-us ver-mik″-yoo-lah′-ris) Gr. *enteron*, intestine, and *bios*, life; L. *vermiculus*, a small worm.

* *Necator americanus* (nee-kay′-tor ah-mer″-ih-cah′-nus) L. *necator*, a murderer, and *americanus*, of New World origin.

* *Ancylostoma duodenale* (an″-kih-los′-toh-mah doo-oh-den-ah′-lee) Gr. *amkylos*, curved or hooked, *stoma*, mouth, and *duodenale*, the intestine.

MICROBITS 23.2
When Parasites Attack the "Wrong" Host

Not only do we humans host a number of our own parasites, but occasionally, worms indigenous to other animals attempt to take up residence in us. When this happens, the parasite cannot complete the life cycle in its usual fashion, but in making the attempt, it can cause discomfort and even death in some cases.

One such parasite is *Echinococcus granulosus,* a tiny intestinal tapeworm that causes **hydatid cyst disease.** The parasite's usual definitive host is the dog, though it can be transmitted to other carnivores and herbivores in egg and larval stages. Humans become involved through frequent, intimate contact with fur, paws, and tongue, especially while "kissing" their dog. If a person accidently ingests eggs, the larvae hatch and migrate through the body, eventually coming to rest in the liver and lungs. There they form compact, water-filled packets called **hydatid cysts** (L. *hydatos,* water drop). The bursting of these capsules gives rise to an allergic reaction and also proliferates the worm by seeding other sites.

Another disease that arises from invasion by animal worms is **cutaneous larval migrans,** also described as a creeping eruption caused by both dog and cat hookworm *(Ancylostoma).* It is acquired when humans walk barefoot through areas where dogs and cats defecate, enabling the larvae to enter the skin. The larvae cannot invade completely and so remain just beneath the skin, producing lesions that become so itchy and inflamed that the person cannot eat or sleep. Another migratory infection called **visceral larval migrans** is caused by the dog ascarid *Toxocara* that is ingested in egg form with contaminated soil. The larvae hatch in the intestine and migrate extensively through the liver, lungs, and circulation.

Likewise, "swimmer's itch" results when larval schistosomes of birds and small mammals penetrate a short distance into the skin and then die. The disintegrating worm sensitizes the humans and causes intense itching and discomfort.

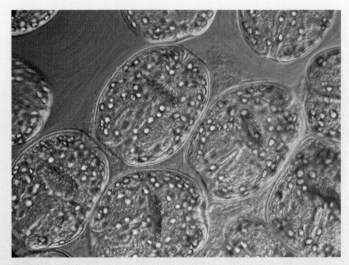

Hydatid cysts from the lung. These fluid-filled spheres are also known as bladder worms.

(Microbits 23.2). The *hook* refers to the adult's oral cutting plates by which it anchors to the intestinal villi and its curved anterior end (figure 23.19). Unlike other intestinal worm infections, hookworm larvae hatch outside the body and infect by penetrating the skin.

Hookworm infestation remains endemic to the tropics and subtropics and is absent in arid, cold regions. In recent times, its worldwide incidence has decreased. In the past, hookworm plagued parts of the southern United States, unfortunately sapping its victims' strength and stigmatizing them as unkempt and ignorant. The decline of the disease can be attributed to such simple changes as improved sanitation (indoor plumbing) and generally better standards of living (being able to afford shoes).

After being deposited into soil by defecation, hookworm eggs hatch into fine *filariform** larvae that instinctively climb onto grass or other vegetation to improve the probability of contact with a host. Ordinarily, the parasite enters sites on bare feet such as hair follicles, abrasions, or the soft skin between the toes, but cases have occurred in which mud splattered on the ankles of people wearing shoes. Infection has even been reported in people handling soiled laundry.

On contact, the hookworm larvae actively burrow into the skin. After several hours, they reach the lymphatic or blood circulation and are immediately carried into the heart and lungs. The larvae proceed up the bronchi and trachea to the throat. Although some larvae are ejected by spitting, most are swallowed with sputum and arrive in the small intestine, where they anchor, feed on tissues, and mature. Eggs first appear in the stool about 6 weeks from the time of entry, and the untreated infection can last about 5 years.

Initial penetration with larvae can evoke a localized dermatitis called "ground itch." The transit of the larvae to the lungs is ordinarily brief, but it can cause symptoms of pneumonia and eosinophilia. The potential for injury is greatest during the intestinal phase, when heavy worm burdens can cause nausea, vomiting, cramps, pain, and bloody diarrhea. Because blood loss is significant, iron-deficient anemia develops, and infants are especially susceptible to hemorrhagic shock. Chronic fatigue, listlessness, apathy, and anemia worsen with chronic and repeated infections.

Strongyloides stercoralis* and Strongyloidiasis** The agent of strongyloidiasis, or threadworm infection, is ***Strongyloides stercoralis.* This nematode is exceptional because of its minute size and its capacity to complete its life cycle either within the human body

---

* **filariform** (fih-lar´-ih-form) L. *filum,* a thread.

* *Strongyloides stercoralis* (stron˝-jih-loy´-deez ster˝-kor-ah´-lis) Gr. *strongylos,* round, and *stercoral,* pertaining to feces.

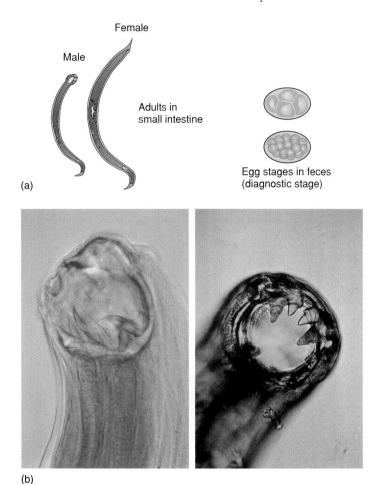

(a)

Female

Male

Adults in
small intestine

Egg stages in feces
(diagnostic stage)

(b)

FIGURE 23.19

The hookworms. **(a)** Adult and egg stages of *Necator americanus*.
(b) Cutting teeth on the mouths of (left) *N. americanus* and (right)
Ancylostoma duodenale.

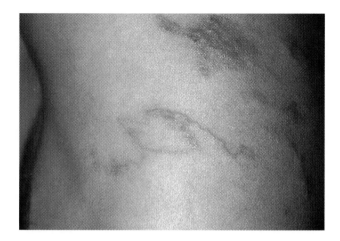

FIGURE 23.20

A patient with disseminated *Strongyloides* infection. Trails under
the skin indicate the migration tracks of the worms.

or outside in moist soil. It shares a similar distribution and life cycle to hookworms and afflicts an estimated 100 to 200 million people worldwide. Infection occurs when soil larvae penetrate the skin.
The worm then enters the circulation, is carried to the respiratory
tract and swallowed, and then enters the small intestine to complete
development. Although adult *S. stercoralis* lay eggs in the gut just
as hookworms do, the eggs can hatch into larvae in the colon and
remain entirely in the host's body to complete the cycle. Eggs can
likewise exit with feces and go through an environmental cycle.
These numerous alternative life cycles greatly increase the chance
of transmission and the likelihood for chronic infection.

The first hint of threadworm infection is usually a red, intensely itchy skin rash at the site of entry. Mild migratory activity in
an otherwise normal person can escape notice, but heavy worm
loads can cause symptoms of pneumonitis and eosinophilia. The
nematode activities in the intestine produce bloody diarrhea, liver
enlargement, bowel obstruction, and malabsorption. In immunocompromised patients there is a risk of severe, disseminated infestation involving numerous organs (figure 23.20). Hardest hit are
AIDS patients, transplant patients on immunosuppressant drugs,
and cancer patients receiving irradiation therapy, who can die if not
treated promptly.

***Trichinella spiralis* and Trichinosis** The life cycle of ***Trichinella
spiralis*** is spent entirely within the body of a mammalian host,
usually a carnivore or omnivore such as a pig, bear, cat, dog, or rat.
In nature, the parasite is maintained in an encapsulated (encysted)
larval form in the muscles of these animal reservoirs and is transmitted when other animals prey upon them. Humans acquire **trichinosis** (also called trichinellosis) by eating the meat of an infected
animal (usually wild or domestic swine or bears), but humans are
essentially dead-end hosts (provided corpses are buried and cannibalism is not practiced!) (figure 23.21).

Because all wild and domesticated mammals appear to be
susceptible to *T. spiralis,* one might expect human trichinellosis to
be common worldwide. But, in actual fact, it is less common in the
more populated countries than it is in the United States (estimated
at 100,000 new cases a year) and in Europe. This distribution appears to be related to regional or ethnic customs of eating raw or
rare pork dishes or wild animal meats. Bear meat is the source of up
to one-third of the cases in the United States. Home or small-scale
butchering enterprises that do not carefully inspect pork can spread
the parasite, although commercial pork can also be a source. Practices such as tasting raw homemade pork sausage for proper sea-
soning or serving rare pork or pork-beef mixtures have been responsible for sporadic outbreaks.

The cyst envelope is digested in the stomach and small intestine, which liberates the larvae. After burrowing into the intestinal
mucosa, the larvae reach adulthood and mate. An unusual aspect of
this worm's life cycle is that males die and are expelled in the feces,
while females incubate eggs and shed live larvae. These penetrate
the intestine and enter the lymphatic channels and blood. All tissues
are at risk for invasion, but final development occurs when the
coiled larvae are encysted in the skeletal muscle (figure 23.21). At
maturity, the cyst is about 1 mm long and can be observed by careful inspection of meat. Although larvae can deteriorate over time,
they have also been known to survive for years.

* *Trichinella spiralis* (trik″-ih-nel′-ah spy-ral′-is) A little, spiral-shaped hair worm.

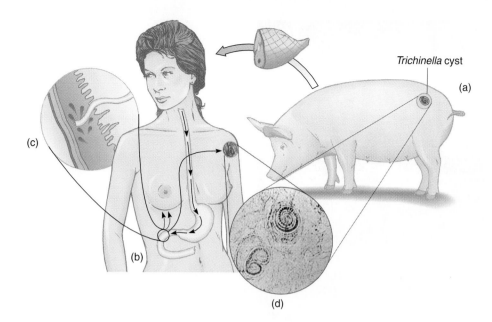

Trichinella cyst

(a)

(b)

(c)

(d)

FIGURE 23.21

The cycle of trichinosis. **(a)** The infective stage is larvae encysted in animal muscle. **(b)** When ingested, the cyst hatches and matures in the intestinal lining. **(c)** Offspring of adult worms burrow through the intestine and penetrate the circulation. **(d)** They eventually form cysts in skeletal muscle that can remain for years. Inset shows a biopsy of human skeletal muscle infected with the coiled larvae of *Trichinella spiralis.*

(d inset) *Source: Koneman et al.,* Diagnostic Microbiology, *4th ed., 1992. Reprinted by permission from J. B. Lippincott Co.*

The first symptoms of trichinosis mimic influenza or viral fevers, with diarrhea, nausea, abdominal pains, fever, and sweating. The second phase, brought on by the mass migration of larvae and their entrance into muscle, produces puffiness around the eyes, intense muscle and joint pain, shortness of breath, and pronounced eosinophilia. The most serious life-threatening manifestations are heart and brain involvement. Although the symptoms eventually subside, a cure is not available once the larvae have encysted in muscles.

The most effective preventive measures for trichinosis are to adequately cook, freeze, or smoke pork and wild meats.

Tissue Nematodes

As opposed to the intestinal worms previously described, tissue nematodes complete their life cycle in human blood, lymphatics, or skin. The most important members of this category are filarial worms, nematodes with elongate, filamentous bodies. These worms also share their transmission mode (biting arthropods) and disease pattern (slow, chronic, deforming).

Filarial Nematodes and Filariasis (Cycle E)

The filarial parasites are threadlike worms that can live in host tissues for years. They are characterized by the production of tiny **microfilarial** larvae that circulate in the blood and lymphatics. Their anatomy is relatively simple, but they have a complex biphasic life cycle that alternates between humans and blood-sucking mosquito or fly vectors. The species responsible for most **filariases** are *Wuchereria bancrofti,* the cause of elephantiasis, *Onchocerca volvulus,* the agent of river blindness, and *Loa loa,* the eye worm.

Wuchereria bancrofti and Bancroftian Filariasis *Wuchereria bancrofti***** causes bancroftian filariasis, a common affliction in tropical regions where its vectors live (female *Culex, Anopheles,*

and *Aedes* mosquitoes). The microfilariae are highly attuned to these vectors, such that they swarm to the skin capillary circulation to coincide with the times the mosquito feeds. The microfilariae complete development in the flight muscles of the mosquito, and the resultant larvae gather in its mouthparts. Infection begins when a mosquito bite deposits the infective larvae into the skin. These larvae penetrate the lymphatic vessels, where they develop into long-lived, reproducing adults. Mature female worms expel huge numbers of microfilariae directly into the circulation throughout their life cycle (figure 23.22*a*).

The first symptoms of larval infestation in humans are inflammation and dermatitis in the skin near the mosquito bite, lymphadenitis, and testicular pain (in males). The most conspicuous and bizarre sign of bancroftian filariasis is the chronic swelling of the scrotum or legs called **elephantiasis.***** This edema is caused by inflammation and blockage of the main lymphatic channels, which prevent return of lymph to the circulation and cause large amounts of fluid to accumulate in the extremities (figure 23.22*b*). The condition can predispose to ulceration and infection. In extreme cases, the patient can become physically incapacitated and suffer from the stigma of having to carry a massive growth. This condition occurs only in individuals in endemic regions after one to five decades of chronic infection. Acute infection in a young, healthy person is usually subclinical.

Onchocerca volvulus* and River Blindness**   *Onchocerca volvulus* is a filarial worm transmitted by small biting vectors called **black flies.** These voracious flies often attack in large numbers, and it is not uncommon to be bitten several hundred times a day. Onchocerciasis is found in rural settlements along rivers bordered by overhanging vegetation, which are typical breeding sites of the black fly. The life and transmission cycle is similar to that of

---

* *Wuchereria bancrofti* (voo″-kur-ee′-ree-ah bang-krof′-tee) After Otto Wucherer and J. Bancroft.

* elephantiasis (el″-eh-fan-ty′-ah-sis) Gr. *elephas,* elephant.

* *Onchocerca volvulus* (ong″-koh-ser′-kah volv′-yoo-lus) Gr. *onkos,* hook, and *kerkos,* tail; L. *volvere,* to twist.

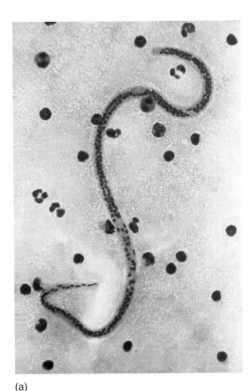

(a)

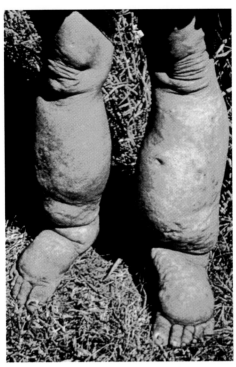

(b)

FIGURE 23.22

Bancroftian filariasis. **(a)** The appearance of a larva in a blood smear. **(b)** Pronounced elephantiasis in a Costa Rican woman.

Wuchereria, although the larvae are deposited into the bite wound and develop into adults in the immediate subcutaneous tissues, where disfiguring nodules form within one to two years after initial contact. Microfilariae given off by the adult female migrate via the blood to many locations, but especially the eyes. It is the blood phase that infects other feeding black flies.

Some cases of onchocerciasis result in a severe itchy rash that can last for years. As worms degenerate, escaping antigens provoke inflammatory and granulomatous lesions in the skin. The other, more serious complication is **river blindness,** arising from infestation of the eye. The worms eventually invade the entire eye, producing much inflammation and permanent damage to the retina and optic nerve. In regions of high prevalence, it is not unusual for an ophthalmologist to see microfilariae wiggling in the anterior chamber during a routine eye checkup. Microfilariae die in several months, but adults can exist for up to 15 years in nodules. River blindness has been a serious problem in many areas of Africa. In some villages, nearly half of the residents are affected by the disease. A campaign to eradicate onchocerciasis is currently underway, using *ivermectin,* a potent antifilarial drug, and insecticides to control the black flies.

Loa loa: **The African Eye Worm**   Another nematode that makes disturbing migratory visits across the eye and under the skin is *Loa loa* (loh´-ah loh´-ah). This worm is larger and more apparent than *O. volvulus.* The agent occurs in parts of western and central Africa and is spread by the bites of dipteran flies. Its life cycle is similar to *Onchocerca* in that the larvae and adults remain more or less confined to the subcutaneous tissues. The first signs of disease are itchy, marble-sized lesions called *calabar swellings* at the site of

penetration. These nodules are transient, and the worm lives for only a year or so, but the infestation is often recurrent. Because the nematode is temperature-sensitive, it is drawn to the body surface in response to warmer temperature and is frequently felt slithering beneath the skin or observed beneath the conjunctiva. One method of treatment is to carefully pull the worm from a small hole in the conjunctiva following application of local anesthesia. Some infections can be suppressed with high doses of diethylcarbamazine.

Dracunculus medinensis* **and Dracontiasis**   Dracontiasis, caused by the guinea worm, once caused untold misery in residents of India, the Middle East, and central Africa. The parasite is carried by *Cyclops,* an arthropod commonly found in still water. When people ingest water from a contaminated community pond or open well, the larvae are freed in the intestine and penetrate into subcutaneous tissues where they form highly irritating lesions.

An intense effort by the WHO aimed to eradicate the disease by converting ponds to wells, filtering water through a fine nylon sieve, and treating the water with an insecticide. So successful was this strategy that, as of 2001, dracontiasis has been eliminated from all parts of the world.

THE TREMATODES, OR FLUKES

The fluke (ME *floke,* flat) is named for the characteristic ovoid or flattened body found in most adult worms, especially liver flukes such as *Fasciola hepatica.* Blood flukes, also called schistosomes,

---

* *Dracunculus medinensis* (drah-kung´-kyoo-lus meh-dih-nen´-sis) The little dragon of Medina. Also called guinea worm.

have a more cylindrical body plan. The name **trematode** (Gr. *trema,* a hole) comes from the muscular sucker containing a mouth (hole) at the fluke's anterior end (see figure 5.34*b*). Flukes have digestive, excretory, neuromuscular, and reproductive systems, but lack circulatory and respiratory systems. Except in the blood flukes, both male and female reproductive systems occur in the same individual and occupy a major portion of the body. The trematode life cycles are closely adapted to agricultural practices such as flood irrigating and using raw sewage as fertilizer. In human fluke cycles, animals such as snails or fish are usually the intermediate hosts, and humans are the definitive hosts.

Blood Flukes: The Schistosomes (Cycle D)

Schistosomiasis has afflicted humans for thousands of years. Ancient Egyptian writings that described males "menstruating" were probably referring to blood in the urine caused by renal schistosomiasis. The disease is caused by *Schistosoma* mansoni, S. japonicum,* or *S. haematobium,*[4] species that are morphologically and geographically distinct but share similar life cycle, transmission methods, and general disease manifestations. The disease occurs in 73 countries located in Africa, South America, the Middle East, and the Far East. Schistosomiasis is a major public health problem, probably affecting 20 million people at any one time, worldwide. Recent increases in its occurrence in Africa have been attributed to new dams on the Nile River, which have provided additional habitat for snail hosts.

The schistosome parasite demonstrates intimate adaptations to both humans and certain species of freshwater snails required to complete its life cycle. The cycle begins when infected humans release eggs into irrigated fields or ponds either by deliberate fertilization with excreta or by defecating or urinating directly into the water. The egg hatches in the water and gives off an actively swimming ciliated larva called a **miracidium*** (figure 23.23*a*), which instinctively swims to a snail and burrows into a vulnerable site, shedding its ciliated covering in the process. In the body of the snail, the miracidium multiplies and transforms into a larger, fork-tailed swimming larva called a **cercaria*** (figure 23.23*b*). Cercariae are given off by the thousands into the water by infected snails.

Upon contact with a human wading or bathing in water, cercariae attach themselves to the skin by ventral suckers and penetrate hair follicles. They pass into small blood and lymphatic vessels and are carried to the liver. Here, the schistosomes achieve sexual maturity, and the male and female worms remain permanently entwined to facilitate mating (figure 23.23*c*). In time, the pair migrates to and lodges in small blood vessels at specific sites. *Schistosoma mansoni* and *S. japonicum* end up in the mesenteric venules of the small intestine; *S. haematobium* enters the venous plexus of the bladder. While attached to these intravascular sites, the worms feed upon blood, and the female lays eggs that are eventually voided in feces or urine.

The first symptoms of infestation are itchiness in the vicinity of cercarial entry, followed by fever, chills, diarrhea, and cough. The most severe consequences, associated with chronic infections,

---

4. The species are named after P. Manson; Japan; and Gr. *haem,* blood, and *obe,* to like.

* *Schistosoma* (skis″-toh-soh′-mah) Gr. *schisto,* split, and *soma,* body.

* miracidium (my″-rah-sid′-ee-um) Gr. *meirakidion,* little boy.

* cercaria (sir-kair′-ee-uh) Gr. *kerkos,* tail.

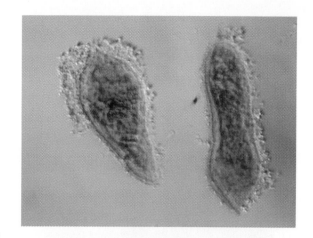

(a)

(b)

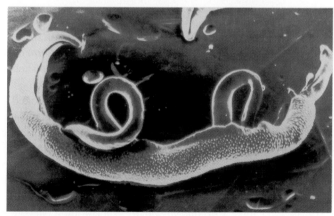

(c)

FIGURE 23.23

Stages in the life cycle of *Schistosoma.* **(a)** The miracidium phase, which infects the snail. **(b)** The cercaria phase, which is released by snails and burrows into the human host. **(c)** An electron micrograph of normal mating position of adult worms. The male worm holds the female in a groove on his ventral surface.

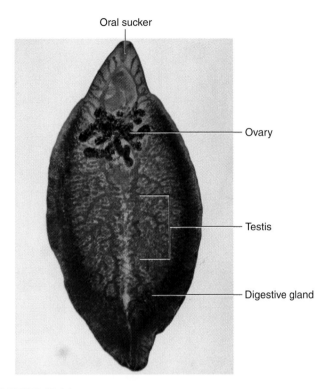

Oral sucker

Ovary

Testis

Digestive gland

FIGURE 23.24

 Fasciola hepatica, the sheep liver fluke (2×).

are hepatomegaly and liver disease, splenomegaly, bladder obstruction, and blood in the urine. Occasionally, eggs carried into the central nervous system and heart create a severe granulomatous response. Adult flukes can live for many years, and by eluding the immune defenses, cause a chronic affliction.

In parts of Africa where this disease is endemic, epidemiologists are attempting a unique form of biological control. They have introduced into local lakes a species of small fish that feeds on snails, which greatly reduces the hosts for cercariae and the possibility for infection.

Liver and Lung Flukes (Cycle D)

Several trematode species that infest humans may be of zoonotic origin. The Chinese liver fluke, **Opisthorchis (Clonorchis) sinensis,*** completes its sexual development in mammals such as cats, dogs, and swine. Its intermediate development occurs in snail and fish hosts. Humans ingest cercariae in inadequately cooked or raw freshwater fish and crustaceans. Larvae hatch and crawl into the bile duct, where they mature and shed eggs into the intestinal tract. Feces containing eggs are passed into standing water that harbors the intermediate snail host. The cycle is complete when infected snails release cercariae that invade fish or invertebrates living in the same water.

The liver fluke **Fasciola hepatica*** is a common parasite in sheep, cattle, goats, and other mammals and is occasionally transmitted to humans (figure 23.24). Periodic outbreaks in temperate

regions of Europe and South America are associated with eating wild watercress. The life cycle is very complex, involving the mammal as the definitive host, the release of eggs in the feces, the hatching of the egg in water into miracidia, invasion of freshwater snails, development and release of cercariae, encystment of cercariae on a water plant, and ingestion of the cyst by a mammalian host eating the plant. The cysts release young flukes into the intestine that wander to the liver, lodge in the gallbladder, and develop into the adult. Humans develop symptoms of vomiting, diarrhea, hepatomegaly, and bile obstruction only if they are chronically infected by a large number of flukes.

The adult lung fluke **Paragonimus westermani,*** endemic to the Orient, India, and South America, occupies the pulmonary tissues of various reservoir mammals, usually carnivores such as cats, dogs, foxes, wolves, and weasels, and has two intermediate hosts—snails and crustaceans. Humans accidently contract infection by eating undercooked crustaceans. After being released in the intestine, the worms migrate to the lungs, where they can cause cough, pleural pain, and abscess.

CESTODE (TAPEWORM) INFESTATIONS (CYCLE C)

The **cestodes,*** or tapeworms, are among the most extreme parasitic worms, consisting of little more than a tiny holdfast connected to a chain of flattened sacs that contain the reproductive organs. Several anatomical features of the adult emphasize their adaptations to an intestinal existence (figure 23.25a,b). The small **scolex,*** or head of the worm, has suckers or hooklets for clinging to the intestinal wall, but it has no mouth because nutrients are absorbed directly through the body of the worm. The importance of the scolex in anchoring the worm makes it a potential target for antihelminthic drugs. The head is attached by a **neck** to the **strobila,*** a long ribbon composed of individual reproductive segments called **proglottids.*** The contents of the proglottids are dominated by male and female reproductive organs, though each one also contains a primitive neuromuscular system that permits feeble movements. The neck generates the proglottids, tapering from the smaller, newly formed, and least mature segments nearest the neck to the larger, older ones containing ripe eggs at the distal end of the worm. In some worms, the proglottids are shed intact in feces after detaching from the strobila, and in others, the proglottids break open in the intestine and expel free eggs into the feces.

All tapeworms for which humans are the definitive hosts require an intermediate animal host. The principal human species are *Taenia saginata,* the beef tapeworm; *T. solium,* the swine tapeworm; and *Diphyllobothrium latum,* the fish tapeworm (Medical Microfile 23.3). The first two tapeworms cause an important infestation called **taeniasis.** The beef tapeworm follows a cycle C pattern, in which the larval worm is ingested in raw beef, and the pork tapeworm follows both a cycle C and a modified cycle A, in which the eggs are ingested and hatch in the body.

---

* *Opisthorchis sinensis* (oh″-piss-thor′-kis sy-nen′-sis) Gr. *opisthein,* behind, *orchis,* testis, and *sinos,* oriental.

* *Fasciola hepatica* (fah-see′-oh-lah heh-pat′-ih-kah) Gr. *fasciola,* a band, and *hepatos,* liver.

* *Paragonimus westermani* (par″-ah-gon′-ih-mus wes-tur-man′-eye) Gr. *para,* to bear, and *gonimus,* productive; after Westerman, a researcher in flukes.

* cestode (ses′-tohd) L. *caestus,* to strike.

* scolex (skoh′-leks) Gr. *scolos,* worm.

* strobila (stroh-bih′-lah) Gr. *strobilos,* twisted.

* proglottid (proh-glot′-id) Gr. *pro,* before, and *glotta,* the tongue.

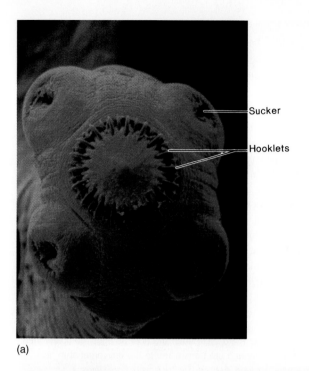

(a)

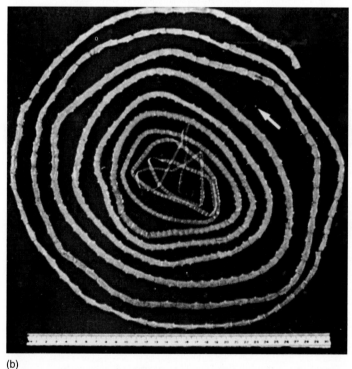

(b)

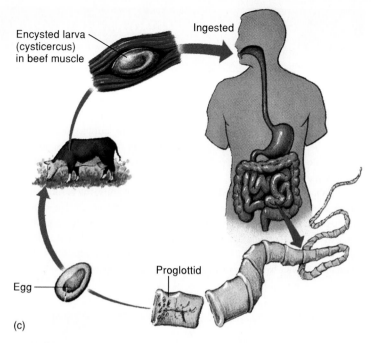

(c)

FIGURE 23.25

🔘 **Tapeworm infestation in humans.** **(a)** Tapeworm scolex showing sucker and hooklets. **(b)** Adult *Taenia saginata*. The arrow points to the scolex; the remainder of the tape, called the strobila, has a total length of 5 meters. **(c)** A generalized diagram of the life cycle of the beef tapeworm *T. saginata*.

23.25*c*). Ultimately, they encyst in the muscles, becoming **cysticerci,** young tapeworms that are the infective stage for humans. When humans ingest a live cysticercus in beef, it is uncoated and flushed into the intestine, where it firmly attaches by the scolex and develops into an adult tapeworm. Humans are not known to acquire infection by ingesting the eggs. For such a large organism, it is remarkable how few symptoms a tapeworm causes. Occasionally, a patient discovers proglottids in his or her stool, and some patients complain of vague abdominal pain and nausea.

Taenia solium differs from *T. saginata* by being somewhat smaller, having a scolex with hooklets and suckers to attach to the intestine, and being infective to humans in both the cysticercus and egg stages. The disease is endemic to regions where pigs eat fecally contaminated food and humans consume raw or partially cooked pork (Southeast Asia, South America, Mexico, and eastern Europe). The cycle involving ingestion of cysticerci is nearly identical to that of the beef tapeworm (figure 23.25*c*).

A different form of the disease, called **cysticercosis,** occurs when humans ingest pork tapeworm eggs from food or water. Although humans are not the usual intermediate hosts, the eggs can still hatch in the intestine, releasing tapeworm larvae that migrate to all tissues. They finally settle down to form peculiar cysticerci, or bladder worms, each of which has a small capsule resembling a little bladder. They do the most harm when they lodge in the heart muscle, eye, or brain, especially if a large number of bladder worms occupy tissue space (figure 23.26). It is not uncommon for patients to exhibit seizures, psychiatric disturbances, and other signs of neurological impairment.

*Taenia saginata** is one of the largest tapeworms, composed of up to 2,000 proglottids and anchored by a scolex with suckers. Taeniasis caused by the beef tapeworm is distributed worldwide but is mainly concentrated in areas where cattle ingest proglottids or eggs from contaminated fields and water, and where raw or undercooked beef is eaten. The eggs hatch in the small intestine, and the released larvae migrate throughout the organs of the cow (figure

---

* Taenia saginata (tee′-nee-ah saj-ih-nah′-tah) L. *taenia,* a flat band, *sagi,* a pouch, and *natare,* to swim.

MEDICAL MICROFILE 23.3
Waiter, There's a Worm in My Fish

Various forms of raw fish or shellfish are part of the cuisine of many cultures. Popular examples are Japanese sashimi and sushi. Although these delicacies are generally safe, fish flesh often harbors various types of helminths. Humans are not usually natural hosts for these parasites, so the worms do not fully invade, but some of their effects can be quite distressing.

One of the common fish parasites transmitted this way is ***Anisakis,*** a nematode parasite of marine vertebrates, fish, and mammals. The symptoms of anisakiasis, usually appearing within a few hours after a person ingests the half-inch larvae, are due to the attempts of the larvae to burrow into the tissues. The first symptom is often a tingling sensation in the throat, followed by acute gastrointestinal pain, cramping, and vomiting that mimic appendicitis. At times, the worms are regurgitated or discovered during surgery, but in many cases, they are probably expelled without symptoms. Another worm that can hide in raw or undercooked freshwater fish is the tapeworm *Diphyllobothrium latum,* common in the Great Lakes, Alaska, and Canada. Because humans serve as its definitive host, this worm can develop in the intestine and cause long-term symptoms.

Students always ask about the safety of sushi and seem uncertain about how to eat it and still avoid parasites. One good safeguard is to patronize a reputable sushi bar, because a competent sushi chef carefully examines the fish for the readily visible larvae. Another is to avoid raw salmon and halibut, which are more often involved than tuna and octopus. Fish can also be freed from live parasites by freezing for 5 days at $-20°C$.

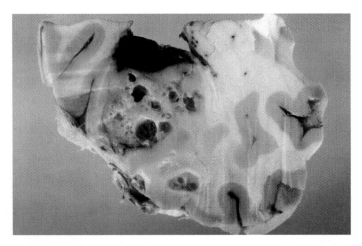

FIGURE 23.26

Cysticercosis caused by *Taenia solium* in the brain of a 13-year-old girl. A section through the cerebrum contains a cluster of encysted larval worms, which gives a spongy or honeycomb appearance.

CHAPTER CHECKPOINTS

Helminths are macroscopic, multicellular worms that have microscopic infective forms: eggs and encysted or free-living larvae. Major divisions of helminths are the nematodes (roundworms) and the platyhelminthes (flatworms). The flatworms are further divided into trematodes (flukes) and cestodes (tapeworms).

Most parasitic helminths have at least two hosts: the definitive host, where sexual reproduction occurs, and the intermediate host, which supports immature or larval forms. The basic patterns of helminth life cycles are characterized by the infectious stage, route of transmission, site of reproduction, and portal of exit. Chemotherapy usually involves drugs that selectively inhibit worm metabolism or weaken their attachment. Preventive measures strive to interrupt the cycle of infection through improved public health conditions.

Nematodes, or roundworms, cause the most severe helminth infestations. They are grouped according to the host site of reproduction as either tissue or intestinal nematodes.

Nematodes of the type A life cycle enter the host as eggs, complete their life cycle in the host, and are transmitted to the environment in new eggs. Examples include *Ascaris, Trichuris,* and *Enterobius.*

Cycle B nematodes enter the host through the skin as larvae. They migrate from the blood to the lungs to the intestine, where they mature and reproduce. Excreted eggs hatch in the environment, producing more larvae. Examples include the hookworms *Necator* and *Ancylostoma* and the threadworm *Strongyloides.*

Ascaris lumbricoides, the largest and most common nematode, is the causative agent of ascariasis, an intestinal infestation in which both larvae and adults can invade other body tissues.

Trichuris trichiura, a tropical nematode, is the causative agent of whipworm infection, an intestinal infestation characterized by ulceration and hemorrhage of the colon.

Enterobius vermicularis, a common nematode in temperate zones, causes pinworm infestation of the intestine and is characterized by the appearance of egg-laying females around the anus.

Trichinella spiralis causes trichinosis, a zoonotic disease associated with eating rare pork. Humans are dead-end hosts that suffer from encystment of larvae in muscle, heart, and brain.

Tissue nematodes include the filarial parasites *Wuchereria* and *Onchocerca,* which cause invasive, debilitating, life-threatening infestations among tropical and subtropical populations.

Trematode pathogens include *Schistosoma, Opisthorchis, Fasciola hepatica,* and *Paragonimus westermani.*

Trematodes exhibit the type D cycle of infection whereby cercarial larvae invade the human host and reproduce in specific organs. Eggs released in the urine or feces hatch into miricidial larvae that invade one or more secondary hosts.

Schistosoma mansonii is a blood fluke that causes a systemic infestation of the bladder and intestine called schistosomiasis.

Cestodes, or tapeworms, are long, chainlike helminths that consist of a feeding attachment to the host and a segmented series of proglottids containing both male and female reproductive structures. They exhibit the type C cycle, in which encysted larvae are ingested in the meat of a secondary animal host. The adult develops and reproduces in the intestine and sheds its eggs in the feces. Cestode pathogens include *Taenia saginata, T. solium, Diphyllobothrium latum,* and *Anisakis.*

CHAPTER CAPSULE WITH KEY TERMS

Parasitology covers **protozoa** and **helminths** (worms) that live on the body of a host. Parasites are spread to humans by other humans, animal hosts, and vectors.

I. **Protozoan Pathogens**

Some protozoans propagate only as **trophozoites** (active feeding stage found in host) whereas others alternate between a trophozoite and a **cyst** (dormant, resistant body). Some have complex life cycles with sexual and asexual phases carried out in more than one host.

A. Infectious Amebas

Entamoeba histolytica is the cause of **amebiasis,** or amebic dysentery, a worldwide human infection that affects approximately 500 million people in the tropics.

1. The protozoan alternates between a trophozoite and a cyst. Cysts released in feces of carriers are spread through unsanitary water and food.

2. Ingested cysts release trophozoites that invade large intestine and cause ulceration and dysentery.

3. Severe cases may occur when the pathogen penetrates into deeper layers; effective drugs are iodoquinol, metronidazole, and chloroquine.

4. Amebic brain infections are caused by *Naegleria fowleri* and *Acanthamoeba,* free-living inhabitants of natural waters. Primary acute meningoencephalitis is acquired through nasal contact with water or traumatic eye damage. Infiltration of brain is usually fatal.

B. Infectious Ciliates

The only important ciliate pathogen is *Balantidium coli,* an occupant of the intestines of domestic animals such as pigs and cattle. It is acquired by humans when cyst-containing food or water is ingested. The trophozoite erodes the intestine and elicits intestinal symptoms.

C. Infectious Flagellates

1. *Trichomonas* is an animal parasite or commensal. The principal human pathogen is *T. vaginalis,* the cause of a very common STD known as trichomoniasis that infects the vagina, cervix, and urethra. Both sex partners can be treated with metronidazole.

2. *Giardia lamblia* is an intestinal flagellate that causes **giardiasis.** Its natural reservoir is animal intestines, and the source of infection is cyst-contaminated fresh water and food. Symptoms are like those of *Entamoeba,* with severe diarrhea that may be chronic in immunodeficient patients. It is preventable by disinfecting water and treatable with quinacrine or metronidazole.

3. Hemoflagellates are protozoans that occur in blood during infection. They cause mostly tropical zoonoses spread by insect vectors. The pathogens have complex life cycles, with various mastigote phases that develop in the insect and human hosts.

a. *Trypanosoma* have tapering, flagellated cells. Two major types of **trypanosomiasis** are geographically isolated and have different blood-feeding vectors endemic to regions where vectors live.

b. *T. brucei* causes **African sleeping sickness,** spread by **tsetse flies** and harbored by reservoir mammalian hosts. The biting fly inoculates skin with the trypanosome, which multiplies in the blood, and damages the spleen, lymph nodes, and brain. Chronic disease symptoms are sleepiness, tremors, paralysis, and coma.

c. *T. cruzi* causes **Chagas disease,** a disease endemic to Latin America. Its cycle is similar to *T. brucei,* except that the **reduviid** (kissing) **bug** is the vector. Infection occurs when bug feces are inoculated into a cutaneous portal. Chronic inflammation occurs in the organs (especially heart and brain).

d. *Leishmania* contains geographically separate species that cause leishmaniasis, a zoonosis of wild animals that is transmitted by the bite of phlebotomine (sand) flies. Infected macrophages carry the pathogen into the skin and bloodstream, giving rise to fever, enlarged organs, and anemia. **Kala azar** is the most severe and fatal form.

D. Apicomplexans are tiny obligate intracellular parasites that have complex life cycles and lack motility in their mature stage. Infectious forms include sporozoites, fecal cysts called oocysts, and tissue cysts. Most diseases are zoonotic and vector-borne.

1. *Plasmodium* is the cause of **malaria;** human is primary host for the asexual phase of parasite; female *Anopheles* mosquito is vector/host for the sexual phase.

a. Malaria is distributed primarily in a belt around the equator, with an estimated 300 to 500 million cases/year worldwide.

b. Infective forms for humans (**sporozoites**) enter blood with mosquito saliva, penetrate liver cells, multiply, and form hundreds of **merozoites.** These multiply in and lyse red blood cells.

c. Symptoms include episodes of chills-fever-sweating, anemia, and organ enlargement. Therapy is chloroquine, quinine, or primaquine.

2. *Toxoplasma gondii* lives naturally in cats that harbor oocysts in the GI tract.

a. **Toxoplasmosis** is acquired by ingesting raw or rare meats containing tissue cysts or by accidently ingesting oocysts from substances contaminated by cat feces.

b. In humans, infection is usually mild and flulike. Immunodepressed (AIDS) patients or fetuses (**fetal toxoplasmosis**) suffer brain and heart damage.

3. *Cryptosporidium* is a vertebrate pathogen that exists in both tissue and oocyst phases. Most cases of **cryptosporidiosis** are caused by handling animals with infections or by drinking contaminated water. Infection causes enteric symptoms.

4. *Isospora* is an uncommon human intestinal parasite transmitted through fecal contamination that causes coccidiosis.

5. *Cyclospora* causes a diarrheal illness when its oocysts are ingested in fecally contaminated produce and water.

II. **Helminth Parasites**

A. The parasitic **helminths,** or worms, are multicellular animals with specialized mouthparts and adaptations such as reduction of organs, protective cuticles, and complex life cycles.

B. Adult worms mate and produce fertile eggs that hatch into **larvae** that mature in several stages to adults. In some worms, the sexes are separate; others are hermaphroditic.

1. Adults live in **definitive host.**

2. Eggs and larvae may develop in the same host, external environment, or **intermediate host.**

3. Types of transmission involve egg-laden feces in soil and water that are infectious when ingested; eggs that go through larval phase in the environment and infect by penetrating

skin; ingestion of animal flesh or plants containing encysted larval worms; or the inoculation of the parasite by an insect vector that is the intermediate host.

4. Helminth pathology arises from worms feeding on and migrating through tissues and accumulation of worms and worm products.

5. Useful antihelminthic drugs control worms by paralyzing their muscles and causing them to be shed or interfering with metabolism and killing them.

6. Other control measures are improved sanitation of food and water, protective clothing, cooking meat adequately, and controlling vectors.

C. Nematodes

Roundworms are filamentous with protective cuticles, circular muscles, a complete digestive tract, and separate sexes with well-developed reproductive organs.

1. Intestinal nematodes: *Ascaris lumbricoides* is a very prevalent species indigenous to humans. Eggs ingested with foods hatch into larvae and burrow through the intestine into the circulation. From there, they travel to the lungs and pharynx and are swallowed. In the intestine, the adult worms complete the reproductive cycle.

2. *Trichuris trichiura,* the **whipworm,** is a small parasite transmitted like *Ascaris* but restricted to the intestine throughout development.

3. *Enterobius vermicularis,* the **pinworm,** is a common childhood infection confined to the intestine. Eggs are picked up from surroundings and swallowed. After hatching, they develop into adults, and the females release eggs that cause anal itching; self-inoculation is common.

4. **Hookworms** have characteristic curved ends and hooked mouths. The two major species, *Necator americanus* (Western Hemisphere) and *Ancylostoma duodenale* (Eastern Hemisphere), share a similar life cycle, transmission, and pathogenesis. Humans shed eggs in feces, which hatch into larvae and burrow into the skin of lower legs. Larvae travel from blood to lungs and are swallowed. Adult worms reproduce in the intestine and complete the cycle.

5. *Strongyloides stercoralis,* the **threadworm,** is a tiny nematode that completes its life cycle in humans or in moist soil. Larvae emerging from soil-borne eggs penetrate the skin and migrate to the lungs, are swallowed, and complete development in the intestine. The parasite can reinfect the same host without leaving the body.

6. *Trichinella spiralis* causes **trichinosis,** a zoonosis in which humans are a dead-end host. Disease is acquired from eating undercooked or raw pork or bear meat containing encysted larvae. Larvae migrate from the intestine to blood vessels, muscle, heart, and brain, where they enter dormancy.

7. **Filarial worms** are long, threadlike worms with tiny larvae (microfilariae) that circulate in blood and reside in various organs; are spread by various biting insects.
   a. *Wuchereria bancrofti* causes **bancroftian filariasis,** a tropical infestation spread by various species of mosquitoes. The vector deposits the larvae which move into lymphatics and develop. Adult females shed microscopic worms into blood. Chronic infection causes blockage of lymphatic circulation and elephantiasis, which manifests as massive swelling in the extremities.
   b. *Onchocerca volvulus* causes **river blindness,** an African disease spread by small, river-associated insects that feed on blood. Inoculated larvae migrate to skin to mature and form nodules. Chronic infections lead to eye involvement and inflammation that can destroy the cornea.
   c. *Loa loa,* the African eye worm, is spread by the bite of small flies. The worm migrates around under the skin and may enter the eye.

D. **Trematodes,** or **flukes,** are flatworms with leaflike bodies bearing suckers; most are hermaphroditic.

1. **Blood flukes** and *Schistosoma* species. **Schistosomiasis** is a prevalent tropical disease. Adult flukes live in humans and release eggs into water. The early larva, or miracidium, develops in the freshwater snail into a second larva, or cercaria. This larva penetrates human skin and moves into the liver to mature; adults migrate to intestine or bladder and shed eggs, giving rise to chronic organ enlargement.

2. **Zoonotic flukes** include the Chinese liver fluke *Opisthorchis sinensis,* which cycles between mammals and snails and fish. Humans are infected when eating fish containing larvae, which invade the liver. *Fasciola hepatica* cycles between herbivores, snails, and aquatic plants. Humans are infected by eating raw aquatic plants; fluke lodges in liver.

E. Cestodes or Tapeworms

These flatworms have long, very thin, ribbonlike bodies (**strobila**) composed of sacs (**proglottids**) and a **scolex** that grips the host intestine. Each proglottid is an independent unit adapted to absorbing food and making and releasing eggs.

1. *Taenia saginata* is the beef tapeworm for which humans are the definitive host. Animals are infected by grazing on land contaminated with human feces. Infection occurs from eating raw beef in which the larval form (**cysticercus**) has encysted. The larva attaches to the small intestine and becomes an adult tapeworm.

2. *T. solium* is the pork tapeworm that infects humans through cysticerci or by ingesting eggs in food or drink (**cysticercosis**). The larvae hatch and encyst in many organs and can do damage.

MULTIPLE-CHOICE QUESTIONS

1. All protozoan pathogens have a ＿＿ phase.
   a. cyst
   b. sexual
   c. trophozoite
   d. blood

2. *Entamoeba histolytica* primarily invades the
   a. liver
   b. large intestine

   c. small intestine
   d. lungs

3. *Giardia* is a/an ＿＿ that invades the ＿＿＿＿.
   a. flagellate, large intestine
   b. ameba, small intestine
   c. ciliate, large intestine
   d. flagellate, small intestine

4. Hemoflagellates are transmitted by
   a. mosquito bites
   b. insect vectors
   c. bug feces
   d. contaminated food

5. *Plasmodium* reproduces sexually in the _____ and asexually in the _____.
   a. liver, red blood cells
   b. mosquito, human
   c. human, mosquito
   d. red blood cell, liver

6. In the exoerythrocytic phase of infection, *Plasmodium* invades the
   a. blood cells
   b. heart muscle
   c. salivary glands
   d. liver

7. An oocyst is found in _____, and a pseudocyst is found in _____.
   a. humans, cats
   b. cats, humans
   c. feces, tissue
   d. tissue, feces

8. A person can acquire toxoplasmosis from
   a. pseudocysts in raw meat
   b. oocysts in air
   c. cleaning out the cat litter box
   d. all of these

9. All adult helminths produce
   a. cysts and trophozoites
   b. scolex and proglottids
   c. fertilized eggs and larvae
   d. hooks and cuticles

10. The _____ host is where the larva develops, and the _____ host is where the adults produce fertile eggs.
    a. intermediate, definitive
    b. definitive, intermediate
    c. secondary, transport
    d. primary, secondary

11. Antihelminthic medications work by
    a. paralyzing the worm
    b. disrupting the worm's metabolism
    c. causing vomiting
    d. both a and b

12. A host defense that is most active in worm infestations is
    a. phagocytes
    b. antibodies
    c. killer T cells
    d. eosinophils

13. Currently, the most common nematode infestation worldwide is
    a. hookworm
    b. ascariasis
    c. pinworm
    d. trichinosis

14. Hookworm diseases are spread by
    a. the feces of humans
    b. mosquito bites
    c. contaminated food
    d. microscopic invertebrates in drinking water

15. Trichinosis can only be spread from human to human by
    a. cannibalism
    b. flies
    c. raw pork
    d. contaminated water

16. The swelling of limbs typical of elephantiasis is due to
    a. allergic reaction to the filarial worm
    b. granuloma development due to inflammation by parasites
    c. lymphatic circulation being blocked by filarial worm
    d. heart and liver failure due to infection

17. **Single Matching.** Match the disease or condition with the causative agent, and indicate for each whether the agent is a protozoan (P) or a helminth (H).

    1. _____ _____ amebic dysentery
    2. _____ _____ Chagas disease
    3. _____ _____ tapeworm
    4. _____ _____ hookworm
    5. _____ _____ African sleeping sickness
    6. _____ _____ pinworm
    7. _____ _____ filariasis
    8. _____ _____ amebic meningoencephalitis
    9. _____ _____ malaria
    10. _____ _____ toxoplasmosis
    11. _____ _____ trichinosis
    12. _____ _____ whipworm
    13. _____ _____ river blindness

    a. *Plasmodium vivax*
    b. *Onchocerca volvulus*
    c. *Enterobius vermicularis*
    d. *Toxoplasma gondii*
    e. *Trichuris trichiura*
    f. *Entamoeba histolytica*
    g. *Necator americanus*
    h. *Trypanosoma brucei*
    i. *Wuchereria bancrofti*
    j. *Trypanosoma cruzi*
    k. *Taenia saginata*
    l. *Naegleria fowleri*
    m. *Trichinella spiralis*

18. **Single Matching.** Match the disease with its primary mode of transmission or acquisition in humans.

    1. _____ amebic dysentery
    2. _____ Chagas disease
    3. _____ cyclosporiasis
    4. _____ toxoplasmosis
    5. _____ giardiasis
    6. _____ tapeworm
    7. _____ filariasis
    8. _____ schistosomiasis
    9. _____ ascariasis
    10. _____ river blindness

    a. eating poorly cooked beef or pork
    b. water or food contaminated with animal wastes containing cysts
    c. water or food contaminated with human feces containing cysts
    d. bite from a reduviid bug
    e. bite from a black fly
    f. water or food contaminated with human feces containing eggs
    g. contact with cats or ingesting rare or raw meat
    h. bite from a mosquito
    i. ingesting fecally contaminated water or produce
    j. freshwater snail vector releases infectious stage

CONCEPT QUESTIONS

1. As a review, compare the four major groups of protozoa according to overall cell structure, locomotion, infective state, and mode of transmission.

2. a. What are the primary functions of the trophozoite and the cyst in the life cycle?
   b. In which diseases is a cyst stage an important part of the infection cycle?

3. a. Why does *Entamoeba* require healthy carriers to complete its life cycle?
   b. What is the role of night soil in transmission of amebiasis?
   c. What is the basic pathology of amebiasis? How and where does it invade?
   d. Why are so many cases asymptomatic?

4. Briefly describe how primary amebic meningoencephalitis is acquired and its outcome.

5. Compare trichomoniasis and amebiasis with respect to transmission, life cycle, and relative hardiness.

6. a. Name ways in which giardiasis and amebic dysentery are similar.
   b. How are they different?

7. a. Compare the infective stages and means of vector transfer in the two types of trypanosomiasis and leishmaniasis.
   b. Are there any stages in the life cycle of the hemoflagellates that occur in all three species?
   c. Which phase or phases is/are never infective to humans? (Hint: Look at table 23.3.)

8. a. What factors cause the symptoms in trypanosomiasis?
   b. In leishmaniasis?

9. a. What is the name of the infective stage of the malaria parasite in humans?
   b. In mosquitoes?
   c. How many times must a female mosquito feed before the parasite can complete the whole cycle?
   d. Where in the human does development take place, and what are the results?
   e. What stage of development is the ring form?
   f. Which events cause the symptoms of malaria?

10. a. Describe the life cycle and host range of *Toxoplasma gondii*.
    b. How are humans infected?
    c. What are the most serious outcomes of infection?

11. Briefly describe the transmission cycle of *Cryptosporidium*, *Sarcocystis*, and *Babesia*.

12. a. In what ways are helminths different from other parasites, and why are there so many parasitic worm infections worldwide?
    b. What is the nature of human defenses against them?

13. a. Outline the five general cycles, and give examples of helminths that exhibit each cycle.
    b. What are the main portals of entry in worm infestations?
    c. How is damage done by parasites?

14. What are some ways that worms adapt to parasitism, and how are these adaptations beneficial to them?

15. How are nematodes, trematodes, and cestodes different from one another?

16. Where in the body do the helminth adults ultimately reside and produce fertile eggs?

17. a. How do adult *Ascaris* get into the intestine?
    b. How do adult hookworms get into the intestine?
    c. How do microfilariae get into the blood?

18. In what ways is trichinosis different from other worm infections?

19. a. Which worms can be found in the eye?
    b. Which worm can cause blindness?

20. a. What are the stages in *Schistosoma* development?
    b. Which organs are affected by schistosomiasis?
    c. What other organs can be invaded by flukes?

21. a. Describe the structure of a tapeworm.
    b. Name several ways in which tapeworms are spread to humans.

22. a. Which helminths are zoonotic?
    b. Which are strictly human parasites?
    c. Which are borne by vectors?
    d. Which can be STDs?
    e. Which cause intestinal symptoms?

CRITICAL-THINKING QUESTIONS

1. Describe some adaptations needed by parasites that enter through the oral cavity.

2. Explain why a person with overt symptoms of intestinal *Entamoeba histolytica* infection is unlikely to transmit infection to others.

3. a. Explain why *Trichomonas vaginalis* is unlikely to be transmitted by casual contact.
   b. What is meant by "ping-pong" infection, and why must both sex partners be treated for trichomoniasis?

4. Explain why only female mosquitoes are involved in malaria and elephantiasis.

5. Which parasitic diseases could conceivably be spread by contaminated blood and needles?

6. For which diseases can one not rely upon chlorination of water as a method of control? Explain why this is true.

7. a. If a person returns from traveling afflicted with trypanosomiasis or leishmaniasis, is he or she generally infective to others?
   b. Why, or why not?

8. Explain why there is no malaria above 6,000 feet in altitude.

9. Describe some strategies to arrive at an effective malaria vaccine.

10. Suggest a possible way to medically circumvent the antigenic switching of trypanosomes.

11. a. Which diseases end up in the intestine from swallowing larval worms?
    b. From swallowing eggs?
    c. From penetration of larvae into skin?

12. a. What one simple act could in time eradicate ascariasis?
    b. Hookworm?
    c. Beef tapeworm?

13. a. To achieve a cure for tapeworm, why must the antihelminthic drug either kill the scolex or slacken its grip?
    b. Which tapeworm is pictured in figure 23.25*a*?
    c. How can you tell?

14. Give some reasons why AIDS patients are so susceptible to certain protozoan diseases.

15. Explain why there is no such thing as a safe form of rare or raw meat.

16. Students sometimes react with horror and distress when they discover that cats and dogs carry parasites to humans. Give an example of a disease for each of these animals that can be spread to humans, and explain how to avoid these diseases and still enjoy your pet.

17. a. Why is it necessary for most parasites to leave their host to complete the life cycle?
    b. What are some ways to prevent completion of the life cycle?
    c. What is the benefit to parasites of having numerous hosts?
    d. What are the disadvantages in having more than one host?

18. East Indian natives habitually chew on betel nuts as an alkaloidal stimulant and narcotic. A side effect is phlegm collection, which is eliminated by frequent spitting. The incidence of *Strongyloides* infections is relatively low in this population. Can you account for this? (No, the betel nuts are not effective antihelminthics.)

19. Give some possible reasons why the incidence of cystic hydatid disease is very high among women and children in hot, arid Turkana, Kenya, where dogs form an integral part of tribal life.

20. In New York City, four Orthodox Jewish patients brought into an emergency clinic with seizures and other neurological symptoms tested positive for antibodies to the pork tapeworm, *Taenia solium.* This was embarrassing to them because they do not eat pork for religious reasons. It also became something of a medical mystery, because there were no indications of how they could have become infected. Later it was shown that the patients had recently employed housekeepers and cooks from Latin America who were free of symptoms yet also seropositive for the worm infection. Use this case study to give a plausible explanation for the transmission of the infection.

INTERNET SEARCH TOPICS

1. Conduct an on-line search for emerging protozoan pathogens. Identify five of the major species and make note of the sources of infection and major risk groups.

2. Look up Chagas disease and research the new treatments available for this disease. What is the current outlook for infected patients?

CHAPTER 24

Introduction to the Viruses of Medical Importance: The DNA Viruses

Viruses are probably the most common infectious agents, though no exact figures exist to support this contention. As a group, they are all obligate parasites and infect not only animals and plants but also other microorganisms. Every time a new virus or viral disease is discovered or a connection is made between a virus and a once-unexplained disease, our understanding of these remarkable infectious particles is increased, and we are reminded of their power. This chapter covers viral diseases of the viral groups most significant to humans and their epidemiology, pathology, and methods of control, with special emphasis on DNA viruses. Chapter 25 covers RNA viruses.

Chapter Overview

- Among microbial groups, viruses cause the majority of infections.
- Viruses are extremely small parasites composed of a nucleic acid core surrounded by a protein capsid.
- Viruses invade host cells by using specific receptors, and they induce the cell machinery to synthesize, assemble, and release new viruses.
- Virus infections vary from mild to severe in effect, local to systemic in scope, and may have the potential for latency and oncogenicity.
- DNA viruses of humans exist in the enveloped or naked state and can carry double-stranded or single-stranded DNA.
- Poxviruses are large, complex DNA viruses that cause smallpox and molluscum contagiosum.
- Herpesviruses are enveloped DNA viruses that are known for being latent in host cells after the initial infection.
- Examples of common herpesvirus diseases are cold sores, genital herpes, chickenpox, shingles, mononucleosis, and roseola.
- Herpesviruses also cause cancers such as lymphoma and sarcoma.
- One cause of hepatitis is hepatitis B virus, an enveloped virus that damages the liver and may cause cancer.
- Nonenveloped DNA viruses are the adenoviruses, one cause of colds, and papillomaviruses, the agents of common skin and genital warts.
- Parvoviruses are single-stranded DNA viruses and are implicated in a respiratory infection of children.

Viruses in Infection and Disease

Viruses are the smallest parasites with the simplest biological structure, being essentially DNA or RNA molecules surrounded by a protein coat that depend on the host cell for their qualities of life.

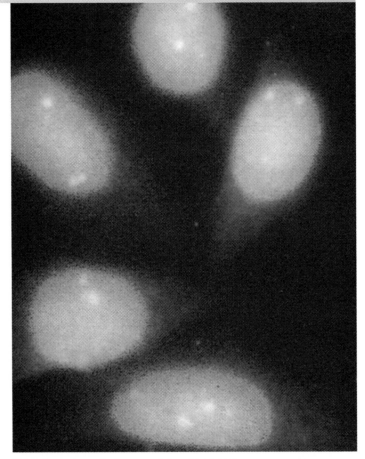

Molecular probes with fluorescent dyes highlight the presence of human papillomavirus (HPV) in the nucleus of infected cells. These fluorescent tags cause integrated viral DNA to appear as small glowing spots.

Viruses have special adaptations for entering a cell and instructing its genetic and molecular machinery to produce and release new viruses. Animal viruses are divided into families based on the nature of the nucleic acid (DNA or RNA), the type of capsid, and whether or not an envelope is present. All DNA viruses are double-stranded except for the parvoviruses, which have single-stranded DNA. All RNA viruses are single-stranded except for the double-stranded reoviruses. The genome of RNA viruses can be further described as segmented (consisting of more than one molecule) or nonsegmented (consisting of a single molecule). The envelope is derived from host cell membranes as the virus buds off the nuclear envelope or cell membrane, and it contains spikes that interact with the host cell (see tables 24.1 and 25.1 for summaries of virus structure).

IMPORTANT MEDICAL CONSIDERATIONS IN VIRAL DISEASES

Target Cells

Because viruses have specific receptors for interacting with host cells, their infectiousness is limited to a particular host or cell type in most cases. Viruses infect most types of tissues, including the nervous system (polio and rabies), liver (hepatitis), respiratory tract (influenza, colds), intestine (polio), skin and mucous membranes (herpes), and immune system (AIDS). Most DNA viruses are assembled and budded off the nucleus, whereas most RNA viruses multiply in and are released from the cytoplasm. The presence of assembling and completed viruses in the cell gives rise to various telltale disruptions called cytopathic effects. Most cells productively infected with viruses are destroyed, which accounts for the sometimes severe pathology and loss of function.

Scope of Infections

Viral diseases range from very mild or asymptomatic infections such as colds to destructive and life-threatening diseases such as rabies and AIDS. Many are so-called childhood infections that occur primarily in the young and are readily transmitted through droplets. Although we tend to consider these infections as self-limited and inevitable, even measles, mumps, chickenpox, and rubella can manifest severe complications. Many viruses are strictly human in origin, but an increasing number (hantavirus, Ebola virus, and viral encephalitis) are zoonoses transmitted by vectors.

The course of viral diseases starts with invasion at the portal of entry by a few virions and a primary infection. In some cases (influenza, colds), the viruses replicate locally and disrupt the tissue, and in others (mumps, polio), the virus enters the blood and invades tissues far from the initial infection. Common manifestations of virus infections include rashes, fever, muscle aches, respiratory involvement, and swollen glands. Many diseases of unknown etiology are thought to have a viral basis. Examples include type I diabetes, multiple sclerosis, and chronic fatigue syndrome (see Medical Microfile 24.2).

Protection in viral infections arises from the combined actions of interferon, neutralizing antibodies, and cytotoxic T cells. Because of the effective immune response to viruses, infections frequently result in lifelong immunities, and many viruses are adaptable to vaccines.

Latency and Oncogenicity

Most DNA viruses and a few RNA viruses can become permanent residents of the host cell. In some cases, the latent virus alternates between periods of relative inactivity and recurrent infections. Viruses can persist in a more permanent state by splicing their DNA into a site in the host's DNA. This process sometimes transforms the cell into a cancer cell. This potential for oncogenesis is more likely with DNA viruses, though retroviruses, which can convert their RNA to DNA, are known to cause cancers. This property of combining with host DNA has been a drawback in developing live, attenuated virus vaccines for them.

Teratogenicity, Congenital Defects, and Viral Diagnosis

Several viruses can cross the placenta from an infected mother to the embryo or fetus. Their infection of the fetus often causes de-velopmental disturbances and permanent defects in the child that are present at birth (congenital). Among viruses with known **teratogenic*** effects are rubella, cytomegalovirus, and adenovirus. Another risk for infants involves infection at the time of birth, as occurs with hepatitis B and herpes simplex.

Viral diseases are diagnosed by a variety of clinical and laboratory methods, including symptoms, isolation in cell or animal culture, and serological testing for antibodies. The use of antigen detection assays is on the increase, especially those based on monoclonal antibodies. Rapid nucleic acid probes directed against DNA or RNA are another option for several viruses. These have the advantage of being so sensitive that they could conceivably detect the viral nucleic acid in a single infected cell. Review figure 6.23 for a summary of identification techniques.

CHAPTER CHECKPOINTS

Viruses are particles of parasitic DNA or RNA. Most DNA viruses are double-stranded, except for the single-stranded parvoviruses. RNA viruses are single-stranded except for the double-stranded reoviruses. Viral infectivity requires a specific viral receptor site on the host cell, but most host tissues possess receptor sites for one or more viruses.

Viral diseases vary in severity, depending on virulence of the virus and age, health, and habitat of the human host. Lifelong immunity develops to some but not all viral agents. Virus infection can be diagnosed by overt symptoms, cultures, antigen detection, and nucleic acid probes.

Most DNA and some RNA viruses can cause chronic infections and combine with the host genome. They also have the potential to activate host oncogenes.

Survey of DNA Virus Groups

The DNA viruses that cause human diseases are placed into six groups based upon the existence of an envelope, the nature of the genome (double-stranded or single-stranded), size, and target cells (table 24.1). The enveloped DNA viruses include the poxviruses, herpesviruses, and hepadnaviruses. The nonenveloped group includes the adenoviruses, papovaviruses, and parvoviruses.

Enveloped DNA Viruses

POXVIRUSES: CLASSIFICATION AND STRUCTURE

Poxviruses produce eruptive skin pustules called **pocks** or **pox** (figure 24.1), which leave small, depressed scars (pockmarks) upon healing. The poxviruses are distinctive because they are the largest and most complex of the animal viruses (see table 24.1), they have the largest genome of all viruses, and they multiply in the cytoplasm in well-defined sites called factory areas, which appear as inclusion bodies in infected cells.

---

* **teratogenic** (tur″-ah-toh-jen′-ik) Gr. *teratos,* monster. Any process that causes physical defects in the embryo or fetus.

TABLE 24.1

DNA Virus Families

Enveloped	Nonenveloped

Enveloped

Poxviridae: **Smallpox, molluscum contagiosum**

Complex structure, lack capsid

Surface tubules · Lateral body · Envelope · Outer membrane · Nucleosome · Core membrane

100 nm

Herpesviridae: **Cold sores, mononucleosis**

Bud off nucleus, tend to become latent

Hepadnaviridae: **Hepatitis B**

Unusual genome containing both double- and single-stranded DNA

Nonenveloped

Adenoviridae: **Common cold, keratoconjunctivitis**

Papovaviridae: **Common and genital warts, leucoencephalopathy**

Parvoviridae: **Erythema infectiosum**

Unusual single-stranded DNA genome

Legend
Genome strandedness

Double (DS) DNA · Single (SS) DNA

Source: Poxviridae *from Buller et al., National Institute of Allergy & Infectious Diseases, Department of Health & Human Services.*

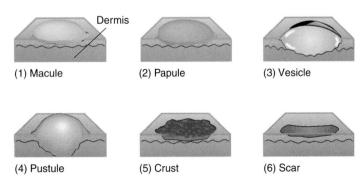

(1) Macule   (2) Papule   (3) Vesicle
(4) Pustule   (5) Crust   (6) Scar

Dermis

FIGURE 24.1

Stages in pock development. Because the infection occurs at the skin-forming level (dermis), these pocks will leave a scar.

Among the best-known poxviruses are **variola,*** the agent of **smallpox,** and vaccinia, a closely related virus used in vaccinations. Other members are of interest to the veterinarian for their role in diseases of domestic or wild animals. A common feature of all poxvirus infections is a specificity for the cytoplasm of epidermal cells and subcutaneous connective tissues. In these sites, they produce the typical lesions and also tend to stimulate cell growth that can lead to tumor formation.

Smallpox: A Perspective

Largely through the World Health Organization's comprehensive global efforts, smallpox is now a disease of the past. It is one of the

* variola (ver-ee-oh′-lah) L. *varius,* varied, mottled.

HISTORICAL HIGHLIGHTS   24.1
The Ultimate End to an Ancient Scourge

The thirty-third World Health Assembly "declares solemnly that the world and all its peoples have won freedom from smallpox . . . an unprecedented achievement in the history of public health . . ." (Resolution 33-3, May 8, 1980, Geneva, Switzerland).

A concerted vaccination effort against smallpox was first instigated by the United Nations' World Health Organization in 1959. Even seven years later, large numbers of cases were still being reported in Africa, Southeast Asia, Indonesia, and Brazil. In 1966 the WHO launched an even more intensive effort. A medical team of 100,000 people began a meticulous village-by-village campaign to identify and report population reservoirs and to isolate and vaccinate specific high-risk groups or individuals. Over the next 10 years, this strategy constantly diminished the numbers of cases, and in 1977, the last natural case was reported in Somalia. The main reasons for this program's success are that only humans with active cases of smallpox are infectious, and variola is not a latent virus nor is it harbored by healthy carriers. These features made it easier to trace the disease and vaccinate in communities where it was prevalent.

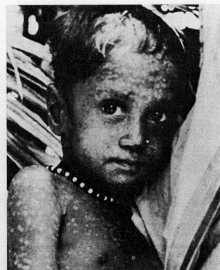

A child in Bangladesh with the last recorded case of variola major (the more serious form of smallpox) in 1975.

Presently, only the United States and the Federation of Russia have stocks of viable variola virus. A great deal of controversy has grown up around the fate of these stocks. At first, the countries had agreed to completely destroy all cultures of the virus after its genome was mapped. Since that time, researchers have protested the loss of the virus, since it could still provide important information on poxvirus structure and function. For the time being, at least, the virus will be retained for further study.

Ultimately, the victory over smallpox has demonstrated that cooperative efforts and a program of vaccination can succeed, even on a planet of over 6 billion people. Hoping for similar success with other agents, the WHO has expanded its drive for immunization against six major childhood killers—measles, polio, diphtheria, whooping cough, tetanus, and tuberculosis. So far, the disease showing greatest promise for total eradication is polio, which has now been declared officially eradicated from the Americas and is almost completely controlled in the rest of the world.

---

few infectious agents that now exists only in government laboratories. (See Historical Highlights 24.1 for a discussion of this stunning achievement.) At one time, smallpox was ranked as one of the deadliest infectious diseases, leaving no civilization untouched. When introduced by explorers into naive populations (those never before exposed to the disease) such as Native Americans and Hawaiians, smallpox caused terrible losses and was instrumental in destroying those civilizations.

Disease Manifestations   Exposure to smallpox usually occurred through inhalation of droplets or skin crusts. Infection was associated with fever, malaise, prostration, and later, a rash that began in the pharynx, spread to the face, and progressed to the extremities. Initially, the rash was macular, evolving in turn to papular, vesicular, and pustular before eventually crusting over, leaving nonpigmented sites pitted with scar tissue (figure 24.1). The two principal forms of smallpox were variola minor and variola major. Variola major was a highly virulent form that caused toxemia, shock, and intravascular coagulation. People who survived any form of smallpox nearly always developed lifelong immunity.

Smallpox Vaccination   The smallpox vaccine uses a single drop of vaccinia virus punctured into the skin with a double-pronged needle or jetgun. In a successful take, a large, scablike pustule with

a reddened periphery develops at the infection site, where it leaves a permanent scar. The vaccine protects people for up to 10 years against variola as well as other poxvirus infections. Smallpox vaccinations have been discontinued for most populations not only because smallpox no longer poses a threat, but also because of the slight risk of vaccinia complications, especially in immunodeficient and allergic people. Vaccination is being reintroduced as a way to control an epidemic of monkeypox in Africa. The vaccinia virus remains an important vehicle for developing genetically engineered vaccines for other diseases (see chapter 16).

Other Poxvirus Diseases

An unclassified poxvirus causes a skin disease called **molluscum contagiosum*** (figure 24.2*a*). This disease is distributed throughout the world, with highest incidence occurring in certain Pacific islands. In endemic regions, it is primarily an infection of children and is transmitted by direct contact and fomites. Skin lesions take the form of smooth, waxy nodules on the face, trunk, and limbs. The infection can also be spread by sexual intercourse and is most common in sexually active young people in the United States. The

---

* molluscum contagiosum (mah-lusk´-uhm kahn″-taj-ee-oh´-sum) L. *molluscus*, soft. In reference to the soft, rounded, cutaneous papules.

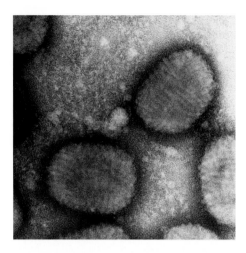

(a)

(b)

FIGURE 24.2

Molluscum contagiosum. **(a)** Virus particles have a typical poxvirus structure with prominent surface tubules. These viruses fill the cytoplasm of infected epithelial cells, causing a prominent mass called a molluscum body. **(b)** A sexually acquired infection. The skin eruption takes the form of small, waxy papules in the genital region.

lesions of the STD form are small, smooth macules that occur on the genital area and thighs (figure 24.2b). They may be spread to other body sites through scratching and self-inoculation. AIDS patients suffer from an atypical version of the disease, which attacks the skin of the face and forms tumorlike growths.

Treatment requires destruction of the virus by freezing (cryotherapy), electric cautery, and chemical agents applied directly to the lesions.

Many mammalian groups host some sort of poxvirus infection. Cowpox, rabbitpox, monkeypox, mousepox, camelpox, and elephantpox are among the types that occur. (But chickenpox is a herpesvirus infection, not a poxvirus, and it does not occur in chickens!) However, only monkeypox and cowpox appear to be capable of infecting humans symptomatically.

The fact that humans are also susceptible to monkeypox virus has produced a new concern in parts of central and western Africa. For the past several years, outbreaks have affected thousands of people, predominantly children. Historically, this disease occurred sporadically in humans having contact with monkeys, squirrels, or rats—the traditional vectors. But nearly three out of four of these latest cases were acquired within communities and families. This changing pattern is thought to be partly due to the discontinuation of smallpox vaccination that had formerly protected against monkeypox as well. Noting that the attack and mortality rates have surpassed any other recorded outbreaks also suggests that the virus is becoming more virulent for humans. The disease manifestations are very similar to smallpox, with skin pocks (see figure 24.1), fever, and swollen lymph nodes. Because of the potential for this emerging disease to spread rapidly through susceptible populations, health authorities have reinstated a vaccine program in regions with greatest risk and are keeping close track of new cases.

Although cowpox is an eruptive cutaneous disease that can develop on cows' udders and teats, cows are not the actual reservoirs of this virus. Other mammals such as rodents, cats, and even zoo animals can carry it. Human infection is rare and usually confined to the hands, although the face and other cutaneous sites can be involved.

THE HERPESVIRUSES: COMMON, PERSISTENT HUMAN VIRUSES

Herpesvirus* was named for the tendency of some herpes infections to produce a creeping rash. It is the common name for a large family whose members include herpes simplex 1 and 2 (HSV), the cause of fever blisters and genital infections; herpes zoster (VZV), the cause of chickenpox and shingles; cytomegalovirus (CMV), which affects the salivary glands and other viscera; Epstein-Barr virus (EBV), associated with infection of the lymphoid tissue; and some recently identified viruses (herpesvirus-6, -7, and -8). Prominent features of the family are its tendency toward viral latency and recurrent infections. This behavior often involves incorporation of viral nucleic acid into the host genome and carries with it the potential for oncogenesis. Two of the viruses in this group have a well-documented association with cancers, and others have suspected involvement.

Virtually everyone becomes infected with a herpesvirus at some time, usually without adverse effect. The clinical complications of latency and recurrent infections become more severe with advancing age, cancer chemotherapy, or other conditions that compromise the immune defenses. The herpesviruses are among the most common and serious opportunists among AIDS patients. Nearly 95% of this group will experience recurrent bouts of skin, mucous membrane, intestinal, and eye disease from these viruses.

The herpesviruses are among the larger viruses, with a diameter ranging from about 150 nm to 200 nm (figure 24.3). They are enclosed within a loosely fitting envelope that contains glycoprotein spikes. Like other enveloped viruses, herpesviruses are prone to deactivation by organic solvents or detergents and are unstable outside the host's body. The icosahedral capsid houses a core of double-stranded DNA that winds around a proteinaceous spindle in some viruses. Replication occurs primarily within the nucleus, and viral release is usually accompanied by cell lysis.

---

* **herpes** (her′-peez) Gr. *herpein*, to creep.

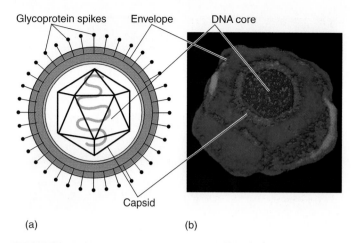

Glycoprotein spikes Envelope DNA core

Capsid

(a)   (b)

FIGURE 24.3

 Herpesviruses. **(a)** A schematic of the general structure of herpesvirus alongside **(b)** a color-enhanced view of cytomegalovirus.

General Properties of Herpes Simplex Viruses

Humans are susceptible to two varieties of **herpes simplex viruses (HSVs):** HSV-1, characterized by lesions on the oropharynx, and HSV-2, with lesions on the genitalia (see table 24.2 for a summary of major characteristics and differences). Even though other mammals can be experimentally infected, humans appear to be the only natural reservoir for herpes simplex virus.

Epidemiology of Herpes Simplex

Herpes simplex infection occurs globally in all seasons and among all age groups. Because these viruses are relatively sensitive to the

TABLE 24.2

Comparative Epidemiology and Pathology of Herpes Simplex, Types 1 and 2

	HSV-1	HSV-2
Usual Etiologic Agent Of:	Herpes labialis Ocular herpes Gingivostomatitis Pharyngitis	Herpes genitalis★
Transmission	Close contact, usually of face	Sexual or close contact
Latency	Occurs in trigeminal ganglion	Occurs primarily in sacral ganglia
Skin Lesions	On face, mouth	On internal, external genitalia, thighs, buttocks
Complications		
Whitlows	Among personnel working on oral cavity	Among obstetric, gynecological personnel
Neonatal encephalitis	Causes up to 30% of cases★★	Causes most cases

★The other herpes simplex type can be involved in this infection, though not as commonly.
★★Due to mothers infected genitally by HSV-1 or contamination of the neonate by oral lesions.

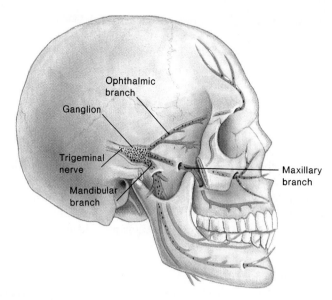

FIGURE 24.4

The site of latency and routes of recurrence in herpes simplex, type 1. Following primary infection, the virus invades the trigeminal, or fifth cranial, nerve and is harbored in its ganglion. Various provocative stimuli cause it to migrate back to the skin surface by the mandibular, maxillary, or ophthalmic branch.

environment, transmission is promoted by direct exposure to secretions containing the virus. People with active lesions are the most significant source of infection, though studies indicate that genital herpes can be transmitted even when no lesions are present. Occasionally cases have been traced to contact with moist secretions on inanimate objects.

Primary infections tend to be age-specific: HSV-1 frequently occurs in infancy and early childhood, and by adulthood, most people exhibit some serological evidence of infection. On the other hand, primary infection with HSV-2 occurs most frequently between the ages of 14 and 29, a pattern that reflects the sexual route of transmission. In fact, genital herpes is one of the most common STDs in the United States with an estimated 1 million new cases each year. Instances of HSV-1 genital infections and HSV-2 infections of the oral cavity do occur, probably from autoinoculation with contaminated hands or oral sexual contact.

The Nature of Latency and Recurrent Attacks In approximately 20% to 50% of primary infections, herpes simplex viruses enter the distal regions of sensory neurons and travel to the dorsal root ganglia. Type 1 HSV enters primarily the trigeminal, or fifth cranial, nerve, which has extensive innervations in the oral region (figure 24.4). Type 2 HSV usually becomes latent in the ganglion of the lumbosacral spinal nerve trunk. The exact details of this unusual mode of latency are not yet fully understood, but the virus remains inside the neuron in a nonproliferative stage for a variable time. Recurrent infection is triggered by various stimuli such as fever, UV radiation, stress, or mechanical injury, all of which reactivate the virus. The virus migrates to the body surface and produces a local skin or membrane lesion, often in the same site as a previous infection. The number of such attacks can range from one to dozens each year.

The Spectrum of Herpes Infection and Disease

Herpes simplex infections, often called simply herpes, usually target the mucous membranes. The virus enters cracks or cuts in the membrane surface and then multiplies in basal and epithelial cells in the immediate vicinity. This results in inflammation, edema, cell lysis, and a characteristic thin-walled vesicle. The main diseases of herpes simplex viruses are facial herpes (oral, optic, and pharyngeal), genital herpes, neonatal herpes, and disseminated disease.

Type 1 Herpes Simplex in Children and Adults **Herpes labialis,** otherwise known as **fever blisters** or **cold sores,** is the most common recurrent HSV-1 infection. Vesicles usually crop up on the mucocutaneous junction of the lips or on adjacent skin (figure 24.5*a*). A few hours of tingling or itching precede the formation of one or more vesicles at the site. Pain is most acute in the earlier stage as vesicles ulcerate; lesions crust over in 2 or 3 days and heal completely in a week.

Although most herpetic infections remain localized, an occasional manifestation known as herpetic *gingivostomatitis* can occur (figure 24.5*b*). This infection of the oropharynx strikes young children most frequently. Inflammation of the oral mucosa can involve the gums, tongue, soft palate, and lips, sometimes with ulceration and bleeding. A common complication in adolescents is pharyngitis, a syndrome marked by sore throat, fever, chills, swollen lymph nodes, and difficulty in swallowing.

Herpetic keratitis (also called ocular herpes) is an infective inflammation of the eye in which a latent virus travels into the ophthalmic rather than the mandibular branch of the trigeminal nerve. Preliminary symptoms are a gritty feeling in the eye, conjunctivitis, sharp pain, and sensitivity to light. Some patients develop characteristic branched or opaque corneal lesions as well. In 25% to 50% of cases, keratitis is recurrent and chronic and can interfere with vision.

Type 2 Herpes Infections Type 2 herpes infection usually coincides with sexual maturation and an increase in sexual encounters. **Genital herpes** (herpes genitalis) starts out with malaise, anorexia, fever, and bilateral swelling and tenderness in the groin. Then, intensely sensitive vesicles break out in clusters on the genitalia, perineum, and buttocks (figure 24.6). The chief symptoms are urethritis, painful urination, cervicitis, and itching. After a period of days to weeks, the vesicles ulcerate and develop a light-gray exudate before healing. Bouts of recurrent genital herpes are usually less severe than the original infection and are triggered by menstruation, stress, and concurrent bacterial infection.

Herpes of the Newborn Although HSV infections in healthy adults are annoying, unpleasant, and painful, only rarely are they life-threatening. However, in the neonate and fetus (figure 24.7), HSV infections are very destructive and can be fatal. Most cases occur when infants are contaminated by the mother's reproductive tract immediately before or during birth, but they have also been traced to hand transmission from the mother's lesions to the baby. Because type 2 is more often associated with genital infection, it is more frequently involved; however, type 1 infection has similar complications. In infants whose disease is confined to the mouth, skin, or eyes, the mortality rate is 30%, but disease affecting the central nervous system has a 50–80% mortality rate.

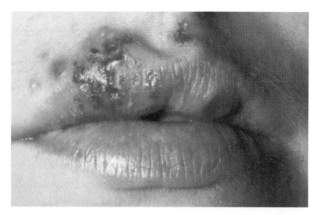

(a)

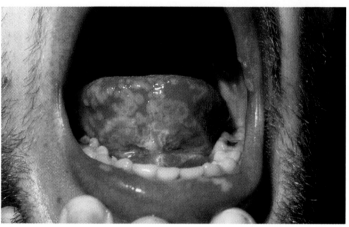

(b)

FIGURE 24.5

Type I herpes simplex. **(a)** Recurrent lesions of herpes labialis (cold sore, fever blister). Tender, itchy papules erupt on the perioral region and progress to vesicles that burst, drain, and scab over. These sores and fluid are highly infectious and should not be touched. **(b)** Primary herpetic gingivostomatitis involving the entire oral mucosa, tongue, cheeks, and lips.

Vesicles

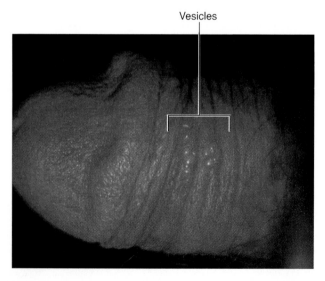

FIGURE 24.6

Genital herpes. Vesicles start out separate, but become confluent and ulcerate into painful red erosions so tender that inspection is difficult.

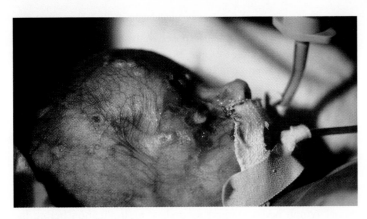

FIGURE 24.7

Neonatal herpes simplex. This premature infant was born with the classic "cigarette burn" pattern of HSV infection. Babies can be born with the lesions or develop them one to two weeks after birth.

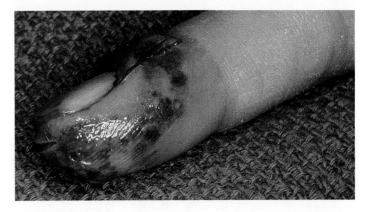

FIGURE 24.8

Herpetic whitlows. These painful, deep-set vesicles can become inflamed and necrotic and are difficult to treat. They occur most frequently in dental and medical personnel or people who carelessly touch lesions on themselves or others.

Because of the danger of herpes to fetuses and newborns and also the increase in the number of cases of genital herpes, it is now standard procedure to screen pregnant women for the herpesvirus early in their prenatal care. Women with a history of recurrent infections must be constantly monitored for any signs of viral shedding, especially in the last 4 weeks of pregnancy. If no evidence of recurrence is seen, vaginal birth is indicated, but any evidence of an outbreak at the time of delivery necessitates a cesarian section.

Miscellaneous Herpes Infections When either form of herpesvirus enters a break in the skin, a local infection can develop. This route of infection is most often associated with occupational exposure or severely damaged skin. A hazard for health care workers who handle patients or their secretions without hand protection is a disease called herpetic **whitlow.*** Workers in the fields of obstetrics, gynecology, dentistry, and respiratory therapy are probably at greatest risk for contracting it. Whitlows are deep-set, usually occur on one finger, and are extremely painful and itchy (figure 24.8). Afflicted personnel should not work with patients until the whitlow has healed, usually in 2 to 3 weeks.

Life-Threatening Complications Although herpes simplex encephalitis is a rare complication of type 1 infection, it is probably the most common sporadic form of viral encephalitis in the United States. The infection disseminates along nerve pathways to the brain or spinal cord. The effects on the CNS begin with headache and stiff neck and can progress to mental disturbances and coma. The fatality rate in untreated cases is 70%. Patients with underlying immunodeficiency are more prone to severe, disseminated herpes infection than are immunocompetent patients. Of greatest concern are patients receiving organ grafts, cancer patients on immunosuppressive therapy, those with congenital immunodeficiencies, and AIDS patients.

Diagnosis, Treatment, and Control of Herpes Simplex

Small, painful, vesiculating lesions on the mucous membranes of the mouth or genitalia, lymphadenopathy, and exudate are typical diagnostic symptoms of herpes simplex. Further diagnostic support is available by examining scrapings from the base of such lesions stained with Giemsa, Wright, or Papanicolaou (Pap) methods (figure 24.9). The presence of multinucleate cells, giant cells, and intranuclear eosinophilic inclusion bodies can help establish herpes infection. This method, however, will not distinguish among HSV-1, HSV-2, or other herpesviruses, which require more specific subtyping.

Laboratory culture and specific tests are essential for diagnosing immunosuppressed and neonatal patients with severe, disseminated herpes infection. A specimen of tissue or fluid is introduced into a primary cell line such as monkey kidney or human embryonic kidney tissue cultures and is then observed for cytopathic effects within 24 to 48 hours. Direct tests on specimens or cell cultures using fluorescent antibodies or DNA probes can differentiate among HSV-1, HSV-2, and closely related herpesviruses. Serological analysis is useful for primary infection but is inconclusive for recurrent illness, because the antibody titer usually does not increase.

Several agents are available for treatment. Acyclovir (Zovirax) is the most effective therapy developed to date that is nontoxic and highly specific to HSV. Famciclovir and valacyclovir are alternate drugs. Topical medications applied to genital and oral lesions cut the length of infection and reduce viral shedding. Systemic therapy is available for more serious complications such as herpes keratitis and disseminated herpes. New evidence indicates that a daily dose of oral acyclovir taken for a period of 6 months to one year can be effective in preventing recurrent genital herpes.

Over-the-counter cold sore medications containing menthol, camphor, and local anesthetics lessen pain and may protect against secondary bacterial infections, but they probably do not affect the progress of the viral infection. Some protection in suppressing cold sores can be obtained from the amino acid lysine, taken orally in the earliest phases of recurrence.

---

* whitlow (hwit′-loh) MI *white*, flaw. An abscess on the distal portion of a finger.

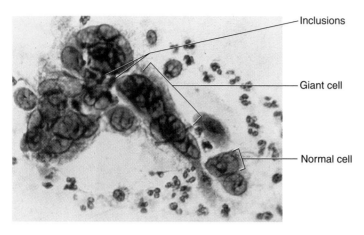

Inclusions

Giant cell

Normal cell

FIGURE 24.9

Direct cytologic diagnosis of herpesvirus infection. A direct Papanicolaou (Pap) smear of a cervical scraping shows typical enlarged (multinucleate giant) cells and intranuclear inclusions. This technique is not specific for HSV, but most other herpesviruses do not infect the reproductive mucosa.

Although the barrier protection afforded by condoms can diminish the prospects of spreading genital herpes sexually, people with active infection should avoid any sexual contact. Mothers with cold sores should observe extreme care in handling their newborns; infants should never be kissed on the mouth; and hospital attendants with active oral herpes infection must be barred from the newborn nursery. Medical and dental workers who deal directly with patients can reduce their risk of exposure by wearing gloves.

The Biology of Varicella–Zoster Virus

Another herpesvirus causes both **varicella*** (commonly known as **chickenpox**) and a recurrent infection called **herpes zoster,*** or **shingles.*** Because the same virus causes two forms of disease, it is known by a composite name, **varicella-zoster virus (VZV).** At one time, the differences in clinical appearance and time separation between varicella and herpes zoster were sufficient to foster the notion that they were caused by different infectious agents. But about 100 years ago, the observation that young children acquired chickenpox from family members afflicted with shingles gave the first hint that the two diseases were related. More recently, doctors followed the diseases in a single patient using Koch's postulates and demonstrated that the same virus caused both diseases and that zoster is a reactivation of a latent varicella virus (figure 24.10).

Epidemiologic Patterns of VZV Infection Humans are the only natural hosts for varicella-zoster virus. The virus is harbored in the respiratory tract, but it is communicable from both respiratory droplets and the fluid of active skin lesions. Infected persons are most infectious a day or two prior to the development of the

rash. The dried scabs are not infectious because the virus is unstable and loses infectivity when exposed to the environment. Patients with shingles are a source of infection for nonimmune children, but the attack rate is significantly lower than with exposure to chickenpox. Only in rare instances will a child acquire chickenpox more than once, and even a subclinical case can result in long-term immunity. Immunity protects against reinfection with chickenpox but not an attack of shingles.

Varicella (Chickenpox) The respiratory epithelium serves as the chief portal of entry and the initial site of viral replication for the chickenpox virus, which initially produces no symptoms. Regional lymph nodes support secondary multiplication before the virus is detectable in the blood. After an incubation period of 10 to 20 days, the first symptoms to appear are fever and an abundant rash that begins on the scalp, face, and trunk and radiates in sparse crops to the extremities. Skin lesions progress quickly from macules and papules to itchy vesicles that encrust and drop off, usually healing completely but sometimes leaving a tiny pit or scar (figure 24.10a). Lesions number from a few to hundreds and are more abundant in adolescents and adults than in young children.

Herpes Zoster (Shingles) In an unknown percentage of individuals, recuperation from varicella is associated with the entry of the varicella-zoster virus into the sensory endings that innervate dermatomes, regions of the skin supplied by the cutaneous branches of nerves, especially the thoracic and trigeminal nerves (figure 24.10b). From here it becomes latent in the ganglia and may reemerge with its characteristic asymmetrical distribution on the skin of the trunk or head (figure 24.10c).

Shingles develops abruptly after reactivation by such stimuli as X-ray treatments, immunosuppressive and other drug therapy, surgery, or developing malignancy. The virus is believed to migrate down the ganglion to the skin, where multiplication resumes and produces crops of tender, persistent vesicles. Inflammation of the ganglia and the pathways of intercostal nerves can cause pain and tenderness (radiculitis) that can last for several months. Involvement of cranial nerves can lead to eye inflammation and ocular and facial paralysis.

Diagnosis, Treatment, and Control The cutaneous manifestations of varicella and shingles are sufficiently characteristic for ready clinical recognition. Supportive evidence for a diagnosis of varicella usually comes from recent close contact with a source or with an active case of VZV infection. A diagnosis of shingles is supported by the distribution pattern of skin lesions and confirmed from multinucleate giant cells in stained smears prepared from vesicle scrapings. However, unequivocal identification or differentiation of herpes simplex from herpes zoster is best done with fluorescent antibody detection of viral antigen in skin lesions, DNA probe analysis, and culture.

Uncomplicated varicella is self-limited and requires no therapy aside from alleviation of discomfort. Secondary bacterial infection that can cause dangerous complications (especially streptococcal and staphylococcal) is prevented by application of anesthetic and antimicrobic ointments. The use of aspirin, though, is contraindicated because of the apparent risk of Reye syndrome in

---

* varicella (var″-ih-sel′-ah) The Latin diminutive for smallpox or variola.

* zoster (zahs′-tur) Gr. zoster, girdle.

* shingles (shing′-gulz) L. cingulus, cinch. (Both zoster and shingles refer to the beltlike encircling nature of the rash.)

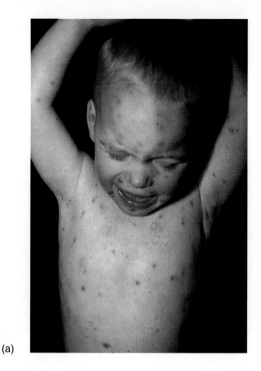

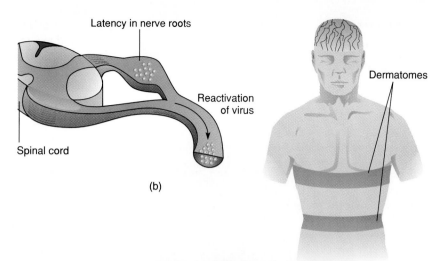

(a)

(b)

Latency in nerve roots

Spinal cord

Reactivation of virus

Dermatomes

(c)

FIGURE 24.10

The relationship between varicella (chickenpox) and zoster (shingles) and the clinical appearance of each. **(a)** First contact with the virus (usually in childhood) results in a macular, papular, vesicular rash distributed primarily on the face and trunk. **(b)** The virus becomes latent in the dorsal ganglia of nerves that supply dermatomes of mid-thoracic nerves and the cranial nerves that supply facial regions. **(c)** The clinical appearance of shingles. Early symptoms are acute pain in the nerve root and redness of the dermatome, followed by a vesicular-papular rash on the chest and back that is usually asymmetrical and does not cross the midline of the body.

children and young adults. Drugs that can be beneficial for systemic disease are intravenous acyclovir or famciclovir and high-dose interferon. Zoster immune globulin (ZIG) and varicella-zoster immune globulin (VZIG) are passive immunotherapies that do not entirely prevent disease but can diminish its complications. A live, attenuated vaccine is currently recommended for routine childhood vaccination. It is given once to a child between one and two years of age and updated by boosters given at age 11 or 12 and to adults. One advantage of this vaccine will possibly be protection against shingles. Because some parents regard the disease as mild and inevitable, they purposely expose their children to infected individuals as a means of "immunizing" them. This practice is not recommended because of the potential for viral latency.

The Cytomegalovirus Group

Another group of herpesviruses, the **cytomegaloviruses* (CMVs),** are named for their tendency to produce giant cells with nuclear and cytoplasmic inclusions (figure 24.11). These viruses, also termed salivary gland virus and cytomegalic inclusion virus, are among the most ubiquitous pathogens of humans.

---

* cytomegalovirus (sy″-toh-may″-gah-loh-vy′-rus) Gr. *cyto*, cell, and *megale*, big, great.

Epidemiology of CMV Disease The testing of human populations has indicated that antibodies to CMV developed during a prior infection are very common. For instance, 50% of women of childbearing age had CMV antibodies, and nearly 10% of newborns reveal some evidence of infection as well, making CMV the most prevalent fetal infection. Other groups with a high carrier rate are IV drug abusers and male homosexuals. Cytomegalovirus is transmitted in saliva, respiratory mucus, milk, urine, semen, cervical secretions, and feces. Transmission usually involves intimate exposure such as sexual contact, vaginal birth, transplacental infection, blood transfusion, and organ transplantation. As with other herpesviruses, CMV is commonly carried in a latent state in various tissues.

Infection and Disease Most healthy adults and children with primary CMV infection are asymptomatic. However, three groups that develop a more virulent form of disease are fetuses, newborns, and immunodeficient adults. The incidence of **congenital CMV** infection is high, occurring in about 20% of pregnancies in which the mother has concurrent infection. Newborns with clinical disease exhibit enlarged liver and spleen, jaundice, capillary bleeding, microcephaly, and ocular inflammation. In some cases, damage can

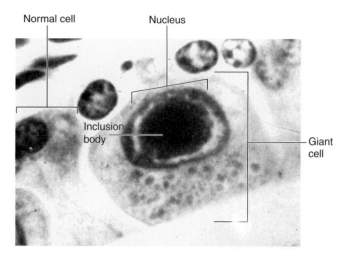

FIGURE 24.11

Identifying cytomegaloviruses. Infection of lung cells with cytomegalovirus shows the cellular enlargement and distortion of the nucleus by a large inclusion (1,500×), called an "owl's eye" effect.

be so severe and extensive that death follows within a few days or weeks. Many babies who survive develop long-term neurological sequelae, including hearing and visual disturbances and even mental retardation. In the United States, about 5,000 babies a year are affected. **Perinatal CMV infection,** which occurs after exposure to the mother's vagina, is chiefly asymptomatic, although pneumonitis and a mononucleosis-like syndrome can develop during the first 3 months after birth.

Cytomegalovirus mononucleosis* is a syndrome characterized by fever and lymphocytosis, somewhat similar to the disease caused by Epstein-Barr virus. Although it can occur in children, it is chiefly an illness of adults. **Disseminated cytomegalovirus** infection can precipitate a medical crisis. It is a common opportunist of AIDS patients, in whom it produces overwhelming systemic disease, with fever, severe diarrhea, hepatitis, pneumonia, and high mortality. Invasion of the retina can lead to blindness. Most patients undergoing kidney transplantation and half of those receiving bone marrow develop CMV pneumonitis, hepatitis, myocarditis, meningoencephalitis, hemolytic anemia, and thrombocytopenia.

Diagnosis, Treatment, and Prevention   In neonates, CMV infection must be differentiated from toxoplasmosis, rubella, and herpes simplex; in adults, it must be distinguished from Epstein-Barr infection and hepatitis. The cytomegalovirus can be isolated from specimens taken from virtually all organs as well as from epithelial tissue. Cell enlargement and prominent inclusions in the cytoplasm and nucleus are suggestive of CMV. The virus can be cultured and tested with monoclonal antibody against the early nuclear antigen. Direct ELISA tests and DNA probe analysis are also useful in diagnosis. Testing serum for antibodies may fail to diagnose infection in neonates and the immunocompromised, so it is less reliable.

* mononucleosis (mah″-noh-noo″-klee-oh′-sis) Gr. *mono,* one, *nucleo,* nucleus, and *sis,* state.

Drug therapy is reserved for immunosuppressed patients. The three main drugs are ganciclovir, valacyclovir, and foscarnet, which have toxic side effects and cannot be administered for long periods. The development of a vaccine is hampered by the lack of an animal that can be infected with human cytomegalovirus. One crucial concern is whether vaccine-stimulated antibodies would be protective, since patients already seropositive can become reinfected; another concern is that a vaccine could be oncogenic.

Epstein–Barr Virus

Although the lymphatic disease known as infectious mononucleosis was first described more than a century ago, its cause was finally discovered through a series of accidental events starting in 1958, when Michael Burkitt discovered in African children an unusual malignant tumor (Burkitt lymphoma) that appeared to be infectious. Later, Michael Epstein and Yvonne Barr cultured a virus from tumors that showed typical herpesvirus morphology. Proof that the two diseases had a common cause was completed when a laboratory technician accidently acquired mononucleosis while working with the Burkitt lymphoma virus. The **Epstein-Barr virus (EBV)** shares morphological and antigenic features with other herpesviruses, and in addition, it contains a circular form of DNA that is readily spliced into the host cell DNA, thus transforming cultured lymphocytes.

Epidemiology of Epstein–Barr Virus   This virus is ubiquitous, with a firmly established niche in human lymphoid tissue and salivary glands. Salivary excretion occurs in about 20% of otherwise normal individuals. Direct oral contact and contamination with saliva are the principal modes of transmission, although transfer through blood transfusions and organ transplants is also possible.

The nature of infection depends on the patient's age at first exposure, socioeconomic level, geographic region, and genetic predisposition. In less-developed regions of the world, infection usually occurs in early childhood. The early and chronic infection means that children living in Africa have higher rates of Burkitt lymphoma. Likewise, residents of certain areas of Asia show high levels of nasopharyngeal carcinoma, another tumor linked to EBV.

In industrialized countries, exposure to EBV is usually postponed until adolescence and young adulthood. This delay is generally ascribed to improved sanitation and less frequent and intimate contact with other children. Because nearly 70% of college-age Americans have never had EBV infection, this population is vulnerable to infectious mononucleosis, often dubbed the "kissing disease" or "mono." However, by mid-life, 90% to 95% of all people, regardless of geographic region, show serological evidence of prior or current infection.

Diseases of EBV   The epithelium of the oropharynx is the portal of entry for EBV during the primary infection. From this site, the virus moves to the parotid gland, which is the major area of viral replication and latency. After extended dormancy, the virus is reactivated by some ill-defined mechanism. In some people, the entire course of infection and latency is asymptomatic.

The symptoms of **infectious mononucleosis** are sore throat, high fever, and cervical lymphadenopathy, which develop after a long incubation period (30–50 days). Many patients also have a

MEDICAL MICROFILE   24.2
Viral Infection: A Connection to Chronic Fatigue Syndrome?

Beginning in 1985, physicians began to see increased numbers of patients with what was apparently a new disease. Patients reported having a cluster of nonspecific symptoms—primarily disabling fatigue, along with some combination of memory loss, sore throat, tender muscles, joint pain, enlarged lymph nodes, and headaches. These cases could not be tied to other medical conditions and were chronic, lasting 6 months or longer. Some cases occurred in small epidemics, whereas others were sporadic. Because most patients had a high titer of antibodies to Epstein-Barr virus, it was initially thought to be a form of mononucleosis caused by this virus. The CDC later concluded that although an infection may have been involved, it was probably not EB virus.

The disease has since been termed **chronic fatigue syndrome (CFS)** or chronic fatigue immunodeficiency syndrome, and is now an accepted diagnosis, but only after other diseases have been ruled out. Although it is not reportable, the prevalence in the United States has been estimated at 500,000 to 800,000 cases. It is most common in Caucasian women between 30 and 50 years of age. At various times, a number of

viruses have been suspected as being the triggers or cofactors in the condition. Because of their potential latency and the fact that they can test positive for these viruses, other herpesviruses, including HHV-6 and CMV are considered possible candidates. Another suspected cause has been an enterovirus such as coxsackievirus, which is known to cause post-viral fatigue syndrome. One problem with all of these possible agents is that none of them is present in all CFS patients.

Most medical experts now believe that the syndrome is multifactorial. Besides the fatigue, patients manifest signs of immune dysfunction, including abnormal levels of CD4 T cells, low levels of natural killer cells, and overactive inflammatory responses. One theory is that the lowered immunity causes reactivation of latent viruses and recurrent infections, which contribute to the symptoms. The reactivated viruses would also further weaken the immune system by keeping it in a cycle of damage and exhaustion. Treatment with interferon and ampligen have been beneficial in some patients, but so far, there is no universal remedy.

---

gray-white exudate in the pharynx (figure 24.12), a skin rash, and enlarged spleen and liver. A notable sign of mononucleosis is sudden leukocytosis, consisting initially of infected B cells and later T cells. The strong, cell-mediated immune response is decisive in controlling the infection and preventing complications. Some people can become chronically infected with EBV, but whether it is a factor in a debilitating illness is controversial (Medical Microfile 24.2).

Tumors and Other Complications Associated with EBV
Burkitt lymphoma is a B-cell malignancy that usually develops in the jaw and grossly swells the cheek (figure 24.13). Most cases are reported in central African children 4 to 8 years old, and a small number are seen in North America. The prevalence in Africa may be associated with chronic coinfections with other diseases, such as malaria, that weaken the immune system and predispose the body to tumors. One theory to explain the neoplastic effect is that persistent EBV infection in young children gives rise to transformed B lymphocytes. At the same time, chronic infections overwhelm the T-cell effector response needed to control the overgrowth of these malignant B-cell clones.

Unlike Burkitt lymphoma, **nasopharyngeal carcinoma** is a malignancy of epithelial cells that occurs in older Chinese and African men. Although much about this disease is still to be explained, high antibody titer to EBV is a consistent finding. It is not known how the virus invades epithelial cells (not its usual target), why the incidence of malignancy is so high in particular areas, or to what extent dietary carcinogens are involved.

In general, any person with an immune deficiency (especially of T cells) is highly susceptible to Epstein-Barr virus. Organ transplant patients carry a high risk because their immune systems are weakened by measures to control rejection. The oncogenic poten-

tial of the virus is another concern, because it is currently believed to cause some types of lymphatic tumors in kidney transplant patients and is commonly isolated from AIDS-related lymphomas. AIDS patients also develop *hairy leukoplakia,* white adherent plaques on the tongue due to invasion of the epithelium by EBV. Lesions can be reduced by oral acyclovir therapy.

Diagnosis, Treatment, and Prevention   Because clinical symptoms in EBV infection are common to many other illnesses, laboratory diagnosis is necessary. A differential blood count that shows

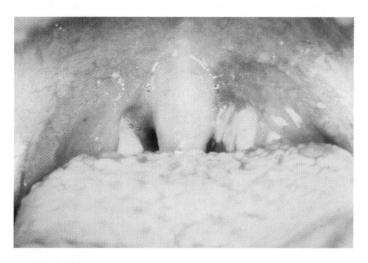

FIGURE 24.12

Infectious mononucleosis. This view of the pharynx shows the swollen tonsils and throat tissues, a gray-white exudate, and rash on the palate that can be of some use in diagnosis.

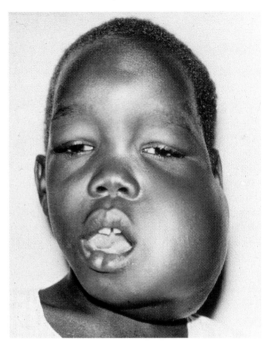

FIGURE 24.13

EBV-associated tumor. Burkitt lymphoma, a malignancy of B cells commonly associated with chronic Epstein-Barr infection.

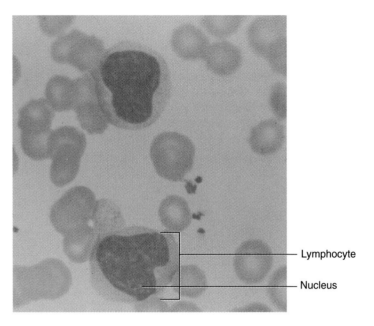

FIGURE 24.14

Epstein-Barr virus in the blood smear of a patient with infectious mononucleosis. Note the abnormally large lymphocytes containing indented nuclei with light discolorations.

lymphocytosis, neutropenia, and large, atypical lymphocytes with lobulated nuclei and vacuolated cytoplasm is suggestive of EBV infection (figure 24.14). Serological assays to detect antibody against the capsid and DNA are helpful, as are direct viral antigen tests using probes and labeled antibodies.

The usual treatments for infectious mononucleosis are directed at symptomatic relief of fever and sore throat. Specific antiviral measures for disseminated disease that have shown some promise are intravenous gamma globulin, interferon, acyclovir, and monoclonal antibodies. Burkitt lymphoma requires systemic chemotherapy with an anticancer drug such as cyclophosphamide or vincristine. These measures, along with surgical removal of the tumor, have considerably reduced the mortality rate.

Other Herpesviruses and the Cancer Connection

The latest addition to the herpesvirus family is **human herpesvirus-6 (HHV-6)**[1], also known as human T-lymphotropic virus. It was originally isolated from infected lymphocytes and has characteristics similar to EBV, but it is genetically distinct from the other herpesviruses. The virus can enter and replicate in T lymphocytes, macrophages, and salivary gland tissues. It is probably transmitted by close contact with saliva and other secretions. It is among the most common herpesviruses, with up to 95% prevalence in tested human populations. It is the cause of **roseola** (also known as roseola infantum), an acute febrile disease in babies between 2 and 12 months of age. It begins with fever that can spike as high as 105°F (40°C) and is followed a few days later by a faint maculopapular rash over the neck, trunk, and buttocks. The disease is usually self-limited, and the children recover spontaneously with support care.

Infected adults present with mononucleosis-like symptoms and may develop lymphadenopathy and hepatitis. Patients with renal or bone marrow transplants often acquire HHV-6 infections that suppress the function of the grafted tissues and are possibly a factor in rejection. Because of the association of HHV-6 with cases of brain infection and encephalitis, it has been considered as a potential cause of chronic neurological disease. Researchers have recently established a link between this virus and multiple sclerosis (MS). Over 70% of MS patients have positive serological signs of HHV-6 infection, and the brain lesions of many of them contain the virus. Although this is not conclusive proof of etiology, it is the first crucial step in connecting this disease to a virus.

The fact that herpesviruses are persistent, can develop latency in the cell nucleus, and have demonstrated oncogenesis has spurred an active search for the roles of these viruses in cancers of unknown cause. A number of clinical studies have disclosed a significant relationship between HHV-6 and Hodgkin lymphoma, oral carcinoma, and certain T-cell leukemias. It must be emphasized that isolating a virus or its antigens from a tumor is not sufficient proof of causation and further evidence is required, but it is highly suggestive of some role in cancer. Another human herpesvirus, HHV-7, has recently been characterized. It is very closely related to HHV-6, and it causes similar diseases in children and adults.

The latest herpesvirus was isolated from Kaposi sarcoma (a common tumor of AIDS patients). This virus, termed *Kaposi sarcoma-associated herpesvirus* (KSHV), or *herpesvirus-8,* was not found in other normal tissues or in patients without sarcoma. The genetic mechanisms and pathology behind this cancer are an ongoing focus of study. Some researchers have discovered an additional

---

1. An alternate system for naming herpesviruses assigns them a number. Thus, 1 and 2 are HSV, 3 is VZV, 4 is EBV, 5 is CMV, 6 and 7 are HHV, and 8 is KSHV.

link between KSHV and multiple myeloma, a relatively common cancer of the blood.

Other mammals are also susceptible to herpesvirus infections. In cases of simian herpesviruses, monkeys can pass the infection to humans, often with fatal results.

HEPADNAVIRUSES: UNUSUAL ENVELOPED DNA VIRUSES

One group of enveloped DNA viruses, called *hepadnaviruses,** is quite unlike any other so far discovered. These viruses have never been grown in tissue culture and have an unusual genome containing both double- and single-stranded DNA (figure 24.15). Hepadnaviruses show a decided tropism for the liver, where they persist and can be a factor in liver cell carcinoma. One member of the group, **hepatitis B virus (HBV),** causes a common form of hepatitis, and other members cause hepatitis in woodchucks, ground squirrels, and Peking ducks.

General Considerations: What Is Viral Hepatitis?

When certain viruses infect the liver, they cause **hepatitis,** an inflammatory disease marked by necrosis of hepatocytes and a mononuclear response that swells and disrupts the liver architecture. This pathologic change interferes with the liver's excretion of bile pigments such as bilirubin into the intestine. When bilirubin, a greenish-yellow pigment, accumulates in the blood and tissues, it causes **jaundice,** a yellow tinge in the skin and eyes.

---

* hepadnavirus (hep″-ah-dee″-en-ay-vy′-rus) Gr. *hepatos,* liver, plus DNA and virus.

* jaundice (jon′-dis) Fr. *jaunisse,* yellow. Also called icterus.

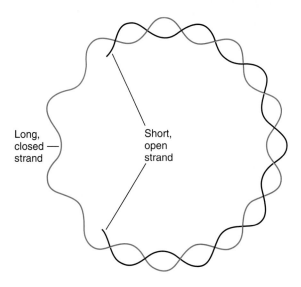

FIGURE 24.15

The nature of the DNA strand in hepatitis B virus. Like the genomes of most DNA viruses, the genome of HBV is circular, but unlike them, it is not a completely double-stranded molecule. A section of the molecule is composed of single-stranded DNA.

Characteristics of the three principal viruses responsible for human hepatitis are summarized in table 24.3. Hepatitis A virus (HAV) is a nonenveloped, single-stranded RNA enterovirus that is the etiologic agent of infectious hepatitis (see chapter 25). In general, HAV disease is far milder, shorter-term, and less virulent than the other forms. Another important agent is *hepatitis C virus*

TABLE 24.3

Principal Morphological and Pathologic Features of the Major Hepatitis Viruses

Property/Diseases	HAV	HBV	HCV
Biology			
Nucleic acid	RNA	DNA	RNA
Size	27 nm	42 nm	Various
Protein coat	–	+	+
Cell culture	+	–	–
Envelope	–	+	+
Synonyms	Infectious hepatitis, yellow jaundice	Serum hepatitis	Post-transfusional hepatitis
Epidemiology	Endemic and epidemic	Endemic	Endemic
Reservoir	Active infections	Chronic carrier	Chronic carrier
Transmission	Oral-fecal; water- or food-borne	Overt inoculation from blood, serum; close contact	Inoculation from blood, serum; intimate contact
Incubation Period	2–7 weeks	1–6 months	2–8 weeks
Symptoms	Fever, GI tract disorder	Fever, rash, arthritis	Similar to HBV
Jaundice	1 in 10	Common	Common
Onset/Duration	Acute, short	Gradual, chronic	Acute to chronic
Complications	Uncommon	Chronic active hepatitis, hepatic cancer	Chronic inflammation, cirrhosis
Availability of Vaccine	+	+	–
Diagnostic Tests to Differentiate	+	+	+

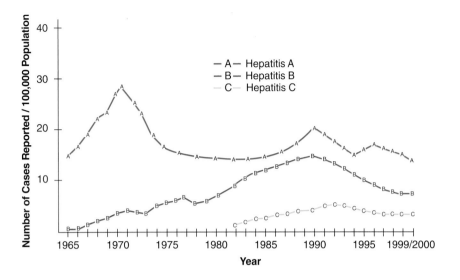

FIGURE 24.16

The comparative incidence of viral hepatitis in the United States from 1965 to 2000.

Source: Data from the Centers for Disease Control and Prevention, Atlanta, GA.

(HCV), an RNA virus in the flavivirus family that causes most cases of infusion hepatitis. HCV is involved in a chronic liver infection that can go undiagnosed, later leading to severe liver damage. It is spread primarily by exposure to blood, and is considered one of the most common unreported agents of hepatitis (an estimated 3 million cases nationally). Another agent is hepatitis E (HEV), a newly identified RNA virus that causes a disease similar to HAV. It is associated with poor hygiene and sanitation and is spread oro-fecally.

These hepatitis viruses have no relationship to hepatitis B virus (HBV), the only DNA virus that causes hepatitis. The incidence of the major forms is presented in figure 24.16. An unusual virus, hepatitis D, or delta agent, is a defective RNA virus that cannot produce infection unless a cell is also infected with HBV. Hepatitis D virus has a nucleocapsid structure that is similar to viroids, and it invades host cells by "borrowing" the outer receptors of HBV. When HBV infection is accompanied by the delta agent, the disease becomes more severe and is more likely to progress to permanent liver damage.

Hepatitis B Virus and Disease

Because a cell culture system for propagating hepatitis B virus is not yet available, by far the most significant source of information about its morphology and genetics has been obtained from viral fragments, intact virions, and antibodies, which are often abundant in patients' blood. One of these blood-borne pieces, the Dane particle, has been suspected as the infectious virus particle, and modern serological and electron microscope studies support that view. Blood-borne viral components are shown in figure 24.17.

Epidemiology of Hepatitis B An important factor in the transmission pattern of hepatitis B virus is that it multiplies exclusively in the liver, which continuously seeds the blood with viruses. Electron microscopic studies have revealed up to 10^7 virions per milliliter of infected blood. Even a minute amount of blood (10^{-6} to 10^{-7} ml) can transmit infection. The abundance of circulating virions is so high and the minimal dose so low that such simple practices as sharing a toothbrush or a razor can transmit infection. Over

the past 10 years, HBV has also been detected in semen and vaginal secretions, and it is currently believed to be transmitted by these fluids in certain populations. Spread of the virus by means of close contact in families or institutions is also well documented.

Hepatitis B is an ancient disease that has been found in all populations, though the incidence and risk are highest among people living under crowded conditions, drug addicts, the sexually promiscuous, and certain occupations, including people who conduct medical procedures involving blood or blood products such as serum (Medical Microfile 24.3). The incidence of HBV infection and carriage among drug addicts who routinely share needles is very high. Homosexual males constitute another high-risk group because such practices as anal intercourse can traumatize

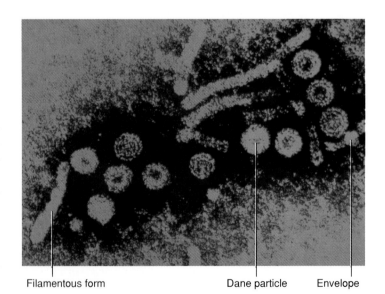

Filamentous form   Dane particle   Envelope

FIGURE 24.17

Hepatitis B. Blood from hepatitis B virus–infected patients presents an array of particles, including Dane particles (the intact virus), the envelope (or surface shell), and filaments that are polymers of the envelope antigen.

Hepatitis B is a risk to patients and health care workers alike. Obvious gross contact, in the form of blood transfusions or injections of human plasma, clotting factors, serum, and other products, is a classic route of infection. The virus remains infective for days in dried blood, for months when stored in serum at room temperature, and for decades if frozen. Although it is not inactivated after 4 hours of exposure to 60°C, boiling for the same period can destroy it. Disinfectants containing chlorine, iodine, and glutaraldehyde show potent anti-hepatitis B activity. It should be noted, however, that HBV infection from transfusion is quite rare now because of blood tests for HBV antigens and processing techniques that inactivate any HBV present.

Such diverse procedures as kidney dialysis, reused needles, and acupuncture needles are also known to transmit the virus. Outbreaks of hepatitis B have developed in people vaccinated with a special needle-less gun. These cases were traced to inadequate disinfection of the casing of the gun nozzle, which must be reused continuously between patients. Even cosmetic manipulations such as tattooing and ear or nose piercing can expose a person to infection. The only reliable method for destroying HBV on reusable instruments is autoclaving.

In regard to occupational exposure, nonimmunized people working with blood, plasma, sera, or materials contaminated with even small amounts of those substances are very vulnerable to infection. Practically any health occupation can be involved, but hematologists, phlebotomists, dialysis personnel, dental workers, surgical staff, and emergency room attendants must take particular precautions. Accidents involving broken blood vials, needlesticks, improper handling of instruments, and splashed blood and serum are common causes of HBV infection in clinical workers.

Other hepatitis viruses, especially HCV, are also closely tied to blood transmission. A blood test can screen blood and organ donations to prevent this source of infection. Currently, because of detection and control measures, all forms of viral hepatitis are declining in the United States.

membranes and permit transfer of virus into injured tissues. Heterosexual intercourse can also spread infection, but is less likely to do so. Infection of the newborn by chronic carrier mothers occurs readily and predisposes to development of the carrier state and increased risk of liver cancer in the child. The role of mosquitoes, which can harbor the virus for several days after biting infected persons, has been well-documented in the tropics and in some parts of the United States.

The tropism of HBV for the liver becomes a significant epidemiologic complication in some people (5–10% of HBV-positive people). Persistent or chronic carriers harbor the virus for up to 6 months, and some continue to shed the virus for many years. Worldwide, 300 million people are estimated to carry the virus, but it is most prevalent in Africa, Asia, and the western Pacific and is least prevalent in North America and Europe.

Pathogenesis of Hepatitis B Virus   The hepatitis B virus enters the body through a break in the skin or mucous membrane or by injection into the bloodstream. Eventually, it reaches the liver cells (hepatocytes), where it multiplies and releases viruses into the blood during an incubation period of 4 to 24 weeks (7 weeks average). Surprisingly, the majority of those infected exhibit few overt symptoms and eventually develop an immunity to HBV. But some people experience malaise, fever, chills, anorexia, abdominal discomfort, and diarrhea. In people with more severe disease, the symptoms and the aftermath of hepatic damage vary widely. Fever, jaundice, rashes, and arthritis are common reactions (figure 24.18), and a smaller number of patients develop glomerulonephritis and arterial inflammation. Complete liver regeneration and restored function occur in most patients. However, a small number of patients develop chronic liver disease in the form of necrosis or cirrhosis (permanent liver scarring and loss of tissue).

The association of HBV with **hepatocellular carcinoma**[2] is based on these observations: (1) Certain hepatitis B antigens are found in malignant cells and are often detected as integrated components of the host genome; (2) persistent carriers of the virus are more likely to develop this cancer; and (3) people from areas of the world with a high incidence of hepatitis B (Africa and the Far East) are more frequently affected. This connection is probably a result of infection early in life and the long-term carrier state. In general, people with chronic hepatitis are 200 times more likely to develop liver cancer, though the exact role of the virus is still the object of molecular analysis.

Diagnosis and Management of Hepatitis B   Hepatitis symptoms are so similar that differential diagnosis on this basis is unlikely. A careful examination of possible risk factors could distinguish between hepatitis B and A. For example, HBV infection is associated with occupational exposure, drug abuse, relatively long incubation period, and insidious onset. Serological tests can detect either virus antigen or antibodies. Recent developments in radioimmunoassay and ELISA testing permit detection of the surface antigen very early in the infection. These same tests are essential for screening blood destined for transfusions, semen in sperm banks, and organs intended for transplant. Antibody tests are most valuable in patients who are negative for the antigen.

Mothers who are carriers of HBV are highly likely to transmit infection to their newborns during delivery. It is not currently known whether transplacental HBV infection occurs. Unfortunately, infection with this virus early in life is a known risk factor for chronic liver disease. In light of the increasing incidence of HBV-positive pregnant women, routine testing of all pregnant

---

2. HOC. Also called hepatoma, a primary malignant growth of hepatocytes.

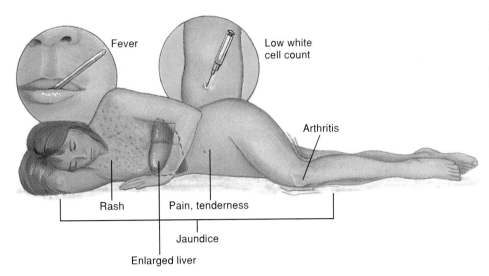

Fever

Low white cell count

Arthritis

Rash

Pain, tenderness

Jaundice

Enlarged liver

FIGURE 24.18

The clinical features of hepatitis B.
Among the most prominent symptoms are fever, jaundice, and intestinal discomfort. The liver is also enlarged and tender.

women is now an important prenatal screen. It has been added to the standard procedures to detect infections that are especially harmful to fetuses and neonates, known collectively as STORCH (see chapter 13).

Mild cases of hepatitis B are managed by symptomatic treatment and supportive care. Chronic infection can be controlled with recombinant interferon, which stops virus multiplication and prevents liver damage in up to 60% of patients. Passive immunization with hepatitis B immune globulin (HBIG) gives significant protection to people who have been exposed to the virus through needle puncture, broken blood containers, or skin and mucosal contact with blood. Other groups for whom prophylaxis is highly recommended are the sexual contacts of actively infected people or carriers and neonates born to infected mothers.

Since 1981, the primary prevention for HBV infection is vaccination. The most widely used vaccines (Recombivax, Energix) contain the pure surface antigen cloned in yeasts. They are given in 3 doses over 18 months, with occasional boosters. An alternate vaccine is Heptavax, made from the purified sterile antigen extracted from carrier blood. It is used mainly in people who are allergic to yeasts. Vaccination is a must for medical and dental workers and students, patients receiving multiple transfusions, immunodeficient persons, and cancer patients. The vaccine is also now strongly encouraged for all newborns as part of a routine immunization schedule.

CHAPTER CHECKPOINTS

Poxviruses are the largest and most complex of all viruses. They infect the skin, producing pustular lesions. Smallpox, which is now eradicated, molluscum contagiosum, monkeypox, and cowpox are the only poxviruses that infect humans.

Herpesviruses are large enveloped viruses that cause recurrent Infections and are potentially carcinogenic. Herpes simplex 1 and 2 infect the skin and mucous membranes but can cause encephalitis in neonates.

The varicella-zoster virus is the causative agent of chickenpox and shingles. Cytomegalovirus (CMV) causes systemic disease in fetuses, neonates, and Immunosuppressed individuals and a type of mononucleosis.

Epstein-Barr virus is the causative agent of Burkitt lymphoma and infectious mononucleosis. Herpesvirus-6 causes roseola and may be a factor in multiple sclerosis.

Other herpesviruses are implicated in cancers such as Kaposi sarcoma.

Hepadnaviruses are the causative agents of liver disease in many animals, including humans. Hepatitis B virus causes the most serious form of human hepatitis. It is spread by direct contact, fomites, and mosquitoes. It is potentially lethal in a small percentage of cases, either by direct liver damage or by causing liver cancer. Other forms of viral hepatitis are caused by RNA viruses.

Nonenveloped DNA Viruses

THE ADENOVIRUSES

During a study seeking the cause of the common cold, a new virus was isolated from the *adenoids** of young children. It turned out that this virus, termed **adenovirus,** was not the sole agent of colds but one of several others (described in chapter 25). It was also the first of about 80 strains of nonenveloped, double-stranded DNA viruses classified as adenoviruses, including 30 types associated with human infection. Besides infecting lymphoid tissue, adenoviruses have a preference for the respiratory and intestinal epithelia and the conjunctiva. Adenoviruses produce aggregations of incomplete and assembled viruses that usually lyse the cell (figure 24.19). In cells that do not lyse, the viral DNA may be harbored latently in the nucleus. Experiments show that adenoviruses are oncogenic in animals, but whether they are oncogenic in humans is not known.

---

* adenoid (ad′-eh-noyd) Gr. *adeno,* gland, and *eidos,* form. A lymphoid tissue (tonsil) in the nasopharynx that sometimes becomes enlarged.

Nuclear membrane   Viral inclusions

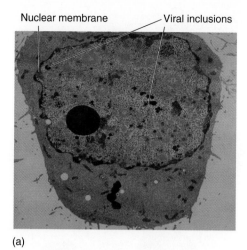

(a)

Mass of virions

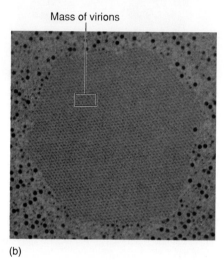

(b)

FIGURE 24.19

Distinctive adenovirus structures. **(a)** Cellular and nuclear alterations in adenovirus infection. Granules of early virus parts accumulating in the nucleus cause inclusions to form (19,000×). **(b)** Magnification of the nucleus shows a crystalline inclusion containing mature virions (35,000×).

Epidemiology of Adenoviruses

Adenoviruses are spread from person to person by means of respiratory and ocular secretions. Most cases of conjunctival infection involve preexisting eye damage and contact with contaminated sources such as swimming pools, dusty working places (shipyards and factories), and unsterilized optical instruments. Infection by adenovirus usually occurs by age 15 in most people, and certain individuals become chronic respiratory carriers. Although the reason is obscure, military recruits are at greater risk for infection than the general population, and respiratory epidemics are common in armed services installations. Civilians, even of the same age group and living in close quarters, seldom experience such outbreaks.

Respiratory, Ocular, and Miscellaneous Diseases of Adenoviruses

The patient infected with an adenovirus is typically feverish, with acute rhinitis, cough, and inflammation of the pharynx, enlarged cervical lymph nodes, and a macular rash (somewhat reminiscent

of rubella). Certain adenoviral strains produce an acute follicular lesion of the conjunctiva. Usually one eye is affected by a watery exudate, redness, and partial closure. A deeper and more serious complication is *keratoconjunctivitis,* an inflammation of the conjunctiva and cornea.

Acute hemorrhagic cystitis in children has been traced to an adenovirus. This self-limiting illness of 4 to 5 days duration is marked by hematuria (blood in the urine), frequent and painful urination with occasional episodes of fever, bedwetting, and suprapubic pain. Because the virus apparently infects the intestinal epithelium and can be isolated from some diarrheic stools, adenovirus is also suspected as an agent of infantile gastroenteritis.

Severe cases of adenovirus infection can be treated with interferon in the early stages. An inactivated polyvalent vaccine prepared from viral antigens is an effective preventive measure, especially for military recruits.

PAPOVAVIRUSES: PAPILLOMA, POLYOMA, VACUOLATING VIRUSES

Letters from the terms **pa**pilloma,* **po**lyoma,* and simian **va**cuolating* viruses gave rise to the term **papovavirus.** This family contains two subtypes: the **papillomaviruses,** which are responsible for human and animal papillomas, and the **polyomaviruses,** which include several human (BK and JC viruses) and animal pathogens.

Epidemiology and Pathology of the Human Papillomaviruses

A papilloma is a benign, squamous epithelial growth commonly referred to as a **wart,** or **verruca,** and caused by one of 40 different strains of **human papillomavirus (HPV).** Some types are specific for the mucous membranes; others invade the skin. The appearance and seriousness of the infection vary somewhat from one anatomical region to another. Painless, elevated, rough growths on the fingers and occasionally on other body parts are called **common,** or **seed, warts** (figure 24.20*a*). These commonly occur in children and young adults. **Plantar warts** are deep, painful papillomas on the soles of the feet; flat warts are smooth, skin-colored lesions that develop on the face, trunk, elbows, and knees. A special form of verruca known as **genital warts** is a prevalent STD and is linked to some types of cancer. Warts are transmissible through direct contact with a wart or contaminated fomites, and they can also spread on the same person by autoinoculation. The incubation period varies from 2 weeks to more than a year. Papillomas are common in all sexes, races, and geographic regions.

Genital Warts: An Insidious Papilloma Genital warts are the latest concern among young, sexually active people. Although this disease is not new, case reports have increased to such an extent that it is now regarded as the most common STD in the United States, with over 6 million new cases per year. There are probably in excess

---

* **papilloma** (pap″-il-oh′-mah) L. *papilla,* pimple, and Gr. *oma,* tumor. A benign, nodular, epithelial tumor.

* **polyoma** (paw″-lee-oh′-mah) L. *poly,* many. A glandular tumor caused by a virus.

* **vacuolating** (vak′-yoo-oh-lay″-ting) L. *vacuus,* empty. The process of forming small spaces or vacuoles in the cytoplasm of the cell.

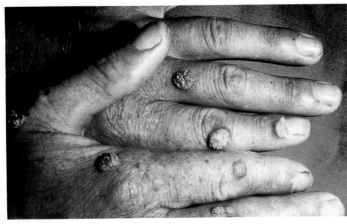

(a)

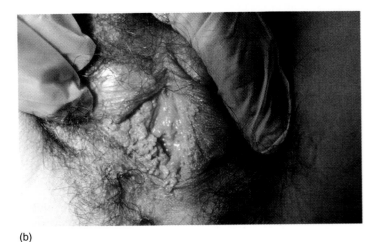

(b)

FIGURE 24.20

 Human papillomas. **(a)** Common warts. Virus infection induces a neoplastic change in the skin and benign growths on the fingers, face, or trunk. One must avoid picking, scratching, or otherwise piercing the lesions to avoid spread and secondary infections. **(b)** Chronic genital warts (condylomata acuminata) that have extended into the labial, perineal, and perianal regions.

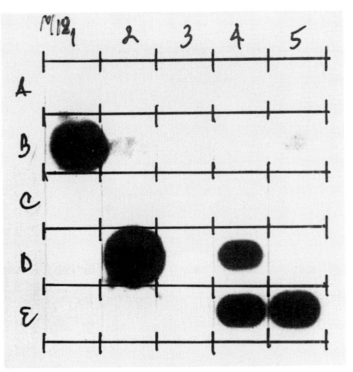

FIGURE 24.21

DNA hybridization or blot technique for detecting human papillomavirus. DNA is extracted from cervical specimens and blotted onto a special carrier. Labeled DNA probes specific to types 16 and 18 are reacted with the samples and exposed to X-ray film. Results show that patients B1, D2, D4, E4, and E5 carry HPV DNA for these two types and have greater risk for developing cervical cancer.

From Koneman et al., Diagnostic Microbiology, *4th ed., 1992, fig. 20.20, p. 1037 © J. B. Lippincott Company.*

of 30 million carriers of one of the five types of HPV associated with genital warts (types 6, 11, 16, 18, and 31). The virus invades the external and internal genital membranes, especially of the vagina and the head of the penis. Wart morphology ranges from tiny, flat, inconspicuous bumps to extensive, branching, cauliflower-like masses called **condylomata acuminata*** (figure 24.20*b*). Infection is also common in the cervix, urethra, and anal skin.

The most disturbing aspect of the sudden increase in HPV infection is its strong association with cancer of the reproductive tract, especially the cervix and penis. Two viral types—16 and 18—can be isolated from a majority of female genital cancers and are known to convert some forms of genital warts into malignant tumors. Susceptibility to this infection and its potential for cancer can be greatly reduced by early detection and treatment of both the patient and the sexual partner. Such early diagnosis depends on thorough inspection of the genitals and, in women, a Papanicolaou smear to screen for abnormal cervical cells.

Diagnosis, Treatment, and Prevention The warts caused by papillomaviruses are usually distinctive enough to permit reliable clinical diagnosis without much difficulty. However, a biopsy and histological examination can help clarify ambiguous cases. Sensitive DNA probes can detect HPV by means of a special hybridization probe (figure 24.21) or directly in infected cells (see chapter opening photo). These techniques can determine the type of HPV and assess the patient's risk for developing cervical cancer. Although most common warts regress over time, they cause sufficient discomfort and cosmetic concern to require treatment. Treatment strategies for all types of warts include direct chemical application of podophyllin and physical removal of the affected skin or membrane by cauterization, freezing, or laser surgery. Immunotherapy with interferon is effective in many cases. Because treatments may not completely destroy the virus, warts can recur.

* condylomata acuminata (kahn″-dee-loh′-mah-tah ah-kyoo″-min-ah′-tah) Gr. *kondyloma,* knob, and L. *acuminatus,* sharp, pointed. Not to be confused with condyloma lata, a broad, flat lesion of syphilis.

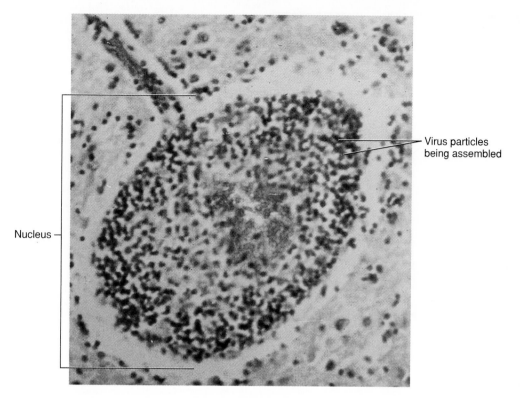

Nucleus

Virus particles
being assembled

FIGURE 24.22

Brain pathology in progressive multifocal leukoencephalopathy. Target cells are oligodendrocytes, shown here with a giant swollen nucleus and masses of assembling virus particles.

The Polyomaviruses

In the mid-1950s the search for a leukemia agent in mice led instead to the discovery of a new virus capable of inducing an assortment of tumors—hence its name, polyomavirus. Among the most important human polyomaviruses are the **JC virus** and **BK virus.**[3] Although many polyomaviruses are fully capable of transforming host cells and inducing tumors in experimental animals, none causes cancer in its natural host.

Epidemiology and Pathology   On the basis of serological surveys, it appears that infections by JC virus and BK virus are commonplace throughout the world. Since the majority of infections are asymptomatic or mild, not a great deal is known about their mode of transmission, portal of entry, or target cells, but urine and respiratory secretions are suspected as possible sources of the viruses. The main complications of infection occur in patients who have cancer or have been immunosuppressed. **Progressive multifocal leukoencephalopathy*** (**PML**) is an uncommon but generally fatal infection by JC virus that attacks accessory brain cells and gradually demyelinizes certain parts of the cerebrum (figure 24.22). Infection with BK viruses, usually associated with renal transplants, causes complications in urinary function. By the time PML is diagnosed, the patient is irreversibly immunocompromised, and extensive brain damage has already occurred. However, BK infection can be prevented by treating renal transplant patients with human leukocyte interferon.

Nonenveloped Single-Stranded DNA Viruses: The Parvoviruses

The **parvoviruses*** (**PVs**) are unique among the viruses in having single-stranded DNA molecules. They are also notable for their extremely small diameter (18–26 nm) and genome size. Parvoviruses are indigenous to and can cause disease in several mammalian groups—for example, distemper in cats, an enteric disease in adult dogs, and a potentially fatal cardiac infection in puppies.

The most important human parvovirus is B19, the cause of erythema infectiosum (fifth disease), a common infection in children (figure 24.23). Often the infection goes unnoticed, though the child may have a low-grade fever and a bright red rash on the cheeks. This same virus can be more dangerous in children with immunodeficiency or sickle-cell anemia, because it destroys red blood stem cells. If a pregnant woman contracts infection and transmits it to the fetus, a severe and often fatal anemia can result. One type of parvovirus, the adeno-associated virus (A-AV), is another example of a defective virus that cannot replicate in the host cell without the function of a helper virus (in this case, an adenovirus). The possible impact of this coinfection on humans is still obscure.

---

3. The initials signify the cancer patients from whom the viruses were first isolated.

* **leukoencephalopathy** (loo″-koh-en-sef″-uh-lop′-uh-thee) Gr. *leuko*, white, *cephalos*, brain, and *pathos*, disease.

---

* **parvovirus** (par″-voh-vy′-rus) L. *parvus*, small.

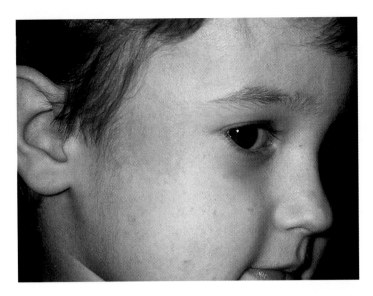

FIGURE 24.23

Typical "slapped face" rash of human parvovirus (B19). Confluent red rash on the cheek gives this child a "slapped face" appearance. This is one symptom of erythema infectiosum, a common childhood ailment caused by human parvovirus (B19).

CHAPTER CHECKPOINTS

Adenoviruses produce inflammatory infections in epithelial cells of the conjunctiva, intestine, respiratory tract, and lymph nodes.

Papillomaviruses infect the skin and mucous membranes, producing warts. Genital warts are the leading STD in the United States.

The polyomaviruses transform infected host cells into tumors, which can be fatal in immunosuppressed patients.

Parvoviruses are extremely small, resistant, single-stranded DNA viruses that cause diseases in cats, dogs, and humans. The most important human parvovirus, B19, is the cause of erythema infectiosum.

CHAPTER CAPSULE WITH KEY TERMS

I. **General Characteristics of Animal Viruses and Human Viral Diseases**

A. Viruses are minute parasitic particles consisting of DNA or RNA genomes packaged within a protein capsid; they invade host cells and appropriate the cell's machinery for mass production of new virions, both in cytoplasm and nucleus; viruses exit the host cell by lysis or budding; budded viruses leave with an envelope; actively infected host cells are usually destroyed.

B. Viruses attack a variety of host and cell types; individual viruses are relatively host/cell specific, due to the need for receptor recognition; target cells include nervous system, blood, liver, skin, and intestine.

C. Diseases range from mild and self-limited to fatal in effects; symptoms are local, systemic, and depend on the exact tissue target; identification and diagnosis are by culture, microscopy, genetic probes, and serology; immunity is both humoral and cell-mediated; limited number of drugs available for treatment.

D. DNA and some RNA viruses persist in inactive state (latent) in host cells; some latent viruses cause recurrent infections, and others integrate into host genome and cause cancerous transformation; some viruses are **teratogenic** and cause congenital defects; zoonotic viruses transmitted by vectors cause severe human disease.

II. **DNA Viruses**

DNA viruses are enveloped or nonenveloped nucleocapsids; most have double-stranded DNA; parvoviruses have single-stranded DNA.

A. Enveloped Viruses

1. *Poxviruses:* Large, complex viruses; produce skin lesions called **pox** (pocks).

a. **Variola** is the agent of **smallpox,** the first disease to be eliminated from the world through vaccination; last smallpox case reported in 1977; infection leads to large deep pustules that may scar; vaccination with vaccinia virus is no longer done routinely, except to protect certain populations against animal poxviruses such as monkeypox.

b. **Molluscum contagiosum,** an unclassified poxvirus, causes waxy nodules on skin; transmitted by direct contact; may be an STD in adults. Humans can also be infected by *monkeypox virus,* which causes a disease similar to smallpox.

2. *Herpesviruses:* Persistent latent viruses that cause recurrent infections and may be involved in neoplastic transformations; attack skin, mucous membranes, and glands.

a. **Herpes simplex virus (HSV):** Types 1 and 2 both create lesions of skin and mucous membranes; migrate into nerve ganglia; are reactivated; cause disseminated disease in newborn infants and immunodeficient patients; cause **whitlows** on fingers of medical and dental personnel. **HSV-1** is transmitted by close contact and droplets and mainly infects lips (cold sores, fever blisters), eyes, and oropharynx. **HSV-2** is sexually transmitted and usually affects the genitalia. Complication in newborns infected at birth with either virus is encephalitis. Infections are treated with some form of cyclovir; controlled by barriers and care in handling secretions.

b. **Varicella-Zoster Virus (VZV):** This herpesvirus causes both **chickenpox (varicella)** and **shingles (zoster);** chickenpox is the primary infection, zoster is later recurrence of infection by latent virus. Chickenpox is transmitted through droplets; symptoms are fever, poxlike papular rash. In zoster, the virus has migrated into spinal nerves and is reactivated by surgery, cancer, or other stimuli; causes painful lesions on skin of trunk or head; severe disease treated with acyclovir or famciclovir;

immune globulin used to relieve symptoms; vaccine now available.

c. **Cytomegalovirus (CMV):** One of the most common herpesvirus; spread through close contact with body fluids; congenital infection affects the liver, spleen, brain, and eyes; also causes mononucleosis-like syndrome with fever and leukocytosis; disseminated disease is common in AIDS and transplant patients; antiviral drugs may be of some benefit.

d. **Epstein-Barr Virus (EBV):** A herpesvirus of the lymphoid and glandular tissue; transmitted by saliva; 95% of all persons develop some form of infection; cause of **mononucleosis,** a "kissing disease" marked by sore throat, fever, swollen lymphoid tissue, leukocytosis; virus is oncogenic; also causes **Burkitt lymphoma,** a malignancy of the B lymphocytes that swells cheek or abdomen and is prevalent in African children; and causes **nasopharyngeal carcinoma,** cancer in older men in China, Africa. Disseminated EBV infection is a complication of AIDS and transplant patients.

e. **Herpesvirus-6** is the cause of **roseola,** an acute febrile disease in children; virus may be involved in multiple sclerosis and lymphoma. **Herpesvirus-8** is the oncogenic viral agent of Kaposi sarcoma, a tumor of AIDS patients.

3. *Hepadnaviruses:* One cause of *viral hepatitis,* an inflammatory disease of liver cells that may result from several different viruses.

a. **Hepatitis B virus (HBV)** causes **serum hepatitis;** virus has strong affinity for liver cells, is carried and shed into blood; only a tiny infectious dose is required.

b. The virus is resistant to heat and disinfectants; risks include contact with blood, drug addiction, and homosexuality.

c. Infection is marked by fever, intestinal disturbance, **jaundice;** chronic carriage may lead to liver disease and liver cancer.

d. Managed by serum globulin, interferon, and vaccines. *Delta agent* is an associated virus that accompanies HBV and causes a more severe disease.

III. **Nonenveloped Viruses**
   A. *Adenoviruses:* Common infectious agents of lymphoid organs, respiratory tract, eyes; spread through close contact with secretions; diseases include common cold with fever and rash; **keratoconjunctivitis,** a severe eye infection, **cystitis,** an acute urinary infection.
   B. *Papovaviruses:* Papilloma- and polyomaviruses.
      1. **Human papillomaviruses (HPVs)** cause skin tumors called **papillomas, verrucas,** or **warts;** spread by close contact with infected skin or fomites. **Common warts** are rough, painless lesions on hands; **plantar warts** are painful, flat, benign tumors on feet, trunk. **Genital warts** are verrucas that start as tiny bumps on membranes or skin of genitals; a very common STD; may progress to large masses called **condylomata acuminata;** chronic infection associated with cervical and penile cancer; treatment by surgery and interferon.
      2. **Polyomaviruses** include animal tumor viruses and human viruses. **JC virus** causes **progressive multifocal leukoencephalopathy,** a slow destruction of the brain in cancer and immunodeficient patients; **BK virus** causes complications in renal transplants.
   C. *Single-Stranded DNA Viruses:* Parvoviruses are the tiniest viruses; cause severe disease in several mammals; **human parvovirus (B19)** causes *erythema infectiosum,* a mild respiratory infection of children that can be dangerous to fetus.

MULTIPLE-CHOICE QUESTIONS

1. Which of the following is *not* caused by a poxvirus?
   a. molluscum contagiosum
   b. smallpox
   c. cowpox
   d. chickenpox

2. In general, zoonotic viral diseases are _____ in humans.
   a. mild, asymptomatic
   b. self-limited
   c. severe, systemic
   d. not contagious

3. Which virus was used in smallpox vaccination?
   a. variola   c. cowpox
   b. vaccinia   d. varicella

4. Herpes simplex 1 causes _____, and herpes simplex 2 causes _____.
   a. cold sores, genital herpes
   b. fever blisters, cold sores
   c. canker sores, fever blisters
   d. shingles, stomatitis

5. Herpetic whitlows are _____ infections of the _____.
   a. pox, oral cavity
   b. CMV, lymph nodes
   c. EBV, skin
   d. herpes, fingers

6. Neonatal herpes simplex is usually acquired through
   a. contaminated mother's hands
   b. other babies in the newborn nursery
   c. crossing the placenta
   d. an infected birth canal

7. _____ is an effective treatment for herpes simplex lesions.
   a. Amantadine   c. Acyclovir
   b. Interferon   d. Cortisone

8. Which herpes virus is commonly associated with a dangerous fetal infection?
   a. herpes simplex   c. EBV
   b. herpes zoster   d. CMV

9. Varicella and zoster are caused by
   a. the same virus
   b. two different strains of VZV
   c. herpes simplex and herpes zoster
   d. CMV and VZV

10. A common sign of hepatitis is
    a. liver cancer   c. anemia
    b. jaundice   d. bloodshot eyes

11. A virus associated with chronic liver infection and cancer is
    a. hepatitis C   c. hepatitis A
    b. hepatitis B   d. delta agent

12. Benign epithelial growths on the skin of fingers are called
    a. polyomas
    c. whitlows
    b. verrucas
    d. pox

13. Parvoviruses are unique because they contain _____.
    a. a double-stranded RNA genome
    b. reverse transcriptase
    c. a single-stranded DNA genome
    d. an envelope without spikes

14. Adenoviruses are the agents of
    a. hemorrhagic cystitis
    b. keratoconjunctivitis
    c. common cold
    d. all of these

15. **Multiple Matching.** Match the virus with the disease or condition. (Some viruses may match more than one condition.)

    _____ herpes simplex 1
    _____ variola
    _____ polyoma
    _____ varicella
    _____ herpes simplex 2
    _____ adenovirus
    _____ papilloma
    _____ Epstein-Barr
    _____ cytomegalovirus
    _____ parvovirus
    _____ herpesvirus-6

    a. gingivostomatitis
    b. condylomata acuminata
    c. chickenpox
    d. erythema infectiosum
    e. fever blisters
    f. mononucleosis
    g. smallpox
    h. genital herpes
    i. leukoencephalopathy
    j. roseola
    k. cancer
    l. keratoconjunctivitis

CONCEPT QUESTIONS

1. Outline 10 medically important characteristics of viruses.

2. a. What accounts for the affinity of viruses for certain hosts or tissues?
   b. What accounts for the symptoms of viral diseases?
   c. Why are some pathologic states caused by viruses so much more damaging and life-threatening than others?

3. Outline the target organs and general symptoms of DNA viral infections.

4. a. What is the general rule governing the cell locations where DNA and RNA viruses multiply?
   b. What factors determine whether latency occurs or not?
   c. In what special way can people carry DNA viruses?

5. a. Briefly describe the epidemiology and symptoms of smallpox.
   b. Why was it such a great killer?
   c. What is a pock, and how is it formed?
   d. What was the basis for the smallpox vaccine?
   e. How was it protective?

6. a. What symptoms characterize molluscum contagiosum?
   b. For what other infection in this chapter could it be mistaken?
   c. Explain why humans would suddenly become more susceptible to monkeypox.

7. What are the common characteristics of herpesviruses?

8. a. Compare the two types of human herpes simplex virus according to the types of diseases they cause, body areas affected, and complications.
   b. Why is neonatal infection accompanied by such severe effects?
   c. Is HSV-1 as severe as HSV-2 in this condition?
   d. What causes recurrent attacks?

9. a. Under what circumstances does one get chickenpox and shingles?
   b. Where do the lesions of shingles occur, and what causes them to appear there?
   c. How are ZIG and VZIG used?

10. a. What are the main target organs of CMV?
    b. How is it transmitted?
    c. What group of people is at highest risk for serious disease?

11. a. What are the main diseases associated with Epstein-Barr virus?
    b. What appears to be the pathogenesis involved in Burkitt's lymphoma?
    c. Why is mononucleosis a disease primarily of college-age Americans?

12. a. Define viral hepatitis, and briefly describe the three main types with regard to causative agent, common name, severity, and mode of acquisition.
    b. What is the effect of the delta agent?

13. a. What is unusual about the genome of hepatitis B virus?
    b. What is the usual source of the virus for study?
    c. What is important about the virus with regard to infectivity and transmission to others?
    d. What groups are most at risk for developing hepatitis B?

14. a. What is the course of hepatitis B infection from portal of entry to portal of exit?
    b. Describe some serious complications.

15. a. What is the nature of the vaccines for hepatitis B?
    b. Who should receive them?

16. a. What are the principal diseases of adenoviruses?
    b. How are they spread?

17. a. Compare the locations and other characteristics of the three types of warts.
    b. What type is most serious and why?
    c. How are warts cured?

18. Name the most important human parvovirus and the disease it causes.

19. a. Which DNA viruses are zoonotic?
    b. Which are intestinal viruses?
    c. Which are spread by blood and blood products?
    d. Which are spread by respiratory droplets, kissing, and other nonsexual intimate contact?

20. a. For which DNA viruses are there effective vaccines?
    b. Who receives them?

CRITICAL-THINKING QUESTIONS

1. a. Which DNA viruses have been linked to cancer?
   b. Why are DNA viruses more likely to cause cancer than RNA viruses?
   c. What are some mechanisms that could instigate cancer formation by viruses?
   d. Why is there a danger in using live attenuated DNA viruses for vaccines?

2. a. Why is isolating a virus from a tumor not indicative of its role in disease?
   b. What important indicators could verify that various cancers are caused by a herpesvirus?

3. a. What features of variola virus made it possible to eradicate smallpox?
   b. Are any other viruses in this chapter possible candidates for this sort of achievement? Why or why not?

4. Discuss the pros and cons of salvaging or destroying the smallpox virus cultures.

5. a. Can you think of some reasons that herpes simplex and zoster viruses are carried primarily in nerve trunks?
   b. What mechanism might account for reactivation of these viruses by various traumatic events?

6. Describe several measures health care workers can take to avoid whitlows.

7. Explain why a baby whose mother has genital herpes is not entirely safe even with a cesarian birth.

8. a. What specific host defenses do immunodeficient, cancer, and AIDS patients lack that make them so susceptible to the viral diseases in this chapter?
   b. Given the ubiquity of most of these viruses, what can health care workers do to prevent transmission of infection to other patients and themselves?

9. What are the medical consequences of transfusing blood labeled Dane (+)?

10. a. Name two different defective viruses that require another virus for function.
    b. Explain how they might interrelate with the host virus.

11. Explain this statement: One acquires chickenpox from others, but one acquires shingles from oneself.

12. A man wants to divorce his wife because he believes her genital herpes could only have been acquired sexually. Her doctor says she has type 1. Can you help them resolve this problem?

13. Weigh the following observation: High titers of antibodies for EBV are found in a leukemia patient; a chronically tired, ill businessman; a healthy military recruit; and an AIDS patient. Comment on the probable significance of antibodies to EBV in human serum.

14. Give some pros and cons of exposing children to chickenpox to infect them and provide subsequent immunity.

15. a. How would you decontaminate a vaccination gun or acupuncture tools used in a series of patients?
    b. What type of virus is most likely to be transmitted by vaccination guns and acupuncture tools?

INTERNET SEARCH TOPICS

1. Access a search engine to locate information on the prevalence of herpes simplex 2, genital warts, and molluscum contagiosum as STDs in the United States.

2. Look in on-line medical sources for up-to-date information on the involvement of DNA viruses in cancers.

The RNA Viruses of Medical Importance

The RNA viruses are a remarkably diverse group of microbes with a variety of morphological and genetic adaptations and extreme and novel biological characteristics. Viruses are assigned to one of 13 families on the basis of their envelope, capsid, and the nature of their RNA genome (table 25.1). RNA viruses are etiologic agents in a number of serious and prevalent human diseases, including influenza, AIDS, hepatitis, and viral encephalitis. Many experts believe that these viruses and dozens of others have the potential of becoming the most important disease threats of the future (Spotlight on Microbiology 25.1).

Chapter Overview

- RNA viruses are responsible for a large number of human infections, and are especially prominent causes of emerging diseases.
- RNA viruses may be enveloped or not; most contain single-stranded RNA, but one group has double-stranded.
- The influenza viruses are highly variable viruses that cause influenza, an epidemic respiratory disease.
- Miscellaneous respiratory infections caused by enveloped RNA viruses are parainfluenza, mumps, measles, and croup.
- Zoonotic RNA viruses are spread by animal vectors:
  - Bunyaviruses (hantaviruses) are spread by rodents and are involved in serious respiratory infections.
  - Arenaviruses are carried by rodents and can cause hemorrhagic fevers.
  - The arthropod-borne arboviruses are the agents of yellow and dengue fever.
  - Lyssavirus (rabies virus) is a fatal brain infection acquired from contact with mammals.
- Retroviruses are unusual viruses that can use single-stranded RNA to synthesize double-stranded DNA using a reverse transcriptase enzyme.
- One prominent example is the human immunodeficiency virus (HIV), the cause of acquired immunodeficiency syndrome (AIDS), a devastating infection of the white blood cells.
  - AIDS is spread through sexual intercourse and contact with blood.
  - The destruction of specific immune function by HIV leads to opportunistic infections, cancers, and metabolic dysfunction.
- Picornaviruses are very small nonenveloped RNA viruses that may be spread through close contact and orofecally in food and water.

A virologist checks a culture while working in a Level 4 biohazard facility at the CDC. He is completely encased in a "space suit" equipped with its own air supply that protects against accidental exposure to the deadliest pathogens.

- Examples of picornaviruses are:
  - enteric viruses, including poliovirus, an infection of the central nervous system that can cause paralysis, and hepatitis A, an acute form of liver infection;
  - rhinoviruses and coxsackieviruses, which are major agents of the common cold syndrome.
- Rotavirus is the only double-stranded RNA virus and a common cause of viral gastroenteritis.
- Prions are very unusual infectious proteins transmitted by infectious tissues and secretions. They cause slow, progressive neurological diseases called spongiform encephalopathies (Creutzfeldt-Jakob disease).

TABLE 25.1

RNA Virus Families and Representative Viruses/Diseases

Enveloped	Nonenveloped

Segmented, Single-Stranded, Negative-Sense Genome★
  Orthomyxoviridae: Influenza
  Bunyaviridae: California encephalitis virus
        Hantavirus hemorrhagic fever
  Arenaviridae: Hemorrhagic fevers
        Lassa fever virus
        Argentine hemorrhagic
          fever virus

Nonsegmented, Single-Stranded, Negative Sense
  Paramyxoviridae: Mumps virus
        Measles virus
        Respiratory syncytial virus

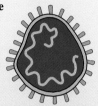

  Rhabdoviridae: Rabies virus
        Vesicular stomatitis virus

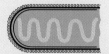

  Filoviridae: Ebola fever virus
        Marburg virus

Nonsegmented, Single-Stranded, Positive Sense
  Togaviridae: Rubella virus
        Western and eastern equine
          encephalitis
  Flaviviridae: Dengue fever virus
        Yellow fever virus
  Coronaviridae: Common cold virus

Single-Stranded, Positive Sense, Reverse Transcriptase
  Retroviridae: AIDS (HIV)
        T-cell leukemia virus
        Hairy-cell leukemia virus

Nonsegmented, Single-Stranded, Positive-Sense Genome
  Picornaviridae: Polio virus
        Hepatitis A virus
        Rhinoviruses

  Caliciviridae: Norwalk agent

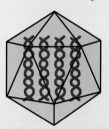

Segmented, Double-Stranded, Positive Sense, Double Capsid
  Reoviridae: Tick fever virus
        Rotavirus gastroenteritis

Legend	Nucleocapsid type: Helical	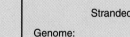

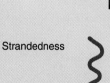

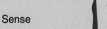

Icosahedral

Genome: Strandedness

Single (SS)   Double (DS)

Number of molecules

Segmented   Nonsegmented

Sense   (+)   (−)

SPOTLIGHT ON MICROBIOLOGY 25.1
Life in the Hot Zone

Although it is true that we have seemingly vanquished smallpox and polio in some areas of the world, these successes are overshadowed by an emerging viral threat. Over the past 30 years, the world has played host to at least 33 outbreaks of serious virus diseases. At least once a year for the past 5 years, a new virus has emerged out of nowhere or a familiar virus has suddenly erupted in unusual outbreaks.

A large number of these emerging diseases are zoonotic in origin and often become more pathogenic in humans than in their natural hosts. Outbreaks have been traced to viruses of traditional concern (influenza and rabies); viruses that are changing their geographic distribution (dengue fever virus) or jumping hosts (monkeypox); and previously unreported forms (hantavirus). The newest occurrences are the Nipah virus in Malaysia, which is transmitted from fruit bats to pigs to humans, and the West Nile virus, which has begun spreading from Africa to the Middle East and North America by means of migrating birds. These "hitchhiking" viruses are picked up by mosquitoes, which can infect humans.

Because of their deadly potential, a great deal of attention has been focused on viruses dubbed "hot agents." These viruses are so lethally infectious that they cannot be safely handled without Level 4 biosafety precautions, including special space suits for protection (see chapter opening

Red Cross workers prepare an Ebola virus victim for burial in Zaire, using high-level containment procedures. Over a thousand Africans have died from this disease.

photo). One of these—the **Ebola virus,** named for a river in Zaire—is fraught with particular dangers. This virus and its German relative, the Marburg virus, belong to a category of viruses called *filoviruses,* characterized by a unique, thready appearance (see figure 6.1). The Ebola virus has furious virulence, as described in this account:

> In this they are like HIV, which also destroys the immune system, but unlike the onset of HIV, the attack by Ebola is explosive. As Ebola sweeps through you, your immune system fails, and you seem to lose your ability to respond to viral attack. . . . Your mouth bleeds, and you bleed around your teeth . . . literally every opening in the body bleeds, no matter how small. Ebola does in ten days what it takes AIDS ten years to accomplish.*

This disease has erupted several times since 1976. It is thought to originate from monkeys and is spread through contact with body fluids. With no effective treatments, the mortality rate is 80–90%.

Other microbiological mysteries involve an unusual viruslike agent found in cattle in England that causes bovine spongiform encephalopathy, or "mad cow disease." This disease, which is transmissible among cows and sheep, causes overwhelming deterioration of the brain, frenzy, and death. A new variant of human Creutzfeldt-Jakob disease was traced to consumption of food products contaminated with the bovine agent. Nearly 200 people in Europe have died from this disease. So great was the concern for a massive epidemic, that millions of cattle and sheep were ordered destroyed and incinerated. All exports of British beef have been banned, and several countries have ceased the commercial sale of beef.

A newly isolated pathogen called Borna disease virus (BDV) is commonly associated with neurological abnormalities in horses, rats, and monkeys. This zoonosis can be transmitted to many hosts, including humans, where it apparently causes neuropsychiatric abnormalities such as schizophrenia. The blood of some psychiatric patients contains antibodies against this virus 600 times higher than normal controls.

The fears fostered by these emerging viruses are the possibility of amplification in the population, leading to rapid global transmission, and the unknown, uncontrollable nature of their origins.

** **Excerpt from** The Hot Zone by Richard Preston, copyright 1994, Random House.*

Enveloped Segmented Single-Stranded RNA Viruses

THE BIOLOGY OF ORTHOMYXOVIRUSES: INFLUENZA

Orthomyxoviruses are spherical particles with an average diameter of 80–120 nm. They are covered with a lipoprotein envelope that is studded with glycoprotein spikes acquired during viral maturation (see figures 25.1a and 6.19). The two glycoproteins that make up the spikes of the envelope and contribute to virulence are hemagglutinin (HA) and neuraminidase (NA). Because **hemagglutinin**

spikes combine with a specific carbohydrate molecule found in all eucaryotic cell membranes, the virus has the capacity to bind and clump a variety of animal cells. The name hemagglutinin is derived from its particular agglutinating action on red blood cells (see figure 16.6) that is the basis for viral assays used to identify several antigenic types. Hemagglutinin contributes to infectivity by binding to host cell receptors of the respiratory mucosa, a process that facilitates viral penetration. **Neuraminidase** has the principal effect of hydrolyzing the protective mucous coating of the respiratory tract, assisting in viral budding and release, keeping viruses from sticking together, and participating in host cell fusion.

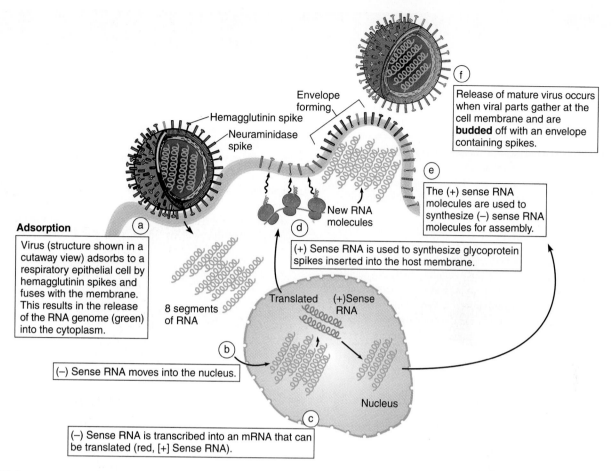

Envelope
forming

Hemagglutinin spike

Neuraminidase
spike

f Release of mature virus occurs
when viral parts gather at the
cell membrane and are
budded off with an envelope
containing spikes.

New RNA
molecules

e The (+) sense RNA
molecules are used to
synthesize (−) sense RNA
molecules for assembly.

d (+) Sense RNA is used to synthesize glycoprotein
spikes inserted into the host membrane.

Adsorption

a Virus (structure shown in a
cutaway view) adsorbs to a
respiratory epithelial cell by
hemagglutinin spikes and
fuses with the membrane.
This results in the release
of the RNA genome (green)
into the cytoplasm.

8 segments
of RNA

Translated   (+)Sense
RNA

b (−) Sense RNA moves into the nucleus.

Nucleus

c (−) Sense RNA is transcribed into an mRNA that can
be translated (red, [+] Sense RNA).

FIGURE 25.1

Stages in cell invasion and disruption by the influenza virus.

The genome of the influenza virus is known for its extreme variability. It is subject to constant genetic changes that alter its antigenic expression and create new strains of the virus with a different epidemiologic pattern. Depending upon the degree and timing of the change, these patterns are called antigenic shift or antigenic drift. **Antigenic shift** is a major alteration in antigenicity occurring when genome segments from different viral strains recombine. A wholly new strain can emerge in a single recombinant event, such as when a human is simultaneously infected by both a human and an animal strain of the virus. By contrast, **antigenic drift** is a minor change in antigenicity that is caused by small mutations in a single virus strain. As the antigenic variations accumulate over time, the gradual trend is toward a new viral strain (Medical Microfile 25.2).

Epidemiology of Influenza A

Influenza is an acute, highly contagious respiratory illness afflicting people of all ages and marked by seasonal regularity and pandemics at predictable intervals. One of its first names was an Italian derivative of *un influenza di freddo,* an "influence of mystery." This name has remained and is now contracted to the vernacular "flu." Periodic outbreaks of respiratory illness that were probably influenza have been noted throughout recorded history. Of the six pandemics occurring in 1900, 1918, 1946, 1957, 1968, and 1977,

the outbreak of 1918 took the greatest toll—at least 20 million deaths. Even today, deaths from influenza or influenzal pneumonia rank among the top 10 causes of death in the United States. Of the three types of influenza viruses, the most virulent, type A, is most commonly associated with human disease.

To characterize each virus strain, virologists have developed a special nomenclature that chronicles the influenza virus type, the presumed animal of origin, the location, and the year of origin. For example, A/duck/Ukraine/63 identifies the strain isolated from domestic or migratory fowl; likewise, A/seal/Mass/1/80 is the causative agent of seal influenza. From this convention come such expressions as "swine flu" and "Hong Kong flu."

Mode of Influenza Transmission

Inhalation of virus-laden aerosols and droplets constitutes the major route of influenza infection, although fomites can play a secondary role. Transmission is greatly facilitated by crowding and poor ventilation in classrooms, barracks, nursing homes, and military installations in the late fall and winter. In the tropics, infection occurs throughout the year, without a seasonal distribution. Occupational contact with ducks, other poultry, and swine constitutes a special high-risk category for disease, The overall mortality rate for influenza A is about 0.1% of cases. Deaths are most common among elderly persons and small children.

MEDICAL MICROFILE 25.2
Community Immunity, Genetic Change, and Influenza

Every fall, an announcement comes from the CDC to expect outbreaks of influenza with exotic names such as Hong King, Australian, or Russian. How can medical experts predict which types of flu to expect in a given year? In order to understand these predictions, we must first examine how changes in the influenza virus influence its epidemiologic pattern. The most important factors in the evolution and fate of influenza virus are the degree of community resistance, antigenic drift, and antigenic shift.

Community Immunity

The concept of community, or herd, immunity was introduced in chapter 16. It is defined as the collective resistance of individuals in a population, acquired by active infection, vaccination, and transplacental antibody transfer. Herd immunity works most effectively when the virus strain has recently passed through a population, and it is least effective when a new virus strain appears or is imported from another population.

Antigenic Drift and New Strains of Virus

The influenza viruses undergo constant genetic variations and antigenic change. The viral components that are most subject to change are the hemagglutinin and neuraminidase spikes, which determine infectivity and induce protective antibodies. Accumulated changes can create viral variants so antigenically different that antibodies from a prior infection are no longer protective. The results are lapses in individual and herd immunity and epidemics.

Antigenic Shift and Coinfections with Influenza Virus

An event that increases the possibility of antigenic shift is the simultaneous infection of one individual by two different influenza strains. Although the influenza viruses are common in humans, they have also been found in swine and birds. It is possible for humans to become infected with animal strains and vice versa. Occasionally, coinfection of the same animal by two antigenic variants gives rise to completely new viral strains quite different from either original. If such an altered strain is introduced into either animal or human populations, it will spread rapidly because neither group has protective antibodies. Experts have traced the pandemics of 1918, 1957, 1968, and 1977 to strains of a virus that came from pigs (swine flu).

Scientists at the Armed Forces Institute of Pathology have analyzed tissue specimens from American soldiers who died in the 1918 pandemic. They were able to isolate enough virus genetic material to determine that genes for spike proteins most closely resemble those of swine. This important finding finally laid to rest the question of where this virulent virus originated, verifying that it was the United States.

More recently, a small but potentially serious outbreak of influenza occurred in China that was due to the chicken influenza virus. This was the first reported instance of avian influenza in humans.

Recurrence of Pandemics

Influenza pandemics are accelerated by the contagiousness of the virus and widespread migrations or travel. They also occur in cycles, with episodes spaced about a decade apart. By compiling extensive serological profiles of populations and the influenza strains to which they are immune, the CDC has a scientific basis for predicting expected epidemic years and the probable strains involved. One advantage of forecasting the expected viral subtypes is that the most vulnerable people can be vaccinated ahead of time with the appropriate virus strain(s).

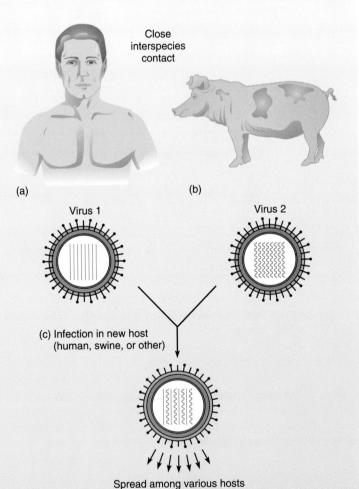

Coinfection of human and pig strains in a single animal. During viral multiplication, the eight-segmented genome of virus 1 can reassort with the eight-segmented genome of virus 2. Theoretically, recombination could produce 256 (2^8) different genome combinations. Only one is shown here.

Infection and Disease

The influenza virus binds primarily to ciliated cells of the respiratory mucosa (figure 25.1). Infection causes the rapid shedding of these cells along with a load of viruses. Stripping the respiratory epithelium to the basal layer eliminates protective ciliary clearance and leads to severe inflammation and irritation. The illness is further aggravated by fever, headache, myalgia, pharyngeal pain, shortness of breath, and repetitive bouts of coughing. The viruses tend to remain in the respiratory tract, and viremia is rare. As the normal ciliated columnar epithelium is restored in a week or two, the symptoms usually abate.

Patients with emphysema or cardiopulmonary disease, along with very young, elderly, or pregnant patients, are more susceptible to serious complications. Most often, the weakened host defenses predispose patients to secondary bacterial infections, especially pneumonia caused by *Streptococcus pneumoniae* and *Staphylococcus aureus*. Although less common than bacterial pneumonia, primary influenzal virus pneumonia is a dangerous complication that can cause rapid death of people even in their prime. Other complications are bronchitis, meningitis, and neuritis.

Diagnosis, Treatment, and Prevention of Influenza

Except during epidemics, influenza must be differentiated from other acute respiratory diseases. Rapid immunofluorescence tests can detect influenza antigens in a pharyngeal specimen (see figure 6.23). Serological testing to screen for antibody titer and virus isolation in chick embryos or kidney cell culture are other specific diagnostic methods. Genetic analysis is valuable for determining the origin of the virus.

Treatments used mainly to control symptoms include fluids, bed rest, and nonaspirin pain relievers and anti-inflammatory drugs. Aspirin is not indicated because it significantly increases the risk for **Reye syndrome,** a disease that strikes the brain, liver, and kidney and is characterized by the fatty degeneration of those organs. Amantadine and rimantidine, which specifically block the uncoating of the influenza virus in the cell, are sometimes prescribed to abate the course of influenza, especially in more severe cases. If given early, drugs can reduce the length of the disease and its symptoms as well as the spread of the virus. Because the drugs show some toxicity for the CNS, they must be used with caution.

The standard vaccine contains dead viruses grown in embryonated eggs. They have an overall effectiveness of 70–90%. An attenuated virus vaccine administered nasally appears to be more effective than the existing vaccine, and it requires smaller doses that could even be administered at home. Influenza vaccination is recommended primarily for certain high-risk groups, such as the chronically ill and the elderly, and for people with a high degree of exposure to the public. For complete protection, it is necessary to immunize against all the different strains of influenza viruses expected in a given year.

A major sequela associated with influenza vaccines is **Guillain-Barré syndrome,** a neurological complication that can arise in approximately one in 100,000 vaccine recipients. This syndrome appears to be an autoimmunity induced by viral proteins and marked by varying degrees of demyelination of the peripheral nervous system, leading to weakness and sensory loss. Most patients recover function, but the disease can also be debilitating and fatal.

BUNYAVIRUSES AND ARENAVIRUSES

Between them, the **bunyavirus*** and **arenavirus*** groups contain hundreds of viruses whose normal hosts are arthropods or rodents. Although none of the viruses are natural parasites of humans, they can be transmitted zoonotically and periodically cause epidemics in human populations. Bunyaviruses are discussed in a later section along with other arboviruses because they are transmitted primarily by insects and ticks. Some of the more important bunyavirus diseases are California encephalitis, spread by mosquitoes; Rift Valley fever, an African disease transmitted by sand flies; and Korean hemorrhagic fever, associated with hantavirus and transmitted by rodents. A newly identified hantavirus, Sin Nombre, is the first American member of this group, and is also an emerging pathogen (Spotlight on Microbiology 25.3).

Four major arenavirus diseases are Lassa fever, found in parts of Africa; Argentine hemorrhagic fever, typified by severe weakening of the vascular bed, hemorrhage, and shock; Bolivian hemorrhagic fever; and lymphocytic choriomeningitis, a widely distributed infection of the brain and meninges. The viruses are closely associated with a rodent host and are continuously shed by the rodent throughout its lifetime. Transmission of the virus to humans is through aerosols and direct contact with the animal or its excreta. These diseases vary in symptomology and severity from mild fever and malaise to complications such as hemorrhage, renal failure, cardiac damage, and shock syndrome. The lymphocytic choriomeningitis virus can cross the placenta and infect the fetus. Pathologic effects include hydrocephaly, blindness, deafness, and mental retardation. So dangerous are these viruses to laboratory workers that the highest level of containment procedures and sterile techniques must be used in handling them.

CHAPTER CHECKPOINTS

The enveloped, segmented, single-stranded RNA virus group includes the Orthomyxoviridae, the Bunyaviridae, and the Arenaviridae.

Segmented viruses such as influenza (flu) virus undergo genetic changes that can create new strains and cause failure in immunity. In **antigenic drift,** small mutations accumulate. In the variation called **antigenic shift,** two genetic variants of the flu virus infect a host and combine to form a third genetically different strain.

Like all enveloped RNA viruses, orthomyxoviruses have two unique glycoprotein virulence factors (receptors) in the envelope, **hemagglutinin** and **neuraminidase.** These assist in viral penetration and release from the host cell. Infection damages the respiratory epithelium and may lead to pneumonia. Control is with amantidine and yearly vaccination.

Bunyaviruses, grouped among the arboviruses because many of them are spread by insects, cause encephalitis and hemorrhagic fevers.

Arenaviruses are typically spread by rodents. They also cause fatal encephalitis and/or hemorrhagic fevers. Specific diseases caused by bunyaviruses include hantavirus pulmonary syndrome and hemorrhagic fever; arenaviruses cause Lassa fever and California encephalitis.

---

* bunyavirus (bun′-yah-vy″-rus) From Bunyamwera, an area of Africa where this type of virus was first isolated.

* arenavirus (ah-ree′-nah-vy″-rus) L. *arena,* sand. In reference to the appearance of virus particles in micrographs.

SPOTLIGHT ON MICROBIOLOGY 25.3
Hunting for Hantaviruses

A bunyavirus with increasing importance is the **hantavirus.** This virus was first associated with a severe kidney disease called Korean hemorrhagic fever reported during the Korean War and later was isolated from a field mouse taken near the Hantaan River in Korea. Similar hantaviruses caused outbreaks of a severe kidney disease in Russia and Scandinavia. Eventually, over 10 recognized strains of this virus have been reported worldwide, all carried by rodents such as field mice and voles. The case rate is currently estimated at 200,000 people, and it appears to be increasing throughout Europe and Asia.

American virologists first isolated strains of hantavirus from rodents in the 1980s, but could not link them to any disease outbreaks. The first American cases came to light suddenly in 1993 in the Four Corners area of New Mexico, when a cluster of unusual cases was reported to the CDC. All victims of the infections were healthy young people who were struck down with high fever, lung edema, and pulmonary failure. Most of them died within a few days. Similar cases were reported sporadically in Arizona, Colorado, and other western states.

Using PCR technology, virologists at the CDC identified a new virus they named *Sin Nombre* ("no name") virus, and the disease was termed hantavirus pulmonary syndrome (HPS). Comprehensive studies of the local rodent population revealed that the virus is carried by deer and harvest mice. It is probably spread by dried animal droppings and urine that have become airborne when the rodent nests are uncovered or removed. The disease occurs sporadically at a rate of 25 to 50 cases per year, but with a high (33%) mortality rate. So far, few specific treatments have been developed. Because of the widespread rodent reservoir and the potential for human contact, this is considered an emerging disease.

Enveloped Nonsegmented Single-Stranded RNA Viruses

PARAMYXOVIRUSES

The important human paramyxoviruses are *Paramyxovirus* (parainfluenza and mumps viruses), *Morbillivirus* (measles virus), and *Pneumovirus* (respiratory syncytial virus), all of which are readily transmitted through respiratory droplets. The envelope of a paramyxovirus possesses HN spikes and F glycoprotein spikes that allow it to infect neighboring cells by a novel mechanism. First, the cell membrane of an infected cell is modified by insertion of the spikes. The HN spikes immediately bind an uninfected neighboring cell, and in the presence of F spikes, the two cells permanently fuse. A chain reaction of multiple cell fusions then produce a *syncytium,** or **multinucleate giant cell**, with cytoplasmic inclusion bodies that is a diagnostically useful cytopathic effect (figure 25.2).

Epidemiology and Pathology of Parainfluenza

One type of infection by *Paramyxovirus,* called **parainfluenza,** is as widespread as influenza but usually more benign. It is spread by droplets and respiratory secretions that are inhaled or inoculated into the mucous membranes by contaminated hands. Parainfluenzal respiratory disease is seen most frequently in children, most of whom have been infected by the age of 6. Babies lacking passive antibodies are particularly susceptible, and they develop more severe symptoms. The usual effects of parainfluenza are minor upper respiratory symptoms (a cold), bronchitis, bronchopneumonia, and laryngotracheobronchitis (croup). *Croup** manifests as labored and noisy breathing accompanied by a hoarse cough that is most common in infants and young children.

Diagnosis, Treatment, and Prevention of Parainfluenza Presenting symptoms typical of a cold are often sufficient to presume respiratory infection of viral origin. Determining the actual viral agent, however, is difficult and usually unnecessary in older children and adults whose infection is usually self-limited and benign. Primary infection in infants can be severe enough to be life-threatening. So far, no specific chemotherapy is available, but supportive treatment with immune serum globulin or interferon can be of benefit.

Mumps: Epidemic Parotitis

Another infection caused by *Paramyxovirus* is **mumps** (Old English for lump or bump). So distinctive are the pathologic features of this disease that Hippocrates clearly characterized it several hundred years B.C. as a self-limited, mildly epidemic illness associated with painful swelling at the angle of the jaw (figure 25.3). Also called **epidemic parotitis,** this infection typically targets the parotid salivary glands, but it is not limited to this region. The mumps virus bears morphological and antigenic characteristics similar to the parainfluenza virus and has only a single serological type.

Epidemiology and Pathology of Mumps Humans are the exclusive natural hosts for the mumps virus. It is communicated primarily through salivary and respiratory secretions. Infection occurs worldwide, with epidemic increases in the late winter and early spring in temperate climates. High rates of infection arise among crowded populations or communities with poor herd immunity. Most cases occur in children under the age of 15, and as many as 40% are subclinical. Because lasting immunity follows any form of mumps infection, no long-term carrier reservoir exists in the population. The incidence of mumps has been reduced in the United States to around 300 cases per year.

---

* syncytium (sin-sish′-yum) Gr. *syn,* together, and *kytos,* cell.

* croup (kroop) Scot. *kropan,* to cry aloud.

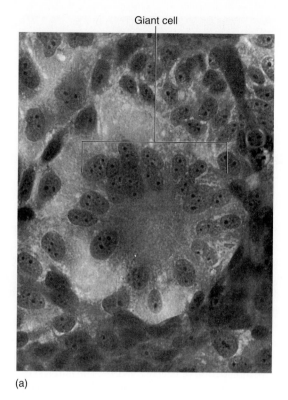

Giant cell

(a)

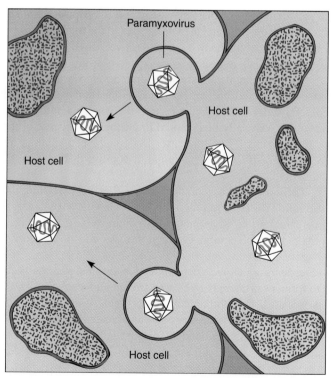

Paramyxovirus

Host cell

Host cell

Host cell

(b)

FIGURE 25.2

The effects of paramyxoviruses. **(a)** When they infect a host cell, paramyxoviruses induce the cell membranes of adjacent cells to fuse into large multinucleate giant cells, or syncytia. **(b)** This fusion allows direct passage of viruses from an infected cell to uninfected cells by communicating membranes. Through this means, the virus evades antibodies.

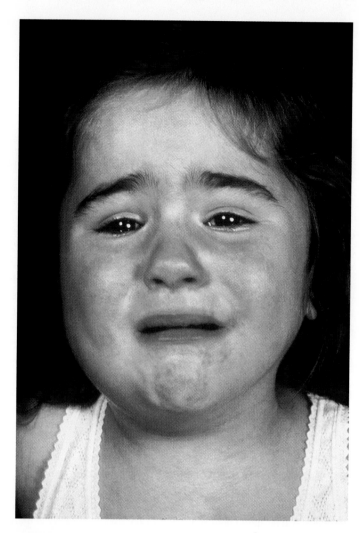

FIGURE 25.3

The external appearance of swollen parotid glands in mumps (parotitis). Usually both sides are affected, though parotitis affecting one side (as shown here) occasionally develops.

After an average incubation period of 2 to 3 weeks, symptoms of fever, nasal discharge, muscle pain, and malaise develop. These may be followed by inflammation of the salivary glands (especially the parotids), producing the classic gopherlike swelling of the cheeks on one or both sides (figure 25.3). This swelling of the gland and its duct can induce considerable discomfort. Viral multiplication in salivary glands is followed by invasion of other organs, especially the testes, ovaries, thyroid gland, pancreas, meninges, heart, and kidney. Despite the invasion of multiple organs, the prognosis for most infections is complete, uncomplicated recovery, with permanent immunity.

Complications in Mumps In 20% to 30% of young adult males, mumps infection localizes in the epididymis and testis, usually on one side only. The resultant syndrome of orchitis and epididymitis may be rather painful, but no permanent damage usually occurs. The popular belief that mumps readily causes sterilization of adult males is still held, despite medical evidence to the contrary. Perhaps this notion has been reinforced by the tenderness that

TABLE 25.2
The Two Forms of Measles

	Synonyms	Etiology	Primary Patient	Complications	Skin Rash	Koplik's Spots
Measles	Rubeola, red measles	Paramyxovirus; *Morbillivirus*	Child	SSPE,* pneumonia	Present	Present
German Measles	Rubella, 3-day measles, rubeola	Togavirus: *Rubivirus*	Child/fetus	Congenital defects	Present	Absent

Subacute sclerosing panencephalitis.

continues long after infection and by the partial atrophy of one testis that occurs in about half the cases. Permanent sterility due to mumps is very rare.

In mumps pancreatitis, the virus replicates in beta cells and pancreatic epithelial cells. Viral meningitis, characterized by fever, headache, and stiff neck, appears 2 to 10 days after the onset of parotitis, lasts for 3 to 5 days, and then dissipates, leaving few or no adverse side effects. Another rare event is infection of the inner ear that leads to deafness.

Diagnosis, Treatment, and Prevention of Mumps Mumps can be tentatively diagnosed in a child with swollen parotid glands and a known exposure 2 or 3 weeks previously. Since parotitis is not always present, and the incubation period can range from 7 to 23 days, a practical diagnostic alternative is to perform a direct fluorescent test for viral antigen or an ELISA test on the serum.

The general pathology of mumps is mild enough that symptomatic treatment to relieve fever, dehydration, and pain is usually adequate. A live, attenuated mumps vaccine given routinely as part of the MMR vaccine at 12 to 15 months of age is a powerful and effective control agent. A separate single vaccine is available for adults who require protection. Although the antibody titer is lower than that produced by wild mumps virus, protection often lasts a decade.

Measles: *Morbillivirus* Infection

Measles, an acute disease caused by *Morbillivirus,* is also known as **red measles*** and **rubeola.*** Unfortunately, German measles (rubella), an entirely unrelated viral infection, is also called rubeola in some countries. Some differentiating criteria for these two forms of measles are summarized in table 25.2. Despite a tendency to think of measles as a mild childhood illness, several recent outbreaks have reminded us that measles can kill babies and young children, and that it is the eighth most frequent cause of death worldwide.

Epidemiology of Measles Measles is one of the most contagious infectious diseases, transmitted principally by respiratory aerosols. Epidemic spread is favored by crowding, low levels of herd immunity, malnutrition, and inadequate medical care. Occasional outbreaks of measles have been linked to the lack of immunization in children or the failure of a single dose of vaccine in many children. There is no reservoir other than humans, and a

person is infectious during the periods of incubation, prodromium, and the skin rash, but not usually during convalescence. Only relatively large, dense populations of susceptible individuals can sustain the continuous chain necessary for transmission. In the United States, the incidence of measles is sporadic, usually less than 100 cases per year.

Infection, Disease, and Complications of Measles The measles virus invades the mucosal lining of the respiratory tract during an incubation period of nearly 2 weeks. The initial symptoms are sore throat, dry cough, headache, conjunctivitis, lymphadenitis, and fever. In a short time, unusual oral lesions called *Koplik's spots* appear as a prelude to the characteristic red maculopapular *exanthem** that erupts on the head and then progresses to the trunk and extremities, until most of the body is covered (figure 25.4). The rash gradually coalesces into red patches that fade to brown.

In a small number of cases, children develop laryngitis, bronchopneumonia, and bacterial secondary infections such as otitis media and sinusitis. Children afflicted with leukemia or thymic deficiency are especially predisposed to pneumonia because of their lack of the natural T-cell defense. Undernourished children may have severe diarrhea and abdominal discomfort that adds to their debilitation.

The most serious complication is **subacute sclerosing panencephalitis (SSPE),*** a progressive neurological degeneration of the cerebral cortex, white matter, and brain stem. Its incidence is approximately one case in a million measles infections, and it afflicts primarily male children and adolescents. The pathogenesis of SSPE appears to involve a defective virus, one that has lost its ability to form a capsid and be released from an infected cell. Instead, it spreads unchecked through the brain by cell fusion and gradually destroys neurons and accessory cells, and breaks down myelin. The disease is known for profound intellectual and neurological impairment. The course of the disease invariably leads to coma and death in a matter of months or years.

Diagnosis, Treatment, and Prevention of Measles The patient's age, a history of recent exposure to measles, and the season of the year are all useful epidemiologic clues for diagnosis. Clinical

---

* measles (mee'-zlz) Dutch *maselen,* spotted.

* rubeola (roo-bee'-oh-lah) L. *ruber,* red.

* exanthem (eg-zan'-thum) Gr. *exanthema,* to bloom or flower. An eruption or rash of the skin.

* subacute sclerosing panencephalitis (sub-uh-kewt' sklair-oh'-sing pan" -en-cef" -uh-ly'-tis) Gr. *skleros,* hard, *pan,* all, and *enkephalos,* brain.

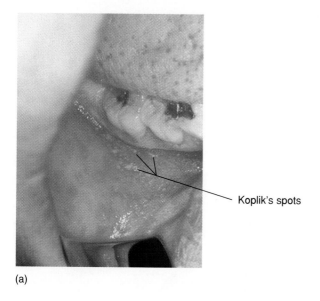

(a)

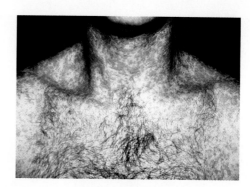

(b)

FIGURE 25.4

🔘 **Signs and symptoms of measles.** **(a)** Koplik's spots, named for the physician who described them in the last century, are tiny white lesions with a red border that form on the cheek membrane adjacent to the molars just prior to eruption of the skin rash. **(b)** A fully developed measles rash.

characteristics such as dry cough, sore throat, conjunctivitis, lymphadenopathy, fever, and especially the appearance of Koplik's spots and a rash are presumptive of measles.

Treatment relies on reducing fever, suppressing cough, and replacing lost fluid. Complications require additional remedies to relieve neurological and respiratory symptoms and to sustain nutrient, electrolyte, and fluid levels. Therapy includes antibiotics for bacterial complications and doses of immune globulin.

Vaccination with attenuated viral vaccine administered by subcutaneous injection achieves immunity that persists for about 20 years. It must be stressed that because the virus is live, it can cause an atypical infection sometimes accompanied by a rash and fever. Measles immunization is recommended for all healthy children at the age of 12 to 15 months (MMR vaccine, with mumps and rubella), and a booster is given before a child enters school. A single antigen vaccine (Meruvax) is also available for older patients who require protection against measles alone. As a general rule, anyone who has had the measles is considered protected.

Respiratory Syncytial Virus: RSV Infections

As its name indicates, **respiratory syncytial virus (RSV)**, also called *Pneumovirus*, infects the respiratory tract and produces giant multinucleate cells. Outbreaks of droplet-spread RSV disease occur regularly throughout the world, with peak incidence in the winter and early spring. Children 6 months of age or younger are especially susceptible to serious disease of the respiratory tract. Approximately 5 in 1,000 newborns are affected, making RSV the most prevalent cause of respiratory infection in this age group. An estimated 100,000 children are hospitalized with RSV infection each year in the United States. The mortality rate is highest for children who have complications such as prematurity, congenital disease, and immunodeficiency. Infection in older children and adults is manifested as a common cold.

The epithelia of the nose and eye are the principal portals of entry, and the nasopharynx is the main site of RSV replication. The first symptoms of primary infection are fever that lasts for 3 days, rhinitis, pharyngitis, and otitis. Infections of the bronchial tree and lung parenchyma give rise to symptoms of croup that include acute bouts

of coughing, wheezing, *dyspnea,** and abnormal breathing sounds (rales). Adults and older children with RSV infection can experience cough and nasal congestion but are frequently asymptomatic.

Diagnosis, Treatment, and Prevention of RSV Diagnosis of RSV infection is more critical in babies than in older children or adults. The afflicted child is conspicuously ill, with signs typical of pneumonia and bronchitis. The best diagnostic procedures are those that demonstrate the viral antigen directly from specimens using direct and indirect fluorescent staining, ELISA testing, and DNA probes.

A highly beneficial new treatment is the administration of RSV immunoglobulin. This serum is obtained from people with high RSV antibody titers, and it greatly reduces complications and the need for hospitalization. An antiviral drug, ribavirin (virazole), can also be administered as an inhaled aerosol. Other supportive measures include drugs to reduce fever, assisting pulmonary ventilation, and treating secondary bacterial infection if present.

RHABDOVIRUSES

The most conspicuous *rhabdovirus** is the **rabies*** virus, genus *Lyssavirus.** The particles of this virus have a distinctive bullet-like appearance, round on one end and flat on the other. Additional features are a helical nucleocapsid and spikes that protrude through the envelope (figure 25.5). The family contains approximately 60 different viruses, but only the rabies lyssavirus infects humans.

Epidemiology of Rabies

Rabies is a slow, progressive zoonotic disease characterized by a fatal meningoencephalitis. It is distributed nearly worldwide, except for 34 countries that have remained rabies-free by practicing

---

* dyspnea (dysp′-nee-ah) Gr. *dyspnoia,* difficulty in breathing.

* rhabdovirus (rab′-doh-vy″-rus) Gr. *rhabdos,* rod. In reference to its bullet, or bacillary, form.

* rabies (ray′-beez) L. *rabidus,* rage or fury.

* *Lyssavirus* (lye′-suh-vy″-rus) Gr. *lyssa,* madness.

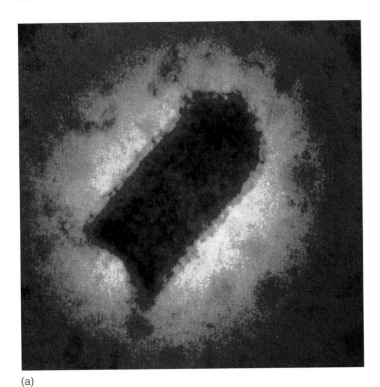

(a)

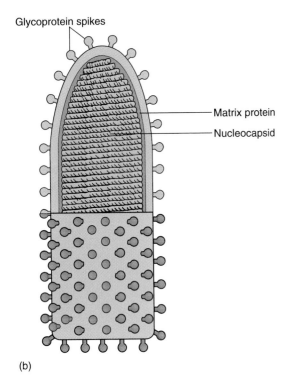

(b)

FIGURE 25.5

The structure of the rabies virus. **(a)** Color-enhanced virion shows internal serrations, which represent the tightly coiled nucleocapsid. **(b)** A schematic model of the virus, showing its major features.

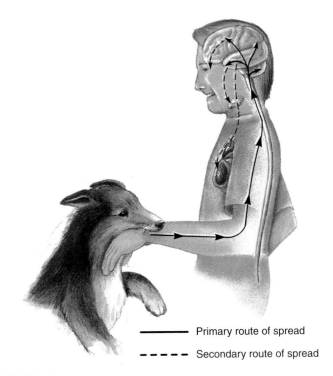

——————— Primary route of spread

- - - - - Secondary route of spread

FIGURE 25.6

A pathologic picture of rabies. After an animal bite, the virus spreads to the nervous system and multiplies. From there, it moves to the salivary glands and other organs.

worldwide total for human rabies is estimated at about 30,000 cases, but only a tiny number of these cases occur in the United States. Most U.S. cases of rabies occur in wild animals (about 5,000–6,000 cases per year), while dog rabies has declined.

The epidemiology of animal rabies in the United States varies both chronologically and geographically. The most common wild animal reservoir host has changed from foxes to skunks to raccoons. Regional differences in the dominant reservoir also occur. Bats, skunks, and bobcats are the most common carriers of rabies in California, raccoons are the predominant carriers in the East, and coyotes dominate in Texas.

Infection and Disease

Infection with rabies virus typically begins when an infected animal's saliva enters a puncture site (figure 25.6). Occasionally, the virus is inhaled or inoculated orally. The rabies virus remains up to a week at the trauma site, where it multiplies. The virus then gradually enters nerve endings and advances toward the ganglia, spinal cord, and brain. Viral multiplication throughout the brain is eventually followed by migration to such diverse sites as the eye, heart, skin, and oral cavity. The infection cycle is completed when the virus replicates in the salivary glands and is shed into the saliva. Clinical rabies proceeds through several distinct stages that inevitably end in death.

Clinical Phases of Rabies

The average incubation period of rabies is 1 to 2 months or more, depending upon the wound site, its severity, and the inoculation dose. The incubation period is shorter in facial, scalp, or neck

rigorous animal control. The primary wild reservoirs of the virus are wild mammals such as canines, skunks, raccoons, badgers, cats, and bats that can spread the infection to domestic dogs and cats. Both wild and domestic mammals can spread the disease to humans through bites, scratches, and inhalation of droplets. The annual

wounds because of greater proximity to the brain. The prodromal phase begins with fever, nausea, vomiting, headache, fatigue, and other nonspecific symptoms. Some patients continue to experience pain, burning, prickling, or tingling sensations at the wound site.

In the form of rabies termed *furious,* the first acute signs of neurological involvement are periods of agitation, disorientation, seizures, and twitching. Spasms in the neck and pharyngeal muscles lead to severe pain upon swallowing. As a result, attempts to swallow or even the sight of liquids bring on *hydrophobia* (fear of water). Throughout this phase, the patient is fully coherent and alert. With the *dumb* form of rabies, a patient is not hyperactive but paralyzed, disoriented, and stuporous. Ultimately, both forms progress to the coma phase, resulting in death from cardiac or respiratory arrest. Until recently, humans were never known to survive rabies. But three patients have now recovered after receiving intensive, long-term treatment.

Diagnosis and Management of Rabies

When symptoms appear after a rabid animal attack, the disease is readily diagnosed. But the diagnosis can be obscured when contact with an infected animal is not clearly defined or when symptoms are absent or delayed. Anxiety, agitation, and depression can pose as a psychoneurosis; muscle spasms resemble tetanus; and encephalitis with convulsions and paralysis mimics a number of other viral infections. Often the disease is diagnosed at autopsy. Criteria indicative of rabies are intracellular inclusions (Negri bodies) in nervous tissue (figure 25.7), identification of rabies virus in saliva or brain tissue, and demonstration of rabies virus antigens in specimens of the brain, serum, cerebrospinal fluid, or cornea using immunofluorescent methods.

Rabies Prevention and Control   A bite from a wild or stray animal demands assessment of the animal, meticulous care of the wound, and a specific treatment regimen. A wild mammal, especially a skunk, raccoon, fox, or coyote that bites without provocation, is presumed to be rabid, and therapy is immediately commenced. If the animal is captured, brain samples and other tissue are examined for verification of rabies. Healthy domestic animals are observed closely for signs of disease and sometimes quarantined. Preventive therapy is initiated if any signs of rabies appear.

After an animal bite, the wound should be scrupulously washed with soap or detergent and water, followed by debridement and application of an antiseptic such as alcohol or peroxide. Rabies is one of the few infectious diseases for which a combination of passive and active postexposure immunization is indicated and successful. Initially the wound is infused with human rabies immune globulin (HRIG) to impede the spread of the virus, and globulin is also injected intramuscularly to provide immediate systemic protection. A full course of vaccination is started simultaneously. The current vaccine of choice is the **human diploid cell vaccine (HDCV).** This potent inactivated vaccine is cultured in human embryonic fibroblasts. The routine in postexposure vaccination entails intramuscular or intradermal injection on the 1st, 3rd, 7th, 14th, 28th, and 60th days, with two boosters. High-risk groups such as veterinarians, animal handlers, laboratory personnel, and travelers should receive three doses to protect against possible exposure.

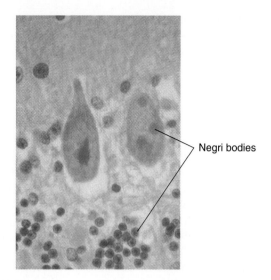

Negri bodies

FIGURE 25.7

Diagnosis of rabies.   Neuronal cells of the cerebellum exhibit distinct intracytoplasmic inclusions called Negri bodies. Although typical of rabies, they are usually not found until the brain is examined in an autopsy.

Control measures such as vaccination of domestic animals, elimination of strays, and strict quarantine practices have helped reduce the virus reservoir. Several countries have tested a new genetically engineered oral vaccine made with a vaccinia virus that carries the gene for the rabies virus surface antigen. The vaccine is incorporated into bait placed in the habitats of wild reservoir species such as skunks and raccoons. General use of this vaccine in the United States has been started in limited regions.

CHAPTER CHECKPOINTS

The enveloped, nonsegmented, single-stranded group of RNA viruses includes the Paramyxoviridae and the Rhabdoviridae.

Paramyxoviruses cause respiratory infections characterized by the formation of multinucleate giant cells. *Paramyxovirus, Morbillivirus,* and *Pneumovirus* are the principal disease agents in this group. They cause mumps, red measles, and respiratory syncytial virus infections through aerosol transmission.

The most serious of the Rhabdoviridae is the *Lyssavirus,* which causes rabies, a slow progressive, usually fatal, zoonotic disease of the CNS acquired by bites and other close contact with infected mammals.

Other Enveloped RNA Viruses

CORONAVIRUSES

Coronaviruses[1] are relatively large RNA viruses with distinctive, widely spaced spikes on their envelopes. These viruses are common in domesticated animals and are responsible for epidemic respiratory, enteric, and neurological diseases in pigs, dogs, cats, and poultry. Thus far, two types of human coronaviruses have been

---

1. Named for the resemblance of the viral spikes to a crown of thorns.

characterized. One of these (HCV) is an etiologic agent of the common cold (see Medical Microfile 25.6). The same virus is also thought to cause some forms of viral pneumonia and myocarditis. The other human virus is very closely associated with the intestine and may have a role in some human enteric infections. Coronaviruses have also been isolated from the brain, but their role in human neurological disease is uncertain at this time.

RUBIVIRUS, THE AGENT OF RUBELLA

Togaviruses are nonsegmented, single-stranded RNA viruses with a loose envelope.[2] There are several important members, including *Rubivirus,* the agent of rubella, and certain arboviruses. **Rubella,** or German measles, was first recognized as a distinct clinical entity in the mid-eighteenth century and was considered a benign childhood disease until its *teratogenic* effects were discovered. It was first observed that cataracts often developed in neonates born to mothers who had contracted rubella in the first trimester. Epidemiologic investigations over the next generation confirmed the link between rubella and many other congenital defects as well.

Epidemiology of Rubella

Rubella is an endemic disease with worldwide distribution. Infection is initiated through contact with respiratory secretions and occasionally urine. The virus is shed during the prodromal phase and up to a week after the rash appears. Because the virus is only moderately communicable, close living conditions are required for its spread. Although epidemics and pandemics of rubella once regularly occurred in 6- to 9-year cycles, the introduction of vaccination has essentially stopped this pattern in the United States. Most cases are reported among adolescents and young adults in military training camps, colleges, and summer camps. The greatest concern is that nonimmune women of childbearing age might be caught up in this cycle, raising the prospect of congenital rubella.

Infection and Disease

Two clinical forms of rubella can be distinguished. Postnatal infection develops in children or adults, and congenital (prenatal) infection of the fetus is expressed in the newborn as various types of birth defects.

Postnatal Rubella During an incubation period of 2 to 3 weeks, the virus multiplies in the respiratory epithelium, infiltrates local lymphoid tissue, and enters the bloodstream. Early symptoms include malaise, mild fever, sore throat, and lymphadenopathy. The rash of pink macules and papules first appears on the face and progresses down the trunk and toward the extremities, advancing and resolving in about 3 days. Adult rubella is often accompanied by joint inflammation and pain rather than a rash. Except for an occasional complication, postnatal rubella is generally mild and produces lasting immunity.

Congenital Rubella Transmission of the rubella virus to a fetus in utero can result in a serious complication called **congenital rubella** (figure 25.8). The mother is able to transmit the virus even if she is asymptomatic. Fetal injury varies according to the time of

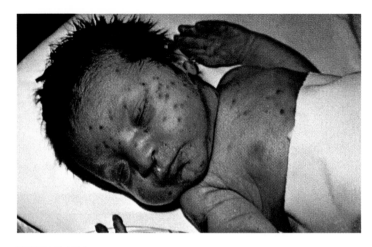

FIGURE 25.8

An infant born with congenital rubella can manifest a papular and purpurotic rash.

Courtesy Kenneth Schiffer, from AJDC 118:25, July 1969. © American Medical Association.

infection. It is generally accepted that infection in the first trimester is most likely to induce miscarriage or multiple permanent defects in the newborn such as cardiac abnormalities, ocular lesions, deafness, and mental and physical retardation. Less drastic sequelae that usually resolve in time are anemia, hepatitis, pneumonia, carditis, and bone infection.

Diagnosis of Rubella

Rubella mimics other diseases and is often asymptomatic, so it should never be diagnosed on clinical grounds alone. The confirmatory methods of choice are serological testing and virus isolation systems. IgM tests can determine recent infection, and a rising titer is a clear indicator of continuing rubella infection. Several serological tests for IgM are used, including complement fixation, ELISA, fluorescent assay, and hemagglutination. The latex-agglutination card, a simple, miniaturized test, is a more rapid method for IgM assay.

Prevention of Rubella

Postnatal rubella is generally benign and requires only symptomatic treatment. Because no specific therapy for rubella is available, most control efforts are directed at maintaining herd immunity with attenuated rubella virus vaccine. This vaccine is usually given to children in the combined form (the MMR vaccination) at 12 to 15 months and a booster at 4 or 6 years of age. The use of rubella vaccine alone varies according to the age and sex of the individual and whether pregnancy is anticipated or already in progress (Medical Microfile 25.4).

Arboviruses: Viruses Spread by Arthropod Vectors

Vertebrates are hosts to more than 400 viruses transmitted primarily by arthropods. For simplicity's sake, these viruses are lumped together in a loose grouping called the **arboviruses** (**ar**thropod-**bo**rne **viruses**). The major arboviruses pathogenic to humans are togaviruses *(Alphavirus),* flaviviruses *(Flavivirus),* some bunyaviruses *(Bunyavirus* and *Phlebovirus),* and reoviruses *(Orbivirus).*

---

2. Reminiscent of a toga.

MEDICAL MICROFILE 25.4
Prevention of Congenital Rubella

Preventing congenital rubella is best accomplished by vaccinating all children at an early age so that girls are already immune by the time of sexual maturity. Unfortunately, the vaccine does not always provide long-term protection, and many children are not vaccinated at all. A second important safety net is routine testing for rubella antibodies in women prior to marriage. A titer of 1:8 or greater is interpreted as being protective.

The current recommendation for nonpregnant, antibody-negative women is immediate immunization. Because the vaccine contains live virus and a teratogenic effect is theoretically possible, the vaccine is administered on the condition that the patient uses contraception for 3 months afterward. Under no circumstances is the vaccine to be given to a pregnant woman. The management of verified rubella in a pregnant woman depends upon whether maternal infection occurred early or late. Therapeutic abortion may be indicated if maternal infection occurred in the first trimester and monitoring of the fetus discloses severe damage.

The chief vectors are blood-sucking arthropods such as mosquitoes, ticks, flies, and gnats. Most types of illness caused by these viruses are mild, undifferentiated fevers, though some cause severe encephalitides (plural of encephalitis) and life-threatening hemorrhagic fevers. Although these viruses have been assigned taxonomic names, they are more often known by their common names, which are based on geographic location and primary clinical profile.

EPIDEMIOLOGY OF ARBOVIRUS DISEASE

Wherever there are arthropods there are also arboviruses, so collectively, their distribution is worldwide (figure 25.9). The vectors and viruses tend to be clustered in the tropics and subtropics, but many temperate zones report periodic epidemics. A given arbovirus type may have very restricted distribution, even to a single isolated region, but some range over several continents, and others can spread along with their vectors.

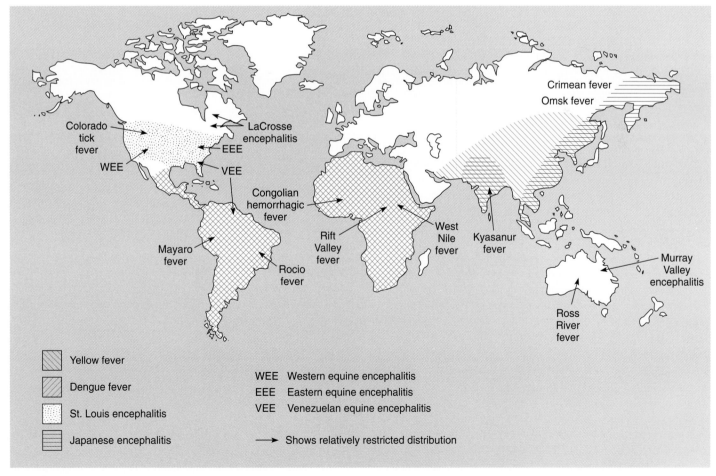

FIGURE 25.9

Worldwide distribution of major arboviral diseases.

FIGURE 25.10

Variations on the arbovirus-vector-human cycle. **(a)** The arthropod vector lives in a habitat separate from humans and spreads the virus among various wild, warm-blooded vertebrates and even to their offspring through transovarial passage. Humans intrude into this cycle by happening on the vector in the wild. **(b)** The infected person then returns to human habitation and serves as a source of viruses for related insects living in the community that may act as biological vectors, or **(c)** wild animal reservoirs and vectors bring the virus into human populations, especially through migration or the encroachment of human habitation into wild areas.

The Influence of the Vector

As one would expect, the activity and distribution of arboviruses are closely tied to the ecology of the vectors. Factors that weigh most heavily are the longevity of the arthropod, the availability of food and breeding sites, and climatic influences such as temperature and humidity. Most arthropod vectors feed on the blood of hosts, which infects them for varying time periods. Infections show a peak incidence when the arthropod is actively feeding and reproducing, usually from late spring through early fall. Warm-blooded vertebrates also maintain the virus during the cold and dry seasons. Humans can serve as dead-end, accidental hosts, as in equine encephalitis and Colorado tick fever, or they can be a maintenance reservoir, as in dengue fever and yellow fever. The general patterns of arbovirus transmission involving vectors are illustrated in figure 25.10.

Arboviral diseases have a great impact on humans. Although exact statistics are unavailable, it is accepted knowledge that millions of people acquire infections each year and thousands of them die. The uncertain nature of host and viral cycles frequently results in sudden, unexpected epidemics, sometimes with previously unreported viruses. Travelers and military personnel entering endemic areas are at special risk because, unlike the natives of that region, they have no immunity to the viruses.

GENERAL CHARACTERISTICS OF ARBOVIRUS INFECTIONS

Although the arboviruses have many inherently interesting characteristics, our coverage here will be limited to (1) the chief manifestations of disease in humans, (2) aspects of the most important North American viruses, and (3) methods of diagnosis, treatment, and control.

Febrile Illness and Encephalitis

One type of disease elicited by arboviruses is an acute, undifferentiated fever, often accompanied by rash. These infections, typified by dengue fever and Colorado tick fever, are usually mild and leave no sequelae. Prominent symptoms are fever, prostration, headache, myalgia, orbital pain, muscle aches, and joint stiffness. About midway through the illness, a maculopapular or petechial rash can erupt over the trunk and limbs.

When the brain, meninges, and spinal cord are invaded, symptoms of viral encephalitis occur. The most common American types are eastern equine, St. Louis, and California encephalitis. The viruses cycle between wild animals (primarily birds) and mosquitoes or ticks, but humans are not usually reservoir hosts. The disease begins with an arthropod bite, the release of the virus into the tissues, and its replication in nearby lymphatics. Prolonged viremia establishes the virus in the brain, where inflammation can cause swelling and damage to the brain, nerves, and meninges. Symptoms are extremely variable and can include coma, convulsions, paralysis, tremor, loss of coordination, memory deficits, changes in speech and personality, and heart disorders. In some cases, survivors experience some degree of permanent brain damage. Young children and the elderly are most sensitive to injury by arboviruses.

Colorado tick fever (CTF) is the most common tick-borne viral fever in the United States. Restricted in its distribution to the Rocky Mountain states, it occurs sporadically during the spring and summer, corresponding to the time of greatest tick activity and human forays into the wilderness. Two hundred to three hundred endemic cases are reported each year, and deaths are rare.

Western equine encephalitis (WEE) occurs sporadically in the western United States and Canada, first in horses and later in humans. The mosquito that carries the virus emerges in the early summer when irrigation begins in rural areas and breeding sites are abundant. The disease is extremely dangerous to infants and small children, with a case fatality rate of 3% to 7%.

Eastern equine encephalitis (EEE) is endemic to an area along the eastern coast of North America and Canada. The usual pattern is sporadic cases, but occasional epidemics can occur in humans and horses. High periods of rainfall in the late summer increase the chance of an outbreak, and disease usually appears first in horses and caged birds. The case fatality rate can be very high (70%).

A disease known commonly as **California encephalitis** is caused by two different viral strains. The California strain occurs occasionally in the western states and has little impact on humans. The LaCrosse strain is widely distributed in the eastern United States and Canada and is a prevalent cause of viral encephalitis in North America. Children living in rural areas are the primary target group, and most of them exhibit mild, transient symptoms. Fatalities are rare.

St. Louis encephalitis (SLE) is the most common of all American viral encephalitides. Cases appear throughout North and South America, but epidemics occur most often in the midwestern and southern states. Inapparent infection is very common, and the total number of cases is probably thousands of times greater than the 50 to 100 reported each year. The seasons of peak activity are spring and summer, depending on the region and species of mosquito. In the east, mosquitoes breed in stagnant or polluted water in urban and suburban areas during the summer. In the west, mosquitoes frequent spring floodwaters in rural areas.

Hemorrhagic Fevers

Certain arboviruses, principally the yellow fever and dengue fever viruses, cause capillary fragility and disrupt the blood clotting system, which leads to localized bleeding and shock. These hemorrhagic syndromes, which are also attended by fever and pain, are caused by a variety of viruses, are carried by a variety of vectors, and are distributed globally. Reservoir animals are usually small mammals, although yellow fever and dengue fever can be harbored in the human population.

The best-known arboviral disease is **yellow fever.** Although this disease was at one time cosmopolitan in its distribution, mosquito control measures have eliminated it in many countries, in-

cluding the United States. Two patterns of transmission occur in nature (see figure 25.10). One is an urban cycle between humans and the mosquito *Aedes aegypti,* which reproduces in standing water in cities. The other is a sylvan cycle, maintained between forest monkeys and mosquitoes. Most cases in the Western Hemisphere occur in the jungles of Brazil, Peru, and Colombia during the rainy season. Infection begins acutely with fever, headache, and muscle pain. In some patients the disease progresses to oral hemorrhage, nosebleed, vomiting, jaundice, and liver and kidney damage with significant mortality rates.

Dengue* fever is caused by a flavivirus and is also carried by *Aedes* mosquitoes. Although mild infection is the usual pattern, a form called dengue hemorrhagic shock syndrome can be lethal. This disease is also called "breakbone fever" because of the severe pain it induces in the muscles and joints. The illness is endemic to Southeast Asia and India, and several epidemics have occurred in South and Central America, the Caribbean, and Mexico. The Pan American Health Organization has reported an ongoing epidemic of dengue fever in the Americas that has increased to 200,000 cases with 5,500 deaths.

Concern over the possible spread of the disease to the United States led the CDC to survey mosquito populations in states along the southern border. In several states, they discovered the Asian tiger mosquito *(Aedes albopictus),* a potential vector of dengue fever virus. Contact between infected human carriers could establish the virus in these mosquito populations. So far, no cases of hemorrhagic dengue fever have been reported in the United States, and it is hoped that continued mosquito abatement will block its spread. One group of researchers is currently testing a method to genetically engineer mosquitoes so that they are resistant to infection and no longer capable of spreading the virus.

DIAGNOSIS, TREATMENT, AND CONTROL OF ARBOVIRUS INFECTION

Except during epidemics, detecting arboviral infections can be difficult. The patient's history of travel or contact with vectors, along with serum analysis, are highly supportive of a diagnosis. Treatment of the various arbovirus diseases relies completely on support measures to control fever, convulsions, dehydration, shock, and edema.

The most reliable arbovirus vaccine is a live, attenuated yellow fever vaccine that provides relatively long-lasting immunity. Vaccination is a requirement for travelers in the tropics and can be administered to entire populations during epidemics. Live, attenuated vaccines for WEE and EEE are administered to laboratory workers, veterinarians, ranchers, and horses.

Most of the control safeguards for arbovirus disease are aimed at the arthropod vectors. Mosquito abatement by eliminating breeding sites and broadcasting insecticides has been highly effective in restricted urban settings. In the case of sylvan vectors, the only real preventive is to restrict human contact with the vector through insect repellents, protective clothing, and by remaining indoors at night, when mosquitoes are most active.

---

* dengue (den'-gha) A corruption of Sp. "dandy" fever. Other synonyms are dengue shock syndrome and stiffneck fever.

Enveloped Single-Stranded RNA Viruses with Reverse Transcriptase: Retroviruses

Retroviruses are among the most unusual viruses for the following reasons: They have known oncogenic abilities; they cause dire, often fatal diseases; and they are capable of altering a host's DNA in profound ways. The most familiar of these viruses, called *lentiviruses,* is the **human immunodeficiency virus (HIV),** which causes **acquired immunodeficiency syndrome (AIDS).** HIV is only one of a whole array of retroviruses associated with cancer and other diseases. Judging from what is currently known about these viruses, they will continue to influence medicine, research, politics, and economics for decades to come.

THE BIOLOGY OF THE RETROVIRUSES

Imagine the surprised group of researchers who discovered the revolutionary properties of retroviruses. Perhaps the most outstanding characteristic is that they come already equipped with an unusual enzyme called **reverse transcriptase** that catalyzes the replication of double-stranded DNA from single-stranded RNA. Because these viruses reverse the order of replication, they are termed *retro*viruses (Latin for reversal). This property is apparently found only in them and certain DNA viruses.

The association of retroviruses with their hosts can be so intimate that viral genes are permanently integrated into the host genome. In fact, as the technology of DNA probes for detecting retroviral genes is employed, it becomes increasingly evident that retroviruses may be integral parts of chromosomes. So striking is their resemblance to transposons (so-called jumping genes) that some biologists think retroviruses may even have originated from host cells. Not only can this retroviral DNA be incorporated into the host genome as a provirus that can be passed on to progeny cells, but some retroviruses also transform cells, regulate certain host genes, and convert normal cells into cancer cells (see chapter 17).

NOMENCLATURE AND CLASSIFICATION

Most retroviruses isolated to date cause slow infections that lead to leukemia, tumors, anemia, immune dysfunction, and even neurological diseases in birds and mammals. The first human retroviruses isolated belong to the lentivirus group. This includes the T-cell lymphotropic viruses I and II (HTLV I and HTLV II) and HTLV III. Types I and II are associated with human leukemia or lymphoma; type III was given the unified name, HIV. There are two types, HIV-1, which is the dominant form in most of the world, and HIV-2, which exists primarily in western Africa.

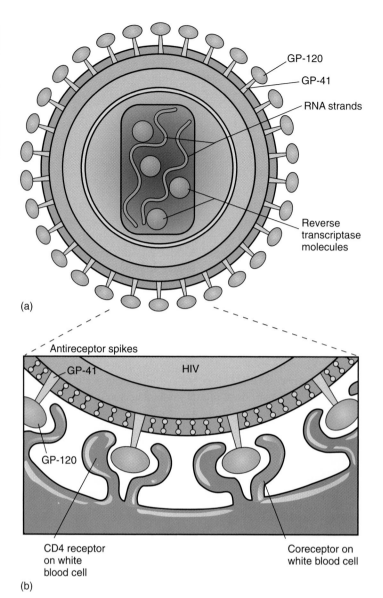

(a)

(b)

FIGURE 25.11

The general structure of HIV. **(a)** The envelope contains two types of glycoprotein (GP) spikes, two identical RNA strands, and several molecules of reverse transcriptase encased in a protein coating. **(b)** The snug attachment of HIV glycoprotein antireceptors (GP-41 and GP-120) to their specific receptors on a human cell membrane. These receptors are CD4 and a coreceptor called CCR-5 (fusin) that permit docking with the host cell and fusion with the cell membrane.

STRUCTURE AND BEHAVIOR OF RETROVIRUSES

HIV and other retroviruses display structural features typical of enveloped RNA viruses (figure 25.11*a*). The outermost component is a lipid envelope with transmembrane glycoprotein spikes and knobs that mediate viral adsorption to the host cell. Although a few retroviruses are nonspecific, most can infect only host cells that present the required receptors. In the case of the AIDS virus, the target cells possess CD4 receptors and other coreceptors that permit the virus to attach and penetrate several types of leukocytes and tissue cells (figure 25.11*b*).

Acquired Immunodeficiency Syndrome (AIDS)

HISTORICAL BACKGROUND

The sudden emergence of AIDS in the early 1980s focused an enormous amount of public attention, research studies, and financial resources on the virus and its disease.

The first cases of AIDS were seen by physicians in Los Angeles, San Francisco, and New York City. They observed clusters of young male patients with one or more of a complex of symptoms: severe pneumonia caused by *Pneumocystis carinii* (ordinarily a harmless fungus); a rare vascular cancer called *Kaposi sarcoma;** sudden weight loss; swollen lymph nodes; and general loss of immune function. Another common feature was that all of these young men were homosexuals. Early hypotheses attempted to explain the disease as a consequence of the homosexual life-style or a result of immune suppression by chronic drug abuse or infections. Soon, however, cases were reported in nonhomosexual patients who had been transfused with blood or blood products. Eventually, virologists at the Pasteur Institute in France isolated a novel retrovirus. That this was clearly a communicable infectious disease left little doubt, and the medical community termed it **acquired immunodeficiency syndrome,** or **AIDS.**

Once HIV was isolated, researchers developed a functioning line of T cells that remained alive after infection with the virus. This development was soon followed by the creation of sensitive tests for detecting the virus and the antibody against it. These tests, in turn, made the diagnosis of AIDS and the screening of donor blood possible.

The Origins of AIDS

There are no conclusive answers to the question of where and how HIV originated. Significant insight into this question has come from studies comparing the genetics of HIV with various African monkey viruses. The following outline is a proposed family tree for the primate lentiviruses:

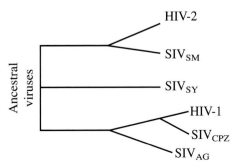

It now appears that HIV-1 and -2 are not as closely related to each other as they are to simian immunodeficiency viruses (SIV). HIV-1 shares a heritage with SIV_{CPZ}, a virus of chimpanzees. Researchers have discovered a new, atypical strain of HIV that is genetically most similar to a strain of SIV from chimpanzees. HIV-2 is allied with SIV_{SM}, a virus from sooty Mangabey monkeys. The two HIVs apparently originated from separate evolutionary events. Distant relatives are SIV_{SY} (Sykes monkey) virus and SIV_{AG} (African Green monkey) virus. All of these monkey viruses are apparently of ancient origin and do not cause severe disease in their natural hosts.

The first well-documented case of AIDS occurred in an African man in 1959. Samples of his blood yielded genetic material from an early version of HIV. This finding has helped to clarify the time frame for the emergence of HIV. It is theorized that it must have first appeared in the early 1950s, after being transmitted by original simian hosts, but we will probably never know the exact time and the series of events involved.

HIV probably remained in small isolated villages, causing sporadic cases and mutating into more virulent strains that were readily transmitted from human to human. When this pattern was coupled with changing social and sexual mores and increased immigration and travel, a pathway was opened up for rapid spread of the virus to the rest of the world.

EPIDEMIOLOGY OF AIDS

Statistical Patterns

AIDS has been reported in every country, and parts of Africa and Asia are especially devastated by it (Spotlight on Microbiology 25.5). Estimates of the number of individuals currently infected with the virus range from 35 to 40 million worldwide, with approximately 1 million (the range is 800,000 to 2 million) in the United States. A large number of these people have not yet begun to show symptoms because they are in the latent phase of the disease.

AIDS first became a notifiable disease at the national level in 1984 (figure 25.12) and has continued in an epidemic pattern, although the number of AIDS cases has decreased since 1994. Approximately 750,000 people have been diagnosed with AIDS over the past 20 years, and over 50% of them have died. New therapies have lengthened the lives of many patients and slowed the death rate dramatically. Despite this improvement, AIDS is still the second leading cause of death in men 25 to 44 years of age and the tenth most frequent overall cause of death nationally.

Other epidemiologic statistics for the United States show that AIDS is more common among males than females and that blacks account for the most cases, followed by whites, Hispanics, and other groups. The average adult patient is about 35 years of age. If one observes specific sociological groups, however, the sexual, racial, and age statistics can be quite different. A majority of AIDS patients are concentrated in large metropolitan centers of New York, California, Florida, New Jersey, and Texas.

---

* Kaposi sarcoma (kah′-poh-seez sahr-koh′-mah) Named for a Viennese dermatologist who first described the condition in the late 1800s.

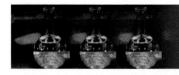

SPOTLIGHT ON MICROBIOLOGY 25.5
The Global Outlook on AIDS

AIDS in the rest of the world shows the same spectrum of illness as is seen in the United States, but it differs in epidemiology. In Africa, the case rate in men and women is about equal, and the mode of transmission there is primarily by sexual contact between men and women, and to a lesser extent via drug abuse and pregnancy. Africans are hardest hit, with 4 million new cases of HIV infection every year and over 25 million people living with the virus. Deaths from AIDS are also having a staggering effect on families and communities. By 2002, Africa will have 12 million orphaned children, and many communities will be missing an entire generation from 20 to 40 years of age.

Statistics from other countries point to a growing pandemic that involves large parts of Europe, Latin America, India, and Southeast Asia. The most recent figures for the year 2000 alone tabulate 36.1 million HIV-infected people, with 3 million deaths worldwide. There has been an alarming rise in new infections in Eastern Europe and the former Soviet Union. In a single year the number of new infections has nearly doubled from 400 thousand to 800 thousand cases. AIDS continues to be prevalent in Asia, with nearly 6 million cases, and in Latin America, with 1.5 million. Unfortunately the newer drug therapies are far beyond the reach of most of the world's AIDS patients, since they average $12,000 to $15,000 per patient per year—a figure far beyond the medical resources of these countries.

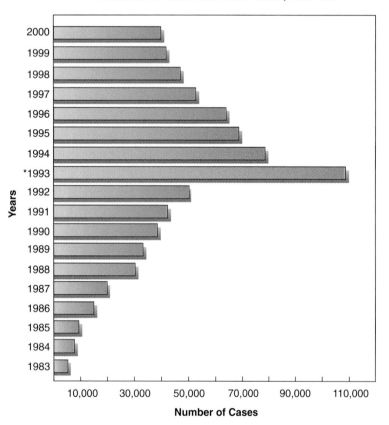

Prevalence of AIDS in the United States, 1983–2000

FIGURE 25.12

Total reported cases of AIDS from 1983 to 2000, based on data from the CDC. *This increase in 1993 was due to several changes in the case definition of AIDS introduced that year.

Data in part from Summary of Notifiable Diseases, United States, *and* Morbidity and Mortality Weekly Report, The Centers for Disease Control and Prevention, Atlanta, GA.

THE CLINICAL DEFINITION OF AIDS

The following sections will cover HIV definitions, the importance of the serological state of patients, the groups most commonly involved, and behavior or events that increase a person's risk.

AIDS is defined as a severe immunodeficiency disease arising from infection with HIV and accompanied by some of the following symptoms: life-threatening opportunistic infections, persistent fever, unusual cancers, chronically swollen lymph nodes, extensive weight loss, chronic diarrhea, and neurological disorders.

At some point during the infection, the patient tests positive for antibodies to HIV. Seropositivity means only that an infection with HIV has taken place; it does not indicate that the disease exists. However, it is presumed that the antibody-positive person is infectious. For further ramifications of this test, see the later section on diagnosis.

The CDC devised a standard system for staging HIV infection and AIDS (table 25.3). This scheme places an HIV-infected person into a particular clinical category based on the state of the

TABLE 25.3

Classification System for HIV Infection and Expanded AIDS Case Definition for Adolescents and Adults★

CD4⁺ T-cell Category	Clinical Category		
	(A) Asymptomatic, Acute (Primary) HIV Infection or PGL★★★	(B) Symptomatic, Not (A) or (C) Conditions	(C) Symptomatic AIDS-Indicator Conditions★★
(1) ≥500/μl	A1	B1	C1
(2) 200–499/μl	A2	B2	C2
(3) <200/μl AIDS-indicator T-cell count	A3	B3	C3

★Persons with AIDS-indicator conditions (category C) as well as those with CD4⁺ T-lymphocyte counts < 200/μl (category A3 or B3) became reportable as AIDS cases in the United States and territories, effective January 1, 1993.

★★Includes opportunistic infections, cancers, wasting, dementia (see figure 25.17).

★★★PGL = persistent generalized lymphadenopathy.

Source: Data from Morbidity and Mortality Weekly Report, *vol. 41. December 18, 1992. The Centers for Disease Control and Prevention, Atlanta, GA.*

immune system and the degree and type of symptoms. It more accurately reflects the spectrum of conditions that occur from asymptomatic infection to HIV disease and early and advanced stages of AIDS. Patients in categories C1, 2, 3 and B2 and 3 are classified as having overt AIDS.

Case Descriptions According to Group, in Order of Prevalence

The following categories summarize the primary exposure and risk categories for HIV infection and AIDS in the United States.

I. Homosexual or Bisexual Males   Male homosexuals still account for the majority of AIDS cases (45%) in the United States, though the rate of infection appears to be declining. This group has increased susceptibility due to anal sex, which is known to lacerate the rectal mucosa, and can provide an entrance for viruses from semen into the blood. Studies have shown that the passive anal partner in these encounters is the more likely of the two to become infected. In addition, bisexual men provide an avenue for the virus to females and the general heterosexual population.

II. Intravenous Drug Users   In large metropolitan areas such as New York City, as many as 60% of intravenous drug addicts can be HIV carriers. It is especially pronounced in "shooting galleries" where needles are shared. Infection from contaminated needles is growing more rapidly than for any other mode of transmission (30% of cases) and is another significant factor in the spread of AIDS to the heterosexual population. A fairly large segment of IV users fall into other risk categories. So concerned are public health officials about this trend that campaigns to hand out sterile needles or disinfection kits to drug users have begun in some cities.

III. Heterosexual Sex Partners of HIV Carriers   In most parts of the world, heterosexual intercourse is the primary mode of transmission. In the United States about 12% of AIDS cases arise from unprotected sexual intercourse with an AIDS-infected partner of the opposite sex. These include drug addicts and bisexual men, prostitutes, and spouses of people who have received transfusions or blood factors. The virus spreads readily in either direction, and multiple sexual encounters increase the chance of infection, although even a single encounter has been known to transmit the virus. The overall rate of heterosexual infection has increased dramatically in the past few years in adolescent and young adult women. Currently, one-third of new cases occur in women.

IV. Blood Transfusions and Blood Products   Now that donated blood is routinely tested for antibodies to the AIDS virus, transfusions are no longer considered a serious risk. Because there can be a lag period of a few weeks to several months before antibodies appear in an infected person (see figure 25.16), it is remotely possible to be infected through donated blood. To further reduce this risk, blood banks thoroughly screen potential donors. Rarely, organ transplants can carry HIV, so they too should be tested. Other blood products (serum, coagulation factors) were once implicated in AIDS. Thousands of hemophilics died from the disease in the 1980s and 1990s. It is now standard practice to heat-treat or sterilize any therapeutic blood products to destroy all viruses.

V. Inapparent or Unknown Risk Factors   About 6% of all AIDS cases occur in people without apparent risk factors. This does not mean that there is some other, unknown route of spread. When the CDC has carefully followed up such cases, a majority of the patients have admitted to prior contact with a prostitute or a history of other STDs. Factors such as patient denial, unavailability, death, or uncooperativeness make it impossible to explain every case.

VI. Congenital and Neonatal AIDS   The majority of mothers that account for neonatal infection are young IV drug addicts or sex partners of drug addicts. The chance of an HIV-positive mother infecting her fetus is 1 in 3, and infected babies develop the disease more rapidly than adults. Recent evidence indicates that treatment of infected mothers with a combination of anti-AIDS drugs can significantly prevent the infection of fetuses. As a result of improved education and treatment, the number of children with AIDS is decreasing in the United States.

VII. Risks Involving Medical and Dental Personnel   Health care personnel are not considered a high-risk group, though several hundred medical and dental workers are known to have acquired AIDS or become antibody-positive as a result of clinical accidents. A health care worker involved in an accident in which gross inoculation with blood occurs (a needlestick) has about one chance in 500 of becoming infected.

It must be emphasized that transmission of HIV in either direction will not occur through casual contact or routine nursing procedures and that basic guidelines and universal precautions for infection control (see "Introduction to Identification Techniques," page 540) were designed to give full protection for both worker and patient.

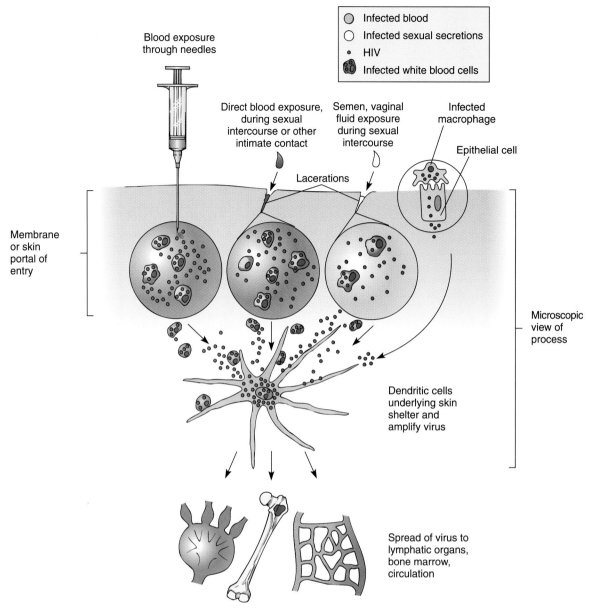

FIGURE 25.13
Primary sources and suggested routes of infection in AIDS.

Modes of AIDS Transmission

It is now clear that the mode of HIV transmission is exclusively through two forms of contact: sexual intercourse and transfer of blood or blood products (figure 25.13). The mode of transmission is rather similar to that of hepatitis B virus, except that the AIDS virus does not survive for as long outside the host, and it is far more sensitive to heat and disinfectants. In general, AIDS is spread only by direct and rather specific routes involving intimate contact with an infectious dose. An infection can take place only if the virus crosses the body's epithelial barriers into the fluid compartments. Because the blood of HIV-infected people often harbors high levels of free virus and infected leukocytes, any form of intimate contact involving transfer of blood (trauma, injection) can be a potential source of infection. Semen and vaginal secretions also harbor free virus and

infected white blood cells, thus they are significant factors in sexual transmission. The virus can be isolated from urine, tears, sweat, and saliva, but in such small numbers (less than 1 virus per cubic centimeter) that they are not considered sources of infection. Because milk contains significant numbers of leukocytes, neonates can become infected through nursing.

Misconceptions: How AIDS Is NOT Transmitted   Inaccurate notions and misconceptions rather than established facts have added to the fear of AIDS. Early on it was mistakenly believed that inhaling droplets, shaking hands, handling fomites, and sharing public facilities, food, and swimming pools were possible ways to become infected. Extensive studies have indicated that, even within families, infection of healthy persons has occurred only after known exposure to blood. Transmission by blood-sucking insects such as mosquitoes is also very unlikely since they are not biological vectors of the AIDS virus and do not harbor it in infectious dose quantities. Epidemiologists at the CDC are convinced that if any of these modes of transmission existed, they would have become evident by now.

THE EVOLVING PICTURE OF AIDS PATHOLOGY

Understanding the pathology of AIDS has been aided to an extent by animal models. Studies of macaque monkeys with a simian form of AIDS that is very similar to the human disease have helped outline some of the mechanisms by which retroviruses suppress the immune system. Although chimpanzees can be infected with HIV, they fail to develop the disease. Much promise lies in studying the colonies of transgenic mice that have been implanted with human immune systems. Preliminary findings show that when these special mice are infected with HIV, they develop some symptoms of AIDS.

Initial Infection of Target Cells

The proposed events in transmission and infection are summarized in figure 25.13. It appears that after HIV enters a mucous membrane or the skin through injection, trauma, or direct passage, it travels to dendritic cells, a type of macrophage living beneath the skin. In the dendritic cells, it grows and is shed from the cells without killing them. It is amplified by multiplying in macrophages in the skin, lymph organs, bone marrow, and blood. One of the great ironies of HIV is that it infects and destroys many of the very cells needed to destroy it, including the helper (T4 or CD4) class of lymphocytes, monocytes, macrophages, and even B lymphocytes. The virus is adapted to docking onto its host cell's surface receptors (see figure 25.11). Its main docking point is CD4, but other coreceptors are involved. It binds to CCR-5 on macrophages, which is ordinarily a receiver for cytokine signals. Using the CXCR/4 (fusin) receptor on T cells, it induces viral fusion with the cell membrane and creates syncytia.

Once inside the cell, its transcriptase makes the RNA into DNA. Although initially it can produce a lytic infection, in many cells, it enters a latent period in the nucleus of the host cell (figure 25.14), which accounts for the lengthy course of the disease that most often follows. It is during this time that the viral DNA becomes integrated into the nucleus.

Stages of HIV Infection and Disease

The scope of HIV infection shows a continuous progression from initial contact to full-blown AIDS (figure 25.15). The initial infection is often attended by vague, mononucleosis-like symptoms that soon disappear. This is followed by a period of asymptomatic infection (HIV disease) with an incubation period that varies in length from 2 to 15 years (the average is about 10). The first antibodies to HIV are usually detectable in serum by the second month following infection, but they are ineffective in neutralizing the virus once it enters the latent period (figure 25.16).

Disease can initially present with fever, swollen lymph glands (persistent lymphadenopathy), fatigue, diarrhea, weight loss, and neurological syndromes, as well as opportunistic infections and neoplasms.

Contrary to what is commonly thought, not everyone who becomes infected or is antibody-positive develops AIDS. About 5% of people who are antibody-positive remain free of disease, indicating that functioning immunity to the virus can develop. Any person who remains healthy despite HIV infection is termed a *non-progressor.* When subjected to study, some non-progressors have been found to lack the cytokine receptors that HIV requires. Others are infected by a weakened virus mutant. Because of the newer therapies (discussed in a later section), there has been a significant reduction in the number of HIV-positive people progressing to AIDS and a significant increase in long-term AIDS survivors.

The Primary Effects: Harm to T Cells and the Brain

In order to understand the outcome of AIDS infection, we must think in terms of an impaired immune system. Over the long period of chronic infection, the infected cells play host to the virus in either a latent or an active phase. Cells with active viruses release large quantities into the circulation and are killed in the process (see figure 25.14c). The death of T cells and other white blood cells results in extreme leukopenia and loss of essential T4-memory clones and stem cells. The viruses also cause the formation of giant T-cell and other cell syncytia, which allow the spread of viruses directly from cell to cell, followed by mass destruction of the syncytia. As the dose of viruses entering the system increases, a vicious cycle occurs, involving a greater infection rate and death of cells and greater viral burden. Because the T4 lymphocytes have multiple cytotoxic, helper, and inducer functions, their destruction or disability paves the way for invasion by opportunistic agents and cancer. When the T4 cell levels fall below 200 cells/mm^3 (200/μl) of blood, symptoms of AIDS appear (figure 25.16).

Targets of the AIDS virus are not limited to the immune system. The central nervous system is involved in a large proportion of cases. It is believed that infected macrophages cross the blood-brain barrier and release viruses, which then invade nervous tissues. Studies have indicated that some of the viral envelope proteins can have a direct toxic effect on the brain's glial cells and other cells. Other research has shown that some peripheral nerves become demyelinated and the brain becomes inflamed.

Secondary Effects That Define AIDS

Opportunistic Infections Among the earliest symptoms of AIDS are debilitating, potentially fatal, and often multiple opportunistic infections (figure 25.17). In general, the infections involve fungi, protozoa, viruses, or bacteria that establish a foothold in the absence of adequate cytotoxic immunity. The most frequent sites of attack are the respiratory, digestive, and nervous systems.

Pneumocystis carinii **pneumonia (PCP),** caused by a widespread fungus, is the leading cause of morbidity and mortality in AIDS patients (see chapter 22). Its symptoms include fever, cough, shortness of breath, thickening of the lung epithelium, and fluid infiltration that compromises respiration. Other fungi most frequently involved in AIDS complications are *Cryptococcus neoformans,* a cause of meningitis; *Candida albicans,* which attacks the oral, pharyngeal, and esophageal mucosa; and *Aspergillus* species, which infect the lungs. Protozoan infections common in AIDS

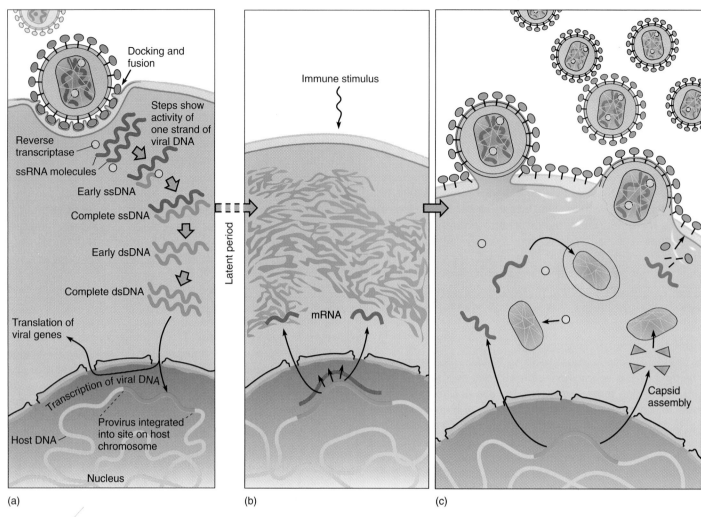

(a)

The virus is adsorbed and endocytosed, and the twin RNAs are uncoated. Reverse transcriptase catalyzes the synthesis of a single complementary strand of DNA (ssDNA). This single strand serves as a template for synthesis of a double strand (ds) of DNA. In latency, dsDNA is inserted into the host chromosome as a provirus.

(b)

After a latent period, various immune activators stimulate the infected cell, causing reactivation of the provirus genes and production of viral mRNA.

(c)

HIV mRNA is translated by the cell's synthetic machinery into virus components (capsid, reverse transcriptase, spikes), and the viruses are assembled. Budding of mature viruses lyses the infected cell.

FIGURE 25.14

 The general multiplication cycle of HIV.

patients are toxoplasmosis of the brain (caused by *Toxoplasma gondii*) and *Cryptosporidium* diarrhea (see chapter 23).

Common herpesviruses also afflict the AIDS patient (see chapter 24). Cytomegalovirus produces a disseminated pneumonia and can infect the brain, eyes, and liver as well. Herpes simplex and herpes zoster can reactivate to produce severe ulceration of the skin, lips, genitalia, and anus. Infections with bacteria such as *Mycobacterium* are particularly problematic and severe. Tuberculosis in AIDS patients is now at epidemic levels. For reasons that are obscure, several non-tuberculous mycobacteria (*Mycobacterium avium* complex) show a strong affinity for AIDS patients, growing in massive numbers in the bone marrow, liver, spleen, and lymph nodes (see chapter 19).

AIDS-Associated Cancers　The predisposition of AIDS patients for tumors is due to the loss of natural cancer-killing T cells along with the intrusion of microbes that can cause cancer. Kaposi sarcoma (KS) accounts for most cancers seen clinically. This sarcoma, which is caused by a newly identified herpesvirus (see chapter 24), is a nodular purple lesion that develops from endothelial cells in blood vessels of the skin, intestine, and mucous membranes (figure 25.18). Other effects of KS are lymphadenopathy, weight loss, hemorrhage, perforation, and intestinal obstruction. Additional cancers common to AIDS patients are epithelial carcinomas of the skin, mouth, and rectum, and lymphomas originating from B lymphocytes.

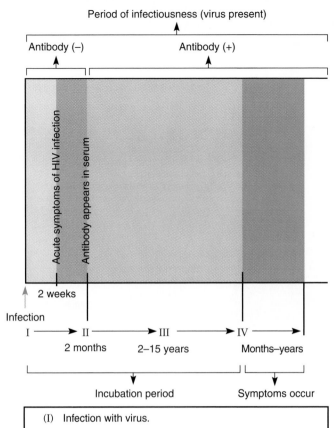

Period of infectiousness (virus present)

Antibody (−)   Antibody (+)

Acute symptoms of HIV infection

Antibody appears in serum

2 weeks

Infection

I → II → III → IV →

2 months   2–15 years   Months–years

Incubation period   Symptoms occur

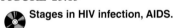

(I) Infection with virus.

(II) Appearance of antibodies in standard HIV tests.

(III) Asymptomatic HIV disease, which can encompass an extensive time period.

(IV) Overt symptoms of AIDS include some combination of opportunistic infections, cancers, and general loss of immune function.

FIGURE 25.15

Stages in HIV infection, AIDS.

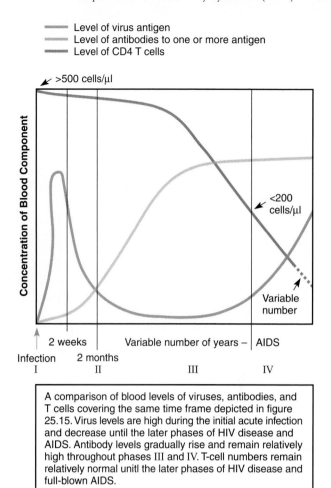

Level of virus antigen
Level of antibodies to one or more antigen
Level of CD4 T cells

>500 cells/µl

Concentration of Blood Component

<200 cells/µl

Variable number

2 weeks   Variable number of years –   AIDS

Infection   2 months

I   II   III   IV

A comparison of blood levels of viruses, antibodies, and T cells covering the same time frame depicted in figure 25.15. Virus levels are high during the initial acute infection and decrease until the later phases of HIV disease and AIDS. Antibody levels gradually rise and remain relatively high throughout phases III and IV. T-cell numbers remain relatively normal unitl the later phases of HIV disease and full-blown AIDS.

FIGURE 25.16

Dynamics of virus antigen, antibody, and T cells in circulation.

Miscellaneous Conditions AIDS patients experience several nonspecific, disease-related symptoms, many of which appear to involve severe immune deregulation, hormone imbalances, and metabolic disturbances. Pronounced wasting of body mass is a consequence of weight loss, diarrhea, and poor nutrient absorption. Protracted fever, fatigue, sore throat, and night sweats are significant and debilitating. Some of the most virulent complications are neurological. Lesions occur in the brain, meninges, spinal column, and peripheral nerves. Patients with nervous involvement show some degree of withdrawal, persistent memory loss, spasticity, sensory loss, and progressive *AIDS dementia.* Both a rash and generalized lymphadenopathy in several chains of lymph nodes are presenting symptoms in many AIDS patients.

The Effects of Coinfections

From the first months of its discovery, medical science sought to identify contributing cofactors in AIDS besides life-style or transfusions. One relationship that seems certain is the effect of coinfections, especially STDs. Genital ulcers such as those present in chlamydia, syphilis, and warts increase susceptibility to AIDS because breaks in the mucous membrane provide both a route of exit and entry for the virus. AIDS patients frequently have concurrent infections with Epstein-Barr virus (mononucleosis), cytomegalovirus, syphilis, tuberculosis, hepatitis B, and mycoplasma (see chapter 21). The severe immune suppression accompanying AIDS appears to enhance the pathogenicity and virulence of both participants and escalates damage.

Testing for HIV Infection

Most HIV testing is based on detection of antibodies specific to the virus in serum or other fluids. This allows for the rapid, inexpensive screening of large numbers of samples. Testing proceeds at two levels. The initial screen tests include the older *ELISA* and newer latex agglutination and oral swab tests. Although these tests are rather accurate, around 1% may give false-positive results and require a follow-up, more specific test called the *Western blot* analysis. This detects several different anti-HIV antibodies and can usually rule out a false-positive result (see figure 16.9).

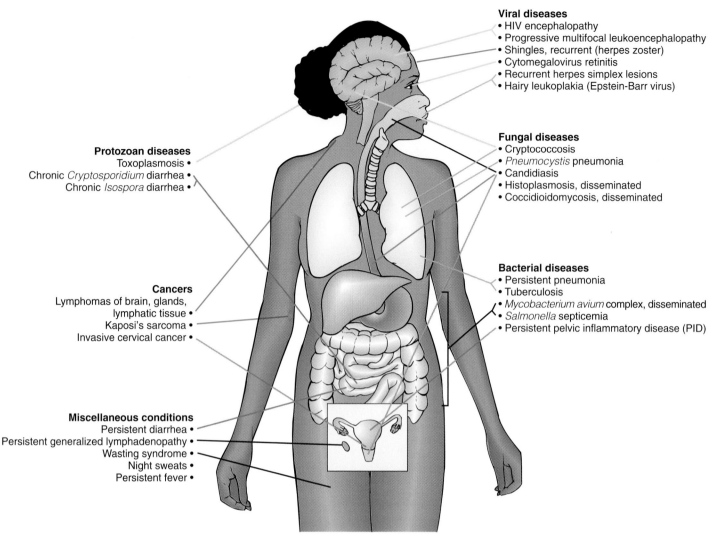

Viral diseases
• HIV encephalopathy
• Progressive multifocal leukoencephalopathy
• Shingles, recurrent (herpes zoster)
• Cytomegalovirus retinitis
• Recurrent herpes simplex lesions
• Hairy leukoplakia (Epstein-Barr virus)

Protozoan diseases
Toxoplasmosis •
Chronic *Cryptosporidium* diarrhea •
Chronic *Isospora* diarrhea •

Fungal diseases
• Cryptococcosis
• *Pneumocystis* pneumonia
• Candidiasis
• Histoplasmosis, disseminated
• Coccidioidomycosis, disseminated

Cancers
Lymphomas of brain, glands, lymphatic tissue •
Kaposi's sarcoma •
Invasive cervical cancer •

Bacterial diseases
• Persistent pneumonia
• Tuberculosis
• *Mycobacterium avium* complex, disseminated
• *Salmonella* septicemia
• Persistent pelvic inflammatory disease (PID)

Miscellaneous conditions
Persistent diarrhea •
Persistent generalized lymphadenopathy •
Wasting syndrome •
Night sweats •
Persistent fever •

FIGURE 25.17

Opportunistic infections and other diseases used in the expanded case definition of AIDS. One or more of these conditions, in combination with serological tests and CD4 T-cell counts, establishes a system to classify the clinical stage of the disease.

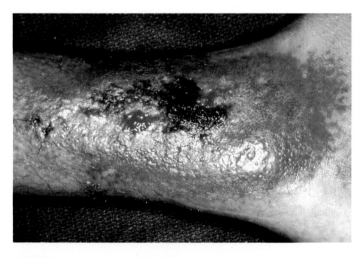

FIGURE 25.18

Kaposi sarcoma lesion on the arm. The flat, purple tumors occur in almost any tissue and are frequently multiple.

Another inaccuracy can be false-negative results, which occur in as many as one in four patients who are indeed infected with HIV. Such silent infections are due to variations in the onset of antibody production. To rule out the possibility of a false-negative, people who test negative but may have been exposed should seek a second test at a later date.

Because an infected person has viruses in his blood from the very earliest stages, a more reliable diagnostic approach would be direct discovery of HIV, its genes, or antigens in blood and other specimens. This form of testing can eliminate the false tests inherent in antibody analysis and are a particular boon in monitoring transfusion blood. Several rapid, specific, very sensitive tests using monoclonal antibodies are now available for research purposes, for diagnosing infection in newborns, and as a means of tracing viral load.

AIDS Therapy

So far, there is no cure for AIDS. None of the therapies do more than prolong life or diminish symptoms. Treatment for AIDS patients focuses on supportive care and drugs to control both

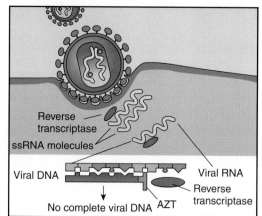

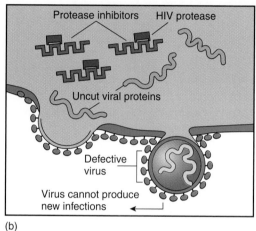

(a)

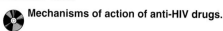

FIGURE 25.19

Mechanisms of action of anti-HIV drugs.

A prominent group of drugs (AZT, ddI, 3TC) are nucleoside analogs that inhibit reverse transcriptase. They are inserted in place of the natural nucleotide by reverse transcriptase but block further action of the enzyme and synthesis of viral DNA.

(b)

Protease inhibitors plug into the active sites on HIV protease. This enzyme is necessary to cut elongate HIV protein strands and produce functioning smaller protein units. Because the enzyme is blocked, the proteins remain uncut, and abnormal defective viruses are formed.

(c)

Integrase inhibitors are a new class of experimental drugs that attach to the enzyme required to splice the dsDNA from HIV into the host genome. This will prevent formation of the provirus and block future virus multiplication in that cell.

opportunistic and HIV infections. Drugs used to control PCP are pentamidine, taken in oral or inhaled form, and sulfamethoxazole-trimethoprim (SxT). Drugs that have some control over CMV infections are ganciclovir and foscarnet. Disseminated fungal infections are treated with fluconazole, and Kaposi sarcoma responds to alpha interferon and human chorionic gonadotropin.

Drugs to inhibit infection and replication by the AIDS virus have been an intense focus of research. More than 200 drugs are in development or clinical trials, and approximately 60 drugs have been formally approved by the Food and Drug Administration for HIV therapy. The first effective drugs were the synthetic nucleoside analogs (reverse transcriptase inhibitors) azidothymidine (AZT), Didanosine (ddI), Epivir (3TC), and Stavudine (d4T). They interrupt the HIV multiplication cycle by mimicking the structure of actual nucleosides and being added to the DNA by reverse transcriptase. Because these drugs lack all of the correct binding sites for further DNA synthesis, viral replication and the viral cycle are terminated (figure 25.19a). Other reverse transcriptase inhibitors that are not nucleosides are nevirapine and sustiva, both of which bind to the enzyme and restructure it. Another important class of

drugs are the protease inhibitors (figure 25.19b) that block the action of an HIV enzyme involved in the final assembly and maturation of the virus. Examples of these drugs include crixivan, norvir, and agenerase.

A regimen that has proved to be most effective in controlling AIDS and inevitable drug resistance is *HAART*, short for highly active anti-retroviral therapy. By combining two reverse transcriptase inhibitors and one protease inhibitor in a cocktail, the virus is interrupted in two different phases of its cycle. The therapy has been successful in reducing viral load to undetectable levels and facilitating the improvement of immune function. It has also reduced the rate of virus drug resistance and the incidence of AIDS deaths. Patients who are HIV-positive but asymptomatic can remain healthy with this therapy as well. The primary drawbacks are high cost, toxic side effects, drug failure due to patient noncompliance, and the inability to completely eradicate the virus.

Several drug companies are testing a new class of anti-AIDS drugs called integrase inhibitors. These would act at the stage of viral integration into the host DNA molecule, thereby adding a third target to block (figure 25.19c).

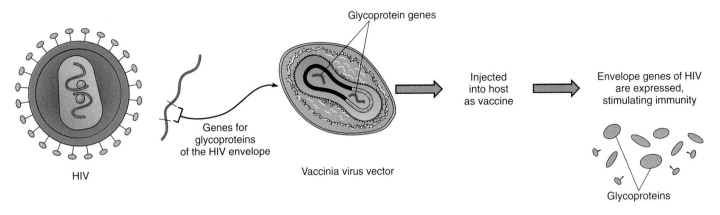

FIGURE 25.20

"Trojan horse," or viral vector technique, a novel technique for making an AIDS vaccine. The part of the HIV genome coding for envelope glycoproteins is inserted into a carrier virus. This hybrid virus replicates and expresses the HIV genes when it is injected into a host. Figure is not shown to scale.

Additional drugs under consideration are fusion inhibitors that prevent attachment of HIV to the host cell, ribozymes that inhibit translation of the viral RNA, and interferon that prevents virus replication. Some patients have acquired untested but potentially useful drugs from countries with less stringent drug policies than the United States. To ease this demand for alternative treatments, the FDA has relaxed its usual regulations to allow for compassionate use of experimental AIDS drugs among patients.

AIDS Protection

With AIDS, a person's sexual history takes on crucial importance. Epidemiologists cannot overemphasize the need to screen prospective sex partners and to follow a monogamous sexual life-style. Measures to prevent direct contact with HIV are an important consideration. There is universal agreement that "safe sex" practices can be very effective. These include the use of condoms with an antimicrobic spermicide and avoidance of anal sex. The use of a condom alone is thought to reduce the risk of infection by several million times. Although avoiding intravenous drugs is an obvious deterrent, many drug addicts do not choose this option. In such cases, risk can be decreased by not sharing syringes and needles or by disinfecting the syringe and needle with hypochlorite solution before use. Anonymous AIDS testing to protect the identity of possible HIV-positive people is currently under way to screen certain populations, such as hospital, STD, and drug treatment patients, military recruits, and newborns. There is some indication that immediate doses of anti-AIDS drugs can deter infection in health care workers accidently exposed to blood during accidents.

Will There Be an AIDS Vaccine?

From the very first years of the AIDS epidemic, the potential for a vaccine has been regarded warily, because the virus presents many seeming insurmountable problems. For example, it becomes latent in cells; its cell surface antigens mutate rapidly; and, although it does elicit immune responses, it is apparently not completely controlled by them. In view of the great need for a vaccine, none of those facts has stopped the medical community from moving ahead.

Most of the 70 vaccines in various stages of development and testing are based on recombinant viruses and viral envelope or core antigens. The principal difficulties in human trials lie in determining the safety of the vaccine in volunteers and determining whether it stimulates effective cytotoxic T cells and neutralizing antibodies.

One vaccine (AIDSVAX) that is currently being tested in Thailand is based on the surface *gp*120 antigen cloned in *E. coli*. Trials are expected to finish up in 2002, and the vaccine will possibly be available within 3 years. Another company has used the viral vector technique to produce a recombinant canarypox virus engineered to contain only HIV envelope genes (figure 25.20). This virus is presently undergoing trials on human volunteers in Australia. Yet another vaccine contains a cloned plasmid of HIV DNA prepared as previously described in figure 16.2. It is currently being tested in parts of Kenya, Africa.

The most controversial vaccine has been tested in monkeys using an SIV model with live, attenuated viruses. This technique has shown promise in animal tests, but is currently considered too controversial to test on humans. It could expose humans to mutation and cancer, or the virus could back-mutate to a more virulent form. Unless its risk can be reduced, it is an unlikely candidate at the present time. Given the challenges of immunizing against such a difficult target as HIV, and the questions as to whether any of these vaccines will be effective, even the most optimistic scientists do not expect a workable vaccine to be available sooner than 2007.

Other Retroviral Diseases in Humans

Other human retroviruses are implicated in various transmissible, fatal lymphocyte malignancies: HTLV I causes adult T-cell leukemia (ATL), a type of lymphosarcoma; HTLV II causes hairy-cell leukemia. The possible mechanisms whereby retroviruses stimulate cancer were covered in chapter 17. One hypothesis says that the virus carries an oncogene that, when spliced into a host's chromosome and triggered by various carcinogens, can immortalize the cell and deregulate the cell division cycle. No oncogenes have been

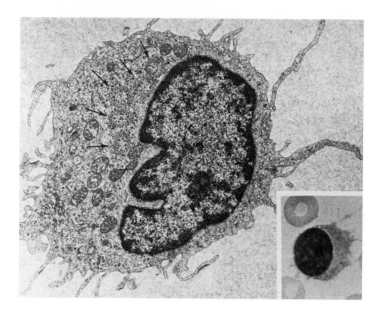

FIGURE 25.21

Appearance of hairy-cell leukemia, a neoplasm of B lymphocytes caused by HTLV II.   Long hairlike processes are evident in an electron microscope section. Arrows indicate cytopathic effects suggestive of viral infection.

From I. Katayarma, C.Y. Li, and L.T. Yam, "Ultrastructure Characteristics of the Hairy Cells of Leukemic Reticuloendotheliosis," American Journal of Pathology, *67:361, 1972. © American Society for Investigative Pathology.*

discovered in HTLV I and II, but they do have genes that code for regulatory proteins. One of HTLV's genetic targets seems to be the gene and receptor for interleukin-2, a potent stimulator of T cells.

EPIDEMIOLOGY AND PATHOLOGY OF THE HUMAN LEUKEMIA VIRUSES

Adult T-cell leukemia (ATL) was first described by physicians working with a cluster of patients in southern Japan. Later, a similar clinical disease was described in Caribbean immigrants. In time, it was shown that these two diseases and a cancer called Sézary-cell leukemia were the same disease. ATL was linked to HTLV I when the virus was isolated from patients' T cells. Although more common in Japan, Europe, and the Caribbean, a small number of cases occur in the United States. The disease is not highly transmissible; studies among families show that repeated close or intimate contact is required. Because the virus is thought to be transferred in infected blood cells, blood transfusions and blood products are potential agents of transmission. Intravenous drug users could spread it through needle sharing.

ATL is chronic and invariably fatal in its course. Symptoms are extreme, persistent lymphocytosis, with large, atypical lymphocytes. One variant of ATL, mycosis fungoides or cutaneous T-cell lymphoma, is accompanied by dermatitis, with thickened, scaly, ulcerative, or tumorous skin lesions. Other complications are lymphadenopathy and dissemination of the tumors to the lung, spleen, and liver. Many types of treatments with drugs, radiation, or some combination have been tried, but the long-term prognosis remains poor. Concerned about the possible transmission in transfusions, health officials are currently screening blood for this virus.

Hairy-cell leukemia (HCL) is a rare form of cancer that has been traced to HTLV II infection. HCL derives its name from the appearance of the afflicted lymphocytes, which have fine cytoplasmic projections or pseudopod-like extensions (figure 25.21). The disease is more common in males than in females and is probably spread through blood and shared syringes and needles. Unlike ATL, this form of leukemia presents with overall leukopenia but with an increased number of neoplastic B lymphocytes. The most serious complication is the infiltration of the spleen, liver, and bone marrow, which interferes with hematopoiesis. Splenectomy, bone marrow transplants, and antineoplastic drugs have been pursued as treatments, but alpha interferon produces the highest rate of remission.

CHAPTER CHECKPOINTS

AIDS is caused by the human immunodeficiency virus (HIV). It is the most significant viral pandemic of this century, with 30–40 million estimated cases worldwide and approximately 1 to 1.5 million cases in the United States. It is acquired through direct contact with blood or sexual fluids in sufficient quantity.

Those at greatest risk for infection include homosexual or bisexual males, IV drug users, heterosexual partners of AIDS carriers, recipients of blood products or tissue transplants, and neonates from infected mothers.

HIV first infects macrophages, where it multiplies and is shed, infecting lymphocytes and other cells that possess the CD4 receptor site. The most significant of these are the T4 (helper) cells that coordinate the immune responses to most infections. Following infection, the virus often enters a dormant stage lasting 2 to 15 years.

AIDS causes overwhelming opportunistic infections resulting from the destruction of T4 lymphocytes and stem cells. Infected macrophages also carry the virus to the CNS, causing neurological degeneration. Opportunistic infections involve fungi, protozoa, bacteria, and other viruses as well as Kaposi sarcoma, an AIDS-associated cancer.

HIV infection is diagnosed by several tests. The most accurate is the Western blot test, which detects several different antibodies to HIV. There is no cure for AIDS, but improved supportive care has increased the life expectancy for many patients. It is difficult to develop drugs that attack the virus itself because of its high mutation rate.

Protection against HIV includes following safe sex practices and taking increased precautions when handling needles and body fluids. Research continues for an effective AIDS vaccine, but many years of research and testing are necessary.

Other retroviruses include the human T-cell leukemia viruses (HTLV I and II). They have been identified as the causative agents of ATL (adult T-cell leukemia) and HCL (hairy-cell leukemia).

Nonenveloped Nonsegmented Single-Stranded RNA Viruses: Picornaviruses and Caliciviruses

As suggested by the prefix, *picorna* viruses are named for their small (pico) size and their RNA core (figure 25.22). Important representatives include **Enterovirus** and **Rhinovirus,** which are responsible for a broad spectrum of human neurological, enteric, and

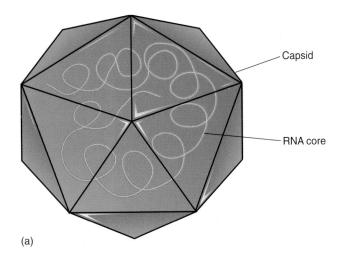

Capsid

RNA core

(a)

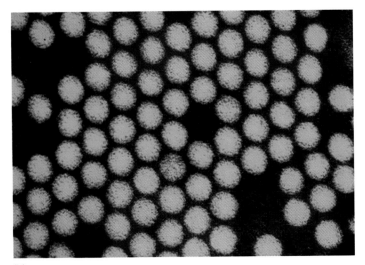

(b)

FIGURE 25.22

Typical structure of a picornavirus. **(a)** A poliovirus, a type of picornavirus that is one of the simplest and smallest viruses (30 nm). It consists of an icosahedral capsid shell around a tightly packed molecule of RNA. **(b)** A crystalline mass of stacked poliovirus particles in an infected host cell.

TABLE 25.4

Selected Representatives of the Human Picornaviruses

Genus	Representative	Primary Diseases
Enterovirus	Poliovirus	Poliomyelitis
	Coxsackievirus A	Focal necrosis, myositis
	Coxsackievirus B	Myocarditis of newborn
	Echovirus	Aseptic meningitis, enteritis, others
	Enterovirus 72	Hepatitis A
Rhinovirus	Rhinovirus	Common cold
Cardiovirus	Cardiovirus	Encephalomyocarditis
Aphthovirus	Aphthovirus	Foot-and-mouth disease (in cloven-foot animals)

Delano Roosevelt suffered its complications. Those of us who experienced the epidemics at the middle of this century will never forget the lasting images of polio: crippling paralysis, braces, the iron lung, and the March of Dimes. Through the modest donations of millions of schoolchildren and families, enough dimes were collected to underwrite increased research that eventually led to effective vaccines.

other illnesses (table 25.4), and *Cardiovirus,* which infects the brain and heart in humans and other mammals. The following discussion on human picornaviruses centers upon the poliovirus and other related enteroviruses, the hepatitis A virus, and the human rhinoviruses (HRVs).

POLIOVIRUS AND POLIOMYELITIS

Poliomyelitis* (polio) is an acute enteroviral infection of the spinal cord that can cause neuromuscular paralysis. Because it often affects small children, it is also called infantile paralysis. No civilization or culture has escaped the devastation of polio. Drawings made by ancient civilizations hint at its long history (figure 25.23), and such famous persons as Sir Walter Scott and President Franklin

* poliomyelitis (poh″-lee-oh-my″-eh-ly′-tis) Gr. *polios,* gray, *myelos,* medulla, and *itis,* inflammation.

FIGURE 25.23

Polio in an ancient civilization. This stone tablet from the Eighteenth Dynasty in Egypt depicts Siptah bearing the signs of paralytic polio.

Infection and Disease

After being ingested, polioviruses adsorb to receptors of mucosal cells in the oropharynx and intestine (figure 25.24). Here, they multiply in the mucosal epithelia and lymphoid tissue. Multiplication results in large numbers of viruses being shed into the throat and feces, and some of them leak into the blood.

Most infections are contained as a short-term, mild viremia. Some persons develop mild nonspecific symptoms of fever, headache, nausea, sore throat, and myalgia. If the viremia persists, viruses can be carried to the central nervous system through its blood supply. The virus then spreads along specific pathways in the spinal cord and brain. Being *neurotropic,** the virus infiltrates the motor neurons of the anterior horn of the spinal cord, though it can also attack spinal ganglia, cranial nerves, and motor nuclei (figure 25.25). Nonparalytic disease involves the invasion but not the destruction of nervous tissue. It gives rise to muscle pain and spasm, meningeal inflammation, and vague hypersensitivity.

Paralytic Disease

Invasion of motor neurons causes various degrees of flaccid paralysis over a period of a few hours to several days. Depending on the level of damage to motor neurons, paralysis of the muscles of the legs, abdomen, back, intercostals, diaphragm, pectoral girdle, and bladder can result. In rare cases of **bulbar poliomyelitis,** the brain stem, medulla, or even cranial nerves are affected. This situation leads to loss of control of cardiorespiratory regulatory centers, requiring mechanical respirators. In time, the unused muscles begin to atrophy, growth is slowed, and severe deformities of the trunk and limbs develop. Common sites of deformities are the spine, shoulder, hips, knees, and feet. Because motor function, but not sensation, are compromised, the crippled limbs are often very painful.

In recent times a condition called post-polio syndrome (PPS) has been diagnosed in long-term survivors of childhood infection. PPS manifests as a progressive muscle deterioration that develops in about 25–50% of patients several decades after their original polio attack.

Diagnosis of Polio

Polio is mainly suspected when epidemics of neuromuscular disease occur in the summer in temperate climates. Poliovirus can usually be isolated by inoculating cell cultures with stool or throat washings in the early part of the disease. The stage of the patient's infection can also be demonstrated by testing serum samples for the type and amount of antibody.

Measures for Treatment and Prevention of Polio

Treatment of polio rests largely on alleviating pain and suffering. During the acute phase, muscle spasm, headache, and associated discomfort can be alleviated by pain-relieving drugs. Respiratory failure may require artificial ventilation maintenance. Prompt physical therapy to diminish crippling deformities and to retrain muscles is recommended after the acute febrile phase subsides.

The mainstay of prevention is vaccination as early in life as possible, usually in four doses starting at about 2 months of age. Adult candidates for immunization are travelers and members of

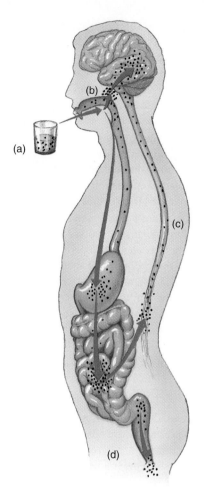

FIGURE 25.24

The stages of infection and pathogenesis of poliomyelitis.
(a) First, the virus is ingested and carried to the throat and intestinal mucosa. **(b)** The virus then multiplies in the tonsils. Small numbers of viruses escape to the regional lymph nodes and blood. **(c)** The viruses are further amplified and cross into certain nerve cells of the spinal column and central nervous system. **(d)** Last, the intestine actively sheds viruses.

Epidemiology of Poliomyelitis

The poliovirus has a naked capsid (see figure 25.22*a*) that confers chemical stability and resistance to acid, bile, and detergents. By this means, the virus survives the gastric environment and other harsh conditions, which contributes to its ease in transmission.

Sporadic cases of polio can break out at any time of the year, but its incidence is more pronounced during the summer and fall. The virus is passed within the population through food, water, hands, objects contaminated with feces, and mechanical vectors. The efforts of a WHO campaign called National Vaccination Days have significantly reduced the global incidence of polio. It is now predicted that all of the last wild polioviruses will be eradicated by 2005. The last remaining cases of infection are confined to a few pockets in Africa, India, and parts of the Middle East. There have been no cases of wild polio in the Americas since 1991.

---

* **neurotropic** (nu″-roh-troh′-pik) Having an affinity for the nervous system.

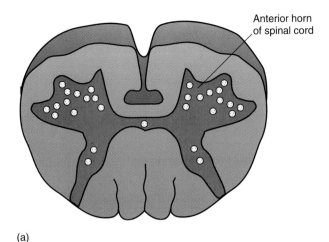

Anterior horn
of spinal cord

(a)

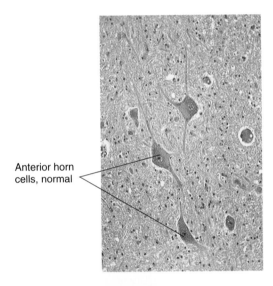

Anterior horn
cells, normal

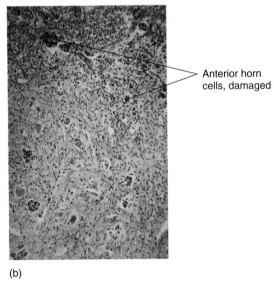

Anterior horn
cells, damaged

(b)

FIGURE 25.25

Targets of poliovirus. **(a)** A cross section of the spinal column indicates the areas most damaged in spinal poliomyelitis. **(b)** The anterior horn cells of a monkey before (top) and after (bottom) poliovirus infection.

the armed forces. The two forms of vaccine currently in use are inactivated poliovirus vaccine (IPV), known as the Salk[3] vaccine, and oral poliovirus vaccine (OPV), known as the Sabin[4] vaccine. Both are prepared from animal cell cultures and are trivalent (combinations of the three serotypes of the poliovirus). Both vaccines are effective, but one may be favored over the other under certain circumstances.

For many years, the Sabin vaccine was used in the United States because it is easily administered by mouth, but it is not free of medical complications. It contains an attenuated virus that can multiply in vaccinated people and be spread to others. In very rare instances, the attenuated virus reverts to a neurovirulent strain that causes disease rather than protects against it. Numerous instances of paralytic polio have occurred among children with hypogamma-globulinemia who have been mistakenly vaccinated. There is a tiny risk (about one case in 4 million) that an unvaccinated family member will acquire infection and disease from a vaccinated child. For these reasons public health officials have revised their recommendations and now favor IPV for the first four doses, followed by OPV for later boosters. Now that the WHO goal of polio eradication is near completion, it may be possible to discontinue polio vaccination, as has been done for smallpox.

NONPOLIO ENTEROVIRUSES

Several viruses related to poliovirus commonly cause transient, nonfatal infections. Such viruses as **coxsackieviruses*** A and B, *echoviruses,** and nonpolio enteroviruses are similar to the poliovirus in many of their epidemiologic and infectious characteristics. The incidence of infection is highest from late spring to early summer in temperate climates and is most frequent in infants and persons living under unhygienic circumstances.

Specific Types of Enterovirus Infection

Enteroviral infections are either subclinical or fall into the category of "undifferentiated febrile illness," characterized by fever, myalgia, and malaise. Symptoms are usually mild and last only a few days. The initial phase of infection is intestinal, after which viruses enter the lymph and blood and disseminate to other organs. The outcome of this infection largely depends on the organ affected. An overview of the more severe complications follows.

Important Complications

Children are more prone than adults to lower respiratory tract illness, namely bronchitis, bronchiolitis, croup, and pneumonia. All ages, however, are susceptible to the "common cold syndrome" of enteroviruses (Medical Microfile 25.6). *Pleurodynia** is an acute disease characterized by recurrent sharp, sudden intercostal or abdominal pain accompanied by fever and sore throat. Although nonpolio enteroviruses are less virulent than the polioviruses, rare

---

3. Named for Dr. Jonas Salk, who developed the vaccine in 1954.

4. Named for Dr. Albert Sabin, who had the idea to develop an oral attenuated vaccine in the 1960s.

  * coxsackievirus (kok-sak´-ee-vy″-rus) Named for Coxsackie, New York, where the viruses were first isolated.

  * echovirus (ek´-oh-vy″-rus) An acronym for enteric cytopathic human orphan virus.

  * pleurodynia (plur″-oh-din´-ee-ah) Gr. *pleura,* rib, side, and *odyne,* pain.

cases of coxsackievirus and echovirus paralysis, aseptic meningitis, and encephalitis occur. Even in severe childhood cases involving seizures, ataxia, coma, and other central nervous system symptoms, recovery is usually complete.

Eruptive skin rashes (exanthems) that resemble the rubella rash are other manifestations of enterovirus infection. Coxsackievirus can cause a peculiar pattern of lesions on the hands, feet, and oral mucosa (hand-foot-mouth disease), along with fever, headache, and muscle pain. Acute hemorrhagic conjunctivitis is an abrupt inflammation associated with subconjunctival bleeding, serous discharge, painful swelling, and sensitivity to light (figure 25.26). Spread of the virus to the heart in infants can cause extensive damage to the myocardium, leading to heart failure and death in nearly half of the cases. Heart involvement in older children and adults is generally less serious, with symptoms of chest pain, altered heart rhythms, and pericardial inflammation.

Hepatitis A Virus and Infectious Hepatitis

One enterovirus that reacts primarily with the intestinal tract is the **hepatitis A virus** (**HAV;** enterovirus 72), the cause of infectious, or short-term, hepatitis. Although this virus is not related to the hepatitis B virus discussed in chapter 24, it shares its tropism for liver cells. Otherwise, the two viruses are different in almost every respect (see table 24.3). The hepatitis A virus is a cubical picornavirus relatively resistant to heat but sensitive to formalin, chlorine, and ultraviolet radiation. There appears to be only one major serotype of this virus.

Epidemiology of Hepatitis A   Hepatitis A virus is spread through the oral-fecal route, but the details of transmission vary from one area to another. In general, the disease is associated with deficient personal hygiene and lack of public health measures. In

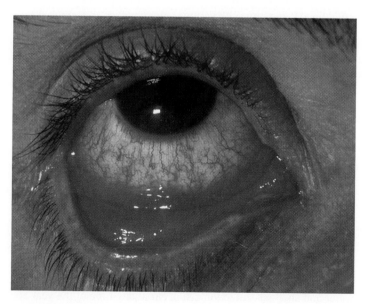

FIGURE 25.26

Acute hemorrhagic conjunctivitis caused by an enterovirus. In the early phase of disease, the eye is severely inflamed, with the sclera bright red owing to subconjunctival hemorrhage. Later, edema of the lids causes complete closure of the eye.

countries with inadequate sewage control, most outbreaks are associated with fecally contaminated water and food. The United States has a yearly reported incidence of 15,000 to 20,000 cases. Most of these result from close institutional contact, unhygienic food handling, eating shellfish, sexual transmission, or travel to other countries. Occasionally, hepatitis A can be spread by blood or blood products. In developing countries, children are the most common targets, because exposure to the virus tends to occur early in life, whereas in North America and Europe, more cases appear in adults. Because the virus is not carried chronically, the principal reservoirs are asymptomatic, short-term carriers or people with clinical disease.

The Course of HAV Infection   Swallowed hepatitis A virus multiplies in the small intestine during an incubation period of 2 to 6 weeks. The virus is then shed in the feces, and in a few days, it enters the blood and is carried to the liver. Most infections are either subclinical or accompanied by vague, flulike symptoms. In more overt cases, the presenting symptoms are loss of appetite, nausea, diarrhea, fever, pain and discomfort in the region of the liver, and darkened urine. Jaundice is present in only about one in 10 cases. Occasionally, hepatitis A occurs as a fulminating disease and causes liver damage, but this manifestation is quite rare. Because the virus is not oncogenic and does not predispose to liver cancer, complete uncomplicated recovery results.

Diagnosis and Control of Hepatitis A   A patient's history, liver and blood tests, serological tests, and viral identification all play a role in diagnosing hepatitis A and differentiating it from the other forms of hepatitis. Certain liver enzymes are elevated, and leukopenia is common. Diagnosis is aided by detection of anti-HAV IgM antibodies produced early in the infection and tests to identify HA antigen or virus directly in stool samples.

There is no specific treatment for hepatitis A once the symptoms begin. Patients who receive immune serum globulin early in the disease usually experience milder symptoms than patients who do not receive it. Prevention of hepatitis A is based primarily on immunization. An inactivated viral vaccine (HAVRAX) is currently approved and an oral vaccine based on an attenuated strain of the virus is in development. Pooled immune serum globulin is also recommended for travelers and armed forces personnel planning to enter endemic areas, contacts of known cases, and occupants of daycare centers and other institutions during epidemics. Control of this disease can be improved by sewage treatment, hygienic food handling and preparation, and adequate cooking of shellfish.

Human Rhinovirus (HRV)

The **rhinoviruses*** are a large group of picornaviruses (more than 110 serotypes) associated with the **common cold** (see Medical Microfile 25.6). Although the majority of characteristics are shared with other picornaviruses, two distinctions set rhinoviruses apart. First, they are sensitive to acidic environments such as that of the stomach, and second, their optimum temperature of multiplication is not normal body temperature, but 33°C, the average temperature in the human nose.

---

* **rhinovirus** (ry'-noh-vy"-rus) Gr. *rhinos,* nose.

The common cold touches the lives of humans more than any other viral infection, afflicting at least half the population every year and accounting for millions of hours of absenteeism from work and school. The reason for its widespread distribution is not that it is more virulent or transmissible than other infections, but that symptoms of colds are linked to hundreds of different viruses and viral strains. Among the known causative viruses, in order of importance, are rhinoviruses (which cause about half of all colds), paramyxoviruses, enteroviruses, coronaviruses, reoviruses, and adenoviruses. A given cold can be caused by a single virus type or it can result from a mixed infection.

The name implies a relationship with cold weather or drafts. But studies in which human volunteers with wet heads or feet chilled or exposed to moist, frigid air have failed to support such a link. Most colds occur in the late autumn, winter, and early spring—all periods of colder weather—but this seasonal connection has more to do with being confined in closed spaces with carriers than with temperature.

The most significant single factor in the spread of colds is contamination of hands with mucous secretions. The portal of entry is the mucous membranes of the nose and eyes. The most common symptom is a nasal discharge, and the least common is fever, except in infants and children.

Finding a cure for the common cold has been a long-standing goal of medical science. This quest is not motivated by the clinical nature of a cold, which is really a rather benign infection. A more likely reason to search for a "magic cold bullet" is productivity in the workplace and schools, as well as the potential profits from a truly effective cold drug.

One nonspecific approach has been to destroy the virus outright and halt its spread. Special facial tissues impregnated with mild acid have been marketed for use during the cold season. After years of controversy, a recent study has finally shown that taking megadoses of vitamin C at the onset of a cold can be beneficial. Zinc lozenges may also help retard the onset of cold symptoms.

The important role of natural interferon in controlling many cold viruses has led to the testing and marketing of a nasal spray containing recombinant interferon. Another company is currently developing a novel therapy based on monoclonal antibodies. These antibodies are raised to the site on the human cell (receptor) to which the rhinovirus attaches. In theory, these antibodies should occupy the cell receptor, competitively inhibit viral attachment, and prevent infection. Experiments in chimpanzees and humans showed that, when administered intranasally, this antibody preparation delayed the onset of symptoms and reduced their severity.

Portrait of Rhinovirus Virologists have subjected rhinovirus (type 14) to a detailed structural analysis. The results of this study provided a striking three-dimensional view of its molecular surface and also explained why immunity to the rhinovirus has been so elusive. The capsid subunits are of two types: protuberances (knobs), which are antigenically diverse among the rhinoviruses, and indentations (pockets), which exist only in two forms (figure 25.27). Because the antigens on the surface are the only ones accessible to the immune system, a successful vaccine would have to contain hundreds of different antigens, so it is not practical. Unfortunately, the recessed antigens are too inaccessible for immune actions. These factors make the development of a vaccine unlikely, but it could provide a basis for developing drugs to block the host receptor (Medical Microfile 25.6).

Epidemiology and Infection of Rhinoviruses Rhinovirus infections occur in all areas and all age groups at all times of the year. Epidemics caused by a single type arise on occasion, but usually many strains circulate in the population at one time. As mutations increase and herd immunity to a given type is established, newer types predominate. Children are the most successful disseminators of colds and often transmit the virus to the rest of a family. Infected respiratory membranes shed the virus for several days to weeks. People acquire infection from contaminated hands and fomites, and to a lesser extent from droplet nuclei. Although other animals have rhinoviruses, interspecies infections have never been reported. After an incubation period of 1 to 3 days, the patient can experience some combination of headache, chills, fatigue, sore throat, cough, a

mild nasal drainage, and atypical pneumonia. Natural host defenses and nasal antibodies have a beneficial local effect on the infection, but immunity is short-lived.

Control of Rhinoviruses The classic therapy is to force fluids and relieve symptoms with various cold remedies and cough syrups that contain nasal decongestants, alcohol, antihistamines, and analgesics. The actual effectiveness of most of these remedies (of which there are hundreds) is rather questionable. Owing to the extreme diversity of rhinoviruses, prevention of colds through vaccination will probably never be a medical reality. Some protection is afforded by such simple measures as handwashing and care in handling nasal secretions.

CALICIVIRUSES

Caliciviruses* are an ill-defined group of enteric viruses found in humans and mammals. The best-known human pathogen is the **Norwalk agent,** named for an outbreak of gastroenteritis that occurred in Norwalk, Ohio, from which a new type of virus was isolated. The Norwalk virus is now believed to cause one-third of all cases of viral gastroenteritis. It is transmitted orofecally in schools, camps, cruise ships, and nursing homes and through contaminated water and shellfish. Infection can occur at any time of the year and in people of all ages. Onset is acute, accompanied by nausea, vomiting, cramps, diarrhea, and chills; recovery is rapid and complete.

---

* calicivirus (kal′-ih-sih-vy″-rus) L. *calix,* the cup of a flower. These viruses have cup-shaped surface depressions.

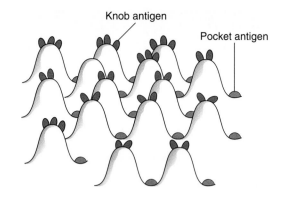

Knob antigen

Pocket antigen

(a)

Antibody   Antibody

(b)

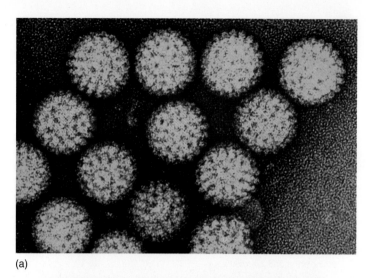

(a)

FIGURE 25.27

Structure of a rhinovirus. **(a)** The surface of a rhinovirus is composed of knobs and pockets. The antigens on the knobs are extremely variable in shape among the scores of viral strains, and those in the pockets do not vary among the strains. **(b)** The knobs are readily accessible to the immune system, and antibodies formed against them will inactivate the virus. But to be fully protected against colds, one would have to produce 100 different kinds of antibodies. Although antibodies against pocket antigens would be universal, the antigens are too inaccessible to react with antibodies.

Nonenveloped Segmented Double-Stranded RNA Viruses: Reoviruses

Reoviruses have an unusual double-stranded RNA genome and both an inner and an outer capsid (see table 25.1). Two of the best-studied viruses of the group are *Rotavirus* and *Reovirus*. Named for its wheel-shaped capsomer, *Rotavirus** (figure 25.28*a*) is a significant cause of diarrhea in newborn humans, calves, and piglets. Because the viruses are transmitted via fecally contaminated food, water, and fomites, disease is prevalent in areas of the world with poor sanitation. Globally, *Rotavirus* is the primary viral cause of mortality and morbidity resulting from diarrhea, accounting for nearly 50% of all the cases. The effects of infection vary with the nutritional state, general health, and living conditions of the infant. Babies from 6 to 24 months of age lacking maternal antibodies have the greatest risk for fatal disease. These children present with symptoms of watery diarrhea, fever, vomiting, dehydration, and shock. The intestinal mucosa can be damaged in a way that chronically compromises nutrition, and long-term or repeated infections

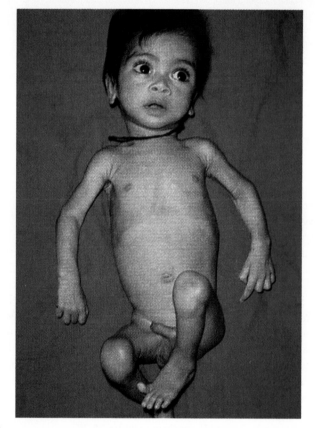

(b)

FIGURE 25.28

Diagnosing gastroenteritis. **(a)** A sample of feces from a child with gastroenteritis as viewed by electron microscopy. Note the unique "spoked-wheel" morphology that can be used to identify *Rotavirus*. **(b)** An infant suffering from chronic rotavirus gastroenteritis shows evidence of malnutrition, failure to thrive, and stunted growth.

can retard growth (figure 25.28*b*). In the United States, rotavirus infection is relatively common, but its course is generally mild. Children are treated as in cholera with oral replacement fluid and electrolytes. The live, attenuated oral vaccine had to be discontinued because it caused too many cases of intestinal blockage.

---

* *Rotavirus* (roh'-tah-vy"-rus) L. *rota,* wheel.

Reovirus* is not considered a significant human pathogen. Adults who were voluntarily inoculated with the virus developed symptoms of the common cold. The virus has been isolated from the feces of children with enteritis and from people suffering from an upper respiratory infection and rash, but most infections are asymptomatic. Although a link to biliary, meningeal, hepatic, and renal syndromes has been suggested, a causal relationship has never been proved.

CHAPTER CHECKPOINTS

Picornaviruses cause a wide variety of viral infections, including polio, aseptic meningitis, the common cold, hepatitis A, and coxsackievirus diseases of the heart and eyes. Major groups include *Enterovirus, Rhinovirus,* and *Cardiovirus.*

Reoviruses are unusual in possessing both double-stranded RNA and two capsids. The two major groups are rotaviruses and reoviruses. *Rotavirus* is a significant intestinal pathogen for young children worldwide. *Reovirus* causes an upper respiratory infection similar to the common cold.

Slow Infections by Unconventional Viruslike Agents

Diseases such as subacute sclerosing panencephalitis and progressive multifocal leukoencephalopathy, arising from persistent viral infections of the CNS, have been described in previous sections. They are characterized by a long incubation period, and they can progress over months or years to a state of severe neurological impairment. Whereas these slow infections are caused by known viruses with conventional morphology, there is another group of CNS infections for which no typical virus has been isolated.

The **spongiform encephalopathies*** are transmissible, uniformly fatal, chronic infections of the nervous system caused by unusual biological entities that some virologists call unconventional viruses, or virinos (see chapter 6). Other virologists have concluded that these diseases are caused by infectious proteins called **prions**,[5] that have extremely atypical chemical and physical properties (table 25.5). The agents cause a degeneration in the brain tissue marked by the formation of vacuoles in nerve cells and a spongy appearance of the cerebrum (figure 25.29).

The diseases of humans known to be caused by unconventional infectious agents are Creutzfeldt-Jakob disease, Gerstmann-Straussler-Scheinker syndrome (GSS), and kuru.

Creutzfeldt-Jakob disease (CJD) has sporadic, worldwide distribution and a familial pattern of inheritance. It appears to be transmitted by intimate contact with tissues of an infected patient. A few cases have been acquired from contaminated growth hormone and grafts. Symptoms include altered behavior, dementia, memory loss, impaired senses, delirium, and premature senility.

5. A contraction of **p**roteinaceous **i**nfectious particle.

* *Reovirus* (ree′oh-vy″-rus) An acronym for **re**spiratory **e**nteric **o**rphan **virus.**

* spongiform encephalopathies (spunj′-ih-form en-sef″-uh-lop′-uh-theez)

TABLE 25.5

Properties of the Agents of Spongiform Encephalopathies

Very resistant to chemicals, radiation, and heat (can withstand autoclaving)

Do not present virus morphology in electron microscopy of infected brain tissue

Not integrated into nucleic acid of infected host cells

Proteinaceous, filterable

Do not elicit inflammatory reaction or cytopathic effects in host

Do not elicit antibody formation in host

Responsible for vacuoles and abnormal fibers forming in brain of host

Transmitted only by intimate contact with infected tissues and secretions

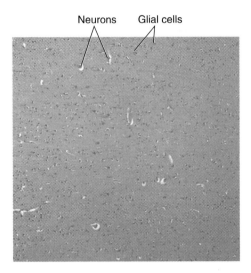

(a)

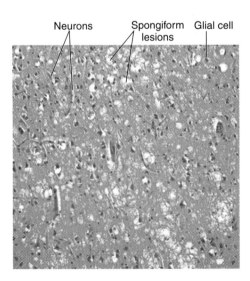

(b)

FIGURE 25.29

The microscopic effects of spongiform encephalopathy. (a) Normal cerebral cortex section, showing neurons and glial cells. **(b)** Section of cortex in CJD patient shows numerous round holes, producing a "spongy" appearance. This destroys brain architecture and causes massive loss of neurons and glial cells.

Uncontrollable muscle contractions continue until death, which usually occurs within one year of diagnosis. An autopsy of the brain shows the spongiform lesions as well as tangled protein fibers (neurofibrillary tangles) and enlarged astroglial cells (figure 25.29*b*). Because the infectious agent is not readily destroyed by usual methods, great care must be taken in handling tissues from CJD patients in a medical setting.

A newly diagnosed form of CJD, known as variant CJD, is caused by ingesting meat from cattle infected with bovine spongiform encephalopathy. Hundreds of human cases have been detected in Europe since 1995, and this has given rise to mass cattle extermination in England. This is the first evidence that a prion disease could spread from animals to humans.

Another well-studied prion disease is **kuru,** an endemic encephalopathy once prevalent among the natives of the New Guinea highlands. It has nearly disappeared since cannibalism was discontinued there. Members of these tribes performed mourning rituals that entailed eating the brains of dead relatives. Because women were the principal participants in these ceremonies, they were the most common victims.

After an incubation period of 20 to 30 years, the parts of the brain controlling gait, posture, speech, and eye movement are adversely affected, leading to symptoms of unsteadiness along with shivering tremor (the tribal word for this is *kuru*). As the victims' health gradually deteriorates, they experience profound loss of locomotion and mental deterioration.

Explaining how a protein particle that lacks nucleic acid can cause a transmissible infection has been a challenge. The most likely theory is that the normal prion protein (Prp) found in mammalian brains has undergone a mutation that alters its structure. Once this happens, the abnormal Prp itself becomes catalytic and able to spontaneously convert normal Prp into the abnormal form. This becomes a self-propagating "chain reaction" that creates a massive accumulation of altered Prp, leading to plaques, spongiform damage, and severe loss of brain function.

A range of animal neurological diseases have a similar pathologic picture to that of CJD. Examples are scrapie in sheep and bovine spongiform encephalopathy. It has been suggested that all spongiform syndromes be lumped together in the category of transmissible virus dementia (TVD).

CHAPTER CHECKPOINTS

Long-term degenerative and incurable CNS infections that cannot be linked to any known viral agent are thought to be caused by *prions,* infectious agents with atypical biological properties that are unusually resistant to destruction. Creutzfeldt-Jakob disease and kuru are two of several TVDs (transmissible virus dementias) identified to date.

CHAPTER CAPSULE WITH KEY TERMS

The RNA Viruses

I. **Enveloped Single-Stranded Viruses**
  A. **Orthomyxoviruses: Influenza viruses** have a segmented genome and glycoprotein spikes (hemagglutinin, neuraminidase); are subject to *antigenic shift* and *drift* that creates new viral subtypes. Influenza A is spread in epidemics and pandemics from human to human and from animal (pigs, poultry) to human. Clinical disease is marked by respiratory symptoms, fever, aches, sore throat; secondary pneumonia is frequent cause of death; long-term complications are Guillain-Barré syndrome and Reye syndrome; controlled with anti-flu drugs and vaccination.
  B. **Bunyaviruses** are zoonotic viruses widely associated with insects and rodents. One type is the *Hantavirus;* cause of Korean hemorrhagic fever worldwide and *hantavirus pulmonary syndrome,* a newly described disease in the United States; spread by dried animal excreta.
  C. **Paramyxoviruses:** Nonsegmented genome; glycoprotein spikes facilitate development of **multinucleate giant cells;** are spread through droplets.
    1. *Paramyxovirus* causes **parainfluenza,** a common, mild form of influenza or coldlike illness, mostly in children; **mumps** or **parotitis,** inflammatory infection of salivary glands, sometimes testes; occurs often in children; can cause deafness; prevented by live, attenuated vaccine (MMR).
    2. *Morbillivirus,* the viral agent of **measles** or **rubeola,** is highly contagious respiratory infection that produces a maculopapular rash over the body and in the oral cavity (**Koplik's spots**); usually mild, but can threaten life in a small number of children; other complication is progressive

brain disease called **SSPE;** attenuated vaccine administered in combination (MMR) or alone.
    3. *Pneumovirus,* or **respiratory syncytial virus (RSV),** causes **croup,** an acute respiratory syndrome in newborns that is a problem in hospital nurseries; treated with aerosol ribavirin.
  D. **Rhabdoviruses:** The main human pathogen is bullet-shaped *Lyssavirus,* the cause of **rabies,** a nearly 100% fatal infection encephalitis; a zoonosis carried by wild carnivores; contracted through an animal bite or scratch or inhaled; virus enters nerves and is carried to brain and salivary glands; symptoms include hyperactivity or paralysis, also hydrophobia; death results from cardiac or respiratory arrest; treatment requires cleaning of wound, passive immunization with rabies immune globulin, and active immunization with **human diploid cell vaccine.**

II. **Other Enveloped Viruses**
  A. **Coronaviruses:** Have distinctive crown of spikes; human coronavirus (HCV) is one cause of the common cold.
  B. **Togaviruses: Rubella** (German measles) **virus** is a **teratogenic** virus that crosses the placenta and causes developmental disruptions; virus is spread through contact with droplets; period of respiratory symptoms is followed by rash; infection during pregnancy can lead to congenital rubella with severe heart and nervous system damage; control by live, attenuated vaccine (MMR), preferably given in childhood or prior to pregnancy.

III. **Zoonotic Viruses**
  A. **Arenaviruses:** Spread by rodent secretions, droppings; example is Lassa fever, a severe and often fatal African hemorrhagic fever.
  B. **Arboviruses:** Loose grouping of 400 viruses spread by arthropod vectors and harbored by numerous birds and mammals; include

togaviruses, bunyaviruses, and reoviruses; distribution is determined by vector location; cause febrile infections (tick-borne fever), mosquito-borne encephalitis in horses and humans (**WEE, EEE**) and in humans and birds (St. Louis encephalitis); hemorrhagic fevers include **yellow fever** and **dengue fever,** sometimes deadly mosquito-borne diseases.

C. Retroviruses

Retroviruses are enveloped, single-stranded RNA viruses with **reverse transcriptase** that converts their single-stranded RNA into double-stranded DNA genome; insertable into host cell, so viruses are latent and can be oncogenic.

1. *Human immunodeficiency virus (HIV, types 1 and 2):* Causes **acquired immunodeficiency syndrome (AIDS),** an HIV antibody-positive state with some combination of opportunistic infections, fever, weight loss, chronic lymphadenitis, neoplasms, diarrhea, brain dysfunction. HIV appears to have originated in Africa; AIDS occurs as a pandemic and is the sixth prevalent cause of death.

2. Virus occurs in blood, semen, and vaginal secretions; is transmitted through homosexual and heterosexual contact and shared needles; risk groups or behaviors include (1) male homosexuals or bisexuals practicing unsafe sex, (2) intravenous drug abusers, (3) heterosexual partners of drug users and bisexuals, (4) blood transfusion and organ transplant patients, (5) persons with unknown risk factors, (6) fetuses or newborns of infected mothers.

3. HIV attacks cells with CD4 receptors (macrophages, T cells, and other lymphocytes); virus becomes latent; long-term infection brings the levels of T-helper cells down.

4. First signs of AIDS are opportunistic infections: *Pneumocystis carinii* **pneumonia (PCP)** and serious opportunistic diseases caused by protozoa, fungi, mycobacteria, and viruses; **Kaposi sarcoma** and lymphoma cancers; coinfections with HIV and other agents can hasten the destruction and weakening of immunities; the death rate is nearly 50%.

5. Treatment for AIDS includes drug cocktails that combine **azidothymidine (AZT),** ddI, 3TC, and d4T and protease inhibitors; drugs must be given for secondary infections; prevention measures are safe sexual practices; numerous vaccines are in clinical trials.

6. *Other retroviruses and diseases:* HTLV I causes **adult T-cell leukemia** or mycosis fungoides, a fatal cancer; HTLV II causes another cancer with high mortality rate, **hairy-cell leukemia.**

IV. Nonenveloped RNA Viruses

A. **Picornaviruses** are the smallest human viruses.

1. **Enteroviruses:** Spread primarily by the oral-fecal route. **Poliovirus,** cause of **polio;** virus can spread through close contact, contaminated food, water; occasionally leads to paralysis caused by infection and destruction of spinal neurons; bulbar type requires life support; disease controlled through vaccination with inactivated polio vaccine (IPV) and oral polio vaccine (OPV).

2. **Coxsackievirus** causes respiratory infection, hand-foot-mouth disease, conjunctivitis.

3. **Hepatitis A virus** causes short-term, mild hepatitis; spread through close institutional contact, contaminated food or shellfish; disease is initially flulike, followed by liver discomfort and diarrhea; prevented by immunization with vaccines and gamma globulin.

4. **Rhinovirus:** Most prominent cause of common cold; spread by droplets; confines itself to respiratory membranes; virus exists in over 110 forms; symptoms are nasal drainage, cough, sneezing, sore throat; treatment is symptomatic.

5. **Calicivirus: Norwalk agent,** a common cause of viral gastroenteritis.

V. Double-Stranded RNA Viruses

Reoviruses: Most important one is *Rotavirus,* the cause of severe infantile diarrhea; most common cause of viral enteric disease worldwide; children can die from effects of diarrhea, dehydration, shock.

VI. Unconventional Viruslike Agents

Prions have unusual properties: very resistant, proteinaceous, cause progressive neurological diseases; diseases are the **spongiform encephalopathies,** named for brain's spongy appearance; **Creutzfeldt-Jakob disease,** a slow deterioration of the brain, is acquired through close contact with patient's tissues and eating beef from infected cattle. Pathology appears to be caused by the buildup of an abnormal protein in the brain.

MULTIPLE-CHOICE QUESTIONS

1. Which receptors of the influenza virus are responsible for binding to the host cell?
   a. hemagglutinin   c. Type A
   b. neuraminidase   d. capsid proteins

2. The primary site of attack in influenza is the
   a. small intestine
   b. respiratory epithelium
   c. skin
   d. meninges

3. Infections with _____ virus cause the development of multinucleate giant cells.
   a. rabies   c. pneumo
   b. influenza   d. corona

4. Which virus is responsible for Korean hemorrhagic fever?
   a. Ebola virus   c. arenavirus
   b. hantavirus   d. flavivirus

5. For which disease is an exanthem (skin rash) *not* a symptom?
   a. measles   c. Coxsackievirus infection
   b. rubella   d. parainfluenza

6. A common, highly diagnostic sign of measles is
   a. viremia   c. sore throat
   b. red rash   d. Koplik's spots

7. Viruses that cause serious diseases in infants are _____.
   a. Mumps, calicivirus
   b. Respiratory syncytial virus, rotavirus
   c. Coxsackievirus, HTLV II
   d. Bunyavirus, cardiovirus

8. Rabies virus has an average incubation period of
   a. 2 to 3 weeks   c. 4 to 5 days
   b. 1 to 2 years   d. 1 to 2 months

9. For which disease are active and passive immunization given simultaneously?
   a. influenza
   b. yellow fever
   c. measles
   d. rabies

10. What property of the retroviruses enables them to integrate into the host genome?
    a. the RNA they carry
    b. presence of glycoprotein receptors
    c. presence of reverse transcriptase
    d. a positive-sense genome

11. Which of the following cells is a target of HIV infection?
    a. dendritic cells
    b. monocytes
    c. helper T cells
    d. all of these can support infection

12. Which of the following conditions is *not* associated with AIDS?
    a. *Pneumocystis carinii* pneumonia
    b. Kaposi sarcoma
    c. dementia
    d. adult T-cell leukemia

13. Polio and hepatitis A viruses are _____ viruses.
    a. arbo
    b. enteric
    c. cold
    d. syncytial

14. Rhinoviruses are the most common cause of
    a. conjunctivitis
    b. gastroenteritis
    c. hand-foot-mouth disease
    d. the common cold

15. Which of the following is *not* a characteristic of the agents of spongiform encephalopathies?
    a. highly resistant
    b. associated with tangled protein fibers in the brain
    c. are naked fragments of RNA
    d. cause chronic transmissible disease

16. **Multiple Matching.** Select all descriptions that fit each type of virus.
    ____ influenza virus
    ____ rhabdovirus
    ____ paramyxovirus
    ____ reovirus
    ____ mumps virus
    ____ measles virus
    ____ poliovirus
    ____ retrovirus
    ____ hantavirus

    a. enveloped
    b. double-stranded RNA
    c. single-stranded RNA
    d. genome in segments
    e. icosahedral capsid
    f. helical nucleocapsid
    g. positive-sense genome
    h. negative-sense genome

17. **Multiple Matching.** Match the virus with its primary target organ or site of attack. Some viruses may attack more than one organ.
    ____ HIV
    ____ mumps virus
    ____ measles virus
    ____ hepatitis A virus
    ____ hantavirus
    ____ western equine encephalitis virus
    ____ enterovirus
    ____ poliovirus
    ____ rabies virus
    ____ rubella virus

    a. brain
    b. parotid gland
    c. upper respiratory tract
    d. white blood cells
    e. kidney
    f. spinal cord
    g. liver
    h. intestine
    i. heart
    j. eye

18. **Matching.** Use the same list of viruses as in question 17 and match each one with its primary mode of transmission from the list below. Some viruses may have more than one mode.
    a. respiratory droplets
    b. sexual transmission
    c. ingestion (oral-fecal)
    d. arthropod bite
    e. contact with mammals
    f. blood transfusion
    g. fomites

CONCEPT QUESTIONS

1. a. Describe the structure and functions of the hemagglutinin and neuraminidase spikes in influenza virus.
   b. What is unusual about the genome of influenza virus?
   c. Use drawings and words to explain the concepts of antigenic drift and antigenic shift, and explain their impact on the epidemiology of this disease.
   d. How does the CDC anticipate which flu strains will predominate each year?
   e. How do the names for the types of flu originate?

2. a. Explain the course of infection and disease in influenza.
   b. What are the complications?
   c. How is the vaccine prepared, and for which groups is it indicated?

3. a. Give examples of bunyaviruses and arenaviruses.
   b. Briefly describe the nature of the diseases they cause.
   c. Describe their modes of transmission.

4. a. Describe the steps in the production of multinucleate giant cells during a viral infection.
   b. Which viruses have this effect, and what impact does cell fusion have on the spread of infection?

5. a. Describe the progress of measles symptoms.
   b. What is the cause of death in measles, and what are the most severe complications of measles infections?
   c. Describe methods of treatment and prevention for measles.

6. a. Describe the epidemiologic cycle in rabies.
   b. Which animals in the United States are most frequently involved as carriers?
   c. Describe the route of infection and the virus' pathologic effects.
   d. Why is rabies so uniformly fatal?
   e. Describe the indications for pre- and postexposure rabies treatment.
   f. What is given in postexposure treatment that is not given in preexposure treatment of rabies?
   g. What makes the latest vaccine less traumatic than earlier ones?

7. a. What is a teratogenic virus?
   b. Which RNA viruses have this potential?
   c. In what ways is rubella different from red measles?
   d. What is the protocol to prevent congenital rubella?

8. a. What are the principal carrier arthropods for arboviruses?
   b. How is the cycle of the virus maintained in the wild?
   c. Describe the symptoms of the encephalitis type of infection (WEE, EEE) and the hemorrhagic fevers (yellow fever).

9. a. What are retroviruses, and how are they different from other viruses?
   b. Give examples of the three principal human retroviruses and the diseases they cause.
   c. Give a comprehensive definition of AIDS, and define categories of HIV disease as characterized by the CDC.

10. a. Briefly explain the activities most likely to spread AIDS and what factors increase the relative risk among certain populations.
    b. Explain why bone marrow transplants are a transmission concern.
    c. Discuss some ways that HIV has *not* been documented to spread.

11. a. Discuss whether AIDS is an all-or-none disease or a series of gradual pathologic changes.
    b. Differentiate between HIV infection, HIV disease, and AIDS.
    c. What are the primary target tissues and effects over time?
    d. Explain why only certain T cells are targeted by HIV and what it does to them.
    e. How is the brain affected by infection?
    f. What major factors cause the long latency period in AIDS?
    g. What does it mean to be a nonprogressor?

12. a. Describe the changes over time in virus antigen levels, antibody levels, and CD4 T cells in the blood of an HIV-infected individual.
    b. Relate these changes to the progress of the disease, the infectiousness of the person, and the effectiveness of the immune response to the virus.

13. a. List the major opportunistic bacterial, fungal, protozoan, and viral diseases used to define AIDS.
    b. Describe four other secondary diseases or conditions that accompany AIDS.
    c. Explain which effects HIV has that create vulnerability to these particular pathogens and the other conditions.

14. a. If a person is seropositive for HIV, does this mean that the patient has AIDS?
    b. What is the most sensible interpretation of a seronegative result for a monogamous person who has not had a blood transfusion or taken IV drugs?
    c. What is a logical interpretation of a seronegativity in a person who has participated in high-risk behavior?
    d. What is meant by the "window" regarding antibody presence in the blood?

15. a. How do AZT and other nucleotide analogs control the AIDS virus? How do protease inhibitors work?
    b. Are there any actual cures for AIDS?
    c. What is the rationale behind the new "drug cocktail" treatment?
    d. Explain several strategies for vaccine development; include in your answer the primary drawbacks to development of an effective vaccine.

16. a. Describe the epidemiology and progress of polio infection and disease.
    b. What causes the paralysis and deformity?
    c. Compare and contrast the two types of vaccines.
    d. What characteristics of enteric viruses cause them to be readily transmissible?

17. a. Describe some common ways that hepatitis A is spread.
    b. What are its primary effects on the body?
    c. How is it different from hepatitis B?

18. a. What is the common cold, and which groups of viruses can be involved in its transmission?
    b. Why is it so difficult to control?
    c. Discuss the similarities in the symptoms of influenza, parainfluenza, mumps, measles, rubella, RSV, and rhinoviruses.

19. a. For which of the RNA viruses are vaccines available?
    b. Which viruses will probably never have a vaccine developed for them and why?
    c. For which of the RNA viruses are there specific drug treatments?

20. Provide the formal genus name for each of the English vernacular names of viruses listed in multiple-choice matching questions 16 and 17.

CRITICAL-THINKING QUESTIONS

1. a. Explain the relationship between herd immunity and the development of influenzal pandemics.
   b. Why will herd immunity be lacking if antigenic shift or drift occurs?

2. a. Explain how the antibody content of a patient's sera can be used to predict which strain of influenza predominated during a previous epidemic.
   b. Explain the precise way that cross-species influenza infections give rise to new and different strains of virus.

3. a. Why do infections such as mumps, measles, polio, rubella, and RSV regularly infect children and not adults?
   b. Measles is considered to be a highly contagious infection. Explain how it is possible to acquire measles from a person who has only been exposed to this infection.

4. AIDS is thought to have originated from simian viruses, and later spread from human populations to the rest of the world.
   a. Give some possible ways that monkeys may have transmitted viruses to humans.
   b. Can you think of a plausible series of events that could explain how HIV came to be so widespread?
   c. Explain why there can be such a discrepancy between the total number of reported cases of AIDS and the projected number of cases estimated by health authorities.
   d. Using the statistics in figure 25.12, estimate the total number of cases occurring between 1983 and 2000 and show the percent decrease occurring since the peak incidence.

5. Trace the likely possible transmission pathways of HIV (source in first person, mode of infection in second person, portals of exit and entry) for each of the following:
   a. From one infected homosexual man to another homosexual man.
   b. From one drug user to another drug user.
   c. From a bisexual man to his wife.
   d. From a prostitute to a contact.
   e. From a monogamous man to his monogamous wife of 50 years.
   f. From one teenager to another.
   g. From a mother to a child.

6. a. Provide a possible explanation for the people with AIDS who are long-term survivors or those with HIV infection who are non-progressors.
   b. Is it possible to become immune to HIV? Explain what brings you to this conclusion.

7. a. Explain the characteristics of the HIV virus and infection that make it such a difficult target for the immune system and for drugs.
   b. Why are coinfections with other microbes so devastating to AIDS patients?

8. a. Explain why a direct antigen test for HIV infection would be preferable to a serological test for antibodies.
   b. What factors have prevented the development of an effective AIDS vaccine?
   c. Give a plausible explanation for the fact that live, attenuated (or even dead) HIV vaccines pose a risk.

9. a. What precautions can a person take to prevent himself or herself from contracting HIV infection?
   b. How can a health care worker prevent possible infection?

10. a. If wild type polio had disappeared from the Western Hemisphere by 1991, how do you explain the 5 cases of polio reported in the United States in 1996?
    b. Provide an explanation for giving the Salk (IPV) vaccine for the first four doses, followed later by Sabin (OPV) vaccine.

11. Various over-the-counter cold remedies control or inhibit inflammation and depress the symptoms of colds. Is this beneficial or not? Support your answer.

12. a. Explain how bovine spongiform encephalopathy can be transmitted to humans.
    b. Make a flowchart to explain the mechanism of how prions can cause a progressive effect, even though they cannot divide by themselves.

13. Case study 1. Three months after having corneal surgery, a patient reported to his physician with fever and numbness of the face. His throat was tight, and he had difficulty swallowing; he was afraid of showering because of the painful spasms it caused. Within hours, he became paralyzed, lapsed into a coma, and died. Physicians were stumped as to the cause. Can you offer some ideas? What therapy should have been given?

14. Case study 2. Late in the spring, a young man from rural Idaho developed fever, loss of memory, difficulty in speech, convulsions, and tremor and lapsed into a coma. He tested negative for bacterial meningitis and had no known contact with dogs or cats. He survived but had long-term mental retardation. What are the possible diseases he might have had, and how would he have contracted them?

15. Case study 3. Biopsies from the liver and intestine of an otherwise asymptomatic 35-year-old male show masses of acid-fast bacilli throughout the tissue. Explain the pathology going on here and provide a preliminary diagnosis.

INTERNET SEARCH TOPICS

1. Look up the latest vaccination guidelines for polio, influenza, measles, mumps, and rubella.

2. Access information on integrase and fusion-inhibiting drugs used in treating HIV infection and AIDS.

Environmental and Applied Microbiology

The last thirteen chapters (13–25) have focused on the somewhat narrow field of pathogenic microbiology, necessarily overlooking many other interesting and useful aspects of microbiology and microorganisms. This last chapter emphasizes microbial activities that help maintain, sustain, and control the life support systems on the earth. This subject is explored from the standpoints of (1) the natural roles of microorganisms in the environment and their contributions to the ecological balance, including soil, water, and mineral cycles, and (2) the artificial applications of microbes in the food, medical, biochemical, drug, and agricultural industries.

Chapter Overview

- Microorganisms contribute in profound ways to the earth's structure and function, and therefore, to the survival of other life forms.
- Microbes also play significant roles in practical endeavors related to agriculture, food production, industrial processes, and waste treatment.
- Microorganisms exist in complex ecological associations that include both living and nonliving components of the environment.
- Microbes have adapted to specific habitats and niches, from which they derive food, energy, shelter, and other essential components of the biosphere.
- Microbes occupy many levels of a community structure, from producers to consumers to decomposers, and they form close partnerships with other microbes, animals, and plants.
- The actions of microbes acting as decomposers or mineralizers are primarily responsible for maintaining and cycling the biologically important elements, such as carbon, nitrogen, and phosphorus, that exist only in certain reservoirs.
- As energy passes through various levels of an ecosystem, large amounts of it are lost. Fortunately, energy can constantly be replenished, largely through the actions of photosynthetic organisms that convert the sun's energy into organic compounds.
- The soil forms complex ecosystems that support a wide variety of microorganisms, many of which decompose dead organisms and foster the growth of plants.
- Aquatic ecosystems are influenced by temperature, sun, geography, and the water cycle interacting to create numerous adaptive zones for microorganisms.
- Water quality is greatly dependent on its microbial and chemical content. Water is made safe by treatment methods that remove pathogenic microbes and toxic wastes.

A site at the Oak Ridge National Laboratory where the Department of Energy is testing several new genetically engineered bacteria that could bioremediate toxic contaminants from soil and water.

- Biotechnology includes the purposeful use of microorganisms to create industrial, agricultural, nutritional, or medical products through a process of fermentation.
- Food fermentations are used to make a variety of milk products (cheeses, yogurt), alcoholic beverages (beer, wine, spirits), and pickles.
- Foods must be properly handled and preserved through the use of temperature, chemicals, or radiation to protect consumers against various types of food poisoning.
- Large-scale industrial fermentations employ microbial metabolism to manufacture drugs, hormones, enzymes, vaccines, and vitamins.

Ecology: The Interconnecting Web of Life

The study of microbes in their natural habitats is known as **environmental,** or **ecological, microbiology;** the study of the practical uses of microbes in food processing, industrial production, and biotechnology is known as **applied microbiology.** Separating this chapter into two disciplines is an organizational convenience, since the two areas actually overlap to a considerable degree—largely because most natural habitats have been altered by human activities. Human intervention in natural settings has changed the earth's warming and cooling cycles, increased wastes in soil, polluted water, and altered some of the basic relationships between microbial, plant, and animal life. Now that humans are also beginning to release totally new, genetically recombined microbes into the environment and to alter the genes of plants, animals, and even themselves, what does the future hold? Although this question may be imponderable, we know one thing for certain: Microbes—the most vast and powerful resource of all—will be silently working in nature.

In chapter 7 we first touched upon the widespread distribution of microorganisms and their adaptations to most habitats of the world, from extreme to temperate. Regardless of their exact location or type of adaptation, microorganisms necessarily are exposed to and interact with their environment in complex and extraordinary ways. The science that studies interactions between microbes and their environment and the effects of those interactions on the earth is **microbial ecology.** Unlike studies that deal with the pathologic effects of a single organism or its individual characteristics in the laboratory, ecological studies are aimed at the interactions taking place between organisms and their environment at many levels at any given moment. Therefore, ecology is a broad-based science that merges many subsciences of biology as well as geology, chemistry, and engineering.

Ecological studies take into account both the biotic and the abiotic components of an organism's environment. **Biotic*** factors include any other living or once-living organisms[1] such as symbionts sharing an organism's habitat, parasites, or food substrates. **Abiotic*** factors include any nonliving surroundings such as the atmosphere, soil, water, temperature, and light. A collection of organisms together with its surrounding physical and chemical factors is defined as an **ecosystem** (figure 26.1).

THE ORGANIZATION OF ECOSYSTEMS

The earth initially may seem like a random, chaotic place, but it is actually an incredibly organized, well-tuned machine. Ecological relationships exist at several levels, ranging from the entire earth all the way down to a single organism (figure 26.1). The most all-encompassing of these levels, the **biosphere,** includes the thin envelope of life (about 14 miles deep) that surrounds the earth's surface. This global ecosystem comprises the **hydrosphere** (water), the **lithosphere** (a few miles into the soil), and the **atmosphere** (a few miles into the air). The biosphere maintains or creates

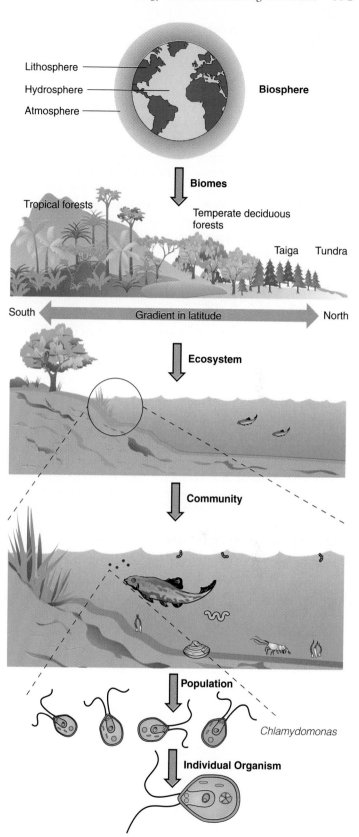

FIGURE 26.1

 Levels of organization in an ecosystem, ranging from the biosphere to the individual organism.

---

1. Biologists make a distinction between nonliving and dead. A nonliving thing has never been alive, whereas a dead thing was once alive but no longer is.

* biotic (by-ah′-tik) Gr. *bios,* living.

* abiotic (ay″-by-ah′-tik) Gr. *a,* not.

the conditions of temperature, light, gases, moisture, and minerals required for life processes. The biosphere can then be naturally subdivided into terrestrial and aquatic realms. The terrestrial realm is usually distributed into particular climatic regions called **biomes** (by'-ohmz), each of which is characterized by a dominant plant form, altitude, and latitude. Particular biomes include grassland, desert, mountain, and tropical rain forest. The aquatic biosphere is generally divisible into freshwater and marine realms.

Biomes and aquatic ecosystems are generally composed of mixed assemblages of organisms that live together at the same place and time and that usually exhibit well-defined nutritional or behavioral interrelationships. These clustered associations are called **communities.** Although most communities are identified by their easily visualized dominant plants and animals, they also contain a complex assortment of bacteria, fungi, algae, protozoa, and even viruses. The basic units of community structure are **populations,** groups of organisms of the same species. The organizational unit of a population is the individual organism, and each organism, in turn, has its own levels of organization (organs, tissues, cells).

Ecosystems are generally balanced, with each organism existing in its particular habitat and niche. The **habitat** is the physical location in the environment to which an organism has adapted. In the case of microorganisms, the habitat is frequently a *microenvironment,* where particular qualities of oxygen, light, or nutrient content are somewhat stable. The **niche** is the overall role that a species (or population) serves in a community. This includes such activities as nutritional intake (what it eats), position in the community structure (what eats it), and rate of population growth. A niche can be broad (such as scavengers that feed on nearly any organic food source) or narrow (microbes that decompose cellulose in forest litter or that fix nitrogen).

ENERGY AND NUTRITIONAL FLOW IN ECOSYSTEMS

All living things must obtain nutrients and a usable form of energy from the abiotic and biotic environments. The energy and nutritional relationships in ecosystems can be described in a number of

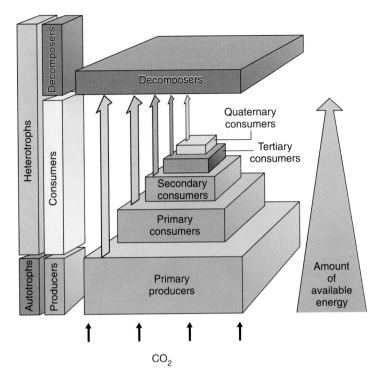

FIGURE 26.2

A trophic and energy pyramid. The relative size of the blocks indicates the number of individuals that exist at a given trophic level. The arrow on the right indicates the amount of usable energy from producers to top consumers. Both the number of organisms and the amount of usable energy decrease with each trophic level. Decomposers are an exception to this pattern, but only because they can feed from all trophic levels (colored arrows).

convenient ways. A **food chain** or **energy pyramid** provides a simple summary of the general trophic (feeding) levels, designated as producers, consumers, and decomposers, and traces the flow and quantity of available energy from one level to another (figure 26.2). It is worth noting that microorganisms are the only living things that exist at all three major trophic levels. The nutritional roles of microorganisms in ecosystems are summarized in table 26.1.

TABLE 26.1

The Major Roles of Microorganisms in Ecosystems

Role	Description of Activity	Examples of Microorganisms Involved
Primary producers	Photosynthesis	Algae, cyanobacteria, sulfur bacteria
	Chemosynthesis	Chemolithotrophic bacteria in thermal vents
Consumers	Predation	Free-living protozoa that feed on algae and bacteria; some fungi that prey upon nematodes
Decomposers	Degradation of plant and animal matter and wastes	Soil saprobes (primarily bacteria and fungi) that degrade cellulose, lignin, and other complex macromolecules
	Mineralization of organic nutrients	Soil bacteria that reduce organic compounds to inorganic compounds such as CO_2 and minerals
Cycling agents for biogeochemical cycles	Recycling compounds containing carbon, nitrogen, phosphorus, sulfur	Specialized bacteria that transform elements into different chemical compounds to keep them cycling from the biotic to the abiotic and back to the biotic phases of the biosphere
Parasites	Living and feeding on hosts	Viruses, bacteria, protozoa, fungi, and worms that play a role in population control

Food Chain

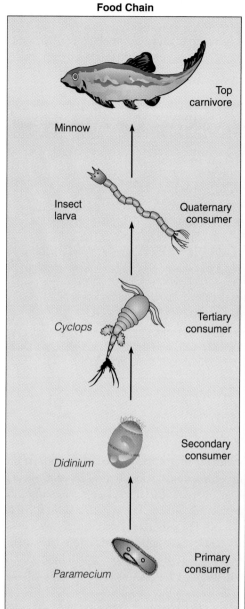

Food Web

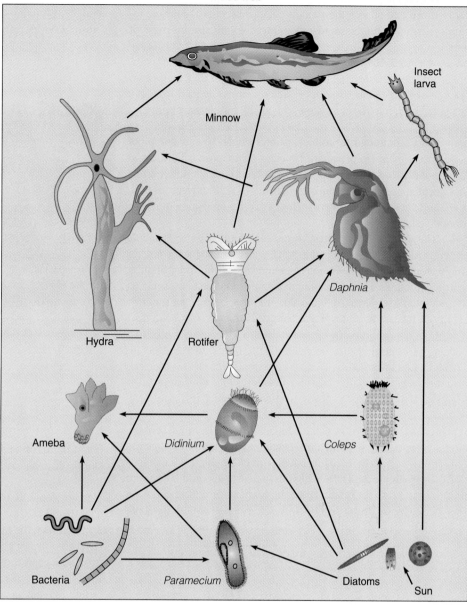

FIGURE 26.3

Food chain. A food chain is the simplest way to present specific feeding relationships among organisms, but it may not reflect the total nutritional interactions in a community.

FIGURE 26.4

Food web. More complex trophic patterns are accurately depicted by a food web, which traces the multiple feeding options that exist for most organisms. Note: Arrows point toward the consumers. Compare this pattern of feeding with the chain in figure 26.3.

Life would not be possible without **producers,** because they provide the fundamental energy source that drives the trophic pyramid. Producers are the only organisms in an ecosystem that can produce organic carbon compounds such as glucose by assimilating (fixing) inorganic carbon (CO_2) from the atmosphere. Such organisms are also termed autotrophs. Most producers are photosynthetic organisms, such as plants and cyanobacteria, that convert the sun's energy into the energy of chemical bonds. Photosynthesis was introduced in chapter 7, and the chemical steps in the process are covered with the carbon cycle later in this chapter. A small but important amount of CO_2 assimilation is brought about by unusual bacteria called lithotrophs. The metabolism of these organisms derives energy from oxidation-reduction reactions of simple inorganic compounds such as sulfides and hydrogen. In certain ecosystems (see thermal vents, Spotlight on Microbiology 7.5), lithotrophs are the sole supporters of the energy pyramid.

Consumers feed on other living organisms and obtain energy from bonds present in the organic substrates they contain. The category includes animals, protozoa, and a few bacteria and fungi. A pyramid usually has several levels of consumers, ranging from *primary consumers* (grazers or herbivores), which consume producers; to *secondary consumers* (carnivores), which feed on primary consumers; to *tertiary consumers,* which feed on secondary consumers; and up to *quaternary consumers* (usually the last level), which feed on tertiary consumers. Figures 26.3 and 26.4 show specific organisms at these levels.

Decomposers, primarily microbes inhabiting soil and water, break down and absorb the organic matter of dead organisms, including plants, animals, and other microorganisms. Because of their biological function, decomposers are active at all levels of the food pyramid. Without this important nutritional class of saprobes, the biosphere would stagnate and die. The work of decomposers is to reduce organic matter into inorganic minerals and gases that can be cycled back into the ecosystem, especially for the use of primary producers. This process, also termed *mineralization,* is so efficient that almost all biological products can be reduced by some type of decomposer. Numerous bacterial species decompose cellulose and lignin, which are complex polysaccharides from plant cell walls that account for the vast bulk of detritus in soil and water. Both fungi and bacteria decompose such complex animal compounds as keratin and chitin.

The pyramid illustrates several limitations of ecosystems with regard to energy. Unlike nutrients, which can be passed among trophic levels, recycled, and reused, energy does not cycle. Maintenance of complex interdependent trophic relationships such as those shown in figures 26.3 and 26.4 requires a constant input of energy at the producer level. As energy is transferred to the next level, a large proportion (as high as 90%) of the energy will be lost in a form (primarily heat) that cannot be fed back into the system. Thus, the amount of energy available decreases at each successive trophic level. This energy loss also decreases the actual number of individuals that can be supported at each successive level.

A concrete image of a feeding pathway can be provided by a food chain. Although it is a somewhat simplistic way to describe a community feeding pattern, a food chain helps identify the specific types of organisms that are present at a given trophic level in a natural setting (figure 26.3). Feeding relationships in communities are more accurately represented by a multichannel food chain, or a **food web** (figure 26.4). A food web reflects the actual nutritional structure of a community. It can help to identify feeding patterns typical of herbivores (plant eaters), carnivores (flesh eaters), and omnivores (feed on both plants and flesh).

ECOLOGICAL INTERACTIONS BETWEEN ORGANISMS IN A COMMUNITY

Whenever complex mixtures of organisms associate, they develop various dynamic interrelationships based on nutrition and shared habitat. These relationships, some of which were described in chapter 7, include mutualism, commensalism, parasitism, competition, synergism (cross-feeding), predation, and scavenging.

Mutually beneficial associations *(mutualism)* such as protozoans living in the termite intestine are so well evolved that the two members require each other for survival. In contrast, *commensalism* is one-sided and independent. Although the action of one microbe favorably affects another, the first microbe receives no benefit. Many commensal unions involve *cometabolism,* meaning that the waste products of the first microbe are useful nutrients for the second one. In *synergism,* two organisms that are usually independent cooperate to break down a nutrient neither one could have metabolized alone. *Parasitism* is an intimate relationship whereby a parasite derives its nutrients and habitat from a host that is usually harmed in the process. In *competition,* one microbe gives off antag-

onistic substances that inhibit or kill susceptible species sharing its habitat. A *predator* is a form of consumer that actively seeks out and ingests live prey (protozoa that prey on algae and bacteria). *Scavengers* are nutritional jacks-of-all-trades; they feed on a variety of food sources, ranging from live cells to dead cells and wastes.

CHAPTER CHECKPOINTS

The study of ecology includes both living (biotic) and nonliving (abiotic) components of the earth. Microbial ecology includes ecological microbiology, the study of microbes in their natural habitats, and applied microbiology, their utilization for commercial purposes.

Ecosystems are organizations of living populations in specific habitats. Each ecosystem requires a continuous outside source of energy for survival and a nonliving habitat consisting of soil, water, and air.

A living community is composed of populations that show a pattern of energy and nutritional relationships called a food web. Microorganisms are essential producers and decomposers in any ecosystem.

The relationships between populations in a community are described according to the degree of benefit or harm they pose to one another. These relationships include mutualism, commensalism, predation, parasitism, synergism, scavenging, and competition.

THE NATURAL RECYCLING OF BIOELEMENTS

Ecosystems are open with regard to energy because the sun is constantly infusing them with a renewable source. In contrast, the bioelements and nutrients that are essential components of protoplasm are supplied exclusively by sources somewhere in the biosphere and are not being continually replenished from outside the earth. In fact, the lack of a required nutrient in the immediate habitat is one of the chief factors limiting organismic and population growth. Because of the finite source of life's building blocks, the long-term sustenance of the biosphere requires continuous **recycling** of elements and nutrients. Essential elements such as carbon, nitrogen, sulfur, phosphorus, oxygen, and iron are cycled through biological, geologic, and chemical mechanisms called **biogeochemical cycles** (figure 26.5). Although these cycles vary in certain specific characteristics, they share several general qualities, as summarized in the following list:

- All elements ultimately originate from a nonliving, long-term reservoir in the atmosphere, sedimentary rocks, or water. They cycle in pure form (N_2) or as compounds (PO_4).
- Elements make the rounds between a free, inorganic state in the abiotic environment and a combined, organic state in the biotic environment.
- Recycling maintains a necessary balance of nutrients in the biosphere so that they do not build up or become unavailable.
- Cycles are complex systems that rely upon the interplay of producers, consumers, and decomposers. Often the waste products of one organism become a source of energy or building material for another.

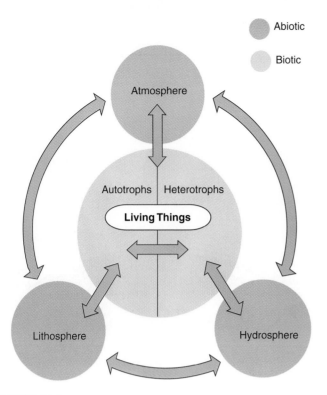

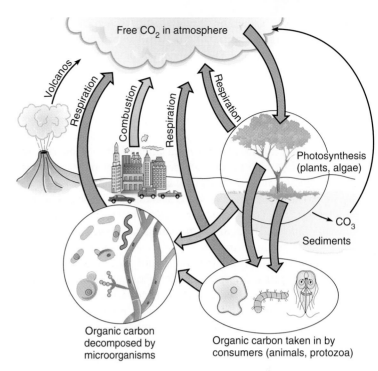

FIGURE 26.5

The general pattern of biogeochemical cycles. Essential elements constantly recycle between their biotic form in protoplasm and their abiotic (inorganic) form in soil, water, or the atmosphere. Although all living things are involved in these cycles, plants and animals have a somewhat general effect on recycling, while microorganisms act upon and alter the forms of elements in the environment.

FIGURE 26.6

The carbon cycle. This cycle traces carbon from the CO_2 pool in the atmosphere to the primary producers (green), where it is fixed into protoplasm. Organic carbon compounds are taken in by consumers (blue) and decomposers (yellow) that produce CO_2 through respiration and return it to the atmosphere (pink). Combustion of fossil fuels and volcanic eruptions also add to the CO_2 pool. Some of the CO_2 is carried into inorganic sediments by organisms that synthesize carbonate (CO_3) skeletons. In time, natural processes acting on exposed carbonate skeletons can liberate CO_2.

- All organisms participate directly in recycling, but only certain categories of microorganisms have the metabolic pathways for converting inorganic compounds from one nutritional form to another.

The English biologist, James Lovelock, has postulated a concept called the **Gaia** (guy′-uh) **hypothesis** after the mythical Greek goddess of earth. This hypothesis proposes that the biosphere contains a diversity of habitats and niches favorable to life because living things have made it that way. Not only does the earth shape the character of living things, but living things shape the character of the earth. After all, we know that the chemical compositions of the aquatic environment, the atmosphere, and even the soil would not exist as they do without the actions of living things. Organisms are also very active in evaporation and precipitation cycles, formation of mineral deposits, and rock weathering.

For billions of years, microbes have played prominent roles in the formation and maintenance of the earth's crust, the development of rocks and minerals, and the formation of fossil fuels. This revolution in understanding the biological involvement in geologic processes has given rise to a new field called *geomicrobiology.*

In the next several sections we examine how, jointly and over a period of time, the varied microbial activities affect and are themselves affected by the abiotic environment.

ATMOSPHERIC CYCLES

The Carbon Cycle

Because carbon is the fundamental atom in all biomolecules and accounts for at least one-half of the dry weight of protoplasm, the **carbon cycle** is more intimately associated with the energy transfers and trophic patterns in the biosphere than are other elements. Besides the enormous organic reservoir in the bodies of organisms, carbon also exists in the gaseous state as carbon dioxide (CO_2) and methane (CH_4) and in the mineral state as carbonate (CO_3). In general, carbon is recycled through ecosystems via photosynthesis (carbon fixation), respiration, and fermentation of organic molecules, limestone decomposition, and methane production. A convenient starting point from which to trace the movement of carbon is with carbon dioxide, which occupies a central position in the cycle and represents a large common pool that diffuses into all parts of the ecosystem (figure 26.6). As a general rule, the cycles of oxygen and hydrogen are closely allied to the carbon cycle.

The principal users of the atmospheric carbon dioxide pool are photosynthetic autotrophs (phototrophs) such as plants, algae, and cyanobacteria. An estimated 165 billion tons of organic material per year are produced by terrestrial and aquatic photosynthesis. A smaller amount of CO_2 is used by chemosynthetic autotrophs such as methane bacteria. A review of the general equation for photosynthesis in figure 26.7a will reveal that phototrophs use energy from the sun to fix CO_2 into organic compounds such as glucose that can be used in synthesis and respiration. Photosynthesis is also the primary means by which the atmospheric supply of O_2 is regenerated.

Just as photosynthesis removes CO_2 from the atmosphere, other modes of generating energy, such as respiration and fermentation, return it. Recall in the general equation for aerobic respiration (chapter 8) that, in the presence of O_2, organic compounds such as glucose are degraded completely to CO_2 and H_2O, with the release of energy. Carbon dioxide is also released by anaerobic respiration, by certain types of fermentation reactions, and by phototrophs, which must respire in the absence of light.

A small but important phase of the carbon cycle involves certain limestone deposits composed primarily of calcium carbonate ($CaCO_3$). Limestone is produced when marine organisms such as mollusks, corals, protozoans, and algae form hardened shells by combining carbon dioxide and calcium ions from the surrounding water. When these organisms die, the durable skeletal components accumulate in marine deposits. As these immense deposits are gradually exposed by geologic upheavals or receding ocean levels, various decomposing agents liberate CO_2 and return it to the CO_2 pool of the water and atmosphere.

The complementary actions of photosynthesis and respiration, along with other natural CO_2-releasing processes such as limestone erosion and volcanic activity, have maintained a relatively stable atmospheric pool of carbon dioxide. Recent figures show that this balance is being disturbed as humans burn *fossil fuels* and other organic carbon sources. Fossil fuels, including coal, oil, and natural gas, were formed through millions of years of natural biologic and geologic activities. Humans are so dependent upon this energy source that, within the past 25 years, the proportion of CO_2 in the atmosphere has steadily increased from 32 ppm to 36 ppm. Although this increase may seem slight and insignificant, most scientists now feel it has begun to disrupt the delicate temperature balance of the biosphere (Microbits 26.1).

Compared with carbon dioxide, **methane** gas (CH_4) plays a secondary part in the carbon cycle, though it can be a significant product in anaerobic ecosystems dominated by **methanogens** (methane producers). In general, when methanogens reduce CO_2 by means of various oxidizable substrates, they give off CH_4. The practical applications of methanogens are covered in a subsequent section on sewage treatment, and their contribution to the greenhouse effect is discussed in Microbits 26.1.

Photosynthesis: The Earth's Lifeline

With few exceptions, the energy that drives all life processes comes from the sun, but this source is directly available only to the cells of photosynthesizers. In the terrestrial biosphere, green plants are the primary photosynthesizers, and in aquatic ecosystems, where 80–90% of all photosynthesis occurs, algae and cyanobacteria fill this role. Other photosynthetic procaryotes are green sulfur, purple sulfur, and purple nonsulfur bacteria.

The anatomy of photosynthetic cells is adapted to trapping sunlight, and their physiology effectively uses this solar energy to produce high-energy glucose from low-energy CO_2 and water. Pho-

Summary equation:

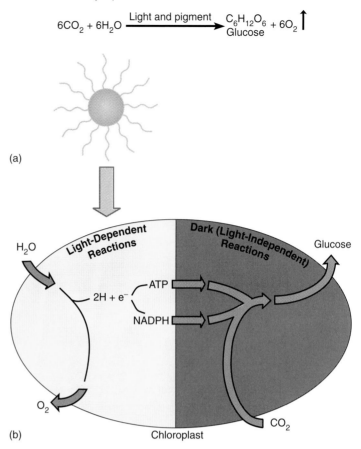

$$6CO_2 + 6H_2O \xrightarrow[\text{Glucose}]{\text{Light and pigment}} C_6H_{12}O_6 + 6O_2 \uparrow$$

FIGURE 26.7

Overview of photosynthesis. **(a)** The equation for photosynthesis. Note that it does not indicate the order of reactions but merely summarizes the initial reactants and final products. **(b)** The general reactions of photosynthesis, divided into two phases called light reactions and dark reactions. The light reactions require light to activate chlorophyll pigment and use the energy given off during activation to split an H_2O molecule into oxygen and hydrogen, producing ATP and NADPH. The dark reactions, which occur either with or without light, utilize ATP and NADPH produced during the light reactions to fix CO_2 into organic compounds such as glucose.

tosynthetic organisms achieve this remarkable feat through a series of reactions, given in the following summary:

- trapping visible light rays by means of special pigments attached to intracellular membranes;
- excitement of atoms in the pigment molecule and release of electrons;
- splitting water to release oxygen and hydrogen ions;
- producing ATP with energy extracted from electrons and hydrogen ions; and
- fixing CO_2 into carbohydrates using ATP.

Photosynthesis proceeds in two phases: the **light-dependent reactions,** which proceed only in the presence of sunlight, and the **dark** (light-independent) **reactions,** which proceed regardless of the lighting conditions (light or dark). The summary equation and overall reactions are depicted in figure 26.7b.

MICROBITS   26.1
Greenhouse Gases, Fossil Fuels, Cows, Termites, and Global Warming

The sun's radiant energy does more than drive photosynthesis; it also helps maintain the stability of the earth's temperature and climatic conditions. As radiation impinges on the earth's surface, much of it is absorbed, but a large amount of the infrared (heat) radiation bounces back into the upper levels of the atmosphere. For billions of years, the atmosphere has been insulated by a layer of gases (primarily CO_2, CH_4, water vapor, and nitrous oxide, N_2O) formed by natural processes such as respiration, decomposition, and biogeochemical cycles. This layer traps a certain amount of the reflected heat yet also allows some of it to escape into space. As long as the amounts of heat entering and leaving are balanced, the mean temperature of the earth will not rise or fall in an erratic or life-threatening way. Although this phenomenon, called the **greenhouse effect,** is popularly viewed in a negative light, it must be emphasized that its function for eons has been primarily to foster life.

The greenhouse effect has recently been a matter of concern because *greenhouse gases* appear to be increasing at a rate that could disrupt the temperature balance. In effect, a denser insulation layer will trap more heat energy and gradually heat the earth. The amount of CO_2 released collectively by respiration, anaerobic microbial activity, fuel combustion, and volcanic activity has increased more than 30% since the beginning of the industrial era. By far the greatest increase in CO_2 production results from human activities such as combustion of fossil fuels, burning forests to clear agricultural land, and manufacturing. Deforestation has the added impact of removing large areas of photosynthesizing plants that would otherwise consume some of the CO_2.

Originally, experts on the greenhouse effect were concerned primarily about increasing CO_2 levels, but it now appears that the other greenhouse gases combined may have a slightly greater contribution than CO_2, and they, too, are increasing. One of these gases, methane (CH_4), released from the gastrointestinal tract of ruminant animals such as cattle, goats, and sheep, has doubled over the past century. The gut of termites also harbors wood-digesting and methane-producing bacteria. Even the human intestinal tract can support methanogens. Other greenhouse gases such as nitrous oxide and sulfur dioxide (SO_2) are also increasing through automobile and industrial pollution.

There is not yet complete agreement as to the extent and effects of global warming. It has been documented that the mean temperature of the earth has increased by 1.6°C since 1860. If the rate of increase continues, by 2020 a rise in the average temperature of 4° to 5°C will begin to melt the polar ice caps and raise the levels of the ocean 2 to 3 feet. Some experts predict more serious effects, including massive flooding of coastal regions, changes in rainfall patterns, expansion of deserts, and long-term climatic disruptions. Early warning signs of global warming are appearing in the Antarctic, where the landmass is breaking up at an increased rate, and in the mass melting of glaciers in many other parts of the world.

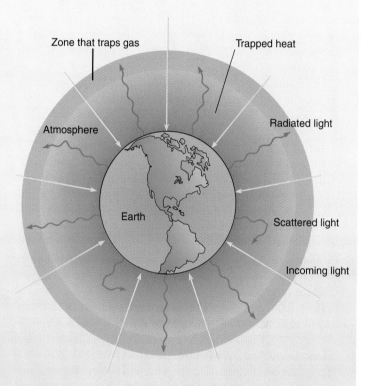

The greenhouse effect. Some of the light radiated or reflected from the earth's surface is trapped in a dense layer of atmospheric gases. Heat radiation trapped in this zone helps maintain the temperature of the atmosphere. Any increase in the density of this gas layer could lead to increased entrapment of heat and a gradual rise in the ambient temperature of the earth.

The Functions of Green Machines   Solar energy is delivered in discrete energy packets called **photons** (also called quanta) that travel as waves. The wavelengths of light operating in photosynthesis occur in the visible spectrum between 400 nm (violet) and 700 nm (red). As this light strikes photosynthetic pigments, some wavelengths are absorbed, some pass through, and some are reflected. The activity that has greatest impact on photosynthesis is the absorbance of light by photosynthetic pigments. These include the **chlorophylls,** which are green; **carotenoids,** which are yellow, orange, or red; and **phycobilins,** which are red or blue-green.[2] By far the most important of these pigments are the chlorophylls, which contain a *photocenter* that consists of a magnesium atom held in the center of a complex ringed structure called a *porphyrin.* As we will see, the chlorophyll molecule harvests the energy of photons and converts it to electron (chemical) energy. The accessory photosynthetic pigments such as carotenes trap light energy and shuttle it to chlorophyll.

---

2. The color of the pigment corresponds to the wavelength of light it reflects.

The detailed biochemistry of photosynthesis is beyond the scope of this text, but we will provide an overview to account for the primary process as it occurs in green plants, algae, and cyanobacteria (figure 26.8). Many of the basic activities (electron transport and phosphorylation) are biochemically similar to certain pathways of respiration previously covered in chapter 8.

Light Reactions The same systems that carry the photosynthetic pigments are also the sites for the light reactions. They occur in the *thylakoid* membranes of the granum in chloroplasts (figure 26.8*a*) and in specialized lamellae of the cell membranes in procaryotes (see chapter 4). These systems exist as two separate complexes called *photosystem I* (P700) and *photosystem II* (P680)[3] (figure 26.8*c*). Both systems contain chlorophyll and they are simultaneously activated by light, but the reactions in photosystem II help drive photosystem I. Together the systems form a powerful scheme to receive light, transport electrons, pump hydrogen ions, and form ATP and NADPH.

When photons enter the photocenter of the P680 system (PS II), the magnesium atom in chlorophyll becomes excited and releases 2 electrons. The loss of electrons from the photocenter has two major effects: (1) It creates a vacancy in the chlorophyll molecule forceful enough to split an H_2O molecule into hydrogens (H^+) (electrons and hydrogen ions) and oxygen (O_2). This splitting of water, termed **photolysis,** is the ultimate source of the O_2 gas that is an important product of photosynthesis. The electrons released from the lysed water regenerate the photosystem for its next reaction with light. (2) Electrons generated by the first photoevent are immediately boosted through a series of carriers (cytochromes) to the P700 system. At this same time, hydrogen ions accumulate in the internal space of the thylakoid complex, thereby producing an electrochemical gradient.

The P700 (PS I) system has been activated by light so that it is ready to accept electrons generated by the PS II. The electrons it receives are passed along a second transport chain to a complex that uses electrons and hydrogen ions to reduce NADP to NADPH. (Recall that reduction in this sense entails the addition of electrons and hydrogens to a substrate.) As noted in chapter 8, a reduced coenzyme such as NADPH is another source of chemical energy and can be used in synthesis.

A second energy reaction involves synthesis of ATP by a chemiosmotic mechanism similar to that shown in figure 8.23. Channels in the thylakoids of the granum actively pump H^+ into the inner chamber, producing a charge gradient. ATP synthase located in this same thylakoid uses the energy from H^+ transport to phosphorylate ADP to ATP. Because it occurs in light, this process is termed **photophosphorylation.** Both NADPH and ATP are released into the stroma of the chloroplast, where they participate in the Calvin cycle.

Dark Reactions The subsequent photosynthetic reactions that do not require light occur in the chloroplast stroma or the cytoplasm of cyanobacteria. These dark reactions use energy produced by the light phase to synthesize glucose by means of a cyclic scheme called the *Calvin cycle* (figure 26.9). The cycle begins at the point where CO_2 is combined with a doubly phosphorylated 5-carbon acceptor molecule called ribulose-1,5-bisphosphate (RuBP). This process, called **carbon fixation,** generates a 6C intermediate compound that immediately splits into two, 3-carbon molecules of 3-phosphoglyceric acid (PGA). The subsequent steps use the ATP and NADPH generated by the photosystems to form high-energy intermediates. First, ATP adds a second phosphate to 3-PGA and produces 1,3-bisphosphoglyceric acid (BPG). Then, during the same step, NADPH contributes its hydrogen to BPG, and one high-energy phosphate is removed. These events give rise to glyceraldehyde-3-phosphate (PGAL). This molecule and its isomer dihydroxyacetone phosphate (DHAP) are key molecules in hexose synthesis leading to fructose and glucose. You may notice that this pathway is very similar to glycolysis, except that it runs in reverse (see figure 8.19). Bringing the cycle back to regenerate RuBP requires PGAL and several steps not depicted in figure 26.9.

Other Mechanisms of Photosynthesis The **oxygenic,** or oxygen-releasing, photosynthesis that occurs in plants, algae, and cyanobacteria is the dominant type on the earth. Other photosynthesizers such as green and purple bacteria possess bacteriochlorophyll, which is more versatile in capturing light. They have only a cyclic photosystem I, which routes the electrons from the photocenter to the electron carriers and back to the photosystem again. This pathway generates a relatively small amount of ATP, and it may not produce NADPH. As photolithotrophs, these bacteria use H_2, H_2S, or elemental sulfur rather than H_2O as a source of electrons and reducing power. As a consequence, they are **anoxygenic** (non-oxygen-producing), and many are strict anaerobes. Bacteria that use sulfur compounds such as H_2S deposit sulfur residues either intracellularly (*Chromatium,* figure 4.35) or extracellularly, depending on the species. The sulfur bacteria are typically isolated from oxygen-deficient environments, such as swamps or the bottom of shallow ponds and lakes, where they play a central role in the biogeochemical cycle of sulfur.

The Nitrogen Cycle

Nitrogen (N_2) gas is the most abundant component of the atmosphere, accounting for nearly 79% of air volume. As we will see, this extensive reservoir in the air is largely unavailable to most organisms. Only about 0.03% of the earth's nitrogen is combined (or fixed) in some other form such as nitrates (NO_3), nitrites (NO_2), ammonium ion (NH_4^+), and organic nitrogen compounds (proteins, nucleic acids).

The **nitrogen cycle** is relatively more intricate than other cycles because it involves such a diversity of specialized microbes to maintain the flow of the cycle. In many ways, it is actually more of a nitrogen "web" because of the array of adaptations that occur. Higher plants can utilize NO_3^- and NH_4^+; animals must receive nitrogen in organic form from plants or other animals; and microorganisms vary in their source, using NO_2^-, NO_3^-, NH_4^+, N_2, and organic nitrogen. The cycle includes four basic types of reactions: nitrogen fixation, ammonification, nitrification, and denitrification (figure 26.10).

---

3. The numbers refer to the wavelength of light to which each system is most sensitive.

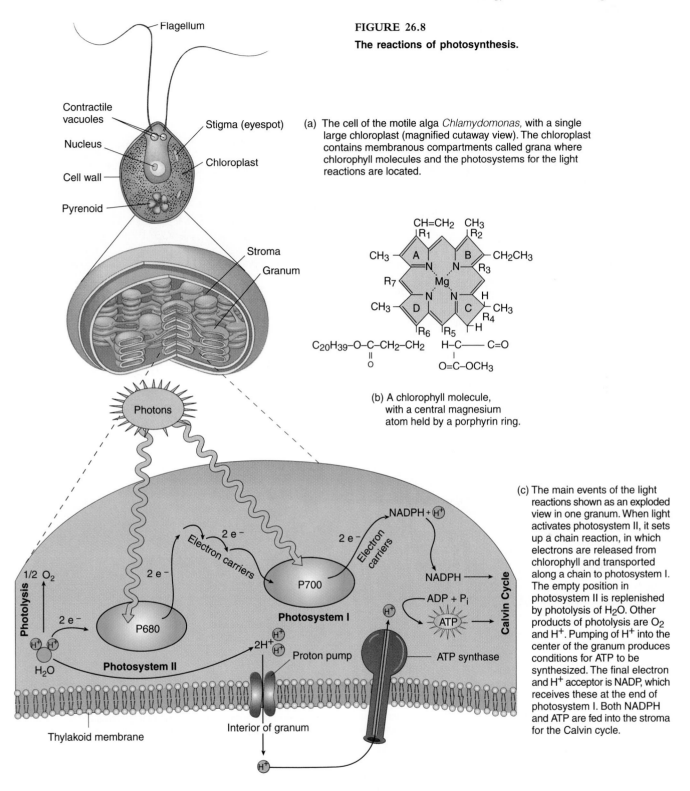

FIGURE 26.8

The reactions of photosynthesis.

(a) The cell of the motile alga *Chlamydomonas,* with a single large chloroplast (magnified cutaway view). The chloroplast contains membranous compartments called grana where chlorophyll molecules and the photosystems for the light reactions are located.

(b) A chlorophyll molecule, with a central magnesium atom held by a porphyrin ring.

(c) The main events of the light reactions shown as an exploded view in one granum. When light activates photosystem II, it sets up a chain reaction, in which electrons are released from chlorophyll and transported along a chain to photosystem I. The empty position in photosystem II is replenished by photolysis of H_2O. Other products of photolysis are O_2 and H^+. Pumping of H^+ into the center of the granum produces conditions for ATP to be synthesized. The final electron and H^+ acceptor is NADP, which receives these at the end of photosystem I. Both NADPH and ATP are fed into the stroma for the Calvin cycle.

Nitrogen Fixation   The biosphere is most dependent upon the only process that can remove N_2 from the air and convert it to a form usable by living things. This process, called **nitrogen fixation,** is the beginning step in the synthesis of virtually all nitrogenous compounds. Nitrogen fixation is brought about primarily by nitrogen-fixing bacteria in soil and water, though a small amount is formed through nonliving processes such as lightning. The nitrogen fixers have developed a unique enzyme system capable of breaking the triple bonds of the N_2 molecule and reducing the N atoms, an anaerobic process that requires the expenditure of considerable ATP. The primary product of nitrogen fixation is the ammonium ion, NH_4^+. Nitrogen-fixing bacteria live free or in a symbiotic relationship with plants. Among the common free-living nitrogen fixers are the aerobic *Azotobacter* and *Azospirillum* and certain members of the anaerobic genus *Clostridium.* Other free-living nitrogen fixers are the cyanobacteria *Anabaena* and *Nostoc.*

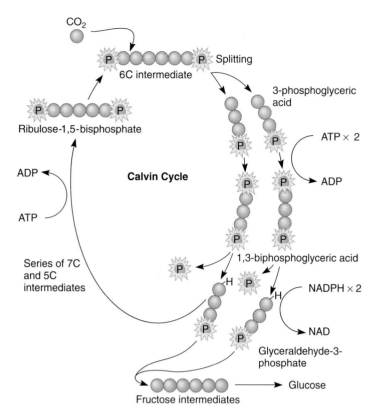

FIGURE 26.9

The Calvin cycle. The main events of the reactions in photosynthesis that do not require light. It is during this cycle that carbon is fixed into organic form using the energy (ATP and NADPH) released by the light reactions. The end product, glucose, can be stored as complex carbohydrates, or it can be used in various amphibolic pathways to produce other carbohydrate intermediates or amino acids.

Root Nodules: Natural Fertilizer Factories A significant symbiotic association occurs between *rhizobia** (bacteria in the genera *Rhizobium, Bradyrhizobium,* and *Azorhizobium*) and *legumes** (plants such as soybeans, peas, alfalfa, and clover that characteristically produce seeds in pods). The infection of legume roots by these gram-negative, motile, rod-shaped bacteria causes the formation of special nitrogen-fixing organs, called **root nodules** (figure 26.11). Nodulation begins when rhizobia colonize specific sites on root hairs. From there, the bacteria invade deeper root cells and induce the cells to form tumorlike masses. The bacterium's enzyme system supplies a constant source of reduced nitrogen to the plant, and the plant furnishes nutrients and energy for the activities of the bacterium. The legume uses the NH_4^+ to aminate (add an amino group to) various carbohydrate intermediates and thereby synthesize amino acids and other nitrogenous compounds that are used in plant and animal synthesis (see figure 8.28).

Plant–bacteria associations have great practical importance in agriculture, because an available source of nitrogen is often a limiting factor in the growth of crops. The self-fertilizing nature of

* rhizobia (ry-zoh′-bee-uh) Gr. *rhiza,* root, and *bios,* to live.

* legume (leg′-yoom) L. *legere,* to gather.

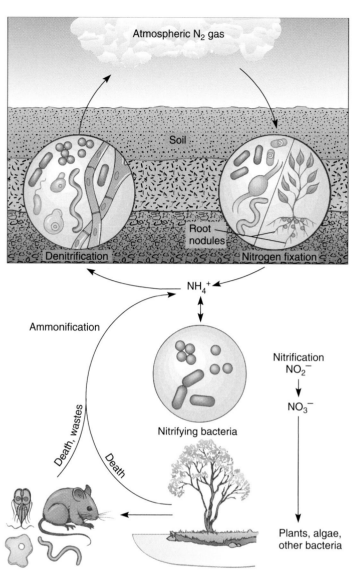

FIGURE 26.10

The simplified events in the nitrogen cycle. In nitrogen fixation, gaseous nitrogen (N_2) is acted on by nitrogen-fixing bacteria, which give off ammonia (NH_4^+). Ammonia is converted to nitrite (NO_2^-) and nitrate (NO_3^-) by nitrifying bacteria in nitrification. Plants, algae, and bacteria use nitrates to synthesize nitrogenous organic compounds (proteins, amino acids, nucleic acids). Organic nitrogen compounds are used by animals and other consumers. In ammonification, nitrogenous macromolecules from wastes and dead organisms are converted to NH_4^+ by ammonifying bacteria. NH_4^+ can be either directly recycled into nitrates or returned to the atmospheric N_2 form by denitrifying bacteria (denitrification).

legumes makes them valuable food plants in areas with poor soils and in countries with limited resources. It has been shown that crop health and yields can be improved by inoculating legume seeds with pure cultures of rhizobia, because the soil is often deficient in the proper strain for forming nodules (figure 26.12).

Ammonification, Nitrification, and Denitrification In another part of the nitrogen cycle, nitrogen-containing organic matter

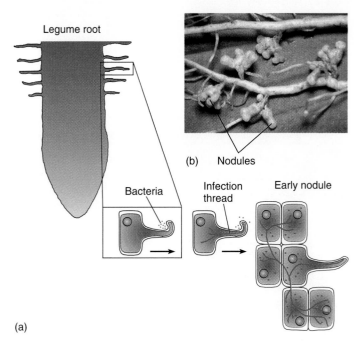

(b) Nodules

(a)

FIGURE 26.11

Nitrogen fixation through symbiosis. **(a)** Events leading to formation of root nodules. Cells of the bacterium *Rhizobium* attach to a legume root hair and cause it to curl. Invasion of the legume root proper by *Rhizobium* initiates the formation of an infection thread that spreads into numerous adjacent cells. The presence of bacteria in cells causes nodule formation. **(b)** Mature nodules that have developed in a sweet clover plant.

is decomposed by various bacteria (*Clostridium, Proteus,* for example) that live in the soil and water. Organic detritus consists of large amounts of protein and nucleic acids from dead organisms and nitrogenous animal wastes such as urea and uric acid. The decomposition of these substances splits off amino groups and produces NH_4^+. This process is thus known as **ammonification.** The ammonium released can be reused by certain plants or converted to other nitrogen compounds, as discussed next.

The oxidation of NH_4^+ to NO_2^- and NO_3^- is a process called **nitrification.** It is an essential conversion process for generating the form of nitrogen (NO_3) that is most useful to living things. This reaction occurs in two phases and involves lithotrophic bacteria in soil and water. In the first phase, certain gram-negative genera such as *Nitrosomonas* and *Nitrosococcus* oxidize NH_4^+ to NO_2^- as a means of generating energy. Nitrite is rapidly acted upon by a second group of nitrifiers, including *Nitrobacter* and *Nitrococcus,* which perform the final oxidation of NO_2^- to NO_3^-. Nitrates can be assimilated into protoplasm by a variety of organisms (plants, fungi, and bacteria); a large number of bacteria also use it as a source of oxygen.

The nitrogen cycle is complete when nitrogen compounds are returned to the reservoir in the air by a reaction series that converts NO_3^- through intermediate steps to atmospheric nitrogen. The first step, which involves the reduction of nitrate to nitrite, is so common that hundreds of different bacterial species can do it. Other bacteria such as *Bacillus, Pseudomonas, Spirillum,* and *Thiobacillus* can carry out this **denitrification process** to completion as follows:

(a)   (b)

FIGURE 26.12

Inoculating legume seeds with *Rhizobium* bacteria increases the plant's capacity to fix nitrogen. The legumes in **(a)** were inoculated and are healthy. The poor growth and yellowish color of the uninoculated legumes in **(b)** indicate a lack of nitrogen.

$$NO_3^- \rightarrow NO_2^- \rightarrow NO \rightarrow N_2O \rightarrow N_2$$

SEDIMENTARY CYCLES

The Sulfur Cycle

The sulfur cycle resembles the nitrogen cycle except that sulfur originates from natural sedimentary deposits in rocks, oceans, lakes, and swamps rather than from the atmosphere. Sulfur exists in the elemental form (S) and as hydrogen sulfide gas (H_2S), sulfate (SO_4), and thiosulfate (S_2O_3). Most of the oxidations and reductions that convert one form of inorganic sulfur compound to another are accomplished by bacteria. Plants and many microorganisms can assimilate only SO_4, and animals must have an organic source. Organic sulfur occurs in the amino acids cystine, cysteine, and methionine, which contain sulfhydryl (—SH) groups and form disulfide (S—S) bonds that contribute to the stability and configuration of proteins.

One of the most remarkable contributors to the cycling of sulfur in the biosphere is the genus *Thiobacillus.** These gram-negative, motile rods flourish in mud, sewage, bogs, mining drainage, and brackish springs that can be inhospitable to organisms that require complex organic nutrients. But the metabolism of these specialized lithotrophic bacteria is adapted to extracting energy by oxidizing elemental sulfur, sulfides, and thiosulfate. One species, *T. thiooxidans,* is so efficient at this process that it secretes large amounts of sulfuric acid into its environment, as shown by the following equation:

$$Na_2S_2O_3 + H_2O + O_2 \rightarrow Na_2SO_4 + H_2SO_4 \text{ (sulfuric acid)} + 4S$$

The marvel of this bacterium is its ability to create and survive in the most acidic habitats on the earth. It also plays an essential

* *Thiobacillus* (thigh''-oh-bah-sil'-us) Gr. *theion,* sulfur. A type of non-photosynthetic sulfur bacteria.

part in the phosphorus cycle, and its relative, *T. ferrooxidans,* participates in the cycling of iron. Other bacteria that can oxidize sulfur to sulfates are the photosynthetic sulfur bacteria mentioned in the section on photosynthesis.

The sulfates formed from oxidation of sulfurous compounds are assimilated into protoplasm by a wide variety of organisms. The sulfur cycle reaches completion when inorganic and organic sulfur compounds are reduced. Bacteria in the genera *Desulfovibrio* and *Desulfuromonas* anaerobically reduce sulfates to hydrogen sulfide and water as the final step in electron transport. Sites in mud and moist soil where they live usually emanate a strong, rotten-egg stench from H_2S and have a blackened film of iron on the surface. Organic sulfur in amino acids can also be reduced to H_2S by a number of saprobic bacteria.

The Phosphorus Cycle

Phosphorus is an integral component of DNA, RNA, and ATP, and all life depends upon a constant supply of it. It cycles between the abiotic and biotic environments almost exclusively as inorganic phosphate (PO_4) rather than its elemental form (figure 26.13). The chief inorganic reservoir is phosphate rock, which contains the insoluble compound $Ca_5(PO_4)_3F$. Before it can enter biological systems, this mineral must be *phosphatized*—converted into more soluble PO_4^{-3} by the action of acid. Phosphate is released naturally when the sulfuric acid produced by *Thiobacillus* dissolves phosphate rock. Soluble phosphate in the soil and water is the principal source for autotrophs, which fix it onto organic molecules and pass it on to heterotrophs in this form. Organic phosphate is returned to the pool of soluble phosphate by decomposers, and it is finally cycled back to the mineral reservoir by slow geologic processes such as sedimentation. Because the low phosphate content of many soils can limit productivity, phosphate is added to soil to increase agricultural yields. The excess run-off of fertilizer into the hydrosphere is often responsible for overgrowth of aquatic pests (see eutrophication in a subsequent section on aquatic habitats).

The involvement of microbes in cycling elements and compounds has both negative and positive effects. For instance, certain species participate in creating toxic compounds (Spotlight on Microbiology 26.2); while others can be used to clean up pesticides (see Spotlight on Microbiology 26.3) and mine minerals.

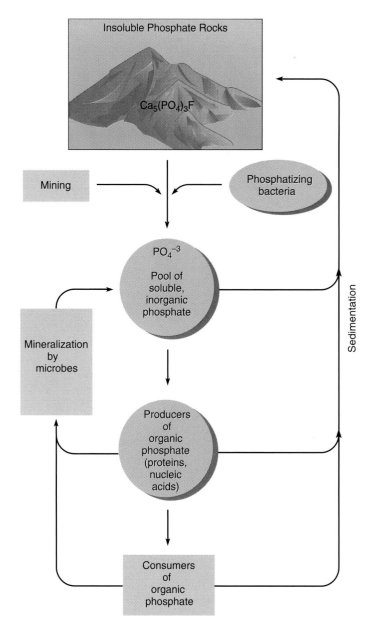

FIGURE 26.13

The phosphorus cycle. The pool of phosphate existing in sedimentary rocks is released into the ecosystem either naturally by erosion and microbial action or artificially by mining and the use of phosphate fertilizers. Soluble phosphate (PO_4^{-3}) is cycled through producers, consumers, and decomposers back into the soluble pool of phosphate, or it is returned to sediment in the aquatic biosphere.

CHAPTER CHECKPOINTS

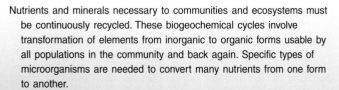
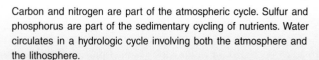

Nutrients and minerals necessary to communities and ecosystems must be continuously recycled. These biogeochemical cycles involve transformation of elements from inorganic to organic forms usable by all populations in the community and back again. Specific types of microorganisms are needed to convert many nutrients from one form to another.

Elements of critical importance to all ecosystems that cycle through various forms are carbon, nitrogen, sulfur, phosphorus, and water.

Carbon and nitrogen are part of the atmospheric cycle. Sulfur and phosphorus are part of the sedimentary cycling of nutrients. Water circulates in a hydrologic cycle involving both the atmosphere and the lithosphere.

The sun is the primary energy source for most surface ecosystems. Photosynthesis captures this energy and utilizes it for carbon fixation by producer populations. Producers include plants, algae, cyanobacteria, and certain bacterial species.

SPOTLIGHT ON MICROBIOLOGY 26.2
Biogeochemical Cycles and Bioamplification of Pollutants

Understanding the action of biogeochemical cycles takes on increased significance when we examine the increasing pollution of the air, soil, and water by humans. The scale of human activity can escalate the natural cycling rate of such toxic elements as arsenic, chromium, lead, and mercury. Many of the hundreds of thousands of synthetic chemicals introduced into the environment over the past hundred years can also get caught up in cycles. Some of these chemicals will be converted into less harmful substances, but others, such as DDT and heavy metals, persist and flow along with nutrients into all levels of the biosphere. If such a pollutant accumulates in living tissue and is not excreted, it can be **bioamplified,** or concentrated and accumulated by living things, through the natural trophic flow of the ecosystem. Microscopic producers such as bacteria and algae begin the accumulation process. With each new level of the food chain, the consumers gather an increasing amount of the chemical, until the top consumers can contain toxic levels.

The long-term effects of *bioaccumulation* are best demonstrated using the heavy metal mercury as a model. Mercury compounds are used in household antiseptics and disinfectants, agriculture, and industry. Elemental mercury precipitates proteins by attaching to functional groups, and it interferes with the bonding of uracil and thymine. But the elemental form is not as toxic as the organic mercurials such as ethyl or methyl mercury.

The severity of this toxicity was first demonstrated by an incident in Minamata Bay, Japan. A local plastics manufacturer had drained large quantities of mercury into the bay. In the aquatic sediments, certain mercury-resistant bacteria *(Desulfovibrio)* metabolized the mercury compounds to the highly toxic organic forms. Motile plankton that fed upon these bacteria accumulated the mercury and passed it on to the next consumer level until eventually, fish and shellfish had accumulated high levels of methyl mercury. These foods were a regular part of the diet of local Japanese fishermen and their families. Many adults and older children in the community suffered nerve and brain damage, and fetuses developed severe birth defects. Of the 130 people who were poisoned, 47 died. Since then, no other large-scale cases of toxicity have been traced to eating fish contaminated with methyl mercury. However, recent studies have disclosed increased mercury content in fish taken from oceans and freshwater lakes in North America, and even in canned tuna.

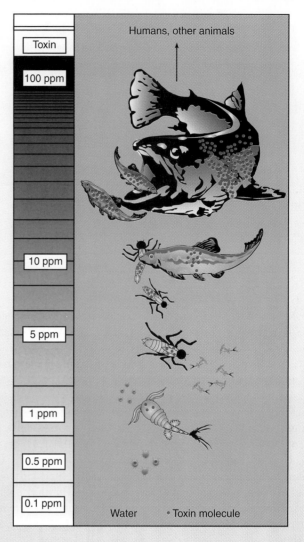

The mechanism for bioamplification of toxic compounds in the food chain.

SOIL MICROBIOLOGY: THE COMPOSITION OF THE LITHOSPHERE

Descriptions such as "soiled" or "dirty" may suggest that soil is an undesirable, possibly harmful substance. To some people, soil can also appear to be a somewhat homogenous, inert substance. At the microscopic level, however, soil is a dynamic ecosystem that supports complex interactions between numerous geologic, chemical, and biological factors. This rich region, called the lithosphere, teems with microbes, serves a dynamic role in biogeochemical cycles, and is an important repository for organic detritus and dead terrestrial organisms.

The abiotic portion of soil is a composite of mineral particles, water, and atmospheric gas. The development of soil begins when

geologic sediments are mechanically disturbed and exposed to weather and microbial action. A mixture of microbes commonly involved in the early colonization and breakdown of rocks are unusual organisms called **lichens*** (figure 26.14). Lichens are complex symbiotic associations between a fungus and a cyanobacterium or green alga. The alga or cyanobacterium feeds itself and the fungus through photosynthesis and could actually live a solitary existence. The fungus is largely dependent upon the algae or cyanobacteria for its survival, although it does cooperate in absorbing water and minerals for the entire lichen complex. Most lichens that are early colonists or pioneers of rock are of the compact crustose form.

* lichens (ly′-kenz) Gr. *leichen,* a tree moss.

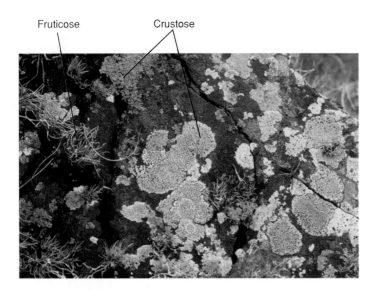

FIGURE 26.14

Rocks covered with patches of crustose and fruticose lichens.
Crustose lichens are flat and spreading, whereas fruticose lichens are branching and shrubby. Such microbial associations begin the process of soil formation by invading and breaking up the rocks into smaller particles.

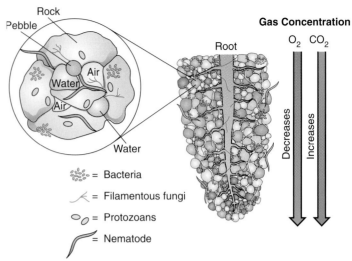

FIGURE 26.15

The structure of the rhizosphere and the microhabitats that develop in response to soil particles, moisture, air, and gas content.

Rock decomposition releases various-sized particles ranging from rocks, pebbles, and sand grains to microscopic morsels that lie in a loose aggregate (figure 26.15). The porous structure of soil creates various-sized pockets or spaces that provide numerous microhabitats. Some spaces trap moisture and form a liquid phase in which mineral ions and other nutrients are dissolved. Other spaces trap air that will provide gases to soil microbes, plants, and animals. Because both water and air compete for these pockets, the water content of soil is directly related to its oxygen content. Water-saturated soils contain less oxygen, and dry soils have more. Gas tensions in soil can also vary vertically. In general, the concentration of O_2 decreases and that of CO_2 increases with the depth of soil. Aerobic and facultative organisms tend to occupy looser, drier soils, whereas anaerobes are adapted to waterlogged, poorly aerated soils.

Within the superstructure of the soil are varying amounts of **humus,*** the slowly decaying organic litter from plant and animal tissues. This soft, crumbly mixture holds water like a sponge. It is also an important habitat for microbes that decompose the complex litter and gradually recycle nutrients. The humus content varies with climate, temperature, moisture and mineral content, and microbial action. Warm, tropical soils have a high rate of humus production and microbial decomposition. Because nutrients in these soils are swiftly released and used up, they do not accumulate. Fertilized agricultural soils in temperate climates build up humus at a high rate and are rich in nutrients. The very low content of humus and moisture in desert soils greatly reduces its microbial flora, rate of decomposition, and nutrient content. Bogs are likewise nutrient-poor due to a slow rate of decomposition of the humus caused by high acid content and lack of oxygen. Humans can artificially increase the amount of humus by mixing plant refuse and animal

wastes with soil and allowing natural decomposition to occur, a process called *composting.*

Living Activities in Soil

The rich culture medium of the soil supports a fantastic array of microorganisms (bacteria, fungi, algae, protozoa, and viruses). A gram of moist loam soil with high humus content can have a count as high as 10 billion, each competing for its own niche and microhabitat. Some of the most distinctive biological interactions occur in the **rhizosphere,** the zone of soil surrounding the roots of plants that contains associated bacteria, fungi, and protozoa (figure 26.15). Plants interact with soil microbes in a truly synergistic fashion. Studies have shown that a rich microbial community grows in a *biofilm* around the root hairs and other exposed surfaces. Their presence stimulates the plant to exude growth factors such as carbon dioxide, sugars, amino acids, and vitamins. These nutrients are released into fluid spaces, where they can be readily captured by microbes. Bacteria and fungi likewise contribute to plant survival by releasing hormonelike growth factors and protective substances. They are also important in converting minerals into forms usable by plants. We saw numerous examples in the nitrogen, sulfur, and phosphorus cycles.

We previously observed that plants can form close symbiotic associations with microbes to fix nitrogen. Other mutalistic partnerships between plant roots and microbes are **mycorrhizae.*** These associations occur when various species of basidiomycetes, ascomycetes, or zygomycetes attach themselves to the roots of vascular plants (figure 26.16). The plant feeds the fungus through photosynthesis, and the fungus sustains the relationship in several ways. By extending its mycelium into the rhizosphere, it helps anchor the plant and increases the surface area for capturing water

---

* humus (hyoo'-mus) L., earth.

* mycorrhizae (my"-koh-ry'-zee) Gr. *mykos*, fungus, and *rhiza*, root.

FIGURE 26.16

Mycorrhizae, symbiotic associations between fungi and plant roots, favor the absorption of water and minerals from the soil.

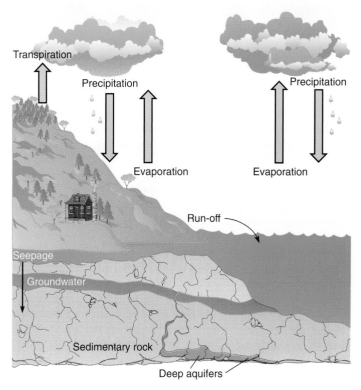

FIGURE 26.17

The hydrologic cycle.   The largest proportion of water cycles through evaporation, transpiration, and precipitation between the hydrosphere and the atmosphere. Other reservoirs of water exist in the groundwater or deep storage aquifers in sedimentary rocks. Plants add to this cycle by releasing water through transpiration, and heterotrophs release it through respiration.

from dry soils and minerals from poor soils. Plants with mycorrhizae can inhabit severe habitats more successfully than plants without them.

The topsoil, which extends a few inches to a few feet from the surface, supports a host of burrowing animals such as nematodes, termites, and earthworms. Many of these animals are decomposer-reducer organisms that break down organic nutrients through digestion and also mechanically reduce or fragment the size of particles so that they are more readily mineralized by microbes. Aerobic bacteria initiate the digestion of organic matter into carbon dioxide and water and generate minerals such as sulfate, phosphate, and nitrate, which can be further degraded by anaerobic bacteria. Fungal enzymes increase the efficiency of soil decomposition by hydrolyzing complex natural substances such as cellulose, keratin, lignin, chitin, and paraffin.

The soil is also a repository for agricultural, industrial, and domestic wastes such as insecticides, herbicides, fungicides, manufacturing wastes, and household chemicals. Spotlight on Microbiology 26.3 explores the problem of soil contamination and the feasibility of harnessing indigenous soil microbes to break down undesirable hydrocarbons and pesticides.

AQUATIC MICROBIOLOGY

Water is the dominant compound on the earth. It occupies nearly three-fourths of the earth's surface. In the same manner as minerals, the earth's supply of water is continuously cycled between the hydrosphere, atmosphere, and lithosphere (figure 26.17). The **hydrologic cycle** begins when surface water (lakes, oceans, rivers) exposed to the sun and wind evaporates and enters the vapor phase of the atmosphere. Living things contribute to this reservoir by various activities. Plants lose moisture through transpiration (evaporation through leaves), and all aerobic organisms give off water during respiration. Airborne moisture accumulates in the atmosphere, most conspicuously as clouds.

Water is returned to the earth through condensation or precipitation (rain, snow). The largest proportion of precipitation falls back into surface waters, where it circulates rapidly between running water and standing water. Only about 2% of water seeps into the earth or is bound in ice, but these are very important reservoirs. Surface water collects in extensive subterranean pockets produced by the underlying layers of rock, gravel, and sand. This process forms a deep **groundwater** source called an **aquifer.** The water in aquifers circulates very slowly and is an important replenishing source for surface water. It can resurface through springs, geysers, and hot vents, and it is also tapped as the primary supply for one-fourth of all water used by humans.

Although the total amount of water in the hydrologic cycle has not changed over millions of years, its distribution and quality have been greatly altered by human activities. Two serious problems have arisen with aquifers. First, as a result of increased well-drilling, land development, and persistent local droughts, the aquifers in many areas have not been replenished as rapidly as they have been depleted. As these reserves are used up, humans will have to rely on other delivery systems such as pipelines, dams, and reservoirs, which can further disrupt the cycling of water. Second, because water picks up materials when falling through air or percolating through the ground, aquifers are also important collection points for pollutants. As we will see, the proper management of water resources is one of the greatest challenges of the new century.

SPOTLIGHT ON MICROBIOLOGY 26.3
Bioremediation: The Pollution Solution?

The soil and water of the earth have long been considered convenient repositories for solid and liquid wastes. Humans have been burying solid wastes for thousands of years, but the process has escalated in the past 50 years. Every year, about 300 metric tons of pollutants, industrial wastes, and garbage are deposited into the natural environment. Often this dumping is done with the mistaken idea that naturally occurring microbes will eventually biodegrade (break down) waste material.

Landfills currently serve as a final resting place for hundreds of castoffs from an affluent society, including yard wastes, paper, glass, plastics, wood, textiles, rubber, metal, paints, and solvents. This conglomeration is dumped into holes and is covered with soil. Although it is true that many substances are readily biodegradable, materials such as plastics and glass are not. Successful biodegradation also requires a compost containing specific types of microorganisms, adequate moisture, and oxygen. The environment surrounding buried trash provides none of these conditions. Large, dry, anaerobic masses of plant materials, paper, and other organic materials will not be successfully attacked by the aerobic microorganisms that dominate in biodegradation. As we continue to fill up hillsides with waste, the future of these landfills is a prime concern. One of the most serious of these concerns is that they will be a source of toxic compounds that seep into the ground and water.

Pollution of groundwater, the primary source of drinking water for 100 million people in the United States, is an increasing problem. Because of the extensive cycling of water through the hydrosphere and lithosphere, groundwater is often the final collection point for hazardous chemicals released into lakes, streams, oceans, and even garbage dumps. Many of these chemicals are pesticide residues from agriculture (dioxin, selenium, 2,4-D), industrial hydrocarbon wastes (PCBs), and hydrocarbon solvents (benzene, toluene). They are often hard to detect, and, if detected, are hard to remove.

For many years, polluted soil and water were simply sealed off or dredged and dumped in a different site, with no attempt to get rid of the pollutant. But now, with greater awareness of toxic wastes, many Americans are adopting an attitude known as NIMBY (not in my backyard!), and environmentalists are troubled by the long-term effects of contaminating the earth.

In a search for solutions, waste management has turned to **bioremediation**—using microbes to break down or remove toxic wastes in water and soil. Some of these waste-eating microbes are natural soil and water residents with a surprising capacity to decompose even artificial substances. Because the natural, unaided process occurs too slowly, most cleanups are accomplished by commercial bioremediation services that treat the contaminated soil with oxygen, nutrients, and water to increase the rate of microbial action. Through these actions, levels of pesticides such as 2,4-D can be reduced to 96% of their original levels, and solvents can be reduced from one million parts per billion (ppb) to 10 ppb or less (see figure 1.2e). Bacteria are also being used to help break up and digest oil spills.

Among the most important bioremedial microbes are species of *Pseudomonas* and *Bacillus* and various toxin-eating fungi. The current quest in bioremediation is for "super bugs" genetically engineered to convert toxic chemicals into CO_2 or other less harmful residues.

So far, about 35 recombinant microbes have been created for bioremediation. Species of *Rhodococcus* and *Burkholderia* have been engineered to decompose PCBs, and certain forms of *Pseudomonas* now contain genes for detoxifying heavy metals, carbon tetrachloride, and naphthalene. With over 3,000 toxic waste sites in the United States alone, the need for effective bioremediation is a top priority (see chapter opening photo).

(a)

(b)

Can microbes rescue us from our polluted world? **(a)** A Santa Monica, California shoreline receives a heavy run-off of raw sewage. **(b)** Solid wastes collect on a beach in the northeastern United States.

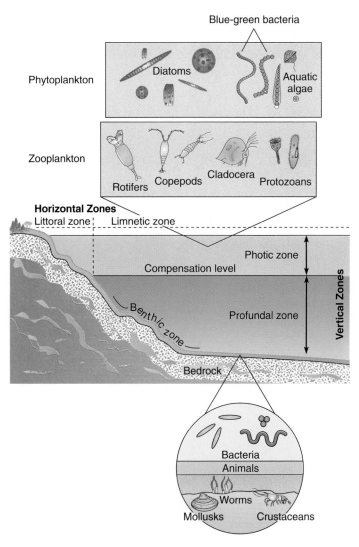

Blue-green bacteria

Phytoplankton

Diatoms

Aquatic algae

Zooplankton

Rotifers   Copepods   Cladocera   Protozoans

Horizontal Zones

Littoral zone | Limnetic zone

Photic zone

Compensation level

Benthic zone

Profundal zone

Vertical Zones

Bedrock

Bacteria
Animals
Worms
Mollusks   Crustaceans

FIGURE 26.18

Stratification in a freshwater lake.   Depth changes in this ecosystem create significant zones that differ in light penetration, temperature, and community structure.

The Structure of Aquatic Ecosystems

Surface waters such as ponds, lakes, oceans, and rivers differ to a considerable extent in size, geographic location, and physical and chemical character. Although an aquatic ecosystem is composed primarily of liquid, it is predictably structured, and it contains significant gradients or local differences in composition. Factors that contribute to the development of zones in aquatic systems are sunlight, temperature, aeration, and dissolved nutrient content. These variations create numerous macro- and microenvironments for communities of organisms. An example of zonation can be seen in the schematic section of a freshwater lake in figure 26.18.

A lake is stratified vertically into three zones, or strata. The uppermost region, called the **photic zone,** extends from the surface to the lowest limit of sunlight penetration. Its lower boundary (the compensation depth) is the greatest depth at which photosynthesis can occur. Directly beneath the photic zone lies the **profundal* zone,** which extends from the edge of the photic zone to the lake sediment. The sediment itself, or **benthic* zone,** is composed of or-

ganic debris and mud, and it lies directly on the bedrock that forms the lake basin. The horizontal zonation includes the shoreline, or **littoral* zone,** an area of relatively shallow water. The open, deeper water beyond the littoral zone is the **limnetic* zone.**

Marine Environments   The marine profile resembles that of a lake, although special characteristics set it apart. The ocean exhibits extreme variations in salinity, depth, temperature, hydrostatic pressure, and mixing. It contains a unique zone where the river meets the sea called an *estuary.* This region fluctuates in salinity, is very high in nutrients, and supports a specialized microbial community. It is often dominated by salt-tolerant species of *Pseudomonas* and *Vibrio.* Another important factor is the tidal and wave action that subjects the coastal habitat to alternate periods of submersion and exposure. The deep ocean, or **abyssal zone,** is poor in nutrients, chiefly because of the absence of sunlight for photosynthesis, and its tremendous depth (up to 10,000 m) makes it oxygen-poor and cold (average temperature 4°C). This zone supports communities with extreme adaptations, including halophilic, psychrophilic, barophilic, and anaerobic.

Aquatic Communities   The freshwater environment is a site of tremendous microbiological activity. Microbial distribution is associated with sunlight, temperature, oxygen levels, and nutrient availability. The photic zone, including littoral and limnetic, is the most productive self-sustaining region because it contains large numbers of **plankton,*** a floating microbial community that drifts with wave action and currents. A major member of this assemblage is the **phytoplankton,** containing a variety of photosynthetic algae and cyanobacteria (figure 26.18). The phytoplankton provide nutrition for **zooplankton,** microscopic consumers such as protozoa and invertebrates that filter feed, prey, or scavenge. The plankton supports numerous other trophic levels such as larger invertebrates and fish. With its high nutrient content, the benthic zone also supports an extensive variety and concentration of organisms, including aquatic plants, aerobic bacteria, and anaerobic bacteria actively involved in recycling organic detritus.

Larger bodies of standing water develop gradients in temperature or thermal stratification, especially during the summer (figure 26.19). The upper region, called the *epilimnion,* is warmest, and the deeper *hypolimnion* is cooler. Between these is a buffer zone, the *thermocline,* that ordinarily prevents the mixing of the two. Twice a year, during the warming cycle of spring and the cooling cycle of fall, temperature changes in the water column break down the thermocline and cause the water from the two strata to mix. Mixing disrupts the stratification and creates currents that bring nutrients up from the sediments. This process, called *upwelling,* is associated with increased activity by certain groups of microbes and is one explanation for the periodic emergence of *red tides* in oceans (figure 26.20) caused by toxin-producing dinoflagellates. A recent outbreak of fish and human disease on the eastern

---

* profundal (proh-fun′-dul) L. *pro,* before, and *fundus,* bottom.

* benthic (ben′-thik) Gr. *benthos,* depth of the sea.

* littoral (lit′-or-ul) L. *litus,* seashore.

* limnetic (lim-neh′-tik) Gr. *limne,* marsh.

* plankton (plang′-tun) Gr. *planktos,* wandering.

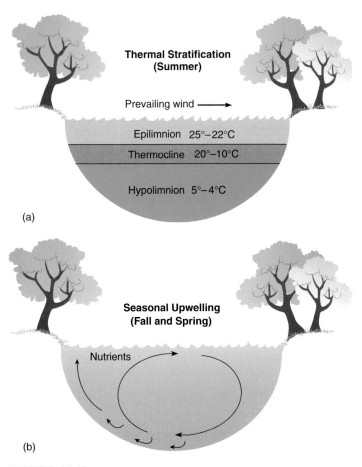

**Thermal Stratification
(Summer)**

Prevailing wind ⟶

Epilimnion 25°–22°C

Thermocline 20°–10°C

Hypolimnion 5°–4°C

(a)

**Seasonal Upwelling
(Fall and Spring)**

Nutrients

(b)

FIGURE 26.19

Profiles of a lake. **(a)** During summer, a lake becomes stabilized into three major temperature strata. **(b)** During fall and spring, cooling or heating of the water disrupts the temperature strata and causes upwelling of nutrients from the bottom sediments.

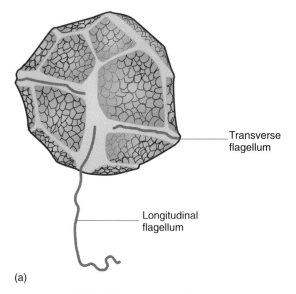

Transverse flagellum

Longitudinal flagellum

(a)

(b)

FIGURE 26.20

Red tides. **(a)** Single-celled red algae called dinoflagellates (*Gymnodinium* shown here) bloom in high-nutrient, warm water and impart a noticeable red color to it, as shown in **(b).** These algae produce a potent muscle toxin that can be concentrated by shellfish through filtration-feeding. When humans eat clams, mussels, or oysters that contain the toxin, they develop paralytic shellfish poisoning. People living in coastal areas are cautioned not to eat shellfish during those months of the year associated with red tides (sometimes characterized as months that lack the letter *r*).

seaboard has been attributed to the overgrowth of certain species of these algae in polluted water (see chapter 5).

Because oxygen is not very soluble in water and is rapidly used up by the plankton, its concentration forms a gradient from highest in the epilimnion to lowest in the benthic zone. In general, the amount of O_2 that can be dissolved is dependent on temperature. Warmer strata on the surface tend to carry lower levels of this gas. Of all the characteristics of water, the greatest range occurs in nutrient levels. Pure water recently carried from melting snow into cold mountain ponds and lakes is lowest in nutrients. Such nutrient-deficient aquatic ecosystems, called **oligotrophic,*** support very few microorganisms and often are virtually sterile. Species that can make a living on such starvation rations are *Hyphomicrobium* and *Caulobacter* (see figure 4.37). These bacteria have special stalks that capture even minuscule amounts of hydrocarbons present in oligotrophic habitats. At one time it was thought that viruses were present only in very low levels in aquatic habitats, but researchers have reported finding bacterial virus levels of 125 million particles per milliliter in a water sample from a pristine lake in Germany. Most of these viruses pose no danger to humans, but as parasites of

bacteria, they appear to be a natural control mechanism for these populations.

At the other extreme are waters overburdened with organic matter and dissolved nutrients. Some nutrients are added naturally through seasonal upwelling and disasters (floods or typhoons), but the most significant alteration of natural waters comes from effluents from sewage, agriculture, and industry that contain heavy loads of organic debris or nitrate and phosphate fertilizers. The addition of excess quantities of nutrients to aquatic ecosystems, termed *eutrophication,** often wreaks havoc on the communities

---

* oligotrophic (ahl″-ih-goh-trof′-ik) Gr. *oligo,* small, and *troph,* to feed.

* eutrophication (yoo″-troh-fih-kay′-shun) Gr. *eu,* good.

FIGURE 26.21

Heavy surface growth of algae and cyanobacteria in a eutrophic pond.

involved. The sudden influx of abundant nutrients along with warm temperatures encourages a heavy surface growth of algae called a bloom (figure 26.21). This heavy algal mat effectively shuts off the O_2 supply to the lake. The oxygen content below the surface is further depleted by aerobic heterotrophs that actively decompose the organic matter. The lack of oxygen greatly disturbs the ecological balance of the community. It causes massive die-offs of strict aerobes (fish, invertebrates), and only anaerobic or facultative microbes will survive. Another serious complication are blooms of certain cyanobacteria that produce the toxin *microcystin*. Several epidemics of fatal toxicosis have decimated populations of birds and fish, and several outbreaks in humans have been linked to toxic blooms.

Water Management to Prevent Disease
Microbiology of Drinking Water Supplies   We do not have to look far for overwhelming reminders of the importance of safe water. Worldwide epidemics of cholera have killed thousands of people, and an outbreak of *Cryptosporidium* in Wisconsin affecting 370,000 people was traced to a contaminated municipal water supply. In a large segment of the world's population, the lack of sanitary water is responsible for billions of cases of diarrheal illness that kills 2 million children each year. In the United States, nearly 1 million people develop water-borne illness every year.

Good health is dependent upon a clean, **potable** (drinkable) **water supply.** This means the water must be free of pathogens, dissolved toxins, and disagreeable turbidity, odor, color, and taste. As we shall see, water of high quality does not come easily, and we must look to microbes as part of the problem and part of the solution.

Through ordinary exposure to air, soil, and effluents, surface waters usually acquire harmless, saprobic microorganisms. But along its course, water can also pick up pathogenic contaminants. Among the most prominent water-borne pathogens of recent times are the protozoans *Giardia* and *Cryptosporidium;* the bacteria *Campylobacter, Salmonella, Shigella, Vibrio,* and *Mycobacterium;*

and hepatitis A and Norwalk viruses. Some of these agents (especially encysted protozoans) can survive in natural waters for long periods without a human host, whereas others are present only transiently and are rapidly lost. The microbial content of drinking water must be continuously monitored to ensure that the water is free of infectious agents.

Attempting to survey water for specific pathogens can be very difficult and time consuming, so most assays of water purity are more focused on detecting fecal contamination. High fecal levels can mean the water contains pathogens and is consequently unsafe to drink. Thus, wells, reservoirs, and other water sources can be analyzed for the presence of various **indicator bacteria.** These generally nonpathogenic species are intestinal residents of birds and mammals, and they are readily identified using routine lab procedures.

Enteric bacteria most useful in the routine monitoring of microbial pollution are *coliforms* and enteric *streptococci*, which survive in natural waters but do not multiply there. Finding them in high numbers thus implicates recent or high levels of fecal contamination. Environmental Protection Agency standards for water sanitation are based primarily on the levels of coliforms, which are described as gram-negative, lactose-fermenting, gas-producing bacteria such as *Escherichia coli, Enterobacter,* and *Citrobacter.* Fecal contamination of marine waters that poses a risk for gastrointestinal disease is more readily correlated with gram-positive cocci, primarily in the genus *Enterococcus.* Occasionally, coliform bacteriophages and reoviruses (the Norwalk virus) are good indicators of fecal pollution, but their detection is more difficult and more technically demanding.

Water Quality Assays   A rapid method for testing the total bacterial levels in water is the **standard plate count.** In this technique, a small sample of water is spread over the surface of a solid medium. The numbers of colonies that develop provide an estimate of the total viable population without differentiating coliforms from other species. This information is particularly helpful in evaluating the effectiveness of various water purification stages. Another general indicator of water quality is the level of dissolved oxygen it contains. It is established that water containing high levels of organic matter and bacteria will have a lower oxygen content because of consumption by aerobic respiration.

Coliform Enumeration   Water quality departments employ some standard assays for routine detection and quantification of coliforms. With the **most probable number (MPN)** procedure, coliforms are detected by a series of *presumptive, confirmatory,* and *completed* tests (figure 26.22). The presumptive test involves a series of three subsets of fermentation tubes, each containing different amounts of lactose or lauryl tryptose broth. Each subset contains five tubes that hold an inverted Durham tube to collect gas produced by fermentation. The three subsets are inoculated with water samples of 10, 1.0, and 0.1 ml, respectively. After 24 hours of incubation, the tubes are evaluated for gas production. A positive test for gas formation is presumptive evidence of coliforms; negative for gas means no coliforms. The number of positive tubes in each subset is tallied, and this set of numbers is applied to a statistical table to estimate the most likely or probable concentration of

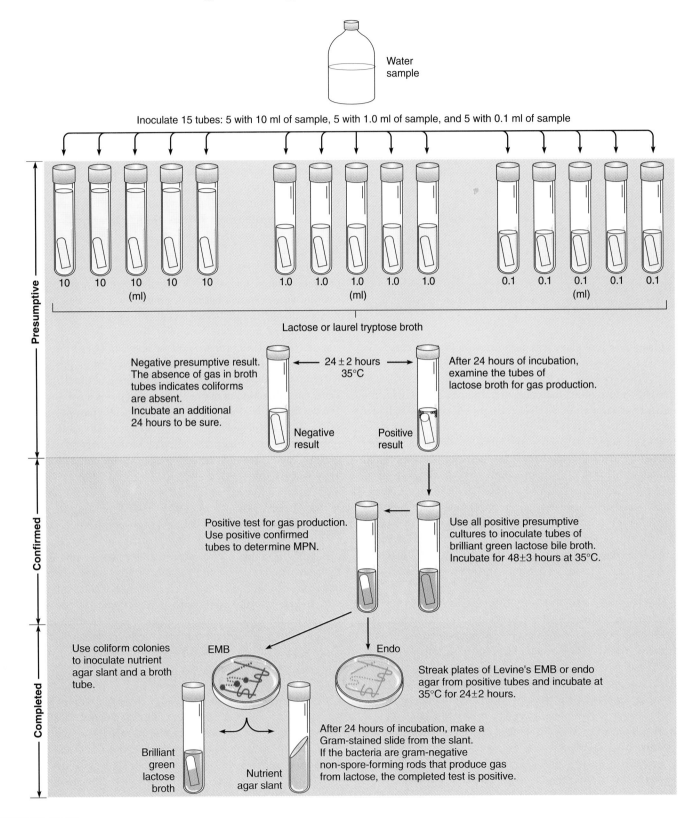

FIGURE 26.22

The most probable number (MPN) procedure for determining the coliform content of a water sample. In the presumptive test, each set of five tubes of broth is inoculated with a water sample reduced by a factor of 10. After incubation, the sets of tubules are examined and rated for gas production (for instance, 0 means no tubes with gas, 1 means one tube with gas, 2 means two tubes with gas). Applying this result to the MPN table in appendix C will indicate the probable number of cells present in 100 ml of the water sample. Confirmation of coliforms can be achieved by confirmatory tests on additional media, and complete identification can be made through selective and differential media and Gram staining.

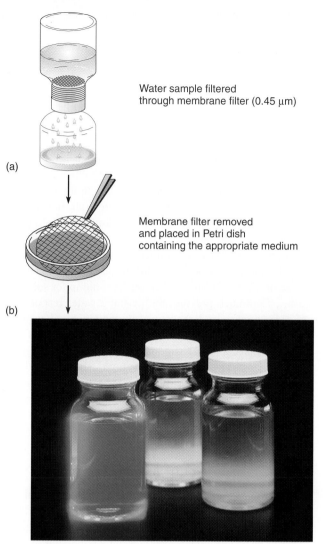

(a)

Water sample filtered
through membrane filter (0.45 µm)

Membrane filter removed
and placed in Petri dish
containing the appropriate medium

(b)

(c)

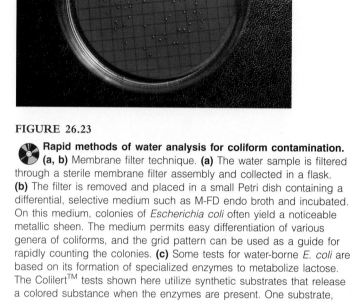

FIGURE 26.23

Rapid methods of water analysis for coliform contamination.
(a, b) Membrane filter technique. **(a)** The water sample is filtered
through a sterile membrane filter assembly and collected in a flask.
(b) The filter is removed and placed in a small Petri dish containing a
differential, selective medium such as M-FD endo broth and incubated.
On this medium, colonies of *Escherichia coli* often yield a noticeable
metallic sheen. The medium permits easy differentiation of various
genera of coliforms, and the grid pattern can be used as a guide for
rapidly counting the colonies. **(c)** Some tests for water-borne *E. coli* are
based on its formation of specialized enzymes to metabolize lactose.
The Colilert™ tests shown here utilize synthetic substrates that release
a colored substance when the enzymes are present. One substrate,
called methyl umbelliferyl gluconuride (MUG), contains a fluorescent
dye that is released when glucuronidase acts on the lactose (blue bottle
on left). A second test uses a synthetic galactose that releases a yellow
dye when galactosidase breaks bonds on lactose (bottle on right). The
center bottle is an indicator of no reaction (negative).

*(c) Source: Courtesy of IDEXX Company. Colilert is a trademark name of the IDEXX
Company.*

coliforms (see appendix C, table A.1). A confirmatory test for coliforms is made by inoculating another broth from one of the positive tubes. The test is completed by final isolation of the coliform
species on selective and differential media, Gram staining the isolate, and reconfirming gas production.

The MPN technique has some notable drawbacks. For one
thing, it requires several days to complete. Another problem is that it
does not differentiate between coliforms that are commonly found
in soil and water *(Enterobacter)* and true *fecal coliforms* that live
primarily in the intestines of animals *(Escherichia)*. Assays for fecal
coliforms require a more specific test such as membrane filtration.

The **membrane filter method** for water analysis is faster,
requires fewer steps and media, is less expensive, is more portable,
and can process larger quantities of water. This method is more
suitable for dilute fluids, such as drinking water, that are relatively
free of particulate matter, and it is less suitable for water containing
heavy microbial growth or debris. This technique is related to the
method described in chapter 11 for sterilizing fluids by filtering out
microbial contaminants, except that in this system, the filter containing the trapped microbes is the desired end product. The steps
in membrane filtration are diagrammed in figure 26.23*a,b*. After fil-

tration, the membrane filter is placed in a small Petri dish containing selective broth. After incubation, fecal coliform colonies can be
counted and often presumptively identified by their distinctive
characteristics on these media. Other rapid techniques for water
quality analysis take advantage of substrates that release colored
substances into the medium in the presence of lactose-processing
enzymes commonly found in coliforms (figure 26.23*c*).

When a test is negative for coliforms, the water is considered
generally fit for human consumption. But even slight coliform levels are allowable under some circumstances. For example, municipal waters can have a maximum of 4 coliforms per 100 ml; private
wells can have an even higher count. There is no acceptable level
for fecal coliforms, enterococci, viruses, or pathogenic protozoans
in drinking water. Waters that will not be consumed but are used for
fishing or swimming are permitted to have counts of 70 to 200 coliforms per 100 ml. If the coliform level of recreational water reaches
1,000 coliforms per 100 ml, health departments usually bar its usage.

Water and Sewage Treatment   Most drinking water comes
from rivers, aquifers, and springs. Only in remote, undeveloped, or
high mountain areas is this water used in its natural form. In most

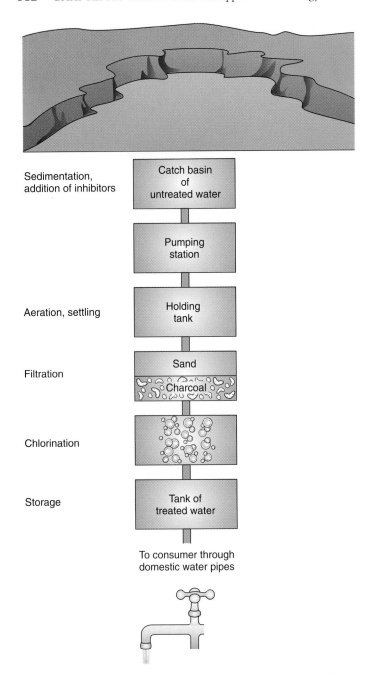

FIGURE 26.24

🔘 The major steps in water purification as carried out by a modern municipal treatment plant.

ppm). Sedimentation to remove large particulate matter is also encouraged during this storage period.

Next, the water is pumped to holding ponds or tanks, where it undergoes further settling, aeration, and filtration. The water is filtered first through sand beds or pulverized diatomaceous earth to remove residual bacteria, viruses, and protozoans, and then through activated charcoal to remove undesirable organic contaminants. Pipes coming from the filtration beds collect the water in storage tanks. The final step in treatment is chemical disinfection by bubbling chlorine gas through the tank until it reaches a concentration of 1–2 ppm, but some municipal plants use chloramines (see chapter 11) for this purpose. A few pilot plants in the United States are using ozone or peroxide for final disinfection, but these methods are expensive and cannot sustain an antimicrobic effect over long storage times. The final quality varies, but most tap water has a slight odor or taste from disinfection.

In many parts of the world, the same water that serves as a source of drinking water is also used as a dump for solid and liquid wastes. Continued pressure on the finite water resources may require reclaiming and recycling of contaminated water such as **sewage.** Sewage is the used wastewater draining out of homes and industries that contains a wide variety of chemicals, debris, and microorganisms. The dangers of typhoid, cholera, and dysentery linked to the unsanitary mixing of household water and sewage have been a threat for centuries. In current practice, some sewage is treated to reduce its microbial load before release, but a large quantity is still being emptied raw (untreated) into the aquatic environment primarily because heavily contaminated waters require far more stringent and costly methods of treatment than are currently available to most cities.

Sewage contains large amounts of solid wastes, dissolved organic matter, and toxic chemicals that pose a health risk. To remove all potential health hazards, treatment typically requires three phases: The primary stage separates out large matter; the secondary stage reduces remaining matter and can remove some toxic substances; and the tertiary stage completes the purification of the water (figure 26.25). Microbial activity is an integral part of the overall process. The systems for sewage treatment are massive engineering marvels (figure 26.26).

In the **primary phase** of treatment, floating bulkier materials such as paper, plastic waste, and bottles are skimmed off. The remaining smaller, suspended particulates are allowed to settle. Sedimentation in settling tanks usually takes 2 to 10 hours and leaves a mixture rich in organic matter. This aqueous residue is carried into a **secondary phase** of active microbial decomposition, or biodegradation. In this phase, a diverse community of natural bioremediators (bacteria, algae, and protozoa) aerobically decompose the remaining particles of wood, paper, fabrics, petroleum, and organic molecules. This forms a suspension of material called *sludge* that tends to settle out and slow the process. To hasten aerobic decomposition of the sludge, most processing plants have systems to *activate* it by injecting air, mechanically stirring it, and recirculating it. A large amount of organic matter is mineralized into sulfates, nitrates, phosphates, carbon dioxide, and water. Certain volatile gases such as hydrogen sulfide, ammonia, nitrogen, and methane may also be released. Water from this process is siphoned off and carried to the **tertiary phase,** which involves further filtering and chlorinating prior to

cities, it must be treated before it is supplied to consumers. Water supplies such as deep wells that are relatively clean and free of contaminants require less treatment than those from surface sources laden with wastes. The stepwise process in water purification as carried out by most cities is shown in figure 26.24. Treatment begins with the impoundment of water in a large reservoir such as a dam or catch basin that serves the dual purpose of storage and sedimentation. The access to reservoirs is controlled to avoid contamination by animals, wastes, and run-off water. In addition, overgrowth of cyanobacteria and algae that add undesirable qualities to the water is prevented by pretreatment with copper sulfate (0.3

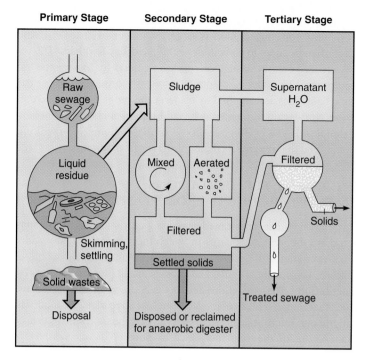

Primary Stage   Secondary Stage   Tertiary Stage

FIGURE 26.25

The primary, secondary, and tertiary stages in sewage treatment.

FIGURE 26.26

Sewage treatment plant. This plant occupies hundreds of acres and can process several hundred million gallons of water a day.

discharge. Such reclaimed sewage water is usually used to water golf courses and parks, rather than for drinking.

In some cases, the solid waste that remains after aerobic decomposition is harvested and reused. Its rich content of nitrogen, potassium, and phosphorus makes it a useful fertilizer. But if the waste contains large amounts of nondegradable or toxic substances, it must be disposed of properly. In many parts of the world, the sludge, which still contains significant amounts of simple, but useful organic matter, is used as a secondary source of energy. Fur-

ther digestion is carried out by microbes in sealed chambers called bioreactors, or **anaerobic digesters.** The digesters convert components of the sludge to swamp gas, primarily methane with small amounts of hydrogen, carbon dioxide, and other volatile compounds. Swamp gas can be burned to provide energy to run the sewage processing facility itself or to power small industrial plants. Considering the mounting waste disposal and energy shortage problems, this technology should gain momentum. (For another microbiological answer to disposal of liquid and solid waste, see Spotlight on Microbiology 26.3.)

CHAPTER CHECKPOINTS

The lithosphere, or soil, is an ecosystem in which mineral-rich rocks are decomposed to organic humus, the base for the soil community. Soil ecosystems vary according to the kinds of rocks and amount of water, air, and nutrients present. The rhizosphere is the most ecologically active zone of the soil.

Aquatic ecosystems are classified as the photic and profundal zones according to the amount of light penetration. The shoreline is the littoral zone, and the soil at the bottom of the water column constitutes the benthic zone. The food web of the aquatic community is built on phytoplankton and zooplankton. The nature of the aquatic community varies with the temperature, depth, minerals, and amount of light present in each zone.

Aquatic ecosystems are readily contaminated by chemical pollutants and pathogens because of industry, agriculture, and improper disposal of human wastes.

Significant water-borne pathogens include protozoans, bacteria, and viruses. *Giardia* and *Cryptosporidium* are the most significant protozoan pathogens. *Campylobacter*, *Salmonella*, and *Vibrio* are the most significant bacterial pathogens. Hepatitis A and Norwalk virus are the most significant viral pathogens.

Water quality assays assess the most probable number of microorganisms in a water sample and screen for the presence of enteric pathogens using *E.coli* as the indicator organism.

Wastewater or sewage is treated in three stages to remove organic material, microorganisms, and chemical pollutants. The primary phase removes physical objects from the wastewater. The secondary phase removes the organic matter by biodegradation. The tertiary phase disinfects the water and removes chemical pollutants.

Applied Microbiology and Biotechnology

Never underestimate the power of the microbe.

—Jackson W. Foster

The profound and sweeping involvement of microbes in the natural world is inescapable. Although our daily encounters with them usually go unnoticed, human and microbial life are clearly intertwined on many levels. It is no wonder that long ago humans realized the power of microbes and harnessed them for specific metabolic tasks. The practical applications of microorganisms in manufacturing products or carrying out a particular decomposition process belong to the large and diverse area of **biotechnology.** Biotechnology has an ancient history, dating back nearly 6,000 years to those first observant humans who discovered that grape juice left sitting produced wine or that bread dough properly infused with a starter would rise. Today, biotechnology has become a fertile ground for hundreds of applications in industry, medicine, agriculture, food sciences, and environmental protection, and it has even come to include the genetic alterations of microbes and other organisms.

Most biotechnological systems involve the actions of bacteria, yeasts, molds, and algae that have been selected or altered to synthesize a certain food, drug, organic acid, alcohol, or vitamin. Many such food and industrial end products are obtained through **fermentation,** a general term used here to refer to the mass, controlled culture of microbes to produce desired organic compounds. It also includes the use of microbes in sewage control, pollution control, and metal mining. A single section cannot cover this diverse area of microbiology in its entirety, but we will touch on some of its more important applications in food technology and industrial processes.

Microorganisms and Food

All human food—from vegetables to caviar to cheese—comes from some other organism, and rarely is it obtained in a sterile, uncontaminated state. Food is but a brief stopover in the overall scheme of biogeochemical cycling. This means that microbes and humans are in direct competition for the nutrients in food, and we must be constantly aware that microbes' fast growth rates give them the winning edge. Somewhere along the route of procurement, processing, or preparation, food becomes contaminated with microbes from the soil, the bodies of plants and animals, water, air, food handlers, or utensils. The final effects depend upon the types and numbers of microbes and whether the food is cooked or preserved. In some cases, specific microbes can even be added to food to obtain a desired effect. The effects of microorganisms on food can be classified as beneficial, detrimental, or neutral to humans, as summarized by the following outline:

Beneficial Effects
Food is fermented or otherwise chemically changed by the addition of microbes or microbial products to alter or improve flavor, taste, or texture.

Microbes can serve as food.

Detrimental Effects
Food poisoning or food-borne illness

Chemical in origin: Pesticides, food additives

Biological in origin: Living things or their products

Infection: Bacterial, protozoan, worm

Intoxication: Bacterial, fungal

Food spoilage

Growth of microbes makes food unfit for consumption; adds undesirable flavors, appearance, and smell; destroys food value

Neutral Effects
The presence or growth of microbes that do not harm or change the nature of the food

As long as food contains no harmful substances or organisms, its suitability for consumption is largely a matter of taste. But what tastes like bouquet to some may seem like decay to others. The test of whether certain foods are edible is guided by experience and preference. The flavors, colors, textures, and aromas of many cultural delicacies are supplied by bacteria and fungi. Poi, pickled cabbage, Norwegian fermented fish, and Limburger cheese are notable examples. If you examine the foods of most cultures, you will find some food that derives its delicious flavor from microbes.

MICROBIAL FERMENTATIONS IN FOOD PRODUCTS FROM PLANTS

In contrast to measures aimed at keeping unwanted microbes out of food are the numerous cases in which they are deliberately added and encouraged to grow. Common substances such as bread, cheese, beer, wine, yogurt, and pickles are the result of **food fermentations.** These reactions actively encourage biochemical activities that impart a particular taste, smell, or appearance to food. The microbe or microbes can occur naturally on the food substrate as in sauerkraut, or they can be added as pure or mixed samples of known bacteria, molds, or yeasts called **starter cultures.** Many food fermentations are synergistic, with a series of microbes acting in concert to convert a starting substrate to the desired end product. Because large-scale production of fermented milk, cheese, bread, alcoholic brews, and vinegar depends upon inoculation with starter cultures, considerable effort is spent selecting, maintaining, and preparing these cultures and excluding contaminants that can spoil the fermentation. Most starting raw materials are of plant origin (grains, vegetables, beans) and, to a lesser extent, of animal origin (milk, meat).

Bread
Microorganisms accomplish three functions in bread making: (1) **leavening** the flour-based dough, (2) imparting flavor and odor, and (3) conditioning the dough to make it workable. Leavening is achieved primarily through the release of gas to produce a porous and spongy product. Without leavening, bread dough remains dense, flat, and hard. Although various microbes and leavening agents can be used, the most common ones are various strains of the baker's yeast *Saccharomyces cerevisiae*. Other gas-forming microbes such as coliform bacteria, certain *Clostridium* species, heterofermentative lactic acid bacteria, and wild yeasts can be employed, depending on the type of bread desired.

Yeast metabolism requires a source of fermentable sugar such as maltose or glucose. Because the yeast respires aerobically in bread dough, the chief products of maltose fermentation are carbon dioxide and water rather than alcohol (the main product in beer and wine). Yeast activity can be modified by adjusting the size of the inoculum, the sugar level, temperature, and salt concentration. Other contributions to bread texture come from kneading, which incorporates air into the dough, and from microbial enzymes, which break down flour proteins (gluten) and give the dough elasticity.

Besides carbon dioxide production, bread fermentation generates other volatile organic acids and alcohols that impart delicate flavors and aromas. These are especially well developed in home-baked bread, which is leavened more slowly than commercial bread. Yeasts and bacteria can also impart unique flavors, depending upon the culture mixture and baking techniques used. The pungent flavor of rye bread, for example, comes in part from starter cultures of lactic acid bacteria such as *Lactobacillus plantarum, L. brevis, L. bulgaricus, Leuconostoc mesenteroides,* and *Streptococcus thermophilus*. Sourdough bread gets its unique tang from *Lactobacillus sanfrancisco* (of course!).

Beer and Other Alcoholic Beverages

The production of alcoholic beverages takes advantage of another useful property of yeasts. By fermenting carbohydrates in fruits or grains anaerobically, they produce ethyl alcohol, as shown by this equation (see figure 8.25):

$$C_6H_{12}O_6 \rightarrow 2C_2H_5OH + 2CO_2$$
(Yeast + Sugar = Ethanol + Carbon dioxide)

Depending upon the starting materials and the processing method, alcoholic beverages vary in alcohol content and flavor. The principal types of fermented beverages are malt liquors, wines, and spirit liquors.

The earliest evidence of beer brewing appears in ancient tablets by the Sumerians and Babylonians around 6,000 B.C. The starting ingredients for both ancient and present-day versions of beer, ale, stout, porter, and other variations are water, malt (barley grain), hops, and special strains of yeasts. The steps in brewing include malting, mashing, adding hops, fermenting, aging, and finishing (figure 26.27).

For brewer's yeast to convert the carbohydrates in grain into ethyl alcohol, the barley must first be sprouted and softened to make its complex nutrients available to yeasts. This process, called **malting,*** releases amylases that convert starch to dextrins and maltose, and proteases that digest proteins. Other sugar and starch supplements added in some forms of beer are corn, rice, wheat, soybeans, potatoes, and sorghum. After the sprouts have been separated, the remaining malt grain is dried and stored in preparation for mashing.

The malt grain is soaked in warm water and ground up to prepare a **mash.** Sugar and starch supplements are then introduced to the mash mixture, which is heated to a temperature of about 65° to 70°C. During this step, the starch is hydrolyzed by amylase, and simple sugars are released. Heating this mixture to 75°C stops the activity of the enzymes. Solid particles are next removed by settling and filtering. **Wort,*** the clear fluid that comes off, is rich in dissolved carbohydrates. It is boiled for about 2.5 hours with **hops,** the dried scales of the female flower of *Humulus lupulus* (figure 26.28) to extract the bitter acids and resins that give aroma and flavor to the finished product.[4] Boiling also caramelizes the sugar and imparts a golden or brown color, destroys any bacterial contaminants that can destroy flavor, and concentrates the mixture. The filtered and cooled supernatant is then ready for the addition of yeasts and fermentation.

---

4. This substance called humulus provides some rather interesting side effects besides flavor. It is a moderate sedative, a diuretic, and a mild antiseptic.

* **malting** (mawlt´-ing) Gr. *malz,* to soften.

* **wort** (wurt) O.E. *wyrt,* a spice.

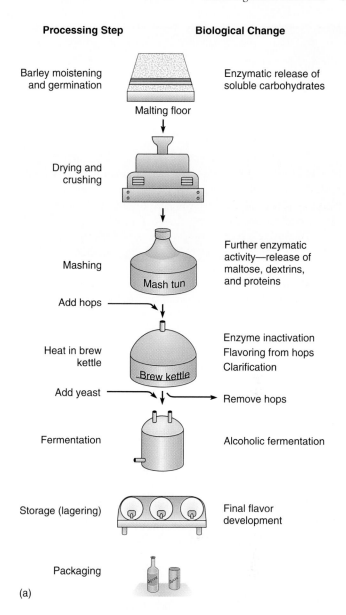

Processing Step	Biological Change
Barley moistening and germination	Enzymatic release of soluble carbohydrates
Malting floor	
Drying and crushing	
Mashing	Further enzymatic activity—release of maltose, dextrins, and proteins
Mash tun	
Add hops	
Heat in brew kettle	Enzyme inactivation / Flavoring from hops / Clarification
Brew kettle	
Add yeast	Remove hops
Fermentation	Alcoholic fermentation
Storage (lagering)	Final flavor development
Packaging	

(a)

(b)

FIGURE 26.27

Brewing basics. **(a)** The general stages in brewing beer, ale, and other malt liquors. **(b)** Fermentation tanks at a commercial brewery.

FIGURE 26.28

Hops. Female flowers of hops, the herb that gives beer some of its flavor and aroma.

Fermentation begins when wort is inoculated with a species of *Saccharomyces* that has been specially developed for beer making. Top yeasts such as *Saccharomyces cerevisiae* function at the surface and are used to produce the higher alcohol content of *ales.* Bottom yeasts such as *S. uvarum (carlsbergensis)* function deep in the fermentation vat and are used to make other beers. In both cases, the initial inoculum of yeast starter is aerated briefly to promote rapid growth and increase the load of yeast cells. Shortly thereafter, an insulating blanket of foam and carbon dioxide develops on the surface of the vat and promotes anaerobic conditions. During 8 to 14 days of fermentation, the wort sugar is converted chiefly to ethanol and carbon dioxide. The diversity of flavors in the finished product is partly due to the release of small amounts of glycerol, acetic acid, and esters. Fermentation is self-limited, and it essentially ceases when a concentration of 3% to 6% ethyl alcohol is reached.

Freshly fermented, or "green," beer is **lagered,*** meaning it is held for several weeks to months in vats near 0°C. During this maturation period, yeast, proteins, resin, and other materials settle, leaving behind a clear, mellow fluid. Lager beer is subjected to a final cold filtration or pasteurization step to remove any residual yeasts that could spoil it. Finally, it is carbonated with carbon dioxide collected during fermentation and packaged in kegs, bottles, or cans.

Wine

Wine is traditionally considered any alcoholic beverage arising from the fermentation of grape juice, but practically any fruit can be rendered into wine. The essential starting point is the preparation of **must,** the juice given off by crushed fruit that is used as a substrate for fermentation. In general, grape wines are either *white* or *red.* The color comes from the skins of the grapes, so white wine is prepared either from white-skinned grapes or from red-skinned grapes that have had the skin removed. Red wine comes from the red- or purple-skinned varieties. Steps in making wine include must preparation (crushing), fermentation, storage, and aging (figure 26.29).

_____

* lagered (law′-gurd) Gr. *lager,* to store or age.

For proper fermentation, must should contain 12% to 25% glucose or fructose, so the art of wine making begins in the vineyard. Grapes are harvested when their sugar content reaches 15% to 25%, depending on the type of wine to be made. Grapes from the field carry a mixed microflora on their surface called the *bloom* that can serve as a source of wild yeasts. Some wine makers allow these natural yeasts to dominate, but many wineries inoculate the must with a special strain of *Saccharomyces cerevisiae,* variety *ellipsoideus.* To discourage yeast and bacterial spoilage agents, wine makers sometimes treat grapes with sulfur dioxide or potassium metabisulfite. The inoculated must is thoroughly aerated and mixed to promote rapid aerobic growth of yeasts, but when the desired level of yeast growth is achieved, anaerobic alcoholic fermentation is begun.

The temperature of the vat during fermentation must be carefully controlled to facilitate alcohol production. The length of fermentation varies from 3 to 5 days in red wines and from 7 to 14 days in white wines. The initial fermentation yields ethanol concentrations reaching 7% to 15% by volume, depending upon the type of yeast, the source of the juice, and ambient conditions. The fermented juice (raw wine) is decanted and transferred to large vats to settle and clarify. Before the final aging process, it is flash-pasteurized to kill microorganisms and filtered to remove any remaining yeasts and sediments. Wine is aged in wooden casks for varying time periods (months to years), after which it is bottled and stored for further aging. During aging, nonmicrobial changes produce aromas and flavors (the bouquet) characteristic of a particular wine.

The fermentation processes discussed thus far can only achieve a maximum alcoholic content of 17%, because concentrations above this level inhibit the metabolism of the yeast. The fermentation product must be **distilled** to obtain higher concentrations such as those found in liquors. During distillation, heating the liquor separates the more volatile alcohol from the less volatile aqueous phase. The alcohol is then condensed and collected. The alcohol content of distilled liquors is rated by *proof,* a measurement that is usually two times the alcohol content. Thus, 80 proof vodka contains 40% ethyl alcohol.

Distilled liquors originate through a process similar to wine making, though the starting substrates can be extremely diverse. In addition to distillation, liquors can be subjected to special treatments such as aging to provide unique flavor or color. Vodka, a colorless liquor, is usually prepared from fermented potatoes, and rum is distilled from fermented sugarcane. Assorted whiskeys are derived from fermented grain mashes; rye whiskey is produced from rye mash, and bourbon from corn mash. Brandy is distilled grape, peach, or apricot wine.

Other Fermented Plant Products

Fermentation provides an effective way of preserving vegetables, as well as enhancing flavor with lactic acid and salt. During pickling fermentations, vegetables are immersed in an anaerobic salty solution (brine) to extract sugar and nutrient-laden juices. The salt also disperses bacterial clumps, and its high osmotic pressure inhibits proteolytic bacteria and sporeformers that can spoil the product.

Sauerkraut is the fermentation product of cabbage (figure 26.30). Cabbage is washed, wilted, shredded, salted, and packed tightly into a fermentation vat. Weights cover the cabbage mass and

Processing Step   **Biological Change**

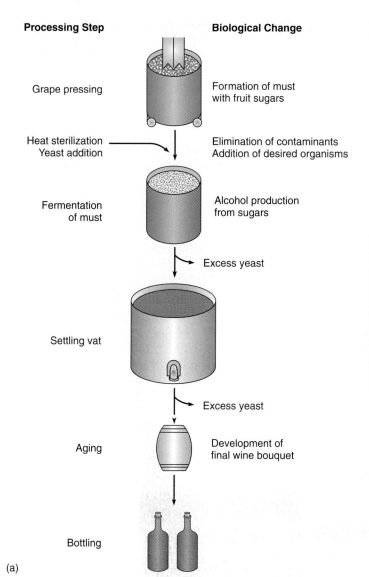

Grape pressing — Formation of must with fruit sugars

Heat sterilization
Yeast addition — Elimination of contaminants
Addition of desired organisms

Fermentation of must — Alcohol production from sugars

Excess yeast

Settling vat

Excess yeast

Aging — Development of final wine bouquet

Bottling

(a)

(b)

FIGURE 26.29

Wine making. **(a)** General steps in wine making. **(b)** Wine fermentation vats in a large commercial winery.

squeeze out its juices. The fermentation is achieved by natural cabbage microflora or by an added culture. The initial agent of fermentation is *Leuconostoc mesenteroides,* which grows rapidly in the brine and produces lactic acid. It is followed by *Lactobacillus plantarum,* which continues to raise the acid content to as high as 2% (pH 3.5) by the end of fermentation. The high acid content restricts the growth of spoilage microbes.

Fermented cucumber pickles come chiefly in salt (sour, sweet-sour, mixed pickles, and relished) and dill varieties. Salt pickles are prepared by washing immature cucumbers, placing them in barrels of brine, and allowing them to ferment for 6 to 9 weeks. The brine can be inoculated with *Pediococcus cerevisiae* and *Lactobacillus plantarum* to avoid unfavorable qualities caused by natural microflora and to achieve a more consistent product. Fermented dill pickles are prepared in a somewhat more elaborate fashion, with the addition of dill herb, spices, garlic, onion, and vinegar.

Natural vinegar is produced when the alcohol in fermented plant juice is oxidized to acetic acid, which is responsible for the pungent odor and sour taste. Although a reasonable facsimile of vinegar could be made by mixing about 4% acetic acid and a dash

of sugar in water, this preparation would lack the traces of various esters, alcohol, glycerin, and volatile oils that give natural vinegar its pleasant character. Vinegar is actually produced in two stages. The first stage is similar to wine or beer making, in which a plant juice is fermented to alcohol by *Saccharomyces.* The second stage involves an aerobic fermentation carried out by acetic acid bacteria in the genera *Acetobacter* and *Gluconbacter.* These bacteria oxidize the ethanol in a two-step process, as shown here:

$$2C_2H_5OH + \tfrac{1}{2} O_2 \rightarrow CH_3CHO + H_2O$$
Ethanol   Acetaldehyde

$$CH_3CHO + \tfrac{1}{2} O_2 \rightarrow CH_3COOH$$
Acetaldehyde   Acetic acid

The abundance of oxygen necessary in commercial vinegar making is furnished by exposing inoculated raw material to air by arranging it in thin layers in open trays, allowing it to trickle over loosely packed beechwood twigs and shavings, or aerating it in a large vat. Different types of vinegar are derived from substrates such as apple cider (cider vinegar), malted grains (malt vinegar), and grape juice (wine vinegar).

Processing Step

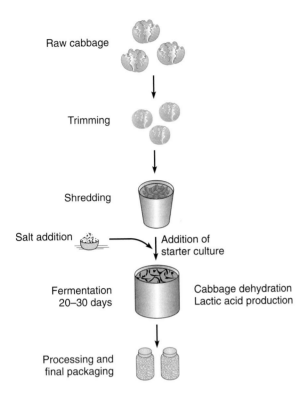

Raw cabbage

Trimming

Shredding

Salt addition → Addition of starter culture

Fermentation 20–30 days — Cabbage dehydration Lactic acid production

Processing and final packaging

FIGURE 26.30

Steps in the preparation of sauerkraut.

MICROBES IN MILK AND DAIRY PRODUCTS

Milk has a highly nutritious composition. It contains an abundance of water and is rich in minerals, protein (chiefly casein), butterfat, sugar (especially lactose), and vitamins. It starts its journey in the udder of a mammal as a sterile substance, but as it passes out of the teat, it is inoculated by the animal's normal flora. Other microbes can be introduced by milking utensils. Because milk is a nearly perfect culture medium, it is highly susceptible to microbial growth. When raw milk is left at room temperature, a series of bacteria ferment the lactose, produce acid, and alter the milk's content and texture (figure 26.31a). This progression can occur naturally, or it can be induced, as in the production of cheese and yogurt.

In the initial stages of milk fermentation, lactose is rapidly attacked by *Streptococcus lactis* and *Lactobacillus* species (figure 26.31b). The resultant lactic acid accumulation and lowered pH cause the milk proteins to coagulate into a solid mass called the **curd.** Curdling also causes the separation of a watery liquid called **whey** on the surface. Curd can be produced by microbial action or by an enzyme, **rennin** (casein coagulase), which is isolated from the stomach of unweaned calves.

Cheese

Since 5000 B.C., various forms of cheese have been produced by spontaneous fermentation of cow, goat, or sheep milk. Present-day, large-scale cheese production is carefully controlled and uses only freeze-dried samples of pure cultures (figure 26.32). These are first inoculated into a small quantity of pasteurized milk to form an ac-

FIGURE 26.31

Microbes at work in milk products. **(a)** Litmus milk is a medium used to indicate pH and consistency changes in milk resulting from microbial action. The first tube is an uninoculated, unchanged control. The second tube has a white, decolorized zone indicative of litmus reduction. The third tube has become acidified (pink), and its proteins have formed a loose curd. In the fourth tube, digestion of milk proteins has caused complete clarification or peptonization of the milk. The fifth tube shows a well-developed solid curd overlaid by a clear fluid, the whey. **(b)** Chart depicting spontaneous changes in the number and type of microorganism and the pH of raw milk as it incubates.

Chart from Philip L. Carpenter, Microbiology, *3rd ed., copyright © 1972 by Holt, Rinehart and Winston, Inc. Reprinted by permission of the publisher.*

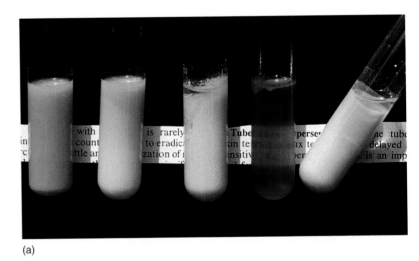

(a)

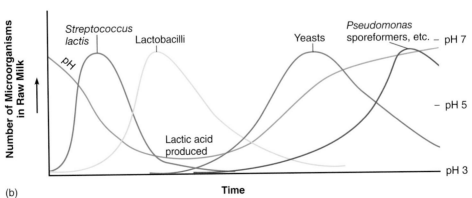

(b)

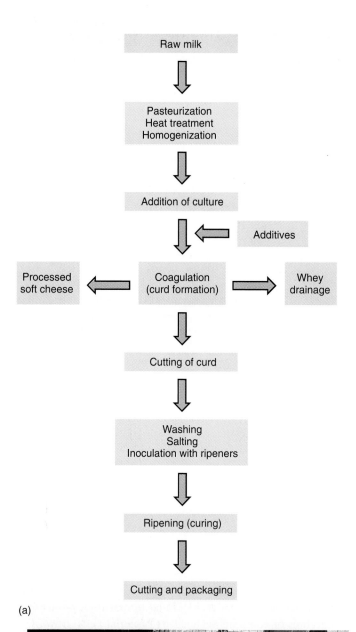

(a)

(b)

FIGURE 26.32

Cheese making. **(a)** A flow diagram that summarizes the main events in cheese making. **(b)** The curd-cutting stage in the making of cheddar cheese.

tive starter culture. This amplified culture is subsequently inoculated into a large vat of milk, where rapid curd development takes place. Such rapid growth is desired because it promotes the overgrowth of the desired inoculum and prevents the activities of undesirable contaminants. Rennin is usually added to increase the rate of curd formation.

After its separation from whey, the curd is rendered to produce one of the 20 major types of soft, semisoft, or hard cheese. The composition of cheese is varied by adjusting water, fat, acid, and salt content. Cottage and cream cheese are examples of the soft, more perishable variety. After light salting and the optional addition of cream, they are ready for consumption without further processing. Other cheeses acquire their character from "ripening," a complex curing process involving bacterial, mold, and enzyme reactions that develop the final flavor, aroma, and other features characteristic of particular cheeses.

The distinctive traits of soft cheeses such as Limburger, Camembert, and Liederkranz are acquired by ripening with a reddish-brown mucoid coating of yeasts, micrococci, and molds. The microbial enzymes permeate the curd and ferment lipids, proteins, carbohydrates, and other substrates. This process leaves assorted acids, salts, aldehydes, ketones, and other by-products that give the finished cheese powerful aromas and delicate flavors. Semisoft varieties of cheese such as Roquefort, blue, or Gorgonzola are infused and aged with a strain of *Penicillium roqueforti* mold. Hard cheeses such as Swiss, cheddar, and Parmesan develop a sharper flavor by aging with selected bacteria. The pockets in Swiss cheese come from entrapped carbon dioxide formed by *Propionibacterium,* which is also responsible for its bittersweet taste.

Other Fermented Milk Products

Yogurt is formed by the fermentation of milk by *Lactobacillus bulgaricus* and *Streptococcus thermophilus.* These organisms produce organic acids and other flavor components and can grow in such numbers that a gram of yogurt regularly contains 100 million bacteria. Live cultures of *Lactobacillus acidophilus* are an important additive to acidophilus milk, which is said to benefit digestion and to help maintain the normal flora of the intestine. Fermented milks such as kefir, koumiss, and buttermilk are a basic food source in many cultures.

MICROORGANISMS AS FOOD

At first, the thought of eating bacteria, molds, algae, and yeasts may seem odd or even unappetizing. We do eat their macroscopic relatives such as mushrooms, truffles, and seaweed, but we are used to thinking of the microscopic forms as agents of decay and disease, or at most, as food flavorings. The consumption of microorganisms is not a new concept. In Germany during World War II, it became necessary to supplement the diets of undernourished citizens by adding yeasts and molds to foods. At present, most countries are able to produce enough food for their inhabitants, but in the future, countries with exploding human populations and dwindling arable land may need to consider microbes as a significant source of protein, fat, and vitamins. Several countries already commercially mass-produce food yeasts, bacteria, and in a few cases, algae. Although eating microbes has yet to win total public acceptance, their

TABLE 26.2
Major Forms of Bacterial Food Poisoning

Disease	Microbe	Foods Involved	Comments
Food Intoxications: Caused by Ingestion of Foods Containing Preformed Toxins			
Staphylococcal enteritis	*Staphylococcus aureus*	Custards, cream-filled pastries, ham, dressings	Very common; symptoms come on rapidly; usually nonfatal
Botulism	*Clostridium botulinum*	Home-canned or poorly preserved low-acid foods	Recent cases involved vacuum-packed foods; can be fatal
Perfringens enterotoxemia	*Clostridium perfringens*	Inadequately cooked meats	Vegetative cells produce toxin within the intestine
Bacillus cereus enteritis	*Bacillus cereus*	Reheated rice, potatoes, puddings, custards	Mimics staphylococcal enteritis; usually self-limited
Food Infections: Caused by Ingestion of Live Microbes That Invade the Intestine			
Salmonellosis	*Salmonella typhimurium* and *S. enteriditis*	Poultry, eggs, dairy products, meats	Very common; can be severe and life-threatening
Shigellosis	Various *Shigella* species	Unsanitary cooked food; fish, shrimp, potatoes, salads	Carriers and flies contaminate food; the cause of bacillary dysentery
Vibrio enteritis	*Vibrio parahaemolyticus*	Raw or poorly cooked seafoods	Microbe lives naturally on marine animals
Listeriosis	*Listeria monocytogenes*	Poorly pasteurized milk, cheeses	Most severe in fetuses, newborns, and the immunodeficient
Campylobacteriosis	*Campylobacter jejuni*	Raw milk; raw chicken, shellfish and meats	Very common; animals are carriers of other species
Escherichia enteritis	*Escherichia coli*	Contaminated raw vegetables, cheese	Various strains can produce infantile and traveler's diarrhea
	E coli 0157:H7	Raw or rare beef, vegetables, water	The cause of hemolytic uremic syndrome (see chapter 20)

use as feed supplements for livestock is increasing. A technology that shows some promise in increasing world food productivity is **single-cell protein (SCP).** This material is produced from waste materials such as molasses from sugar refining, petroleum by-products, and agricultural wastes. In England, an animal feed called **pruteen** is produced by mass culture of the bacterium *Methylophilus methylotrophus.* **Mycoprotein,** a product made from the fungus *Fusarium graminearum,* is also sold there. The filamentous texture of this product makes it a likely candidate for producing meat substitutes for human consumption.

Health food stores carry bottles of dark green pellets or powder that are a culture of a spiral-shaped cyanobacterium called *Spirulina.* This microbe is harvested from the surface of lakes and ponds, where it grows in great mats. In some parts of Africa and Mexico, *Spirulina* has become a viable alternative to green plants as a primary nutrient source. It can be eaten in its natural form or added to other foods and beverages. The microbe has high productivity and requires very little area or resources to grow. *Spirulina* is nontoxic, high in protein, and has a mild, vegetable-like flavor.

MICROBIAL INVOLVEMENT IN FOOD-BORNE DISEASES

Diseases caused by ingesting food are usually referred to as **food poisoning,** and although this term is often used synonymously with microbial food-borne illness, not all food poisoning is caused by

microbes or their products. Several illnesses are caused by poisonous plant and animal tissues or by ingesting food contaminated by pesticides or other poisonous substances. Table 26.2 summarizes the major types of bacterial food-borne disease.

Food poisoning of microbial origin can be divided into two general categories[5] (figure 26.33). **Food intoxication** results from the ingestion of exotoxin secreted by bacterial cells growing in food. The absorbed toxin disrupts a particular target such as the intestine (if an enterotoxin) or the nervous system (if a neurotoxin). The symptoms of intoxication vary from bouts of vomiting and diarrhea (staphylococcal intoxication) to severely disrupted muscle function (botulism). In contrast, **food infection** is associated with the ingestion of whole, intact microbial cells that target the intestine. In some cases, they infect the surface of the intestine, and in others, they invade the intestine and other body structures. Most food infections manifest some degree of diarrhea and abdominal distress (see Medical Microfile 20.2). It is important to realize that disease symptoms in food infection can be initiated by toxins but that the toxins are released by microbes growing in the infected tissue rather than in the food.

Reports of food poisoning are escalating in the United States and worldwide. Outbreaks attributed to common pathogens

---

5. Although these categories are useful for clarifying the general forms of food poisoning, some diseases are transitional. For example, perfringens intoxication is caused by a toxin, but the toxin is released in the intestinal lumen rather than in the food.

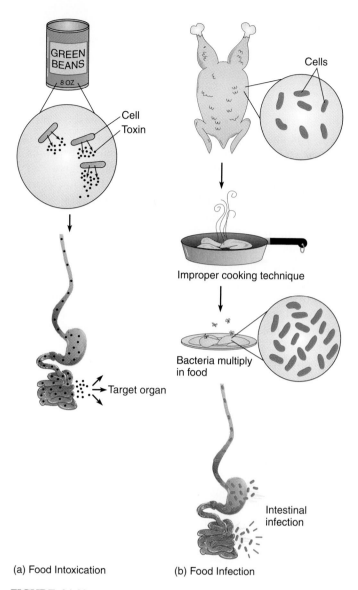

(a) Food Intoxication   (b) Food Infection

FIGURE 26.33

Food-borne illnesses of microbial origin. **(a)** Food intoxication. The toxin is produced by microbes growing in the food. After the toxin is ingested, it acts upon its target tissue and causes symptoms. **(b)** Food infection. The infectious agent comes from food or is introduced into it through poor food processing and storage. After the cells are ingested, they invade the intestine and cause symptoms of gastroenteritis.

(*Salmonella, E. coli, Vibrio,* hepatitis A, and protozoa) have doubled in the past 20 years. The CDC estimates that several million people suffer each year from some form of food infection. A major factor in this changing pattern is the mass production and distribution of processed food such as raw vegetables, fruits, and meats. Improper handling can lead to gross contamination of these products with soil or animal wastes.

Many reported food poisoning outbreaks occur where contaminated food has been served to large groups of people,[6] but most cases probably occur in the home and are not reported. The most common food-borne illness in the United States is staphylococcal food intoxication. Other relatively common agents of food-associated disease include *Campylobacter, Salmonella, Clostrid-*

ium perfringens, and *Shigella.* Food poisoning by *Clostridium botulinum, Bacillus cereus,* and *Vibrio* is less common. The detailed pathology of the diseases listed in table 26.2 was covered in chapters 18, 19, 20, and 21.

PREVENTION MEASURES FOR FOOD POISONING AND SPOILAGE

It will never be possible to avoid all types of food-borne illness because of the ubiquity of microbes in air, water, food, and the human body. But most types of food poisoning require the growth of microbes in the food. In the case of food infections, an infectious dose (sufficient cells to initiate infection) must be present, and in food intoxication, enough cells to produce the toxin must be present. Thus, food poisoning or spoilage can be prevented by proper food handling, preparation, and storage (see Microbits 20.4). The methods shown in figure 26.34 are aimed at preventing the incorporation of microbes into food, removing or destroying microbes in food, and keeping microbes from multiplying.

Preventing the Incorporation of Microbes into Food Most agricultural products such as fruits, vegetables, grains, meats, eggs, and milk are naturally exposed to microbes. Vigorous washing reduces the levels of contaminants in fruits and vegetables, whereas meat, eggs, and milk must be taken from their animal source as aseptically as possible. Aseptic techniques are also essential in the kitchen. Contamination of foods by fingers can be easily remedied by handwashing and proper hygiene, and contamination by flies or other insects can be stopped by covering foods or eliminating pests from the kitchen. Care and common sense also apply in managing utensils. It is important to avoid cross-contaminating food by using the same cutting board for meat and vegetables without disinfecting it between uses. The subject of cutting board safety is discussed in Microbits 26.4.

Preventing the Survival or Multiplication of Microbes in Food Since it is not possible to eliminate all microbes from certain types of food by clean techniques alone, a more efficient approach is to preserve the food by physical or chemical methods. Hygienically preserving foods is especially important for large commercial companies that process and sell bulk foods and must ensure that products are free from harmful contaminants. Regulations and standards for food processing are administered by two federal agencies: the Food and Drug Administration (FDA) and the United States Department of Agriculture (USDA).

Temperature and Food Preservation Heat is a common way to destroy microbial contaminants or to reduce the load of microorganisms. Commercial canneries preserve food in hermetically sealed containers that have been exposed to high temperatures over a specified time period. The temperature used depends upon the type of food, and it can range from 60°C to 121°C, with exposure times ranging from 20 minutes to 115 minutes. The food is usually processed at a thermal death time (TDT; see chapter 11) that will destroy the main spoilage organisms and pathogens but will not

---

6. One-third of all reported cases result from eating restaurant food.

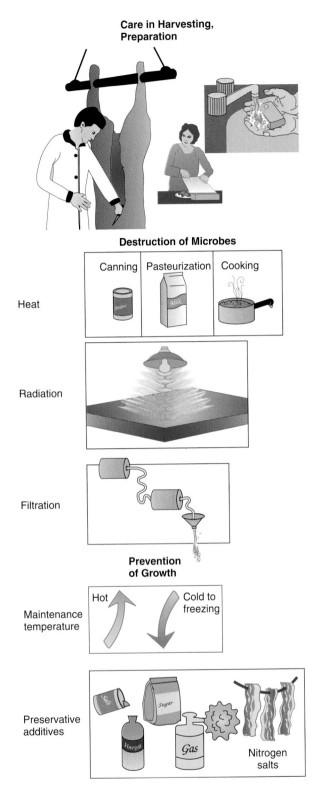

Care in Harvesting, Preparation

Destruction of Microbes

	Canning	Pasteurization	Cooking
Heat			

Radiation

Filtration

Prevention of Growth

	Hot	Cold to freezing
Maintenance temperature		

Preservative additives — Salt, Sugar, Vinegar, Gas, Nitrogen salts

FIGURE 26.34

The primary methods to prevent food poisoning and food spoilage.

ous enough to sterilize the food completely, but some only render the food "commercially sterile," which means it contains live bacteria that are unable to grow under normal conditions of storage.

Another use of heat is **pasteurization,** usually defined as the application of heat below 100°C to destroy nonresistant bacteria and yeasts in liquids such as milk, wine, and fruit juices. The heat is applied in the form of steam, hot water, or even electrical current. The most prevalent technology is the *high-temperature short-time (HTST),* or flash method, using extensive networks of tubes that expose the liquid to 72°C for 15 seconds (figure 26.35). An alternative method, ultrahigh-temperature (UHT) pasteurization, steams the product until it reaches a temperature of 134°C for at least one second. Although milk processed this way is not actually sterile, it is often marketed as sterile, with a shelf life of up to 3 months. Older methods involve large bulk tanks that hold the fluid at a lower temperature for a long time, usually 62.3°C for 30 minutes.

Cooking temperatures used to boil, roast, or fry foods can render them free or relatively free of living microbes if carried out for sufficient time to destroy any potential pathogens. A quick warming of chicken or an egg is inadequate to kill microbes such as *Salmonella.* In fact, any meat is a potential source of infectious agents and should be adequately cooked. Because most meat-associated food poisoning is caused by nonsporulating bacteria, heating the center of meat to at least 80°C and holding it there for 30 minutes is usually sufficient to kill pathogens. Roasting or frying food at temperatures of at least 200°C or boiling it will achieve a satisfactory degree of disinfection (see guidelines in Microbits 20.4).

Any perishable raw or cooked food that could serve as a growth medium must be stored to prevent the multiplication of bacteria that have survived during processing or handling. Because most food-borne bacteria and molds that are agents of spoilage or infection can multiply at room temperature, manipulation of the holding temperature is a useful preservation method (figure 26.36). A good general directive is to store foods at temperatures below 4°C or above 60°C.

Regular refrigeration reduces the growth rate of most mesophilic bacteria by ten times, although some psychrotrophic microbes can continue to grow at a rate that causes spoilage. This factor limits the shelf life of milk, because even at 7°C, a population could go from a few cells to a billion in 10 days. Pathogens such as *Listeria monocytogenes* and *Salmonella* can also continue to grow in refrigerated foods. Freezing is a longer-term method for cold preservation. Foods can be either slow-frozen for 3 to 72 hours at −15°C to −23°C or rapidly frozen for 30 minutes at −17°C to −34°C. Because freezing cannot be counted upon to kill microbes, rancid, spoiled, or infectious foods will still be unfit to eat after freezing and defrosting. *Salmonella* is known to survive several months in frozen chicken and ice cream, and *Vibrio parahaemolyticus* can survive in frozen shellfish. For this reason, frozen foods should be defrosted rapidly and immediately cooked or reheated. However, even this practice will not prevent staphylococcal intoxication if the toxin is already present in the food before it is heated.

Foods such as soups, stews, gravies, meats, and vegetables that are generally eaten hot should not be maintained at warm or room temperatures, especially in settings such as cafeterias, banquets, and picnics. The use of a hot plate, chafing dish, or hot water

alter the nutrient value or flavor of the food. For example, tomato juice must be heated to between 121°C and 132°C for 20 minutes to ensure destruction of the spoilage agent *Bacillus coagulans.* Likewise, green beans must be heated to 121°C for 20 minutes to destroy pathogenic *Cl. botulinum.* Most canning methods are rigor-

MICROBITS 26.4
Wood or Plastic: On the Cutting Edge of Cutting Boards

Inquiring cooks have long been curious to have the final word on which type of cutting board is the better choice for food safety. When the USDA first recommended plastic cutting boards in the 1980s, it seemed the logical, reasonable choice. After all, plastic is nonabsorbent and easy to clean, presumably making it less likely to harbor bacteria and other microorganisms on its surface than wood is. But this recommendation was never based on evidence from scientific tests. Recently, two separate research groups turned their attention to this important kitchen question. What emerged from these studies came as rather a surprise—the two groups came up with exactly opposite conclusions.

First came the study by a team of microbiologists from the University of Wisconsin. They experimented with hardwood chopping blocks and acrylic plastic boards inoculated with pathogens such as *Salmonella, Escherichia coli,* and *Listeria monocytogenes.* One of the most unexpected results was that the wooden boards actually killed 99.9% of the bacteria within a few minutes. The team concluded from the lack of viable cells that wood must contain some antibacterial substances, although they were unable to isolate them. The plastic boards did not similarly reduce the numbers of pathogens and they failed to live up to expectations in other ways. For instance, they continued to harbor bacteria if left unwashed for a given time period. If they were scored from extensive use, even after scrubbing with soap and water, they still held live bacteria. In contrast, even heavily used wooden boards did not grow microorganisms and had a far lower bacterial count. The Wisconsin researchers concluded that the grounds for advocating plastic are questionable and that wood is as safe as plastic, if not superior to it.

In the other study, researchers from the Food and Drug Administration performed an electron microscope study of wood. They found that pathogens such as *E. coli* 0157:H7 and *Campylobacter* became trapped in the porous spaces of wooden boards and were able to survive for 2 hours to several days, depending on the moisture content of the wood. They continue to recommend the use of plastic because bacteria trapped in wood would be difficult to remove and could be released during use.

What is a chef to do? Although these contradictory studies seem not to provide a definitive answer, they can serve to emphasize an important point. The solution still exists in simple, commonsense guidelines that are the crux of good kitchen practices. It is apparent that both boards can be safe if properly handled and their limitations are taken into account. All boards should be scrubbed with soap and hot water and disinfected between uses, especially if meats, poultry, or fish have been cut on

them. Plastic boards should be replaced if their surface has become too roughened with use, and wooden boards must not be left moist for any period of time. A new line of plastic boards containing antibacterial triclosan is expected to improve the safety of cutting boards.

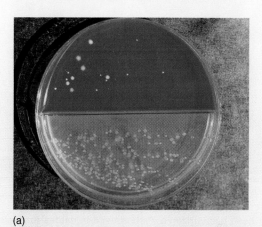

(a)

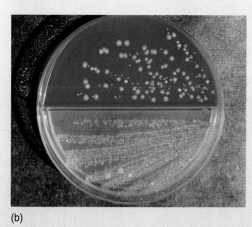

(b)

Double-sided plates of blood agar (top) and MacConkey agar (bottom) after swabbing with samples from cutting boards. The boards were equally contaminated with a fresh chicken carcass, and the samples were taken 10 minutes later. Results appear in **(a)** for the wooden board and in **(b)** for the plastic board. Note that, in this case, the wooden board yielded significantly fewer colonies on both types of media.

bath will maintain foods above 60°C, well above the incubation temperature of food-poisoning agents.

As a final note about methods to prevent food poisoning, remember the simple axiom: "When in doubt, throw it out."

Radiation Food industries have led the field of radiation sterilization and disinfection (see chapter 11). Ultraviolet (nonionizing) lamps are commonly used to destroy microbes on the surfaces of foods or utensils, but they do not penetrate far enough to sterilize bulky foods or food in packages. Food preparation areas are often

equipped with UV radiation devices that are used to destroy spores on the surfaces of cheese, breads, and cakes and to disinfect packaging machines and storage areas.

Food itself is usually sterilized by gamma or cathode radiation because these ionizing rays can penetrate denser materials. Gamma rays are more dangerous because they originate from a radioactive source such as cobalt-60, but the technology has sophisticated chambers that protect workers from the cobalt. It must also be emphasized that this method does not cause the targets of irradiation to become radioactive.

FIGURE 26.35

A modern flash pasteurizer, a system used in dairies for high-temperature short-time (HTST) pasteurization.

Photo taken at Alta Dena Dairy, City of Industry, California.

Concerns have been raised about the possible secondary effects of radiation that could alter the safety and edibility of foods. Experiments over the past 30 years have demonstrated some side reactions that affect flavor, odor, and vitamin content, but it is currently thought that irradiated foods are relatively free of toxic by-products. The government has currently approved the use of radiation in sterilizing beef, pork, poultry, fish, spices, grain, and some fruits and vegetables (see figure 11.9). Less than 10% of these products are sterilized this way, but outbreaks of food-borne illness have increased its desirability for companies and consumers. It also increases the shelf life of perishable foods, thus lowering their cost.

Other Forms of Preservation The addition of chemical preservatives to many foods can prevent the growth of microorganisms that could cause spoilage or disease. Preservatives include natural chemicals such as salt (NaCl) or table sugar and artificial substances such as ethylene oxide. The main classes of preservatives are organic acids, nitrogen salts, sulfur compounds, oxides, salt, and sugar.

Organic acids, including lactic, benzoic, and propionic acids, are among the most widely used preservatives. They are added to baked goods, cheeses, pickles, carbonated beverages,

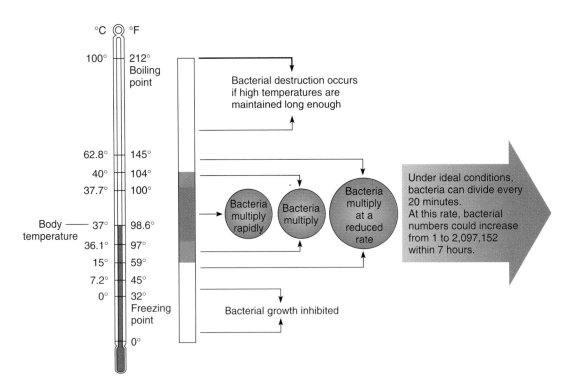

FIGURE 26.36

Temperatures favoring and inhibiting the growth of microbes in food. Most microbial agents of disease or spoilage grow in the temperature range of 15°–40°C. Preventing unwanted growth in foods in long-term storage is best achieved by refrigeration or freezing (4°C or lower). Preventing microbial growth in foods intended to be consumed warm in a few minutes or hours requires maintaining the foods above 60°C.

From Ronald Atlas, Microbiology: Fundamentals and Applications, *2nd ed., © 1998, p. 475. Reprinted by permission of Prentice Hall, Upper Saddle River, New Jersey.*

jams, jellies, and dried fruits to reduce spoilage from molds and some bacteria. Nitrites and nitrates are used primarily to maintain the red color of cured meats (hams, bacon, and sausage). By inhibiting the germination of *Clostridium botulinum* spores, they also prevent botulism intoxication, but their effects against other microorganisms are limited. Sulfite prevents the growth of undesirable molds in dried fruits, juices, and wines and retards discoloration in various foodstuffs. Ethylene and propylene oxide gases disinfect various dried foodstuffs. Because they are carcinogenic and can leave a residue in foods, their use is restricted to fruit, cereals, spices, nuts, and cocoa.

The high osmotic pressure contributed by hypertonic levels of salt plasmolyzes bacteria and fungi and removes moisture from food, thereby inhibiting microbial growth. Salt is commonly added to brines, pickled foods, meats, and fish. However, it does not retard the growth of pathogenic halophiles such as *Staphylococcus aureus,* which grows readily even in 7.5% salt solutions. The high sugar concentrations of candies, jellies, and canned fruits also exert an osmotic preservative effect. Other chemical additives that function in preservation are alcohols and antibiotics. Alcohol is added to flavoring extracts, and antibiotics are approved for treating the carcasses of chickens, fish, and shrimp.

Food can also be preserved by **desiccation,** a process that removes moisture needed by microbes for growth by exposing the food to dry, warm air. Solar drying was traditionally used for fruits and vegetables, but modern commercial dehydration is carried out in rapid-evaporation mechanical devices. Drying is not a reliable microbicidal method, however. Numerous resistant microbes such as micrococci, coliforms, staphylococci, salmonellae, and fungi survive in dried milk and eggs, which can subsequently serve as agents of spoilage and infections.

CHAPTER CHECKPOINTS

The use of microorganisms for practical purposes to benefit humans is called biotechnology.

Microorganisms can compete with humans for the nutrients in food. Their presence in food can be beneficial, detrimental, or of neutral consequence to human consumers.

Food fermentation processes utilize bacteria or yeast to produce desired components such as alcohols and organic acids in foods and beverages. Beer, wine, yogurt, and cheeses are examples of such processes.

Some microorganisms are used as a source of protein. Examples are single-cell protein, mycoprotein, and *Spirulina*. Microbial protein could replace meat as a major protein source.

Food poisoning can be an intoxication caused by microbial toxins produced as by-products of microbial decomposition of food. Or it can be a food infection when pathogenic microorganisms in the food attack the human host after being consumed.

Heat, radiation, chemicals, and drying are methods used to limit numbers of microorganisms in food. The type of method used depends on the nature of the food and the type of pathogens or spoilage agents it contains.

General Concepts in Industrial Microbiology

Virtually any large-scale commercial enterprise that enlists microorganisms to manufacture consumable materials is part of the realm of **industrial microbiology.** Here the term pertains primarily to bulk production of organic compounds such as antibiotics, hormones, vitamins, acids, solvents (table 26.3), and enzymes (table 26.4). Many of the processing steps involve fermentations similar to those described in food technology, but industrial processes usually occur on a much larger scale, produce a specific compound, and involve numerous complex stages. The aim of industrial microbiology is to produce chemicals that can be purified and packaged for sale or for use in other commercial processes. Thousands of tons of organic chemicals worth several billion dollars are produced by this industry every year. To create just one of these products, an industry must determine which microbes, starting compounds, and growth conditions work best. The research and development involved requires an investment of 10 to 15 years and billions of dollars.

The microbes used by fermentation industries are mutant strains of fungi or bacteria that selectively synthesize large amounts of various metabolic intermediates, or **metabolites.** Two basic kinds of metabolic products are harvested by industrial processes: (1) *Primary metabolites* are produced during the major metabolic pathways and are essential to the microbe's function. (2) *Secondary metabolites* are by-products of metabolism that may not be critical to the microbe's function (figure 26.37). In general, primary products are compounds such as amino acids and organic acids synthesized during the logarithmic phase of microbial growth, and secondary products are compounds such as vitamins, antibiotics, and steroids synthesized during the stationary phase (see chapter 7). Most strains of industrial microorganisms have been chosen for their high production of a particular primary or secondary metabolite. Certain strains of yeasts and bacteria can produce 20,000 times more metabolite than a wild strain of that same microbe.

Industrial microbiologists have several tricks to increase the amount of the chosen end product. First, they can manipulate the growth environment to increase the synthesis of a metabolite. For instance, adding lactose instead of glucose as the fermentation substrate increases the production of penicillin by *Penicillium.* Another strategy is to select microbial strains that genetically lack a feedback system to regulate the formation of end products, thus encouraging mass accumulation of this product. Many syntheses occur in sequential fashion, wherein the waste products of one organism become the building blocks of the next. During these *biotransformations,* the substrate undergoes a series of slight modifications, each of which gives off a different by-product (figure 26.38). The production of an antibiotic such as tetracycline requires several microorganisms and 72 separate metabolic steps.

FROM MICROBIAL FACTORIES TO INDUSTRIAL FACTORIES

Industrial fermentations begin with microbial cells acting as living factories. When exposed to optimum conditions, they multiply in massive numbers and synthesize large volumes of a desired

TABLE 26.3

Industrial Products of Microorganisms

Chemical	Microbial Source	Substrate	Applications
Pharmaceuticals			
Bacitracin	*Bacillus subtilis*	Glucose	Antibiotic effective against gram-positive bacteria
Cephalosporins	*Cephalosporium*	Glucose	Antibacterial antibiotic, broad spectrum
Pencillins	*Penicillium chrysogenum*	Lactose	Antibacterial antibiotics, broad and narrow spectrum
Erythromycin	*Streptomyces*	Glucose	Antibacterial antibiotic, broad spectrum
Tetracycline	*Streptomyces*	Glucose	Antibacterial antibiotic, broad spectrum
Amphotericin B	*Streptomyces*	Glucose	Antifungal antibiotic
Vitamin B_{12}	*Pseudomonas*	Molasses	Dietary supplement
Riboflavin	*Asbya*	Glucose, corn oil	Animal feed supplement
Steroids (hydrocortisone)	*Rhizopus, Cunninghamella*	Deoxycholic acid, Stigmasterol	Treatment of inflammation, allergy; hormone replacement therapy
Food Additives and Amino Acids			
Citric acid	*Aspergillus, Candida*	Molasses	Acidifier in soft drinks; used to set jam; candy additive; fish preservative; retards discoloration of crabmeat; delays browning of sliced peaches
Lactic acid	*Lactobacillus, Bacillus*	Whey, corncobs, cottonseed; from maltose, glucose, sucrose	Acidifier of jams, jellies, candies, soft drinks, pickling brine, baking powders
Xanthan	*Xanthomonas*	Glucose medium	Food stabilizer; not digested by humans
Acetic acid	*Acetobacter*	Any ethylene source, ethanol	Food acidifer; used in industrial processes
Glutamic acid	*Corynebacterium, Arthrobacter, Brevibacterium*	Molasses, starch source	Flavor enhancer monosodium glutamate (MSG)
Lysine	*Corynebacterium*	Casein	Dietary supplement for cereals
Miscellaneous Chemicals			
Ethanol	*Saccharomyces*	Beet, cane, grains, wood, wastes	Additive to gasoline (gasohol)
Acetone	*Clostridum*	Molasses, starch	Solvent for lacquers, resins, rubber, fat, oil
Butanol	*Clostridium*	Molasses, starch	Added to lacquer, rayon, detergent, brake fluid
Gluconic acid	*Aspergillus, Gluconobacter*	Corn steep, any glucose source	Baking powder, glass-bottle washing agent, rust remover, cement mix, pharmaceuticals
Glycerol	Yeast	By-product of alcohol fermentation	Explosive (nitroglycerine)
Dextran	*Klebsiella, Acetobacter, Leuconostoc*	Glucose, molasses, sucrose	Polymer of glucose used as adsorbents, blood expanders, and in burn treatment; a plasma extender; used to stabilize ice cream, sugary syrup, candies
Thuricide insecticide	*Bacillus thuringiensis*	Molasses, starch	Used in biocontrol of caterpillars, moths, loopers, and hornworm plant pests

product. Producing appropriate levels of growth and fermentation requires cultivation of the microbes in a carefully controlled environment. This process is basically similar to culturing bacteria in a test tube of nutrient broth. It requires a sterile medium containing appropriate nutrients, protection from contamination, provisions for introduction of sterile air or total exclusion of air, and a suitable temperature and pH (figure 26.39).

Many commercial fermentation processes have been worked out on a small scale in a lab and then *scaled up* to a large commercial venture. An essential component for scaling up is a **fermentor,** a device in which mass cultures are grown, reactions take place, and product develops. Some fermentors are large tubes, flasks, or vats, but most industrial types are metal cylinders with built-in

mechanisms for stirring, cooling, monitoring, and harvesting product (figure 26.40). Fermentors are made of materials that can withstand pressure and are rust-proof, nontoxic, and leakproof. They range in holding capacity from small, 5-gallon systems used in research labs to larger, 5,000- to 100,000-gallon vessels (see figure 26.39), and in some industries, to tanks of 250 million to 500 million gallons.

For optimum yield, a fermentor must duplicate the actions occurring in a tiny volume (a test tube) on a massive scale. Most microbes performing fermentations have an aerobic metabolism, and the large volumes make it difficult to provide adequate oxygen. Fermentors have a built-in device called a *sparger* that aerates the medium to promote aerobic growth. Paddles *(impellers)* located in

TABLE 26.4

Industrial Enzymes and Their Uses

Enzyme	Source	Application
Amylase	*Aspergillus, Bacillus, Rhizopus*	Flour supplement, desizing textiles, mash preparation, syrup manufacture, digestive aid, precooked foods, spot remover in dry cleaning
Catalase	*Micrococcus, Aspergillus*	To prevent oxidation of foods; used in cheese production, cake baking, irradiated foods
Cellulase	*Aspergillus, Trichoderma*	Liquid-coffee concentrates, digestive aid, increase digestibility of animal feed, degradation of wood or wood by-products
Glucose oxidase	*Aspergillus*	Removal of glucose or oxygen that can decolorize or alter flavor in food preparations as in dried egg products; glucose determination in clinical diagnosis
Hyaluronidase	Various bacteria	Medical use in wound cleansing, preventing surgical adhesions
Keratinase	*Streptomyces*	Hair removal from hides in leather preparation
Lipase	*Rhizopus*	Digestive aid and to develop flavors in cheese and milk products
Pectinase	*Aspergillus, Sclerotina*	Clarifies wine, vinegar, syrups, and fruit juices by degrading pectin, a gelatinous substance; used in concentrating coffee
Penicillinase	*Bacillus*	Removal of penicillin in research
Proteases	*Aspergillus, Bacillus, Streptomyces*	To clear and flavor rice wines, process animal feed, remove gelatin from photographic film, recover silver, tenderize meat, unravel silkworm cocoon, remove spots
Rennet	*Mucor*	To curdle milk in cheese making
Streptokinase	*Streptococcus*	Medical use in clot digestion, as a blood thinner
Streptodornase	*Streptococcus*	Promotes healing by removing debris from wounds and burns

the central part of the fermentor increase the contact between the microbe and the nutrients by vigorously stirring the fermentation mixture. Their action also maintains its uniformity.

SUBSTANCE PRODUCTION

The general steps in mass production of organic substances in a fermentor are illustrated in figure 26.41. These can be summarized as: (1) introduction of microbes and sterile media into the reaction

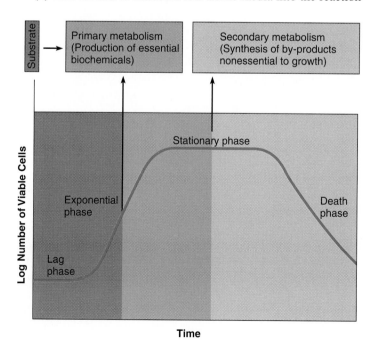

FIGURE 26.37

The origins of primary and secondary microbial metabolites harvested by industrial processes.

chamber, (2) fermentation, (3) *downstream processing* (recovery, purification, and packaging of product), and (4) removal of waste. All phases of production must be carried out aseptically and monitored (usually by computer) for rate of flow and quality of product. The starting raw substrates include crude plant residues, molasses, sugars, fish and meat meals, and whey. Additional chemicals can be added to control pH or to increase the yield. In *batch fermentation,* the substrate is added to the system all at once and taken through a limited run until product is harvested. In *continuous feed systems,* nutrients are continuously fed into the reactor and the product is siphoned off throughout the run.

Ports in the fermentor allow the raw product and waste materials to be recovered from the reactor chamber when fermentation is complete. The raw product is recovered by settling, precipitation, centrifugation, filtration, or cell lysis. Some products come from this process ready to package, whereas others require further purification, extraction, concentration, or drying. The end product is usually in a powder, cake, granular, or liquid form that is placed in sterilized containers. The waste products can be siphoned off to be used in other processes or discarded, and the residual microbes and nutrients from the fermentation chamber can be recycled back into the system or removed for the next run.

Fermentation technology for large-scale cultivation of microbes and production of microbial products is versatile. Table 26.3 itemizes some of the major pharmaceutical substances, food additives, and solvents produced by microorganisms.

Pharmaceutical Products

Health care products derived from microbial biosynthesis are antibiotics, hormones, vitamins, and vaccines. The first mass-produced antimicrobic was penicillin, which came from *Penicillium chrysogenum,* a mold first isolated from a cantaloupe in Wisconsin. The current strain of this species has gone through

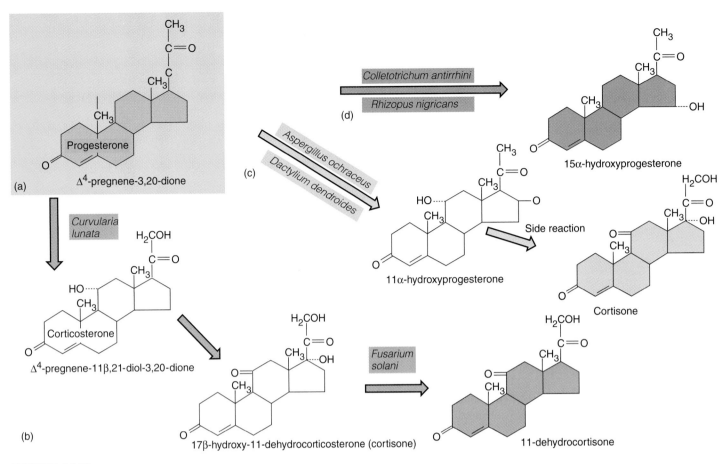

FIGURE 26.38

An example of biotransformation by microorganisms in the industrial production of steroid hormones. Desired hormones such as cortisone and progesterone can require several steps and microbes, and rarely can a single microbe carry out all the required synthetic steps. **(a)** Beginning with starting compound Δ⁴-pregnene-3,20-dione, **(b)** two species of mold can produce 11-dehydrocortisone in three steps. **(c)** Two other species of mold can convert this starting compound to cortisone in two steps, and **(d)** two species of mold can produce 15α-hydroxyprogesterone in one step.

40 years of selective mutation and screening to increase its yield. (The original wild *P. chrysogenum* synthesized 60 mg/ml of medium, and the latest isolate yields 85,000 mg/ml.) The semisynthetic penicillin derivatives are produced by introducing the assorted side-chain precursors to the fermentation vessel during the most appropriate phase of growth. These experiences with penicillin have provided an important model for the manufacture of other antibiotics (see table 26.3).

Several steroid hormones used in therapy are produced industrially (see figure 26.38). Corticosteroids of the adrenal cortex, cortisone and cortisol (hydrocortisone), are invaluable for treating inflammatory and allergic disorders, and female hormones such as progesterone or estrogens are the active ingredients in birth control pills. For years, the production of these hormones was tedious and expensive because it involved purifying them from slaughterhouse animal glands or chemical syntheses. In time, it was shown that, through biotransformation, various molds could convert a precursor compound called diogenin into cortisone. By the same means, stigmasterol from soybean oil could be transformed into progesterone.

Some vaccines are also adaptable to mass production through fermentation. Vaccines for *Bordetella pertussis, Salmonella typhi,*

FIGURE 26.39

A cell culture vessel used to mass-produce pharmaceuticals. Such elaborate systems require the highest levels of sterility and clean techniques.

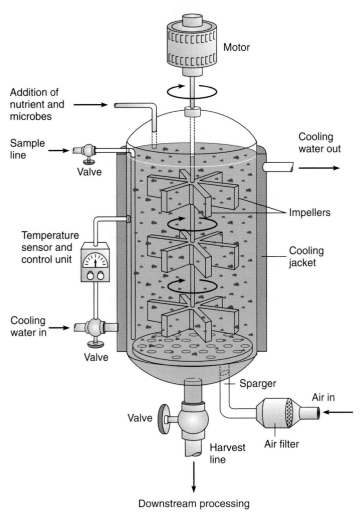

FIGURE 26.40

A schematic diagram of an industrial fermentor for mass culture of microorganisms. Such instruments are equipped to add nutrients and cultures, to remove product under sterile or aseptic conditions, and to aerate, stir, and cool the mixture automatically.

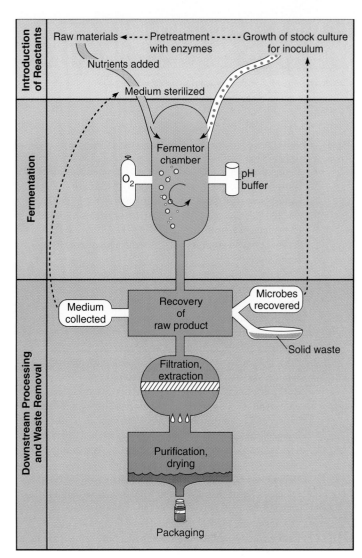

FIGURE 26.41

The general layout of a fermentation plant for industrial production of drugs, enzymes, fuels, vitamins, and amino acids.

Vibrio cholerae, and *Mycobacterium tuberculosis* are produced in large batch cultures. *Corynebacterium diphtheriae* and *Clostridium tetani* are propagated for the synthesis of their toxins, from which toxoids for the DT vaccines are prepared.

Miscellaneous Products

An exciting innovation has been the development and industrial production of natural *biopesticides* using *Bacillus thuringiensis.* During sporulation, these bacteria produce intracellular crystals that can be toxic to certain insects. When the insect ingests this endotoxin, its digestive tract breaks down and it dies, but the material is relatively nontoxic to other organisms. Commercial dusts are now on the market to suppress caterpillars, moths, and worms on various agricultural crops and trees. A strain of this bacterium is also being considered to control the mosquito vector of malaria and the black fly vector of onchocerciasis.

Enzymes are critical to chemical manufacturing, the agriculture and food industries, textile and paper processing, and even

laundry and dry cleaning. The advantage of enzymes is that they are very specific in their activity and are readily produced and released by microbes. Mass quantities of proteases, amylases, lipases, oxidases, and cellulases are produced by fermentation technology (see table 26.4). The wave of the future appears to be custom-designing enzymes to perform a specific task by altering their amino acid content. Other compounds of interest that can be mass-produced by microorganisms are amino acids, organic acids, solvents, and natural flavor compounds to be used in air fresheners and foods.

CHAPTER CHECKPOINTS

Industrial microbiology refers to the bulk production of any organic compound derived from microorganisms. Currently these include antibiotics, hormones, vitamins, acids, solvents, and enzymes.

CHAPTER CAPSULE WITH KEY TERMS

I. Microbial Ecology

Microbial ecology deals with the interaction between the environment and microorganisms. The environment is composed of **biotic** (living or once-living) and **abiotic** (nonliving) components. The combination of organisms and the environment make up an **ecosystem.**

A. Ecosystem Organization
1. Living things inhabit only that area of the earth called the **biosphere,** which is made up of the **hydrosphere** (water), the **lithosphere** (soil), and the **atmosphere** (air).
2. The biosphere consists of terrestrial ecosystems of **biomes** and aquatic ecosystems.
3. Biomes contain **communities,** assemblages of coexisting organisms.
4. Communities consist of **populations,** groups of like organisms of the same species.
5. The space within which an organism lives is its **habitat;** its role in community dynamics is its **niche.**

B. Energy and Nutrient Flow
1. Organisms consume nutrients and derive energy.
2. Their collective trophic status relative to one another is summarized in a **food** or **energy pyramid.**
3. At the beginning of the chain or pyramid are **producers**—photosynthetic or lithotrophic organisms that synthesize large, complex organic compounds from small, simple inorganic molecules.
4. The levels above producer are occupied by **consumers,** organisms that feed upon other organisms.
5. **Decomposers** are consumers that obtain nutrition from the remains of dead organisms and help recycle and **mineralize** nutrients.
6. Organisms are usually not related in a linear **food chain** fashion; a cross-linked **food web** better reflects their ability to use alternative nutritional sources.

C. Microbial Interrelationships include:
1. **Mutualism,** a beneficial, dependent, two-way partnership.
2. **Commensalism,** a one-sided relationship that benefits one partner while neither benefiting nor harming the other.
3. **Cometabolism,** the by-product of one microbe becomes a nutrient for another.
4. **Synergism,** two microbes degrade a substrate cooperatively.
5. **Parasitism,** a host is harmed by the activities of its passenger.
6. **Competition,** the inhibition of one organism by another organism.
7. **Predation,** the seeking and ingesting of live prey.

D. Biogeochemical Cycles
1. The processes by which bioelements and essential building blocks of protoplasm are **recycled** between the biotic and abiotic environments are called **biogeochemical cycles.**
2. They require microorganisms that can remove elements and compounds from a nonliving inorganic reservoir and return them to the food web.
3. The **Gaia hypothesis** is a philosophical explanation for the self-sustaining and self-regulating nature of organisms and the environment.

E. Atmospheric Cycles
Key compounds in the **carbon cycle** include carbon dioxide, methane, and carbonates.
1. Carbon is **fixed** when autotrophs (photosynthesizers) add carbon dioxide to organic carbon compounds.
2. This is later returned to carbon dioxide by respiration and burning fossil fuels. Volcanic activity is one of several natural ways in which carbonates revert to carbon dioxide.
3. Anaerobic metabolism of **methane** by **methanogens** also contributes to carbon recycling.
4. The accumulation of a heat-trapping layer of carbon dioxide, methane, nitrous oxide, and water vapor in the upper atmosphere contributes to the **greenhous effect,** which gives rise to global warming.
5. Photosynthesis takes place in two stages—**light reactions** (light-dependent) and **dark reactions** (light-independent).
   a. During the light reactions, **photons** of solar energy are absorbed by **chlorophyll, carotenoid,** and **phycobilin** pigments in **thylakoid** membranes in chloroplasts or in specialized membrane lamellae. Captured light energy fuels a series of two photosystems that split water by **photolysis** and release oxygen gas and electrons to drive **photophosphorylation.** In this process, released light energy is used to synthesize ATP and NADPH.
   b. Dark reactions occur during the **Calvin cycle,** which uses ATP to fix carbon dioxide to a carrier molecule, ribulose-1,5-bisphosphate, and convert it to glucose in a multistep process.
   c. The type of photosynthesis commonly found in plants, algae, and cyanobacteria is called **oxygenic** because it liberates oxygen. **Anoxygenic** photosynthesis occurs in photolithotrophs, which do not produce oxygen.
6. The **nitrogen cycle** requires four processes and several types of microbes.
   a. In **nitrogen fixation,** atmospheric N_2 gas (the primary reservoir) is combined with oxygen or hydrogen to form NO_2^-, NO_3^-, or NH_4^+ salts. Microbial fixation is achieved both by free-living soil and symbiotic bacteria (*Rhizobium*) that help form **root nodules** in association with legumes.
   b. **Ammonification** is a stage in the degradation of nitrogenous organic compounds (proteins, nucleic acids) by bacteria to ammonium.
   c. Some bacteria **nitrify** NH_4^+ by converting it to NO_2^- and then to NO_3^-, which can be incorporated into protoplasm by still other microbes.
   d. **Denitrification** is a multistep microbial conversion of various nitrogen salts back to atmospheric N_2.

F. Cycles in the Lithosphere
1. In the **sulfur cycle,** environmental sulfurous compounds are converted into useful substrates and returned to the inorganic reservoir through the action of microbes. The major inorganic forms of sulfur are S, H_2S, SO_4, and S_2O_3, and the chief organic forms are in certain amino acids.
2. The chief compound in the **phosphorus cycle** is phosphate (PO_4) found in certain mineral rocks. Microbial action on this reservoir makes it available to be incorporated into organic phosphate forms such as DNA, RNA, and ATP. Phosphate is often a limiting factor in ecosystems.
3. Microorganisms often cycle and help **bioamplify** (accumulate) heavy metals and other toxic pollutants that have been added to habitats by human activities.
4. Soil is a dynamic, complex ecosystem that accommodates a vast array of microbes, animals, and plants coexisting among rich organic debris, water and air spaces, and minerals.

a. The breakdown of rocks and the release of minerals are due in part to the activities of acid-producing bacteria and pioneering **lichens,** symbiotic associations between fungi and cyanobacteria or algae.

b. **Humus** is the rich, moist layer of soil containing plant and animal debris being decomposed by microorganisms.

c. An important habitat is the **rhizosphere,** a zone of soil around plant roots, which nurtures the growth of microorganisms.

d. **Mycorrhizae** are specialized symbiotic organs formed between specialized fungi and certain plant roots.

G. Cycles in the Hydrosphere

1. The surface water, atmospheric moisture, and groundwater are linked through a **hydrologic cycle** that involves evaporation and precipitation. Living things contribute to the cycle through respiration and transpiration.

2. Water that percolates into the earth accumulates in deep underground **aquifers.**

3. The diversity and distribution of water communities are related to sunlight, temperature, aeration, and dissolved nutrients.

   a. The three strata occurring vertically are the **photic zone,** the **profundal zone,** and the **benthic zone;** horizontal sections are the **littoral zone** and the **limnetic zone.** Features of the marine environment include tidal and wave action and estuary, intertidal, and abyssal zones that harbor unique microorganisms.

   b. **Phytoplankton** and **zooplankton** drifting in the photic zone constitute a microbial community that supports the aquatic ecosystem.

   c. Temperature gradients in still waters cause stratified layers to form; the **epilimnion** and **hypolimnion** meet at an interface called a **thermocline.**

   d. Currents brought on by temperature changes cause **upwelling** of nutrient-rich benthic sediments and outbreaks of abundant microbial growth (red tides).

   e. **Oligotrophic** waters contain few nutrients and support a sparser population.

   f. Waters artificially spiked with high levels of nutrients lead to **eutrophication** and explosive aquatic overgrowth of algae and other plants.

4. Water analysis:

   a. Providing **potable water** is central to prevention of water-borne disease.

   b. Water is constantly surveyed for certain **indicator bacteria** (coliforms and enterococci) that signal fecal contamination.

   c. Assays for possible water contamination include the **standard plate count,** the **most probable number (MPN)** procedure, and **membrane filter tests** to enumerate coliforms.

5. Water and sewage treatment:

   a. Drinking water is rendered safe by a purification process that involves storage, sedimentation, settling, aeration, filtration, and disinfection.

   b. Sewage or used wastewater can be processed to remove solid matter, dangerous chemicals, and microorganisms.

   c. Treatment occurs in three phases. Microbes biodegrade the waste material or sludge. Solid wastes are further processed in **anaerobic digesters** that can provide combustible methane and fertilizer.

   d. Future solutions may involve **bioremediation,** the degradation of pollutants by microbial activities.

II. **Applied Microbiology and Biotechnology**

Biotechnology is the practical application of microbiology in the manufacture of food, industrial chemicals, drugs, and other products. Many of these processes use mass, controlled microbial **fermentations** and bioengineered microorganisms.

A. Food Microbiology

Microbes and humans compete for the rich nutrients in food. Although some microbes are present on food as harmless contaminants, others create favorable flavors and nutrients or unfavorable reactions.

1. Fermentations in foods: Microbes can impart desirable aroma, flavor, or texture to foods. Bread, alcoholic beverages, some vegetables, and some dairy products are infused with **starter cultures** of pure microbial strains to yield the necessary fermentation products.

   a. Baker's yeast, *Saccharomyces cerevisiae,* is used to **leaven** bread dough by giving off CO_2; bacteria contribute additional flavors.

   b. Beer making involves the following steps: Barley is sprouted (**malted**) to generate digestive enzymes and then dried; malt is transformed to **mash** and heated with hops to produce a **wort.** Wort is fermented by yeast to a concentration of 3% to 6% alcohol and **lagered** in large tanks before it is carbonated and packaged.

   c. Wine is started by fermentation of **must** (fruit juices) to yield 7% to 15% alcohol. Whiskey, vodka, brandy, and other alcoholic beverages are **distilled** to increase their alcohol content.

   d. Vegetable products, including sauerkraut, pickles, and soybean derivatives, can be pickled in the presence of salt or sugar.

   e. Vinegar is produced by fermenting plant juices first to alcohol and then to acetaldehyde and acetic acid.

   f. Most dairy products are produced by microbes acting on nutrients in milk. In cheese production, milk proteins are coagulated with **rennin** to form solid **curd** that separates from the watery **whey.** Different cheeses are obtained by varying water, fat, acid, and salt content and by aging with bacteria and yeasts; yogurt and buttermilk are also processed with live cultures.

   g. Mass-cultured microbes such as yeasts, molds, and algae can serve as food. **Single-cell protein,** pruteen, and mycoprotein are currently added to animal feeds; in some countries, large colonies of the cyanobacterium *Spirulina* are harvested to be used as food supplements.

2. Food-borne disease: Some microbes cause spoilage and **food poisoning.**

   a. In **food intoxication,** damage is done when the toxin is produced in food, is ingested, and then acts on a specific body target; examples are staphylococcal and clostridial poisoning.

   b. In **food infection,** the intact microbe does damage when it is ingested and invades the intestine; examples are salmonellosis and shigellosis.

3. Precautions for food:

   a. Microbial growth that leads to spoilage and food poisoning can be avoided by high temperature and pressure treatment (canning) and **pasteurization** for disinfecting milk and other heat-sensitive beverages.

   b. Refrigeration and freezing inhibit microbial growth.

   c. Irradiation sterilizes or disinfects foods for longer-term storage.

d. Alternative preservation methods include additives (salt, nitrites), treatment with ethylene oxide gas, and drying.

B. Industrial Microbiology

1. Industrial microbiology involves the large-scale commercial production of organic compounds such as antibiotics, vitamins, amino acids, enzymes, and hormones using specific microbes in carefully controlled fermentation settings.

2. Microbes are chosen for their production of a desired **metabolite;** several different species can be used to **biotransform** raw materials in a stepwise series of metabolic reactions.

3. Fermentations are conducted in massive culture devices called **fermentors** that have special mechanisms for adding nutrients, stirring, oxygenating, altering pH, cooling, monitoring, and harvesting product.

4. Plant organization includes downstream processing (purification and packaging).

MULTIPLE-CHOICE QUESTIONS

1. Which of the following is *not* a major subdivision of the biosphere?
   a. hydrosphere
   b. lithosphere
   c. stratosphere
   d. atmosphere

2. A/an _____ is defined as a collection of populations sharing a given habitat.
   a. biosphere
   b. community
   c. biome
   d. ecosystem

3. The quantity of available nutrients _____ from the lower levels of the energy pyramid to the higher ones.
   a. increases
   b. decreases
   c. remains stable
   d. cycles

4. Photosynthetic organisms convert the energy of _____ into chemical energy.
   a. electrons
   b. protons
   c. photons
   d. hydrogen atoms

5. Which of the following is considered a greenhouse gas?
   a. CO_2
   b. CH_4
   c. N_2O
   d. all of these

6. The Calvin cycle operates during which part of photosynthesis?
   a. only in the light
   b. only in the dark
   c. in both light and dark
   d. only during photosystem I

7. Root nodules contain _____, which can _____.
   a. *Azotobacter,* fix N_2
   b. *Nitrosomonas,* nitrify NH_3
   c. rhizobia, fix N_2
   d. *Bacillus,* denitrify NO_3

8. Which element(s) has/have an inorganic reservoir that exists primarily in sedimentary deposits?
   a. nitrogen
   b. phosphorus
   c. sulfur
   d. b and c

9. The floating assemblage of microbes, plants, and animals that drifts on or near the surface of large bodies of water is the _____ community.
   a. abyssal
   b. benthic
   c. littoral
   d. plankton

10. An oligotrophic ecosystem would be most likely to exist in a/an
    a. ocean
    b. high mountain lake
    c. tropical pond
    d. polluted river

11. Which of the following does not vary predictably with the depth of the aquatic environment?
    a. dissolved oxygen
    b. temperature
    c. penetration by sunlight
    d. salinity

12. Which of the following would be *least* accurate in detecting coliform bacteria in a water sample?
    a. the presumptive MPN test
    b. the standard plate count
    c. the membrane filter method
    d. the confirmatory MPN test

13. Milk is usually pasteurized by
    a. the high-temperature short-time method
    b. ultrapasteurization
    c. batch method
    d. electrical currents

14. The dried, presprouted grain that is soaked to activate enzymes for beer is
    a. hops
    b. malt
    c. wort
    d. mash

15. Substances given off by yeasts during fermentation are
    a. alcohol
    b. carbon dioxide
    c. organic acids
    d. all of these

16. Which of the following is added to facilitate milk curdling during cheese making?
    a. lactic acid
    b. salt
    c. *Lactobacillus*
    d. rennin

17. A food maintenance temperature that generally will prevent food poisoning is
    a. below 4°C
    b. above 60°C
    c. room temperature
    d. both a and b

18. Secondary metabolites of microbes are formed during the _____ phase of growth.
    a. exponential
    b. stationary
    c. death
    d. lag

CONCEPT QUESTIONS

1. a. Present in outline form the levels of organization in the biosphere. Define the term *biome*.
   b. Compare autotrophs and heterotrophs; producers and consumers.
   c. Where in the energy and trophic schemes do decomposers enter?
   d. Compare the concepts of habitat and niche using *Chlamydomonas* (figure 26.1) as an example.

2. a. Using figures 26.3 and 26.4, point out specific examples of producers; primary, secondary, and tertiary consumers, herbivores; primary, secondary, and tertiary carnivores; and omnivores.
    b. What is mineralization, and which organisms are responsible for it?

3. Using specific examples, differentiate between commensalism, mutualism, parasitism, competition, and scavenging.

4. a. Outline the general characteristics of a biogeochemical cycle.
    b. What are the major sources of carbon, nitrogen, phosphorus, and sulfur?

5. a. In what major forms is carbon found? Name three ways carbon is returned to the atmosphere.
    b. Name a way it is fixed into organic compounds.
    c. What form is the least available for the majority of living things?
    d. Which form is produced during anaerobic reduction of CO_2?
    e. Describe the relationship between photosynthesis and respiration with regard to CO_2 and O_2.

6. a. Tell whether each of the following is produced during the light or dark reactions of photosynthesis: O_2, ATP, NADPH, and glucose.
    b. When are water and CO_2 consumed?
    c. What is the function of chlorophyll and the photosystems?
    d. What is the fate of the ATP and NADH?
    e. Compare oxygenic with nonoxygenic photosynthesis.

7. a. Outline the steps in the nitrogen cycle from N_2 to organic nitrogen and back to N_2.
    b. Describe nitrogen fixation, ammonification, nitrification, and denitrification.
    c. What form of nitrogen is required by plants? By animals?

8. a. Outline the phosphorus and sulfur cycles.
    b. What are the most important inorganic and organic phosphorus compounds? How is phosphorus made available to plants?
    c. What are the most important inorganic and organic sulfur compounds?
    d. What are the roles of microorganisms in these cycles?

9. a. Describe the structure of the soil and the rhizosphere.
    b. What is humus?
    c. Compare and contrast root nodules with mycorrhizae.

10. a. Outline the modes of cycling water through the lithosphere, hydrosphere, and atmosphere.
    b. What are the roles of precipitation, condensation, respiration, transpiration, surface water, and aquifers?

11. a. Map a section through the structure of a lake.
    b. What types of organisms are found in plankton?
    c. Why is phytoplankton not found in the benthic zone?
    d. What would you expect to find there?

12. a. What causes the formation of the epilimnion, hypolimnion, and thermocline?
    b. What is upwelling?
    c. In what ways are red tides and eutrophic algal blooms similar and different?

13. a. Why must water be subjected to microbiological analysis?
    b. What are the characteristics of good indicator organisms, and why are they monitored rather than pathogens?
    c. Give specific examples of indicator organisms and water-borne pathogens.
    d. Describe two methods of water analysis.
    e. What are the principles behind the most probable number test?
    f. Describe the three phases of sewage treatment.
    g. What is activated sludge?

14. a. Explain the meaning of fermentation from the standpoint of industrial microbiology.
    b. Describe five types of fermentations.

15. a. What is the main criterion regarding the safety of foods?
    b. Differentiate between food infection and food intoxication.
    c. When are microbes on food harmless?
    d. What are the two most common food intoxications?
    e. What are the three most common food infections?
    f. Summarize the major methods of preventing food poisoning and spoilage.

16. a. Which microbes are used as starter cultures in bread, beer, wine, cheeses, and sauerkraut?
    b. Outline the steps in beer making.
    c. List the steps in wine making.
    d. What are curds and whey, and what causes them?

17. a. Describe the aims of industrial microbiology.
    b. Differentiate between primary and secondary metabolites.
    c. Describe a fermentor.
    d. How is it scaled up for industrial use?
    e. What are specific examples of products produced by these processes?

CRITICAL-THINKING QUESTIONS

1. a. What factors cause energy to decrease with each trophic level?
    b. How is it possible for energy to be lost and the ecosystem to still run efficiently?
    c. Are the nutrients on the earth a renewable resource? Why, or why not?

2. Describe the course of events that could result in a parasitic relationship evolving into a mutualistic one.

3. Give specific examples from biogeochemical cycles that support the Gaia hypothesis.

4. Biologists can set up an ecosystem in a small, sealed aquarium that continues to function without maintenance for years. Describe the minimum biotic and abiotic components it must contain to remain balanced and stable.

5. a. Is the greenhouse effect harmful under ordinary circumstances?
    b. What occurrence has made it dangerous to the global ecosystem?
    c. What could each person do on a daily basis to cut down on the potential for disrupting the delicate balance of the earth?

6. Outline a possible genetic engineering method to produce root nodules in nonleguminous plants. (Hint: See chapter 10.)

7. a. If we are to rely on microorganisms to biodegrade wastes in landfills, aquatic habitats, and soil, list some ways that this process could be made more efficient.
    b. Since elemental poisons (heavy metals) cannot be further degraded even by microbes, what is a possible fate of these metals?
    c. What, if anything, can be done about this form of pollution?

8. Why are organisms in the abyssal zone of the ocean necessarily halophilic, psychrophilic, barophilic, and anaerobic?

9. a. What happens to the nutrients that run off into the ocean with sewage and other effluents?
   b. Why can high mountain communities usually dispense with water treatment?

10. Every year supposedly safe municipal water supplies cause outbreaks of enteric illness.
    a. How in the course of water analysis and treatment might these pathogens be missed?
    b. What kinds of microbes are they most likely to be?

11. Describe three important potential applications of bacteria that can make methane.

12. a. Describe four food-preparation and food-maintenance practices in your own kitchen that could expose people to food poisoning and explain how to prevent them.
    b. What is a good general rule to follow concerning suspicious food?

13. a. Describe a controlled experiment that might solve the contradiction between which cutting board is safer to use in the kitchen.
    b. What mistakes might the researchers have made in the experiments described in Microbits 26.4?

14. a. What is the purpose of boiling the wort in beer preparation?
    b. What are hops used for?
    c. If fermentation of sugars to produce alcohol in wine is anaerobic, why do wine makers make sure that the early phase of yeast growth is aerobic?

15. Predict the differences in the outcome of raw milk that has been incubated for 48 hours versus pasteurized milk that has been incubated for the same length of time.

16. How are cometabolism and biotransformations of microorganisms harnessed in industrial microbiology?

17. a. Outline the steps in the production of an antibiotic, starting with a newly discovered mold.
    b. Review chapter 10 and describe several ways that recombinant DNA technology can be used in biotechnology processes.

INTERNET SEARCH TOPICS

1. Look up information on bioremediation. Name three new technologies in this field involving microorganisms.

2. Find information on the uses of microorganisms to mine metals such as gold and copper. How can bacteria extract metals from ore?

3. Locate <http://www.fightbac.org> on the Internet. What are the most important factors in reducing food-borne illnesses?

Exponents

Dealing with concepts such as microbial growth often requires working with numbers in the billions, trillions, and even greater. A mathematical shorthand for expressing such numbers is with exponents. The exponent of a number indicates how many times (designated by a superscript) that number is multiplied by itself. These exponents are also called common *logarithms*, or logs. The following chart, based on multiples of 10, summarizes this system.

Number	Quantity	Exponential Notation*	Number Arrived at By:	One Followed By:
1	One	10^0	Numbers raised to zero power are equal to one	No zeros
10	Ten	10^1**	10×1	One zero
100	Hundred	10^2	10×10	Two zeros
1,000	Thousand	10^3	$10 \times 10 \times 10$	Three zeros
10,000	Ten thousand	10^4	$10 \times 10 \times 10 \times 10$	Four zeros
100,000	Hundred thousand	10^5	$10 \times 10 \times 10 \times 10 \times 10$	Five zeros
1,000,000	Million	10^6	10 times itself 6 times	Six zeros
1,000,000,000	Billion	10^9	10 times itself 9 times	Nine zeros
1,000,000,000,000	Trillion	10^{12}	10 times itself 12 times	Twelve zeros
1,000,000,000,000,000	Quadrillion	10^{15}	10 times itself 15 times	Fifteen zeros
1,000,000,000,000,000,000	Quintillion	10^{18}	10 times itself 18 times	Eighteen zeros

Other large numbers are sextillion (10^{21}), septillion (10^{24}), and octillion (10^{27}).

*The proper way to say the numbers in this column is 10 raised to the nth power, where n is the exponent. The numbers in this column can also be represented as 1×10^n, but for brevity, the $1 \times$ can be omitted.

**The exponent 1 is usually omitted in numbers at this level.

Converting Numbers to Exponent Form

As the chart shows, using exponents to express numbers can be very economical. When simple multiples of 10 are used, the exponent is always equal to the number of zeros that follow the 1, but this rule will not work with numbers that are more varied. Other large whole numbers can be converted to exponent form by the following operation: First, move the decimal (which we assume to be at the end of the number) to the left until it sits just behind the first number in the series (example: 3568. = 3.568). Then count the number of spaces (digits) the decimal has moved; that number will be the exponent. (The decimal has moved from 8. to 3., or 3 spaces.) In final notation, the converted number is multiplied by 10 with its appropriate exponent: 3568 is now 3.568×10^3.

Rounding Off Numbers

The notation in the previous example has not actually been shortened, but it can be reduced further by rounding off the decimal fraction to the nearest thousandth (three digits), hundredth (two digits), or tenth (one digit). To round off a number, drop its last digit and either increase the one next to it or leave it as it is. If the number dropped is 5, 6, 7, 8, or 9, the subsequent digit is increased by one (rounded up); if it is 0, 1, 2, 3, or 4, the subsequent digit remains as is. Using the example of 3.528, removing the 8 rounds off the 2 to a 3 and produces 3.53 (two digits). If further rounding is desired, the same rule of thumb applies, and the number becomes 3.5 (one digit). Other examples of exponential conversions are shown on the following page:

Number	Is the Same As	Rounded Off, Placed in Exponent Form
16,825.	$1.6825 \times 10 \times 10 \times 10 \times 10$	1.7×10^4
957,654.	$9.57654 \times 10 \times 10 \times 10 \times 10 \times 10$	9.58×10^5
2,855,000.	$2.855000 \times 10 \times 10 \times 10 \times 10$ $\times 10 \times 10$	2.86×10^6

Negative Exponents

The numbers we have been using so far are greater than 1 and are represented by positive exponents. But the correct notation for numbers less than 1 involves negative exponents (10 raised to a negative power, or 10^{-n}). A negative exponent says that the number has been divided by a certain power of 10 (10, 100, 1,000). This usage is handy when working with concepts such as pH that are based on very small numbers otherwise needing to be represented by large decimal fractions—for example, 0.003528. Converting this and other such numbers to exponential notation is basically similar to converting positive numbers, except that you work from left to right and the exponent is negative. Using the example of 0.003528, first convert the number to a whole integer followed by a decimal fraction and keep track of the number of spaces the decimal point moves (example: $0.003528 = 3.528$). The decimal has moved three spaces from its original position, so the finished product is 3.528×10^{-3}. Other examples are:

Number	Is the Same As	Rounded Off, Express with Exponents
0.0005923	$\dfrac{5.923}{10 \times 10 \times 10 \times 10}$	5.92×10^{-4}
0.00007295	$\dfrac{7.295}{10 \times 10 \times 10 \times 10 \times 10}$	7.3×10^{-5}

Significant Events in Microbiology

Date	Discovery/People Involved	Date	Discovery/People Involved
1546	Italian physician Girolamo Fracastoro suggests that invisible organisms may be involved in disease.	1876–1877	German bacteriologist Robert Koch★ studies anthrax in cattle and implicates the bacterium *Bacillus anthracis* as its causative agent.
1590	Zaccharias Janssen, a Dutch spectacle maker, invents the first compound microscope.	1881	Pasteur develops a vaccine for anthrax in animals.
1660	Englishman Robert Hooke explores various living and nonliving matter with a compound microscope that uses reflected light.		Koch introduces the use of pure culture techniques for handling bacteria in the laboratory.
			Walther and Fanny Hesse introduce agar-agar as a solidifying gel for culture media.
1668	Francesco Redi, an Italian naturalist, conducts experiments that demonstrate the fallacies in the spontaneous generation theory.	1882	Koch identifies the causative agent of tuberculosis.
1676	Antonie van Leeuwenhoek, a Dutch linen merchant, uses a simple microscope of his own design to observe bacteria and protozoa.	1884	Koch outlines his postulates.
			Elie Metchnikoff,★ a Russian zoologist, lays groundwork for the science of immunology by discovering phagocytic cells.
1776	An Italian anatomist, Lazzaro Spallanzani, conducts further convincing experiments that dispute spontaneous generation.		The Danish physician Hans Christian Gram devises the Gram stain technique for differentiating bacteria.
1796	English surgeon Edward Jenner introduces a vaccination for smallpox.	1885	Pasteur develops a special vaccine for rabies.
1838	Phillipe Ricord, a French physician, inoculates 2,500 human subjects to demonstrate that syphilis and gonorrhea are two separate diseases.	1887	Julius Petri, a German bacteriologist, adapts two plates to form a container for holding media and culturing microbes.
1839	Theodor Schwann, a German zoologist, and Matthias Schleiden, a botanist, formalize the theory that all living things are composed of cells.	1890	A German, Emil von Behring,★ and a Japanese, Shibasaburo Kitasato, demonstrate the presence of antibodies in serum that neutralize the toxins of diphtheria and tetanus.
1847–1850	The Hungarian physician Ignaz Semmelweis substantiates his theory that childbed fever is a contagious disease transmitted to women by their physicians during childbirth.	1892	A Russian, D. Ivanovski, is the first to isolate a virus (the tobacco mosaic virus) and show that it could be transmitted in a cell-free filtrate.
1853–1854	John Snow, a London physician, demonstrates the epidemic spread of cholera through a water supply contaminated with human sewage.	1895	Jules Bordet,★ a Belgian bacteriologist, discovers the antimicrobial powers of complement.
1857	French bacteriologist Louis Pasteur shows that fermentations are due to microorganisms and originates the process now known as pasteurization.	1898	R. Ross★ and G. Grassi demonstrate that malaria is transmitted by the bite of female mosquitoes.
			Germans Friedrich Loeffler and P. Frosch discover that "filterable viruses" cause foot-and-mouth disease in animals.
1858	Rudolf Virchow, a German pathologist, introduces the concept that all cells originate from preexisting cells.	1899	Dutch microbiologist Martinus Beijerinck further elucidates the viral agent of tobacco mosaic disease and postulates that viruses have many of the properties of living cells and that they reproduce within cells.
1861	Louis Pasteur completes the definitive experiments that finally lay to rest the theory of spontaneous generation.		
1867	The English surgeon Joseph Lister publishes the first work on antiseptic surgery, beginning the trend toward modern aseptic techniques in medicine.	1900	The American physician Walter Reed and his colleagues clarify the role of mosquitoes in transmitting yellow fever.
1869	Johann Miescher, a Swiss pathologist, discovers in the cell nucleus the presence of complex acids, which he terms nuclein (DNA, RNA).		An Austrian pathologist, Karl Landsteiner,★ discovers the ABO blood groups.

continued

★ *These scientists were awarded Nobel prizes for their contributions to the field.*

Date	Discovery/People Involved	Date	Discovery/People Involved
1903	American pathologist James Wright and others demonstrate the presence of antibodies in the blood of immunized animals.	1959–1960	Gerald Edelman★ and Rodney Porter★ determine the structure of antibodies.
1905	Syphilis is shown to be caused by *Treponema pallidum,* through the work of German bacteriologists Fritz Schaudinn and E. Hoffman.	1972	Paul Berg★ develops the first recombinant DNA in a test tube.
1906	August Wasserman, a German bacteriologist, develops the first serologic test for syphilis.	1973	Herb Boyer and Stanley Cohen clone the first DNA using plasmids.
	Howard Ricketts, an American pathologist, links the transmission of Rocky Mountain spotted fever to ticks.	1975	A technique for making monoclonal antibodies is developed by Cesar Milstein, Georges Kohler, and Niels Kai Jerne.
1908	The German Paul Ehrlich★ becomes the pioneer of modern chemotherapy by developing salvarsan to treat syphilis.	1979	Genetically engineered insulin is first synthesized by bacteria.
1910	An American pathologist, Francis Rous,★ discovers viruses that can induce cancer.	1982	Development of first hepatitis B vaccine using virus isolated from human blood.
1915–1917	British scientist F. Twort and French scientist F. D'Herelle independently discover bacterial viruses.	1983	Isolation and characterization of human immunodeficiency virus (HIV) by Luc Montagnier of France and Robert Gallo of the United States.
1928	Frederick Griffith lays the foundation for modern molecular genetics by his discovery of transformation in bacteria.		The polymerase chain reaction is invented by Kerry Mullis.★
1929	A Scottish bacteriologist, Alexander Fleming,★ discovers and describes the properties of the first antibiotic, penicillin.	1987	The molecular genetics of antibody genes is worked out by Susumu Tonegawa.★
1933–1938	Germans Ernst Ruska★ and B. von Borries develop the first electron microscope.		First release of recombinant strain of *Pseudomonas* to prevent frost formation on strawberry plants.
1935	Gerhard Domagk,★ a German physician, discovers the first sulfa drug and paves the way for the era of antimicrobic chemotherapy.	1989	Cancer-causing genes called oncogenes are characterized by J. Michael Bishop, Robert Huber, Hartmut Michel, and Harold Varmus.
	Wendell Stanley★ is successful in inducing tobacco mosaic viruses to form crystals that still retain their infectiousness.	1990	First clinical trials in gene therapy testing.
			Vaccine for *Haemophilus influenzae,* a cause of meningitis, is introduced.
1941	Australian Howard Florey★ and Englishman Ernst Chain★ develop commercial methods for producing penicillin; this first antibiotic is tested and put into widespread use.	1991	Development of transgenic animals to synthesize human hemoglobin.
1944	Oswald Avery, Colin MacLeod, and Maclyn McCarty show that DNA is the genetic material.	1994	Human breast cancer gene isolated.
	Joshua Lederberg★ and E. L. Tatum★ discover conjugation in bacteria.	1995	First bacterial genome fully sequenced, for *Haemophilus influenzae.*
	The Russian Selman Waksman★ and his colleagues discover the antibiotic streptomycin.	1996	Dr. David Ho develops a "cocktail" of drugs to treat AIDS.
1953	James Watson,★ Francis Crick,★ Rosalind Franklin, and Maurice Wilkins★ determine the structure of DNA.	1997	Medical researchers at Case Western University construct human artificial chromosomes (HACs).
1954	Jonas Salk develops the first polio vaccine.		Scottish researchers clone first mammal (a sheep) from adult nuclei.
1957	Alick Isaacs and Jean Lindenmann discover the natural antiviral substance interferon.	1999	Heat-loving bacteria discovered 2 miles beneath earth in African gold mine.
	D. Carleton Gajdusek★ discovers the underlying cause of slow virus diseases.	2000	A rough version of the human genome is mapped.
			250-million-year-old bacterium unearthed by Pennsylvania team.

Methods for Testing Sterilization and Germicidal Processes

Most procedures used for microbial control in the clinical setting are either physical methods (heat, radiation) or chemical methods (disinfectants, antiseptics). These procedures are so crucial to the well-being of patients and staff that their effectiveness must be monitored in a consistent and standardized manner. Particular concerns include the time required for the process, the concentration or intensity of the antimicrobic agent being used, and the nature of the materials being treated. The effectiveness of an agent is frequently established through controlled microbiological analysis. In these tests, a biological indicator (a highly resistant microbe) is exposed to the agent, and its viability is checked. If this known test microbe is destroyed by the treatment, it is assumed that the less resistant microbes have also been destroyed. Growth of the test organism indicates that the sterilization protocol has failed. The general categories of testing include:

> **Heat**   The usual form of heat used in microbial control is steam under pressure in an autoclave. Autoclaving materials at a high temperature (121°C) for a sufficient time (15–40 minutes) destroys most bacterial endospores. To monitor the quality of any given industrial or clinical autoclave run, technicians insert a special ampule of *Bacillus stearothermophilus,* a spore-forming bacterium with extreme heat resistance (figure A.1). After autoclaving, the ampule is incubated at 56°C (the optimal temperature for this species) and checked for growth.

Radiation   The effectiveness of ionizing radiation in sterilization is determined by placing special strips of dried spores of *Bacillus sphaericus* (a common soil bacterium with extreme resistance to radiation) into a packet of irradiated material.

Filtration   The performance of membrane filters used to sterilize liquids may be monitored by adding *Pseudomonas diminuta* to the liquid. Because this species is a tiny bacterium that can escape through the filter if the pore size is not small enough, it is a good indicator of filtrate sterility.

Gas sterilization   Ethylene oxide gas, one of the few chemical sterilizing agents, is used on a variety of heat-sensitive medical and laboratory supplies. The most reliable indicator of a successful sterilizing cycle is the sporeformer *Bacillus subtilis,* variety *niger.*

Germicidal assays   United States hospitals and clinics commonly employ more than 250 products to control microbes in the environment, on inanimate objects, and on patients. Controlled testing of these products' effectiveness is not usually performed in the clinic itself, but by the chemical or pharmaceutical manufacturer. Twenty-five standardized *in vitro* tests are currently available to assess the effects of germicides. Most of them use a non-spore-forming pathogen as the biological indicator.

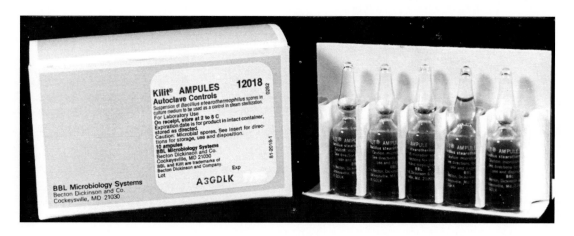

FIGURE A.1

The sterilizing conditions of autoclave runs may be tested by indicator bacteria in small broth ampules. One ampule is placed in the center of the autoclave chamber along with the regular load. After the process, the ampule is incubated to test for survival.

The Phenol Coefficient (PC)

The older disinfectant phenol has been the traditional standard by which other disinfectants are measured. In the PC test, a water-soluble phenol-based (phenolic) disinfectant (amphyl, lysol) is tested for its bactericidal effectiveness as compared with phenol. These disinfectants are serially diluted in a series of *Salmonella choleraesuis, Staphylococcus aureus,* or *Pseudomonas aeruginosa* cultures. The tubes are left for 5, 10, or 15 minutes and then sub-cultured to test for viability. The PC is a ratio derived by comparing the following data:

$$PC = \frac{\text{Greatest dilution of the phenolic that kills test bacteria in 10 min but not in 5 min}}{\text{Greatest dilution of phenol giving same result}}$$

The general interpretation of this ratio is that chemicals with lower phenol coefficients have greater effectiveness. The primary disadvantage of the PC test is that, because it is restricted to phenolics, it is inappropriate for the vast majority of clinical disinfectants.

Use Dilution Test

An alternative test with broader applications is the use dilution test. This test is performed by drying the test culture (one of those species used in the PC test) onto the surface of small stainless steel cylindrical carriers. These carriers are then exposed to varying concentrations of disinfectant for 10 minutes, removed, rinsed, and placed in tubes of broth. After incubation, the tubes are inspected for growth. The smallest concentration of disinfectant that kills the test organism on 10 carrier pieces is the correct dilution for use.

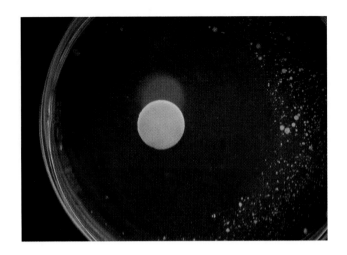

FIGURE A.2

A filter paper disc containing hydrogen peroxide produces a broad clear zone of inhibition in a culture of *Staphylococcus aureus.*

Filter Paper Disc Method

A quick measure of the inhibitory effects of various disinfectants and antiseptics can be achieved by the filter paper disc method. First, a small (1/2-inch) sterile piece of filter paper is dipped into a disinfectant of known concentration. This is placed on an agar medium seeded with a test organism (*S. aureus* or *P. aeruginosa*), and the plate is incubated. A zone devoid of growth (zone of inhibition) around the disc indicates the capacity of the agent to inhibit growth (figure A.2). Like the antimicrobic sensitivity test (fig. 12.21, p. 373), this test measures the minimum inhibitory concentration of the chemical. In general, chemicals with wide zones of inhibition are effective even in high dilutions.

TABLE A.1

Most Probable Number Chart for Evaluating Coliform Content of Water

The Number of Tubes in Series, with Positive Growth at Each Dilution							
10 ml	1 ml	0.1 ml	MPN* per 100 ml of Water	10 ml	1 ml	0.1 ml	MPN* per 100 ml of Water
0	1	0	0.18	5	0	1	3.1
1	0	0	0.20	5	1	0	3.3
1	0	0	0.40	5	1	1	4.6
2	0	0	0.45	5	2	0	4.9
2	0	1	0.68	5	2	1	7.0
2	2	0	0.93	5	2	2	9.5
3	0	0	0.78	5	3	0	7.9
3	0	1	1.1	5	3	1	11.0
3	1	0	1.1	5	3	2	14.0
3	2	0	1.4	5	4	0	13.0
4	0	0	1.3	5	4	1	17.0
4	0	1	1.7	5	4	2	22.0
4	1	0	1.7	5	4	3	28.0
4	1	1	2.1	5	5	0	24.0
4	2	0	2.2	5	5	1	35.0
4	2	1	2.6	5	5	2	54.0
4	3	0	2.7	5	5	3	92.0
5	0	0	2.3	5	5	4	160.0

*Most probable number of cells per sample of 100 ml.

Classification of Major Microbial Disease Agents by System Affected, Site of Infection, and Routes of Transmission

Tables A.2 through A.8 provide a cross reference to the organismic approach of presenting pathogens used in this text. The tables list the primary organs and systems that serve as targets for infectious agents and the sources and routes of infection. Many infectious agents affect both the system they first enter and others. (For example, polio infects the gastrointestinal tract and also inflicts major damage on the central nervous system.) This reference highlights agents that have major effects on more than one system with a star (★). The chapter that contains detailed information on the listings is shown in parentheses in the far right column of each table.

■ Respiratory Tract (A.2)	■ Multiple Systems (Circulatory and Lymphatic) (A.6)
■ Gastrointestinal Tract (A.3)	■ Nervous System (A.7)
■ Skin and Wounds (A.4)	■ Eye and Ear (A.8)
■ STDs and GU Tract (A.5)	

TABLE A.2

Infections of the Respiratory Tract (RT)

Name of Microbe	Name of Disease	Specific Targets of Infection	Source/Mode of Acquisition	(Ch)
Bacterial pathogens				
Streptococcus pyogenes★	Streptococcal pharyngitis	Upper RT membranes	Droplets	(18)
Viridans streptococci	Sinusitis	Upper RT and saliva	Normal flora	(18)
Bacillus anthracis★	Pulmonary anthrax	Upper and lower RT	Spores inhaled from animals/products	(19)
Corynebacterium diphtheriae★	Diphtheria	Upper RT membranes; also skin	Droplets	(19)
Haemophilus influenzae★	Pharyngitis	Upper RT membranes	Droplets	(20)
Streptococcus pneumoniae★	Pneumococcal pneumonia; otitis media, sinusitis meningitis	Lower RT, lungs; meninges	Droplets, mucus	(18)
Mycobacterium tuberculosis★	Tuberculosis	Lower RT, lungs	Droplets; droplet nuclei	(19)
Legionella pneumophila	Legionnaire disease Pontiac fever	Lower RT, lungs	Aerosols from aquatic habitats	(20)
Brucella spp.★	Brucellosis (undulant fever)	Respiratory tract	Aerosols from infected animals	(20)
Bordetella pertussis	Whooping cough	Upper, lower RT	Droplets	(20)
Klebsiella pneumoniae *Pseudomonas aeruginosa*	Opportunistic pneumonias	Lower RT, lungs	Droplets, normal flora, aerosols, fomites	(20)
Yersinia pestis★	Pneumonic plague	Lower RT, lungs	Droplets, animal aerosols	(20)
Mycoplasma pneumoniae	Primary atypical pneumonia	Lower RT, lungs	Droplets, mucus	(21)

continued

TABLE A.2 *continued*

Infections of the Respiratory Tract (RT)

Name of Microbe	Name of Disease	Specific Targets of Infection	Source/Mode of Acquisition	(Ch)
Bacterial pathogens *continued*				
Chlamydia pneumoniae	Chlamydial pneumonitis	Lower RT, lungs	Droplets, mucus	(21)
Chlamydia psittaci	Ornithosis (parrot fever)	Lower RT, lungs	Aerosols, droplets from birds	(21)
Coxiella burnetii	Q (for query) fever	Lower RT, lungs	Ticks, aerosols from animals, dust	(21)
Nocardia brasiliensis	Nocardiosis (a form of pneumonia)	Lower RT, lungs	Spores inhaled from environment	(19)
Fungal pathogens (chapter 22)				
Histoplasma capsulatum★	Histoplasmosis	Lower RT, lungs	Spores inhaled in dust	
Blastomyces dermatitidis★	Blastomycosis	Lower RT, lungs	Spores inhaled	
Coccidioides immitis★	Coccidioidomycosis (Valley fever)	Lower RT, lungs	Spores inhaled	
Cryptococcus neoformans★	Cryptococcosis	Lower RT, lungs	Yeasts inhaled, can be normal flora	
Sporothrix schenckii★	Sporothix pneumonia	Lower RT, lungs	Spores on vegetation	
Pneumocystis carinii	Pneumocystis pneumonia (PCP)	Lower RT, lungs	Spores widespread, normal flora	
Aspergillus fumigatus★	Aspergillosis	Lower RT, lungs	Spores widespread in dust, air	
Parasitic (protozoan, helminth) pathogens (chapter 23)				
Paragonimus westermani	Adult lung fluke	Lung tissue; intestine	Larvae ingested in raw or under-cooked crustacea	
Viral pathogens				
Influenza viruses	Influenza A, B, C	Mucosal cells of trachea, bronchi	Inhalation of droplets, aerosols from active cases	(25)
Hantavirus	Hantavirus pulmonary syndrome	Lung epithelia	Inhalation of airborne rodent excrement	(25)
Respiratory syncytial pneumovirus	RSV pneumonitis, bronchiolitis	Nasopharynx; upper and lower RT	Droplets and secretions from active cases	(25)
Parainfluenza virus Paramyxovirus	Parainfluenza	Upper, lower RT	Direct hand contact with secretions; inhalation of droplets	(25)
Adenovirus	Common cold	Nasopharynx; sometimes lower RT	Close contact with secretions from person with active infection; hand to nose	(24)
Rhinovirus	Common cold	Same as above		(25)
Coronavirus	Common cold	Same as above		(25)
Coxsackievirus	Common cold	Same as above		(25)
Mumps virus★	See table A.3			
Rubella virus★	See table A.4			
Measles virus★	See table A.4			
Chickenpox virus (VZV)★	See table A.4			

TABLE A.3

Diseases of the Gastrointestinal Tract (Food Poisoning and Intestinal, Gastric, Oral Infections)

Name of Microbe	Name of Disease	Specific Targets of Disease	Source/Mode of Acquisition	(Ch)
Bacterial pathogens				
Staphylococcus aureus	Staph food intoxication	Small intestine	Foods contaminated by human carriers	(18)
Clostridium perfringens	Perfringens food poisoning	Small intestine	Rare meats; toxin produced in the body	(19)
Clostridium difficile	Antibiotic-associated colitis	Large intestine	Spores ingested; can be normal flora; caused by antibiotic therapy	(19)
Listeria monocytogenes	Listeriosis	Small intestine	Water, soil, plants, animals, milk, food infection	(19)
Escherichia coli	Traveler's diarrhea 0157: H7 enteritis	Small intestine	Food, water, rare or raw beef	(20)
Salmonella typhi★	Typhoid fever	Small intestine	Human carriers; food, drink	(20)
Salmonella species	Food infection, enteric fevers	Small intestine	Cattle, poultry, rodents; fecally contaminated foods	(20)
Shigella spp.	Bacillary dysentery	Large intestine	Human carriers; food, flies	(20)
Yersinia enterocolitica *Y. paratuberculosis*	Yersiniosis (gastroenteritis)	Small intestine, lymphatics	A zoonosis; food and drinking water	(20)
Vibrio cholerae	Epidemic cholera	Small intestine	Natural waters; food contaminated by human carriers	(21)
Helicobacter pylori	Gastritis, gastric intestinal ulcers	Esophagus, stomach, duodenum	Humans, cats (?)	(21)
Campylobacter jejuni	Enteritis	Jejunum segment of small intestine	Water, milk, meat, food infection	(21)
Streptococcus mutans	Dental caries	Tooth surface	Normal flora; saliva	(21)
Actinomyces israelii★	Actinomycosis	Oral membranes, cervicofacial area	Normal flora; enter damaged tissue	(19)
Fusobacterium, *Treponema, Bacteroides*	Various forms of gingivitis, periodontitis	Gingival pockets	Normal flora; close contact	(21)
Fungal pathogens (chapter 22)				
Aspergillus flavus	Aflatoxin poisoning	Liver	Toxin ingested in food	
Candida albicans★	Esophagitis	Esophagus	Normal flora	
Parasitic (protozoan, helminth) pathogens (chapter 23)				
Entamoeba histolytica★	Amebic dysentery	Caecum, appendix, colon, rectum	Food, water-contaminated fecal cysts	
Giardia lamblia (intestinalis)	Giardiasis	Duodenum, jejunum	Ingested cysts from animal, human carriers	
Balantidium coli	Balantidiosis	Large intestine	Swine, contaminated food	
Cryptosporidium spp.	Cryptosporoidosis	Small and large intestine	Animals, natural waters, contaminated food	
Toxoplasma gondii★	Toxoplasmosis	Lymph nodes	Cats, rodents, domestic animals; raw meats	
Enterobius vermicularis	Pinworm	Rectum, anus	Eggs ingested from fingers, food	
Ascaris lumbricoides	Ascariasis	Worms burrow into intestinal mucosa	Fecally contaminated food and water containing eggs	
Trichuris trichiura	Trichiuriasis, whipworm	Worms invade small intestine	Food containing eggs	
Trichinella spiralis★	Trichinosis	Larvae penetrate intestine	Pork and bear meat containing encysted larvae	
Necator americanus★ *Ancylostoma duodenale*★	Hookworms	Large and small intestine	Larvae deposited by feces into soil; burrow into feet	
Fasciola hepatica	Sheep liver fluke	Liver and intestine	Watercress	
Opisthorchis sinensis	Chinese liver fluke	Liver and intestine	Raw or undercooked fish or shellfish	
Taenia saginata	Beef tapeworm	Large intestine	Rare or raw beef or pork containing larval forms	
T. solium	Pork tapeworm	Pork larva can migrate to brain		

continued

TABLE A.3 *continued*

Diseases of the Gastrointestinal Tract (Food Poisoning and Intestinal, Gastric, Oral Infections)

Name of Microbe	Name of Disease	Specific Targets of Disease	Source/Mode of Acquisition	(Ch)
Viral pathogens				
Herpes simplex virus, type 1	Cold sores, fever blisters, gingivostomatitis	Skin, mucous membranes of oral cavity	Close contact with active lesions of human	(24)
Mumps virus★ (*Paramyxovirus*)	Parotitis	Parotid salivary gland	Close contact with salivary droplets	(25)
Hepatitis A virus	Infectious hepatitis	Liver	Food, water contaminated by human feces; shellfish	(25)
Rotavirus	Rotavirus diarrhea	Small intestinal mucosa	Fecally contaminated food, water, fomites	(25)
Norwalk agent (enterovirus)	Gastroenteritis	Small intestine	Close contact; oral-fecal, water, shellfish	(25)
Poliovirus★	Poliomyelitis	Intestinal mucosa, systemic	Food, water, mechanical vectors	(25)

TABLE A.4

Infections of Skin and Skin Wounds

Name of Microbe	Name of Disease	Specific Targets of Infection	Source/Mode of Transmission	(Ch)
Bacterial pathogens				
Staphylococcus aureus★	Folliculitis Furuncles (boils) Carbuncles Bullous impetigo Scalded skin syndrome	Glands, follicles, glabrous skin	Close contact, nasal and oral droplets; associated with poor hygiene, tissue injury	(18)
Streptococcus pyogenes★	Pyoderma (impetigo) Erysipelas Necrotizing fasciitis	Skin and underlying tissues	Contact with human carrier; invasion of tiny wounds in skin	(18)
Bacillus anthracis★	Cutaneous anthrax	Necrotic lesion in skin (eschar)	Spores from animal secretions, products, soil	(19)
Clostridium perfringens★	Gas gangrene	Skin, muscle, connective tissue	Introduction of spores into damaged tissue	(19)
Propionibacterium acnes	Acne	Sebaceous glands	Normal flora; bacterium is lipophilic	(19)
Erysipelothrix rhusiopathiae	Erysipeloid	Epidermal, dermal regions of skin	Normal flora of pigs, other animals; enters in injured tissue	(19)
Mycobacterium leprae★	Leprosy (Hansen disease)	Skin, underlying nerves	Long-term contact with infected person; skin is inoculated	(19)
Mycobacterium marinum	Swimming pool granuloma	Superficial skin	Skin scraped against contaminated surface	(19)
Pasteurella multocida	Pasteurellosis	Localized skin abscess	Animal bite or scratch	(20)
Bartonella henselae	Cat-scratch disease	Subcutaneous tissues, lymphatic drainage	Cat's claws carry bacteria, inoculate skin	(21)
Fungal pathogens (chapter 22)				
Blastomyces dermatitidis	Blastomycosis	Subcutaneous	Accidental inoculation of skin with conidia	
Trichophyton Microsporum Epidermophyton	Dermatomycoses: tineas, ringworm, athlete's foot	Cutaneous tissues (epidermis, nails, scalp, hair)	Contact with infected humans, animals, spores in environment	
Sporothrix schenckii★	Sporotrichosis	Epidermal, dermal, subcutaneous	Puncture of skin with sharp plant materials	
Madurella	Madura foot, mycetoma	Subcutaneous tissues, bones, muscles	Traumatic injury with spore-infested objects	

continued

TABLE A.4 *continued*

Infections of Skin and Skin Wounds

Name of Microbe	Name of Disease	Specific Targets of Infection	Source/Mode of Transmission	(Ch)
Fungal pathogens (chapter 22) *continued*				
Cladosporium *Fonsecaea*	Chromoblastomycosis	Deep subcutaneous tissues of legs, feet	Penetration of skin with contaminated objects	
Candida albicans★	Onychomycosis	Skin, nails	Chronic contact with moisture, warmth	
Malassezia furfur	Tinea versicolor	Superficial epidermis	Normal flora of skin	
Parasitic (protozoan, helminth) pathogens (chapter 23)				
Leishmania★ *tropica* *L. mexicana*	Cutaneous leishmaniasis	Local infection of skin capillaries	Sand fly salivary gland harbors infectious stage; transmitted through bite	(23)
Dracunculus medinensis	Dracontiasis	Starts in intestine, migrates to skin	Microscopic aquatic arthropod harbors larval form of guinea worm; is ingested	
Viral pathogens				
Varicella-zoster virus★	Chickenpox Shingles	Skin of face and trunk; becomes systemic and latent in nerves	Spread by respiratory droplets; recurrences from latent infection	(24)
Herpesvirus-6	Roseola infantum	Generalized skin rash	Close contact with droplets	
Parvovirus B19	Erythema infectiosum	Skin of cheeks, trunk	Close contact with droplets	
Measles virus★ (*Morbillivirus*)	Measles, rubeola	Skin on head, trunk, extremities	Contact with respiratory aerosols from active case	(25)
Rubella virus★ (*Rubivirus*)	Rubella	Skin on face, trunk, limbs	Respiratory droplets, urine of active cases	(25)
Human papillomavirus (See also table A.5)	Warts, verruca	Skin of hands, feet, body	Contact with warts, fomites	(24)
Molluscum contagiosum	See table A.5			(24)
Herpes simplex, type 2	See table A.5			(24)

TABLE A.5

Sexually Transmitted Diseases (STDs) and Infections of the Genitourinary (GU) Tract

Name of Microbe	Name of Disease	Specific Targets of Infection	Source/Mode of Acquisition	(Ch)
Bacterial pathogens				
Neisseria gonorrhoeae★	Gonorrhea	Vagina, urethra	Mucous secretions; STD	(18)
Haemophilus ducreyi	Chancroid	Genitalia, lymph nodes	Skin lesions; STD	(20)
Treponema pallidum★	Syphilis	Genitalia; mucous membranes	Chancres, skin lesions; STD	(21)
Chlamydia trachomatis★	Chlamydiosis (non-gonococcal urethritis), lymphogranuloma venereum	Vagina, urethra, lymph nodes	Mucous discharges; STD	(21)
Escherichia coli★	Urinary tract infection (UTI)	Urethra, bladder	Normal flora; intestinal tract	(20)
Leptospira interrogans★	Leptospirosis	Kidney, liver	Animal urine; contact with contaminated water, soil	(21)
Gardnerella vaginalis	Bacterial vaginosis	Vagina, occasionally penis	Mixed infection; normal flora; intimate contact	(20)
Fungal pathogens (chapter 22)				
Candida albicans	Yeast infections, candidiasis	Vagina, vulva, penis, urethra	Discharge from infected membranes; STD	
Parasitic (protozoan, helminth) pathogens (chapter 23)				
Trichomonas vaginalis	Trichomoniasis	Vagina, vulva, urethra	Genital discharges, direct contact; STD	

continued

TABLE A.5 *continued*
Sexually Transmitted Diseases (STDs) and Infections of the Genitourinary (GU) Tract

Name of Microbe	Name of Disease	Specific Targets of Infection	Source/Mode of Acquisition	(Ch)
Viral pathogens				
Molluscum contagiosum poxvirus	Molluscum contagiosum	Skin, mucous membranes	Pox lesions; STD	(24)
Herpes simplex, type 2	Genital herpes	Genitalia, perineum	Contact with vesicles, shed skin cells; STD	(24)
Human papillomavirus	Genital warts, condylomata	Membranes of the vagina, penis	Direct contact with warts; some fomites; STD	(24)
Hepatitis B virus★	Hepatitis B	Liver	Semen, vaginal fluids, blood; STD	(24)
Human immunodeficiency virus (HIV)★	AIDS	White blood cells, brain cells	Blood, semen, vaginal fluids; STD	(25)

TABLE A.6
Infections Affecting Multiple Systems (Circulatory and Lymphatic)

Name of Microbe	Name of Disease	Specific Targets of Disease	Source/Mode of Acquisition	(Ch)
Bacterial pathogens				
Staphylococcus aureus★	Toxic shock syndrome	Vagina, uterus, kidney, liver, blood	Normal flora; grow in environment created by tampons	(18)
	Osteomyelitis	Interior of long bones	Focal from skin infection	
Streptococcus pyogenes★	Scarlet fever	Skin, brain	Toxemia; complication of strep pharyngitis	(18)
	Rheumatic fever	Heart valves	Autoimmune reaction to strep toxins	
	Glomerulonephritis	Kidney	Immune reaction blocks filtration apparatus	
Streptococcus spp. (viridans)	Subacute endocarditis	Lining of heart and valves	Normal flora invades during dental work; colonizes heart	(18)
Corynebacterium diphtheriae★	Diphtheria	Heart muscle, nerves, occasionally skin	Droplets from carriers; local infection leads to toxemia	(19)
Mycobacterium tuberculosis★	Disseminated tuberculosis	Lymph nodes, kidneys, bones, brain, sex organs	Reactivation of latent lung infection	(19)
Yersinia pestis★	Plague	Lymph nodes, blood	Bite of infected flea	(20)
Francisella tularensis	Tularemia Rabbit fever	Skin, lymph nodes, eyes, lungs	Blood-sucking arthropods (ticks, mosquitoes); direct contact with animals and meat; aerosols	(20)
Borrelia recurrentis	Borreliosis, relapsing fever	Liver, heart, spleen, kidneys, nerves	Human lice; influenced by poor hygiene, crowding	(21)
B. hermsii	Same as above	Same as above	Soft tick bite; reservoir is wild rodents	(21)
B. burgdorferi	Lyme disease	Skin, joints, nerves, heart	Larval *Ixodes* ticks; reservoir is deer and rodents	(21)
Rickettsia prowazekii	Epidemic typhus	Blood vessels in many organs	Human lice, bites, feces, crowding, lack of sanitation	(21)
R. rickettsii	Rocky Mountain spotted fever	Cardiovascular, central nervous, skin	Wood and dog ticks; picked up in mountain or rural habitats	(21)
R. typhi	Endemic typhus	Similar to epidemic typhus	Rat flea vector; rodent reservoir; occupational contact	(21)
Coxiella burnetii★	See table A.2			

Fungal pathogens (chapter 22)
Many pathogenic fungi (*Histoplasma, Coccidioides, Blastomyces,* and *Cryptococcus*) have a systemic component to their infection, especially in immunocompromised patients. Because their primary infection involves the lungs, information on those fungi is listed in table A.2.

continued

TABLE A.6 *continued*

Infections Affecting Multiple Systems (Circulatory and Lymphatic)

Name of Microbe	Name of Disease	Specific Targets of Disease	Source/Mode of Acquisition	(Ch)
Parasitic (protozoan, helminth) pathogens (chapter 23)				
Leishmania spp.	Systemic leishmaniasis (kala-azar)	Spleen, liver, lymph nodes	Bite by sand (phlebotamine) fly; reservoir is dogs, rodents, wild carnivores	
Plasmodium spp.	Malaria	Liver, blood, kidney	Female *Anopheles* mosquito; human reservoir, mainly tropical	
Toxoplasma gondii★	Toxoplasmosis	Pharynx, lymph nodes, nervous system	Cysts from cats and other animals are ingested; many hosts	
Strongyloides stercoralis	Strongyloidiasis Threadworm	Intestine, trachea, lungs, pharynx, subcutaneous; migrate through systems	Larvae deposited into soil or water by feces; burrow into skin	
Schistosoma spp.	Schistosomiasis, blood fluke	Liver, lungs, blood vessels in intestine, bladder	Larval forms develop in aquatic snails; penetrate skin of bathers	
Wuchereria bancrofti	Filariasis Elephantiasis	Lymphatic circulation, blood vessels	Various mosquito species carry larval stage; human reservoir	
Viral pathogens				
Cytomegalovirus (CMV)	CMV mononucleosis, congenital CMV	Lymph glands, salivary glands	Human saliva, milk, urine, semen, vaginal fluids; intimate contact	(24)
Epstein-Barr virus	Infectious mononucleosis	Lymphoid tissue, salivary glands	Oral contact, exposure to saliva, fomites	(24)
Hepatitis B virus★	See table A.5			
Flavivirus	Yellow fever	Liver, blood vessels, kidney, mucous membranes	Various mosquito species; human and monkey reservoirs	(24) (25)
	Dengue fever (breakbone fever)	Skin, muscles, and joints	Several mosquito vectors	(25)
Filovirus	Ebola fever	Every organ and tissue except muscle and bone	Close contact with blood and secretions of victim; monkey reservoir?	(25)
Arenavirus	Lassa fever	Blood vessels, kidney, heart	Inhalation of airborne virus from rodent excreta	(25)
HIV★	See table A.5			
HTLV I	Adult T-cell leukemia Sézary-cell leukemia	T cells, lymphoid tissue	Blood, blood products, sexual intercourse, IV drugs	(25)
HTLV II	Hairy-cell leukemia	B cells	Blood, shared needles	(25)

TABLE A.7
Diseases of the Nervous System

Name of Microbe	Name of Disease	Specific Targets of Disease	Source/Mode of Acquisition	(Ch)
Bacterial pathogens				
Neisseria meningitidis★	Meningococcal meningitis	Membranes covering brain, spinal cord	Contact with respiratory secretions or droplets of human carrier	(18)
Haemophilus influenzae★	Acute meningitis	Meninges, brain	Contact with carrier; normal flora of throat	(20)
Clostridium tetani	Tetanus	Inhibitory neurons in spinal column	Spores from soil introduced into wound; toxin formed	(19)
Clostridium botulinum★	Botulism	Neuromuscular junction	Canned vegetables, meats; an intoxication	(19)
Mycobacterium leprae★	Leprosy (Hansen disease)	Peripheral nerves of skin	Close contact with infected person; enters skin	(19)
Fungal pathogens (chapter 22)				
Cryptococcus neoformans★	Cryptococcal meningitis	Brain, meninges	Pigeon droppings; inhalation of yeasts in air and dust	
Parasitic (protozoan, helminth) pathogens (chapter 23)				
Naegleria fowleri	Primary acute meningoencephalitis	Brain, spinal cord	Acquired while swimming in fresh, brackish water; ameba invades nasal mucosa	
Acanthamoeba	Meningoencephalitis	Eye, brain	Ameba lives in fresh water; can invade cuts and abrasions	
Trypanosoma brucei	African trypanosomiasis, sleeping sickness	Brain, heart, many organs	Bite of tsetse fly, vector	
Trypanosoma cruzi	South American trypanosomiasis; Chagas disease	Nerve ganglia, muscle	Feces of reduviid bug rubbed into bite	
Viral pathogens (chapters 24 and 25)				
Poliovirus★	Poliomyelitis	Anterior horn cells; motor neurons	Contaminated food, water objects	
Rabies virus	Rabies	Brain, spinal nerves, ganglia	Saliva of infected animal enters bite wound	
Encephalitis viruses	St. Louis encephalitis Western equine encephalitis Eastern equine encephalitis	Brain, meninges, spinal cord	Bite of infected mosquito	
Bunyavirus	California encephalitis	Brain, meninges	Mosquito bite; reservoir is wild birds	
HIV★ (see table A.5)	AIDS, dementia	Destroys brain cells, peripheral nerves	Blood, semen, vaginal fluids; STD	
JC polyomavirus	Progressive multifocal leukoencephalopathy	Oligodendrocytes in cerebrum	Not well defined	
Prions	Spongiform encephalopathies Creutzfeldt-Jakob disease	Brain, neurons	Intimate contact with infected tissues	

Answers to Multiple-Choice Questions and Selected Matching Questions

Chapter 1
1. c
2. d
3. d
4. c
5. d
6. a
7. c
8. b
9. d
10. a
11. a
12. c
13. b
14. d
15. 1st col:
3, 7, 4, 2
2nd col:
8, 5, 6, 1

Chapter 2
1. c
2. c
3. b
4. c
5. a
6. e
7. a
8. d
9. a
10. c
11. c
12. a
13. b
14. c
15. c
16. b
17. b
18. d
19. b
20. c
21. c
22. c
23. a
24. d

Chapter 3
1. b
2. b
3. c
4. d
5. a
6. d
7. b
8. b

9. b
10. c
11. a
12. b

Chapter 4
1. d
2. a
3. c
4. a
5. c
6. b
7. c
8. d
9. b
10. d
11. c
12. d
13. c
14. b
15. a

Chapter 5
1. b
2. d
3. b
4. d
5. a
6. b
7. c
8. d
9. b
10. b
11. d
12. a
13. d
14. b
15. c
16. Matching:
b, e, c, h, g,
j, i, d, a, f

Chapter 6
1. c
2. d
3. d
4. b
5. d
6. a
7. a
8. d
9. b
10. b
11. c
12. d

13. d
14. a

Chapter 7
1. c
2. a
3. a
4. c
5. c
6. b
7. a
8. a
9. b
10. b
11. c
12. c
13. c
14. c
15. b
16. a

Chapter 8
1. b
2. a
3. c
4. d
5. d
6. b
7. b
8. c
9. b
10. b
11. a
12. c
13. a
14. d
15. b
16. c
17. c
18. c
19. c
20. Matching:
c, a, b, c,
b, a, c, c

Chapter 9
1. b
2. e
3. b
4. b
5. c
6. b
7. c
8. a
9. b

10. a
11. b
12. a
13. d
14. b
15. d
16. b
17. Matching:
e/h, f, b, g,
e, c, a, i,
c/e/h

Chapter 10
1. c
2. c
3. b
4. c
5. a
6. b
7. b
8. c
9. c
10. d
11. d
12. Matching:
h, c, f, a,
g, b, e, d

Chapter 11
1. d
2. c
3. b
4. a
5. c
6. b
7. b
8. d
9. c
10. b
11. d
12. c
13. d
14. a
15. b
16. c

Chapter 12
1. b
2. c
3. b
4. a
5. d
6. b
7. c
8. c

9. a
10. d
11. a
12. c
13. c
14. b

Chapter 13
1. a
2. d
3. b
4. d
5. c
6. d
7. b
8. c
9. c
10. c
11. c
12. c
13. a
14. a
15. d

Chapter 14
1. b
2. b
3. d
4. b
5. b
6. c
7. b
8. a
9. c
10. d
11. c
12. c
13. c
14. d
15. d

Chapter 15
1. a
2. c
3. a
4. c
5. c
6. c
7. a
8. c
9. a
10. c
11. a
12. c
13. b

14. e
15. Matching:
IgG bghi
IgA ac
IgD be
IgE bj
IgM dfi

Chapter 16
1. b
2. c
3. d
4. b
5. c
6. c
7. d
8. b
9. c
10. a
11. a
12. d
13. c
14. Matching:
aei, adh,
cdh, bdh,
adh, bdh,
cj, ch, agh,
adh, ce

Chapter 17
1. d
2. d
3. d
4. c
5. b
6. c
7. b
8. a
9. d
10. d
11. d
12. a
13. d
14. c
15. d
16. d

Chapter 18
1. d
2. a
3. c
4. d
5. b
6. c
7. b

8. d
9. c
10. c
11. a
12. c
13. d
14. a
15. Matching:
h, j, d, l, m,
a, f, i, e, k,
b, c

Chapter 19
1. b
2. c
3. b
4. d
5. c
6. c
7. d
8. c
9. d
10. b
11. c
12. d
13. c
14. d
15. a
16. a
16. Matching:
1. a
2. b
3. c
4. b
5. c
6. d
7. bf
8. d
9. a

Chapter 20
1. b
2. b
3. d
4. b
5. c
6. d
7. d
8. b
9. d
10. d
11. c
12. d
13. a, c, d, e, h
14. all but c

15. Matching:
l, i, h, a, d,
g, k, e, j,
m, f, c, b

Chapter 21
1. b
2. b
3. c
4. d
5. d
6. e
7. d
8. c
9. b
10. d
11. a
12. d
13. c
14. d
15. d
16. a
17. b
21. Matching:
b, c, f, f, c,
d, a, d, g, a,
e, g, g

Chapter 22
1. c
2. a
3. a
4. d
5. d
6. a
7. a
8. a
9. c
10. c
11. c
12. c
13. d
14. b
15. Matching:
b, g, d, e, c,
a, f

Chapter 23
1. c
2. b
3. d
4. b
5. b
6. d
7. c

8. d
9. c
10. a
11. d
12. d
13. b
14. a
15. a
16. c
17. Matching:
f, j, k, g, h,
c, i, l, a, d,
m, e, b

Chapter 24
1. d
2. c
3. b
4. a
5. d
6. d
7. c
8. d
9. a
10. b
11. b
12. b
13. c
14. d
15. Matching:
a/e, g, i, c,
h, l, b/k,
f/k, i, d, j

Chapter 25
1. a
2. b
3. c
4. b
5. d
6. d
7. b
8. d
9. d
10. c
11. d
12. d
13. b
14. c
15. c
17. Matching:
4, 2, 3, 7,
3, 1, 8, 6,
1, 3

Chapter 26
1. c
2. b
3. b
4. c
5. d
6. c
7. c
8. d
9. d
10. b
11. c
12. b
13. a
14. b
15. d
16. d
17. d
18. b

TABLE A.8

Infections of the Eye and Ear

Name of Microbe	Name of Disease	Specific Targets of Infection	Source/Mode of Transmission	(Ch)
Bacterial pathogens				
Neisseria gonorrhoeae★	Ophthalmia neonatorum	Conjunctiva, cornea, eyelid	Vaginal mucus enters eyes during birth	(18)
Streptococcus pneumoniae★	Otitis media	Middle ear	Nasopharyngeal secretions forced into eustachian tube	(18)
Haemophilus ducreyi	Pinkeye	Conjunctiva, sclera	Contaminated fingers, fomites	(20)
Chlamydia trachomatis★	Inclusion conjunctivitis, trachoma	Conjunctiva, inner eyelid, cornea	Infected birth canal, contaminated fingers	(21)
Fungal pathogens (chapter 22)				
Aspergillus fumigatus	Aspergillosis	Eyelids, conjunctiva	Contact lens contaminated with spores	
Fusarium solani	Corneal ulcer, keratitis	Cornea	Eye is accidentally scratched, contact lens is contaminated	
Parasitic (protozoan, helminth) pathogens (chapter 23)				
Onchocerca volvulus	Onchocerciasis, river blindness	Eye, skin, blood	Black flies inject larvae into bite	
Loa loa	African eye worm	Subcutaneous tissues, conjunctiva, cornea	Bites by dipteran flies	
Viral pathogens				
Coxsackievirus	Hemorrhagic conjunctivitis	Conjunctiva, sclera, eyelid	Contact with fluids, secretions of carrier	(25)

Glossary

A

abiogenesis The belief in spontaneous generation as a source of life.

abiotic Nonliving factors such as soil, water, temperature, and light that are studied when looking at an ecosystem.

ABO blood group system Developed by Karl Landsteiner in 1904; the identification of different blood groups based on differing isoantigen markers characteristic of each blood type.

abscess An inflamed, fibrous lesion enclosing a core of pus.

abyssal zone The deepest region of the ocean; a sunless, high-pressure, cold, anaerobic habitat.

acid-fast A term referring to the property of mycobacteria to retain carbol fuchsin even in the presence of acid alcohol. The staining procedure is used to diagnose tuberculosis.

acidic A solution with a pH value above 7 on the pH scale.

actinomycetes A group of filamentous, funguslike bacteria.

active immunity Immunity acquired through direct stimulation of the immune system by antigen.

active site The specific region on an apoenzyme that binds substrate. The site for reaction catalysis.

activated lymphocyte A T or B cell that has received an immune stimulus, such as antigens or cytokines, and is undergoing rapid synthesis and proliferation.

active transport Nutrient transport method that requires carrier proteins in the membranes of the living cells and the expenditure of energy.

acute Characterized by rapid onset and short duration.

acyclovir A synthetic purine analog that blocks DNA synthesis in certain viruses, particularly the herpes simplex viruses.

adenine (A) One of the nitrogen bases found in DNA and RNA, with a purine form.

adenosine triphosphate (ATP) A nucleotide that is the primary source of energy to cells.

adenovirus Nonenveloped DNA virus; means of transmission is human-to-human via respiratory and ocular secretions.

adhesion The process by which microbes gain a more stable foothold at the portal of entry; often involves a specific interaction between the molecules on the microbial surface and the receptors on the host cell.

adjuvant In immunology, a chemical vehicle that enhances antigenicity, presumably by prolonging antigen retention at the injection site.

aerobe A microorganism that lives and grows in the presence of free gaseous oxygen (O_2).

aflatoxin From *Aspergillus flavus* toxin, a mycotoxin that typically poisons moldy animal feed and can cause liver cancer in humans and other animals.

agammaglobulinemia Also called hypogammaglobulinemia. The absence of or severely reduced levels of antibodies in serum.

agar A polysaccharide found in seaweed and commonly used to prepare solid culture media.

agglutination The aggregation by antibodies of suspended cells or similar-sized particles (agglutinogens) into clumps that settle.

agglutinin A specific antibody that cross-links agglutinogen, causing it to aggregate.

agglutinogen An antigenic substance on cell surfaces that evokes agglutinin formation against it.

agranulocyte One form of leukocyte (white blood cells), having globular, non-lobed nuclei and lacking prominent cytoplasmic granules.

AIDS Acquired immunodeficiency syndrome. The complex of signs and symptoms characteristic of the late phase of human immunodeficiency virus (HIV) infection.

allele A gene that occupies the same location as other alternative (allelic) genes on paired chromosomes.

allergen A substance that provokes an allergic response.

allergy The altered, usually exaggerated, immune response to an allergen. Also called hypersensitivity.

alloantigen An antigen that is present in some but not all members of the same species.

allograft Relatively compatible tissue exchange between nonidentical members of the same species. Also called homograft.

allosteric Pertaining to the altered activity of an enzyme due to the binding of a molecule to a region other than the enzyme's active site.

amastigote The rounded or ovoid nonflagellated form of the *Leishmania* parasite.

Ames test A method for detecting mutagenic and potentially carcinogenic agents based upon the genetic alteration of nutritionally defective bacteria.

amination The addition of an amine ($—NH_2$) group to a molecule.

amino acids The building blocks of protein. Amino acids exist in 20 naturally occurring forms that impart different characteristics to the various proteins they compose.

aminoglycoside A complex group of drugs derived from soil actinomycetes that impairs ribosome function and has antibiotic potential. Example: streptomycin.

ammonification Phase of the nitrogen cycle in which ammonia is released from decomposing organic material.

amphibolism Pertaining to the metabolic pathways that serve multiple functions in the breakdown, synthesis, and conversion of metabolites.

amphipathic Relating to a compound that has contrasting characteristics, such as hydrophilic-hydrophobic or acid-base.

amphitrichous Having a single flagellum or a tuft of flagella at opposite poles of a microbial cell.

amplicon DNA strand that has been primed for replication during polymerase chain reaction.

anabolism The energy-consuming process of incorporating nutrients into protoplasm through biosynthesis.

anaerobe A microorganism that grows best, or exclusively, in the absence of oxygen.

analog In chemistry, a compound that closely resembles another in structure.

anamnestic In immunology, an augmented response or memory related to a prior stimulation of the immune system by antigen. It boosts the levels of immune substances.

anaphylaxis The unusual or exaggerated allergic reaction to antigen that leads to severe respiratory and cardiac complications.

anion A negatively charged ion.

anneal Characteristic of changing binding properties in response to heating and cooling.

antagonism Relationship in which microorganisms compete for survival in a common environment by taking actions that inhibit or destroy another organism.

anthrax A zoonotic disease of herbivorous livestock. The anthrax bacillus is a facultative parasite and can infect humans in a number of ways. In its most virulent form, it can be fatal.

antibiotic A chemical substance from one microorganism that can inhibit or kill another microbe even in minute amounts.

antibody A large protein molecule evoked in response to an antigen that interacts specifically with that antigen.

anticodon The trinucleotide sequence of transfer RNA that is complementary to the trinucleotide sequence of messenger RNA (the codon).

antigen Any cell, particle, or chemical that induces a specific immune response by B cells or T cells and can stimulate resistance to an infection or a toxin. *See* immunogen.

antigen binding site Specific region at the ends of the antibody molecule that recognize specific antigens. These sites have numerous shapes to fit a wide variety of antigens.

antigenic determinant The precise molecular group of an antigen that defines its specificity and triggers the immune response.

antigenic drift Minor antigenic changes in the influenza A virus due to mutations in the spikes' genes.

antigenic shift Major changes in the influenza A virus due to recombination of viral strains from two different host species.

antihistamine A drug that counters the action of histamine and is useful in allergy treatment.

anti-idiotype An anti-antibody that reacts specifically with the idiotype (variable region or antigen-binding site) of another antibody.

antimetabolite A substance such as a drug that competes with, substitutes for, or interferes with a normal metabolite.

antimicrobic A special class of compounds capable of destroying or inhibiting microorganisms.

anti-oncogene A gene that is responsible for regulating the function of a proto-oncogene. The interaction of these two genes keeps the cell division cycle operating normally.

antiseptic A growth-inhibiting agent used on tissues to prevent infection.

antiserum Antibody-rich serum derived from the blood of animals (deliberately immunized against infectious or toxic antigen) or from people who have recovered from specific infections.

antitoxin Globulin fraction of serum that neutralizes a specific toxin. Also refers to the specific antitoxin antibody itself.

apoenzyme The protein part of an enzyme, as opposed to the nonprotein or inorganic cofactors.

apoptosis The genetically programmed death of cells that is both a natural process of development and the body's means of destroying abnormal or infected cells.

aquifer A subterranean water-bearing stratum of permeable rock, sand, or gravel.

arbovirus Arthropod-borne virus, including togaviruses, reoviruses, flaviviruses, and bunyaviruses. These viruses generally cause mild, undifferentiated fevers and occasionally cause severe encephalitides and hemorrhagic fever.

arthrospore A fungal spore formed by the septation and fragmentation of hyphae.

Arthus reaction An immune complex phenomenon that develops after repeat injection. This localized inflammation results from aggregates of antigen and antibody that bind, complement, and attract neutrophils.

artificial chromosome A large packet of DNA from yeasts (YAC), bacteria (BAC), or humans (HAC) that can be used to carry, transfer, or analyze isolated foreign DNA.

ascospore A spore formed within a saclike cell (ascus) of Ascomycota following nuclear fusion and meiosis.

ascus Special fungal sac in which haploid spores are created.

asepsis A condition free of viable pathogenic microorganisms.

asymptomatic An infection that produces no noticeable symptoms even though the microbe is active in the host tissue.

atomic number (AN) A measurement that reflects the number of protons in an atom of a particular element.

atomic weight The average of the mass numbers of all the isotopic forms for a particular element.

atom The smallest particle of an element to retain all the properties of that element.

atopy Allergic reaction classified as type I, with a strong familial relationship; caused by allergens such as pollen, insect venom, food, and dander; involves IgE antibody; includes symptoms of hay fever, asthma, and skin rash.

ATP synthase A unique enzyme located in the mitochondrial cristae and chloroplast grana that harnesses the flux of hydrogen ions to the synthesis of ATP.

attenuate To reduce the virulence of a pathogenic bacterium or virus by passing it through a non-native host or by long-term subculture.

autoantibody An "anti-self" antibody having an affinity for tissue antigens of the subject in which it is formed.

autoantigen Molecules that are inherently part of self but are perceived by the immune system as foreign.

autograft Tissue or organ surgically transplanted to another site on the same subject.

autoimmune disease The pathologic condition arising from the production of antibodies against autoantigens. Example: rheumatoid arthritis. Also called autoimmunity.

autosome A chromosome of somatic cells as opposed to a sex chromosome of gametes.

autotroph A microorganism that requires only inorganic nutrients and whose sole source of carbon is carbon dioxide.

axenic A sterile state such as a pure culture. An axenic animal is born and raised in a germ-free environment. *See* gnotobiotic.

axial filament A type of flagellum (called an endoflagellum) that lies in the periplasmic space of spirochetes and is responsible for locomotion. Also called periplasmic flagellum.

azole Five-membered heterocyclic compounds typical of histidine, which are used in antifungal therapy.

B

bacillus Bacterial cell shape that is cylindrical (longer than it is wide).

bacteremia The presence of viable bacteria in circulating blood.

bacterial chromosome A circular body in bacteria that contains the primary genetic material. Also called nucleoid.

bactericide An agent that kills bacteria.

bacteriocin Proteins produced by certain bacteria that are lethal against closely related bacteria and are narrow spectrum compared with antibiotics; these proteins are coded and transferred in plasmids.

bacteriophage A virus that specifically infects bacteria.

bacteriostatic Any process or agent that inhibits bacterial growth.

barophile A microorganism that thrives under high (usually hydrostatic) pressure.

basement membrane A thin layer (1–6 μm) of protein and polysaccharide found at the base of epithelial tissues.

basic A solution with a pH value below 7 on the pH scale.

basidiospore A sexual spore that arises from a basidium. Found in basidiomycota fungi.

basidium A reproductive cell created when the swollen terminal cell of a hypha develops filaments (sterigmata) that form spores.

basophil A motile polymorphonuclear leukocyte that binds IgE. The basophilic cytoplasmic granules contain mediators of anaphylaxis and atopy.

bdellovibrio A bacterium that preys on certain other bacteria. It bores a hole into a specific host and inserts itself between the protoplast and the cell wall. There it elongates before subdividing into several cells and devouring the host cell.

benign tumor A self-contained mass within an organ that does not spread into adjacent tissue.

benthic zone The sedimentary bottom region of a pond, lake, or ocean.

beta-lactamase An enzyme secreted by certain bacteria that cleaves the beta-lactam ring of penicillin and cephalosporin and thus provides for resistance against the antibiotic. *See* penicillinase.

beta oxidation The degradation of long-chain fatty acids. Two-carbon fragments are formed as a result of enzymatic attack directed against the second or beta carbon of the hydrocarbon chain. Aided by coenzyme A, the fragments enter the tricarboxylic acid (TCA) cycle and are processed for ATP synthesis.

binary fission The formation of two new cells of approximately equal size as the result of parent cell division.

binomial system Scientific method of assigning names to organisms that employs two names to identify every organism—genus name plus species name.

bioamplication The concentration or accumulation of a pollutant in living tissue through the natural flow of an ecosystem.

biochemistry The study of organic compounds produced by (or components of) living things. The four main categories of biochemicals are carbohydrates, lipids, proteins, and nucleic acid.

bioenergetics The study of the production and use of energy by cells.

bioethics The study of biological issues and how they relate to human conduct and moral judgment.

biofilm A complex association that arises from a mixture of microorganisms growing together on the surface of a habitat.

biogenesis Belief that living things can only arise from others of the same kind.

biomes Particular climate regions in a terrestrial realm.

bioremediation The use of microbes to reduce or degrade pollutants, industrial wastes, and household garbage.

biosphere Habitable regions comprising the aquatic (hydrospheric), soil-rock (lithospheric), and air (atmospheric) environments.

biotechnology The use of microbes or their products in the commercial or industrial realm.

biotic Living factors such as parasites, food substrates, or other living or once-living organisms that are studied when looking at an ecosystem.

blast cell An immature precursor cell of B and T lymphocytes. Also called a lymphoblast.

blocking antibody The IgG class of immunoglobulins that competes with IgE antibody for allergens, thus blocking the degranulation of basophils and mast cells.

B lymphocyte (B cell) A white blood cell that gives rise to plasma cells and antibodies.

booster The additional doses of vaccine antigen administered to increase an immune response and extend protection.

botulin *Clostridium botulinum* toxin. Ingestion of this potent exotoxin leads to flaccid paralysis.

bradykinin An active polypeptide that is a potent vasodilator released from IgE-coated mast cells during anaphylaxis.

broad spectrum A word to denote drugs that affect many different types of bacteria, both gram-positive and gram-negative.

Brownian movement The passive, erratic, nondirectional motion exhibited by microscopic particles. The jostling comes from being randomly bumped by submicroscopic particles, usually water molecules, in which the visible particles are suspended.

brucellosis A zoonosis transmitted to humans from infected animals or animal products; causes a fluctuating pattern of severe fever in humans as well as muscle pain, weakness, headache, weight loss, and profuse sweating. Also called undulant fever.

bubo The swelling of one or more lymph nodes due to inflammation.

budding *See* exocytosis.

bulla A large, bubblelike vesicle in a region of separation between the epidermis and the subepidermal layer. The space is usually filled with serum and sometimes with blood.

C

calculus Dental deposit formed when plaque becomes mineralized with calcium and phosphate crystals. Also called tartar.

Calvin cycle The recurrent photosynthetic pathway characterized by CO_2 fixation and glucose synthesis. Also called the dark reactions.

cancer Any malignant neoplasm that invades surrounding tissue and can metastasize to other locations. A carcinoma is derived from epithelial tissue, and a sarcoma arises from proliferating mesodermal cells of connective tissue.

capsid The protein covering of a virus's nucleic acid core. Capsids exhibit symmetry due to the regular arrangement of subunits called capsomers. *See* icosahedron.

capsomer A subunit of the virus capsid shaped as a triangle or disc.

capsule In bacteria, the loose, gel-like covering or slime made chiefly of simple polysaccharides. This layer is protective and can be associated with virulence.

carbohydrate A compound containing primarily carbon, hydrogen, and oxygen in a 1:2:1 ratio.

carbuncle A deep staphylococcal abscess joining several neighboring hair follicles.

carcinoma Cancers originating in epithelial tissue.

carotenoid Yellow, orange, or red photosynthetic pigments.

carrier A person who harbors infections and inconspicuously spreads them to others. Also, a chemical agent that can accept an atom, chemical radical, or subatomic particle from one compound and pass it on to another.

caseous lesion Necrotic area of lung tubercle superficially resembling cheese. Typical of tuberculosis.

catabolism The chemical breakdown of complex compounds into simpler units to be used in cell metabolism.

catalyst A substance that alters the rate of a reaction without being consumed or permanently changed by it. In cells, enzymes are catalysts.

cation A positively charged ion.

cecum The intestinal pocket that forms the first segment of the large intestine. Also called the appendix.

cell-mediated The type of immune responses brought about by T cells, such as cytotoxic, suppressor, and helper effects.

cellulose A long, fibrous polymer composed of β-glucose; one of the most common substances on earth.

cephalosporins A group of broad-spectrum antibiotics isolated from the fungus *Cephalosporium.*

cercaria The free-swimming larva of the schistosome trematode that emerges from the snail host and can penetrate human skin, causing schistosomiasis.

cestode The common name for tapeworms that parasitize humans and domestic animals.

chancre The primary sore of syphilis that forms at the site of penetration by *Treponema pallidum.* It begins as a hard, dull red, painless papule that erodes from the center.

chancroid A lesion that resembles a chancre but is soft and is caused by *Haemophilus ducreyi.*

chemical bond A link formed between molecules when two or more atoms share, donate, or accept electrons.

chemoautotroph An organism that relies upon inorganic chemicals for its energy and carbon dioxide for its carbon. Also called a chemolithotroph.

chemoheterotroph Microorganisms that derive their nutritional needs from organic compounds.

chemokine Chemical mediators (cytokines) that stimulate the movement and migration of white blood cells.

chemostat A growth chamber with an outflow that is equal to the continuous inflow of nutrient media. This steady-state growth device is used to study such events as cell division, mutation rates, and enzyme regulation.

chemotaxis The tendency of organisms to move in response to a chemical gradient (toward an attractant or to avoid adverse stimuli).

chemotherapy The use of chemical substances or drugs to treat or prevent disease.

chemotroph Organism that oxidizes compounds to feed on nutrients.

chimera A product formed by the fusion of two different organisms.

chitin A polysaccharide similar to cellulose in chemical structure. This polymer makes up the horny substance of the exoskeletons of arthropods and certain fungi.

chloroplast An organelle containing chlorophyll that is found in photosynthetic eucaryotes.

cholesterol Best-known member of a group of lipids called steroids. Cholesterol is commonly found in cell membranes and animal hormones.

chromatic aberration Deviant focus of magnification due to refraction of the colored wavelengths that make up white light.

chromatin The genetic material of the nucleus. Chromatin is made up of nucleic acid and stains readily with certain dyes.

chromophore The chemical radical of a dye that is responsible for its color and reactivity.

chromosome The tightly coiled bodies in cells that are the primary sites of genes.

chronic Any process or disease that persists over a long duration.

cilium (plural: *cilia*) Eucaryotic structure similar to flagella that propels a protozoan through the environment.

class In the levels of classification, the division of organisms that follows phylum.

clonal selection theory A conceptual explanation for the development of lymphocyte specificity and variety during immune maturation.

clone A colony of cells (or group of organisms) derived from a single cell (or single organism) by asexual reproduction. All units share identical characteristics. Also used as a verb to refer to the process of producing a genetically identical population of cells or genes.

cloning host An organism such as a bacterium or a yeast that receives and replicates a foreign piece of DNA inserted during a genetic engineering experiment.

coagulase A plasma-clotting enzyme secreted by *Staphylococcus aureus.* It contributes to virulence and is involved in forming a fibrin wall that surrounds staphylococcal lesions.

coccobacillus An elongated coccus; a short, thick, oval-shaped bacterial rod.

coccus A spherical-shaped bacterial cell.

codon A specific sequence of three nucleotides in mRNA (or the sense strand of DNA) that constitutes the genetic code for a particular amino acid.

coenzyme A complex organic molecule, several of which are derived from vitamins (e.g., nicotinamide, riboflavin). A coenzyme operates in conjunction with an enzyme. Coenzymes serve as transient carriers of specific atoms or functional groups during metabolic reactions.

cofactor An enzyme accessory. It can be organic, such as coenzymes, or inorganic, such as Fe^{+2}, Mn^{+2}, or Zn^{+2} ions.

cold sterilization The use of nonheating methods such as radiation or filtration to sterilize materials.

coliform A collective term that includes normal enteric bacteria that are gram-negative and lactose-fermenting.

colinear Having corresponding parts of a molecule arranged in the same linear order as another molecule, as in DNA and mRNA.

colony A macroscopic cluster of cells appearing on a solid medium, each arising from the multiplication of a single cell.

colostrum The clear yellow early product of breast milk that is very high in secretory antibodies. Provides passive intestinal protection.

commensalism An unequal relationship in which one species derives benefit without harming the other.

communicable infection Capable of being transmitted from one individual to another.

community The interacting mixture of populations in a given habitat.

competitive inhibition Control process that relies on the ability of metabolic analogs to control microbial growth by successfully competing with a necessary enzyme to halt the growth of bacterial cells.

complement In immunology, serum protein components that act in a definite sequence when set in motion either by an antigen-antibody complex or by factors of the alternative (properdin) pathway.

complementary DNA (cDNA) DNA created by using reverse transcriptase to synthesize DNA from RNA templates.

compounds Molecules that are a combination of two or more different elements.

concentration The expression of the amount of a solute dissolved in a certain amount of solvent. It may be defined by weight, volume, or percentage.

conidia Asexual fungal spores shed as free units from the tips of fertile hyphae.

conjugation In bacteria, the contact between donor and recipient cells associated with the transfer of genetic material such as plasmids. Can involve special (sex) pili. Also a form of sexual recombination in ciliated protozoans.

conjunctivitis Sometimes called "pinkeye." A *Haemophilus* infection of the subconjunctiva that is common among children, is easily transmitted, and is treated with antibiotic eyedrops.

constitutive enzyme An enzyme present in bacterial cells in constant amounts, regardless of the presence of substrate. Enzymes of the central catabolic pathways are typical examples.

contagious Communicable; transmissible by direct contact with infected people and their fresh secretions or excretions.

contaminant An impurity; any undesirable material or organism.

convalescence Recovery; the period between the end of a disease and the complete restoration of health in a patient.

corepressor A molecule that combines with inactive repressor to form active repressor, which attaches to the operator gene site and inhibits the activity of structural genes subordinate to the operator.

covalent bond A chemical bond formed by the sharing of electrons between two atoms.

crista The infolded inner membrane of a mitochondrion that is the site of the respiratory chain and oxidative phosphorylation.

culture The visible accumulation of microorganisms in or on a nutrient medium. Also, the propagation of microorganisms with various media.

curd The coagulated milk protein used in cheese making.

cutaneous Second level of skin, including the stratum corneum and occasionally the upper dermis.

cyst The resistant, dormant, but infectious form of protozoans. Can be important in spread of infectious agents such as *Entamoeba histolytica* and *Giardia lamblia*.

cysticercus The larval form of certain *Taenia* species, which typically infest muscles of mammalian intermediate hosts. Also called bladderworm.

cystine An amino acid, HOOC—CH(NH$_2$)—CH$_2$—S—S—CH$_2$—CH(NH$_2$)COOH. An oxidation product of two cysteine molecules in which the —SH (sulfhydryl) groups form a disulfide union. Also called dicysteine.

cytochrome A group of heme protein compounds whose chief role is in electron and/or hydrogen transport occurring in the last phase of aerobic respiration.

cytokine A chemical substance produced by white blood cells and tissue cells that regulates development, inflammation, and immunity.

cytopathic effect The degenerative changes in cells associated with virus infection. Examples: the formation of multinucleate giant cells (Negri bodies), the prominent cytoplasmic inclusions of nerve cells infected by rabies virus.

cytoplasm Dense fluid encased by the cell membrane; the site of many of the cell's biochemical and synthetic activities.

cytosine (C) One of the nitrogen bases found in DNA and RNA, with a purine form.

cytotoxic Having the capacity to destroy specific cells. One class of T cells attacks cancer cells, virus-infected cells, and eucaryotic pathogens. *See* killer T cells.

D

death phase End of the cell growth due to lack of nutrition, depletion of environment, and accumulation of wastes. Population of cells begins to die.

debridement Trimming away devitalized tissue and foreign matter from a wound.

decomposition The breakdown of dead matter and wastes into simple compounds, that can be directed back into the natural cycle of living things.

decontamination The removal or neutralization of an infectious, poisonous, or injurious agent from a site.

deduction Problem-solving process in which an individual constructs a hypothesis, tests its validity by outlining particular events that are predicted by the hypothesis, and then performs experiments to test for those events.

definitive host The organism in which a parasite develops into its adult or sexually mature stage. Also called the final host.

degerm To physically remove surface oils, debris, and soil from skin to reduce the microbial load.

degranulation The release of cytoplasmic granules, as when cytokines are secreted from mast cell granules.

dehydration synthesis During the formation of a carbohydrate bond, the step in which one carbon molecule gives up its OH group and the other loses the H from its OH group, thereby producing a water molecule. This process is common to all polymerization reactions.

denaturation The loss of normal characteristics resulting from some molecular alteration. Usually in reference to the action of heat or chemicals on proteins whose function depends upon an unaltered tertiary structure.

dendritic cell A large, antigen-processing cell characterized by long, branchlike extensions of the cell membrane.

denitrification The end of the nitrogen cycle when nitrogen compounds are returned to the reservoir in the air.

dental caries A mixed infection of the tooth surface that gradually destroys the enamel and may lead to destruction of the deeper tissue.

deoxyribose A 5-carbon sugar that is an important component of DNA.

deoxyribonucleic acid (DNA) The nucleic acid often referred to as the "double helix." DNA carries the master plan for an organism's heredity.

desensitization *See* hyposensitization.

desiccation To dry thoroughly. To preserve by drying.

desquamate To shed the cuticle in scales; to peel off the outer layer of a surface.

diapedesis The migration of intact blood cells between endothelial cells of a blood vessel such as a venule.

differential medium A single substrate that discriminates between groups of microorganisms on the basis of differences in their appearance due to different chemical reactions.

differential stain A technique that utilizes two dyes to distinguish between different microbial groups or cell parts by color reaction.

diffusion The dispersal of molecules, ions, or microscopic particles propelled down a concentration gradient by spontaneous random motion to achieve a uniform distribution.

dimorphic In mycology, the tendency of some pathogens to alter their growth form from mold to yeast in response to rising temperature.

dinoflagellate A marine algae whose toxin can cause food poisoning. Overgrowth of this algae is responsible for the phenomenon known as "red tide."

diphtheria Infection by *Corynebacterium diphtheriae*. It is transmitted by human carriers or contaminated milk, and the primary infection is in the upper respiratory tract. Several forms of this infection may be fatal if untreated. Vaccination on the recommended schedule can prevent infection.

diplococcus Spherical or oval-shaped bacteria, typically found in pairs.

diploid Somatic cells having twice the basic chromosome number. One set in the pair is derived from the father, and the other from the mother.

disaccharide A sugar containing two monosaccharides. Examples: sucrose (fructose + glucose).

disinfection The destruction of pathogenic nonsporulating microbes or their toxins, usually on inanimate surfaces.

division In the levels of classification, an alternate term for phylum.

domain In the levels of classification, the broadest general category to which an organism is assigned. Members of a domain share only one or a few general characteristics.

droplet nuclei The dried residue of fine droplets produced by mucus and saliva sprayed while sneezing and coughing. Droplet nuclei are less than 5 μm in diameter (large enough to bear a single bacterium and small enough to remain airborne for a long time) and can be carried by air currents. Droplet nuclei are drawn deep into the air passages.

dysentery Diarrheal illness caused by exotoxins.

dyspnea Difficulty in breathing.

E

ecosystem A collection of organisms together with its surrounding physical and chemical factors.

ectoplasm The outer, more viscous region of the cytoplasm of a phagocytic cell such as an ameba. It contains microtubules, but not granules or organelles.

eczema An acute or chronic allergy of the skin associated with itching and burning sensations. Typically, red, edematous, vesicular lesions erupt, leaving the skin scaly and sometimes hyperpigmented.

edema The accumulation of excess fluid in cells, tissues, or serous cavities. Also called swelling.

electrolyte Any compound that ionizes in solution and conducts current in an electrical field.

electromagnetic radiation A form of energy that is emitted as waves and is propagated through space and matter. The spectrum extends from short gamma rays to long radio waves.

electron A negatively charged subatomic particle that is distributed around the nucleus in an atom.

electrophoresis The separation of molecules by size and charge through exposure to an electrical current.

electrostatic Relating to the attraction of opposite charges and the repulsion of like charges. Electrical charge remains stationary as opposed to electrical flow or current.

element A substance comprising only one kind of atom that cannot be degraded into two or more substances without losing its chemical characteristics.

ELISA Abbreviation for **e**nzyme-**l**inked **i**mmuno**s**orbent **a**ssay, a very sensitive serological test used to detect antibodies in diseases such as AIDS.

emerging disease Newly identified diseases that are becoming more prominent.

encystment The process of becoming encapsulated by a membranous sac.

endemic disease A native disease that prevails continuously in a geographic region.

endergonic reaction A chemical reaction that occurs with the absorption and storage of surrounding energy. Antonym: exergonic.

endocarditis An inflammation of the lining and valves of the heart. Often caused by infection with pyogenic cocci.

endocytosis The process whereby solid and liquid materials are taken into the cell through membrane invagination and engulfment into a vesicle.

endoenzyme An intracellular enzyme, as opposed to enzymes that are secreted.

endogenous Originating or produced within an organism or one of its parts.

endoplasmic reticulum An intracellular network of flattened sacs or tubules with or without ribosomes on their surfaces.

endospore A small, dormant, resistant derivative of a bacterial cell that germinates under favorable growth conditions into a vegetative cell. The bacterial genera *Bacillus* and *Clostridium* are typical sporeformers.

endosymbiosis Relationship in which a microorganism resides within a host cell and provides a benefit to the host cell.

endotoxin A bacterial intracellular toxin that is not ordinarily released (as is exotoxin). Endotoxin is composed of a phospholipid-polysaccharide complex that is an integral part of gram-negative bacterial cell walls. Endotoxins can cause severe shock and fever.

energy of activation The minimum energy input necessary for reactants to form products in a chemical reaction.

enriched medium A nutrient medium supplemented with blood, serum, or some growth factor to promote the multiplication of fastidious microorganisms.

enteric Pertaining to the intestine.

enteroinvasive Predisposed to invade the intestinal tissues.

enteropathogenic Pathogenic to the alimentary canal.

enterotoxin A bacterial toxin that specifically targets intestinal mucous membrane cells. Enterotoxigenic strains of *Escherichia coli* and *Staphylococcus aureus* are typical sources.

enveloped virus A virus whose nucleocapsid is enclosed by a membrane derived in part from the host cell. It usually contains exposed glycoprotein spikes specific for the virus.

enzyme A protein biocatalyst that facilitates metabolic reactions.

enzyme induction One of the controls on enzyme synthesis. This occurs when enzymes appear only when suitable substrates are present.

enzyme repression The inhibition of enzyme synthesis by the end product of a catabolic pathway.

eosinophil A leukocyte whose cytoplasmic granules readily stain with red eosin dye.

epidemic A sudden and simultaneous outbreak or increase in the number of cases of disease in a community.

epidemiology The study of the factors affecting the prevalence and spread of disease within a community.

epimastigote The trypanosomal form found in the tsetse fly or reduviid bug vector. Its flagellum originates near the nucleus, extends along an undulating membrane, and emerges from the anterior end.

erysipelas An acute, sharply defined inflammatory disease specifically caused by hemolytic *Streptococcus*. The eruption is limited to the skin but can be complicated by serious systemic symptoms.

erysipeloid An inflammation resembling erysipelas but caused by *Erysipelothrix*, a gram-positive rod. The self-limited cellulitis that appears at the site of an infected wound, usually the hand, comes from handling contaminated fish or meat.

erythema An inflammatory redness of the skin.

erythroblastosis fetalis Hemolytic anemia of the newborn. The anemia comes from hemolysis of Rh-positive fetal erythrocytes by anti-Rh maternal antibodies. Erythroblasts are immature red blood cells prematurely released from the bone marrow.

erythrogenic toxin An exotoxin produced by lysogenized group A strains of β-hemolytic streptococci that is responsible for the severe fever and rash of scarlet fever in the nonimmune individual. Also called a pyrogenic toxin.

eschar A dark, sloughing scab that is the lesion of anthrax and certain rickettsioses.

essential nutrient Any ingredient such as a certain amino acid, fatty acid, vitamin, or mineral that cannot be formed by an organism and must be supplied in the diet. A growth factor.

ester bond A covalent bond formed by reacting carboxylic acid with an OH group:

$$(R—\overset{\overset{\displaystyle O}{\|}}{C}—O—R')$$

Olive and corn oils, lard, and butter fat are examples of triacylglycerols—esters formed between glycerol and three fatty acids.

estuary The intertidal zone where a river empties into the sea.

ethylene oxide A potent, highly water-soluble gas invaluable for gaseous sterilization of heat-sensitive objects such as plastics, surgical and diagnostic appliances, and spices. Potential hazards are related to its carcinogenic, metagenic, residual, and explosive nature. Ethylene oxide is rendered nonexplosive by mixing with 90% CO_2 or fluorocarbon.

etiologic agent The microbial cause of disease; the pathogen.

eucaryotic cell A cell that differs from a procaryotic cell chiefly by having a nuclear membrane (a well-defined nucleus), membrane-bounded subcellular organelles, and mitotic cell division.

eutrophication The process whereby dissolved nutrients resulting from natural seasonal enrichment or industrial pollution of water cause overgrowth of algae and cyanobacteria to the detriment of fish and other large aquatic inhabitants.

evolution Scientific principle that states that living things change gradually through hundreds of millions of years, and these changes are expressed in structural and functional adaptations in each organism. Evolution presumes that those traits which favor survival are preserved and passed on to following generations, and those traits which do not favor survival are lost.

exanthem An eruption or rash of the skin.

exergonic A chemical reaction associated with the release of energy to the surroundings. Antonym: endergonic.

exfoliative toxin A poisonous substance that causes superficial cells of an epithelium to detach and be shed. Example: staphylococcal exfoliatin. Also called an epidermolytic toxin.

exocytosis The process that releases enveloped viruses from the membrane of the host's cytoplasm.

exoenzyme An extracellular enzyme chiefly for hydrolysis of nutrient macromolecules that are otherwise impervious to the cell membrane. It functions in saprobic decomposition of organic debris and can be a factor in invasiveness of pathogens.

exogenous Originating outside the body.

exon A stretch of eucaryotic DNA coding for a corresponding portion of mRNA that is translated into peptides. Intervening stretches of DNA that are not expressed are called introns. During transcription, exons are separated from introns and are spliced together into a continuous mRNA transcript.

exotoxin A toxin (usually protein) that is secreted and acts upon a specific cellular target. Examples: botulin, tetanospasmin, diphtheria toxin, and erythrogenic toxin.

exponential Pertaining to the use of exponents, numbers that are typically written as a superscript to indicate how many times a factor is to be multiplied. Exponents are used in scientific notation to render large, cumbersome numbers into small workable quantities.

extremophiles Organisms capable of living in harsh environments, such as extreme heat or cold.

F

facultative Pertaining to the capacity of microbes to adapt or adjust to variations; not obligate. Example: The presence of oxygen is not obligatory for a facultative anaerobe to grow. *See* obligate.

family In the levels of classification, a mid-level division of organisms that groups more closely related organisms than previous levels. An order is divided into families.

fastidious Requiring special nutritional or environmental conditions for growth. Said of bacteria.

feedback inhibition Temporary end to enzyme action caused by an end product molecule binding to the regulatory site and preventing the enzyme's active site from binding to its substrate.

fermentation The extraction of energy through anaerobic degradation of substrates into simpler, reduced metabolites. In large industrial processes, fermentation can mean any use of microbial metabolism to manufacture organic chemicals or other products.

fertility (F′) factor Donor plasmid that allows synthesis of a pilus in bacterial conjugation. Presence of the factor is indicated by F^+, and lack of the factor is indicated by F^-.

filament A helical structure composed of proteins that is part of bacterial flagella.

filariasis Illnesses caused by infection by nematodes. These illnesses include river blindness, elephantiasis, and eye worm.

fixation In microscopic slide preparation of tissue sections or bacterial smears, fixation pertains to rapid killing, hardening, and adhesion to the slide, while retaining as many natural characteristics as possible. Also refers to the assimilation of inorganic molecules into organic ones, as in carbon or nitrogen fixation.

flagellum A structure that is used to propel the organism through a fluid environment.

flora Beneficial or harmless resident bacteria commonly found on and/or in the human body.

fluid mosaic model A conceptualization of the molecular architecture of cellular membranes as a bilipid layer containing proteins. Membrane proteins are embedded to some degree in this bilayer, where they float freely about.

fluorescence The property possessed by certain minerals and dyes to emit visible light when excited by ultraviolet radiation. A fluorescent dye combined with specific antibody provides a sensitive test for the presence of antigen.

folliculitis An inflammatory reaction involving the formation of papules or pustules in clusters of hair follicles.

fomite Virtually any inanimate object an infected individual has contact with that can serve as a vehicle for the spread of disease.

food pyramid A triangular summation of the consumers, producers, and decomposers in a community and how they are related by trophic, number, and energy parameters.

formalin A 37% aqueous solution of formaldehyde gas; a potent chemical fixative and microbicide.

fructose One of the carbohydrates commonly referred to as sugars. Fructose is commonly fruit sugars.

functional group In chemistry, a particular molecular combination that reacts in predictable ways and confers particular properties on a compound. Examples: —COOH, —OH, —CHO.

furuncle A boil; a localized pyogenic infection arising from a hair follicle.

G

Gaia hypothesis The concept that biotic and abiotic factors sustain suitable conditions for one another simply by their interactions. Named after the mythical Greek goddess of earth.

GALT Abbreviation for **g**ut-**a**ssociated **l**ymphoid **t**issue. Includes Peyer's patches.

gamma globulin The fraction of plasma proteins high in immunoglobulins (antibodies). Preparations from pooled human plasma containing normal antibodies make useful passive immunizing agents against pertussis, polio, measles, and several other diseases.

gas gangrene Disease caused by a clostridial infection of soft tissue or wound. The name refers to the gas produced by the bacteria growing in the tissue. Unless treated early, it is fatal. Also called myonecrosis.

gastroenteritis Inflammation of the lining of the stomach and intestine. May be caused by infection, intestinal disorders, or food poisoning.

gene A site on a chromosome that provides information for a certain cell function. A specific segment of DNA that contains the necessary code to make a protein or RNA molecule.

gene probe Short strands of single-stranded nucleic acid that hybridize specifically with complementary stretches of nucleotides on test samples and thereby serve as a tagging and identification device.

gene therapy The introduction of normal functional genes into people with genetic diseases such as sickle-cell anemia and cystic fibrosis. This is usually accomplished by a virus vector.

generation time Time required for a complete fission cycle—from parent cell to two new daughter cells. Also called doubling time.

genetic engineering A field involving deliberate alterations (recombinations) of the genomes of microbes, plants, and animals through special technological processes.

genetics The science of heredity.

genome The complete set of chromosomes and genes in an organism.

genotype The genetic makeup of an organism. The genotype is ultimately responsible for an organism's phenotype, or expressed characteristics.

genus In the levels of classification, the second most specific level. A family is divided into several genera.

germicide An agent lethal to non-endospore-forming pathogens.

giardiasis Infection by the *Giardia* flagellate. Most common mode of transmission is contaminated food and water. Symptoms include diarrhea, abdominal pain, and flatulence.

gingivitis Inflammation of the gum tissue in contact with the roots of the teeth.

gluconeogenesis The formation of glucose (or glycogen) from noncarbohydrate sources such as protein or fat. Also called glyconeogenesis.

glucose One of the carbohydrates commonly referred to as sugars. Glucose is characterized by its 6-carbon structure.

glycan A polysaccharide.

glycerol A 3-carbon alcohol, with three OH groups that serve as binding sites.

glycocalyx A filamentous network of carbohydrate-rich molecules that coats cells.

glycogen A glucose polymer stored by cells.

glycolysis The energy-yielding breakdown (fermentation) of glucose to pyruvic or lactic acid. It is often called anaerobic glycolysis because no molecular oxygen is consumed in the degradation.

glycosidic bond A bond that joins monosaccharides to form disaccharides and polymers.

gnotobiotic Referring to experiments performed on germ-free animals.

Golgi apparatus An organelle of eucaryotes that participates in packaging and secretion of molecules.

gonococcus Common name for *Neisseria gonorrhoeae,* the agent of gonorrhea.

Gram stain A differential stain for bacteria useful in identification and taxonomy. Gram-positive organisms appear purple from crystal violet-mordant retention, whereas gram-negative organisms appear red after loss of crystal violet and absorbance of the safranin counterstain.

grana Discrete stacks of chlorophyll-containing thylakoids within chloroplasts.

granulocyte A mature leukocyte that contains noticeable granules in a Wright stain. Examples: neutrophils, eosinophils, and basophils.

granuloma A solid mass or nodule of inflammatory tissue containing modified macrophages and lymphocytes. Usually a chronic pathologic process of diseases such as tuberculosis or syphilis.

greenhouse effect The capacity to retain solar energy by a blanket of atmospheric gases that redirects heat waves back toward the earth.

growth factor An organic compound such as a vitamin or amino acid that must be provided in the diet to facilitate growth. An essential nutrient.

guanine (G) One of the nitrogen bases found in DNA and RNA in the purine form.

gumma A nodular, infectious granuloma characteristic of tertiary syphilis.

H

habitat The environment to which an organism is adapted.

halophile A microbe whose growth is either stimulated by salt or requires a high concentration of salt for growth.

Hansen disease A chronic, progressive disease of the skin and nerves caused by infection by a mycobacterium that is a slow-growing, strict parasite. Hansen disease is the preferred name for leprosy.

H antigen The flagellar antigen of motile bacteria. *H* comes from the German word *hauch,* which denotes the appearance of spreading growth on solid medium.

haploid Having a single set of unpaired chromosomes, such as occurs in gametes and certain microbes.

hapten An incomplete or partial antigen. Although it constitutes the determinative group and can bind antigen, hapten cannot stimulate a full immune response without being carried by a larger protein molecule.

hay fever A form of atopic allergy marked by seasonal acute inflammation of the conjunctiva and mucous membranes of the respiratory passages. Symptoms are irritative itching and rhinitis.

helical Having a spiral or coiled shape. Said of certain virus capsids and bacteria.

helminth A term that designates all parasitic worms.

helper T cell A class of thymus-stimulated lymphocytes that facilitate various immune activities such as assisting B cells and macrophages. Also called a T helper cell.

hemagglutinin A molecule that causes red blood cells to clump or agglutinate. Often found on the surfaces of viruses.

hemolysin Any biological agent that is capable of destroying red blood cells and causing the release of hemoglobin. Many bacterial pathogens produce exotoxins that act as hemolysins.

hemolytic disease Incompatible Rh factor between mother and fetus causes maternal antibodies to attack the fetus and trigger complement-mediated lysis in the fetus.

hemolyze When red blood cells burst and release hemoglobin pigment.

hemopoiesis The process by which the various types of blood cells are formed, such as in the bone marrow.

hepadnavirus Enveloped DNA viruses with a predisposition to affect the liver. Hepatitis B is the most serious form.

hepatocyte A liver cell.

heredity Genetic inheritance.

herd immunity The status of collective acquired immunity in a population that reduces the likelihood that nonimmune individuals will contract and spread infection. One aim of vaccination is to induce herd immunity.

hermaphroditic Containing the sex organs for both male and female in one individual.

herpes zoster A recurrent infection caused by latent chickenpox virus. Its manifestation on the skin tends to correspond to dermatomes and to occur in patches that "girdle" the trunk. Also called shingles.

herpetic keratitis Corneal or conjunctival inflammation due to herpesvirus type 1.

heterophile antigen An antigen present in a variety of phylogenetically unrelated species. Example: red blood cell antigens and the glycocalyx of bacteria.

heterotroph An organism that relies upon organic compounds for its carbon and energy needs.

hexose A 6-carbon sugar such as glucose and fructose.

hierarchies Levels of power. Arrangement in order of rank.

histamine A cytokine released when mast cells and basophils release their granules. An important mediator of allergy, its effects include smooth muscle contraction, increased vascular permeability, and increased mucus secretion.

histiocyte Another term for macrophage.

histone Proteins associated with eucaryotic DNA. These simple proteins serve as winding spools to compact and condense the chromosomes.

histoplasmin The antigenic extract of *Histoplasma capsulatum,* the causative agent of histoplasmosis. The preparation is used in skin tests for diagnosis and for conducting surveys to determine the geographic distribution of the fungus.

HLA An abbreviation for **h**uman **l**eukocyte **a**ntigens. This closely linked cluster of genes programs for cell surface glycoproteins that control immune interactions between cells and is involved in rejection of allografts. Also called the major histocompatibility complex (MHC).

holoenzyme An enzyme complete with its apoenzyme and cofactors.

hops The ripe, dried fruits of the hop vine *(Humulus lupulus)* that is added to beer wort for flavoring.

host Organism in which smaller organisms or viruses live, feed, and reproduce.

host range The limitation imposed by the characteristics of the host cell on the type of virus that can successfully invade it.

human immunodeficiency virus (HIV) A retro virus that causes acquired immunodeficiency syndrome (AIDS).

humoral immunity Protective molecules (mostly B lymphocytes) carried in the fluids of the body.

hybridization A process that matches complementary strands of nucleic acid (DNA-DNA, RNA-DNA, RNA-RNA). Used for locating specific sites or types of nucleic acids.

hybridoma An artificial cell line that produces monoclonal antibodies. It is formed by fusing (hybridizing) a normal antibody-producing cell with a cancer cell, and it can produce pure antibody indefinitely.

hydatid cyst A sac, usually in liver tissue, containing fluid and larval stages of the echinococcus tapeworm.

hydration The addition of water as in the coating of ions with water molecules as ions enter into aqueous solution.

hydrogen bond A weak chemical bond formed by the attraction of forces between molecules or atoms—in this case, hydrogen and either oxygen or nitrogen. In this type of bond, electrons are not shared, lost, or gained.

hydrologic cycle The continual circulation of water between hydrosphere, atmosphere, and lithosphere.

hydrolase An enzyme that catalyzes the cleavage of a bond with the additions of —H and —OH (parts of a water molecule) at the separation site.

hydrolysis A process in which water is used to break bonds in molecules. Usually occurs in conjunction with an enzyme.

hydrophilic The property of attracting water. Molecules that attract water to their surface are called hydrophilic.

hydrophobic The property of repelling water. Molecules that repel water are called hydrophobic.

hypertonic Having a greater osmotic pressure than a reference solution.

hyphae The tubular threads that make up filamentous fungi (molds). This web of branched and intertwining fibers is called a mycelium.

hypogammaglobulinemia An inborn disease in which the gamma globulin (antibody) fraction of serum is greatly reduced. The condition is associated with a high susceptibility to pyogenic infections.

hyposensitization A therapeutic exposure to known allergens designed to build tolerance and eventually prevent allergic reaction.

hypothesis A tentative explanation of what has been observed or measured.

hypotonic Having a lower osmotic pressure than a reference solution.

I

icosahedron A regular geometric figure having 20 surfaces that meet to form 12 corners. Some virions have capsids that resemble icosahedral crystals.

immune complex reaction Type III hypersensitivity of the immune system. It is characterized by the reaction of soluble antigen with antibody, and the deposition of the resulting complexes in basement membranes of epithelial tissue.

immune surveillance The continual function of macrophages, cytotoxic T cells, and natural killer cells in identifying and destroying cancer cells within the body.

immunity An acquired resistance to an infectious agent due to prior contact with that agent.

immunoassays Extremely sensitive tests that permit rapid and accurate measurement of trace antigen or antibody.

immunocompetence The ability of the body to recognize and react with multiple foreign substances.

immunodeficiency Immune function is incompletely developed, suppressed, or destroyed.

immunogen Any substance that induces a state of sensitivity or resistance after processing by the immune system of the body.

immunoglobulin The chemical class of proteins to which antibodies belong.

immunology The study of the system of body defenses that protect against infection.

immunopathology The study of disease states associated with overreactivity or underreactivity of the immune response.

IMViC Abbreviation for four identification tests: **i**ndole production, **m**ethyl red test, **V**oges-Proskauer test (**i** inserted to concoct a wordlike sound), and **c**itrate as a sole source of carbon. This test was originally developed to distinguish between *Enterobacter aerogenes* (associated with soil) and *Escherichia coli* (a fecal coliform).

incidence In epidemiology, the number of new cases of a disease occurring during a period.

inclusion A relatively inert body in the cytoplasm such as storage granules, glycogen, fat, or some other aggregated metabolic product.

incubate To isolate a sample culture in a temperature-controlled environment to encourage growth.

inducible enzyme An enzyme that increases in amount in direct proportion to the amount of substrate present.

induction Process by which an individual accumulates data or facts and then formulates a general hypothesis that accounts for those facts.

infection The entry, establishment, and multiplication of pathogenic organisms within a host.

infectious disease The state of damage or toxicity in the body caused by an infectious agent.

inflammation A natural, nonspecific response to tissue injury that protects the host from further damage. It stimulates immune reactivity and blocks the spread of an infectious agent.

inoculation The implantation of microorganisms into or upon culture media.

inorganic chemicals Molecules that lack the basic framework of the elements of carbon and hydrogen.

interferon Naturally occurring polypeptides produced by fibroblasts and lymphocytes that can block viral replication and regulate a variety of immune reactions.

interleukin A macrophage agent (interleukin-1, or IL-1) that stimulates lymphocyte function. Stimulated T cells release yet another interleukin (IL-2), which amplifies T-cell response by stimulating additional T cells. T helper cells stimulated by IL-2 stimulate B-cell proliferation and promote antibody production.

intoxication Poisoning that results from the introduction of a toxin into body tissues through ingestion or injection.

intron The segments on split genes of eucaryotes that do not code for polypeptide. They can have regulatory functions. *See* exon.

in utero Literally means "in the uterus"; pertains to events or developments occurring before birth.

in vitro Literally means "in glass," signifying a process or reaction occurring in an artificial environment, as in a test tube or culture medium.

in vivo Literally means "in a living being," signifying a process or reaction occurring in a living thing.

iodophor A combination of iodine and an organic carrier that is a moderate-level disinfectant and antiseptic.

ion An unattached, charged particle.

ionic bond A chemical bond in which electrons are transferred and not shared between atoms.

ionization The aqueous dissociation of an electrolyte into ions.

ionizing radiation Radiant energy consisting of short-wave electromagnetic rays (X ray) or high-speed electrons that cause dislodgment of electrons on target molecules and create ions.

irradiation The application of radiant energy for diagnosis, therapy, disinfection, or sterilization.

isograft Transplanted tissue from one monozygotic twin to the other; transplants between highly inbred animals that are genetically identical.

isolation The separation of microbial cells by serial dilution or mechanical dispersion on solid media to achieve a clone or pure culture.

isotonic Two solutions having the same osmotic pressure such that, when separated by a semipermeable membrane, there is no net movement of solvent in either direction.

isotope A version of an element that is virtually identical in all chemical properties to another version except that their atoms have slightly different atomic masses.

J

jaundice The yellowish pigmentation of skin, mucous membranes, sclera, deeper tissues, and excretions due to abnormal deposition of bile pigments. Jaundice is associated with liver infection, as with hepatitis B virus and leptospirosis.

J chain A small molecule with high sulfhydryl content that secures the heavy chains of IgM to form a pentamer and the heavy chains of IgA to form a dimer.

K

Kaposi sarcoma A malignant or benign neoplasm that appears as multiple hemorrhagic sites on the skin, lymph nodes, and viscera and apparently involves the metastasis of abnormal blood vessel cells. It is a clinical feature of AIDS.

keratitis Inflammation of the cornea.

keratoconjunctivitis Inflammation of the conjunctiva and cornea.

killer T cells A T lymphocyte programmed to directly affix cells and kill them. *See* cytotoxic.

kingdom In the levels of classification, the second division from more general to more specific. Each domain is divided into kingdoms

Koch's postulates A procedure to establish the specific cause of disease. In all cases of infection: (1) The agent must be found; (2) inoculations of a pure culture must reproduce the same disease in animals; (3) the agent must again be present in the experimental animal; and (4) a pure culture must again be obtained.

Koplik's spots Tiny red blisters with central white specks on the mucosal lining of the cheeks. Symptomatic of measles.

L

labile In chemistry, molecules or compounds that are chemically unstable in the presence of environmental changes.

lacrimation The secretion of tears, especially in profusion.

lactose One of the carbohydrates commonly referred to as sugars. Lactose is commonly found in milk.

lactose (*lac*) operon Control system that manages the regulation of lactose metabolism. It is composed of three DNA segments, including a regulator, a control locus, and a structural locus.

lager The maturation process of beer, which is allowed to take place in large vats at a reduced temperature.

lagging strand The newly forming 5′ DNA strand that is discontinuously replicated in segments (Okazaki fragments).

lag phase The early phase of population growth during which no signs of growth occur.

latency The state of being inactive. Example: a latent virus or latent infection.

leading strand The newly forming 3′ DNA strand that is replicated in a continuous fashion without segments.

Legionnaire disease Infection by *Legionella* bacterium. Weakly gram-negative rods are able to survive in aquatic habitats. Some forms may be fatal.

leaven To lighten food material by entrapping gas generated within it. Example: the rising of bread from the CO_2 produced by yeast or baking powder.

leprosy *See* Hansen disease.

lesion A wound, injury, or some other pathologic change in tissues.

leukocidin A heat-labile substance formed by some pyogenic cocci that impairs and sometimes lyses leukocytes.

leukocytes White blood cells. The primary infection-fighting blood cells.

leukocytosis An abnormally large number of leukocytes in the blood, which can be indicative of acute infection.

leukopenia A lower than normal leukocyte count in the blood that can be indicative of blood infection or disease.

leukotriene An unsaturated fatty acid derivative of arachidonic acid. Leukotriene functions in chemotactic activity, smooth muscle contractility, mucous secretion, and capillary permeability.

L form L-phase variants; wall-less forms of some bacteria that are induced by drugs or chemicals. These forms can be involved in infections.

ligase An enzyme required to seal the sticky ends of DNA pieces after splicing.

limnetic zone The deep-water region beyond the shoreline.

lipid A term used to describe a variety of substances that are not soluble in polar solvents such as water, but will dissolve in nonpolar solvents such as benzene and chloroform. Lipids include triglycerides, phospholipids, steroids, and waxes.

lipase A fat-splitting enzyme. Example: Triacylglycerol lipase separates the fatty acid chains from the glycerol backbone of triglycerides.

lipopolysaccharide A molecular complex of lipid and carbohydrate found in the bacterial cell wall. The lipopolysaccharide (LPS) of gram-negative bacteria is an endotoxin with generalized pathologic effects such as fever.

listeriosis Infection with *Listeria monocytogenes,* usually due to eating contaminated dairy products, poultry, or meat. It is usually mild in healthy adults, but can produce severe symptoms in neonates and immunocompromised adults.

lithoautotroph Bacteria that rely on inorganic minerals to supply their nutritional needs. Sometimes referred to as chemoautotrophs.

lithotroph An autotrophic microbe that derives energy from reduced inorganic compounds such as N_2S.

littoral zone The shallow region along a shoreline.

loci A site on a chromosome occupied by a gene.

log phase Maximum rate of cell division during which growth is geometric in its rate of increase. Also called exponential growth phase.

lophotrichous Describing bacteria having a tuft of flagella at one or both poles.

lumen The cavity within a tubular organ.

lymphadenitis Inflammation of one or more lymph nodes. Also called lymphadenopathy.

lymphatic system A system of vessels and organs that serve as sites for development of immune cells and immune reactions. It includes the spleen, thymus, lymph nodes, and GALT.

lymphocyte The second most common form of white blood cells.

lymphokine A soluble substance secreted by sensitized T lymphocytes upon contact with specific antigen. About 50 types exist, and they stimulate inflammatory cells: macrophages, granulocytes, and lymphocytes. Examples: migration inhibitory factor, macrophage activating factor, chemotactic factor.

lyophilization Freeze-drying; the separation of a dissolved solid from the solvent by freezing the solution and evacuating the solvent under vacuum. A means of preserving the viability of cultures.

lyse To burst.

lysin A complement-fixing antibody that destroys specific targeted cells. Examples: hemolysin and bacteriolysin.

lysis The physical rupture or deterioration of a cell.

lysogeny The indefinite persistence of bacteriophage DNA in a host without bringing about the production of virions. A lysogenic cell can revert to a lytic cycle, the process that ends in lysis.

lysosome A cytoplasmic organelle containing lysozyme and other hydrolytic enzymes.

lysozyme An enzyme that attacks the bonds on bacterial peptidoglycan. It is a natural defense found in tears and saliva.

M

macromolecules Large, molecular compounds assembled from smaller subunits, most notably biochemicals.

macrophage A white blood cell derived from a monocyte that leaves the circulation and enters tissues. These cells are important in nonspecific phagocytosis and in regulating, stimulating, and cleaning up after immune responses.

macroscopic Visible to the naked eye.

macule A small, discolored spot or patch of skin that is neither raised above nor depressed below the surrounding surface.

malignant tumor Cancerous growth characterized by uncontrolled growth or abnormal cells within normal tissue.

malt The grain, usually barley, that is sprouted to obtain digestive enzymes and dried for making beer.

maltose One of the carbohydrates referred to as sugars. A fermentable sugar formed from starch.

Mantoux test An intradermal screening test for tuberculin hypersensitivity. A red, firm patch of skin at the injection site greater than 10 mm in diameter after 48 hours is a positive result that indicates current or prior exposure to the TB bacillus.

mapping Determining the location of loci and other qualities of genomic DNA.

marker Any trait or factor of a cell, virus, or molecule that makes it distinct and recognizable. Example: a genetic marker.

mash In making beer, the malt grain is steeped in warm water, ground up, and fortified with carbohydrates to form mash.

mass number (MN) Measurement that reflects the number of protons and neutrons in an atom of a particular element.

mast cell A nonmotile connective tissue cell implanted along capillaries, especially in the lungs, skin, gastrointestinal tract, and genitourinary tract. Like a basophil, its granules store mediators of allergy.

mastitis Inflammation of the breast or udder.

matrix The dense ground substance between the cristae of a mitochondrion that serves as a site for metabolic reactions.

matter All tangible materials that occupy space and have mass.

medium (plural: *media*) A nutrient used to grow organisms outside of their natural habitats.

meiosis The type of cell division necessary for producing gametes in diploid organisms. Two nuclear divisions in rapid succession produce four gametocytes, each containing a haploid number of chromosomes.

memory cell The long-lived progeny of a sensitized lymphocyte that remains in circulation and is genetically programmed to react rapidly with its antigen.

meningitis An inflammation of the membranes (meninges) that surround and protect the brain. It is often caused by bacteria such as *Neisseria meningitidis* (the meningococcus) and *Haemophilus influenzae.*

merozoite The motile, infective stage of an apicomplexan parasite that comes from a liver or red blood cell undergoing multiple fission.

mesophile Microorganisms that grow at intermediate temperatures.

mesosome The irregular invagination of a bacterial cell membrane that is more prominent in gram-positive than in gram-negative bacteria. Although its function is not definitely known, it appears to participate in DNA replication and cell division; in certain cells, it appears to play a role in secretion.

messenger RNA A single-stranded transcript that is a copy of the DNA template that corresponds to a gene.

metabolic analog Enzyme that mimics the natural substrate of an enzyme and vies for its active site.

metabolites Small organic molecules that are intermediates in the stepwise biosynthesis or breakdown of macromolecules.

metabolism A general term for the totality of chemical and physical processes occurring in a cell.

metachromatic Exhibiting a color other than that of the dye used to stain it.

metastasis In cancer, the dissemination of tumor cells so that neoplastic growths appear in other sites away from the primary tumor.

methanogens Methane producers.

MHC Major histocompatibility complex. *See* HLA.

MIC Abbreviation for **m**inimum **i**nhibitory **c**oncentration. The lowest concentration of antibiotic needed to inhibit bacterial growth in a test system.

microaerophile An aerobic bacterium that requires oxygen at a concentration less than that in the atmosphere.

microfilaments Cellular cytoskeletal element formed by thin protein strands that attach to cell membrane and form a network though the cytoplasm. Responsible for movement of cytoplasm.

microsopic Invisible to the naked eye.

miliary tuberculosis Rapidly fatal tuberculosis due to dissemination of mycobacteria in the blood and formation of tiny granules in various organs and tissues. The term *miliary* means resembling a millet seed.

mineralization The process by which decomposers (bacteria and fungi) convert organic debris into inorganic and elemental form. It is part of the recycling process.

minimum inhibitory concentration (MIC) The smallest concentration of drug needed to visibly control microbial growth.

miracidium The ciliated first-stage larva of a trematode. This form is infective for a corresponding intermediate host snail.

mitochondrion A double-membrane organelle of eucaryotes that is the main site for aerobic respiration.

mitosis Somatic cell division that preserves the somatic chromosome number.

mixed culture A container growing two or more different, known species of microbes.

molecule A distinct chemical substance that results from the combination of two or more atoms.

monoclonal antibody An antibody produced by a clone of lymphocytes that respond to a particular antigenic determinant and generate identical antibodies only to that determinant. *See* hybridoma.

monocyte A large mononuclear leukocyte normally found in the lymph nodes, spleen, bone marrow, and loose connective tissue. This type of cell makes up 3% to 7% of circulating leukocytes.

monomer A simple molecule that can be linked by chemical bonds to form larger molecules.

monosaccharide A simple sugar such as glucose that is a basic building block for more complex carbohydrates.

monotrichous Describing a microorganism that bears a single flagellum.

morbidity A diseased condition; the relative incidence of disease in a community.

mordant A chemical that fixes a dye in or on cells by forming an insoluble compound and thereby promoting retention of that dye. Example: Gram's iodine in the Gram stain.

morphology The study of organismic structure.

mortality rate Total number of deaths in a population attributable to a particular disease.

most probable number (MNP) Coliform test used to detect the concentration of contaminants in water supplies.

motility Self-propulsion.

must Juices expressed from crushed fruits that are used in fermentation for wine.

mutagen Any agent that induces genetic mutation. Examples: certain chemical substances, ultraviolet light, radioactivity.

mutation A permanent inheritable alteration in the DNA sequence or content of a cell.

mutualism Organisms living in an obligatory, but mutually beneficial, relationship.

mycelium The filamentous mass that makes up a mold. Composed of hyphae.

mycetoma A chronic fungal infection usually afflicting the feet, typified by swelling and multiple draining lesions. Example: maduromycosis or Madura foot.

mycoplasma The smallest, self-replicating microorganisms. Mycoplasma naturally lack a cell wall. Most species are parasites of animals and plants.

mycorrhizae Various species of fungi adapted in an intimate, mutualistic relationship to plant roots.

mycosis Any disease caused by a fungus.

mycotoxicosis Illness resulting from eating poisonous fungi.

myonecrosis Another name for gas gangrene.

N

NAD/NADH Abbreviations for the oxidized/reduced forms of nicotinamide adenine dinucleotide, an electron carrier. Also known as the vitamin niacin.

nanobes Cell-like particles, found in sediments and other geologic deposits, that some scientists speculate are the smallest bacteria. Short for nanobacteria.

narrow-spectrum Denotes drugs that are selective and limited in their effects. For example, they inhibit either gram-negative or gram-positive bacteria, but not both.

necrosis A pathologic process in which cells and tissues die and disintegrate.

negative feedback Enzyme regulation of metabolism by the end product of a multienzyme system that blocks the action of a "pacemaker" enzyme at or near the beginning of the pathway.

negative stain A staining technique that renders the background opaque or colored and leaves the object unstained so that it is outlined as a colorless area.

nematode A common name for helminths called roundworms.

neoplasm A synonym for tumor.

neurotropic Having an affinity for the nervous system. Most likely to affect the spinal cord.

neutron An electrically neutral particle in the nuclei of all atoms except hydrogen.

neutralization The process of combining an acid and a base until they reach a balanced proportion, with a pH value close to 7.

neutrophil A mature granulocyte present in peripheral circulation, exhibiting a multilobular nucleus and numerous cytoplasmic granules that retain a neutral stain. The neutrophil is an active phagocytic cell in bacterial infection.

niche In ecology, an organism's biological role in or contribution to its community.

night soil An archaic euphemism for human excrement collected for fertilizing crops.

nitrification Phase of the nitrogen cycle in which ammonium is oxidized.

nitrogen base A ringed compound of which pyrimidines and purines are types.

nitrogen fixation A process occurring in certain bacteria in which atmospheric N_2 gas is converted to a form (NH_4) usable by plants.

nomenclature A set system for scientifically naming organisms, enzymes, anatomical structures, etc.

noncoliforms Lactose-negative enteric bacteria in normal flora.

noncommunicable An infectious disease that does not arrive through transmission of an infectious agent from host to host.

nonpolar A term used to describe an electrically neutral molecule formed by covalent bonds between atoms that have the same or similar electronegativity.

non-self Molecules recognized by the immune system as containing foreign markers, indicating a need for immune response.

nonsense codon A triplet of mRNA bases that does not specify an amino acid but signals the end of a polypeptide chain.

normal flora The native microbial forms that an individual harbors.

nosocomial infection An infection not present upon admission to a hospital but incurred while being treated there.

nucleocapsid In viruses, the close physical combination of the nucleic acid with its protective covering.

nucleoid The basophilic nuclear region or nuclear body that contains the bacterial chromosome.

nucleolus A granular mass containing RNA that is contained within the nucleus of a eucaryotic cell.

nucleosome Structure in the packaging of DNA. Formed by the DNA strands wrapping around the histone protein to form nucleus bodies arranged like beads on a chain.

nucleus The central core of an atom, composed of protons and neutrons.

nucleotide A composite of a nitrogen base (purine or pyrimidine), a 5-carbon sugar (ribose or deoxyribose), and a phosphate group; a unit of nucleic acids.

numerical aperture In microscopy, the amount of light passing from the object and into the object in order to maximize optical clarity and resolution.

nutrient Any chemical substance that must be provided to a cell for normal metabolism and growth. Macronutrients are required in large amounts, and micronutrients in small amounts.

O

obligate Without alternative; restricted to a particular characteristic. Example: An obligate parasite survives and grows only in a host; an obligate aerobe must have oxygen to grow; an obligate anaerobe is destroyed by oxygen.

Okazaki fragment In replication of DNA, a segment formed on the lagging strand in which biosynthesis is conducted in a discontinuous manner dictated by the $5' \rightarrow 3'$ DNA polymerase orientation.

oligodynamic action A chemical having antimicrobial activity in minuscule amounts. Example: Certain heavy metals are effective in a few parts per billion.

oligonucleotides Short pieces of DNA or RNA that are easier to handle than long segments.

oligotrophic Nutrient-deficient ecosystem.

oncogene A naturally occurring type of gene that when activated can transform a normal cell into a cancer cell.

oncology The study of neoplasms, their cause, disease characteristics, and treatment.

oocyst The encysted form of a fertilized macrogamete or zygote; typical in the life cycles of apicomplexan parasites.

operator In an operon sequence, the DNA segment where transcription of structural genes is initiated.

operon A genetic operational unit that regulates metabolism by controlling mRNA production. In sequence, the unit consists of a regulatory gene, inducer or repressor control sites, and structural genes.

opportunistic In infection, ordinarily nonpathogenic or weakly pathogenic microbes that cause disease primarily in an immunologically compromised host.

opsonization The process of stimulating phagocytosis by affixing molecules (opsonins such as antibodies and complement) to the surfaces of foreign cells or particles.

orbitals The pathways of electrons as they rotate around the nucleus of an atom.

order In the levels of classification, the division of organisms that follows class. Increasing similarity may be noticed among organisms assigned to the same order.

organelle A small component of eucaryotic cells that is bounded by a membrane and specialized in function.

organic chemicals Molecules that contain the basic framework of the elements carbon and hydrogen.

ornithosis Worldwide zoonosis carried in the latent state in wild and domesticated birds; caused by *Chlamydia psittacosis*.

osmophile A microorganism that thrives in a medium having high osmotic pressure.

osmosis The diffusion of water across a selectively permeable membrane in the direction of lower water concentration.

oxidation In chemical reactions, the loss of electrons by one reactant.

oxidation-reduction Redox reactions, in which paired sets of molecules participate in electron transfers.

oxidative phosphorylation The synthesis of ATP using energy given off during the electron transport phase of respiration.

P

palindrome A word, verse, number, or sentence that reads the same forward or backward. Palindromes of nitrogen bases in DNA have genetic significance as transposable elements, as regulatory protein targets, and in DNA splicing.

palisades The characteristic arrangement of *Corynebacterium* cells resembling a row of fence posts and created by snapping.

pandemic A disease afflicting an increased proportion of the population over a wide geographic area (often worldwide).

pannus The granular membrane occurring on the cornea in trachoma.

papilloma Benign, squamous epithelial growth commonly referred to as a wart.

papule An elevation of skin that is small, demarcated, firm, and usually conical.

parasite An organism that lives on or within another organism (the host), from which it obtains nutrients and enjoys protection. The parasite produces some degree of harm in the host.

parenteral Administering a substance into a body compartment other than through the gastro-intestinal tract, such as via intravenous, subcutaneous, intramuscular, or intramedullary injection.

paroxysmal Events characterized by sharp spasms or convulsions; sudden onset of a symptom such as fever and chills.

passive immunity Specific resistance that is acquired indirectly by donation of preformed immune substances (antibodies) produced in the body of another individual.

passive transport Nutrient transport method that follows basic physical laws and does not require direct energy input from the cell.

pasteurization Heat treatment of perishable fluids such as milk, fruit juices, or wine to destroy heat-sensitive vegetative cells, followed by rapid chilling to inhibit growth of survivors and germination of spores. It prevents infection and spoilage.

pathogen Any agent, usually a virus, bacterium, fungus, protozoan, or helminth, that causes disease.

pathogenicity The capacity of microbes to cause disease.

pathology The structural and physiological effects of disease on the body.

pellicle A membranous cover; a thin skin, film, or scum on a liquid surface; a thin film of salivary glycoproteins that forms over newly cleaned tooth enamel when exposed to saliva.

pelvic inflammatory disease (PID) An infection of the uterus and fallopian tubes that has ascended from the lower reproductive tract. Caused by gonococci and chlamydias.

penicillinase An enzyme that hydrolyzes penicillin; found in penicillin-resistant strains of bacteria.

penicillins A large group of naturally occurring and synthetic antibiotics produced by *Penicillium* mold and active against the cell wall of bacteria.

pentose A monosaccharide with five carbon atoms per molecule. Examples: arabinose, ribose, xylose.

peptidase An enzyme that can hydrolyze the peptide bonds of a peptide chain.

peptide bond The covalent union between two amino acids that forms between the amine group of one and the carboxyl group of the other. The basic bond of proteins.

peptidoglycan A network of polysaccharide chains cross-linked by short peptides that forms the rigid part of bacterial cell walls. Gram-negative bacteria have a smaller amount of this rigid structure than do gram-positive bacteria.

perinatal In childbirth, occurring before, during, or after delivery.

periodontal Involving the structures that surround the tooth.

periplasmic space The region between the cell wall and cell membrane of the cell envelopes of gram-negative bacteria.

peritrichous In bacterial morphology, having flagella distributed over the entire cell.

petechiae Minute hemorrhagic spots in the skin that range from pinpoint- to pinhead-sized.

pertussis Infection by *Bordetella pertussis*. A highly communicable disease that causes acute respiratory syndrome. Pertussis can be life-threatening in infants, but vaccination on the recommended schedule can prevent infection. Also called whooping cough.

Peyer's patches Oblong lymphoid aggregates of the gut located chiefly in the wall of the terminal and small intestine. Along with the tonsils and appendix, Peyer's patches make up the gut-associated lymphoid tissue that responds to local invasion by infectious agents.

pH The symbol for the negative logarithm of the H ion concentration; p (power) or $[H^+]_{10}$. A system for rating acidity and alkalinity.

phage A bacteriophage; a virus that specifically parasitizes bacteria.

phagocytosis A type of endocytosis in which the cell membrane actively engulfs large particles or cells into vesicles.

phagolysosome A body formed in a phagocyte, consisting of a union between a vesicle containing the ingested particle (the phagosome) and a vacuole of hydrolytic enzymes (the lysosome).

phenetic Based on phenotype, or expression of traits.

phenotype The observable characteristics of an organism produced by the interaction between its genetic potential (genotype) and the environment.

phlebotomine Pertains to a genus of very small midges or blood-sucking (phlebotomous) sand flies and to diseases associated with those vectors such as kala-azar, Oroya fever, and cutaneous leishmaniasis.

phospholipid A class of lipids that compose a major structural component of cell membranes.

phosphorylation Process in which inorganic phosphate is added to a compound.

photic zone The aquatic stratum from the surface to the limits of solar light penetration.

photoautotroph An organism that utilizes light for its energy and carbon dioxide chiefly for its carbon needs.

photolysis Literally, splitting water with light. In photosynthesis, this step frees electrons and gives off O_2.

photosynthesis A process occurring in plants, algae, and some bacteria that traps the sun's energy and converts it to ATP in the cell. This energy is used to fix CO_2 into organic compounds.

phototrophs Microbes that use photosynthesis to feed.

phylum In the levels of classification, the third level of classification from general to more specific. Each kingdom is divided into numerous phyla. Sometimes referred to a division.

physiology The study of the function of an organism.

pili Small, stiff filamentous appendages in gram-negative bacteria that function in DNA exchange during bacterial conjugation.

pinocytosis The engulfment, or endocytosis, of liquids by extensions of the cell membrane.

plankton Minute animals (zooplankton) or plants (phytoplankton) that float and drift in the limnetic zone of bodies of water.

plaque In virus propagation methods, the clear zone of lysed cells in tissue culture or chick embryo membrane that corresponds to the area containing viruses. In dental application, the filamentous mass of microbes that adheres tenaciously to the tooth and predisposes to caries, calculus, or inflammation.

plasma The carrier fluid element of blood.

plasma cell A progeny of an activated B cell that actively produces and secretes antibodies.

plasmids Extrachromosomal genetic units characterized by several features. A plasmid is a double-stranded DNA that is smaller than and replicates independently of the cell chromosome; it bears genes that are not essential for cell growth; it can bear genes that code for adaptive traits; and it is transmissible to other bacteria.

pleomorphism Normal variability of cell shapes in a single species.

pluripotential Stem cells having the developmental plasticity to give rise to more than one type. Example: undifferentiated blood cells in the bone marrow.

pneumonia An inflammation of the lung leading to accumulation of fluid and respiratory compromise.

pneumococcus Common name for *Streptococcus pneumoniae,* the major cause of bacterial pneumonia.

pneumonic plague The acute, frequently fatal form of pneumonia caused by *Yersinia pestis.*

point mutation A change that involves the loss, substitution, or addition of one or a few nucleotides.

polar Term to describe a molecule with an asymmetrical distribution of charges. Such a molecule has a negative pole and a positive pole.

poliomyelitis An acute enteroviral infection of the spinal cord that can cause neuromuscular paralysis.

polyclonal In reference to a collection of antibodies with mixed specificities that arose from more than one clone of B cells.

polymer A macromolecule made up of a chain of repeating units. Examples: starch, protein, DNA.

polymerase An enzyme that produces polymers through catalyzing bond formation between building blocks (polymerization).

polymerase chain reaction (PCR) A technique that amplifies segments of DNA for testing. Using denaturation, primers, and heat-resistant DNA polymerase, the number can be increased several million-fold.

polymorphonuclear leukocytes (PMNLs) White blood cells with variously shaped nuclei. Although this term commonly denotes all granulocytes, it is used especially for the neutrophils.

polymyxin A mixture of antibiotic polypeptides from *Bacillus polymyxa* that are particularly effective against gram-negative bacteria.

polypeptide A relatively large chain of amino acids linked by peptide bonds.

polyribosomal complex An assembly line for mass production of proteins composed of a chain of ribosomes involved in mRNA transcription.

polysaccharide A carbohydrate that can be hydrolyzed into a number of monosaccharides. Examples: cellulose, starch, glycogen.

porin Transmembrane proteins of the outer membrane of gram-negative cells that permit transport of small molecules into the periplasmic space but bar the penetration of larger molecules.

potable Describing water that is relatively clear, odor-free, and safe to drink.

portal of entry Characteristic route of entry for an infectious agent; typically a cutaneous or membranous route.

portal of exit Characteristic route through which a pathogen departs from the host organism.

pox The thick, elevated pustular eruptions of various viral infections. Also called pocks.

prevalence The total cumulative number of cases of a disease in a certain area and time period.

primary response The first response of the immune system when exposed to an antigen.

primary structure Initial protein organization described by type, number, and order of amino acids in the chain. The primary structure varies extensively from protein to protein.

primers Synthetic oligonucleotides of known sequence that serve as landmarks to indicate where DNA amplification will begin.

prion A concocted word to denote "proteinaceous infectious agent"; a cytopathic protein associated with the slow-virus spongiform encephalopathies of humans and animals.

probes Small fragments of single-stranded DNA (RNA) that are known to be complementary to the specific sequence of DNA being studied.

procaryotic cell Small cells, lacking special structures such as a nucleus and organelles. All procaryotes are microorganisms.

prodromium A short period of mild symptoms occurring at the end of the period of incubation. It indicates the onset of an infection.

proglottid The egg-generating segment of a tapeworm that contains both male and female organs.

promastigote A morphological variation of the trypanosome parasite responsible for leishmaniasis.

promoter Part of an operon sequence. The DNA segment that is recognized by RNA polymerase as the starting site for transcription.

promoter region The site composed of a short signaling DNA sequence that RNA polymerase recognizes and binds to commence transcription.

properdin pathway A normal serum protein involved in the alternate complement pathway that leads to nonspecific lysis of bacterial cells and viruses.

prophage A lysogenized bacteriophage; a phage that is latently incorporated into the host chromosome instead of undergoing viral replication and lysis.

prophylactic Any device, method, or substance used to prevent disease.

prostaglandin A hormonelike substance that regulates many body functions. Prostaglandin comes from a family of organic acids containing 5-carbon rings that are essential to the human diet.

protease inhibitors Drugs that act to prevent the assembly of functioning viral particles.

protein Predominant organic molecule in cells, formed by long chains of amino acids.

proton An elementary particle that carries a positive charge. It is identical to the nucleus of the hydrogen atom.

proto-oncogene A gene that regulates the onset of mitosis.

protoplast A bacterial cell whose cell wall is completely lacking and that is vulnerable to osmotic lysis.

protozoa A group of single-celled, eucaryotic organisms.

pseudohypha A chain of easily separated, spherical to sausage-shaped yeast cells partitioned by constrictions rather than by septa.

pseudopods Protozoan appendage responsible for motility. Also called "false feet."

pseudomembrane A tenacious, noncellular mucous exudate containing cellular debris that tightly blankets the mucosal surface in infections such as diphtheria and pseudomembranous enterocolitis.

pseudopodium A temporary extension of the protoplasm of an ameboid cell. It serves both in ameboid motion and for food gathering (phagocytosis).

psychrophile A microorganism that thrives at low temperature (0°–20°C), with a temperature optimum of 0°–15°C.

pure culture A container of microbial cells whose identity is known.

purine A nitrogen base that is an important encoding component of DNA and RNA. The two most common purines are adenine and guanine.

purpura A condition characterized by bleeding into skin or mucous membrane, giving rise to patches of red that darken and turn purple.

pus The viscous, opaque, usually yellowish matter formed by an inflammatory infection. It consists of serum exudate, tissue debris, leukocytes, and microorganisms.

pyogenic Pertains to pus formers, especially the pyogenic cocci: pneumococci, streptococci, staphylococci, and neisseriae.

pyrimidine Nitrogen bases that help form the genetic code on DNA and RNA. Uracil, thymine, and cytosine are the most important pyrimidines.

pyrimidine dimer The union of two adjacent pyrimidines on the same DNA strand, brought about by exposure to ultraviolet light. It is a form of mutation.

pyrogen A substance that causes a rise in body temperature. It can come from pyrogenic microorganisms or from polymorphonuclear leukocytes (endogenous pyrogens).

Q

Q fever A disease first described in Queensland, Australia, initially dubbed Q for "query" to denote a fever of unknown origin. Q fever is now known to be caused by a rickettsial infection.

quaternary structure Most complex protein structure characterized by the formation of large, multiunit proteins by more than one of the polypeptides. This structure is typical of antibodies and some enzymes that act in cell synthesis.

quats A byword that pertains to a family of surfactants called quaternary ammonium compounds. These detergents are only weakly microbicidal and are used as sanitizers and preservatives.

Quellung test The capsular swelling phenomenon; a serological test for the presence of pneumococci whose capsules enlarge and become opaque and visible when exposed to specific anticapsular antibodies.

quinolone A class of synthetic antimicrobic drugs with broad-spectrum effects.

R

rabies The only rhabdovirus that infects humans. Zoonotic disease characterized by fatal meningoencephalitis.

rad A unit of measure for absorbed dose of ionizing radiation.

radiation Electromagnetic waves or rays, such as those of light given off from an energy source.

radioactive isotopes Unstable isotopes whose nuclei emit particles of radiation. This emission is called radioactivity or radioactive decay. Three naturally occurring emissions are alpha, beta, and gamma radiation.

radioimmunoassay (RIA) A highly sensitive laboratory procedure that employs radioisotope-labeled substances to measure the levels of antibodies or antigens in the serum.

reactants Molecules entering or starting a chemical reaction.

real image An image formed at the focal plane of a convex lens. In the compound light microscope, it is the image created by the objective lens.

receptor In intercellular communication, cell surface molecules involved in recognition, binding, and intracellular signaling.

recombinant DNA A technology, also known as genetic engineering, that deliberately modifies the genetic structure of an organism to create novel products, microbes, animals, plants, and viruses.

recombination A type of genetic transfer in which DNA from one organism is donated to another.

redox Denoting an oxidation-reduction reaction.

reduction In chemistry, the gain of electrons.

reemerging disease Previously identified disease that is increasing in occurrence.

refraction In optics, the bending of light as it passes from one medium to another with a different index of refraction.

regulator DNA segment that codes for a protein capable of repressing an operon.

rennin The enzyme casein coagulase, which is used to produce curd in the processing of milk and cheese.

reovirus Respiratory enteric orphan virus. Virus with a double-stranded RNA genome and both an inner capsid and an outer capsid. Not a significant human pathogen.

replication In DNA synthesis, the semiconservative mechanisms that ensure precise duplication of the parent DNA strands.

replication fork The Y-shaped point on a replicating DNA molecule where the DNA polymerase is synthesizing new strands of DNA.

replicon A piece of DNA capable of replicating. Contains an origin of replication.

reportable disease Those diseases that must be reported to health authorities by law.

repressor The protein product of a repressor gene that combines with the operator and arrests the transcription and translation of structural genes.

reservoir In disease communication, the natural host or habitat of a pathogen.

resident flora The deeper, more stable microflora that inhabit the skin and exposed mucous membranes, as opposed to the superficial, variable, transient population.

resistance (R) factor Plasmids, typically shared among bacteria by conjugation, that provide resistance to the effects of antibiotics.

resolving power The capacity of a microscope lens system to accurately distinguish between two separate entities that lie close to each other. Also called resolution.

respiratory chain In cellular respiration, a series of electron-carrying molecules that transfers energy-rich electrons and protons to molecular oxygen. In transit, energy is extracted and conserved in the form of ATP.

restriction endonuclease An enzyme present naturally in cells that cleaves specific locations on DNA. It is an important means of inactivating viral genomes, and it is also used to splice genes in genetic engineering.

reticuloendothelial system Also known as the mononuclear phagocyte system, it pertains to a network of fibers and phagocytic cells (macrophages) that permeates the tissues of all organs. Examples: Kupffer cells in liver sinusoids, alveolar phagocytes in the lung, microglia in nervous tissue.

retrovirus A group of RNA viruses (including HIV) that have the mechanisms for converting their genome into a double strand of DNA that can be inserted on a host's chromosome.

reverse transcriptase The enzyme possessed by retroviruses that carries out the reversion of RNA to DNA—a form of reverse transcription.

Reye syndrome A sudden, usually fatal neurological condition that occurs in children after a viral infection. Autopsy shows cerebral edema and marked fatty change in the liver and renal tubules.

Rh factor An isoantigen that can trigger hemolytic disease in newborns due to incompatibility between maternal and infant blood factors.

rhabdovirus Family of bullet-shaped viruses that includes rabies.

rhinovirus A picornavirus associated with the common cold. Transmission is through human-to-human contact, and symptoms typically are short-lived. Most effective control is effective hand washing and care in handling nasal secretions.

rhizobia Bacteria that live in plant roots and supply supplemental nitrogen that boosts plant growth.

rhizosphere The zone of soil, complete with microbial inhabitants, in the immediate vicinity of plant roots.

ribose A 5-carbon monosaccharide found in RNA.

ribonucleic acid (RNA) The nucleic acid responsible for carrying out the hereditary program transmitted by an organism's DNA.

ribosome A bilobed macromolecular complex of ribonucleoprotein that coordinates the codons of mRNA with tRNA anticodons and, in so doing, constitutes the peptide assembly site.

rickettsias Medically important family of bacteria, commonly carried by ticks, lice, and fleas. Significant cause of important emerging diseases.

ringworm A superficial mycosis caused by various dermatophytic fungi. This common name is actually a misnomer.

RNA polymerase Enzyme process that translates the code of DNA to RNA.

rolling circle An intermediate stage in viral replication of circular DNA into linear DNA.

rosette formation A technique for distinguishing surface receptors on T cells by reacting them with sensitized indicator sheep red blood cells. The cluster of red cells around the central white blood cell resembles a little rose blossom and is indicative of the type of receptor.

rotavirus Virus with a double-stranded RNA genome and both an inner capsid and an outer capsid. Transmitted by fecal contamination, it is common in areas with poor sanitation. Rotavirus typically causes diarrheal disease and can be fatal.

rough endoplasmic reticulum (RER) Microscopic series of tunnels that originates in the outer membrane of the nuclear envelope and is used in transport and storage. Large numbers of ribosomes, partly attached to the membrane, give the rough appearance.

rubella Commonly known as German measles. Rubella is caused by *Rubivirus*, a member of the togavirus family. Postnatal rubella is generally a mild condition; congenital rubella poses a risk of birth defects and results when virus passes from infected mother to fetus.

S

saccharide Scientific term for sugar. Refers to a simple carbohydrate with a sweet taste.

salmonelloses Illnesses caused by infection of the noncoliform *Salmonella* pathogens. The cause of typhoid fever, salmonella food poisoning, and gastroenteritis.

salpingitis Inflammation of the fallopian tubes.

sanitize To clean inanimate objects using soap and degerming agents so that they are safe and free of high levels of microorganisms.

saprobe A microbe that decomposes organic remains from dead organisms. Also known as a saprophyte or saprotroph.

sarcina A cubical packet of 8, 16, or more cells; the cellular arrangement of the genus *Sarcina* in the family Micrococcaceae.

sarcoma A fleshy neoplasm of connective tissue. Growth, usually highly malignant, is of mesodermal origin.

satellite phenomenon A type of commensal relationship in which one microbe produces growth factors that favor the growth of its dependent partner. When plated on solid media, the dependent organism appears as surrounding colonies. Example: *Staphylococcus* surrounded by its *Haemophilus* satellite.

schistosomiasis Infection by blood fluke, often as a result of contact with contaminated water in rivers and streams. Symptoms include fever, chills, diarrhea, and cough. Infection may be chronic.

schizogony A process of multiple fission whereby first the nucleus divides several times, and subsequently the cytoplasm is subdivided for each new nucleus during cell division.

scientific method Principles and procedures for the systematic pursuit of knowledge, involving the recognition and formulation of a problem, the collection of data through observation and experimentation, and the formulation and testing of a hypothesis.

scolex The anterior end of a tapeworm characterized by hooks and/or suckers for attachment to the host.

SCP Abbreviation for single-cell protein, a euphemistic expression for microbial protein intended for human and animal consumption.

sebaceous glands The sebum- (oily, fatty) secreting glands of the skin.

secondary immune response The more rapid and heightened response to antigen in a sensitized subject due to memory lymphocytes.

secondary infection An infection that compounds a preexisting one.

secondary structure Protein structure that occurs when the functional groups on the outer surface of the molecule interact by forming hydrogen bonds. These bonds cause the amino acid chain to either twist, forming a helix, or to pleat into an accordion pattern called a β-pleated sheet.

secretion The process of actively releasing cellular substances into the extracellular environment.

secretory antibody The immunoglobulin (IgA) that is found in secretions of mucous membranes and serves as a local immediate protection against infection.

selective media Nutrient media designed to favor the growth of certain microbes and to inhibit undesirable competitors.

self Natural markers of the body that are recognized by the immune system.

self-limited Applies to an infection that runs its course without disease or residual effects.

semiconservative replication In DNA replication, the synthesis of paired daughter strands, each retaining a parent strand template.

semisolid media Nutrient media with a firmness midway between that of a broth (a liquid medium) and an ordinary solid medium; motility media.

sensitizing dose The initial effective exposure to an antigen or an allergen that stimulates an immune response. Often applies to allergies.

sepsis The state of putrefaction; the presence of pathogenic organisms or their toxins in tissue or blood.

septic shock Blood infection resulting in a pathological state of low blood pressure accompanied by a reduced amount of blood circulating to vital organs. Endotoxins of all gram-negative bacteria can cause shock, but most clinical cases are due to gram-negative enteric rods.

septicemia Systemic infection associated with microorganisms multiplying in circulating blood.

septum A partition or cellular cross wall, as in certain fungal hyphae.

sequela A morbid complication that follows a disease.

sequencing Determining the actual order and types of bases in a segment of DNA.

serology The branch of immunology that deals with *in vitro* diagnostic testing of serum.

seropositive Showing the presence of specific antibody in a serological test. Indicates ongoing infection.

serotonin A vasoconstrictor that inhibits gastric secretion and stimulates smooth muscle.

serotyping The subdivision of a species or subspecies into an immunologic type, based upon antigenic characteristics.

serum The clear fluid expressed from clotted blood that contains dissolved nutrients, antibodies, and hormones but not cells or clotting factors.

serum sickness A type of immune complex disease in which immune complexes enter circulation, are carried throughout the body, and are deposited in the blood vessels of the kidney, heart, skin, and joints. The condition may become chronic.

sex pilus A conjugative pilus.

sexually transmitted disease (STD) Infections resulting from pathogens that enter the body via sexual intercourse or intimate, direct contact.

shigellosis An incapacitating dysentery caused by infection with *Shigella* bacteria.

sign Any abnormality uncovered upon physical diagnosis that indicates the presence of disease. A sign is an objective assessment of disease, as opposed to a symptom, which is the subjective assessment perceived by the patient.

smooth endoplasmic reticulum (SER) A microscopic series of tunnels lacking ribosomes that functions in the nutrient processing function of a cell.

sodoku A Japanese term pertaining to rat bite fever.

solution A mixture of one or more substances (solutes) that cannot be separated by filtration or ordinary settling.

solvent A dissolving medium.

Southern blot A technique that separates fragments of DNA using electrophoresis and identifies them by hybridization.

species In the levels of classification, the most specific level of organization.

spheroplast A gram-negative cell whose peptidoglycan, when digested by lysozyme, remains intact but is osmotically vulnerable.

spike A receptor on the surface of certain enveloped viruses that facilitates specific attachment to the host cell.

spirillum A type of bacterial cell with a rigid spiral shape and external flagella.

spirochete A coiled, spiral-shaped bacterium that has endoflagella and flexes as it moves.

spongiform encephalopathies Transmissible, fatal, chronic infections of the nervous system caused by unconventional viruses now identified as prions (infectious proteins).

spontaneous generation Early belief that living things arose from vital forces present in nonliving, or decomposing, matter.

sporangium A fungal cell in which asexual spores are formed by multiple cell cleavage.

spore A differentiated, specialized cell form that can be used for dissemination, for survival in times of adverse conditions, and/or for reproduction. Spores are usually unicellular and may develop into gametes or vegetative organisms.

sporotrichosis A subcutaneous mycosis caused by *Sporothrix schenckii*; a common cause is the prick of a rose thorn.

sporozoite One of many minute elongated bodies generated by multiple division of the oocyst. It is the infectious form of the malarial parasite that is harbored in the salivary gland of the mosquito and inoculated into the victim during feeding.

sporulation The process of spore formation.

start codon The nucleotide triplet AUG that codes for the first amino acid in protein sequences.

starter culture The sizeable inoculation of pure bacterial, mold, or yeast sample for bulk processing, as in the preparation of fermented foods, beverages, and pharmaceuticals.

stasis A state of rest or inactivity; applied to nongrowing microbial cultures. Also called microbistasis.

stationary growth phase Survival mode in which cells either stop growing or grow very slowly.

stem cells Pluripotent, undifferentiated cells.

sterile Completely free of all life forms, including spores and viruses.

sterilization Any process that completely removes or destroys all viable microorganisms, including viruses, from an object or habitat. Material so treated is sterile.

strain In microbiology, a set of descendants cloned from a common ancestor that retain the original characteristics. Any deviation from the original is a different strain.

streptolysin A hemolysin produced by streptococci.

stroma The matrix of the chloroplast that is the site of the dark reactions.

structural gene A gene that codes for the amino acid sequence (peptide structure) of a protein.

subacute Indicates an intermediate status between acute and chronic disease.

subcellular vaccine A vaccine against isolated microbial antigens rather than against the entire organism.

subclinical A period of inapparent manifestations that occurs before symptoms and signs of disease appear.

subculture To make a second-generation culture from a well-established colony of organisms.

subcutaneous The deepest level of the skin structure.

substrate The specific molecule upon which an enzyme acts.

sucrose One of the carbohydrates commonly referred to as sugars. Common table or cane sugar.

sulfonamide Antimicrobial drugs that interfere with the essential metabolic process of bacteria and some fungi.

superantigens Bacterial toxins that are potent stimuli for T cells and can be a factor in diseases such as toxic shock.

superinfection An infection occurring during antimicrobic therapy that is caused by an overgrowth of drug-resistant microorganisms.

superoxide A toxic derivative of oxygen; (O_2^-).

suppressor T cell A class of T cells that inhibits the actions of B cells and other T cells.

surfactant A surface-active agent that forms a water-soluble interface. Examples: detergents, wetting agents, dispersing agents, and surface tension depressants.

sylvatic Denotes the natural presence of disease among wild animal populations. Examples: sylvatic (sylvan) plague, rabies.

symbiosis An intimate association between individuals from two species; used as a synonym for mutualism.

symptom The subjective evidence of infection and disease as perceived by the patient.

syncytium A multinucleated protoplasmic mass formed by consolidation of individual cells.

syndrome The collection of signs and symptoms that, taken together, paint a portrait of the disease.

synergism The coordinated or correlated action by two or more drugs or microbes that results in a heightened response or greater activity.

syngamy Conjugation of the gametes in fertilization.

syphilis A sexually transmitted bacterial disease caused by the spirochete *Treponema pallidum*.

systemic Occurring throughout the body; said of infections that invade many compartments and organs via the circulation.

T

tartar *See* calculus.

taxa Taxonomic categories.

taxonomy The formal system for organizing, classifying, and naming living things.

temperate phage A bacteriophage that enters into a less virulent state by becoming incorporated into the host genome as a prophage instead of in the vegetative or lytic form that eventually destroys the cell.

teratogenic Causing abnormal fetal development.

tetanospasmin The neurotoxin of *Clostridium tetani,* the agent of tetanus. Its chief action is directed upon the inhibitory synapses of the anterior horn motor neurons.

tetracyclines A group of broad-spectrum antibiotics with a complex 4-ring structure.

tertiary structure Protein structure that results from additional bonds forming between functional groups in a secondary structure, creating a three-dimensional mass.

tetanus A neuromuscular disease caused by infection with *Clostridium tetani.* Usual portals of entry include puncture wounds, burns, umbilical stumps, frostbite sites, and crushed body parts. Vaccination repeated at the recommended times can prevent infection. Also called lockjaw.

tetrads Groups of four.

theory A collection of statements, propositions, or concepts that explains or accounts for a natural event.

therapeutic index The ratio of the toxic dose to the effective therapeutic dose that is used to assess the safety and reliability of the drug.

thermal death point The lowest temperature that achieves sterilization in a given quantity of broth culture upon a 10-minute exposure. Examples: 55°C for *Escherichia coli,* 60°C for *Mycobacterium tuberculosis,* and 120°C for spores.

thermal death time The least time required to kill all cells of a culture at a specified temperature.

thermocline A temperature buffer zone in a large body of water that separates the warmer water (the epilimnion) from the colder water (the hypolimnion).

thermoduric Resistant to the harmful effects of high temperature.

thermophile A microorganism that thrives at a temperature of 50°C or higher.

thrush *Candida albicans* infection of the oral cavity.

thylakoid Vesicles of a chloroplast formed by elaborate folding of the inner membrane to form "discs." Solar energy trapped in the thylakoids is used in photosynthesis.

thymine (T) One of the nitrogen bases found in DNA, but not in RNA. Thymine is in a pyrimidine form.

tincture A medicinal substance dissolved in an alcoholic solvent.

tinea Ringworm; a fungal infection of the hair, skin, or nails.

titer In immunochemistry, a measure of antibody level in a patient, determined by agglutination methods.

T lymphocyte (T cell) A white blood cell that is processed in the thymus gland and is involved in cell-mediated immunity.

topoisomerases Enzymes that can add or remove DNA twists and thus regulate the degree of supercoiling.

toxemia An abnormality associated with certain infectious diseases. Toxemia is caused by toxins or other noxious substances released by microorganisms circulating in the blood.

toxigenicity The tendency for a pathogen to produce toxins. It is an important factor in bacterial virulence.

toxin A specific chemical product of microbes, plants, and some animals that is poisonous to other organisms.

toxoid A toxin that has been rendered nontoxic but is still capable of eliciting the formation of protective antitoxin antibodies; used in vaccines.

trace elements Micronutrients (zinc, nickel, and manganese) that occur in small amounts, and are involved in enzyme function and maintenance of protein structure.

tracheostomy A surgically created emergency airway opening into the trachea.

trachoma Strain of *Chlamydia trachomatis* that attacks mucous membranes of the eye, genitourinary tract, and lungs in humans.

transcription mRNA synthesis; the process by which a strand of RNA is produced against a DNA template.

transduction The transfer of genetic material from one bacterium to another by means of a bacteriophage vector.

transfer RNA (tRNA) A transcript of DNA that specializes in converting RNA language into protein language.

transformation In microbial genetics, the transfer of genetic material contained in "naked" DNA fragments from a donor cell to a competent recipient cell.

transgenic technology Introduction of foreign DNA into cells or organisms. Used in genetic engineering to create recombinant plants, animals, and microbes.

transients In normal flora, the assortment of superficial microbes whose numbers and types vary depending upon recent exposure. The deeper-lying residents constitute a more stable population.

translation Protein synthesis; the process of decoding the messenger RNA code into a polypeptide.

transposon A DNA segment with an insertion sequence at each end, enabling it to migrate to another plasmid, to the bacterial chromosome, or to a bacteriophage.

trematode A fluke or flatworm parasite of vertebrates.

tricarboxylic acid cycle (TCA or Krebs cycle) The second pathway of the three pathways that complete the process of primary catabolism. Also called the citric acid cycle.

trichinosis Infection by the *Trichinella spiralis* parasite, usually caused by eating the meat of an infected animal. Early symptoms include fever, diarrhea, nausea, and abdominal pain that progress to intense muscle and joint pain and shortness of breath. In the final stages, heart and brain function are at risk, and death is possible.

trichomoniasis Sexually transmitted disease caused by infection by the trichomonads, a group of protozoa. Symptoms include urinary pain and frequency, and foul-smelling vaginal discharge in females or recurring urethritis, with a thin milky discharge, in males.

triglyceride A type of lipid composed of a glycerol molecule bound to three fatty acids.

trophozoite A vegetative protozoan (feeding form) as opposed to a resting (cyst) form.

trypomastigote The infective morphological stage transmitted by the tsetse fly or the reduviid bug in African trypanosomiasis and Chagas disease.

tubercle In tuberculosis, the granulomatous well-defined lung lesion that can serve as a focus for latent infection.

tuberculin A glycerinated broth culture of *Mycobacterium tuberculosis* that is evaporated and filtered. Formerly used to treat tuberculosis, tuberculin is now used chiefly for diagnostic tests.

tularemia Infection by *Francisella tularensis.* A zoonotic disease of mammals common to the northern hemisphere. Occasionally called rabbit fever. Portal of entry and symptoms are varied.

turbid Cloudy appearance of nutrient solution in a test tube due to growth of microbe population.

tyndallization Fractional (discontinuous, intermittent) sterilization designed to destroy spores indirectly. A preparation is exposed to flowing steam for an hour, and then the mineral is allowed to incubate to permit spore germination. The resultant vegetative cells are destroyed by repeated steaming and incubation.

typhoid fever Form of salmonelloses. It is highly contagious. Primary symptoms include fever, diarrhea, and abdominal pain. Typhoid fever can be fatal if untreated.

typhus Rickettsia infection characterized by high fever, chills, frontal headache, muscular pain, and a generalized rash within seven days of infection. In more severe cases, a personality change, low urine output, hypotension, and gangrene can cause complications. Mortality is high in older adults.

U

uncoating The process of removal of the viral coat and release of the viral genome by its newly invaded host cell.

undulant fever *See* brucellosis.

universal donor In blood grouping and transfusion, a group O individual whose erythrocytes bear neither agglutinogen A nor B.

Universal Precautions (UP) Center for Disease Control guidelines for health care workers regarding the prevention of disease transmission when handling patients and body substances.

uracil (U) One of the nitrogen bases in RNA, but not in DNA. Uracil is in a pyrimidine form.

V

vaccine Originally used in reference to inoculation with the cowpox or vaccinia virus to protect against smallpox. In general, the term now pertains to injection of whole microbes (killed or attenuated), toxoids, or parts of microbes as a prevention or cure for disease.

vacuoles In the cell, membrane-bounded sacs containing fluids or solid particles to be digested, excreted, or stored.

valence The combining power of an atom based upon the number of electrons it can either take on or give up.

variable region The antigen binding fragment of an immunoglobulin molecule, consisting of a combination of heavy and light chains whose molecular conformation is specific for the antigen.

variolation A hazardous, outmoded process of deliberately introducing smallpox material scraped from a victim into the nonimmune subject in the hope of inducing resistance.

VDRL A flocculation test that detects syphilis antibodies. An important screening test. The abbreviation stands for Venereal Disease Research Laboratories.

vector An animal that transmits infectious agents from one host to another, usually a biting or piercing arthropod like the tick, mosquito, or fly. Infectious agents can be conveyed mechanically by simple contact or biologically whereby the parasite develops in the vector.

vector$_2$ A genetic element such as a plasmid or a bacteriophage used to introduce genetic material into a cloning host during recombinant DNA experiments.

vegetative In describing microbial developmental stages, a metabolically active feeding and dividing form, as opposed to a dormant, seemingly inert, nondividing form. Examples: a bacterial cell versus its spore; a protozoan trophozoite versus its cyst.

vehicle An inanimate material (solid object, liquid, or air) that serves as a transmission agent for pathogens.

verruca A flesh-colored wart. This self-limited tumor arises from accumulations of growing epithelial cells. Example: papilloma warts.

vesicle A blister characterized by a thin-skinned, elevated, superficial pocket inflated with serum.

vibrio A curved, rod-shaped bacterial cell.

viremia The presence of a virus in the bloodstream.

virion An elementary virus particle in its complete morphological and thus infectious form. A virion consists of the nucleic acid core surrounded by a capsid, which can be enclosed in an envelope.

viroid An infectious agent that, unlike a virion, lacks a capsid and consists of a closed circular RNA molecule. Although known viroids are all plant pathogens, it is conceivable that animal versions exist.

virtual image In optics, an image formed by diverging light rays; in the compound light microscope, the second, magnified visual impression formed by the ocular from the real image formed by the objective.

virus Microscopic, acellular agent composed of nucleic acid surrounded by a protein coat.

virulence In infection, the relative capacity of a pathogen to invade and harm host cells.

vitamins A component of coenzymes critical to nutrition and the metabolic function of coenzyme complexes.

W

wart An epidermal tumor caused by papillomaviruses. Also called a verruca.

Western blot test A procedure for separating and identifying antigen or antibody mixtures by two-dimensional electrophoresis in polyacrylamide gel, followed by immune labeling.

wheal A welt; a marked, slightly red, usually itchy area of the skin that changes in size and shape as it extends to adjacent area. The reaction is triggered by cutaneous contact or intradermal injection of allergens in sensitive individuals.

whey The residual fluid from milk coagulation that separates from the solidified curd.

white piedra A fungus disease of hair, especially of the scalp, face, and genitals, caused by *Trichosporon beigelii*. The infection is associated with soft, mucilaginous, white-to-light-brown nodules that form within and on the hair shafts.

whitlow A deep inflammation of the finger or toe, especially near the tip or around the nail. Whitlow is a painful herpes simplex virus infection that can last several weeks and is most common among health care personnel who come in contact with the virus in patients.

whooping cough *See* pertussis.

Widal test An agglutination test for diagnosing typhoid.

wild type The natural, nonmutated form of a genetic trait.

wort The clear fluid derived from soaked mash that is fermented for beer.

X

xenograft The transfer of a tissue or an organ from an animal of one species to a recipient of another species.

Y

yaws A tropical disease caused by *Treponema pertenue* that produces granulomatous ulcers on the extremities and occasionally on bone, but does not produce central nervous system or cardiovascular complications.

yellow fever Best-known arbovirus. Yellow fever is transmitted by mosquitoes. Its symptoms include fever, headache, and muscle pain that can proceed to oral hemorrhage, nosebleeds, vomiting, jaundice, and liver and kidney damage.

Z

zoonosis An infectious disease indigenous to animals that humans can acquire through direct or indirect contact with infected animals.

zygospore A thick-walled sexual spore produced by the zygomycete fungi. It develops from the union of two hyphae, each bearing nuclei of opposite mating types.

Credits

Photographs

Chapter 1

Opener: Courtesy Russell Vreeland, Biology Dept., West Chester University; **1.1:** © NASA/Photo Researchers, Inc.; **1.2a:** © Doug Sokell/Tom Stack & Associates; **1.2b:** Courtesy Tom Volk; **1.2c:** © Fred McConnaughey/Photo Researchers, Inc.; **1.2d:** © C.G. Van Dyke; **1.3a:** © Science VU/SIM, NBS/Visuals Unlimited; **1.3b:** © Corale L. Brierley/Visuals Unlimited; **1.3c:** Courtesy Affymetrix; **1.3d:** USDA/Agricultural Research Service; **1.3e:** Courtesy of General Electric Research and Development Center; **1.5c(lower left):** © Sinclair Stammers/SPL/Photo Researchers, Inc.; **1.5c(top left):** © A.M. Siegelman/Visuals Unlimited; **1.5d:** © Carolina Biological Supply/Phototake; **1.5d(center):** © T.E. Adams/Visuals Unlimited; **1.5d-3:** Courtesy Tom Volk; **1.7:** © Bettmann/Corbis; **1.8a, 1.8b-1:** © Kathy Park Talaro/Visuals Unlimited; **1.8b-2:** © Science/ VU/ Visuals Unlimited; **Microfile 1.3a–c:** Brian Smale/© 1998. Reprinted with permission of Discover Magazine; **1.9a:** © The Bettman Archives; **1.12, 1.13:** © Bettmann/Corbis.

Chapter 2

Opener: Courtesy T. Mark Harrison/UCLA; **2.6d:** Kathy Park Talaro; **2.10a–c:** © John W. Hole, Jr.; **Microbits 2.3:** © Don Fawcett/Visuals Unlimited; **2.22d:** From A.S. Moffat "Nitrogenase Structure Revealed," *Science,* 250:1513, 12/14/90. Copyright 1990 by the AAAS. Photo by M.M. Georgiadis and D.C. Rees, Caltech.

Chapter 3

Opener: Courtesy Alan Raumbach/CHROMagar; **3.3b,f:** Kathy Park Talaro; **Microbits 3.1:** Courtesy Charles River Lab; **3.3, 3.4b, 3.5b, 3.6b:** © Kathy Park Talaro; **3.7a:** © A. M. Siegelman/Visuals Unlimited; **3.7b:** © Photograph courtesy of Becton Dickinson Microbiology Systems; **3.9a,b, 3.10a,b:** © Kathy Park Talaro; **3.11:** Courtesy of Harold J. Benson; **3.12a–c:** Kathy Park Talaro; **3.13:** © Kathy Park Talaro/Visuals Unlimited; **3.14:** Courtesy Leica, Inc.; **3.19a:** © Carolina Biological Supply/Phototake, NYC; **3.19b,c:** © Abbey/Visuals Unlimited; **3.20:** © E.C.S. Chan/Visuals Unlimited; **3.21a:** © George J. Wilder/Visuals Unlimited; **3.21b:** © Abbey/Visuals Unlimited; **3.22:** © Molecular Probes, Inc.; **3.23c:** © William Ormerod/Visuals Unlimited; **Microbits 3.2:** Courtesy Randal Feenstra, IBM Thomas, J. Watson Research Center, Yorktown Heights, NY; **3.24a:** Courtesy Cynthia Goldsmith, Charles D. Humphrey and Luanne Elliott/Centers for Disease Control and Prevention; **3.24b:** © W.L. Dentler, University of Kansas/Biological Photo Service; **3.25:** © Dennis Kunkel/CNRI/Phototake, NYC; **3.26a–1:** Kathy Park Talaro; **3.26a–2:** Courtesy Harold Bensen; **3.26b:** Kathy Park Talaro; **3.26c–2(both):** © Jack Bostrack/Visuals Unlimited; **3.26c–3:** © Manfred Kage/Peter Arnold, Inc.; **3.26d–1:** © A.M. Siegelman/Visuals Unlimited; **3.26d–2:** Kathy Park Talaro.

Chapter 4

Opener: Courtesy Kit Pogliano/UCSD; **4.2a:** Courtesy Julius Adler; **4.3a:** Courtesy of Dr. Jeffrey C. Burnham; **4.3b:** From Reichelt and Baumann, *Arch. Microbiology,* 94:283–330 © Springer-Verlag, 1973; **4.3c:** From Noel R. Krieg in *Bacteriological Reviews,* March 1976, Vol. 40(1):87 fig. 7; **4.3d:** From Preer et al, *Bacteriological Reviews,* June 1974, 38(2): 121, fig. 7. © ASM; **4.7c:** Courtesy Stanley F. Hayes, Rocky Mountain Laboratories, NIAID, NIH; **4.8a,b:** Courtesy of Dr. S. Knutton from D.R. Lloyd and S. Knutton, *Infection and Immunity,* January 1987, p. 86–92. © ASM; **4.9:** © L. Caro/SPL/Photo Researchers, Inc.; **4.12a:** Courtesy Harriet & Ephrussi-Taylor, Rockefeller University Press; **4.12b:** © John D. Cunningham/Visuals Unlimited; **Spotlight 4.3:** Courtesy Heide N. Schulz/Max Planck Institute for Marine Microbiology; **4.13:** © Science VU-Charles W. Stratton/Visuals Unlimited; **4.15a:** © S.C. Holt/Biological Photo Service; **4.15b:** © T. J. Beveridge/Biological Photo Service; **4.18:** © E.S. Anderson/Photo Researchers, Inc.; **4.20:** © Paul W Johnson/John Sieburth/Biological Photo Service; **4.21b:** © Cabisco/Visuals Unlimited; **4.21c:** © Lee D. Simon/Photo Researchers, Inc.; **4.23a,b:** © David M. Phillips/Visuals Unlimited; **4.23c:** From *Microbiological Reviews,* 55(1):25, fig.2b, March 1991. Courtesy of Jorge Benach; **4.23d:** © R.G. Kessel-G. Shih/Visuals Unlimited; **4.24a:** © A. M. Siegelman/Visuals Unlimited; **4.24b:** From Braude, *Infectious Diseases and Microbiology,* 2E, fig. 3, Page 257 © Saunders College Publishing Company; **4.27a,b:** Courtesy of Analytab Products, a division of Sherwood Medical; **4.28e:** Courtesy Steven J. Norris; **4.30:** © Science VU-M.D. Maser/Visuals Unlimited; **4.32:** From Baca and Paretsky, *Microbiological Reviews,* 47(2):133, fig. 16, June 1983 © ASM; **4.33:** © David M. Phillips/Visuals Unlimited; **4.34a:** Courtesy John Waterbury, WHOI; **4.34b,c:** © T.E. Adams/Visuals Unlimited; **4.35a:** From *ASM News,* 53(2), Feb. 1987, American Society for Microbiology. Photo by H. Kaltwasser; **4.35b:** © Paul W. Johnson/Biological Photo Service; **4.36b:** Courtesy Dr. David Graham, Univ. of Illinois, Urbana; **4.38a:** © Kathy Park Talaro; **4.38b:** From Kessel and Cohen, "Ultrastructure of Square Bacteria From a Brine Pool in Southern Sinai" in *Journal of Bacteriology,* 150(2):851–860, 1982. American Society for Microbiology.

Chapter 5

Opener: From: Wood and Bousfield: "Common Objects of the Microscope," 1900 George Routledge IV, London; **5.1a,b:** © Andrew Knoll; **5.3a:** © Michael Webb/Visuals Unlimited; **5.4:** Courtesy of R.G. Garrison, Ph. D. From *Fungal Dimorphism with Emphasis on Fungi Pathogenic for Human,* P.J. Szaniszlo (ed). Reprinted with permission of Plenum Publishing Corp.; **5.5:** © Don W. Fawcett/Visuals Unlimited; **5.6b:** © Science VU/Visuals Unlimited; **5.12b:** © Don W. Fawcett/Visuals Unlimited; **5.15a:** © David M. Phillips/Visuals Unlimited; **5.16a,b:** Courtesy Dr. Judy A. Murphy, San Joaquin Delta College, Dept. of Microscopy, Stockton, CA.; **5.17a:** © John D. Cunningham/Visuals Unlimited; **5.17b:** © Everett S. Beneke/Visuals Unlimited; **Spotlight 5.2:** Courtesy Gregory M. Filip, Oregon State University; **5.23:** © Kathy Park Talaro/Visuals Unlimited; **5.24a:** © A. M. Siegelman/Visuals Unlimited; **5.24b:** © John D. Cunningham/Visuals Unlimited; **Medical Microbiology fig. D:** © Fred Marsik/Visuals Unlimited; **5.25:** From Robert Simmons, *ASM News,* 1991, Vol. 57, #8, p. 400 © ASM; **Medical Microbiology fig. F:** Courtesy Wadsworth Center, New York State Department of Health; **5.26b:** © T. E. Adams/Visuals Unlimited; **5.26c:** © Jan Hinsch/Photo Researchers, Inc.; **5.26d:** Courtesy Howard Glasgow, Dept. Botany, North Carolina State University; **5.28b:** © David M. Phillips/Visuals Unlimited; **5.30c:** © Biophoto Associates/Science Source/Photo Researchers, Inc.; **5.31b:** Michael Riggs et al, *Infection and Immunity,* Vol. 62, #5, May 1994, p. 1931 © ASM.

Chapter 6

Opener: © AFP Photo/Doug Kanter; **6.1a:** © K. G. Murti/Visuals Unlimited; **6.1b:** © CDC/Phototake, NYC; **6.1c:** © A. B. Dowsette/SPL/Photo Researchers, Inc.; **6.2a:** Reprinted from Schaffer et al, *Proceedings of the National Academy of Science,* 41:1020, 1955; **6.2b:** © Omikron/ Photo Researchers, Inc.; **6.5b:** © Dennis Kunkel/Phototake, NYC; **6.5d:** © K. G. Murti/Visuals Unlimited; **6.6c:** © Science VU-NIH, R. Feldman/Visuals Unlimited; **6.7a:** © Boehringer Ingelheim International GMBH; **6.7b:** Kathy Park Talaro; **6.8b:** Courtesy of Harold Fisher, University of Rhode Island; **6.11b, 6.13:** © Lee D. Simon/Photo Researchers Inc.; **6.18:** © K. G. Murti/Visuals Unlimited; **6.19b:** © Christ Bjornberg/Photo Researchers, Inc.; **6.20a:** © Patricia Barber/Custom Medical Stock; **6.20b:** © Science VU-Charles W. Stratton/Visuals Unlimited; **6.21a:** Photo by Ted Heald, State of Iowa Hygienic Laboratory; **6.22a:** © E. S. Chan/Visuals Unlimited; **6.22b,c:** Courtesy Jack W. Frankel; **Microfile 6.1:** Science VU/ Wayside/Visuals Unlimited; **6.23a:** © Carroll M. Weiss/ Camera M. D. Studios; **6.23b–1:** © A. M. Siegelman/Visuals Unlimited; **6.23b–2:** © Science VU/CDC/Visuals Unlimited; **6.23c–1:** © K.G. Murti/Visuals Unlimited; **6.23c–2:** Courtesy Fred Williams, U.S. Environmental Protection Agency; **6.23d:** Courtesy of Reference Laboratory, A PCC Laboratory.

Chapter 7

Opener: Courtesy Mary Ann Cunningham; **Spotlight 7.1a:** © Corale Brierley/Visuals Unlimited; **Spotlight 7.1b:** Courtesy E. E. Adams, Montana State University/National Science Foundation, Office of Polar Programs; **7.1a:** © Ralph Robinson/Visuals Unlimited; **7.1b:** Courtesy Jack Jones, US EPA; **7.10a:** © Pat Armstrong/Visuals Unlimited; **7.10b:** © Philip Sze/Visuals Unlimited; **Spotlight 7.4:** © Michael Milstein; **7.11a:** Courtesy of Sheldon Manufacturing, Inc.; **Spotlight 7.5a:** Courtesy of Dr. Hans-Dieter Gortz, Zoologisches Institute, Münster, Germany; **Spotlight 7.5b:**

© Mike Abbey/Visuals Unlimited; **Spotlight 7.5c:** Bauchop et al, *Applied Microbiology,* Vol. 30, #4, p. 673, October 1975 © ASM; **7.13:** © Science VU-Fred Marsik/Visuals Unlimited; **7.17a:** © Kathy Park Talaro/Visuals Unlimited; **7.19b:** Courtesy of Howard Shapiro.

Chapter 8

Opener: Courtesy Nicholas C. DeMello/UCLA; **Microbits 8.2:** Courtesy Heinz W. Pley, Kevin M. Flaherty & David B. McKay; **8.24b:** Courtesy Donald F. Parsons.

Chapter 9

Opener: © A. Barrington Brown/Photo Researchers, Inc.; **9.3:** © K. G. Murti/Visuals Unlimited; **Hist. Highlights 9.1:** © Lawrence Livermore Laboratory/SPL/Custom Medical Stock Photo; **9.17c:** Courtesy Steven McKnight and O. L. Miller, Department of Biology, University of Virginia.

Chapter 10

Opener: © Jean Claude Levy/Phototake, NYC; **10.3b:** © Kathy Park Talaro; **10.4e:** From Koneman et al, *Diagnostic Microbiology,* 4/e, fig. 20.20, p. 1037 © J.B. Lippincott Company; **Table, fig. 10.3a:** Courtesy David M. Stalker; **Table, fig. 10.3b:** Courtesy Richard Shade, Purdue University; **Table, 10.4(left),** Courtesy R. L. Brinster; **Table, 10.4(right):** © Karen Kasmauski/Matrix; **10.13:** Courtesy Brigid Hogan, Howard Hughes Medical Institute, Vanderbilt University; **10.16:** Courtesy of Dr. Michael Baird, Lifecodes Corporation; **Microbits 10.2:** © Marilyn Humphries/Impact Visuals.

Chapter 11

Opener: © David Corio/Discover Magazine; **Hist. Highlights 11.1:** © Bettmann/Corbis; **11.5a:** © Science VU/Visuals Unlimited; **11.6:** © Raymond B. Otero/Visuals Unlimited; **11.9b:** © Leonard Lessin/Peter Arnold, Inc.; **11.12b:** © Fred Hossler/Visuals Unlimited; **11.16:** © Kathy Park Talaro/Visuals Unlimited; **Hist. Highlights 11.3a:** © David M. Phillips/Visuals Unlimited; **Hist. Highlights 11.3b:** Kathy Park Talaro; **11.18a:** © Raymond B. Otero/Visuals Unlimited.

Chapter 12

Opener: © Bettmann/Corbis; **12.3e,f:** Reprinted with permission of Robert L. Perkins from A. S. Kleiner & R. L. Perkins, *Journal of Infectious Diseases,* 122:4, 1970, © University of Chicago Press; **Microbits 12.2.1A:** © Kathy Park Talaro/Visuals Unlimited; **Microbits 12.2.1B:** © Kathy Park Talaro; **12.15:** © Cabisco/Visuals Unlimited; **12.19:** © Kenneth E. Greer/Visuals Unlimited; **12.22:** Courtesy ABBiodisk, Solna Sweden.

Chapter 13

Opener: © AP Photo/Robyn Beck, Pool; **13.5:** © David M. Phillips/Visuals Unlimited; **13.8:** © Science Vu/Visuals Unlimited; **13.22:** © Marshall W. Jennison; **Hist. Highlights 13.5:** © The Bettman Archive.

Chapter 14

Opener: © Manfred Kage/Peter Arnold, Inc.; **14.3b:** © Ellen R. Dirksen/Visuals Unlimited; **14.12c:** Courtesy Steve Kunkel; **14.21a:** © David M. Phillips/Visuals Unlimited; **14.24e:** Reprinted from Wendell F. Rosse et al, "Immune Lysis of Normal Human and Parozysmal Nocturnal Hemogloginuria (PNH) Red Blood Cells," *Journal of Experimental Medicine,* 123:969, 1966. Rockefeller University Press.

Chapter 15

Opener: © Lennart Nilsson, "The Body Victorious" Bonnier Fakta; **15.14b:** © R. Feldmann/Rainbow; **15.22:** © Boehringer Ingelheim International GMBH.

Chapter 16

Opener: James Gillroy, British, 1757–1815, "The Cow-Pock", engraving, 1802, William McCallin McKee Memorial Collection, 1928. 1407. © 1991, The Art Institute of Chicago. All Rights Reserved.; **16.7a:** From Gillies and Dodds, *Bacteriology Illustrated,* 5/E, fig. 13, page 24. Reprinted by permission of the publisher Churchill Livingstone; **16.9:** Courtesy of Philip Markham, Advanced BioScience Laboratories, Inc.; **16.13a:** From Miguel Pedraza et al, *Laboratory Medicine,* Vol. 14, #1, January 1983; **16.13b:** From F. Latel et al., *Journal of Clinical Microbiology,* 23(6):1018, American Society of Microbiology.

Chapter 17

Opener: © Runk/Schoenberger/Grant Heilman Photography; **17.2b:** © SPL/Photo Researchers, Inc.; **17.2c:** © David M. Phillips/Visuals Unlimited; **17.5:** © Kenneth E. Greer/Visuals Unlimited; **17.6a:** © STU/Custom Medical Stock Photo; **17.10a,b:** © Stuart I. Fox; **17.15a:** © Kenneth E. Greer/Visuals Unlimited; **17.15b:** © SIU/Visuals Unlimited; **17.18a:** © Kathy Park Talaro; **17.18b:** © Kenneth E. Greer/Visuals Unlimited; **Spotlight 17.3(top):** © Renee Lynn/Photo Researchers, Inc.; **Spotlight i7.3(lower):** © Walter H. Hodge/Peter Arnold, Inc.; **17.22:** Reprinted from R. Kretschmer, *New England Journal of Medicine,* 279:1295, 1968. Copyright © 1968 Massachusetts Medical Society. All rights reserved.; **Spotlight 17.5:** Courtesy of Baylor College of Medicine, Public Affairs.

Chapter 18

Opener: From C.F.J. Rayner, A.D. Jackson, A. Rutman, A. Dewar, T. J. Mitchell, P.W. Andrew, P.J. Cole, and R. Wilson, *Infection and Immunity,* Vol. 63, No. 2, February 1995, pp. 442–447, fig. 5. © 1995 ASM.; **18.1:** © David M. Phillips/Visuals Unlimited; **18.2:** © Kathy Park Talaro/Visuals Unlimited; **18.3b,c:** © Carroll H. Weiss/Camera M.D. Studios; **18.4b:** © Science VU-Charles W. Stratton/Visuals Unlimited; **18.5a:** Courtesy National Institute Slide Bank/The Wellcome Centre for Medical Sciences; **18.5b:** From Braude, *Infections Diseases and Medical Microbiology,* 2/E, fig. 3, page 1320 © Saunders College Publishing; **18.6a:** © Raymond B. Otero/Visuals Unlimited; **18.7:** Courtesy Analytab Products; **18.8:** © Kathy Park Talaro; **18.9a:** © G.W. Willis, MD/Biological Photo Service; **18.9b:** © L.M. Pope and D. R. Grote/Biological Photo Service; **18.11a,b:** © Kenneth E. Greer/Visuals Unlimited; **18.12:** From Farrar WE, Woods MJ, Innes JA: *Infectious Diseases: Text and Color Atlas,* ed 2. London, Mosby Europe, 1992.; **18.14a:** Courtesy Dr. David Schlaes; **18.14b(both):** Diagnostic Products Corporation; **18.15a:** Courtesy of Dr. David Schlaes; **18.16:** © Carroll H. Weiss/Camera M.D. Studios; **18.17:** © A. M. Siegelman/Visuals Unlimited; **18.19:** © Custom Medical Stock Photo; **18.20:** © Raymond B. Otero/Visuals Unlimited; **18.21:** © G. Musil/Visuals Unlimited; **18.25:** From James Bingham, *Pocket Guide for Clinical Medicine.* © 1984 Williams and Wilkins Co., Baltimore; **18.26:** © George J. Wilder/Visuals Unlimited; **18.28:** © Kenneth E. Greer/Visuals Unlimited; **18.29:** © Kathy Park Talaro/Visuals Unlimited.

Chapter 19

Opener: © Arthur C. Aufderheide, MD; **19.1a:** © A. M. Siegelman/Visuals Unlimited; **19.1b:** © George J. Wilder/Visuals Unlimited; **19.1c:** © John D. Cunningham/Visuals Unlimited; **19.2a,b:** © Science VU-Charles W. Stratton/Visuals Unlimited; **19.3a:** From M.A. Boyd et al, *Journal of Medical Microbiology,* 5:459, 1972. Reprinted by permission of Longarm Group, Ltd. © Pathological Society of Great Britain and Ireland; **19.4:** © Science VU-Charles W. Stratton/Visuals Unlimited; **19.5:** © Los Angeles Times Photo/Rolando Otero; **19.6a:** Courtesy of Dr. Fred E. Pittman; **19.6b:** Reprinted from Farrar and Lambert, *Pocket Guide for Nurses: Infectious Diseases.* © 1984, Williams and Wilkins, Baltimore; **19.8:** Courtesy of Dr. T. F. Sellers, Jr.; **19.10:** © Kenneth E. Greer/Visuals Unlimited; **19.11:** © George J. Wilder/Visuals Unlimited; **19.13:** Centers for Disease Control; **19.14:** © A. M. Siegelman/Visuals Unlimited; **19.15a:** From Gillies and Dodds, *Bacteriology Illustrated,* 5th ed., fig. 45, p. 58. Reprinted by permission of Churchill Livingstone; **19.15b:** From *ASM News,* Vol. 59(12), Dec. 1993, courtesy of Pascal Meylan.; **19.16a:** © John D. Cunningham/Visuals Unlimited; **19.17:** © Science VU-Charles W. Stratton/Visuals Unlimited; **19.19:** © CNRI/SPL/Photo Researchers, Inc.; **19.20:** © Elmer Koneman/Visuals Unlimited; **19.21a:** © Kenneth E. Greer/Visuals Unlimited; **19.21b:** © Science VU/Visuals Unlimited; **19.22:** © Kenneth E. Greer/Visuals Unlimited; **19.23:** © NMSB/Custom Medical Stock Photo; **19.24:** World Health Organization; **19.25, 19.26a:** © Kenneth E. Greer/Visuals Unlimited; **19.26b:** © Science Vu-Charles W. Stratton/Visuals Unlimited; **19.27:** From J. Walter Wilson, *Fungous Diseases of Man,* Plate 26, © 1965, The Regents of University of California.

Chapter 20

Opener: Courtesy Jerry Shenep, St. Jude's Children's Hospital; **20.1:** © Science VU/Visuals Unlimited; **20.2:** © Bottone E. J. and Perez A.A., II, The Mount Sinai Hospital New York; **20.3:** © Kathy Park Talaro/Visuals Unlimited; **20.5:** © Jack M. Bostrack/Visuals Unlimited; **20.6:** Centers for Disease Control; **20.9a:** © Kathy Park Talaro/Visuals Unlimited; **20.9b:** © Kathy Park Talaro; **Spotlight 20.3:** Courtesy Susan Broadaway and Barry Pyle, Montana State University; **20.12:** © John D. Cunningham/Visuals Unlimited; **20.13:** © Elmer Koneman/Visuals Unlimited; **20.14:** © A. M. Siegelman/Visuals Unlimited; **20.18, 20.20:** © Science VU-Charles W. Stratton/Visuals Unlimited; **20.21:** From Gillies and Dodds, *Bacteriology Illustrated,* 5/e, fig. 78 (right), p. 104. Reprinted by permission of Churchill Livingstone; **20.23:** Reprinted from Farrar and Lambert, *Pocket Guide for Nurses: Infectious Diseases,* © 1984, Williams and Wilkins, Baltimore. Photo by Dr. M. Tapert.

Chapter 21

Opener: Courtesy Didier Raoult; **21.1b:** © Science VU-Charles R. Stratton/Visuals Unlimited; **21.1c,d:** © Science VU-Charles W. Stratton/Visuals Unlimited; **21.2:** © Custom Medical Stock Photo; **21.3:** © Carroll H. Weiss/Camera M.D. Studios; **21.4:** © Science VU/Visuals Unlimited; **21.5:** © Kenneth E. Greer/Visuals Unlimited; **21.6a,b:** © Science VU/CDC/Visuals Unlimited; **21.7:** © Science VU/CDC/Visuals Unlimited; **21.8a:** From Strickland, *Hunter's Tropical Medicine,* 6/E, © Saunders College Publishing; **21.8b:** © Science VU/Visuals Unlimited; **21.9:** © Science VU-Fred Marsik/Visuals Unlimited; **21.12:** © Centers for Disease Control/Peter Arnold, Inc.; **21.13a,b:** From Jacob S. Teppema, "In Vivo Adherence and Colonization of Vibro Cholerae Strains That Differ in Hemagglutinating Activity and Motility", *Journal of Infection and Immunity*, 55(9): 2093–2102, Sept. 1987. Reprinted by permission of American Society for Microbiology.; **Microfile 21.2:** © AP/Wide World Photos; **21.15:** From R.R. Colwell and D.M. Rollins, "Viable but Nonculturable Stage of Campylobacter jejuni and Its Role in Survival in the Natural Aquatic Environment," *Applied and Environmental Microbiology*, 52(3):531–538, 1986. Reprinted by permission of American Society for Microbiology.; **Microbits 21.3:** Courtesy of James G. Fox from *Infection and Immunity*, June 1994, pp. 2367–2374. © American Society for Microbiology; **21.16a:** Reprinted from R.L. Anaker et al, "Details of the Untrastructure of Rickettsia Prowazekii Grown in the Chick Volk Sac," *Journal of Bacteriology*, 94(1): 260, 1967. American Society of Microbiology; **21.16b:** From Bruce Merrell et al., "Morphological and Cell Association Characteristics of Rochalimaea quintana: Comparison of the Vole and Fuller Strains," *Journal of Bacteriology*, 135 (2):633–640, 1978. Reprinted by permission from the American Society of Microbiology; **Microbits 21.4b:** © Science VU/Visuals Unlimited; **Microbits 21.4c:** © A.M. Siegelman/Visuals Unlimited; **21.19:** Dept. of Health and Human Resources, Courtesy of Dr. W. Burgdorfer; **21.20:** From McCaul and Williams, "Development Cycle of C. Burnetii," *Journal of Bacteriology*, 147:1063, 1981. Reprinted by permission of American Society for Microbiology; **21.21:** © Kenneth E. Greer/Visuals Unlimited; **21.23a:** Armed Forces Institute of Pathology; **21.23b:** © Science VU-Bascom Palmer Institute/Visuals Unlimited; **21.24:** © Harvey Blank/Camera M.D. Studios; **21.25a:** © Science VU/CDC/Visuals Unlimited; **21.25b:** © Science VU-AFIP/Visuals Unlimited; **21.26a:** Photo courtesy Duncan C. Krause, from *Trends in Microbiology*, Vol. 6, no. 1, Jan. 1998, fig. 1, pg. 16, © Elsevier Sciences L.T.D.; **21.30a:** © R. Gottsegen/Peter Arnold, Inc.; **21.30b:** © Stanley Flegler/Visuals Unlimited; **21.31:** © Science VU-Max A. Listgarten/Visuals Unlimited; **21.33:** © Science VU-Max A. Listgarten/Visuals Unlimited.

Chapter 22

Opener: © William R. Harrison; **22.4:** Courtesy Reinhard Ruchel, Univ. Göttingen, Germany; **22.6a:** © Elmer W. Koneman/Visuals Unlimited; **22.6b:** © Everett S. Beneke/Visuals Unlimited; **22.7b:** © A.M. Siegelman/Visuals Unlimited; **22.8d:** © Science VU-Charles Sutton/Visuals Unlimited; **22.9:** Reprinted from J. Walter Wilson, *Fungous Diseases of Man*, Plate 2, © 1965 The Regents of the University of California; **22.10:** From Rippon, *Medical Mycology*, 2/E, fig. 17.18, page 419 © Saunders College Publishing; **22.11a:** © Elmer W. Koneman/Visuals Unlimited; **22.11b:** © A.M. Siegelman/Visuals Unlimited; **22.12:** Reprinted by permission of Upjohn Company from E.S. Beneke, et al, *Human Mycoses*, 1984; **22.13:** From Elmer W. Koneman and Roberts, *Practical Laboratory Mycology*, 1985, pgs. 133, 134, © Williams and Wilkins Co., Baltimore; **22.14:** © HarkisanRaj/Visuals Unlimited; **22.15:** © Everett S. Beneke/Visuals Unlimited; **22.16:** Reprinted by permission of Upjohn Company from E.S. Beneke, et al, *Human Mycoses*, 1984.; **22.17a,c:** From Elmer W. Koneman and Roberts, *Practical Laboratory Mycology*, 1985, pgs. 133, 134, © Williams and Wilkins Co., Baltimore; **22.17b:** © A.M. Siegelman/Visuals Unlimited; **Microfile 22.2:** Courtesy Toshio Kanbe; **22.18a:** © Everett S. Beneke/Visuals Unlimited; **22.18b, 22.19a:** © Kenneth E. Greer/Visuals Unlimited; **22.19b:** Reprinted from J. Walter Wilson, *Fungous Diseases of Man*, Plate 42(middle right), © 1965, The Regents of the University of California; **22.20:** © Carroll

H. Weiss/Camera M.D. Studios; **22.21a,b, 22.22a:** © Everett S. Beneke/Visuals Unlimited; **22.22b:** © Science VU/CDC/Visuals Unlimited; **22.23a:** © Raymond B. Otero/Visuals Unlimited; **22.23b:** Courtesy Glenn S. Bulmer; **22.23c:** © Elmer Koneman/Visuals Unlimited; **22.24:** Courtesy Gordon Love; **22.25:** Reprinted from J. Walter Wilson, *Fungous Diseases of Man*, Plate 21, © 1965, The Regents of the University of California.; **22.26:** © A.M. Siegelman/Visuals Unlimited; **22.27a:** © Everett S. Beneke/Visuals Unlimited; **22.27b:** © Science VU-AFIP/Visuals Unlimited; **22.28:** Abarca ML. Taxonomía e identificación de especies implicadas en la aspergilosis nosocomial (Taxonomy and identification of the species involved in nosocomial aspergillosis). Rev Iberoam Micol 2000; 17: S79–S84]; **22.29:** From Rippon, *Medical Mycology*, 2/E, fig. 25.4b, page 620 © Saunders College Publishing. Photo by L. Calkins; **22.30a,b:** © M.F. Brown/Visuals Unlimited.

Chapter 23

Opener: Ellen Li et al, *Infection and Immunity*, Nov. 1994, Vol. 62 #11, fig. 5c, p. 5116. © American Society for Microbiology; **23.2:** © Science VU-Charles W. Stratton/Visuals Unlimited; **23.4:** © Science VU-David John/Visuals Unlimited; **23.7b:** Courtesy Stanley Erlandsen; **23.9a:** Reprinted from Katz, Despommier, and Dwadz, "Parasitic Diseases", Springer-Verlag. Photo by T. Jones; **23.10b:** Armed Forces Institute of Pathology; **23.12:** Centers for Disease Control; **23.14a,b:** Courtesy of Dr. Henry W. Murray. From *Harrison's Principles of Internal Medicine*, 12/ed., McGraw-Hill, Inc. Reproduced with permission of the publisher; **23.15a:** © Cecil H. Fox/Photo Researchers, Inc.; **23.16:** Courtesy Ynes R. Ortega; **23.18a:** © Lauritz Jensen/Visuals Unlimited; **23.18b:** © A.M. Siegelman/Visuals Unlimited; **Microbits 23.2:** © Lauritz Jensen/Visuals Unlimited; **23.19b1:** © R. Calentine/Visuals Unlimited; **23.19b2:** © Science VU-Fred Marsik/Visuals Unlimited; **23.20:** © Carroll H. Weiss/Camera M.D. Studios; **23.22a,b:** © Science VU-Fred Marsik/Visuals Unlimited; **23.23a:** © Cabisco/Visuals Unlimited; **23.23b:** Courtesy Harvey Blankespoor; **23.23c:** © Science VU/Visuals Unlimited; **23.24:** © A.M. Siegelman/Visuals Unlimited; **23.25a:** © Stanley Flegler/Visuals Unlimited; **23.25b:** From Katz et al., "Parasitic Diseases", Springer/Verlag; **23.26:** © Science VU-Charles W. Stratton/Visuals Unlimited.

Chapter 24

Opener: Courtesy NEN Life Science Products; **Hist. Highlights 24.1:** World Health Organization; **24.2a:** Courtesy Derek Garbellini; **24.2b:** © Charles Stoer/Camera M.D. Studios; **24.3b:** © Boehringer Ingeheim International Gmbh; **24.5a:** © Carroll H. Weiss/Camera M.D. Studios; **24.5b:** © Kenneth E. Greer/Visuals Unlimited; **24.6, 24.7:** © Carroll H. Weiss/Camera M.D. Studios; **24.8:** © Kathy Park Talaro; **24.9:** © Science VU-Charles W. Stratton/Visuals Unlimited; **24.10a:** © John D. Cunningham/Visuals Unlimited; **24.10c:** © Carroll H. Weiss/Camera M.D. Studios; **24.11:** From J.S. Nelson and J.P. Wyatt, *Medicine*, 38:22 © 1959 Williams & Wilkins; **24.12:** © Science VU-Charles W. Stratton/Visuals Unlimited; **24.13:** Centers for Disease Control; **24.14:** Courtesy Barbara O'Connor; **24.17:** © Science VU-NIH/Visuals Unlimited; **24.19a:** © David M. Phillips/Visuals Unlimited; **24.19b:** © K.G. Murti/Visuals Unlimited; **24.20a,b:** © Kenneth E. Greer/Visuals Unlimited; **24.21:** From Koneman et al, *Diagnostic Microbiology*, 4/e, fig. 20.20, p. 1037 © J.B. Lippincott Company; **24.22:** © Science VU-Charles W. Stratton/Visuals Unlimited; **24.23:** © Dr. P. Marazzi/SPL/Photo Researchers, Inc.

Chapter 25

Opener: © Robin Thomas; **Spotlight 25.1:** © Patrick Robert/Corbis Sygma; **25.2a:** From Exeen M. Morgan and Fred Rapp, "Measles Virus and Its Associates Disease," *Bacteriological Reviews*, 41(3):636–666, 1977. Reprinted by permission of American Society for Microbiology; **25.3:** © Biophoto Associates/Photo Researchers, Inc.; **25.4a:** Centers for Disease Control; **25.4b:** © Kenneth E. Greer/Visuals Unlimited; **25.5a:** © Boehringer Ingelheim International Gmbh; **25.7:** © Science VU-AFIP/Visuals Unlimited; **25.8a:** Reprinted by permission of Dr. Kenneth A. Schiffer from Cooper et al, *Am. J. Dis. Child*, 110:419, © 1965, American Medical Association; **25.18:** © Science VU/Visuals Unlimited; **25.21:** From I. Katayarma, C.Y. Li, and L.T. Yam, "Ultrastructure Characteristics of the Hairy Cells of Leukemic Reticuloen-dotheliosis," *American Journal of Pathology*, 67:361, 1972. Reprinted by permission of the American Society for Investigative Pathology; **25.22b:** © Science VU-CDC/Visuals Unlimited; **25.23:** © Corbis/Bettmann; **25.25b-1:** © Cabiso/Visuals Unlimited; **25.25b-2:** © Science VU-Charles W. Stratton/Visuals Unlimited; **25.26:** © Carroll H. Weiss/Camera M. D. Studios; **25.28a:** © K.G. Murti/Visuals Unlimited; **25.28b:** © From Farrar and Lambert, *Pocket Guide for Nurses: Infectious Diseases*, 1984. Williams and Wilkins, Baltimore. Reprinted by permission of Dr. William Edmund Farrar, Jr.; **25.29a,b:** Image from WebPath, courtesy of Edward C. Klatt MD.

Chapter 26

Opener: Courtesy Oak Ridge National Laboratory; **26.11b:** © John D. Cunningham/Visuals Unlimited; **26.12a,b:** © Sylvan Wittwer/Visuals Unlimited; **26.14, 26.16:** © John D. Cunningham/Visuals Unlimited; **Spotlight 26.3a:** © Frank Hanna/Visuals Unlimited; **Spotlight 26.3b:** © John D. Cunningham/Visuals Unlimited; **26.20b:** © Carleton Ray/Photo Researchers, Inc.; **26.21:** © John Cunningham/Visuals Unlimited; **26.23b:** © Kathy Park Talaro/Visuals Unlimited; **26.23c:** Courtesy IDEXX Laboratories, Inc., Westbrook, Maine, U.S.A.; **26.27b:** © Vance Henry/Nelson Henry; **26.28:** © John D. Cunningham/Visuals Unlimited; **26.29b:** © Kevin Schafer/Peter Arnold, Inc.; **26.31a:** © Kathy Park Talaro/Visuals Unlimited; **26.32b:** © Joe Munroe/Photo Researchers, Inc.; **Box 26.4a,b:** © Kathy Park Talaro; **26.26:** Courtesy Donald Klein; **26.35:** © Kathy Park Talaro/Visuals Unlimited; **26.39:** © J.T. MacMillan.

Line Art

3.23: Courtesy William A. Jensen; **4.37:** Reprinted with permission from *Nature*, Vol. 196, pp. 1189–1192. Copyright © 1962 Macmillan Magazines Ltd.; **6.8a:** From Westwood, et al., *Journal of Microbiology*, 34:67, 1964. Reprinted by permission of The Society for General Microbiology, United Kingdom; **11.5b:** From John J. Perkins, Principles and Methods of Sterilization in Health Science, 2e, 1969. Courtesy of Charles C. Thomas, Springfield, Illinois; **11.5b:** From Nolte, et al., *Oral Microbiology*, 4e. © 1982 Mosby. Reprinted by permission of Gloria-Mae Nolte; **15.18:** From *Immunology III*, Copyright © 1985 W. B. Saunders and Co. Philadelphia, PA. Reprinted by permission of Joseph A. Bellanti; **20.10:** Reprinted courtesy of Becton, Dickinson and Company.

Index

Page numbers in boldface indicate illustrations and tables.

Notifiable Diseases—Summary of Reported Cases, United States, 1989–2000

Disease	1989	1990	1991	1992	1993
AIDS	33,722	41,595	43,627	45,472	103,691
Amebiasis	3,217	3,328	2,989	2,942	2,970
Aseptic meningitis	10,274	11,852	14,526	12,223	12,848
Botulism, total (including wound and unsp.)	89	92	114	91	97
Foodborne	23	23	27	21	27
Infant	60	65	81	66	65
Brucellosis	95	85	104	105	120
Chancroid	4,692	4,212	3,476	1,886	1,399
Chlamydia¶			..........................**..........................		
Cholera	–	6	26	103	18
Diphtheria	3	4	5	4	–
Encephalitis, primary	981	1,341	1,021	774	919
Post – infectious	88	105	82	129	170
Escherichia coli O157:H7			..........................**..........................		
Gonorrhea	733,151	690,169	620,478	501,409	439,673
Granuloma inguinale	7	97	29	6	19
Haemophilus influenzae, invasive		..........................**..........................	2,764	1,412	1,419
Hansen disease (leprosy)	163	198	154	172	187
Hepatitis A	35,821	31,441	24,378	23,112	24,238
Hepatitis B	23,419	21,102	18,003	16,126	13,361
Hepatitis, C/non-A, non-B††	2,529	2,553	3,582	6,010	4,786
Hepatitis, unspecified	2,306	1,671	1,260	884	627
Legionellosis	1,190	1,370	1,317	1,339	1,280
Leptospirosis	93	77	58	54	51
Lyme disease		..........................**..........................	9,465	9,895	8,257
Lymphogranuloma venereum	189	277	471	302	285
Malaria	1,277	1,292	1,278	1,087	1,411
Measles (rubeola)	18,193	27,786	9,643	2,237	312
Meningococcal disease	2,727	2,451	2,130	2,134	2,637
Mumps	5,712	5,292	4,264	2,572	1,692
Pertussis (whooping cough)	4,157	4,570	2,719	4,083	6,586
Plague	4	2	11	13	10
Poliomyelitis, paralytic§§	11	6	10	6	4
Psittacosis	116	113	94	92	60
Rabies, animal	4,724	4,826	6,910	8,589	9,377
Rabies, human	1	1	3	1	3
Rheumatic fever, acute	144	108	127	75	112
Rocky Mountain spotted fever	623	651	628	502	456
Rubella (German measles)	396	1,125	1,401	160	192
Rubella, congenital syndrome	3	11	47	11	5
Salmonellosis, excluding typhoid fever	47,812	48,603	48,154	40,912	41,641
Shigellosis	25,010	27,077	23,548	23,931	32,198
Syphilis, primary and secondary	44,540	50,223	42,935	33,973	26,498
Total, all stages	110,797	134,255	128,569	112,581	101,259
Tetanus	53	64	57	45	48
Toxic-shock syndrome	400	322	280	244	212
Trichinosis	30	129	62	41	16
Tuberculosis	23,495	25,701	26,283	26,673	25,313
Tularemia	152	152	193	159	132
Typhoid fever	460	552	501	414	440
Varicella (chickenpox)***	185,441	173,099	147,076	158,364	134,722

Source: Data from *Morbidity and Mortality Weekly Report.*

*The total number of acquired immunodeficiency syndrome (AIDS) cases includes all cases reported to the Division of HIV/AIDS Prevention, Surveillance, and Epidemiology, National Center for HIV, STD, and TB Prevention (NCHSTP) through December, 2000.

†No longer nationally notifiable.

§Cases were updated through the Division of Sexually Transmitted Diseases Prevention, NCHSTP, as of June, 2000.

¶Chlamydia refers to genital infections caused by *C. trachomatis.*

**Not previously nationally notifiable.

††Anti-HCV antibody test was available as of May 1990.

§§Numbers may not reflect changes based on retrospective case evaluations or late reports (see *MMWR* 1986;35:180 – 2).

¶¶Cases were updated through the Division of Tuberculosis Elimination, NCHSTP, as of May, 2000.

***Varicella was taken off the nationally notifiable disease list in 1991. Many states continue to report these cases to CDC.